REFRIGERATION & AIR CONDITIONING TECHNOLOGY

6th Edition

William C. Whitman
William M. Johnson
John A. Tomczyk
Eugene Silberstein

DELMAR
CENGAGE Learning

Australia Brazil Canada Mexico Singapore Spain United Kingdom United States

**Refrigeration and Air Conditioning
Technology, Sixth Edition**
William C. Whitman, William M.
Johnson, John A. Tomczyk, and
Eugene Silberstein

Vice President, Career and Professional
Editorial: Dave Garza

Director of Learning Solutions:
Sandy Clark

Senior Acquisitions Editor: James Devoe

Managing Editor: Larry Main

Senior Product Manager: John Fisher

Editorial Assistant: Thomas Best

Vice President, Career and Professional
Marketing: Jennifer McAvey

Marketing Director: Deborah S. Yarnell

Marketing Manager: Kevin Rivenburg

Marketing Coordinator: Mark Pierro

Production Director: Wendy Troeger

Production Manager: Stacy Masucci

Content Project Manager: Cheri Plasse

Art Director: Bethany Casey

Technology Project Manager:
Christopher Catalina

Production Technology Analyst:
Thomas Stover

For product information and technology assistance, contact us at
**Professional & Career Group Customer Support,
1-800-648-7450**

For permission to use material from this text or product,
submit all requests online at **cengage.com/permissions.**
Further permissions questions can be e-mailed to
permissionrequest@cengage.com.

Library of Congress Cataloging-in-Publication Data

Refrigeration & air conditioning technology /
William C. Whitman . . . [et al.]. — 6th ed.
  p. cm.
  ISBN 1-4283-1936-0
  1. Refrigeration and refrigerating machinery. 2. Air conditioning.
I. Whitman, William C. II. Title: Refrigeration and air conditioning
technology.
  TP492.W6 2008
  621.5'6—dc22

2008000465

ISBN-13: 978-1428319363

ISBN-10: 1428319360

Delmar
5 Maxwell Drive
Clifton Park, NY 12065-2919
USA

Cengage Learning products are represented in Canada by Nelson
Education, Ltd.

For your lifelong learning solutions, visit **delmar.cengage.com**

Visit our corporate website at **cengage.com.**

Notice to the Reader
Publisher does not warrant or guarantee any of the products described herein or perform any independent analysis in connection with any of the product information contained herein. Publisher does not assume, and expressly disclaims, any obligation to obtain and include information other than that provided to it by the manufacturer. The reader is expressly warned to consider and adopt all safety precautions that might be indicated by the activities described herein and to avoid all potential hazards. By following the instructions contained herein, the reader willingly assumes all risks in connection with such instructions. The publisher makes no representations or warranties of any kind, including but not limited to, the warranties of fitness for particular purpose or merchantability, nor are any such representations implied with respect to the material set forth herein, and the publisher takes no responsibility with respect to such material. The publisher shall not be liable for any special, consequential, or exemplary damages resulting, in whole or part, from the readers' use of, or reliance upon, this material.

Printed in the United States of America
2 3 4 5 08

BRIEF TABLE OF CONTENTS

Contents

CD in Back of Book Contains:

Chapter A Alternative Heating

Section 9 Domestic Appliances

Appendix C Section 608

Appendix D Refrigerant Numbering System

Appendix E Refrigerant Changeover Guidelines

Video Clips

Preface

*R*efrigeration and Air Conditioning Technology is designed and written for students in vocational-technical schools and colleges, community colleges, and apprenticeship programs. The content is in a format appropriate for students who are attending classes full-time while preparing for their first job, for students attending classes part-time while preparing for a career change, or for those working in the field who want to increase their knowledge and skills. Emphasis throughout the text is placed on the practical applications of the knowledge and skills technicians need to be productive in the refrigeration and air-conditioning industry. The contents of this book can be used as a study guide to prepare for the Environmental Protection Agency (EPA) mandatory technician certification examinations. It can be used in the HVACR field or closely related fields by students, technicians, installers, contractor employees, service personnel, and owners of businesses.

This text is also an excellent study guide for the Industry Competency Exam (ICE), the North American Technician Excellence (NATE), the HVAC Excellence, the Refrigeration Service Engineers Society (RSES), the United Association (UA) STAR certification, and the Heating, Air Conditioning, and Refrigeration Distributors International (HARDI) voluntary HVACR technician certification and home study examinations.

The book is also written to correspond to the National Skill Standards for HVACR technicians. Previous editions of this text are often carried to the job site by technicians and used as a reference for service procedures. "Do-it-yourselfers" will find this text valuable for understanding and maintaining heating and cooling systems.

As general technology has evolved, so has the refrigeration and air-conditioning industry. A greater emphasis is placed on digital electronic controls and system efficiency. Every central split cooling system manufactured in the United States today must have a Seasonal Energy Efficiency Ratio (SEER) rating of at least 13. This energy requirement was mandated by federal law as of January 23, 2006. SEER is calculated on the basis of the total amount of cooling (in Btu) the system will provide over the entire season, divided by the total number watt-hours it will consume. Higher SEER ratings reflect a more efficient cooling system. Air-conditioning and refrigeration technicians are responsible for following procedures to protect our environment, particularly with regard to the handling of refrigerants. Technician certification has become increasingly important in the industry.

Global warming has become a major environmental issue. When HVACR systems are working correctly and efficiently, they will greatly reduce energy consumption and greenhouse gases. Organizations like the Green Mechanical Council (GreenMech) are advocates for the HVACR industry and assist the industry in meeting with government, educational, industry, and labor interests to find solutions to the world's global-warming problem. GreenMech has created a scoring system designed to help engineers, contractors, and consumers know the "green value" of each mechanical installation. The "green value" encompasses the system's energy efficiency, pollution output, and sustainability. Realtors, building inspectors, builders, and planning and zoning officials will now have some knowledge about and guidance on how buildings and mechanical systems are performing. Green buildings and green mechanical systems are becoming increasingly popular in today's world as a way to curb global warming.

TEXT DEVELOPMENT

This text was developed to provide the technical information necessary for a technician to be able to perform satisfactorily on the job. It is written at a level that most students can easily understand. Practical application of the technology is emphasized. Terms commonly used by technicians and mechanics have been used throughout to make the text easy to read and to present the material in a practical way. Many of these key terms are also defined in the glossary. This text is updated regularly in response to market needs and emerging trends. Refrigeration and air-conditioning instructors have reviewed each unit. A technical review takes place before a revision is started and also during the revision process.

Illustrations and photos are used extensively throughout the text. Full-color treatment of most photos and illustrations helps amplify the concepts presented.

No prerequisites are required for using this text. It is designed to be used by beginning students, as well as by those with training and experience.

ORGANIZATION

Considerable thought and study has been devoted to the organization of this text. Difficult decisions had to be made to provide text in a format that would meet the needs of varied institutions. Instructors from different areas of the country and from various institutions were asked for their ideas regarding the organization of the instructional content.

The text is organized so that after completing the first four sections, students may concentrate on courses in refrigeration or air conditioning (heating and/or cooling). If the objective is to complete a whole program, the instruction may proceed until the sequence scheduled by the school's curriculum is completed.

New in This Edition

INTRODUCTORY MATERIALS

- A new section on the "green awareness" movement and global warming. Coverage includes key organizations and their present and future goals for how to accomplish green buildings and green mechanical rooms throughout the United States.
- A update on technician certification and on the key organizations that support either voluntary or mandatory technician certification.
- The new federal mandate of January 23, 2006 (requiring that every central split cooling system manufactured in the United States today have a Seasonal Energy Efficiency Ratio [SEER] rating of 13) is mentioned in the introductory material and is covered in more detail in appropriate chapters.
- The "Career Opportunities" section has been expanded to include a comprehensive list of career opportunities available in the HVACR field.

UNIT UPDATES

Most units have been updated to include advances or changes in technology, procedures, and/or equipment. The authors have added over 250 new images to this edition to emphasize the practical-application approach to the book. The following units have received major content additions and revisions for the sixth edition.

UNIT 1 Heat and Pressure

- More discussion of heat and pressure

UNIT 3 Refrigeration and Refrigerants

- Expanded coverage on pressure/enthalpy diagrams and system applications
- Detailed descriptions of "heat of work" and "heat of compression"

UNIT 4 General Safety Practices

New coverage on and photos illustrating the following topics:
- Cylinder safety
- Modern eye protection
- Pressure regulators
- Pressure relief valves
- Cylinder storage
- Electrical disconnects
- Emergency stop buttons
- Building signage
- First aid kits and placement
- Fire extinguishers and categories
- Refrigerant-specific leak detectors
- Evacuation plans
- Building directories
- Eye wash stations
- Emergency call stations

UNIT 5 Tools and Equipment

- Many new photos
- Discussion about how to choose and buy tools and instruments

UNIT 6 Fasteners

- New discussion about safety and fasteners

UNIT 7 Tubing and Piping

- Expanded coverage on tubing bending
- New coverage on new-generation mechanical piping connection systems

UNIT 8 System Evacuation

- Updated photos illustrating tools and instrumentation

UNIT 10 System Charging

- Expanded coverage on service valves
- Expanded discussion about the use of graduated charging cylinders

UNIT 14 Automatic Control Components and Applications

Updated photos illustrating the following:
- Transformers
- Heat anticipators
- Bimetal motor thermostats
- Oil safety controls

UNIT 15 Troubleshooting Basic Controls

- Expanded step-by-step electrical circuit evaluation

UNIT 16 Advanced Automatic Controls—Direct Digital Controls (DDC) and Pneumatics

New technical coverage on Direct Digital Control (DDC) systems. Many photos are included. The following systems and components are discussed:
- Controlled output devices
- Controlled environments
- Signal converters
- Control points
- Memory
- Control system components
- Active sensors
- Passive sensors
- Digital and analog inputs and outputs
- Set points
- Open control loops
- Closed control loops

- Feedback loops
- Control systems
- Control agents
- Controlled mediums
- DDC control system responsibilities

UNIT 17 Types of Electric Motors

- Additional information about motor slip and slip calculations

UNIT 18 Application of Motors

- Additional information about motor and blower speeds with respect to pulley sizes

UNIT 20 Troubleshooting Electric Motors

- Expanded coverage on the use of the belt tension gage
- Expanded coverage on motor evaluation methods. Instructs the technician on how to evaluate motors by measuring winding resistance

UNIT 21 Evaporators and the Refrigeration System

- Coverage on the 13 SEER federal mandate
- New photo illustrating the foaming of a compressor's sight glass
- Extensive coverage on the aluminum parallel-flow, flat-plate-and-fin evaporator (including many photos)

UNIT 22 Condensers

- New coverage on water-regulating valves (including photos)
- New, extensive coverage on the aluminum parallel-flow, flat-plate-and-fin condenser (including many photos)
- New coverage on cooling towers (including photos)
- New photos illustrating variable-frequency drives (VFDs)
- New, extensive coverage on low-ambient head pressure control valves (including many photos)

UNIT 23 Compressors

- New photos of and extensive coverage on discus compressors
- New photos of and extensive coverage on scroll compressors
- New photos of and coverage on the two-step modulating scroll compressor
- New coverage on digital capacity control for scroll compressors (including photos)
- New coverage on scroll compressor protection (including photos)
- New coverage on the high-efficiency, oil-free centrifugal compressor (including photos)

UNIT 25 Special Refrigeration System Components

- New diagrams and coverage on an automatic pumpdown system that will not short cycle the compressor while being pumped down

UNIT 30 Electric Heat

- New Service Calls

UNIT 32 Oil Heat

This expanded unit includes the following additions:
- New content on oil storage tanks
- New content on oil deaerators
- Expanded information about primary controls
- Expanded information about testing and evaluating oil-fired systems
- New content on oil-line vacuums and their effects on system and equipment operation
- New information about how to calculate the desired oil-line vacuum
- New information about how to compare one-pipe and two-pipe oil delivery systems

UNIT 33 Hydronic Heat

This expanded unit includes the following additions:
- Discussion of geothermal heat pumps as a heat source for hydronic heating systems
- More in-depth information about radiant heating systems (including installation types and applications)
- New content on primary–secondary pumping
- Expanded calculations relating the water flow, temperature differential, and Btu output of terminal units
- New information about multitemperature heating systems
- New information about outdoor reset

UNIT 34 Indoor Air Quality

- New discussion about filter applications
- Two new mold charts

UNIT 35 Comfort and Psychrometrics

- Expanded psychrometric chart

UNIT 36 Refrigeration Applied to Air Conditioning

- Discussion about attic ventilation
- Discussion about choosing equipment for different humidity locations
- Discussion about selecting equipment for different outside weather design conditions and maintaining equipment efficiencies

UNIT 37 Air Distribution and Balance

Additional coverage on system zoning that includes the following topics:
- Zoning using a single-speed blower
- Zoning using a variable-speed blower
- Adding zones to an existing system
- Airflow mathematical calculations

UNIT 39 Controls

- New photos of a solid-state anti-short-cycle timer, rooftop-unit air conditioner with parallel scroll compressors

UNIT 41 Troubleshooting

- Discussion about approach temperatures and temperature splits
- Information added to some of the illustrations
- Many new illustrations

UNIT 43 Air Source Heat Pumps

- Discussion of the Coefficient of Performance (COP) of heat pump systems
- Expanded discussion of heat pump defrost modes
- Expanded content on electrical strip heater operation in the emergency-heat, second-stage-heat, and defrost modes.

UNIT 45 Domestic Refrigerators

- Many new photographs

UNIT 46 Domestic Freezers

- Some new photographs

UNIT 47 Room Air Conditioners

- Some revised art
- Two new photographs

UNIT 48 High-Pressure, Low-Pressure, and Absorption Chilled-Water Systems

- Discussion about changing load conditions and equipment unloading
- Discussion about cooling for manufacturing processes
- More on variable-frequency drives (VFDs)
- Saving energy using building core heat to heat the perimeter of the building
- The use of subcooling in large systems
- Electronic expansion valves for large systems
- The use of electronic starters
- Power failure, low-voltage, voltage unbalance, and phase-reversal protection
- Several new photos

UNIT 50 Operation, Maintenance, and Troubleshooting of Chilled-Water Air-Conditioning Systems

- New discussion about and photos of many types of refrigerant, water, and steam valves
- Valve service information
- The servicing of pumps and strainers in water and steam systems
- Refrigerant safety in equipment rooms
- New photos and illustrations

How to Use the Text and Supplementary Materials

This text may be used as a classroom text, as a learning resource for an individual student, as a reference text for technicians on the job, or as a homeowner's guide. An instructor may want to present the unit objectives, briefly discuss the topics included, and assign the unit to be read. The instructor then may want to discuss the material with the students. This can be followed by the students completing the review questions, which can later be reviewed in class. The lecture outline provided in the *Instructor's Guide* may be utilized in this process. Lab assignments may be made at this time, followed by the students completing the lab review questions.

The instructor *e.resource CD* may be used to access a computerized test bank for end-of-unit review questions, teaching tips, PowerPoint® presentations, and more. The CD bound into the back of this book contains Section 9 (units 45–46) titled "*Domestic Appliances*," as well as a unit titled "*Alternative Heating (Stoves, Fireplace Inserts, Solar),*" appendices, and two video-clip samples from the 24-video DVD series that accompanies this text.

Features of the Text

Objectives

- *Objectives* are listed at the beginning of each unit. The objective statements are kept clear and simple to give students direction.

Safety Checklists

- A *Safety Checklist* is presented at the beginning of each unit, when applicable, immediately following the Objectives. This checklist emphasizes the importance of safety and is included in units where "hands-on" activities are discussed.
- *Safety* is emphasized throughout the text. In addition to the safety checklist at the beginning of most units, safety precautions and techniques are highlighted in red throughout the text. It would be impossible to include a safety precaution for every conceivable circumstance that may arise, but an attempt has been made to be as thorough as possible. The overall message is to work safely whether in a school shop, laboratory, or on the job and to use common sense.

Troubleshooting

Practical troubleshooting procedures are an important feature of this text. Practical component and system troubleshooting suggestions and techniques are provided.

Preventive Maintenance

Preventive Maintenance Procedures are contained in many units and relate specifically to the equipment presented in that particular unit. Technicians can provide some routine preventive maintenance service when on other types of service calls as well as when on strictly maintenance calls. The preventive maintenance procedures provide valuable information for the new or aspiring technician and homeowner, as well as for those technicians with experience.

HVAC Golden Rules

Golden Rules for the refrigeration and air-conditioning technician give advice and practical hints for developing good customer relations. These "golden rules" appear in appropriate units.

Service Technician Calls

In many units, practical examples of service technician calls are presented in a down-to-earth situational format. These are realistic service situations in which technicians may find themselves. In many instances the solution is provided in the text, and in others the reader must decide what the best solution should be. These solutions are provided in the *Instructor's Guide*.

Recovery/Recycling/Reclaiming Retrofitting

Discussions relating to *recovery, recycling, reclaiming, retrofitting,* or other *environmental issues* are highlighted in orange throughout the text. In addition, one complete unit on refrigerant management is included—Unit 9, "Refrigerant and Oil Chemistry and Management— Recovery, Recycling, Reclaiming, and Retrofitting."

Green Awareness

As previously mentioned, global warming stemming from the uncontrolled rate of greenhouse gas emissions is a major global environmental issue. Buildings are the major source of demand for energy and materials. Because of this, buildings are the major source of greenhouse gases that are the by-products of the energy and materials. At the time of this writing, there are approximately 5 million commercial buildings and 125 million housing units in the United States. Surprisingly, almost every one of their mechanical systems is obsolete. Discussions relating to the green awareness movement (for example, lowering energy costs, reducing operating and maintenance costs, increasing productivity, and decreasing the amount of generated pollution) will be highlighted in green throughout the text.

Summary

The *Summary* appears at the end of each unit prior to the Review Questions. It can be used to review the unit and to stimulate class discussion.

Review Questions

Review Questions follow the summary in each unit and can help to measure the student's knowledge of the unit. There are a variety of question types—multiple choice, true/false, short answer, short essay, and fill-in-the-blank.

Diagnostic Charts

Diagnostic Charts are included at the end of many units. These charts include material on troubleshooting and diagnosis.

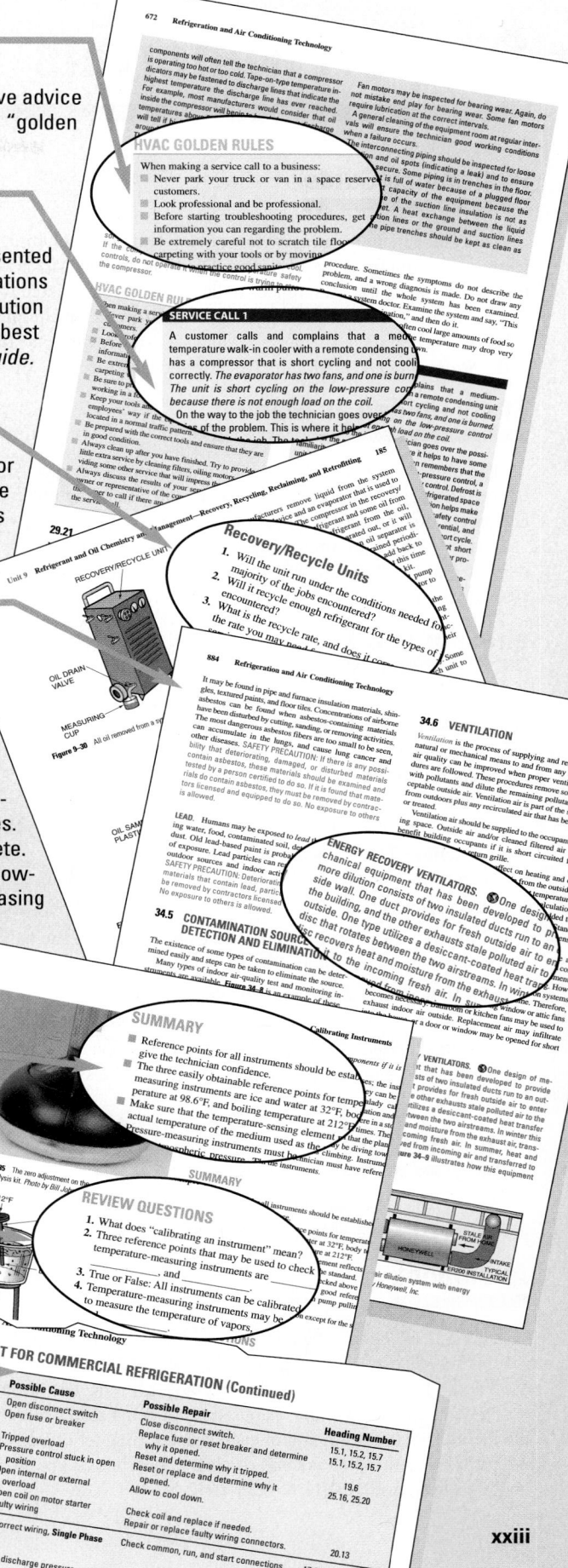

Support Materials

Instructor's Guide

This guide includes an overview of each text unit, including a summary description, a list of objectives, and important safety notes. Diagnoses for service technician calls that are not solved in the text are provided. References to lab exercises associated with each unit's study are included. "Special Notes to Instructors" specify how to create an equipment "problem" for students to resolve during certain lab exercises. Answers to the review questions in the text and to all questions in the *Study Guide/Lab Manual* (review and lab exercises) are provided. ISBN: 1-4283-1938-7.

Study Guide/Lab Manual

The *Study Guide/Lab Manual* includes a unit overview, key terms, and a unit review test. Each lab includes a general introduction to the lab, including objectives, text references, tools, materials, and safety precautions. A series of practical exercises is provided for the student to complete in a "hands-on" lab environment. Maintenance instructions are given for the workstation and tools. A reference to the "Special Note to Instructors" in the *Instructor's Guide* describes how to create a system "problem" to be solved in the lab. ISBN: 1-4283-1937-9.

e.resource CD

This educational resource creates a truly electronic classroom. It is a CD-ROM containing tools and instructional resources that enrich the classroom and make the instructor's preparation time shorter. The elements of the *e.resource* link directly to the text to provide a unified instructional system. With the *e.resource* you can spend your time teaching, not preparing to teach. ISBN: 1-4283-1939-5.

Features contained in the *e.resource* include the following:

- **Syllabus.** This is the standard course syllabus for this textbook, providing a summary outline for teaching HVACR.
- **Teaching Tips.** Teaching hints provide the basis for a lecture outline to present concepts and material. Key points and concepts can be graphically highlighted for student retention.
- **Lecture Outlines.** Each unit has key topics outlined, and the key concepts that should be covered for each topic.
- **PowerPoint Presentation.** These slides provide the basis for a lecture outline to present concepts and material. Key points and concepts can be graphically highlighted for student retention.
- **Optical Image Library.** This database of key images (all in full color) taken from the text can be used in lecture presentations, transparencies, tests and quizzes, and PowerPoint presentations.
- **Computerized Test Bank.** Over 1,000 questions of varying levels of difficulty are provided in true/false, multiple choice, fill-in-the-blank, and short answer formats so you can assess student comprehension. This versatile tool enables the instructor to manipulate the data to create original tests.

Video DVD Set

A six-DVD video set addressing over 120 topics covered in the text is available. Each DVD will contain four 20-minute videos. To order the six-DVD set, reference ISBN: 1-4180-7283-4.

Audiobook

This is a collection of audio files covering every chapter in *Refrigeration and Air Conditioning Technology, Sixth Edition.* The audio files are organized into "A" head groupings (comparable to songs), which allow content to be accessed within the chapter. Once downloaded, MP3 audio files can be accessed on portable MP3 players or on PCs with standard media programs.

Students can listen to the chapter content being read while they follow along and look at the illustrations. References to page numbers are included at the beginning of each chapter. Chapter objectives, boxed features, figure and photo captions, and end-of-chapter elements are included as well (but not end-of-chapter questions). The audio files will not replace the book, since the artwork and photos are essential and must be viewed. ISBN: 1-4283-1942-5.

About the Authors

Bill Whitman

Bill Whitman graduated from Keene State College in Keene, New Hampshire, with a bachelor's degree in Industrial Education. He received his master's degree in School Administration from St. Michael's College in Winooski, Vermont.

After instructing drafting courses for 3 years, Mr. Whitman became the Director of Vocational Education for the Burlington Public Schools in Burlington, Vermont, a position he held for 8 years. He spent 5 years as the Associate Director of Trident Technical College in Charleston, South Carolina. Mr. Whitman was the head of the Department of Industry for Central Piedmont Community College in Charlotte, North Carolina, for 18 years.

Bill Johnson

Bill Johnson graduated from Southern Polytechnic with an associate's degree in Gas Fuel Technology and Refrigeration. He worked for the State of North Carolina's Weights and Measures Department; Coosa Valley Vocational and Technical Institute in Rome, Georgia; and the Trane Company of North Carolina. He also owned and operated an air-conditioning, heating, and refrigeration business for 10 years. He has unlimited licenses for North Carolina in heating, air conditioning, and refrigeration. Mr. Johnson taught heating, air-conditioning, and refrigeration installation, service, and design for 15 years at Central Piedmont Community College in Charlotte, North Carolina, and was instrumental in standardizing the heating, air-conditioning, and refrigeration curriculum for the state community college system. He is a member of Refrigeration Service Engineers Society (RSES). He has a series of articles on the Web site for the *Air Conditioning, Heating, Refrigeration News.* These articles are service situation calls for technicians.

John Tomczyk

John Tomczyk received his associate's degree in Refrigeration, Heating and Air-Conditioning Technology from Ferris State University in Big Rapids, Michigan; his bachelor's degree in Mechanical Engineering from Michigan State University in East Lansing, Michigan; and his master's degree in Education from Ferris State University. Mr. Tomczyk has worked in refrigeration, heating, and air-conditioning service; project engineering; and technical writing consultation for both the academic and industrial fields. His technical articles have been featured in the *Refrigeration News, Service and Contracting Journal,* and *Engineered Systems Journal.* He writes monthly for the *Air Conditioning, Heating, Refrigeration News* and is coauthor of an EPA-approved *Technician Certification Program Manual.* Mr. Tomczyk also is the author of the book *Troubleshooting and Servicing Modern Air Conditioning and Refrigeration Systems.* He is currently a professor in the Refrigeration, Heating, and Air-Conditioning Technology program at Ferris State University with 23 years of teaching experience. Mr. Tomczyk is a member of Refrigeration Service Engineers Society (RSES).

Eugene Silberstein

Eugene Silberstein has been involved in all aspects of the HVACR industry over the past 26 years—from field technician and system designer to company owner, teacher, administrator, consultant, and author. Mr. Silberstein is the author of *Heat Pumps* and *Residential Construction Academy: HVAC.* Mr. Silberstein is presently the Director and Professor of the HVAC/R Associates of Applied Sciences program on the Michael J. Grant Campus of Suffolk County Community College in Brentwood, New York. Mr. Silberstein is a member of the American Society of Heating, Refrigeration, and Air Conditioning Engineers (ASHRAE) and the Refrigeration Service Engineers Society (RSES).

Acknowledgments

The authors thank the following individuals, companies, and universities for their valuable contributions, to this text:

James DeVoe, Senior Acquisitions Editor, for his work with the authors and the publisher to produce a workable text that is both economical and comprehensive. His perpetual energies and insistence for the best possible product have resulted in a quality text that is usable both in schools and in the field.

John Fisher, Senior Developmental Editor, for his work with the authors and publisher to ensure the accuracy of the details in this text. His professional, thorough, and enthusiastic handling of the text manuscript will provide the student with a well-presented, usable product.

Cheri Plasse, Art and Design Specialist, for a great job of making sure the pages, artwork, and photographs are well presented and easy to follow. Her professional skills and talents are appreciated.

The late **Ed Bottum, Sr.,** President of Refrigeration Research, Inc., in Brighton, Michigan, for supplying much of the history and many photographs for the time line found in the introduction of this text. Mr. Bottum's historical collection of refrigeration items and artifacts in Brighton, Michigan, has been designated a National Historic Site by the *American Society of Mechanical Engineers* (ASME).

Ferris State University, Big Rapids, Michigan, for permission to use their building and HVACR applications laboratories to take digital photographs. These digital photographs have certainly enhanced many units in this book.

Bill Litchy, Training Materials Manager, Scotsman Ice Systems, for his valued technical consultation and professional guidance in Unit 27, "Commercial Ice Machines," in this edition. Technical literature, photographs, and illustrations from the Scotsman Company have greatly enhanced this book.

Danny Moore, Director of Technical Support, Hoshizaki America, Inc., for his technical assistance in Unit 27, "Commercial Ice Machines." His published technical articles on the topics of water and ice quality and water filtration and treatment surely enhanced the quality of this unit.

Mitch Rens, Service Publications Manager, Manitowoc Ice Inc., for his valued technical assistance and professional consultation in Unit 27, "Commercial Ice Machines." Technical literature and photographs from Manitowoc have made the unit current and applicable.

Rex Ambs, Manager of GeoFurnace Heating and Cooling, LLC, and CoEnergies, LLC, in Traverse City, Michigan, for his assistance with the enhancements of Unit 44, "Geothermal Heat Pumps." He supplied detailed technical information and digital photographs on waterless, earth-coupled, closed-loop geothermal heat pump technology.

Jim Holstine, Manager of GeoFurnace Heating and Cooling, LLC, and CoEnergies, LLC, in Traverse City, Michigan, for his technical assistance in Unit 44, "Geothermal Heat Pumps."

Dennis Weston and **Tom Kiessel,** Managers of CoEnergies, LLC, in Traverse City, Michigan, for supplying many digital photographs of waterless, earth-coupled, closed-loop geothermal heat pump systems used in Unit 44, "Geothermal Heat Pumps."

Roger McDow, Senior Instructional Lab Facilitator, Central Piedmont Community College in Charlotte, North Carolina, for assisting the setup of the photographic sessions for all editions of this book. He organized and provided many tools and controls for photography, which have provided an invaluable educational experience for students.

Tony Young, Emerson Climate Technologies, Inc. for making his company's valuable technical literature and photographs available for use in this edition.

Modine Manufacturing Company, 1500 DeKoven Ave., Racine, Wisconsin, 53403, for the use of both their technical literature and photographs of their all-aluminum Micro Channel coil technology.

Dan Mason, Danfoss Turbocore Compressors, Inc., for making his company's technical literature and photographs available for use in this edition.

Sporlan Division, Parker Hannifin Corporation, for making their technical literature and photographs available for use in this edition.

John Levey of Oilheat Associates in Wantagh, New York, for his assistance in compiling and reviewing material for Unit 32, "Oil Heat," as well as providing new images for use in the text.

A special thanks to the family members and close relatives of the authors for their help and patience while this edition was being developed.

The contributions of the following reviewers of the 4th edition text are gratefully acknowledged:

George Gardianos, Lincoln Technical Institute, Mahwah, New Jersey

Raymond Norris, Central Missouri State University, Warrensburg, Missouri

Arthur Gibson, Erwin Technical Center, Tampa, Florida

Robert J. Honer, New England Institute of Technology at Palm Beach, West Palm Beach, Florida

Richard McDonald, Santa Fe Community College, Gainesville, Florida

Joe Moravek, Lee College, Baytown, Texas

Neal Broyles, Rolla Technical Institute, Rolla, Missouri

John B. Craig, Sheridan Vocational Technical Center, Hollywood, Florida

Rudy Hawkins, Kentucky Tech–Jefferson Campus, Louisville, Kentucky

Richard Dorssom, N.S. Hillyard AVTS, St. Joseph, Missouri

Robert Ortero, School of Cooperative Technical Education, New York, New York

John Sassen, Ranken Technical College, St. Louis, Missouri

Billy W. Truitt, Worcester Career and Technology Center, Newark, Maryland

George M. Cote, Erwin Technical Center

Marvin Maziarz, Niagara County Community College, Sanborn, New York

Greg Skudlarek, Minneapolis Community and Technical College, Minneapolis, Minnesota

Chris Rebecki, Baran Institute, Windsor and West Haven, Connecticut

Wayne Young, Midland College, Midland, Texas

Darren M. Jones, Meade County Area Technology Center, Brandenburg, Kentucky

Keith Fuhrman, Del Mar College West, Corpus Christi, Texas

Eugene Dickson, Indian River Community College, Fort Pierce, Florida

Mark Davis, New Castle School of Trades, Pulaski, Pennsylvania

Gene Silberstein, Apex Technical School, New York, New York

Phil Coulter, Durham College, Skills Training Center (Whitby Campus), Whitby, Ontario, Canada

Larry W. Wyatt, Advanced Tech Institute, Virginia Beach, Virginia

Bob Kish, Belmont Technical College, St. Clairsville, Ohio

John Pendleton, Central Texas College, Killeen, Texas

Richard Wirtz, Columbus State Community College, Columbus, Ohio

Brad Richmand, ACCA, Washington, District of Columbia

Greg Perakes, Tennessee Technology Center at Murfreesboro, Murfreesboro, Tennessee

Thomas Schafer, Macomb Community College, Warren, Michigan

Larry Penar, Refrigeration Service Engineers Society (RSES), Des Plaines, Illinois

Johnnie O. Bellamy, Eastfield College Continuing Education, Mesquite, Texas

John Corbitt, Eastfield College Continuing Education, Mesquite, Texas

Dick Shaw, ACCA, Washington, D.C.

Hugh Cole, Gwinnett Technical Institute, Lawrenceville, Georgia

Norman Christopherson, San Jose City College, San Jose, California

The contributions of the following reviewers of the 5th and 6th editions are gratefully acknowledged:

Barry Burkan, Apex Technical School, New York, New York

Victor Cafarchia, El Camino College, Torrance, California

Cecil W. Clark, American Trades Institute, Dallas, Texas

Bill Litchy, Training Materials Manager, Scotsman Ice Systems, Vernon Hills, Illinois

Danny Moore, Director of Technical Support, Hoshizaki America, Peachtree City, Georgia

Lawrence D. Priest, Tidewater Community College, Virginia Beach, Virginia

Mitch Rens, Service Publications Manager, Manitowoc Ice, Inc., Manitowoc, Wisconsin

Terry M. Rogers, Midlands Technical College, West Columbia, South Carolina

Russell Smith, Athens Technical College, Athens, Georgia

Avenue for Feedback

The authors would appreciate feedback from students and/or instructors. They can be reached through Delmar Cengage Learning in Clifton Park, New York, or through the following e-mail addresses:

William C. Whitman

William M. Johnson
billj@carolina.rr.com

John A. Tomczyk
tomczykj@tucker-usa.com

Eugene Silberstein
aquavirg29@aol.com

Introduction

Refrigeration, as used in this text, relates to the cooling of air or liquids, thus providing lower temperatures to preserve food, cool beverages, make ice, and for many other applications. Air conditioning includes space cooling, heating, humidification, dehumidification, air filtration, and ventilation to condition the air and improve indoor air quality.

History of Refrigeration and Air Conditioning (Cooling)

Most evidence indicates that the Chinese were the first to store natural ice and snow to cool wine and other delicacies. Evidence has been found that ice cellars were used as early as 1000 B.C. in China. Early Greeks and Romans also used underground pits to store ice, which they covered with straw, weeds, and other materials to provide insulation and preserve it over a long period.

Ancient people of Egypt and India cooled liquids in porous earthen jars. These jars were set in the dry night air, and the liquids seeping through the porous walls evaporated to provide the cooling. Some evidence indicates that ice was produced due to the vaporization of water through the walls of these jars, radiating heat into the night air.

In the eighteenth and nineteenth centuries, natural ice was cut from lakes and ponds in the winter in northern climates and stored underground for use in the warmer months. Some of this ice was packed in sawdust and transported to southern states to be used for preserving food. In the early twentieth century, it was still common in the northern states for ice to be cut from ponds and then stored in open ice houses. This ice was insulated with sawdust and delivered to homes and businesses.

In 1823, Michael Faraday discovered that certain gases under constant pressure will condense when they cool. In 1834, Jacob Perkins, an American, developed a closed refrigeration system using liquid expansion and then compression to produce cooling. He used ether as a refrigerant, a hand-operated compressor, a water-cooled condenser, and an evaporator in a liquid cooler. He was awarded a British patent for this system. In Great Britain during the same year, L. W. Wright produced ice by the expansion of compressed air.

In 1842, Florida physician John Gorrie placed a vessel of ammonia atop a stepladder, letting the ammonia drip, which vaporized and produced cooling. This basic principle is used in air conditioning and refrigeration today.

In 1856, Australian inventor James Harrison, an emigrant from Scotland, used an ether compressor. He used ammonia on an experimental basis but used ether in equipment that was previously constructed.

In 1858, a French inventor, Ferdinand Carré, developed a mechanical refrigerator using liquid ammonia in a compression machine. He made blocks of ice. Generally, mechanical refrigeration was first designed to produce ice.

In 1875, Raoul Pictet of Switzerland used sulfur dioxide as a refrigerant. Sulfur dioxide is also a lubricant and could be used as the refrigerant and as a lubricant for the compressor. This refrigerant was used frequently after 1890 and was used on British ships into the 1940s.

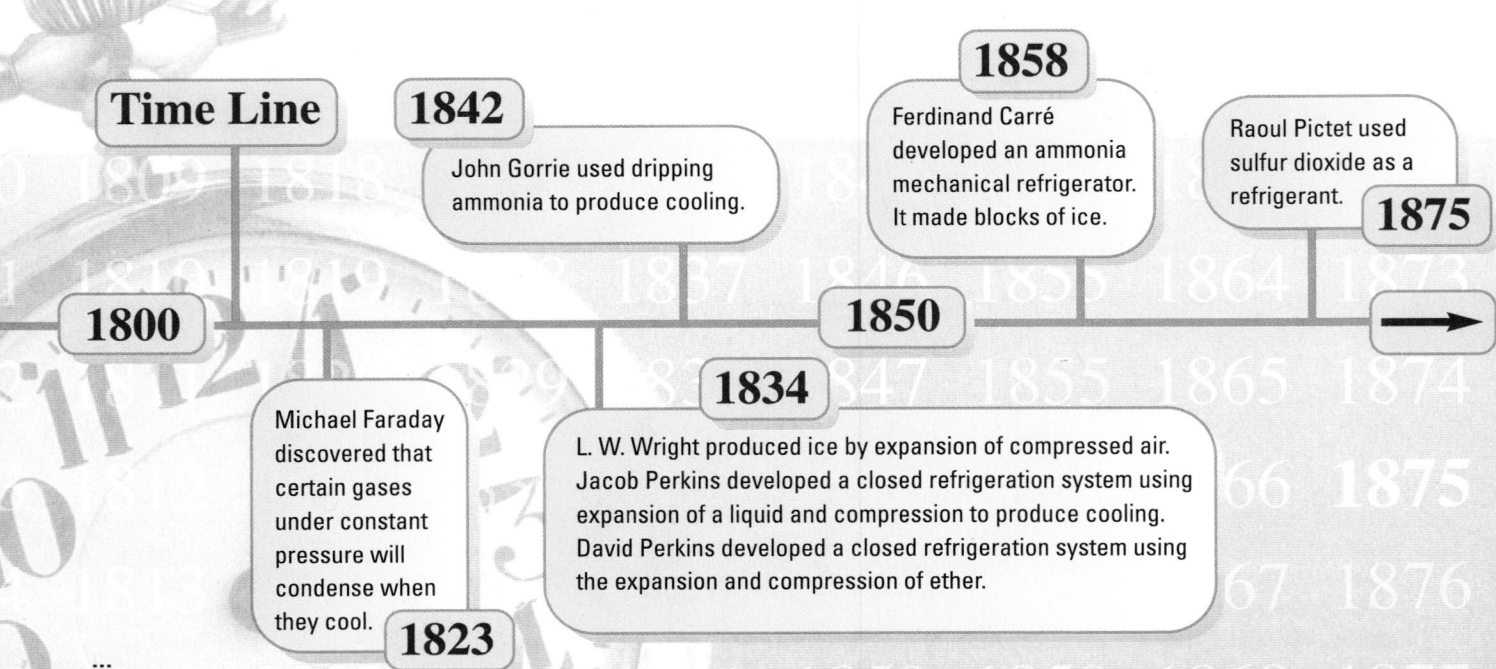

Time Line

1842 — John Gorrie used dripping ammonia to produce cooling.

1858 — Ferdinand Carré developed an ammonia mechanical refrigerator. It made blocks of ice.

1875 — Raoul Pictet used sulfur dioxide as a refrigerant.

1800

1850

1823 — Michael Faraday discovered that certain gases under constant pressure will condense when they cool.

1834 — L. W. Wright produced ice by expansion of compressed air. Jacob Perkins developed a closed refrigeration system using expansion of a liquid and compression to produce cooling. David Perkins developed a closed refrigeration system using the expansion and compression of ether.

In 1881, Gustavus Swift worked to develop refrigeration railcars that were used at that time, and in 1890, Michael Cudahy improved refrigeration railcars.

In 1894, the Audiffren-Singrün refrigeration machine was patented by a French priest and physicist, Father Marcel Audiffren. Its original design was for cooling liquids, such as wine, for the monks. Its compressor design is of the Scotch-yoke type. Most units used sulfur dioxide as the refrigerant.

In 1902, Willis Carrier, the "father of air conditioning," designed a humidity control to accompany a new air-cooling system. He pioneered modern air conditioning. In 1915, he, along with other engineers, founded Carrier Engineering, now known as Carrier Corporation.

In 1918, the name of Electro Automatic Refrigeration Corporation was changed to Kelvinator. This is also when the first of the Kelvinator household units were sold. The refrigerator was a remote-split type in which the condensing unit was installed in the basement and connected to an evaporator in a converted icebox in the kitchen.

Guardian Refrigerator Company developed a refrigerator they called the "Guardian." General Motors purchased Guardian in 1919 and developed the refrigerator they named Frigidaire. In 1929, refrigerator sales topped 800,000. The average price fell from $600 in 1920 to $169 in 1939.

In 1923, Nizer introduced a water-cooled compressor and condensing unit for ice cream cabinets. This unit was considered the first ice cream unit for the market. Nizer soon merged into the Kelvinator Company.

In 1923–1926, Savage Arms were among the first automatically controlled commercial units to appear. The Savage Arms compressor had no seals, no pistons, and no internal moving parts. A mercury column compressed the refrigerant gas as the entire unit rotated. The compressor was practically noiseless.

In 1928, Paul Crosley introduced an absorption-type refrigeration machine so that people could have refrigeration in rural areas where there was no electricity. Ammonia and water, charged by generating the system over a kerosene burner, could lower the inside temperature to 43°F or less. Ice cubes actually could be made for a period of about 36 hours, depending on the room temperature.

Many different refrigerants have been developed over the years. The refrigerant R-12, a chlorofluorocarbon (CFC), was developed in 1931 by Thomas Midgley of Ethyl Corporation and C. F. Kettering of General Motors. It was produced by DuPont. By the 1930s refrigeration was well on its way to being used extensively in American homes and commercial establishments.

In 1939, the Copeland Company introduced the first successful semi-hermetic (Copelametic) field-serviceable compressor. Three engineering changes made these compressors successful:

1. Cloth-insulated motor windings were replaced with Glyptal insulation.
2. Neoprene insulation replaced porcelain enamel in the electric terminals.
3. Valves were redesigned to improve efficiency.

In 1974, two professors from the University of California, Sherwood Rowland and Mario Molina, presented the "ozone theory." This hypothesis stated that released CFC refrigerants were depleting the earth's protective ozone layer. Scientists conducted high-altitude studies and concluded that CFCs were linked to ozone depletion.

Representatives from the United States, Canada, and more than 30 other countries met in Montreal, Canada, in September, 1987, to try to solve the problems of released refrigerants and the effect they had on ozone depletion. This meeting was known as the Montreal Protocol. This Protocol was ratified by 100 nations in 1989 and mandated a global production freeze on CFCs that froze their production levels back to 1986 levels. The Protocol also froze production of HCFCs to 1986 levels, which was to start in 1992. In addition, the Protocol placed a tax-rate

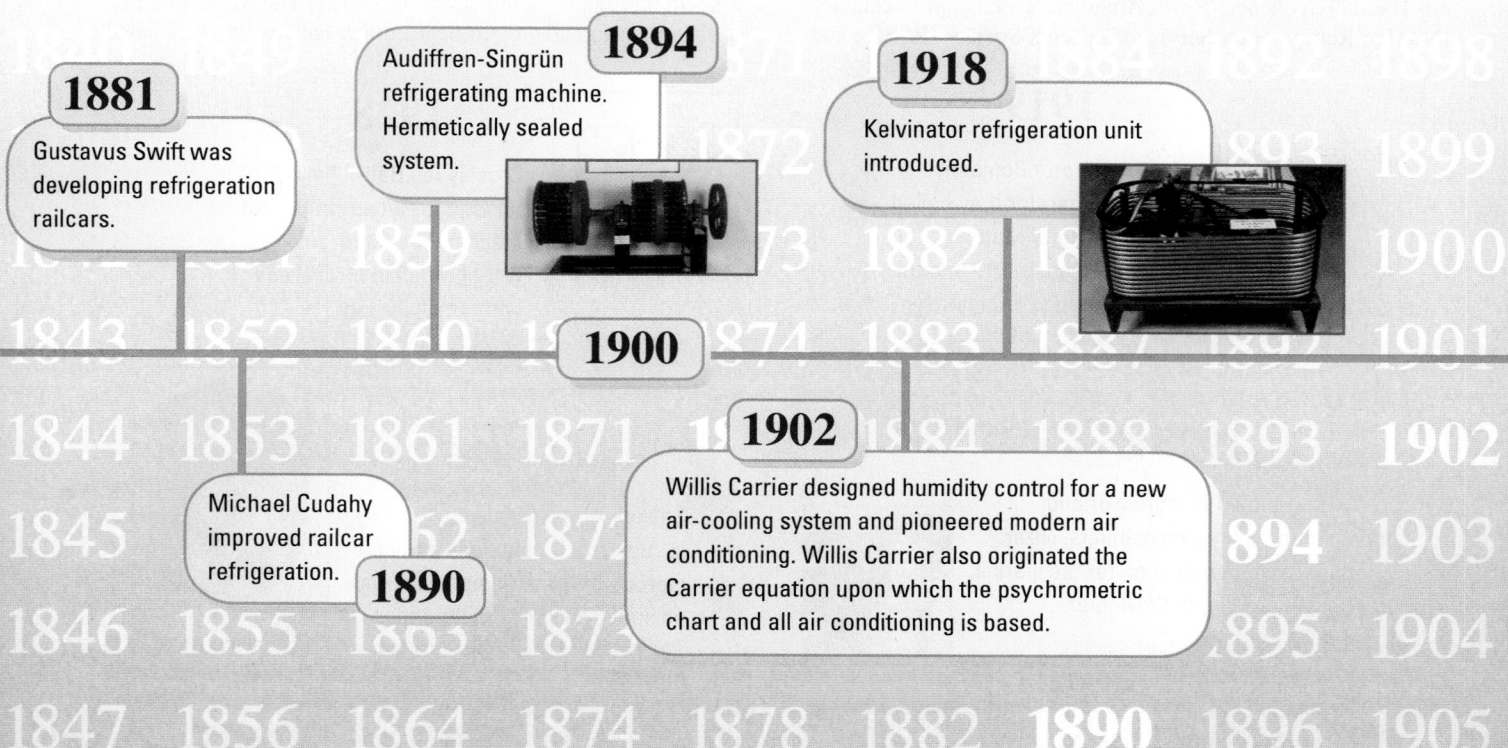

1881 Gustavus Swift was developing refrigeration railcars.

1894 Audiffren-Singrün refrigerating machine. Hermetically sealed system.

1918 Kelvinator refrigeration unit introduced.

1900

1890 Michael Cudahy improved railcar refrigeration.

1902 Willis Carrier designed humidity control for a new air-cooling system and pioneered modern air conditioning. Willis Carrier also originated the Carrier equation upon which the psychrometric chart and all air conditioning is based.

schedule on CFC refrigerants. As research on ozone depletion continues today, reassessments and updates to the Montreal Protocol also continue. At the time of writing this edition, the most current updates are as follows:

- 1990 (November)—President George H. W. Bush signed the Clean Air Act amendments which initiated production freezes and bans on certain refrigerants.
- 1992 (July)—The EPA made it against the law to intentionally vent CFC and HCFC refrigerants into the atmosphere.
- 1993—The EPA mandated the recycling of CFC and HCFC refrigerants.
- 1994 (November)—The EPA mandated a technician certification program deadline. Working HVACR technicians had to be EPA-certified by this date.
- 1995 (November)—The EPA made it against the law to intentionally vent alternative refrigerants (HFCs and all refrigerant blends) into the atmosphere.
- 1996—The EPA made it against the law to manufacture CFC refrigerants.
- 1996—The EPA put into place a gradual HCFC production phaseout schedule, which will totally phase out the production of HCFC refrigerants by the year 2030.
- 1998 (June)—The EPA proposed new regulations on recovery/recycling standards, equipment leak rates, and alternative refrigerants.

In 1997, the Kyoto Protocol was introduced, with the objective of reducing worldwide global-warming gas emissions. The greenhouse effect, or global warming, had become a major environmental issue.

From 1997–2000, voluntary HVACR technician certification became a major focus.

From 1998–2008, the AC&R Safety Coalition; Air Conditioning and Refrigeration Institute (ARI); Heating, Air Conditioning, and Refrigeration Distributors International (HARDI); Carbon Monoxide Safety Association (COSA); Green Mechanical Council; HVAC Excellence; North American Technician Excellence (NATE); Refrigeration Service Engineers Society (RSES); and the United Association of Journeymen and Apprentices (UA) became important players in voluntary HVACR technician certification and home study examinations.

R-410A, an efficient and chlorine-free HFC-based refrigerant blend for residential and light-commercial air-conditioning applications is used with the scroll compressor for greater efficiencies.

Today, every central split cooling system manufactured in the United States must have a Seasonal Energy Efficiency Ratio (SEER) rating of at least 13. This energy requirement was mandated by federal law as of January 23, 2006.

In 2007, global warming became a major environmental issue. A scoring system was designed to help engineers, contractors, and consumers know the "green value" of each mechanical installation. The "green value" encompasses the system's energy efficiency, pollution output, and sustainability. Green buildings and green mechanical systems are becoming increasingly popular in today's world as a way to curb global warming.

Green Awareness

As mentioned previously, global warming stemming from the uncontrolled rate of greenhouse gas emissions is a major global environmental issue. Most of the sun's energy reaches the earth as visible light. After passing through the atmosphere, part of this energy is absorbed by the earth's surface and is converted into heat energy. The earth, warmed by the sun, radiates heat energy back into the atmosphere toward space. Naturally occurring gases and lower atmospheric pollutants such as CFCs, HCFCs, HFCs, carbon dioxide, carbon monoxide, water vapor, and many other chemicals absorb, reflect, and/or refract the earth's infrared radiation and prevent it from escaping the lower atmosphere. Carbon dioxide, occurring mainly from the burning of fossil fuels, is the major global-warming gas today. Humans are chiefly responsible for many of these greenhouse gases. This process slows the earth's heat loss, making the earth's surface warmer than it would be if this heat energy had passed unobstructed through the atmosphere into space. The warmer earth's surface then radiates more heat until a balance is established

1919

Guardian Refrigerator Company developed a refrigerator they called the Guardian. General Motors purchased Guardian in 1918 and the name was changed to Frigidaire.

1928

Crosley Icy Ball—An absorption-type refrigerator machine provided refrigeration in rural areas without electricity.

1910

1923

Nizer water-cooled compressor and condenser for ice cream cabinets. Nizer made the first ice cream unit for the market. Nizer soon merged into Kelvinator.

1926

Savage Arms ice cream unit introduced. It contained a unique mercury column compressor with no seals, pistons, or lubrication.

between incoming and outgoing energy. This warming process is called *global warming* or the *greenhouse effect.*

Over 70% of the earth's fresh water supply is either in ice cap or glacier form. Scientists are concerned that these ice caps or glaciers will melt if the average earth temperature rises too much, thereby causing increased water levels. Scientific consensus is that we must limit the rise in global temperatures to less than 2 degrees centigrade above pre-industrial levels to avoid disastrous impacts. At a rise of 2 degrees centigrade, millions of people will likely be displaced from their homes because of rising water levels. Food production will decline, rivers will become too warm to support fish, coral reefs will die, snow packs will decrease and threaten water supplies, weather will become unpredictable and extreme, and many plant and animal species will die and become extinct.

Nineteen of the hottest 20 years on record have occurred since 1980. Atmospheric carbon dioxide levels are now at their highest. Half of the world's oil is gone and other natural resources are dwindling. The average American uses 142 gallons of water per day, and in some regions of the country water is drying up. Because of this, slowing the growth rate of greenhouse gas emissions and then reversing it has become a global effort.

Buildings are the major source of demand for energy and materials. Because of this, buildings are the major source of greenhouse gases that are the by-products of the energy and materials. At the time of this writing, there are approximately 5 million commercial buildings and 125 million housing units in the United States. Surprisingly, almost every one of their mechanical systems is obsolete. It is these global-warming scares, the rising price of fuels, the scarcity of clean water, and the ever-growing waste stream that demands improvements in our homes and businesses today. Trained contractors, with the help of the government, installers, builders, manufacturers, and educators, must renovate and improve the efficiency of these buildings and mechanical systems.

In the United States, buildings account for
- 36% of total energy used.
- 65% of electrical consumption.
- 30% of greenhouse gas emissions.
- 30% of raw materials used.
- 30% of waste output (136 millions tons annually).
- 12% of potable water consumption.

Organizations like the Green Mechanical Council (Green-Mech) and the United States Green Building Council (USGBC) are setting goals to use fewer fossil fuels in existing and new buildings. Some of these goals are listed below:
- All new buildings, developments, and major renovation projects must be designed to use one-half of the fossil fuel energy they would typically consume.
- The fossil fuel reduction standard for all new buildings must be increased to
  - 60% in 2010.
  - 70% in 2015.
  - 80% in 2020.
  - 90% in 2025.
- By 2030, new buildings must be carbon-neutral. Carbon-neutral means that they cannot use any fossil-fuel greenhouse-gas-emitting energy to operate.
- Joint efforts must be made to change existing building standards and codes to reflect these targets.

Builders can accomplish these goals by choosing proper siting, building forms, glass properties and locations, and material selection, and by incorporating natural heating, cooling, ventilating, and lighting strategies. Renewable energy sources such as solar, wind, biomass, and other carbon-free sources can operate equipment within the building.

The Leadership in Energy and Environmental Design (LEED) is a voluntary national rating system for developing high-performance, sustainable buildings. It is referred to as the LEED Green Building Rating System. It was established by the

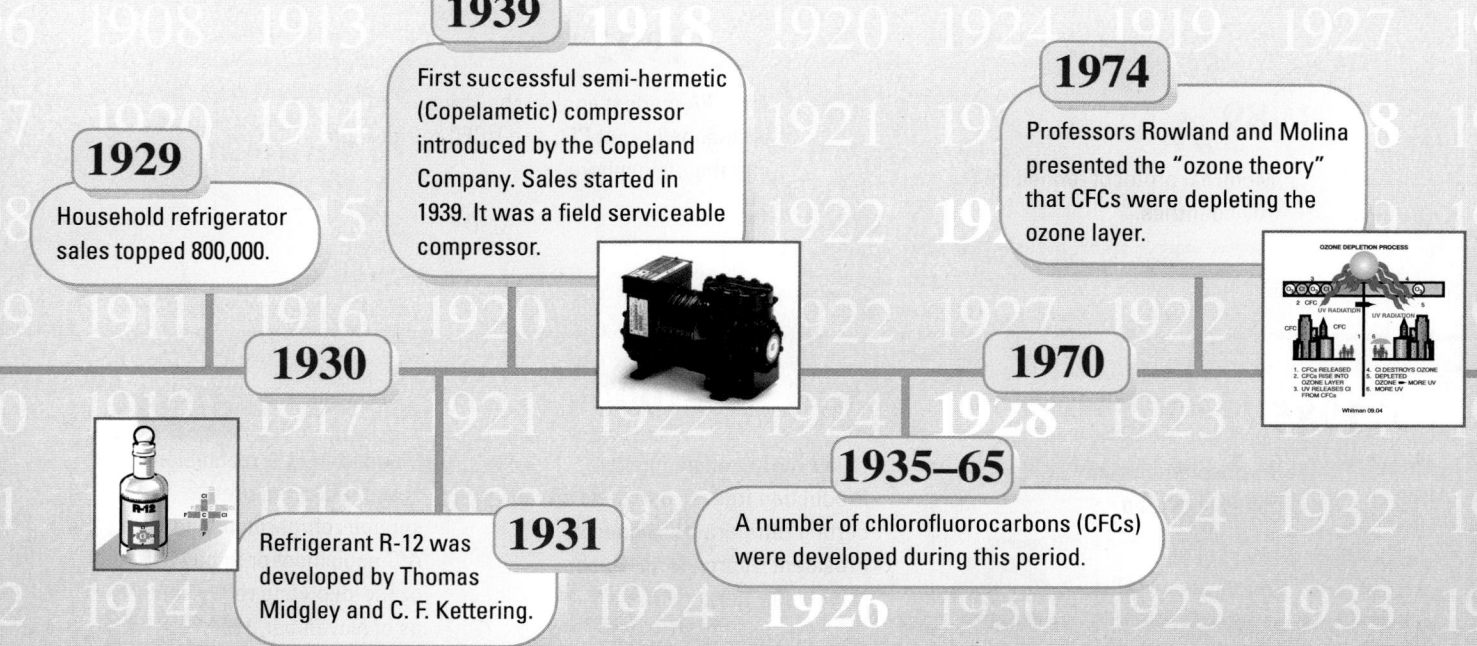

1939

First successful semi-hermetic (Copelametic) compressor introduced by the Copeland Company. Sales started in 1939. It was a field serviceable compressor.

1974

Professors Rowland and Molina presented the "ozone theory" that CFCs were depleting the ozone layer.

1929

Household refrigerator sales topped 800,000.

1930

1970

1931

Refrigerant R-12 was developed by Thomas Midgley and C. F. Kettering.

1935–65

A number of chlorofluorocarbons (CFCs) were developed during this period.

USGBC in 1999 and is widely recognized as the third-party verification system and guideline for measuring what constitutes a green building.

A LEED-certified building means that it has achieved at least a minimum standard as judged in six categories:

1. Sustainable sites
2. Water efficiency
3. Energy and atmosphere (HVAC systems)
4. Materials and resources
5. Indoor environmental quality
6. Innovation and design process

Points are awarded in each individual category, depending on how the building meets the category's requirements. In order to be LEED certified, a building must receive a minimum of 26 points out of the 69 points available. The energy and atmosphere category deals with the HVAC systems and consists of one-third of the total LEED points. This category addresses the amount of energy the HVAC system consumes, the environmental impact of generating this energy, and the ozone depletion potential of the refrigerant used in the HVAC system.

There are four levels of LEED certification. They are as follows:

1. Certified (26–32 points)
2. Silver (33–38 points)
3. Gold (39–51 points)
4. Platinum (52–69 points)

The USGBC membership, which is composed of every sector of the building industry and consists of over 9000 organizations, developed and continue to refine LEED. LEED addresses all building types including new construction, commercial interiors, core and shell, operation and maintenance, homes, neighborhoods, campuses, schools, health care, laboratories, and lodging. LEED promotes expertise in green building by offering project certification, professional accreditation, and training. LEED emphasizes state-of-the-art strategies for sustainable site development, water savings, energy efficiency, material selection, and indoor environmental quality. According to the United Nations World Commission on Environment and Development, a sustainable design "meets the needs of the present without compromising the ability of future generations to meet their own needs." Companies looking to "go green," or incorporate sustainable design into their facilities, want products that help them lower energy costs, reduce operating and maintenance costs, increase productivity, and decrease the amount of pollution that is generated. Sustainable buildings typically have lower annual costs for energy, water, maintenance/repair, and other operating expenses.

The green awareness movement isn't just a temporary "buzz word" that will fade away with time. It is a movement that will be rapidly gaining momentum in the coming years. If contractors want to remain competitive, they must obtain the necessary training with regard to green building and LEED certification.

History of Home and Commercial Heating

Human beings' first exposure to fire was probably when lightning or another natural occurrence, such as a volcanic eruption, ignited forests or grasslands. After overcoming the fear of fire, early humans found that placing a controlled fire in a cave or other shelter could create a more comfortable living environment. Fire was often carried from one place to another. Smoke was always a problem, however, and methods needed to be developed for venting it outside. An example seen in later years was that of Native Americans venting smoke through holes at the peak of their tepees, and some of these vents were constructed with a vane that could be adjusted to prevent downdrafts.

Fireplaces were common in Europe and North America and were vented through chimneys. Early stoves were found to be more efficient than fireplaces. These early stoves were constructed of a type of firebrick, ceramic materials, or iron.

In the mid-eighteenth century, a jacket for the stove and a duct system were developed. The stove could then be located at the lowest place in a structure; the air in the jacket around the

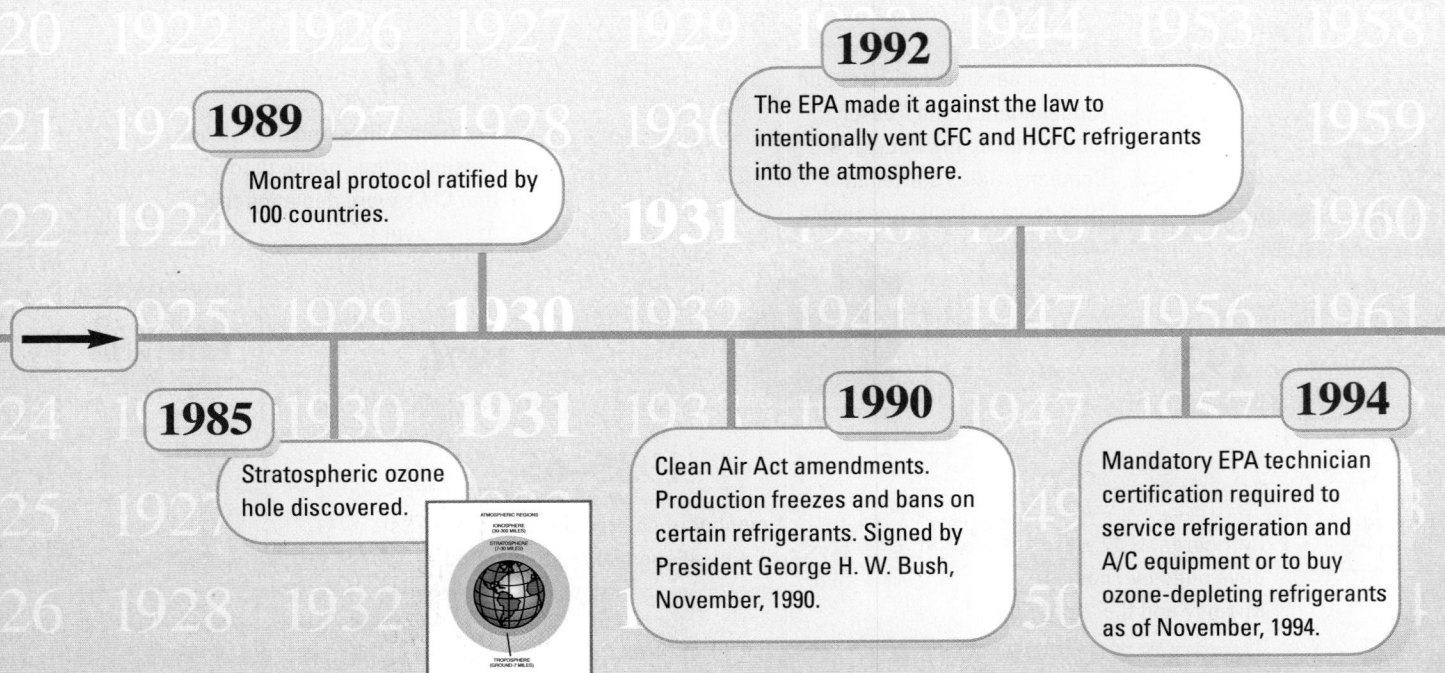

1989
Montreal protocol ratified by 100 countries.

1992
The EPA made it against the law to intentionally vent CFC and HCFC refrigerants into the atmosphere.

1985
Stratospheric ozone hole discovered.

1990
Clean Air Act amendments. Production freezes and bans on certain refrigerants. Signed by President George H. W. Bush, November, 1990.

1994
Mandatory EPA technician certification required to service refrigeration and A/C equipment or to buy ozone-depleting refrigerants as of November, 1994.

stove was heated and would rise through the duct system and grates into the living area. This was the beginning of the development of circulating warm air heating systems.

Boilers that heated water were developed, and this water was circulated through pipes in duct systems. The water heated the air around the pipes. This heated air passed into the rooms to be heated. Radiators were then developed. The heated water circulated by convection through the pipes to the radiators, and heat was passed into the room by radiation. These early systems were forerunners of modern hydronic heating systems.

Career Opportunities

With the advancement of technology that is being spurred on by the need for increased energy efficiencies, the HVACR industry is rapidly changing. The career opportunities available in HVACR for those who have acquired formal technical training coupled with field experience is unlimited. Schools that provide excellent technical training in HVACR are becoming easier to identify through HVACR program accreditation. As new equipment becomes more technically challenging and the existing workforce continues to age, the available employment positions will continue to outnumber applicants for the foreseeable future. This shortfall in available competent HVACR service technicians is being addressed through the cooperative efforts of educational institutions, labor unions, employers, and manufacturers. Many organizations offer apprenticeship opportunities that can lead to high-income positions. Manufacturers are also teaming up with select educational institutions across North America to help develop the next generation of HVACR technician.

People in the United States expect to be comfortable. In cold weather, they expect to be able to go inside and be warm, and in the warmer climates, they expect to be able to go inside and be cool. They expect beverages to be cold when they want them to be and their food to be properly preserved.

Many buildings are constructed so that the quality of the air must be controlled by specialized equipment. The condition of the air must be controlled in many manufacturing processes. Heating and air-conditioning systems control the temperature, humidity, and total air quality in residential, commercial, industrial, and other types of buildings. Refrigeration systems are used to store and transport food, medicine, and other perishable items. Refrigeration and air-conditioning technicians design, sell, install, or maintain these systems. Many contractors and service companies specialize in commercial refrigeration. The installation and service technicians employed by these companies install and service refrigeration equipment in supermarkets, restaurants, hotels/motels, flower shops, and many other types of retail and wholesale commercial businesses.

Other contractors and service companies may specialize in air conditioning. Many specialize in residential-only or commercial-only installation and service; others may install and service both residential and commercial equipment up to a specific size. Air conditioning may include cooling, heating, humidifying, dehumidifying, or air cleaning. The heating equipment may include gas, oil, electric, or heat pumps. The number of each type of installation will vary from one part of the country to another, depending on the climate and availability of the heat source. The heating equipment may be a space-heating (air distribution) type or a hydronic furnace. The hydronic furnace heats water and pumps it to the space to be heated, where one of many types of heat exchangers transfers the heat to the air.

Technicians may specialize in installation or service of this equipment, or they may be involved with both. Other technicians may design installations or work in the sales area. Sales representatives may be in the field selling equipment to contractors,

1995

Unlawful to intentionally vent alternative refrigerants (HFCs and all refrigerant blends) into the atmosphere as of November, 1995.

1997

Kyoto Protocol introduced, intended to reduce worldwide global warming gases. Global warming has become a major environmental issue.

1998–2005

R-410A, an efficient and chlorine-free HFC-based refrigerant blend for residential and light-commercial air-conditioning applications, is used with the scroll compressor for greater efficiencies.

1995

2006

"Minimum 13 SEER" required for 1 1/2- to 5-ton unitary equipment and split/ packaged air conditioners and heat pumps.

1996

Gradual HCFC production phaseout schedule with the total phaseout of HCFC refrigerants in the year 2030. Phaseout of CFC refrigerants.

1997–2000

Voluntary HVACR technician certification became a major focus.

2007

"Green awareness"—Green mechanical systems and green buildings become increasingly popular as a way to curb global warming and conserve energy.

businesses, or homeowners; others may work in wholesale supply stores. Other technicians may represent manufacturers, selling equipment to wholesalers and large contractors.

Many opportunities exist for technicians to be employed in the industry or by companies owning large buildings. These technicians may be responsible for the operation of air-conditioning equipment, or they may be involved in the service of this equipment.

Opportunities also exist for employment in servicing household refrigeration and room air conditioners. This would include refrigerators, freezers, and window or through-the-wall air conditioners.

Opportunities are available for employment in a field often called transport refrigeration. This includes servicing refrigeration equipment on trucks or on large containers hauled by trucks and ships.

Most modern houses and other buildings are constructed to keep outside air from entering, except through planned ventilation. Consequently, the same air is circulated through the building many times. The quality of this air may eventually cause a health problem for people spending many hours in the building. This indoor air quality (IAQ) presents another opportunity for employment in the air-conditioning field. Technicians clean filters and ducts, take air measurements, check ventilation systems, and perform other tasks to help ensure healthy air quality.

Other technicians work for manufacturers of air-conditioning equipment. These technicians may be employed to assist in equipment design, in the manufacturing process, or as equipment salespersons.

Following is a list of many career opportunities in the HVACR field:

- Field service technician
- Service manager
- Field supervisor
- Field installer
- Journeyman
- Project manager
- Job foreman
- Application engineer
- Controls technician
- Draftsperson
- Contractor
- Lab technician
- Inspector
- Facilities technician
- Instructor
- Educational administrator
- Inside/outside sales
- Sales manager
- Research and development
- Estimator

Technician Certification Programs

History. Even though mandatory technician certification programs are in place today, the EPA originally did not consider them as its lead option. As a matter of fact, the EPA initially thought private incentives would ensure that technicians were properly trained in refrigerant recycling and recovery. The EPA also stated that it would play an important role through a voluntary technician certification program by recognizing those who provide and participate in voluntary technician training programs that meet certain minimum standards. The EPA also thought that a mandatory certification program would be an administrative burden for it. The EPA then requested public comments on a mandatory versus voluntary technician certification program. More than 18,000 comments were in favor of a mandatory program, and only 142 were in favor of a voluntary program. Most of the 18,000 in favor of the mandatory certification program were major trade organizations and technicians themselves. Manufacturers of recovery and recycling equipment, along with environmental organizations, also supported mandatory certifications. They believed it would increase compliance with venting, recovery, and recycling laws and the general safe handling of refrigerants. The following were reasons given by those favoring mandatory technician certification:

- Improve refrigerant leak detection techniques
- Promote awareness of problems relating to venting, recovery, and recycling of refrigerants
- Improve productivity and cost savings through proper maintenance practices
- Ensure environmentally safe service practices
- Gain more consumer trust
- Receive more liability protection
- Ensure that equipment is properly maintained
- Educate technicians on how to effectively contain and conserve refrigerants
- Create uniform and enforceable laws
- Foster more fair competition in the regulated community

With these comments in mind, the EPA decided that mandatory technician certification would increase fairness by ensuring that all technicians are complying with today's rules. The EPA also said that a mandatory certification program would also enhance the EPA's ability to enforce today's rules by providing a tool to use against intentional noncompliance: the ability to revoke the technician's certification. The EPA then created a mandatory technician certification program that mandated that all technicians be certified after November 14, 1994.

All technicians now must pass an examination administered by an approved EPA testing organization in the private sector in order to purchase refrigerant and to work on equipment that contains refrigerant. *Technicians* for mandatory certification are defined as installers, contractor employees, in-house service personnel, and anyone else who installs, maintains, or repairs equipment that might reasonably have the opportunity to release CFCs or HCFCs into the atmosphere. The EPA created three separate technician certification types:

- Small appliances
- High- and very high-pressure appliances
- Low-pressure appliances

Persons who successfully pass a *core of questions* on stratospheric ozone protection and legislation and also pass one of the three certification types will be certified in that type. If all three certification types are passed, a person will be *universally* certified. To this date, the EPA is not requiring recertification. However, it will be the technicians' responsibility to keep updated on new technologies and governmental rule changes. By creating certification types, the EPA allowed technicians to be tested on information concerning equipment and service practices that the technicians primarily service and maintain.

Although training programs are beneficial, participation in a training program is not required by today's rule. If training programs are requested by technicians, they will be administered by the private sector to create price-competitive training programs. Many national educational and trade organizations such as the AC&R Safety Coalition; Air Conditioning and Refrigeration Institute (ARI); Air Conditioning Contractors of America (ACCA); Heating, Air Conditioning, and Refrigeration Distributors International (HARDI); Carbon Monoxide Safety Association (COSA); Educational Standards Corporation (ESCO); Environmental Protection Agency (EPA); Ferris State University (FSU); Green Mechanical Council (GreenMech); HVAC Excellence; North American Technician Excellence (NATE); Refrigeration Service Engineers Society (RSES); and the United Association of Journeymen and Apprentices (UA) have developed training and/or testing programs. These programs are specifically intended to help technicians comply with the July 1, 1992, refrigerant venting law. Unit 9 of this text, "Refrigerant and Oil Chemistry and Management—Recovery, Recycling, Reclaiming, and Retrofitting," will give more detailed information on the EPA's mandatory technician certification program, including details on the specific types of certification tests and specifications. For a complete list of EPA approved certifying organizations, contact the EPA hotline at 1-800-296-1996.

Certification Programs. Technician certification programs can be divided into two categories. They are:

- ***Mandatory technician certification programs***
- ***Voluntary technician certification programs***

Mandatory technician certification programs are covered in the preceding paragraphs and in Unit 9. Voluntary technician certification programs are becoming popular because they are industry led and are much more comprehensive in nature when compared to mandatory certification programs. They give technicians an educational opportunity from the beginning to the end of their careers. These programs allow technicians to become recognized at their level of expertise. They also allow technicians to excel to higher levels of competence. Their diverse nature allows for almost every avenue of the industry to be covered. Voluntary certification testing is based on the courses taken for each level, with an outline and roadmap on what material will be covered on the test and where to find the material.

Why Technicians Should Become Certified. As mentioned earlier, mandatory technician certification legally allows the technician to purchase ozone-depleting refrigerants and work on equipment that contains refrigerant. Some advantages of having both mandatory and voluntary technician certifications are that

- customers tend to ask for certified technicians because of their reliability and good workmanship.
- equipment manufacturers develop faith in certified technicians and have a sense of well-being when they know the job has been accomplished by a certified technician.
- higher standards are set on the job by certified technicians, giving them more respect, recognition, trust, higher pay, and a higher quality of life in the long run.
- employers would rather hire a certified technician, because they know certified technicians care more about their reputation, customer relations, and overall professionalism.
- certification gives the technician a status symbol for other technicians to work up to.
- certified technicians have proven technical proficiencies with measured capabilities.

National Skill Standards

The National Skill Standards (NSS), as interpreted by the Vocational-Technical Education Consortium of States (VTECS) for Heating, Air Conditioning, and Refrigeration Technicians, were funded by the U.S. Department of Education as part of 22 projects from the National Skill Standards Board from 1992 to 1998. The NSS were created by a joint effort of committees composed of heating, air-conditioning, and refrigeration industry professionals. These skill standards not only help technicians identify the skills and knowledge needed for their occupation but also assess their weaknesses and/or needs for additional training.

Skill standards are often described as workplace behaviors, technical skills, and the general body of knowledge required of technicians to be successful, productive, and competitive in today's workforce. As HVACR manufacturers increase the efficiency and sophistication of their equipment, technicians require additional updated information as well as a sound technical skills base to maintain, install, and service this equipment. An increased number of environmental regulations for more energy efficient and environmentally friendly HVACR equipment have also created a new knowledge and skills base for technicians to learn and use on their jobs.

Although it is difficult to provide all users of this text with information on the vast array of issues covered in the NSS, the authors have made every effort to do so. Hopefully, both private and public institutions as well as industry will use our comprehensive book to provide both students and workers with the competencies needed for successful employment and advancement in the ever-changing and growing technical HVACR field.

The NSS are divided into three main areas with subdivisions as follows:

- **Core Knowledge**
  - *Communications*
  - *Mathematics*
  - *Science*
- **Occupational-Specific Skills**
  - *Core skills*
  - *Occupational-specific skills*
- **Workplace Behaviors**
  - *Ethics*
  - *Environment*
  - *Communications*
  - *Professionalism*
  - *Problem solving*

The *core skills* consist of:
- Safety and environment
- Electrical principles
- Electric motors
- Controls
- Refrigeration principles and practices
- Heating principles and practices
- Air-conditioning principles and practices
- Piping principles and practices

The *occupational-specific skills* consist of:
- Residential and light-commercial heating
- Residential and light-commercial air conditioning
- Residential and light-commercial heat pumps
- Commercial conditioned-air systems
- Commercial refrigeration

For more detailed information on the NSS, go to http://www. nssb.org.

Customer Relations

Customer relations are extremely important to a service business and consequently to a service technician. Without customers there will be no business and no income. The technician is a major factor in acquiring and keeping customers. This is true whether work is performed at a residence, an office, a restaurant, or a store, or whether the technician is an inside or outside salesperson for a distributor or contractor. The HVACR business and technician are dependent on the customers. All technicians should be concerned with the quality of their work because customers have the right to insist on quality. If they have had a previous unsatisfactory experience, customers may have some doubt as to whether they will get the quality service for which they are paying. As professionals, technicians should strive to provide the best workmanship possible. Quality work will prove beneficial to the technician, to the company, and to the consumer. Customers depend on the technician for their comfort and air quality at home and at the office.

First Impressions. The impression the technician makes on the customer is very important, and the **first impression** is the most important. The first impression begins with the technician **arriving on time.** Most customers feel that their time is valuable. If the technician is going to be delayed, the customer should be called and given an explanation. An appointment should be scheduled for either later that day or another time convenient for the customer. The customer affected by a delay should be given priority in scheduling a makeup appointment. If the service call is an emergency, all efforts should be made to arrive as soon as possible.

When arriving, **do not park in or block the customer's driveway** unless necessary. If carrying equipment or having to make several trips to the vehicle, ask permission to park in the driveway. The customer may suggest another location. **Ensure that the service vehicle is kept in a neat, clean, and orderly manner.** This will help to make a good impression and provide better working conditions for the technician.

Remember the customer's name and use it frequently, preferably with Ms., Mrs., or Mr. *Sir* or *Ma'am* may also be used when appropriate. When meeting a customer, be prepared to **shake hands.** In many cases it may be appropriate to let the customer initiate the handshake. Your handshake should be firm and accompanied with a smile. A handshake that is too limp may give the impression of weakness; one that is too strong may indicate an overbearing type of person. Not all people like to shake hands. The technician should be friendly and always have a smile.

Appearance. Another major factor in first impressions and maintaining good customer relations is **appearance.** Included is the following:

Hair—brushed or combed, neatly trimmed. Male technicians should be clean shaven or have a neatly trimmed beard or mustache. Female technicians with longer hair may wish to contain it in a ponytail.

Clothing—neat and clean. For most uniforms, ensure that the shirttail is tucked in. A clean and neat uniform will help to make the appropriate presentation. If you have an ID badge, wear it in plain sight.

Personal hygiene—Cleanliness is important. Hands should be washed and clean. A shower before going to bed or before going to work should be a regular habit. Your appearance and personal hygiene are major indicators of your personality and the quality of work you offer.

After arriving at the customer's address take a minute or two to get organized. You may have a clipboard with material to organize and review. Think about what you are going to say and do when meeting the customer. Do not flip a cigarette butt to the ground outside the truck or on the way to the house or other location. There should be **no smoking** while making the service call.

After arriving at the house but before entering, put on your shoe covers.

Do not use the customer's phone for personal calls, and do not use the customer's bathroom.

Communication Skills. The technician must be able to describe the service that can be provided. However, the technician must not monopolize the conversation. A big part of communicating is **listening.** Most people like to be listened to and the more you listen to the customer about the problems involved with the system, the easier it will be to diagnose. **Courtesy** and a **show of respect** for the customer should be evident at all times. The training and high skill level of the technician should also be evident as a result of the conversation and being able to answer and ask questions. Telling people how capable and skilled you are is often a turnoff. Remember to smile often. Ask pertinent questions and do not interrupt when the customer is answering. **Never say anything to discredit a competing business.**

Conflicts and Arguments. Conflicts and arguments with customers should be avoided at all costs. When you are dealing with an angry customer, you are dealing with an emotional customer. Listen until the customer is finished before replying. A complaint may be an opportunity to solve a problem. The customer should feel assured that the technician is competent and that the work will be done properly and in a timely manner. Never be **critical** of a customer, even in a joking manner. People hate to be criticized. It is very important to be friendly.

Even when angry, most customers are good individuals. They may have had a bad experience or may be disappointed, frustrated, and upset. Angry customers may have reviewed what they want to say and will not feel right until they have said it to a willing listener. Be sympathetic, listen carefully, and try to determine why the customer is so upset. Do not take it personally. Do not reply until the customer is definitely finished with the complaint and then try to concentrate on the solution. Ask the customer what you as the technician can do to help resolve the problem. If you can resolve the problem, do it. If you must report it to your supervisor, let the customer know that you will do this right away and will get back to them immediately if possible.

After listening carefully to the customer and resolving any complaint to the extent possible, you should be ready to start the troubleshooting process.

The Service Call. After arriving and introducing yourself, it is important to ask as many questions as needed to have a clear understanding of the problem. These questions will help to assure the customer that you are capable of solving the problem. During the service procedure you may need to talk with the customer to explain what you have found and to indicate the parts needed and possibly state the approximate costs if they may be higher than expected. If you must leave the job for any reason, tell the customer the reason and when you will return. You may need to go for parts or to another job emergency, but the customer needs to be informed. An informed customer is less likely to become angry or to complain. Keep the customer informed of all unusual circumstances. Double-check all your work. Clean the work site when finished and protect the customer's property from damage.

After the service work is completed, tell the customer what you found to be wrong, indicate that it has been corrected, and demonstrate when possible by turning the unit on while explaining how the problem was corrected. Customers deserve to know what they are paying for. All discussions should be in terms the customer will understand. Before leaving, billing information should be given to the customer. This should include a description of the work done and the costs.

The Technician as a Salesperson. A good technician is also a good salesperson. All options to resolve a problem should be presented in an honest and fair manner. Provide estimates and work orders in writing. A customer is buying not only service or equipment but also a solution to a problem. The customer may be offered an option not necessarily needed but should not be "talked into it." The sale and installation of a new system is not always the best option for a customer. If a customer feels that he or she was talked into something that was not needed, there is a good chance that the transaction will end the relationship between the customer and the company.

A company may have written recommendations for guiding the technician in presenting options. For instance if a unit is "x" years old and the repair will cost "x" amount, a recommendation to replace the unit or system may be appropriate.

In summary:

First impressions are very important.

Appearance and personal hygiene are major indicators of your personality and the quality of work you offer.

As a technician, you must be a good communicator.

Conflicts and arguments with customers should be avoided at all costs.

When making a service call, ask as many questions as needed to have a clear understanding of the problem.

If you must leave the job site, tell the customer why you are leaving and when you will return.

As a salesperson, you should be honest and fair to the customer, to yourself, and to your employer.

SECTION 1

Theory of Heat

Heat and Pressure

OBJECTIVES

After studying this unit, you should be able to

- define temperature.
- make conversions between Fahrenheit and Celsius scales.
- describe molecular motion at absolute zero.
- define the British thermal unit.
- describe heat flow between substances of different temperatures.
- explain the transfer of heat by conduction, convection, and radiation.
- discuss sensible heat, latent heat, and specific heat.
- state atmospheric pressure at sea level and explain why it varies at different elevations.
- describe two types of barometers.
- explain psig and psia as they apply to pressure measurements.

Figure 1–1 These are the tools of our trade; they are used to measure temperature and pressure.

1.1 HOW WE USE HEAT AND PRESSURE

The term *heat* can be applied to events in our lives that refer to our comfort, food, weather, and many other things. How does *cold* figure in? Cold will be discussed later as the absence of heat. We say, "It is hot outside," or "The coffee is hot; the ice cream is cold." In the field of refrigeration and air-conditioning technology, we must really understand what heat is and how it works for or against us and be able to explain it to other people.

Pressure is another term that we use in our everyday lives that must be understood. We talk about atmospheric pressure when we talk about the weather. We talk about tire pressure on our bikes and cars.

Both of these terms, temperature and pressure, will be used daily in conversation about our industry. **Figure 1–1** shows the instruments that we will use to measure temperature and pressure.

1.2 TEMPERATURE

Temperature can be thought of as a description of the level of heat and also may be referred to as heat intensity. Both heat level and heat intensity should not be confused with the amount of heat, or heat content. Heat can also be thought of as energy in the form of molecules in motion. Everything in the universe is made up of molecules. A molecule is the smallest portion that an element or substance can be divided

into. For example, we all know that water is made of hydrogen and oxygen (H_2O), which is two molecules of hydrogen and one molecule of oxygen combined together. All of these molecules vibrate and create heat while vibrating. The starting point of temperature is, therefore, the starting point of molecular motion.

Most people know that the freezing point of water is 32 degrees Fahrenheit (32°F) and that the boiling point is 212 degrees Fahrenheit (212°F), **Figure 1–2**. These points are commonly indicated on a thermometer, which is an instrument that measures temperature.

Early thermometers were of glass-stem types and operated on the theory that when the substance in the bulb was heated it would expand and rise in the tube, **Figure 1–3**. Mercury and alcohol are still commonly used today for this application.

We must qualify the statement that water boils at 212°F. Pure water boils at 212°F when standard atmospheric conditions exist. Standard conditions are sea level with the barometer reading 29.92 in. Hg (14.696 psia). This qualification concerns the relationship of the earth's atmosphere to the boiling point and will be covered in detail later in this section in the discussion on pressure. The statement that water boils at 212°F at standard conditions is important because these are standard conditions that will be applied to actual practice in later units.

Pure water has a freezing point of 32°F. Obviously the temperature can go lower than 32°F, but the question is, how much lower?

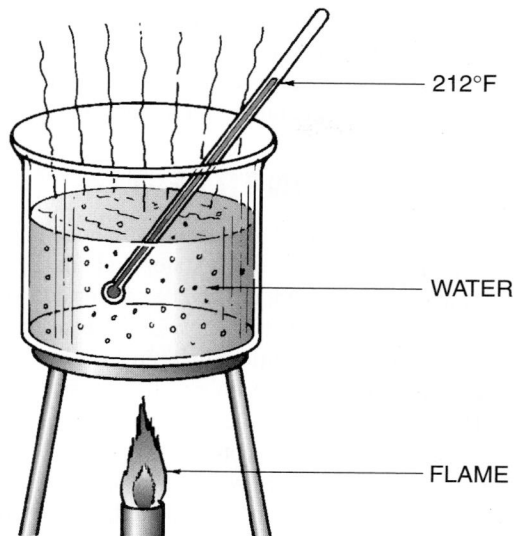

212°F

WATER

FLAME

Figure 1–2   The water in the container increases in temperature because the molecules move faster as heat is applied. When the water temperature reaches 212°F, boiling will occur. The bubbles in the water are small steam cells that are lighter than water and rise to the top.

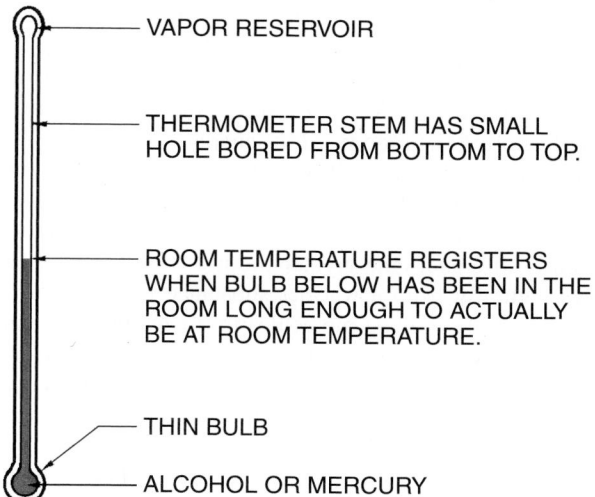

VAPOR RESERVOIR

THERMOMETER STEM HAS SMALL HOLE BORED FROM BOTTOM TO TOP.

ROOM TEMPERATURE REGISTERS WHEN BULB BELOW HAS BEEN IN THE ROOM LONG ENOUGH TO ACTUALLY BE AT ROOM TEMPERATURE.

THIN BULB

ALCOHOL OR MERCURY

Figure 1–3   A glass stem thermometer.

The theory is that molecular motion stops at −460°F. This is theoretical because molecular motion has never been totally stopped. The complete stopping of molecular motion is expressed as absolute zero. This has been calculated to be −460°F. Scientists have actually come within a few degrees of causing substances to reach absolute zero. **Figure 1–4** is an illustration of some levels of heat (molecular motion) shown on a thermometer scale.

The Fahrenheit scale of temperature is used in the English measurement system by the United States, one of only a few countries in the world that uses this system. The Celsius scale of temperature measurement is used in the International System of Units (SI) or *metric* system used by most other countries.

As the United States develops trade opportunities with the rest of the world, it may become necessary to use the metric

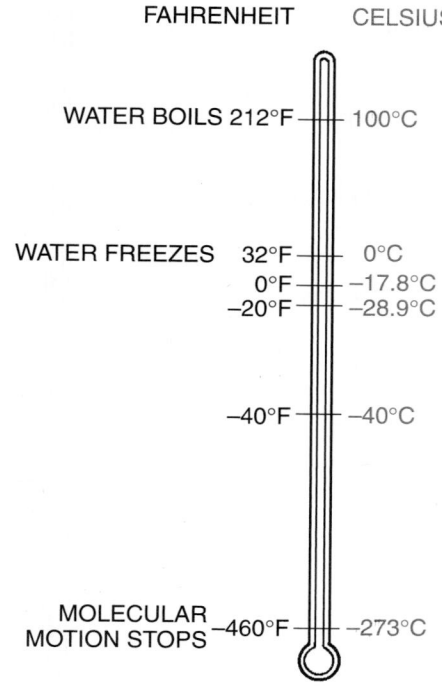

FAHRENHEIT   CELSIUS

WATER BOILS 212°F ── 100°C

WATER FREEZES 32°F ── 0°C
0°F ── −17.8°C
−20°F ── −28.9°C

−40°F ── −40°C

MOLECULAR
MOTION STOPS −460°F ── −273°C

Figure 1–4   Fahrenheit scale compared with Celsius scale.

system. **Figure 1–4** illustrates a thermometer with some important Fahrenheit and Celsius equivalent temperatures. **Figure 1–5** illustrates a thermometer showing more equivalent temperatures. See the Temperature Conversion Table in the appendix for conversion of temperatures. Formulas can also be used to make conversions. Use of the table and formulas is discussed at the end of this unit.

Most of the terms of measurement in this book are in English terms because at this time these are the terms most often used by technicians in this industry in the United States.

Temperature has been expressed in everyday terms up to this point. It is equally important in the air-conditioning, heating, and refrigeration industry to describe temperature in terms engineers and scientists use. Performance ratings of equipment are established using *absolute* temperature. Equipment is rated to establish criteria for comparing equipment performance. Using these ratings, different manufacturers can make comparisons with other manufacturers regarding their products. We can use the equipment rating to evaluate these comparisons. The Fahrenheit absolute scale is called the *Rankine* scale (named for its inventor, W. J. M. Rankine), and the Celsius absolute scale is known as the *Kelvin* scale (named for the scientist Lord Kelvin). Absolute temperature scales begin where molecular motion starts; they use 0 as the starting point. For instance, 0 on the Fahrenheit absolute scale is called absolute zero or 0° Rankine (0°R). Similarly, 0 on the Celsius absolute scale is called absolute zero or 0 Kelvin (0°K), **Figure 1–6**.

The Fahrenheit/Celsius and the Rankine/Kelvin scales are used interchangeably to describe equipment and fundamentals of this industry. Memorization is not required. A working knowledge of these scales and conversion formulas

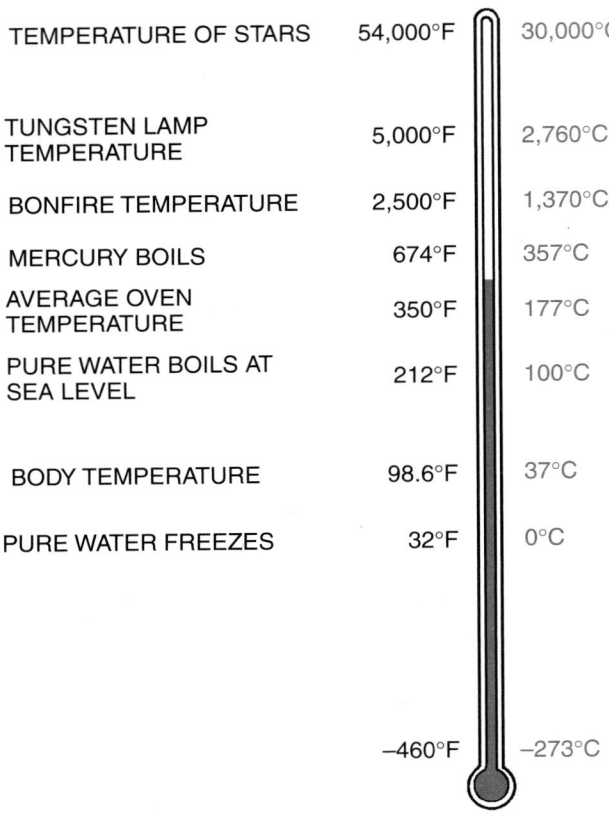

TEMPERATURE OF STARS	54,000°F	30,000°C
TUNGSTEN LAMP TEMPERATURE	5,000°F	2,760°C
BONFIRE TEMPERATURE	2,500°F	1,370°C
MERCURY BOILS	674°F	357°C
AVERAGE OVEN TEMPERATURE	350°F	177°C
PURE WATER BOILS AT SEA LEVEL	212°F	100°C
BODY TEMPERATURE	98.6°F	37°C
PURE WATER FREEZES	32°F	0°C
	−460°F	−273°C

Figure 1–5 Civilization is generally exposed to a comparatively small range of temperatures.

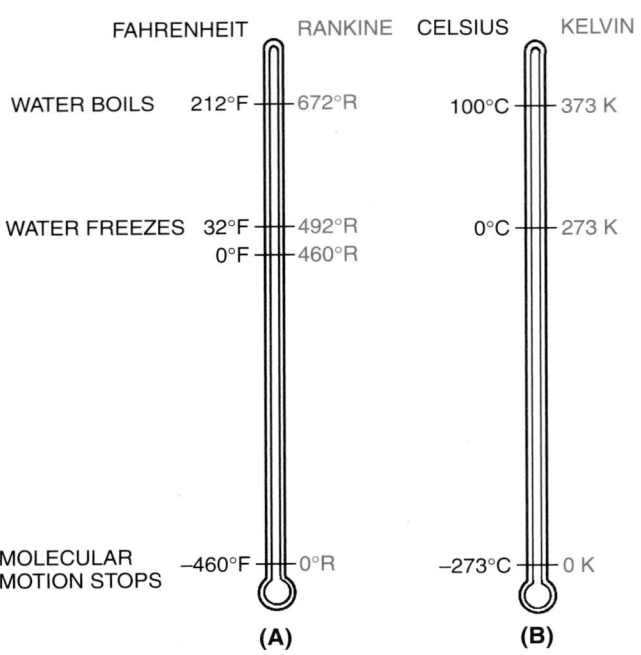

Figure 1–6 **(A)** A Fahrenheit and Rankine thermometer. **(B)** A Celsius and Kelvin thermometer.

and a ready reference table are more practical. **Figure 1–6** shows how these four scales are related. The world we live in accounts for only a small portion of the total temperature spectrum.

Our earlier statement that temperature describes the level of heat, heat intensity, or molecular motion can now be explained. As a substance receives more heat, its molecular motion, and therefore its temperature, increases, **Figure 1–2**.

1.3 INTRODUCTION TO HEAT

The laws of thermodynamics can help us to understand what heat is all about. One of these laws states that energy can be neither created nor destroyed, but can be converted from one form to another. This means that most of the heat the world experiences is not being continuously created but is being converted from other forms of energy like fossil fuels (gas and oil). This heat can also be accounted for when it is transferred from one substance to another.

Temperature describes the level of heat with reference to no heat. The term used to describe the quantity of heat or heat content is known as the *British thermal unit (Btu)*. This term explains how much heat is contained in a substance. The rate of heat consumption can be determined by adding time.

The Btu is defined as the amount of heat required to raise the temperature of 1 lb of water 1°F. For example, when 1 lb of water (about 1 pint) is heated from 68°F to 69°F, 1 Btu of heat energy is absorbed into the water, **Figure 1–7**. To actually measure how much heat is absorbed in a process like this, we need an instrument of laboratory quality. This instrument is called a *calorimeter*. Notice the similarity to the word "calorie," the food word for energy.

When a temperature difference exists between two substances, heat transfer will occur. Temperature difference is the driving force behind heat transfer. The greater the temperature difference, the greater the heat transfer rate. Heat flows naturally from a warmer substance to a cooler substance. Rapidly moving molecules in the warmer substance give up some of their energy to the slower moving molecules in the cooler substance. The warmer substance cools because the molecules have slowed. The cooler substance becomes warmer because the molecules are moving faster.

The following example illustrates the difference in the quantity of heat compared with the level of heat. One tank of water weighing 10 lb (slightly more than 1 gallon [gal]) is heated to a temperature level of 200°F. A second tank of water weighing 100,000 lb (slightly more than 12,000 gal) is heated to 175°F. The 10-lb tank will cool to room temperature much faster than the 100,000-lb tank. The temperature difference of 25°F is not much, but the cool-down time is much longer for the 100,000-lb tank due to the quantity of water, **Figure 1–8**.

A comparison using water may be helpful in showing the heat intensity level versus the quantity of heat. A 200-ft deep well would not contain nearly as much water as a large lake with a water depth of 25 ft. The depth of water (in feet) tells us the level of water, but it in no way expresses the quantity (gallons) of water.

In practical terms, each piece of heating equipment is rated according to the amount of heat it will produce. If the equipment had no such rating, it would be difficult for a buyer to intelligently choose the correct appliance.

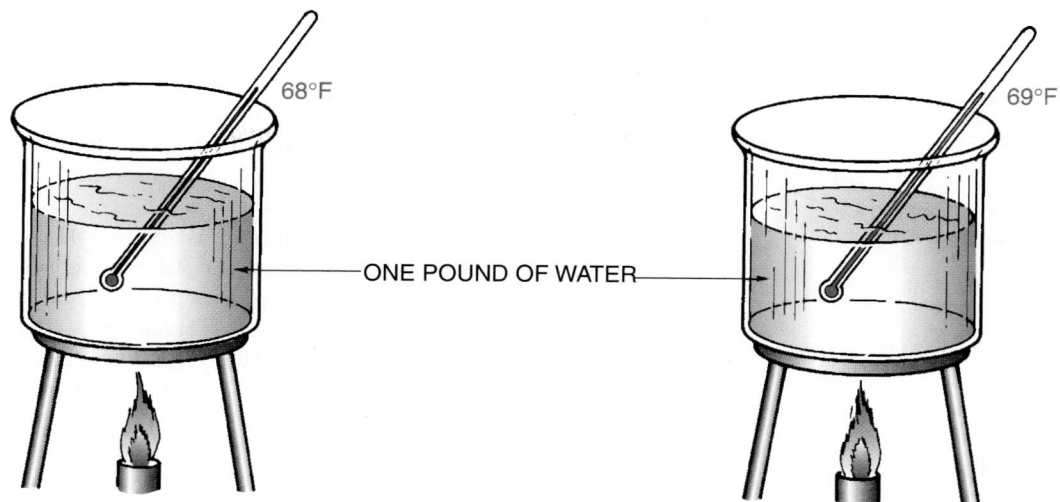

Figure 1–7   One British thermal unit (Btu) of heat energy is required to raise the temperature of 1 lb (pound) of water from 68°F to 69°F.

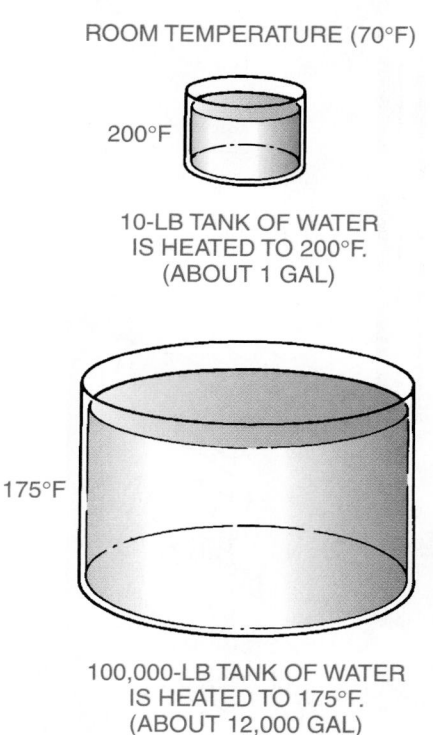

ROOM TEMPERATURE (70°F)

200°F

10-LB TANK OF WATER
IS HEATED TO 200°F.
(ABOUT 1 GAL)

175°F

100,000-LB TANK OF WATER
IS HEATED TO 175°F.
(ABOUT 12,000 GAL)

Figure 1–8   The smaller tank will cool to room temperature first because a smaller quantity of heat exists, even though its temperature or heat intensity level is higher.

A gas or oil furnace used to heat a home has the rating permanently printed on a nameplate. Either furnace would be rated in Btu per hour, which is a *rate* of energy consumption. Later, this rate will be used to calculate the amount of fuel required to heat a house or a structure. For now, it is sufficient to say that if one needs a 75,000-Btu/h furnace to heat a house on the coldest day, a furnace rated at 75,000 Btu/h should be chosen. If not, the house will begin to get cold any time the heat loss of the house exceeds the furnace output in Btu/h.

In the metric, or SI, system of measurement, the term *joule (J)* is used to express the quantity of heat. Because a joule is very small, metric units of heat in this industry are usually expressed in kilojoules (kJ) or 1000 joules. One Btu equals 1.055 kJ.

The term *gram (g)* is used to express weight in the metric system. Again, this is a small quantity of weight, so the term kilograms (kg) is often used. One pound equals 0.45359 kg.

The amount of heat required to raise the temperature of 1 kg of water 1°C is equal to 4.187 kJ.

Cold is a comparative term used to describe lower temperature levels. Because all heat is a positive value in relation to no heat, cold is not a true value. It is really an expression of comparison. When a person says it is cold outside, it is in relation to the normal expected temperature for the time of year or to the inside temperature. Cold has no number value and is used by most people as a basis of comparison only. Cold is sometimes referred to as a level of heat absence.

1.4   CONDUCTION

Conduction heat transfer can be explained as the energy actually traveling from one molecule to another. As a molecule moves faster, it causes others to do the same. For example, if one end of a copper rod is in a flame, the other end gets too hot to handle. The heat travels up the rod from molecule to molecule, **Figure 1–9.**

Conduction heat transfer is used in many heat transfer applications that are experienced regularly. Heat is transferred by conduction from the hot electric burner on the cookstove to the pan of water. It then is transferred by conduction into the water. Note that there is an orderly explanation for each step.

Heat does not conduct at the same rate in all materials. Copper, for instance, conducts at a different rate from iron. Glass is a very poor conductor of heat.

Touching a wooden fence post or another piece of wood on a cold morning does not give the same sensation as touching a car fender or another piece of steel. The piece of steel

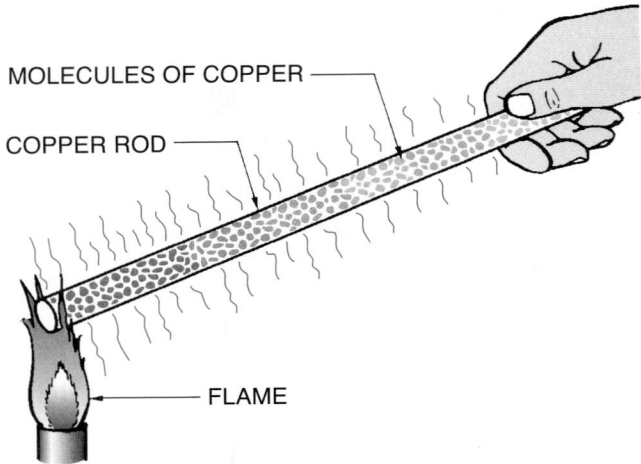

Figure 1–9   The copper rod is held in the flame only for a short time before heat is felt at the far end.

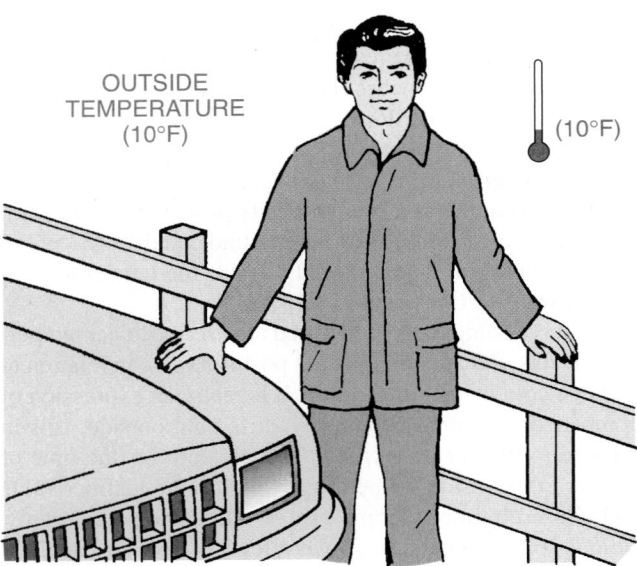

Figure 1–10   The car fender and the fence post are actually the same temperature, but the fender feels colder because the metal conducts heat away from the hand faster than the wooden fence post.

feels much colder. Actually the steel is not colder; it just conducts heat out of the hand faster, **Figure 1–10.**

The different rates at which various materials conduct heat have an interesting similarity to the conduction of electricity. As a rule, substances that are poor conductors of heat are also poor conductors of electricity. For instance, copper is one of the best conductors of electricity and heat, and glass is one of the poorest conductors of both. Glass is actually used as an insulator of electrical current flow.

1.5   CONVECTION

Convection heat transfer is used to move heat from one location to another by means of currents set up in a fluid medium. The most common fluid mediums in the heating and air-

conditioning trades are air and water. When heat is moved, it is normally transferred into some substance that is readily movable, such as air or water. Many large buildings have a central heating plant where water is heated and pumped throughout the building to the final heated space. Notice the similarity of the words "convection" and "convey" (to carry from one place to another).

A forced air gas furnace is an example of forced convection. Room air is drawn into the return air of the furnace by the fan. This air is forced out the fan outlet over the furnace heat exchanger, which exchanges heat to the air from a gas flame. The air is then forced into the ductwork and distributed into the various rooms in the structure. **Figure 1–11**

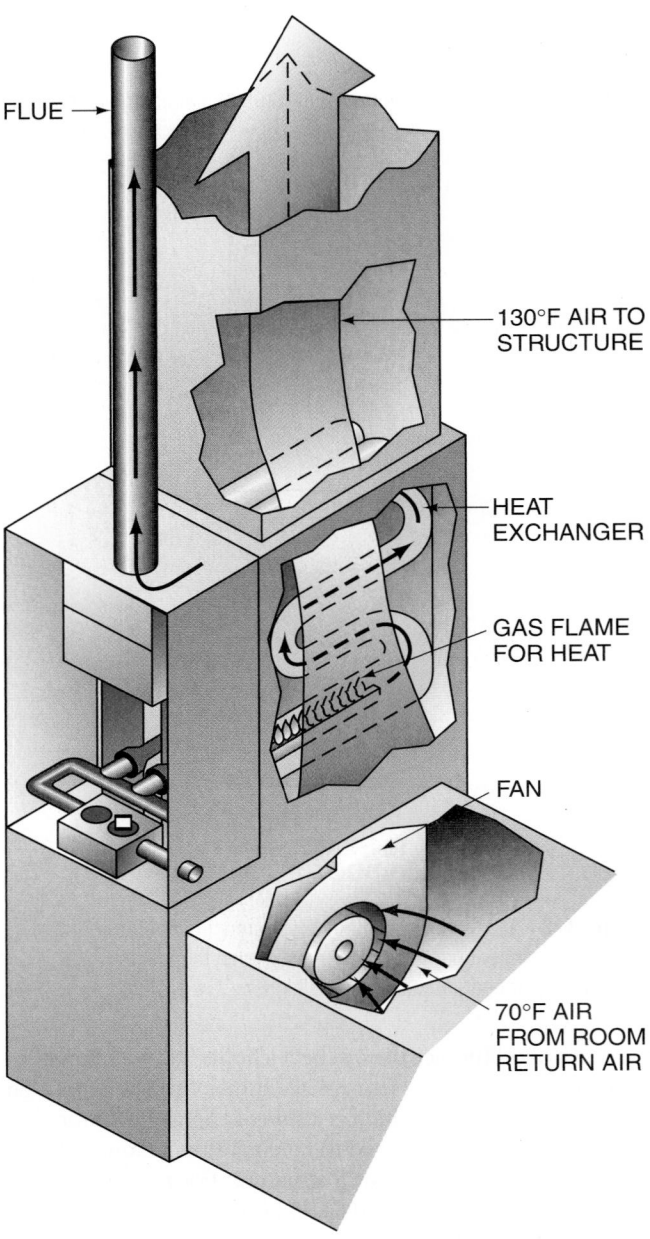

Figure 1–11   Air from the room enters the fan at 70°F. The fan forces the air across the hot heat exchanger and out into the structure at 130°F.

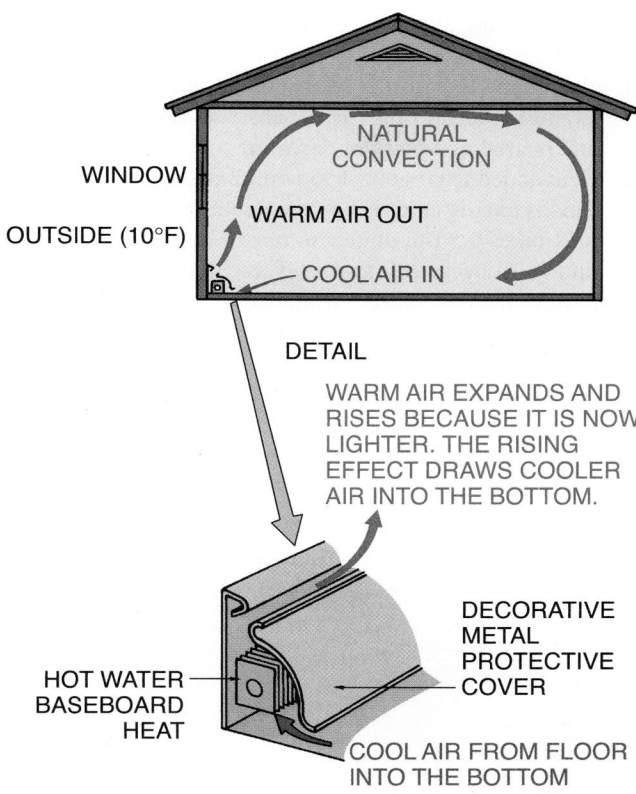

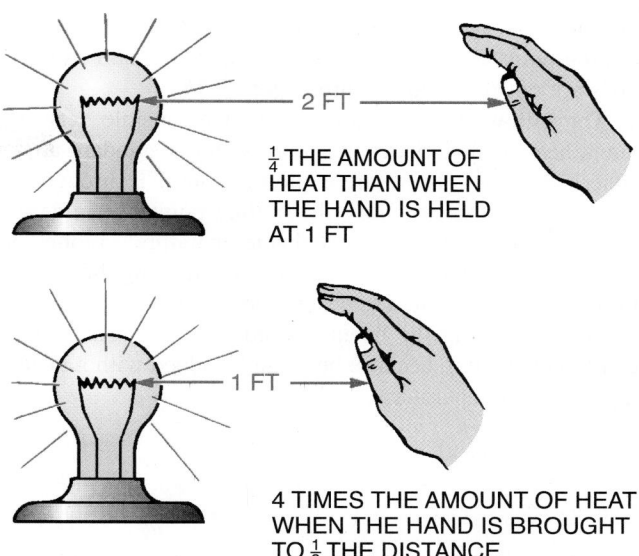

Figure 1–13 The intensity of the radiant heat diminishes by the square of the distance.

Figure 1–12 Natural convection occurs when heated air rises and cool air takes its place.

shows this process in which 70°F room air is entering the furnace; 130°F air is leaving; and the fan is creating the pressure difference to force the air into the various rooms. The fan provides the *forced convection.*

Another example of heat transfer by convection is that when air is heated, it rises; this is called *natural* convection. When air is heated, it expands, and the warmer air becomes less dense or lighter than the surrounding unheated air. This principle is applied in many ways in the air-conditioning industry. Baseboard heating units are an example. They are normally installed on the outside walls of buildings and use electricity or hot water as the heat source. When the air near the floor is heated, it expands and rises. This heated air is displaced by cooler air around the heater, which sets up a natural convection current in the room, **Figure 1–12.**

1.6 RADIATION

Radiation heat transfer can best be explained by using the sun as an example of the source. The sun is approximately 93 million miles from the earth's surface, yet we can feel its intensity. The sun's surface temperature is extremely hot compared with anything on earth. Heat transferred by radiation travels through space without heating the space and is absorbed by the first solid object that it encounters. Radiation is the only type of heat transfer that can travel through a vacuum, such as space, because radiation is not dependent on matter as a medium of heat transfer. This is impossible with

convection and conduction because they require some form of matter, like air or water, to be the transmitting medium. The earth does not experience the total heat of the sun because heat transferred by radiation diminishes by the inverse of the square of the distance traveled. In practical terms, this means that every time the distance is doubled, the heat intensity decreases by one-fourth. If you hold your hand close to a light bulb, for example, you feel the heat's intensity, but if you move your hand twice the distance away, you feel only one-fourth of the heat intensity, **Figure 1–13.** Keep in mind that, because of the inverse-square-of-the-distance explanation, radiant heat does not transfer the actual temperature or heat quantity value. If it did, the earth would be as hot as the sun.

Electric heaters that glow red hot are practical examples of radiant heat. The electric heater coil glows red hot and radiates heat into the room. It does not heat the air, but it warms the solid objects that the heat rays encounter. Any heater that glows has the same effect.

1.7 SENSIBLE HEAT

Heat level or heat intensity can readily be measured when it changes the temperature of a substance (remember the example of changing 1 lb of water from 68°F to 69°F). This change in the heat level can be measured with a thermometer. When a change of temperature can be registered, we know that the level of heat or heat intensity has changed; this is called *sensible heat.*

1.8 LATENT HEAT

Another type of heat is called *latent* or *hidden* heat. In this process heat is known to be added, but no temperature rise is noticed. An example is heat added to water while it is boiling in an open container. Once water is brought to the boiling point, adding more heat only makes it boil faster; it does not raise the temperature, **Figure 1–14.**

The following example describes the sensible heat and latent heat characteristics of 1 lb of water at standard atmospheric pressure. These are explored from 0°F through the temperature range to above the boiling point. Examine the chart in **Figure 1–15** and notice that temperature is plotted on the left margin, and heat content is plotted along the bottom of the chart. We see that as heat is added the temperature will rise except during the latent- or hidden-heat process. This chart is interesting because heat can be added without causing a rise in temperature.

The following statements should help you to understand the chart.

1. Water is in the form of ice at point 1 where the example starts. Point 1 is *not* absolute zero. It is 0°F and is used as a point of departure.
2. Heat added from point 1 to point 2 is sensible heat. This is a registered rise in temperature. Note that it only takes 0.5 Btu of heat to raise 1 lb of ice 1°F. We know this because it took only 16 Btu of sensible heat to raise the temperature from 0°F to 32°F.
3. When point 2 is reached, the ice is at its highest temperature of 32°F. This means that if more heat is added, it will be known as latent heat and will start to

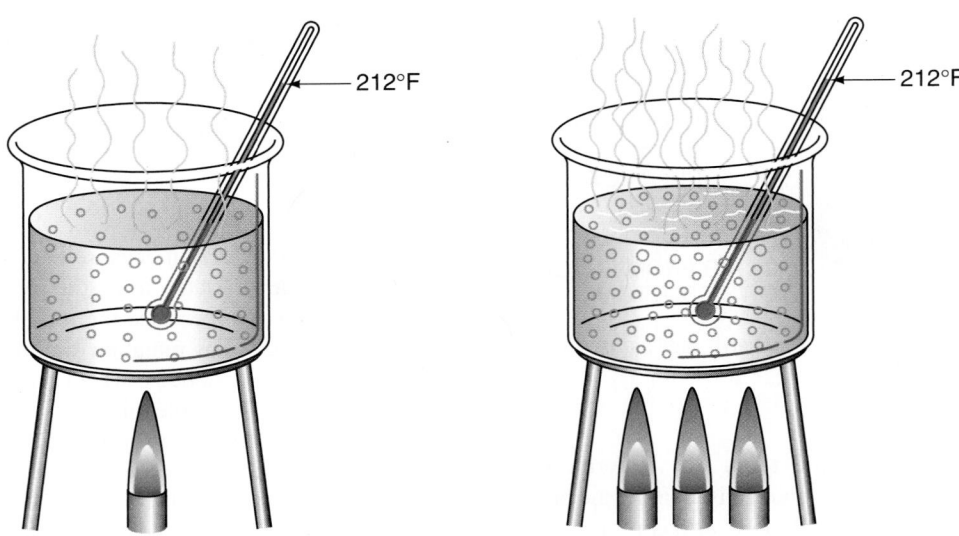

Figure 1–14   Adding three times as much heat only causes the water to boil faster. The water does not increase in temperature.

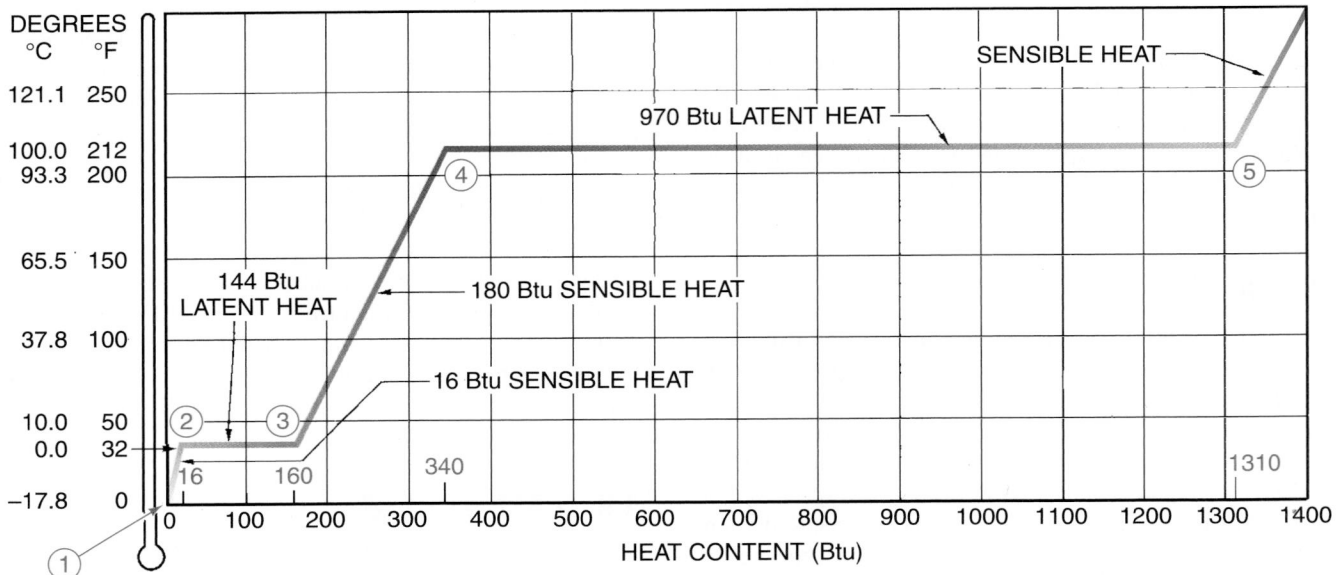

Figure 1–15   The heat/temperature graph for 1 lb of water at atmospheric pressure explains how water responds to heat. An increase in sensible heat causes a rise in temperature. An increase in latent heat causes a change of state, for example, from solid ice to liquid water.

SUBSTANCE	SPECIFIC HEAT Btu/lb/°F	SUBSTANCE	SPECIFIC HEAT Btu/lb/°F
ALUMINUM	0.224	BEETS	0.90
BRICK	0.22	CUCUMBERS	0.97
CONCRETE	0.156	SPINACH	0.94
COPPER	0.092	BEEF, FRESH	
ICE	0.504	LEAN	0.77
IRON	0.129	FISH	0.76
MARBLE	0.21	PORK, FRESH	0.68
STEEL	0.116	SHRIMP	0.83
WATER	1.00	EGGS	0.76
SEA WATER	0.94	FLOUR	0.38
AIR	0.24 (AVERAGE)		

Figure 1–16   The specific heat table shows how much heat is required to raise the temperature of several different substances 1°F.

melt the ice but not raise the temperature. Adding 144 Btu of heat will change the 1 lb of ice to 1 lb of water. Removing any heat will cool the ice below 32°F.

4. When point 3 is reached, the substance is now 100% water. Adding more heat causes a rise in temperature. (This is sensible heat.) Removal of any heat at point 3 results in some of the water changing back to ice. This is known as removing latent heat because there is no change in temperature.

5. Heat added from point 3 to point 4 is sensible heat; when point 4 is reached, 180 Btu of heat will have been added from point 3: 1 Btu/lb/°F temperature change for water.

6. Point 4 represents the 100% saturated liquid point. The water is saturated with heat to the point that the removal of any heat causes the liquid to cool off below the boiling point. Heat added is identified as latent heat and causes the water to boil and to start changing to a vapor (steam). Adding 970 Btu makes the 1 lb of liquid boil to point 5 and become a vapor.

7. Point 5 represents the 100% saturated vapor point. The water is now in the vapor state. Heat removed would be latent heat and would change some of the vapor back to a liquid. This is called *condensing the vapor.* Any heat added at point 5 is sensible heat; it raises the vapor temperature above the boiling point. Heating the vapor temperature above the boiling point is called *superheating.* Any water vapor with a temperature above the boiling point of 212°F is called a superheated vapor in this example. Superheat will be important in future studies. Note that in the vapor state it takes only 0.5 Btu to heat the water vapor (steam) 1°F. The same was true while water was in the ice (solid) state.

SAFETY PRECAUTION: When examining these principles in practice, be careful because the water and steam are well above body temperature, and you could be seriously burned.

1.9   SPECIFIC HEAT

We now realize that different substances respond differently to heat. When 1 Btu of heat energy is added to 1 lb of water, it changes the temperature 1°F. This only holds true for water.

When other substances are heated, different values occur. For instance, we noted that adding 0.5 Btu of heat energy to either ice or steam (water vapor) caused a 1°F rise per pound while in these states. They heated, or increased the temperature, at twice the rate. Adding 1 Btu would cause a 2°F rise. This difference in heat rise is known as *specific heat.*

Specific heat is the amount of heat necessary to raise the temperature of 1 lb of a substance 1°F. Every substance has a different specific heat. Note that the specific heat of water is 1 Btu/lb/°F. See **Figure 1–16** for the specific heat of some other substances.

1.10   SIZING HEATING EQUIPMENT

Specific heat is significant because the amount of heat required to change the temperatures of different substances is used to size equipment. Recall the example of the house and furnace earlier in this unit.

The following example shows how this would be applied in practice. A manufacturing company may need to buy a piece of heating equipment to heat steel before it can be machined. The steel may be stored outside in the cold at 0°F and need preheating before machining. The temperature desired for the machining is 70°F. How much heat must be added to the steel if the plant wants to machine 1000 lb/h?

The steel is coming into the plant at a fixed rate of 1000 lb/h, and that heat has to be added at a steady rate to stay ahead of production. **Figure 1–16** gives a specific heat of 0.116 Btu/lb/°F for steel. This means that 0.116 Btu of heat energy must be added to 1 lb of steel to raise its temperature 1°F.

$$Q = \text{Weight} \times \text{Specific Heat} \times \text{Temperature Difference}$$

where Q = quantity of heat needed. Substituting in the formula, we get

$$Q = 1000 \text{ lb/h} \times 0.116 \text{ Btu/lb/°F} \times 70\text{°F}$$

$$Q = 8120 \text{ Btu/h required to heat the steel for machining.}$$

The previous example has some known values and a value to be found. The known information is used to find the unknown value with the help of the formula. The formula can be used when adding heat or removing heat and is often used in heat-load calculations for sizing both heating and cooling equipment.

1.11 PRESSURE

Pressure is defined as force per unit of area. This is normally expressed in pounds per square inch (psi). Simply stated, when a 1-lb weight rests on an area of 1 square inch (1 in^2), the pressure exerted downward is 1 psi. Similarly, when a 100-lb weight rests on a 1-in^2 area, 100 psi of pressure is exerted, **Figure 1–17**.

When you swim below the surface of the water, you feel a pressure pushing inward on your body. This pressure is caused by the weight of the water and is very real. A different sensation is felt when flying in an airplane without a pressurized cabin. Your body is subjected to less pressure instead of more, yet you still feel uncomfortable.

It is easy to understand why the discomfort under water exists. The weight of the water pushes in. In the airplane, the reason is just the reverse. There is less pressure high in the air than down on the ground. The pressure is greater inside your body and is pushing out.

Water weighs 62.4 pounds per cubic foot (lb/ft^3). A cubic foot (7.48 gal) exerts a downward pressure of 62.4 lb/ft^2 when it is in its actual cube shape, **Figure 1–18**. How much weight is then resting on 1 in^2? The answer is simply calculated. The bottom of the cube has an area of 144 in^2 (12 in. $\times$ 12 in.)

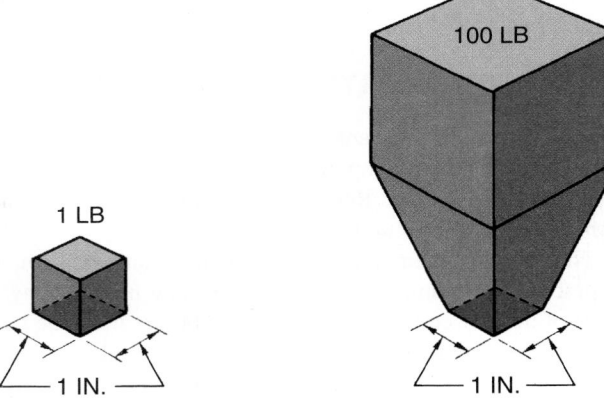

Figure 1–17   Both weights are resting on a 1-square-inch (1-in^2) surface. One weight exerts a pressure of 1 psi, the other a pressure of 100 psi.

WATER IN CUBIC
FOOT CONTAINER

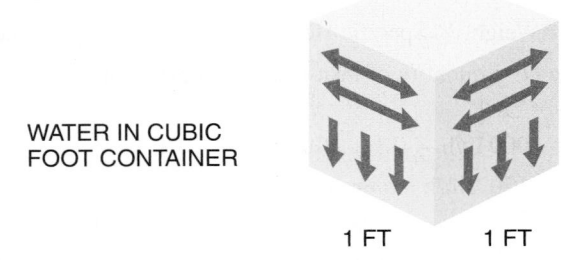

1 FT   1 FT

Figure 1–18   One cubic foot (1 ft^3) of water (7.48 gal) exerts its pressure outward and downward. 1 ft^3 of water weighs 62.4 lb spread over 1 ft^2.

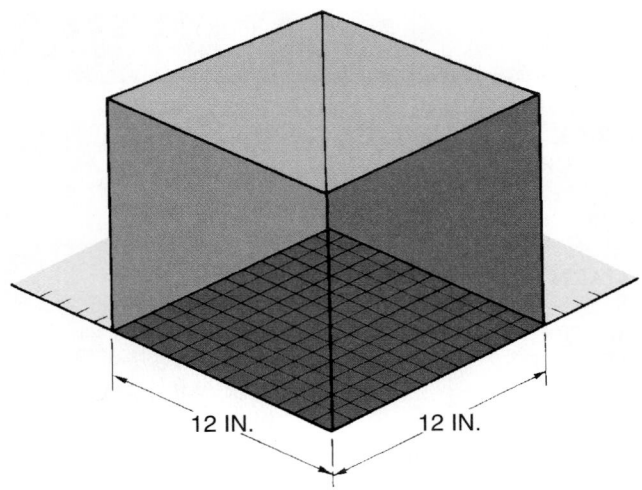

12 IN.   12 IN.

Figure 1–19   One cubic foot (1 ft^3) of water exerts a downward pressure of 62.4 lb/ft^2 on the bottom surface area of a cube.

sharing the weight. Each square inch has a total pressure of 0.433 lb (62.4 ÷ 144) resting on it. Thus, the pressure at the bottom of the cube is 0.433 psi, **Figure 1–19**.

1.12 ATMOSPHERIC PRESSURE

The sensation of being underwater and feeling the pressure of the water is familiar to many people. The earth's atmosphere is like an ocean of air that has weight and exerts pressure. The earth's surface can be thought of as being at the bottom of this ocean of air. Different locations are at different depths. For instance, there are sea-level locations such as Miami, Florida, or mountainous locations such as Denver, Colorado. The atmospheric pressures at these two locations are different. For now, we will assume that we live at the bottom of this ocean of air.

The atmosphere that we live in has weight just as water does, but not as much. Actually the earth's atmosphere exerts a weight or pressure of 14.696 psi at sea level. This is known as a standard condition.

Atmospheric pressure can be measured with an instrument called a *barometer*. The barometer is a glass tube about 36 in. long that is closed on one end and filled with mercury. It is then inserted open-side down into a puddle of mercury and held upright. The mercury will try to run down into the puddle, but it will not all run out. The atmosphere is pushing down on the puddle, and a vacuum is formed in the top of the tube. The mercury in the tube will fall to 29.92 in. at sea level when the surrounding atmospheric temperature is 70°F, **Figure 1–20**. This is a standard that is used for comparison in engineering and scientific work. If the barometer is taken up higher, such as on a mountain, the mercury column will start to fall. It will fall about 1 in./1000 ft of altitude. When the barometer is at standard conditions and the mercury drops, it is called a low-pressure system by the weather forecaster; this means the weather is going to change. Listen closely to the weather report, and the weather forecaster will make these terms more meaningful.

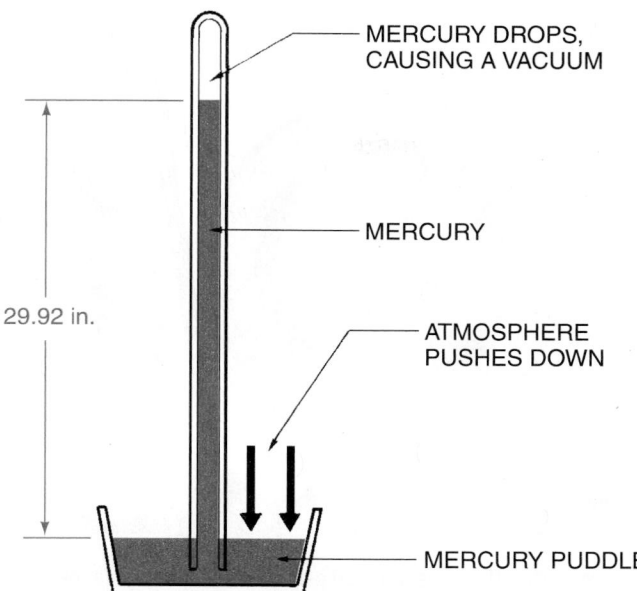

Figure 1–20   Mercury (Hg) barometer.

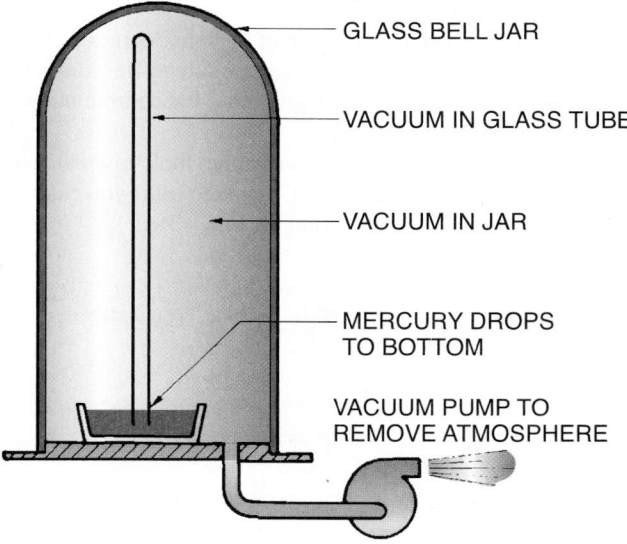

Figure 1–21   When the mercury barometer is placed in a closed glass jar (known as a bell jar) and the atmosphere is removed from the jar, the column of mercury drops to the level of the puddle in the dish.

If the barometer is placed inside a closed jar and the atmosphere evacuated, the mercury column falls to a level with the puddle in the bottom, **Figure 1–21**. When the atmosphere is allowed back into the jar, the mercury again rises because a vacuum exists above the mercury column in the tube.

The mercury in the column has weight and counteracts the atmospheric pressure of 14.696 psi at standard conditions. A pressure of 14.696 psi then is equal to the weight of a column of mercury (Hg) 29.92 in. high. The expression "inches of mercury" thus becomes an expression of pressure and can be converted to pounds per square inch. The conversion factor

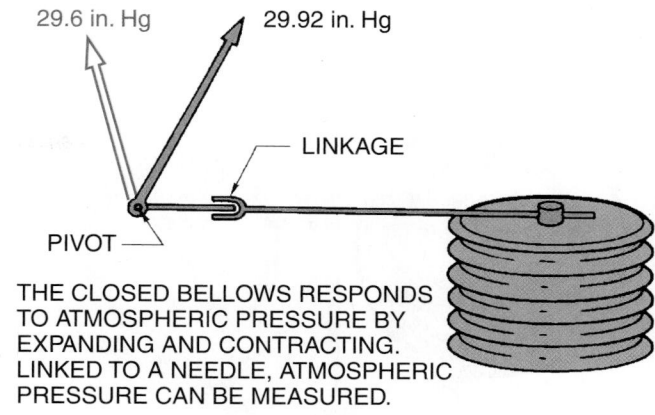

Figure 1–22   The aneroid barometer uses a closed bellows that expands and contracts with atmospheric pressure changes.

is 1 psi = 2.036 in. Hg (29.92 ÷ 14.696); 2.036 is often rounded off to 2 (30 in. Hg ÷ 15 psi).

Another type of barometer is the *aneroid* barometer. This is a more practical instrument to transport. Atmospheric pressure has to be measured in many places, so instruments other than the mercury barometer had to be developed for field use, **Figure 1–22**.

1.13   PRESSURE GAGES

Measuring pressures in a closed system requires a different method—the Bourdon tube, **Figure 1–23**. The *Bourdon tube* is linked to a needle and can measure pressures above and below atmospheric pressure. A common tool used in the refrigeration industry to take readings in the field or shop is a combination of a low-pressure gage (called the *low-side gage*) and a high-pressure gage (called the *high-side gage*),

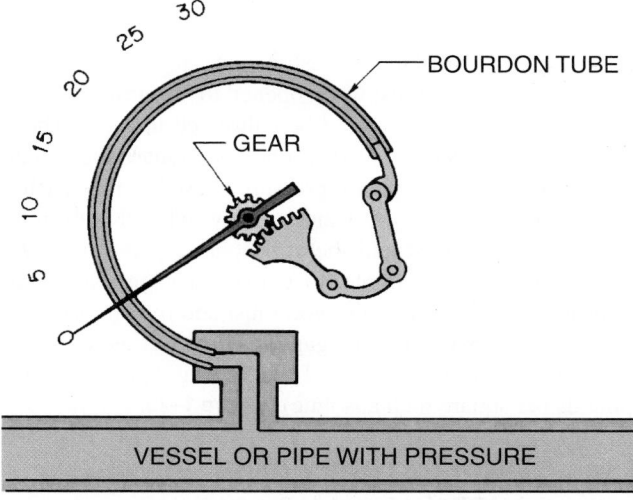

Figure 1–23   The Bourdon tube is made of a thin substance such as brass. It is closed on one end, and the other end is fastened to the pressure being checked. When pressure increases, the tube tends to straighten out. When attached to a needle linkage, pressure changes are indicated.

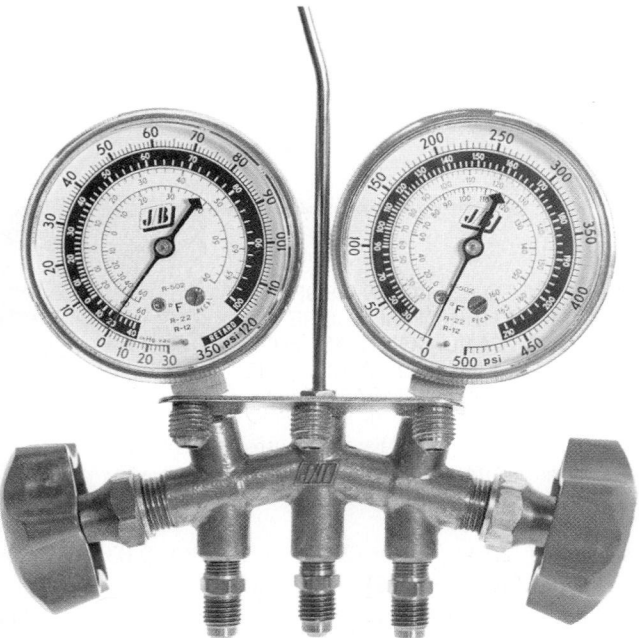

Figure 1–24 The gage on the left is called a compound gage because it reads below atmospheric pressure in inches Hg (mercury) and above atmospheric pressure in psig. The right-hand gage reads high pressure up to 500 psig. *Photo by Bill Johnson*

Figure 1–24. The gage on the left reads pressures above and below atmospheric pressure. It is called a *compound gage.* The gage on the right will read up to 500 psi and is called the *high-pressure (high-side) gage.*

SAFETY PRECAUTION: Working with temperatures that are above or below body temperature can cause skin and flesh damage. Proper protection, such as gloves and safety glasses, **must be used.** Pressures that are above or below the atmosphere's pressure can cause bodily injury. A vacuum can cause a blood blister on the skin. Pressure above atmospheric can pierce the skin or inflict damage when blowing air lifts small objects like filings.

These gages read 0 psi when opened to the atmosphere. If they do not, then they should be calibrated to 0 psi. These gages are designed to read pounds-per-square-inch gage pressure (psig). Atmospheric pressure is used as the starting or reference point. If you want to know what the absolute pressure is, you must add the atmospheric pressure to the gage reading. For example, to convert a gage reading of 50 psig to absolute pressure, you must add the atmospheric pressure of 14.696 psi to the gage reading. Let us round off 14.696 to 15 for this example. Then 50 psig + 15 = 65 psia (pounds per square inch absolute), **Figure 1–25.**

1.14 TEMPERATURE CONVERSION— FAHRENHEIT AND CELSIUS

You may find it necessary to convert specific temperatures from Fahrenheit to Celsius or from Celsius to Fahrenheit. This conversion can be done by using the Temperature Conversion Table in the appendix or by using a formula.

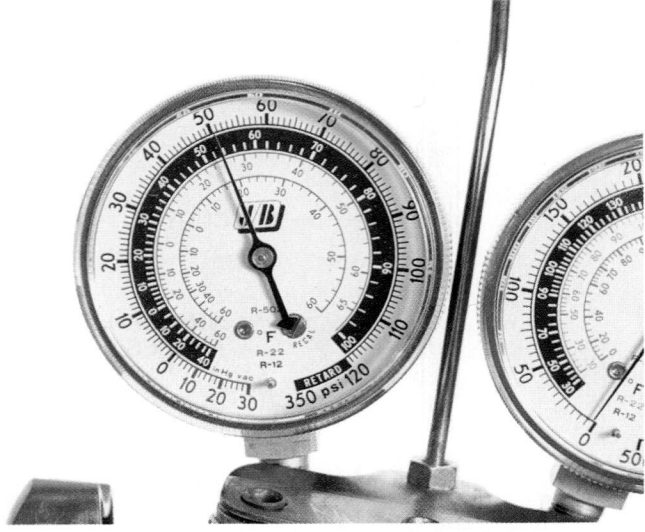

Figure 1–25 This gage reads 50 psig. To convert this gage reading to psia, add 50 psig to 15 psi (atmospheric pressure) for a sum of 65 psia. *Photo by Bill Johnson*

Using the table:

To convert a room temperature of 78°F to degrees Celsius, move down the column labeled Temperature to Be Converted until you find 78. Look to the right under the column marked °C, and you will find 25.6°C.

To convert 36°C to degrees Fahrenheit, look down the column labeled Temperature to Be Converted until you find 36. Look to the left and you will find 96.8°F.

Using formulas:

$$°C = \frac{°F - 32°}{1.8} \qquad °F = (1.8 \times °C) + 32°$$

$$\text{or} \qquad\qquad \text{or}$$

$$°C = \frac{5}{9}(°F - 32°) \quad °F = \left(\frac{9}{5} \times °C\right) + 32°$$

To convert a room temperature of 75°F to degrees C:

$$°C = \frac{75° - 32°}{1.8} = 23.9°$$

$$°C = 23.9°F$$

$$75°F = 23.9°C$$

To convert a room temperature of 25°C to degrees F:

$$°F = (1.8 \times 25°) + 32° = 77°$$

$$°F = 77°$$

$$25°C = 77°F$$

1.15 PRESSURE MEASURED IN METRIC TERMS

Pressure, like temperature, can be expressed in metric terms. Remember, pressure is an expression of force per unit of area. In the past, several terms have been used by different countries to express the measurement of pressure, but the present standard metric expression for pressure is the term newton per square meter (N/m^2). Pressure in English measurement is expressed in pounds per square inch (psi). It is difficult to compare pounds per square inch and newton per square meter. To make this comparison easier, the newton per square meter has been given the name pascal in honor of the scientist and mathematician Blaise Pascal. The standard metric term for pressure is the kilopascal (kPa) or 1000 pascal. One psi is equal to 6890 pascal, or 6.89 kPa. To convert psi to kPa, simply multiply the number of psi by 6.89. Barometric pressure in metric terms is measured in mm Hg (millimeters of mercury). A standard atmosphere is 760 mm Hg or 101.3 kPa.

SUMMARY

- Thermometers measure temperature. Four temperature scales are Fahrenheit, Celsius, Fahrenheit absolute (Rankine), and Celsius absolute (Kelvin).
- Molecules in matter are constantly moving. The higher the temperature, the faster they move.
- The British thermal unit (Btu) describes the quantity of heat in a substance. One Btu is the amount of heat necessary to raise the temperature of 1 lb of water 1°F.
- The transfer of heat by conduction is the transfer of heat from molecule to molecule. As molecules in a substance move faster and with more energy, they cause others near them to do the same.
- The transfer of heat by convection is the actual moving of heat in a fluid (vapor state or liquid state) from one place to another.
- Radiant heat is a form of energy that does not depend on matter as a medium of transfer. Solid objects absorb the energy, become heated, and transfer the heat to the air.
- Sensible heat causes a rise in temperature of a substance.
- Latent (or hidden) heat is that heat added to a substance causing a change of state and not registering on a thermometer. For example, heat added to melting ice causes ice to melt but does not increase the temperature.
- Specific heat is the amount of heat (measured in Btu) required to raise the temperature of 1 lb of a substance 1°F. Substances have different specific heats.
- Pressure is the force applied to a specific unit of area. The atmosphere around the earth has weight and therefore exerts pressure.
- Barometers measure atmospheric pressures in inches of mercury. Two barometers used are the mercury and the aneroid.
- Gages have been developed to measure pressures in enclosed systems. Two common gages used in the air-conditioning, heating, and refrigeration industry are the compound gage

and the high-pressure gage. The compound gage reads pressures both above and below atmospheric pressure.
- The metric term kilopascal (kPa) is used to express pressure in the refrigeration and air-conditioning field.

REVIEW QUESTIONS

1. Temperature is defined as
   A. how hot it is.
   B. the level of heat.
   C. how cold it is.
   D. why it is hot.
2. State the standard conditions for water to boil at 212°F.
3. List four types of temperature scales.
4. Under standard conditions, water freezes at _____°C.
5. Molecular motion stops at _____°F.
6. One British thermal unit will raise the temperature of _____ lb of water _____°F.
7. In which direction does heat flow?
   A. From a cold substance to a cold substance
   B. Up
   C. Down
   D. From a warm substance to a cold substance
8. Describe heat transfer by conduction.
9. A rise in sensible heat causes
   A. a rise in a thermometer.
   B. a fall in a thermometer.
   C. no change in a thermometer.
   D. ice to melt.
10. Latent heat causes
    A. a rise in a thermometer.
    B. temperature to rise.
    C. a change of state.
    D. temperature to fall.
11. Describe how heat is transferred by convection.
12. Describe how heat is transferred by radiation.
13. Specific heat is the amount of heat necessary to raise the temperature of 1 lb of a _____ 1°F.
14. Atmospheric pressure at sea level under standard conditions is _____ inches of mercury (Hg) or _____ pounds per square inch absolute (psia).
15. Describe the difference between a mercury and an aneroid barometer.
16. Pressure inside a Bourdon tube pressure gage causes the Bourdon tube to straighten or curl?
17. To change from psig to psia you must add _____ to psig.
18. Convert 80°F to Celsius.
19. Convert 22°C to Fahrenheit.
20. Convert 70 psig to kPa (kilopascal).

UNIT 2

Matter and Energy

OBJECTIVES

After studying this unit, you should be able to

- define matter.
- list the three states in which matter is commonly found.
- define density.
- discuss Boyle's Law.
- state Charles' Law.
- discuss Dalton's Law as it relates to the pressure of different gases.
- define specific gravity and specific volume.
- state two forms of energy important to the air-conditioning (heating and cooling) and refrigeration industry.
- describe work and state the formula used to determine the amount of work in a given task.
- define horsepower.
- convert horsepower to watts.
- convert watts to British thermal units.

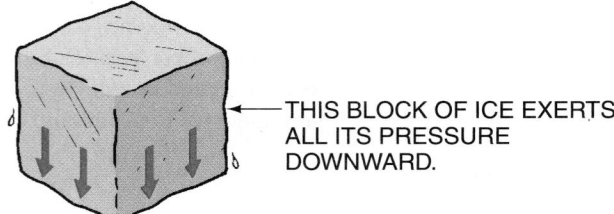

THIS BLOCK OF ICE EXERTS ALL ITS PRESSURE DOWNWARD.

Figure 2–1   Solids exert all their pressure downward. The molecules of solid water have a great attraction for each other and hold together.

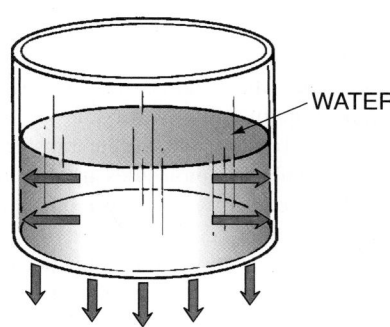

WATER

Figure 2–2   The water in the container exerts pressure outward and downward. The outward pressure is what makes water seek its own level. The water molecules still have a small amount of adhesion to each other. The pressure is proportional to the depth.

2.1   MATTER

Matter is commonly explained as a substance that occupies space and has mass. The weight comes from the earth's gravitational pull. Matter is made up of atoms. Atoms are very small parts of a substance and may combine to form molecules. Atoms of one substance may be combined chemically with those of another to form a new substance. When molecules are formed they cannot be broken down any further without changing the chemical nature of the substance. Matter also exists in three states: *solids, liquids,* and *gases.* The heat content and pressure determine the state of matter. For instance, water is made up of molecules containing atoms of hydrogen and oxygen. Two atoms of hydrogen and one atom of oxygen are in each molecule of water. The chemical expression of these molecules is H_2O.

Water in the solid state is known as ice. It exerts all of its force downward—it has weight. The molecules of the water are highly attracted to each other, **Figure 2–1.**

When the water is heated above the freezing point, it begins to change to a liquid state. The molecular activity is higher, and the water molecules have less attraction for each other. Water in the liquid state exerts a pressure outward and downward. Because water pressure is proportional to its depth, water seeks a level surface where its pressure equals that of the atmosphere above it, **Figure 2–2.**

Water heated above the liquid state, 212°F at standard conditions, becomes a vapor. In the vapor state the molecules have even less attraction for each other and are said to travel at random. The vapor exerts pressure more or less in all directions, **Figure 2–3.**

The study of matter leads to the study of other terms that help to understand how different substances compare with each other.

2.2   MASS AND WEIGHT

Mass is the property of matter that responds to gravitational attraction. *Weight* is the force that matter (solid, liquid, or gas) applies to a supporting surface when it is at rest.

Weight is not a property of matter but is dependent on the gravitational attraction. The stronger the force of gravity, the more an object will weigh. The earth has stronger gravitational attraction forces than the moon. This is why objects weigh more on the earth than on the moon.

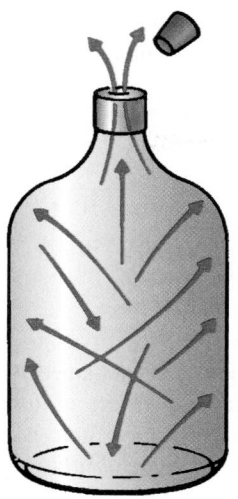

Figure 2–3   Gas molecules travel at random. When a container with a small amount of gas pressure is opened, the gas molecules seem to repel each other and fly out.

All solid matter has mass. A liquid such as water is said to have mass. The air in the atmosphere has weight and mass. When the atmosphere is evacuated out of a jar, the mass is removed, and a vacuum is created.

2.3   DENSITY

The *density* of a substance describes its mass-to-volume relationship. The mass contained in a particular volume is the density of that substance. In the British system of units, volume is measured in cubic feet. Sometimes it is advantageous to compare different substances according to weight per unit volume. Water, for example, has a density of 62.4 lb/ft^3. Wood floats on water because the density (weight per volume) of wood is less than the density of water. In other words, it weighs less per cubic foot. Iron, on the other hand, sinks because it is denser than water. **Figure 2–4** lists some typical densities.

2.4   SPECIFIC GRAVITY

Specific gravity is a unitless number because it is the density of a substance divided by the density of water. Water is simply used as the standard comparison. The density of water is

SUBSTANCE	DENSITY lb/ft^3	SPECIFIC GRAVITY
ALUMINUM	171	2.74
BRASS (RED)	548	8.78
COPPER	556	8.91
GOLD	1208	19.36
ICE @ 32°F	57.5	0.92
TUNGSTEN	1210	19.39
WATER	62.4	1
MARBLE	162	2.596

Figure 2–4   Table of density and specific gravity.

62.4 lb/ft^3. So, the specific gravity of water is 62.4 lb/ft^3 ÷ 62.4 lb/ft^3 = 1. Notice that the units cancel because of the division. The density of red brass is 548 lb/ft^3. Its specific gravity is then 548 lb/ft^3 ÷ 62.4 lb/ft^3 = 8.78. **Figure 2–4** lists some typical specific gravities of substances.

2.5   SPECIFIC VOLUME

Specific volume compares the volume that each pound of gas occupies. It specifies that there must be only 1 lb of the gas. Its units are ft^3/lb. This differs from total volume whose units are simply ft^3. One pound of clean dry air has a total volume of 13.33 ft^3 at standard atmospheric conditions. Its specific volume would then be 13.33 ft^3/lb. Hydrogen has a specific volume of 179 ft^3/lb under the same conditions. Because more cubic feet of hydrogen exist per pound, it has a higher specific volume, thus is lighter than air. Although both are gases, the hydrogen has a tendency to rise when mixed with air.

Specific volume and density are considered inverses of one another. This means that specific volume = 1 ÷ density and that density = 1 ÷ specific volume. If one knows the specific volume of a substance, its density can be calculated and vice versa. For example, the specific volume of dry air is 13.33 ft^3/lb. Its density would then be 1 ÷ 13.33 ft^3/lb = .075 lb/ft^3. Notice that even the units are inverse to one another. Substances with high specific volumes can be said to have low densities. Also, substances with high densities can be said to have low specific volumes.

The specific volume of air is a factor in determining the fan horsepower needed in air-conditioning work. As an example, a low specific volume of air requires a higher horsepower blower motor, and a high specific volume of air requires a lower horsepower blower motor.

Natural gas is explosive when mixed with air, but it is lighter than air and has a tendency to rise like hydrogen. Propane gas is another frequently used heating gas and has to be treated differently from natural gas because it is heavier than air. Propane has a tendency to fall and collect in low places and to cause potential danger from ignition.

The specific volumes of various gases that are pumped is valuable information that enables the engineer to choose the size of the compressor or vapor pump to do a particular job. The specific volumes for vapors vary according to the pressure the vapor is under. An example is refrigerant-22, which is a common refrigerant used in residential air-conditioning units. At 3 psig about 2.5 ft^3 of gas must be pumped to move 1 lb of gas. At the standard design condition of 70 psig, only 0.48 ft^3 of gas needs to be pumped to move 1 lb of the same gas. A complete breakdown of specific volume can be found in the properties of saturated and superheated conditions for liquid and/or vapor in engineering manuals for any refrigerant.

2.6   GAS LAWS

It is necessary to have a working knowledge of gases and how they respond to pressure and temperature changes. Several scientists made significant discoveries many years ago.

A simple explanation of some of the gas laws developed by these scientists may help you understand the reaction of gases and the pressure/temperature/volume relationships in various parts of a refrigeration system. Whenever using pressure or temperature in an equation like the gas laws, one has to use the absolute scales of pressure (psia) and temperature (Rankine or Kelvin), or the solutions to these equations will be meaningless. Absolute scales use zero as their starting points, because zero is where molecular motion actually begins.

Boyle's Law

Robert Boyle, a citizen of Ireland, developed in the early 1600s what has come to be known as Boyle's Law. He discovered that when pressure is applied to a volume of air that is contained, the volume of air becomes smaller and the pressure greater. Boyle's Law states that *the volume of a gas varies inversely with the absolute pressure, provided the temperature remains constant.* For example, if a cylinder with a piston at the bottom and enclosed at the top were filled with air and the piston moved halfway up the cylinder, the pressure of the air would double, **Figure 2–5.** That part of the law pertaining to the temperature remaining constant keeps Boyle's Law from being used in practical situations. This is because when a gas is compressed some heat is transferred to the gas from the mechanical compression, and when gas is expanded heat is given up. However, this law, when combined with another, will make it practical to use.

The formula for Boyle's Law is as follows:

$$P_1 \times V_1 = P_2 \times V_2$$

where P_1 = original absolute pressure
V_1 = original volume
P_2 = new pressure
V_2 = new volume

For example, if the original pressure was 40 psia and the original volume 30 in^3, what would the new volume be if the pressure were increased to 50 psia? We are determining the new volume. The formula would have to be rearranged so we could find the new volume.

$$V_2 = \frac{P_1 \times V_1}{P_2}$$

$$V_2 = \frac{40 \times 30}{50}$$

$$V_2 = 24 \text{ in}^3$$

Charles' Law

In the 1800s, a French scientist named Jacques Charles made discoveries regarding the effect of temperature on gases. Charles' Law states that *at a constant pressure, the volume of a gas varies directly as to the absolute temperature, and at a constant volume, the pressure of a gas varies directly with the absolute temperature.* Stated in a different form, when a gas is heated and if it is free to expand, it will do so, and the volume will vary directly as to the absolute temperature. If a gas is confined in a container that will not expand and it is heated, the pressure will vary directly with the absolute temperature.

This law can also be stated with formulas. Two formulas are needed because one part of the law pertains to pressure and temperature and the other part to volume and temperature.

This formula pertains to volume and temperature:

$$\frac{V_1}{T_1} = \frac{V_2}{T_2}$$

where V_1 = original volume
V_2 = new volume
T_1 = original temperature
T_2 = new temperature

If 2000 ft^3 of air is passed through a gas furnace and heated from 75°F room temperature to 130°F, what is the volume of the air leaving the heating unit? See **Figure 2–6.**

$V_1 = 2000 \text{ ft}^3$
$T_1 = 75°F + 460° = 535°R \text{ (absolute)}$
$V_2 = \text{unknown}$
$T_2 = 130°F + 460° = 590°R$

We must mathematically rearrange the formula so that the unknown is alone on one side of the equation.

$$V_2 = \frac{V_1 \times T_2}{T_1}$$

$$V_2 = \frac{2000 \text{ ft}^3 \times 590°R}{535°R}$$

$$V_2 = 2205.6 \text{ ft}^3$$

The air expanded when heated.
The following formula pertains to pressure and temperature:

$$\frac{P_1}{T_1} = \frac{P_2}{T_2}$$

where P_1 = original pressure
T_1 = original temperature
P_2 = new pressure
T_2 = new temperature

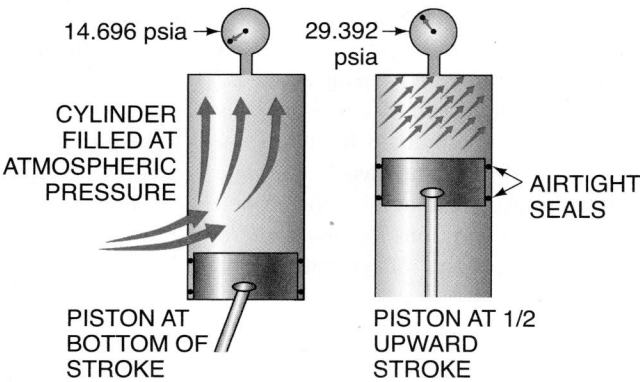

14.696 psia →   29.392 → psia

CYLINDER
FILLED AT
ATMOSPHERIC
PRESSURE

AIRTIGHT
SEALS

PISTON AT
BOTTOM OF
STROKE

PISTON AT 1/2
UPWARD
STROKE

Figure 2–5   Absolute pressure in a cylinder doubles when the volume is reduced by half.

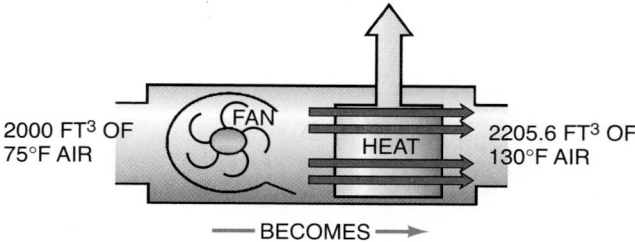

Figure 2–6   Air expands when heated.

If a large natural gas tank holding 500,000 ft³ of gas is stored at 70°F in the spring and the temperature rises to 95°F in the summer, what would the pressure be if the original pressure was 25 psig in the spring?

$$P_1 = 25 \text{ psig} + 14.696 \text{ (atmospheric pressure) or}$$
$$39.696 \text{ psia}$$
$$T_1 = 70°F + 460°R \text{ or } 530°R \text{ (absolute)}$$
$$P_2 = \text{unknown}$$
$$T_2 = 95°F + 460° \text{ or } 555°R \text{ (absolute)}$$

Again the formula must be rearranged so that the unknown is on one side of the equation by itself.

$$P_2 = \frac{P_1 \times T_2}{T_1}$$

$$P_2 = \frac{39.696 \text{ psia} \times 555°R}{530°R}$$

$$P_2 = 41.57 \text{ psia} - 14.696 = 26.87 \text{ psig}$$

General Law of Perfect Gas

A general gas law, often called the General Law of Perfect Gas, is a combination of Boyle's and Charles' Laws. This combination law is more practical because it includes temperature, pressure, and volume.

The formula for this law can be stated as follows:

$$\frac{P_1 \times V_1}{T_1} = \frac{P_2 \times V_2}{T_2}$$

where P_1 = original pressure
V_1 = original volume
T_1 = original temperature
P_2 = new pressure
V_2 = new volume
T_2 = new temperature

For example, 20 ft³ of gas is being stored in a container at 100°F and a pressure of 50 psig. This container is connected by pipe to one that will hold 30 ft³ (a total of 50 ft³), and the gas is allowed to equalize between the two containers. The temperature of the gas is lowered to 80°F. What is the pressure in the combined containers?

$$P_1 = 50 \text{ psig} + 14.696 \text{ or } 64.696$$
$$V_1 = 20 \text{ ft}^3$$
$$T_1 = 100° + 460°R = 560°R$$
$$P_2 = \text{unknown}$$
$$V_2 = 50 \text{ ft}^3$$
$$T_2 = 80°F + 460° \text{ or } 540°R$$

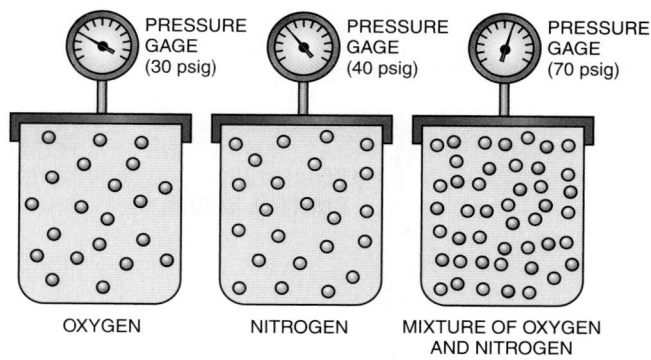

Figure 2–7   Dalton's Law of Partial Pressures. The total pressure is the sum of the individual pressures of each gas.

The formula is mathematically rearranged to solve for the unknown P_2:

$$P_2 = \frac{P_1 \times V_1 \times T_2}{T_1 \times V_2}$$

$$P_2 = \frac{64.696 \times 20 \times 540}{560 \times 50}$$

$$P_2 = 24.95 \text{ psia} - 14.696 = 10.26 \text{ psig}$$

Dalton's Law

In the early 1800s, John Dalton, an English mathematics professor, made the discovery that the atmosphere is made up of several different gases. He found that each gas created its own pressure and that the total pressure was the sum of each. Dalton's Law states that *the total pressure of a confined mixture of gases is the sum of the pressures of each of the gases in the mixture.* For example, when nitrogen and oxygen are placed in a closed container, the pressure on the container will be the total pressure of the nitrogen as if it were in the container by itself added to the oxygen pressure in the container by itself, **Figure 2–7.**

2.7   ENERGY

Using energy properly to operate equipment is a major goal of the air-conditioning and refrigeration industry. Energy in the form of electricity drives the motors; heat energy from the fossil fuels of natural gas, oil, and coal heats homes and industry. What is this energy and how is it used?

The only new energy we get is from the sun heating the earth. Most of the energy we use is converted to usable heat from something already here (e.g., fossil fuels). This conversion from fuel to heat can be direct or indirect. An example of direct conversion is a gas furnace, which converts the gas flame to usable heat by combustion. The gas is burned in a combustion chamber, and heat from combustion is transferred to circulated air by conduction through the heat exchanger wall of thin steel. Some gas furnaces may have a condensing heat exchanger. This will be explained in the unit on gas heat. The heated air is then distributed throughout the heated space, **Figure 2–8.**

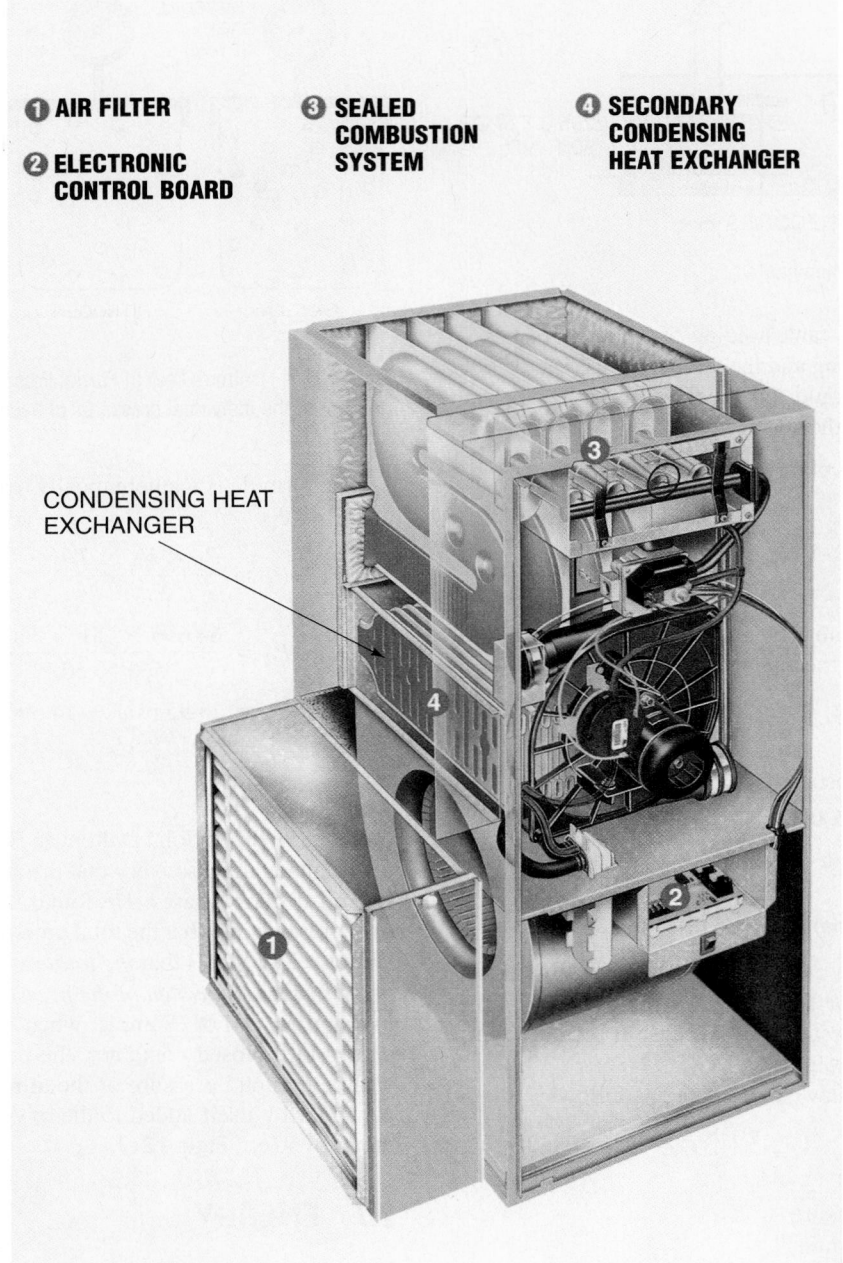

❶ **AIR FILTER**

❷ **ELECTRONIC CONTROL BOARD**

❸ **SEALED COMBUSTION SYSTEM**

❹ **SECONDARY CONDENSING HEAT EXCHANGER**

CONDENSING HEAT EXCHANGER

Figure 2–8   A high-efficiency furnace with burners on top of the heat exchanger. *Courtesy Bryant Heating and Cooling Systems*

An example of indirect conversion is a fossil-fuel power plant. Gas may be used in the power plant to produce the steam that turns a steam turbine generator to produce electricity. The electricity is then distributed by the local power company and consumed locally as electric heat, **Figure 2–9.**

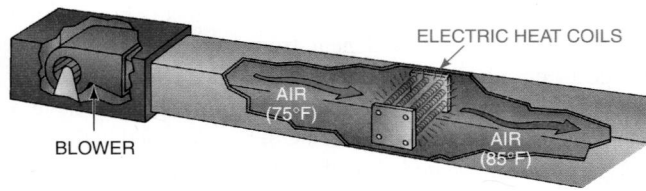

ELECTRIC HEAT COILS

AIR (75°F)

AIR (85°F)

BLOWER

Figure 2–9   An electric heat airstream.

2.8   CONSERVATION OF ENERGY

The preceding leads to the law of conservation of energy. This law states that *energy is neither created nor destroyed but can be converted from one form to another.* It can then be said that energy can be accounted for.

Most of the energy we use is a result of the sun supporting plant growth for thousands of years. Fossil fuels come from decayed vegetable and animal matter covered by earth and rock during changes in the earth's surface. This decayed matter is in various states, such as gas, oil, or coal, depending on the conditions it was subjected to in the past, **Figure 2–10.** The energy stored in fossil fuels is called chemical energy because a chemical reaction is needed to release the energy.

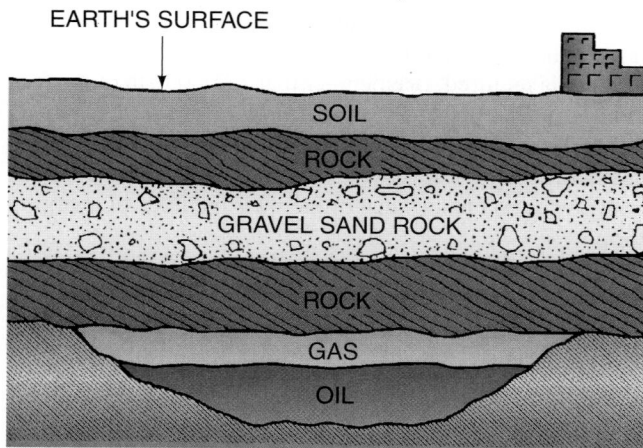

Figure 2–10   Gas and oil deposits settle into depressions.

2.9   ENERGY CONTAINED IN HEAT

Temperature is a measure of the degree of heat or the heat intensity, but not necessarily the amount of heat. Heat is a form of energy because of the motion of molecules. To be specific, heat is thermal energy. If two substances of different temperatures are moved close to each other, heat from the substance with the higher temperature will flow to the one with the lower temperature, **Figure 2–11(A)**. Because molecular motion does not stop until −460°F, energy is still available in a substance even at very low temperatures. This energy is in relationship to other substances that are at lower temperatures. For example, if two substances at very low temperatures are moved together, heat will transfer from the warmer substance to the colder one. In **Figure 2–11(B)** a substance at −200°F

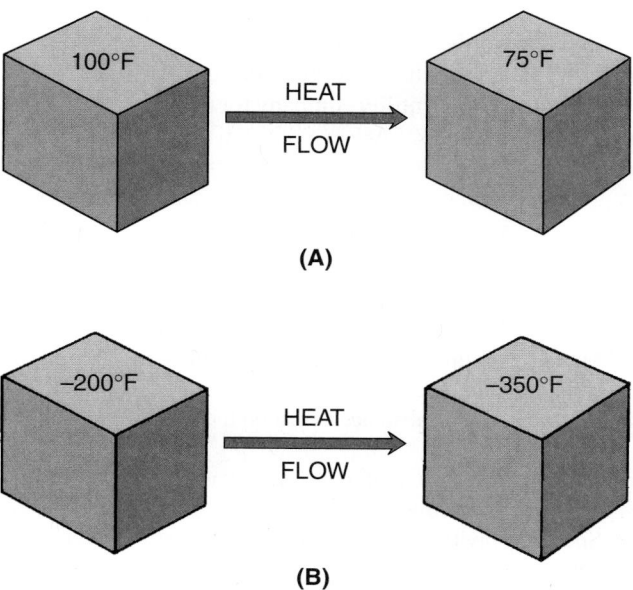

Figure 2–11   (A) If two substances of different temperatures are moved close to each other, heat from the substance with the higher temperature will flow to the one with the lower temperature. **(B)** Heat energy is still available at these low temperatures and will transfer from the warmer substance to the colder substance.

is placed next to a substance at −350°F. As we discussed earlier, the warmer substance gives up heat (energy) to the cooler substance. The energy used by home and industry is not at these low levels.

Most of the heat energy used in homes and industry is provided from fossil fuels, but some comes from electrical energy. As shown in **Figure 2–9**, electron flow in high-resistance wire causes the wire to become hot, thus heating the air. A moving airstream is then passed over the heated wire, allowing heat to be transferred to the air by conduction and moved to the heated space by forced convection (the fan).

2.10   ENERGY IN MAGNETISM

Magnetism is another method of converting electron flow to usable energy. Electron flow is used to develop magnetism to turn motors. The motors turn the prime movers—fans, pumps, compressors—of air, water, and refrigerant. In **Figure 2–12** an electric motor turns a water pump to boost the water pressure from 20 to 60 psig. This takes energy. The energy in this example is purchased from the power company.

The preceding examples serve only as an introduction to the concepts of chemical energy, heat energy, and electrical energy. Each subject will later be covered in detail. For now it is important to realize that any system furnishing heating or cooling uses energy.

2.11   PURCHASE OF ENERGY

Energy must be transferred from one owner to another and accounted for. This energy is purchased as a fossil fuel or as electric power. Energy purchased as a fossil fuel is normally purchased by the unit. Natural gas is an example. Natural gas flows through a meter that measures how many cubic feet have passed during some time span, such as a month. Fuel oil is normally sold by the gallon, coal by the ton. Electrical energy is sold by the kilowatt-hour or kWh. The amount of heat each of these units contains is known, so a known amount of heat is purchased. Natural gas, for instance, has a heat content of about 1000 Btu/ft^3; whereas the heat content of coal varies from one type of coal to another.

Figure 2–12   Electrical energy used in an electric motor is converted to work to boost water pressure to force circulation.

2.12 ENERGY USED AS WORK

Energy purchased from electrical utilities is known as electric power. *Power* is the rate of doing work. *Work* can be explained as a force moving an object in the direction of the force. It is expressed by this formula:

$$\text{Work} = \text{Force} \times \text{Distance}$$

For instance, when a 150-lb man climbs a flight of stairs 100 ft high (about the height of a 10-story building), he performs work. But how much? The amount of work in this example is equivalent to the amount of work necessary to lift this man the same height. We can calculate the work by using the preceding formula.

$$\text{Work} = 150 \text{ lb} \times 100 \text{ ft}$$
$$= 15,000 \text{ ft-lb}$$

Notice that no time limit has been added. This example can be accomplished by a healthy man in a few minutes. But if the task were to be accomplished by a machine such as an elevator, more information is necessary. Do we want to take seconds, minutes, or hours to do the job? The faster the job is accomplished, the more power is required.

2.13 POWER

Power is the rate of doing work. An expression of power is *horsepower (hp)*. Many years ago it was determined that an average good horse could lift the equivalent of 33,000 lb to a height of 1 ft in 1 min, which is the same as 33,000 ft-lb/min or 1 hp. This describes a rate of doing work because time has been added. Keep in mind that lifting 330 lb to a height of 100 ft in 1 min or lifting 660 lb 50 ft in 1 min is the same amount of work. As a point of reference, the fan motor in the average furnace can be rated at 1/2 hp. See **Figure 2–13** for an illustration of the horse lifting 1 hp.

When the horsepower is compared with the man climbing the stairs, the man would have to climb the 100 ft in less than 30 sec to equal 1 hp. That makes the task seem even

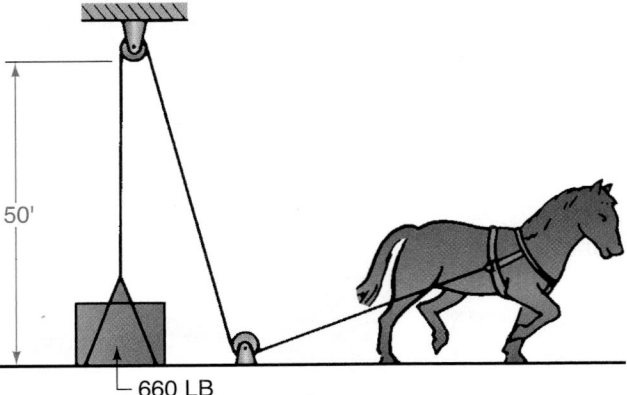

Figure 2–13 When a horse can lift 660 lb to a height of 50 ft in 1 min, it has done the equivalent of 33,000 ft-lb of work in 1 min, or 1 hp.

harder. A 1/2-hp motor could lift the man 100 ft in 1 min if only the man were lifted. The reason is that 15,000 ft-lb of work is required. (Remember that 33,000 ft-lb of work in 1 min equals 1 hp.)

Our purpose in discussing these topics is to help you understand how to use power effectively and to understand how power companies determine their methods of charging for power.

2.14 ELECTRICAL POWER—THE WATT

The unit of measurement for electrical power is the *watt (W)*. This is the unit used by the power company. When converted to electrical energy, 1 hp = 746 W; that is, when 746 W of electrical power is properly used, the equivalent of 1 hp of work has been accomplished.

Fossil-fuel energy can be compared with electrical energy, and one form of energy can be converted to the other. There must be some basis, however, for conversion so that one fuel can be compared with another. The examples we use to illustrate this comparison will not take efficiencies into account. Efficiencies for the various fuels will be covered in the section on applications for each fuel. Some examples of conversions follow.

1. **Converting electric heat rated in kilowatts (kW) to the equivalent gas or oil heat rated in Btu.** Suppose that we want to know the capacity in Btu for a 20-kW electric heater (a kilowatt is 1000 watts). 1 kW = 3413 Btu.

   20 kW × 3413 Btu/kW = 68,260 Btu of heat energy

2. **Converting Btu to kW.** Suppose that a gas or oil furnace has an output capacity of 100,000 Btu/h. Since 3413 Btu = 1 kW, we have

   100,000 Btu ÷ 3413 Btu/kW = 29.3 kW

In other words, a 29.3-kW electric heat system would be required to replace the 100,000-Btu/h furnace.

Contact the local utility company for rate comparisons between different fuels. SAFETY PRECAUTION: Any device that consumes power, such as an electric motor or gas furnace, is potentially dangerous. These devices should only be handled or adjusted by experienced people.

SUMMARY

- Matter takes up space, has mass, and can be in the form of a solid, a liquid, or a gas.
- The weight of a substance at rest on the earth is proportional to its mass.
- In the British system of units, density is the weight of a substance per cubic foot.
- Specific gravity is the term used to compare the density of various substances.
- Specific volume is the amount of space a pound of a vapor or a gas will occupy.
- Boyle's Law states that the volume of a gas varies inversely with the absolute pressure, provided the temperature remains constant.

▓ Charles' Law states that at a constant pressure, the volume of a gas varies directly as to the absolute temperature, and at a constant volume the pressure of a gas varies directly with the absolute temperature.

▓ Dalton's Law states that the total pressure of a confined mixture of gases is the sum of the pressures of each of the gases in the mixture.

▓ Electrical energy and heat energy are two forms of energy used in this industry.

▓ Fossil fuels are purchased by the unit. Natural gas is metered by the cubic foot; oil is purchased by the gallon; and coal is purchased by the ton. Electricity is purchased from the electric utility company by the kilowatt-hour (kWh).

▓ Work is the amount of force necessary to move an object: Work = Force × Distance.

▓ Horsepower is the equivalent of lifting 33,000 lb to a height of 1 ft in 1 min or some combination totaling the same.

▓ Watts are a measurement of electrical power. One horsepower equals 746 W.

▓ 3.413 Btu = 1 W. 1 kW (1000 W) = 3413 Btu.

REVIEW QUESTIONS

1. Matter is a substance that occupies space and has
   A. color.
   B. texture.
   C. temperature.
   D. mass.

2. What are the three states in which matter is commonly found?

3. _____ is the term used for water when it is in the solid state.

4. In what direction does a solid exert force?

5. In what direction does a liquid exert force?

6. Vapor exerts pressure in what direction?
   A. Outward
   B. Upward
   C. Downward
   D. All of the above

7. Define density.

8. Define specific gravity.

9. Describe specific volume.

10. Why does an object weigh less on the moon than on the earth?

11. The density of tungsten is 1210 lb/ft^3. What would be its specific volume?

12. The specific volume of red brass is .001865 ft^3/lb. What would be its density?

13. Aluminum has a density of 171 lb/ft^3. What would be its specific gravity?

14. Four pounds of a gas occupies 10 ft^3. What would be its total volume, density, and specific gravity?

15. Why is information regarding the specific volume of gases important to the designer of air-conditioning, heating, and refrigeration equipment?

16. Whose law states that the volume of a gas varies inversely with the absolute pressure, as long as the temperature remains constant?
   A. Charles'
   B. Boyle's
   C. Newton's
   D. Dalton's

17. At a constant pressure how does a volume of gas vary with respect to the absolute temperature?

18. Describe Dalton's Law as it relates to a confined mixture of gases.

19. What are the two types of energy most frequently used or considered in this industry?

20. How were fossil fuels formed?

21. _____ is the time rate of doing work.

22. State the formula for determining the amount of work accomplished in a particular task.

23. If an air-conditioning compressor weighing 300 lb had to be lifted 4 ft to be mounted on a base, how many ft-lb of work must be accomplished?

24. Describe horsepower and list the three quantities needed to determine horsepower.

25. How many watts of electrical energy are equal to 1 hp?

26. How many Btu of heat can be produced by 4 kWh of electricity?

27. How many Btu/h would be produced in a 12-kW electric heater?

28. What unit of energy does the power company charge the consumer for?

29. If a 30-ft^3 volume of air at 10 psig is compressed to 25 ft^3 at a constant temperature, what would be the new pressure in psig?

30. If 3000 ft^3 of air is crossing an evaporator coil and is cooled from 75°F to 55°F, what would be the volume of air in ft^3 exiting the evaporator coil?

31. A gas is compressed inside a compressor's cylinder. When the piston is at its bottom dead center, the gas is initially at 10 psig, 65°F, and 10.5 in^3. After compression and when the piston is at top dead center, the gas is 180°F and occupies 1.5 in^3. What would be the new pressure of the gas in psig?

UNIT 3

Refrigeration and Refrigerants

OBJECTIVES

After studying this unit, you should be able to

- discuss applications for high-, medium-, and low-temperature refrigeration.
- describe the term *ton of refrigeration.*
- describe the basic refrigeration cycle.
- explain the relationship between pressure and the boiling point of water or other liquids.
- describe the function of the evaporator or cooling coil.
- explain the purpose of the compressor.
- list the compressors normally used in residential and light commercial buildings.
- discuss the function of the condensing coil.
- state the purpose of the metering device.
- list four characteristics to consider when choosing a refrigerant for a system.
- list the designated colors for refrigerant cylinders for various types of refrigerants.
- describe how refrigerants can be stored or processed while refrigeration systems are being serviced.
- plot a refrigeration cycle for refrigerants (R-22, R-12, R-134a, and R-502) on a pressure/enthalpy diagram.
- plot a refrigeration cycle on a pressure/enthalpy diagram for refrigerant blends R-404A and R-410A.
- plot a refrigeration cycle on a pressure/enthalpy diagram for a refrigerant blend (R-407C) that has a noticeable temperature glide.

SAFETY CHECKLIST

✔ Areas in which there is the potential for refrigerant leaks should be properly ventilated.
✔ Extra precautions should be taken to ensure that no refrigerant leaks occur near an open flame.
✔ Refrigerants are stored in pressurized containers and should be handled with care. Goggles with side shields and gloves should be worn when checking pressures and when transferring refrigerants from the container to a system or from the system to an approved container.

3.1 INTRODUCTION TO REFRIGERATION

This unit is an introduction to refrigeration and refrigerants. The term refrigeration is used here to include both the cooling process for preserving food and comfort cooling (air conditioning).

Preserving food is one of the most valuable uses of refrigeration. The rate of food spoilage gets slower as molecular motion slows. This retards the growth of bacteria that causes food to spoil. Below the frozen hard point, food-spoiling bacteria stop growing. The frozen hard point for most foods is considered to be 0°F. The food temperature range between 35°F and 45°F is known in the industry as medium temperature; below 0°F is considered low temperature. These ranges are used to describe many types of refrigeration equipment and applications. Refrigeration systems that operate to produce warmer temperatures are referred to as high-temperature refrigeration systems. Comfort cooling systems, commonly referred to as air-conditioning systems, are used, for example, to cool our homes and commercial spaces and are classified as high-temperature refrigeration systems.

For many years dairy products and other perishables were stored in the coldest room in the house, the basement, the well, or a spring. In the South, temperatures as low as 55°F could be reached in the summer with underground water. This would add to the time that some foods could be kept. Ice in the North and to some extent in the South was placed in "ice boxes" in kitchens. The ice melted when it absorbed heat from the food in the box, cooling the food, **Figure 3–1.**

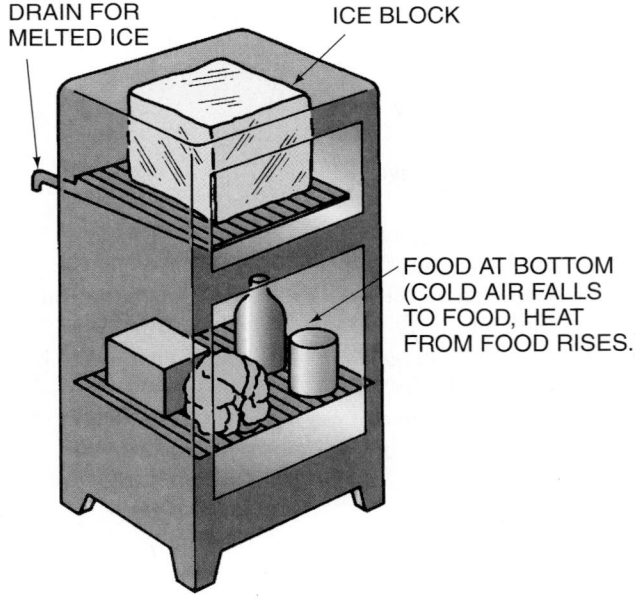

DRAIN FOR MELTED ICE

ICE BLOCK

FOOD AT BOTTOM (COLD AIR FALLS TO FOOD, HEAT FROM FOOD RISES.)

Figure 3–1 Ice boxes were made of wood at first, then metal. The boxes were insulated with cork. If a cooling unit were placed where the ice is, this would be a refrigerator.

In the early 1900s, ice was manufactured by mechanical refrigeration and sold to people with ice boxes, but still only the wealthy could afford it.

Also in the early 1900s, some companies manufactured the household refrigerator. Like all new items, it took a while to become popular. Now, most houses have a refrigerator with a freezing compartment. Modern refrigerators have become state-of-the-art appliances—and some models even include automatic beverage and ice dispensers, built-in television screens, and connections to the World Wide Web.

Frozen food was just beginning to become popular about the time World War II began. Because most people did not have a freezer at this time, central frozen food locker plants were established so that a family could have its own locker. Food that is frozen fresh is appealing because it stays fresh for a longer period of time. Refrigerated and frozen foods are so common now that most people take them for granted.

3.2 REFRIGERATION

Refrigeration **is the process of removing heat from a place where it is not wanted and transferring that heat to a place where it makes little or no difference.** In the average household, the room temperature from summer to winter is normally between 70°F and 90°F. The temperature inside the refrigerator fresh food section should be about 35°F. Heat flows naturally from a warm level to a cold level. Therefore, heat in the room is trying to flow into the refrigerator, and it does through the insulated walls, the door when it is opened, and warm food placed in the refrigerator, **Figure 3–2, Figure 3–3,** and **Figure 3–4.** For this reason, to

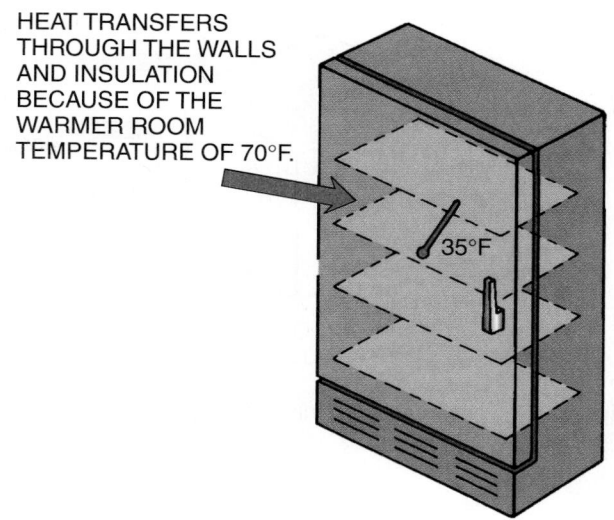

Figure 3–3   Heat transfers through the walls into the box by conduction. The walls have insulation, but this does not stop the heat leakage completely.

Figure 3–4   Warm food that is placed in the refrigerator adds heat to the refrigerator and is also considered heat leakage. This added heat has to be removed or the temperature inside the refrigerator will rise.

increase the efficiency of the unit, it is always best to allow food to cool down to room temperature before placing it in the refrigerator.

3.3 RATING REFRIGERATION EQUIPMENT

Refrigeration equipment must also have a capacity rating system so that equipment of different manufacturers and models can be compared. The method for rating refrigeration

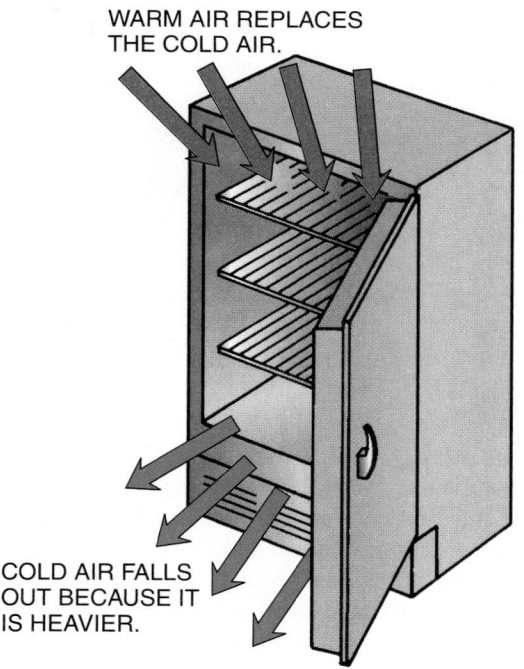

Figure 3–2   The colder air falls out of the refrigerator because it is heavier than the warmer air located outside the refrigerator. The cooler air is replaced with warmer air at the top. This is referred to as heat leakage.

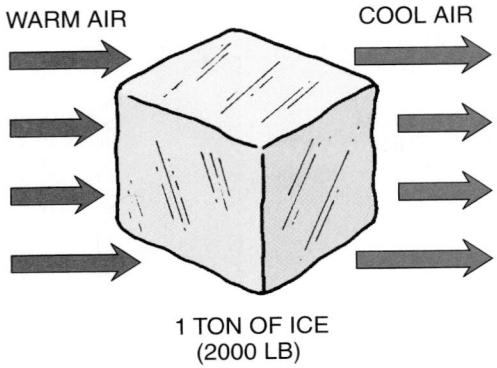

Figure 3–5 Ice requires 144 Btu/lb to melt. Melting 1 ton of ice requires 288,000 Btu (2000 lb × 144 Btu/lb = 288,000 Btu).

equipment capacity goes back to the days of using ice as the source for removing heat. It takes 144 Btu of heat energy to melt 1 lb of ice at 32°F. This same figure is also used in the capacity rating of refrigeration equipment.

The term for this capacity rating is the ton. *One ton of refrigeration* is the amount of heat required to melt 1 ton of ice in a 24-hour period. Previously, we saw that it takes 144 Btu of heat to melt 1 lb of ice. It would then take 2000 times that much heat to melt a ton of ice (2000 lb = 1 ton):

$$144 \text{ Btu/lb} \times 2000 \text{ lb} = 288,000 \text{ Btu}$$

When accomplished in a 24-hour period, it is known as 1 ton of refrigeration. The same rules apply when removing heat from a substance. For example, an air conditioner that has a 1-ton capacity will remove 288,000 Btu/24 h or 12,000 Btu/h (288,000 ÷ 24 = 12,000) or 200 Btu/min (12,000 ÷ 60 = 200), **Figure 3–5.**

3.4 THE REFRIGERATION PROCESS

The refrigerator has to pump the heat up the temperature scale from the 35°F or 0°F refrigeration compartments to the 70°F room. The components of the refrigerator are used to accomplish this task, **Figure 3–6.** The heat leaking into the refrigerator raises the air temperature but does not normally raise the temperature of the food an appreciable amount. If it did, the food would spoil. When the temperature inside the refrigerator rises to a predetermined level, the refrigeration system comes on and pumps the heat out.

The process of pumping heat out of the refrigerator could be compared to pumping water from a valley to the top of a hill. It takes just as much energy to pump water up the hill as it does to carry it. A water pump with a motor accomplishes work. If a gasoline engine, for instance, were driving the pump, the gasoline would be burned and converted to work energy. An electric motor uses electric power as work energy, **Figure 3–7.** Refrigeration is the process of moving heat from an area of lower temperature into an area or medium with higher temperature. This takes energy that must be purchased.

Following is an example using a residential window air-conditioning system to explain the basics of refrigeration. Residential air conditioning, whether a window unit or a

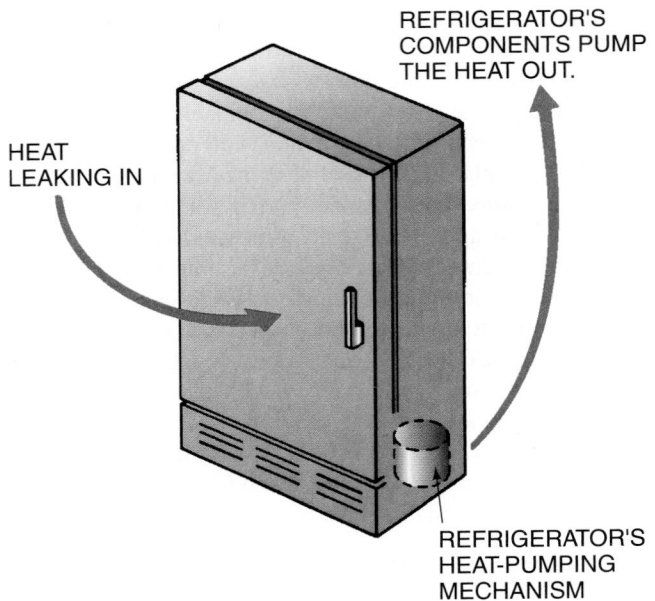

Figure 3–6 Heat that leaks into the refrigerator from any source must be removed by the refrigerator's heat-pumping mechanism. The heat has to be pumped from the cool, 35°F interior of the refrigerator to the warmer, 70°F air in the room in which the refrigerator is located.

Figure 3–7 Power is required to pump water uphill. The same is true for pumping heat up the temperature scale from a 35°F box temperature to a 70°F room temperature.

central system, is considered to be high-temperature refrigeration and is used for comfort cooling. The residential system can be seen from the outside, touched, and listened to for examples that will be given.

The refrigeration concepts utilized in the residential air conditioner are the same as those in the household refrigerator. It pumps the heat from inside the house to the outside of the house, and the household refrigerator pumps heat from the refrigerator into the kitchen. In addition, heat leaks into the house just as heat leaks into the refrigerated compartments in the refrigerator.

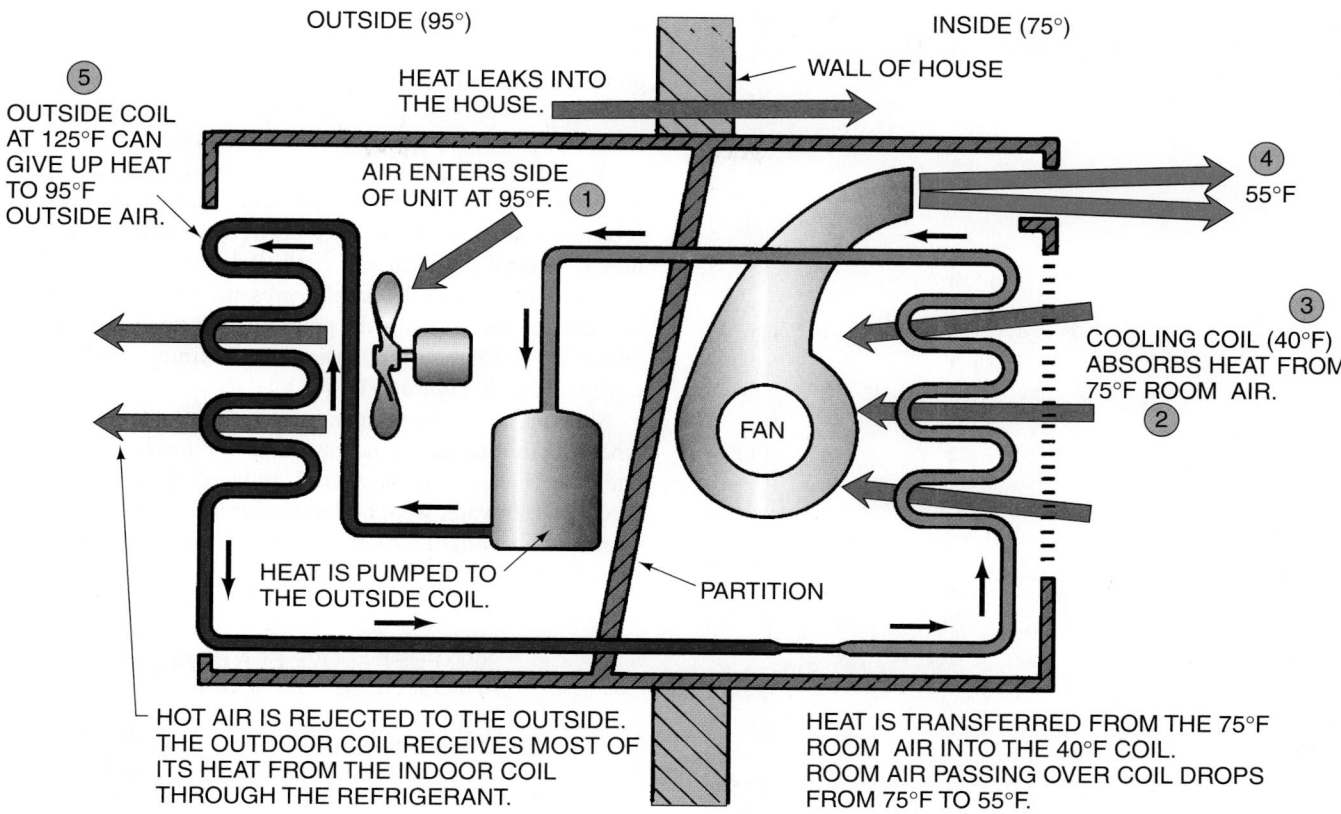

OUTSIDE (95°)   HEAT LEAKS INTO   INSIDE (75°)
THE HOUSE.   WALL OF HOUSE

⑤
OUTSIDE COIL
AT 125°F CAN
GIVE UP HEAT
TO 95°F
OUTSIDE AIR.

AIR ENTERS SIDE
OF UNIT AT 95°F. ①

④
55°F

③
COOLING COIL (40°F)
ABSORBS HEAT FROM
75°F ROOM AIR.

②

FAN

HEAT IS PUMPED TO
THE OUTSIDE COIL.

PARTITION

HOT AIR IS REJECTED TO THE OUTSIDE.
THE OUTDOOR COIL RECEIVES MOST OF
ITS HEAT FROM THE INDOOR COIL
THROUGH THE REFRIGERANT.

HEAT IS TRANSFERRED FROM THE 75°F
ROOM AIR INTO THE 40°F COIL.
ROOM AIR PASSING OVER COIL DROPS
FROM 75°F TO 55°F.

Figure 3–8 A window air-conditioning unit.

When heat enters the house, it must be removed. The heat is transferred outside by the air-conditioning system. The cold air in the house is recirculated air. Room air at approximately 75°F goes into the air-conditioning unit, and air at approximately 55°F comes out. This is the same air with some of the heat removed, **Figure 3–8.**

The following example illustrates this concept. The following statements are also guidelines to some of the design data used throughout the air-conditioning field.

1. The outside design temperature is 95°F.
2. The inside design temperature is 75°F.
3. The design cooling coil temperature is 40°F. This coil transfers heat from the room into the refrigeration system. Notice that with a 75°F room temperature and a 40°F cooling coil temperature, heat will transfer from the room air into the refrigerant in the coil.
4. This heat transfer makes the air leaving the coil and entering the fan about 55°F. The air exits the fan also at 55°F.
5. The outside coil temperature is 125°F. This coil transfers heat from the system to the outside air. Notice that when the outside air temperature is 95°F and the coil temperature is 125°F, heat will be transferred from the system to the outside air.

Careful examination of **Figure 3–8** shows that heat from the house is transferred into the refrigeration system through the inside coil and transferred to the outside air from the refrigeration system through the outside coil. The air-conditioning system is actually pumping the heat out of the

house. The system capacity must be large enough to pump the heat out of the house faster than it is leaking back in so the occupants will not become uncomfortable.

3.5 TEMPERATURE AND PRESSURE RELATIONSHIP

To understand the refrigeration process, we must go back to **Figure 1–15** (heat/temperature graph), where water was changed to steam. Water boils at 212°F at 29.92 in. Hg pressure. This suggests that water has other boiling points. The next statement is one of the most important in this text. You may wish to memorize it. **The boiling point of water can be changed and controlled by controlling the vapor pressure above the water.** Understanding this concept is necessary because water is used as the heat transfer medium in the following example. The next few paragraphs are important for understanding refrigeration.

The *temperature/pressure relationship* correlates the vapor pressure and the boiling point of water and is the basis for controlling the system's temperatures. So, if we are able to control the pressures in a refrigeration or air-conditioning system, we will be able to control the temperatures that the system will maintain.

Pure water boils at 212°F at sea level when the barometric pressure is at the standard value of 29.92 in. Hg. This condition exerts an atmospheric pressure of 14.696 psia (0 psig) on the water's surface. This reference point is on the last line of the table in **Figure 3–9.** Also see **Figure 3–10,** showing the container

WATER TEMPERATURE °F	ABSOLUTE PRESSURE	
	lb/in² (psia)	in. Hg
10	0.031	0.063
20	0.050	0.103
30	0.081	0.165
32	0.089	0.180
34	0.096	0.195
36	0.104	0.212
38	0.112	0.229
40	0.122	0.248
42	0.131	0.268
44	0.142	0.289
46	0.153	0.312
48	0.165	0.336
50	0.178	0.362
60	0.256	0.522
70	0.363	0.739
80	0.507	1.032
90	0.698	1.422
100	0.950	1.933
110	1.275	2.597
120	1.693	3.448
130	2.224	4.527
140	2.890	5.881
150	3.719	7.573
160	4.742	9.656
170	5.994	12.203
180	7.512	15.295
190	9.340	19.017
200	11.526	23.468
210	14.123	28.754
212	14.696	29.921

Figure 3–9   The boiling point of water. The temperature at which water will boil at a specified pressure can be found on the temperature/pressure chart for water.

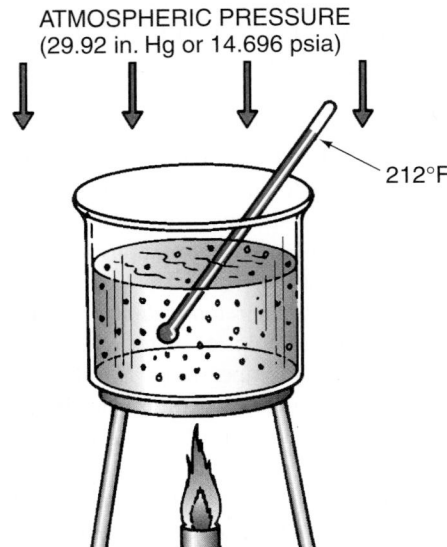

Figure 3–10   Water boils at 212°F when the atmospheric pressure is 29.92 in. Hg.

of water boiling at sea level at atmospheric pressure. When this same pan of water is taken to a mountaintop, the boiling point changes, **Figure 3–11,** because the thinner atmosphere causes a reduction in pressure (about 1 in. Hg/1000 ft). In Denver, Colorado, for example, which is about 5000 ft above sea level, the atmospheric pressure is approximately 25 in. Hg. Water boils at 203.4°F at that pressure. This makes cooking foods such as potatoes and dried beans more difficult because they now need more time to cook. But, by placing the food in a closed container that can be pressurized, such as a pressure cooker, and allowing the pressure to go up to about 15 psi above atmosphere (or 30 psia), the boiling point can be raised to 250°F, **Figure 3–12.**

Studying the water temperature/pressure table reveals that whenever the pressure is increased, the boiling point increases, and that whenever the pressure is reduced, the boiling point is reduced. If water were boiled at a temperature low enough to absorb heat out of a room, we could have comfort cooling (air conditioning).

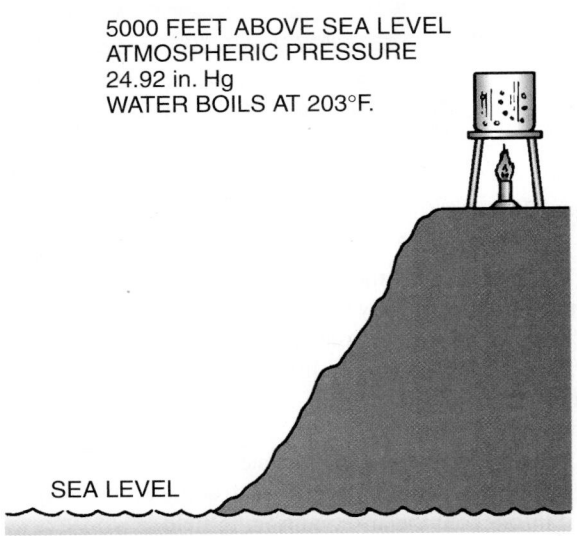

Figure 3–11   Water boils at 203°F when the atmospheric pressure is 24.92 in. Hg.

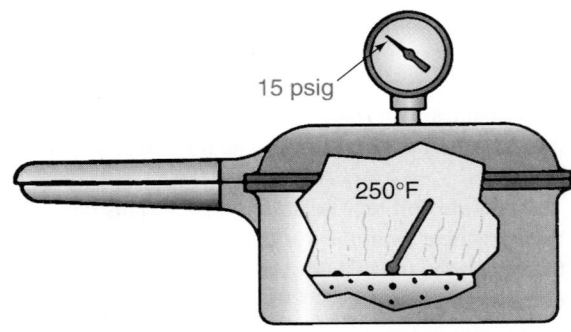

Figure 3–12   The water in a pressure cooker boils at 250°F. As heat is added, the water boils to make vapor. Since the vapor cannot escape, the vapor pressure rises to 15 psig. The water boils at a temperature higher than 212°F because the pressure in the vessel has risen to a level above atmospheric pressure.

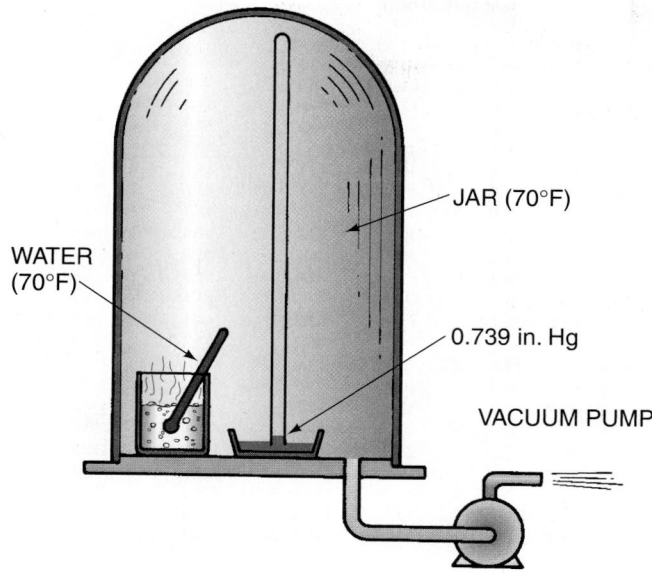

Figure 3–13   The pressure in the bell jar is reduced to 0.739 in. Hg. The boiling temperature of the water is reduced to 70°F because the pressure is 0.739 in. Hg (0.363 psia).

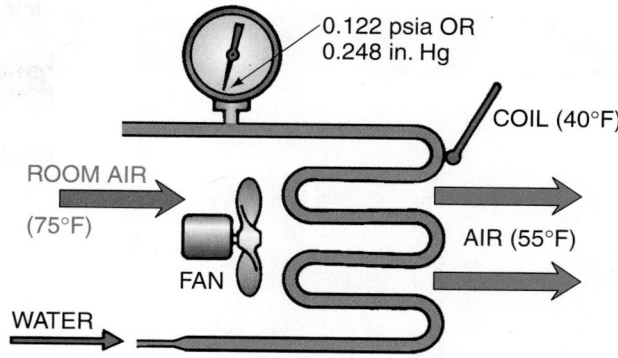

Figure 3–14   The water is boiling at 40°F because the pressure is 0.122 psia or 0.248 in. Hg. The room air is 75°F and gives up heat to the 40°F coil.

Let us place a thermometer in the pan of pure water, put the pan inside a bell jar with a barometer, and start the vacuum pump. Suppose the water is at room temperature (70°F). When the pressure in the jar reaches the pressure that corresponds to the boiling point of water at 70°F, the water will start to boil and vaporize. This point is 0.739 in. Hg (0.363 psia). **Figure 3–13** illustrates the container in the jar. These figures can be found in the table in **Figure 3–9.**

Notice that, in **Figure 3–9,** the temperatures are listed in the left-hand column and the pressures are found in the body of the chart, hence the name temperature/pressure chart (or table). Sometimes, however, the information is presented differently. On some charts, the pressures are located in the left-hand column and the temperatures are found in the body of the chart. These charts or tables are referred to as pressure/temperature charts. Although the information presented is the same, the way that it is presented is different. Make certain that you know which type of chart you are using—or inaccurate calculations and system conclusions may result.

If we were to lower the pressure in the jar to correspond to a temperature of 40°F, this new pressure of 0.248 in. Hg (0.122 psia) will cause the water to boil at 40°F. The water is not hot even though it is boiling. The thermometer in the pan indicates this. If the jar were opened to the atmosphere, the water would be found to be cold. Also, if the jar were opened to the atmosphere, the pressure in the jar would rise and the boiling process would stop.

Now let us circulate this water boiling at 40°F through a cooling coil. If room air were passed over it, it would absorb heat from the room air. Because this air is giving up heat to the coil, the air leaving the coil is cold. **Figure 3–14** illustrates the cooling coil.

When water is used in this way, it is called a *refrigerant.* **A *refrigerant* is a substance that can be changed readily**

to a vapor by boiling it and then changed to a liquid by condensing it. The refrigerant must be able to make this change repeatedly without altering its characteristics. Water is not normally used as a refrigerant in small applications for reasons that will be discussed later. We used it in this example because most people are familiar with its characteristics.

To explore how a real refrigeration system works, we will use refrigerant-22 (R-22) in the following examples because it is commonly used in residential air conditioning. See **Figure 3–15** for the temperature/pressure relationship chart for several refrigerants including R-22. This chart is like that for water but at different temperature and pressure levels. Take a moment to become familiar with this chart; observe that temperature is in the left column, expressed in °F, and pressure is to the right expressed in psig. Find 40°F in the left column, read to the right, and notice that the gage reading is 68.5 psig for R-22. What does this mean in usable terms? It means that when R-22 liquid is boiled at a vapor pressure of 68.5 psig, it will boil at a temperature of 40°F. When air is passed over the coil, it will cool just as in the example of water.

The pressure and temperature of a refrigerant will correspond when both liquid and vapor are present under two conditions:

1. When the change of state (boiling or condensing) is occurring.
2. When the refrigerant is at equilibrium (i.e., no heat is added or removed).

In both conditions 1 and 2, the refrigerant is said to be *saturated.* When a refrigerant is saturated, both liquid and vapor can exist simultaneously. When they do, both the liquid and the vapor will have the same saturation temperature. NOTE: The liquid and vapor being at the same temperature is not true for some of the newer blended refrigerants that have a temperature glide. Temperature glide will be covered in Unit 9. This saturation temperature depends on the pressure of the liquid/vapor mixture. This pressure is called the saturation pressure. The higher the pressure, the higher the saturation temperature of the liquid and vapor mixture. The lower the pressure, the lower the saturation temperature.

Suppose that a cylinder of R-22 is allowed to set in a room until it reaches the room temperature of 75°F. It will then be

TEMPERATURE °F	REFRIGERANT 12	22	134a	502	404A	410A
−60	19.0	12.0		7.2	6.6	0.3
−55	17.3	9.2		3.8	3.1	2.6
−50	15.4	6.2		0.2	0.8	5.0
−45	13.3	2.7		1.9	2.5	7.8
−40	11.0	0.5	14.7	4.1	4.8	9.8
−35	8.4	2.6	12.4	6.5	7.4	14.2
−30	5.5	4.9	9.7	9.2	10.2	17.9
−25	2.3	7.4	6.8	12.1	13.3	21.9
−20	0.6	10.1	3.6	15.3	16.7	26.4
−18	1.3	11.3	2.2	16.7	18.2	28.2
−16	2.0	12.5	0.7	18.1	19.6	30.2
−14	2.8	13.8	0.3	19.5	21.1	32.2
−12	3.6	15.1	1.2	21.0	22.7	34.3
−10	4.5	16.5	2.0	22.6	24.3	36.4
−8	5.4	17.9	2.8	24.2	26.0	38.7
−6	6.3	19.3	3.7	25.8	27.8	40.9
−4	7.2	20.8	4.6	27.5	30.0	42.3
−2	8.2	22.4	5.5	29.3	31.4	45.8
0	9.2	24.0	6.5	31.1	33.3	48.3
1	9.7	24.8	7.0	32.0	34.3	49.6
2	10.2	25.6	7.5	32.9	35.3	50.9
3	10.7	26.4	8.0	33.9	36.4	52.3
4	11.2	27.3	8.6	34.9	37.4	53.6
5	11.8	28.2	9.1	35.8	38.4	55.0
6	12.3	29.1	9.7	36.8	39.5	56.4
7	12.9	30.0	10.2	37.9	40.6	57.8
8	13.5	30.9	10.8	38.9	41.7	59.3
9	14.0	31.8	11.4	39.9	42.8	60.7
10	14.6	32.8	11.9	41.0	43.9	62.2
11	15.2	33.7	12.5	42.1	45.0	63.7

TEMPERATURE °F	REFRIGERANT 12	22	134a	502	404A	410A
12	15.8	34.7	13.2	43.2	46.2	65.3
13	16.4	35.7	13.8	44.3	47.4	66.8
14	17.1	36.7	14.4	45.4	48.6	68.4
15	17.7	37.7	15.1	46.5	49.8	70.0
16	18.4	38.7	15.7	47.7	51.0	71.6
17	19.0	39.8	16.4	48.8	52.3	73.2
18	19.7	40.8	17.1	50.0	53.5	75.0
19	20.4	41.9	17.7	51.2	54.8	76.7
20	21.0	43.0	18.4	52.4	56.1	78.4
21	21.7	44.1	19.2	53.7	57.4	80.1
22	22.4	45.3	19.9	54.9	58.8	81.9
23	23.2	46.4	20.6	56.2	60.1	83.7
24	23.9	47.6	21.4	57.5	61.5	85.5
25	24.6	48.8	22.0	58.8	62.9	87.3
26	25.4	49.9	22.9	60.1	64.3	90.2
27	26.1	51.2	23.7	61.5	65.8	91.1
28	26.9	52.4	24.5	62.8	67.2	93.0
29	27.7	53.6	25.3	64.2	68.7	95.0
30	28.4	54.9	26.1	65.6	70.2	97.0
31	29.2	56.2	26.9	67.0	71.7	99.0
32	30.1	57.5	27.8	68.4	73.2	101.0
33	30.9	58.8	28.7	69.9	74.8	103.1
34	31.7	60.1	29.5	71.3	76.4	105.1
35	32.6	61.5	30.4	72.8	78.0	107.3
36	33.4	62.8	31.3	74.3	79.6	108.4
37	34.3	64.2	32.2	75.8	81.2	111.6
38	35.2	65.6	33.2	77.4	82.9	113.8
39	36.1	67.1	34.1	79.0	84.6	116.0
40	37.0	68.5	35.1	80.5	86.3	118.3
41	37.9	70.0	36.0	82.1	88.0	120.5

TEMPERATURE °F	REFRIGERANT 12	22	134a	502	404A	410A
42	38.8	71.4	37.0	83.8	89.7	122.9
43	39.8	73.0	38.0	85.4	91.5	125.2
44	40.7	74.5	39.0	87.0	93.3	127.6
45	41.7	76.0	40.1	88.7	95.1	130.0
46	42.6	77.6	41.1	90.4	97.0	132.4
47	43.6	79.2	42.2	92.1	98.8	134.9
48	44.6	80.8	43.3	93.9	100.7	136.4
49	45.7	82.4	44.4	95.6	102.6	139.9
50	46.7	84.0	45.5	97.4	104.5	142.5
55	52.0	92.6	51.3	106.6	114.6	156.0
60	57.7	101.6	57.3	116.4	125.2	170.0
65	63.8	111.2	64.1	126.7	136.5	185.0
70	70.2	121.4	71.2	137.6	148.5	200.8
75	77.0	132.2	78.7	149.1	161.1	217.6
80	84.2	143.6	86.8	161.2	174.5	235.4
85	91.8	155.7	95.3	174.0	188.6	254.2
90	99.8	168.4	104.4	187.4	203.5	274.1
95	108.2	181.8	114.0	201.4	219.2	295.0
100	117.2	195.9	124.2	216.2	235.7	317.1
105	126.6	210.8	135.0	231.7	253.1	340.3
110	136.4	226.4	146.4	247.9	271.4	364.8
115	146.8	242.7	158.5	264.9	290.6	390.5
120	157.6	259.9	171.2	282.7	310.7	417.4
125	169.1	277.9	184.6	301.4	331.8	445.8
130	181.0	296.8	198.7	320.8	354.0	475.4
135	193.5	316.6	213.5	341.2	377.1	506.5
140	206.6	337.2	229.1	362.6	401.4	539.1
145	220.3	358.9	245.5	385.9	426.8	573.2
150	234.6	381.5	262.7	408.4	453.3	608.9
155	249.5	405.1	280.7	432.9	479.8	616.2

VACUUM (in. Hg) – RED FIGURES
GAGE PRESSURE (psig) – BOLD FIGURES

Figure 3–15   This chart shows the temperature/pressure relationship in in. Hg vacuum, or psig. Pressures for R-404A and R-410A are an average liquid and vapor pressure.

in equilibrium because no outside forces are acting on it. The cylinder and its partial liquid, partial vapor contents will now be at the room temperature of 75°F. The temperature and pressure chart indicates a pressure of 132 psig, **Figure 3–15.** This temperature/pressure chart is also referred to as a saturation chart because it contains saturation temperatures for different saturation pressures.

Suppose that the same cylinder of R-22 is moved into a walk-in cooler and allowed to reach the room temperature of 35°F and attain equilibrium. The cylinder will then reach a new pressure of 61.5 psig because while it is cooling off to 35°F, the vapor inside the cylinder is reacting to the cooling effect by partially condensing; therefore, the pressure drops.

If we move the cylinder (now at 35°F) back into the warmer room (75°F) and allow it to warm up, the liquid inside it reacts to the warming effect by boiling slightly and creating vapor. Thus, the pressure gradually increases to 132 psig, which corresponds to 75°F.

If we move the cylinder (now at 75°F) into a room at 100°F, the liquid again responds to the temperature change by slightly boiling and creating more vapor. As the liquid boils and makes vapor, the pressure steadily increases (according to the temperature/pressure chart) until it corresponds to the liquid temperature. This continues until the contents of the cylinder reach the pressure, 196 psig, corresponding to 100°F, **Figure 3–16, Figure 3–17,** and **Figure 3–18.**

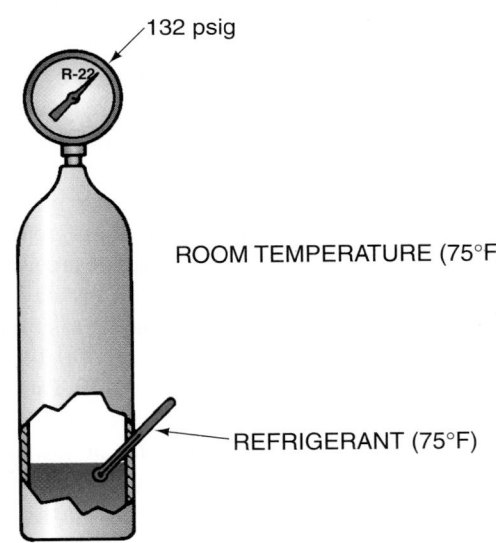

132 psig

ROOM TEMPERATURE (75°F)

REFRIGERANT (75°F)

Figure 3–16   The cylinder of R-22 is left in a 75°F room until it and its contents are at room temperature. The cylinder contains a partial liquid, partial vapor mixture. When both reach room temperature, they are in equilibrium and no further temperature changes will occur. At this time, the cylinder pressure, 132 psig, will correspond to the ambient, or surrounding temperature of 75°F. The liquid and vapor refrigerant are said to be saturated at 75°F.

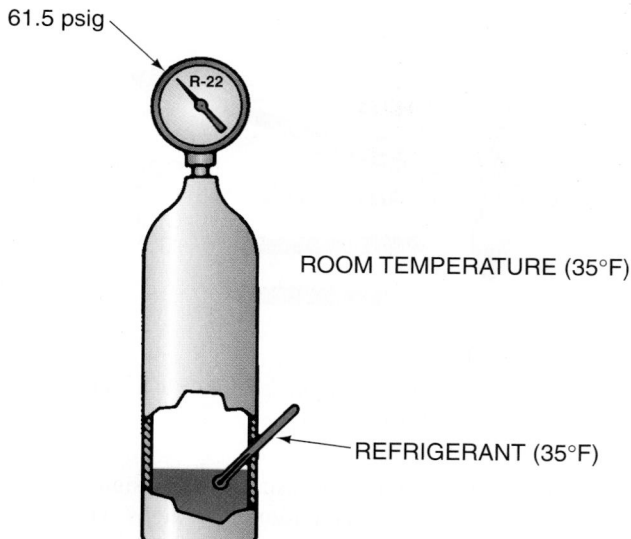

61.5 psig

R-22

ROOM TEMPERATURE (35°F)

REFRIGERANT (35°F)

Figure 3–17   The cylinder of R-22 is moved into a 35°F walk-in cooler and left until the cylinder and its contents are at the same temperature as the cooler. As the refrigerant in the cylinder cools, some of the vapor will condense into a liquid, reducing the vapor pressure in the cylinder. Once the partial liquid, partial vapor mixture reaches the cooler temperature of 35°F, they are in equilibrium and no further temperature changes will occur. At this time, the cylinder pressure, 61.5 psig, will correspond to the cooler temperature of 35°F. The liquid and vapor refrigerant are now saturated at 35°F.

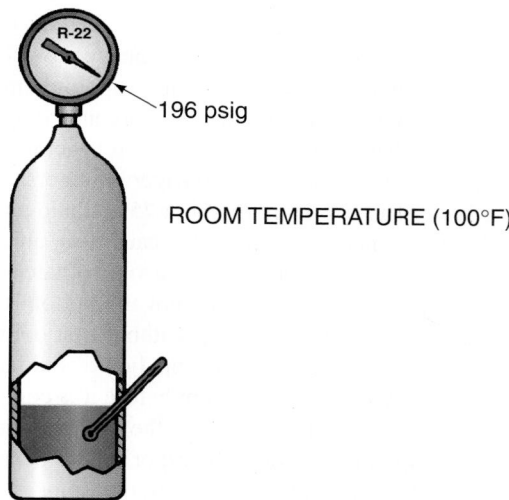

R-22

196 psig

ROOM TEMPERATURE (100°F)

Figure 3–18   The cylinder of R-22 is moved to a 100°F room and allowed to reach the equilibrium point of 100°F and 196 psig. The rise in pressure is due to the fact that, at the higher temperature, some of the liquid refrigerant vaporizes. This causes the vapor pressure to increase. The liquid and vapor are still saturated but are now at a higher saturation temperature.

In fact, the vapor that was generated because of the increase in temperature is referred to as *vapor pressure*. Vapor pressure is the pressure exerted on a saturated liquid. Any time saturated vapor and liquid are together, vapor pressure is generated. Vapor pressure acts equally in all directions,

and this action is what the pressure gage reads on a refrigeration or air-conditioning system. As the temperature of the liquid/vapor mixture increases, vapor pressure increases. As the temperature of the liquid/vapor mixture decreases, vapor pressure decreases.

Further study of the temperature/pressure chart shows that when the pressure is lowered to atmospheric pressure, R-22 boils at about −41°F. ♻ Do not perform the following exercises because allowing refrigerant to intentionally escape to the atmosphere is against the law! These are mentioned here for illustration purposes only. ♻ If the valve on the cylinder of R-22 were opened slowly and the vapor allowed to escape to the atmosphere, the pressure loss of the vapor would cause the liquid remaining in the cylinder to boil and drop in temperature. Whenever any liquid boils, heat is absorbed in the process, which causes a cooling effect. The heat in this case came from the R-22 liquid in the cylinder. Soon the pressure in the cylinder would be down to atmospheric pressure, and it would frost over and become −41°F. We are assuming that the vapor valve at the top of the R-22 cylinder is large enough to allow the R-22 vapor to escape freely. This way, the vapor will leave the cylinder at the same rate the R-22 liquid is boiling in the cylinder.

Now, let's say that we took one end of a gage hose and connected it to the liquid port of a refrigerant (R-22) cylinder and we directed the other end to a cup. If the liquid valve were opened very slowly, liquid refrigerant would flow from the cylinder, through the gage hose, and into the cup. The liquid refrigerant would accumulate in the cup and you might notice the liquid boiling. If a thermometer were placed in the cup of boiling refrigerant, the thermometer would register a reading of −41°F, **Figure 3–19.** You can double-check these numbers by looking up the saturation temperature for R-22 at 0 psig on the temperature/pressure chart. ♻ Again, do not perform the preceding experiments. ♻

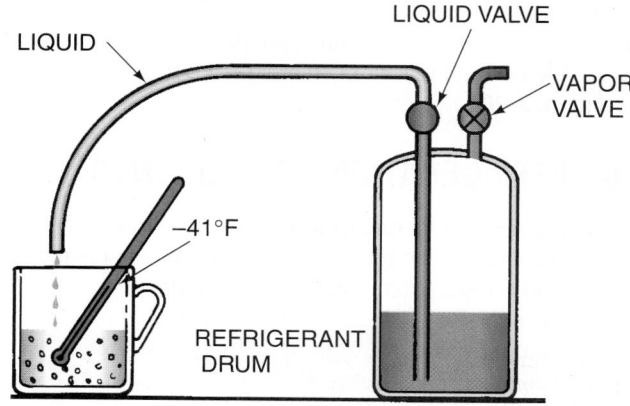

LIQUID

LIQUID VALVE

VAPOR VALVE

−41°F

REFRIGERANT DRUM

Figure 3–19   When the gage hose is attached to the liquid line valve on an R-22 cylinder and liquid is allowed to trickle out of it and into the cup, the liquid will collect in the cup. The liquid R-22 will continue to boil at a temperature of −41°F until all the liquid has vaporized. ♻ Do not perform this experiment because it is illegal to intentionally vent or release refrigerants to the atmosphere. ♻

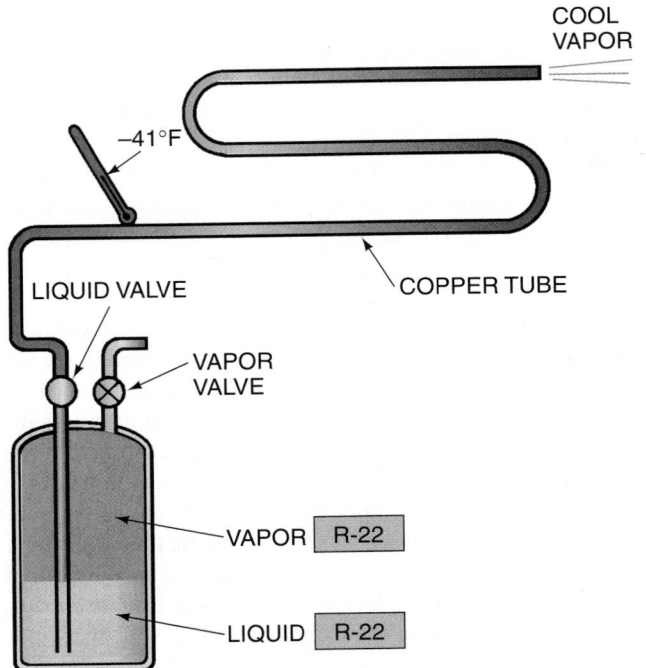

Figure 3–20   When the tubing is attached to the liquid line valve on an R-22 cylinder and liquid is allowed to trickle into the tubing, the liquid will boil at a temperature of −41°F at atmospheric pressure. ♻ Do not perform this experiment as it is illegal to intentionally vent or release refrigerants to the atmosphere. ♻

A crude but effective demonstration has been used in the past to show how air can be cooled. ♻ Do not perform this experiment because intentionally venting refrigerant to the atmosphere is illegal. ♻ A long piece of copper tubing is fastened to the liquid tap on the refrigerant cylinder, and liquid refrigerant is allowed to trickle into the tube while air passes over it. The tube has a temperature of −41°F, corresponding to atmospheric pressure, because the refrigerant is escaping out of the end of the tube at atmospheric pressure. If the tube were coiled up and placed in an airstream, it would cool the air, **Figure 3–20.**

3.6   REFRIGERATION COMPONENTS

By adding some components to the system, these problems can be eliminated. The four major components that make up mechanical refrigeration systems are the following:

1. The evaporator
2. The compressor
3. The condenser
4. The refrigerant metering device

3.7   THE EVAPORATOR

The *evaporator* absorbs heat into the system. When the refrigerant is boiled at a lower temperature than that of the substance to be cooled, it absorbs heat from the substance. The boiling temperature of 40°F was chosen in the previous

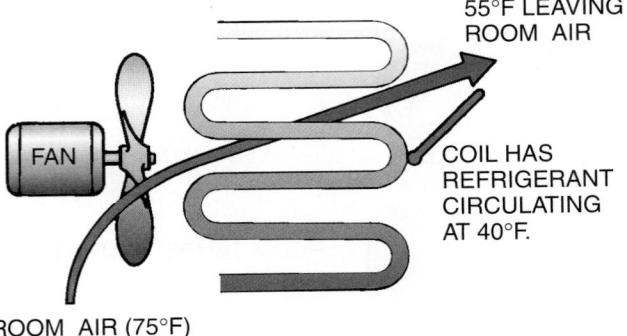

Figure 3–21   The evaporator is operated at 40°F to be able to absorb heat from the 75°F air.

air-conditioning examples because it is the design temperature normally used for air-conditioning systems. The reason is that the ideal room temperature is close to 75°F, which readily gives up heat to a 40°F coil. The 40°F temperature is also well above the freezing point of the coil. See **Figure 3–21** for the coil-to-air relationships.

The evaporator can be thought of as a "heat sponge." Just as a dry kitchen sponge absorbs liquid from a spill because the water content of the sponge is low, the evaporator is able to absorb heat because the temperature of the coil is lower than the temperature of the medium being cooled. Continuing our analogy, the sponge will stop absorbing liquid if it becomes completely soaked. This is why we have to squeeze the sponge to remove the absorbed water. Looking back at our evaporator, if we just keep adding heat to the coil, its temperature will rise and the amount of heat it can absorb will drop. So, the heat that is absorbed into the evaporator must be removed later on to allow the system to continue to operate.

Let us see what happens as the R-22 refrigerant passes through the evaporator coil. The refrigerant enters the coil from the bottom as a mixture of about 75% liquid and 25% vapor. These percentages can change because they are system and application dependent and will be a topic of later discussion. Evaporators are usually fed from the bottom to help ensure that no liquid leaves the top without first boiling off to a vapor. If they were fed from the top, liquid could quickly drop by its own weight to the bottom before it is completely boiled off to a vapor. This protects the compressor from any liquid refrigerant. In addition, vapor is less dense than liquid and, as the liquid refrigerant boils, the vapor has a tendency to rise. Since the refrigerant enters the coil from the bottom and leaves from the top, the direction of refrigerant flow is from the bottom to the top. This is the same direction as the direction of the rising vapor. If the refrigerant entered the evaporator from the top and left through the bottom, the direction of flow would be opposite to that of the rising vapor. This would cause the refrigerant to slow down.

The mixture is tumbling and boiling as it flows through the tubes, with the liquid being turned to vapor all along the coil because heat is being added to the coil from the air, **Figure 3–22.** About halfway down the coil, the mixture becomes more vapor than liquid. The purpose of the evaporator is to boil all of the liquid into a vapor just before the end of

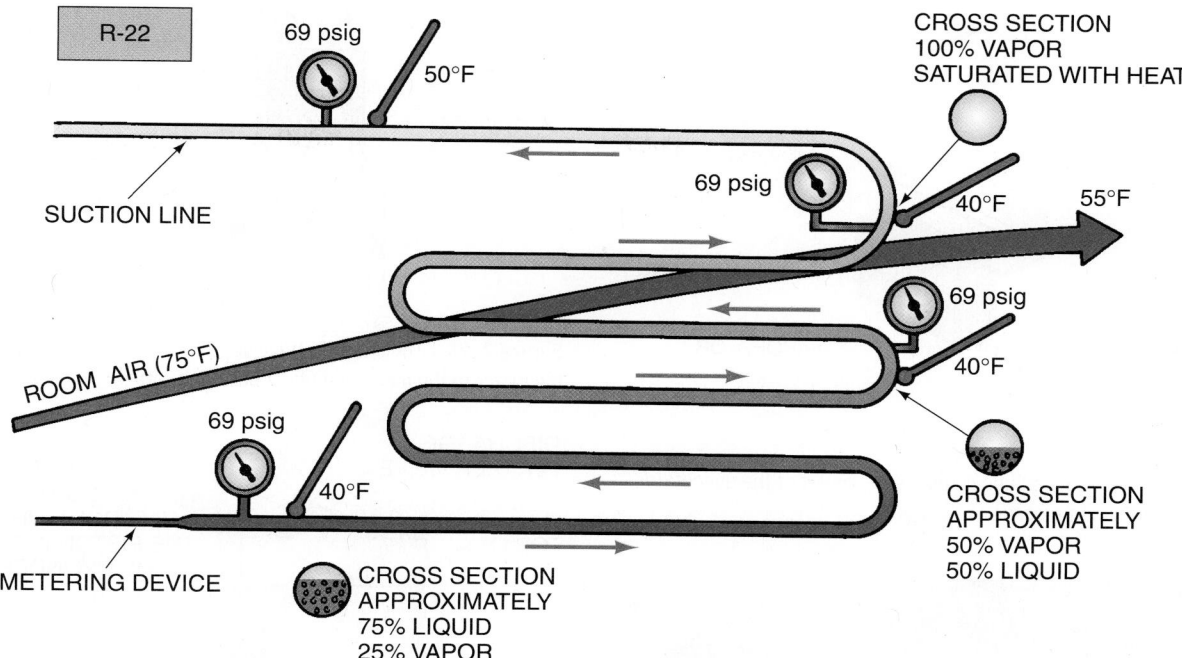

Figure 3–22   The evaporator absorbs heat into the refrigeration system by boiling the refrigerant at a temperature that is lower than the temperature of the room air passing over it. The 75°F room air readily gives up heat to the 40°F evaporator by conduction.

the coil. This occurs approximately 90% of the way through the coil, leaving pure vapor. At the point where the last droplet of liquid vaporizes, we have what is called a saturated vapor. This is the point where the vapor would start to condense if heat were removed from it or become superheated if any heat were added to it. ***When a vapor is superheated, it no longer corresponds to a temperature/pressure relationship. Because no liquid remains to boil off to vapor, no more vapor pressure can be generated when heat is added. The vapor will now take on sensible heat when heated, and its temperature will rise, but the pressure will remain unchanged.*** Superheat is considered insurance for the compressor because it ensures that no liquid leaves the evaporator and enters the compressor. When there is superheat, there is no liquid.

To summarize, the three main functions of the evaporator are to:

1. absorb heat from the medium being cooled.
2. allow the heat to boil off the liquid refrigerant to a vapor in its tubing bundle.
3. allow the heat to superheat the refrigerant vapor in its tubing bundle.

Evaporators have many design configurations. But for now just remember that they absorb the heat into the system from the substance to be cooled. The substance may be solid, liquid, or gas, and the evaporator has to be designed to fit the condition. See **Figure 3–23** for a typical evaporator. Once absorbed into the system, the heat is now contained in the refrigerant and is drawn into the compressor through the suction line. The suction line simply connects the evaporator to

Figure 3–23   A typical refrigeration evaporator. *Courtesy Ferris State University. Photo by Eugene Silberstein*

the compressor and provides a path for refrigerant vapor to travel, **Figure 3–24.** The suction line typically passes through unrefrigerated spaces and the surrounding air temperature is much higher than the temperature of the refrigerant line. Heat from the surrounding air can, therefore, be readily absorbed into the refrigeration system, making the system work much harder than needed. For this reason, the suction line is insulated, **Figure 3–25.** A well-insulated suction line helps the system operate as efficiently as possible.

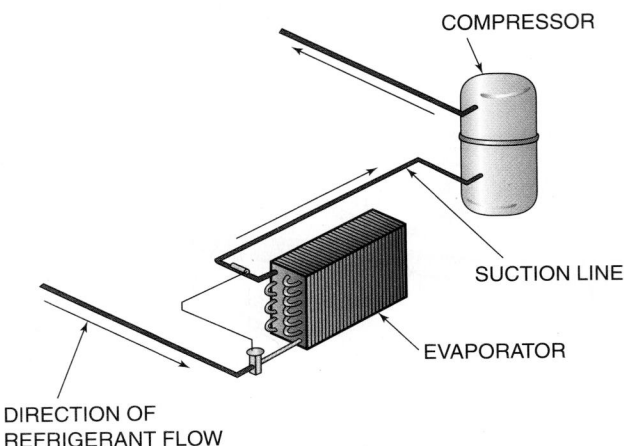

Figure 3–24 The suction line connects the outlet of the evaporator to the inlet of the compressor.

Figure 3–25 Suction lines should be well insulated to increase system efficiency and capacity. *Courtesy Ferris State University. Photo by John Tomczyk*

3.8 THE COMPRESSOR

The compressor is the heart of the refrigeration system. It pumps heat through the system in the form of heat-laden refrigerant. A compressor can be considered a *vapor pump.* It reduces the pressure on the low-pressure side of the system, which includes the evaporator, and increases the pressure on the high-pressure side of the system. This pressure difference is what causes the refrigerant to flow through the system. All compressors in refrigeration systems perform this function by compressing the vapor refrigerant. This compression can be accomplished in several ways with different types of compressors. The most common compressors used in residential and light commercial air conditioning and refrigeration are the *reciprocating,* the *rotary,* and the *scroll.*

The reciprocating compressor uses a piston in a cylinder to compress the refrigerant, **Figure 3–26.** Valves, usually reed or flapper valves, ensure that the refrigerant flows in the correct direction, **Figure 3–27.** This compressor is known as a positive displacement compressor. Positive displacement compressors increase the pressure of the refrigerant by

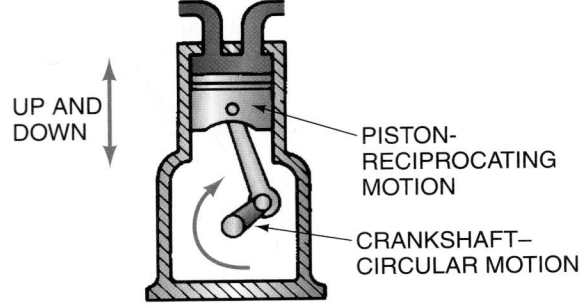

Figure 3–26 The crankshaft converts the circular, rotating motion of the motor to the reciprocating, or back-and-forth, motion of the piston.

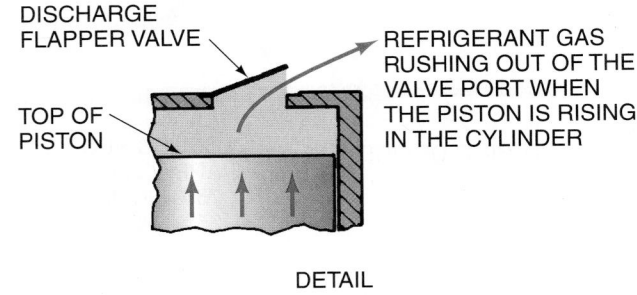

DETAIL

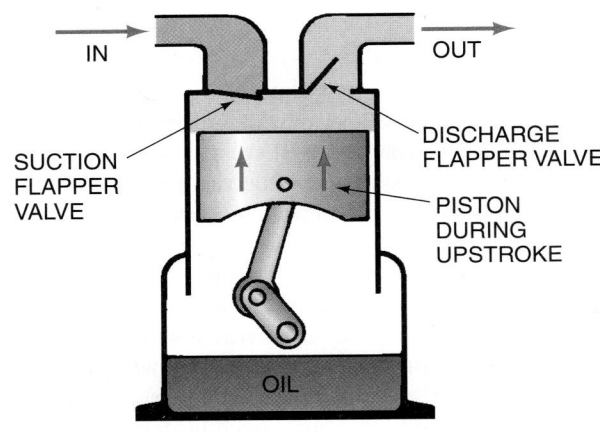

Figure 3–27 Flapper valves and compressor components.

physically decreasing the volume of the container that is holding the refrigerant. In the case of the reciprocating compressor, the volume is decreased as the piston moves up in the cylinder. When the cylinder is filled with vapor, it must be emptied as the compressor turns, or damage will occur. For many years, it was the most commonly used compressor for systems up to 100 hp. Newer and more efficient designs of compressors are now being used.

The rotary compressor is also a positive displacement compressor and is used for applications that are typically in the small equipment range, such as window air conditioners, household refrigerators, and some residential air-conditioning systems. These compressors are extremely efficient and have few moving parts, **Figure 3–28.** This compressor uses a rotating drumlike piston that squeezes the refrigerant vapor out the discharge port. These compressors are typically very

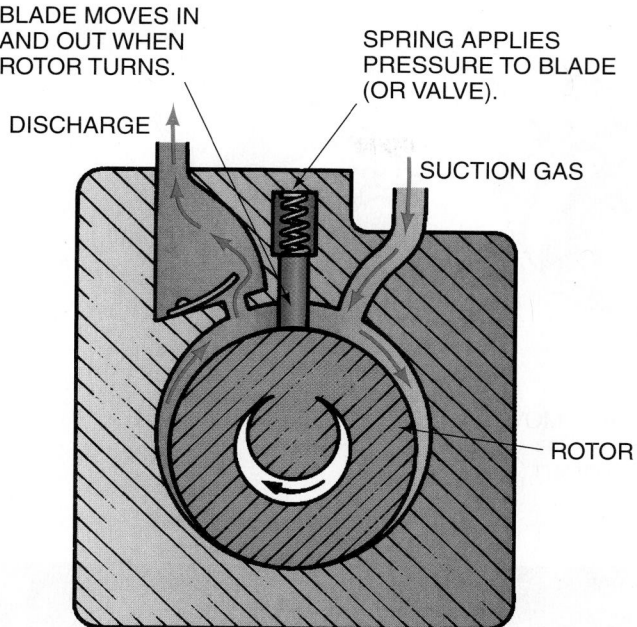

Figure 3–28   A rotary compressor with motion in one direction and no backstroke.

small compared with the same capacity of reciprocating compressors.

The scroll compressor is one of the latest compressors to be developed and has an entirely different working mechanism. It has a stationary part that looks like a coil spring and a moving part that matches and meshes with the stationary part, **Figure 3–29.** The movable part orbits inside the stationary part and squeezes the vapor from the low-pressure side to the high-pressure side of the system between movable and stationary parts. Several stages of compression are taking place in the scroll at the same time, making it a very smooth running compressor with few moving parts. The scroll is sealed on the bottom and top with the rubbing action and at the tip with a tip seal. These sealing surfaces prevent refrigerant from the high-pressure side from pushing back to the low-pressure side while running. It is a positive displacement compressor with a limitation. It is positive displacement until too much pressure differential builds up; then the scrolls are capable of moving apart, and high-pressure refrigerant can blow back through the compressor and prevent overload. This ability to move the mating scrolls apart makes it more forgiving if a little liquid refrigerant enters the compressor's inlet. Thus, compressor damage is less likely to occur.

Recent advances in computer-aided manufacturing techniques have enabled the scroll compressor to gain popularity in air conditioning and high-, medium-, and low-temperature refrigeration applications. The capacity of the scroll compressor can be controlled by the size and wall height of the orbiting and stationary (spirals) scrolls.

Large commercial systems use other types of compressors because they must move much more refrigerant vapor through the system. The centrifugal compressor is used in large air-conditioning systems. It is much like a large fan and is not positive displacement, **Figure 3–30.** The centrifugal

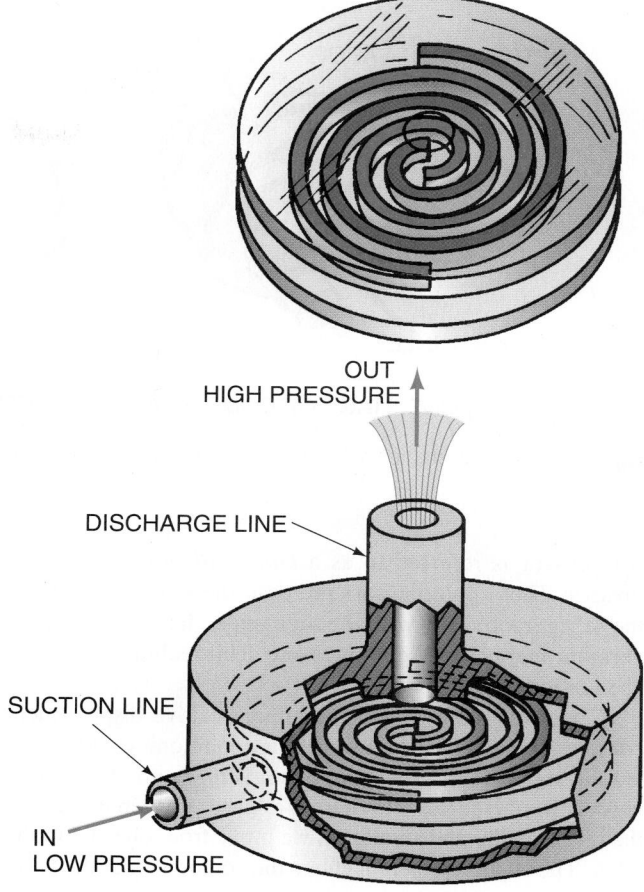

Figure 3–29   An illustration of the operation of a scroll compressor mechanism.

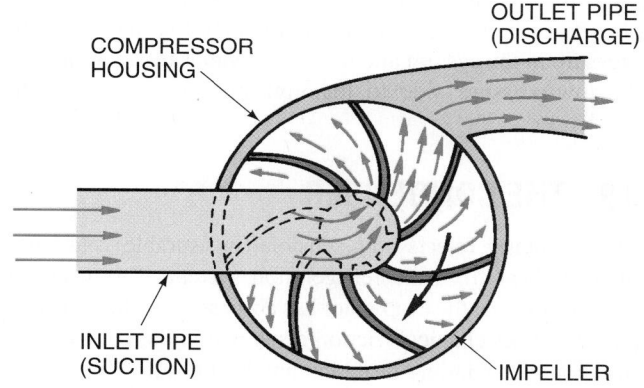

THE TURNING IMPELLER IMPARTS CENTRIFUGAL FORCE ON THE REFRIGERANT, FORCING THE REFRIGERANT TO THE OUTSIDE OF THE IMPELLER. THE COMPRESSOR HOUSING TRAPS THE REFRIGERANT AND FORCES IT TO EXIT INTO THE DISCHARGE LINE. THE REFRIGERANT MOVING TO THE OUTSIDE CREATES A LOW PRESSURE IN THE CENTER OF THE IMPELLER WHERE THE INLET IS CONNECTED.

Figure 3–30   An illustration of the operation of a centrifugal compressor mechanism.

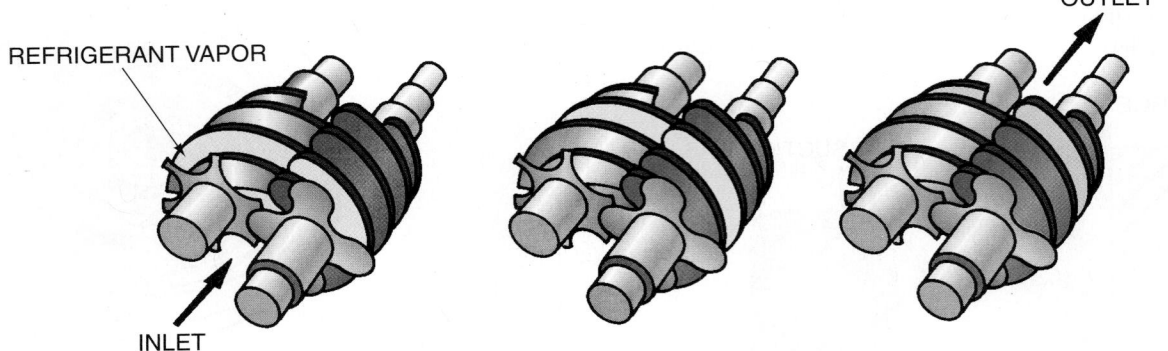

ONE STAGE OF COMPRESSION AS REFRIGERANT MOVES THROUGH SCREW COMPRESSOR

Figure 3–31   An illustration of the internal working mechanism of a screw compressor.

compressor is referred to as a *kinetic displacement compressor*. These compressors increase the kinetic energy of the refrigerant vapor with the high-speed fan and then convert this energy to a higher pressure. This technology is also used in jet engines.

The screw compressor is another positive displacement compressor and is used for larger air-conditioning and refrigeration applications. The compressor is made up of two nesting "screws," **Figure 3–31.** There is a space between the screws that gets smaller and smaller as we move from one end to the other. The vapor refrigerant enters the compressor at the point where the screw spacing is the widest. As the refrigerant flows between these rotating screws, the volume is reduced and the pressure of the vapor increases. The screw compressor is popular for use in low-temperature refrigeration applications.

The important thing to remember is that a compressor performs the same function no matter what the type is. For now it can be thought of as a component that increases the pressure in the system and moves the vapor refrigerant from the low-pressure side to the high-pressure side into the condenser.

3.9   THE CONDENSER

The *condenser* rejects both sensible (measurable) and latent (hidden) heat from the refrigeration system. This heat can come from what the evaporator has absorbed, any heat of compression or mechanical friction generated in the compression stroke, motor-winding heat, and any heat absorbed by superheating the refrigerant in the suction line before it enters the compressor.

The condenser receives the hot gas after it leaves the compressor through the short pipe between the compressor and the condenser; this pipe is called the hot gas line, **Figure 3–32.** This line is also referred to as the *discharge line.* The hot gas is forced into the top of the condenser coil by the compressor. The discharge gas from the compressor is a high-pressure, high-temperature, superheated vapor. The temperature of the hot gas from the compressor can be in the 200°F range and will change depending on the surrounding temperatures and

Figure 3–32   The discharge line connects the outlet of the compressor to the inlet of the condenser. *Courtesy Ferris State University. Photo by Eugene Silberstein*

the system application. The refrigerant at the outlet of the compressor DOES NOT follow a temperature/pressure relationship. This is because the refrigerant is 100% vapor and superheated. The high-side pressure reading of 278 psig in **Figure 3–33** corresponds to the 125°F condenser saturation temperature, which is the temperature at which the refrigerant will begin to condense from a vapor into a liquid. The

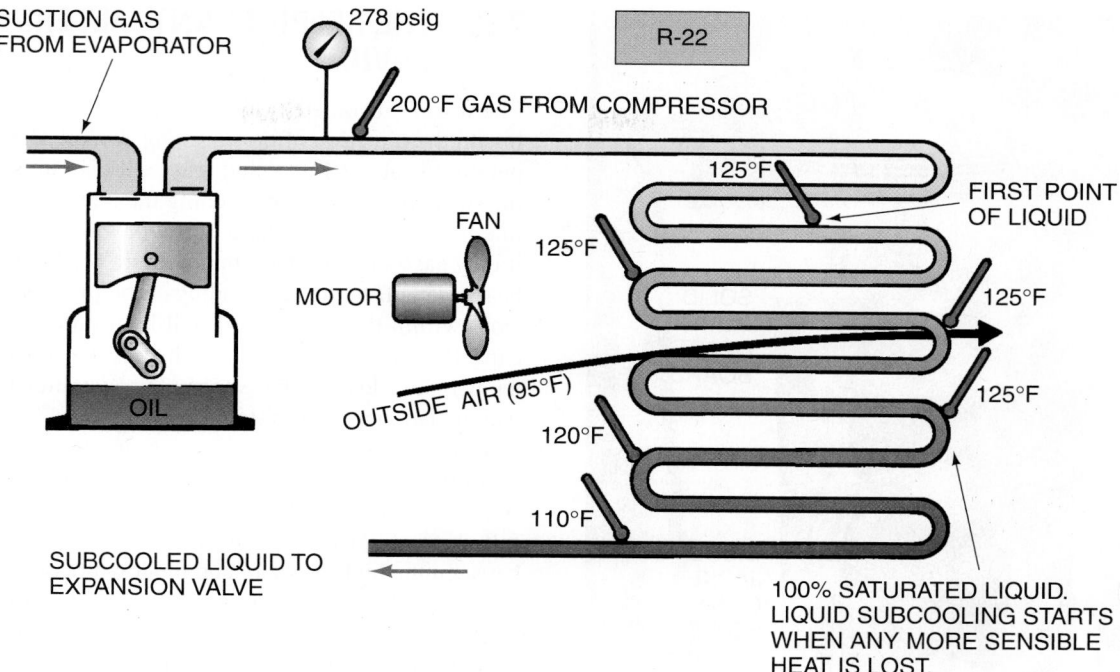

SUCTION GAS FROM EVAPORATOR

278 psig

R-22

200°F GAS FROM COMPRESSOR

FAN

MOTOR

OIL

OUTSIDE AIR (95°F)

125°F

125°F

FIRST POINT OF LIQUID

125°F

125°F

120°F

110°F

SUBCOOLED LIQUID TO EXPANSION VALVE

100% SATURATED LIQUID. LIQUID SUBCOOLING STARTS WHEN ANY MORE SENSIBLE HEAT IS LOST.

Figure 3–33   Subcooled liquid at the outlet of the condenser.

condenser is where the vapor pressure of 278 psig is actually coming from for the high-side gage to read. This vapor pressure is often referred to as head pressure, high-side pressure, discharge pressure, or condensing pressure in the heating, ventilation, air-conditioning, and refrigeration (HVACR) industry. The 200°F hot gas temperature coming out of the compressor is superheated by 75°F (200°F − 125°F) and again cannot have a temperature/pressure relationship. Therefore, the hot vapor must first cool, or desuperheat, 75 degrees before it can begin to condense.

The gas leaving the compressor, flowing through the discharge line, and entering the condenser is so hot compared with the surrounding air that a heat exchange between the hot gas and the surrounding air begins to occur immediately. The surrounding air that is being passed over the condenser is 95°F as compared with the near 200°F gas entering the condenser. As the gas moves through the condenser, it begins to give up sensible heat to the surrounding air. This causes a drop in gas temperature. The gas keeps cooling off, or *desuper-heating,* until it reaches the condensing temperature of 125°F. Then, the change of state begins to occur. The change of state begins slowly at first with small amounts of vapor changing to liquid and gets faster as the vapor-liquid mixture moves through the condenser. This change of state from vapor to liquid happens at the condensing saturation temperature and pressure of 125°F and 278 psig, respectively. This change of state is a latent heat process, meaning that even though heat in Btu is being rejected from the refrigerant, the temperature is staying constant at 125°F. NOTE: This constant temperature process occurring during the change of state does not occur for some of the newer blended refrigerants that have a temperature glide.

When the condensing refrigerant gets about 90% of the way through the condenser, the refrigerant in the pipe becomes pure saturated liquid. If any more heat is removed from the 100% saturated liquid, the liquid will now go through a sensible-heat rejection process because no more vapor is left to condense. This causes the liquid to drop below the condensing saturation temperature of 125°F. Liquid cooler than the condensing saturation temperature is called *subcooled* liquid, **Figure 3–33**.

Three important things may happen to the refrigerant in the condenser:

1. The hot gas from the compressor is desuperheated from the hot discharge temperature to the condensing temperature. Remember, the condensing temperature determines the head pressure. This is sensible heat transfer.

2. The refrigerant is condensed from a vapor to a liquid. This is latent heat transfer.

3. The liquid refrigerant temperature may then be lowered below the condensing temperature, or subcooled. The refrigerant can usually be subcooled to between 10°F and 20°F below the condensing temperature, but is system dependent, **Figure 3–33**. This is a sensible heat transfer.

Many types of condensing devices are available. The condenser is the component that rejects the heat out of the refrigeration system. The heat may have to be rejected into a solid, liquid, or gas substance, and a condenser can be designed to do the job. Very often, the condenser and the compressor are incorporated into a single piece of equipment called a *condensing unit.* **Figure 3–34** shows some typical condensing units.

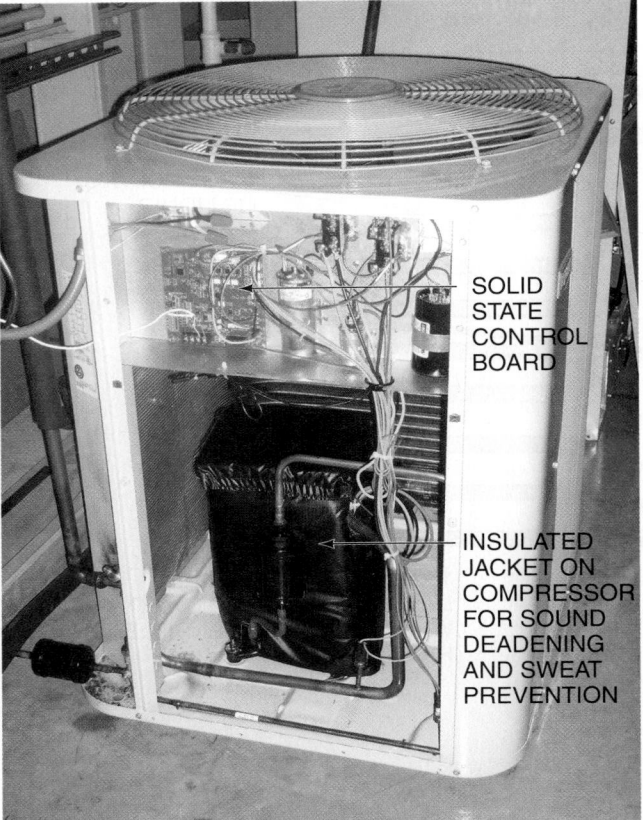

(A)

(B)

Figure 3–34   **(A)** Typical air-cooled condensing unit found on a split-type central air-conditioning system. **(B)** An air-cooled hermetic condensing unit from a refrigeration system. *(A) Courtesy Ferris State University. Photo by Eugene Silberstein. (B) Photo by Bill Johnson*

By incorporating both of these components into a single unit, system installation becomes an easier task. The suction line carries low-pressure, low-temperature superheated vapor to the condensing unit and the liquid line carries high-pressure, high-temperature subcooled liquid from the condensing unit to the metering device.

3.10   THE REFRIGERANT METERING DEVICE

The warm subcooled liquid is now moving down the liquid line in the direction of the *metering device*. The liquid temperature is about 110°F and may still give up some heat to the surroundings before reaching the metering device. This line may be routed under a house or through a wall where it may easily reach a new temperature of about 105°F. Any heat given off to the surroundings is helpful because it came from within the system and will help improve the system capacity and efficiency. It also brings the subcooled liquid temperature closer to the evaporator's saturation temperature, which also increases system capacity.

One type of metering device is a simple fixed-size (*fixed-bore*) type known as an *orifice*. It is a small restriction of a fixed size in the line, **Figure 3–35.** This device holds back the full flow of refrigerant and is one of the dividing points between the high-pressure and the low-pressure sides of the system. Only pure liquid must enter it. The pipe leading to the orifice may be the size of a pencil, and the precision-drilled hole in the orifice may be the size of a very fine sewing needle. As you can see from the figure, the liquid flow is greatly restricted here. For an R-22 system, the liquid refrigerant entering the orifice is at a pressure of 278 psig; the refrigerant leaving the orifice is a *mixture* of about 75% liquid and 25% vapor at a new pressure of 70 psig and a new temperature of 41°F, **Figure 3–35.** Two questions usually arise at this time:

1. Why did approximately 25% of the liquid change to a gas?
2. How did the mixture of 100% pure liquid go from about 105°F to 41°F in such a short space?

These questions can be answered by using a garden hose under pressure, in which the water coming out feels cooler, **Figure 3–36.** The water actually is cooler because some of it evaporates and turns to mist. This evaporation takes heat out of the rest of the water and cools it down. Now when the high-pressure subcooled refrigerant passes through the orifice, it does the same thing as the water in the hose; it changes pressure (278 psig to 70 psig), and some of the refrigerant flashes to a vapor (called *flash gas*). At this point, the refrigerant is a saturated mixture of liquid and vapor at 70 psig. Since the refrigerant is saturated, it follows the

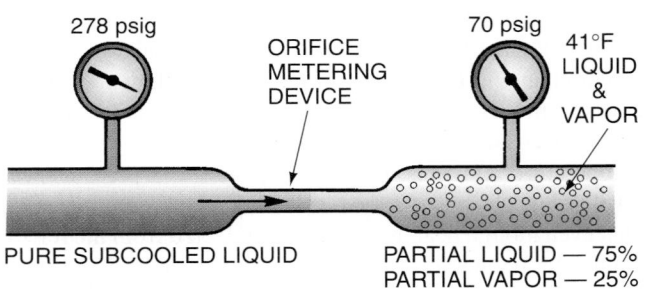

Figure 3–35   A fixed-bore (orifice) metering device on an R-22 system.

Figure 3–36   A person squeezing the end of a garden hose.

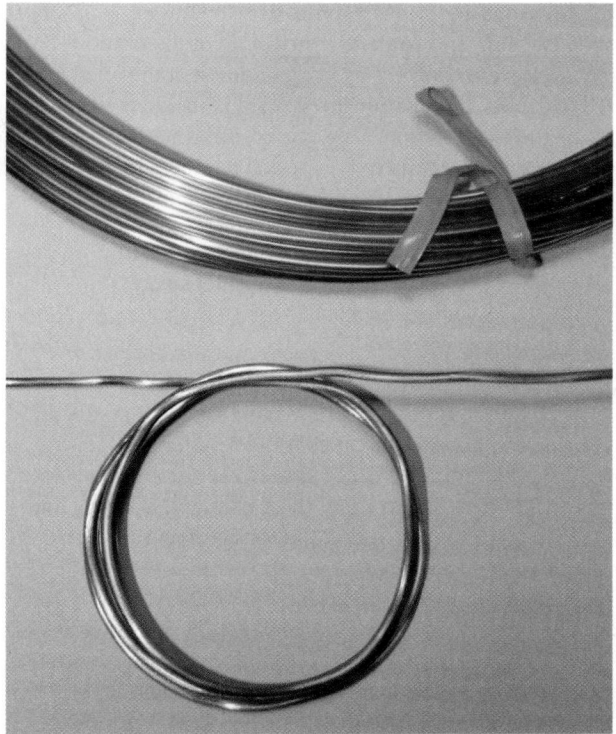

(A)

temperature/pressure relationship for that refrigerant—hence, the 41°F refrigerant temperature at the outlet of the metering device. In addition, the evaporating refrigerant further cools the liquid refrigerant that remains as the refrigerant flows through the metering device. *Flash gas* at the exit of the metering device is considered a loss to the system's capacity. This is because less liquid is now available to boil to vapor in the evaporator when cooling the refrigerated space. Because of this, flash gas should be kept to a minimum. This quick drop in pressure in the metering device lowers the boiling point or saturation temperature of the liquid leaving the metering device.

Several types of metering devices are available for many applications. They will be covered in detail in later units. See **Figure 3–37** for some examples of the various types of metering devices.

3.11   REFRIGERATION SYSTEM AND COMPONENTS

The basic components of the mechanical, vapor-compression refrigeration system have been described according to function. These components must be properly matched for each specific application. For instance, a low-temperature compressor cannot be applied to a high-temperature application because of the pumping characteristics of the compressor. Some equipment can be mixed and matched successfully by using the manufacturer's data, but only someone with considerable knowledge and experience should do so.

Following is a description of a matched system correctly working at design conditions. Later we will explain malfunctions and adverse operating conditions.

A typical air-conditioning system operating at a design temperature of 75°F inside temperature has a relative humidity

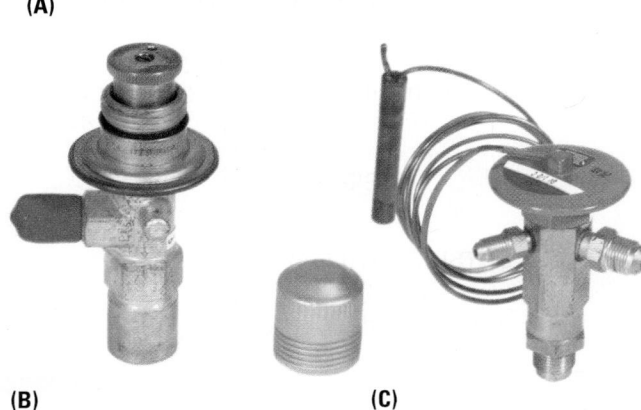

(B)   **(C)**

Figure 3–37   Metering devices. **(A)** Capillary tube. **(B)** Automatic expansion valve. **(C)** Thermostatic expansion valve. *(A) Courtesy Ferris State University. Photo by John Tomczyk. (B) and (C) Photos by Bill Johnson*

(moisture content of the conditioned room air) of 50%. These conditions are to be maintained inside the house. The air in the house gives up heat to the refrigerant. The humidity factor has been brought up at this time because the indoor coil is also responsible for removing some of the moisture from the air to keep the humidity at an acceptable level. This is known as *dehumidifying*.

Moisture removal requires considerable energy. Approximately the same amount (970 Btu) of latent heat removal is required to condense a pound of water vapor from the air as to condense a pound of steam. All air-conditioning systems must have a method for dealing with this moisture after it has turned to a liquid. Some units drip, some drain the liquid into plumbing waste drains, some use a slinger ring on the condenser fan, and some use the liquid at the outdoor coil to help the system capacity by evaporating it at the condenser.

Remember that part of the system is inside the house and part of the system is outside the house. The numbers in the description correspond to the circled numbers in **Figure 3–38**.

1. A mixture of 75% liquid and 25% vapor leaves the metering device and enters the evaporator.
2. The mixture is saturated R-22 at a pressure of 69 psig, which corresponds to a 40°F boiling point. It is important to remember that *the pressure is 69 psig because the evaporating refrigerant is boiling at 40°F.*
3. The mixture tumbles through the tube in the evaporator with the liquid evaporating from the 75°F inside air's heat and humidity load as it moves along.
4. When the mixture is about halfway through the coil, the refrigerant is composed of about 50% liquid and 50% vapor and is still at the same temperature and pressure because a change of state is taking place. Remember that this is a latent heat transfer.
5. The refrigerant is now 100% vapor. In other words, it has reached the 100% *saturation* point of the vapor. Recall the example using the saturated water table; the water reached various points where it was saturated with heat. We say it is saturated with heat because if any heat is removed at this point, some of the vapor changes back to a liquid; if any heat is added, the temperature of the vapor rises. This rise in temperature of the vapor makes

it a *superheated* vapor. (Superheat is sensible heat.) At point 5, the saturated vapor is still at 40°F and still able to absorb more heat from the 75°F room air.

6. Pure vapor now exists that is normally superheated about 10°F above the saturation temperature. Examine the line in **Figure 3–38** at this point, and you will see that the temperature is about 50°F. NOTE: To arrive at the correct superheat reading at this point, take the following steps.
   A. Note the suction pressure or evaporating pressure reading from the suction, or low-side, gage: 69 psig.
   B. Convert the suction pressure reading to suction or evaporating temperature using the temperature/pressure chart for R-22: 40°F.
   C. Use a suitable thermometer to record the actual temperature of the suction line at the outlet of the evaporator: 50°F.
   D. Subtract the saturated suction temperature from the actual suction line temperature: 50°F − 40°F = 10°F of superheat.

This vapor is said to be heat laden because it contains the heat removed from the room air. The heat was absorbed into the vaporizing refrigerant that boiled off to this vapor as it traveled through the evaporator. The vapor superheats another 10°F and is now 60°F as it travels down the suction line to the compressor. The superheat that is picked up only in the evaporator is referred to as

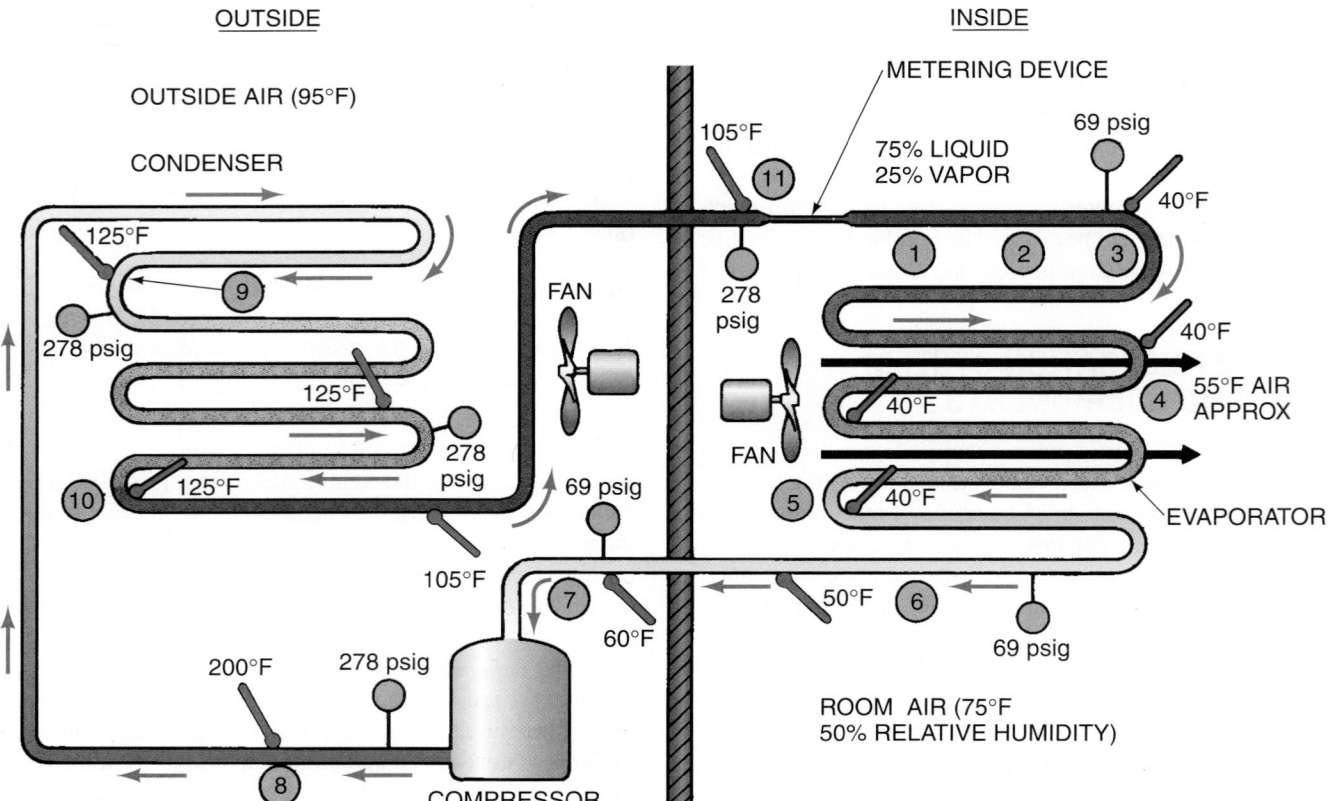

Figure 3–38   A typical R-22 air-conditioning system showing temperatures and airflow. Red indicates warm/hot refrigerant; blue indicates cool/cold refrigerant.

evaporator superheat, while the total superheat that is picked up in the evaporator and the suction line is referred to as system superheat. To calculate the system superheat, we will take the suction line temperature (from item C, above) close to the compressor's inlet instead of at the outlet of the evaporator. In this example, the evaporator superheat is 10°F (50°F − 40°F) and the system superheat is 20°F (60°F − 40°F). System superheat is often referred to as compressor superheat.

7. The vapor is drawn into the compressor by its pumping action, which creates a low-pressure suction. When the vapor left the evaporator, its temperature was about 50°F with 10°F of superheat above the saturated boiling temperature of 40°F. As the vapor moves along toward the compressor, it is contained in the suction line. This line is usually copper and should be insulated to keep it from drawing heat into the system from the surroundings and to prevent it from sweating. However, it still picks up some heat. Because the suction line carries vapor, any heat that it picks up will quickly raise the temperature. Remember that it does not take much sensible heat to raise the temperature of a vapor. Depending on the length of the line and the quality of the insulation, the suction line temperature may be 60°F at the compressor inlet.

8. Highly superheated gas leaves the compressor through the *hot gas line* on the high-pressure side of the system. This line normally is very short because the condenser is usually close to the compressor. On a hot day the hot gas line may be close to 200°F with a pressure of 278 psig. Because the saturated temperature corresponding to 278 psig is 125°F, the hot gas line has about 75°F (200°F − 125°F) of superheat that must be removed before condensing can occur. Because the line is so hot and a vapor is present, the line will give up heat readily to the surroundings. The surrounding air temperature is 95°F.

9. The superheat has been removed and the refrigerant has cooled down to the 125°F condensing temperature. Point 9 is the point in the system where the desuperheating vapor has just cooled to a temperature of 125°F; this is referred to as a 100% saturated vapor point. If any heat is taken away or rejected, liquid will start to form. As more heat is rejected, the remaining saturated vapor will continue to condense to saturated liquid at the condensing temperature of 125°F. Now notice that the coil temperature is corresponding to the high-side pressure of 278 psig and 125°F. The high-pressure reading of 278 psig is due to the refrigerant condensing at 125°F. In fact, 278 psig is the *vapor pressure* that the vapor is exerting on the liquid while it is condensing. Remember, it is vapor pressure that the gage reads. The condensing conditions are arrived at by knowing the efficiency of the condenser. In this example we use a standard condenser, which has a condensing temperature about 30°F higher than the surrounding air used to absorb heat from the condenser. In this example, 95°F outside air is used to absorb the heat—so 95°F + 30°F = 125°F condensing temperature.

Some condensers will condense at 25°F above the surrounding air; these are high-efficiency condensers, and the high-pressure side of the system will be operating at a lower pressure. Condensing temperatures and pressures are also dependent on the heat load given to the condenser to reject. The higher the heat load, the higher the condensing temperature and corresponding pressure will be. It can also be concluded that, as the outside temperature rises, the operating pressure on the high side of the system will also rise. For example, if the outside ambient (surrounding) temperature rises to 105°F, the condenser saturation temperature on our sample system would rise to about 135°F.

10. The refrigerant is now 100% liquid at the saturated temperature of 125°F. As the liquid continues along the coil, the air continues to cool the liquid to below the actual condensing temperature. The liquid may go as much as 20°F below the condensing temperature of 125°F before it reaches the metering device. Any liquid at a temperature below the condensing temperature of 125°F is called subcooled liquid. In this example, the liquid is cooled to 105°F before it reaches the metering device. The liquid now has 20°F (125°F − 105°F) of subcooling.

11. The liquid refrigerant reaches the metering device through a pipe, usually copper, from the condenser. This liquid line is often field installed and not insulated. Since the temperature of the liquid line is warmer than the temperature of the surrounding air, keeping the liquid line uninsulated allows some additional heat to be rejected by the refrigerant to the surrounding air. This helps increase the operating efficiency of the system. Since the liquid line may be long, and depending on the distance between the condenser and the metering device, the amount of additional heat being rejected may be significant. Heat given up here is leaving the system, and that is good. The refrigerant entering the metering device may be as much as 20°F cooler than the condensing temperature of 125°F, so the liquid line entering the metering device may be 105°F. The refrigerant entering the metering device is 100% subcooled liquid. In the short distance of the metering device's orifice (a pinhole about the size of a small sewing needle), the subcooled liquid is changed to a mixture of about 75% saturated liquid and 25% saturated vapor. The percent of liquid to vapor leaving the metering device is both system and application dependent. The 25% vapor is known as *flash gas* and is used to cool the remaining 75% of the liquid down to 40°F, the boiling temperature of the evaporator. Flash gas is considered a system loss because it is cooling the liquid temperature down to the 40°F evaporating temperature. This wasted cooling cannot be performed in the evaporator in cooling the inside air and removing humidity. The only way to minimize flash gas is to get the subcooled liquid temperature entering the metering device closer to the evaporating temperature.

The refrigerant has now completed one cycle and is ready to go around again. It should be evident that a refrigerant does the same thing over and over, changing from a liquid to a vapor in the evaporator and back to liquid form in the condenser. The expansion device meters the flow to the evaporator, and the compressor pumps the refrigerant out of the evaporator.

The following statements briefly summarize the refrigeration cycle:

1. **The evaporator absorbs heat into the system.**
2. **The condenser rejects heat from the system.**
3. **The compressor pumps the heat-laden vapor.**
4. **The expansion device meters the flow of refrigerant.**

3.12 REFRIGERANTS

Previously we have used water and R-22 as examples of refrigerants. Although many products have the characteristics of a refrigerant, only a few will be covered here. More detailed information will be provided in Unit 9, "Refrigerant and Oil Chemistry and Management—Recovery, Recycling, Reclaiming, and Retrofitting."

The following four refrigerants either can no longer be manufactured or have phaseout dates in the near future:

R-12—Used primarily in medium- and high-temperature refrigeration applications. Manufacturing and importing banned as of January 1, 1996.

R-22—Used primarily in residential, commercial, and industrial air-conditioning applications and in some commercial and industrial refrigeration. R-22 is subject to phaseout in new equipment in 2010, and total production will be phased out in 2020.

R-500—Used primarily in older air-conditioning applications and some commercial refrigeration. Manufacturing and importing banned as of January 1, 1996.

R-502—Used primarily in low-temperature refrigeration applications. Manufacturing and importing banned as of January 1, 1996. An *azeotropic* refrigerant blend that has no *temperature glide* and behaves like a *pure compound*.

The following are some of the newer, more popular long-term replacement refrigerants:

R-134a—Has properties very similar to R-12. Used primarily in medium- and high-temperature refrigeration applications, refrigerators and freezers, and automotive air conditioning. A replacement for R-12 but not a direct drop-in replacement because retrofitting is required. The ester-based lubricants used with R-134a are not compatible with the oils typically used on R-12 systems.

R-404A—Is replacing R-502 in low- and medium-temperature refrigeration applications. Has slightly higher working pressures than R-502. A *near-azeotropic* refrigerant blend with a small temperature glide.

R-407C—Has similar properties to R-22. Replacing R-22 in residential and commercial air-conditioning applications. Can be used as a retrofit refrigerant for R-22 but has a large temperature glide and fractionation potential. A near-azeotropic refrigerant blend. R-407C, like R-134a, operates with ester-based lubricants that are not compatible with the oils typically used on R-22 systems. R-407C is a great choice when the condensing unit of a system must be replaced, as most of the refrigeration oil is contained in the compressor.

R-410A—Is a near-azeotropic refrigerant blend replacing R-22 in residential and commercial air-conditioning applications. Has much higher operating pressures than R-22. Has special safety concerns. Has a very small temperature glide and is not recommended as a retrofit refrigerant. SAFETY PRECAUTION: Never add R-410A to a system that was manufactured for use with R-22. R-22 system components are typically not manufactured to accept the higher operating pressures that are present in R-410A systems. For example, the shells of R-410A compressors are manufactured with thicker steel than are the shells of R-22 compressors.

R-507—Is replacing R-502 in low- and medium-temperature refrigeration applications. Has slightly higher pressures and capacity than R-404A. An azeotropic refrigerant blend.

♻ As will be seen later in this unit, the choice of refrigerant is becoming more important because of environmental issues. It has been thought for many years that the common refrigerants were perfectly safe to use. New discoveries have shown that some of the common refrigerants, R-12, R-500, R-502, and R-22, may be causing damage to the ozone layer in the stratosphere, 7 to 30 miles above the earth's surface. Refrigerants are also being blamed for global warming effects that take place in the troposphere, 0 to 7 miles above the earth. ♻

3.13 REFRIGERANTS MUST BE SAFE

A refrigerant must be safe to protect people from sickness or injury, even death, if the refrigerant should escape from its system. For instance, it could be a disaster to use ammonia for the air-conditioning system in a public place even though it is an efficient refrigerant from many standpoints.

Modern refrigerants are nontoxic, and equipment is designed to use a minimum amount of refrigerant to accomplish its job. A household refrigerator or window air conditioner, for example, normally uses less than 2 lb of refrigerant, yet for years almost 1 lb of refrigerant was used as the propellant in a 16-oz aerosol can of hair spray. SAFETY PRECAUTION: Because refrigerants are heavier than air, proper ventilation is important. For example, if a leak in a large container of refrigerant should occur in a basement, the oxygen could be displaced by the refrigerant and a person could be overcome. Avoid open flame when a refrigerant is present. When refrigeration equipment or cylinders are located in a room with an open gas flame, such as a pilot light on a gas water heater or furnace, the equipment must be kept leak free. If the refrigerant escapes and gets to the flame, the flame will sometimes burn an off-blue or blue-green. This means the flame is giving off a toxic and corrosive gas that will deteriorate any steel in the vicinity and burn the eyes and nose and severely hamper the breathing of anyone in the room. The refrigerants themselves will not burn.

3.14 REFRIGERANTS MUST BE DETECTABLE

A good refrigerant must be readily detectable. The first leak detection device that can be used for some large leaks is listening for the hiss of the escaping refrigerant. **Figure 3–39(A)**. This is not the best way in all cases as some leaks may be so small they may not be heard by the human ear. However, many leaks can be found in this way. There is an ultrasonic leak-detecting device on the market that detects leaks by sound, **Figure 3–39(B)**. This leak detector enables the technician to hear the sound of the fluid moving through the piping circuit toward the leak location. The pitch of the emitted sound changes as the sensor is brought closer and closer to the leak.

Soap bubbles are a practical and yet simple leak detector. Commercially prepared products that blow large elastic types of bubbles are used by many service technicians, **Figure 3–39(C)**. These are valuable when it is known that a leak is in a certain area. Soap-bubble solution can be applied with a brush to the tubing joint to see exactly where the leak is. When refrigerant lines are below freezing temperature, a small amount of antifreeze can be added to the bubble solution. Leaking refrigerant will cause bubbles, **Figure 3–39(D)**. At times a piece of equipment can be submerged in water to watch for bubbles. This is effective when it can be used.

The halide leak detector, **Figure 3–39(E)**, is available for use with acetylene or propane gas. It operates on the principle that when the refrigerant is allowed in an open flame in the presence of glowing copper, the flame will change color, **Figure 3–39(F)**. The halide leak detector should be used in well-ventilated areas, because the resulting blue or green flame indicates that toxic gas is being produced. Although the amounts produced can be very small, the gas is an irritant—corrosive as well as toxic.

The electronic leak detector in **Figure 3–39(G)** is battery operated, is small enough to be easily carried, and has a flexible probe. Some residential air-conditioning equipment has refrigerant charge specifications that call for half-ounce accuracy. The electronic leak detectors are capable of detecting leak rates down to a quarter of an ounce per year.

Another system uses a high-intensity ultraviolet lamp, **Figure 3–39(H)**. An additive is induced into the refrigerant system. The additive will show as a bright yellow-green glow under the ultraviolet lamp at the source of the leak. The area can be wiped clean with a general-purpose cleaner after the leak has been repaired, and the area can be reinspected. The additive can remain in the system. Should a new leak be suspected at a later date it will still show the yellow-green

color under the ultraviolet light. This system will detect leaks as small as a quarter of an ounce per year.

3.15 THE BOILING POINT OF THE REFRIGERANT

The boiling point of the refrigerant should be low at atmospheric pressure so that low temperatures may be obtained without going into a vacuum. For example, R-502 can be boiled as low as $-50°F$ before the boiling pressure goes into a vacuum; whereas R-12 can be boiled only down to $-21°F$ before it goes into a vacuum. Water would have to be boiled at 29.67 in. Hg vacuum just to boil at 40°F. NOTE: When using the compound gage below atmospheric pressure, the scale reads in reverse of the inches of mercury absolute scale. It starts at atmospheric pressure and counts down to a perfect vacuum, called inches of mercury vacuum. When possible, design engineers avoid using refrigerants that boil below 0 psig. This is one reason why R-502 was a good choice for a low-temperature system. When a system operates in a vacuum and a leak occurs, the atmosphere is pulled inside the system instead of the refrigerant leaking out of the system.

3.16 PUMPING CHARACTERISTICS

The pumping characteristics have to do with how much refrigerant vapor is pumped per amount of work accomplished. Water was disqualified as a practical refrigerant for small equipment partly for this reason. One pound of water at 40°F has a vapor volume of 2445 ft^3 compared to about 0.6 ft^3 for R-22. As we will see later on, system capacity is directly related to the number of pounds of refrigerant that are circulated through the system per unit of time. This is expressed in units called pounds per minute, or lb/min. Since the vapor volume of 40°F water is so high, the compressor would have to move 2445 ft^3 of vapor to move a single pound. Thus, the compressor would have to be very large for a water system.

Modern refrigerants meet all of these requirements better than any of the older types. **Figure 3–40** presents the temperature/pressure chart for the refrigerants we have discussed.

3.17 POPULAR REFRIGERANTS AND THEIR IMPORTANT CHARACTERISTICS

The American National Standards Institute (ANSI) and the American Society of Heating, Refrigerating and Air-Conditioning Engineers (ASHRAE) are responsible for naming refrigerants and identifying their characteristics. **Figure 3–41** on page 44 is an organized list of some of the more popular refrigerants and their characteristics. For more detailed information on refrigerants, refer to Unit 9, "Refrigerant and Oil Chemistry and Management—Recovery, Recycling, Reclaiming, and Retrofitting."

♻ As mentioned earlier, environmental issues like ozone depletion and global warming have forced many refrigerants to have phaseout dates for manufacturing. However, these

(A)

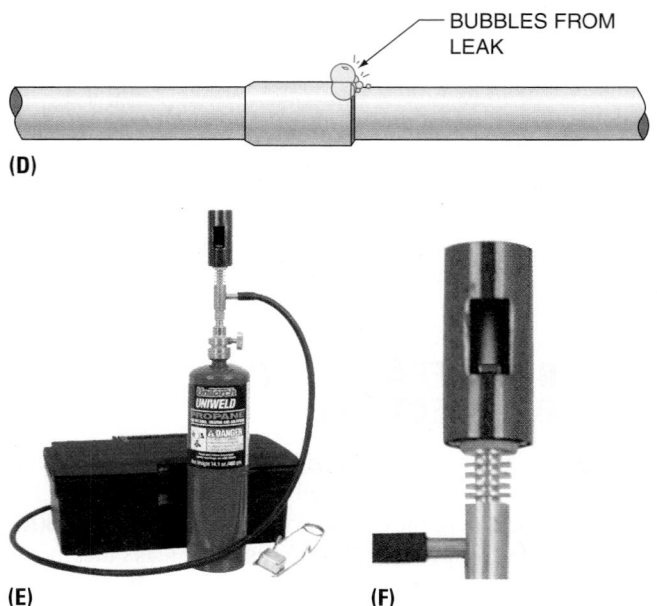

(D)

(E)   **(F)**

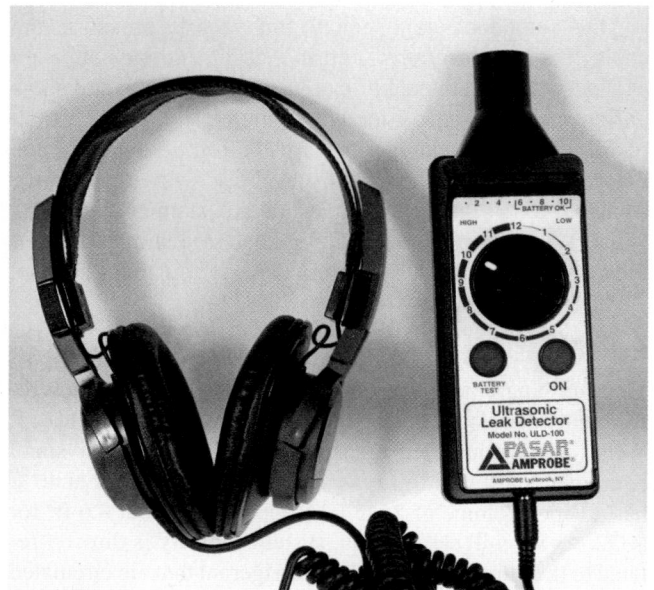

(B)

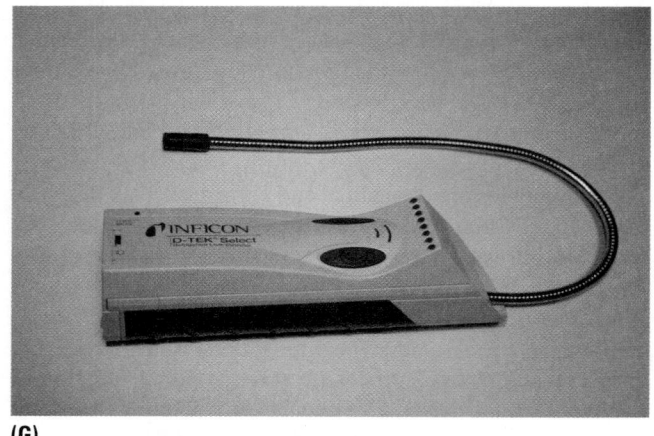

(G)

(C)

(H)

Figure 3–39 Common methods used for leak detection. **(A)** Listening for audible leaks. **(B)** Ultrasonic leak detector. **(C)** Soap-bubble solution. **(D)** Soap-bubble solution causes bubbles to form when a leak is present. **(E)** Halide torch. **(F)** The flame on the halide leak detector turns green when a leak is present. **(G)** Electronic leak detector. **(H)** Ultraviolet leak detection equipment. **(B)** *Photo by Bill Johnson.* **(E)** *Courtesy Uniweld Products.* **(F)** *Courtesy Uniweld Products.* **(G)** *Courtesy Ferris State University. Photo by John Tomczyk.* **(H)** *Courtesy Spectronics Corporation*

BUBBLES FROM LEAK

TEMPERATURE °F	__ 12	REFRIGERANT 22	134a	502	404A	410A
−60	19.0	12.0		7.2	6.6	0.3
−55	17.3	9.2		3.8	3.1	2.6
−50	15.4	6.2		0.2	0.8	5.0
−45	13.3	2.7		1.9	2.5	7.8
−40	11.0	0.5	14.7	4.1	4.8	9.8
−35	8.4	2.6	12.4	6.5	7.4	14.2
−30	5.5	4.9	9.7	9.2	10.2	17.9
−25	2.3	7.4	6.8	12.1	13.3	21.9
−20	0.6	10.1	3.6	15.3	16.7	26.4
−18	1.3	11.3	2.2	16.7	18.2	28.2
−16	2.0	12.5	0.7	18.1	19.6	30.2
−14	2.8	13.8	0.3	19.5	21.1	32.2
−12	3.6	15.1	1.2	21.0	22.7	34.3
−10	4.5	16.5	2.0	22.6	24.3	36.4
−8	5.4	17.9	2.8	24.2	26.0	38.7
−6	6.3	19.3	3.7	25.8	27.8	40.9
−4	7.2	20.8	4.6	27.5	30.0	42.3
−2	8.2	22.4	5.5	29.3	31.4	45.8
0	9.2	24.0	6.5	31.1	33.3	48.3
1	9.7	24.8	7.0	32.0	34.3	49.6
2	10.2	25.6	7.5	32.9	35.3	50.9
3	10.7	26.4	8.0	33.9	36.4	52.3
4	11.2	27.3	8.6	34.9	37.4	53.6
5	11.8	28.2	9.1	35.8	38.4	55.0
6	12.3	29.1	9.7	36.8	39.5	56.4
7	12.9	30.0	10.2	37.9	40.6	57.8
8	13.5	30.9	10.8	38.9	41.7	59.3
9	14.0	31.8	11.4	39.9	42.8	60.7
10	14.6	32.8	11.9	41.0	43.9	62.2
11	15.2	33.7	12.5	42.1	45.0	63.7
12	15.8	34.7	13.2	43.2	46.2	65.3
13	16.4	35.7	13.8	44.3	47.4	66.8
14	17.1	36.7	14.4	45.4	48.6	68.4
15	17.7	37.7	15.1	46.5	49.8	70.0
16	18.4	38.7	15.7	47.7	51.0	71.6
17	19.0	39.8	16.4	48.8	52.3	73.2
18	19.7	40.8	17.1	50.0	53.5	75.0
19	20.4	41.9	17.7	51.2	54.8	76.7
20	21.0	43.0	18.4	52.4	56.1	78.4
21	21.7	44.1	19.2	53.7	57.4	80.1
22	22.4	45.3	19.9	54.9	58.8	81.9
23	23.2	46.4	20.6	56.2	60.1	83.7
24	23.9	47.6	21.4	57.5	61.5	85.5
25	24.6	48.8	22.0	58.8	62.9	87.3
26	25.4	49.9	22.9	60.1	64.3	90.2
27	26.1	51.2	23.7	61.5	65.8	91.1
28	26.9	52.4	24.5	62.8	67.2	93.0
29	27.7	53.6	25.3	64.2	68.7	95.0
30	28.4	54.9	26.1	65.6	70.2	97.0
31	29.2	56.2	26.9	67.0	71.7	99.0
32	30.1	57.5	27.8	68.4	73.2	101.0
33	30.9	58.8	28.7	69.9	74.8	103.1
34	31.7	60.1	29.5	71.3	76.4	105.1
35	32.6	61.5	30.4	72.8	78.0	107.3
36	33.4	62.8	31.3	74.3	79.6	108.4
37	34.3	64.2	32.2	75.8	81.2	111.6
38	35.2	65.6	33.2	77.4	82.9	113.8
39	36.1	67.1	34.1	79.0	84.6	116.0
40	37.0	68.5	35.1	80.5	86.3	118.3
41	37.9	70.0	36.0	82.1	88.0	120.5
42	38.8	71.4	37.0	83.8	89.7	122.9
43	39.8	73.0	38.0	85.4	91.5	125.2
44	40.7	74.5	39.0	87.0	93.3	127.6
45	41.7	76.0	40.1	88.7	95.1	130.0
46	42.6	77.6	41.1	90.4	97.0	132.4
47	43.6	79.2	42.2	92.1	98.8	134.9
48	44.6	80.8	43.3	93.9	100.7	136.4
49	45.7	82.4	44.4	95.6	102.6	139.9
50	46.7	84.0	45.5	97.4	104.5	142.5
55	52.0	92.6	51.3	106.6	114.6	156.0
60	57.7	101.6	57.3	116.4	125.2	170.0
65	63.8	111.2	64.1	126.7	136.5	185.0
70	70.2	121.4	71.2	137.6	148.5	200.8
75	77.0	132.2	78.7	149.1	161.1	217.6
80	84.2	143.6	86.8	161.2	174.5	235.4
85	91.8	155.7	95.3	174.0	188.6	254.2
90	99.8	168.4	104.4	187.4	203.5	274.1
95	108.2	181.8	114.0	201.4	219.2	295.0
100	117.2	195.9	124.2	216.2	235.7	317.1
105	126.6	210.8	135.0	231.7	253.1	340.3
110	136.4	226.4	146.4	247.9	271.4	364.8
115	146.8	242.7	158.5	264.9	290.6	390.5
120	157.6	259.9	171.2	282.7	310.7	417.4
125	169.1	277.9	184.6	301.4	331.8	445.8
130	181.0	296.8	198.7	320.8	354.0	475.4
135	193.5	316.6	213.5	341.2	377.1	506.5
140	206.6	337.2	229.1	362.6	401.4	539.1
145	220.3	358.9	245.5	385.9	426.8	573.2
150	234.6	381.5	262.7	408.4	453.3	608.9
155	249.5	405.1	280.7	432.9	479.8	616.2

VACUUM (in. Hg) – RED FIGURES
GAGE PRESSURE (psig) – BOLD FIGURES

Figure 3–40 This chart shows the temperature/pressure relationship in in. Hg vacuum, or psig. Pressures for R-404A and R-410A are an average liquid and vapor pressure.

refrigerants can still be used if recovered or recycled, or if they are in an operating refrigeration or air-conditioning system. Environmental issues and phaseout dates have made many refrigerants very expensive due to the heavy taxes imposed. Alternative (environmentally friendly) refrigerants have entered the market because of these issues. It is now illegal to intentionally vent to the atmosphere any refrigerant. Stiff fines of up to $32,500 and/or imprisonment can follow. Because of this, mandatory technician certification programs have educated HVACR personnel on environmental issues, alternative refrigerants, and legislation issues. ♻

3.18   REFRIGERANT CYLINDER COLOR CODES

Each type of refrigerant is contained in a cylinder or drum that has a designated color. Following are the colors for some of the most frequently used refrigerants.

R-407B	Cream	R-114	Dark blue
R-407C	Chocolate	R-500	Yellow
R-410A	Rose	R-502	Orchid
R-11	Orange	R-717	Silver
R-12	White	R-409A	Tan
R-22	Green	R-123	Light gray
R-113	Purple	R-401A	Coral red
R-134a	Light blue	R-401B	Mustard yellow
R-401C	Aqua	R-404A	Orange
R-402A	Light brown	R-406A	Light gray-green
R-402B	Green-brown	R-407A	Bright green

Some equipment manufacturers color code their compressors to indicate the type of refrigerant used in the system. **Figure 3–42** shows refrigerant containers for some newer refrigerants.

3.19   RECOVERY, RECYCLE, OR RECLAIM OF REFRIGERANTS

♻ It is mandatory for technicians to recover and sometimes recycle refrigerants during installation and servicing operations to help reduce emissions of chlorofluorocarbons (CFCs), hydrochlorofluorocarbons (HCFCs), and hydrofluorocarbons (HFCs) to the atmosphere. Examples of recovery equipment are shown in **Figure 3–43**. Many larger systems can be fitted with receivers or dump tanks into which the refrigerant can be pumped and stored while the system is serviced. However, in smaller capacity systems it is not often feasible to provide these components. Recovery units or other storage devices may be necessary. Most recovery and/or recycle units that have been developed to date vary in their technology and capabilities so manufacturers' instructions must be followed carefully when using this equipment. Unit 9 includes a detailed description of the recovery, recycle, and reclaim of refrigerants. ♻

ANSI/ASHRAE designation	Safety* classification	Empirical formula	Molecular formula	Components/weight percentages Chemical name		Cylinder color
R-11	A1	CFC	CCl_3F	Trichlorofluoromethane		Orange
R-12	A1	CFC	CCl_2F_2	Dichlorodifluoromethane		White
R-13	A1	CFC	$CClF_3$	Chlorotrifluoromethane		Light Blue
R-14	A1	PFC	CF_4	Tetrafluoromethane		Mustard
R-22	A1	HCFC	$CHClF_2$	Chlorodifluoromethane		Light Green
R-23	A1	HFC	CHF_3	Trifluoromethane		Light Gray-Blue
R-32	A2	HFC	CH_2F_2	Difluoromethane		White/Red Stripe
R-113	A1	CFC	$CCl_2F\text{-}CClF_2$	1,1, 2-Trichloro-1, 2, 2-trifluoroethane		Dark Purple (Violet)
R-114	A1	CFC	$CClF_2\text{-}CClF_2$	1, 2-Dichloro-1,1, 2, 2-tetrafluoroethane		Dark Blue (Navy)
R-115	A1	CFC	$CClF_2\text{-}CF_3$	Chloropentafluoroethane		White/Red Stripe
R-116	A1	PFC	$CF_3\text{-}CF_3$	Hexafluoroethane		Dark Gray (Battleship)
R-123	B1	HCFC	$CHCl_2\text{-}CF_3$	2, 2-Dichloro-1,1,1-trifluoroethane		Light Gray-Blue
R-124	A1	HCFC	$CHClF\text{-}CF_3$	2-Chloro-1,1,1, 2-tetrafluoroethane		Dark Green
R-125	A1	HFC	$CHF_2\text{-}CF_3$	Pentafluoroethane		Medium Brown (Tan)
R-134a	A1	HFC	$CH_2F\text{-}CF_3$	1,1,1, 2-Tetrafluoroethane		Light Sky Blue
R-143a	A2	HFC	$CH_3\text{-}CF_3$	1,1,1-Trifluoroethane		White/Red Stripe
R-152a	A2	HFC	$CH_3\text{-}CHF_2$	1,1-Difluoroethane		White/Red Stripe
R-290	A3	HC	$CH_3\text{-}CH_2\text{-}CH_3$	Propane		White
R-500	A1	CFC	$CCl_2F_2/CH_3\text{-}CHF_2$	R-12/R-152a	73.8/26.2	Yellow
R-502	A1	CFC	$CHClF_2/CClF_2\text{-}CF_3$	R-22/R-115	48.8/51.2	Light Purple (Lavender)
R-503	A1	CFC	$CHF_3/CClF_3$	R-23/R-13	40.1/59.9	Blue-Green (Aqua)
R-507	A1/A1	HFC	$CHF_2\text{-}CF_3/CH_3\text{-}CF_3$	R-125/R-143a	50/50	Blue-Green (Teal)
R-717	B2		NH_3	Ammonia		Silver
R-401A	A1/A1	HCFC	$CHClF_2/CH_3\text{-}CHF_2/CHClF\text{-}CF_3$	R-22/R-152a/R-124	53/13/34	Coral Red
R-401B	A1/A1	HCFC	$CHClF_2/CH_3\text{-}CHF_2/CHClF\text{-}CF_3$	R-22/R-152a/R-124	61/11/28	Yellow-Brown (Mustard)
R-401C	A1/A1	HCFC	$CHClF_2/CH_3\text{-}CHF_2/CHClF\text{-}CF_3$	R-22/R-152a/R-124	33/15/52	Blue-Green (Aqua)
R-402A	A1/A1	HCFC	$CHF_2\text{-}CF_3/CH_3\text{-}CH_2\text{-}CH_3/CHClF_2$	R-125/R-290/R-22	60/02/38	Light Brown (Sand)
R-402B	A1/A1	HCFC	$CHF_2\text{-}CF_3/CH_3\text{-}CH_2\text{-}CH_3/CHClF_2$	R-125/R-290/R-22	38/02/60	Green-Brown (Olive)
R-403A	A1/A1	HCFC	$CH_3\text{-}CH_2\text{-}CH_3/CHClF_2/CF_3\text{-}CF_2\text{-}CF_3$	R-290/R-22/R-218	05/75/20	Light Purple
R-404A	A1/A1	HFC	$CHF_2\text{-}CF_3/CH_3\text{-}CF_3/CH_2F\text{-}CF_3$	R-125/R-143a/R-134a	44/52/04	Orange
R-406A	A1/A2	HCFC	$CHClF_2/CH(CH_3)_3/CH_3\text{-}CClF_2$	R-22/R-600a/R-142b	55/04/41	Light Gray-Green
R-407A	A1/A1	HFC	$CH_2F_2/CHF_2\text{-}CF_3/CH_2F\text{-}CF_3$	R-32/R-125/R-134a	20/40/40	Bright Green
R-407B	A1/A1	HFC	$CH_2F_2/CHF_2\text{-}CF_3/CH_2F\text{-}CF_3$	R-32/R-125/R-134a	10/70/20	Cream
R-407C	A1/A1	HFC	$CH_2F_2/CHF_2\text{-}CF_3/CH_2F\text{-}CF_3$	R-32/R-125/R-134a	23/25/52	Medium Brown
R-408A	A1/A1	HCFC	$CHF_2\text{-}CF_3/CH_3\text{-}CF_3/CHClF_2$	R-125/R-143a/R-22	07/46/47	Medium Purple
R-409A	A1/A1	HCFC	$CHClF_2/CHClF\text{-}CF_3/CH_3\text{-}CClF_2$	R-22/R-124/R-142b	60/25/15	Mustard Brown (Tan)
R-410A	A1/A1	HFC	$CH_2F_2/CHF_2\text{-}CF_3$	R-32/R-125	50/50	Rose

Figure 3–41 A list of refrigerants and some of their characteristics. *NOTE: Safety classifications are covered in Unit 4, "General Safety Practices."

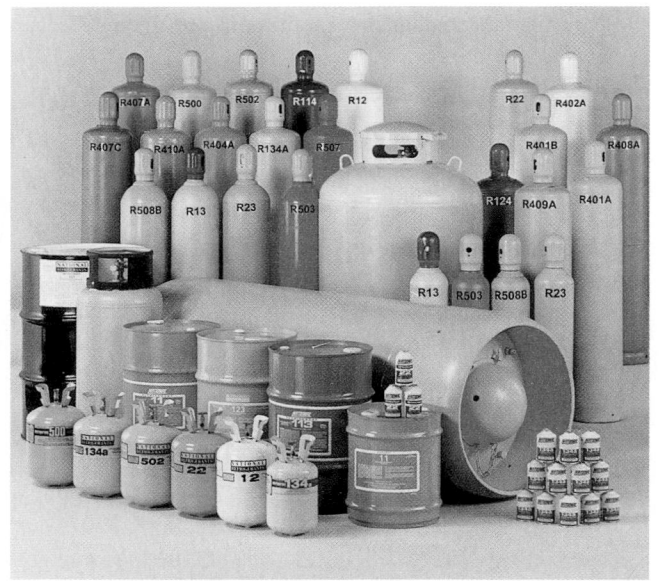

Figure 3–42 Color-coded refrigerant cylinders and drums for some of the newer refrigerants. *Courtesy National Refrigerants, Inc.*

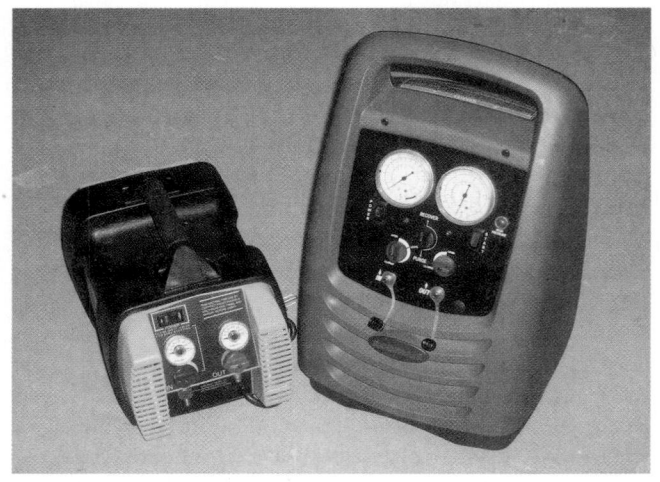

Figure 3–43 Recovery units. *Photo by Eugene Silberstein*

3.20 PLOTTING THE REFRIGERANT CYCLE

A graphic picture of the refrigerant cycle may be plotted on a pressure/enthalpy diagram. This diagram plots pressure on the left-hand side of the diagram and enthalpy on the bottom of the diagram, **Figure 3–44.** *Enthalpy* describes how much heat a substance contains with respect to an accepted reference point. Quite often, people refer to enthalpy as total heat, but this is not absolutely accurate because it refers to the heat content above the selected reference point. Refer to **Figure 1–15,** the heat/temperature graph for water. We used 0°F as the starting point of heat for water, knowing that you can really remove more heat from the water (ice) and lower the temperature below 0°F. We described the process as the amount of heat added starting at 0°F. This heat is called enthalpy. The pressure/enthalpy diagram is a similar diagram, is available for all refrigerants, and is sometimes referred to as a p-e (pressure/enthalpy) or p-h (pressure/heat) chart. Since different refrigerants have different

characteristics, properties, and temperature/pressure relationships, the pressure/enthalpy chart for each refrigerant is different. The pressure/enthalpy chart is used to plot the complete refrigeration cycle as a continuous loop.

The selected reference point for measuring the heat content in a refrigerant is −40°F. On the pressure/enthalpy chart in **Figure 3–45,** it can be seen that the enthalpy or heat content, in Btu/lb (along the bottom of the chart), has a value of 0 Btu/lb when the refrigerant is a saturated liquid at −40°F. The heat content for temperature readings below −40°F saturated liquid are indicated as being negative. As we move from left to right on the chart, the heat content per pound of refrigerant increases. As we move from right to left, the heat content per pound of refrigerant decreases.

As you inspect the pressure/enthalpy chart in **Figure 3–44,** you will notice that there is a horseshoe-shaped curve toward the center of the chart. This curve is called the saturation curve and contains the same information that is contained on the temperature/pressure chart that was discussed earlier.

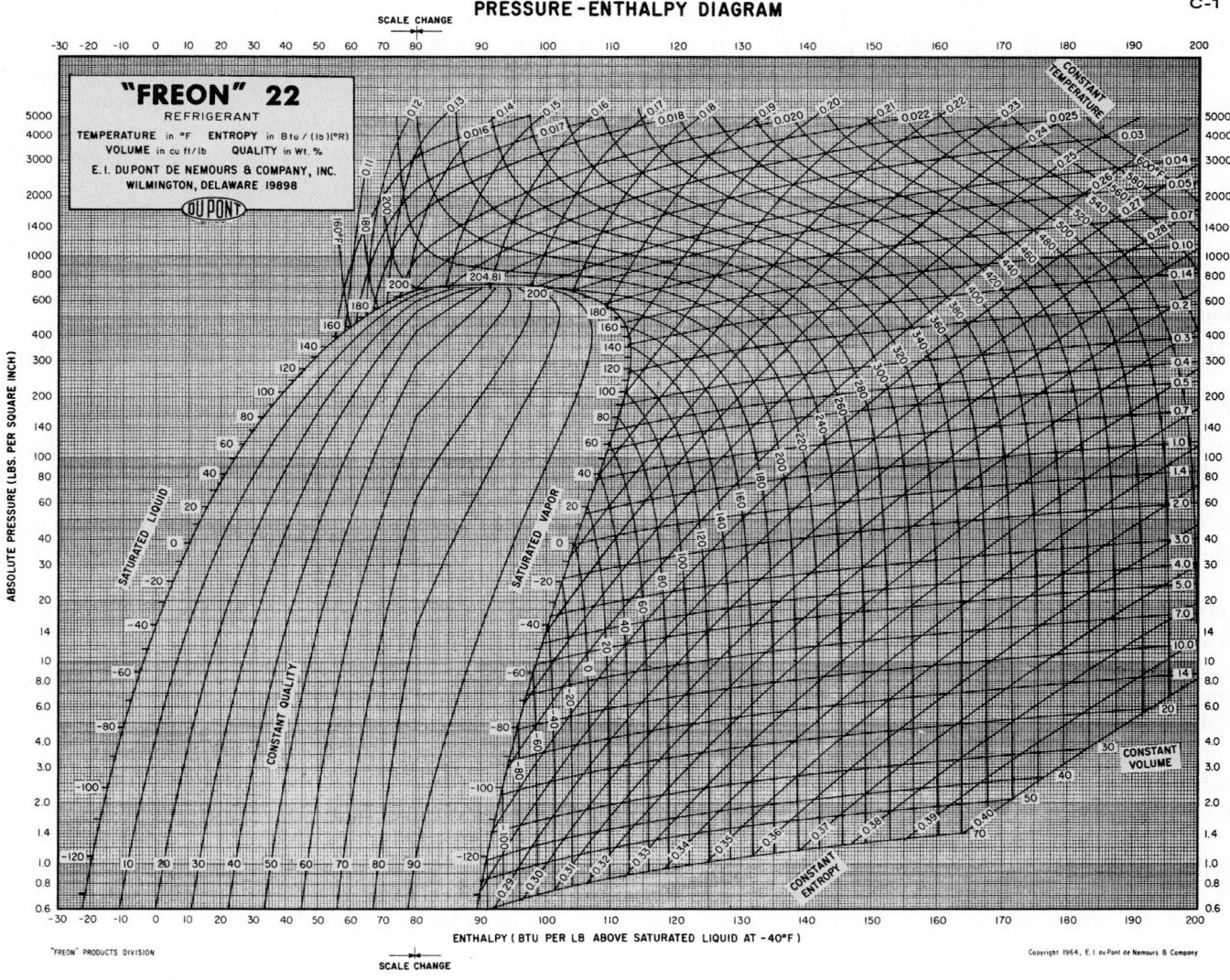

Figure 3–44   The pressure/enthalpy chart relates system operating pressure and temperatures to the heat content of the refrigerant in Btu/lb. *Courtesy E. I. DuPont*

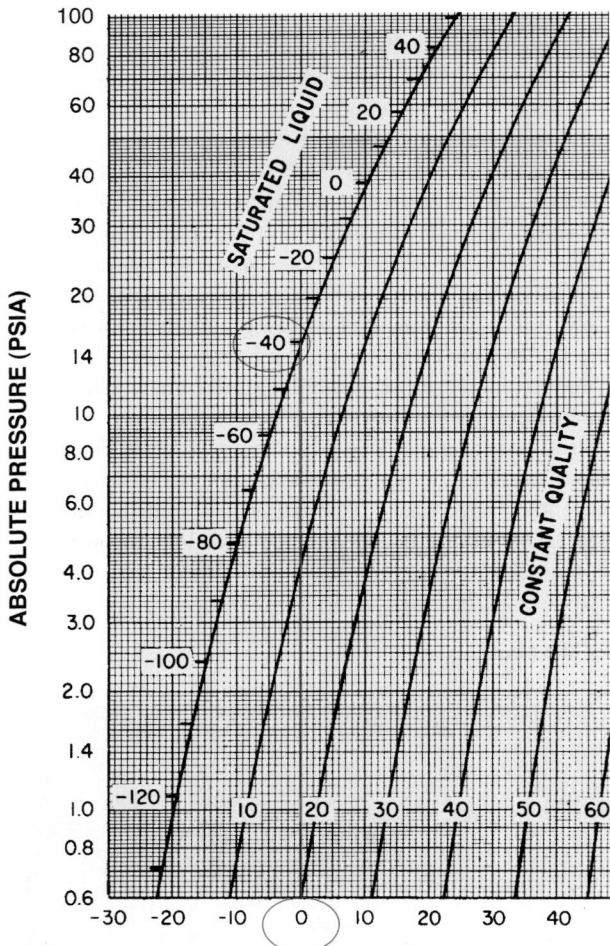

Figure 3–45   The reference point used for measuring heat content is saturated liquid at −40°F. *Courtesy E. I. DuPont*

The only difference is that the pressures on the pressure/enthalpy chart are expressed in absolute pressures, psia, instead of gage pressures, psig.

Any time a plot falls on, or under, this curve, the refrigerant is saturated and will have a corresponding temperature and pressure relationship. Two saturation curves exist; the one on the left is the saturated liquid curve. If heat is added, the refrigerant will start changing state to a vapor. If heat is removed, the liquid will be subcooled. The right-hand curve is the saturated vapor curve. If heat is added, the vapor will superheat. If heat is removed, the vapor will start changing state to a liquid. Notice that the saturated liquid and vapor curves touch at the top. This is called the critical temperature, or pressure. Above this point, the refrigerant will not condense. It is a vapor regardless of how much pressure is applied.

The area between the saturated liquid and saturated vapor curve, inside the horseshoe-shaped curve, is where the change of state occurs. Any time a plot falls between the saturation curves, the refrigerant is in the partial liquid, partial vapor state. The slanted, near-vertical lines between the saturated liquid and saturated vapor lines are the constant quality lines and describe the percentage of vapor to liquid in the mixture between the saturation points. There are nine of these lines

under the saturation curve and each of these lines represents 10%. The saturated liquid line on the left side of the curve represents 0% vapor and 100% liquid, while the saturated vapor line on the right side of the saturation curve represents 100% vapor and 0% liquid. The nine constant quality lines that are located under the saturation curve are labeled, from left to right, 10 through 90. These numbers represent the percentage of quality. Percentage of *quality* means percentage of vapor. This means that if a point falls on the 20% constant quality line, it would be 20% vapor and 80% liquid. If the plot is closer to the saturated liquid curve, there is more liquid than vapor. If the plot is closer to the saturated vapor curve, there is more vapor than liquid. For example, let's find a point on the chart at 40°F (on the saturated liquid curve) and 30 Btu/lb (along the bottom), **Figure 3–46**. This point is inside the horseshoe-shaped curve, and the refrigerant is 90% liquid and 10% vapor. **Figure 3–47** summarizes in skeletal form the important regions, points, and lines of the pressure/enthalpy diagram. For practical reasons, we will use only a few of these skeletal forms for illustrating the functions of the refrigeration cycle.

A refrigeration cycle is plotted in **Figure 3–48** on page 49. The system to be plotted is an air-conditioning system using R-22. The system is operating at 130°F condensing temperature (296.8 psig or 311.5 psia discharge pressure) and an evaporating temperature of 40°F (68.5 psig or 83.2 psia suction pressure). The cycle plotted has no subcooling, and 10°F of superheat as the refrigerant leaves the evaporator with another 10°F of superheat absorbed by the refrigerant in the suction line as it returns to the compressor. The compressor is an air-cooled compressor, and the suction gas enters the suction valve adjacent to the cylinders. Assume that the hot gas line is very short and its heat of rejection is negligible. The following five steps summarize the basic refrigeration cycle (refer to **Figure 3–48**):

1. Refrigerant R-22 enters the expansion device as a saturated liquid at 311.5 psia (296 psig) and 130°F, point A. The heat content is 49 Btu/lb entering the expansion valve and 49 Btu/lb leaving the expansion valve. It can be seen in **Figure 3–48** that the metering device is represented by a vertical line. It can therefore be concluded that, even though the temperature and pressure of the refrigerant drops as it flows through the metering device, the heat content of the refrigerant remains the same. It is important that heat content (Btu/lb) is not confused with temperature. Note that the temperature of the liquid refrigerant before the valve is 130°F, and the temperature leaving the valve is 40°F. The temperature drop can be accounted for by observing that we have 100% liquid entering the valve and about 70% liquid leaving the valve. About 30% of the liquid has changed to a vapor (called flash gas). During the process of evaporating, heat is absorbed from the remaining liquid, lowering its temperature to 40°F. Remember, this flash gas does not contribute to the net refrigeration effect. The *net refrigeration effect (NRE)* is expressed in Btu/lb and is the quantity of

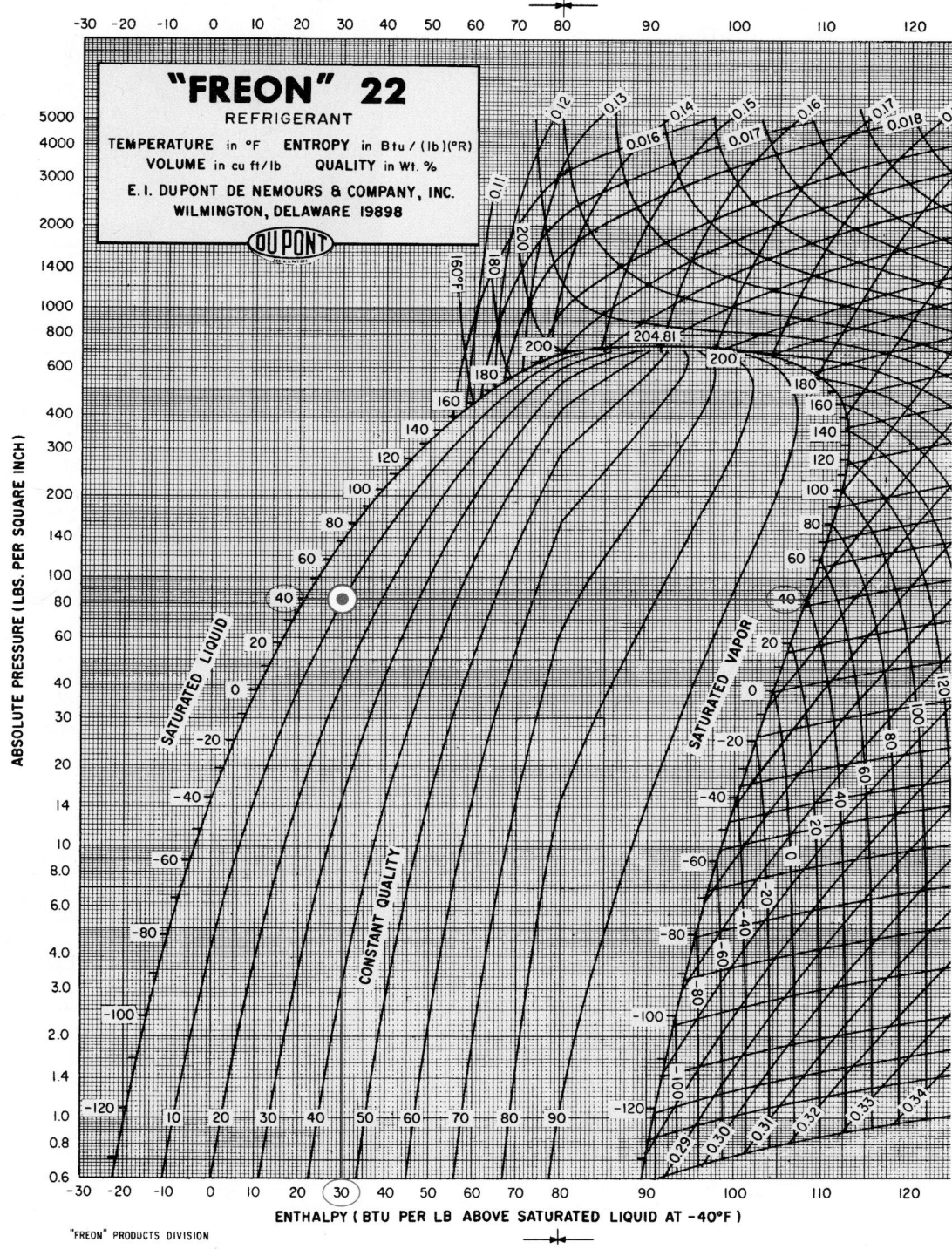

Figure 3–46 The quality of saturated R-22 at 40°F and with a heat content of 30 Btu/lb is 10% vapor and 90% liquid. *Courtesy E. I. DuPont*

heat that each pound of refrigerant absorbs from the refrigerated space to produce useful cooling.

Flash gas occurs because the refrigerant entering the evaporator from the metering device (130°F) must be cooled to the evaporating temperature (40°F) before the remaining liquid can evaporate in the evaporator and produce useful cooling as part of the net refrigeration effect. The heat needed to flash the liquid

came from the liquid itself and not from the conditioned space. No enthalpy is gained or lost in this process, which further explains why the expansion line from point A to point B happened at a constant enthalpy. This expansion process is called *adiabatic expansion* because it happened at a constant enthalpy. Adiabatic processes, by definition, result in temperature and pressure changes with no change in heat content.

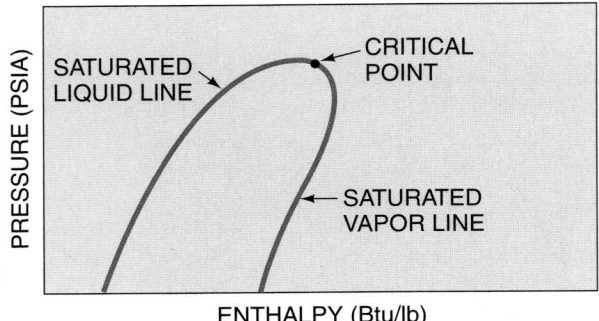

ENTHALPY (Btu/lb)

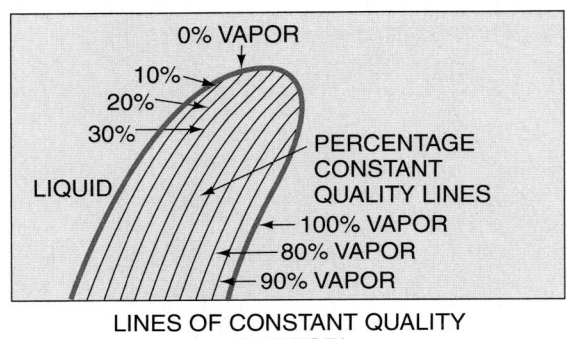

LINES OF CONSTANT QUALITY
(%VAPOR)

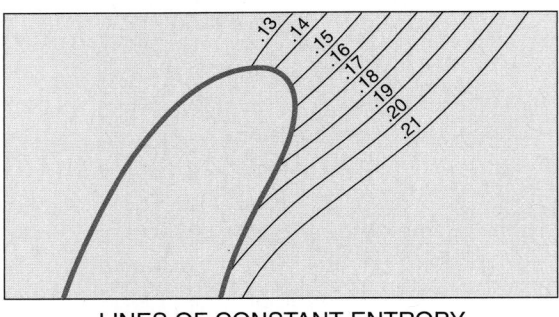

LINES OF CONSTANT ENTROPY
(Btu/lb/°R)

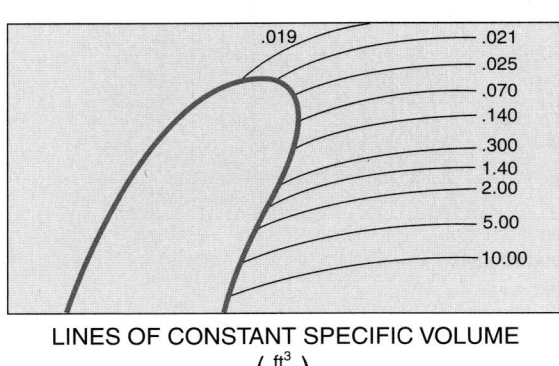

LINES OF CONSTANT SPECIFIC VOLUME
$\left(\dfrac{ft^3}{lb}\right)$

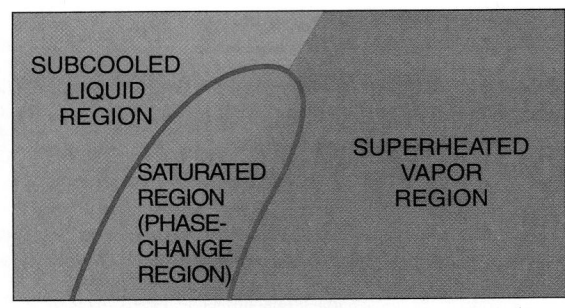

SUBCOOLED–SATURATED–SUPERHEATED
REGIONS

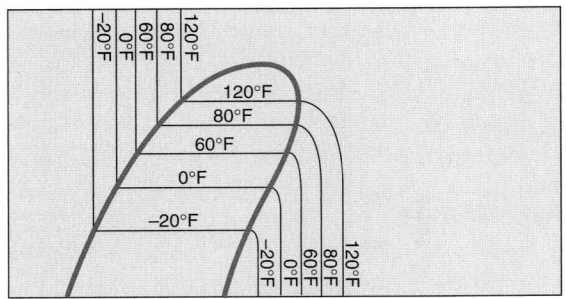

LINES OF CONSTANT TEMP.
(ISOTHERMS)

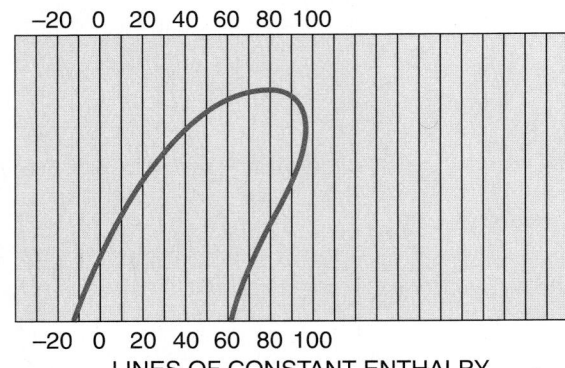

LINES OF CONSTANT ENTHALPY

LINES OF CONSTANT PRESSURE
(ISOBARS)

Figure 3–47 Skeletal pressure/enthalpy diagrams.

2. Usable refrigeration starts at point B where the refrigerant has a heat content of 49 Btu/lb. As heat is added to the refrigerant in the evaporator, the refrigerant gradually changes state to a vapor. Notice that, as we move toward the right from point B in a horizontal

line, the pressure remains unchanged but the heat content increases. The source of this increase in enthalpy is the heat content in the air being cooled by the evaporator. All liquid has changed to a vapor when it reaches the saturated vapor curve, and a small

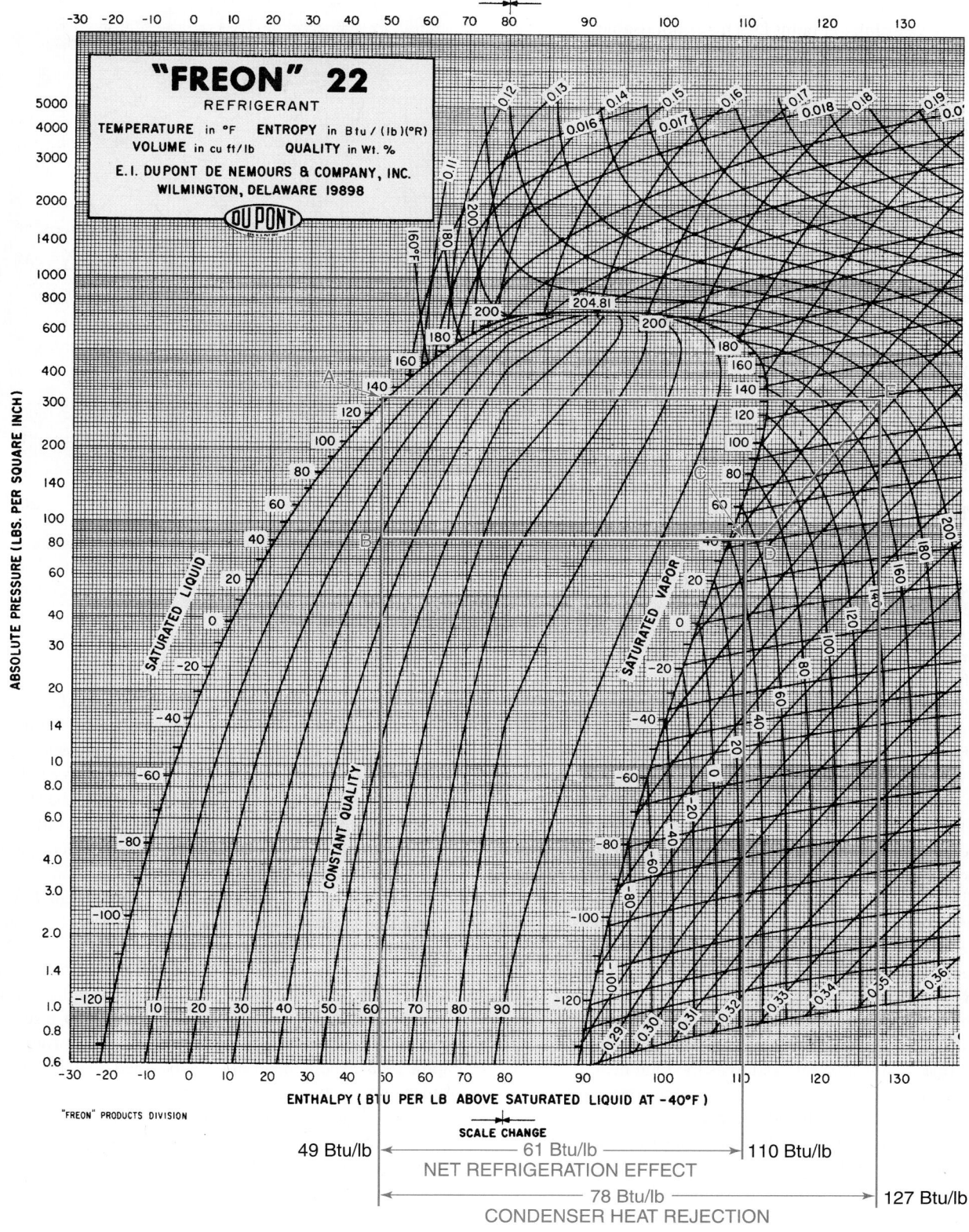

Figure 3–48 A refrigeration cycle plotted on a pressure/enthalpy chart. *Courtesy E. I. DuPont*

amount of heat is added to the refrigerant in the form of superheat (10°F). When the refrigerant leaves the evaporator at point C, it contains about 110 Btu/lb. This is a net refrigeration effect of 61 Btu/lb

(110 Btu/lb − 49 Btu/lb = 61 Btu/lb) of refrigerant circulated. The net refrigeration effect, NRE, is the same as usable refrigeration, the heat actually extracted from the conditioned space. About 10 more Btu/lb are

absorbed into the suction line before reaching the compressor inlet at point D. This is not usable refrigeration because the heat does not come from the conditioned space, but it is heat that must be pumped by the compressor and rejected by the condenser.

3. The refrigerant enters the compressor at point D and leaves the compressor at point E. No heat has been added in the compressor except heat of compression, because the compressor is air cooled. Some of the heat of compression will conduct to the head of the compressor and be rejected to the surroundings. The refrigerant enters the compressor cylinder from the suction line. (A fully hermetic compressor with a suction-cooled motor would not plot out just like this. We have no way of knowing how much heat is added by the motor so we do not know what the temperature of the suction gas entering the compressor cylinder would be for a suction-cooled motor. Manufacturers obtain their own figures for this using internal thermometers during testing.) Notice that the line that represents the compressor is sloped up and to the right. This indicates that both the heat content and pressure of the refrigerant are increasing.

The line that represents the compressor is drawn parallel to another set of lines on the chart that are referred to as lines of constant *entropy*, **Figure 3–47.** Entropy, in our case, represents the compression process and the relationship among the system characteristics of heat content, absolute pressure, and absolute temperature. These lines of constant entropy indicate that, during the compression process, the changes in pressure and temperature are predictable. The units of constant entropy are Btu/lb/°R, where °R is an absolute temperature (as mentioned in Unit 1).

4. The refrigerant leaves the compressor at point E and contains about 127 Btu/lb. At point E, the refrigerant is now at the outlet of the compressor and traveling in the discharge line toward the condenser. This condenser must reject 78 Btu/lb (127 Btu/lb − 49 Btu/lb = 78 Btu/lb), called the heat of rejection. Remember that the condenser must reject all of the heat that is absorbed in the evaporator and suction line as well as the heat generated and concentrated in the compressor during the compression process. Therefore, the heat of rejection is also referred to as the total heat of rejection, THOR, because the condenser must reject all of the heat introduced to the system.

At point E we can also determine the temperature at the outlet of the compressor. We can do this by looking at the position of the point with respect to the lines of constant temperature, which are the downward-sloped, curved lines on the right-hand side of the saturation curve, **Figure 3–49.** The temperature of the discharge gas is about 190°F (see the constant temperature lines for temperature of superheated gas). When the hot gas leaves the compressor it contains the maximum amount of heat that must be rejected by the condenser.

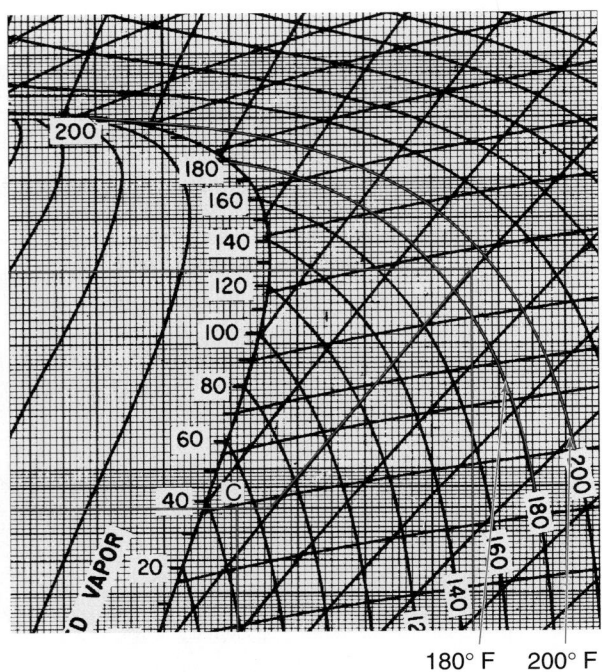

Figure 3–49 The compressor discharge temperature can be found by locating point E and its relation to the constant temperature lines. The compressor discharge temperature shown here is 190°F. *Courtesy E. I. DuPont*

5. The refrigerant enters the condenser at point E as a highly superheated gas. The refrigerant condensing temperature is 130°F and the hot gas leaving the compressor is 190°F, so it contains 60°F (190°F − 130°F = 60°F) of superheat. The condenser will first remove the superheat down to the condensing temperature, which falls on the saturated vapor line. (This process is, once again, referred to as desuperheating.) Then, the condenser will condense the refrigerant to a liquid at 130°F for reentering the expansion device at point A, the saturated liquid line, for another trip around the cycle.

The refrigerant cycle in the previous example can be improved by removing some heat from the condensed liquid by subcooling it. This can be seen in **Figure 3–50,** a scaled-up diagram. The same conditions are used in this figure as in **Figure 3–48,** except the liquid is subcooled 20°F (from 130°F condensing temperature to 110°F liquid). The system then has a net refrigeration effect of 68 Btu/lb instead of 61 Btu/lb. This is an increase in capacity of about 11%. Notice the liquid leaving the expansion valve is only about 23% vapor instead of the 30% vapor in the first example. Also notice that the heat content at the outlet of the metering device is 42 Btu/lb, so the NRE of 68 Btu/lb was determined by subtracting the heat content of 42 Btu/lb at the inlet of the evaporator from the heat content of 110 Btu/lb at the outlet of the evaporator (110 Btu/lb − 42 Btu/lb = 68 Btu/lb). This is where the capacity is gained. Less capacity is lost to flash gas, because the subcooled liquid temperature at 110°F is now a bit closer to the evaporator temperature of 40°F.

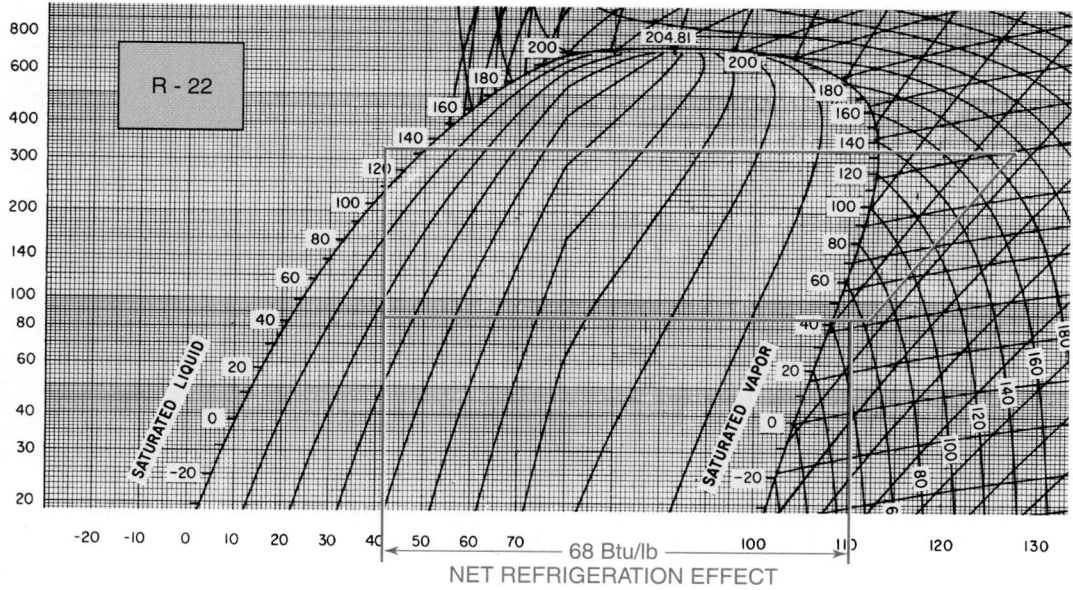

Figure 3–50   Adding subcooling increases the net refrigeration effect of the system. *Courtesy E. I. DuPont*

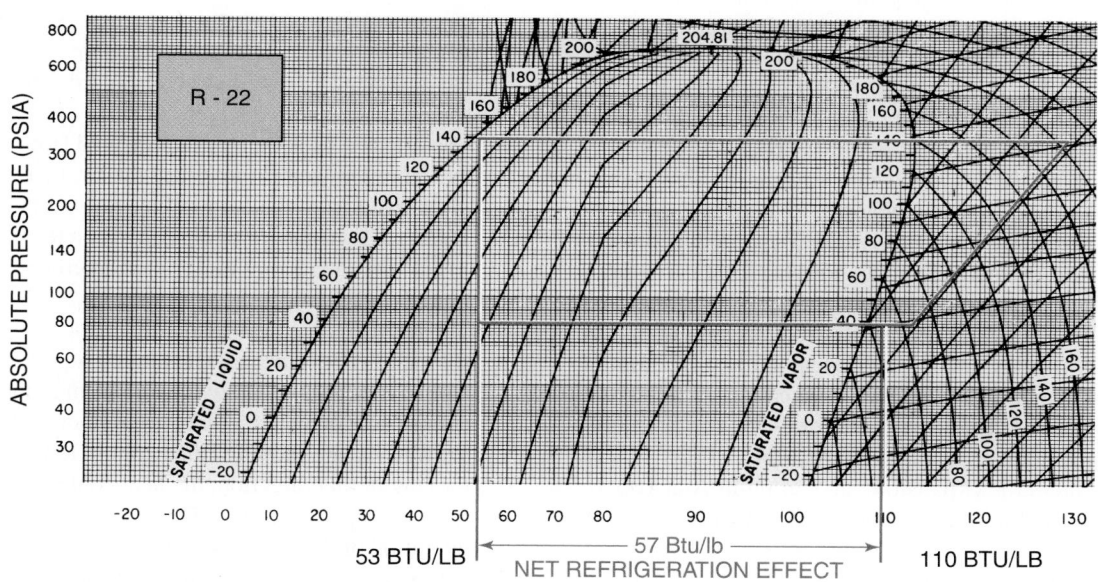

Figure 3–51   A rise in system head pressure will result in a reduction of system capacity. *Courtesy E. I. DuPont*

Other conditions may be plotted on the pressure/enthalpy diagram. For example, suppose the head pressure is raised due to a dirty condenser, **Figure 3–51.** Using the first example, **Figure 3–48,** and raising the condensing temperature to 140°F (337.2 psig or 351.9 psia), we see the percentage of liquid leaving the expansion valve to be about 64% (36% flash gas) with a heat content of 53 Btu/lb. Using the same heat content leaving the evaporator, 110 Btu/lb, we have a net refrigeration effect of 57 Btu/lb. This is a net refrigeration effect reduction of about 7% from the original example, which contained 49 Btu/lb at the same point. This shows the importance of keeping condensers clean.

Figure 3–52 shows how increased superheat affects the first system in **Figure 3–48.** The suction line has not been insulated and absorbs heat. The suction gas may leave the evaporator at 50°F and rise to 75°F before entering the compressor. Notice the high discharge temperature (about 200°F). This is approaching the temperature that will cause oil to break down and form acids in the system. Most compressors must not exceed 250°F. The compressor must pump more

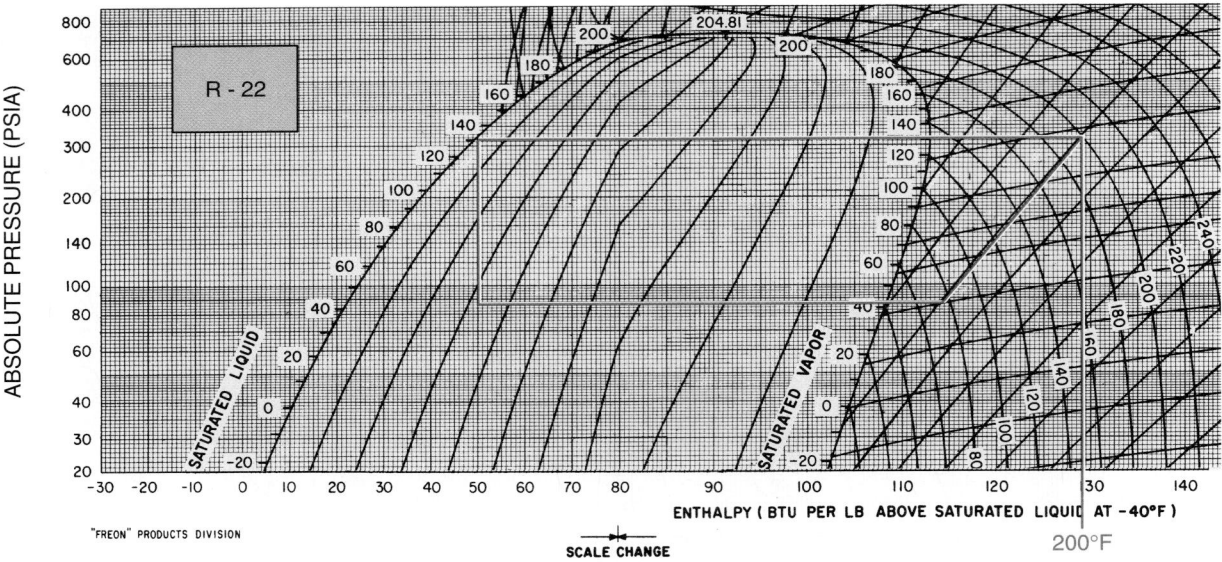

Figure 3–52   An increase in superheat results in an increase in compressor discharge temperature. *Courtesy E. I. DuPont*

refrigerant to accomplish the same refrigeration effect, and the condenser must reject more heat.

R-12 has been the most popular refrigerant for medium-temperature applications for many years. Environmental issues have caused the manufacturers to explore some different refrigerants for the replacement of R-12, namely, R-134a, R-22, and the newer refrigerant blends. R-134a has a zero ozone depletion potential but has oil compatibility problems and cannot be easily retrofitted into existing R-12 systems. R-134a also suffers in capacity when used in low-temperature applications. Environmental issues will be covered in Unit 9.

The following sequence shows how R-12 performs in a typical medium-temperature application. We will use an evaporator temperature of 20°F and a condensing temperature of 115°F. This lower condensing temperature is common for medium- and low-temperature applications. Follow the description in the illustration in **Figure 3–53**. The pressure/enthalpy explanation appears in **Figure 3–54.**

1. Liquid enters the expansion valve at point A at 100°F. Note that the liquid is subcooled 15°F to 100°F, **Figure 3–54.**
2. Partial liquid and partial vapor leave the expansion valve at point B (16% vapor and 84% liquid). The heat content per pound is 31 Btu/lb.
3. Vapor refrigerant leaves the evaporator at point C with 10° of superheat and a heat content of 82 Btu/lb. Note that the net refrigeration effect is 51 Btu/lb (82 − 31 = 51).
4. The refrigerant enters the compressor at point D at a temperature of 50°F. The refrigerant has a total of 30° of superheat considering what was picked up in the evaporator and the suction line. The heat content at the inlet of the compressor is 84 Btu/lb. This means that, in the suction line, each pound of refrigerant picks up 2 Btu/lb (84 Btu/lb − 82 Btu/lb) that must later be rejected by the condenser.

5. The refrigerant is compressed along the line between point D and E, where it leaves the compressor at a temperature of 160°F. Note the lower discharge temperature of 160°F. This is because we are operating at a lower head (condensing) pressure, which causes the heat of compression to be lowered. This is accomplished with a larger condenser.
6. The refrigerant is then desuperheated from point E to the saturated vapor line. Since the refrigerant is discharged from the compressor at a temperature of 160°F and will begin to condense at a temperature of 115°F, the refrigerant must be desuperheated 45°F (160°F − 115°F).
7. The refrigerant is now gradually condensed from the saturated vapor line to the saturated liquid line at 115°F. Notice that, as we follow the line from point E (the compressor outlet) to point A (the inlet of the metering device), we are moving from right to left. Since we are moving in that direction, the heat content of the refrigerant is decreasing. This is consistent with the operation of the condenser, which is the system component that is responsible for rejecting system heat.
8. The liquid is now subcooled from the saturated liquid line to point A. The process then repeats itself.

Following is an example of the same system using R-22, for medium-temperature application. Follow the description in **Figure 3–55** and compare with the previous example.

1. Refrigerant enters the expansion valve at point A at 100°F, subcooled 15°F from the condensing temperature of 115°F, just like the example for R-12.
2. The refrigerant leaves the expansion valve at point B at 28% vapor and 72% liquid with a heat content of 38 Btu/lb. It then travels through the evaporator.
3. The refrigerant leaves the evaporator at point C in the vapor state with 10° of superheat, a heat content of 108 Btu/lb, and a net refrigeration effect of 70 Btu/lb.

20 psig 34.7 psia
146.5 psig 161.2 psia
50°F
160°F
CONDENSING TEMPERATURE 115°F
CONDENSING UNIT
30°F
82 Btu/lb
EVAPORATING TEMPERATURE 20°F
EVAPORATOR
100°F
29% VAPOR
71% LIQUID

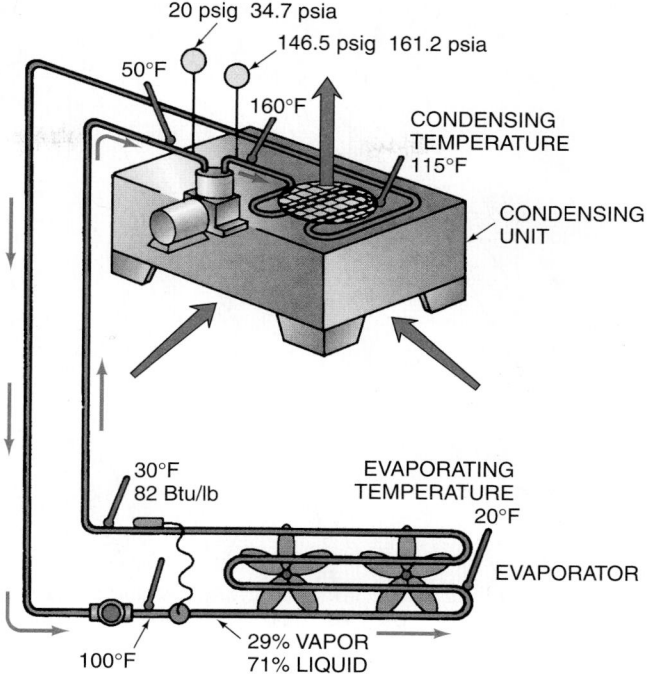

Figure 3–53 A medium-temperature refrigeration system showing operating temperatures and pressures.

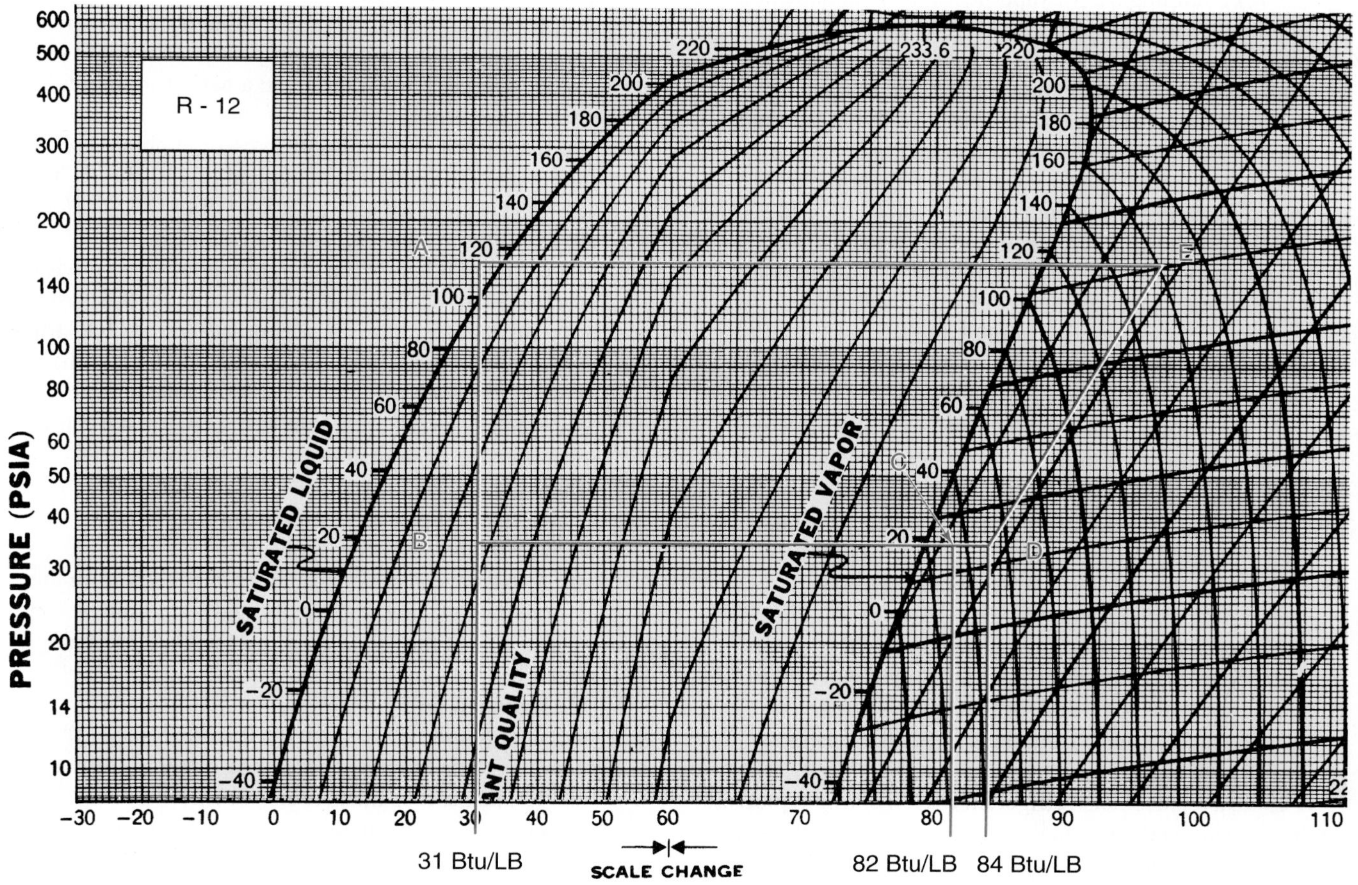

R - 12
PRESSURE (PSIA)
SATURATED LIQUID
SATURATED VAPOR
SCALE CHANGE
31 Btu/LB
82 Btu/LB
84 Btu/LB

Figure 3–54 An R-12 medium-temperature refrigeration system. *Courtesy E. I. DuPont*

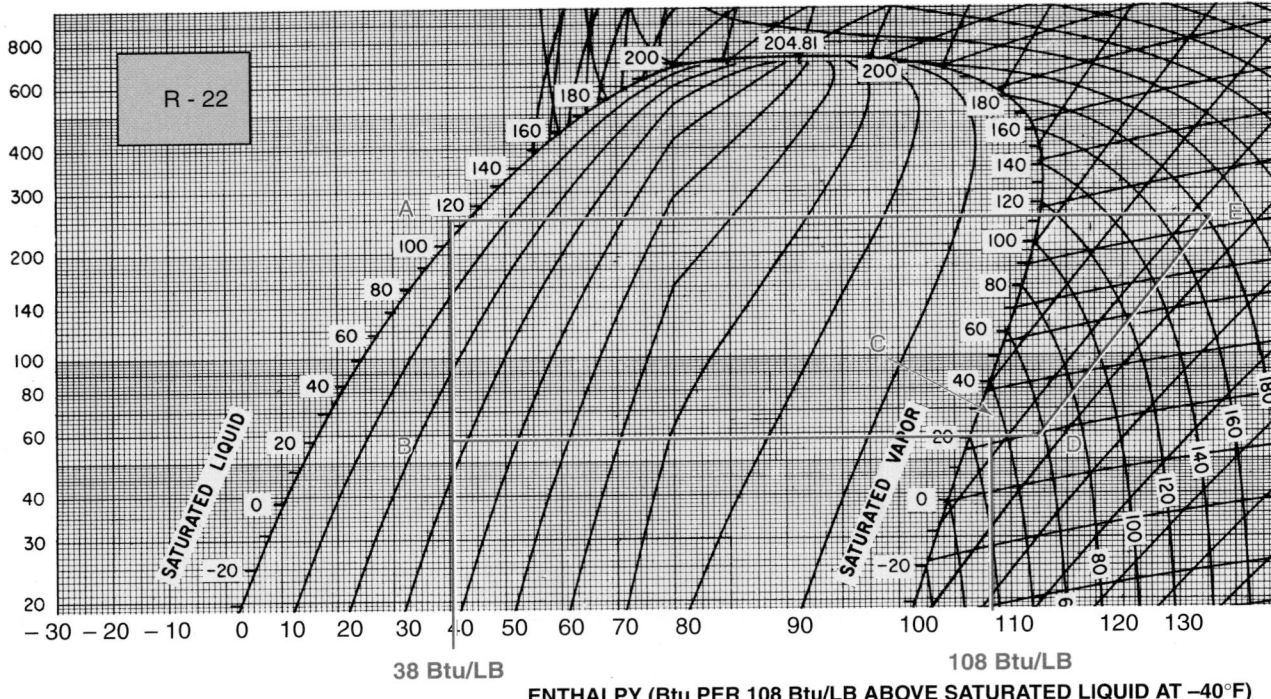

Figure 3–55   An R-22 medium-temperature refrigeration system. *Courtesy E. I. DuPont*

4. Refrigerant vapor enters the compressor at point D at 57°F, containing 37° of superheat. The heat content at the inlet of the compressor, point D, is 113 Btu/lb.

5. The vapor refrigerant is compressed on the line from D to E and leaves the compressor at point E at a temperature of about 180°F. *This is 20°F higher than the temperature for similar conditions for R-12 and is one of the main differences in the refrigerants. R-22 has a much higher discharge temperature than R-12.* As the condensing temperature of the application becomes higher, the temperature rises. At some point, the designer must decide to either use a different refrigerant or change the application.

6. The heat content at the outlet of the compressor, point E, is 136 Btu/lb. Since the refrigerant entered the compressor with a heat content of 113 Btu/lb, the amount of heat added to the refrigerant during the compression process is 23 Btu/lb (136 Btu/lb − 113 Btu/lb). This is referred to as the heat of work (HOW) for the compressor.

7. There was an additional 5 Btu/lb added to the refrigerant in the suction line (113 Btu/lb − 108 Btu/lb). The additional heat added to the system outside of the evaporator is referred to as the heat of compression (HOC) and, in this case, is equal to 28 Btu/lb (23 Btu/lb from the compression process + 5 Btu/lb from the suction line).

R-134a, a replacement refrigerant for R-12, would plot out on the pressure/enthalpy chart as follows. Use **Figure 3–56** to follow this example.

1. Refrigerant enters the expansion valve at point A at 105°F, subcooled 10° from 115°F.

2. Refrigerant leaves the expansion valve at point B with a heat content of 47 Btu/lb. The quality is 33% vapor with 67% liquid. It then travels through the evaporator.

3. Vapor refrigerant leaves the evaporator at point C with a heat content of 109 Btu/lb and a net refrigeration effect of 62 Btu/lb (109 − 47 = 62).

4. The refrigerant enters the compressor at point D with a superheat of 40° and a heat content of 114 Btu/lb. The vapor is compressed along the line from D to E. *Note the lower discharge temperature of 160°F.*

5. The refrigerant leaves the compressor at point E with a heat content of 130 Btu/lb. The HOW for this system is 16 Btu (130 Btu/lb − 114 Btu/lb) and the HOC for this system is 21 Btu/lb [(130 Btu/lb − 114 Btu/lb) + (114 Btu/lb − 109 Btu/lb)]. The HOC can also be determined by subtracting the heat content at point C from the heat content at point E (130 Btu/lb − 109 Btu/lb).

6. The refrigerant is then desuperheated 45°F from point E to the saturated vapor line.

7. The refrigerant is now gradually condensed from the saturated vapor line to the saturated liquid line at 115°F. This is the *latent heat of condensation* being rejected.

8. The saturated liquid is now subcooled from the saturated liquid line to point A. It is subcooled 10° (115°F − 105°F) until it enters the metering device at 105°F. The process then repeats itself.

Another comparison of refrigerants may be made using low temperature as the application. Here we will see much higher discharge temperatures and see why some decisions about various refrigerants are made. We will use a condensing

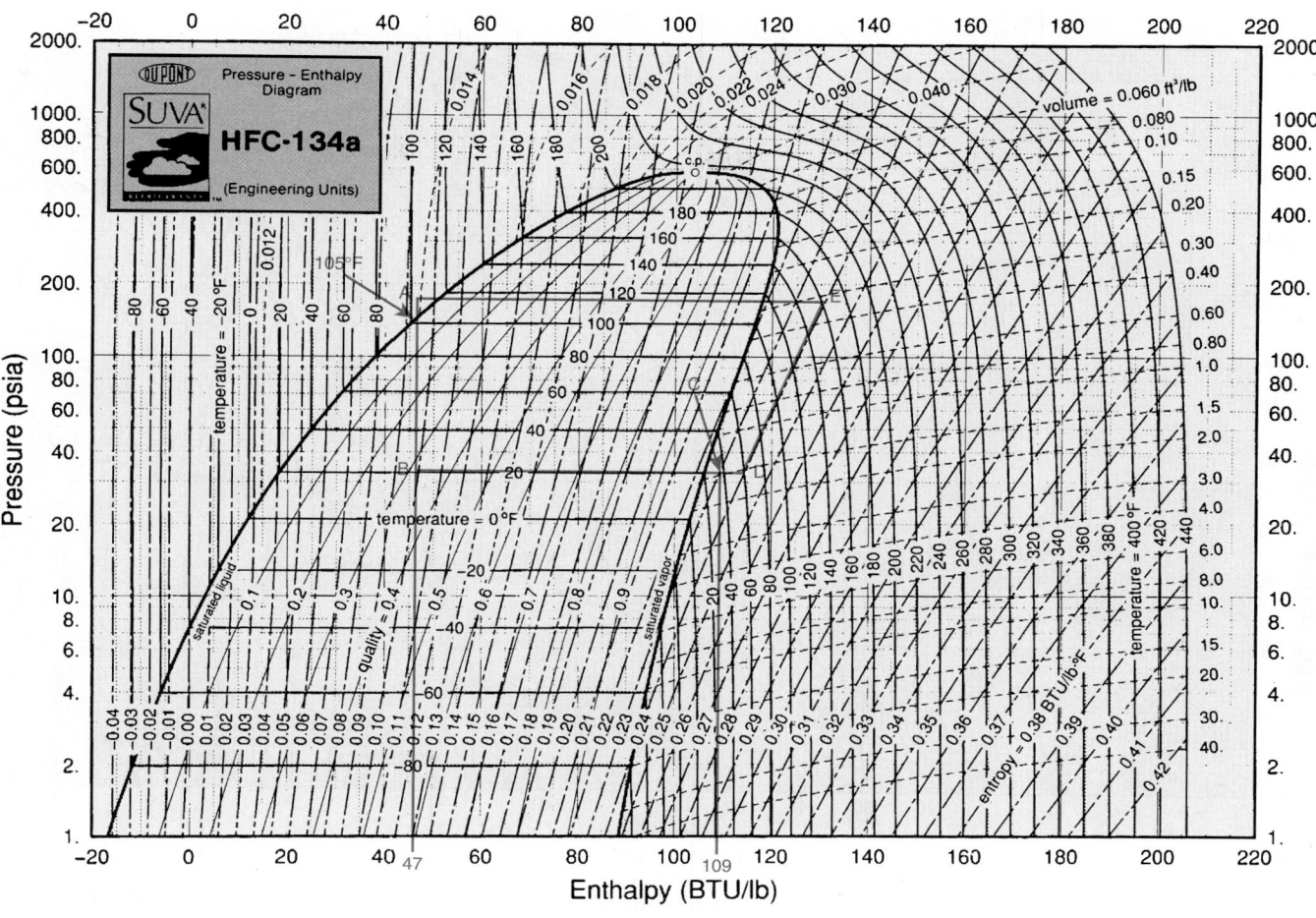

Figure 3–56   An R-134a pressure/enthalpy diagram. *Courtesy E. I. DuPont*

temperature of 115°F and an evaporator temperature of −20°F and compare R-12 to R-502, then to R-22.

Figure 3–57 shows a low-temperature R-12 refrigeration system on a pressure/enthalpy chart. Follow the plot below.

1. The refrigerant enters the expansion valve at point A at 105°F, subcooled 10° from 115°F, like the above examples.

2. The refrigerant leaves the expansion valve at point B at −20°F to be boiled to a vapor in the evaporator. NOTE: The pressure for R-12 boiling at −20°F is 0.6 psig, very close to atmospheric pressure. If the boiling point were any lower, the low side of the system would be in a vacuum. This is one of the disadvantages of R-12 and R-134a as refrigerants for low-temperature application. For this system to maintain a room temperature of about 0°F, the coil temperature can be only 20°F below the return air temperature (0°F room or return air temperature −20°F temperature difference = −20°F coil temperature), **Figure 3–58**. If the thermostat were to be turned down too low, the low-pressure side of this system would be in a vacuum. NOTE: The percentage of liquid to vapor mixture is different for low-temperature than for medium-temperature applications. In this example, we have 39% vapor

and 61% liquid leaving the expansion valve. The difference is that extra flash gas is required to lower the remaining liquid to the lower temperature of −20°F. The heat content of the refrigerant at the inlet of the evaporator is 32 Btu/lb.

3. At point C, the refrigerant leaves the evaporator in the vapor state at a temperature of −10°F, with 10° of superheat and a heat content of 76 Btu/lb. The NRE for this evaporator is 44 Btu/lb.

4. The vapor refrigerant enters the compressor at point D with 30° of superheat at a temperature of 10°F and is compressed along the line to point E. The heat content at point D is 78 Btu/lb.

5. Notice that the discharge temperature is only 170°F, a very cool discharge temperature. The refrigerant leaves the compressor at point E with a heat content of 98 Btu/lb. The refrigerant is then desuperheated 55°F from point E to the saturated vapor line. It is left as an exercise to confirm that the HOW for this system is 20 Btu/lb and that the HOC for this system is 22 Btu/lb.

6. The refrigerant is now gradually condensed from the saturated vapor line to the saturated liquid line at 115°F. This is the *latent heat of condensation* being rejected.

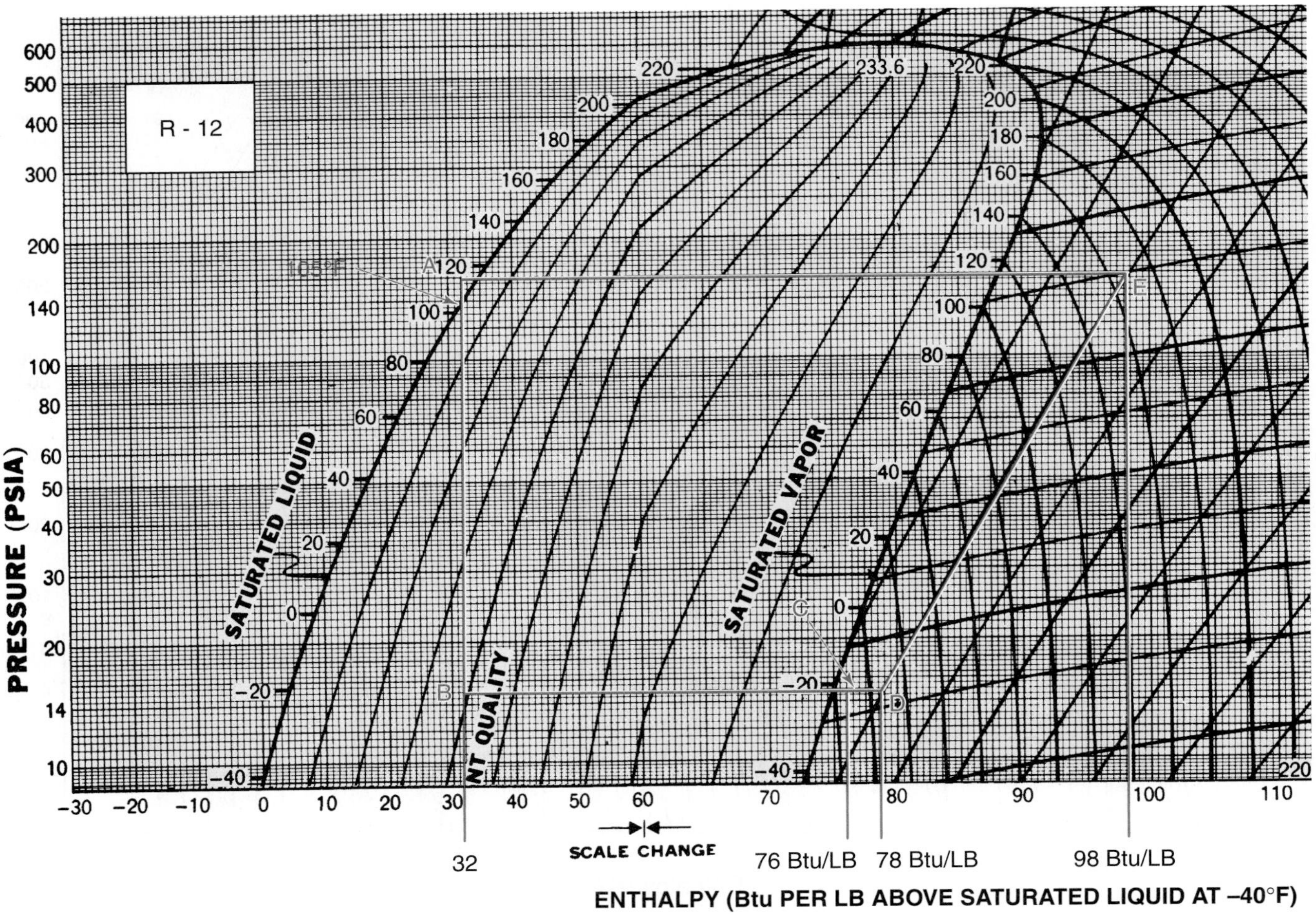

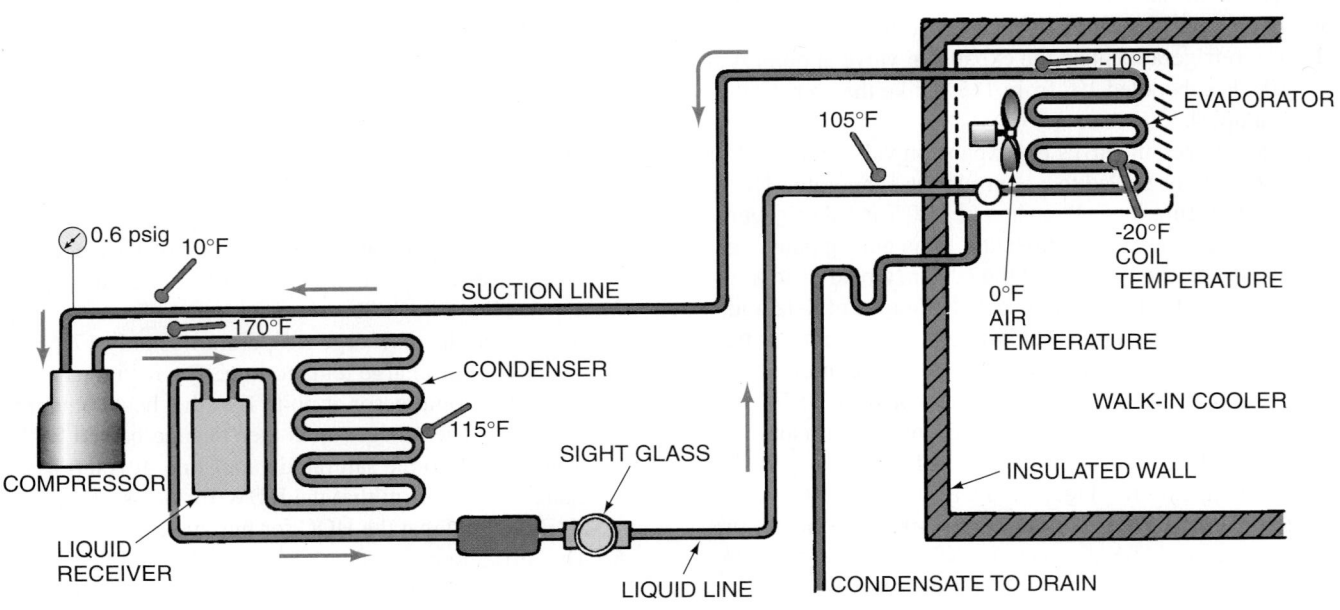

Figure 3–57 An R-12 low-temperature refrigeration system. *Courtesy E. I. DuPont*

Figure 3–58 The refrigerant is boiling at −20°F. The room air temperature is 0°F.

7. The saturated liquid is now subcooled from the saturated liquid line to point A. It is subcooled 10° (115°F − 105°F) until it enters the metering device at 105°F. The process then repeats itself.

R-502 has been used for many low-temperature applications to prevent the system from operating in a vacuum on the low-pressure side. Follow the same conditions using R-502 in **Figure 3–59**.

1. Refrigerant enters the expansion valve at 105°F, subcooled from 115°F and 48% vapor, 52% liquid. The heat content is 38 Btu/lb.

2. The remaining vapor is evaporated and 10° of superheat is added by the time the vapor reaches point C. The heat content rises to 78 Btu/lb, and the NRE is 40 Btu/lb.

3. The vapor enters the compressor at point D and is compressed to point E where the vapor leaves the compressor at 160°F.

4. The refrigerant is then desuperheated, condensed, and subcooled from point E to A.

5. Notice that the suction pressure is 15.3 psig (30 psia) for R-502 while boiling at −20°F. R-502 will not go into a vacuum until the temperature is −50°F. This refrigerant is very good for low-temperature applications because of this. It also has a very acceptable discharge gas temperature.

As an exercise, verify the following:
- Heat content at the inlet of the compressor is 80 Btu/lb.
- Heat content at the outlet of the compressor is 98 Btu/lb.
- Heat of work, HOW, is 18 Btu/lb.
- Heat of compression, HOC, is 20 Btu/lb.

R-22 is being used for many low-temperature applications because R-502 has been phased out due to the CFC/ozone depletion issue. R-22 has the problem of high discharge gas temperature. **Figure 3–60** shows a plot of the low-temperature application above using R-22. Use the following sequence.

1. The refrigerant enters the expansion valve at point A at 105°F, subcooled from 115°F, just as in the preceding problem. At the inlet of the evaporator, the refrigerant has a heat content of 40 Btu/lb. The refrigerant leaves the evaporator at point C with a heat content of 104 Btu/lb. The temperature at the outlet of the evaporator is −10°F, so the evaporator is operating with 10 degrees of superheat. The NRE for this evaporator is 64 Btu/lb (104 Btu/lb − 40 Btu/lb).

2. The refrigerant is evaporated to a vapor at −20°F and enters the compressor at point D at a temperature of 20°F. Notice that R-22 boils at 10.1 psig at −20°F. It is in a positive pressure. Evaporators using R-22 can be operated down to −41°F before the suction pressure goes into a vacuum. The heat content at the inlet of the compressor is 110 Btu/lb.

3. The vapor is compressed along line D to E where it leaves the compressor at point E at a temperature of 240°F. Remember, R-12 had a discharge temperature of 170°F at the same condition and R-502 a discharge temperature of 160°F. R-22 is much hotter. A 240°F

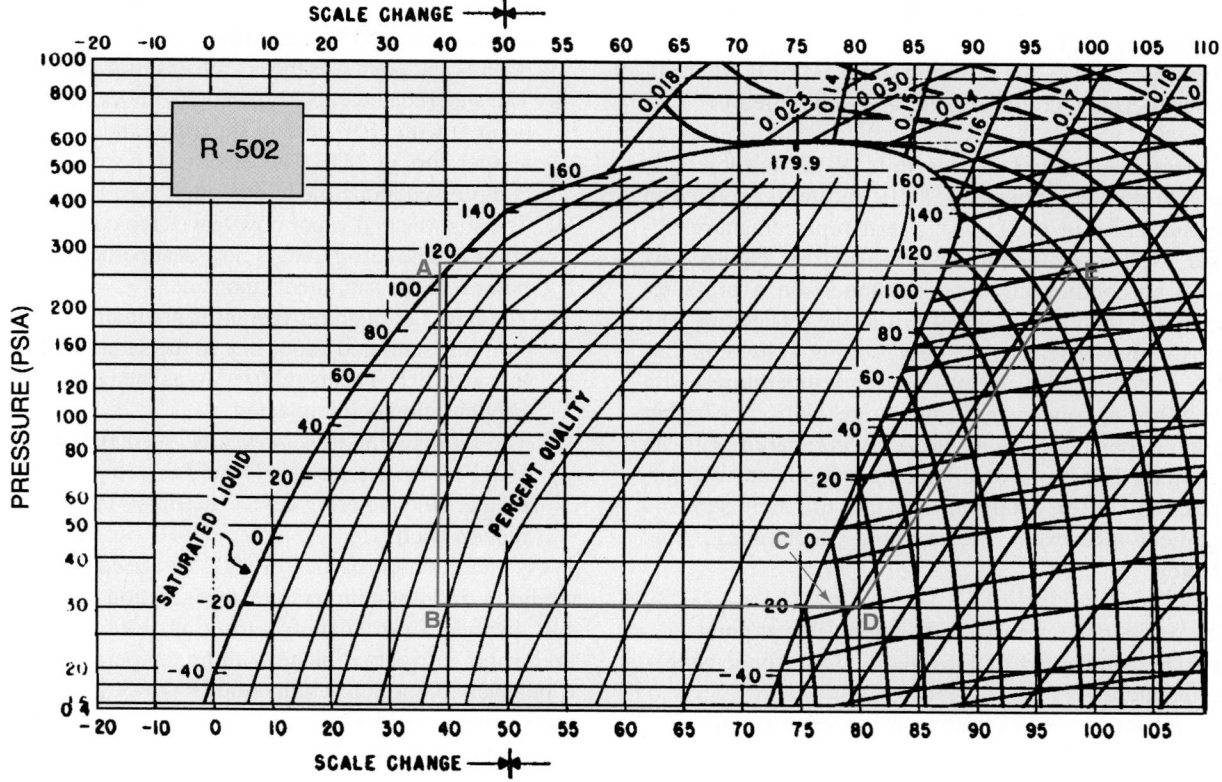

Figure 3–59 An R-502 low-temperature refrigeration system. *Courtesy E. I. DuPont*

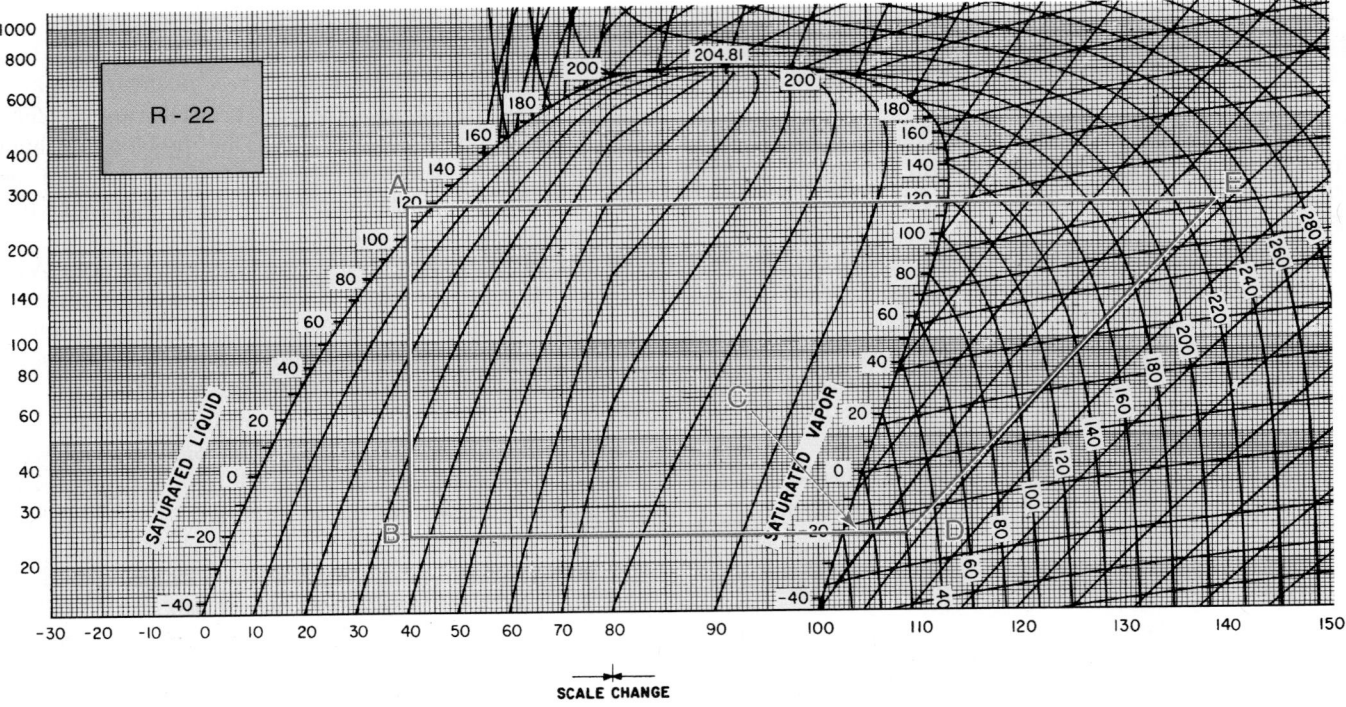

Figure 3–60 An R-22 low-temperature refrigeration system. *Courtesy E. I. DuPont*

discharge temperature can be worked with but is close to being too high. Any increase in discharge temperature due to a dirty condenser can cause serious system problems. The heat content at the outlet of the compressor is 138 Btu/lb, the HOW is 28 Btu/lb, and the HOC is 34 Btu/lb.

4. The vapor leaves the compressor at point E where it is desuperheated, condensed, and subcooled to point A.

As mentioned earlier in this unit, R-22 is subject to phaseout in new equipment in 2010, and total production will be phased out in 2020. A long-term replacement refrigerant for R-22 in new equipment is R-410A. **Figure 3–61** shows an air-conditioning system plotted on a pressure/enthalpy chart incorporating R-410A as the refrigerant. The system is a high-efficiency air-conditioning system operating with a 45°F (130-psig) evaporating temperature and a 115°F (390-psig) condensing temperature. Notice the higher pressures associated with R-410A as compared with an R-22 system. For more detailed information on refrigerants, refer to Unit 9, "Refrigerant and Oil Chemistry and Management—Recovery, Recycling, Reclaiming, and Retrofitting."

Follow the steps below.

1. The refrigerant enters the expansion valve at point A at 105°F, subcooled 10°F from the 115°F condensing temperature.

2. The refrigerant leaves the expansion valve at point B at 45°F (27% vapor and 73% liquid) to be totally evaporated in the evaporator. Because R-410A has

such a small temperature glide (0.3°F), it can be ignored in most air-conditioning applications involving pressure/temperature relationships.

3. At point C, the refrigerant leaves the evaporator as 100% vapor with 10°F of evaporator superheat and at a temperature of 55°F. The difference in enthalpy between point B and point C is the net refrigeration effect.

4. The superheated vapor now enters the compressor at point D with 30°F of total superheat and at a temperature of 75°F. The superheated vapor is now compressed along the line to point E to 180°F.

5. The superheated vapor now leaves the compressor at point E. The refrigerant is now desuperheated from point E to the saturated vapor line.

6. The now saturated refrigerant is gradually condensed from the saturated vapor line to the saturated liquid line at 115°F. This is referred to as the latent heat of condensation being rejected.

7. The saturated liquid is now subcooled 10°F (115°F − 105°F) from the saturated liquid line to point A where it enters the metering device at 105°F. The process then repeats itself.

As mentioned earlier, R-502 is used primarily in medium- and low-temperature refrigeration applications. It was banned from manufacturing in 1996. R-502 is an azeotropic refrigerant blend with no temperature glide, and it behaves like a pure compound. R-404A is its long-term replacement refrigerant and has a slightly higher working pressure than R-502. **Figure 3–62** shows a low-temperature refrigeration system plotted on a pressure/enthalpy chart

Enthalpy (BTU/lb)

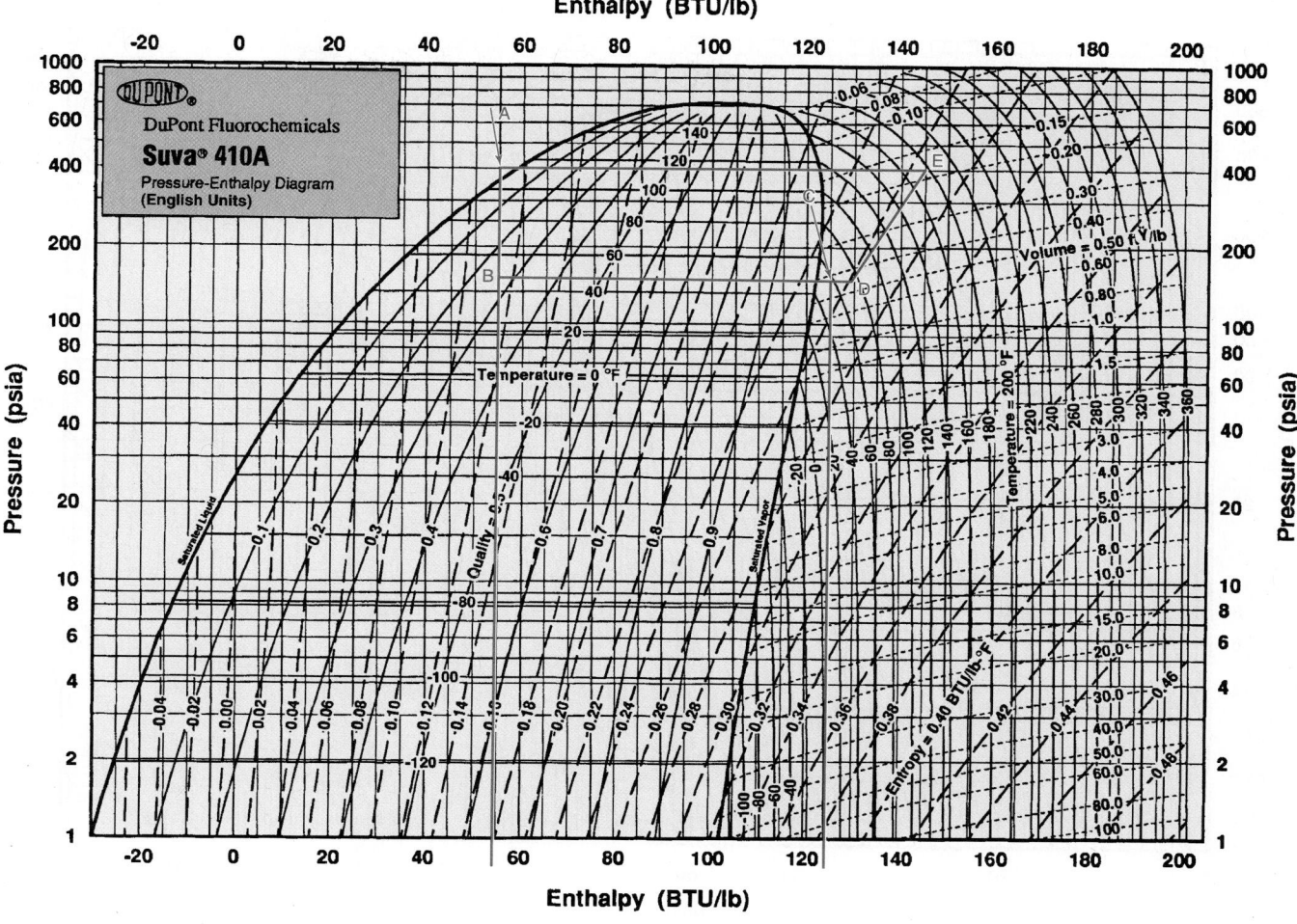

Figure 3–61 An R-410A air-conditioning system. *Courtesy E. I. DuPont*

incorporating R-404A as the refrigerant. The system is a commercial freezer operating with an evaporating temperature of −20°F (16.7-psig) and a 115°F (290-psig) condensing temperature.

Follow the sequence below.

1. The refrigerant enters the expansion valve at point A at 105°F, subcooled 10°F from the 115°F condensing temperature.

2. The refrigerant leaves the expansion valve at point B at −20°F (55% vapor and 45% liquid) to be totally evaporated in the evaporator. At point C, the refrigerant leaves the evaporator as 100% vapor with 10°F of evaporator superheat and at a temperature of −10°F. The difference in enthalpy between point B and point C is the net refrigeration effect.

3. The superheated vapor enters the compressor at point D with 30°F of total superheat and at a temperature of 10°F. The superheated vapor is now compressed along the line to point E to 170°F.

4. The superheated vapor leaves the compressor at point E. The refrigerant is now desuperheated from point E to the saturated vapor line.

5. The now saturated refrigerant is gradually condensed from the saturated vapor line to the saturated liquid line at 115°F. This is referred to as the latent heat of condensation being rejected.

6. The saturated liquid is subcooled 10°F (115°F − 105°F) from the saturated liquid line to point A where it enters the metering device at 105°F. The process then repeats itself.

Pressure/enthalpy diagrams are useful for showing the refrigerant cycle for the purpose of establishing the various conditions around the system. They are partially constructed from properties of refrigerant tables. **Figure 3–63** is a page from a typical table for R-22. Column 1 is the temperature corresponding to the pressure columns for the saturation temperature.

Column 5 lists the specific volume for the saturated vapor refrigerant in cubic feet per pound. For example, at 60°F, the compressor must pump 0.4727 ft³ of refrigerant to circulate 1 lb of refrigerant in the system. The specific volume along with the net refrigeration effect help the engineer determine the compressor's pumping capacity. The example in **Figure 3–48** using R-22 had a net refrigeration effect of

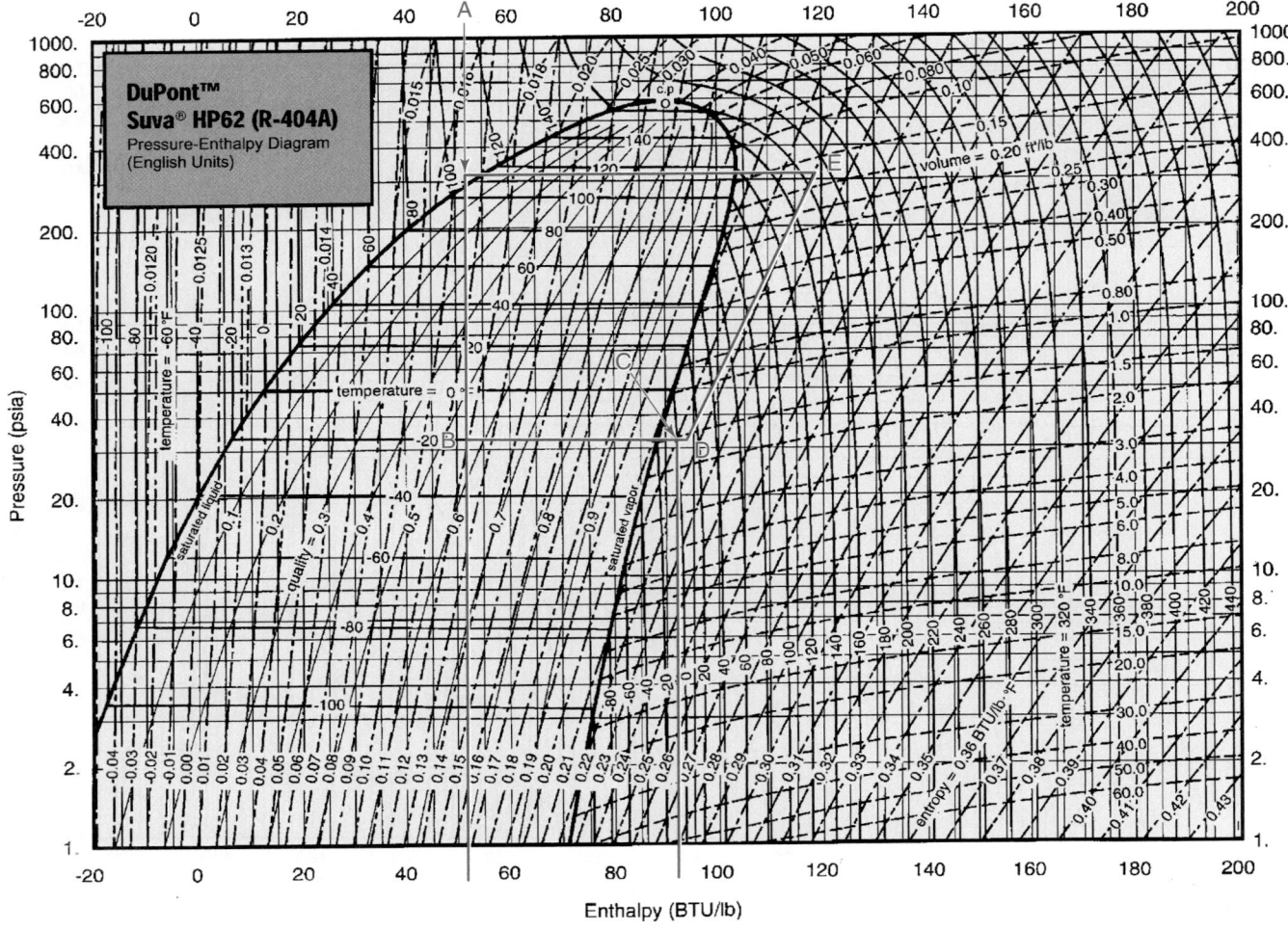

Figure 3–62   An R-404A low-temperature refrigeration system. *Courtesy E. I. DuPont*

61 Btu/lb of refrigeration circulated. If we had a system needing to circulate enough refrigerant to absorb 36,000 Btu/h (3 tons of refrigeration) we would need to circulate 590.2 lb of refrigerant per hour (36,000 Btu/h divided by 61 Btu/lb = 590.2 lb/h). If the refrigerant entered the compressor at 60°F, the compressor must move 275 ft^3 of refrigerant per hour (590.2 lb/h × 0.46523 ft^3/lb = 275 ft^3/h). There is a slight error in this calculation because the 0.46523 ft^3/h is for saturated refrigerant, and the vapor is superheated entering the compressor. Superheat tables are available, but will only complicate this calculation more and there is very little error. Many compressors are rated in cubic feet per minute so this compressor would need to pump 4.58 ft^3/min (275 ft^3/h ÷ 60 min/h = 4.58 ft^3/min).

The density portion of the table tells the engineer how much a particular volume of liquid refrigerant will weigh at the rated temperature. For example, 1 ft^3 of R-22 weighs 76.773 lb when the liquid temperature is 60°F. This is important for determining the weight of refrigerant in components, such as evaporators, condensers, and receivers.

The enthalpy portion of the table (total heat) lists the heat content of the liquid and vapor and the amount of latent heat

required to boil 1 lb of liquid to a vapor. For example, at 60°F, saturated liquid refrigerant would contain 27.172 Btu/lb compared with 0 Btu/lb at −40°F. It would require 82.54 Btu/lb to boil 1 lb of 60°F saturated liquid to a vapor. The saturated vapor would then contain 109.712 Btu/lb total heat (27.172 + 82.54 = 109.712).

The entropy column is of no practical value except on the pressure/enthalpy chart where it is used to plot the compressor discharge temperature.

These charts and tables are not normally used in the field for troubleshooting but are for engineers to use to design equipment. They help the technician understand the refrigerants and the refrigerant cycle.

Different refrigerants have different temperature/pressure relationships and enthalpy relationships. These all must be considered by the engineer when choosing the correct refrigerant for a particular application. A complete study of each refrigerant and its comparison to all other refrigerants is helpful, but you do not need to understand the complete picture to successfully perform in the field. A complete study and comparison is beyond the scope of this book.

TEMP.	PRESSURE		VOLUME cu ft/lb		DENSITY lb/cu ft		ENTHALPY Btu/lb			ENTROPY Btu/(lb X °R)		TEMP.
°F	PSIA	PSIG	LIQUID v_f	VAPOR v_g	LIQUID $1/v_f$	VAPOR $1/v_g$	LIQUID h_f	LATENT h_{fg}	VAPOR h_g	LIQUID s_f	VAPOR s_g	°F
10	47.464	32.768	0.012088	1.1290	82.724	0.88571	13.104	92.338	105.442	0.02932	0.22592	10
11	48.423	33.727	0.012105	1.1077	82.612	0.90275	13.376	92.162	105.538	0.02990	0.22570	11
12	49.396	34.700	0.012121	1.0869	82.501	0.92005	13.648	91.986	105.633	0.03047	0.22548	12
13	50.384	35.688	0.012138	1.0665	82.389	0.93761	13.920	91.808	105.728	0.03104	0.22527	13
14	51.387	36.691	0.012154	1.0466	82.276	0.95544	14.193	91.630	105.823	0.03161	0.22505	14
15	52.405	37.709	0.012171	1.0272	82.164	0.97352	14.466	91.451	105.917	0.03218	0.22484	15
16	53.438	38.742	0.012188	1.0082	82.051	0.99188	14.739	91.272	106.011	0.03275	0.22463	16
17	54.487	39.791	0.012204	0.98961	81.938	1.0105	15.013	91.091	106.105	0.03332	0.22442	17
18	55.551	40.855	0.012221	0.97144	81.825	1.0294	15.288	90.910	106.198	0.03389	0.22421	18
19	56.631	41.935	0.012238	0.95368	81.711	1.0486	15.562	90.728	106.290	0.03446	0.22400	19
20	57.727	43.031	0.012255	0.93631	81.597	1.0680	15.837	90.545	106.383	0.03503	0.22379	20
21	58.839	44.143	0.012273	0.91932	81.483	1.0878	16.113	90.362	106.475	0.03560	0.22358	21
22	59.967	45.271	0.012290	0.90270	81.368	1.1078	16.389	90.178	106.566	0.03617	0.22338	22
23	61.111	46.415	0.012307	0.88645	81.253	1.1281	16.665	89.993	106.657	0.03674	0.22318	23
24	62.272	47.576	0.012325	0.87055	81.138	1.1487	16.942	89.807	106.748	0.03730	0.22297	24
25	63.450	48.754	0.012342	0.85500	81.023	1.1696	17.219	89.620	106.839	0.03787	0.22277	25
26	64.644	49.948	0.012360	0.83978	80.907	1.1908	17.496	89.433	106.928	0.03844	0.22257	26
27	65.855	51.159	0.012378	0.82488	80.791	1.2123	17.774	89.244	107.018	0.03900	0.22237	27
28	67.083	52.387	0.012395	0.81031	80.675	1.2341	18.052	89.055	107.107	0.03958	0.22217	28
29	68.328	53.632	0.012413	0.79604	80.558	1.2562	18.330	88.865	107.196	0.04013	0.22198	29
30	69.591	54.895	0.012431	0.78208	80.441	1.2786	18.609	88.674	107.284	0.04070	0.22178	30
31	70.871	56.175	0.012450	0.76842	80.324	1.3014	18.889	88.483	107.372	0.04126	0.22158	31
32	72.169	57.473	0.012468	0.75503	80.207	1.3244	19.169	88.290	107.459	0.04182	0.22139	32
33	73.485	58.789	0.012486	0.74194	80.089	1.3478	19.449	88.097	107.546	0.04239	0.22119	33
34	74.818	60.122	0.012505	0.72911	79.971	1.3715	19.729	87.903	107.632	0.04295	0.22100	34
35	76.170	61.474	0.012523	0.71655	79.852	1.3956	20.010	87.708	107.719	0.04351	0.22081	35
36	77.540	62.844	0.012542	0.70425	79.733	1.4199	20.292	87.512	107.804	0.04407	0.22062	36
37	78.929	64.233	0.012561	0.69221	79.614	1.4447	20.574	87.316	107.889	0.04464	0.22043	37
38	80.336	65.640	0.012579	0.68041	79.495	1.4697	20.856	87.118	107.974	0.04520	0.22024	38
39	81.761	67.065	0.012598	0.66885	79.375	1.4951	21.138	86.920	108.058	0.04576	0.22005	39
40	83.206	68.510	0.012618	0.65753	79.255	1.5208	21.422	86.720	108.142	0.04632	0.21986	40
41	84.670	69.974	0.012637	0.64643	79.134	1.5469	21.705	86.520	108.225	0.04688	0.21968	41
42	86.153	71.457	0.012656	0.63557	79.013	1.5734	21.989	86.319	108.308	0.04744	0.21949	42
43	87.655	72.959	0.012676	0.62492	78.892	1.6002	22.273	86.117	108.390	0.04800	0.21931	43
44	89.177	74.481	0.012695	0.61448	78.770	1.6274	22.558	85.914	108.472	0.04855	0.21912	44
45	90.719	76.023	0.012715	0.60425	78.648	1.6549	22.843	85.710	108.553	0.04911	0.21894	45
46	92.280	77.584	0.012735	0.59422	78.526	1.6829	23.129	85.506	108.634	0.04967	0.21876	46
47	93.861	79.165	0.012755	0.58440	78.403	1.7112	23.415	85.300	108.715	0.05023	0.21858	47
48	95.463	80.767	0.012775	0.57476	78.280	1.7398	23.701	85.094	108.795	0.05079	0.21839	48
49	97.085	82.389	0.012795	0.56532	78.157	1.7689	23.988	84.886	108.874	0.05134	0.21821	49
50	98.727	84.031	0.012815	0.55606	78.033	1.7984	24.275	84.678	108.953	0.05190	0.21803	50
51	100.39	85.69	0.012836	0.54698	77.909	1.8282	24.563	84.468	109.031	0.05245	0.21785	51
52	102.07	87.38	0.012856	0.53808	77.784	1.8585	24.851	84.258	109.109	0.05301	0.21768	52
53	103.78	89.08	0.012877	0.52934	77.659	1.8891	25.139	84.047	109.186	0.05357	0.21750	53
54	105.50	90.81	0.012898	0.52078	77.534	1.9202	25.429	83.834	109.263	0.05412	0.21732	54
55	107.25	92.56	0.012919	0.51238	77.408	1.9517	25.718	83.621	109.339	0.05468	0.21714	55
56	109.02	94.32	0.012940	0.50414	77.282	1.9836	26.008	83.407	109.415	0.05523	0.21697	56
57	110.81	96.11	0.012961	0.49606	77.155	2.0159	26.298	83.191	109.490	0.05579	0.21679	57
58	112.62	97.93	0.012982	0.48813	77.028	2.0486	26.589	82.975	109.564	0.05634	0.21662	58
59	114.46	99.76	0.013004	0.48035	76.900	2.0818	26.880	82.758	109.638	0.05689	0.21644	59
60	116.31	101.62	0.013025	0.46523	76.773	2.1154	27.172	82.540	109.712	0.05745	0.21627	60
61	118.19	103.49	0.013047	0.46523	76.644	2.1495	27.464	82.320	109.785	0.05800	0.21610	61
62	120.09	105.39	0.013069	0.45788	76.515	2.1840	27.757	82.100	109.857	0.05855	0.21592	62
63	122.01	107.32	0.013091	0.45066	76.386	2.2190	28.050	81.878	109.929	0.05910	0.21575	63
64	123.96	109.26	0.013114	0.44358	76.257	2.2544	28.344	81.656	110.000	0.05966	0.21558	64

Figure 3–63 A portion of the R-22 properties table. *Courtesy E. I. DuPont*

3.21   PLOTTING THE REFRIGERANT CYCLE FOR BLENDS WITH NOTICEABLE TEMPERATURE GLIDE (ZEOTROPIC BLENDS)

Temperature glide occurs when the refrigerant blend has many temperatures as it evaporates or condenses at a given pressure. The pressure/enthalpy diagram for refrigerants with temperature glide differs from that for refrigerants that do not have temperature glide. **Figure 3–64** illustrates a skeletal pressure/enthalpy diagram for a refrigerant blend with temperature glide. Notice that the lines of constant temperature (isotherms) are not horizontal, but are angled downward as they travel from saturated liquid to saturated vapor. As you follow the lines of constant pressure (isobars) straight across from saturated liquid to saturated vapor, more than one isotherm will be intersected. This illustrates that for any one pressure (evaporating or condensing), there will be a range of temperatures associated with it, and the blend is said to have temperature glide. For example, in **Figure 3–64,** for a constant pressure of 130 psia, the temperature glide is 10°F (70°F − 60°F). The isobar of 130 psia intersects both the 60°F and the 70°F isobars as it travels from saturated liquid to saturated vapor.

Figure 3–65 illustrates an actual pressure/enthalpy diagram for refrigerant R-407C, a zeotropic refrigerant blend with very similar properties to R-22 in air-conditioning equipment. It is a blend of R-32, R-125, and R-134a with a fairly large temperature glide (10°F) in the ranges of air-conditioning applications. Notice the isotherms being angled downward from left to right showing temperature glide.

The terms *near-azeotropic* blends and *zeotropic blends* are used interchangeably in the HVACR industry because they both exhibit temperature glide and fractionation. However, zeotropes have temperature glide to a larger extent than near-azeotropic blends. Refer to Unit 9 for more in-depth information on refrigerants, refrigerant blends, temperature glide, and fractionation. Charging methods for refrigerant blends will be covered in Unit 10, "System Charging." SAFETY PRECAUTION: All refrigerants that have been discussed in this text are stored in pressurized containers and should be handled with care. Consult your instructor or supervisor about the use and handling of these refrigerants. Goggles and gloves should be worn while transferring the refrigerants from the container to the system.

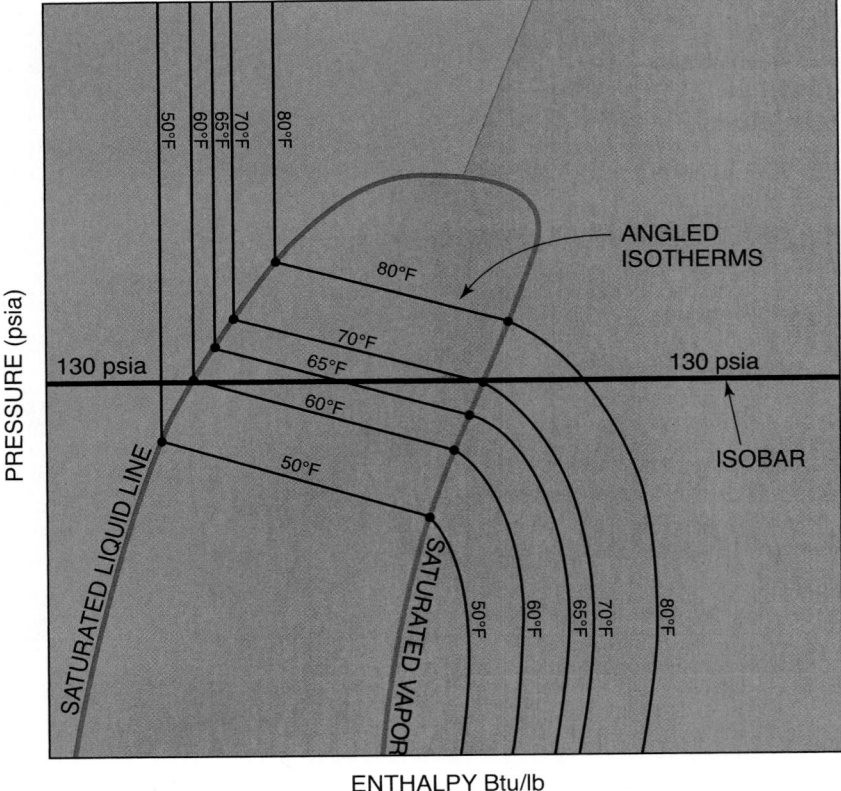

Figure 3–64   A skeletal pressure/enthalpy diagram of a refrigerant blend with a noticeable temperature glide (near-azeotropic blend). Notice the angled isotherms.

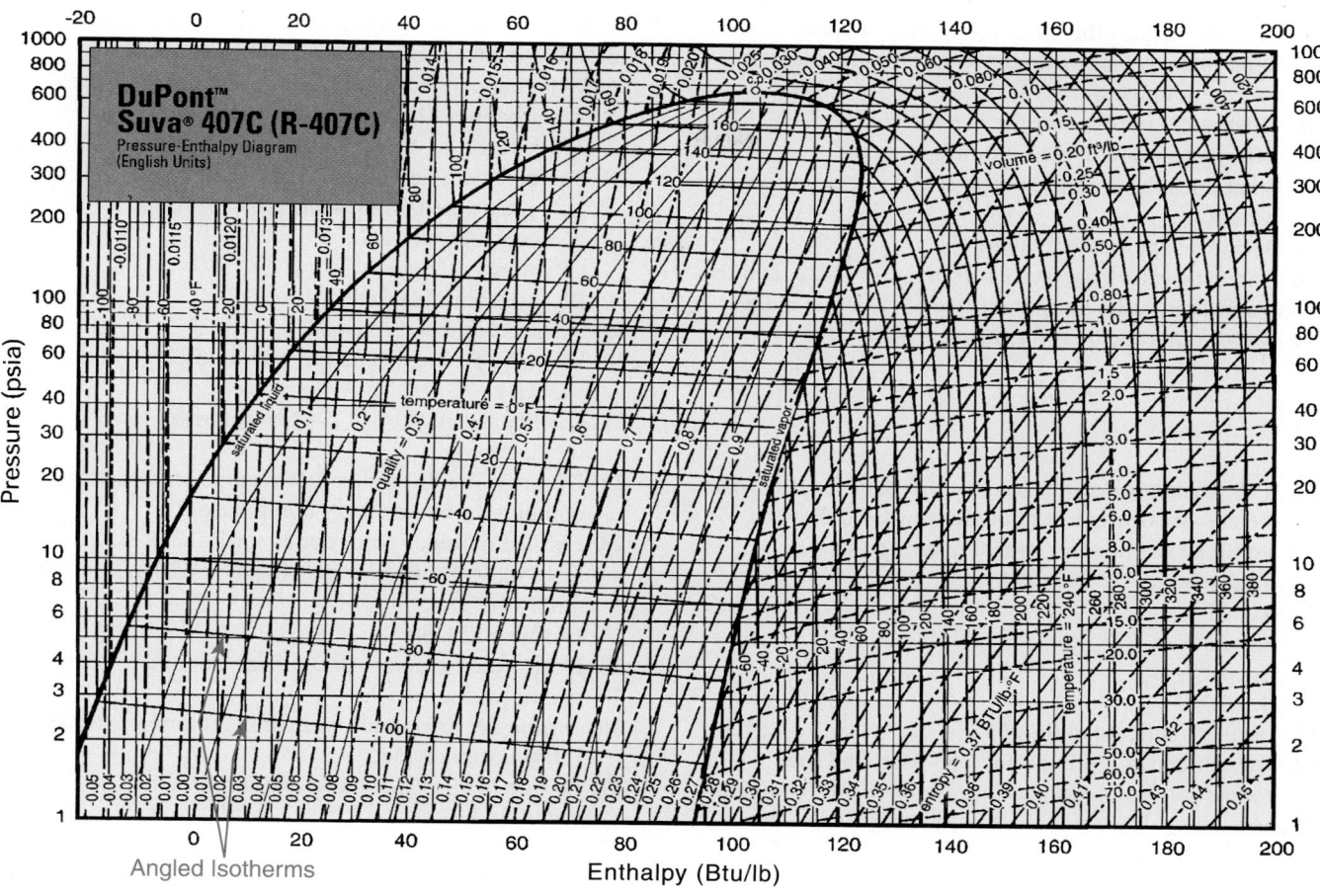

Figure 3–65 A pressure/enthalpy diagram for refrigerant R-407C showing the angled isotherms. *Courtesy E. I. DuPont*

SUMMARY

■ Bacterial growth that causes food spoilage slows at low temperatures.

■ Product temperatures above 45°F and below room temperature are considered high-temperature refrigeration.

■ Product temperatures between 35°F and 45°F are considered medium-temperature refrigeration.

■ Product temperatures from 0°F to −10°F are considered low-temperature refrigeration.

■ Refrigeration is the process of removing heat from a place where it is not wanted and transferring it to a place where it makes little or no difference.

■ One ton of refrigeration is the amount of heat necessary to melt 1 ton of ice in a 24-hour period. It takes 288,000 Btu to melt 1 ton of ice in a 24-hour period or 12,000 Btu in 1 h or 200 Btu in 1 min.

■ The relationship of the vapor pressure and the boiling point temperature is called the temperature/pressure relationship.

■ A compressor can be considered a vapor pump. It lowers the pressure in the evaporator to the desired temperature and increases the pressure in the condenser to a level where the vapor may be condensed to a liquid.

■ The liquid refrigerant moves from the condenser to the metering device where it again enters the evaporator.

■ Refrigerants have a definite chemical makeup and are usually designated with an "R" and a number for field identification.

■ A refrigerant must be safe, must be detectable, must be environmentally friendly, must have a low boiling point, and must have good pumping characteristics.

■ Refrigerant cylinders are color coded to indicate the type of refrigerant they contain.

■ Refrigerants should be recovered or stored while a refrigeration system is being serviced, then recycled, if appropriate, or sent to a manufacturer to be reclaimed.

■ Pressure/enthalpy diagrams may be used to plot refrigeration cycles.

REVIEW QUESTIONS

1. Name three reasons why ice melts in an icebox:
   _____, _____, _____.

2. What are the approximate temperature ranges for low-, medium-, and high-temperature refrigeration applications?

3. One ton of refrigeration is
   A. 1200 Btu.
   B. 12,000 Btu/h.
   C. 120,000 Btu.
   D. 120,000 Btu/h.

4. Describe briefly the basic refrigeration cycle.
5. What is the relationship between pressure and the boiling point of liquids?
6. What is the function of the evaporator in a refrigeration or air-conditioning system?
7. What does the compressor do in the refrigeration system?
8. Define a superheated vapor.
9. The evaporating pressure is 76 psig for R-22, and the evaporator outlet temperature is 58°F. What is the evaporator superheat for this system?
   A. 13°F
   B. 74°F
   C. 18°F
   D. 17°F
10. If the evaporating pressure was 76 psig for R-22 and the compressor inlet temperature was 65°F, what would be the total superheat entering the compressor?
    A. 11°F
    B. 21°F
    C. 10°F
    D. 20°F
11. Define a subcooled liquid.
12. The condensing pressure is 260 psig, and the condenser outlet temperature is 108°F for R-22. By how many degrees is the liquid subcooled in the condenser?
    A. 12°F
    B. 42°F
    C. 7°F
    D. 14°F
13. The condensing pressure is 260 psig, and the metering device inlet temperature is 100°F for R-22. What is the total subcooling in this system?
    A. 15°F
    B. 20°F
    C. 25°F
    D. 30°F

14. What is meant by a saturated liquid and vapor?
15. What is meant by desuperheating a vapor?
16. What happens to the refrigerant in the condenser?
17. What happens to refrigerant heat in the condenser?
18. The metering device
    A. cycles the compressor.
    B. controls subcooling.
    C. stores refrigerant.
    D. meters refrigerant.
19. What is adiabatic expansion?
20. Describe flash gas and tell how it affects system capacity.
21. Quality means _____ when referring to a refrigerant.
22. Describe the difference between a reciprocating compressor and a rotary compressor.
23. List the cylinder color codes for R-12, R-22, R-502, R-134a, R-11, R-401A, R-402B, R-410A, R-404A, and R-407C.
24. Define enthalpy.
25. Define a pure compound refrigerant and give two examples.
26. Define net refrigeration effect as it applies to the refrigeration cycle.
27. Define heat of work and explain how it is computed.
28. Define heat of compression and explain how it is computed.
29. Explain the relationship between HOW and HOC.
30. Define flash gas and explain how it applies to the net refrigeration effect of the refrigeration cycle.
31. Define temperature glide as it pertains to a refrigerant blend.
32. Define a zeotropic refrigerant blend and give an example.
33. Define a near-azeotropic refrigerant blend and give two examples.

SECTION 2

Safety, Tools and Equipment, Shop Practices

UNIT 4
General Safety Practices

OBJECTIVES

After studying this unit, you should be able to

- describe proper procedures for working with pressurized systems and vessels, electrical energy, heat, cold, rotating machinery, and chemicals; for moving heavy objects; and for utilizing proper ventilation.
- work safely, avoiding safety hazards.

The heating, air-conditioning, and refrigeration technician works close to many potentially dangerous situations: liquids and gases under pressure, electrical energy, heat, cold, chemicals, rotating machinery, moving heavy objects, and areas needing ventilation. The job must be completed in a manner that is safe for the technician and the public.

This unit describes some general safety practices and procedures with which all technicians should be familiar. Whether in a school laboratory, a shop, a commercial service shop, or a manufacturing plant, always be familiar with the location of emergency exits and first aid and eye wash stations. When you are on the job, be aware of how you would get out of a building or particular location should an emergency occur. Know the location of first aid kits and first aid equipment. Many other more specific safety practices and tips are given within units where they may be applied. Always use common sense and be prepared.

All tools of the trade have limitations. The technician should make it a practice to know what these limitations are by reading the manufacturer's literature. If you have the feeling that you are going beyond the manufacturer's limitations, you probably are and should not proceed. Improper use of tools can cause property damage and personal injury.

Safety goggles are one of the most overlooked safety devices in the industry. Flying debris can cause an eye injury before a person can react, so it is good practice to wear goggles anytime the possibility of injury exists.

Even though good work practices are used, accidents do happen. It would be a good practice for every technician to take a class in cardiopulmonary resuscitation (CPR). Classes are offered by many schools and the Red Cross.

Technicians should always be professional and not play on the job; countless mistakes have been made while playing around. It may seem like fun, until someone gets hurt. Remember, you may involve someone else in the problem if you are working at their place of business or home when an accident occurs.

4.1 PRESSURE VESSELS AND PIPING

Pressure vessels and piping are part of many systems that are serviced by refrigeration and air-conditioning technicians. For example, a cylinder of R-22 in the back of an open truck with the sun shining on it may have a cylinder temperature of 110°F on a summer day. The temperature/pressure chart indicates that the pressure inside the cylinder at 110° is 226 psig. This pressure reading means that the cylinder has a pressure of 226 lb for each square inch of surface area. A large cylinder may have a total area of 1500 in². This gives a total inside pressure (pushing outward) of

$$1500 \text{ in}^2 \times 226 \text{ psi} = 339,000 \text{ lb}$$

This is equal to

$$339,000 \text{ lb} \div 2000 \text{ lb/ton} = 169.5 \text{ tons}$$

See **Figure 4–1.** This pressure is well contained and will be safe if the cylinder is protected. Do not drop it. SAFETY PRECAUTION: Move the cylinder only while a protective cap is on it, if it is designed for one, **Figure 4–2.** Cylinders too large to be carried should be moved chained to a cart, **Figure 4–3.** If a refrigerant cylinder ever falls off the moving cart, the protective cap will protect the valve from breaking off and becoming a projectile. The cylinder will have a tendency to rock back and forth when it falls because of the momentum of the dense liquid contained within it. It is this rocking action that

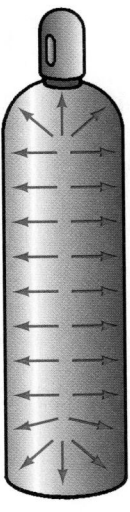

Figure 4–1 Pressure exerted across the entire surface area of a refrigerant cylinder.

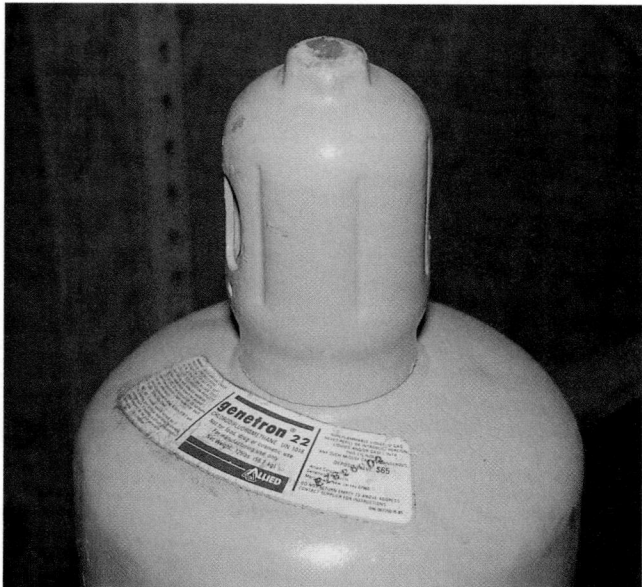

Figure 4–2   A refrigerant cylinder with a protective cap. *Photo by Bill Johnson*

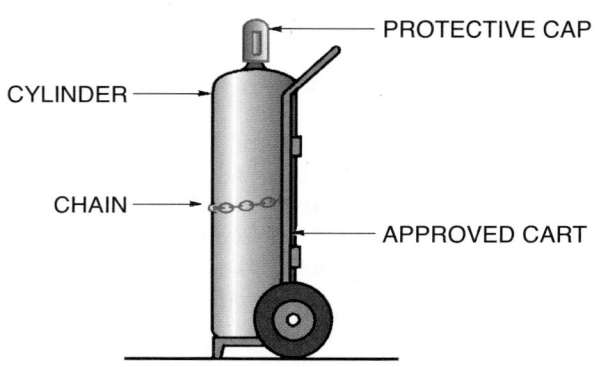

Figure 4–3   Pressurized cylinders should be chained to and moved safely on an approved cart. The protective cap must be secured.

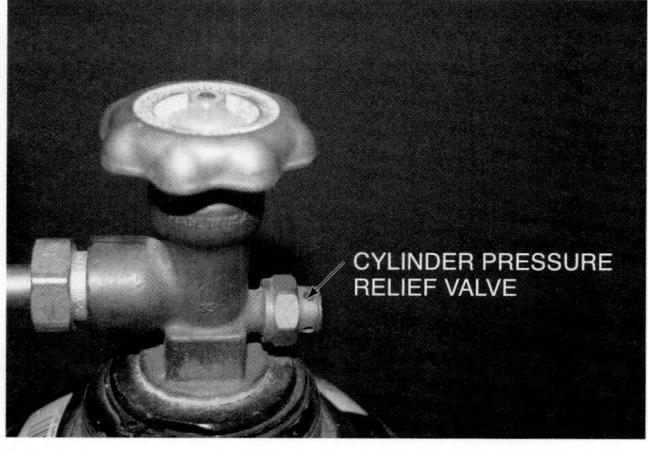

(A)

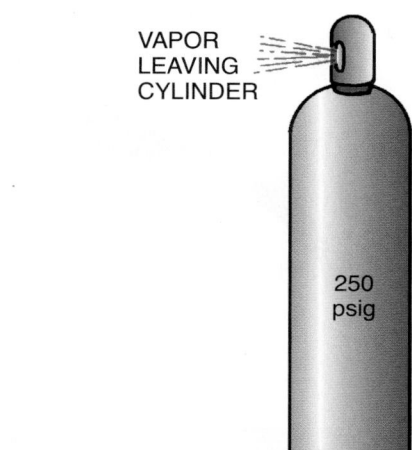

(B)

Figure 4–4   (A) Cylinder pressure relief valve. **(B)** The pressure relief valve will reduce pressure in the cylinder when overpressure occurs. *(A) Courtesy Ferris State University. Photo by John Tomczyk*

may cause the valve to break off. If the cylinder accidentally falls, the protective cap will also prevent the vaporization of all the liquid in the cylinder, which can displace the air we breathe and cause death or personal injury. The air is displaced because the refrigerant vapor is denser than the air.

The pressure in this cylinder can be thought of as a potential danger. It will not become dangerous unless it is allowed to escape in an uncontrolled manner. The cylinder has a relief valve at the top in the vapor space, **Figure 4–4(A)**. If the pressure builds up to the relief valve setting, it will start relieving vapor. As the vapor pressure is relieved, the liquid in the cylinder will begin to vaporize and absorb heat from the surrounding liquid, giving a cooling effect. This will reduce the pressure in the cylinder, **Figure 4–4(B)**. Relief valve settings are set at values above the worst typical operating conditions and are typically more than 400 psig.

The refrigerant cylinder has a fusible plug made of a material with a low melting temperature. The plug will melt and blow out if the cylinder gets too hot. This prevents the

cylinder from bursting and injuring personnel and property around it. SAFETY PRECAUTION: Refrigerant cylinders should be stored and transported in the upright position to keep the pressure relief valve in contact with the vapor space, not the liquid inside the cylinder. Some technicians may apply heat to refrigerant cylinders while charging a system to keep the pressure from dropping in the cylinder. This is an extremely dangerous practice. It is recommended for the above purpose that the cylinder of refrigerant be set in a container of warm water with a temperature no higher than 90°F, **Figure 4–5**.

You will be taking pressure readings of refrigeration and air-conditioning systems. Liquid R-22 boils at −41°F when it is released to the atmosphere. If you are careless and get this refrigerant on your skin or in your eyes, it will quickly cause frostbite. Keep your skin and eyes away from any liquid refrigerant. When you attach gages and take refrigerant pressure readings or transfer refrigerant into or out of a system, wear gloves and side-shield goggles. **Figure 4–6(A)** is a photo of protective eye goggles. SAFETY PRECAUTION: If a

Figure 4–5   A refrigerant cylinder in warm water (not warmer than 90°F).

(A)

(B)

Figure 4–6   **(A)** Protective eye side-shield goggles. **(B)** Modern wraparound, tinted safety glasses. **(A)** *Photo by Bill Johnson.* **(B)** *Courtesy Ferris State University. Photo by John Tomczyk*

leak develops and refrigerant is escaping, the best thing to do is stand back and look for a valve with which to shut it off. Do not try to stop it with your hands.

Released refrigerant, and any particles that may become airborne with the blast of vapor, can harm the eyes. Wear approved goggles, which are vented to keep them from

Figure 4–7   Color-coded refrigerant recovery cylinders. The bodies of the cylinders are gray; the tops are yellow. *Courtesy White Industries*

fogging over with condensation and which help keep the operator cool, **Figure 4–6(A).** Modern wraparound, tinted safety glasses may also be worn to protect the eyes from particles and refrigerant, **Figure 4–6(B).**

SAFETY PRECAUTION: All cylinders into which refrigerant is transferred should be approved by the Department of Transportation as recovery cylinders. **Figure 4–7** shows cylinders with the color code indicating they are approved as recovery cylinders—gray bodies with yellow tops. It is against the law to transport refrigerant in nonapproved cylinders.

In addition to the pressure potential inside a refrigerant cylinder, there is tremendous pressure potential inside nitrogen and oxygen cylinders. They are shipped at pressures of 2500 psig. These cylinders must not be moved unless the protective cap is in place. They should be chained to and moved on carts designed for this purpose, **Figure 4–3.** Dropping the cylinder without the protective cap may result in breaking the valve off the cylinder. The pressure inside can propel the cylinder like a balloon full of air that is turned loose, **Figure 4–8.**

Nitrogen must also have its pressure regulated before it can be used, **Figure 4–9.** The pressure in the cylinder is too great to be connected to a system. If a person allowed nitrogen under cylinder pressure to enter a refrigeration system, the pressure could burst some weak point in the system. This could be particularly dangerous if it were at the compressor shell. **Figure 4–9** shows nitrogen tank pressure of approximately 2200 psi being regulated down to about 130 psi. Notice the downstream safety pressure relief that protects the system and the operator.

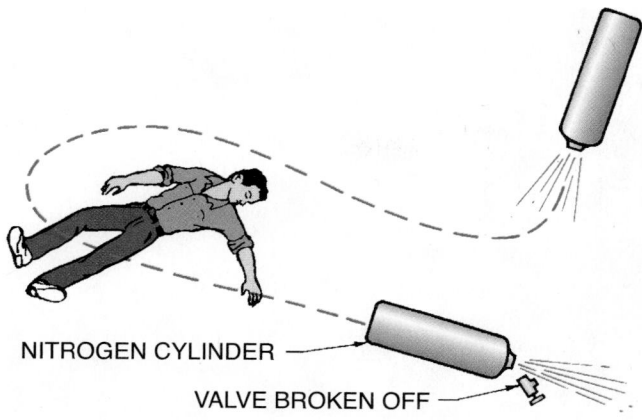

Figure 4–8   When a cylinder valve is broken off, the cylinder becomes a projectile until pressure is exhausted.

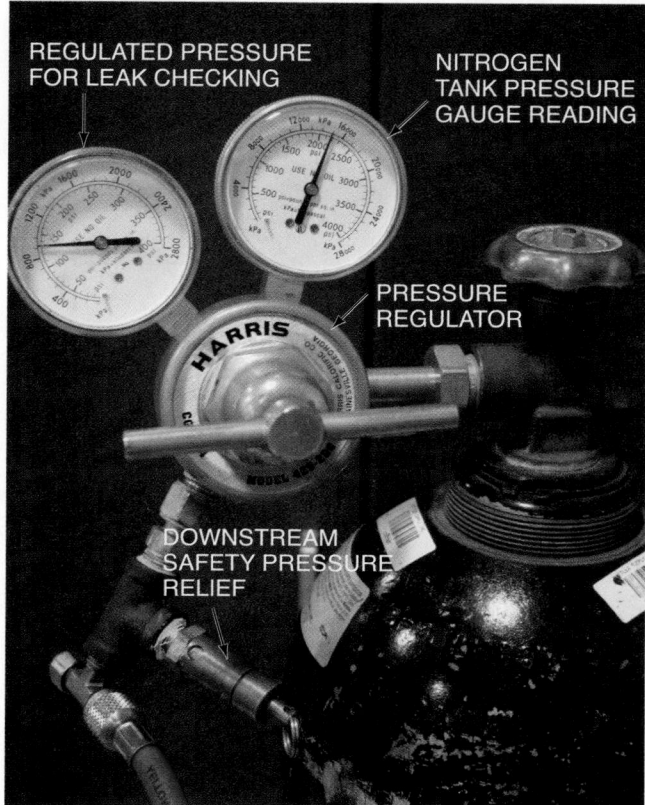

Figure 4–9   A nitrogen tank with pressure regulator and safety downstream pressure relief. *Courtesy Ferris State University. Photo by John Tomczyk*

Because of this high pressure, oxygen also must be regulated. In addition, all oxygen lines must be kept absolutely oil free. Oil residue in an oxygen regulator connection may cause an explosion. The explosion may blow the regulator apart in your hand. SAFETY PRECAUTION: Oxygen is under a great deal of pressure but should NEVER be used as a pressure source to pressurize lines or systems. If there is oil or oil residue in the system, adding oxygen to pressurize will lead to an explosion.

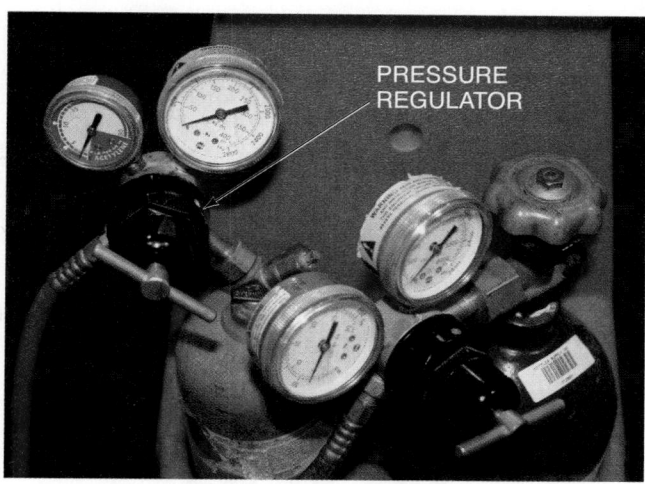

Figure 4–10   Pressure regulators on oxygen and acetylene tanks. *Courtesy Ferris State University. Photo by John Tomczyk*

Oxygen is often used with acetylene. Acetylene cylinders are not under the same high pressure as nitrogen and oxygen but must be treated with the same respect because acetylene is highly explosive. A pressure-reducing regulator must be used, **Figure 4–10.** Always use an approved cart, chain the cylinder to the cart, and secure the protective cap, **Figure 4–3.** All stored cylinders must be supported so they will not fall over, and they must be separated as required by code, **Figure 4–11(A** and **B).**

4.2   ELECTRICAL HAZARDS

Electrical shocks and burns are ever-present hazards. It is impossible to troubleshoot all circuits with the power off, so you must learn safe methods for troubleshooting "live" circuits. As long as electricity is contained in the conductors and the device is where it is supposed to function, there is nothing to fear. When uncontrolled electrical flow occurs (e.g., if you touch two live wires), you are very likely to get hurt.

SAFETY PRECAUTION: Electrical power should always be shut off at the distribution or entrance panel and locked out in an approved manner when installing equipment. Specific requirements furnished by the Occupational Safety and Health Administration (OSHA) give the details for working safety conditions, including electrical lock-out and tag procedures. Whenever possible the power should be shut off and locked out when servicing the equipment. Electrical panels are furnished with a place for a padlock for the purpose of lockout, **Figure 4–12(A).** To avoid someone turning the power on, you should keep the panel locked when you are out of sight and keep the only key with you. Emergency STOP controls can be installed in the electrical lines for added safety, **Figure 4–12(B).** Do not ever think that you are good enough or smart enough to work with live electrical power when it is not necessary.

However, at certain times tests have to be made with the power on. Extreme care should be taken when making these

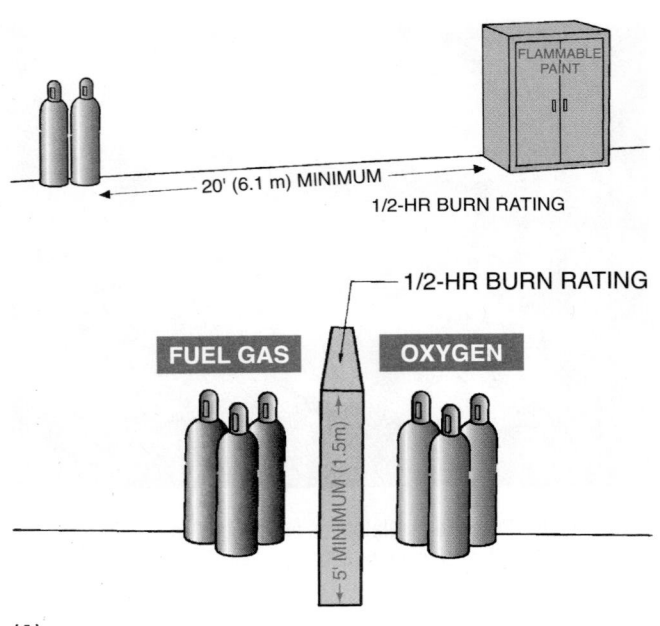

(A)

(B)

(A)

(B)

Figure 4–11 **(A)** Pressurized cylinders must be chained when stored. The minimum safe distance between stored fuel gas cylinders and any flammable materials is 20 ft or a wall 5 ft high. **(B)** Refrigerant cylinders chained to a wall for safety. **(B)** *Courtesy Ferris State University. Photo by John Tomczyk*

Figure 4–12 **(A)** Electrical disconnect "locks out" with padlock. **(B)** Emergency stop disconnect for electrical safety. **(A)** *Courtesy Ferris State University. Photo by John Tomczyk.* **(B)** *Courtesy Ferris State University. Photo by John Tomczyk*

tests. Ensure that your hands touch only the meter probes and that your arms and the rest of your body stay clear of all electrical terminals and connections. You should know the voltage in the circuit you are checking. Make sure that the range selector on the test instrument is set properly before using it. SAFETY PRECAUTION: Do not stand in a wet or damp area when making these checks. Use only proper test equipment, and make sure it is in good condition. Wear heavy shoes with an insulating sole and heel. Intelligent and competent technicians take all precautions. Many areas display danger signs warning that proper eyewear and footwear is required in that area, **Figure 4–13.**

Electrical Shock

Electrical shock occurs when you become part of the circuit. Electricity flows through your body and can damage your heart—stop it from pumping—resulting in death if it is not restarted quickly. It is a good idea to take a first aid course

Figure 4–13 Signage requiring safety eyewear and footwear. *Courtesy Ferris State University. Photo by John Tomczyk*

that includes CPR methods for lifesaving. SAFETY PRECAUTION: To prevent electrical shock, do not become a conductor between two live wires or a live (hot) wire and ground. The electricity must have a path to flow through. The higher the voltage that you may be working with, the greater the potential to become injured. In this industry, we work with circuits that may have voltages at 460 V or up to 560 V. Always know the power supply you are working with and act

accordingly. Do not let your body be the path. **Figure 4–14** is a wiring diagram showing situations in which the technician is part of a circuit.

SAFETY PRECAUTION: The technician should use only properly grounded power tools connected to properly grounded circuits. Caution should be taken when using portable electric tools. These are handheld devices with electrical energy inside just waiting for a path to flow through. Some portable electric tools are constructed with metal frames. These should all have a grounding wire in the power cord. The grounding wire protects the operator. The tool will work without it, but it is not safe. If the motor inside the tool develops a loose connection and the frame of the tool becomes electrically hot, the third wire, rather than your body, will carry the current, and a fuse or breaker will interrupt the circuit, **Figure 4–15.**

In some instances technicians may use the three- to two-prong adapters at job sites because the wall receptacle may have only two connections and their portable electric tool has a three-wire plug, **Figure 4–16.** This adapter has a third wire that must be connected to a ground for the circuit to give protection. If this wire is fastened under the wall plate screw and this screw terminates in an ungrounded box nailed to a wooden wall, you are not protected, **Figure 4–17.** Ensure that the third or ground wire is properly connected to a ground.

Alternatives to the older style of metal portable electric tools are the plastic-cased and the battery-operated tool. In the plastic-cased tool the motor and electrical connections are insulated within the tool. This is called a double-insulated

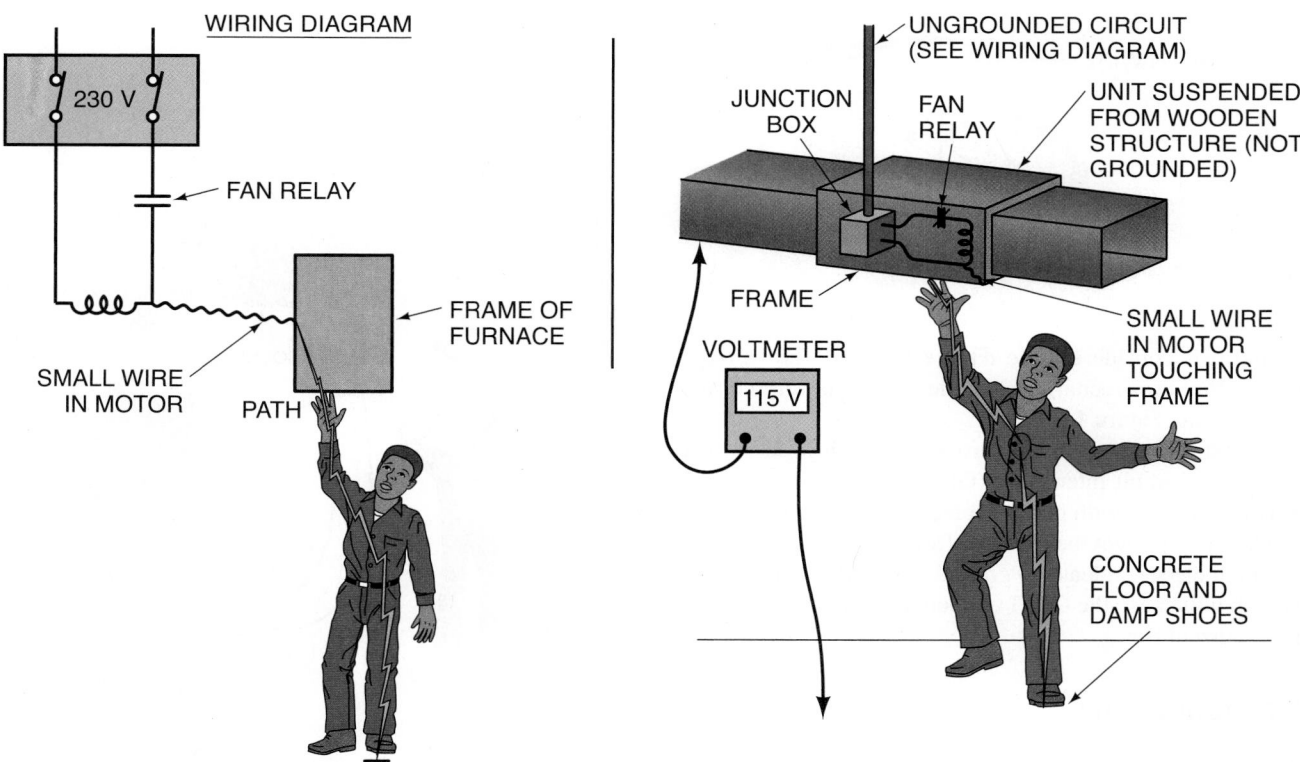

Figure 4–14 This figure shows ways that the technician can become part of the electrical circuit and receive an electrical shock.

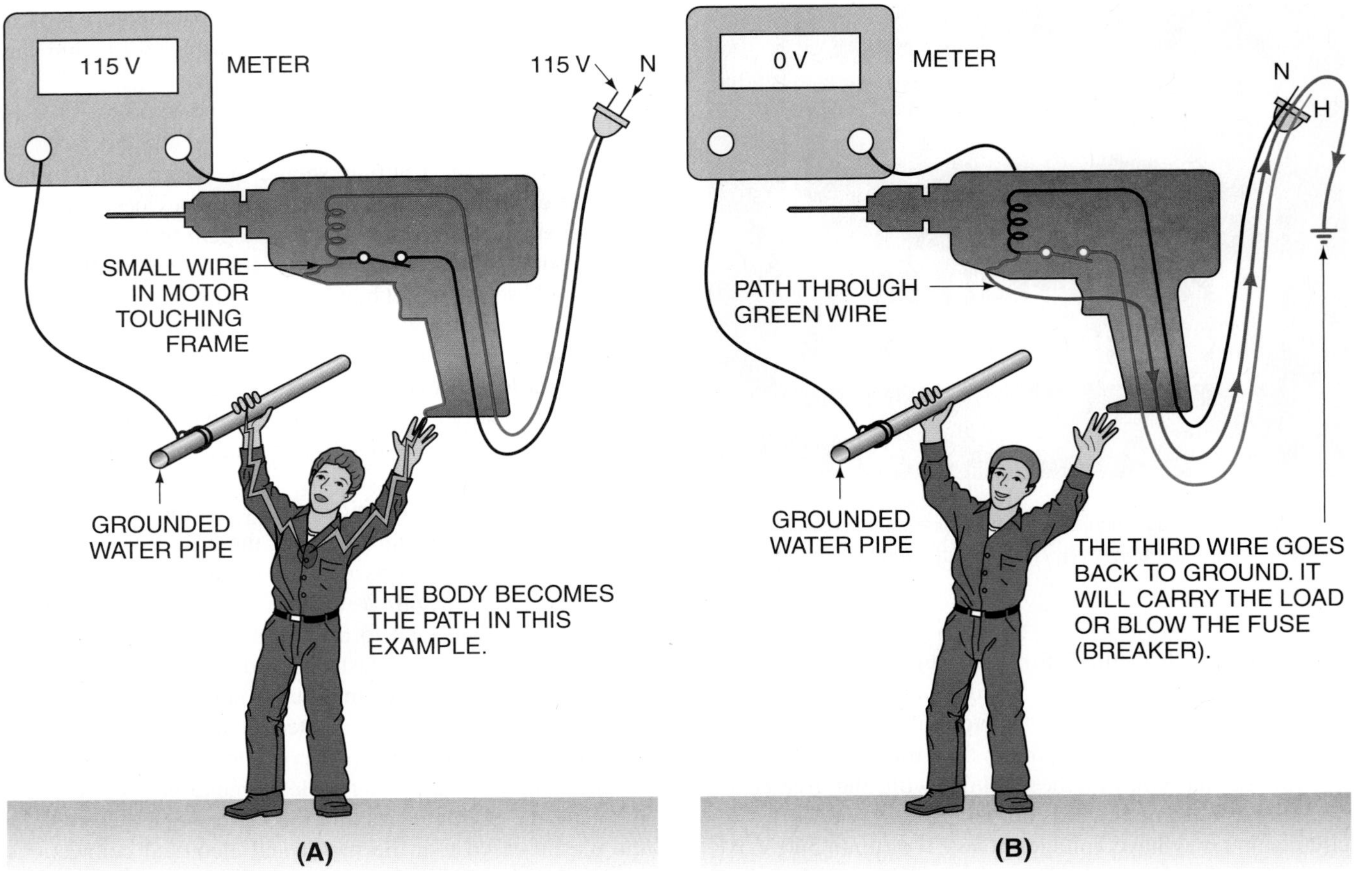

115 V   METER   115 V   N

SMALL WIRE
IN MOTOR
TOUCHING
FRAME

GROUNDED
WATER PIPE

THE BODY BECOMES
THE PATH IN THIS
EXAMPLE.

(A)

0 V   METER   N   H

PATH THROUGH
GREEN WIRE

GROUNDED
WATER PIPE

THE THIRD WIRE GOES
BACK TO GROUND. IT
WILL CARRY THE LOAD
OR BLOW THE FUSE
(BREAKER).

(B)

Figure 4–15   **(A)** An electrical circuit to ground from the metal frame of a drill. **(B)** Metal frame of a drill properly grounded.

GREEN WIRE

Figure 4–16   A three- to two-prong adapter.

tool and it is considered safe, **Figure 4–18(A)**. The battery-operated tool uses rechargeable batteries and is very convenient and safe, **Figure 4–18(B)**.

The extension cord in **Figure 4–19** is plugged into a ground fault circuit interrupter (GFCI) receptacle. It is recommended for use with portable electric tools. These are designed to help protect the operator from shock as they detect very small electrical leaks to ground. A small electrical current leak will cause the GFCI to open the circuit, preventing further current flow.

Electrical Burns

Do not wear jewelry (rings and watches) while working on live electric circuits because they can cause shock and possible burns.

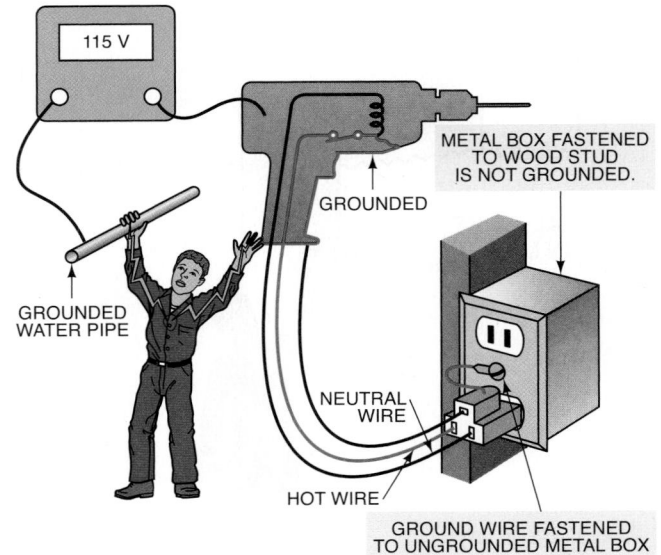

115 V

GROUNDED

METAL BOX FASTENED
TO WOOD STUD
IS NOT GROUNDED.

GROUNDED
WATER PIPE

NEUTRAL
WIRE

HOT WIRE

GROUND WIRE FASTENED
TO UNGROUNDED METAL BOX

Figure 4–17   The wire from the adapter is intended to be fastened under the screw in the duplex wall plate. However, this will provide no protection if the outlet box is not grounded.

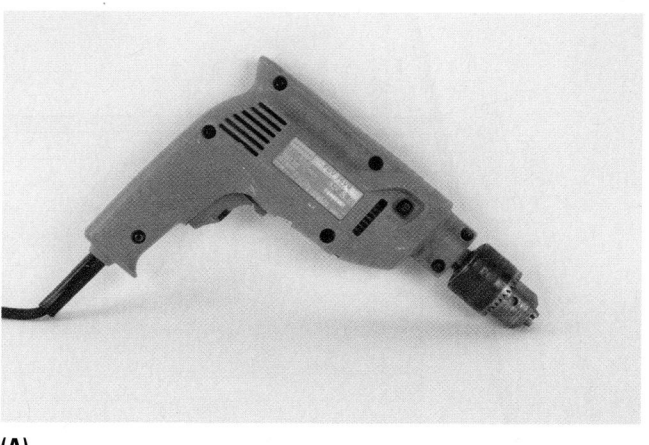

(A)

(B)

Figure 4–18 **(A)** A double-insulated electric drill. **(B)** A battery-operated electric drill. *Photos by Bill Johnson*

Figure 4–19 An extension cord with a ground fault circuit interrupter receptacle. *Photo by Bill Johnson*

Never use a screwdriver or other tool in an electrical panel when the power is on. Electrical burns can come from an electrical arc, such as in a short circuit to ground when uncontrolled electrical energy flows. For example, if a screwdriver slipped while you were working in a panel and the blade completed a circuit to ground, the potential flow of electrical energy is tremendous. When a circuit has a resistance of 10 Ω and is operated on 120 V, it would, using Ohm's law, have a current flow of

$$I = \frac{E}{R} = \frac{120 \text{ V}}{10 \text{ Ω}} = 12 \text{ A}$$

If this example is calculated again with less resistance, the current will be greater because the voltage is divided by a smaller number. If the resistance is lowered to 1 Ω, the current flow is then

$$I = \frac{E}{R} = \frac{120 \text{ V}}{1 \text{ Ω}} = 120 \text{ A}$$

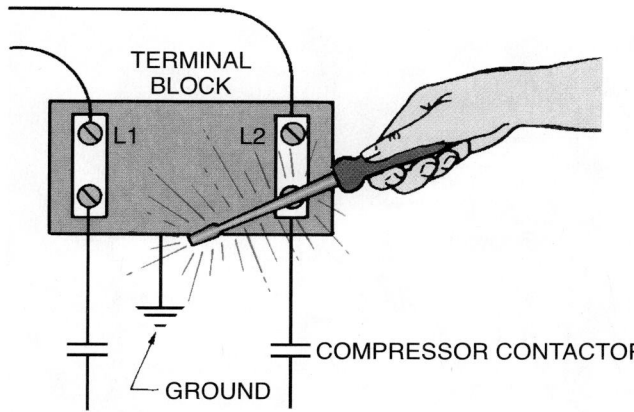

Figure 4–20 This wiring illustration shows a short circuit caused by the slip of a screwdriver.

If the resistance is reduced to 0.1 Ω, the current flow will be 1200 A. By this time the circuit breaker will trip, but you may have already incurred burns or an electrical shock, **Figure 4–20.** Current flow of 0.015 ampere or less through the body can prove fatal. It is a good idea to keep a first aid kit in a permanent location for any type of injury, **Figure 4–21(A** and **B).**

Ladder Safety

Nonconducting ladders should be used on all jobs and should be the type furnished with service trucks. Two types of nonconducting ladders are those made of wood and those made of fiberglass. Nonconducting ladders work as well as aluminum ladders; they are just heavier. They are also much safer. SAFETY PRECAUTION: A technician may raise a ladder into a power line or place it against a live electrical hazard without realizing it. Chances should not be taken. When the technician is standing on a nonconducting ladder, it will provide protection from electrical shock to ground. However, it will not provide protection between two or more electrical conductors.

(A)

(B)

Figure 4–21 **(A)** First aid kit mounted on a wall. **(B)** Contents of a general first aid kit. **(A)** *Courtesy Ferris State University. Photo by John Tomczyk.* **(B)** *Courtesy Ferris State University. Photo by John Tomczyk*

There are ladder safety practices other than those involving electricity. When a ladder is used for access to an upper landing area such as a roof, the ladder should extend at least 3 ft above the landing surface. Also, the horizontal distance from the top support to the foot of the ladder should be approximately one-fourth the working length of the ladder, **Figure 4–22.** Ensure that ladders are placed on a stable, level surface. They should not be used on slippery surfaces, and they should be provided with slip-resistant feet. If there is any question, secure the ladder at the feet. Ladders should be used only for their designed purpose and should not be loaded with more weight than they were designed to hold.

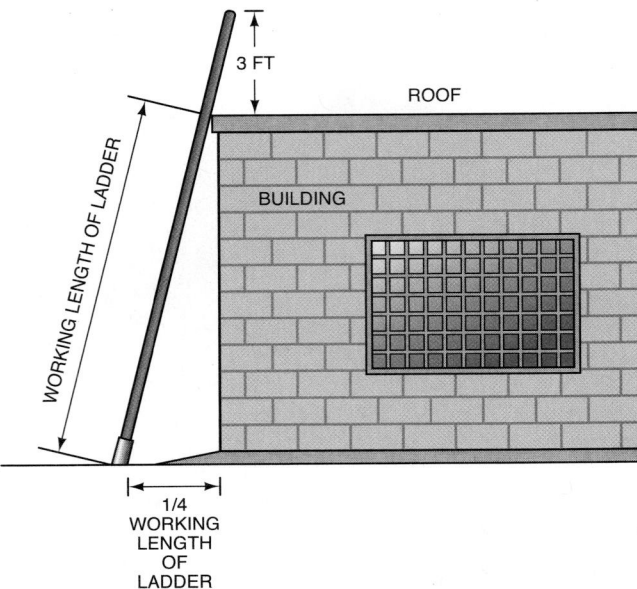

Figure 4–22 The side rails of a portable ladder should extend 3 ft above the upper landing surface. The angle of the ladder should be such that the horizontal space at the bottom is about one-fourth the distance of the working length of the ladder.

Use other methods for raising loads to the roof when the weight is excessive. Ladders should be maintained free of oil, grease, and any other slipping hazards.

Working from heights on rooftops or from ladders has the potential for great danger. Rooftops have different pitches or slants. When the angle becomes steep, the technician should use a safety belt and be tied off in such a manner that a fall would be stopped. Technicians should always work in pairs when doing rooftop work or when working on a high ladder. In many cases the ladder should be secured at the top, particularly if it is to be used for several climbs. SAFETY PRECAUTION: Never use the cross bracing on the far side of stepladders for climbing. The far side is only safe for climbing if it is provided with steps.

4.3 HEAT

The use of *heat* requires special care. A high concentration of heat comes from torches. Torches are used for many things, including soldering, brazing, or welding. Many combustible materials may be in the area where soldering is required. For example, the refrigeration system in a restaurant may need repair. The upholstered restaurant furniture, grease, and other flammable materials must be treated as carefully as possible. SAFETY PRECAUTION: When soldering or using concentrated heat, you should have a fire extinguisher close by, and you should know exactly where it is and how to use it. Learn to use a fire extinguisher *before* the fire occurs. A fire extinguisher should always be included as a part of the service tools and equipment on a service truck, **Figure 4–23(A).**

(A)

(B)

Figure 4–23 **(A)** A typical fire extinguisher. **(B)** Fire extinguisher details showing the types of fires it should be used on. **(A)** *Photo by Bill Johnson.* **(B)** *Courtesy Ferris State University. Photo by John Tomczyk*

Many times, fire extinguishers are designed for specific types of fires. Type A encompasses trash, wood, and paper. Type B is for liquids on fire and Type C is for electrical equipment fires. However, some fire extinguishers are rated for all three types of fires, Types A, B, and C, **Figure 4–23(B).** These extinguishers are usually the dry-chemical type. Make sure you know what type of fire extinguisher you are using on a fire or the fire can spread and serious personal injury and property damage can occur.

Figure 4–24 A shield used when brazing or soldering. *Courtesy of Thermadyne Industries, Inc.*

Figure 4–25 A fire-retardant spray.

SAFETY PRECAUTION: When a solder connection must be made next to combustible materials or a finished surface, use a shield of noncombustible material for insulation, **Figure 4–24.** A fire-resistant spray may also be used to decrease the flammability of wood if a torch must be used nearby, **Figure 4–25.** This spray retardant should be used with an appropriate shield. It is also often necessary to use a shield when soldering within an equipment cabinet, for example, when an ice-maker compressor is changed or when a drier must be soldered in line. A shield should be used to protect any wires that may be close by.

SAFETY PRECAUTION: Never solder tubing lines that are sealed. Service valves or Schrader ports should be open before soldering is attempted. When heat from a torch is added to a sealed tubing line, a small amount of pressure will build up inside the tubing and may bubble through the hot molten solder. This will cause leaks in the solder joint.

SAFETY PRECAUTION: Hot refrigerant lines, hot heat exchangers, and hot motors can burn your skin and leave a permanent scar. Care should be used while handling them.

SAFETY PRECAUTION: Working in very hot weather or in hot attics can be very hard on the body. Technicians should be aware of how their body and the bodies of those around them are reacting to the working conditions. Watch for signs of overheating, such as someone's face turning very red or someone who has stopped sweating. Get them out of the heat and cool them off. Overheating can be life threatening; call for emergency help.

4.4 COLD

Cold can be as harmful as heat. Liquid refrigerant can freeze your skin or eyes instantaneously. But long exposure to cold is also harmful. Working in cold weather can cause frostbite. Wear proper clothing and waterproof boots, which also help protect against electrical shock. A cold, wet technician will not always make decisions based on logic. Make it a point to stay warm. SAFETY PRECAUTION: Be aware of the wind chill when working outside. The wind can quickly take the heat out of your body, and frostbite will occur. Waterproof boots not only protect your feet from water and cold but help to protect you from electrical shock hazard when you are working in wet weather. However, do not depend on these boots for protection against electrical hazards.

Low-temperature freezers are just as cold in the middle of the summer as in the winter. Cold-weather gear must be used when working inside these freezers. For example, an expansion valve may need changing, and you may be in the freezer for more than an hour. It is a shock to the system to step from the outside where it may be 95°F or 100°F into a room that is 0°F. If you are on call for any low-temperature applications, carry a coat and gloves and wear them in cold environments.

4.5 MECHANICAL EQUIPMENT

Rotating equipment can damage body and property. Motors that drive fans, compressors, and pumps are among the most dangerous because they have so much power. SAFETY PRECAUTION: If a shirt sleeve or coat were caught in a motor drive pulley or coupling, severe injury could occur. Loose clothing should never be worn around rotating machinery, **Figure 4–26**. Even a small electric hand drill can wind a necktie up before the drill can be shut off.

SAFETY PRECAUTION: When starting an open motor, stand well to the side of the motor drive mechanism. If the coupling or belt were to fly off the drive, it would fly outward in the direction of rotation of the motor. All set screws or holding mechanisms must be tight before a motor is started, even if the motor is not connected to a load. All wrenches must be away from a coupling or pulley. A wrench or nut thrown from a coupling can be a lethal projectile, **Figure 4–27**.

SAFETY PRECAUTION: When a large motor, such as a fan motor, is coasting to a stop, do not try to stop it. If you try to stop the motor and fan by gripping the belts, the momentum of the fan and motor may pull your hand into the pulley and under the belt, **Figure 4–28**.

Figure 4–26 SAFETY PRECAUTION: Never wear a necktie or loose clothing when using or working around rotating equipment.

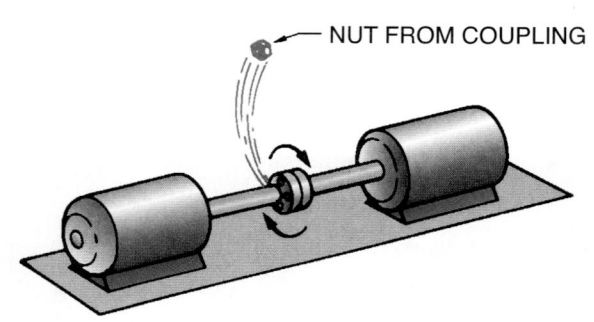

Figure 4–27 SAFETY PRECAUTION: Ensure that all nuts are tight on couplings and other components.

Figure 4–28 SAFETY PRECAUTION: Never attempt to stop a motor or other mechanism by gripping the belt.

SAFETY PRECAUTION: Never wear jewelry while working on a job that requires much movement. A ring may be caught on a nail head, or a bracelet may be caught on the tailgate of a truck as you jump down, **Figure 4–29**.

SAFETY PRECAUTION: When using a grinder to sharpen tools, remove burrs, or for other reasons, always use a face shield, **Figure 4–30**. Most grinding stones are made for grinding ferrous metals, cast iron, steel, stainless steel, and others.

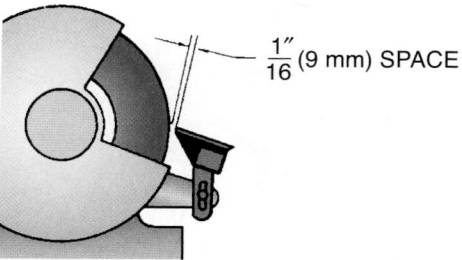

Figure 4–31   Keep the tool rest on a grinder adjusted properly.

Figure 4–29   Jewelry can catch on nails or other objects and cause injury.

Figure 4–32   When installing a grinding stone, ensure that it is compatible with the grinder.

Figure 4–31. As the stone wears down, keep the tool rest adjusted to this setting. A grinding stone must not be used on a grinder that turns faster than the stone's rated maximum revolutions per minute (rpm), as it may explode, **Figure 4–32.**

4.6   MOVING HEAVY OBJECTS

Heavy objects must be moved from time to time. Think out the best and safest method for moving these objects. Do not just use muscle power. Special tools can help you move equipment. When equipment must be installed on top of a building, a crane or even a helicopter can be used, **Figure 4–33**. Do not try to lift heavy equipment by yourself; get help from another person and use tools and equipment designed for that purpose. A technician without proper equipment is limited.

SAFETY PRECAUTION: When you must lift, use your legs, not your back, and wear an approved back brace belt, **Figure 4–34.** Some available tools are a pry bar, a lever truck, a refrigerator hand truck, a lift gate on the pickup truck, and a portable dolly, **Figure 4–35.**

When moving large equipment across a carpeted or tiled floor or across a gravel-coated roof, first lay down some plywood. Keep the plywood in front of the equipment as it is moved along. When equipment has a flat bottom, such as a package air conditioner, short lengths of pipe may be used to move the equipment across a solid floor, **Figure 4–36.**

Figure 4–30   Use a face shield when grinding. *Photo by Bill Johnson*

However, other stones are made for nonferrous metals such as aluminum, copper, or brass. Use the correct grinding stone for the metal you are grinding. The tool rest should be adjusted to approximately 1/16 in. from the grinding stone,

Figure 4–33 A helicopter lifting air-conditioning equipment to a roof.

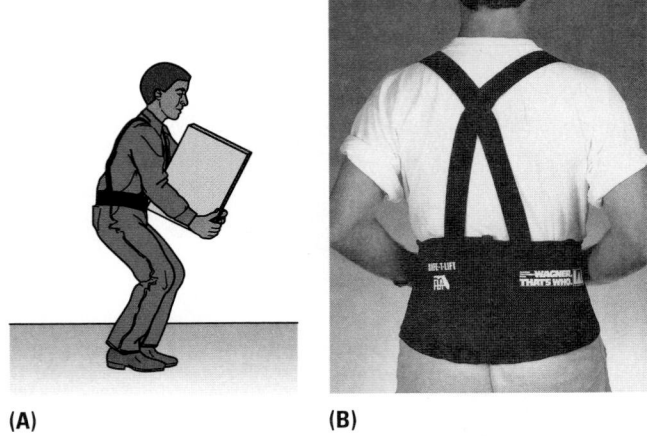

(A)   **(B)**

Figure 4–34 SAFETY PRECAUTION: **(A)** Use your legs, not your back, to lift objects. Keep your back straight. **(B)** Use a back belt brace. **(B)** *Courtesy Wagner Products Corp.*

(A)

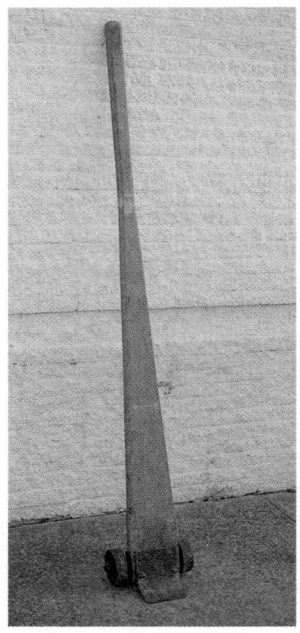

(B)

(C)

(D)

(E)

Figure 4–35 **(A)** A pry bar. **(B)** A lever truck. **(C)** A hand truck. **(D)** The lift gate on a pickup truck. **(E)** A portable dolly. *Photos by Bill Johnson*

Figure 4–36   Moving equipment using short lengths of pipe as rollers.

4.7   REFRIGERANTS IN YOUR BREATHING SPACE

Fresh refrigerant vapors and many other gases are heavier than air and can displace the oxygen in a closed space. Because of this, low-level ventilation should be used. Proper ventilation must be used at all times to prevent the technician from being overcome by the lack of oxygen. Should you be in a close space and the concentration of refrigerant becomes too great, you may not notice it until it is too late. Your symptoms would be a dizzy feeling, and your lips may become numb. If you should feel this way, move quickly to a place with fresh air.

SAFETY PRECAUTION: Proper ventilation should be set up in advance of starting a job. Fans may be used to push or pull fresh air into a confined space where work must be performed. Cross ventilation can help prevent a buildup of fumes, **Figure 4–37**.

Special leak detectors with alarms are required in some installations. These detectors sound an alarm well in advance of harmful buildup. **Figure 4–38(A)** shows a refrigerant-specific leak detector for HCFC-123 (R-123) that will sniff out certain parts per million (PPM) of refrigerant. If the programmed PPM level is reached, the detector will sound an alarm and color-coded lights will flash. **Figure 4–38(B)** shows a written warning on an equipment room's entry door for added safety. **Figure 4–38(C)** is a refrigerant-specific leak

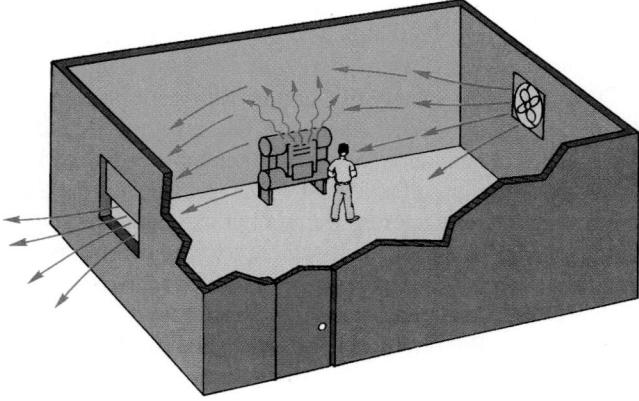

Figure 4–37   Cross ventilation with fresh air will help prevent fumes from accumulating.

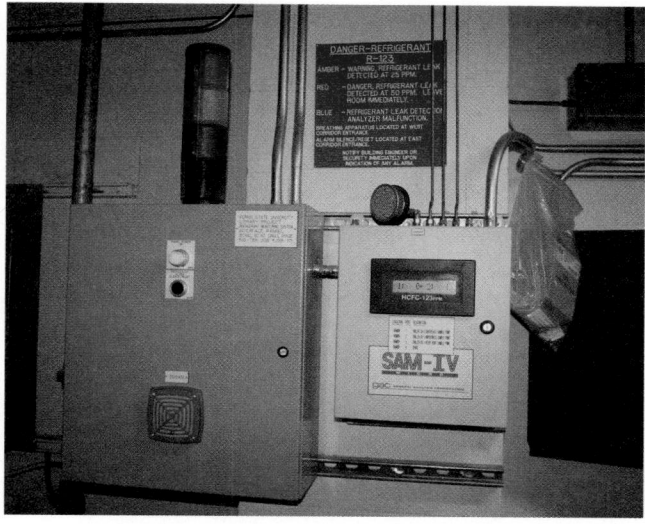

(A)

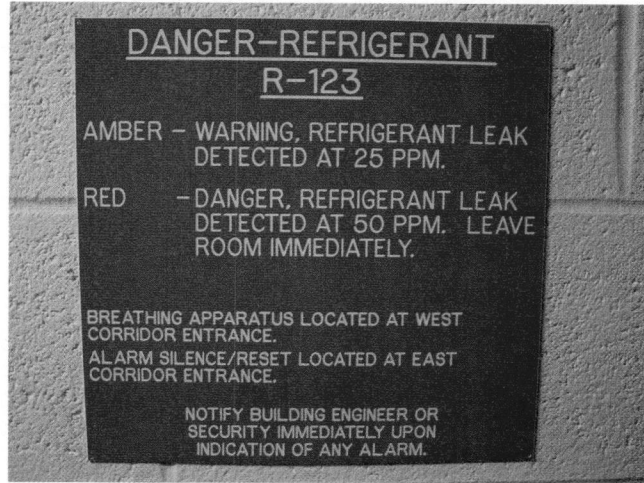

(B)

(C)

Figure 4–38   **(A)** Refrigerant-specific leak detector for R-123 with alarm lights. **(B)** Written warning about R-123 on equipment room's door. **(C)** Refrigerant-specific leak detector for R-134a with warning lights. *(B) Courtesy Ferris State University. Photo by John Tomczyk. (C) Courtesy Ferris State University. Photo by John Tomczyk*

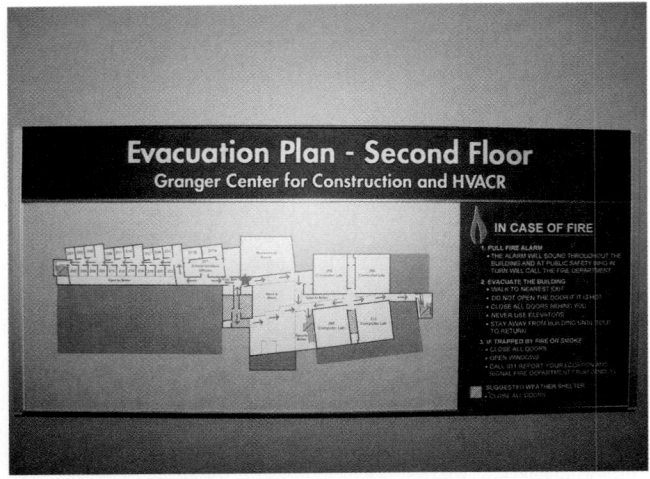

(A)

(B)

Figure 4–39 **(A)** Evacuation plan on building's interior wall for safety. **(B)** Building directory on building's interior wall for safety. *(A) Courtesy Ferris State University. Photo by John Tomczyk. (B) Courtesy Ferris State University. Photo by John Tomczyk*

detector for HFC-134a (R-134a) located in an equipment room that houses a certain amount of R-134a either in a working refrigeration system's receiver or in storage cylinders. **Figure 4–39(A)** shows an evacuation plan for the second floor of a building in case of a refrigerant leak, fire, or any other emergency. Building directories are also important for human safety because they let the occupants know where main entrances, stairs, elevators, and restrooms are located. Room configurations and locations are also covered in building directories, **Figure 4–39(B)**.

When the leak-detector alarm sounds, take proper precautions, which may require special breathing gear. Turn on ventilation.

Refrigerant vapors that have been heated during soldering or because they have passed through an open flame (for example, when there is a leak in the presence of a fire) are dangerous. They are toxic and may cause harm. They have a

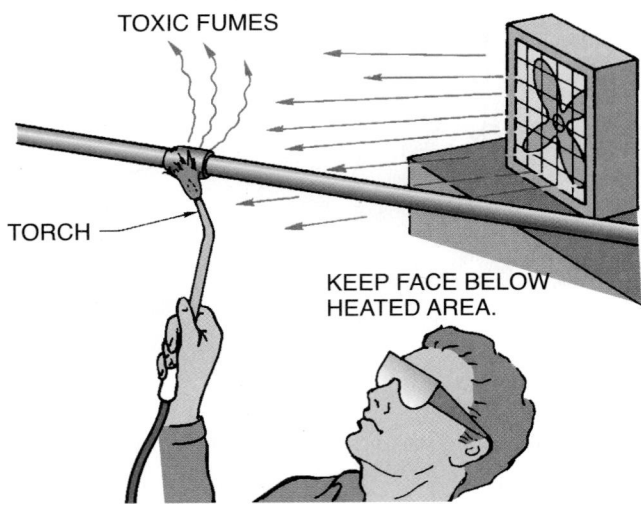

Figure 4–40 Keep your face below the heated area and ensure that the area is well ventilated.

LOWER TOXICITY	HIGHER TOXICITY	
A3	B3	HIGHER FLAMMABILITY
A2	B2	LOWER FLAMMABILITY
A1	B1	NO FLAME PROPAGATION

INCREASING TOXICITY →
INCREASING FLAMMABILITY ↑

Figure 4–41 ASHRAE's Standard 34-1992 refrigerant safety group classifications.

strong odor. SAFETY PRECAUTION: If you are soldering in a close place, keep your head below the rising fumes and ensure that you have plenty of ventilation, **Figure 4–40.**

The American Society of Heating, Refrigerating and Air-Conditioning Engineers (ASHRAE) has developed Standard 34-1992 for refrigerant toxicity and flammability. **Figure 4–41** is a matrix outlining Standard 34-1992, and the following explains the Standard's letter and number classifications.

Class A—For refrigerants where toxicity levels have not been identified at concentrations less than or equal to 400 ppm (parts per million).

Class B—For refrigerants where evidence of toxicity exists at concentrations below 400 ppm.

Class 1—For refrigerants that do not show flame propagation when tested in 65°F air at 14.7 psia.

(A)

(B)

(C)

Figure 4–42 **(A)** Wall sign showing the presence of an eye wash station for safety. **(B)** Eye wash station. **(C)** Emergency body and eye wash station. *(A) Courtesy Ferris State University. Photo by John Tomczyk.* *(B) Courtesy Ferris State University. Photo by John Tomczyk.* *(C) Courtesy Ferris State University. Photo by John Tomczyk*

Class 2—For refrigerants with a lower flammability limit of more than 0.00625 lb/ft^3 when tested at 70°F and 14.7 psia. They also have a heat of combustion of less than 8174 Btu/lb.

Class 3—For refrigerants with high flammability. Their lower flammability limit is less than or equal to 0.00625 lb/ft^3 when tested at 70°F and 14.7 psia. They have a heat of combustion greater than or equal to 8174 Btu/lb.

4.8 USING CHEMICALS

Chemicals are often used to clean equipment such as air-cooled condensers and evaporators. They are also used for water treatment. The chemicals are normally simple and mild, except for some harsh cleaning products used for water treatment. SAFETY PRECAUTION: These chemicals should be handled according to the manufacturer's directions. Do not get careless. If you spill chemicals on your skin or splash them in your eyes, follow the manufacturer's directions and go to a doctor. It is a good idea to read the entire label before starting a job. It is hard to read the first aid treatment for eyes after your eyes have been damaged.

Refrigerant and oil from a motor burnout can be harmful. The contaminated refrigerant and oil may be hazardous to your skin, eyes, and lungs because they contain acid. SAFETY PRECAUTION: Keep your distance if a line is subject to rupture or any amount of this refrigerant is allowed to escape.

If any harmful chemicals or refrigerants accidentally get on a person's body, there should be emergency eye and body wash facilities readily available. Signs directing a person to the wash facilities should be posted and clearly marked, **Figure 4–42(A, B,** and **C).** Emergency call boxes can

Figure 4–43 Emergency call station. *Courtesy Ferris State University. Photo by John Tomczyk*

also be located throughout the building for added safety. By simply pushing the red button on the call box, the proper authority will be notified that help is needed in a specific location, **Figure 4–43.**

SUMMARY

- The technician must use every precaution when working with pressures, electrical energy, heat, cold, rotating machinery, and chemicals, and when moving heavy objects.
- Technicians encounter safety situations involving pressure while working with pressurized systems and vessels.
- Electrical energy is present whenever you troubleshoot energized electrical circuits. Be careful. Turn off electrical power, if possible, when working on an electrical component. Lock the panel or disconnect the box and keep the only key in your possession.
- Refrigerant-specific leak detectors that sense parts per million (PPM) of refrigerant are used in equipment rooms for safety.

- Some fire extinguishers are designed for specific types of fires, and some can be used for many types: trash, wood, paper, liquid, and electrical fires.
- Evacuation plans and building directories mounted in a conspicuous place on the building's interior walls are an important safety feature.
- Heat is encountered during soldering and when heating systems are being worked on.
- Liquid R-22 refrigerant boils at −41°F at atmospheric pressure and will cause frostbite.
- Rotating equipment such as fans and pumps can be dangerous and should be treated with caution.
- When moving heavy equipment, use correct techniques, choose appropriate tools and equipment, and wear a back brace belt.
- Chemicals are used for cleaning and water treatment and must be handled with care.

REVIEW QUESTIONS

1. Where would a technician encounter freezing temperatures when working with liquid refrigerant?
2. Which of the following can happen to a full nitrogen cylinder when the top is broken off?
   A. Nitrogen will slowly leak out.
   B. There will be a loud bang.
   C. The tank will act like a rocket and move very fast.
   D. It will make a sound like a washing machine.
3. Why must nitrogen be used only with a regulator when it is charged into a system?
4. What can happen when oil is mixed with oxygen under pressure?
5. Trying to stop escaping liquid refrigerant with your hands will result in _____.
6. Electrical shock to the body affects the _____.
7. Electrical energy passing through the body causes two types of injury, _____ and _____.
8. How can a technician prevent burning his or her surroundings while soldering in a tight area?
9. What safety precautions should be taken when starting a large electric motor whose coupling is disconnected?
10. The third (green) wire on an electric drill is the _____ wire.
11. Describe how heavy equipment can be moved across a rooftop.
12. What special precautions should be taken before using chemicals to clean a condenser?
13. When a hermetic motor burns while in use, _____ are produced.
14. What refrigerant safety group classification has the highest flammability and the highest toxicity?
15. What do the letters A, B, and C mean on a fire extinguisher?

UNIT 5

Tools and Equipment

OBJECTIVES

After studying this unit, you should be able to

- describe hand tools used by the air-conditioning, heating, and refrigeration technician.
- describe equipment used to install and service air-conditioning, heating, and refrigeration systems.

SAFETY CHECKLIST

✔ Tools and equipment should be used only for the job for which they were designed. Other use may damage the tool or equipment and may be unsafe for the technician.

Air-conditioning, heating, and refrigeration technicians must be able to properly use hand tools and specialized equipment relating to this field. Technicians must use the tools and equipment intended for the job. Using the correct tool for the job is more efficient, saves time, and often is more safe. Accidents often can be prevented by using the correct tool.

This unit contains a brief description of most of the general and specialized tools and equipment that technicians use in this field. Some of these are described in more detail in other units as they apply to specific tasks.

5.1 GENERAL HAND TOOLS

Figure 5–1 through **Figure 5–6** illustrate many general hand tools used by service technicians.

In general, it is a good idea to buy the best tools that you can afford. Good hand tools are usually more expensive and will usually last longer. For example, an inexpensive screwdriver will often have an undersized handle and will be hard to grip. Also, the bit on the end will not be made of hard, tempered steel and will deform easily. The better choice would be a screwdriver with a comfortable handle and a very hard tip, which is under great stress when the technician has a very tight screw to work with. Good hand tools will usually carry the guarantee of replacement if they are broken during regular use.

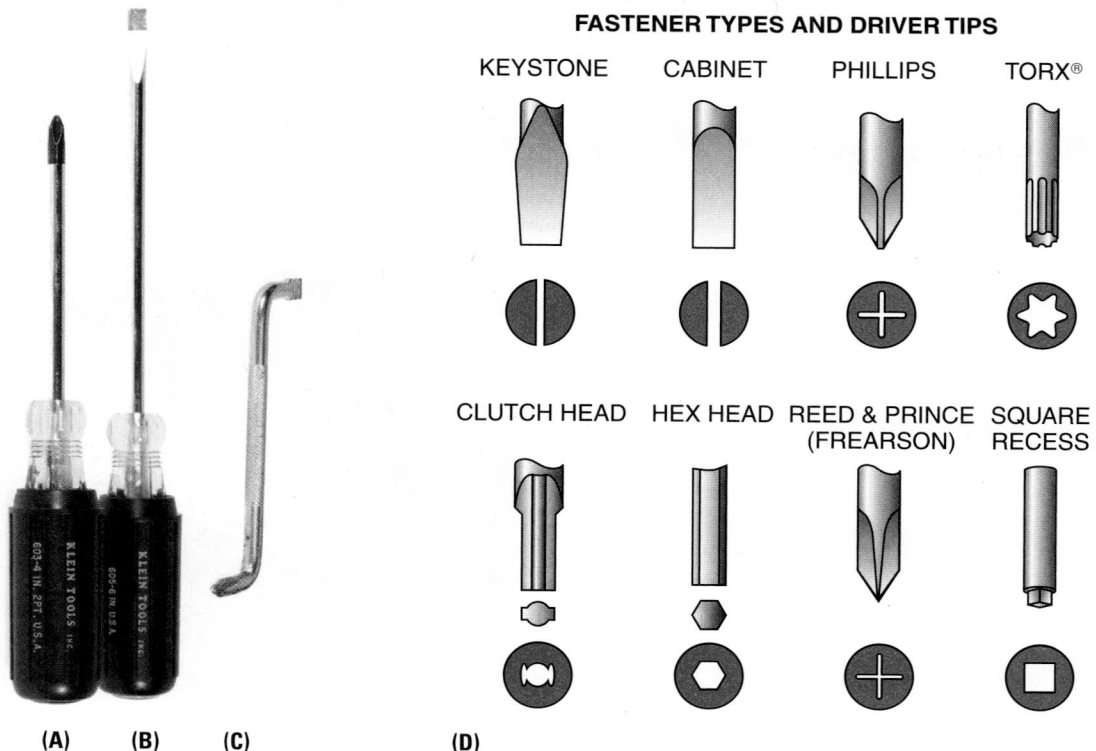

FASTENER TYPES AND DRIVER TIPS

KEYSTONE CABINET PHILLIPS TORX®

CLUTCH HEAD HEX HEAD REED & PRINCE (FREARSON) SQUARE RECESS

(A)   (B)   (C)   (D)

Figure 5–1   Screwdrivers. **(A)** Phillips tip. **(B)** Straight or slot blade. **(C)** Offset. **(D)** Standard screwdriver bit types. *Photos **(A)**, **(B)**, and **(C)** by Bill Johnson. **(D)** Courtesy Klein Tools*

(A)

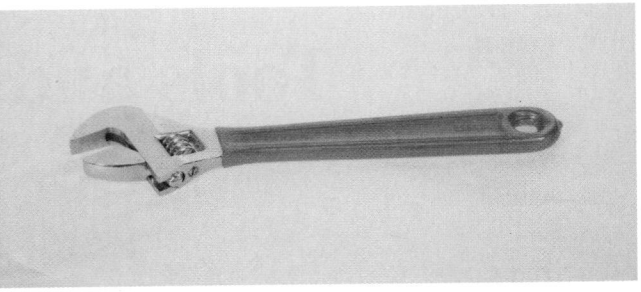

(E)

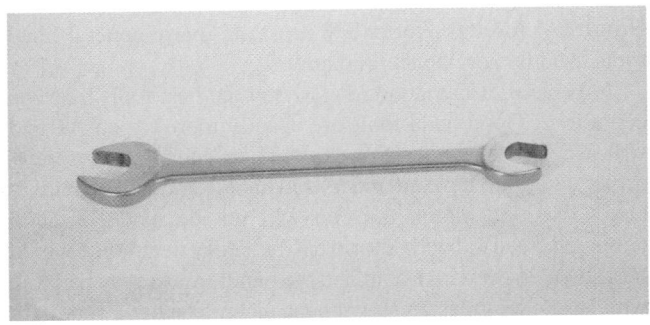

(B)

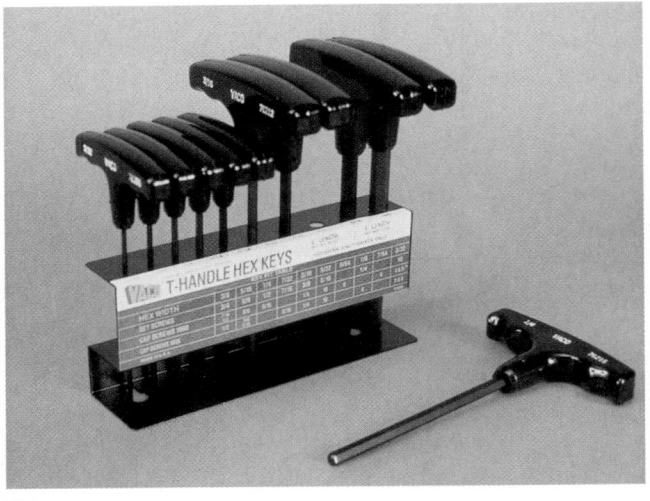

(F)

(G)

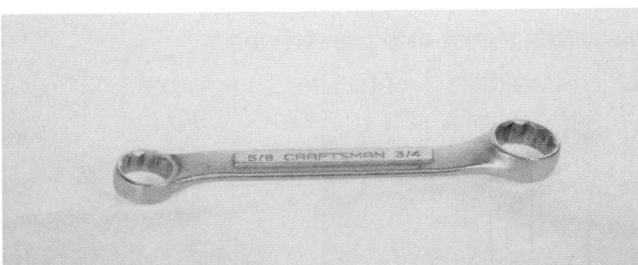

(C)

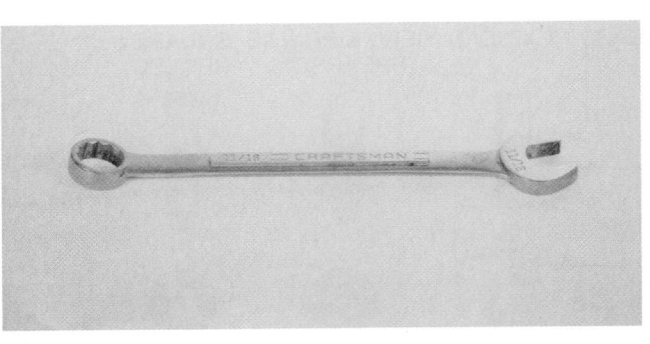

(D)

(H)

Figure 5–2   Wrenches. **(A)** Socket with ratchet handle. **(B)** Open end. **(C)** Box end. **(D)** Combination. **(E)** Adjustable open end. **(F)** Ratchet box. **(G)** Pipe. **(H)** T-Handle hex keys. *Photos **(A)** through **(G)** by Bill Johnson. **(H)** Courtesy Klein Tools*

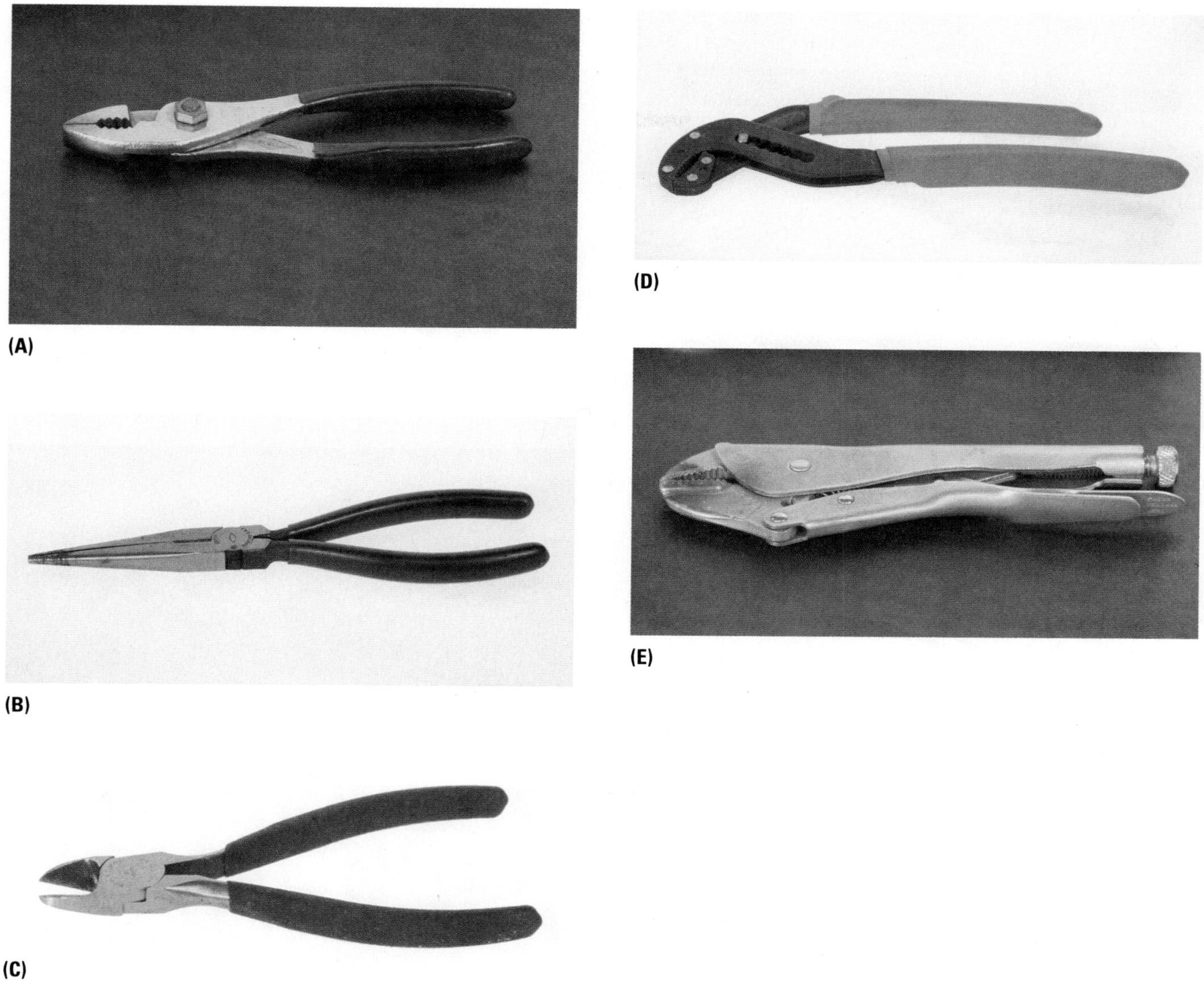

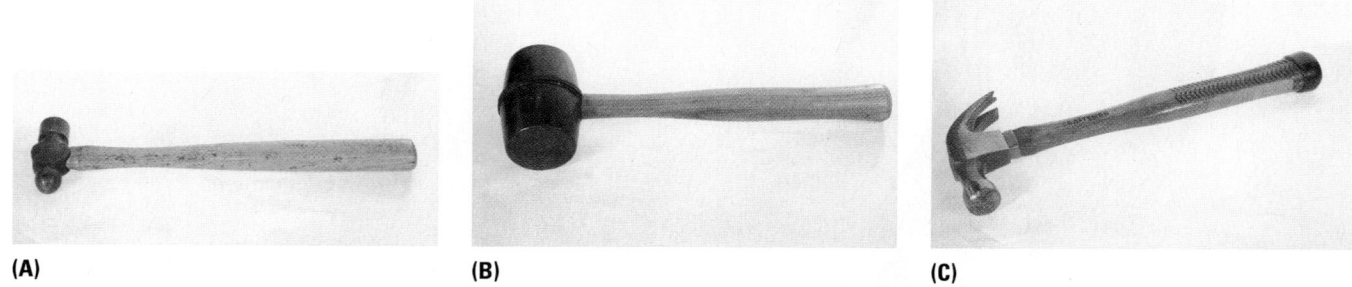

Figure 5–3 Pliers. **(A)** General purpose. **(B)** Needle nose. **(C)** Side cutting. **(D)** Slip joint. **(E)** Locking. *Photos by Bill Johnson*

Figure 5–4 Hammers. **(A)** Ball peen. **(B)** Soft head. **(C)** Carpenter's claw. *Photos by Bill Johnson*

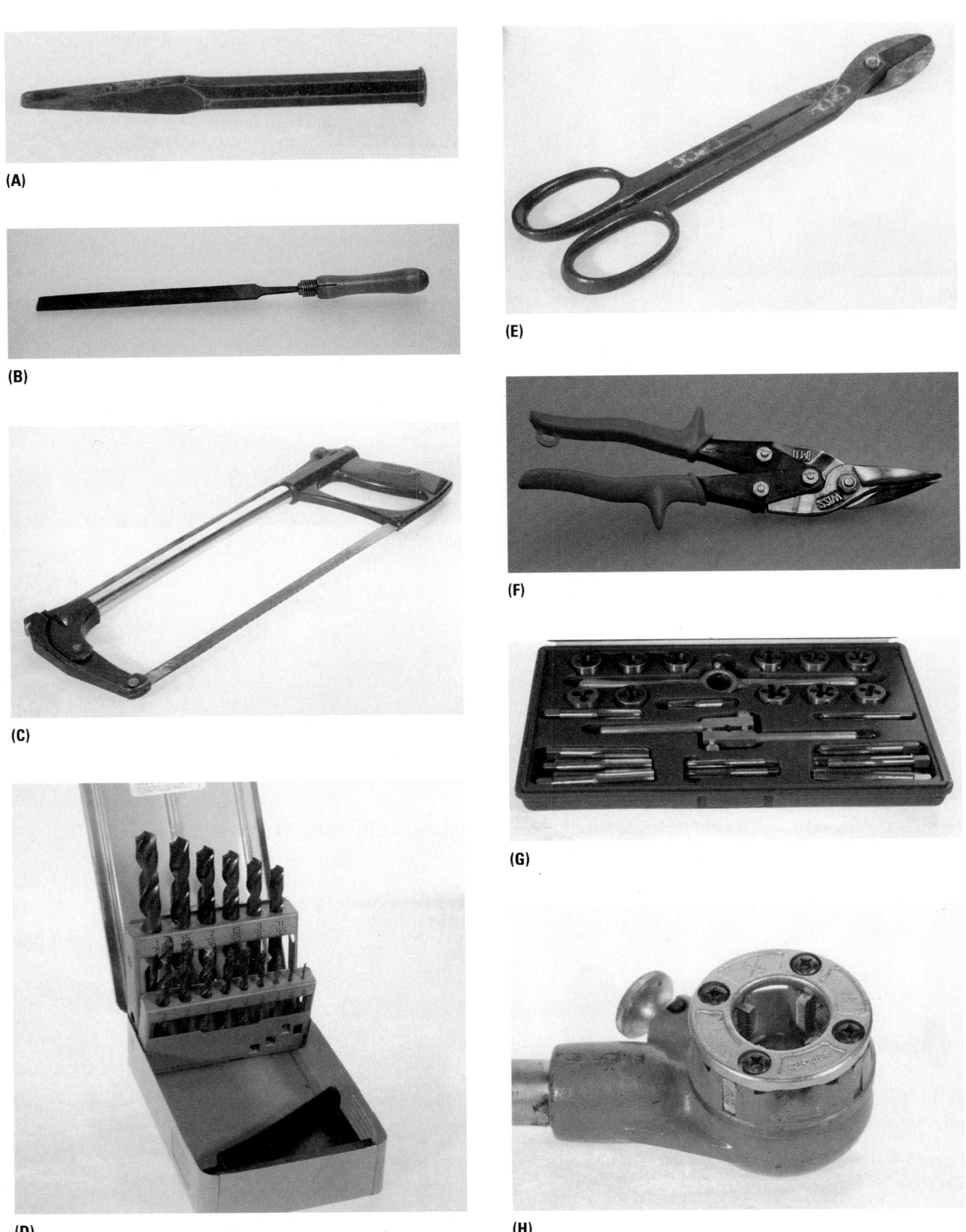

Figure 5–5 General metal-cutting tools. **(A)** Cold chisel. **(B)** File. **(C)** Hacksaw. **(D)** Drill bits. **(E)** Straight metal snips. **(F)** Aviation metal snips. **(G)** Tap and die set. A tap is used to cut an internal Tread. A die is used to cut an external thread. **(H)** Pipe-threading die. *Photos by Bill Johnson*

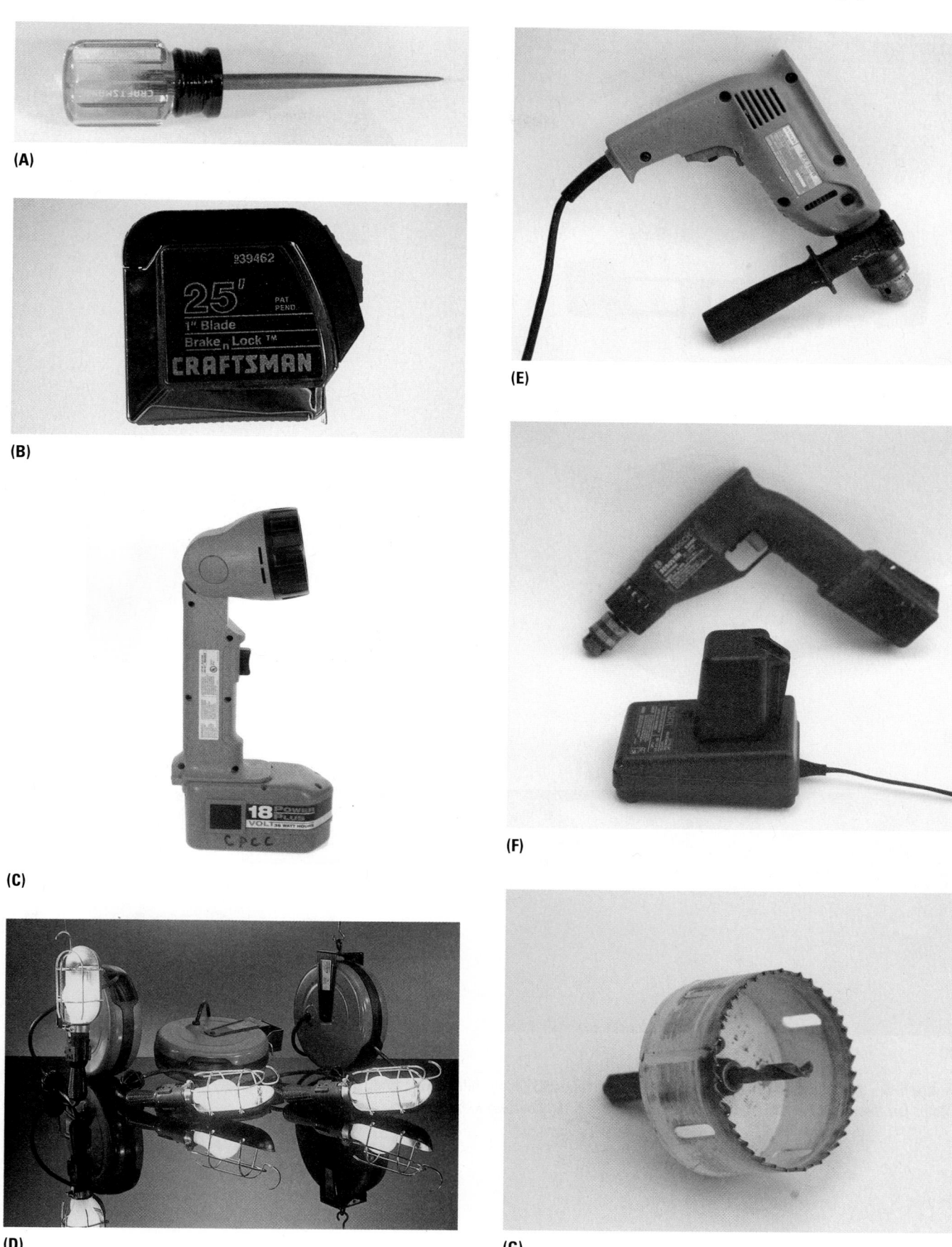

Figure 5–6   Other general-purpose tools. **(A)** Awl. **(B)** Rule. **(C)** Flashlight. **(D)** Extension cord/lights. **(E)** Portable electric drill, cord type. **(F)** Portable electric drill, cordless. **(G)** Hole saw.

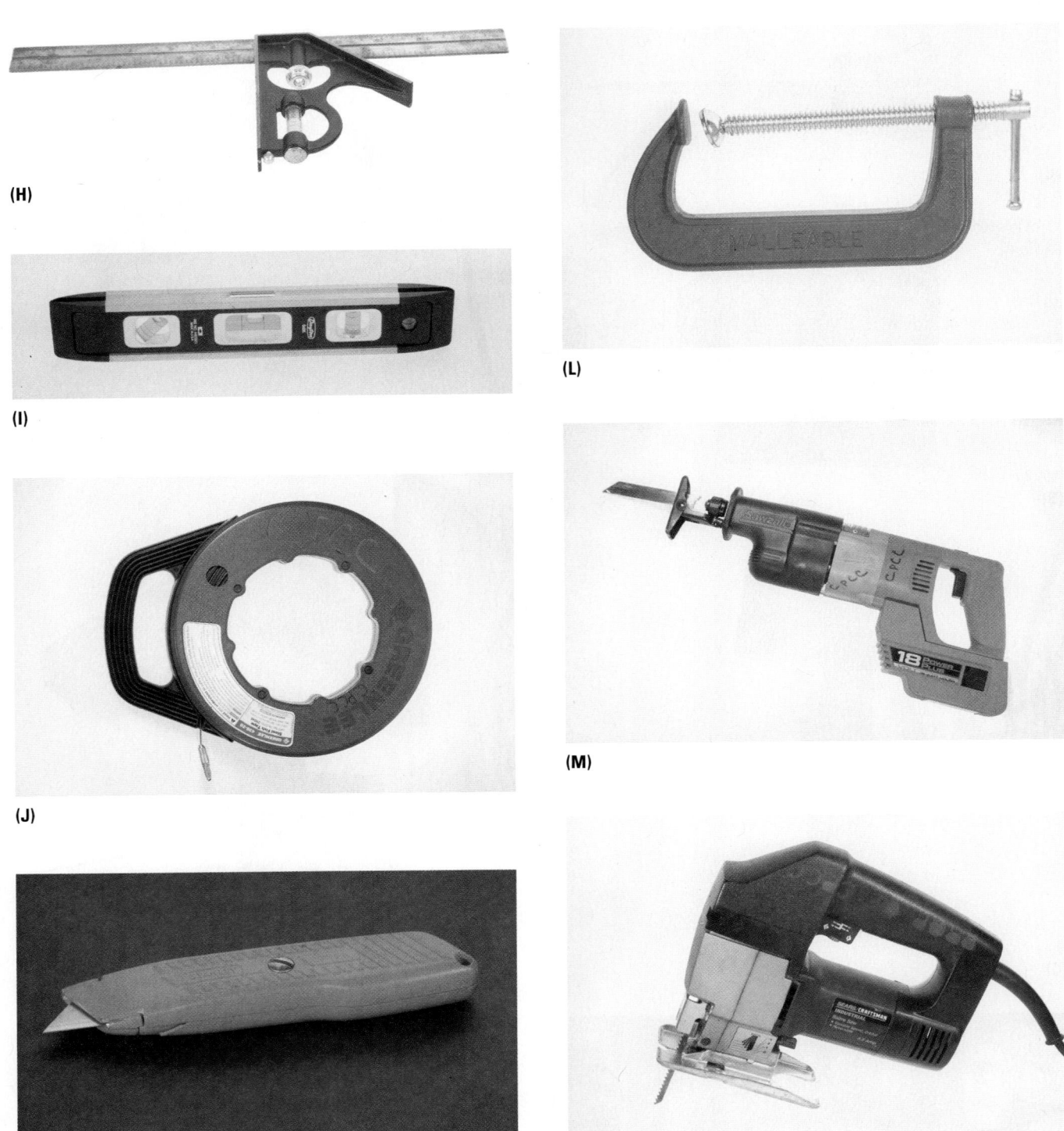

(H)

(I)

(J)

(K)

(L)

(M)

(N)

Figure 5–6   (*continued*) **(H)** Square. **(I)** Levels. **(J)** Fish tape. **(K)** Utility knife. **(L)** C-Clamp. **(M)** Reciprocating saw. **(N)** Jigsaw. *Photos **(A)** through **(C)** and **(E)** through **(N)** by Bill Johnson. **(D)** Courtesy Klein Tools*

PORTABLE ELECTRIC DRILLS. Portable electric drills are used extensively by refrigeration and air-conditioning technicians. They are available in cord type (115 V) or cordless (battery operated), **Figure 5–6(E)** and **Figure 5–6(F)**. If the drill is a cord type and not double-insulated, it must have a three-prong plug and be used only in a grounded receptacle. Drills that have variable speed and reversing options are recommended. Many technicians prefer the battery-operated type for its convenience and safety. Be sure that the batteries are kept charged.

Power tools are much like hand tools. Quality costs more. Many tools are made for the consumer market; the consumer picks the tool up once a year and does a light-duty job with it. A technician, however, must use these tools under all kinds of conditions—wet, dry, cold, and hot. The tools must be used often and will begin to let you down when you need them most. You should buy the best tools you can and mark them with a distinctive mark that will indicate whom the tool belongs to. At job sites there are often several technicians working in the same area using the same types of tools. Make sure you know your own tools.

5.2 SPECIALIZED HAND TOOLS

The following tools are regularly used by technicians in the air-conditioning, heating, and refrigeration field.

NUT DRIVERS. Nut drivers have a socket head and are used primarily to drive hex head screws from panels on air-conditioning, heating, and refrigeration cabinets. They are available with hollow shafts, solid shafts, and extra long or stubby shafts, **Figure 5–7.** The hollow shaft allows the screw to protrude into the shaft when it extends beyond the nut.

AIR-CONDITIONING AND REFRIGERATION REVERSIBLE RATCHET BOX WRENCHES. Ratchet box wrenches, which allow the ratchet direction to be changed by pushing the lever-like button near the end of the wrench, are used with air-conditioning and refrigeration valves and fittings. Two openings on each end of the wrench allow it to be used on four sizes of valve stems or fittings, **Figure 5–8.**

FLARE NUT WRENCH. The flare nut wrench is used like a box end wrench. The opening at the end of the wrench allows it to be slipped over the tubing. The wrench can then be placed over the flare nut to tighten or loosen it, **Figure 5–9.**

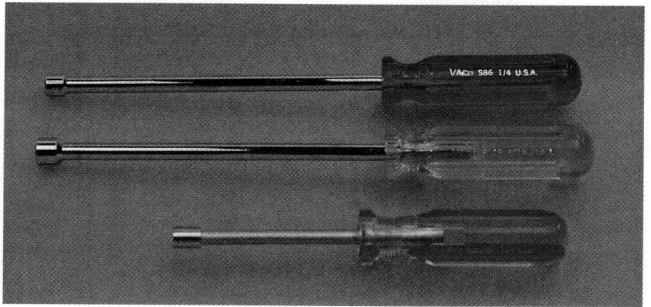
Figure 5–7 Assorted nut drivers. *Photo by Bill Johnson*

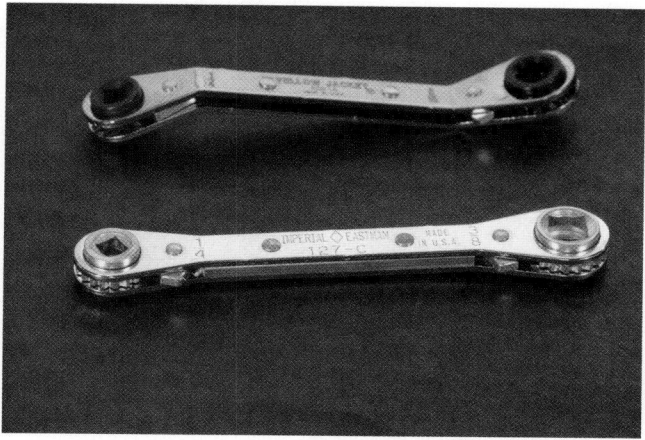
Figure 5–8 Air-conditioning and refrigeration reversible ratchet box wrenches. *Photo by Bill Johnson*

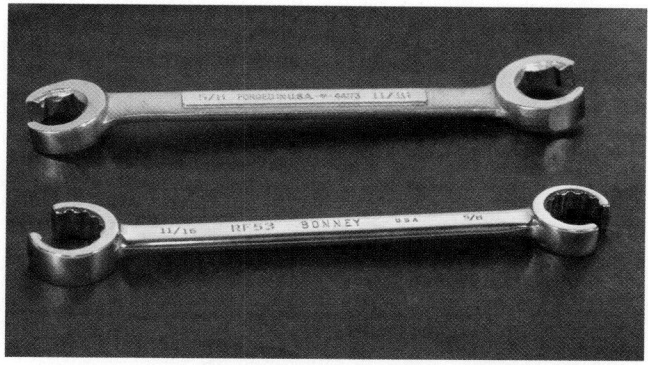
Figure 5–9 Flare nut wrenches. *Photo by Bill Johnson*

Adjustable wrenches should not be used because they may round off the corners of fittings made of soft brass. Do not use a pipe to extend the handle on a wrench for more leverage. Fittings can be loosened by heating or by using a penetrating oil.

WIRING AND CRIMPING TOOLS. Wiring and crimping tools are available in many designs. **Figure 5–10** illustrates a combination tool for crimping solderless connectors, stripping wire, cutting wire, and cutting small bolts. This figure also illustrates an automatic wire stripper. To use this tool, insert the wire into the proper strip-die hole. The length of the strip is determined by the amount of wire extending beyond the die away from the tool. Hold the wire in one hand and squeeze the handles with the other. Release the handles and remove the stripped wire.

INSPECTION MIRRORS. Inspection mirrors are usually available in rectangular or round shapes, with fixed or telescoping handles, some more than 30 in. long. The mirrors are used to inspect areas or parts in components that are behind or underneath other parts, **Figure 5–11.**

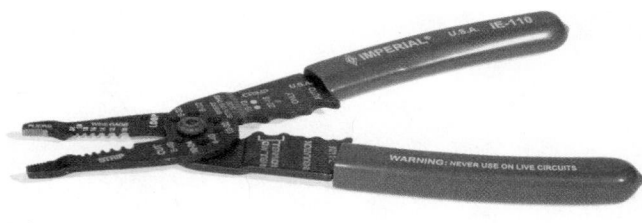

(A)

(B)

Figure 5–10   Wire stripping and crimping tools. **(A)** A combination crimping and stripping tool for crimping solderless connectors, stripping wire, cutting wire, and cutting small bolts. **(B)** An automatic wire stripper. *Photos by Bill Johnson*

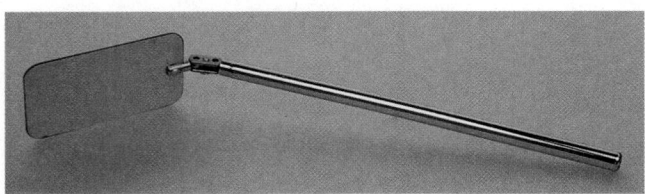

Figure 5–11   An inspection mirror. *Photo by Bill Johnson*

Figure 5–12   A stapling tacker. *Photo by Bill Johnson*

STAPLING TACKERS.   Stapling tackers are used to fasten insulation and other soft materials to wood; some types may be used to secure low-voltage wiring, **Figure 5–12.**

5.3   TUBING TOOLS

The following tools are used to install tubing. Most will be described more fully in other units in the book.

TUBE CUTTER.   Tube cutters are available in different sizes and styles. The standard tube cutter is shown in **Figure 5–13.** They are also available with a ratchet-feed mechanism. The cutter opens quickly to insert the tubing and slides to the cutting position. Some models have a flare cut-off groove that reduces the tube loss when removing a cracked flare. Many models also include a retractable reamer to remove inside burrs and a filing surface to remove outside burrs. **Figure 5–14** illustrates a cutter for use in tight spaces where standard cutters do not fit.

INNER–OUTER REAMERS.   Inner–outer reamers use three cutters to both ream the inside and trim the outside edges of tubing, **Figure 5–15.**

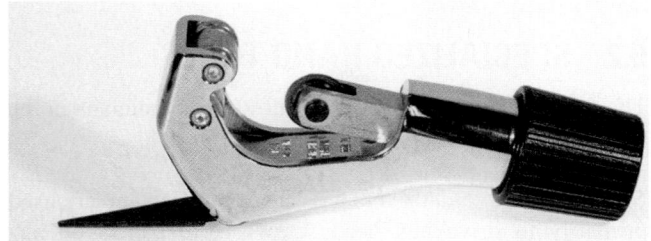

Figure 5–13   A tube cutter. *Photo by Bill Johnson*

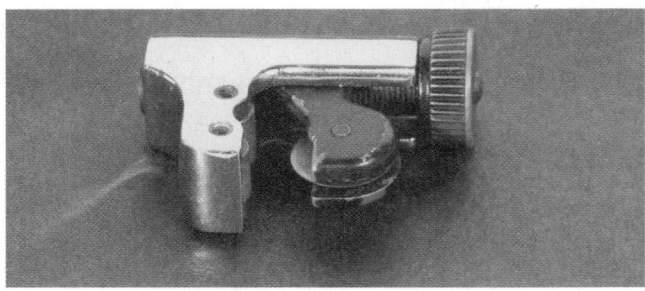

Figure 5–14   A small tube cutter used in tight places. *Photo by Bill Johnson*

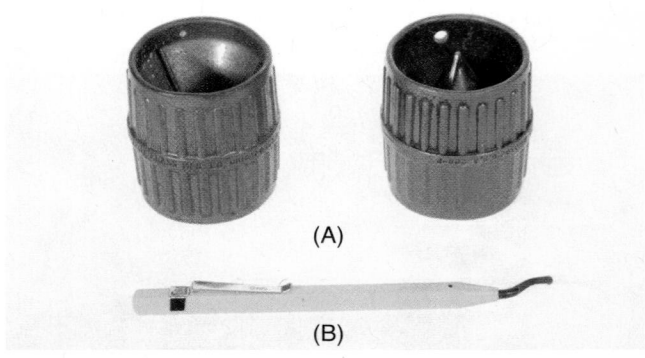

(A)

(B)

Figure 5–15   **(A)** Inner–outer reamers. **(B)** Deburring tool. *Photos by Bill Johnson*

FLARING TOOLS. The flaring tool has a flaring bar to hold the tubing, a slip-on yoke, and a feed screw with flaring cone and handle. Several sizes of tubing can be flared with this tool, **Figure 5–16.**

SWAGING TOOLS. Swaging tools are available in the punch type and lever type, **Figure 5–17.**

TUBE BENDERS. Three types of tube benders may be used: the spring type, lever type, and, to a lesser extent, gear type. **Figure 5–18** shows spring and lever types. These tools are used for bending soft copper and aluminum.

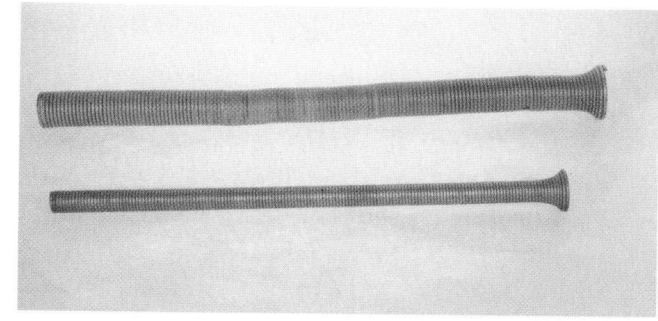

(A)

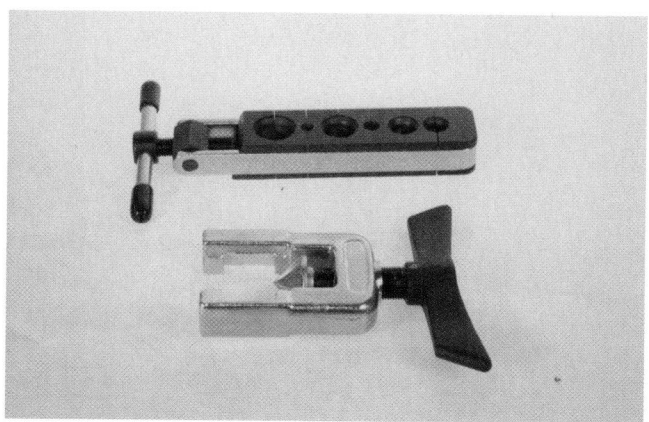

Figure 5–16 This flaring tool has a flaring bar, a yoke, and a feed screw with a flaring cone. *Photo by Bill Johnson*

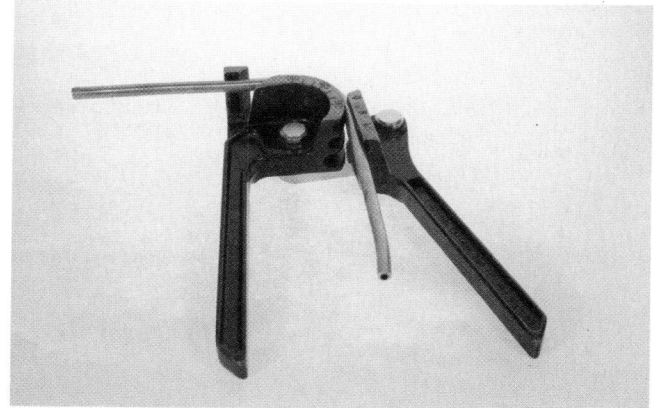

(B)

Figure 5–18 Tube benders. **(A)** Spring type. **(B)** Lever type. *Photos by Bill Johnson*

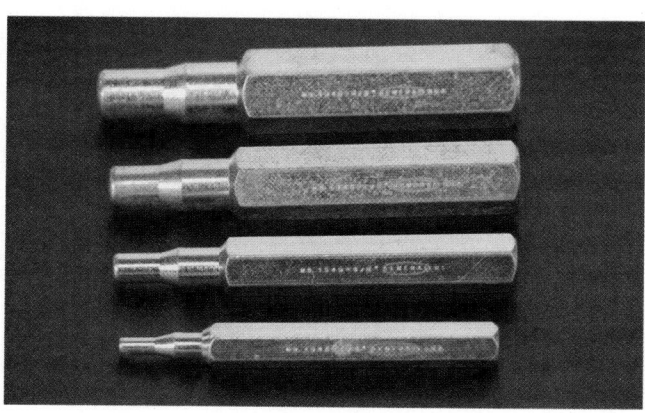

(A)

Figure 5–19 Tube brushes. *Courtesy Shaefer Brushes*

TUBE BRUSHES. Tube brushes clean the inside and outside of tubing and the inside of fittings. Some types can be turned by hand or by an electric drill, **Figure 5–19.**

PLASTIC TUBING SHEAR. A plastic tubing shear cuts plastic tubing and non-wire-reinforced plastic or synthetic hose, **Figure 5–20.**

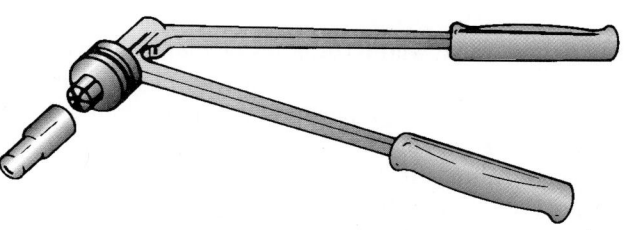

(B)

Figure 5–17 A swaging tool. **(A)** Punch type. **(B)** Lever type. **(A)** *Photo by Bill Johnson*

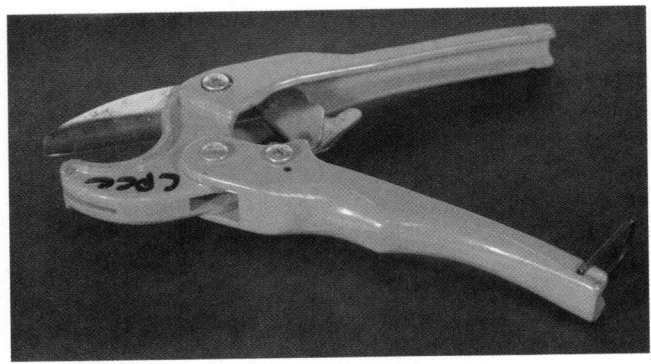

Figure 5–20   A plastic tubing shear. *Photo by Bill Johnson*

Figure 5–22   A metalworker's hammer. *Photo by Bill Johnson*

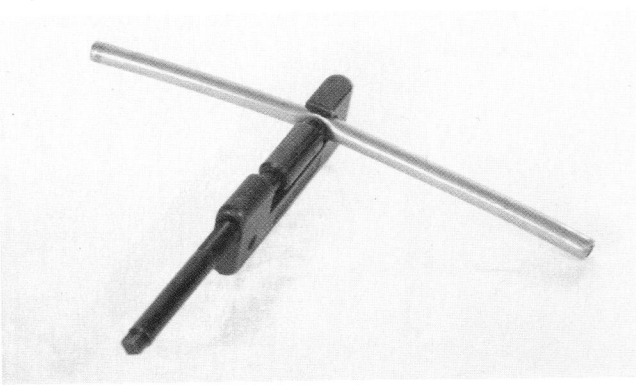

(A)

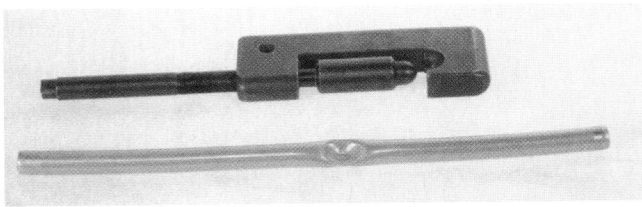

(B)

Figure 5–21   A tubing pinch-off tool **(A)–(B)**. *Photos by Bill Johnson*

TUBING PINCH-OFF TOOL.   A tubing pinch-off tool is used to pinch shut the short stub of tubing often provided for service, such as the service stub on a compressor. This tool is used to pinch shut this stub before sealing it by soldering, **Figure 5–21**.

METALWORKER'S HAMMER.   A metalworker's hammer is used to straighten and form sheet metal for ductwork, **Figure 5–22**.

5.4   SPECIALIZED SERVICE AND INSTALLATION EQUIPMENT

GAGE MANIFOLD.   One of the most important of all pieces of refrigeration and air-conditioning service equipment is the *gage manifold*. This equipment normally includes the *compound gage* (low-pressure and vacuum), the *high-pressure gage*, and the *manifold*, valves, and hoses. The gage manifold

(A)

(B)

Figure 5–23   Gage manifolds. **(A)** A two-valve gage manifold with three hoses. **(B)** A four-valve gage manifold with four hoses. *Courtesy Robinair SPX Corporation*

may be two-valve with three hoses or four-valve with four hoses. The four-valve design has separate valves for the vacuum, low-pressure, high-pressure, and refrigerant cylinder connections, **Figure 5–23**.

Gage manifolds that are capable of handling the higher pressures of some refrigerant blends must be used when working with these blends, **Figure 5–24**. SAFETY PRECAUTION: Care should be taken that all components, including hoses, are designed for use with these blends.

Electronic manifolds are also available that can provide more information than those indicated above. The manifold in

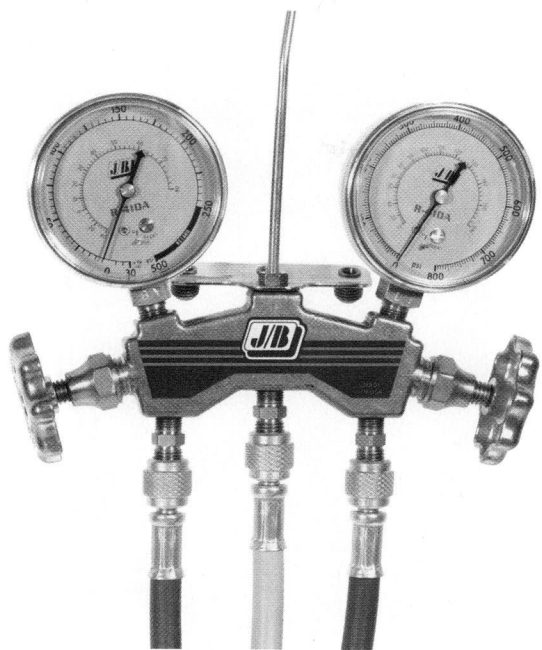

Figure 5–24   A gage manifold used with R-410A. *Photo by Bill Johnson*

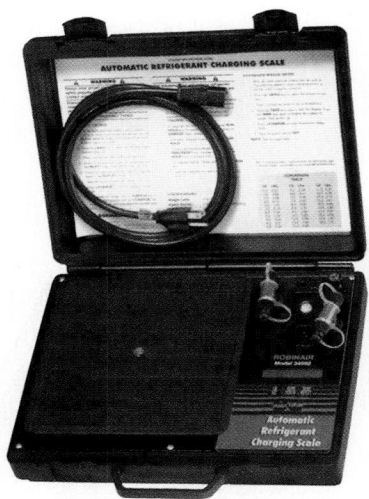

Figure 5–26   A programmable charging meter or scale. *Courtesy Robinair SPX Corporation*

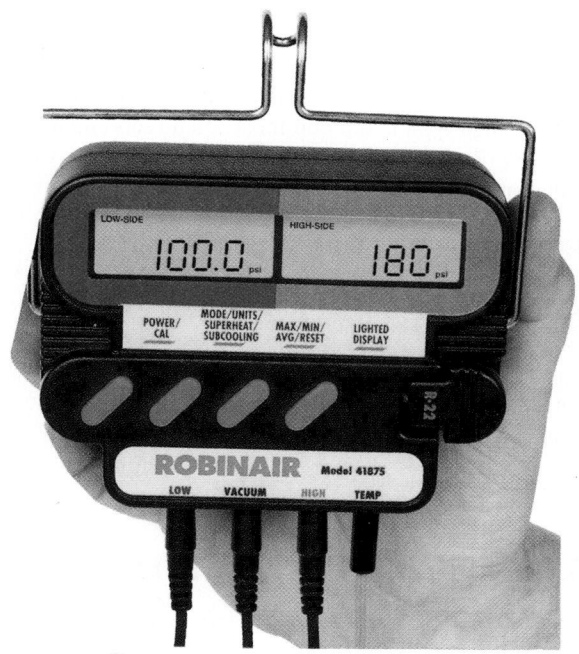

Figure 5–25   An electronic gage manifold. *Courtesy Robinair SPX Corporation*

Figure 5–25 can measure system pressures and temperatures, calculate superheat and subcooling, and measure vacuum level in microns. It may also be used as a digital thermometer.

PROGRAMMABLE CHARGING METER OR SCALE. These types of scales or meters will allow a technician to accurately charge refrigerant by weight. This usually can be done manually or automatically. The amount of refrigerant to be charged into the system can be programmed. **Figure 5–26** is an example of a programmable charging meter or scale.

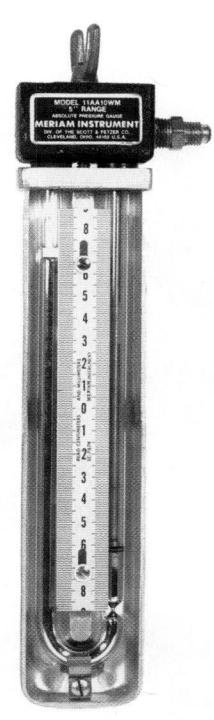

Figure 5–27   A U-tube mercury manometer. *Photo by Bill Johnson*

U-TUBE MERCURY MANOMETER. A U-tube mercury manometer shows the level of vacuum while a refrigeration system is being evacuated. It is accurate to approximately 0.5 mm Hg, **Figure 5–27.**

ELECTRONIC THERMISTOR VACUUM GAGE. An electronic thermistor vacuum gage measures the vacuum when a refrigeration, air-conditioning, or heat pump system is being evacuated. It measures a vacuum down to about 50 microns or 0.050 mm Hg (1000 microns = 1 mm), **Figure 5–28.**

VACUUM PUMP. Vacuum pumps designed specifically for servicing air-conditioning and refrigeration systems remove

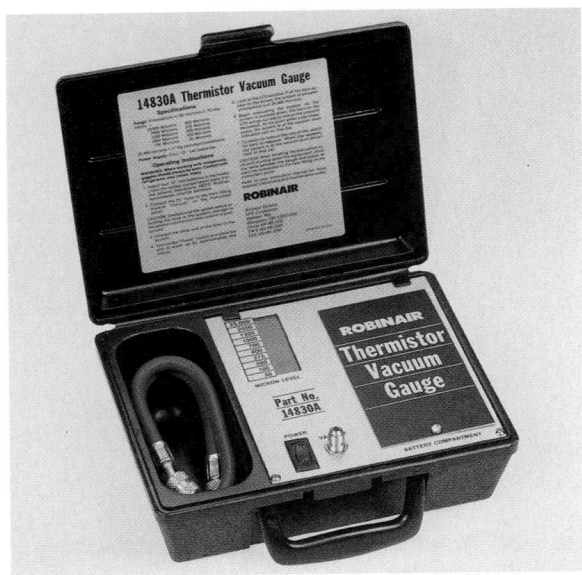

Figure 5–28   An electronic thermistor vacuum gage. *Courtesy Robinair SPX Corporation*

Figure 5–29   Typical vacuum pumps. *Courtesy Robinair SPX Corporation*

the air and noncondensable gases from the system. This is called evacuating the system and is necessary because the air and noncondensable gases take up space, contain moisture, and cause excessive pressures. **Figure 5–29** shows photos of two vacuum pumps.

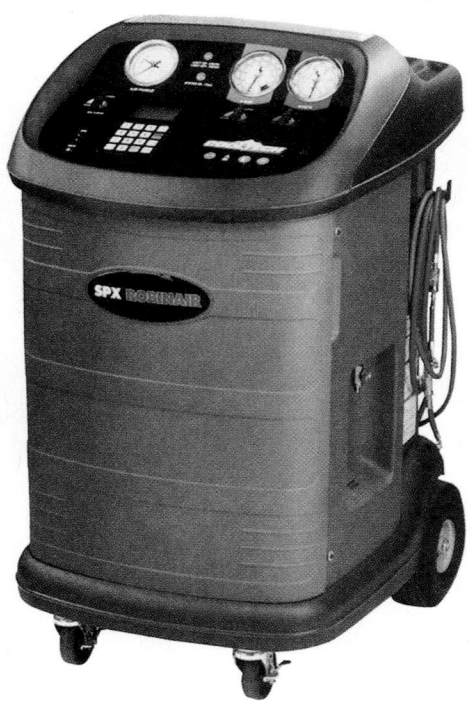

Figure 5–30   A refrigerant recovery and recycling station. *Courtesy Robinair Division, SPX Corporation*

Vacuum pumps may be either one- or two-stage types. The two-stage type is the one most often recommended for use. It will reach a lower vacuum for a better system vacuum in situations where moisture is an issue.

REFRIGERANT RECOVERY RECYCLING STATION.   ♲ It is illegal to intentionally vent refrigerant to the air. **Figure 5–30** is a photo of a typical recovery recycling station. The refrigerant from a refrigerator or air-conditioning system is pumped into a cylinder or container at this station where it is stored until it can be either charged back into the system if it meets the requirements or transferred to another approved container for transportation to a refrigerant reclaiming facility. ♲

5.5 REFRIGERANT LEAK DETECTORS

HALIDE LEAK DETECTOR.   A halide leak detector, **Figure 5–31,** detects refrigerant leaks. It is used with acetylene or propane gas. When the detector is ignited, the flame heats a copper disc. Air for the combustion is drawn through the attached hose. The end of the hose is passed over or near fittings or other areas where a leak may be suspected. If there is a leak, the refrigerant will be drawn into the hose and contact the copper disc. This breaks down the halogen refrigerants into other compounds and changes the color of the flame. The colors range from green to purple, depending on the size of the leak.

ELECTRONIC LEAK DETECTORS.   Electronic leak detectors contain an element sensitive to a particular refrigerant. The device may be battery- or AC-powered and often has a

Figure 5–31   A halide leak detector used with acetylene. *Photo by Bill Johnson*

pump to suck in the gas and air mixture. A ticking signal that increases in frequency and intensity as the probe "homes in" on the leak is used to alert the operator. Many also have varying sensitivity ranges that can be adjusted, **Figure 5–32.**

Many modern leak detectors have selector switches for detecting all three refrigerant types: CFC, HCFC, or HFC. HCFCs have less chlorine than CFCs, and the sensitivity has to be changed by the selector switch. HFCs do not contain chlorine but do contain fluorine. When the selector switch is set to HFC, the sensitivity of the leak detector is increased, and the more elusive fluorine can be detected.

ADDITIVE USED WITH A HIGH-INTENSITY ULTRAVIOLET LAMP.   With this system, **Figure 5–33,** an additive is induced into the refrigerant system. This additive shows up as a bright

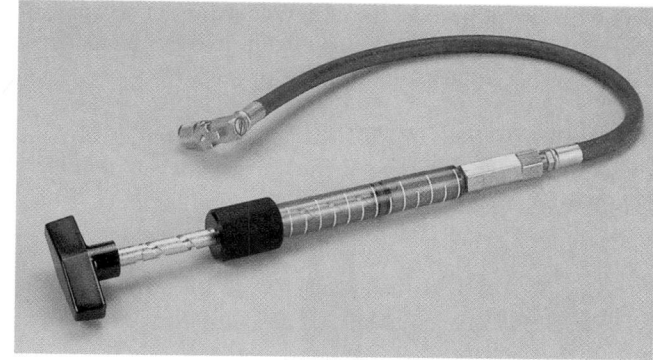

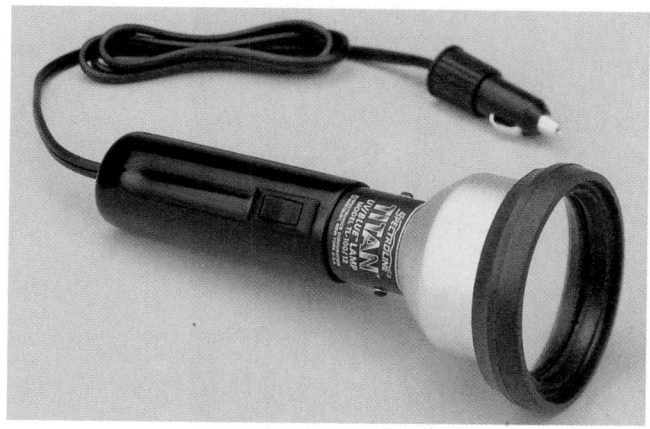

Figure 5–33   A fluorescent refrigerant leak detection system using an additive with a high-intensity ultraviolet lamp. *Courtesy Spectronics Corporation, Westbury, NY*

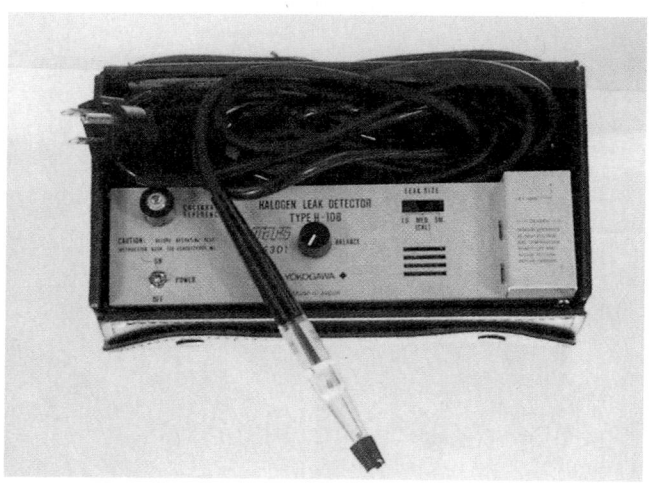

(A)

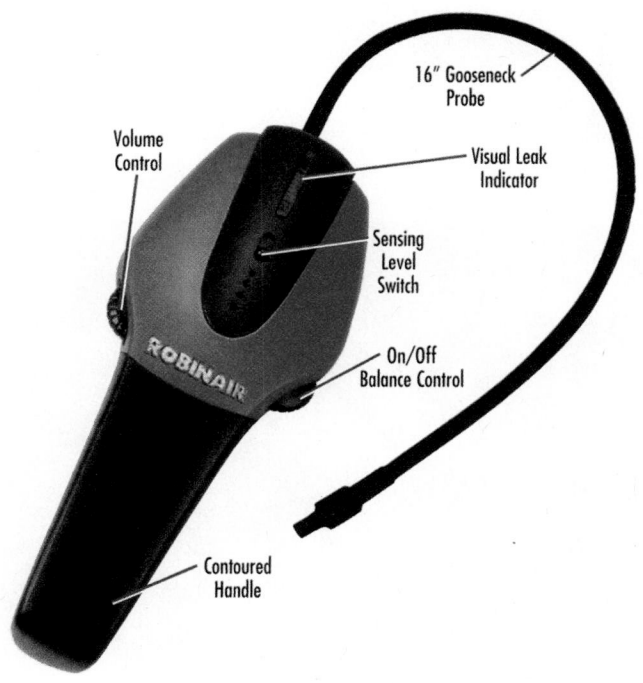

(B)

Figure 5–32   **(A)** AC-powered leak detector. **(B)** Heated diode type. *(A) Photo by Bill Johnson. (B) Courtesy Robinair SPX Corporation*

yellow-green glow under the ultraviolet lamp at the source of the leak. The additive can remain in the system should a new leak be suspected at a later date.

ULTRASOUND LEAK DETECTOR. These detectors use the sound from the escaping refrigerant to detect the leak, **Figure 5–34**.

THERMOMETERS. Thermometers range from simple pocket styles to infrared, electronic, and recording types. Pocket styles may be mercury (or alcohol) column glass-stem, dial-indicator, or digital-thermometer types, **Figure 5–35**. Other thermometers are shown in the following figures: electronic, **Figure 5–36(A)**, infrared, **Figure 5–36(B)**, and recording types, **Figure 5–37**.

(A)

(B)

Figure 5–34 Ultrasound leak detectors. *(A) Courtesy Robinair SPX Corporation. (B) Courtesy Amprobe Instruments*

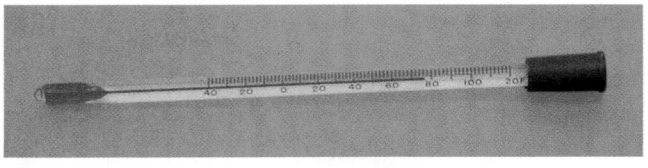

(A)

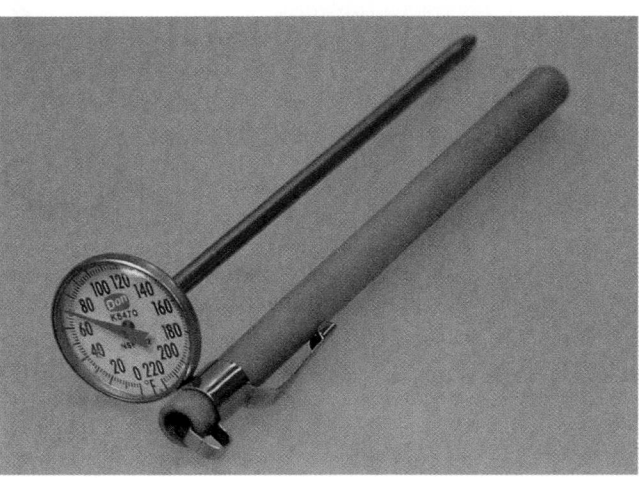

(B)

(C)

Figure 5–35 Thermometers. **(A)** Glass stem. **(B)** Pocket dial indicator. **(C)** Pocket digital indicators. *Photos **(A)** and **(B)** by Bill Johnson. (C) Courtesy UEi*

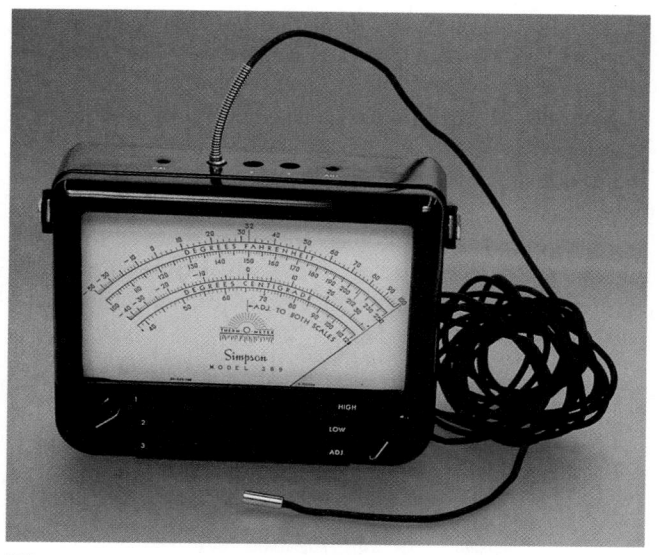

(A)

(B)

Figure 5–36 **(A)** An electronic thermometer. **(B)** An infrared thermometer. *(A) Photo by Bill Johnson. (B) Courtesy UEi*

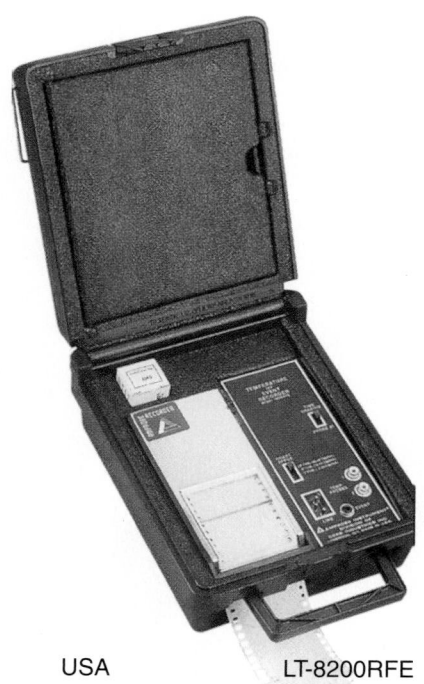

USA   LT-8200RFE

Figure 5–37 A recording thermometer. *Courtesy Amprobe Instruments*

Figure 5–38 A condenser or evaporator coil fin straightener. *Photo by Bill Johnson*

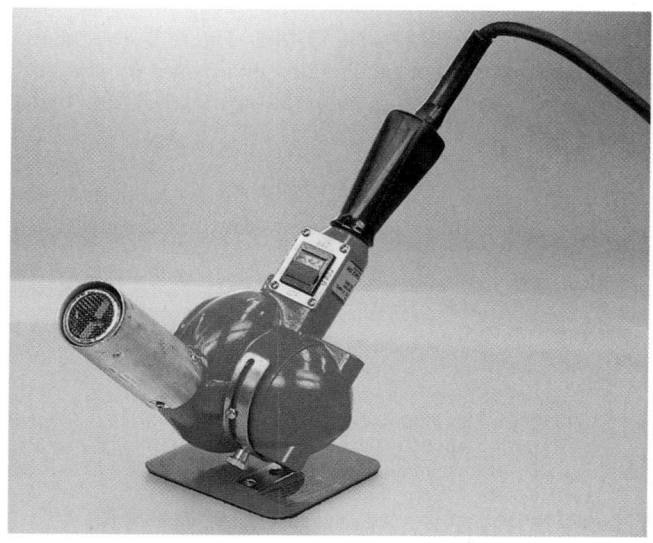

Figure 5–39 A heat gun. *Photo by Bill Johnson*

FIN STRAIGHTENERS. Fin straighteners are available in different styles. **Figure 5–38** is an example of one type capable of straightening condenser and evaporator coil fins having spacing of 8, 9, 10, 12, 14, and 15 fins per inch.

HEAT GUNS. Heat guns are needed in many situations to warm refrigerant, melt ice, and do other tasks. **Figure 5–39** is a heat gun carried by many technicians.

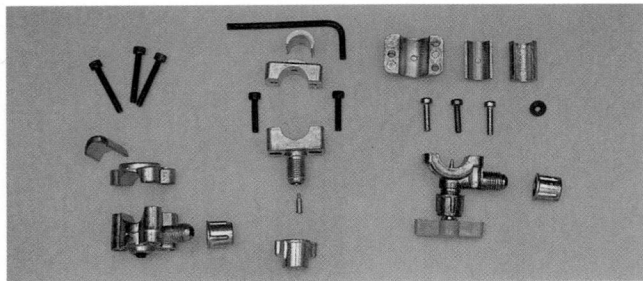

(A)

(B)

Figure 5–40   A tubing piercing valve **(A)–(B)**. *Photos by Bill Johnson*

HERMETIC TUBING PIERCING VALVES.   Piercing valves are an economical way to tap a line for charging, testing, or purging hermetically sealed units. A valve such as the one in **Figure 5–40** is clamped to the line. Often these are designed so that the valve stem is turned until a sharp needle pierces the tubing. Then the stem is backed off so that the service operations can be accomplished. These valves should be installed following the manufacturer's directions.

Piercing valves come as either gasket type or solder type. The gasket type is used only for temporary system access because the gaskets will get brittle and deteriorate in time when exposed to the atmosphere and high temperatures. Solder-type piercing valves are used for permanent access to a system because they do not employ gaskets.

COMPRESSOR OIL CHARGING PUMP.   **Figure 5–41** is an example of a compressor oil charging pump used for charging refrigeration compressors with oil while they are under pressure.

SOLDERING AND WELDING EQUIPMENT.   Soldering and welding equipment is available in many styles, sizes, and qualities.

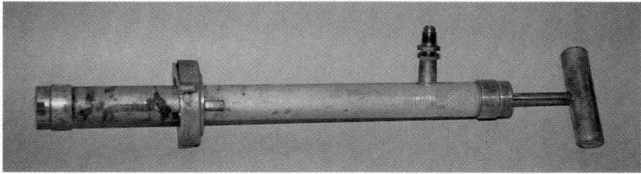

Figure 5–41   A compressor oil charging pump. *Photo by Bill Johnson*

A *soldering gun* is used primarily to solder electrical connections. It does not produce enough heat for soldering tubing, **Figure 5–42.**

Figure 5–43 shows a *propane gas torch* with a disposable propane gas tank. This is an easy torch to use. The flame adjusts easily and can be used for many soldering operations.

Air-acetylene units provide sufficient heat for soldering and brazing. They consist of a torch, which can be fitted with several sizes of tips, and a regulator, hoses, and an acetylene tank, **Figure 5–44.**

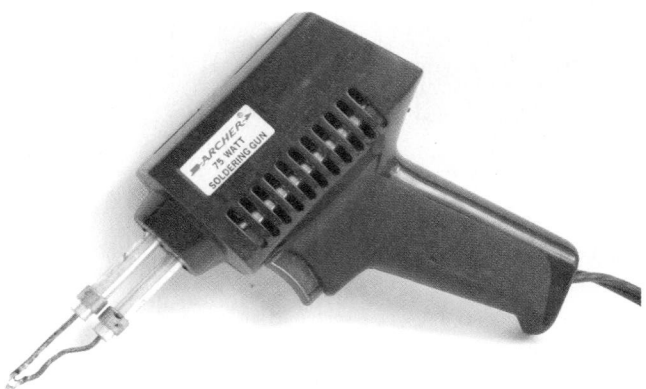

Figure 5–42   A soldering gun. *Photo by Bill Johnson*

Figure 5–43   A propane gas torch with a disposable propane gas tank. *Photo by Bill Johnson*

Figure 5–44 An air-acetylene unit. *Photo by Bill Johnson*

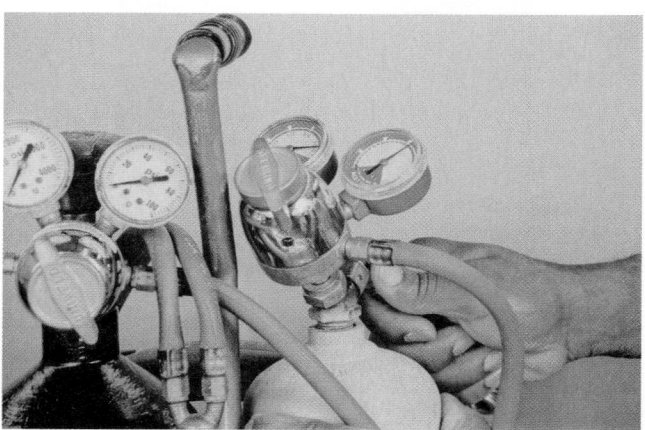

Figure 5–45 An oxyacetylene unit. *Photo by Bill Johnson*

Oxyacetylene welding units are used for certain solder-ing, brazing, and welding applications. They consist of a torch, regulators, hoses, oxygen tanks, and acetylene gas tanks. When the oxygen and acetylene gases are mixed in the proper proportion, a very hot flame is produced, **Figure 5–45.**

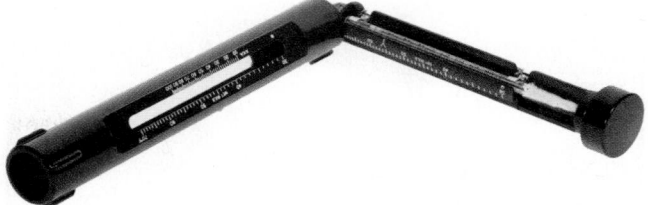

Figure 5–46 A sling psychrometer. *Photo by Bill Johnson*

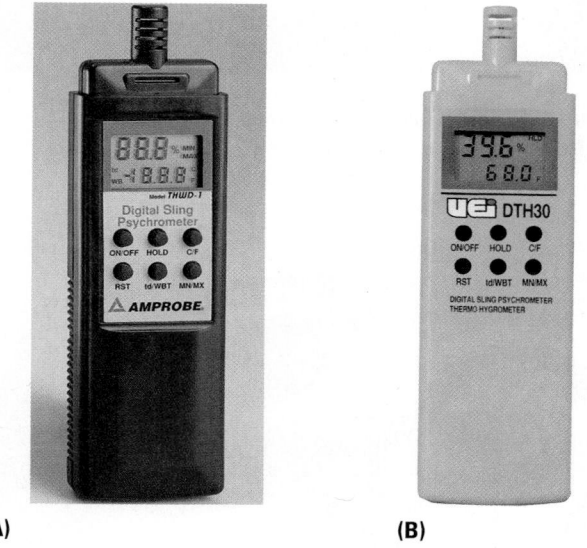

(A)   **(B)**

Figure 5–47 Two styles of digital sling psychrometers. **(A)** *Courtesy Amprobe.* **(B)** *Courtesy UEi*

SLING PSYCHROMETER. The sling psychrometer uses the wet-bulb/dry-bulb principle to obtain relative humidity read-ings quickly. Two thermometers, a dry bulb and a wet bulb, are whirled together in the air. Evaporation will occur at the wick of the wet-bulb thermometer, giving it a lower temper-ature reading. The difference in temperature depends on the humidity in the air. The drier the air, the greater the differ-ence in the temperature readings because the air absorbs more moisture, **Figure 5–46.** Most manufacturers provide a scale so that the relative humidity can easily be determined. Before you use it, be sure that the wick is clean and wet (with pure water, if possible).

Digital sling psychrometers have been developed. These will provide the readings just mentioned. **Figure 5–47** shows two styles of digital sling psychrometers.

MOTORIZED PSYCHROMETER. A motorized psychrometer is often used when many readings are taken over a large area.

NYLON STRAP FASTENER. **Figure 5–48** shows the installa-tion tool used to install nylon strap clamps around flexible duct. This tool automatically cuts the clamping strap off flush when a preset tension is reached.

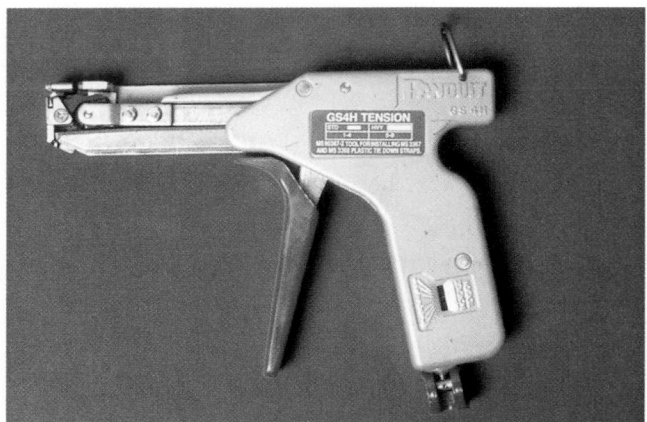

Figure 5–48   A tool used to install nylon strap clamps around flexible duct. *Photo by Bill Johnson*

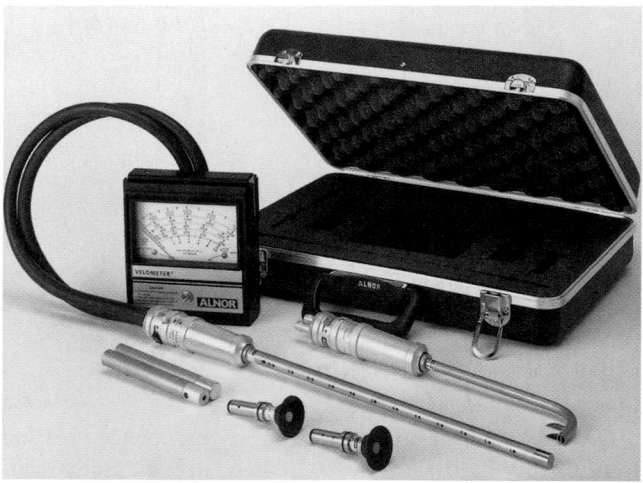

Figure 5–49   An air velocity measuring kit. *Courtesy Alnor Instrument Company*

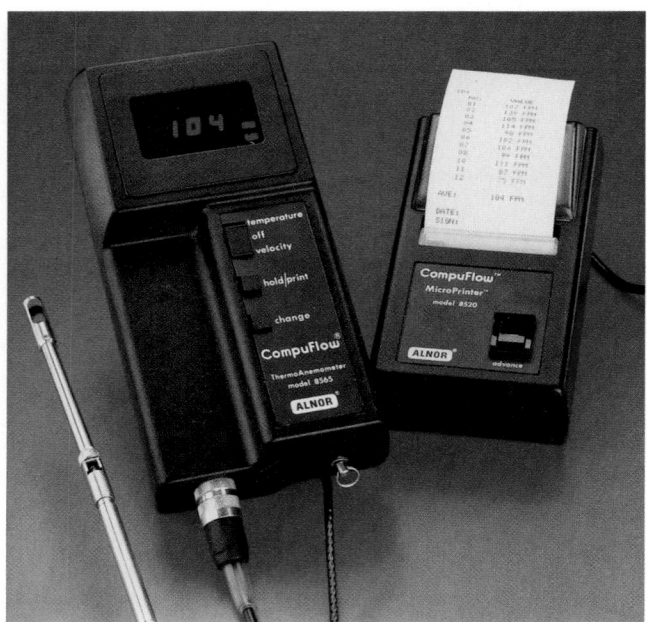

Figure 5–50   An air velocity measuring instrument with a microprocessor. *Courtesy Alnor Instrument Company*

Figure 5–51   An air balancing meter. *Courtesy Alnor Instrument Company*

AIR VELOCITY MEASURING INSTRUMENTS. Air velocity measuring instruments are necessary to balance duct systems, check fan and blower characteristics, and make static pressure measurements. They measure air velocity in feet per minute. **Figure 5–49** shows an air velocity measuring kit. This type of instrument makes air velocity measurements from 50 to 10,000 ft/min. **Figure 5–50** shows an air velocity instrument incorporating a microprocessor that will take up to 250 readings across hood openings or in a duct and will display the average air velocity and temperature readings when needed.

AIR BALANCING METER. The air balancing meter shown in **Figure 5–51** eliminates the need to take multiple readings. The meter will make readings directly from exhaust or supply grilles in ceilings, floors, or walls, and read out in cubic feet per minute (cfm).

CARBON DIOXIDE (CO_2) AND OXYGEN (O_2) INDICATORS. CO_2 and O_2 indicators are used to make flue-gas analyses to determine combustion efficiency in gas or oil furnaces, **Figure 5–52.**

CARBON MONOXIDE (CO) INDICATOR. A CO indicator is used to take flue-gas samples in natural gas furnaces to determine the percentage of CO present, **Figure 5–53.**

COMBUSTION ANALYZER. This combustion analyzer, **Figure 5–54(A)**, measures and displays concentrations in flue gases of oxygen (O_2), carbon monoxide (CO) (a poisonous

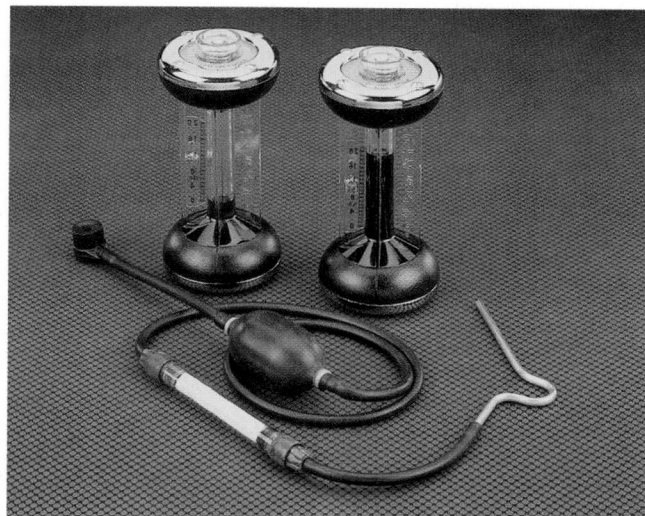

Figure 5–52 Carbon dioxide and oxygen indicators. *Courtesy Bacharach, Inc. Pittsburgh, PA USA*

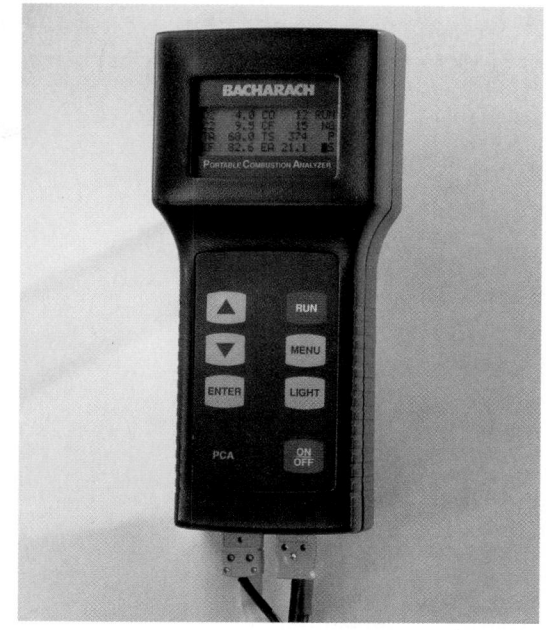

(A)

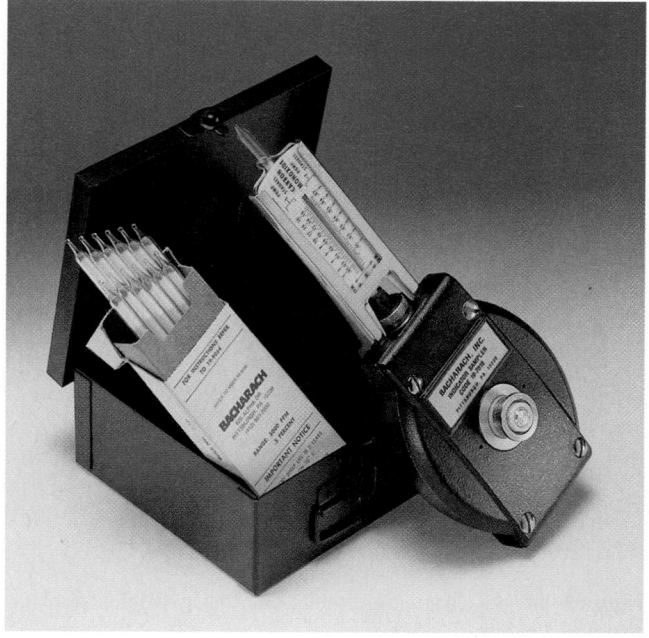

Figure 5–53 A carbon monoxide indicator. *Courtesy Bacharach, Inc. Pittsburgh, PA USA*

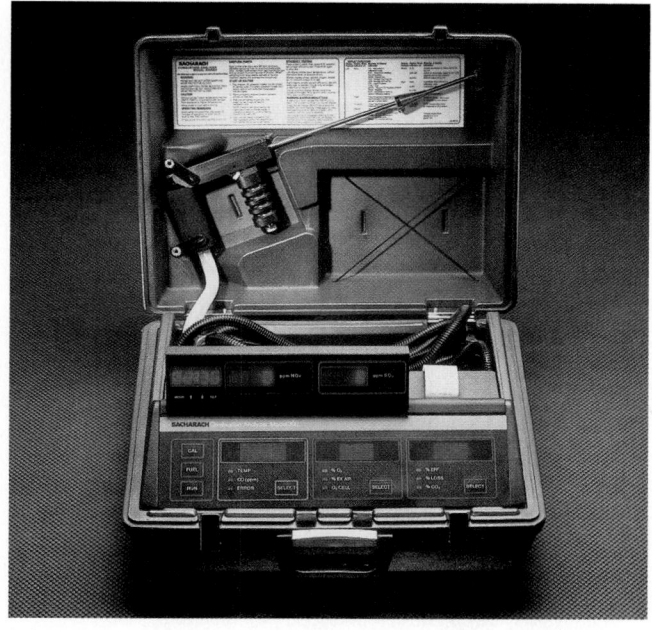

(B)

Figure 5–54 Combustion analyzers **(A)–(B)**. *Courtesy Bacharach, Inc. Pittsburgh, PA USA*

gas), oxides of nitrogen (NO_x), and sulfur dioxide (SO_2), and also measures and displays stack gas temperature. It computes and displays combustion efficiency, stack loss, and excess air.

Figure 5–54(B) is a portable analyzer that measures and displays flue gas oxygen, stack temperature, draft, oxides of nitrogen, and carbon monoxide.

COMBUSTIBLE GAS LEAK DETECTOR. This gas leak detector has an audible alarm and visual leak detection by light-emitting diode (LED) indicators. A red LED light on the tip illuminates the search area, **Figure 5–55.**

DRAFT GAGE. A draft gage is used to check the pressure of the flue gas in gas and oil furnaces to ensure that the flue gases are moving up the flue at a satisfactory speed, **Figure 5–56.** The flue gas is normally at a slightly negative pressure.

VOLT-OHM-MILLIAMMETER (VOM). A VOM, often referred to as a multimeter, is an electrical instrument that measures voltage (volts), resistance (ohms), and current (milli-amperes). These instruments have several ranges in each

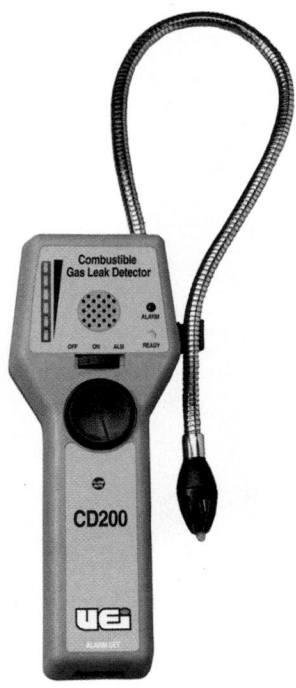

Figure 5–55 A combustible gas leak detector. *Courtesy UEi*

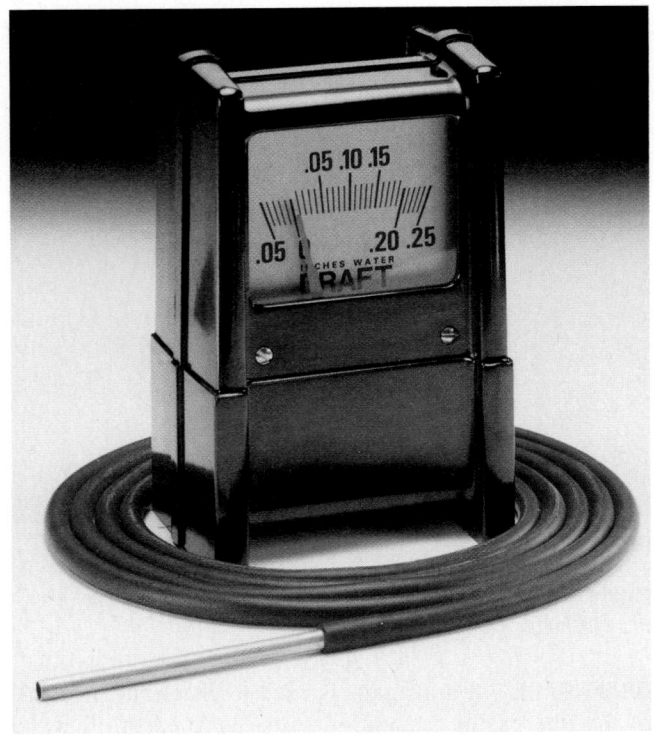

Figure 5–56 A draft gage. *Courtesy Bacharach, Inc. Pittsburgh, PA, USA*

mode. They are available in many types, ranges, and quality, either with a regular dial (analog) readout or a digital readout. **Figure 5–57** illustrates both an analog and digital type. If you purchase one, be sure to select one with the features and ranges used by technicians in this field. The voltmeter that

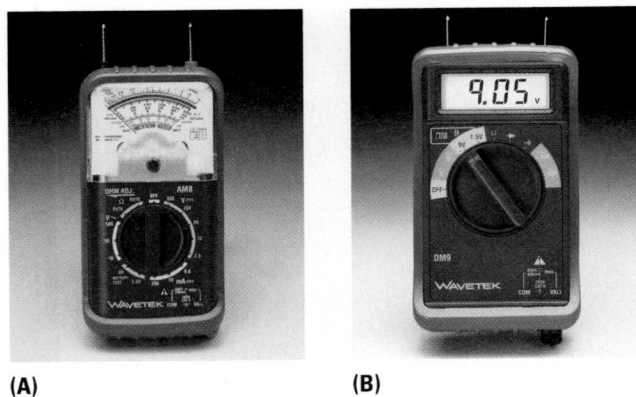

(A)   **(B)**

Figure 5–57 Volt-ohm-milliammeter (VOM). **(A)** An analog type. **(B)** A digital type. *Courtesy Wavetek*

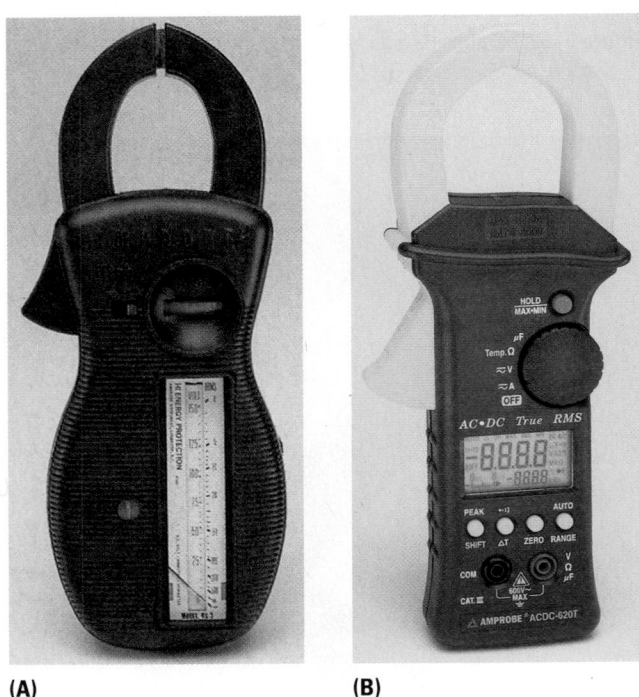

(A)   **(B)**

Figure 5–58 AC clamp-on ammeters. **(A)** Analog. **(B)** Digital. *Courtesy Amprobe*

you use is very important. Your life may depend on knowing what voltages you are working with. Be sure to get the best one you can afford. Inexpensive ones are tempting.

AC CLAMP-ON AMMETER. An AC clamp-on ammeter is a versatile instrument. It is also called clip-on, tang-type, snap-on, or other names. Some can also measure voltage or resistance or both. Unless you have an ammeter like this, you must interrupt the circuit to place the ammeter in the circuit. With this instrument you simply clamp the jaws around a single conductor, **Figure 5–58**.

MEGOHMMETER. A megohmmeter is used for measuring very high resistances. This particular device can measure up to 4000 megohms, **Figure 5–59**.

Figure 5–59   This megohmmeter will measure very high resistances. *Reproduced with permission of Fluke Corporation*

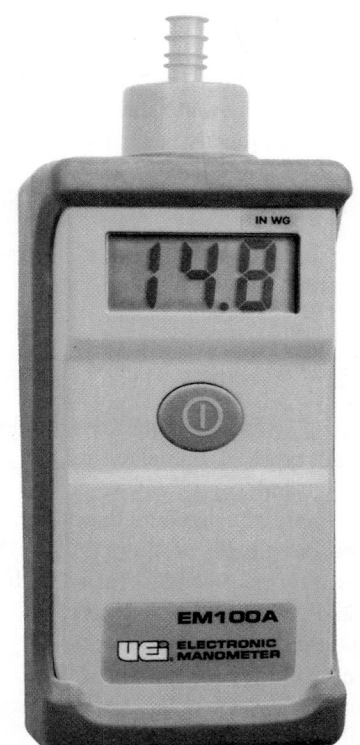

Figure 5–61   A digital electronic manometer. *Courtesy UEi*

U-TUBE WATER MANOMETER.   A U-tube water manometer displays natural gas and propane gas pressures during servicing or installation of gas furnaces and gas-burning equipment, **Figure 5–60(A)**.

INCLINED WATER MANOMETER.   An inclined water manometer determines air pressures in very low-pressure systems, such as to 0.1 in. of water column. These are used to analyze airflow in air-conditioning and heating air-distribution systems, **Figure 5–60(B)**.

DIGITAL ELECTRONIC MANOMETER.   This instrument is used with tubing that, when connected to an air-distribution system, can measure from −20 in. to +20 in. water gage, **Figure 5–61**.

A technician will use many other tools and equipment, but the ones covered in this unit are the most common.

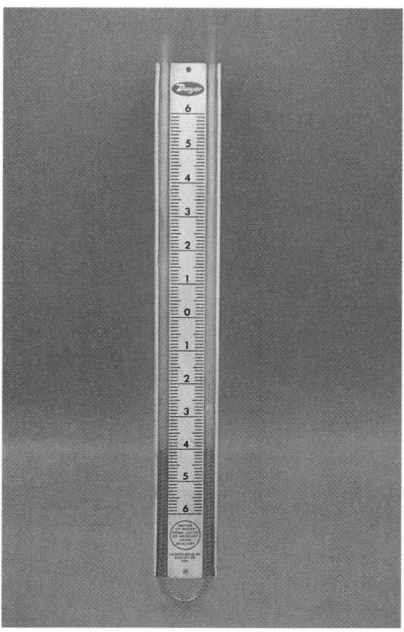

(A)

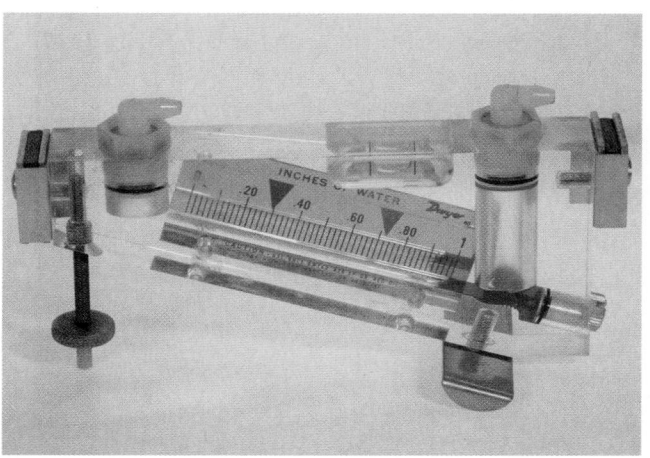

(B)

Figure 5–60   Manometers. **(A)** Water U-tube. **(B)** Inclined. *Photos by Bill Johnson*

SUMMARY

■ Air-conditioning, heating, and refrigeration technicians should be familiar with available hand tools and equipment.

■ Technicians should properly use hand tools and specialized equipment.

■ SAFETY PRECAUTION: These tools and this equipment should be used only for the job for which they were designed. Other use may damage the tool or equipment and may be unsafe for the technician.

REVIEW QUESTIONS

1. _____ and _____ are two types of screwdriver bits.
2. Five different types of wrenches are _____, _____, _____, _____, and _____.
3. Give an example of a use for a nut driver.
4. List four different types of pliers.
5. Describe two types of benders for soft tubing.
6. A wiring and crimping tool is used to
   A. crimp tubing.
   B. assemble tubing.
   C. cut and strip wire.
   D. install wire staples.
7. A flaring tool is used to
   A. cut copper tubing.
   B. flare a sheet metal joint.
   C. put a flare on the end of copper tubing.
   D. provide a yoke for bending brazing filler metal.
8. A vacuum pump is used to
   A. remove contaminants from a refrigeration system.
   B. vacuum dust before installing the refrigeration system compressor.
   C. vacuum the area before using an air-acetylene unit.
   D. prepare an area for taking a temperature reading.
9. Two types of refrigerant leak detectors are _____ and _____.
10. List the components usually associated with a gage manifold unit.
11. List three types of thermometers.
12. What type of soldering is a soldering gun normally used for?
13. An air-acetylene unit is used for _____ and _____.
14. What is a sling psychrometer used for?
15. When are air velocity measuring instruments used?
16. Describe the purpose of CO_2 and O_2 indicators.
17. What is the purpose of a draft gage?
18. List three electrical measurements that a VOM can make.
19. Why would a technician use an electronic charging scale?
20. Why is it necessary to recover refrigerant?

UNIT 6 · Fasteners

OBJECTIVES

After studying this unit, you should be able to

- identify common fasteners used with wood.
- identify common fasteners used with sheet metal.
- write and explain a typical tapping screw dimension.
- identify typical machine screw heads.
- write and explain each part of a machine screw thread dimension.
- describe a fastener used in masonry.
- describe hanging devices for piping, tubing, and duct.
- describe solderless terminals and screw-on wire connectors.

SAFETY CHECKLIST

- ✔ When staples are used to fasten a wire in place, do not hammer them too tight as they may damage the wire.
- ✔ When using fasteners, be sure that all materials are strong enough for the purpose for which they are being used.
- ✔ Wear goggles whenever drilling holes and when cleaning out holes.
- ✔ Do not use powder actuated systems without the proper training.

As an air-conditioning (heating and cooling) and refrigeration technician, you need to know about different types of fasteners and the various fastening systems so that you will use the right fastener or system for the job and securely install all equipment and materials.

6.1 NAILS

Probably the most common fastener used in wood is the nail. Nails are available in many styles and sizes. Common nails are large, flat-headed wire nails with a specific diameter for each length, **Figure 6–1(A)**. A finishing nail, **Figure 6–1(B)**, has a very small head so that it can be driven below the surface of the wood; it is used where a good finish is desired.

These nails are sized by using the term *penny*, which is abbreviated by the letter *d*, **Figure 6–1(A)**. For instance, an 8d common nail will describe the shape and size. These nails can be purchased plain or with a coating, usually zinc or resin. The coating protects against corrosion and helps to prevent the nail from working out of the wood.

A roofing nail is used to fasten shingles and other roofing materials to the roof of a building. It has a large head to keep

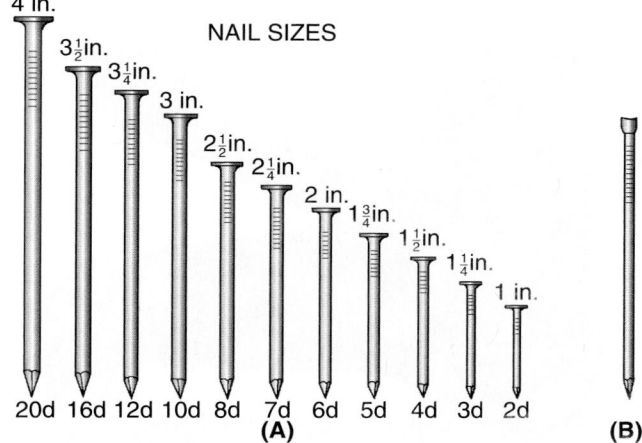

Figure 6–1 **(A)** Common nails. **(B)** A finishing nail.

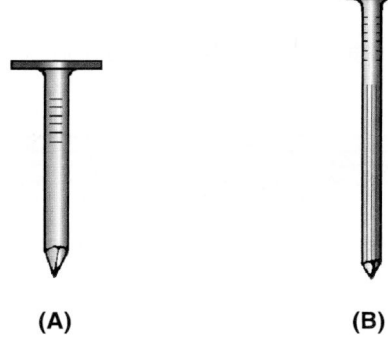

Figure 6–2 **(A)** A roofing nail. **(B)** A masonry nail.

the shingles from tearing and pulling off. A roofing nail can sometimes be used on strapping to hang ductwork or tubing, **Figure 6–2(A)**. Masonry nails are made of hardened steel and can be driven into masonry, **Figure 6–2(B)**.

6.2 STAPLES AND RIVETS

STAPLES. A staple, **Figure 6–3**, is a fastener made somewhat like a nail but shaped in a U with a point on each end. Staples are available in different sizes or for different uses. One use is to fasten wire in place. The staples are simply hammered in over the wire. SAFETY PRECAUTION: Never hammer them too tight because they can damage the wire. Other types of staples may be fastened in place with a stapling tacker, **Figure 6–4**. By depressing the handle a staple can be driven through paper, fabric, or insulation into the wood. *Outward*

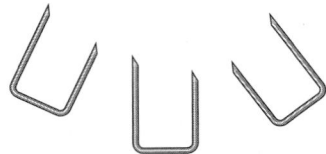

Figure 6–3 Staples used to fasten wire to wood.

Figure 6–4 A stapling tacker. *Photo by Bill Johnson*

Figure 6–5 A staple that is clinched outward (used to fasten soft materials together).

clinch tackers will anchor staples inside soft materials. The staple legs spread outward to form a strong tight clinch, **Figure 6–5.** They can be used to fasten insulation around heating or cooling pipes and ducts and to install ductboard. Other models or tackers are used to staple low-voltage wiring to wood.

RIVETS. Pin rivets or blind rivets are used to join two pieces of sheet metal. They are actually hollow rivets assembled on a pin, often called a *mandrel.* **Figure 6–6** illustrates a *pin rivet assembly,* how it is used to join sheet metal, and a riveting tool or gun. The rivets are inserted and set from only one side of the metal. Thus, they are particularly useful when there is no access to the back side of the metal being fastened.

To use a pin rivet, drill a hole in the metal the size of the rivet diameter. Insert the pin rivet in the hole with the pointed end facing out. Then place the nozzle of the riveting tool over the pin and squeeze the handles. Jaws grab the pin and pull the head of the pin into the rivet, expanding it and forming a head on the other side. Continue squeezing the handles until the head on the reverse side forms a tight joint.

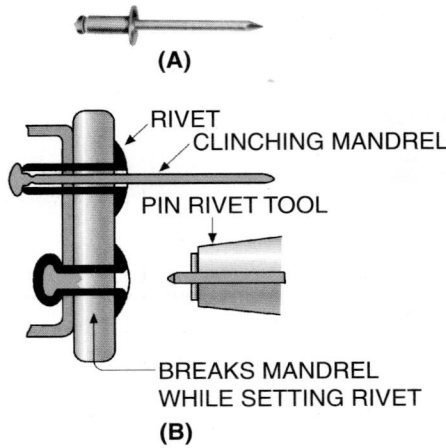

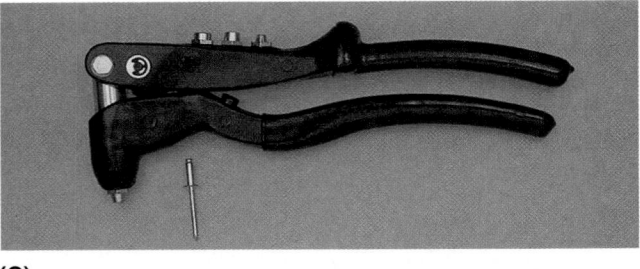

(C)

Figure 6–6 **(A)** A pin rivet assembly. **(B)** Using pin rivets to fasten sheet metal. **(C)** A riveting gun. *(A) and (B) Courtesy Duro Dyne Corp. (C) Photo by Bill Johnson*

When this happens, the pin breaks off and is ejected from the riveting tool.

This rivet can only be removed by drilling it out. To drill it out, use a drill bit that is the same size as the rivet. When the rivet is drilled out, the back side of the rivet will fall on the inside of the sheet metal pieces that are fastened together. SAFETY PRECAUTION: If you are working on an electrical panel, be sure the power is off. Another rivet of the same size can be used to refasten the panels together.

6.3 THREADED FASTENERS

We will describe only a few of the most common threaded fasteners.

WOOD SCREWS. Wood screws are used to fasten many types of materials to wood. They generally have a flat, round, or oval head, **Figure 6–7.** Flat and oval heads have an

WOOD SCREW STYLES

Figure 6–7 **(A)** A flat head wood screw. **(B)** A round head wood screw. **(C)** An oval head wood screw.

angle beneath the head. Holes for these screws should be countersunk so that their angular surface will be recessed. Wood screws can have a straight slotted head or a Phillips recessed head.

Wood screw sizes are specified by length (in inches) and shank diameter (a number from 0 to 24). The larger the number, the larger the diameter or gage of the shank.

For example, a No. 8 screw is smaller than a No. 6 screw. The numbering system described here is used for smaller screws until a screw reaches the diameter of 1/4 inch—and then the system changes to fractions of an inch. Larger sizes of wood-type screws (screws with a tapered shaft) are called lag screws. They often come with a hexagon or a square head and are tightened with a wrench, not a hand driver. These are typically used to fasten timber or large pieces of wood together.

TAPPING SCREWS.

Tapping or sheet metal screws are used extensively by service technicians. These screws may have a straight slot, a Phillips head, or a straight slot and hex head, **Figure 6–8.** To use a tapping screw, drill a hole into the sheet metal the approximate size of the root diameter of the thread. Then turn the screw into the hole with a conventional screwdriver or electric drill with screwdriver bit.

Figure 6–9 shows a tapping screw, often called a self-drilling screw, that can be turned into sheet metal with an electric drill using a chuck similar to the one shown. This screw has a special point used to start the hole. It can be used with light to medium gages of sheet metal.

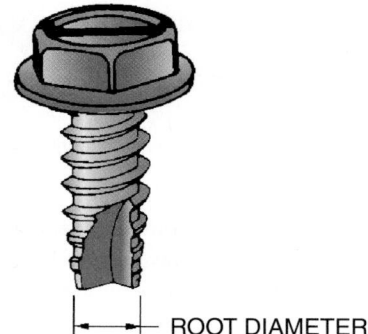

Figure 6–8   A tapping screw.

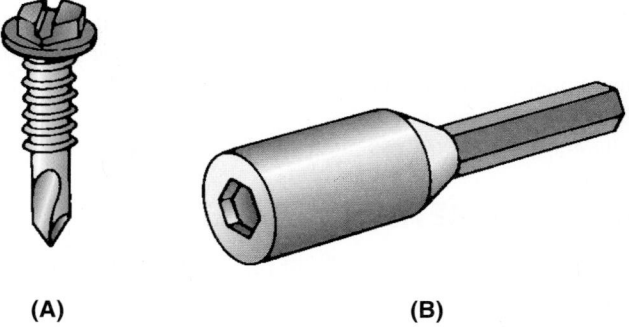

(A)   **(B)**

Figure 6–9   **(A)** A self-drilling screw. Note the special point used to start the hole. **(B)** A drill chuck.

Tapping screws are identified or labeled in a specific way. The following is a typical identifying label for tapping screws:

6−20 × 1/2 Type AB, Slotted hex
6(0.1380) = Outside thread diameter
20 = Number of threads per inch
1/2 = Length of the screw
Type AB = Type of point
Slotted hex = Type of head

Other items can be included, but they are very technical in nature. A thread dimension may not include all of the features listed. The diameter can be a number, a fraction, or a decimal fraction. The type of point is often not included.

MACHINE SCREWS.

Many types of machine screws exist. They normally are identified by their head styles. Some of these are indicated in **Figure 6–10.**

A thread dimension or label is indicated as follows:

5/16–18 UNC −2
5/16 = Outside thread diameter
18 = Number of threads per inch
UNC = Unified thread series (Unified National Coarse)
2 = Class of fit (the amount of play between the internal and external threads)

SET SCREWS.

Set screws have points as illustrated in **Figure 6–11.** They have square heads, hex heads, or no heads. The headless type may be slotted for a screwdriver, or it may have a hexagonal or fluted socket. These screws are used to keep a pulley from turning on a shaft and for other similar applications.

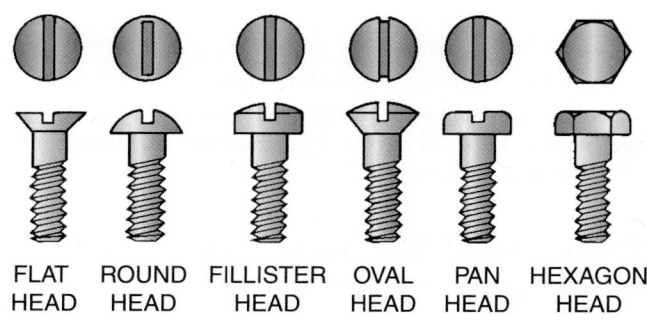

FLAT HEAD   ROUND HEAD   FILLISTER HEAD   OVAL HEAD   PAN HEAD   HEXAGON HEAD

Figure 6–10   Machine screws.

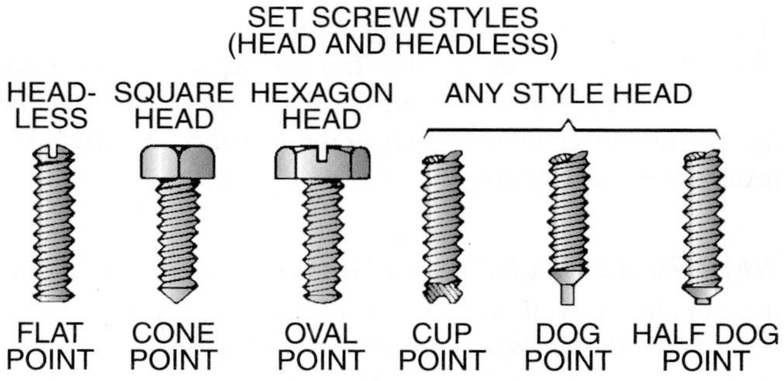

SET SCREW STYLES
(HEAD AND HEADLESS)

HEADLESS   SQUARE HEAD   HEXAGON HEAD   ANY STYLE HEAD

FLAT POINT   CONE POINT   OVAL POINT   CUP POINT   DOG POINT   HALF DOG POINT

Figure 6–11   Styles of set screws.

FASTENER TYPES AND DRIVER TIPS

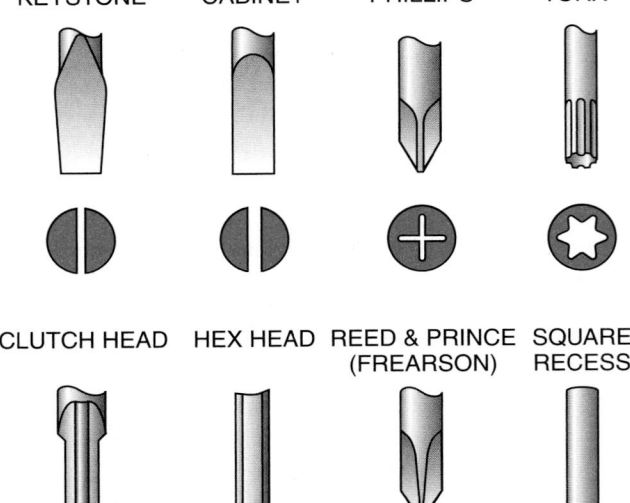

Figure 6–12 Several fastener heads and their driver tips. *Courtesy Klein Tools*

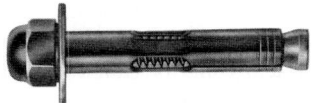

Figure 6–13 An anchor shield with a screw. *Courtesy Rawlplug Company, Inc.*

The set screw that is used to hold a pulley tight on a shaft is made of hardened steel and the pulley shaft is made of soft steel. When the set screw is tightened onto the shaft, it indents the shaft with its tip.

OTHER FASTENER HEADS. **Figure 6–12** is an illustration of several fastener heads and their driver tips.

ANCHOR SHIELDS. Anchor shields with bolts or screws are used to fasten objects to masonry or, in some instances, hollow walls. **Figure 6–13** illustrates a multipurpose steel anchor bolt used in masonry material. Drill a hole in the masonry the size of the sleeve. Tap the sleeve and bolt into the hole with a hammer. Turn the bolt head. This expands the sleeve and secures the bolt in the masonry. A variety of head styles are available.

WALL ANCHOR. A *hollow wall anchor,* **Figure 6–14,** can be used in plaster, wallboard, gypsum board, and similar materials. Once the anchor has been set, the screw may be removed as often as necessary without affecting the anchor.

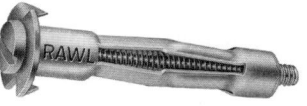

Figure 6–14 A hollow wall anchor. *Courtesy Rawlplug Company, Inc.*

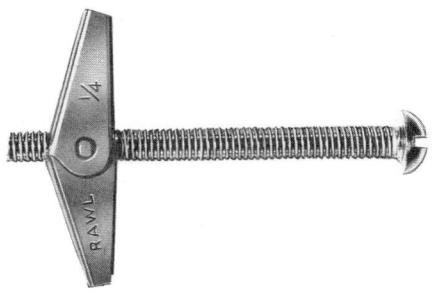

Figure 6–15 A toggle bolt. *Courtesy Rawlplug Company, Inc.*

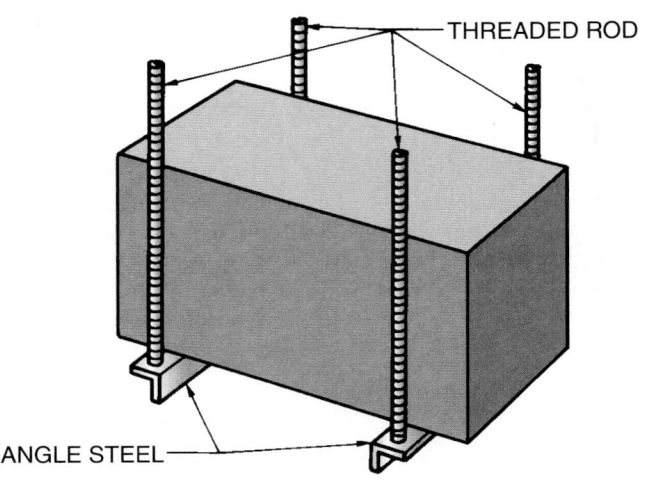

Figure 6–16 Using threaded rod and angle steel to make a hanger.

TOGGLE BOLTS. *Toggle bolts* provide a secure anchoring in hollow tiles, building block, plaster over lath, and gypsum board. Drill a hole in the wall large enough for the toggle in its folded position to go through. Push the toggle through the hole and use a screwdriver to turn the bolt head. You must maintain tension on the toggle or it will not tighten, **Figure 6–15.**

THREADED ROD AND ANGLE STEEL. Threaded rod and angle steel can be used to custom-make hangers for pipes or components such as an air handler, **Figure 6–16.** SAFETY PRECAUTION: Be sure that all materials are strong enough to adequately support the equipment.

6.4 CONCRETE FASTENERS

SCREW FASTENERS FOR CONCRETE. A concrete screw fastener may be used in poured concrete, concrete block, or brick, **Figure 6–17.** A hole must first be drilled in the base material. It is important to use the drill diameter recommended

Figure 6–17 A concrete screw fastener.

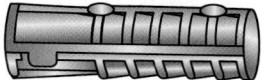

Figure 6–18 A lag shield anchor.

Figure 6–19 A powder load.

Figure 6–20 A cotter pin.

Figure 6–21 A wire pipe hook.

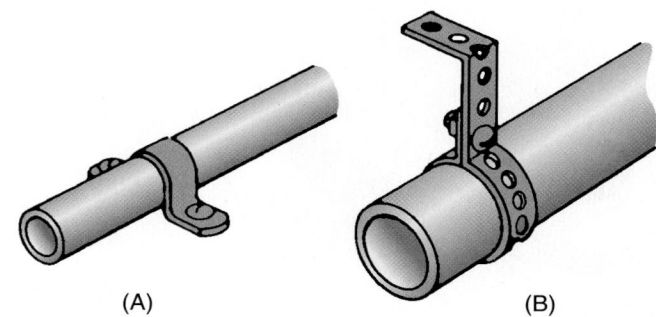

(A)   (B)

Figure 6–22 **(A)** A pipe strap. **(B)** A perforated strap.

by the manufacturer for each screw size. Drill the hole slightly deeper than the length of the screw that will extend into the base material. Blow out the dust and other material from the hole to the extent possible. SAFETY PRECAUTION: Wear safety goggles during this drilling and cleaning process. The screw may then be installed with an electric screwdriver or drill using a screwdriver bit or drill chuck, depending on the type of head on the screw.

LAG SHIELD ANCHORS.   A two-part shield has tapered internal threads for part of its length, **Figure 6–18.** These shields are generally used with lag screws. For installation, drill a hole of the diameter recommended by the manufacturer for the size of shield being used. The hole should be drilled as deep as the length of the shield plus approximately 1/2 inch. Clean dust or loose concrete material out of the hole. SAFETY PRECAUTION: Wear goggles when drilling and cleaning. Tap in the shield with a hammer until it is flush with the surface. Position the material to be fastened with a hole over the shield and turn in the lag screw.

POWDER ACTUATED FASTENER SYSTEMS.   These systems provide another method of fastening materials to concrete. SAFETY PRECAUTION: This type of system is very dangerous. Operators must have specific training and in many instances must be licensed. It is only mentioned here to make installers aware that there is such a system. A powder load, **Figure 6–19,** is used in a powder actuated tool (PAT). When actuated, it forces a pin, threaded stud, or other fastener into the concrete. SAFETY PRECAUTION: Do not try this without the training and the license to operate this equipment.

Many other types of fasteners are used to fasten material to concrete. Those indicated above are but a few examples.

6.5   OTHER FASTENERS

COTTER PINS.   *Cotter pins,* **Figure 6–20,** are used to secure objects on pins or shafts. The cotter pin is inserted through the hole in the pin, and the ends spread to retain it.

PIPE HOOK.   A wire pipe hook is a hardened steel wire bent into a U with a point at an angle on both ends. The pipe or tubing rests in the bottom of the U, and the pointed ends are driven into wooden joists or other wood supports, **Figure 6–21.**

PIPE STRAP.   A pipe strap is used for fastening pipe and tubing to joists, ceilings, or walls, **Figure 6–22(A).** The arc on the strap should be approximately the same as the outside diameter of the pipe or tubing. These straps are normally fastened with round head screws. Notice that the underside of the head of a round head screw is flat and provides good contact with the strap.

PERFORATED STRAP.   A perforated strap may also be used to support pipe and tubing. Round head stove bolts with nuts fasten the straps to themselves, and round head wood screws fasten the straps to wood supports, **Figure 6–22(B).**

NYLON STRAP.   To fasten round flexible duct to a sheet metal collar, apply the inner liner of the duct with a sealer and clamp it with a nylon strap. A special tool is manufactured to install this strap, which applies the correct tension and cuts the strap off. Now position the insulation and vapor barrier over the collar, which can then be secured with another nylon strap. Apply duct tape over this strap and duct end to further seal the system. **Figure 6–23** illustrates this procedure.

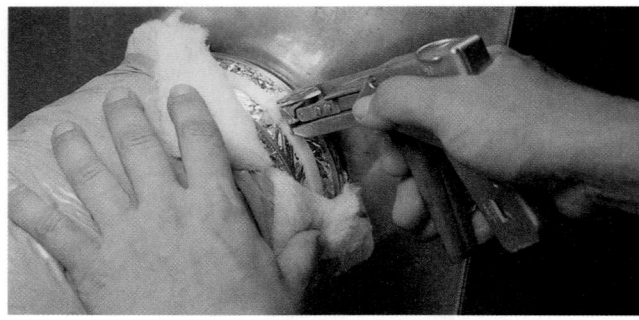

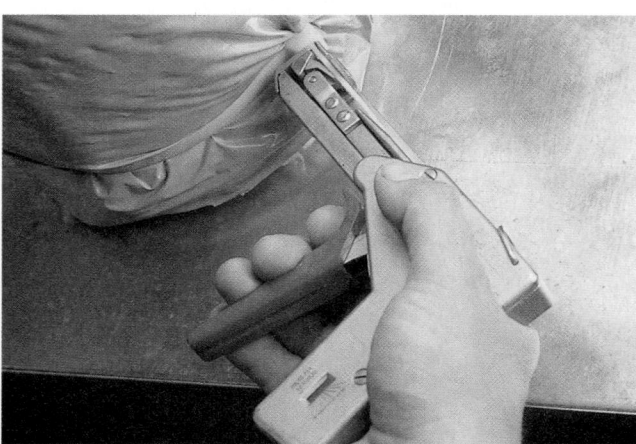

Figure 6–23   A system for fastening round flexible duct to a sheet metal collar. *Courtesy Panduit Corporation*

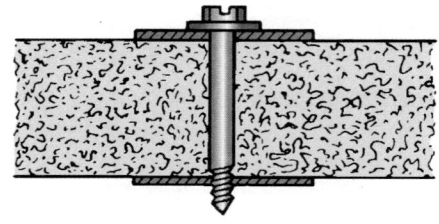

Figure 6–25   A system for fastening fiberglass duct.

**PRIMARY TYPES OF
SOLDERLESS TERMINALS**

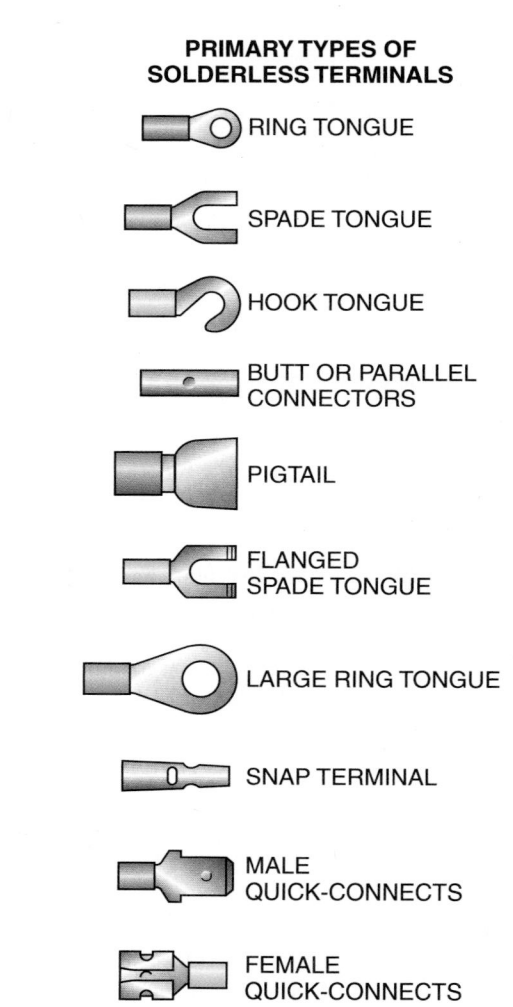

RING TONGUE

SPADE TONGUE

HOOK TONGUE

BUTT OR PARALLEL
CONNECTORS

PIGTAIL

FLANGED
SPADE TONGUE

LARGE RING TONGUE

SNAP TERMINAL

MALE
QUICK-CONNECTS

FEMALE
QUICK-CONNECTS

Figure 6–26   Typical solderless terminals. *Courtesy Klein Tools*

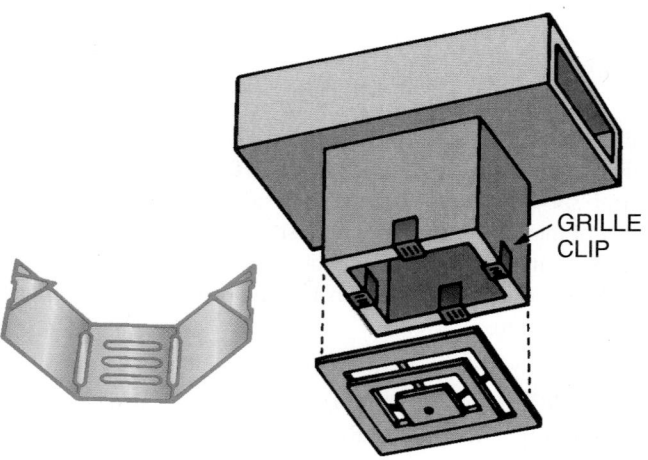

Figure 6–24   Fastening a grille to fiberglass duct with a grille clip.

GRILLE CLIPS.   Grille clips fasten grilles to the ends of fiberglass duct. The clips are bent around the end of the duct, with the points pushed into the sides of the duct, **Figure 6–24.** Screws fasten the grilles to these clips.

Figure 6–25 illustrates a system for fastening damper regulators, controls, and other components to fiberglass duct. It consists of a drill screw, a head plate, and a backup plate.

SOLDERLESS TERMINALS.   *Solderless terminals* are used to fasten stranded wire to various terminals or to connect two lengths of stranded wire together. Various styles of these terminals are shown in **Figure 6–26.**

Figure 6–27(A) shows a typical terminal before the wire has been inserted. To connect wire to a terminal, strip a short end of the insulation. A stripping and cutting tool similar to the one in **Figure 6–28(A)** may be used. Insert wire into the terminal as indicated in **Figure 6–27(B).** On most terminals a double crimp can be made using a tool similar to that shown in **Figure 6–28(B).** One crimp is made over the bare wire and one over the insulation.

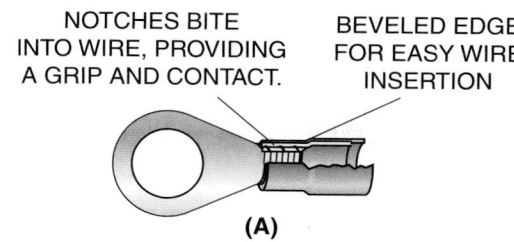

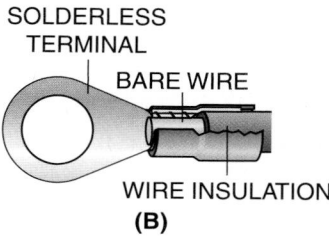

Figure 6–27 **(A)** The cutaway view of a typical solderless connector. **(B)** A connector with wire inserted. *Courtesy Klein Tools*

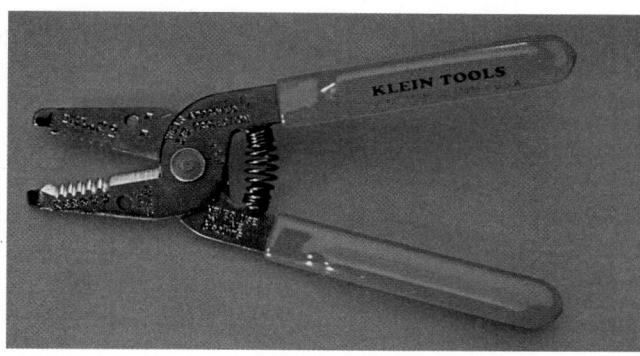

(A)

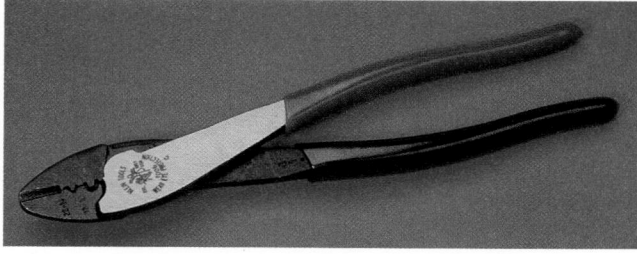

(B)

Figure 6–28 **(A)** A wire stripping and cutting tool. **(B)** A crimping tool. *Courtesy Klein Tools*

SCREW-ON WIRE CONNECTORS. Screw-on wire connectors, **Figure 6–29,** are used to connect two or more wires together. They come in various sizes for different sizes and numbers of wires. Be sure to follow the manufacturer's instructions when installing these connectors. Following is a summary of one or more manufacturers' instructions:
1. Strip each wire to the depth of the connector.
2. Insert the wires into the connector and twist it on, **Figure 6–30.**

Figure 6–29 Screw-on wire connectors. *Courtesy Klein Tools*

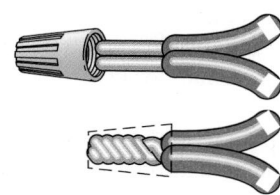

Figure 6–30 Installing screw-on wire connectors. *Courtesy Klein Tools*

3. The screw action will twist the wires together, pressing them into the threads of the connector and forming a strong bond.
SAFETY PRECAUTION: Ensure that the connector completely covers the bare wire. Refrigeration and air-conditioning technicians normally will use these connectors on control voltage or other low-voltage circuits unless licensed for higher voltage applications. It is important that the proper size connector is used.

Many fasteners not described here are highly specialized. More are developed each year. The intent of this unit is to depict the more common ones and to encourage you to keep up to date by discussing new fastening techniques with your supplier.

SUMMARY

- Technicians need to use a broad variety of fasteners.
- Some of these fasteners are nails, screws, staples, tapping screws, and set screws.
- Anchor shields and other devices are often used with screws to secure them in masonry walls or hollow walls.
- Typical hangers used with pipe, tubing, and ductwork are wire pipe hooks, pipe straps, perforated steel straps, and custom hangers made from threaded rod and angle steel.
- Other specialty fasteners are available for flexible duct and fiberglass duct.
- Solderless terminals are used to fasten wire to terminals.
- Screw-on wire connectors are used to fasten two or more lengths of wire together.

REVIEW QUESTIONS

1. Three types of nails are _____, _____, and _____.
2. The term _____ is used to indicate the size of a common nail.
3. What is the abbreviation of the term in question 2?

4. An outward clinch tacker is used to
   A. bend wire on a pipe hook.
   B. install a pin rivet.
   C. drive roofing nails.
   D. spread the ends of a staple outward.
5. Wood screws generally have
   A. a flat head
   B. a round head.
   C. an oval head.
   D. any of the above.
6. What are masonry nails made from?
7. Describe two types of staples.
8. Describe the procedure for fastening two pieces of metal together with a pin rivet.
9. Write a typical dimension for a tapping screw. Explain each part of the dimension.
10. Sketch three types of machine screw heads.
11. Write a typical machine screw thread dimension and explain each part of the dimension.
12. Describe the procedure for fastening two pieces of sheet metal together with a tapping screw.
13. Hollow rivets are assembled in a pin that is often called
    A. a mandrel.
    B. a nozzle.
    C. an anchor.
    D. a clinch.
14. Describe two types of tapping screws.
15. Three types of set screw points are _____, _____, and _____.
16. How are anchor shields used?
17. What is a pipe strap? What is it used for?
18. What is a grille clip used for?
19. Describe the procedure for fastening three wires together using the screw-on wire connectors.

OBJECTIVES

After studying this unit, you should be able to

- list the different types of tubing used in heating, air-conditioning, and refrigeration applications.
- describe two common ways of cutting copper tubing.
- list procedures used for bending tubing.
- discuss procedures used for soldering and brazing tubing.
- describe two methods for making flared joints.
- state procedures for making swaged joints.
- describe procedures for preparing and threading steel pipe ends.
- list four types of plastic pipe and describe uses for each.
- describe alternative, mechanical methods for joining pipe sections.

SAFETY CHECKLIST

- ✔ Use care while reaming tubing; the burr can stick in your hand.
- ✔ Be careful while using a hacksaw; the blade is sharp.
- ✔ Do not allow your bare skin near the flame of a torch or near material that has just been soldered, brazed, or welded; the 5000°F and higher temperatures can cause severe burns.
- ✔ Always wear eye protection while performing any task where particles may be in the air.
- ✔ Use extra caution with the oxyacetylene heat source. SAFETY PRECAUTION: Do not allow any oil around the fittings or allow it to enter the hoses or regulators.
- ✔ Be aware of flammable materials in the area where you are soldering or brazing.
- ✔ Pipe cutting and threading creates burrs that can cut your skin and otherwise be dangerous; use caution.
- ✔ Do not breathe excessive amounts of the adhesive used for plastic fittings.

7.1 PURPOSE OF TUBING AND PIPING

🌐The correct size, layout, and installation of tubing, piping, and fittings helps to keep a refrigeration or air-conditioning system operating properly and efficiently and prevents refrigerant loss.🌐 The piping system provides passage for the refrigerant to the evaporator, the compressor, the condenser, and the expansion valve. It also provides the way for oil to drain back to the compressor. Tubing, piping, and fittings are used in numerous applications, such as fuel lines for oil and gas burners and water lines for hot water

heating. The tubing, piping, and fittings must be of the correct material and the proper size; the system must be laid out properly and installed correctly.

SAFETY PRECAUTION: Careless handling of the tubing and poor soldering or brazing techniques may cause serious damage to system components. You must keep contaminants, including moisture, from air-conditioning and refrigeration systems.

7.2 TYPES AND SIZES OF TUBING

Copper tubing is generally used for plumbing, heating, and refrigerant piping. Steel and wrought iron pipe are used for gas piping and frequently for hot water heating. Plastic pipe is often used for waste drains, condensate drains, water supplies, water-source heat pumps, and venting high-efficiency gas furnaces.

Copper tubing is available as soft- or hard-drawn copper. Soft copper tubing, **Figure 7–1,** may be bent or used with elbows, tees, and other fittings. Hard-drawn tubing is not intended to be bent and is used only with fittings to obtain the necessary configurations.

This tubing is available in four standard weights: K, L, M and DWV. K-type copper is heavy duty with a thick tubing wall. L-type copper is considered medium weight and is the

Figure 7–1 Soft-drawn copper tubing typically comes in 50-foot rolls. Note that the ends of the rolls are capped to keep the interior of the tubing clean.

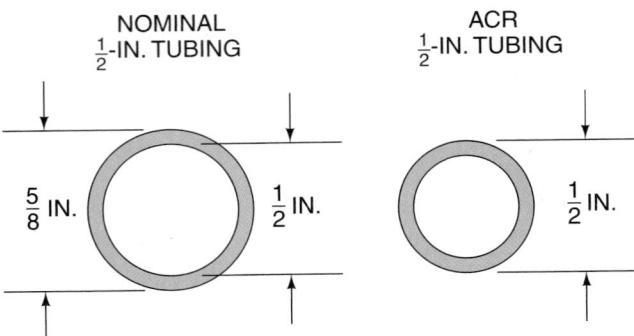

NOMINAL
$\frac{1}{2}$-IN. TUBING

ACR
$\frac{1}{2}$-IN. TUBING

$\frac{5}{8}$ IN. $\frac{1}{2}$ IN.

$\frac{1}{2}$ IN.

Figure 7–2   Copper tubing (nominal) used for plumbing and heating is sized by its inside diameter. ACR tubing is sized by its outside diameter.

type used most frequently. M is a lightweight copper and is not commonly used in our industry. DWV copper stands for drain, waste, and vent and is also a lightweight copper not typically used in the HVACR industry as far as refrigerant-carrying applications are concerned.

Copper tubing used for refrigeration or air conditioning is called air-conditioning refrigeration (ACR) tubing and is sized by its outside diameter (OD). Therefore, if a specification calls for 1/2-in. ACR tubing, the OD would measure 1/2 in., **Figure 7–2.**

Tubing used for plumbing and heating applications, called nominal-sized tubing, is sized by its inside diameter (ID). The OD for most applications in the plumbing and heating industry will be 1/8 in. larger than the size indicated. A 1/2-in. nominal-sized tubing would have an OD of 5/8 in., **Figure 7–2.** Copper tubing normally is available in diameters from 3/16 in. to greater than 6 in.

Soft copper tubing is normally available in 25-ft or 50-ft rolls and in diameters from 3/16 in. to 3/4 in. It can be special ordered in 100-ft lengths. ACR tubing is capped on each end to keep it dry and clean inside and often has a charge of nitrogen to keep it free of contaminants. Proper practice should be used to remove tubing from the coil. Never uncoil the tubing from the side of the roll. Place it on a flat surface and unroll it, **Figure 7–3.** Cut only what you need and recap the ends, **Figure 7–1.**

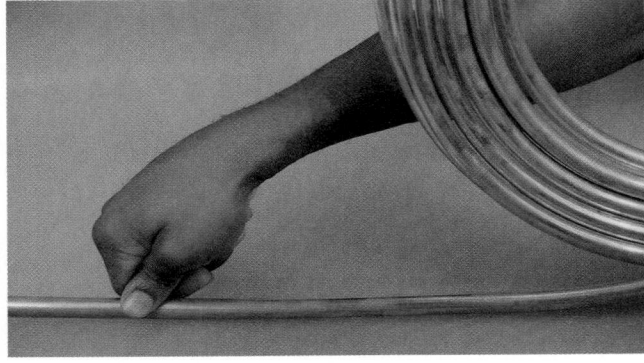

Figure 7–3   A roll of soft-drawn copper tubing. Place on a flat surface and unroll. *Photo by Bill Johnson*

Do not bend or straighten the tubing more than necessary because it will harden. This is called *work hardening*. Work-hardened tubing can be softened by heating and allowing it to cool slowly. This is called *annealing*. When annealing, do not use a high concentrated heat in one area, but use a flared flame over 1 ft at a time. Heat to a cherry red and allow to cool slowly.

Hard-drawn copper tubing is available in 20-ft lengths and in larger diameters than soft copper tubing. Be as careful with hard-drawn copper as with soft copper, and recap the ends when the tubing is not used.

7.3   TUBING INSULATION

ACR tubing is often insulated on the low-pressure side of an air-conditioning or refrigeration system between the evaporator and compressor to keep the refrigerant from absorbing excess heat, **Figure 7–4.** Insulation also prevents condensation from forming on the lines. The closed-cell structure of this insulation eliminates the need for a vapor barrier. The insulation may be purchased separately from the tubing, or it may be factory installed. If you install the insulation, it is easier, where practical, to apply it to the tubing before assembling the line. The ID of the insulation is usually powdered to allow easy slippage even around most bends. You can buy adhesive to seal the ends of the insulation together, **Figure 7–5.**

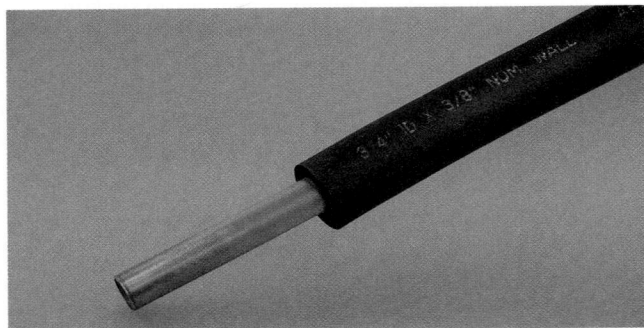

Figure 7–4   ACR tubing with insulation. *Photo by Bill Johnson*

Figure 7–5   When joining two ends of tubing insulation, use an adhesive intended for this purpose. *Photo by Bill Johnson*

Figure 7–6 A typical line set. *Photo by Bill Johnson*

(A)

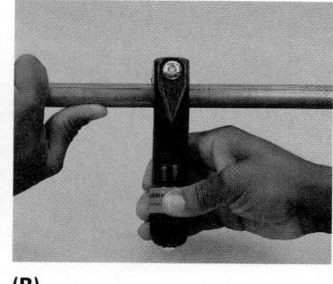
(B)

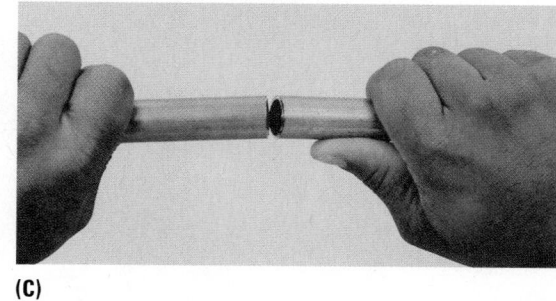
(C)

Figure 7–7 The proper procedure for using a tubing cutter **(A)–(C)**. *(A) Photo by Eugene Silberstein. (B) and (C) Photos by Bill Johnson*

For existing lines, or when it is impractical to insulate before installing the tubing, the insulation can be slit with a sharp utility knife and snapped over the tubing. All seams must be sealed with adhesive. Do not use tape.

Do not stretch tubing insulation because the wall thickness of the insulation will be reduced, which will reduce the insulation effectiveness, and the adhesive may fail to hold.

7.4   LINE SETS

Tubing can be purchased as line sets. One type of line set is precharged with refrigerant, sealed on both ends, and may be obtained with insulation installed on the suction line. These line sets normally will have quick-connect fittings on each end for quicker and cleaner field installation, **Figure 7–6**.

Another type of line set is charged with nitrogen, has an insulated suction line, and may be purchased with various types of fittings. This type may be cut to size, and other types of fittings may be used. Before completing the installation, the nitrogen will need to be evacuated.

7.5   CUTTING TUBING

Tubing is normally cut with a tube cutter or a hacksaw. The tube cutter is most often used with soft tubing and smaller diameter hard-drawn tubing. A hacksaw may be used with larger diameter hard-drawn tubing. To cut the tubing with a tube cutter, follow these steps, as shown in **Figure 7–7**.

A. Place the tubing in the cutter and align the cutting wheel with the cutting mark on the tube. Tighten the adjusting screw until a moderate pressure is applied to the tubing. When properly tightened, the cutter should be held in place on the tubing and there should be no cutter wheel indentations in the copper tubing. Overtightening the cutter on the tubing will result in oblong or out-of-round edges on the tubing that can cause refrigerant leaks.

B. Revolve the cutter around the tubing, keeping a moderate pressure applied to the tubing by gradually turning the adjusting screw.

C. Continue until the tubing is cut. *Do not apply excessive pressure because it may break the cutter wheel and constrict the opening in the tubing.*

When the cut is finished, the excess material (called a *burr*) pushed into the pipe by the cutter wheel must be removed, **Figure 7–8**. Burrs cause turbulence and restrict the fluid or vapor passing through the pipe. When removing burrs from the end of a cut pipe section, make every attempt to prevent the burrs from falling into the pipe. Burrs that remain in the system can cause restrictions to refrigerant flow and affect system operation and performance.

To cut the tubing with a hacksaw, make the cut at a 90° angle to the tubing. A fixture may be used to ensure an accurate cut, **Figure 7–9**. After cutting, ream the tubing and file the end. Remove all chips and filings, making sure that no debris or metal particles get into the tubing. Since the process of using a hacksaw under normal conditions produces a large amount of copper filings and burrs, cutting piping materials with this method is not preferable when using a tubing cutter is an option. Given certain conditions, however, it may be the only choice available.

7.6   BENDING TUBING

In an effort to keep refrigerant in the system, the number of solder joints and mechanical fittings should be kept to a minimum. Each solder joint, fitting, and mechanical connection represents a potential leak. One way to reduce the number of solder joints in a piping system is to bend the tubing material into the desired configuration. Typically, only soft tubing

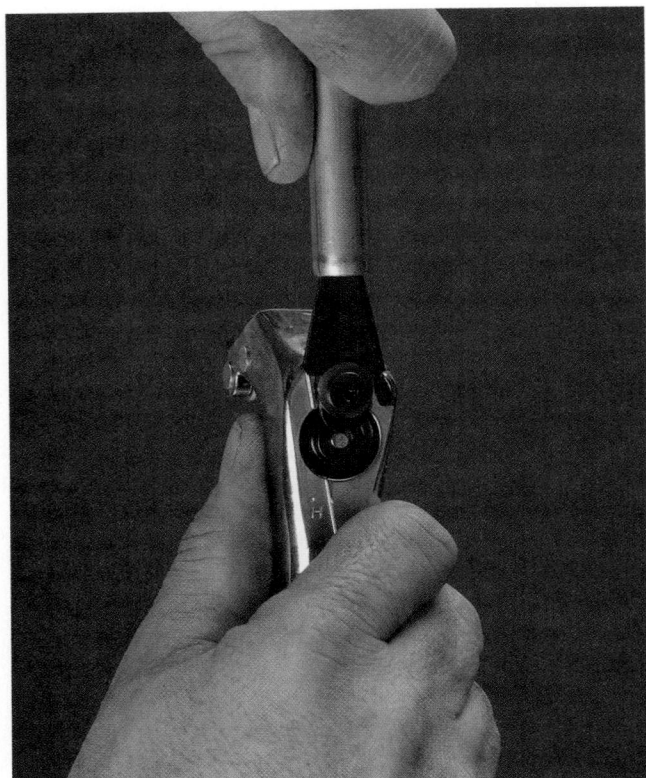

Figure 7–8   Removing burrs from inside the tubing.

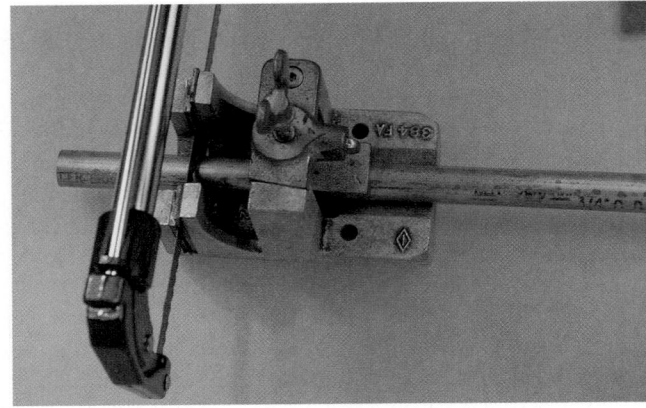

(A)

(B)

Figure 7–9   The proper procedure for using a hacksaw **(A)–(B)**.
Photos by Bill Johnson

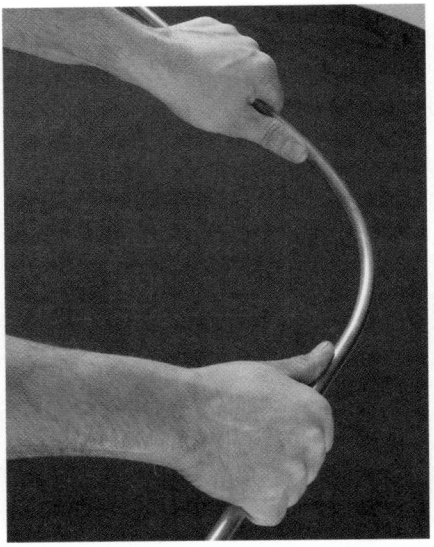

(A)

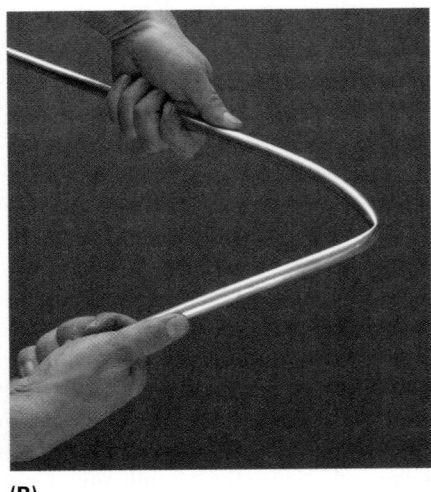

(B)

Figure 7–10   **(A)** Small-diameter tubing can be bent by hand.
(B) Larger tubing can easily kink when bent by hand.

should be bent. Use as large a radius bend as possible, **Figure 7–10(A)**. All areas of the tubing must remain round. *Do not allow it to flatten or kink,* **Figure 7–10(B).** Carefully bend the tubing, gradually working around the radius.

Tube bending springs may be used to help make the bend, **Figure 7–11.** They can be used either inside or outside the tube. They are available in different sizes for different diameters of tubing, **Figure 7–12.** To remove the spring after the bend, you might have to twist it. If you use a spring on the OD, bend the tube before flaring so that the spring may be removed.

Lever-type tube benders, **Figure 7–13,** which are available in different sizes, are used to bend soft copper, aluminum and thin-walled steel tubing. There are many different types of benders available; what follows, however, is a brief explanation of the operation of the lever-type bender.

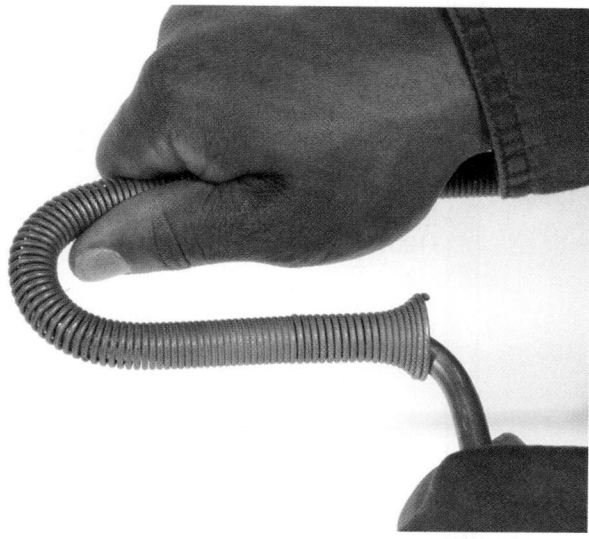

Figure 7–11 Tube bending springs are used to help prevent kinking the tubing while bending. Be sure to use the proper size. *Photo by Bill Johnson*

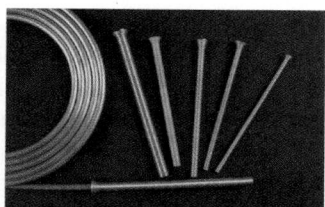

Figure 7–12 Different-sized bending springs.

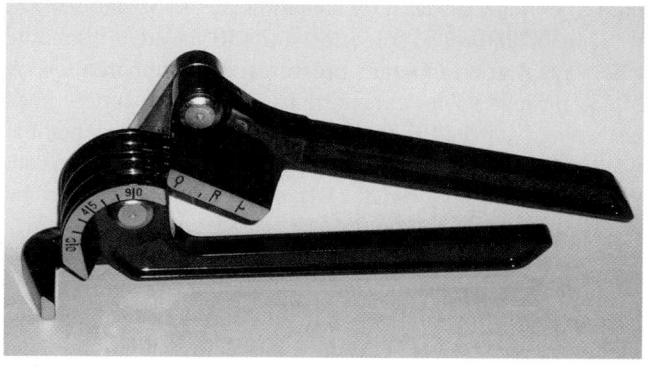

Figure 7–13 Lever-type tubing bender. *Photo by Eugene Silberstein*

The lever-type bender is made up of two parts, the body and the lever, **Figure 7–14.** The body of the bender will have markings on it that represent the amount of bend, in degrees, that will be put into a tubing section if the section is bent up to that point. The lever also has markings. The lever markings indicate the proper position of the lever when the bend is started as well as the proper position for the tubing when the bend is started. The tubing position in the bender depends on whether the piping is extending from the left or the right side of the bender.

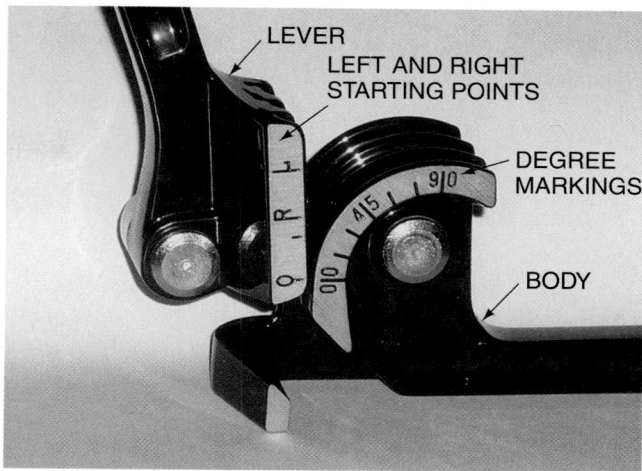

Figure 7–14 Markings on the lever and body of the lever-type tubing bender. *Photo by Eugene Silberstein*

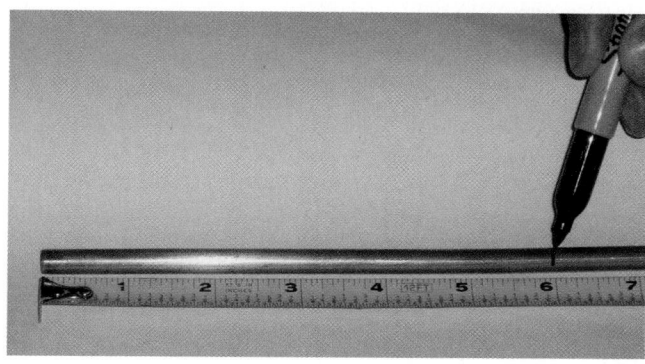

Figure 7–15 Mark tubing at the desired point. *Photo by Eugene Silberstein*

Let's say we wanted to put a 90-degree bend in a piece of soft-drawn copper tubing so that the length from the unbent end of the tubing to the center of the bent end is 6 inches. In this case, we would first measure and mark the tubing at the 6-inch point, **Figure 7–15.** Once marked, we insert the tubing into the bender and line up the mark with the "L," which indicates that the measured portion of tubing is extending from the left side of the bender, **Figure 7–16.** We also want to make certain that the "0" on the lever lines up with the "00" on the body of the bender. This ensures that the lever is in the proper position prior to bending.

While holding the handle on the body of the bender, we pull on the lever until the "0" on the lever is lined up with the "90" on the body. This ensures that the bend will be 90 degrees, **Figure 7–17.**

If, however, we wanted to make a 45-degree bend in the tubing instead of a 90-degree bend, the lever would be positioned so that the "0" on the lever is lined up with the "45" on the body, **Figure 7–18.** Be sure to practice on smaller, scrap sections of tubing to gain confidence in creating tubing bends at the desired positions. It takes a little time to get it just right, but the results will be well worth the effort.

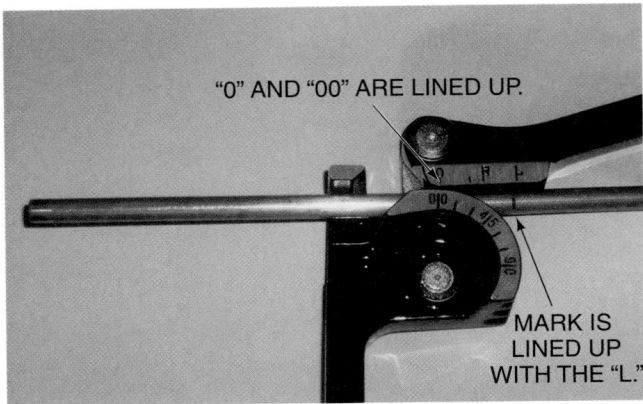

Figure 7–16 Proper tubing and lever positions at the beginning of the bending process. *Photo by Eugene Silberstein*

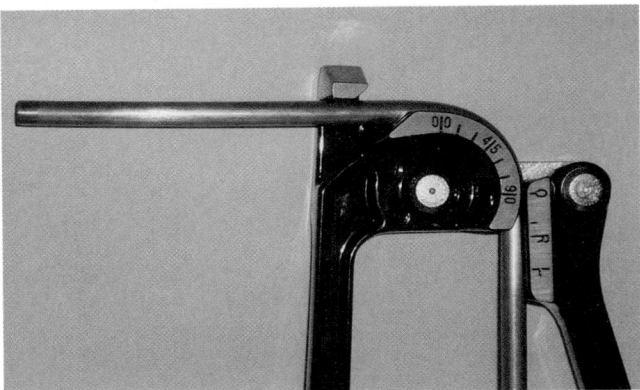

Figure 7–17 Completing a 90-degree bend. *Photo by Eugene Silberstein*

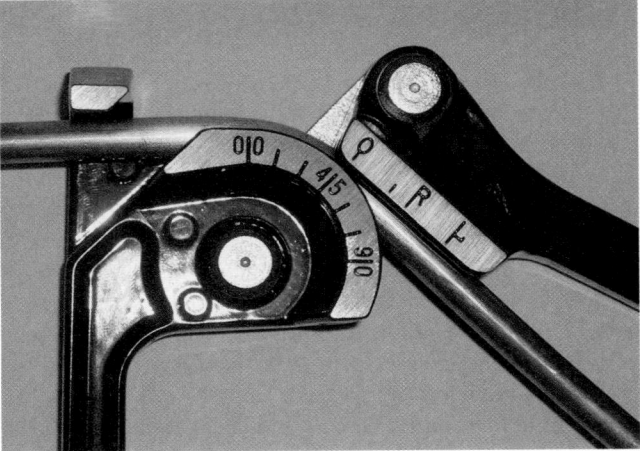

Figure 7–18 A 45-degree bend. *Photo by Eugene Silberstein*

7.7 SOLDERING AND BRAZING PROCESSES

Soldering is a process used to join piping and tubing to fittings. It is used primarily in plumbing and heating systems utilizing copper and brass piping and fittings. Soldering,

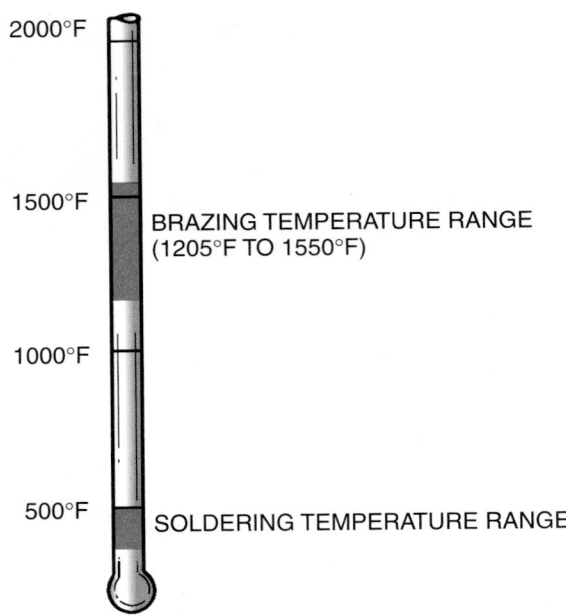

Figure 7–19 Temperature ranges for soldering and brazing.

often called soft soldering, is done at temperatures under 800°F, usually in the 361°F to 500°F range, **Figure 7–19.**

When soldering or brazing, there is a gap between the two surfaces that are fastened together. For example, when two lengths of copper tubing are fastened together, a fitting or coupling may be placed between them into which each pipe slides. A clearance is required for the tubing to slide into the coupling or fitting. A filler metal must be used to fill this clearance. The filler metal actually adheres to the tubing and fitting to be joined as it fills the gap. Only the filler metal melts. Two common filler materials used for soldering are 50/50 and 95/5.

The 50/50 tin-lead solder, 50% tin-50% lead, is a suitable filler metal for moderate pressures and temperatures. A 50/50 tin-lead solder cannot be used for water supply lines because of its lead content. For higher pressures potable water supplies, or where greater joint strength is required, 95/5 tin-antimony solder, 95% tin-5% antimony, can be used.

Brazing, requiring higher temperatures, is often called silver brazing or hard soldering and is similar to soldering. It is used to join tubing and piping in air-conditioning and refrigeration systems. Do not confuse this with welding brazing. In brazing processes, temperatures over 800°F are used. The differences in temperature are necessary due to the different combinations of alloys used in the filler metals.

Brazing filler metals suitable for joining copper tubing are alloys containing 15% to 60% silver (BAg) or copper alloys containing phosphorous (BCuP). Brazing filler metals are sometimes referred to as hard solders or silver solders.

In soldering and brazing, the base metal (the piping, tubing, and/or fitting) is heated to the melting point of the filler material. *The piping and tubing must not melt.* When two close-fitting, clean, smooth metals are heated to the point where the filler metal melts, this molten metal is drawn into the close-fitting space by *capillary attraction*

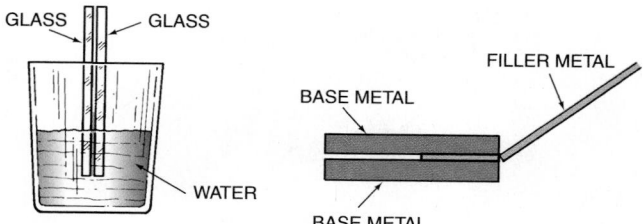

Figure 7–20 Two examples of capillary attraction. On the left are two pieces of glass spaced closely together. When the pieces of glass are inserted in water, capillary attraction draws the water into the space between them. The water molecules have a greater attraction for the glass than they do for each other. On the right is an illustration showing molten filler material being drawn into the space between the two pieces of base metal. The molecules in the filler material have a greater attraction for the base metal than they do for each other. These molecules work their way along the joint, first "wetting" the base metal and then filling the joint.

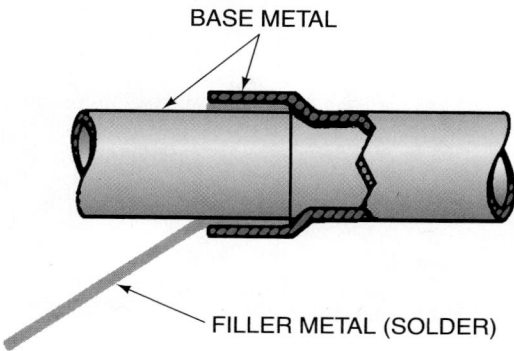

Figure 7–21 The molten solder in a soldered joint will be absorbed into the surface pores of the base metal.

Figure 7–22 A propane torch with a typical soldering tip. *Photo by Bill Johnson*

(see **Figure 7–20** for an explanation of this). If the soldering is properly done, the molten solder will be absorbed into the pores of the base metal, adhere to all surfaces, and form a bond, **Figure 7–21.**

7.8 HEAT SOURCES FOR SOLDERING AND BRAZING

Propane, butane, air-acetylene, or oxyacetylene torches are the most commonly used sources of heat for soldering or brazing. A *propane* or *butane* torch can be ignited easily and adjusted to the type and size of joint being soldered. Various tips are available, **Figure 7–22.**

An *air-acetylene* unit is a type of heat source used often by air-conditioning and refrigeration technicians. It usually consists of a B tank of acetylene gas, a regulator, a hose, and a torch, **Figure 7–23.** Various sizes of standard tips are available for a unit like this. The smaller tips are used for small-diameter tubing and for high-temperature applications. A high-velocity tip may be used to provide more concentration of heat. **Figure 7–24** illustrates a popular high-velocity tip.

Follow these procedures when setting up, igniting, and using an air-acetylene unit, **Figure 7–25.**

Figure 7–23 A typical air-acetylene torch setup. *Photo by Bill Johnson*

A. Before connecting the regulator to the tank, quickly open and close the tank valve slightly to blow out any dirt that may be lodged at the valve. SAFETY PRECAUTION: Stand back from the tank and wear safety goggles when blowing dirt from the tank valve.

B. Connect the regulator with hose and torch to the tank. Be sure that all connections are tight.

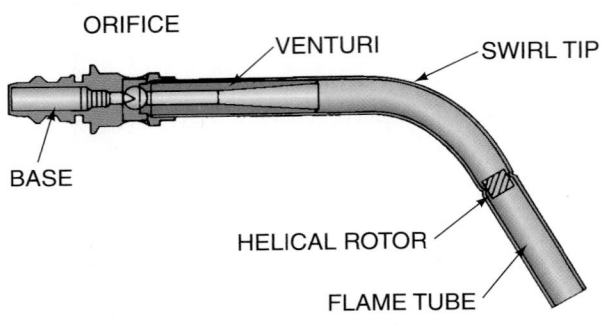

Figure 7–24 A high-velocity torch tip. *Courtesy Thermadyne Industries, Inc.*

(A)

(B)

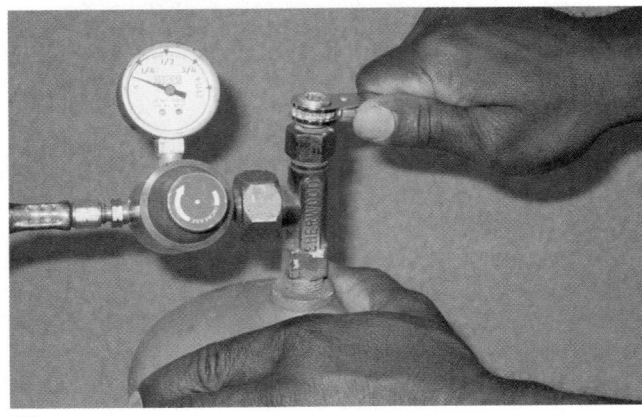

(C)

(D)

(E)

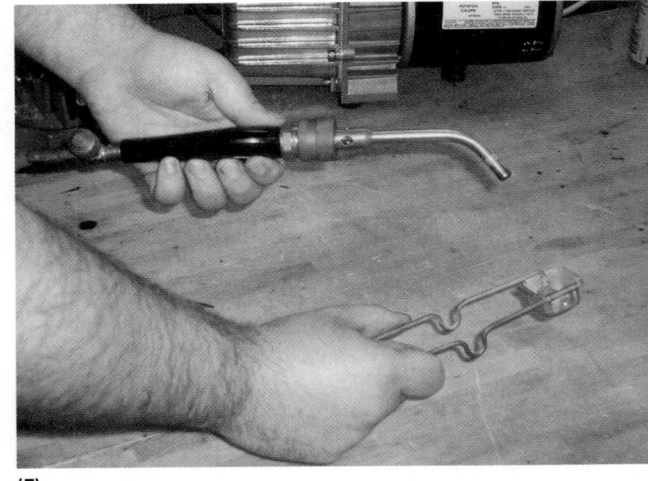

(F)

(G)

Figure 7–25 The proper procedures for setting up, lighting, and using an air-acetylene torch. *Photos (A), (B), (C), (D), and (G) by Bill Johnson. Photos (E) and (F) by Eugene Silberstein*

C. Open the tank valve one-half turn. SAFETY PRECAUTION: There is no need to open the tank valve more than one-half turn as this will give adequate flow. Leave the wrench on the tank valve stem. The valve can then be quickly closed in the event of an emergency.

D. Adjust the regulator valve to about midrange.

E. Using a soap-bubble solution, check for leaks on the hose, regulator, and torch handle.

F. Open the needle valve on the torch slightly, and ignite the gas with the spark lighter. SAFETY PRECAUTION: Do not use matches or cigarette lighters.

G. Adjust the flame using the needle valve at the handle so that there will be a sharp inner flame and a blue outer flame. After each use, shut off the valve on the tank and bleed off the acetylene in the hose by opening the valve on the torch handle. Bleeding the acetylene from the hoses relieves the pressure when the hoses are not in use.

Oxyacetylene torches may be preferred by some technicians, particularly when brazing large-diameter tubing or other applications requiring higher temperatures. The addition of pure oxygen produces a much hotter flame. SAFETY PRECAUTION: This equipment can be extremely dangerous when not used properly. It is necessary that you thoroughly understand proper instructions for using oxyacetylene equipment before attempting to use it. When you first begin to use the equipment, use it only under the close supervision of a qualified person.

Following is a brief description of oxyacetylene welding and brazing equipment. Oxyacetylene brazing and welding processes use a high-temperature flame. Oxygen is mixed with acetylene gas to produce this high heat. The equipment includes oxygen and acetylene cylinders, oxygen and acetylene pressure regulators, hoses, fittings, safety valves, torches, and tips, **Figure 7–26.**

The regulators each have two gages, one to register tank pressure and the other to register pressure to the torch. The pressures indicated on these regulators are in pounds per square inch gage (psig). **Figure 7–27** is a photo of an oxygen regulator and an acetylene regulator. These regulators must

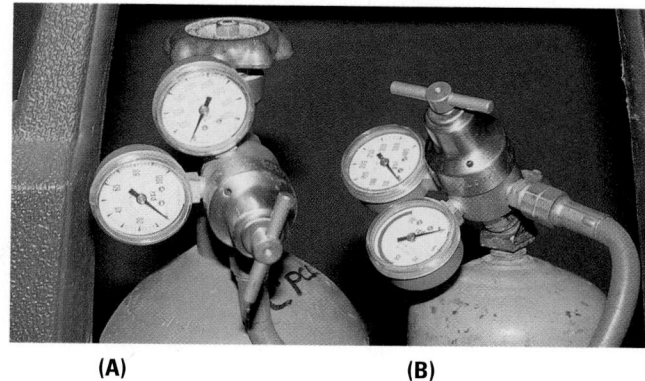

(A) **(B)**

Figure 7–27 **(A)** An oxygen regulator. **(B)** An acetylene regulator. *Photo by Bill Johnson*

be used only for the gases and service for which they are intended. SAFETY PRECAUTION: All connections must be free from dirt, dust, grease, and oil. Oxygen can produce an explosion when in contact with grease or oil. A reverse flow valve should be used somewhere in the hoses. These valves allow the gas to flow in one direction only and prevent the two gases from mixing in the hose, which could be very dangerous. **Figure 7–28(A)** shows the reverse flow check valve. Follow instructions when attaching these valves because some are designed to attach to the hose connection on the torch body and some to the hose connection on the regulator.

The red hose is attached to the acetylene regulator with left-handed threads, and the green hose is attached to the oxygen regulator with right-handed threads.

The torch body is then attached to the hoses and the appropriate tip to the torch, **Figure 7–28(A).** Many tip sizes and

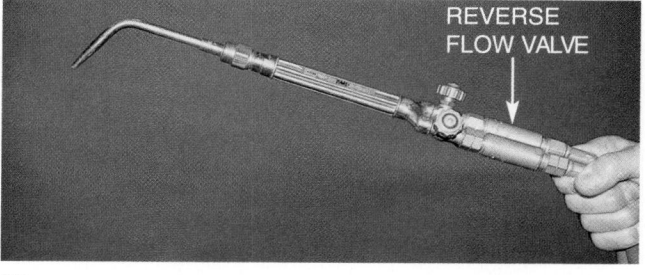

(A)

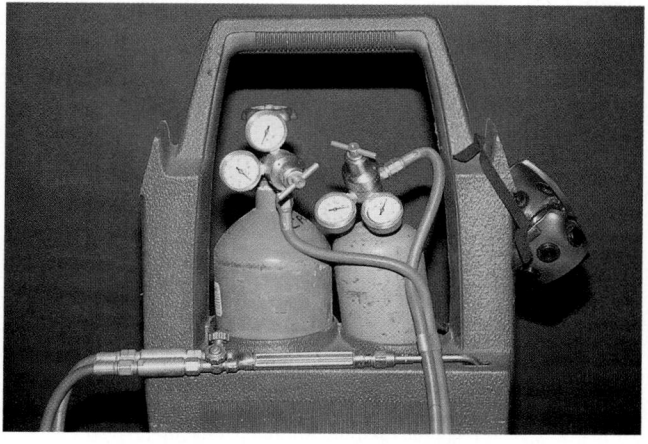

Figure 7–26 Oxyacetylene equipment. *Photo by Bill Johnson*

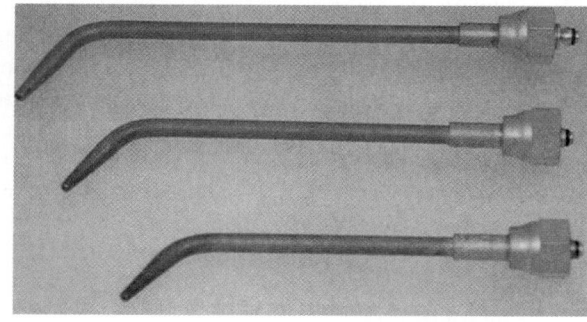

(B)

Figure 7–28 **(A)** Reverse flow check valve with an oxyacetylene torch body and tip. **(B)** An assortment of oxyacetylene torch tips. *Photos by Bill Johnson*

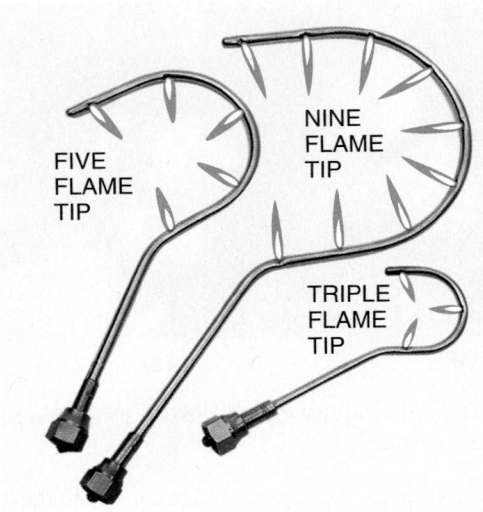

Figure 7–29 A tip that can be used for heating the entire circumference of a fitting at the same time. *Courtesy Uniweld Products, Inc.*

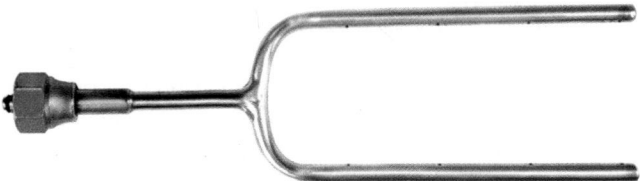

Figure 7–30 A tip that can be used for removing and replacing the four-way reversing valve on a heat pump system. *Courtesy Uniweld Products, Inc.*

styles are available, **Figure 7–28(B).** Tips have been developed that provide heat around the tubing and fittings. This allows the entire circumference of the tubing to be heated at one time, **Figure 7–29.** Another tip is available to remove and replace a four-way or reversing valve for a heat pump, **Figure 7–30.** All three of the copper fittings from this valve can be heated simultaneously, and the tubing can be pulled from the valve when the brazing alloys reach the proper temperature. You should learn through your specialized training in how to use this equipment which tips to use for particular applications.

The following procedure may be used to set up an oxygen acetylene torch system for use. SAFETY PRECAUTION: Always turn the fuel gas (acetylene) on or off first. With the regulators and hoses fastened to the tank and the torch tip and the T handles on the regulators turned counterclockwise so that no pressure will go to the hoses (they may feel loose):

1. Turn on the acetylene cylinder valve one-half turn so that pressure is introduced to the regulator. SAFETY PRECAUTION: Always stand to the side of the regulator. If the regulator were to blow out, it would likely blow toward the T handle. The wrench should remain on the stem of the cylinder valve so the valve can be quickly turned off in case of an emergency. You should see the pressure register on the acetylene cylinder gage, **Figure 7–31.**

Figure 7–31 Pressure registers on the acetylene cylinder gage. *Photo by Bill Johnson*

2. Slowly turn on the oxygen cylinder valve to the oxygen regulator. SAFETY PRECAUTION: Again, stand to the side. You should see the pressure register on the oxygen cylinder pressure gage, **Figure 7–32.**

3. Slightly open the valve on the torch handle that controls the acetylene (the red hose). With this valve slightly open, adjust the T handle on the acetylene regulator until the gage on the red hose reads 5 psig, **Figure 7–33.** SAFETY PRECAUTION: Now shut off the valve at the torch. This part of the torch is ready.

4. Slightly open the valve on the torch handle that controls the oxygen (the green hose). With the valve slightly open, adjust the T handle on the oxygen regulator until the gage reads 10 psig, **Figure 7–34.** SAFETY PRECAUTION: Now shut off the valve at the torch. This part of the torch is ready.

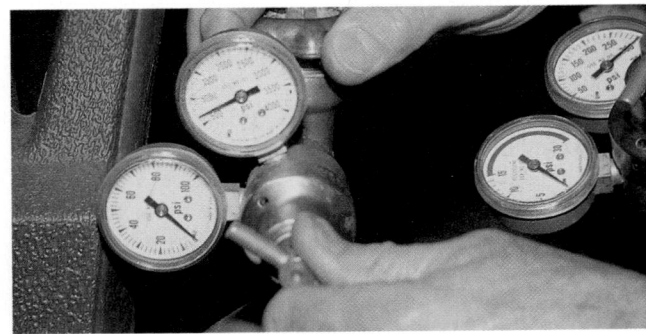

Figure 7–32 Pressure registers on the oxygen cylinder gage. *Photo by Bill Johnson*

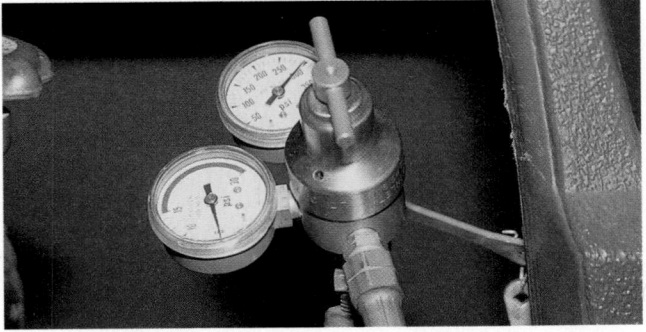

Figure 7–33 The acetylene regulator gage indicates 5 psig. *Photo by Bill Johnson*

Figure 7–34 This oxygen regulator gage indicates 10 psig. *Photo by Bill Johnson*

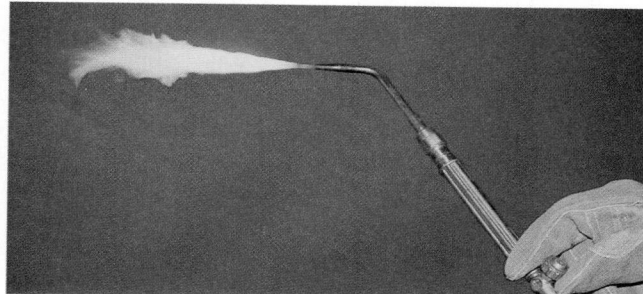

Figure 7–35 An acetylene flame.

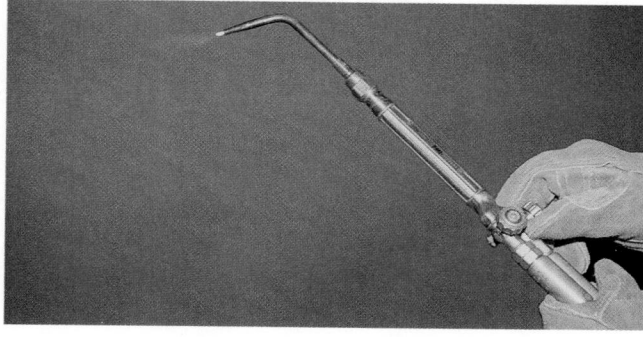

Figure 7–36 An oxygen-acetylene (oxyacetylene) flame.

The following procedure may be used to light the torch for use. SAFETY PRECAUTION: Before lighting, always point the torch tip away from you or any flammable substance. Make sure that you point the torch away from the cylinders and hoses. Failure to do so could result in tragedy. While holding the torch handle in your left hand:

1. Slightly open the valve on the torch handle on the acetylene side and allow it to flow a moment to get any air out of the line.
2. Using only an approved lighter, light the fuel gas, acetylene. You will have a large yellow-orange flame that smokes, **Figure 7–35.**
3. Slightly open the valve on the torch handle on the oxygen side. The flame will begin to clear up and turn blue, **Figure 7–36.** You will have to adjust the valves until you get the flame you want. The flame should be blue and be setting firmly on the torch tip, not blowing away from it.

To shut the system down, follow these procedures:

1. Shut off the fuel gas (acetylene) valve at the torch first.
2. Shut off the oxygen valve at the torch.
3. Turn off the cylinder valve on the fuel gas (acetylene).
4. Open the valve on the torch handle on the fuel gas side to relieve the pressure on the hose. Both regulators will lose their pressure to 0 psig.
5. Turn the T handle on the fuel gas regulator counter-clockwise until it appears to be loose. This side of the system has now been bled of pressure. Shut off the valve on the torch handle.
6. Turn off the cylinder valve on the oxygen cylinder.
7. Open the valve on the torch handle on the oxygen side to relieve the pressure on the hose. Both regulators will lose their pressure to 0 psig. This side of the system has been bled of pressure. Turn off the torch handle valve on the torch handle.
8. Turn the T handle on the oxygen regulator counter-clockwise until it feels loose.

The system is now ready for storage. SAFETY PRECAUTION: If the oxygen acetylene system is to be transported, it must be moved while in the preceding mode using an approved hand cart or carrier with the tanks secured. If it is to be transported on the road in a truck, caps must be used on the cylinders.

A neutral flame should be used for most operations. See **Figure 7–37** for photos of a neutral flame, a carbonizing

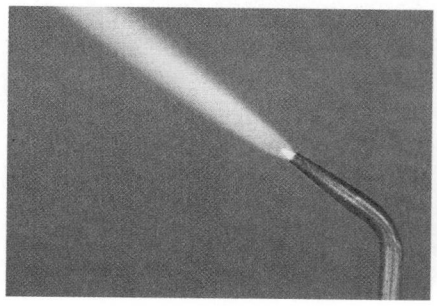

(A)

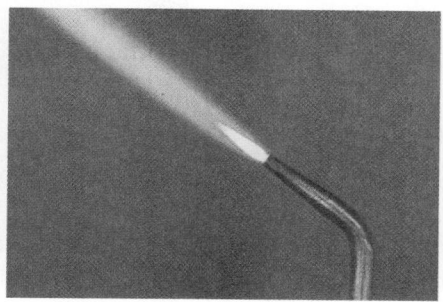

(B)

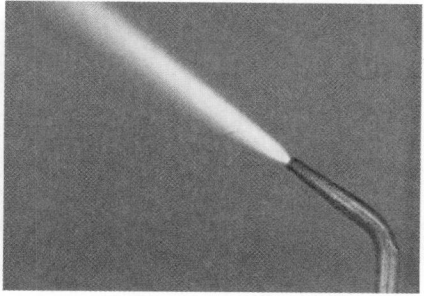

(C)

Figure 7–37 **(A)** A neutral flame. **(B)** A carbonizing flame. **(C)** An oxidizing flame. *Photos by Bill Johnson*

flame (too much acetylene), and an oxidizing flame (too much oxygen).

SAFETY PRECAUTION: As stated previously there are many safety precautions that should be followed when using oxyacetylene equipment. Be sure that you follow all of these precautions. You should become familiar with them during your training in how to use this equipment.

7.9 SOLDERING TECHNIQUES

The mating diameters of tubing and fittings are designed or sized to fit together properly. For good capillary attraction there should be a space between the metals of approximately 0.003 in. After the tubing has been cut to size and deburred, you must do the following for good joint soldering:

1. Clean mating parts of the joint.
2. Apply a flux to the male connection.
3. Assemble the tubing and fitting.
4. Heat the joint and apply the solder.
5. Wipe the joint clean.

CLEANING. The end of the copper tubing and the inside of the fitting must be absolutely clean. Even though these surfaces may look clean, they may contain fingerprints, dust, or oxidation. A fine sand cloth, a cleaning pad, or a special wire brush may be used. When the piping system is for a hermetic compressor, the sand cloth should be a nonconducting approved type, **Figure 7–38.**

FLUXING. Apply flux soon after the surfaces are cleaned. For soft soldering, flux may be a paste, jelly, or liquid. Apply the flux with a *clean* brush or applicator. Do *not* use a brush that has been used for any other purpose. Apply the flux only to the area to be joined and avoid getting it into the piping system. The flux minimizes oxidation while the joint is being heated. It also helps to float dirt or dust out of the joint. Too much flux will make it easier for the molten filler material to flow into the pipe itself, possibly restricting flow through the pipe. Be sure to apply flux sparingly and to only the male end of the joint. This will help ensure that the filler material remains in the space between the mating surfaces.

ASSEMBLY. Soon after the flux is applied, assemble and support the joint so that it is straight and will not move while being soldered.

HEATING AND APPLYING SOLDER. When soldering, heat the tubing near the fitting first for a short time. Then move the torch from the tubing to the fitting. Keep moving the torch from the tubing to the fitting. Heat the entire joint, not just one side. Keep moving the torch to spread the heat evenly and do not overheat any area. Do not point the flame into the fitting socket. Hold the torch so that the inner cone of the flame just touches the metal. After briefly heating the joint, touch the solder to the joint. If it does not readily melt,

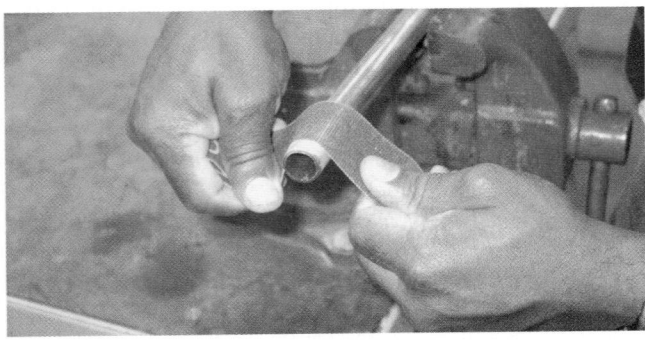

(A)

(B)

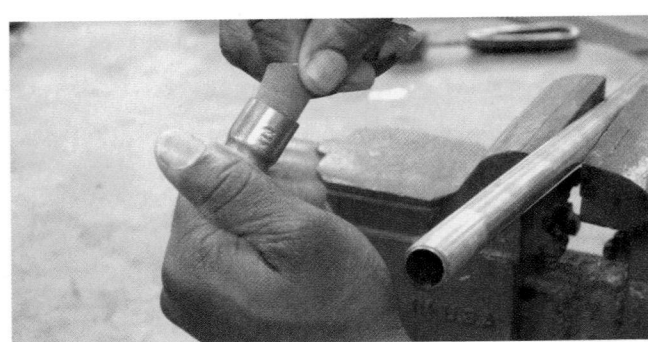

(C)

(D)

Figure 7–38 Cleaning and fluxing. **(A)** Clean the tube with sand cloth. **(B)** Clean the fitting with a tubing brush. **(C)** Clean the fitting with sand cloth. **(D)** Apply flux to the male portion of the fitting. *Photos by Bill Johnson*

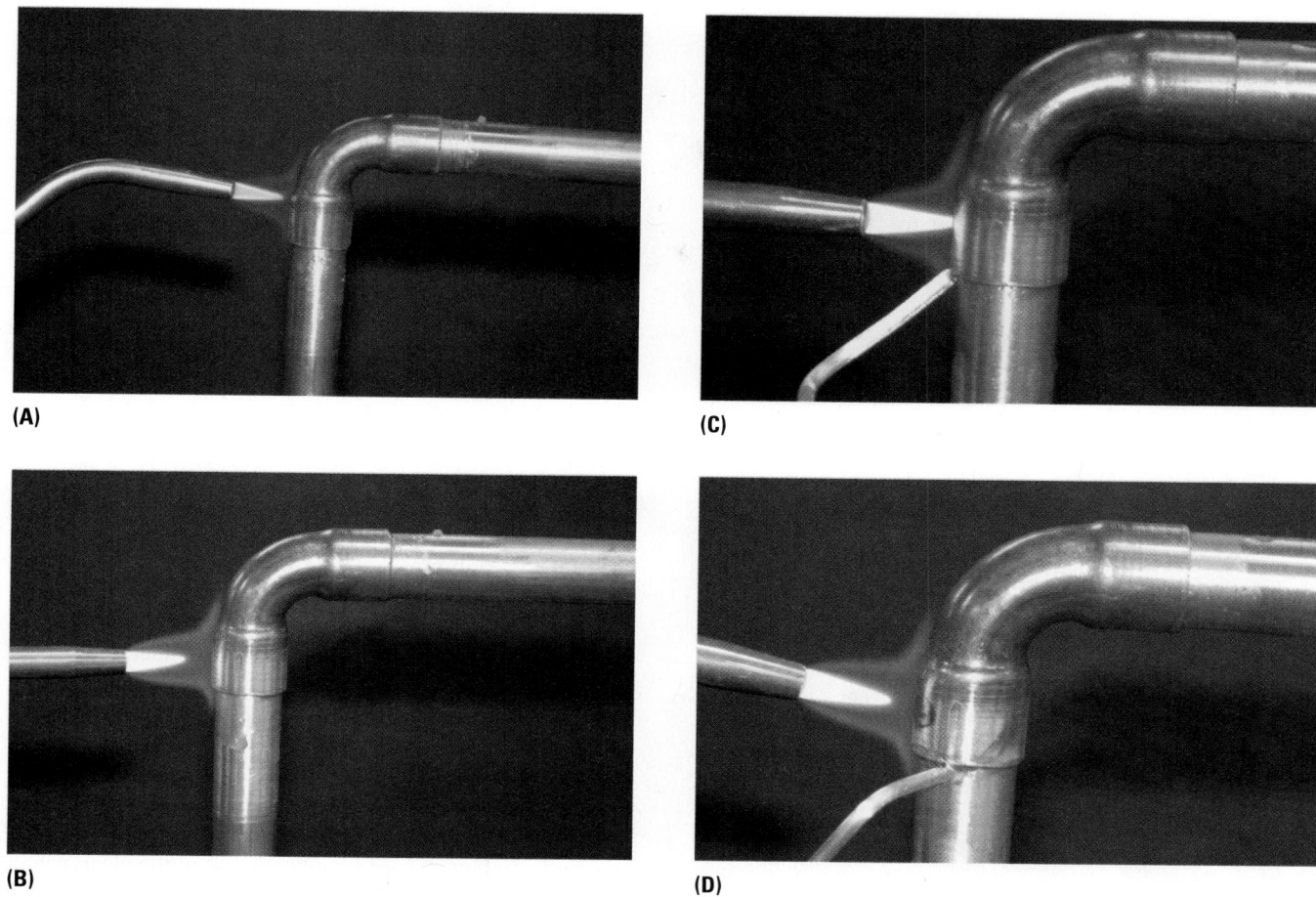

(A)

(B)

(C)

(D)

Figure 7–39   The proper procedures for heating a joint and applying solder. **(A)** Start by heating the tubing. **(B)** Keep the flame moving. Do not point the flame into the end of the fitting. **(C)** Touch the solder to the joint to check for proper heat. Do not melt the solder with the flame. **(D)** When the joint is hot enough, solder will flow. *Photos by Bill Johnson*

remove it and continue heating the joint. Continue to test the heat of the metal with the solder. Do *not* melt the solder with the flame; use the heat in the metal. When the solder flows freely from the heat of the metal, feed enough solder in to fill the joint. Do not use excessive solder. **Figure 7–39** is a step-by-step procedure for heating the joint and applying solder.

For horizontal joints apply the filler metal first at the bottom, then to the sides, and finally on the top, making sure the operations overlap, **Figure 7–40**. On vertical joints it does not matter where the filler is first applied.

WIPING.   While the joint is still hot, you may wipe it with a rag to remove excess solder. This is not necessary for producing a good bond, but it improves the appearance of the joint. Be careful not to break the joint when wiping.

7.10   BRAZING TECHNIQUES

CLEANING.   The cleaning procedures for brazing are similar to those for soldering. The brazing flux is applied with a brush to the cleaned area of the tube end. Avoid getting flux inside the piping system. Some brands of silver or copper-

phosphorous alloys do not require extensive cleaning or flux when brazing copper to copper. This is because, at the higher temperatures required for brazing, the mating surfaces are cleaned by the torch flame. Follow the instructions from the filler-material manufacturer.

APPLYING HEAT FOR BRAZING.   *Before you heat the joint, it is good practice to introduce nitrogen into the system to purge the air (containing oxygen) and reduce the possibility of oxidation to a minimum. A pressure of 1 to 2 psig is enough to purge the piping.* Apply heat to the parts to be joined with an air-acetylene or oxyacetylene torch. Heat the tube first, beginning about 1 in. from the edge of the fitting, sweeping the flame around the tube. It is very important to keep the flame in motion and not overheat any one area. Then switch the flame to the fitting at the base of the cup. Heat uniformly, sweeping the flame from the fitting to the tube. Apply the filler rod or wire at a point where the tube enters the socket. When the proper temperature is reached, the filler metal will flow readily by capillary attraction into the space between the tube and the fitting. As in soldering, do not heat the rod or wire itself. The temperature of the metal at the joint should be hot enough to melt the filler

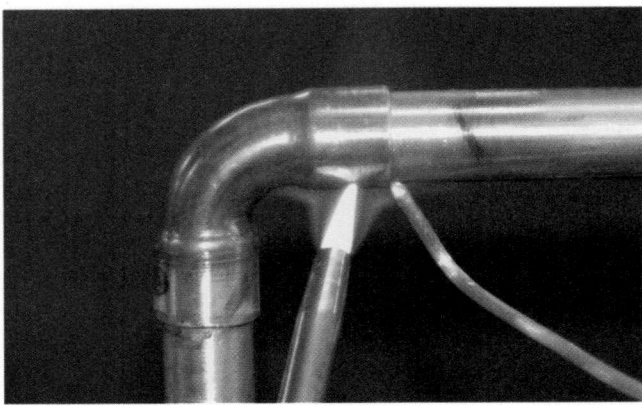

(A)

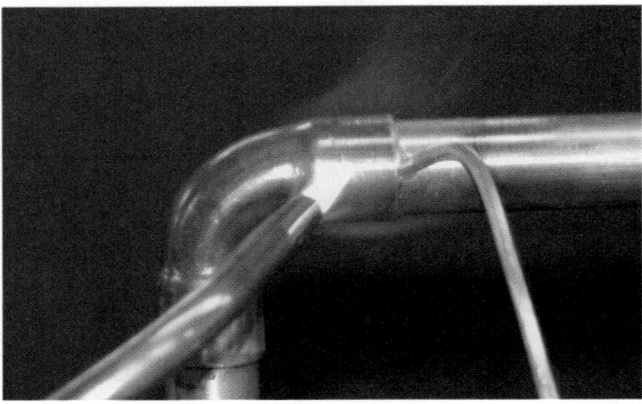

(B)

(C)

Figure 7–40 When making horizontal soldered or brazed joints, **(A)** apply the filler material at the bottom, **(B)** then to the two sides, and **(C)** finally on the top, making certain that the operations overlap. *Photos by Bill Johnson*

metal. When the joint is at the correct temperature, it will be cherry red in color. The procedures are the same as those used for soldering except for the materials used and the higher heat applied. The flux used in the brazing process will cause oxidation.

7.11 PRACTICAL SOLDERING AND BRAZING TIPS

LOW TEMPERATURE. When soldering, the surfaces between the male and female parts being joined must be clean. Clean surfaces are necessary to ensure leak-free connections. It takes much longer to prepare surfaces for soldering than it does to actually make the soldered joint. If a connection to be soldered is cleaned in advance, it may need a touch-up cleaning when it is actually soldered. Copper oxidizes, and iron or steel begins to rust immediately. Some fluxes may be applied after cleaning to prevent oxidation and rust until the tubing and fittings are ready for soldering.

Only the best solders should be used for low-temperature solder connections for refrigeration and air conditioning. For many years, systems were soldered successfully with 95/5 solder. If the soldered connection is completed in the correct manner, 95/5 can still be used. It should never be used on the high-pressure side of the system close to the compressor. The high temperature of the discharge line and the vibration will very likely cause it to leak.

A better choice for low-temperature solder than 95/5 would be a solder with a high strength and low temperature. Low-temperature solders with a silver content offer greater strength with low melting temperatures.

One of the problems with most low-temperature solders is that the melting and flow points are too close together. This is evident when you are trying to use the solder; it flows too fast, and you have a hard time keeping it in the clearance in the joint. Some of the silver-type low-temperature solders have a wider melt and flow point and are easier to use. They also have the advantage of being more elastic during the soldering procedure. This allows gaps between fittings to be filled more easily.

HIGH-TEMPERATURE (BRAZING). Several choices of high-temperature brazing materials are available. Some include a high silver content (45% silver) and must always be used with a flux. Some high-temperature brazing materials have been developed that do not have high silver content (15% silver). These may not require flux when making copper-to-copper connections; however, flux may be used. Other brazing materials have been developed with no silver content. Manufacturers are continually developing new soldering and brazing filler alloys. Your experience will soon help you choose which you prefer for a particular application.

DIFFERENT JOINTS. The type of joint dictates the solder or brazing materials used. All connections are not copper-to-copper. Some may be copper-to-steel, copper-to-brass, or brass-to-steel. These are called dissimilar metal connections. Examples follow.

1. **A copper suction line to a steel compressor or connection.** The logical choice is to use 45% silver content because of the strength of the connection combined with the high melting temperature.
2. **A copper suction line to a brass accessory valve.** The best choice would be 45% silver from a strength

standpoint. Another choice would be a low-melting temperature solder with a silver content that has a high strength and a low melting temperature. The valve body will not have to be heated to the high melting temperature of the 45% silver solder.

3. **A copper liquid line to a steel filter drier.** Although 45% silver brazing material is a good choice, it requires a lot of heat. A low-temperature solder with silver content may be the best choice. It also gives you the option of easily removing the drier for replacement at a later date.

4. **A large copper suction line connection using hard-drawn tubing.** A high-temperature brazing material with low silver content is the choice of many technicians, but you may not want to take the temper out of the hard-drawn tubing. A low-temperature solder with a silver content will give the proper strength, and the low melting temperature will not take the temper from the pipe.

Any time copper tubing is heated to a temperature that causes the tubing to become red-hot, heavy oxidation occurs on the inside and the outside of the pipe. This can be noted by the black scale that accumulates on the outside of a pipe when it is brazed. The pipe looks the same on the inside, where there is black scale. The oxygen that is in the air inside the pipe causes this black scale to form. Displacing the oxygen with dry nitrogen can prevent this. All technicians should

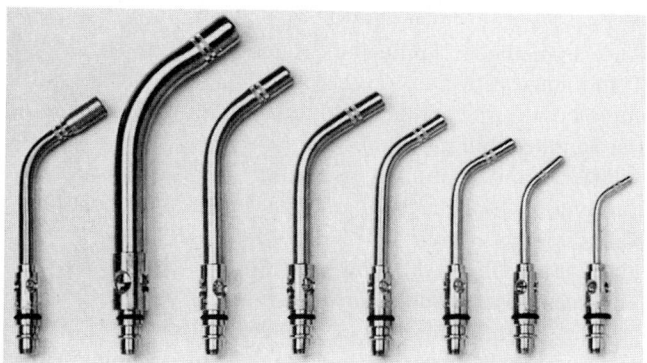

sweep the system with dry nitrogen when high-temperature connections are made. When a system is made of many connections, the scale can amount to a considerable quantity. It is trapped in either the oil, the filter driers, or the metering device— whichever it reaches first.

CHOICE OF HEAT FOR SOLDERING AND BRAZING. Air-acetylene or oxyacetylene units may be used as the heat source for soldering and brazing. Air-acetylene combinations that use the twist-tip method for mixing the air and acetylene may be used for both low- and high-temperature soldering and brazing. An air-acetylene torch has a flame temperature of 5589°F. The correct tip must be used for the solder type and pipe size. **Figure 7–41** is a photo and table of several tip sizes for different pipe sizes and solder combinations.

Another heat source is MAPP™ gas. It is a composite gas that is similar in nature to propane and may be used with air. The flame temperature of MAPP gas is 5301°F. It does not get as hot as air-acetylene but is supplied in lighter containers, **Figure 7–42.**

SOLDERING AND BRAZING TIPS.   Examples follow.
1. Clean all surfaces to be soldered or brazed.
2. Keep filings, burrs, and flux from inside the pipe.
3. When making upright soldered joints, apply heat to the top of the fitting, causing the filler to rise.
4. When soldering or brazing fittings of different weights, such as soldering a copper line to a large brass valve body, most of the heat should be applied to the large mass of metal, the valve.
5. Do not overheat the connections. The heat may be varied by moving the torch closer to or farther from the joint. Once heat is applied to a connection, it should not be completely removed because the air moves in and oxidation occurs.
6. Do not apply excessive amounts of low-temperature solders to a connection. It is a good idea to mark the length of solder you intend to use with a bend,

Tip No.	Tip Size		Gas Flow		Copper Tubing Size Capacity			
			@ 14 psi	(0.9 Bar)	Soft Solder		Silver Solder	
	in.	mm	ft³/hr	m³/hr	in.	mm	in.	mm
A-2	3/16	4.8	2.0	.17	1/8-1/2	3-15	1/8-1/4	3-10
A-3	1/4	6.4	3.6	.31	1/4-1	5-25	1/8-1/2	3-12
A-5	5/16	7.9	5.7	.48	3/4-1 1/2	20-40	1/4-3/4	10-20
A-8	3/8	9.5	8.3	.71	1-2	25-50	1/2-1	15-30
A-11	7/16	11.1	11.0	.94	1 1/2-3	40-75	7/8-1 5/8	20-40
A-14	1/2	12.7	14.5	1.23	2-3 1/2	50-90	1-2	30-50
A-32*	3/4	19.0	33.2	2.82	4-6	100-150	1 1/2-4	40-100
MSA-8	3/8	9.5	5.8	.50	3/4-3	20-40	1/4-3/4	10-20

ACETYLENE TORCH TIPS

*Use with large tank only.
NOTE: For air conditioning, add 1/8 inch for type L tubing.

Figure 7–41   Different tip sizes may be used for different pipe sizes and solder combinations. *Courtesy Thermadyne Industries, Inc.*

Figure 7–42 A MAPP™ gas kit. *Courtesy Thermadyne Industries, Inc.*

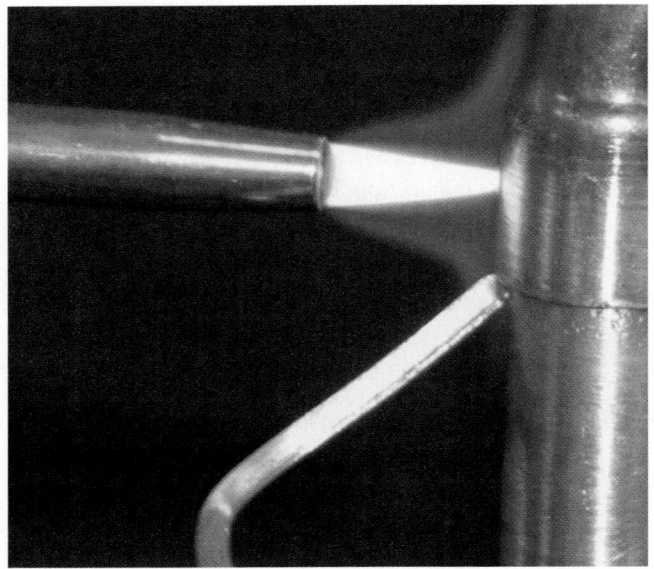

Figure 7–43 Make a bend in the end of the solder so you will know when to stop. *Photo by Bill Johnson*

Figure 7–43. When you get to the bend, stop or you will overfill the joint, and the excess may be in the system.

7. When using flux with high-temperature brazing material, always chip the flux away when finished.

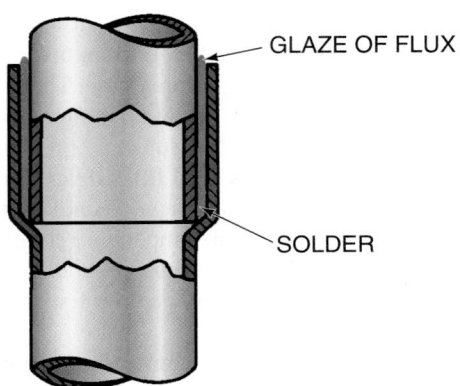

Figure 7–44 Flux used with high-temperature brazing materials will form a glaze that looks like glass.

SAFETY PRECAUTION: Wear eye protection. Flux is hard and appears like glass on the brazed connection, **Figure 7–44.** This hard substance may cover a leak and be blown out later.

8. When using any flux that will corrode the pipe, such as some fluxes for low-temperature solders, wash the flux off the connection or corrosion will occur. If this is not done, it will soon look like a poor job.

9. Talk to the expert at a supply house for your special solder needs.

7.12 MAKING FLARE JOINTS

Another method of joining tubes and fittings is the flare joint. This joint uses a flare on the end of the tubing against an angle on a fitting and is secured with a flare nut behind the flare on the tubing, **Figure 7–45.**

The flare on the tubing can be made with a screw-type flaring tool. To make the flare on the end of the tube, use the following procedure:

1. Cut the tube to the right length.
2. Ream to remove all burrs and clean all residue from the tubing.
3. Slip the flare nut or coupling nut over the tubing with the threaded end facing the end of the tubing.
4. Clamp the tube in the flaring block, **Figure 7–46(A).** Adjust it so that the tube is slightly above the block (about one-third of the total height of the flare).
5. Place the yoke on the block with the tapered cone over the end of the tube. Many technicians use a drop or two of refrigerant oil to lubricate the inside of the flare while it is being made, **Figure 7–46(B).**
6. Turn the screw down firmly, **Figure 7–46(C).** Tighten and loosen the screw several times during the flaring

Figure 7–45 Components of a flare joint. *Photo by Bill Johnson*

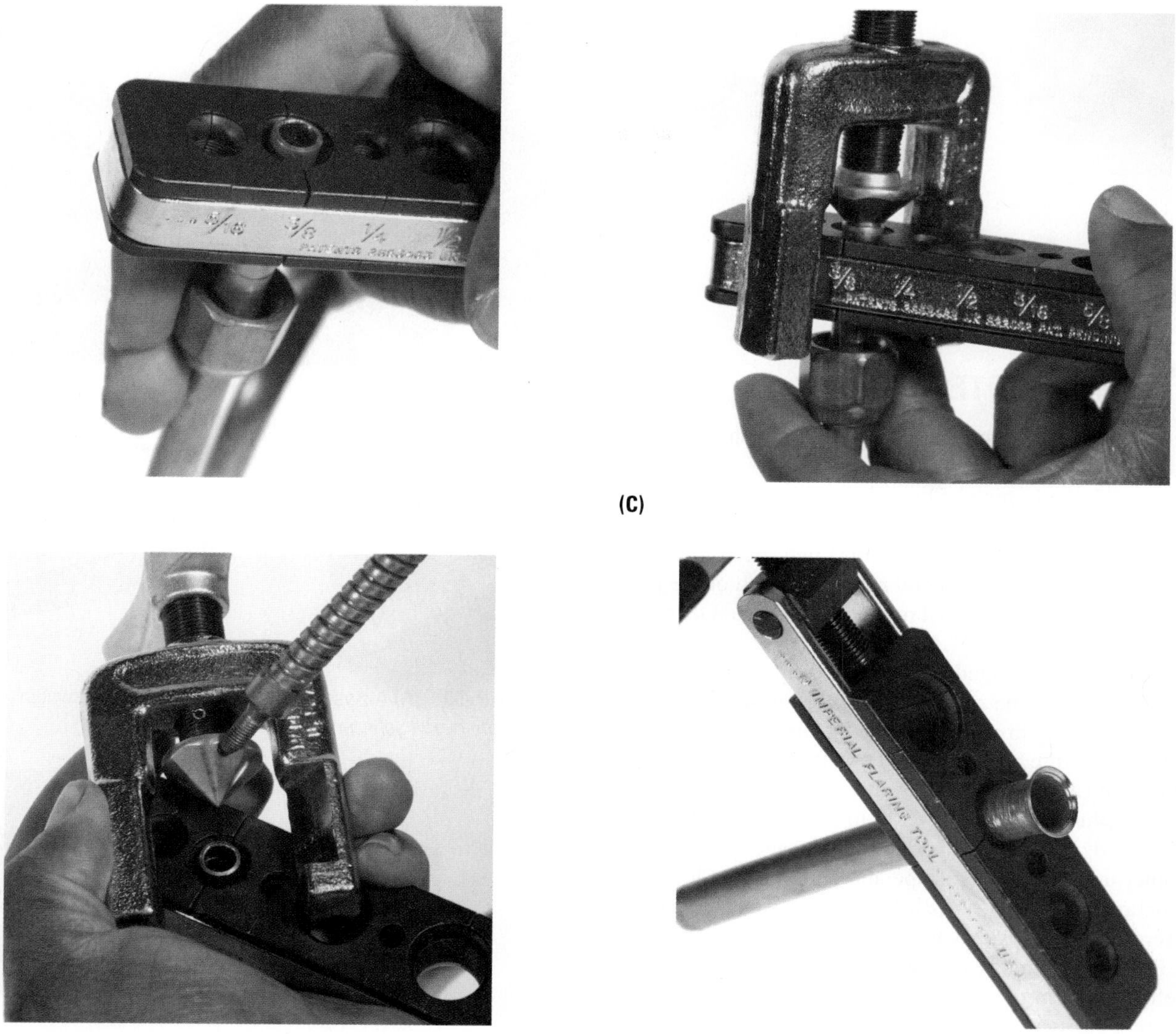

(A)

(B)

(C)

(D)

Figure 7–46   The proper procedure for making a flare joint using a screw-type flaring tool **(A)–(D)**. *Photos by Bill Johnson*

process to prevent the work from hardening. Continue until the flare is completed.
7. Remove the tubing from the block, **Figure 7–46(D)**. Inspect for defects. If you find any, cut off the flare and start over.
8. Assemble the joint.

7.13   MAKING A DOUBLE-THICKNESS FLARE

A double-thickness flare provides more strength at the flare end of the tube. To make the flare is a two-step operation. Either a punch and block or combination flaring tool is used. **Figure 7–47** illustrates the procedure for making double-thickness flares with the combination flaring tool.

Many fittings are available to use with a flare joint. Each of the fittings has a 45° angle on the end that fits against the flare on the end of the tube, **Figure 7–48**.

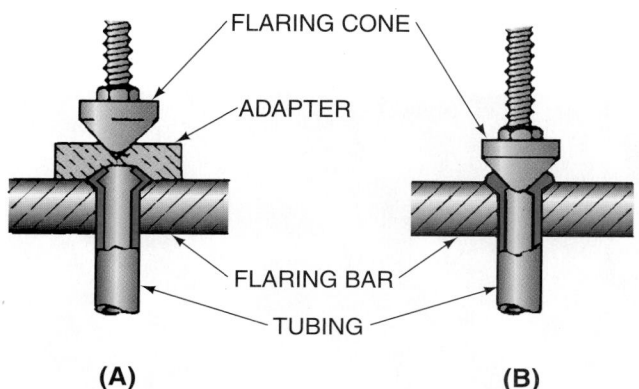

(A)   **(B)**

Figure 7–47   Procedure for making a double-thickness flare. **(A)** Place the adapter of the combination flaring tool over the tubing in the block. Screw down to bell out the tubing. **(B)** Remove the adapter, place a cone over the tubing, and screw down to form the double flare.

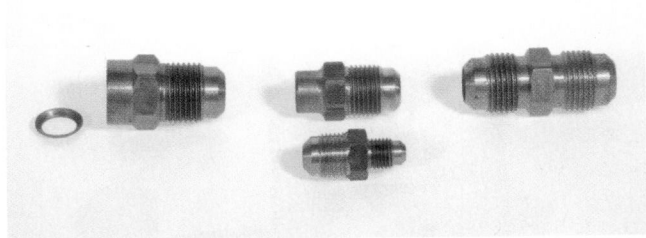

Figure 7–48   Flare fittings. *Photo by Bill Johnson*

7.14 SWAGING TECHNIQUES

Swaging is not as common as flaring, but you should know how to make a swaged joint.

Swaging is the joining of two pieces of copper tubing of the same diameter by expanding or stretching the end of one piece to fit over the other so the joint may be soldered or brazed, **Figure 7–49**. As a general rule, the length of the joint that fits over the other is equal to the approximate OD of the tubing.

You can make a swaged joint by using a punch or a lever-type tool to expand the end of the tubing, **Figure 7–50**.

Place the tubing in a flare block or an anvil block that has a hole equal to the size of the OD of the tubing. The tube should extend above the block by an amount equal to the OD of the tube plus approximately 1/8 in., **Figure 7–51(A)**. Place the correct-size swaging punch in the tube and strike it with a hammer until the proper shape and length of the joint has been obtained, **Figure 7–51(B)**. Follow the same procedure with screw-type or level-type tools. A drop or two of refrigerant oil on the swaging tool will help but must be cleaned off before soldering. Assemble the joint. The tubing should fit together easily.

Always inspect the tubing after swaging to see whether there are cracks or other defects. If any are seen or suspected, cut off the swage and start over.

SAFETY PRECAUTION: Field fabrication of tubing is not done under factory-clean conditions, so you need to be observant and careful that no foreign materials enter the

(A)

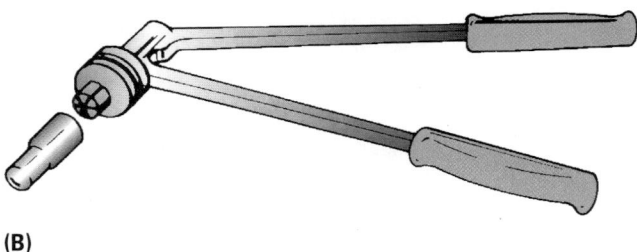

(B)

Figure 7–50   **(A)** A swaging punch. **(B)** A lever-type swaging tool. *(A) Photo by Bill Johnson*

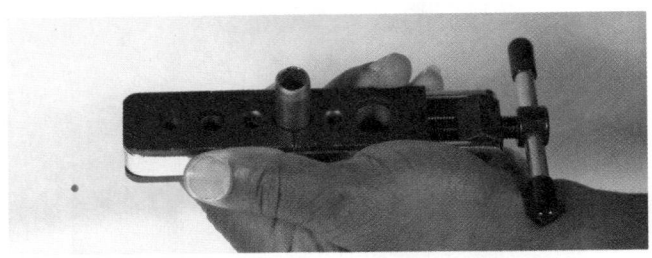

(A)

(B)

Figure 7–51   Making a swaged joint. **(A)** A tube secured in a block. **(B)** Striking the swage punch to expand the metal. *Photos by Bill Johnson*

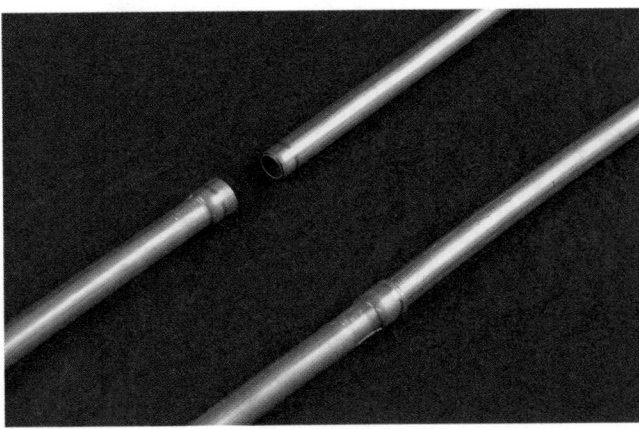

Figure 7–49   A swage connection.

tubing. When the tubing is applied to air conditioning or refrigeration, it should be remembered that any foreign matter will cause problems. Utmost care must be taken.

Here are some tips to help ensure that the swage is correct, neat, and straight:

- If using a swage punch, rotate the flaring block 180 degrees every three or four strikes. This will keep the tubing on the other side of the block straight as the swage is formed.
- Do not tap the swage punch from side to side with the hammer. This will expand the swage too much and the space between the mating surfaces will be too large.
- Avoid placing the flaring block in a vise to hold the tubing while striking the swage punch.

7.15 STEEL AND WROUGHT IRON PIPE

The terms "steel pipe," "wrought steel," and "wrought iron pipe" are often used interchangeably and incorrectly. When you want wrought iron pipe, specify "genuine wrought iron" to avoid confusion.

When manufactured, the steel pipe is either seam welded or produced without a seam by drawing hot steel through a forming machine. This pipe may be painted, left black, or coated with zinc (galvanized) to help resist rusting.

Steel pipe is often used in plumbing, hydronic (hot water) heating, and gas heating applications. The size of the pipe is referred to as the *nominal* size. For pipe sizes 12 in. or less in diameter, the nominal size is approximately the size of the ID of the pipe. For sizes larger than 12 in. in diameter, the OD is considered the nominal size. The pipe comes in many wall thicknesses but is normally furnished in standard, extra-strong, and double-extra-strong sizes. **Figure 7–52** is a cross section showing the different wall thicknesses of a 2-in. pipe. Steel pipe is normally available in 21-ft lengths.

7.16 JOINING STEEL PIPE

Steel pipe is joined (with fittings) either by welding or by threading the end of the pipe and using threaded fittings. The two types of American National Standard pipe threads are tapered pipe and straight pipe. In this industry only the tapered threads are used because they produce a tight joint and help prevent the pressurized gas or liquid in the pipe from leaking.

Pipe threads have been standardized. Each thread is V-shaped with an angle of 60°. The diameter of the thread has a taper of 3/4 in./ft or 1/16 in./in. There should be approximately seven perfect threads and two or three imperfect threads for each joint, **Figure 7–53**. Perfect threads must not be nicked or broken, or leaks may occur.

Thread diameters refer to the approximate ID of the steel pipe. The nominal size then will be smaller than the actual diameter of the thread. **Figure 7–54** shows the number of threads per inch for some pipe sizes. A thread dimension is written as follows: first the diameter, then the number of threads per inch, then the letters *NPT*, **Figure 7–55**.

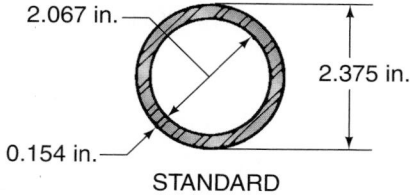

STEEL PIPE NOMINAL SIZE 2 in.

2.067 in. — 2.375 in. — 0.154 in.

STANDARD

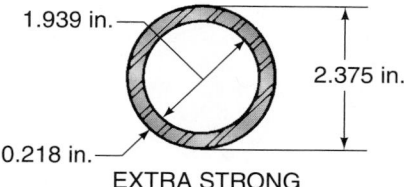

1.939 in. — 2.375 in. — 0.218 in.

EXTRA STRONG

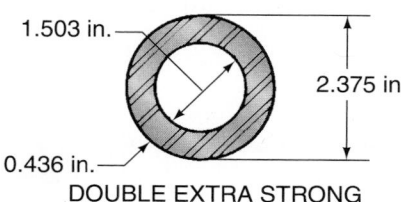

1.503 in. — 2.375 in. — 0.436 in.

DOUBLE EXTRA STRONG

Figure 7–52 A cross section of standard, extra-strong and double-extra-strong steel pipe.

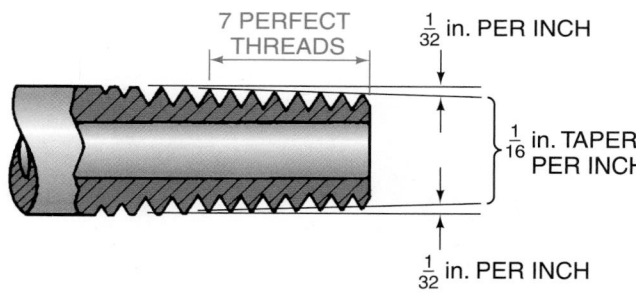

7 PERFECT THREADS

$\frac{1}{32}$ in. PER INCH

$\frac{1}{16}$ in. TAPER PER INCH

$\frac{1}{32}$ in. PER INCH

Figure 7–53 The cross section of a pipe thread.

PIPE SIZE (INCHES)	THREADS PER INCH
$\frac{1}{8}$	27
$\frac{1}{4}$, $\frac{3}{8}$	18
$\frac{1}{2}$, $\frac{3}{4}$	14
1 to 2	$11\frac{1}{2}$
$2\frac{1}{2}$ to 12	8

Figure 7–54 Number of threads per inch for some common pipe sizes.

You also need to be familiar with various fittings. Some common fittings are illustrated in **Figure 7–56**.

Four tools are needed to cut and thread pipe:

- A hacksaw (use one with 18 to 24 teeth per inch) or pipe cutter is generally used to cut the pipe, **Figure 7–57**.

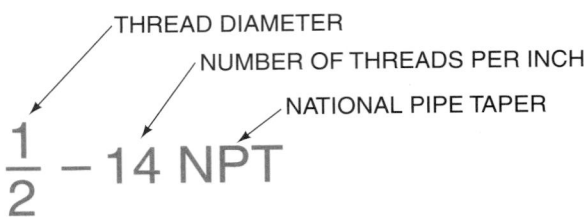

THREAD DIAMETER
NUMBER OF THREADS PER INCH
NATIONAL PIPE TAPER

$\frac{1}{2}$ – 14 NPT

Figure 7–55 A thread specification.

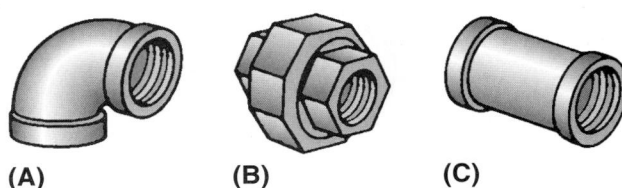

(A)   **(B)**   **(C)**

Figure 7–56 Steel pipe fittings. **(A)** A 90° elbow. **(B)** A union. **(C)** A coupling.

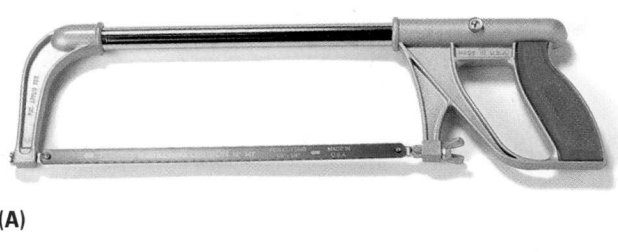

(A)

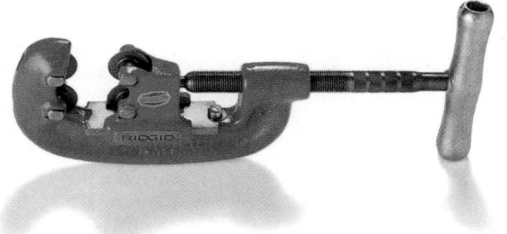

(B)

Figure 7–57 **(A)** A hacksaw. **(B)** Three- and four-wheel pipe cutters. *Courtesy Ridge Tool Company*

A pipe cutter is best because it makes a square cut, but there must be room to swing the cutter around the pipe.

■ A *reamer* removes burrs from the inside of the pipe after it has been cut. The burrs must be removed because they restrict the flow of the fluid or gas, **Figure 7–58.**

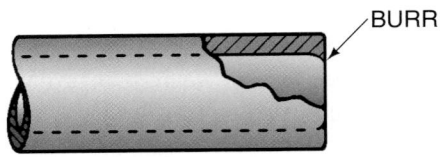

BURR

(A)

(B)

Figure 7–58 **(A)** A burr inside a pipe. **(B)** Using a reamer to remove a burr. *Photo by Bill Johnson*

Figure 7–59 A three-way fixed die-type pipe threader. *Courtesy Ridge Tool Company*

■ A threader is also known as a *die*. Most threading devices used in this field are fixed-die threaders, **Figure 7–59.**

■ Holding tools such as the chain vise, yoke vise, and pipe wrench, **Figure 7–60,** are also needed.

When large quantities of pipe are cut and threaded regularly, special machines can be used. These machines are not covered in this text.

CUTTING. The pipe must be cut square to be threaded properly. If there is room to revolve a pipe cutter around the pipe, you can use a one-wheel cutter. Otherwise, use one with more than one cutting wheel. Hold the pipe in the chain vise or yoke vise if it has not yet been installed. Place the cutting wheel directly over the place where the pipe is to be cut. Adjust the cutter with the T handle until all the rollers or cutters contact the pipe. Apply moderate pressure with the T handle and rotate the cutter around the pipe. Turn the handle about one-quarter turn for each revolution around the pipe. SAFETY PRECAUTION: Do not apply too much pressure because it will cause a large burr inside the pipe and excessive wear of the cutting wheel, **Figure 7–61.**

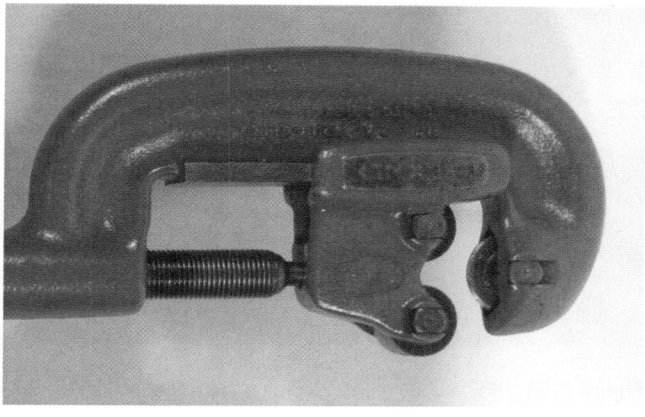

Figure 7–61 A pipe cutter. *Photo by Bill Johnson*

Figure 7–62 Cutting steel pipe with a hacksaw and a holding fixture. *Photo by Bill Johnson*

Figure 7–60 **(A)** A tri-stand with a chain vise. **(B)** A tri-stand with a yoke vise. **(C)** A pipe wrench. *Courtesy Ridge Tool Company*

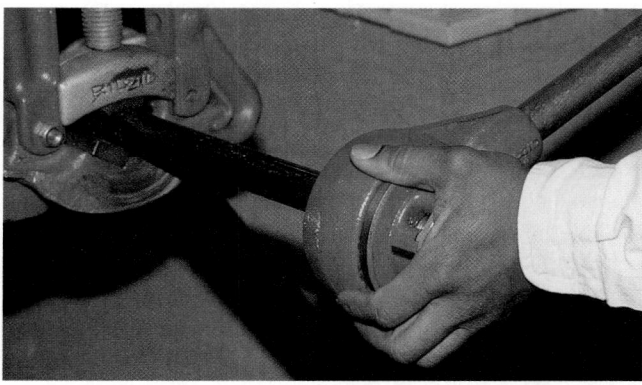

Figure 7–63 Threading pipe. *Photo by Bill Johnson*

To use a hacksaw, start the cut gently, using your thumb to guide the blade or use a holding fixture, **Figure 7–62**. SAFETY PRECAUTION: Keep your thumb away from the teeth. A hacksaw will only cut on the forward stroke. Do not apply pressure on the backstroke. Do not force the hacksaw or apply excessive pressure. Let the saw do the work.

REAMING. After the pipe is cut, put the reamer in the end of the pipe. Apply pressure against the reamer and turn clockwise. Ream only until the burr is removed, **Figure 7–58(B)**.

THREADING. To thread the pipe, place the die over the end and make sure it lines up square with the pipe. Apply cutting oil on the pipe and turn the die once or twice. Then reverse the die approximately one-quarter turn. Rotate the die one or two more turns, and reverse again. Continue this procedure and apply cutting oil liberally until the end of the pipe is flush with the far side of the die, **Figure 7–63**.

Figure 7–64   Holding the pipe and turning the fitting with pipe wrenches. *Photo by Bill Johnson*

USE MODERATE AMOUNT OF DOPE.

LEAVE TWO END THREADS BARE.

Figure 7–65   Applying pipe dope.

7.17   INSTALLING STEEL PIPE

When installing steel pipe, hold or turn the fittings and pipe with pipe wrenches. These wrenches have teeth set at an angle so that the fitting or pipe will be held securely when pressure is applied. Position the wrenches in opposite directions on the pipe and fitting, **Figure 7–64.**

When assembling the pipe, use the *correct* pipe thread dope on the male threads. Do not apply the dope closer than two threads from the end of the pipe, **Figure 7–65;** otherwise, it might get into the piping system.

All state and local codes must be followed. You should continually familiarize yourself with all applicable codes.

NOTE: The technician should also realize the importance of installing the pipe size specified. A designer has carefully studied the entire system and has indicated the size that will deliver the correct amount of gas or fluid. Pipe sizes other than those specified should never be substituted without permission of the designer.

7.18   PLASTIC PIPE

Plastic pipe is used for many plumbing, venting, and condensate applications. You should be familiar with the following types.

ABS (ACRYLONITRILEBUTADIENE STYRENE).   *ABS* is used for water drains, waste, and venting. It can withstand heat to 180°F without pressure. Use a solvent cement to join ABS with ABS; use a transition fitting to join ABS to a

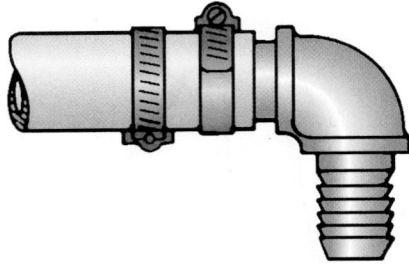

Figure 7–66   The position of clamps on PE pipe.

metal pipe. ABS is rigid and has good impact strength at low temperatures.

PE (POLYETHYLENE).   *PE* is used for water, gas, and irrigation systems. It can be used for water-supply and sprinkler systems and water-source heat pumps. PE is not used with a hot water supply, although it can stand heat with no pressure. It is flexible and has good impact strength at low temperatures. It normally is attached to fittings with two hose clamps. Place the screws of the clamps on opposite sides of the pipe, **Figure 7–66.**

PVC (POLYVINYL CHLORIDE).   *PVC* can be used in high-pressure applications at low temperatures. It can be used for water, gas, sewage, certain industrial processes, and in irrigation systems. It is a rigid pipe with a high impact strength. PVC can be joined to PVC fittings with a solvent cement, or it can be threaded and used with a transition fitting for joining to metal pipe.

CPVC (CHLORINATED POLYVINYL CHLORIDE).   *CPVC* is similar to PVC except that it can be used with temperatures up to 180°F at 100 psig. It is used for both hot and cold water supplies and is joined to fittings in the same manner as PVC.

The following and **Figure 7–67** describe how to prepare PVC or CPVC for joining.

A. Cut the end square with a plastic tubing shear, a hacksaw, or a tube cutter. The tube cutter should have a special wheel for plastic pipe.

B. Deburr the pipe inside and out with a knife or half-round file.

C. Clean the pipe end. Apply primer if required and cement to both the outside of the pipe and the inside of the fitting. (One-step primer/cement products are available for some applications. Follow instructions on primer and cement containers.)

D. Insert the pipe all the way into the fitting. Turn approximately one-quarter turn to spread the cement and allow it to set (dry) for about 1 min.

Schedule #80 PVC and CPVC can be threaded. A regular pipe thread die can be used. NOTE: Do not use the same die for metal and plastic pipe. The die used for metal will become too dull to be used for plastic. The plastic pipe die must be kept very sharp. Always follow the manufacturer's directions when using any plastic pipe and cement.

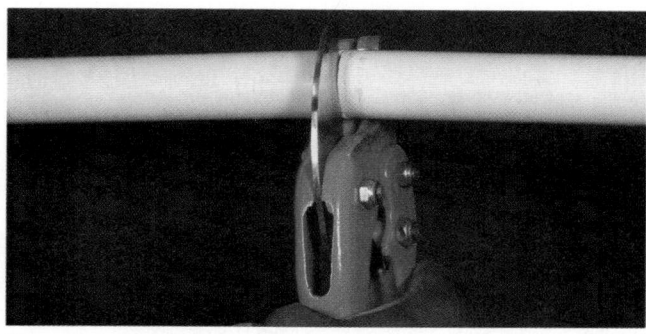

(A)

(B)

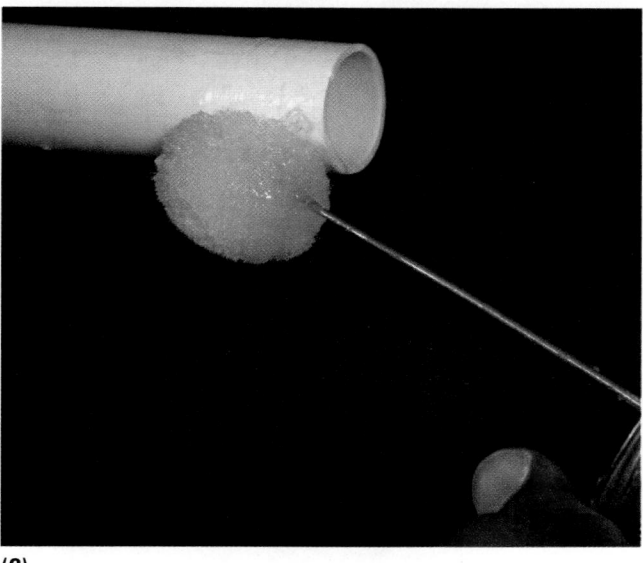

(C)

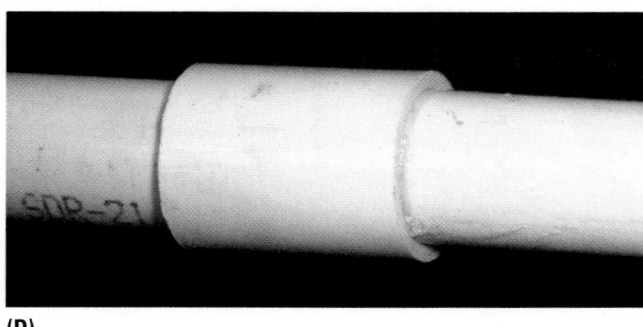

(D)

Figure 7–67 Cutting and joining PVC or CPVC pipe **(A)–(D)**. *Photos by Bill Johnson*

7.19 ALTERNATIVE MECHANICAL PIPING CONNECTIONS

In addition to the piping-connection methods already mentioned in this unit, there are a number of other alternative methods that have become popular. Although each method is unique in that each is intended for specific purposes and applications, the methods are also similar in that they all require special fittings and/or equipment for making the necessary connections. Quite often, the equipment and the required fittings are far more expensive than traditional pipe fittings and connections, but the manufacturers of such product lines stress that these products result in substantial time savings that can easily translate into reliable and profitable piping systems. It is not the intention of this text to determine the reliability of these items, but to present them to the reader for the sake of comprehensiveness. Please refer to the manufacturer's literature for important information regarding applications, operating pressure ratings, and other safety-related issues.

One such mechanical piping connection is shown in **Figure 7–68.** This connector can be used to connect two sections of copper tubing as well as to join a copper tubing section to one made of a different material, such as aluminum. There are adapters available that enable the service technician to join together tubing sections of different sizes.

The mechanics of the joint involves the insertion of the ends of the tube sections to be joined into the coupling/connection device. The coupler is then pressed together to form the connection. **Figure 7–69** shows the connection before (top) and after (bottom) the fitting has been compressed. **Figure 7–70** shows an illustration of how an installation tool is used to compress the ends of the fitting.

Two other mechanical methods are intended to eliminate the use of soldering. As with the previous method, they both use special fittings to make the connection, but neither is intended for use on refrigerant-carrying piping circuits. The fitting shown in **Figure 7–71** is intended to receive sections of pipe that are simply pushed into them. The tabs located around the interior circumference of the fitting, along with the gasket material inside the fitting, provide the seal, **Figure 7–72.**

The connection shown in **Figure 7–73** is a press-on fitting. The fitting is placed on the pipe section and is then squeezed

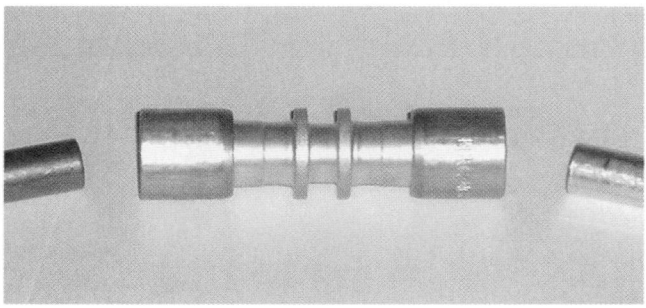

Figure 7–68 Fitting used to mechanically join two tubing sections. *Photo by Eugene Silberstein*

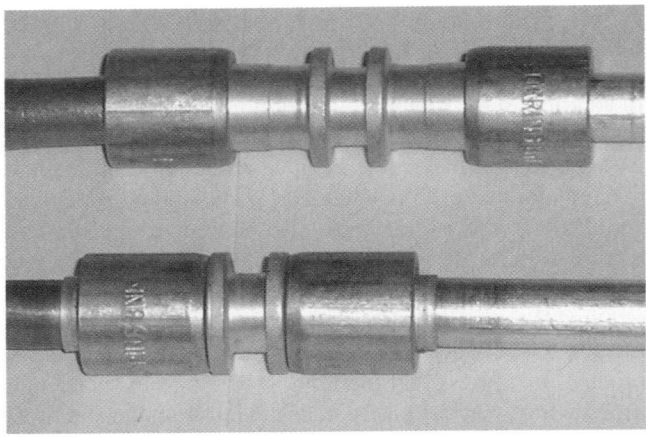

Figure 7–69 Tubing connector before (top) and after (bottom) compression. *Photo by Eugene Silberstein*

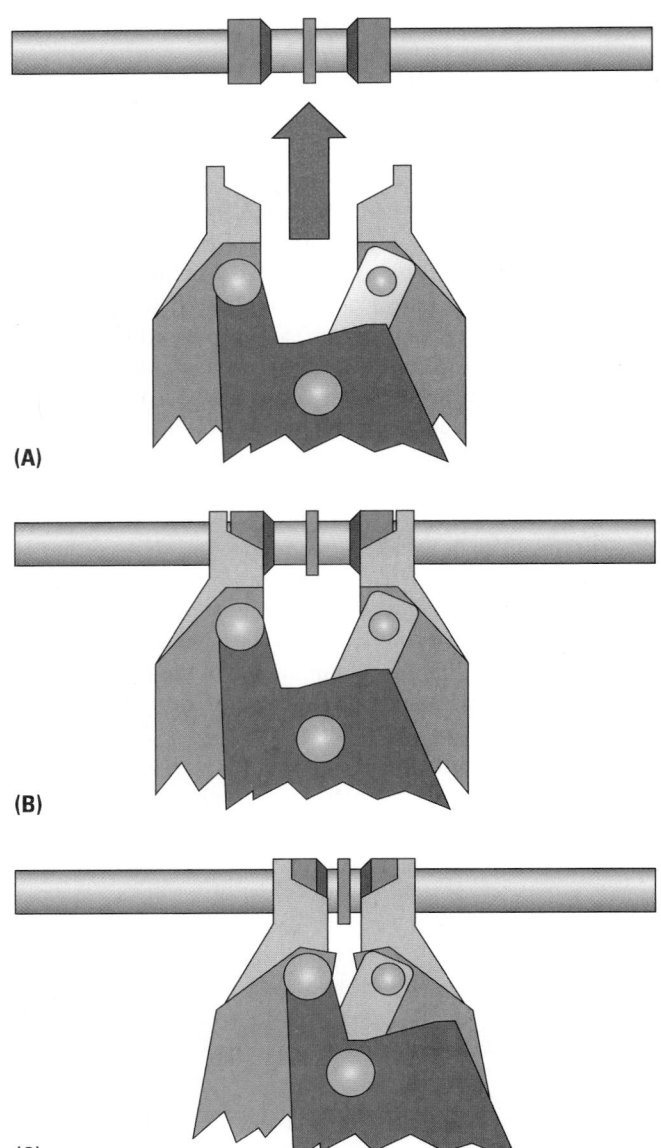

(A)

(B)

(C)

Figure 7–70 Process of compressing fitting onto two piping sections **(A)–(C)**.

Figure 7–71 Push-on-type fitting with internal gasket and metal teeth for grabbing onto the pipe section. *Photo by Eugene Silberstein*

Figure 7–72 Pipe section pushed into the fitting. *Photo by Eugene Silberstein*

Figure 7–73 Cutaway view of press-type fitting installed on pipe. *Courtesy Viega*

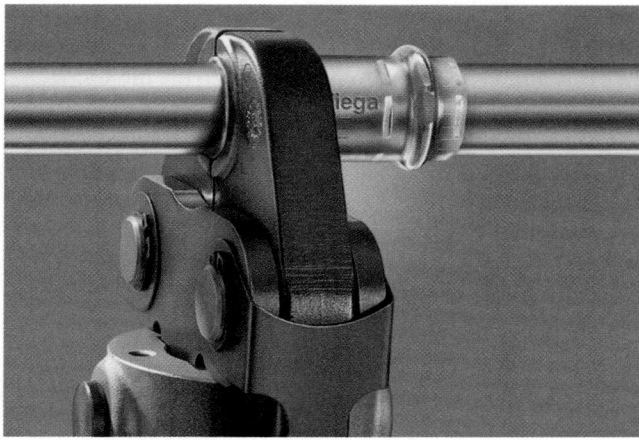

Figure 7–74　Crimping the press-type fitting to join two pipe sections. *Courtesy Ridge Tool Company*

Figure 7–75　Various jaw sizes used for installing press-type fittings on pipes of different sizes. *Courtesy Ridge Tool Company*

onto the pipe using a special tool, **Figure 7–74.** Internal gaskets in the fitting help reduce the possibility of leaks. The crimping tool that is used has interchangeable jaws, **Figure 7–75,** so that the same tool can be used on a number of different pipe sizes. As mentioned at the beginning of this section, please check with the individual manufacturer for information on pressure ratings and applications. Also, when installing mechanical fittings, be sure to follow the manufacturer's instructions carefully.

SUMMARY

- *The use of correct tubing, piping, and fittings—as well as their proper installation—is necessary in order for a refrigeration or air-conditioning system to operate properly. Careless handling of the tubing and poor soldering or brazing techniques may cause serious damage to the components of the system.*
- Copper tubing is generally used for plumbing, heating, and refrigerant piping.
- Copper tubing is available in soft- or hard-drawn copper.
- ACR tubing can be purchased as line sets.
- Tubing may be cut with a hacksaw or tubing cutter.

- Soft tubing may be bent. Tube-bending springs or lever-type benders may be used, or the bend can be made by hand.
- Soldering and brazing fasten tubing and fittings together.
- Air-acetylene units are frequently used for soldering and brazing.
- Oxyacetylene equipment is also used, particularly for brazing requiring higher temperatures.
- The flare joint is another method of joining tubing and fittings.
- The soldered swaged joint is a method used to fasten two pieces of copper tubing together.
- Steel pipe is used in plumbing, hydronic heating, and gas heating applications.
- Steel pipe is joined with threaded fittings or by welding.
- ABS, PE, PVC, and CPVC are four types of plastic pipe; each has a different use.
- Alternative mechanical piping-connection methods can be used for certain applications.

REVIEW QUESTIONS

1. The standard weight of copper tubing used most frequently in the heating and air-conditioning industry is
   A. K.
   B. L.
   C. DWV.
   D. M.
2. The size of 1/2 in. would refer to the _____ with regard to copper tubing used in plumbing and heating.
   A. ID
   B. OD
   C. length
   D. fitting length
3. The size of 1/2 in. would refer to the _____ with regard to ACR copper tubing.
   A. ID
   B. OD
   C. length
   D. fitting length
4. Tubing used for air-conditioning installations is usually insulated at
   A. the low-pressure side.
   B. the discharge line.
   C. the high-pressure side from the condenser to the metering device.
   D. all of the above.
5. In what size rolls is soft copper tubing normally available?
6. Why are some ACR tubing lines insulated?
7. Describe procedures for bending soft copper tubing.
8. Describe the procedure for cutting tubing with a tube cutter.
9. What type of solder is suitable for moderate temperatures and pressures?
10. What are elements that make up brazing filler metal alloys?

11. Describe steps to make a good soldered copper tubing/fitting joint.
12. Describe the procedures for making a flare for joining copper tubing.
13. Brazing is done at _____ for soldering.
    A. the same temperature as that used
    B. lower temperatures than those used
    C. higher temperatures than those used
    D. temperatures that can be higher or lower than those used
14. A common filler material used for brazing is composed of
    A. 50/50 tin-lead.
    B. 95/5 tin-antimony.
    C. 15% to 60% silver.
    D. cast steel.
15. When soldering, a flux is used to
    A. minimize oxidation while the joint is being heated.
    B. allow the fitting to fit easily onto the tubing.
    C. keep the filler metal from dropping on the floor.
    D. help the tubing and fitting to heat faster.
16. What should you do if you see or suspect a crack in a flared joint?
17. What are some uses of steel pipe?
18. List the procedures used when setting up, igniting, and using an air-acetylene unit.
19. Describe the procedure for preparing and threading the end of steel pipe.
20. Describe each part of the thread dimension 1/4–18 NPT.
21. List four types of plastic pipe.
22. Describe three alternative mechanical piping methods.

UNIT 8

System Evacuation

After studying this unit, you should be able to

- describe a standing pressure test.
- choose a leak detector for a particular type of leak.
- describe a deep vacuum.
- describe two different types of evacuation.
- describe two different types of vacuum measuring instruments.
- choose a proper high-vacuum pump.
- list some of the proper evacuation practices.
- describe a high-vacuum single evacuation.
- describe a triple evacuation.

SAFETY CHECKLIST

- ✔ Care should be used while handling any of the products from a contaminated system because acids may be encountered.
- ✔ Do not place your hand over any opening that is under high vacuum because the vacuum may cause a blood blister on the skin.
- ✔ Wear goggles and gloves while transferring refrigerant.
- ✔ Do not allow mercury from any instrument to escape. It is a hazardous material.

8.1 RELIABLE AND EFFICIENT SYSTEMS

The previous unit explained how refrigeration system piping is connected. These systems must be connected to be as leak free as possible. All systems leak because many of the metals and their connections are slightly porous. When they are viewed under an electron microscope, cracks and faults can be readily seen. When a system runs for years without showing signs of leaks, it is just less porous than one that does not run as long. Technicians use good piping and connection practices to minimize leaks on the portions of the systems they work with. Many systems have run efficiently for 50 or more years with the original refrigerant charge. The system is leaking, but at a rate that is not detectable and does not affect the efficiency. Many systems that are sold today have a critical refrigerant charge down to 1/2 oz and will not run efficiently with a loss of charge of more than 1/2 oz. These systems operate for years at good efficiency levels when correctly connected because the leak rate is so slow. This is because of careful piping and system assembly practices. The original assembly must be done correctly as well as any field service connections.

Today, systems that are assembled under factory conditions are less likely to have detectable leaks unless there is a material malfunction. 🌐 Installation and service personnel must learn good field piping and assembly practices to ensure that systems will run efficiently as long as possible. Many field-installed systems last for 50 or more years because of good practices. It all begins with technicians knowing what they are doing when they are installing systems. After the installation or service, a proper leak test must be performed. 🌐

8.2 STANDING PRESSURE TEST

The technician can start with a standing pressure test for a newly assembled system to see if it will hold pressure. At the same time, the system can be checked with a sensitive leak detector.

The technician should assemble the system first and then give it a visual test to make sure that all connections look as they should.
- The solder connections should have no gaps.
- All flanges and threaded connectors should be tight.
- All control valves must be installed in the correct direction and correctly set.
- All valve covers must be on.

Several leak-checking procedures can be conducted. The first test is to charge a trace amount of refrigerant into the system for the purpose of leak detection. Most technicians start at 0 psig and let R-22 into the system until the pressure reads about 10 psig. This practice is accepted by the Environmental Protection Agency (EPA) because there must be some refrigerant in the system to detect a leak. This refrigerant is being used only to detect leaks, not as a refrigerant. R-22 is the only refrigerant approved for this purpose. When the trace refrigerant is allowed into the system, dry nitrogen is then pushed into the system to a practical test pressure. The low-pressure side of most systems has a working pressure of 150 psig. Some of the new refrigerants have a higher working pressure; therefore, the system components have a higher working pressure. The technician should check each system for the proper working pressure before pressurizing it. The components on the low-pressure side of the system are the evaporator, suction line, and compressor. The only parts of the compressor that are considered to be on the high-pressure side are the compressor head and the discharge piping, most of which is internal to the compressor. The compressor is considered a low-pressure device because the compressor shell is

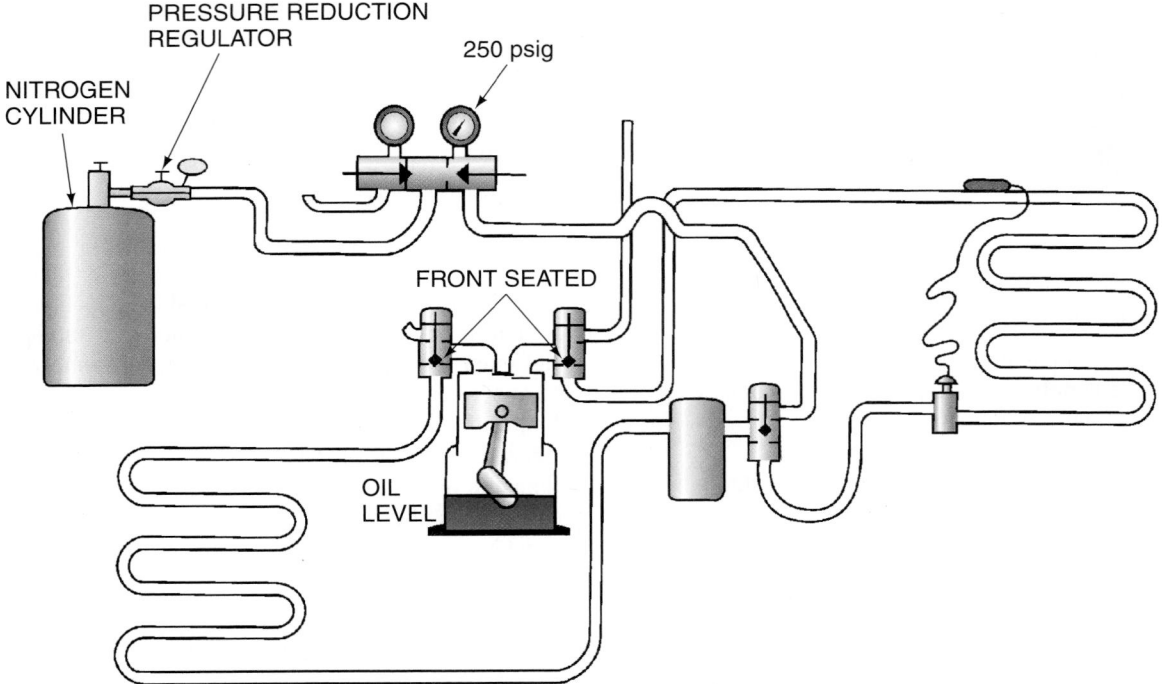

Figure 8–1   Isolating the compressor to pressurize the condenser and evaporator.

on the low-pressure side of the system. The high-pressure side of the system consists of the discharge line to the condenser, the condenser, and the liquid line. The high-pressure side test pressure for most refrigerants is at least 350 psig when the compressor shell is not pressurized. Systems using R-410A may have working pressures up to 450 psig. When the compressor has service valves, it can be isolated, **Figure 8–1.**

The pressure used to pressurize a complete system must not exceed the lowest test pressure for the system. Once pressurized, the system should be allowed to stand for about 10 min to allow pressures in the system to equalize. At this time, the technician should listen to each connection for an obvious leak, **Figure 8–2.** Once the pressures have equalized, the high-pressure gage should be marked. The technician

should tap the gage to settle the needle down and then mark the place, **Figure 8–3.** SAFETY PRECAUTION: Never use oxygen or compressed air as they can cause explosive gases to accumulate.

The standing pressure test should be left on the equipment for as long as practical. A 1 1/2-ton package unit that has a small system may need to stand for 1 hour. A larger system, such as a 10-ton split system, should be allowed to stand overnight, or for 24 hours, if possible. The longer the system stands under pressure without a drop in pressure, the more confidence the technician may have that the system is leak-free. If there is a drop in pressure, a very close leak test should be conducted unless there is a significant drop in the ambient (surrounding) temperature.

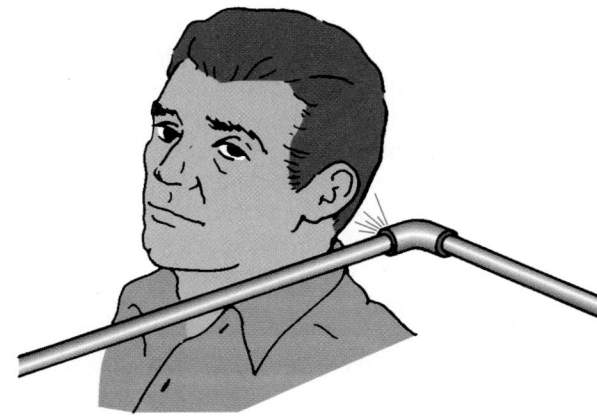

Figure 8–2   Listen for leaks.

Figure 8–3   When using a pressure gage for a standing leak test, tap the gage lightly to make sure the needle is free, then mark the gage.

8.3 **LEAK DETECTION METHODS**

The most basic but effective method is to just listen for the rush of nitrogen and refrigerant from the connections. Leaks can often be heard but cannot be pinpointed. When there is a leak this large, you can often find it by wetting the back of your hand and moving it around where you suspect the leak. The leak will spray vapor on your wet hand and the evaporation will cool your hand and help locate the leak.

Halide leak detectors like the one shown in **Figure 8–4** have been around for a long time. This is considered a primitive detector by many people, but it still has its place. It is powered by a propane or acetylene gas flame which creates a draft that draws a sample of whatever is in the vicinity of the sample tube. A sample of refrigerant, for example, passes over a glowing copper element and causes the flame to change color. This change in flame color can best be viewed where there is very little light. This detector is not very effective in bright sunlight. It can be used only with refrigerants that have chlorine as part of their formula, such as CFCs and HCFCs. It cannot be used in an explosive atmosphere, such as around gasoline or other flammables, because of the open flame. The halide detector is capable of detecting a leak as small as about 1 oz per year. The best application is in the case of a suspected fairly large leak because of the relatively low sensitivity.

Ultrasonic leak detectors have earphones, a microphone, and an amplification system for listening for the high-pitched sound of a leak. These detectors would be very good in a place where there is a lot of refrigerant in the air, **Figure 8–5.**

Ultraviolet leak detectors work by means of a dye that is charged into the system. The dye mixes with the oil in the system that circulates with the refrigerant. This dye glows under black light when it escapes through the leak, **Figure 8–6.** This technology was developed for the auto industry to find leaks in automatic transmissions. It is very effective, particularly for older systems that may be oil soaked from prior leaks.

Dye-saturated oil is a leak detection system that causes a red dye residue to seep out wherever there is a leak and oil escapes. This dye is normally charged into the system with a special factory-charged refrigerant cylinder.

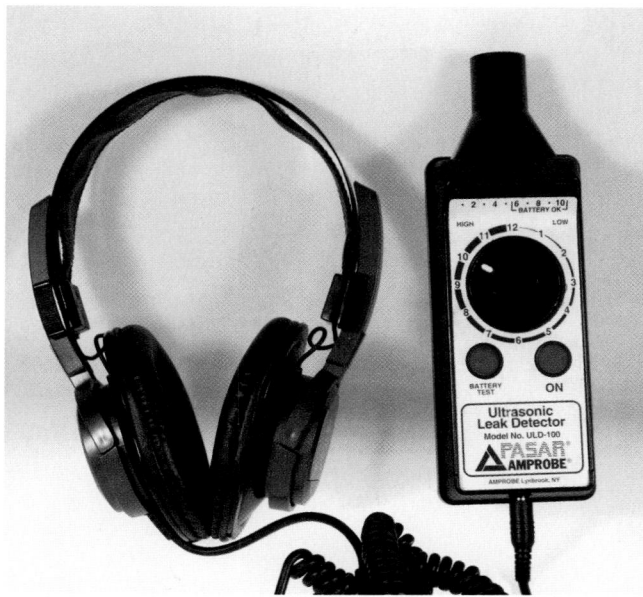

Figure 8–5   An ultrasonic leak detector that listens for leaks. *Photo by Bill Johnson*

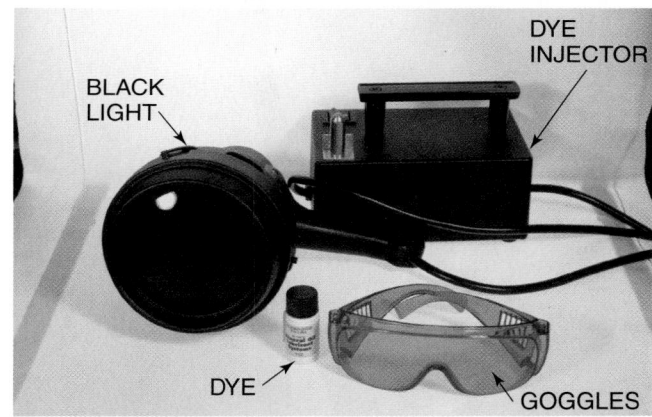

Figure 8–6   An ultraviolet leak detector that uses black light and a special dye. *Photo by Bill Johnson*

Before allowing any dye into a system, you should consult with the compressor manufacturer to make sure the dye is approved, particularly while the system is under warranty. Most dyes are lubricant specific, meaning a different type of dye has to be used with mineral oil, alkylbenzene, and ester-type lubricants. If the compressor is out of warranty, it belongs to the customer.

Electronic leak detectors, the most accurate and probably the most often used detectors, will detect leaks down to about 1/4 oz per year, **Figure 8–7.** These detectors are great for certain applications. When there is a lot of refrigerant in the air, electronic detectors may have a hard time locating where to start unless they have a feature that allows them to correct for background refrigerant. The sensitivity of these instruments can be their greatest asset or their greatest liability, depending on whether the technician takes the time to understand

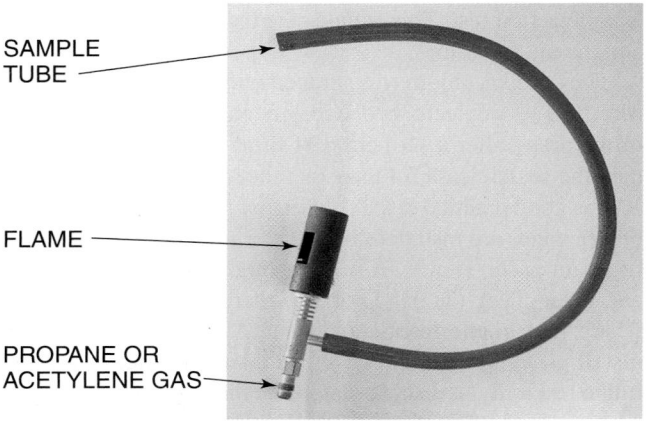

Figure 8–4   A halide torch for leak detection. *Photo by Bill Johnson*

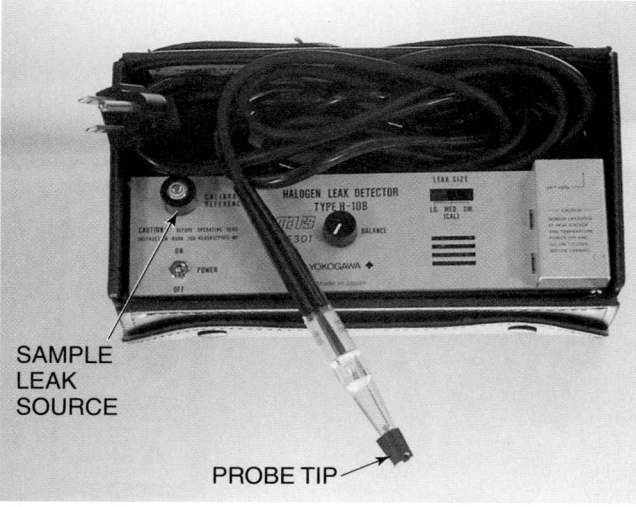

SAMPLE
LEAK
SOURCE

PROBE TIP

(A)

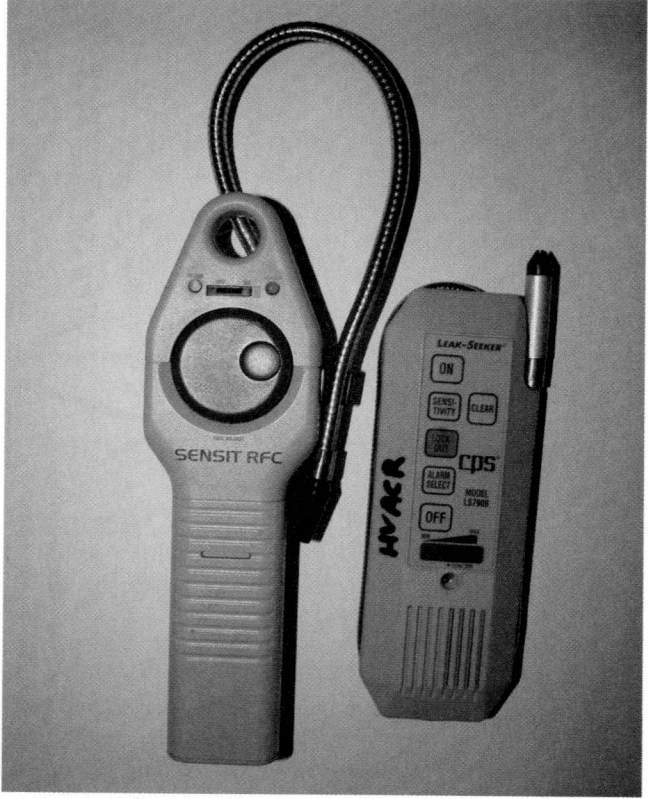

(B)

Figure 8–7   (A) An electronic leak detector. **(B)** Modern electronic leak detectors for use on CFC, HCFC, and HFC refrigerants. **(A)** *Photo by Bill Johnson.* **(B)** *Courtesy Ferris State University. Photo by John Tomczyk*

8.4   LEAK DETECTION TIPS

Leak detection may be performed on a split system that is new and has not been operating or as a service call where the system has been operating and a leak is suspected.

When a split system is new, the field-installed portion of the system should be leak checked (as described previously) using nitrogen, a trace refrigerant, and a standing leak check.

Leak checking a system that has been operating is a different situation. This system may or may not have refrigerant in it. If it is cooling at all, at least some liquid refrigerant is in the system and nitrogen should not be added because it cannot be recovered with the refrigerant. Before nitrogen is added, the system must meet certain vacuum level requirements. To leak check a system that has refrigerant in it, the technician should shut down the compressor and let the evaporator fan run. This will boost the low-pressure side of the system up to the temperature of the conditioned space. For low-temperature systems the pressure may not be high enough for an adequate leak test. For an air-conditioning application the indoor pressure may correspond to the indoor temperature, and this may be high enough for a leak test. Otherwise, for either situation, the refrigerant should be properly recovered and the system pressured with a trace refrigerant, R-22, and nitrogen for a high-pressure leak test. This can be very time consuming but will get the best results.

When there is a persistent leak in a split system, the following procedures can be used. Always do a visual check to look for fresh oil or dust spots where dust collects on oil. Leaks are often caused by vibration or temperature. For example, the discharge line of the compressor is a high-vibration area. It may also be an area where there are great temperature changes. The gas line on a heat pump is at high pressure and temperature in the winter and low pressure and temperature in the summer. Look at any threaded connections, such as quick-connect fittings, where expansion and contraction due to temperature difference may occur.

The system parts may be isolated by disconnecting the liquid line and the suction line at the outdoor unit and pressurizing the entire low-pressure side of the system. This would include the greater part of the field-installed system as well as the evaporator, **Figure 8–8.** This portion of the system could be pressurized to 250 psig for a standing pressure and leak detection test. This isolates the leak between the indoor and the outdoor unit.

The outdoor unit may be draped with plastic and pressurized. The leak detector probe may be inserted under the edge of the drape after a set period of time. If a leak is detected, then the search is on for the exact location.

It is good practice to leak check any gage connection ports before gages are installed. First, determine if there is any refrigerant in the system. If it is cooling, it has some refrigerant. Leak check the gage ports. If there is a leak, you know where part of the refrigerant went. Many technicians will install gages, pressurize the system, and will not be able to find a leak only to discover later that the gage ports were the problem in the first place. SAFETY PRECAUTION: When servicing Schrader valve ports, use only an approved cap, the

them. Most of these detectors have very small air pumps that pull a sample of air, or air and refrigerant, over a sensor. These sensors are sensitive to moisture. They must not come in contact with water, such as in drain lines or on a wet coil.

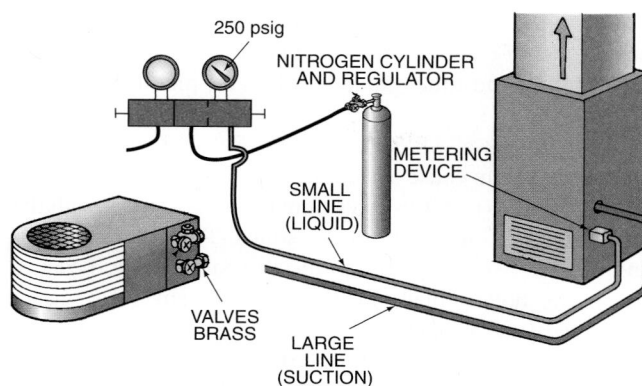

Figure 8–8   The low-pressure side of the system is isolated and pressurized to 250 psig.

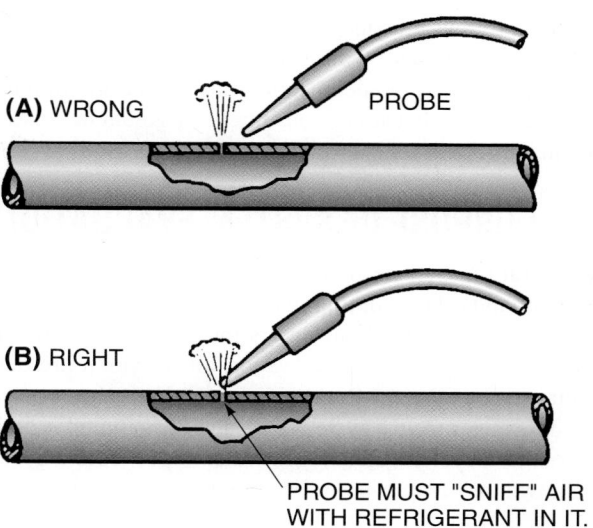

Figure 8–9   **(A)** The electronic leak detector is not sensing the small pinhole leak because it is spraying past the detector's sensor. **(B)** The sensor will detect a refrigerant leak.

one that came with it. If you screw a brass cap down too tight, you may create a real problem by crushing the delicate seat in the valve down to where the valve core may not ever be removed.

Many technicians use a special soap to pinpoint leaks. The soap has a special additive that keeps it from evaporating and makes it much more elastic so that larger bubbles are developed. Some bubble solutions have a microfoamer additive which forms a small cocoon of foam (resembling shaving cream) at the leak. Some technicians mix mild dishwashing soap with glycerin to give the soap elasticity. Remember, it takes time to blow a bubble at 1/2 oz a year, which is a very small leak. Bubbles do not work well on hot surfaces, such as the compressor discharge line. NOTE: The technician must perform a thorough cleanup after using soap for leak detecting. The soap residue will turn the copper pipe green and it will start to corrode. Painted surfaces will start to peel if not cleaned. Carefully wiping surfaces with a wet rag or rinsing them with a hose will usually remove the residual soap.

Refrigerant is heavier than air and has a tendency to fall away from the leak source. It is important to start at the highest point and work downward. The technician can check an evaporator coil by stopping the evaporator fan, letting the system stand for a few minutes, and then placing the detector probe in the vicinity of the condensate drain line connection and letting it remain there for about 10 min. (The technician must make sure the probe does not contact water in the drain line.) If it sounds the refrigerant leak alarm, the test should be repeated. If the alarm sounds again, there is a leak in the evaporator section. Removing panels and going over the connections one at a time may find the leak.

Technicians should not rush when trying to locate a leak. Leak detection with any device is a slow process. It takes time for the instrument to respond, so the probe must be moved very slowly—about 1 in. every 2 sec is a good rate of speed. The probe tip must be directly over the leak to detect a very small leak, **Figure 8–9.** If the technician moves too fast, the probe may pass over the leak or indicate a leak that is several inches back.

8.5   REPAIRING LEAKS

The technician must know when to actually repair a leak. For a very small leak in a small piece of equipment, it is usually not as economical to repair the leak as it is to add refrigerant. Therefore, the technician must know when to add refrigerant and when to repair a leak.

The following EPA guidelines must be followed while making these decisions:

- Systems containing less than 50 lb of refrigerant do not require repair.
- Industrial process and commercial refrigeration equipment and systems containing more than 50 lb of refrigerant require repair of substantial leaks. Substantial leaks are leaks that lose more than 35% of the refrigerant charge in a year. For example, a system with 100 lb of refrigerant may be allowed to lose 35 lb per year, and this must be documented by the owner. This is an average of 2.9 lb per month (35/12 = 2.9). When this leak rate is reached, the owner has 30 days to make a repair or to set up a retrofit or replacement program to be fulfilled within a year.
- Comfort cooling chillers and all other equipment have an allowable leak rate of 15% per year. The latest revision of the regulations states that the leak needs only to be "reduced" to a level below the required percentage.

The technician must put together a system that does not leak as an initial installation. The equipment should run leak free and correctly at least through the warranty period, usually the first year. If a leak develops after that, it is out of warranty and the responsibility of the owner.

Most residential systems hold less than 10 lb of refrigerant and do not require repair. Many technicians just "top off the charge" each year without really looking for the leak.

This is not good practice. Often the leak is in the Schrader valve connection, and with just a little effort this leak could be repaired. Many homeowners think that air-conditioning equipment actually uses or consumes refrigerant because the technician adds it each spring.

8.6 PURPOSE OF SYSTEM EVACUATION

Refrigeration systems are designed to operate with only refrigerant and oil circulating inside them. When systems are assembled or serviced, air enters the system. Air contains oxygen, nitrogen, and water vapor, all of which are detrimental to the system. Removing air and/or other noncondensable gases from a system with a vacuum pump is called *degassing* a system. Removing water vapor from a system is known as *dehydration.* In the HVACR industry, the process of removing both air and water vapor is referred to as *evacuation.*

Degassing + Dehydration = Evacuation

These gases cause two problems. The nitrogen, contained in air, is called a noncondensable gas. It will not condense in the condenser and move through like the liquid; instead, it will occupy condenser space that would normally be used for condensing refrigerants. This will cause a rise in head pressure, resulting in an increase in discharge temperatures and compression ratios, which cause unwanted inefficiencies. **Figure 8–10** illustrates a condenser with noncondensable vapors inside. The other gases cause chemical reactions that produce acids in the system. Acids in the system cause deterioration of the system's parts, copper plating of the running gear, and the breakdown of motor insulation.

These acids can be very mild to quite strong. Electroplating in a refrigeration system is much like electroplating of any other kind. It requires electrical current, acid, and dissimilar metals. In the refrigeration system, we have copper, brass, cast iron, and steel with aluminum in some systems. The electroplating in refrigeration systems seems to only plate from copper to steel—copper from the pipes and steel on the crankshaft and bearing surfaces. The bearing surfaces have close tolerances. When a small amount of copper is plated on the steel, the bearings become tight and bind. The breakdown of electric motor insulation will cause electrical short circuits and either ground or short the motor from phase to phase.

Air contains about 20% oxygen. Because noncondensables in a system cause system head pressures and discharge temperatures to rise, this oxygen in the air will react with the refrigeration oil to form organic solids. This reaction of oil and oxygen usually occurs at the discharge valve, because this is the hottest place in the entire system. Also, when moisture (water vapor), heat, and refrigerant are present in a system, acids will start to form after a short period of time. Refrigerants, such as R-12, R-502, R-22, R-134a, and the newer blends, contain either chlorine or fluorine and will *hydrolyze* (a chemical reaction) with water, forming hydrochloric and hydrofluoric acids and more water. After acids form, motor windings may deteriorate, and metal corrosion and sludge can occur. Sludge is a tightly bound mixture of water, acid, and oil.

Moisture + Acid + Oil = Sludge

These acids may remain in the system for years without showing signs of a problem. Then, the motor will burn out, or copper will be deposited on the crankshaft from the copper in the system, causing the crankshaft to become slightly oversized and causing binding to occur. This will cause the rubbing surfaces to score and become worn prematurely. 🌐To avoid corrosion and sludge problems in refrigeration systems, moisture must be kept out through good service practices and effective preventive maintenance.🌐 Sludge and corrosion cause expansion devices, filter driers, and strainers to plug and malfunction. 🌐The only sure way to rid the cooling system of moisture is to use good evacuation procedures through the use of a high vacuum pump. However, once sludge is formed, standard cleanup procedures with oversize driers specified for sludge removal must be used. 🌐 Vacuum pumps are not designed to remove solids such as sludge. Deep vacuum procedures will not take the place of liquid-line or suction-line driers because the vacuum pump will not remove sludge or solids. This can be accomplished only with the correct filtration.

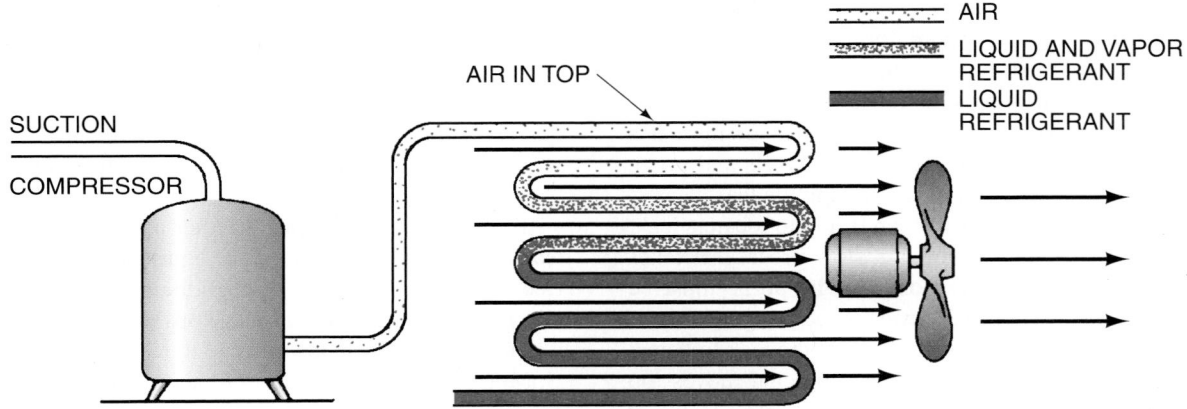

Figure 8–10   A condenser containing noncondensable gases.

These noncondensable gases must be removed from the system if it is to have a normal life expectancy. Many systems have been operated for years with small amounts of these products inside them, but they will not last or give the reliability the customer pays for.

Noncondensable gases are removed by vacuum pumps after the system is leak checked. The pressure inside the system must be reduced to an almost perfect vacuum for this to be accomplished.

8.7   THEORY INVOLVED WITH EVACUATION

To *pull a vacuum* means to lower the pressure in a system below the atmosphere's pressure. The atmosphere exerts a pressure of 14.696 psia (29.92 in. Hg) at sea level. Vacuum is commonly expressed in millimeters of mercury (mm Hg). The atmosphere will support a column of mercury 760 mm (29.92 in.) high. To pull a perfect vacuum in a refrigeration system, for example, the pressure inside the system must be reduced to 0 psia (29.92 in. Hg vacuum) to remove all of the atmosphere. This would represent a perfect vacuum, which has never been achieved.

A compound gage is often used to indicate the vacuum level. A compound gage starts at 0 in. Hg vacuum and reduces to 30 in. Hg vacuum. When the term in. Hg "vacuum" is used, it is applied to the compound gage. When the term in. Hg is used without "vacuum," it is applied to a manometer or barometer. Gages used for refrigeration have scales graduated in in. Hg vacuum.

We can use the bell jar in **Figure 8–11** to describe a typical system evacuation. A refrigeration system contains a volume of gas like the bell jar. The only difference is that a refrigeration system is composed of many small chambers

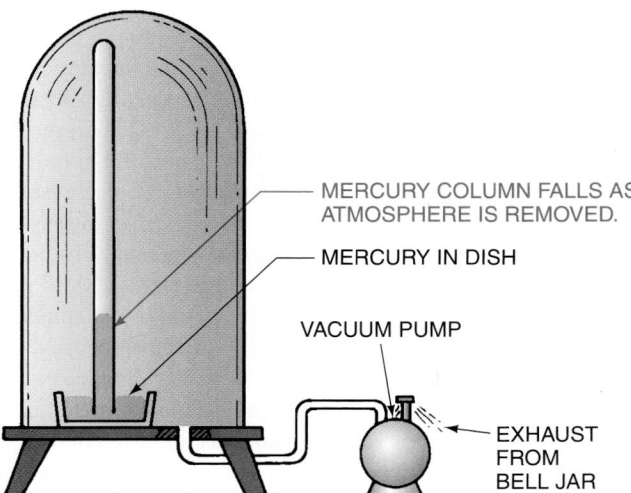

Figure 8–11 The mercury barometer in this bell jar illustrates how the atmosphere will support a column of mercury. As the atmosphere is removed from the jar, the column of mercury will begin to fall. If all of the atmosphere could be removed, the mercury would be at the bottom of the column.

ATMOSPHERIC PRESSURES, ABSOLUTE VALUES				COMPOUND GAGE READING in. Hg VACUUM	SATURATION POINTS of H₂O (BOILING—CONDENSING) °F
psia	in. Hg	mm Hg	microns		
14.696	29.921	759.999	759,999	00.000	212.00
14.000	28.504	724.007	724,007	1.418	209.56
13.000	26.468	672.292	672,292	3.454	205.88
12.000	24.432	620.577	620,577	5.490	201.96
11.000	22.396	568.862	568,862	7.526	197.75
10.000	20.360	517.147	517,147	9.617	193.21
9.000	18.324	465.432	465,432	11.598	188.28
8.000	16.288	413.718	413,718	13.634	182.86
7.000	14.252	362.003	362,003	15.670	176.85
6.000	12.216	310.289	310,289	17.706	170.06
5.000	10.180	258.573	258,573	19.742	162.24
4.000	8.144	206.859	206,859	21.778	152.97
3.000	6.108	155.144	155,144	23.813	141.48
2.000	4.072	103.430	103,430	25.849	126.08
1.000	2.036	51.715	51,715	27.885	101.74
0.900	1.832	46.543	46,543	28.089	98.24
0.800	1.629	41.371	41,371	28.292	94.38
0.700	1.425	36.200	36,200	28.496	90.08
0.600	1.222	31.029	31,029	28.699	85.21
0.500	1.180	25.857	25,857	28.903	79.58
0.400	0.814	20.686	20,686	29.107	72.86
0.300	0.611	15.514	15,514	29.310	64.47
0.200	0.407	10.343	10,343	29.514	53.14
0.100	0.204	5.171	5,171	29.717	35.00
0.000	0.000	0.000	0.000	29.921	—

NOTE: psia × 2.035 966 = in. Hg   psia × 51.715 = mm Hg   psia × 51.715 = microns

Figure 8–12 Pressure and temperature relationships for water below atmospheric pressure.

connected by piping. These chambers include the cylinders of the compressor, which may have a reed valve partially sealing it from the system. When the atmosphere is removed from the bell jar, it is often called *pulling a vacuum* or *degassing the bell jar*. When the noncondensables are removed from the refrigeration equipment, the process is often called *pulling a vacuum*. As the atmosphere is pulled out of the bell jar, the barometer inside the jar changes, **Figure 8–11**. The standing column of mercury begins to drop. When the column drops down to 1 mm, only a small amount of the atmosphere is still in the jar (1/760 of the original volume). **Figure 8–12** shows comparative scales for the saturation points of water. There will be more in this unit on how to apply the vacuum to both degassing and dehydration of a system. For now, realize that the compound pressure gage on the refrigeration manifold is only an indicator as far as vacuum is concerned. More accurate methods of measuring vacuum must be used.

When pulling a vacuum on a refrigeration system, the technician should attach the vacuum pump to the high- and the low-pressure sides of the system. This will prevent refrigerant from being trapped in one side of the system, such as when only one line is used. All large systems provide gage ports on both the high and low sides of the system.

8.8   MEASURING THE VACUUM

When the pressure in the bell jar is reduced to 1 mm Hg, the mercury column is hard to see, so another pressure measurement called the micron is used (1000 microns = 1 mm Hg; 1 in. Hg = 25,400 microns). Microns are measured with electronic instruments.

Accurately measuring and proving a vacuum in the low micron range can be accomplished with an electronic instrument, such as a thermocouple or thermistor vacuum gage. **Figure 8–13** shows typical electronic vacuum gages. Several companies manufacture an electronic vacuum gage, so the choice may be made by asking a reliable supply house

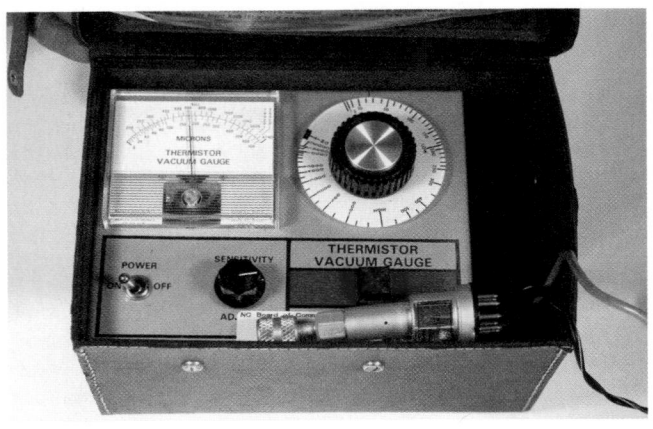

(A)

(B)

Figure 8–13 **(A)** An analog micron gage for deep vacuum measurements. **(B)** Modern electronic vacuum gages for measuring a deep vacuum. **(A)** Photo by Bill Johnson. **(B)** Courtesy Ferris State University. Photo by John Tomczyk

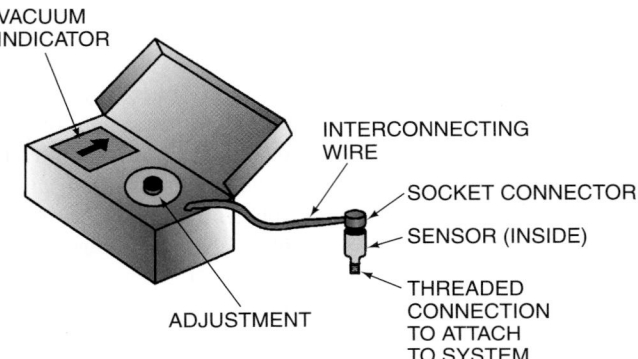

Figure 8–14 Components of an electronic vacuum gage.

which gage seems to be the most popular. Micron gage displays can be electronic analog, digital, or *light-emitting diode (LED).*

The electronic vacuum gage is attached to the system so it can measure the pressure inside the system. It has a separate sensor that attaches to the system with wires connecting it to the instrument, **Figure 8–14.** The sensor portion of the instrument must not be exposed to system pressure, so it is advisable to have a valve between it and the system. The sensor may also be installed using gage lines where it is disconnected from the system before refrigerant is added, **Figure 8–15.** *The sensor should always be located in an upright position so that any oil in the system will not drain into the sensor,* **Figure 8–16.**

When the electronic sensor is used, the vacuum pump should be operated for some period of time; when the gage-manifold gage begins to drop, indicating a deep vacuum, the valve to the sensor should be opened and the instrument turned on. It is pointless to turn the instrument on until a fairly deep vacuum is achieved, some value below 25 in. Hg as indicated on the manifold gage.

When the vacuum gage reaches the vacuum desired, usually about 250 microns, the vacuum pump should then be valved off and the reading marked. The instrument reading

may rise for a very short time, about 1 min, then it should stabilize. When the reading stabilizes, the true system reading can be recorded. If it continues to rise, either a leak or moisture is present and boiling to create pressure.

One of the advantages of the electronic vacuum gage is the rapid response to pressure rise. It will respond quickly to very small pressure rises. The smaller the system, the faster the rise. A very large system will take time to reach new pressure levels. These pressure differences may be seen instantly on the electronic instrument. *Be sure you allow for the first rise in pressure mentioned above before deciding a leak is present.*

One of the most practical and accurate methods of measuring a deep vacuum today is with a thermistor micron gage, which is a temperature-sensitive resistor that changes electrical resistance with a change in its temperature. Thermistors used in vacuum measurements are negative-coefficient thermistors, meaning that as their temperature increases, their electrical resistance decreases and vice versa. The electronic thermistor is mounted somewhere in the vacuum line, or a sensing tube is mounted in the vacuum line. They are heat-sensing devices in that the sensing element (thermistor) generates heat. The heat flow away from the thermistor changes as the surrounding vapors are removed from the refrigeration system. This results in a decrease in the thermistor's surrounding vapor pressure. As gases are removed from the system, less heat is swept away from the thermistor. This is because of the deeper vacuum (less gas molecules) around the thermistor. The thermistor's temperature now increases, thus decreasing its resistance because of its negative temperature coefficient. This change in its temperature and resistance will be indicated on a meter calibrated in microns of mercury. As soon as all of the moisture in the system is vaporized, the vapor pressure and heat taken away from the thermistor will continue decreasing, thus decreasing the micron measurement even further. The more vaporization of water or degassing of unwanted gases from the system, the more time it will take to reach a low micron level when using the same size vacuum pump. Again, it is the measurement from the meter that determines when an

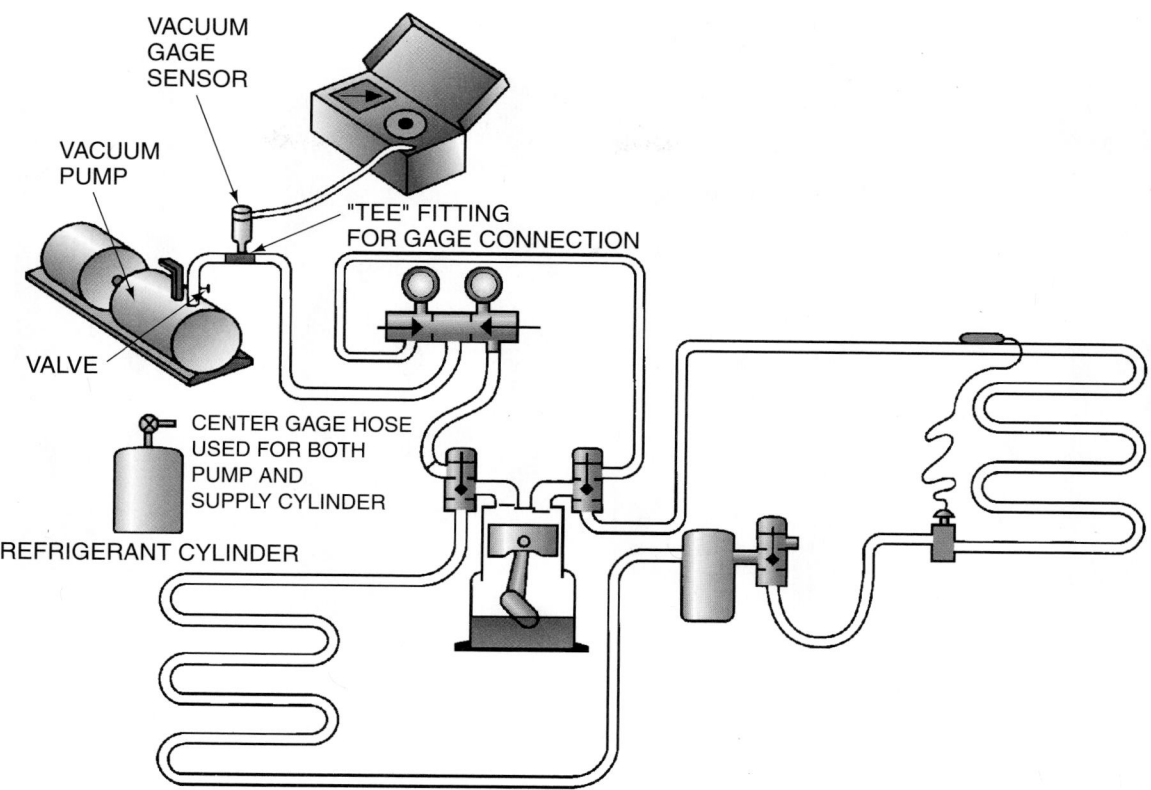

Figure 8–15   An electronic vacuum gage sensor is located near the vacuum pump. Notice that it can be disconnected at the "tee" fitting to prevent pressure from entering the sensor.

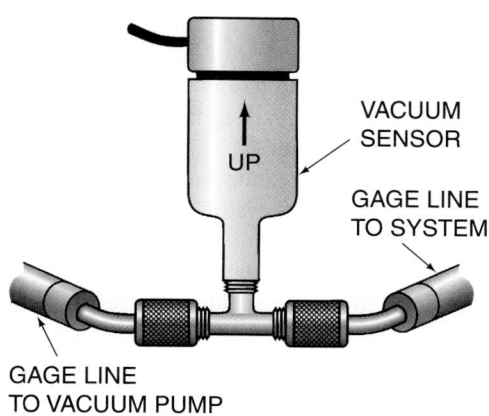

Figure 8–16   The electronic vacuum sensor must be mounted upright to prevent oil from entering the sensor.

evacuation is complete. Electronic thermistor vacuum gages are rugged and portable so the service technician can use them on the job. They can be accurate to 1 micron.

Another vacuum gage often used in refrigeration work is the U-tube manometer, **Figure 8–17(A),** which has a glass gage closed on one side and uses mercury as an indicator. The two columns of mercury balance each other. The atmosphere has been removed from one side of the mercury column so that the instrument has a standing column of about 5 in. Hg. This device can be used for fairly accurate reading

down to about 1 mm Hg. Since the columns of mercury are only about 5 in. different in height, this gage starts indicating at about 25 in. Hg vacuum below atmospheric pressure. When the gage is attached to the system and the vacuum pump is started, the gage will not read until the vacuum reaches about 25 in. Hg vacuum, **Figure 8–17(B).** Then the gage will gradually fall until the two columns of mercury are equal. At this time, the vacuum in the system is between 1 mm Hg and a perfect vacuum. The instrument cannot be read much closer than that because the eye cannot see any better, **Figure 8–17(C).**

Using a special valve arrangement helps a technician check the system pressure and the vacuum pump's capability. This arrangement allows the technician to isolate the vacuum pump and the sensor by closing the valve closest to the system so just the sensor can be evacuated. If the vacuum pump cannot develop enough vacuum, it can be shown at this point. By closing the valve closest to the vacuum pump, the sensor and the system are isolated to check the system pressure, **Figure 8–18.** These tests can be useful for determining vacuum pump operation and system pressures for a standing vacuum test.

The vacuum pump used to evacuate systems can often become contaminated with whatever is evacuated from a system. The vacuum pump has an oil sump where all contaminants seem to settle. It can contain acids and moisture. The oil in the vacuum pump must be changed on a regular basis for good pump performance. When a vacuum pump

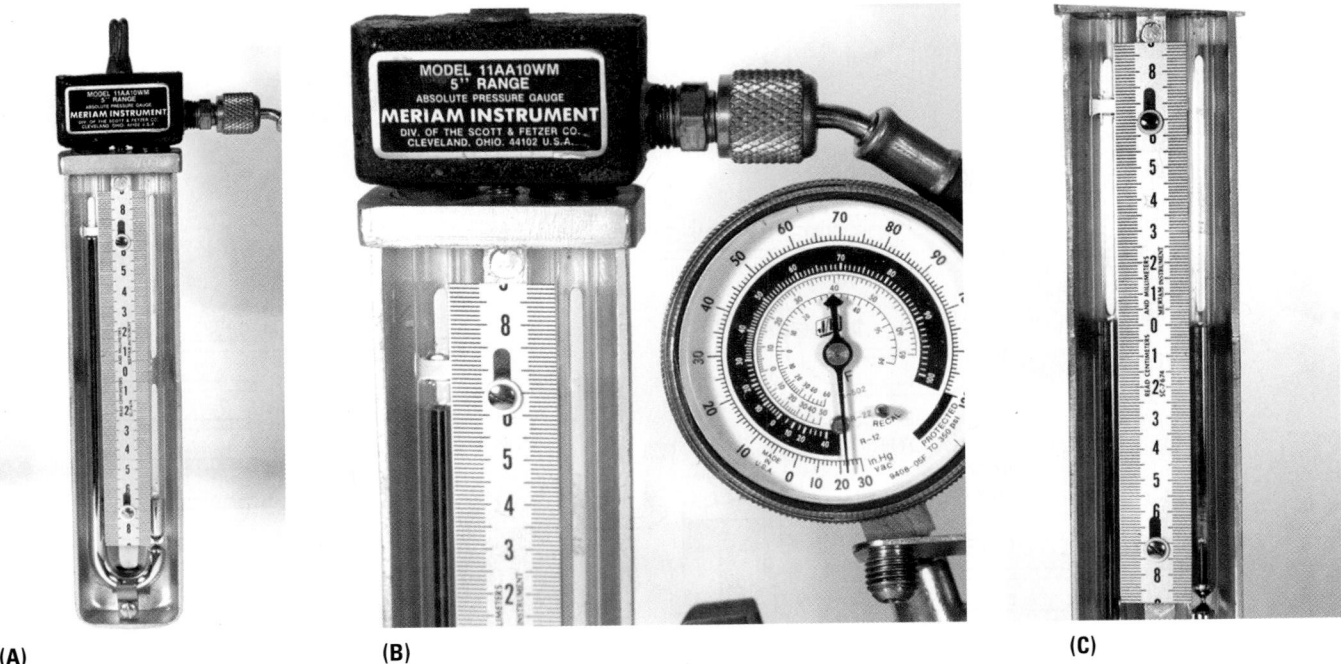

(A)   **(B)**   **(C)**

Figure 8–17   The mercury U-tube manometer is at various stages of evacuation. **(A)** The column on the left is a closed column with no atmosphere above it. **(B)** When the column on the right is connected to a system that is below atmosphere at a very low vacuum, the column on the left will fall, and the column on the right will rise. This will not start until the vacuum on the right is about 25 in. Hg vacuum or 5 in. Hg absolute. See the text for an explanation of the difference. As the atmosphere is pulled out of the right-hand column, the column rises. **(C)** At a perfect vacuum, it will rise to be exactly parallel to the left-hand column. This is very hard to see. *Photo by Bill Johnson*

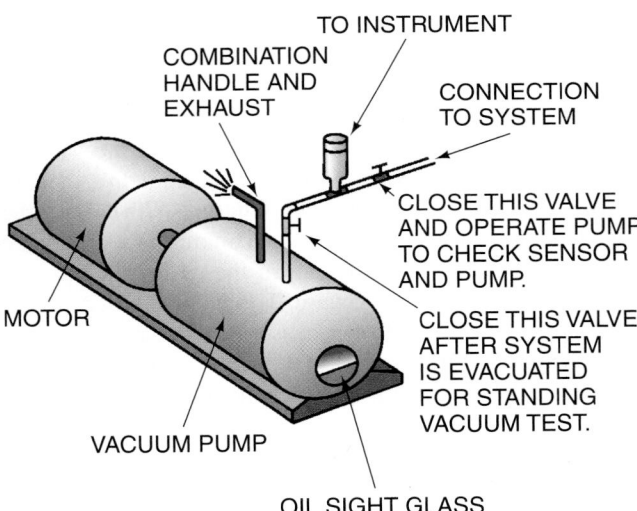

Figure 8–18   Valve arrangements for the vacuum pump and sensor.

will not reduce the pressure to the correct levels for a deep vacuum, the first suspect is a leak in the piping. When a valve arrangement such as the one shown in **Figure 8–18** is used, the piping is reduced to a minimum and reduces the checkpoints. If this valve and piping arrangement has been used on a regular basis and is known to be in good shape, the next thing that should be checked is the vacuum pump oil. It should be clean and clear. If it is cloudy, moisture is likely

to be present. The pump should then be serviced by changing the oil. NOTE: Only approved vacuum pump oil should be used for the best performance for your vacuum pump. Any other oil will have a boiling point that is too high, and deep vacuums cannot be achieved. Be sure to dispose of the old oil in the correct manner.

With the new oil in the pump, run the pump long enough for the pump to get up to operating temperature. It will usually be too hot to touch or hold. Drain the oil again. With fresh oil in the pump, repeat the vacuum test. If it is still not satisfactory, change the oil again and observe what the oil looks like. If it is discolored, add new oil and run a new test. The pump may need several changes to get the contaminants out of the entire pump. Fortunately, most vacuum pumps use only small quantities of oil, so you do not have much expense or much oil to dispose of.

Vacuum pump oil has a very low boiling point, and it is difficult to keep it from being contaminated while in storage. If your vacuum pump is in the truck, the vacuum pump is open to the atmosphere, and a certain amount of contamination will occur as moisture from the atmosphere enters the pump. In the daytime when the sun is out, the air temperature rises, and the pump gets up to daytime temperatures. At night when the air cools down, the pump cools, and this moist air migrates into the pump, **Figure 8–19**. If the pump is allowed to sweat due to weather changes, even more moisture will enter the pump, **Figure 8–20**. Oil stored in a can that is not properly sealed goes through the same process. All of this

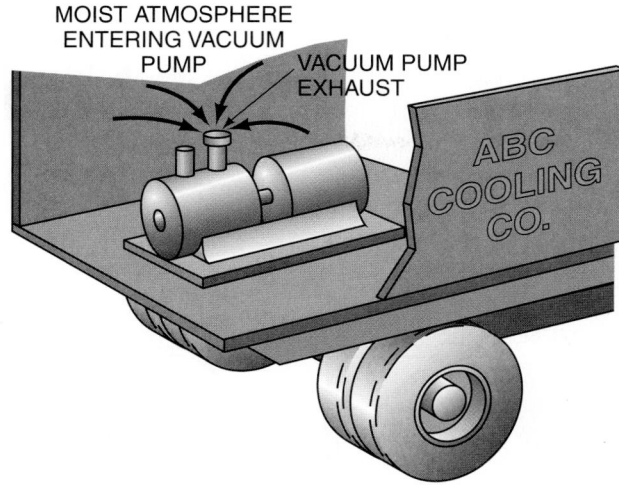

MOIST ATMOSPHERE ENTERING VACUUM PUMP

VACUUM PUMP EXHAUST

ABC COOLING CO.

Figure 8–19 When the vacuum pump cools down during the night, moist atmospheric air is pulled into the inside of the pump.

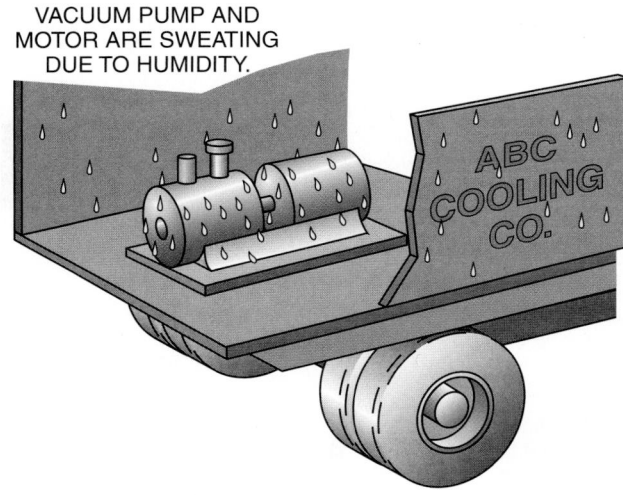

VACUUM PUMP AND MOTOR ARE SWEATING DUE TO HUMIDITY.

ABC COOLING CO.

Figure 8–20 When the vacuum pump sweats on the outside, it sweats on the inside. This contaminates the oil.

means that to keep a vacuum pump in top condition for very deep vacuums, the pump should not be charged with fresh oil until it is time to use it. This is often not practical as you move from job to job, so frequent oil changes are required. Always cap the top of the vacuum pump oil container and vacuum pump as soon as possible to prevent atmospheric moisture from migrating to the oil.

8.9 RECOVERING REFRIGERANT

♻ Before evacuating a system to remove contaminants or noncondensable gases, if there is or has been refrigerant in that system, the technician must remove the refrigerant. This must be done with EPA-approved recovery equipment. The amount of vacuum to be achieved in removing this refrigerant depends on the size of the system, the type of refrigerant,

and whether or not the recovery equipment was manufactured before or after November 15, 1993. The recovery of refrigerant is discussed in more detail in Unit 9, "Refrigerant and Oil Chemistry and Management—Recovery, Recycling, Reclaiming, and Retrofitting." ♻

8.10 THE VACUUM PUMP

A vacuum pump capable of removing the atmosphere down to a very low vacuum is necessary. The vacuum pumps usually used in the refrigeration field are manufactured with rotary compressors. The pumps that produce the lowest vacuums are two-stage rotary vacuum pumps, **Figure 8–21.** These vacuum pumps are capable of reducing the pressure in a leak-free vessel down to 0.1 micron. It is not practical to pull a vacuum this low in a field-installed system because the refrigerant oil in the system will boil slightly and create a vapor. The usual vacuum required by manufacturers is approximately 500 microns, although some may require a vacuum as low as 250 microns.

Two-stage vacuum pumps are nothing but two single-stage vacuum pumps in series. They are almost always of the rotary style because rotary vacuum pumps do not require any head clearance like piston-type pumps. This allows them to have a much higher volumetric efficiency. Two-stage vacuum pumps can pull much lower vacuums because their second stage experiences a much lower intake pressure. This lower pressure results from the exhaust of the first-stage vacuum pump being exhausted into the intake of the second-stage vacuum pump instead of to the atmospheric pressure. This gives the first-stage pump less back pressure and, therefore, a higher efficiency. The second-stage vacuum pump begins pulling at a lower pressure, so it can pull to a lower vacuum, as low as 0.1 microns. Two-stage vacuum pumps have the best track record because they

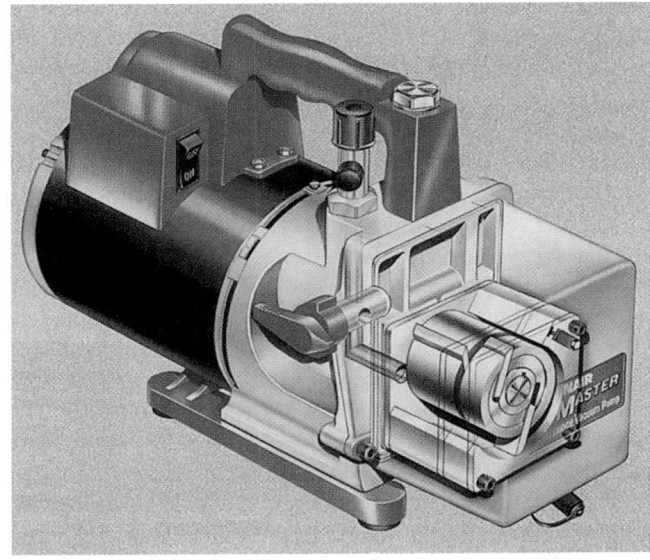

Figure 8–21 A two-stage rotary vacuum pump. *Courtesy Robinair SPX Corporation*

consistently pull lower vacuums and are much more efficient when removing moisture.

When moisture is in a system, a very low vacuum will cause the moisture to boil to a vapor. This vapor will be removed by the vacuum pump and exhausted to the atmosphere. Small amounts of moisture can be removed this way, but it is not practical to remove large amounts with a vacuum pump because of the large amount of vapor produced by boiling water. For example, 1 lb of water (about a pint) in a system will turn to 867 ft³ of vapor if boiled at 70°F.

8.11 DEEP VACUUM

The *deep vacuum* method involves reducing the pressure in the system to about 50 to 250 microns. When the vacuum reaches the desired level, the vacuum pump is valved off, and the system is allowed to stand for some time period to see whether the pressure rises. If the pressure rises and stops at some point, a material such as water is boiling in the system. If this occurs, continue evacuating. If the pressure continues to rise, there is a leak, and the atmosphere is seeping into the system. In this case the system should be pressured and leak checked again.

When a system pressure is reduced to 50 to 250 microns and the pressure remains constant, no noncondensable gas or moisture is left in the system. Reducing the system pressure to 250 microns is a slow process because when the vacuum pump pulls the system pressure to below about 5 mm (5000 microns), the pumping process slows down. The technician should have other work planned and let the vacuum pump run. Most technicians plan to start the vacuum pump as early as possible and finish other work while the vacuum pump does its work.

Some technicians leave the vacuum pump running all night because of the time involved in reaching a deep vacuum. Then the vacuum should be at the desired level the next morning. This is a good practice if some precautions are taken. Be sure to read the vacuum pump manufacturer's instructions regarding the length of running time for their pump. Some of the modern pumps may not have long running-time capabilities. When the vacuum pump pulls a vacuum, the system becomes a large volume of low pressure with the vacuum pump between this volume and the atmosphere, **Figure 8–22**. If the vacuum pump shuts off during the night from a power failure, the vacuum pump's oil may be sucked into the refrigeration system because of a strong pressure difference. If the power is restored and it starts back up, it will be without adequate lubrication and could be damaged. The oil is pulled out of the vacuum pump by the large vacuum volume, **Figure 8–23**. This can be prevented by installing a large solenoid valve in the vacuum line entering the vacuum pump and wiring the solenoid valve coil in parallel with the vacuum pump motor. The solenoid valve should have a large port to keep from restricting the flow, **Figure 8–24**. This will be discussed in more detail later in this unit. Now, if the power fails, or if someone disconnects the vacuum pump (a good possibility at a construction site), the vacuum will not be lost, and the vacuum pump will not lose its lubrication.

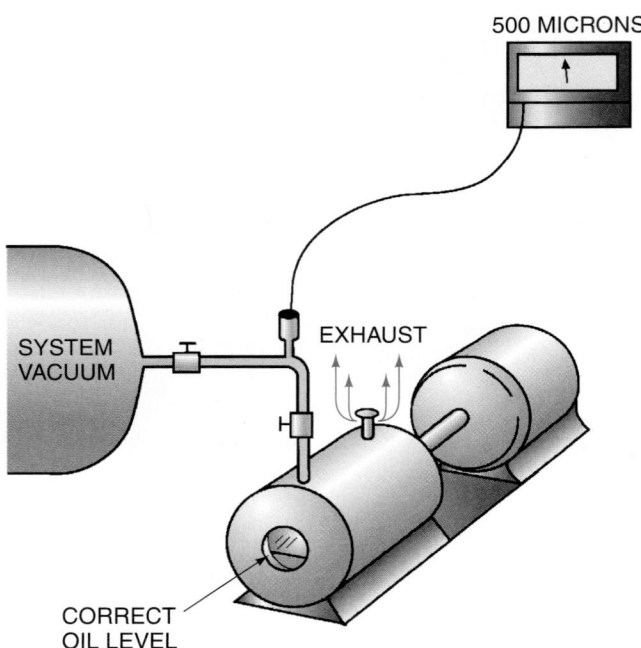

Figure 8–22 This vacuum pump has pulled a vacuum on a large system.

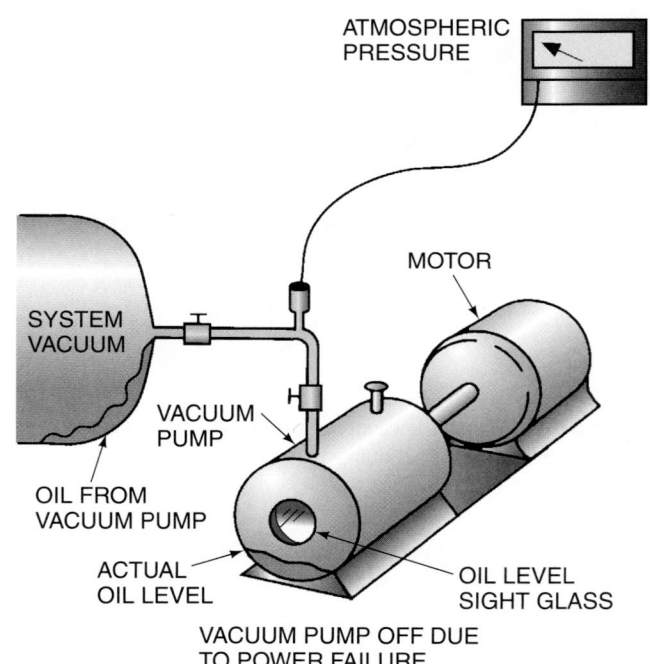

Figure 8–23 This system in a vacuum has pulled the oil out of the vacuum pump.

8.12 MULTIPLE EVACUATION

Multiple evacuation is used by many technicians for removing the atmosphere to the lowest level of contamination. Multiple evacuation is normally accomplished by evacuating a system to a low vacuum, about 1 or 2 mm, and then allowing a small amount of refrigerant to bleed into the system.

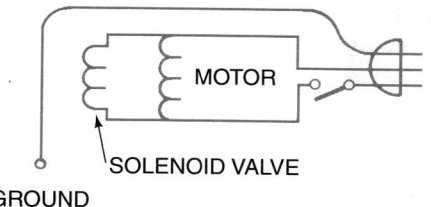

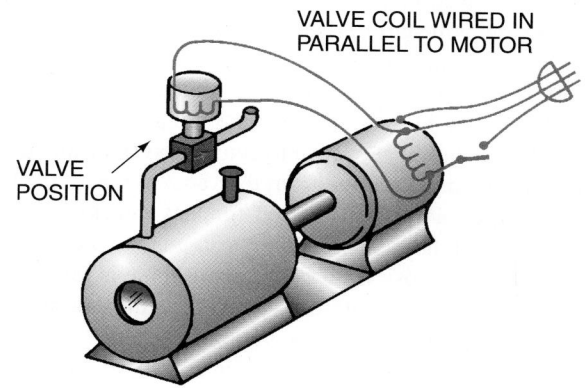

Figure 8–24   A vacuum pump with a solenoid valve in the inlet line. Note the direction of the arrow on the solenoid valve. The valve is installed to prevent flow from the pump. It must be installed in this direction.

The system is then evacuated until the vacuum is again reduced to 1 mm Hg. The following is a detailed description of a multiple evacuation. This one is performed three times and called a *triple evacuation*. **Figure 8–25** is a diagram of the valve attachments.

1. Attach a U-tube mercury manometer to the system. The best place is as far from the vacuum port as possible. For example, on a refrigeration system, the pump may be attached to the suction and discharge service valves and the U-tube manometer or micron gage to the liquid receiver valve port. Then start the vacuum pump.

2. Let the vacuum pump run until the indicator reaches 1 mm. The mercury manometer should be positioned vertically to take accurate readings. Lay a straightedge across the manometer to help determine the column heights, **Figure 8–26.**

3. Allow a small amount of refrigerant or nitrogen to enter the system until the vacuum is about 20 in. Hg. This must be indicated on the manifold gage because the mercury in the mercury gage will rise to the top and give no indication. **Figure 8–27** shows the manifold gage reading. This small amount of refrigerant vapor or nitrogen will fill the system and absorb and mix with other vapors.

4. Open the vacuum pump valve and start the vapor moving from the system again. Let the vacuum pump run until the vacuum is again reduced to 1 mm Hg. Then repeat step 3.

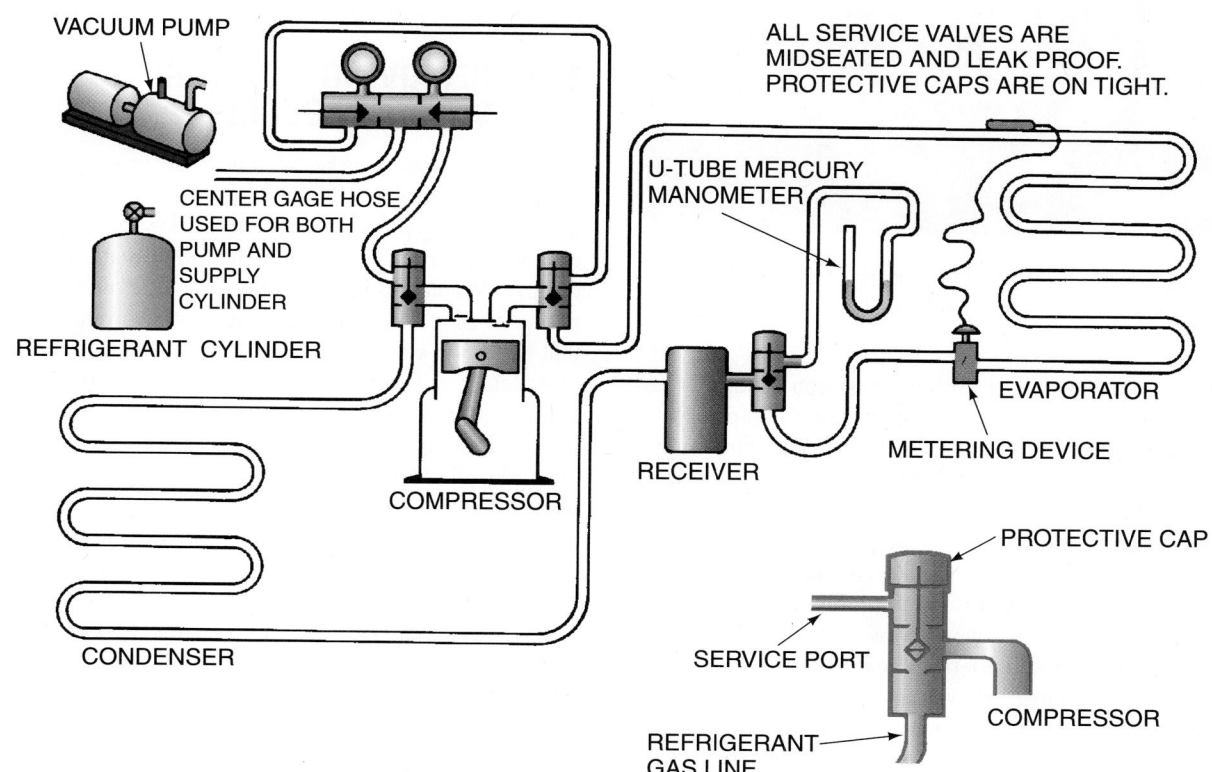

Figure 8–25   This system is ready for multiple evacuation. Notice the valve arrangements and where the U-tube manometer is installed.

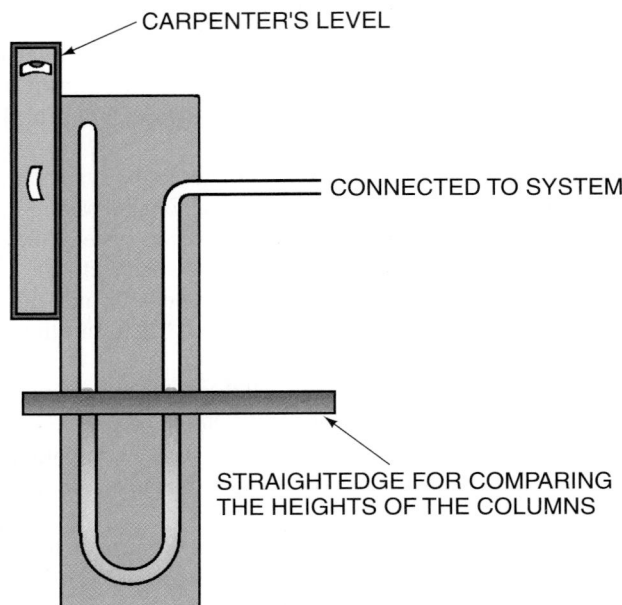

Figure 8–26 A mercury U-tube manometer being read as closely as possible. The manometer is positioned vertically, and a straightedge is used to compare the two columns of mercury.

Figure 8–27 Manifold gage reading 20 in. Hg vacuum. *Courtesy Ferris State University. Photo by John Tomczyk*

5. When the refrigerant or nitrogen has been added to the system the second time, open the vacuum pump valve and again remove the vapor. Operate the vacuum pump for a long time during this third pulldown. It is best to operate the vacuum pump until the manometer columns are equal. Some technicians call this *flat out*.

6. When the vacuum has been pulled the third time, allow refrigerant to enter the system until the system is about 2 psig above the atmosphere. Now remove the mercury manometer (it should not be exposed to system pressure) and charge the system.

The electronic vacuum gage may be used for triple evacuation by using the valve arrangement mentioned above to isolate the system between evacuations. Again, remember that the advantage of the electronic vacuum gage is that it

responds to pressure changes in the system very quickly. If a leak is present, it will tell you much faster than the U-tube manometer.

8.13 LEAK DETECTION WHILE IN A VACUUM

We mentioned that if a leak is present in a system, the vacuum gage will start to rise if the system is still in a vacuum, indicating a pressure rise. The vacuum gage will rise very fast. NOTE: Many technicians use this as an indicator that a leak is still in the system, but this is not a recommended leak test procedure. It allows air to enter the system, and the technician cannot determine from the vacuum where the leak is. Also, when a vacuum is used for leak checking it is only proving that the system will not leak under a pressure difference of 14.696 psi. If all of the atmosphere is removed from the system, only the atmosphere under atmospheric pressure is trying to get back into the system.

When checking for a leak using a vacuum, the technician is using a reverse pressure (the atmosphere trying to get into the system) of only 14.696 psi. The system, when fully loaded, typically has an operating pressure of 350 psig + 14.696 psi = 364.696 psia for an R-22 air-cooled condenser on a very hot day, **Figure 8–28.**

Using a vacuum for leak checking also may hide a leak. For example, if a pin-sized hole is in a solder connection that has a flux buildup on it, the vacuum will tend to pull the flux into the pinhole and may even hide it to the point where a deep vacuum can be achieved, **Figure 8–29.** Then when pressure is applied to the system, the flux will blow out of the pinhole, and a leak will appear.

8.14 REMOVING MOISTURE WITH A VACUUM

Removing moisture with a vacuum is the process of using the vacuum pump to remove moisture from a refrigeration system. Two kinds of moisture are in the system, vapor and liquid. When the moisture is in the vapor state, it is easy to remove. When it is in the liquid state, it is much more difficult to remove. The example given earlier in this unit shows that 867 ft³ of vapor at 70°F must be pumped to remove 1 lb of water. This is not a complete explanation because as the vacuum pump begins to remove the moisture, the water will boil, and the temperature of the trapped water will drop. For example, if the water temperature drops to 50°F, 1 lb of water will then boil to 1702 ft³ of vapor that must be removed. This is a pressure level in the system of 0.362 in. Hg or 9.2 mm Hg (0.362 × 25.4 mm/in.). The vacuum level is just reaching the low ranges. As the vacuum pump pulls lower, the water will boil more (if the vacuum pump has the capacity to pump this much vapor), and the temperature will decrease to 36°F. The water will now create a vapor volume of 2837 ft³. This is a vapor pressure in the system of 0.212 in. Hg or 5.4 mm Hg (0.212 × 25.4 mm/in.). This illustrates that lowering the pressure level creates more vapor. It takes a large vacuum pump to pull moisture out of a system.

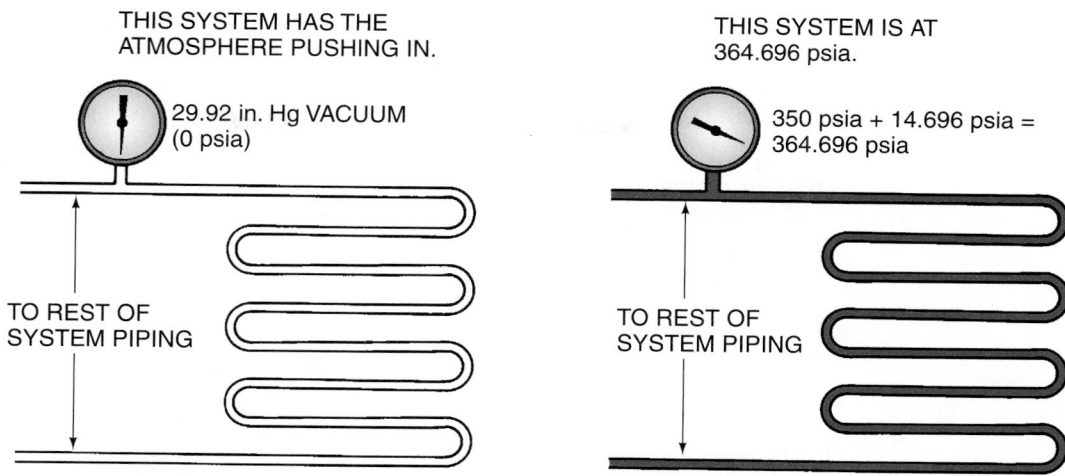

Figure 8–28   These two systems are being compared with each other under different pressure situations. One is evacuated and has the atmosphere under atmospheric pressure trying to get into the system. The other has 350 psig + 14.696 psia. The one under the most pressure is under the most stress. Using a vacuum as a leak test does not give the system a proper leak test.

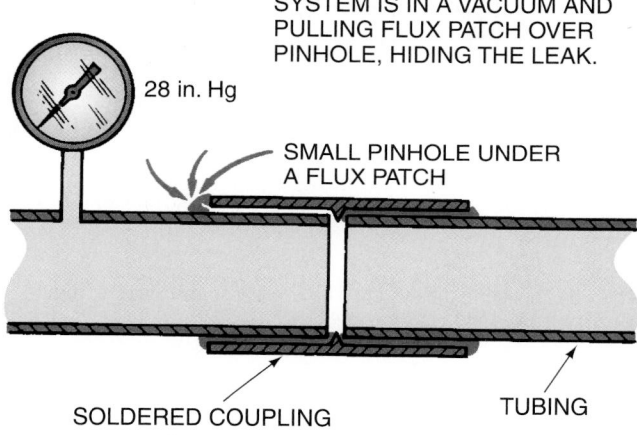

Figure 8–29   This system was leak checked under a vacuum.

(See **Figure 8–30** for the relationships between temperature, pressure, and volume.)

If the system pressure is reduced further, the water will turn to ice and be even more difficult to remove. If large amounts of moisture must be removed from a system with a vacuum pump, the following procedure will help.

1. Use a large vacuum pump. If the system is flooded, for example if a water-cooled condenser pipe ruptures from freezing, a 5-cfm (cubic feet per minute) vacuum pump is recommended for systems up to 10 tons. If the system is larger, a larger pump or a second pump should be used.

2. Drain the system in as many low places as possible. Remove the compressor and pour the water and oil from the system. **Do not add the oil back until the system is ready to be started, after evacuation. If you add it earlier, the oil may become wet and hard to evacuate.**

TEMPERA-TURE		SPECIFIC VOLUME OF WATER VAPOR	ABSOLUTE PRESSURE		
°C	°F	ft³/lb	lb/in.²	kPa	in. Hg
−12.2	10	9054	0.031	0.214	0.063
−6.7	20	5657	0.050	0.345	0.103
−1.1	30	3606	0.081	0.558	0.165
0.0	32	3302	0.089	0.613	0.180
1.1	34	3059	0.096	0.661	0.195
2.2	36	2837	0.104	0.717	0.212
3.3	38	2632	0.112	0.772	0.229
4.4	40	2444	0.122	0.841	0.248
5.6	42	2270	0.131	0.903	0.268
6.7	44	2111	0.142	0.978	0.289
7.8	46	1964	0.153	1.054	0.312
8.9	48	1828	0.165	1.137	0.336
10.0	50	1702	0.178	1.266	0.362
15.6	60	1206	0.256	1.764	0.522
21.1	70	867	0.363	2.501	0.739
26.7	80	633	0.507	3.493	1.032
32.2	90	468	0.698	4.809	1.422
37.8	100	350	0.950	6.546	1.933
43.3	110	265	1.275	8.785	2.597
48.9	120	203	1.693	11.665	3.448
54.4	130	157	2.224	15.323	4.527
60.0	140	123	2.890	19.912	5.881
65.6	150	97	3.719	25.624	7.573
71.1	160	77	4.742	32.672	9.656
76.7	170	62	5.994	41.299	12.203
82.2	180	50	7.512	51.758	15.295
87.8	190	41	9.340	64.353	19.017
93.3	200	34	11.526	79.414	23.468
98.9	210	28	14.123	97.307	28.754
100.0	212	27	14.696	101.255	29.921

Figure 8–30   This partial temperature/pressure relationship table for water shows the specific volume of water vapor that must be removed to remove a pound of water from a system.

3. Apply as much heat as possible without damaging the system. If the system is in a heated room, the room may be heated to 90°F without fear of damaging the room and its furnishings or the system, **Figure 8–31**. If part of the system is outside, use a heat lamp, **Figure 8–32**. The entire system, including the interconnecting piping, must be heated to a warm temperature, or the water will boil to a vapor where the heat is applied and condense

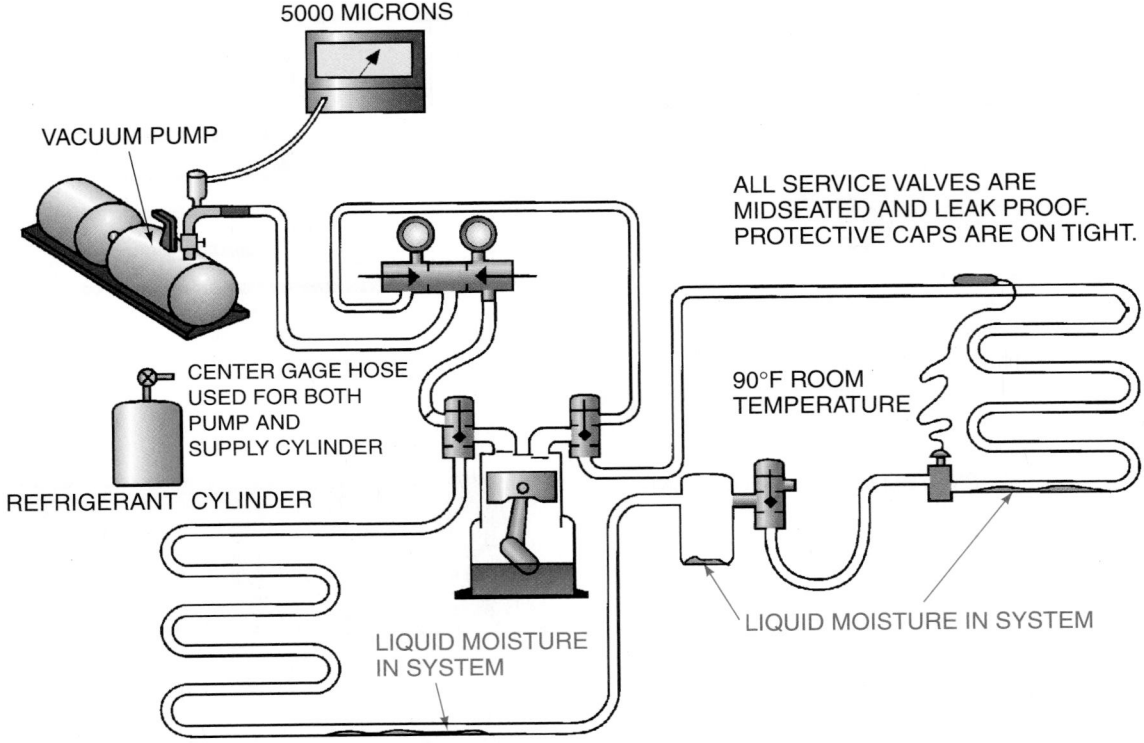

Figure 8–31   When a system has moisture in it and is being evacuated, heat may be applied to the system. This will cause the water to turn to vapor, and the vacuum pump will remove it.

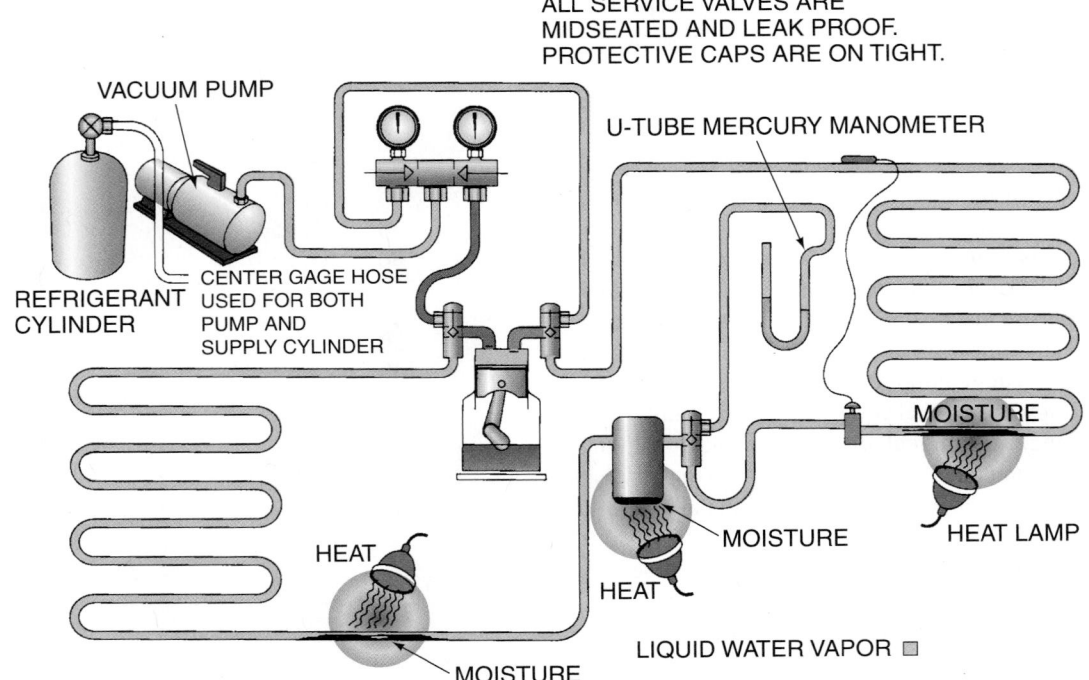

Figure 8–32   When heat is supplied to a large system with components inside and outside, the entire system must be heated. If not, the moisture will condense where the system is cool.

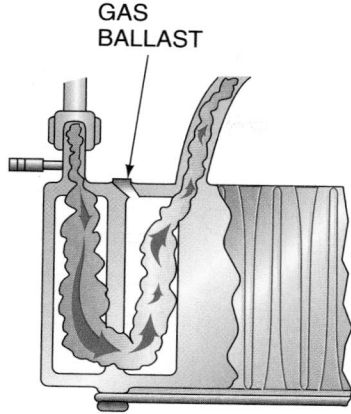

Figure 8–33 This vacuum pump has a feature called gas ballast, which allows a small amount of atmosphere to enter between the first and second stage. This prevents some moisture from condensing in the pump to cause oil contamination. *Courtesy Robinair Division, SPX Corporation*

where the system is cool. For example, if you know water is in the evaporator inside the structure and you apply heat to the evaporator, the water will boil to a vapor. If it is cool outside, the water vapor may condense outside in the condenser piping. The water is only being moved around.

4. Start the vacuum pump and observe the oil level in it. As moisture is removed, some of it will condense in the vacuum pump crankcase. Some vacuum pumps have a feature called *gas ballast* that introduces some atmosphere between the first and second stages of the two-stage pump, **Figure 8–33.** This prevents some of the moisture from condensing in the crankcase. Regardless of the vacuum pump, watch the oil level. NOTE: The water will displace the oil and raise the oil out of the pump. Soon, water may be the only lubricant in the vacuum pump crankcase, and damage may occur to the vacuum pump. They are *very* expensive and should be protected.

8.15 GENERAL EVACUATION PROCEDURES

Some general rules apply to deep vacuum and multiple evacuation procedures. If the system is large enough or if you must evacuate the moisture from several systems, you can construct a cold trap to use in the field. The *cold trap* is a refrigerated volume in the vacuum line between the wet system and the vacuum pump. When the water vapor passes through the cold trap, the moisture freezes to the walls of the trap, which is normally refrigerated with dry ice (CO_2), a commercially available product. The trap is heated, pressurized, and drained periodically to remove the moisture, **Figure 8–34.** NOTE: The cold-trap container must be able to withstand atmospheric pressure, 14.696 psi pushing in, when in a deep vacuum. A light-duty can collapse. The cold trap can save a vacuum pump.

Noncondensable gases and moisture may be trapped in a compressor and are as difficult to release as a vapor that can be pumped out of the system. A compressor has small chambers, such as cylinders, which may contain air or moisture. Only the flapper valves sit on top of these chambers, but there is no reason for the air or water to move out of the cylinder while it is under a vacuum. At times it is advisable to start the compressor after a vacuum has been tried. This is easy to do with the triple evacuation method. When the first vacuum has been reached, nitrogen can be charged into the system until it reaches atmospheric pressure. The compressor can then be started for a few seconds. All chambers should be flushed at this time. NOTE: Do not start a hermetic compressor while it is in a deep vacuum. Motor damage may occur. **Figure 8–35** is an example of vapor trapped in the cylinder of a compressor.

Water can be trapped in a compressor under the oil. The oil has surface tension, and the moisture may stay under it even under a deep vacuum. During a deep vacuum, the oil surface tension can be broken with vibration, such as the vibration that occurs when striking the compressor housing with a soft face hammer. Any kind of movement that causes the oil's surface to shake will work, **Figure 8–36.** Applying heat to the compressor crankcase will also release the water, **Figure 8–37.**

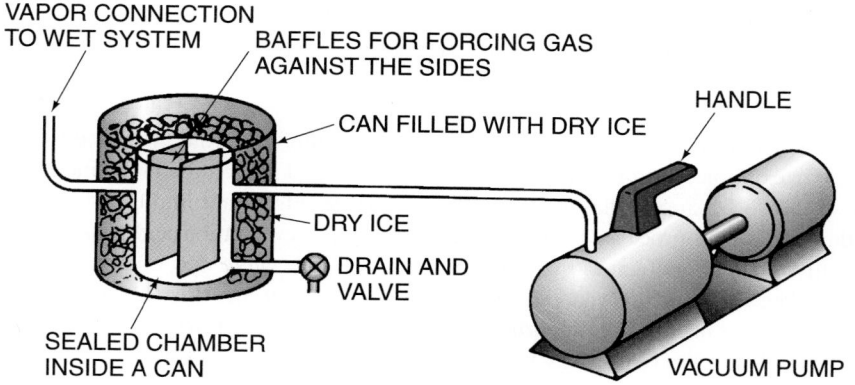

Figure 8–34 A cold trap.

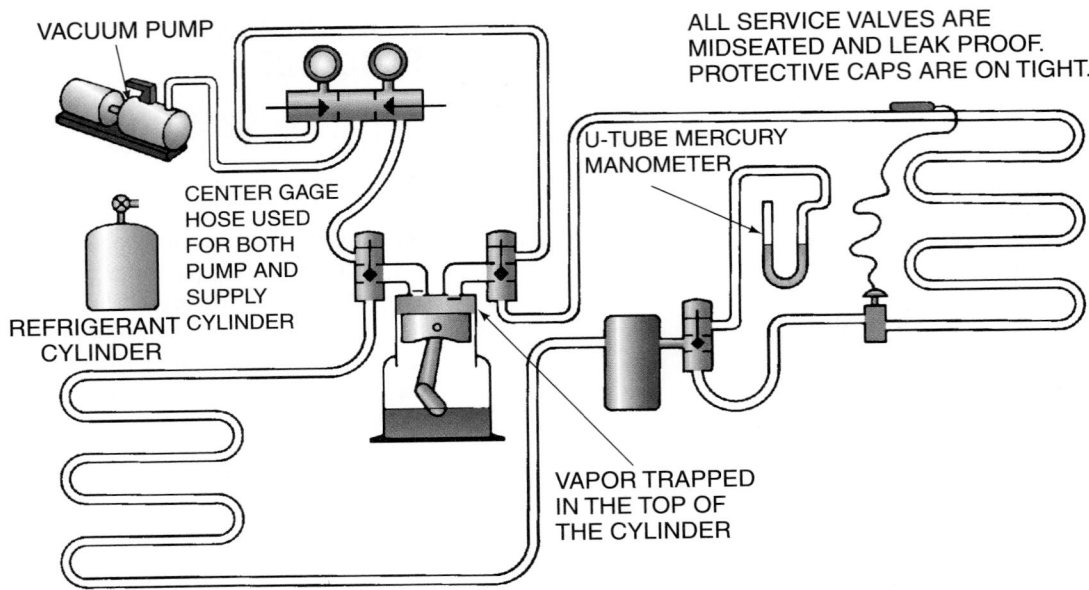

Figure 8–35   Vapor trapped in the cylinder of the compressor.

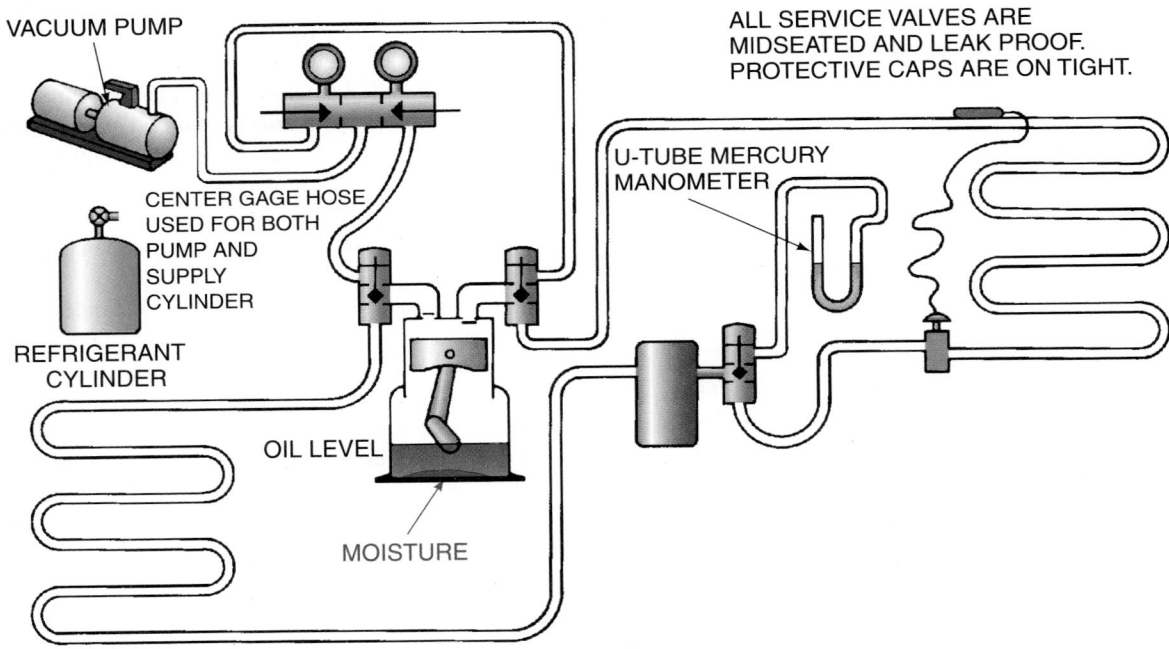

Figure 8–36   A compressor with water under the oil in the crankcase.

The technician who evacuates many systems must use time-saving procedures. For example, a typical gage manifold may not be the best choice because it has very small valve ports that slow the evacuation process, **Figure 8–38**. However, some gage manifolds are manufactured with large valve ports and a special large hose for the vacuum pump connection, **Figure 8–39**. The gage manifold in **Figure 8–40** has four valves and four hoses. The extra two valves are used to control the refrigerant and the vacuum pump lines. When

using this manifold, you need not disconnect the vacuum pump and switch the hose line to the refrigerant cylinder to charge refrigerant into the system. When the time comes to stop the evacuation and charge refrigerant into the system, close one valve and open the other, **Figure 8–40**. This is a much easier and cleaner method of changing from the vacuum line to the refrigerant line.

When a gage line is disconnected from the vacuum pump, as in a three-hose manifold, air is drawn into the gage hose.

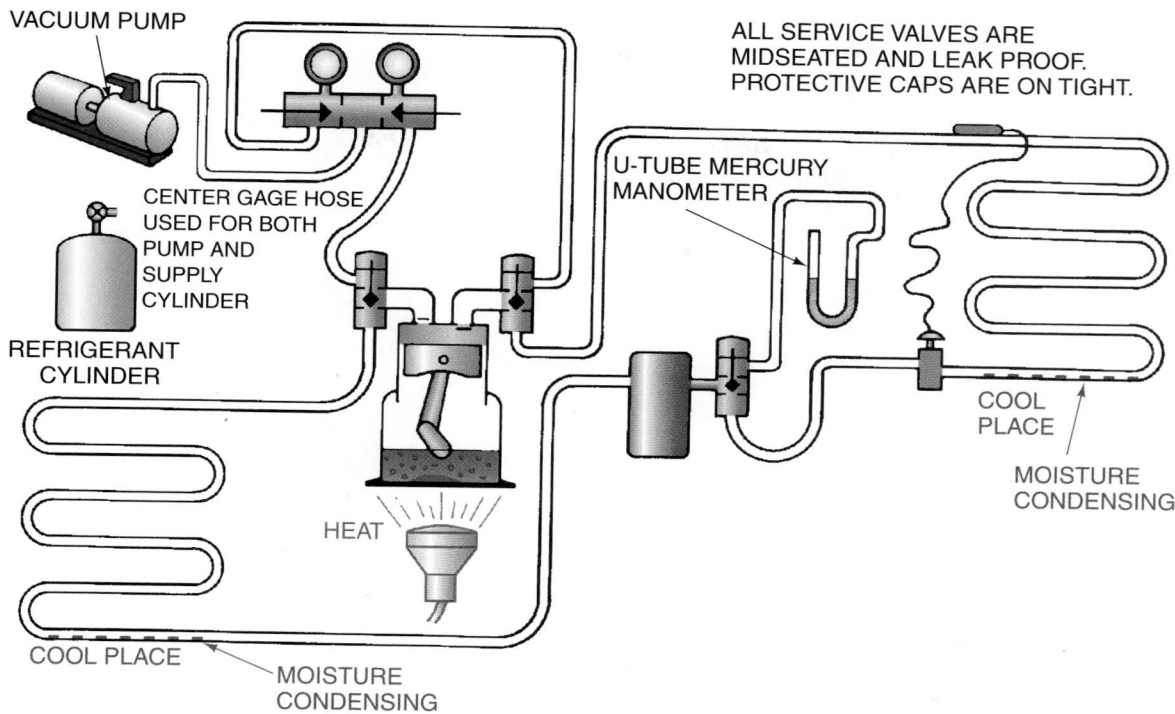

Figure 8–37   Heat is applied to the compressor to boil water under the oil.

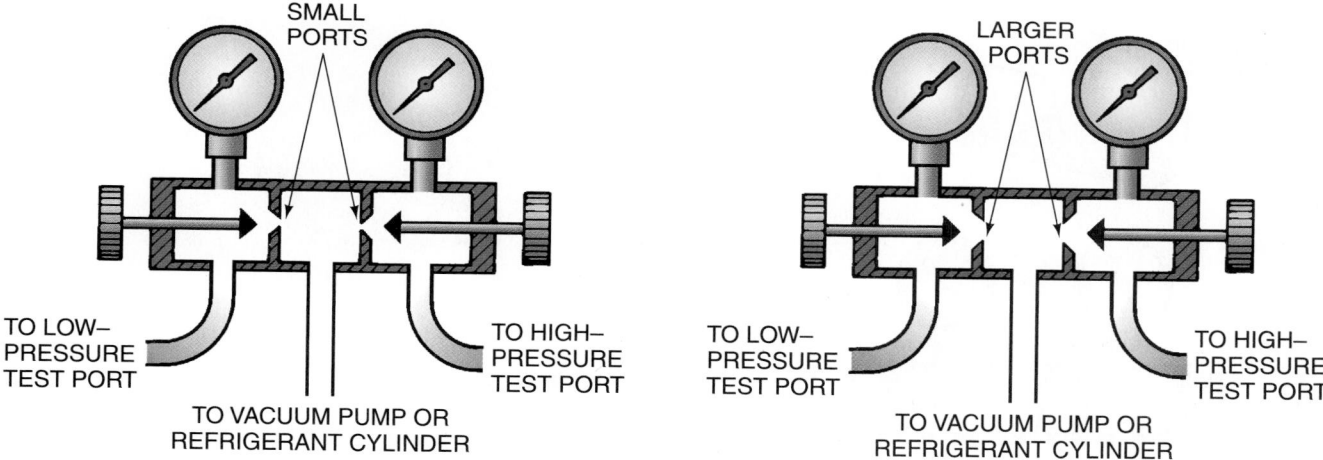

Figure 8–38   A gage manifold with small ports.

Figure 8–39   A gage manifold with large ports.

This air must be purged from the gage hose at the top, near the manifold. It is impossible to get all of the air out of the manifold because some will be trapped and pushed into the system, **Figure 8–41.**

Most gage manifolds have valve stem depressors in the ends of the gage hoses. The depressors are used for servicing systems with Schrader access valves. These valves are much like the valve and stem on an automobile tire. These depressors are a restriction to the evacuation process. When a vacuum pump pulls down to the very low ranges (1 mm Hg),

these valve depressors slow the vacuum process considerably. Many technicians erroneously use oversized vacuum pumps and undersized connectors, because they do not realize that the vacuum can be pulled much faster with large connectors. The valve depressors can be removed from the ends of the gage hoses, and adapters can be used when valve depression is needed. **Figure 8–42** shows one of these adapters. **Figure 8–43** is a small valve that can be used on the end of a gage hose; it will even give the technician the choice as to when the valve stem is depressed.

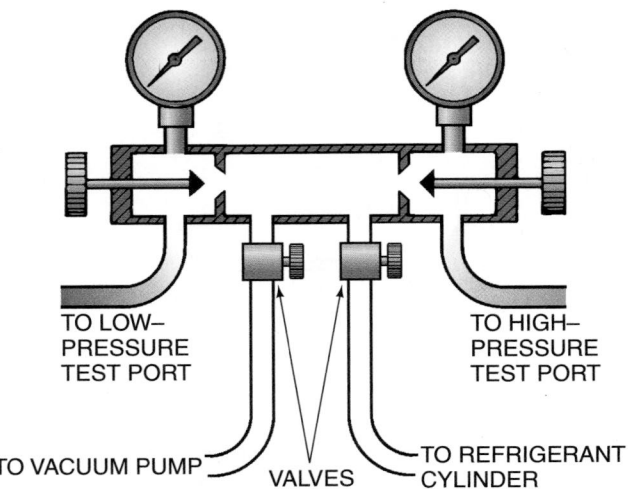

Figure 8–40   A manifold with four valves and four gage hoses.

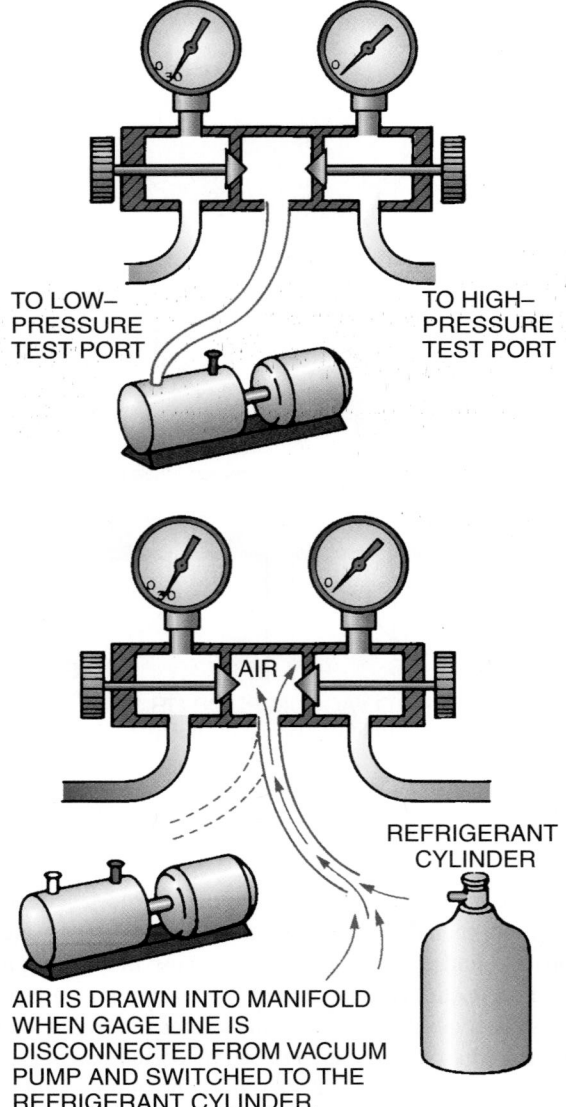

AIR IS DRAWN INTO MANIFOLD
WHEN GAGE LINE IS
DISCONNECTED FROM VACUUM
PUMP AND SWITCHED TO THE
REFRIGERANT CYLINDER.

Figure 8–41   This piping diagram shows how air is trapped in the gage manifold.

Figure 8–42   This gage adapter can be used instead of the gage depressors that are normally in the end of the gage lines. The adapters may be used for gage readings. *Photo by Bill Johnson*

Figure 8–43   This small valve can also be used for controlled gage readings. When the technician wants to read a pressure in a Schrader port, this adapter valve may be used by turning the valve handle. *Courtesy Ferris State University. Photo by John Tomczyk*

8.16   SYSTEMS WITH SCHRADER VALVES

A system with Schrader valves for gage ports will take much longer to evacuate than a system with service valves. The reason is that the valve stems and the depressors act as very small restrictions. An alternative is to remove the valve stems during evacuation and replace them when evacuation is finished. A system with water to be removed will take a great deal of time to evacuate if there are Schrader valve stems in the service ports. These valve stems are designed to be removed for replacement, so they can also be removed for evacuation, **Figure 8–44.**

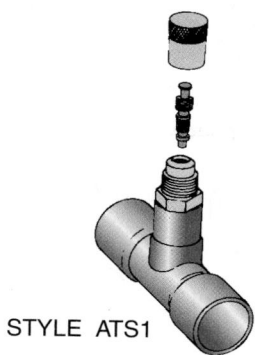

STYLE ATS1

Figure 8–44   A Schrader valve assembly. *Courtesy J/B Industries*

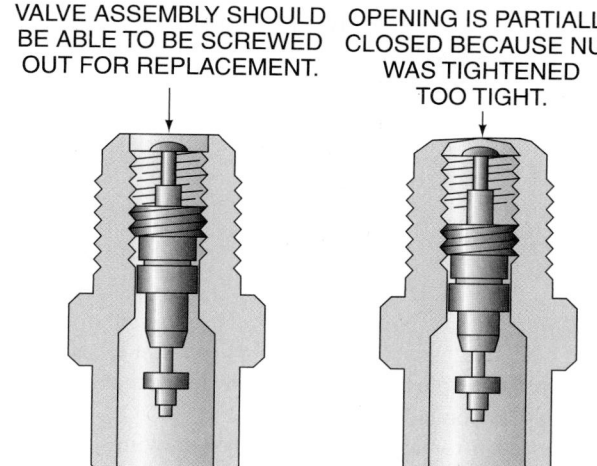

VALVE ASSEMBLY SHOULD BE ABLE TO BE SCREWED OUT FOR REPLACEMENT.

OPENING IS PARTIALLY CLOSED BECAUSE NUT WAS TIGHTENED TOO TIGHT.

Figure 8–45   This fitting has a sensitive sealing shoulder on top. The sealing cap that comes with it has a neoprene gasket that is soft. If a brass flare cap is used and tightened down tight, the top of the fitting will become distorted, and the valve stem cannot be removed.

A special tool, called a field service valve, can be used to replace Schrader valve stems under pressure, or it can be used as a control valve during evacuation. The tool has a valve arrangement that allows the technician to evacuate a system through it with the stem backed out of the Schrader valve. The stem is replaced when the evacuation is completed.

Schrader valves are shipped with a special cap, which is used to cover the valve when it is not in use. This cover has a soft gasket and should be the only cover used for Schrader valves. If a standard brass flare cap is used and overtightened, the Schrader valve top will be distorted and valve stem service will be difficult, if it can be done at all, **Figure 8–45**.

8.17   GAGE MANIFOLD HOSES

The standard gage manifold uses flexible hoses with connectors on the ends. These hoses sometimes get pinhole leaks, usually around the connectors, which may leak while under a vacuum but not be evident when the hose has pressure inside it. The reason is that the hose swells when pressurized.

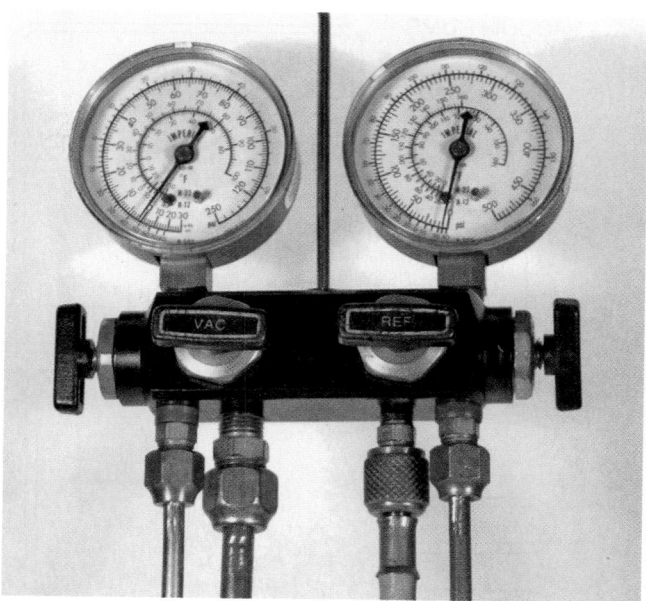

Figure 8–46   A gage manifold with copper gage lines and a large vacuum pump line. *Photo by Bill Johnson*

If you have trouble while pulling a vacuum and you cannot find a leak, substitute soft copper tubing for the gage lines, **Figure 8–46**.

8.18   SYSTEM VALVES

For a system with many valves and piping runs, perhaps even multiple evaporators, check the system's valves to see whether they are open before evacuation. A system may have a closed solenoid valve. The valve may trap air in the liquid line between the expansion valve and the solenoid valve, **Figure 8–47**. This valve must be opened for complete evacuation. It may even need a temporary power supply to operate its magnetic coil. Some solenoid valves have a screw on the bottom to manually jack the valve open, **Figure 8–48**.

8.19   USING DRY NITROGEN

Maintaining good workmanship practices while assembling or installing a system can make system evacuation an easier task. ⊙When piping is field installed, sweeping dry nitrogen through the refrigerant lines can keep the atmosphere pushed out and clean the pipe. It is relatively inexpensive to use a dry nitrogen setup, and using it saves time and money.⊙

When a system has been open to the atmosphere for some time, it needs evacuation. The task can be quickened by sweeping the system with dry nitrogen before evacuation. **Figure 8–49** shows how this is done.

8.20   CLEANING A DIRTY SYSTEM

The technician should be aware of how to use a vacuum to clean a dirty system. Several types of contaminants may be in a system. Water and air have been discussed, but they are

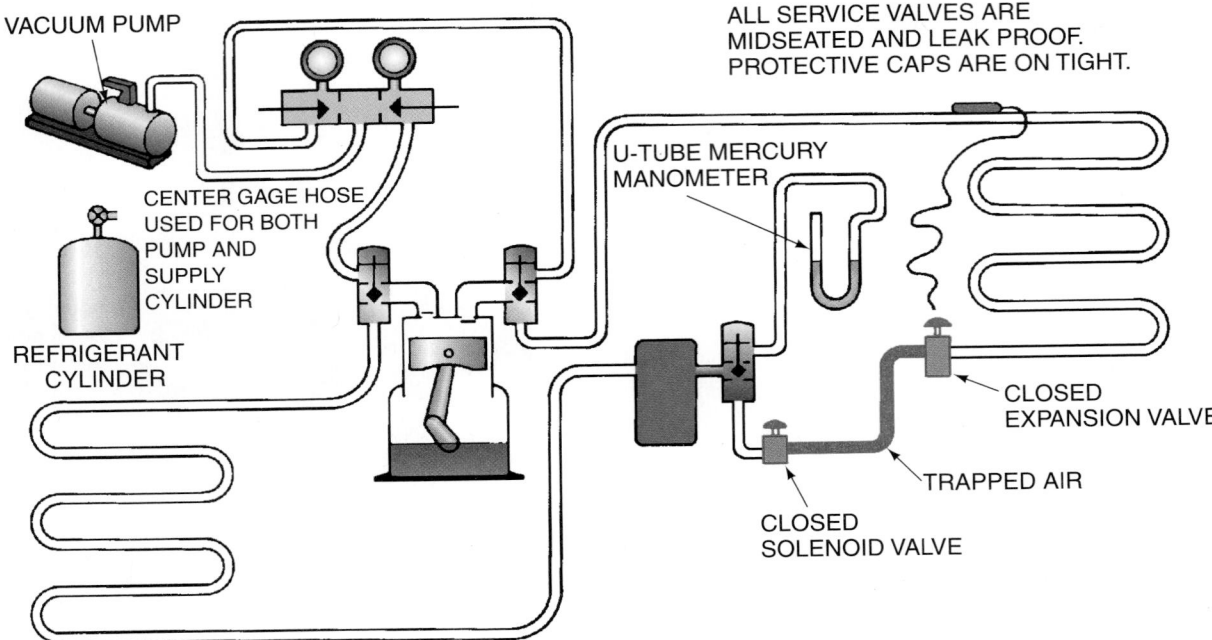

Figure 8–47 A closed solenoid valve trapping air in the system liquid line.

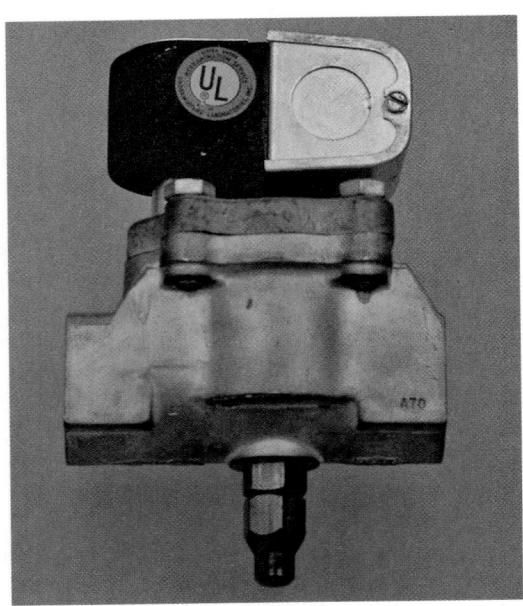

Figure 8–48 A manual opening stem for a solenoid valve. *Photo by Bill Johnson*

not the only contaminants that can form in a system. The hermetic motor inside a sealed system is the source of heat in a motor-burn circumstance. This heat source can heat the refrigerant and oil to temperatures that will break down the oil and refrigerants to acids, soot (carbon), and sludge that cannot be removed with a vacuum pump. SAFETY PRECAUTION: Whenever working with a burned-out system, extreme caution must be used when handling and/or operating the system. Adequate ventilation, butyle gloves, and safety glasses are required to protect the technician from acids. The oil

from a burnout can cause serious skin irritation and possible burns. In some cases, the fumes are toxic. Let us use a bad motor-burn example to demonstrate how most manufacturers would expect you to clean a system.

Suppose a 5-ton air-conditioning system with a fully hermetic compressor were to have a severe motor burn while running. When a motor burn occurs while the compressor is running, the soot and sludge from the hot oil move into the condenser, **Figure 8–50.** The following steps will be taken to clean the system.

1. ♻ *The refrigerant must be recovered from the system. This process is discussed in detail in Unit 9.* ♻
2. When the refrigerant is removed from the system, the compressor can then be changed for the new one. It will not be connected until later.
3. As mentioned before, contamination is in the system. It is in the vapor, liquid, and solid states. The vacuum pump can remove only the vapor state substances so other methods may be used to sweep some of the contaminants from the system.
4. Dry nitrogen may be used to push some of the contaminants out of the system by simply attaching the nitrogen regulator to one of the lines and allowing it to blow out the other. SAFETY PRECAUTION: Do not exceed the system working pressure with the nitrogen. Without the compressor in the system, you can safely use 250 psig for high-pressure refrigerant systems. Because the contaminants are known to be in the condenser, the line may be disconnected before the expansion device where a liquid-line filter drier can be installed later. Nitrogen may then be purged through the liquid line toward the compressor and discharged out the compressor discharge line before the

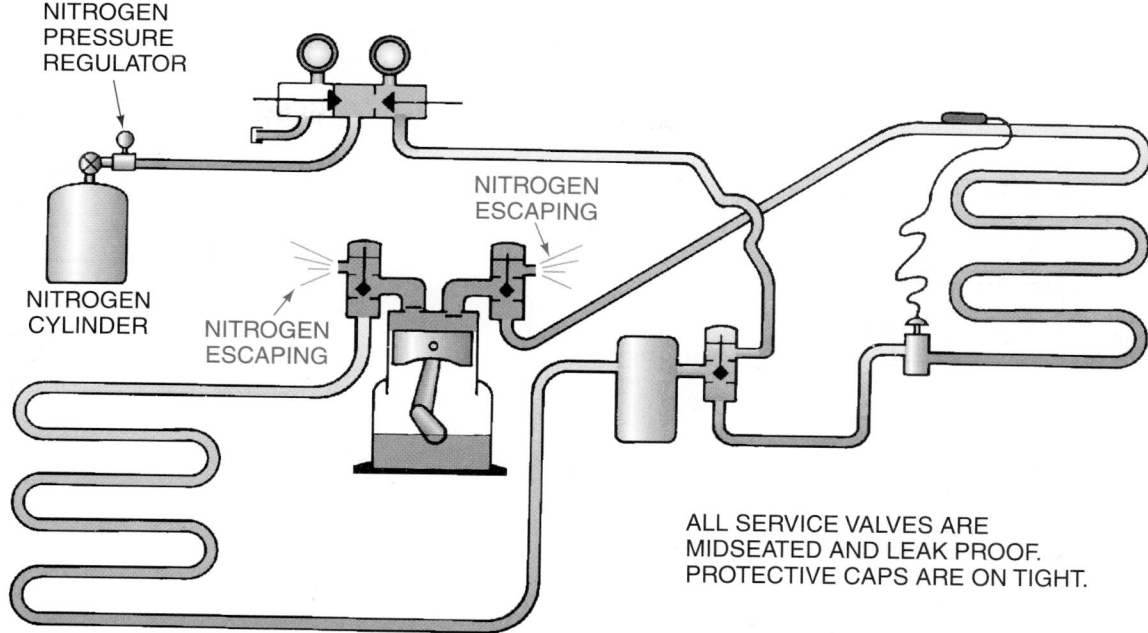

Figure 8–49   The technician is using dry nitrogen to sweep this system before evacuation.

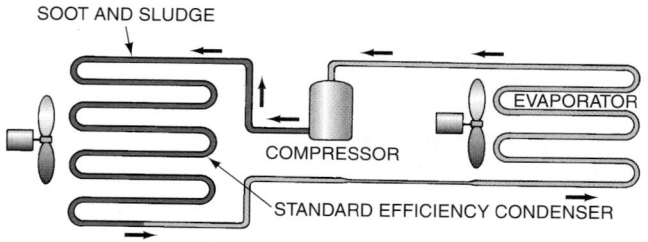

Figure 8–50   Soot and sludge move into the condenser when a motor burn occurs while the compressor is running.

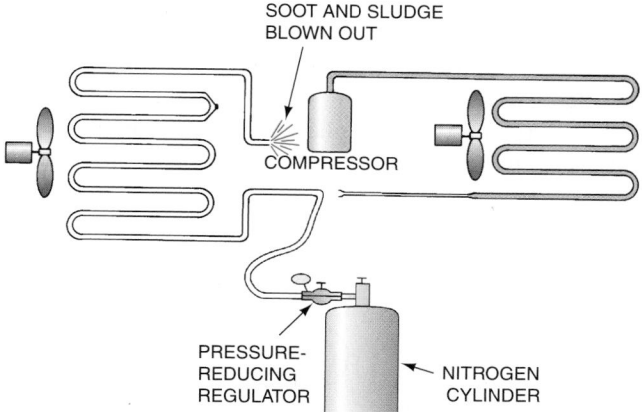

Figure 8–51   Purging contaminants from the condenser.

compressor is connected, **Figure 8–51**. This will push all loose contaminants out at this point.

5. The nitrogen cylinder may then be connected to the expansion-device side of the liquid line, and this line may be purged toward the compressor suction line.

The velocity of the refrigerant will be reduced because of the expansion device, but this is all that can be done without disconnecting the expansion device. If it is a capillary tube system, there will be several connections and this will not be practical.

6. The system has been purged as much as possible, but contaminants are still in the system, as solids and liquids (soot and contaminated oil). The compressor is now connected to the system with a suction-line filter drier installed just before the compressor. This will prevent any contamination from entering the new compressor on start-up. Purging the system with dry nitrogen also ensures maximum capacity from the filter drier because some of the contaminants have already been pushed out of the system.

7. A liquid-line filter drier is installed just before the metering device to prevent contamination from restricting the refrigerant flow. Oversized (temporary) suction-line and liquid-line driers are often used before the standard (permanent) driers are installed.

8. The system is leak checked and ready to evacuate with a vacuum pump. If you have a choice of an old used or a new vacuum pump, use the old one, because contamination may be pulled through the pump.

9. Evacuate the system to a low vacuum, 250 microns, or triple evacuate (whichever you usually do), then charge the system with refrigerant.

10. Start the system and keep it running as long as practical to circulate the refrigerant through the filter driers. The refrigerant in the system is the best solvent you can find to break loose the contaminants that will be trapped in the filter driers. The pressure drop across the suction-line drier may be monitored to see whether it is gathering particles and beginning

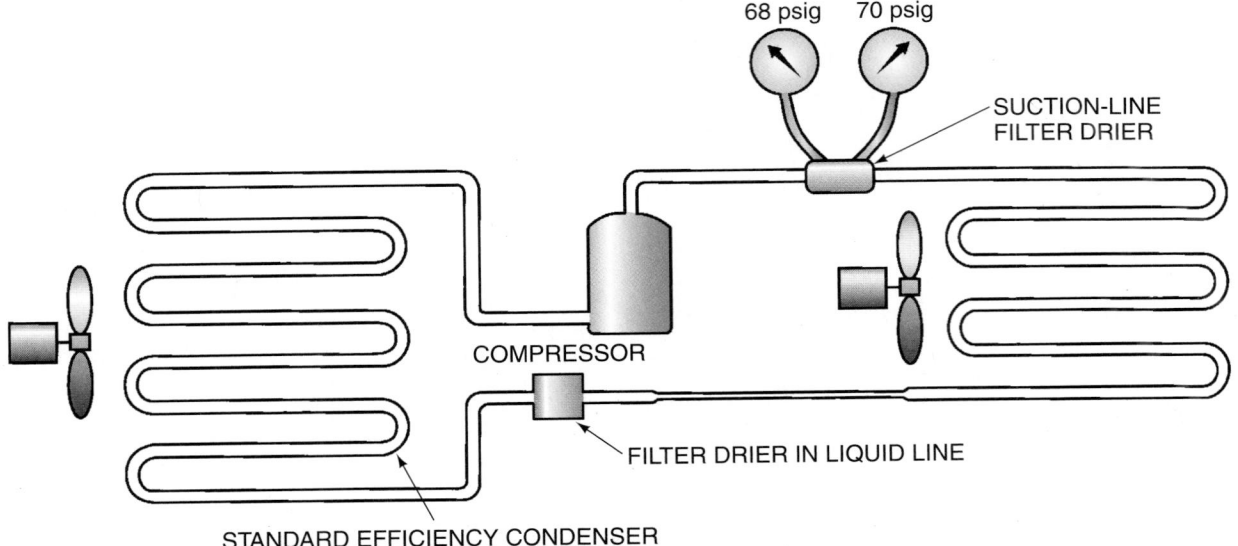

Figure 8–52   Checking the pressure drop across a suction-line filter drier.

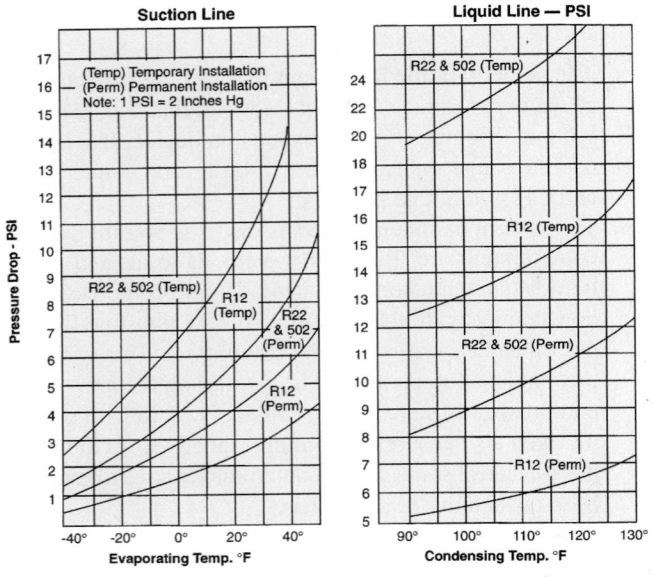

Figure 8–53   The maximum recommended pressure drop for filter driers. *Courtesy Copeland Corporation*

to become restricted, **Figure 8–52.** The manufacturer's literature will tell you what maximum pressure drop is allowed. If it becomes restricted beyond the manufacturer's recommendation, pump the system down and change the drier. If it does not become restricted, just leave it in the system; it will not hurt anything, **Figure 8–53.**

11. Change the oil in the vacuum pump while it is hot. Run it for 30 min and change it again to be sure all contamination has been removed from the pump crankcase. Many technicians will do a great job of cleaning the refrigeration system and neglect the vacuum pump, which is very expensive and just as important as the system compressor.

12. Check the pressure drop across the liquid-line filter drier by measuring the temperature in and out of the drier. If there is a temperature drop, there is a pressure drop, **Figure 8–54.**

Filter drier manufacturers have done a good job of developing filter media that will remove acid, moisture, and carbon

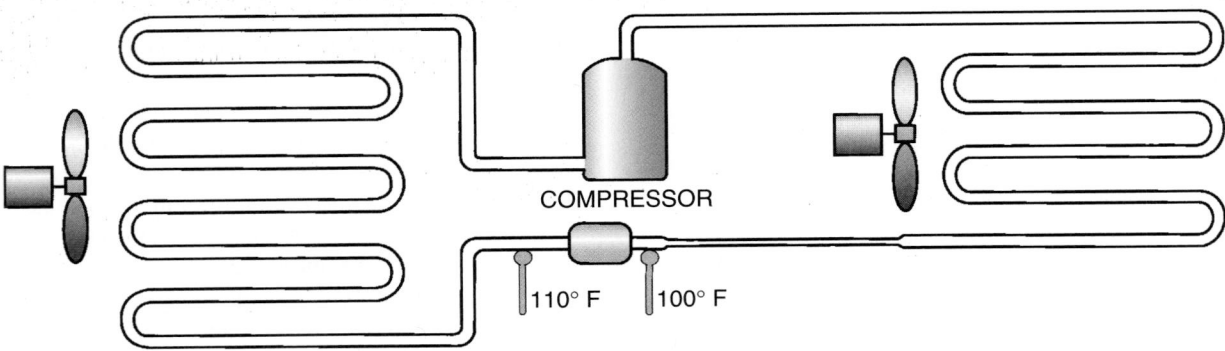

Figure 8–54   The 10°F temperature differential indicates a pressure drop; the drier should be changed.

sludge. They claim that nothing can be created inside a system with only oil and refrigerant that the filter driers will not remove.

SUMMARY

■ Nitrogen and R-22 (used as a trace refrigerant for finding leaks) are commonly used to leak check a system.

■ Using R-22 as a trace refrigerant is approved, because it is not used as an operating refrigerant.

■ The entire system must not be pressurized higher than the lowest system test pressure.

■ While under pressure, the system can be leak checked using several different methods: halide, soap, ultrasonic, and electronic.

■ It is not required by the Environmental Protection Agency to repair systems that contain less than 50 lb of refrigerant. For industrial processes equipment and commercial refrigeration equipment, systems that contain more than 50 lb of refrigerant must be repaired when the leak rate is more than 35% of their total charge in 1 year. An annual leak rate of 15% is allowed for comfort cooling chillers and all other equipment.

■ Noncondensable gases and moisture are common foreign matter that get into systems during assembly and repair. They must be removed.

■ Evacuation using low vacuum levels removes noncondensable gases and involves pumping the system to below atmospheric pressure.

■ Vapors will be pumped out by the vacuum pump. Liquids must boil to be removed with a vacuum.

■ Water makes a large volume of vapor when boiled at low pressure levels. It should be drained from a system if possible.

■ The two common vacuum gages are the U-tube manometer and the electronic micron gage.

■ Pumping a vacuum may be quickened with large unrestricted lines.

■ Good workmanship and piping practice along with a dry nitrogen setup will lessen the evacuation time.

■ When noncondensables are left in a system, mild acids (hydrochloric and hydrofluoric) will slowly form and deteriorate the system by attacking the motor windings and causing copper plating on the crankshaft.

■ The oxygen in the air is the real problem when air is allowed to enter a system.

■ Nitrogen in a system will cause excess head pressure because it takes up condensing space.

■ The best system pressure test is a standing pressure test, normally 150 psig.

■ The only advantage in using a vacuum to test for leaks is that vacuum instruments quickly respond to leaks. The vacuum leak test only proves that the system piping will prevent atmosphere from entering the system (14.969 psi).

■ When a vacuum pump is allowed to run unattended, the system becomes a large vacuum reservoir—and if the vacuum pump were to be shut off, the oil will be pulled into the system. If the vacuum pump is then restarted, it will be operating without lubrication.

■ Special valve arrangements allow the technician to check the vacuum gage and pump and also check the system pressure.

■ System Schrader valve cores may be removed to quicken the vacuum on a system; then the cores may be replaced before the system is put back in operation.

■ Using dry nitrogen to sweep a contaminated system will help in evacuating a dirty system.

■ *Do not forget to clean the vacuum pump after evacuating a contaminated system.*

REVIEW QUESTIONS

1. The low-pressure side of the system must not be pressurized to more than 150 psig because of
   A. the type of refrigerant.
   B. a dry-nitrogen danger.
   C. the low-side test pressure.
   D. atmospheric pressure.

2. True or False: The high-pressure side working pressure for most systems is 700 psig.

3. The _____ is both the greatest asset and the greatest liability of an electronic leak detector.

4. Why should oxygen or compressed air never be used to pressurize a system?

5. List some of the foreign matter that may enter a refrigeration system.

6. True or False: A vacuum pump can remove any type of foreign matter that enters a refrigeration system.

7. Air in a refrigeration system causes which of the following problems?
   A. Acid buildup
   B. Moisture
   C. Copper plating
   D. All of the above

8. The only two products that should be circulating in a refrigeration system are _____ and _____.

9. For a technician to remove moisture from a refrigeration system using a vacuum pump, it must be in a _____ state.

10. Are the best vacuum pumps one stage or two stage?

11. Name the two types of vacuum procedures used by technicians.

12. What can be done to get water out from under the oil level in a compressor?

13. When evacuating a refrigeration system that has solenoid valves, what special procedure must be followed?

14. Name two common vacuum gages.

15. _____ microns are in 1 mm Hg.

16. What gas is commonly used to sweep a refrigeration system to push out any air in the system?

17. What is the advantage of using large gage lines for evacuation?

18. At _____ in. Hg, water boils at 60°F.

19. When a vacuum is pulled on the water in question 18, the water vapor created for each pound of water removed is _____ cubic feet.

UNIT 9

Refrigerant and Oil Chemistry and Management—Recovery, Recycling, Reclaiming, and Retrofitting

OBJECTIVES

After studying this unit you should be able to

- describe ozone depletion and global warming.
- discuss how CFCs deplete the earth's ozone layer.
- differentiate between CFCs, HCFCs, HFCs, and HCs.
- discuss popular refrigerants (including R-410A) and their applications.
- discuss refrigerant blends.
- discuss temperature glide and fractionation as it applies to refrigerant blends.
- discuss refrigerant oils and their applications.
- discuss EPA regulations as they relate to refrigerants.
- define the terms *recover, recycle,* and *reclaim.*
- describe methods of recovering refrigerants, including active and passive methods.
- identify a DOT-approved recovery cylinder.

SAFETY CHECKLIST

- ✔ Gloves and goggles should be worn any time a technician transfers refrigerant from one container to another.
- ✔ Transfer refrigerants into only DOT-approved containers.
- ✔ When retrofitting refrigerants and oils, always follow retrofit guidelines.
- ✔ Never use R-22 or any other refrigerant's service equipment on R-410A equipment. Service equipment must be rated to handle the higher operating pressures of R-410A.
- ✔ ASHRAE Standard 15 requires the use of room sensors and alarms to detect R-123 refrigerant leaks.
- ✔ All refrigeration systems are required to have some sort of pressure relief device, which should be vented to the outdoors to prevent the buildup of excessive system pressures.
- ✔ A high-pressure reading in a recovery tank is an indication of air or other noncondensables in the tank.
- ✔ Never overheat the system's tubing while the system is pressurized. This can cause a blowout, and serious injury can occur. When certain refrigerants are exposed to either high temperatures, a flame, or a glowing hot metal, they can decompose into hydrochloric and/or hydrofluoric acids and phosgene gas.
- ✔ Never energize the compressor when under a deep vacuum. This can cause a short in the motor windings at the Fusite terminals or at a weak spot in the windings, which can cause compressor damage.
- ✔ Always check the nameplate of the system into which dry nitrogen will enter. Never exceed the design low-side factory test pressure on the nameplate. Never pressurize a system with oxygen or compressed air because dangerous pressures can be generated from a reaction with the oil as it oxidizes.
- ✔ When using dry nitrogen, always use a pressure relief valve and pressure regulator on the nitrogen tank.

9.1 REFRIGERANTS AND THE ENVIRONMENT

As the earth's population grows and demands for comfort and newer technologies increase, more and more chemicals are produced and used in various combinations. Many of these chemicals reach the earth's atmospheric layers, producing different types of pollution. Some of these chemicals make up the refrigerants that cool or freeze our food and cool our homes, office buildings, stores, and other buildings in which we live and work. Many of the refrigerants we use were developed in the 1930s.

While contained within a system, the refrigerants are stable, are not pollutants, and produce no harmful effects to the earth's atmospheric layers. Over the years, however, quantities have leaked from systems and have been allowed to escape by the willful purging or venting of systems during routine service procedures.

9.2 OZONE DEPLETION

Ozone is a gas that is found in both the stratosphere and troposphere, **Figure 9–1.** Ozone is a form of oxygen gas whose molecule consists of three oxygen atoms (O_3), **Figure 9–2(A).** The standard oxygen molecule that we breathe is (O_2), **Figure 9–2(B).** Stratospheric ozone, which resides 7 to 30 miles above the earth, is considered good ozone because it prevents harmful ultraviolet (UV-B) radiation from reaching the earth. Tropospheric ozone, however, is considered unwanted ozone because it is a pollutant.

Stratospheric ozone is formed by the reaction that occurs when sunlight interacts with oxygen. This ozone shelters the earth from UV-B radiation; this protection is essential for human health. Stratospheric ozone accounts for more than 90% of the earth's total ozone and is rapidly being depleted by man-made chemicals containing chlorine, including refrigerants such as CFCs and HCFCs.

ATMOSPHERIC REGIONS

IONOSPHERE
(30–300 MILES)

STRATOSPHERE
(7–30 MILES)

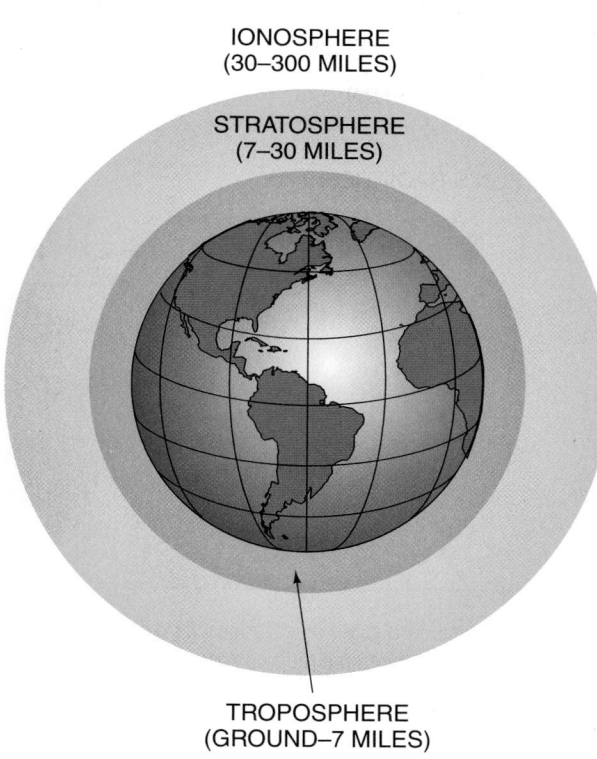

TROPOSPHERE
(GROUND–7 MILES)

Figure 9–1 Atmospheric regions.

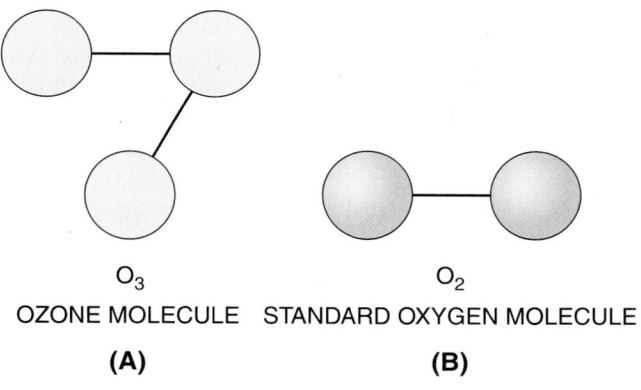

O_3 O_2
OZONE MOLECULE   STANDARD OXYGEN MOLECULE

(A)   **(B)**

Figure 9–2 **(A)** An ozone molecule (O_3) and **(B)** a standard oxygen molecule (O_2).

The troposphere, or lower atmosphere that extends upward from ground level to about 7 miles, holds only about 10% of the total ozone. The troposphere contains 90% of the atmosphere and is well mixed by weather patterns. Sunlight acting on chemicals in the troposphere causes bad ozone. This ozone has a bluish color when seen from earth and may irritate mucous membranes when inhaled. A popular term for tropospheric ozone pollution is *smog*.

Once stratospheric ozone is formed, much of it is easily destroyed by the sun's ultraviolet radiation, which is very powerful and has a frequency almost as high as that of X-rays. The sun's radiation will break down the ozone (O_3)

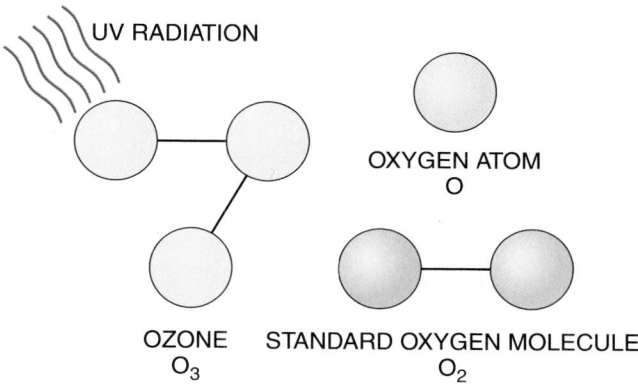

Figure 9–3 The breaking up of an ozone molecule by ultraviolet radiation.

OZONE DEPLETION PROCESS

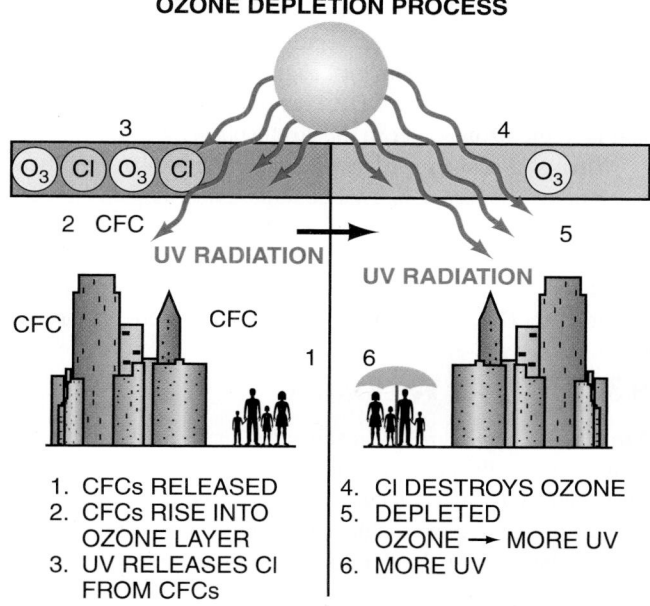

1. CFCs RELEASED
2. CFCs RISE INTO OZONE LAYER
3. UV RELEASES Cl FROM CFCs
4. Cl DESTROYS OZONE
5. DEPLETED OZONE → MORE UV
6. MORE UV

Figure 9–4 The ozone depletion process. *Courtesy U.S. EPA*

molecule into a standard oxygen (O_2) molecule and an elemental free oxygen (O) molecule, **Figure 9–3**. At the same time, more ozone is produced through photosynthesis and the bonding of standard oxygen (O_2) with free oxygen (O). Ozone is constantly being made and destroyed in the stratosphere. This balance has been going on for millions of years. However, man-made chlorine-containing chemicals have knocked this delicate process out of balance in the last couple of decades.

Once in the stratosphere, ultraviolet radiation will break off a chlorine atom (Cl) from the CFC or HCFC molecule, **Figure 9–4**. The chlorine atom destroys many ozone molecules. One chlorine atom (Cl) can destroy up to 100,000 ozone molecules. The depleted ozone molecules in the stratosphere let more UV-B radiation reach the earth. This causes an increase in skin cancers, an increase in the frequency of cataracts in humans and animals, a weakening of

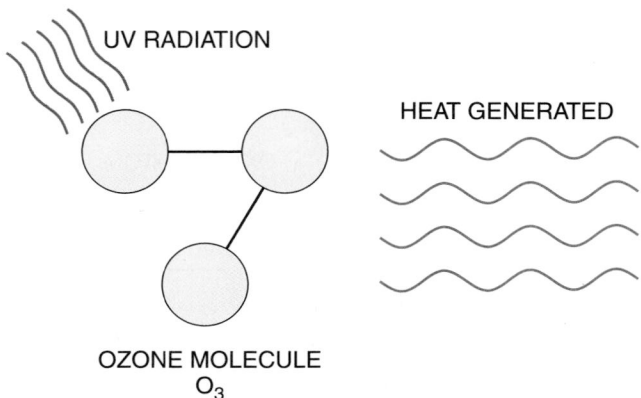

Figure 9–5 Heat generated from the breakup of an ozone molecule.

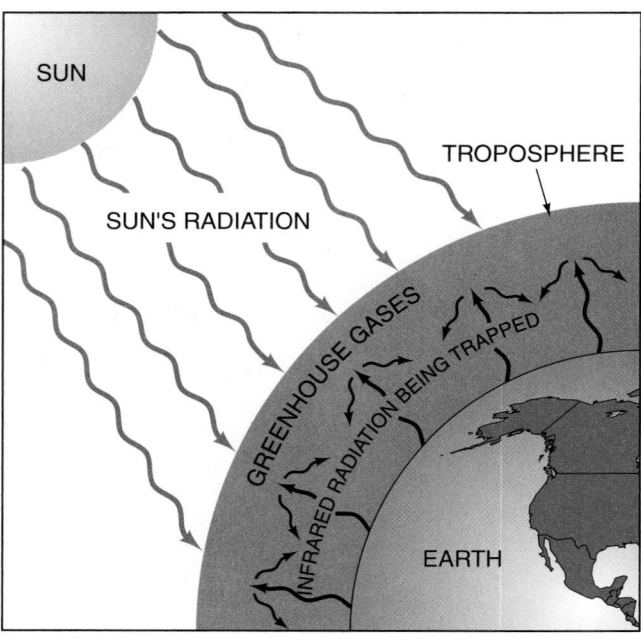

Figure 9–6 Greenhouse gases in the troposphere absorbing, refracting, and reflecting infrared radiation given off by the earth. This causes the greenhouse effect, or global warming.

the human immune system, and a decrease in plant and marine life. An index called the *ozone depletion potential (ODP)* has been used for regulatory purposes under the United Nations Environment Programme (UNEP) Montreal Protocol. The higher the ODP, the more damaging the chemical is to the ozone layer in the stratosphere.

When stratospheric ozone intercepts ultraviolet light, heat is generated, **Figure 9–5**. This generated heat is the force behind stratospheric winds, which affect weather patterns on earth. By changing the amount of ozone, or even its distribution in the stratosphere, the temperature of the stratosphere can be affected. This can seriously affect weather here on earth.

9.3 GLOBAL WARMING

Most of the sun's energy reaches the earth as visible light. After passing through the atmosphere part of this energy is absorbed by the earth's surface and, in the process, is converted into heat energy. The earth, now warmed by the sun, radiates heat energy back into the atmosphere toward space. Naturally occurring gases and tropospheric pollutants like CFCs, HCFCs, and HFCs—as well as carbon dioxide, water vapor, and many other chemicals—absorb, reflect, and/or refract the earth's infrared radiation and prevent it from escaping the lower atmosphere, **Figure 9–6**. These gases are often referred to as *greenhouse gases.* This process slows the earth's heat loss, making the earth's surface warmer than it would be if this heat energy had passed unobstructed through the atmosphere into space. The warmer earth's surface, in turn, radiates more heat until a balance is established between incoming and outgoing energy. This warming process, caused by the atmosphere's absorption of the heat energy radiated from the earth's surface, is called the *greenhouse effect,* or *global warming.*

The *direct effects* of global warming are created by chemicals that are emitted directly into the atmosphere. These direct emissions are measured by an index referred to as the refrigerant's *global warming potential (GWP).* Refrigerants leaking from a refrigeration or air-conditioning system contribute to global warming. These direct effects of global warming are measured by comparing them to carbon dioxide (CO_2), which has a GWP of 1.

Carbon dioxide is the number-one contributor to global warming. People have increased the amount of carbon dioxide in the atmosphere in excess of 25% of its usual amounts. Most of this increase is through the combustion of fossil fuels. This combustion of fossil fuels is caused by the need for electricity in our modern world. In fact, much of this electricity is generated to power refrigeration and air-conditioning equipment. ⊕The more efficient the HVACR equipment is, the less electrical energy that is needed. Even refrigeration or air-conditioning equipment with a relatively small charge of refrigerant that never leaks out may have a great impact on global warming if the equipment is undercharged or overcharged. The equipment would be very inefficient under these conditions, and the carbon dioxide generated from the longer run times created by these inefficiencies would contribute to global warming more than the refrigerant leaking out. This is an example of an *indirect effect* of global warming. ⊕

In the HVACR industry, these indirect effects of global warming are what concern scientists. Most of the newer refrigerants introduced to the HVACR industry are more energy efficient than the ones they are replacing. However, some are not as efficient. Even though some of the newer refrigerants do not contribute to ozone depletion, they still can contribute to the *direct* and *indirect* effects of global warming. An example of this would be the newer refrigerant HFC-134a. It has a 0 ozone depletion index and contributes to global warming directly via a refrigerant leak and indirectly through being used in an inefficient system with a longer run time that is caused by the refrigerant leak.

The *total equivalent warming impact (TEWI)* takes into consideration both the direct and indirect global warming effects of refrigerants. Refrigerants with the lowest global warming and ozone depletion potentials will have the lowest TEWI. Using HFC and HCFC refrigerants in the place of CFCs will also reduce the TEWI. All of the newer refrigerant alternatives introduced have a much lower TEWI.

9.4 REFRIGERANTS

Most refrigerants are made from two molecules, methane and ethane, **Figure 9–7**. These two molecules simply contain hydrogen (H) and carbon (C) and are referred to as pure *hydrocarbons (HCs)*. Pure hydrocarbons were at one time considered good refrigerants, but because of their flammability were not used after the 1930s in any large scale. However, hydrocarbons are making a comeback in Europe where they are used in domestic refrigerators. They are also making a comeback in the United States as a small percentage of the mixture in refrigerant blends for nonflammable applications. An example of a refrigerant that is not methane- or ethane-based is ammonia. Ammonia contains only nitrogen and hydrogen (NH_3) and is not an ozone-depleting refrigerant.

Any time some of the hydrogen atoms are removed from either the methane or ethane molecule and replaced with either chlorine or fluorine, the new molecule is said to be either chlorinated, fluorinated, or both. Abbreviations are used to describe refrigerants chemically and to make it simpler for technicians to differentiate between them. The following are some common abbreviations:

- Chlorofluorocarbons (CFCs)
- Hydrochlorofluorocarbons (HCFCs)
- Hydrofluorocarbons (HFCs)
- Hydrocarbons (HCs)

SAFETY PRECAUTION: Most refrigerants are heavier than air. Because of this, they can displace air and cause someone to suffocate. Always vacate and ventilate the area if large refrigerant leaks occur. In fact, some sort of self-contained breathing apparatus should be worn if you have to work in the area of a large refrigerant leak. Inhalation of large amounts of refrigerant can cause cardiac arrest and eventually death.

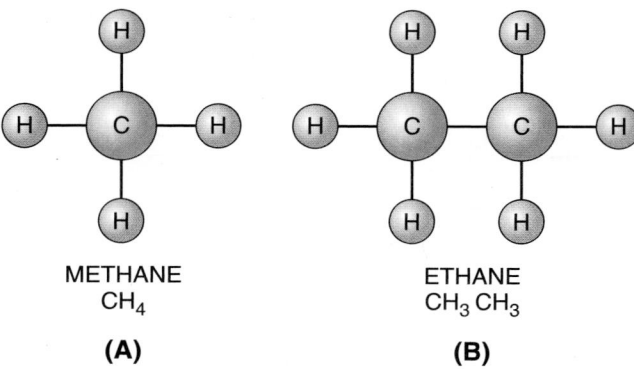

METHANE
CH_4

(A)

ETHANE
CH_3 CH_3

(B)

C = CARBON
H = HYDROGEN

Figure 9–7 Most refrigerants originate from **(A)** methane and **(B)** ethane molecules. Both molecules are hydrocarbons.

9.5 CFC REFRIGERANTS

The CFCs contain chlorine, fluorine, and carbon and are considered the most damaging because their molecules are not destroyed as they reach the stratosphere. CFC molecules have a very long life when exposed to the atmosphere because of their stable chemical structure. This allows them to be blown up to the stratosphere by atmospheric winds where they react with ozone molecules and cause destruction. CFCs also contribute to global warming. As of July 1, 1992, it became illegal to intentionally vent CFC refrigerants into the atmosphere.

Following is a chart of CFCs showing the chemical name and formula for each:

Chlorofluorocarbons (CFCs)

Refrigerant No.	Chemical Name	Chemical Formula
R-11	Trichlorofluoromethane	CCl_3F
R-12	Dichlorodifluoromethane	CCl_2F_2
R-113	Trichlorotrifluoroethane	CCl_2FCClF_2
R-114	Dichlorotetrafluoroethane	$CClF_2CClF_2$
R-115	Chloropentafluoroethane	$CClF_2CF_3$

All of the refrigerants in the CFC group were phased out of manufacturing at the end of 1995. One refrigerant that is important to us is R-12, because it is commonly used for residential and light commercial refrigeration and for centrifugal chillers in some commercial buildings. The other refrigerants are used in other applications. For example, R-11 is used for many centrifugal chillers in office buildings and also as an industrial solvent to clean parts. R-113 is used in smaller commercial chillers in office buildings and also as a cleaning solvent. R-114 has been used in some household refrigerators in the past and in centrifugal chillers for marine applications. R-115 is part of the blend of refrigerants that make up R-502 used for low-temperature refrigeration systems.

9.6 HCFC REFRIGERANTS

The second group of refrigerants in common use is the HCFC group. These refrigerants contain hydrogen, chlorine, fluorine, and carbon. They have a small amount of chlorine in them, but also have hydrogen in the compound that makes them less stable in the atmosphere. These refrigerants have much less potential for ozone depletion because they tend to break down in the atmosphere, releasing the chlorine before it reaches and reacts with the ozone in the stratosphere. However, the HCFC group is scheduled for a total phaseout by the year 2030. HCFC-22 (R-22) is an exception, with an earlier phaseout date for new equipment in 2010 and a total production phaseout in 2020 under the *Montreal Protocol*. Total phaseout means no production and no importing of the refrigerants. As of July 1, 1992, it became illegal to intentionally vent HCFC refrigerants to the atmosphere. HCFCs do have some global-warming potential, but it is much lower than most CFCs.

This group of refrigerants includes (to mention a few):

Hydrochlorofluorocarbons (HCFCs)

Refrigerant No.	Chemical Name	Chemical Formula
R-22	Chlorodifluoromethane	$CHClF_2$
R-123	Dichlorotrifluoroethane	$CHCl_2CF_3$
R-124	Chlorotetrafluoroethane	$CHClFCF_3$
R-142b	Chlorodifluoroethane	CH_3CClF_2

9.7   HFC REFRIGERANTS

The third group of refrigerants is the HFC group. HFC molecules contain no chlorine atoms and will not deplete the earth's protective ozone layer. HFCs contain hydrogen, fluorine, and carbon atoms. HFCs do have small global-warming potentials. HFCs are the long-term replacements for many CFC and HCFC refrigerants. It became unlawful to intentionally vent HFC refrigerants to the atmosphere on November 15, 1995. The original plan was to replace R-12 with R-134a, but this plan is complicated by the fact that R-134a is not compatible with any oil left in an R-12 system. It is also not compatible with some of the gaskets in the R-12 systems. An R-12 system, however, may be converted to R-134a by modifying the system. This task must be performed by an experienced technician because a complete oil change is necessary. The residual oil left in the system pipes must be removed, and all system gaskets must be checked for compatibility. To be sure the compressor materials are compatible, they must be checked by contacting the manufacturer.

Following is a list of refrigerants in the HFC group:

Hydrofluorocarbons (HFCs)

Refrigerant No.	Chemical Name	Chemical Formula
R-125	Pentafluoroethane	CHF_2CF_3
R-134a	Tetrafluoroethane	CH_2FCF_3
R-23	Trifluoromethane	CHF_3
R-32	Difluoromethane	CH_2F_2
R-125	Pentafluoroethane	CHF_2CF_3
R-143a	Trifluoroethane	CH_3CF_3
R-152a	Difluoroethane	CH_3CHF_2
R-410A	Blend	$CH_2F_2/CHF_2\text{-}CF_3$

9.8   HC REFRIGERANTS

The fourth group of refrigerants is the hydrocarbon (HC) group. These refrigerants have no fluorine or chlorine in their molecule; thus, they have a 0 ODP. They contain nothing but hydrogen and carbon. They do, however, contribute to global warming. Hydrocarbons are used as stand-alone refrigerants in Europe, but are not used as stand-alone refrigerants in the United States because they are flammable. Small percentages of hydrocarbons are used in many refrigerant blends in the United States but are not flammable when mixed in such small percentages. Some popular hydrocarbons (HC) include propane, butane, methane, and ethane.

9.9   NAMING REFRIGERANTS

Instead of their complex chemical names, an easier way to refer to refrigerants is by number. The DuPont Company has created an easy method of naming refrigerants, depending upon their chemical composition.

1. The first digit from the right is the number of fluorine atoms.
2. The second digit from the right is the number of hydrogen atoms plus 1.
3. The third digit from the right is the number of carbon atoms minus 1. If this number is zero, omit it.

Example 1

HCFC-123 $CHCl_2CF_3$ Dichlorotrifluoroethane

Number of fluorine atoms	= 3
Number of hydrogen atoms +1	= 2
Number of carbon atoms −1	= 1

Thus HCFC-123

The number of chlorine atoms is found by subtracting the sum of fluorine, bromine, and hydrogen atoms from the total number of atoms that can be connected to the carbon atom(s).

Example 2

CFC-12 CCL_2F_2 Dichlorodifluoromethane

Number of fluorine atoms	= 2
Number of hydrogen atoms +1	= 1
Number of carbon atoms −1	= 0 (omit)

Thus CFC-12

The small letters "a,b,c,d, . . ." at the end of the numbering system represent how symmetrical the molecular arrangement is. An example is R-134a and R-134. These are completely different refrigerants with completely different properties. Both have the same number of atoms of the same elements but differ in how they are arranged. These molecules are said to be isomers of one another. In the following diagram, notice that the R-134 molecule has perfect symmetry. As isomers of this molecule start to get less and less symmetrical, the small letters a,b,c,d, . . . are assigned to their ends.

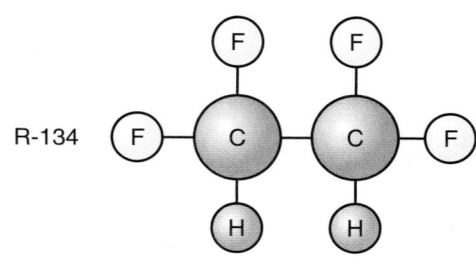

ISOMERS

MOST SYMMETRICAL

R-134

ISOMERS

NEXT SYMMETRICAL

R-134a

9.10 REFRIGERANT BLENDS

Ozone depletion and global warming issues have caused a need for much research on *refrigerant blends.* Refrigerant blends have been used successfully in both comfort cooling and refrigeration.

Blends can have as many as four refrigerants mixed together to give properties and efficiencies similar to the refrigerants they will replace. HCFC-based blends are short-term replacements for many CFC and HCFC refrigerants, whereas HFC-based blends are long-term replacements.

Azeotropes or azeotropic mixtures are blended refrigerants of two or more liquids. When mixed together, they behave like R-12 and R-22 when phase changing from liquid to vapor. Only one boiling point and/or one condensing point exists for each given system pressure. **Figure 9–8** illustrates a temperature/pressure graph of an azeotropic refrigerant blend showing the refrigerants boiling at one temperature for a given pressure through the length of the heat exchanger (evaporator). Pure compounds or single-component refrigerants like R-12 and R-22 also exhibit this behavior as they evaporate and condense. Azeotropic blends are not new to the HVACR industry. R-500 and R-502 are both azeotropic blends that have been used in the industry for years. R-500 is made up of 73.8% R-12 and 26.2% R-152a (by weight). R-500 is used in some cooling systems. R-502 is made up of 48.8% R-22 and 51.2% R-115. It has been an excellent low-temperature refrigerant since the 1960s.

However, many refrigerant blends are not azeotropic blends; they are near-azeotropic mixtures. A *refrigerant mixture* is a blend of two or more refrigerants that can still separate into individual refrigerants. These blends act differently because two or three molecules instead of one are present in any one sample of liquid or vapor; more than one molecule is always present (for example, sugar water and salt water). This is where a difference arises in temperature/pressure relationships of the near-azeotropic blends versus refrigerants like CFC-12, HCFC-22, HFC-134a, and azeotropic blends.

Near-azeotropic blends experience a temperature glide. Temperature glide occurs when the blend has many temperatures as it evaporates and condenses at a given pressure. **Figure 9–9** illustrates a temperature/pressure graph of a near-azeotropic refrigerant blend showing the refrigerants boiling at many temperatures (temperature glide) for a given pressure through the length of the heat exchanger. In fact, as the near-azeotropic refrigerant blend changes phase from liquid to vapor and back, more of one component in the blend will transfer to the other phase faster than the rest. Zeotropic refrigerant blends also exhibit these properties, but to a greater extent. This is different than refrigerants like R-12, R-134a, and R-22, which, as shown in previous units, evaporate and condense at one temperature for a given pressure.

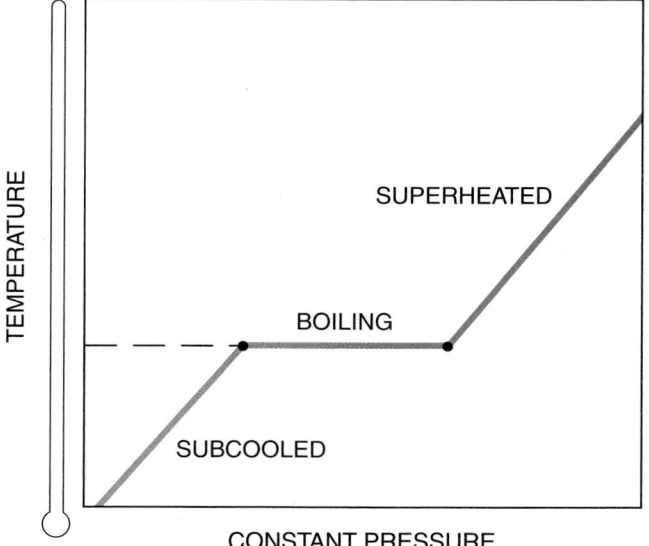

Figure 9–8 An azeotropic refrigerant blend showing only one temperature for a given pressure as it boils.

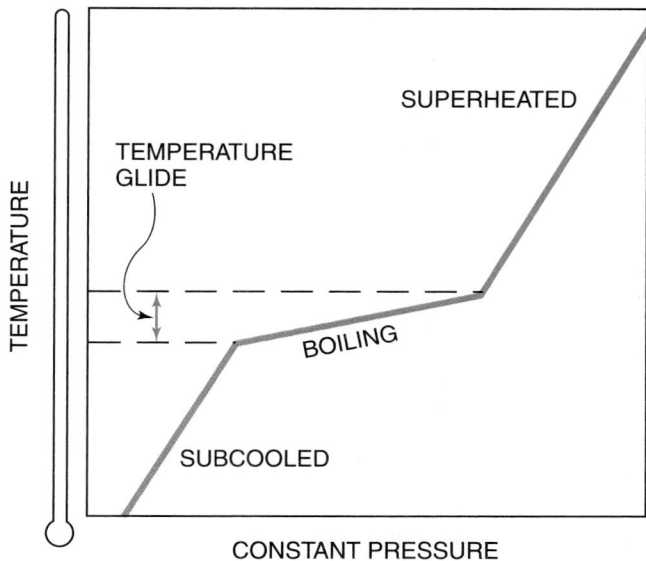

Figure 9–9 A near-azeotropic refrigerant blend showing temperature glide as it boils at a constant pressure.

Near-azeotropic refrigerant blends may also experience *fractionation.* Fractionation occurs when one or more of the refrigerants in the blend will condense or evaporate at different rates than the other refrigerants. This different rate is caused by the slightly varied vapor pressures of each refrigerant in the blend. Fractionation will be applied in detail in Unit 10, "System Charging."

Zeotropes, or zeotropic blends, also have temperature glide and fractionation but to a larger extent than near-azeotropic blends. The terms zeotropic blends and near-azeotropic blends are used interchangeably in the industry because they both have temperature glide and fractionation. Charging methods for refrigerant blends will be covered in Unit 10, "System Charging."

"R" Numbers for the Blends

- The 400-series blends are the near-azeotropic (zeotropic) blends. If a blend has an R-410 designation, it is a near-azeotropic blend. The 10 indicates that it is the tenth one commercially produced. They all have temperature glide and can fractionate.
- The 500-series blends represent the azeotropic blends. R-502 would represent an azeotropic blend. The 2 means that it is the second one produced on the market.
- Blends can also be represented by the percentage of each refrigerant that makes up the blend. The refrigerant with the lowest boiling point at atmospheric pressure will be named first. For example, a blend of 20% (R-12) and 80% (R-22) would be represented as R-22/12 (80/20). Because R-22 has the lowest boiling point, it is listed first.
- Blends can also have capital letters at their ends. The capital letters at the end of R-401A, R-401B, and R-401C mean that the same three refrigerants make up these near-azeotropic blends, but their individual percentages differ.

9.11 POPULAR REFRIGERANTS AND THEIR COMPATIBLE OILS

The following are some of the more popular refrigerants used in the HVACR industry.

HFC-134a (R-134a) is the alternative refrigerant of choice to replace CFC-12 in many medium- and high-temperature stationary refrigeration and air-conditioning applications and automotive air-conditioning applications. Its safety classification is A1, **Figure 9–10.** R-134a is a pure compound, not a refrigerant blend. The reason for this is that R-134a has similar temperature/pressure and capacity relationships as compared with R-12, **Figure 9–11.** R-134a also suffers small capacity losses when used as a low-temperature refrigerant. R-134a contains no chlorine in its molecule and thus has a zero ODP. Its GWP is 0.27. Polarity differences between commonly used organic mineral oils and HFC refrigerants make R-134a insoluble and thus incompatible with

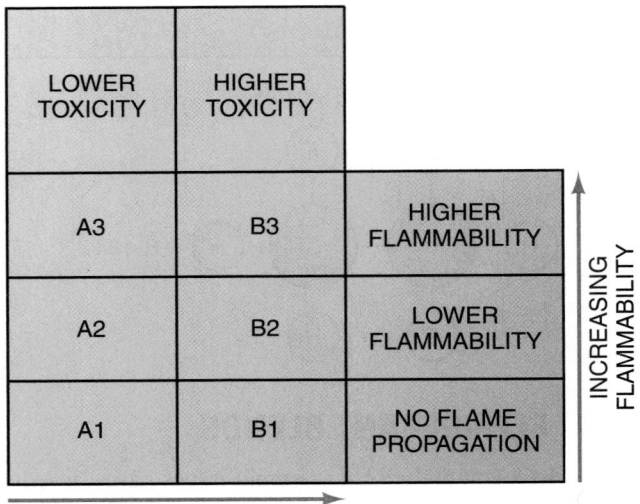

Figure 9–10 ASHRAE's Standard 34-1992 refrigerant safety group classifications.

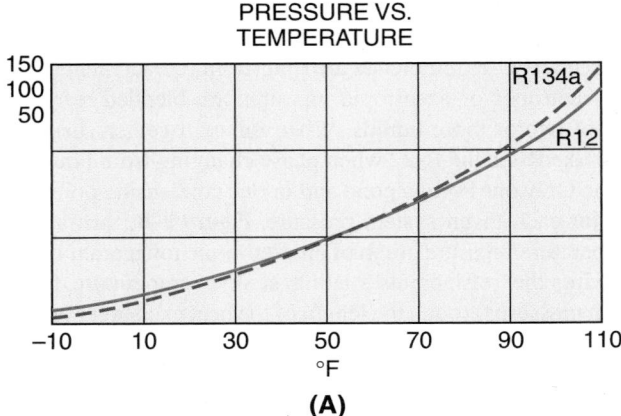

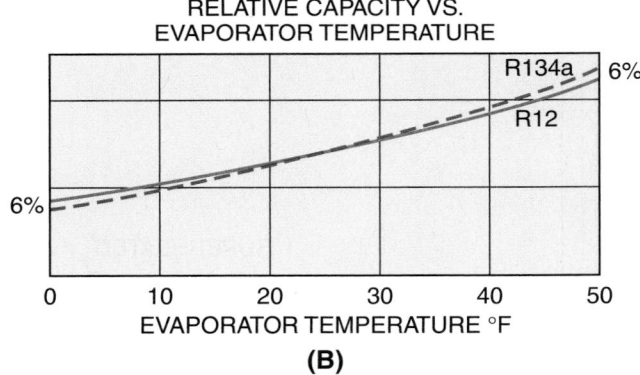

Figure 9–11 **(A)** Pressure/temperature and **(B)** capacity curves for R-134a. *Courtesy Tecumseh Products Company*

mineral oils used in many refrigeration and air-conditioning applications today. R-134a systems must employ synthetic *polyol ester (POE)* or *polyalkylene glycol (PAG)* lubricants. R-134a is also replacing R-12 and R-500 in many centrifugal chiller applications. Ester-based lubricants are also required in these applications because of R-134a's incompatibility with organic mineral oils. In many centrifugal chiller applications, efficiencies have improved but not without some reductions in capacity. SAFETY PRECAUTION: R-134a is not a direct drop-in refrigerant for R-12. Retrofit guidelines must be followed carefully when converting a system from R-12 to R-134a. Appendix IV is an example of a retrofit guideline from a major manufacturer.

HCFC-22 (R-22) is a pure compound, not a refrigerant blend. Its safety classification is A1. R-22 is considered an interim replacement refrigerant because it has a current phaseout date for new equipment in 2010 and a total production phaseout in 2020. R-22 has been the primary refrigerant used in commercial and residential air-conditioning systems for many years. R-22 has also been used in large industrial centrifugal chiller air-conditioning applications. Industrial cooling processes also use R-22 in their equipment. It is currently replacing R-502 in low-temperature commercial supermarket refrigeration. However, because of its high discharge temperatures in low-temperature commercial refrigeration applications, these systems require compound compression systems with interstage cooling. R-407C and R-410A are two refrigerant blends that will be the long-term replacement for R-22 in air-conditioning applications.

HFC-407C (R-407C) is a near-azeotropic refrigerant blend of R-32, R-125, and R-134a. R-407C is a long-term replacement refrigerant for R-22 in residential and commercial air-conditioning and refrigeration applications. Its safety classification is A1. It is being used by manufacturers in new equipment and can also be used as a retrofit refrigerant blend replacing R-22. Always follow retrofit guidelines when retrofitting from R-22 to R-407C. R-407C has a large temperature glide (10°F) over air-conditioning and refrigeration temperature ranges. R-407C also has fractionation potential. POE lubricant must be used with R-407C.

HFC-410A (R-410A) is a near-azeotropic refrigerant blend of R-32 and R-125. R-410A is a higher-efficiency, long-term replacement refrigerant for R-22 in new residential and light commercial air-conditioning applications. Its safety classification is A1/A1. Retrofitting an R-22 system to an R-410A system is not recommended because R-410A systems operate at much higher pressures (60% higher) in air-conditioning applications than a conventional R-22 system. If an air-conditioning system is operating at a 45°F evaporating temperature and a 110°F condensing temperature, the corresponding pressures would be as follows:

R-410A................................130 psig evaporating pressure
                                      365 psig condensing pressure
R-22....................................76 psig evaporating pressure
                                      226 psig condensing pressure

Figure 9–12 illustrates pressure/temperature comparisons of R-410A with R-22 in tabular and graphical format. Gage manifold sets, hoses, recovery equipment, and recovery storage cylinders all have to be designed to handle the higher pressures of R-410A. Different gage line connections are used for the two refrigerants, preventing the technician from making a mistake. In fact, the gage and manifold set for

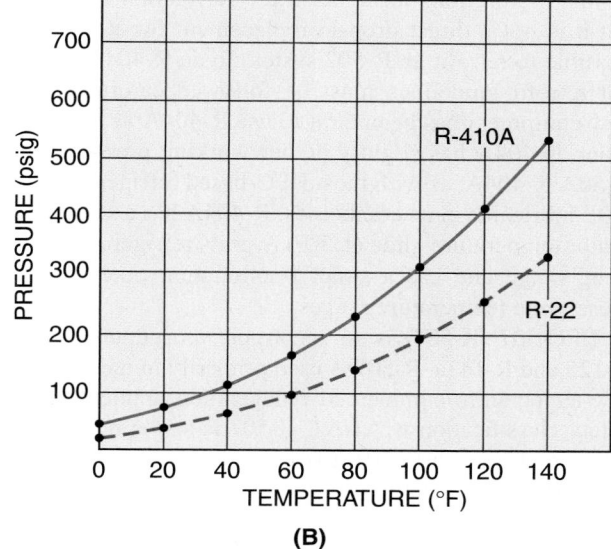

TEMP.	PRESSURE (psig)	
°F	R-410A	R-22
0	48.3	24.0
20	78.4	43.0
40	118.2	68.5
60	169.9	101.6
80	235.4	143.6
100	317.1	195.9
120	417.4	259.9
140	539.1	337.2

(A)

(B)

Figure 9–12   Pressure/temperature comparison of R-410A with R-22 in **(A)** tabular and **(B)** graphical format.

R-410A are required to range up to 800 psig on the high side and 250 psig on the low side with a 550-psig low-side retard. All R-410A hoses must have a service rating of 800 psig. Even filter driers for R-410A must be rated and compatible for the higher pressures. SAFETY PRECAUTION: Never use R-22 or any other refrigerant service equipment on R-410A equipment. Service equipment must be rated to handle the higher operating pressures of R-410A. Air-conditioning equipment has been redesigned, and safer service tools have been introduced to handle the higher pressures of R-410A. As with all refrigerants, safety glasses and gloves must always be worn when handling R-410A. If an R-22 system has to be retrofitted, it is recommended to use R-407C as the retrofit refrigerant because of its similar properties to R-22. Because R-410A is a near-azeotropic refrigerant blend, it has a very small evaporator temperature glide (0.2°F) and has negligible fractionation potential. R-410A systems use POE lubricant in their crankcases.

HFC-417A (R-417A) is a near-azeotropic refrigerant blend of R-134a, R-125, and R-600. R-600 is the hydrocarbon butane. Its safety classification is A1/A1. R-417A is used primarily in new and existing R-22 systems utilizing direct-expansion metering devices. R-417A also can be used as an R-22 direct drop-in replacement refrigerant. Its applications range from large tonnage industrial refrigeration to commercial packaged air-conditioning units. It is a non-ozone-depleting HFC-based refrigerant blend with a low global-warming impact. It has a moderate temperature glide. R-417A is compatible with all standard refrigeration oils. R-417A operating pressures are slightly lower than R-22 operating pressures.

HFC-404A (R-404A) is a near-azeotropic refrigerant blend of R-125, R-143a, and R-134a. Its safety classification is A1/A1. R-404A is used primarily in the medium- and low-temperature commercial refrigeration markets. R-404A is the long-term substitute refrigerant for R-502, but it is not a direct drop-in replacement for R-502. It is possible to retrofit an R-502 system to an R-404A system, but retrofit guidelines must be followed carefully. Most new equipment has been built to use R-404A as the refrigerant. R-404A has slightly higher working pressures than R-502. R-404A, as with most HFC-based refrigerants, uses POE lubricants in its crankcases. R-404A has a small evaporator temperature glide (1.5°F) over its refrigeration operating range and has a small fractionation potential over these same temperature ranges.

HFC-507 (R-507) is an azeotropic refrigerant blend of R-125 and R-143a. R-507 is used primarily in medium- and low-temperature commercial refrigeration applications. Its safety classification is A1/A1. R-507 is also a replacement refrigerant for R-502, but it is not a direct drop-in replacement. Retrofit guidelines must be followed carefully when retrofitting an R-502 system to an R-507 system. Most new equipment is rated to handle both R-507 and R-404A interchangeably. R-507 has slightly higher pressures than both R-502 and R-404A. R-507 has slightly higher capacities as compared with R-404A. Because R-507 is an azeotropic blend of refrigerants, it has negligible temperature glide and fractionation potential. R-507 requires POE lubricant in the compressor's crankcase.

The following are some of the refrigerants still in use in today's refrigeration and air-conditioning systems. However, because of current or near-future production bans, these refrigerants are, or soon will be, no longer manufactured. Even though it is unlawful to produce these refrigerants, they can still be used when recovered or reclaimed out of an existing system, as long as certain Environmental Protection Agency (EPA) guidelines are followed. More detailed information on recovering and reclaiming refrigerants is provided later in this unit.

CFC-11 (R-11) is a low-pressure refrigerant used in centrifugal chillers. Its safety classification is A1. R-11 has been in the HVACR industry since 1932. Centrifugal chillers employing R-11 as their refrigerant have low operating costs and are one of the most inexpensive ways to supply large amounts of chilled water to large commercial and industrial buildings for air-conditioning purposes. In 1996, environmental concerns about ozone depletion and global warming forced a production ban on R-11. R-11 centrifugal chiller systems operate in a vacuum on their low sides. Because of this, any leaks will introduce air into the system. Efficient purge systems have

been designed to rid these systems of unwanted air. R-11 has an interim replacement refrigerant of R-123. However, a retrofit procedure is required when retrofitting an R-11 system to an R-123 system. SAFETY PRECAUTION: Always consult with the original equipment manufacturer for retrofit guidelines when retrofitting any system. Newer chiller systems may come with R-123, R-134a, or ammonia. Lithium bromide and water absorption systems may also be used in newer chiller applications.

CFC-12 (R-12) has been widely used in the HVACR industry since it was invented in 1931. Its safety classification is A1. R-12 has been used extensively in small- and large-scale refrigeration and air-conditioning system applications. Systems ranging in size from small hermetic systems to large centrifugal positive-pressure chiller applications have incorporated R-12 as their refrigerant. Its main applications have been automotive air conditioning; high-, medium-, and low-temperature refrigeration; and other commercial and industrial refrigeration and air-conditioning applications. In 1996, environmental concerns about ozone depletion and global warming forced a production ban on R-12. R-134a has been the most popular refrigerant to replace R-12 in medium- and high-temperature stationary refrigeration and air-conditioning systems. R-134a is also the refrigerant replacing R-12 in automotive air-conditioning systems. However, 134a is not a direct drop-in replacement for R-12 in any application. Many interim refrigerant blends that are listed here have also entered the market to replace R-12 with minimal retrofit. Always refer to the manufacturer's retrofit guidelines and procedures when retrofitting any system.

CFC-502 (R-502) is an azeotropic refrigerant blend of R-22 and R-115. Its safety classification is A1. R-502 was first manufactured in 1961. In 1996, environmental concerns about ozone depletion and global warming forced a production ban on R-502 also. Long-term replacement refrigerants for R-502 include R-404A and R-507. R-22 has been used as an interim replacement for R-502, but it usually requires a compound compression system with intercoolers for low-temperature refrigeration applications. As mentioned earlier, R-22 is considered an interim replacement refrigerant because it has a current phaseout date for new equipment in 2010 and a total production phaseout in 2020. R-502 has been a popular low-temperature refrigerant for low-temperature refrigeration applications. One of the main advantages of R-502 over R-22 in low-temperature refrigeration is its ability to operate in positive pressure even with a −40°F evaporating temperature. Also, because of its relatively higher pressures, R-502 operates with lower compression ratios than R-12 in low-temperature refrigeration applications. Lower compression ratios lead to improved capacity as compared with R-12 and R-22 systems. R-502 also has much lower discharge temperatures as compared with R-22. Because of these advantages, R-502 can perform low-temperature refrigeration with single-stage compression and relatively inexpensive compressors.

HCFC-123 (R-123) is an interim replacement refrigerant for R-11 in low-pressure centrifugal chiller applications. The American Society of Heating, Refrigerating and

Air-Conditioning Engineers (ASHRAE) has given R-123 a safety rating of B1, meaning there is evidence of toxicity at concentrations below 400 ppm, **Figure 9–10.** This means that lower levels are allowed for persons exposed to R-123 in the course of normal daily operations in machine rooms and service conditions. Special precautions for R-123 and machine room requirements are covered in ASHRAE Standard 15, "Safety Code for Mechanical Refrigeration." SAFETY PRECAUTION: ASHRAE Standard 15 requires the use of room sensors and alarms to detect R-123 refrigerant leaks. The machine room shall activate an alarm and mechanical ventilation equipment before the concentration of R-123 exceeds certain limits. When retrofitting an R-11 system to an R-123 system, retrofit guidelines must be followed. R-123 systems may require new seals, gaskets, and other system components to prevent leaks and get maximum performance and capacity.

HCFC-401A (R-401A) is a near-azeotropic refrigerant blend of R-22, R-152a, and R-124. Its safety classification is A1/A1. R-401A is intended for use as a retrofit refrigerant for R-12 or R-500 systems with only minor modifications. It is especially designed for systems with evaporator temperatures ranging from 10°F to 20°F, because the pressure/temperature curve closely matches that of R-12 in these ranges. There is a drop in capacity at lower temperatures as compared with R-12 systems. An 8°F temperature glide in the evaporator is typical when R-401A is being used. There is an increase in discharge pressure and a slight increase in discharge temperature as compared with R-12. An oil retrofit to an *alkylbenzene* lubricant is recommended when retrofitting from R-12 to R-401A.

HCFC-402A (R-402A) is a near-azeotropic refrigerant blend of R-22, R-125, and R-290 (propane). The hydrocarbon propane is added to enhance oil solubility and circulation with the refrigerant for oil-return purposes. Its safety classification is A1/A1. R-402A is intended for use as an R-502 retrofit refrigerant with minimal retrofitting. R-402A will have 25- to 40-psi increases in discharge pressure over R-502 but will not significantly increase the discharge temperature. An oil retrofit to an alkylbenzene lubricant is recommended when retrofitting from R-502 to R-402A.

HCFC-402B (R-402B) is a near-azeotropic refrigerant blend of R-22, R-125, and R-290 that contains more R-22 and less R-125 than R-402A does. Its safety classification is A1/A1. R-402B is intended for use as an R-502 ice-machine retrofit refrigerant because of its lower discharge pressures and higher discharge temperatures as compared with R-502. The higher discharge temperatures make for quicker and more efficient hot gas defrosts in ice-machine applications. An oil retrofit to an alkylbenzene lubricant is recommended when retrofitting from R-502 to R-402B.

HCFC-409A (R-409A) is a near-azeotropic refrigerant blend of R-22, R-142b, and R-124. Its safety classification is A1/A1. R-409A was designed for retrofitting R-12 systems having an evaporator temperature between 10°F and 20°F. R-409A systems usually experience a 13°F temperature glide in the evaporator. R-409A has been used in retrofitting R-12 and R-500 direct-expansion refrigeration and

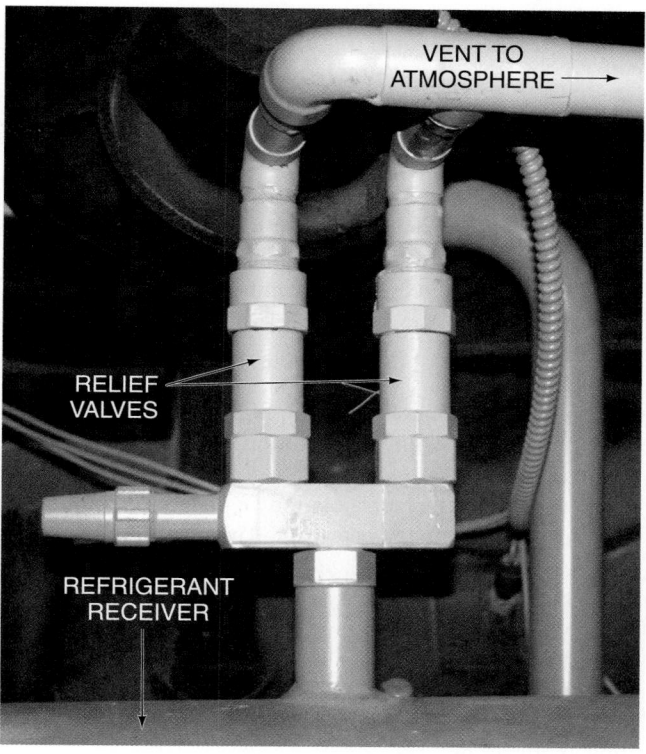

Figure 9–13 Dual pressure relief valves installed in parallel. *Courtesy Ferris State University. Photo by John Tomczyk*

air-conditioning systems. R-409A operates with higher discharge pressures and temperatures than R-12. R-409A has a similar capacity as that of R-12 down to −30°F evaporator temperatures. R-409A mixes well with mineral oil lubricants down to 0°F, so oil retrofitting may not be necessary above this temperature. However, an alkylbenzene lubricant should be used in place of the mineral oil lubricant in evaporator temperatures below 0°F.

SAFETY PRECAUTION: Always consult with the original equipment manufacturer for retrofit guidelines when retrofitting any system.

SAFETY PRECAUTION: All refrigeration systems are required to have some sort of pressure relief device, which should be vented to the outdoors to prevent the buildup of excessive system pressures. This is usually accomplished through the use of a pressure relief valve. Pressure relief valves should always be installed in parallel with one another, never in series, **Figure 9–13.**

9.12 REFRIGERANT OILS AND THEIR APPLICATIONS

Today, almost as many new oils are on the market as there are refrigerants. Let the compressor manufacturers specify what oil to use in any application. If you are unsure of what oil to use, please call the compressor manufacturer or use their specification data along with their performance curve literature.

9.13 OIL GROUPS

Oils fall into three basic groups. The groupings are
- animal.
- vegetable.
- mineral.

Neither animal nor vegetable oils are good refrigeration or air-conditioning lubricants. However, mineral oils are proven lubricants for both refrigeration and air-conditioning applications. The three groups of mineral oils are
- paraffinic.
- naphthenic.
- aromatic.

Of the three mineral oil groupings, naphthenic oils are used most often with refrigeration and air-conditioning applications.

Synthetic oils for refrigeration applications have been used successfully for many years. Three popular synthetic oils are
- alkylbenzenes.
- glycols.
- esters.

HCFC-based blends work best with alkylbenzene lubricants. Today, alkylbenzenes are used quite often in refrigeration applications.

The HFC-based blends perform best with an ester lubricant. Manufacturers are using a synthetic ester-based lubricant for many HFC-based refrigerant blends. Extensive oil flushing is required when retrofitting a CFC/mineral oil system to an HFC/ester oil system.

Mineral oils do not mix completely with the HCFC-based refrigerant blends. HCFC-based refrigerant blends are soluble with mineral oil up to a 20% concentration. Mineral oil systems retrofitted with HCFC-based refrigerant blends will not require a lot of oil flushing.

As mentioned earlier, a popular glycol-based lubricant is *polyalkylene glycol.* PAGs are used in automotive applications. PAG lubricants attract and hold moisture readily. PAGs will not tolerate chlorine. When retrofitting a mineral oil system to a PAG system, a labor-intensive retrofit is involved.

Ester-based oils are also synthetic oils. Ester-based oils are wax-free oils and are used with HFC refrigerants like R-134a and HFC-based refrigerant blends. R-134a is a non-ozone-depleting refrigerant with properties very similar to R-12 at medium- and high-temperature applications, **Figure 9–11.** Notice the increase in capacity at the higher temperature applications, and the decrease in capacity at lower temperatures. R-134a is not compatible with current refrigerant mineral oils. Systems will have to come from the factory as virgin systems with ester oil, or the technician will have to go through a step-by-step flushing process in an effort to eliminate the residual mineral oil. The system cannot contain more than 1% to 5% mineral oil after retrofitting. **Figure 9–14** shows a refractometer that a technician can use in the field to make sure the system has no more than 1% to 5% residual mineral oil left in it. SAFETY PRECAUTION: Even though the EPA has ruled that refrigeration oils are not

Figure 9–14 A refractometer used in the field for checking the percentage of residual oil left in the system. *Courtesy Nu-Calgon Wholesaler, Inc.*

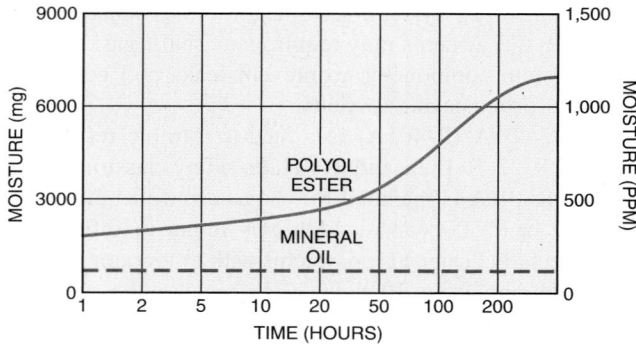

Figure 9–15 The moisture-absorbing capability of mineral oil versus polyol ester lubricant. *Courtesy Copeland Corporation*

hazardous waste unless they are contaminated, disposing of used oil in a careless manner is against the law. Check with your local waste disposal codes when disposing of refrigeration oil.

Polyol esters, ester-based lubricants, have become very popular recently. POEs do, however, absorb atmospheric moisture more rapidly than mineral oils, **Figure 9–15.** POEs not only absorb moisture more readily but also hold it more tightly. Refrigeration-system evacuation procedures may require longer times because of this. Containers of POE oil, therefore, should not be exposed to the atmosphere any longer than possible to avoid moisture contamination. Many manufacturers are using metal oil containers for POE oils instead of plastic ones to prevent moisture penetration through the walls of the oil container. **Figure 9–16** lists the appropriate oils for many popular refrigerants. However, always consult with the compressor or condensing unit manufacturer when unsure of what oil to use.

Retrofit guidelines have been written by some of the larger compressor and chemical manufacturers for changing a system over to a newer alternative refrigerant and its appropriate oil. Always consult with the compressor manufacturer before retrofitting any system. The appendices contain a specific step-by-step refrigerant changeover guideline (CFC-12 to HFC-134a) from the Copeland Corporation. Here are some general retrofitting steps to follow when retrofitting from a CFC/mineral oil system:
- Establish baseline performance with the CFC system.
- Recover the refrigerant.

CFC Refrigerants	Appropriate Lubricant		
	Mineral Oil	Alkylbenzene	Polyol Ester
R-11	+		
R-12	+	+	
R-13	+	+	
R-113	+	+	
R-114	+	+	
R-115	+	+	
R-500	+	+	
R-502	+	+	
R-503	+	+	

HCFC Refrigerants	Appropriate Lubricant		
	Mineral Oil	Alkylbenzene	Polyol Ester
R-22	+	+	
R-123	+	+	
R-124		+	
R-401A	+	+	
R-401B			
R-401C			
R-402A		+	
R-402B	+	+	
R-403A			
R-403B			
R-405A			
R-406A	+		
R-408A		+	
R-409A	+	+	

HFC Refrigerants	Appropriate Lubricant		
	Mineral Oil	Alkylbenzene	Polyol Ester
R-23			+
R-32			+
R-125			+
R-134a			+
R-143a			+
R-152a			+
R-404A			+
R-407A			+
R-407B			+
R-407C			+
R-410A			+
R-410B			+
R-507			+

+ Good Suitability   Applications with Limitations

NOTE: Always consult the compressor manufacturer for the appropriate lubricant.

Figure 9–16 A list of refrigerants with their appropriate lubricants. POE = Polyol ester, AB = Alkylbenzene, MO = Mineral oil, and PAG = Polyalkylene glycol (automotive applications).

- Drain the mineral oil from the system.
- Charge the system with the new lubricant.
- Change filter driers.
- Evacuate the system.
- Charge the system with CFC refrigerant.
- Start up and run the system for at least 24 hours up to two weeks maximum.
- Recover or isolate the refrigerant.
- Drain the new lubricant and test for residual mineral oil content down to 5% or less.
- Repeat this oil-flushing process until residual mineral oil is down to 5% or below, or per the manufacturer's guidelines.
- Evacuate the old CFC refrigerant.
- Make any last equipment or system modifications.
- Leak test the system one final time.
- Charge the system with the new alternative refrigerant.

9.14 REGULATIONS

The United States, Canada, and more than 30 other countries met in Montreal, Canada, in September 1987 to try to solve the problems of released refrigerants. This conference is known as the Montreal Protocol, and those attending agreed to reduce by 50% by 1999 the production of refrigerants believed to be harmful ozone-destroying chemicals. This would begin when ratified by at least eleven countries representing 60% of the global production of these chemicals. Additional meetings have been held since 1987 and further reduction of these refrigerants has been agreed on. The use of refrigerants as we have known them has changed to some extent and will change dramatically in the near future.

The United States Clean Air Act Amendments of 1990 regulate the use and disposal of CFCs and HCFCs. The EPA is charged with implementing the provisions of this legislation.

This act affects the technician servicing refrigeration and cooling equipment. According to the Clean Air Act of 1990, technicians may not "knowingly vent or otherwise release or dispose of any substance used as a refrigerant in such an appliance in a manner which permits such substance to enter the environment." This prohibition became effective July 1, 1992. Severe fines and penalties are provided for, including prison terms. The first level of the act is the power of the EPA to obtain an injunction against the offending party that prohibits the discharge of refrigerant to the atmosphere. The second level is a $27,500 fine per day and a prison term not exceeding five years. Also, rewards of up to $10,000 are offered to persons furnishing information that leads to the conviction of a person in violation.

To help police the illegal discharge of refrigerants, a reward may be given to any person who furnishes information leading to the conviction of a person willfully venting refrigerant into the environment. New legislation made it illegal to manufacture CFC refrigerants after January 1, 1996.

9.15 RECOVER, RECYCLE, OR RECLAIM

You should become familiar with these three terms: *recover, recycle,* and *reclaim*. These are three terms the industry and its technicians must understand.

Technicians will have to contain all refrigerants except those used to purge lines and tools of the trade, such as the gage manifold or any device used to capture and save the refrigerant. The Clean Air Act Amendments of 1990 charged the EPA to issue regulations to reduce the use and emissions of refrigerants to the lowest achievable level. The law states that the technician shall not "knowingly vent or otherwise release or dispose of any substance used as a refrigerant in such an appliance in a manner which permits such substance to enter the environment. De minimus releases associated with good faith attempts to recapture and recycle or safely dispose of any such substance shall not be subject to prohibition set forth in the preceding sentence." This prohibition became effective July 1, 1992. *De minimus* releases means the minimum amount possible under the circumstances.

Some situations requiring the removal of refrigerant from a component or system exist when the following occurs:

1. A compressor motor is burned.
2. A system is being removed to be replaced. It cannot be disposed of at a salvage yard with refrigerant in the system.
3. A repair must be made to a system, and the refrigerant cannot be pumped using the system compressor and captured in the condenser or receiver.

When refrigerant must be removed, the technician must study the system and determine the best procedure for removal. The system may have several different valve configurations. The compressor may or may not be operable. Sometimes the system compressor may be used to assist in refrigerant recovery. We will consider only the refrigerants and systems covered in this text. The following situations where refrigerant must be removed from a system may be encountered by the technician in the field:

1. The compressor will run, and the system has no service valves or access ports; for example, when a unit like a domestic refrigerator, freezer, or window unit is being discarded.
2. The compressor will *not* run, and the system has no access ports; for example, a burnout on a domestic refrigerator, freezer, or window unit.
3. The compressor will run, and the unit has service ports (Schrader ports) only, as in a central air conditioner.
4. The compressor will *not* run, and the unit has service ports (Schrader ports) only, as in a central air conditioner.
5. The compressor will run, and the system has some service valves; for example, a residential air-conditioning system with liquid- and suction-line isolation valves in the line set.
6. The compressor will *not* run, and the system has some service valves; for example, a residential air-conditioning system with liquid- and suction-line isolation valves in the line set.
7. The compressor will run, and the system has a complete set of service valves; for example, a refrigeration system.
8. The compressor will *not* run, and the system has a complete set of service valves; for example, a refrigeration system.
9. A heat pump with any combination of service valves.

When the technician begins to try to recover the refrigerant in these cases, a certain amount of knowledge and experience is necessary. The condition of the refrigerant in the system must also be considered. It may be contaminated with air, another refrigerant, nitrogen, acid, water, or motor burn particulates.

Cross contamination to other systems must never be allowed because it can cause damage to the system into which the contaminated refrigerant is transferred. This damage may occur slowly and may not be known for long periods of time. Warranties of compressors and systems may be voided if contamination can be proven. Food loss from inoperative refrigeration systems and money loss in places of business when the air-conditioning system is off can be a result.

The words *recover, recycle,* and *reclaim* may be used to mean the same thing by many technicians, but they refer to three entirely different procedures.

RECOVER REFRIGERANT. "To remove refrigerant in any condition from a system and store it in an external container without necessarily testing or processing it in any way" is the EPA's definition of recovering refrigerant. This refrigerant may be contaminated with air, another refrigerant, nitrogen, acid, water, or motor-burn particulates. In fact, an Industry Recycling Guideline (IRG-2) states that the only way recovered refrigerant can change hands from one owner to another is when it is brought up (cleaned or known to be cleaned) to the Air Conditioning and Refrigeration Institute (ARI) Standard 700 purity levels by an independent EPA-certified testing agency. It must also remain in the contractor's custody and control. If the refrigerant is to be put back into the "same owner's" equipment, the recycled refrigerant must meet the maximum contaminant levels listed in **Figure 9–17**. The purity levels in **Figure 9–17** can be reached by field recycling and testing, followed by documentation. Another option is to send the refrigerant to a certified refrigerant reclaimer. The following are four options for handling recovered refrigerant:

- Charge the recovered refrigerant back into the same owner's equipment without recycling it.
- Recycle the recovered refrigerant and charge it back into the same system or back into another system owned by the same person or company.
- Recycle the refrigerant to ARI 700 standards before reusing it in another owner's equipment. The refrigerant must remain in the original contractor's custody and control at all times from recovery through recycling and reuse.
- Send the recovered refrigerant to a certified reclaimer.

CONTAMINANTS	LOW-PRESSURE SYSTEMS	R-12 SYSTEMS	ALL OTHER SYSTEMS
Acid content (by wt.)	1.0 ppm	1.0 ppm	1.0 ppm
Moisture (by wt.)	20 ppm	10 ppm	20 ppm
Noncondensable gas (by vol.)	N/A	2.0%	2.0%
High-boiling residues oil (by vol.)	1.0%	0.02%	0.02%
Chlorides by silver nitrate test	no turbidity	no turbidity	no turbidity
Particulates	visually clean	visually clean	visually clean
Other refrigerants	2.0%	2.0%	2.0%

Figure 9–17   Maximum contaminant levels of recycled refrigerant in the same owner's equipment.

If the technician is removing the refrigerant from the system and it is operable, it may be charged back into the system. For example, assume a system has no service valves, and a leak occurs. The remaining refrigerant may be recovered and reused in this system only. It may not be sold to another customer without meeting the ARI 700 specification standard. Analysis of the refrigerant to the ARI 700 standard may be performed only by a qualified chemical laboratory. This test is expensive and is usually performed only on larger volumes of refrigerant. Some owners of multiple equipment may reuse refrigerant in another unit they own if they are willing to take the chance with their own equipment. It is up to the technician to use an approved, clean refrigerant cylinder for recovery to avoid cross contamination.

RECYCLE REFRIGERANT. "To clean the refrigerant by oil separation and single or multiple passes through devices, such as replaceable core filter driers, which reduce moisture, acidity and particulate matter" is the EPA's definition of recycling refrigerant. "This term usually applies to procedures implemented at the job site or at a local service shop." If it is suspected that refrigerant is dirty with certain contaminants, it may be recycled, which means to filter and clean. Filter driers may be used only to remove acid, particles, and moisture from a refrigerant. In some cases, the refrigerant will need to pass through the driers several times before complete cleanup is accomplished. Purging air from the system may be done only when the refrigerant is contained in a separate container or a recycling unit.

A form of recycling has been used for many years. When a motor burn occurs and the compressor has service valves, the valves are front seated, the compressor is changed, and the only refrigerant lost is the refrigerant vapor in the compressor. This is only a small amount of vapor. Driers are used to clean the refrigerant while running the new system compressor. Often the system is started, pumped down into the condenser and receiver, and acid-removing filter driers are added to the suction line to further protect the compressor, **Figure 9–18.** This recycle process occurs within the equipment, and normal operation resumes.

A recycling unit removes the refrigerant from the system, cleans it, and returns it to the system. This is accomplished

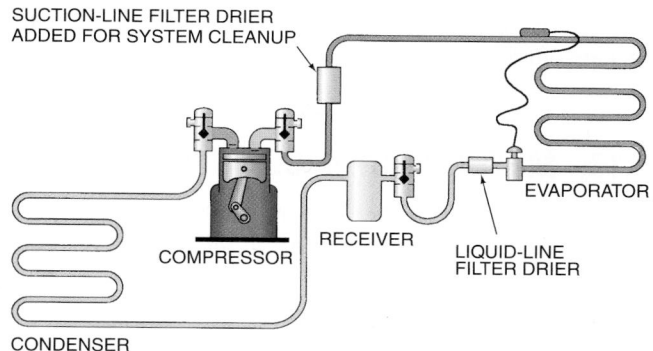

Figure 9–18   Filter driers in the system can be used to recycle the refrigerant within the system.

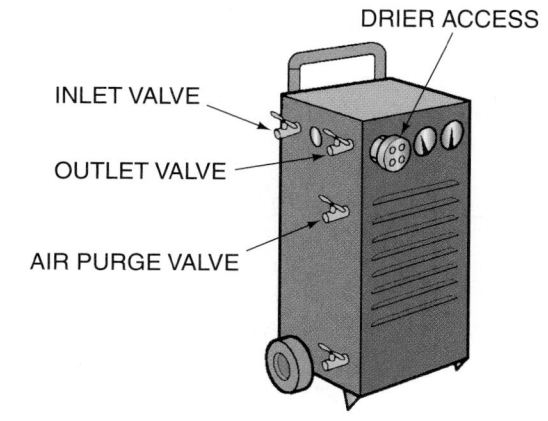

Figure 9–19   The recovery/recycle unit has an air purge valve from which accumulated air may be vented.

by moving the refrigerant to the recovery cylinder. Then using the two valves on the cylinder, the recycle unit circulates the refrigerant around the unit and its filtration system. Some units have an air purge that allows any air introduced into the system to be purged before transferring the refrigerant back into the system, **Figure 9–19.**

RECLAIM REFRIGERANT. "To process refrigerant to new product specifications by means which may include distillation" is the EPA's definition of reclaiming refrigerant. "It will require chemical analysis of the refrigerant to determine that appropriate product specifications are met. This term usually implies the use of processes or procedures available only at a reprocessing or manufacturing facility." This refrigerant is recovered at the job site, stored in approved cylinders, and shipped to the reprocess site where it is chemically analyzed and declared that it can be reprocessed (reclaimed). It is then reprocessed and shipped as reclaimed refrigerant because it meets the ARI 700 standard. At this time there is no way of testing to the ARI 700 standard in the field.

When reclaimed refrigerant is sold, it is not subject to an excise tax because this tax applies to only newly manufactured refrigerant. This tax saving is an incentive to save and reclaim old refrigerant.

9.16 METHODS OF RECOVERY

Several methods of recovery for refrigerants are used in the systems mentioned in this text. All of these refrigerants are thought of as high-pressure refrigerants. They are also called low-boiling-point refrigerants. Some of the CFC refrigerants, such as R-11, are low-pressure refrigerants (high boiling point). It boils at 74.9°F at atmospheric pressure. This refrigerant is very hard to remove from a system because of its high boiling point. A very low vacuum must be pulled to boil it all from the system. This is different from R-12, which has the highest boiling point of the commonly used refrigerants in small systems. It boils at −21.62°F at atmospheric pressure. That means if the temperature of the system is above −21.62°F, the R-12 will boil and can be removed from the system easily. Refrigerant 134a has a boiling point of −15.7°F. The boiling points of some refrigerants at atmospheric pressure are as follows:

Refrigerant	Boiling Point (°F)
R-11	+74.9
R-12	−21.62
R-22	−41.36
R-123	+82.2 (to replace R-11)
R-125	−55.3
R-134a	−15.7 (to replace R-12)
R-500	−28.3
R-502	−49.8
R-410A	−61 (to replace R-22)
R-404A	−51 (to replace R-502)
R-402A	−54
R-402B	−51

The following chart shows that most of the common refrigerants discussed in this text have high vapor pressures and will be relatively easy to remove from the systems. The vapor pressures at a room temperature of 70°F would be as follows:

Refrigerant	Vapor Pressure at 70°F
R-12	70 psig
R-22	121 psig
R-125	158 psig
R-134a	71 psig
R-500	85 psig
R-502	137 psig
R-410A	201 psig
R-404A	148 psig
R-402A	158 psig
R-402B	147 psig

The boiling point is important because the EPA has ruled on how low the pressure in a system must be reduced to accomplish the recovery guideline.

PASSIVE RECOVERY. Passive recovery or system-dependent recovery is when the service technician uses the internal pressure of the system or the system's compressor to aid in the refrigerant recovery process. If the system has an inoperative compressor and passive recovery is to be used, the refrigerant must be recovered from both the low side and high side of the system and achieve the recovery efficiency requirements. Many times, a seasoned service technician will gently tap the side of the compressor's shell with a soft wood, rubber, or leather mallet. This action will agitate the crankcase oil and cause any refrigerant dissolved in it to vaporize and be recovered more easily. Because the compressor is not working, the service technician can also use the aid of a vacuum pump as long as its discharge is not into a pressurized container, but to the atmosphere only. If the compressor is working, the technician can recover refrigerant only from the high side of the system. All manufacturers of appliances containing Class I (CFC) or Class II (HCFC) refrigerants must supply a service aperture or process stub on the system. Also, nonpressurized containers must be used to capture the refrigerant when a passive recovery process is used, **Figure 9–35.** Passive refrigerant recovery is used often on small appliances that contain 5 lb or less of refrigerant in their systems. Small appliances and their recovery techniques are covered in this unit under Section 9.18, "Recovering Refrigerant from Small Appliances."

ACTIVE RECOVERY. Active refrigerant recovery is the method used by technicians the greater part of the time. Refrigerant is removed through the use of a certified self-contained recovery machine. The refrigeration or air-conditioning system's compressor is not used with an active recovery method. These machines are usually capable of both the recovery and recycling of refrigerant. To speed the recovery of refrigerant in an active recovery process, always recover at the highest possible ambient temperature to cause a higher pressure in the system being recovered. SAFETY PRECAUTION: A high-pressure reading in a recovery tank is an indication of air or other noncondensables in the tank. Dangerous pressures could result. After the tank has cooled to the ambient temperature, a temperature/ pressure relationship with the ambient temperature will indicate what the recovery tank pressure should be. Before starting a recovery, it is always important to know what type of refrigerant is in the system to be recovered. Unintentional mixing of refrigerants can also cause high pressures in a recovery tank. Always make sure that refrigerants are not mixed together in a recovery tank. The following are some suggestions for how to avoid mixing refrigerants in a recovery tank:

■ Use pressure/temperature charts for refrigerant in the recovery cylinder.
■ Make sure the recovery machine has a self-clearing feature.
■ Dedicate a lubricant-specific recovery unit to specific refrigerants.
■ Dedicate a certain refrigerant cylinder to a specific refrigerant and mark the cylinder appropriately.
■ Use commercially available refrigerant identifiers and air-detection devices on the recovery tank.
■ Keep appropriate records of refrigerant in inventory.

Effective August 12, 1993, all companies have had to certify to the EPA that they have recovery equipment to perform adequate recovery of the refrigerant for the systems they service.

Equipment manufactured before November 15, 1993, must meet qualifications set for this time period, **Figure 9–20.**

Equipment that is manufactured after November 15, 1993 must be approved by an ARI-approved, third-party laboratory to meet ARI Standard 740.

As of November 15, 1993, the following is the guide used by the EPA for evacuation of equipment except small appliances and motor vehicle air conditioning:

1. HCFC-22 appliance, or isolated component of such appliance, normally containing less than 200 lb of refrigerant, 0 psig.
2. HCFC-22 appliance, or isolated component of such appliance, normally containing more than 200 lb of refrigerant, 10 in. Hg vacuum.
3. Other high-pressure appliance, or isolated component for such appliance, normally containing less than 200 lb of refrigerant (R-12, R-500, R-502, R-114), 10 in. Hg vacuum.
4. Other high-pressure appliance, or isolated component of such appliance, normally containing 200 lb or more of refrigerant (R-12, R-500, R-502, R-114), 15 in. Hg vacuum.
5. Very high-pressure appliance (R-13, R-503), 0 psig.
6. Low-pressure appliance (R-11, R-113, R-123), 25 mm Hg absolute, which is 29 in. Hg vacuum, **Figure 9–20.**

The exception to the preceding would be when the system could not be reduced to these pressure levels; for example, when the system has a large leak, there is no need to pull the system full of air.

Required Levels of Evacuation for Air-Conditioning, Refrigeration, and Recovery/Recycling Equipment (Except for small appliances, MVACs, and MVAC-like equipment) Inches of Hg Vacuum		
Type of Air-Conditioning or Refrigeration Equipment	**Using Recovery or Recycling Equipment Manufactured before November 15, 1993**	**Using Recovery or Recycling Equipment Manufactured on or after November 15, 1993**
HCFC-22 equipment, or isolated component of such equipment, normally containing less than 200 pounds of refrigerant.	0	0
HCFC-22 equipment, or isolated component of such equipment, normally containing 200 pounds or more of refrigerant.	4	10
Other high-pressure equipment, or isolated component of such equipment, normally containing less than 200 pounds of refrigerant.	4	10
Other high-pressure equipment, or isolated component of such equipment, normally containing 200 pounds or more of refrigerant.	4	15
Very high-pressure equipment.	0	0
Low-pressure equipment.	25	29

NOTE: MVAC = Motor Vehicle Air Conditioning

Figure 9–20   Required levels of evacuation for air-conditioning, refrigeration, and recovery/recycling equipment. *Courtesy U.S. EPA*

Technicians are required to be certified to purchase refrigerant after November 14, 1994. The EPA defines a *technician* as: "Any person who performs maintenance, service, or repair that could reasonably be expected to release Class I (CFC) or Class II (HCFC) substances into the atmosphere, including but not limited to installers, contractor employees, in-house service personnel, and, in some cases, owners." There are four different certification areas for the different types of service work. A person must take an EPA-approved proctored test and pass both the core section and either the Type I, Type II, or Type III certification sections to become certified. Certification types are discussed in the following paragraphs. A person who passes the core and certification Types I, II, and III is automatically Universally Certified. Consult with your company's training official, industry chapters, or local wholesale supply houses for training information, testing dates, and test sites. Once a technician is certified, it will be the certified technician's responsibility to comply with any changes in the law.

TYPE I CERTIFICATION. Small Appliance—Manufactured, *charged and hermetically sealed with 5 lb or less of refrigerant.* Includes refrigerators, freezers, room air conditioners, *package terminal heat pumps,* dehumidifiers, under-the-counter ice makers, vending machines, and drinking-water coolers.

TYPE II CERTIFICATION. High-Pressure Appliance—Uses refrigerant with a boiling point between −50°C (−58°F) and 10°C (50°F) at atmospheric pressure. Includes 12, 22, 114, 500, and 502 refrigerant. Replacement refrigerants for these refrigerants are also included.

TYPE III CERTIFICATION. Low-Pressure Appliance—Uses refrigerant with a boiling point above 10°C (50°F) at atmospheric pressure. Includes 11, 113, 123, and their replacement refrigerants.

UNIVERSAL CERTIFICATION. Certified in: Type I, II, and III. It is believed that technician certification should ensure that the technician knows how to handle the refrigerant in a safe manner without exhausting it to the atmosphere.

The refrigerant may be removed from the system as either a vapor or a liquid or in the partial liquid and vapor state. It must be remembered that lubricating oil is also circulating in the system with the refrigerant. If the refrigerant is removed in the vapor state, the oil is more likely to stay in the system. This is desirable from two standpoints:

1. If the oil is contaminated, it may have to be handled as a hazardous waste. Much more consideration is necessary, and a certified hazardous-waste technician must be available.
2. If the oil remains in the system, it will not have to be measured and replaced back to the system. Time may be saved in either case, and time is money. The technician should pay close attention to the management of the oil from any system.

Figure 9–21 Approved Department of Transportation (DOT) cylinders. *Courtesy National Refrigerants, Inc.*

This cylinder must be clean and in a deep vacuum before the recovery or recycle process is started. It is suggested that the cylinder be evacuated to 1000 microns (1 mm Hg) before recovery is started. Refrigerant from a previous job will be removed with this vacuum. All lines to the cylinder must be purged of air before connections are made.

SAFETY PRECAUTION: Refrigerant must be transferred only to approved refrigerant cylinders. These cylinders are approved by the Department of Transportation (DOT). Never use the DOT 39 disposable cylinders. The approved cylinders are recognizable by their color and valve arrangement. They are yellow on top with gray bodies and a special valve that allows liquid or vapor to be added to or removed from the cylinder, **Figure 9–21.**

9.17 MECHANICAL RECOVERY SYSTEMS

Many mechanical recovery systems are available. Some are sold for the purpose of recovering refrigerant only. Some are designed to recover and recycle, and a few sophisticated ones claim to have reclaim capabilities. ARI is the organization that will certify the equipment specifications, and these machines will be expected to perform to ARI 740 specifications.

For recovery and recycle equipment to be useful in the field, it must be portable. Some of the equipment is designed to be used in the shop where refrigerant is brought in for recycling. But for the equipment to be useful in the field, it must be able to be easily moved to the rooftop, where many systems are located. This presents a design challenge because the typical unit must be able to both recover and recycle in the field for it to be effective. It must also be small enough to be hauled in the technician's truck. Units for use on rooftops should not weigh more than approximately 50 lb because anything heavier may be too much for the technician to carry up a ladder. A heavy unit may require two technicians where only one would normally be needed.

Some technicians may overcome this weight problem by using long hoses to reach from the rooftop to the unit, which may be left on the truck or on the top floor of a building, **Figure 9–22**. This is not often used because the technician must recover the refrigerant from the long hose. The long hose also slows the recovery process. Many units have wheels so they may be rolled to the top floor, **Figure 9–23**.

RECOVERY UNIT

Figure 9–22   The refrigerant is recovered from this unit through a long hose to the unit on the truck. This is a slower process.

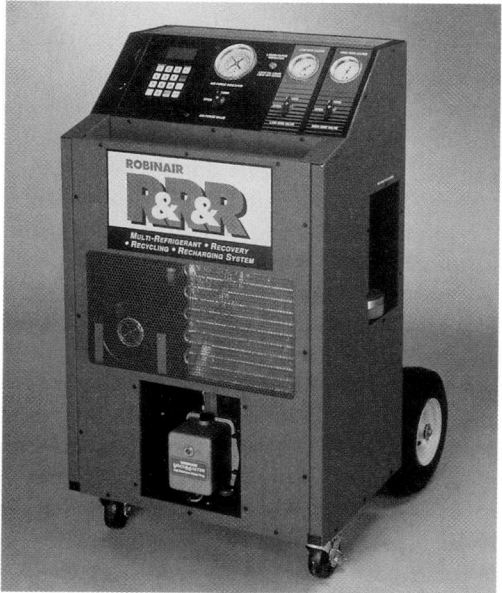

Figure 9–23   The recovery/recycle unit is heavy and has wheels to make it easy to move around. *Courtesy of Robinair SPX Corporation*

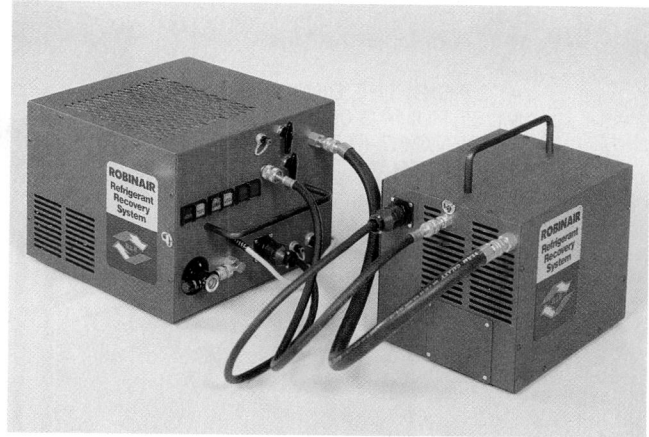

Figure 9–24   This modular unit reduces a heavy unit to two manageable pieces. *Courtesy of Robinair SPX Corporation*

Manufacturers are meeting this portability challenge in several ways. Some have developed modular units that may be carried to the rooftops or remote jobs as separate components, **Figure 9–24**. Some are reducing the size of the units by limiting the internal components and using them for recovery only. In this case, the refrigerant may be recovered to a cylinder, taken to the shop for recycling, and then returned to the job. However, every time the refrigerant is transferred from one container to another, refrigerant will be lost.

Manufacturers are attempting to provide a unit that is small as well as practical, yet will still perform properly. The task is to remove the refrigerant from the system. They are trying to remove vapor, liquid, and vapor/liquid mixtures and deal with oil that may be suspended in the liquid refrigerant. Many are accomplishing this challenge by using a small hermetic compressor in the unit to pump the refrigerant vapor. As the compressors become larger, they become heavier. Larger condensers are also needed, which adds to the weight. Rotary compressors may also pull a deeper vacuum on the system with less effort than a reciprocating compressor. Deeper vacuums may be required in the future for removing more of the system refrigerant. All of these compressors in recovery units also pump oil, so some method of oil return is needed to return the oil to the recovery unit compressor crankcase.

The *fastest* method of removing refrigerant from a system is to take it out in the liquid state. It occupies a smaller volume per pound of refrigerant. If the system is large enough, it may have a liquid receiver where most of the charge is collected. Many systems, however, are small, and if the compressor is not operable it cannot pump the refrigerant.

The *slowest* method of removing refrigerant is to remove it in the vapor state. When removed in the vapor state using a recovery/recycle unit, the unit will remove the vapor faster if the hoses and valve ports are not restricted and if a greater pressure difference can be created. The warmer the system is, the warmer and denser the vapor is. The compressor in the recovery/recycle unit will be able to pump more pounds in a minute. As the vapor pressure in the system is reduced, the vapor becomes less dense, and the unit capacity is reduced.

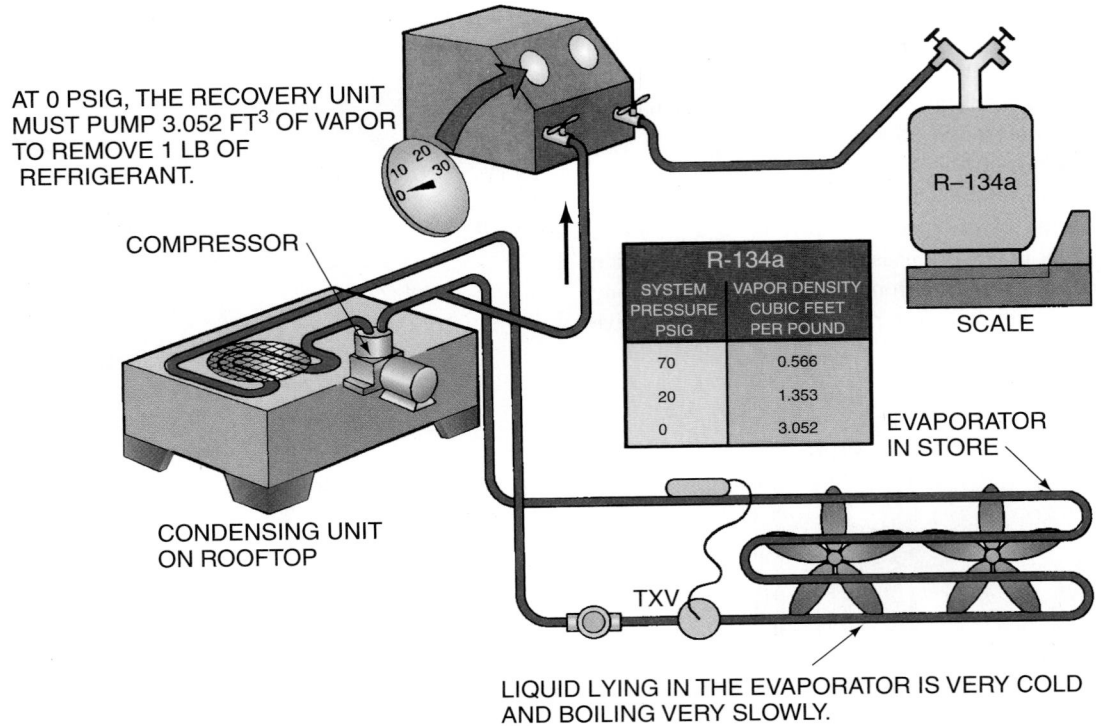

Figure 9–25   Refrigerant vapor has less weight per cubic foot as the pressure is reduced.

It will take more time as the system pressure drops. For example, when removing R-134a from a system, if saturated vapor is removed at 70 psig, only 0.566 ft³ of refrigerant vapor must be removed to remove 1 lb of refrigerant. When the pressure in the system is reduced to 20 psig, 1.353 ft³ of vapor must be removed to remove 1 lb of refrigerant. When the pressure is reduced to 0 psig, there is still refrigerant in the system, but 3.052 ft³ of refrigerant vapor must be removed to remove 1 lb of refrigerant, **Figure 9–25.** The unit recovery rate will slow down as the refrigerant pressure drops because the compressor pumping rate is a constant.

The technician must also be able to rely on having equipment that will operate under the conditions for which it will be needed. It may be 100°F on the rooftop. The unit will be functioning under some extremely hot conditions. The system pressure may be very high due to the ambient temperature, which may tend to overload the unit compressor. A crankcase pressure regulator is sometimes used in recovery/recycle units to prevent overloading the compressor with high suction pressure. The recovery unit may have to condense the refrigerant using the 100°F air across the condenser, **Figure 9–26.** The condenser on the unit must have enough capacity to condense this refrigerant without overloading the compressor.

Recovery/recycling unit manufacturers give the typical pumping rate for their equipment in their specifications. This is the rate at which the equipment is capable of removing refrigerant. The rate is generally 2 to 6 lb/min. The rate will be more constant and faster while refrigerant is being removed in the liquid state. More factors must be considered

when removing vapor. When the system pressure is high, the unit will be able to remove vapor relatively fast. When the system pressure is lowered, the rate of vapor removal will slow down. With a small compressor, when the system vapor pressure reaches about 20 psig, the rate of removal will become very slow.

The compressor in the recovery/recycle unit is a vapor pump. Liquid refrigerant cannot be allowed to enter the compressor because it will cause damage. Different methods may be used to prevent this from occurring. Some manufacturers use a push-pull method of removing refrigerant. This is accomplished by connecting the liquid-line fitting on the unit to the liquid-line fitting on the cylinder. A connection is then made from the unit discharge back to the system. When the unit is started, vapor is pulled out of the recovery cylinder from the vapor port and condensed by the recovery unit. A very small amount of liquid is pushed into the system where it flashes to a vapor to build pressure and push liquid into the receiving cylinder, **Figure 9–27.** A sight glass is used to monitor the liquid, and the recovery cylinder is weighed to prevent overfilling. When it is determined that no more liquid may be removed, the suction line from the recovery unit is reconnected to the vapor portion of the system, and the unit discharge is fastened to the refrigerant cylinder, **Figure 9–28.** The unit is started, and the remaining vapor is removed from the system and condensed in the cylinder.

A crankcase pressure-regulating valve is often installed in the suction line to the compressor in the recovery/recycle

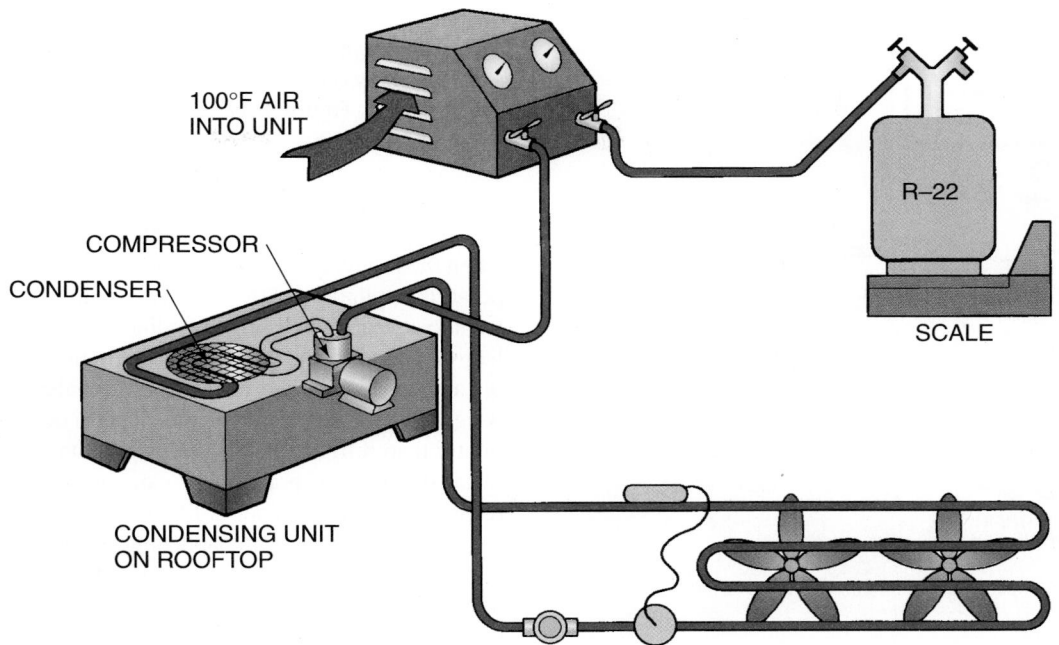

Figure 9–26 The unit will have to condense the refrigerant using 100°F air on the rooftop. The condenser must be large enough to keep the pressure low enough that the compressor is not overloaded.

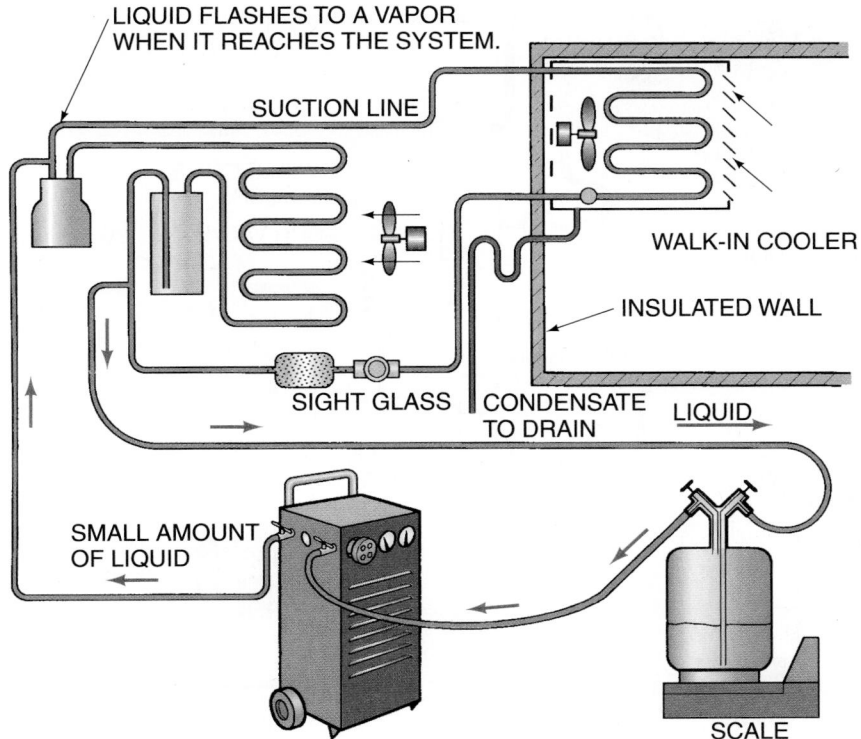

Figure 9–27 This unit uses a push-pull method of removing liquid refrigerant from a system. Vapor is pulled out of the recovery cylinder, creating a low pressure in the cylinder. A small amount of condensed liquid is allowed back into the system to build pressure and push liquid into the cylinder.

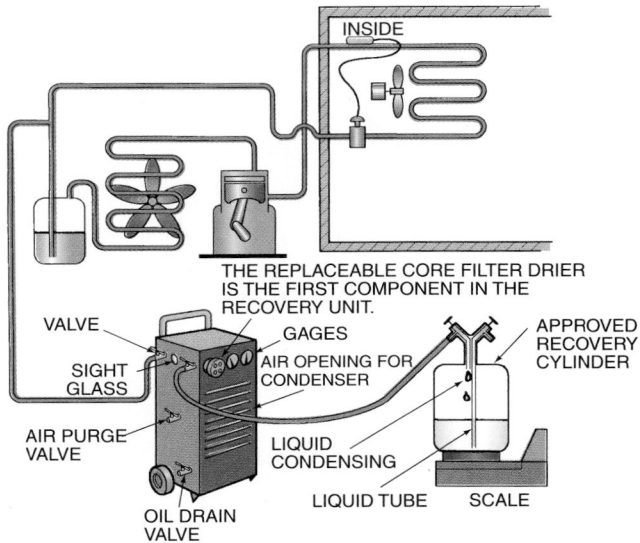

Figure 9–28 The hoses are reconnected to remove the system's vapor.

unit to protect it from overloading and offers some liquid refrigerant protection, **Figure 9–29.** When liquid is removed, it contains oil. The oil will be in the refrigerant in the cylinder. When this liquid refrigerant is charged back into the system,

the oil will go with it. Oil that may move through the recovery unit with any small amount of liquid will be stopped by the oil separator. Any oil removed must be accounted for to add back to the system, **Figure 9–30.** An acid check of the system's oil should be performed if there is any question as to its quality, such as after a motor burn, **Figure 9–31.**

The DOT-approved cylinder used for some of the recovery/recycle systems has a float and switch that will shut the unit off when the liquid in the cylinder reaches a certain level. At this point, the cylinder is 80% full. This switch helps prevent the technician from overfilling the cylinder with refrigerant. It also helps compensate for oil that may be in the refrigerant. When oil and refrigerant are added to a cylinder, the amount of weight that the cylinder can hold changes because oil is much lighter than refrigerant. If a cylinder contains several pounds of oil due to poor practices, it can easily be overfilled with refrigerant if the cylinder is filled only by weight. The float switch prevents this from happening, **Figure 9–32.** NOTE: The recovery cylinder must be clean, empty, and in a vacuum of about 1000 microns before recovery is started. Other methods besides a float device are being used to make sure recovery cylinders are not overfilled. They include having electronic thermistors built into the recovery cylinder's wall and simply weighing the recovery cylinder before and after the refrigerant recovery process. Always make sure the recovery tank is free of oil when using the weighing method.

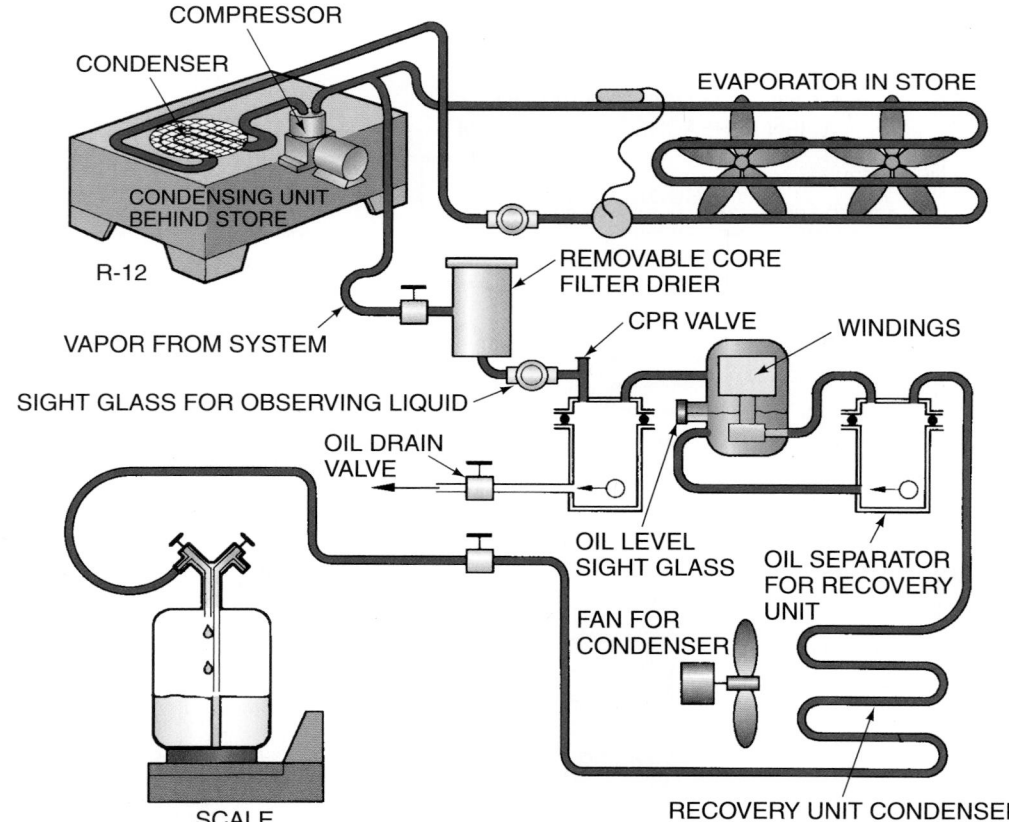

Figure 9–29 A system with some of the possible features incorporated, including a crankcase pressure-regulating valve for protecting the compressor.

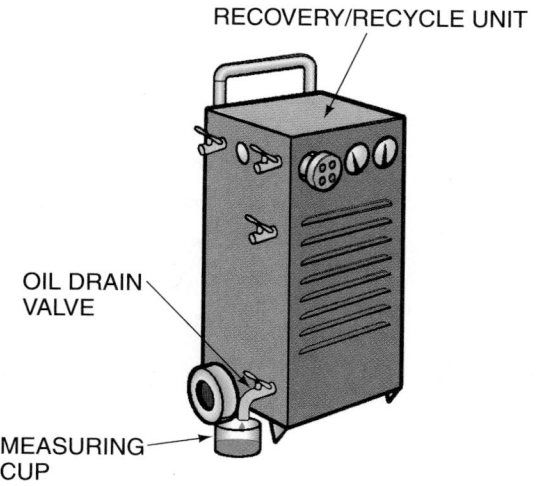

RECOVERY/RECYCLE UNIT

OIL DRAIN VALVE

MEASURING CUP

Figure 9–30   All oil removed from a system must be accounted for.

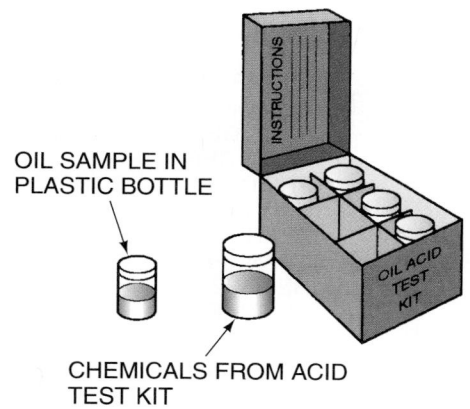

OIL SAMPLE IN PLASTIC BOTTLE

CHEMICALS FROM ACID TEST KIT

Figure 9–31   An acid test may be performed on the refrigerant oil in the field.

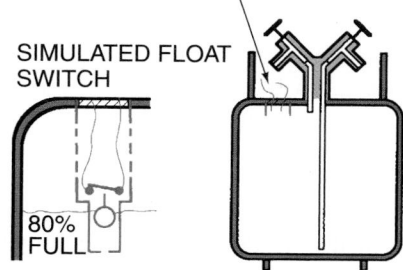

WIRES FOR CONTROLLING CIRCUIT

SIMULATED FLOAT SWITCH

80% FULL

WHEN LIQUID RISES IN THE CYLINDER, THE BALL RISES AND OPENS THE SWITCH. THIS MAY TURN ON A LIGHT, OR PREFERABLY STOP THE UNIT TO PREVENT OVERFILLING.

Figure 9–32   This DOT-approved cylinder has a float switch to prevent overfilling.

Some manufacturers remove liquid from the system through a metering device and an evaporator that is used to boil the liquid to a vapor. The compressor in the recovery/recycle unit moves the liquid refrigerant and some oil from the system and evaporates the refrigerant from the oil, **Figure 9–33**. The oil must then be separated out, or it will collect in the unit compressor. Usually an oil separator is used. The oil from the separator must be drained periodically and measured to know how much oil to add back to the system if it is to be put back in service. Oil at this time may be checked for acid content using an acid test kit.

All of the recovery/recycle unit compressors will pump oil out the discharge line and should have an oil separator to direct the oil back to the unit compressor crankcase.

Manufacturers may use other features to accomplish the same things. Keep in mind that manufacturers are trying to incorporate features in their equipment to make it lightweight, dependable, and efficient. Follow the manufacturers' instructions; they know the capabilities of their equipment.

The unit to be purchased should be chosen carefully. Some questions should be asked when determining which unit to purchase.

Recovery Unit

1. Will it run under the conditions needed for the majority of the jobs encountered? For example, will it operate properly on the rooftop in hot or cold weather?
2. Will it recover both liquid and vapor if you need these features?
3. What is the pumping rate in the liquid and vapor state for the refrigerants you need to recover?
4. Are the drier cores standard, and can they be purchased locally?
5. Is the unit portability such that you can conveniently perform your service jobs?
6. Is the recovery unit oil-less? If oil-less, it can be used on systems with different lubricants such as mineral oils, alkylbenzenes, esters, and glycols. Oil changes and oil contamination are no longer a problem with an oil-less recovery unit, **Figure 9–34**.

Recovery/Recycle Units

1. Will the unit run under the conditions needed for the majority of the jobs encountered?
2. Will it recycle enough refrigerant for the types of jobs encountered?
3. What is the recycle rate, and does it correspond with the rate you may need for the size of the systems you service?
4. Are the drier cores standard, and can they be purchased locally?
5. Is the unit portable enough for your jobs and transportation availability?

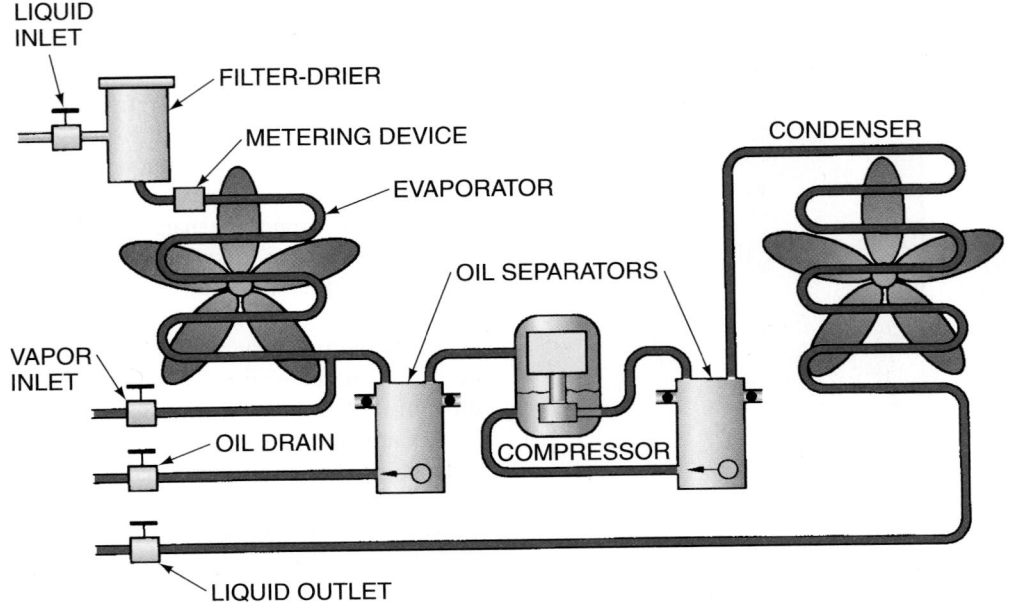

Figure 9–33   This system removes either liquid or vapor.

Figure 9–34   An oil-less recovery unit. *Courtesy Ferris State University. Photo by John Tomczyk*

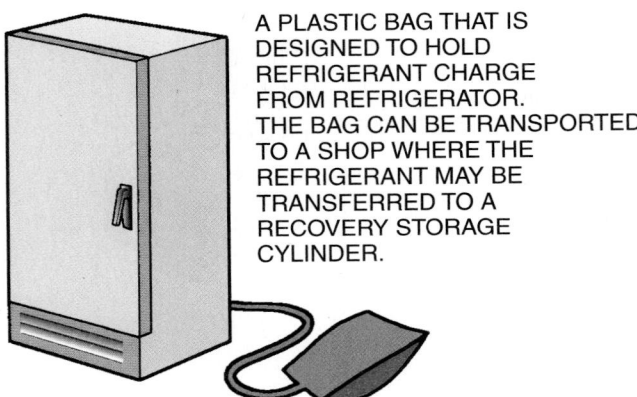

Figure 9–35   Removing refrigerant to a special plastic bag.

9.18   RECOVERING REFRIGERANT FROM SMALL APPLIANCES

A small appliance is defined as a system that is manufactured, charged, and hermetically sealed with 5 lb or less of refrigerant. This includes refrigerators, freezers, room air conditioners, packaged terminal heat pumps, dehumidifiers, under-the-counter ice makers, vending machines, and drinking-water coolers.

Recovering refrigerant from small appliances may be an easier job than from larger systems because not as much refrigerant is involved. Recovery in small appliances can be either *active* or *passive recovery*. *Active recovery* is recovering refrigerant with the use of a recovery machine that has its own built-in compressor. *Passive recovery* uses the refrigeration system's compressor or internal vapor pressure for the recovery process. However, when a service technician uses passive recovery, the refrigerant must be recovered in a nonpressurized container, **Figure 9–35.** No matter what kind of recovery device is being used, an environmentally conscious service technician should do a daily leak check with a refrigerant leak detector on the recovery device. To speed up the recovery process, a technician can:

- Put the recovery cylinder in cold water to lower its vapor pressure.
- Heat the refrigeration system to cause the vapor pressure inside the system to increase. Heating the compressor's crankcase will surely release a lot of refrigerant dissolved in the oil. Heat lamps, electric blankets, and heat guns work best. Use of the system's defrost heater will also suffice. Never use an open flame for heating the system.
- Make sure the ambient temperature around the system being recovered is as high as possible. This will ensure high vapor pressures inside the system.

Figure 9–36 A Schrader valve with core and cap. *Photo by John Tomczyk*

■ Softly tap the compressor's crankcase with a soft wood, rubber, or leather mallet to agitate the crankcase oil and release refrigerant.

LOW-LOSS FITTINGS. Any recovery equipment manufactured after November 15, 1993, must have low-loss fittings on the system. **Figure 9–41** is an example of a low-loss fitting. There are many different types of low-loss fittings on the market today. Low-loss fittings can be either manually closed off or automatically closed off when a service technician takes them off a pressurized refrigeration system or recovery device.

SCHRADER VALVES. A lot of small appliances have Schrader valves on their systems, **Figure 9–36.** ♻ To prevent refrigerant leaks, Schrader valves must be periodically examined for defects like bends, cracks, or loose cores and caps. ♻ If leaks do occur in the Schrader valve, replace the valve core and make sure the cap is on finger tight. Schrader valve caps help prevent leaks and also help prevent the valve core from accumulating dirt and dust. Accidental depression of the valve core can also be prevented by placing a cap on the valve.

PIERCING VALVES. Many service technicians gain access to small appliances by using a piercing valve, **Figure 9–37.** With piercing valves, a sharp pin is driven through the system's tubing. When the needle is retracted, access is gained to the system. Piercing valves come in both clamp-on type and solder-on type. The clamp-on type piercing valves are used where temporary system access is needed, such as a recovery and charging process. All clamp-on type piercing valves should be removed when service is finished because their soft rubber gaskets can rot over time and cause leaks. It is a good idea to leak check these valves before the system's tubing is pierced. Piercing valves are not recommended for use on steel tubing.

Solder-on-type piercing valves are used for permanent access to a system. Once the valve is soldered onto the system's tubing, leak check the assembly before piercing the system's tubing. This can be done by pressurizing the assembly.

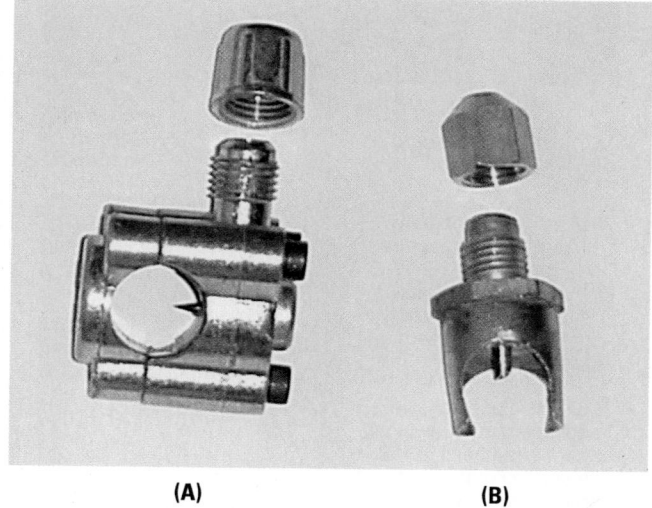

     (A)   **(B)**

Figure 9–37 **(A)** A gasket-type piercing valve. **(B)** A solder-on-type piercing valve. *Photo by John Tomczyk*

SAFETY PRECAUTION: Never heat the system's tubing to cherry red. This can cause a blowout, and serious injury can occur. When certain refrigerants are exposed to high temperatures, a flame, or a glowing hot metal, they can decompose into hydrochloric and/or hydrofluoric acids and phosgene gas. A technician who accesses a system and finds 0 psig pressure should not continue on with a recovery process. A pressure reading of 0 psig could mean that there is a leak in the system or that the line has not been successfully pierced. If the system leaked and is not at 0 psig, noncondensables like air with water vapor are in the system and the recovery device will only take in these contaminants.

When accessing a small appliance for recovery purposes, if the compressor is inoperative, both high- and low-side access is recommended to speed up the recovery process and to make sure the required vacuum levels are reached. **Figure 9–38** lists recovery requirements for small appliances. A vacuum pump can assist in a passive recovery with an inoperative compressor, but the refrigerant must be recovered in a nonpressurized container as in **Figure 9–35.** This is because vacuum pumps cannot handle pumping against anything but atmospheric pressure. Too large a vacuum pump will freeze any moisture in the system because it causes a speedy drop in system pressure. The technician should then introduce dry nitrogen into the system to increase pressure and avoid moisture freezing. SAFETY PRECAUTION: Never energize the compressor when under a deep vacuum. This can result in a short in the motor windings at the Fusite terminals or at a weak spot in the windings and cause compressor damage. For accuracy in pulling a vacuum with a vacuum pump, the system's vacuum gauge should be located as close to the system tubing and as far away from the vacuum pump as possible. To determine how long it will take to pull a deep vacuum (500 microns), the vacuum pump's capacity and the diameter of the service hose coming from the vacuum pump are two important factors. Make sure the vacuum pump is turned off or valved off when

Recovery Efficiency Requirements for Small Appliances*		
Recovery Efficiencies Required	**Recovered Percentages**	**Inches of Mercury Vacuum**
For active and passive equipment manufactured after November 15, 1993, for service or disposal of small appliances with an operative compressor on the small appliance.	90%	4*
For active and passive equipment manufactured after November 15, 1993, for service or disposal with an inoperative compressor on the small appliance.	80%	4*
For grandfathered active and passive equipment manufactured before November 15, 1993, for service or disposal with or without an operating compressor on the small appliance.	80%	4*

*ARI 740 Standards
*NOTE: Small appliances are products that are fully manufactured, charged, and hermetically sealed in a factory and that have 5 lb or less of refrigerant.

Figure 9–38 Small-applicance recovery requirements.

checking the final vacuum. Never recover from a small appliance in the following situations:

- Systems having sulfur dioxide, methyl chloride, or methyl formate.
- The pressure in the system to be recovered is 0 psig.
- Refrigerants like water, hydrogen, or ammonia, which are found in recreational vehicles and absorption systems, are present.
- A strong odor of acid is smelled, as in a compressor burnout.
- Nitrogen is mixed with R-22 as a trace gas for electronic leak detection, or nitrogen is used as a holding charge.

♲ Even though it is not necessary to repair a leak on a small appliance, every service technician should find and repair the leak because of environmental concerns, which include ozone depletion and global warming. Also, a service technician cannot avoid recovering refrigerant by adding nitrogen to a system that already has a refrigerant charge in it. The system must first be evacuated to the levels in **Figure 9–38** before nitrogen can be added.♲

SAFETY PRECAUTION: Always check the nameplate of the system that dry nitrogen will enter. Never exceed the design low-side factory test pressure on the nameplate, **Figure 9–39**. When using dry nitrogen, always use a pressure relief valve and pressure regulator on the nitrogen tank because of its high pressures (over 2000 psi). Pressure relief valves must

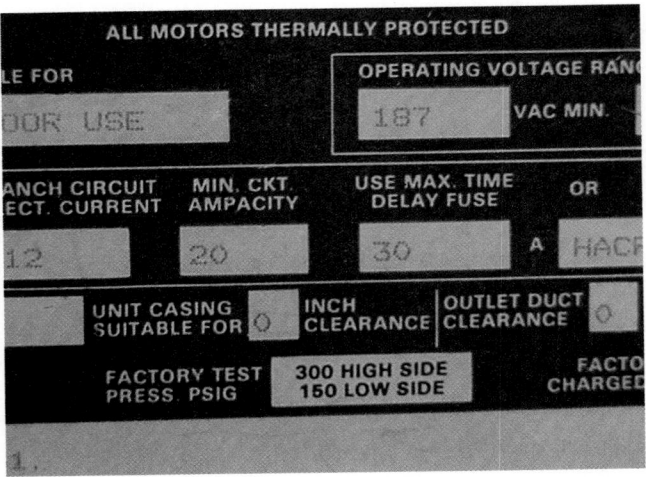

Figure 9–39 A condensing unit nameplate showing factory test pressures. *Photo by John Tomczyk*

never be installed in series. Never pressurize a system with oxygen or compressed air, because dangerous pressures can be generated from a reaction with the oil as it oxidizes.

RECOVERY RULES AND REGULATIONS. Effective August 12, 1993, contractors must certify to the EPA that the equipment they are using is capable of accomplishing the recovery requirements of ARI Standard 740, listed in **Figure 9–38**. Also, any recovery equipment manufactured after November 15, 1993 and used on small appliances must be certified by an EPA-approved lab. Within 20 days of opening a new business, you must register or certify to the EPA that you have recovery equipment that is able to recover at least 80% of the refrigerant charge or able to pull a 4-in. Hg vacuum.

DISPOSAL OF SMALL APPLIANCES. Before disposing of a small appliance, you must do the following:
- Recover the refrigerant from the appliance.
- Possess a signed statement that the refrigerant has been recovered, with the date of recovery.

Refrigerant bags are available for recovery of refrigerant from small appliances. These bags are plastic and will hold the charge of several refrigerators, **Figure 9–35**. The technician must have enough room in the service truck to haul the bag. When the bag is full, it may be taken to the shop and the refrigerant transferred into the reclaim cylinder for later pickup by or shipment to the reclaim company.

9.19 RECLAIMING REFRIGERANT

Because refrigerant must meet the ARI 700 standard before it is resold or reused in another customer's equipment, all refrigerant for reclaim will need to be sent to a reclaim company where it can be checked and received (provided it is in good enough condition to be used). Many service companies will not try to recycle refrigerant in the field because of the complications and the possibility of cross contamination of products. Reclaiming the refrigerant is another option.

Refrigerant that meets the ARI 700 specification can be purchased from reclaim companies, and it does not have the added tax. Since the refrigerant cannot be vented, the only other choice is to recover it to a central location and place it in a cylinder that may be shipped to a reclaim company. The reclaim companies will accept only large quantities of refrigerant so the service shop will need a cylinder of the correct size for each of the refrigerants that will be shipped. Reclaim companies will be accepting refrigerant in cylinders of 100 lb and larger. Some companies are buying recovered refrigerant and accumulating it to be reclaimed. Only DOT-approved cylinders with the proper labels can be used, **Figure 9–40.**

The refrigerant in the reclaim cylinders must be of one type of refrigerant. The refrigerants cannot be allowed to be mixed, or the reclaim company cannot reclaim it. As mentioned before, mixed refrigerants cannot be separated.

When the reclaim company receives refrigerant that cannot be reclaimed, it must be destroyed. The method of destroying refrigerant is to incinerate it in such a manner that the fluorine in the refrigerant is captured. This is an expensive process, so technicians must be careful not to mix the refrigerants.

9.20   REFRIGERANTS AND TOOLS IN THE FUTURE

All refrigerants in the future will have to be environmentally friendly. Hopefully, all refrigerants will someday have 0 ODPs and very low GWPs. It will pay the technician of the future to stay up to date on all developments as they occur. You should join local trade organizations that are watching industry trends. For example, in the future there will be changes in the tools used to handle refrigerant.

Figure 9–40   Returnable refrigerant cylinder labels for recovered refrigerant. *Courtesy National Refrigerants, Inc.*

Figure 9–41 A quick-couple or quick-connect to help reduce refrigerant loss. *Courtesy of Robinair SPX Corporation*

Gage lines that will not leak refrigerant are necessary. Many older gage lines seep refrigerant while connected. Gage lines are also required to have valves to reduce the amount of refrigerant lost when they are disconnected from Schrader connections. These valves are located in the end of the hoses and hold the refrigerant in the hose when disconnected, **Figure 9–41**.

Each technician should make an effort to know what the latest information and technology is and practice the best techniques available.

SUMMARY

- CFC refrigerants must not be vented to the atmosphere.
- Ozone is a form of oxygen, with each molecule consisting of three atoms of oxygen (O_3). The oxygen we breathe contains two atoms (O_2).
- The ozone layer, located in the stratosphere 7 to 30 miles above the earth's surface, protects the earth from harmful ultraviolet rays from the sun, which can cause damage to human beings, animals, and plants.
- Ozone Depletion Potentials (ODPs) have been established by the Montreal Protocol.
- Global warming is a process that takes place in the lower atmosphere, or troposphere.
- Both direct and indirect effects of global warming exist.
- Global warming potentials (GWPs) are given to refrigerants.
- Chlorofluorocarbons (CFCs) contain chlorine, fluorine, and carbon atoms.
- Hydrochlorofluorocarbons (HCFCs) contain hydrogen, chlorine, fluorine, and carbon atoms.
- Hydrofluorocarbons (HFCs) contain hydrogen, fluorine, and carbon atoms.
- Some hydrocarbons (HCs) are entering the markets as part of refrigerant blends.
- Both zeotropic and azeotropic refrigerant blends have become very popular as replacements for both CFC and HCFC refrigerants.
- Both near-azeotropic and zeotropic blends exhibit temperature glide and fractionation when they evaporate and condense.
- Refrigerant oils have been researched extensively because of the refrigerant transition.
- Synthetic oils have gained popularity and are being used with a lot of the newer refrigerants.
- Refrigerant and oil retrofitting guidelines have been established for service purposes because of the refrigerant and oil transition.
- Hydrogen atoms help the HCFCs break down before they reach the stratosphere and react with the ozone.

- The EPA has established many regulations governing the handling of refrigerants.
- To recover refrigerant is "to remove refrigerant in any condition from a system and store it in an external container without necessarily testing or processing it in any way."
- To recycle refrigerant is "to clean the refrigerant by oil separation and single or multiple passes through devices, such as replaceable core filter-driers, which reduce moisture, acidity, and particulate matter. This term usually applies to procedures implemented at the job site or at a local service shop."
- To reclaim refrigerant is "to process refrigerant to new product specifications by means which may include distillation. It will require chemical analysis of the refrigerant to determine that appropriate product specifications are met. This term usually implies the use of processes or procedures available only at a reprocessing or manufacturing facility."
- Active recovery involves recovering refrigerant through the use of a certified self-contained recovery machine.
- Passive recovery or system-dependent recovery is when the service technician uses the internal pressure of the system and/or the system's compressor to aid in the refrigerant recovery process.
- Recovery of small appliances involves the use of different methods and different evacuation levels.
- Refrigerant can be transferred only into Department of Transportation (DOT)–approved cylinders and tanks.
- Manufacturers have developed many types of equipment to recover and recycle refrigerant.

REVIEW QUESTIONS

1. What is the difference between the oxygen we breathe and the ozone located in the earth's stratosphere?
2. Describe the difference between good and bad ozone.
3. Explain the process of ozone depletion resulting from the breaking up of a CFC molecule in the stratosphere.
4. Which refrigerants have a 0 ODP?
   - A. HFCs
   - B. HCFCs
   - C. CFCs
   - D. Both A and B
5. An increase in the natural greenhouse effect that leads to heating of the earth is called _____.
6. True or False: Only CFCs and HCFCs contribute to global warming.
7. Explain what a global warming potential (GWP) is.
8. Describe the differences between global warming and ozone depletion.
9. Describe the difference between a direct and an indirect global-warming potential.
10. What measurement or index takes into consideration both the indirect and direct global-warming effects of refrigerants?
    - A. Total equivalent warming impact (TEWI)
    - B. Greenhouse effect number
    - C. Global-warming number
    - D. Greenhouse effect index

11. Describe the differences between CFCs, HCFCs, HFCs, and HCs.
12. CFCs are more harmful to the stratospheric ozone layer than HCFCs because they contain _____.
13. True or False: HFCs do not damage the ozone layer but still contribute to global warming.
14. Describe the difference between an azeotropic refrigerant blend and a near-azeotropic or zeotropic refrigerant blend.
15. Why are there more safety precautions with R-410A than with other popular refrigerants like R-22 and R-404A?
16. The three most popular synthetic oils in the refrigeration and air-conditioning industry are _____, _____, and _____.
17. What is the name of the conference that was held in Canada in 1987 to attempt to solve the problems of released refrigerants?
18. What agency of the federal government is charged with implementing the United States Clean Air Act Amendments of 1990?
19. Explain the differences between an active and a passive recovery method.

20. Are federal excise taxes on reclaimed refrigerants the same as those for new refrigerants?
21. To recover refrigerant means to _____.
22. To recycle refrigerant means to _____.
23. To reclaim refrigerant means to _____.
24. Explain what is meant by a nonpressurized recovery container and briefly explain which applications it is used for.
25. What is the name of the federal agency that must approve containers into which refrigerant may be transferred?
26. What is a quick-couple or quick-connect and where are they used?
27. The standard that a refrigerant must meet after it has been reclaimed and before it can be sold to another owner is
    A. ANSI 7000.
    B. ANSI 700.
    C. EPA 7000.
    D. ARI 700.
28. Briefly describe the two types of piercing valves used in the industry and list some advantages and disadvantages of both.

OBJECTIVES

After studying this unit, you should be able to

■ describe how refrigerant is charged into systems in the vapor and liquid states.

■ describe system charging using two different weighing methods.

■ state the advantage of using electronic scales for weighing refrigerant into a system.

■ describe two types of charging devices.

■ charge refrigerant blends incorporating a temperature glide and fractionation potential.

■ charge fixed orifice, capillary tube, and piston (short tube) air-conditioning systems using charging charts and curves.

■ use the subcooling method of charging to charge air-conditioning and heat pump systems that incorporate a thermostatic expansion valve (TXV) as the metering device.

■ use modern temperature/pressure charts that incorporate dew point and bubble point values for calculating subcooling and superheat amounts.

SAFETY CHECKLIST

✔ Never use concentrated heat from a torch to apply heat to a refrigerant cylinder. Use a gentle heat such as from a tub of warm water no hotter than 90°F.

✔ Do not charge liquid refrigerant into the suction line of a system unless it is vaporized before it reaches the compressor. Liquid refrigerant must not enter the compressor.

✔ Liquid refrigerant is normally added in the liquid line and then only under the proper conditions.

✔ If a device is used to flash liquid refrigerant into the suction line, 100% vapor must enter the compressor. Liquid refrigerant can cause severe damage to any compressor.

10.1 CHARGING A REFRIGERATION SYSTEM

Charging a system refers to the adding of refrigerant to a refrigeration system. 🌐 The correct charge must be added for a refrigeration system to operate as it was designed to, and this is not always easy to do. 🌐 Each component in the system must have the correct amount of refrigerant. The refrigerant may be added to the system in the vapor or liquid states by weighing, measuring, or using operating pressure charts. When using system operating charts and tables to determine the correct system charge, a number of factors should be taken into account. These factors include, but are not limited to, high- and low-side saturation pressures, ambient temperatures, evaporator superheat, condenser subcooling, compressor amperage draw, and temperature differentials across heat exchange surfaces.

This unit describes how to add the refrigerant correctly and safely in the vapor state and liquid state. The correct charge for a particular system will be discussed in the unit where that system is covered. Heat pump charging, for example, is covered in Unit 43, which discusses heat pumps.

10.2 VAPOR REFRIGERANT CHARGING

The *vapor refrigerant charging* of a system is accomplished by allowing vapor to move out of the vapor space of a refrigerant cylinder and into the refrigerant system. When the system pressures are lower than the pressure in the refrigerant tank, for example, when the system has just been evacuated or when the system is out of refrigerant, refrigerant can be added to both the high- and low-pressure sides of the system. This is because the pressure on both sides of the system is lower than the pressure in the refrigerant tank. Remember that higher pressure will flow toward a lower pressure. When the system is operating, however, refrigerant is added to the low side of the system. This is because the compressor raises the pressure of the refrigerant on the high side of the system and lowers the pressure on the low side of the system. The high-side pressure of an operating refrigeration system is typically higher than the pressure of the refrigerant in the tank. For example, an R-134a system may have a head pressure of 185 psig on a 95°F day (this is determined by taking the outside temperature of 95°F and adding 30°F, which gives a condensing temperature of 125°F, or 185 psig for R-134a). The cylinder is exposed to the same ambient temperature of 95°F but has a pressure of only 114 psig, **Figure 10–1**. It should be noted that, if valve B in **Figure 10–1** is opened, refrigerant will flow toward the refrigerant tank. If the tank is a "one-trip" tank that is used only for new, "virgin" refrigerant, the refrigerant will not enter the tank. This is because one-trip tanks are manufactured with check valves that only allow refrigerant to leave the tank, not enter it. If the tank is a reusable tank, there are no check valves and system refrigerant can enter the tank as long as the system pressure is higher than the pressure in the tank.

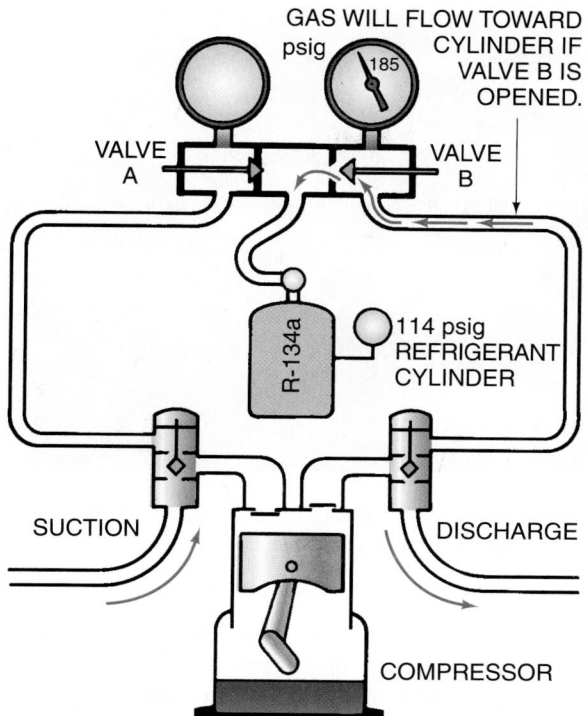

Figure 10–1 This refrigerant cylinder has a pressure of 114 psig. The high side of the system has a pressure of 185 psig. Pressure in the system will prevent the refrigerant in the cylinder from flowing into the system.

The low-side pressure in an operating system is much lower than the cylinder pressure if the cylinder is warm. For example, on a 95°F day, the cylinder of saturated R-134a will have a pressure of 114 psig, but the evaporator pressure may be only 20 psig. Refrigerant will easily move into the system from the cylinder, **Figure 10–2**. In the winter, however, the cylinder may have been in the back of the truck all night, and its pressure may be lower than the low side of the system, **Figure 10–3**. In this case the cylinder will have to be warmed to get refrigerant to move from the cylinder to the system. It is a good idea to have a cylinder of refrigerant stored in the equipment room of large installations. The cylinder will always be there in case you have no refrigerant in the truck, and the cylinder will be at room temperature even in cold weather.

When vapor refrigerant is pulled out of a refrigerant cylinder, the vapor pressure is reduced, and the liquid boils to replace the vapor that is leaving. This vapor pressure reduction causes a lower saturation temperature for both the liquid and the vapor in the cylinder because of the associated temperature/pressure relationship. As more and more vapor is released from the cylinder, the liquid in the bottom of the cylinder continues to boil, and its temperature decreases. If enough refrigerant is released, the cylinder pressure will decrease to the low-side pressure of the system. Heat will have to be added to the liquid refrigerant to keep the pressure up. SAFETY PRECAUTION: *Never* use concentrated heat from a torch. Gentle heat, such as the heat from a tub of warm water, is safer. The water temperature should not

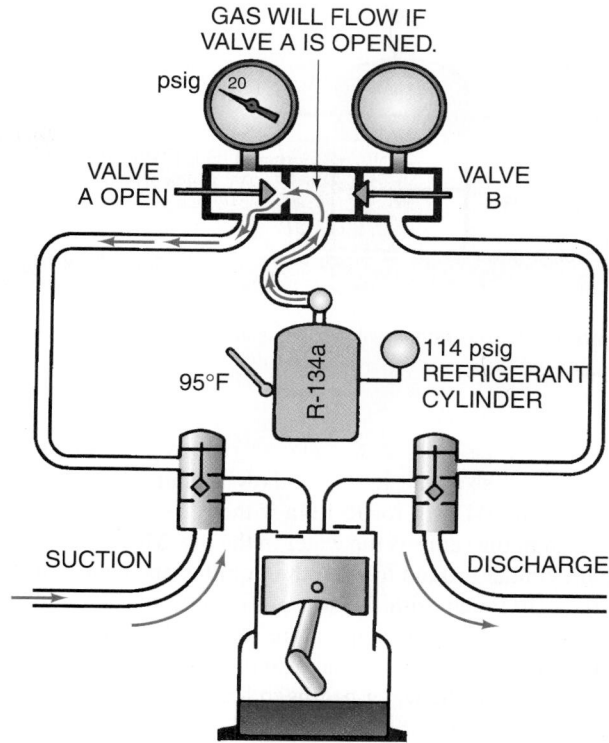

Figure 10–2 The temperature of the cylinder is 95°F. The pressure inside the cylinder is 114 psig. The low-side pressure is 20 psig.

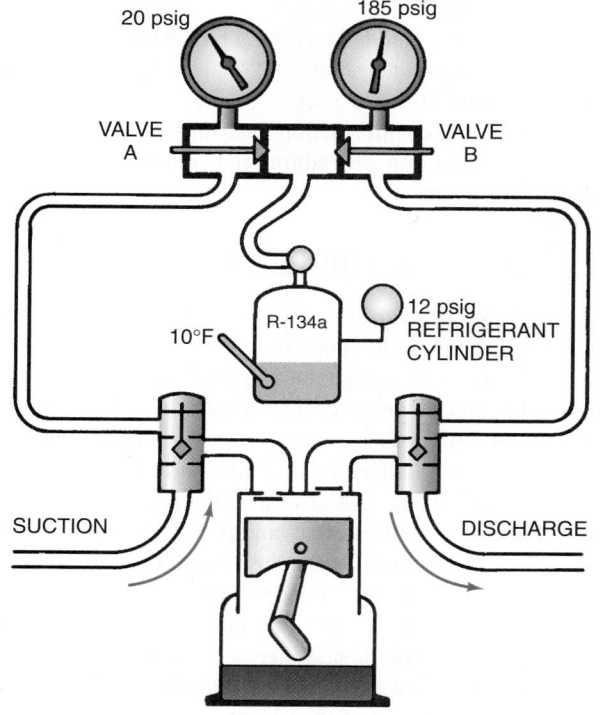

Figure 10–3 The refrigerant in this cylinder is at a low temperature and pressure because it has been in the back of the service truck all night in cold weather. The cylinder pressure is 12 psig, which corresponds to 10°F. The pressure in the system is 20 psig.

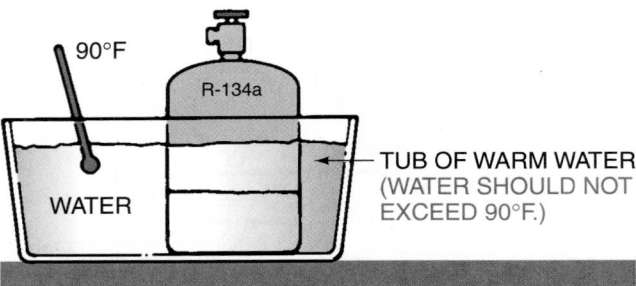

Figure 10–4 This refrigerant cylinder is in warm water to keep the pressure up.

Figure 10–5 Liquid-line service valve located on the refrigerant receiver. *Photo by Eugene Silberstein*

exceed 90°F. Electric heating blankets manufactured for this purpose are also safe to use. This will maintain a cylinder pressure of 104 psig for R-134a if the refrigerant temperature is kept the same as the water in the tub. Move the refrigerant cylinder around to keep the liquid in the center of the cylinder in touch with the warmth outside of the cylinder, **Figure 10–4**. A water temperature of 90°F is a good one to work with because it is the approximate temperature of the human hand. If the water begins to feel warm to the hand, it is getting too hot.

The larger the volume of the liquid refrigerant in the bottom of the cylinder, the longer the cylinder will maintain the pressure. When large amounts of refrigerant must be charged into a system, use the largest cylinder available. For example, do not use a 25-lb cylinder to charge 20 lb of refrigerant into a system if a 125-lb cylinder is available.

If you are planning on charging vapor into a system that is operating with a blended refrigerant, care must be taken to ensure that the blend does not fractionate, or separate into its component parts. Refer to Section 10.8 in this unit as well as Unit 9 for important information regarding blended refrigerants and the charging procedures that must be used when working with them.

10.3 LIQUID REFRIGERANT CHARGING

Liquid refrigerant charging of a system is normally accomplished in the liquid line. For example, when a system is out of refrigerant, liquid refrigerant can be charged into the *king valve* on the liquid line or receiver, **Figure 10–5**. If the system is in a vacuum or has been evacuated, you can connect to the liquid valve of a cylinder of refrigerant, and liquid refrigerant may be allowed to enter the system until it has nearly stopped. The liquid will enter the system and move toward the evaporator and the condenser. When the system is started, the refrigerant is about equally divided between the evaporator and the condenser, and no danger of liquid flooding into the compressor exists, **Figure 10–6**. When technicians are charging with liquid refrigerant, they do not reduce the cylinder's vapor pressure or temperature. When large amounts of refrigerant are needed, adding the liquid through the king valve is preferable to other methods because it saves time.

The king valve is used to help technicians perform a number of system service tasks that include taking pressure

readings on the high side of the system and trapping all of the system refrigerant in the receiver. In the past, king valves were used primarily as safety valves on ammonia and sulfur dioxide systems. They were located outside the mechanical equipment rooms. By positioning the stem on the king valve so that the liquid line was sealed off, refrigerant could not leave the receiver, causing the remaining system refrigerant to become stored in the receiver. This could be done from outside the equipment room so that the technician would not have to enter a toxic environment in the event of a refrigerant leak. Stems that are positioned this way are referred to as front seated. The king valve has four positions: back seated, cracked-off-the-back-seat, midseated, and front seated, **Figure 10–7**.

The back-seated position, **Figure 10–7(A)**, is the normal operating position of the valve. Refrigerant flows freely from the receiver to the liquid line, while the service port is closed off. This is the port where our high-side gage hose is connected. When the valve is cracked-off-the-back-seat, the service port is now open to the system and the technician can take operating pressure readings from the high side of the system, **Figure 10–7(B)**. While in the cracked-off-the-back-seat position, refrigerant still flows freely from the receiver to the liquid line. The midseated position, **Figure 10–7(C)**, is used for performing standing pressure tests and for system evacuation. The front-seated position, **Figure 10–7(D)**, is used for system pumpdown. In this position, the refrigerant flows to the receiver, but cannot leave it. This traps all of the system refrigerant in the receiver, enabling the technician to service the low-pressure side of the system.

When a system has a king valve, it may be front seated while the system is operating, and the low-side pressure of the system will drop. This happens because the rest of the liquid line, expansion valve, evaporator, suction line, and compressor are now being starved of refrigerant. Liquid from the cylinder may be charged into the system at this time through an extra charging port. The liquid from the cylinder is actually feeding the expansion device. *Be careful not to overcharge the system,*

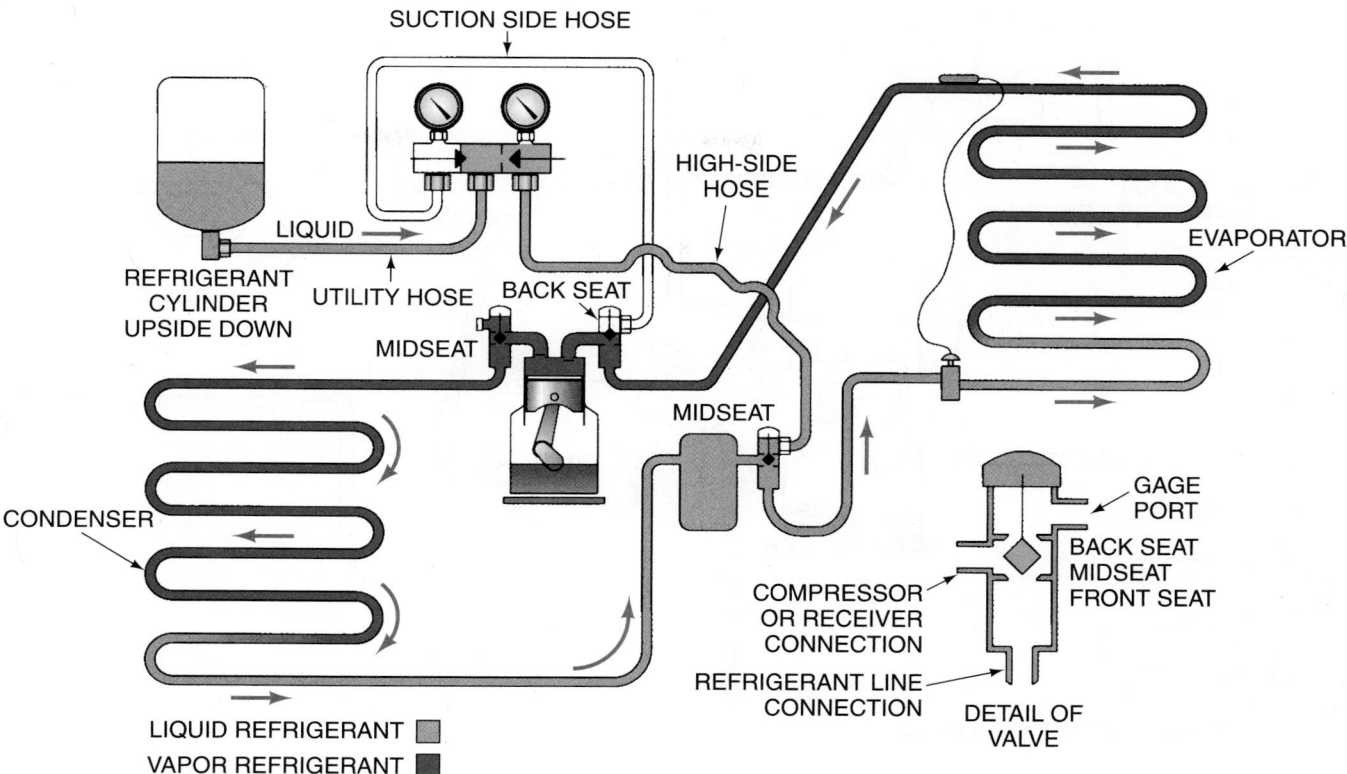

Figure 10–6 This system is being charged while it has no refrigerant in it. The liquid refrigerant moves toward the evaporator and the condenser when doing this. No liquid refrigerant will enter the compressor.

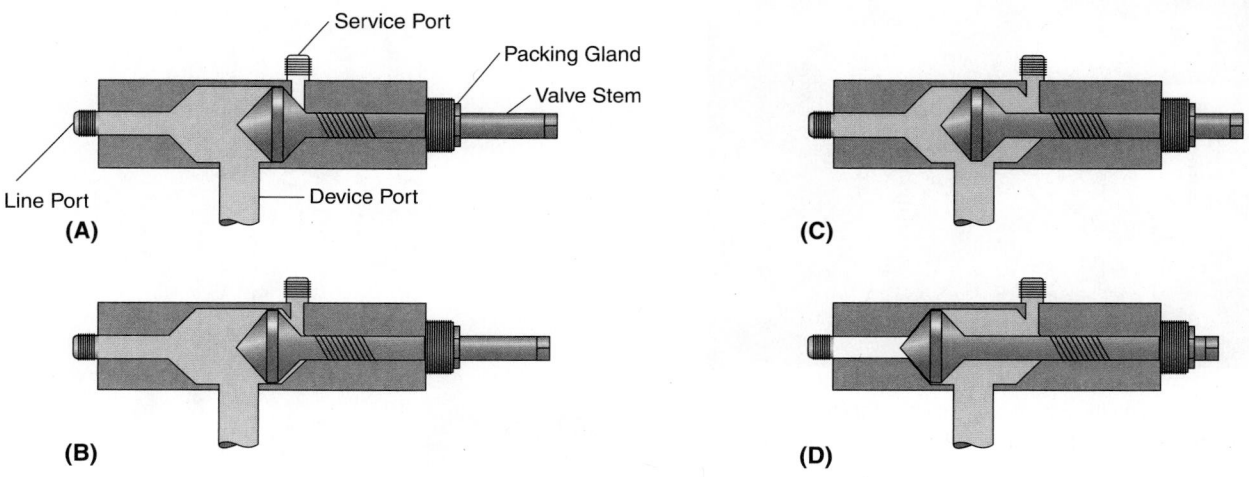

Figure 10–7 Cross-sectional view of a service valve. **(A)** Back-seated position. **(B)** Cracked-off-the-back-seat position. **(C)** Midseated position. **(D)** Front-seated position.

Figure 10–8. The low-pressure control may have to be by-passed during charging to keep it from shutting the system off. Be sure to remove the electrical bypass when charging is completed, **Figure 10–9.** NOTE: Every manufacturer cautions against charging liquid refrigerant into the suction line of a compressor. Precautions must be taken when charging liquid into the suction line of a system. Liquid must be vaporized once it leaves the charging cylinder to prevent any liquid from entering the compressor. Liquid refrigerant must not enter the compressor as damage can result if the compressor attempts to compress liquids, which are not compressible. When working with blended refrigerants, note that the refrigerant must leave the cylinder as a liquid in order to prevent fractionation.

Some commercially available charging devices allow the cylinder liquid line to be attached to the suction line for charging a system that is running. They are orifice-metering devices that are actually a restriction between the gage manifold and

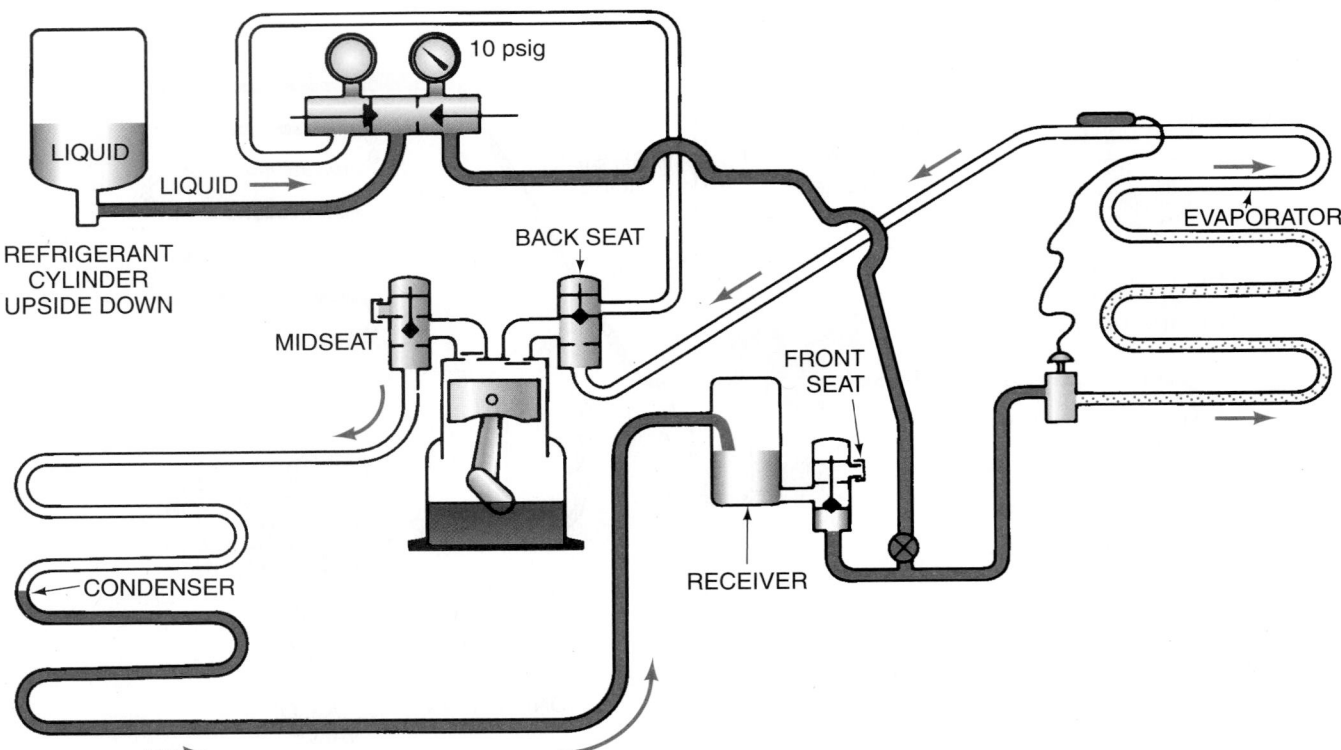

Figure 10–8 This system is being charged by front seating the king valve and allowing liquid refrigerant to enter the liquid line.

Figure 10–9 Bypassing the low-pressure control for charging purposes. *Courtesy Ferris State University. Photo by John Tomczyk*

the system's suction line, **Figure 10–10.** They meter liquid refrigerant into the suction line where it flashes into a vapor. The same thing may be accomplished using the gage manifold valve, **Figure 10–11.** The pressure in the suction line is maintained at a pressure of not more than 10 psig higher than the system suction pressure. This will meter the liquid refrigerant into the suction line as a vapor. During the process of throttling the liquid refrigerant through the gage manifold, the portion of the gage manifold where the center hose is connected should be warm, while the portion of the manifold where the low-side hose is connected should be cold. The temperature difference that exists between the center and

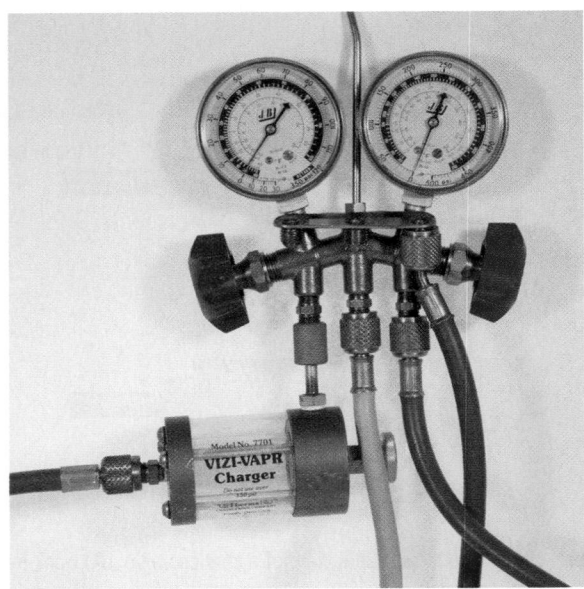

Figure 10–10 A charging device in the gage line between the liquid refrigerant in the cylinder and the suction line of the system. *Courtesy Ferris State University. Photo by John Tomczyk*

low-side connections on the gage manifold is due to the fact that the vaporizing liquid will absorb heat from the manifold. NOTE: The gage manifold valve should be used only on compressors where the suction gas passes over the motor windings. This will boil any small amounts of liquid refrigerant that may reach the compressor. If the lower compressor housing

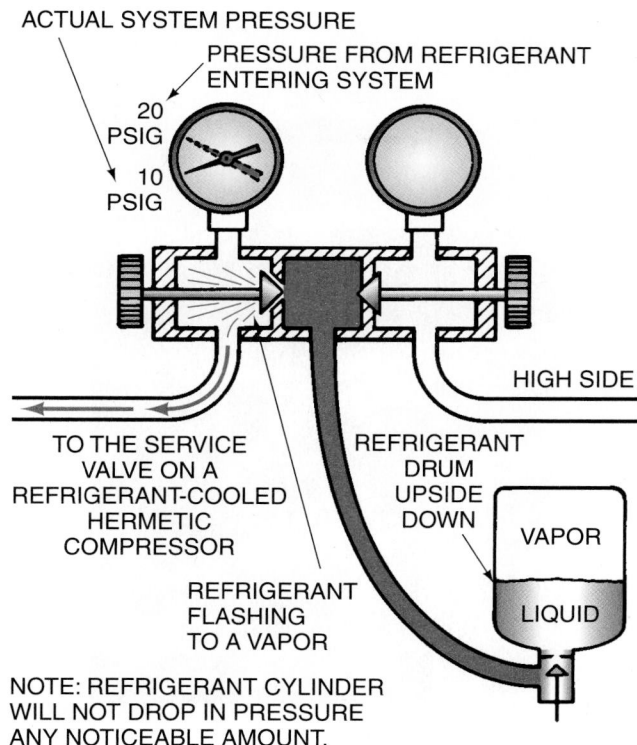

ACTUAL SYSTEM PRESSURE

PRESSURE FROM REFRIGERANT ENTERING SYSTEM

20 PSIG

10 PSIG

HIGH SIDE

TO THE SERVICE VALVE ON A REFRIGERANT-COOLED HERMETIC COMPRESSOR

REFRIGERANT DRUM UPSIDE DOWN

VAPOR

LIQUID

REFRIGERANT FLASHING TO A VAPOR

NOTE: REFRIGERANT CYLINDER WILL NOT DROP IN PRESSURE ANY NOTICEABLE AMOUNT.

Figure 10–11 A manifold gage set used to accomplish the same purpose as that in **Figure 10–10.**

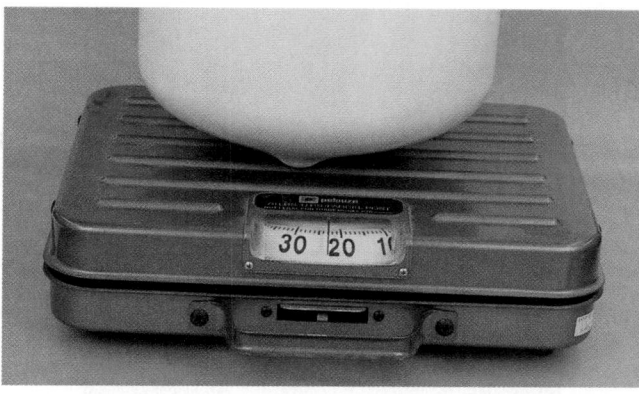

Figure 10–12 A scale for measuring refrigerant. *Photo by Bill Johnson*

becomes cold, stop adding liquid. This method should be performed only under the supervision of an experienced person. Throttling liquid refrigerant from a charging cylinder, however, must be done when charging certain refrigerant blends to avoid fractionation. This throttling must be done regardless of whether the compressor is refrigerant cooled or air cooled.

When a measured amount of refrigerant must be charged into a system, it may be weighed into the system or measured in by using a graduated charging cylinder. Package systems such as air conditioners and refrigerated cases will have the recommended charge printed on the nameplate. Attempting to weigh refrigerant into the system when it contains a partial charge will lead to inaccurate charging and improper system operation. The remaining charge must first be recovered and then the system must be properly evacuated before the nameplate charge can be added to the system.

10.4 WEIGHING REFRIGERANT

Weighing refrigerant may be accomplished using various scales. Bathroom and other inaccurate scales should *not* be used. **Figure 10–12** shows an accurate dial scale graduated in pounds and ounces. Secure the scales (make sure they are portable) in the truck to keep the mechanism from shaking and knocking the device out of calibration. Dial scales can be difficult to use, as the next example shows.

Suppose 28 oz of refrigerant is needed. A dial scale weighs the cylinder of refrigerant at 24 lb 4 oz. As refrigerant runs

into the system, cylinder weight decreases. Determining the final cylinder weight is not easy for some technicians.

The calculated final cylinder weight is

$$24 \text{ lb } 4 \text{ oz} - 28 \text{ oz} = 22 \text{ lb } 8 \text{ oz}$$

To determine this, 24 lb 4 oz was converted to ounces:

$$24 \text{ lb} \times 16 \text{ oz/lb} = 384 \text{ oz} + 4 \text{ oz} = 388 \text{ oz}$$

Now subtract 28 oz from 388 oz:

$$388 \text{ oz} - 28 \text{ oz} = 360 \text{ oz}$$

This is the final cylinder weight. Because scales do not read in ounces, you must convert to pounds and ounces.

$$360 \text{ oz} \div 16 \text{ oz/lb} = 22.5 \text{ lb}$$
$$= 22 \text{ lb } 8 \text{ oz}$$

Since it is easy to make a mistake in this calculation, electronic scales are often used, **Figure 10–13(A)**. These are very accurate but more expensive than dial-type scales. These scales can be adjusted to zero with a full cylinder, **Figure 10–13(B)**, so as refrigerant is added to the system the scales read a positive value. For example: If 28 oz of refrigerant is needed in a system, put the refrigerant cylinder on the scale and set the scale at 0. As the refrigerant leaves the cylinder, the scale counts upward. When 28 oz is reached, the refrigerant flow can be stopped. This is a time-saving feature that avoids the cumbersome calculations involved with the dial scale. Electronic scales are available with different features. Some of the more desirable models have an automatic charging feature, **Figure 10–14,** that enables the technician to program the amount of refrigerant that will be charged into the system. A solid-state microprocessor automatically stops the charging process once the desired amount of refrigerant has been introduced to the system.

10.5 USING CHARGING DEVICES

Graduated cylinders, **Figure 10–15,** are often used to add refrigerant to smaller air-conditioning and refrigeration systems, such as domestic refrigerators and window or through-the-wall air conditioners, that have their total refrigerant charge indicated on the appliance's nameplate. These cylinders are clear so that you can observe the liquid level in the cylinder.

(A)

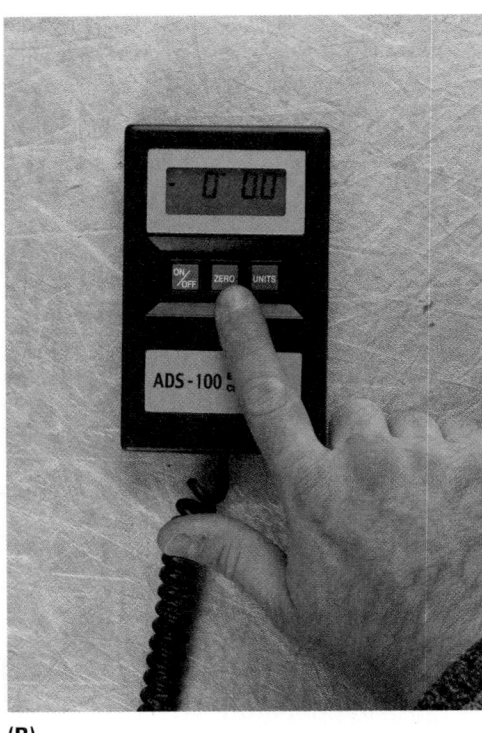

(B)

Figure 10–13   (A) Digital scale used to measure the amount of refrigerant added to or removed from the system. **(B)** The scale's display can be zeroed out to reduce the number of manual calculations.

Figure 10–14 Automatic charging scales can be programmed to charge a precise quantity of refrigerant into the system. *Courtesy Ferris State University. Photo by Eugene Silberstein*

There is a pressure gage located at the top of the graduated charging cylinder, from which we read the saturation pressure and temperature of the refrigerant. The saturation temperature gives us the temperature of the surrounding ambient air, which we need to properly prepare the cylinder for use. The volume of liquid refrigerant changes as the surrounding air temperature changes, so the cylinder must be calibrated to the ambient temperature before each use. The amount of liquid refrigerant in the cylinder, in pounds

Figure 10–15 A graduated charging cylinder. *Photo by Bill Johnson*

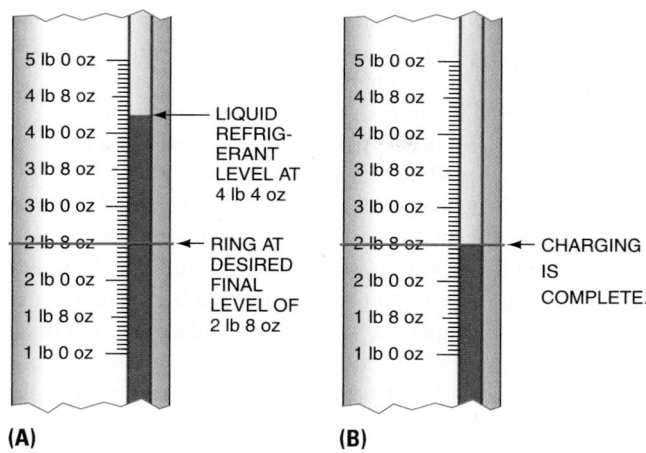

(A)   **(B)**

Figure 10–16   (A) The ring is set to desired refrigerant level once the charging process has been completed. **(B)** When the refrigerant level is even with the position of the ring, the system has been charged.

and ounces, can be read from the graduated scale on the sides of the cylinder.

Once calibrated, the amount of liquid refrigerant in the cylinder is recorded. The amount of refrigerant that must be added to the system is then subtracted from the amount in the cylinder. The result is the amount of refrigerant that must remain in the cylinder after the charging process has been completed. This level is marked with a sliding ring, **Figure 10–16(A)**. The final liquid level inside the cylinder must be calculated much like the previous example, but it is not as complicated.

Suppose a graduated cylinder has 4 lb 4 oz of R-134a in the cylinder at 100 psig. Turn the dial to 100 psig and record the level of 4 lb 4 oz. The system charge of 28 oz is subtracted from the 4 lb 4 oz as follows:

4 lb × 16 oz/lb = 64 oz + the remaining 4 oz = 68 oz

Then

$$68 \text{ oz} - 28 \text{ oz} = 40 \text{ oz}$$

is the final cylinder level.

$$40 \text{ oz} \div 16 \text{ oz/lb} = 2.5 \text{ lb} = 2 \text{ lb } 8 \text{ oz}$$

When the charging process is completed, the level of the liquid refrigerant in the graduated charging cylinder will be level with the sliding ring, **Figure 10–16(B)**.

Some graduated cylinders have heaters in the bottom to keep the refrigerant temperature from dropping when vapor is pulled from the cylinder or when it is being used in a cold surrounding temperature.

When selecting a graduated cylinder for charging purposes, be sure you select one that is large enough for the systems with which you will be working. It is difficult to use a cylinder twice for one accurate charge. When charging systems with more than one type of refrigerant, you need a charging cylinder for each type of refrigerant. You will also not overcharge the customer or use the wrong amount of refrigerant if you closely follow the preceding methods.

Other available refrigerant charging devices may make it more convenient to charge refrigerant into a system.

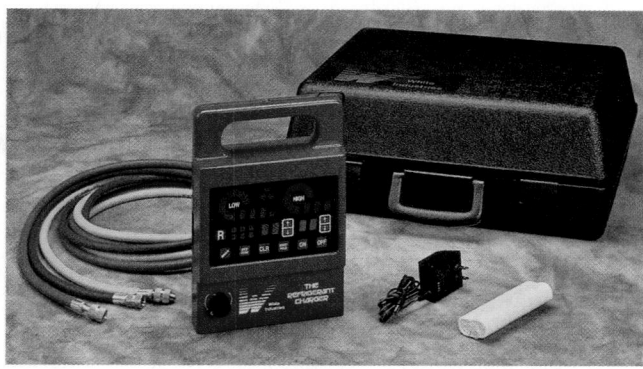

Figure 10–17   A refrigerant charger. *Courtesy White Industries*

Figure 10–17 is an example of one of these. This device can be set to charge a system using pressure or weight. A predetermined amount of refrigerant can be charged in 1-oz increments. This type of device also has many other features. Be sure to follow the manufacturer's instructions.

10.6   USING CHARGING CHARTS

Some manufacturers will supply charts or curves to assist the service technician in correctly charging air-conditioning and heat pump units. These are called *charging charts* or *charging curves,* **Figure 10–18**. Manufacturers may vary the style and type of superheat charging chart or curve that they offer. Many times the type of charging chart depends on what type of metering device the equipment is incorporating. In the HVACR industry, charging charts or curves that utilize superheat measurements, such as those shown in **Figure 10–18,** are typically used in conjunction with fixed-bore metering

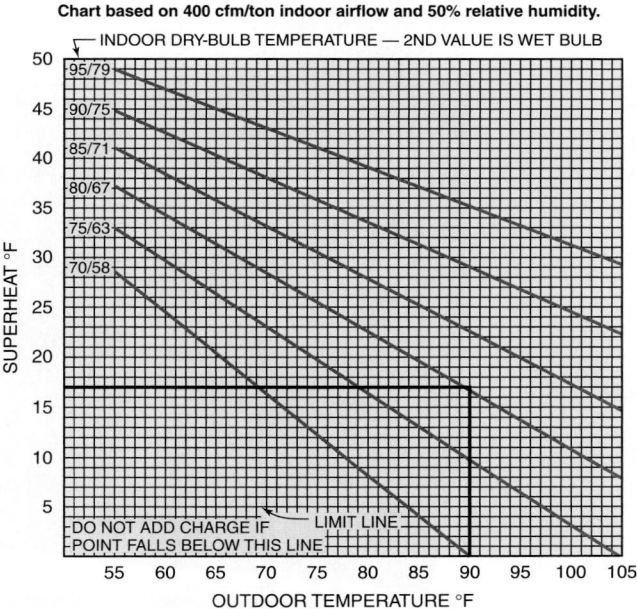

Figure 10–18   A charging curve used for charging a split-type air-conditioning system that incorporates a capillary tube metering device.

devices such as capillary tubes, fixed orifices, and pistons (short-tube devices). Charging curves that rely on subcooling measurements are typically used on systems that are equipped with thermostatic expansion valves. The same basic underlying principles hold for just about all charging curves or charts. However, the safest and most accurate way to charge any unit is to follow the manufacturer's charging instructions on the unit or in a manual supplied with the unit.

The charging curve in **Figure 10–18** is based on 400 cfm/ ton of airflow at 50% relative humidity across the evaporator coil. To use a charging curve to charge a split air-conditioning system that incorporates a capillary tube for a metering device, follow these steps:

Step 1 Measure the indoor dry-bulb temperature (DBT). This is the return air at the air handler. NOTE: Use wet-bulb temperature (WBT) if the percentage of relative humidity is above 70% or below 20%.

Step 2 Measure the outdoor DBT at the outdoor unit. This is the temperature of the air coming into the condenser.

Step 3 Measure the suction pressure at the service valve on the outdoor condensing unit and convert to a temperature using a temperature/ pressure chart.

Step 4 Measure the suction-line inlet temperature on the suction line near the service valve on the outdoor condensing unit, **Figure 10–19.**

Step 5 Calculate the amount of superheat by subtracting the temperature obtained in step 3 from the temperature obtained in step 4.

Step 6 Find the intersection where the outdoor temperature and the indoor temperature meet and read the degrees superheat.

The superheat that is being calculated here is NOT the evaporator superheat. Since the suction-line temperature reading is being taken from a point near the compressor, the sensible heat being picked up in the suction line is included in the calculation. This superheat calculation is often referred to as the *system superheat.* If the superheat of the system is more than 5°F above what the charging chart reads, add refrigerant into the low side of the operating system until the superheat is within 5°F of the chart.

If the superheat of the system is more than 5°F below what the chart reads, recover some refrigerant from the system until the superheat is within 5°F of the chart.

NOTE: Always let the system run for at least 15 min after adding refrigerant to or recovering refrigerant from the system before recalculating system superheat. To avoid fractionation in the charging cylinder when using a refrigerant blend that has a temperature glide and fractionation potential, liquid refrigerant has to be taken from the charging cylinder and then throttled into the low side of the system as a vapor while the system is running. For a review of refrigerant blends with temperature glide and fractionation potential, refer to Unit 9.

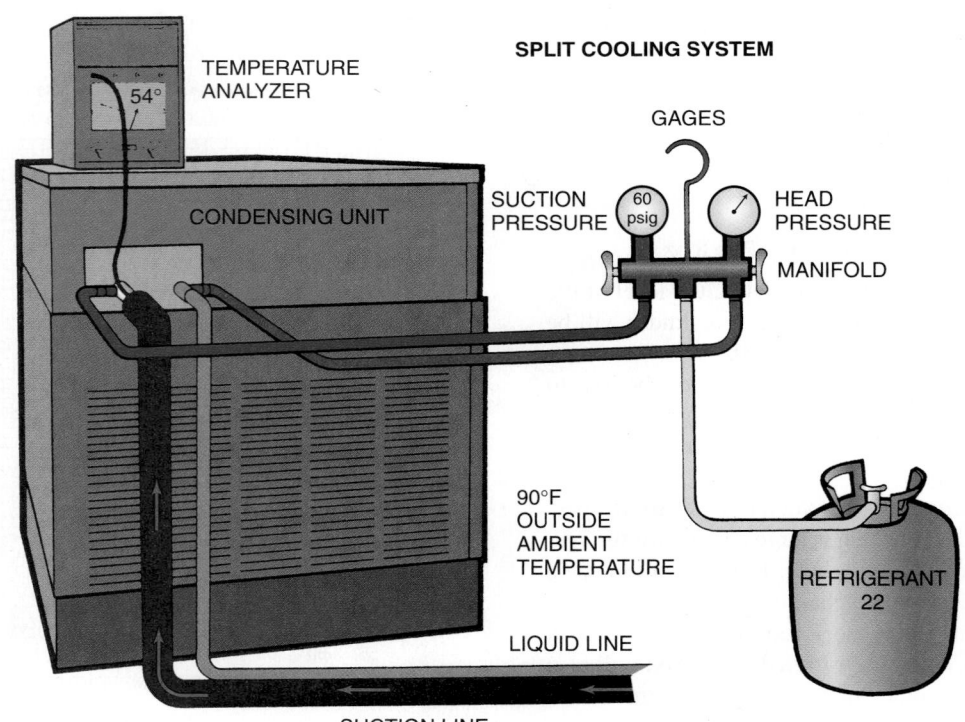

Figure 10–19   A split-type air-conditioning system using a capillary tube metering device and R-22 as the refrigerant.

Example

R-22 capillary tube or fixed orifice system, **Figure 10–19**

Indoor DBT = 80°F

Outdoor DBT = 90°F

Suction pressure at service valve of condensing unit = 60 psig or 34°F using a P/T chart

Suction-line temperature at service valve = 54°F

System superheat is (54°F – 34°F) = 20°F

The intersection of the 90°F outdoor temperature with the 80°F indoor temperature indicates that there should be about 17°F of superheat. Our system has 20°F of superheat. This is within 5°F of the superheat chart, so the system is fully charged. The technician would not have to add any refrigerant to the system.

The theory behind these charging curves is simple. Let us take the charging curve in **Figure 10–18** as an example. As you move to the right on the bottom axis, the outdoor temperature rises. Notice that for a constant indoor DBT or WBT (lines that slant downward from left to right), as you increase the outdoor ambient temperature the operating compressor superheat decreases. The reason for this is that there is now more head pressure on the high side of the system pushing the subcooled liquid out of the condenser's bottom through the liquid line and the capillary tube metering device. This will force more refrigerant into the evaporator and give less superheat. This is why some air-conditioning systems incorporating capillary tubes or fixed orifice metering devices will flood or slug liquid at the compressor on a day with hot outdoor ambient temperatures when the devices are overcharged. The superheat curve will prevent this from occurring if it is followed properly.

Referring to **Figure 10–18,** if we assume a constant indoor DBT of 75°F across the evaporator coil, and increase the outdoor DBT from 70°F to 105°F, we can see that the operating compressor superheat will fall from 23°F to 0°F. This is caused by the hotter outdoor ambient temperature, which results in higher head pressures pushing more liquid through the capillary tube and into the evaporator. So, it is normal for the system to run 23°F of compressor superheat when the outdoor ambient temperature is 70°F. Do not add any refrigerant to this system, because if the outdoor ambient temperature climbed to 95°F later in the day, the system's compressor would slug or flood.

NOTE: For the manufacturer's charging curve in **Figure 10–18,** if the relative humidity is above 70% or below 20%, use WBTs instead of DBTs across the evaporator coil to compensate for the varying latent (moisture) loads.

If the outdoor temperature stays constant and the indoor DBT or indoor WBT increases, the operating superheat will increase. This loading of the indoor coil with either sensible or latent heat or both will cause a more rapid vaporization of refrigerant in the evaporator. This will cause high compressor superheats. This is a normal occurrence. Many technicians will add refrigerant in this case and overcharge the system. It is completely normal for a capillary or fixed orifice metering device system to run high superheat at high evaporator loadings.

Again referring to **Figure 10–18,** as the outdoor ambient temperature stays constant at 95°F, and the indoor DBT across the evaporator coil rises from 75°F to 95°F, the operating superheat will rise from 6°F to 33°F. At a 95°F indoor air DBT and a 95°F outdoor air DBT, the superheat should normally be 33°F according to the chart. These conditions create an inefficient system with an inactive evaporator. This is the greatest disadvantage of a fixed orifice metering device. However, this is the only way to prevent slugging and flooding of refrigerant with the varying indoor and outdoor loads that air-conditioning systems often experience.

Although charging capillary tube, fixed orifice, and piston (short-tube) systems in air-conditioning applications follow much the same underlying idea, it is strongly recommended that the technician consult with the manufacturers of air-conditioning systems to use their exact methods of charging following their charging curves and tables. Some manufacturers use different curves and tables for different models of their equipment. Some manufacturers have eliminated the need for a WBT because of custom-made charging curves that represent their laboratory tests on the equipment.

Figure 10–20 is a charging chart for a 5-ton, unitary air-conditioning system using a capillary tube as a metering device. This charging chart was affixed to the side of the unit by the manufacturer to help the service technician in servicing and charging the unit. This type of chart is much easier to use than the one in **Figure 10–18.** There are only three variables on this chart. They are the outdoor temperature, suction-line temperature, and suction pressure. The service technician must first establish what the outdoor ambient temperature is and then read the suction-pressure gage at the condensing unit's service valve. With these two conditions known and the charging chart handy, the service

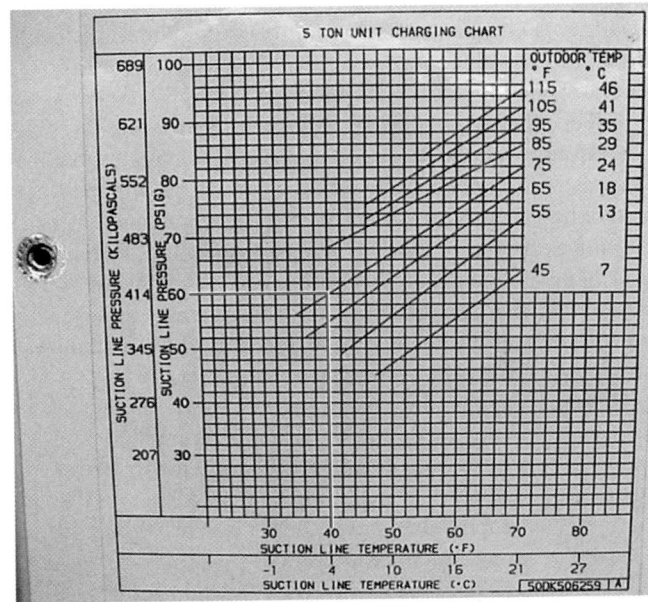

Figure 10–20 A charging chart from the side of a 5-ton unitary air-conditioning system. *Courtesy Ferris State University. Photo by John Tomczyk*

technician can determine what the appropriate suction-line temperature entering the condensing unit at the service valve should be. If the suction-line temperature is high, add some refrigerant. If the suction-line temperature is low, recover some refrigerant.

Example: R-22

Outdoor temperature = 75°F
Suction pressure = 60 psig
Suction-line temperature at the condensing unit's service
  valve should be 40°F (see **Figure 10–20**).

If the suction-line temperature measured at the service valve was 60°F, the service technician would have to add refrigerant. If the suction-line temperature measured at the service valve was 32°F, the service technician would have to recover some refrigerant from the unit. Always give the unit an ample amount of time (15 to 20 min) to find a new equilibrium when adding or subtracting refrigerant and before using the charging chart again. It is much easier to add the correct charge than to remove refrigerant to the correct charge level.

10.7   SUBCOOLING METHOD OF CHARGING FOR TXV SYSTEMS

The previous two charging charts illustrated were for fixed restrictor-type metering devices such as capillary tubes, fixed orifices, or piston (short-tube) devices. The main advantage of fixed orifice metering devices is their inexpensive cost and simplicity. However, some air-conditioning units, especially high-efficiency units, come with thermostatic expansion valves (TXVs) as their metering devices. When systems have a TXV as a metering device, the TXV will control a constant amount of evaporator superheat and keep the evaporator active under both high- and low-heat loadings. TXV systems also keep a constant superheat in the evaporator with changing outdoor ambient temperatures. For a review of TXVs, refer to Unit 24, "Expansion Devices."

When TXVs are used as metering devices in air-conditioning applications, a refrigerant charging method known as the "subcooling method" must be applied. The subcooling method involves measuring the amount of sub-cooling exiting the outdoor condensing unit. This subcooling will include condenser subcooling and a small part of the liquid line that the outdoor condensing unit employs. To use the subcooling method to charge a cooling system that employs a TXV, follow these steps:

Step 1 Operate the system for at least 15 min before taking any pressure or temperature measurements. If the outdoor temperature is below 65°F, make sure the condensing temperature is at least 105°F. This would be a minimum pressure of 340 psig for R-410A and 210 psig for R-22. This can be done by covering up part of the outdoor condenser's coil and restricting some of its airflow.

Step 2 Measure the liquid-line (condensing) pressure at the outdoor condensing unit's service valve using a pressure gage. Convert the liquid-line pressure to a saturation temperature using a temperature/pressure chart.

Step 3 Measure the liquid-line temperature at the outdoor condensing unit's service valve where the liquid-line pressure measurement was taken.

Step 4 Refer to the unit's data plate or technical manual to find the required subcooling value for the unit. In the absence of any data plate or manual, use a subcooling value of 10°F to 15°F.

Step 5 Subtract the subcooling value (read from the data plate or technical manual) from the saturation temperature (corresponding to the liquid-line pressure in step 2) to determine the desired liquid-line temperature at the outdoor condensing unit.

Step 6 If the measured liquid-line temperature is higher than the value in step 5, add refrigerant to the unit. If the liquid-line temperature is lower than the value in step 5, recover refrigerant from the unit. (Only a 3°F to 4°F tolerance of the desired temperature is necessary when using this method.) Adding or removing refrigerant from the cooling unit should be done in small increments, allowing at least 15 min before the next reading of pressures and temperatures. This allows the cooling unit to reach a new equilibrium.

Refer to **Figure 10–21,** which shows a high-efficiency, split air-conditioning system incorporating R-410A as the refrigerant. This system has a TXV as the metering device. The technician lets the air-conditioning system run for 15 min and takes a pressure reading (390 psig) at the liquid service valve on the outdoor condensing unit. Through the use of a temperature/pressure chart for R-410A, the service technician converts this pressure to a saturation temperature (115°F). The technician then measures the liquid-line temperature (110°F) at the same place the pressure reading was taken at the service valve of the outdoor condensing unit. Referring to the data plate specifications on the unit, the technician finds out that this unit should have 15°F of subcooling when charged properly. The technician then subtracts this 15°F value received from the data plate from the saturation temperature of 115°F to get the desired liquid-line temperature (100°F) at the outdoor condensing unit. See the following equation:

$$\begin{array}{ccc} \text{Saturation} & \text{Data plate} & \text{Desired liquid-line} \\ \text{temperature} - \text{subcooling value} = \text{temperature} \\ 115°F & - \quad 15°F \quad = \quad 100°F \end{array}$$

The technician then compares the measured liquid-line temperature of 110°F with the desired liquid-line temperature of 100°F. Since the measured liquid-line temperature is higher than the desired liquid-line temperature, the service technician adds refrigerant to the unit while it is running. Because R-410A is a near-azeotropic refrigerant blend,

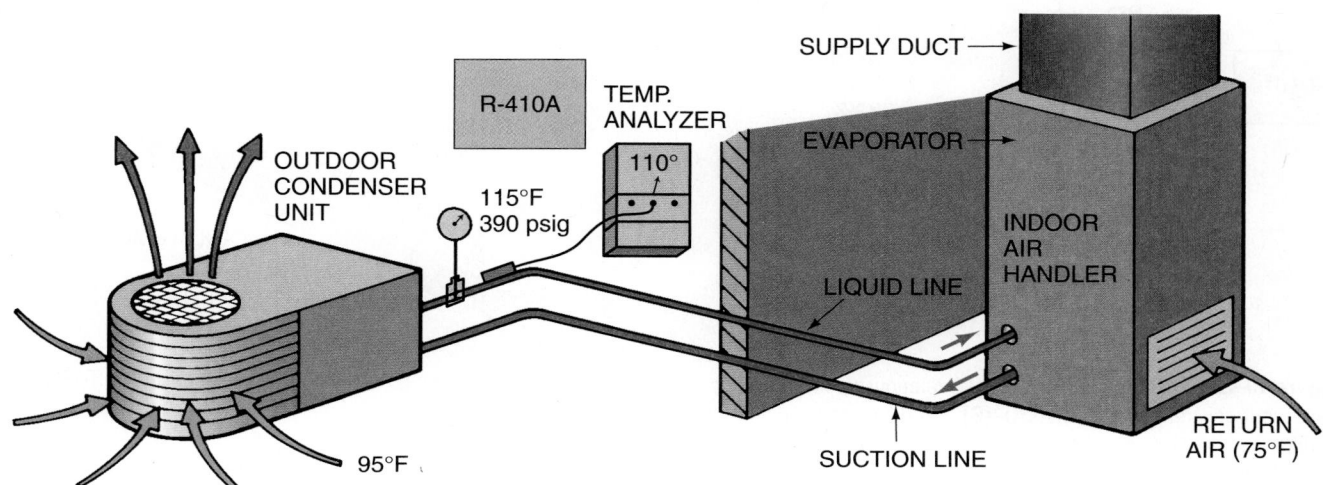

Figure 10–21   The subcooling charging method for a high-efficiency split R-410A air-conditioning system with a TXV metering device.

refrigerant must leave the charging cylinder as a liquid to prevent fractionation. It must then be throttled (vaporized) before it reaches the compressor. The technician must wait at least 15 min for this added refrigerant charge to stabilize in the system. The pressure and temperature measuring process must then be repeated until the measured temperature and desired temperature at the liquid line of the outdoor condensing unit are within 3°F to 4°F of one another.

10.8   CHARGING NEAR-AZEOTROPIC (ZEOTROPIC) REFRIGERANT BLENDS

Refrigerant blends and their characteristics were covered in Unit 9, "Refrigerant and Oil Chemistry and Management—Recovery, Recycling, Reclaiming, and Retrofitting." A more in-depth look at refrigerant blends and their characteristics is needed when charging them into a refrigeration or air-conditioning system.

As mentioned in Unit 9, refrigerant blends are nothing more than two or more refrigerants blended to create another refrigerant. This new refrigerant will have properties completely different than the refrigerants that were blended. Two azeotropic blends used in the HVACR industry for many

years are CFC-502 and CFC-500. These blends consist of azeotropic mixtures of HCFC-22/CFC-115 and CFC-12/HFC-152a, respectively. However, because of the environmental concerns of ozone depletion and global warming, both of these blends were phased out in 1996.

Azeotropic blends are mixtures of two or more liquids that when mixed together behave as commonly known refrigerants like R-12, R-22, and R-134a. Each system pressure will have only one boiling point and/or one condensing point. For a given evaporator or condensing saturation pressure, only one corresponding temperature exists.

However, many refrigerant blends in use today are called *near-azeotropic* blends or *zeotropes*. These blends act very differently when they evaporate and/or condense. Two or three molecules instead of one exist in each sample of liquid. A difference exists in the temperature/pressure relationships of these blends as compared with azeotropic blends and other refrigerants, such as CFC-12, HCFC-22, or HFC-134a.

Zeotropic blends have a *temperature glide* when they boil and condense. Temperature glide occurs when the refrigerants that make up the blend have different temperatures when they evaporate and condense at a given pressure. The liquid and vapor temperatures actually will be different for one

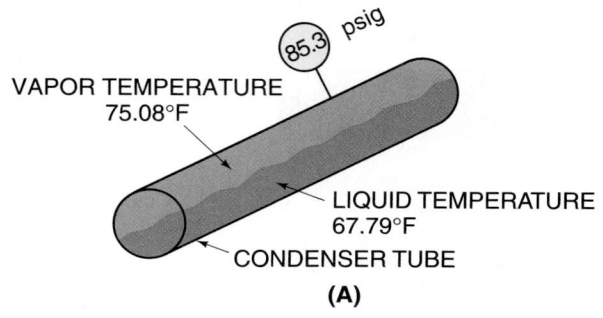

	°F	PRESSURE PER SQUARE INCH (psig)
V	75.08	85.3
L	67.79	85.3
V	76.30	87.3
L	69.04	87.3
V	77.51	89.3
L	70.27	89.3
V	78.69	91.3
L	71.48	91.3
V	79.86	93.3
L	72.67	93.3
V	81.02	95.3
L	73.85	95.3

R-401A

V = VAPOR
L = LIQUID
(B)

Figure 10–22 The temperature/pressure relationship for R-401A, a near-azeotropic refrigerant blend **(A)–(B)**. R-401A is made up of R-22, R-152a, and R-124 with percentages of 53%, 13%, and 34%, respectively.

	°F	PRESSURE PER SQUARE INCH (psig)
V	79.0	85.3
L	79.0	85.3
V	80.5	87.3
L	80.5	87.3
V	81.7	89.3
L	81.7	89.3
V	82.6	91.3
L	82.6	91.3
V	83.8	93.3
L	83.8	93.3
V	85.0	95.3
L	85.0	95.3

R-134a

V = VAPOR
L = LIQUID
(A)

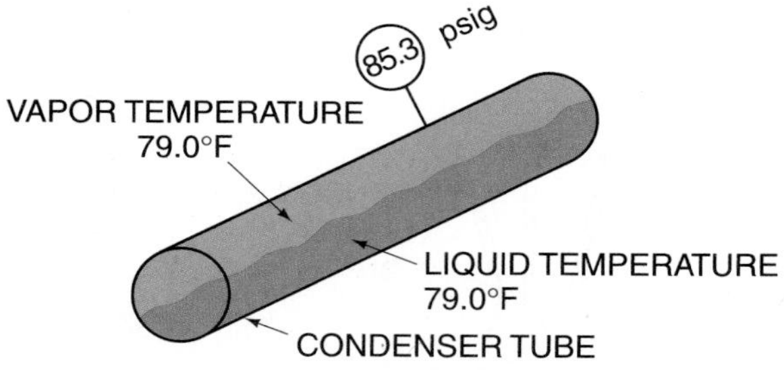

Figure 10–23 The temperature/pressure relationship of R-134a **(A)–(B)**.

given pressure. **Figure 10–22(A)** shows a zeotropic blend (R-410A) as it is phase changing in a condenser tube. Notice that for a given pressure of 85.3 psig, the liquid and vapor temperature are different. An examination of **Figure 10–22(B)** clearly shows that the liquid phase and vapor phase of this blend have two distinct temperatures for one given pressure. In fact, at a condensing pressure of 85.3 psig the liquid temperature (L) is 67.79°F and the vapor temperature (V) is 75.08°F. Its condensing *temperature glide* range is then from 75.08°F to 67.79°F. Its condensing temperature glide would be 7.29°F (75.08°F − 67.79°F). This means that the condensing process in the condenser is taking place through a range of temperatures. The condensing temperature would be an average of the condensing temperature glide range or (75.08°F + 67.79°F) ÷ 2 = 71.43°F. Temperature glide happens only when near-azeotropic (zeotropic) blends phase change.

One temperature no longer corresponds to one pressure as with refrigerants such as R-12, R-22, or R-134a. Temperature glide may range from 2°F to 14°F in refrigeration and air-conditioning applications depending on the specific blend. Usually, at the higher temperature applications like air conditioning, the temperature glide will be greater. Most of the blends exhibit low temperature glides with little effect on the refrigeration system. However, because of temperature

glide, different ways for calculating superheat and subcooling will have to be learned. These methods will be covered in later units. Examples of some near-azeotropic (zeotropic) refrigerant blends are R-401A, R-401B, R-402A, R-402B, R-404A, R-407C, and R-410A.

Even though R-410A is listed as a near-azeotropic refrigerant blend, it has a very low temperature glide and can be ignored in air-conditioning applications.

Refrigerants like R-12, R-22, R-134a, or azeotropic refrigerant blends like R-500 and R-502, however, boil and condense at constant temperatures for each pressure. Notice in **Figure 10–23,** R-134a has both a liquid and vapor temperature of 79.0°F at 85.3 psig as it phase changes in the condenser tube. This is because R-134a has no temperature glide.

New pressure/temperature charts on the market make it easier for the service technician to check a system for the right amount of refrigerant charge when dealing with near-azeotropic refrigerant blends that have a temperature glide. For example, the pressure/temperature chart in **Figure 10–24** instructs the service technician who is checking for a superheat value on the system to use *dew point* values. Dew point is the temperature at which saturated vapor first starts to condense. If the service technician is checking for a subcooling value, the chart instructs the technician to use the *bubble point* values only. Bubble point is the temperature where the saturated liquid starts to boil off its first bubble of vapor.

PRESSURE-TEMPERATURE CHART

PSIG	Pink — MP39 or 401A (X)	Sand — HP80 or 402A (L)	Orange — HP62 or 404A (S)	Green — KLEA 60 or 407A	Lt. Brown — 9000 or KLEA 66 407C (N)	Brown — FX-56 or 409A
5*	-23	-59	-57	-45	-40	-22
4*	-22	-58	-56	-43	-39	-20
3*	-20	-56	-54	-42	-37	-19
2*	-19	-55	-53	-41	-36	-17
1*	-17	-54	-52	-39	-35	-16
0	-16	-53	-51	-38	-34	-15
1	-13	-50	-48	-36	-31	-12
2	-11	-48	-46	-33	-29	-9
3	-9	-45	-43	-31	-27	-7
4	-6	-43	-41	-29	-24	-5
5	-4	-41	-39	-27	-22	-2
6	-2	-39	-37	-25	-20	0
7	0	-37	-35	-23	-18	2
8	2	-36	-33	-21	-17	4
9	4	-34	-32	-20	-15	6
10	6	-32	-30	-18	-13	8
11	8	-30	-28	-16	-12	9
12	9	-29	-27	-15	-10	11
13	11	-27	-25	-13	-8	13
14	13	-26	-23	-12	-7	14
15	14	-24	-22	-10	-5	16
16	16	-23	-20	-9	-4	17
17	17	-21	-19	-8	-3	19
18	19	-20	-18	-6	-1	20
19	20	-19	-16	-5	0	22
20	21	-17	-15	-4	1	23
21	23	-16	-14	-2	3	25
22	24	-15	-12	-1	4	26
23	25	-14	-11	0	5	27
24	27	-12	-10	1	6	29
25	28	-11	-9	2	8	30
26	29	-10	-8	4	9	31
27	30	-9	-6	5	10	32
28	32	-8	-5	6	11	34
29	33	-7	-4	7	12	35
30	34	-6	-3	8	13	36
31	35	-5	-2	9	14	37
32	36	-4	-1	10	15	38
33	37	-2	0	11	16	39
34	38	-1	1	12	17	40
35	39	0	2	13	18	41
36	40	0	3	14	19	43
37	42	1	4	15	20	44
38	43	2	5	16	21	45 30
39	44	3	6	17	22	46 31
40	45 34	4	7	18	23	47 32
42	46 36	6	8	21	25	48 34
44	48 38	8	10	23	26	50 36
46	50 40	10	12	28	28	38
48	42	11	14	30	30	39
50	44	13	16	26	31	41
52	45	14	17	28	33	43
54	47	16	19	29	34	45
56	49	18	20	31	36	46
58	50	19	22	32	37	48
60	52	20	23	33	39	50
62	53	22	25	35	40	51
64	55	23	26	36	42	53
66	56	25	27	38	43	54
68	58	26	29	39	44	56
70	59	27	30	40	46 36	57
72	61	29	31 32	41 31	47 36	58
74	62	30	33 32	43 32	48 37	60
76	64	31	34 33	44 34	49 38	61
78	65	32	35 34	45 35	39	63
80	66	34 31	37 36	46 36	41	64
85	69	34	40 37	49 39	44	67
90	73	40 37	42 42	42	44 46	70
95	76	42 43	45 44	48 47	45 47	73
100	78	45 43	48 47	47	49 52	76
105	81	48 45	50 48	50	54 57	79
110	84	50 48	52 52	53	55 59	82
115	87	50	55	55	59 62	84
120	89	53 53	57 57	57	62 64	87
125	92	55 55	59 60	60	64	89
130	94	57	62 60	64	66	92
135	96	60	64 62	64	69	94
140	99	62	66 62	66	71	96
145	101	64 66	68 70	68	73	99
150	103	66	70	70	75	101
155	105	68 68	72	72	77	103
160	108	70	74	74	79	105
165	110	72	76	76	81	107
170	112	74	78	78	82	109
175	114	75	80	80	84	111
180	116	77	82	81	86	113
185	117	79	83	83	88	115
190	119	81	85	85	90	117
195	121	82	87	88	91	119
200	123	84	88	88	93	121
205	125	86	90	90	95	123
210	127	87	92	91	96	124
220	130	91	95	94	99	128
230	133	94	98	97	102	131
240	136	97	101	100	105	134
250	140	99	104	103	108	137
260	143	102	107	106	111	141
275	147	106	111	110	115	145
290	151	110	115	114	119	149
305	155	114	118	117	123	153
320	159	118	122	121	126	157
335	163	121	126	124	130	161
350	167	125	129	128	133	165
365	170	128	132	131	137	169

Temperature, °F. REFRIGERANT – (Sporlan Code). Each refrigerant column carries both a BUBBLE POINT and a DEW POINT scale (vertical labels within the chart).

*Inches mercury below one atmosphere

Figure 10–24   A pressure/temperature chart for refrigerant blends that have temperature glides. *Courtesy Sporlan Valve Company*

Example

A service technician takes the following reading on a low-temperature refrigeration system incorporating R-404A as the refrigerant.

Head pressure = 250 psig
Suction pressure = 33 psig
Condenser outlet temperature = 90°F
Evaporator outlet temperature = 10°F

The technician wants to find out the condenser subcooling and the evaporator superheat amounts. According to the pressure/temperature chart in **Figure 10–24,** the corresponding condensing temperature for a 250-psig head pressure is 104°F. Notice that the chart will let the technician use only the bubble point temperature for the pressure/temperature relationship. This way there is no confusing a liquid or vapor temperature from the chart when subcooling or superheat readings are taken. The condenser subcooling amount would then be:

$$\begin{array}{ccc} \text{(Condensing} & \text{(Condenser outlet} & \text{Condenser} \\ \text{temperature)} - \text{temperature)} = \text{subcooling} \\ (104°F) - (90°F) = (14°F) \end{array}$$

To find the evaporator superheat, the technician would again refer to the chart in **Figure 10–24** and find that the corresponding evaporating temperature for a 33-psig suction pressure is 0°F. Notice again that the chart will let the technician use only the dew point temperature for the pressure/temperature relationship. Again, this ensures that there is no way of confusing a liquid or vapor temperature from the chart when subcooling or superheat readings are taken. The evaporator superheat amount would then be:

$$\begin{array}{ccc} \text{(Evaporator outlet} & \text{(Evaporating} & \text{Evaporator} \\ \text{temperature)} - \text{temperature)} = \text{superheat} \\ (10°F) - (0°F) = (10°F) \end{array}$$

The pressure/temperature chart in **Figure 10–24** is much easier and less confusing for the service technician to use than the chart in **Figure 10–22(B),** which lists both a liquid temperature and a vapor temperature for the same pressure.

Another property of zeotropic refrigerant blends is *fractionation*. Because zeotropic blends are made up of different refrigerants and mix together as several molecules, they act differently than refrigerants like R-12, R-22, R-134a, and azeotropic blends when phase changing in the evaporator and condenser. Fractionation occurs when some of the blend will condense or evaporate before the rest of the blend will. Phase changes actually will happen at different rates within the same blend. The many different refrigerants in the blend all have vapor pressures that are slightly different, which causes them to condense and evaporate at different rates.

If a zeotropic blend is to leak (vaporizing), the refrigerant in the blend with the lowest boiling point (highest vapor pressure) will leak faster than the other refrigerants in the blend. The refrigerant in the blend with the next lowest boiling point (next higher vapor pressure) will also leak out, but at a slower rate than the refrigerant with the lowest boiling point. Many blends are made up of three refrigerants (ternary blends) resulting in three different leakage rates at once. This causes problems because once the leak is found and repaired, the refrigerant blend that is left in the system after the leak will not have the same percentages of the refrigerants that it had when it was initially charged. Fractionation happens only when the leak occurs as a vapor leak; it will not occur when the system is leaking pure liquid. This is because vapor pressures are involved, and you need liquid and vapor to coexist for fractionation to occur.

Fractionation also happens when refrigerant vapor is taken out of a charging cylinder when the technician is charging with a zeotropic blend. The charging or recharging of a refrigeration system using a near-azeotropic or zeotropic blend should be done with liquid refrigerant. As long as virgin liquid refrigerant is used in recharging a system that has leaked, capacity losses from leak fractionation will be small and harmless. Remove liquid only from the charging cylinder to ensure that the proper blend of percentages or composition enters the refrigeration system. If vapor is removed from the charging cylinder, fractionation will occur. However, once the right amount of liquid is removed from the charging cylinder to another holding cylinder, the refrigerant can be charged as vapor as long as all of the refrigerant vapor is put in the system. This will ensure that the proper percentages of the blend enter the system.

Refrigerant cylinders containing zeotropic refrigerant blends often, but not always, have dip tubes that run to the bottom of the cylinder to allow liquid to be removed from the cylinder when upright, **Figure 10–25.** This should ensure that the technician will always remove liquid from the cylinder

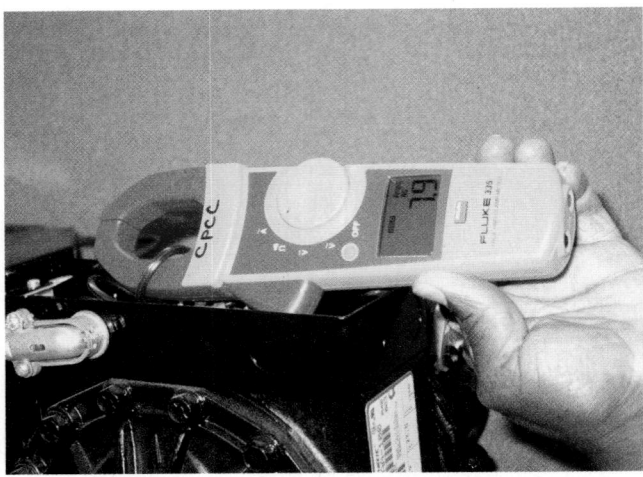

Figure 10–26   Taking a current reading at the compressor. *Photo by Bill Johnson*

when charging a system. The cylinder has to be tipped upside down to remove vapor. Of course you cannot look inside the refrigerant tank to determine if there is a dip tube in the cylinder or not. Refer to the side of the refrigerant tank for information regarding the correct positions of the tank for liquid or vapor removal.

When adding liquid refrigerant to the suction line of a system that is running, the liquid, once out of the charging cylinder, has to be restricted and vaporized into the system to avoid any damage to the compressor. Some sort of a restricting valve must be used to make sure that the liquid refrigerant is vaporized before it enters the compressor. **Figure 10–10** and **Figure 10–11** show examples. Keep an ammeter on the compressor's power lead and monitor the running current of the compressor when charging this way, **Figure 10–26.** These precautions avoid overloading the compressor and creating nuisance overload trips caused from refrigerant vapors that are too dense when they are entering the compressor. If the current increases too much over the nameplate rating current, discontinue charging until the current stabilizes at a lower reading. You could also try to restrict the liquid down further with the manifold gage set, which would introduce the refrigerant into the system more slowly. This would result in vapors that are less dense entering the compressor, which would create a lower current draw. The compressor's three-way suction service valve can also be cracked just off its back seat, which will create a throttling effect and help vaporize any liquid refrigerant trying to enter the compressor.

SUMMARY

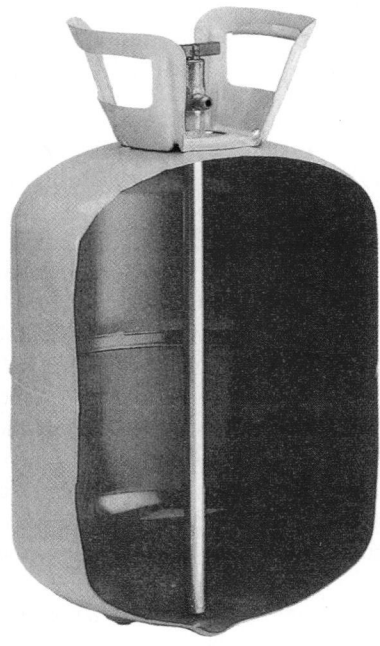

Figure 10–25   A cylinder of a near-azeotropic refrigerant blend showing a liquid dip tube extending to the bottom of the cylinder while the cylinder is in the upright position. This allows the charging of liquid only while the cylinder is in the upright position. *Courtesy Worthington Cylinders*

- ■ Refrigerant may be added to the refrigeration system in the vapor state or the liquid state under proper conditions.
- ■ When refrigerant is added in the vapor state, the refrigerant cylinder will lose pressure as the vapor is pushed out of the cylinder.
- ■ Liquid refrigerant is normally added in the liquid line and only under the proper conditions.

- Liquid refrigerant must come out of the charging cylinder when the technician is charging with certain refrigerant blends. This will avoid fractionation. Once out of the charging cylinder, this liquid must be restricted and vaporized to avoid liquid refrigerant from entering the compressor.
- Liquid refrigerant must never be allowed to enter the compressor.
- Refrigerant is measured into systems using weight and volume.
- It can be difficult to add refrigerant with dial scales because the final cylinder weight must be calculated. The scales are graduated in pounds and ounces.
- Electronic scales may have a cylinder-emptying feature that allows the scales to be adjusted to zero with a full cylinder of refrigerant on the platform.
- Graduated cylinders use the volume of the liquid refrigerant. This volume varies at different temperatures. These may be dialed onto the cylinder for accuracy.
- Near-azeotropic or zeotropic blends have entered the refrigeration and air-conditioning industry with the onset of the ozone-depletion and global-warming scares.
- Zeotropic blends can exhibit temperature glide (when phase changing) and fractionation (when vapor leaking or charging as a vapor).
- All zeotropic blends must come out of the charging cylinder as a liquid to avoid fractionation.
- Some sort of restricting device must be used with zeotropic blends when charging to vaporize the liquid refrigerant before it enters the compressor.
- Manufacturer charging charts and curves can assist the service technician in correctly charging an accurate amount of refrigerant into an air-conditioning or heat pump system.
- The subcooling method of refrigerant charging is used on air-conditioning and heat pump systems incorporating a thermostatic expansion valve for a metering device.
- Manufacturers have designed new pressure/temperature charts that make it easier for service technicians to check a system for the right amount of refrigerant charge when dealing with near-azeotropic refrigerant blends that have a temperature glide.

REVIEW QUESTIONS

1. How is liquid refrigerant added to the refrigeration system when the system is out of refrigerant?
2. How is the refrigerant cylinder pressure kept above the system pressure when a system is being charged with vapor from a cylinder?
3. Why does the refrigerant pressure decrease in a refrigerant cylinder while charging with vapor?
4. The main disadvantage of a dial scale is _____.
5. What type of equipment normally has the refrigerant charge printed on the nameplate?
6. The _____ feature of a digital electronic scale makes them useful for refrigerant charging.
7. How is refrigerant pressure maintained in a graduated cylinder?

8. How does a graduated cylinder account for the volume change due to temperature changes?
9. You must remember _____ when purchasing a charging cylinder.
10. What methods besides weighing and measuring are used for charging systems?
11. Two or more refrigerants blended together to create another refrigerant is a
    A. refrigerant blend.
    B. refrigerant solution.
    C. refrigerant slurry.
    D. refrigerant slush.
12. Define temperature glide and give examples of where it happens in a refrigeration system.
13. A portion of a refrigerant blend that is evaporating or condensing before the rest of the same blend is called
    A. temperature glide.
    B. phase glide.
    C. enthalpy change.
    D. fractionation.
14. What is the main difference between a zeotropic and an azeotropic refrigerant blend?
15. What is meant by "restricting" liquid refrigerant into a refrigeration system when charging?
16. What causes fractionation to happen in certain blends of refrigerants?
17. When using the charging chart in **Figure 10–18,** what happens to the compressor superheat when the outdoor ambient temperature increases? Explain why.
18. When using the charging chart in **Figure 10–18,** what happens to the compressor superheat when the outdoor ambient temperature stays constant but the indoor DBT or WBT increases? Explain why.
19. When using the charging chart in **Figure 10–20,** if the outdoor ambient temperature is 85°F and the suction pressure is 74 psig, what should the suction-line temperature at the condensing unit's service valve be?
20. When should a service technician use the subcooling method of charging air-conditioning and heat pump systems?
21. A service technician is charging an air-conditioning system that has a TXV as the metering device using the subcooling method of charging. The low-side saturation temperature is 110°F and the unit's technical manual specifies 10°F of liquid subcooling. What would the desired liquid-line temperature be at the condensing unit?
22. A technician takes the following readings on an R-404A refrigeration system:
    Head pressure = 205 psig
    Suction pressure = 30 psig
    Condenser outlet temperature = 80°F
    Evaporator outlet temperature = 10°F
    Using the pressure/temperature chart in **Figure 10–24,** find the condenser subcooling and evaporator superheat amounts.

Calibrating Instruments

OBJECTIVES

After studying this unit you should be able to

- describe instruments used in heating, air conditioning, and refrigeration.
- test and calibrate a basic thermometer at the low- and high-temperature ranges.
- check an ohmmeter for accuracy.
- describe the comparison test for an ammeter and a voltmeter.
- describe procedures for checking pressure instruments above and below atmospheric pressure.
- check flue-gas analysis instruments.

11.1 THE NEED FOR CALIBRATION

The service technician cannot always see or hear what is occurring within a machine or piece of equipment. Instruments such as multimeters, pressure gages, and temperature testers are used to help. Therefore, these instruments must be reliable. Although the instruments should be calibrated when manufactured, they do not always remain in calibration. They may need to be checked before the technician uses them and on a periodic basis. Even if they are perfectly calibrated, the instruments may not remain so due to use and because of conditions such as moisture and vibration. Instruments may be transported in a truck over rough roads to the job site. The instrument stays in the truck through extremes of hot and cold weather. The instrument compartment may sweat (moisture due to condensation of humidity from the air) and cause the instruments to become damp. All of these things cause stress to the instruments. They may not stay in calibration over a long period of time.

Technicians should always take proper care of tools and equipment because they are very dependent on them. There is no substitute for good common sense, careful use, and proper storage of these tools and various pieces of equipment. SAFETY PRECAUTION: Some of these instruments are used to check for voltage to protect the technician from electrical shock. These instruments must function correctly for safety's sake.

11.2 CALIBRATION

Some instruments cannot be calibrated. *Calibration* means the changing of the instrument's output or reading to correspond to a standard or correct reading. For example, if a speedometer shows 55 mph for an automobile actually traveling at 60 mph, the speedometer is out of calibration. If the speedometer can be changed to read the correct speed, it can be calibrated.

Some instruments are designed for field use and will stay calibrated longer. The electronic instruments with digital readout features may stay in calibration better than the analog (needle) type and be more appropriate for field use, **Figure 11–1**.

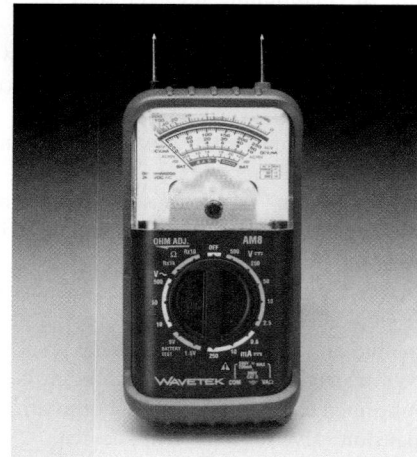

(A)

(B)

Figure 11–1 **(A)** An analog meter. **(B)** A digital meter. *Courtesy Wavetek*

This unit deals with the most common instruments used for troubleshooting. These instruments measure temperature, pressure, voltage, amperage, and resistance; check for refrigerant leaks; and make flue-gas analyses. To check and calibrate instruments, you must have reference points. Some instruments can be readily calibrated; some must be returned to the manufacturer for calibration; and some cannot be calibrated at all. We recommend that whenever you buy an instrument, you check some readings against known values. If the instrument is not within the standards the manufacturer states, return it to the supplier or manufacturer. Save the box the instrument came in as well as the directions and warranty. They can save you much time.

11.3 TEMPERATURE-MEASURING INSTRUMENTS

Temperature-measuring instruments measure the temperature of vapors, liquids, and solids. Air, water, and refrigerant in copper lines are the common substances measured for temperature level. Regardless of the medium to be measured, the methods for checking the accuracy of the instruments are similar.

Refrigeration technicians must have thermometers that are accurate from −50°F to 50°F to measure the refrigerant lines and the inside of coolers. Higher temperatures are experienced when measuring ambient temperatures, such as when the operating pressures for the condenser are being examined. Heating and air-conditioning technicians must measure air temperatures from 40°F to 150°F and liquid temperatures as high as 220°F for normal service. This can require a wide range of instruments. For temperatures above 250°F—for example, flue-gas analysis in gas- and oil-burning equipment—special thermometers are used. The thermometer is included in the flue-gas analysis kit, **Figure 11–2.**

In the past, most technicians relied on glass-stem mercury or alcohol thermometers. These are easy to use for measuring

fluid temperature when the thermometer can be inserted into the fluid, but they are difficult to use for measuring the temperature of solids. They are being replaced by the electronic thermometer, which is very popular. Electronic thermometers are simple, economical, and accurate, **Figure 11–3** and **Figure 11–4.** Both the analog and digital versions are adequate. Although the digital instrument costs more, it retains accuracy for a longer time under rough conditions.

The pocket-type dial thermometer is often used for field readings, **Figure 11–5.** It is not intended to be a laboratory-grade instrument. The scale on this unit goes from 0°F to 220°F in a very short distance, so obtaining accurate measurements is more difficult. The distance the needle must move to travel from the bottom of the scale to the top is only

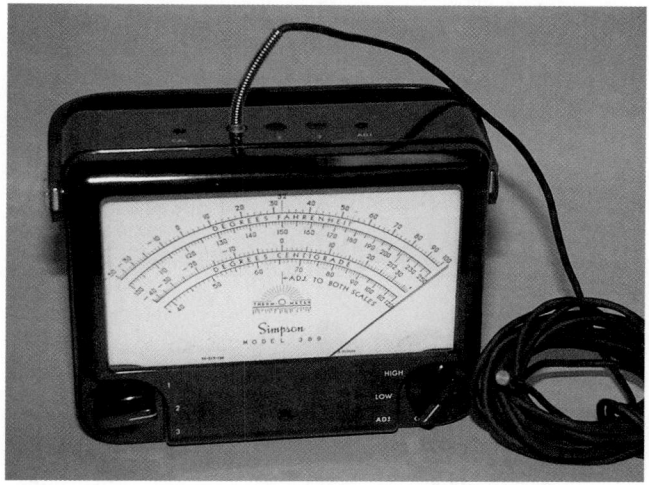

Figure 11–3   An analog-type electronic thermometer. *Photo by Bill Johnson*

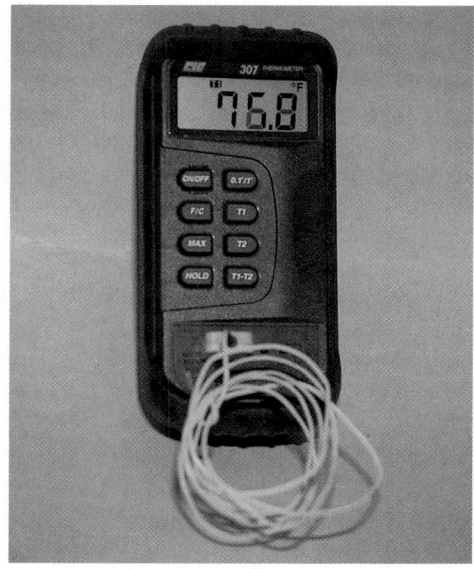

Figure 11–4   A digital-type electronic thermometer. *Photo by Bill Johnson*

Figure 11–2   The thermometer is included in the flue-gas analysis kit. Note the high temperature range of the thermometer. *Courtesy Ferris State University. Photo by Eugene Silberstein*

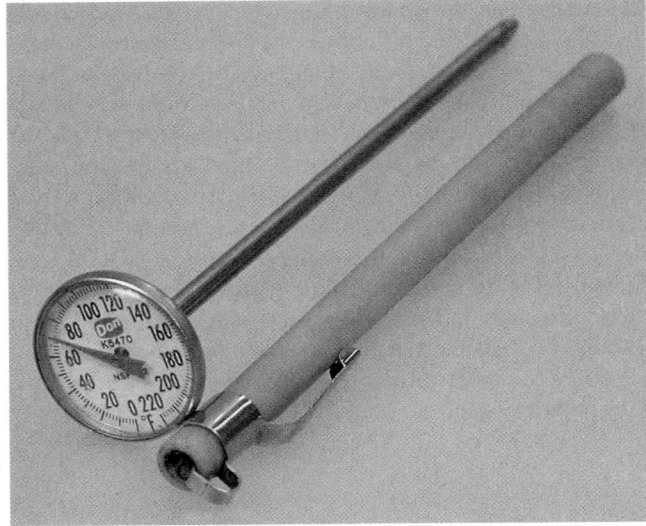

Figure 11–5   A pocket dial-type thermometer. *Photo by Bill Johnson*

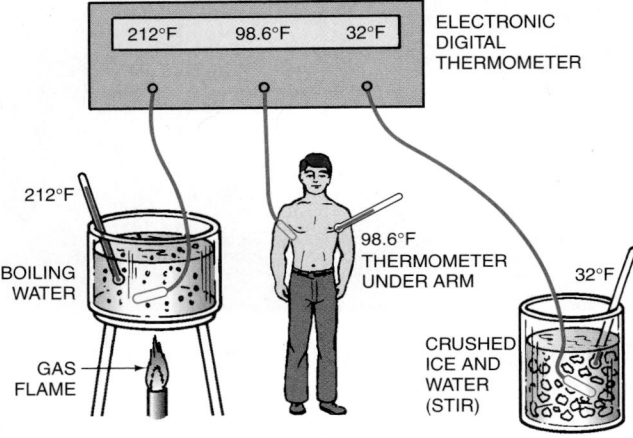

Figure 11–6   Three reference points that a service technician may use.

about 2.5 in. (the circumference of the dial). This would be like having an automobile speedometer on which 1.25 in. indicates a range of 0 to 100 mph. The speedometer dial would be so narrow that the width of the needle would cover several miles per hour, and the driver could not be sure of the actual speed.

Three reference points are easily obtainable for checking temperature-measuring instruments: 32°F (ice and water), 98.6°F (body temperature), and 212°F (boiling point of water), **Figure 11–6**. The reference points should be close to the temperature range in which you are working. When using any of these as a reference for checking the accuracy of a temperature-measuring device, remember that a thermometer indicates the temperature of the sensing element. Many technicians make the mistake of thinking that the sensing element indicates the temperature of the medium being checked. This is not necessarily the case. Many inexperienced technicians merely set a thermometer lead on a copper line and read the

temperature, but the thermometer temperature-sensing element has more contact with the surrounding air than with the copper line, **Figure 11–7**. For an accurate reading, the thermometer must be in contact with the medium to be measured for a long enough time for the sensor to become the same temperature as the medium.

One method of checking temperature instruments is to submerge the instrument-sensing element into a known temperature condition (such as ice and water while the change of state is occurring) and allow the sensing element to reach the known temperature. The following method checks an electronic thermometer with four plug-in leads that can be moved from socket to socket.

1. Fasten the four leads together as shown in **Figure 11–8** and **Figure 11–9**. Something solid can be fastened with them so that they can be stirred in ice and water.

Figure 11–7   The technician must remember that the temperature-sensing element of a thermometer indicates the temperature of the sensing element. *Photo by Bill Johnson*

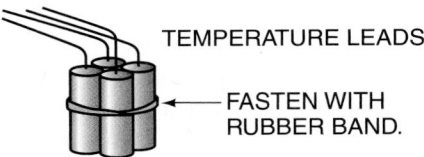

Figure 11–8   The four leads to the temperature tester are fastened at the ends so they can all be submerged in water at the same time.

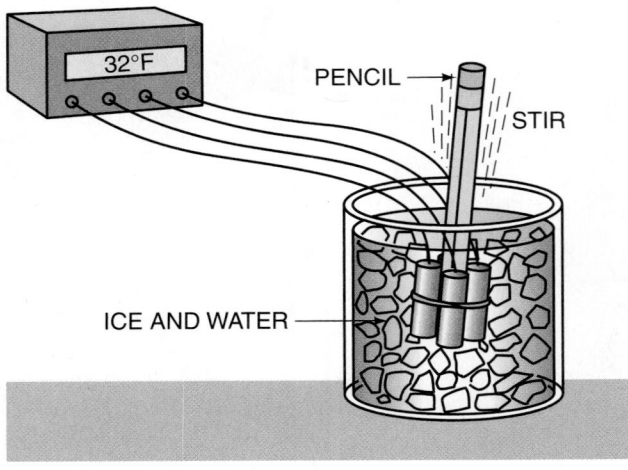

Figure 11–9   A pencil is fastened to the group to give the leads some rigidity so that they can be stirred in the ice and water. Note that the ice must reach the bottom of the container.

2. For a low-temperature check, crush about a quart of ice, preferably made from pure water. If pure water is not available, make sure the water has no salt or sugar because either one changes the freezing point. You must crush the ice very fine (wrap it in a towel and pound it with a hammer), or there may be warm spots in the mixture.

3. Pour enough water, pure if possible, over the ice to almost cover the ice. **Do not cover the ice completely with water,** or it will float and may be warmer on the bottom of the mixture. The ice must reach to the bottom of the vessel.

4. Stir the temperature leads in the mixture of ice and water, where the change of state is taking place, for at least 5 min. The leads must have enough time to reach the temperature of the mixture.

5. If the leads vary, note which leads are in error and by how much. The leads should be numbered, and the temperature differences marked on the instrument case, or the leads can be marked with their error.

6. For a high-temperature check, put a pan of water on a stove-top heating unit and bring the water to a boil. Make sure the thermometers you are checking are capable of registering the temperature of boiling water. If they are, immerse the sensors in the boiling water. NOTE: Do not let them touch the bottom of the pan or they may conduct heat directly from the pan bottom to the lead, **Figure 11–10.** Stir the thermometers for at least 5 min and check the readings. It is not critical that the thermometers be accurate to a perfect 212°F, because with products at these temperature levels, a degree or two one way or the other does not make a big difference. If any lead reads more than 4°F from 212°F, mark it defective. Remember that water boils at 212°F at sea level under standard conditions. Any altitude above sea level will make a slight difference. If you are more than 1000 ft above sea level, we highly recommend that you use a laboratory glass thermometer as a standard and that you do not rely on the boiling water temperature being correct.

Accuracy is more important in the lower temperature ranges where small temperature differences are measured. A 1°F error does not sound like much until you have one lead that is off +1°F and another off −1°F and then you try to take an accurate temperature drop across a water heat exchanger that only has a 10°F drop. You have a built-in 20% error, **Figure 11–11.**

If a digital thermometer with leads that cannot be moved from socket to socket is used, there may be an adjustment in the back of the instrument for each lead. **Figure 11–12** shows how this thermometer can be calibrated.

Glass thermometers often cannot be calibrated because the graduations are etched on the stem. If the graduations are printed on the back of the instrument, the back may be adjustable. A laboratory-grade glass thermometer is certified as to its accuracy, and it may be used as a standard for field instruments. It is a good investment for calibrating electronic thermometers, **Figure 11–13.**

Many dial-type thermometers have built-in means for making adjustments. These instruments may be tested for accuracy as we have described and calibrated, if possible. If not, the dial may need to be marked, **Figure 11–14.**

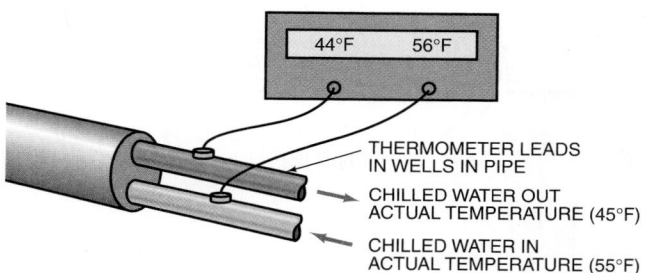

Figure 11–11 A thermometer with two leads that are each accurate to within 1°F.

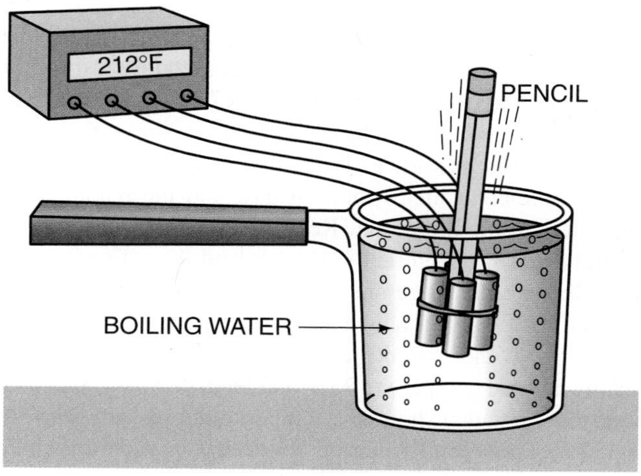

Figure 11–10 A high-temperature test for the accuracy of four temperature leads.

Figure 11–12 Calibration of a thermometer with leads that cannot be moved or relocated within the instrument. *Photo by Bill Johnson*

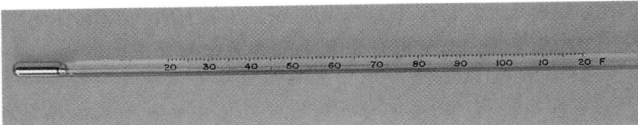

Figure 11–13   A laboratory-grade glass thermometer. *Photo by Bill Johnson*

Figure 11–14   Large dial-type thermometers may be calibrated with the calibration screw. *Photo by Bill Johnson*

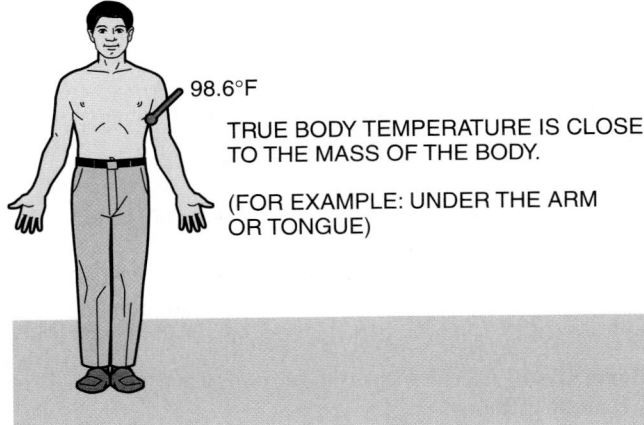

98.6°F

TRUE BODY TEMPERATURE IS CLOSE TO THE MASS OF THE BODY.

(FOR EXAMPLE: UNDER THE ARM OR TONGUE)

Figure 11–15   Using the human body as a standard.

Body temperature may also be used as a standard when needed. Remember that the outer extremities, such as the hands, are not at body temperature. The body is, on average, 98.6°F in the main blood flow, next to the trunk of the body, **Figure 11–15**.

11.4   PRESSURE TEST INSTRUMENTS

Pressure test instruments register pressures above and below atmospheric pressure. The gage manifold and its construction were discussed earlier. The technician must be able to rely on these gages and have some reference points with which to check them periodically. This is particularly necessary when there is reason to doubt gage accuracy. This instrument is used frequently and is subject to considerable abuse, **Figure 11–16**.

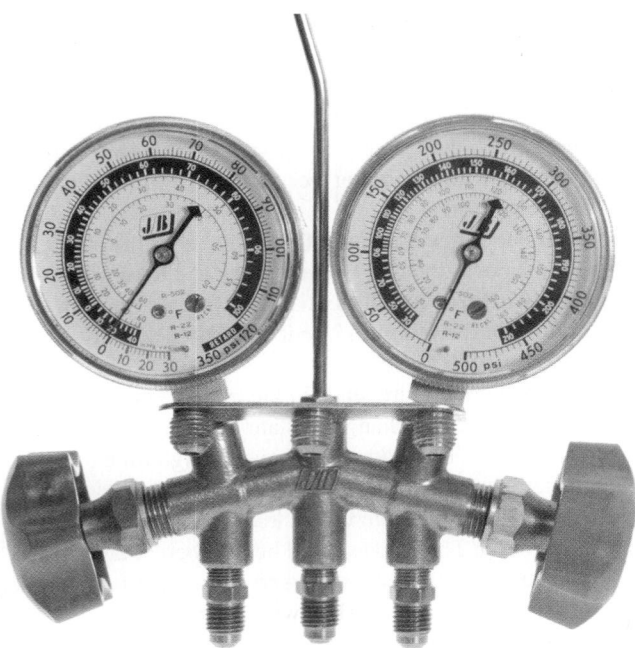

Figure 11–16   A gage manifold. *Photo by Bill Johnson*

Gage readings can be taken from a cylinder of refrigerant, and the pressure/temperature relationship can be compared in the following manner. The gage manifold should be opened to the atmosphere, and both gages should be checked to see that they read 0 psig. *It is impossible to determine a correct gage reading if the gage is not set at 0 at the start of the test.* Connect the gage manifold to a cylinder of fresh new refrigerant that has been in a room at a fixed temperature for a long time. Purge the gage manifold of air. Using an old cylinder may lead to errors due to cylinder pollution of air or another refrigerant. If the cylinder pressure is not correct due to refrigerant contaminants, it *cannot* be used to check the gages. The cylinder pressure is the standard and must be reliable. Most new virgin and disposable refrigerant cylinders have check valves that allow the refrigerant to leave the cylinder—but nothing can enter it. It is not likely that one of these cylinders will contain any pollution. A 1-lb cylinder may be purchased and kept at a fixed temperature just for the purpose of checking gages, **Figure 11–17**.

The refrigerant should have a known pressure if the cylinder temperature is known. Typically, the cylinder is left in a temperature-controlled office all day, and the readings are taken late in the afternoon. Keep the cylinder out of direct sunlight. If the refrigerant is R-134a and the office is 75°F, the cylinder and refrigerant should be 75°F if they have been left in the office for a long enough time. When the gages are connected to the cylinder and purged of any air, the gage reading should compare to 75°F and read 78.7 psig. (Refer to **Figure 3–15**, which shows the temperature/pressure chart for R-134a.) The cylinder can be connected in such a manner that both gages (the low and high sides) may be checked at the same time, **Figure 11–18**. The same test can be performed with R-22, and the reading will be higher. The reading for

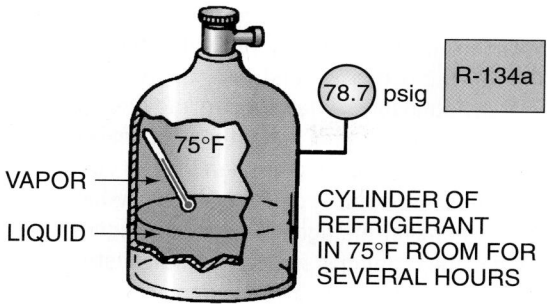

Figure 11–17   This refrigerant cylinder has been left at a known temperature for long enough that the temperature of the refrigerant is the same as the temperature of the room. When the gage manifold is attached to this cylinder, the pressure inside can be compared with the corresponding temperature on a temperature/pressure chart.

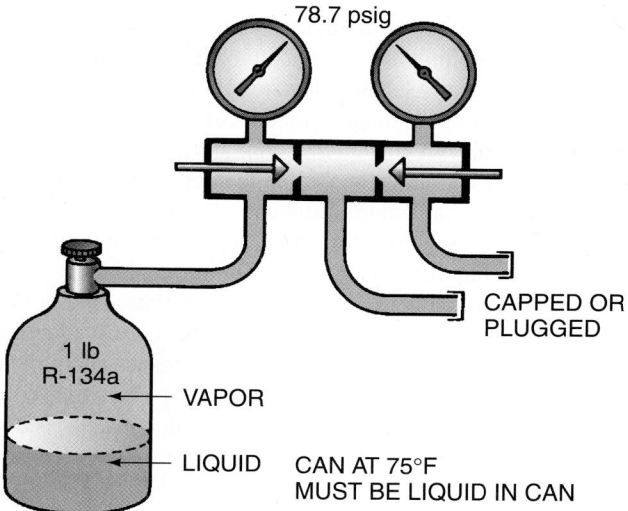

Figure 11–18   Both gages are connected to the cylinder so they can be checked at the same time.

R-22 at 75°F should be 132 psig. Performing the test with both refrigerants checks the gages at two different pressure ranges.

Checking the low-side gage in a vacuum is not as easy as checking the gages above atmosphere because we have no readily available known vacuum. One method is to open the gage to atmosphere and make sure that it reads 0 psig. Then connect the gage to a two-stage vacuum pump and start the pump. When the pump has reached its lowest vacuum, the gage should read 30 in. Hg (29.92 in. Hg vacuum), **Figure 11–19.** NOTE: The vacuum pump will not make as much noise at low vacuum as it will at a pressure close to atmospheric pressure—so an experienced technician can tell from the sound of the pump when it is in a deep vacuum. If the gage is correct at atmospheric pressure and at the bottom end of the scale, you can assume that it is correct in the middle of the scale. If vacuum readings closer than this are needed for monitoring a system that runs in a vacuum, you should buy a larger, more accurate vacuum gage, **Figure 11–20.**

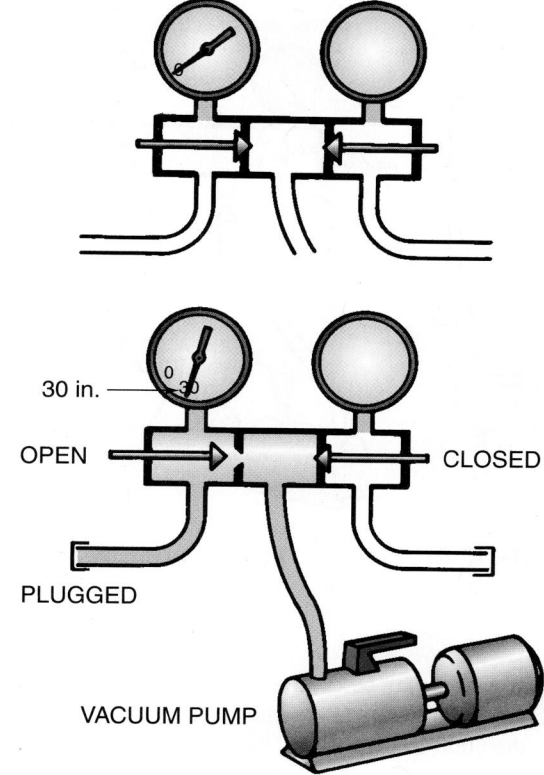

Figure 11–19   When evacuated, the gage should read 30 in. Hg vacuum. If this is the case, all points between 0 psig and 30 in. Hg should be correct.

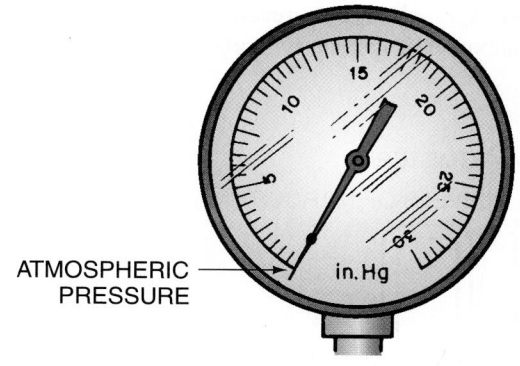

Figure 11–20   A large gage is more accurate for monitoring systems that operate in a vacuum because the needle moves farther from 0 psig to 30 in. Hg.

The mercury manometer and the electronic micron gage may be checked in the following manner. This field test is not 100% accurate, but it is sufficient to tell the technician whether the instruments are within a working tolerance or not.

1. Prepare a two-stage vacuum pump for the lowest vacuum that it will pull. Change the oil to improve the pumping capacity. Connect a gage manifold to the vacuum pump with the mercury manometer and the micron gage as shown in **Figure 11–21.** The low-pressure gage, the micron gage, and the mercury

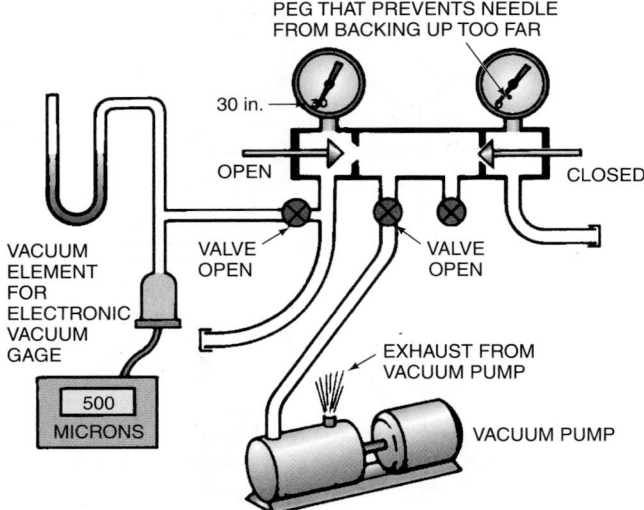

Figure 11–21   The setup for checking a mercury U-tube manometer and an electronic micron gage.

manometer may be compared at the same time. Start the vacuum pump. If the micron gage has readings in the 5000-micron range, the mercury manometer and the micron gage may be compared at this point. Remember, 1 mm Hg = 1000 microns, so 5 mm Hg = 5000 microns. The low-side manifold gage will read 30 in. Hg, and you cannot easily distinguish movement below 1 mm on the mercury manometer.

2. When the vacuum pump has evacuated the manifold and gages, observe the readings. If the mercury manometer is reading *flat out*—both columns of

mercury are at the same level when compared (be sure the instrument is perfectly vertical)—the micron gage should read between 0 and 1000 microns.

NOTE: If any atmosphere has seeped into the left column of the mercury manometer, when the vacuum is pulled the right column will rise higher than the left column. This indicates more than a perfect vacuum, which is not possible. Whenever the right column rises higher than the left column, the reading is *wrong*. With the instrument at atmospheric pressure, check the manometer for a bubble on top of the left column.

It is difficult to compare the mercury manometer closer than this because it is hard to compare the columns. If the mercury columns are flat out and the micron gage is still reading high, you should send the micron gage to the manufacturer for calibration.

NOTE: If the vacuum pump will not pull the mercury manometer and the micron gage down to a very low level—flat out on the mercury manometer and 50 microns on the micron gage—either the vacuum pump is not pumping or the connections are leaking. The connection can be replaced with copper lines if you suspect a leak. If the vacuum still will not pull down, connect the micron gage directly to the vacuum pump with the shortest possible connection and see whether the vacuum pump will pull the gage down. If it will not, check the gage on another pump to see which is not performing, the gage or the pump.

It is difficult to get instruments to correlate exactly in a vacuum. It is also difficult to determine which instrument is correct and whether the vacuum pump is evacuating the system. If the evacuation happens too quickly to be observed, a volume, such as an empty refrigerant cylinder, can be used to slow the pulldown time, **Figure 11–22.**

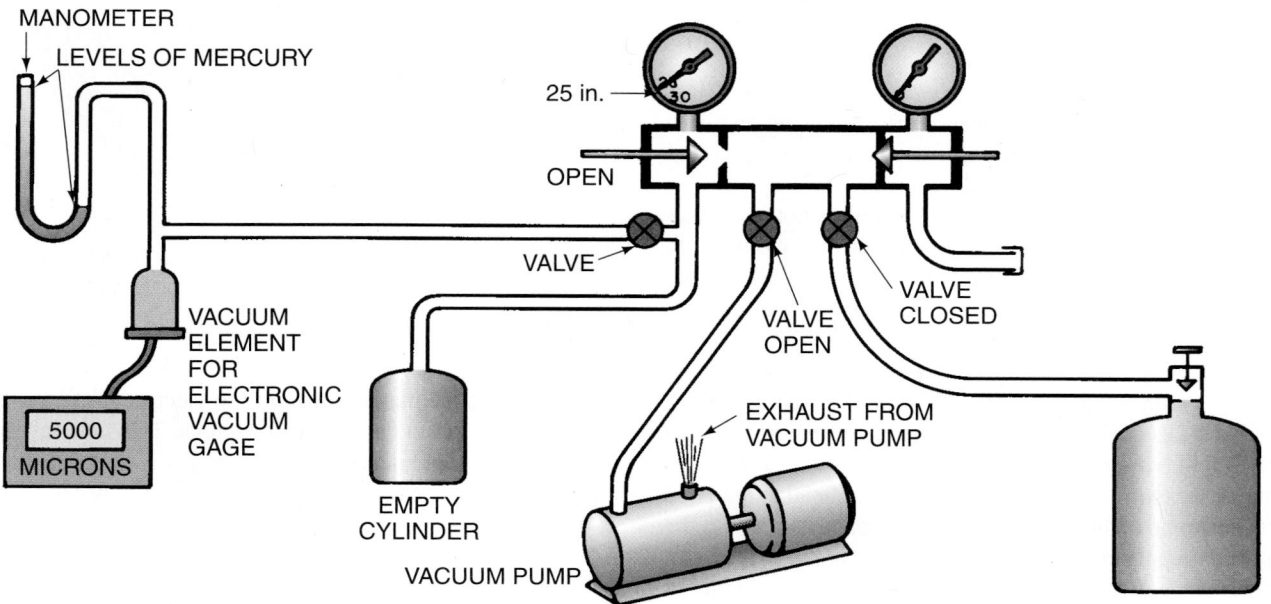

Figure 11–22   If the evacuation in **Figure 11–21** happens too fast, an additional volume, such as an empty refrigerant cylinder, may be connected to slow the evacuation speed.

11.5   ELECTRICAL TEST INSTRUMENTS

Electrical test instruments are not as easy to calibrate; however, they may be checked for accuracy. The technician must know that the ohm scale, the volt scale, and the ammeter scale are correct. The milliamp scale on the meter is seldom used and must be checked by the manufacturer or compared with another meter.

Many grades of electrical test instruments exist. When electrical testing procedures must be relied on for accuracy, it pays to buy a good quality instrument. If you do not need accuracy from an electrical test instrument, a less expensive one may be satisfactory. You should periodically (at least once a year) check electrical test instruments for accuracy. One way to check these instruments is to compare the instrument reading against known values.

The ohmmeter feature of a volt-ohm-milliammeter (VOM) can be checked by obtaining several high-quality resistors of a known resistance at an electronic supply house. Get different values of resistors, so you can test the ohmmeter at each end and at the midpoint of every scale on the meter, **Figure 11–23**. Always start the test by a zero adjustment check of the ohmmeter. NOTE: If the meter is out of calibration to the point that it cannot be brought to the zero adjustment, check the batteries and change if necessary. If the instrument will not read zero with fresh batteries, check the leads. The leads must be connected perfectly, particularly with digital meters. You may want to use the alligator clips on the end to ensure a perfect connection. Leads can become frayed internally. Good meter leads cost a little more, but they have many strands of very small wire for flexibility and will last longer and have better reliability. If the meter will not check at zero, send it to the experts. Do not try to repair it yourself. Be sure to start each test with a zero adjustment check of the ohmmeter.

The volt scale is not as easy to check as the ohm scale. A friend at the local power company or technical school may allow you to compare your meter to a high-quality bench meter. This is recommended at least once a year or whenever you suspect your meter readings are incorrect, **Figure 11–24**. It is satisfying to know your meter is correct when you call the local power company to report that the meter has a low-voltage reading for a particular job.

The clamp-on ammeter is used most frequently for amperage checks. This instrument clamps around one conductor in an electrical circuit. Like the voltmeter, it can be compared with a high-quality bench meter, **Figure 11–25**. Some amount of checking can be done by using Ohm's law

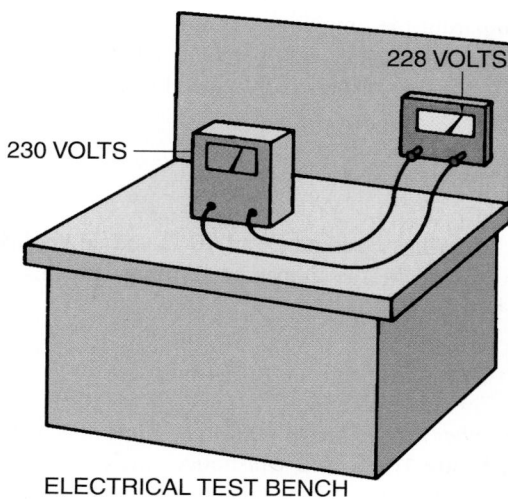

Figure 11–24   A VOM being compared with a quality bench meter in the voltage mode.

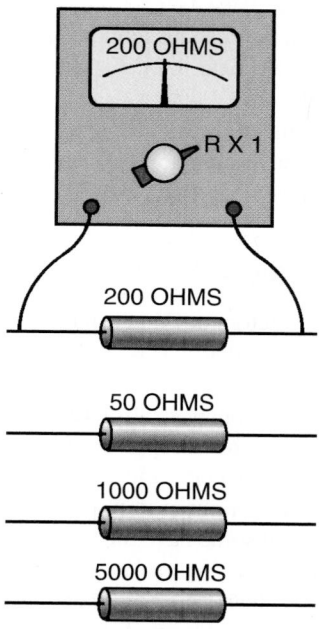

Figure 11–23   Checking an ohmmeter at various ranges with resistances of known values.

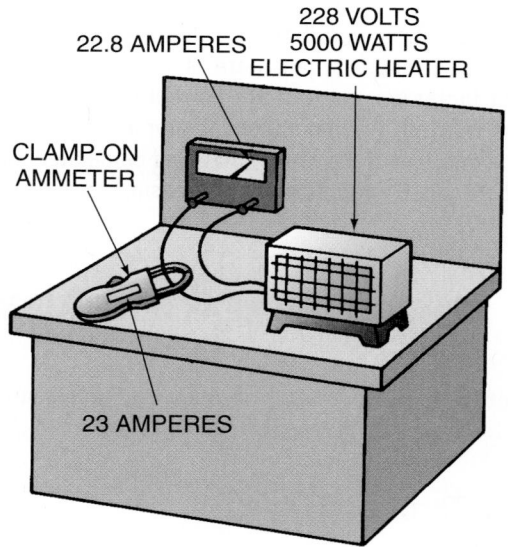

Figure 11–25   A clamp-on ammeter being compared with a quality bench ammeter.

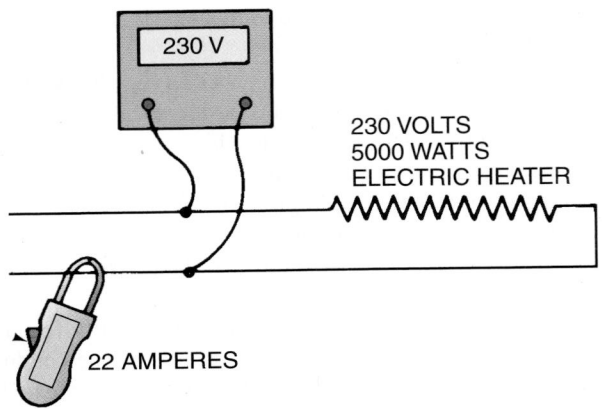

Figure 11–26 Using an electric heater to check the calibration of an ammeter.

and comparing the amperage reading with a known resistance heater circuit. For example, Ohm's law states that current (I) is equal to voltage (E) divided by resistance (R or Ω for ohms):

$$I = \frac{E}{R}$$

If a heater has a resistance of 10 Ω and an applied line voltage of 228 V, the amperage (current) on this circuit should be

$$I = \frac{228 \text{ V}}{10 \text{ }\Omega} = 22.8 \text{ A}$$

Remember to read the voltage at the same time as the amperage, **Figure 11–26.** You will notice small errors because the resistance of the electric heaters will change when they get hot. The resistance will be greater, and the exact ampere reading will not compare precisely to the calculated one. But the purpose is to check the ammeter. If it is off by more than 10%, send it to the repair shop.

NOTE: It is important to note that Ohm's law, as written, will hold for DC-powered circuits that are used to operate resistive-type loads. When there is an AC power supply, a power factor must be used to correct the calculated values. For example, it is not uncommon for a 115-volt circuit to produce one-half of the power that was calculated. Refer to Unit 12, "Basic Electricity and Magnetism," for more on this topic.

11.6   REFRIGERANT LEAK DETECTION DEVICES

Two refrigerant detection devices are commonly used: the halide torch and the electronic leak detector.

Halide Torch

The halide torch cannot be calibrated, but it can be checked to make sure that it will detect leaks. It must be maintained to be reliable. It will detect a leak rate of about 7 oz per year.

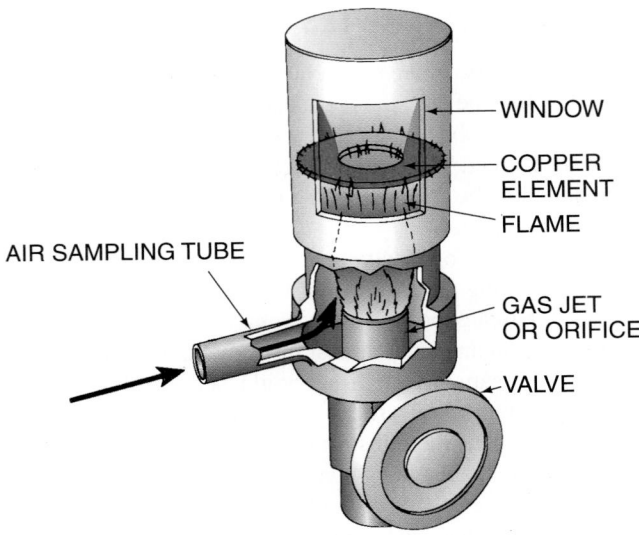

Figure 11–27 A halide torch for detecting refrigerant leaks. The halide torch flame changes color from blue to green when chlorine is present. Chlorine is a component part of CFC and HCFC refrigerants. The glowing copper element makes this color change possible.

The halide torch uses the primary air port to draw air into the burner through a flexible tube. If there is any chlorine-containing (CFC or HCFC) refrigerant in this air sample, as it passes over a copper element, the color of the flame changes from the typical blue of a gas flame to a green color, **Figure 11–27.** A large leak will extinguish the flame of the halide torch. The halide torch cannot be used with HFC refrigerants.

The maintenance on this torch consists of keeping the tube clear of debris and keeping a copper element in the burner head. If the sampling tube becomes restricted, the flame may burn yellow. You can place the end of the sampling tube close to your ear and hear the rushing sound of air being pulled into the tube. If you cannot hear it, or if it burns yellow, clean the tube, **Figure 11–28.**

The copper element is replaceable. If you cannot find an element, you can make a temporary one out of a piece of copper wire, **Figure 11–29.**

Electronic Leak Detectors

Electronic leak detectors are much more sensitive than the halide torch (they can detect leak rates of about 1/4 oz per year) and are widely used. The electronic leak detector samples air; if the air contains refrigerant, the detector either sounds an alarm or lights the probe end. These devices are manufactured in both battery- and 120-V AC-powered units. Some units may have a pump to pull the sample across the sensing element, and some have the sensing element located in the head of the probe, **Figure 11–30.**

Some electronic leak detectors have an adjustment that will compensate for background refrigerant. Some equipment rooms may have many small leaks, and a small amount of refrigerant may be in the air all the time. The electronic

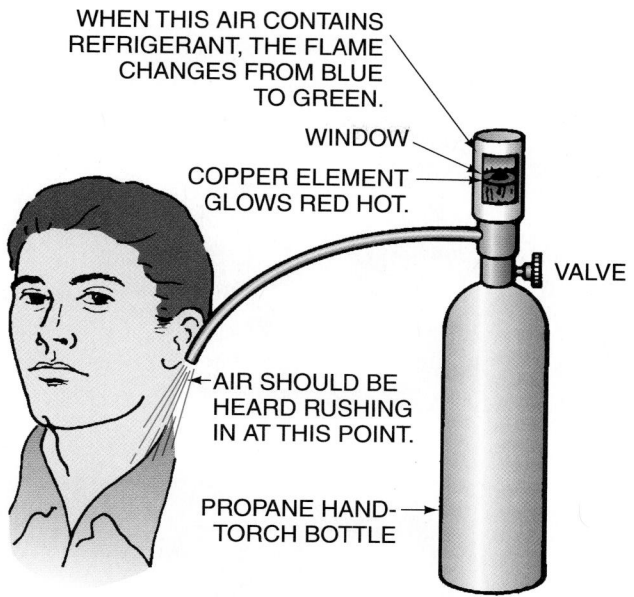

WHEN THIS AIR CONTAINS REFRIGERANT, THE FLAME CHANGES FROM BLUE TO GREEN.

WINDOW

COPPER ELEMENT GLOWS RED HOT.

VALVE

AIR SHOULD BE HEARD RUSHING IN AT THIS POINT.

PROPANE HAND-TORCH BOTTLE

Figure 11–28 A rushing sound may be heard at the end of the sampling tube if the halide torch is pulling in air and working properly.

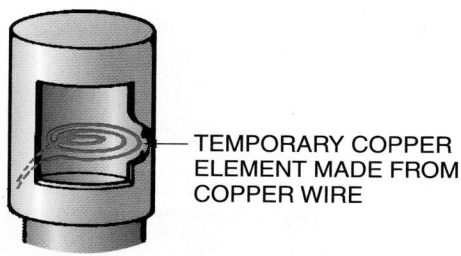

TEMPORARY COPPER ELEMENT MADE FROM COPPER WIRE

Figure 11–29 A temporary copper element.

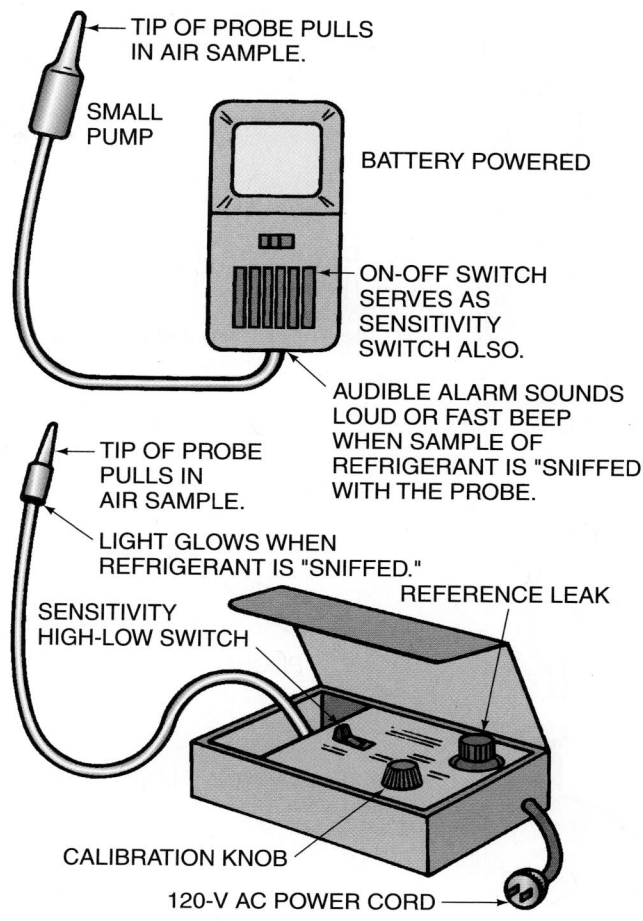

TIP OF PROBE PULLS IN AIR SAMPLE.

SMALL PUMP

BATTERY POWERED

ON-OFF SWITCH SERVES AS SENSITIVITY SWITCH ALSO.

AUDIBLE ALARM SOUNDS LOUD OR FAST BEEP WHEN SAMPLE OF REFRIGERANT IS "SNIFFED" WITH THE PROBE.

TIP OF PROBE PULLS IN AIR SAMPLE.

LIGHT GLOWS WHEN REFRIGERANT IS "SNIFFED."

SENSITIVITY HIGH-LOW SWITCH

REFERENCE LEAK

CALIBRATION KNOB

120-V AC POWER CORD

Figure 11–30 Electronic leak detectors.

leak detector will indicate all the time unless it has a feature to account for this background refrigerant.

No matter what style of leak detector you use, you must be confident that the detector will actually detect a leak. Remember that the detector only indicates what it samples. If the detector is in the middle of a refrigerant cloud and the sensor is sensing air, it will not sound an alarm or light up. For example, a pinhole leak in a pipe can be passed by with the probe of a leak detector. The sensor is sensing air next to the leak, not the leak itself, **Figure 11–31.**

Some manufacturers furnish a sample refrigerant container of R-11. The container has a pinhole in the top with a calibrated leak, **Figure 11–32.** The refrigerant will remain in the container for a long time if the lid is replaced after each use.

NOTE: Never spray pure refrigerant into the sensing element. Damage will occur. If you do not have a reference leak canister, a gage line under pressure from a refrigerant cylinder can be loosened slightly, and the refrigerant can be fanned to the electronic leak-detector sensing element. In doing it this way, air is mixed with the refrigerant to dilute it, **Figure 11–33.**

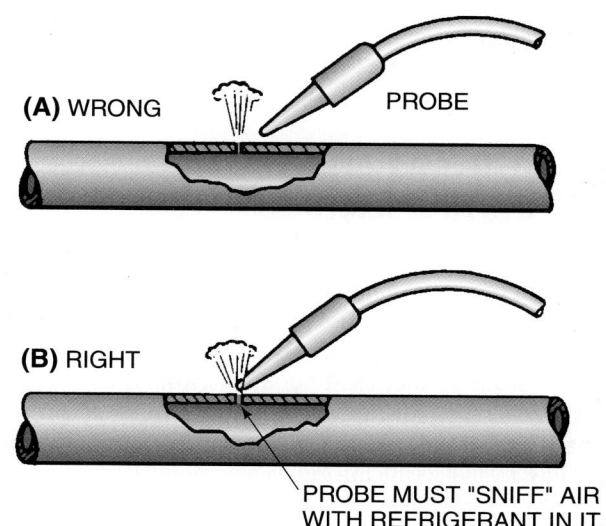

(A) WRONG   PROBE

(B) RIGHT

PROBE MUST "SNIFF" AIR WITH REFRIGERANT IN IT.

Figure 11–31 **(A)** This electronic leak detector is not sensing refrigerant escaping from the small pinhole in the tubing. This is because the refrigerant is spraying past the detector's sensor. **(B)** In this position, the sensor will detect the leaking refrigerant.

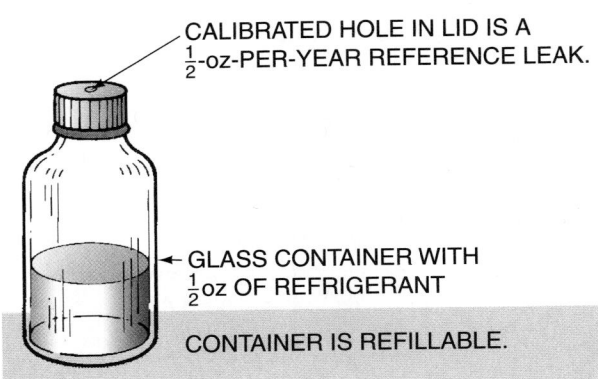

Figure 11–32 This small refrigerant vial serves as a reference leak.

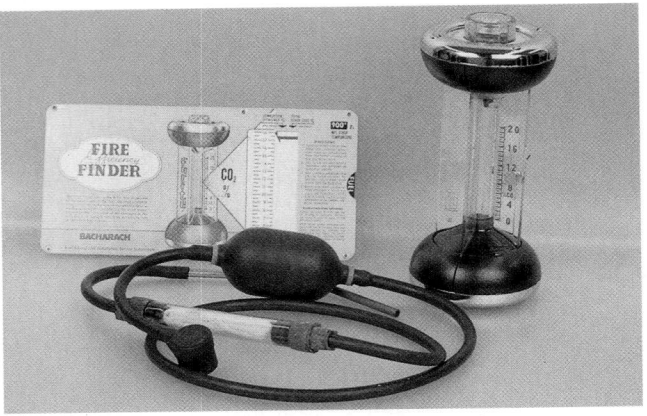

(A)

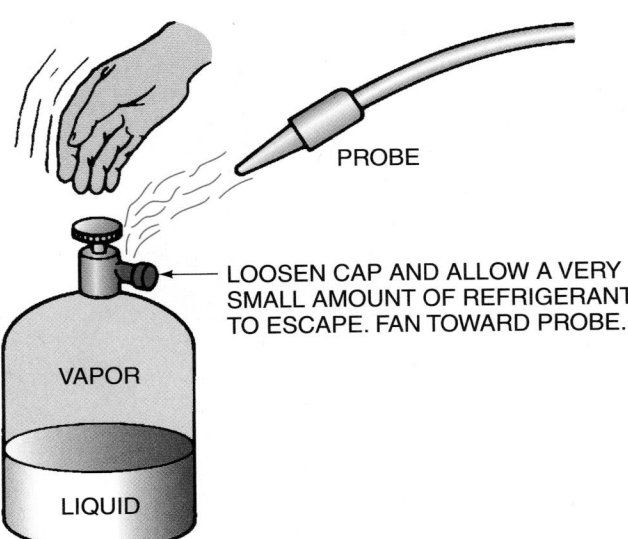

Figure 11–33 If a reference-leak vial is not available, a small amount of refrigerant may be allowed to escape and mix with the air. This sample may be fanned to dilute the concentration and direct it toward the sensing probe on the leak detector.

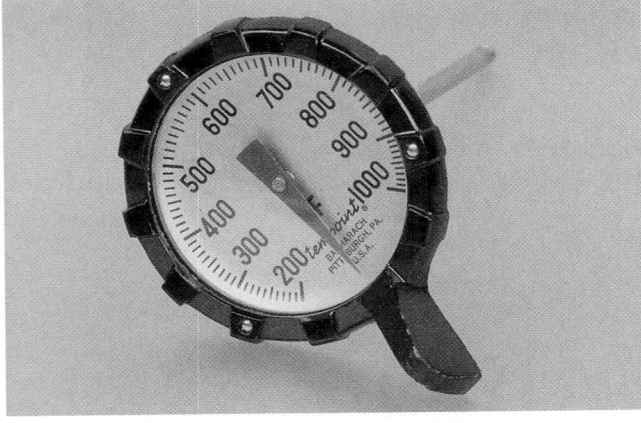

(B)

11.7 FLUE-GAS ANALYSIS INSTRUMENTS

Flue-gas analysis instruments analyze fossil fuel–burning equipment products of combustion, such as oil and gas furnaces. These instruments are normally sold in kit form with a carrying case. SAFETY PRECAUTION: There are chemicals in the flue-gas kit that must not be allowed to contact any tools or other instruments. This chemical is intended to stay in the container or the instrument that uses it. The instrument has a valve at the top that is a potential leak source. It should be checked periodically. It is best to store and transport the kit in the upright position so that if the valve does develop a leak, the chemical will not run out of the instrument, **Figure 11–34**.

These are precision instruments that cost money and deserve care and attention. The draft gage is sensitive to a very

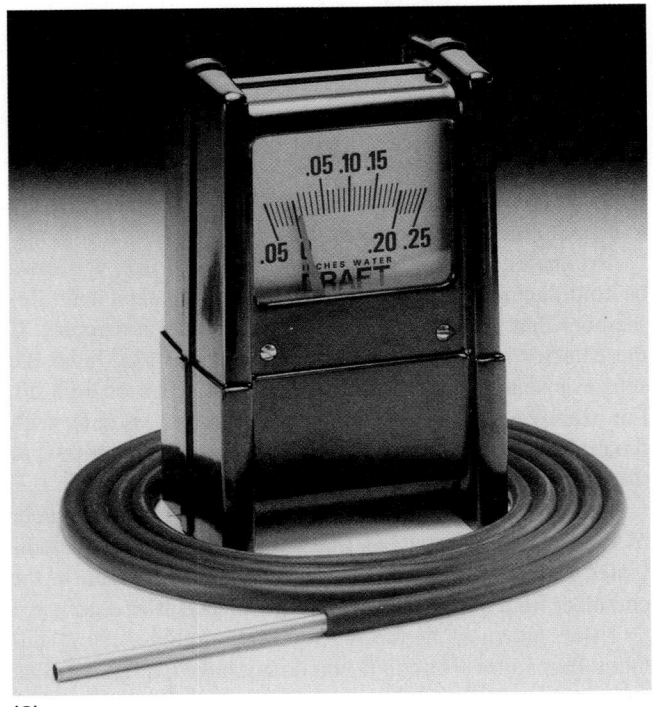

(C)

Figure 11–34 A flue-gas analysis kit. **(A)** CO_2 tester. **(B)** Thermometer. **(C)** Draft gage. **(A)**, **(B)** *Photos by Bill Johnson.* **(C)** *Courtesy Bacharach, Inc., Pittsburgh, PA USA*

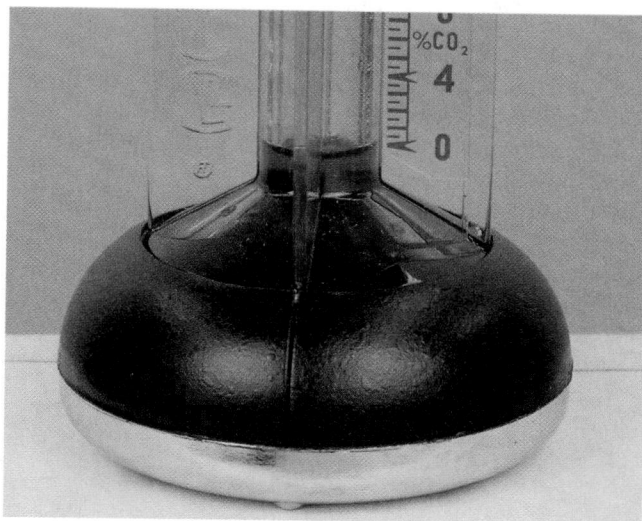

Figure 11–35 The zero adjustment on the sliding scale of the flue-gas analysis kit. *Photo by Bill Johnson*

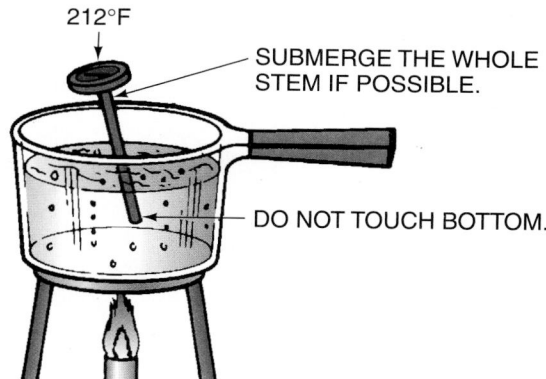

212°F

SUBMERGE THE WHOLE STEM IF POSSIBLE.

DO NOT TOUCH BOTTOM.

Figure 11–36 A flue-gas thermometer being checked in boiling water.

fine degree. This kit should not be hauled in the truck except when you intend to use it.

A calibration check of these instruments is not necessary. The chemicals in the analyzer should be changed according to the manufacturer's suggestions. These instruments are direct-reading instruments and cannot be calibrated. The only adjustment is the zero adjustment on the sliding scale, **Figure 11–35,** which is adjusted at the beginning of the test. This is done by venting the sample chamber to the atmosphere. The fluid should fall. If it gets dirty, change it.

The thermometer in the kit is used to register very high temperatures, up to 1000°F. There is no easy reference point except the boiling point of water, 212°F. This may be used as the reference, even though it is near the bottom of the scale, **Figure 11–36.**

11.8 GENERAL MAINTENANCE

Any instrument with a digital readout will have batteries. They must be maintained. *Buy the best batteries available; inexpensive batteries may cause problems. A good battery*

will not leak acid on the instrument components if it is left unattended and goes dead.

The instruments you use extend your senses; the instruments, therefore, must be maintained so that they can be believed. Airplane pilots sometimes have a malady called vertigo. They become dizzy and lose their orientation and relationship with the horizon. Suppose the pilot were in a storm and being tossed around, even upside down at times. The pilot can be upside down and have a sensation that the plane is right side up and climbing. The plane may be diving toward the earth while the pilot thinks it is climbing. Instruments must be believable, and the technician must have reference points to have faith in the instruments.

SUMMARY

- Reference points for all instruments should be established to give the technician confidence.
- The three easily obtainable reference points for temperature-measuring instruments are ice and water at 32°F, body temperature at 98.6°F, and boiling temperature at 212°F.
- Make sure that the temperature-sensing element reflects the actual temperature of the medium used as the standard.
- Pressure-measuring instruments must be checked above and below atmospheric pressure. There are no good reference points below the atmosphere, so a vacuum pump pulling a deep vacuum is used as the reference.
- Flue-gas analysis kits need no calibration except for the sliding scale on the sample chamber.

REVIEW QUESTIONS

1. What does "calibrating an instrument" mean?
2. Three reference points that may be used to check temperature-measuring instruments are _____, _____, and _____.
3. True or False: All instruments can be calibrated.
4. Temperature-measuring instruments may be designed to measure the temperature of vapors, _____, and _____.
5. Glass stem thermometers are being replaced by _____ thermometers.
6. True or False: To check the temperature leads of an electronic thermometer for low temperatures, they should be stirred in a container of crushed ice, water, and a tablespoon of salt.
7. Accuracy of a thermometer is more important in the _____ temperature ranges.
8. True or False: Pressure test instruments measure pressures only above atmospheric pressure.
9. A pressure gage can be checked by
   A. connecting it to an operating refrigeration system and comparing the pressure with the manufacturer's recommended pressure.
   B. connecting it to a cylinder of refrigerant with a known temperature and comparing the pressure reading with that on a temperature/pressure relationship chart.

    C. connecting it to an operating system and comparing the pressure reading with the subcooling temperature.

    D. connecting it to an operating system and comparing the reading with that of a megohmmeter.

10. A vacuum pump at a low vacuum will make _____ noise than at a higher vacuum.

    A. more

    B. less

    C. the same

11. The mercury manometer or the _____ _____ _____ may be used for accurate vacuum readings.

12. What should a gage manifold reading indicate when opened to the atmosphere?

13. Describe how the ohmmeter feature of a VOM can be checked.

14. Describe how amperage readings are taken with a clamp-on ammeter.

15. Electronic leak detectors are _____ [more or less] sensitive than a halide torch.

16. Good-quality electronic leak detectors can detect leak rates down to approximately _____ ounce(s) per year.

17. What are three instruments that a flue-gas analysis kit might contain?

SECTION 3

Basic Automatic Controls

UNIT 12

Basic Electricity and Magnetism

OBJECTIVES

After studying this unit, you should be able to

- describe the structure of an atom.
- identify atoms with a positive charge and atoms with a negative charge.
- explain the characteristics that make certain materials good conductors and others good insulators.
- describe how magnetism is used to produce electricity.
- state the differences between alternating current and direct current.
- list the units of measurement for electricity.
- explain the differences between series and parallel circuits.
- state Ohm's law.
- state the formula for determining electrical power.
- describe a solenoid.
- explain inductance.
- describe the construction of a transformer and the way a current is induced in a secondary circuit.
- describe how a capacitor works.
- describe a sine wave.
- state the reasons for using proper wire sizes.
- describe the physical characteristics and the function of several semiconductors.
- describe procedures for making electrical measurements.

SAFETY CHECKLIST

- ✔ Do not make any electrical measurements without specific instructions from a qualified person.
- ✔ Use only electrical conductors of the proper size to avoid overheating and possibly fire.
- ✔ Electrical circuits must be protected from current overloads. These circuits are normally protected with fuses or circuit breakers.
- ✔ Extension cords used by technicians to provide electrical power for portable power tools and other devices should be protected with ground fault circuit interrupters.
- ✔ When servicing equipment, the electrical service should be shut off at a disconnect panel whenever possible, the disconnect panel locked, and the only key kept by the technician.

12.1 STRUCTURE OF MATTER

To understand the theory of how an electric current flows, you must understand something about the structure of matter. Matter is made up of atoms, which are the smallest amounts of a substance that can exist. Atoms are made up of protons, neutrons, and electrons. Protons and neutrons are located at the center (or nucleus) of the atom. Protons have a positive charge. Neutrons have no charge and have little or no effect as far as electrical characteristics are concerned. Electrons have a negative charge and travel around the nucleus in orbits. The number of electrons in an atom is the same as the number of protons. Electrons in the same orbit are the same distance from the nucleus but do not follow the same orbital paths, **Figure 12–1.**

The hydrogen atom is a simple atom to illustrate because it has only one proton and one electron, **Figure 12–2.** Not

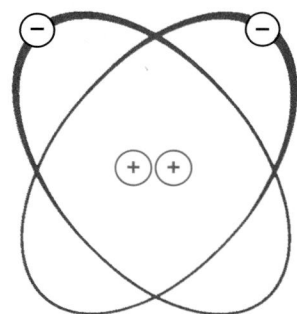

Figure 12–1   The orbital paths of electrons.

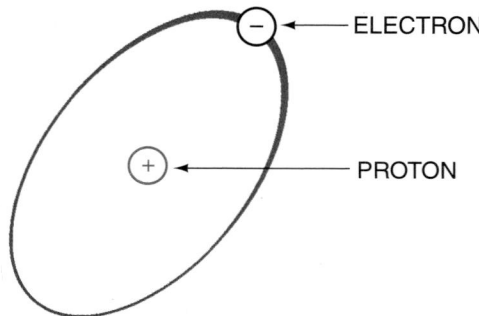

Figure 12–2   A hydrogen atom with one electron and one proton.

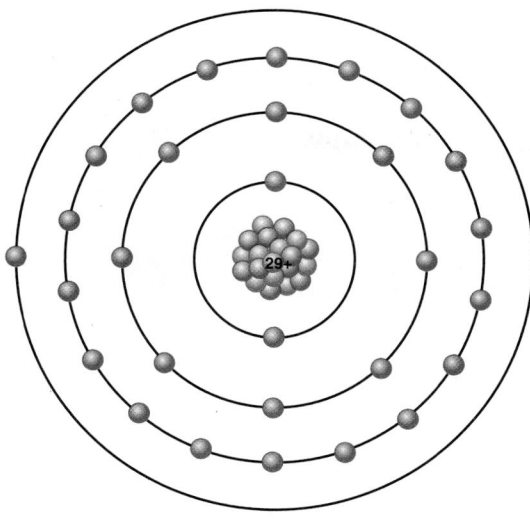

Figure 12–3 A copper atom with 29 protons and 29 electrons.

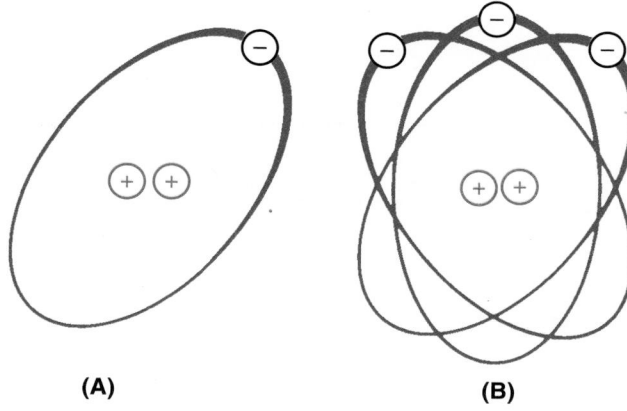

(A)   **(B)**

Figure 12–4 **(A)** This atom has two protons and one electron. It has a shortage of electrons and a positive charge. **(B)** This atom has two protons and three electrons. It has an excess of electrons and a negative charge.

all atoms are as simple as the hydrogen atom. Most wiring used to conduct an electrical current is made of copper. **Figure 12–3** illustrates a copper atom, which has 29 protons and 29 electrons. Some electron orbits are farther away from the nucleus than others. As can be seen, 2 travel in an inner orbit, 8 in the next, 18 in the next, and 1 in the outer orbit. It is this single electron in the outer orbit that makes copper a good conductor of electricity.

The goal of an atom is to become as stable as possible. This involves having either a completely full or a completely empty outermost shell. Since the copper atom has only one electron in its outer shell, the atom can become stable if it is able to give up this outermost, or *valence,* electron. For this reason, the copper atom, and other metallic elements and substances as well, are eager to give up their "loose" electrons. Since electricity is described as the flow of electrons, metallic substances facilitate current flow by not holding onto their valence electrons tightly and by allowing them to move freely from atom to atom.

12.2 MOVEMENT OF ELECTRONS

When sufficient energy or force is applied to an atom, the outer electron (or electrons) becomes free and moves. If it leaves the atom, the atom will contain more protons than electrons. Protons have a positive charge. This means that this atom will have a positive charge, **Figure 12–4(A).** The atom the electron joins will contain more electrons than protons, so it will have a negative charge, **Figure 12–4(B).**

Like charges repel each other, and unlike charges attract each other. An electron in an atom with a surplus of electrons (negative charge) will be attracted to an atom with a shortage of electrons (positive charge). An electron entering an orbit with a surplus of electrons will tend to repel an electron already there and cause it to become a free electron.

12.3 CONDUCTORS

Good conductors are those with few electrons in the outer orbit. Three common metals—copper, silver, and gold—are good conductors, and each has one electron in the outer orbit. These are considered to be free electrons because they move easily from one atom to another.

Other metallic substances, such as mercury and aluminum, are also good conductors, even though they have two and three electrons in their outermost electron shells, respectively. Good conductors of electricity are typically also good conductors of heat.

12.4 INSULATORS

Atoms with several electrons in the outer orbit are poor conductors. These electrons are difficult to free, and materials made with these atoms are considered to be insulators. Glass, rubber, and plastic are examples of good insulators.

12.5 ELECTRICITY PRODUCED FROM MAGNETISM

Electricity can be produced in many ways, for example, from chemicals, pressure, light, heat, and magnetism. The electricity that air-conditioning and heating technicians are more involved with is produced by a generator using magnetism.

Magnets are common objects with many uses. Magnets have poles usually designated as the north (N) pole and the south (S) pole. They also have fields of force. **Figure 12–5** shows the lines of the field of force around a permanent bar magnet. This field causes the like poles of two magnets to repel each other and the unlike poles to attract each other.

If a conductor, such as a copper wire, is passed through this field and crosses these lines of force, the outer electrons in the atoms in the wire are freed and begin to move from

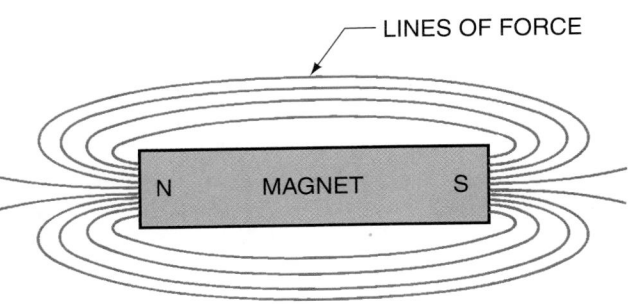

Figure 12–5   A permanent magnet with lines of force.

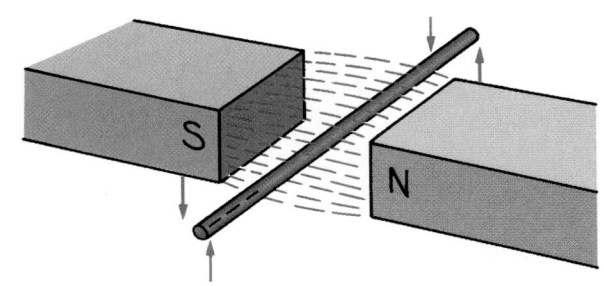

Figure 12–6   The movement of wire up and down cuts the lines of force, generating voltage, which in turn facilitates current flow.

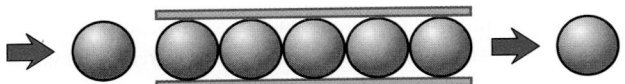

Figure 12–7   A tube filled with golf balls; when one ball is pushed in, one ball is pushed out.

atom to atom. This movement of electrons is considered to be an electrical current. They will move in one direction. It does not matter if the wire moves or if the magnetic field moves. It is only necessary that the conductor cross through the lines of force, **Figure 12–6.**

This movement of electrons in one direction produces the electric current. The current is an impulse transferred from one electron to the next. For example, if you pushed a golf ball into a tube already filled with golf balls, one would be ejected instantly from the other end, **Figure 12–7.** Electric current travels in a similar manner at a speed of 186,000 miles/sec. The electrons do not travel through the wire at this speed, but the repelling and attracting effect causes the current to do so.

An electrical generator has a large magnetic field and many turns of wire crossing the lines of force. A large magnetic field will produce more current than a smaller magnetic field, and many turns of wire passing through a magnetic field will produce more current than a few turns of wire. The magnetic force field for generators is usually produced by electromagnets. Electromagnets have similar characteristics to permanent magnets and are discussed later in this unit. **Figure 12–8** shows a simple generator.

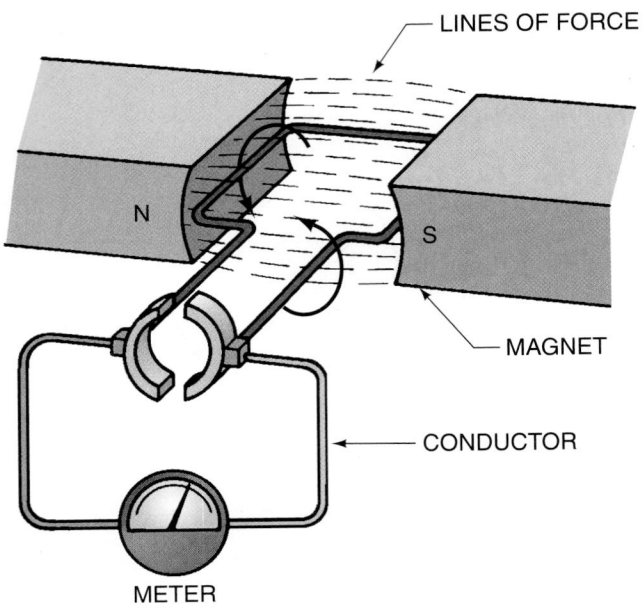

Figure 12–8   A simple generator.

12.6   DIRECT CURRENT

Direct current (DC) travels in one direction. Because electrons have a negative charge and travel to atoms with a positive charge, DC is considered to flow from negative to positive.

DC voltage is generated primarily by chemical reactions and is commonly known as the battery-power source used in wireless and portable devices. We can also create direct current from alternating current (AC) power sources by utilizing solid-state or electronic circuitry.

12.7   ALTERNATING CURRENT

Alternating current (AC) is continually and rapidly reversing. The charge at the power source (generator) is continually changing direction; thus, the current continually reverses itself. For several reasons, most electrical energy generated for public use is AC. It is much more economical to transmit electrical energy long distances in the form of AC. The voltage of this type can be readily changed so that it has many more uses. DC still has many applications, but it is usually obtained by changing AC to DC or by producing the DC locally where it is to be used.

12.8   ELECTRICAL UNITS OF MEASUREMENT

Electromotive force (emf) or voltage (V) is used to indicate the difference of potential in two charges. When an electron surplus builds up on one side of a circuit and a shortage of electrons exists on the other side, a difference of potential or emf is created. The unit used to measure this force is the *volt.*

The *ampere* is the unit used to measure the quantity of electrons moving past a given point in a specific period of time (electron flow rate). *Amperage* and *ampere(s)* are sometimes termed "amp(s)" for short. The abbreviation for ampere is A.

All materials oppose or resist the flow of an electrical current to some extent. In good conductors this opposition or resistance is very low. In poor conductors the resistance is high. The unit used to measure resistance is the ohm. A resistance of 1 ohm is present when a force of 1 volt causes a current of 1 ampere to flow in an electric circuit.

Volt = Electrical force or pressure (V)
Ampere = Quantity of electron flow rate (A)
Ohm = Resistance to electron flow (Ω)

12.9 THE ELECTRICAL CIRCUIT

An electrical circuit must have a power source, a conductor to carry the current, and a load or device to use the current. There is also generally a means for turning the electrical current flow on and off. **Figure 12–9** shows an electrical generator for the source, a wire for the conductor, a light bulb for the load, and a switch for opening and closing the circuit.

The generator produces the current by passing many turns of wire through a magnetic field. The wire or conductor provides the path for the electricity to flow to the bulb and complete the circuit. The electrical energy is converted to heat and light energy at the bulb element. The switch is used to open and close the circuit. When the switch is open, no current will flow. When it is closed, the bulb element will produce heat and light because current is flowing through it.

12.10 MAKING ELECTRICAL MEASUREMENTS

In the circuit illustrated in **Figure 12–9,** electrical measurements can be made to determine the voltage (emf) and current (amperes). In making the measurements, **Figure 12–10,**

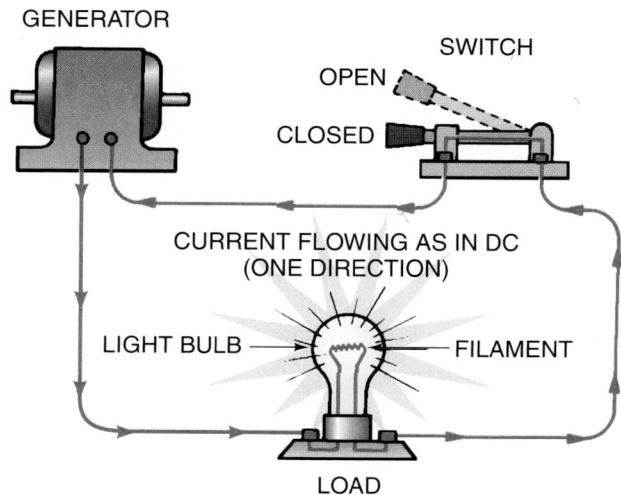

Figure 12–9 An electric circuit.

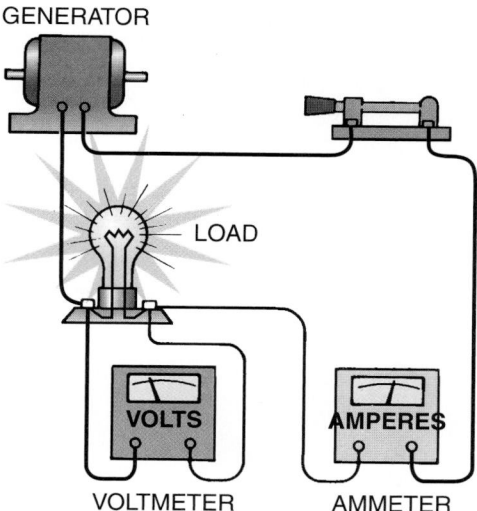

Figure 12–10 Voltage is measured across the load (in parallel). Amperage is measured in series.

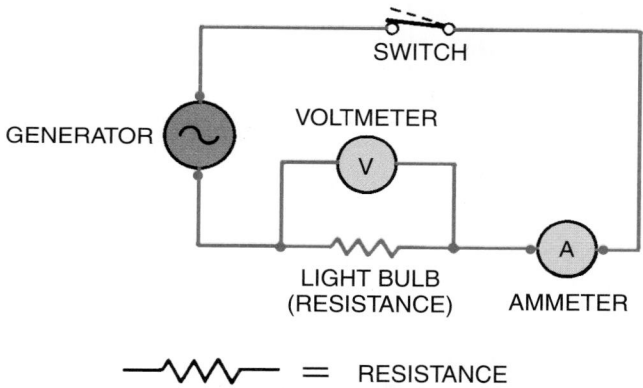

Figure 12–11 The same circuit as **Figure 12–10,** illustrated with symbols.

the voltmeter is connected across the terminals of the bulb without interrupting the circuit. The ammeter is connected directly into the circuit so that all the current flows through it. **Figure 12–11** illustrates the same circuit using symbols. Voltmeters are used to measure the potential difference between two points in an electric circuit and are therefore connected in parallel to the portion of the circuit between those points. The internal resistance of the voltmeter is very high so that the meter does not alter the current flow or the potential difference between the points being evaluated.

Ammeters are used to measure the intensity of current flow in a circuit or a particular branch in a circuit, so they are wired in series with the circuit being evaluated. Since we do not want the meter to affect the circuit operation, current, or voltage, the internal resistance of an ammeter is very low.

Often a circuit will contain more than one resistance or load. These resistances may be wired in series or in parallel, depending on the application or use of the circuit. **Figure 12–12** shows three loads in series. This is shown

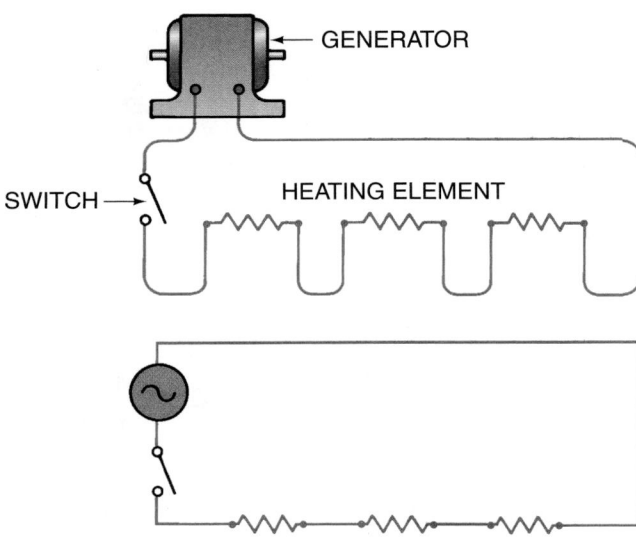

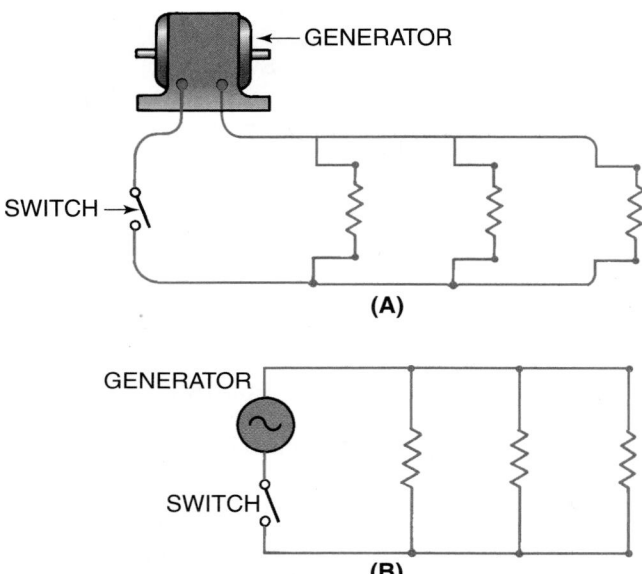

Figure 12–12 Multiple resistances (small heating elements) in series.

Figure 12–13 **(A)** Multiple resistances in parallel. **(B)** Three resistances in parallel using symbols.

pictorially and with symbols. **Figure 12–13** illustrates three loads wired in parallel.

In circuits where devices are wired in series, all of the current passes through each load. When two or more loads are wired in parallel, the current is divided among the loads. This is explained in more detail later. Power-passing devices such as switches are wired in series with the devices or loads they are controlling. Most resistances or loads (power-consuming devices) that air-conditioning and heating technicians work with are wired in parallel.

Figure 12–14 illustrates how a voltmeter is connected for each of the resistances. The voltmeter is in parallel with each resistance. The ammeter is also shown and is wired in series in the circuit. An ammeter has been developed that can be

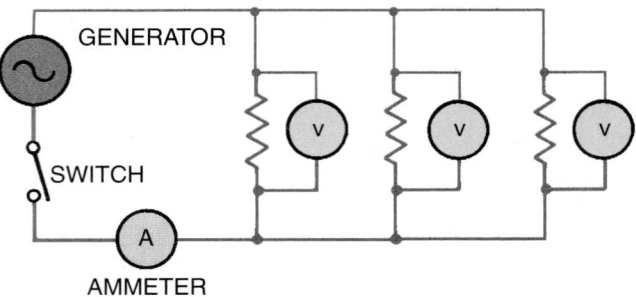

Figure 12–14 Voltage readings are taken across the resistances in the circuit.

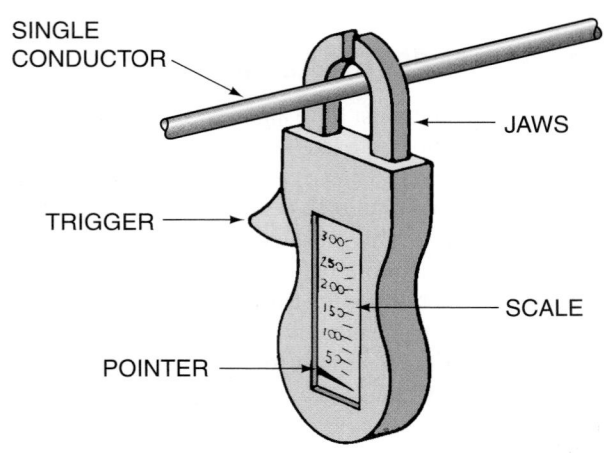

Figure 12–15 A clamp-on ammeter. *Courtesy Amprobe Instrument Division, Core Industries*

clamped around a single conductor to measure amperes, **Figure 12–15**. This is convenient because it is often difficult to disconnect the circuit to connect the ammeter in series. This type of ammeter, usually called a clamp-on type, is discussed in Section 12.20, "Electrical Measuring Instruments."

12.11   OHM'S LAW

During the early 1800s the German scientist George S. Ohm did considerable experimentation with electrical circuits and particularly with regard to resistances in these circuits. He determined that a relationship exists among the factors in an electrical circuit. This relationship is called Ohm's law. The following describes this relationship. Letters are used to represent the different electrical factors.

$$E \text{ or } V = \text{Voltage (emf)}$$
$$I = \text{Amperage (current)}$$
$$R = \text{Resistance (load)}$$

The voltage equals the amperage times the resistance:

$$E = I \times R$$

The amperage equals the voltage divided by the resistance:

$$I = \frac{E}{R}$$

The resistance equals the voltage divided by the amperage:

$$R = \frac{E}{I}$$

Figure 12–16 shows a convenient way to remember these formulas. The symbol for ohms is Ω.

In **Figure 12–17** the resistance of the heating element can be determined as follows:

$$R = \frac{E}{I} = \frac{120}{3} = 40 \ \Omega$$

In **Figure 12–18** the voltage across the resistance can be calculated as follows:

$$E = I \times R = 2 \times 60 = 120 \ \text{V}$$

Figure 12–19 indicates the voltage to be 120 V and the resistance to be 20 Ω. The formula for determining the current flow is

$$I = \frac{E}{R} = \frac{120}{20} = 6 \ \text{A}$$

In series circuits with more than one resistance, simply add the resistances together to find the total resistance of the circuit. In **Figure 12–20** there is only one path for the current

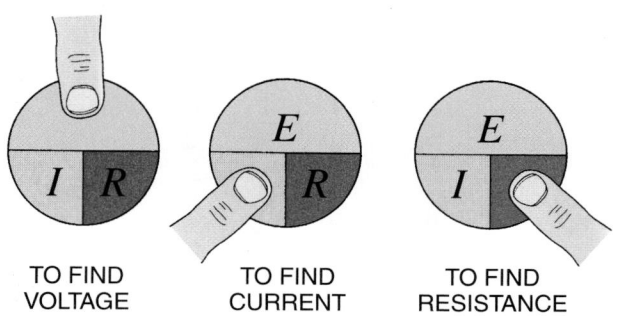

TO FIND VOLTAGE   TO FIND CURRENT   TO FIND RESISTANCE

Figure 12–16   To determine the formula for the unknown quantity, cover the letter that represents the unknown.

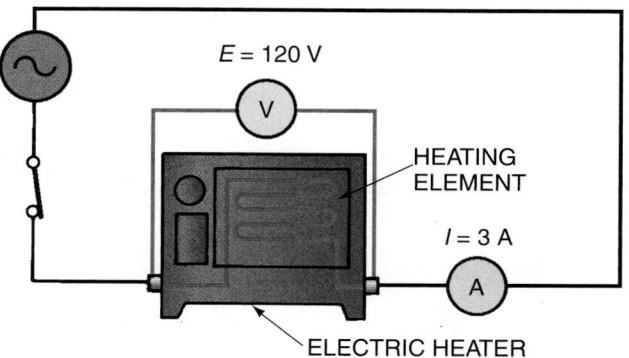

E = 120 V

HEATING ELEMENT

I = 3 A

ELECTRIC HEATER

Figure 12–17   The resistance of the heating element can be determined by using the voltage and amperage indicated.

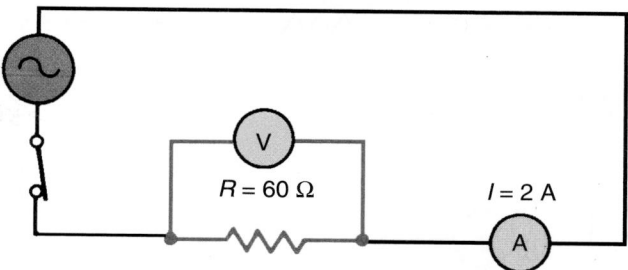

R = 60 Ω   I = 2 A

Figure 12–18   The voltage across the resistance can be calculated using the information in this diagram.

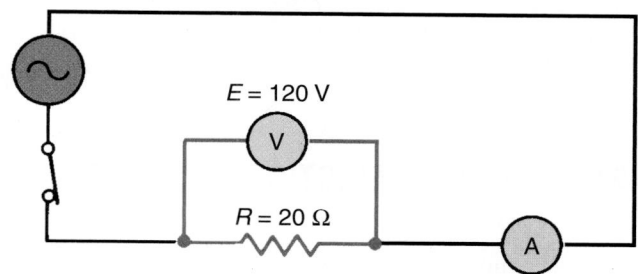

E = 120 V

R = 20 Ω

Figure 12–19   Voltage is 120 V and the resistance is 20 Ω. The amperage of the circuit can be determined using this information.

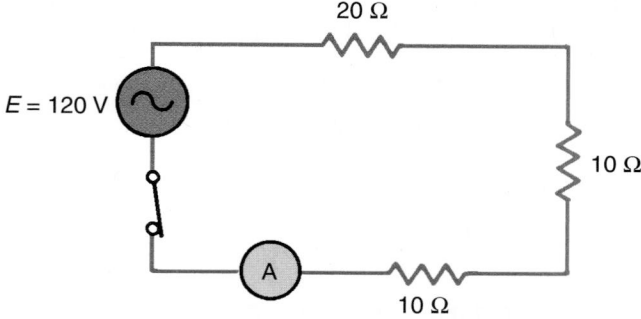

20 Ω

E = 120 V

10 Ω

10 Ω

Figure 12–20   The current in a series circuit has only one possible path to follow.

to follow (a series circuit), so the resistance is 40 Ω (20 Ω + 10 Ω + 10 Ω). The amperage in this circuit will be

$$I = \frac{E}{R} = \frac{120}{40} = 3 \ \text{A}$$

The resistances, in ohms, of the individual loads can be determined by disconnecting the resistance to be measured from the circuit and reading the ohms from an ohmmeter, as illustrated in **Figure 12–21**. NOTE: Ohmmeters are to be used only on circuits that are deenergized. Also, be sure that the component being evaluated is disconnected from the circuit to avoid inaccurate or false readings.

SAFETY PRECAUTION: Do not use any electrical measuring instruments without specific instructions from a qualified person. Follow the manufacturer's instructions. The use of electrical measuring instruments is discussed in Section 12.20.

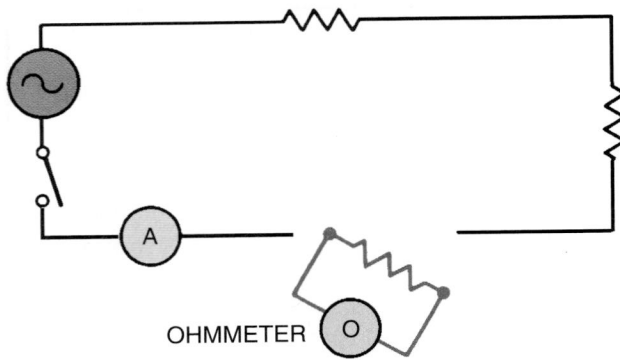

Figure 12–21 To determine the resistance of a circuit component, turn off the power and disconnect the component from the circuit. Then, check the component with an ohmmeter.

12.12 CHARACTERISTICS OF SERIES CIRCUITS

In series circuits
- the voltage is divided across the different resistances. The mathematical formula is as follows:

$$E_{total} = E_1 + E_2 + E_3 + \cdots$$

- the total current flows through each resistance or load. The mathematical formula is as follows:

$$I_{total} = I_1 = I_2 = I_3 = \cdots$$

- the resistances are added together to obtain the total resistance. The formula for calculating total resistance in a series circuit is as follows:

$$R_{total} = R_1 + R_2 + R_3 + \cdots$$

12.13 CHARACTERISTICS OF PARALLEL CIRCUITS

In parallel circuits
- the total voltage is applied across each resistance. The mathematical formula is as follows:

$$E_{total} = E_1 = E_2 = E_3 = \cdots$$

- the current is divided between the different loads according to their individual resistances, and the total current is equal to the sum of the currents in each branch. The mathematical formula is as follows:

$$I_{total} = I_1 + I_2 + I_3 + \cdots$$

- the total resistance is less than the value of the smallest resistance. As more and more resistive paths or branches are added to the parallel circuit, the total resistance of the circuit gets lower and lower.

Calculating the total resistance in a parallel circuit requires a different procedure than simply adding them together as in a series circuit. A parallel circuit allows current flow along two or more paths at the same time. This type of circuit applies equal voltage to all loads. The general formula used to determine total resistance in parallel circuits is as follows:

For two resistances

$$R_{total} = \frac{R_1 \times R_2}{R_1 + R_2}$$

For more than two resistances

$$R_{total} = \frac{1}{\dfrac{1}{R_1} + \dfrac{1}{R_2} + \dfrac{1}{R_3} + \cdots}$$

The total resistance of the circuit in **Figure 12–22** is determined as follows:

$$R_{total} = \frac{1}{\dfrac{1}{10} + \dfrac{1}{20} + \dfrac{1}{30}}$$

$$= \frac{1}{0.1 + 0.05 + 0.033}$$

$$= \frac{1}{0.183}$$

$$= 5.46 \ \Omega$$

Notice that the total resistance of 5.46 Ω is lower than the lowest individual resistance in the circuit. Another way to look at the resistance calculation is as follows:

$$\frac{1}{R_{total}} = \frac{1}{R_1} + \frac{1}{R_2} + \frac{1}{R_3}$$

$$\frac{1}{R_{total}} = \frac{1}{10} + \frac{1}{20} + \frac{1}{30}$$

$$\frac{1}{R_{total}} = \frac{6}{60} + \frac{3}{60} + \frac{2}{60}$$

$$\frac{1}{R_{total}} = \frac{11}{60}$$

$$R_{total} = \frac{60}{11}$$

$$R_{total} = 5.46 \ \Omega$$

To determine the total current draw, use Ohm's law:

$$I = \frac{E}{R} = \frac{120}{5.46} = 22 \ A$$

We can confirm this value for the total current in the circuit by determining the current flow through the individual loads

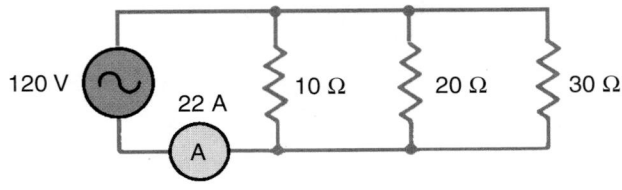

Figure 12–22 The total resistance of this circuit can be determined using the information in this diagram.

or resistances and then adding them up. The current flow through the 10 Ω resistance is determined here:

$$I_{10} = E \div R$$
$$I_{10} = 120 \text{ V} \div 10 \text{ } \Omega = 12 \text{ A}$$

The current flow through the 20 Ω resistance is determined here:

$$I_{20} = E \div R$$
$$I_{20} = 120 \text{ V} \div 20 \text{ } \Omega = 6 \text{ A}$$

The current flow through the 30 Ω resistance is determined here:

$$I_{30} = E \div R$$
$$I_{30} = 120 \text{ V} \div 30 \text{ } \Omega = 4 \text{ A}$$

By adding these three individual branch currents together, we get

$$I_{\text{total}} = I_1 + I_2 + I_3$$
$$I_{\text{total}} = 12 \text{ A} + 6 \text{ A} + 4 \text{ A}$$
$$I_{\text{total}} = 22 \text{ A}$$

12.14 ELECTRICAL POWER

Electrical power (*P*) is measured in watts. A *watt (W)* is the power used when 1 ampere flows with a potential difference of 1 volt. Therefore, power can be determined by multiplying the voltage times the amperes flowing in a circuit.

$$\text{Watts} = \text{Volts} \times \text{Amperes}$$

or

$$P = E \times I$$

The consumer of electrical power pays the electrical utility company according to the number of kilowatts (kW) used for a certain time span usually billed as kilowatt hours (kWh). A kilowatt is equal to 1000 W. To determine the power being consumed, divide the number of watts by 1000:

$$\text{P (in kW)} = \frac{E \times I}{1000}$$

In the circuit shown in **Figure 12–22,** the power consumed can be calculated as follows:

$$P = E \times I$$
$$P = 120 \text{ V} \times 22 \text{ A} = 2640 \text{ W}$$
$$\text{kW} = P \div 1000$$
$$\text{kW} = 2640 \text{ W} \div 1000 = 2.64 \text{ kW}$$

12.15 MAGNETISM

Magnetism was briefly discussed previously in the unit to point out how electrical generators are able to produce electricity. Magnets are classified as either permanent or temporary. Permanent magnets are used in only a few applications that air-conditioning and refrigeration technicians would work with. Electromagnets, a type of temporary magnet, are used in many electrical components of air-conditioning and refrigeration equipment.

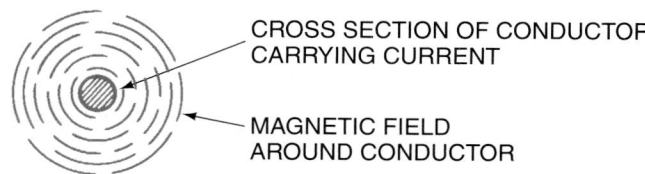

Figure 12–23 This cross section of a wire shows a magnetic field around the conductor.

Figure 12–24 The magnetic field around a loop of wire. This is a stronger field than that around a straight wire.

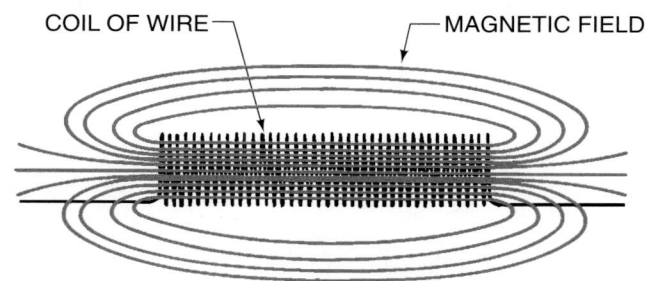

Figure 12–25 There is a stronger magnetic field surrounding wire formed into a coil.

A magnetic field is generated around a wire whenever an electrical current is flowing through it, **Figure 12–23.** If the wire or conductor is formed into a loop, the strength of the magnetic field will be increased, **Figure 12–24.** If the wire is wound into a coil, a stronger magnetic field will be created, **Figure 12–25.** This coil of wire carrying an electrical current is called a *solenoid.* This solenoid or electromagnet will attract or pull an iron bar into the coil, **Figure 12–26.** If an iron bar is inserted permanently in the coil, the strength of the magnetic field will be increased even more.

This magnetic field can be used to generate electricity and to cause electric motors to operate. The magnetic attraction can also cause motion, which is used in many controls and switching devices, such as solenoids, relays, and contactors, **Figure 12–27(A). Figure 12–27(B)** is a cutaway view of a solenoid.

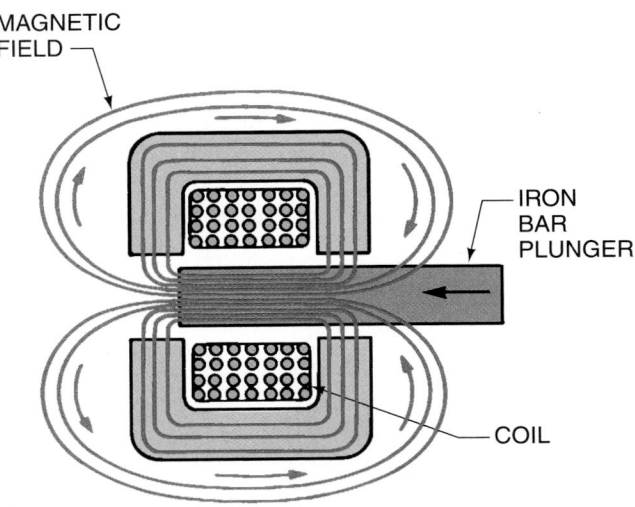

Figure 12–26   When current flows through a coil, the iron bar will be pulled into it.

(A)

(B)

(C)

Figure 12–27(A)   **(A)** A solenoid. **(B)** A relay. **(C)** Contactors. *Photos (A) and (B) by Bill Johnson. (C) Courtesy Honeywell, Inc.*

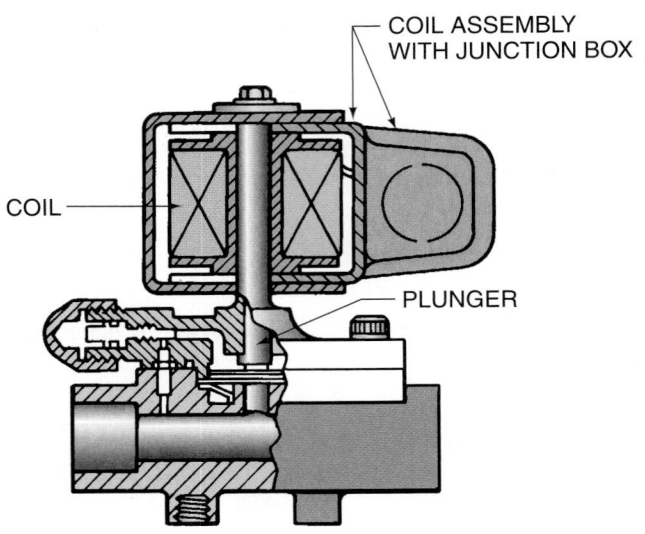

Figure 12–27(B)   Cutaway view of a solenoid. *Courtesy Parker Hannifin Corporation*

12.16   INDUCTANCE

As mentioned previously, when voltage is applied to a conductor and current flows, a magnetic field is produced around the conductor. In an AC circuit the current is continually changing direction. This causes the magnetic field to continually build up and immediately collapse. When these lines of force build up and collapse, they cut through the wire or conductor and produce an emf or voltage. This voltage opposes the existing voltage in the conductor.

In a straight conductor, this induced voltage is very small and is usually not considered, **Figure 12–28.** However, if a conductor is wound into a coil, these lines of force overlap and reinforce each other, **Figure 12–29.** This does develop an emf or voltage that is strong enough to provide opposition to the existing voltage. This opposition is called *inductive reactance* and is a type of resistance in an AC circuit. Coils,

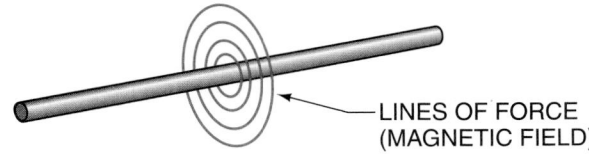

Figure 12–28   A straight conductor with a magnetic field.

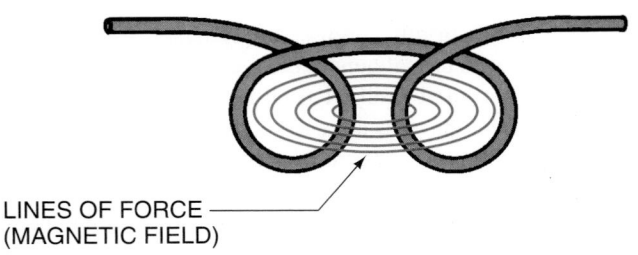

Figure 12–29   A conductor formed into a coil with lines of force.

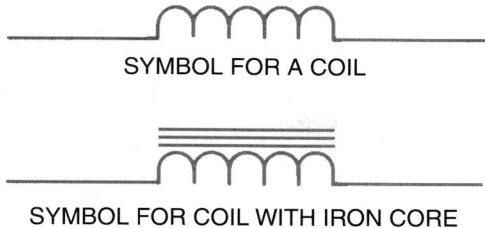

SYMBOL FOR A COIL

SYMBOL FOR COIL WITH IRON CORE

Figure 12–30 Electrical symbols for a coil.

Figure 12–31(B) A transformer. *Photo by Bill Johnson*

chokes, and transformers are examples of components that produce inductive reactance. **Figure 12–30** shows symbols for the electrical coil.

12.17 TRANSFORMERS

Transformers are electrical devices that produce an electrical current in a second circuit through electromagnetic induction.

In **Figure 12–31(A)**, a voltage applied across terminals A-A will produce a magnetic field around the steel or iron core. This is AC causing the magnetic field to continually build up and collapse as the current reverses. This will cause the magnetic field around the core in the second winding to cut across the conductor wound around it. An electric voltage is induced in the second winding. If this induced voltage is then connected to a separate, otherwise complete, circuit, current will flow in that circuit.

Transformers, **Figure 12–31(B)**, have a primary winding, a core usually made of thin plates of steel laminated together, and a secondary winding. There are step-up and step-down transformers. A step-down transformer contains more turns of wire in the primary winding than in the secondary winding.

The voltage at the secondary is directly proportional to the number of turns of wire in the secondary (as compared to the number of turns in the primary) windings. For example, **Figure 12–32** is a transformer with 1000 turns in the primary and 500 turns in the secondary. A voltage of 120 V is applied, and the voltage induced in the secondary is 60 V. Actually the voltage is slightly less due to some loss into the air of the magnetic field and because of resistance in the wire.

A step-up transformer has more windings in the secondary than in the primary. This causes a larger voltage to be induced into the secondary. In **Figure 12–33,** with 1000 turns in the primary, 2000 in the secondary, and an applied voltage of 120 V, the voltage induced in the secondary is double, or

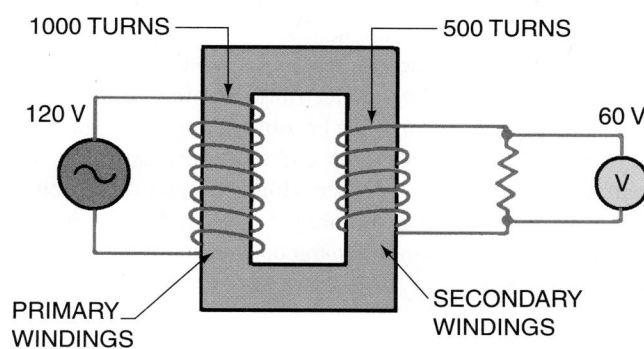

Figure 12–32 A step-down transformer.

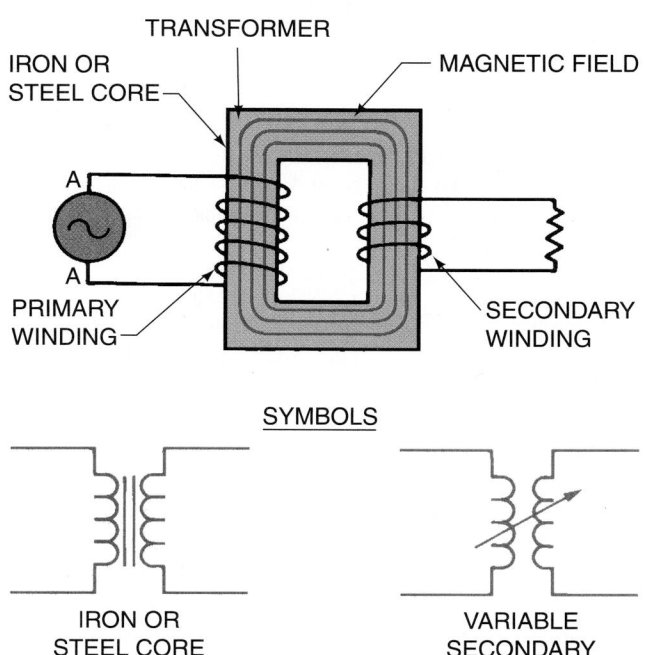

Figure 12–31(A) Voltage applied across terminals produces a magnetic field around an iron or steel core.

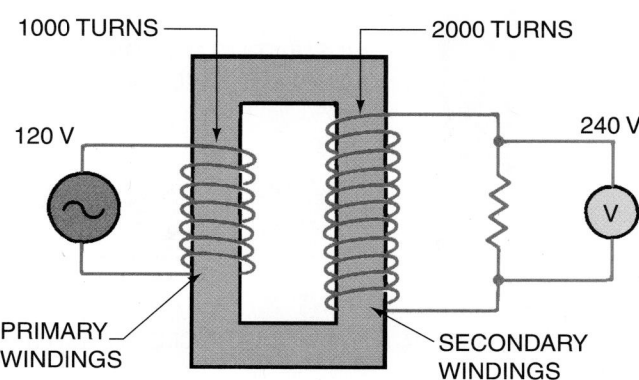

Figure 12–33 A step-up transformer.

approximately 240 V. Step-up transformers are commonly used to generate sparks that ignite air/fuel mixtures for heating applications.

The same power (watts) is available at the secondary as at the primary (except for a slight loss). If the voltage is reduced to one-half that at the primary, the current capacity nearly doubles.

Step-up transformers are used at generating stations to increase the voltage to produce more efficiency in delivering the electrical energy over long distances to substations or other distribution centers. At the substation the voltage is reduced for further distribution. To reduce the voltage, a step-down transformer is used. At a residence, the voltage may be reduced to 240 V or 120 V. Further step-down transformers may be used with air-conditioning and heating equipment to produce the 24 V commonly used in thermostats and other control devices.

12.18   CAPACITANCE

A device in an electrical circuit that allows electrical energy to be stored for later use is called a *capacitor*. A simple capacitor is composed of two plates with insulating material between them, **Figure 12–34**. This insulating material is referred to as a dielectric and can be air, paper, or other nonconductive material. The capacitor can store a charge of electrons on one plate. When the plate is fully charged in a DC circuit, no current will flow until there is a path back to the positive plate, **Figure 12–35**. When this path is available, the electrons will flow to the positive plate until the negative plate no longer has a charge, **Figure 12–36** and **Figure 12–37**. At this point both plates are neutral.

In an AC circuit, the voltage and current are continually changing direction. As the electrons flow in one direction, the capacitor plate on one side becomes charged. As this current and voltage reverses, the charge on the capacitor becomes greater than the source voltage, and the capacitor begins to discharge. It is discharged through the circuit, and the opposite plate becomes charged. This continues through each AC cycle.

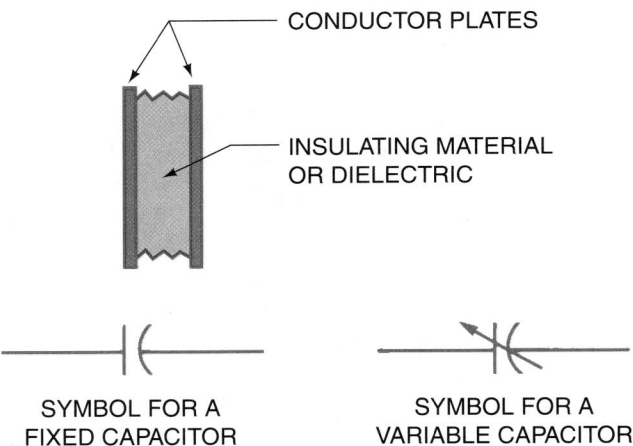

Figure 12–34   A capacitor.

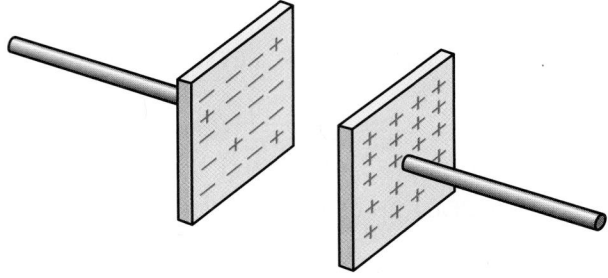

Figure 12–35   A charged capacitor.

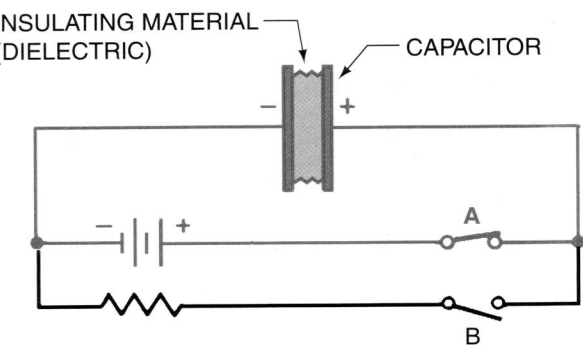

Figure 12–36   Electrons will flow to the negative plate from the battery. The negative plate will charge until the capacitor has the same potential difference as the battery.

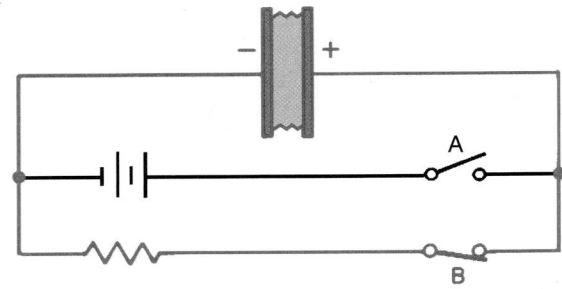

Figure 12–37   When the capacitor is charged, switch A is opened, switch B is closed, and the capacitor discharges through the resistor to the positive plate. The capacitor is then completely discharged and has no charge.

A capacitor has *capacitance,* which is the amount of charge that can be stored. The capacitance is determined by the following physical characteristics of the capacitor:

1. Distance between the plates
2. Surface area of the plates
3. Dielectric material between the plates

Capacitors are rated in farads. However, farads represent such a large amount of capacitance that the term microfarad is normally used. A microfarad is one millionth (0.000001) of a farad. Capacitors can be purchased in ranges up to several hundred microfarads. The symbol for micro is the Greek letter μ (mu) and the symbol for farad is the capital letter F (μ F).

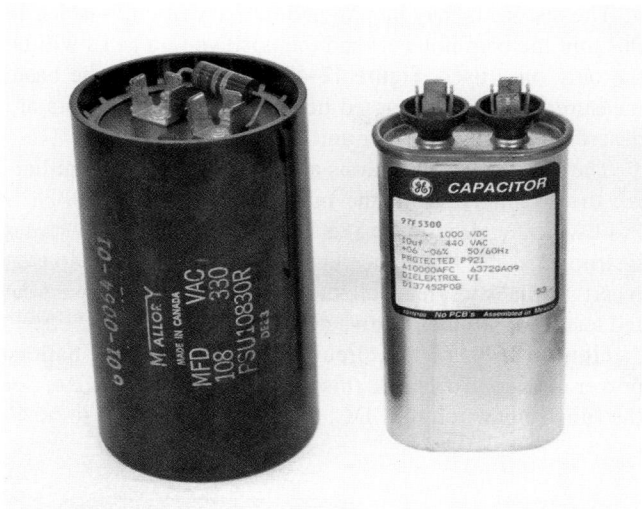

Figure 12–38   A start capacitor, left, and a run capacitor, right. *Photo by Bill Johnson*

A capacitor opposes current flow in an AC circuit similar to a resistor or to inductive reactance. This opposition or type of resistance is called *capacitive reactance*. The capacitive reactance depends on the frequency of the power supply and the capacitance of the capacitor.

Two types of capacitors used frequently in the air-conditioning and refrigeration industry are the start and run capacitors used on electric motors, **Figure 12–38.**

12.19   IMPEDANCE

We have learned that there are three types of opposition to current flow in an AC circuit. There is pure resistance, inductive reactance, and capacitive reactance. The total effect of these three is called impedance. The voltage and current in a circuit that has only resistance are in phase with each other. The voltage leads the current across an inductor and lags behind the current across a capacitor. **Figures 12–57** and **12–58** in Section 12.21 illustrate these conditions. Inductive reactance and capacitive reactance can cancel each other. Impedance is a combination of the opposition to current flow produced by these characteristics in a circuit.

12.20   ELECTRICAL MEASURING INSTRUMENTS

A multimeter is an instrument that measures voltage, current, resistance, and, on some models, temperature. It is a combination of several meters and can be used for making AC or DC measurements in several ranges. It is the instrument used most often by heating, refrigeration, and air-conditioning technicians.

A multimeter often used is the *volt-ohm-milliammeter (VOM)* such as the one in **Figure 12–39.** This meter is used to measure AC and DC voltages, DC, and resistance. AC amperage may be measured with some meters when used with an AC ammeter adapter, **Figure 12–40.** Meters have

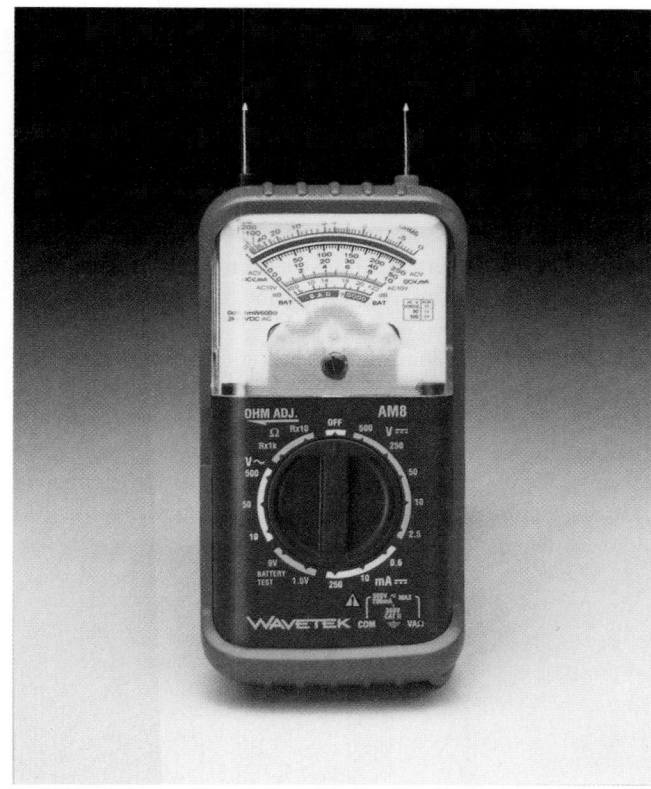

Figure 12–39   A volt-ohm-milliammeter (VOM). *Courtesy Wavetek*

Figure 12–40   An ammeter adapter for a VOM. *Courtesy Amprobe*

different features depending on the manufacturer. Most meters have a function switch and a range switch. We will use a typical meter with examples to explain how the technician uses the meter.

The function switch, located at the left side of the lower front panel, **Figure 12–41**, has −DC, +DC, and AC positions.

The range switch is in the center of the lower part of the front panel, **Figure 12–41**. It may be turned in either direction to obtain the desired range. It also selects the proper position for making AC measurements when the AC clamp-on adapter is being used.

The "zero ohm" control on the right of the lower panel is used to adjust the meter to compensate for the aging of the meter's batteries, **Figure 12–42**. When the meter is ready to operate, the pointer must read zero. If the pointer is off zero, use a screwdriver to turn the screw clockwise or counterclockwise until the pointer is set exactly at zero.

The test leads may be plugged into any of eight jacks. In this unit the common (−) and the positive (+) jacks will be the only ones used, **Figure 12–43**. Only a few of the basic measurements are discussed here. Other measurements are described in detail in other units.

The following instructions are intended only to familiarize the technician with the meter and procedure. SAFETY PRECAUTION: Do not make any measurements without instructions and approval from an instructor or supervisor. Insert the black test lead into the common (−) jack. Insert the red test lead into the positive (+) jack.

Figure 12–44 is a DC circuit with 15 V from the battery power source. To check this voltage with the VOM, set the function switch to +DC. Set the range switch to 50 V,

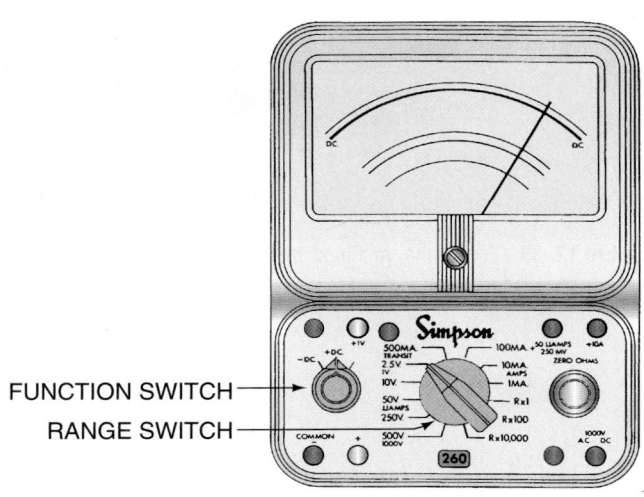

Figure 12–41 The function switch and range switch on a VOM. *Courtesy Simpson Electric Co.*

FUNCTION SWITCH
RANGE SWITCH

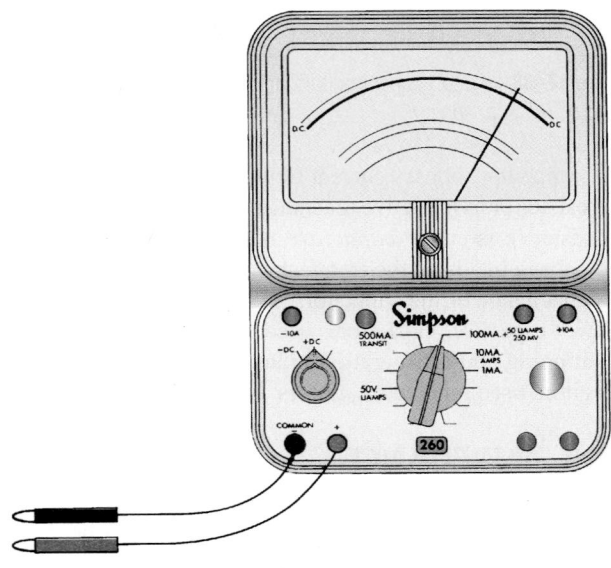

Figure 12–43 A VOM showing the − and + jacks. *Courtesy Simpson Electric Co.*

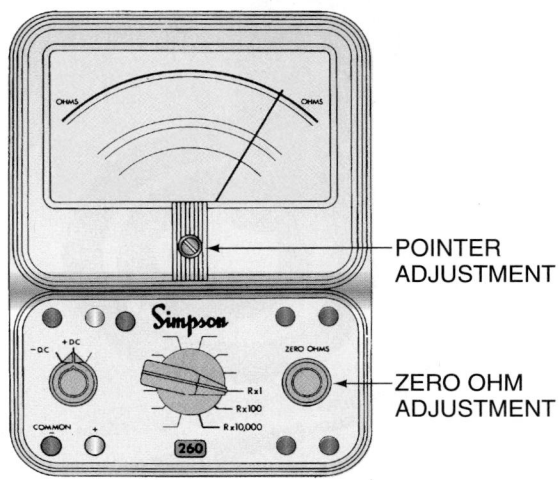

POINTER ADJUSTMENT

ZERO OHM ADJUSTMENT

Figure 12–42 Zero ohm adjustment and pointer adjustment. *Courtesy Simpson Electric Co.*

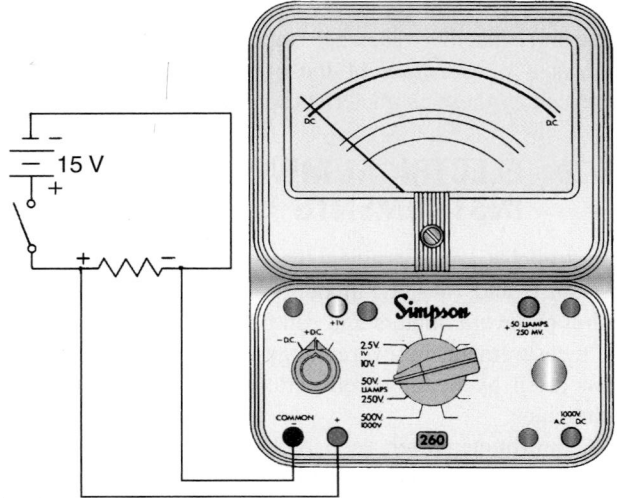

Figure 12–44 A VOM with the function switch set at + DC and range switch set at 50 V. *Courtesy Simpson Electric Co.*

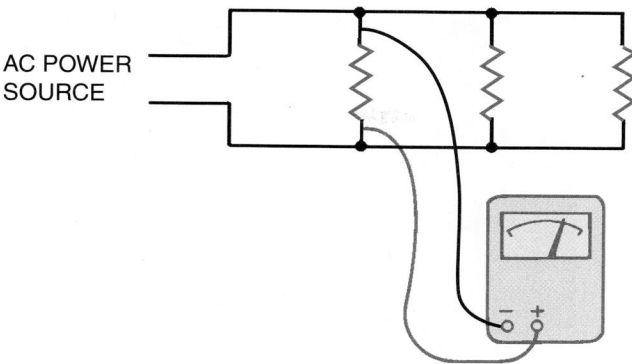

Figure 12–45 Connecting test leads across a load to check voltage.

Figure 12–44. If you are not sure about the magnitude of the voltage, always set the range switch to the highest setting. After measuring, you can set the switch to a lower range, if necessary, to obtain a more accurate reading. Be sure the switch in the circuit is open. Now connect the black test lead to the negative side of the circuit, and connect the red test lead to the positive side, as indicated in **Figure 12–44.** Note that the meter is connected across the load (in parallel). Close the switch and read the voltage from the DC scale.

To check the voltage in the AC circuit in **Figure 12–45,** follow the steps listed:

1. Turn off the power.
2. Set the function switch to AC.
3. Set the range switch to 500 V.
4. Plug the black test lead into the common (−) jack and the red test lead into the positive (+) jack.
5. Connect the test leads across the load as shown in **Figure 12–45.**
6. Turn on the power. Read the red scale marked AC. Use the 0 to 50 figures and multiply the reading by 10.
7. Turn the range switch to 250 V. Read the red scale marked AC and use the black figures immediately above the scale.

To determine the resistance of a load, disconnect the load from the circuit. Make sure all power is off while doing this. If the load is not entirely disconnected from the circuit, the voltage from the internal battery of the ohmmeter may damage solid-state components. These solid-state components may be part of an electronic control board or microprocessor.

1. Make the zero ohm adjustment in the following manner:
   A. Turn the range switch to the desired ohms range.
      Use $R \times 1$ for 0 to 200 Ω
      Use $R \times 100$ for 200 to 20,000 Ω
      Use $R \times 10,000$ for above 20,000 Ω
   B. Plug the black test lead into the common (−) jack and the red test lead into the positive (+) jack.
   C. Connect the test leads to each other.
   D. Rotate the zero ohm control until the pointer indicates zero ohms. (If the pointer cannot be adjusted to zero, replace the batteries.)

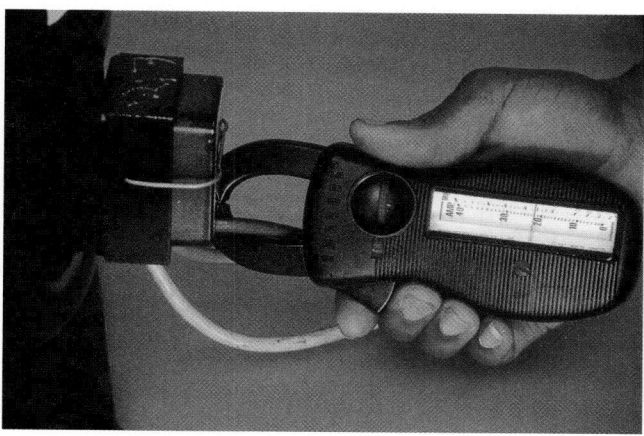

Figure 12–46 Measuring amperage by clamping the jaws of the meter around the conductor. *Photo by Bill Johnson*

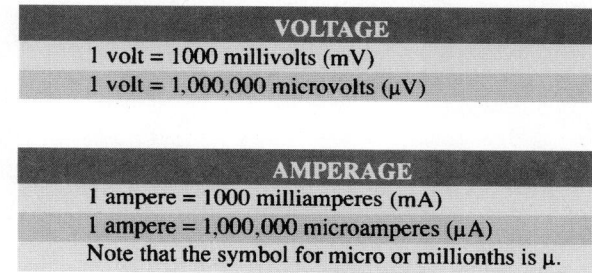

VOLTAGE
1 volt = 1000 millivolts (mV)
1 volt = 1,000,000 microvolts (µV)

AMPERAGE
1 ampere = 1000 milliamperes (mA)
1 ampere = 1,000,000 microamperes (µA)
Note that the symbol for micro or millionths is µ.

Figure 12–47 Units of voltage and amperage.

2. Disconnect the ends of the test leads and connect them to the load being tested.
3. Set the function switch at either −DC or +DC.
4. Observe the reading on the ohms scale at the top of the dial. (Note that the ohms scale reads from right to left.)
5. To determine the actual resistance, multiply the reading by the factor at the range switch position.

The ammeter has a clamping feature that can be placed around a single wire in a circuit, and the current flowing through the wire can be read as amperage from the meter, **Figure 12–46.** SAFETY PRECAUTION: Do not perform any of the preceding or following tests in this unit without approval from an instructor or supervisor. These instructions are simply a general orientation to meters. Be sure to read the operator's manual for the particular meter available to you.

It is often necessary to determine voltage or amperage readings to a fraction of a volt or ampere, **Figure 12–47.**

Many styles and types of meters are available for making electrical measurements. **Figure 12–48** shows some of these meters. Many modern meters come with digital readouts, **Figure 12–49.**

Electrical troubleshooting is taken one step at a time. **Figure 12–50** is a partial wiring diagram of an oil burner fan circuit showing the process. The fan motor does not operate because of an open motor winding. The following voltage

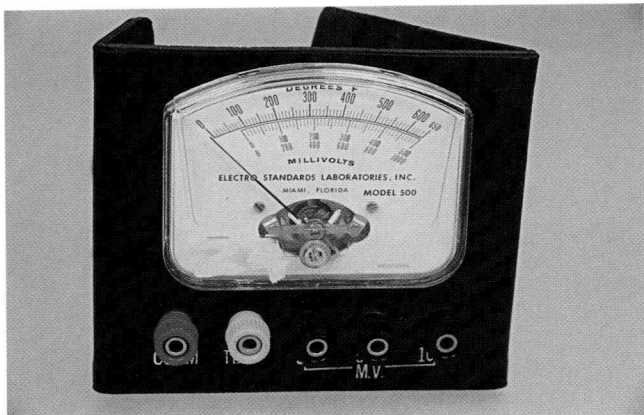

(A)

(B)

(C)

Figure 12–48 Meters used for electrical measurements. **(A)** DC millivoltmeter. **(B)** Multimeter (VOM). **(C)** Digital clamp-on ammeter. **(A)** *Photo by Bill Johnson.* **(B)** *Courtesy Wavetek.* **(C)** *Courtesy Amprobe*

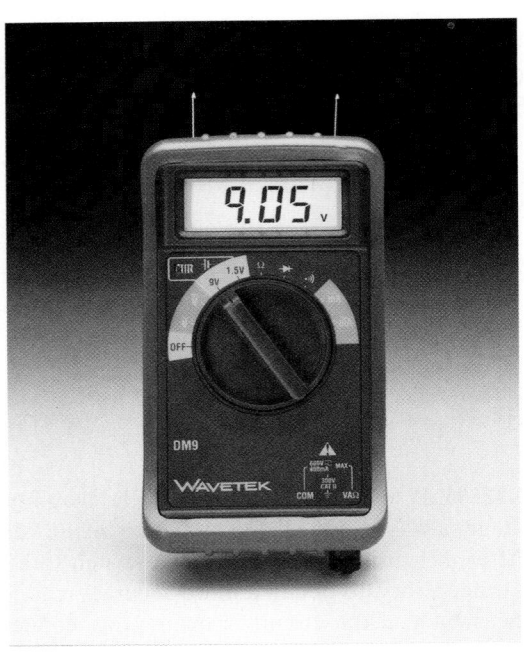

Figure 12–49 A typical VOM with digital readout/display. *Courtesy Wavetek*

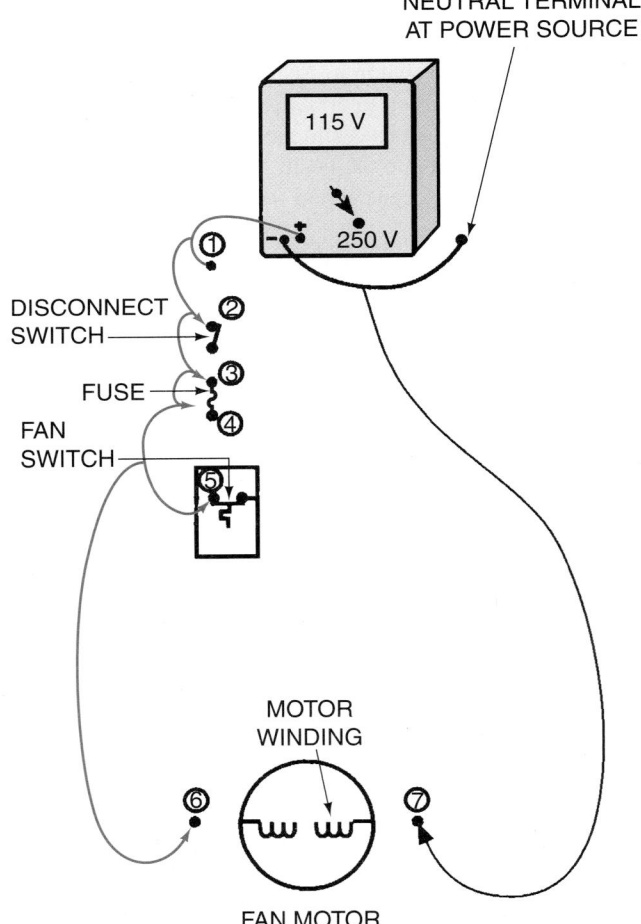

Figure 12–50 A partial diagram of an oil burner fan motor circuit.

checks could be made by the technician to determine where the failure is.

1. This is a 120-V circuit. The technician would set the VOM range selector switch to 250 V AC. The neutral (− or common) meter lead is connected to a neutral terminal at the power source.
2. The positive (+) meter probe is connected to a line-side power source terminal, **Figure 12–50 (1).** The meter should read 115 V, indicating that power is at the source.
3. Then the positive lead is connected to the line-side terminal of the disconnect switch **(2).** There is power.
4. The positive lead is connected to the line side of the fuse **(3).** There is power.
5. The positive lead is connected to the load side of the fuse **(4).** There is power.
6. The positive lead is connected to the load side of the fan switch **(5).** There is power.
7. The positive lead is connected to the line side of the motor terminal **(6).** There is power.
8. To ensure there is power through the conductor and terminal connections on the neutral or ground side, connect the neutral meter probe to the neutral side of the motor **(7).** Leave the positive probe on the line side of the motor. There should be a 115-V reading, indicating that current is flowing through the neutral conductor.

In all of the above checks, there is 115 V, but the motor does not run. It would be appropriate to conclude that the motor is defective.

The motor winding can be checked with an ohmmeter. The motor winding must have a measurable resistance for it to function properly. To check this resistance the technician may do the following, **Figure 12–51:**

1. Turn off the power source to the circuit.
2. Disconnect the terminals on the motor from the circuit.
3. Set the meter selector switch to ohms $R \times 1$.
4. Touch the meter probes together and adjust the meter to 0 ohms.
5. Touch one meter probe to one motor terminal and the other to the other terminal. The meter reads infinity (∞). This is the same reading you would get by holding the meter probes apart in the air. There is no circuit through the windings indicating that they are open.

Most clamp-on-type ammeters do not read accurately in the lower amperage ranges such as in the 1-ampere-or-below range. However, a standard clamp-on ammeter can be modified to produce an accurate reading. For instance, transformers are rated in volt-amperes (VA). A 40-VA transformer is often used in the control circuit of combination heating and cooling systems. These transformers produce 24 V and can carry a maximum of 1.66 A. This is determined as follows:

$$\text{Output in amperes} = \frac{\text{VA rating}}{\text{Voltage}}$$

$$I = 40/24$$

$$= 1.66 \text{ A}$$

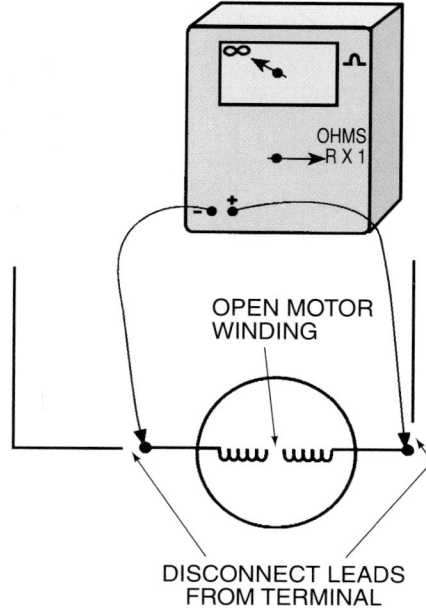

Figure 12–51   Checking an electric motor winding with an ohmmeter.

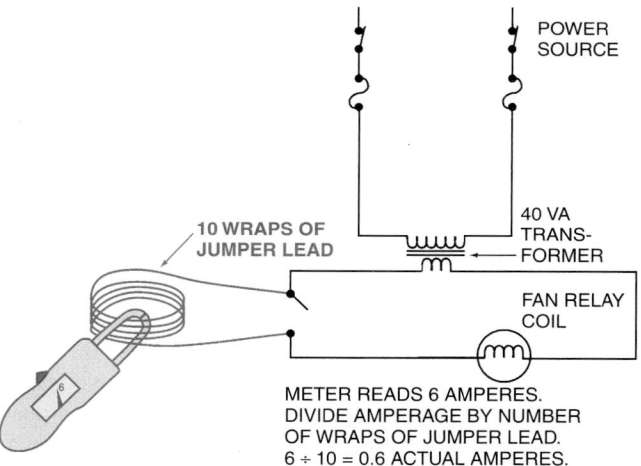

Figure 12–52   An illustration showing the use of a 10-wrap multiplier with an ammeter to obtain accurate, low-amperage readings.

Figure 12–52 illustrates how the clamp-on ammeter can be used with 10 wraps of wire to multiply the amperage reading by 10. To determine the actual amperage, divide the amperage indicated on the meter by 10.

12.21   SINE WAVES

As mentioned earlier, alternating current continually reverses direction. An oscilloscope is an instrument that measures the amount of voltage over a period of time and can display this voltage on a screen. This display is called a wave form. Many types of wave forms exist, but we will discuss

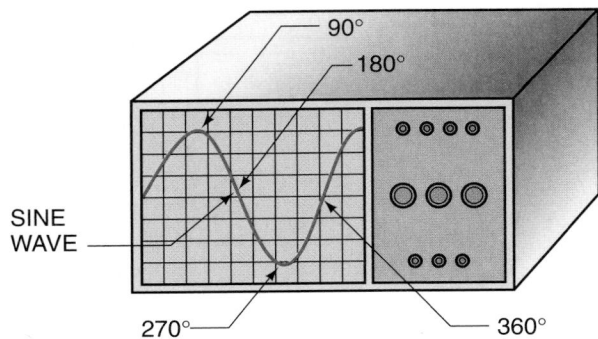

Figure 12–53 A sine wave displayed on an oscilloscope.

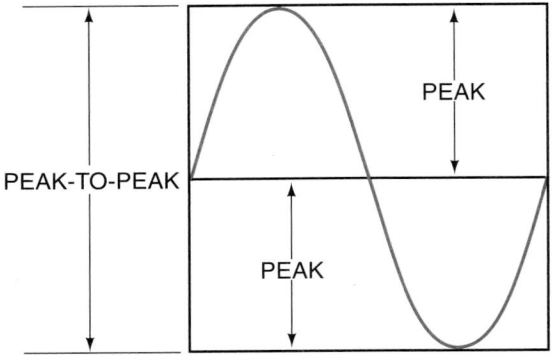

Figure 12–54 Peak and peak-to-peak AC voltage values.

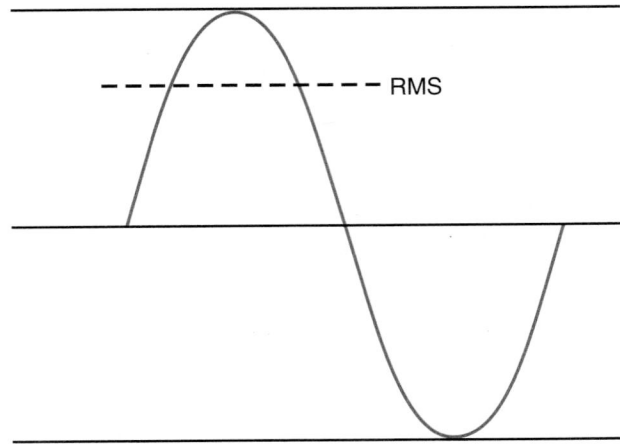

Figure 12–55 The root-mean-square (RMS) or effective voltage value.

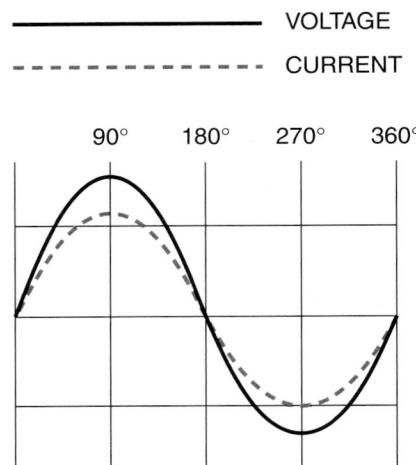

Figure 12–56 This sine wave represents both the voltage and current in phase with each other in a pure resistive circuit.

the one most refrigeration and air-conditioning technicians are involved with, the sine wave, **Figure 12–53.**

The sine wave displays the voltage of one cycle through 360°. In the United States and Canada, the standard voltage in most locations is produced with a frequency of 60 cycles per second. Frequency is measured in hertz. Therefore, the standard frequency in this country and Canada is 60 Hz.

Figure 12–53 is a sine wave as it would be displayed on an oscilloscope. At the 90° point the voltage reaches its peak (positive); at 180° it is back to 0; at 270° it reaches its negative peak; and at 360° it is back to 0. If the frequency is 60 Hz, this cycle would be repeated 60 times every second. The sine wave is a representation of a trigonometric function of an alternating current cycle.

Figure 12–54 shows the peak and peak-to-peak values. As the sine wave indicates, the voltage is at its peak value briefly during the cycle. Therefore, the peaks of the peak-to-peak values are not the effective voltage values.

The effective voltage is the RMS voltage, **Figure 12–55.** The RMS stands for root-mean-square value. This is the alternating current value measured by most voltmeters and ammeters. The RMS voltage is 0.707 × the peak voltage. If the peak voltage were 170 V, the effective voltage measured by a voltmeter would be 120 V (170 V × 0.707 = 120.19 V).

Sine waves can illustrate a cycle of an AC electrical circuit that contains only a pure resistance, for example, a circuit with electrical heaters. In a pure resistive circuit such as

this, the voltage and current will be in phase. This is illustrated with sine waves such as in **Figure 12–56.** Notice that the voltage and current reach their negative and positive peaks at the same time.

Sine waves can also illustrate a cycle of an AC electrical circuit that contains a fan relay coil. This coil will produce an inductive reactance. The sine wave will show the current lagging the voltage in this circuit, **Figure 12–57.**

Figure 12–58 shows a sine wave illustrating a cycle of an AC electrical circuit that has a capacitor producing a pure capacitive circuit. In this case the current leads the voltage.

12.22 WIRE SIZES

All conductors have some resistance. The resistance depends on the conductor material, the cross-sectional area of the conductor, the temperature of the conductor, and the length of the conductor. A conductor with low resistance carries a current more easily than a conductor with high resistance.

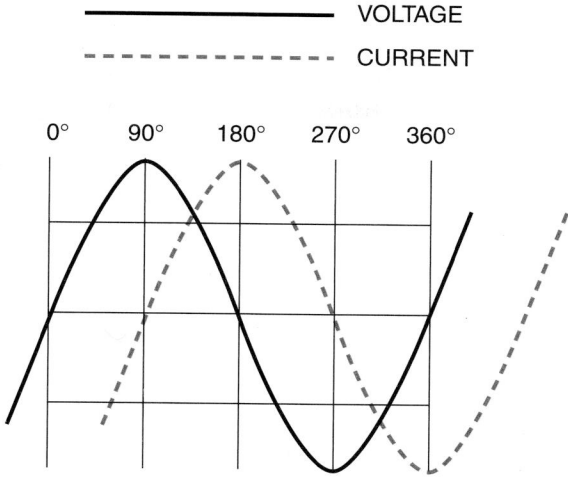

Figure 12–57 This sine wave represents the current lagging the voltage (out of phase) by 90 degrees in an inductive circuit.

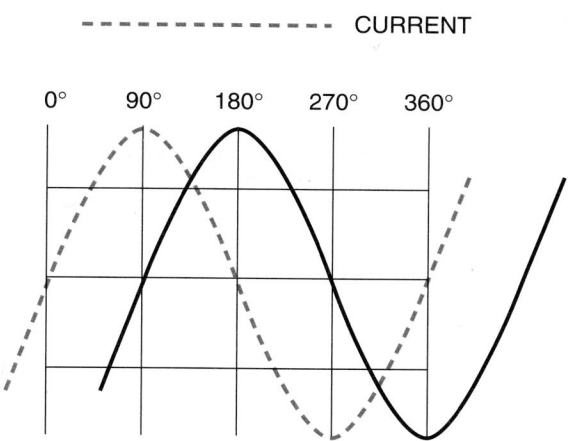

Figure 12–58 This sine wave represents a capacitive AC cycle with the voltage lagging the current by 90 degrees. The voltage and current are out of phase.

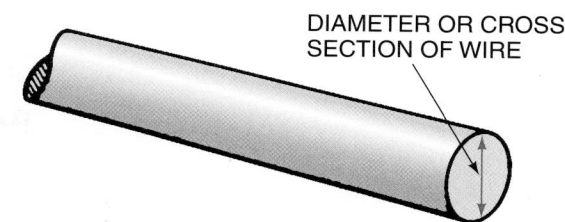

Figure 12–59 Cross section of a wire.

WIRE SIZE	
	60°C (140°F)
	Types TW, UF
AWG or kemil	
COPPER	
18	—
16	—
14*	20
12*	25
10*	30
8	40
6	55
4	70
3	85
2	95
1	110

*Small Conductors. Unless specifically permitted in (e) through (g), the overcurrent protection shall not exceed 15 amperes for No. 14, 20 amperes for No. 12, and 30 amperes for No. 10 copper.

Figure 12–60 This section of the *National Electrical Code®* shows an example of how a wire is sized for only one type of conductor. It is not the complete and official position, but is a representative example.

The proper wire (conductor) size must always be used. The size of a wire is determined by its diameter or cross section, **Figure 12–59.** A large-diameter wire has lower resistance and has more current-carrying capacity than a small-diameter wire. SAFETY PRECAUTION: If a wire is too small for the current passing through, it will overheat and possibly burn the insulation and could cause a fire. Standard copper wire sizes are identified by American Standard Wire Gauge (AWG) numbers and measured in circular mils. A circular mil is the area of a circle 1/1000 in. in diameter. Temperature is also considered because resistance increases as temperature increases. Increasing wire-size numbers indicate *smaller* wire diameters and greater resistance. For example, number 12 wire size is smaller than number 10 wire size and has less current-carrying capacity.

The technician should not establish the size of a conductor nor install the wire unless licensed to do so. The technician should, however, be able to recognize an undersized conductor and bring it to the attention of a qualified person. As mentioned previously, an undersized wire may cause voltage to drop, breakers or fuses to trip, and conductors to overheat.

The conductors are sized by their amperage-carrying capacity. This is called *ampacity.* **Figure 12–60** contains a small part of a chart from the *National Electrical Code® (NEC®)* and a partial footnote for one type of conductor. This chart as shown and the footnote should not be used when determining a wire size. It is shown here only to familiarize you with the way in which it is presented in the *NEC.* The footnote actually reduces the amount of amperes for the number 12

General Data   4TWX4042A1000A

OUTDOOR UNIT ①②	4TWX4042A1000A
SOUND RATING (DECIBELS) ②⑨	77/75
POWER CONNS.—V/PH/HZ ③	208/230/1/60
MIN. BRCH. CIR. AMPACITY	25
BR. CIR. } MAX. (AMPS)	40
PROT. RTG. } MIN. (AMPS)	40
COMPRESSOR	CLIMATUFF® -SCROLL
NO. USED - NO. SPEEDS	1 - 1
VOLTS/PH/HZ	208/230/1/60
R.L. AMPS ⑦ - L.R. AMPS	19.2 - 104
FACTORY INSTALLED	
START COMPONENTS ⑧	NO
INSULATION/SOUND BLANKET	YES
COMPRESSOR HEAT	YES
OUTDOOR FAN — TYPE	PROPELLER
DIA. (IN.) - NO. USED	27.6 - 1
TYPE DRIVE - NO. SPEEDS	DIRECT - 2
CFM @ 0.0 IN. W.G. ④	4200
NO. MOTORS - HP	1 - 1/6
MOTOR SPEED R.P.M.	825
VOLTS/PH/HZ	200/230
F.L. AMPS	1.4
OUTDOOR COIL — TYPE	SPINE FIN™
ROWS - F.P.I.	1 - 24
FACE AREA (SQ. FT.)	27.81
TUBE SIZE (IN.)	5/16
REFRIGERANT CONTROL	EXPANSION VALVE
REFRIGERANT	
LBS.— R-410A (O.D. UNIT) ⑤	7/04 -LB/OZ
FACTORY SUPPLIED	YES
LINE SIZE - IN. O.D. GAS ⑥	3/4
LINE SIZE - IN. O.D. LIQ. ⑥	3/8
FCCV	
RESTRICTOR ORIFICE SIZE	0.071
DIMENSIONS	H X W X D
OUTDOOR UNIT - CRATED (IN.)	53.4 X 35.1 X 38.7
UNCRATED	SEE OUTLINE DWG.
WEIGHT	
SHIPPING (LBS.)	315
NET (LBS.)	267

OUTDOOR UNIT WITH HEAT PUMP COILS

① CERTIFIED IN ACCORDANCE WITH THE AIR-SOURCE UNITARY HEAT PUMP EQUIPMENT CERTIFICATION PROGRAM WHICH IS BASED ON A.R.I. STANDARD 210/240.

② RATED IN ACCORDANCE WITH A.R.I. STANDARD 270.

③ CALCULATED IN ACCORDANCE WITH NATIONAL ELECTRIC CODE. ONLY USE HACR CIRCUIT BREAKERS OR FUSES.

④ STANDARD AIR - DRY COIL - OUTDOOR

⑤ THIS VALUE APPROXIMATE. FOR MORE PRECISE VALUE SEE UNIT NAMEPLATE AND SERVICE INSTRUCTION.

⑥ MAX. LINEAR LENGTH: 80 FT WITH RECIPROCATING COMPRESSOR - 60 FT WITH SCROLL. MAX. LIFT - SUCTION 60 FT; MAX LIFT - LIQUID 60 FT. FOR GREATER LENGTH REFER TO REFRIGERANT PIPING SOFTWARE PUB. NO. 32-3312-01.

⑦ THE VALUE SHOWN FOR COMPRESSOR RLA ON THE UNIT NAMEPLATE AND ON THIS SPECIFICATION SHEET IS USED TO COMPUTE MINIMUM BRANCH CIRCUIT AMPACITY AND MAXIMUM FUSE SIZE. THE VALUE SHOWN IS THE BRANCH CIRCUIT SELECTION CURRENT.

⑧ NO MEANS NO START COMPONENTS
YES MEANS QUICK START KIT COMPONENTS
PTC MEANS POSITIVE TEMPERATURE COEFFICIENT STARTER.

⑨ RATED IN ACCORDANCE WITH ARI STANDARD 270/SECTION 5.3.6.

SPLIT SYSTEM

Figure 12–61 This specification sheet for one unit shows the electrical data that are used to size the wire to the unit. *Courtesy Trane*

and number 14 wire listed in the *NEC* table in **Figure 12–60.** The reason is that number 12 and number 14 wire are used in residential houses where circuits are often overloaded unintentionally by the homeowner. The footnote exception simply adds more protection.

The following is an example of a procedure that might be used for calculating the wire size for the outdoor unit of a heat pump, **Figure 12–61.**

This outdoor unit is a 3½-ton unit (the 42A in the model number means 42,000 Btu/hr or 3½ tons). The electrical data show that the unit compressor uses 19.2 FLA (full-load amperage) and the fan motor uses 1.4 FLA for a total of 20.6 FLA. The specifications round this up to an ampacity of 25. In other words, the conductor must be sized for 25 amperes. According to the *NEC*, the wire size would be 10. The specifications for the unit indicate that a maximum fuse or breaker size would be 40 A. This difference allows for the compressor's locked-rotor amperage (LRA) draw at

start-up. If the circuit were wired and protected to accommodate only 10 A of current, the circuit breaker would trip every time the compressor started. Some manufacturers give the ampacity and the recommended wire size in the directions or printed on the unit nameplate.

When you find a unit that has low voltage while operating, you should first check the voltage at the entrance panel to the building. If the voltage is correct there and low at the unit, there is either a loose connection, there is undersized wire, or the wire run is too long.

12.23 CIRCUIT PROTECTION DEVICES

SAFETY PRECAUTION: Electrical circuits *must* be protected from current overloads. If too much current flows through the circuit, the wires and components will overheat, resulting in damage and possible fire. Circuits are normally protected with fuses or circuit breakers.

Fuses

A *fuse* is a simple device used to protect circuits from overloading and overheating. Most fuses contain a strip of metal that has a higher resistance than the conductors in the circuit. This strip also has a relatively low melting point. Because of its higher resistance, it will heat up faster than the conductor. When the current exceeds the rating on the fuse, the strip melts and opens the circuit. Fuses are one-time devices and must be replaced when the strip or element melts. The cause for the circuit overload must be identified and corrected before replacing the fuse.

PLUG FUSES. Plug fuses have either an Edison base or a type S base, **Figure 12–62(A)**. Edison-base fuses are used in older installations and can be used for replacement only. Type S fuses can be used only in a type S fuse holder specifically designed for the fuse; otherwise an adapter must be used, **Figure 12–62(B)**. Each adapter is designed for a specific amperage rating, and these fuses cannot be interchanged. The amperage rating determines the size of the adapter. Plug fuses are rated up to 125 V and 30 A.

DUAL-ELEMENT PLUG FUSES. Many circuits have electric motors as the load or part of the load. Motors draw more current when starting and can cause a plain (single-element) fuse to burn out or open the circuit. Dual-element fuses are frequently used in this situation, **Figure 12–63**. One element

in the fuse will melt when there is a large overload, such as a short circuit. The other element will melt and open the circuit when there is a smaller current overload lasting more than a few seconds. This allows for the larger starting current of an electric motor.

CARTRIDGE FUSES. For 230-V to 600-V service up to 60 A, the ferrule cartridge fuse is used, **Figure 12–64(A)**. From 60 A to 600 A, knife-blade cartridge fuses can be used, **Figure 12–64(B)**. A cartridge fuse is sized according to its amperage rating to prevent a fuse with an inadequate rating from being used. Many cartridge fuses have an arc-quenching material around the element to prevent damage from arcing in severe short-circuit situations, **Figure 12–65**.

Circuit Breakers

A circuit breaker can function as a switch as well as a means for opening a circuit when a current overload occurs. Most modern installations in houses and many commercial and industrial installations use circuit breakers rather than fuses for circuit protection.

Circuit breakers use two methods to protect the circuit. One is a bimetal strip that heats up with a current overload and trips the breaker, opening the circuit. The other is a magnetic coil that causes the breaker to trip and open the circuit when there is a short circuit or other excessive current overload in a short time, **Figure 12–66**.

Figure 12–62 **(A)** A type S base plug fuse. **(B)** A type S fuse adapter. *Reprinted with permission by Bussman Division, McGraw-Edison Company*

Figure 12–63 A dual-element plug fuse. *Courtesy Cooper Bussman Division*

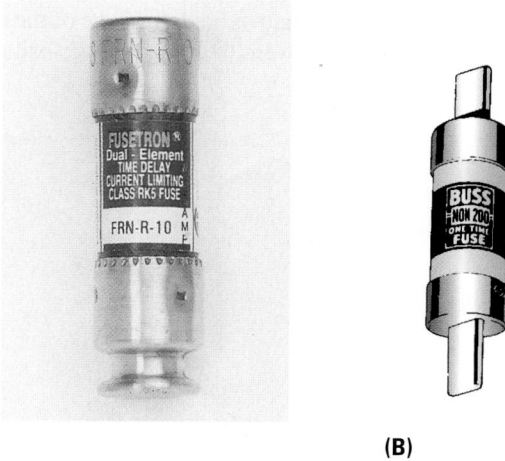

(A)   (B)

Figure 12–64 **(A)** A ferrule-type cartridge fuse. **(B)** A knife-blade cartridge fuse. *Courtesy Cooper Bussman Division*

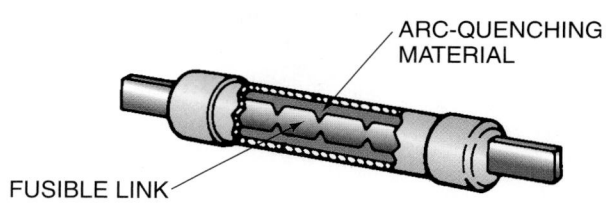

Figure 12–65 A knife-blade cartridge fuse with arc-quenching material.

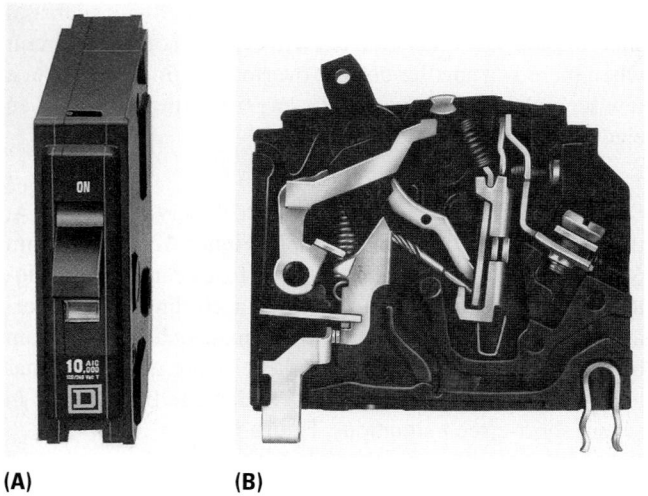

(A)   **(B)**

Figure 12–66 **(A)** A circuit breaker. **(B)** Cutaway view of the breaker. *Courtesy Square D Company*

Ground Fault Circuit Interrupters (GFCI)

SAFETY PRECAUTION: Ground fault circuit interrupters (GFCI) help protect individuals against shock, in addition to providing current overload protection. The GFCI, **Figure 12–67**, detects even a very small current leak to a ground. Under certain conditions this leak may cause an electrical shock. This small leak, which may not be detected by a conventional circuit breaker, will cause the GFCI to open the circuit.

Circuit protection is essential for preventing the conductors in the circuit from being overloaded. If one of the circuit power-consuming devices were to cause an overload due to

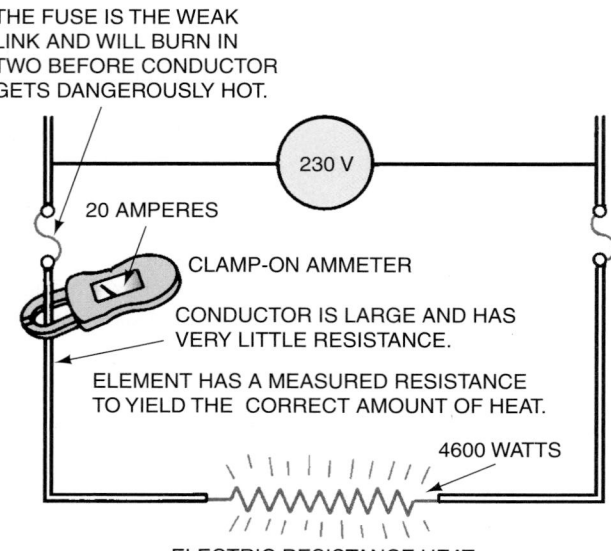

THE FUSE IS THE WEAK LINK AND WILL BURN IN TWO BEFORE CONDUCTOR GETS DANGEROUSLY HOT.

230 V

20 AMPERES

CLAMP-ON AMMETER

CONDUCTOR IS LARGE AND HAS VERY LITTLE RESISTANCE.

ELEMENT HAS A MEASURED RESISTANCE TO YIELD THE CORRECT AMOUNT OF HEAT.

4600 WATTS

ELECTRIC RESISTANCE HEAT

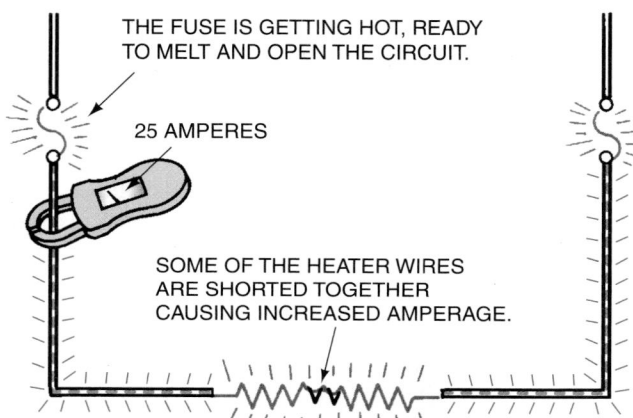

THE FUSE IS GETTING HOT, READY TO MELT AND OPEN THE CIRCUIT.

25 AMPERES

SOME OF THE HEATER WIRES ARE SHORTED TOGETHER CAUSING INCREASED AMPERAGE.

Figure 12–68 Fuses protect the circuit.

a short circuit within its coil, the circuit protector would stop the current flow before the conductor became overloaded and hot. Remember, a circuit consists of a power supply, the conductor, and the power-consuming device. The conductor must be sized large enough that it does not operate beyond its rated temperature, typically 140°F (60°C) while in an ambient of 86°F (30°C). For example, a circuit may be designed to carry a load of 20 A. As long as the circuit is carrying up to its amperage, overheating is not a potential hazard. If the amperage in the circuit is gradually increased, the conductor will begin to become hot, **Figure 12–68**. Proper understanding of circuit protection is a lengthy process. More details can be obtained from the *NEC* and from further study of electricity.

12.24 SEMICONDUCTORS

The development of what are commonly called semiconductors or solid-state components has caused major changes in the design of electrical devices and controls.

Figure 12–67   A ground fault circuit interrupter. *Courtesy Square D Company*

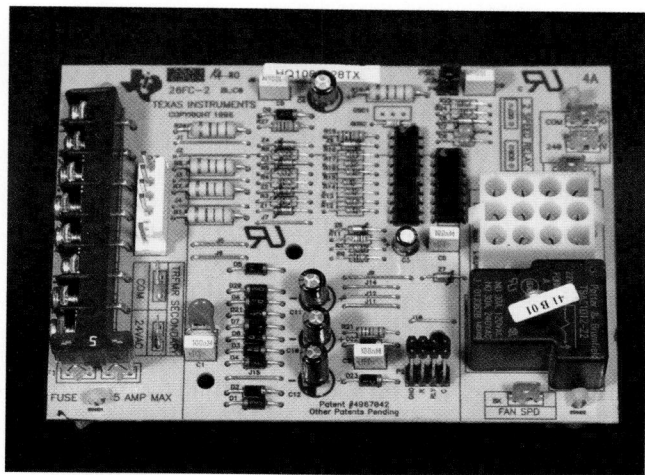

Figure 12–69 A circuit board with semiconductors. *Photo by Bill Johnson*

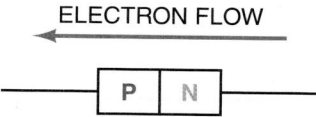

Figure 12–70 A pictorial diagram of a diode.

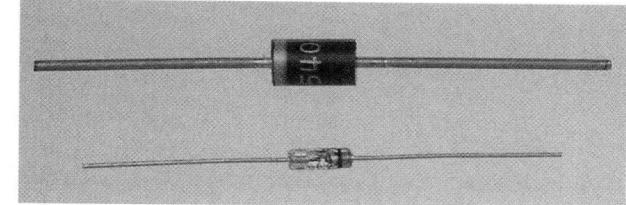

Figure 12–71 Typical diodes. *Photo by Bill Johnson*

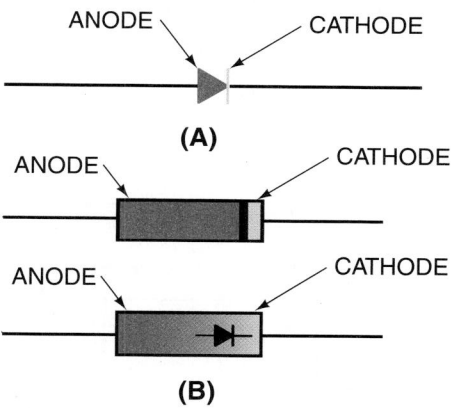

Figure 12–72 **(A)** The schematic symbol for a diode. **(B)** Identifying markings on the diode.

Semiconductors are generally small and lightweight and can be mounted in circuit boards, **Figure 12–69.** In this unit we describe some of the individual solid-state devices and some of their uses. Refrigeration and air-conditioning technicians do not normally replace solid-state components on circuit boards. They should have some knowledge of these components, however, and should be able to determine when one or more of the circuits in which they are used are defective. In most cases, when a component is defective the entire board will need to be replaced. Often these circuit boards can be returned to the manufacturer or sent to a company that specializes in repairing or rebuilding them.

Semiconductors are usually made of silicon or germanium. Semiconductors in their pure form, as their name implies, do not conduct electricity very well. However, for semiconductors to be of value they must conduct electricity in some controlled manner. To accomplish this an additional substance, often called an impurity, is added to the crystal-like structures of the silicon or germanium. This is called doping. One type of impurity produces a hole in the material where an electron should be. Because the hole replaces an electron (which has a negative charge), it results in the material having fewer electrons or a net result of a positive charge. This is called a P-type material. If a material of a different type is added to the semiconductor, an excess of electrons is produced; the material has a negative charge and is called an N-type material.

When a voltage is applied to a P-type material, electrons fill these holes and move from one hole to the next still moving from negative to positive. However, this makes it appear that the holes are moving in the opposite direction (from positive to negative) as the electrons move from hole to hole.

N-type material has an excess of electrons that move from negative to positive when a voltage is applied.

Solid-state components are made from a combination of N-type and P-type substances. The manner in which the materials are joined together, the thickness of the materials, and other factors determine the type of solid-state component and its electronic characteristics.

DIODES. Diodes are simple solid-state devices. They consist of P-type and N-type material connected together. When this combination of P- and N-type material is connected to a power source one way, it will allow current to flow and is said to have forward bias. When reversed it is said to have reverse bias, and no current will flow. **Figure 12–70** is a drawing of a simple diode. **Figure 12–71** is a photo of two types of diodes. One of the connections on the diode is called the cathode and the other the anode, **Figure 12–72.** If the diode is to be connected to a battery to have forward bias (current flow), the negative terminal on the battery should be connected to the cathode, **Figure 12–73.** Connecting the negative terminal of the battery to the anode will produce reverse bias (no current flow), **Figure 12–74.**

CHECKING A DIODE. A diode may be tested by connecting an ohmmeter across it. The diode must be removed from the circuit. The negative probe should be touched to the cathode

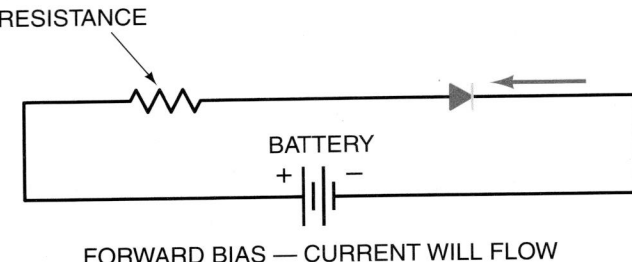

FORWARD BIAS — CURRENT WILL FLOW

Figure 12–73 A simple diagram with a diode indicating forward bias.

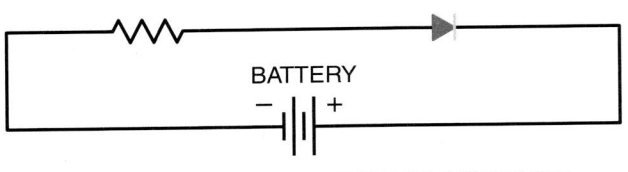

REVERSE BIAS — CURRENT WILL NOT FLOW

Figure 12–74 A circuit with a diode indicating reverse bias.

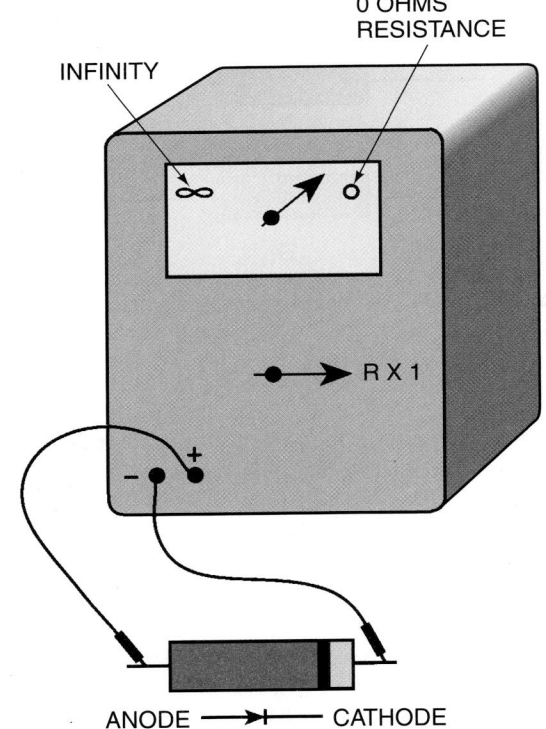

ANODE ———▷|——— CATHODE

Figure 12–75 Checking a diode.

and the positive to the anode. With the selector switch on $R \times 1$, the meter should show a small resistance, indicating that there is continuity, **Figure 12–75**. Reverse the leads. The meter should show infinity, indicating there is no continuity. A diode should show continuity in one direction and not in the other. If it shows continuity in both directions, it is defective. If it does not show continuity in either direction, it is defective.

RECTIFIER. A diode can be used as a solid-state rectifier, changing AC to DC. The term diode is normally used when rated for less than 1 A. A similar component rated above 1 A is called a rectifier. A rectifier allows current to flow in one direction. Remember that AC flows in first one direction and then reverses, **Figure 12–76(A)**. The rectifier allows the AC to flow in one direction but blocks it from reversing, **Figure 12–76(B)**. Therefore, the output of a rectifier circuit is in one direction or direct current. **Figure 12–77** illustrates a rectifier circuit. This is called a half wave rectifier because it allows only that part of the AC moving in one direction to pass through. **Figure 12–77(A)** shows the AC before it is rectified and **Figure 12–77(B)** shows the AC after it is rectified. Full wave rectification can be achieved by using a more complicated circuit such as in **Figure 12–78**. This is a full wave bridge rectifier commonly used in this industry.

SILICON-CONTROLLED RECTIFIER. Silicon-controlled rectifiers (SCRs) consist of four semiconductor materials bonded together. These form a PNPN junction, **Figure 12–79(A)**;

AC VOLTAGE WAVEFORM

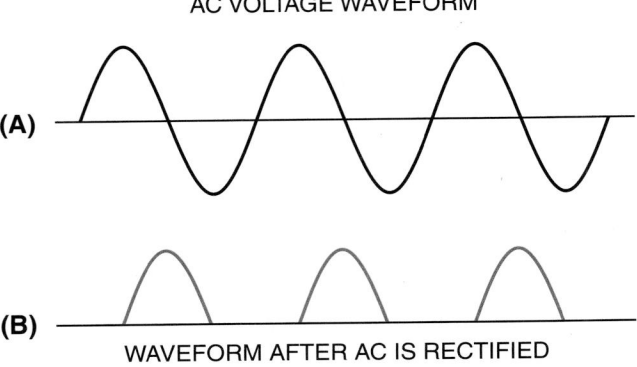

WAVEFORM AFTER AC IS RECTIFIED

Figure 12–76 **(A)** A full wave AC waveform. **(B)** A half wave DC waveform.

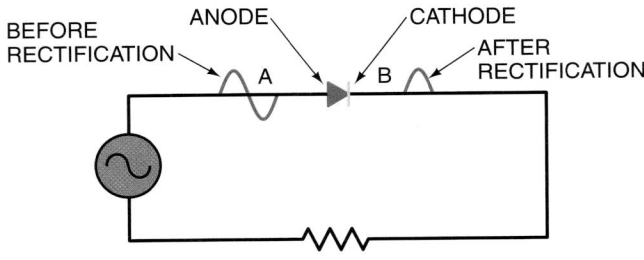

Figure 12–77 A diode rectifier circuit.

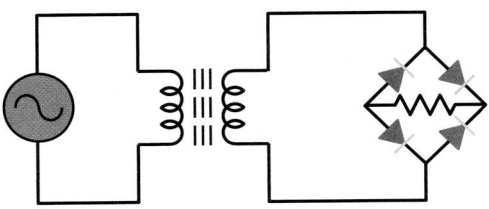

Figure 12–78 A full wave bridge rectifier.

SILICON-CONTROLLED RECTIFIER
(SCR)

PICTORIAL

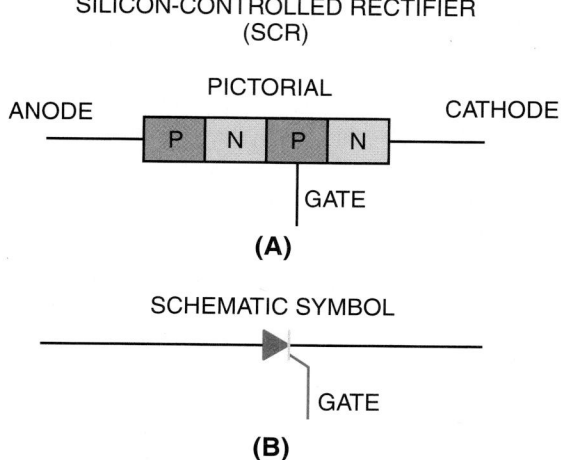

(A)

SCHEMATIC SYMBOL

(B)

Figure 12–79 Pictorial and schematic drawings of a silicon-controlled rectifier.

Figure 12–80 A typical silicon-controlled rectifier. *Photo by Bill Johnson*

Figure 12–79(B) illustrates the schematic symbol. Notice that the schematic is similar to the diode except for the gate. The SCR is used to control devices that may use large amounts of power. The gate is the control for the SCR. These devices may be used to control the speed of motors or to control the brightness of lights. **Figure 12–80** shows a photo of a typical SCR.

Checking the Silicon-Controlled Rectifier

The SCR can also be checked with an ohmmeter. Ensure that the SCR is removed from a circuit. Set the selector switch on the meter to $R \times 1$. Zero the meter. Fasten the negative lead from the meter to the cathode terminal of the SCR and the positive lead to the anode, **Figure 12–81**. If the SCR is good, the needle should not move. This is because the SCR has not fired to complete the circuit. Use a jumper to connect the gate terminal to the anode. The meter needle should now show continuity. If it does not, you may not have the cathode and anode properly identified. Reverse the leads and change the jumper to the new suspected anode. If it fires, you had the anode and cathode reversed. When the jumper is removed, the SCR will continue to conduct if the meter has enough capacity to keep the gate closed. If the meter were to show current flow without firing the gate, the SCR is defective.

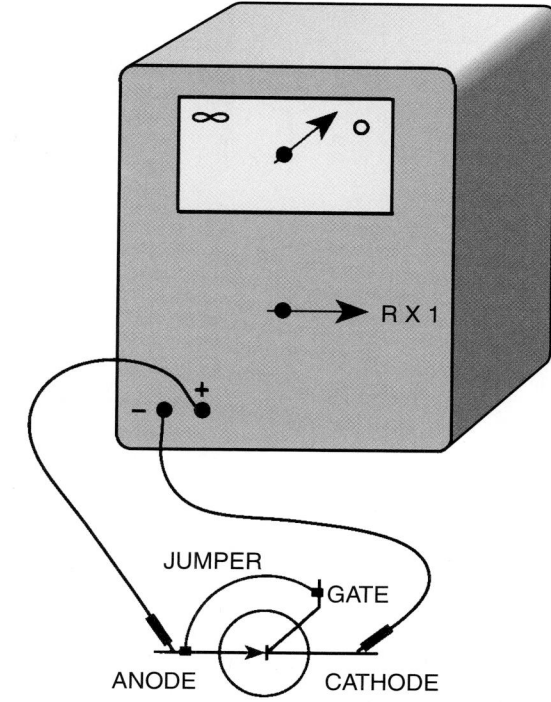

Figure 12–81 Checking a silicon-controlled rectifier.

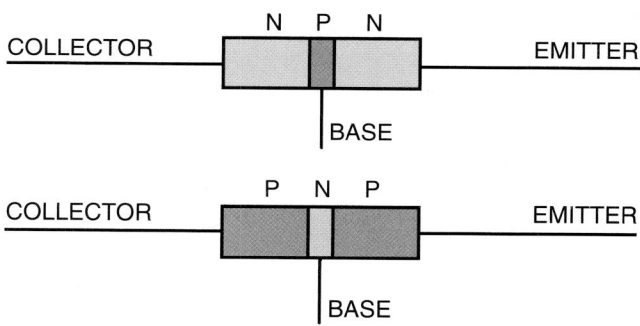

Figure 12–82 Pictorial drawings of NPN and PNP transistors.

If the gate will not fire after the jumper and leads are attached correctly, the SCR is defective.

TRANSISTORS. Transistors are also made of N-type and P-type semiconductor materials. Three pieces of these materials are sandwiched together. Transistors are either NPN or PNP types. **Figure 12–82** shows diagrams, and **Figure 12–83** shows the schematic symbols, for the two types. As the symbols show, each transistor has a base, a collector, and an emitter. In the NPN type the collector and the base are connected to the positive; the emitter is connected to the negative. The PNP transistor has a negative base and collector connection and a positive emitter. The base must be connected to the same polarity as the collector to provide forward bias. **Figure 12–84** is a photo of typical transistors.

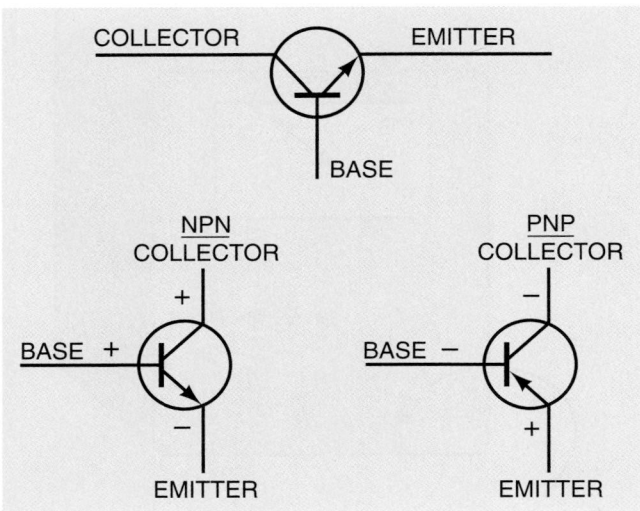

Figure 12–83   Schematic drawings of NPN and PNP transistors.

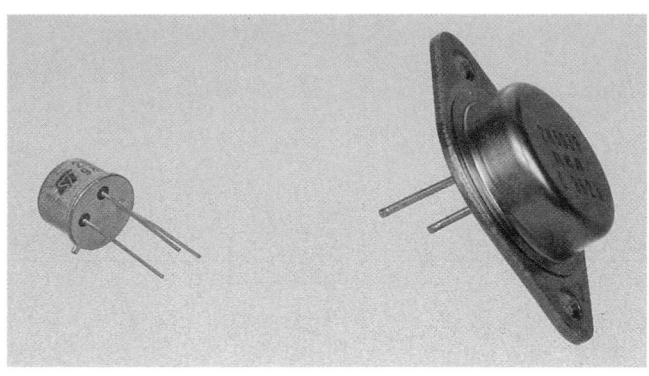

Figure 12–84   Typical transistors. *Photo by Bill Johnson*

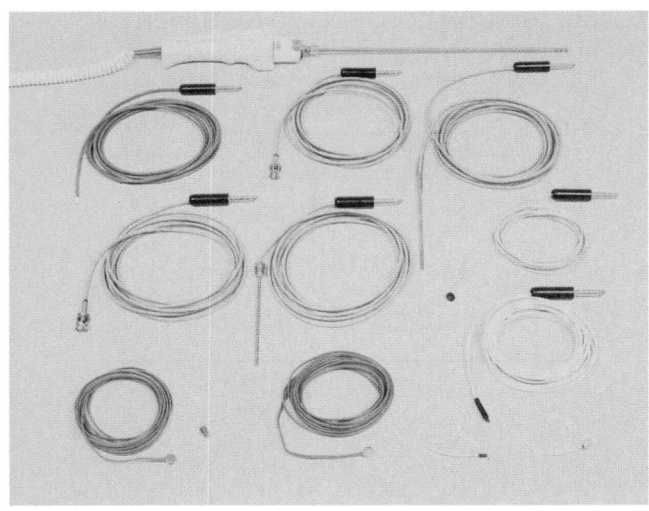

Figure 12–85   Typical thermistors. *Courtesy Omega Engineering*

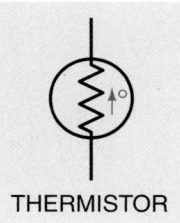

THERMISTOR

Figure 12–86   The schematic symbol for a thermistor.

The transistor may be used as a switch or a device to amplify or increase an electrical signal. One application of a transistor used in an air-conditioning control circuit would be to amplify a small signal to provide enough current to operate a switch or relay.

Current will flow through the base emitter and through the collector emitter. The base-emitter current is the control, and the collector-emitter current produces the action. A very small current passing through the base emitter may allow a much larger current to pass through the collector-emitter junction. A small increase in the base-emitter junction can allow a much larger increase in the current flow through the collector emitter.

THERMISTORS.   A thermistor is a type of resistor that is sensitive to temperature, **Figure 12–85.** The resistance of a thermistor changes with a change in temperature. Two types of thermistors exist. A positive temperature coefficient (PTC) thermistor causes the resistance of the thermistor to increase when the temperature increases. A negative temperature coefficient (NTC) thermistor causes the resistance to decrease

with an increase in temperature. **Figure 12–86** illustrates the schematic symbol for a thermistor.

An application of a thermistor is to provide motor overload protection. The thermistor is imbedded in the windings of a motor. When the winding temperature exceeds a predetermined amount, the thermistor changes in resistance. This change in resistance is detected by an electronic circuit that causes the motor circuit to open.

Another application is to provide start assistance in a PSC (permanent split-capacitor) electric motor. This thermistor is known as a positive temperature coefficient device. It allows full voltage to reach the start windings during start-up of the motor. The thermistor heats during start-up and creates resistance, turning off power to the start winding at the appropriate time. This does not give the motor the starting torque that a start capacitor does, but it is advantageous in some applications because of its simple construction and lack of moving parts.

DIACS.   The diac is a two-directional electronic device. It can operate in an AC circuit and its output is AC. It is a voltage-sensitive switch that operates in both halves of the AC waveform. When a voltage is applied, it will not conduct (or will remain off) until the voltage reaches a predetermined level. Let us assume this predetermined level to be 24 V. When the voltage in the circuit reaches 24 V, the diac will begin to conduct or fire. Once it fires it will continue to conduct even at a lower voltage. Diacs are designed to have a higher cut-in voltage and lower cut-out voltage. If the cut-in voltage is 24 V, let us assume that the cut-out voltage is 12 V. In this case the diac will continue to operate until the voltage drops below 12 V—at which time it will cut off.

Figure 12–87 shows two schematic symbols for a diac. **Figure 12–88** illustrates a diac in a simple AC circuit. Diacs are often used as switching or control devices for triacs.

TRIACS. A triac is a switching device that will conduct on both halves of the AC waveform. The output of the triac is AC. **Figure 12–89** shows the schematic symbol for a triac. Notice that it is similar to the diac but has a gate lead. A pulse supplied to the gate lead can cause the triac to fire or to conduct. Triacs were developed to provide better AC switching. As was mentioned previously, diacs often provide the pulse to the gate of the triac. A common use of a triac is as a motor speed control for AC motors.

HEAT SINKS. Some solid-state devices may appear a little different from others of the same type. This is because of the differing voltages and current they are rated to carry and the purpose for which they are designed and used. Some produce much more heat than others and will operate only at the rating specified if kept within a certain temperature range.

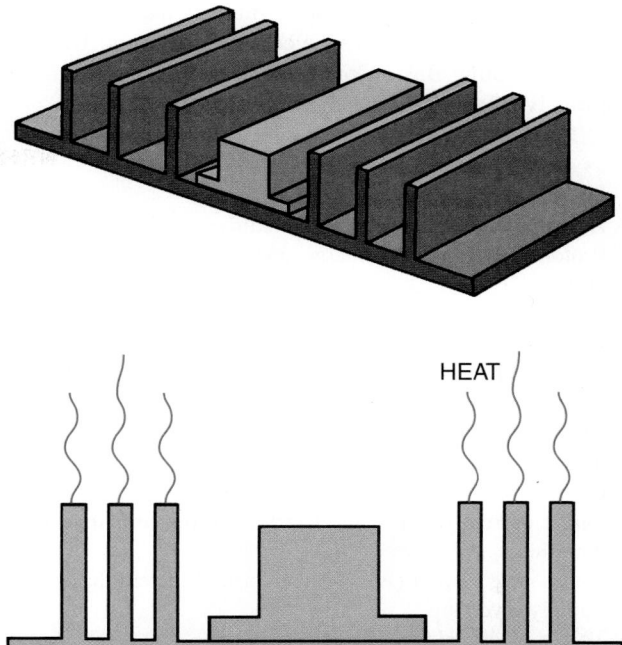

Figure 12–90 One type of heat sink.

Figure 12–87 Schematic symbols for a diac.

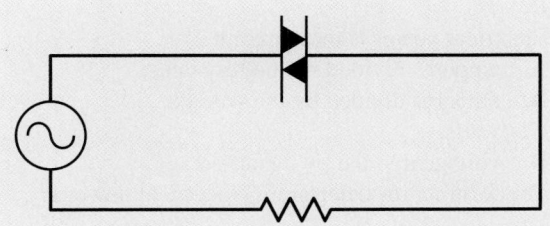

Figure 12–88 A simple diac circuit.

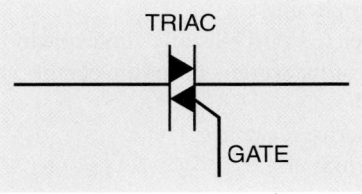

Figure 12–89 The schematic symbol for a triac.

Heat that could change the operation of or destroy the device must be dissipated. This is done by adhering the solid-state component to an object called a heat sink that has a much greater surface area, **Figure 12–90**. Heat will travel from the device to this object with the large surface area allowing the excess heat to be dissipated into the surrounding air.

Practical Semiconductor Applications for Technicians

This has been a short introduction into semiconductors. Brief descriptions in previous paragraphs outlined procedures for checking certain semiconductors. These tests were described to show how semiconductors may be checked in the field. However, it is seldom practical to remove these devices from a printed circuit board and perform the test. Other instruments may be used in the shop or laboratory to check these same components while they are still in the circuit. Refrigeration and air-conditioning technicians will be more involved with checking the input and output of circuit board circuits than with checking individual electronic components. The preceding information has been provided so that you will have some idea of the purpose of these control devices. You are encouraged to pursue the study of these components because more and more solid-state electronics will be used in the future to control the systems you will be working with. Each manufacturer will have a control sequence procedure that you must use to successfully check their controls. Make it a practice to attend seminars and factory schools in your area to increase your knowledge with regard to all segments of this ever-changing field.

Appendix B is an Electrical Symbols Chart that will be helpful as you learn to read electrical drawings. Unit 15, "Troubleshooting Basic Controls," contains many exercises that involve reading electrical schematic drawings.

SUMMARY

- Atoms are made up of protons, neutrons, and electrons.
- Protons have a positive charge, and electrons have a negative charge.
- Electrons travel in orbits around the protons and neutrons.
- When a surplus of electrons is in an atom, it has a negative charge. When a deficiency of electrons exists, the atom has a positive charge.
- Electricity can be produced by using magnetism. A conductor cutting through magnetic lines of force produces electricity.
- Direct current (DC) is an electrical current moving in one direction.
- Alternating current (AC) is an electrical current that is continually reversing.
- Volt = electrical force or pressure.
- Ampere = quantity of electron flow.
- Ohm = resistance to electron flow.
- Voltage (E) = Amperage (I) × Resistance (R). This is Ohm's law.
- In series circuits the voltage is divided across the resistances, the total current flows through each resistance, and the resistances are added together to obtain the total resistance.
- In parallel circuits the total voltage is applied across each resistance, the current is divided between the resistances, and the total resistance is less than that of the smallest resistance.
- Electrical power is measured in watts, $P = E \times I$.
- Inductive reactance is the resistance caused by the magnetic field surrounding a coil in an AC circuit.
- A step-up transformer increases the voltage and decreases the current. A step-down transformer decreases the voltage and increases the current.
- A capacitor in an AC circuit will continually charge and discharge as the current in the circuit reverses.
- A capacitor has capacitance, which is the amount of charge that can be stored.
- Impedance is the opposition to current flow in an AC circuit from the combination of resistance, inductive reactance, and capacitive reactance.
- A multimeter often used is the VOM (volt-ohm-milliammeter).
- A sine wave displays the voltage of one AC cycle through 360°.
- SAFETY PRECAUTION: Properly sized conductors must be used. Larger wire sizes will carry more current than smaller wire sizes without overheating.
- SAFETY PRECAUTION: Fuses and circuit breakers are used to interrupt the current flow in a circuit when the current is excessive.
- Semiconductors in their pure state do not conduct electricity well, but when they are doped with an impurity, they form an N-type or P-type material that will conduct in one direction.
- Diodes, rectifiers, transistors, thermistors, diacs, and triacs are examples of semiconductors.

REVIEW QUESTIONS

1. The _____ is that part of an atom that moves from one atom to another.
   A. electron
   B. proton
   C. neutron
2. When this part of an atom moves to another atom, the losing atom will have a _____ charge.
   A. negative
   B. positive
   C. neutral
3. State the differences between AC and DC.
4. Describe how a meter would be connected in a circuit to measure the voltage at a light bulb.
5. Describe how an amperage reading would be taken using a clamp-on or clamp-around ammeter.
6. Describe how the total resistance in a series circuit is determined.
7. Ohm's law for determining voltage is _____.
8. Ohm's law for determining amperage is _____.
9. Ohm's law for determining resistance is _____.
10. Sketch a circuit with three loads wired in parallel.
11. Describe the characteristics of the voltage, amperage, and resistances when there is more than one load in a parallel circuit.
12. If there were a current flowing of 5 amperes in a 120-volt circuit, what would the resistance be?
    A. 25 ohms
    B. 24 ohms
    C. 600 ohms
    D. 624 ohms
13. If the resistance in a 120-volt circuit was 40 ohms, what would the current be in amperes?
    A. 4800
    B. 48
    C. 4
    D. 3
14. Electrical power is measured in
    A. amperes divided by the resistance.
    B. amperes divided by the voltage.
    C. watts.
    D. voltage divided by the amperage.
15. The formula for determining electrical power is _____.
16. Describe how a step-down transformer differs from a step-up transformer.
17. What are the three types of opposition to current flow that impedance represents?
18. Why is it important to use a properly sized wire in a particular circuit?
19. What does forward bias on a diode mean?
20. The unit of measurement for the charge a capacitor can store is the
    A. inductive reactance.
    B. microfarad.
    C. ohm.
    D. joule.

UNIT 13

Introduction to Automatic Controls

OBJECTIVES

After studying this unit, you should be able to

- define bimetal.
- make general comparisons between different bimetal applications.
- describe the rod and tube.
- describe fluid-filled controls.
- describe partial liquid, partial vapor-filled controls.
- distinguish between the bellows, diaphragm, and Bourdon tube.
- discuss the thermocouple.
- explain the thermistor.

13.1 TYPES OF AUTOMATIC CONTROLS

The heating, air-conditioning, and refrigeration field requires many types and designs of automatic controls to stop or start equipment. Modulating controls that vary the speed of a motor or those that open and close valves that vary fluid flow are found less frequently in the residential and light-commercial range of equipment. Controls also provide protection to people and equipment.

Controls can be classified in the following categories: electrical, mechanical, electromechanical, and electronic. Pneumatic controls are discussed in Unit 16.

Electrical controls are electrically operated and normally control electrical devices. Mechanical controls are operated by pressure and temperature to control fluid flow. Electromechanical controls are driven by pressure or temperature to provide electrical functions, or they are driven by electricity to control fluid flow. Electronic controls use electronic circuits and devices to perform the same functions that electrical and electromechanical controls perform.

The automatic control of a system is intended to maintain stable or constant conditions with a controllable device. This can involve protection of people and equipment. The system must regulate itself within the design boundaries of the equipment. If the system's equipment is allowed to operate outside of its design boundaries, the equipment components may be damaged.

In this industry, the job is to control space or product condition by controlling temperature, humidity, and cleanliness.

13.2 DEVICES THAT RESPOND TO THERMAL CHANGE

Automatic controls in this industry usually provide some method of controlling temperature. Temperature control is used to maintain space or product temperature and to protect equipment from damaging itself. When used to control space or product temperature, the control is called a *thermostat;* when used to protect equipment, it is known as a safety device. A good example of both of these applications can be found in a household refrigerator. The refrigerator maintains the space temperature in the fresh food section at about 35°F. When food is placed in the box and stored for a long time, it becomes the same temperature as the space, **Figure 13–1.** If the space temperature is allowed to go much below 35°F, the food begins to freeze. Foods such as eggs, tomatoes, and lettuce are not good after freezing.

The refrigerator is often able to maintain this fresh food compartment condition for 15 or 20 years without failure. The frozen food compartment is another situation and is discussed later. Without automatic controls, the owner of the refrigerator would have to anticipate the temperature in the food compartment and get up in the middle of the night and

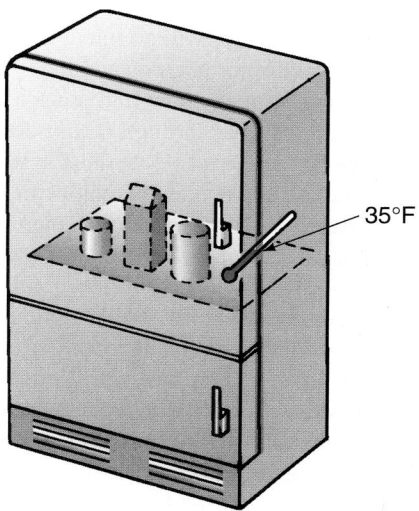

Figure 13–1 The household refrigerator maintains a specific temperature inside the box.

(A)

(B)

Figure 13–2   **(A)** Compressor overload device. **(B)** Overload heater. *Courtesy Ferris State University. Photos by John Tomczyk*

turn it on or off to maintain the temperature. The thermostat stops and starts the refrigeration cycle thousands of times over the course of 20 years to maintain the proper conditions in the refrigerator.

The refrigerator compressor has a protective device to keep it from overloading and damaging itself, **Figure 13–2(A)**. This overload is an automatic control designed to function on the rare occasion that an overload or power problem may cause damage to the compressor. One such occasion is when the power goes off and comes right back on while the refrigerator is running. The overload will stop the compressor for a cool-down period until it is ready to go back to work again without overloading and damaging itself. This overloading happens because of the refrigerator's low-torque, fractional-horsepower compressor motor trying to start against unequal system pressures. The motor will go into a locked rotor situation and draw locked rotor amperage (LRA). The overload will then open the electrical circuit and protect the compressor. After the overload goes through a cool-down period, it will reset and close the circuit. Hopefully, system pressures will be nearing equalization, and the motor can start safely.

The system overload may be electrical, thermal, or a combination of both. An electrical overload exists when the electrical circuit is drawing too much current for which it is designed, for numerous reasons. Too much current may be due to an electrical problem such as a short circuit, improper electrical connections, or a faulty electrical component.

A thermal system overload may occur if a specific electrical or mechanical component is not cooled properly, resulting in overheating. Excess electrical current is converted to heat energy. System overloads may also be mechanically related, causing too much current, which in turn causes excess heat. An example would be a restricted airflow through a condensing coil of a central air conditioner resulting in high head pressure, requiring the compressor motor to work harder. **Figure 13–2(B)** shows a small heater located under the bimetallic element of a compressor overload. This heater will warp the bemetallic element and open the circuit when either an electrical or thermal overload occurs.

Some common automatically controlled devices are
- household refrigerators' fresh and frozen food compartments.
- residential and office cooling and heating systems.
- water heater temperature controls.
- electric oven temperature controls.
- garbage disposal overload controls.
- fuse and circuit breakers that control current flow in electrical circuits in a home.

Figure 13–3 shows two examples of automatic controls.

Automatic controls used in the air-conditioning and refrigeration industry are devices that monitor temperature and its changes. Some controls respond to temperature changes and are used to monitor electrical overloads by temperature changes in the wiring circuits. This response is usually a change of dimension or electrical characteristic in the control-sensing element.

13.3   THE BIMETAL DEVICE

The *bimetal* device is probably the most common device used to detect thermal change. In its simplest form, the device consists of two unlike metal strips, attached back to back, each with different rates of expansion and contraction, **Figure 13–4**. Brass and steel are commonly used. When the device is heated, the brass expands faster than the steel, and the device is warped out of shape. This warping action is a known dimensional change that can be attached to an electrical component or valve to stop, start, or modulate electrical current or fluid flow, **Figure 13–5**. This control is limited in its application by the amount of warp it can accomplish with a temperature change. For instance, when the bimetal is fixed on one end and heated, the other end moves a certain amount per degree of temperature change, **Figure 13–6**.

To obtain enough travel to make the bimetal practical over a wider temperature range, add length to the bimetal strip. When adding length, the bimetal strip is normally coiled into a circle, shaped like a hairpin, wound into a helix, or formed into a worm shape, **Figure 13–7**. The movable end of the coil

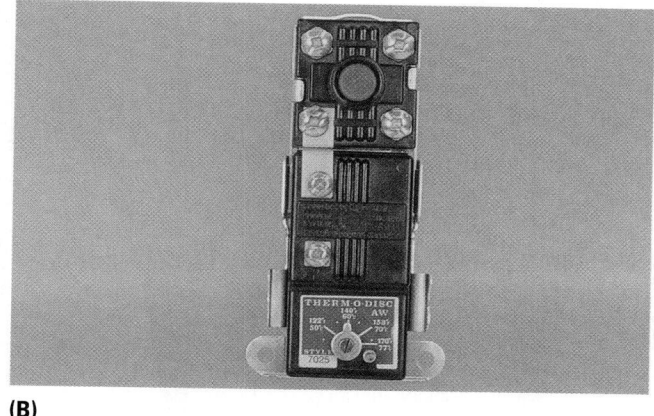

(A)   **(B)**

Figure 13–3   These controls operate household appliances. **(A)** Refrigerator thermostat. **(B)** Water heater thermostat. *Photos by Bill Johnson*

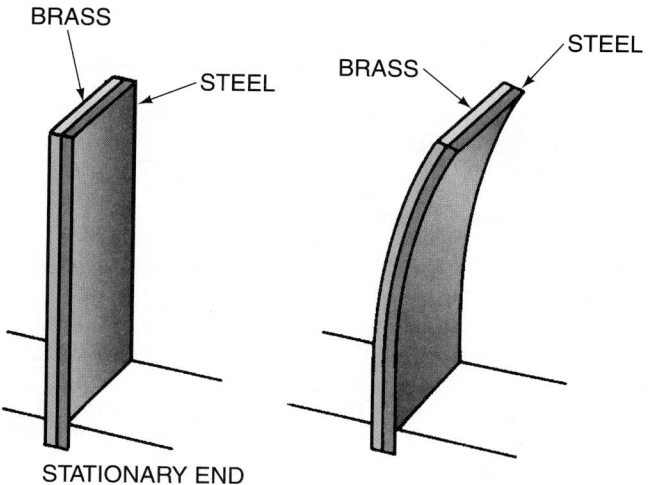

Figure 13–4   A basic bimetal strip made of two unlike metals such as brass and steel fastened back to back.

or helix can be attached to a pointer to indicate temperature, a switch to stop or start current flow, or a valve to modulate fluid flow. One of the basic control applications is shown in **Figure 13–8.**

ROD AND TUBE.   The *rod and tube* is another type of control that uses two unlike metals and the difference in thermal expansion and contraction. The rod and tube can be described more accurately as the rod in tube. It has an outer tube of metal with a high expansion rate and an internal rod of metal with a low expansion rate, **Figure 13–9.** This control has been used for years in the residential gas water heater. The tube is inserted into the tank and provides very accurate sensing of the water temperature. As the tank water temperature changes, the tube pushes the rod and opens or closes the gas valve to start or stop the heat to the water in the tank, **Figure 13–10.**

Another application for the rod and tube is the heat motor-type motorized valve. A small heater is wrapped around the

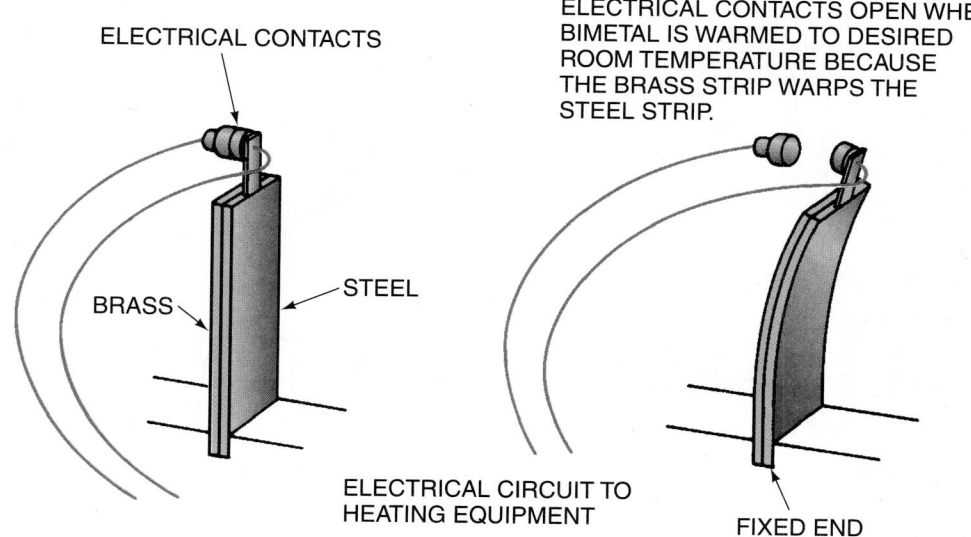

Figure 13–5   A basic bimetal strip used as a heating thermostat.

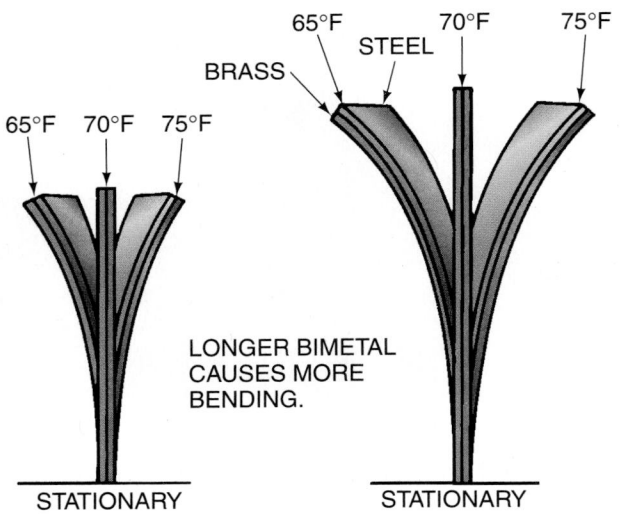

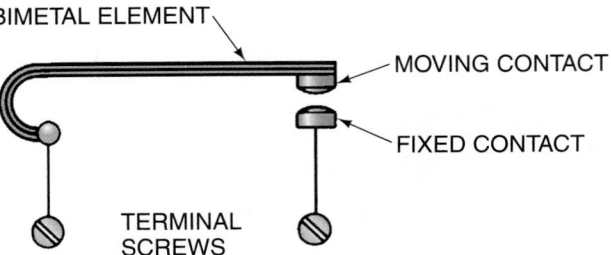

Figure 13–8 Movement of the bimetal due to changes in temperature opens or closes electrical contacts.

Figure 13–6 This bimetal is straight is at 70°F. The brass side contracts more than the steel side when cooled, causing a bend to the left. The brass side expands faster than the steel side on a temperature rise, causing a bend to the right. This bend reflects a predictable amount per degree of temperature change. The longer the strip, the greater the bend.

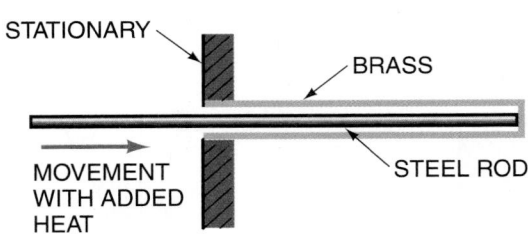

Figure 13–9 The rod and tube.

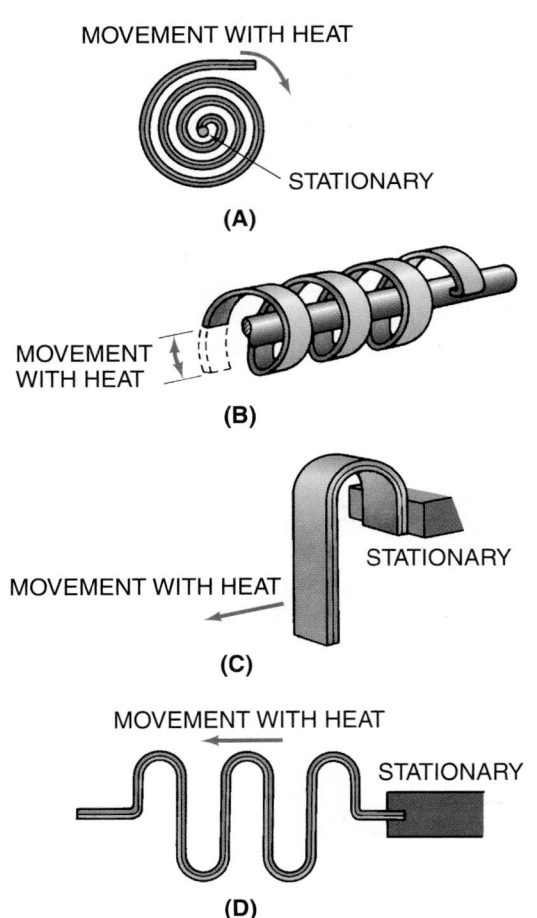

Figure 13–7 Adding length to the bimetal. **(A)** Coiled. **(B)** Wound into helix. **(C)** Hairpin shape. **(D)** Formed into worm shape.

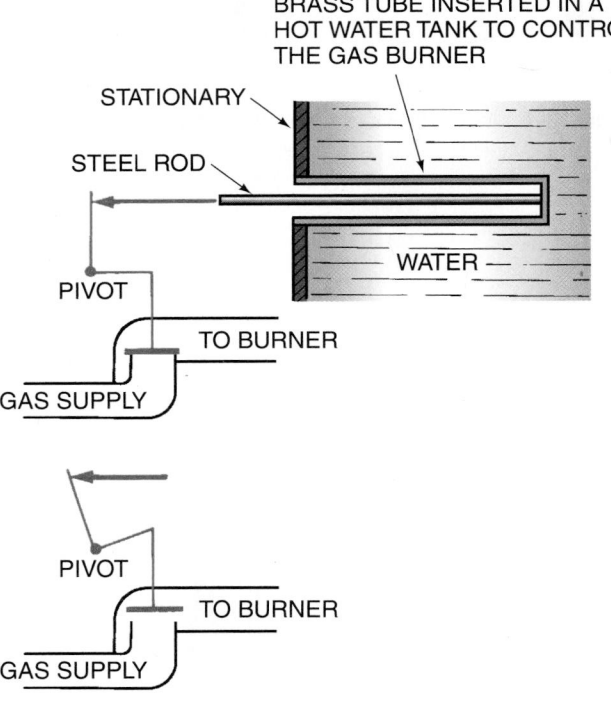

Figure 13–10 The rod-and-tube type of control consists of two unlike metals—with the fastest expanding metal in a tube normally inserted in a fluid environment such as a hot water tank.

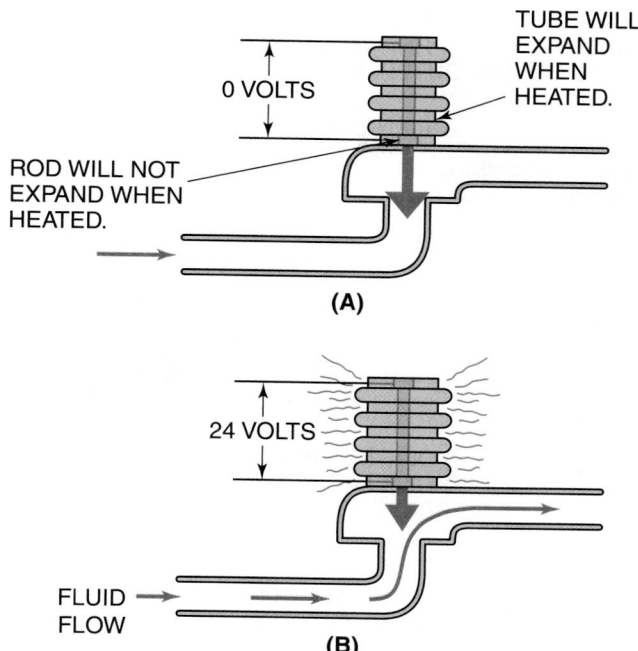

Figure 13–11 This heat motor valve is used to control the flow of water. **(A)** When it receives a 24-V signal, an electrical heater heats the tube and it expands, pulling the rod. **(B)** When the rod moves, it pulls the seat off the valve.

tube portion of the mechanism. When voltage is applied, it heats the tube, which expands to pull the rod and open the valve for water flow. When the heat is removed, the tube will shrink and close the valve, **Figure 13–11**.

SNAP-DISC. The *snap-disc* is another type of bimetal used in some applications to sense temperature changes. This control is treated apart from the bimetal because of its snap characteristic that gives it a quick open-and-close feature. Some sort of snap-action feature has to be incorporated into all controls that stop and start electrical loads, **Figure 13–12**.

13.4 CONTROL BY FLUID EXPANSION

Fluid expansion is another method of sensing temperature change. In Unit 1, a mercury thermometer was described as a bulb with a thin tube of mercury rising up a stem. As the mercury in the bulb is heated or cooled, it expands or contracts and either rises or falls in the stem of the thermometer. The level of the mercury in the stem is based on the temperature of the mercury in the bulb, **Figure 1–3**. The reason for the rising and falling is that the mercury in the tube has no place else to go. When the mercury in the bulb is heated and expands, it has to rise up in the tube. When it is cooled, it naturally falls down the tube. This same idea can be used to transmit a signal to a control that a temperature change is occurring.

The liquid rising up in the transmitting tube has to act on some device to convert the rising liquid to usable motion.

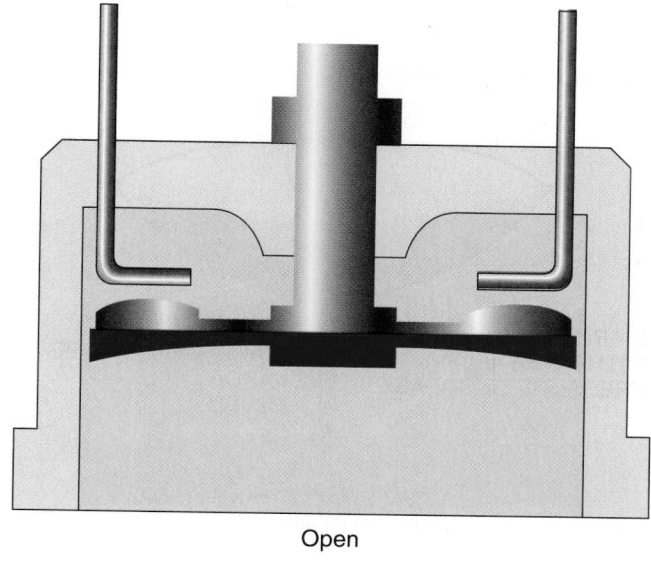

(A)

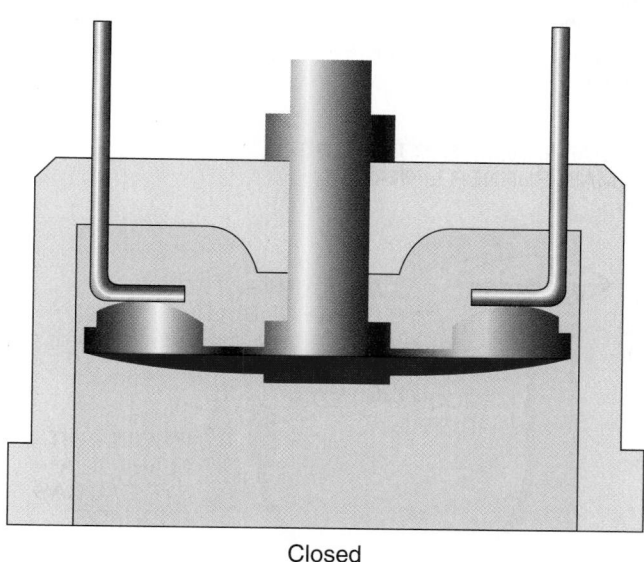

(B)

Figure 13–12 The snap-disc is another variation of the bimetal concept. It is usually round and fastened on the outside. When heated, the disc snaps to a different position. **(A)** Open circuit. **(B)** Closed circuit.

One device used is a diaphragm. A *diaphragm* is a thin, flexible metal disc with a large area. It can move in and out with pressure changes underneath it, **Figure 13–13**.

When a bulb is filled with liquid and connected to a diaphragm with piping, the bulb temperature can be transmitted to the diaphragm by the expanding liquid. In **Figure 13–14** the bulb is filled with mercury and placed in the pilot light flame on a gas furnace to ensure that a pilot light is present to ignite the gas burner before the main gas valve is opened. The entire mercury-filled tube and mercury-filled diaphragm are sensitive to temperature changes. The pilot light is very hot so only the sensing bulb is located in the pilot light flame.

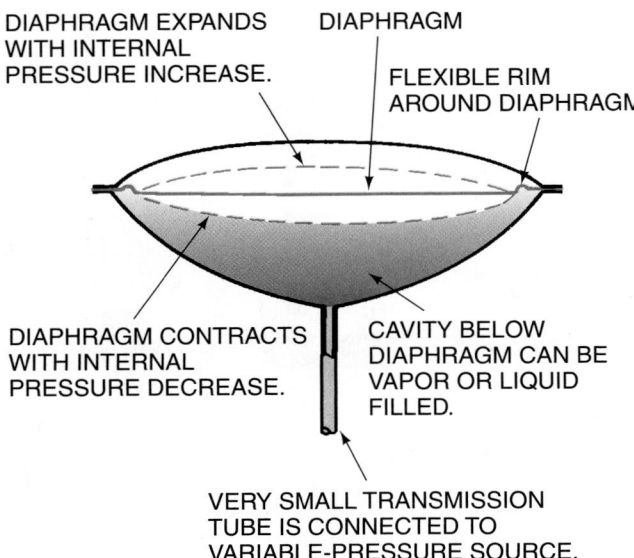

DIAPHRAGM EXPANDS WITH INTERNAL PRESSURE INCREASE.

DIAPHRAGM

FLEXIBLE RIM AROUND DIAPHRAGM

DIAPHRAGM CONTRACTS WITH INTERNAL PRESSURE DECREASE.

CAVITY BELOW DIAPHRAGM CAN BE VAPOR OR LIQUID FILLED.

VERY SMALL TRANSMISSION TUBE IS CONNECTED TO VARIABLE-PRESSURE SOURCE.

Figure 13–13 The diaphragm is a thin, flexible, moveable membrane (brass, steel, or other metal) used to convert pressure changes to movement. This movement can stop, start, or modulate controls.

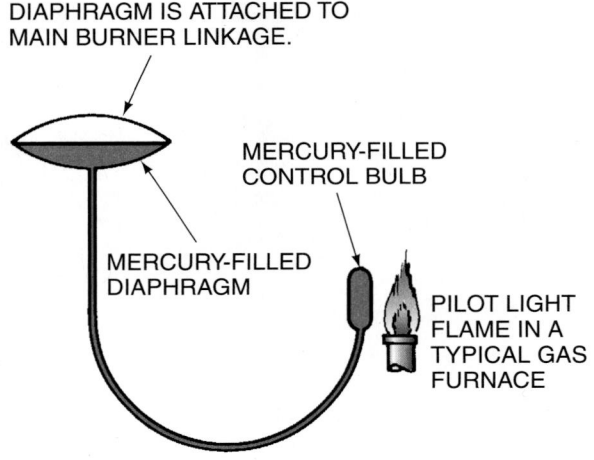

DIAPHRAGM IS ATTACHED TO MAIN BURNER LINKAGE.

MERCURY-FILLED CONTROL BULB

MERCURY-FILLED DIAPHRAGM

PILOT LIGHT FLAME IN A TYPICAL GAS FURNACE

Figure 13–14 The bulb in or near the flame is mercury filled. The heated bulb causes the mercury to expand, move up the transmission tube, and flex out the diaphragm. This proves that the pilot flame is present to ignite the main burner.

To maintain more accurate control at the actual bulb location, you can use a bulb partially filled with a liquid that will boil and make a vapor, which is then transmitted to the diaphragm at the control point, **Figure 13–15.** You should realize that the liquid will respond to temperature change much more than will the vapor, which is used to transmit the pressure.

The following example using a walk-in refrigerated box with R-134a describes how this can work. Refer to **Figure 3–15** for the temperature/pressure chart for R-134a. The inside temperature is maintained by cutting the refrigeration system off when the box temperature reaches 35°F and

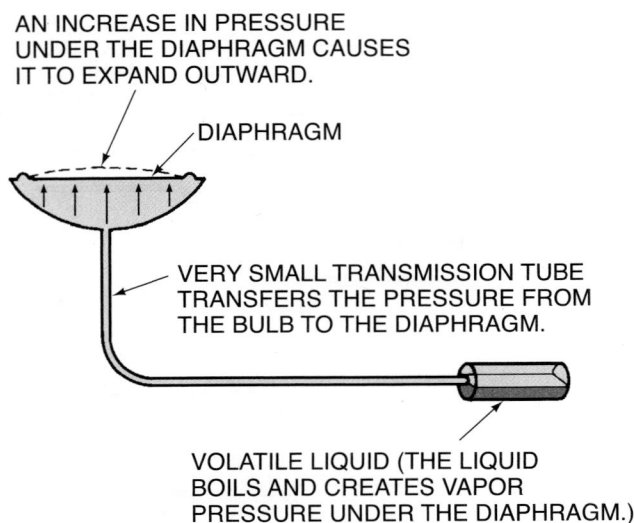

AN INCREASE IN PRESSURE UNDER THE DIAPHRAGM CAUSES IT TO EXPAND OUTWARD.

DIAPHRAGM

VERY SMALL TRANSMISSION TUBE TRANSFERS THE PRESSURE FROM THE BULB TO THE DIAPHRAGM.

VOLATILE LIQUID (THE LIQUID BOILS AND CREATES VAPOR PRESSURE UNDER THE DIAPHRAGM.)

Figure 13–15 A large bulb is partially filled with a volatile liquid, one that boils and creates vapor pressure when heated. This causes an increase in vapor pressure, which forces the diaphragm to move outward. When cooled, the vapor condenses and the diaphragm moves inward.

starting it when the box temperature reaches 45°F. A control with a remote bulb is used to regulate the space temperature. The bulb is located inside the box, and the control is located outside the box so it can be adjusted.

For illustration purposes, a pressure gage is installed in the bulb to monitor the pressures inside the bulb as the temperature changes. **Figure 13–16** presents a progressive explanation of this example. At the point that the unit needs to be cycled off, the bulb temperature is 35°F. This corresponds to a pressure of 30 psig for R-134a. A control mechanism can be designed to open an electrical circuit and stop it at this point. When the cooler temperature rises to 45°F, it is time to restart the unit. At 45°F for R-134a, the pressure inside the control is 40 psig, and the same mechanism can be designed to close the electrical circuit and start the refrigeration system.

The diaphragm has a limited amount of travel but a great deal of power during this movement. The travel of the liquid-filled control is limited to the expansion of the liquid in the bulb for the temperature range within which it is working. When more travel is needed, another device, called the *bellows,* can be used. The bellows is much like an accordion. It has a lot of internal volume with a lot of travel, **Figure 13–17.** The bellows is normally used with a vapor inside it instead of a liquid.

A remote bulb partially filled with liquid may also be used to indicate temperature. The device uses a Bourdon tube that straightens out with an increase in vapor pressure. This movement moves a needle on a calibrated dial to indicate temperature, **Figure 13–18.**

The partially filled bulb control is widely used in this industry because it is reliable, simple, and economical. This control has been in the industry since it began, and it has many configurations. **Figure 13–19** illustrates some remote

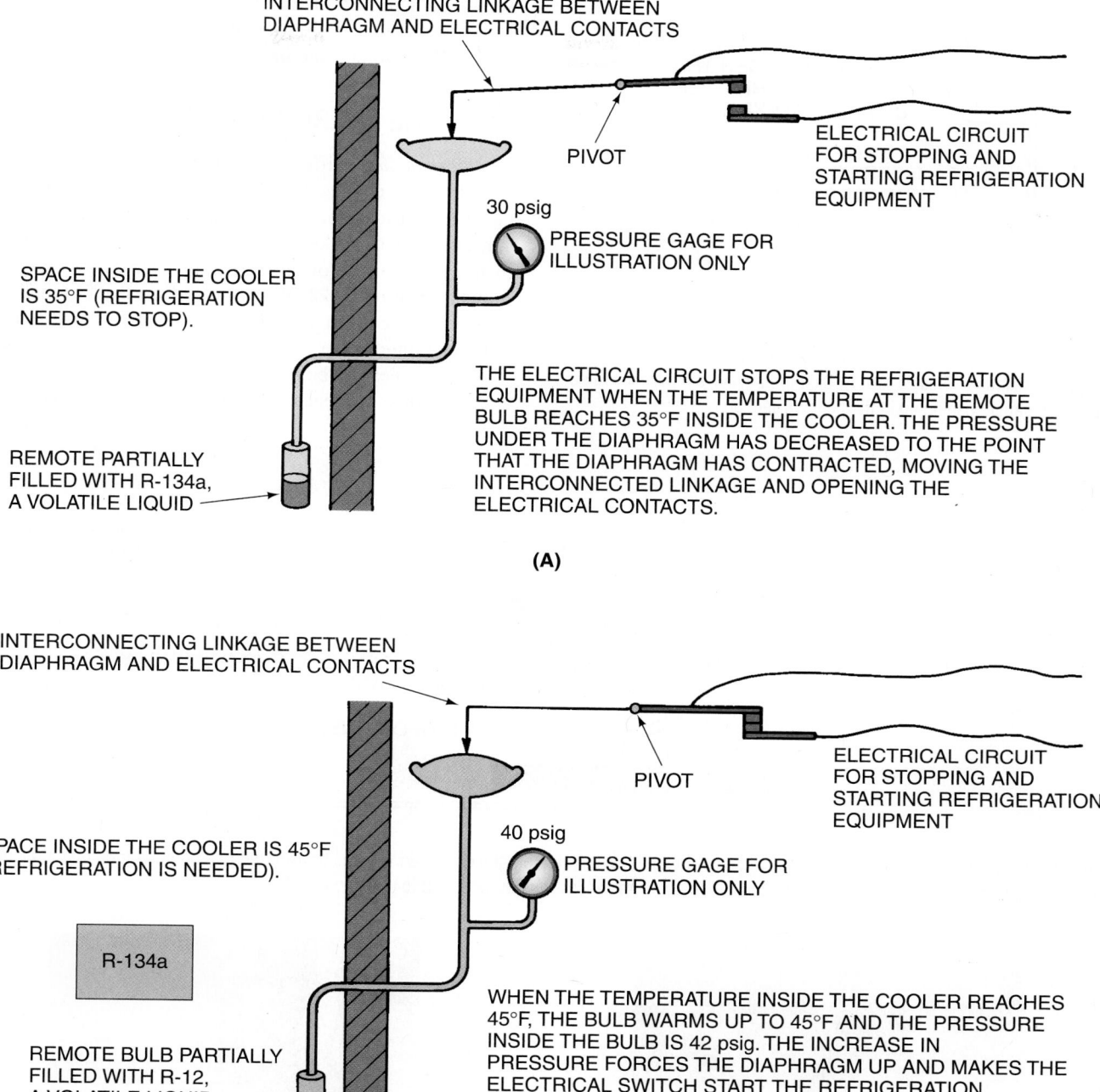

INTERCONNECTING LINKAGE BETWEEN
DIAPHRAGM AND ELECTRICAL CONTACTS

PIVOT

ELECTRICAL CIRCUIT
FOR STOPPING AND
STARTING REFRIGERATION
EQUIPMENT

30 psig

PRESSURE GAGE FOR
ILLUSTRATION ONLY

SPACE INSIDE THE COOLER
IS 35°F (REFRIGERATION
NEEDS TO STOP).

THE ELECTRICAL CIRCUIT STOPS THE REFRIGERATION
EQUIPMENT WHEN THE TEMPERATURE AT THE REMOTE
BULB REACHES 35°F INSIDE THE COOLER. THE PRESSURE
UNDER THE DIAPHRAGM HAS DECREASED TO THE POINT
THAT THE DIAPHRAGM HAS CONTRACTED, MOVING THE
INTERCONNECTED LINKAGE AND OPENING THE
ELECTRICAL CONTACTS.

REMOTE PARTIALLY
FILLED WITH R-134a,
A VOLATILE LIQUID

(A)

INTERCONNECTING LINKAGE BETWEEN
DIAPHRAGM AND ELECTRICAL CONTACTS

PIVOT

ELECTRICAL CIRCUIT
FOR STOPPING AND
STARTING REFRIGERATION
EQUIPMENT

40 psig

PRESSURE GAGE FOR
ILLUSTRATION ONLY

SPACE INSIDE THE COOLER IS 45°F
(REFRIGERATION IS NEEDED).

R-134a

WHEN THE TEMPERATURE INSIDE THE COOLER REACHES
45°F, THE BULB WARMS UP TO 45°F AND THE PRESSURE
INSIDE THE BULB IS 42 psig. THE INCREASE IN
PRESSURE FORCES THE DIAPHRAGM UP AND MAKES THE
ELECTRICAL SWITCH START THE REFRIGERATION
EQUIPMENT.

REMOTE BULB PARTIALLY
FILLED WITH R-12,
A VOLATILE LIQUID

(B)

Figure 13–16   A remote bulb transmits pressure to the diaphragm on the basis of the temperature in the cooler.

bulb thermostats. The Bourdon tube is often used in the same manner as the diaphragm and the bellows to monitor fluid expansion.

The mechanisms described here are simplified and are not practical because each electrical switch must have a snap action built into it. As explained earlier, an electrical arc occurs when the electrical circuit is closed or opened, but especially when opened. Electrons are flowing and will try to keep flowing when the switch is opened, creating the arc, **Figure 13–20.** This arc is a small version of an electric welding arc. Every time a switch is opened, wear takes place. It can be said that every switch has a lifetime. That lifetime is a certain number of openings; after that the switch will fail. The faster the opening, the longer the switch will last.

Many types of snap-action mechanisms are used. Some switches use an over-center device to quicken the opening.

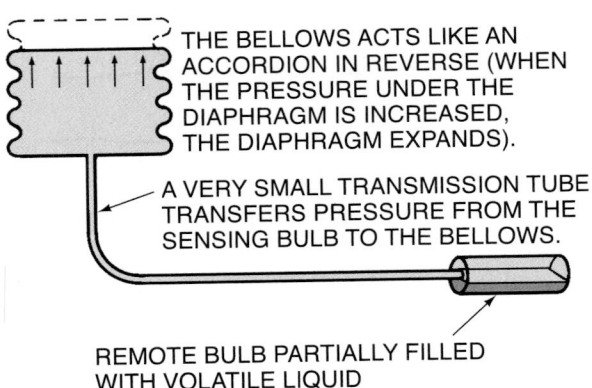

THE BELLOWS ACTS LIKE AN ACCORDION IN REVERSE (WHEN THE PRESSURE UNDER THE DIAPHRAGM IS INCREASED, THE DIAPHRAGM EXPANDS).

A VERY SMALL TRANSMISSION TUBE TRANSFERS PRESSURE FROM THE SENSING BULB TO THE BELLOWS.

REMOTE BULB PARTIALLY FILLED WITH VOLATILE LIQUID

Figure 13–17 The bellows is applied where more movement per degree is desirable. This control would normally have a partially filled bulb with vapor pushing up in the bellows section.

Some use a puddle of mercury to create the quick action. Some may use a magnet to make the action faster. See **Figure 13–21** for some examples of snap-action mechanisms.

13.5 THE THERMOCOUPLE

The *thermocouple* differs from other methods of controlling with thermal change because it does not use expansion; instead, it uses electrical principles. The thermocouple consists of two unlike metals formed together on one end (usually wire made of unlike metals such as iron and constantan), **Figure 13–22**. When heated on the fastened end, an electrical current flow is started due to the difference in temperature in the two ends of the device, **Figure 13–23**. Thermocouples can be made of many different unlike metal combinations, and each one has a different characteristic.

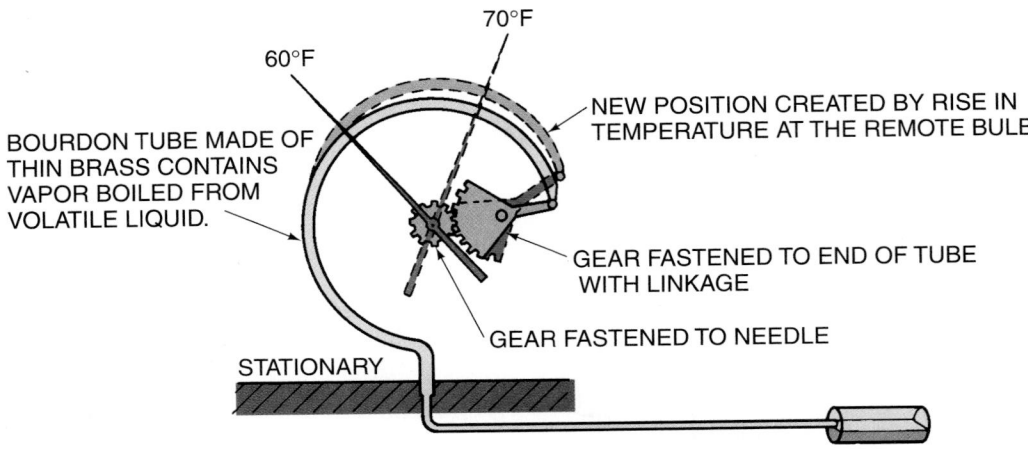

70°F

60°F

NEW POSITION CREATED BY RISE IN TEMPERATURE AT THE REMOTE BULB

BOURDON TUBE MADE OF THIN BRASS CONTAINS VAPOR BOILED FROM VOLATILE LIQUID.

GEAR FASTENED TO END OF TUBE WITH LINKAGE

GEAR FASTENED TO NEEDLE

STATIONARY

Figure 13–18 This remote bulb is partially filled with liquid. When heated, the expanded vapor is transmitted to a Bourdon tube that straightens out with an increase in vapor pressure. A decrease in pressure causes the Bourdon tube to curl inward.

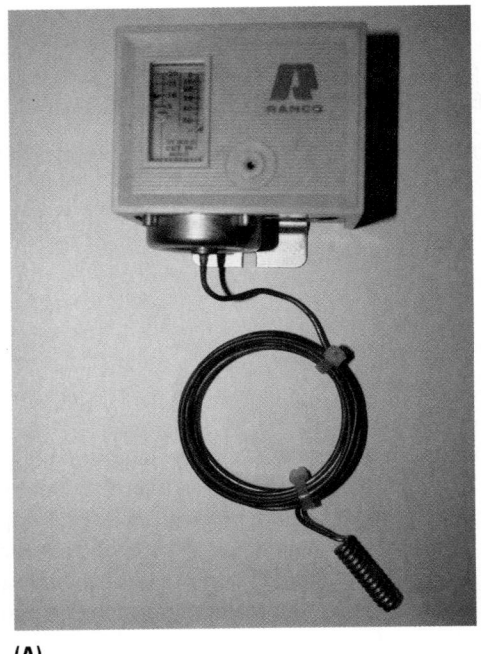

(A)

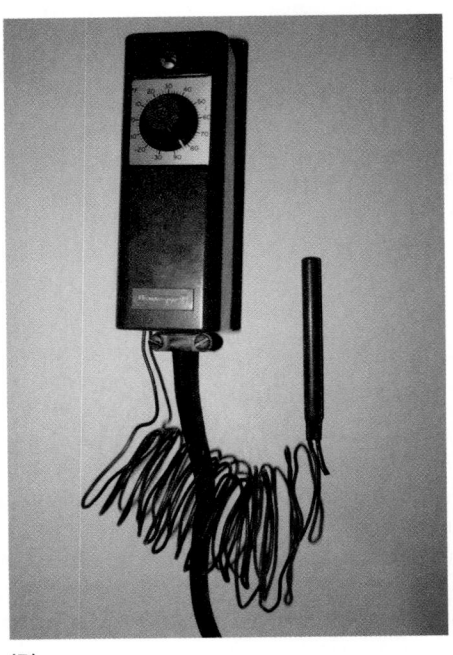

(B)

Figure 13–19 Remote bulb refrigeration temperature controls **(A)–(B)**. *Courtesy Ferris State University. Photos by John Tomczyk*

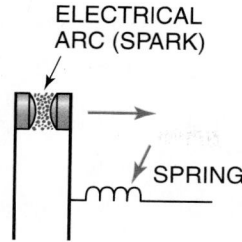

Figure 13–20 The arc damage between the contacts when they open is in proportion to the speed with which they open.

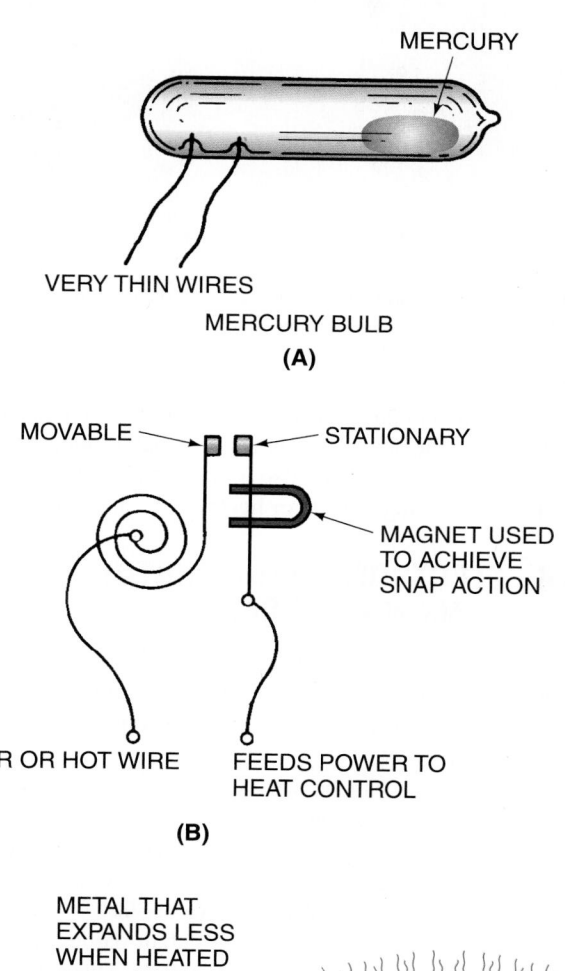

MERCURY

VERY THIN WIRES

MERCURY BULB

(A)

MOVABLE — STATIONARY

MAGNET USED TO ACHIEVE SNAP ACTION

R OR HOT WIRE   FEEDS POWER TO HEAT CONTROL

(B)

METAL THAT EXPANDS LESS WHEN HEATED

METAL THAT EXPANDS MORE WHEN HEATED

(C)

Figure 13–21 Devices used to create snap action in switches. **(A)** Puddle of mercury. **(B)** Magnet. **(C)** Over-center device.

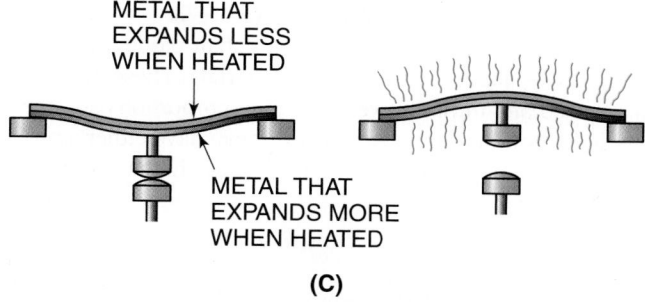

Figure 13–22 A thermocouple used to detect a pilot light in a gas-burning appliance. *Courtesy Robertshaw Controls Company*

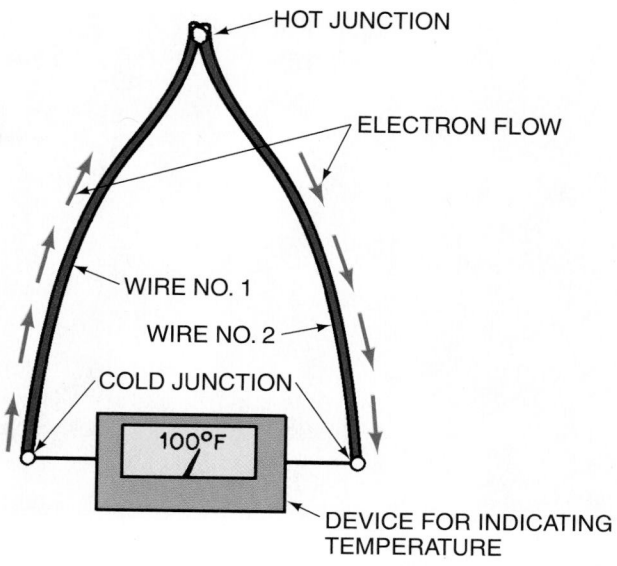

HOT JUNCTION

ELECTRON FLOW

WIRE NO. 1

WIRE NO. 2

COLD JUNCTION

100°F

DEVICE FOR INDICATING TEMPERATURE

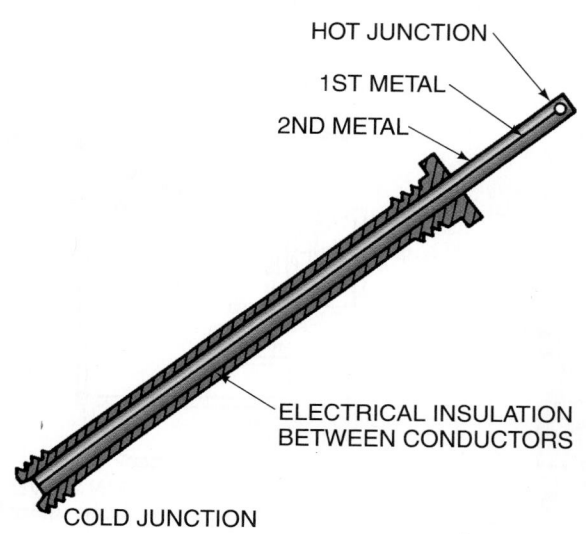

HOT JUNCTION

1ST METAL

2ND METAL

ELECTRICAL INSULATION BETWEEN CONDUCTORS

COLD JUNCTION

Figure 13–23 The thermocouple on the top is made of two wires of different metals welded on one end. It is used to indicate temperature. The thermocouple on the bottom is a rigid device and is used to detect a pilot light.

Each thermocouple has a *hot* junction and a *cold* junction. The hot junction, as the name implies, is at a higher temperature level than the cold junction. This difference in temperature is what starts the current flowing. Heat will cause an electrical current to flow in one direction in one metal and in the opposite direction in the other. When these metals are connected, they make an electrical circuit, and current will flow when heat is applied to one end of the device.

Tables and graphs for various types of thermocouples show how much current flow can be expected from a thermocouple under different conditions for hot and cold junctions. The current flow in a thermocouple can be monitored by an electronic circuit and used for many temperature-related applications, such as a thermometer, a thermostat to stop or start a process, or a safety control, **Figure 13–24.**

Figure 13–24 This thermocouple senses whether the gas furnace pilot light is on. *Photo by Bill Johnson*

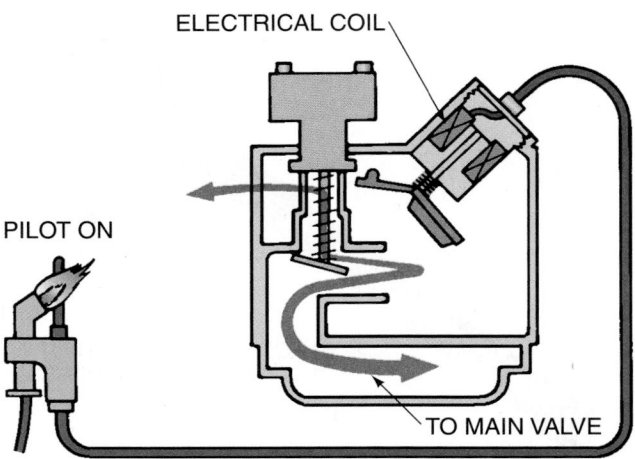

Figure 13–25 A thermocouple and control circuit used to detect a gas flame. When the flame is lit, the thermocouple generates an electrical current. This energizes an electromagnet that holds the gas valve open. When the flame is out, the thermocouple stops generating electricity and the valve closes. Gas is not allowed to flow.

The thermocouple has been used extensively for years in gas furnaces to detect the pilot light flame for safety purposes. This application is beginning to be phased out because it works best with standing pilot light systems. A gradual change in design that is resulting in the use of intermittent pilots (these are pilot lights that are extinguished each time the burner goes out and relit each time the room thermostat calls for heat) and other newer types of ignition—is causing this phaseout. This thermocouple application has an output voltage of about 20 to 30 millivolts (DC), all that is needed to control a safety circuit to prove there is flame, **Figure 13–25.**

Thermocouples ganged together to give more output are called *thermopiles,* **Figure 13–26.** The thermopile is used on some gas-burning equipment as the only power source. This

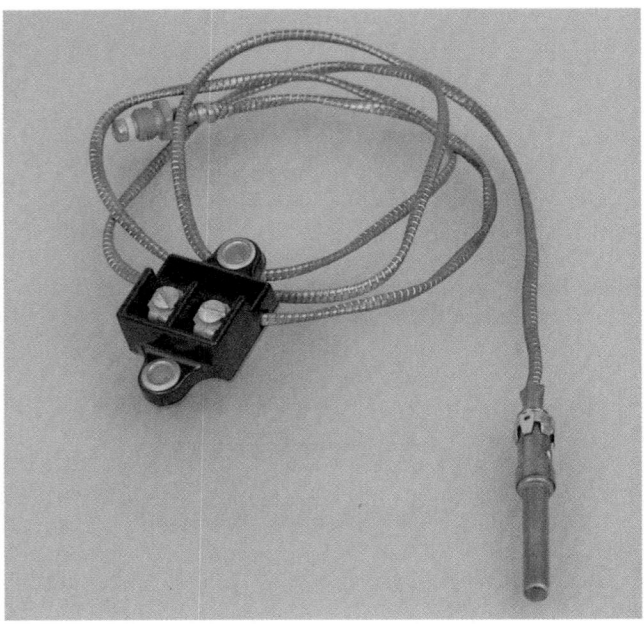

Figure 13–26 A thermopile consists of a series of thermocouples in one housing. *Photo by Bill Johnson*

type of equipment has no need for power other than the control circuit, so the power supply is very small (about 500 millivolts). The thermopile has also been used in remote areas to operate radios using the sun or heat from a small fire.

13.6 ELECTRONIC TEMPERATURE-SENSING DEVICES

The *thermistor* is an electronic solid-state device known as a semiconductor and requires an electronic circuit to utilize its capabilities. It varies its resistance to current flow based on its temperature.

The thermistor can be very small and will respond to small temperature changes. The changes in current flow in the device are monitored by special electronic circuits that can stop, start, and modulate machines or provide a temperature readout, **Figure 13–27** and **Figure 13–28.**

Thermistors are usually made of cobalt oxide, nickel, or manganese, and from a few other materials. These materials are mixed and milled in very accurate proportions and then hardened for durability. A thermistor can have either a positive or negative coefficient of resistance. If the thermistor

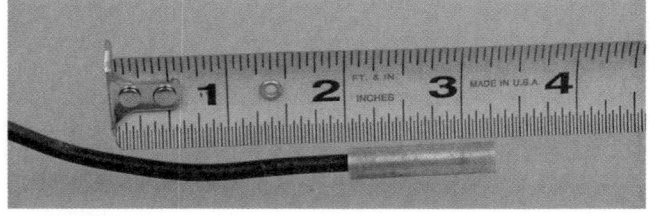

Figure 13–27 A thermometer probe using a thermistor to measure temperature. *Photo by Bill Johnson*

Figure 13–28 A thermistor application. *Photo by Bill Johnson*

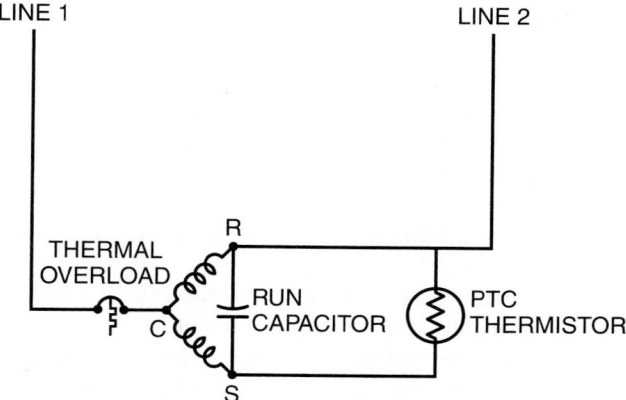

Figure 13–29 A PTC thermistor wired in parallel with a run capacitor. This increases motor starting torque on this permanent split-capacitor (PSC) motor.

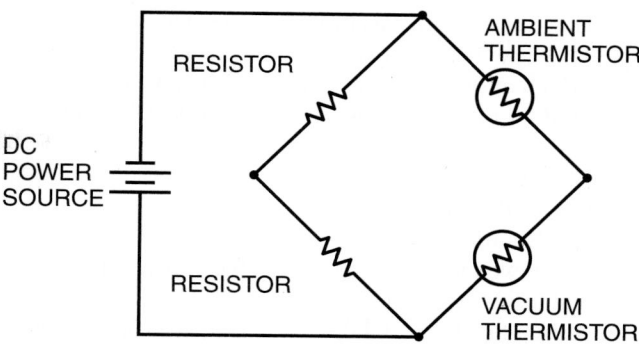

Figure 13–30 A bridge circuit showing thermistors used for micron readings when a vacuum is being measured.

increases its resistance as the temperature increases, it has a positive temperature coefficient (PTC) of resistance. If it decreases its resistance as the temperature increases, it has a negative temperature coefficient (NTC) of resistance. Also, some thermistors can double their resistance with as little as one degree of temperature change. Thermistors are used in solid-state motor protectors, motor-starting relays, room thermostats, electronic expansion valves, duct sensors, micron gages for vacuum pumps, heat pump controls, and many other control schemes. **Figure 13–29** shows how a PTC thermistor assists in starting a *permanent split-capacitor (PSC)* motor. At the instant of start, the PTC thermistor has a very low resistance. This allows a large inrush of current to help increase the starting torque of the motor. However, after a fraction of a second, the thermistor's resistance increases because its temperature increases from the high starting current. Now the thermistor is like an open circuit, and the motor is running like a normal PSC motor.

Thermistors can also be used in a bridge circuit when measuring the vacuum in a refrigeration system, **Figure 13–30**. One of the thermistors is placed in the vacuum, while the other is placed in the outside ambient. As the vacuum gets deeper and deeper, the heat cannot be transferred away as easily from the thermistor placed in the vacuum, because fewer molecules are surrounding it. This difference in resistance of the two thermistors can now be sensed by the bridge circuit and calibrated into microns for vacuum measurement.

SUMMARY

- A bimetal element is two unlike metal strips such as brass and steel fastened back to back.
- Bimetal strips warp with temperature changes and can be used to stop, start, or modulate electrical current flow and fluid flow when used with different mechanical, electrical, and electronic helpers.
- The travel of the bimetal can be extended by coiling it. The helix, worm shape, hairpin shape, and coil are the names given on the extended bimetal.
- The rod and tube is another version of the bimetal.
- Fluid expansion is used in the thermometer to indicate temperature and to operate controls that are totally liquid filled.
- The temperature/pressure relationship is applied to some controls that are partially filled with liquid.
- The diaphragm is used to move the control mechanism when either liquid or vapor pressure is applied to it.
- The diaphragm has very little travel but much power.
- The bellows is used for more travel and is normally vapor filled.
- The Bourdon tube is sometimes used like the diaphragm or bellows.
- The thermocouple generates electrical current flow when heated at the hot junction.
- This current flow can be used to monitor temperature changes to stop, start, or modulate electrical circuits.
- The thermistor is an electronic device that varies its resistance to electrical current flow on the basis of temperature changes.
- Positive temperature coefficient (PTC) thermistors increase in resistance as the temperature increases. Negative temperature coefficient (NTC) thermistors decrease in resistance when the temperature increases.

REVIEW QUESTIONS

1. What is a device that consists of two unlike metal strips that are attached together and that have a different rate of expansion and contraction?
   A. Thermistor
   B. Thermocouple
   C. Diaphragm
   D. Bimetal
2. What are some applications of the bimetal strip?

3. How can the bimetal strip be extended to have more stroke per degree of temperature change?

4. Which of the following is a bimetallic type of control used for many years in residential gas water heaters?
   A. Snap disc
   B. Rod and tube
   C. Coiled bimetal
   D. Bourdon tube

5. What two metals can be used in the bimetal or the rod and tube?

6. A thin, flexible metal disc with a large surface area is a _____.

7. Name two characteristics of a diaphragm.

8. What fluid can be used in a totally liquid-filled bulb type of control?

9. What is one application for the totally liquid-filled control?

10. An accordion-type device with a large internal volume that has a lot of travel is referred to as a _____.

11. What is the difference between a bellows and a diaphragm?

12. Which type of fluid is normally found in a bellows?

13. How is the temperature/pressure relationship used to understand the partially filled remote-bulb type of control?

14. Which of the following is a device consisting of two unlike metals fastened together on one end that, when heated on the fastened end, produces a DC current?
   A. Thermocouple
   B. Thermistor
   C. Diaphragm
   D. Bimetal

15. How can a thermocouple be used to verify that a gas flame is present?

16. What is used with a thermocouple to control mechanical devices?

17. Thermocouples are normally made of _____ and _____ metals.

18. Define a thermistor.

19. What must be used with a thermistor to control a mechanical device or machine?

20. What does a thermistor do that makes it different from a thermocouple?

21. What do the terms PTC and NTC mean with regard to thermistors?

22. Draw and explain how a PTC thermistor assists with increasing the starting torque of a motor.

23. Two PTC thermistors can be used in an electronic circuit to measure _____ when pulling a vacuum.

UNIT 14

Automatic Control Components and Applications

OBJECTIVES

After studying this unit, you should be able to

- ■ discuss space temperature control.
- ■ describe the mercury control bulb.
- ■ describe system overshoot and temperature swing.
- ■ describe the difference between low- and high-voltage controls.
- ■ name components of low- and high-voltage controls.
- ■ name two ways motors are protected from high temperature.
- ■ describe the difference between a diaphragm and a bellows control.
- ■ state the uses of pressure-sensitive controls.
- ■ describe a high-pressure control.
- ■ describe a low-pressure control.
- ■ discuss the range and differential of a control.
- ■ describe pressure transducers.
- ■ describe a pressure relief valve.
- ■ describe the functions of mechanical and electromechanical controls.

SAFETY CHECKLIST

- ✔ The subbase for a line-voltage thermostat is attached to an electrical outlet box. If an electrical arc due to overload or short circuit occurs, it is enclosed in the conduit or box. Do not consider this as a low-voltage device.
- ✔ The best procedure to use when a hot compressor is encountered is to shut off the compressor with the space temperature thermostat "off" switch and return the next day. This gives it time to cool and keeps the crankcase heat on.
- ✔ If the above is not possible or feasible and you must cool the compressor quickly, turn the power off, and cover electrical circuits with plastic before you use water to cool the compressor. Do not stand in this water when you make electrical checks.
- ✔ When used as safety controls, pressure controls should be installed in such a manner that they are not subject to being valved off by the service valves.
- ✔ The high-pressure control is a method of providing safety for equipment and surroundings. It should not be tampered with.

- ✔ A gas pressure switch is a safety control and should never be bypassed except for troubleshooting and then only by experienced service technicians.
- ✔ A pressure relief valve is a safety device that is factory set and should not be changed or tampered with.

14.1 RECOGNITION OF CONTROL COMPONENTS

This unit describes how various controls look and operate. Recognizing a control and understanding its function and what component it influences are vitally important and will eliminate confusion when you read a diagram or troubleshoot a system. The ability to see a control on a circuit diagram and then recognize it on the equipment will be easier after a description and illustration are studied.

14.2 TEMPERATURE CONTROLS

Temperature is controlled in many ways for many reasons. For instance, space temperature is controlled for comfort. The motor-winding temperature in a compressor motor is controlled to prevent overheating and damage to the motor. The motor could overheat, and damage could occur from the very power that operates it. Both types of controls require some device that will sense temperature rise with a sensing element and make a known response. The space temperature example used the control as an operating control, whereas the motor temperature example serves as a safety device.

The space temperature application has two different actions depending on whether winter heating or summer cooling is needed. In winter, the control must break a circuit to stop the heat when conditions are satisfied. In summer, the conditions are reversed; the control must make a circuit and start the cooling on the basis of a temperature rise. The heating thermostat opens on a rise in temperature, and the cooling thermostat closes on a rise in temperature. NOTE: In both cases the control was described as *functioning on a rise*. This terminology is important because it is used in this industry.

The motor temperature cut-out has the same circuit action as the heating thermostat. It opens the circuit on a rise in motor temperature and stops the motor. The heating thermostat and a motor-winding thermostat may make the same move under the same conditions, but they do not physically resemble each other, **Figure 14–1.**

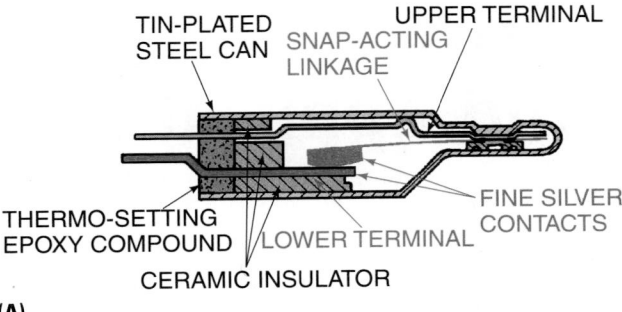

(A)

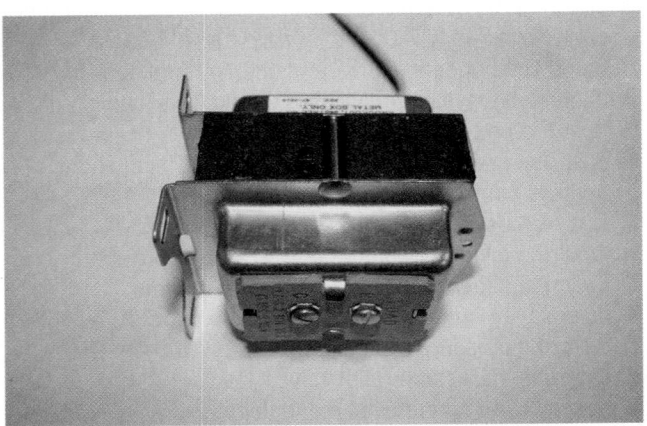

(B)

Figure 14–1 Both thermostats open on a rise in temperature, but they serve two different purposes. **(A)** The motor-winding thermostat measures the temperature of iron and copper while in close contact with the motor windings. **(B)** The space thermostat measures air temperature from random air currents. *Photo by Bill Johnson*

Another difference between the two thermostats is the medium to be detected. The motor-winding thermostat must be in close contact with the motor winding. It is fastened to the winding itself. The space temperature thermostat is mounted on a wall with the control components suspended in air under the decorative cover. The thermostat relies on random air currents passing over it.

Another important design concept is the current-carrying characteristics of the various controls. In the space temperature application the stopping or starting of a heating system, such as the gas or oil furnace, involved stopping and starting low-voltage (24 V) components and high-voltage (115 V or 230 V) components. The gas or oil furnace normally has a low-voltage gas valve or relay and a high-voltage fan motor.

No firm rule states that one voltage or the other is all that is used in any specific application. However, the stopping and starting of a 3-hp compressor requires a larger switching mechanism than a simple gas valve. A 3-hp compressor could require a running current of 18 A and a starting current of 90 A, whereas a simple gas valve might use only 1/2 A. If the bimetal were large enough to carry the current for a 3-hp compressor, the control would be so large that it would be slow to respond to air temperature changes. This is one

reason for using low-voltage controls to stop and start high-voltage components.

Residential systems usually have low-voltage control circuits. There are four reasons for this:

1. Economy
2. Safety
3. More precise control of relatively still air temperature
4. In many states a technician does not need an electrician's license to install and service low-voltage wiring.

The low-voltage thermostat receives its voltage from the residence power supply that is reduced to 24 V with a small transformer usually furnished with the equipment, **Figure 14–2**.

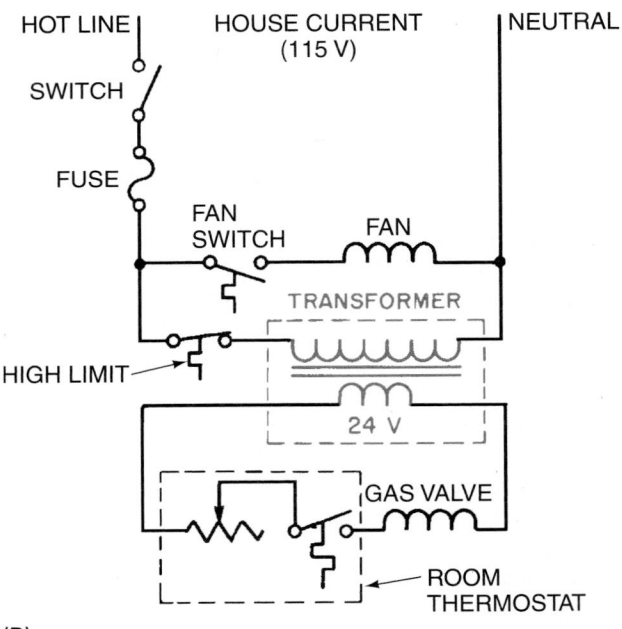

(A)

(B)

Figure 14–2 **(A)** A typical transformer used in residential and light commercial buildings to change 115 V to 24 V (control voltage). **(B)** The typical transformer circuit used in residences and light commercial buildings to change 115 V to 24 V (control voltage). **(A)** *Courtesy Ferris State University. Photo by John Tomczyk.* **(B)** *Photo by Bill Johnson*

14.3 SPACE TEMPERATURE CONTROLS, LOW VOLTAGE

The low-voltage space temperature control (thermostat) normally regulates other controls and does not carry much current—seldom more than 2 A. The thermostat consists of the following components.

ELECTRICAL CONTACT TYPE. The mercury bulb is probably the most popular component used to make and break the electric circuit in low-voltage thermostats. The mercury bulb is inside the thermostat, **Figure 14–3.** It consists of a glass bulb filled with an inert gas (a gas that will not support oxidation) and a small puddle of mercury free to move from one end to the other. The principle is to be able to make and break a small electrical current in a controlled atmosphere. When an electrical current is either made or broken, a small arc is present. The arc is hot enough to cause oxidation in the vicinity of the

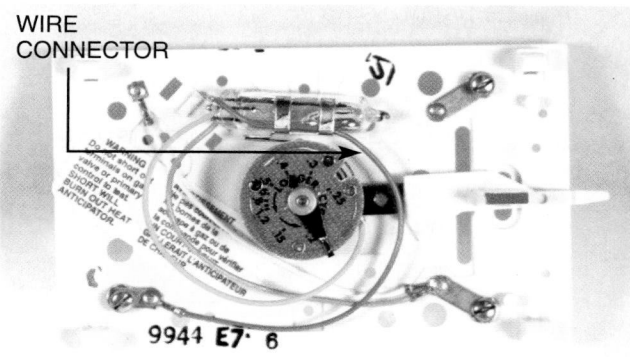

(A)

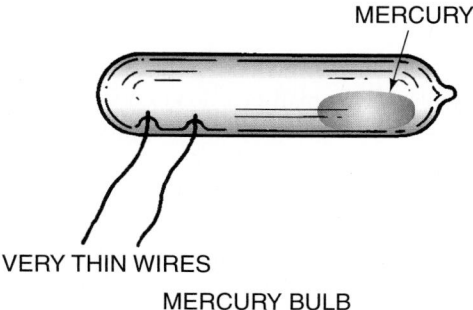

(B)

Figure 14–3 **(A)** A wall thermostat with the cover off and the mercury bulb exposed. **(B)** A detail of the mercury bulb. Note the very fine wire that connects the bulb to the circuit. The bulb is attached to the movable end of a bimetal coil. When the bimetal tips the bulb, the mercury flows to the other end and closes the circuit by providing the contact between the two wires. The wires are fastened to contacts inside the glass bulb, which is filled with an inert gas. This inert gas helps keep the contacts from pitting and burning up. *Photo by Bill Johnson*

arc. When this arc occurs inside the bulb with an inert gas, where there is no oxygen, there is no oxidation.

The mercury bulb is fastened to the movable end of the bimetal, so it is free to rotate with the movement of the bimetal. The wire that connects the mercury bulb to the electrical circuit is very fine to prevent drag on the movement of the bulb. The mercury cannot be in both ends of the bulb at the same time, so when the bimetal rolls the mercury bulb to a new position, the mercury rapidly makes or breaks the electric current flow. This is called *snap* or *detent action*.

Two other types of contacts in the low-voltage thermostat use conventional contact surfaces of silver-coated steel contacts. One is simply an open set of contacts, usually with a protective cover, **Figure 14–4.** The other is a set of silver-coated steel contacts enclosed in a glass bulb. Both of these contacts use a magnet mounted close to the contact to achieve detent or snap action.

HEAT ANTICIPATOR. The *heat anticipator* is a small resistor, which is usually adjustable, located close to the bimetal sensing element, and used to cut off the heating equipment prematurely, **Figure 14–5.** Consider that an oil or a gas furnace has

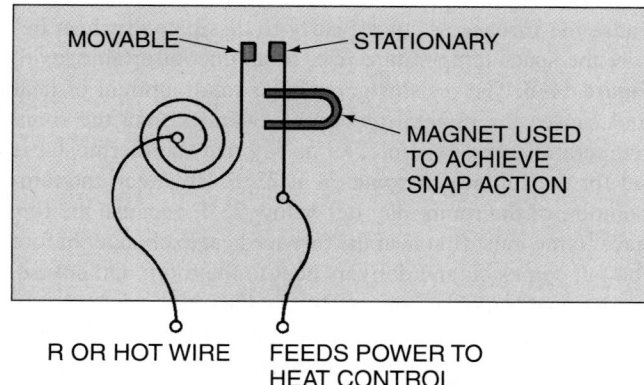

Figure 14–4 A low-voltage thermostat illustrated with open contacts.

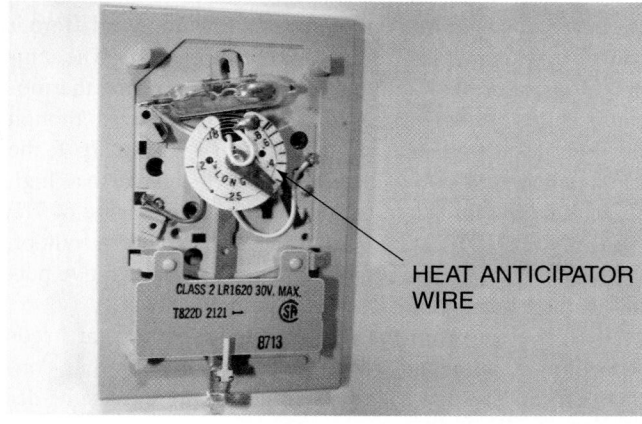

Figure 14–5 The heat anticipator as it appears in an actual thermostat. *Courtesy Ferris State University. Photo by John Tomczyk*

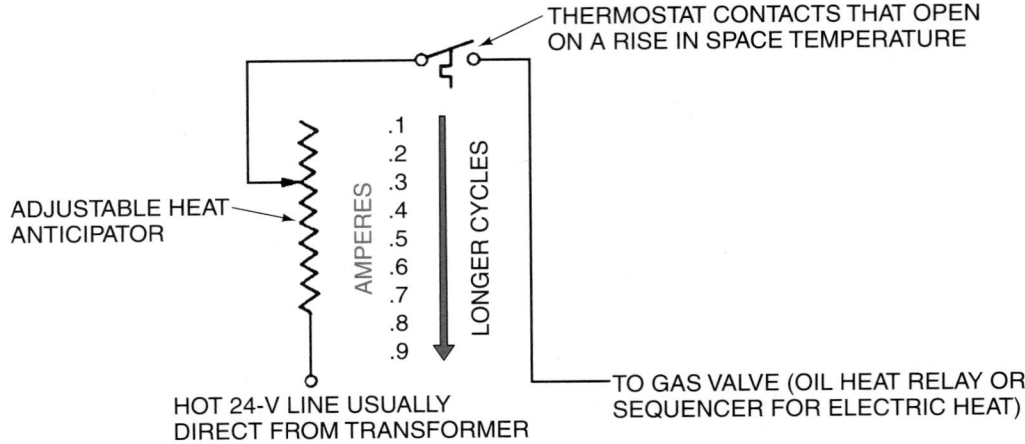

Figure 14–6   The heat anticipator is usually a wirewound, slide-bar-type variable resistor.

heated a home to just the right cut-off point. The furnace itself could weigh several hundred pounds and be hot from running a long time. When the combustion is stopped in the furnace, it still contains a great amount of heat. The fan is allowed to run to dissipate this heat. The heat left in the furnace is enough to drive the house temperature past the comfort point.

The heat anticipator is located inside the thermostat to cause the furnace to cut off early to dissipate this heat before the space temperature rises to an uncomfortable level, **Figure 14–6**. The resistor gives off a small amount of heat and causes the bimetal to become warmer than the room temperature. For example, let us say that the thermostat is set for the furnace to come on at 73°F. However, the temperature of the room may fall below 73°F because the furnace flame must first heat the furnace heat exchanger before the fan comes on and delivers heat to the room. Otherwise, objectionable colder air would be delivered to the room. The temperature drop of the heated room below the set point of the thermostat is referred to as *system lag.* System lag may cause the room to drop in temperature to 71°F before heated air is delivered to the room by the fan.

If the thermostat has a differential of 3°F, the thermostat will open its contacts when its bimetal reaches 76°F (73°F + 3°F). However, the fan will still blow the residual heat from the heat exchanger into the room for a few minutes before it shuts off. This may cause the room to reach 78°F. The temperature rise of the room above the set point of the thermostat is called *system overshoot.* As you can see, even though the thermostat was set to close at 73°F and open at 76°F, the room temperature could fall as low as 71°F and rise to as high as 78°F. This causes a large room *temperature swing* of 7°F (78°F minus 71°F) from system lag and system overshoot. **Figure 14–7,** which has been exaggerated for illustrative purposes, illustrates system lag and system overshoot.

It is obvious from the figure that the thermostat needs something to anticipate system overshoot and open its contacts earlier. This lets the residual heat in the furnace be delivered to the heated space. This is where the heating anticipator comes into play. The heating anticipator actually fools the bimetal of the thermostat into opening its contacts

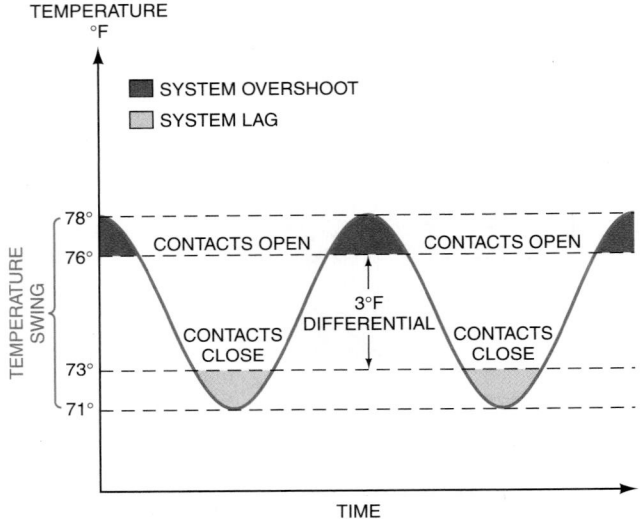

Figure 14–7   Unanticipated thermostat cycles showing system lag and system overshoot causing a large temperature swing.

sooner by placing a small heat load on it. This causes the thermostat's bimetal to be always slightly higher than the room temperature when the thermostat contacts are closed. The thermostat will thus open sooner, and system overshoot will be reduced.

The heat anticipator is wired in series with the mercury bulb heating contacts so that it is energized anytime the thermostat contacts are closed. For the thermostat to control properly, the heat given off from the anticipator must be the same regardless of the current flow through it. Ohm's law says that power in watts is $W = I^2R$. For different values of I (current), different values of R (resistance) will be required to produce the same amount of heat in watts in the anticipator. The designer of the thermostat has precalculated how much resistance will produce the right amount of heat for the proper anticipation.

Because the adjustable anticipator setting is calibrated in amperage to compensate for varying amperage draw,

Figure 14–6, the heating technician must determine what the current flow is through the heating control circuit, thereby determining the current flow through the heating anticipator when the furnace is in a *steady-state condition.* Steady-state conditions are met when the ignition process has shut off and the furnace has been running for about one minute. The most accurate way to measure the current through the heating anticipator is with an AC ammeter measuring current between the R and W terminals of the thermostat's subbase, **Figure 14–8.** When the steady-state current is found, adjust the heating anticipator to that current.

If the heating anticipator is ever set at a higher amperage setting than what it is supposed to be, the smaller resistance in the anticipator would dissipate less heat to the thermostat's bimetal, and greater system overshoot would occur.

Heating anticipators do not eliminate system overshoot and large temperature swings; they simply make them smaller and less objectionable for human comfort. Anytime directions come with a thermostat showing how to set the anticipator, follow them.

Electronic or microelectronic thermostats have different anticipation circuits. Usually these thermostats have a factory built-in microprocessor with software that will determine the cycle rate. Another method is a factory-installed thermistor wired into an electronic circuit that will determine the cycle rate.

THE COLD ANTICIPATOR. The cooling system needs to be started just a few minutes early to allow the air-conditioning system to get up to capacity when needed. If the air-conditioning system were not started until it was needed, it would be 5 to 15 min before it would be producing to full capacity. This is enough time to cause a temperature rise in the conditioned space. The cold anticipator causes the system to start a few minutes early, to anticipate this capacity lag. This is normally a high-resistance fixed resistor that is not adjustable in the field, **Figure 14–9.** The cold anticipator is wired in parallel with the mercury bulb cooling contacts and is energized during the off cycle, **Figure 14–10.** A small

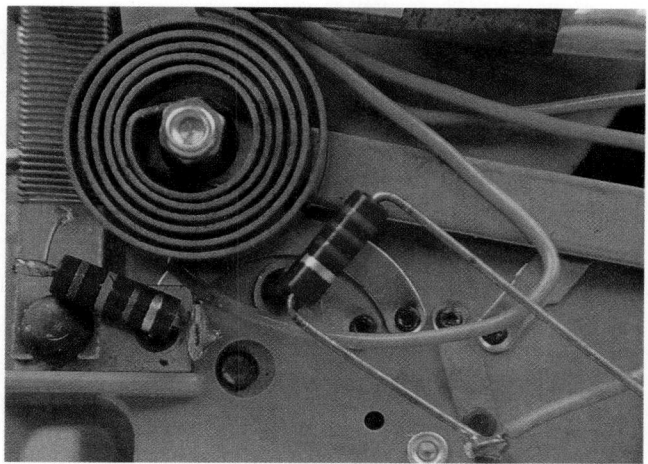

Figure 14–9 The cold anticipator is usually a fixed resistor similar to resistors found in electronic circuitry. It is a small round device with colored bands to denote the resistance and wattage. *Photo by Bill Johnson*

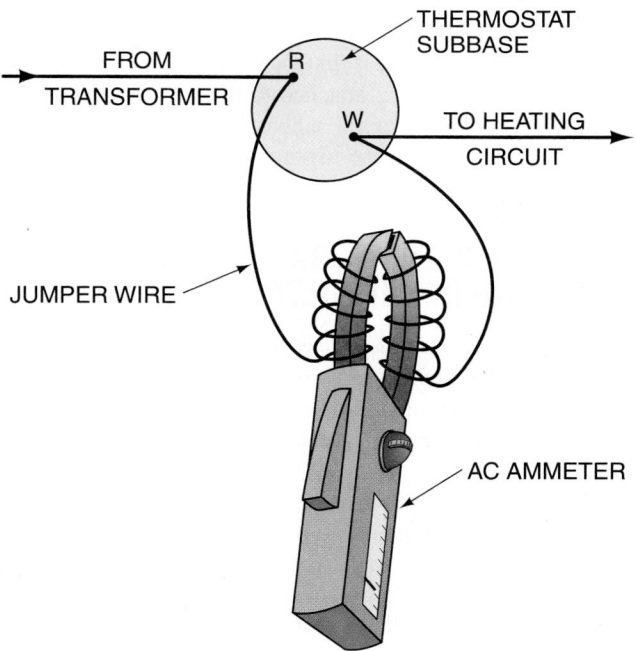

Figure 14–8 An AC ammeter measuring current through the heating circuit. Often, 10 turns of wire wrapped around an ammeter will give a more accurate reading. Do not forget to divide by 10. Modern digital ammeters can read small amperages without the wire turns.

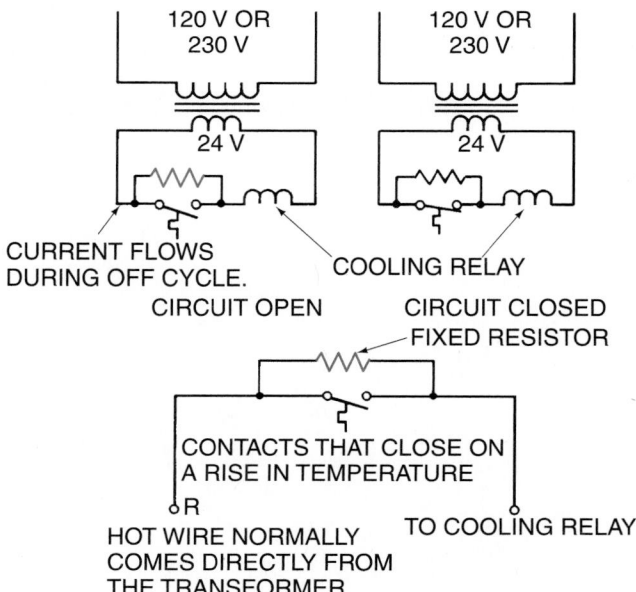

Figure 14–10 The cold anticipator is wired in parallel with the cooling contacts of the thermostat. This allows the current to flow through it during the off cycle.

heat load is put on the thermostat's bimetal by the anticipator when the thermostat's contacts are opened. This will fool the bimetal and keep it at a temperature above the room temperature. This causes the thermostat contacts to close a few minutes early, anticipating the cooling system's lag.

Because the cold anticipator has a high resistance, it will drop most of the voltage when in series with another power-consuming device like the cooling relay coil in **Figure 14–10**. This will prevent the cooling relay coil from being energized when the thermostat contacts are open. The cooling anticipator is, however, giving off heat for the thermostat's bimetal to sense when the thermostat is open. When the thermostat's contacts close, the cooling anticipator is shunted out of the circuit, which allows the cooling relay to be energized and cooling to begin.

THERMOSTAT COVER. The thermostat cover is intended to be decorative and protective. A thermometer is usually mounted on it to indicate the surrounding (ambient) temperature. The thermometer is functionally separate from any of the controls and would serve the same purpose if it were hung on the wall next to the thermostat. Thermostats come in many shapes, **Figure 14–11.**

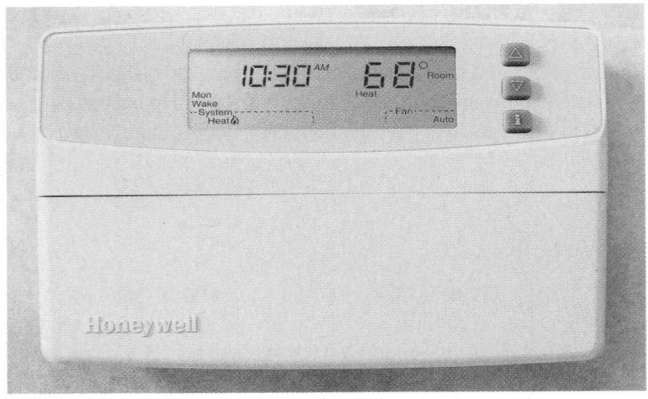

(A)

(B)

Figure 14–11 Thermostat decorative covers. *Courtesy Honeywell*

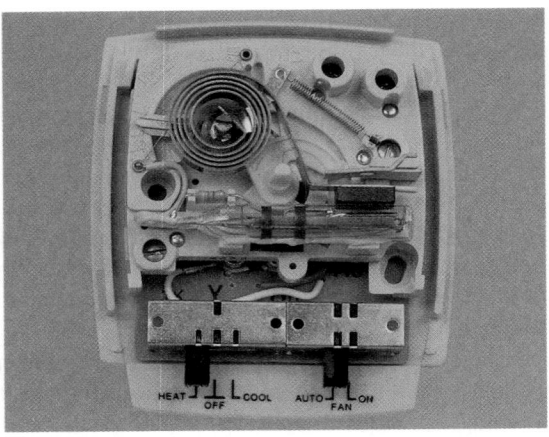

Figure 14–12 A thermostat assembly. *Photo by Bill Johnson*

THERMOSTAT ASSEMBLY. The *thermostat assembly* houses the thermostat components already mentioned and is normally mounted on a subbase fastened to the wall. This assembly could be called the brain of the system. In addition to mercury bulbs and anticipators, the thermostat assembly includes the movable levers that adjust the temperature. These levers or indicators normally point to the set point or desired temperature, **Figure 14–12.** When the thermostat is functioning correctly, the thermometer on the front will read the same as the set point.

THE SUBBASE. The *subbase,* which is usually separate from the thermostat, contains the selector switching levers, such as the FAN-ON switch or the HEAT-OFF-COOL switch, **Figure 14–13.** The subbase is important because the thermostat mounts on it. The subbase is first mounted on the wall, then the interconnecting wiring is attached to the subbase. When the thermostat is attached to the subbase, the electrical connections are made between the two components. Often, cooling anticipators are located on the subbase of the thermostat.

14.4 SPACE TEMPERATURE CONTROLS, HIGH (LINE) VOLTAGE

Sometimes it is desirable to use line-voltage thermostats to stop and start equipment. Some self-contained equipment does not need a remote thermostat. To add a remote thermostat would be an extra expense and would only result in more potential trouble.

The window air conditioner is a good example of this. It is self-contained and needs no remote thermostat. If it had a remote thermostat, it would not be a plug-in device but would require an installation of the thermostat. Note that the remote-bulb type of thermostat is normally used with the bulb located in the return airstream, **Figure 14–14.** The fan usually runs all of the time and keeps a steady stream of room return air passing over the bulb. This gives more sensitivity to this type of application.

The concept of the line-voltage thermostat is used in many types of installations. The household refrigerator,

SOME OF THE FIELD WIRING CONNECTORS

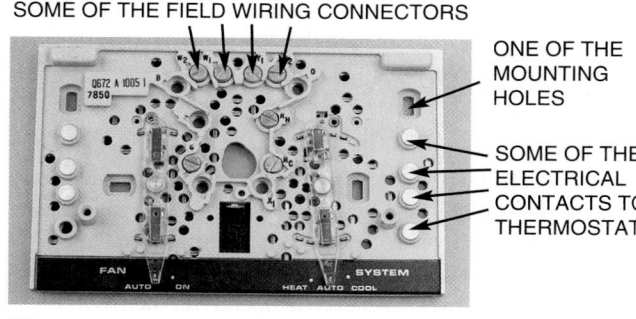

ONE OF THE MOUNTING HOLES

SOME OF THE ELECTRICAL CONTACTS TO THERMOSTAT

(A)

(B)

Figure 14–13 **(A)** The subbase normally mounts on the wall, and the wiring is fastened inside the subbase on terminals. These terminals are designed in such a way as to allow easy wire makeup. When the thermostat is screwed down onto the subbase, electrical connections are made between the two. The subbase normally contains the selector switches, such as FAN ON-AUTO and HEAT-OFF-COOL. **(A)** *Photo by Bill Johnson.* **(B)** *Courtesy Ferris State University. Photo by John Tomczyk*

THERMOSTAT

AIR OUT

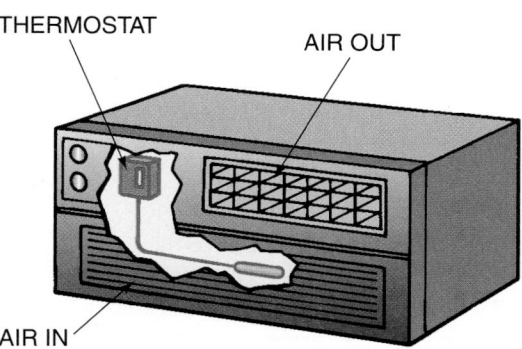

AIR IN

Figure 14–14 The window air conditioner has a line-voltage thermostat that starts and stops the compressor. When the selector switch is turned to cool, the fan comes on and stays on. The thermostat cycles the compressor only.

reach-in coolers, and free-standing package air-conditioning equipment are just a few examples. All of these have something in common—the thermostat is a heavy-duty type and may not be as sensitive as the low-voltage type. When replacing these thermostats, an exact replacement or one recommended by the equipment supplier should be used.

The line-voltage thermostat must be matched to the voltage and current that the circuit is expected to use. For example, a reach-in cooler for a convenience store has a compressor and fan motor to be controlled (stopped and started). The combined running-current draw for both is 16.2 A at 115 V (15.1 A for the compressor and 1.1 A for the fan motor). The locked-rotor amperage for the compressor is 72 A. (Locked-rotor amperage is the inrush current that the circuit must carry until the motor starts turning. The inrush is not considered for the fan because it is so small.) This is a 3/4-hp compressor motor. A reliable, long-lasting thermostat must be chosen to operate this equipment. A control rated at 20 A running current and 80 A locked-rotor current is selected, **Figure 14–15.**

Usually, thermostats are not rated at more than 25 A because the size becomes prohibitive. This limits the size of the compressor that can be started directly with a line-voltage thermostat to about 1 1/2 hp on 115 V or 3 hp on 230 V. Remember, the same motor would draw exactly half of the current when the voltage is doubled.

When larger current-carrying capacities are needed, a motor starter is normally used with a line-voltage thermostat or a low-voltage thermostat.

Line-voltage thermostats usually consist of the following components.

Figure 14–15 The line-voltage thermostat used to stop and start a compressor that draws up to 20 A full-load current has an inrush current up to 80 A. *Courtesy Ferris State University. Photo by John Tomczyk*

THE SWITCHING MECHANISM. Some line-voltage thermostats use mercury as the contact surface, but most switches are a silver-coated base metal. The silver helps conduct current at the contact point. This silver contact point takes the real load in the circuit and is the component in the control that wears first, **Figure 14–16.**

THE SENSING ELEMENT. The sensing element is normally a bimetal, a bellows, or a liquid-filled remote bulb, **Figure 14–17,** located where it can sense the space temperature. If sensitivity is important, a slight air velocity should cross the element. The levers or knobs used to change the adjustment of the thermostat are attached to the main thermostat.

THE COVER. Because of the line voltage inside, the cover is usually attached with some sort of fastener to discourage easy entrance, **Figure 14–18.** If the control is applied to room space temperature, such as a building, a thermometer may be mounted on the cover to read the room temperature. If the control is used to control a box space temperature, such as a reach-in cooler, the cover might be just a plain protective cover.

SUBBASE. When the thermostat is used for room space temperature control, there must be some way to mount it to the wall. A subbase that fits on the electrical outlet box is usually used. The wire leading into a line-voltage control is normally a high-voltage wire routed between points in conduit. The conduit is connected to the box that the thermostat is mounted on. SAFETY PRECAUTION: If an electrical arc due to overload or short circuit occurs, it is enclosed in the conduit box. This reduces fire hazard, **Figure 14–19.**

As mentioned earlier, a room thermostat for the purpose of controlling a space-heating system is the same as a motor temperature thermostat as far as action is concerned. They both open or interrupt the electrical circuit on a rise in temperature. However, the difference in the appearance of the two controls is considerable. One control is designed to sense

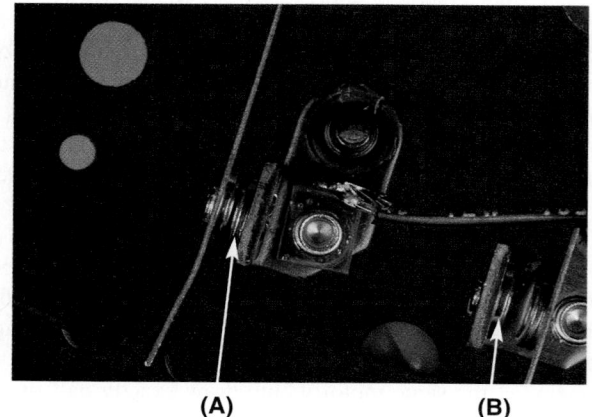

(A)   **(B)**

Figure 14–16   Silver-contact-type line-voltage contacts. **(A)** Closed. **(B)** Open. *Photo by Bill Johnson*

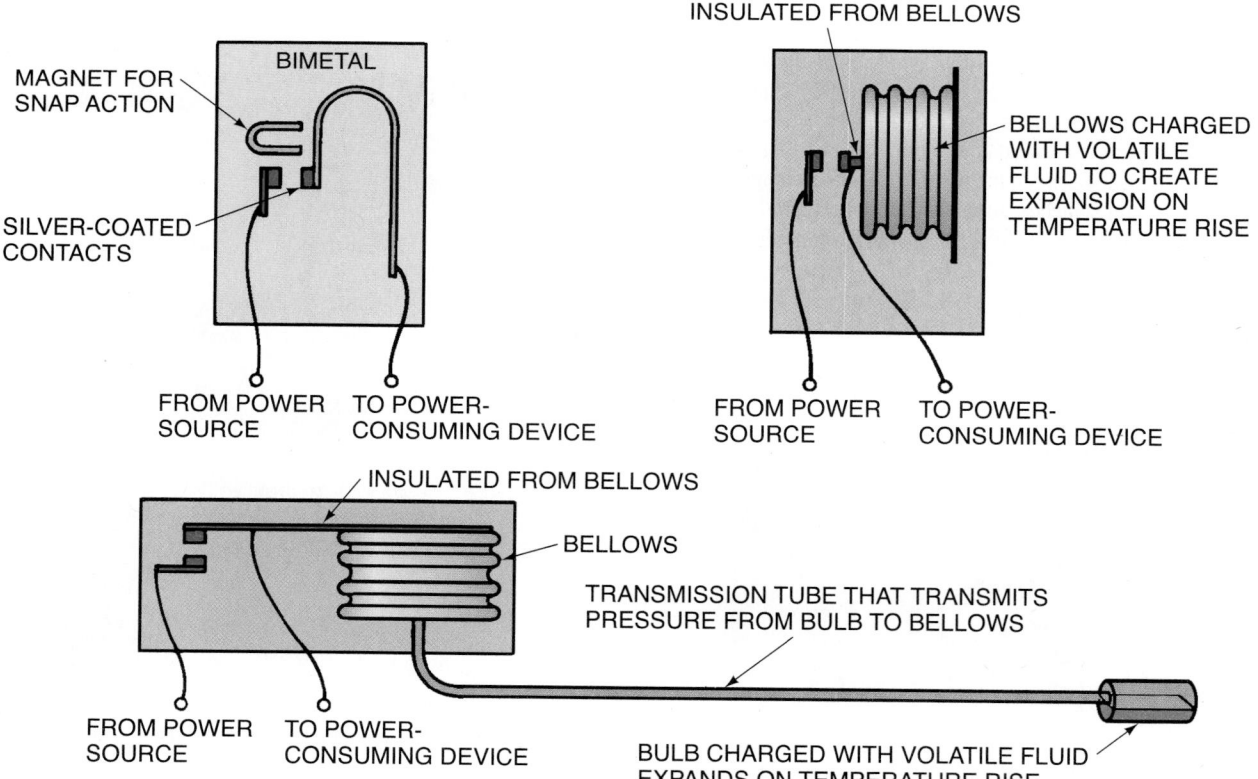

Figure 14–17   Bimetal, bellows, and remote bulb-sensing elements used in line-voltage thermostats. The levers or knobs used to adjust these controls are arranged to apply more or less pressure to the sensing element to vary the temperature range.

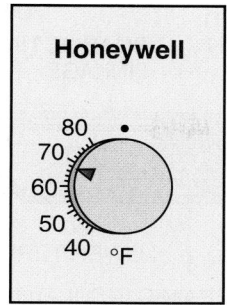

Figure 14–18 The cover for a line-voltage thermostat is usually less decorated than the cover for a low-voltage thermostat.

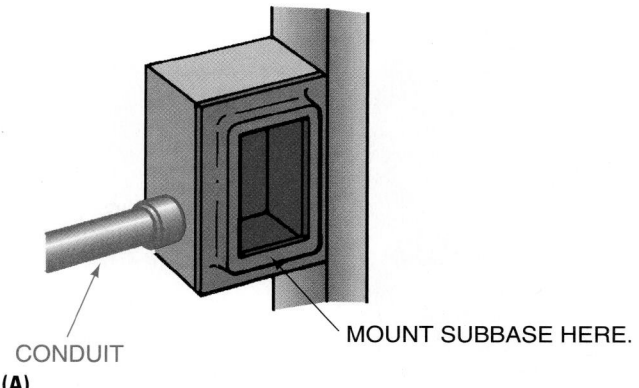

CONDUIT
(A)

MOUNT SUBBASE HERE.

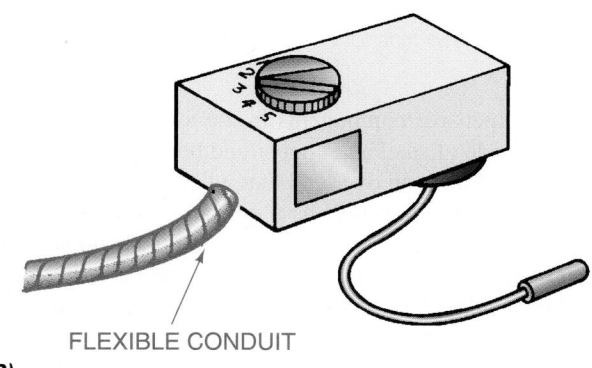

FLEXIBLE CONDUIT
(B)

Figure 14–19 **(A)** The box for a line-voltage, wall-mounted thermostat. **(B)** A remote bulb thermostat with flexible conduit. In both cases the interconnected wiring is covered and protected.

the temperature of slow-moving air, and the other is designed to sense the temperature of the motor winding. The winding has much more mass, and the control must be in much closer contact to get the response needed.

14.5 SENSING THE TEMPERATURE OF SOLIDS

A key point to remember is that any sensing device indicates or reacts to the temperature of the sensing element. A mercury bulb thermometer indicates the temperature of the bulb at the end of the thermometer, not the temperature of the

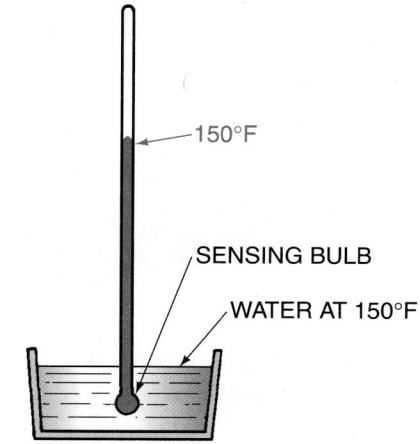

150°F

SENSING BULB

WATER AT 150°F

Figure 14–20 The mercury thermometer rises or falls on the basis of the expansion or contraction of the mercury in the bulb. The bulb is the sensing device.

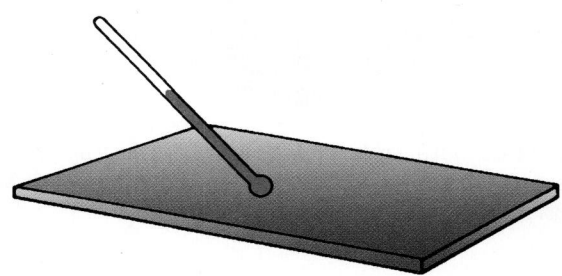

Figure 14–21 Only a fractional part of the mercury bulb thermometer can touch the surface of the flat iron plate at one time. This leaves most of the bulb exposed to the surrounding (ambient) temperature.

substance it is submerged in. If it stays in the substance long enough to attain the temperature of the substance, an accurate reading will be achieved. **Figure 14–20** illustrates the mercury thermometer.

To accurately determine the temperature of solids, the sensing element must assume the temperature of the substance to be sensed as soon as practical. It can be difficult to get a round mercury bulb close enough to a flat piece of metal so that it senses only the temperature of the metal. Only a fractional part of the bulb will touch the flat metal at any time, **Figure 14–21**. This leaves most of the area of the bulb exposed to the surrounding (ambient) air.

Placing insulation over the thermometer to hold it tightly on the plate and to shield the bulb from the ambient air helps give a more accurate reading. Sometimes a gum-type substance is used to hold the thermometer's bulb against the surface and insulate it from the ambient air, **Figure 14–22**. A well in which to insert the thermometer may be necessary for a more permanent installation, **Figure 14–23**.

Most sensing elements will have similar difficulties in reaching the same temperature of the substance to be sensed. Some sensing elements are designed to fit the surface to be sensed. The external motor temperature-sensing element is a good example. It is manufactured flat to fit close to the motor housing, **Figure 14–24**. One reason it can be made flat

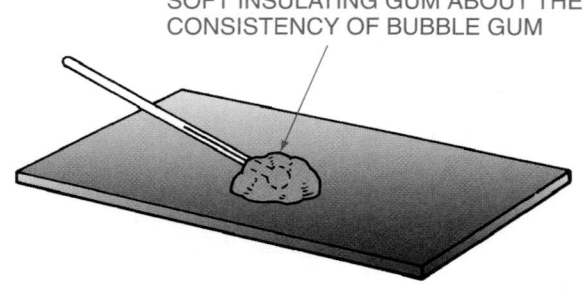

Figure 14–22   An example of temperature sensing often used for field readings. The insulating gum holds the bulb against the metal and insulates it from the ambient air.

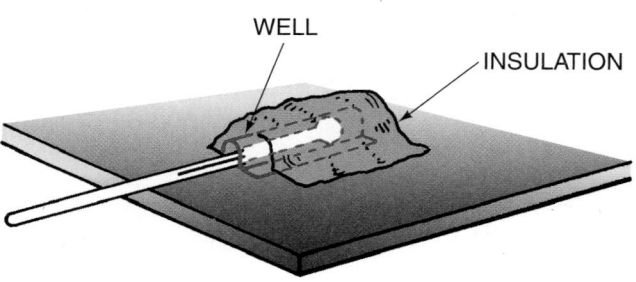

Figure 14–23   An example showing how a well for a thermometer is designed. The well is fastened to the metal plate so that heat will conduct both into and out of the bulb.

Figure 14–24   A motor temperature-sensing thermostat. *Courtesy Ferris State University. Photo by John Tomczyk*

is that it is a bimetal. This control is normally mounted inside the terminal box of the motor or compressor to shield it from the ambient temperature.

The protection of electric motors is important in this industry because motors are the prime movers of refrigerant, air, and water. Motors, especially the compressor motor, are

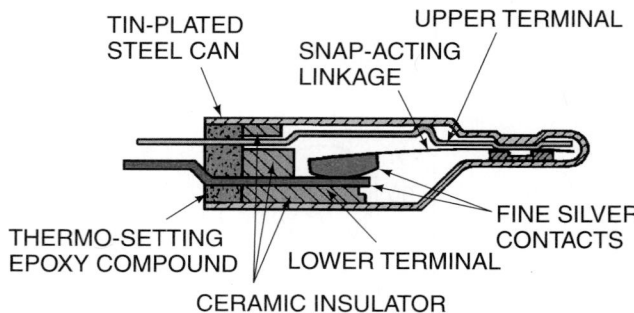

Figure 14–25   A bimetal motor temperature protection device.

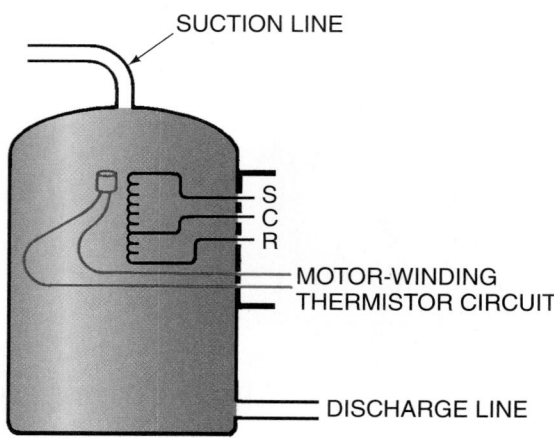

Figure 14–26   A terminal box on the side of the compressor. It has more terminals than the normal common, run, and start of the typical compressor. The extra terminals are for internal motor protection.

the most expensive components in the system. Motors are normally made of steel and copper and need to be protected from heat and from an overload that will cause heat.

All motors build up heat as a normal function of work. The electrical energy that passes to the motor is intended to be converted to magnetism and work, but some of it is converted to heat. All motors have some means of detecting and dissipating this heat under normal or design conditions. If the heat becomes excessive, it must be detected and dealt with. Otherwise, the motor will overheat and be damaged.

Motor high-temperature protection is usually accomplished with some variation of the bimetal or the thermistor. The bimetal device can either be mounted on the outside of the motor, usually in the terminal box, or be embedded in the windings themselves, **Figure 14–25.** The section on motors covers in detail how these devices actually protect the motors.

The thermistor type is normally embedded in the windings. This close contact with the windings gives fast, accurate response, but it also means that the wires must be brought to the outside of the compressor. This involves extra terminals in the compressor terminal box, **Figure 14–26. Figure 14–27** and **Figure 14–28** show wiring diagrams of temperature protection devices.

Because of its size and weight, a motor can take a long time to cool after overheating. If the motor is an open type, a

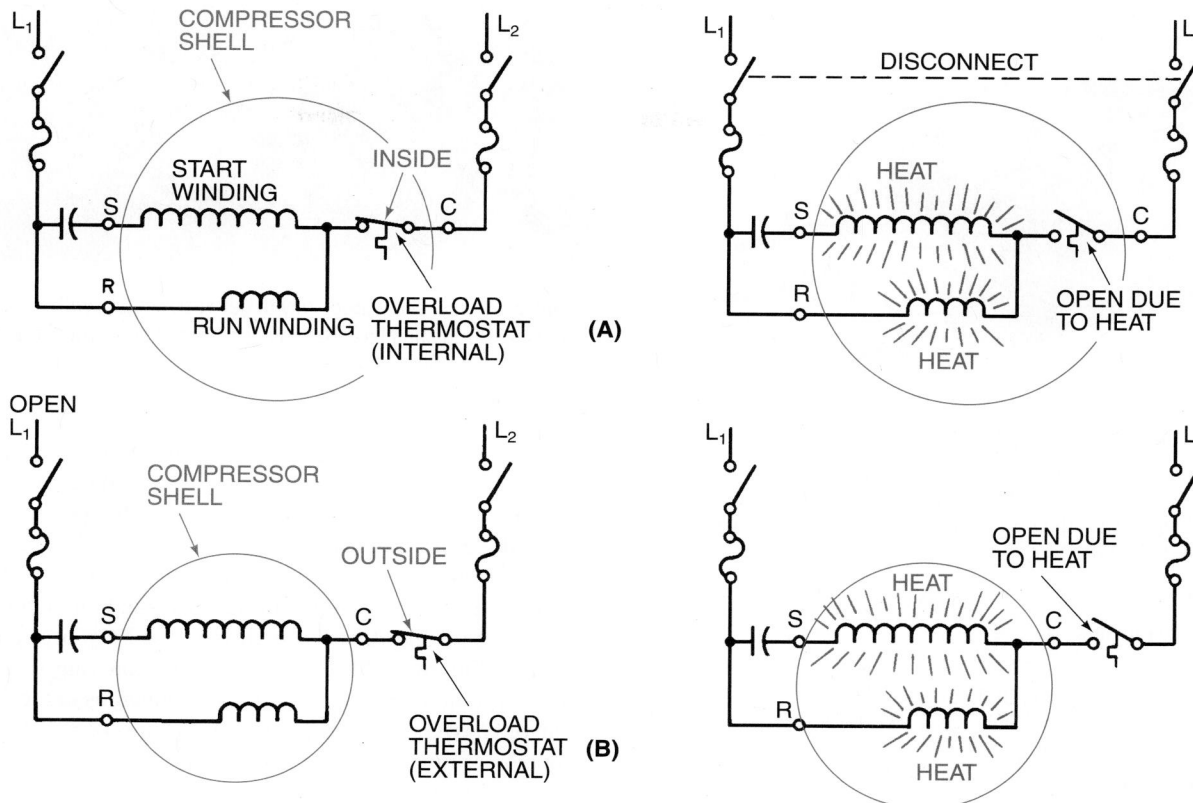

Figure 14–27   Line-voltage bimetal sensing devices under hot and normal operating conditions. **(A)** Permanent split-capacitor motor with internal protection. Note that a meter would indicate an open circuit if the ohm reading were taken at the C terminal to either start or run if the overload thermostat were to open. There is still measurable resistance between start and run. **(B)** The same motor, except that the motor protection is on the outside. Note that it is easier to troubleshoot the overload, but it is not as close to the winding for fast response.

Figure 14–28   The thermistor type of temperature-monitoring device uses an electronic monitoring circuit to check the temperature at the thermistor. When the temperature reaches a predetermined high, the monitor interrupts at the circuit to the contactor coil and stops the compressor.

fan or moving airstream can be devised to cool it more quickly. If the motor is inside a compressor shell, it may be suspended from springs inside the shell. This means that the actual motor and the compressor may be hot and hard to cool even though the shell does not feel hot. **Figure 14–29** shows a method often used to cool a hot compressor. There is a vapor space between the outside of the shell and the actual heat source, **Figure 14–30**. SAFETY PRECAUTION: The unit must have time to cool. If you are in a hurry, set up a fan, or even cool the compressor with water, but do not allow water to get into the electrical circuits. Turn the power off and cover electrical circuits with plastic before using water to cool a hot compressor. Be careful when restarting. Use an ammeter to look for overcurrent and gages to determine the charge level of the equipment. Most hermetic compressors are cooled by the suction gas. If there is an undercharge, there will be undercooling (or overheating) of the motor.

SAFETY PRECAUTION: The best procedure when a hot compressor is encountered is to shut off the compressor with the space temperature thermostat switch and return the next day. This gives it ample time to cool and keeps the crankcase heat on. Many service technicians have diagnosed an open winding in a compressor that was only hot, and they later discovered that the winding was open because of internal thermal protection.

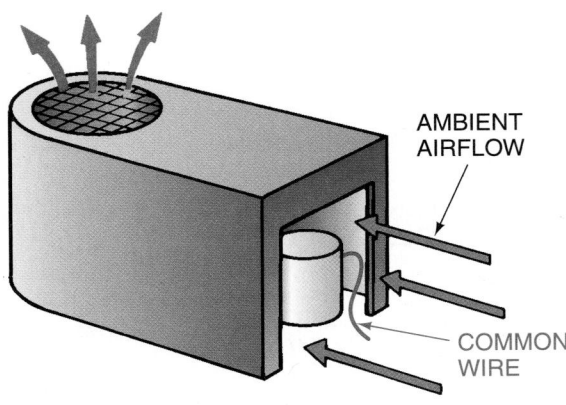

THE COMPRESSOR COMPARTMENT DOOR IS REMOVED AND THE FAN IS STARTED. THIS MAY BE ACCOMPLISHED BY REMOVING THE COMPRESSOR'S COMMON WIRE AND SETTING THE THERMOSTAT TO CALL FOR COOLING.

Figure 14–29 Ambient air is used to cool the compressor motor in the fan compartment.

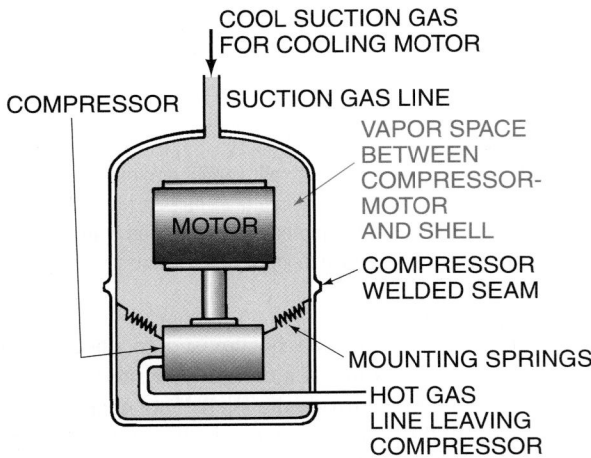

Figure 14–30 The compressor and motor are suspended in vapor space in the compressor shell. The vapor conducts heat slowly.

14.6 MEASURING THE TEMPERATURE OF FLUIDS

The term *fluid* applies to both the liquid and vapor states of matter. Liquids are heavy and change temperature very slowly. The sensing element must be able to reach the temperature of the medium to be measured as soon as practical. Because liquids are contained in vessels (or pipes), the measurement can be made either by contact with the vessel or by some kind of immersion. When a temperature is detected from the outside of the vessel by contact, care must be taken that the ambient temperature does not affect the reading.

A good example of this is the sensing bulb used in refrigeration work for measuring the performance of the thermostatic expansion valve. The sensing bulb must sense refrigerant gas temperature accurately to keep liquid from entering the suction line. Often the technician will strap

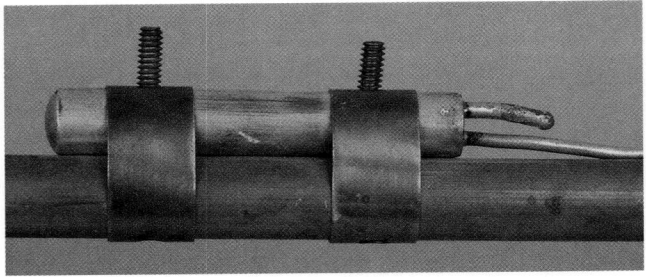

Figure 14–31 The correct way to get good contact with a sensing bulb and the suction line. The bulb is mounted on a straight portion of the line with a strap that holds it securely against the line. *Photo by Bill Johnson*

the sensing bulb to the suction line in the correct location but will fasten it to the line incorrectly. The technician may forget to insulate the bulb from the ambient temperature when it needs to be insulated. In this case the bulb is sensing the ambient temperature and averaging it in with the line temperature. Make sure when mounting a sensing bulb to a line for a contact reading that the bulb is in the very best possible contact, **Figure 14–31**. Brass or copper mounting straps, not plastic, should be used to secure sensing bulbs to the suction line or anywhere else freezing temperatures occur. Brass and copper are much stronger than plastic and can withstand the freezing and thawing cycles. This will prevent breakage from expansion when moisture gets in between the suction line and the sensing bulb and freezes.

When a temperature reading is needed from a larger pipe in a permanent installation, different arrangements are made. A well can be welded into the pipe during installation so that a thermometer or controller-sensing bulb can be inserted into it. The well must be matched to the sensing bulb to get a good contact fit, or you will not get an accurate reading.

Sometimes it is desirable to remove the thermometer from the well and insert an electronic thermometer with a small probe for troubleshooting purposes. The well inside diameter is much larger than the probe. The well can be packed so that the probe is held firm against the well for an accurate reading, **Figure 14–32**.

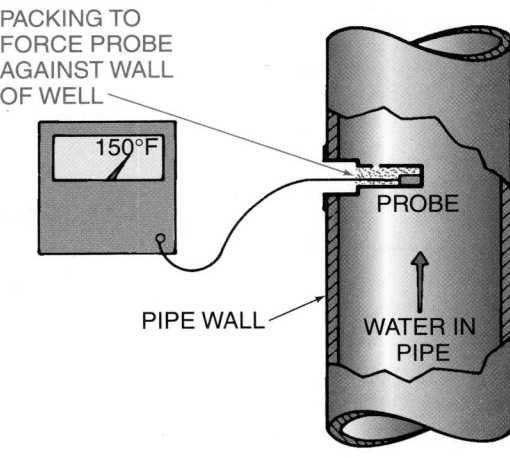

Figure 14–32 The well is packed to get the probe against the wall of the well.

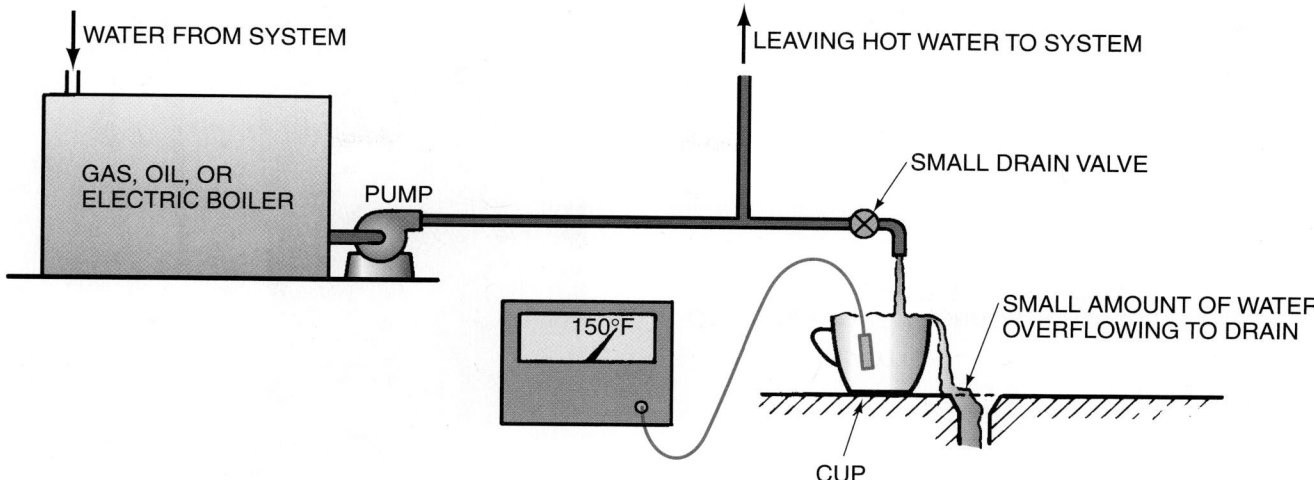

Figure 14–33 A method for obtaining a leaving-water temperature reading from a small boiler.

Another method for obtaining an accurate temperature reading in a water circuit for test purposes is to use one of the valves in a water line to give constant bleed. For instance, if the leaving-water temperature of a home boiler is needed and the thermometer in the well is questionable, try the following procedure. Place a small container under the drain valve in the leaving-water line. Allow a small amount of water from the system to run continuously into the container. An accurate reading can be obtained from this water. This is not a long-term method for detecting temperature because it requires a constant bleeding of water, but it is an effective field method, **Figure 14–33.**

14.7 SENSING TEMPERATURE IN AN AIRSTREAM

Sensing temperature in fast-moving airstreams such as ductwork and furnace heat exchangers is usually done by inserting the sensing element into the actual airstream. A bimetal device such as the flat-type snap-disc or the helix coil is usually used. **Figure 14–34** shows a snap-disc.

14.8 THINGS TO REMEMBER ABOUT SENSING DEVICES

Sensing devices are not mysterious; something, such as a bimetal device or a thermistor, reacts to the temperature change. Look over any temperature-sensing control and study

it, and you will usually be able to determine the way it operates. If you are still confused, consult a catalog or the supplier of the control.

14.9 PRESSURE-SENSING DEVICES

Pressure-sensing devices are normally used when measuring or controlling the pressure of refrigerants, air, gas, and water. These are sometimes strictly pressure controls, or they can be used to operate electrical switching devices. The terms *pressure control* or *pressure switch* are often used interchangeably in the field. The actual application of the component should indicate if the device is used to control a fluid or to operate an electrical switching device.

Some applications for pressure-operated controls are as follows:

1. Pressure switches are used to stop and start electrical loads, such as motors, **Figure 14–35.**
2. Pressure controls contain a bellows, a diaphragm, or a Bourdon tube to create movement when the pressure inside it is changed. Pressure controls may be attached to switches or valves, **Figure 14–36.**

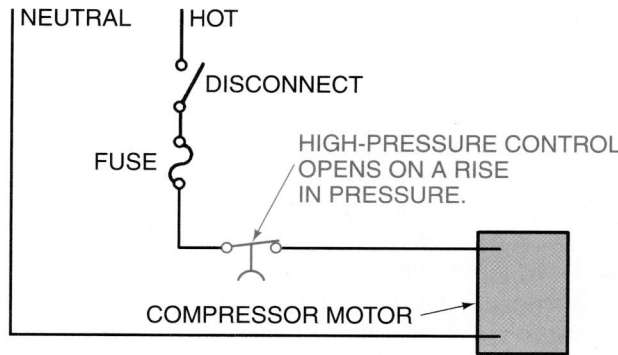

Figure 14–35 The electrical circuit of a refrigeration compressor with a high-pressure control. This control has a normally closed circuit that opens on a rise in pressure.

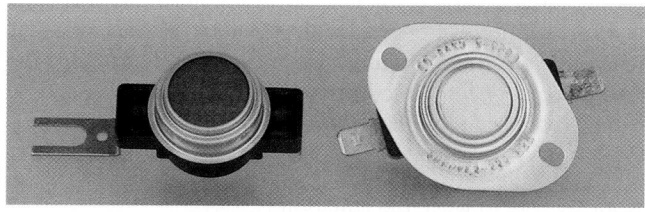

Figure 14–34 A snap-disc, which can be used to sense air temperature in ductwork and furnaces. *Photo by Bill Johnson*

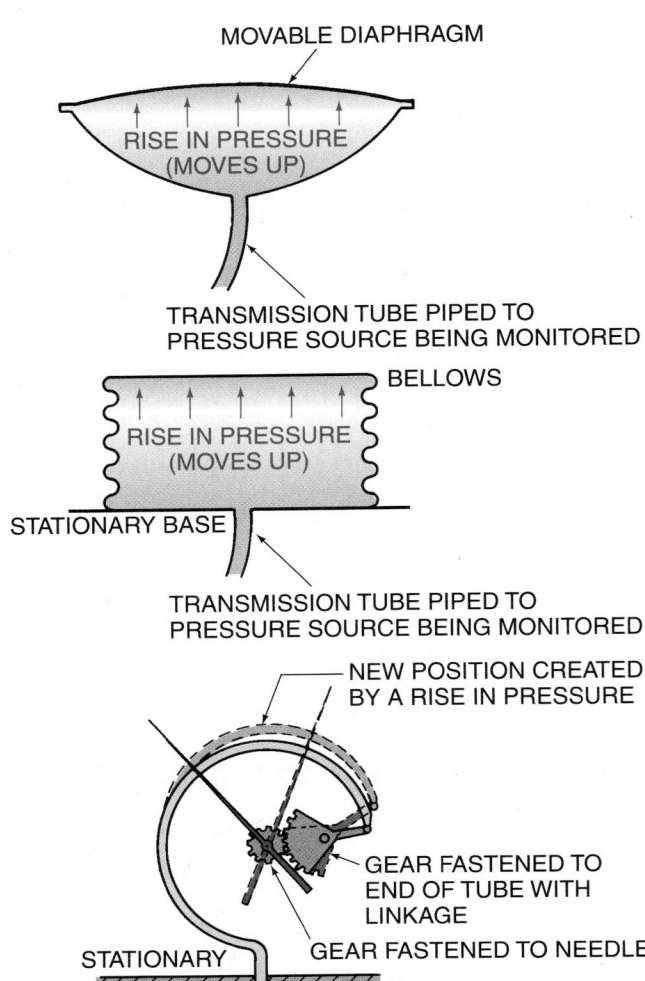

Figure 14–36 The moving part of most pressure-type controls.

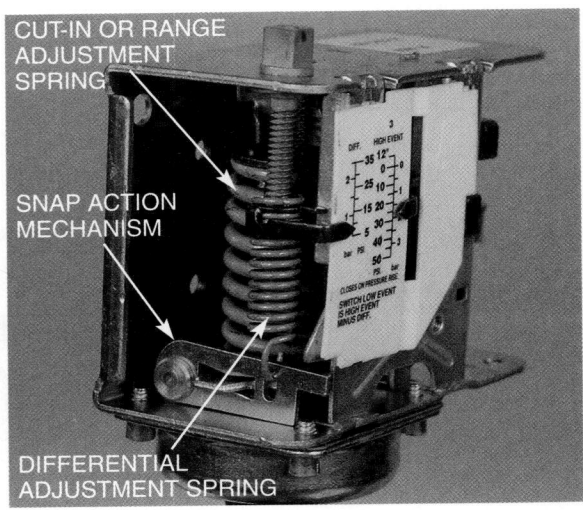

Figure 14–37 Snap action over-center device. *Photo by Bill Johnson*

3. When used as a switch, the bellows, Bourdon tube, or diaphragm is attached to the linkage that operates the electrical contacts. When used as a valve, it is normally attached directly to the valve.

4. The electrical contacts are the components that actually open and close the electrical circuit.

5. The electrical contacts either open or close with snap action on a rise in pressure, **Figure 14–37**.

6. The pressure control can either open or close on a rise in pressure. This opening and closing action can control water or other fluids, depending on the type.

7. The pressure control can sense a pressure differential and be designed to open or close a set of electrical contacts, **Figure 14–38**.

8. The pressure control can be an operating-type control or a safety-type control, **Figure 14–39**.

9. It can operate either at low pressures (even below atmospheric pressure) or high pressures, depending on the design of the control mechanism.

10. Pressure controls can sometimes be recognized by the small pipe running to them for measuring fluid pressures.

11. Pressure switches are manufactured to handle control voltages or line currents that start a compressor up to about 3 hp maximum. The refrigeration industry is the only industry that uses the high-current draw controls.

12. The high-pressure and low-pressure controls in refrigeration and air-conditioning equipment are the two most widely used pressure controls in this industry.

13. Some pressure switches are adjustable (**Figure 14–40**), and some are not.

14. Some controls are automatic reset, and some are manual reset, **Figure 14–41**.

15. In some pressure controls the high-pressure and the low-pressure controls are built into one housing. They are called dual-pressure controls, **Figure 14–42**.

16. Pressure controls are usually located near the compressor on air-conditioning and refrigeration equipment.

17. SAFETY PRECAUTION: When used as safety controls, pressure controls should be installed in such a manner that they are not subject to being valved off by the service valves.

18. The point or pressure setting at which the control interrupts the electrical circuit is known as the *cut-out*. The point or pressure setting at which the electrical circuit is made is known as the *cut-in*. The difference in the two settings is known as the *differential*.

Both pressure and temperature controls incorporate cut-out, cut-in, and differential adjustments in the HVACR field. As mentioned earlier, the cut-out of a control is where the control interrupts or opens the electrical circuit. The cut-in is where the electrical circuit closes, and the differential is the difference between the cut-in and cut-out points. This logic can be represented in the following formula.

$$\text{Cut-in} - \text{Cut-out} = \text{Differential}$$

Another way to write the equation is:

$$\text{Cut-in} - \text{Differential} = \text{Cut-out}$$

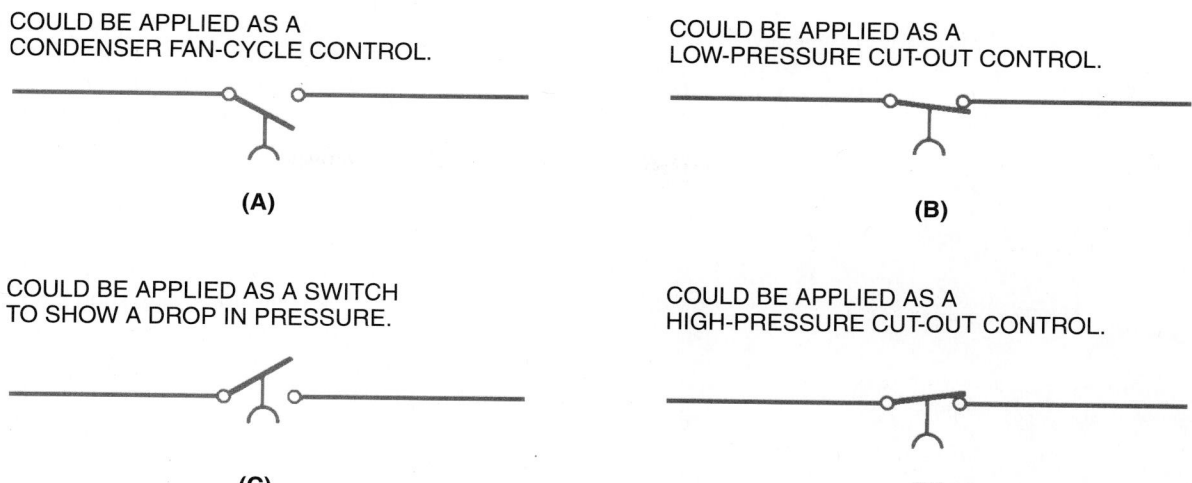

COULD BE APPLIED AS A
CONDENSER FAN-CYCLE CONTROL.

(A)

COULD BE APPLIED AS A
LOW-PRESSURE CUT-OUT CONTROL.

(B)

COULD BE APPLIED AS A SWITCH
TO SHOW A DROP IN PRESSURE.

(C)

COULD BE APPLIED AS A
HIGH-PRESSURE CUT-OUT CONTROL.

(D)

Figure 14–38 These symbols show how pressure controls connected to switches appear on control diagrams. Symbols A and B indicate that the circuit will make a rise in pressure. Symbol A indicates the switch is normally open when the machine does not have power to the electrical circuit. B shows that the switch is normally closed without power. Symbols C and D indicate that the circuit will open on a rise in pressure. C is normally open, and D is normally closed.

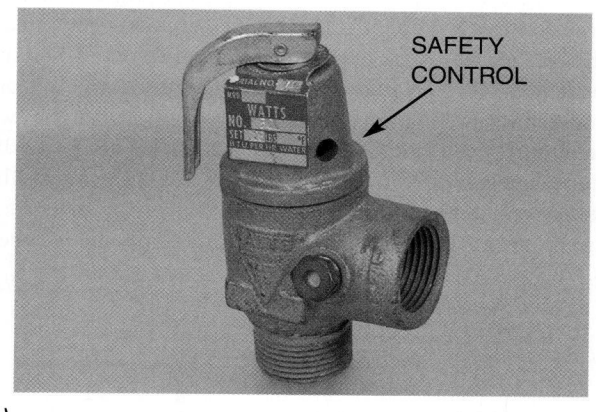

SAFETY
CONTROL

(A)

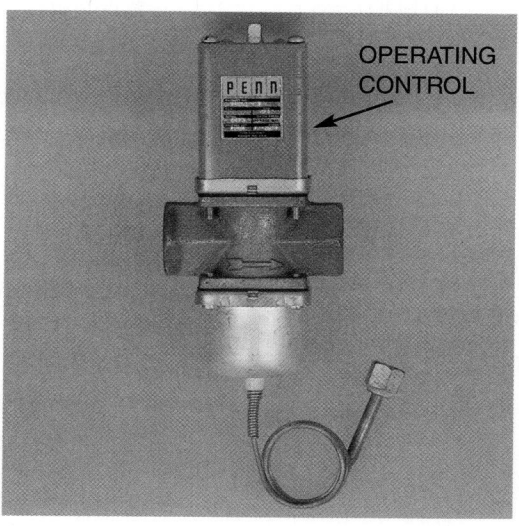

OPERATING
CONTROL

(B)

Figure 14–39 **(A)** A safety control (boiler relief valve). **(B)** An operating control (water-regulating valve). *Photos by Bill Johnson*

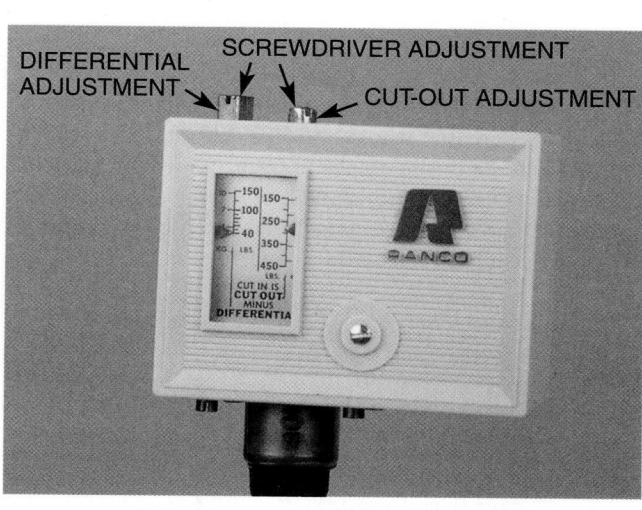

DIFFERENTIAL
ADJUSTMENT

SCREWDRIVER ADJUSTMENT

CUT-OUT ADJUSTMENT

Figure 14–40 A commonly used high-pressure control. *Photo by Bill Johnson*

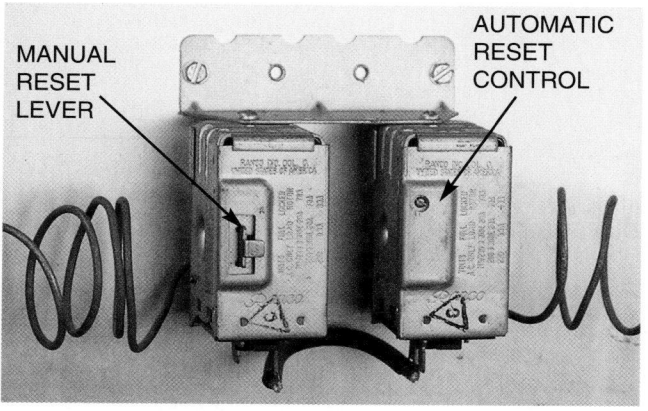

MANUAL
RESET
LEVER

AUTOMATIC
RESET
CONTROL

Figure 14–41 The control on the left is a manual reset control. The one on the right is automatic reset. Note the push lever on the left-hand control. *Photo by Bill Johnson*

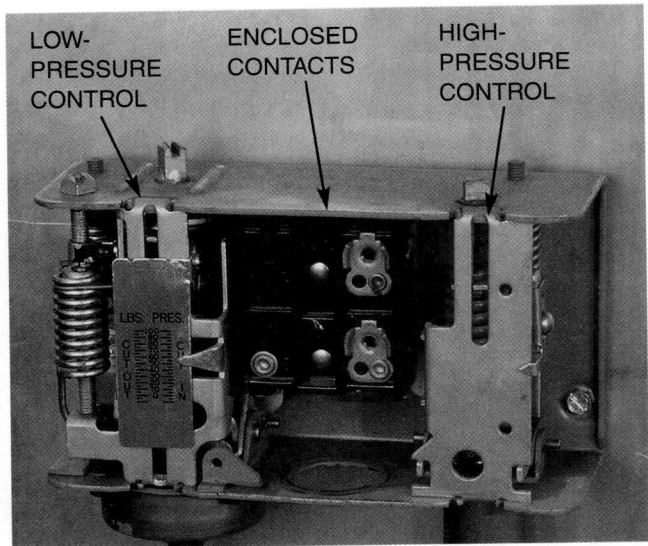

Figure 14–42 The control has two bellows acting on one set of contacts. Either control can stop the compressor. *Photo by Bill Johnson*

If a pressure control has a cut-in of 20 psi and a cut-out of 5 psi, the differential would be 15 psi (20 psi − 5 psi). **Figure 14–43** shows this in graphical form. Notice that this control language can be used for both pressure and temperature controls. Some control manufacturers refer to the cut-out as the "low event" and the cut-in as the "high event."

As you can see, the differential controls the pressure or temperature difference between the cut-in and cut-out settings. The differential adjustment is usually not easily accessible to the owner or operator when making a temperature or pressure change with the control. It is usually protected or locked up by some sort of mechanism that takes time and knowledge to access. In most cases, only a service technician makes a differential adjustment to the control. If the differential is set too low, the compressor motor or other power-consuming devices it is controlling may short cycle. Short cycling can overheat motor windings and prematurely wear and pit electrical contacts. On the other hand, if the differential is set too high, the motor may never shut off and may draw unwanted high power. Some differential controls can be nonadjustable and built into the control as a fixed differential. Most high-pressure controllers and inexpensive box thermostats are nonadjustable.

Changing the cut-in setting without changing the differential will automatically change the cut-out setting, **Figure 14–44**. The *range* of the control is increased and so is the average box temperature. In the figure, the beginning range was from 5°F to 20°F. By changing the cut-in adjustment of the control to cut in at 25°F, the cut-out point is automatically moved to 10°F because of the differential staying constant at 15°F. Notice that both the cut-in and cut-out temperatures are higher, but the difference between the two temperatures (differential) does not change. This gives the control a new range, from 10°F to 25°F. It also increases the average box temperature. The range provides for the correct minimum and maximum pressures or temperatures when

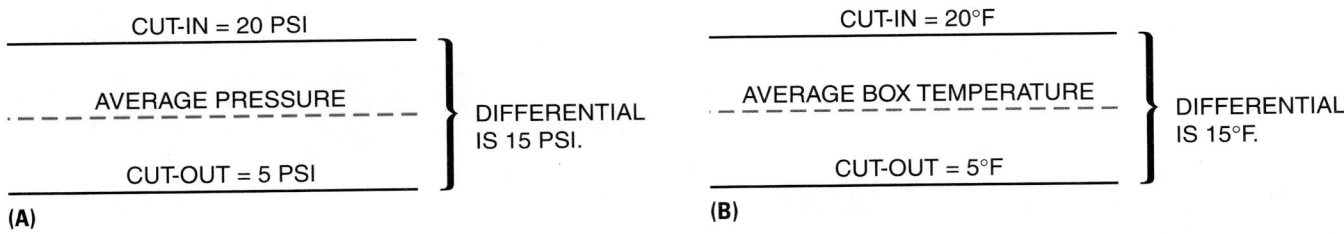

Figure 14–43 Cut-in, cut-out, and differential points shown in a graphical format for **(A)** a pressure controller and **(B)** a temperature controller.

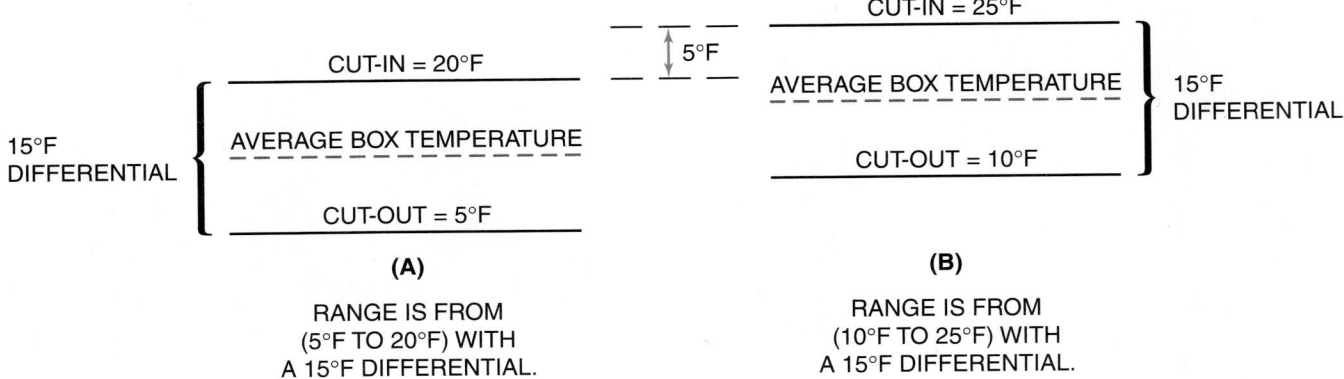

Figure 14–44 A change in the cut-in adjustment by 5°F in **(B)** will automatically change the cut-out setting by 5°F. This changes the range. Notice that the differential is the same in both examples **(A)** and **(B)**.

controlled automatically by a controller. There must be a range of temperatures to control because it is very difficult to control the box temperature of a refrigerated case at one exact temperature.

The cut-in adjustment of a control is often referred to as the range adjustment because changing the cut-in causes both the cut-out and cut-in (range of the control) to change, as in **Figure 14–44**. If the differential of the control is changed, the range of the control will also be changed. This is because the differential is the distance between the two settings.

The range (cut-in) adjustment spring always acts directly on the control bellows that operates the temperature- or pressure-control electrical switch. The differential adjustment or spring is usually set to the side of the bellows and adds resistance to the mechanism that the control's bellows is acting on. This action makes it either easier or harder for the electrical contact points to open or close, **Figure 14–37**. Applications and more detailed examples of cut-in, cut-out, and differential settings will be covered in Unit 25, "Special Refrigeration System Components."

Some manufacturers have electronic pressure controls on the market, **Figure 14–45**. They often have small microprocessors on board that are hardwired to an electronic pressure transducer. These versatile electronic controls have easy-to-read digital liquid crystal displays. Some of their features include the following:

- Selectable pressure ranges
- Anti–short cycling devices
- Remote pressure transducer (easily mounted)
- Wiring harness (no capillary tubes)
- Lockable touch pads (prevents tampering)
- Ability to open contacts on increase or decrease in pressure

14.10   PRESSURE TRANSDUCERS

Pressure transducers are devices that measure pressure by converting a sensed pressure signal to an electronic signal to be processed by a microprocessor. Two modern pressure transducers are shown in **Figure 14–46**. Most pressure transducers today are capacitive-type pressure transducers. They are usually constructed of a durable, compact housing containing two closely spaced parallel plates, **Figure 14–47**. The parallel plates are generally made of some sort of metal and are electrically isolated from one another. However, the plates can be made of other materials. One of the metallic plates is stationary. The other acts like a diaphragm that slightly flexes when pressure is applied to its surface. A slight mechanical flexing of the movable plate caused by the applied pressure will change the gap between the parallel plates. This creates a capacitor that varies its capacitance as more or less pressure is applied to the flexible plate or diaphragm. This varying capacitor, or change in capacitance, is sensed by solid-state linear comparative circuitry that is usually linked to a microprocessor. The solid-state circuit analyzes the signal and then amplifies its proportional output signal. Capacitive-type pressure transducers can sense pressures in the ranges of 0.1 in. of water column (WC) to 10,000 psig. Older and less accurate pressure transducers can operate on the strain gage principle or simply off a mechanical linkage mechanism.

Figure 14–48 shows a working pressure transducer connected to the common suction line of a parallel compressor system. The pressure transducer will sense the common suction pressure in the suction line and convert this signal to a voltage signal for a microprocessor to analyze. This will cycle different size compressors on and off to more accurately match the heat loads on the refrigerated cases in the store. There is also solid-state circuitry in the head of the transducer as shown in **Figure 14–49**. The transducer usually has three wires connected to it. Two wires are its input voltage, which is usually a DC voltage, and the other wire is the output signal, which can also (but not always) be a DC voltage. The three wires usually originate from a microprocessor board,

Figure 14–45   An electronic pressure control with a connecting pressure transducer. *Photo by John Tomczyk*

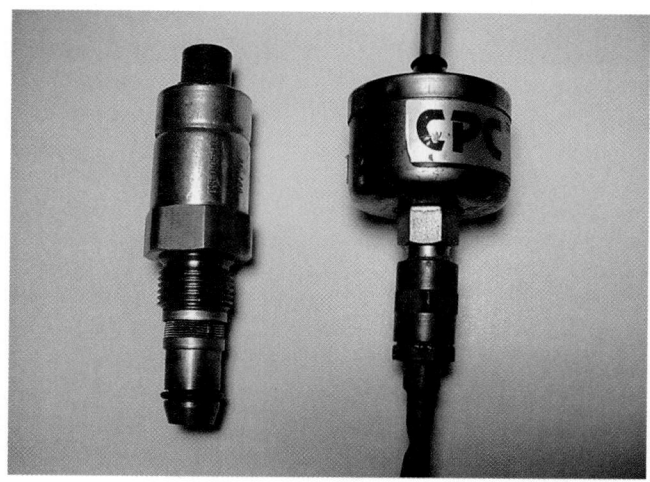

Figure 14–46   Two modern pressure transducers. *Photo by John Tomczyk*

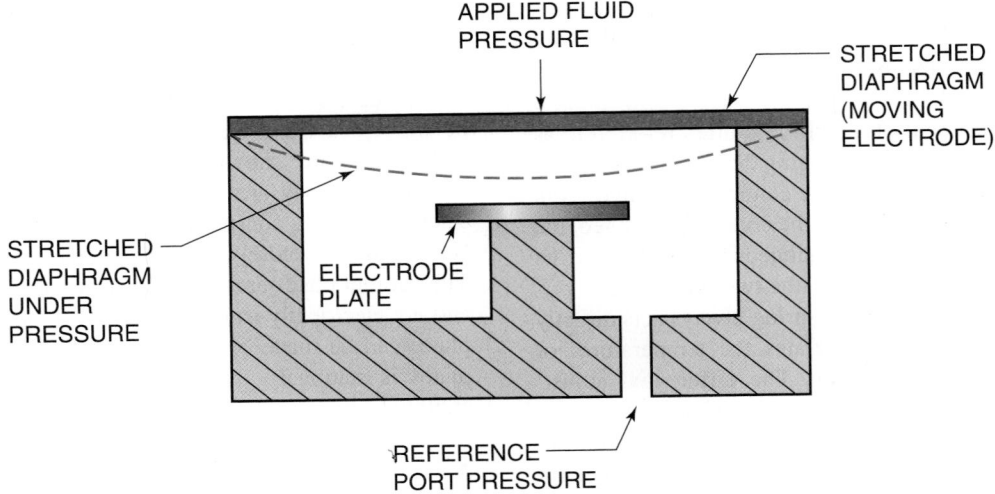

Figure 14–47 The cross section of a capacitive-type pressure transducer showing a movable diaphragm, which creates a variable capacitance in a circuit.

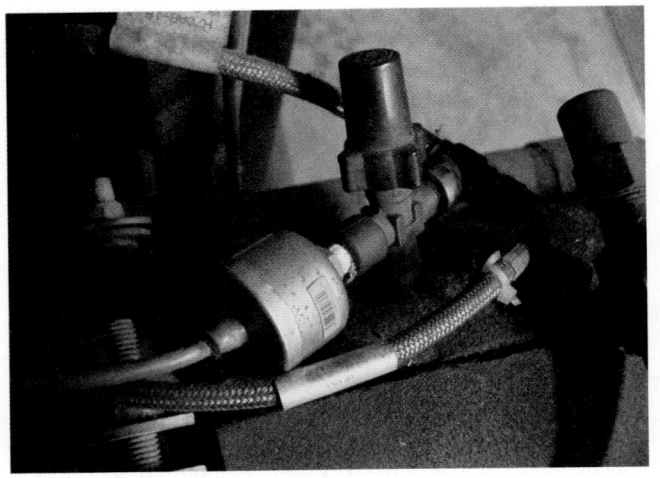

Figure 14–48 A working pressure transducer on a parallel compressor rack's suction line. *Photo by John Tomczyk*

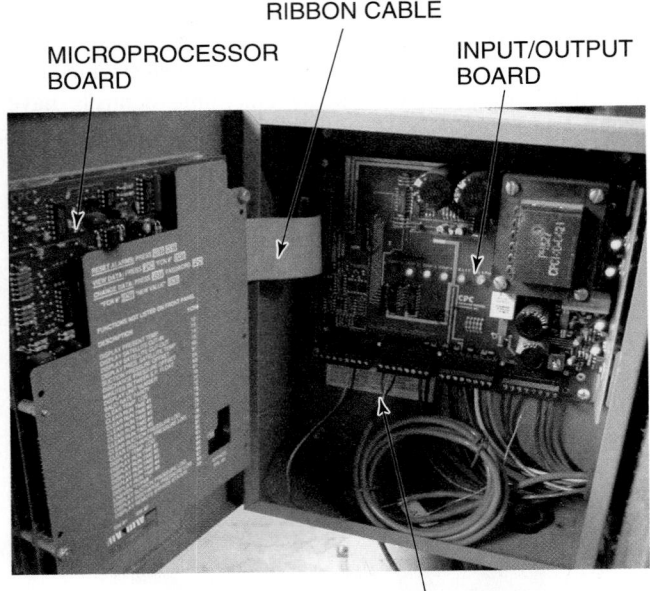

Figure 14–50 A microprocessor board ribbon-cabled to an input/output board. *Photo by John Tomczyk*

which is ribbon-cabled to an input/output board, **Figure 14–50.** The transducer senses pressure and converts the signal to an output voltage signal, which is processed by the microprocessor board. This signal is then sent to the input/output board to operate relays to either start or stop power-consuming devices.

14.11 HIGH-PRESSURE CONTROLS

The high-pressure control (switch) on an air conditioner stops the compressor if the pressure on the high-pressure side becomes excessive. This control appears in the wiring diagram as a normally closed control that opens on a rise in

Figure 14–49 Solid-state circuitry in the head of the pressure transducer. *Photo by John Tomczyk*

(A)

(B)

Figure 14–51 High-pressure controls. **(A)** Automatic reset.
(B) Manual reset. *Photos by Bill Johnson*

compressor continues to operate, very high pressures will occur. SAFETY PRECAUTION: The high-pressure control is one method of ensuring safety for the equipment and the surroundings. Some compressors are strong enough to burst a pipe or a container. Often, on semihermetic compressors, pressures can elevate high enough to blow out a head gasket or a valve plate gasket. The results can be a large refrigerant loss, not to mention a spoiled product. The overload device in the compressor offers some protection in this event, but it is really a secondary device because it is not directly responding to the pressures. The motor overload device may also be a little slow to respond.

14.12 LOW-PRESSURE CONTROLS

The low-pressure control (switch) is used in the air-conditioning field as a low-charge protection, **Figure 14–52**. In commercial refrigeration applications, low-pressure controls are used to cycle compressors on and off in response to evaporator pressures that indirectly control box temperatures. They can also be used in ice makers to initiate the harvest cycle. When the equipment loses some of the refrigerant from the system, the pressure on the low-pressure side of the system will fall. Manufacturers may have a minimum pressure under which they will not allow the equipment to operate. This is the point at which the low-pressure control cuts off the compressor. Manufacturers use different settings based on their requirements, so their recommendations should be followed.

The recent popularity of the capillary tube as a metering device has caused the low-pressure control to be reconsidered as a standard control on all equipment. The capillary tube metering device equalizes pressure during the off cycle and causes the low-pressure control to short cycle the compressor if it is not carefully applied. The capillary tube is a fixed-bore metering device and has no shutoff valve action.

pressure. The manufacturer may determine the upper limit of operation for a particular piece of equipment and furnish a high-pressure cut-out control to ensure that the equipment does not operate above these limits, **Figure 14–51**. High-pressure controls on HVACR equipment usually have a built-in differential of about 50 psig. For example, if the control opens at a pressure of 400 psig, it will usually reset itself (close contacts) at 350 psig.

A reciprocating compressor is known as a positive displacement device. When it has a cylinder full of vapor, it is going to pump out the vapor or stall. If a condenser fan motor on an air-cooled piece of equipment burns out and the

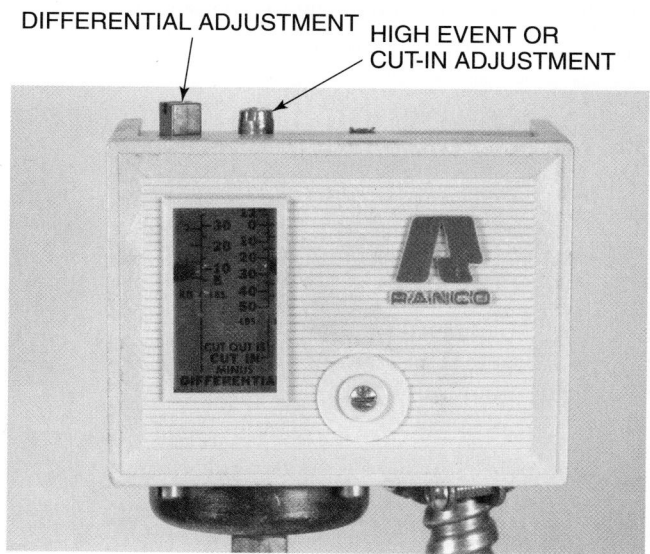

DIFFERENTIAL ADJUSTMENT   HIGH EVENT OR
CUT-IN ADJUSTMENT

Figure 14–52 The low-pressure control. *Photo by Bill Johnson*

To prevent this short cycle, some equipment comes with a time-delay circuit that will not allow the compressor to restart for a predetermined time period.

It is undesirable to operate a system without an adequate charge for two good reasons:

1. The compressor motors used most commonly in this industry, particularly in the air-conditioning field, are cooled by the refrigerant. Without this cooling action, the motor will build up heat when the charge is low. The motor temperature cut-out is used to detect this condition. It often takes the place of the low-pressure cut-out by sensing motor temperature.

2. If the refrigerant escaped from the system through a leak in the low side of the system, the system can operate until it goes into a vacuum. When a vessel is in a vacuum, the atmospheric pressure is greater than the vessel pressure. This causes the atmosphere to be pushed into the system. Technicians often say "pulled air into the system." The reference point the technician uses is atmospheric pressure. This air in the system is sometimes hard to detect; if it is not removed it will be pumped through the evaporator and the compressor. Because air is a noncondensable gas, it will not condense as refrigerant will. The air will now take up valuable volume in the condenser and cause unwanted high head pressures. These higher head pressures cause high-compression ratios and low efficiencies. The compressor's discharge temperature will now elevate to a point where oil may break down. Also, the combination of oxygen and moisture in the air, excess heat, and the refrigerant will create an oil sludge and acid formations.

14.13 OIL PRESSURE SAFETY CONTROLS

The oil pressure safety control (switch) is used to ensure that the compressor has oil pressure when operating, **Figure 14–53**. This control is used on larger compressors and has a different sensing arrangement than the high- and low-pressure controls. The high- and low-pressure controls are

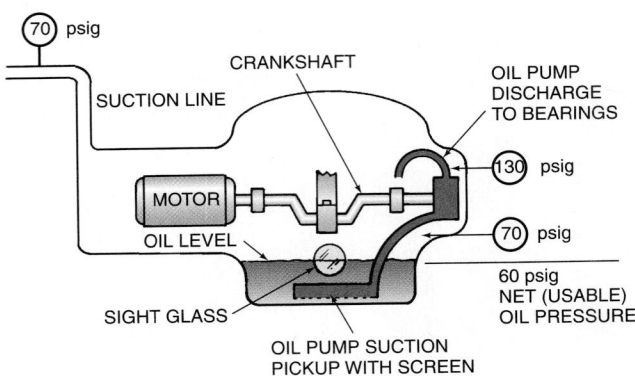

Figure 14–54 The oil pump suction is actually the suction pressure of the compressor. This means that the true oil pump pressure is the oil pump discharge pressure less the compressor suction pressure. For example, if the oil pump discharge is 130 psig and the compressor suction pressure is 70 psig, the net oil pressure is 60 psig. This is usable oil pressure.

single-diaphragm or single-bellows controls, because they are comparing atmospheric pressure with the pressures inside the system. Atmospheric pressure can be considered a constant for any particular locality because it does not vary more than a small amount.

The oil pressure safety control is a pressure differential control. This control measures a difference in pressure to establish that positive oil pressure is present. A study of the compressor will show that the compressor crankcase (this is where the oil pump suction inlet is located) is the same as the compressor suction pressure, **Figure 14–54**. The suction pressure will vary from the off or standing reading to the actual running reading. For example, when a system is using R-22 as the refrigerant, the pressures may be similar to the following: 125 psig while standing, 70 psig while operating, and 20 psig during a low-charge situation.

A plain low-pressure cut-out control would not function at all of these levels, so a control had to be devised that would sensibly monitor pressures at all of these conditions.

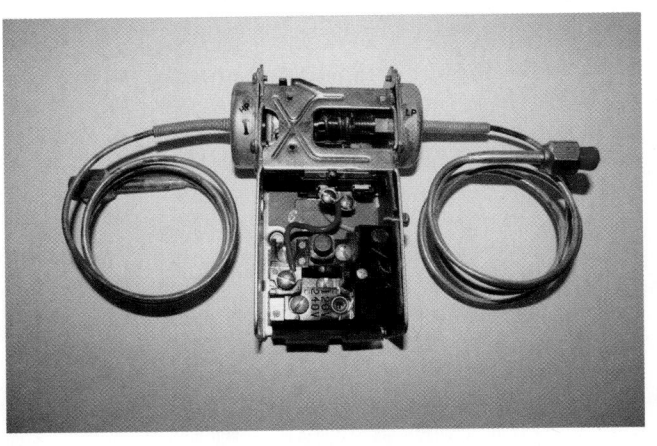

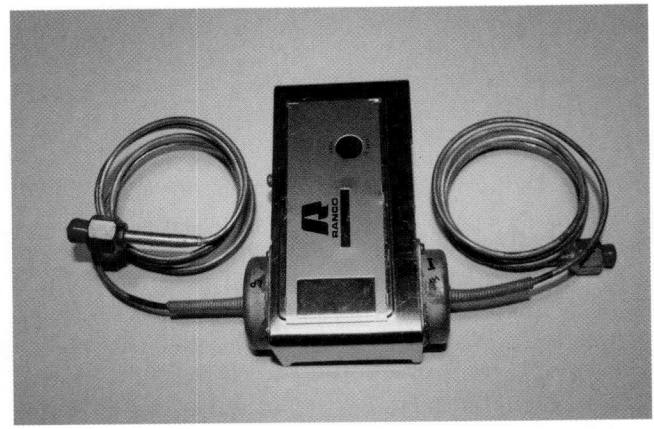

(A)

(B)

Figure 14–53 Two views of an oil pressure safety control. This control satisfies two requirements: how to measure net oil pressure effectively and how to get the compressor started to build oil pressure. **(A)** and **(B)**, Courtesy Ferris State University. Photos by John Tomczyk

Most compressors need at least 30 psig of actual oil pressure for proper lubrication. This means that whatever the suction pressure is, the oil pump discharge pressure has to be at least 30 lb above the oil pump inlet pressure, because the oil pump inlet pressure is the same as the suction pressure. For example, if the suction pressure is 70 psig, the oil pump outlet pressure must be 100 psig for the bearings to have a net oil pressure of 30 psig. This difference in the suction pressure and the oil pump outlet pressure is called the *net oil pressure.*

The basic low-pressure control has the pressure under the diaphragm or bellows and the atmospheric pressure on the other side of the diaphragm or bellows. The atmospheric pressure is considered a constant because it does not vary more than a small amount. The oil pressure control uses a double bellows—one bellows opposing the other to detect the net or actual oil pressure. The pump inlet pressure is under one bellows, and the pump outlet pressure is under the other bellows. These bellows are opposite each other either physically or by linkage. The bellows with the most pressure is the oil pump outlet, and it overrides the bellows with the least amount of pressure. This override reads out in net pressure and is attached to a linkage that can stop the compressor when the net pressure drops for a predetermined time.

Because the control needs a differential in pressure to allow power to pass to the compressor, it must have some means for allowing the compressor to get started. Remember, there is no pressure differential until the compressor starts to turn, because the oil pump is attached to the compressor crankshaft. There is a time delay built into the control to allow the compressor to get started and to prevent unneeded cut-outs when oil pressure may vary for only a moment. This time delay is normally about 90 sec. It is accomplished with a heater circuit and a bimetal device or electronically. NOTE: The manufacturer's instructions should be consulted when working with any compressor that has an oil safety control, **Figure 14–55.**

ELECTRONIC OIL SAFETY
CONTROL PRESSURE
TRANSDUCER

Figure 14–56 A pressure transducer of an electronic oil safety controller that senses net oil pressure. *Photo by John Tomczyk*

Oil safety controllers can also incorporate a pressure transducer to sense the combination of oil pump discharge pressure and suction (crankcase) pressure, **Figure 14–56.** The pressure transducer has two separate ports for sensing crankcase pressure and oil pump discharge pressure. The subtraction, or difference, between these two pressures (net oil pressure) is accomplished by the transducer. As seen in **Figure 14–57,** the pressure transducer is connected to an electronic controller by wires. The transducer is actually mounted directly into the oil pump. The pressure transducer then transforms a pressure signal to an electrical signal for the electronic controller to process. One advantage of an electronic oil safety controller over a mechanical bellows-type controller is that it eliminates capillary tubes; thus,

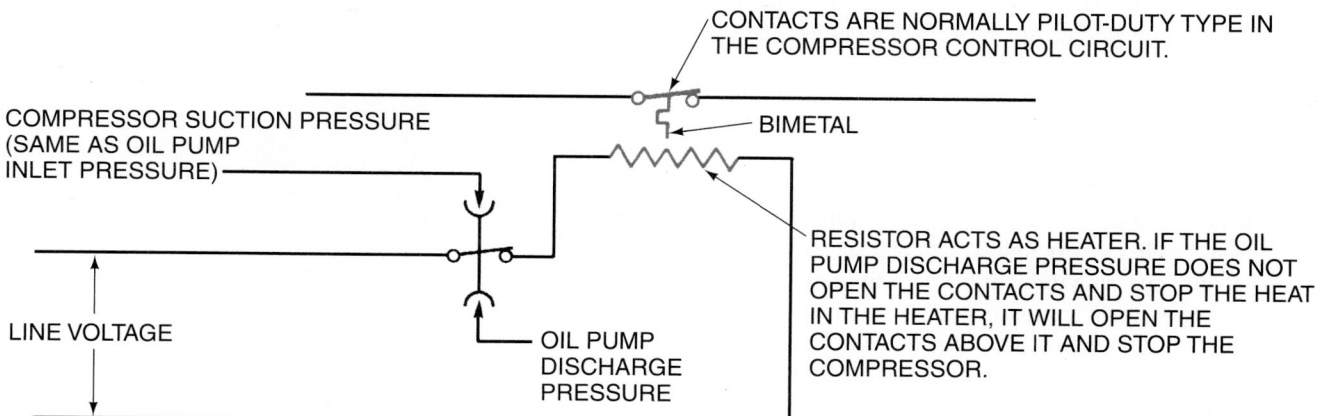

Figure 14–55 The oil pressure control is used for lubrication protection of the compressor. The oil pump that lubricates the compressor is driven by the compressor crankshaft. Therefore, a time delay is necessary to allow the compressor to start up and build oil pressure. This time delay is normally 90 sec. The time delay is accomplished with either a heater circuit heating a bimetal or an electronic circuit.

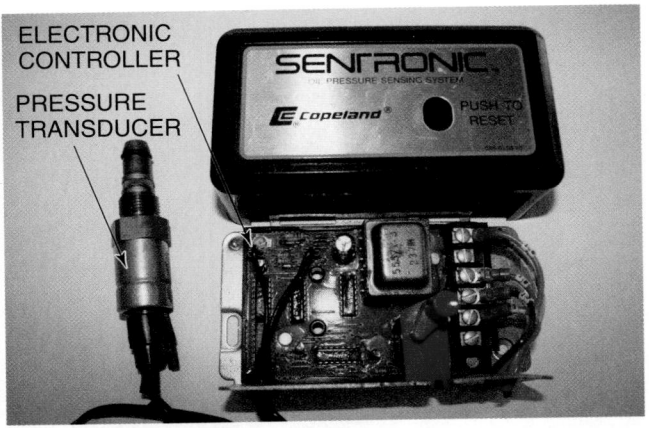

Figure 14–57 The pressure transducer connected by wires to the electronic oil safety controller. *Courtesy Ferris State University. Photo by John Tomczyk*

Figure 14–58 An air pressure differential switch used to detect ice on a heat pump outdoor coil. *Photo by Bill Johnson*

there is less chance for refrigerant leaks. Also, the electronic clock and circuitry are much more accurate and reliable. A more detailed explanation of both the mechanical bellows and the electronic type of oil safety controllers, with accompanying circuitry, will be covered in Unit 25, "Special Refrigeration System Components."

14.14 AIR PRESSURE CONTROLS

Air pressure controls (switches) are used in the following applications.

1. Heat pumps have air pressure drop across the outdoor coil due to ice buildup. When the coil has a predetermined pressure drop across it, there is an ice buildup that justifies a defrost. When used in this manner, the control is an operating control, **Figure 14–58.** Not all manufacturers use air pressure as an indicator for defrost.

   Air pressure switch sensors need a large diaphragm to sense the very low pressures in air systems. Mounted on the outside of the diaphragm is a small switch (microswitch) capable only of stopping and starting control circuits.

2. When electric heat is used in a remote duct as terminal heat, an air switch (sail switch) is sometimes used to ensure that the fan is passing air through the duct before the heat is allowed to be energized. This is a safety switch and should be treated as such, **Figure 14–59.** This switch has a very lightweight and sensitive sail. When air is passing through the duct from the fan, the sail is blown to an approximate horizontal position. In this position it allows the heat source to be activated. When the fan stops, the sail rises, ensuring that the heat source is turned off.

3. Air switches or sail switches are also used to prove that the combustion blower motor is operating before ignition and/or that the main gas valve is allowed to be energized.

Figure 14–59 A sail switch detects air movement. *Photo by Bill Johnson*

14.15 GAS PRESSURE SWITCHES

Gas pressure switches are similar to air pressure switches but usually smaller. They detect the presence of gas pressure in gas-burning equipment before burners are allowed to ignite. SAFETY PRECAUTION: This is a safety control and should never be bypassed except for troubleshooting and then only by experienced service technicians.

14.16 DEVICES THAT CONTROL FLUID FLOW AND DO NOT CONTAIN SWITCHES

The pressure relief valve can be considered a pressure-sensitive device. It is used to detect excess pressure in any system that contains fluids (water or refrigerant). Boilers, water heaters, hot water systems, refrigerant systems, and gas systems all use pressure relief valves, **Figure 14–60.** This device can be either pressure and temperature sensitive (called a P & T valve) or just pressure sensitive. The P & T valves are normally associated with water heaters. The pressure-sensitive type will be the only type considered here.

The pressure relief valve can usually be recognized by its location. It can be found on all boilers at the high point on the boiler. SAFETY PRECAUTION: It is a safety device that must be treated with the utmost respect. Some of these

(A)

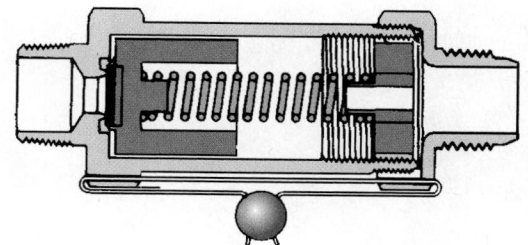

REFRIGERANT RELIEF VALVE
(B)

Figure 14–60 Pressure relief valves. *(A) Photo by Bill Johnson.* *(B) Courtesy Superior Valve Company*

valves have levers on top that can be raised to check the valve for flow. Most of the valves have a visible spring that pushes the seat downward toward the system's pressure. A visual inspection will usually reveal that the principle of the valve is for the spring to hold the seat against the valve body. The valve body is connected to the system, so, in effect, the spring is holding the valve seat down against the system pressure. When the system pressure becomes greater than the spring tension, the valve opens and relieves the pressure and then reseats. SAFETY PRECAUTION: The pressure setting on a relief valve is factory set and should never be tampered with. Normally the valve setting is sealed or marked in some manner so that it can be seen if it is changed.

Most relief valves are automatically reset. After they relieve, they automatically seat back to the original position. A seeping or leaking relief valve should be replaced. SAFETY PRECAUTION: Never adjust or plug the valve. This could be extremely dangerous.

14.17 WATER PRESSURE REGULATORS

Water pressure regulators are common devices used to control water pressure. Two types of valves are commonly used: the pressure-regulating valve for system pressure, and the pressure-regulating valve for head pressure control on water-cooled refrigeration systems. Air-conditioning and refrigeration systems operate at lower pressures and temperatures when water cooled, but the water circuit needs a special kind of maintenance. At one time water-cooled equipment was widely used. Air-cooled equipment has now become the dominant type because it is easy to maintain. Water valves, although still used on equipment, are therefore not as common as they were on air-conditioning and refrigeration equipment.

Water pressure–regulating valves are used in two basic applications:

1. They reduce supply water pressure to the operating pressure in a hot water heating system. This type of valve has an adjustment screw on top that increases the tension on the spring regulating the pressure, **Figure 14–61.** A great many boilers in hydronic systems in homes or businesses use circulating hot water at about 15 psig. The supply water may have a working pressure of 75 psig or more and have to be reduced to the system working pressure. If the supply pressure were to be allowed into the system, the boiler pressure relief valve would open. The water-regulating valve is installed in the supply water makeup line that adds water to the system when some leaks out. Most water-regulating valves have a manual valve arrangement that allows the service technician to remove the valve from the system and service it without stopping the system. A manual feedline is also furnished with most systems to allow the system to be filled by bypassing the water-regulating valve. SAFETY PRECAUTION: Care must be taken not to overfill the system.

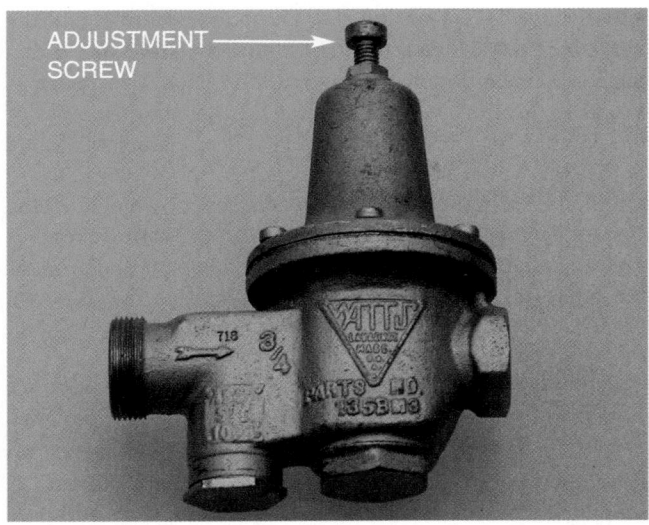

Figure 14–61 Adjustable water-regulating valve for a boiler. *Photo by Bill Johnson*

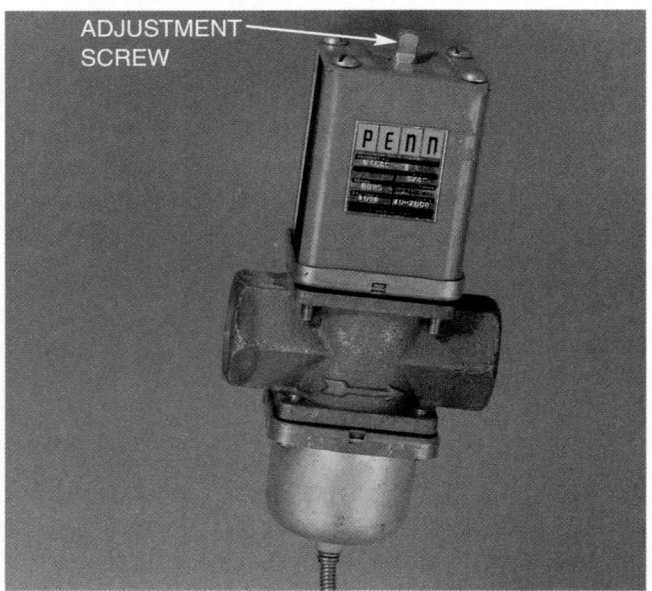

Figure 14–62 This valve maintains a constant head pressure for a water-cooled system during the changing of water temperatures and pressures. *Photo by Bill Johnson*

2. The water-regulating valve controls water flow to the condenser on water-cooled equipment for head pressure control. This valve takes a pressure reading from the high-pressure side of the system and uses this to control the water flow to establish a predetermined head pressure, **Figure 14–62.** For example, if an ice maker were to be installed in a restaurant where the noise of the fan would be objectionable, a water-cooled ice maker may be used. This might be a "wastewater type of system" where the cooling water is allowed to go down the drain. In the winter the water may be very cool, and in the summer the water may get warm. In

other words, the ice maker may have a winter need and a summer need. The water-regulating valve would modulate the water in both cases to maintain the required head pressure. There is an added benefit because when the system is off for any reason, the head pressure is reduced and the water flow is stopped until needed again. All of this is accomplished without an electrical connection.

14.18 GAS PRESSURE REGULATORS

The gas pressure regulator is used in all gas systems to reduce the gas transmission pressure to usable pressure at the burner. The gas pressure at the street in a natural gas system could be 5 psig—and the burners could be manufactured to burn gas at 3.5 in. WC (this is a pressure-measuring system that indicates how high the gas pressure can push a column of water). SAFETY PRECAUTION: This pressure must be reduced to the burner's design, and this is done with a pressure-reducing valve that acts like the water pressure–regulating valve in the boiler system, **Figure 14–63.** There is a regulator on top of the valve that qualified personnel can use to adjust the valve.

When bottled gas is used, the tank can have as much pressure as 150 psig needing to be reduced to the burner-design pressure of 11 in. WC. For this pressure reduction, the regulator is normally located on the tank, **Figure 14–64.**

14.19 MECHANICAL CONTROLS

An example of a mechanical control is a water pressure–regulating valve. This valve is used to maintain a preset water pressure in a boiler circuit and has no electrical contacts or connections, **Figure 14–61.** It acts independently from any of the other controls, yet the system depends on it as part of the team of controls that make the system function trouble free. The technician must be able to recognize and understand the function of each mechanical control. Doing this is

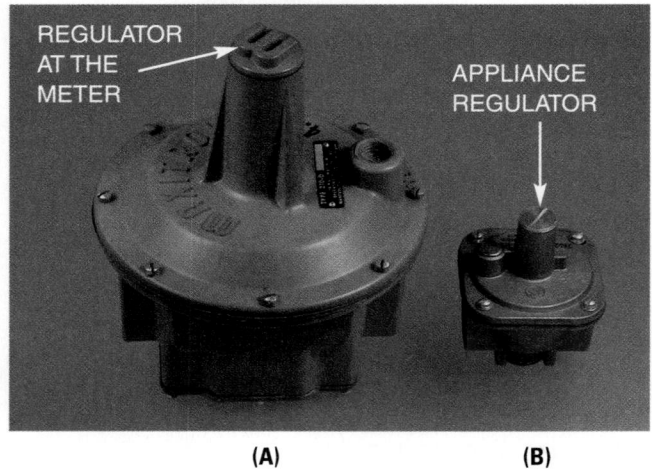

(A)   **(B)**

Figure 14–63 Gas pressure regulators. **(A)** Gas pressure regulator at the meter. **(B)** Gas pressure regulator at the appliance. *Photo by Bill Johnson*

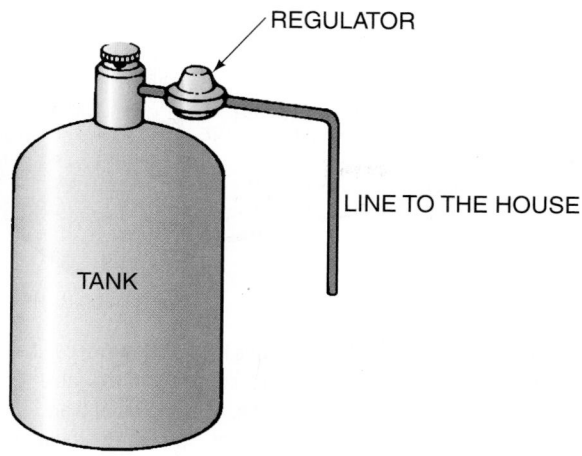

Figure 14–64 The gas pressure regulator on a bottled-gas system. This regulator is at the tank.

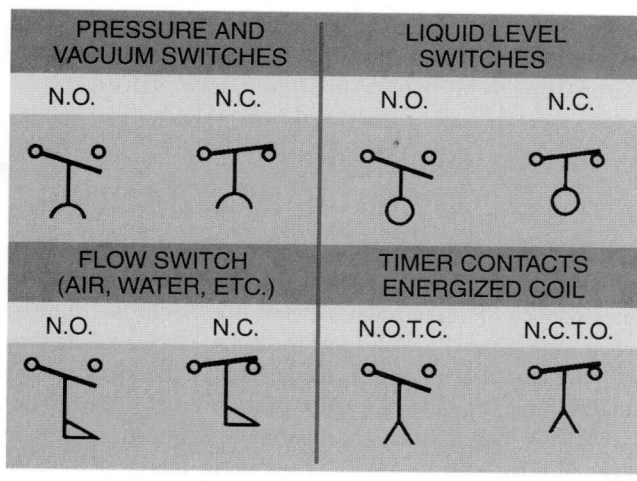

Figure 14–66 Table of electromechanical control symbols.

more difficult than for electrical controls because the control diagram does not always describe mechanical controls as well as it does electrical controls.

14.20 ELECTROMECHANICAL CONTROLS

Electromechanical controls convert a mechanical movement into some type of electrical activity. A high-pressure switch is an example of an electromechanical control. The switch contacts are the electrical part, and the bellows or diaphragm is the mechanical part. The mechanical action of the bellows is transferred to the switch to stop a motor when high pressures occur, **Figure 14–65**. Electromechanical controls normally appear on the electrical print with a symbol adjacent to the electrical contact describing what the control does, **Figure 14–66**. These symbols are supposed to be standard; however, old equipment, installed long before standardization was considered, is still in service. In some instances you need imagination and experience to understand the intent of the manufacturer.

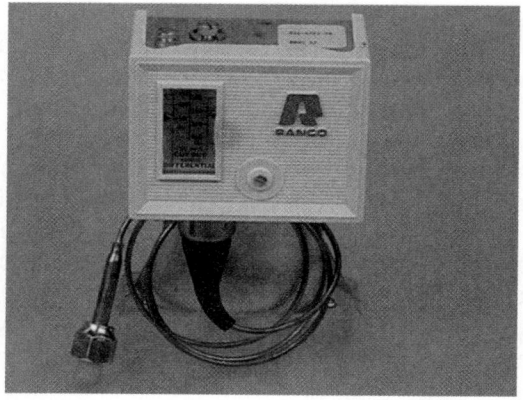

Figure 14–65 This high-pressure control is considered an electromechanical control. *Photo by Bill Johnson*

14.21 MAINTENANCE OF MECHANICAL CONTROLS

Mechanical controls are used to control the flow of fluids such as water, refrigerant liquids and vapors, or natural gas. These fluids flowing through the controls must be contained within the system. The controls typically are designed with diaphragms, bellows, and gaskets that are subject to leakage after being used for long periods of time.

Water is one of the most difficult substances to contain and is likely to leak through the smallest opening. All water-regulating valves should be inspected for leaks by looking for wet spots or rust streaks. Water circulating in a system will also leave mineral deposits in the piping and valve sets or mechanisms. For example, a water-regulating valve for a hot water system, **Figure 14–61,** is subject to problems that may cause water leaks or control set point drift. Some of these valves are made from brass to prevent corrosion and some are made of cast iron. The cast-iron ones are less expensive but may rust and bind the moving parts. The valve also has a flexible diaphragm that moves every time the valve functions. This diaphragm is subject to deterioration from flexing and age. If the diaphragm leaks, water will escape to the outside and drip on the floor. Mineral deposits or rust inside a valve can prevent the valve from moving up and down to feed water. This might overfeed or underfeed the system.

Pressure relief valves on boilers are constructed of material that will not corrode because this control is a safety control. These valves must be set free to ensure that they will not stick shut and not function. An explosion may occur because of a valve that is stuck shut. Many technicians pull the lever on the top of safety relief valves from time to time and let them relieve a small amount of steam or hot water. This may be considered a good practice because it ensures that the valve port is free from deposits that may stop it from functioning. On some occasions, the valve may not seat back when the lever is released and the valve may seep. Usually pulling the lever again will clear this up by blowing out any trash in the valve seat. If it cannot be stopped from leaking, the valve must be replaced. The fact that this procedure may start a leak is the reason many technicians do not "test relieve" valves.

14.22 MAINTENANCE OF ELECTROMECHANICAL CONTROLS

Electromechanical controls have both a mechanical and an electrical action. Many of the same procedures used in the maintenance of the mechanical controls should be followed when water is involved. Inspect for leaks. Electromechanical pressure controls often are connected to the system with small tubing similar to capillary tubes used as metering devices. This tube is usually copper and is the cause of many leaks in refrigeration systems because of misapplication or poor installation practices. For example, a low- or high-pressure control may be mounted close to a compressor and the small control tube routed to the compressor to sense low or high pressure. The control must be mounted securely to the frame and the control tube routed in such a manner that it does not touch the frame or any other component. The vibration of the compressor will vibrate the tube and rub a hole in the control tube or one of the refrigerant lines if they are not kept isolated.

The electrical section of the control will usually have a set of contacts or a mercury switch to stop and start some component in the system. The electrical contacts are often enclosed and cannot be viewed so visual inspection is not possible. You should look for frayed wires or burned wire insulation adjacent to the control when a problem occurs.

The mercury in mercury switches may be viewed through the clear glass enclosure. If the mercury becomes dark, the tube is allowing oxygen to enter the space where the mercury makes and breaks the electrical contact and oxidation occurs. In this case the switch should be replaced. It is possible that the switch will conduct across the oxidation and the switch will then appear to be closed at all times. This could prevent a boiler from shutting off, allowing overheating to occur.

14.23 SERVICE TECHNICIAN CALLS

SERVICE CALL 1

A customer calls indicating that the boiler in the equipment room at a motel has hot water running out and down the drain all the time. Another service company has been performing service at the motel for the last few months. *The problem is that the water-regulating valve (boiler water feed) is out of adjustment. Water is seeping from the boiler's pressure relief valve,* **Figure 14–67.** The technician arrives, goes to the boiler room, and notices that, as the customer said, water is seeping out of the relief valve pipe that terminates in the drain. This is heated water, causing a very inefficient situation. The technician looks at the temperature and pressure on the boiler gage. The needle is in the red; the pressure is 30 psig. No wonder the valve is relieving the pressure; it is too high. The boiler is hot, but

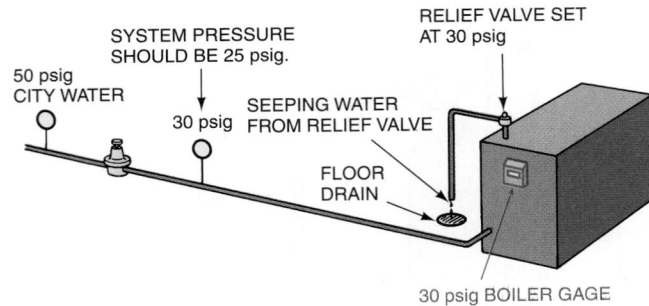

Figure 14–67 This illustration indicates how one control can cause another to look defective. The water pressure–regulating valve adjustment has been changed by mistake, allowing more than operating pressure into the system. The water pressure–regulating valve is designed to regulate the water pressure down to 25 psig. The valve is necessary to allow water to automatically keep the system full should a water loss develop due to a leak.

the technician cannot tell if it is too hot. The burner is not operating, so the technician decides that this may be a pressure problem instead of a temperature problem.

The technician looks for and finds the water pressure–regulating valve, which is in the supply water line entering the boiler. Actually, this valve keeps the supply water pressure from reaching the boiler and feeds water into the system when there is a loss of water due to a leak. The supply water pressure could easily be 75 psig or more. The boiler working pressure is 30 psig with a relief valve setting of 30 psig. If the supply water were allowed directly into the boiler, it would push the pressure past the design pressure of the system. The design pressure of the system must be less than 30 psig. If the pressure were to be above this, it would cause the pressure relief valve to relieve the pressure and dump water from the system until the pressure was lowered.

The technician shuts off the water valve leading to the boiler makeup system and bleeds some water from the system until the pressure at the boiler is down to 15 psig and then adjusts the water-regulating valve for a lower pressure. The supply water valve is then opened and water is allowed to enter the system. The technician can hear water entering the system through the water-regulating valve. In a few minutes, the water stops entering. The pressure in the system is now adjusted to 18 psig and all is well.

The problem in this example is one in which a mechanical automatic control is causing another control to show that the system has problems. If you did not examine the problem, you might think that the relief valve is defective because water is seeping out of it. This is not the case; the relief valve is doing its job. The water-regulating valve is at fault.

The technician shows the motel manager what the problem was and completes the paperwork before leaving for another call.

SERVICE CALL 2

A customer in a service station calls and says the hot water boiler in the equipment room is blowing water out from time to time. A technician is quickly dispatched to the job because of the potential hazard. *The problem is the gas valve will not close. The boiler is overheating. The relief valve is relieving water periodically.* The season is mild, so the boiler has a small load on it.

Because the relief valve is relieving intermittently, you may believe there is a periodic pressure buildup in the boiler. If the problem were the same as in Service Call 1, where a constant, bleeding relief occurred, the water-regulating valve or the relief valve might be the suspect. This is not the case.

The technician arrives and goes straight to the boiler and reads the gage; it is in the red and about to relieve again. The technician then looks at the flame in the burner section and notices that it is not a full flame; it is small. It is decided that the gas valve may be stuck open, allowing fuel to enter the burner all the time.

The in-line gas hand valve is shut off to put out the flame so the technician can discover the problem.

A check with a voltmeter shows no voltage at the gas valve terminals, so the thermostat is trying to shut the gas off. The valve must be stuck open.

The technician obtains a new gas valve and installs it. The boiler is started up and allowed to run. The boiler is cold after being off for some time for the gas valve change so it takes awhile for it to get up to temperature. The set point of the thermostat is 190°F and the gas valve shuts off when this temperature is reached.

These service calls are examples of how a trained technician might approach these service problems. Technicians develop their own methods of approach to problems. The point is that no diagram was consulted. A technician either must have prior knowledge of the system or must be able to look at the system and make accurate deductions based on a knowledge of basic principles.

Each control has distinguishing features to give it a purpose. There are hundreds of controls and dozens of manufacturers (some of whom are out of business). If you find a control with no information as to what it does, consult the manufacturer. If this is impossible, try to find someone who has experience with the equipment. In general, when the control is examined in the proper perspective, the design parameters will help. For example: What is the maximum temperature or pressure that is practical? What is the minimum temperature or pressure that is practical? Ask yourself these questions when you run across a strange control.

The following service calls do not have the solution described. The solution can be found in the *Instructor's Guide*.

SERVICE CALL 3

A residential customer calls and reports that the heating system was heating until early this morning, but for no apparent reason it stopped. The furnace seems to be hot, but no heat is coming out of the air outlets. This is a gas furnace located in the basement.

The technician talks to the home owner about the problem and from the description believes the problem may be with the fan. The technician first checks to see if there is power at the furnace and notices that although the furnace is hot to the touch, no heat is coming out of the duct. Further investigation shows that the fan is not running.

The technician knows there is power to the furnace because it is hot and the burner ignites from time to time. The problem must be in the fan circuit.

A voltage check shows that power is leaving the fan switch going to the fan motor, but it is still not running. See the diagram in **Figure 14–68** for direction.

What is the problem and the recommended solution?

SERVICE CALL 4

A customer calls and says the central air-conditioning system in their small office building will not cool.

The service technician arrives and turns the thermostat to COOL and turns the thermostat indicator to call for cooling. The indoor fan motor starts. This indicates that the control voltage is operating and will send power on as requested for cooling. Only recirculated air is coming out of the ducts.

The technician checks the outside unit and notices that it is not running.

The control panel cover is removed to see what controls will actually prevent the unit from starting. This will give some indication as to what the possible problem may be. Use the diagram in **Figure 14–69**.

Pressure gages are applied to determine the system pressures. Most air-conditioning systems use R-22. If the system were standing at 80°F, the system should have 144 psig of pressure, according to the temperature/pressure chart for R-22 at 80°F.

NOTE: Care should be used in the standing pressure comparison. Part of the unit is inside, at 75°F air, and part of the unit is outside. Some of the liquid refrigerant will go to the coolest place if the system valves will allow it to.

The gage reading may correspond to a temperature between the indoor and outdoor temperature but not less than the cooler temperature. Many modern systems use fixed-opening metering devices that allow the system pressures to equalize and the refrigerant to migrate to the coolest place in the system.

In this example the pressure in the system is 60 psig, the refrigerant is R-22, and the temperature is about 80°F. A typical pressure control setup may call for the control to

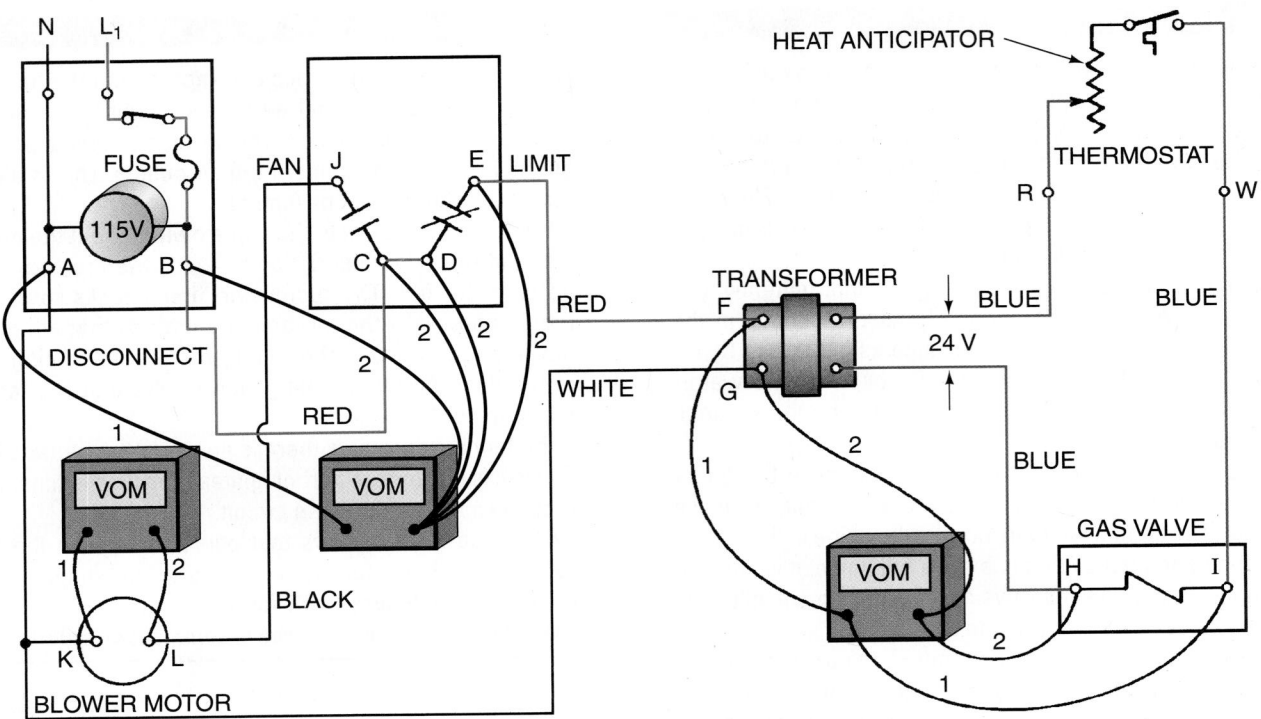

Figure 14–68   This diagram can be used for Service Call 3.

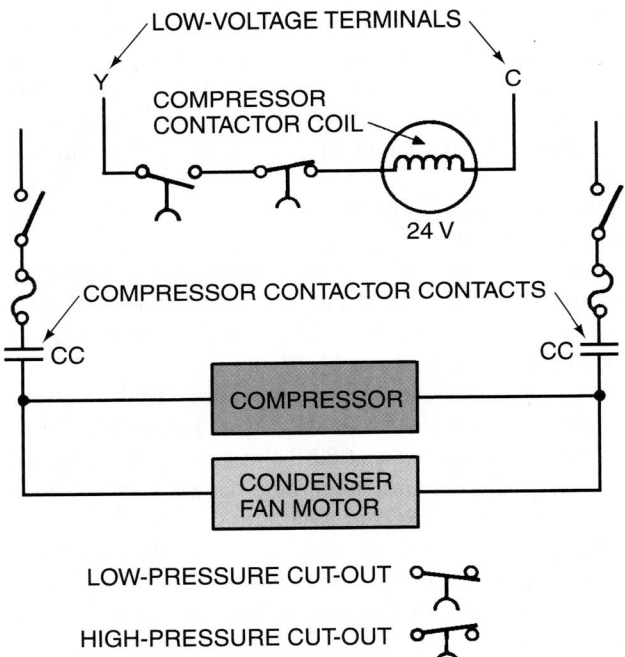

Figure 14–69   This diagram can be used for Service Call 4.

interrupt the compressor when the system pressure gets down to 20 psig and to make the control and start the compressor when the pressure rises back to a temperature corresponding to 70°F.

What is the problem and the recommended solution?

The foregoing examples are used as typical service problems to show how a service technician might arrive at the correct diagnosis, using the information at hand.

Always save and study the manufacturer's literature. There is no substitute for having access to the manufacturer's specifications.

SUMMARY

▨ Space temperature controls can be either low-voltage or line-voltage types.

▨ Low voltage is normally applied to residential heating and cooling controls.

▨ Heating thermostats normally have a heat anticipator circuit in series with the thermostat contacts.

▨ Cooling thermostats may have a cooling anticipator in parallel with the thermostat contacts.

▨ Line-voltage thermostats are normally rated up to 20 A of current because they are used to switch high-voltage current.

▨ Low-voltage thermostats normally will carry only 2 A of current.

▨ To get a correct temperature reading of a flat or round surface, the sensing element (either mercury bulb, remote bulb, or bimetal) must be in good contact with the surface and insulated from the ambient air.

▨ Some installations have wells in the substance to be sensed in which the sensor can be placed.

▨ Motors have both internal and external types of motor temperature-sensing devices.

▨ The internal type of motor temperature-sensing device can be either a bimetal or a thermistor inserted inside the motor windings.

▣ Some sensing elements are inserted into the fluid stream. Examples are the fan or limit switch on a gas, oil, or electric furnace.

▣ All temperature-sensing elements change in some manner with a change in temperature: the bimetal warps, the thermistor changes resistance, and the thermocouple changes voltage.

▣ Pressure controls normally are either diaphragm, Bourdon tube, or bellows operated.

▣ Pressure controls can operate at high pressures, low pressures, and even below atmospheric pressure (in a vacuum), and can detect differential pressures.

▣ SAFETY PRECAUTION: Gas pressure must be reduced before it enters a house or place of business to be burned in an appliance.

▣ Some purely mechanical controls are water-regulating valves, pressure relief valves, and expansion valves.

▣ Some electromechanical controls are low-pressure cut-out, high-pressure cut-out, and thermostats.

▣ Keep and study manufacturers' literature.

REVIEW QUESTIONS

1. Why is a low-voltage thermostat normally more accurate than a high-voltage thermostat?
2. The three kinds of switching mechanism used in low-voltage thermostats are _____, _____, and _____.
3. What does a heat anticipator do?
4. What does a cold anticipator do?
5. Define system overshoot when dealing with thermostats.
6. True or False: The temperature drop below the thermostat's set point is referred to as system lag.
7. What is meant by the steady-state condition with regard to setting a heating anticipator?
8. Describe the method a service technician should take when measuring the current that normally travels through a heating anticipator.
9. Explain how microelectronic or electronic thermostats handle system overshoot and system lag.
10. Describe how a bimetal functions.
11. What component steps down the voltage to the low-voltage value?
12. Low voltage is desirable for residential control voltages because of _____, _____, _____, and _____.
13. What is the maximum amperage usually encountered by a low-voltage thermostat?
14. A heating thermostat _____ on a rise in temperature.
15. A cooling thermostat _____ on a rise in temperature.
16. What two types of switches are normally found in the subbase of a low-voltage thermostat?
17. When is a line-voltage thermostat used?
18. The three sensing elements that can be used with a line-voltage thermostat are _____, _____, and _____.
19. What is the maximum amperage generally encountered in a line-voltage thermostat?

20. Why is a line-voltage thermostat mounted on an electrical box that has conduit connected to it?
21. Why do motors build up heat?
22. The two types of motor temperature-sensing devices are _____ and _____.
23. What is the main precaution for an externally mounted motor temperature protector?
24. What is the principle of operation for most externally mounted overload protection devices?
25. Name a method for speeding up the cooling of an open motor.
26. Describe two methods for cooling an overheated compressor.
27. How can an electronic thermometer be used in a well to obtain an accurate response?
28. A temperature-sensing device used in measuring airflow is the _____.
29. How does a thermocouple change with a temperature change?
30. Name two methods used to convert pressure changes into action.
31. Name two actions that can be obtained with a pressure change.
32. True or False: Pressures below atmospheric pressure cannot be detected.
33. Name one function of the low-pressure control.
34. Which is true of a high-pressure control?
    A. It closes on a rise in pressure.
    B. It opens on a fall in pressure.
    C. It is located on the evaporator.
    D. It protects the compressor from high pressures.
35. Name two types of water pressure control.
36. Why should a safety control not be adjusted if it has a seal to prevent tampering?
37. True or False: If the differential of a motor control is set too small, the motor may short cycle.
38. True or False: A change in the differential setting will automatically change the cut-in point of the control.
39. True or False: A change in the differential of a control will always change the range of the control.
40. True or False: Changing the cut-out adjustment without changing the differential adjustment will automatically change the cut-in setting.
41. Explain what is meant by the range of a temperature control or pressure control.
42. If a high-pressure control has a cut-out of 500 psig and a cut-in of 450 psig, what would be the differential of the control?
43. Explain the function of a pressure transducer and give an example of where one would be used in the HVACR industry.
44. List two advantages an electronic oil safety controller has over a mechanical bellows oil safety controller.

UNIT 15

Troubleshooting Basic Controls

OBJECTIVES

After studying this unit, you should be able to

- describe and identify power- and non-power-consuming devices.
- describe how a voltmeter is used to troubleshoot electrical circuits.
- identify some typical problems in an electrical circuit.
- describe how an ammeter is used to troubleshoot an electrical circuit.
- describe how a voltmeter is used to troubleshoot an electrical circuit.
- recognize the components in a heat-cool electrical circuit.
- follow the sequence of electrical events in a heat-cool electrical circuit.
- differentiate between a pictorial and a line-type electrical wiring diagram.

SAFETY CHECKLIST

- ✔ When troubleshooting electrical malfunctions, disconnect electrical power unless power is necessary to make appropriate checks. Lock and tag the panel where the disconnect is made and keep the only available key on your person.
- ✔ When checking continuity or resistance, turn off power and disconnect both leads from the component being checked.
- ✔ Ensure that meter probes come in contact with only terminals or other contacts intended.
- ✔ Make certain that all electrical connections are tight.
- ✔ Make certain that all unused wires are properly capped to prevent them from coming into contact with other conductors or with the metal casing of the equipment.

15.1 INTRODUCTION TO TROUBLESHOOTING

Each control must be evaluated as to its function (major or minor) in the system. Recognizing the control and its purpose requires that you understand its use in the system. Studying the purpose of a control before you take action will save a great deal of troubleshooting time. Look for pressure lines going to pressure controls and temperature-activated elements on temperature controls. See if the control stops and starts a motor, opens and closes a valve, or provides some other function. As mentioned previously, controls are either electrical, mechanical, or a combination of electrical and mechanical. (Electronic controls are considered the same as electrical controls for now.)

Electrical devices can be considered as power consuming or non–power consuming as far as their function in the circuit is concerned. Power-consuming devices use power and have either a magnetic coil or a resistance circuit. Power-consuming devices are more commonly referred to as *loads*. Non-power-consuming devices pass power on to power-consuming devices. Non-power-consuming, or power-passing, devices are commonly referred to as *switches*. They are used to simply pass power to the loads that are ultimately being controlled.

Both power-consuming and power-passing devices are wired in series with the power supply. Multiple loads are often wired in parallel with each other and the switches that are used to control each load are connected in series with that load. If multiple switches are to be used to control the operation of a single load, the switches are wired in series with each other and in series with the load being controlled.

The concepts of the load and switch can be easily clarified using a simple light bulb circuit with a switch. The light bulb in the circuit is the load and actually consumes the power, whereas the switch passes the power to the light bulb. The object is to wire the light bulb to both sides (hot and neutral) of the power supply, **Figure 15–1.** This will then ensure that a complete circuit will be established. The switch is a power-passing device and is wired in series with the light bulb. For any power-consuming device to consume power it must have a potential voltage across it. A potential voltage is the voltage indicated on a voltmeter between two power legs (such as line 1 to line 2) of a power supply. This can be thought of as electrical pressure in a home; for example, the light bulb has to be wired from the hot leg (the wire that has a fuse or breaker in it) to the neutral leg (the grounded wire that actually is wired to the earth ground).

The following statements may help you understand electron flow in an electrical circuit. Some liberties are taken in explaining the electrical circuits in this unit, but this has been done to aid your understanding of the concepts involved.

1. It does not make any difference which way the current flows in an alternating current (AC) circuit when the object is to get the electrons (current) to a power-consuming device and complete the circuit.
2. The electrons may pass through many power-passing devices before getting to the power-consuming device(s).

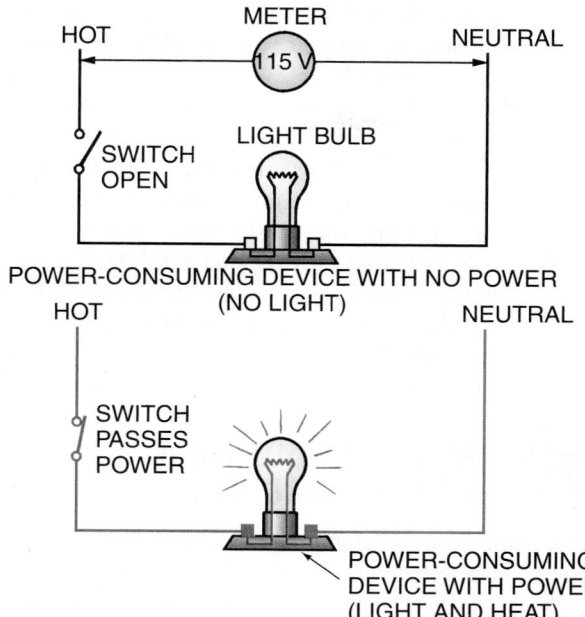

Figure 15–1 A power-consuming device (light bulb) and a non-power-consuming device (switch). The switch passes power to the light bulb.

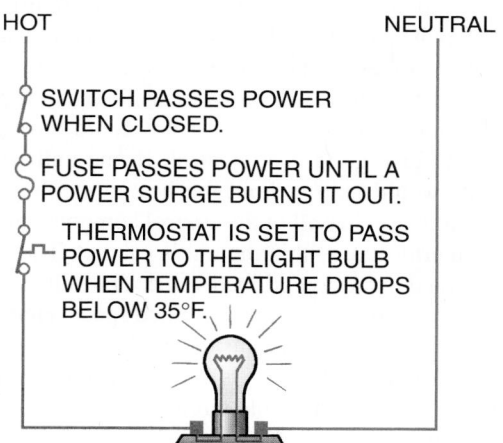

Figure 15–2 When closed, the switch passes power to the fuse, then to the thermostat, and on to the light bulb.

3. Devices that do not consume power pass power.
4. Devices that pass power can be either safety devices or operating devices.

Suppose the light bulb mentioned above was to be used to add heat to a well pump to keep it from freezing in the winter. To prevent the bulb from burning all the time, a thermostat is installed. A fuse must also be installed in the line to protect the circuit, and a switch must be wired to allow the circuit to be serviced, **Figure 15–2**. Note that the power now must pass through three power-passing devices to reach the light bulb. These three switches are connected in series with each other. The fuse is a safety control, and the thermostat turns the light bulb on and off. The switch is not a control but

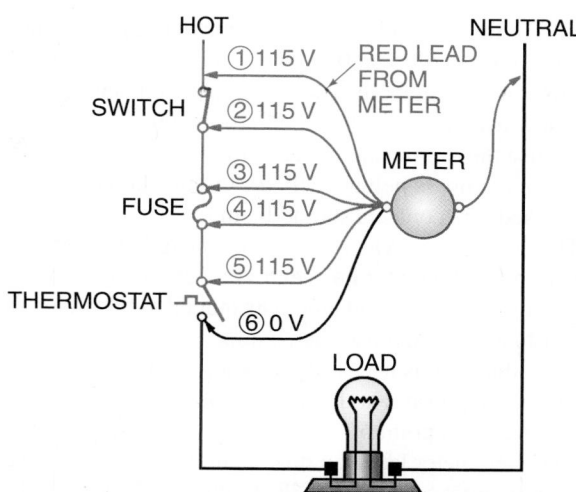

Figure 15–3 The troubleshooting procedure is to establish the main power supply of 115 volts from the hot line to the neutral line. When this power supply is verified, the lead on the hot side is moved down the circuit toward the light bulb. The voltage is established at all points. When point 6 is reached, no voltage is present. The thermostat contacts are open.

a service convenience device. Notice that all switches are on one side of the circuit. Let us suppose that the light bulb does not light up when it is supposed to. Which component is interrupting the power to the bulb: the switch, the fuse, or the thermostat? Or is the light bulb itself at fault? In this example the thermostat (switch) is open, **Figure 15–3**.

15.2 ▶ **TROUBLESHOOTING A SIMPLE CIRCUIT**

To troubleshoot the circuit in the previous paragraph, a voltmeter may be used in the following manner. Turn the voltmeter selector switch to a voltage setting higher than the voltage supply. In this case the supply voltage should be 115 V, so the 250-V scale is a good choice. Follow the diagram in **Figure 15–3** as you read the following.

1. Place the red lead of the voltmeter on the hot line and the black lead on the neutral. The meter will read 115 V.
2. Place the red lead, the lead being used to find and detect power in the hot line, on the load side of the switch. (The "load" side of the switch is the side of the switch that the load is connected to. The other side of the switch, where the line is connected, is the "line" side of the switch.) The black lead should remain in contact with the neutral line. The meter will read 115 V.
3. Place the red lead on the line side of the fuse. The meter will read 115 V.
4. Place the red lead on the load side of the fuse. The meter will read 115 V.
5. Place the red lead on the line side of the thermostat. The meter will read 115 V.

6. Place the red lead on the load side of the thermostat. The meter will read 0 V. There is no power available to energize the bulb. The thermostat contacts are open. Now ask the question, "Is the room cold enough to cause the thermostat contacts to make?" If the room temperature is below 35°F, the circuit should be closed; the contacts should be made.

NOTE: The red lead was the only one moved. It is important to note that if the meter had read 115 V at the light bulb connection when the red lead was moved to this point, then further tests should have been made.

Another step is necessary to reach the final conclusion. Let us suppose that the thermostat is good and 115 V is indicated at point 6.

7. The red meter lead can now be moved to the terminal on the light bulb, **Figure 15–4**. Suppose it reads 115 V. Now, move the black lead to the light bulb terminal on the right. If there is no voltage, the neutral wire is open between the source and the bulb.

If there is voltage at the light bulb and it will not burn, the bulb is defective, **Figure 15–5**. *When there is a power supply*

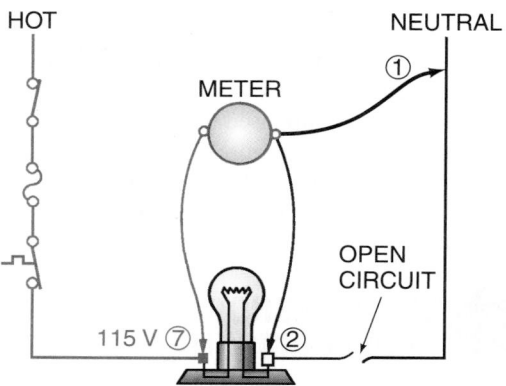

Figure 15–4 The thermostat contacts are closed, and 115 volts is the measurement when one lead is on the neutral and one lead is on the light bulb terminal (see position 7). When the black lead is moved to the light bulb at position 2, there is no voltage reading. The neutral line is open.

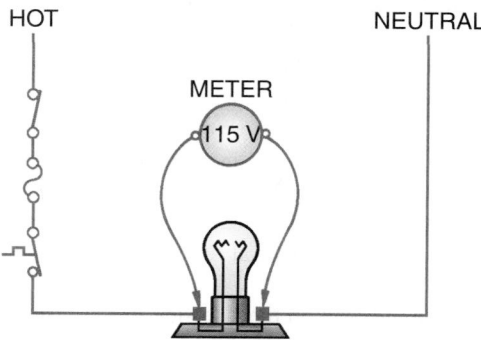

Figure 15–5 Power is available at the light bulb. The hot line on the left side and the neutral on the right side complete the circuit. The light bulb filament is burned out, so there is no current path through the bulb and no current flows through the circuit.

and a path to flow through, current will flow. The light bulb filament is measurable resistance and should be the path for the current to flow through.

The power supply used in the preceding example was 115 V with neutral as one side of the circuit. Neutral and earth ground are usually the same thing. With 115-V circuits, the technician need have only a ground source, such as a water pipe, for one side of the meter. When troubleshooting 208-V or higher circuits, one meter lead must be placed on the other side of the line feeding the control circuit. This lead may be labeled line 1 and line 2, line 2 and line 3, or line 1 and line 3.

15.3 ▶ TROUBLESHOOTING A COMPLEX CIRCUIT

The following example is a progressive circuit that would be typical of a combination heating-cooling unit. This circuit is not standard because each manufacturer may design its circuits differently. They may vary but will be similar in nature.

The unit in this example is a package air conditioner with 1 1/2 tons cooling and 5 kW of electric heat. This unit resembles a window air conditioner because all of the components are in the same cabinet. The unit can be installed through a wall or on a roof, and the supply and return air can be ducted to the conditioned space. The reason for using this unit is that it comes in many sizes, from 1 1/2 tons of cooling (18,000 Btu/h) to very large systems. The unit is popular with shopping centers because a store could have several roof units to give good zone control. If one unit were off, other units would help to hold the heating-cooling conditions. The unit also has all of the control components, except the room thermostat, within the unit's cabinet. The thermostat is mounted in the conditioned space. The unit can be serviced without disturbing the conditioned space, **Figure 15–6**.

The first thing we will consider is the thermostat. The thermostat is not standard but is a version used for illustration purposes. This is a usable circuit designed to illustrate troubleshooting.

This thermostat is equipped with a selector switch for either HEAT or COOL. When the selector switch is in the HEAT position and heat is called for, the *electric heat relay*

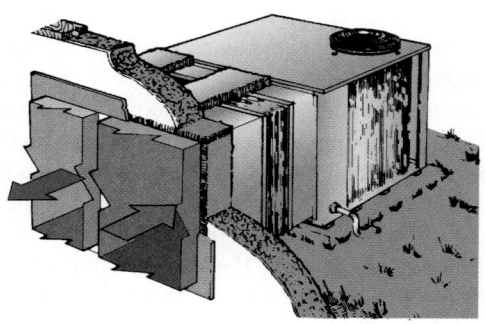

Figure 15–6 A small package air conditioner. *Courtesy Climate Control*

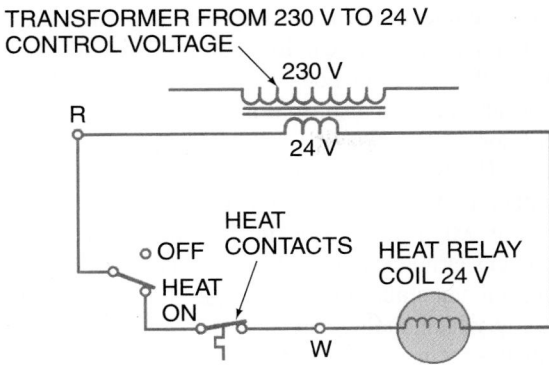

Figure 15–7   A simple thermostat with a HEAT OFF/HEAT ON position. The selector switch is closed, calling for heating. When the heat relay is energized, the heating system starts.

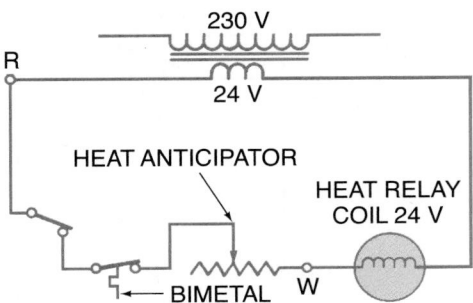

NOTE: THE HEAT ANTICIPATOR IS A VARIABLE RESISTANCE THAT IS DESIGNED TO CREATE A SMALL AMOUNT OF HEAT NEAR THE BIMETAL. THIS FOOLS THE THERMOSTAT INTO CUTTING OFF EARLY. THE FAN WILL CONTINUE TO RUN AND DISSIPATE THE HEAT REMAINING IN THE HEATERS.

Figure 15–8   This is the same diagram as shown in **Figure 15–7**, but a heat anticipator has been added in series with the thermostat contacts. The selector switch is set for heat and the thermostat contacts are closed, calling for heat.

is energized. The fan must run in the heating cycle. It will be started and run in the high-voltage circuit. This is discussed later in this unit.

Figure 15–7 is the beginning of the control explanation. This is the simplest of thermostats, having only a set of heating contacts and a selector switch. Follow the power from the R terminal through the selector switch to the thermostat heat contacts and on to the W terminal. The W terminal destination is not universal but is common. When these heat contacts are closed, we can say that the thermostat is calling for heat. These contacts pass power to the heat relay coil. Power for the other side of the heat relay coil comes directly from the control transformer. When this coil is energized (24 V), the heat should come on. This coil is going to close a set of contacts in the high-voltage circuit to pass power to the electric heat element.

It should be noted that the circuit in **Figure 15–7** has two switches and one load. The two switches are the manual heat switch and the other is the set of heat contacts. The manual heat switch is opened and closed by the operator of the equipment, hence the term "manual." The other switch, the set of heat contacts, is an automatic switch that opens and closes its contacts depending on the temperature sensed by the contacts. In order for the heating system to be energized, both the manual switch and the automatic temperature-sensing switch must be in the closed position. If the occupied space is very cold but the manual heat switch is off, the heating system will not operate. Similarly, if the manual heat switch is on and the space is warm, the system will not operate. As with any circuit that has multiple switches connected in series with each other, all of the switches must be in the closed position in order to pass power to the load being controlled.

In **Figure 15–8** the heat anticipator is added to the circuit. The heat anticipator is in series with the heat relay coil and has current passing through it at any time the heat relay coil is energized. This current passing through the anticipator creates a small amount of heat in the vicinity of the thermostat bimetal sensing element. This causes the bimetal to break the thermostat contacts early to dissipate the heat in the heat exchanger. This is done because the fan will blow hot air into the heated space for a short time after the heat relay is

deenergized while cooling the heat exchanger off. If the contacts to the heater did not open until the space temperature was actually up to the cut-off point of the room thermostat, the extra heat would overheat the space; this is referred to as a system overshoot. The fan to move the heat is stopped and started in the high-voltage circuit. This is discussed later in this unit.

The heat anticipator is often a variable resistor and must be adjusted to the actual system. The current must be matched to the current used by the heat relay coil. All thermostat manufacturers explain how this is done in the installation instructions for the specific thermostat. Heat anticipators are covered in detail in Unit 14, "Automatic Control Components and Applications."

Figure 15–9 illustrates how a fan-starting circuit can be added to a thermostat. Notice the addition of the G terminal. It is used to start the indoor fan. The G terminal designation is not universal but is common. Follow the power from the R terminal through the fan selector switch to the G terminal and on to the indoor fan relay coil. Power is supplied to the other side of the coil directly from the control transformer. When this coil has power (24 V), it closes a set of contacts in the high-voltage circuit and starts the indoor fan.

Figure 15–10 completes the circuitry in this thermostat by adding a Y terminal for cooling. Again, this is not the only letter used for cooling, but it is also common. Follow the power from the R terminal down to the cool side of the selector switch through to the contacts and on to the Y terminal. When these contacts are closed, power will pass through them and go two ways:

- One path will be through the fan AUTO switch to the G terminal and on to start the indoor fan. The indoor fan has to run in the cooling cycle.
- The other way is straight to the cool relay. When the cool relay is energized (24 V), it closes a set of contacts in the high-voltage circuit to start the cooling cycle.

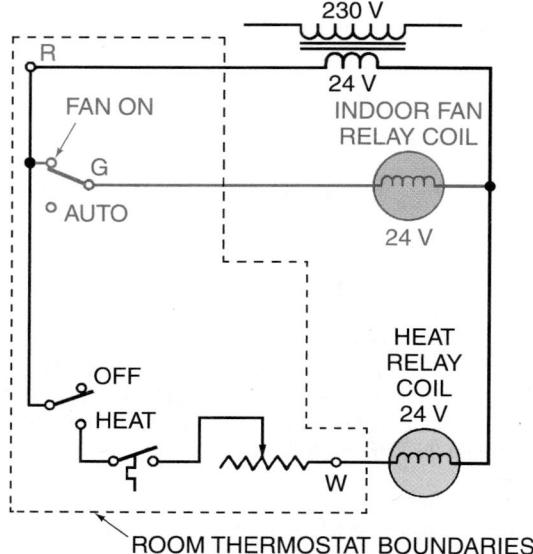

Figure 15–9 The addition of the fan relay to the thermostat allows the owner to switch the fan on for continuous operation. The fan must be operating during the heating cycle. Automatic fan operation will be explained later in this unit.

When the fan is switched to ON, the indoor fan will run all the time, even when the system switch is in the OFF position. This thermostat is equipped with a FAN ON or FAN AUTO position, so the fan can be switched on manually for air circulation when desired. When the fan switch is in the AUTO position, the fan will start on a call for cooling.

When the selector switch is in the cooling mode, the cooling system will start on a call for cooling. The indoor fan will start through the AUTO mode on the fan selector switch. The outdoor fan must run also, so it is wired in parallel with the compressor.

NOTE: The cooling anticipator is in parallel with the cooling contacts. This means that current will flow through this anticipator when the thermostat is satisfied (or when the thermostat's circuit is open). The cool anticipator is normally a fixed, nonadjustable type. It causes the cooling cycle to start early, which enables the system to be up to capacity at the time the room warms to above the thermostat set point.

15.4 ▶ **TROUBLESHOOTING THE THERMOSTAT**

The thermostat is an often misunderstood, frequently suspected component during equipment malfunction. But it does a straightforward job of monitoring temperature and distributing the power leg of the 24-V circuit to the correct components to control the temperature. Service technicians should remind themselves as they approach the job that "one power leg enters the thermostat and it distributes power where called." Every technician needs a technique for checking the thermostat for circuit problems.

It should be restated that the thermostat is nothing more than a set of switches. There is a manual selector switch that selects the mode of operation, which can be OFF, COOL, or HEAT, and the automatic temperature-sensing component. When the conditions are correct, the thermostat will pass

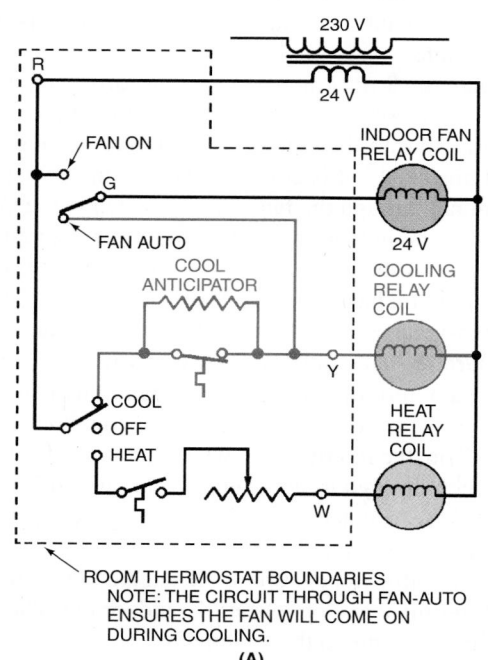

(A)

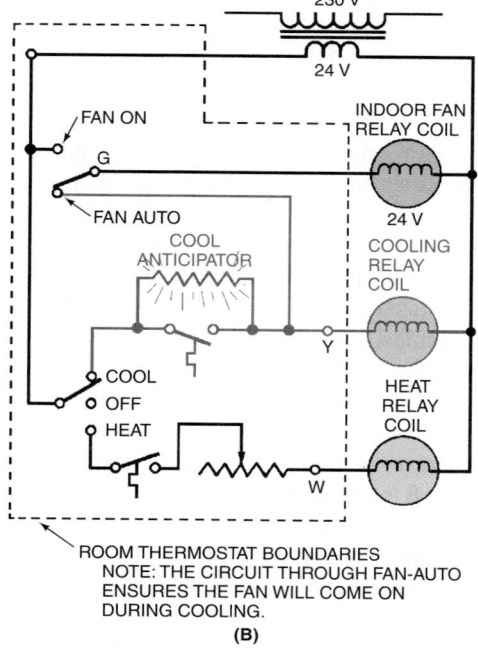

(B)

Figure 15–10 **(A)** Cooling has been added to **Figure 15–9.** **(B)** The cooling circuit has a cooling anticipator in parallel with the cooling contacts. Current flows through the cooling anticipator during the off cycle. NOTE: This is one of the few times that there are two loads connected in series with each other.

power through the device onto the system components to be controlled. When the system is calling for heat, power will be passed to the heat relay. When the system is calling for cooling, power will be passed to the cooling relay and the indoor fan relay. If power enters the thermostat and does not pass through the device, either the thermostat is not calling for the desired mode of operation or there is a problem with the control.

One way to troubleshoot a thermostat on a heating and cooling system is to first turn the fan selector switch to the FAN ON position and see if the fan starts. If the fan does not start, the problem may be with the control voltage. Control voltage must be present for the thermostat to operate. Assume for a moment that the thermostat will cause the fan to come on when turned to FAN ON but will not operate the heat or cooling cycles. The next step may be to take the thermostat off the subbase and jump the circuit out manually with an *insulated jumper.* Jump from R to G, and the fan should start. Jump from R to W, and the heat should come on. Jump from R to Y, and the cooling should come on. If the circuit can be made to operate without the thermostat but not with it, the thermostat is defective, **Figure 15–11.** One does not have to be afraid to jump these circuits out one at a time because only one leg of power comes to the thermostat. Some more sophisticated thermostats may have a clock or indicator lights and may have the common side of the circuit wired to the thermostat. Caution must be used for these thermostats. If all circuits were jumped out at one time in this thermostat, the only thing that would happen is that the heating and cooling would run at the same time. Of course, this should not be allowed to continue.

We have just covered what happens in the basic thermostat. The next step is to move into the high-voltage circuit and see how the thermostat's actions actually control the fan, cooling, and heating. The progressive circuit will again be used.

The first thing to remember is that the high voltage (230 V) is the input to the transformer. Without high voltage there is no low voltage. When the service disconnect is closed, the potential voltage between line 1 and line 2 is the power supply to the primary of the transformer. The primary input then induces power to the secondary winding of the transformer.

Figure 15–12 illustrates the high-voltage operation of the fan circuit. The power-consuming device is the fan motor in the high-voltage circuit. The fan relay contact is in the line 1 circuit. It should be evident that the fan relay contacts have to be closed to pass power to the fan motor. These contacts close when the fan relay coil is energized in the low-voltage control circuit. Although not drawn together, the fan relay coil in the low-voltage circuit causes the fan relay contacts in the line-voltage circuit to open and close. The magnetic field that is generated when current flows through the fan relay coil is the force that causes the relay contacts to close.

Figure 15–13 illustrates the addition of the electric heat element and the electric heat relay. The electric heat element and the fan motor are the power-consuming devices in the high-voltage circuit. In the low-voltage circuit when the HEAT-COOL selector switch is moved to the HEAT position,

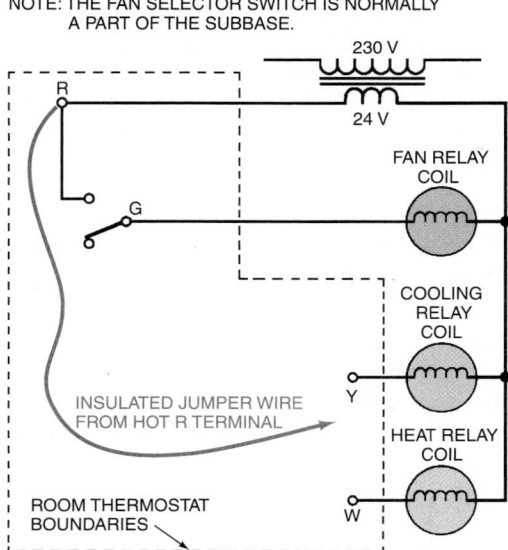

Figure 15–11   Terminal designation in a heat-cool thermostat subbase. The letters R, G, Y, and W are common designations. A jumper wire placed between R and G should start the indoor fan motor. A jumper wire placed between R and Y should start the cooling cycle. A jumper wire placed between R and W should start the heating cycle.

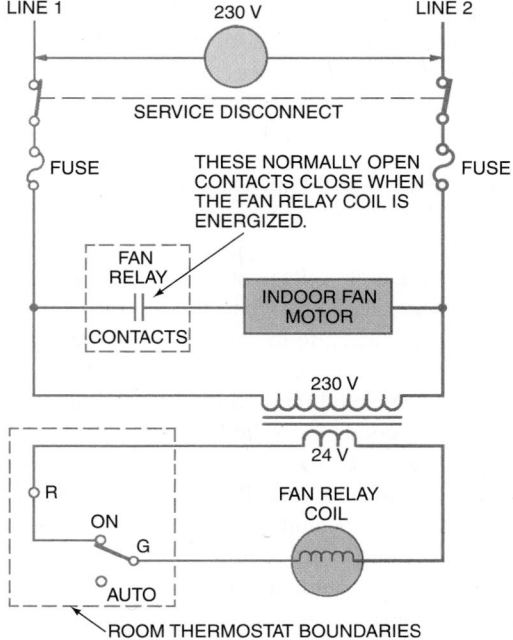

Figure 15–12   When the fan switch is turned to the ON position, a circuit is completed to the fan relay coil. When this coil is energized, it closes the fan relay contacts, passing power to the indoor fan motor.

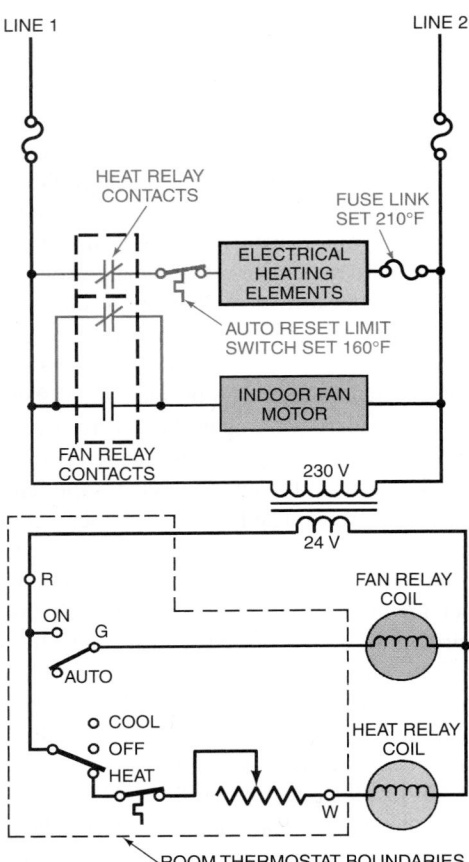

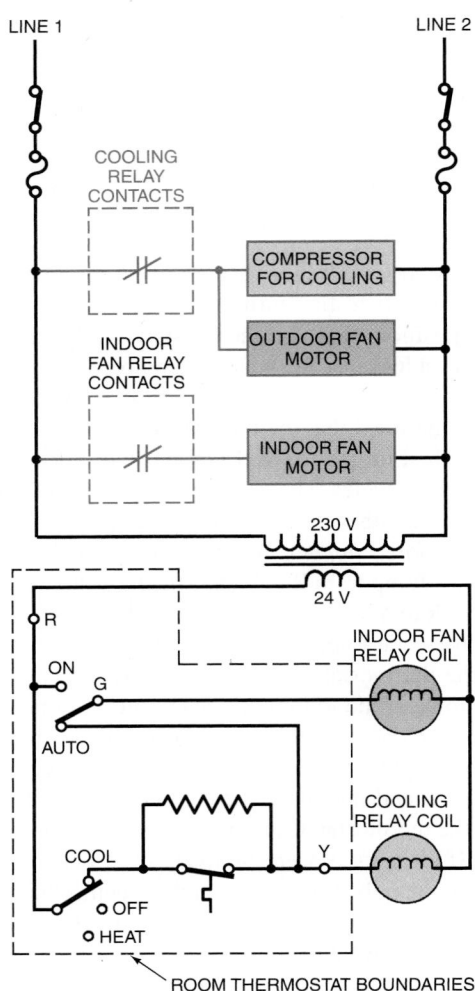

Figure 15–13 When the selector switch is in the HEAT position and the thermostat contacts close, the heat relay coil is energized. This action closes two sets of contacts in the heat relay. One set starts the fan because they are wired in parallel with the fan relay contacts. The other set passes power to the auto reset limit switch (set at 160°F). Power reaches the other side of the heater element through the fusible link. When power reaches both sides of the heating element, the heating system is functioning.

Figure 15–14 Addition of cooling to the system. When the thermostat selector switch is set to COOL and the thermal contacts close, the cooling relay coil is energized, closing the high-voltage contacts. These high-voltage contacts feed power to the compressor and condenser fan motor. Notice that the circuit in the thermostat starts the indoor fan motor through the AUTO position of the fan selector switch.

power is passed to the thermostat contacts. When there is a call for heat, these contacts close and pass power through the heat anticipator to the heat relay coil. This closes the two sets of contacts in the high-voltage circuit. One set starts the fan, and the other set passes power to the limit switch and then through the limit switch to the line 1 side of the heater. This limit switch (165°F) is the primary overheat safety control. The circuit is completed through the fuse link in line 2. The fuse link is the secondary overheat safety control (210°F).

In **Figure 15–14** cooling or air conditioning has been added to the circuit. For this illustration the heating circuit was removed to reduce clutter and confusion. Three components have to operate in the cooling mode: the indoor fan, the compressor, and the outdoor fan. The compressor and the outdoor fan are wired in parallel and, for practical purposes, can be thought of as one component.

The line 2 side of the circuit goes directly to the three power-consuming components. Power for the line 1 side of the circuit comes through two different relays. The cooling relay contacts start the compressor and outdoor fan motor; the

indoor fan relay starts the indoor fan motor. In both cases, the relays pass power to the components on a call from the thermostat.

When the thermostat selector switch is set to the COOL position, power is passed to the contacts in the thermostat. When these contacts are closed, power passes to the cooling relay coil. When the cooling relay coil is energized, the cooling relay contacts in the high-voltage circuit are closed. This passes power to the compressor and outdoor fan motor.

The indoor fan must run whenever there is a call for cooling. When power passes through the contacts in the thermostat, a circuit is made through the AUTO side of the selector switch to energize the indoor fan relay coil and start the indoor fan. Note that if the fan selector switch is in the FAN ON position, the fan would run all the time and would still be on for cooling.

15.5 ▶ TROUBLESHOOTING AMPERAGE IN THE LOW-VOLTAGE CIRCUIT

Transformers are rated in *volt-amperes,* commonly called VA. This rating can be used to determine whether the transformer is underrated or drawing too much current.

For example, it is quite common to use a 40-VA transformer for the low-voltage power source on a combination cooling and heating piece of equipment. This tells the technician that at 24 V, the maximum amperage the transformer can be expected to carry is 1.66 A. This is determined as follows:

$$\frac{40\ VA}{24\ V} = 1.66\ A$$

Many clamp-on ammeters will not readily measure such a low current with any degree of accuracy, so arrangements must be made to determine an accurate current reading. Use a jumper wire and coil it 10 times (called a 10-wrap amperage multiplier). Place the ammeter's jaws around the 10-wrap loop and place the jumper in series with the circuit, **Figure 15–15. The reading on the ammeter will have to be divided by 10.** For instance, if the cooling circuit coil amperage reads 7 A, it is really only carrying 0.7 A. Many digital clamp-on ammeters can read small amperages accurately and may not need a 10-wrap coil of wire around them. Refer to Unit 5, "Tools and Equipment," for an illustration of a modern digital clamp-on ammeter.

Some ammeters have attachments for readings in ohms and volts. The volt attachments can be helpful for voltage readings, but the ohm attachment should be checked to make sure that it will read in the range of ohms needed. Some of the ohmmeter attachments will not read very high resistance. For instance, it may show an open circuit on a high-voltage coil that has a considerable amount of resistance.

15.6 ▶ TROUBLESHOOTING VOLTAGE IN THE LOW-VOLTAGE CIRCUIT

The volt-ohm-milliammeter (VOM) has the capability of checking continuity, milliamps, and volts. The most common applications are checking voltage and continuity. The volt scale can be used to check for the presence of voltage, and the ohmmeter scale can be used to check for continuity. A high-quality meter will give accurate readings for each measurement.

Remember that any power-consuming device in the circuit must have the correct voltage applied to it in order to operate. The voltmeter, when set on the correct scale, can be applied to any power-consuming device for the voltage check by placing one probe on one side of the device and the other probe on the other side of the device. When power is present at both points, the component should function. If not, a check of continuity through the device is the next step, using the ohmmeter. Again keep in mind that when voltage is present and a path is provided, current will flow. If the flowing current does not cause the load to do its job, it is defective and will have to be changed.

The voltmeter cannot be used to much advantage at the actual thermostat because the thermostat is a closed device. When the thermostat is removed from the subbase, the method discussed earlier of jumping the thermostat terminals is usually more effective. With the thermostat removed from the subbase and the subbase terminals exposed, the voltmeter can be used in the following manner.

Turn the voltmeter selector switch to the scale just higher than 24 V. Attach the voltmeter lead to the hot leg feeding the thermostat, **Figure 15–16.** This terminal is sometimes labeled R or V; no standard letter or number is used. When the other

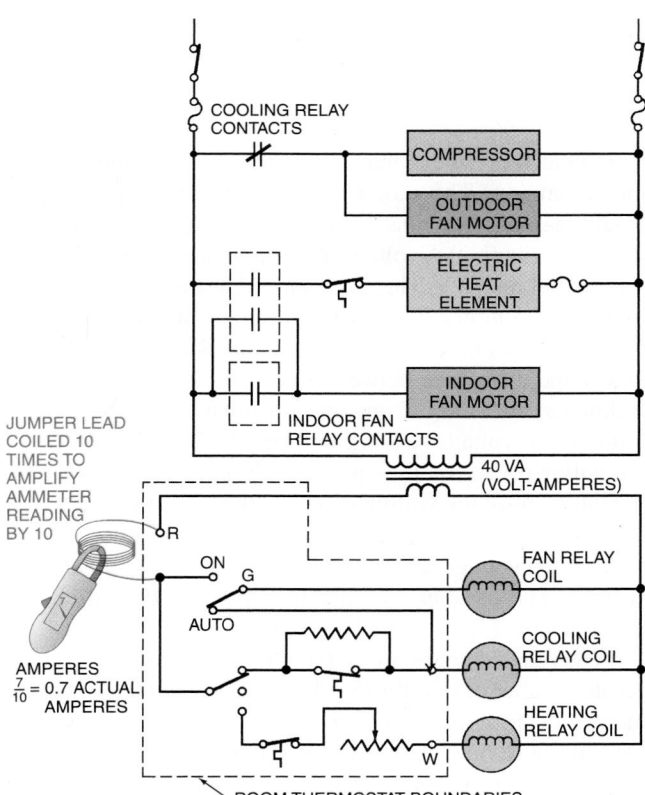

Figure 15–15   A clamp-on ammeter is used to measure the current draw in the 24-V control circuit.

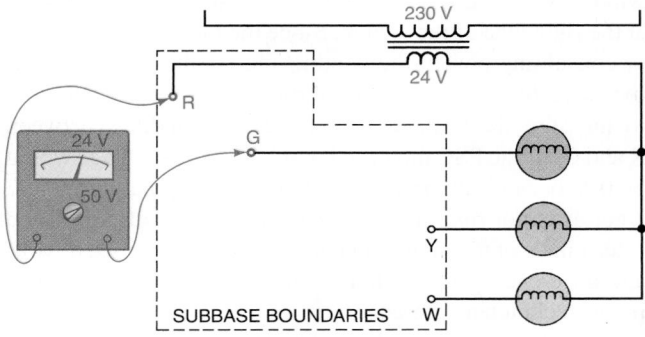

Figure 15–16   The VOM can be used at the thermostat location with the thermostat removed from the subbase.

lead is placed on the other terminals one at a time, the circuits will be verified.

For example, with one lead on the R terminal and one on the G terminal, the terminal normally used for the fan circuit, a voltage reading of 24 V will be read. The meter is reading one side of the line straight from the transformer and the other side of the line through the coil on the fan relay. In this case, the fan relay coil is not energized and simply acts as a coil of wire. It will now let the other line of the transformer feed through to the G terminal of the thermostat's subbase. This phenomenon is sometimes referred to as *voltage feedback*. The voltmeter now will be reading each side of the transformer or 24 V. When the meter probe is moved to the circuit assigned to cooling (this terminal could be lettered Y), 24 V is read. When the meter probe is moved to the circuit assigned to heating (sometimes the W terminal), 24 V again appears on the meter. The fact that the voltage reads through the coil in the respective circuits is evidence that a complete circuit is present.

15.7 ► ELECTRICALLY TROUBLESHOOTING SWITCHES AND LOADS

Service technicians often encounter electrical problems when troubleshooting equipment. These problems are often nothing but electrical switches that are either opened or closed. However, when power-consuming devices (electrical loads) are in series with these switches it can complicate matters. Most of the time electrical switches are in series with one another and are relatively simple to troubleshoot electrically. It is when these electrical switches are in parallel with one another that it gets more complicated.

Figure 15–17 shows a pressure switch in series with a motor (electrical load or power-consuming device). In this case the electrical load is a permanent split-capacitor (PSC) motor. The potential difference (voltage) between line 1 and line 2 is 230 V. This means that if the two leads of a voltmeter were placed between line 1 and line 2, a voltage of 230 V would be read. If a technician measured the voltage across the open switch, a voltage of 230 V would also be read on the voltmeter. This happens because line 1 ends at the left side of the open switch, and line 2 simply bleeds through the run winding of the motor, through the closed overload, and ends at the right side of the switch. Since the motor is not running or consuming power, the windings are nothing but conductive wire for line 2 to bleed through. If a technician were to measure the voltage across the run winding (between R and C) of the PSC motor in **Figure 15–17,** the voltage would be 0 V because the motor winding is simply passing line 2 when it is not running. Line 2 would be at both the R and C terminals of the motor, and the potential difference or voltage difference between line 2 and line 2 is 0 V. However, if the technician references either the R or C terminal to ground, the voltage would be 115 V because line 2 relative to ground is 115 V.

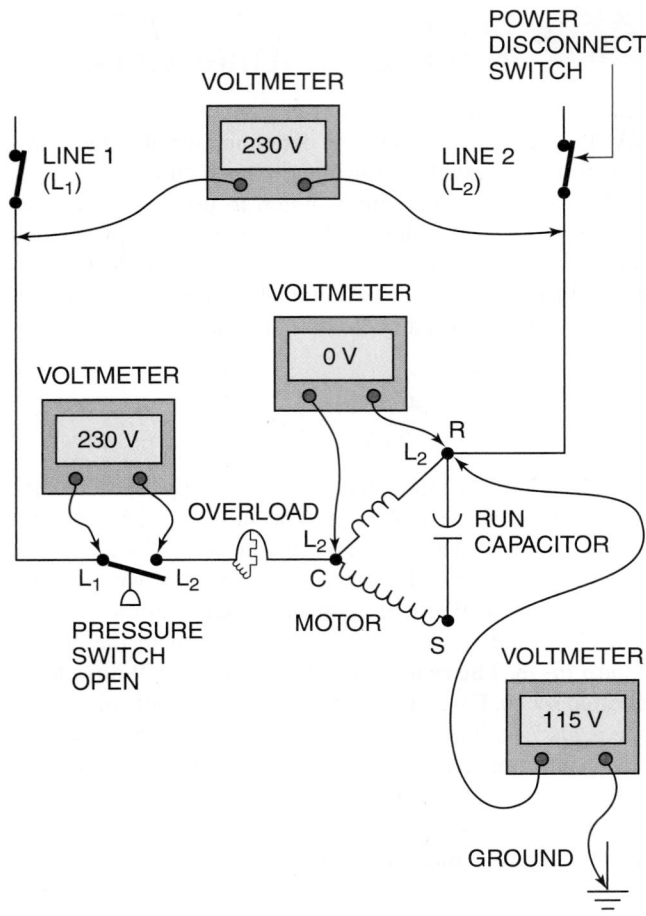

Figure 15–17   Voltage readings for a switch in series with a motor.

In **Figure 15–18** the pressure switch is closed and the motor is running. The motor is now consuming power and would read 230 V across its run winding (terminals R and C) while it is running. A voltage measurement across the closed switch will read 0 V. This is because line 1 ends at the C terminal of the motor when it is running. The switch experiences line 1 at both its terminals, and the potential difference or voltage difference between line 1 and line 1 is 0 V. If a technician measures from one terminal of the switch to ground, the voltage reading will be 115 V. In **Figure 15–17** the voltmeter across the open switch reads 230 V, whereas in **Figure 15–18** the voltmeter across the closed switch reads 0 V. It is incorrect to conclude that open switches always read some voltage and closed switches always read zero voltage. The parallel switches in **Figure 15–19** will clarify this concept.

When the switches are in parallel, the technician faces a greater challenge. **Figure 15–19** shows two switches that are in parallel but at the same time in series with a motor (power-consuming device). The top switch is closed, but the bottom switch is open. A voltmeter across terminals A and B of the top switch will read 0 V because it is measuring a potential difference between line 1 and line 1. Since the motor is

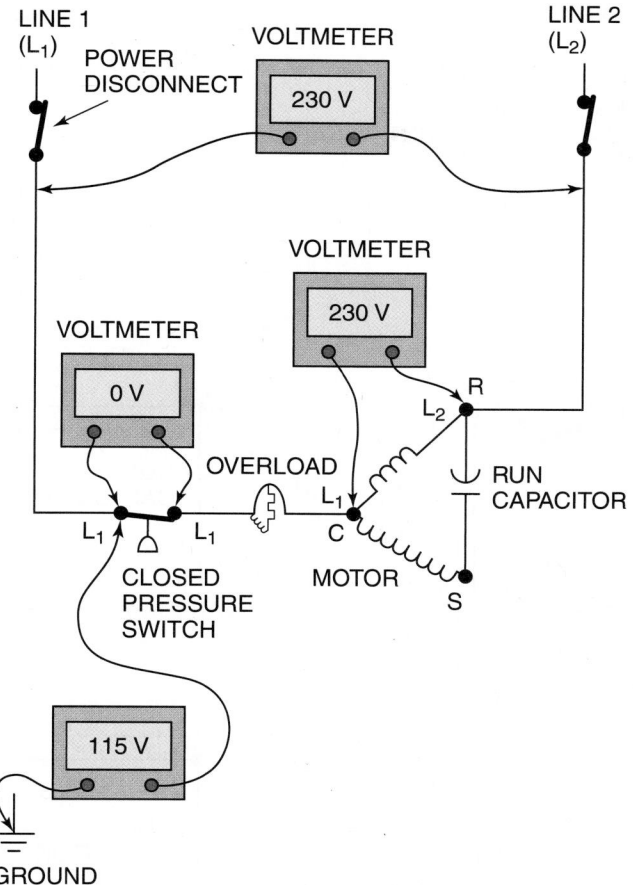

Figure 15–18   Voltage readings for a closed switch in series with an operating motor.

running, it has 230 V applied across the run winding (C and R terminals) of the PSC motor. However, a voltmeter across terminals C and D of the bottom switch will also read 0 V. This happens because line 1 actually extends to the common terminal of the motor when it is running. This would make terminals C and D of the bottom switch both line 1, and the voltage difference between line 1 and line 1 is 0 V. Actually points A, B, C, and D are all line 1. This example is a scenario where both an open switch and a closed switch read 0 V. *When doing electrical troubleshooting with a voltmeter, technicians must ask themselves where line 1 and line 2 are, not whether the switch is open or closed.* Measuring across the same line with a voltmeter, whether it be line 1 or line 2, will always give 0 V. Measuring from line 1 to line 2 will always give the total circuit voltage, which in these examples was 230 V.

There will be times when a technician must switch to an ohmmeter and shut the electrical power off in order to solve the problem. In **Figure 15–20,** what would be the voltage between R and C if the run winding between R and C opened, causing the motor to stall and draw locked-rotor amperage (LRA)? This is, of course, before the overload has opened. Notice that a voltmeter placed across the R and C terminals

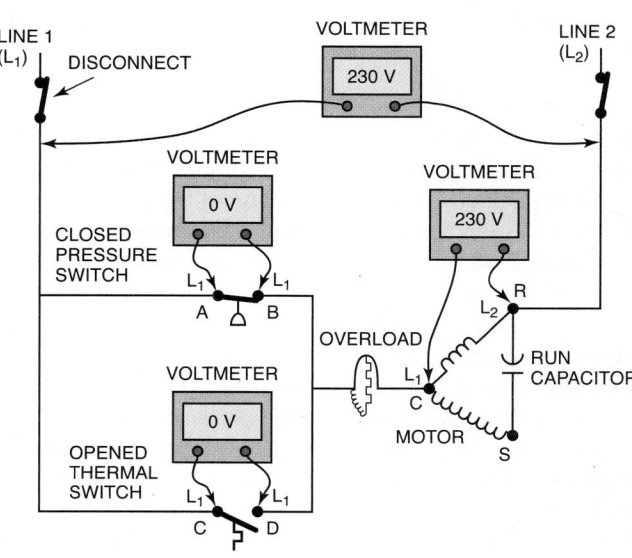

Figure 15–19   Voltage readings for two switches wired in parallel with each other. These switches are both wired in series with the operating motor.

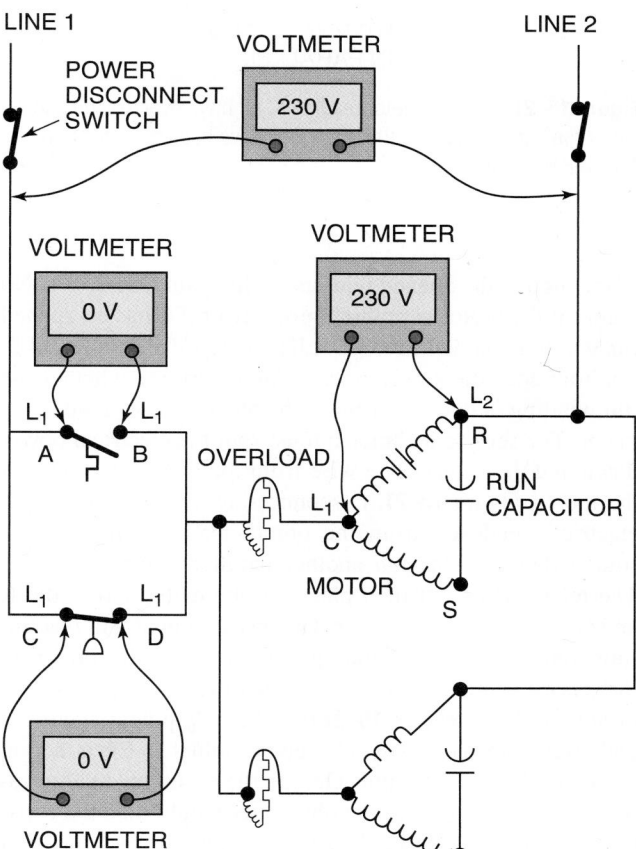

Figure 15–20   A voltmeter reading 230 V across the run winding of a motor even when the run winding is open.

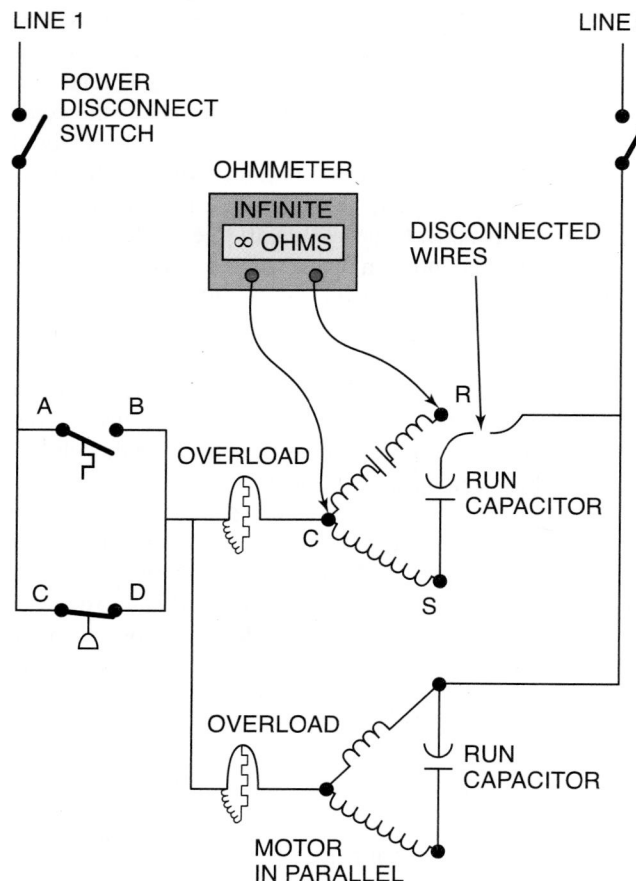

Figure 15–21 An ohmmeter reading of infinity ∞ across an open motor run winding. Notice that the wires have been disconnected from the R terminal.

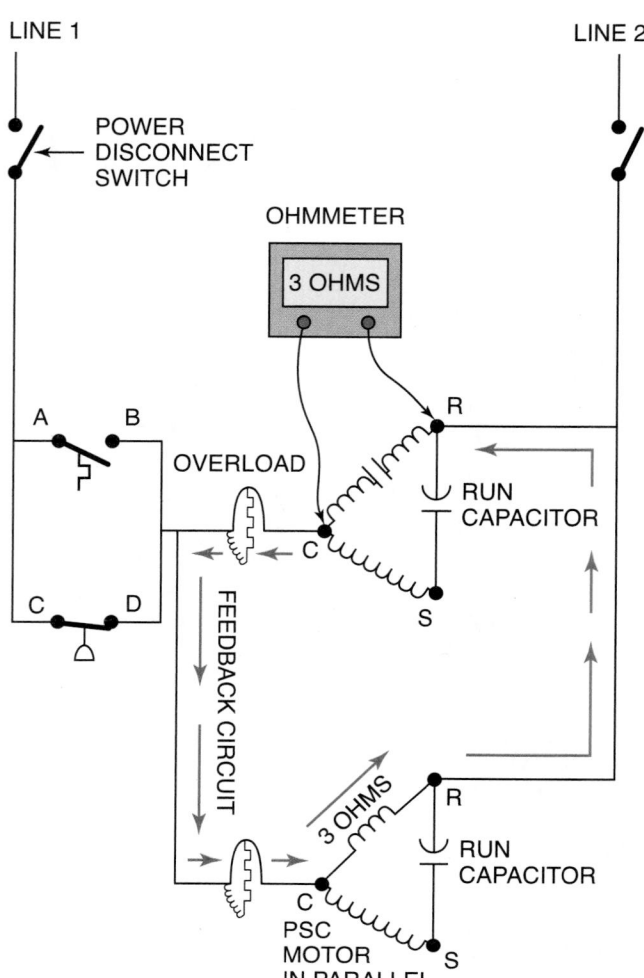

Figure 15–22 A feedback circuit from the ohmmeter's battery when a wire is not disconnected from the motor being tested. The ohmmeter will now measure the 3 ohms of the run winding of the PSC motor that is wired in parallel with the motor being tested.

of the motor (the opened winding) will again read 230 V. No matter if the motor is running properly or if it has an opened run winding, the voltage will still read 230 V across R and C. So, how does the service technician determine whether the run winding is opened or not? The answer is, with an ohmmeter. The service technician must shut the electrical power down and disconnect one wire from the R or C terminal of the motor, **Figure 15–21**. Disconnecting a wire will prevent electrical feedback from the ohmmeter's internal voltage source (battery) through another parallel electrical circuit. The technician must then place an ohmmeter across the R and C terminals of the motor. The measurement will read infinite ohms (∞) if the winding is open. This is the only way the service technician can tell whether the winding is opened or not. Notice in **Figure 15–21** that the only potential current path exists between terminal C and terminal R. Even the run capacitor has been disconnected. In the event that the run capacitor had a short in it, the ohmmeter would read the resistance in the path through the run capacitor and the start winding, instead of the infinite resistance of the open path between terminals C and R. **Figure 15–22** shows a feedback circuit from the ohmmeter's battery if a wire is not disconnected from the motor terminals. In this case, the ohmmeter

reading would be 3 ohms. This resistance comes from the resistance of the run winding in the feedback circuit of the PSC motor in parallel with it. This would fool the technician into thinking the winding was still good and not opened.

Figure 15–23 is a 230-V, single-phase electrical schematic of a typical commercial refrigeration system. The diagram includes a timer assembly with a defrost termination solenoid (DTS), evaporator fans, defrost heaters, a temperature-activated defrost termination/fan delay (DTFD) switch, a low-pressure control (LPC), a high-pressure control (HPC), a compressor contactor assembly, and a compressor/potential relay assembly. Although this diagram may look a little complicated, it is actually multiple series circuits connected to a single power supply. We will now examine each of these branch circuits individually. The system is drawn in the refrigeration mode and shows where line 1 (L₁) is in relation to line 2 (L₂) for ease of understanding the measured voltages.

The first load we will look at in this circuit is the timer motor. It can be seen from **Figure 15–24** that the timer motor

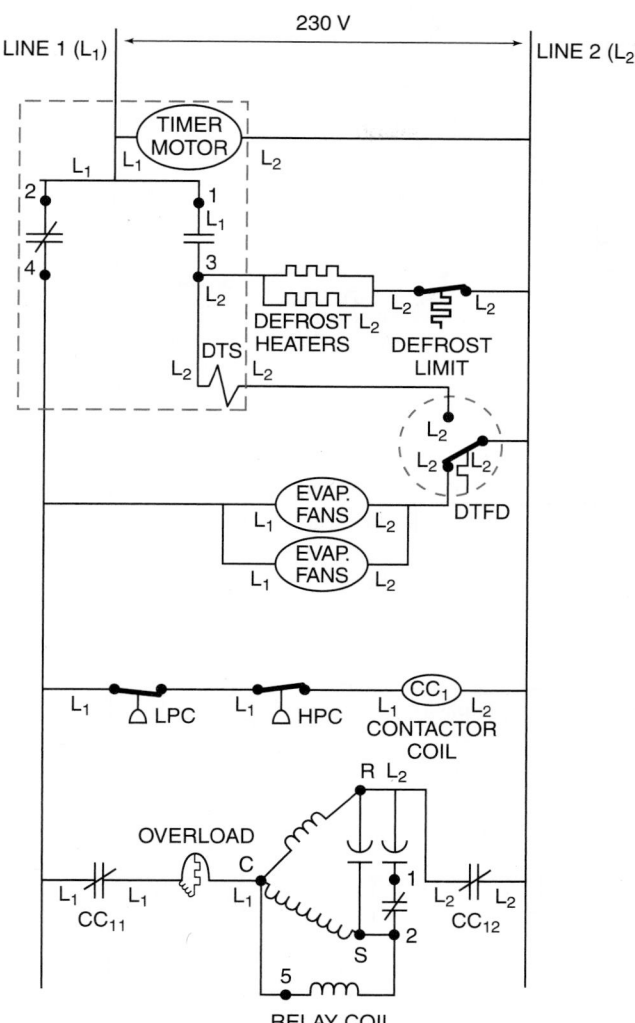

Figure 15–23 A 230-volt electrical schematic of a typical refrigeration circuit showing relative positions of line 1 (L_1) and line 2 (L_2).

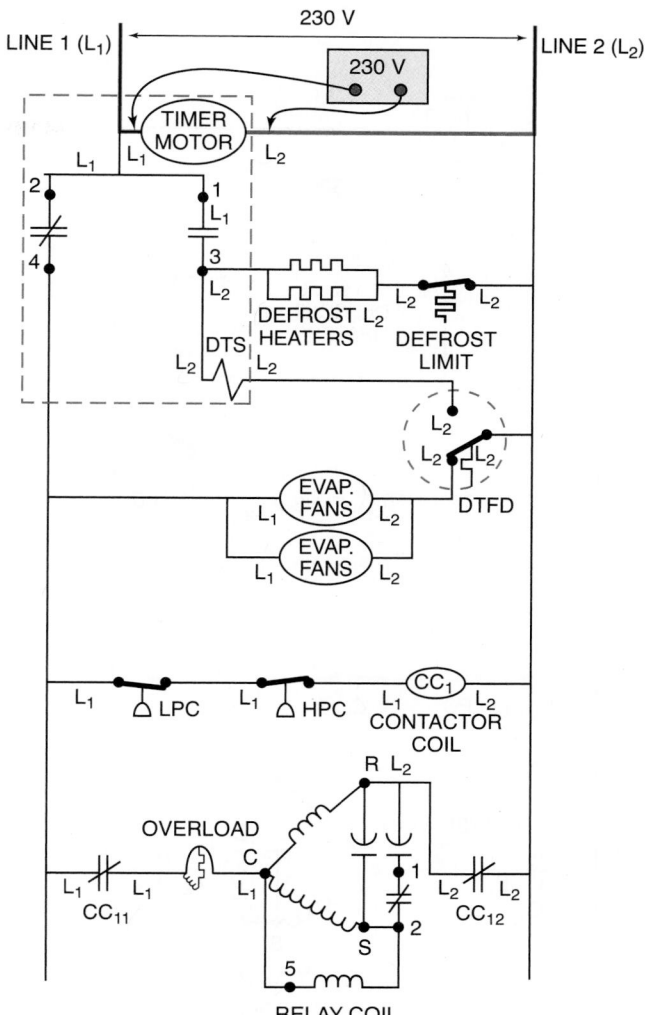

Figure 15–24 The timer motor circuit.

is connected directly to the 230-volt power supply. As long as the power to the system is on, there should be a reading of 230 volts across the timer motor. If the power is on and there are not 230 volts being supplied to the timer motor, there is a problem with the wiring that is connecting the power supply to the timer motor. If there are 230 volts being supplied to the timer motor—and the motor is not operating—there is a problem with the timer motor.

The next circuit we will examine is the defrost heater circuit, **Figure 15–25**. By starting at L_1 and tracing a path through the defrost heaters and back to L_2, we can see that there are a number of switches in series with the heaters. The first switch is the set of contacts that is located between contact points 1 and 3 on the timer, which is the component that is outlined with the dashed line. When the system is operating in the refrigeration mode, the heater contacts are in the open position, as we do not want them to be energized when the system is providing refrigeration. When the system goes into defrost, these contacts will close and the heaters will

become energized as long as the second switch in the circuit and all interconnecting wiring is correct. The second switch in the circuit is the defrost limit switch. This is a safety device and will remain closed as long as the temperature it senses is below the safe operational limit. If the temperature rises too high, the limit switch will open and the heaters will be turned off. A reading of 230 volts across the heaters indicates that the heaters are being powered. An amperage reading must then be taken to ensure that the heaters are indeed energized. If the heaters are being powered but the amperage draw of the heaters is 0 A, there is a problem with the heaters. A positive amperage reading in the heater circuit indicates that the heaters are energized.

The next circuit controls the operation of the defrost termination solenoid (DTS), **Figure 15–26**. Just as with the defrost heaters, the defrost solenoid is controlled by a set of time-controlled contacts as well as set of temperature-controlled contacts. It just so happens that the set of time-controlled contacts is the same exact set that is used to control the defrost

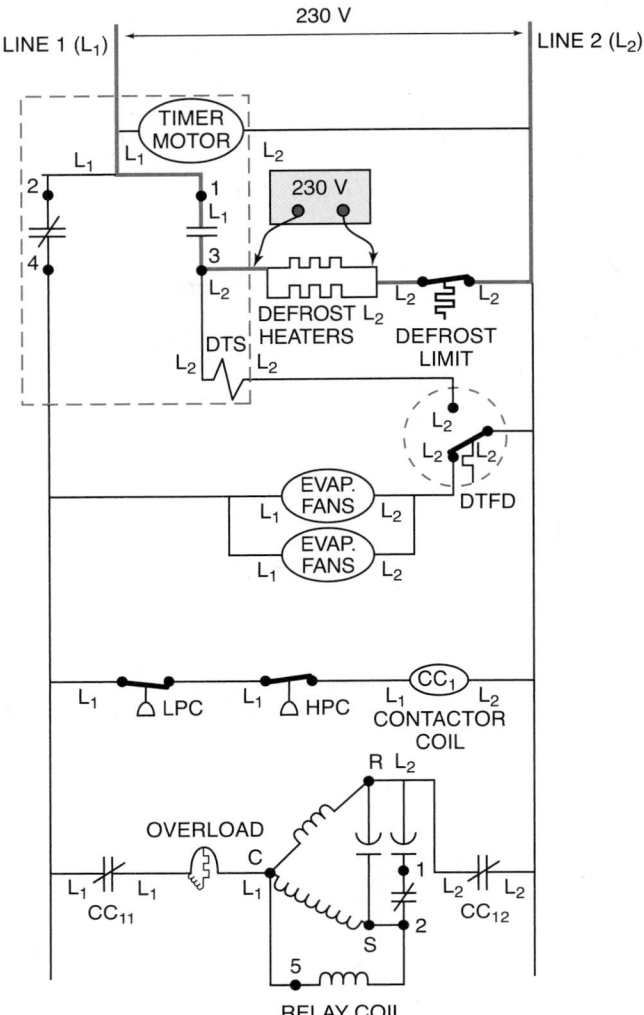

Figure 15–25 The defrost heater circuit.

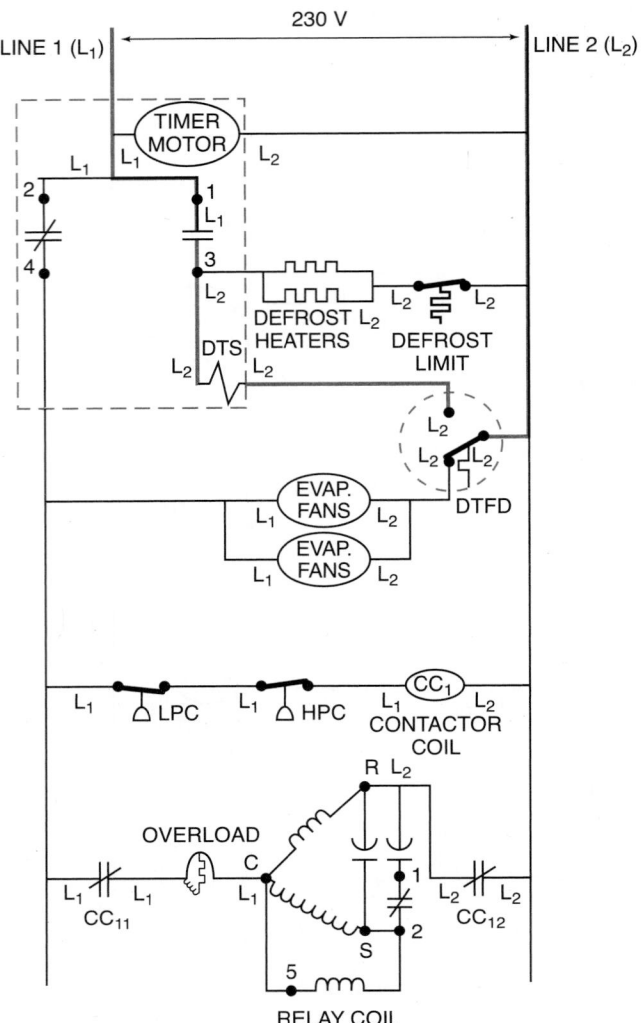

Figure 15–26 The defrost termination solenoid circuit.

heaters; namely, the contacts between terminals 1 and 3 on the defrost timer. The temperature-controlled switch performs two functions:

■ It takes the system out of defrost if all of the evaporator frost has been removed.

■ It controls the evaporator fan motors so they can only operate when the evaporator is cold enough to provide refrigeration.

In order for the defrost termination solenoid to be energized, the system must be in defrost and the temperature sensed by the defrost termination thermostat must be warm enough. When the DTS is energized, the closed contacts on the timer will open and the open contacts will close, putting the system back into the refrigeration mode.

The defrost termination solenoid is also often referred to as a "Z-coil." Taking amperage readings in the Z-coil circuit is difficult, as the coil is never energized for more than a fraction of a second. Here's why (let's say the system is in defrost):

■ The set of contacts between terminal 1 and 3 is closed.

■ There is frost on the evaporator coil, so the defrost termination/fan delay switch (DTFD) is in the position shown in **Figure 15–26.**

■ So, the Z-coil circuit is not energized.

■ When the frost is melted, the DTFD changes its position and energizes the DTS.

■ The instant the DTS is energized, it opens the closed contacts on the defrost timer and closes the open contact.

■ This opens the very contacts that are energizing the DTS in the first place.

For this reason, it is possible to take voltage readings, but amperage readings will not prove to be very useful for troubleshooting purposes.

We will now take a look at the evaporator fan circuit, **Figure 15–27.** There are two control devices that cycle the evaporator fans on and off. One is a set of contacts on the defrost timer that is located between terminals 2 and 4 on the timer. These contacts are in the closed position when the system is in the refrigeration mode. The other control device is the defrost termination/fan delay switch. When the evaporator is

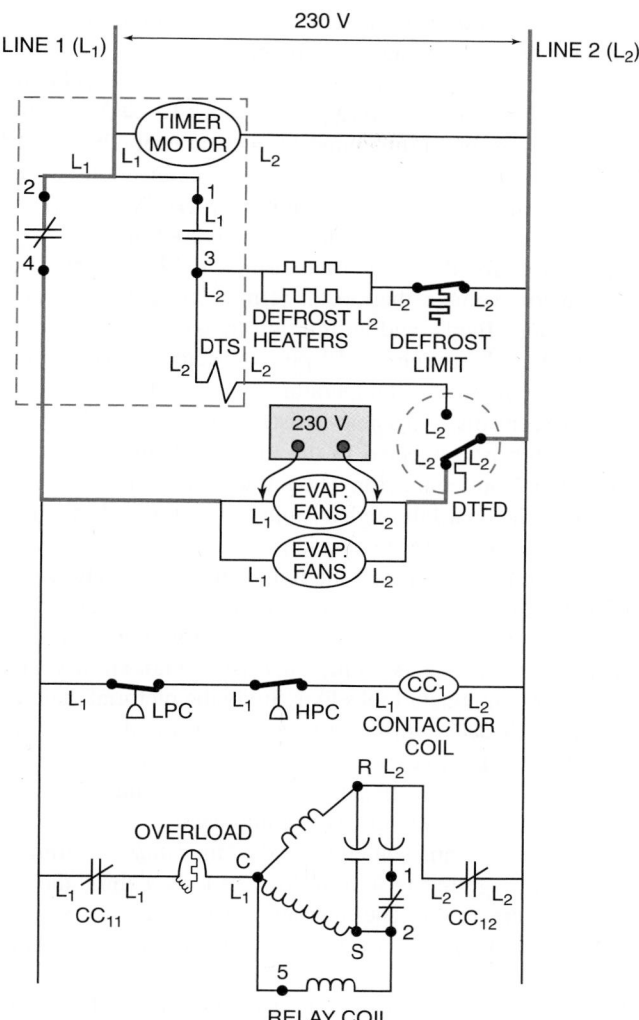

Figure 15–27   The evaporator fan motor circuit.

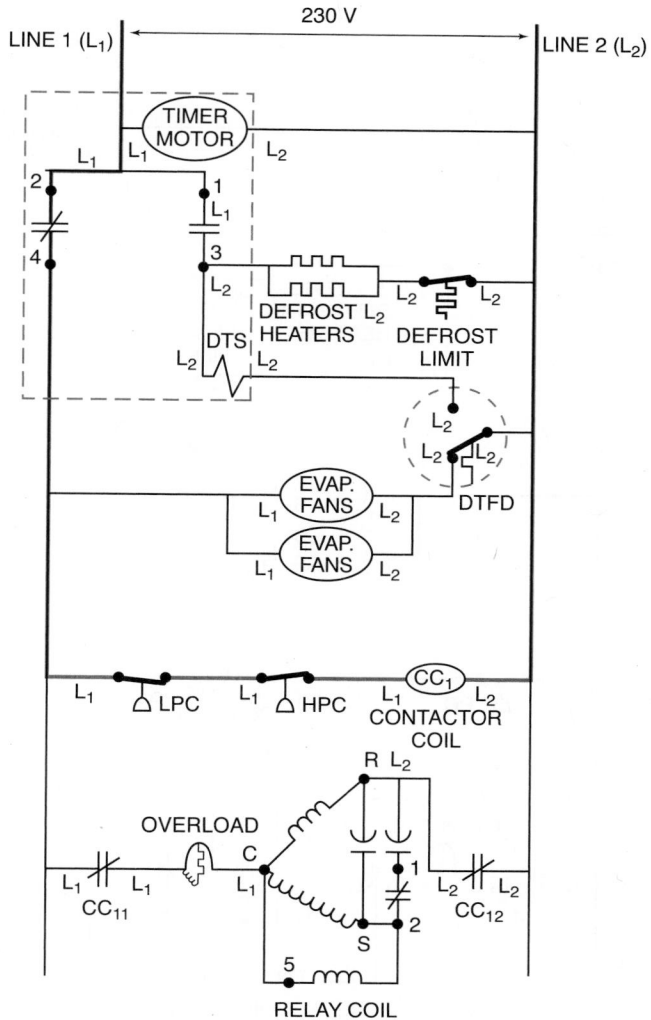

Figure 15–28   The contactor coil circuit.

cold enough for refrigeration, the DTFD will close the contacts in series with the evaporator fan motors. If there is no voltage being supplied to the evaporator fan motors, the technician should check for voltage across the timer contacts 2 and 4 as well as for voltage across the DTFD. A reading of line voltage across contacts 2 and 4 indicates that the system is not in the refrigeration mode, while a line-voltage reading across the DTFD is an indication that the evaporator coil temperature is too high for refrigeration.

The next part of the wiring diagram is the circuit that energizes and deenergizes the compressor contactor coil, **Figure 15–28**. In this circuit, the contactor coil is the load and there are three switches that are wired in series with this coil. The three power-passing devices are the

- set of defrost timer contacts between terminals 2 and 4.
- low-pressure control.
- high-pressure control.

So, in order for the contactor coil to be energized, the system's defrost timer must be calling for refrigeration and the system pressures must be within the safe or desired operating range. If the system pressure is too low, the low-pressure

control will open and deenergize the coil. If the high-side pressure is too high, the high-pressure switch will open and deenergize the coil, as well. If there is line voltage being supplied to the contactor coil, the voltage readings across each of the three switches in the circuit will be 0 volts. A reading of 0 volts across the contactor coil most likely indicates that one of the three switches is in the open position. For example, a 230-volt reading across contacts 2 and 4 on the defrost timer indicates that the contacts are open and that the defrost timer is not calling for refrigeration.

The last circuit in this diagram is the compressor circuit, **Figure 15–29**. Once again the load is wired in series with a number of switches. In this case, there are four sets of contacts that must be closed in order for the compressor to operate. The four switches are the

- set of defrost timer contacts between terminals 2 and 4.
- contactor contacts CC_{11}.
- contactor contacts CC_{12}.
- compressor overload.

We already know that, in order for contacts 2 and 4 to be closed, the system has to be calling for refrigeration. The

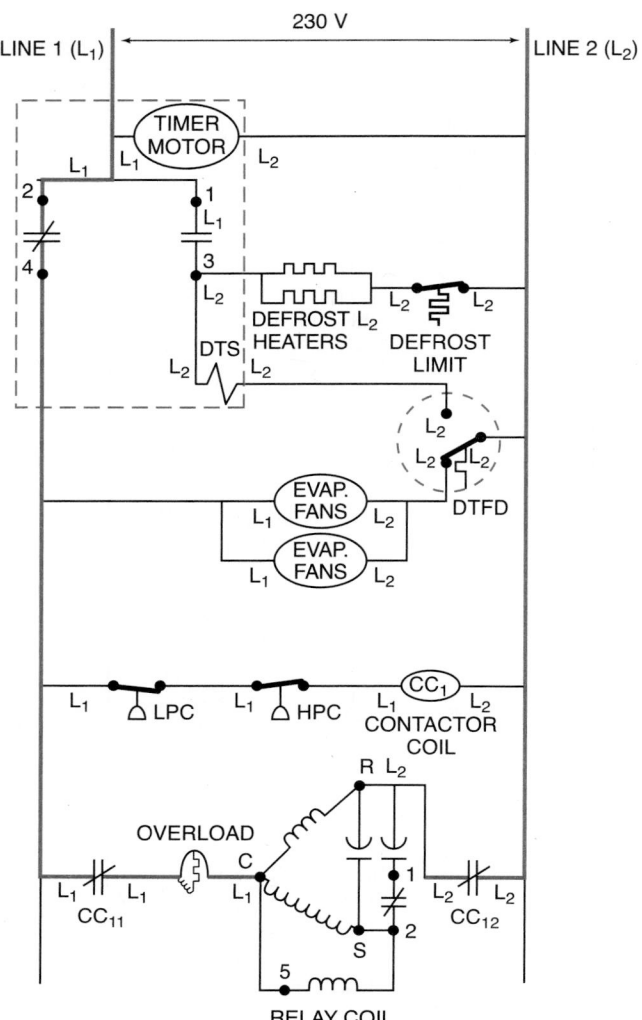

Figure 15–29 The compressor circuit.

two sets of contactor contacts, CC_{11} and CC_{12}, are controlled by the contactor coil in the circuit branch we just discussed, **Figure 15–28.** The magnetic field generated by the contactor coil will cause these contacts to close. The last switch is the overload, which is a safety device. This device will open if there is excessive current flow in that circuit branch or if the compressor overheats. A voltage reading of 230 volts across the overload indicates that there is an overcurrent, overload, or overheating condition. The cause for this condition must be identified and corrected to put the system back into operation.

15.8 PICTORIAL AND LINE DIAGRAMS

The previous examples were all on troubleshooting the basic circuit. This is recommended to start with. The control circuits that use circuit boards work much the same way. The same rules apply with more emphasis on manufacturers' directions. The service technician must have a good mental picture of the circuit or a good diagram to work from. Two distinct types of diagrams are furnished with equipment: the

pictorial wiring diagram and the *line wiring diagram.* Equipment may have one diagram or both.

The pictorial diagram is used to locate the different components in the circuit, **Figure 15–30. Figure 15–30(A)** is a pictorial diagram of an integrated furnace control (IFC) board for a two-stage gas furnace. It contains connections for an electronic air cleaner (EAC), a humidifier (HUM), line 1 and neutral, fan connections for heating and cooling, secondary connections for the 24-V transformer, a 12-pin connector plug, a low-voltage thermostat terminal board, flame and power status lights, and an ignition transformer. Also built into the board through the 12-pin connector is a flame sensor, limit, rollout, overtemperature switches, high- and low-pressure controls, and a heat-assisted limit control. A blower door switch is connected to the integrated furnace control through a 2-pin connector. The induced draft motor is connected to the board through a 3-pin connector. This diagram is organized just as the technician would see it with the panel door open. For example, if the flame and power status lights are in the upper left-hand corner of the diagram, they will be in the upper left-hand corner of the control board when the door is opened. This is useful when you do not know what a particular component looks like. Study the pictorial diagram in **Figure 15–30(A)** until you find a control and then locate it in the actual integrated furnace control board in **Figure 15–30(B)**. Notice that the pictorial diagram also gives wire colors to further verify the components.

The line diagram, sometimes called the *ladder* diagram, is the easier diagram to use to follow the logic behind the circuitry, **Figure 15–31.** The diagram normally can be studied briefly, and the circuit functions will become obvious. Notice that the power-consuming devices are to the right of the diagram. Switches and controls are usually to the left in a ladder diagram. Most of the time, a technician must know what the sequence of events is for the IFC in order to follow the logic when the furnace is called on to ignite. A technical manual from the manufacturer will sometimes have the sequence of events in it, which will assist the technician in troubleshooting. These manuals often come with the furnace, or the homeowner may have one from the day of purchase. If the technician does not have such a manual, an Internet search or phone call to the furnace manufacturer may be needed before going any further in troubleshooting. Often, one of the diagrams on the inside of the front cover of the furnace will assist the service technician in figuring out the sequence of events for the furnace.

Pictorial and line diagrams are an example of the way most manufacturers illustrate the wiring in their equipment. Each manufacturer has its own way to illustrate points of interest. The only standard the industry seems to have established is the symbols used to illustrate the various components, but even these tend to vary from one manufacturer to another.

Anyone studying electrical circuits could benefit by first using a colored pencil for each circuit. A skilled person can divide every diagram into circuits. The colored pencil will allow an unskilled person to make a start in dividing the circuits into segments.

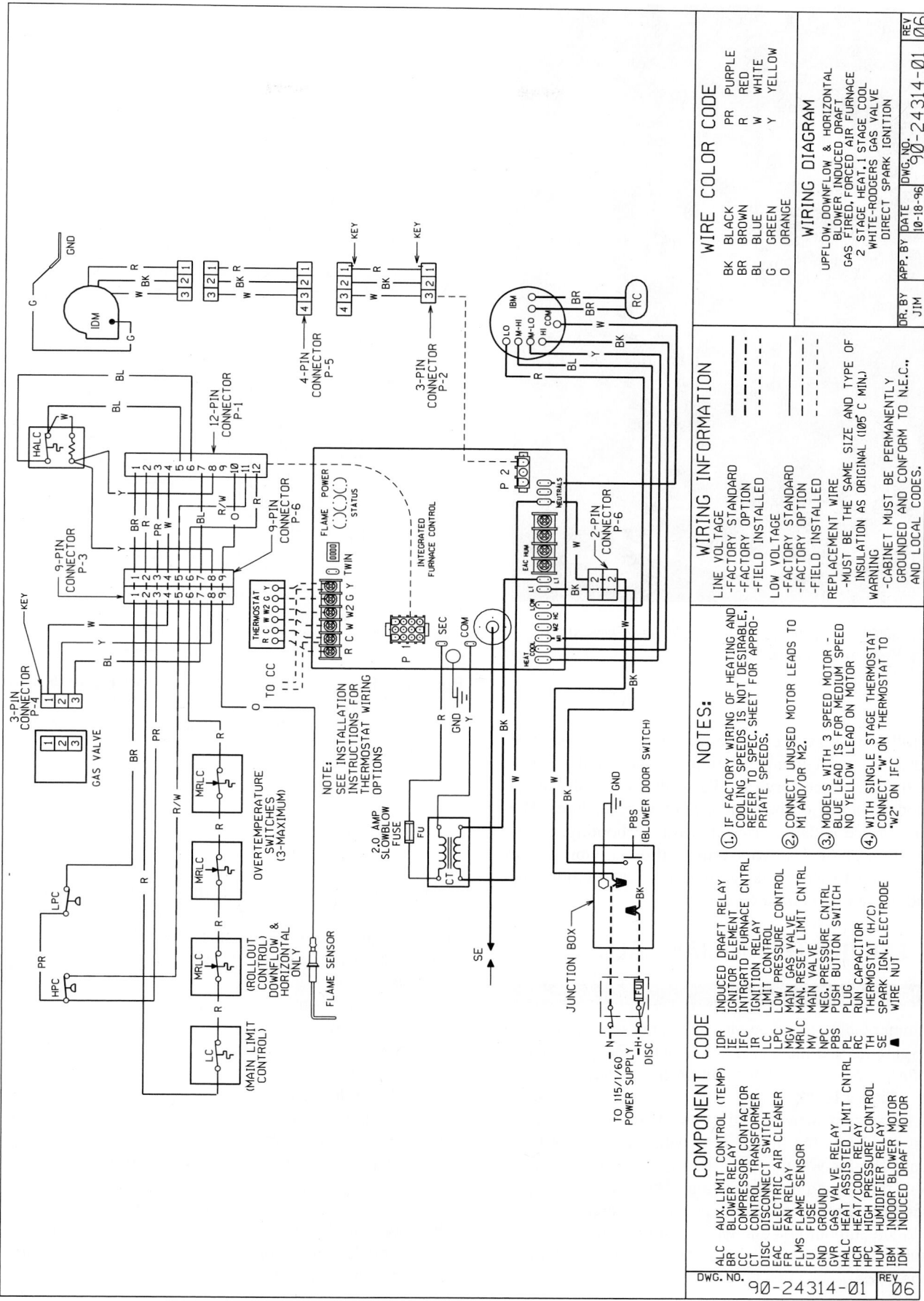

Figure 15-30(A)   A pictorial diagram showing the relative positions of the components as they actually appear. *Courtesy Rheem Manufacturing Company*

Figure 15–30(B)   An actual circuit board. *Courtesy Ferris State University. Photo by John Tomczyk*

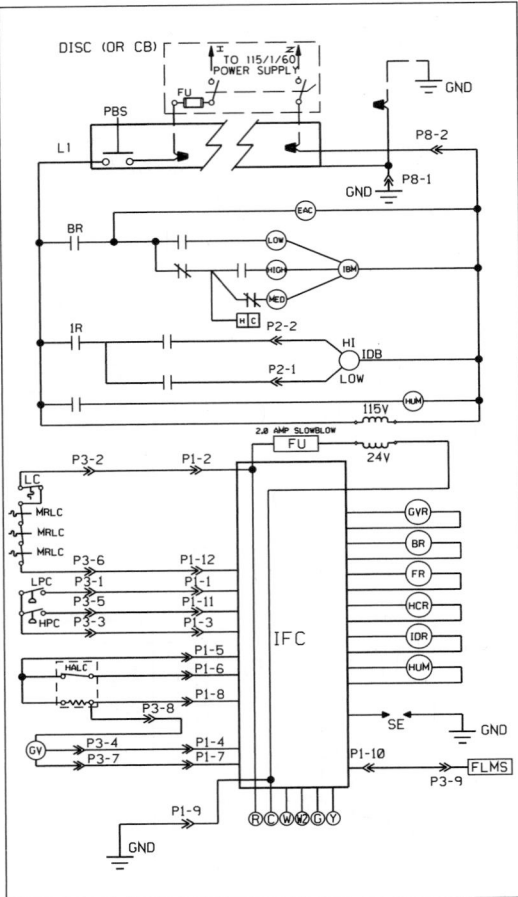

Figure 15–31   Line diagram of the circuit shown in **Figure 15–30(A)**. Notice the arrangements of the components. The right side has no switches. The power goes to the integrated furnace controller (IFC) and then to the power-consuming devices.

The following service calls are examples of actual service situations. The problem and solution are stated at the beginning of the first three service calls so that you will have a better understanding of the troubleshooting procedures. The last four service calls describe the troubleshooting procedures up to a point. Use your critical thinking abilities or class discussion to determine the solutions. The solutions can be found in the *Instructor's Guide*.

15.9   SERVICE TECHNICIAN CALLS

SERVICE CALL 1

A customer with a package air conditioner calls and tells the dispatcher that the unit is not cooling. *The problem is that the control transformer has an open circuit in the primary*, **Figure 15–32**.

The technician arrives and goes to the indoor thermostat and turns the indoor fan switch to the FAN ON position. The fan will not start. This is an indication that possibly there is no control voltage. There must be high voltage before there is low voltage. The technician then goes to the outdoor unit and with the meter range selector switch set on 250 V, a check for high voltage shows there is power.

The cover is removed from the control compartment (usually this compartment can be found because the low-voltage wires enter close by). A check for high voltage at the primary of the transformer shows that there is power.

A check for power at the secondary of the low-voltage transformer indicates no power. The transformer must be defective.

To prove this, the main power supply is turned off. One lead of the low-voltage transformer is removed and the ohmmeter feature of the VOM is used to check the transformer for continuity. *Here the technician finds the primary circuit is open*. A new transformer must be installed.

The technician installs the new transformer and turns the power on. The system starts and runs correctly. The paperwork is handled with the customer, and the technician goes to the next call.

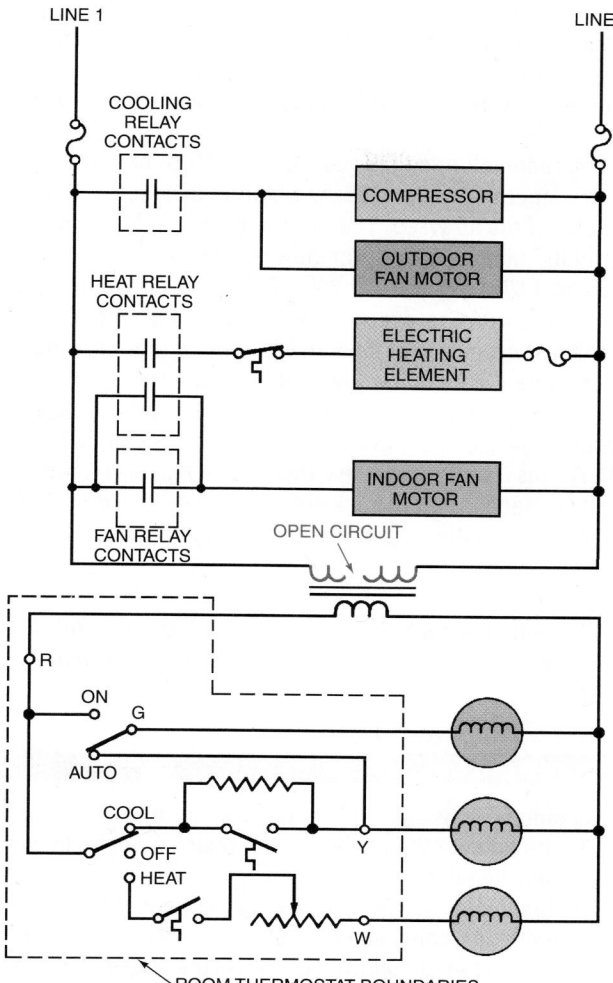

Figure 15–32   The control transformer has an open circuit.

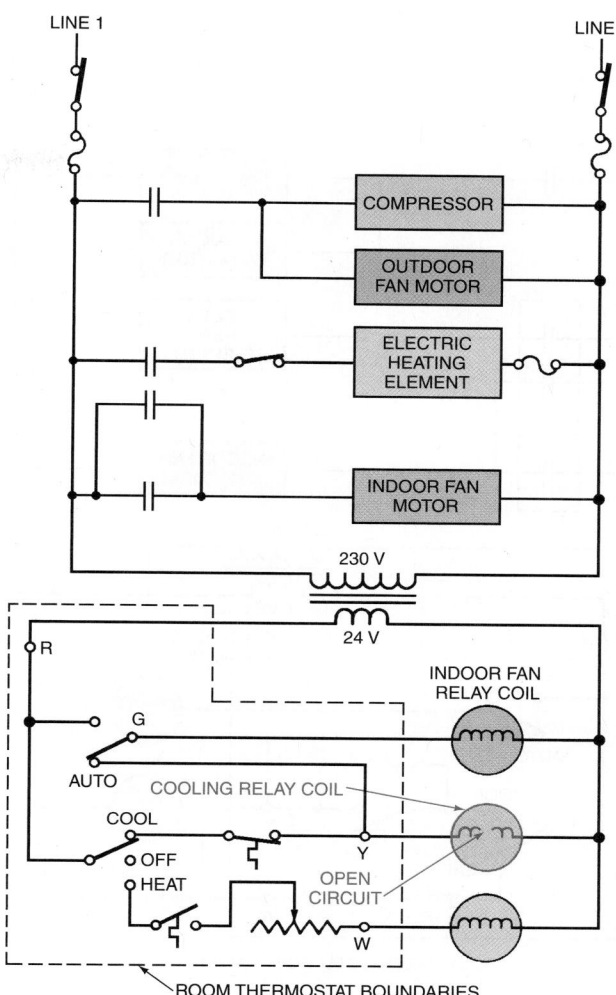

Figure 15–33   The cooling relay has an open circuit and will not close the cooling relay contacts. Notice that the indoor fan motor remains operational.

SERVICE CALL 2

A customer calls with a "no cooling" complaint. *The problem is that the cooling relay coil is open,* **Figure 15–33.**

The technician arrives, goes to the thermostat, and tries the fan circuit. The fan switch is moved to FAN ON, and the indoor fan starts. Then the thermostat is switched to the COOL position, and the fan runs. This means that the power is passing through the thermostat, so the problem is probably not there. The thermostat is left in position to call for cooling while the technician goes to the outdoor unit.

Since the primary for the low-voltage transformer comes from the high voltage, the power supply is established.

A look at the diagram shows the technician that the only requirement for the cooling relay to close its contacts is that its coil be energized. A check for 24 V at the cooling relay coil shows 24 volts. The coil must be defective. To be sure, the technician turns off the power and removes one of the leads on the coil. The ohmmeter is used to check the coil. There is no continuity. It is open.

The technician changes the whole contactor (it is less expensive to change the contactor from stock on the truck than to go to a supply house and get a coil), turns the power on, and starts the unit. It runs correctly. All paperwork is handled, and the technician leaves.

SERVICE CALL 3

A customer complains of no heat. *The problem is the heat relay has a shorted coil that has overloaded the transformer and burned it out,* **Figure 15–34.**

The technician arrives, goes to the thermostat, and tries the indoor fan. It will not run. There is obviously no current flow to the fan. This could be a problem in either the high- or low-voltage circuits.

The technician goes to the outdoor unit to check for high voltage and finds there is power. A check for low voltage shows there is none. The transformer must be defective.

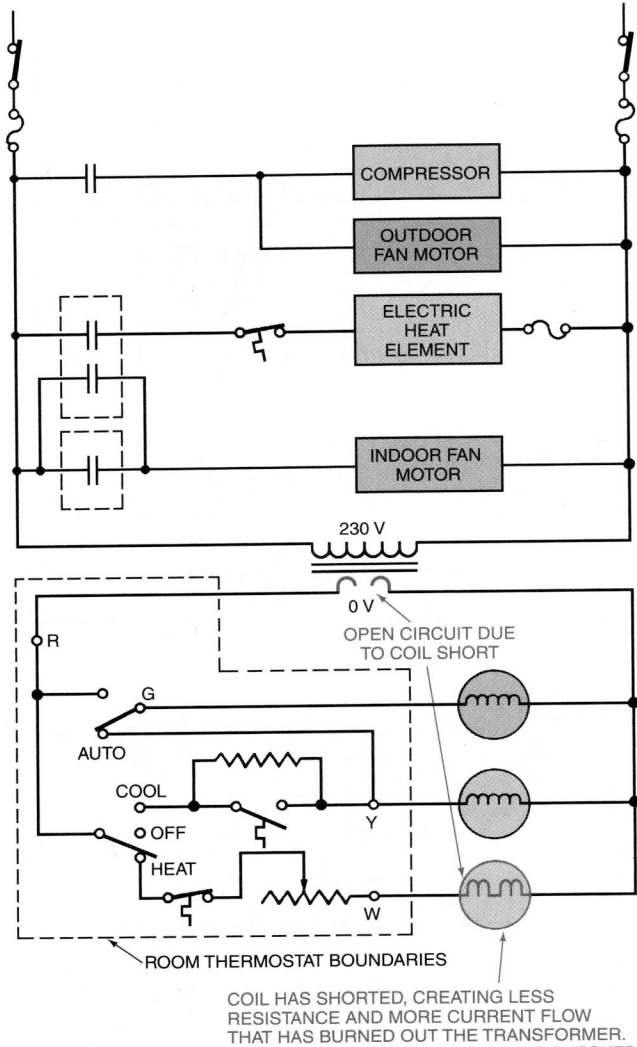

Figure 15–34 A shorted heat relay causes the transformer to burn out.

It is removed and has a burnt smell. It is checked with the VOM and found to have an open secondary.

When the transformer is changed and the system is switched to HEAT, the system does not come on. The technician notices the transformer is getting hot. It appears to be overloaded. The technician turns off the power before a second transformer is damaged. The problem is to find out which circuit is overloading. Since the system was in the heating mode, the technician assumes for the moment it is the heat relay. The technician realizes that if too much current is allowed to pass through the heat anticipator in the room thermostat, it will burn also, and a thermostat will have to be replaced.

The technician goes to the room thermostat and turns it off to take it out of the circuit and then goes to the outdoor unit where the low-voltage terminal block is located. An ammeter is applied to the transformer circuit leaving the transformer. A 10-wrap coil of wire is installed to amplify the current reading. The power is turned on, and no amperage is recorded. The problem must not be a short in the wiring.

The technician then jumps from R to W to call for heat. The amperage goes to 25 A, which is divided by 10 because of the 10 wrap. The real amperage is 2.5 A in the circuit. This is a 40-VA transformer with a rated amperage of 1.67 A (40/24 = 1.666). The heat relay is pulling enough current to overheat the transformer and burn it up in a short period of time. It is lucky that the room thermostat heat anticipator was not burned also.

The technician changes the heat relay and tries the circuit again, and the average is only 5 A, divided by 10, or 0.5 A. This is correct for the circuit. The jumper is removed, and the panels are put back in place. The system is started with the room thermostat and it operates normally.

The following service calls do not have the solution described. The solution can be found in the *Instructor's Guide*.

SERVICE CALL 4

A customer calls and indicates that the air conditioner quit running after the cable TV repairman had been under the house installing TV cable.

The technician arrives and enters the house. It is obvious that the air conditioner is not running. The technician can hear the fan running, but it is hot in the house. A check outside shows that the condensing unit is not running. The technician then proceeds through the following checklist. Use the diagram in **Figure 15–35** to follow the checkout procedure.

1. Low voltage is checked at the condenser, and there is none.
2. The technician then goes to the room thermostat and removes it from the subbase and jumps from R to Y; the condensing unit does not start.
3. Leaving the jumper in place, the technician then goes to the condenser, checks voltage at the contactor, and finds no voltage.

What is the problem and the recommended solution?

SERVICE CALL 5

A shoe store manager calls and reports that the heat is off. This store has an electric heat package unit on the roof.

The technician arrives, talks to the customer, and uses the following checkout procedure to determine the problem. You can follow the procedure using the diagram in **Figure 15–36**.

1. The store manager says the system worked well until about an hour ago when the store seemed to be getting cold.

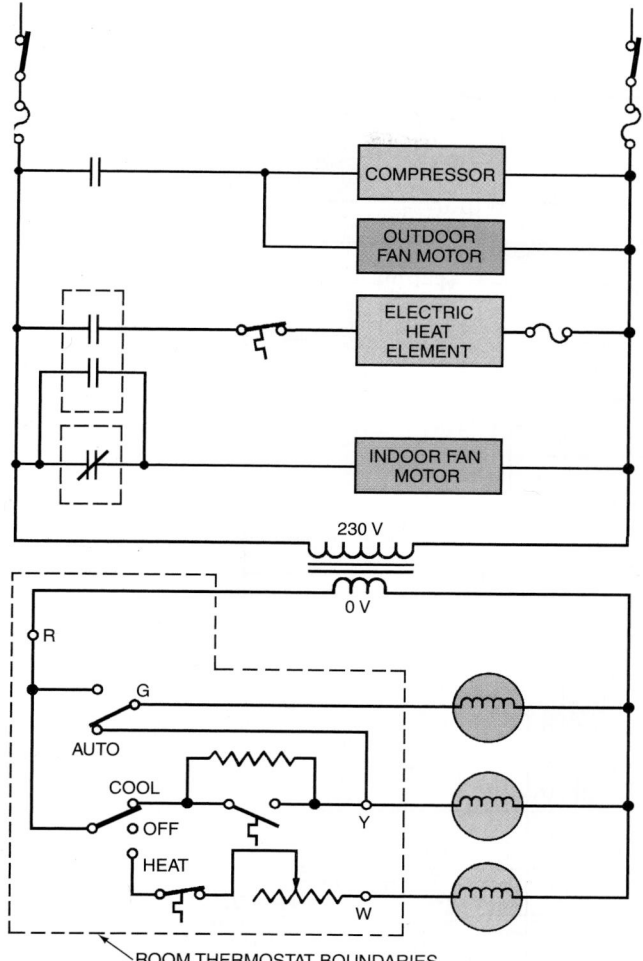

Figure 15–35 Use this diagram for the discussion of Service Call 4.

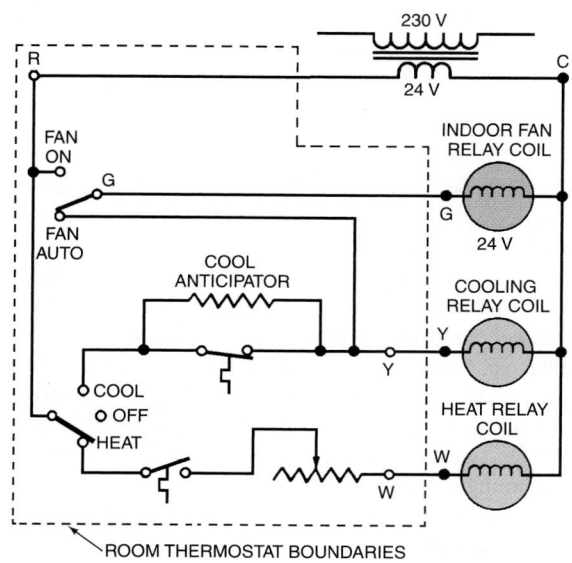

Figure 15–36 Use this diagram for the discussion of Service Call 5.

2. The technician notices that the set point on the room thermostat is 75°F, and the store temperature is 65°F.
3. The technician then goes to the roof and removes the low-voltage control panel. The voltage from C to R is 24 V and the voltage from C to W (the heat terminal) is 0 V. There is a 24-V power supply, but the heat is not operating.
4. The room thermostat is removed from the subbase, and a jumper is applied from R to W. The heating system starts.

What is the problem and the recommended solution?

SERVICE CALL 6

A customer calls on a very cold day and explains that the gas furnace just shut off and the house is getting cold. The system has a printed circuit board, **Figure 15–37.**

The technician arrives and checks the room thermostat. It is set at 73°F, and the house temperature is 65°F. The thermostat is calling for heat. The fan is running, but there is no heat.

The technician goes to the furnace and removes the front door. This stops the furnace fan because of the door switch (see 9G at the top of the diagram). The door switch is taped down so the technician can troubleshoot the circuit.

The technician places one meter lead on SEC-2 (the common side of the circuit) and the other on SEC-1 and finds that there is 24 V. The meter lead on SEC-1 is then moved to the R terminal, and 24 V is found there. The lead is then moved to G_H, where the signal from the room thermostat passes power to the circuit board, and there is power.

The technician then removes the lead from G_H and places it on the GAS-1 terminal where the technician finds 24 V.

What is the problem and the recommended solution?

SERVICE CALL 7

A service technician is called to a small market where the customer is complaining of products spoiling in a dairy case. The technician takes the box temperature of the dairy case with a thermometer and finds it to be 47°F. The case is supposed to be holding product at 35°F to 39°F. While measuring the dairy case's temperature, the service technician notices that there is no airflow on one side of the case. A quick inspection of the evaporator section and evaporator fans indicates that one of the three PSC evaporator fan motors is not operating.

The three fans are wired in parallel and operate off 230 V, **Figure 15–38.** The run winding of the inoperative PSC fan motor has burned open. The technician suspects that the inoperative fan motor probably has an open winding. The service technician then opens the power disconnect switch and places an ohmmeter

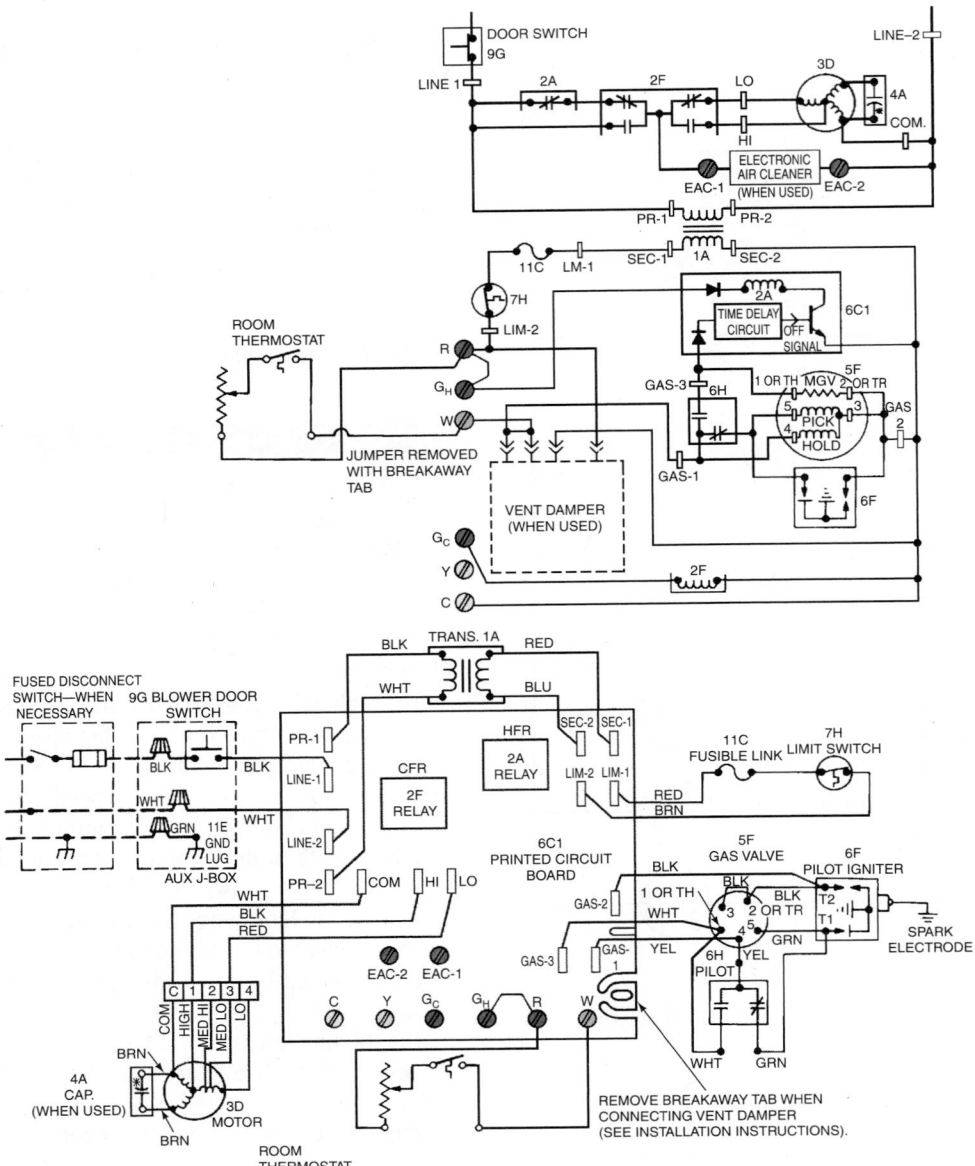

Figure 15–37 Use this diagram for the discussion of Service Call 6.

across the run winding of the inoperative fan motor. The ohmmeter reads 2 ohms.

Did the service technician take the right steps when ohmmeter troubleshooting the open winding in the inoperative evaporator fan motor? Explain your answer.

SUMMARY

- Each control must be evaluated as to its purpose in a circuit.
- Electrical controls are divided into two categories: power-consuming and power-passing (non–power consuming).
- One method of understanding a circuit is that the power supply is the potential difference between two power legs.
- Devices or controls that pass power are known as safety, operating, or control devices.

- The light bulb controlled by a thermostat with a fuse in the circuit is an example of an operating device and a safety device in the same circuit.
- The voltmeter may be used to follow the circuit from the beginning to the power-consuming device.
- The three separate power-consuming circuits in the low-voltage control of a typical heating and cooling fan unit are the heat circuit, cool circuit, and fan circuit. The selector switch for heating and cooling decides which function will operate.
- The fan relay, cooling relay, and heat relay are all power-consuming devices.
- The low-voltage relays start the high-voltage power-consuming devices.
- The voltmeter is used to trace the actual voltage at various points in the circuit.
- The ohmmeter is used to check for continuity in a circuit.
- The ammeter is used to detect current flow.

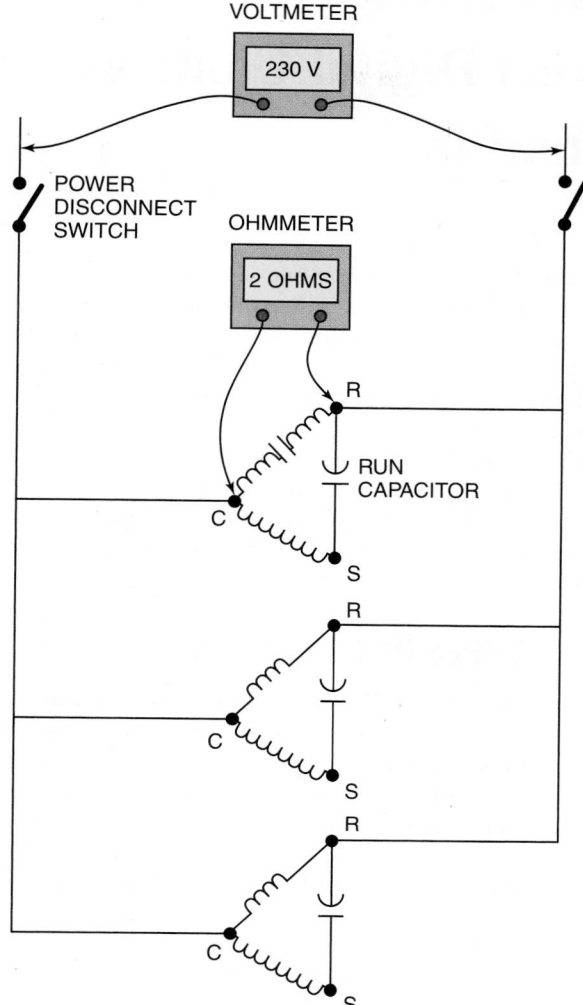

Figure 15–38 The fan circuit for a dairy case showing three PSC motors in parallel with one another.

- The pictorial diagram has wire colors and destinations printed on it. It shows the actual locations of all components.
- The line or ladder diagram is used to trace the circuit and understand its purpose.

REVIEW QUESTIONS

1. Name three types of automatic controls.
2. Two categories of electric controls are _____ and _____.

3. What circuit is energized with the thermostat Y circuit?
4. The component in the system that is energized when the G terminal is energized is the _____.
5. The _____ terminal must be energized in the thermostat for the heat to start.
6. The hot circuit in the thermostat is the _____ terminal.
7. Is the heat anticipator in **series** or **parallel** with the thermostat contacts?
8. The cooling anticipator is wired
   A. parallel with the cooling contact.
   B. next to the liquid line.
   C. in series with the cooling contact.
   D. inside the indoor unit.
9. Which anticipator is adjustable, **heat** or **cool**?
10. True or False: The indoor fan must always run during the cooling cycle.
11. True or False: When the control transformer is not working, the indoor fan will not run in the FAN ON position.
12. Which of the following should be turned off while working on the compressor circuit?
   A. The indoor unit fan
   B. The unit disconnect
   C. The thermostat
   D. The main power supply to the building
13. The _____ diagram is used to locate different components from the diagram to the unit.
14. To trace the circuits in the unit, the _____ diagram is used.
15. Describe how the amperage in a low-voltage circuit can be measured.
16. Explain why an ohmmeter, and not a voltmeter, should be used to assist the service technician in finding an opened run winding on a motor.
17. True or False: When using a voltmeter for troubleshooting, any time the technician measures across the same line (whether it be line 1 to line 1 or line 2 to line 2), the voltage will always be zero.
18. True or False: Open switches will always read voltage.
19. Explain why a service technician must always know where line 1 and line 2 are in relation to one another when voltage troubleshooting a schematic wiring diagram.

OBJECTIVES

After studying this unit, you should be able to

- recognize advanced control terminology.
- demonstrate control applications.
- describe electronic control circuits.
- describe pneumatic control circuits.
- discuss a control loop.
- understand direct digital control systems.
- discuss sensitivity or gain in controls.

16.1 CONTROL APPLICATIONS

When HVACR started, there was only manual control of all the components in a unit. For example, coal furnaces required a person called a "stoker" to put the coal in at the correct time to heat a structure. When the furnace was overfired, it was too hot. The windows had to be opened to release some of the heat. If the furnace was underfired, it took time to get the temperature up to a comfortable level. The first big improvement was an automatic stoker, which was a conveyor belt operated with a timer. Every so many minutes, it started up and ran for a prescribed length of time, dumping coal into the furnace. This method had no connection to the actual temperature of the heated space, but it eliminated a person from the process. As time passed, the process was improved with a thermostat in the conditioned space. Then it was further improved with a thermostat outside that anticipated heating needs. All of these methods of control would be considered primitive and unacceptable by today's standards. Today we expect the temperature to be well within the bounds of comfort, summer and winter. This can be accomplished with modern systems and a small number of technicians, even in large buildings.

If you walked through a large modern building with the best control systems and noticed warm and cool spots, you would be correct to think that the temperature level should be much more even. Many of these buildings have adequate systems and control arrangements but do not have a staff of people who understand how to operate the controls. The control system must be thoroughly understood to be set up to operate correctly. Even after the system is set up to operate and control temperature correctly, the controls may drift out of calibration from time to time and have to be readjusted. Buildings are designed to be divided into zones and into different styles and sizes of office space. The walls are easily removed and new ones may be installed. Many times this is done with no regard to the heating and cooling

system. It is not unusual to find the return air inlet in one office, the supply air outlet in another office, and the thermostat in yet another office. Sometimes the construction people lay the thermostat in the ceiling while they are building walls and forget to remount it. All of this leads to unbalanced systems.

The accomplished technician should be able to look at the building blueprints and interpret the intentions of the control designer. Many control systems are not designed to control as well as expected, but most will be satisfactory if set up correctly.

16.2 TYPES OF CONTROL SYSTEMS

The very first control systems only stopped and started the components to be controlled. These controls were discussed in Unit 14. Control technology has been expanded not only to switch components on and off but also to modulate. An example of modulation is the throttle on an automobile. Can you imagine a car that only had on and off as controls? You would be either accelerating at full speed or decelerating. The accelerator is properly used to slowly move up to the required speed and to slow down by applying less pressure to the gas pedal. Modern control systems can accomplish these same functions with automatic controls.

Control systems can be broken down into three basic components. These components make up a *control loop.* All control systems can be subdivided into control loops, no matter how many controls there are.

- The sensor measures the change in conditions.
- The controller controls an output signal to the controlled device. The sensor and controller are often in the same device.
- The controlled device reacts to stop, start, or modulate the flow of water, air, or refrigerant to the system or the conditioned space.

The sensor is the device that responds to a change in conditions. In this field, the sensors respond to temperature, velocity of air, humidity, water level, and pressure. **Figure 16–1** shows the inside of a pneumatic room thermostat. The sensor sends a signal to the controller, which in turn sends the correct signal to the device to be controlled. The controller may be built into the sensor.

The controls that control the temperature in a home will serve as a simple example of a basic control loop. The wall thermostat may be located in the hall close to the return air where it senses the space temperature. When the temperature rises above the set point, for example to 75°F, the bimetal in

THERMOSTAT 1

Figure 16–1 Typical pneumatic thermostat. *Photo by Bill Johnson*

the thermostat will move and cause the mercury contacts to close. This is the action of the sensor coupled to the controller. When the mercury contacts close, a 24-V signal goes to the compressor contactor (the controlled device), which passes power to the compressor and starts to cool the house. It will continue to run until the mercury contacts open the circuit to the compressor contactor.

The technician must understand that all systems can be divided into "control loops" and that they can be identified. A large building will have a control print that shows the basic control loops as individual loops, **Figure 16–2.** The legend that shows abbreviated words and symbols is at the beginning of the mechanical print section. You may drive by a 200-room motel and think you could never understand what goes on inside to operate the air-conditioning and heating systems. When you look further, you will find that it is actually one plan that has been implemented 200 times. Each room is practically identical as far as the utilities are concerned. They each have the same sink, toilet, lights, and air-conditioning system. When you learn one of them, you have learned all of them because it is a control loop for a portion of the system. The control loop in an office building may be much the same. When you learn one area, you have a basic understanding of the next one.

The basic types of controls being installed today to control large systems are pneumatic and digital electronic. These systems can also be divided into control loops for the purpose of design, installation, and troubleshooting. Pneumatic control systems were used before electronics. These controls are mechanical in nature and use air to transmit signals from the sensor controller to the controlled device. Electronic controls are quickly taking the place of pneumatic controls and

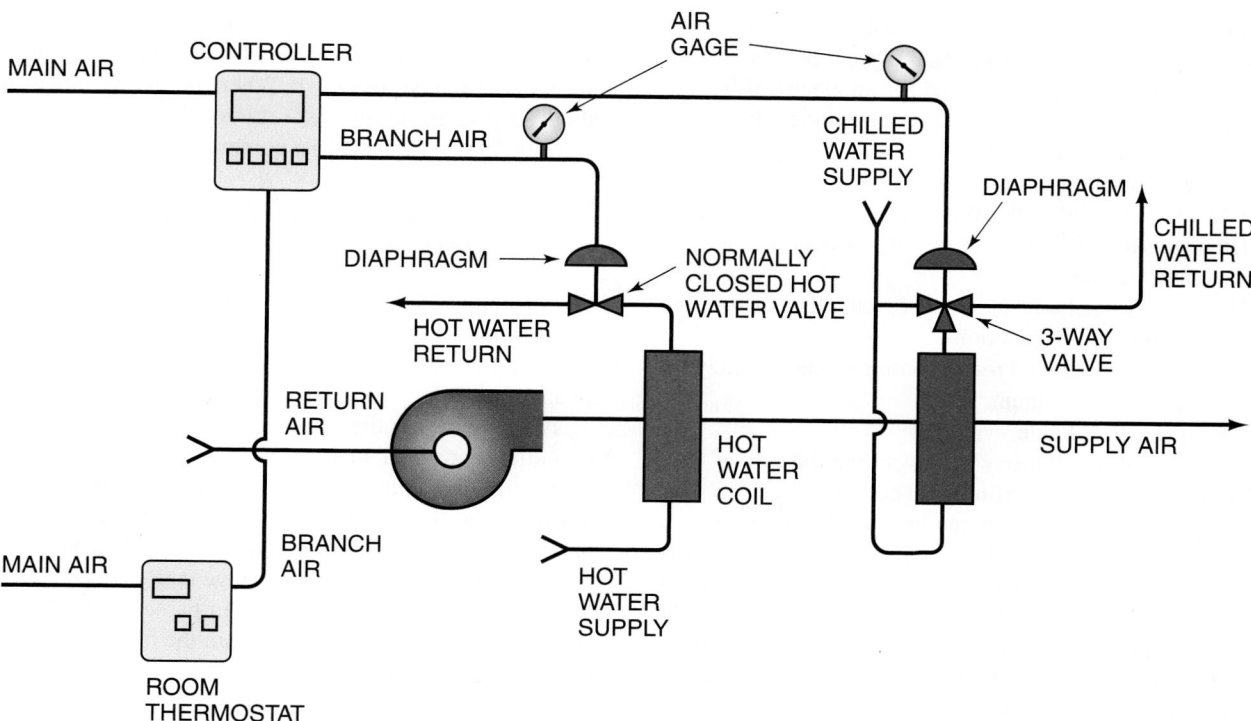

Figure 16–2 A building is divided into control loops for each area. One area may be larger than another, but the control loop may be the same.

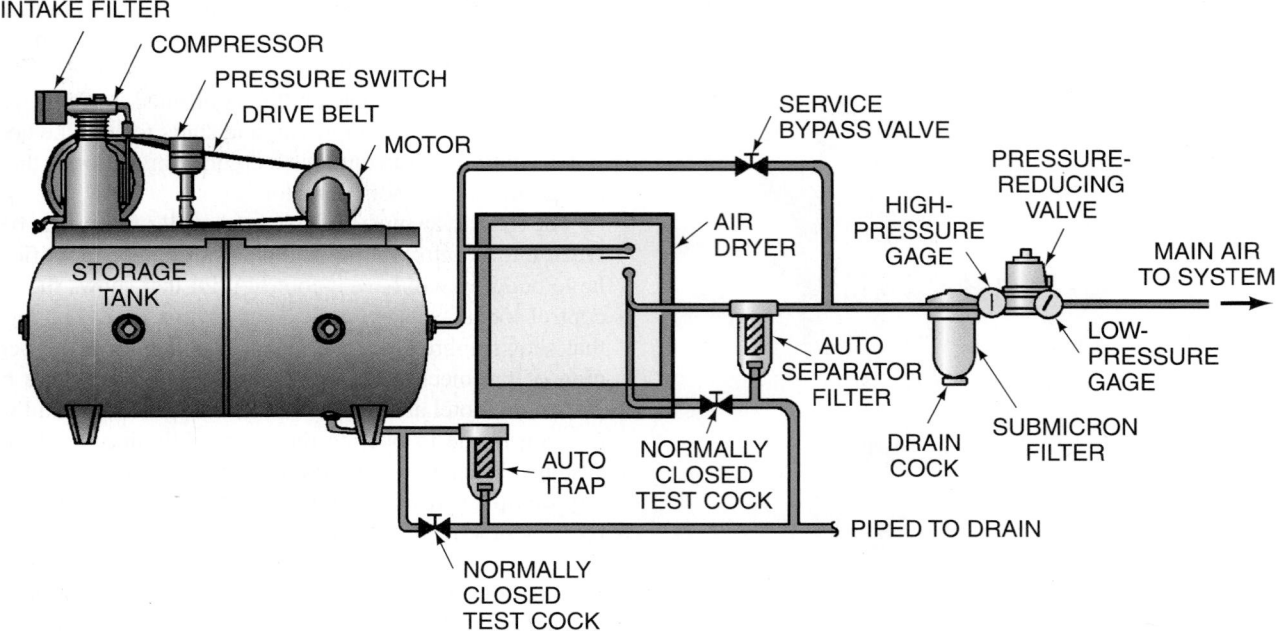

Figure 16–3 An air compressor system for pneumatic control system. *Courtesy Honeywell*

are the controls of the future. However, many installations that are already set up and operating with pneumatic controls will be used for many years. Electronic control systems can be integrated into the pneumatic systems for the best of both control systems.

16.3 PNEUMATIC CONTROLS

Many buildings and industrial installations were supplied with pneumatic controls in the past. These controls are simple and easy for a good mechanic to master. The control signal is air and is actually run in small pipe, typically 3/8 in. or 1/4 in. This eliminates the need for an electrical license and expertise in installing control wiring. Some advantages of pneumatic controls are as follows:

- They are simple to install and understand. A person with a mechanical background can soon adapt to them.
- They are safe and explosion-proof where combustible materials are used.
- They can be easily adapted to provide either on–off or modulating control functions.
- They are versatile, and many control sequences may be accomplished with them.
- They are very reliable.
- The installation cost may be lower than the cost of running long electrical lines and conduit.

The starting point for understanding pneumatic controls is to understand the air system that supplies the power to operate the controls. Pneumatic controls must have a permanent air supply that is **clean** and **dry.** Most systems use a dedicated air compressor and tank with a rather elaborate filtering system and drying system to supply the air. The actual control portion of the system will have many components with small air passages called orifices. They are so small that

a dirt particle or a drop of water cannot be tolerated. These systems use air continuously because some of the controls have a constant bleed rate with air escaping all the time. **Figure 16–3** shows a drawing of a typical dedicated single-compressor system for the air supply. Some systems have two compressors of the same capacity so that there is always a backup compressor. Either compressor can supply the total air requirements for the entire system by running only about half the time. The second compressor operates when there is a great demand.

Most system compressors provide 100 to 125 psig of air pressure to the storage tank, and the compressors are controlled with a pressure switch. When the air pressure drops to about 75 psig, the compressor starts; when the pressure rises to about 100 or 125 psig, the compressor stops.

16.4 CLEANING AND DRYING CONTROL AIR

The cleaning of the air starts with the prefilter to the compressor. It is much like the air filter on an automobile. It may be a dry-type filter that filters all entering air. This must be kept clean, or pressure will drop, causing excess running times.

The air entering the compressor usually has moisture, or humidity, entrained in it. Most parts of the country have some humidity, and many states have high humidity, particularly in the summer. When air with high humidity is compressed from atmospheric pressure to 100 psig, the outlet air from the compressor into the storage tank may be 100% saturated with moisture. The air often passes through a small coil that has air passing over it from the blades on the compressor flywheel. This coil will condense some of the moisture to a liquid, where it will quickly drop to the bottom of the storage tank. The storage tank normally will have a

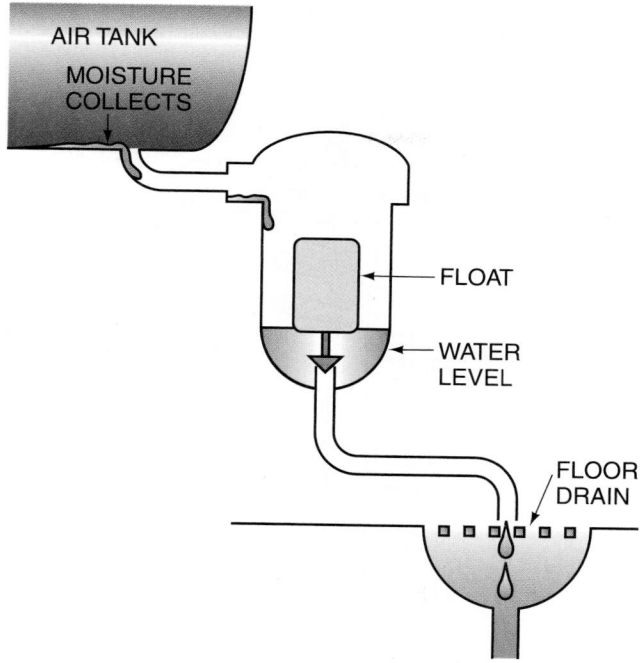

Figure 16–4   As water accumulates in the float chamber, the float rises and drains the water to the floor drain.

float-operated automatic bleed valve that will drain water from the tank, **Figure 16–4.**

The remaining air in the storage tank normally will still have too much moisture to be used for control air, so it must be further dried. Usually one of four types of drier is used:

■ Local water supply and coil
■ Chilled water coil
■ Desiccant driers
■ Refrigerated air drier

The local water supply may be used if it is cold enough to remove the correct amount of moisture from the air. This is not likely with most applications.

Chilled water from the refrigeration system may be used in some installations. However, if the chiller is not operating, the air may not be dried enough, and problems may occur.

Desiccant driers use some of the same products that refrigerant driers use—silica gel or molecular sieve as an adsorbent, which removes moisture much like table salt attracts moisture. The desiccant may be regenerated by means of heat, so there must be two driers—one in operation and one being regenerated at all times. Regeneration may be accomplished with heated dry air. Using some portion of the dried air from the system that is in operation may be part of the process.

Refrigerated air driers are probably the most common, **Figure 16–5.** These use a small refrigerated system that passes the compressed air through a refrigerated heat exchanger, which is operated below the dew point temperature of the compressed air. This condenses much of the moisture from the air, where it is automatically drained from the system with a float-type drain device. When the water level rises in the float chamber, a valve is opened, and the water goes down the drain.

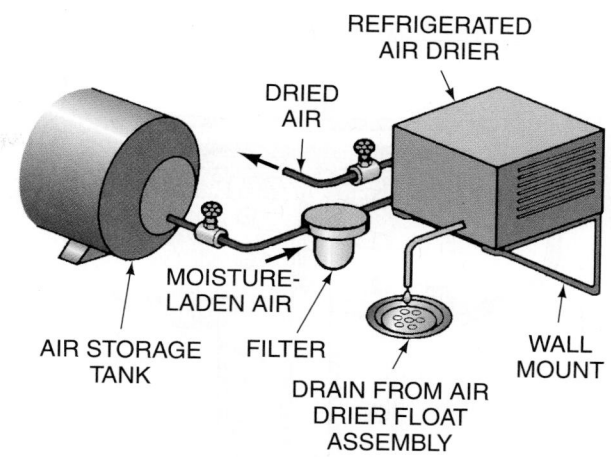

Figure 16–5   A small refrigerated air drier.

When the air is dried, it often passes through another filter that is so fine that oil vapor or liquid cannot pass through it. This is a high-density filter to prevent any leftover particles from entering the control system. The system has virtually a zero tolerance for contaminants.

The actual control systems require an air supply that is usually no more than 20 psig. A pressure-reducing regulator is located in the line after the air-drying device. This pressure reducer or regulator reduces the air pressure from about 100 psig to 20 to 30 psig, **Figure 16–3.** A relief valve is either in the pressure reducer or in the downstream side of the system to ensure that the control air pressure does not exceed about 30 psig. The reason for this is that the controls themselves are not rated for high pressure. High pressures will destroy the control components.

16.5   CONTROL COMPONENTS

The 20-psig control air is used to position various components in the system. We will use a valve actuator assembly as an example. Remember that we will be working with an air pressure of 20 psig; that is, 20 pounds per square inch. Main air pressure is used in this example. This is important because we will use the 20 psig to open a valve with a large diaphragm that has a diameter of 4 in. This diaphragm has an area of 12.56 in^2 (area = $\pi \times$ r^2 or 3.14 $\times$ 2 $\times$ 2 = 12.56 in^2). When we apply 20 psig of pressure to the top of the diaphragm, we have 251.2 lb of force (12.56 in^2 $\times$ 20 psi = 251.2 pounds of force), **Figure 16–6.** The diaphragm is held in the upward location with a strong spring, and the air above the diaphragm overcomes the spring. Therefore, when the air pressure is released, the diaphragm will be repositioned to the upward location.

The diaphragm and its plunger can be fastened to any device that we want to move with the air pressure, such as a steam or water valve or a set of air dampers, **Figure 16–7.** We can move a rather large load with 251.2 lb of energy. The air pressure, which is the force, can be piped to the various components using small copper or plastic tubing. The 20-psig main air lines are typically run using 3/8-in. lines, and the

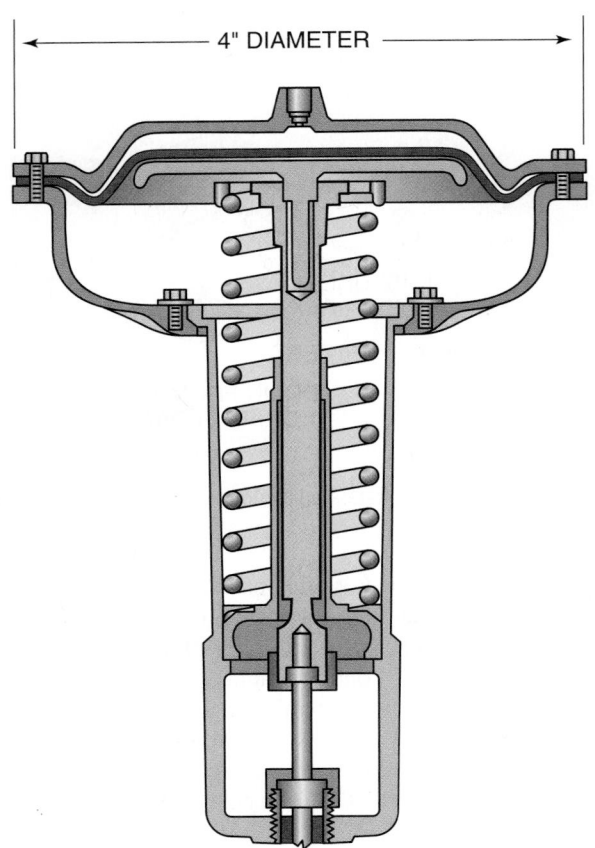

Figure 16–6 This diaphragm assembly can exert a pressure of up to 250 pounds because it has 20 psi of air pressure exerted on a 4-in. diaphragm.

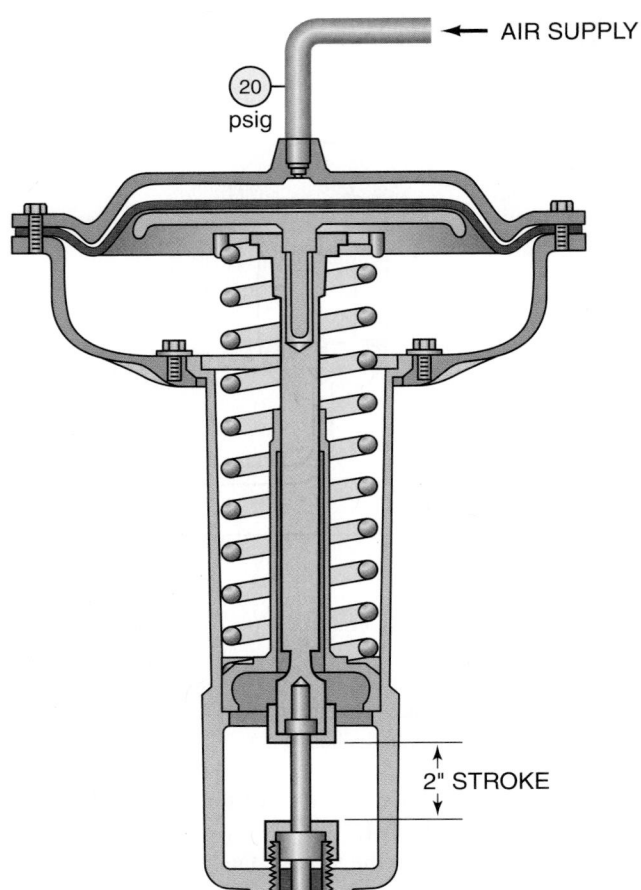

Figure 16–7 This diaphragm and plunger can be fastened to any number of devices that need to be positioned to control conditions.

branch control pressures are run with 1/4-in. lines. If you have ever watched a service station operator push an air valve and lift a 4000-lb car with the air lift, it is the same principle only with about 100 to 125 psig of air pressure.

The pneumatic control components—the sensors such as thermostats, humidistats, pressure controls, relays, and switches—perform the same functions that other controls do. The control components to be controlled are valves, dampers, pumps, compressors, and fans. These components work to control the conditioned space. There are as many combinations of controls in various systems as there are systems. We will outline some basic processes.

The system in **Figure 16–8** can control the room temperature in many increments, depending on the sensitivity of the thermostat. This is a basic control loop where the room conditions change and cause a change in the room thermostat, which changes the outlet air, which changes the room thermostat. This is called a closed loop because all changes happen within the loop. This control setup is found in many office buildings and motels. If you change the thermostat and then hear air movement, it is a pneumatic control system. The system uses a normally closed diaphragm valve that uses air pressure to overcome the spring to allow hot water to pass to the coil. Many applications use a normally open valve that is held closed by air pressure. When the room conditions call for

heat, it allows air pressure to bleed off the diaphragm and open the valve. This is called reverse acting. It has the advantage that if the systems fail for any reason, the building will overheat rather than have no heat. This can be considered a safety feature in cold climates that prevents building freezing conditions over a holiday.

The following is an example of how a room thermostat controls chilled water flow by modulating it to the room's needs. The output pressure positions the valve for the correct chilled water flow through the coil to maintain even temperatures in the conditioned space. Using electric controls, only a solenoid valve may be used to divert water around the chilled water coil or through the coil, **Figure 16–9**. Pneumatic controls can be used to position the valve for the correct water flow by modulating the air pressure to the valve that controls the water. The valve is the controlled device; the pneumatic device is the controller (controlled by the sensor). This high-air-volume valve may be located some distance from the room thermostat. A device called a *pilot positioner* is used to control the large volume of air. The pilot positioner receives the signal from the thermostat and regulates the air pressure to the air cylinder, which modulates the water valve anywhere from fully closed to fully open. The pilot positioner uses main air pressure, 20 psig, to move an air cylinder piston. The piston is held in the retracted position by a

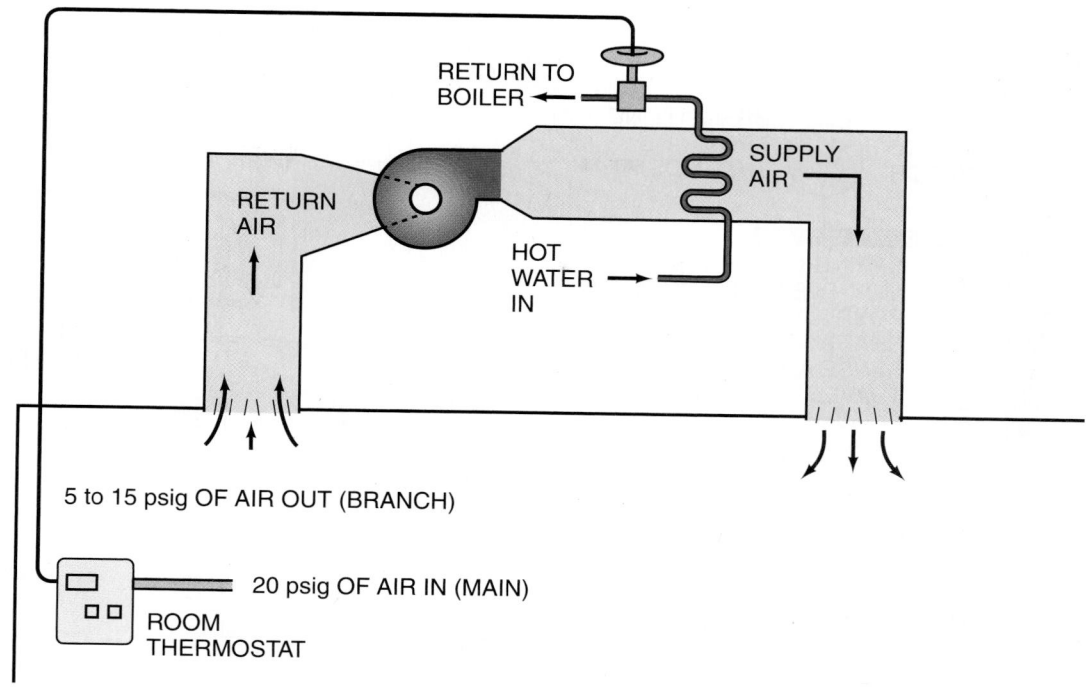

Figure 16–8   This control loop is used to control room temperature.

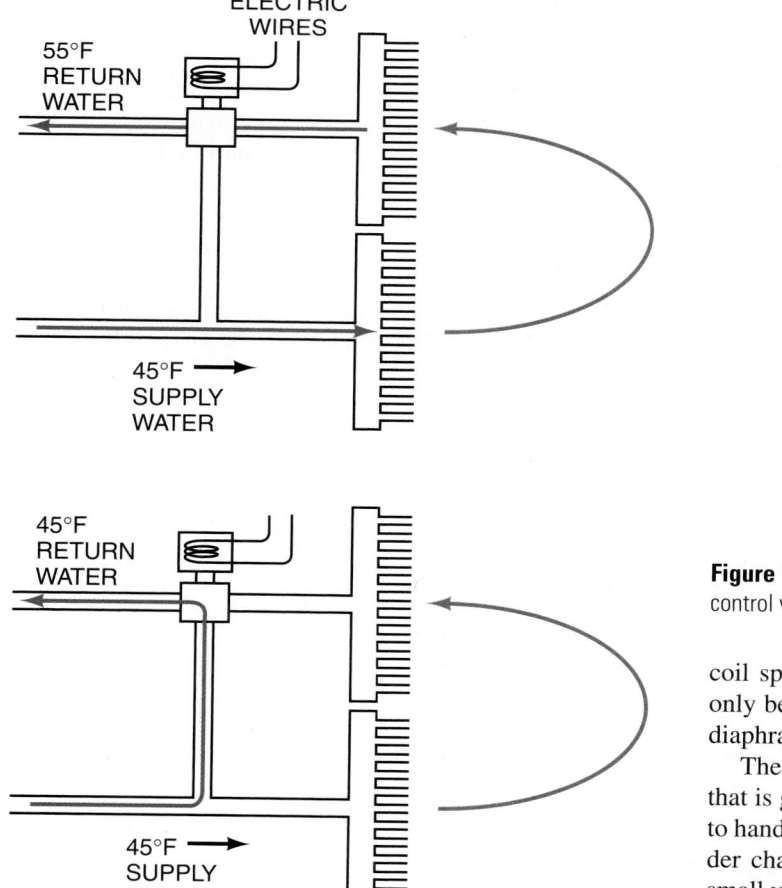

Figure 16–9   The electrical solenoid can only provide on–off control. This causes a swing in room temperature.

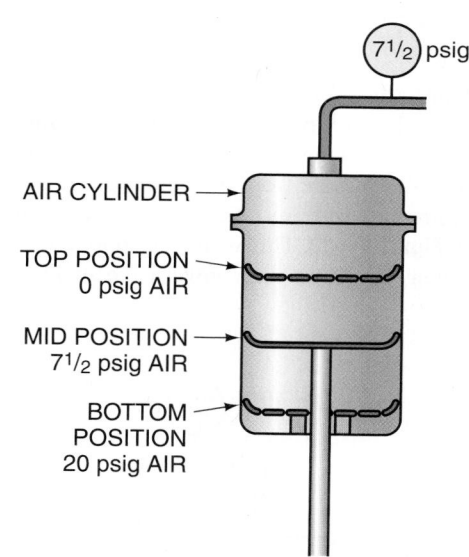

Figure 16–10   The air-operated cylinder is commonly used to control valves and dampers.

coil spring within the air cylinder body, and the shaft can only be moved with air pressure on the opposite side of the diaphragm, **Figure 16–10.**

The pilot positioner has a main air supply (the largest line) that is governed by the internal diaphragm, which enables it to handle the large volume of air required to fill the air cylinder chamber, **Figure 16–11.** The thermostat can now use a small volume of air that can be piped a greater distance from the pilot positioner. **Figure 16–12** shows the pilot positioner and the air cylinder it operates. For example, suppose that the room thermostat calls for maximum cooling. It will provide

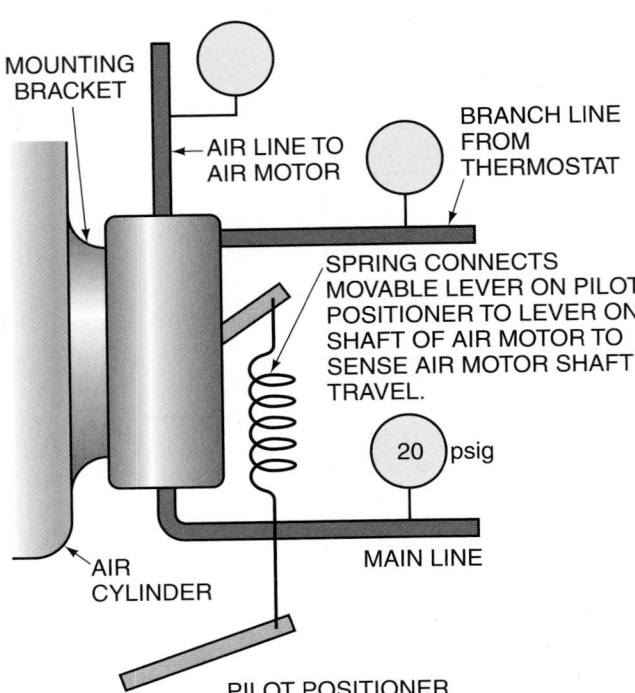

Figure 16–11 The pilot positioner in this system allows the thermostat to handle a very small amount of air from the pilot positioner, which governs a high volume of air to the cylinder.

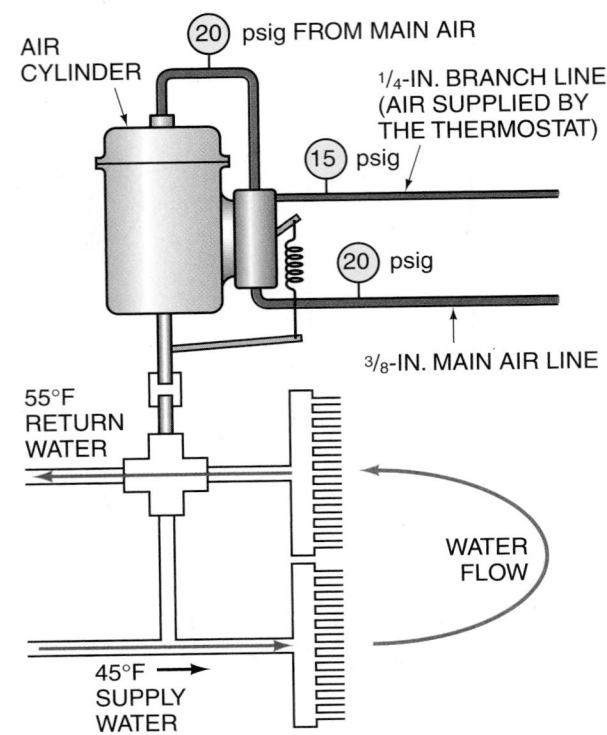

Figure 16–12 The room thermostat is supplying 15 psig of air to the pilot positioner, causing full stroke of the air motor. This will result in full water flow through the coil, providing maximum cooling. The room temperature is 78°F and the thermostat is set for 75°F.

15 psig of air to the pilot positioner, and full water flow will occur through the coil, providing maximum cooling. When the room temperature gets close to the set point, for example 75°F, the air pressure from the output of the thermostat, the branch line pressure into the pilot positioner, would be 7 1/2 psig, **Figure 16–13**. This would be supplied to the pilot positioner that would, in turn, govern the main air to the

valve. The spring would move the valve to halfway, and 50% flow through the coil would result, **Figure 16–14**.

When the room thermostat is satisfied at about 72°F, the valve will be positioned to completely bypass water around the coil and back into the return water to the chiller,

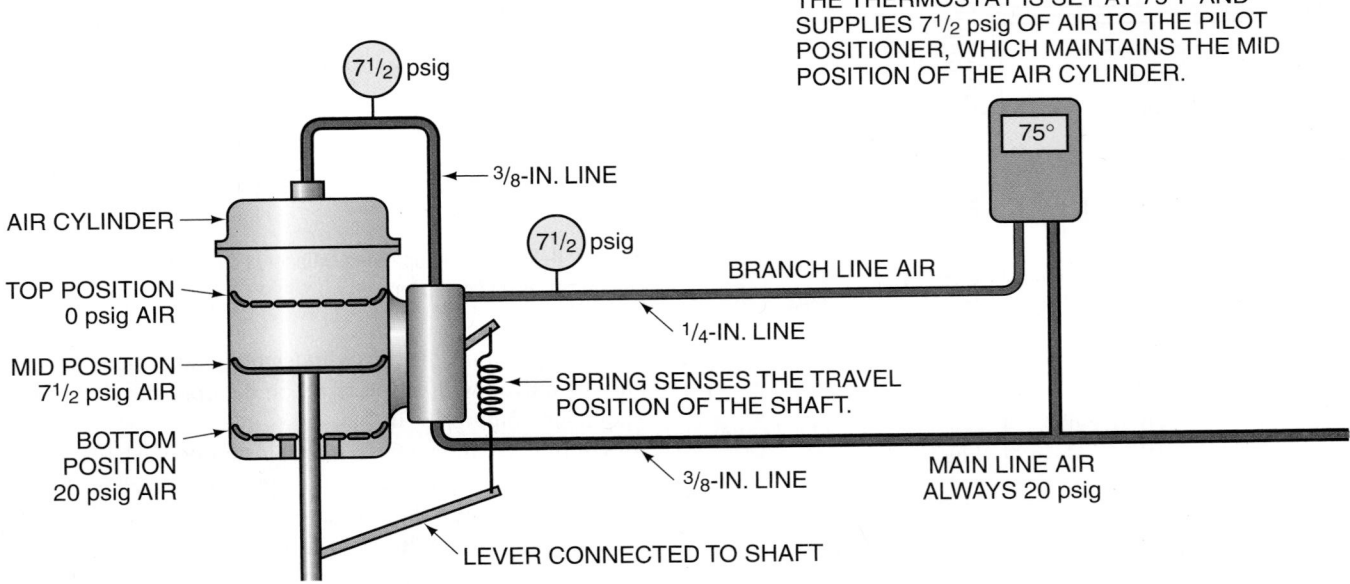

AIR CYLINDER CONTROLLED WITH ROOM THERMOSTAT AND PILOT POSITIONER

Figure 16–13 This shows the entire system.

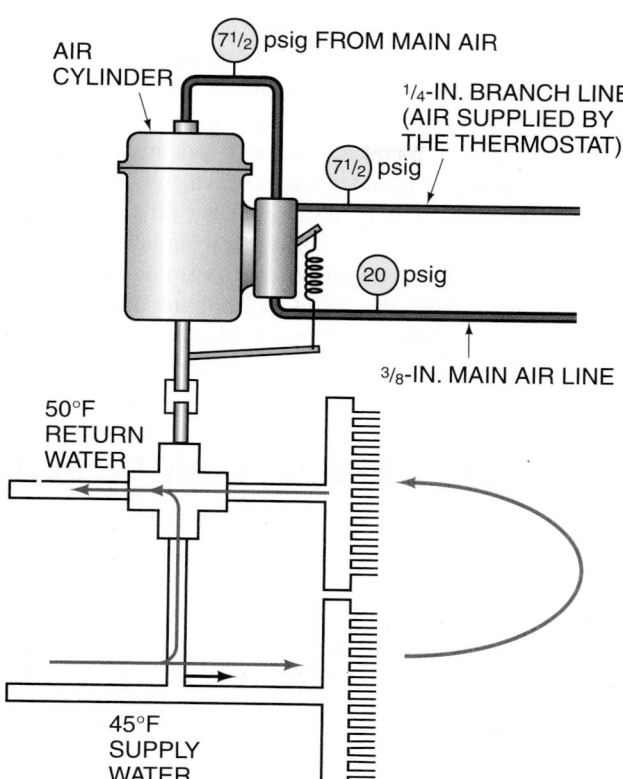

Figure 16–14   The room thermostat is set to control the room temperature at 75°F. The room is at 75°F so the thermostat is sending 7 1/2 psig of air pressure to the pilot positioner. This positions the valve to allow 50% water flow through the coil.

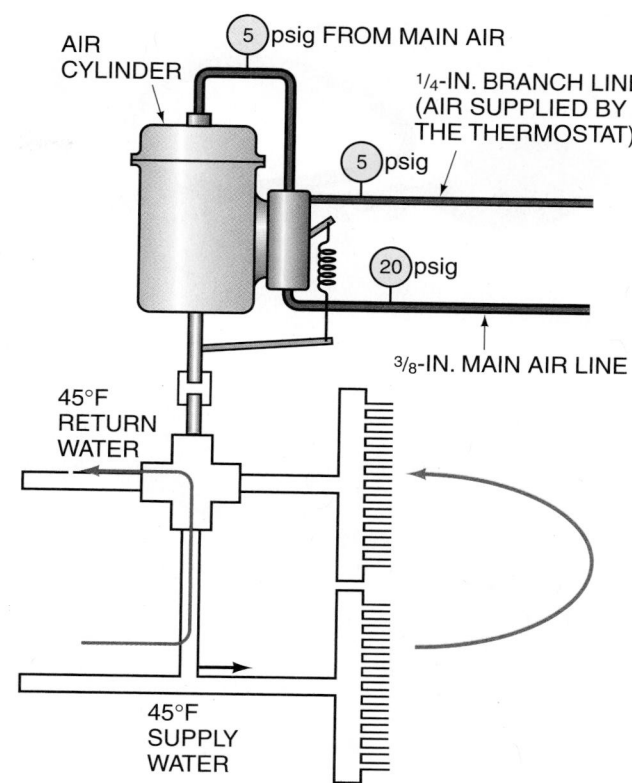

Figure 16–15   The room temperature is now 72°F. The room thermostat is satisfied and has an output of 5 psig. The valve is now bypassing all water around the coil. If the room temperature drops much more, heating will be required. A hot water coil may then be allowed to provide heat in a similar manner.

Figure 16–15. This allows the thermostat to control the water flow to the room's needs, rather than on–off flow. This is called *modulating flow.* Much like the accelerator on an automobile, it modulates the flow of fuel.

Notice in the preceding explanation that at 75°F we had 50% flow and a branch line pressure of 7 1/2 psig, and at 72°F we had no flow. The pressure should be 5 psig. When the temperature reaches 78°F we should have 15 psig branch line pressure and full flow. We have a temperature difference of 78°F − 72°F = 6°F and an air pressure change of 10 psig. The *sensitivity,* or *gain* as it is sometimes called, is 1.67 psig per °F (10 psig/6°F = 1.67 psig per °F temperature change). When there is 1°F temperature change, the thermostat will change the branch line pressure 1.67 psig. This variable is called for on the description for the application. If the sensitivity is too great, the control may have a tendency to hunt, back and forth, as if it cannot find equilibrium or a steady state. If the sensitivity is too little, there will be a noticeable difference in the room temperature that will not be much better than on–off control. The job application dictates what the sensitivity should be; the previous example is typical for cooling applications. The three-way valve is used in the cooling application to maintain a constant water flow through the chiller. If a building has 20 chilled water coils similar to the one in the example and they all bypass water, there is still

water flow through the chiller, and the chiller will shut off due to low load.

The tools of the pneumatic technician are a pressure gage that can be plugged into the various control points, an Allen wrench to make adjustments, and a good pocket thermometer, **Figure 16–16.** The pressure gage has a quick-connect fitting that will fit the particular brand of pneumatic control, so you will have to contact the manufacturer for this device. Many technicians carry one for each manufacturer.

The technician may be called on to check a room thermostat for calibration. The technician uses the pocket thermometer to detect the temperature in the vicinity (ambient air temperature) of the thermostat by leaving it in the air long enough to adjust to the air temperature. The air pressure at the thermostat is then recorded. Suppose the air temperature is 75°F, the thermostat is set for 75°F, and the air pressure is 3 psig. The thermostat indicates that cooling should be occurring, but the air pressure does not call for cooling. The thermostat can be adjusted to call for 7 1/2 psig at this condition and cooling will start. The thermostat has drifted out of calibration.

Many different types of pneumatic controls and arrangements are possible, but they all operate using the principles described. Pneumatic controls can be recognized by the air pressure line going to them. When they change conditions

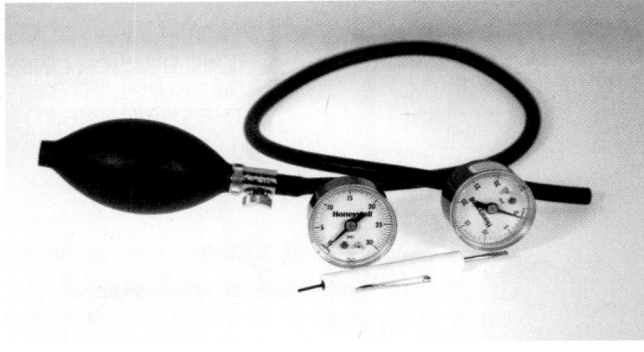

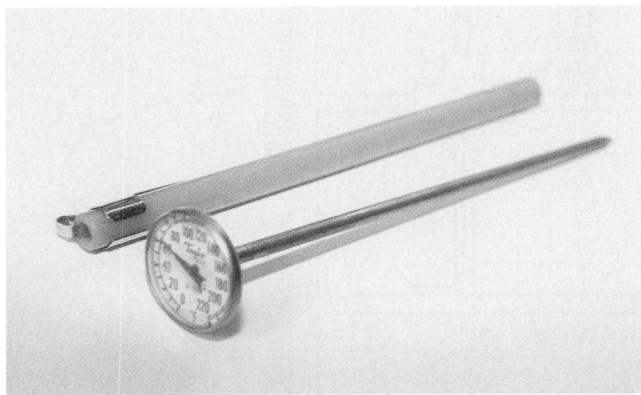

Figure 16–16 A thermometer, a gage, and an Allen wrench are the tools used by the pneumatic control technician to make control adjustments. *Photo by Bill Johnson*

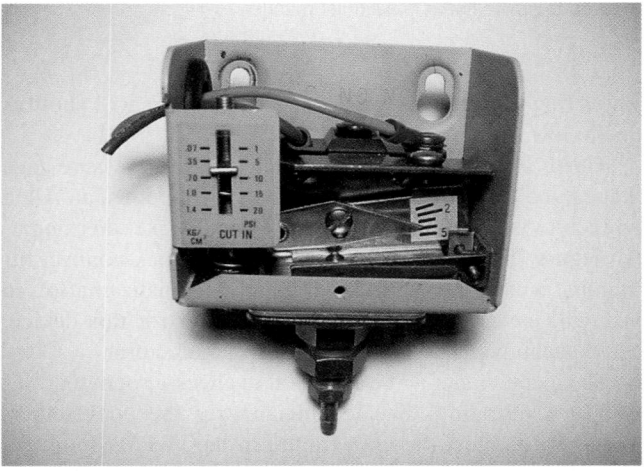

Figure 16–17 This switch changes an air pressure difference to an electrical signal. It can be thought of as a pressure switch, only operating at low pressures. *Courtesy Ferris State University. Photo by John Tomczyk*

or positions, you will often hear the air bleeding from the control.

Pneumatic controls may be used to switch electrical circuits, or electrical circuits may be used to switch pneumatic control circuits on or off. The pneumatic device that is used to switch electrical circuits is called a pneumatic-electric (PE) device. **Figure 16–17** shows a PE switch that is much like a

pressure control except that it operates at much lower pressures. The electric device used to switch pneumatic devices on and off is called an electric-pneumatic (EP) device.

Technicians who have pneumatic controls in their maintenance program should attend any pneumatic control schools available, try to watch a seasoned technician, and ask a lot of questions. The control sequence provided by the building blueprint is invaluable for showing how the building controls are supposed to function. Control manufacturers are very cooperative with regard to supplying duplicate diagrams and material on their controls. Many times, the building blueprint is lost and the original control company or architect will have a control blueprint on file.

16.6 DIRECT DIGITAL CONTROLS (DDCs)

Electronic controls have taken over the majority of today's control systems. These controls, commonly called *direct digital controls (DDCs),* have the advantage of being able to operate on very low voltages. The control wiring may be installed without an electrical license in some states. DDCs can also be modulated like pneumatic controls. They are much more compact than pneumatic controls; however, they do the same sort of duty. They may need to open, close, or modulate a water valve, steam valve, or an air damper in an airstream. They can also modulate motor speeds as can be seen in Unit 17.

DDC is a control process in which a microprocessor, acting as a digital controller, constantly updates a *database* of internal information. Through *sensors,* the microprocessor updates this information by watching and monitoring information from an external *controlled environment* or conditioned space. The microprocessor then continually produces corrections and outputs this corrected information back to the controlled environment. The controlled environment could be controlled by an array of *controlled output devices* such as dampers, fans, variable-frequency drives (VFDs), cooling coil valves, heating coil valves, relays, and motors—to name a few. The sensors can sense many things like pressure, temperature, light, humidity, velocity, or flow rate. They can also do a status check to see if something is energized or running or not. A conventional control system (pneumatic, electric, or electromechanical) works very similarly to a DDC system except that digital (binary) information can be processed much faster and is much more accurate at all times and under all conditions. Binary information consists of nothing but zeros and ones (0 or 1), and often represents one of two elements in opposition to each other, such as on/off, high/low, or yes/no.

The DDC system actually changes or conditions a signal in three steps, **Figure 16–18.** The following is an explanation of the three steps:

- The input information from sensors is converted from analog (modulating) to a digital (on/off) binary format. *Signal converters,* usually housed internally inside the controller, convert the information from either analog to digital (A/D) or digital to analog (D/A) format, **Figure 16–18.** This allows the microprocessor or digital controller to be able to read the incoming data.

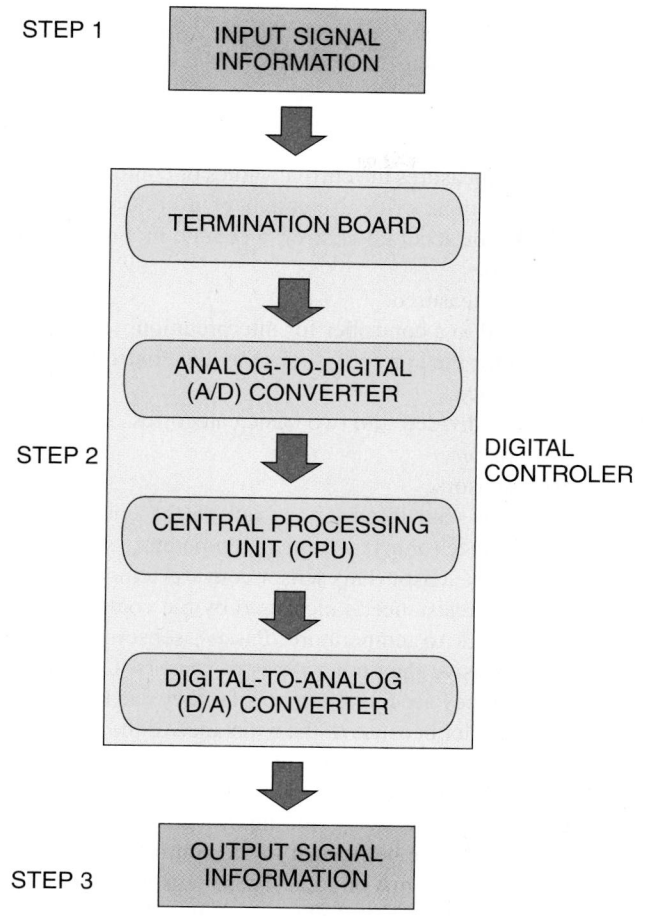

STEP 1

STEP 2

DIGITAL CONTROLER

STEP 3

Figure 16–18   Three-step procedure for conditioning a signal on a DDC control system.

■ Control commands (algorithms) stored in the digital controller's central processing unit (CPU) are performed by the microprocessor with this newly converted digital data.

■ The correct output response of the microprocessor or digital controller is reconverted from digital data to analog data for analog output devices. Digital data is the output of the digital controller when the controlled output device is digital. This allows the appropriate controlled output devices to recognize the controller's output signals and respond accordingly.

Figure 16–19 shows this process in graphical form. The digital controller (microprocessor) will address each *control point* in the system through an input/output (I/O) card. The signal then passes through an analog/digital converter which gives the signal a binary value. This allows the digital controller to understand the input signal and will allow the internal instructions (*algorithms*) stored in its memory to act upon this information. The controller's output response is converted from digital to analog format through a digital-to-analog converter. Notice that the signal is now back into its original analog format. The signal can now be sent to the controlled devices. Today, however, control output devices can receive signals as either analog or digital. So, digital controllers can be programmed to send the output signal that is appropriate to the device under control.

The microprocessor or digital controller must remember its tasks, what information is available for its use, what this information does, and the results of what it has done. The digital controller must also remember a number of rules and mathematical functions for controlling the system. *Memory* is the part of the digital controller where this information is

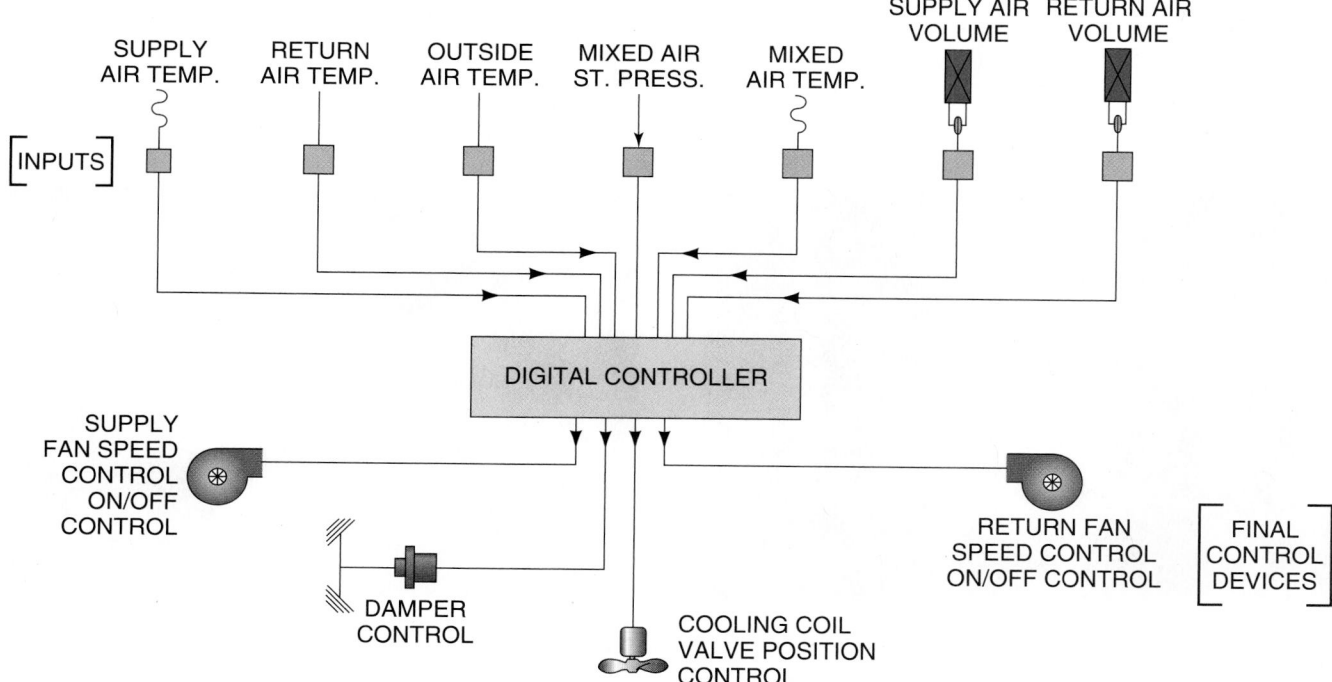

Figure 16–19   Digital controller showing inputs and final control devices.

stored. An algorithm (program of instructions) resides in the memory of the digital controller and serves as the instructions or step-by-step operations sequence for how the digital controller should act (take action) on the data received from the inputs. The controlled output devices can now implement this action and make changes to the controlled environment. Some of the major advantages a DDC control system has over a traditional control system are the following:

- Speed and accuracy
- Ability to control many control sequences simultaneously
- Ability to manage many control loops simultaneously
- Repeatable control of system set points
- Operating functions of the controller can be adjusted through software
- Ability to automatically adapt to changing conditions in the controlled environment

DDC controls are being used worldwide to reduce energy costs. Current trends in DDC technology are toward smaller, modular, stand-alone DDC panels, **Figure 16–20**. Today, with microprocessor-based controls located right on the equipment being controlled, component failure is not risked by the entire building's HVACR system. These systems are said to be intelligent, self-compensating, self-maintaining, and self-diagnosing.

Basically, a control system, no matter how complex, consists of the following:

- Sensors
- Controller
- Controlled devices

Sensors are devices that can measure some type of environmental parameter or a controlled variable and convert this parameter to a value that the controller can understand. Sensors can measure parameters such as the following:

- Temperature
- Pressure
- Humidity
- Light
- Position
- Moisture
- Velocity
- Enthalpy
- Conductivity

Since a sensor measures the current values of controlled variables, it is the most critical element of a control system, **Figure 16–21**. For accurate control, a change in a controlled variable must be

- sensed or measured.
- transmitted to a controller for interpretation.
- acted on for the correct response by the controlled output device.

Sensors can be divided into two basic categories. They are
- active sensors.
- passive sensors.

Passive sensors send information back to the controller in terms of resistance (ohms). A typical temperature sensor is a 1000-ohm nickel sensor. This sensor converts temperature to resistance. The resistance is measured by the controller and is translated back to temperature. Passive sensors are typically less expensive than other types of sensors. One disadvantage is that they are limited in how far they can be located from the controller because of the resistance of the wire connecting the sensor to the controller. *Active sensors* send information back to the controller in terms of milliamps (mA) or volts. The sensor converts the input signal at the sensor to mA or volts before being sent to the controller. The controller then measures mA or volts. These sensors are not sensitive to the length of wire to the controller. Typical output signals sent back to the controller from the active sensors are 0–10 volts or 4–20 mA.

Sensors have both binary (digital) and analog inputs to the controller. Digital inputs report back to the controller as one of two opposing conditions. Examples of digital inputs are the following:

- Yes/no
- Open/closed

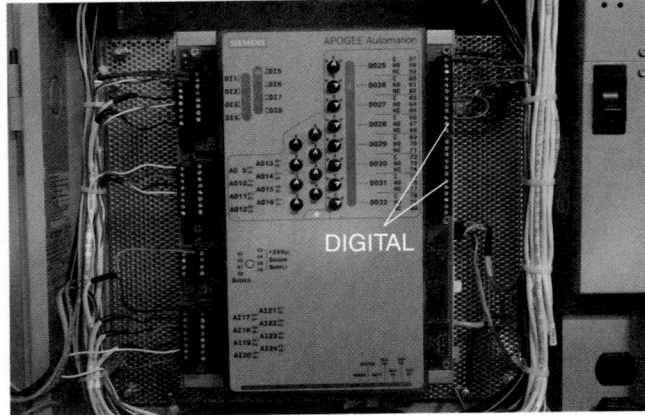

Figure 16–20   Small, modular, stand-alone DDC controller. DI = digital inputs, DO = digital outputs; AI = analog inputs, AO = analog outputs. *Courtesy Ferris State University. Photo by John Tomczyk*

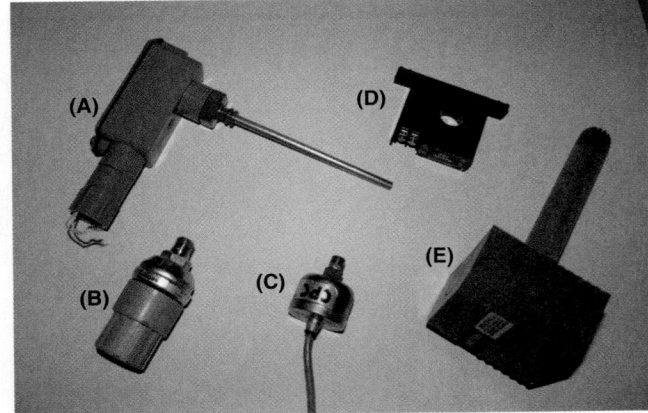

Figure 16–21   Sensors for a DDC controller. **(A)** Thermistor (anolog). **(B)** and **(C)** Pressure transducers (analog). **(D)** Current sensor (digital). **(E)** Humidity sensor (analog). *Courtesy Ferris State University. Photo by John Tomczyk*

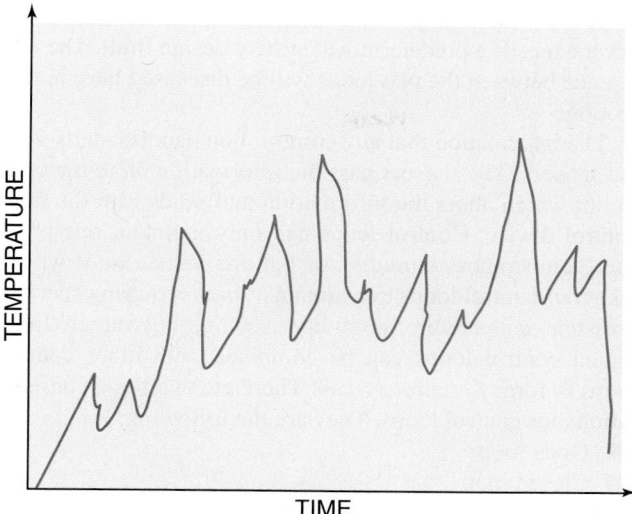

Figure 16–22   Analog, or modulating, input from a temperature sensor.

- On/off
- Start/stop

In the HVACR field, digital or binary inputs can be a set of contacts that are either open or closed or a motor that is either on or off. They can be used as indicators of status, proof, safety, and limits.

Analog inputs have a range of values over any given time period, **Figure 16–22.** Examples of analog inputs are the following:

- Temperature
- Pressure
- Humidity
- Flow
- Light

In the HVACR field, analog inputs can measure room temperature, discharge air temperature, return air temperature, mixed air temperature, outdoor air temperature, chilled water temperature, hot water temperature, static pressure, chilled water flow, outdoor air humidity, return air humidity, or light levels—to name a few. Probably the most recognized analog inputs to a controller are thermistors and resistance temperature devices (RTDs). Thermistor sensors measure temperature using silicone instead of a metal. However, the thermistor does not have a linear relationship between resistance and temperature, **Figure 16–23(A).** So, the digital controller has to be able to handle this type of sensor. RTDs, on the other hand, have a more linear and predictable output resistance whenever their temperature changes. The controller will have to analyze a much more linear temperature/resistance-to-voltage relationship, **Figure 16–23(B).**

Controlled output devices can be either analog or digital. When analog, the final control device has a range of operation such as "full open to full close" or "full speed to stop." Analog output devices are typically operated with either a 0–10 volt or a 4–20 mA signal range. The controller has to be able to handle one of these two outputs or both. Typical analog outputs in the HVACR field are valves,

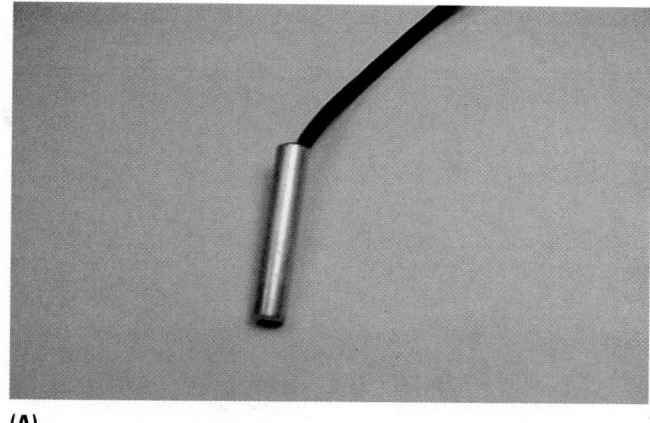

(A)

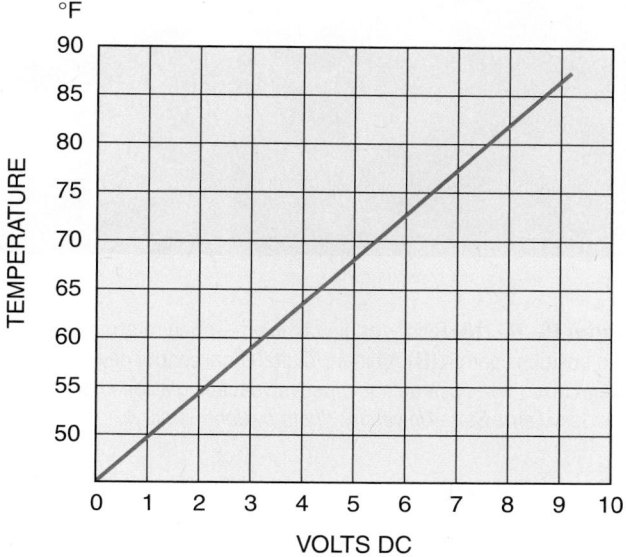

(B)

Figure 16–23   **(A)** A thermistor used as a temperature sensor for the controller's input. **(B)** This temperature-voltage graph shows the relationship of temperature to voltage, simplified. **(A)** *Courtesy Ferris State University. Photo by John Tomczyk*

dampers, and VFDs (variable-frequency drives) that control motor speed.

Digital or binary output devices have two states of operation—such as on/off, open/closed, start/stop, and yes/no, **Figure 16–24(A and B).** Binary outputs are simply an electronic switch (triac) that is operated by the controller. Examples of binary outputs in the HVACR fields are fans (start/stop), pumps (start/stop), direct-expansion (DX) cooling (on/off), and gas heat (on/off)—to name a few.

Any time there is a group of control devices interconnected in such a way that they control a single process at a certain desired condition, a *control loop* is said to exist. An example of this is a group of sensors, controllers, valve actuators, dampers, VFDs, and signal converters. Processes controlled by a control loop in the HVACR fields are usually comfort related. Certain control loops can maintain temperature, pressure, lighting, flow rates, or humidity at a certain

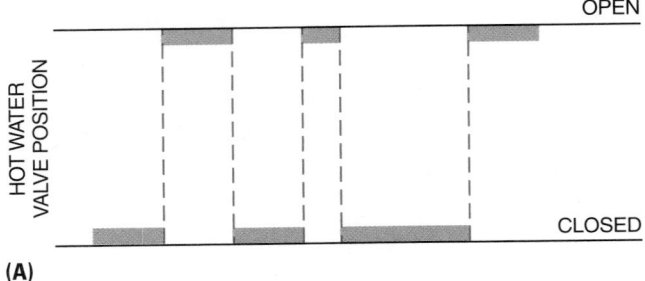

(A)

(B)

Figure 16–24 **(A)** Binary, or digital, input—often referred to as a two-position control. **(B)** A binary (digital) final control device. An open/closed two-position hot water valve feeding a coil in ductwork. *Courtesy Ferris State University. Photo by John Tomczyk*

set point. Others simply make sure a process or control point never exceeds a predetermined system design limit. The difference between the two loops will be discussed later in this section.

The information that any control loop handles starts with the sensors. The sensors pass the information on to the controller which alters the information and sends it to the final control device. Control loops can only maintain one pressure, temperature, humidity, or lighting value. So, it would take four control loops to maintain a space requiring specific pressure, temperature, humidity, and lighting values. Individual control loops can be combined with other control loops to form a *control system.* There are two basic configurations for control loops. They are the following:

- Open loop
- Closed loop

Open- and closed-loop configurations do not describe how a loop's control devices are connected, but how the loop is connected to the HVACR process. It is actually the sensor's location that determines whether the loop is open or closed.

In an *open-loop control configuration,* the sensor is located upstream of the location where the *control agent* causes a change in the *controlled medium,* **Figure 16–25.** The control agent is the fluid that transfers energy or mass to the controlled medium. It is the control agent (such as hot or cold water) that will flow through a final control device (water valves). The controlled medium is what absorbs or releases the energy or mass transferred to or from the HVACR process. Controlled mediums can be air in a duct or water in a pipe. In **Figure 16–25,** the control agent is the hot water in the coil, the controlled medium is the air in the duct, the control point is the outdoor air temperature, and the final control device is the hot water valve. **Figure 16–26(A, B, and C)** shows

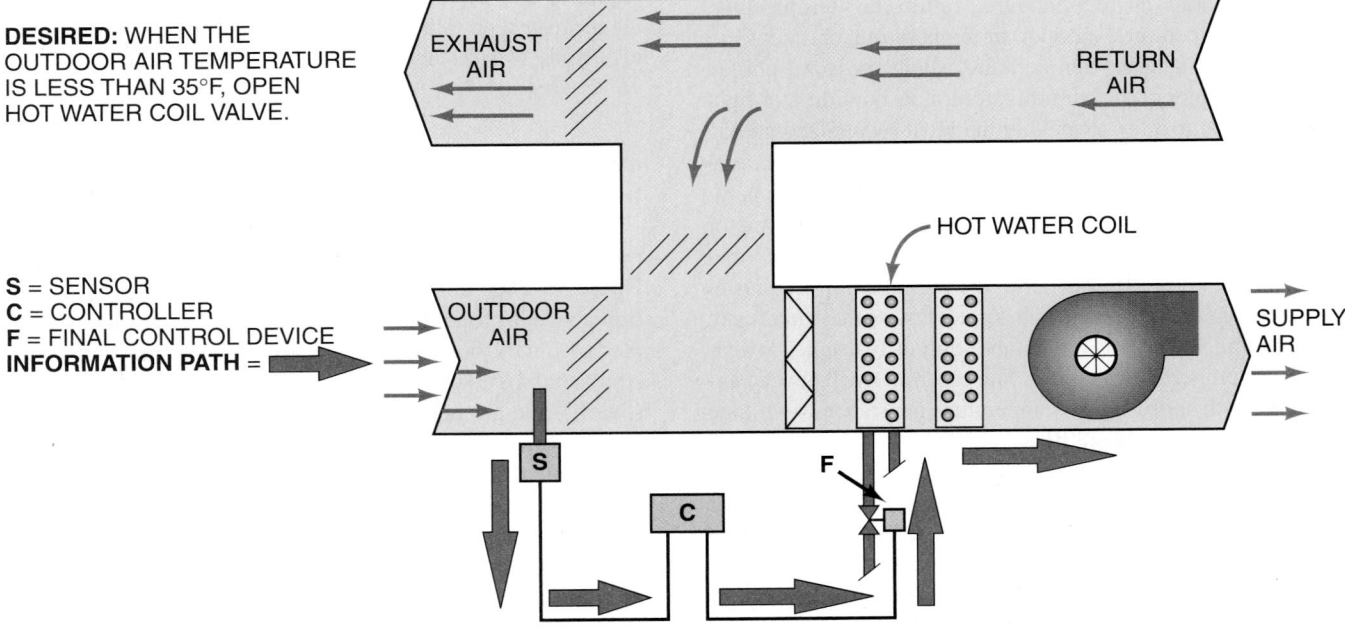

Figure 16–25 An open-loop control configuration.

(A)

(B)

(C)

Figure 16–26   (A) Mixing valve acting as an (analog) final control device. **(B)** Return air damper acting as an (analog) final control device. **(C)** Two variable-frequency drives (VFDs) acting as (analog) final control devices for supply-and-return air fans. *Courtesy Ferris State University. Photos by John Tomczyk*

three analog (modulating) final control devices. The flow of information in an open-loop configuration travels in only one direction. Because the sensor is located upstream of the control agent, it never realizes the results of the final control device's (hot water valve's) action. Open-loop control systems

cannot maintain a balance between the energy transferred to or from the system and the process's heating or cooling loads. The sensor does not monitor a control point like supply air temperature, but can monitor a variable like the outdoor air temperature in **Figure 16–25.** The sensor is simply making sure that if the outdoor air temperature is ever less than 40 degrees, the hot water valve will open. This opening or closing of the hot water valve would be considered a binary or two-position response, and is simply preventing the controlled variable (outdoor air temperature) from exceeding a predetermined design limit.

A *closed-loop control configuration* can maintain the controlled variable, like the air in the duct in **Figure 16–27,** at its set point because the sensor in now located in the air being heated or cooled (controlled medium). The sensor is located downstream of the water coils or point where energy or mass is being introduced. The sensor can now sense of what the final control device (water valve) did and make adjustments. This is called a *feedback loop.* The feedback loop actually "closes" the information loop, causing the flow of information to be in a circle. This is where the term "closed loop" originates. This type of control loop can now maintain a controlled medium at a set point—whether the control is a modulating (analog) or digital (binary) type. With the modulating closed-loop control in **Figure 16–27,** there can now be a balance between the HVACR process load and the energy transferred from the controlled agent. Refer to Section 24.24, "Step-Motor Expansion Valves," and Section 24.25, "Algorithms and PID Controllers," in Unit 24, "Expansion Devices," for an example of a closed-loop control configuration in which you can apply the principles you have just learned about DDC. These sections are applications of DDC applied to an expansion valve for a commercial refrigeration system used in today's supermarket applications.

DDC control systems in large buildings or supermarkets can be responsible for the following:

- Total building energy management
- HVACR functions
- Building lighting
- Fire protection
- Temperature setback
- Refrigeration

DDCs use a typical personal computer or laptop computer to assimilate and distribute information to small computers used in the system such as control boards. An advantage is that the technician can control the system from wherever there is a telephone line. Security codes prevent others from accessing the system. A technician who gets a call about a building temperature being out of control can go to a phone line and view the entire building conditions, what is running and what is not running, and what the conditions are in all parts of the system. The corrective action can often be entered from a remote location. This gives building maintenance personnel some extra freedom and eliminates some overtime calls for the owners. Small computers that are part of the DDC system for each section of the building are tied to the main personal computer. There may be multiple sensors and controllers for each portion of the building feeding

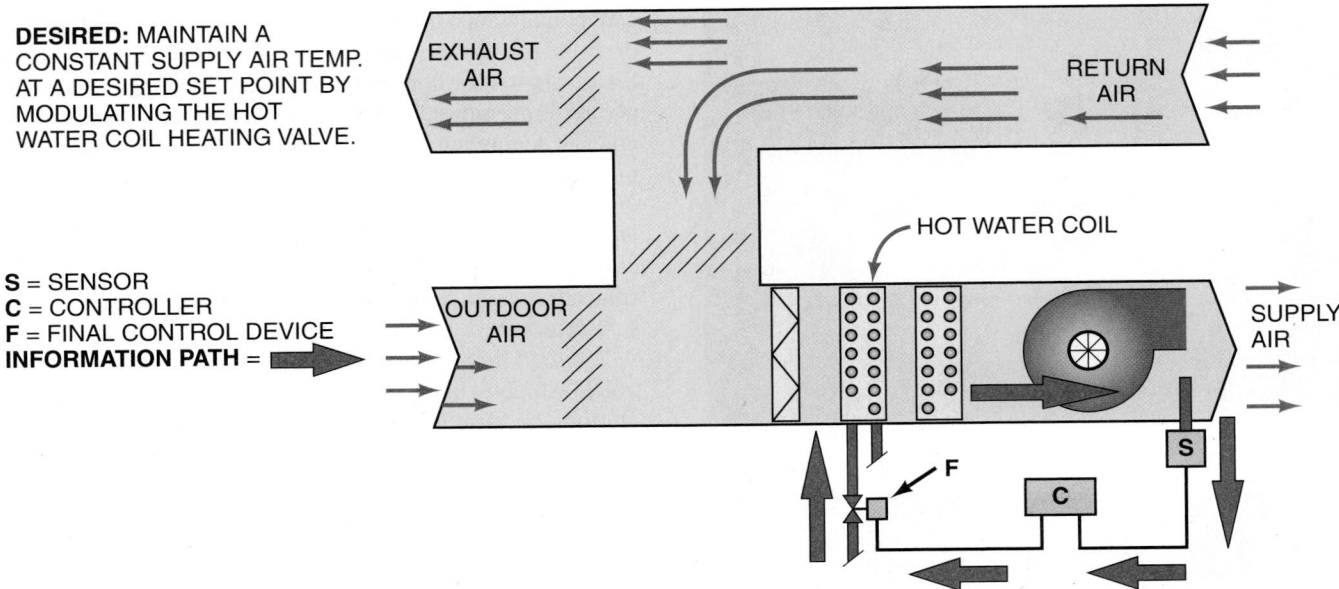

DESIRED: MAINTAIN A CONSTANT SUPPLY AIR TEMP. AT A DESIRED SET POINT BY MODULATING THE HOT WATER COIL HEATING VALVE.

S = SENSOR
C = CONTROLLER
F = FINAL CONTROL DEVICE
INFORMATION PATH =

Figure 16–27   A closed-loop control configuration.

information to the individual control computers. Reliable sensors and their control wiring are the greatest challenge.

🌐Building management may want to perform what is known as *load shedding* or *duty cycling* to improve electrical power costs. Electrical load shedding is using management methods to reduce the power consumption of a building.🌐 For example, suppose the air conditioner is turned off at exactly 5:00 PM rather than 5:30 PM. The people exiting the building may notice that the space temperature is rising, but it does not matter because they are leaving. Duty cycling is shutting down nonessential equipment during high peak loads. This could be shutting down the water coolers, decorative water fountains, ice makers, water heaters, and other nonessential devices to temporarily reduce the electrical load. Remember, every Btu that is allowed into the building must be removed by the air-conditioning system in the summer. It pays to shut them off when they are not needed.

Power companies use a system called *demand metering* to charge commercial buildings for their electricity. Demand metering bills the owner at the power rate for the month based on the highest usage for some time span. This rate is set by the highest power consumption for this period of time, usually 15 or 30 min. Power companies try to collect in a way that charges for the demand on their systems. To understand how demand billing works, suppose a building's air-conditioning system had not been operated for a 30-day period, and on the next day, it became very hot toward the last part of the day. If the system operated at maximum capacity for the last 30 min of the last day of the month, the building owners would be billed for the entire month at that rate, even the previous 30 days. A sharp operator or a computer-analyzed control and power system would discover this event and not allow the system to operate at full

capacity or would shed some nonessential loads. The power bill for the month would be greatly reduced. In fact, a building's power consumption can be monitored by the minute for the entire year with a well-designed system.

🌐A system can be designed to stop on time at night and start early in the morning. This allows the equipment to bring the building conditions to temperature at a lower power consumption as opposed to running the equipment at full load for a short period of time.🌐 The pull-down time in the summer and the heat-up time in the winter can be lengthened to reduce energy consumption. The computer system can keep a record of the pull-down time for the building and allow for extra time or less time as needed during future events. It keeps a history of how the system responds to specific condition changes and works to make it more efficient. Weather patterns can be tracked to predict fast temperature changes that may be needed and phase them in at lower energy cost. Different solar zone requirements can be predicted and adjusted for a building. The energy requirements are different for the east side of a building than for the west side at different times of day because of the changing position of the sun. In actuality, there are so many variations that can be accomplished with DDC and a good computer system that they cannot all be mentioned here.

Figure 16–28 is a simple example of a closed-loop control system that controls a chilled water coil to a conditioned space using DDC. The thermostat in the return air from the conditioned space is checking the air temperature. Suppose we want to maintain the temperature at 75°F. The digital signal voltage range is from 0 to 10 V and we want to have a sensitivity (often called gain in electronics) of 1.7 V per degree of temperature change. The voltage should read 5 V at 75°F. The three-way mixing valve should be bypassing about

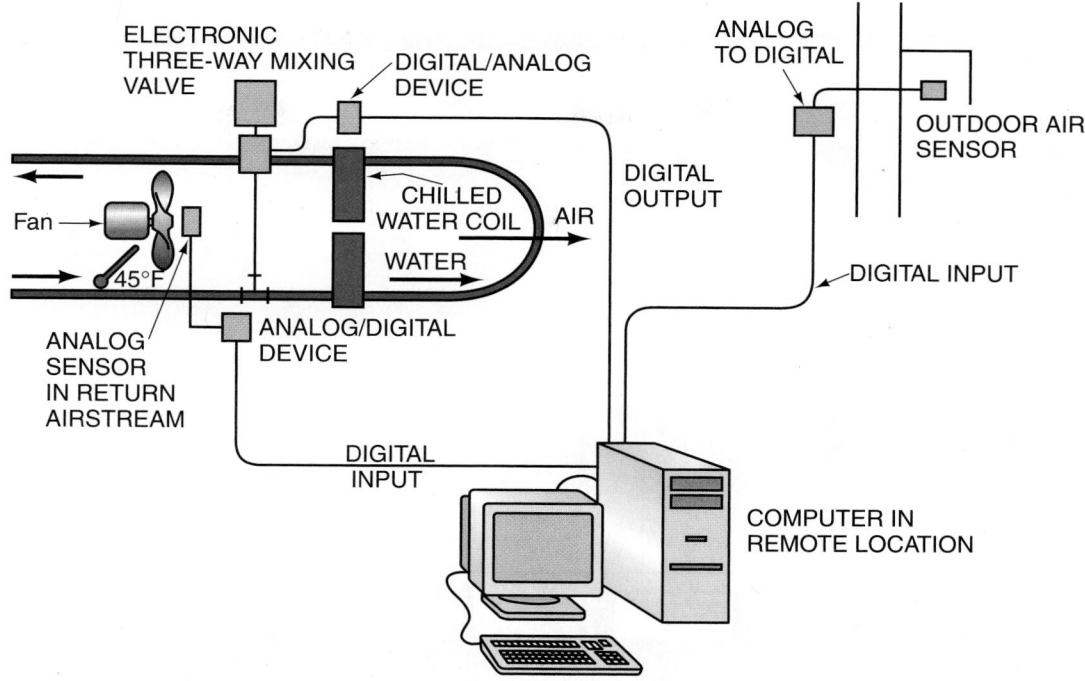

Figure 16–28   The analog signal from the return air sensor is changed to a digital signal and analyzed by the computer to allow more or less chilled water to flow through the chilled water coil to maintain space temperature.

50% of the water, and 50% should be passing through the coil to cool the air. This should result in about a 5°F rise in the water temperature if it has a 10°F rise at full load. The coil is operating at about 50% capacity. If the return air temperature goes up to 76°F, the voltage will rise to 6.7 V and the three-way mixing valve will begin to allow more chilled water through the coil, bypassing less water. When the load begins to reduce, the sensor will alert the computer to tell the valve to bypass more water. The technician can use the DC voltage feature of the voltmeter to monitor these voltages and make adjustments. The technician can check the voltage at various points to determine how the controller is supposed to react, and make changes accordingly.

The reason for using a three-way mixing valve for this system is because the central water chiller must have a constant flow through it. It is not like the boiler and hot water coil used in the pneumatic example. If too many chilled water coils were to shut down using straight-through valves instead of bypassing the water, the chiller would be in danger of approaching freezing conditions that could result in damage. It is customary to use mixing valves for chilled water circuits.

There is a lot of versatility in these control arrangements, particularly for large buildings, **Figure 16–29.** These control advantages are beginning to move into the residential systems because the commercial systems have been proven to work. Electronic control components, like computers, are becoming more affordable to the average person.

Many old buildings are still outfitted with pneumatic controls, and it is not cost-effective to replace them. This is what makes the DDC system so versatile. DDC systems can be meshed in with the older systems for less cost than a complete

building update. Both control systems are reliable and work well together when the system is designed correctly. Again, it is best to have the designer's prints to know what is really expected of a system.

16.7 RESIDENTIAL ELECTRONIC CONTROLS

Technicians have been dealing with electronic controls in residential control systems for many years, often without knowing it. The residential oil burner control board is an example. It has been used for many years. Some technicians seem to be scared when the word *electronics* is mentioned, but it is not necessary to have the knowledge of a radio or TV technician to work with the electronic controls in the HVACR industry. We do not check out the individual components like the TV technician; we work with complete circuit boards. We use manufacturers' diagnostic systems or treat the circuit board as a switch in many cases. Many systems use error messages at the thermostat or blinking light-emitting diodes (LEDs) as codes for problems. Manufacturers' literature has become vital for troubleshooting. Attending manufacturers' service schools will help you understand their goals with equipment and controls.

Electronics are used in virtually every type of equipment today. The electronic thermostat often uses thermistors to control temperature. Recall the discussion about thermistors from Unit 12. A thermistor changes resistance with a change in temperature. Thermistors are very small and respond quickly to a temperature change. They can send a different signal for a different temperature. **Figure 16–30** shows a

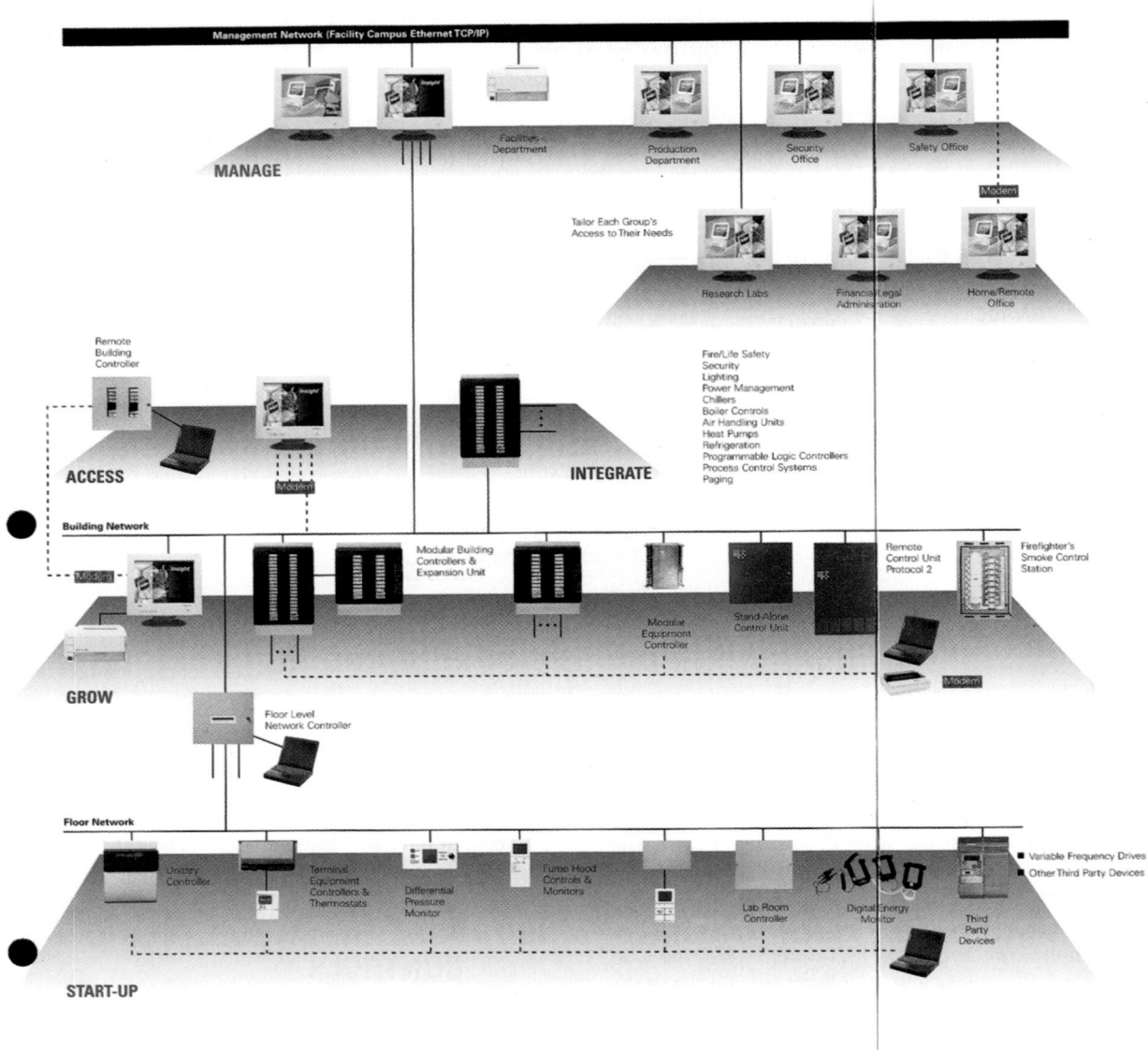

Figure 16–29 This pictorial shows the layout of a typical system. Notice that there is a start-up phase for when a building is first built. Then there is a growth phase where new equipment may be added as a building expands. The "Integrate" part of the figure shows that different types of systems may be operated with the system. The "Access" part shows that the system can be accessed away from the site. The "Manage" portion will allow off-property management of the system by an outside company. *Courtesy Siemens Building Technologies, Inc.*

temperature and resistance graph for a thermistor. The thermistor can be tested with a quality ohmmeter. Just record the temperature of the actual element with an accurate thermometer and check out the plot on a graph for that thermistor. The change in temperature of the thermistor can be used in electronic circuits to monitor temperatures throughout the system and send information to the circuit board that contains a computer chip, a small computer. It can, in turn, respond and cause a change in conditions. Space temperature is not the only temperature that is important in modern systems. Other temperatures that may be important are the following:

■ Discharge-line temperature
■ Motor-winding temperature
■ Return air temperature

■ Suction-line temperature
■ Liquid-line temperature
■ Flue-gas temperature
■ Outdoor temperature
■ Supply air temperature
■ Indoor temperature

Figure 16–31 shows an electronic thermostat that performs many functions other than just turning the equipment on and off. This thermostat can be used with a two-stage cooling or heating system and is fully programmable for different events 7 days per week. 🌐 It is suitable for a working family that wants to change the space temperature at night and during the day while at work, and then have the space temperature brought back to the comfortable set point while at home.

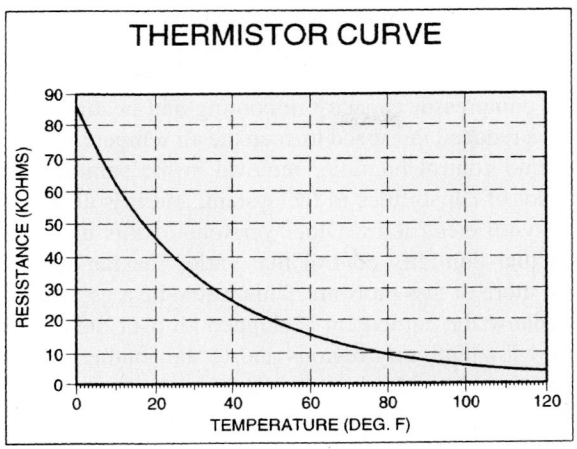

A91431

Figure 16–30 This is a temperature-resistance graph for a thermistor. The technician can use this graph, a thermometer, and an ohmmeter to test a thermistor for accuracy.

It can be programmed to stay at the comfortable set point for weekends and to go to setback functions for vacations. This is typical of many versatile thermostats on the market.

Some of the functions are:

- 3-minute compressor off time to prevent compressor short cycling.
- 15-minute cycle timer to prevent the heating system from short cycling.
- 15-minute stage timer to prevent the second-stage heat from starting up too soon.
- 3-minute minimum on timer for staged heat.
- heat/cool set points of desired temperature.
- auto changeover from heat to cool and back.
- emergency heat mode for heat pumps.
- power on check.
- low voltage.
- error codes for thermostat operation, line-voltage drops, and outdoor temperature sensor problems.
- outdoor temperature smart recovery that tells the system when to restart the equipment after night or day setback.

1 **Easy operation** of the mode button selects between OFF, HEAT, COOL and AUTO operations. Heat pump thermostat models also include an EMERGENCY HEAT mode.

2 **Airflow is monitored** by the fan button. You can use this function to choose between ON or AUTO fan operations.

3 **Simple maintenance reminder** of the clean filter indicator ensures that you keep your system operating at peak performance and efficiency.

4 **Added convenience** is offered by the outdoor temperature sensor. This optional feature displays the outdoor temperature on the LCD readout.

Backlit LCD displays large, easy-to-read numbers with a back-lighting feature that is activated by the touch of a button.

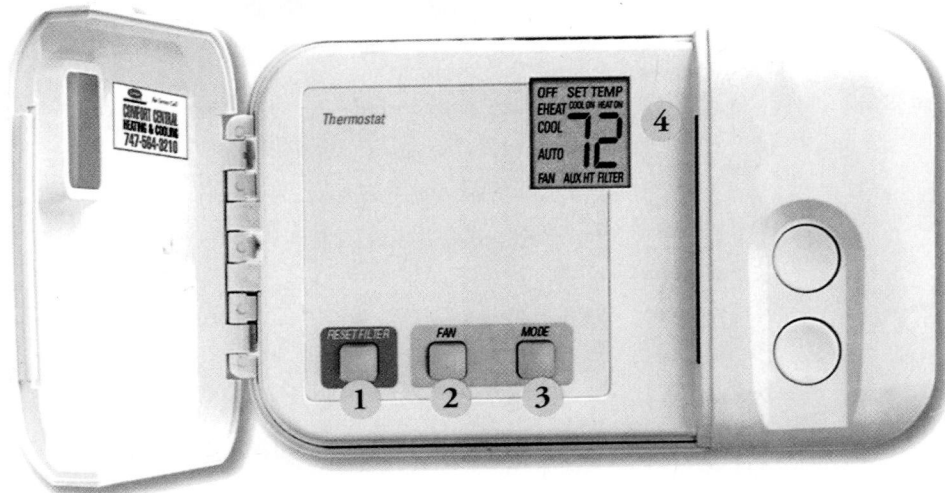

Figure 16–31 The electronic thermostat can perform many functions that could not be accomplished with earlier thermostats. *Courtesy Carrier Corporation*

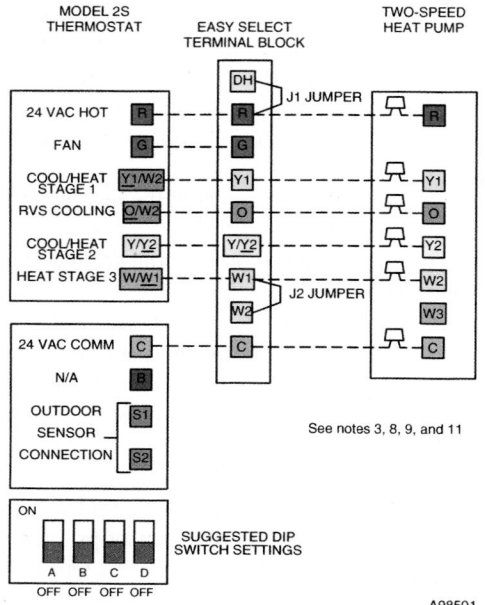

MODEL 2S
THERMOSTAT

EASY SELECT
TERMINAL BLOCK

TWO-SPEED
HEAT PUMP

See notes 3, 8, 9, and 11

SUGGESTED DIP
SWITCH SETTINGS

A98501

Figure 16–32 This diagram shows the field control wiring diagram and the DIP switches that must be set up for the various configurations that the thermostat can accomplish. *Courtesy Carrier Corporation*

The thermostat has some small switches in the back that are called dual in-line pair (DIP) switches. These allow the technician to set the thermostat up for the particular piece of equipment it is controlling. Using DIP switches and following the wiring diagram for the particular installation will give the customer the desired type of control over the equipment. **Figure 16–32** is the wiring diagram and DIP switch setup for a two-speed heat pump operated with a variable-speed fan

coil. Notice the two-speed heat pump (this is achieved with a combination two- and four-pole compressor motor) and variable-speed fan coil. This allows the system to operate at 50% compressor capacity in cooling and heating, and it utilizes a reduced fan speed to keep the air temperature up in winter and control humidity removal in the summer. This adds a lot of capabilities to the system, and it is all accomplished with electronics. Other versions of this thermostat have actual humidity control that varies the fan speed to remove more or less moisture. This would be a great advantage because the equipment is shipped all over the country, to desert low-humidity regions and to the southern coastal regions where humidity is very high.

The outdoor unit for a heat pump also has some electronics that the technician will be interested in. The circuit board in the outdoor unit receives a signal from the thermostat asking it to either heat or cool, and this may be a high or low speed. When all is well, there is nothing to worry about. Suppose the technician arrives at the job because of a service request for no heat. What should be done first?

■ Go to the thermostat and see if the indoor fan will run. If not, check the power supply.

■ Make sure there is power to the indoor unit, which may be under the house or in the attic.

■ When power is established, make sure the indoor fan will run; then go to the outdoor unit and verify power to the unit. Make sure the thermostat is calling for it to operate.

When all signs seem to call for operation at the outdoor unit and it will not run, remove the cover and look at what the manufacturer recommends. **Figure 16–33** is a table called an LED control function light code that shows the status lights for this particular manufacturer's equipment. This chart is typical of what the technician can expect. During the service call, the

NOTE: A signal (code) is not sent through the L lead to thermostat unless a failure has occurred.

LED Control Function Light Code		
CODE	**DEFINITION**	*
Constant flash No pause	No demand Stand by	9
1 flash w/pause	Low-speed operation	8
2 flashes w/pause	High-speed operation	7
3 flashes w/pause	Outdoor ambient thermistor failure	6
4 flashes w/pause	Outdoor coil thermistor failure	5
3 flashes pause 4 flashes	Thermistor out of range†	4
5 flashes w/pause	Pressure switch trip (LM1/LM2)	3
6 flashes w/pause	Compressor PTC's out of limit	2
Constant light No pause No flash	Board failure	1

* Function light signal order of importance in case of multiple signal request; 1 is most important.
† Check both thermistors to determine which is faulty.

Figure 16–33 This chart shows the control function light codes for this particular system. *Courtesy Carrier Corporation*

technician can look at the status lights and see what is supposed to be happening. If the equipment is supposed to be running and is not running, more investigation is in order.

The manufacturer will supply a trouble chart to take the technician to the next step. Manufacturers are working to design more efficient equipment with unique features. Installers and service technicians must keep up to date. Most contractors try to work with one or two brands of equipment so that their installers and service technicians can be efficient with those systems.

SUMMARY

- Automatic controls are a big part of all modern buildings.
- The industry technician should be able to look at and interpret building blueprints and their control circuits.
- Modern control systems use modulating controls as well as on–off control systems.
- Control systems can be subdivided into control loops for better understanding.
- A basic control loop has a sensor, controller, and something to be controlled.
- In a home air-conditioning system the thermostat is the sensor, the contactor is the controller, and the compressor (and fans) is the controlled device.
- Pneumatic controls are typically used in commercial applications and are mechanical in nature. They use air pressure instead of electricity to position the system devices.
- Pneumatic controls are explosion-proof, simple, safe, and reliable.
- The air that operates pneumatic controls is typically 0 to 20 psig and must be clean and dry.
- Air pressure is amplified for more power using diaphragms of a large size.
- Diaphragm mechanisms with a plunger may be used to operate dampers and valves to modulate the flow of fluids (air, water, and refrigerant) to control room and building conditions.
- Since pneumatic controls use air, they can often be heard repositioning as the air bleeds from the control.
- A pilot positioner is often used on the controlled device to handle large volumes of air needed to fill or exhaust a large diaphragm.
- The tools of the pneumatic technician are an air gage, thermometer, and special wrench (usually Allen type).
- Direct digital control (DDC) is a control process in which a microprocessor, acting as a digital controller, constantly updates a database of internal information. Through sensors, the microprocessor updates this information by watching and monitoring information from an external controlled environment or conditioned space. The microprocessor then continually produces corrections and outputs this corrected information back to the controlled environment.
- Direct digital controls, called DDCs, are an electronic version of pneumatic controls, only they are much more versatile.
- Computers are used to create logic with these control systems, and many more features and control points can be coordinated together.

- DDCs and pneumatic controls may be used together; therefore, older systems can be updated with DDC systems.
- Modern electronic controls have been developed, tried, and proven in commercial systems and are now being used for many residential applications.
- The residential technician does not have to be an electronic technician to troubleshoot the system. Many of these systems have built-in diagnostics.
- The electronics control temperature, humidity, air, and refrigerant flow in residential systems.

REVIEW QUESTIONS

1. True or False: Pneumatic controls are explosion-proof.
2. Pneumatic controls are operated by
   A. water.
   B. electricity.
   C. electronics.
   D. air.
3. One advantage of a modern control system is
   A. simplicity.
   B. faster response.
   C. that they can modulate.
   D. that they are less expensive.
4. A control loop contains _____, _____, and _____.
5. Pneumatic air must:
   A. be clean and dry.
   B. have very high pressure.
   C. encompass a large volume.
   D. be hot.
6. The air supplied to a diaphragm is 20 psig, and the diaphragm has a diameter of 6 in. How much pressure can this diaphragm apply to a plunger?
7. When a large volume of air is needed at a diaphragm, the thermostat may feed branch line pressure to the _____.
8. Which are the typical tools that a pneumatic control technician would use?
   A. Gage manifold and nitrogen cylinder
   B. Flue-gas analyzer and thermometer
   C. Flaring tools, adjustable wrench, and screwdriver
   D. Pressure gage, Allen wrench, and pocket thermometer
9. True or False: Electronic controls operate on either low voltages or milliamperes.
10. An analog control signal
    A. has an infinite number of steps of control.
    B. is very limited in the number of steps of control.
    C. makes a hissing noise when the position is changed.
    D. is not used in this industry.
11. Another term used to describe sensitivity in electronic controls is
    A. temperature.
    B. gain.
    C. RPM.
    D. pressure.

12. One of the advantages of DDCs is
    A. total building control management.
    B. that they are simple.
    C. that no air compressor is needed.
    D. that they operate on high voltage.
13. True or False: Electronic controls have been used in the residential industry for a long time.
14. One of the major components of a residential electronic control circuit that measures temperature is a
    A. thermometer.
    B. thermistor.
    C. capacitor.
    D. filter.
15. True or False: Diagnostic light-emitting diodes are often used to alert the technician of problems.

16. Describe a final control device and give three examples of these devices used in a DDC system in today's HVACR industry.
17. Describe what a sensor does in a DDC system used in today's HVACR industry, and list five examples of sensors.
18. What is the difference between an active and a passive sensor used in DDC control systems?
19. A control system consists of what three basic components?
20. List six advantages that DDC control systems have over traditional control systems.
21. Describe the difference between the analog and digital signals in a DDC control system.
22. Describe the difference between an open-loop and a closed-loop control configuration in a DDC system.

SECTION 4

Electric Motors

Types of Electric Motors

OBJECTIVES

After studying this unit, you should be able to

- describe the different types of open single-phase motors used to drive fans, compressors, and pumps.
- describe the applications of the various types of motors.
- state which motors have high starting torque.
- list the components that cause a motor to have a higher starting torque.
- describe a multispeed permanent split-capacitor motor and indicate how the different speeds are obtained.
- explain the operation of a three-phase motor.
- describe a motor used for a hermetic compressor.
- explain the motor terminal connections in various compressors.
- describe the different types of compressors that use hermetic motors.
- describe the use of variable-speed motors.

17.1 USES OF ELECTRIC MOTORS

Electric motors are used to turn the prime movers of air, water, and refrigerant. The prime movers are the fans, pumps, and compressors, **Figure 17–1**. Several types of motors, each with its particular use, are available. For example, some applications need motors that will start under heavy loads and still develop their rated work horsepower under a continuous running condition. Some motors run for years in dirty operating conditions, and others operate in a refrigerant atmosphere. These are a few of the typical applications of motors in this industry. The technician must understand which motor is suitable for each job so that effective troubleshooting can be accomplished and, if necessary, the motor replaced by the proper type. But the basic operating principles of an electric motor first must be understood. Although many types of electric motors are used, most motors operate on similar principles.

17.2 PARTS OF AN ELECTRIC MOTOR

Electric motors have a *stator* with windings, a *rotor, bearings, end bells, housing,* and some means to hold these parts in the proper position, **Figure 17–2** and **Figure 17–3**.

The stator is a winding that, when energized, will generate a magnetic field—because there will be current flowing through it. The rotor is the rotating portion of the motor and is made of iron or copper bars bound on the ends with aluminum. The rotor is not wired to the power source like the stator is.

(A)

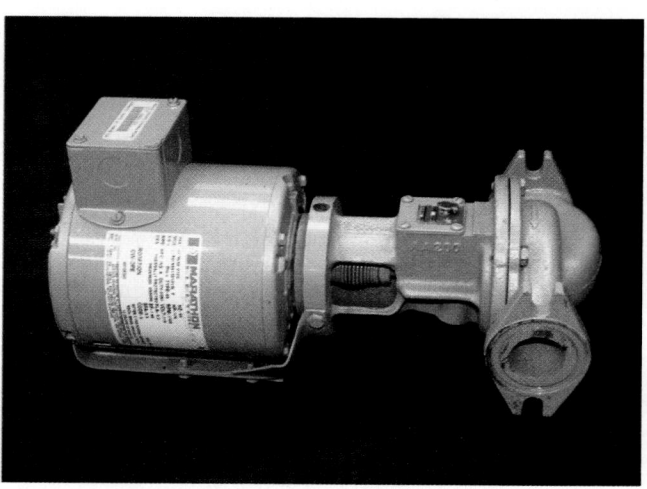

(B)

Figure 17–1   **(A)** Fans move air. **(B)** Pumps move water. *(A) and (B) Courtesy W. W. Grainger, Inc.*

Figure 17–2   A cutaway of an electric motor. *Courtesy Century Electric, Inc.*

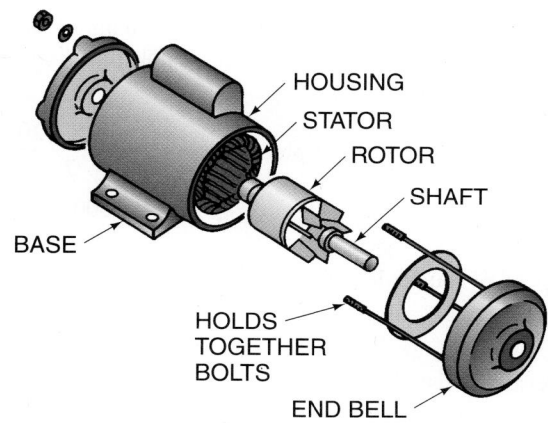

Figure 17–3   Individual electric motor parts.

The motor shaft is connected to the rotor; therefore, when the rotor turns, so does the motor shaft. The motor's bearings allow for low-friction rotation of the rotor. Bearings help reduce the friction and the heat generated by the motor's moving and rubbing parts and surfaces. The motor housing, end bells, and base hold the motor components in place and provide a means for securely mounting the motor itself.

17.3   ELECTRIC MOTORS AND MAGNETISM

Electricity and magnetism are used to create the rotation in an electric motor to drive the fans, pumps, and compressors. Magnets are known to have two different electrical poles, north and south. Unlike poles of a magnet attract each other, and like poles repel each other. If a stationary horseshoe magnet were placed with its two poles (north and south) at either end of a free-turning magnet as in **Figure 17–4,** one pole of the free-rotating magnet would line up with the opposite pole of the horseshoe magnet. If the horseshoe magnet were an electromagnet and the wires on the battery were reversed, the poles of this magnet would reverse, and the

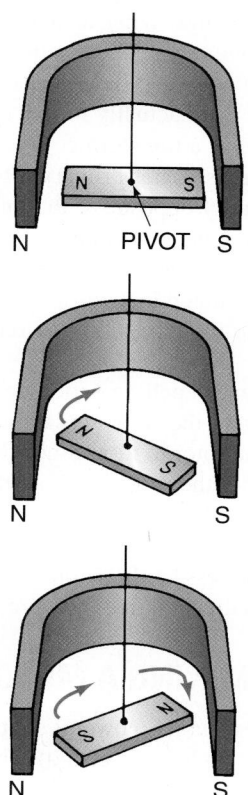

Figure 17–4   Poles (north and south) on a rotating magnet will line up with the opposite poles on a stationary magnet.

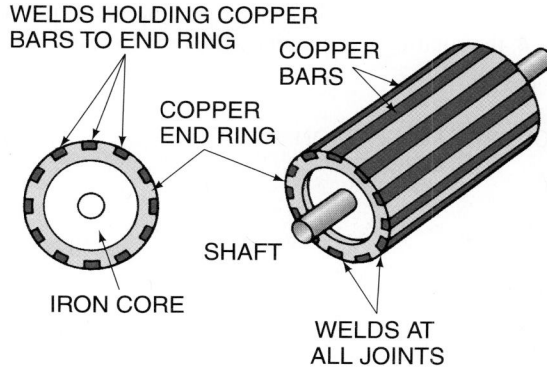

Figure 17–5   A simple sketch of a squirrel cage rotor.

poles on the free magnet would be repelled, causing it to rotate until the unlike poles again were lined up. This is the basic principle of electric motor operation. In this example, the horseshoe magnet acts as the stator, and the free-rotating magnet acts as the rotor.

In a two-pole split-phase motor the stator has two poles with insulated wire windings called the *run windings.* When an electrical current is applied, these poles become an electromagnet with the polarity changing constantly. In normal 60-cycle operation the polarity changes 60 times per second.

The rotor may be constructed of bars, **Figure 17–5.** This type is called a *squirrel cage rotor.* The rotor is positioned between the run windings. When an alternating current (AC) is applied to these windings, a magnetic field is produced in the windings and a magnetic field is also induced in the rotor. The bars in the rotor actually form a coil. This is similar to the field induced in a transformer secondary by the magnetic field in the transformer primary. The field induced in the rotor has a polarity opposite to that in the run windings. The opposite poles of the run winding are wound in different directions. If one pole is wound clockwise, the opposite pole would be wound counterclockwise. This would set up opposite polarities for the attraction and repulsion forces that cause rotation.

The attracting and repelling action between the poles of the run windings and the rotor sets up a rotating magnetic field and causes the rotor to turn. Since this is AC reversing 60 times per second, the rotor turns, in effect "chasing" the changing polarity in the run windings. The motor will continue to rotate as long as power is supplied. The motor starting method determines the direction of motor rotation. The motor will run equally well in either direction.

17.4 DETERMINING A MOTOR'S SPEED

In the United States, our AC power is supplied at a frequency of 60 hertz (Hz, or cycles per second). This means that, in a one-second time interval, 60 complete sine waves are generated. From this, we can conclude that one cycle is completed in 1/60 (one-sixtieth) of a second. During this very brief instant, the direction of current flow changes two times, **Figure 17–6**. This means that every second, the current changes direction 120 times.

Referring back to our horseshoe magnet example, **Figure 17–4,** the rotating bar magnet will come to rest when the

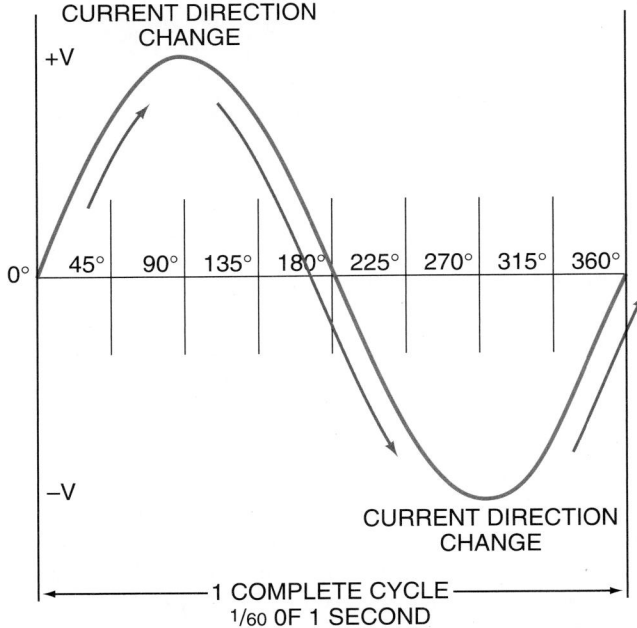

Figure 17–6 The direction of current flow changes twice during each cycle.

north pole of the bar magnet is aligned with the south pole of the horseshoe magnet and when the south pole of the bar magnet is aligned with the north pole of the horseshoe magnet. So, in order for the bar magnet to turn 360 degrees, or one full rotation, the horseshoe magnet must change the polarity of its poles twice. We are discussing AC power; therefore, during each sine wave, the current changes direction twice each cycle. So, the rotor of a motor with two poles, as in our bar magnet example, will turn 60 times every second, or 3600 times a minute. The more poles a motor has, the slower the motor will turn.

The following formula can be used to determine the synchronous speed (without load) of motors in rpm, or rotations per minute.

$$S \text{ (rpm)} = \frac{\text{Frequency} \times 120}{\text{Number of poles}}$$

NOTE: The magnetic field builds and collapses twice each second (each time it changes direction); the 120 consists of a time conversion that changes seconds into minutes. It also consists of a conversion from the motor's pole pairs (poles/2) to the actual number of poles.

$$\text{Speed of two-pole split-phase motors} = \frac{60 \times 120}{2}$$
$$= 3600 \text{ rpm}$$

$$\text{Speed of four-pole split-phase motors} = \frac{60 \times 120}{4}$$
$$= 1800 \text{ rpm}$$

The speeds listed above are the calculated speeds and not the actual speeds at which the motor will turn. Internal motor component friction and the mechanisms or drives to which the motor is connected add a certain amount of resistance to rotation. This resistance is referred to as the motor *slip* and is often expressed as a percentage. Motor slip represents the difference between the motor's calculated and actual speeds. Our 3600 rpm motor actually turns at a speed of about 3450 rpm. We can calculate slip as follows:

Motor slip (%)

$$= \frac{\text{Calculated motor speed} - \text{Actual motor speed}}{\text{Calculated motor speed}} \times 100$$

In this example we have the following:

$$\text{Motor slip (\%)} = \frac{3600 - 3450}{3600} \times 100$$

$$\text{Motor slip (\%)} = \frac{150}{3600} \times 100$$

$$\text{Motor slip (\%)} = 4.17\%$$

It is left as an exercise for the reader to verify that the motor slip for an 1800-rpm motor that is turning at 1725 rpm is also 4.17%. Excessive motor slip can be an indication of improper motor lubrication, defective motor bearings, poorly adjusted belts (too tight), or an improperly selected motor for the desired application (overloaded).

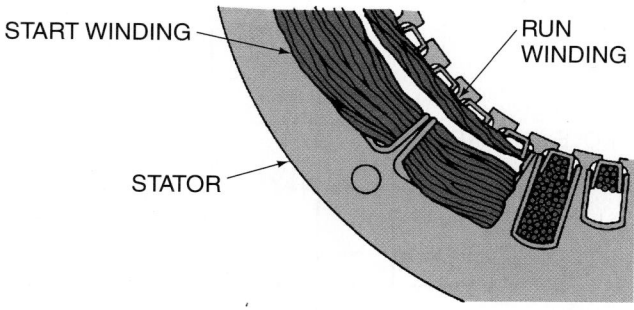

Figure 17–7   The placement of the start and run windings inside the stator.

17.5   START WINDINGS

The preceding explanation does not explain how to start the motor rotation, which is accomplished by using a separate motor winding called the start winding. The start winding is wound next to the run winding but is a few electrical degrees out of phase with the run winding. This design is much like the pedals on a bicycle—if both pedals were at the same angle, it would be very hard to get started. The start winding is in the circuit only for as long as it takes the rotor to get close to its rated speed, and it then is electrically disconnected, **Figure 17–7**. The start windings have more turns than the run windings and are wound with a smaller diameter wire. Opposite poles of the start winding are wound in different directions. If one pole is wound clockwise, then the opposite pole would be wound counterclockwise. This would set up opposite polarities for the attraction and repulsion forces that cause rotation. This produces a larger magnetic field and greater resistance, which helps the rotor to start turning and determines the direction in which it will turn. This happens as a result of these windings being located between the run windings. It changes the phase angle between the voltage and the current in these windings.

We have just described a two-pole split-phase *induction motor*, which is rated to run at 3600 revolutions per minute (rpm). It actually turns at a slightly slower speed when running under full load. When the motor approaches its normal speed, the start winding is often removed from the circuit, leaving only the run windings energized. Many split-phase motors have four poles and run at 1800 rpm.

17.6   STARTING AND RUNNING CHARACTERISTICS

Two major considerations of electric motor applications are the starting and running characteristics. A motor applied to a refrigeration compressor must have a high starting *torque*—it must be able to start under heavy starting loads. Torque is the twisting power of the motor shaft. The starting torque is the power to turn the shaft from the stopped position. The motor must also have enough torque to operate with the load of the application. Some motors must have a great deal of

starting torque to turn the motor at the stopped position but do not need a great deal of torque to maintain speed. For example, systems equipped with capillary tubes can typically use compressors with low starting torque as the system pressures can equalize during the off cycle. Some thermostatic expansion valve (TXV) systems require greater starting torque as there may be a pressure differential across the compressor even when the compressor is not operating.

Motors are rated with two different current ratings, full-load or rated-load amperes (FLA or RLA) and locked-rotor amperes (LRA). Locked-rotor amperage is often referred to as inrush current. When a motor is still, it takes a great deal of torque to get it turning, particularly if the motor has a load on it at start-up. Typically, the LRA for a motor is about five times the FLA/RLA. For example, suppose a compressor operates on 25 A when at full load. This same compressor will pull about 125 A on start-up for just a moment. As the motor gets up to speed, the amperage will reduce to FLA, 25 A. Since this inrush of current is of a short duration, it does not figure in the wire sizing for the motor. A refrigeration compressor may have a head pressure of 155 psig and a suction pressure of 5 psig and still be required to start in systems where the pressures do not equalize. The pressure difference of 150 psig is the same as saying that the compressor has a starting resistance of 150 psi of piston area. If this compressor has a 1-in. diameter piston, the area of the piston is 0.78 in^2 ($A = \pi r^2$ or $3.14 \times 0.5 \times 0.5$). This area multiplied by the pressure difference of 150 psi is the starting resistance for the motor (117 lb). This is similar to a 117-lb weight resting on top of the piston when it tries to start, **Figure 17–8**.

To start a small fan, a motor does not need as much starting torque. The motor must simply overcome the friction needed to start the fan moving. There is no pressure difference because the pressures equalize when the fan is not running, **Figure 17–9**. The high starting load, called starting torque, puts more load on the compressor as compared with the light load of the fan. These are two different motor applications and require two different types of motors. These applications will be discussed one at a time in this unit as the different types of motors are discussed.

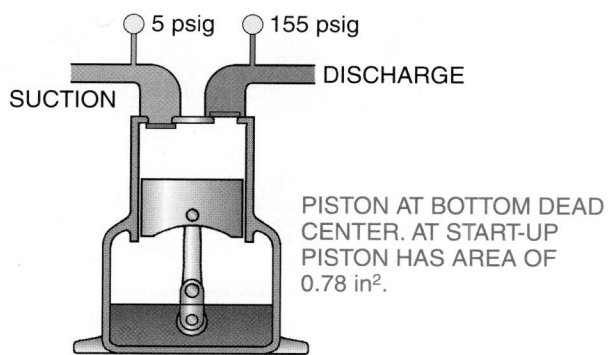

Figure 17–8   A compressor with a high-side pressure of 155 psig and a low-side pressure of 5 psig.

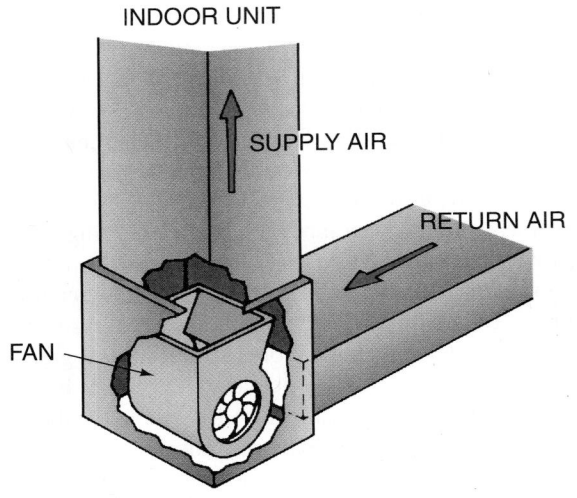

Figure 17–9 This fan has no pressure difference to overcome while starting. When the fan stops, the air pressure equalizes.

17.7 ELECTRICAL POWER SUPPLIES

The power company furnishes the power to the customer and determines what type of power supply they have. The power company moves the power across the country by means of high-voltage transmission lines to a transformer that reduces the power to the voltage needed at the consumer level. Residences are typically furnished with a single-phase (φ is the symbol often used for phase) power supply. The power pole to the house may look like the one in **Figure 17–10.** Several homes may be furnished from the same power-reducing transformer. The power-reducing transformer has three wires that pass through the meter base and on to the house electrical panel where the power is distributed through the circuit protectors (circuit breakers or fuses), to the various circuits in the house, and then to the power-consuming equipment, **Figure 17–11.**

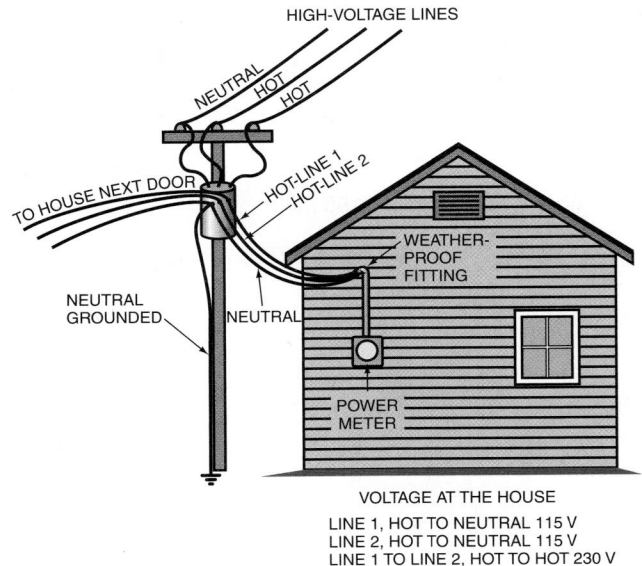

VOLTAGE AT THE HOUSE
LINE 1, HOT TO NEUTRAL 115 V
LINE 2, HOT TO NEUTRAL 115 V
LINE 1 TO LINE 2, HOT TO HOT 230 V

Figure 17–10 A single-phase power supply.

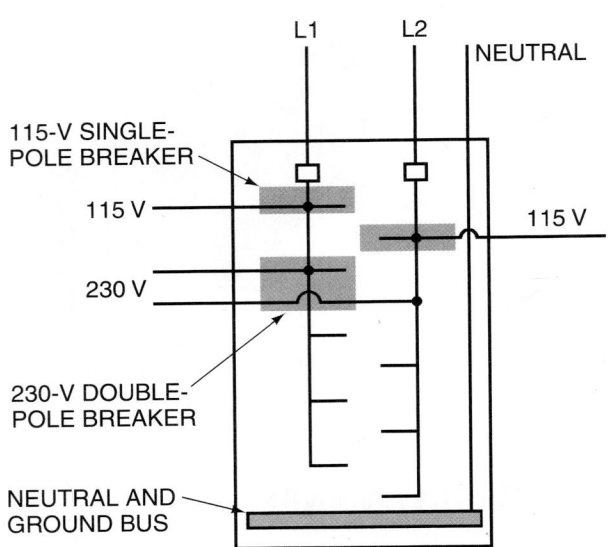

Figure 17–11 A wiring diagram of a main circuit breaker panel for a typical residence.

The power panel in the house would look like the illustration in **Figure 17–11.** Notice that it will furnish both 115-V and 230-V service. Typically the 230-V service is used for the electric dryer, electric range, electric oven, many electric heat devices, and the air conditioning. All other appliances are typically operated from the 115-V circuits.

Commercial buildings and factories use equipment large enough that they require three-phase power. A three-phase power supply may furnish several different voltage options:
A. 115-V single-phase for common appliances
B. 230-V single-phase for heavy-duty appliances
C. 230-V three-phase for large loads such as electric heat or motors
D. 460-V three-phase for large loads such as electric heat or motors
E. 277-V single-phase for lighting circuits (may be obtained between 460 V and neutral with certain systems)
F. 560-V three-phase for special industrial systems

A typical three-phase system wiring from the pole may look like the illustration in **Figure 17–12.** Again, this is furnished by the power company for this application. This building could very well have been supplied with a 230-V three-phase power supply. One of the reasons for using 460-V three-phase is that it reduces the wire size to all components in the building, which reduces the materials cost and the labor cost for installation. The same load using 230 V would use twice the current flow, so the wire would have to be sized up to twice the current-carrying capacity. Many of the components in the building will need to operate from 115 V and maybe even 230 V so a step-down transformer is used for equipment such as small fans, computers, and office equipment. The step-down transformer enables the building to take advantage of the benefits of 460 V, 277 V, 230 V, and 115 V in the same system, **Figure 17–13.**
SAFETY PRECAUTION: A technician must be very careful with any live power circuits. 460-V circuits are particularly

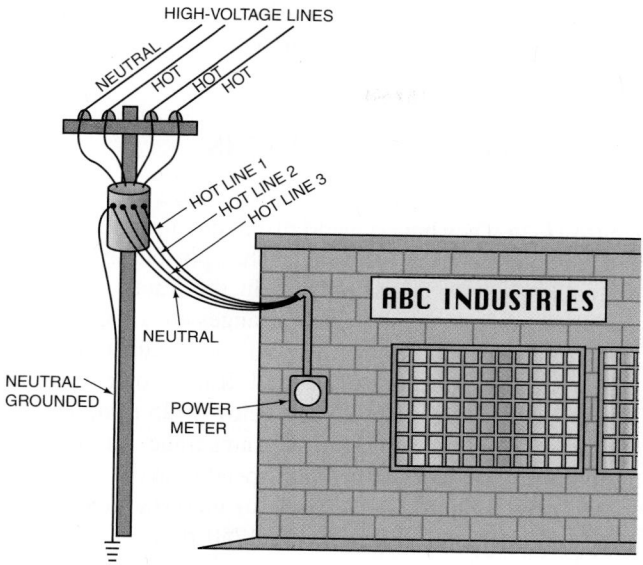

VOLTAGE SUPPLY TO BUILDING

LINE 1, 2, OR 3 TO NEUTRAL = 277 V
LINE 1 TO LINE 2 = 460 V
LINE 1 TO LINE 3 = 460 V
LINE 2 TO LINE 3 = 460 V

Figure 17–12   A 460-V, three-phase power supply to a commercial building.

dangerous, and the technician must take extra care while servicing them.

Following is a description of some motors currently used in the heating, air-conditioning, and refrigeration industry. The electrical characteristics, not the working conditions, are emphasized. Some older motors are still in operation, but they are not discussed in this text.

17.8   SINGLE-PHASE OPEN MOTORS

The power supply for most *single-phase* motors is either 115 V or 208 V to 230 V. A home furnace uses a power supply of 115 V, whereas the air conditioner outside uses a power supply of 230 V. A commercial building may have either 208 V, 230 V, or 460 V, depending on the power company. Some single-phase motors are dual voltage. The motor has two run windings and one start winding. The two run windings have the same resistance, and the start winding has a high resistance. The motor will operate with the two run windings in parallel in the low-voltage mode. When it is required to run in the high-voltage mode, the technician changes the numbered motor leads according to the manufacturer's instructions. This wires the run windings in series with each other and delivers an effective voltage of 115 V to each winding. The motor windings are actually only 115 V because they operate only on 115 V, no matter which mode they are in. The technician can change the wiring connections at the motor terminal box, **Figure 17–14**.

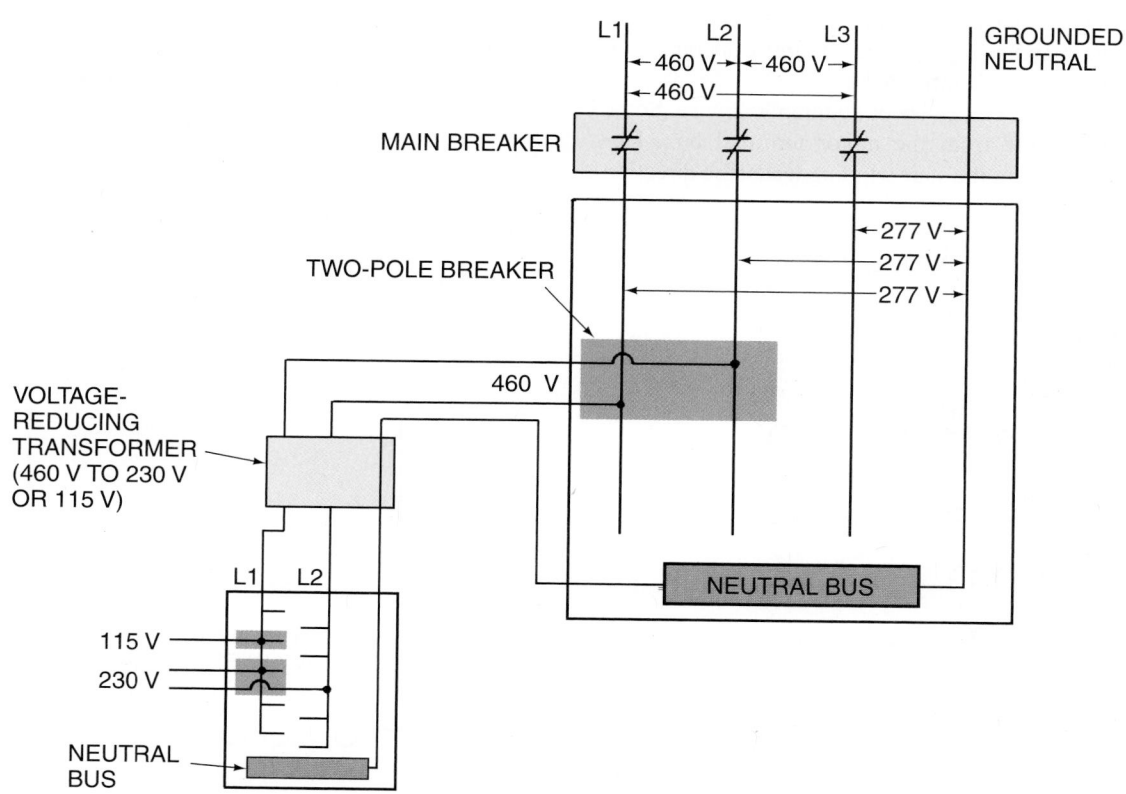

Figure 17–13   When a building has a 460-V power supply, it will have a step-down transformer to provide 115-volt circuits to power office machines and small appliances.

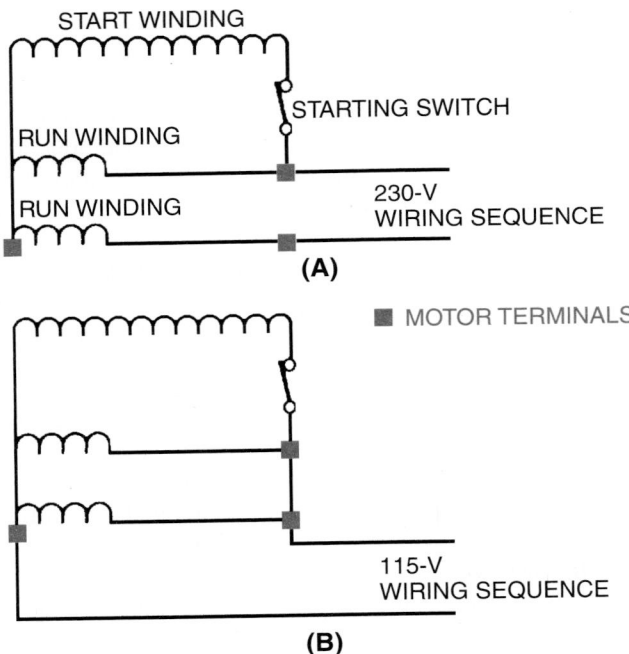

(A)

■ MOTOR TERMINALS

(B)

Figure 17–14 The wiring diagram of a dual-voltage motor. This motor is designed to operate at either 115 V or 230 V, depending on how the motor is wired in the field. **(A)** A 230-V wiring diagram. **(B)** A 115-V wiring diagram.

Some commercial and industrial installations may use a 460-V power supply for large motors. The 460 V may be reduced to a lower voltage to operate the small motors. The smaller motors may be single-phase and must operate from the same power supply, **Figure 17–13.**

A motor can rotate clockwise or counterclockwise. Some motors are reversible from the motor terminal box, **Figure 17–15.** By reversing the start winding leads, the direction

DUAL-VOLTAGE MOTOR

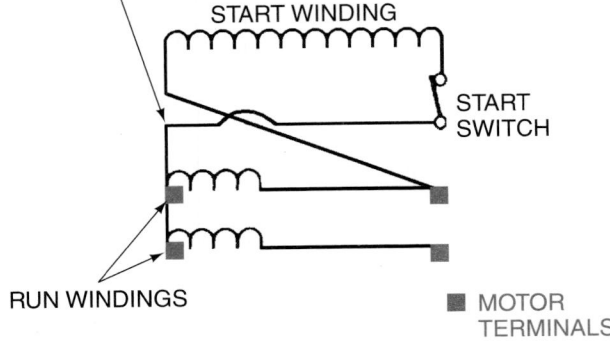

Figure 17–15 The direction of rotation of a single-phase motor can be reversed by changing the connections in the motor terminal box. This will change the direction of current flow through the start winding.

of current flow through the start winding is reversed, changing the charges of the poles. This will cause the motor to rotate in the opposite direction.

17.9 SPLIT-PHASE MOTORS

Split-phase motors have two distinctly different windings, **Figure 17–16.** They have a medium amount of starting torque and a good operating efficiency. The split-phase motor is normally used for operating fans in the fractional horsepower range. Its normal operating ranges are 1800 rpm and 3600 rpm. An 1800-rpm motor will normally operate at 1725 rpm to 1750 rpm under a load. If the motor is loaded to the point where the speed falls below 1725 rpm, the current draw will climb above the rated amperage. Motors rated at 3600 rpm will normally slip in speed to about 3450 rpm to 3500 rpm, **Figure 17–17.** Some of these motors are designed to operate at either speed, 1750 rpm or 3450 rpm.

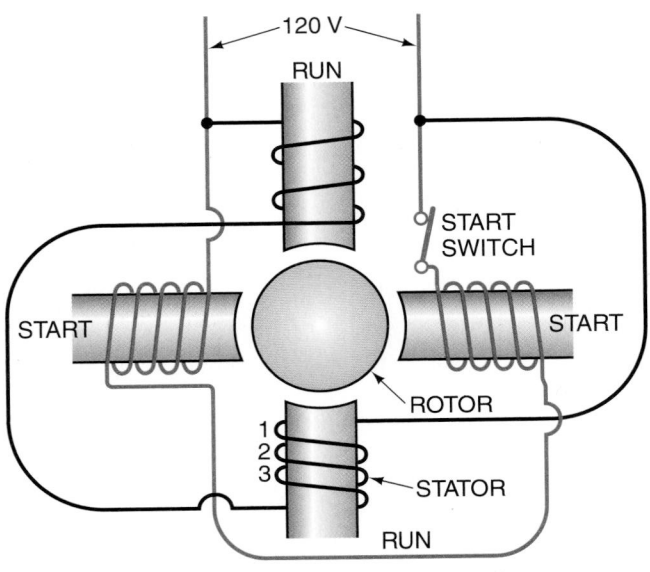

LINE WIRING DIAGRAM

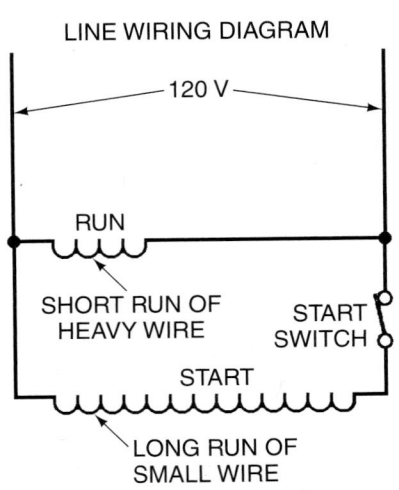

Figure 17–16 Diagram of the start and run windings.

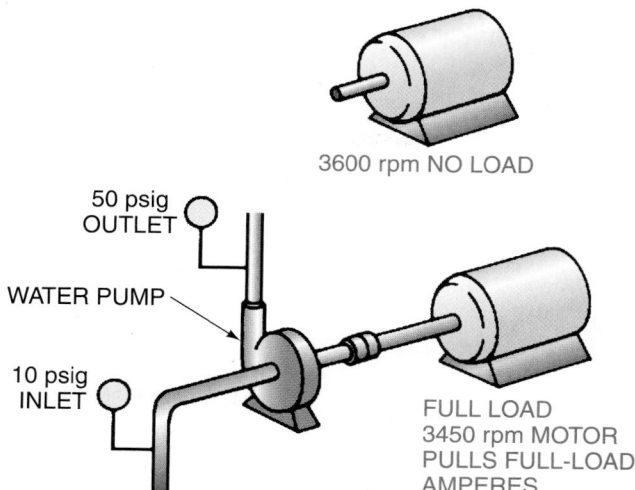

Figure 17–17   Motor speed changes depend on the load on the motor.

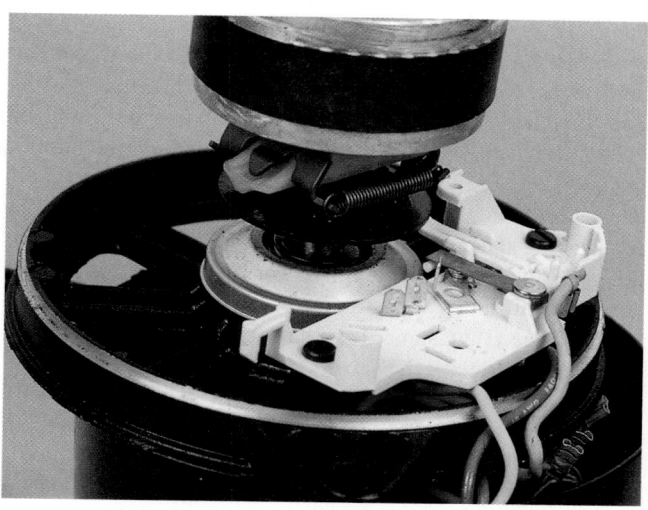

Figure 17–18   The centrifugal switch is located at the end of the motor. *Photo by Bill Johnson*

The speed of the motor is determined by the number of motor poles and by the method of wiring the motor poles. The technician can change the speed of a two-speed motor at the motor terminal box.

17.10   THE CENTRIFUGAL SWITCH

All split-phase motors have a start and run winding. The start windings must be disconnected from the circuit within a very short period of time, or they will overheat. Several methods are used to disconnect the start windings. With open motors, the centrifugal switch is the most common; however, electronic start switches are sometimes used.

The *centrifugal switch* is used to disconnect the start winding from the circuit when the motor reaches approximately 75% of the rated speed. Motors described here are those that run in the atmosphere. (Hermetic motors that run in the refrigerant environment are discussed later in this unit.) When a motor is started in the air, the arc from the centrifugal switch will not harm the atmosphere. (It will harm the refrigerant, so there must be no arc in a refrigerant atmosphere.)

The centrifugal switch is a mechanical device attached to the end of the shaft with weights that will sling outward when the motor reaches approximately 75% speed. For example, if the motor has a rated speed of 1725 rpm, at 1294 rpm (1725 × 0.75) the centrifugal weights will change position and open a switch to remove the start winding from the circuit. This switch is under a fairly large current load, so a spark will occur. If the switch fails to open its contacts and remove the start winding, the motor will draw too much current, and the overload device will cause it to stop. When the motor is deenergized, the motor will slow down and the centrifugal switch will close its contacts in preparation for the next motor starting attempt.

The more the switch is used, the more its contacts will burn from the arc. If this type of motor is started many times,

the first thing that will likely fail will be the centrifugal switch. This switch makes an audible sound when the motor starts and stops, **Figure 17–18**.

17.11   THE ELECTRONIC RELAY

The *electronic relay* is used with some motors to open the start windings after the motor has started. This is a solid-state device designed to open the start winding circuit when the design speed has been obtained. Other devices are also used to perform this function; they are described with hermetic motors.

17.12   CAPACITOR-START MOTORS

The capacitor-start motor is the same basic motor as the split-phase motor, **Figure 17–19**. It has two distinctly different windings for starting and running. However, a start

Figure 17–19   A capacitor-start motor. *Courtesy W. W. Grainger, Inc.*

capacitor is wired in series with the start windings to give the motor more starting torque. **Figure 17–20** shows voltage and current cycles in an induction motor. In an inductive circuit the current *lags* the voltage. In a capacitive circuit the current *leads* the voltage. The amount by which the current leads or lags the voltage is the *phase angle*. A capacitor is chosen to make the phase angle such that it is most efficient for starting the motor, **Figure 17–21**. This capacitor is not designed to be used while the motor is running, and it must be switched out of the circuit soon after the motor starts. This is done at the same time the windings are taken out of the circuit, and with the same switch.

Once the start winding and the start capacitor have been removed from the circuit, the motor will continue to operate with only the run winding energized. The rotating motor shaft and rotor will create, or induce, the needed imbalance in the magnetic field to keep the motor turning. For this reason, capacitor-start motors are often referred to as CSIR

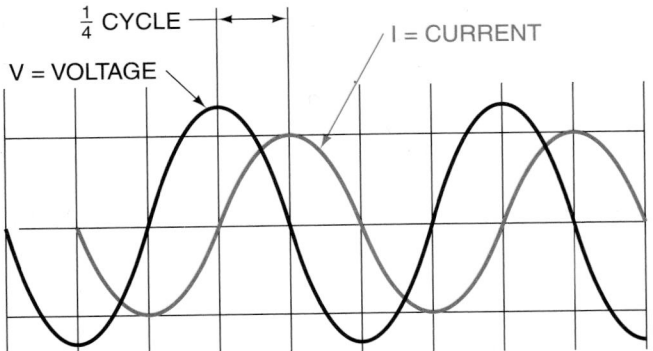

Figure 17–20   The current lags the voltage in an inductive circuit.

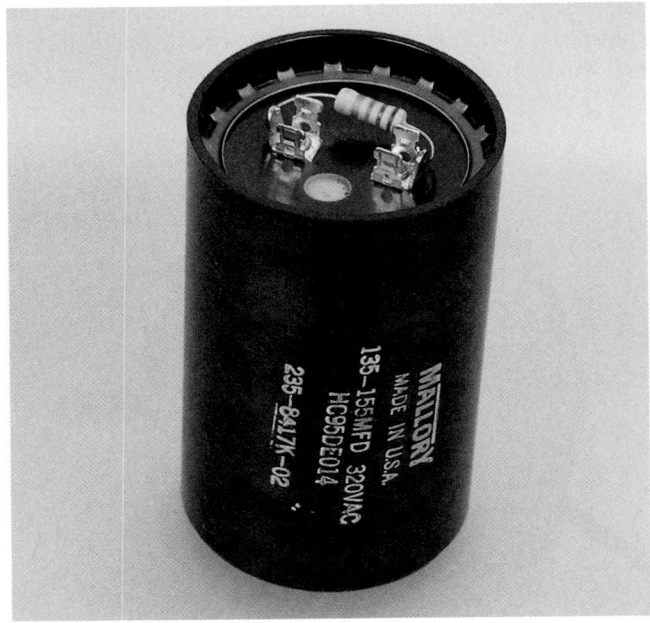

Figure 17–21   A start capacitor. *Photo by Bill Johnson*

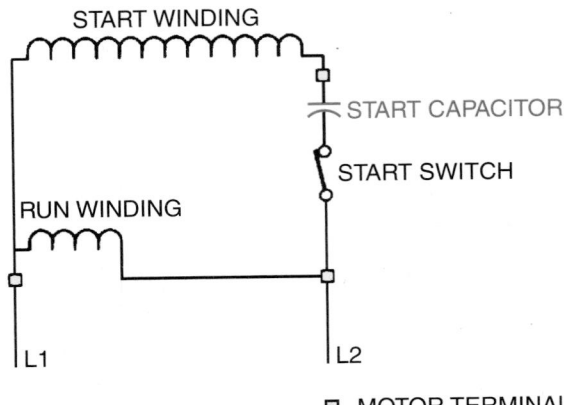

Figure 17–22   Wiring diagram of a capacitor-start motor.

motors. The CSIR stands for capacitor-start-induction-run. A simplified wiring diagram for a CSIR motor is shown in **Figure 17–22**.

17.13   CAPACITOR-START, CAPACITOR-RUN MOTORS

Capacitor-start, capacitor-run motors are much the same as the split-phase motors. A run capacitor is wired into the circuit to provide the most efficient phase angle between the current and voltage when the motor is running. The run capacitor is in the circuit at any time the motor is running. Both the run and start capacitors are wired in series with the start winding but are in parallel to one another, **Figure 17–23**. The microfarad rating of the run capacitor is much lower than that of the start capacitor for any given motor application. The capacitances of two capacitors in parallel add their values the same as resistors do in series. If the run capacitor has a capacitance of 10 microfarads and the start capacitor has a capacitance of 110 microfarads, their total capacitance would add to be 120 microfarads. During start-up, this

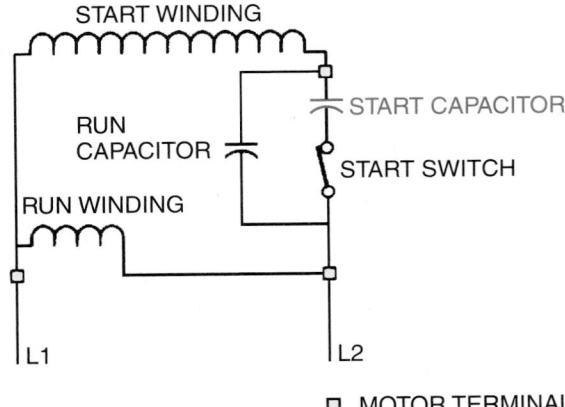

Figure 17–23   Wiring diagram of a capacitor-start, capacitor-run motor. The start capacitor is in the circuit only during motor start-up, whereas the run capacitor is in the circuit whenever the motor is energized.

added capacitance in series with the start winding causes a greater phase angle between the run and start winding, which gives the motor more starting torque. When the start switch opens, the start capacitor is taken out of the circuit. However, the run capacitor and start winding stay in the circuit. The run capacitor stays in series with the start winding for extra running torque. The capacitor, being in series with the start winding during the running mode, also limits the current through the start winding so the winding will not get hot. This motor is actually a permanent split-capacitor (PSC) motor when running (PSC motors will be discussed in the next section). The start capacitor is used for nothing but added starting torque. These motors are specially designed to have the start winding energized whenever the motor is energized, so be sure to use and wire the motor according to the manufacturer's recommendations and guidelines.

If a run capacitor fails because of an open circuit within the capacitor, the motor may start, but the running amperage will be about 10% too high, and the motor will get hot if operated at full load. The capacitor-start, capacitor-run motor is one of the most efficient motors used in refrigeration and air-conditioning equipment. It is normally used with belt-drive fans and compressors. Belt-driven motor and blower assemblies offer a great deal of resistance to the motor when it initially starts. This is due to the fact that the motor must turn the pulley, which is connected to a belt, which is then connected to a pulley and blower assembly. The added starting torque is needed to help overcome this resistance. Once the motor starts, however, keeping the motor turning does not require the same amount of torque.

17.14   PERMANENT SPLIT-CAPACITOR MOTORS

The PSC motor has windings very similar to the split-phase motor, **Figure 17–24,** but it does not have a start capacitor. Instead it uses one run capacitor wired into the circuit in a way similar to the run capacitor in the capacitor-start,

capacitor-run motor. The run capacitor is connected to the L2 sides of the start and run windings. The PSC motor does not utilize a start switch as both the start and run windings, along with the run capacitor, are in the active electric circuit whenever the motor is energized. This is the simplest split-phase motor. It is very efficient and has no moving parts for the starting of the motor; however, the starting torque is very low so the motor can be used only in low-starting-torque applications, **Figure 17–25.**

A multispeed motor can be identified by the many wires at the motor electrical connections, **Figure 17–26** and **Figure 17–27.** As the resistance of the motor winding decreases, the speed of the motor increases. When more resistance is wired into the circuit, the motor speed decreases. Most manufacturers use this motor in the fan section in air-conditioning and heating systems. Motor speed can be changed by switching the wires. Earlier systems used a capacitor-start, capacitor-run motor and a belt drive, and air volumes were adjusted by varying the drive pulley diameter. Many PSC motors are manufactured in 2-, 4-, 6-, and 8-pole designs. The speed in rpm of these motors depends on the

Figure 17–25   This open-type PSC motor can be used to turn a fan. *Courtesy Universal Electric Company*

Figure 17–24   A permanent split-capacitor (PSC) motor. *Courtesy Universal Electric Company*

Figure 17–26   A multispeed PSC motor. *Photo by Bill Johnson*

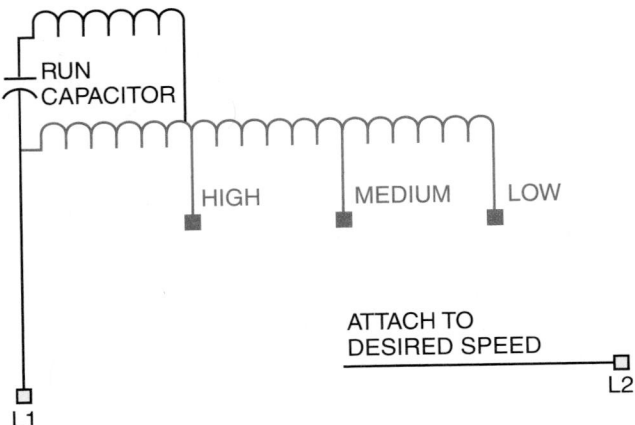

Figure 17–27 This diagram shows how the windings of a three-speed PSC motor are configured. As the winding resistance increases, the motor speed decreases.

frequency of the AC source and the number of poles wired in the motor. The higher the frequency, the faster the motor. The more poles there are, the slower the rpm of the motor. For 60-hertz operation, the rated or synchronous speeds would be 3600, 1800, 1200, and 900 rpm, respectively. However, due to motor slip, the actual speeds would be 3450, 1725, 1075, and 825 rpm.

The PSC motor may be used to obtain slow fan speeds during the winter heating season for higher leaving-air temperatures with gas, oil, and electric furnaces. The fan speed can be increased by switching to a different resistance in the winding using a relay. This will provide more airflow in summer to satisfy cooling requirements, **Figure 17–28.**

The PSC fan motor also has another advantage over the split-phase motor for fan applications. It has a very soft start-up. When a split-phase motor is used for a belt-drive application, the motor, belt, and fan must get up to speed very quickly, which often creates a start-up noise. The PSC motor starts up very slowly, gradually getting up to speed. This is very desirable if the fan is close to the return air inlet to the duct system.

17.15 SHADED-POLE MOTORS

The shaded-pole motor has very little starting torque and is not as efficient as the PSC motor, so it is used only for light-duty applications. These motors have small starting windings or shading coils at the corner of each pole that help the motor start by providing an induced current and a rotating field, **Figure 17–29.** It is an economical motor from the standpoint of initial cost. The shaded-pole motor is normally manufactured in the fractional horsepower range. For years it has been used in air-cooled condensers to turn the fans, **Figure 17–30.** The location of the shading coil on the pole face determines the direction of rotation for a shaded-pole motor. On most shaded-pole motors, rotation can be reversed by disassembling the motor and turning the stator over. This moves the shaded coil to the opposite side that the pole faces and reverses rotation.

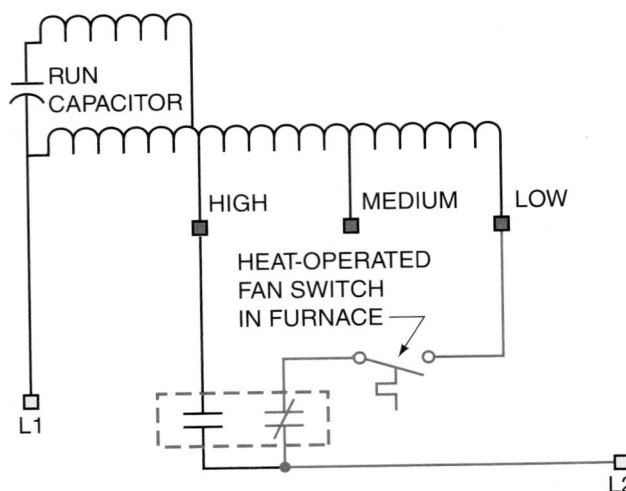

FAN RELAY: WHEN ENERGIZED, SUCH AS IN COOLING, THE FAN CANNOT RUN IN THE LOW-SPEED MODE. WHEN DEENERGIZED, THE FAN CAN START IN THE LOW-SPEED MODE THROUGH THE CONTACTS IN THE HEAT-OPERATED FAN SWITCH.

IF THE FAN SWITCH AT THE THERMOSTAT IS ENERGIZED WHILE THE FURNACE IS HEATING, THE FAN WILL MERELY SWITCH FROM LOW TO HIGH. THIS RELAY PROTECTS THE MOTOR FROM TRYING TO OPERATE AT 2 SPEEDS AT ONCE.

Figure 17–28 Diagram showing how a multispeed PSC motor is wired to operate at low speed (lower air volume) in the winter and high speed (higher air volume) in the summer.

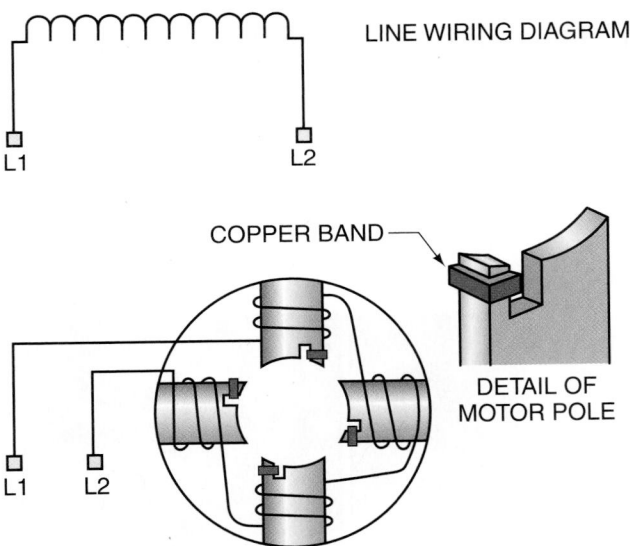

Figure 17–29 Wiring diagram of a shaded-pole motor.

17.16 THREE-PHASE MOTORS

Three-phase motors have some characteristics that make them popular for many applications from about 1 hp into the thousands of horsepower. These motors are very efficient and

Figure 17–30   A shaded-pole motor. *Photo by Bill Johnson*

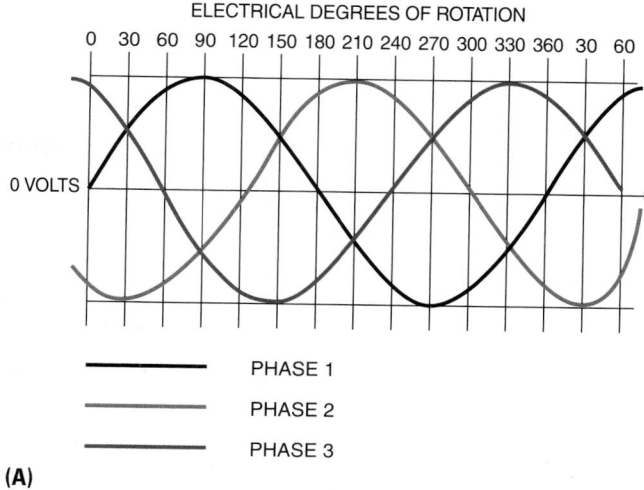

(A)

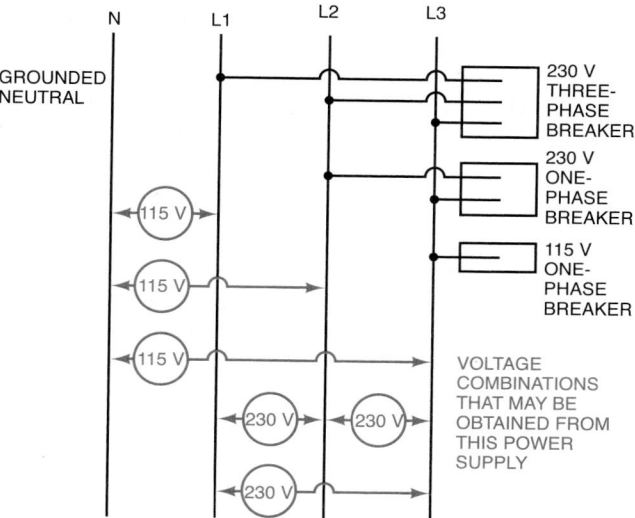

Figure 17–31   Wiring diagram of a three-phase power supply.

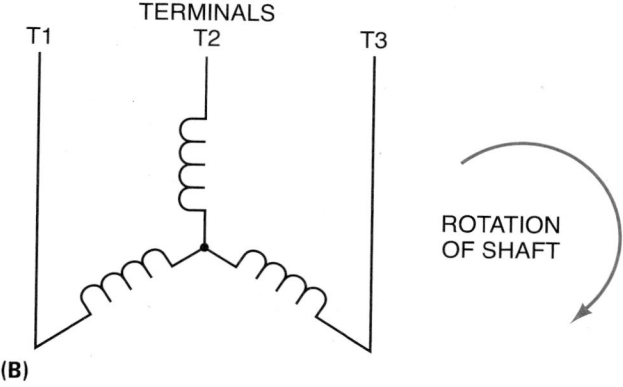

(B)

Figure 17–32   **(A)** Diagram of a three-phase power supply.
(B) Diagram of a typical, single-speed, three-phase motor.

require no start assist for high-torque applications. Three-phase motors are used mainly on commercial equipment. The building power supply must have three-phase power available. (Three-phase power is seldom found in a home.) Three-phase motors have no starting windings or capacitors. They can be thought of as having three single-phase power supplies, **Figure 17–31.** Each of the phases can have either two or four poles. A 3600-rpm motor will have three sets, each with two poles (total of six), and an 1800-rpm motor will have three sets, each with four poles (total of 12). Each phase changes the direction of current flow at different times but always in the same order. A three-phase motor has high starting torque because of the three phases of current that operate the motor. At any given part of the rotation of the motor, one of the windings is in position for high torque. This makes starting large fans and compressors very easy, **Figure 17–32.**

Three-phase power supplies are made up of three single-phase power supplies that are generated 120 electrical

degrees out of phase with each other. As can be seen in **Figure 17–32(A),** phase number 1, indicated by the black sine wave, starts at "0" electrical degrees of rotation and has a potential of zero volts. At 90 electrical degrees, phase number 1 has reached its maximum potential, and at 360 electrical degrees it has completed a full cycle. It can be seen that phases 2 and 3 start at 120 and 240 electrical degrees of rotation, respectively. All three of these phases are 120 degrees out of phase with each other. It should be noted that 120 degrees after the start of phase number 3 (which is at 360 electrical degrees), phase number 1 is ready to start another cycle.

The three-phase motor rpm also slips to about 1750 rpm and 3450 rpm when under full load. The motor is not normally available with dual speed; it is either an 1800-rpm or a 3600-rpm motor.

The rotation of a three-phase motor may be changed by switching any two motor leads, **Figure 17–33.** This rotation must be carefully observed when three-phase fans are used. If a fan rotates in the wrong direction, it will move only about half as much air. If this occurs, reverse the motor leads, and the fan will turn in the correct direction.

Motors have other characteristics that must be considered when selecting a motor for a particular application: for

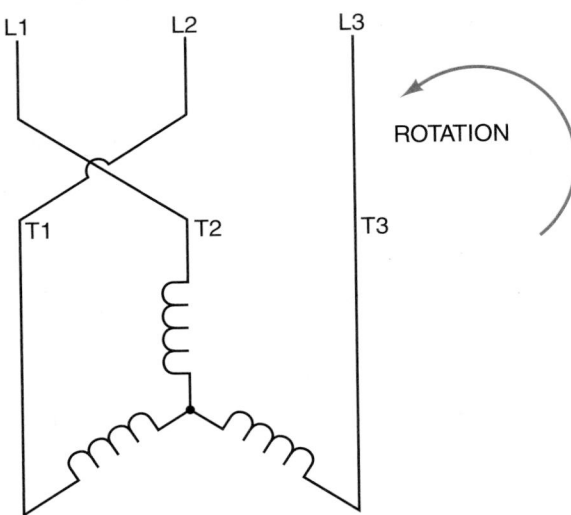

Figure 17–33 A wiring diagram of a three-phase motor. The direction of rotation of the motor can be reversed by switching any two power leads.

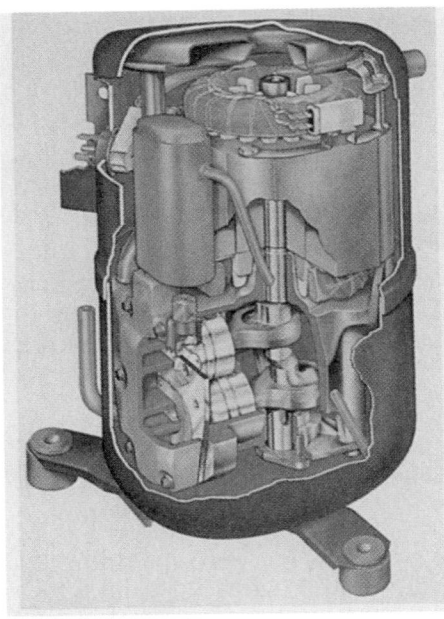

Figure 17–34 A typical motor for a hermetic compressor. *Courtesy Tecumseh Products Company*

example, the motor mounting. Is the motor solidly mounted to a base, or is there a flexible mount to minimize noise?

The sound level is another factor. Will the motor be used where ball bearings would make too much noise? If so, sleeve bearings should be used.

Still another factor is the operating temperature of the motor surroundings. A condenser fan motor that pulls the air over the condenser coil and then over the motor requires a motor that will operate in a warmer atmosphere. The best advice is to replace any motor with an exact replacement whenever possible.

17.17 SINGLE-PHASE HERMETIC MOTORS

The wiring in a single-phase hermetic motor is similar to that in a split-phase motor. It has start and run windings, each with a different resistance. The motor runs with the run winding and may have a start assist to open the circuit to the start windings. A run capacitor is often used to improve running efficiency. A hermetic motor is designed to operate in a refrigerant, usually vapor, atmosphere. It is undesirable for liquid refrigerant to enter the shell—as, for example, by an overcharge of refrigerant. Single-phase hermetic compressors usually are manufactured up to 5 hp, **Figure 17–34.** If more capacity is needed, multiple systems or larger three-phase units are used.

Hermetic compressor motor materials must be compatible with the refrigerant and oil circulating in the system. The coatings on the windings, the materials used to tie the motor windings, and the papers used as wedges must be of the correct material. The motor is assembled in a dry, clean atmosphere.

Hermetic motors are started in much the same way as the other motors described. The start windings must be removed from the circuit when the motor nears normal operation. The

start windings are not removed from the circuit in the same way as for an open motor because the windings are in a refrigerant atmosphere. Open single-phase motors are operated in air, and a spark is allowed when the start winding is disconnected. This cannot be allowed in a hermetic motor because the spark will deteriorate the refrigerant. Special devices determine when the compressor motor is drawing the desired current or running at the correct speed to disconnect the start winding from outside the compressor shell.

Because the hermetic motor is enclosed in refrigerant, the motor leads must pass through the compressor shell to the outside. A terminal box on the outside houses the three motor terminals, **Figure 17–35,** one for the run winding, one for the start winding, and one for the line common to the run and

Figure 17–35 The motor terminal box on the outside of the compressor shell. *Photo by Bill Johnson*

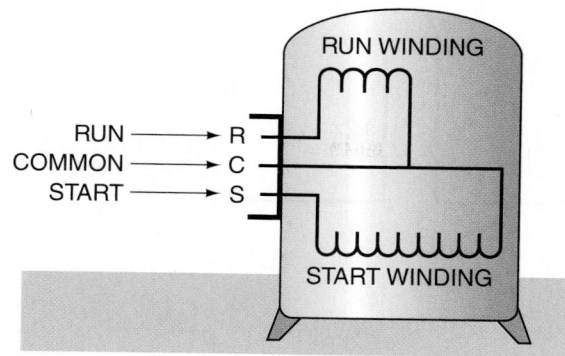

Figure 17–36 A wiring diagram of the terminal and internal wiring connections on a single-phase compressor.

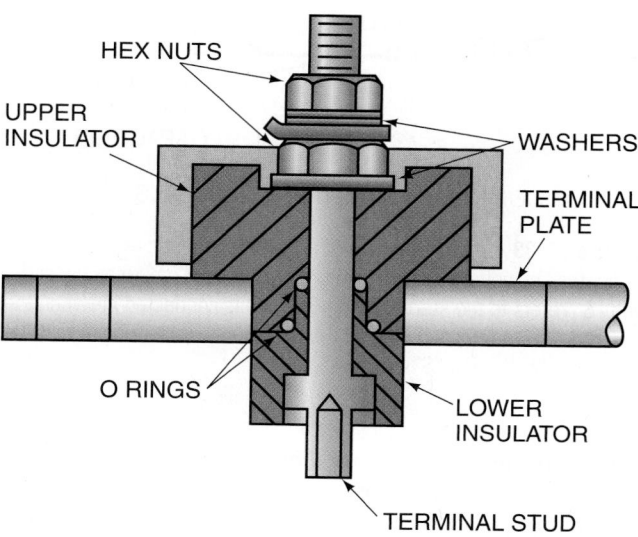

Figure 17–37 These motor terminals use neoprene O rings as the insulator between the terminals and the compressor housing. *Courtesy Trane Company*

start windings. See **Figure 17–36** for a wiring diagram of a three-terminal compressor. The start winding has more resistance than the run winding.

The motor leads are insulated from the steel compressor shell. Neoprene was the most popular insulating material for years. However, if the motor terminal becomes too hot, due to a loose connection, the neoprene may eventually become brittle and possibly leak, **Figure 17–37**. Many compressors now use a ceramic material to insulate the motor leads.

17.18   THE POTENTIAL RELAY

Potential relays or "voltage" relays are used with single-phase capacitor-start, capacitor-run motors that need relatively high starting torque. Their main function is to assist in starting the motor. These relays are also called potential magnetic relays, or simply PMRs.

Potential relays consist of a high-resistance coil and a set of normally closed contacts. The coil is wired between terminals 2 and 5, and the normally closed contacts are wired

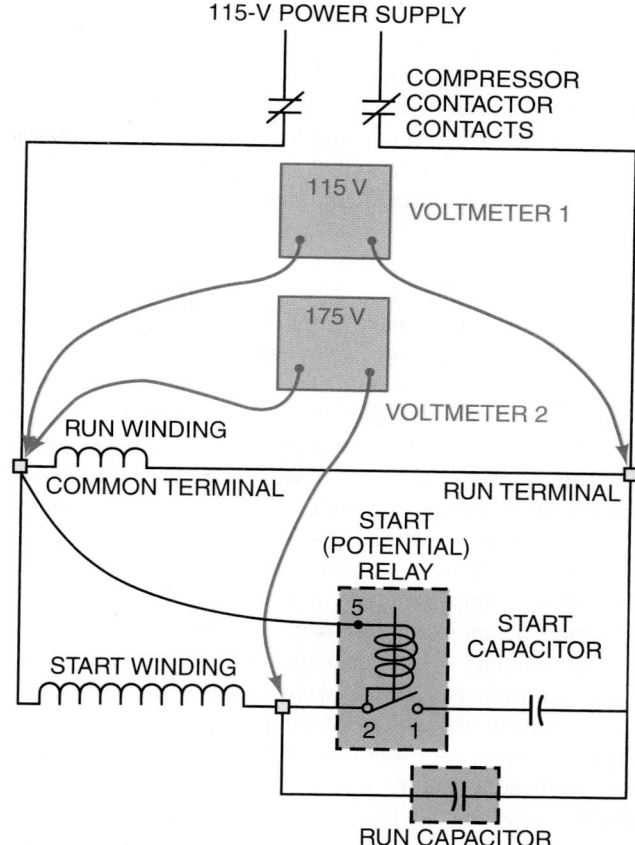

Figure 17–38 Diagram illustrating the higher induced voltage that is measured across the start winding in a typical motor.

between terminals 1 and 2, **Figure 17–38**. Other terminal designations on the relay act only as wire connections and are sometimes referred to as accommodation terminals. These dummy terminals are often used for wiring such devices as condenser fans or capacitors.

When power is applied through the compressor's contactor contacts, both the run and start windings are energized, and the motor's rotor starts to turn. The large metal mass of the motor's rotor turning at high speeds with motor windings in close proximity has a voltage-generating effect. The generated voltage has more of an effect across the start winding because it is wound with a longer and thinner wire than the run winding. This generated voltage is often referred to as induced voltage, or *back electromotive force (BEMF)*. This BEMF opposes line voltage and can be measured with a voltmeter across the start winding. This BEMF is higher than line voltage and can range over 400 V AC with a 230-V circuit, **Figure 17–38**. All motors have different magnitudes of BEMF.

The BEMF generated across the start winding causes a small current to flow in the start winding and potential relay coil because they are in the same circuit. When the BEMF has built up to a high enough value—referred to as "pickup voltage"—the contacts between terminals 1 and 2 will be picked up or "opened." This will take the start capacitor out of the circuit. The start winding is still in the BEMF circuit keeping the relay coil energized while the motor is running at full speed. Pickup voltage usually occurs at about 3/4 motor speed. The run capacitor causes a small phase shift between the run and start winding for better running torque. It also limits the current through the start winding so overheating does not occur during the running cycle.

When the compressor contactor contacts open, line voltage is taken away from the motor. The motor rotor now decreases in speed; thus the BEMF generated across the start winding decreases in magnitude. The relay coil now sees a decreased BEMF and no longer can generate enough magnetism in its iron core to keep the contacts between terminals 1 and 2 open. The contacts thus return to their originally closed position by spring power as the motor coasts to a stop. The voltage where the contacts return to their original closed position is called the "drop-out voltage." One benefit of using the PMR as a starting relay is that the start winding can be brought back into the active electric circuit if the motor should encounter a temporary load increase. If, during motor operation, the motor slows down, the induced voltage across the start winding will drop and the normally closed contacts between terminals 1 and 2 will close. This will bring the start capacitor back into the circuit and provide additional torque to get the motor back up to speed. Once the induced voltage rises (indicating that the motor is operating at a satisfactory speed), the contacts will open to remove the start capacitor from the circuit.

Because these motors generate different BEMF values, service technicians must realize that potential relays must be sized to each individual compressor. The pickup voltage needed to open the contacts is one of the considerations for sizing the relay. The actual current-carrying capacity of the relay contacts, the drop-out voltage, and the continuous coil voltage are other important considerations when sizing the relay. Consult with a service manual, supply house, or the compressor manufacturer on selecting the correct potential relay for a specific compressor.

17.19 THE CURRENT RELAY

Current relays are used on single-phase, fractional horsepower motors requiring low starting torque. Their main function is to assist in starting the motor. Current relays, also referred to as current magnetic relays, or CMRs, are often seen when capillary tubes or fixed orifices are used as metering devices, because systems employing capillary tubes or fixed orifice metering devices equalize pressure during their off cycles. This causes a lower starting torque condition than systems that do not equalize pressure during their off cycles, as with the conventional thermostatic-expansion or automatic-expansion metering devices. Some typical applications for current relays include domestic refrigeration compressors, drinking fountains, small window air conditioners, small ice makers, and small unitary supermarket display cases.

Current starting relays consist of a low-resistance coil (1 ohm or less) and a set of normally open contacts. The coil is short in length and large in diameter. In fact, the current

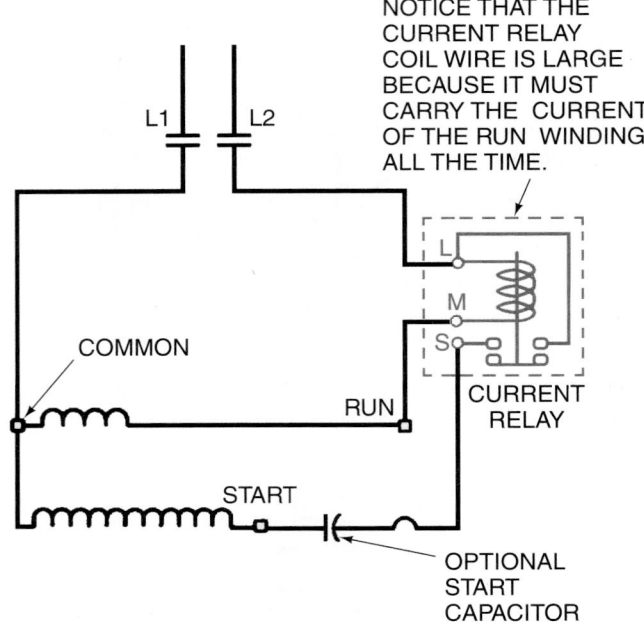

NOTICE THAT THE CURRENT RELAY COIL WIRE IS LARGE BECAUSE IT MUST CARRY THE CURRENT OF THE RUN WINDING ALL THE TIME.

□ COMPRESSOR MOTOR TERMINALS

Figure 17–39 Wiring diagram of a current magnetic relay. The "L" indicates line voltage, the "M" refers the main, or run, winding, and the "S" refers to the start winding. The coil is connected between the "L" and "M" terminals, whereas the relay contacts are connected between the "L" and "S" terminals.

relay may always be identified by the size of the wire in the relay coil. This wire is large because it must carry the full-load current of the motor. The coil is wired between terminals L and M, and the contacts usually are wired between terminals L and S, **Figure 17–39**. The standard designation is L for line, S for start winding, and M for main winding.

When power is applied, both the run (main) winding and relay coil are at LRA. This happens because the relay coil is in series with the run winding. For this reason the relay coil is designed to be of very small resistance; otherwise it would drop too much voltage and interfere with the run winding's power consumption needs. The start winding will not experience the LRA because of the contacts between L and S being normally open. Once the relay coil experiences LRA, a strong electromagnetic field is generated around it. This will make an electromagnet out of the iron core that the coil surrounds. This magnetic force will close the contacts between L and S and energize the start winding. The motor's rotor then starts to turn, **Figure 17–40**.

Once the start winding circuit is closed, the motor will quickly accelerate in speed. After the motor starts, the current draw of the run winding will decrease because of a BEMF generated in its winding. It is this reduced current draw that will decrease the magnetism in the iron core of the relay. Now spring pressure or gravity forces the contacts between L and S back to their normally open position because the magnetic field generated by the lower FLA/RLA is not strong enough to keep the relay contacts closed. When power

Figure 17–40 The current magnetic relay is easily identified by the large wire used to wind the holding coil. *Photo by Bill Johnson*

is taken away from the circuit, the motor will coast to a stop, and the contacts will already be open and waiting for the next run cycle.

These starting methods are used on many compressors with split-phase motors that need high starting torque. If a system has a capillary tube metering device or a fixed-bore orifice metering device, the pressures will equalize during the off cycle, and a high starting torque compressor may not be necessary.

17.20 POSITIVE TEMPERATURE COEFFICIENT START DEVICE

The PSC motor may not need any start assistance when conditions are well within the design parameters. If it does need start assistance, a potential relay and start capacitor may be added to provide additional torque, or a positive temperature coefficient (PTC) device may be added. The PTC is a thermistor that has no resistance to current flow when the unit is off. Remember, a thermistor changes resistance with a change in temperature. When the unit is started, the current flow through the PTC causes it to heat very fast and create a high resistance in its circuit. This changes the phase angle of the start windings. It will not give a motor the starting torque that a start capacitor will, but it is advantageous because it has no moving parts. The PTC is wired in parallel with the run capacitor and acts like a short across the run capacitor during starting. This provides full-line voltage to the start windings during starting. **Figure 17–41(A)** shows how a PTC device works in a circuit. **Figure 17–41(B)** shows a PTC device.

17.21 TWO-SPEED COMPRESSOR MOTORS

Two-speed compressor motors are used by some manufacturers to control the capacity required from small compressors. For example, a residence or small office building may have a 5-ton air-conditioning load at the peak of the season

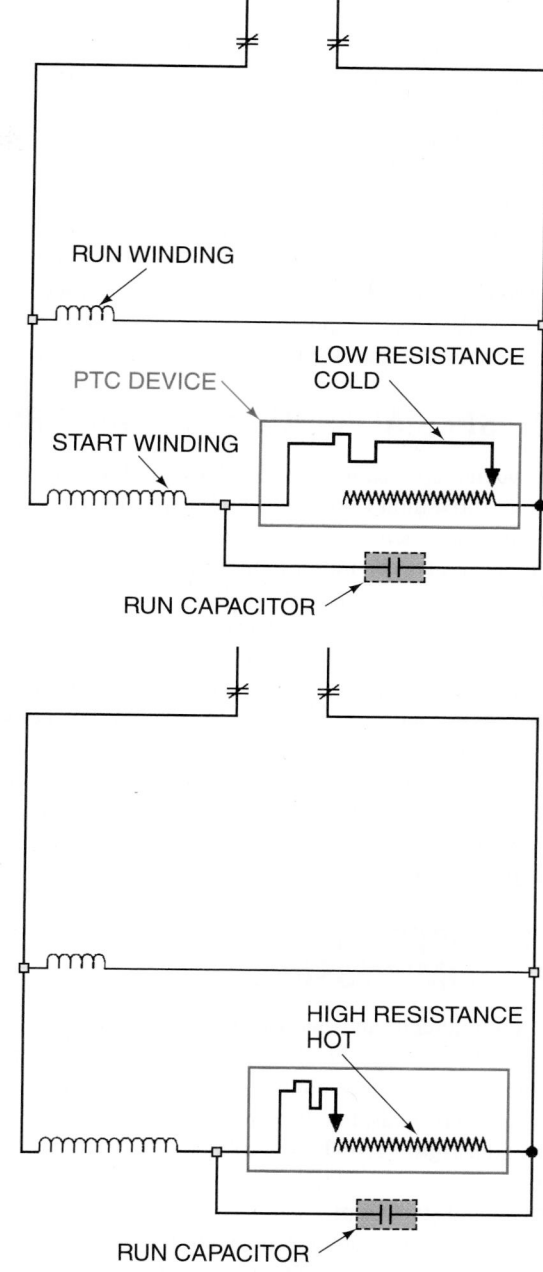

(A)

(B)

Figure 17–41 A positive temperature coefficient (PTC) device. *Photo by Bill Johnson*

and a 2 1/2-ton load as a minimum. Capacity control is desirable for this application. A two-speed compressor may be used to accomplish capacity control. Two-speed operation is obtained by wiring the compressor motor to operate as a two-pole motor or a four-pole motor. The automatic changeover is accomplished with the space temperature thermostat and the proper compressor contactor for the proper speed. For all practical purposes, this can be considered two motors in one compressor housing. One motor turns at 1800 rpm, the other at 3600 rpm. The compressor uses either motor, based on the capacity needs. This compressor has more than three motor terminals to operate the two motors in the compressor.

17.22 SPECIAL APPLICATION MOTORS

Some special application single-phase motors may have more than three motor terminals and not be two-speed motors. Some manufacturers design an auxiliary winding in the compressor to give the motor more efficiency. These motors are normally in the 5-hp and smaller range. Other special motors may have the winding thermostat wired through the shell with extra terminals. A large three-phase compressor motor may have a winding thermostat for each winding. This calls for three thermostats within the compressor housing. Using a combination of four small terminals, they can be wired in series where any one of them will shut the compressor off. If one of them fails, the extra terminals would allow the technician to still use the others for some motor protection, **Figure 17–42.**

17.23 THREE-PHASE MOTOR COMPRESSORS

Large commercial and industrial installations will have three-phase power for the air-conditioning and refrigeration equipment. Three-phase compressor motors normally have three motor terminals, but the resistance across each winding is the same, **Figure 17–43.** As explained earlier, three-phase motors have a high starting torque and, consequently, should experience no starting problems.

Welded hermetic compressors were limited to 7 1/2 tons for many years but are now being manufactured in sizes up to about 50 tons. The larger welded hermetic compressors are traded to the manufacturer when they fail and are remanufactured. These must be cut open for service, **Figure 17–44.**

Serviceable hermetic compressors of the reciprocating type are manufactured in sizes up to about 125 tons, **Figure 17–45.** These compressors may have dual-voltage motors for 208-V to 230-V or 460-V operation, **Figure 17–46.** These compressors are normally rebuilt or remanufactured when the motor fails, so an overhaul is considered when a motor fails. The compressors may be rebuilt in the field or traded for remanufactured compressors. Most large metropolitan areas will have companies that can rebuild the compressor to the proper specifications. The difference in rebuilding and remanufacturing is that one is done by an independent rebuilder and the other by the original manufacturer or an authorized rebuilder.

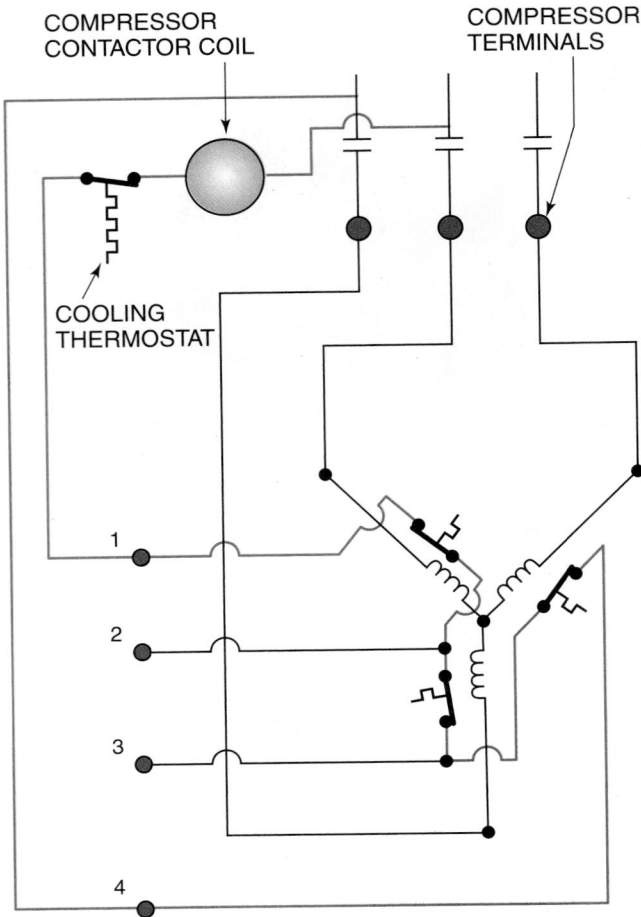

COMPRESSOR WINDING THERMOSTAT TERMINALS. THESE ARE SMALL COMPARED WITH THE MOTOR TERMINALS.

Figure 17–42 This compressor has extra winding thermostat terminals. In the event that one winding thermostat goes bad, the others can still be used.

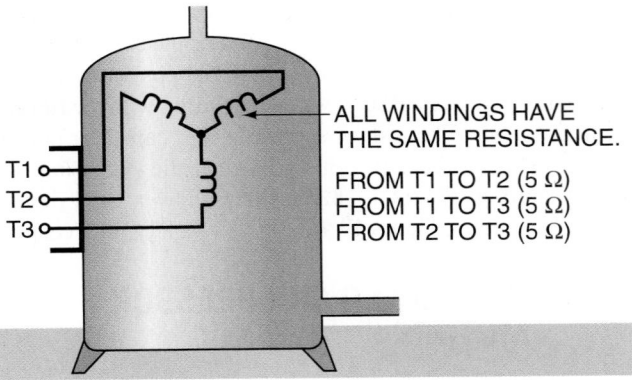

Figure 17–43 A three-phase compressor motor with three leads for the three windings. Unlike a single-phase motor, the resistance of all the three windings is the same.

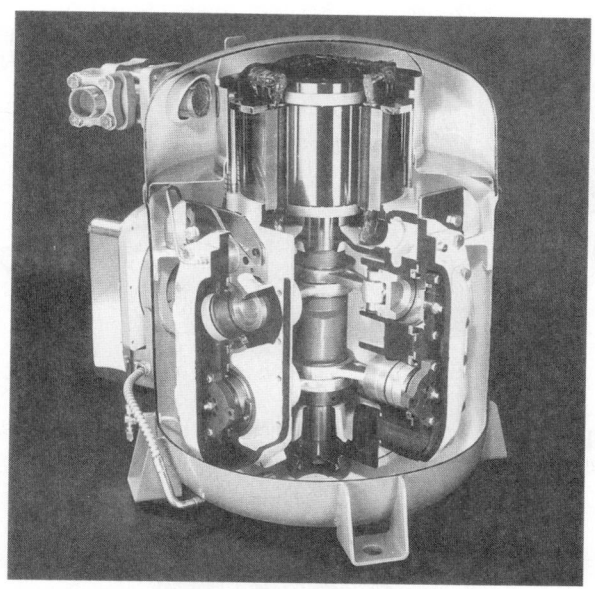

Figure 17–44 A large welded hermetic compressor. *Courtesy Trane Company*

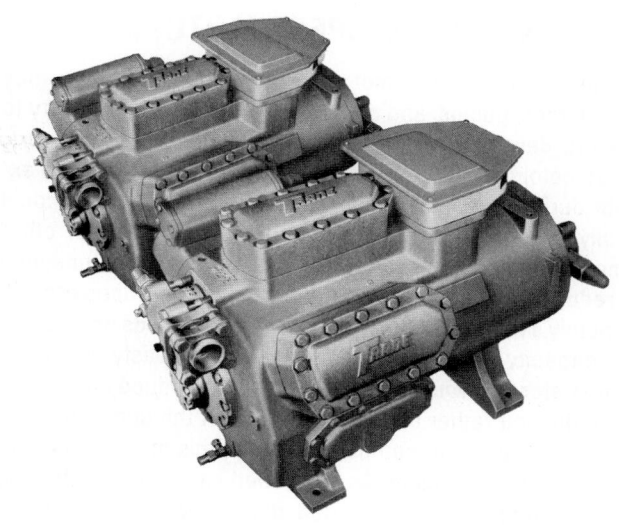

Figure 17–45 Serviceable compressors. *Courtesy Trane Company*

MOTOR TERMINAL BOX
1
2
3
4
5
6
7
8
9
460-V THREE-PHASE
L1
L2
L3
208/230-V THREE-PHASE
TWO CONTACTORS
½-SEC. TIME DELAY
THE MOTOR WINDINGS ARE IN SERIES
AT 460 V.

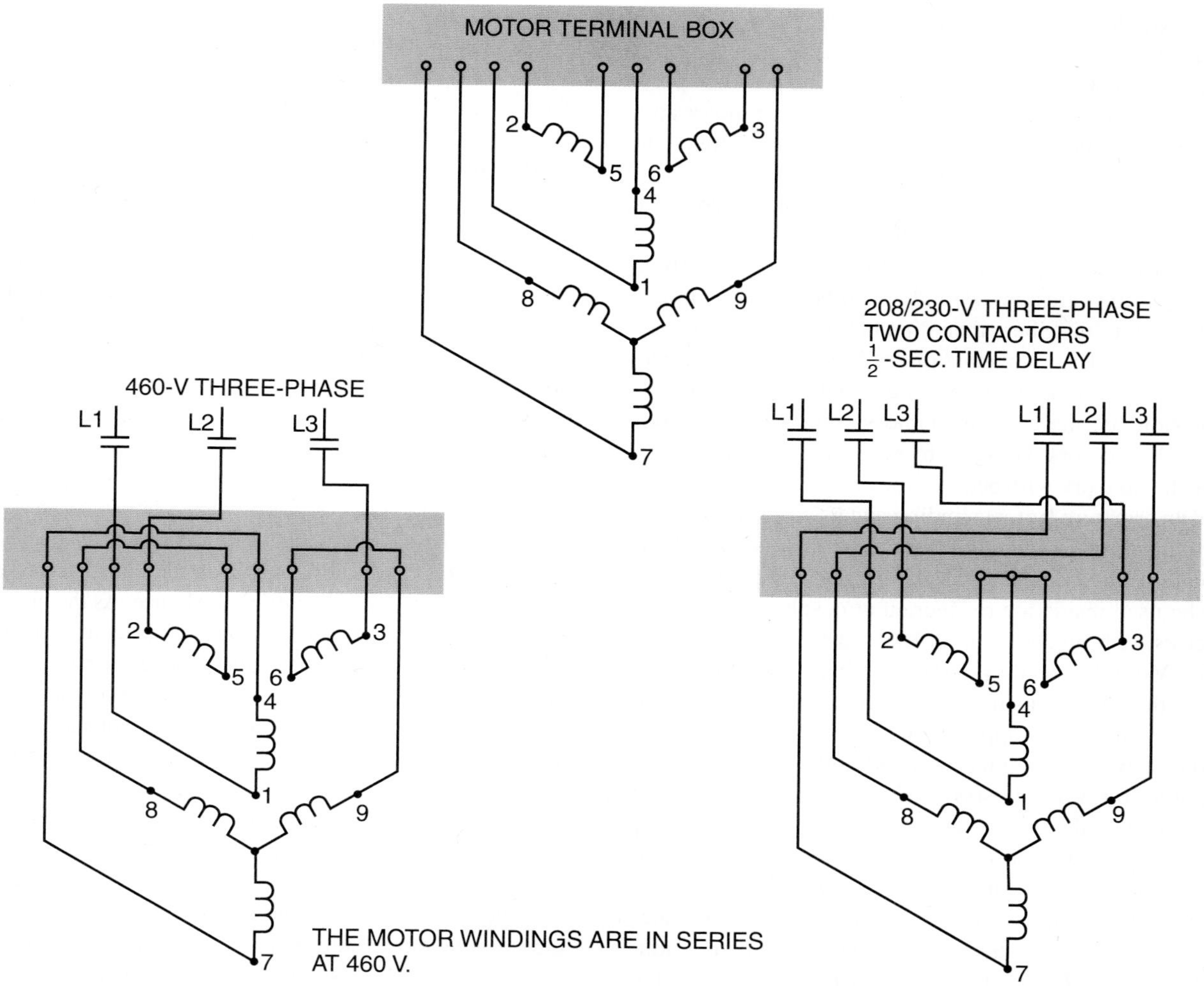

Figure 17–46 A dual-voltage compressor motor wiring diagram. This motor may be used for either 208 V/230 V or 460 V.

17.24 VARIABLE-SPEED MOTORS

🌐 The desire to control motors to provide a greater efficiency for the fans, pumps, and compressors has led the industry to explore development of and use of variable-speed motors. Most motors do not need to operate at full speed and load except during the peak temperature of the season and could easily satisfy the heating or air-conditioning load at other times by operating at a slower speed. When the motor speed is reduced, the power to operate the motor reduces proportionately. For example, if a home or building needs only 50% of the capacity of the air-conditioning unit to satisfy the space temperature, it would be advantageous to reduce the capacity of the unit rather than stop and restart the unit. When the power consumption can be reduced in this manner, the unit becomes more efficient. As mentioned earlier in this chapter, when motors are initially started, they draw locked-rotor amperage, which can be five to seven times greater than the full-load or rated-load amperage of the motor. By limiting the number of times the motor starts, the rate of power consumption is reduced. In addition, having (for example) a compressor motor operate at a lower speed for a longer period of time will result in more even cooling of an occupied space. 🌐

The frequency (cycles per second) of the power supply and the number of poles determine the speed of a conventional motor. New motors are being used that can operate at different speeds with the use of electronic circuits. Several methods are used to vary the frequency of the power supply depending on the type of motor. The compressor motor and the fan motors may be controlled through any number of speed combinations based on the needs of the system.

Unit 12, "Basic Electricity and Magnetism," covered some of the fundamentals of AC electricity. All of the electricity furnished in the United States is AC and the current frequency is 60 cycles per second, or 60 hertz. AC is much more efficient to distribute than direct current (DC). AC can be converted to DC by means of rectifiers. Many applications for motors in the industry need variable-speed motors, and traditionally, the only available variable-speed motors have been DC motors. These motors are typically much more complicated to work with because they have both a field (stator) and an armature (rotor) winding and use brushes to carry power to the rotor. Brushes are carbon connectors that rub on the armature and create an arc. The brushes on the armature cannot be used inside the refrigerant atmosphere; therefore, DC motors with brushes cannot be used for hermetic compressors. Variable-speed systems have been used in our industry in the past by using open-drive DC motors, but the hermetic compressor application is the most common method of driving a compressor. In addition, there are many applications for variable-speed drive, such as pumps and fans and fractional horsepower motors.

The two types of motors that may be found in equipment today are the squirrel cage induction motor and the ECM (electronically commutated) DC motor. Instead of brushes rubbing on an armature, the motor is electronically commutated, **Figure 17–47**. This motor would be applied to a fan application as it is a fractional horsepower motor.

Figure 17–47   An electronically commutated motor (ECM). *Courtesy General Electric*

Figure 17–48   This power transistor is basically an electronic switch that does not arc when it changes position and that has the ability to turn on and off very quickly.

Variable-speed AC motor operation can now be accomplished with electronic components. The electronic components can switch power on and off in microseconds without creating an arc; therefore, open-type contacts are not necessary for the switching purposes. The no-arc components have virtually no wear, have a long life cycle, and are reliable. **Figure 17–48** shows a transistor that is an electronic switch with no contacts to arc.

The air-conditioning load on a building varies during the season and during each day. The central air-conditioning system in a house or other building would have many of the same operating characteristics. Let us use a house as an example. Starting at noon, the outside temperature may be 95°F and the system may be required to run at full load all the time to remove heat as fast as it is entering the house. As the house cools off in the evening, the unit may start to cycle off and then back on, based on the space temperature. Remember, every time the motor stops and restarts, there is wear at the contactor contacts and a burden is put on the motor in the form of starting up. Motor inrush current stresses the bearings and windings of the motor. Most motor bearing wear occurs in the first few seconds of start-up because the bearings are not lubricated until the motor is turning. It would be best not to ever turn the motor off and instead just keep it running at a reduced capacity.

When an air conditioner shuts off, there is normally a measurable temperature rise before it starts back up. This is very noticeable in systems that only stop and start. The humidity rises during this period. In the winter, the typical gas

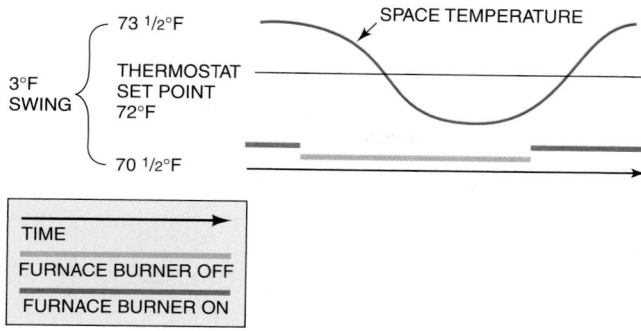

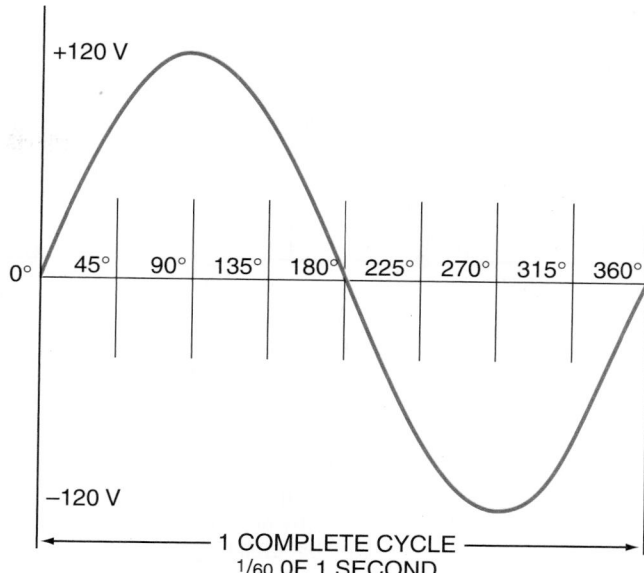

Figure 17–49 This typical ON–OFF furnace thermostat has a 3°F temperature swing during its cycle. This is enough temperature swing to be noticeable to the occupants of the space.

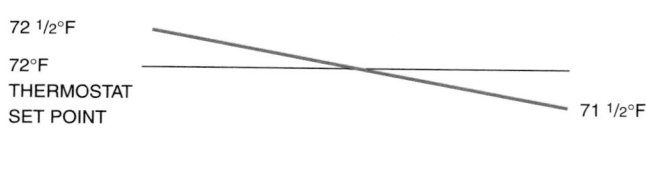

Figure 17–50 This furnace is controlled by a variable-speed fan motor and firing rate. It has only a 1°F temperature swing during its cycle. This will not be noticeable.

Figure 17–51 This sine wave is generated in one-sixtieth (1/60) of a second.

or oil furnace does the same thing. It starts up and runs until the thermostat is satisfied and then it shuts off. There is a measurable temperature rise before it shuts off and a measurable temperature drop before it starts back up. The actual space temperature at the thermostat location may look like the graph in **Figure 17–49.** The temperature graph may look more like the one in **Figure 17–50** when variable-speed motor controls are used along with variable firing rate for a furnace. The same temperature curve profile would be true for the cooling season, a more flat profile with less temperature and humidity variations.

Variable-speed motors can level out the air-conditioner operation by running for longer periods of time. You may notice in any building when the system thermostat is satisfied and the unit shuts off. Suppose we could just keep the unit running at a reduced capacity that matched the building load. If we could gently ramp the motor speed down as the load reduces and then ramp it up as the load increases, the temperature and the humidity would be more constant in the summer. This can be accomplished with modern electronics and variable-speed motor drives.

For example, a light bulb turns on and off 120 times per second when used with 60-cycle current furnished to a typical household. **Figure 17–51** shows a sine wave of the current that is furnished to the home. The voltage goes from 0 V to 120 V to 0 V and back to 120 V 60 times per second; thus it is a 60-cycle current. Current to the light bulb element is interrupted 120 times in that second. The light from the bulb comes from the glowing element. It does not cool off

between the cycles, so you do not notice it. Electronic devices can detect any part of the sine wave and switch off the current at any place in time.

If a switch could be installed to make even more interruptions in the current to the light bulb, it would glow at a lower light output. **Figure 17–52** shows a simple rectifier in the form of a diode installed in the line that will interrupt the current flow to the light bulb on half of the sine wave cycle. The light bulb will now glow at half of its light output. This diode does not draw any current or have any moving parts but interrupts half of the current flow to the light bulb. The light bulb not only uses half power, it also gives off half as much light. The diode is like a check valve in a water circuit; it allows current flow only in one direction. The term *alternating current* tells you that current flows in two directions, one direction in one part of the cycle and the other direction in the other part of the cycle. When the current tries to reverse for the other half of the cycle, it stops it from flowing. It saves power and gives the owner some control over the light output.

AC electric motors can be controlled in a similar manner with electronic circuits. AC motor speed is directly proportional to the cycles per second, hertz. If the cycles per second are varied, the motor speed will vary. The voltage must also be varied in proportion to the cycles per second for the motor to remain efficient at all speeds. Once the voltage is converted to DC and filtered, it then goes through an inverter to change it back to AC that is controllable. Actually, this is still pulsating DC. The reason for all of this is to be able to change the frequency (cycles per second) and voltage at the same time. As the frequency is reduced, the voltage must also be reduced at the same rate. For example, if you have a 3600-rpm, 230-V, 60-cycle-per-second motor and you want to reduce the speed to 1800 rpm, the voltage and the cycles must be reduced in proportion. By reducing the voltage to 115 V and the frequency to 30 cycles per second, the motor will have the

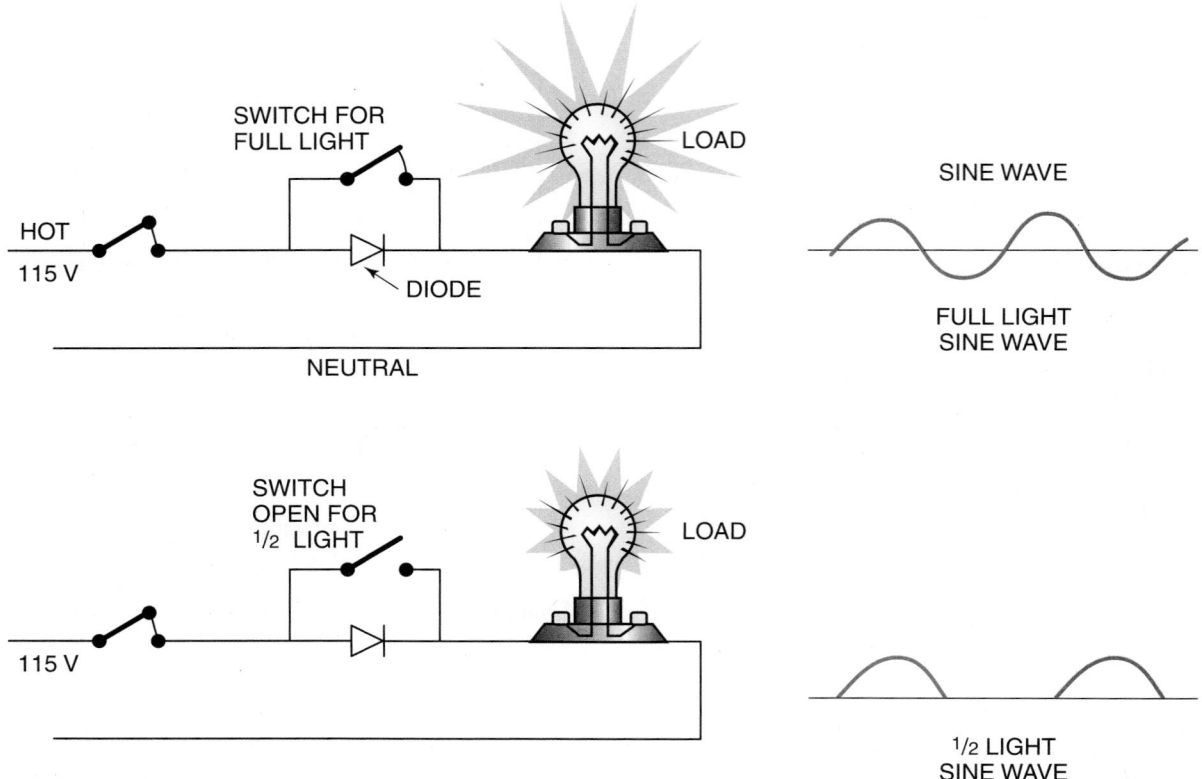

Figure 17–52   This light dimmer uses a diode to cut the voltage in half at low light.

correct power to operate at 1800 rpm. It will not lose its torque characteristics. The motor will also operate on about half the power requirement of the full-load operation. If you reduce only the voltage, the motor will overheat.

AC that is supplied from the power company is very hard to regulate for usage at the motor, so it must be altered to make the process easier and more stable. The process involves changing the incoming AC voltage to DC. This is accomplished with a device called a converter or rectifier. This is much like a battery charger that converts AC to 14 V DC to charge an automobile battery. This DC voltage is actually known as pulsating DC voltage. The DC voltage is then filtered using capacitors to create a more pure DC voltage.

A simplified diagram of a circuit used to accomplish this voltage conversion is shown in **Figure 17–53**. There are many components in an actual system, but the technician should think about the system one component at a time for the purposes of understanding and troubleshooting it.

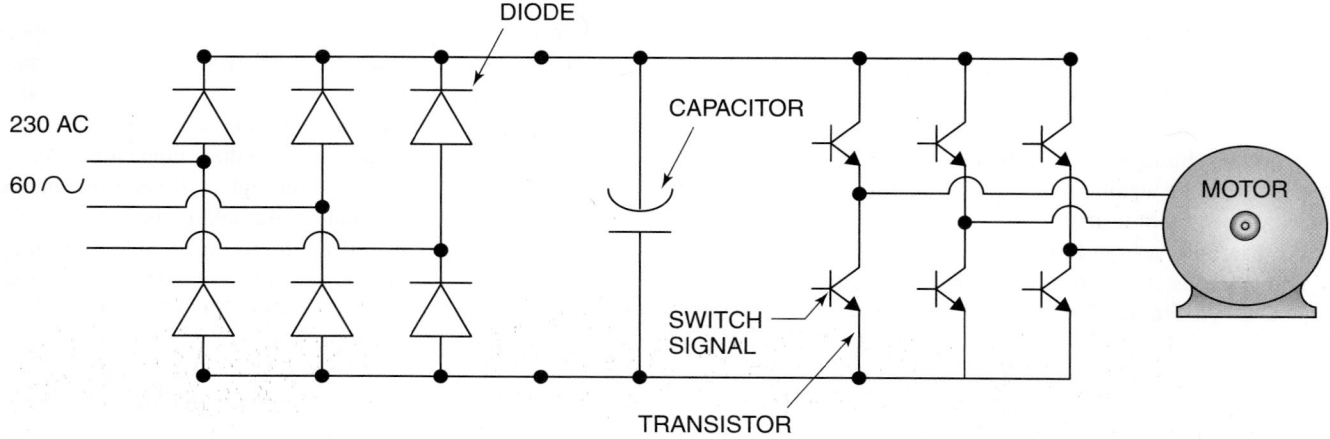

Figure 17–53   A simple diagram of a variable-speed motor drive.

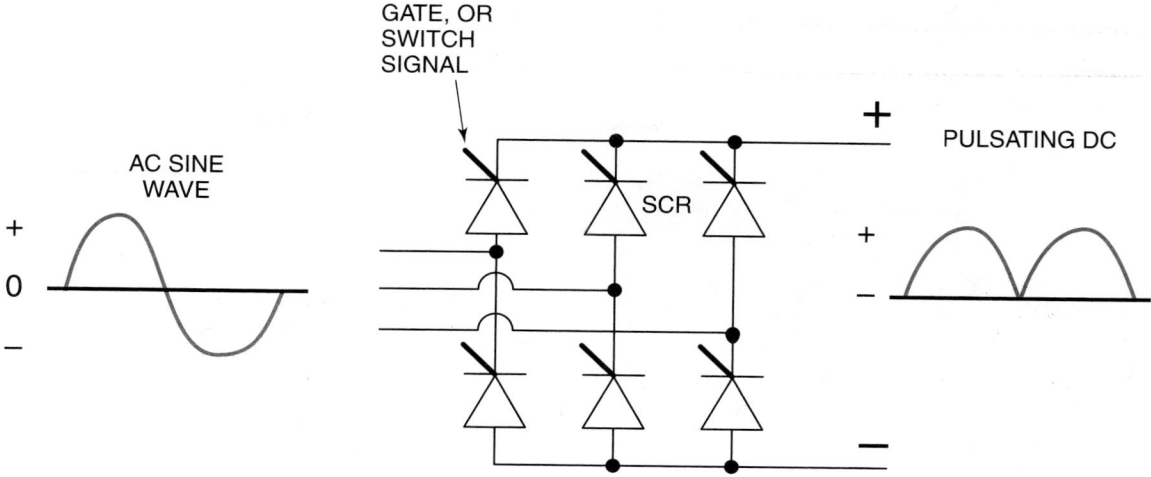

Figure 17–54 The full wave rectifier moves the entire sine wave to the positive side of the graph.

17.25 DC CONVERTERS (RECTIFIERS)

There are two basic types of converters, the *phase-controlled rectifier* and the *diode bridge rectifier.*

The phase-controlled rectifier receives AC from the power company and converts it to variable voltage DC. This is done by using silicon-controlled rectifiers (SCRs) and transistors that can be turned off and back on in microseconds. **Figure 17–54** shows the wave form of AC that enters the phase-controlled rectifier, which is furnished by the power company, and the DC current that leaves the device. Notice the connection for turning these diodes or transistors on and off. The DC voltage leaving this rectifier is varied within the rectifier to coincide with the motor speed. The frequency of the power will be varied to the required motor speed in the inverter, which is between the converter and the motor. Remember, the voltage and the frequency must be changed for an efficient motor speed adjustment.

The other component in the system is a capacitor bank to smooth out the DC voltage. The rectifier turns AC into a pulsating DC voltage that looks like all of the AC voltage is on one side of the sine curve. The voltage looks more like pure DC voltage when it leaves the capacitor bank, **Figure 17–55.** This type of capacitor bank is used for any rectifier to create a better DC profile.

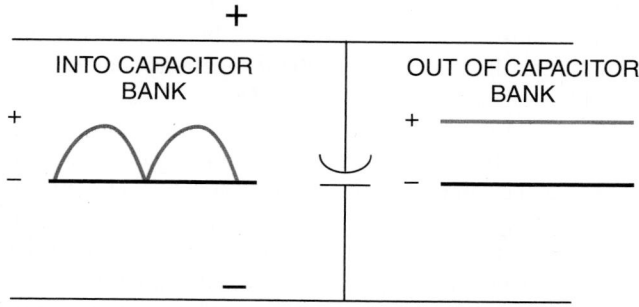

Figure 17–55 Pulsating DC enters the capacitor bank and straight-line DC leaves.

The diode bridge rectifier is a little different in that the DC voltage is not regulated in the rectifier. The diodes used in this rectifier are not controllable. It is a constant pure DC voltage after it has been filtered through the capacitor bank. The voltage and the frequency will be adjusted at the inverter of the system. The diode bridge rectifier has no connection for switching the diodes on and off.

17.26 INVERTERS

Inverters produce the correct frequency to the motor for the desired speed. Conventional motor speeds are controlled by the number of poles, and the frequency is a constant 60 Hz. Inverters can actually control motor speeds down to about 10% of their rated speed at 60 Hz and up to about 120% of their rated speed by adjusting the hertz to above the 60-Hz standard.

There are different types of inverters. A common one is a *six-step* inverter, and there are two variations. One controls voltage and the other controls current. The six-step inverter has six switching components, two for each phase of a three-phase motor. This inverter receives regulated voltage from the converter, such as the phase-controlled inverter, and the frequency is regulated in the inverter.

The voltage-controlled six-step converter has a large capacitor source at the output of the DC bus that maintains the output voltage, **Figure 17–56.** Notice the controllers are transistors that can be switched on and off.

The current-controlled six-step inverter also has the voltage controlled at the input. It uses a large coil, often called a choke, in the DC output bus, **Figure 17–57.** This helps stabilize the current flow in the system.

The *pulse-width modulator* (PWM) inverter receives a fixed DC voltage from the converter, and then it pulses the voltage to the motor. At low speeds, the pulses are short; at high speeds, the pulses are longer. The PWM pulses are sine coded to where they are narrower at the part of the cycle close to the ends. This makes the pulsating signal look more like a sine wave to the motor. **Figure 17–58** shows the signal the motor receives. This motor speed can be controlled very closely.

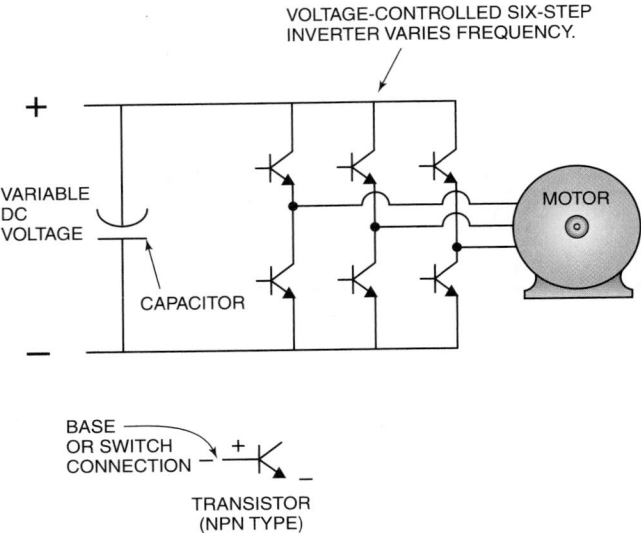

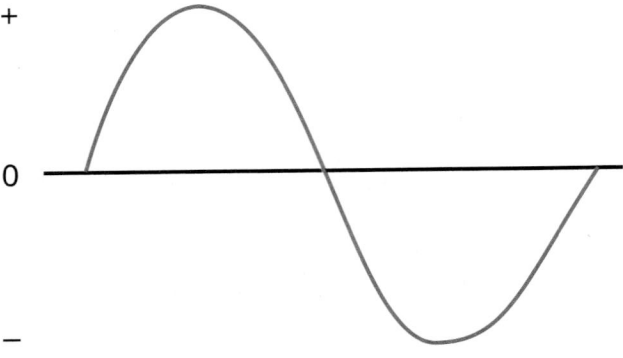

Figure 17–56 When the control system (computer) sends a signal to the base connection on the transistor, the transistor turns on and allows current to flow through the device. When the signal is dropped, the transistor turns off.

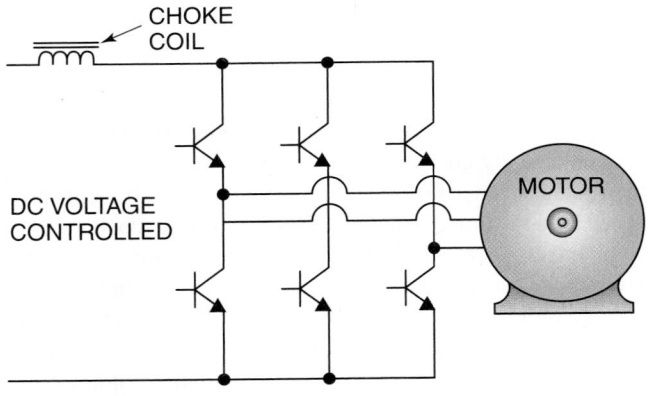

Figure 17–57 The choke coil stabilizes current flow.

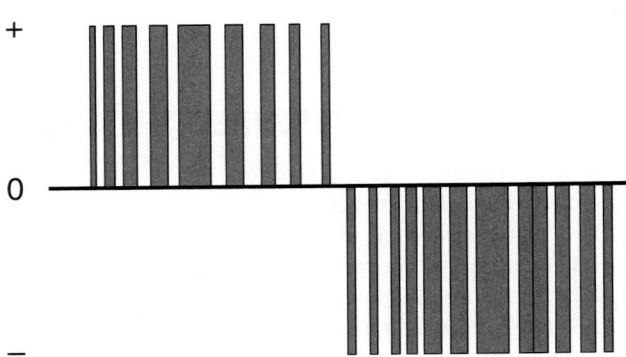

Figure 17–58 Sine-coded, pulse-width modulation.

17.27 ELECTRONICALLY COMMUTATED MOTORS

The ECM is used for applications for open-drive fans in sizes 1 hp and smaller, **Figure 17–47.** It is reliable and has an energy savings that is worth the difference over time. The technology has been developed to electronically commutate a DC motor where there is no need for brushes, as mentioned earlier. The DC motor has always been more efficient than the AC motor, but in the past the armature had to be magnetized with DC power to the armature with the brushes. The armature can be commutated (magnetized) now with permanent magnets that are attached to the armature with reliable high-tech adhesives.

This motor is factory calibrated to suit the piece of equipment it is applied to. A speed, torque, airflow, and external static pressure relationship is programmed into the motor at the factory for that particular application. This can help the air handler supply the correct airflow for different field conditions. For example, when a typical system has restricted air filters or when supply registers are closed, there is a reduced airflow. With the ECM, the motor simply speeds up to accommodate the reduced airflow and stabilizes at the correct airflow. This is accomplished with the relationship of airflow to torque, speed, and external static pressure set up by the manufacturer.

The ECM can also be applied to any system that can furnish the correct 24-V signal to the motor for load demand. This technology has provided variable-speed motors for gas furnaces, air-conditioning systems, oil burner systems, and heat pumps. It is also used in the refrigeration industry for variable loads in refrigerated cases, **Figure 17–59.**

The ECM is actually a two-piece motor, the controls and the motor section. If the technician suspects that the motor is defective, the controls can be removed and the motor may be checked using an ohmmeter in the conventional manner. If the motor shows to be electrically sound, a test module (furnished by the manufacturer) can be fastened to the motor and operated manually at variable speeds, **Figure 17–60.** When the motor will operate correctly with the test module and will not operate without it, the ECM electronic portion of the motor

Figure 17–59 This is a very small motor that may be used for a refrigerated case. *Courtesy General Electric*

Figure 17–60 This analyzer is connected to the motor and control board. The technician can diagnose whether the problem lies within the motor or the board itself. *Courtesy Ferris State University. Photo by John Tomczyk*

can be checked for correct input and tight connections. If the motor cannot be made to operate with the correct signal to the ECM control portion, the ECM controls can be changed and the same motor used. NOTE: Each motor has its own ECM control package, and it is not interchangeable with just any other ECM control package from the same manufacturer. Remember, the control package is programmed for the specific piece of equipment. An exact replacement must be used.

All of these variable-speed motors must have very close control for faults. They all tend to give off a lot of electronic noise, peaks, and valleys to the input supply AC voltage. This is because they are constantly stopping and starting the input current flow to the converter. If a large current flow is stopped in the middle of the entering sine wave cycle, the power furnished by the power company will have a tendency to fluctuate. This will create a spike in voltage. The same thing will happen if a large current load is started, like the inrush current on start-up of a motor. The fluctuations must be

filtered to prevent problems with computers and other electronic equipment in the vicinity. Special precautions are taken by the manufacturer of the equipment and all directions should be followed.

One of the advantages of variable-speed motors is that they can be started at a reduced speed, which will be a reduced inrush current. As discussed earlier, the inrush current is about five times the running current. When a large motor is started in an office building, the inrush current can cause the voltage to the entire building to drop to a lower value. This can cause lights to blink and computers to cycle off if it is extreme. Large motors have methods to soften the start-up. Variable-speed motors have what is called a soft start-up, at a reduced speed. Then the motor speed is ramped up to full load and full current.

The compressor motor is the largest motor in the system. When the load is reduced, the indoor fan and the condenser fan speeds must be reduced in proportion to the load on them. By reducing the indoor fan speed, the humidity in the conditioned space can be controlled. If the indoor fan speed is not reduced when the compressor speed is reduced, the suction pressure will rise and proper humidity removal will not be accomplished. By reducing the indoor fan speed, the suction pressure will be reduced and humidity removal will continue.

The same is true with the condenser fan motor. If the compressor load is reduced, the load on the condenser is reduced. If the condenser fan speed is still operating at full speed, the head pressure will become too low and operating problems may occur.

Each manufacturer will have a troubleshooting procedure that should be followed. As the components become more cost effective, they will be used in smaller equipment. These systems are very efficient. Always check the incoming power supply to any device to make sure it is what it should be before you decide the electronics are the problem. It is possible to have one fuse blown on a three-phase circuit. A quick check with a voltmeter will verify whether the power supply is good.

Some benefits of variable-speed technology are

- power savings.
- load reduction based on demand.
- soft starting of the motor (no LRA).
- better space temperature and humidity control.
- solid-state motor starters without open contacts.
- oversized units for future expansion that can run at part load until the expansion.
- load and capacity matching.

17.28 COOLING ELECTRIC MOTORS

All motors must be cooled because part of the electrical energy input to the motor is given off as heat. Most open motors are air cooled. Hermetic motors may be cooled by air or refrigerant gas, **Figure 17–61.** Small and medium-sized motors are water cooled. An air-cooled motor has fins on the surface to help give the surface more area for dissipating heat. These motors must be located in a moving airstream. To cool properly, refrigerant gas-cooled motors must have an adequate refrigerant charge.

(A)

(B)

Figure 17–61 All motors must be cooled or they will overheat. **(A)** This compressor is cooled by the refrigerant gas that passes over the motor windings. **(B)** This compressor is cooled by air from the fan. The air-cooled motor has fins on the compressor to help dissipate heat. *Courtesy Copeland Corporation*

SUMMARY

- Motors turn fans, compressors, and pumps.
- Some of these applications need high starting torque and good running efficiencies; some need low starting torque with average or good running efficiencies.
- Compressors applied to refrigeration normally require motors with high starting torque.
- Small fans normally need motors that have low starting torque.
- The voltage supplied to a particular installation will determine the motor's voltage. The common voltage for furnace fans is 115 V; 230 V is the common voltage for home air-conditioning systems.
- Common single-phase motors are split-phase, PSC, and shaded-pole.
- When more starting torque is needed, a start capacitor is added to the motor.
- A run capacitor improves the running efficiency of the split-phase motor.
- A centrifugal switch breaks the circuit to the start winding when the motor is up to running speed. The switch changes position with the speed of the motor.
- An electronic switch may be used to interrupt power to the start winding.
- The common rated speed of a single-phase motor is determined by the number of poles or windings in the motor. The common speeds are 1800 rpm, which will slip in speed to about 1725 rpm, and 3600 rpm, which will slip to about 3450 rpm.
- The difference between 1800/3600 and the running speeds of 1750/3450 is known as the slip. Slip is due to the load imposed on the motor while operating.
- Three-phase motors are used for all large applications. They have a high starting torque and a high running efficiency.

- Three-phase power is not available at most residences, so these motors are limited to commercial and industrial installations.
- The power to operate hermetic motors must be conducted through the shell of the compressor by way of insulated motor terminals.
- Since the winding of a hermetic compressor is in the refrigerant atmosphere, a centrifugal switch may not be used to interrupt the power to the start winding.
- A potential relay takes the start winding out of the circuit using BEMF.
- A current relay breaks the circuit to the start winding using the motor's run current.
- The PSC motor is used when high starting torque is not required. It needs no starting device other than the run capacitor.
- The PTC device is used with some PSC motors to give small amounts of starting torque. It has no moving parts.
- When compressors are larger than 5 tons, they are normally three-phase.
- Dual-voltage three-phase compressors are built with two motors wired into the housing.
- Three-phase reciprocating compressors come in sizes up to about 125 tons.
- Variable-speed motors operate at higher efficiencies with varying loads.
- The electronically commutated motor (ECM) is a DC motor that does not have brushes.
- Variable-speed motors can even out the load on the heating and air-conditioning systems by adjusting the equipment speed to the actual load.
- Motor speed can be varied from about 10% of rated speed to about 120% using electronics.
- Electronic switches, silicon-controlled rectifiers (SCRs), and transistors can be turned on and off without making an arc. They are reliable and do not use much power.

- DC converters convert AC power to DC power.
- Capacitors even out the pulsating DC power.
- Inverters create an alternating frequency that can be varied.
- The frequency and the voltage must be changed together for the motor to perform efficiently.

REVIEW QUESTIONS

1. The two popular operating voltages for residences are _____ V and _____ V.
2. When an open motor gets up to speed, the _____ switch takes the start winding out of the circuit.
3. Is the resistance in the start winding **greater** or **less** than the resistance in the run winding?
4. Which of the following devices may be wired into the starting circuit of a motor to improve the starting torque?
   A. Another winding
   B. Start capacitor
   C. Thermostat
   D. Winding thermostat
5. Which of the following devices may be wired into the running circuit of a motor to improve the running efficiency?
   A. A start capacitor
   B. A run capacitor
   C. Both a start and a run capacitor
   D. None of the above
6. Define BEMF.
7. Why is the hermetic compressor motor manufactured from special materials?
8. The two types of motors used for hermetic compressors are _____ and _____.
9. How does power pass through the compressor shell to the motor inside?

10. The two types of relays used to start hermetic compressors are the _____ relay and the _____.
11. A PTC device is used
    A. to start a hermetic compressor.
    B. as a motor protector.
    C. to protect against low oil pressure.
    D. to protect against high oil temperature.
12. True or False: Three-phase motors have low starting torque.
13. True or False: All large motors are three-phase.
14. Why is it desirable to have a two-speed compressor?
15. True or False: All electric motors must be cooled, or they will overheat.
16. True or False: An SCR is an electronic switch.
17. When a transistor is used as a switch
    A. a great deal of power is consumed.
    B. it makes a lot of noise.
    C. it is subject to quick failure.
    D. it does not create an arc.
18. Capacitors are used after the converter to
    A. smooth out the DC current.
    B. suppress the noise.
    C. change the voltage frequency.
    D. reduce the voltage.
19. The converter
    A. changes the frequency of the power.
    B. changes AC to DC.
    C. is a switch.
    D. is used only with DC motors.
20. The inverter
    A. adjusts the frequency.
    B. adjusts the DC voltage.
    C. smoothes out the AC power.
    D. makes a lot of noise.

UNIT 18 Application of Motors

OBJECTIVES

After studying this unit, you should be able to

- identify the proper power supply for a motor.
- describe the application of three-phase versus single-phase motors.
- describe other motor applications.
- explain how the noise level in a motor can be isolated from the conditioned space.
- describe the different types of motor mounts.
- identify the various types of motor drive mechanisms.

SAFETY CHECKLIST

✔ Ensure that electric motor ground straps are connected properly when appropriate so that the motor will be grounded.

✔ Never touch a motor drive belt when it is moving. Be sure that your fingers do not get between the belt and the pulley.

18.1 MOTOR APPLICATIONS

Because electric motors perform so many different functions, choosing the proper motor is necessary for safe and effective performance. Usually the manufacturer or design engineer for a particular job chooses the motor for each piece of equipment. However, as a technician you often will need to substitute a motor when an exact replacement is not available, so you should understand the reasons for choosing a particular motor for a job. For example, when a fan motor burns out in the air-conditioning condensing unit, the correct motor must be obtained, or another failure may occur. NOTE: In an air-cooled condenser, the air is normally pulled through the hot condenser coil and passed over the fan motor. This hot air is used to cool the motor. You must be aware of this, or you may install the wrong motor. The motor must be able to withstand the operating temperatures of the condenser air, which may be as high as 130°F, **Figure 18–1**.

Open motors are discussed in this unit because they are the only ones from which a technician has a choice of selection. Some design differences that influence the application are as follows:

- Power supply
- Work requirements
- Motor insulation type or class
- Bearing types

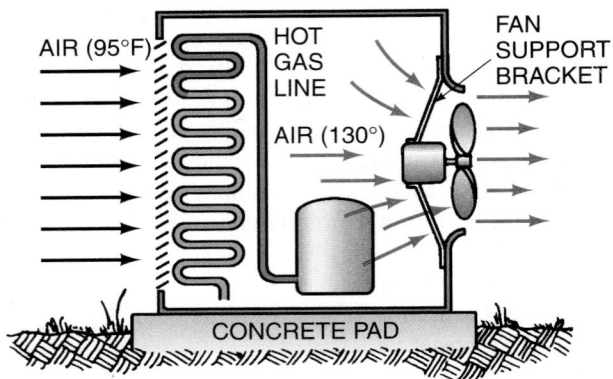

Figure 18–1 This motor is operating in the hot airstream after the air has been pulled through the condenser.

- Mounting characteristics
- Cooling requirements

18.2 THE POWER SUPPLY

The power supply must provide the correct voltage and sufficient current. For example, the power supply in a small shop building may be capable of operating a 5-hp air compressor. But suppose air conditioning is desired. If the air-conditioning contractor prices the job expecting the electrician to use the existing power supply, the electrical service for the whole building may have to be changed. The motor equipment nameplate information and the manufacturer's catalogs provide the needed information for the additional service, but someone must put the whole project together. The installing air-conditioning contractor may have that responsibility. See **Figure 18–2** for a typical

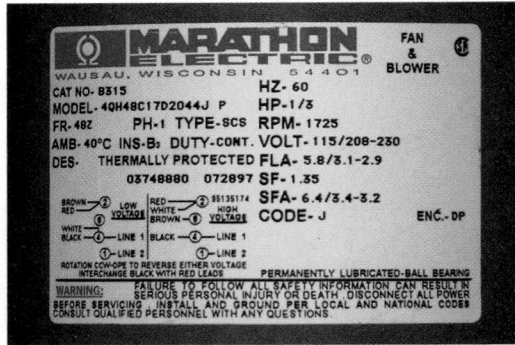

Figure 18–2 A motor nameplate. *Photo by Bill Johnson*

motor nameplate and **Figure 18–3** for a part of a page from a manufacturer's catalog with the electrical data included. **Figure 18–4** is an example of a typical electrical panel rating.

The power supply data contains the following:

1. The voltage (115 V, 208 V, 230 V, 460 V)
2. The current capacity in amperes

TYPE CS • NEMA SERVICE FACTOR • 60 HERTZ • CUSHION BASE					
HP	RPM	Bearings	Overload Protector	Full Load ·(12) Amps	Frame
$\frac{1}{6}$	1800	Sleeve	Auto	4.4	K48
$\frac{1}{4}$	1800	Sleeve	Auto	2.7	K48
		Ball	Auto	2.7	K48
$\frac{1}{3}$	3600	Sleeve	Auto	3.1	L48
	1800	Sleeve	Auto	2.9	L48
		Sleeve	Auto	2.6	J56
		Ball	No	2.9	L48
		Ball	No	2.6	J56
		Ball	Auto	2.6	J56
	1200	Sleeve	No	3.0	K56
$\frac{1}{2}$	3600	Sleeve	Auto	4.0	L48
	1800	Sleeve	No	3.6	J56
		Sleeve	Auto	3.6	J56
		Ball	No	3.6	J56
		Ball	Auto	3.6	J56
$\frac{3}{4}$	3600	Sleeve	Auto	4.6	J56
	1800	Sleeve	No	5.2	K56
		Sleeve	Auto	5.2	K56
		Ball	No	5.2	K56
		Ball	Auto	5.2	K56
1	3600	Sleeve	Auto	6.0	K56
		Ball	No	6.0	K56
	1800	Sleeve	Auto	6.5	L56
		Ball	No	6.5	L56
		Ball	Auto	6.5	L56
$1\frac{1}{2}$	3600	Ball	Auto	8.0	L56
	1800	Ball	Auto	7.5	M56
2	3600	Ball	No	9.5	L56

Figure 18–3 Part of a page from a manufacturer's catalog. *Courtesy Century Electric, Inc.*

MANUFACTURER / MANUFACTURERO
Cutler-Hammer
125 AMPS MAX.
120/240 VAC / VCA 1PH- 3W ; *1 FASE - 3 HILOS*
208Y/120 VAC / VCA 3W ; *3 HILOS*
SEE CASE SIDEWALL FOR FURTHER INFORMATION ;

Figure 18–4 A nameplate from an electrical panel used for a power supply. *Photo by Bill Johnson*

3. The frequency in hertz or cycles per second (60 cps in the United States and 50 cps in many foreign countries)
4. The phase (single or three phase)

Motors and electrical equipment must fit within the system's total electrical capacity, or failures may occur. Voltage and current are most often the inadequate characteristic encountered. The technician can check them, but it is usually preferable to have a licensed electrician make the final calculations.

Voltage

The voltage of an installation is important because every motor operates within a specified voltage range, usually within ±10%. **Figure 18–5** gives the upper and lower limits of common voltages. If the voltage is too low, the motor will draw a high current. For example, if a motor is designed to operate on 230 V but the supply voltage is really 200 V, the motor's current draw will go up. The motor is trying to do its job, but it lacks the power, and it will overheat, **Figure 18–6**.

If the applied voltage is too high, the motor may develop local hot spots within its windings, but it will not always experience high amperage. The high voltage will actually give

208 VOLT-RATED MOTOR	+10%	228.8 VOLTS
	−10%	187.2 VOLTS
230 VOLT-RATED MOTOR	+10%	253 VOLTS
	−10%	207 VOLTS
208–230 VOLT-RATED MOTOR	+10%	253 VOLTS
	−10%	187.2 VOLTS

Figure 18–5 This table shows the maximum and minimum operating voltages of typical motors.

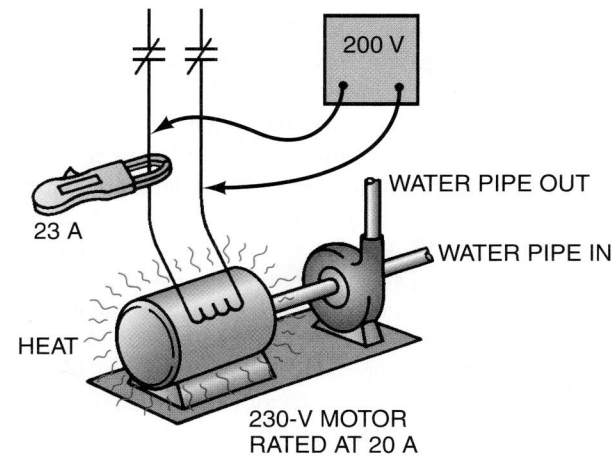

Figure 18–6 A motor operating under a low-voltage condition.

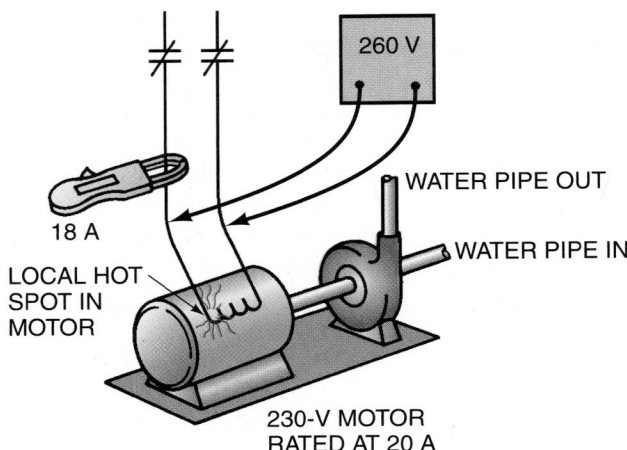

Figure 18–7   A motor rated at 230 V and operating at 260 V.

Motor	Single Phase		3-Phase-Squirrel Cage Induction		
HP	120V	230V	230V	460V	575V
$\frac{1}{6}$	4.4	2.2			
$\frac{1}{4}$	5.8	2.9			
$\frac{1}{3}$	7.2	3.6			
$\frac{1}{2}$	9.8	4.9	2	1.0	0.8
$\frac{3}{4}$	13.8	6.9	2.8	1.4	1.1
1	16	8	3.6	1.8	1.4
$1\frac{1}{2}$	20	10	5.2	2.6	2.1
2	24	12	6.8	3.4	2.7
3	34	17	9.6	4.8	3.9
5	56	28	15.2	7.6	6.1
$7\frac{1}{2}$			22	11.0	9.0
10			28	14.0	11.0

APPROXIMATE FULL-LOAD AMPERAGE VALUES FOR ALTERNATING CURRENT MOTORS

Does not include shaded pole.

Figure 18–8   This chart shows approximate full-load amperage values. *Courtesy BDP Company*

the motor more power than it can use. A 1-hp motor with a voltage rating of 230 V that is operating at 260 V is running above its 10% maximum. This motor may be able to develop 1 1/4 hp at this higher-than-rated voltage, but the windings are not designed to operate at that level. The motor can overheat and eventually burn out if it continually runs overloaded. This can happen *without* drawing excessive current, **Figure 18–7.**

Current Capacity

There are two current ratings for a motor. The full-load amperage (FLA) is the current the motor draws while operating at a full-load condition at the rated voltage. This is also called the run-load amperage or rated-load amperage (RLA). For example, a 1-hp motor will draw approximately 16 A at 115 V or 8 A at 230 V in a single-phase circuit. **Figure 18–8** is a chart that shows approximate amperages for some typical motors.

The other amperage rating that may be given for a motor is the locked-rotor amperage (LRA). The LRA is the amperage the motor draws at start-up before the rotor has started to turn. These two current ratings are available for every motor and are stamped on the motor nameplate for an open motor. By taking amperage readings at start-up and while the motor is operating under load and comparing these readings with those on the motor nameplate, the technician can determine whether the motor is operating within its design parameters. Some compressors do not have both ratings printed. Normally the LRA is about five times the FLA. For example, a motor that has an FLA of 5 A will normally have an LRA of about 25 A. If the LRA is given on the nameplate and the FLA is not given, divide the LRA by 5 to get an approximate FLA or RLA. For example, if a compressor nameplate shows an LRA of 80 A, the approximate FLA is 80/5 = 16 A.

Every motor has a service factor that may be listed in the manufacturer's literature. This service factor is actually reserve horsepower. A service factor of 1.15 applied to a motor means that the motor can operate at 15% over the nameplate

horsepower before it is out of its design parameters. A motor operating with a variable load and above normal conditions for short periods of time should have a larger service factor. If the voltage varies at a particular installation, a motor with a high service factor may be chosen. The service factor is standardized by the National Electrical Manufacturer's Association (NEMA). **Figure 18–9** is a typical manufacturer's chart showing service factors.

Frequency

The frequency in cycles per second (cps) is the frequency of the electrical current that the power company supplies. The technician has no control over this. Most motors are 60 cps in the United States but could be 50 cps in another country. Most 60-cps motors will run on 50 cps, but they will develop only five-sixths of their rated speed (50/60). If you believe that the supply voltage is not 60 cps, contact the local power company. When motors are operated with local generators as the power supply, the generator's speed will determine the frequency. A cps meter is normally mounted on the generator and can be checked to determine the frequency.

Phase

The number of phases of power to be supplied by the power company is determined by the type of equipment to be used at a particular installation. If equipment using three-phase motors is to be used, three-phase power must be supplied. Normally, single-phase power is supplied to residences, and

THREE PHASE • DRIPPROOF				

Type SC Squirrel Cage • Fractional HP
- 60 Hertz
- Ball Bearing
- 40°C Ambient
  Class B Insulation

• NEMA Service Factor
  1/20 thru 1/8 HP—1.40
  1/6 thru 1/3 HP—1.32
  1/2 thru 3/4 HP—1.25
  1 thru 200 HP—1.15

• Versatile 208-430/460 volt motors available in many ratings.

HP	RPM	Volts	Full Load (5) Amps	Frame
Rigid Base				
1/4	1800	200-230/460	0.8	K48
	1200	230/460	0.6	H56
1/3	3600	200-230/460	0.7	B56
	1800	200-230/460	0.8	K48
		208-230/460	0.8	B56
	1200	200-208	1.7	J56
		230/460	0.8	J56
1/2	3600	208-230/460	0.9	B56
	1800	200-208	2.4	B56
		230/460	1.1	B56
		208-230/460	1.1	B56
	1200	200-208	2.0	J56
		230/460	1.0	J56
3/4	3600	208-230/460	1.2	J56
	1800	200-208	3.2	H56
		230/460	1.3	H56
		200-230/460	1.3	H56
	1200	200-208	3.3	J56
		200-208	3.3	M143T
		230/460	1.6	J56
		230/460	1.6	M143T
1	3600	200-208	3.2	J56
		230/460	1.5	J56
	1800	200-208	3.8	J56
		200-230/460	1.7	L143T
		200-230/460	1.7	J56
		575	1.4	L143T
	1200	200-208	3.8	N145T
		230/460	1.9	K56
		230/460	1.9	N145T

Figure 18–9 This chart shows service factors for motors. *Courtesy Century Electric, Inc.*

three-phase power is supplied to commercial and industrial installations when needed. Single-phase motors will operate on two phases of three-phase power, **Figure 18–10**. Three-phase motors will not operate on single-phase power, **Figure 18–11**.

The power company furnishes the power to the respective services to meet the specifications for the particular job. For example, if the job or system calls for a large electrical load, such as with a manufacturing plant or high-rise building, they will consult with the engineers about what is recommended. Power companies rate the furnished voltage according to the nominal voltage they will furnish. These are typically 480 V, 240 V, 208 V, and 120 V. Motors and appliances are typically rated at 460 V, 230 V, 208 V, and 115 V. The technician must work with the motor and appliance rated voltages. Usually the power supply will be close to the power company rating and will still be within the motor's rating.

The technician needs to know the characteristics of the voltage supplied. The supplied voltage depends on the power company's transformer and the way it is connected. The two transformer types commonly used for commercial and industrial applications are the wye or star transformer and the delta transformer, **Figure 18–12**.

The wye transformer is configured in a Y-shaped connection with three windings going to a center tap. The center tap is considered the neutral or grounded neutral leg. The consumer typically will have a power supply of 208 V at the secondary terminals from phase to phase and 115 V (120 V) from any phase to the grounded neutral.

Another option for the wye transformer is to furnish 460 V (480 V) phase to phase and 277 V phase to the grounded neutral. The 277 V to the neutral is often used for the lighting service for the building. Smaller wire sizes can be furnished for the lighting circuits. When 460-V (480 V) and 277-V grounded neutral are used, there must be a further reduction of voltage for any 115-V (120 V) circuits in the

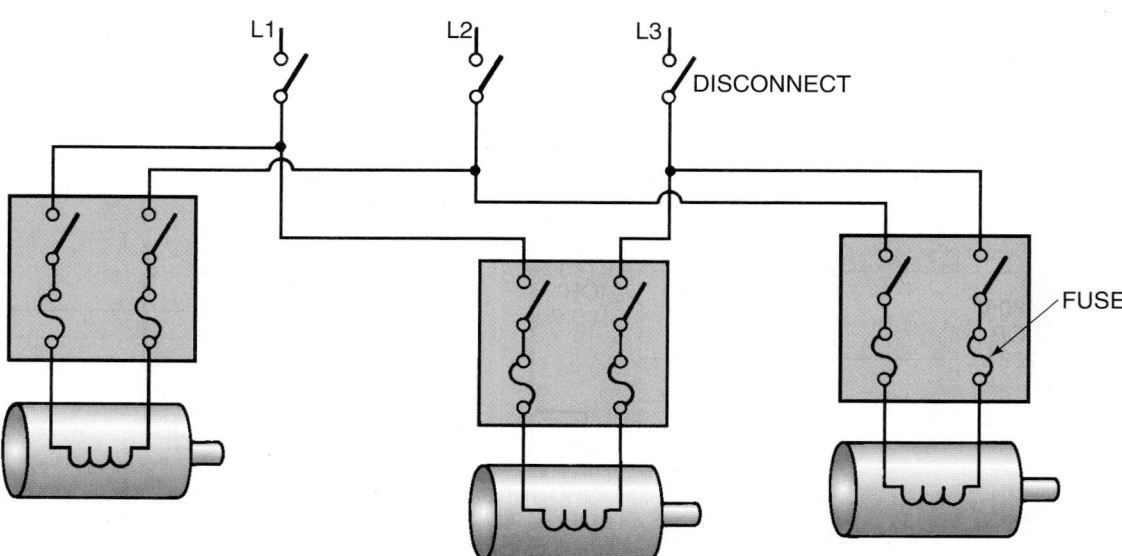

Figure 18–10 This wiring diagram shows how single-phase motors are wired into a three-phase circuit.

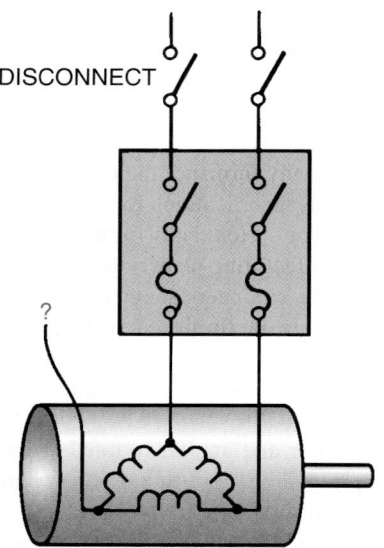

DISCONNECT

?

Figure 18–11 A three-phase motor with three power leads cannot be connected to a single-phase power supply.

building. This is accomplished with step-down transformers at locations near the point of consumption.

The wye transformer connection service is popular because it is easier to balance the load of each phase-to-phase transformer connection. This is necessary to keep the load on the power supply balanced.

The delta transformer service has been popular in some localities in the past. This service furnishes 230 V and can have a "center tapped" leg to furnish 115 V. Notice that this service furnishes 230 V phase to phase, 115 V from phase 1 to neutral, and 115 V from phase 2 to neutral. It is hard to balance the loads on this system because of the two phases that have a relationship to neutral. NOTE: The relationship from the third phase to neutral is not usable. This is often called a "wild leg" or a "high leg" and is identified with orange along the length to prevent confusion. This leg often has 185 V or more to neutral or ground. If a 115 V appliance is connected to this leg, it will be burned out in short order.

The technician is advised to carry and use a quality voltmeter and to always check the power supply voltages to know exactly what is available at the job site.

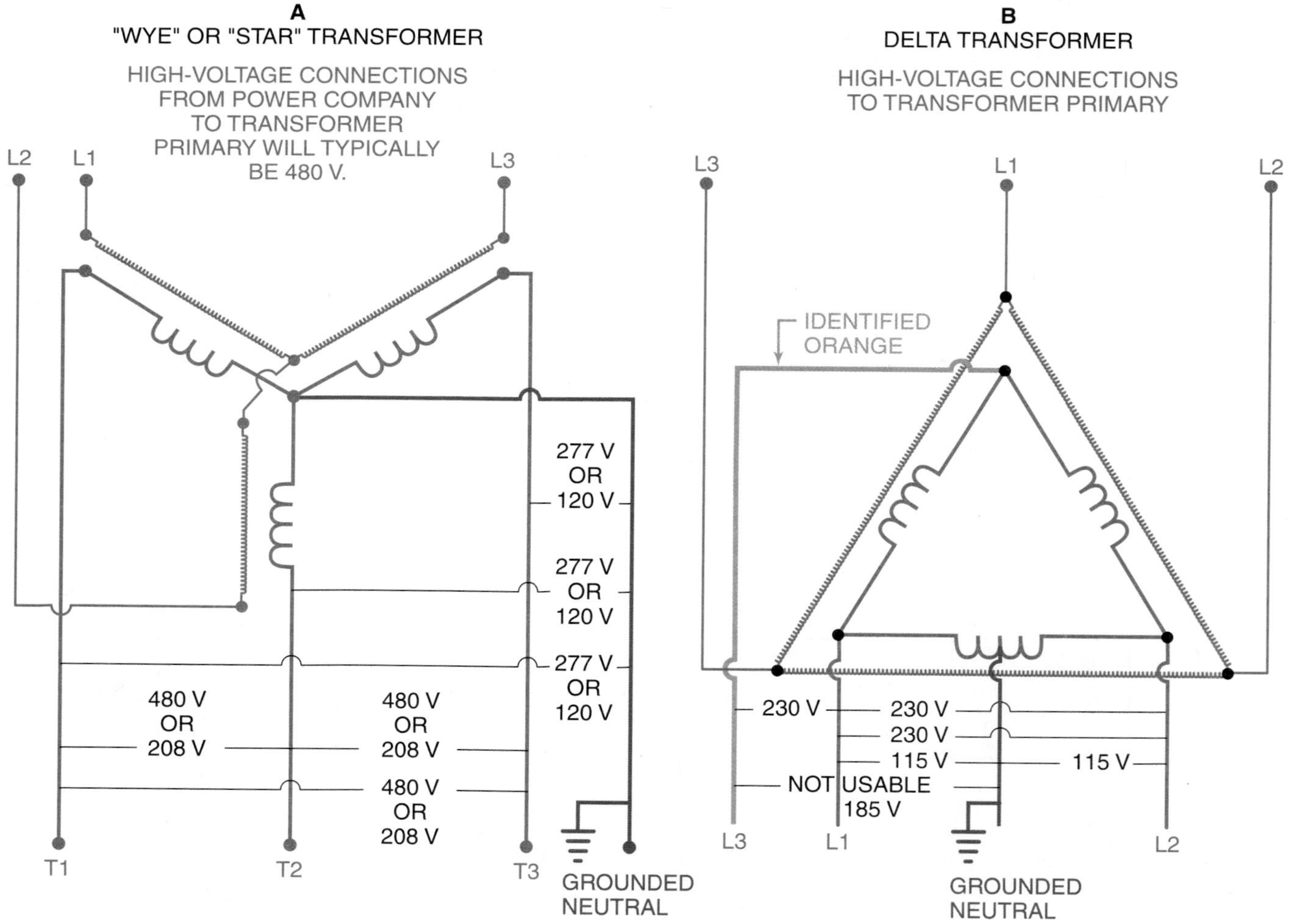

A
"WYE" OR "STAR" TRANSFORMER

HIGH-VOLTAGE CONNECTIONS
FROM POWER COMPANY
TO TRANSFORMER
PRIMARY WILL TYPICALLY
BE 480 V.

L2   L1   L3

277 V OR 120 V

277 V OR 120 V

277 V OR 120 V

480 V OR 208 V   480 V OR 208 V

480 V OR 208 V

T1   T2   T3   GROUNDED NEUTRAL

B
DELTA TRANSFORMER

HIGH-VOLTAGE CONNECTIONS
TO TRANSFORMER PRIMARY

L3   L1   L2

IDENTIFIED ORANGE

230 V   230 V
230 V
115 V   115 V
NOT USABLE
185 V

L3   L1   L2

GROUNDED NEUTRAL

Figure 18–12 Two different transformers for commercial and industrial applications.

18.3 ELECTRIC-MOTOR WORKING CONDITIONS

The motor's working conditions determine which motor is the most economical for a particular installation. An open motor with a centrifugal starting switch (single phase) for the air-conditioning fan may not be used in a room with explosive gases. When the motor's centrifugal switch opens to interrupt the power to the start winding, the gas may ignite. An explosion-proof motor enclosed in a housing must be used, **Figure 18–13.** Local codes should be checked and adhered to. The explosion-proof motor is too expensive to be installed in a standard office building, so a proper choice of motors should be made. A motor operated in a very dirty area may need to be enclosed, giving no ventilation for the motor windings. This motor must have some method to dissipate the heat from the windings, **Figure 18–14.**

Figure 18–15   A drip-proof motor. *Photo by Bill Johnson*

A drip-proof motor should be used where water can fall on it. It is designed to shed water, **Figure 18–15.**

18.4 INSULATION TYPE OR CLASS

The insulation type or class describes how hot the motor can safely operate in a particular ambient temperature condition. The example of the motor used earlier in this unit in an air-conditioning condensing unit is typical. This motor must be designed to operate in a high ambient temperature. Motors are classified by the maximum allowable operating temperatures of the motor windings, **Figure 18–16.**

For many years motors were rated by allowable temperature rise of the motor above the ambient temperature. Many motors still in service are rated this way. A typical motor has a temperature rise of 40°C. If the maximum ambient temperature is 40°C (104°F), the motor should have a winding temperature of 40°C + 40°C = 80°C (176°F). When troubleshooting these motors, the technician may have to convert from Celsius to Fahrenheit or may need a conversion table if the temperature rating is given only in Celsius. If the temperature of the motor winding can be determined, the technician can tell if a motor is running too hot for the conditions. For example, a motor in a 70°F room is allowed a 40°C rise on the motor. The maximum winding temperature is 142°F (70°F = 21°C; 21°C + 40°C rise = 61°C). This is 142°F, **Figure 18–17.**

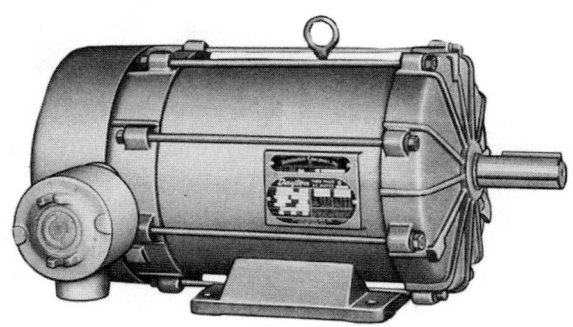

Figure 18–13   An explosion-proof motor. *Courtesy W. W. Grainger, Inc.*

Figure 18–14   A totally enclosed motor. *Photo by Bill Johnson*

Class A	221°F	(105°C)
Class B	266°F	(130°C)
Class F	311°F	(155°C)
Class H	356°F	(180°C)

Figure 18–16   Temperature classifications for typical motors.

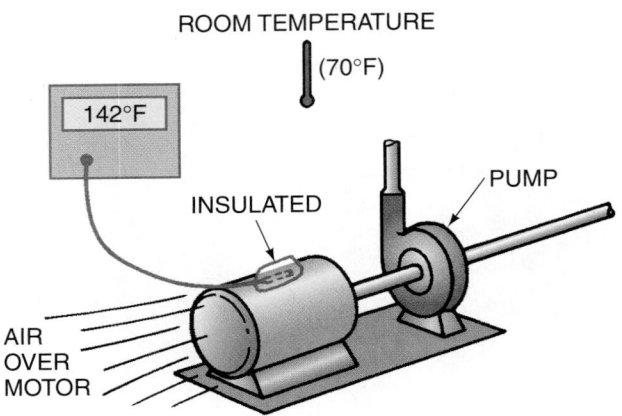

NOTE: CARE MUST BE TAKEN THAT TEMPERATURE TESTER LEAD IS TIGHT ON MOTOR AND NEXT TO MOTOR WINDINGS. THE LEAD MUST BE INSULATED FROM SURROUNDINGS. THE TEMPERATURE CHECK POINT MUST BE AS CLOSE TO THE WINDING AS POSSIBLE.

Figure 18–17 This motor is being operated in an ambient that is equal to the maximum allowable for its insulation with a 40°C rise.

18.5 TYPES OF BEARINGS

Load characteristics and noise level determine the type of bearings that should be selected for the motor. Two common types are the *sleeve* bearing and the *ball* bearing, **Figure 18–18.**

The sleeve bearing is used where the load is light and the noise must be low (e.g., a fan motor on a residential furnace). A ball-bearing motor would probably make excessive noise in the conditioned space. Metal ductwork is an excellent sound carrier. Any noise in the system is carried throughout the entire system. Sleeve bearings are normally used in smaller applications, such as in residential and light commercial air-conditioning systems, for this reason. They are quiet and dependable, but they cannot stand great pressures (e.g., if fan belts are too tight). These motors have either vertical or

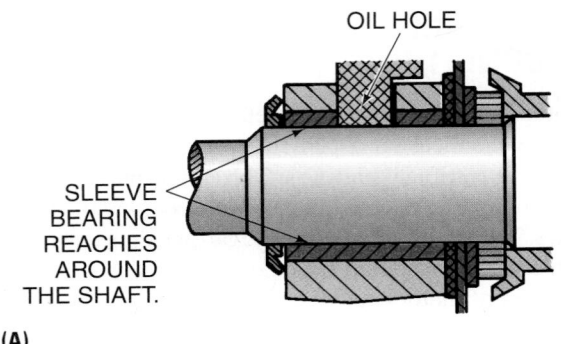

(A)

(B)

Figure 18–18 **(A)** Sleeve bearings. **(B)** Ball bearings. *(B) Courtesy Century Electric, Inc.*

horizontal shaft applications. The typical air-cooled condenser has a vertical motor shaft and is pushing the air out the top of the unit. This results in a downward thrust on the motor bearings, **Figure 18–19.** A furnace fan has a horizontal motor shaft, **Figure 18–20.** These two types may not look very different, but they are. The vertical condenser fan is trying to fly downward into the unit to push air out the top. This puts a real load on the end of the bearing (called the *thrust surface*), **Figure 18–21.**

Sleeve bearings are made from material that is softer than the motor shaft. The bearing must have lubrication—an oil

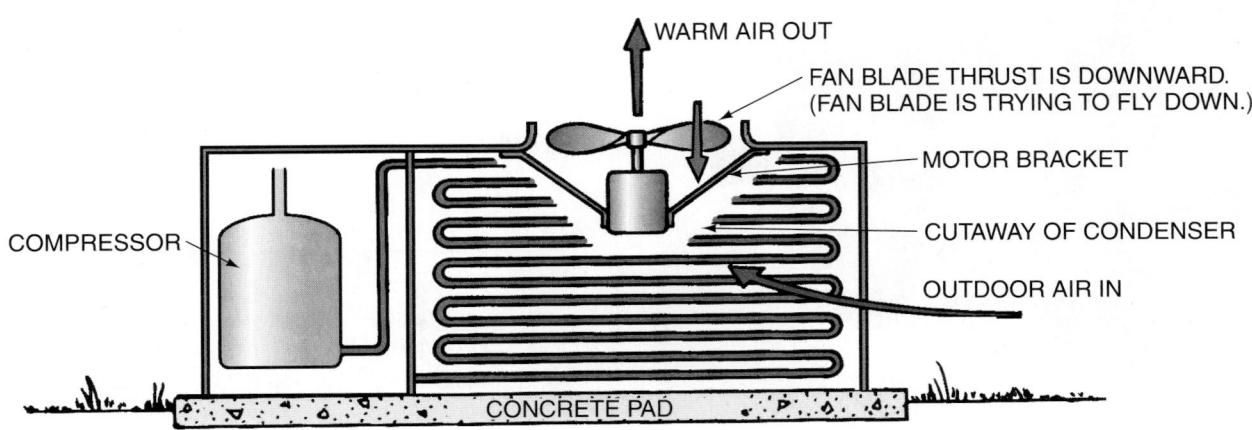

Figure 18–19 This motor is working against two conditions: the normal motor load and the fact that the fan is trying to fly downward while pushing air upward.

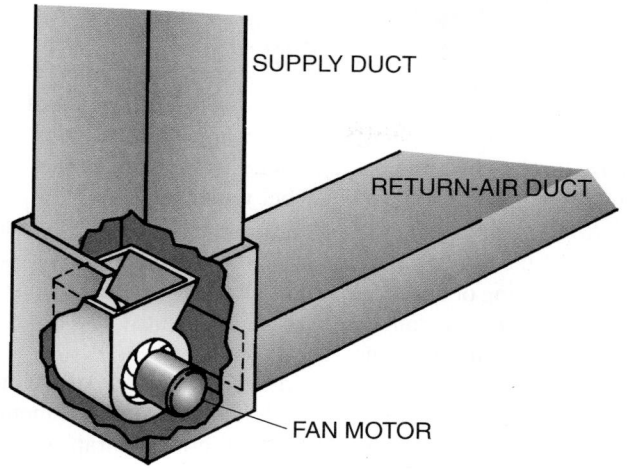

Figure 18–20 A furnace fan motor mounted in a horizontal position.

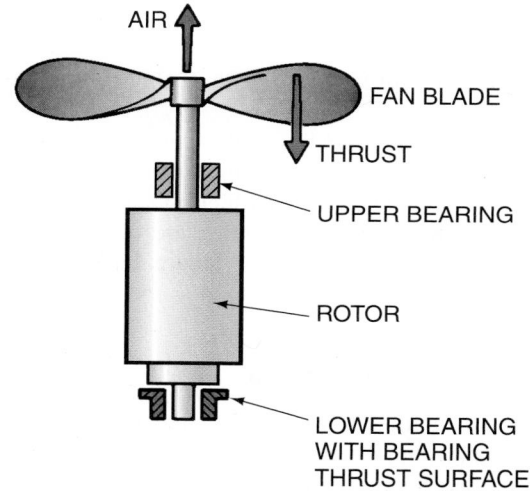

Figure 18–21 The thrust surface on a fan motor bearing.

film between the shaft and the bearing surface. The shaft actually floats on this oil film and never should touch the bearing surface. The oil film is supplied by the lubrication system. Two types of lubrication systems for sleeve bearings are the *oil port* and the *permanently lubricated* bearing.

The oil port bearing has an oil reservoir that is filled from the outside by means of an access port. This bearing must be lubricated at regular intervals with the correct type of oil, which is usually 20-weight nondetergent motor oil or an oil specially formulated for electric motors. If the oil is too thin, it will allow the shaft to run against the bearing surface. If the oil is too thick, it will not run into the clearance between the shaft and the bearing surface. The correct interval for lubricating a sleeve bearing depends on the design and use of the motor. The manufacturer's instructions will indicate the recommended interval. Some motors have large reservoirs and do not need lubricating for years. This is good if there is limited access to the motor.

The permanently lubricated sleeve bearing is constructed with a large reservoir and a wick to gradually feed oil to the

bearing. This bearing truly does not need lubrication until the oil deteriorates. If the motor has been running hot for many hours, the oil will deteriorate and fail. Shaded-pole motors operating in the heat and weather have these bearing systems, and many have operated without failure for years.

Ball-bearing motors are not as quiet as sleeve-bearing motors and are used in locations where their noise levels will not be noticed. Large fan motors and pump motors are normally located far enough from the conditioned space that the bearing noises will not be noticed. These bearings are made of very hard material and usually lubricated with grease rather than oil. Motors with ball bearings generally have permanently lubricated bearings or grease fittings.

Permanently lubricated ball bearings are similar to permanently lubricated sleeve bearings, but they have reservoirs of grease sealed in the bearing. They are designed to last for years with the lubrication furnished by the manufacturer, unless the conditions in which they operate are worse than those for which the motor was designed.

Bearings needing lubrication have grease fittings, so a grease gun can force grease into the bearing. This is often done by hand, **Figure 18–22.** Only an approved grease must be used. In **Figure 18–23** the slotted screw at the bottom of the bearing housing is a *relief screw*. When grease is pumped

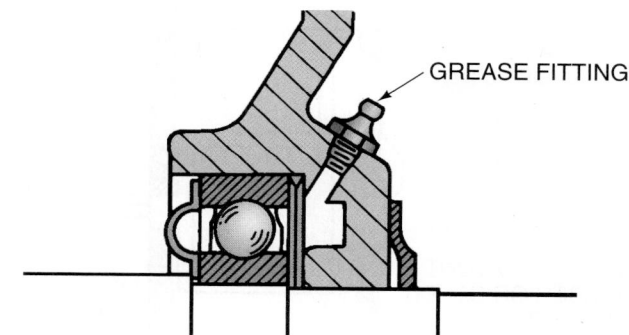

Figure 18–22 A fitting through which the bearing is greased.

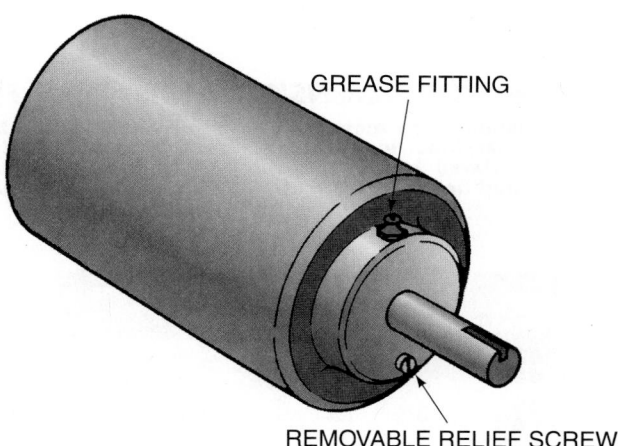

Figure 18–23 A motor bearing using grease for lubrication. Notice the relief plug.

into the bearing, this screw must be removed, or the pressure of the grease may push the grease seal out, and grease will leak down the motor shaft.

Large motors use a type of ball bearing called a *roller bearing,* which has cylindrically shaped rollers instead of balls.

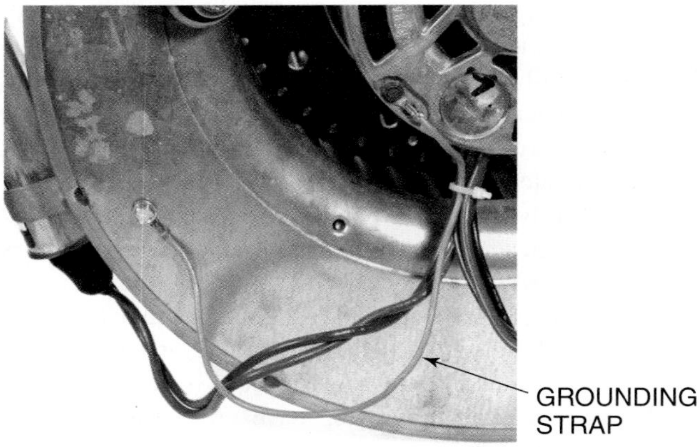

Figure 18–24 The grounding strap to carry current from the frame if the motor has a grounded winding. *Photo by Bill Johnson*

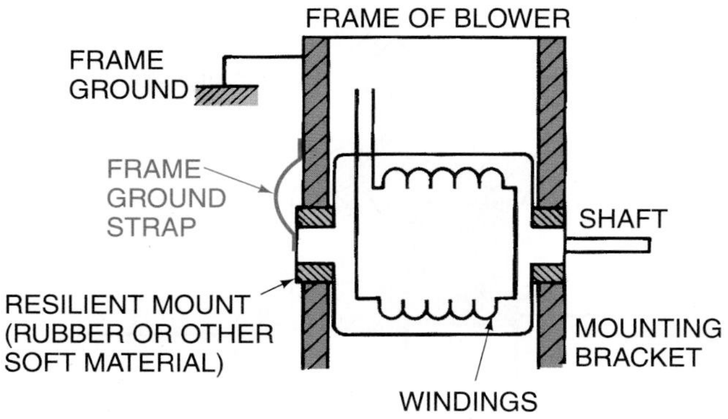

Figure 18–25 This illustration shows how the grounding strap works.

18.6 MOTOR MOUNTING CHARACTERISTICS

Mounting characteristics of a motor determine how it will be secured during its operation. The technician must consider noise level when mounting a motor. Two primary means are rigid mount and resilient, or rubber mount.

Rigid-mount motors are bolted, metal to metal, to the frame of the fan or pump and will transmit any motor noise into the piping or ductwork. Motor hum is an electrical noise, which is different from bearing noise, and must also be isolated in some installations.

Resilient-mount motors use different methods of isolating the motor noise and bearing noise from the metal framework of the system. Notice the ground strap on the resilient-mount motor, **Figure 18–24.** This motor is electrically and mechanically isolated from the metal frame. If the motor were to have a ground (circuit from the hot line to the frame of the motor), the motor frame would be electrically hot without the ground strap. SAFETY PRECAUTION: When replacing a motor always connect the ground strap properly or the motor could become dangerous, **Figure 18–25.**

The four basic mounting styles are *cradle mount, rigid-base mount, end mount* (with tabs or studs or flange mount), and *belly-band-mount,* all of which fit standard dimensions established by NEMA and which are distinguished from each other by *frame numbers.* **Figure 18–26** shows some typical examples.

Cradle-Mount Motors

Cradle-mount motors are used for either direct-drive or belt-drive applications. They have a cradle that fits the motor end housing on each end. The end housing is held down with a bracket, **Figure 18–27.** The cradle is fastened to the equipment or pump base with machine screws, **Figure 18–28.** Cradle-mount motors are available only in the small horsepower range. A handy service feature is that the motor can be removed easily.

MOTOR DIMENSIONS FOR NEMA FRAMES

Standardized motor dimensions as established by the National Electrical Manufacturers Association (NEMA) are tabulated below and apply to all base-mounted motors listed herein which carry a NEMA frame designation.

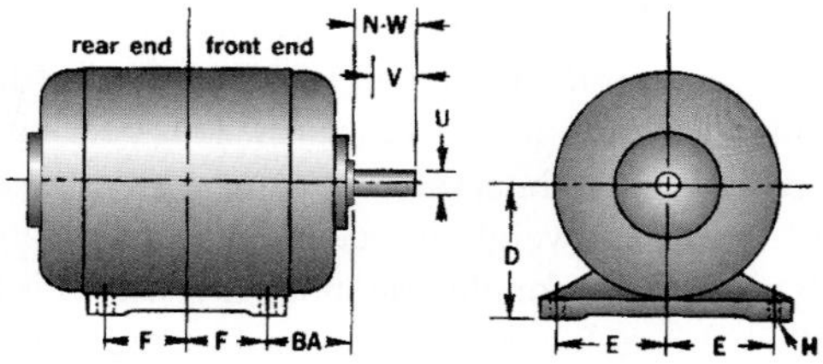

NEMA FRAME	All Dimensions in Inches							V(§) Min.	Key			NEMA FRAME
	D(*)	2E	2F	BA	H	N-W	U		Wide	Thick	Long	
42	2 5/8	3 1/2	1 11/16	2 1/16	9/32 slot	1 1/8	3/8	—	—	21/64 flat	—	42
48	3	4 1/4	2 3/4	2 1/2	11/32 slot	1 1/2	1/2	—	—	29/64 flat	—	48
56 56H	3 1/2	4 7/8	3 3&5(‡)	2 3/4	11/32 slot	1 7/8(†)	5/8(†)	—	3/16(†)	3/16(†)	1 3/8(†)	56 56H
56HZ	3 1/2	**	**	**	**	2 1/4	7/8	2	3/16	3/16	1 3/8	56HZ

Figure 18–26 Dimensions of typical motor frames. *Courtesy W. W. Grainger, Inc.*

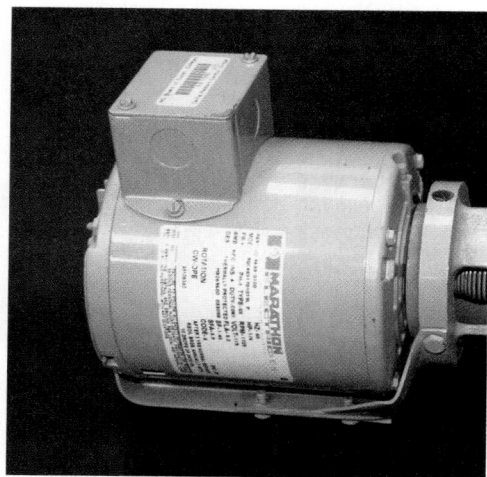

Figure 18–27 A cradle-mount motor. *Courtesy W. W. Grainger, Inc.*

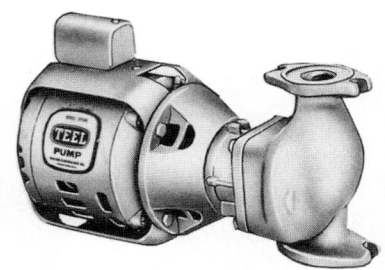

Figure 18–28 A cradle fastened to the base of a pump. *Courtesy W. W. Grainger, Inc.*

Figure 18–29 A rigid-mount motor. *Photo by Bill Johnson*

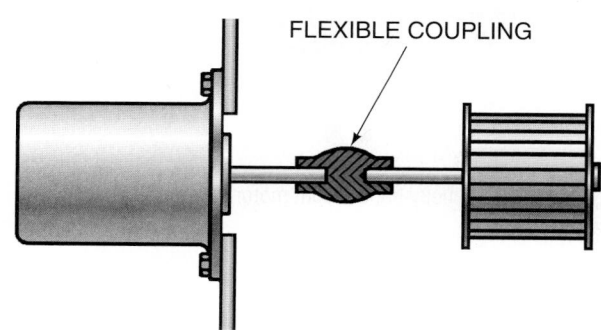

Figure 18–30 A direct-drive motor with flexible coupling.

Rigid-Base-Mount Motors

Rigid-base-mount motors are similar to cradle-mount motors except that the base is fastened to the motor body, **Figure 18–29**. The sound isolation for this motor is in the belt, if one is used, that drives the prime mover. The belt is flexible and dampens motor noise. This motor is often used as a direct drive to turn a compressor or pump. A flexible coupling is used between the motor and prime mover in a direct-drive installation, **Figure 18–30**.

End-Mount Motors

End-mount motors are very small motors mounted to the prime mover with tabs or studs fastened to the motor housing, **Figure 18–31**. Flange-mounted motors have a flange as a part of the motor housing—for example, an oil burner motor, **Figure 18–32**.

Belly-Band-Mount Motors

Belly-band-mount motors have a strap that wraps around the motor to secure it with brackets mounted to the strap. These motors are often used in air-conditioning air handlers.

Figure 18–31 A motor that is end mounted with tabs and studs. *Photo by Bill Johnson*

Figure 18–32 The motor for an oil burner with a flange on the end to hold the motor to the equipment that is being turned. *Photo by Bill Johnson*

Figure 18–33 A belly-band-mount motor. *Courtesy W. W. Grainger, Inc.*

Several universal types of motor kits are belly-band mounted and will fit many different applications. These motors are all direct drive, **Figure 18–33.**

18.7 MOTOR DRIVES

Motor drives are devices or systems that connect a motor to the driven load. For instance, the motor is a driving device, and a fan is a driven component. All motors drive their loads by belts or direct drives through couplings, or the driven component may be mounted on the motor shaft. Gear drives are a form of direct drive and will not be covered in this text because they are used mainly in large industrial applications.

The drive mechanism is intended to transfer the motor's rotating power or energy to the driven device. For example, a compressor motor is designed to transfer the motor's power to the compressor, which compresses the refrigerant vapor and pumps refrigerant from the low side to the high side of the system. Efficiency, speed of the driven device, and noise level are some factors involved in this transfer. It takes

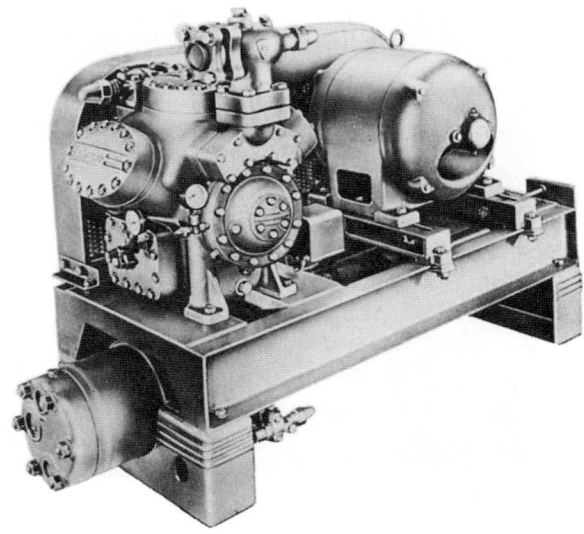

(A)

(B)

Figure 18–34 Motor drive mechanisms. **(A)** Belt drive. **(B)** Direct drive. *Courtesy Carrier Corp.*

energy to turn the belts and pulleys on a belt-drive system in addition to the compressor load. Therefore a direct-drive motor may be better suited for this application. **Figure 18–34** shows an example of both direct and belt drives.

Belt-driven applications have been used for years to drive both fans and compressors. Pulley sizes can be changed, and the speed of the driven device may be changed. This is a versatility feature of the belt-drive type system, **Figure 18–35.** This can be a great advantage if the capacity of a compressor or a fan speed needs to be changed. However, the changes must be made within the capacity of the drive motor.

There is a definite relationship between the speed of the motor and the speed of the blower. This relationship involves the diameters of the drive and of the driven pulleys. The drive pulley is the pulley connected to the motor shaft, while the

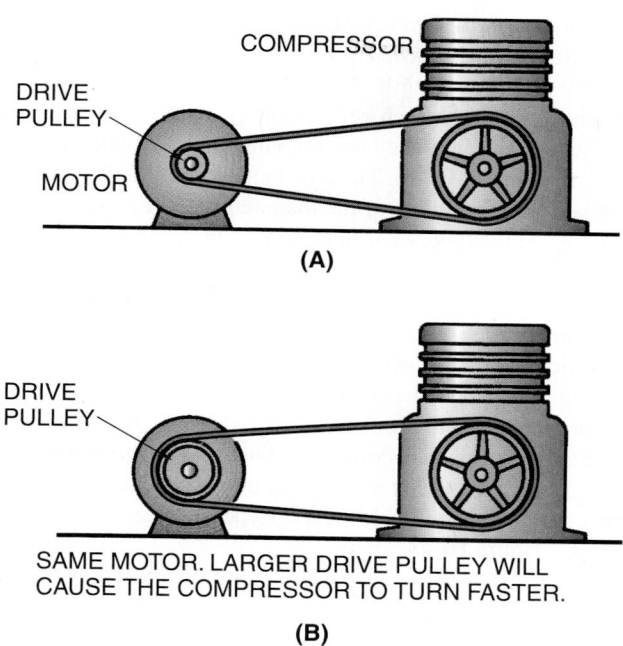

COMPRESSOR

DRIVE PULLEY

MOTOR

(A)

DRIVE PULLEY

SAME MOTOR. LARGER DRIVE PULLEY WILL CAUSE THE COMPRESSOR TO TURN FASTER.

(B)

Figure 18–35   A belt drive.

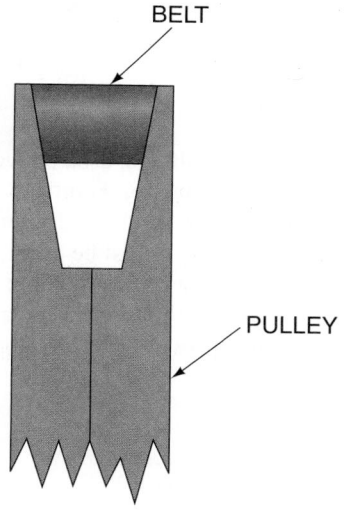

BELT

PULLEY

Figure 18–36   A belt riding at the outer edge of the pulley.

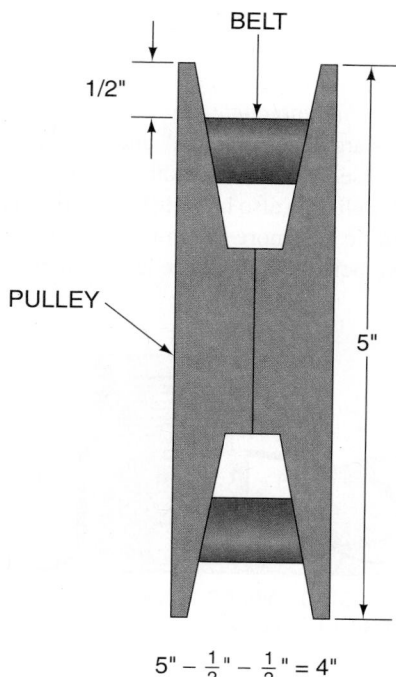

BELT

1/2"

PULLEY

5"

$5" - \frac{1}{2}" - \frac{1}{2}" = 4"$

Figure 18–37   The belt is riding inside the pulley, so adjustments to the calculations need to be made.

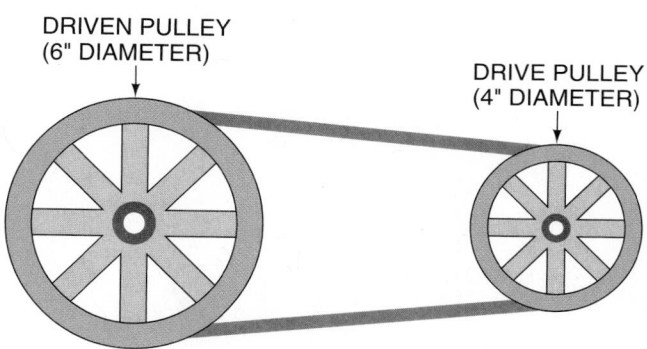

DRIVEN PULLEY (6" DIAMETER)

DRIVE PULLEY (4" DIAMETER)

Figure 18–38   A blower-speed example.

driven pulley is the pulley connected to the blower shaft. The relationship is as follows:

Motor speed × Drive pulley diameter
= Blower speed × Driven pulley diameter

This formula assumes that the belts are positioned at the outermost edge of the pulleys, **Figure 18–36**. If the belts are resting in the pulley grooves with the outermost edge of the belt well within the pulley itself, use this formula instead: the diameter of the pulley minus twice the distance between the top of the belt and the edge of the pulley, **Figure 18–37**.

Consider the pulley arrangement in **Figure 18–38**. The motor is turning at an actual speed of 3000 rpm and has a 4-inch

pulley connected to its shaft. The blower shaft has a 6-inch pulley attached to it. How fast is the blower turning?

From the formula, we get:

Motor speed × Drive pulley diameter
= Blower speed × Driven pulley diameter

3000 rpm × 4 in. = Blower speed × 6 in.

Blower speed = [(3000 rpm)(4 in.)] ÷ 6 in.

Blower speed = 2000 rpm

As long as three of the four values are known, the fourth can be computed using this very useful formula.

Belts are manufactured in different types and sizes. Some have different fibers inside to prevent stretching. SAFETY PRECAUTION: Handle belts carefully during installation. A belt designed for minimum stretch must not be installed by forcing it over the side of the pulley, because it may not stretch enough. Fibers will break and weaken the belt, **Figure 18–39**. Do not get your fingers between the belt and the pulley. Never touch the belt when it is moving.

Belt widths are denoted by "A" and "B." An A-width belt must not be used with a B-width pulley nor vice versa, **Figure 18–40**. Belts can also have different grips, **Figure 18–41**.

When a drive has more than one belt, the belts must be matched. Two belts with the same length marked on the belt

are not necessarily matched. They may not be exactly the same length. A *matched* set of belts means the belts are *exactly* the same length. A set of 42-in. belts *marked* as a matched set means each belt is exactly 42 in. If the belts are not marked as a matched set, one may be 42 1/2 in., and the other may be 41 3/4 in. Thus the belts will not pull evenly—one belt will take most of the load and will wear out first.

Belts and pulleys wear like any moving or sliding surface. When a pulley begins to wear, the surface roughens and wears out the belts. Normal pulley wear is caused by use or running time. Belt slippage will cause premature wear. Pulleys must be inspected occasionally, **Figure 18–42**.

Belts must have the correct tension, or they will cause the motor to operate in an overloaded condition. A belt tension gage should be used to correctly adjust belts to the proper tension. This gage is used in conjunction with a chart that gives the correct tension for different types of belts of various lengths, **Figure 18–43**.

Direct-drive motor applications are normally used with drive motors for fans, pumps, and compressors. Small fans and hermetic compressors are direct drive, but the motor shaft is actually an extended shaft with the fan or compressor on the end, **Figure 18–44**. The technician can do nothing to alter these. When this type is used in an open-drive application, some sort of coupling must be installed between the motor and the driven device, **Figure 18–45**. Some couplings have springs that connect the two coupling halves together to absorb small amounts of vibration from the motor or pump.

A more complicated coupling is used between the motor and a larger pump or a compressor, **Figure 18–46**. This coupling and shaft must be in very close alignment, or vibration will occur. The alignment must be checked to see that the motor shaft is parallel with the compressor or pump shaft. Alignment is a very precise operation and is done by experienced technicians. If two shafts are aligned to within

MOTOR IS ADJUSTED TOWARD COMPRESSOR FOR BELTS TO BE INSTALLED.

Figure 18–39 The correct method for installing belts over a pulley. The adjustment is loosened to the point that the belts may be passed over the pulley.

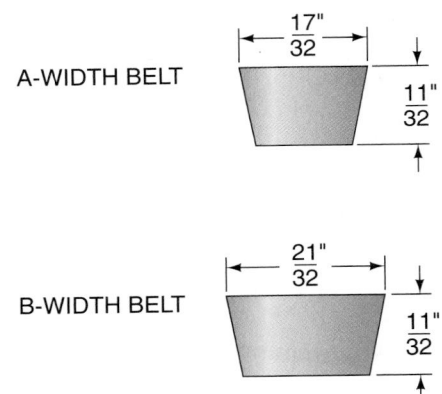

Figure 18–40 A- and B-width belts.

Figure 18–41 This belt with grooves has a tractor-type grip. *Photo by Bill Johnson*

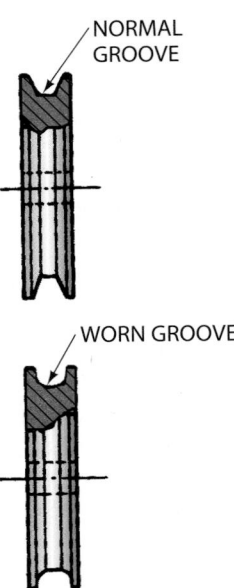

Figure 18–42 A comparison between a normal and a worn pulley.

Figure 18–43   Belt tension gage. *Courtesy Robinair SPX Corporation*

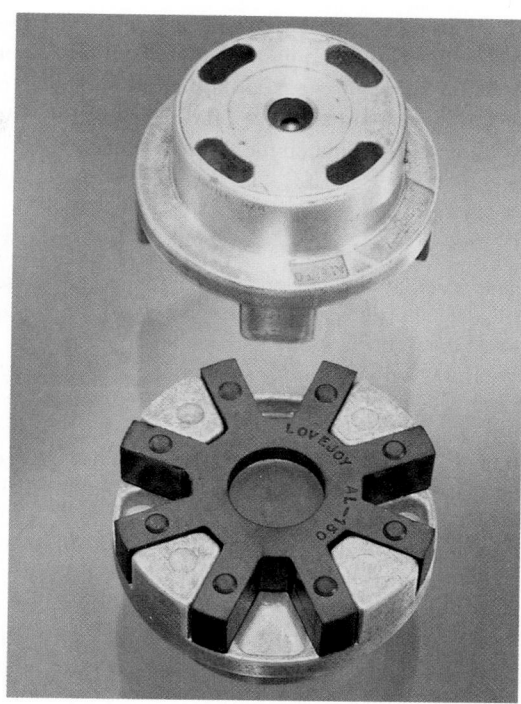

Figure 18–46   A more complicated coupling used to connect larger motors to large compressors and pumps. *Courtesy Lovejoy, Inc.*

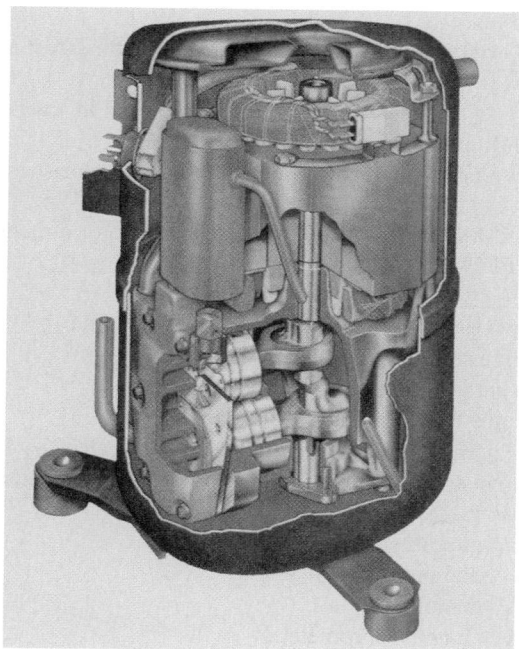

Figure 18–44   A direct-drive compressor. *Courtesy Tecumseh Products Company*

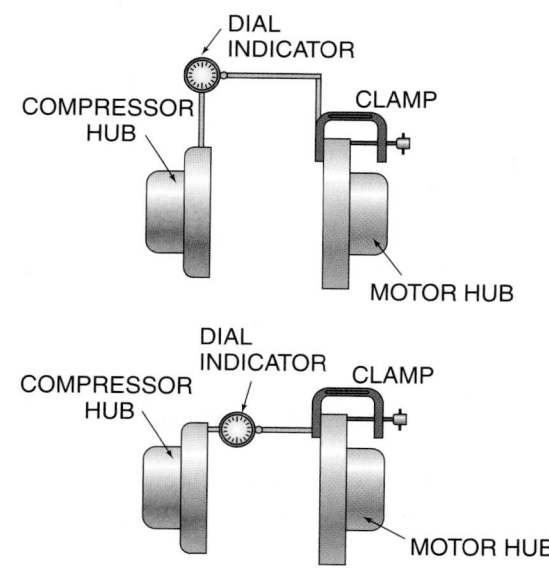

Figure 18–47   The alignment of the two shafts must be very close for the system to operate correctly. *Courtesy Trane Company*

tolerance while the motor and driven mechanism is at room temperature, the alignment must be checked again after the system is run long enough to get the system up to operating temperature. The motor may not expand and move the same distance as the driven mechanism, and the alignment may need to be adjusted to the warm value, **Figure 18–47.**

When a new motor must be installed to replace an old one, try to use an exact replacement. An exact replacement may

Figure 18–45   A small flexible coupling. *Courtesy Lovejoy, Inc.*

be found at a motor supply house or it may have to be obtained from the original equipment manufacturer. When the motor is not a normally stocked motor, you can save much time by taking the old motor to the distributor or supply house so that they can match the motor's specifications.

SUMMARY

- ▓ In many installations only one type of motor can be used.
- ▓ The power supply determines the applied voltage, the current capacity, the frequency, and the number of phases.
- ▓ The working conditions (duty) for a motor specify the atmosphere in which the motor must operate (i.e., wet, explosive, or dirty).
- ▓ Motors are also classified according to the insulation of the motor windings and the motor temperature under which they operate.
- ▓ Each motor has sleeve, ball, or roller bearings. Sleeve bearings are the quietest but will not stand heavy loads.
- ▓ SAFETY PRECAUTION: Motors must be mounted in the fashion designed for the installation.
- ▓ SAFETY PRECAUTION: An exact motor replacement should be obtained whenever possible.
- ▓ The drive mechanism transfers the motor's energy to the driven device (the fan, pump, or compressor).

REVIEW QUESTIONS

1. List four characteristics of an application that would influence the electric-motor selection.
2. Power supply data should include the following information:
   A. _____
   B. _____
   C. _____
   D. _____
3. The voltage range for an electric motor is usually within + or − _____%.
4. If an electric motor is rated for 208 V, the maximum allowable voltage would be _____, and the minimum allowable voltage would be _____.
5. The full-load amperage (FLA) of an electric motor is the amperage that the
   A. motor draws when operating at run-load conditions.
   B. motor draws at start-up.
   C. motor draws when operating at +25% of its rating.
6. In the United States, the frequency in cycles per second (cps) at which an AC electric motor operates under normal conditions is
   A. 50.
   B. 60.
   C. 120.
   D. 208.
7. True or False: Three-phase motors are often used in air-conditioning systems in residences.
8. Two types of bearings commonly used on small motors are _____ and _____.
9. An open single-phase motor with a centrifugal starting switch may not be used in
   A. an air-conditioning condenser fan motor.
   B. an air-conditioning indoor air handler.
   C. an area where explosive gases exist.
   D. a ventilating fan motor.
10. What is meant by the service factor of an electric motor?
11. How does the insulation factor of a motor affect its use?
12. A _____ bearing surface must be used in a vertical shaft motor application.
13. Two primary electric-motor mount characteristics are _____ and _____.
14. NEMA has established standardized motor dimensions. NEMA stands for _____ _____ _____ _____.
15. Two types of motor drives are _____ and _____ drives.
16. What is a matched set of belts?
17. Name the different types of belts.
18. Why must direct-drive couplings be aligned?
19. An instrument often used to check the alignment of two shafts is the
   A. micrometer.
   B. dial indicator.
   C. manometer.
   D. sling psychrometer.
20. Why must resilient-mount motors have a ground strap?
21. A belt-driven blower assembly is equipped with a motor that turns a 5-inch drive pulley at a speed of 3450 rpm. The blower is turning at a speed of 1725 rpm. Determine the size of the driven pulley.

UNIT
19 Motor Controls

OBJECTIVES

After studying this unit, you should be able to

■ describe the differences between a relay, a contactor, and a starter.

■ state how the locked-rotor current of a motor affects the choice of a motor starter.

■ list the basic components of a contactor and starter.

■ compare two types of external motor overload protection.

■ describe conditions that must be considered when resetting safety devices to restart electric motors.

SAFETY CHECKLIST

✔ When a motor is stopped for safety reasons, do not restart it immediately. If possible, determine the reason for the overload condition before restarting.

✔ Conductor wiring should not be allowed to pass too much current, or it will overheat causing conductor failure or fire.

✔ When replacing or servicing line voltage components, ensure that the power is off, that the panel is locked and tagged, and that you have the only key.

19.1 INTRODUCTION TO MOTOR CONTROL DEVICES

This unit concerns those automatic components used to close and open the power-supply circuit to the motor. These devices are called relays, contactors, and starters.

For example, a compressor in a residential central air conditioner is controlled in the following manner. With a temperature rise in the space to be conditioned, the thermostat contacts close and pass low voltage to energize the coil on the compressor contactor. This closes the contacts in the compressor contactor, allowing the applied or line voltage to pass to the compressor motor windings. Relays, contactors, and starters are names given to these motor controls, even though they perform the same function (but with some simple differences). Relays are designed to control loads that draw less than 20 amps of electric current, while contactors are designed to control loads that draw more than 20 amps of current. Motor starters are, simply put, three-pole contactors with motor overload protection built into the device. More details will be presented as we discuss each of these system

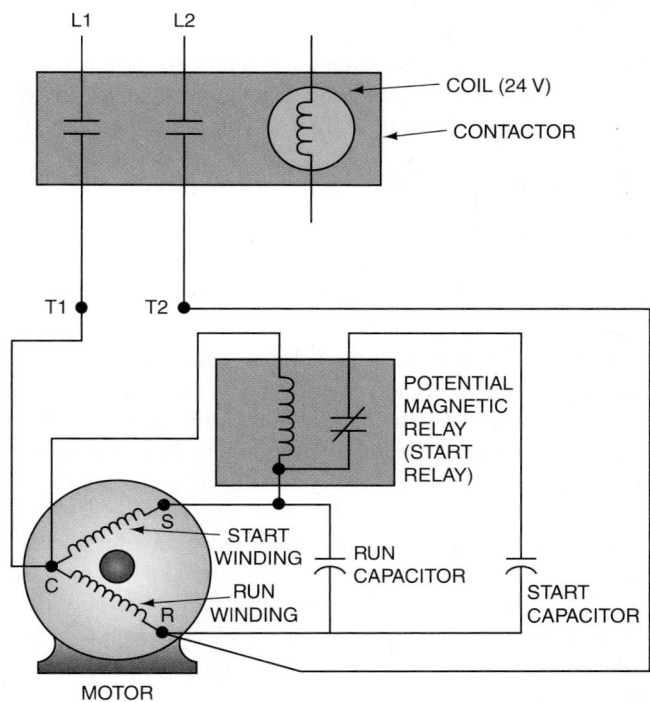

Figure 19–1 Typical motor wiring diagram utilizing a contactor and a potential magnetic relay (PMR) as a motor-starting relay.

components. **Figure 19–1** is a diagram of a typical motor circuit. In this circuit there is a contactor as well as a potential magnetic relay (PMR) that is acting as a motor-starting relay. The contactor passes power to the motor on an external call for motor operation, and the starting relay helps the motor start on a call for operation. Notice the contactor is a major part of the motor circuit even though it is external to the motor windings. Power is fed to the motor windings and starting relay from the contactor.

The size of the motor and the application usually determines the type of switching device (relay, contactor, or starter). For example, a small manual switch can start and stop a handheld hair drier, and a person operates the switch. A large 100-hp motor that drives an air-conditioning compressor must start, run, and stop unattended. It will also consume much more current than the hair drier. The components that start and stop large motors must be more elaborate than those that start and stop small motors.

19.2 RUN-LOAD AND LOCKED-ROTOR AMPERAGE

Electric motors have two current (amperage) ratings: the *run-load amperage (RLA)*, sometimes referred to as the *full-load amperage (FLA)*, and the *locked-rotor amperage (LRA)*. The RLA or FLA is the current drawn while the motor is running. The LRA is the current drawn by the motor just as it begins to start but before the rotor begins to rotate. The LRA of a motor is typically five times higher than the FLA of the motor and can be as high as seven times the FLA. Both currents must be considered when choosing the component (relay, contactor, starter) that passes the line voltage to the motor.

19.3 THE RELAY

The *relay* has a magnetic coil that opens or closes one or more sets of contacts, **Figure 19–2**. It is considered a throwaway device because parts are not available for rebuilding it when it no longer functions properly.

Relays are designed for light duty applications. *Pilot-duty* relays can switch (on and off) larger contactors or starters. Pilot-duty relays for switching circuits are very light duty and are not designed to directly start motors. Relays designed for starting motors are not really suitable as switching relays because they have more resistance in the contacts.

The pilot-duty relay contacts are often made of a fine silver alloy and designed for low-level current switching. Use on a higher load would melt the contacts. Heavier-duty motor switching relays are often made of silver cadmium oxide with a higher surface resistance and are physically larger than pilot-duty relays.

If a relay starts the indoor fan in the cooling mode of a central air-conditioning system, it must be able to withstand the inrush current of the fan motor on start-up, **Figure 19–3**. (A motor normally has a starting current of five times the

Figure 19–3 A throwaway fan relay that may be used to start an evaporator fan motor on a central air-conditioning system. *Photo by Bill Johnson*

running current, but only for a very short time.) Relays are often rated in horsepower: If a relay is rated for a 3-hp motor, it will be able to stand the inrush or locked-rotor current of a 3-hp motor as well as the run current.

A relay may have more than one type of contact configuration. It could have two sets of contacts that close when the magnetic coil is energized, **Figure 19–4**, or it may have two sets of contacts that close and one set that opens when the coil is energized, **Figure 19–5**. A relay with a single set of contacts that close when the coil is energized is called a *single-pole–single-throw, normally open relay* (spst, NO). A relay with two contacts that close and one that opens is called a *triple-pole–single-throw*, with two *normally open* contacts, and one *normally closed contact* (tpst). **Figure 19–6** shows some different relay contact arrangements.

Figure 19–2 General-purpose relay typically used for starting a motor that has a relatively low amperage draw. This relay has a holding coil, one normally open set of contacts, and one normally closed set of contacts. Both sets of contacts change position when the holding coil is energized. *Photo by Bill Johnson*

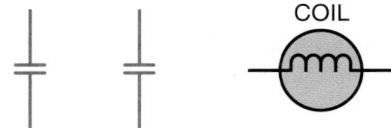

Figure 19–4 A double-pole–single-throw relay with two sets of contacts that close when the coil is energized.

Figure 19–5 A relay with two sets of contacts that close and one set of contacts that opens when the coil is energized. It has two normally open (NO) sets and one normally closed (NC) set.

CONTACT ARRANGEMENT DIAGRAMS

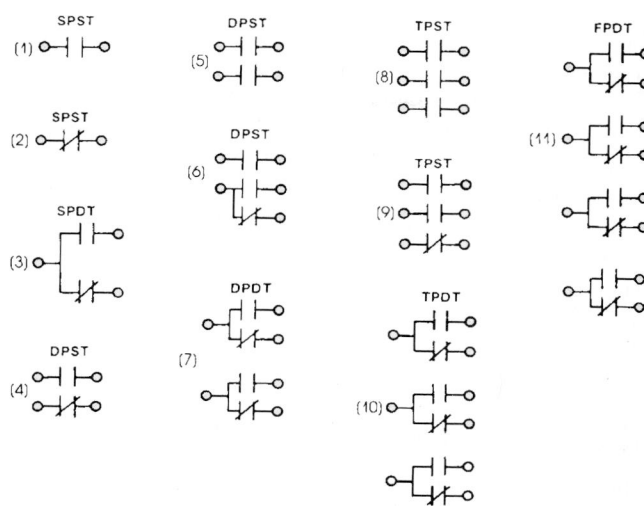

Figure 19–6   Some of the more common configurations of contacts found on relays. *Reproduced courtesy of Carrier Corporation*

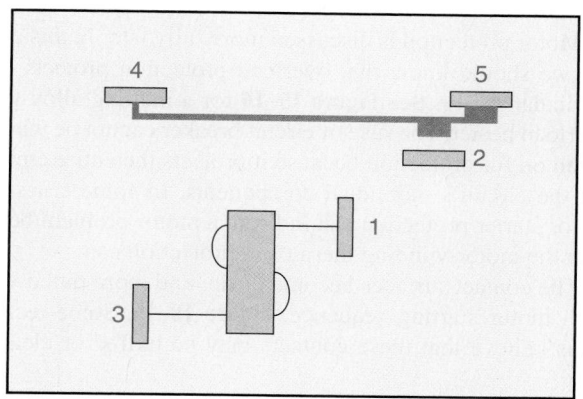

Figure 19–7   Wiring diagram of the relay shown in **Figure 19–2**. The coil is connected between terminals 1 and 3. Terminals 2 and 4 are connected to the NO contacts and terminals 4 and 5 are connected to the NC contacts.

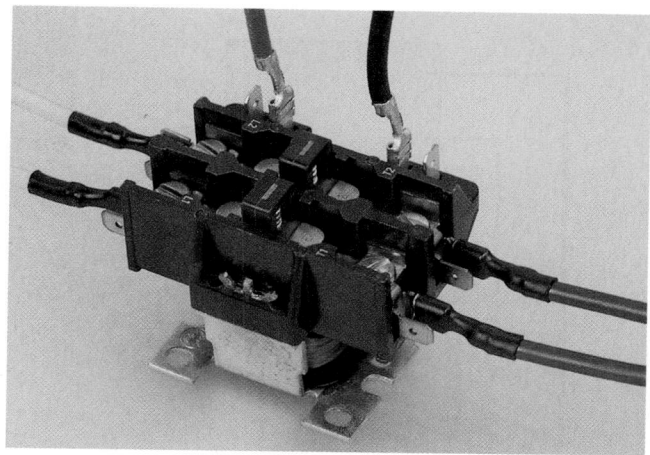

Figure 19–8   The contactor. *Photo by Bill Johnson*

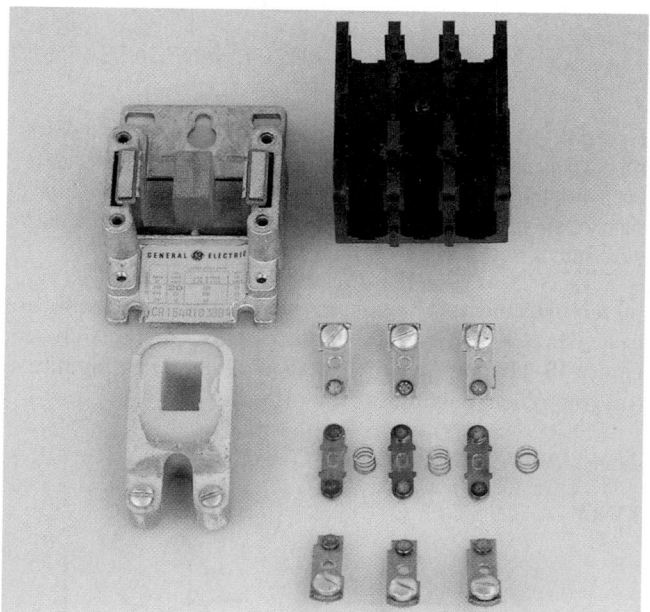

Figure 19–9   The parts of a contactor that can be replaced: the contacts (both moving and stationary), the holding coil, and the contact holding springs. *Photo by Bill Johnson*

Most relays have wiring diagrams printed on the side of the device. These diagrams identify the coil terminals as well as the normally open and normally closed contacts. **Figure 19–7** shows the wiring diagram of the relay shown in **Figure 19–2**. By referring to the diagram and identifying these marked terminals on the relay itself, replacing one relay with another in the event of component failure is not a complicated process. Many relays use terminals 1 and 3 for the coil, terminals 2 and 4 for a normally open set of contacts, and terminals 2 and 5 for a normally closed set of contacts.

19.4   THE CONTACTOR

The *contactor* is a larger version of a relay, **Figure 19–8**. It can be rebuilt if it fails, **Figure 19–9**. A contactor has movable and stationary contacts. The holding coil can be designed for

various operating voltages: 24-V, 115-V, 208-V to 230-V, or 460-V operation.

Contactors can be very small or very large, but all have similar parts. A contactor may have many configurations of contacts, from the most basic, which has just one contact, to more elaborate types with several. The single-contact type contactor is used on many residential air-conditioning units. It interrupts power to one leg of the compressor, which is all that is necessary to stop a single-phase compressor. Sometimes a single-contact contactor is needed to provide crankcase heat to the compressor. A trickle charge of electricity passes through line 2 to the motor windings and crankcase heater when the contacts are open, during the off cycle, **Figure 19–10**. If you use a contactor with two contacts

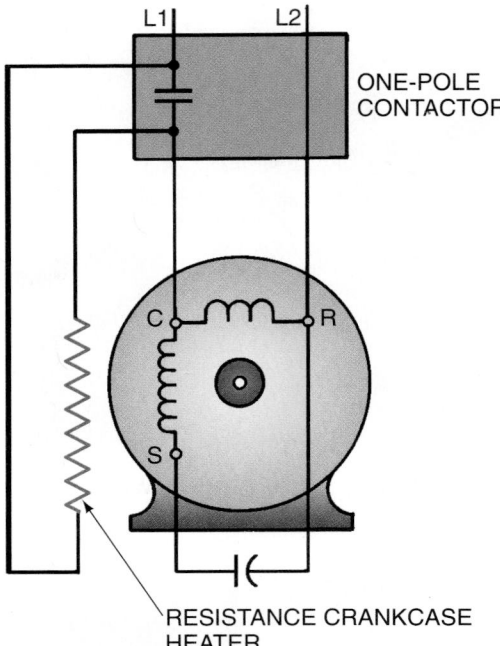

Figure 19–10 This contactor (a single-pole contactor) has only one set of contacts. Only one line of power needs to be broken to stop a motor like the one in the diagram. This is a method commonly used to supply crankcase heat to the compressor during the off cycle. When the contacts close, current will bypass the heater.

for a replacement, the compressor would not have crankcase heat. Once again, an exact replacement is the best choice. **Figure 19–11** shows another method of supplying crankcase (oil sump) heat only during the off cycle.

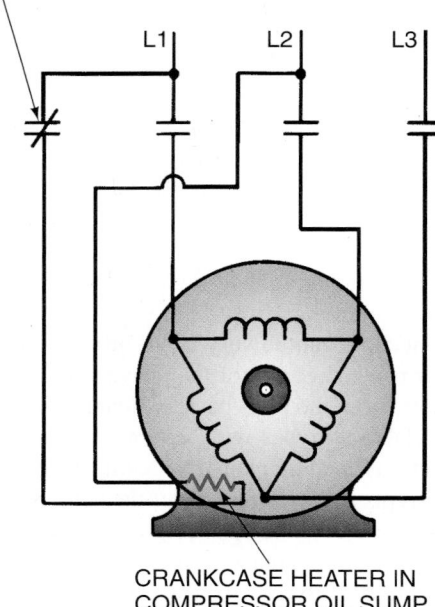

Figure 19–11 A contactor with auxiliary contacts for switching other circuits.

Some contactors have as many as five or six sets of contacts. Generally for large motors there will be three heavy-duty sets to start and stop the motor; the rest (*auxiliary contacts*) can be used to switch auxiliary circuits.

Auxiliary contacts are rated at lower amperages than the main contacts. The auxiliary contacts are typically used to open and close control circuits and to energize and deenergize other smaller system loads that have low current requirements. Auxiliary contacts can be normally open or normally closed and are often mounted to the side of the contactor, where its moving armature can change the position of the auxiliary contacts at the same moment the main contacts change position.

19.5 MOTOR STARTERS

The *starter,* or motor starter as it is sometimes called, is similar to the contactor. In some cases, a contactor may be converted to a motor starter. The motor starter differs from a contactor because it has motor overload protection built into its framework, **Figure 19–12** and **Figure 19–13**. A motor starter may be rebuilt and is available in a wide range of sizes.

Motor protection is discussed more fully later in this unit, but we should know that overload protection protects that particular motor. See **Figure 19–14** for a melting alloy-type overload heater. The fuse or circuit breaker cannot be wholly relied on for protection because it protects the entire circuit, not the circuit's individual components. In some cases the motor starter protection can indicate a motor problem better than the motor-winding thermostat protection can.

The contact surfaces become dirtier and more pitted with each motor starting sequence, **Figure 19–15**. Some technicians believe that these contacts may be buffed or cleaned

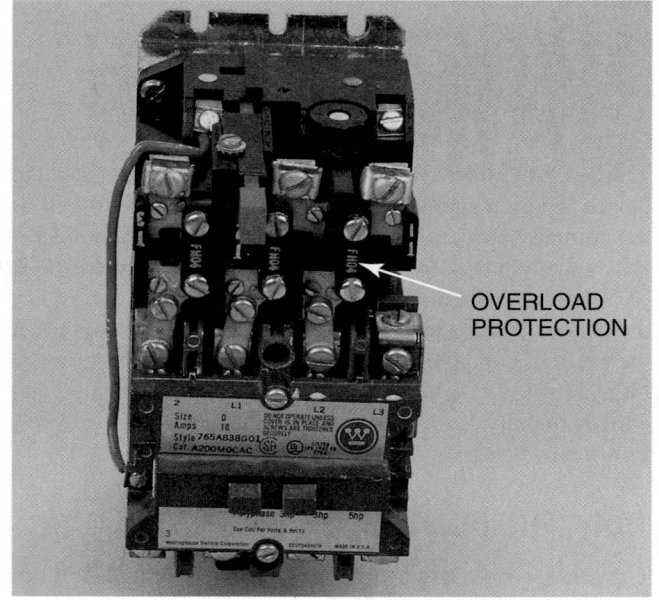

Figure 19–12 The motor starter. *Photo by Bill Johnson*

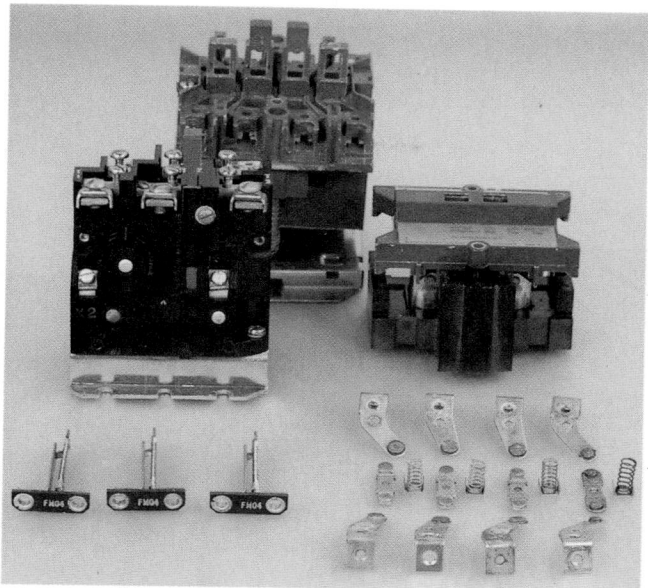

Figure 19–13 Components that may be changed on the motor starter: contacts (both moving and stationary), the springs, the coil, and the overload protection devices (heaters and switches). *Photo by Bill Johnson*

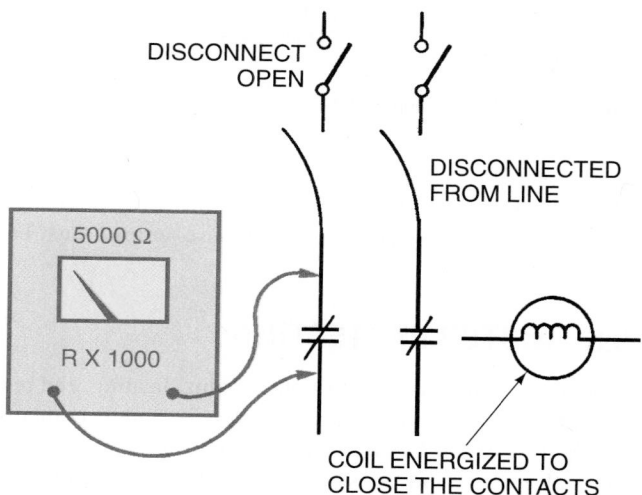

Figure 19–14 A melting alloy-type overload heater. *Courtesy Square D Company*

Figure 19–15 Clean contacts contrasted with a set of dirty, pitted contacts. *Courtesy Square D Company*

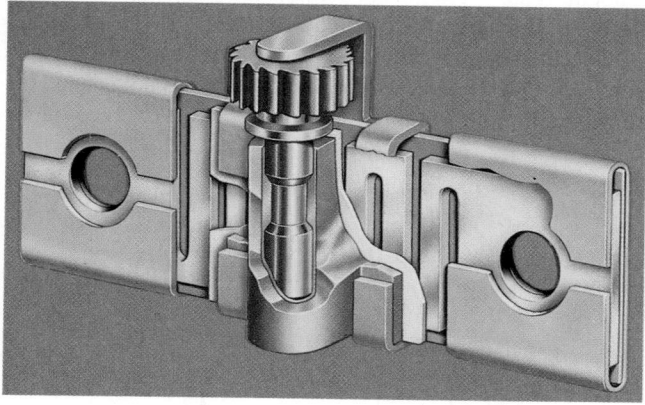

Figure 19–16 Checking contacts with an ohmmeter shows resistance in the contacts caused by dirty surfaces. Make certain that there is no current flowing through the contacts when checking resistance.

with a file or sandpaper. NOTE: Contacts may be cleaned, but this should be done as a temporary measure only. Filing or sanding exposes base metal under the silver plating and speeds its deterioration. Replacing the contacts is the only recommended repair. If this device is a relay, the complete relay must be replaced. If the device is a contactor or a starter, the contacts may be replaced with new ones, **Figure 19–15**. Both movable and stationary contacts and springs hold tension on the contacts.

You can use a voltmeter to check resistance across a set of contacts under full load. When the meter leads are placed on each side of a set of contacts and a measurable voltage exists, there is resistance in the contacts. When the contacts are new and have no resistance, no voltage should be read on the meter. An old set of contacts will have some slight voltage drop and will produce heat, due to resistance in the contact's surface, where the voltage drop occurs. **Figure 19–16** shows an example of how contacts may be checked with an ohmmeter.

Each time a motor starts, the contacts will be exposed to the inrush current (the same as LRA). These contacts are under a tremendous load for this moment. When the contacts open, an arc is caused by the breaking of the electrical circuit. The contacts must make and break as fast as possible. The contacts have a magnet that pulls them together with a mechanism to take up any slack. For example, three large movable contacts in a row may all need to be held equally tight against the stationary contacts. The springs behind the movable contacts keep this pressure even, **Figure 19–17**. If there is a resistance in the contact surface, the contacts may get hot enough to take the tension out of the springs. This

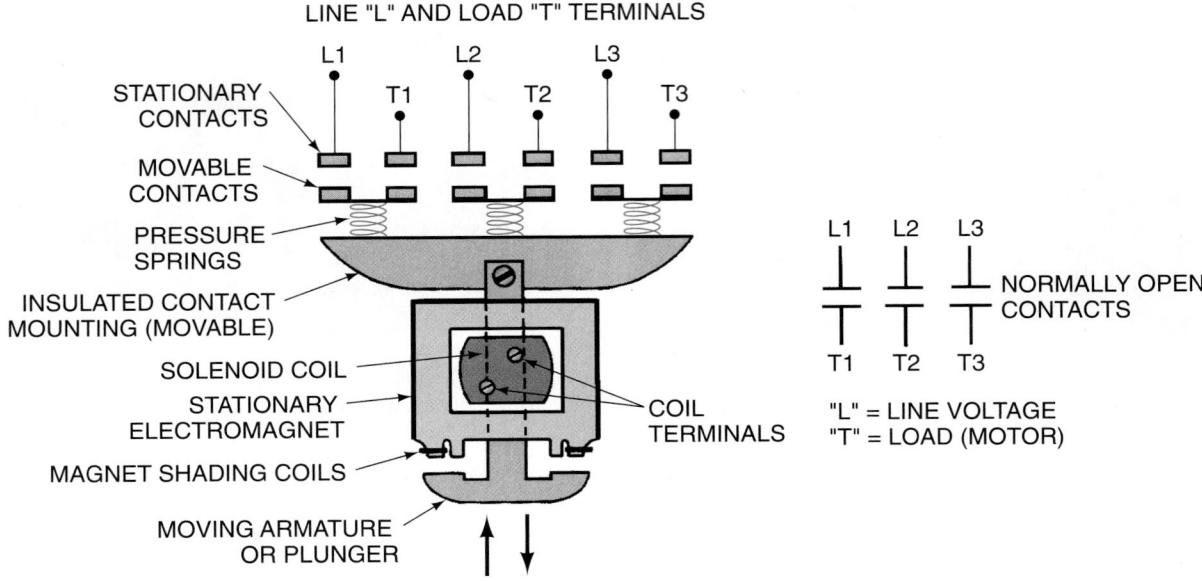

Figure 19–17   When energized, springs will hold the three movable contacts tightly against the stationary contacts.

will make the contact-to-contact pressure even less and make the heat greater. The tension in these springs must be maintained.

19.6   MOTOR PROTECTION

The electric motors used in air-conditioning, heating, and refrigeration equipment are the most expensive single components in the system. These motors consume large amounts of electrical energy, and considerable stress is placed on the motor windings. Therefore, they deserve the best protection possible within the economic boundaries of a well-designed system. The more expensive the motor, the more expensive and elaborate the protection should be.

Fuses are used as circuit (not motor) protectors. SAFETY PRECAUTION: The conductor wiring in the circuit must not be allowed to pass too much current, or it will overheat and cause conductor failures or fires. A motor may be operating at an overloaded condition that would not cause the conductor to be overloaded; hence, the fuse will not open the circuit. Let us use a central air-conditioning system as an example. There are two motors in the condensing unit: the compressor (the largest) and the condenser fan motor (the smallest). In a typical unit the compressor may have a current draw of 23 A, and the fan motor may use 3 A. The fuse protects the total circuit. The circuit will be fused to protect the circuit wiring. If one motor is overloaded, the fuse may not open the circuit, **Figure 19–18**. Each motor should be protected within its own operating range. The condenser fan motor and the compressor will have an inherent motor protector that protects the fan motor at 3 A and the compressor at 23 A.

Motors can operate without harm for short periods at a slight overcurrent condition. The overload protection is designed to disconnect the motor at some current draw value that is slightly more than the FLA value, so the motor can be

operated at its full-load design condition. Time is involved in this value in such a manner that the higher the current value above FLA, the more quickly the overload should react. The amount of the overload and the time are both figured into the design of the particular overload device.

Overload current protection is applied to motors in different ways. Overload protection, for example, is not needed for some small motors that will not cause circuit overheating or

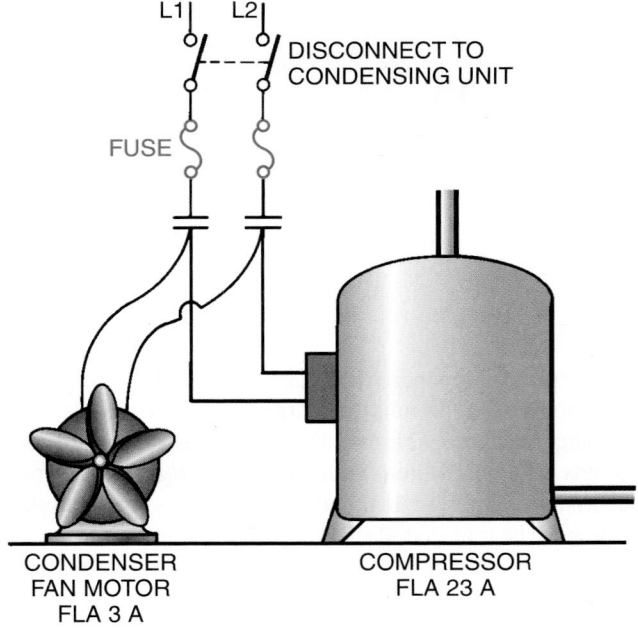

Figure 19–18   Two motors in the same circuit are served by the same conductors and have different overload requirements. The main circuit fuses protect the entire circuit and not the individual circuit components.

Figure 19–19   A small condenser fan motor that does not pull enough amperage at locked rotor amperage to create enough heat to be a problem. This is known as impedance motor protection. *Photo by Bill Johnson*

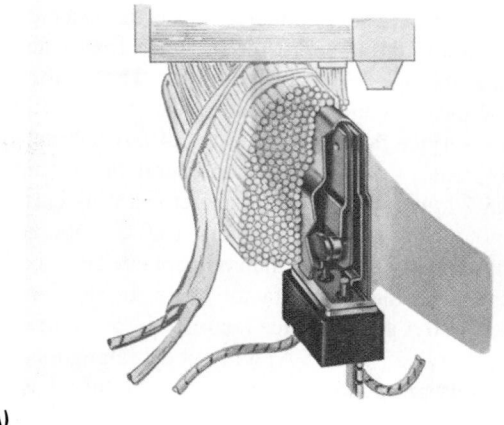

(A)

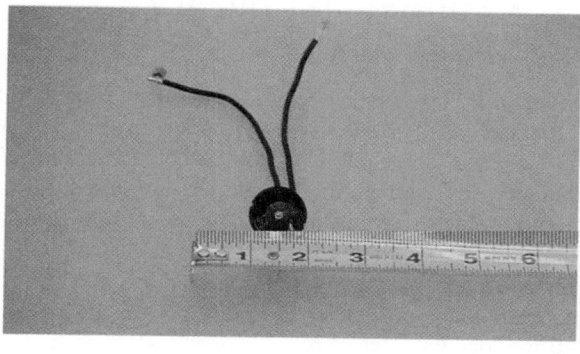

(B)

Figure 19–20   **(A)** An inherent overload protector. **(B)** An external overload protector. **(A)** *Courtesy Tecumseh Products Company.* **(B)** *Photo by Bill Johnson*

will not damage themselves. Some small motors do not have overload protection because they will not consume enough power to damage the motor unless shorted from winding to winding or from the winding to the frame (to ground). See **Figure 19–19** for an example of a small condenser fan motor that does not draw enough amperage at the LRA condition to overheat. This motor does not have overload protection. This motor is described as "impedance protected." It may only generate 10 W of power, the same as a 10-W light bulb. It is not an expensive motor. If the motor fails because of a burnout, the current draw will be interrupted by the circuit protector.

Overload protection is divided into inherent (internal) protection and external protection.

19.7   INHERENT MOTOR PROTECTION

Inherent protection is provided by internal thermal overloads in the motor windings or the thermally activated snap-disc (bimetal), **Figure 19–20**. The same types of devices are used with open and hermetic motors.

19.8   EXTERNAL MOTOR PROTECTION

External protection is often applied to the device passing power to the motor contactor or starter. These devices normally are actuated by current overload and break the circuit to the contactor coil. The contactor stops the motor. When a motor is started with a relay, the motor is normally small and has only internal protection, **Figure 19–21**. Contactors are used to start larger motors, and either inherent protection or external protection is used. Large motors (above 5 hp in air-conditioning, heating, and refrigeration systems) use starters and overload protection built into either the starter or the contactor's circuit.

The value (trip point) and type of the overload protection are normally chosen by the system design engineer or by the

Figure 19–21   A fan motor that is normally started with a relay. Motor protection is internal. *Courtesy W. W. Grainger, Inc.*

manufacturer. The technician checks the overload devices when there is a problem, such as random shutdowns because of an overload tripping. The technician must be able to understand the designer's intent with regard to the motor's operation and the overload device operation because they are closely related in a working system.

Many motors have a service factor; this is the reserve capacity of the motor and is expressed as a percentage above the full-load amperage (FLA) of the motor. The motor can operate above the FLA and within the service factor without harm. Typical service factors are 1.15 to 1.40. A service factor of 1.15 indicates that the motor can operate at an amperage that is 15% higher than the nameplate operating current, whereas a motor with a service factor of 1.40 can operate at amperages that are 40% higher than nameplate. The smaller the motor, the larger the service factor. For example, a motor with an FLA of 10 A and a service factor of 1.25 can operate at 12.5 A (10 A $\times$ 1.25 = 12.5 A) without damaging the motor. The overload protection for a particular motor takes the service factor into account.

19.9 NATIONAL ELECTRICAL CODE® STANDARDS

The *National Electrical Code®* (NEC®) sets the standard for all electrical installations, including motor overload protection. The code book published by the National Electrical Manufacturer's Association should be consulted for any overload problems or misunderstandings that may occur regarding correct selection of the overload device.

The purpose of the overload protection device is to disconnect the motor from the circuit when an overload condition occurs. Detecting the overload condition and opening the circuit to the motor can be done in several ways: by an overload device mounted on the motor starter, or by a separate overload relay applied to a system with a contactor. **Figure 19–22** shows an example of a thermal overload relay.

19.10 TEMPERATURE-SENSING DEVICES

Various sensing devices are used for overcurrent situations. The most popular ones are those sensitive to temperature changes.

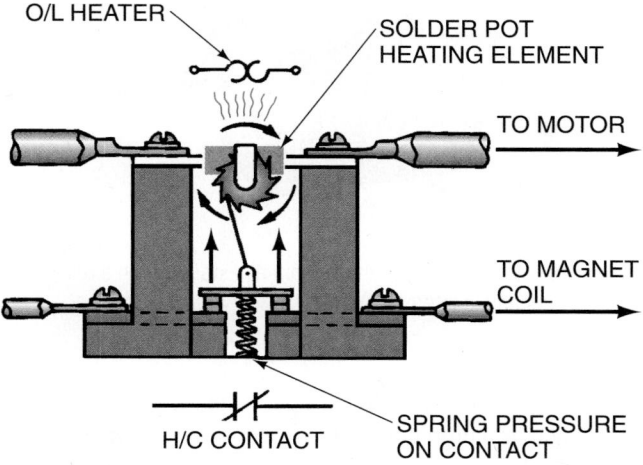

Figure 19–22 An overload using a resistor-type heater that heats a low-temperature solder.

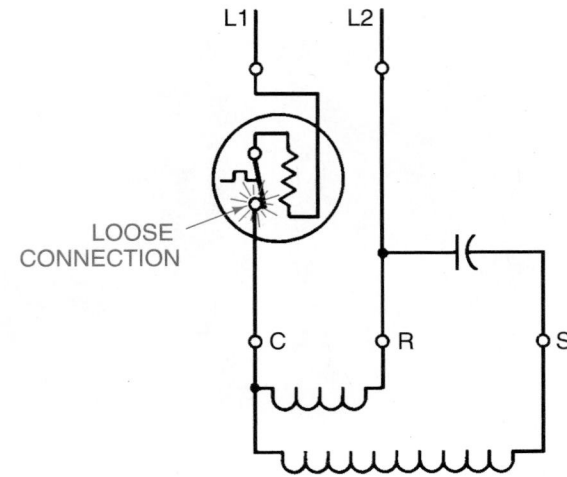

Figure 19–23 Overload is tripping because of a loose connection.

The bimetal element is an example. The line current of the motor passes through a heater (that can be changed to suit a particular motor amperage) that heats a bimetal strip. When the current is excessive, the heater warps the bimetal, opening a set of contacts that interrupt power to the contactor's coil circuit. All bimetal overload devices are designed with snap action to avoid excessive arcing. These thermal-type overloads are sensitive to any temperature and conditions around them. High ambient temperature and loose connections are frequent problems encountered. **Figure 19–23** shows an example of a thermal overload with a loose connection.

A low-melting solder may be used in place of the bimetal. This is called a *solder pot*. The solder will melt from the heat caused by an overcurrent condition. The overload heater is sized for the particular amperage draw of the motor it is protecting. The overload control circuit will interrupt the power to the motor contactor coil and stop the motor in case of overload. The solder melts, and the overload mechanism turns because it is spring loaded. It can be reset when it cools, **Figure 19–22**.

Both of these overload protection devices are sensitive to temperature. The temperature of the heater causes them to function. Heat from any source, even if it has nothing to do with motor overload, makes the protection devices more sensitive. For example, if the overload device is located in a metal control panel in the sun, the heat from the sun may affect the performance of the overload protection device. A loose connection on one of the overload device leads will cause local heat and may cause it to open the circuit to the motor even though there is actually no overload, **Figure 19–24**.

19.11 MAGNETIC OVERLOAD DEVICES

Magnetic overload devices are separate components and are not attached to the motor starter. This component is very accurate and not affected by ambient temperature, **Figure 19–25**. The advantage of this overload device is that it can be located

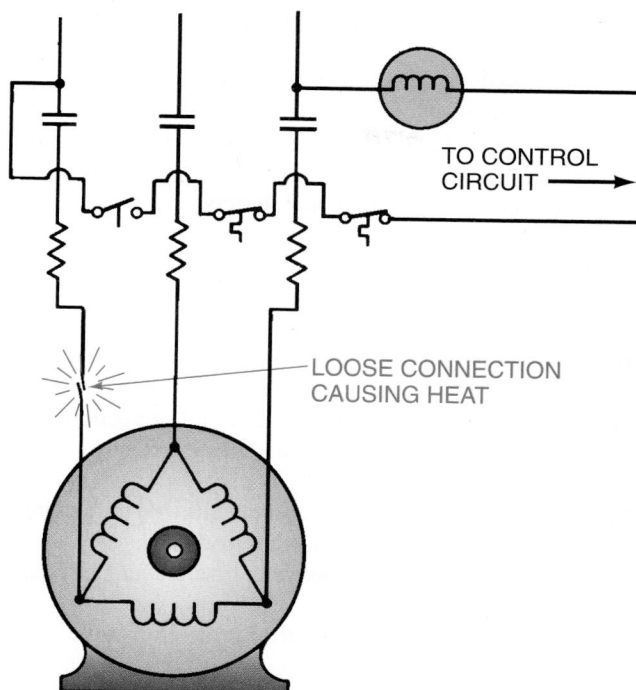

TO CONTROL
CIRCUIT →

LOOSE CONNECTION
CAUSING HEAT

Figure 19–24 Local heat, such as from a loose connection, will influence the thermal-type overload.

Figure 19–25 A magnetic overload device. *Photo by Bill Johnson*

in a hot cabinet on the roof, and the temperature will not affect it. It will shut the motor off at an accurate ampere rating regardless of the temperature.

19.12 RESTARTING THE MOTOR

SAFETY PRECAUTION: When a motor is found stopped for safety reasons, such as an overload, do not restart it immediately. Look around for the possible problem before a restart. When a motor stops because it is overloaded, the overload condition at the instant the motor is stopped is reduced to 0 A. This does *not* mean that the motor should be restarted immediately. The cause of the overload may still exist, and the motor could be too hot and may need to cool.

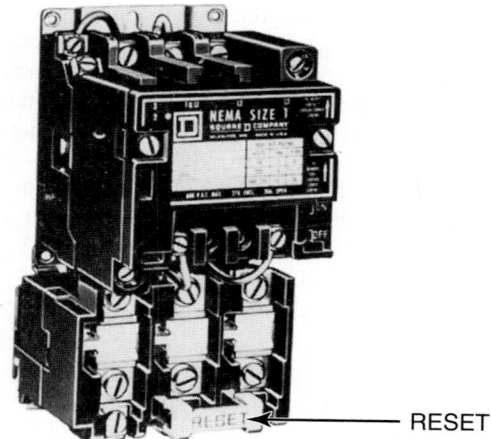

— RESET

Figure 19–26 Motor starters are equipped with overloads that are manually reset. The manual reset feature prevents excessive wear and tear on the motor. *Courtesy Square D Company*

There are various ways of restarting a motor after an overload condition has occurred. Some manufacturers design their control circuits with a manual reset to keep the motor from restarting, and some use a time delay to keep the motor off for a predetermined time. Others use a relay that will keep the motor off until the thermostat is reset. The units that have a manual reset at the overload device may require someone to go to the roof to reset the overload if the unit is located on the roof. See **Figure 19–26** for an example of a manual reset. When the reset is in the thermostat circuit, the protection devices may be reset from the room thermostat. This is convenient, but several controls may be reset at the same time, and the technician may not know which control is being reset. When the manual reset button is pushed and a restart occurs, there is no doubt which control has been reset to start the motor.

Time-delay reset devices keep the unit from short cycling but may reset themselves with a problem condition still existing.

SUMMARY

- The relay, the contactor, and the motor starter are three types of motor starting and stopping devices.
- The relay is used for switching circuits and motor starting.
- Motor starting relays are used for heavier duty jobs than are switching relays.
- Contactors are large relays that may be rebuilt.
- Starters are contactors with motor overload protection built into the framework of the contactor.
- The contacts on relays, contactors, and starters should not be filed or sanded.
- Large motors should be protected from overload conditions through the use of devices other than normal circuit overload protection devices.
- Inherent motor overload protection is provided by sensing devices within the motor.

- External motor overload protection is applied to the current-passing device: the relay, the contactor, or the starter.
- The service factor is the reserve capacity of the motor.
- Bimetal and solder-pot devices are thermally operated.
- Magnetic overload devices are very accurate and not affected by ambient temperature.
- SAFETY PRECAUTION: Most motors should not be restarted immediately after shutdown from an overload condition because they may need time to cool. When possible, determine the reason for the overload condition before restarting the motor.

REVIEW QUESTIONS

1. The recommended repair for a defective relay is to
   A. replace it.
   B. send it to a repair shop.
   C. fix it yourself.
   D. none of the above.
2. What components can be changed on a contactor and a starter for rebuilding purposes?
3. The two types of relays are _____ and _____.
4. The two amperages that influence the choice for replacing a motor starter are _____ and _____.
5. What is the difference between a contactor and a starter?
6. True or False: A contactor can always be converted to a starter.
7. What are the contact surfaces of relays, contactors, and starters made of?
8. What causes an overload protection device to function?
9. Which is not a typical operating voltage used for relays, contactors, and motor starters?
   A. 12 V
   B. 24 V
   C. 115 V
   D. 230 V
10. Why is it not a good idea to file or sand the contactor contacts?
11. Why is it not a good idea to use circuit protection devices to protect large motors from overload conditions?
12. Under what conditions are motors allowed to operate with slightly higher-than-design loads?
13. Describe the difference between inherent and external overload protection.
14. What is the purpose of overload protection at the motor?
15. True or False: A motor can be restarted immediately after it has stopped or been overloaded.

UNIT 20
Troubleshooting Electric Motors

OBJECTIVES

After studying this unit, you should be able to

- describe different types of electric motor problems.
- list common electrical problems in electric motors.
- identify various mechanical problems in electric motors.
- describe a capacitor checkout procedure.
- explain the difference between troubleshooting a hermetic motor problem and troubleshooting an open motor problem.

SAFETY CHECKLIST

- ✔ If it is suspected that a motor has electrical problems, pull the motor disconnect to prevent further damage or an unsafe condition, lock and tag the panel, and keep the only key on your person.
- ✔ Before checking a motor capacitor, short from one terminal to the other with a 20,000-Ω 5-W resistor to discharge the capacitor. This practice is recommended even if the capacitor has its own bleed resistor. Use insulated pliers.
- ✔ When wiring a motor run capacitor be sure to connect the lead that feeds power to the capacitor to the terminal identified for this purpose.
- ✔ Turn the power off before trying to turn the open drive compressor over using a wrench.

20.1 ▶ MOTOR TROUBLESHOOTING

Electric motor problems are either mechanical or electrical. Mechanical problems may appear to be electrical. For example, a bearing dragging in a small permanent split-capacitor (PSC) fan motor may not make any noise. The motor may not start, and it appears to be an electrical problem. The technician must know how to diagnose the problem correctly. This is particularly true with open motors because if the driven component is stuck, a motor may be changed unnecessarily. If the stuck component is a hermetic compressor, the whole compressor must be changed; if it is a serviceable hermetic compressor, the motor can be replaced, or the compressor running gear can be rebuilt.

20.2 MECHANICAL MOTOR PROBLEMS

Mechanical motor problems normally occur in the bearings or the shaft where the drive is attached. The bearings may be tight or worn due to lack of lubrication. Grit can easily get into the bearings of some open motors and cause them to wear. Overtightened belts can also put excessive stress on the motor's bearings, causing them to wear and fail prematurely.

Problems with large motors are not usually repaired by heating, air-conditioning, and refrigeration technicians. They are handled by technicians trained in rebuilding motors and rotating equipment. A motor vibration may require you to seek help from a qualified balancing technician. Explore every possibility to ensure that the vibration is not caused by a field problem, such as a fan loaded with dirt or liquid flooding into a compressor.

Motor bearing failure with roller and ball bearings can often be identified by the bearing noise. When sleeve bearings fail, they normally lock up (will not turn) or sag to the point that the motor is out of its magnetic center. At this point the motor will not start. When motor bearings fail, they can be replaced.

If the motor is small, the motor is normally replaced because it would cost more to change the bearings than to purchase and install a new motor. The labor involved in obtaining bearings and disassembling the motor can take too much time to make a profit. This is particularly true for fractional horsepower fan motors. These small motors almost always have sleeve bearings pressed into the end bells of the motor, and special tools may be needed to remove and install new bearings, **Figure 20–1**.

When deciding whether to repair or replace a motor, a number of factors should be considered. These factors include the availability of a new motor versus the availability of the parts needed to repair the motor, the cost of repairing the motor, the cost of the new motor, the labor involved in repairing as opposed to replacing the motor, and the warranty/guarantee on the new and repaired motors. If the cost of repairing the motor is close to the cost of replacing the motor, for example, it would definitely pay to replace the motor if the new component carries a one-year warranty and the repaired motor carries only a 90-day guarantee.

20.3 REMOVING DRIVE ASSEMBLIES

To remove the motor, you need to remove the pulley, coupling, or fan wheel from the motor shaft. The fit between the shaft and whatever assembly it is fastened to may be very tight. SAFETY PRECAUTION: Removing the assembly from the motor shaft must be done with care. The assembly may have been running on this shaft for years, and it may have

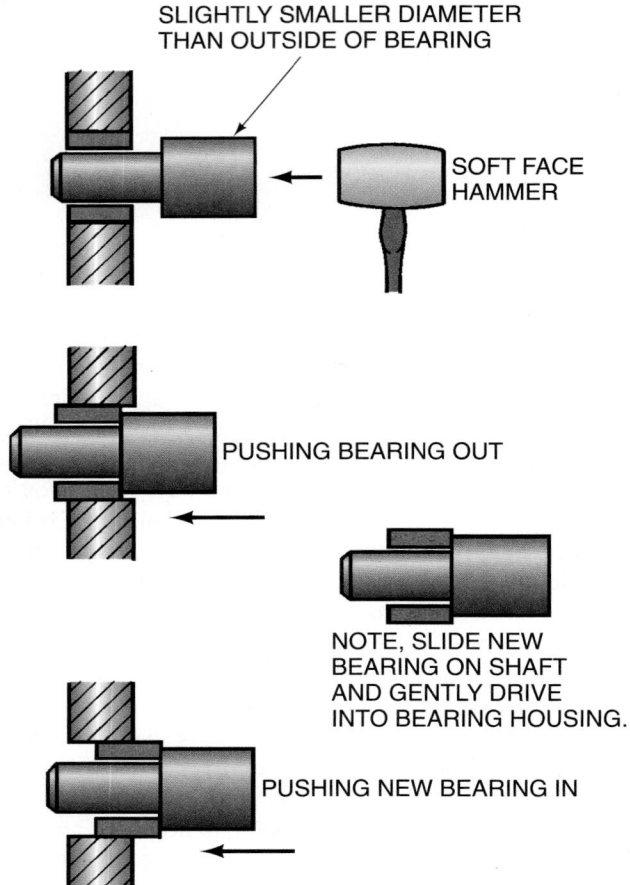

Figure 20–1 A special tool for removing bearings.

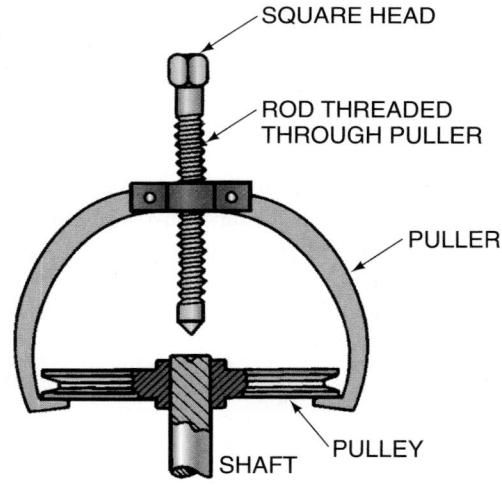

Figure 20–2 A pulley puller.

rust between the shaft and the assembly. You must remove the assembly without damaging it. Special pulley pullers, **Figure 20–2,** will help, but other tools or procedures may be required.

Most assemblies are held to the motor shaft with setscrews threaded through the assembly and tightened

Figure 20–3 A flat spot on the motor shaft where the pulley setscrew is tightened. *Photo by Bill Johnson*

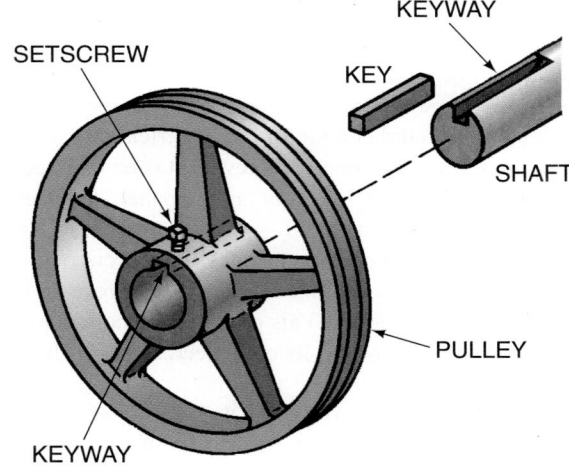

Figure 20–4 A pulley with a groove cut in it that matches a groove in the shaft. A key is placed in these grooves and the setscrew is often tightened down on top of the key.

against the shaft. A flat spot is usually provided on the shaft for the seating of the setscrew and to keep it from damaging the shaft surface, **Figure 20–3.** The setscrew is made of very hard steel, much harder than the motor shaft. When larger motors with more torque are used, a matching keyway is normally machined in the shaft and assembly. This keyway with a key provides a better bond between the motor and assembly, **Figure 20–4.** A setscrew is then often tightened down on the top of the key to secure the assembly to the motor shaft.

Many technicians make the mistake of trying to drive a motor shaft out of the assembly fastened to the shaft. In doing so, they blunt or distort the end of the motor shaft. When it is distorted, the motor shaft will never go through the assembly without damaging it, **Figure 20–5.** The shaft is made from mild steel and can be damaged easily. If the shaft must be driven, you may need to use as the driving tool a similar shaft with a slightly smaller diameter, **Figure 20–6.** Before attempting to remove a belt from a motor shaft, it is a good idea to clean, sand, and remove all debris from the motor

Figure 20–5   A motor shaft damaged from trying to drive it through the pulley. *Photo by Bill Johnson*

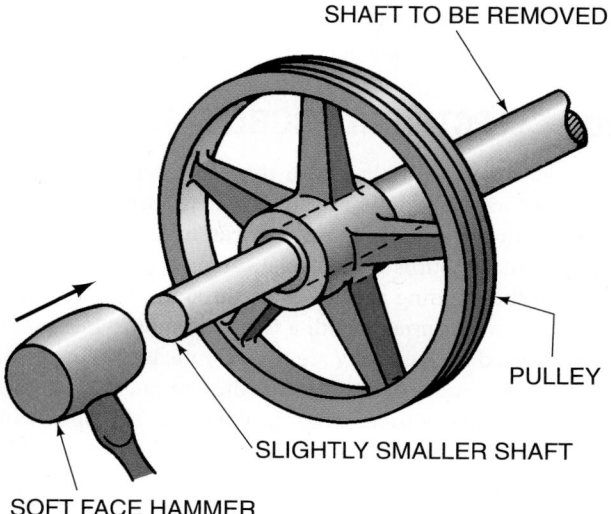

SHAFT TO BE REMOVED

PULLEY

SLIGHTLY SMALLER SHAFT

SOFT FACE HAMMER

Figure 20–6   A shaft driven through the pulley with another shaft as the contact surface. The shaft that is used as the contact surface has a smaller diameter than the original shaft.

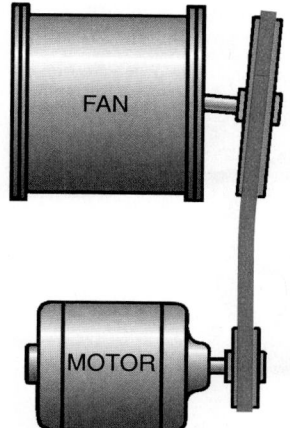

FAN

MOTOR

Figure 20–7   A belt that is too tight.

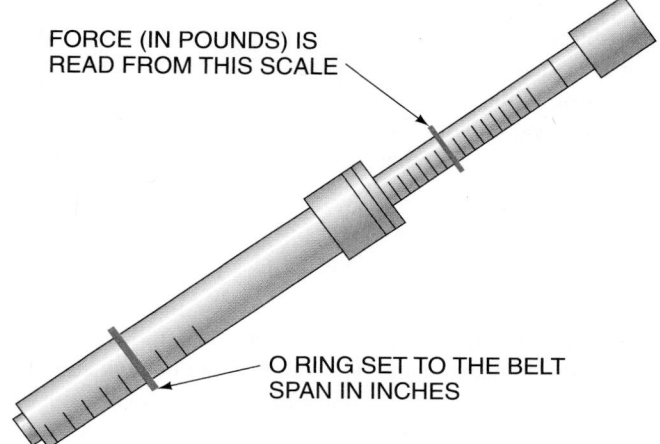

FORCE (IN POUNDS) IS READ FROM THIS SCALE

O RING SET TO THE BELT SPAN IN INCHES

Figure 20–8   A belt tension gage.

shaft. In addition, spraying the shaft with a rust-dissolving solution or penetrating oil will help make the removal process much easier.

20.4   BELT TENSION

Many motors fail because of overtightened belts and incorrect alignment. The technician should be aware of the specifications for the motor belt tension on belt-drive systems. A belt tension gage will ensure properly adjusted belts when the gage manufacturer's directions are followed. Belts that are too tight strain the bearings so that they wear out prematurely, **Figure 20–7. Figure 20–8** shows one type of belt tension gage.

This type of belt tension gage uses two O rings. In order to use this gage, the center-to-center measurement must be taken between the two shafts, **Figure 20–9.** The larger O ring is positioned on the gage so that it is lined up with the corresponding distance, in inches, between the two shafts. A straightedge is then rested across the two pulleys. The gage is then positioned at the center of the belt span between the two pulleys and pushed toward the other end of the pulley until the large O ring is even with the straightedge, **Figure 20–10.** The smaller O ring will register a force, in pounds, on the top

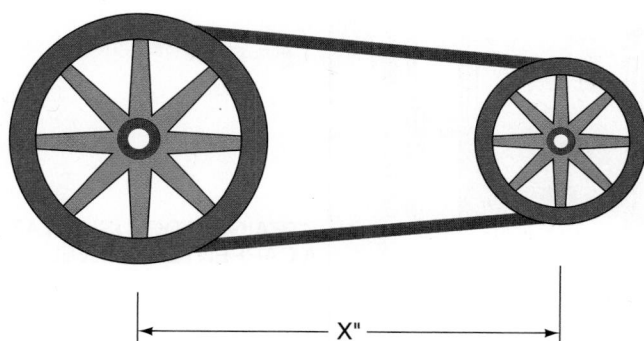

X"

Figure 20–9   The measurement between the centers of the drive and driven shafts.

portion of the gage. By comparing this value with the chart that accompanies the gage, the technician can determine if the belt is too tight or too loose. A low force reading indicates that the belt is too loose, whereas a force reading that is too high indicates that the belt is too tight.

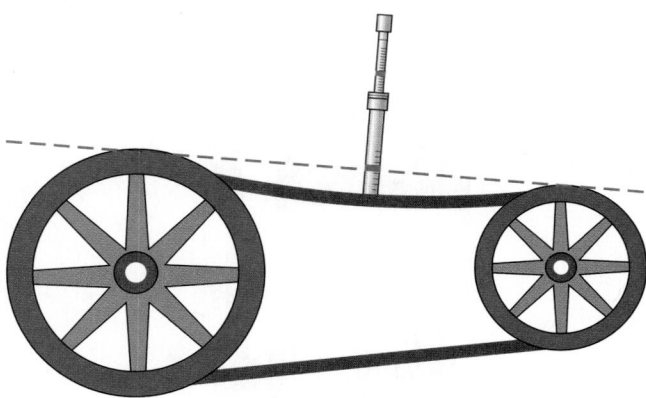

Figure 20–10 Using the belt tension gage.

Belts that are too loose can cause problems—as can those that are too tight. Loose belts have a tendency to slip, and slipping belts are a cause for reduced airflow through the air distribution system. Since motors often rely on the air moving over them to cool the windings, a reduction in airflow can cause motors to overheat. As a result, the internal thermal overload on a motor may open and give the technician reason to believe that there is an electrical problem with the motor.

20.5 PULLEY ALIGNMENT

Pulley alignment is very important. If the drive pulley and driven pulley are not in line, a strain is imposed on the shafts' drive mechanisms. The pulleys may be aligned with the help of a straightedge, **Figure 20–11.** A certain amount of adjustment tolerance is built into the motor base on small

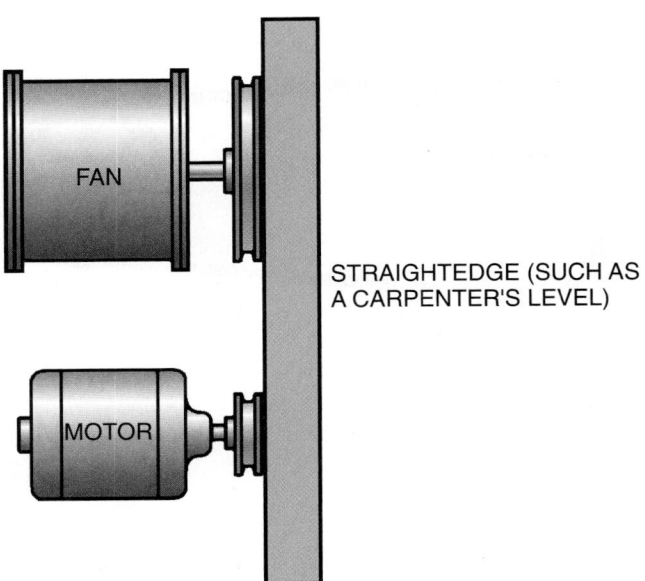

Figure 20–11 Pulleys must be in proper alignment or excessive belt and bearing wear will occur. The pulleys can be aligned using a straightedge.

motors, and this may be enough to allow the motor to be out of alignment if the belt becomes loose. Aligning shafts on some pieces of equipment may not be easy to do, but it must be done, or the motor or drive mechanism will not last.

When mechanical problems occur with a motor, the motor is normally either replaced or taken to a motor repair shop. Bearings can sometimes be replaced in the field by a competent technician, but it is generally better to leave this type of repair to motor experts. When the problem is the pulley or drive mechanism, the air-conditioning, heating, and refrigeration technician may be responsible for the repair. NOTE: Proper tools must be used for motor repair, or shaft and motor damage may occur.

20.6 ELECTRICAL PROBLEMS

Electrical motor problems are the same for hermetic and open motors. Open motor problems are a little easier to understand or diagnose because they can often be seen. When an open motor burns up, this can often be easily diagnosed because the winding may be seen through the end bells. It may also smell burned. With a hermetic motor, instruments must be used because they are the only means of diagnosing problems that are not visible inside the compressor. There are three common electrical motor problems: (1) an open winding, (2) a short circuit from winding to winding, and (3) a short circuit from the winding to ground.

20.7 OPEN WINDINGS

Open windings in a motor can be found with an ohmmeter. There should be a known measurable resistance from terminal to terminal on every motor for it to run when power is applied to the windings. Single-phase motors must have the applied system voltage at the run winding to run and at the start winding during starting, **Figure 20–12. Figure 20–13** is an illustration of a motor with an open start winding.

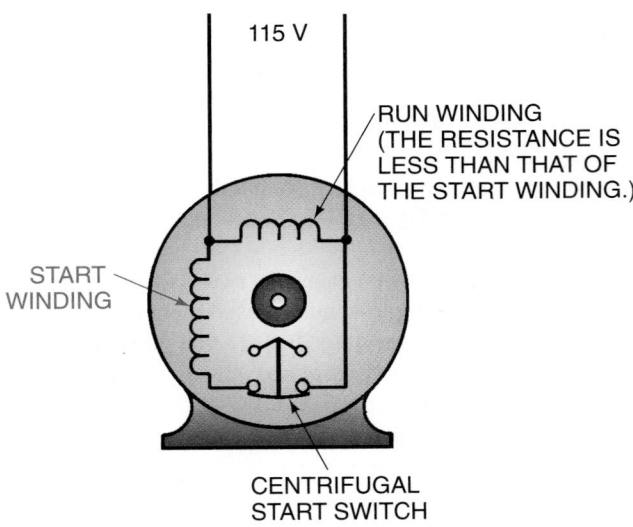

Figure 20–12 A wiring diagram with run and start windings.

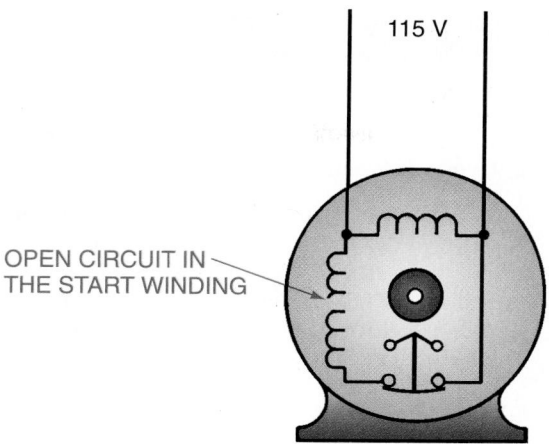

Figure 20–13 A motor with an open winding.

A single-phase motor that is in good working order may have the following resistance readings between the motor terminals:

Common to start: 3 Ω
Common to run: 2 Ω
Run to start: 5 Ω

This situation is shown in **Figure 20–14.**

The resistance reading from the common terminal to the run terminal only provides the technician with the resistance reading of the run winding. The resistance reading from the common terminal to the start terminal only provides the technician with the resistance reading of the start winding. The resistance between the start and run terminals gives the sum of the resistances of both the start and run windings. If the start winding in this example were to open, the resistance readings for the same motor would be as follows:

Common to start: ∞ (infinite resistance)
Common to run: 2 Ω
Run to start: ∞ (infinite resistance)

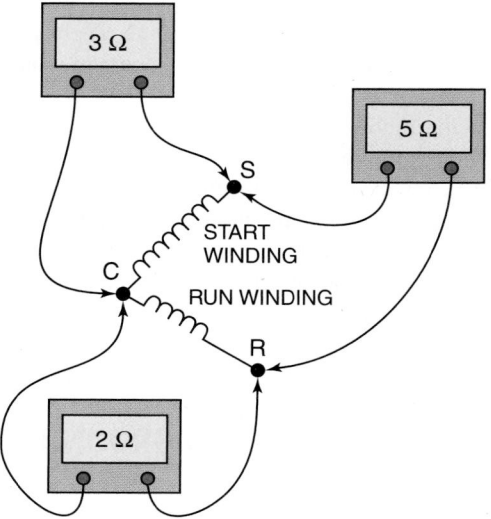

Figure 20–14 Resistance readings between the terminal pairs on a hermetic motor.

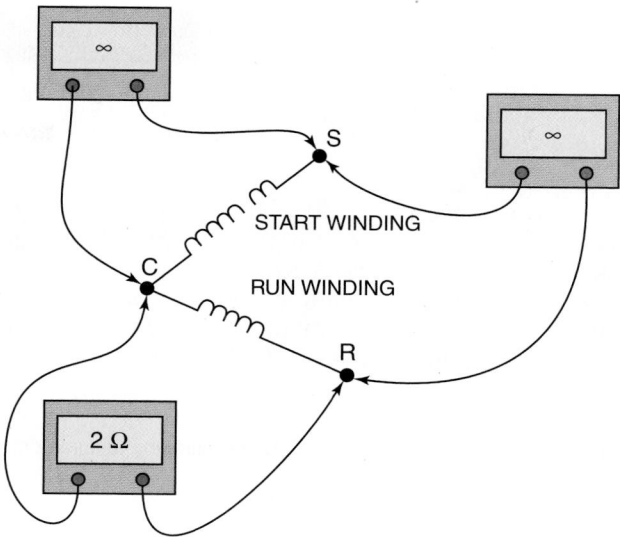

Figure 20–15 Resistance readings between terminal pairs on a hermetic motor with an open start winding.

The open winding would affect the resistance reading of the start winding as well as the resistance reading between the run and start terminals, **Figure 20–15.**

20.8 SHORTED MOTOR WINDINGS

Short circuits in windings occur when the conductors in the winding touch each other where the insulation is worn or in some way defective. This creates a short path through which the electrical energy flows. This path has a lower resistance and increases the current flow in the winding. Although motor windings appear to be made from bare copper wire, they are coated with an insulator to keep the copper wires from touching each other. The measurable resistance mentioned in the previous paragraph is known for all motors. Some motors have a published resistance for their windings. The best way to check a motor for electrical soundness is to *know what the measurable resistance should be for a particular winding and verify it with a good ohmmeter*, **Figure 20–16.** This measurable resistance will be less than the rated value when a motor has short-circuit problems. The decrease in resistance causes the current to rise, which causes motor overload devices to open the circuit and possibly even trips the circuit overload protection. If the resistance does not read within these tolerances, there is a problem with the winding. This table is helpful when you troubleshoot a hermetic compressor. Tables may not be easy to obtain for open motors, and the windings are not as easy to check because the individual windings do not all come out to terminals as they do on a hermetic compressor.

If the decrease in resistance in the windings is in the start winding, the motor may not start. The reason for this is that, in order to start, the motor relies on an imbalance in the magnetic field. If the start winding is partially shorted, the resistance in that winding will be closer to the resistance of the run winding. This will reduce, or possibly even eliminate, the

Compressor Model		Voltage	MOTOR AMPS				FUSE SIZE		Winding Resistance in Ohms
			Full Winding		1/2 Winding		Recommended Max		
			Rated Load	Locked Rotor	Rated Load	Locked Rotor	Fusetron	Std.	
9RA - 0500 - CFB		230/1/60	27.5	125.0			FRN-40	50	Start 1.5 Run 0.40
9RB	TFC	208-230/3/60	22.0	115.0			FRN-25	40	0.51-0.61
9RJ	TFD	460/3/60	12.1	53.0			FRS-15	15	2.22-2.78
9TK	TFE	575/3/60	7.8	42.0			FRS-10	15	3.40-3.96
MRA	FSR	200-240/3/50	17.0	90.0	8.5	58.0	FRN-25	35	0.58-0.69
MRB	FSM	380-420/3/50	9.5	50.0	4.8	32.5	FRS-15	20	1.80-2.15
MRF									

Figure 20–16   Resistances for some typical hermetic compressors. *Courtesy Copeland Corporation*

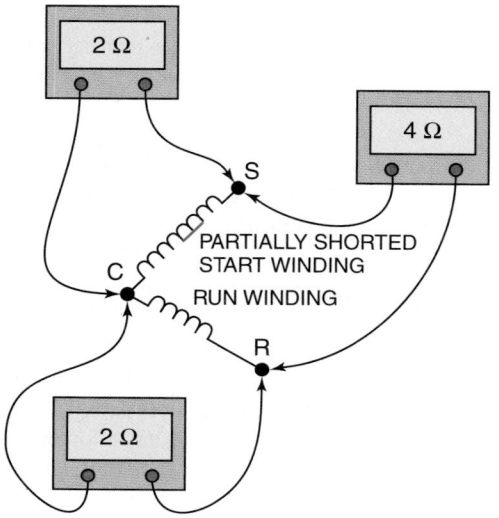

Figure 20–17   Resistance readings between terminal pairs on a hermetic motor with a partially shorted start winding.

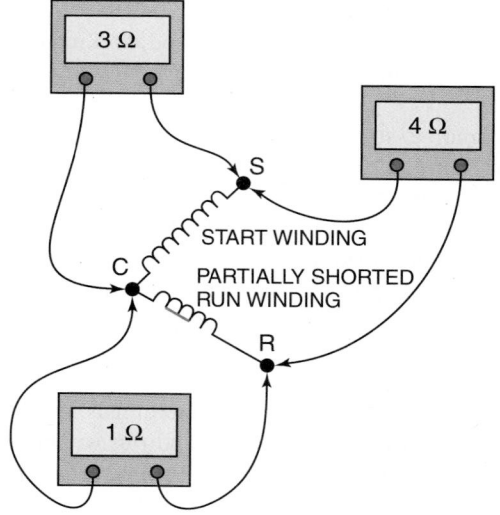

Figure 20–18   Resistance readings between terminal pairs on a hermetic motor with a partially shorted run winding.

magnetic field imbalance that is needed to start the motor, **Figure 20–17.** If the decrease in resistance is in the run winding, the motor may start and draw too much current while running. If the resistance of the run winding is lowered, the imbalance in magnetic fields between the start and the run winding will be increased, and the starting torque of the motor will actually increase, **Figure 20–18,** but the motor may very well overheat. If the motor winding resistance cannot be determined, then it is hard to know whether a motor is overloaded or whether it has a defective winding when only a few of its coils are shortened.

When stretched out, the windings of a motor may be many feet long. This insulated winding wire is wound into coils. The shorting of a winding wire may make it a shorter winding; thus a higher amperage reading and a localized "hot spot" will occur where the windings are shorted together. For example, suppose the total winding length is 50 ft and the factory ohm reading is supposed to be 30 Ω. When it is shorted to 45 ft where two turns are touching due to insulation failure,

the ohm reading may be reduced to 27 Ω. There would be a hot spot where they touch and the amperage would go up slightly.

When the motor is an open motor, the load can be removed. For example, the belts can be removed, or the coupling can be taken apart, and the motor can be started without the load, **Figure 20–19.** If the motor starts and runs correctly without the load, the load may be too great.

Three-phase motors must have the same resistance for each winding. (There are three identical windings.) Otherwise there is a problem. An ohmmeter check will quickly reveal an incorrect winding resistance, **Figure 20–20.**

20.9   SHORT CIRCUIT TO GROUND (FRAME)

A short circuit from winding to ground or to the frame of the motor may be detected with a good ohmmeter. No circuit should be detectable from the winding to ground. The copper

Figure 20–19 The coupling was disconnected between this motor and pump because it was suspected that the motor or pump was locked up and would not turn.

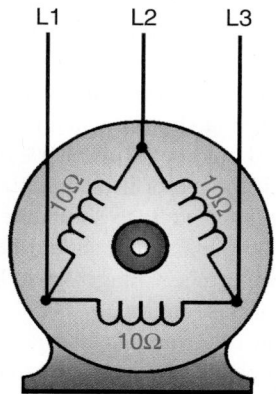

Figure 20–20 A wiring diagram of a three-phase motor. The resistance is the same across all three windings.

suction line on a compressor is a good source for checking to the ground. NOTE: "Ground" or "frame" are interchangeable terms because the frame should be grounded to the earth ground through the building electrical system, **Figure 20–21**.

To check a motor for a ground, use a good ohmmeter with an R × 10,000-Ω scale. Special instruments for finding very high resistances to ground are used for larger, more sophisticated motors. Most technicians use ohmmeters. Top-quality instruments can detect a ground in the 10,000,000 Ω and higher range. A megohmmeter may be used. It has an internal high-voltage direct-current (DC) supply to help create conditions for detecting the ground, **Figure 20–22**. The megohmmeter is often called a Megger and will measure resistances in the millions of ohms.

A typical ohmmeter will detect a ground of about 1,000,000 Ω or less. The rule of thumb is that if an ohmmeter set to the R × 10,000 scale will even move the needle with one lead touching the motor terminal and the other touching a ground (such as a copper suction line), the motor should be started with care. SAFETY PRECAUTION: If the meter needle moves to the midscale area, do *not* start the motor, **Figure 20–23**. When a meter reads a very slight resistance to ground, the windings may be dirty and damp if it is an open motor. Clean the motor, and the ground will probably be eliminated. Some open motors operating in a dirty, damp atmosphere may indicate a slight circuit to ground in damp weather. Air-cooled condenser fan motors are an example. When the motor is started and allowed to run long enough to get warm and dry, the ground circuit may disappear.

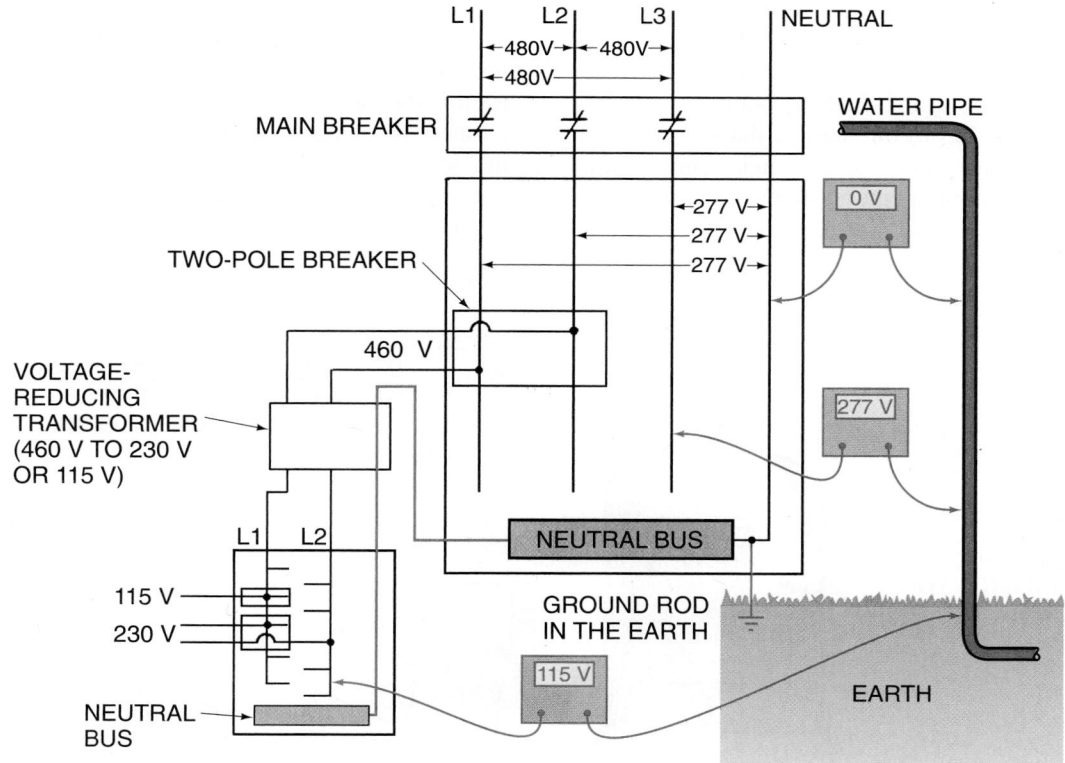

Figure 20–21 This building electrical diagram shows the relationship of the earth ground system to the system's piping.

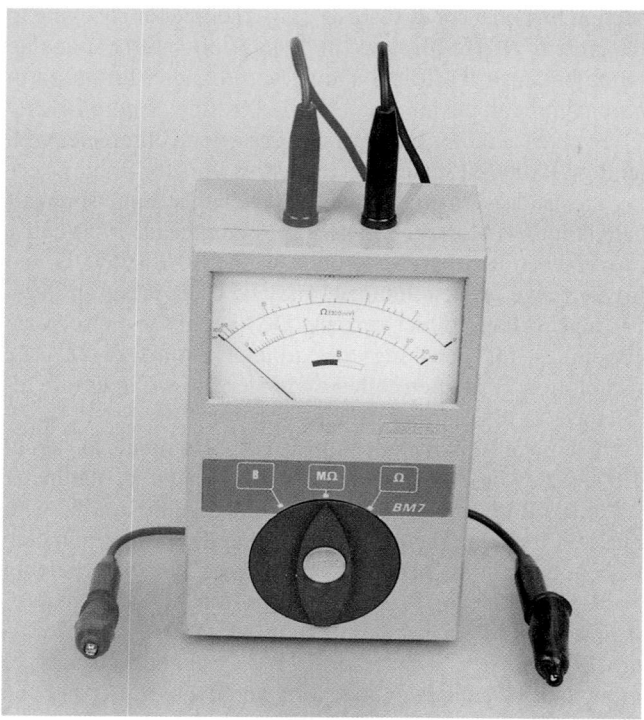

Figure 20–22   A Megger. *Photo by Bill Johnson*

Hermetic compressors may occasionally have a slight ground due to the oil and liquid refrigerant in the motor splashing on the windings. The oil may have dirt suspended in it and show a slight ground. Liquid refrigerant causes this condition to be worse. If the ohmmeter shows a slight ground but the motor starts, run the motor for a little while and check again. If the ground persists, the motor is probably going to fail soon if the system is not cleaned. A suction-line filter drier may help remove particles that are circulating in the system and causing the slight ground.

For troubleshooting electric motors, the ammeter and the voltmeter are the main instruments used. If the resistance is correct, the motor is electrically sound. Other problems may be found using the ammeter.

20.10 SINGLE-PHASE MOTOR STARTING PROBLEMS

Single-phase open electric motors are fairly easy to troubleshoot for starting problems because they can be seen and heard. It is very frustrating to turn the power on to any motor if it fails to start. First, make sure that you really have full power to the motor leads. This entails fastening voltmeter leads to the motor terminals and reading the meter when the power is applied. Low voltage can occur at any place in the electrical system, such as at loose connections, loose fuse holders, undersized wire, or low voltage from the main electrical panel. The only way to be sure is to measure it at the motor terminals. Even at start-up when the motor is pulling LRAs, the voltage should be + or −10% of the rated voltage. For example, if the motor is a combination 208/230-V fan motor, the minimum voltage should be 187 V (208 × 0.90 = 187.2) and the maximum should be 253 V (230 × 1.10 = 253). The motor's voltage must be within these limits during starting or running.

Open motors must be started with the start winding and the run winding receiving power at the same time. The start winding is then disconnected from the circuit as the motor reaches about three-fourths of its rated speed. On some motors, such as capacitor-start, capacitor-run (CSCR) motors, the start winding remains in the circuit through the run capacitor. The run capacitor thus limits the current through the start winding to prevent it from getting too hot and burning up. The run capacitor is a low microfarad capacitor. Disconnecting the start winding can be just as much a part of troubleshooting as operating the motor itself. When the open motor with a centrifugal switch is started and starts to turn, you should hear the audible disconnect of the start winding by the switch. When the motor is stopped, the switch makes

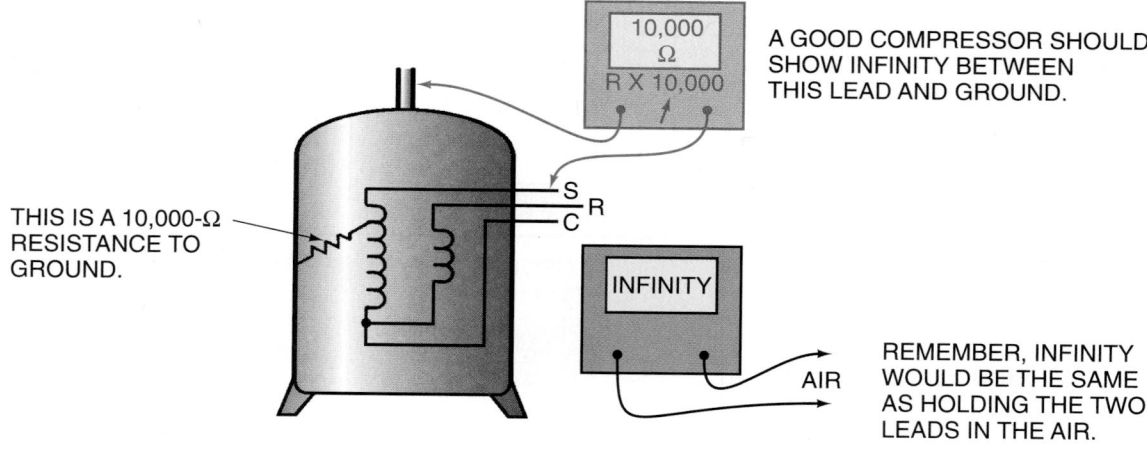

Figure 20–23   A volt-ohmmeter detecting a circuit to ground in a compressor winding.

EACH SHOULD TURN
EASILY BY HAND.

MAKE SURE POWER IS OFF.

Figure 20–24   An example of how a pump or compressor coupling can be disconnected and the component's shaft turned by hand to check for a hard-to-turn shaft or a stuck component. SAFETY PRECAUTION: Turn off the power to the motor and lock out before servicing the motor.

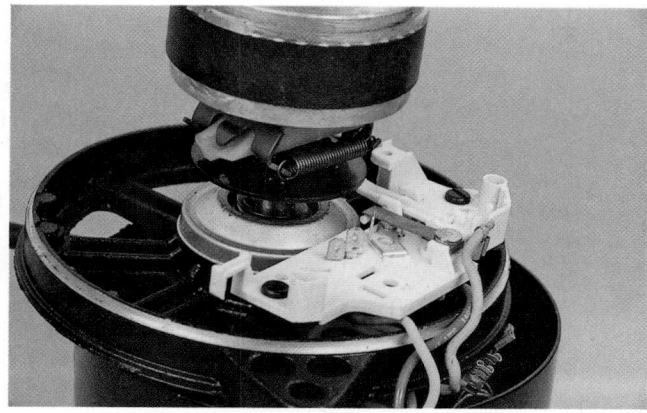

Figure 20–25   The end bell of this motor has been removed to examine the start switch and the windings. *Photo by Bill Johnson*

BLEED RESISTOR

Figure 20–26   A start capacitor with a bleed resistor across the terminals to bleed off the charge during the off cycle. SAFETY PRECAUTION: Make certain that capacitors are completely discharged before you come into contact with the terminals. *Photo by Bill Johnson*

an audible sound when the contacts close. If you hear the centrifugal switch function and the motor still will not start, you must apply meter leads to the start winding in the end bell (start to common terminals) of the motor to make sure full voltage is reaching the start winding during start-up. If the motor still will not start, perform the same test on the run windings (run to common terminals).

The electronic start switch must be checked with a voltmeter applied to the start winding circuit in the end bell of the motor. This is a little more involved than just listening to the centrifugal switch.

The following are symptoms of electric-motor starting problems:

- The motor hums and then shuts off.
- The motor runs for short periods and shuts off.
- The motor will not try to start at all.

The technician must decide whether there is a motor mechanical problem, a motor electrical problem, a circuit problem, or a load problem. SAFETY PRECAUTION: If the motor is an open motor, turn the power off. Lock and tag panel. Try to rotate the motor by hand. If it is a fan or a pump, it should be easy to turn, **Figure 20–24**. If it is a compressor, the shaft may be hard to turn. Use a wrench to grip the coupling when trying to turn a compressor. Be sure the power is off.

If the motor and the load turn freely, examine the motor windings and components. If the motor is humming and not starting, the starting switch may need replacing, or the windings may be burned. If the motor is open, you may be able to visually check them; if you cannot, remove the motor end bell, **Figure 20–25**.

20.11   CHECKING CAPACITORS

Motor capacitors may be checked to some extent with an ohmmeter in the following manner. Turn off the power to the motor and remove one lead of the capacitor. SAFETY PRECAUTION: Short from one terminal to the other with a 20,000-Ω, 5-W resistor to discharge the capacitor in case the capacitor has a charge stored in it. Some start capacitors have a resistor across these terminals to bleed off the charge during the off cycle, **Figure 20–26**. Short across the capacitor anyway by using the above resistor. Use insulated pliers. The resistor may be open and will not bleed the charge from the capacitor. If you place ohmmeter leads across a charged capacitor, the meter movement may be damaged. Set the ohmmeter scale on an analog meter to the R × 10 scale and touch the leads to the capacitor terminals, **Figure 20–27**. If the capacitor is good, the meter needle will go toward 0 and begin to fall back toward infinite resistance. If the leads are left on the capacitor for a long time, the needle will fall back to infinite resistance. If the needle falls part of the way back and will drop no more, the capacitor has an internal short. The internal plate is touching the other internal plate. If the needle will not rise at all, try the R × 100 scale. Try reversing the leads. The ohmmeter is charging the capacitor with its internal battery. This is DC voltage. When the capacitor is charged in one direction, the meter leads must be reversed for the next check. If the capacitor has a bleed resistor, the capacitor will charge toward 0 Ω, then drop back to the value of the resistor.

DISCONNECT BLEED RESISTOR AFTER BLEEDING WITH A 20,000-OHM, 5-W RESISTOR FOR TEST IF DESIRED. WHEN BLEED RESISTOR IS LEFT IN THE CIRCUIT, THE METER NEEDLE WILL RISE FAST AND FALL BACK TO THE BLEED RESISTOR VALUE.

FOR BOTH RUN AND START CAPACITORS

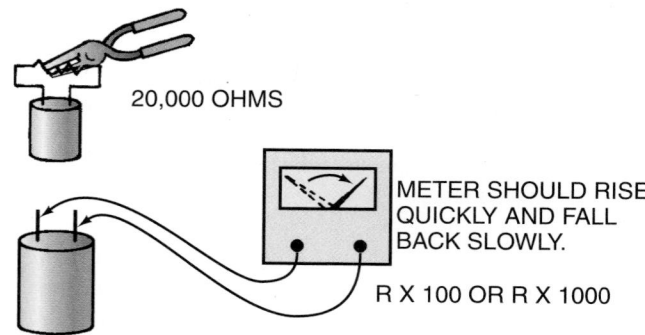

1. FIRST, SHORT THE CAPACITOR FROM POLE TO POLE USING A 20,000-OHM, 5-WATT RESISTOR WITH INSULATED PLIERS.

   20,000 OHMS

2. USING THE R X 100 OR R X 1000 SCALE, TOUCH THE METER'S LEADS TO THE CAPACITOR'S TERMINALS. METER NEEDLE SHOULD RISE FAST AND FALL BACK SLOWLY. IT WILL EVENTUALLY FALL BACK TO INFINITY IF THE CAPACITOR IS GOOD (PROVIDED THERE IS NO BLEED RESISTOR).

   METER SHOULD RISE QUICKLY AND FALL BACK SLOWLY.

   R X 100 OR R X 1000

   START CAPACITOR

3. YOU CAN REVERSE THE LEADS FOR A REPEAT TEST, OR SHORT THE CAPACITOR TERMINALS AGAIN. IF YOU REVERSE THE LEADS, THE METER NEEDLE MAY RISE EXCESSIVELY HIGH AS THERE IS STILL A SMALL CHARGE LEFT IN THE CAPACITOR.

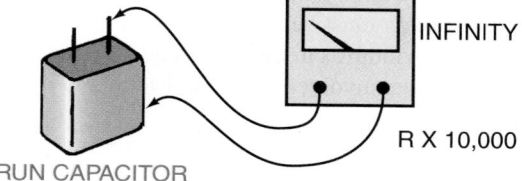

4. FOR RUN CAPACITORS THAT ARE IN A METAL CAN: WHEN ONE LEAD IS PLACED ON THE CAN AND THE OTHER LEAD ON A TERMINAL, INFINITY SHOULD BE INDICATED ON THE METER USING THE R X 10,000 OR R X 1000 SCALE.

   INFINITY

   RUN CAPACITOR

   R X 10,000

Figure 20–27   Procedure for field testing a capacitor.

This simple check with an ohmmeter will not give the capacitance of a capacitor. You need a capacitor tester to find the actual capacitance of a capacitor, **Figure 20–28.** It is not often that a capacitor will change in value, so this capacitor analyzer is used primarily by technicians who need to perform many capacitor checkouts.

20.12   IDENTIFICATION OF CAPACITORS

A run capacitor is contained in a metal can and is oil filled. If this capacitor gets hot due to overcurrent, it will often swell, **Figure 20–29.** The capacitor should then be changed.

SAFETY PRECAUTION: Run capacitors have an identified terminal to which the lead that feeds power to the capacitor should be connected. When the capacitor is wired in this manner, a fuse will blow if the capacitor is shorted to the container. If the capacitor is not wired in this manner and a short occurs, current can flow through the motor winding to ground during the off cycle and overheat the motor, **Figure 20–30.**

The start capacitor is a dry type and may be contained in a shell of paper or plastic. Paper containers are no longer used, but you may find one in an older motor. If this capacitor has

Figure 20–28   A digital capacitor tester. *Courtesy Davis Instruments*

been exposed to overcurrent, it may have a bulge in the vent at the top of the container, **Figure 20–31.** The capacitor should then be changed.

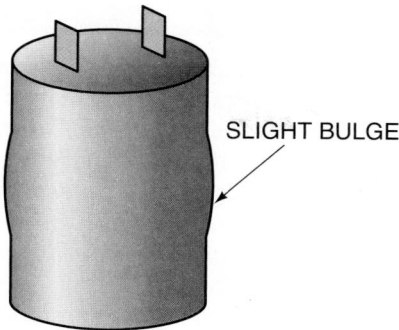

Figure 20–29 This capacitor swelled because of internal pressure.

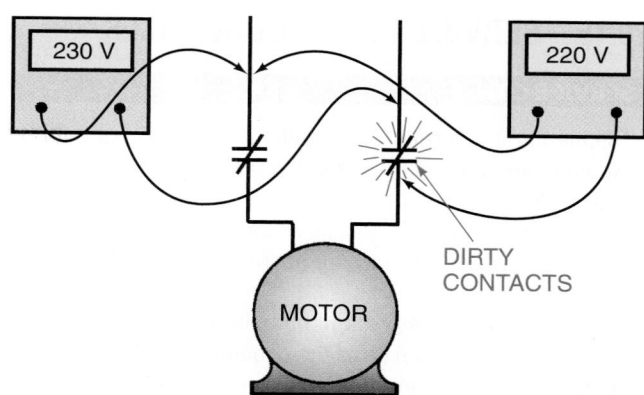

Figure 20–32 When voltmeter leads are applied to both sides of a set of dirty or burnt contacts, the voltmeter will read a voltage drop of 10 V or more.

20.13 WIRING AND CONNECTORS

The wiring and connectors that carry the power to a motor must be in good condition. When a connection becomes loose, oxidation of the copper wire occurs. The oxidation acts as an electrical resistance and causes the connection to heat more, which in turn causes more oxidation. This condition will only get worse. Loose connections will result in low voltages at the motor and in overcurrent conditions. Loose connections will appear the same as a set of dirty or burnt contacts and may be located with a voltmeter, **Figure 20–32.** If a connection is loose enough to create an overcurrent condition, it can often be located by a temperature rise at the loose connection.

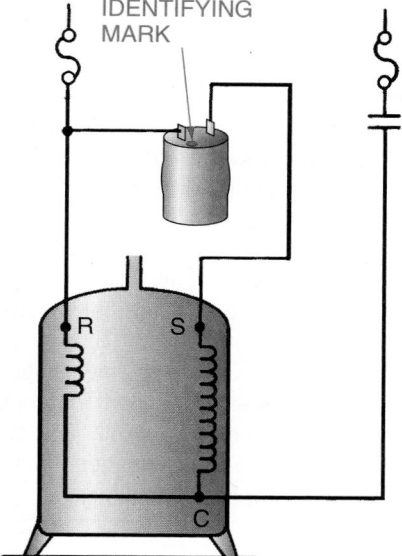

Figure 20–30 The wiring to a run capacitor for proper fuse protection of the circuit.

20.14 **TROUBLESHOOTING HERMETIC MOTORS**

Diagnosing hermetic compressor motor problems differs from diagnosing open motor problems because the motor is enclosed in a shell and cannot be seen; the motor sound level may be dampened by the compressor shell. Motor noises that are obvious in an open motor may become hard to hear in a hermetic compressor.

The motor inside a hermetic compressor can be checked electrically only from the outside. It will have the same problems as an open motor; that is, open circuit, short circuit, or grounded circuit. The technician must give the motor a complete electrical checkout from the outside of the shell; if it cannot be made to operate, the compressor must be changed. A motor checkout includes the starting and running components for a single-phase compressor, such as the run and start capacitors and the start relay.

As mentioned earlier, compressors operate in remote locations and are not normally attended. A compressor can throw a rod, tearing it up, and the motor may keep running until damaged. The compressor parts damage the motor, which may then burn out. The compressor damage is not detected, because the damage cannot be seen. When a compressor is changed because of a bad motor, you should suspect mechanical damage.

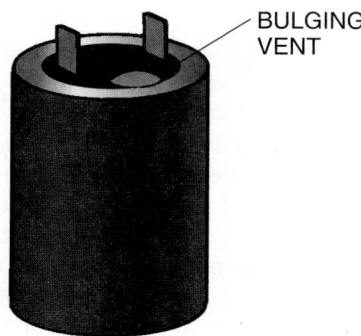

Figure 20–31 When the start capacitor has overheated, the small rubber diaphragm in the top of the component will bulge.

20.15   SERVICE TECHNICIAN CALLS

SERVICE CALL 1

The manager of a retail store calls. *They have no air conditioning and the piping going into the unit is frozen solid. The air handler is in the stock room. The evaporator fan motor winding is open, and the outdoor unit has continued to run without air passing over the evaporator, freezing it solid.*

On arriving at the store and noticing that the fan is not running the technician turns the cooling thermostat to OFF to stop the condensing unit, then turns the fan switch to ON. SAFETY PRECAUTION: The technician then turns off the power, locks and tags the panel, and removes the fan compartment cover. The fan motor body is cool to the touch, indicating the motor has been off for some time. The technician then turns on the power and checks power at the motor terminal block; the motor has power. The power is turned off and locked out. The technician checks the motor winding with an ohmmeter; the motor windings are open.

The fan motor is changed. The unit cannot be started until the coil is allowed to thaw. The technician instructs the store manager to allow the fan to run until air is felt at the outlet in the men's room. This ensures the airflow has started. Then the technician waits 15 min and sets the thermostat to COOL and the fan switch to AUTO. The technician leaves. A call back later in the day indicates that the system is now cooling correctly.

SERVICE CALL 2

A residential customer calls. *There is no heat. The fan motor on the gas furnace has a defective run capacitor. The furnace burner is cycling on and off with the high-limit control. The fan will run for a few minutes each cycle before shutting off from overtemperature. The defective capacitor causes the fan motor current to run about 15% too high.*

The technician arrives and goes under the house where the horizontal furnace is located. The furnace is hot. The technician hears the burner ignite and watches it for a few minutes. The burner shuts off, and about that time the fan comes on. The technician removes the fan compartment door and sees the fan turning very slowly. SAFETY PRECAUTION: The power is shut off. The panel is locked and tagged. The motor is very hot. It seems to turn freely, so the bearings must be normal. The technician removes the run capacitor and uses an ohmmeter for a capacitor check, **Figure 20–27**. The capacitor is open—no circuit or continuity. The capacitor is replaced. The fan compartment door is shut, and power is restored. It is evident from the sound of the motor that it is turning faster than before the capacitor was changed.

Before leaving the job, the technician turns off the power, oils the motor, and replaces the air filters.

SERVICE CALL 3

A retail store customer calls. *There is no air conditioning. The compressor, located outside, starts up, then shuts off. The fan motor is not running. The dispatcher tells the customer to shut the unit off until the technician arrives. The PSC fan motor for the condenser has bad bearings. The motor feels free to turn the shaft, but if power is applied to the motor when it is turning, the motor will stop.*

The technician arrives and talks to the store manager. The technician goes to the outdoor unit and disconnects the power so that the system can be controlled from outside. The room thermostat is set to call for cooling.

The technician goes outside and turns the disconnect on and observes. The compressor starts, but the fan motor does not. SAFETY PRECAUTION: The power is turned off. The panel is locked and tagged. The compressor leads are disconnected at the load side of the contactor so that the compressor will not be in a strain. The technician makes the disconnect switch again and, using the clamp-on ammeter, checks the current going to the fan motor. It is drawing current and trying to turn. The technician does not know at this point if the motor or the capacitor is defective. The power is shut off. The technician spins the motor and turns the power on while the motor is turning. The motor acts like it has brakes and stops. This indicates bad bearings or an internal electrical motor problem.

The motor is changed, and the new motor performs normally even with the old capacitor. The compressor is reconnected, and power is resumed. The system now cools normally. The technician changes the air filter and oils the indoor fan motor before leaving.

The following service calls do not have the solution described. The solution can be found in the *Instructor's Guide.*

SERVICE CALL 4

An insurance office customer calls. *There is no air conditioning. This system has an electric furnace with the air handler (fan section) mounted above the suspended ceiling in the stock room. The condensing unit is on the roof. The low-voltage power supply is at the air handler.*

The technician arrives and goes to the room thermostat. It is set on cooling, and the indoor fan is running. The technician goes to the roof and discovers the breaker is tripped. Before resetting the breaker, the technician decides to find out why the breaker tripped. The cover to the electrical panel to the unit is removed. A voltage check shows that the breaker has all the voltage off. The contactor's 24-V coil is the only thing energized because its power comes from downstairs at the air handler.

The technician checks the motors for a ground circuit by placing one lead on the load side of the contactor and the other on the ground with the meter set on R × 10,000. The meter shows a short circuit to ground (no resistance to ground). See **Figure 20–23** for an example of a motor ground check. From this test, the technician does not know if the problem is the compressor or the fan motor. The wires are disconnected from the line side of the contactor to isolate the two motors. The compressor shows normal, no reading. The fan motor shows 0 Ω resistance to ground.

What is the problem and the recommended solution?

SERVICE CALL 5

A commercial customer calls. *There is no air conditioning in the upstairs office. This is a three-story building with an air handler on each floor and a chiller in the basement.*

The technician arrives and goes to the fan room on the second floor where the complaint is. There is no need to go to the chiller or the other floors because there would be complaints if the chiller was not furnishing cooling to them.

The chilled water coil piping feels cold; the chiller is definitely running and furnishing cold water to the coil. The fan motor is not running. SAFETY PRECAUTION: Because the motor may have electrical problems, the technician proceeds with caution by opening, locking, and tagging the electrical disconnect. The technician pushes the reset button on the fan motor starter and hears the ratchet mechanism reset. The motor will not try to start while the technician is pushing the reset button because the disconnect switch is open. The unit must have been pulling too much current for the overload to trip.

Using an ohmmeter, the technician checks for a ground by touching one lead to a ground terminal and the other to one of the motor leads on the load side of the starter. The meter is set on R × 10,000 and will detect a fairly high resistance to ground. See **Figure 20–23** for an example of a motor ground check. SAFETY PRECAUTION: Any movement of the meter needle when the ohmmeter's leads are connected between ground and the motor lead would indicate a ground—so caution is necessary. If the meter needle moves as much as one-fourth of its scale, the circuit should not be energized until the ground is cleared up, or physical damage may occur. The motor is not grounded. The resistance between each winding is the same, so the motor appears to be normal. The technician then turns the motor over by hand to see whether the bearings are too tight; the motor turns normally.

The technician shuts the fan compartment door and then fastens the clamp-on ammeter to one of the motor leads at the load side of the starter. The electrical disconnect is then closed to start the motor. When it is closed the motor tries to start. It will not start and pulls a high amperage. The motor seems to be normal from an electrical standpoint and turns freely, so the power supply is now suspected.

SAFETY PRECAUTION: The technician quickly pulls the disconnect to the off position and gets an ohmmeter. Each fuse is checked with the ohmmeter. The fuse in L2 is open. The fuse is replaced and the motor is started again. The motor starts and runs normally with normal amperage on all three phases. The question is, why did the fuse blow?

What is the problem and the recommended solution?

SERVICE CALL 6

A customer who owns a truck stop calls. *There is no cooling in the main dining room. The customer does some of the maintenance, so the technician can expect some problem related to this maintenance program.*

The technician arrives and can hear the condensing unit on the roof running; it shuts off about the time the technician gets out of the truck. The customer had just serviced the system by changing the filters, oiling the motors, and checking the belts. Since then, the fan motor has been shutting off. It can be reset by pressing the reset button at the fan contactor, but it will not run long.

The technician and the customer go to the stock room where the air handler is located. The fan is not running, and the condensing unit is shut down from low pressure. The technician fastens the clamp-on ammeter on one of the motor leads and restarts the motor. The motor is pulling too much current on all three phases, and the motor seems to be making too much noise.

What is the problem and the recommended solution?

SUMMARY

- Bearing problems are often caused by belt tension.
- Shaft problems may be caused by the technician while he or she is removing pulleys or couplings.
- Motor balancing problems are normally not handled by the heating, air-conditioning, and refrigeration technician.
- Make sure that vibration is not caused by the system.
- Most electrical problems are open windings, short-circuited windings, or grounded windings.
- The laws of electrical current flow must be used when troubleshooting motors.
- If the motor is receiving the correct voltage and is electrically sound, check the motor components.
- Troubleshooting hermetic compressor motors is different from troubleshooting open motors because hermetic motors are enclosed.

REVIEW QUESTIONS

1. True or False: If a sleeve bearing fails in a small fractional horsepower motor, the motor is generally replaced.
2. An open winding in an electric motor means that
   A. the winding is making contact with the motor frame.
   B. one winding is making contact with another winding.

C. a wire in one winding is broken.

D. the centrifugal switch to the start winding is open.

3. There is a short circuit in an electric motor winding when

A. there is an adverse weather condition.

B. a winding conductor is worn bare and touches another part of the conductor that is also bare.

C. there is a defect in the centrifugal start switch.

D. the motor will not start.

4. An open winding or a shorted winding can be determined with which of the following instruments?

A. Voltmeter

B. Ammeter

C. Ohmmeter

5. A decrease in resistance will cause the amperage to

A. decrease.

B. increase.

C. decrease or increase.

D. none of the above.

6. A _____ may be used to detect resistances in the millions of ohms.

A. megohmmeter

B. voltmeter

C. clamp-on ammeter

D. psychrometer

7. What size of resistor is recommended for shorting across an electrical motor capacitor before checking with an ohmmeter?

8. To determine the capacitance of a capacitor, a _____ should be used.

9. State the physical differences between a run and start capacitor.

10. In a three-phase motor in good operating condition, the resistance

A. should be the same across all three windings.

B. should vary from a lower resistance in the first winding to a higher resistance in the third winding.

C. across each winding is not important.

11. Describe what may happen to an electric motor if the drive belt is too tight.

12. How are the electrical components checked on a hermetic compressor?

13. A loose connection may result in _____ voltage and _____ current at the motor.

14. The amperage can be checked with a _____ ammeter.

15. Electric motor roller and ball bearing failure often can be determined by the _____ .

UNIT 20 DIAGNOSTIC CHART FOR OPEN-TYPE ELECTRIC MOTORS

Electric motors are used to turn the prime movers in all heating and cooling, air-conditioning, and refrigeration systems. The prime movers are the fans, pumps, and compressors in these systems. Compressors are usually hermetically sealed in the system and are discussed in the units covering compressors. This diagnostic chart will cover open-type electric motors, single phase and three phase. When possible, the technician should listen to the customer for possible causes of the problem. Often the customer's comments will result in quickly resolving the problem in the system.

Problem	Possible Cause	Possible Repair	Heading Number
Motor does not attempt to start—makes no sound.	Open disconnect switch	Close disconnect switch	15.1, 15.2, 15.7
	Open fuse or breaker	Replace fuse or reset breaker and determine why it opened	15.1, 15.2, 15.7
	Tripped overload	Reset overload and determine why it tripped	19.6
	Faulty fan switch	Repair or replace switch mechanism	15.4, 15.7
	Faulty wiring	Repair or replace faulty wiring or connectors	20.13
Motor will not start—hums and trips on overload.	Incorrect power supply	Correct the power supply	15.1, 17.7
Single Phase			
A. Shaded-pole and split-phase motors	Tight or dragging bearings	Change motor	20.10
	Tight belt or overloaded	Adjust belt	20.4
	Either winding open	Change motor	20.7
	Start circuit contacts on split phase	Change motor	17.10
	Defective start capacitor	Change capacitor	17.12, 20.11
B. Permanent split-capacitor motor	Defective run capacitor	Change capacitor	17.14, 20.11
	Either winding open		
Three Phase	Incorrect power supply—voltage unbalance	Correct voltage balance	17.7, 48.46
	One phase out—single-phase condition	Correct before starting motor	17.7
	Tight or dragging bearings	Change bearings or motor	20.10
	Tight belt	Adjust belt	20.4
	Overloaded	Reduce load	20.10
Motor starts and runs for short time, then shuts off due to overload.	Defective overload	Check actual load on overload and replace if it trips below specifications	19.6
	System problem—flooding or slugging or low superheat	Determine why and repair low superheat	29.12
	Excess current in overload circuit—check for added load, such as fan or pumps	Correct load to match motor and overload	19.6
	Low voltage	Determine reason and correct	17.7
Three Phase	One phase out—single-phase condition	Correct before starting motor	17.7
	Voltage unbalance	Correct voltage balance	48.46
	Excess load on motor	Fan moving too much air—correct	
		Motor undersized—change to larger motor	18.1
	Fan or pump bearings too tight	Disconnect load and check bearings—repair as needed	20.8

SECTION 5

Commercial Refrigeration

UNIT 21
Evaporators and the Refrigeration System

OBJECTIVES

After studying this unit, you should be able to

- define high-, medium-, and low-temperature refrigeration.
- determine the boiling temperature in an evaporator.
- identify different types of evaporators.
- describe a parallel-flow, plate-and-fin evaporator.
- describe multiple- and single-circuit evaporators.
- describe how a voltmeter is used to troubleshoot an electrical circuit.

SAFETY CHECKLIST

✔ Wear goggles and gloves when attaching or removing gages to transfer refrigerant or to check pressures.
✔ Wear warm clothing when working in a walk-in cooler or freezer.

21.1 REFRIGERATION

Refrigeration is the process of removing heat from a place where it is not wanted and transferring that heat to a place where it makes little or no difference. Commercial refrigeration is similar to the refrigeration that occurs in your household refrigerator. The food that you keep in the refrigerator is stored at a temperature lower than the room temperature. Typically, the fresh food compartment temperature is about 35°F. Heat from the room moves through the walls of the refrigerator from the warm room (typically 75°F) to the cooler temperature in the refrigerator. Heat travels normally and naturally from a warm to a cool medium.

If the heat that is transferred into the refrigerator remains in the refrigerator, it will warm the food products, and spoilage will occur. This heat may be removed from the refrigerator by mechanical means using the refrigeration equipment furnished with the refrigerator. This mechanical means requires energy, or work. **Figure 21–1** shows how this heat is removed with the compression cycle. Because it is 35°F in the box and 75°F in the room, the mechanical energy in the compression cycle actually pumps the heat to a warmer environment from the box to the room.

The heat is transferred into a cold refrigerant coil and pumped with the system compressor to the condenser where it is released into the room. This is much like using a sponge to move water from one place to another. When a dry sponge is allowed to absorb water in a puddle and you take the wet

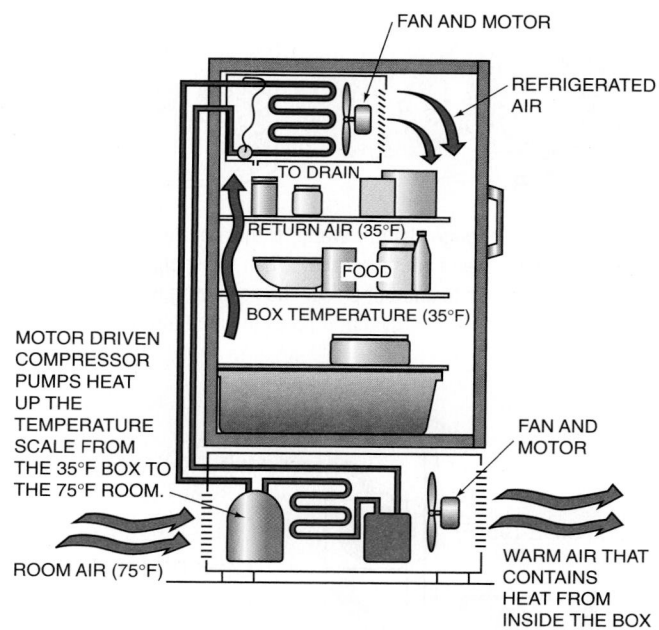

HIGH-PRESSURE/HIGH-TEMPERATURE REFRIGERANT ■
LOW-PRESSURE/LOW-TEMPERATURE REFRIGERANT ■

Figure 21–1 Heat normally flows from a warm place to a colder place. When it is desirable for heat to be removed from a colder to a warmer place, the heat must be moved by force. The compressor in the refrigeration system is the pump that forces the heat up the temperature scale. These compressors are normally driven by electric motors.

sponge to a container and squeeze it, you exert energy, much like a compressor in the refrigeration system, **Figure 21–2**.

Another example of refrigeration is a central air-conditioning system in a residence. It absorbs heat from the home by passing indoor air at about 75°F over a coil that is cooled to about 40°F. Heat will transfer from the room air to the coil cooling the air. This cooled air may be mixed with the room air, lowering its temperature, **Figure 21–3**. This process is called air conditioning, but it is also refrigeration at a higher temperature level than the household refrigerator. It is frequently called high-temperature refrigeration.

Commercial refrigeration differs from domestic (household) refrigeration because it is located in commercial business locations. The food store, fast food restaurant, drug store, flower shop, and food processing plant are only a few of the applications for commercial refrigeration. Some of

Figure 21–2 A sponge absorbs water. The water then can be carried in the sponge to another place. When the sponge is squeezed, the water is rejected to another place. The squeezing of the sponge may be considered the energy that it takes to pump the water.

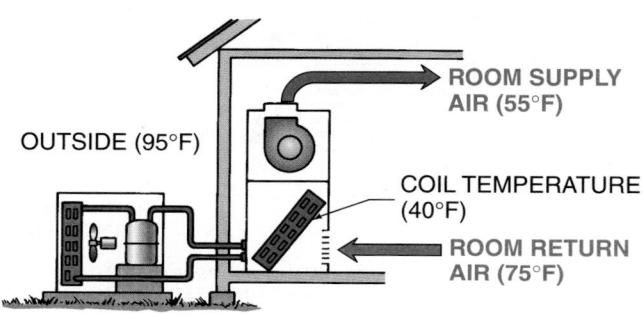

Figure 21–3 An air-conditioning example of refrigeration.

the systems are plug-in appliances, such as a small reach-in ice storage bin at the local convenience store. The refrigeration system is all located within the one unit. Some systems consist of individual boxes with single remote condensing units, and some are complex systems with several compressors in a rack serving several reach-in display cases in a supermarket. Most commercial refrigeration is installed and serviced by a special group of technicians who work only with commercial refrigeration and the food service business.

21.2   TEMPERATURE RANGES OF REFRIGERATION

The temperature ranges for commercial refrigeration may refer to the temperature of the refrigerated box, or the boiling temperature of the refrigerant in the coil. When discussing box temperatures, the following temperatures will illustrate some of the guidelines used in industry.

HIGH-TEMPERATURE APPLICATIONS. High-temperature refrigeration applications will normally provide box temperatures of 47°F to 60°F. The storage of such products as flowers and candy may require these temperatures.

MEDIUM-TEMPERATURE APPLICATIONS. The household refrigerator fresh-food compartment is a good example of medium-temperature refrigeration, which typically ranges from 28°F to 40°F. Many different products are stored at the medium-temperature range. The medium-temperature refrigeration range is above freezing for most products. Few products are stored below 32°F. Items such as eggs, lettuce, and tomatoes lose their appeal if they freeze in a refrigerator.

LOW-TEMPERATURE APPLICATIONS. Low-temperature refrigeration produces temperatures below the freezing point of water, 32°F. One of the higher low-temperature applications is the making of ice, which usually occurs at about 32°F.

Low-temperature food storage applications generally start at 0°F and go as low as −20°F. At this temperature ice cream would be frozen hard. Frozen meats, vegetables, and dairy products are only a few of the foods preserved by freezing. Some foods may be kept for long periods of time and are appetizing when thawed for cooking, provided they are frozen correctly and kept frozen.

21.3   THE EVAPORATOR

The evaporator in a refrigeration system is responsible for absorbing heat into the system from whatever medium is to be cooled. This heat-absorbing process is accomplished by maintaining the evaporator coil at a lower temperature than the medium to be cooled. For example, if a walk-in cooler is to be maintained at 35°F to preserve food products, the coil in the cooler must be maintained at a lower temperature than the 35°F air that will be passing over it. **Figure 21–4** shows the refrigerant in the evaporator boiling at 20°F, which is 15°F lower than the entering air.

The evaporator operating at these low temperatures removes latent and sensible heat from the cooler. Operating at

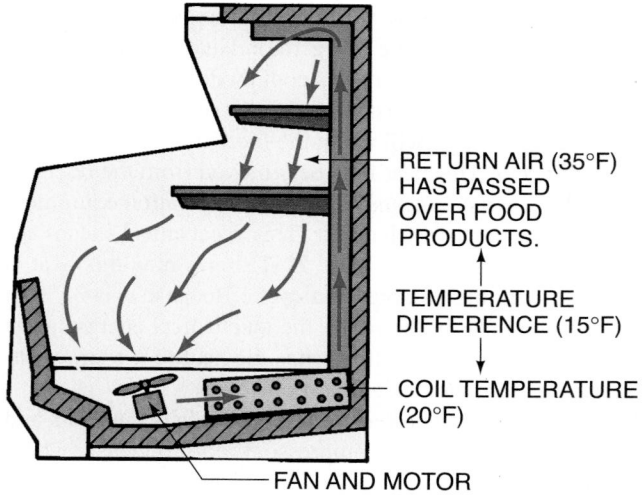

THE COMPRESSOR IS RUNNING — THIS LOWERS COIL TEMPERATURE.

Figure 21–4 The relationship of the coil's boiling temperature to the air passing over the coil while it is operating in the design range.

TEMPERATURE °F	12	22	134a	502	404A	410A
−60	19.0	12.0		7.2	6.6	0.3
−55	17.3	9.2		3.8	3.1	2.6
−50	15.4	6.2		0.2	0.8	5.0
−45	13.3	2.7		1.9	2.5	7.8
−40	11.0	0.5	14.7	4.1	4.8	9.8
−35	8.4	2.6	12.4	6.5	7.4	14.2
−30	5.5	4.9	9.7	9.2	10.2	17.9
−25	2.3	7.4	6.8	12.1	13.3	21.9
−20	0.6	10.1	3.6	15.3	16.7	26.4
−18	1.3	11.3	2.2	16.7	18.2	28.2
−16	2.0	12.5	0.7	18.1	19.6	30.2
−14	2.8	13.8	0.3	19.5	21.1	32.2
−12	3.6	15.1	1.2	21.0	22.7	34.3
−10	4.5	16.5	2.0	22.6	24.3	36.4
−8	5.4	17.9	2.8	24.2	26.0	38.7
−6	6.3	19.3	3.7	25.8	27.8	40.9
−4	7.2	20.8	4.6	27.5	30.0	42.3
−2	8.2	22.4	5.5	29.3	31.4	45.8
0	9.2	24.0	6.5	31.1	33.3	48.3
1	9.7	24.8	7.0	32.0	34.3	49.6
2	10.2	25.6	7.5	32.9	35.3	50.9
3	10.7	26.4	8.0	33.9	36.4	52.3
4	11.2	27.3	8.6	34.9	37.4	53.6
5	11.8	28.2	9.1	35.8	38.4	55.0
6	12.3	29.1	9.7	36.8	39.5	56.4
7	12.9	30.0	10.2	37.9	40.6	57.8
8	13.5	30.9	10.8	38.9	41.7	59.3
9	14.0	31.8	11.4	39.9	42.8	60.7
10	14.6	32.8	11.9	41.0	43.9	62.2
11	15.2	33.7	12.5	42.1	45.0	63.7

TEMPERATURE °F	12	22	134a	502	404A	410A
12	15.8	34.7	13.2	43.2	46.2	65.3
13	16.4	35.7	13.8	44.3	47.4	66.8
14	17.1	36.7	14.4	45.4	48.6	68.4
15	17.7	37.7	15.1	46.5	49.8	70.0
16	18.4	38.7	15.7	47.7	51.0	71.6
17	19.0	39.8	16.4	48.8	52.3	73.2
18	19.7	40.8	17.1	50.0	53.5	75.0
19	20.4	41.9	17.7	51.2	54.8	76.7
20	21.0	43.0	18.4	52.4	56.1	78.4
21	21.7	44.1	19.2	53.7	57.4	80.1
22	22.4	45.3	19.9	54.9	58.8	81.9
23	23.2	46.4	20.6	56.2	60.1	83.7
24	23.9	47.6	21.4	57.5	61.5	85.5
25	24.6	48.8	22.0	58.8	62.9	87.3
26	25.4	49.9	22.9	60.1	64.3	90.2
27	26.1	51.2	23.7	61.5	65.8	91.1
28	26.9	52.4	24.5	62.8	67.2	93.0
29	27.7	53.6	25.3	64.2	68.7	95.0
30	28.4	54.9	26.1	65.6	70.2	97.0
31	29.2	56.2	26.9	67.0	71.7	99.0
32	30.1	57.5	27.8	68.4	73.2	101.0
33	30.9	58.8	28.7	69.9	74.8	103.1
34	31.7	60.1	29.5	71.3	76.4	105.1
35	32.6	61.5	30.4	72.8	78.0	107.3
36	33.4	62.8	31.3	74.3	79.6	108.4
37	34.3	64.2	32.2	75.8	81.2	111.6
38	35.2	65.6	33.2	77.4	82.9	113.8
39	36.1	67.1	34.1	79.0	84.6	116.0
40	37.0	68.5	35.1	80.5	86.3	118.3
41	37.9	70.0	36.0	82.1	88.0	120.5

TEMPERATURE °F	12	22	134a	502	404A	410A
42	38.8	71.4	37.0	83.8	89.7	122.9
43	39.8	73.0	38.0	85.4	91.5	125.2
44	40.7	74.5	39.0	87.0	93.3	127.6
45	41.7	76.0	40.1	88.7	95.1	130.0
46	42.6	77.6	41.1	90.4	97.0	132.4
47	43.6	79.2	42.2	92.1	98.8	134.9
48	44.6	80.8	43.3	93.9	100.7	136.4
49	45.7	82.4	44.4	95.6	102.6	139.9
50	46.7	84.0	45.5	97.4	104.5	142.5
55	52.0	92.6	51.3	106.6	114.6	156.0
60	57.7	101.6	57.3	116.4	125.2	170.0
65	63.8	111.2	64.1	126.7	136.5	185.0
70	70.2	121.4	71.2	137.6	148.5	200.8
75	77.0	132.2	78.7	149.1	161.1	217.6
80	84.2	143.6	86.8	161.2	174.5	235.4
85	91.8	155.7	95.3	174.0	188.6	254.2
90	99.8	168.4	104.4	187.4	203.5	274.1
95	108.2	181.8	114.0	201.4	219.2	295.0
100	117.2	195.9	124.2	216.2	235.7	317.1
105	126.6	210.8	135.0	231.7	253.1	340.3
110	136.4	226.4	146.4	247.9	271.4	364.8
115	146.8	242.7	158.5	264.9	290.6	390.5
120	157.6	259.9	171.2	282.7	310.7	417.4
125	169.1	277.9	184.6	301.4	331.8	445.8
130	181.0	296.8	198.7	320.8	354.0	475.4
135	193.5	316.6	213.5	341.2	377.1	506.5
140	206.6	337.2	229.1	362.6	401.4	539.1
145	220.3	358.9	245.5	385.9	426.8	573.2
150	234.6	381.5	262.7	408.4	453.3	608.9
155	249.5	405.1	280.7	432.9	479.8	616.2

VACUUM (in. Hg) – RED FIGURES
GAGE PRESSURE (psig) – BOLD FIGURES

Figure 21–5 The temperature/pressure chart in inches mercury vacuum, or psig. Pressures for P-404A and B-410 are an average liquid and vapor pressure.

20°F, as in the preceding example, the evaporator will collect moisture from the air in the cooler, latent heat. The removal of sensible heat reduces the food temperature.

21.4 BOILING AND CONDENSING

Two important factors in understanding refrigeration are (1) boiling temperature and (2) condensing temperature. The boiling temperature and its relationship to the system exist in the evaporator. The condensing temperature exists in the condenser and will be discussed in the next unit.

These temperatures can be followed by using the temperature/pressure chart in conjunction with a set of refrigeration pressure gages, **Figure 21–5** and **Figure 21–6.**

21.5 THE EVAPORATOR AND BOILING TEMPERATURE

The *boiling temperature* of the liquid refrigerant determines the coil operating temperature. In an air-conditioning system a 40°F evaporator coil with 75°F air passing over it produces conditions used for air-conditioning or high-temperature refrigeration. Boiling is normally associated with high temperatures and water. Unit 3, "Refrigeration and Refrigerants," discussed the fact that water boils at 212°F at atmospheric pressure. It also discussed the fact that water boils at other temperatures, depending on the pressure. When the pressure is reduced, water will boil at 40°F. This is still boiling—

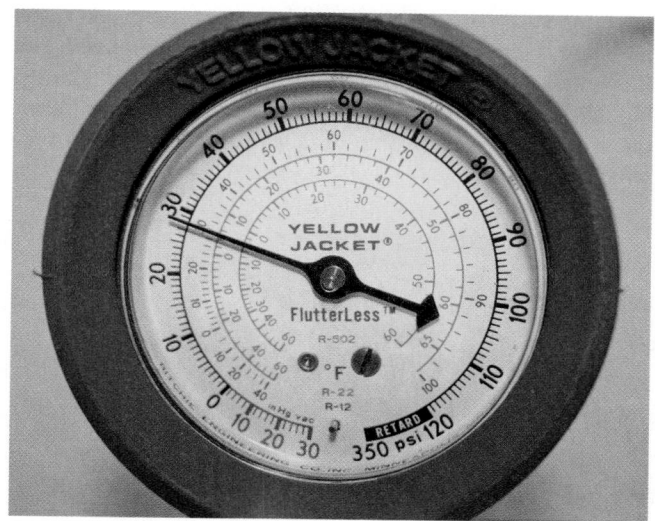

Figure 21–6 Many pressure gages have temperature/pressure relationships printed on the gage. *Photo by John Tomczyk*

changing a liquid to a vapor. In a refrigeration system, the refrigerant may boil at 20°F by absorbing heat from the 35°F food.

The service technician must be able to determine what operating pressures and temperatures are correct for the various systems being serviced under different load conditions. Much of this knowledge comes from experience. When the thermometers and gages are observed, the readings must be evaluated. There can be as many different readings as there are changing conditions.

Guidelines can help the technician know the pressure and temperature ranges at which the equipment should operate. Relationships exist between the entering air temperature and the evaporator for each system. These relationships are similar from installation to installation.

21.6 REMOVING MOISTURE

Dehumidifying the air means to remove the moisture, and this is frequently desirable in refrigeration systems. Moisture removal is similar from one refrigeration system to another. Knowing this relationship can help the technician know what conditions to look for. The load on the coil would rise or fall accordingly as the return air temperature rises or falls. Warmer return air in the box will also have more moisture content, which imposes further load on the coil. If the cooler is warm due to food added to it, the coil would have more heat to remove because it has more load on it. It would be much like boiling water in an open pan on the stove. The water boils at one rate with the burner on medium and at an increased rate with the burner on high. The boiling pressure stays the same in the boiling water in a pan because the pan is open to the atmosphere. When this same boiling process occurs in an enclosed coil, the pressures will rise when the boiling occurs at a faster rate. This causes the operating pressure of the whole system to rise, **Figure 21–7.**

When the evaporator removes heat from air and lowers the temperature of the air, sensible heat is removed. When moisture is removed from the air, latent heat is removed. The moisture is piped to a drain, **Figure 21–8.** Latent heat is called hidden heat because it does not register on a thermometer, but it is heat, like sensible heat, and it must be removed, which takes energy.

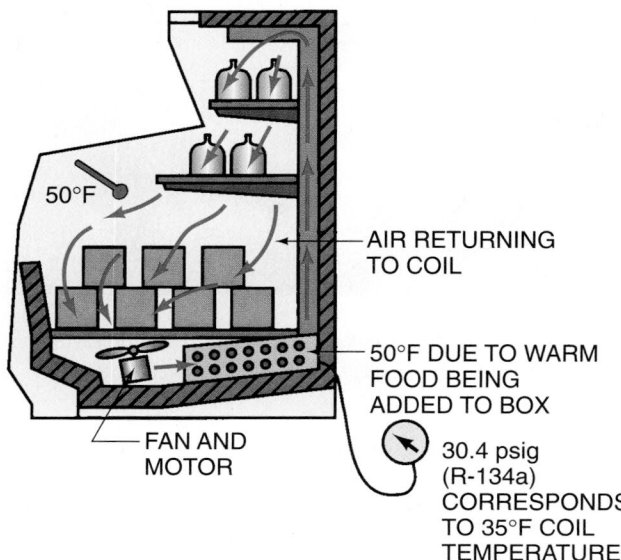

Figure 21–7 The coil-to-air temperature relationship under increased coil load.

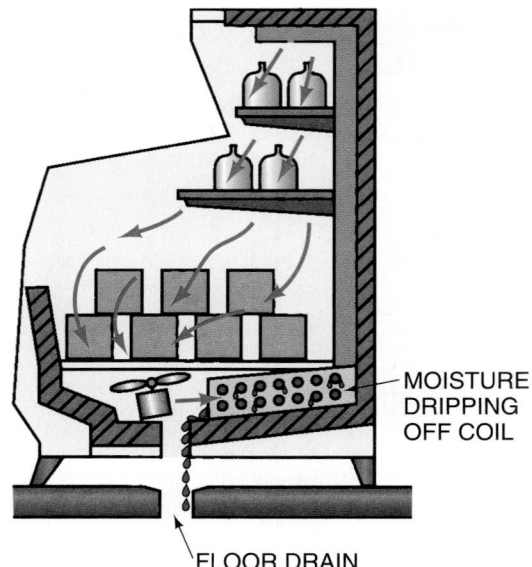

Figure 21–8 The cooling coil condenses moisture from the air.

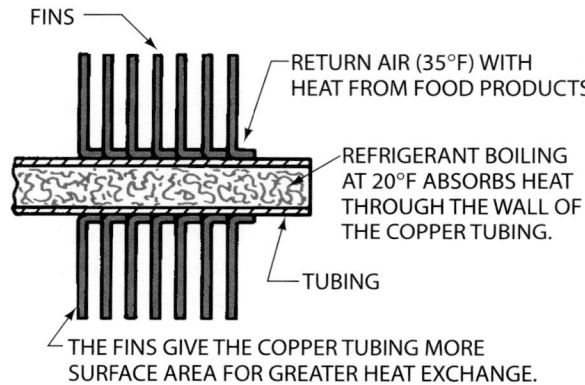

Figure 21–9 The heat exchange relationship between air and refrigerant.

The refrigeration evaporator is a component that absorbs heat from the conditioned space into the refrigeration system. The evaporator can be thought of as the *sponge* of the system. It is responsible for a heat exchange between the conditioned space or product and the refrigerant inside the system. Some evaporators absorb heat more efficiently than others. **Figure 21–9** illustrates this heat exchange between air and refrigerant.

21.7 HEAT EXCHANGE CHARACTERISTICS OF THE EVAPORATOR

The following are conditions that govern the rate of heat exchange:

1. The evaporator *material* through which the heat has to be exchanged. Evaporators may be manufactured from copper, steel, brass, stainless steel, or aluminum.

Corrosion is one factor that determines what material is used. For instance, when acidic materials need to be cooled, copper or aluminum coils would be eaten away. Stainless steel may be used instead, but stainless steel does not conduct heat as well as copper. Some evaporators are even coated with a plastic-like substance to protect the metal underneath from rust or oxidation. This application is often seen in restaurants with smaller commercial, medium-temperature coolers that store salad preparation materials. These materials are acidic because they often have a vinegar base for added flavor and for extended shelf life.

2. The *medium* to which the heat is exchanged. Giving heat up from air to refrigerant is an example. The best heat exchange occurs between two liquids, such as water to liquid refrigerant. This is because liquids are more dense than vapors and usually have a higher specific heat. However, this is not always practical because heat frequently has to be exchanged between air and vapor refrigerant. The vapor-to-vapor exchange is slower than the liquid-to-liquid exchange, **Figure 21–10.**

3. The *film factor.* This is a relationship between the medium giving up heat and the heat-exchange surface. The film factor relates to the velocity of the medium passing over the exchange surface. When the velocity is too slow, the film between the medium and the surface becomes an insulator and slows the heat exchange. The velocity keeps the film to a minimum, **Figure 21–11.** The correct velocity is chosen by the manufacturer.

4. The *temperature difference* between the two mediums in which the heat exchange is taking place. The greater the temperature difference between the evaporator coil and the medium giving up the heat, the faster the heat exchange will occur.

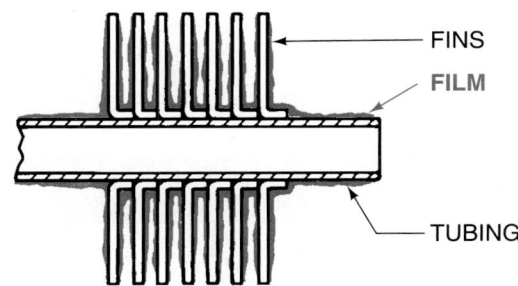

Figure 21–11 One of the deterring factors in a normal heat exchange. The film factor is the film of air or liquid next to the tube in the heat exchanger.

21.8 TYPES OF EVAPORATORS

Numerous types of evaporators are available, and each has its purpose. The first evaporators for cooling air were the natural-convection type. They were actually bare-pipe evaporators with refrigerant circulating through them, **Figure 21–12.** This evaporator was used in early walk-in coolers and was mounted high in the ceiling. It relied on the air being cooled, falling to the floor, and setting up a natural air current. The evaporator had to be quite large for the particular application because the velocity of the air passing over the coil was so slow. It is still occasionally used today.

The use of a blower to force or induce air over the coil improved the efficiency of the heat exchange. This meant that smaller evaporators could be used to do the same job. Design trends in the industry have always been toward smaller, more efficient equipment, **Figure 21–13.**

The expansion of the evaporator surface to a surface larger than the pipe itself gives a more efficient heat exchange. The *stamped evaporator* is a result of the first designs to create a large pipe surface. The stamped evaporator is two pieces of metal stamped with the impression of a pipe passage through it, **Figure 21–14.**

Pipe with fins attached, called a *finned-tube evaporator,* is used today more than any other type of heat exchange between air and refrigerant. This heat exchanger is efficient because the fins are in good contact with the pipe carrying the refrigerant. **Figure 21–15** shows an example of a finned-tube evaporator.

Multiple circuits improve evaporator performance and efficiency by reducing pressure drop inside the evaporator. The pipes inside the evaporator can be polished smooth, but they still offer resistance to the flow of both liquid and vapor refrigerants. The shorter the evaporator is, the less resistance there is to this flow. The "U" bends at the ends of the evaporator also cause a great deal of resistance to the flow of refrigerant. When an evaporator becomes quite long, it is common to cut it off and run another circuit in parallel next to it, **Figure 21–16.**

The evaporator for cooling liquids or making ice operates under the same principles as one for cooling air but is designed differently. It may be strapped on the side of a cylinder with liquid inside, submerged inside the liquid container,

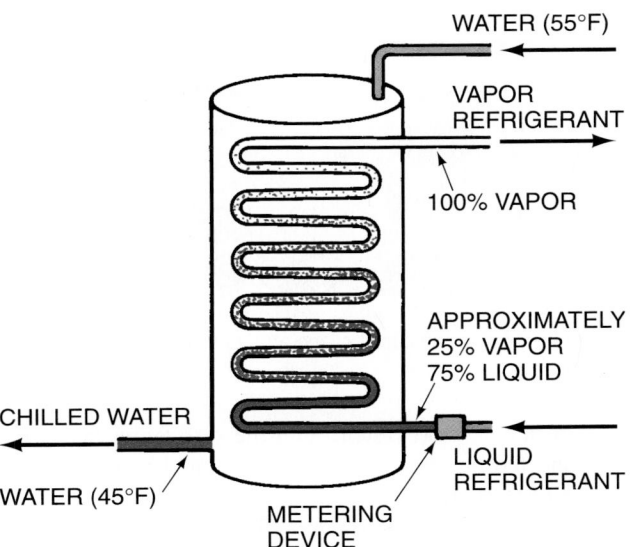

Figure 21–10 The heat exchange relationship between a liquid in a heat exchanger and the refrigerant inside the coil.

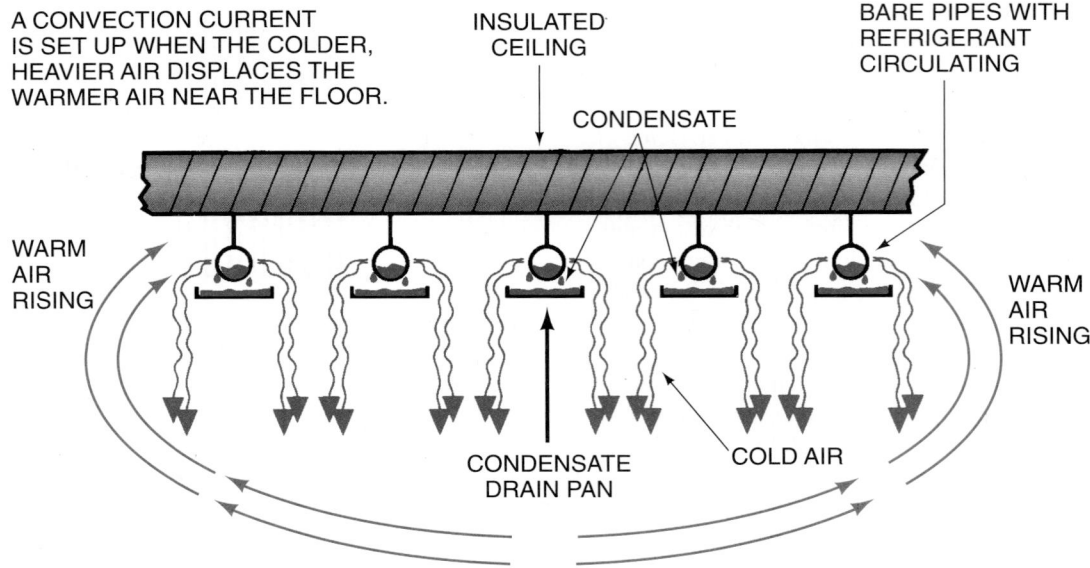

A CONVECTION CURRENT IS SET UP WHEN THE COLDER, HEAVIER AIR DISPLACES THE WARMER AIR NEAR THE FLOOR.

INSULATED CEILING

BARE PIPES WITH REFRIGERANT CIRCULATING

CONDENSATE

WARM AIR RISING

WARM AIR RISING

CONDENSATE DRAIN PAN

COLD AIR

Figure 21–12   A bare-pipe evaporator.

Figure 21–13   A forced-draft evaporator. *Courtesy Bally Case and Cooler, Inc.*

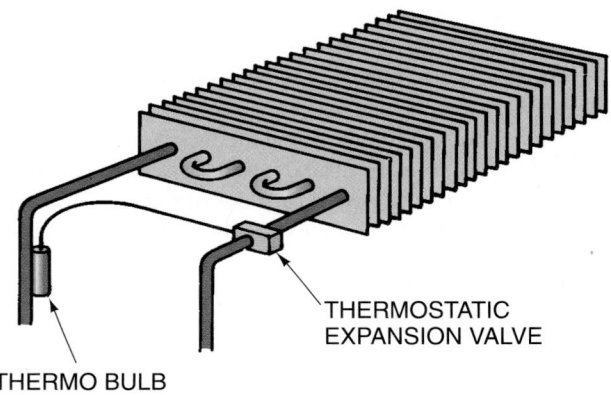

THERMOSTATIC EXPANSION VALVE

THERMO BULB

Figure 21–15   A finned-tube evaporator.

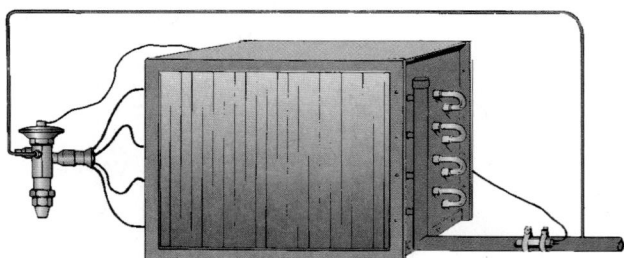

Figure 21–16   A multicircuit evaporator. *Courtesy Sporlan Valve Company*

or be a double-pipe system with the refrigerant inside one pipe and the liquid to be cooled circulated inside an outer pipe, **Figure 21–17.**

Every central split-cooling system manufactured in the United States today must have a Seasonal Energy Efficiency Ratio (SEER) of at least 13. This energy requirement was mandated by federal law as of January 23, 2006. Also, with

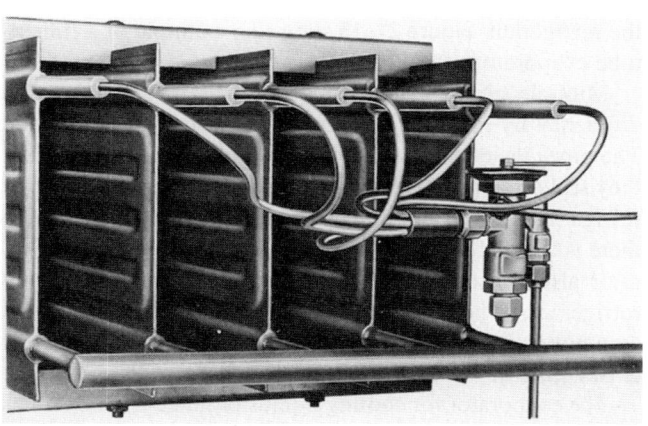

Figure 21–14   A stamped evaporator. *Courtesy Sporlan Valve Company*

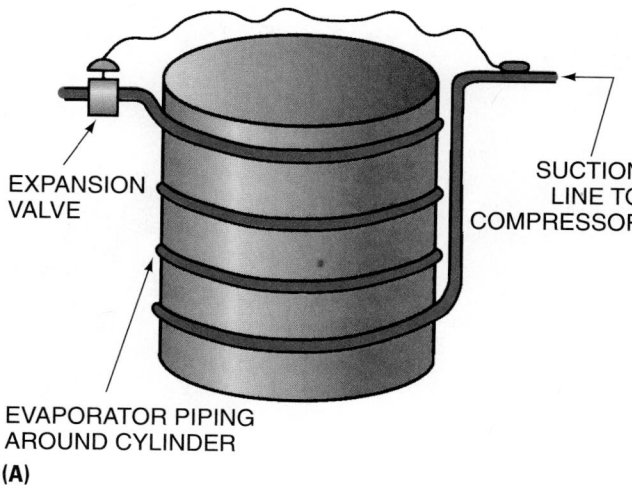

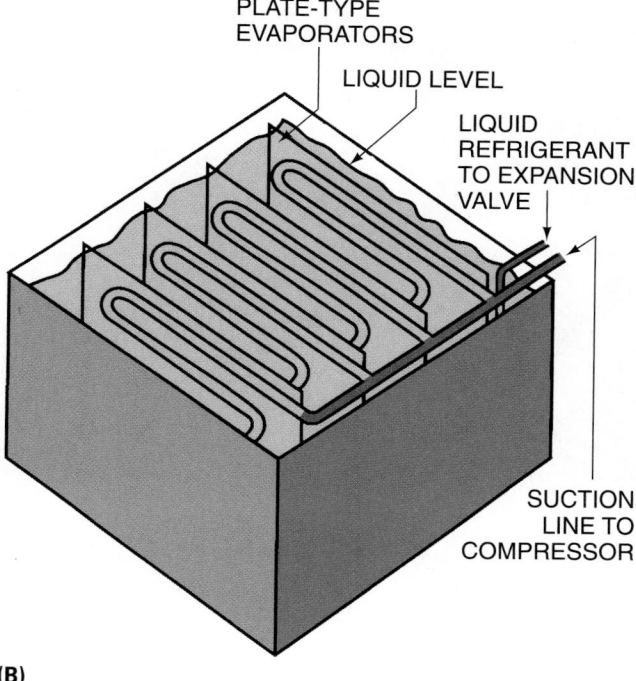

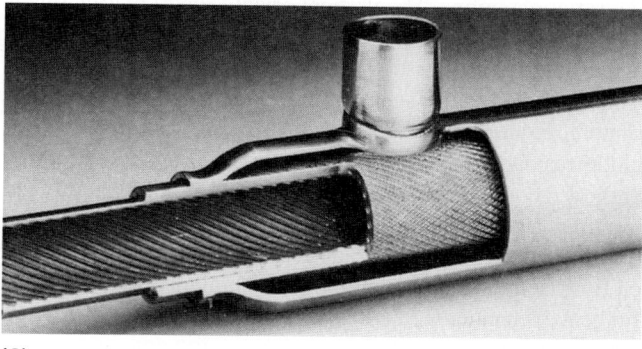

Figure 21–17 Liquid heat exchangers. **(A)** A drum-type evaporator. **(B)** A plate-type evaporator in a tank. **(C)** A pipe-in-pipe evaporator. **(C)** *Courtesy Noranda Metal Industries, Inc.*

the phaseout of R-22 just around the corner, manufacturers of HVACR equipment are looking for energy-efficient methods to apply to their equipment to meet these new energy requirements. The timeline for R-22 is as follows:

- 2010—R-22 use is banned at original equipment manufacturer (OEM) level with a 65% reduction on HCFC production
- 2015—90% reduction on HCFC production
- 2020—total ban on R-22 production

Equipment covered in this federal mandate includes

- unitary equipment from 1 1/2 to 5 tons.
- split/packaged air conditioners and heat pumps.

Equipment not covered includes

- commercial equipment greater than 6 tons.
- space-constrained units smaller than 3 tons (room air conditioners).
- water-source units.

SEER is calculated on the basis of the total amount of cooling (in Btu) the system will provide over the entire season, divided by the total number watt-hours it will consume. Higher SEERs reflect a more efficient cooling system. This federal mandate impacts 95% of the unitary market in the United States, which is about 8 million units at the time of this writing. Because of this new federal mandate of 13 SEER, most air-conditioning and heat pump manufacturers are looking for more efficient evaporator and condenser designs, more efficient compressors and fan motors, and more sophisticated control systems in order to meet this new energy efficiency requirement.

One such evaporator design incorporates an aluminum, parallel-flow, flat-plate-and-fin configuration with small parallel channels inside the flat plate. These plates are flattened, streamlined tubes with each one split into smaller, parallel ports, **Figure 21–18.** Refrigerant will phase change or evaporate from a liquid to a vapor inside the channels in the plate, while strategically shaped fins (extended surfaces) will enhance heat transfer from the air into the evaporator. The plates and fins are bonded or soldered to increase heat transfer and to eliminate any contact resistance (air gaps) that will reduce heat transfer. Headers at the inlet and outlet of the heat exchanger are also bonded to the plates through soldering.

Heat is transferred from the air to the evaporating refrigerant in three steps. The heat transfer steps are as follows:

1. Air side—between the fins and the air to be cooled
2. Heat conduction—between the fins and the tubes
3. Refrigerant side—between the tubes and the evaporating refrigerant

The air side of the heat exchange can be enhanced through fin geometry. The addition of louvres, lances, and rippled edges all increase heat transfer, **Figure 21–19.** The conduction between the fins and the tubes is enhanced through the application of a metallic bond (soldering) that eliminates any air gaps. The refrigerant side of the heat transfer deals with how much surface area of the inside of the tubes will come into contact with the phase-changing refrigerant. This internal surface area is often referred to as a *wetted perimeter.* As the internal surface area of the tubes

(A)

(B)

Figure 21–18 Aluminum parallel-flow, flat-plate-and-fin heat exchanger. *Courtesy Modine Manufacturing Co., Racine, WI*

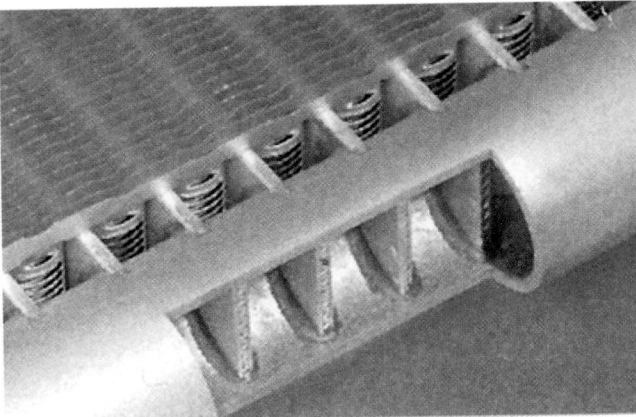

(A)

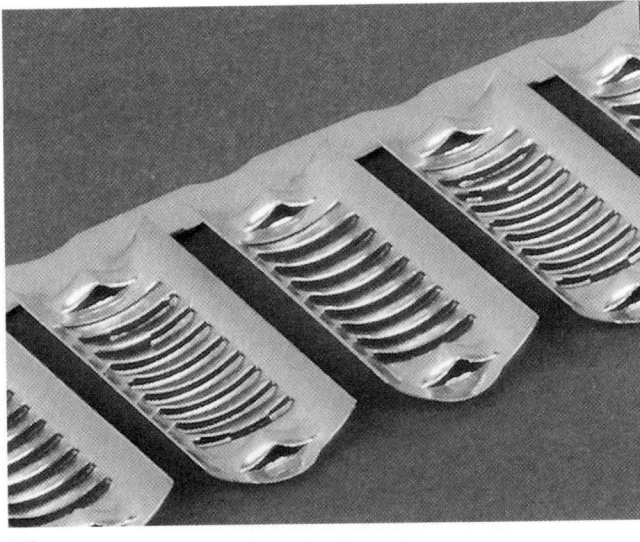

(B)

Figure 21–19 Fin geometry for increasing heat transfer. *Courtesy Modine Manufacturing Co., Racine, WI*

The capacity (tonnage) of the heat exchanger can vary with its height and length. The plates can be oriented vertically for an evaporator application or horizontally for condenser applications. The vertical orientation of the flat plates allows condensate removal to occur naturally, alleviating any water drainage issues from the evaporator, **Figure 21–20.** This technology is being used with condensers as well as evaporators, **Figure 21–21.** Applications in the HVACR field include residential air conditioning, rooftop air conditioning, chillers, geothermal heat pumps, electronic cooling, packaged terminal air conditioners (PTACs), ice machines, beverage dispensers, refrigerated display cases, and food-service refrigeration. Some of the benefits of this parallel-flow, plate-and-fin heat exchanger technology are listed below:

- 🌐 Reduces static pressure through the coil—which means fewer fan watts and less horsepower
- Reduces coil depth for the evaporator and condenser, which leads to easier cleaning and less air-side static pressure

increases, the heat transfer increases. Internal surface area can be increased by

- increasing the number of parallel channels inside the flat plates.
- increasing the number of flat plates (decreasing the spacing between them).

MOUNTING WITH HORIZONTAL HEADERS

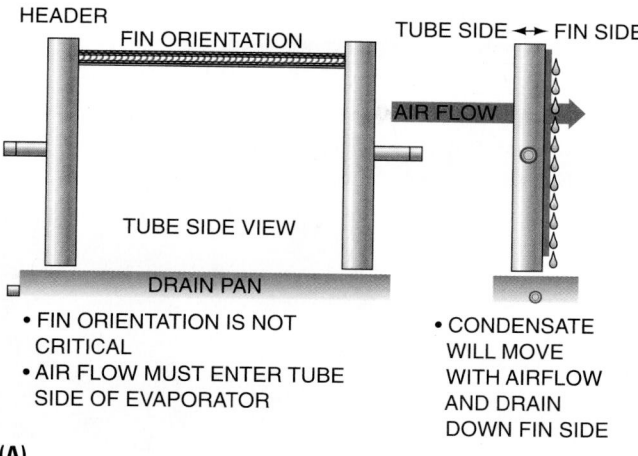

TUBE SIDE VIEW

- FIN ORIENTATION IS NOT CRITICAL
- AIR FLOW MUST ENTER TUBE SIDE OF EVAPORATOR

(A)

- CONDENSATE WILL MOVE WITH AIRFLOW AND DRAIN DOWN FIN SIDE

MOUNTING WITH HORIZONTAL HEADERS

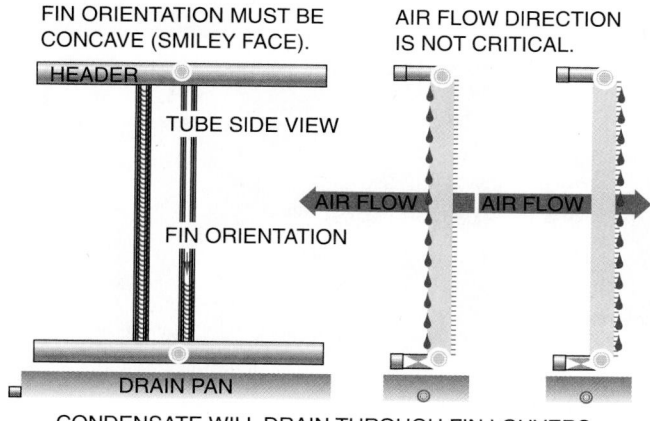

FIN ORIENTATION MUST BE CONCAVE (SMILEY FACE).

AIR FLOW DIRECTION IS NOT CRITICAL.

TUBE SIDE VIEW

CONDENSATE WILL DRAIN THROUGH FIN LOUVERS

(B)

Figure 21–20   Condensate removal for vertical and horizontal header configuration. *Courtesy Modine Manufacturing Co., Racine, WI*

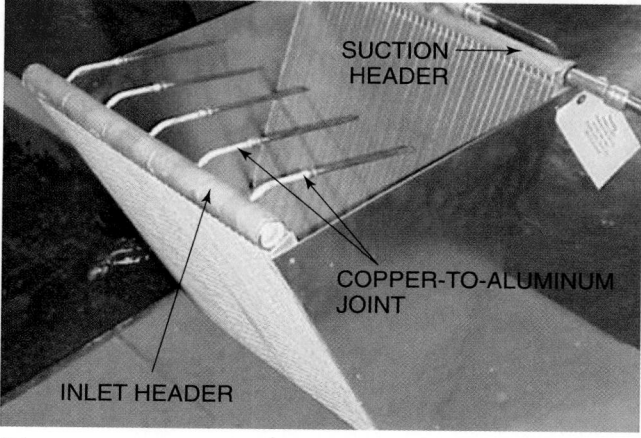

(A)

(B)

Figure 21–21   **(A)** Condenser. **(B)** Evaporator. *Courtesy Modine Manufacturing Co., Racine, WI*

- Reduced internal volume reduces refrigerant charge
- Reduced face area of condenser and evaporator
- Smaller footprint for the condensing unit
- A 30% coil weight and size reduction
- A reduction in packing costs, sizes, and weights
- The all-aluminum coil, header, and fins enhance corrosion resistance
- Lower system costs
- Provides higher system efficiencies than does a round, copper-tube heat exchanger with aluminum fins, **Figure 21–34(B)**
- Lower operating costs 🌐
- Quieter operations

Field repair of leaking heat exchangers, **Figure 21–22,** can be accomplished by

- recovering the refrigerant.
- cleaning the leaking area with a solution.
- brushing the area with a wire brush.

- using a utility knife blade and removing any fins that may be in the local area.
- pulling a vacuum with a vacuum pump.
- applying a two-part epoxy that will be sucked into the flat plate where the leak exists.
- applying heat with an electric blow drier until the epoxy is cured.
- evacuating to a 500-micron vacuum.
- charging with the appropriate refrigerant.

Field cleaning the heat exchangers can be accomplished using the same methods used for a standard, round, copper-tube heat exchanger with aluminum fins:

- Elevate the temperature of the mixed cleaner to 120°F.
- Use a power washer with a broad spray pattern.
- Use nonacidic cleaners (pH < 10.5).
- When the heat exchanger is clean, rinse the coil with clean water.

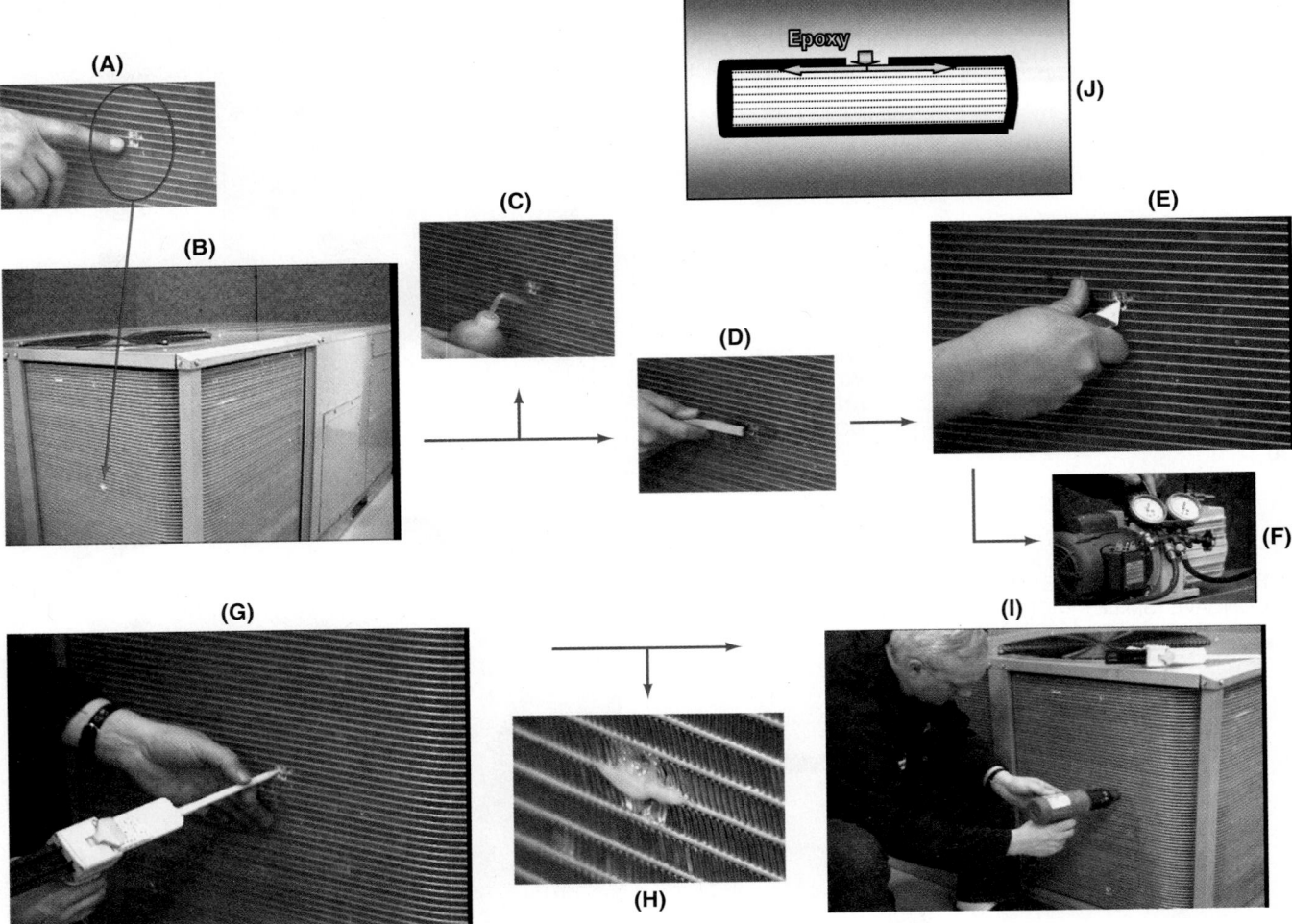

Figure 21–22   Field repair of a leaking heat exchanger. *Courtesy Modine Manufacturing Co., Racine, WI*

21.9 EVAPORATOR EVALUATION

Knowing the design considerations helps in *evaporator evaluation.* When the service technician arrives at the job, it may be necessary to evaluate whether a particular evaporator is performing properly. This can be considered one of the starting points in organized troubleshooting. The evaporator absorbs heat, the compressor pumps it, and the condenser rejects it. The following example pertains to a medium-temperature walk-in box. However, the evaporator evaluation would be about the same for any typical application.

Evaporator Specifications

1. Copper pipe coil
2. Aluminum fins attached to the copper pipe coil
3. Forced draft with a prop-type fan
4. One continual refrigerant circuit
5. R-134a
6. Evaporator to maintain space temperature at 35°F
7. Evaporator clean and in good working condition

First is a description of how the evaporator functions when it is working correctly.

Entering the evaporator is a partial liquid–partial vapor mixture at 20°F and 18.4 psig; it is approximately 65% liquid and 35% vapor. Approximately 35% of the liquid entering the expansion device at the evaporator is changed to a vapor and cools the remaining 65% of the liquid to the evaporator's boiling temperature (20°F). This is accomplished with the pressure drop across the expansion device. When the warm liquid passes through the small opening in the expansion device into the low pressure (18.4 psig) of the evaporator side of the device, some of the liquid flashes to a gas, **Figure 21–23.**

As the partial liquid–partial vapor mixture moves through the evaporator, more of the liquid changes to a vapor. This is called *boiling* and is a result of heat absorbed into the coil from whatever medium the evaporator is cooling. Finally, near the end of the evaporator the liquid is all boiled away to a vapor. At this point the refrigerant is known as *saturated vapor.* This means that the refrigerant vapor is saturated with heat. If any more heat is added to it, it will rise in temperature.

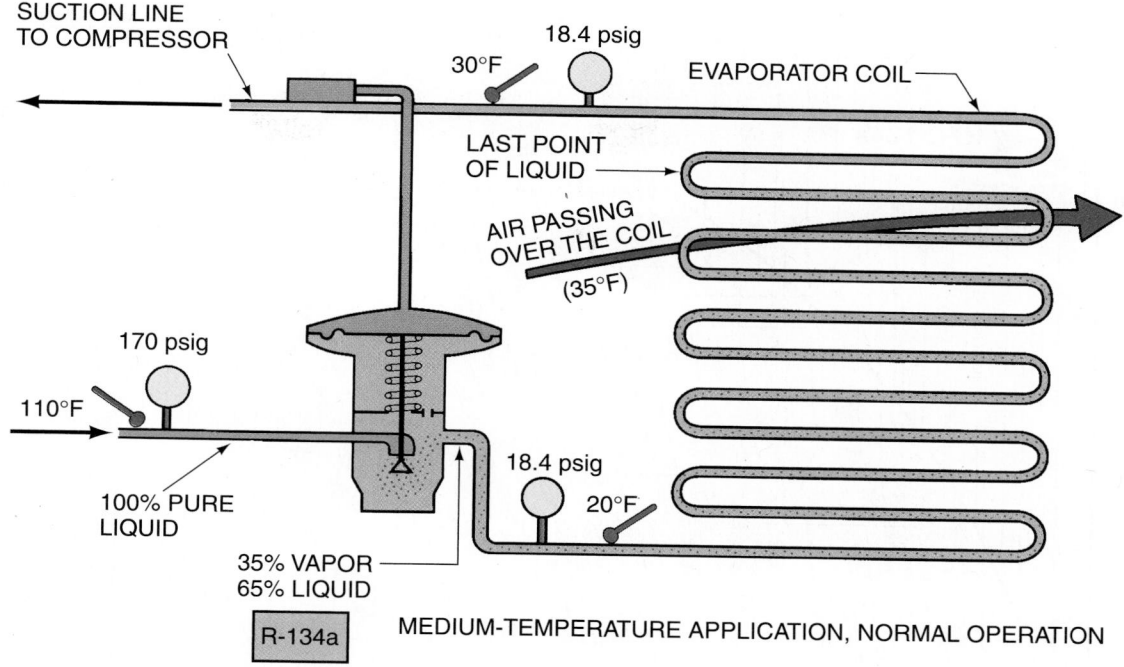

SUCTION LINE
TO COMPRESSOR

18.4 psig

30°F

EVAPORATOR COIL

LAST POINT
OF LIQUID

AIR PASSING
OVER THE COIL
(35°F)

170 psig

110°F

100% PURE
LIQUID

18.4 psig

20°F

35% VAPOR
65% LIQUID

R-134a   MEDIUM-TEMPERATURE APPLICATION, NORMAL OPERATION

Figure 21–23   When the 110°F liquid passes through the expansion valve orifice, some of the liquid flashes to a vapor and cools the remaining liquid to the evaporator temperature at 20°F.

If any heat is taken away from it, it will start changing back to a liquid. This vapor is saturated with heat, but it is still at the evaporating temperature corresponding to the boiling point, 20°F. ⬤This is a most important point in the function of an evaporator because all of the liquid must be boiled away as close to the end of the coil as possible. This is necessary to (1) keep the coil efficiency up and (2) ensure that liquid refrigerant does not leave the evaporator and move into the compressor.⬤ For the evaporator to run efficiently, it must operate as full of liquid as possible without allowing liquid to be present at the outlet of the coil, because the best heat exchange is between the liquid refrigerant and the air passing over the coil.

The pressure/enthalpy chart in **Figure 21–24** shows graphically what happens inside the evaporator for the preceding walk-in cooler example. The refrigerant enters the evaporator at point A (after leaving the expansion valve). The liquid pressure is 18.4 psig and contains 48.7 Btu/lb of heat at this point. Approximately 35% of the liquid was flashed to a vapor passing through the expansion valve. As the liquid proceeds through the evaporator, it is changing to a vapor. It has all changed to a vapor at point B, but the vapor temperature is still 20°F and capable of absorbing heat, in the form of superheat. The vapor temperature starts to rise while it is still in the evaporator until the temperature is 30°F (containing 10°F of superheat). The vapor leaves the evaporator at point C with a heat content of 108.1 Btu/lb. The usable refrigeration in the evaporator is from point A to C where the refrigerant absorbed 59.4 Btu/lb (108.1 Btu/lb − 48.7 Btu/lb) of refrigerant circulated. You only need to know how many Btu/h capacity is needed to determine the amount of refrigerant

that needs to be circulated. For example, if the evaporator needs to have a capacity of 35,000 Btu/h, it must have 673 lb of refrigerant circulate through it per hour (35,000 Btu/h ÷ 59.4 Btu/lb = 589.2 lb/h). This sounds like a lot of refrigerant, but it is only 9.82 lb/min (589.2 lb/h ÷ 60 min/h = 9.82 lb/min). The size of the compressor and system operating conditions determine how much refrigerant can be pumped.

21.10   LATENT HEAT IN THE EVAPORATOR

The latent heat absorbed during the change of state is much more concentrated than the sensible heat that would be added to the vapor leaving the coil. Refer to the example of heat in Unit 1, Section 1.3, that showed how it takes 1 Btu to change the temperature of 1 lb of 68°F water to 69°F water. Section 1.8 also showed that it takes 970 Btu to change 1 lb of 212°F water to 212°F steam. The change of state is where the greatest amount of heat is absorbed into the system. The preceding example showed that 59.4 Btu of heat were absorbed for every 1 lb of refrigerant that was circulated (59.4 Btu/lb). This happened at a boiling temperature of 20°F, without a change in pressure.

21.11   THE FLOODED EVAPORATOR

To get the maximum efficiency from the evaporator heat exchange, some evaporators are operated full of liquid or flooded and are equipped with a device to keep the liquid refrigerant from passing to the compressor. These *flooded evaporators* are specially made and normally use a float

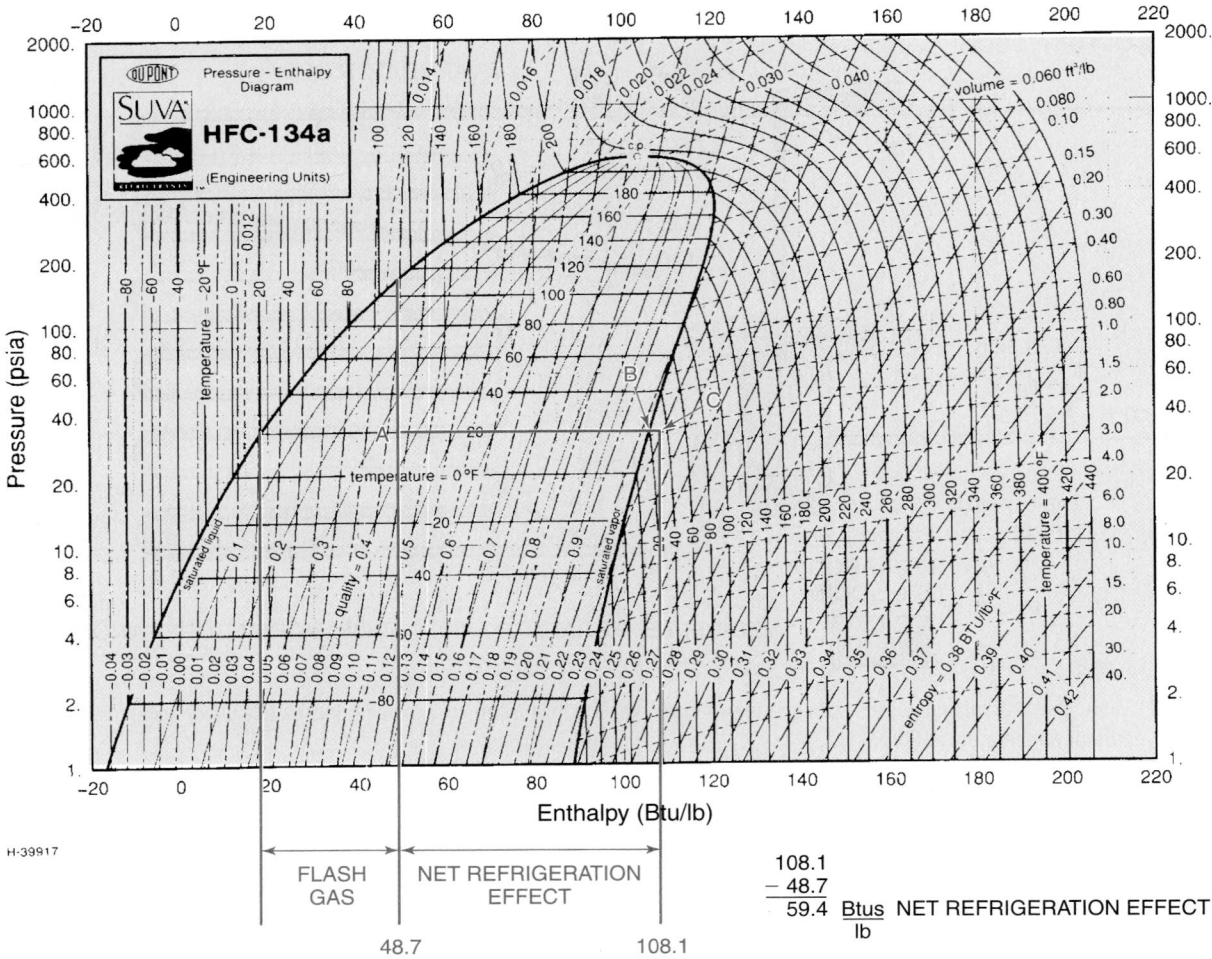

Figure 21–24 The refrigeration effect in the evaporator. *Courtesy E. I. DuPont*

metering device to keep the liquid level as high as possible in the evaporator. This text will not go into detail about this system because it is not a device encountered often. The manufacturer's literature should be consulted for any special application.

When an evaporator is flooded, it would be much like water boiling in a pot with a compressor taking the vapor off the top of the liquid. There would always be a liquid level. If the evaporator is not flooded, that is, when the refrigerant starts out as a partial liquid and boils away to a vapor in the heat exchange pipes, it is known as a *dry-type,* or direct-expansion, evaporator.

21.12 DRY-TYPE EVAPORATOR PERFORMANCE

To check the performance of a dry-type evaporator, the service technician would first make sure that the refrigerant coil is operating with enough liquid inside the coil. To determine whether enough liquid is in the coil, the technician must calculate the evaporator superheat. This is generally done by comparing the boiling temperature of the refrigerant inside the coil with the line temperature leaving the coil. The difference

in temperatures is usually 8°F to 12°F. For example, in the coil pictured in **Figure 21–25,** the superheat in the coil was arrived at by converting the coil pressure (suction pressure) to temperature. In this example, the pressure was 18.4 psig, which corresponds to 20°F. This suction pressure reading is important to the technician because the boiling temperature must be known to arrive at the superheat reading for the coil. In the following example, the evaporator superheat reading is 10°F (30°F − 20°F).

21.13 EVAPORATOR SUPERHEAT

The difference in temperature between the boiling refrigerant temperature and the evaporator outlet temperature is known as evaporator superheat. Superheat is the sensible heat added to the vapor refrigerant after the change of state has occurred. Superheat is the best method of checking to see when a refrigerant coil has a proper level of refrigerant. When a metering device is not feeding enough refrigerant to the coil, the coil is said to be a *starved coil,* and the superheat is greater, **Figure 21–26.** It can be seen from the example that all of the refrigeration takes place at the beginning of the coil. The suction pressure is very low, below freezing, but only a

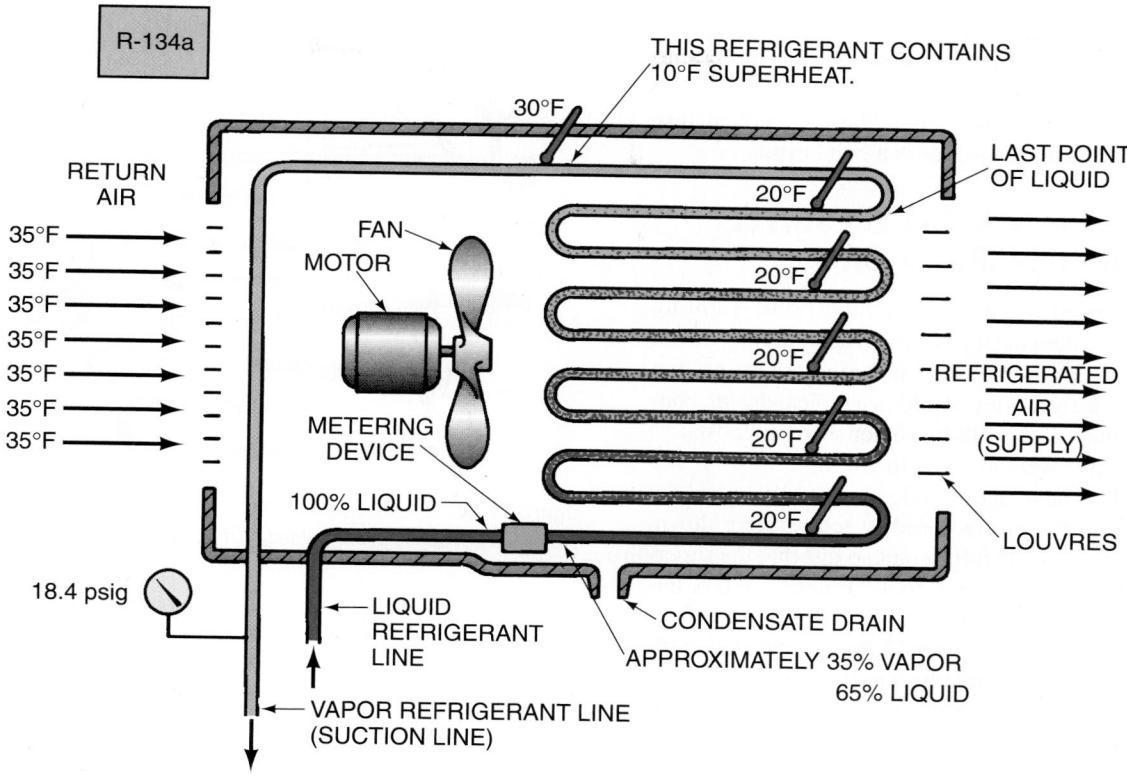

Figure 21–25   The evaporator operating under normal load.

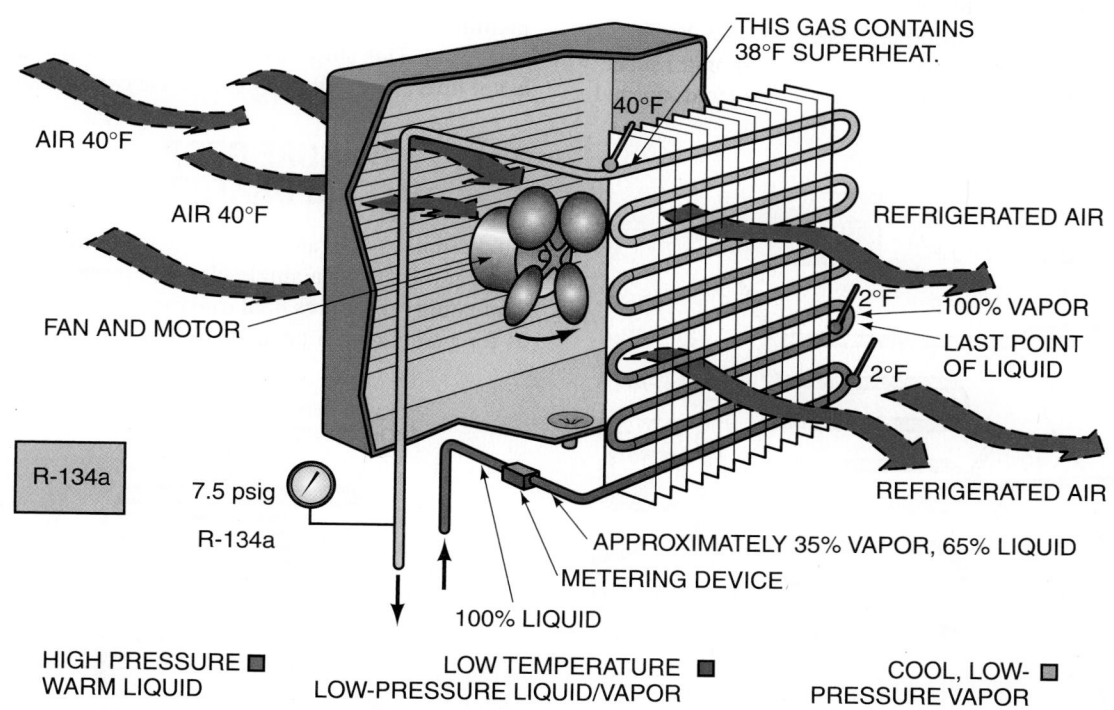

Figure 21–26   A starved evaporator coil showing 38°F (40°F − 2°F) of evaporator superheat.

portion of the coil is being used effectively. This coil would freeze solid, and no air would pass through it. The freeze line would creep upward until the whole coil is a block of ice, and the refrigeration would do no good. The refrigerated box temperature would rise because ice is a good insulator.

21.14 HOT PULLDOWN (EXCESSIVELY LOADED EVAPORATOR)

When the refrigerated space has been allowed to warm up considerably, the system must go through a hot pulldown. On a hot pulldown the evaporator and metering device are not expected to act exactly as they would in a typical design condition. For instance, if a walk-in cooler were supposed to maintain 35°F and it were allowed to warm up to 60°F and have some food or beverages inside, it would take an extended time to pull the air and product temperature down. The coil may be boiling the refrigerant so fast that the superheat may not come down to 8°F to 12°F until the box has cooled down closer to the design temperature. **A superheat reading on a hot pulldown should be interpreted with caution, Figure 21–27.** The superheat reading will be correct only when the coil is at or near design conditions. However, many modern thermostatic expansion valves (TXVs) today have wide temperature control ranges. Some can control evaporator superheat from +20°F to −20°F. They advertise that they do this effectively even when under heavy or light heat loadings of the evaporator. TXVs should control superheat under most normal conditions. However, when a system is under a hot pulldown, the technician should let the system get past this heavy load period and reach a somewhat stabilized condition before trying to calculate an evaporator superheat reading. Hot pulldowns are not considered normal

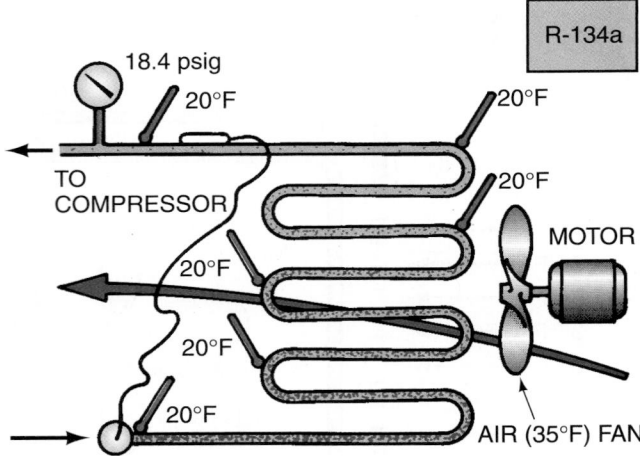

Figure 21–28   The evaporator is flooding because the thermostatic expansion device is not controlling refrigerant flow properly.

conditions, and the technician must be patient when calculating evaporator superheat. It takes time for a TXV to fill out the evaporator with refrigerant even when it is wide open during a hot pulldown.

When a dry-type coil is fed too much refrigerant, the refrigerant does not all change to a vapor. This coil is thought of as a *flooded* coil—flooded with liquid refrigerants, **Figure 21–28. Do not confuse this with a coil flooded by design.**

This is a symptom that can cause real trouble because unless this liquid in the suction line boils to a vapor before it reaches the compressor, compressor damage may occur. **Remember, the evaporator is supposed to boil all of the liquid to a vapor. Therefore, a thermostatic expansion device that is not operating correctly can cause compressor failure.**

21.15 PRESSURE DROP IN EVAPORATORS

Multicircuit evaporators are used when the coil would become too long for a single circuit, **Figure 21–29.** The same evaluating procedures hold true for a multicircuit evaporator as for a single-circuit evaporator.

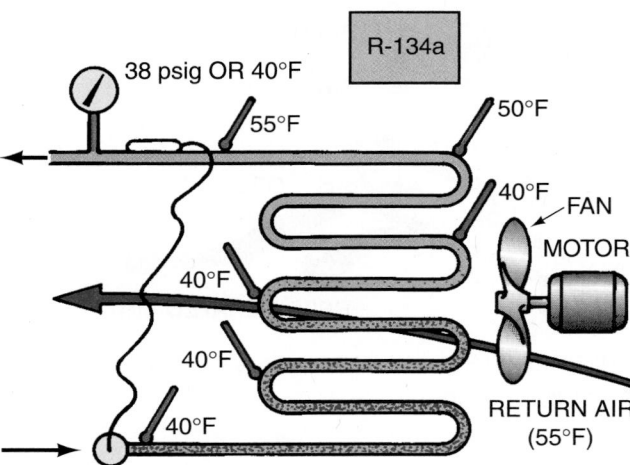

Figure 21–27   Hot pulldown with a coil. This is a medium-temperature evaporator that should be operating at 18.4 psig (R-134a, 20°F). The return air is 55°F instead of 35°F. This causes the pressure in the coil to rise. The warm box boils the refrigerant at a faster rate. The thermostatic expansion valve is not able to feed the evaporator quickly enough to keep the superheat at 10°F. The evaporator has 15°F of superheat.

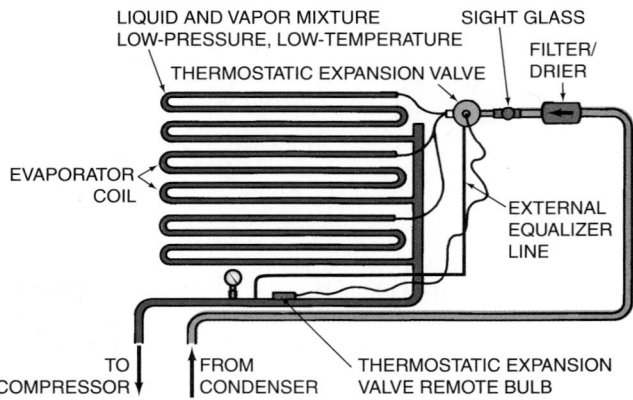

Figure 21–29   A multicircuit evaporator. *Courtesy Sporlan Valve Company*

A dry-type evaporator has to be as full as possible with refrigerant to be efficient. Each circuit should be feeding the same amount of refrigerant. If this needs to be checked, the service technician can check the common pressure tap for the boiling pressure. This pressure can be converted to temperature. Then the temperature will have to be checked at the outlet of each circuit to see whether any circuit is overfeeding or starving, **Figure 21–30** and **Figure 21–31**.

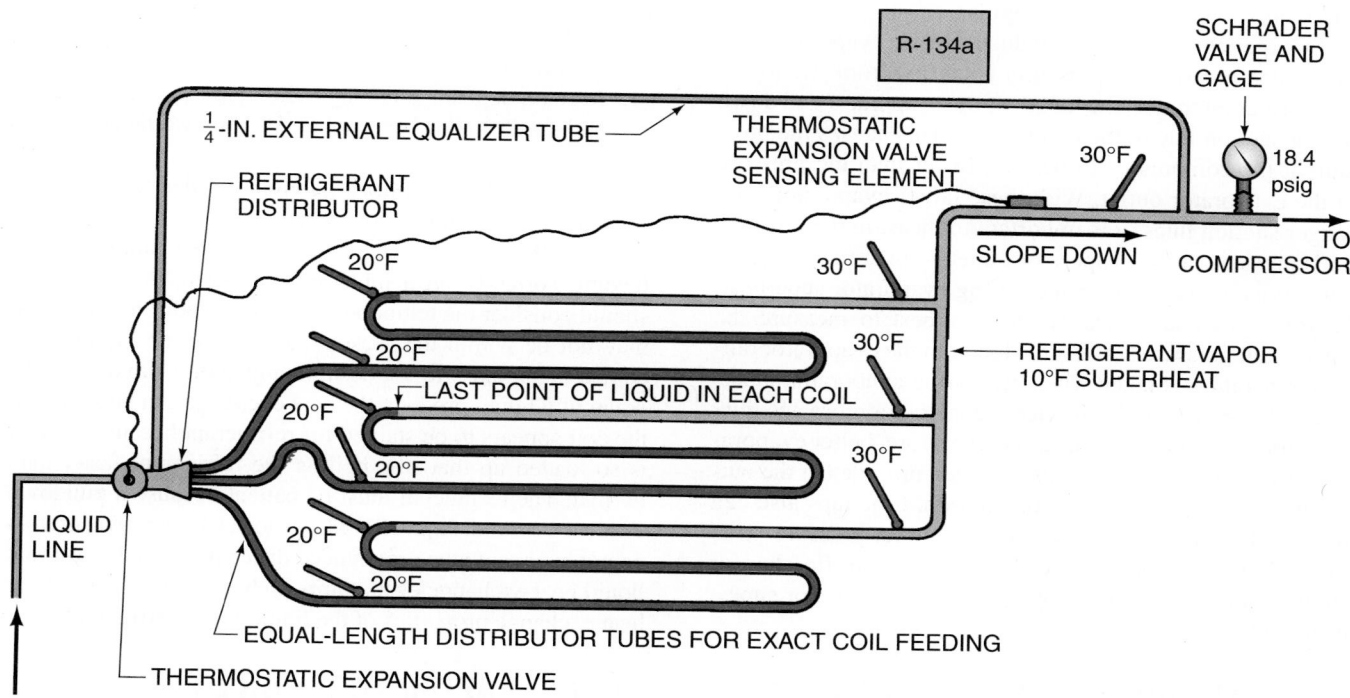

Figure 21–30 The appearance of a multicircuit evaporator on the inside when it is feeding correctly. It resembles several evaporators piped in parallel.

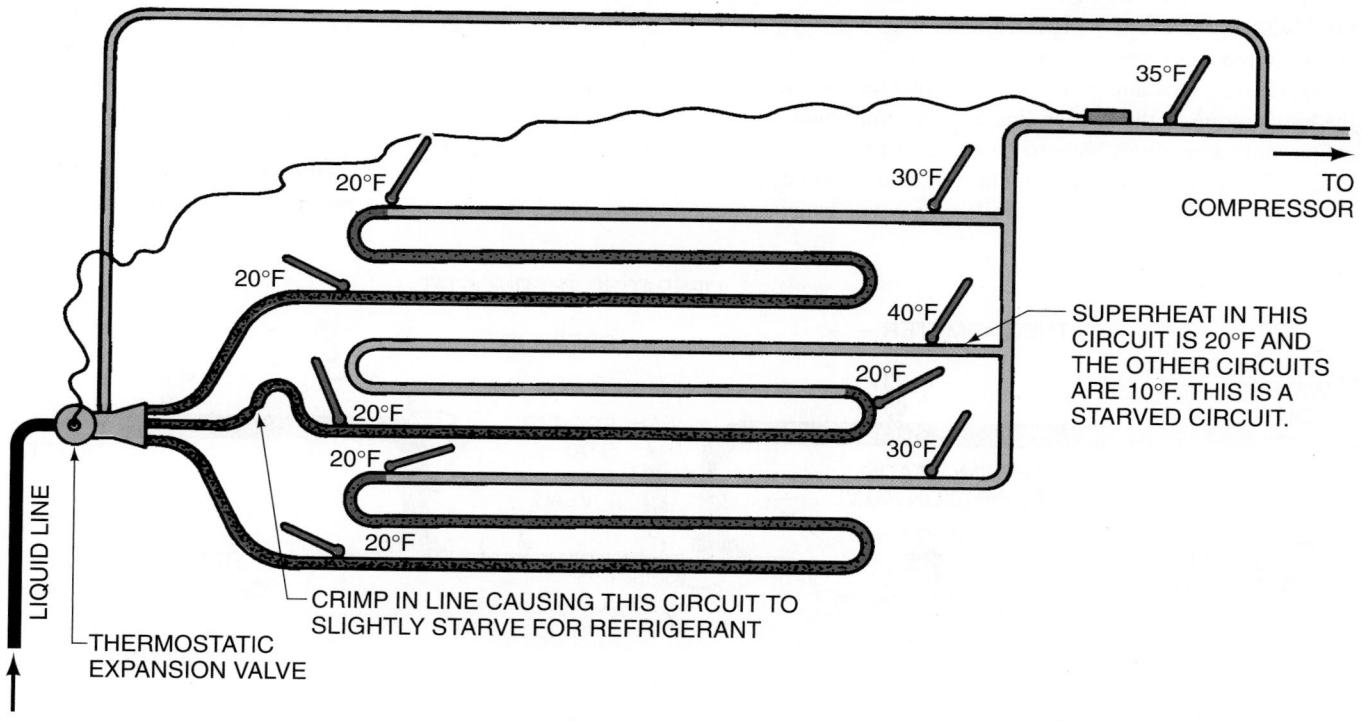

Figure 21–31 The appearance of a multicircuit evaporator on the inside when it is not feeding evenly.

Some reasons for uneven feeding of a multicircuit evaporator are the following:

1. Blocked distribution system
2. Dirty coil
3. Uneven air distribution
4. Coil circuits of different lengths

In larger commercial and industrial-type evaporators, an associated pressure drop usually is caused from friction as the refrigerant travels the length of the evaporator and down a long suction line to the compressor. This causes the pressure at the compressor to be a bit lower than the pressure at the evaporator outlet. With these larger evaporators and longer suction lines, it is important to measure the refrigerant's pressure at the evaporator outlet—not at the compressor service valves—when measuring evaporator superheat, **Figure 21–29** and **Figure 21–30**. It is best to measure the refrigerant pressure at the same location the evaporator outlet temperature is taken when measuring evaporator superheat. This will give the service technician a more accurate evaporator superheat reading and, therefore, better evaporator efficiencies. Schrader taps are often provided at the outlet of larger evaporators for this reason. Line taps also can be used to gain access to evaporator outlet pressure. This method also will protect the compressor from flooding or slugging problems caused by inaccurate evaporator superheat readings.

21.16 LIQUID COOLING EVAPORATORS (CHILLERS)

A different type of evaporator is required for liquid cooling. It functions much like the one for cooling air. They are normally the dry-type expansion evaporator in the smaller systems, **Figure 21–32**. These evaporators have more than one refrigerant circuit to prevent pressure drop in the evaporator. The use of refrigeration gages and some accurate method for checking the temperature of the suction line are very important. These evaporators sometimes have to be checked for performance to see whether they are absorbing heat like they should. They have a normal superheat range similar to air-type evaporators

(8°F to 12°F). When the superheat is within this range and all circuits in a multicircuit evaporator are performing alike, the evaporator is doing its job on the refrigerant side. However, this does not mean that it will cool properly. The liquid side of the evaporator must be clean so that the liquid will come in proper contact with the evaporator.

The following are typical problems on the liquid side of the evaporator:

1. Mineral deposits that may build up on the liquid side and cause a poor heat exchange. They would act like an insulator.
2. Poor circulation of the liquid to be cooled where a circulating pump is concerned.

When the superheat is found to be correct and the coil is feeding correctly on a multicircuit system, the technician should consider the temperature of the liquid. The superheat may not be within the prescribed limits if the liquid to be cooled is not close to the design temperature. On a hot pull-down of a liquid product, the heat exchange can be such that the coil appears to be starved for refrigerant because the coil is so loaded up that it is boiling the refrigerant faster than normal. The technician must be patient because a pulldown cannot be rushed, **Figure 21–33**. Air-to-refrigerant evaporators do not have quite the pronounced difference in pulldown that liquid heat exchange evaporators do because of the excellent heat exchange properties of the liquid to the refrigerant.

21.17 EVAPORATORS FOR LOW-TEMPERATURE APPLICATION

Low-temperature evaporators used for cooling space or product to below freezing are designed differently because they require the coil to operate below freezing.

In an airflow application, the water that accumulates on the coil will freeze and will have to be removed. The design of the fin spacing must be carefully chosen. A very small amount of ice accumulated on the fins will restrict the airflow. Low-temperature coils have fin spacings that are wider than medium-temperature coils, **Figure 21–34**. Other than the air-flow blockage due to ice buildup, these low-temperature

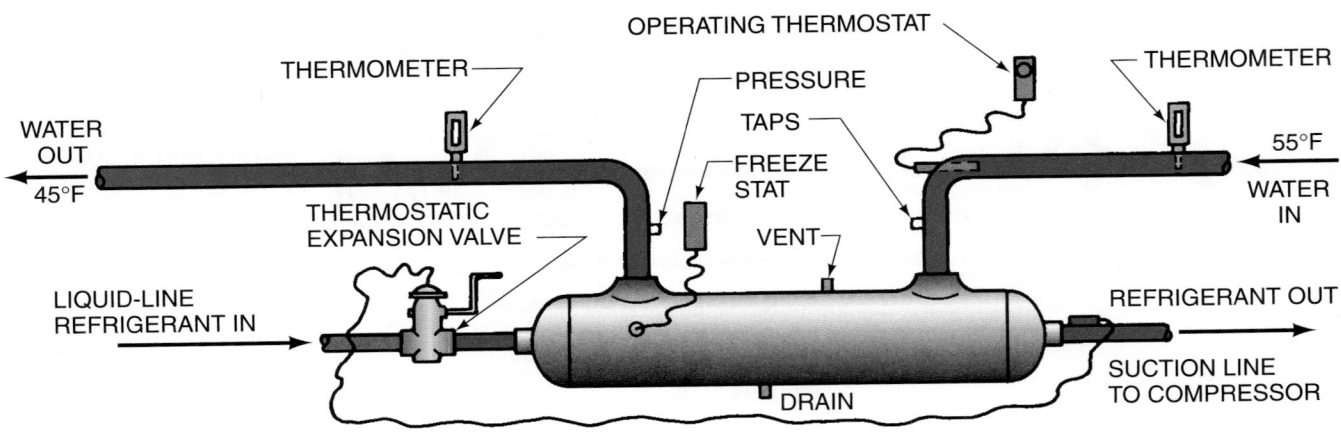

Figure 21–32 An evaporator used for exchanging heat between liquids and refrigerant. Most of these evaporators are of the direct-expansion type.

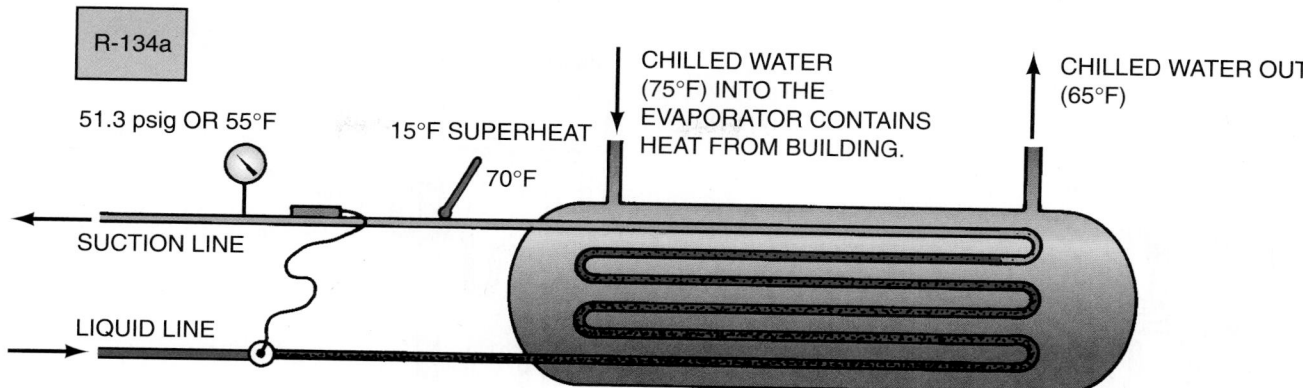

Figure 21–33 A hot pulldown on a liquid evaporator giving up its heat to refrigerant. This evaporator normally has 55°F water in and 45°F water out. The hot pulldown with 75°F water instead of 55°F water boils the refrigerant at a faster rate. The expansion valve may not be able to feed the evaporator quickly enough to maintain 10°F superheat. No conclusions should be made until the system approaches design conditions.

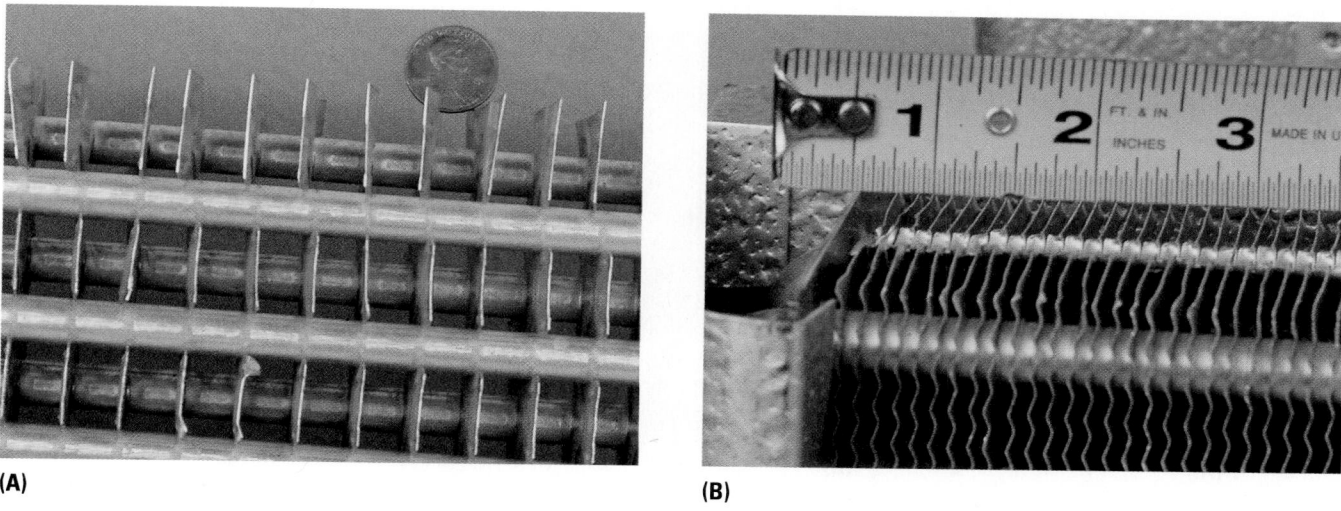

(A)   **(B)**

Figure 21–34 Fin spacing. **(A)** Low-temperature evaporator. **(B)** Medium-temperature evaporator. *Photo by Bill Johnson*

evaporators perform much the same as medium-temperature evaporators. They are normally dry-type evaporators and have one or more fans to circulate the air across the coil. The defrosting of the coil has to be done by raising the coil temperature above freezing to melt the ice. Then the condensate water has to be drained off and kept from freezing.

Defrost is sometimes accomplished with heat from outside the system. Electric heat can be added to the evaporator to melt the ice, but this heat adds to the load of the system and needs to be pumped out after defrost.

21.18 DEFROST OF ACCUMULATED MOISTURE

Defrost can be accomplished with heat from inside the system using the hot gas from the discharge line of the compressor. This is accomplished by routing a hot gas line from the compressor discharge line to the outlet of the expansion valve and installing a solenoid valve to control the flow. When defrost is needed, hot gas is released inside the evaporator, which will quickly melt any ice, **Figure 21–35.**

When the hot gas enters the evaporator, it is likely that liquid refrigerant will be pushed out of the suction line toward the compressor. In fact, when hot gas enters an evaporator and starts to cool as it melts ice or frost, it will soon lose all of its superheat and turn to liquid or condense. This liquid will go to the accumulator and fall to its bottom. Dense saturated vapors will be drawn into the compressor's suction stroke. The compressor will see an increased load from these dense vapors and may draw a higher amp than during the normal running cycle. If the system does not have an accumulator, this condensed liquid may flood the compressor's crankcase, and foaming of the oil in the crankcase may occur, **Figure 21–36.** This can lower the oil level in the crankcase and cause scoring of bearing surfaces in the compressor. This condition is often referred to as *bearing washout.*

Flooding the compressor's crankcase with liquid refrigerant can also cause the foaming refrigerant and oil mixture to pressurize the crankcase. This causes the mixture of liquid and vapor refrigerant and oil foam to be forced through any crevice available, including the compressor's piston rings. The mixture is often pumped into the high side of the system

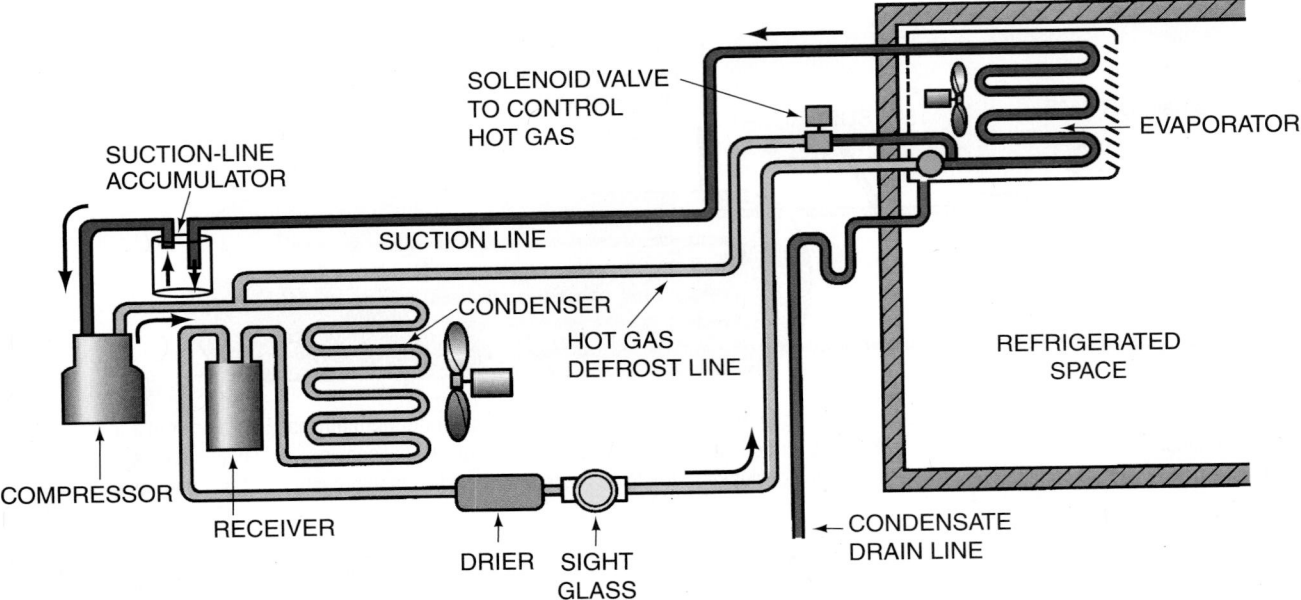

Figure 21–35 Using hot gas to defrost an evaporator.

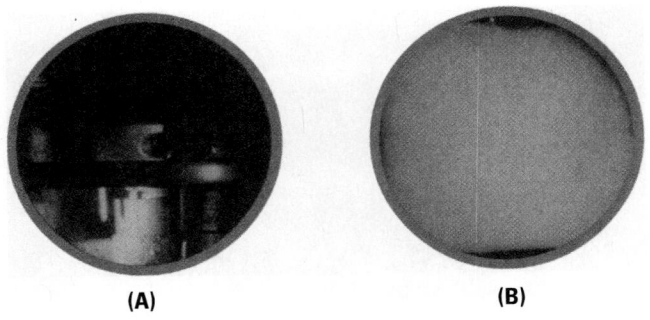

Figure 21–36 **(A)** A clear compressor's oil sight glass. **(B)** A sight glass from a flooding crankcase foams with liquid refrigerant. *Courtesy Tecumseh Products Company*

by the compressor. The compressor's discharge temperature will decrease from the *wet compression* of the rich mixture of refrigerant and oil foam. As soon as this mixture is compressed, it will vaporize and absorb heat away from the cylinder walls. This is what causes a lower-than-normal discharge temperature on the compressor's discharge line. Some manufacturers of systems incorporating hot gas defrost place a thermistor on the discharge line of the compressor to sense this cooler-than-normal discharge temperature while in the defrost mode. The thermistor relays a message to a control circuit to deenergize the hot gas solenoid; closing the hot gas solenoid temporarily prevents any more hot gas from entering the evaporator and thus turning to liquid. As the compressor's discharge temperature rises again, the thermistor senses the rise in temperature and relays a message to the control circuit to continue with hot gas defrost by energizing the hot gas solenoid again.

This sequence of events may continue until defrost is complete. The sequence simply protects the compressor from bearing washout and wet compression while in defrost.

It may also protect the compressor during the refrigeration cycle by shutting down the compressor if the discharge temperature gets too low, indicating wet compression. To prevent this liquid from entering the compressor, often a suction line accumulator will be added to the suction piping, **Figure 21–35.**

🌐 The hot gas defrost system is economical because power does not have to be purchased for defrost using external heat, such as electric heaters that will heat the evaporator. The heat is already in the system. 🌐

Electric defrost is accomplished using electric heating elements located at the evaporator. The compressor is stopped, and these heaters are energized on a call for defrost and allowed to operate until the frost is melted from the coil, **Figure 21–37.** These heaters are often embedded in the actual evaporator fins and cannot be removed if they burn out. Frequently, in the event that the heaters do burn out, hot gas

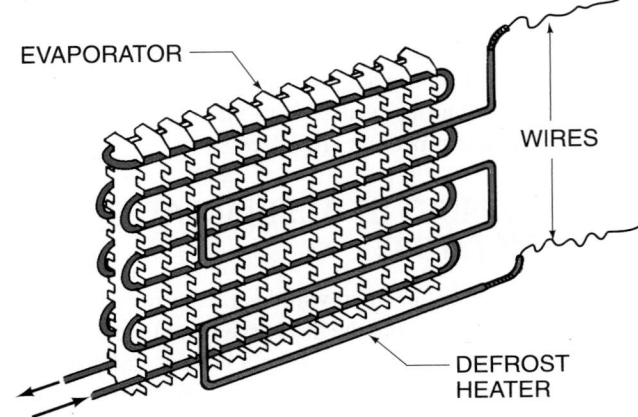

Figure 21–37 A heater used for electric defrost of low-temperature evaporators.

defrost can be added to the system and the electric heat defrost procedures discontinued.

When either system is used for defrost, the evaporator fan is often turned off during defrost; if it is not, two things will happen:

1. The heat from defrost will be transferred directly to the conditioned space.
2. The cold conditioned air will slow down the defrost process.

However, some manufacturers design their open, frozen food cases so that the evaporator fans are left on during defrost. This allows the supply-air and return-air ducts to be defrosted along with the coil. The warm defrost air is discharged out of the supply duct and rises out of the case, having little effect on the product temperature. Fans are always shut off on closed, glass-door cases because of the problem of fogging on glass or mirrored surfaces.

Evaporators applied to some ice-making processes have similar defrost methods. They must have some method of applying heat to the evaporator to free the ice. Sometimes the heat is electric or hot gas. When the evaporator is being used to make ice, the makeup water for the ice maker is sometimes used for defrost.

In summary, when checking an evaporator remember that its job is to absorb heat into the refrigeration system.

SUMMARY

■ Heat travels normally from a warm substance to a cool substance.

■ For heat to travel from a cool substance to a warm substance, work must be performed. The motor that drives the compressor in the refrigeration cycle does this work.

■ The evaporator is the component that absorbs the heat into the refrigeration system.

■ The evaporator must be cooler than the medium to be cooled to have a heat exchange.

■ The refrigerant boils to a vapor in the evaporator and absorbs heat because it is boiling at a low pressure and low temperature.

■ The boiling temperature of the refrigerant in the evaporator determines the evaporator (low-side) pressure.

■ Medium-temperature systems can use off-cycle defrost. The product is above freezing, and the heat from it can be used to cause the defrost.

■ Low-temperature refrigeration must have heat added to the evaporator to melt the ice.

■ Evaporators have the same characteristics for the same types of installations regardless of location.

■ Most refrigeration coils are copper with aluminum fins.

■ The starting point in organized troubleshooting is to determine whether the evaporator is operating efficiently.

■ Checking the superheat is the best method the service technician has for evaluating evaporator performance.

■ Some evaporators are called dry-type because they use a minimum of refrigerant.

■ Dry-type evaporators are also called direct-expansion evaporators.

■ Some evaporators are flooded and use a float to meter the refrigerant. **Superheat checks on these evaporators should be interpreted with caution.**

■ Some evaporators have a single circuit, and some have multiple circuits.

■ Multicircuit evaporators keep excessive pressure drop from occurring in the evaporator.

■ There is a relationship between the boiling temperature of the refrigerant in the evaporator and the temperature of the medium being cooled.

■ The coil normally operates at temperatures from 10°F to 20°F colder than the temperature of the air passing over it.

REVIEW QUESTIONS

1. What is the function of the evaporator in the refrigeration system?
2. Refrigerant in the evaporator
   A. changes from vapor to liquid.
   B. changes from liquid to vapor.
   C. stays in the vapor state.
   D. stays in the liquid state.
3. The heat that is added to a vapor after all of the liquid has boiled away is referred to as _____.
4. What determines the pressure on the low-pressure side of the system?
5. A refrigerant system's evaporator typically runs about _____ degrees of superheat.
6. What does a high superheat indicate?
7. A low evaporator superheat indicates
   A. undercharge.
   B. system restriction.
   C. overcharge.
   D. dirt buildup.
8. Why is a multicircuit evaporator used?
9. Flooded evaporators use a _____ type of expansion device.
10. An evaporator that is not flooded is thought of as what type of evaporator?
11. When an evaporator experiences a heat-load increase, the suction pressure
    A. remains constant.
    B. decreases.
    C. varies up and down.
    D. increases.
12. What is commonly used to defrost the ice from a low-temperature evaporator?
13. A medium-temperature refrigeration box operates within what temperature range?
    A. 28°F to 40°F
    B. 40°F to 60°F
    C. 0°F to −20°F
    D. 0°F to 50°F
14. List seven advantages that an aluminum, parallel-flow, plate/fin evaporator has over a standard, round, copper-tube plate/fin evaporator.

UNIT 22 Condensers

OBJECTIVES

After studying this unit, you should be able to

- explain the purpose of the condenser in a refrigeration system.
- describe differences between the operating characteristics of water-cooled and air-cooled systems.
- describe the basis of the heat exchange in a condenser.
- explain the difference between a tube-within-a-tube coil-type condenser and a tube-within-a-tube serviceable condenser.
- describe the difference between a shell-and-coil condenser and a shell-and-tube condenser.
- describe a wastewater system.
- describe a recirculated water system.
- describe a cooling tower.
- explain the relationship between the condensing refrigerant and the condensing medium for cooling tower systems.
- compare an air-cooled, high-efficiency condenser with a standard condenser.
- describe the operation of head pressure control values.

SAFETY CHECKLIST

✔ Wear goggles and gloves when attaching or removing gages to transfer refrigerant or to check pressures.
✔ Wear warm clothing when working in a walk-in cooler or freezer. A technician does not think properly when chilled.
✔ Keep hands well away from moving fans. Do not try to stop a fan blade as it slows after the power has been turned off.
✔ Do not touch the hot gas line.

22.1 THE CONDENSER

The *condenser* is a heat exchange device similar to the evaporator; it rejects the heat from the system absorbed by the evaporator. This heat is rejected from a hot superheated vapor in the first passes of the condenser. The middle of the condenser rejects latent heat from the saturated vapor, which is in the process of phase changing to a saturated liquid. The last passes of the condenser reject heat from subcooled liquid. This further subcools the liquid to below its condensing temperature. In fact, the three functions of a normal condenser are to *desuperheat, condense,* and *subcool* the refrigerant. When heat is being absorbed into the system, it is at the point

of change of state (liquid to a vapor) of the refrigerant where the greatest amount of heat is absorbed. The same thing, in reverse, is true in the condenser. The point where the change of state (vapor to a liquid) occurs is where the greatest amount of heat is rejected.

The condenser is operated at higher pressures and temperatures than the evaporator and is often located outside. The same principles apply to heat exchange in the condenser as in the evaporator. The materials a condenser is made of and the medium used to transfer heat make a difference in the efficiency of the heat exchanger.

22.2 WATER-COOLED CONDENSERS

The first commercial refrigeration condensers were water cooled. These condensers were crude compared with modern water-cooled devices, **Figure 22–1.** Water-cooled condensers are more efficient than air-cooled condensers and operate at much lower condensing temperatures. Water-cooled equipment comes in several styles. The tube within a tube, the shell and coil, and the shell and tube are the most common.

22.3 TUBE-WITHIN-A-TUBE CONDENSERS

The *tube-within-a-tube* condenser comes in two styles: the coil type and the cleanable type with flanged ends, **Figure 22–2.**

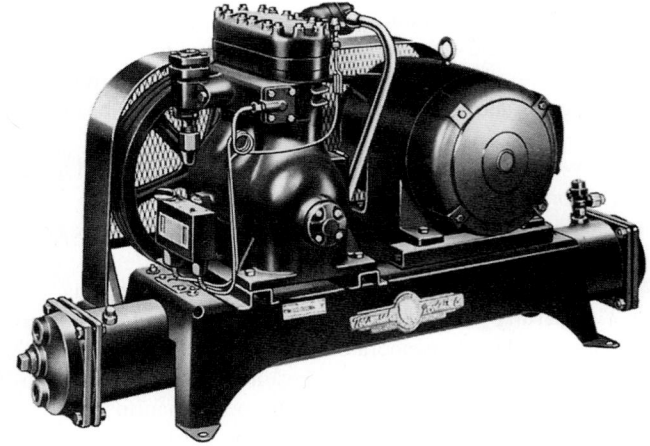

Figure 22–1 An early water-cooled condensing unit. *Courtesy Tecumseh Products Company*

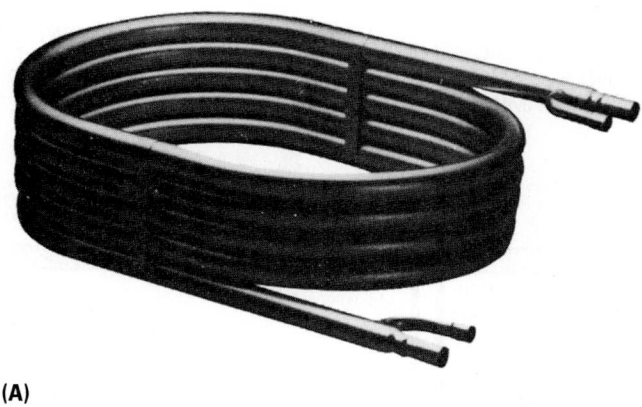

(A)

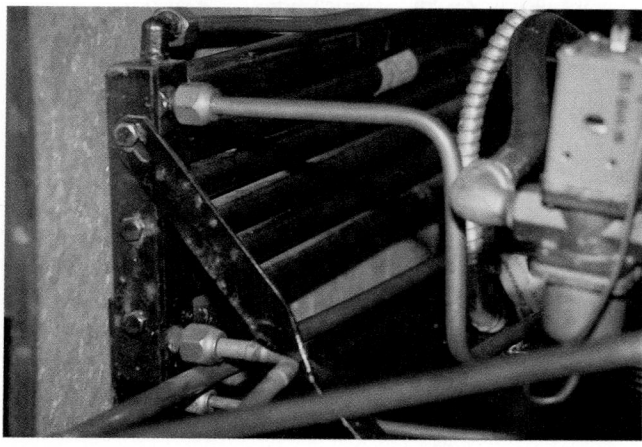

(B)

Figure 22–2 Two types of tube-within-a-tube condensers. **(A)** A pipe within a pipe. **(B)** A flanged type of condenser. The flanged condenser can be cleaned by removing the flanges. Removal of the flanges opens only the water circuit, not the refrigerant circuit. **(A)** *Courtesy Noranda Metal Industries, Inc.* **(B)** *Photo by Bill Johnson*

The tube within a tube that is fabricated into a coil is manufactured by slipping one pipe inside another and sealing the ends in such a manner that the outer tube becomes one container and the inner tube becomes another container, **Figure 22–3.** The two pipes are then formed into a coil to save space. The heat exchange occurs between the fluid inside the outer pipe and the fluid inside the inner pipe, **Figure 22–4.** Because water flows in the inner tube and hot refrigerant flows in the outer tube, the hot refrigerant also can reject some of its heat to the surrounding air. However, most of the heat rejected is from the refrigerant to the water because of the higher density, flow rate, and specific heat of the water.

22.4 MINERAL DEPOSITS

Because water flows through one of the tubes, mineral deposits and scale will form even in the best water. The heat in the vicinity of the discharge gas has a tendency to cause any minerals in the water to deposit onto the tube surface. This is

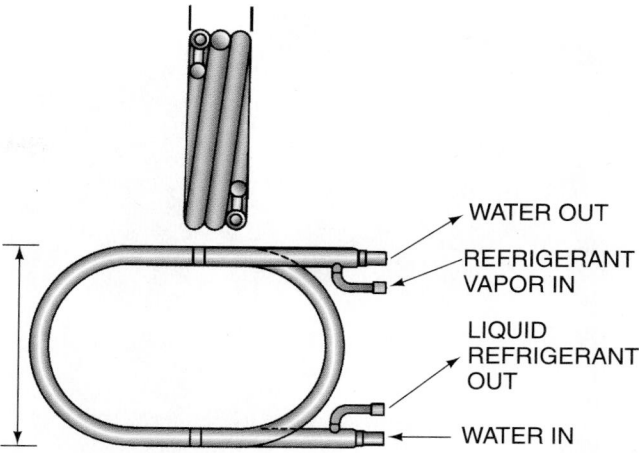

Figure 22–3 A tube-within-a-tube condenser constructed by sliding one tube through the other tube. The tubes are sealed in such a manner that the inside tube is separated from the outside tube. *Courtesy Noranda Metal Industries, Inc.*

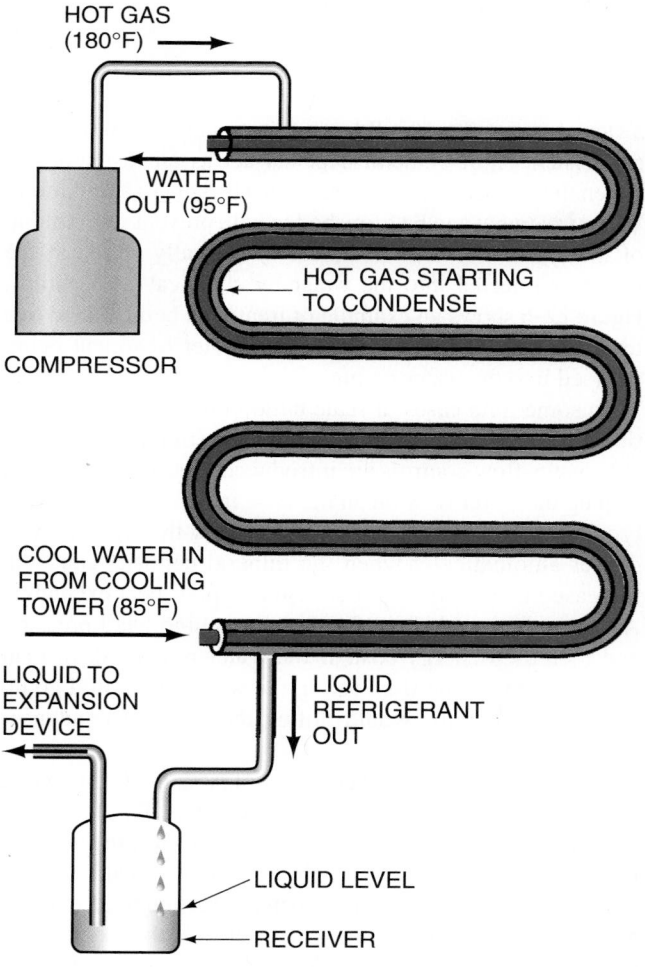

Figure 22–4 Fluid flow through the condenser. The refrigerant is flowing in one direction, water in the other.

Figure 22–5 A method of adding water treatment to a cooling tower. Treatment is being metered at a rate that will last about a month so that the operator will not have to be in attendance at the tower all of the time. There is a constant bleed of the tower water to the drain to keep the water from being overconcentrated with minerals. *Courtesy Calgon Corporation*

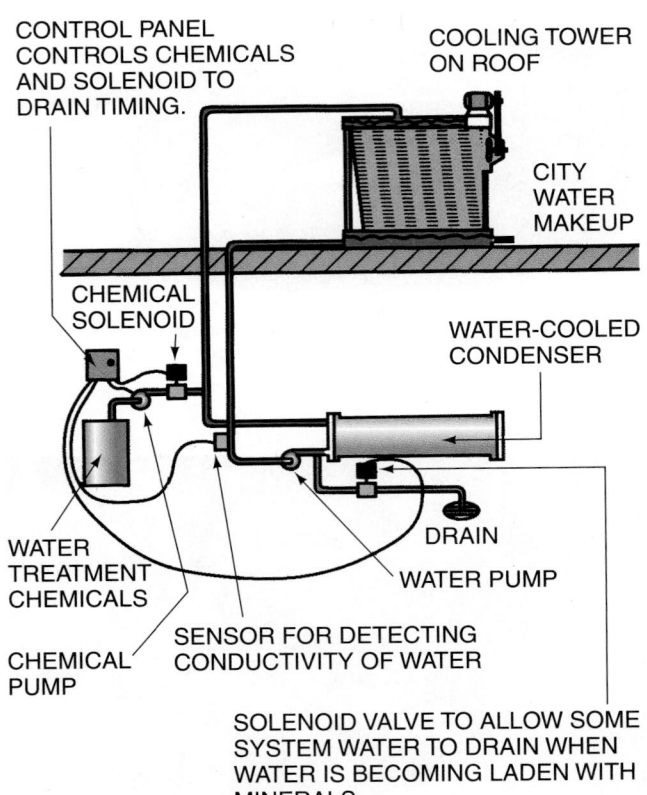

Figure 22–6 An automatic system of feeding the treatment chemicals to the water. It includes an automatic monitoring system that determines when the system actually needs chemicals added. This type is normally used on larger systems because of the economics of the total system cost.

a slow process, but it will happen in time to any water-type condenser. These mineral deposits act as an insulator between the tube and the water and must be kept to a minimum. Water treatment can be furnished to help prevent this buildup of mineral scale. This treatment is normally added at the tower or injected into the water by chemical feed pumps. **Figure 22–5** shows an example of treatment being added to a tower. **Figure 22–6** is an example of water treatment being pumped into the water piping.

In some mild cases of scale buildup more water circulation will improve the heat exchange. Later in this unit, variable water-flow controls are introduced. These controls will step up the water flow on an increase in head pressure. This type of control causes more water to flow through the condenser automatically when the mineral deposits cause an increase in head pressure because of a poor heat exchange. A dirty water-cooled condenser causes high head pressure and increased energy cost. If the water is wasted, instead of being cooled and used again, the water bill would go up before the operator would notice there was a condenser problem.

The tube-within-a-tube condenser that is made into a coil cannot be cleaned mechanically with brushes. This type must be cleaned with chemicals designed not to harm the metal in the condenser. Professional help from a chemical company that specializes in water treatment is recommended when a condenser must be cleaned with chemicals. Condensers of this type are normally made from copper or steel; some special condensers are made of stainless steel or copper and nickel.

22.5 CLEANABLE CONDENSERS

The tube-within-a-tube condenser that is fabricated with flanges on the end can be mechanically cleaned. The flanges can be removed, and the tubes can be examined and brushed with an approved brush, **Figure 22–7**. The flanges and gaskets on this type of condenser are in the water circuit with the refrigerant flowing around the tubes. The refrigerant circuit is not opened to clean the tubes. **Figure 22–8** shows how tubes are cleaned. Consult the manufacturer for the correct brush. Fiber is usually preferable. This is a more expensive type of condenser, but it is serviceable.

22.6 SHELL-AND-COIL CONDENSERS

The shell-and-coil condenser is similar to the tube-within-a-tube coil. It is a coil of tubing packed into a shell that is then closed and welded. Normally the refrigerant gas is discharged into the shell, and the water is circulated in the coil located in the shell. The shell of the condenser serves as a receiver storage tank for the extra refrigerant in the system. This condenser is not mechanically cleanable because the coil is not straight, **Figure 22–9**. It must be cleaned chemically.

Figure 22–7   This condenser is flanged for service. When the flanges are removed, the refrigerant circuit is not disturbed. *Photo by Bill Johnson*

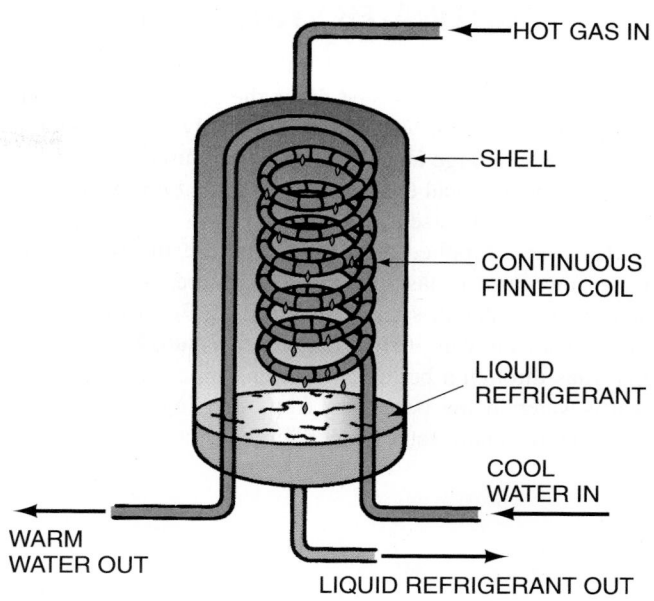

Figure 22–9   The shell-and-coil condenser. The hot refrigerant gas is piped into the shell, and the water is contained inside the tubes.

22.7   SHELL-AND-TUBE CONDENSERS

Shell-and-tube condensers are more expensive than shell-and-coil condensers, but they can be cleaned mechanically with brushes. They are constructed with the tubes fastened into an end sheet in the shell. The refrigerant is discharged into the shell, and the water is circulated through the tubes. The ends of the shell are like end caps (known as *water boxes*) with the water circulating in them, **Figure 22–10**. These end caps can be removed, so the tubes can be inspected and brushed out if needed. The shell acts as a receiver storage tank for extra refrigerant. This is the most expensive condenser and is normally used in larger applications.

The water-cooled condenser is used to remove the heat from the refrigerant. After the heat has been removed from the refrigerant, the heat is now in the water, which gets warmer. Two things can be done at this point: (1) waste the water, or (2) pump the water to a remote place, remove the heat, and reuse the water.

(A)

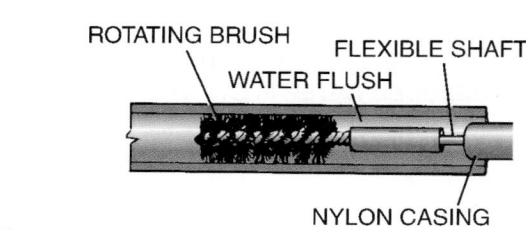

(B)

Figure 22–8   **(A)** Brushes actually being pushed through the water side of the condenser. Use only approved brushes. **(B)** Cutaway of brush. *Courtesy Tools Corporation*

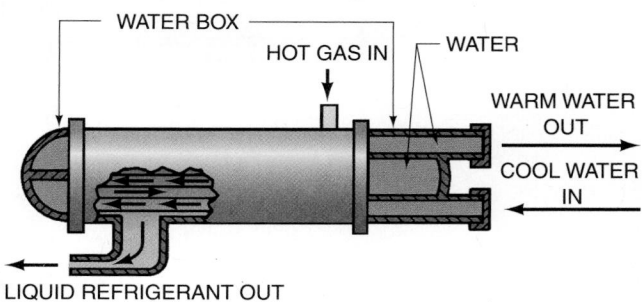

Figure 22–10   Water can be circulated back and forth through the condenser by using the end caps to give the water the proper direction.

22.8 WASTEWATER SYSTEMS

Wastewater systems are just what the name implies. The water is used once, then wasted down the drain, **Figure 22–11.** This is worthwhile if water is free or if only a small amount is used. Where large amounts of water are used, it is probably more economical to save the water, cool it in an outside water tower, and reuse it.

The water supplied to systems that use the water only once and waste it has a broad temperature range. For instance, the water in summer may be 75°F out of the city mains and as low as 40°F in the winter, **Figure 22–12.** Water that runs through a building with long pipe runs may have warm water in the beginning from standing in the pipes; however, the main water temperature may be quite low. This

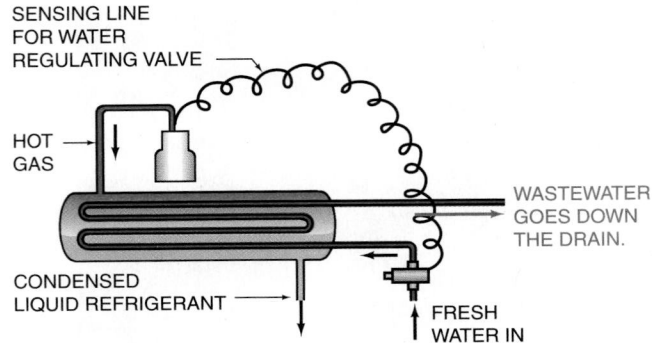

Figure 22–11 The wastewater system is used when water is plentiful at a low cost, such as from a well or lake.

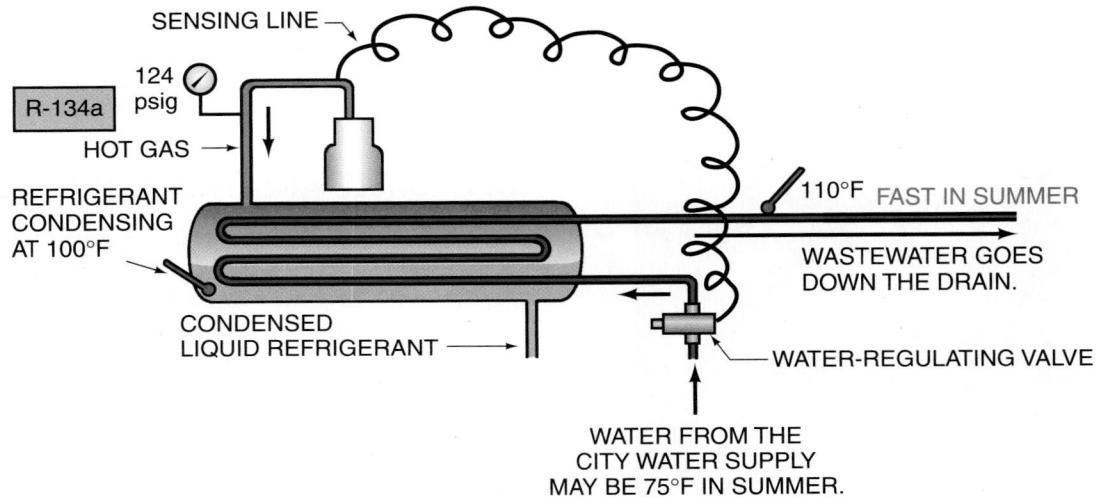

(A)

(B)

Figure 22–12 A wastewater condenser system at two sets of conditions. **(A)** Summer with warm water entering the system. **(B)** Winter with cooler water entering the system.

change in water temperature has an effect on the refrigerant head pressure (condensing temperature). A typical wastewater system with 75°F entering water uses 1.5 gal/min per ton of refrigeration. In the winter, when the water temperature is lower, the required water flow will be lower.

22.9 REFRIGERANT-TO-WATER TEMPERATURE RELATIONSHIP FOR WASTEWATER SYSTEMS

In a wastewater system, the water flow can be varied to suit the need by means of a regulating valve for the water. This valve has a pressure tap that fastens the bellows in the control to the high-pressure side of the system. When the head pressure goes up, the valve opens and allows more water to flow through the condenser to keep the head pressure correct, **Figure 22–13(A)**.

This valve will keep a constant head pressure by letting more or less water into the water-cooled condenser as the head pressure changes. **Figure 22–13(A and B)** shows a direction arrow stamped on the body of the valve to let the service technician know that the valve must be installed so the water flow will follow the arrow's direction. **Figure 22–13(B)** shows that an increase in head pressure will put more pressure on the flexible bellows and push up on the moveable rod to open the valve and let more water through the condenser. This in turn will keep the head pressure constant at the spring's set point. The spring pressure can be adjusted by turning the spring tension adjustment located on top of the valve. A refrigeration service wrench, standard slot screwdriver, or crescent wrench can be used to turn the spring tension adjustment. A spring retainer plate with female threads rides up and down on a threaded stationary stud when the spring tension is adjusted, **Figure 22–13(C)**. An arrow on the top of the valve indicates that a counterclockwise turn direction will increase the head pressure setting by letting in less water. A pressure gage that is connected to the high side of the refrigeration system while the spring tension of the water-regulating valve is being changed will show the head pressure change almost instantaneously. **Figure 22–13(D)** shows the valve's specification plate and the refrigerants the valve is designed to be used with. These valves have a maximum water pressure and refrigerant pressure range stamped on the specification plate for safety and operational reasons. Notice that this valve can handle a water pressure of 150 psig maximum and that its adjustment, or refrigerant pressure, range is from 70 to 260 psig. **Figure 22–13(E)** shows a water-regulating valve operating with a tube-within-a-tube water-cooled condenser. The water-regulating valve can be flushed by prying up on the bottom of the spring with two standard slot screwdrivers. This will rid the valve body of any loose scale deposits, dirt, or sand.

Wastewater systems often have seasonal water inlet temperatures. For example, in the summer, the water mains may be delivering water at 75°F to the water-cooled condenser. However, in the winter months, the water from the mains may be as low as 40°F. **Figure 22–12** illustrates a water-

regulating valve maintaining a constant head pressure of 124 psig for an R-134a refrigeration system. When the condenser water entering temperature is 75°F, the flow of water through the condenser is much greater and faster. This is because the water-regulating valve is more opened to maintain a constant condensing temperature of 100°F (124 psig) for the R-134a refrigeration system. Notice that the water exiting the condenser is 110°F, which is 10°F hotter than the condensing temperature of 100°F. This phenomenon happens because the water exiting the condenser comes in contact with hot superheated discharge gases from the compressor before it exits down the drain.

However, in the winter months when the entering water temperature is 40°F, the water-regulating valve will throttle the water to a reduced flow rate to maintain the same 100°F (124 psig) condensing temperature for this system. Because of this reduced flow rate of colder water, the water will stay in contact with the condenser's internal tubing longer.

SPRING TENSION ADJUSTMENT

(A)

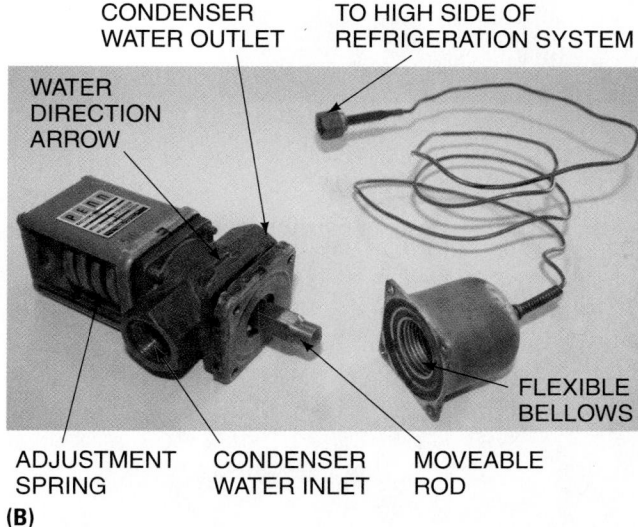

CONDENSER WATER OUTLET

TO HIGH SIDE OF REFRIGERATION SYSTEM

WATER DIRECTION ARROW

FLEXIBLE BELLOWS

ADJUSTMENT SPRING

CONDENSER WATER INLET

MOVEABLE ROD

(B)

Figure 22–13 **(A)** This water-regulating valve controls water flow during different demand periods. **(B)** Internal components of a water-regulating valve. **(A)** Photo by Bill Johnson. **(B)** Courtesy Ferris State University. Photo by John Tomczyk

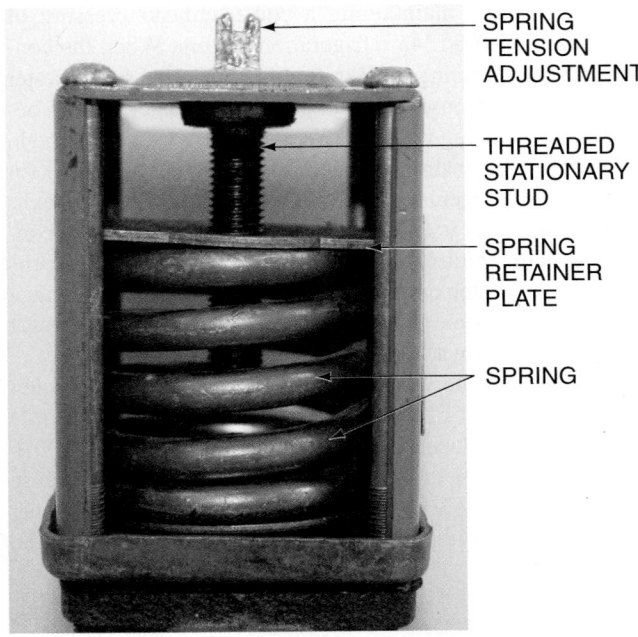

SPRING
TENSION
ADJUSTMENT

THREADED
STATIONARY
STUD

SPRING
RETAINER
PLATE

SPRING

(C)

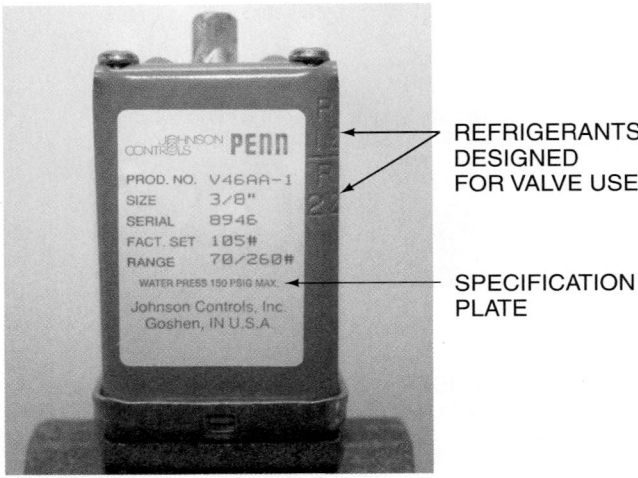

REFRIGERANTS
DESIGNED
FOR VALVE USE

SPECIFICATION
PLATE

(D)

(E)

→ **Figure 22–13** *(continued)* **(C)** Top of a water-regulating valve showing spring, spring adjustment, threaded stud, and retainer plate. **(D)** Specification plates for a water-regulating valve. **(E)** Flushing a water-regulating valve. **(C)** *Courtesy Ferris State University. Photo by John Tomczyk.* **(D)** *Courtesy Ferris State University. Photo by John Tomczyk.* **(E)** *Courtesy Ferris State University. Photo by John Tomczyk*

Notice this has caused the water exiting temperature to be 120°F.

Because of the variable flow and variable temperature of water flowing through a wastewater system during different seasons of the year, there is no refrigerant-to-water temperature relationship or rule of thumb for the technician to go by for these systems. The service technician must simply rely on the water-regulating valve to throttle more or less water at different temperatures through the condenser to maintain a constant head pressure of 124 psig.

22.10 RECIRCULATED WATER SYSTEMS

When the system gets large enough that saving water is a concern, then a system that will recirculate the water is considered. This system uses the condenser to reject the heat to the water just as in the wastewater system. The water is then pumped to an area away from the condenser where the heat is removed from the water, **Figure 22–14**. There is a known temperature relationship between the

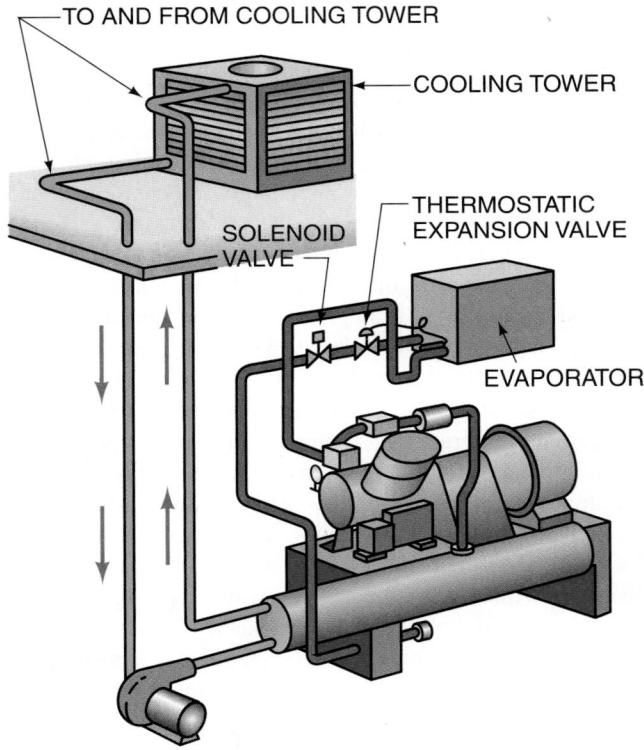

TO AND FROM COOLING TOWER

COOLING TOWER

THERMOSTATIC
EXPANSION VALVE

SOLENOID
VALVE

EVAPORATOR

Figure 22–14 This water-cooled condenser absorbs heat from the refrigerant and pumps the water to a cooling tower at a remote location. The condenser is located close to the compressor, and the tower is on the roof outside the structure.

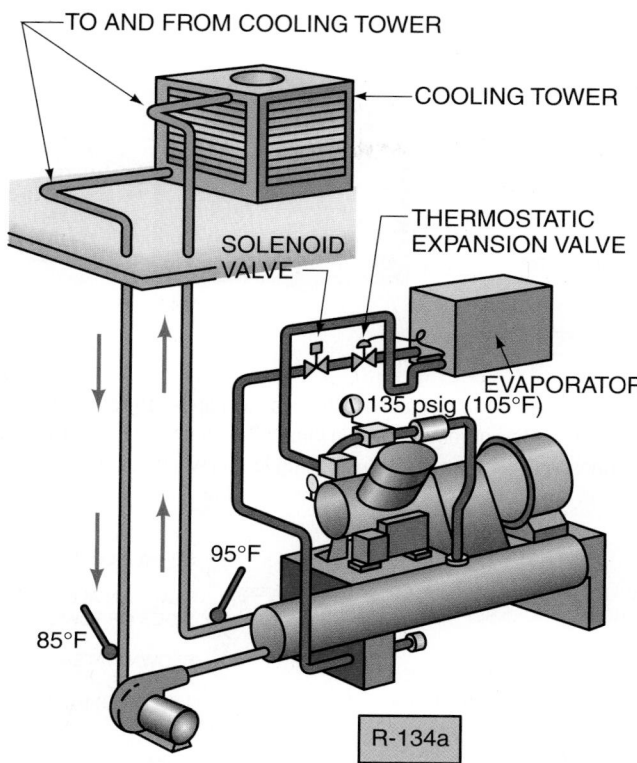

Figure 22–15 The relationship of the condensing refrigerant to the leaving-water temperature.

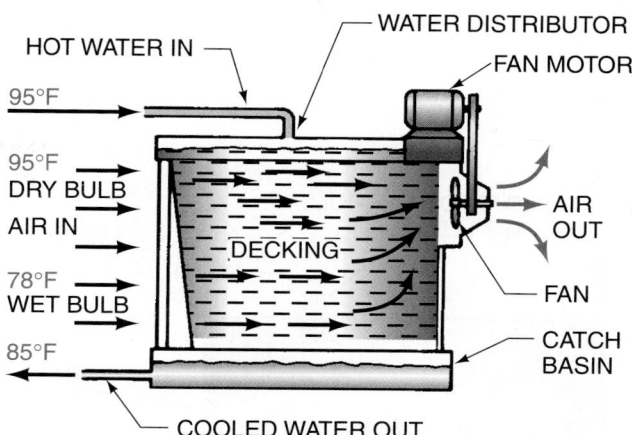

Figure 22–16 The relationship of a forced-draft cooling tower to the ambient air. Cooling tower performance depends on the wet-bulb temperature of the air. This relates to the humidity and the ability of the air to absorb moisture.

water and the refrigerant. The refrigerant will normally condense at a temperature of about 10°F higher than the leaving-water temperature, **Figure 22–15.** A recirculating water system will circulate approximately 3 gal of water per min per ton of refrigeration. The typical design temperature for the entering water is 85°F, and most systems have a 10°F rise across the condenser. Therefore, you could look for 85°F entering and 95°F leaving when the system is under full load.

Since the R-134a refrigerant will condense at a temperature about 10°F higher than the leaving-water temperature, the approximate condensing temperature would be 105°F (95°F + 10°F). The head pressure for R-134a corresponding to this 105°F condensing temperature is 135 psig.

22.11 COOLING TOWERS

The *cooling tower* is a device that passes outside air over the water to remove the system heat from the water. Any cooling tower is limited in capacity to the amount of evaporation that occurs. The evaporation rate is linked to the wet-bulb temperature of the outside air (humidity). Usually a cooling tower can cool the water that returns back to the condenser to within 7°F of the wet-bulb temperature of the outside air, **Figure 22–16.** With 78°F wet-bulb temperature air entering the cooling tower, the water in the tower can be cooled to 85°F. This tower arrangement comes in sizes of about 2 tons of refrigeration and up. Towers can be either (1) natural draft, (2) forced draft, or (3) evaporative.

22.12 NATURAL-DRAFT TOWERS

The *natural-draft tower* does not have a blower to move air through the tower. It is customarily made of some material that the weather will not deteriorate, such as redwood, fiberglass, or galvanized sheet metal.

Because natural-draft towers rely on the natural prevailing breezes to blow through them, they need to be located in the prevailing wind. The water is sprayed into the top of the tower through spray heads, and some of the water evaporates as it falls to the bottom of the tower where the water is collected in a basin. This evaporation takes heat from the remaining water and adds to the capacity of the tower. Water must be made up for the evaporated water. A makeup system using a float assembly connected to the water supply makes up for evaporated water automatically by adding fresh water, **Figure 22–17(A and B).**

The tower location must be chosen carefully. If it is located in a corner between two buildings where the breeze cannot blow through it, higher-than-normal water temperatures will occur, which will cause higher-than-normal head pressures, **Figure 22–18.**

These towers have two weather-related conditions that must be considered.

1. The tower must be operated in the winter on refrigeration systems, and the water will freeze in some climates if freeze protection is not provided. Heat can be added to the water in the basin of the tower; antifreeze will also prevent this, **Figure 22–19.**

2. The water can get cold enough to cause a head pressure drop. A water-regulating valve can be installed to prevent this from happening. Natural-draft towers can be seen on top of buildings as structures that look like they are made of slats. These slats help keep the water from blowing out of the tower, **Figure 22–20.**

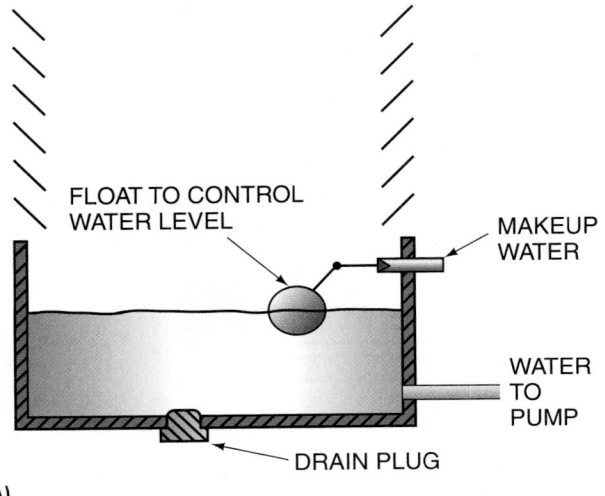

(A)

(B)

Figure 22–17 **(A)** A makeup water system in a cooling tower. Because the cooling tower performance depends partly on evaporation of water from the tower, this makeup is necessary. **(B)** Makeup water with ball-and-water valve assembly. Plastic honeycomb fill material is shown in background. *Courtesy Ferris State University. Photo by John Tomczyk*

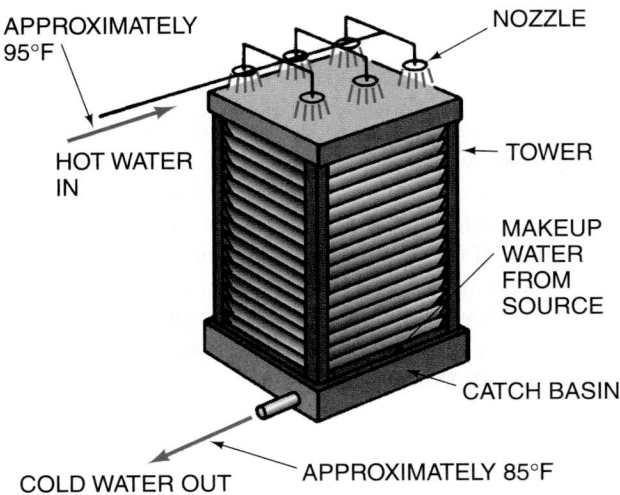

Figure 22–18 A natural-draft cooling tower. (It must be located in the path of prevailing winds.)

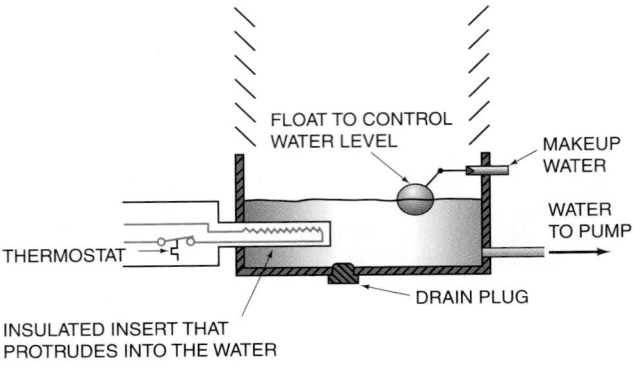

Figure 22–19 The type of heat that may be applied to keep the water in the basin from freezing in winter. This heat can be controlled thermostatically to prevent it from being left on when not needed.

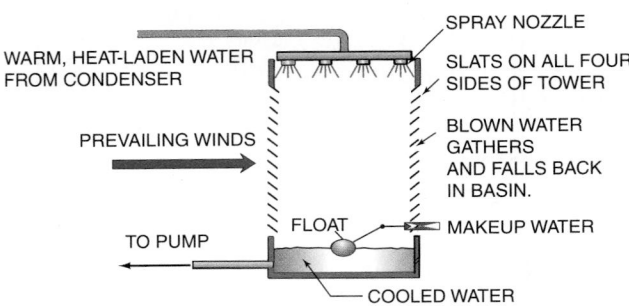

Figure 22–20 Stats on the sides of the natural-draft cooling tower keep the water inside the tower when the wind is blowing.

22.13 FORCED- OR INDUCED-DRAFT TOWERS

Forced-draft or induced-draft cooling towers differ from natural-draft towers because they have a fan to move air over a wetted surface, **Figure 22–21.** They are customarily designed with the warm water from the condenser pumped into a flat basin at the top of the tower. This basin has calibrated holes drilled in it to allow a measured amount of water to pass downward through the fill material, **Figure 22–22(A).** The calibrated holes in the basin at the top of the cooling tower often may become plugged with scale from minerals in the water, **Figure 22–22(B).** This can cause an erratic water flow through the fill material in the cooling tower, thus causing the tower to be less efficient. The fill material is usually redwood or manmade fiber and gives the water surface area for the fan to blow air over to evaporate and cool the water, **Figure 22–23.** Today, modern fill materials used in cooling towers are usually made of a plastic material formed in a honeycomb pattern for higher efficiencies and ease of cleaning, **Figure 22–17(B).** As the water is evaporated, it is replaced with a water makeup system using a float, similar to the natural-draft tower.

Forced- or induced-draft towers can be located almost anywhere because the fan can move the air. They can even be located inside buildings, where the air is brought in and out through ducts, **Figure 22–24.** The tower is fairly enclosed to

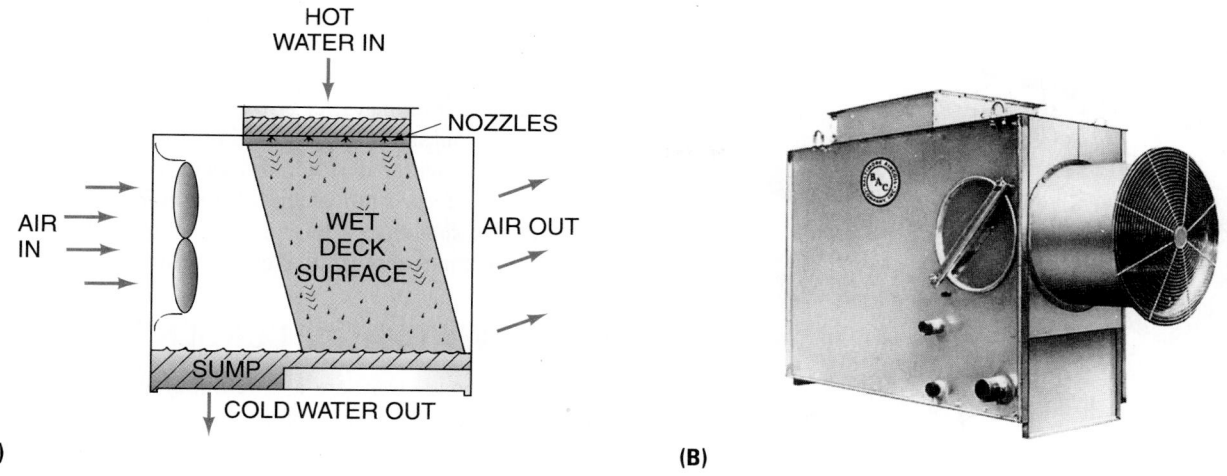

Figure 22–21 **(A)** A forced-draft tower. **(B)** An induced-draft tower. *Courtesy of Baltimore Aircoil Company, Inc.*

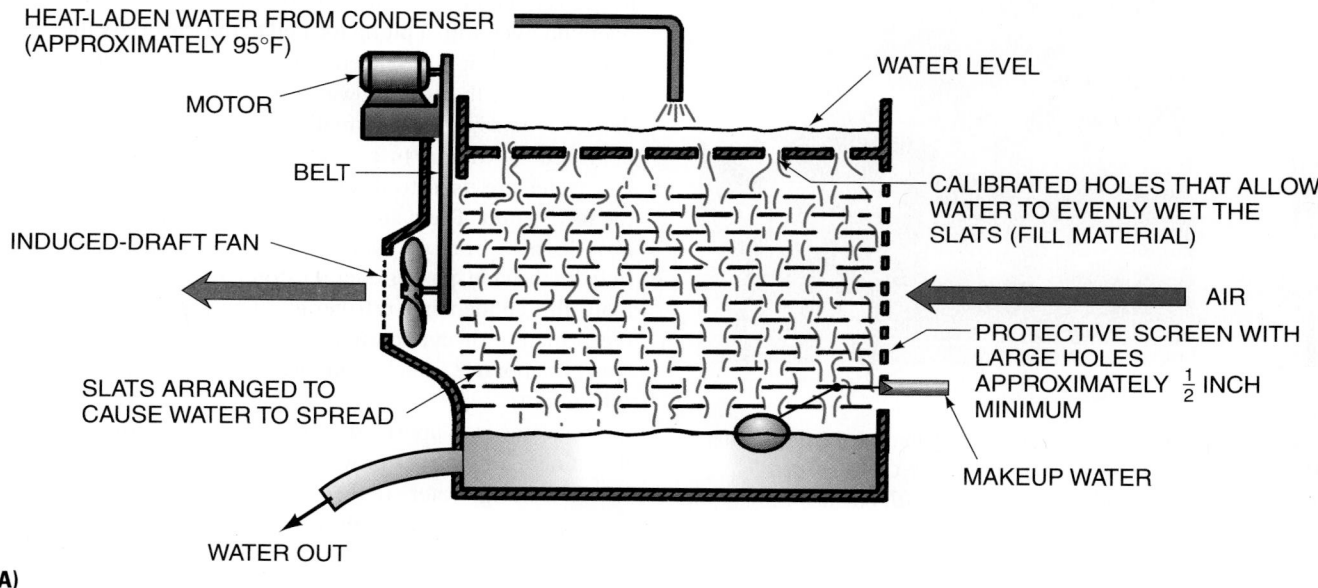

Figure 22–22 **(A)** Calibrated holes at the top of an induced-draft tower. The holes distribute the water over the fill material below. **(B)** Calibrated holes at the top of a cooling tower showing significant mineral deposits blocking water flow to the fill material. *Courtesy Ferris State University. Photo by John Tomczyk*

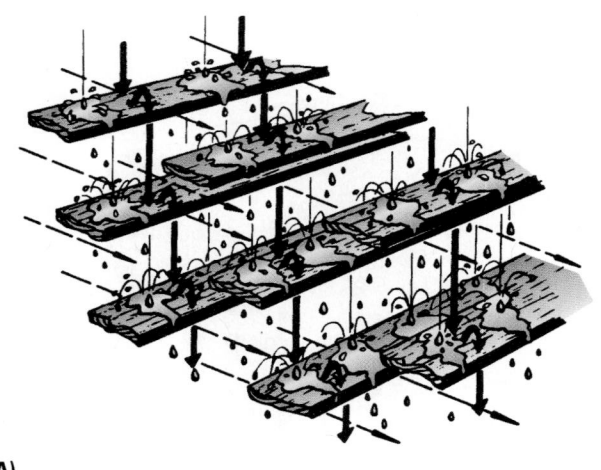

(A)

(B)

Figure 22–23   **(A)** The water trickles down through the fill material. **(B)** Plastic honeycomb fill material in a modern cooling tower. *(A) Courtesy Marley Cooling Tower Company. (B) Courtesy Ferris State University. Photo by John Tomczyk*

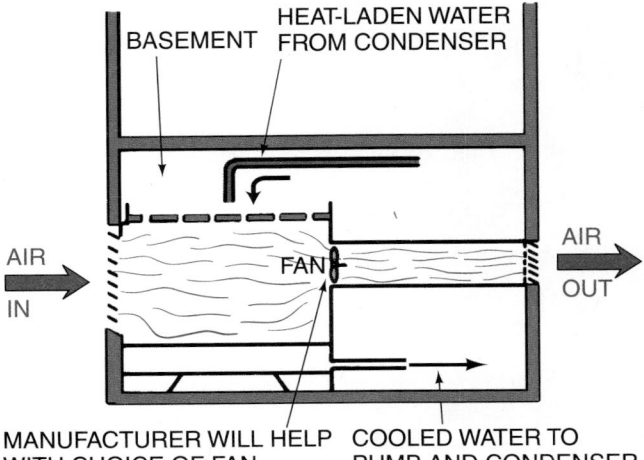

Figure 22–24   An induced-draft tower located inside a building with air ducted to the outside.

the prevailing winds, so no water-regulating valve is normally necessary. The fan can be cycled off and on to control the water temperature and thus control the head pressure. The mass of the water in the tower provides a long cycle between the time the fan starts and stops. Forced-draft towers are small compared with natural-draft towers. They are versatile because of the forced movement of air. NOTE: An induced-draft tower is similar to a forced-draft tower, but the air is pulled, not pushed, across the wetted surface.

22.14   EVAPORATIVE CONDENSERS

Evaporative condensers are a different type altogether because the refrigerant condenser is actually located inside the tower. These types are often confused with cooling towers, **Figure 22–25**. In the cooling towers discussed previously, the condenser containing the refrigerant was remote from the tower, and the water was piped through the condenser to the tower. The evaporative condenser uses the same water over and over with a pump located at the tower. As the water is evaporated, it is replaced with a makeup system using a float, as with the other towers. When the evaporative condenser is used in cold climates, freeze protection must be provided in winter. In an evaporative condenser, both air and water are used to desuperheat, condense, and subcool the refrigerant. The water is sprayed over the condenser and absorbs heat from the warmer refrigerant. The air is pulled in from the side, passes over the condenser, and absorbs heat from the condensing refrigerant. The process is based on the principle of evaporative cooling, because as the water spray evaporates, much more heat can be absorbed from the refrigerant.

As water is evaporated from any cooling tower system, the minerals in the water will become more concentrated in the remaining water. If these minerals are allowed to become overconcentrated, they will begin to deposit on the condenser surface and cause head pressure problems. To prevent this from happening, water must be allowed to escape the system on a continuous basis. This escaping water is called *blowdown.* It is actually water that is allowed to escape down the drain to be made up with fresh water and the float system. It is not unusual for people who do not know the purpose of blowdown to shut off the drain line of what appears to be clean water wasted down the drain. Problems will be the result. Technicians must establish a regular procedure for checking the rate of blowdown. Cooling towers and blowdown are discussed in more detail in Unit 49.

22.15   AIR-COOLED CONDENSERS

Air-cooled condensers use air as the medium into which the heat is rejected. This can be advantageous where it is difficult to use water. The first air-cooled condensers were bare pipe with air from the compressor flywheel blowing over the condenser. The compressors were open drive at this time. To improve the efficiency of the condenser and to make it smaller, the surface area was then extended with fins. The condensers at this time were normally steel with steel fins,

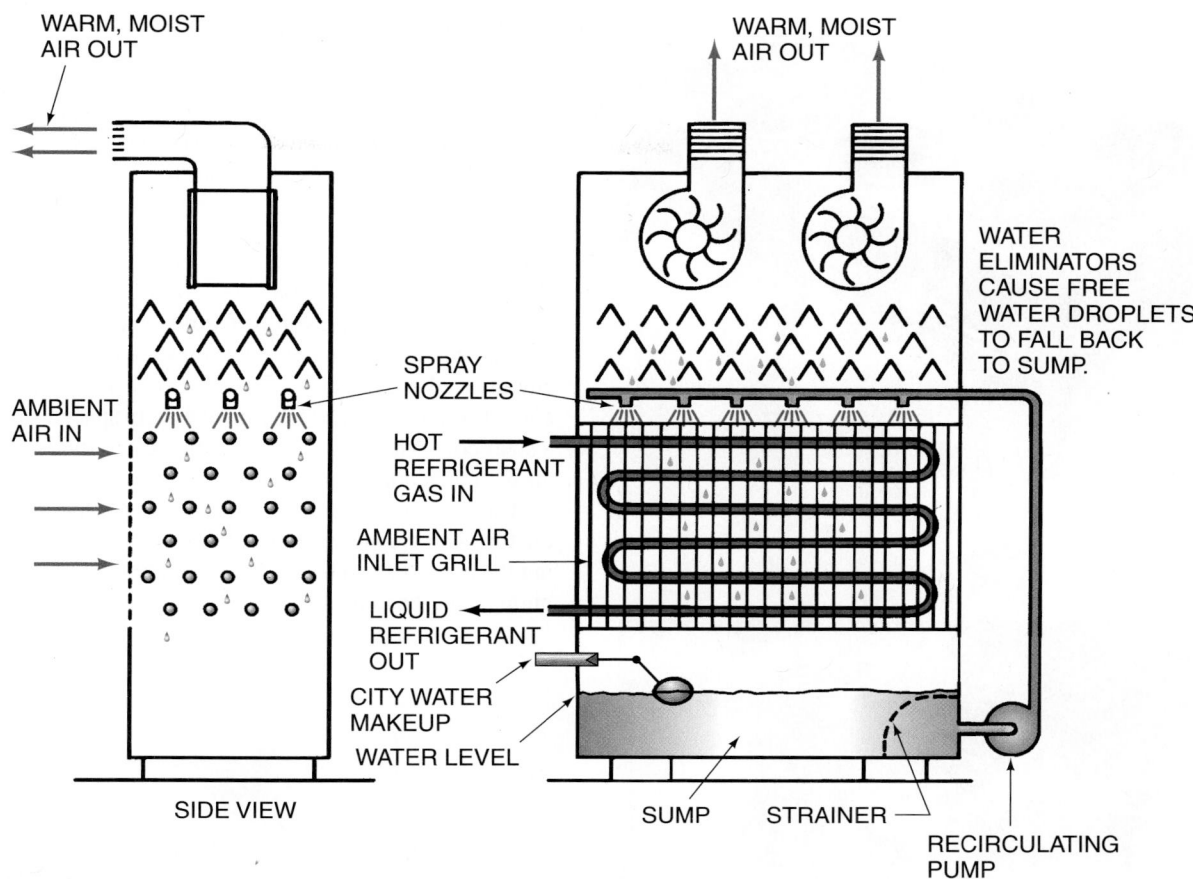

Figure 22–25 Water recirculates in the evaporative condenser. The condenser tubes are in the tower, rather than in a condenser shell located in a building.

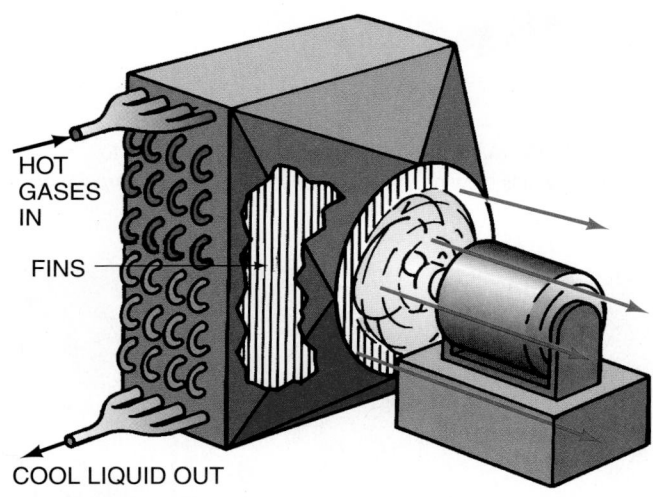

Figure 22–26 Fins designed to give the coil more surface area.

Figure 22–26. These condensers resembled radiators and were sometimes referred to as radiators.

Steel air-cooled condensers are still used in many small installations on refrigeration units. **Figure 22–27(A)** is a photo of a condenser for a large refrigeration system. **Figure 22–27(B and C)** illustrates some newer technology in

condenser and evaporator coil design that has entered the market to improve on system efficiency. One such condenser design incorporates an aluminum, parallel-flow, flat-plate-and-fin configuration that has small parallel channels inside the flat plate. These plates are flattened, streamlined tubes with each one split into smaller, parallel ports. Refrigerant will phase change or condense from a vapor to a liquid inside the channels in the plate, while strategically shaped fins (extended surfaces) will enhance heat transfer from the condenser coil to the air. The plates and fins are bonded or soldered to increase heat transfer and to eliminate any contact resistance (air gaps) that will reduce heat transfer. Headers at the inlet and outlet of the heat exchanger are also bonded to the plates through soldering. Please refer to Unit 21, "Evaporators and the Refrigeration System," for a more detailed explanation about using the parallel-flow, flat-plate-and-fin condenser and evaporator configuration to increase the efficiency of a refrigeration or air-conditioning system.

Air-cooled condensers come in a variety of styles. In some, the air blows horizontally through them, and they are subject to prevailing winds, **Figure 22–28**. In other air-cooled condensers the airflow pattern is vertical. They take air into the bottom and discharge it out the top. The prevailing winds do not affect these condensers to any extent, **Figure 22–29**.

(A)

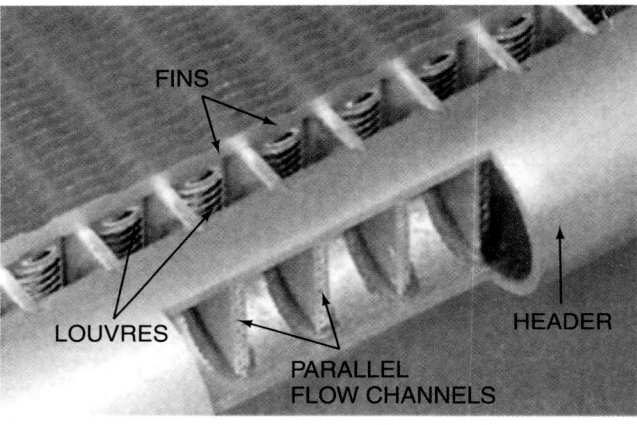

(B)

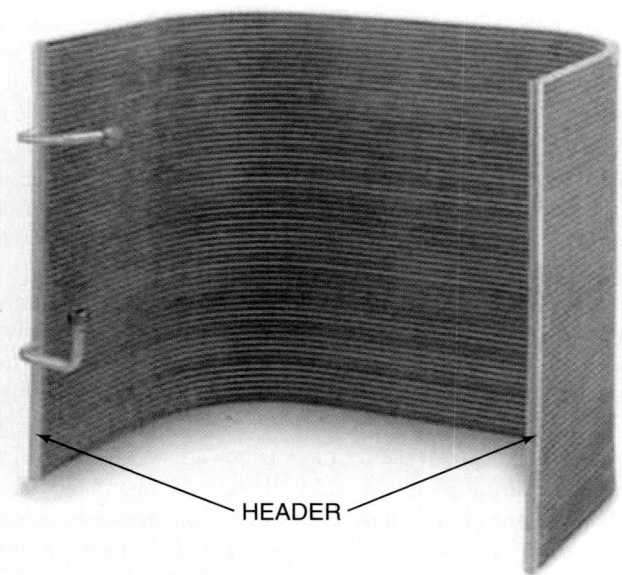

(C)

Figure 22–27 **(A)** This larger refrigeration air-cooled condenser resembles an air-conditioning condenser. **(B)** Aluminum parallel-flow, flat-plate-and-fin heat exchanger. **(C)** Aluminum parallel-flow, flat-plate-and-fin condenser. *(A) Courtesy Heatcraft, Inc., Refrigeration Products Division. (B) Courtesy Modine Manufacturing Co., Racine, WI. (C) Courtesy Modine Manufacturing Co., Racine, WI*

Another style of air-cooled condenser takes the air in the sides and discharges it out the top. This condenser can be affected by prevailing winds, **Figure 22–30.**

The smaller refrigeration systems are often located in the conditioned space, such as a restaurant or store. These air-cooled condensers normally have widely spaced steel fins on a steel coil, which allow more time before the coil will stop up with dust and other airborne material.

The hot gas normally enters the air-cooled condenser at the top. The beginning tubes of the condenser will be receiving the hot gas straight from the compressor. This gas will be highly superheated. (Remember that superheat is heat that is

Figure 22–28 A horizontal air-cooled condenser subject to prevailing winds blowing through it. *Courtesy Copeland Corporation*

Figure 22–29 A condenser with a vertical airflow pattern. Air enters the bottom and is blown out the top. It is unaffected by prevailing winds. *Courtesy Heatcraft, Inc., Refrigeration Products Division*

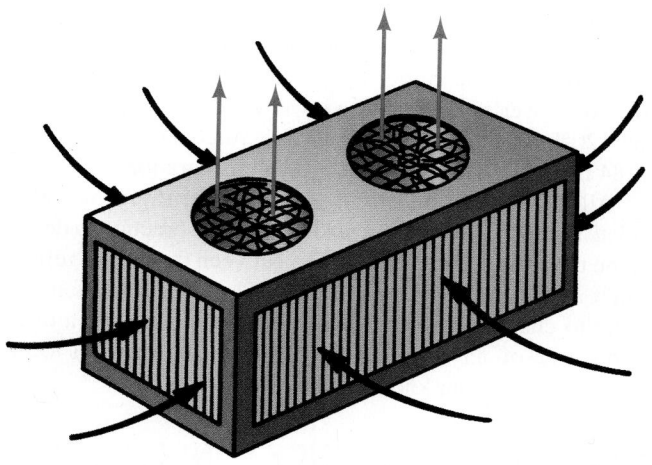

Figure 22–30 This condenser takes the air in from the sides and discharges it out the top. The prevailing winds could affect it.

added to the refrigerant after the change of state in the evaporator.) When the superheated refrigerant from the evaporator reaches the compressor and is compressed, more superheat is added to the gas. Part of the energy applied to the compressor transfers into the refrigerant in the form of heat energy. This additional heat added by the compressor causes the refrigerant leaving the compressor to be heavily heat laden. On a hot day (95°F) the hot gas leaving the compressor could easily reach 250°F. SAFETY PRECAUTION: DO NOT TOUCH the hot gas discharge line from the compressor. It will burn your fingers. The condenser must remove heat from the refrigerant until it reaches the condensing temperature before any condensation can occur.

Air-cooled condensers have a relationship to the temperature of the air passing over them (as the evaporator did in the previous unit). For instance, the refrigerant inside the coil will normally condense at a 30°F higher temperature than the air passing over it (also known as the ambient air). This statement is true for most standard-efficiency condensers that have been in service long enough to have a typical dirt deposit on the fins and tubing. The relationship can be improved by adding condenser surface area. With an outside air temperature of 95°F, the condensing temperature will be about 125°F. With the refrigerant R-134a condensing at 125°F, the head pressure or high-pressure gage should read 184 psig. (See the temperature/pressure chart for R-134a.) This is important because it helps the service technician establish what the head pressure should be.

See **Figure 22–31** for an illustration of the following description of an air-cooled, R-134a condenser located inside a store. It is responsible for rejecting the heat absorbed inside a medium-temperature walk-in cooler. This cooler has reach-in doors typical of those in a convenience store. In this box the beverages are on shelves, and store personnel can stock the shelves from the walk-in portion behind the shelves. The cooler is maintained at 35°F. The outside air temperature is 95°F. The refrigeration system must absorb heat at 35°F and reject the same heat to the outside where the condensing medium is 95°F.

1. The hot gas is entering the condenser at 200°F. The condensing temperature is going to be 30°F warmer than the outside (ambient) air.

2. The outside air temperature is 95°F. The condensing temperature is 95°F + 30°F = 125°F. The refrigerant must be cooled to 125°F before any actual condensing

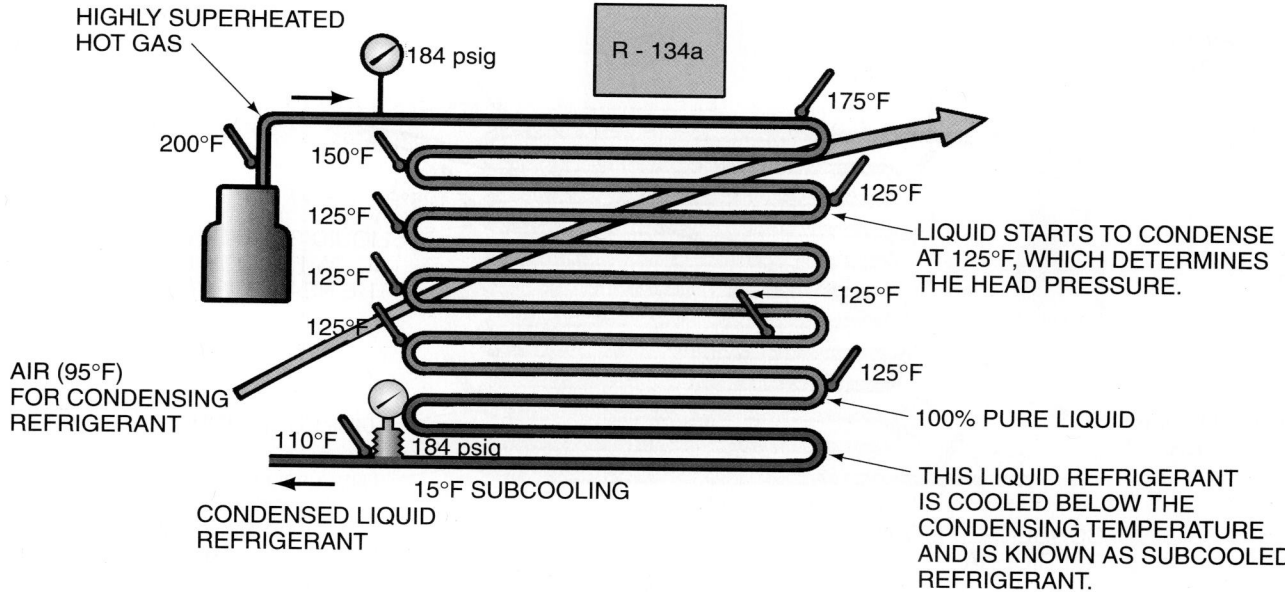

Figure 22–31 The following points are illustrated: (1) The hot gas flowing into the first part of the condenser is highly superheated. The condensing temperature is 125°F, and the discharge gas must be cooled from 200°F to 125°F before any condensing will occur. (2) The best heat exchange is between the liquid and the air on the outside of the coil. More heat is removed during the change of state than while desuperheating the vapor. (3) When the liquid is all condensed and pure liquid is in the coil, the liquid can be subcooled to below the condensing temperature.

occurs. Thus the condenser has to lower the hot gas temperature 75°F (200°F − 125°F) in the first part of the coil. This is called *desuperheating*. It is the first job of the condenser.

3. Partway through the coil the superheat is removed down to the actual condensing temperature of 125°F, and liquid begins to form in the coil. Liquid will continue to form at the condensing temperature of 125°F until all of the vapor has turned to 100% liquid. This is called the *latent heat* of condensation.

4. This liquid can now reject sensible heat and *subcool* to below the 125°F condensing temperature. Any liquid below the 125°F condensing temperature is referred to as a *subcooled liquid*. When the subcooled liquid refrigerant gets to the end of the coil, the condenser tubes will be full of liquid and then drain into the receiver.

5. The liquid in the bottom of the condenser draining into the receiver may cool to 110°F. This would give the liquid 15°F (125°F − 110°F) of subcooling.

In larger commercial and industrial applications in which the condensers are large, a pressure drop will occur as the refrigerant flows through the condenser. This pressure drop is caused from friction. When measuring condenser subcooling in these cases, the refrigerant pressure has to be measured close to the condenser outlet, not at the compressor service valve, **Figure 22–32**. The condenser outlet pressure can be read at the receiver's *king valve* or at the receiver charging valve. In fact, the refrigerant pressure should be taken very near the same place the condenser outlet temperature is taken, or inaccurate condenser subcooling calculations will occur. If the condenser in **Figure 22–32** was a large condenser, and the

pressure was measured at the condenser outlet, the condenser subcooling calculation would be 120°F − 110°F = 10°F of condenser subcooling. The lower pressure of 171 psig at the condenser's outlet gives a lower condensing temperature of 120°F when the condenser pressure drop of 13 psig (184 psig − 171 psig) is taken into consideration.

All air-cooled condensers do not have the same 30°F relationship with the ambient air. It is good practice to determine the temperature relationship between the air and refrigerant at start-up time and record it. Then if the relationship were to change, you would suspect trouble, such as a dirty condenser or an overcharge of refrigerant that has been added without your knowledge.

22.16 HIGH-EFFICIENCY CONDENSERS

Condensing temperatures can be reduced by increasing the condenser surface area. The larger the condenser surface area, the closer the condensing temperature is to the ambient temperature. A perfect condenser would condense the refrigerant at the same temperature as the ambient, but would be so large that it would not be practical. In the previous examples, we have used a temperature difference of 30°F between the ambient and the condensing temperature. **Figure 22–33** shows an example of a condenser condensing at 110°F, 15°F above the ambient temperature. The head pressure for R-134a condensing at 110°F would be 146.4 psig. (R-404A would be 271.4 psig.) ⬤The reduction of head pressure provides for a more efficient system—less power is consumed for the same amount of usable refrigeration.⬤ Condensing temperatures to within 10°F of the ambient have been used on some installations, particularly in extra-low-temperature systems.

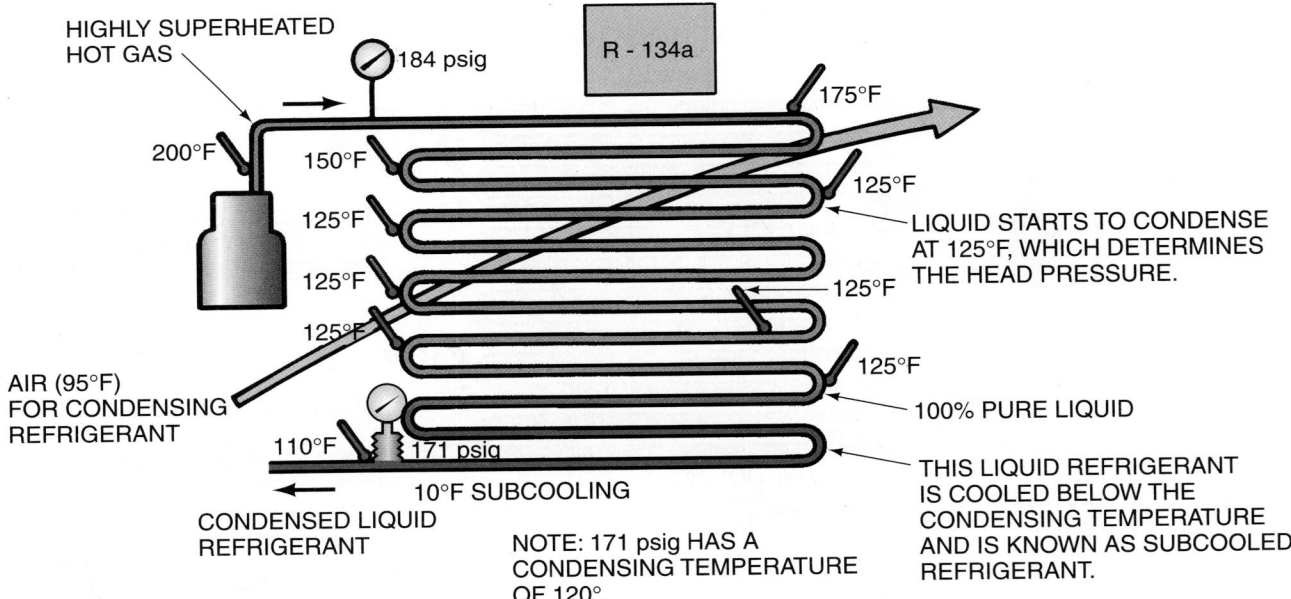

Figure 22–32 The air-cooled condenser has a 13-psig pressure drop from where the hot gas enters to where the subcooled liquid leaves the condenser. This is due to friction loss in the piping and is typical of larger condensers. The subcooling must be measured close to the same pressure location as where the temperature pressure is measured.

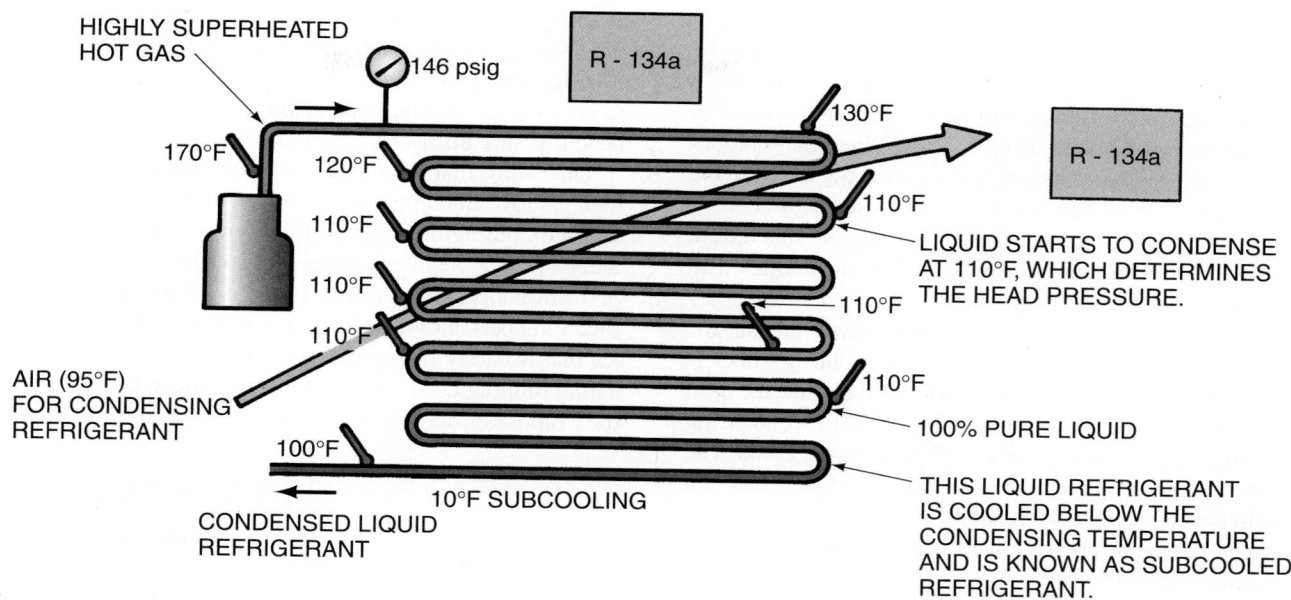

Figure 22–33 The operating conditions for a high-efficiency condenser.

Compressors are affected by a decrease in head pressure and will be discussed in the next unit.

22.17 THE CONDENSER AND LOW AMBIENT CONDITIONS

The foregoing example described how an air-cooled condenser operates on a hot day. An example of a condenser operating under different conditions might be in a supermarket with a small package-display case located inside the store. Because this is a package-display case, the condenser is also located inside the store, **Figure 22–34**. The inlet air to the case is in the store itself and may be quite cool if the air conditioner is operating. The store temperature may be 70°F or even cooler at times. This reduces the operating pressure on the high-pressure side of the system. When the condenser relationship temperature rule is applied, we see that the new condensing temperature would be 70°F + 30°F = 100°F.

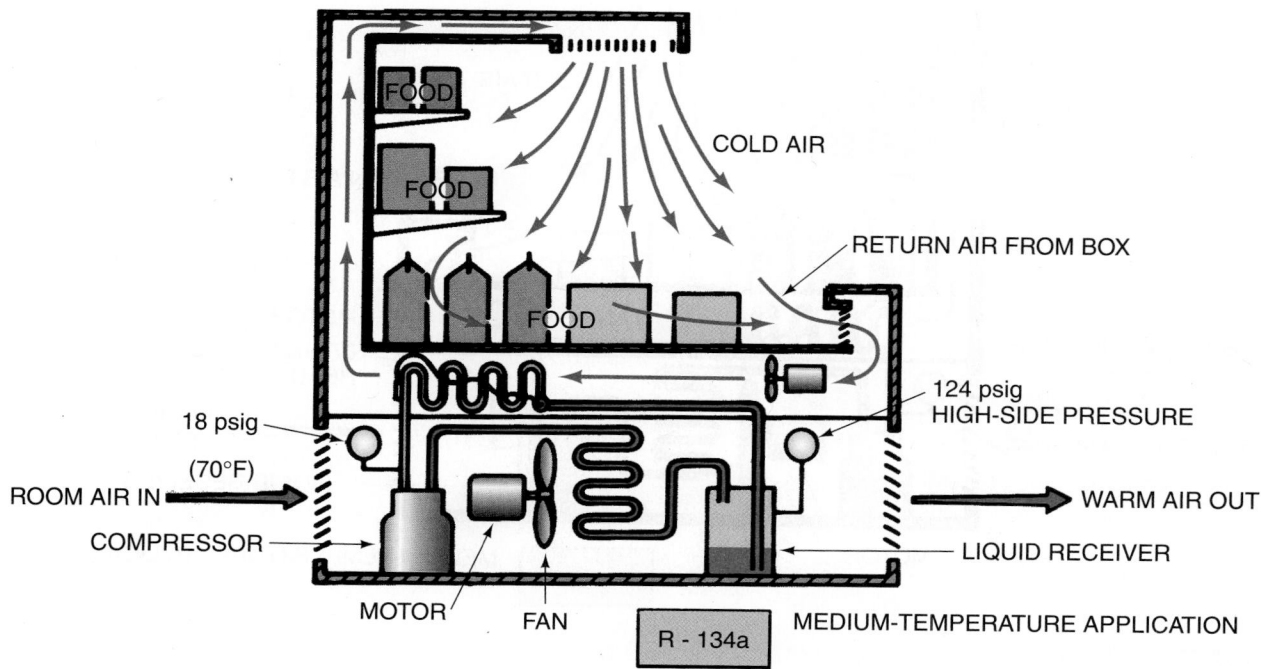

Figure 22–34 The package-display case located inside the store with the compressor and the condenser located inside the cabinet. This is a plug-in, self-contained device; no piping is required.

The head pressure would be 124 psig. This may be enough to affect the performance of the expansion device. This will be covered in more detail later. The low-side pressure for a medium-temperature fixture will be in the neighborhood of 18 psig at the lowest point using R-134a. When the relationship of the low-side pressure of 18 psig is subtracted from the high-side pressure of 124 psig for a condenser with 70°F air passing over it, the difference is 106 psig. Should this fixture be exposed to temperatures lower than 70°F, difficulties with a starved evaporator coil may result.

If the medium-temperature box were moved outside the store, the fixture may quit working to capacity during winter. If the air temperature over the coil drops to 50°F outside, the head pressure is going to drop to 50°F + 30°F = 80°F condensing temperature, which corresponds to 87 psig, **Figure 22–35**. The pressure difference of 87 psig − 18 psig = 69 psig may not be sufficient to feed enough liquid refrigerant through the expansion device to properly feed the coil, **Figure 22–35**. In the figure, notice the pressure and temperature difficulties encountered. The head pressure must be regulated.

Another example of a condenser operating outside the design parameters would be an ice-holding box like those found at service stations and convenience stores. These fixtures hold ice made at another location. They must keep the ice hard in all types of weather and often operate at about 0°F to 20°F inside the box.

It may be 30°F outside the box where the small air-cooled condenser is rejecting the heat. The condenser would be operating at about 30°F + 30°F = 60°F, or at a head pressure of 57 psig, **Figure 22–36**. The evaporator should be operating at about −15°F, or at a suction pressure of 0 psig to maintain 0°F. This gives a pressure difference of 57 psig − 0 psig = 57 psig, which will starve the evaporator. This unit has to run to keep the ice frozen, so something has to be done to get the head pressure up.

One thing that helps prevent problems for some equipment in low ambient conditions is that the load is reduced. The ice-holding box, for example, does not have to run as much at 30°F weather to keep the ice frozen at 0°F. It may perform poorly, but you will not know it unless it must run for a long period of time for some reason, such as if a load of ice that is barely frozen is loaded into it. The unit may have a long running cycle pulling the load temperature down, and the evaporator may freeze solid.

22.18 HEAD PRESSURE CONTROL

Practical methods to maintain the correct workable head pressure automatically and not cause equipment wear are fan-cycle control, variable-frequency drive (VFD) condenser fan motors, dampers, and condenser flooding. Each type has different characteristics and features. Different companies recommend different head pressure control types for their own reasons.

Fan-Cycling Devices

The air-cooled condenser normally has a small fan that passes the air over the condenser. When this fan is cycled off, the head pressure will go up in any kind of weather, provided

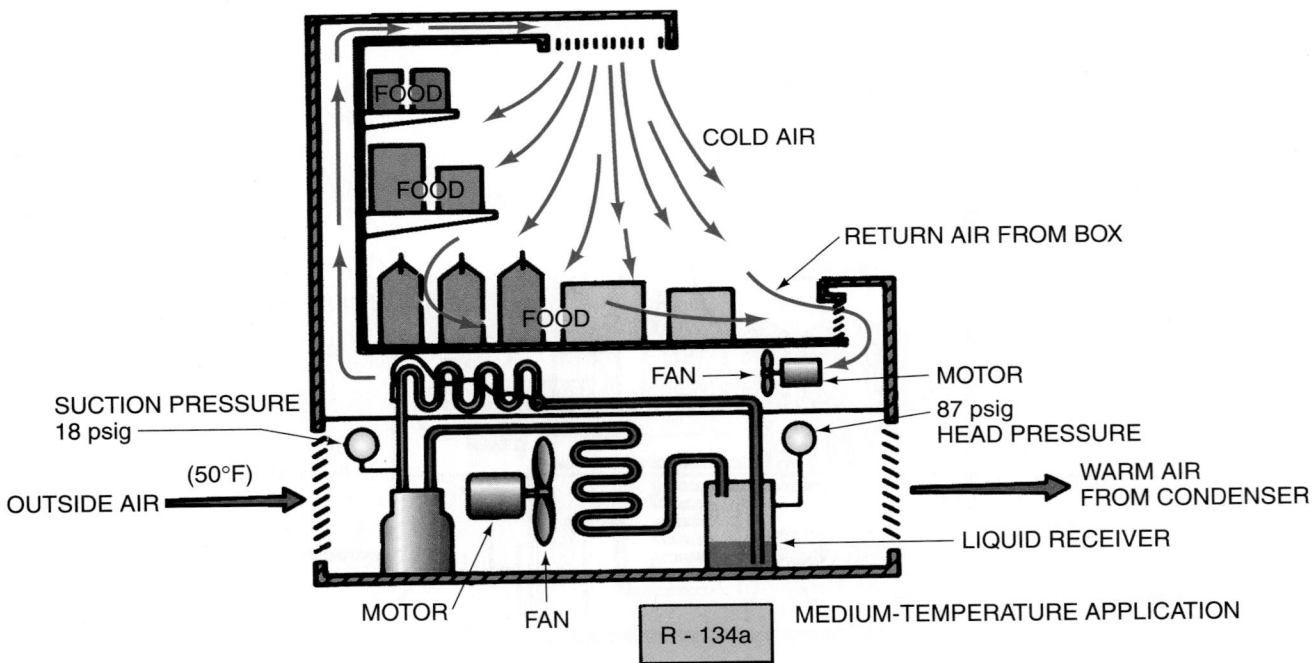

COLD AIR

RETURN AIR FROM BOX

FOOD

FOOD

FOOD

FAN — MOTOR

87 psig
HEAD PRESSURE

SUCTION PRESSURE
18 psig

(50°F)

OUTSIDE AIR

MOTOR   FAN

WARM AIR
FROM CONDENSER

LIQUID RECEIVER

R - 134a

MEDIUM-TEMPERATURE APPLICATION

Figure 22–35 The fixture in **Figure 22–34** is relocated outside the store. Now the condenser is subject to the winter conditions. The performance will fall off if some type of head pressure control is not furnished. The head pressure is so low that it cannot push enough liquid refrigerant through the expansion valve. The evaporator is starved. The unit has a reduced capacity that would be evident if a load of warm food were placed in the box. It might not pull the food temperature down. The coil would ice up because there would not be any off-cycle defrost.

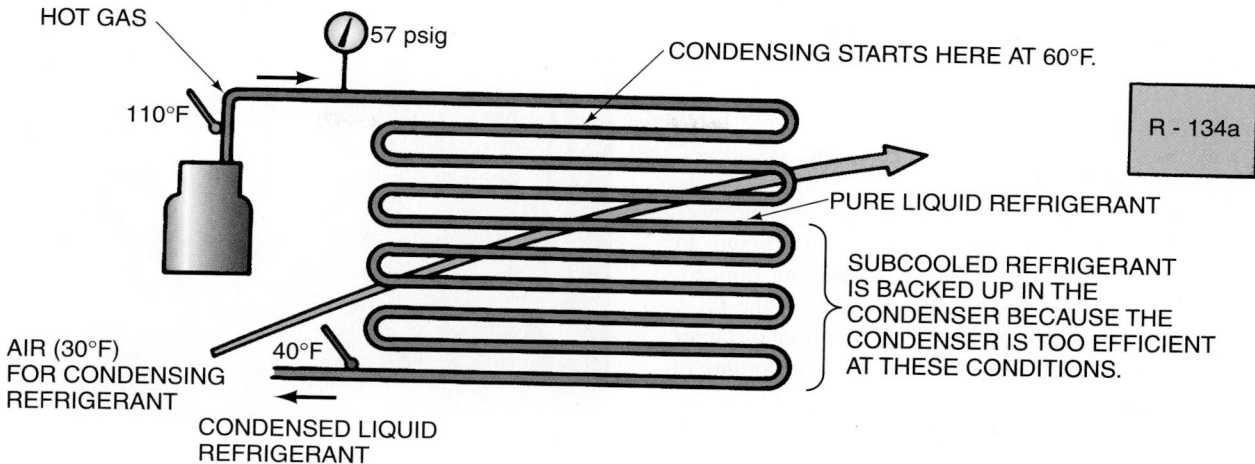

Figure 22–36 An illustration of how the condenser acts in a low ambient condition. Notice that this condenser has much more liquid refrigerant in it and the head pressure is very low. It may be so low that it may starve the expansion device. It relies on this pressure to force liquid refrigerant through its metering device.

the prevailing winds do not take over and do the fan's job. To cycle off the fan, a pressure control can be piped into the high-pressure side of the system that will close a set of electrical contacts on a preset rise in pressure, **Figure 22–37.** The electrical contacts will stop and start the fan motor on pressure changes.

A common setup for R-134a may call for the fan to cut off when the head pressure falls to 135 psig and to restart the fan when the head pressure reaches 190 psig. This setting will not interfere with the summer operation of the system and will give good performance in the winter. The settings are far enough apart to keep the fan from short cycling any more than necessary. When the thermostat calls for the compressor to come on and the ambient air is cold, the head pressure would be so low that the system would never get up to good running capacity because the condenser fan would be moving more cold air over the condenser. The condenser fan-cycling control would keep the condenser fan off until the head pressure is within the correct operating range.

The fan-cycling device is one method of maintaining a correct operating head pressure. It can be added to the system without much expense for the control, and the piping normally does not have to be altered. One problem with this device is that it has a tendency to cause the head pressure to swing up and down as the fan is stopped and started. This can also affect the expansion device operation because the pressure may be 190 psig for part of the cycle and then, when the fan comes on, drop rapidly to 135 psig. This pressure swing causes the expansion device to operate erratically. There are other means for controlling the head pressure at a steady state on air-cooled condensers.

When a condenser has more than one fan, one fan can be put in the lead, and the other fans can be cycled off by temperature, **Figure 22–38.** The lead fan can be cycled off by pressure like a single fan. This can help prevent the pressure swings from being so close together like they are in the single-fan application. For example, when three fans are used, the first will cycle off at approximately 70°F, and the second will

Figure 22–37 A condenser fan-cycling device. This control has the same action (make on a rise in pressure) as the low-pressure control except that it operates at a much higher range. *Photo by Bill Johnson*

Figure 22–38 A multiple-fan condenser. *Courtesy Heatcraft, Inc., Refrigeration Products Division*

cycle off at approximately 60°F. The remaining fan may be controlled by the pressure or by a temperature sensor located on the liquid line.

Variable-Speed Condenser Fan Motors

As mentioned earlier, one problem with cycling condenser fan motors is their tendency to cause the head pressure to swing up and down as the fan motor is started and stopped. This can cause erratic and inconsistent operation of the TXV. However, variable-speed condenser fan motors used on larger condensing units can vary their speed gradually according to head pressure or outside ambient changes. Many of these fan motors are equipped with variable-frequency drives (VFDs) and electronic inverters to control the drives. VFDs operate on the principle that when the frequency or hertz (Hz) of the voltage is changed coming into the fan motor, the speed of the fan motor will also be changed. As the voltage frequency is increased, the motor speed will be increased. As the voltage frequency is decreased, the motor speed will decrease. The formula below governs the speed (rpm) of a motor. Notice that the numerator is the frequency or hertz of the incoming voltage in cycles/sec, multiplied by a factor for converting from minutes to seconds. The denominator is the number of motor pole pairs.

$$\text{rpm} = \frac{(\text{Frequency})(60 \text{ min/hr})}{(\text{Number of motor poles} \div 2)}$$

This simplifies down to:

$$\text{rpm} = \frac{(\text{Frequency})(120)}{(\text{Number of motor poles})}$$

Since we cannot change the number of motor poles using electronics, the frequency is the easiest variable to alter for a change in the rpm of the motor.

Through special electronic circuitry called an inverter, the voltage frequency feeding a condenser fan motor can be altered with changing head pressure. This is done through the refrigeration system's high-side pressure transducer linked to the inverter. Another control scheme is to let changing outdoor ambient temperatures be sensed through a temperature sensor near the condenser's inlet air. Their signals can then be linked to the electronic inverter.

Figure 22–39 shows a VFD for changing the frequency of the voltage signal to a motor. This frequency change can be done in small increments, thus changing the speed of the condenser fan motors in small increments. This type of control will keep a more consistent head pressure as the outside ambient temperature changes. It will also prevent the head pressure from swinging up and down erratically which occurs when the fan motors are cycled on and off. VFDs are covered in more detail in Unit 17, "Types of Electric Motors."

Air Shutters or Dampers

Shutters may be located either at the inlet to the condenser or at the outlet. The air shutter has a pressure-operated piston that pushes a shaft to open the shutters when the head pressure

Figure 22–39 A variable-frequency drive (VFD) used in changing the voltage frequency to a motor. This will change the motor's speed in very small increments. *Courtesy Ferris State University. Photo by John Tomczyk*

Figure 22–40 A condenser with an air shutter. With one fan it is the only control. With multiple fans the other fans can be cycled by temperature with the shutter controlling the final fan. *Courtesy Trane Company*

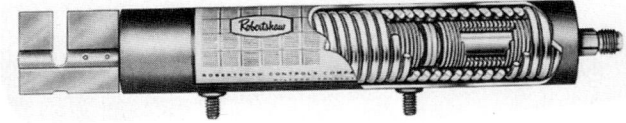

Figure 22–41 A piston-type shutter operator. With this piston the high-pressure discharge gas is on one side of the bellows, and the atmosphere is on the other side. When the head pressure rises to a predetermined point, the shutters begin to open. In the summer the shutters will remain wide open during the running cycle. *Courtesy Robertshaw Controls Company*

rises to a predetermined pressure, **Figure 22–40**. This pressure-operated piston extends to open the shutters when the pressure rises, **Figure 22–41**. Shutters open and close slowly and provide even head pressure control.

When a single fan is used, the shutter is installed over the inlet or over the outlet to the fan. When there are multiple fans, the shutter-covered fan is operated all the time, and the other fans can be cycled off by pressure or temperature. This arrangement can give good head pressure control down to low temperatures.

Condenser Flooding

Flooding the condenser with liquid refrigerant causes the head pressure to rise just as though the condenser were covered with a plastic blanket. It is accomplished by having enough refrigerant in the system to flood the condenser with liquid refrigerant in both mild and cold weather. This calls for a large refrigerant charge and a place to keep it. In addition to the charge there must be a valve arrangement to allow the refrigerant liquid to fill the condenser during both mild and cold weather. This condenser flooding method is designed to maintain the correct head pressure in the coldest weather during start-up and while operating.

One big problem designers face when designing air-cooled refrigeration and air-conditioning equipment is the changing ambient of the changing seasons. A properly designed condenser will operate inefficiently during extremely high ambient conditions. The inefficiencies come from the higher head pressures and thus higher compression ratios associated with high ambient conditions. However, it is the low ambient conditions that cause more serious problems. Moreover, it is a fact that most units will be required to operate at temperatures below their design dry-bulb temperature during the fall, winter, and spring seasons. It is because of these changing ambient conditions that head pressure control valves were designed. Without these valves, head pressures would fluctuate as widely as ambients do during the changing seasons. Both running-cycle and off-cycle problems can occur if head pressure controls are not used.

Low head pressures from low ambients can cause insufficient refrigerant flow rates through metering devices, which in turn can starve evaporators. Low suction pressures, iced coils, short cycling, and inefficient cooling can also result from low head pressures. Also, any system using hot gas defrost or hot gas bypass for compressor capacity control must have a minimum head pressure for proper defrosting and capacity control. With outdoor forced-air condensers, low ambient conditions also cause refrigerant to migrate to the condenser, compressor, and suction line during the off cycle.

Hard starting may also occur in a low ambient condition, causing low head pressures. Often, refrigeration systems are not able to get started because most of the refrigerant has migrated to the colder condenser. The evaporator has a hard time building vapor pressure and may never get to the cut-in pressure of its low-pressure control. Even if the compressor does eventually start by reaching the cut-in pressure, it may soon short cycle. This is caused by low refrigerant flow through the metering device at the lower head pressure in the low ambient condition.

The solution to the low head-pressure problem is to install a pressure-actuated "holdback" valve at the outlet of the condenser, **Figure 22–42(A and B).** This valve will throttle shut

when the condenser pressure reaches a preset minimum for low ambient conditions. This will allow liquid refrigerant in the condenser to be held back and actually flood portions of the condenser. This partial flooding will cause some of the condenser to become inactive and to have a smaller internal volume. At this point, desuperheating and condensation must take place in a smaller-volume condenser. Condensing pressures will rise, thus giving sufficient liquid-line pressures and pressure differences across the thermostatic expansion valve (TXV) for normal system operation in the colder ambient. The valve shown in **Figure 22–42(A)** is referred to as an ORI (open on rise of inlet pressure) valve. It is an inlet-pressure-regulating valve and responds to changes to inlet pressure (condensing pressure) only. A decrease in condensing pressure causes less pressure to act on the bottom of the seat disk. This action throttles the valve more in the closed position and starts to back up liquid refrigerant in the bottom of the condenser. Soon the head pressure will start to rise as a result of the condenser's reduced internal volume. Any increase in inlet (condensing pressure) above the valve setting will tend to open the valve. The condensing pressure is opposed by the force of an adjustable spring that is on top of the seat disc. The valve setting can be changed simply by either increasing or decreasing the tension of the spring with a screwdriver or an Allen wrench. Increasing the spring pressure by turning the fitting clockwise will increase the minimum opening pressure of the valve. The outlet pressure of the valve is then cancelled out and has no bearing on valve movement. This is because the outlet pressure is exerted on top of the bellows and on top of the seat disc simultaneously. Since the effective area of the bellows is equal to the area of the top of the seat disc, the pressures cancel one another and do not effect the valve movement, **Figure 22–43.** Only changes in condensing pressure can throttle the valve either open or closed.

An ORI valve is usually used in conjunction with an ORD (open on rise of differential pressure) valve, **Figure 22–44(A and B).** The ORD valve is located between the discharge line and the receiver inlet, **Figure 22–42(B).** It responds to changes in pressure differences across the valve. The ORD valve is thus dependent on the ORI valve for its operation. The ORD valve will bypass hot compressor discharge gas from the compressor to the receiver inlet when it senses a preset determined pressure difference across the valve. As the ORI valve senses a drop in condenser pressure and starts to throttle shut, a pressure difference is created across the ORD valve. The pressure difference is created from the reduced flow to the receiver because of the throttling action of the ORI valve. One outlet of the ORD valve senses receiver pressure and the inlet side senses discharge pressure from the compressor. When the ORI valve starts to throttle shut, the receiver is still supplying refrigerant to the TXV and receiver pressure will eventually drop. If the receiver pressure drops too low, its ability to keep feeding the liquid line and TXV's liquid will diminish. Something has to keep the receiver pressure up while the ORI valve is throttling liquid from the condenser. This is when the function of the ORD valve comes into play.

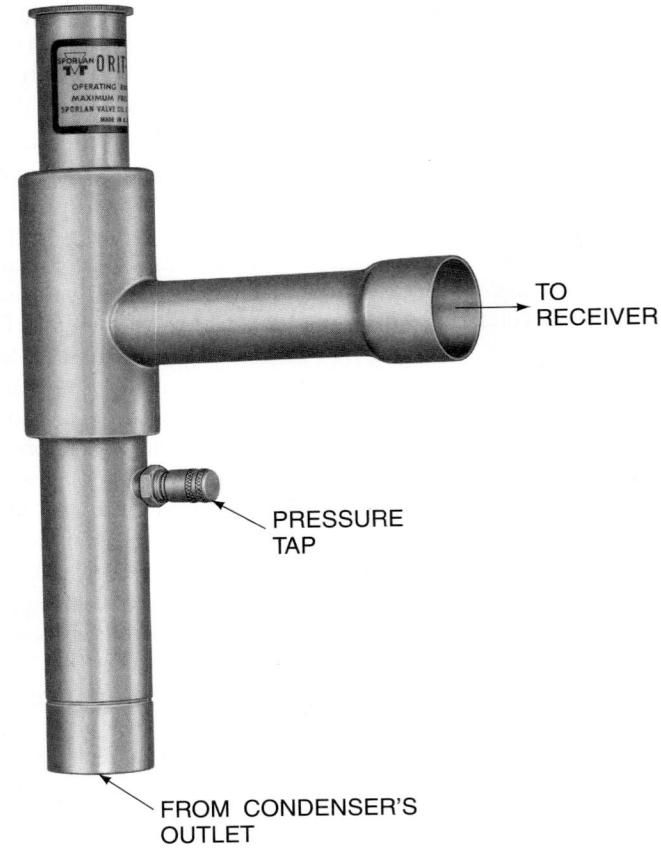

(A)

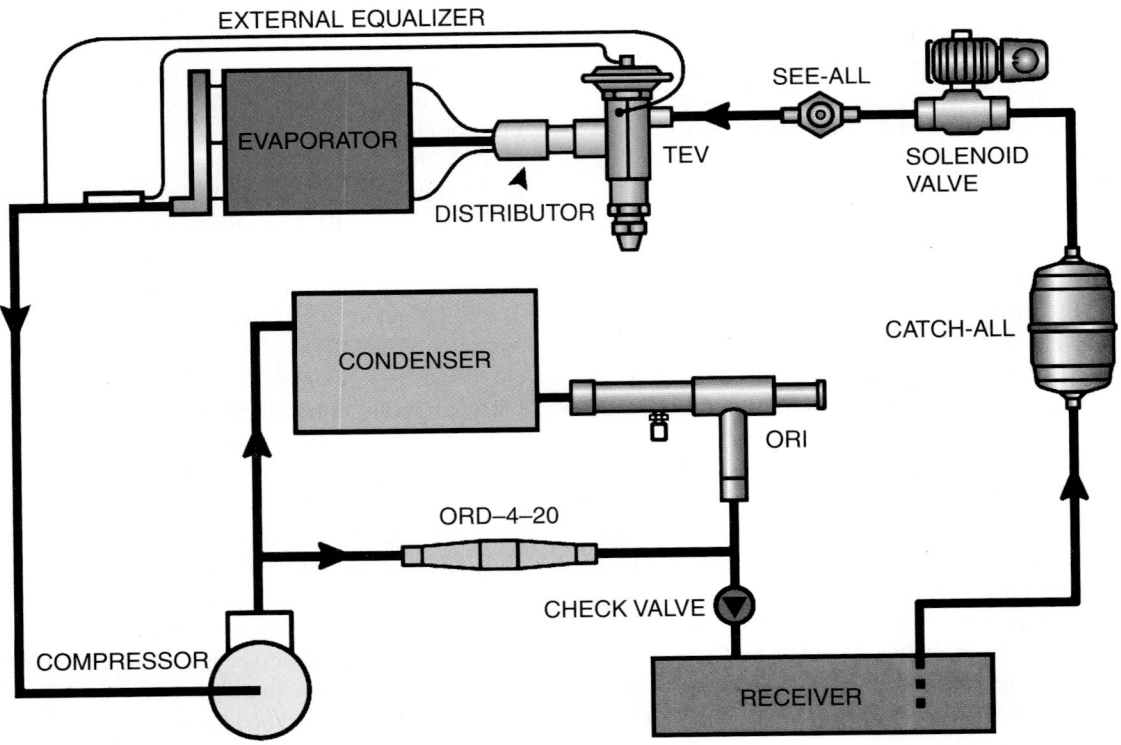

(B)

Figure 22–42   **(A)** Head pressure control valve designed for low ambient conditions. **(B)** Open on rise of inlet pressure (ORI) head pressure control valve shown at the outlet of the condenser. *(A) and (B) Courtesy Sporlan Division, Parker Hannifin Corp.*

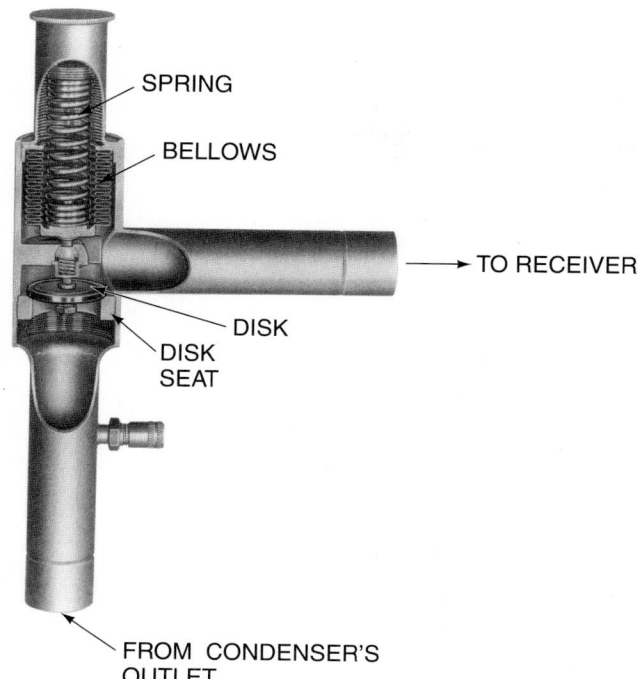

Figure 22–43 Cutaway of an ORI head pressure control valve. *Courtesy Sporlan Division, Parker Hannifin Corp.*

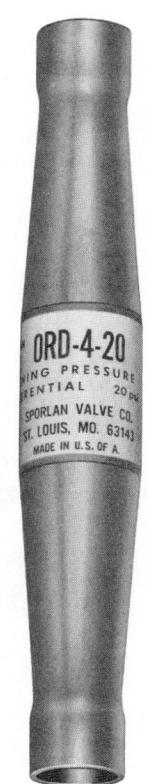

(A)

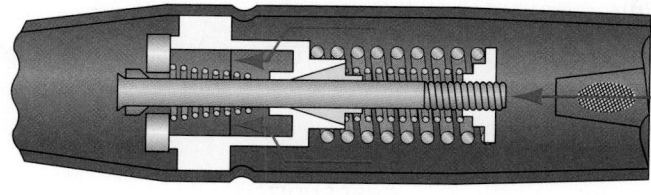

(B)

Figure 22–44 **(A)** Open on rise of differential (ORD) head pressure control valve. **(B)** Cutaway of an ORD head pressure control valve. *(A) and (B) Courtesy Sporlan Division, Parker Hannifin Corp.*

When the ORD valve senses a factory preset pressure difference of 20 psi between the receiver and compressor discharge, it will start to open and bypass hot compressor discharge gas to the receiver's inlet. If the pressure difference across the ORD valve ever reaches 30 psi, the valve will be fully open. The hot gas from the compressor, which flows through the ORD valve, serves to heat up any cold liquid refrigerant being throttled through the ORI valve at the receiver's inlet. This hot gas entering the receiver will also increase the pressure of the receiver and allow it to deliver liquid to the liquid line and TXV when the ORI valve is throttling shut on the outlet of the condenser. Both the ORI and ORD valve will automatically work in conjunction with one another to maintain proper receiver pressure, regardless of the outside ambient conditions. The ORD valve also acts as a check valve to prevent reverse flow from the receiver to the compressor's discharge line during the off cycle.

A combination ORI/ORD nonadjustable valve arrangement can also be used—which simplifies the piping arrangement, **Figure 22–45(A and B).** These valves also limit the flow of refrigerant from the condenser to the receiver, and at the same time regulate the flow of the compressor's hot gas to the receiver. This valve has a round dome at the top that is pressure charged. The pressurized dome will be explained in the paragraphs to come.

Figure 22–46(A and B) is another style of head pressure control valve for low ambient conditions that has a round, pressurized dome at the top that is pressure charged. This valve is often referred to as a low-ambient control (LAC) valve. This dome charge is independent of the refrigerant charge in the actual refrigeration system. The dome charge

will expand and contract in volume and act on an internal diaphragm as the outside ambient changes. This happens because the entire valve is actually located in the condensing unit that is in the outside ambient temperature. This expanding and contracting of the pressure charge will move the diaphragm, which in turn will move a piston in the valve and modulate the valve to either a more open or closed position. When the temperature of the condenser is above 70°F, the refrigerant flow from the compressor is directed by the mixing valve through the condenser and into the receiver. If the outside ambient temperature drops below 70°F, the condensing pressure will fall and the pressure of the liquid coming from the condenser will also fall to a point below that of the bellows in the dome of the valve. This causes a piston to move in the valve and partially restrict the flow of refrigerant leaving the condenser. The condenser will now partially

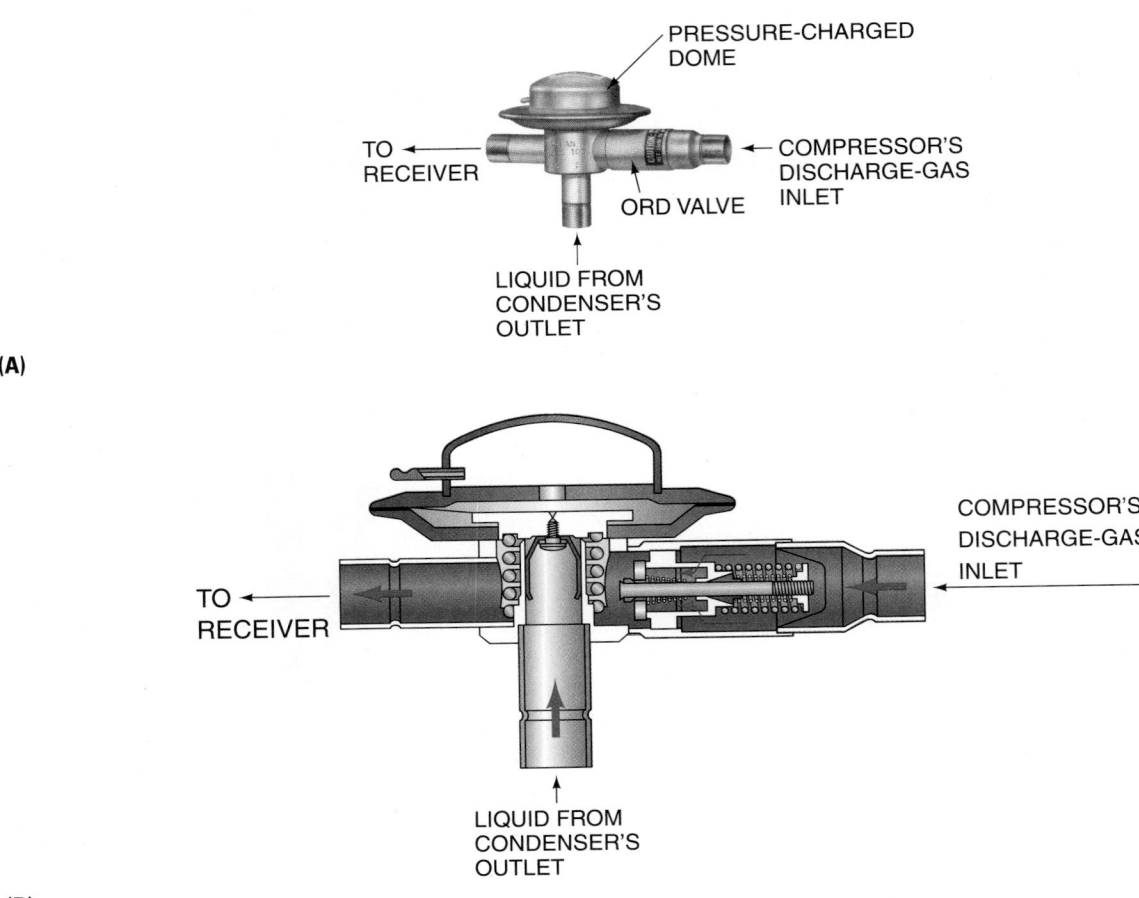

Figure 22–45 **(A)** Combination ORI/ORD head pressure control valve. **(B)** Internal view of a combination ORI/ORD head pressure control valve. *(A)* and *(B)* Courtesy Sporlan Division, Parker Hannifin Corp.

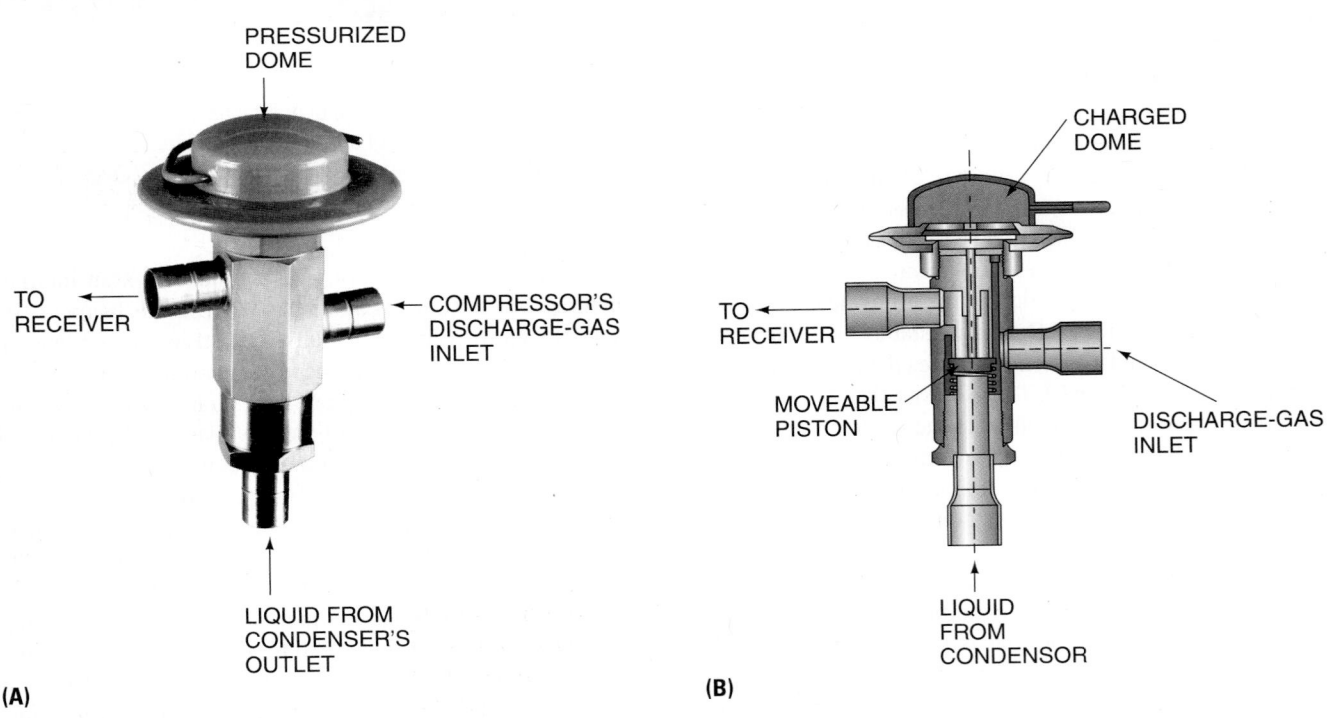

Figure 22–46 **(A)** Low-ambient control (LAC) valve with a pressurized dome. **(B)** Internal view of a low-ambient control (LAC) valve. *(A)* and *(B)* Courtesy Sporlan Division, Parker Hannifin Corp.

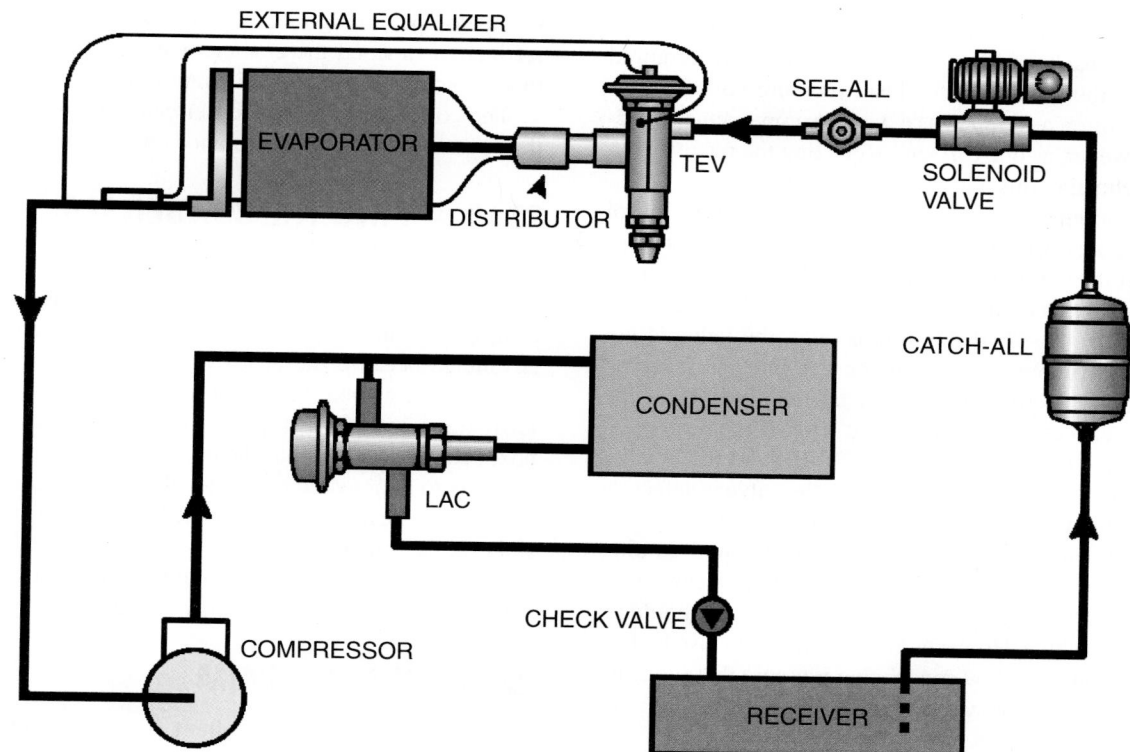

Figure 22–47 System diagram showing location of the low-ambient control (LAC) head pressure control valve. *Courtesy Sporlan Division, Parker Hannifin Corp.*

flood with refrigerant to maintain a certain condensing pressure. At the same time, discharge gas will bypass the condenser and flow directly to the receiver, **Figure 22–47**. This hot, superheated discharge gas going to the receiver now mixes with liquid refrigerant coming from the condenser. This keeps the receiver pressure up and helps keep metering devices fed with refrigerant. In low ambient conditions, these mixing valves will maintain a head pressure of approximately 190 psi for R-502 systems, a head pressure of either 240 psi or 190 psi for R-404A systems (depending on the model of the valve), and a head pressure of approximately 180 psi for R-22 systems. The dome charges can also be custom ordered from the manufacturer to meet specific system requirements. This often allows these head pressure control valves to be used on *floating head pressure* systems. Floating head pressure is covered in Section 22.21 of this unit.

The receivers on systems having head pressure control valves must be sized large enough to hold the normal operating charge plus the additional charge that is necessary to totally flood the condenser in wintertime operations. In fact, the receivers should be sized so that they are about 80% full when they contain the entire system charge. This allows for a 20% vapor head for safety when pumping the system down. Any refrigeration system's receiver should be able to hold all of the system's refrigerant charge and still have a 20% vapor head for safety. **Figure 22–48** shows a refrigeration system with a head pressure control mixing valve used with an oversized receiver. Most manufacturers publish informative charts and tables of system charge recommendations that show the service technician how much refrigerant to add to sys-

| AIR-COOLED | HOT GAS | CONDENSER | OVERSIZED |
| CONDENSER | DISCHARGE | OUTLET | RECEIVER |

Figure 22–48 Low-ambient control (LAC) head pressure control valve shown piped to condenser and receiver. *Courtesy Ferris State University. Photo by John Tomczyk*

tems having head pressure control valves designed to flood the condenser in colder weather. Nowadays, these tables and charts can be accessed on the Internet by a service technician with a personal computer. If the receivers of a refrigeration or air-conditioning system haven't been oversized to accommodate the extra refrigerant needed for flooding the condenser during low ambient conditions, the service technician will have to remove refrigerant every spring to prevent high spring to prevent high head pressures at a design ambient—only to add it back again in the fall when it will be required for flooding the condenser for head pressure control. This technique is often referred to as a winter/summer charge procedure. However, as one can see, oversizing the receiver can save the technician this added labor.

Troubleshooting a low-ambient head pressure control valve with a pressurized dome can be easy if the proper steps are taken. If the outdoor ambient is below 70°F and the head pressure is low, feel the line between the valve and the receiver, **Figure 22–47.** If this line is cold, the valve is not allowing the compressor's discharge gas into the receiver. The valve is defective and should be replaced. This line should be warm at an ambient below 70°F. If the head pressure is low and the ambient is above 70°F—and the line between the valve and the receiver is hot—the valve is defective and should be replaced. The line should be a little warmer than the ambient because of the subcooled liquid coming from the condenser's outlet to the receiver at the warmer outside ambient. Discharge gas flow from the receiver should be shut off by the valve. However, the system could also be low on charge—causing a low condensing pressure that fools the valve. So, the defective valve should only be replaced after verifying that the system charge is correct.

To replace a defective low-ambient control valve with a pressurized dome charge, follow these steps:

- Recover refrigerant.
- Cut the process tube on the valve's charged dome to remove the pressure charge.
- With a torch, heat the valve until the solder liquifies, then remove the valve.
- Wrap the valve's body with a heat sink to prevent damage and silver-solder a new valve in place.
- Install a new filter drier.
- Leak check the system and evacuate to a deep vacuum (500 microns).
- Charge the system and leak check again.

Systems that use condenser flooding for head pressure control have large refrigerant charges, as mentioned earlier. The total charge for the system can be added by following the manufacturer's calculations. When a service technician is working on a unit that has a partial charge, it is very hard to charge the system without some guidelines. When the system is small—for example, under 10 tons—the technician would be advised to recover the charge that is in the system and measure in the charge that is recommended by the manufacturer. Then the system should perform in all temperature conditions.

Large systems will normally have two sight glasses on the liquid receiver, one at the bottom and one at the 80%-full level. The proper procedure for charging one of these systems is to turn off the system until the load builds up and the compressor or compressors can be run at full load. Then block the air over the condenser until the head pressure is that of a 95°F day. This will assure you that there is no excess liquid refrigerant in the condenser. Then charge the system until there is a liquid level in the sight glass at the 80% level of the liquid receiver. If the weather is warm outside, you may have to uncover the condenser if the head pressure

begins to rise too high. Now you know that there is enough refrigerant to flood the condenser in cold weather and that there is not an excess of refrigerant in the condenser.

The cover from the condenser can then be removed and the system should get into balance on its own.

22.19 USING THE CONDENSER SUPERHEAT

Air-cooled condensers have the characteristic of high discharge line temperatures even in winter. This can be used to advantage because the heat can be captured in winter and redistributed as heat for the structure or to heat water. The refrigeration system is rejecting heat out of the refrigerated box, heat that leaked into the box from the store itself. This heat has to be rejected to a place that is unobjectionable. In summer the heat should be rejected outside the store or possibly into a domestic hot-water system, **Figure 22–49.** Water temperatures of 140°F are obtainable for use around the store. This can lead to big energy savings.

22.20 HEAT RECLAIM

🌎In winter the store needs heat. If heat could be rejected inside the store, it could reduce the heating cost. Any heat that is recovered from the system is heat that does not have to be purchased, **Figure 22–50.**🌎

With air-cooled equipment, heat recovery can be accomplished easily. The discharge gas can be passed to the rooftop condenser or to a coil mounted in the ductwork that supplies heat to the store. The condensing temperature of air-cooled equipment is high enough to be used as heat, and the quantity of heat is sizable enough to be important. Some stores in moderate climates will be able to extract enough heat from the refrigeration system to supply the full amount of heat to the store. Heat reclaim is covered in detail in Unit 26, "Application of Refrigeration Systems."

Figure 22–49 A heat exchanger to capture the heat from the highly superheated discharge line and use it to heat domestic water. This line can easily be 210°F and can furnish 140°F water in a supply limited by the size of the refrigeration system. The larger the system, the more heat is available. *Courtesy Noranda Metal Industries, Inc.*

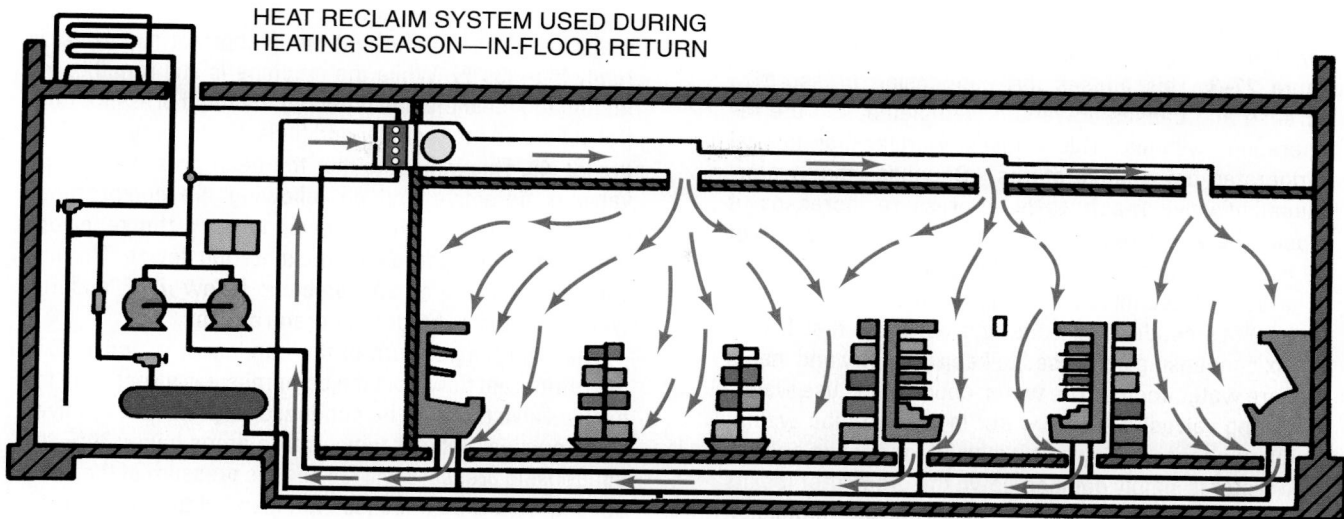

Figure 22–50 A system that can supply heat to the store.

22.21 FLOATING HEAD PRESSURES

Historically, high condensing pressures and temperatures were artificially maintained with head pressure controls in a refrigeration system so that it would function properly at low ambient temperatures. These higher pressures were considered mandatory in order for the metering device to feed the evaporator properly. However, as condensing pressures increase, the refrigeration system draws more electrical power and becomes more inefficient. At today's power costs, these inefficiencies are becoming more and more unacceptable.

In the past, designers of air-conditioning and refrigeration systems picked an outdoor design condition for the system. This outdoor design condition typically was a temperature that would not be reached more than 2% of the time during the life of the equipment. The selection of the condenser was then made on this seldom-reached condition. Today, designers of refrigeration and air-conditioning systems are often trying to attain the lowest possible condensing pressures and temperatures and are designing condensers accordingly. An industry term used for attaining the lowest possible condensing pressure is called *floating head pressure.* Floating head pressure systems simply let the condensing pressures follow the ambient temperature as the seasons swing from summer to winter and back again. This means that in the fall, winter, and spring, the condensing pressures will be at their lowest. It is not uncommon for condensing temperatures to be 30°F to 40°F in the winter months.

Working with metering device suppliers, researchers discovered that *thermostatic expansion valves (TXVs)* would work with much less pressure drop across them than expected in the past, as long as pure liquid was supplied to them. Some newer technology metering devices can operate with as little as a 30-psi pressure drop. 🌐With this new knowledge, designers are allowing condensing pressures to *float* downward with the ambient temperature in the spring, fall, and winter months. This gives the system higher efficiencies with less power consumption. 🌐 In fact, a majority of the outdoor

temperatures in the United States are below 70°F more than they are above. The compressor capacity increases about 6% for every 10°F drop in condensing temperatures. Low ambient controls can still be used on these systems when extreme cold temperatures are experienced. Often, their settings will have to be modified for the specific application and extreme cold ambient conditions.

When head pressures are floated in the fall, winter, and spring, condensing pressures will be lowered. This causes less heat to be available for the heat reclaim coils to heat buildings. However, it is much more efficient to heat a building with a fossil-fuel furnace than by artificially elevating head pressures with head pressure controls on a refrigeration or air-conditioning system and causing inefficiencies. Because of this, fewer heat reclaim systems are being installed.

22.22 CONDENSER EVALUATION

A final note: Do not get lost in the details of the equipment. All of the system's components interact with each other; for example, the condenser operating pressure affects the evaporator. You will be able to draw the correct conclusions with a little experience. Every compression system has a condenser to reject the heat from the system. Examination of the equipment will disclose the condenser, whatever type it may be. The condenser will be hot on air-cooled equipment and warm on water-cooled equipment.

22.23 SERVICE TECHNICIAN CALLS

SERVICE CALL 1

The owner of a small grocery store calls a service technician complaining of high water bills and warm refrigeration temperatures for a few coolers and freezers. *The problem is that two of the helical tube-within-a-tube water-cooled condensers are fouled with mineral deposits and are not transferring the condenser's rejected heat to the water,* **Figure 22–3.** This causes the condensing pressure to increase and causes unwanted inefficiencies in the refrigeration systems. This causes warmer-than-normal refrigerated-box temperatures. Because the water-regulating valve reacts to head-pressure increases, it causes the water-regulating valve to remain open all of the time—wasting water down the drain.

The service technician locates all of the refrigeration equipment's condensing units and observes that five of the six condensing units used to keep the dairy and meats cool are water cooled. The water-cooled units use water-regulating valves to supply water to the five tube-within-a-tube condensers. City water is used for cooling and the used water is simply dumped down the drain after it exits the condenser. The customer explains to the technician that the water bill has always been somewhat high because of his use of water-cooled condensers, but it has gradually been increasing in the last three months. Also, refrigerated-box temperatures have been warmer than usual.

The technician carefully listens and realizes that two of the water-cooled condensers are dumping water down the floor drain very fast. The service technician then measures the temperature difference between the water going into the condenser and the water going out of the condenser on these two units and finds that there is only a 1-degree temperature difference. This tells the service technician that both of the tube-within-a-tube condensers are fouled with mineral deposits and are preventing heat from being transferred from the condenser to the water. Because both of the fouled, helical tube-within-a-tube water-cooled condensers are not cleanable, the service technician tells the customer that he will have to order two new condensers. A week later, the service technician recovers the entire refrigerant charge and replaces both of the noncleanable tube-within-a-tube water-cooled condensers with cleanable shell-and-tube condensers of the same capacity, **Figure 22–22(B).** This will allow the service technician to clean instead of replace the condensers the next time they become fouled with minerals.

SERVICE CALL 2

A store customer is complaining that a walk-in cooler designed to keep meats at 36°F is short cycling and not keeping the refrigerated space below 45°F. The R-134a unit has an outdoor, air-cooled condensing unit. *The problem is that the outdoor ambient is 10°F and that the low-ambient head pressure control mixing valve is defective. The valve is not allowing compressor's discharge gas into the receiver to keep the receiver pressure up to feed the metering device.* The evaporator is thus being starved of refrigerant and is running a low evaporator pressure with high superheats. This causes the low-pressure control to shut the unit off prematurely and short cycle.

A service technician is called. After installing gages on the unit, the technician realizes that the head pressure is really low (50 psig) and that the evaporator superheat is really high (30°F). While the machine is still running, the technician feels the line between the head pressure control valve and the receiver and finds it to be very cold, **Figure 22–48.** This indicates that the head pressure control valve is defective and not allowing the compressor's discharge gas into the receiver to keep the receiver's pressure up in order to feed the metering device. This line should be warm at any ambient below 70°F and get warmer as the outdoor temperature drops.

When the temperature of the condenser is above 70°F, the refrigerant flow from the compressor is directed by the mixing valve through the condenser and into the receiver. If the outside ambient temperature drops below 70°F, the condensing pressure will fall and the pressure of the liquid coming from the condenser will also fall to a point below that of the bellows in the dome of the valve. This causes a piston to move in the valve and partially restrict the flow of refrigerant leaving the condenser. The condenser will now partially flood with refrigerant to maintain a certain condensing pressure. At the same time, discharge gas will bypass the condenser and flow directly to the receiver, **Figure 22–47.** This hot superheated discharge gas going to the receiver now mixes with liquid refrigerant coming from the condenser. This keeps the receiver pressure up and helps keep metering devices fed with refrigerant.

The service technician decides to replace the valve by following these steps:

- Recover refrigerant.
- Cut the process tube on valve's charged dome to remove the pressure charge.
- With a torch, heat the valve until the solder liquifies, then remove the valve.
- Wrap the valve's body with a heat sink to prevent damage and silver-solder a new valve in place.
- Install a new filter drier.
- Leak check the system and evacuate to a deep vacuum (500 microns).
- Charge the system and leak check again.

The system is now started and normal evaporator superheat and system pressures are realized. The refrigerated space pulls down in a matter of one hour. The technician feels the line between the low-ambient control mixing valve and the receiver and realizes that it is warm. This means the valve is allowing the compressor's discharge gas into the receiver to keep the receiver's pressure up in order to feed the metering device. At the same time the condenser will be partially flooded with liquid refrigerant to control head pressure.

SUMMARY

- The condenser is the component that rejects the heat from the refrigeration system.
- The refrigerant condenses to a liquid in the condenser and gives up heat.
- Water is the first medium that heat is rejected into through the water-cooled condenser.

■ There are three types of water-cooled condensers: the tube within a tube, the shell and coil, and the shell and tube.

■ The greatest amount of heat is given up from the refrigerant while the condensing process is taking place.

■ The refrigerant normally condenses about 10°F higher than the leaving condensing medium in a water-cooled condenser that uses a cooling tower.

■ The first job of the condenser is to desuperheat the gas flowing from the compressor.

■ After the refrigerant is condensed to a liquid, the liquid can be further cooled below the condensing temperature. This is called subcooling.

■ When the condensing medium is cold enough to reduce the head pressure to the point that the expansion device will starve the evaporator, a head pressure control must be used, whether air or water is being used as the condensing medium.

■ The three types of cooling towers are natural draft, forced draft, and evaporative.

■ Recirculated water uses evaporation to help the cooling process.

■ When water is evaporated, it will overconcentrate the minerals, and water must be added to the system to keep this from happening.

■ Four types of common head pressure controls are fan-cycling controls, variable-speed fans, shutters, and condenser flooding controls.

■ The relationship of the condensing temperature to the temperature of the air passing over a condenser can help the service technician determine what the high-pressure gage reading should be on air-cooled condensers.

REVIEW QUESTIONS

1. Why do some condensers have to be cleaned with brushes and others with chemicals?
2. The three materials of which condensers are normally made are _____, _____, and _____.
3. Who should be consulted when condenser cleaning is needed?

4. When is the most heat removed from the refrigerant in the condensing process?
5. After heat is absorbed into a condenser medium in a water-cooled condenser, the heat can be deposited in one of two places. What are they?
6. A water-cooling tower capacity is governed by what aspect of the ambient air?
   A. Dry-bulb temperature
   B. Wet-bulb temperature
   C. Enthalpy of the air
   D. Altitude
7. When a standard-efficiency air-cooled condenser is used, the condensing refrigerant will normally be _____ °F higher in temperature than the entering air temperature.
8. Four methods for controlling head pressure in an air-cooled condenser are _____, _____, _____, and _____.
9. The prevailing winds can affect which of the air-cooled condensers?
10. True or False: Air-cooled condensers are much more efficient than water-cooled condensers.
11. The type of condenser that has the lowest operating head pressure is the _____ cooled condenser.
12. Compare the standard conditions of high-efficiency and standard air-cooled condensers.
13. Name two ways in which the heat from an air-cooled condenser can be used for heat reclaim.
14. Briefly describe what is meant by floating head pressure and tell why it is used in refrigeration and air-conditioning systems.
15. Explain how to flush impurities from a water-regulating valve that is on a water-cooled condenser.
16. Explain the basic operation of a combination ORI and ORD low-ambient head pressure control valve.
17. Name two problems a refrigeration system may have if the condensing unit is exposed to a low ambient condition when it is not using a low-ambient head pressure control valve.

OBJECTIVES

After studying this unit, you should be able to

■ explain the function of the compressor in a refrigeration system.
■ discuss compression ratio.
■ describe four different methods of compression.
■ state specific conditions under which a compressor is expected to operate.
■ explain the difference between a hermetic compressor and a semi-hermetic compressor.
■ describe the various working parts of reciprocating and rotary compressors.

SAFETY CHECKLIST

✔ Wear goggles and gloves when attaching or removing gages to transfer refrigerant or to check pressures.
✔ Wear warm clothing when working in a walk-in cooler or freezer.
✔ Do not touch the compressor discharge line with your bare hands.
✔ Be careful not to get your hands or clothing caught in moving parts such as pulleys, belts, or fan blades.
✔ Wear an approved back brace belt when lifting and use your legs, keeping your back straight and upright.
✔ Observe all electrical safety precautions. Be careful at all times and use common sense.

23.1   THE FUNCTION OF THE COMPRESSOR

The *compressor* is considered the heart of the refrigeration system. The term that best describes a compressor is a *vapor pump*. The compressor actually increases the pressure from the suction pressure level to the discharge pressure level. For example, in a low-temperature system the suction pressure for a system that has R-12 as the refrigerant may have a suction pressure of 3 psig and a discharge pressure of 169 psig. The compressor increases the pressure 166 psig (169 − 3 = 166), **Figure 23–1.** A system next to the low-temperature system may have a different pressure increase. It may be a medium-temperature system and have a suction pressure of 21 psig with a discharge pressure of 169 psig. This system has an increase of 148 psig (169 − 21 = 148), **Figure 23–2.**

Compression ratio is the technical expression of pressure difference; it is the high-side *absolute pressure* divided by the low-side absolute pressure. Compression ratio is calculated using absolute pressures. Absolute pressures rather than gage pressures are used when figuring compression ratio in order to keep the calculated compression ratio from becoming a negative number. Absolute pressures keep compression ratios positive and meaningful.

For example, when a compressor is operating with R-12, a head pressure of 169 psig (125°F), and a suction pressure of 2 psig (−16°F), the compression ratio would be:

$$\text{Compression Ratio} = \frac{\text{Absolute Discharge}}{\text{Absolute Suction}}$$

$$CR = \frac{169 \text{ psig} + 14.7 \text{ atmosphere}}{2 \text{ psig} + 14.7 \text{ atmosphere}}$$

$$CR = \frac{183.7}{16.7}$$

$$CR = 11 \text{ to } 1 \text{ or } 11:1$$

A compression ratio of 11:1 would indicate to a service technician that the absolute or true discharge pressure is 11 times as great as the absolute suction pressure.

Let us now assume this same refrigeration system is operating with R-134a as the refrigerant. With the same condensing temperature of 125°F (184.6 psig) and an evaporating temperature of −16°F (0.7 in. Hg vacuum), the compression ratio would be:

$$CR = \frac{184.6 \text{ psig} + 14.7 \text{ atmosphere}}{(29.92 \text{ in. Hg} - 0.7 \text{ in. Hg}) \div 2.036}$$

$$CR = \frac{199.3}{14.35}$$

$$CR = 13.89 \text{ to } 1 \text{ or } 13.89:1$$

NOTE: Because the suction (evaporating) pressure of the R-134a system was in a vacuum of 0.7 in. Hg, a different calculation has to be applied to convert this vacuum reading to absolute pressure in psia. The gage's vacuum measurement of 0.7 in. Hg has to first be converted to in. Hg absolute. This is done by subtracting the vacuum gage reading from standard atmospheric pressure of 29.92 in. Hg. The result is then divided by 2.036 to convert in. Hg to psi. There are 2.036 in. Hg in every 1 psi.

Notice that the compression ratio for the R-134a system is higher than that for the R-12 system even though both systems have the same condensing and evaporating temperatures of 125°F and −16°F, respectively. The compression ratio for the R-134a system is higher, because of the higher condensing pressure and lower evaporating pressure associated with R-134a at these temperature ranges. Either an increase in

R-12   LOW-TEMPERATURE APPLICATION

THE REFRIGERANT ON THE LOW-PRESSURE
SIDE OF THE SYSTEM IS EVAPORATING AT
−13°F AND 3 psig.

THE REFRIGERANT ON THE HIGH-PRESSURE
SIDE OF THE SYSTEM IS CONDENSING AT
125°F AND 169 psig.

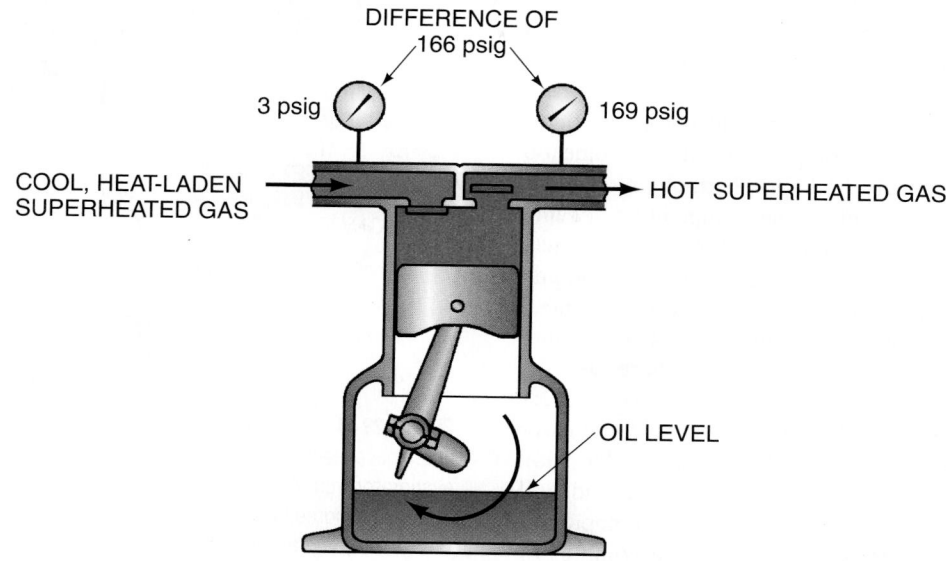

Figure 23–1   The pressure difference between the suction and discharge side of the compressor.

THE REFRIGERANT
IS EVAPORATING AT
20°F AND 21 psig.

NOTE THAT THE REFRIGERANT IN THIS
MEDIUM-TEMPERATURE APPLICATION
CONDENSES AT THE SAME CONDITIONS
AS THE PREVIOUS LOW-TEMPERATURE
APPLICATION.

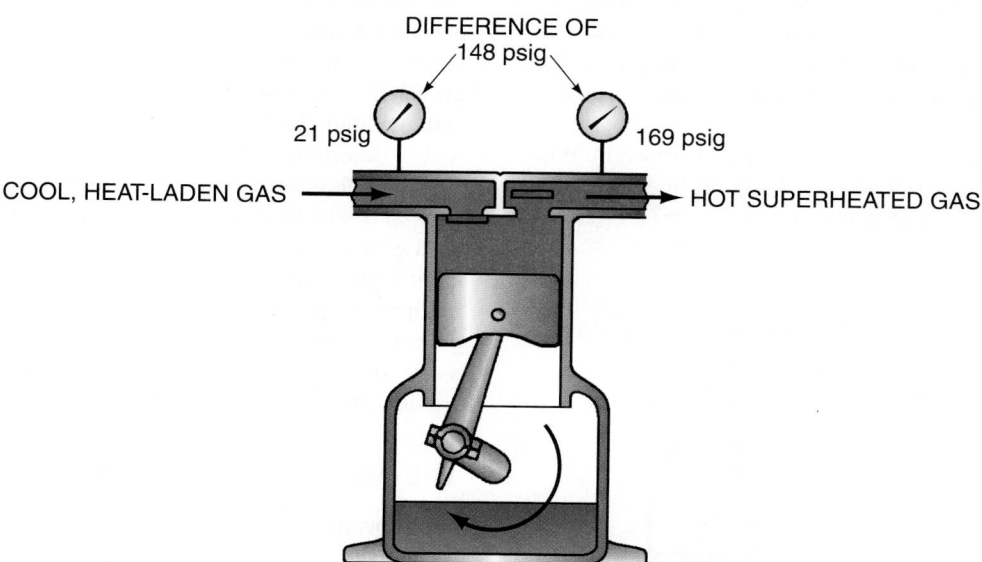

Figure 23–2   A combination of suction and discharge pressure. This is R-12 applied to a medium-temperature system. The compressor only has to lift the suction gas 148 psi.

head pressure or a decrease in suction pressure will cause higher compression ratios. Any time a system has high compression ratios (over 12:1), the compressor's discharge temperature will also be elevated. This is because of the higher heat of compression during the compression stroke associated with higher compression ratios. When compression ratios are high, more energy is needed to raise the pressure of the suction gases to the discharge pressure. This means more heat is generated as heat of compression during the compression stroke.

Compression ratio is used to compare pumping conditions for a compressor. When compression ratios become too high—above approximately 12:1 for a hermetic reciprocating compressor—the refrigerant gas temperature leaving the compressor rises to the point that oil for lubrication may become overheated. Overheated oil may turn to carbon and create acid in the system. Compression ratios can be reduced by two-stage compression. One compressor discharges into the suction side of the second compressor. In **Figure 23–3**, the first-stage compressor has a compression ratio of 3.2:1 (114.7 psia ÷ 35.7 psia), and the second-stage compressor has a compression ratio of 1.6:1 (183.7 psia ÷ 114.7 psia). Both of these compression ratios are acceptable and will result in good compressor efficiencies and low compressor discharge temperatures. However, if only one compressor were used, the compression ratio would be 5.14:1. Even though a compression ratio of 5.14:1 is acceptable, the compressor would have lower efficiencies and higher discharge temperatures than if two compressors were used. The previous example is for illustrative purposes only. Two-stage or compound compression is not generally used until the compression ratio has exceeded 10:1. This is often experienced with low-temperature freezing applications in commercial and industrial storage facilities.

The compressor has cool refrigerant passing through the suction valve to fill the cylinders. This cool vapor contains the heat absorbed in the evaporator. The compressor pumps this heat-laden vapor to the condenser so that it can be rejected from the system.

The vapor leaving the compressor can be very warm. With a discharge pressure of 169 psig, the discharge line at the compressor could easily be 200°F or higher. SAFETY PRECAUTION: Do not touch the compressor discharge line because

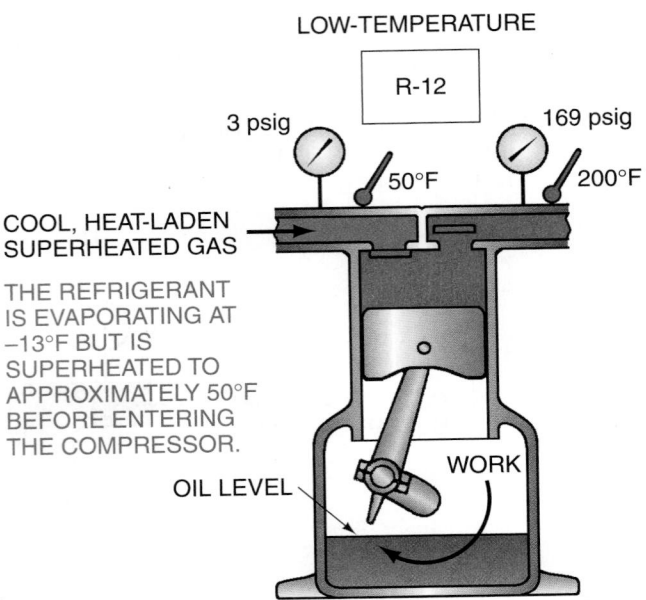

Figure 23–4 The refrigerant entering the compressor is called heat laden because it contains the heat that was picked up in the evaporator from the boiling process. The gas is cool but full of heat that was absorbed at a low pressure and temperature level. When this gas is compressed in the compressor, the heat concentrates. In addition to the heat that was absorbed in the evaporator, the act of compression converts some energy to heat.

you can burn your fingers. The vapor is compressed with the heat from the suction gas concentrated in the gas leaving the compressor, **Figure 23–4.**

23.2 TYPES OF COMPRESSORS

Five major types of compressors are used in the refrigeration and air-conditioning industry. These are the *reciprocating, screw, rotary, scroll,* and *centrifugal.* The reciprocating, **Figure 23–5,** is the compressor used most frequently in small- and medium-sized commercial refrigeration systems and will be described in detail in this unit. The screw compressor, **Figure 23–6,** is used in large commercial and industrial systems and will be described only briefly here because it is used in larger systems described later in this text. The rotary, **Figure 23–7,** and the scroll, **Figure 23–8,** along with the recriprocating, are used in residential and light commercial air conditioning. Centrifugal compressors, **Figure 23–9,** are used extensively for air conditioning in large buildings and are described in Section 48.7.

The Reciprocating Compressor

Reciprocating compressors are categorized by the compressor housing and by the drive mechanisms. The two housing categories are *open* and *hermetic* compressors, **Figure 23–10.** Hermetic refers to the type of housing the compressor is contained in and is divided into two types: fully welded, **Figure 23–11,** and serviceable or semi-hermetic, **Figure 23–10(B).** The drive mechanisms may be either enclosed inside the

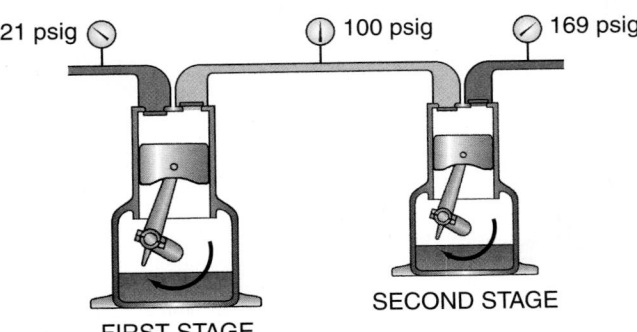

Figure 23–3 Two-stage compression. Notice that the second stage of the compressor is smaller than the first stage.

Figure 23–5   A reciprocating compressor. *Courtesy Copeland Corporation*

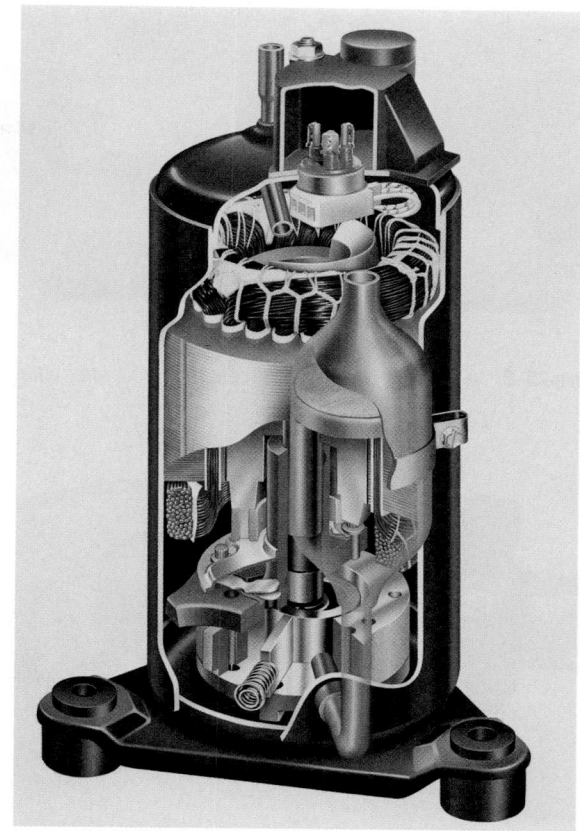

Figure 23–7   A rotary compressor. *Reprinted with permission of Motors and Armatures, Inc.*

Figure 23–6   A screw compressor. *Courtesy Frick Company*

shell or outside the shell. When the compressor is hermetic, the drive mechanism is direct. The compressor and motor shaft are the same shaft inside the compressor shell.

FULLY WELDED HERMETIC COMPRESSORS.   The motor and compressor are contained inside a single shell that is welded closed when a welded hermetic compressor is manufactured, **Figure 23–11.** This unit is sometimes called the *tin can* compressor because it cannot be serviced without cutting open the shell. Characteristics of the fully welded hermetic compressor are as follows:

1. The only access to the inside of the shell is by cutting the shell open.
2. They are only opened by a very few companies that specialize in this type of work. Otherwise, unless the manufacturer wants the compressor back for examination, it is a throwaway compressor.
3. The motor shaft and the compressor crankshaft are one shaft.

Figure 23–8   A scroll compressor. *Courtesy Copeland Corporation*

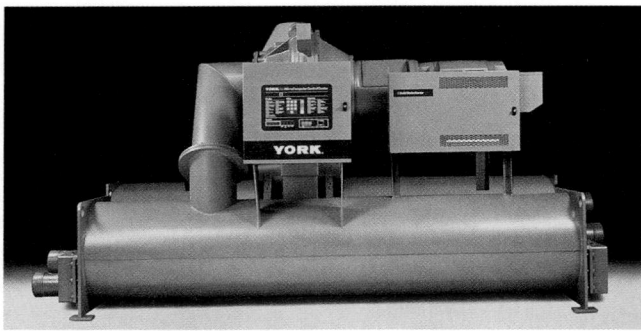

Figure 23–9 A centrifugal compressor. *Courtesy York International Corp.*

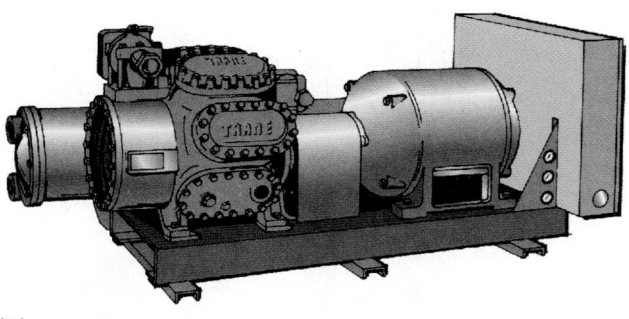

(A)

(B)

Figure 23–10 **(A)** An open-drive compressor. **(B)** A serviceable hermetic compressor. **(A)** *Courtesy Trane Company 2000.* **(B)** *Courtesy Copeland Corporation*

4. It is usually considered a low-side device because the suction gas is vented to the whole inside of the shell, which includes the crankcase. The discharge (high-pressure) line is normally piped to the outside of the shell so that the shell only has to be rated at the low-side working-pressure value.
5. Generally, they are cooled with suction gas.

Figure 23–11 The welded hermetic compressor is used in the smaller compressor sizes, from fractional horsepower to about 50 hp. Most welded hermetic compressors have a few things in common. The suction line is usually piped directly into the shell and is open to the crankcase. The discharge line is normally piped from the compressor inside the shell to the outside of the shell. The compressor shell is typically thought of as a low-side component. *Courtesy Bristol Compressors, Inc.*

6. They usually have a pressure lubrication system, but smaller hermetics are splash lubricated.
7. The combination motor and crankshaft are customarily in a vertical position with a bearing at the bottom of the shaft next to the oil pump. The second bearing is located about halfway on the shaft between the compressor and the motor.
8. The pistons and rods work outward from the crankshaft, so they are working at a 90° angle in relation to the crankshaft, **Figure 23–12.**

SERVICEABLE HERMETIC COMPRESSORS. When a serviceable hermetic (semi-hermetic) compressor is manufactured, the motor and compressor are contained inside a single shell that is bolted together. This unit can be serviced by removing the bolts and opening the shell at the appropriate place, **Figure 23–13.** Following are the characteristics of the serviceable hermetic compressor:

1. The unit is bolted together at locations that facilitate service and repair. Gaskets separate the mating parts connected by the bolts.
2. The housing is normally cast-iron and may have a steel housing fastened to the cast-iron compressor. They are normally heavier than the fully welded type.
3. The motor and crankshaft combination are similar to the motor and crankshaft in the fully welded type except that the crankshaft is usually horizontal.

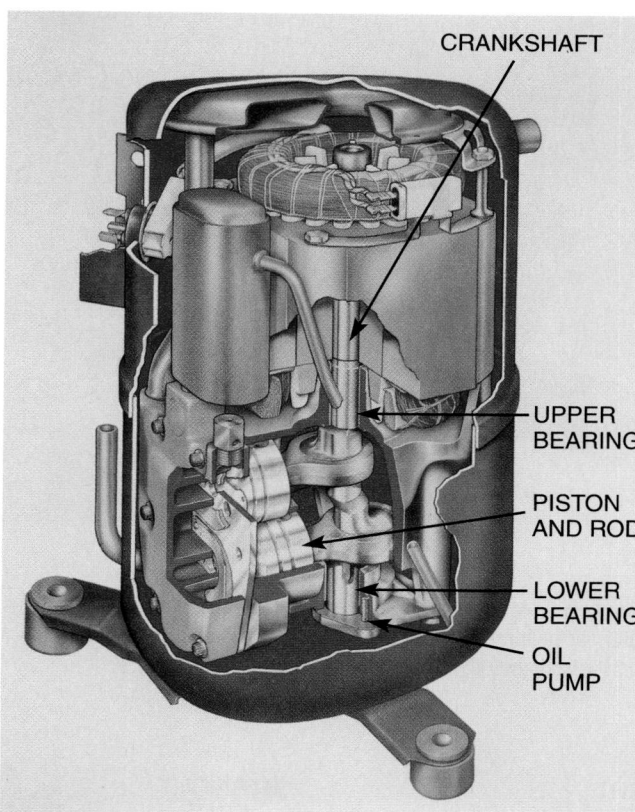

Figure 23–12 The internal workings of a welded hermetic compressor. *Courtesy Tecumseh Products Company*

Figure 23–13 A serviceable hermetic (semi-hermetic) compressor designed in such a manner that it can be serviced in the field. *Courtesy Copeland Corporation*

4. They generally use a splash-type lubrication system in the smaller compressors and a pressure lubrication system in the larger compressors.
5. They are often air cooled and can be recognized by the fins in the casting or extra sheet metal on the outside of the housing that give the shell more surface area.

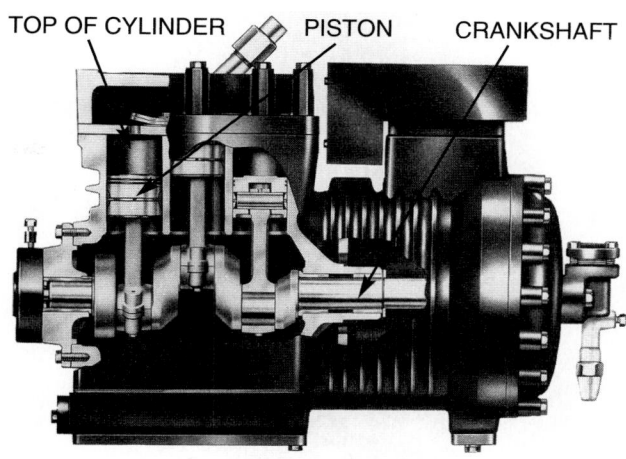

Figure 23–14 The working parts of a serviceable hermetic compressor. The crankshaft is in the horizontal position, and the rods and pistons move in and out from the center of the shaft. The oil pump is on the end of the shaft and draws the oil from the crankcase at the bottom of the compressor. *Courtesy Copeland Corporation*

6. The piston heads are normally at the top or near the top of the compressor and work up and down from the center of the crankshaft, **Figure 23–14**.

OPEN-DRIVE COMPRESSORS. Open-drive compressors are manufactured in two styles: belt drive and direct drive, **Figure 23–15.** Any compressor with the drive on the outside of the casing must have a shaft seal to keep the refrigerant from escaping to the atmosphere. This seal arrangement has not changed much in many years. Open-drive compressors are bolted together and may be disassembled for service to the internal parts.

BELT-DRIVE COMPRESSORS. The belt-drive compressor was the first type of compressor and is still used to some extent. With the belt-drive unit the motor and its shaft are parallel with the compressor's shaft. The motor is beside the compressor. Notice that since the compressor and motor shaft are parallel, there is a sideways pull on both shafts to tighten the belts. This strains both shafts and requires the manufacturer to compensate for this in the shaft bearings. **Figure 23–16** indicates correct and incorrect alignments of belt-drive compressors and motors.

DIRECT-DRIVE COMPRESSORS. The direct-drive compressor differs from the belt-drive in that the compressor shaft is end to end with the motor shaft. These shafts have a coupling between them with a small amount of flexibility. The two shafts have to be in very good alignment to run correctly, **Figure 23–17.**

THE ROTARY SCREW COMPRESSOR. The rotary screw compressor is another mechanical method of compressing refrigerant gas used in larger installations. Instead of a piston and

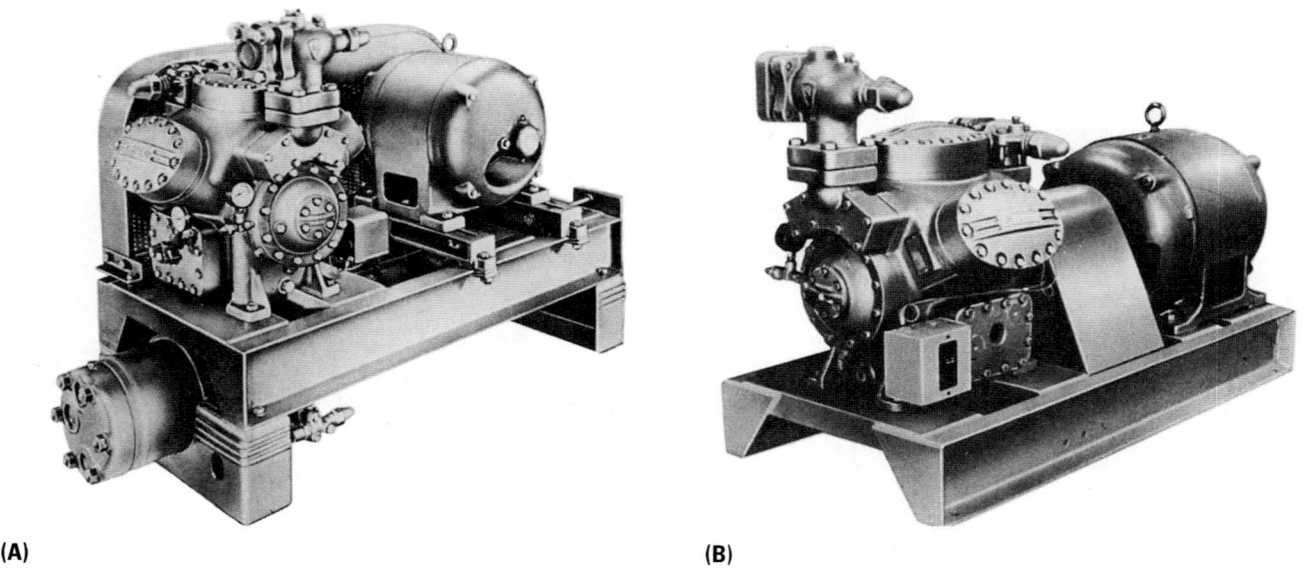

(A)   **(B)**

Figure 23–15 **(A)** A belt-drive compressor. **(B)** A direct-drive compressor. The belt-drive compressor may have different speeds, depending on the pulley sizes. The direct-drive compressor turns at the speed of the motor because it is attached directly to the motor. Common speeds are 1750 rpm and 3450 rpm. The coupling between the motor and the compressor is slightly flexible. *Reproduced courtesy of Carrier Corporation*

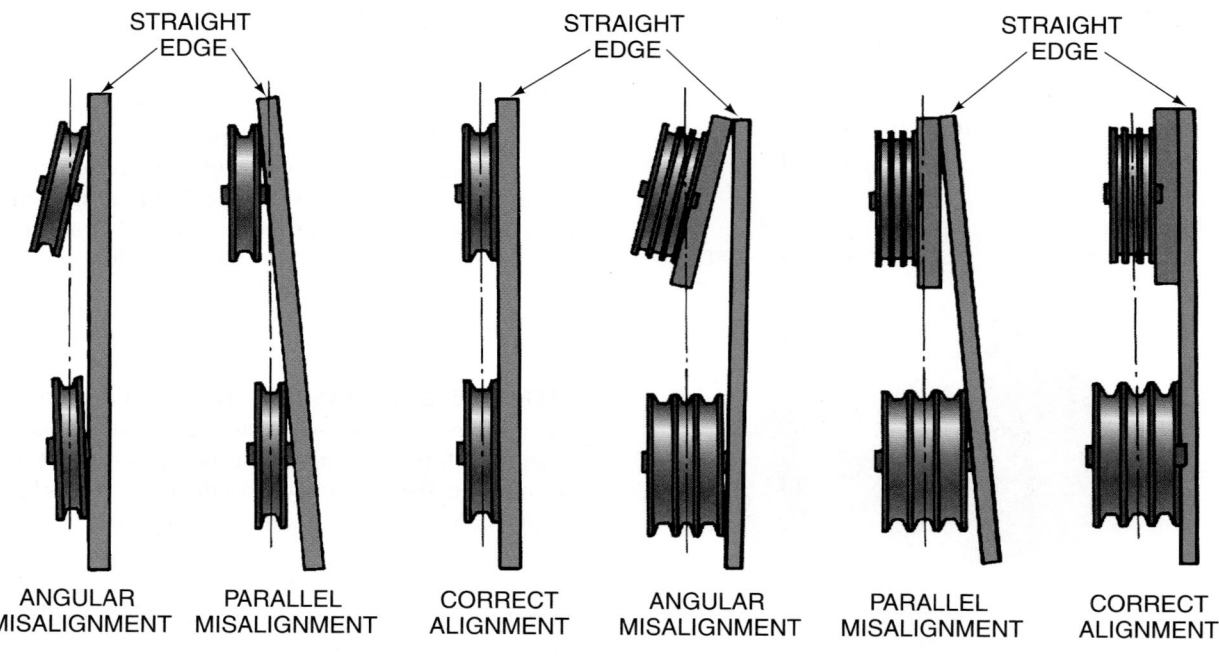

Figure 23–16 Correct and incorrect alignments of a belt-drive compressor with its motor. The belts used on a multiple-belt installation are matched at the factory for the correct length.

cylinder, this compressor uses two matching, tapered, machined, screw-type gears that squeeze the refrigerant vapor from the inlet to the outlet, **Figure 23–18.**

The rotary screw compressor uses an open motor instead of the hermetic design. A shaft seal traps the refrigerant in the compressor housing where the rotating shaft leaves the compressor housing. A flexible coupling is used to connect the motor shaft to the compressor shaft to prevent minor misalignment from causing seal or bearing damage.

Screw-compressor applications are for large systems (see **Figure 23–6**). The refrigerant may be any of the common refrigerants. The operating pressure on the low- and high-pressure sides of the system are the same as for a reciprocating system of like application.

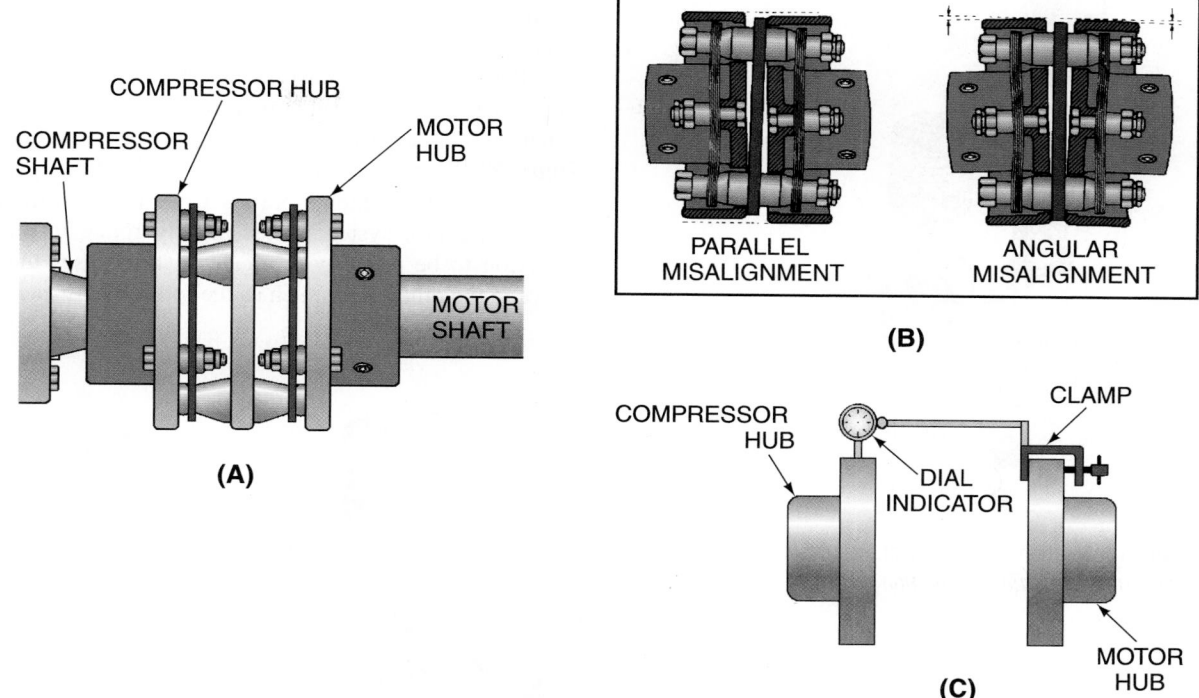

Figure 23–17 **(A)** and **(C)** Alignment of a direct-drive compressor and motor. The alignment must be within the compressor manufacturer's specifications. **(B)** These compressors turn so fast that if the alignment is not correct, a seal or bearing will soon fail. When the correct alignment is attained, the compressor and motor are normally fastened permanently to the base on which they are mounted. The motor and the compressor can be rebuilt in place in larger installations. *Courtesy Trane Company*

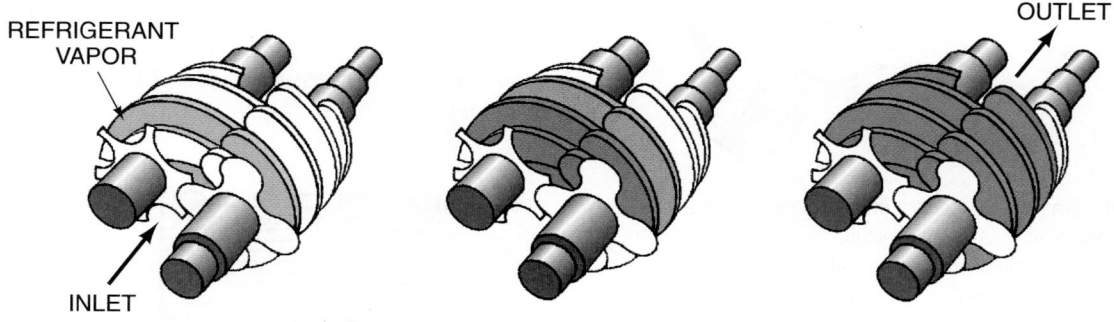

ONE STAGE OF COMPRESSION AS REFRIGERANT MOVES THROUGH THE SCREW COMPRESSOR

Figure 23–18   Tapered machined screw-type gears in a screw compressor.

23.3   RECIPROCATING COMPRESSOR COMPONENTS

Reciprocating compressors house a vast array of moving parts. Most of these component parts have been precision machined with very tight tolerances for proper compressor efficiency and operation.

Following is a list and description of the reciprocating compressor's major component parts.

The Crankshaft

The *crankshaft* of a reciprocating compressor transmits the circular motion of the rods, and the motion is changed to back and forth (reciprocating) for the pistons, **Figure 23–19.** Crankshafts are normally manufactured of cast iron or soft steel. The crankshaft can be cast (molten metal poured into a mold) into a general shape and machined into the exact size and shape. In this case the shaft is made of cast iron,

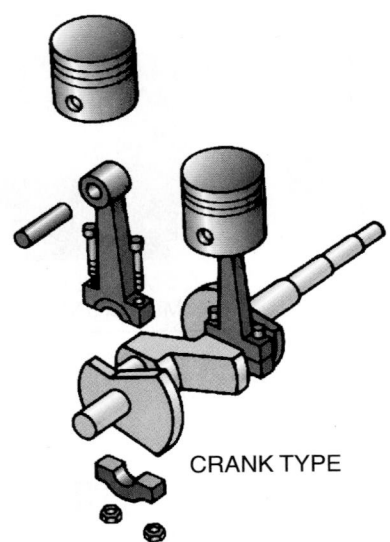

CRANK TYPE

Figure 23–19 The relationship of the pistons, rods, and crankshaft. *Reproduced courtesy of Carrier Corporation*

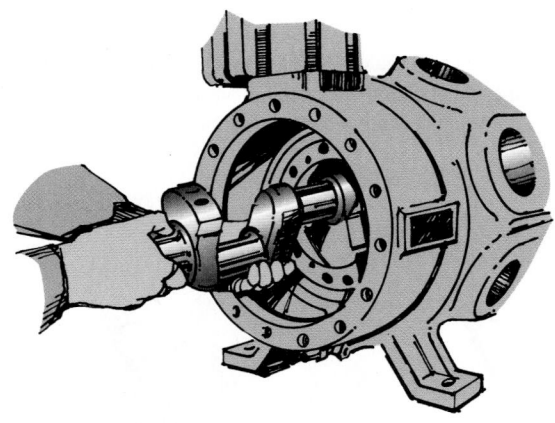

Figure 23–20 Crankshafts are cast into the general shape and machined to the correct shape in a machine shop. This crankshaft is in an open-drive compressor but is typical also for hermetic compressors. *Courtesy Trane Company 2000*

Figure 23–20. This machining process is critical because the throw (the off-center part where the rod fastens) does not turn in a circle (in relationship to the center of the shaft) when placed in a lathe. The machinist must know how to work with this type of setup.

These off-center shafts normally have two main-bearing surfaces in addition to the off-center rod-bearing surfaces: One is on the motor end of the shaft, and one is on the other end. The bearing on the motor end is normally the largest because it carries the greatest load.

Some shafts are straight and have a cam-type arrangement called an *eccentric.* This allows the shaft to be manufactured straight and from steel. The shaft may not be any more durable, but it is easier to machine. The eccentrics can be machined off-center to the shaft to accomplish the reciprocating

action, **Figure 23–21.** Notice that the rod has to be different for the eccentric shaft because the end of the rod has to fit over the large eccentric on the crankshaft.

All of these shafts must be lubricated. The smaller compressors using the splash system may have a catch basin to catch oil and cause it to flow down the center of the shaft, **Figure 23–22.** It is then slung to the outside of the crankshaft surface when the compressor runs. Compressors with splash-type lubrication systems must usually rotate in only one direction to be properly lubricated. This causes the oil to move to the other parts, such as the rods on both ends.

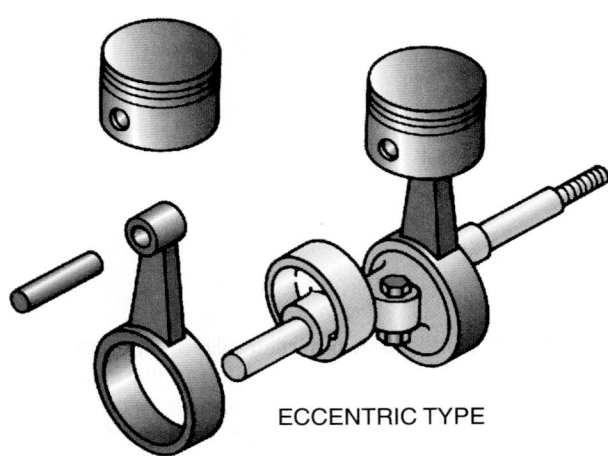

ECCENTRIC TYPE

Figure 23–21 This crankshaft obtains the off-center action with a straight shaft and an eccentric. The eccentric is much like a cam lobe. The rods on this shaft have large bottom throws and slide off the end of the shaft. This means that to remove the rods, the shaft must be taken out of the compressor. *Reproduced courtesy of Carrier Corporation*

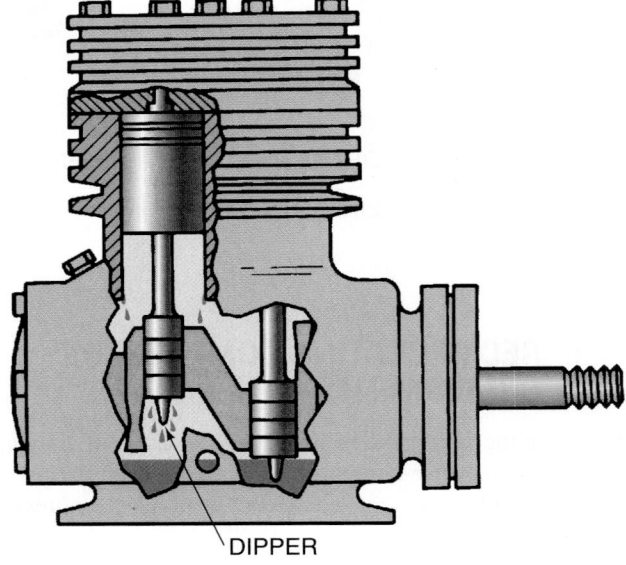

DIPPER

Figure 23–22 A splash lubrication system that splashes oil up onto the parts. *Reproduced courtesy of Carrier Corporation*

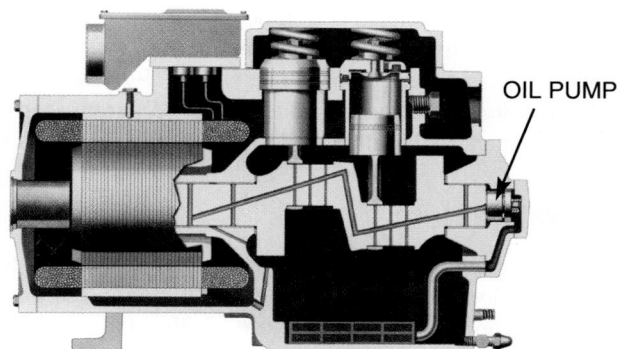

Figure 23–23 A crankshaft drilled for the oil pump to force the oil up the shaft to the rods, then up the rods to the wrist pins. Magnetic elements are sometimes placed along the passage to capture iron filings. *Courtesy Trane Company 2000*

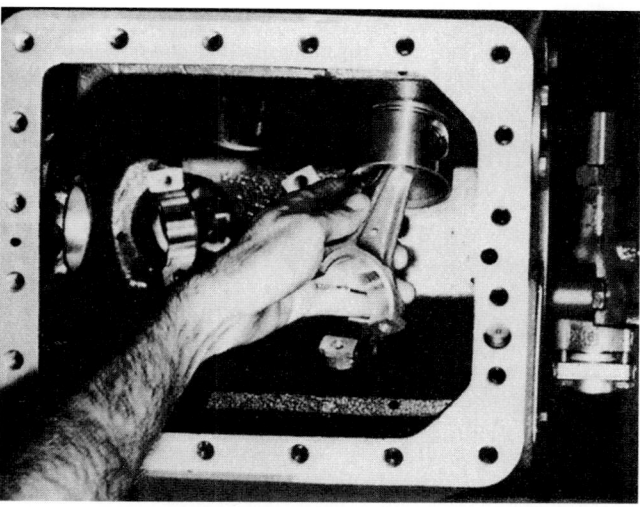

Figure 23–24 The service technician is removing the rods from a compressor. The rods have the split bottoms. *Courtesy Trane Company 2000*

Some of the shafts are drilled and lubricated with a pressure lubrication system. These compressors have an oil pump mounted on the end of the crankshaft that turns with the crankshaft, **Figure 23–23**.

NOTE: There is no pressure lubrication when the compressor first starts. The compressor must be running up to speed before the lubrication system is fully effective.

Because some compressors have vertical shafts and some have horizontal, manufacturers have been challenged to provide proper lubrication where needed. You may have to consult the compressor manufacturer for questions about a specific application.

Connecting Rods

Connecting rods connect the crankshaft to the piston. These rods are normally made in two styles: the type to fit the crankshaft with off-center throws and the type to fit the eccentric crankshaft, **Figure 23–19** and **Figure 23–21**. Rods can be made of several different metals such as iron, brass, and aluminum. The rod design is important because it takes a lot of the load in the compressor. If the crankshaft is connected directly to the motor and the motor is running at 3450 rpm, the piston at the top of the rod is changing directions 6900 times per minute. The rod is the connection between this piston and the crankshaft and is the link between this changing of direction.

The rods with the large holes in the shaft end are for eccentric shafts. They cannot be taken off with the shaft in place. The shaft has to be removed to take the piston out of the cylinder. The rods with the smaller holes are for the off-center shafts, are split, and have rod bolts, **Figure 23–24**. These rods can be separated at the crankshaft, and the rod and piston can be removed with the crankshaft in place.

The rod is small on the piston end and fastens to the piston by a different method. The rod normally has a connector called a *wrist pin* that slips through the piston and the upper end of the rod. This almost always has a *snapring* to keep the wrist pin from sliding against the cylinder wall, **Figure 23–25**.

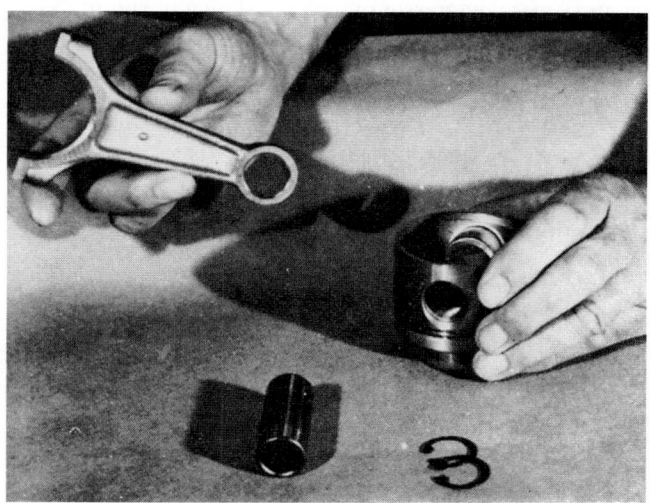

Figure 23–25 The end of the rod that fits into the piston. The wrist pin holds the rod to the piston while allowing the pivot action that takes place at the top of the stroke. The wrist pin is held secure in the piston with snaprings. *Courtesy Trane Company 2000*

The Piston

The *piston* is the part of the cylinder assembly exposed to the high-pressure gas during compression. Pistons have high-pressure gas on top and suction or low-pressure gas on the bottom during the upstroke. They have to slide up and down in the cylinder in order to pump. They must have some method of preventing the high-pressure gas from slipping by to the crankcase. Piston rings like those used in automobile engines are used on the larger pistons. These rings are of two types: compression and oil. The smaller compressors use the oil on the cylinder walls as the seal. A cross-sectional view of these rings can be seen in **Figure 23–26**.

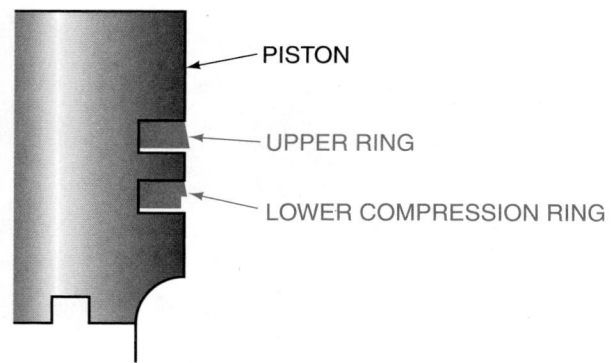

Figure 23–26   The piston rings for a refrigeration compressor resemble the rings used on automobile pistons. *Courtesy Trane Company 2000*

Refrigerant Cylinder Valves

The valves in the top of the compressor determine the direction in which the gas entering the compressor will flow. A cutaway of a compressor cylinder is shown in **Figure 23–27**. These valves are made of very hard steel. The two styles that make up the majority of the valves on the market are the *ring valve* and the *flapper (reed) valve.* They serve both the suction and the discharge ports of the compressor.

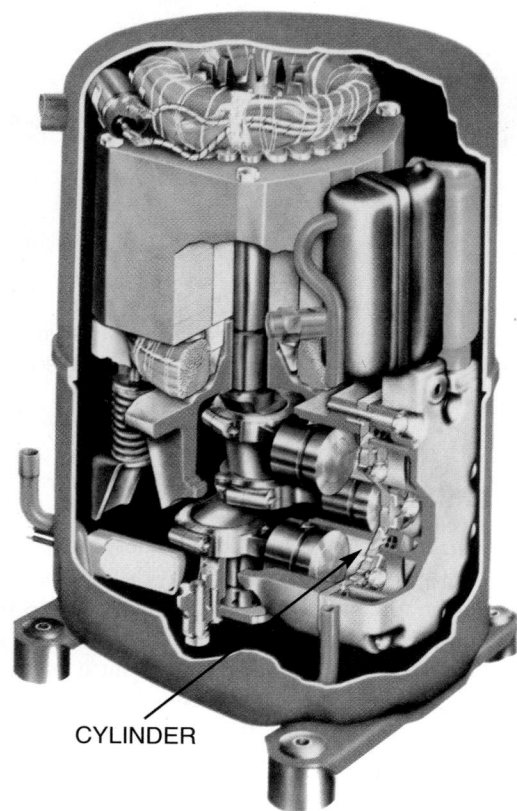

Figure 23–27   This cutaway of a compressor shows a typical cylinder. *Courtesy Tecumseh Products Company*

The ring valve is made in a circle with springs under it. If ring valves are used for the suction and the discharge, the larger one will be the suction valve, **Figure 23–28**.

Flapper valves have been made in many different shapes. Each manufacturer has its own design, **Figure 23–29**.

The Valve Plate

The *valve plate* holds the suction and discharge flapper valves. It is located between the head of the compressor and the top of the cylinder wall, **Figure 23–30**. Many different methods have been used to hold the valves in place without taking up any more space than necessary. The bottom of the plate actually protrudes into the cylinder. Any volume of gas that cannot be pumped out of the cylinder because of the valve design will reexpand on the downstroke of the piston. This makes the compressor less efficient.

Other versions of the crankshaft and valve arrangements exist than those listed here. These other arrangements, however, are not used enough in the refrigeration industry to justify coverage. If you need more information, contact the specific manufacturer.

The Head of the Compressor

The component that holds the top of the cylinder and its assembly together is the *head*. It sets on top of the cylinder and contains the high-pressure gas from the cylinder until it moves into the discharge line. It often contains the suction chamber, separated from the discharge chamber by a partition and gaskets. These heads have many different design configurations and need to accomplish two things. They hold the pressure in and hold the valve plate on the cylinder. They are made of steel in some welded hermetic compressors and of cast iron in a serviceable hermetic type. The cast iron heads may be in the moving airstream and have fins on them to help dissipate the heat from the top of the cylinder, **Figure 23–31**.

Mufflers

Mufflers are used in many fully hermetic compressors to muffle compressor pulsation noise. Audible suction and discharge pulsations can be transmitted into the piping if they are not muffled. Mufflers must be designed to have a low pressure drop and still muffle the discharge pulsations, **Figure 23–32**.

The Compressor Housing

The *housing* holds the compressor and sometimes the motor. It is made of stamped steel for the welded hermetic and of cast iron for the serviceable hermetic.

The welded hermetic compressor is designed so the compressor shell is under low-side pressure and will often have a working pressure of 150 psig. Many compressors manufactured today must use the newer refrigerants that operate under much higher pressure; therefore, compressor shell working pressures are much higher. SAFETY PRECAUTION: The technician should always check the working pressures of all system

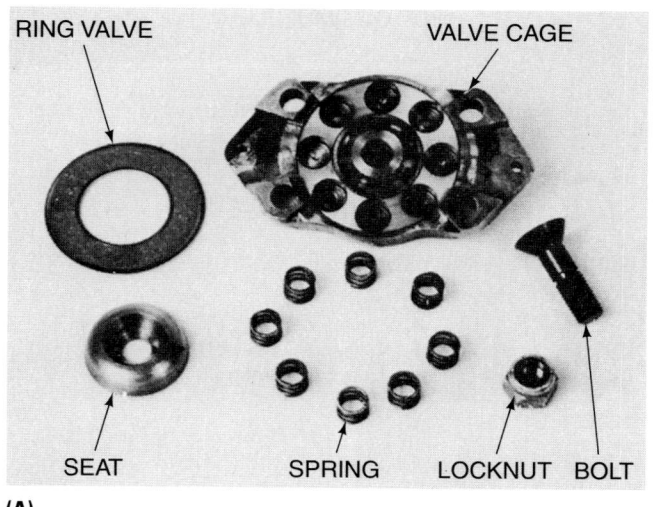

(A)

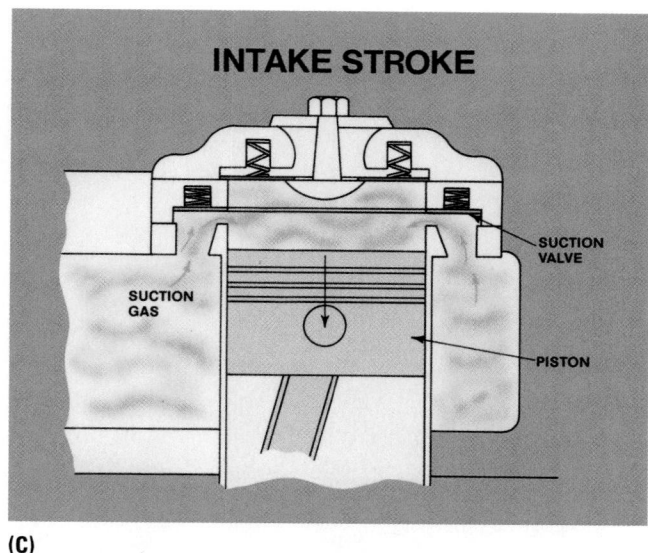

(C)

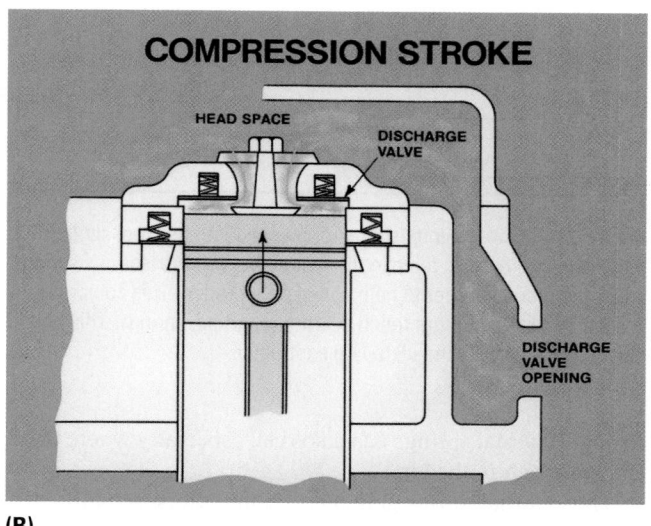

(B)

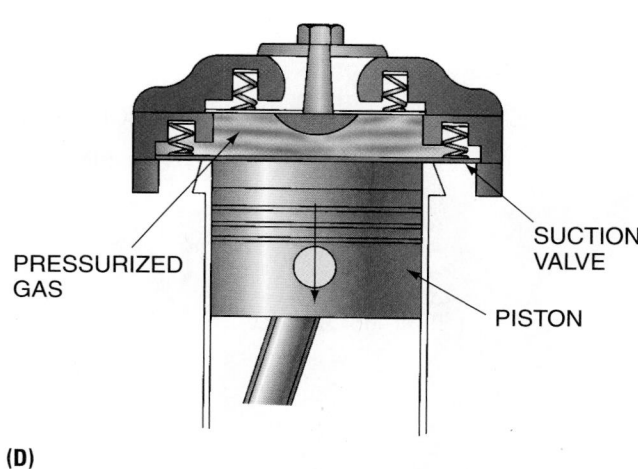

(D)

Figure 23–28   **(A)** Ring valves. **(B)–(D)** They normally have a set of small springs to close them. *Courtesy Trane Company 2000*

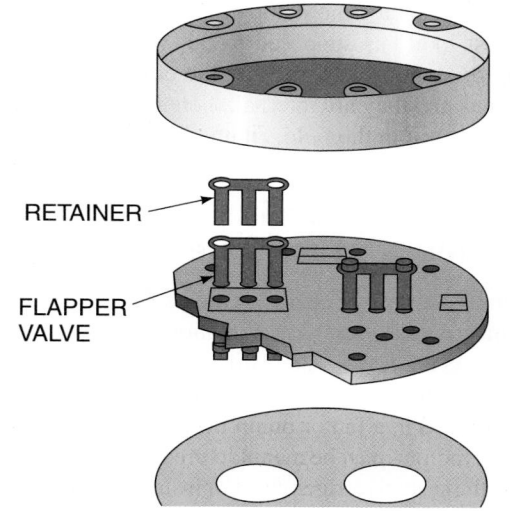

Figure 23–29   Reed or flapper valves held down on one end. This provides enough spring action to close the valve when reverse flow occurs. *Reproduced courtesy of Carrier Corporation*

Figure 23–30   Valve plates typical of those used to hold the valves. They can be replaced or rebuilt if not badly damaged. There is a gasket on both sides. *Reproduced courtesy of Carrier Corporation*

parts before applying high pressures for leak checking. The compressor is mounted inside the shell, and the discharge line is normally piped to the outside of the shell. This means the shell does not need to have a test pressure as high as the high-side pressure. A cutaway of a hermetic compressor

COMPRESSOR HEAD

(A)

FINS   COMPRESSOR HEAD
FINS

(B)

Figure 23–31 Typical compressor heads. **(A)** A suction-cooled compressor. **(B)** Air-cooled compressors that have air-cooled motors must be located in a moving airstream, or overheating will occur. *Courtesy Copeland Corporation*

inside a welded shell and the method used to weld the shell together are shown in **Figure 23–27.**

Two methods are used to mount the compressor inside the shell: *rigid mount* and *spring mount.*

Rigid-mounted compressors were used for many years. The compressor shell was mounted on external springs that had to be bolted tightly for shipment. The springs were supposed to be loosened when installed, **Figure 23–33.** Occasionally they were not, and the compressor vibrated because, without the springs, it was mounted rigidly to the condenser

HOT GAS MUFFLER

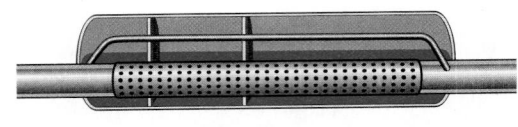

(HORIZONTAL INSTALLATION)

Figure 23–32 A compressor muffler.

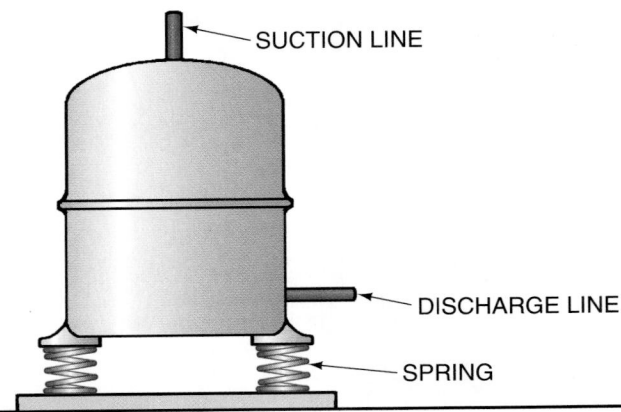

SUCTION LINE

DISCHARGE LINE

SPRING

Figure 23–33 A compressor motor pressed into its steel shell. It requires experience to remove the motor. The compressor has springs under the mounting feet to help eliminate vibration. This compressor is shipped with a bolt tightened down through the springs. This bolt must be loosened by the installing contractor.

casing. External springs can also rust, especially where there is a lot of salt in the air.

The internal spring-mounted compressors actually suspend the compressor from springs inside the shell. These compressors have methods of keeping the compressor from moving too much during shipment. Sometimes a compressor will come loose from one or two of the internal springs. When this happens, the compressor will normally run and pump just like it is supposed to but will make a noise on start-up or shutdown or both. If the compressor comes off the springs and they are internal, there is nothing that can be done to repair it in the field, **Figure 23–34.**

Compressor Motors in a Refrigerant Atmosphere

The compressor motor operating inside the refrigerant atmosphere must have special consideration. Motors for hermetic compressors differ from standard electric motors. The materials used in a hermetic motor are not the same materials that would be used in a fan or pump motor that would run in air. Hermetic motors must be manufactured of materials compatible with the system refrigerants. For instance, rubber cannot be used because the refrigerant would dissolve it. The motors are assembled in a very clean atmosphere and kept dry. When a hermetic motor malfunctions, it cannot be repaired in the field.

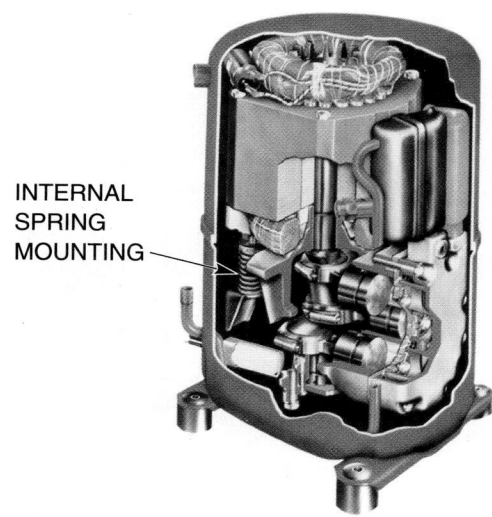

Figure 23–34 A compressor mounted inside the welded hermetic shell on springs. The springs have guides that only allow them to move a certain amount during shipment. *Courtesy Tecumseh Products Company*

MOTOR ELECTRICAL TERMINALS. There must be some conductor to carry the power from the external power supply to the internal motor. The power to operate the compressor must be carried through the compressor housing without the refrigerant leaking. The connection also has to insulate the electric current from the compressor shell. These terminals are sometimes fused glass with a terminal stud through the middle on the smaller compressors. When large terminals are required, the terminals are sometimes placed in a fiber block with an O ring–type seal, **Figure 23–35.**

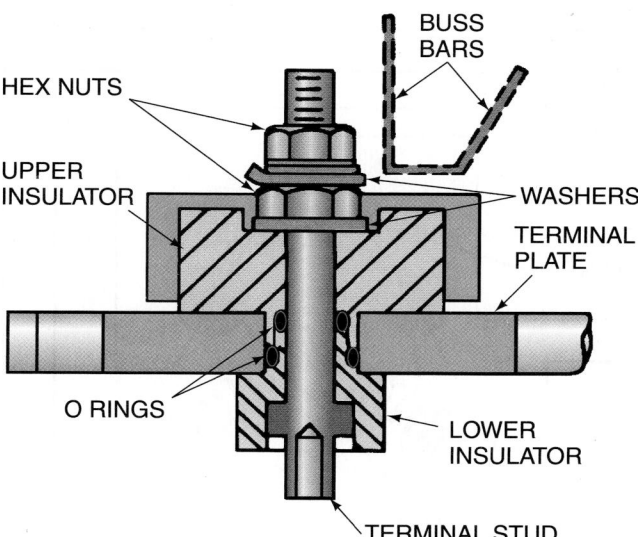

Figure 23–35 Motor terminals. The power to operate the compressor is carried through the compressor shell but must be insulated from the shell. This is a fiber block used as the insulator. O rings keep the refrigerant in. *Courtesy Trane Company*

NOTE: Care must be taken with these terminals (due to loose electrical connections) to prevent overheating. Should the terminal overheat, a leak could occur.

If the terminal block is a fused-glass type, it would be hard to repair. The fused-glass type will stand more heat, but there is a limit to how much heat it can take. Less heat can be tolerated with the O ring and fiber type of terminal board. When the O ring and fiber board are damaged, they can be replaced with new parts. However, refrigerant loss can result before the problem is discovered.

INTERNAL MOTOR PROTECTION DEVICES. Internal overload protection devices in hermetic motors protect the motor from overheating. These devices are embedded in or near the windings and are wired in two different ways. One style breaks the line circuit inside the compressor. Because it is internal and carries the line current, it is limited to smaller compressors. It has to be enclosed to prevent the electrical arc from affecting the refrigerant, **Figure 23–36(A).** If a contact in this line type remains open, the compressor cannot be restarted. The compressor would have to be replaced. **Figure 23–36(B)** illustrates a three-phase internal thermal overload. All three of the overload contacts will open at the same time if a motor overheating problem occurs. This prevents single phasing of the three-phase motor. This overload device is located in the compressor's shell, usually in the motor barrel compartment over the motor windings. **Figure 23–36(C)** is a photo of a three-phase overload protector taken out of a working three-phase motor. **Figure 23–36(D)**

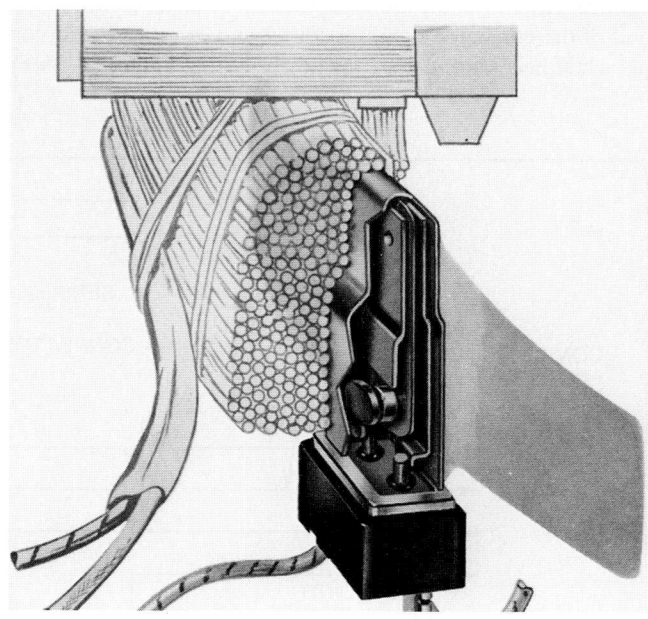

Figure 23–36(A) An internal compressor overload protection device that breaks the line circuit. Because this set of electrical contacts is inside the refrigerant atmosphere, they are contained inside a hermetic container of their own. If the electrical arc were allowed inside the refrigerant atmosphere, the refrigerant would deteriorate in the vicinity of the arc. *Courtesy Tecumseh Products Company*

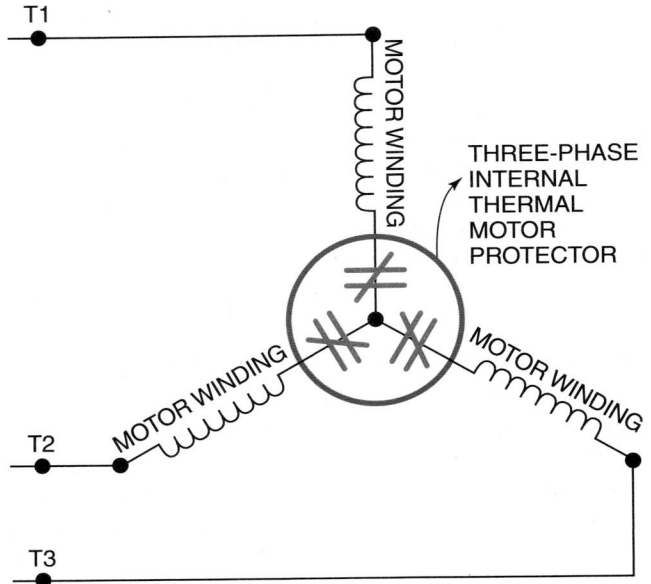

Figure 23–36(B)   A three-phase wye wound motor with internal thermal overload protection. All three internal overload contacts will open at the same time to prevent single phasing of the three-phase motor.

shows a three-phase motor with internal thermistor protection. Thermistors are resistors that vary their resistance according to their temperature. If a motor overheating problem occurs, the thermistors will sense the heat, change resistance, and relay this change in resistance to a solid-state motor protection module. The thermistor circuit is usually a 24-V control circuit. The motor protection module controls the coil of the contactor or motor starter and can turn the motor off when necessary for a cool-down period. **Figure 23–36(E)**

Figure 23–36(C)   A three-phase internal overload protector. *Courtesy Ferris State University. Photo by John Tomczyk*

is a wiring diagram of a typical electronic motor protection package showing the motor contactor and electronic protection module.

Another type of motor overload protection device breaks the control circuit. This is wired from the outside of the compressor to the control circuit. If the pilot-duty type wired to the outside of the compressor were to remain open, an external overload device could be substituted.

THE SERVICEABLE HERMETIC COMPRESSOR.   The serviceable hermetic normally has a cast-iron shell and is considered a low-side device, **Figure 23–37**. Because of the piping arrangement in the head, the discharge gas is contained either under the head or out the discharge line. The motor is rigidly

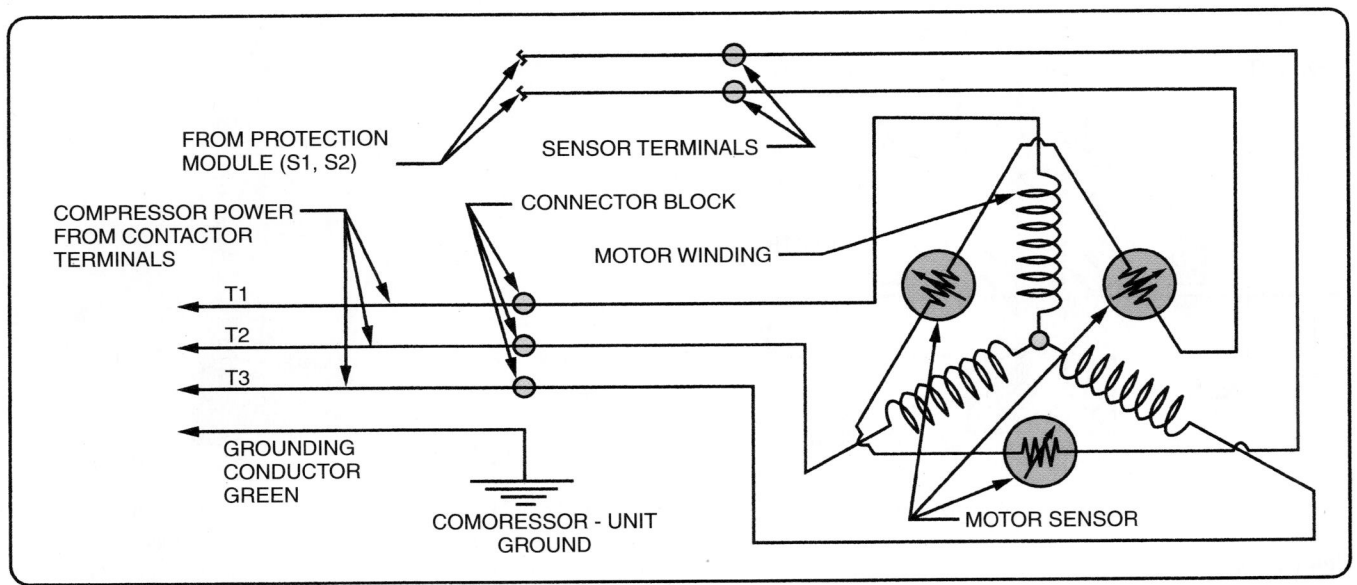

Figure 23–36(D)   A three-phase wye wound motor with internal thermistor protection linked to a solid-state electronic motor protection module. The motor sensors are themistors. *Courtesy Tecumseh Products Company*

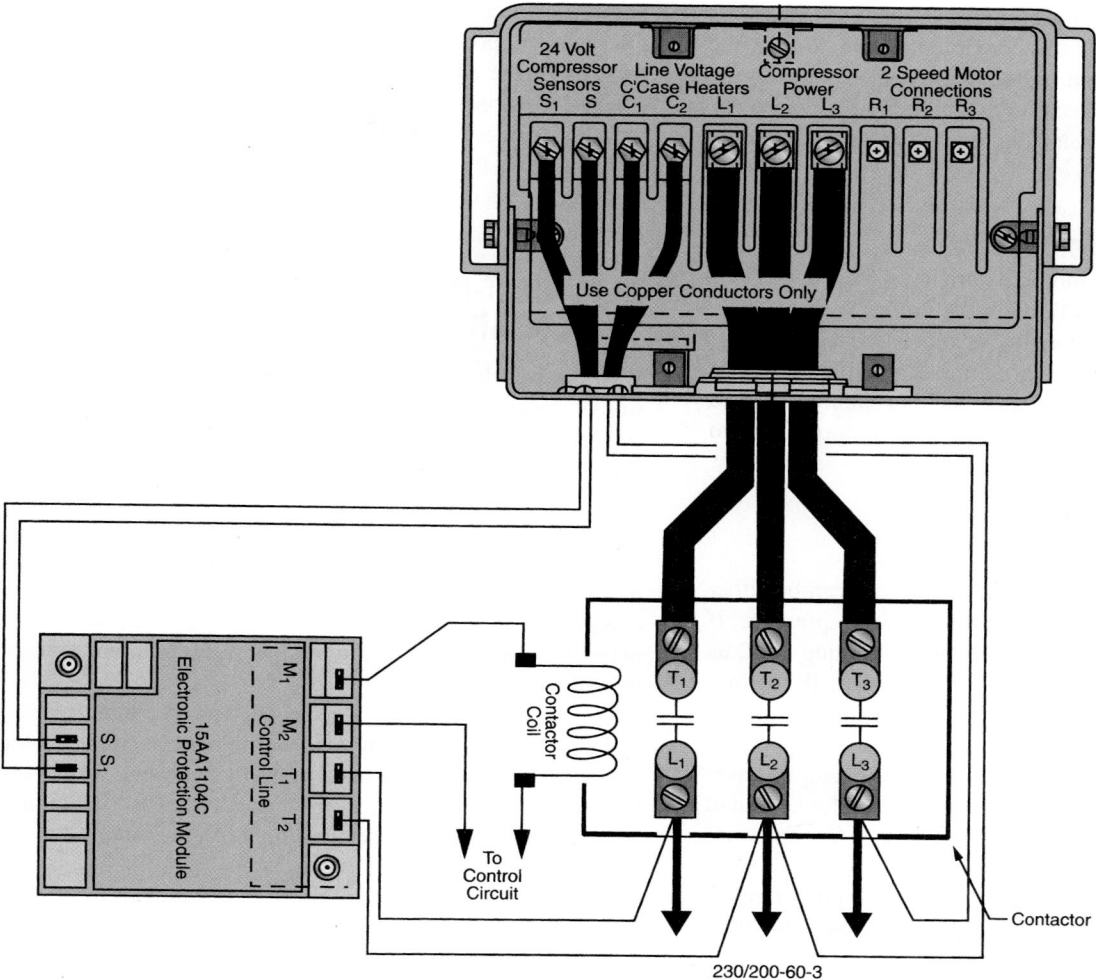

Figure 23–36(E) The wiring diagram of a typical electronic three-phase motor protection package. *Courtesy Tecumseh Products Company*

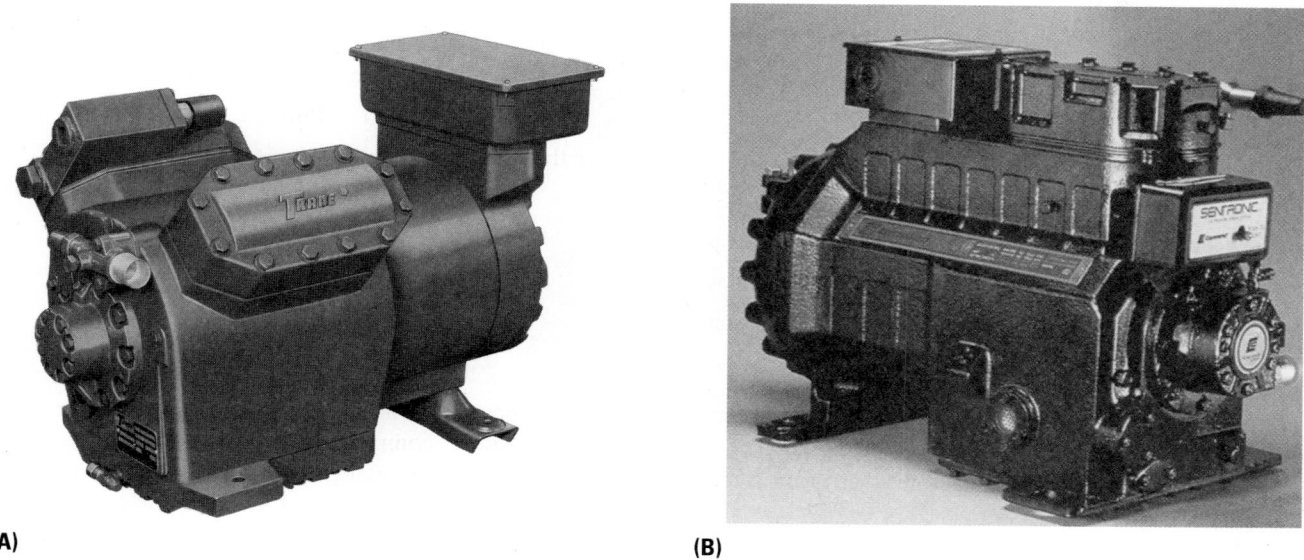

(A)

(B)

Figure 23–37 Serviceable hermetic compressors. *(A) Courtesy Trane Company. (B) Courtesy Copeland Corporation*

mounted to the shell, and the compressor must be externally mounted on springs or other flexible mounts to prevent vibration. The serviceable hermetic is used exclusively in larger compressor sizes because it can be rebuilt. The compressor components are much the same as the components in the welded hermetic type.

OPEN-DRIVE COMPRESSOR. Open-drive compressors are manufactured with the motor external to the compressor shell. The shaft protrudes through the casing to the outside where either a pulley or a coupling is attached. This compressor is normally heavy duty in nature. It must be mounted tightly to a foundation. The motor is either mounted end to end with the compressor shaft or beside the compressor and belts used to turn the compressor.

The Shaft Seal

The pressure inside the compressor crankcase can be either in a vacuum (below atmosphere) or a positive pressure. If the unit were an extra-low-temperature unit using R-12 as the refrigerant, the crankcase pressure could easily be in a vacuum. If the shaft seal were to leak, the atmosphere would enter the crankcase. When the compressor is setting still, it could have a high positive pressure on it. For example, when R-502 or R-404A is used and the system is off for extended periods (the whole system may get up to 100°F in a hot climate), the crankcase pressure may go over 200 psig. The crankcase shaft seal must be able to hold refrigerant inside the compressor under all of these conditions and while the shaft is turning at high speed, **Figure 23–38.**

The shaft seal has a rubbing surface to keep the refrigerant and the atmosphere separated. This surface is normally a carbon or ceramic material rubbing against a steel surface.

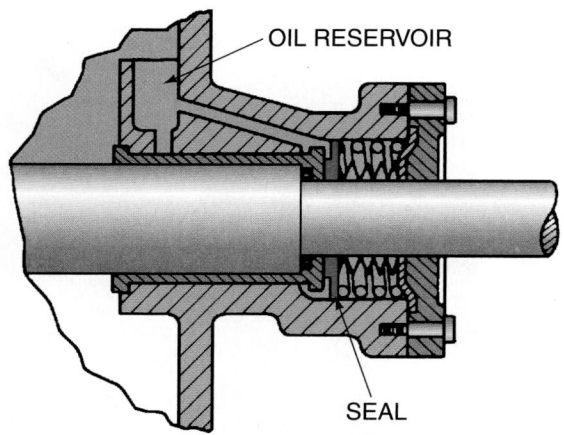

OIL RESERVOIR

SEAL

Figure 23–38 The shaft seal is responsible for keeping the refrigerant inside the crankcase and allowing the shaft to turn at high speed. This seal must be installed correctly. If the seal is installed on a belt-drive compressor, the belt tension is important. If it is installed on a direct-drive compressor, the shaft alignment is important.
Reproduced courtesy of Carrier Corporation

If assembled correctly, these two surfaces can rub together for years and not wear out. This correct assembly normally consists of the shafts being aligned correctly on a system where the bearings are in good working order. The belts must have the correct tension if the unit is a belt drive. If the unit is direct drive, the shafts have to be aligned according to the manufacturer's instructions.

23.4 BELT-DRIVE MECHANISM CHARACTERISTICS

With the belt-drive compressor, the motor is mounted at the side of the compressor and has a pulley on the motor as well as on the compressor. The pulley on the motor is called the *drive pulley,* and the pulley on the compressor is called the *driven pulley.* The drive pulley is sometimes adjustable, which allows the compressor speed to be adjustable. The drive pulley can also be changed to a different size to vary the compressor speed. This can be advantageous when a compressor is too large for a job (too much capacity) and needs to be slowed down to compensate. Different pulleys are shown in **Figure 23–39.**

The compressor can also be speeded up for more capacity if the motor has enough horsepower in reserve (if the motor is not already running at maximum horsepower). If the pulley size change is needed, consult the compressor manufacturer to be sure that the design limits of the compressor are not exceeded.

The formula for determining the size of the drive pulley is determined from information gathered from the existing system. The motor speed is normally fixed, 1725 or 3450 rpm, and the pulley (flywheel) on the compressor is normally a fixed size. The only variable is the drive pulley. A typical problem may be a motor with an rpm of 1725 and a drive pulley of 4 in. is used to drive a compressor at 575 rpm with a pulley size of 12 in. It is desirable to reduce the compressor speed to 500 rpm to reduce the capacity. What would the new drive pulley size be? The formula is

Drive pulley size × Drive pulley rpm =
Driven pulley size × Driven pulley rpm, or

Pulley 1 × rpm 1 = Pulley 2 × rpm 2

Solve the problem for pulley 1, restated as

$$\text{Pulley 1} = \frac{\text{Pulley 2} \times \text{rpm 2}}{\text{rpm 1}}$$

$$\text{Pulley 1} = \frac{12 \text{ in.} \times 500 \text{ rpm}}{1725 \text{ rpm}}$$

Pulley 1 = 3.48 or 3 1/2 in.

Most compressors are not designed to operate over a certain rpm, and the compressor manufacturer has this information. When a particular pulley size is needed, the pulley supplier will help you choose the correct size. Belt sizes also must be calculated. Choosing the correct belts and pulleys is important and should be done by an experienced person.

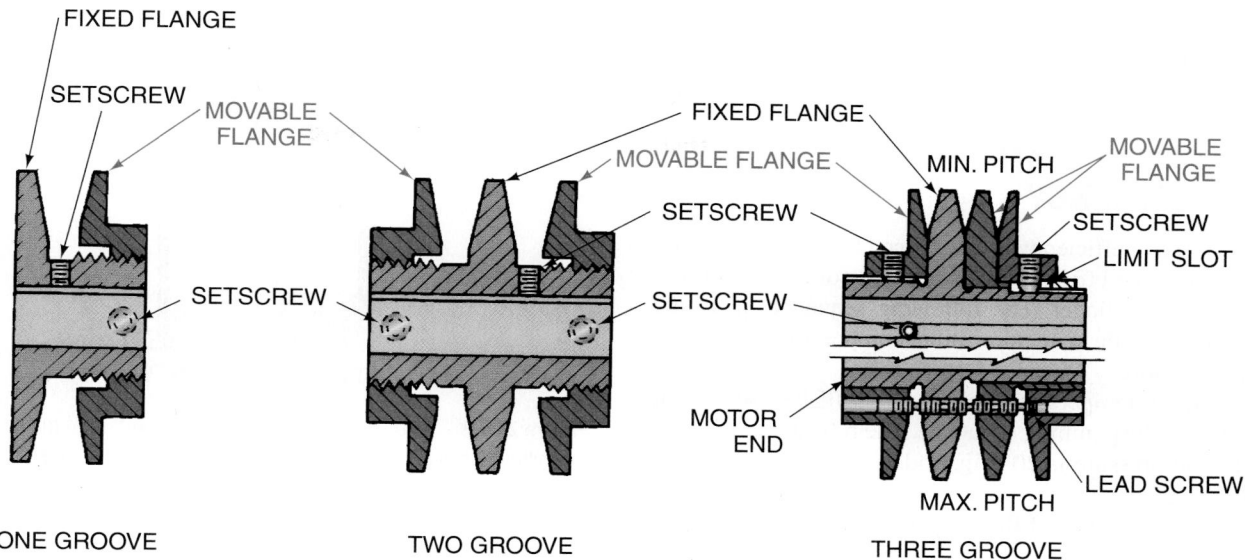

FIXED FLANGE
SETSCREW
MOVABLE FLANGE
SETSCREW
SETSCREW

ONE GROOVE

FIXED FLANGE
MOVABLE FLANGE
SETSCREW
SETSCREW

TWO GROOVE

MIN. PITCH
MOVABLE FLANGE
SETSCREW
LIMIT SLOT
MOTOR END
LEAD SCREW
MAX. PITCH

THREE GROOVE

Figure 23–39 Different pulleys used on belt-drive compressors. They are single-groove and multiple-groove pulleys with different belt widths. Some pulleys are adjustable for changing the compressor speed.

The compressor and motor shafts must be in correct alignment with the proper tension applied to the belts. The motor base and the compressor base must be tightened rigidly so that no dimensions will vary during the operation. Several belt combinations may have to be considered. The compressor drive mechanism may have multiple belts, **Figure 23–40**, or a single belt if the compressor is small. Belts also come in different types. The width of the belt, the grip type, and the material have to be considered.

When multiple belts are used, they have to be bought as matched sets. For example, if a particular compressor and motor drive has four V belts, which are of B width and are 88 in. long, you should order four 88-in. belts of B width that are factory matched for the exact length. These are called *matched belts* and have to be used on multiple-belt installations.

23.5 DIRECT-DRIVE COMPRESSOR CHARACTERISTICS

The direct-drive compressor is limited to the motor speed that the drive motor is turning. In this type of installation the motor and compressor shafts are end to end. These shafts have a slightly flexible coupling between them but must be in very close alignment, or the bearings and seal will fail prematurely, **Figure 23–41**. This compressor and motor combination is mounted on a common rigid base.

Both the motor and compressor are customarily manufactured so that they can be rebuilt in place. Thus, once the shafts are aligned, the shell of the motor and the shell of the compressor can be fastened down and always remain in place. If the motor or compressor has to be rebuilt, the internal parts may be removed for rebuilding, then the shafts will automatically line back up when reassembled.

The reciprocating compressor has not changed appreciably for many years. Manufacturers continuously try to improve

Figure 23–40 A multiple-belt compressor. *Courtesy Tecumseh Products Company*

Figure 23–41 A flexible coupling. An extensive procedure is used to obtain the correct shaft alignment. This must be done, or bearings and seal will fail prematurely. *Courtesy Lovejoy, Inc.*

the motor and the pumping efficiencies. The valve arrangements can make a difference in the pumping efficiency and are being studied for improvement.

23.6 RECIPROCATING COMPRESSOR EFFICIENCY

A compressor's efficiency is determined by the design of the compressor. The efficiency of a compressor starts with the filling of the cylinder. The following sequence of events takes place inside a reciprocating compressor during the pumping action.

A medium-temperature application will be used as an example for the pumping sequence. The refrigerant is R-12, the suction pressure is 20 psig, and the discharge pressure is 180 psig.

1. **Piston at the top of the stroke and starting down.** When the piston has moved down far enough to create less pressure in the cylinder than is in the suction line, the intake flapper valve will open and the cylinder will start to fill with gas, **Figure 23–42.**

2. **Piston continues to the bottom of the stroke.** At this point the cylinder is nearly as full as it is going to get. There is a very slight time lag at the bottom of the stroke as the crankshaft carries the rod around the bottom of the stroke, **Figure 23–43.**

3. **Piston is starting up.** The rod throw is past bottom dead-center, and the piston starts up. When the cylinder is as full as it is going to get, the suction flapper valve closes.

4. **The piston proceeds to the top of the stroke.** When the piston reaches a point that is nearly at the top, the pressure in the cylinder becomes greater than the pressure in the discharge line. If the discharge pressure

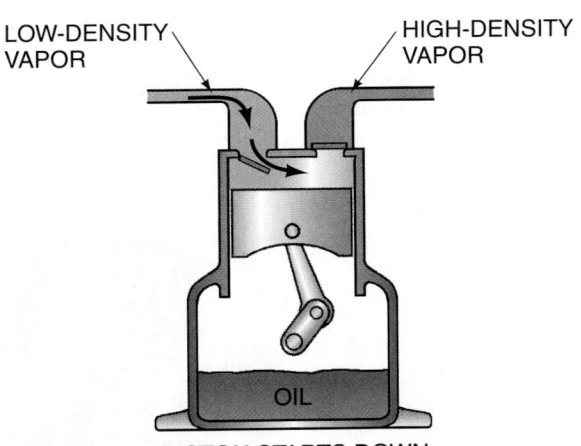

LOW-DENSITY VAPOR

HIGH-DENSITY VAPOR

OIL

PISTON STARTS DOWN

Figure 23–42   An illustration of what happens inside the reciprocating compressor while it is pumping. When the piston starts down, a low pressure is formed under the suction reed valve. When this pressure becomes less than the suction pressure and the valve spring tension, the cylinder will begin to fill. Gas will rush into the cylinder through the suction reed valve.

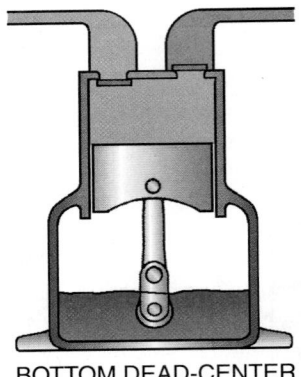

BOTTOM DEAD-CENTER

Figure 23–43   When the piston gets near the bottom of the stroke, the cylinder is nearly as full as it is going to get. There is a short time lag as the crankshaft circles through bottom dead-center, during which a small amount of gas can still flow into the cylinder.

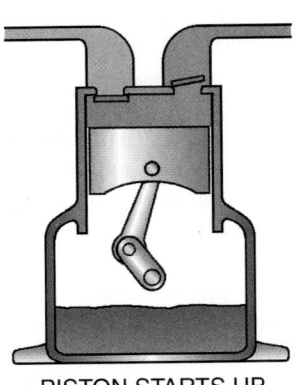

PISTON STARTS UP

Figure 23–44   When the piston starts back up and gets just off the bottom of the cylinder, the suction valve will have closed, and pressure will begin to build in the cylinder. When the piston gets close to the top of the cylinder, the pressure will start to approach the pressure in the discharge line. When the pressure inside the cylinder is greater than the pressure on the top side of the discharge reed valve, the valve will open, and the discharge gas will empty out into the high side of the system.

is 180 psig, the pressure inside the cylinder may have to reach 190 psig to overcome the discharge valve's weight and spring tension, **Figure 23–44.**

5. **The piston is at exactly top dead-center.** This is as close to the top of the head as it can go. There has to be a certain amount of clearance in the valve assemblies and between the piston and the head, or they would touch. This clearance is known as *clearance volume.* The piston is going to push as much gas out of the cylinder as time and clearance volume will allow. There will be a small amount of gas left in the clearance volume, **Figure 23–45.** This gas will be at the discharge pressure mentioned earlier. When the piston starts back down, this gas will reexpand, and the cylinder will not start to fill until the cylinder pressure

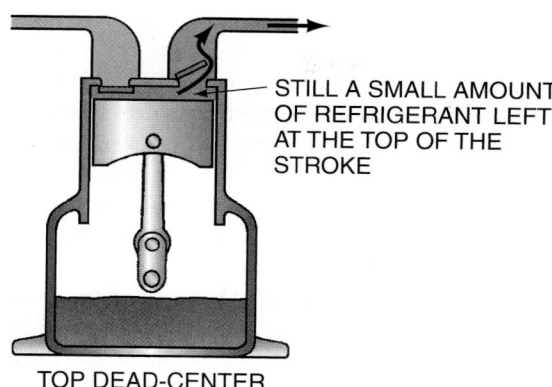

STILL A SMALL AMOUNT OF REFRIGERANT LEFT AT THE TOP OF THE STROKE

TOP DEAD-CENTER

Figure 23–45 A reciprocating compressor cylinder cannot completely empty because of the clearance volume at the top of the cylinder. The manufacturers try to keep this clearance volume to a minimum but cannot completely do away with it.

is lower than the suction pressure of 20 psig. This reexpanded refrigerant is part of the reason that the compressor is not 100% efficient. Valve design and the short period of time the cylinder has to fill at the bottom of the stroke are other reasons the compressor is not 100% efficient.

23.7 DISCUS VALVE DESIGN

The discus valve design allows a closer tolerance inside the compressor cylinder at top dead-center. This closer tolerance gives the compressor more efficiency because of less clearance volume. The discus valve also has a larger bore and allows more gas through the port within a short period of time. **Figure 23–46** shows a discus compressor, a conventional valve design, a discus valve design, and a cutaway view of a discus compressor. Discus compressors are refrigeration compressors used in supermarkets, walk-in coolers and freezers, and industrial applications. They range in horsepower from 5 to 60 hp.

Discus compressors are available with capacity modulation that ranges from 10% to 100%. This provides the ability to match capacity to the load desired for the refrigeration equipment. Capacity modulation reduces the suction pressure and temperature variation of the refrigerated space and provides for a decrease in the compressor's cycling rate. A reduced cycling rate increases the compressor's reliability. Modulation is accomplished with a blocked suction technology by feeding a variable voltage to a solenoid for its open and closed intervals.

23.8 NEW TECHNOLOGY IN COMPRESSORS

Discus compressor technology also offers onboard diagnostics. This technology provides real-time intelligence monitoring, which lets the service technician know the status of what is happening inside the compressor before any major

(A)

DISCHARGE VALVE

INTAKE VALVE

CONVENTIONAL VALVE DESIGN

DISCUS VALVE

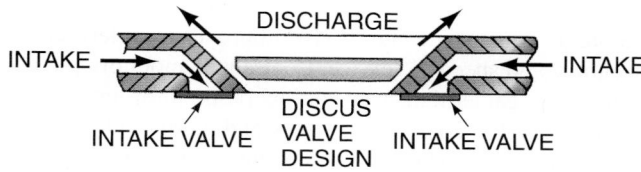

DISCHARGE

INTAKE → → ← ← INTAKE

INTAKE VALVE   DISCUS VALVE DESIGN   INTAKE VALVE

(B)

(C)

Figure 23–46 **(A)** A discus compressor. **(B)** A discus valve design. **(C)** The cutaway view of a discus compressor. *Courtesy Copeland Corporation*

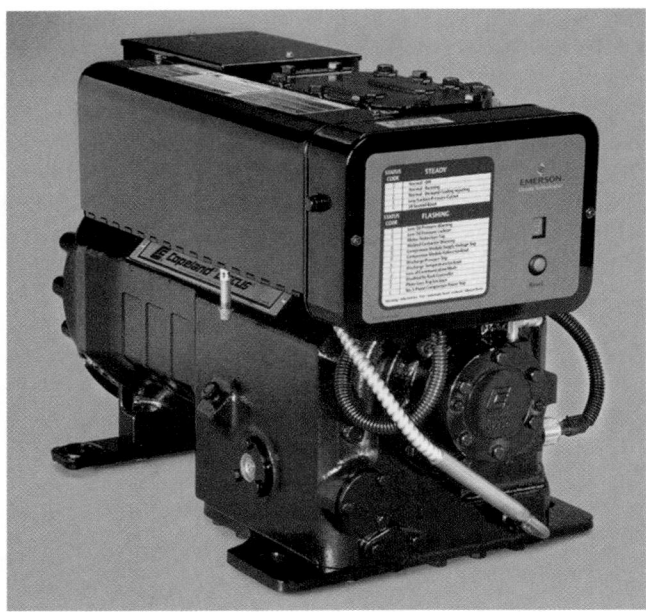

Figure 23–47 An Intelligent Store™ Discus compressor with integrated system electronics and onboard diagnostics. *Courtesy Emerson Climate Technologies*

problems develop. Today, compressor monitoring operations can be centralized. This technology allows service technicians to systematically troubleshoot compressor problems before arriving at the site. It also improves troubleshooting accuracy and speed of service. Centralized monitoring technology can gather data, transmit operating information, and visually display compressor status and alarm codes on the front control box. It also can record and retain a history of the compressor's operating information and past alarms. This technology allows service technicians to be dispatched automatically if an alarm problem exists. **Figure 23–47** shows a modern discus compressor that incorporates onboard diagnostics. This new technology has the following advantages:

- Monitors the compressor's discharge temperature
- Provides contactor protection
- Enables remote diagnostics
- Integrates the compressor's system electronics—including the high- and low-pressure controls, the cooling and temperature control, oil pressure monitoring, motor protection devices, and input/output (I/O) boards
- Reduces the number of brazed joints on the compressor that can develop leaks
- Guarantees consistent field installation because of less wiring requiring fewer components

The Scroll Compressor

The concept of compressing a gas by turning one scroll against another around a common axis isn't new. Scroll compressor technology has been around for 100 years, but it did not become commercially available and cost-effective until

the mid 1980s. Today, it is becoming even more fine-tuned and is available in many more applications. A major hurdle in making the scroll compressor a viable product has been the ability to achieve a balance between the need for high-volume precision manufacturing and the need for consistent high performance and efficiency, low sound levels, and great reliability. In fact, it is the scroll compressor's ability to reach higher levels of Seasonal Energy Efficiency Ratio (SEER) ratings that has helped guide the industry through government-mandated regulations. **Figure 23–48** shows a fifth-generation scroll compressor with its very few moving parts. In fact, every central split-cooling system manufactured in the United States today must have a SEER of at least 13. This energy requirement was mandated by federal law as of January 23, 2006. Also, with the phaseout of R-22 just around the corner, manufacturers of HVACR equipment are looking for energy-efficient methods to apply to their equipment in order to meet these new energy requirements.

SEER is calculated on the basis of the total amount of cooling (in Btu) the system will provide over the entire season, divided by the total number watt-hours it will consume. Higher SEER ratings reflect a more efficient cooling system. This federal mandate will impact 95% of the unitary market in the United States, which is about 8 million units at the time of this writing. To meet this new energy-efficiency requirement of 13 SEER, most air-conditioning and heat pump manufacturers are looking for more efficient evaporator and

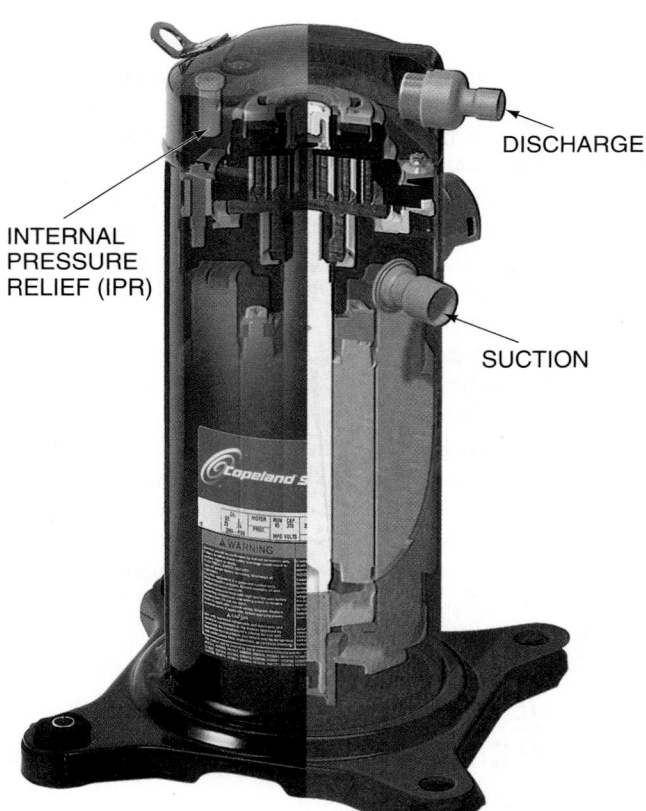

Figure 23–48 Fifth-generation Copeland Scroll Compressor showing very few moving parts. *Courtesy Emerson Climate Technologies*

condenser designs, more efficient compressors (like the scroll compressor), and more efficient fan motors—along with more sophisticated control systems. Refer to Unit 21, "Evaporators and the Refrigeration System," and Unit 22, "Condensers," for more detailed information on evaporator and condenser designs.

Because of the scroll compressor's fewer moving parts, it operates much more quietly than other compressor designs and technologies. Less noise and higher efficiencies have become two key selling points for installing the scroll compressor in residential air-conditioning units.

The next use for the scroll compressor was in commercial cooling equipment. A major development for commercial scroll compressors has been the advent of modulating technology. Modulating technology is able to maintain ideal comfort levels in buildings such as churches, where rooms can be empty for hours and then suddenly fill to capacity. Modulation technology operates in several ways that will be covered later in the unit. Some high-end systems now have communication systems onboard that enable components to talk to each other, **Figure 23–47.**

The next major application for the scroll compressor has been in refrigeration. The scroll compressor's reliability and durability has proved itself in refrigeration applications. It is used in walk-in coolers, reach-ins, and distributed-refrigeration systems. Beyond coolers and freezers, the scroll compressor has found its way to soft-serve ice cream machines, frozen carbonated-beverage machines, and other machine types. Scroll compressors are also being utilized in cryogenic equipment, magnetic resonance imaging (MRI) machines, and even some dental equipment.

Scroll Compressor Operation

Scroll compressor operation is relatively simple. Two spiral-shaped scrolls fit inside one another, **Figure 23–49.** The two mating parts are often referred to as *involute spirals.* One of the spiral-shaped parts stays stationary while the other orbits around the stationary member. The orbiting motion is created from the centers of the journal bearings and the motor being offset, **Figure 23–50.**

It is a true orbital motion, not a rotational motion. This orbiting motion causes continuous crescent-shaped gas pockets to be formed, **Figure 23–51.** The orbiting motion draws gas into the outer pocket and seals it as the orbiting continues. This continuous orbiting motion causes the crescent-shaped gas pocket to become smaller and smaller in volume as it nears the center of the scroll form. The gas pocket is fully compressed and is discharged out of a port of the nonorbiting (fixed) scroll member, **Figure 23–52.**

Several crescent-shaped gas pockets are compressed at the same time, which provides for a smooth and continuous compression cycle. Thus, the scroll compressor conducts its intake, compression, and discharge phases simultaneously.

The scroll compressor always takes a fixed volume of gas at suction pressure and then decreases the same gas volume, which increases its pressure. In fact, during the discharge phase, the scroll compressor compresses the discharge gas

Figure 23–49 Two spiral-shaped scrolls fitting inside one another. *Courtesy Emerson Climate Technologies*

Figure 23–50 A scroll compressor's drive train showing center bearing and motor shaft offset. *Courtesy Ferris State University. Photo by John Tomczyk*

to a zero volume, eliminating any carryover of trapped discharge gas in a clearance volume that is characteristic of piston-type compressors. Because of this, the scroll compressor is often referred to as a "fixed compression ratio" compressor.

Unlike piston-type compressors, the scroll compressor has no reexpansion of discharge gas that can be trapped in a clearance volume. This is why piston-type compressors are often referred to as "variable compression ratio" compressors. Reexpansion of discharge gas contributes to low volumetric efficiencies in piston-like compressors.

1 GAS ENTERS AN OUTER OPENING AS ONE SCROLL ORBITS THE OTHER.

2 THE OPEN PASSAGE IS SEALED AS GAS IS DRAWN INTO THE COMPRESSION CHAMBER.

3 AS ONE SCROLL CONTINUES ORBITING, THE GAS IS COMPRESSED INTO AN INCREASINGLY SMALLER "POCKET."

4 GAS IS CONTINUALLY COMPRESSED TO THE CENTER OF THE SCROLLS, WHERE IT IS DISCHARGED THROUGH PRECISELY MACHINED PORTS AND RETURNED TO THE SYSTEM.

5 DURING ACTUAL OPERATION, ALL PASSAGES ARE IN VARIOUS STAGES OF COMPRESSION AT ALL TIMES, RESULTING IN NEAR-CONTINUOUS INTAKE AND DISCHARGE.

Figure 23–51 Continuous crescent-shaped gas pockets being formed in the scroll. *Courtesy Emerson Climate Technologies*

(A)

(B)

Figure 23–52 **(A)** Nonorbiting or "fixed" scroll member showing discharge port. **(B)** Top side of nonorbiting or "fixed" scroll member showing discharge valve. *Courtesy Ferris State University. Photos by John Tomczyk*

Scroll Compressor Advantages

There are several advantages to the ways in which the two scroll members interact and operate. When liquid refrigerant, oil, or small solid particles enter between the two scrolls, the mating scroll parts can actually move apart in a sideways direction. This is referred to as "radial" movement. This radial movement eliminates high-stress situations and allows for just the right amount of contact force between mating scroll surfaces.

This action allows the compressor to handle some liquid. When a liquid slug is experienced by the compressor, the scroll's mating parts will separate slightly and allow the pressurized gas to vent to suction pressure. This allows the liquid slug to be swept from the mating scroll surfaces to the suction pressure and be vaporized. A gurgling noise may be heard during this process. The compressor may even stop pumping briefly and then restart as the excess liquid is purged out of

the scrolls. Scroll compressors handle liquid better than other compressors, but still can require additional accessories like crankcase heaters and suction-line accumulators for added protection.

As the orbiting scroll orbits, centrifugal forces on the sides of the mating scrolls, along with some lubricating oil, form a seal that prevents gas-pocket leakage. This is often referred to as "flank sealing," a major contributor to the scroll's high efficiency. A small amount of lubricating oil is usually entrained in the suction gases, and along with the centrifugal forces, provides the flank sealing.

Tight up-and-down mating or sealing of each scroll's tips prevents any compressed gas-pocket leakage and adds to efficiency. This up-and-down sealing is often referred to as "axial" sealing. Some scroll manufacturers use tip seals for the axial seal, **Figure 23–53.**

Scroll-tip seals serve the same purpose as the piston rings in a reciprocating piston-type compressor. These tip seals

Figure 23–53 Scroll member showing a tip seal for axial sealing. *Courtesy Ferris State University. Photo by John Tomczyk*

ride on the surface of the opposite scroll and provide a seal so that gases cannot escape between mating scroll parts and the tips of the scrolls.

The scroll compressor requires no reed valves, so it does not have valve losses that contribute to inefficiencies as piston-type compressors do. As mentioned earlier, the scroll compressor causes no reexpansion of discharge gases, which can be trapped in a clearance volume and cause low volumetric efficiencies. This is why the scroll compressor has a very high capacity in high-compression-ratio applications.

A considerable distance separates the scroll compressor's suction and discharge ports or locations. This greatly reduces the transfer of heat between the suction and discharge gases. Because of this, the suction gases will see less heat transferred into them and will have a higher density. This will increase the mass flow rate of refrigerant through the scroll compressor.

Because of the scroll's continuous compression process and the fact that it has no reed valves to create valve noise, the scroll compressor produces very soft gas-pulsation noises and very little vibration as compared with piston-type compressors.

Scroll compressors have a check valve in the discharge chamber that prevents high-side gas from flowing into the low side when the compressor stops. This valve opens by pressure difference when the compressor restarts.

Finally, the scroll compressor's simplicity requires only the stationary and the orbiting scroll to compress gas. Piston-type compressors require about 15 parts to do the same task.

The main reasons scroll compressors are gaining popularity over piston-type compressors in energy efficiency, reliability, and quieter operation are summarized as follows:

- There are no volumetric losses through gas reexpansion as with piston-type compressors.
- The scroll compressor requires no reed valves, so it does not have valve losses that contribute to inefficiencies as piston-type compressors do.
- Separation of suction and discharge gases reduces heat-transfer losses.

- Centrifugal forces within the mating scrolls maintain nearly continuous compression and constant leak-free contact.
- Radial movement eliminates high-stress situations and allows for just the right amount of contact force between mating scroll surfaces. This action allows the compressor to handle some liquid.
- Scroll compressors maintain a continuous compression process and have no reed valves to create valve noise. This creates very soft gas-pulsation noises and very little vibration as compared with piston-type compressors.

Two-Step Capacity Control for Scroll Compressors

A cooling system that has a modulating-capacity-control ability within the compressor and a variable-speed blower motor for the conditioned air will deliver much tighter temperature control and a reduced humidity level within the conditioned space than will a standard cooling system. The overall comfort level for occupants will be much higher because the compressor will run longer at part load to reduce humidity levels and maintain very precise temperature levels.

A new technology in scroll compressor design allows the compressor to have a two-step capacity of either 67% or 100%. A modern, two-step scroll compressor is shown in **Figure 23–54**. An internal unloading mechanism in the scroll compressor opens a bypass port or vent at the end of the first compression pocket. This internal unloading mechanism is a direct-current (DC) solenoid controlled by the second stage of a conditioned space thermostat in either the heating or cooling modes. The DC solenoid, which is controlled by a rectified, external, 24-volt alternating-current (AC) signal initiated by the conditioned space thermostat, moves a slider that covers and uncovers the bypass ports or vents, **Figure 23–55(A and B)**. The compressed gas is then vented into the beginning of a suction pocket within the scroll. When the bypass or vent port is opened by deenergizing the DC solenoid, the effective displacement of the scroll is reduced to 67%. When the DC solenoid is energized by the second stage of the conditioned space thermostat, the bypass ports are blocked or closed and the scroll is at 100% capacity. Again, this opening and closing of the bypass ports or vents is controlled by an internal, electrically operated DC solenoid. The unloading and loading of the two-step scroll compressor is done while the compressor's motor is running—and without cycling the motor on and off.

The compressor's motor is a single-speed, high-efficiency motor that will continue to run while the scroll alternates between the two capacity steps. Whether in the high-capacity (100%) or low-capacity (67%) mode, the two-step scroll operates like a standard scroll compressor.

As mentioned earlier, the internal DC solenoid in the compressor that operates the internal unloading mechanism is energized by the second stage of a conditioned space thermostat. It is expected that the majority of run hours will be at low capacity (unloaded at 67%). It is in this mode that

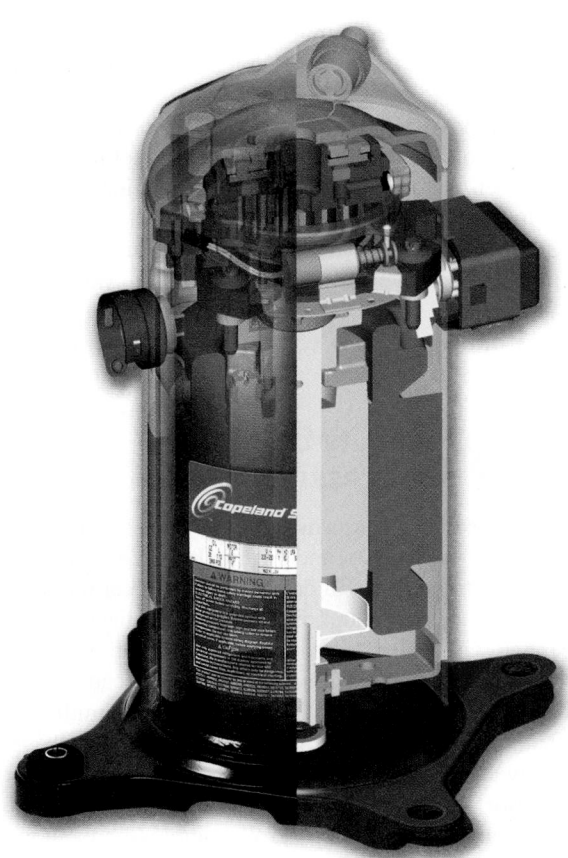

Figure 23–54 A modern, two-step modulating scroll compressor with either 100% or 67% capacity. *Courtesy Emerson Climate Technologies*

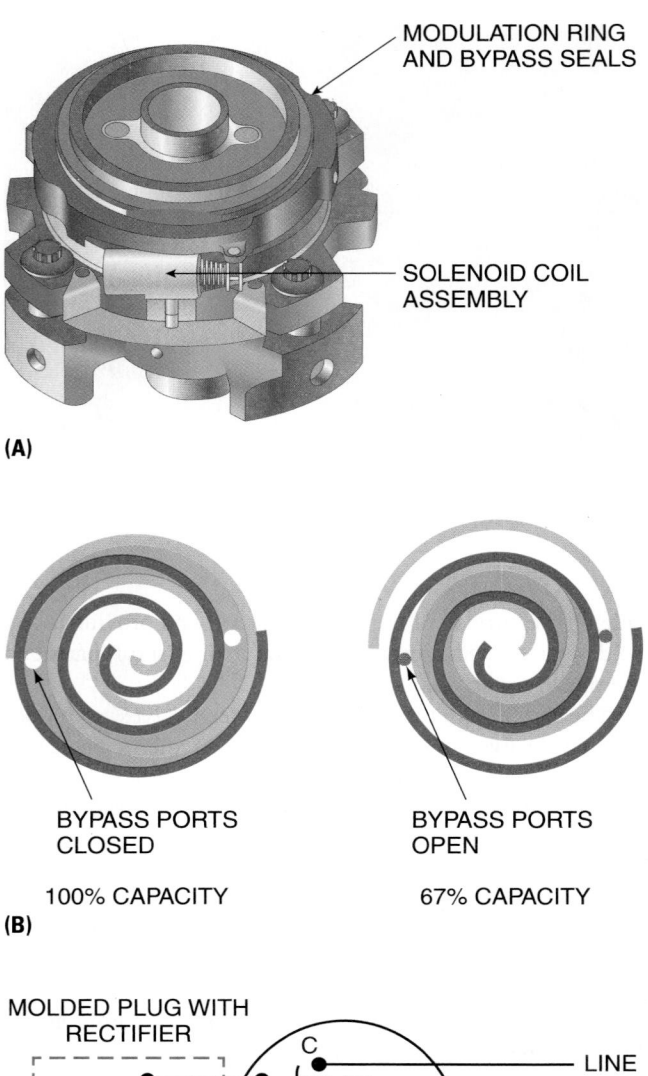

(A)

BYPASS PORTS
CLOSED

BYPASS PORTS
OPEN

100% CAPACITY

67% CAPACITY

(B)

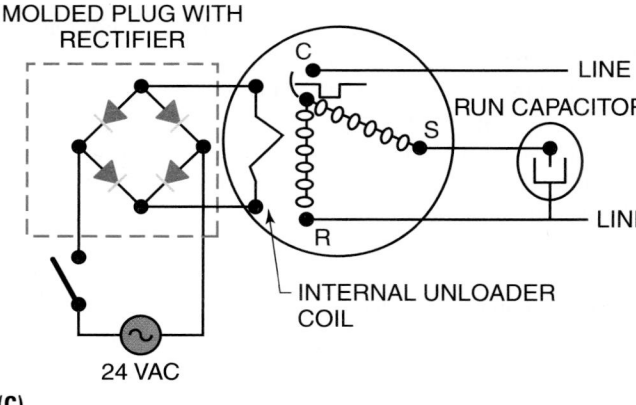

(C)

Figure 23–55 **(A)** Internal unloading mechanism showing modulating ring and solenoid coil assembly. **(B)** Bypass ports shown open and closed. **(C)** Electrical diagram of internal unloading and compressor motor circuit. *(A–C) Courtesy Emerson Climate Technologies*

the solenoid is deenergized. This allows the two-stage thermostat to control capacity through the second stage of the thermostat in both cooling and heating, **Figure 23–55(C).** An extra external electrical connection is made with a molded plug assembly that contains a full wave rectifier to supply direct current to the solenoid unloaded coil, **Figure 23–54.** The rectifier is actually located in the external power plug on the compressor.

Digital Capacity Control for Scroll Compressors

As mentioned in previous paragraphs, compressor capacity control is desirable for optimum system performance when loads vary over a wide range. Compressor capacity control through modulation can reduce power consumption, produce better dehumidification control, and reduce compressor cycling along with smaller compressor starting currents. This is especially important if you are trying to cool an area that has different cooling needs throughout the day. With a digital capacity scroll doing the cooling, constant temperatures of +/−0.5°F can be accomplished in every room, at any time of the day, whether the room is mostly empty or standing-room-only.

If the mating scroll members of a scroll compressor are separated axially, there will be no refrigerant gas compressed and only 10% power usage will be realized. If axial separation of the mating scrolls can be controlled by varying the amount of time they are separated, capacity control can be achieved at percentages between 10% and 100%. The separation of the mating scrolls is achieved by bypassing a controlled amount of discharge gas to the suction side of the compressor through a solenoid valve, **Figure 23–56**. When the pressure in the modulating chamber is lowered by energizing the solenoid valve, discharge gas will be metered

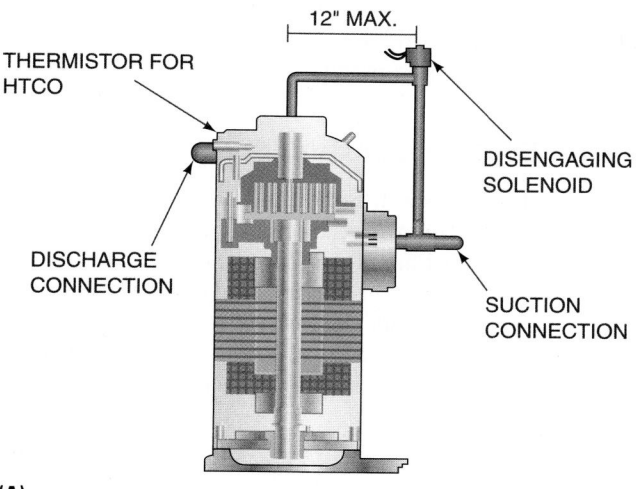

(A)

(B)

Figure 23–56 **(A)** Digital scroll compressor showing a disengaging solenoid valve for bypassing discharge gas to the suction side of the compressor. **(B)** Digital scroll with internal piping and enclosed solenoid. **(A)** and **(B)** *Courtesy Emerson Climate Technologies*

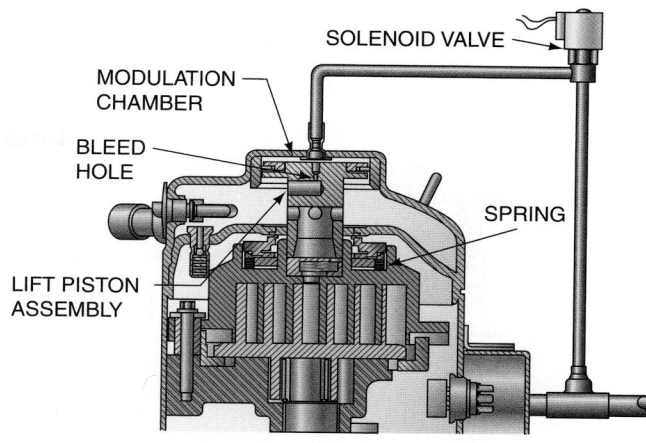

DIGITAL SCROLL COMPRESSOR

Figure 23–57 An internal view of the head of a digital scroll compressor showing modulation chamber, bleed hole, lift piston assembly, spring, and solenoid valve. *Courtesy Emerson Climate Technologies*

through a bleed hole and the scrolls will separate axially, **Figure 23–57**. No flow of refrigerant gas will take place when the mating scrolls are separated. This axial separation will cause the scroll compressor to be unloaded. The scroll compressor is shown in both the loaded and unloaded positions in **Figure 23–58**.

A modulating capacity is achieved by either energizing or deenergizing the solenoid valve. An energized solenoid will unload the compressor by axially separating the scrolls, making the compressor's capacity zero. When the solenoid valve is deenergized, the compressor's capacity is 100%. Solenoid cycle times of between 10 and 30 seconds should be used to minimize solenoid-valve cycling and to make the system more responsive. One complete "cycle time" is a combination of solenoid-valve energized (unloaded) time plus deenergized (loaded) time. It is suggested that the technician never deenergize the solenoid less than 10% of the cycle time; this will ensure that there is enough refrigerant gas flow for motor cooling, since the digital scroll is a refrigerant-cooled compressor.

Example: If you have a 30-second cycle time and the solenoid is deenergized for 20 seconds and then energized for 10 seconds, the resulting capacity will be

$$20 \div 30 = 66.6\%$$

Any normal control parameter (including the surrounding air temperature, humidity, or suction pressure) can unload the compressor. A compressor discharge-line thermistor is required with a cut-out temperature of 280°F. The solenoid valve must have 15 watts of power at the appropriate voltage.

A compressor controller that is used with the digital scroll compressor offers many protective features such as phase control, short-cycling control, amperage and voltage unbalance, and high-amperage monitoring, to name a few. The controller can also supply a variable voltage to the unloading solenoid valve for open/closed time intervals, **Figure 23–59**.

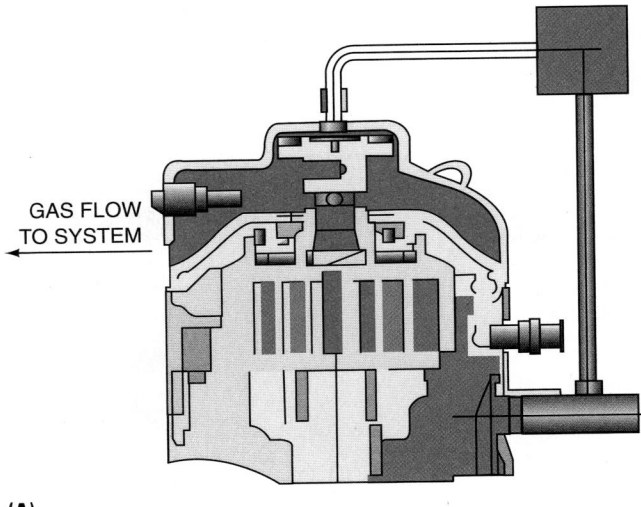

(A)

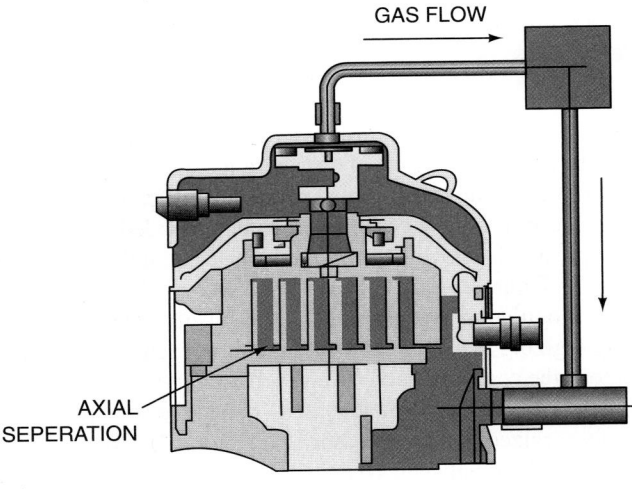

AXIAL
SEPERATION

(B)

Figure 23–58 **(A)** Digital scroll compressor shown in the "loaded" position. The solenoid is deenergized. **(B)** Digital scroll compressor shown in the "unloaded" position. **(A)** and **(B)** Courtesy Emerson Climate Technologies

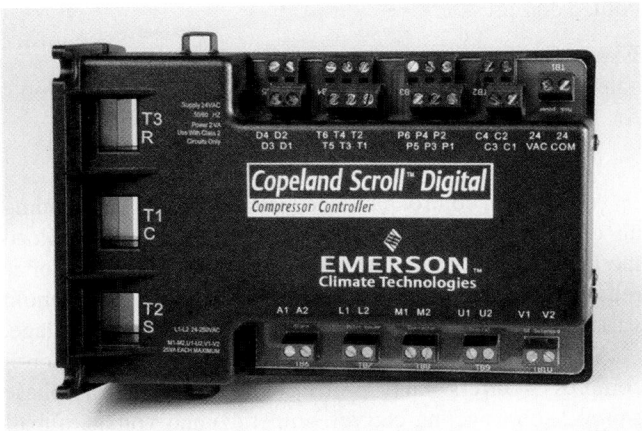

Figure 23–59 A controller used with a digital scroll compressor. Courtesy Emerson Climate Technologies

In summary, the digital scroll compressor can offer the following:

- Ability to hold a precise temperature and humidity level
- Efficient full-load and part-load efficiencies
- Ability to be 30% more efficient than traditional compressor modulation methods
- Less compressor cycling for longer compressor life
- Delivery of maximum comfort, efficiency, and reliability in one compressor

Scroll Compressor Protection

Many modern-day scroll compressors incorporate a variety of internal safety controls which can actuate the internal line-break motor protection, **Figure 23–60.** Some safety features that can be found in the air-conditioning scroll with less than 7 tons of capacity include the following:

- Temperature-operated disc (TOD)—A bimetallic disc that senses compressor discharge temperature and opens at 270°F.

SCROLL COMPRESSOR PROTECTION

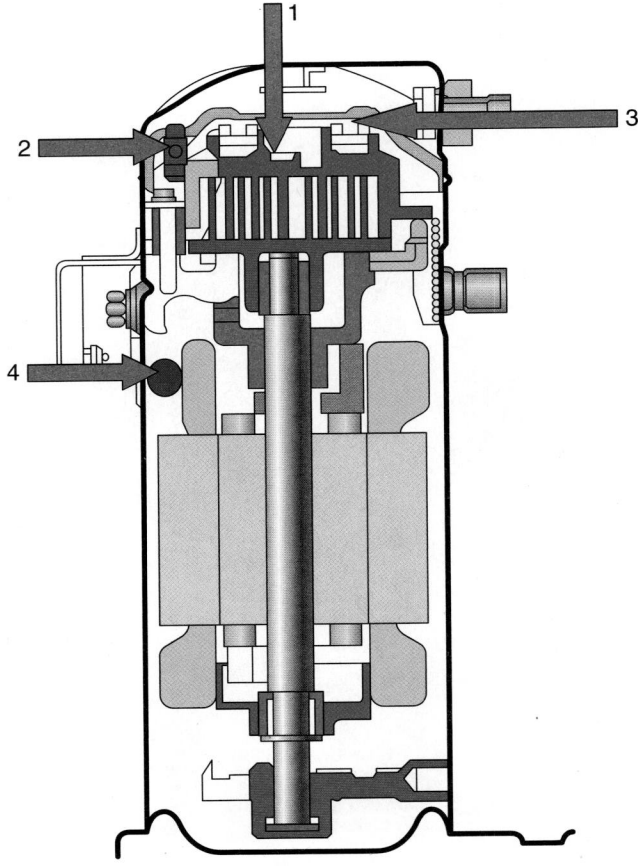

1 = TEMPERATURE OPERATING DISC
2 = INTERNAL PRESSURE RELIEF
3 = FLOATING SEAL
4 = MOTOR PROTECTOR

Figure 23–60 Variety of internal safety controls for a modern-day scroll compressor. Courtesy Emerson Climate Technologies

- Internal pressure relief (IPR)—Opens at an approximate 400 +/−50 psi differential between high- and low-side pressures for R-22, and at a 500 to 625 psi differential for R-410A.
- Floating seal—Separates the high side from the low side. Also prevents the compressor from drawing into a deep vacuum and damaging the Fusite electrical terminal.
- Internal motor protection—Senses both internal temperatures and amperages.

Another innovative scroll compressor protection device is used with compressors to ensure better reliability, diagnostic accuracy, speed of service, and a reduction in the number of callbacks when the technician is systematically trouble-shooting the compressor. The diagnostic controller is installed in the electrical box of a commercial condensing unit or residential unit, or inside a rooftop unit. It is completely self-contained and has no external sensors, **Figure 23–61**. In fact, it actually uses the compressor as a sensor because the compressor's electrical lines run through the device, which acts as a current transformer. It monitors vital information from the scroll compressor that will help pinpoint the root cause of cooling system problems. These common problems include electrical problems, compressor defects, and other general system faults. Phase dropouts, miswiring, and short cycling can also be detected with this device. A flashing light-emitting diode (LED) will quickly communicate the alert code and direct the service technician to the problem.

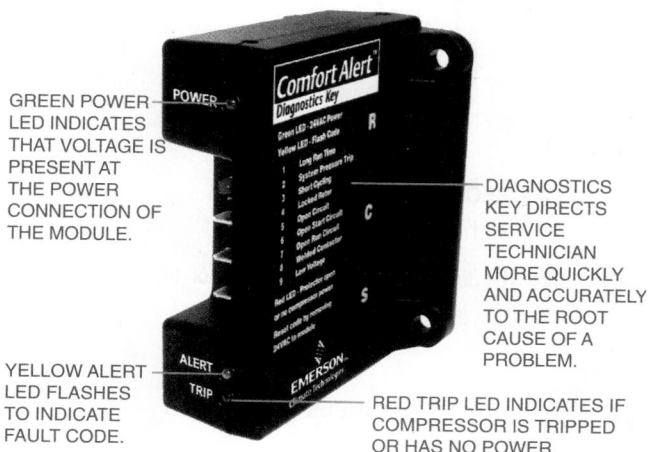

GREEN POWER LED INDICATES THAT VOLTAGE IS PRESENT AT THE POWER CONNECTION OF THE MODULE.

DIAGNOSTICS KEY DIRECTS SERVICE TECHNICIAN MORE QUICKLY AND ACCURATELY TO THE ROOT CAUSE OF A PROBLEM.

YELLOW ALERT LED FLASHES TO INDICATE FAULT CODE.

RED TRIP LED INDICATES IF COMPRESSOR IS TRIPPED OR HAS NO POWER.

Commercial Comfort Alert Diagnostics Codes

Alert code	System condition	Alert indicator blinks	Lockout
Code 2	System pressure trip	2 times	Yes
Code 3	Short-cycling	3 times	Yes
Code 4	Locked rotor	4 times	Yes
Code 5	Open circuit	5 times	No
Code 6	Missing phase	6 times	Yes
Code 7	Reverse phase	7 times	Yes
Code 8	Welded contactor	8 times	No
Code 9	Low voltage	9 times	No

Figure 23–61 Diagnostic controller with no external sensors. *Courtesy Emerson Climate Technologies*

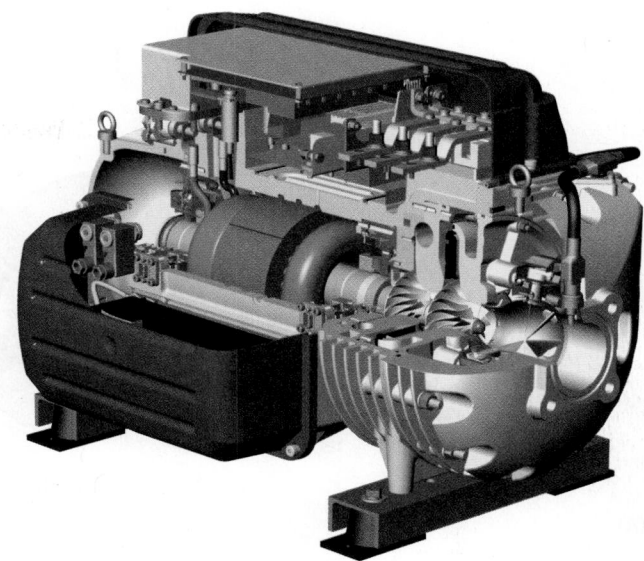

Figure 23–62 Oil-free centrifugal compressor with variable-frequency drive (VFD), onboard digital electronic controls, and magnetically levitated bearings. *Courtesy Danfoss Torbocor Compressors, Inc.*

Centrifugal Compressor

A revolutionary oil-free centrifugal compressor has been designed for the heating, ventilation, air-conditioning, and refrigeration (HVACR) industry. This centrifugal compressor has magnetic bearings, variable speeds, and digital electronic technology for achieving very high efficiencies, **Figure 23–62**.

The rotor shaft and impellers are the only moving parts in this compressor. The shaft acts as the rotor for the permanent-magnet synchronous motor. The rotor shaft has magnetically levitated bearings for frictionless operation, **Figure 23–63(A and B)**. The magnetic bearings and bearing sensors are composed of both permanent magnets and electromagnets. They precisely control the frictionless compressor shaft rotation on a levitated magnetic cushion, **Figure 23–63(B)**. Bearing sensors, located at each magnetic bearing, feed back rotor orbit information in real time to the bearing control system, **Figure 23–64**.

The permanent-magnet synchronous motor is powered by a pulse-width-modulated (PWM) voltage supply. High-speed variable-frequency-drive (VFD) operation allows for high efficiency, compactness, and soft-start capability. Soft starting reduces high in-rush currents at start-up.

The variable-frequency drive has an inverter that converts a DC voltage into an adjustable three-phase AC voltage. Signals from the motor/bearing controller determine the inverter output frequency, torque, and voltage phase—all of which regulate the motor speed. The compressor motor can reduce speed as the condensing temperatures or heat loads are reduced. This allows the variable-speed compressor to optimize energy performance from 100% to 20% of rated capacity.

MAGNETIC BEARINGS LEVITATE THE ROTOR SHAFT

(A)

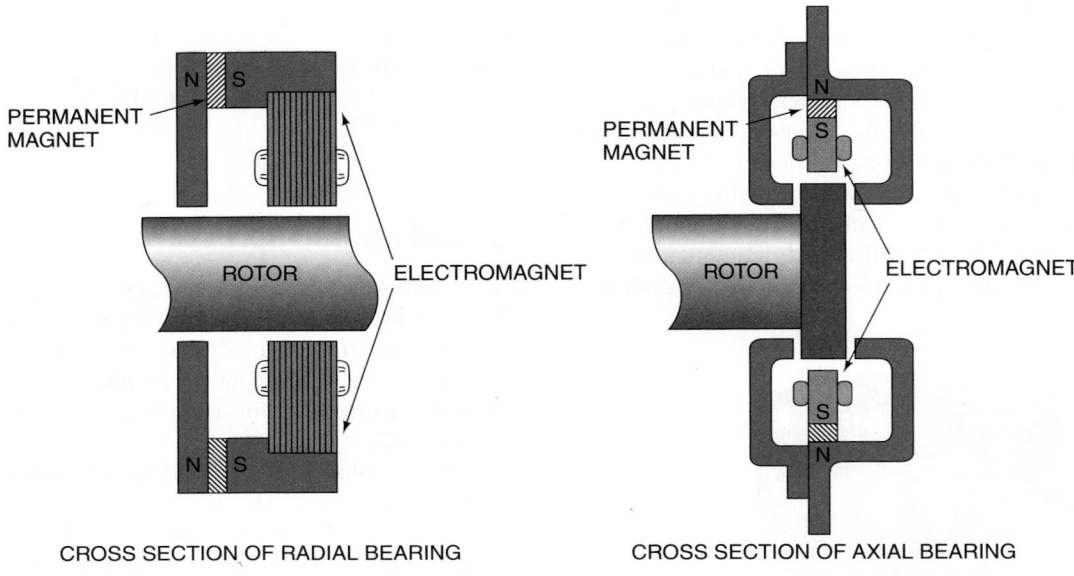

CROSS SECTION OF RADIAL BEARING   CROSS SECTION OF AXIAL BEARING

(B)

Figure 23–63   (A) Rotor shaft, centrifugal impellers, and magnetically levitated bearings. **(B)** Rotor shaft shown cushioned on magetically levitated bearings. *(A) and (B) Courtesy Danfoss Torbocor Compressors, Inc.*

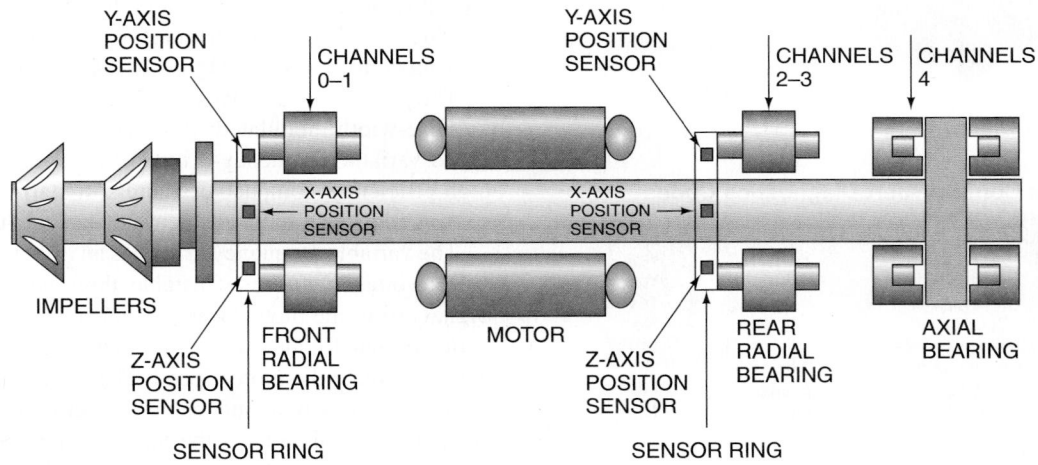

Figure 23–64   Bearing control system and sensors. *Courtesy Danfoss Torbocor Compressors, Inc.*

The compressor has onboard digital control electronics which oversee compressor operation. The electronic package also provides for Web-enabled monitoring access and external control. This provides valuable performance and reliability information. It also saves costs because it replaces the power and control panel functions that are seen on traditional chillers and rooftop packages.

23.9 LIQUID IN THE COMPRESSOR CYLINDER

The piston-type reciprocating compressor is known as a positive displacement device. This means that when the cylinder starts on the upstroke, it is going to empty itself or stall. SAFETY PRECAUTION: If the cylinder is filled with liquid refrigerant that does not compress (liquids are not compressible), something is going to break. Piston breakage, valve breakage, and rod breakage can all occur if a large amount of liquid reaches the cylinder, **Figure 23–65**.

Liquid in compressors can be a problem from more than one standpoint. Large amounts of liquid, called a *slug* of liquid or *liquid slugging,* usually cause immediate damage. Small amounts of liquid floodback can be just as detrimental but with a slower action. When small amounts of liquid enter the compressor, they can cause oil dilution. If the compressor has no oil lubrication protection, this may not be noticed until the compressor fails. One of the aspects of this failure is that marginal oil pressure may cause the compressor to throw a rod. Throwing a rod would be called mechanical failure. This may cause motor damage and burn out the motor. The technician may erroneously diagnose this as an electrical problem. If the compressor is a welded hermetic, the technician may have difficulty diagnosing the problem until another failure occurs. Many electrical failures are caused by mechanical failures. Following are some causes that may lead to *slugging* a compressor with liquid refrigerant, which in turn may lead to electrical failures because of a motor seizure.

- Metering device overfeeding
- Low air circulation over evaporator
- Low heat load
- Evaporator fan motor not working
- Dirty or iced-up evaporator coil
- Defrost heaters or defrost timer not working
- Dirty or blocked evaporator filters

NOTE: Never take a compressor failure for granted. Give the system a thorough checkout on start-up and actively look for a problem.

23.10 SYSTEM MAINTENANCE AND COMPRESSOR EFFICIENCY

The compressor's overall efficiency can be improved by maintaining the correct working conditions. This involves keeping the suction pressure as high as practical and the head pressure as low as practical within the design parameters.

A dirty evaporator will cause the suction pressure to drop. When the suction pressure goes below normal, the vapor that the compressor is pumping becomes less dense and gets thin, sometimes called rarified vapor. The compressor performance decreases. Low suction pressures also cause the high-pressure gases caught in the clearance volume of the compressor to expand more during the piston's downstroke. This gives the compressor a lower volumetric efficiency. These gases have to expand to a pressure just below the suction pressure before the suction valve will open. Because more of the downward stroke is used for reexpansion, less of the stroke can be used for suction. Suction ends when the piston reaches bottom dead-center.

In the past, not much attention has been given to dirty evaporators. Technicians are now beginning to realize that low evaporator pressures cause high compression ratios. For example, if an ice cream storage walk-in cooler is designed to operate at −5°F room temperature, it would result in a coil temperature of about −20°F (using a coil to return air temperature of 15°F). If the refrigerant is R-134a, the suction pressure would be 3.6 psig, slightly above atmospheric pressure. Assume the condenser will operate with a 25°F temperature difference between the outside air and the condensing temperature. If the outside air is 95°F, the condensing temperature would be 120°F (95°F + 25°F). The condensing pressure would then be 171.2 psig. The following calculation would show the compression ratio to be 10:1, **Figure 23–66(A)**.

Compression ratio = Absolute discharge pressure
÷ Absolute suction pressure

CR = [171.2 psig + 15 (atmosphere)]
÷ [3.6 psig + 15 (atmosphere)]

CR = 10.01:1

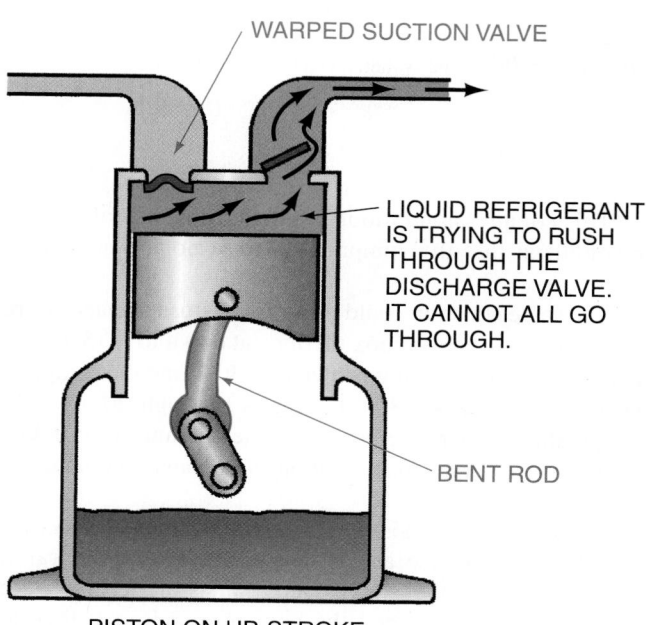

WARPED SUCTION VALVE

LIQUID REFRIGERANT IS TRYING TO RUSH THROUGH THE DISCHARGE VALVE. IT CANNOT ALL GO THROUGH.

BENT ROD

PISTON ON UP-STROKE

Figure 23–65 A cylinder trying to compress liquid. Something has to give.

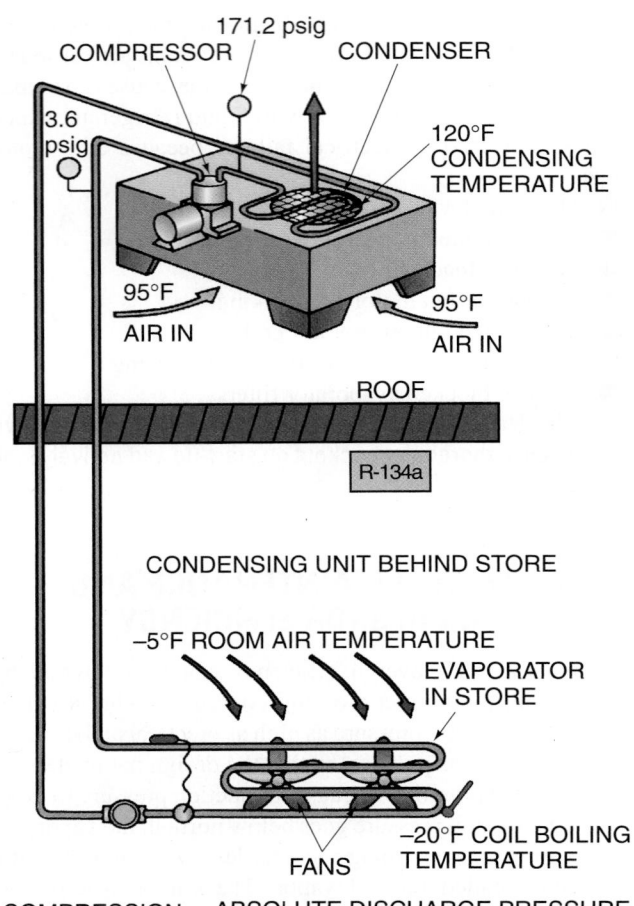

COMPRESSION RATIO $= \dfrac{\text{ABSOLUTE DISCHARGE PRESSURE}}{\text{ABSOLUTE SUCTION PRESSURE}}$

$CR = \dfrac{171.2 \text{ psig} + 15 \text{ (ATMOSPHERE)}}{3.6 \text{ Hg} + 15 \text{ (ATMOSPHERE)}}$

CR = 10.01 TO 1

Figure 23–66(A)   A system operating with a normal compression ratio. Notice that the condenser is large enough to keep the head pressure low.

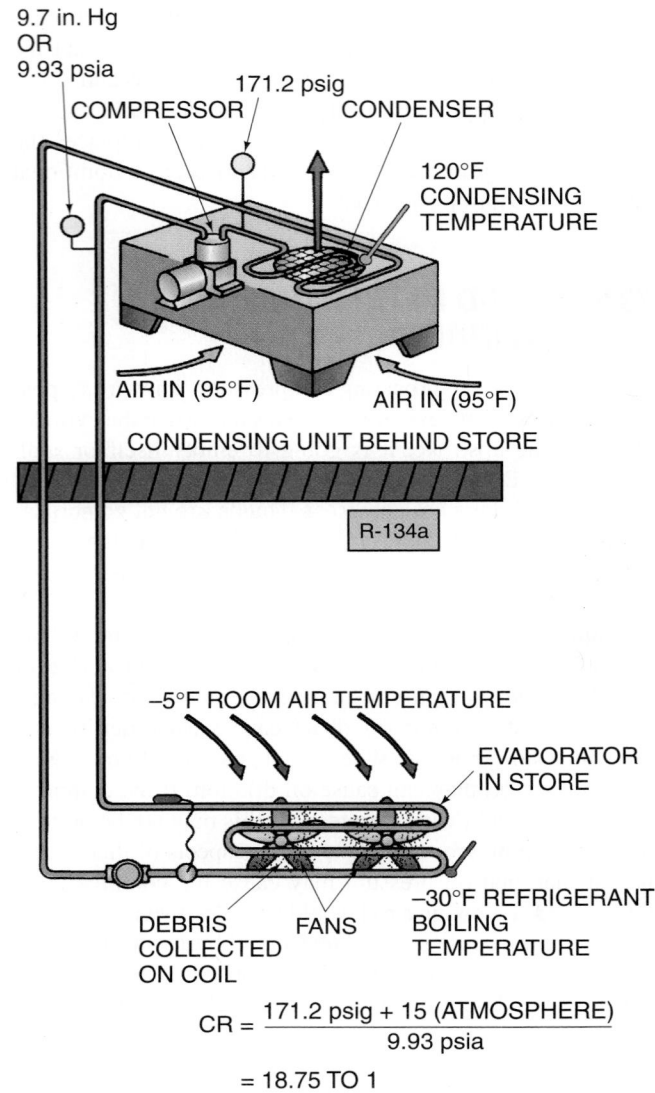

$CR = \dfrac{171.2 \text{ psig} + 15 \text{ (ATMOSPHERE)}}{9.93 \text{ psia}}$

= 18.75 TO 1

Figure 23–66(B)   This system has a dirty evaporator.

If the coil were old and dirty, **Figure 23–66(B),** and the coil temperature reduced to −30°F, assuming the head pressure remained the same, a new refrigerant boiling temperature of −30°F would produce a suction pressure of 9.7 in. Hg vacuum. This is below atmospheric pressure. It is an absolute pressure of 9.93 psia. (This was arrived at by converting inches of mercury vacuum to inches of mercury absolute, then converting this to psia.)

29.92 in. Hg absolute − 9.7 in. Hg vacuum
= 20.22 in. Hg absolute

20.22 in. Hg absolute ÷ 2.036 in. Hg per psi = 9.93 psia

The new compression ratio is 18.75:1

CR = [171.2 + 15 (atmosphere)] ÷ 9.93 psia

CR = 18.75:1

This compression ratio is too high. Most manufacturers recommend that the compression ratio not be more than 12:1.

The same situation could occur if the store manager were to turn the refrigerated box thermostat down to −15°F. The coil temperature would go down to −30°F and the compression ratio would be 18.75:1. This is too high. Many low-temperature coolers are operated at temperatures far below what is necessary. This condition often causes compressor problems.

Dirty condensers also cause compression ratios to rise, but they do not rise as fast as with a dirty evaporator. For example, if the above ice cream storage box were operated at 3.6 psig and the condenser became dirty to the point that the head pressure rose to 190 psig, the compression ratio would be 11.02:1, **Figure 23–66(C).** This is

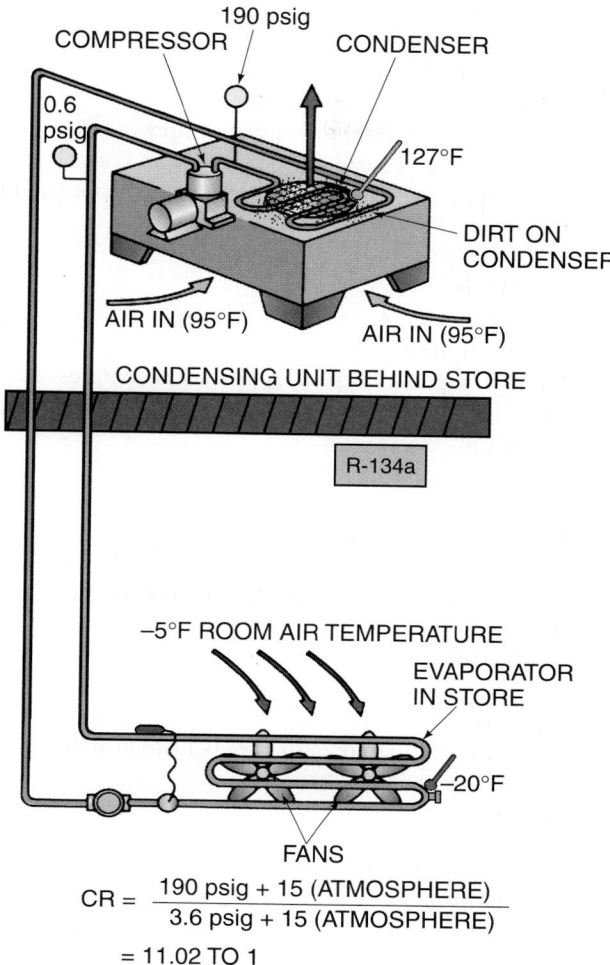

$$CR = \frac{190 \text{ psig} + 15 \text{ (ATMOSPHERE)}}{3.6 \text{ psig} + 15 \text{ (ATMOSPHERE)}}$$

$$= 11.02 \text{ TO } 1$$

Figure 23–66(C) A system with high head pressure because the condenser is dirty.

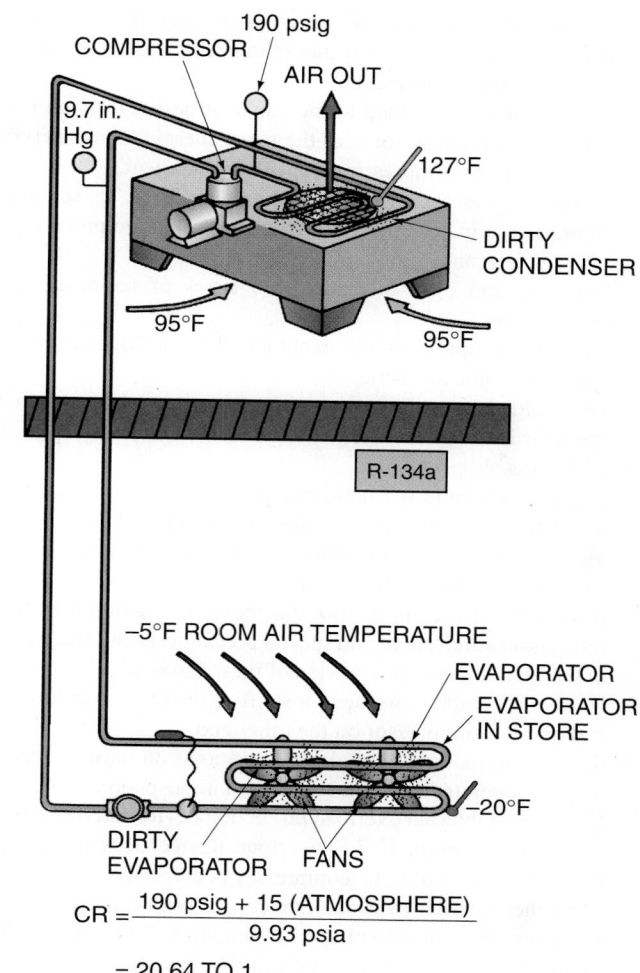

$$CR = \frac{190 \text{ psig} + 15 \text{ (ATMOSPHERE)}}{9.93 \text{ psia}}$$

$$= 20.64 \text{ TO } 1$$

Figure 23–66(D) This system has a dirty condenser and a dirty evaporator.

not good but not as bad as the 18.75:1 ratio in the previous example.

$$CR = [190 \text{ psig} + 15 \text{ (atmosphere)}]$$
$$\div [3.6 \text{ psig} + 15 \text{ (atmosphere)}]$$
$$CR = 11.02:1$$

If both situations were to occur at the same time, as in **Figure 23–66(D),** the compressor would have a compression ratio of 20.64:1. The compressor would probably not last very long under these conditions.

$$CR = [190 \text{ psig} + 15 \text{ (atmosphere)}] \div 9.93 \text{ psia}$$
$$CR = 20.64:1$$

The technician has a responsibility to maintain the equipment to obtain the greatest efficiency and to otherwise protect the equipment with a maintenance program. Clean coils are part of this program. The owner usually does not know the difference.

A dirty condenser makes the head pressure rise. This causes the amount of refrigerant in the clearance volume (at the top of the compressor cylinder) to be greater than the design conditions allow. This makes the compressor efficiency drop. If there is a dirty condenser (high head pressure) and a dirty evaporator (low suction pressure), the compressor will run longer to keep the refrigerated space at the design temperature. The overall efficiency drops. A customer with a lot of equipment may not be aware of this if only part of the equipment is not efficient.

When the efficiency of a compressor drops, the owner is paying more money for less refrigeration. A good maintenance program is economical in the long term.

SUMMARY

- The compressor is a vapor pump.
- The compressor lifts the low-pressure gas from the suction side of the system to the discharge side of the system.

■ The discharge gas can be quite hot because the heat contained in the cool suction gas is concentrated when compressed in the compressor.

■ Additional heat is added to the gas as it passes through the compressor because some of the work energy does not convert directly to compression but converts to heat.

■ Three types of compressors are usually used to achieve compression in commercial refrigeration: the reciprocating, rotary screw, and scroll compressor.

■ Hermetic and open-drive are two types of reciprocating compressors.

■ Hermetic compressors are manufactured as welded hermetic and serviceable hermetic.

■ The welded hermetic compressor must have the shell cut open for it to be serviced, and this work is done by special rebuilding shops only.

■ In the serviceable hermetic compressor the shell is bolted together and can be disassembled in the field.

■ Most reciprocating compressor motors are cooled by suction gas. Some are air cooled.

■ In all hermetic compressors the motor is operating in the refrigerant atmosphere and special precautions must be taken in the manufacture and service of these motors.

■ Hermetic compressors have a shaft with the motor on one end and the compressor on the other end.

■ Special internal overload devices are used on hermetic motors that operate inside the refrigerant atmosphere.

■ One type of internal overload protection device interrupts the actual line current. If this overload device does not close back when it should, the compressor is defective.

■ The other internal overload device is a pilot-duty type that interrupts the control voltage. If something were to happen to this device, an external type could be installed.

■ Some compressors use thermistors in or near the motor windings to sense motor heat. The thermistors are then wired to a solid-state electronic compressor protection module, which will decide when the compressor should be cycled off because of excessive heat.

■ Reciprocating compressors are positive displacement pumps, meaning that when they have a cylinder full of gas or liquid, the cylinder is going to be emptied, or damage will occur.

■ With open compressors the motor is on the outside of the system.

■ The motor can be mounted beside the compressor, the compressor and motor shafts may be side by side, or the motor may be mounted at the end of the compressor shaft with a flexible coupling between them.

■ Shaft-to-motor alignment is very important.

■ Belts for belt-drive applications come in different types. The manufacturer's supplier should be consulted for advice.

■ Discus valve design provides for smaller clearance volume, a greater area through which the gas can flow, and consequently greater efficiency.

■ Continued compressor efficiencies depend to a great extent on a good maintenance program.

REVIEW QUESTIONS

1. Describe the operation of a compressor.
2. Name five types of compressors.
3. True or False: A compressor can compress a liquid.
4. What would be considered a higher-than-normal temperature for a discharge gas line on a reciprocating compressor?
   A. 190°F
   B. 215°F
   C. 250°F
   D. All of the above.
5. Which compressor type uses pistons to compress the gas?
   A. Scroll
   B. Reciprocating
   C. Rotary
   D. Screw
6. Which compressor type uses tapered, machined gear components to trap the gas for compression?
   A. Scroll
   B. Reciprocating
   C. Rotary
   D. Screw
7. Which style of compressor uses belts to turn the compressor?
   A. Open drive
   B. Closed drive
   C. Chassis drive
   D. Hermetic drive
8. What would be done to increase the compressor speed on a belt-driven compressor?
   A. Decrease drive pulley size
   B. Increase drive pulley size
   C. Increase driven pulley size
   D. Both A and C
9. Describe the reciprocating compressor piston, rod, crankshaft, valves, valve plate, head, shaft seal, internal motor overload device, pilot-duty motor overload device, and coupling.
10. Name two things in the design of a reciprocating compressor that control the efficiency of the compressor.
    A. _____
    B. _____
11. At what speeds in rpm does a hermetic compressor normally turn?
    A. 1750
    B. 3450
    C. 4500
    D. Both A and B.
12. What effect does a slight amount of liquid refrigerant have on a compressor over a long period of time?

13. What lubricates the refrigeration compressor?
  A. Refrigerant
  B. Moisture
  C. Desiccants
  D. Oil

14. What can the service technician do to keep the refrigeration system operating at peak efficiency?

15. List the main advantage that a discus compressor's valve plate design has over a conventional piston-type compressor.

16. Briefly describe how the scroll compressor compresses refrigerant gas.

17. List four main advantages a scroll compressor has over a conventional piston-type compressor.

18. What are two main reasons why a compressor should have modulating capacity control?

19. Briefly describe how the digital capacity control for a scroll compressor works.

20. List the three main advantages of having digital capacity on a scroll compressor.

21. Explain how a compressor can be oil free.

UNIT 24
Expansion Devices

OBJECTIVES

After studying this unit, you should be able to

- describe the three most popular types of expansion devices.
- describe the operating characteristics of the three most popular expansion devices.
- describe how the three expansion devices respond to load changes.
- describe the operation of a balanced-port expansion valve.
- describe the operation of a dual-port expansion valve.
- describe how electronic expansion valves and their controllers operate.

SAFETY CHECKLIST

✔ Wear warm clothing when working in a walk-in cooler or freezer.

24.1 EXPANSION DEVICES

The *expansion device,* often called the *metering device,* is the fourth component necessary for the compression refrigeration cycle to function. The expansion device is not as visible as the evaporator, the condenser, or the compressor. Generally, the device is concealed inside the evaporator cabinet and not obvious to the casual observer. It can be either a valve or a fixed-bore device. **Figure 24–1** illustrates an expansion valve installed inside the evaporator cabinet.

The expansion device is one of the division lines between the high side of the system and the low side of the system (the compressor is the other). **Figure 24–2** shows the location of the device. The expansion device is responsible for metering the correct amount of refrigerant to the evaporator. The evaporator performs best when it is as full of liquid refrigerant as possible without leaving any in the suction line. Any liquid refrigerant that enters the suction line may reach the compressor because only a small amount of heat should be added to the refrigerant in the suction line. Later in this section a suction-line heat exchanger for special applications will be discussed, which is used to boil away liquid that may be in the suction line. Usually, liquid in the suction line is a problem.

The expansion device is normally installed in the liquid line between the condenser and the evaporator. The liquid line may be warm to the touch on a hot day and can be followed quite easily to the expansion device where there is a

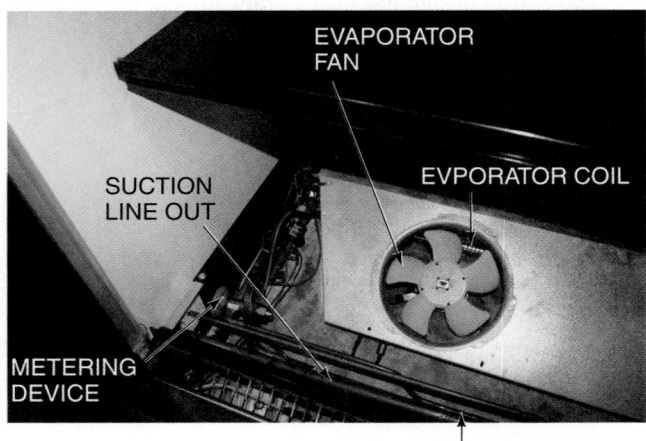

Figure 24–1 Metering devices installed on a refrigerated case. The valve is not out in the open and is not as visible as the compressor, condenser, or evaporator. *Courtesy Ferris State University. Photo by John Tomczyk*

pressure drop and an accompanying temperature drop. For example, on a hot day the liquid line entering the expansion device may be 110°F. If this is a low-temperature cooler using R-134a, the low-side pressure on the evaporator side may be 3 psig at a temperature of −8°F. This is a dramatic temperature drop and can be easily detected when found. The device may be warm on one side and frosted on the other, **Figure 24–3**. Because some expansion devices are valves and some are fixed-bore devices, this change can occur in a very short space—less than an inch on a valve, or a more gradual change on some fixed-bore devices.

Expansion devices come in the following different types: (A) thermostatic expansion valve, (B) automatic expansion valve, and (C) fixed bore, such as the capillary tube, **Figure 24–4.**

24.2 THERMOSTATIC EXPANSION VALVE

The *thermostatic expansion valve* (TXV) meters the refrigerant to the evaporator using a thermal sensing element to monitor the superheat. This valve opens or closes in response to a thermal element. The TXV maintains a constant superheat in the evaporator. Remember, when there is superheat, there is no liquid refrigerant. Excess superheat is *not* desirable, but a small amount is necessary with this valve to

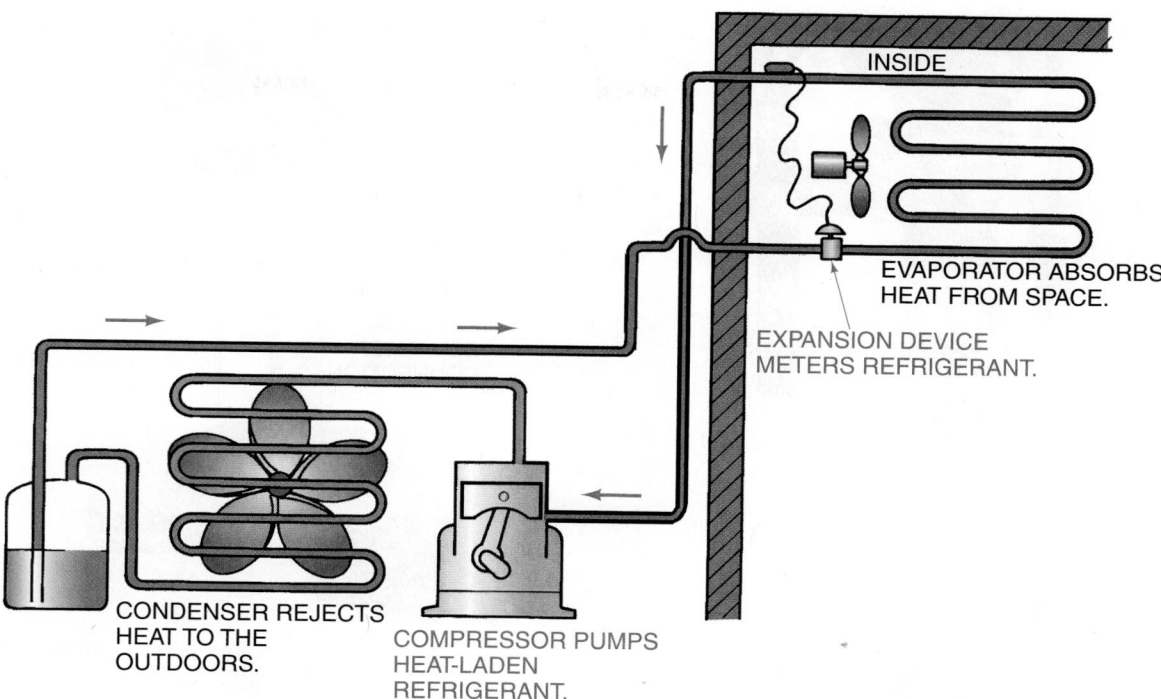

Figure 24–2 The complete refrigeration cycle with the four basic components: compressor, condenser, evaporator, and expansion device.

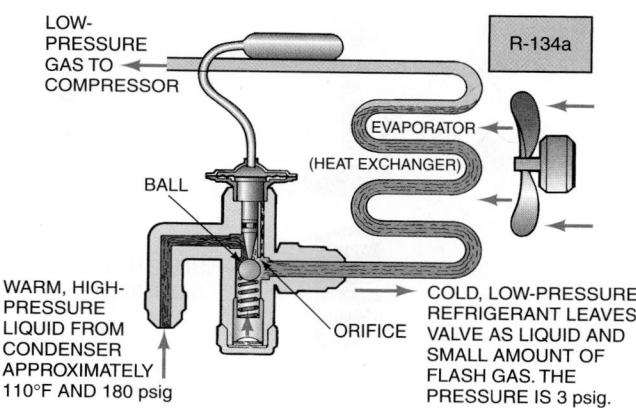

Figure 24–3 The expansion device has a dramatic temperature change from one side to the other. *Courtesy Parker Hannifin Corp.*

ensure that no liquid refrigerant leaves the evaporator. 🌐 The TXV can be adjusted to maintain a low superheat to ensure that the majority of the evaporator surface is being used. This will give the refrigeration system a higher net refrigeration effect and higher capacities and efficiencies. 🌐 The service technician should always check evaporator superheat to make sure the evaporator is fully utilized with refrigerant. When adjusting the TXV, care should be taken to do so slowly. The superheat spring adjustment should be turned in the counterclockwise direction about one half to one full turn at a time. The system should then be allowed to operate for approximately 15 min before making additional adjustments. Evaporator superheat measurements should be made after each adjustment to help ensure that liquid refrigerant does

not leave the evaporator. Recommended evaporator superheat settings are listed below:

- High-temperature evaporators (30°F) 10–12 degrees
- Medium-temperature evaporators (0°–30°F) 5–10 degrees
- Low-temperature evaporators (below 0°F) 2–5 degrees

The lower the evaporator temperature, the more need there is to keep the evaporator active with refrigerant and have less superheat. This is because at the lower evaporator temperatures, system efficiencies and capacities are at their lowest and there is a greater need for the most active evaporator.

24.3 TXV COMPONENTS

The TXV consists of the valve body, diaphragm, needle and seat, spring, adjustment and packing gland, and the sensing bulb and transmission tube, **Figure 24–5**.

24.4 THE VALVE BODY

On common refrigerant systems the *valve body* is an accurately machined piece of solid brass or stainless steel that holds the rest of the components and fastens the valve to the refrigerant piping circuit, **Figure 24–6**. Notice that the valves have different configurations. Some of them are one piece and cannot be disassembled, and some are made so that they can be taken apart.

These valves may be fastened to the system by three methods: flare, solder, or flange. When a valve is installed in a refrigeration system, future service should be considered,

(A)

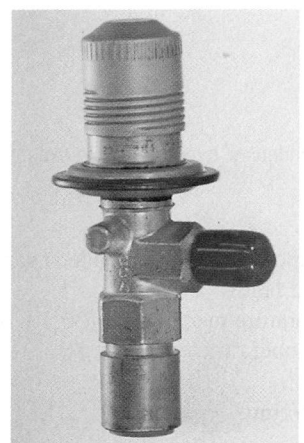

(B)

(C)

Figure 24–4 Three metering devices. **(A)** Thermostatic expansion valve. **(B)** Automatic expansion valve. **(C)** Capillary tube with liquid-line drier. *(A) Courtesy Sporlan Division, Parker Hannifin Corp. (B) Photo by Bill Johnson. (C) Courtesy Parker Hannifin Corp.*

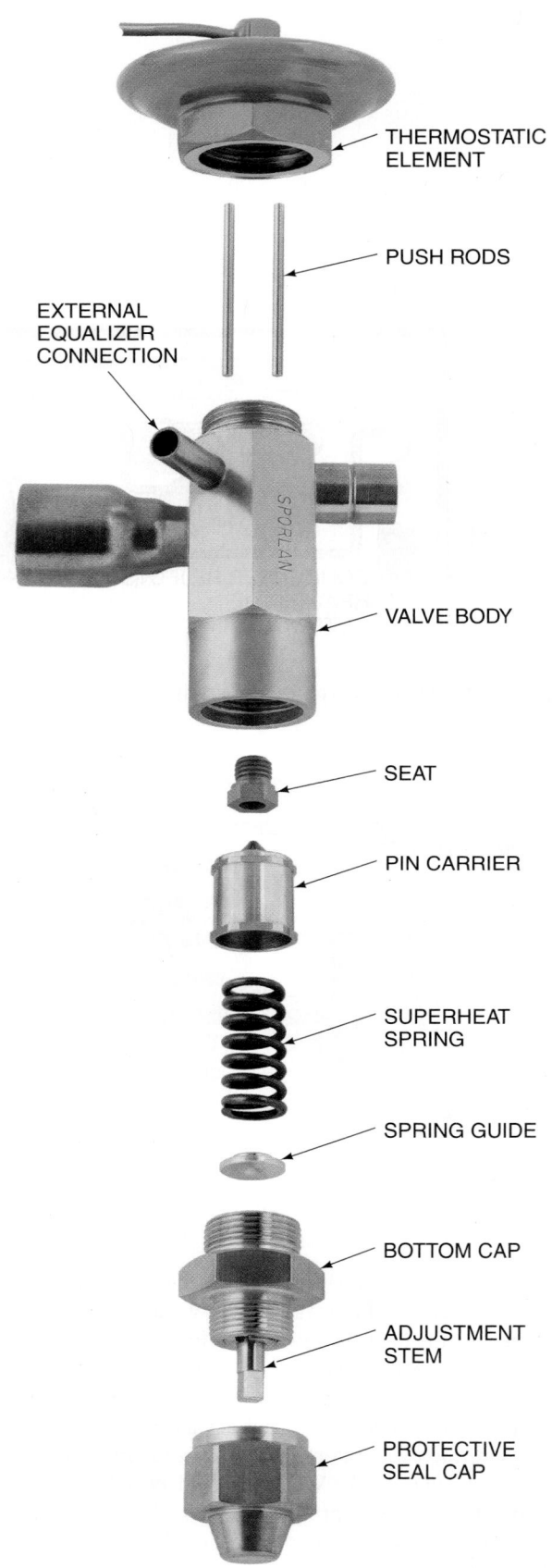

THERMOSTATIC ELEMENT

PUSH RODS

EXTERNAL EQUALIZER CONNECTION

VALVE BODY

SEAT

PIN CARRIER

SUPERHEAT SPRING

SPRING GUIDE

BOTTOM CAP

ADJUSTMENT STEM

PROTECTIVE SEAL CAP

Figure 24–5 An exploded view of the TXV. All parts are visible in the order in which they go together in the valve. *Sporlan Division, Parker Hannifin Corp.*

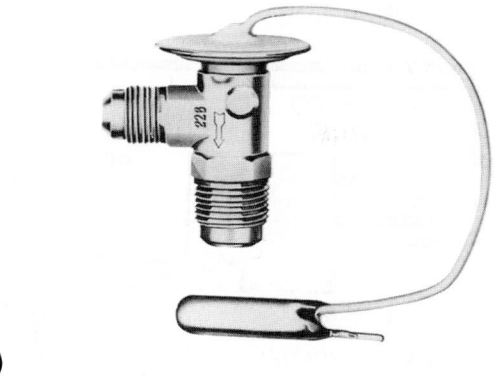

(A)

(B)

Figure 24–6   TXVs have remote sensing elements. **(A)** Some of these valves are one-piece valves that are thrown away when defective. **(B)** Some of the remote sensing elements can be replaced. This feature is particularly good when the valve is soldered into the system. **(A)** *Courtesy Singer Controls Division.* **(B)** *Courtesy Sporlan Division, Parker Hannifin Corp.*

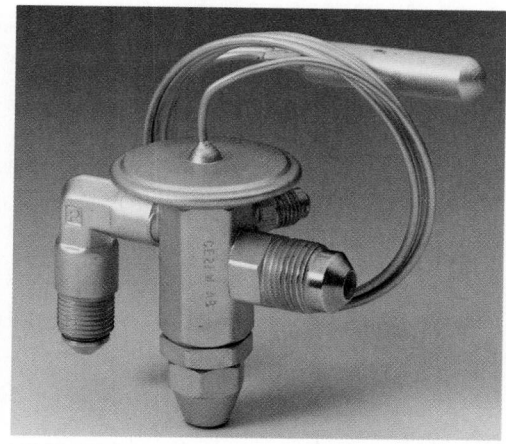

(A)

(B)

Figure 24–7   **(A)** The flare and **(B)** the flange-type valve. Either can be removed from the system and replaced easily when it is installed where it can be reached with wrenches. **(A)** *Courtesy Parker Hannifin Corp.* **(B)** *Courtesy Singer Controls Division*

so a flare connection or a flange-type valve should be used, **Figure 24–7.** If a solder connection is used, a valve that can be disassembled and rebuilt in place is desirable, **Figure 24–8.** The valve often has an inlet screen with a very fine mesh to strain out any small particles that may stop up the needle and seat, **Figure 24–9.**

Some valves have a third connection called an *external equalizer.* This connection is normally a 1/4-in. flare or 1/4-in. solder and is on the side of the valve close to the diaphragm, **Figure 24–10.** As will be discussed later, the evaporator pressure has to be represented under the diaphragm of the expansion valve. When an evaporator has a very long circuit, a pressure drop in the evaporator may occur, and an external equalizer is used. A pressure connection is made at the end of the evaporator, which supplies the evaporator pressure under the diaphragm. Some evaporators have several circuits and a method of distributing the refrigerant that will cause pressure drop between the expansion

Figure 24–8   A solder-type valve that can be disassembled and rebuilt without taking it out the system. This valve can be quite serviceable in some situations. *Courtesy Sporlan Division, Parker Hannifin Corp.*

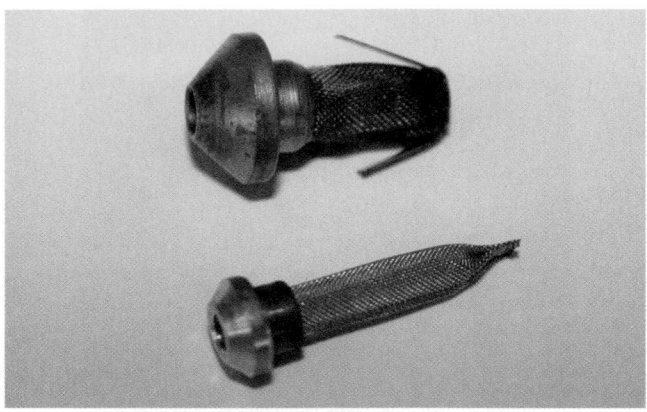

Figure 24–9   Most valves have some sort of inlet screen to strain any small particles out of the liquid refrigerant before they reach the very small opening in the expansion valve. *Courtesy Ferris State University. Photo by John Tomczyk*

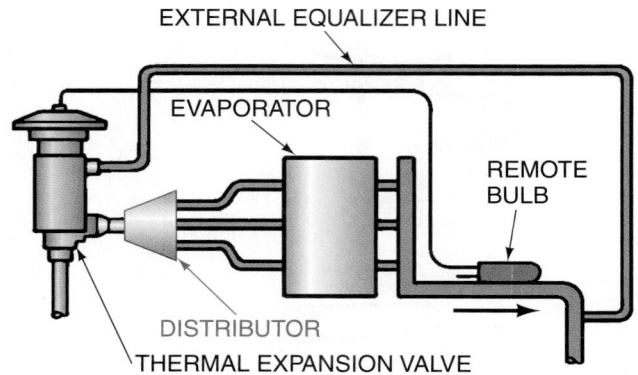

Figure 24–11   An evaporator with circuits and an external equalizer line to account for pressure drop. When an evaporator becomes so large that the length would create pressure drop, the evaporator is divided into circuits, and each circuit must have the correct amount of refrigerant.

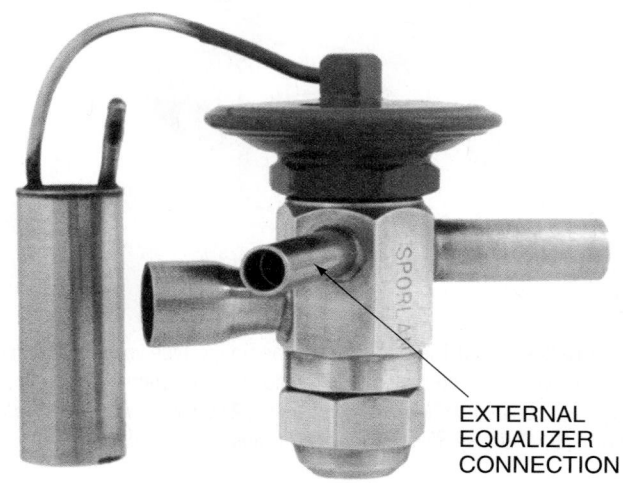

Figure 24–10   The third connection on this expansion valve is called the external equalizer. *Courtesy Sporlan Division, Parker Hannifin Corp.*

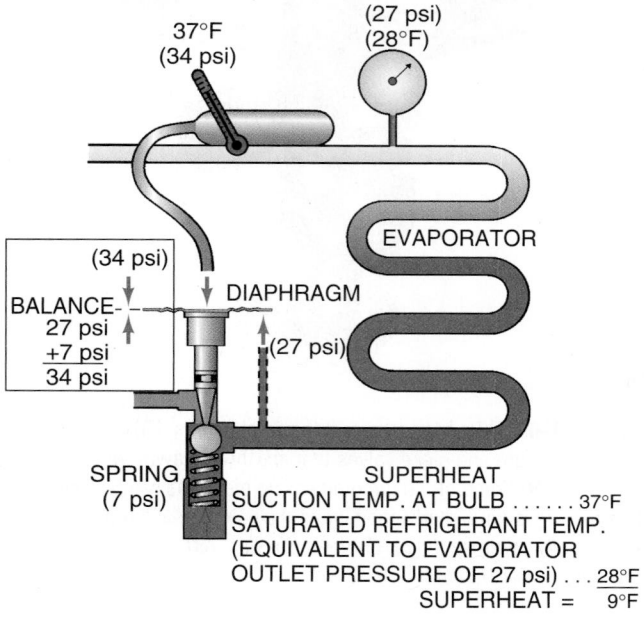

Figure 24–12   The diaphragm in the expansion valve is a thin membrane that has a certain amount of flexibility. It is normally made of a hard metal such as stainless steel. *Courtesy Parker Hannifin Corp.*

valve outlet and the evaporator inlet. This installation must have an external equalizer for the expansion valve to have correct control of the refrigerant, **Figure 24–11**.

24.5   THE DIAPHRAGM

The *diaphragm* is located inside the valve body and moves the needle in and out of the seat in response to system load changes. The diaphragm is made of thin metal and is under the round dome-like top of the valve, **Figure 24–12**.

24.6   NEEDLE AND SEAT

The *needle* and *seat* control the flow of refrigerant through the valve. They are normally made of some type of very hard metal, such as stainless steel, to prevent the refrigerant passing through from eroding the seat. The needle and seat are used in a metering device so that close control of the

refrigerant can be obtained, **Figure 24–13**. Some valve manufacturers have needle and seat mechanisms that can be changed for different capacities or to correct a problem.

The size of the needle and seat determines how much liquid refrigerant will pass through the valve with a specific pressure drop. For example, when the pressure is 170 psig on one side of the valve and 2 psig on the other side of the valve, a measured and predictable amount of liquid refrigerant will pass through the valve. If this same valve were used when the pressure is 100 psig and 3 psig, the valve would not be able to pass as much refrigerant. The conditions that the valve will operate under must be considered when selecting the valve. The manufacturer's manual for a specific valve is the best

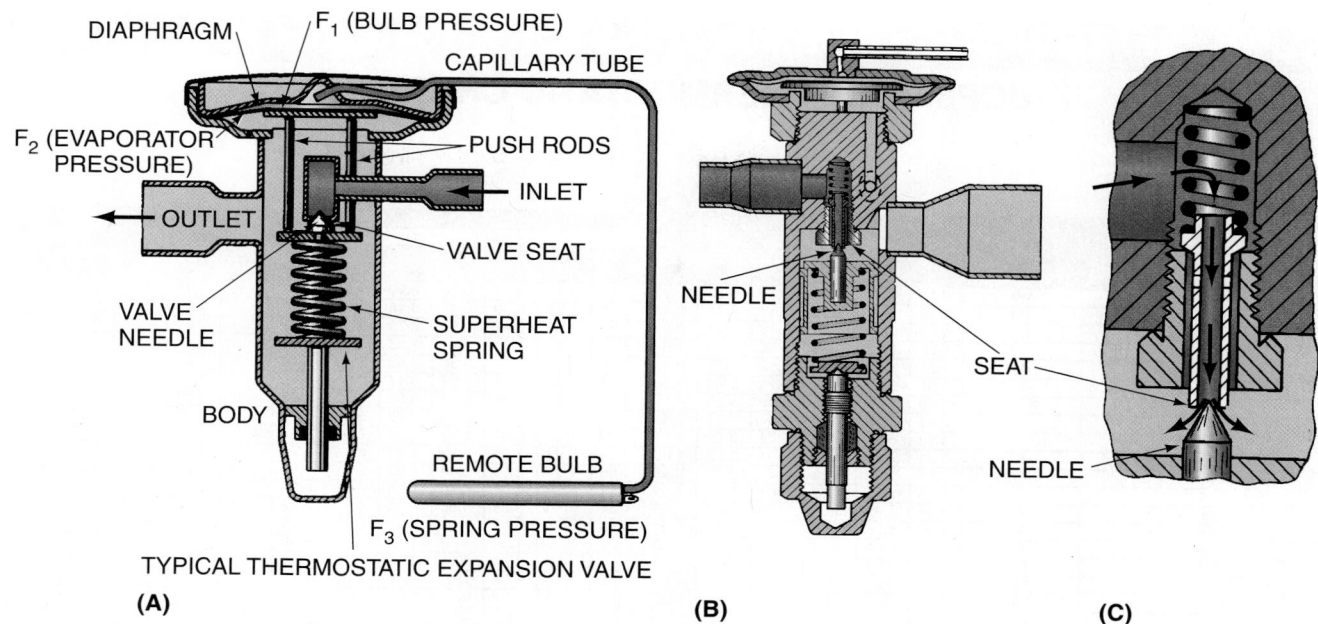

Figure 24–13   Needle and seat devices used in expansion valves **(A)–(C)**. **(A)** *Courtesy Singer Controls Division.* **(B)** and **(C)** *Courtesy Sporlan Valve Company*

place to get the proper information to make these decisions. The pressure difference from one side of the valve to the other is not necessarily the discharge versus the suction pressure. Pressure drop in the condenser and interconnecting piping may be enough to cause a problem if not considered, **Figure 24–14**. Notice in the figure that the actual discharge and suction pressure are not the same as the pressure drop across the expansion valve. The pressure drop across the distributor would be 25 psig (62 psig – 37 psig). The pressure drop across the TXV, not considering the distributor, would be 58 psig (120 psig – 62 psig).

Thermostatic expansion valves are rated in tons of refrigeration at a particular pressure drop condition. The capacity of the system and the working conditions of the system must be known.

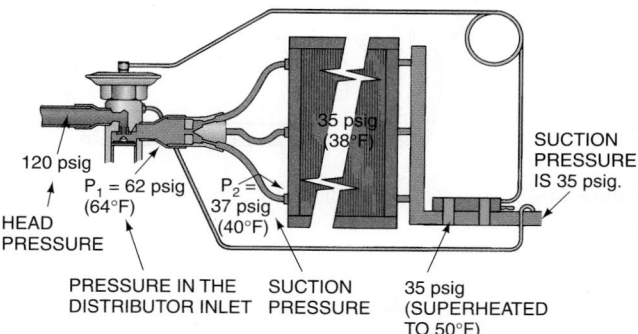

Figure 24–14   Real pressures as they would appear in a system. The pressure drop across the expansion valve is not necessarily the head pressure versus the suction pressure. The pressure drop through the refrigerant distributor is one of the big pressure drops that has to be considered. The refrigerant distributor is between the expansion valve and the evaporator coil. *Courtesy Sporlan Valve Company*

For example, using the manufacturer's catalog data for sizing a TXV in the table in **Figure 24–15**, we see that a 1-ton R-134a TXV is needed in a medium-temperature cooler with a 20°F evaporator (18.4 psig) and a 1-ton (12,000 Btu/h) capacity. This system is going to operate inside the store, which is expected to stay at 75°F year-round. The head pressure is expected to remain constant at 124 psig (100°F). Assuming negligible pressure drops in the associated piping and accessories in the system, the TXV would be operating with a pressure drop across the valve of 105.6 psig (124 psig – 18.4 psig). Since the 100-psig column in the table is closest to the 105.6-psig calculated pressure drop, we will use it. A 1-ton nominal capacity valve has a capacity of 1.34 tons at these conditions.

If the same cooler is moved outside where the temperature is warmer and the condensing pressure rose to 145 psig, the new pressure drop across the TXV would be 126.6 psig (145 psig – 18.4 psig). We will use the 120-psig pressure drop column in the table because it is closest to our 126.6-psig calculated pressure drop across the valve. A 1-ton nominal capacity valve has a capacity of 1.47 tons under this new condition of 120-psig pressure drop across the valve. The increase in capacity of the 1-ton nominal TXV from 1.34 to 1.47 was due to the increase in head pressure, which caused an increase in the pressure drop across the valve.

The preceding two examples of TXV capacity ratings assumed that the subcooled liquid temperature coming into the TXV was 100°F. The bottom of the table in **Figure 24–15** includes liquid correction factors for liquid temperatures other than 100°F. Notice that the correction factor for 100°F liquid is 1.00. Correction factors greater than 1.00 will increase the valve's capacity, and correction factors less than 1.00 will decrease the valve's capacity. In our first example, the nominal 1-ton TXV at a 20°F evaporating temperature

TEV CAPACITIES in TONS for R-134a (HFC)
JCP60 & JC THERMOSTATIC CHARGES

Condition 1   Condition 2

VALVE TYPES	NOMINAL CAPACITY tons	EVAPORATOR TEMPERATURE (°F) 40° PRESSURE DROP ACROSS VALVE (PSI)						20°						0°					
		40	60	80	100	120	140	60	80	100	120	140	160	60	80	100	120	140	160
NI-F-G-EG	1/8	0.12	0.15	0.17	0.19	0.21	0.23	0.13	0.15	0.17	0.18	0.20	0.21	0.12	0.13	0.15	0.16	0.18	0.19
NI-F-G-EG	1/4	0.26	0.31	0.36	0.40	0.44	0.48	0.30	0.35	0.39	0.42	0.46	0.49	0.28	0.32	0.36	0.39	0.43	0.45
NI-F-G-EG	1/2	0.49	0.60	0.70	0.78	0.85	0.92	0.52	0.60	0.67	0.73	0.79	0.85	0.47	0.54	0.60	0.66	0.71	0.76
NI-F-G-EG	1	0.98	1.21	1.39	1.56	1.70	1.84	1.04	1.20	1.34	1.47	1.58	1.69	0.93	1.07	1.20	1.32	1.42	1.52
F-G-EG	1-1/2	1.57	1.93	2.23	2.49	2.73	2.95	1.66	1.91	2.14	2.34	2.53	2.71	1.49	1.72	1.92	2.11	2.28	2.43
F(Ext)-G & EG(Ext)-S	2	1.97	2.41	2.78	3.11	3.41	3.68	2.07	2.39	2.67	2.93	3.17	3.38	1.86	2.15	2.40	2.63	2.84	3.04
C-S	2-1/2	2.46	3.01	3.48	3.89	4.26	4.60	2.59	2.99	3.34	3.66	3.96	4.23	2.33	2.69	3.00	3.29	3.56	3.80
C-S	3	2.95	3.62	4.18	4.67	5.11	5.52	3.11	3.59	4.01	4.40	4.75	5.08	2.79	3.22	3.61	3.95	4.27	4.56
C&S(Ext)	5	4.92	6.03	6.96	7.78	8.52	9.21	4.32	4.98	5.57	6.10	6.59	7.05	3.53	4.08	4.56	4.99	5.39	5.77
S(Ext)	6	5.91	7.23	8.35	9.34	10.2	11.0	5.18	5.98	6.69	7.33	7.91	8.46	4.24	4.89	5.47	5.99	6.47	6.92
H	1-1/2	1.57	1.93	2.23	2.49	2.73	2.95	1.66	1.91	2.14	2.34	2.53	2.71	1.49	1.72	1.92	2.11	2.28	2.43
H	3	2.95	3.62	4.18	4.67	5.11	5.52	3.11	3.59	4.01	4.40	4.75	5.08	2.57	2.97	3.32	3.63	3.92	4.19
H	4	3.94	4.82	5.57	6.22	6.82	7.37	4.14	4.79	5.35	5.86	6.33	6.77	3.42	3.95	4.42	4.84	5.23	5.59
H	5	4.92	6.03	6.96	7.78	8.52	9.21	5.18	5.98	6.69	7.33	7.91	8.46	4.28	4.94	5.53	6.05	6.54	6.99
P-H	8	7.38	9.04	10.4	11.7	12.8	13.8	7.77	8.97	10.0	11.0	11.9	12.7	6.42	7.41	8.29	9.08	9.81	10.5
P-H	12	11.5	14.1	16.3	18.2	19.9	21.5	12.1	14.0	15.6	17.1	18.5	19.8	10.0	11.6	12.9	14.2	15.3	16.4
M	13	12.8	15.7	18.1	20.2	22.2	23.9	13.5	15.6	17.4	19.0	20.6	22.0	10.7	12.4	13.8	15.1	16.4	17.5
M	15	15.3	18.7	21.6	24.1	26.4	28.5	16.1	18.5	20.7	22.7	24.5	26.2	12.8	14.7	16.5	18.1	19.5	20.9
M	20	19.7	24.1	27.8	31.1	34.1	36.8	20.7	23.9	26.7	29.3	31.7	33.8	16.5	19.0	21.3	23.3	25.2	26.9
M	25	24.6	30.1	34.8	38.9	42.6	46.0	25.9	29.9	33.4	36.6	39.6	42.3	20.6	23.8	26.6	29.1	31.5	33.6

Liquid Temperature (°F)	0°	10°	20°	30°	40°	50°	60°	70°	80°	90°	100°	110°	120°	130°	140°
Correction Factor	1.70	1.63	1.56	1.49	1.42	1.36	1.29	1.21	1.14	1.07	1.00	0.93	0.85	0.78	0.71

These factors include corrections for liquid refrigerant density and net refrigerating effect and are based on an average evaporator temperature of 0°F. However they may be used for any evaporator temperature from 40°F to 0°F since the variation in the actual factors across this range is insignificant.

EXAMPLE: 2 ton valve at 20°F evaporator, 80 psi pressure drop, 90°F liquid temperature = 2.39 x 1.07 = 2.56 tons.

Figure 24–15 The manufacturer's table shows the capacity of valves at different pressure drops. Notice that the same valve has different capacities at different pressure drops. The more the pressure drop, the more the capacity a valve has. This is a partial table and not intended for use in the design of a system. *Courtesy Sporlan Valve Company*

and a 100-psig pressure drop across the valve had a capacity of 1.34 tons. This is assuming 100°F liquid temperatures entering the valve. However, if the liquid temperature entering the valve were 80°F, the valve would have a new capacity of 1.34 tons × 1.14 = 1.52 tons. A correction factor of 1.14 from the table was used for the 80°F liquid temperature entering the valve. The colder the liquid temperature entering the TXV, the more capacity in tons the TXV will have. As the liquid temperature gets colder, the less flashing (loss effect) there will be at the entrance of the evaporator in cooling the liquid down to the evaporating temperature. The more liquid subcooling there is, the less loss effects there will be at the evaporator entrance. This will increase the net refrigeration effect and the capacity of the system. Because increased subcooling increases the refrigeration effect of the system, many systems have the suction lines in contact with the liquid line to increase the subcooling effect. This also helps reduce the possibility of liquid refrigerant entering the compressor by adding heat to the refrigerant in the suction line.

24.7 THE SPRING

The *spring* is one of the three forces that act on the diaphragm. It raises the diaphragm and closes the valve by pushing the needle into the seat. When a valve has an adjustment, the adjustment applies more or less pressure to the

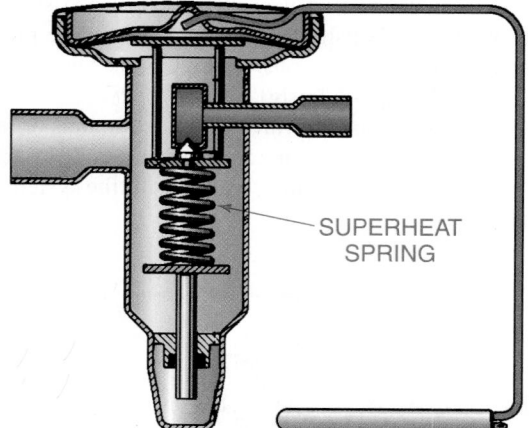

Figure 24–16 The spring used in the TXV. *Courtesy Singer Controls Division*

SUPERHEAT SPRING

spring to change the tension for different superheat settings. The spring is often referred to as the superheat spring. The spring tension is factory set for a predetermined superheat of 8°F to 12°F, **Figure 24–16**.

The adjustment part of this valve can be either a screw slot or a square-headed shaft. Either type is normally covered with a cap to prevent water, ice, or other foreign matter from collecting on the stem. The cap also serves as a backup leak

(A)

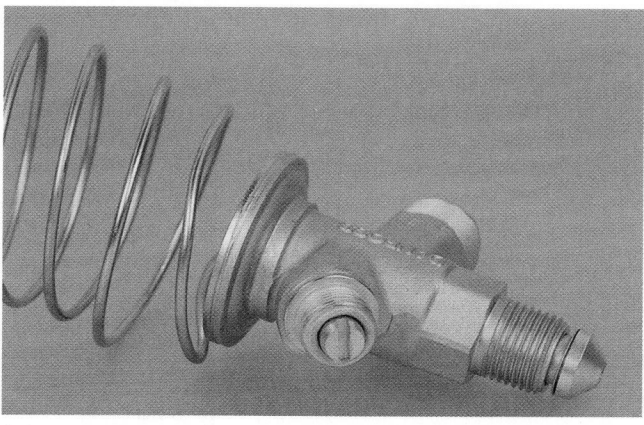

(B)

Figure 24–17 Adjustment stems on expansion valves. Some of them are adjusted with **(A)** a valve wrench and some with **(B)** a screwdriver. *Photos by Bill Johnson*

prevention. Most adjustment stems on expansion valves have a packing gland that can be tightened to prevent refrigerant from leaking. The cap would cover the stem and the gland, **Figure 24–17.** Normally one complete turn of the stem can change the superheat reading significantly depending on the manufacturer. It is good field practice not to adjust the spring unless you are a seasoned technician. If the valve is adjusted improperly, compressor damage can occur.

24.8 THE SENSING BULB AND TRANSMISSION TUBE

The *sensing bulb* and *transmission line* are extensions of the valve diaphragm. The bulb detects the temperature at the end of the evaporator on the suction line and transmits this temperature, converted to pressure, to the top of the diaphragm. The bulb contains a fluid, such as refrigerant, that responds to a temperature/pressure relationship chart just like any refrigerant would. When the suction line temperature goes up, this temperature change occurs inside the bulb. As the temperature of the TXV's sensing bulb increases, the valve will gradually open, letting more refrigerant into the evaporator. This will increase both the evaporating pressure and the compressor's suction pressure, while attempting to

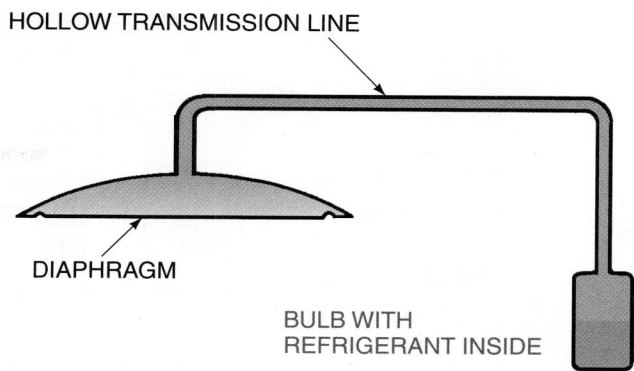

Figure 24–18 An illustration of the diaphragm, the bulb, and the transmission tube.

reduce the amount of superheat in the evaporator. When the pressure changes, the transmission line (which is nothing more than a small-diameter hollow tube) allows the pressure between the bulb and diaphragm to equalize back and forth, **Figure 24–18.**

The seat in the valve is stationary in the valve body, and the needle is moved by the diaphragm. One side of the diaphragm gets its pressure from the bulb, and the other side gets its pressure from the evaporator and the spring. The diaphragm moves up and down in response to three different pressures. These three different pressures act at the proper time to open, close, or modulate the valve needle between open and closed. These three pressures are the *bulb pressure,* the *evaporator pressure,* and the *spring pressure.* They all work as a team to position the valve needle at the correct position for the load conditions at any particular time. **Figure 24–19** is a series of illustrations that show the TXV functions under different load conditions.

24.9 TYPES OF BULB CHARGE

The fluid inside the expansion valve bulb is known as the *charge* for the valve. Four types of charge can be obtained with the TXV: liquid charge, cross liquid charge, vapor charge, and cross vapor charge.

24.10 THE LIQUID CHARGE BULB

The *liquid charge bulb* is a bulb charged with a fluid characteristic of the refrigerant in the system. The diaphragm and bulb are not actually full of liquid. They have enough liquid, however, that they always have some liquid inside them. The liquid will not all boil away. The temperature/pressure relationship is almost a straight line on a graph. When the temperature goes up a degree, the pressure goes up a specific amount and can be followed on a temperature/pressure chart. When this temperature/pressure concept is carried far enough, it can easily be seen that when high temperatures are encountered, high pressures will exist. During defrost the expansion valve bulb may reach high temperatures. This can cause two things to happen. The pressures inside the bulb can cause excessive pressures over the diaphragm, and the valve will open wide

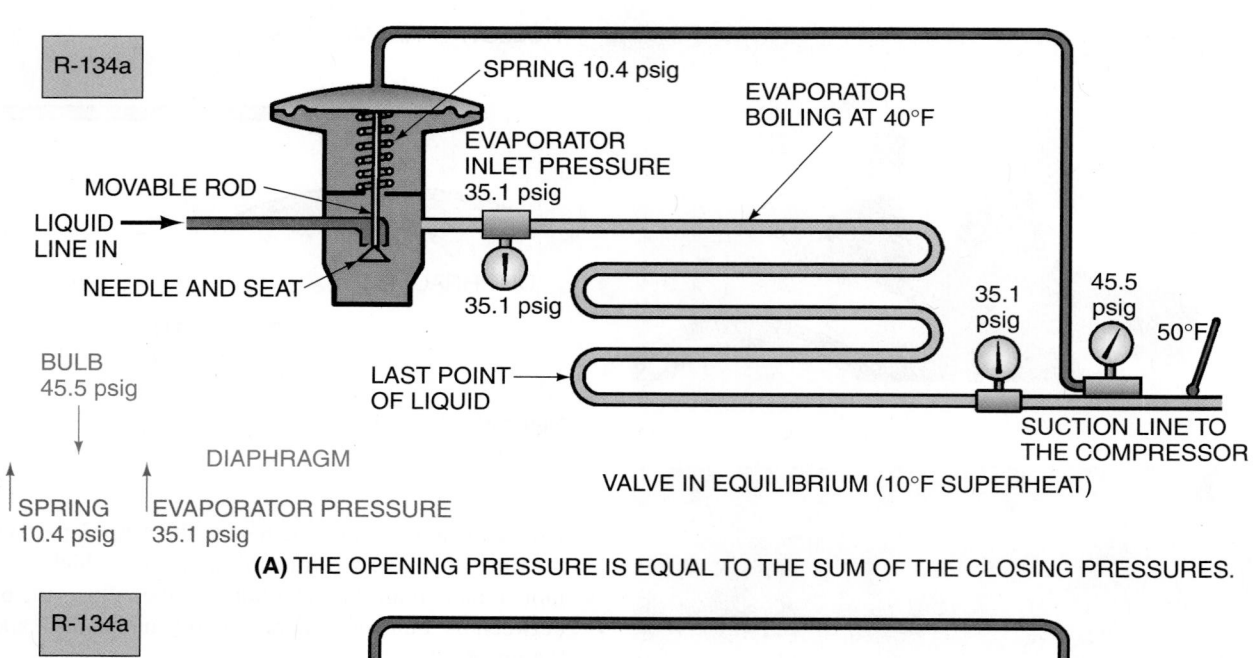

(A) THE OPENING PRESSURE IS EQUAL TO THE SUM OF THE CLOSING PRESSURES.

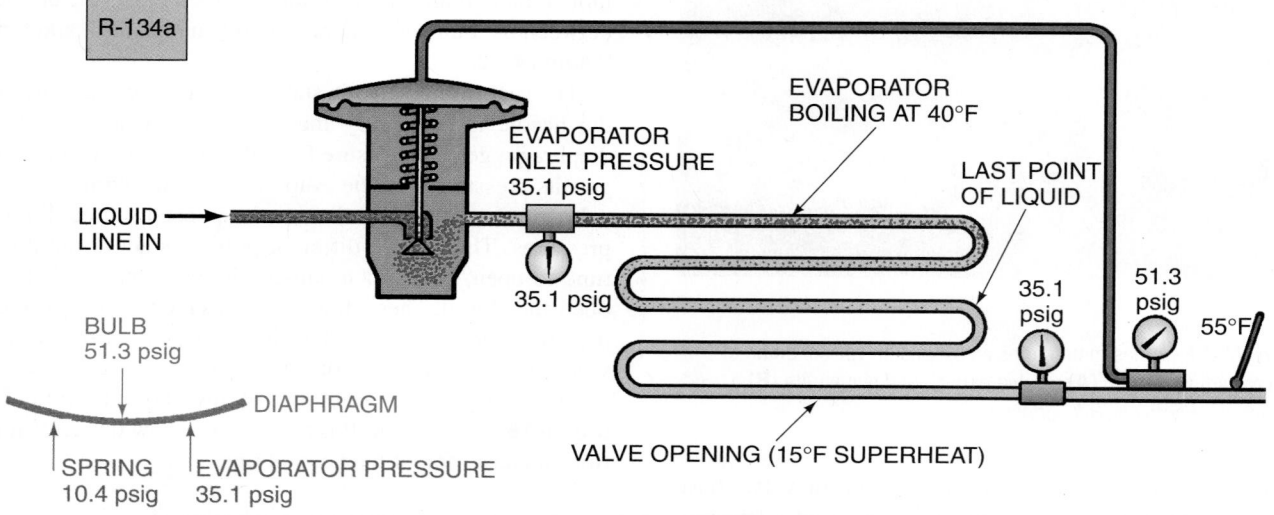

(B) THE OPENING PRESSURE IS GREATER THAN THE SUM OF THE CLOSING PRESSURES.

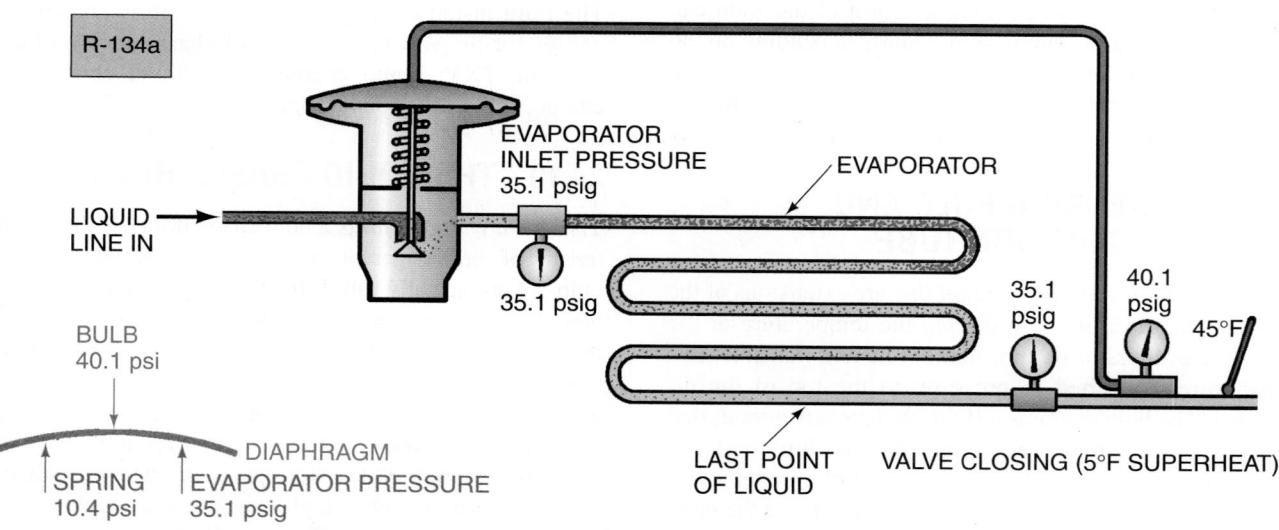

(C) THE OPENING PRESSURE IS LOWER THAN THE CLOSING PRESSURES.

Figure 24–19 All components work together to hold the needle and seat in the correct position to maintain stable operation with the correct preheat. **(A)** Valve in equilibrium. **(B)** Valve opening. **(C)** Valve closing.

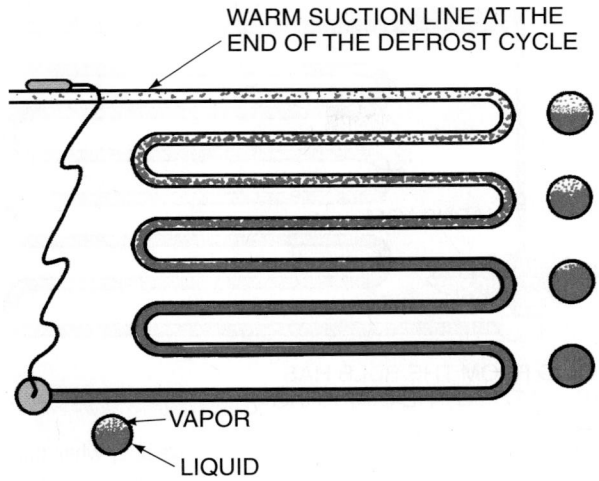

WARM SUCTION LINE AT THE
END OF THE DEFROST CYCLE

VAPOR

LIQUID

Figure 24–20   This can happen at the end of the defrost cycle. The valve is open because the evaporator suction line is warm after the defrost cycle. This allows the liquid in the evaporator to spill over into the suction line before the expansion valve can gain control. The bulb can become quite hot in an extended defrost cycle, and the remaining liquid can move on to the suction line, causing possible compressor damage.

when the bulb gets warm. This will overfeed the evaporator and can cause liquid to flood the compressor when defrost is terminated, **Figure 24–20.** This can cause the service technician or manufacturer to change to the cross liquid charge bulb.

The liquid charge bulb is usually limited to narrow-operating-range applications or to special functions like desuperheating in the high side of the system.

24.11   THE CROSS LIQUID CHARGE BULB

The *cross liquid charge bulb* has a fluid that is different from the system fluid. It does not follow the temperature/pressure relationship. It has a flatter curve and will close the valve faster on a rise in evaporator pressure. Liquid cross charges are good for low-temperature applications. They will usually give high evaporator superheats at high evaporator-heat load conditions. Normal evaporator superheats will be experienced within the operating range of the valve. The valve closes during the off cycle when the compressor shuts off and the evaporator pressure rises. This will help prevent liquid refrigerant from flooding over into the compressor at startup, **Figure 24–21.**

24.12   THE VAPOR CHARGE BULB

The *vapor charge bulb* is actually a valve that has only a small amount of liquid refrigerant in the bulb, **Figure 24–22.** It is sometimes called a *critical charge bulb.* When the bulb temperature rises, more and more of the liquid will boil to a vapor until there is no more liquid. When this point is reached, an increase in temperature will no longer bring an increase in pressure. The pressure curve will be flat, **Figure 24–23(A).** This acts to limit the maximum operating pressure the valve will allow the evaporator to experience. These

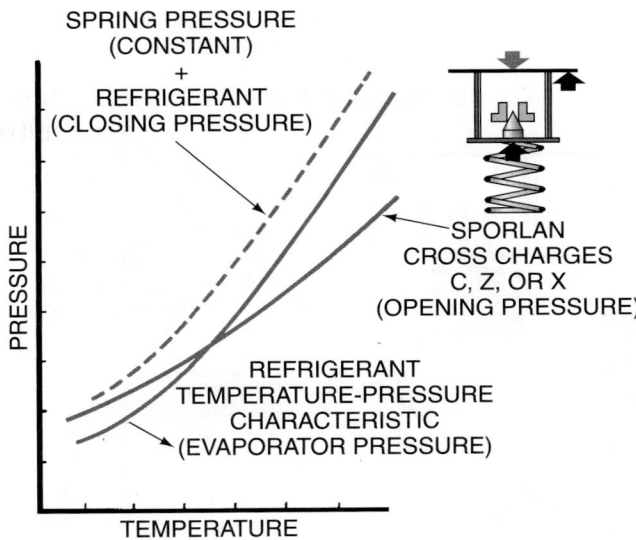

SPRING PRESSURE
(CONSTANT)
+
REFRIGERANT
(CLOSING PRESSURE)

SPORLAN
CROSS CHARGES
C, Z, OR X
(OPENING PRESSURE)

PRESSURE

REFRIGERANT
TEMPERATURE-PRESSURE
CHARACTERISTIC
(EVAPORATOR PRESSURE)

TEMPERATURE

Figure 24–21   This cross charge valve has a flatter temperature/pressure curve. The pressure rise after defrost or at the end of the cycle will have a tendency to close the valve and prevent liquid floodback problems. *Courtesy Sporlan Valve Company*

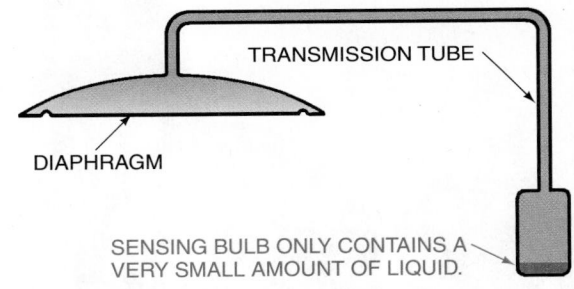

TRANSMISSION TUBE

DIAPHRAGM

SENSING BULB ONLY CONTAINS A
VERY SMALL AMOUNT OF LIQUID.

Figure 24–22   This gas charge bulb is actually a liquid charge bulb with a critical charge. When the bulb reaches a predetermined temperature, the liquid is boiled away, and the pressure will not rise any more.

valves are often referred to as maximum operating pressure (MOP) valves. Notice that the TXV in **Figure 24–23(B)** has a MOP of 15. This means that when the evaporator pressure reaches 15 psig, all of the liquid in the remote bulb will be vaporized and there will be no more opening pressure on the diaphragm of the valve. The evaporator pressure will never go higher than 15 psig. Limiting the evaporator pressure will protect the compressor from an overloaded condition resulting from excessively high pressures. Pressure entering the compressor that is too high will cause the density of suction gases entering the compressor to increase. This in turn causes an increase in the mass flow rate of refrigerant through the compressor. High mass flow rates of refrigerant will overload the compressor by causing high compressor-motor amperage draws. These overloaded conditions can overheat compressor motors and open compressor-motor overload protection devices, which will take the compressor out of service temporarily. When the vapor charge bulb is installed, care must be taken that the valve body does not get colder

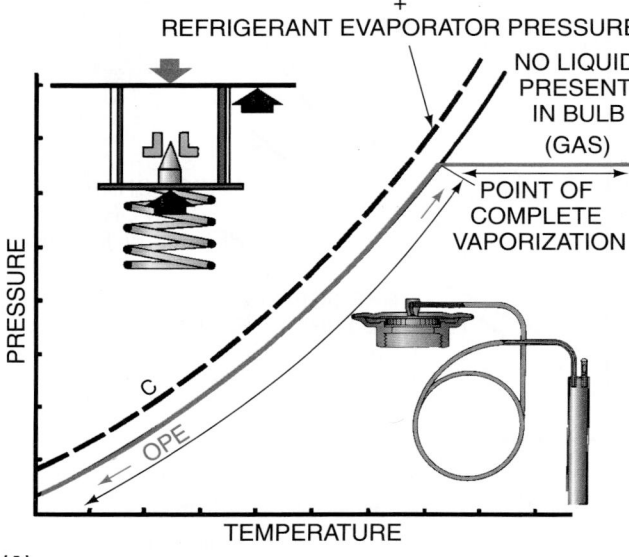

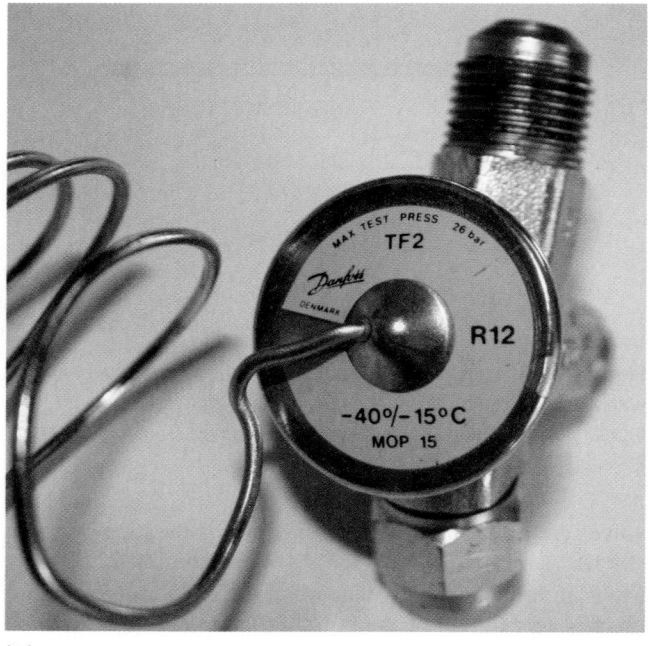

(B)

Figure 24–23 **(A)** This graph shows pressures inside the valve when the sensing bulb is in a hot area. **(B)** TXV showing a maximum operating pressure (MOP) of 15 psi for compressor protection. **(A)** *Courtesy Sporlan Valve Company.* **(B)** *Courtesy Ferris State University. Photo by John Tomczyk*

than the bulb, or the liquid that is in the bulb will condense above the diaphragm. When this happens, the control at the end of the evaporator by the bulb will be lost to the valve diaphragm area, **Figure 24–24**. The valve will be controlled by the temperature of the liquid that is at the diaphragm, and the valve will lose control. Small heaters have been installed at the valve body to keep this from happening.

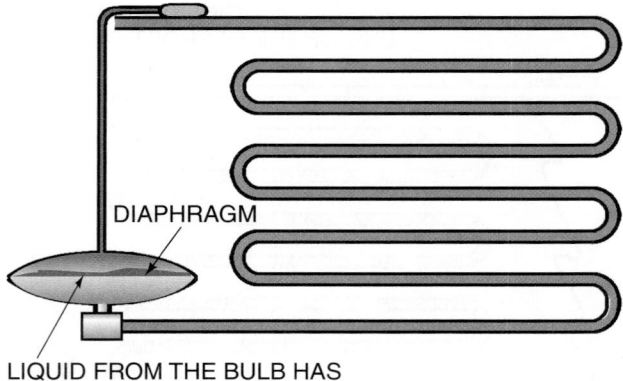

Figure 24–24 The liquid can migrate to the valve body when the body becomes cooler than the bulb. When this happens, the valve body can be heated with warm cloth, and the liquid will move back to the bulb.

24.13 THE CROSS VAPOR CHARGE BULB

The *cross vapor charge bulb* has a similar characteristic to the vapor charged valve, but it has a fluid different from the refrigerant in the system. This gives a different pressure and temperature relationship under different conditions. These special valves are applied to special systems. Manufacturers or suppliers should be consulted when questions are encountered.

24.14 FUNCTIONING EXAMPLE OF A TXV WITH INTERNAL EQUALIZER

When all of these components are assembled, the expansion valve will function as follows. NOTE: This is a liquid-filled bulb.

1. **Normal load conditions.** The valve is operating in equilibrium (no change, stable), **Figure 24–25**. The evaporator is operating at a medium-temperature application and just before the cut-off point. The suction pressure is 18.4 psig, and the refrigerant, R-134a, is boiling (evaporating) in the evaporator at 20°F. The expansion valve is maintaining 10°F of superheat, so the suction-line temperature is 30°F at the bulb location. The bulb has been on the line long enough that it is the same temperature as the line, 30°F. For now, suppose that the liquid in the bulb will be 26.1 psig, corresponding to 30°F. The spring is exerting a pressure equal to the difference in pressure to hold the needle at the correct position in the seat to maintain this condition. The spring pressure in this example is 7.7 psi.
2. **Load changes with food added to the cooler.** See **Figure 24–26(A)**. When food that is warmer than the inside of the cooler is added to the cooler, the load of the evaporator changes. The warmer food warms the air inside the cooler, and the load is added to the refrigeration coil by the air. This warmer air passing

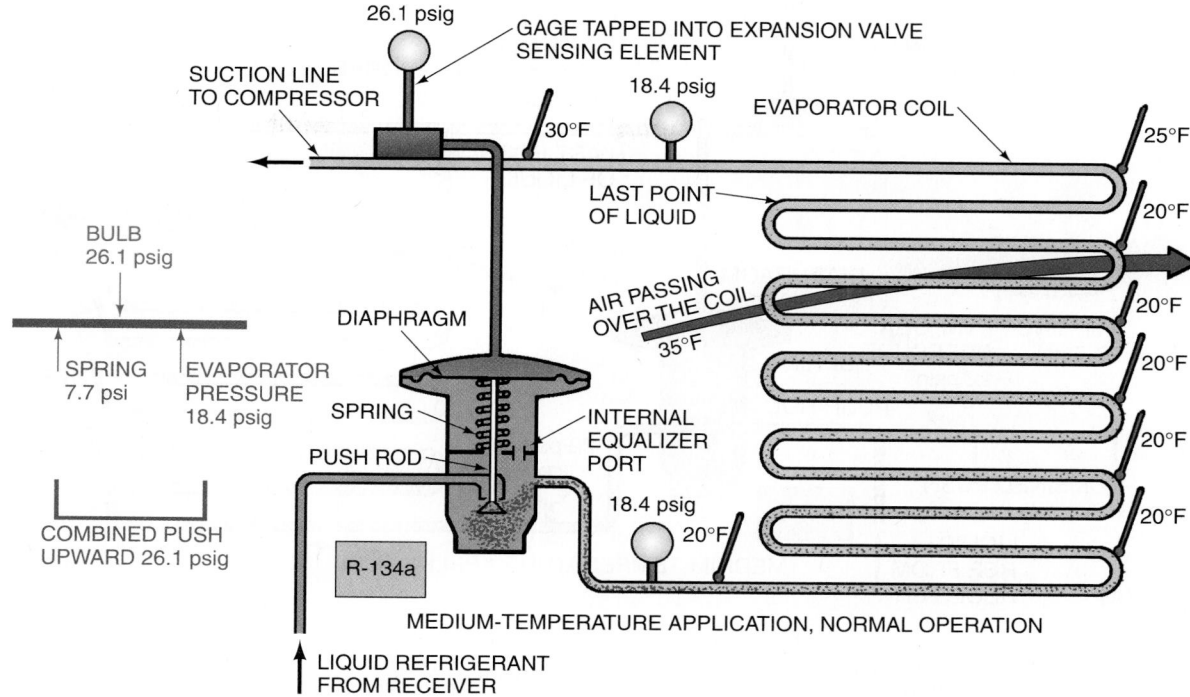

Figure 24–25   A TXV under a normal load condition. The valve is said to be in equilibrium. The needle is stationary.

over the coil causes the liquid refrigerant inside the coil to boil faster. The suction pressure will also rise. The net effect of this condition will be that the last point of liquid in the coil will be farther from the end of the coil than when the coil was under normal working conditions. The coil will start to starve for liquid refrigerant. The TXV will start to feed more refrigerant to compensate for this shortage. This happens because the increased superheat in the evaporator causes the temperature of the remote bulb to rise. Since the pressure in the remote bulb is the only opening pressure in the TXV, the valve will open to feed more refrigerant in the evaporator. The rate of vaporization of the refrigerant in the evaporator will increase and cause the evaporator pressure to increase. In fact, the pressure on the whole low side of the system will increase all the way down to the compressor. When this condition of increased load has gone on for an extended time, the TXV will stabilize in the feeding of refrigerant and reach a new point of equilibrium, where no adjustment occurs.

3. **Load changes with food removed from the cooler.** See **Figure 24–26(B)**. When a large portion of the food is removed from the cooler, the load will decrease on the evaporator coil. There will no longer be enough load to boil the amount of refrigerant that the expansion valve is feeding into the coil, so the expansion valve will start overfeeding the coil. The coil is beginning to flood with liquid refrigerant. The TXV needs to throttle the refrigerant flow. When the TXV has operated at this condition for a period

of time, it will stabilize and reach another point of equilibrium at or near 10°F of evaporator superheat. When the condition exists for a long enough time, the thermostat will stop the compressor because the air in the cooler is reduced to the thermostat cut-out point.

24.15   TXV WITH EXTERNAL EQUALIZERS

Evaporators with pressure drop from the inlet to the outlet are often designed and applied. This pressure drop may be from a distributor located after the expansion valve or because the evaporator has a long piping circuit. An external equalizer is required whenever the pressure drop in the evaporator exceeds

- 3 psig (air-conditioning applications).
- 2 psig (commercial-refrigeration applications).
- 1 psig (low-temperature applications).

    The external equalizer on a TXV is used to compensate, not get rid of, pressure drops from the inlet to the outlet of the evaporator. As mentioned earlier, these pressure drops could be caused from a distributor after the TXV or simply from long evaporator piping circuits with a lot of U bends. Excess pressure drop in a coil where a TXV is used will cause the valve to starve the coil of refrigerant, **Figure 24–27**. When evaporator outlet pressure is sensed, less pressure is on the bottom of the TXV bellows than if evaporator inlet pressure were sensed. ⓢThis lower pressure causes a smaller closing force, so the valve remains more open and fills out more of the evaporator with refrigerant by compensating for the pressure drop through the evaporator.

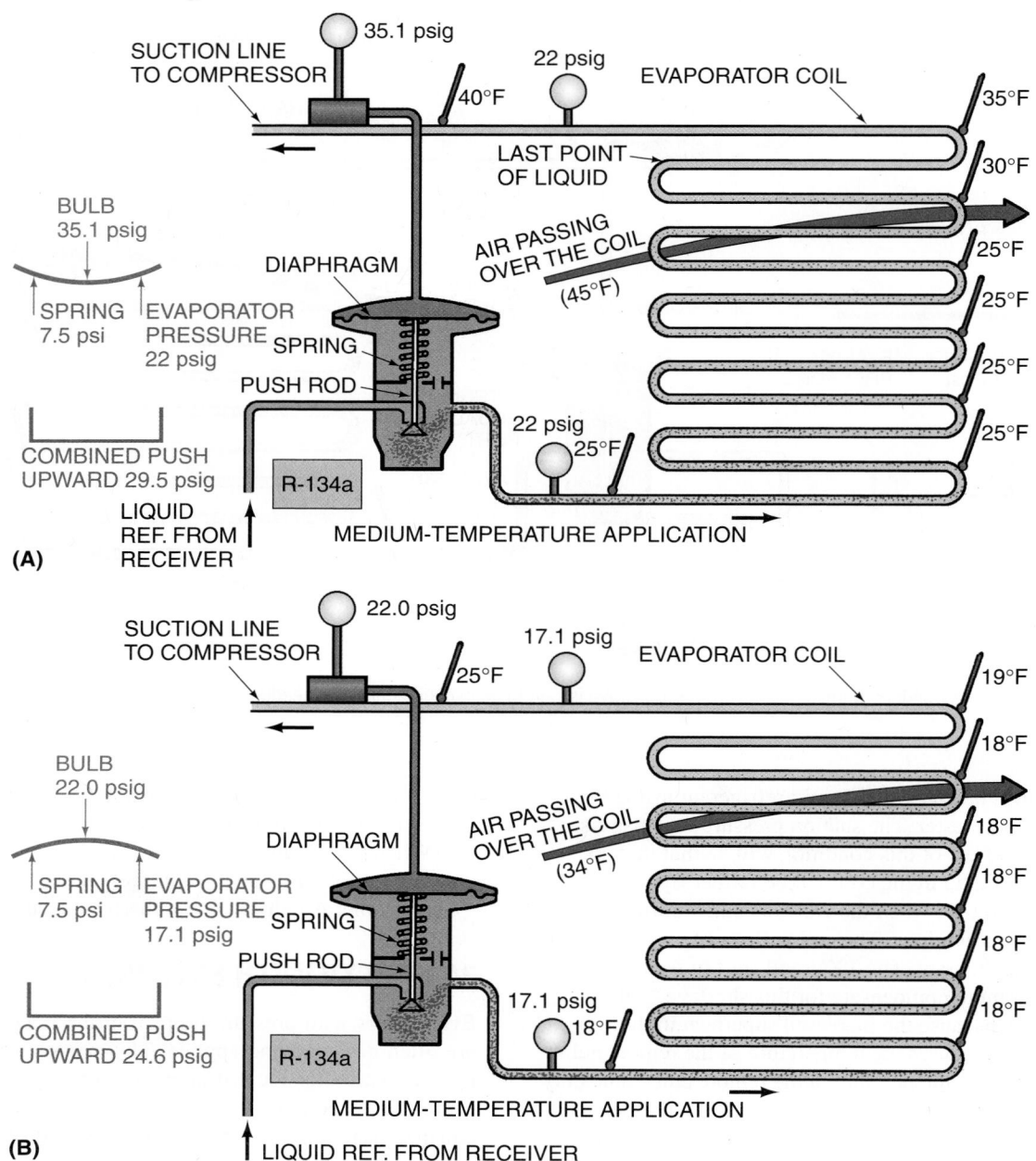

Figure 24–26 **(A)** When food is added to the cooler, the load on the coil goes up. The extra Btu added to the air in the cooler are transferred into the coil, which causes a temperature increase in the suction line. This causes the valve to open and feed more refrigerant. If this condition is prolonged, the valve will reach a new equilibrium point with the needle being steady and not moving. **(B)** When there is a load decrease, such as when some of the food is removed, the refrigeration machine now has prolonged time; the valve will reach a new point of equilibrium at a low load condition. The thermostat will cut off the unit if this condition goes on long enough. The equilibrium point in **(A)** and **(B)** is the valve superheat set point of 10°F superheat. The valve should always return to this 10°F set point with prolonged steady-state operation.

Remember, the evaporator is more efficient when it has the maximum amount of refrigerant without flooding any back to the compressor. 🌐 An expansion valve with an external equalizer was shown in **Figure 24–10.**

The external equalizer line should always be piped to the suction line after the expansion valve sensing bulb to prevent superheat problems from internal leaks in the valve, **Figure 24–28.** If a TXV with an external equalizer line were

to have an internal leak, very small amounts of liquid would vent to the suction line through the external equalizer line. If the sensing bulb for the valve were to have this small amount of liquid touching it, it would throttle toward closed because this would simulate a flooded coil, **Figure 24–29.** Sometimes this condition can be sensed by feeling the line leading to the suction line. It should not become cold as it leaves the expansion valve.

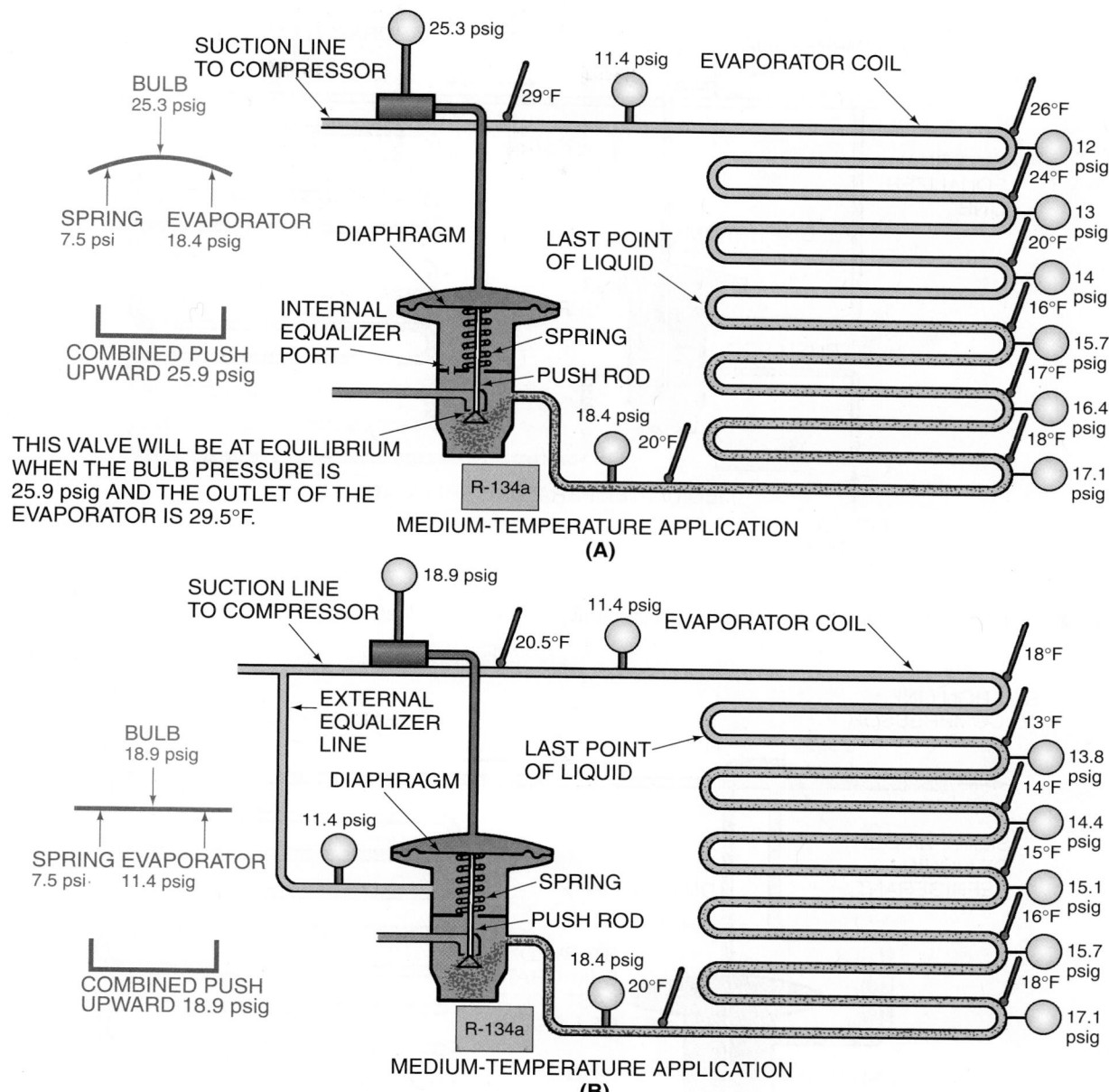

Figure 24–27 (A) This evaporator has 7-psig pressure drop, which is causing a starved evaporator. **(B)** When a TXV with an external equalizer is added, the coil has the correct amount of refrigerant for best coil efficiency.

24.16 TXV RESPONSE TO LOAD CHANGES

The TXV responds to a change in load in the following manner. When the load is increased, for example when a load of food warmer than the cooler (refrigerated box) is placed inside the cooler, the TXV opens, allowing more refrigerant into the coil. The evaporator needs more refrigerant at this time because the increased load is evaporating the refrigerant in the evaporator faster. The suction pressure increases, **Figure 24–26(A)**. When there is a load decrease (e.g., when some of the food is removed from the cooler), the liquid in the evaporator evaporates more slowly, and the suction

pressure decreases. At this time the TXV will throttle back by slightly closing the needle and seat to maintain the correct superheat. **The TXV responds to a load increase by metering more refrigerant into the coil. This causes an increase in suction pressure from the refrigerant boiling at a faster rate.**

24.17 TXV VALVE SELECTION

The TXV must be carefully chosen for a particular application. **Each TXV is designed for a particular refrigerant or group of refrigerants.** Many newer TXVs entering

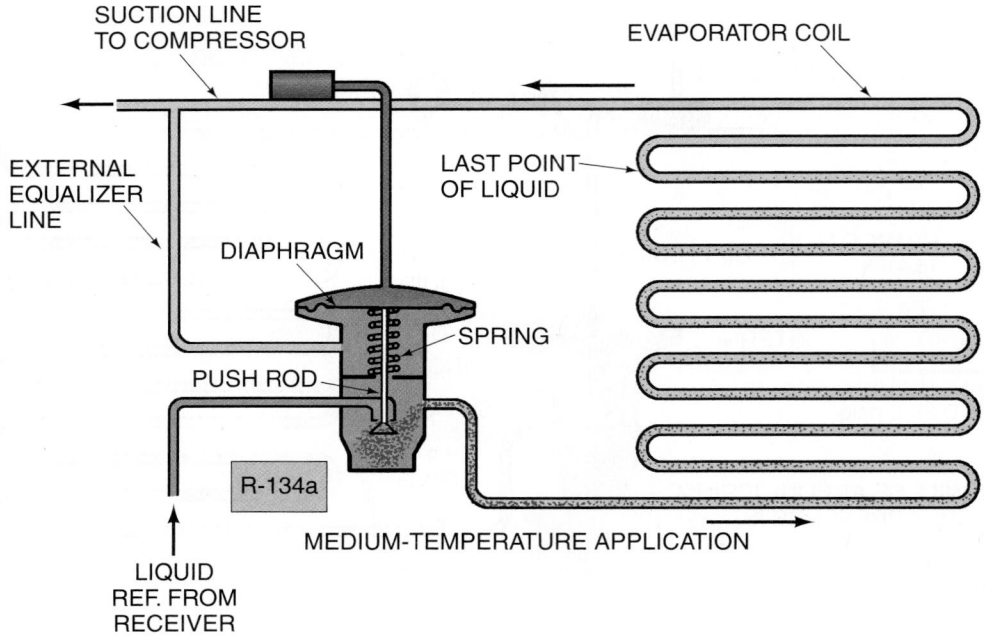

Figure 24–28   This illustration shows the correct method for connecting an external equalizer line.

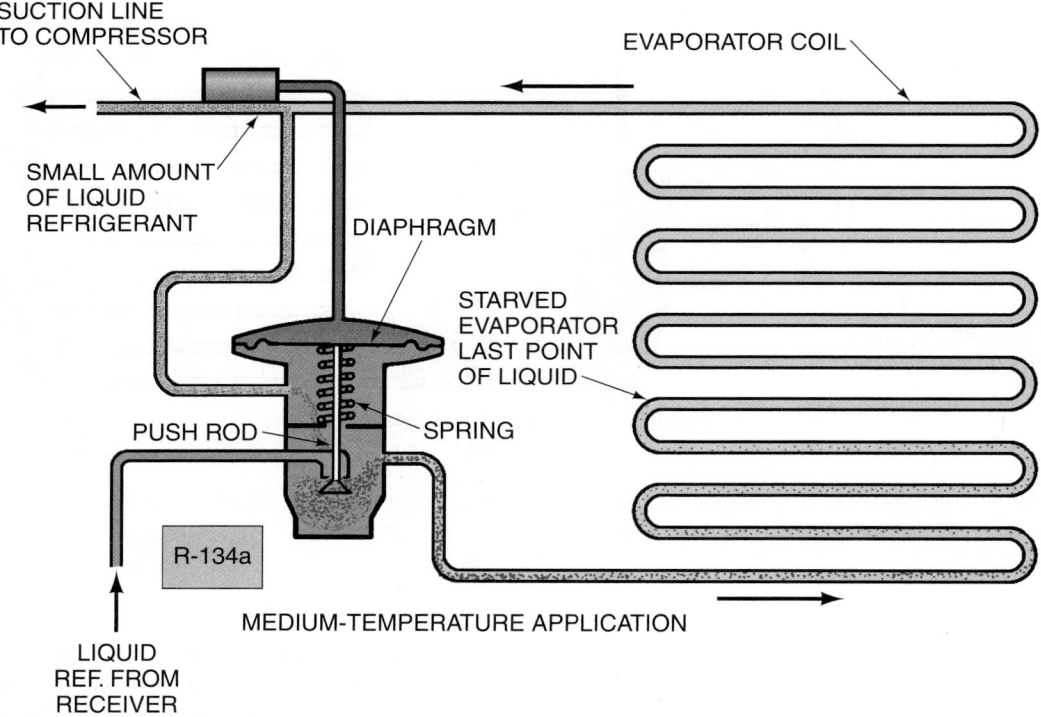

Figure 24–29   When the equalizer line is connected before the sensing bulb, a small leak in the internal parts of the valve will simulate a flooded evaporator.

the market are manufactured to handle some of the newer alternative refrigerants and refrigerant blends. These TXVs are designed to be used with more than one refrigerant. This helps in their selection process and makes it a lot easier for service technicians when retrofitting a system.

The capacity of the system is very important. If the system calls for a 1/2-ton expansion valve and a 1-ton valve is used, the valve will not control correctly because the needle and seat are too large. If a 1/4-ton valve were used, it would not pass enough liquid refrigerant to keep the coil full, and a starving coil condition would occur.

24.18 BALANCED-PORT TXV

The need for systems to operate in low ambient temperatures has led valve manufacturers to develop TXVs that will feed refrigerant at the same rate when the ambient temperatures are low. These valves allow refrigerant flows that do not reduce when the head pressure is low in mild weather. The evaporator will then have the correct amount of refrigerant and be able to operate at design conditions at lower outdoor temperatures. On larger TXVs, the flow of liquid on the face of the needle valve tends to open the valve, which can cause erratic superheat control. On the *balanced-port TXV,* the liquid pressure is actually cancelled out because it acts on equal areas but in opposite directions. **Figure 24–30(A)** illustrates the liquid force on the face of the needle valve. This can be a significant opening force for a larger TXV. This phenomenon, which can overfeed evaporators, happens because of the larger surface area associated with the needle valve faces on larger TXVs. This liquid force or opening force also becomes significant with a large pressure drop across the valve from either high head pressures, low suction pressures, or a combination of both. **Figure 24–30(B)** illustrates how the liquid force on a balanced-port TXV is cancelled out because it acts on equal areas but in opposite directions. If this liquid force is not cancelled out, erratic operation of

the TXV will occur. Also, the valve cannot hold its original superheat setting. This can cause compressor damage if certain system parameters change such as head pressure or suction pressure.

Balanced-port valves have a large orifice or needle valve area as compared with a conventional TXV. They also operate with the needle valve controlling very close to its seat, which provides a very stable control of evaporator superheat with a minimum movement of the valve. This allows a large orifice or needle valve assembly to handle both large and small loads. This type of design can help refrigeration equipment pull down faster. The larger orifice or needle valve assembly can also handle some liquid-line bubbles or flash gas if it exists. This is where the term *balanced port* originates. Balanced-port TXVs should be used if any of the following conditions exist:

- Large varying head pressures
- Larger varying pressure drops across the TXV
- Widely varying evaporator loads
- Very low liquid-line temperatures

You cannot tell a balanced-port expansion valve from a regular expansion valve by its appearance. It will have a different model number, and you can look it up in a manufacturer's catalog to determine the exact type of valve.

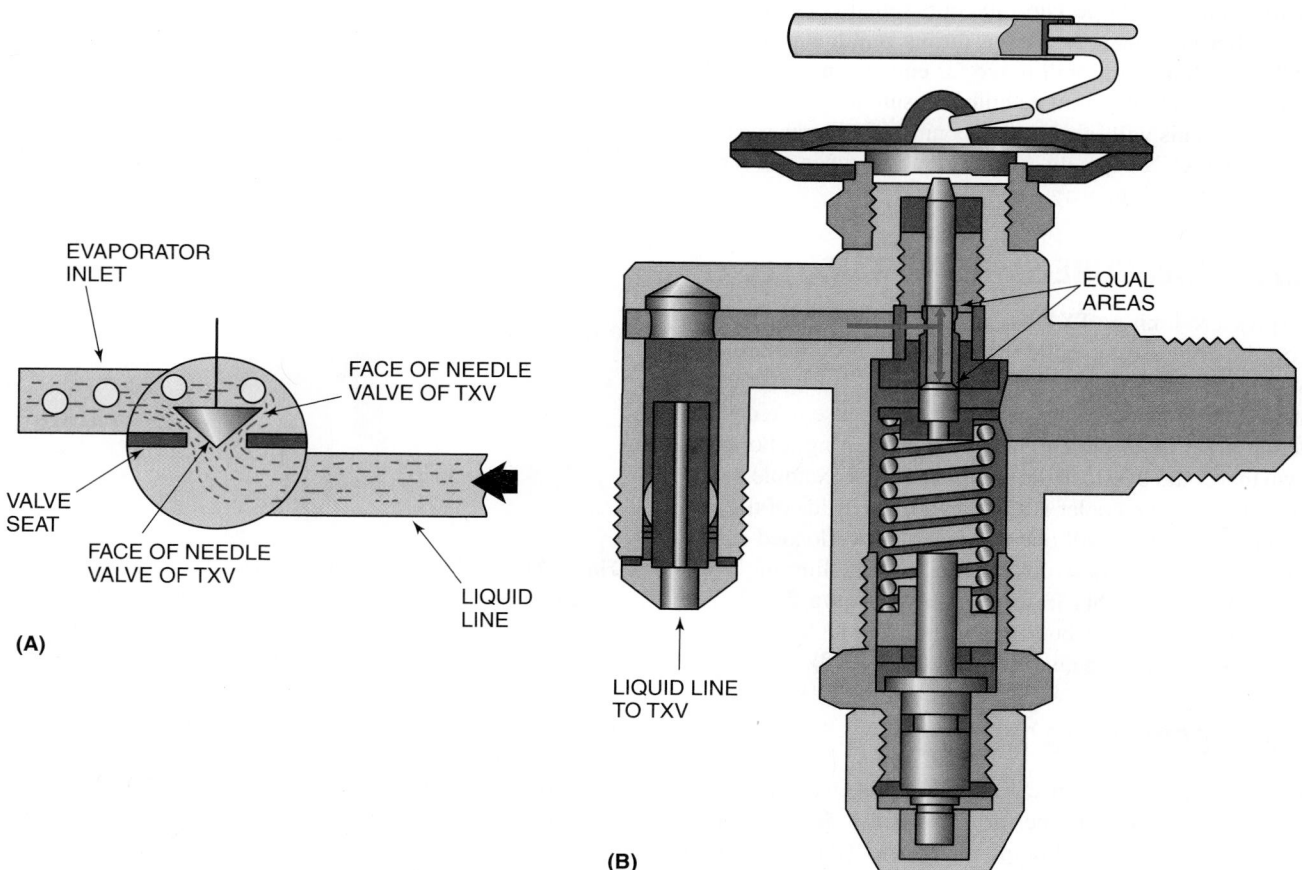

Figure 24–30 **(A)** The liquid force on the face of the needle valve, which can be a significant opening force for the TXV. **(B)** Equal liquid forces, but opposite in direction, which cancel out one another in a balanced-port TXV. **(B)** *Courtesy Copeland Corporation and Emerson Flow Controls*

24.19 DUAL-PORT TXV

There are times when the evaporator experiences excessive heat loadings and could use a larger TXV with a larger port area for a short period. Evaporators experience these excessive heat loadings after defrost periods, after short or long shutdowns (start-ups), or when product is being loaded into the refrigerated area. It is during these periods of time that the refrigeration system must operate in its pulldown mode until the box temperature begins to stabilize to its standard cooling or "holding load." If the TXV is sized to the holding load, it will be undersized at maximum loads. This will cause starved evaporators, high superheats, and long pulldown times after defrost. However, the valve will perform just fine in normal loading positions. If the valve is intentionally oversized to handle maximum loadings, problems will arise where the valve will overfeed and then underfeed trying to find equilibrium at which to control a constant evaporator superheat. This continuous overfeeding and underfeeding is referred to as *hunting*. Hunting can also occur as the outdoor ambient changes, thus changing the condensing pressure and pressure drop across the TXV. The balanced-port TXV is the only valve immune to capacity variations due to head pressure swings as seasons change.

Some newer technology has created a TXV with two independent capacities. It uses a larger port for pulldowns and a smaller port for holding loads, **Figure 24–31**. During the holding load, the large capacity port remains closed with the smaller port open. However, during pulldown loadings, the TXV's diaphragm will move far enough in the opening direction from higher remote bulb pressure to contact a sliding piston. This will initiate the opening of a larger capacity port. The TXV's capacity will be doubled when the larger port is opened all the way.

24.20 PRESSURE-LIMITING TXV

The pressure-limiting TXV has another bellows that will only allow the evaporator to build to a predetermined pressure, and then the valve will throttle the flow of liquid. This valve is desirable on low-temperature applications because it keeps the suction pressure to the compressor down during a hot pulldown that could overload the compressor. For example, when a low-temperature cooler is started with the inside of the box hot, the compressor will operate under an overloaded condition until the box cools down. The pressure-limiting TXV valve will prevent this from happening, **Figure 24–32**. The vapor-charged remote bulb application on TXVs will also have pressure-limiting qualities, **Figure 24–23(B)**.

24.21 SERVICING THE TXV

When any TXV is chosen, care should be taken that the valve is serviceable and will perform correctly. Several things should be considered: (1) type of fastener (flare, solder, or flange), (2) location of the valve for service and performance, and (3) expansion valve bulb location. This valve has moving parts that are subject to wear. When a valve must be

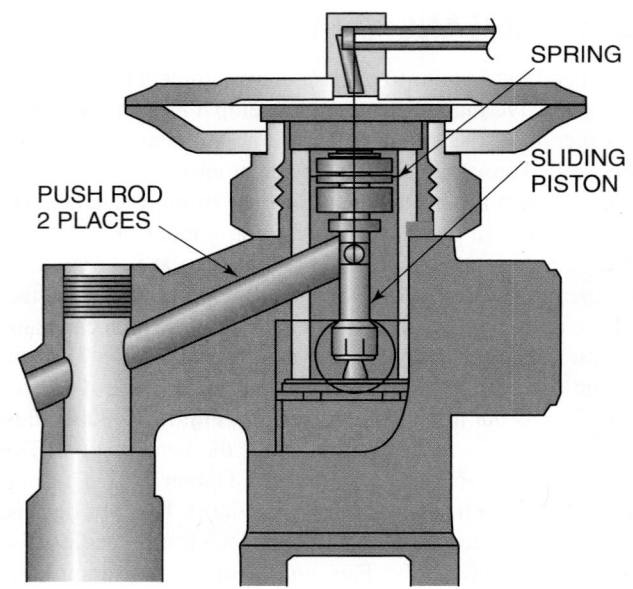

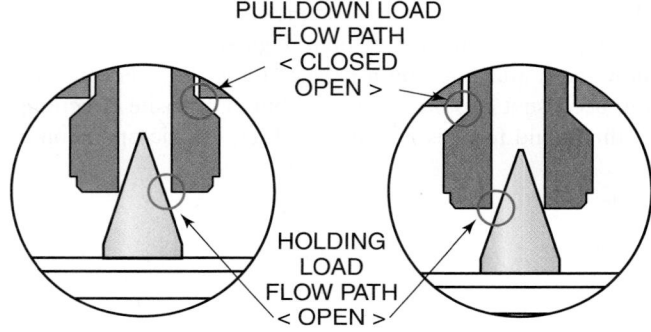

Figure 24–31 A dual-port TXV showing both orifices and the sliding piston. *Courtesy Sporlan Valve Company*

Figure 24–32 A pressure-limiting expansion valve. When the pressure is high in the evaporator, such as during a hot pulldown, this valve will override the thermostatic element with a pressure element and throttle the refrigerant. *Courtesy Sporlan Valve Company*

replaced, the exact replacement is usually best. When this is not possible, a supplier can furnish you with the information needed to change to another valve.

Many technicians (and owners) will adjust the TXV. This may cause problems because it will change pressures and temperatures in a coil. If the valve sensing element is mounted properly and in the correct location where it can sense refrigerant vapor, the valve should work correctly as shipped from the factory. These valves are very reliable and normally do not require any adjustment. If there are signs that the valve has been adjusted, look for other problems, because the valve probably did not need adjustment to begin with. Many manufacturers are producing nonadjustable valves to prevent people from adjusting the valve when not needed.

24.22 SENSING ELEMENT INSTALLATION

Particular care should be taken when installing the expansion valve sensing element. Each manufacturer has a recommended method for this installation, but they are all similar. The valve sensing bulb has to be mounted at the end of the evaporator on the suction line. The recommended location for the thermal bulb on smaller suction lines is near the top of the line. The best location for larger suction lines is near the bottom of the line on a horizontal run where the bulb can be mounted flat and not be raised by a fitting, **Figure 24–33**. The bulb should not be located at the bottom of the line because there will be oil returning that will act as an insulator to the sensing element. The object of the sensing element is to sense the temperature of the refrigerant gas in the suction line. To do this, the line should be very clean and the bulb should be fastened to this line very securely. Normally the manufacturer suggests that the bulb be insulated from the ambient temperature if the ambient is much warmer than the suction-line temperature, because the bulb will be influenced by the ambient temperature as well as by the line temperature.

24.23 THE SOLID-STATE CONTROLLED EXPANSION VALVE

The solid-state controlled expansion valve uses a thermistor as a sensing element to vary the voltage to a heat motor-operated valve (a valve with a bimetal element). This valve normally uses 24 V as the control voltage to operate the valve, **Figure 24–34(A)**.

When voltage is applied to the coil in the valve, the valve opens. Modulation is accomplished by varying the voltage, **Figure 24–34(B)**. The valve is versatile and can be used to accomplish different functions in the system. When the voltage is cut off at the end of the cycle, the valve will close, and the system can be pumped down. If the voltage is allowed to remain on the element, the valve will remain open during the off cycle, and the pressures will equalize.

The thermistor is inserted into the vapor stream at the end of the evaporator. It is very small in mass and will respond quickly to temperature changes.

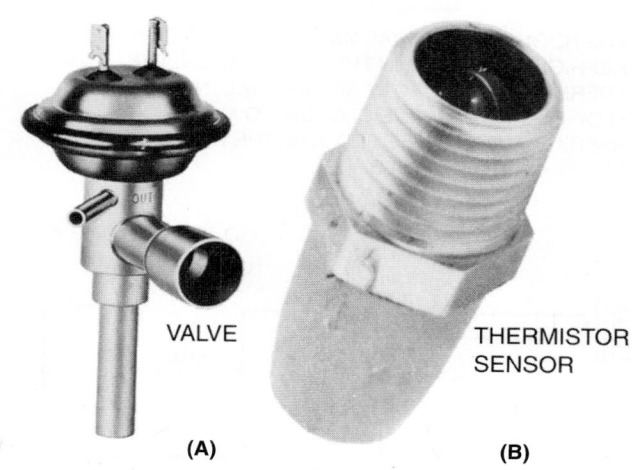

Figure 24–34(A) An expansion valve controlled with a thermistor and a heat motor. *Courtesy Singer Controls Division*

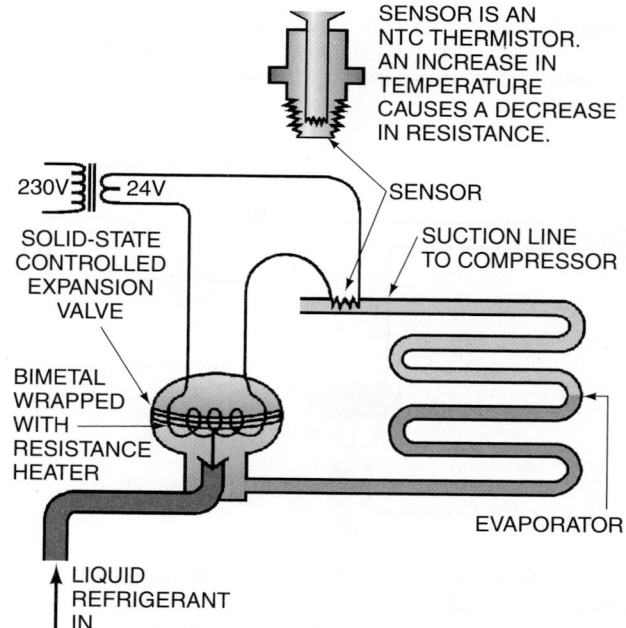

Figure 24–34(B) The temperature of the sensor controls the current flow to the bimetal in the valve. More current flow causes the bimetal to warp and open the valve. Less current flow cools the bimetal and throttles the valve closed.

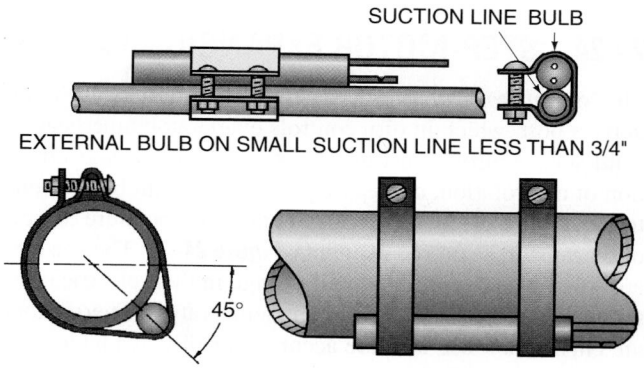

SUCTION LINE BULB

EXTERNAL BULB ON SMALL SUCTION LINE LESS THAN 3/4"

45°

EXTERNAL BULB ON LARGE SUCTION LINE OVER 7/8"

Figure 24–33 The best positions for mounting the expansion valve sensing bulb. *Courtesy ALCO Controls Division, Emerson Electric Company*

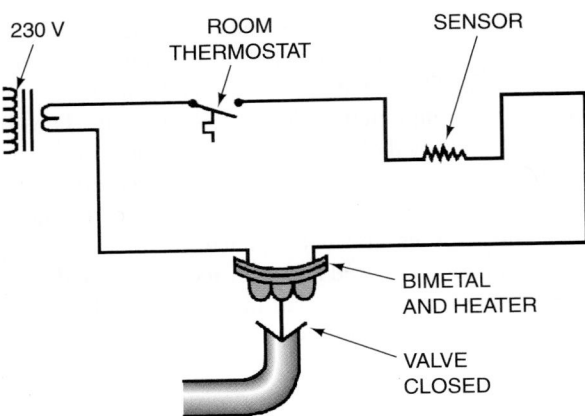

THE ROOM THERMOSTAT MAY BE USED TO CONTROL REFRIGERANT FLOW. IN THIS APPLICATION, THE ROOM THERMOSTAT CONTACTS ARE OPENED, NO CURRENT FLOWS, AND THE VALVE CLOSES. NO MORE REFRIGERANT WILL BE FED INTO THE EVAPORATOR DURING THE OFF CYCLE.

(1)

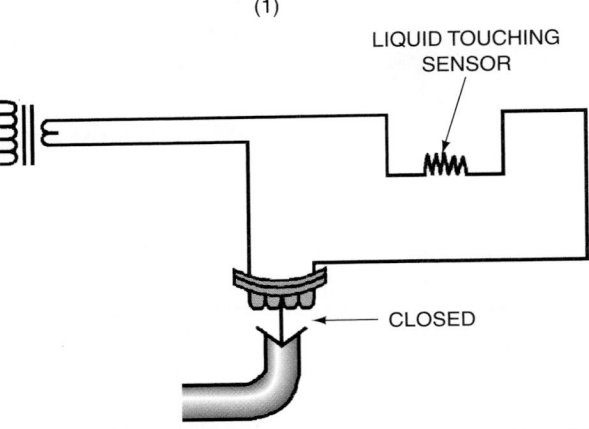

LIQUID REFRIGERANT TOUCHING THE SENSOR WILL COOL IT AND INCREASE THE RESISTANCE, DECREASING THE CURRENT FLOW AND THROTTLING THE VALVE TOWARD CLOSED.

(2)

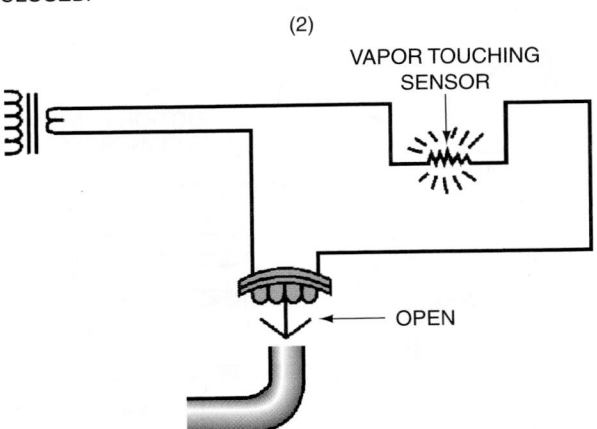

WHEN VAPOR TOUCHES THE SENSOR, THE CURRENT INCREASES, HEATING THE BIMETAL AND OPENING THE VALVE. WITHOUT THE THERMOSTAT SHOWN IN (1), THE SENSOR WOULD BE WARM DURING THE OFF CYCLE, WHICH WOULD OPEN THE VALVE AND EQUALIZE THE PRESSURE.

(3)

Figure 24–34(C) Different applications and actions for the solid-state (thermistor) controlled valve.

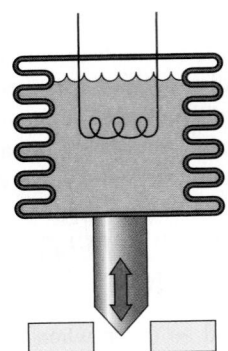

Figure 24–35 An electric heater immersed in a volatile fluid within the heat motor diaphragm. *Courtesy Sporlan Valve Company*

The solid-state controlled expansion valve responds to the change in temperature of the sensing element like the typical TXV except that it does not have a spring. When the thermistor is suspended in dry vapor, it is heated by the current passing through it. This creates a faster response than does merely measuring the vapor temperature, **Figure 24–34(C)**. When the valve opens and saturated vapor reaches the element, the valve begins to close slightly. 🌐This valve controls to a very low superheat, which allows the evaporator to use maximum surface area.🌐

The solid-state controlled expansion valve is unique because refrigerant can flow in either direction through the valve body. It is suitable for heat pump applications because of this and is used on packaged heat pumps where the manufacturer can build a system with only one expansion valve allowing refrigerant to flow in either direction.

Some heat-motor-operated valves use a small electric heater immersed in a volatile fluid within the valve. When the heater is energized by varying the power to it, the heat generated is used to expand the bulb's volatile fluid, which causes movement of the valve, **Figure 24–35**. A thermistor is usually used on the evaporator outlet to vary the amount of heat the heating element transfers to the volatile fluid within the valve. The thermistor is the sensor and it is used to control evaporator superheat.

24.24 STEP-MOTOR EXPANSION VALVES

The *step-motor valve* uses a small motor to control the valve's port, which in turn controls evaporator superheat. A solid-state controller instructs the step motor to rotate a fraction of a revolution, or step, for each signal sent by the controller, **Figure 24–36**. The controller has a temperature sensor, or thermistor, as one of its inputs, **Figure 24–37**. Thermistors are solid-state devices that will change in their electrical resistance in response to a change in temperature. Thermistors are chosen because they are accurate, readily available, and reasonably priced. Once the electrical signal is taken away, the motor will stop. The motor does not rotate continuously. The motor can return the valve to any previous position at any time because the controller can remember the number of steps it has taken. Many of these motors can operate at a rate of 200 steps per second.

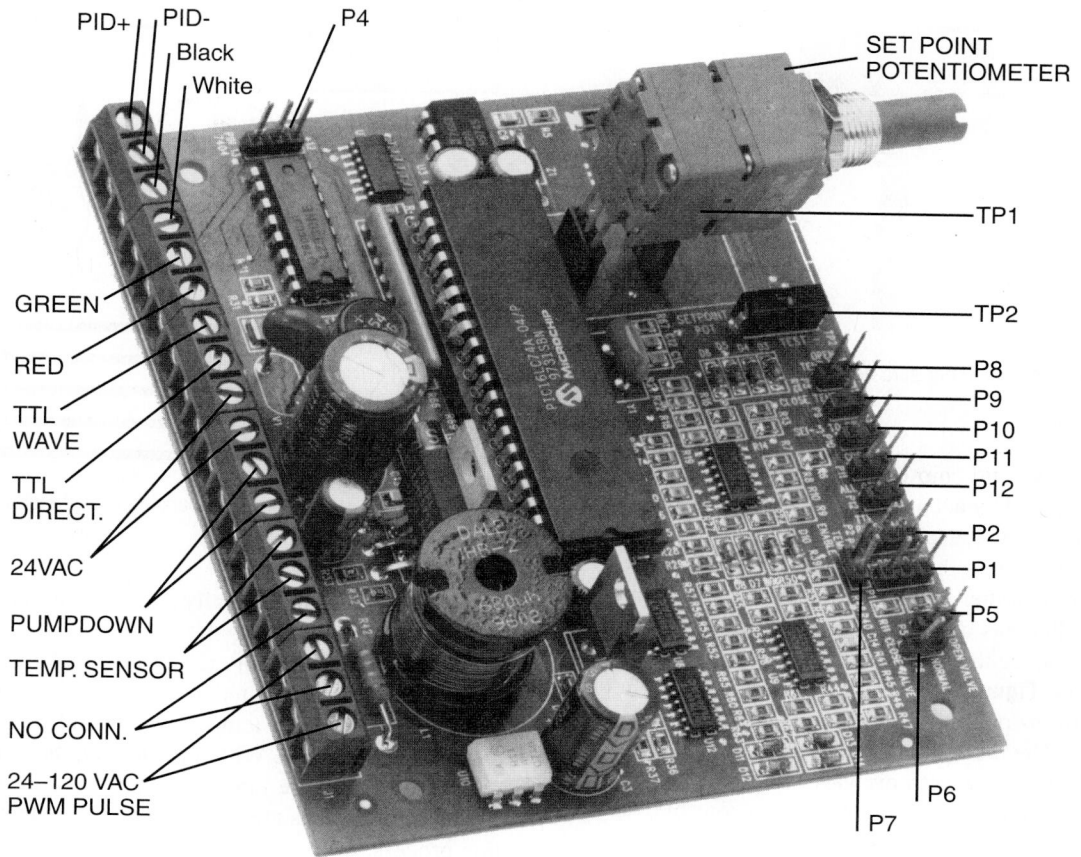

PID+
PID-
Black
White
P4
SET POINT POTENTIOMETER
TP1
TP2
GREEN
RED
TTL WAVE
TTL DIRECT.
24VAC
PUMPDOWN
TEMP. SENSOR
NO CONN.
24–120 VAC PWM PULSE
P8
P9
P10
P11
P12
P2
P1
P5
P6
P7

Figure 24–36 A solid-state step-motor controller. *Courtesy Sporlan Valve Company*

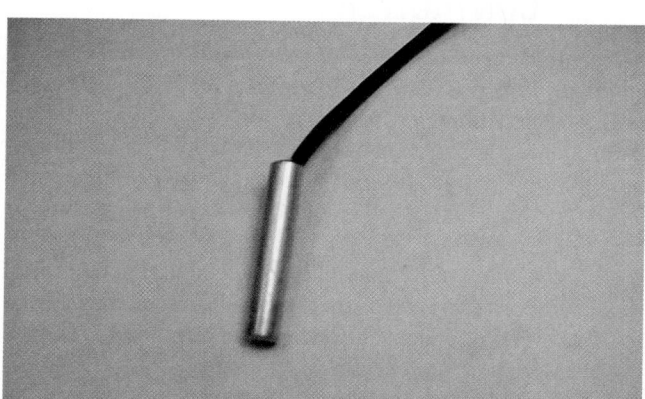

Figure 24–37 A thermistor used as a temperature sensor for the controller input. *Photo by John Tomczyk*

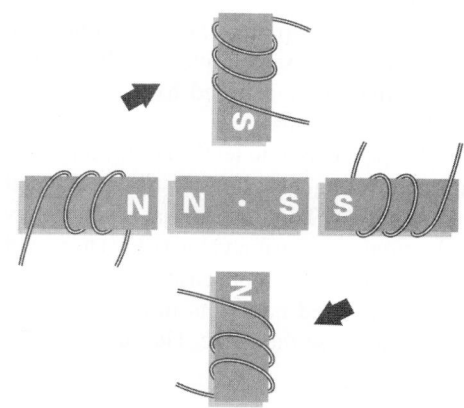

Figure 24–38 Electromagnets placed around a permanent magnet rotor. *Courtesy Sporlan Valve Company*

Most modern step motors used in electronic expansion valve applications use a bipolar motor design. Multiple electromagnets are placed around a permanent magnet acting as the motor's rotor, **Figure 24–38.** There can be as many as 100 electromagnets placed around the rotor. They can have step increments as small as 3.6° in rotation. This rotation comes from a total of 360° divided by 100 magnets. The pin or valve port piston of the electronic step-motor expansion valve can move in a linear motion of 0.0000783 in./step.

The motor is powered by signals that change in polarity. This creates a push/pull action within the motor's magnetic

fields and increases the torque and efficiency of the motor. This allows for a very small motor. The torque, or rotational force, is then translated into linear motion by a digital linear actuator, **Figure 24–39.** The motor and valve is thus a true modulating device. It can be compared to the dimmer switch on a typical household light, which has thousands of steps. It is not a pulsing digital motor or a common light switch that has only two steps, on or off.

The controller consists of many transistors for each switching function. Transistors are solid-state switches that

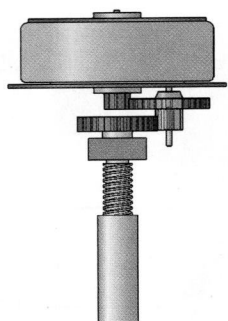

Figure 24–39 A digital linear actuator. *Courtesy Sporlan Valve Company*

are made from a solid chip of silicone. Transistors have no moving parts and they act as relays by using small electrical signals to turn a larger electrical signal on and off. The small electrical signal enters the base of the transistor, which allows flow from emitter to collector. A small microcomputer in the controller has a programmed *algorithm,* which can control or sequence the electrical signal given to the base of each transistor, **Figure 24–40.** An algorithm is simply a program or set of instructions resident in the *microprocessor* of the controller. The transistors are usually turned on and off in pairs, which step the electronic valve in either the open or the closed direction in small increments to control evaporator superheat.

The controls of electronic expansion valves (EEVs) use algorithms, and a set of electronic instructions, along with a *proportional, integral, and derivative (PID) controller.* PID controllers will be explained in the next section. The valve's hardware includes the step motor, the controller, transistors, and the microprocessor. As mentioned earlier, algorithms or sets of instructions, often referred to as *software,* must be resident in the microprocessor.

Feedback loops are used to let the controller know that the process of controlling evaporator superheat needs to be changed or modified. This means that when the controller opens the electronic expansion valve (EEV) too much, which in turn overcools the thermistor sensor on the evaporator outlet, the sensor will feed back this information so the controller can start to close the valve, **Figure 24–41.** The sensor

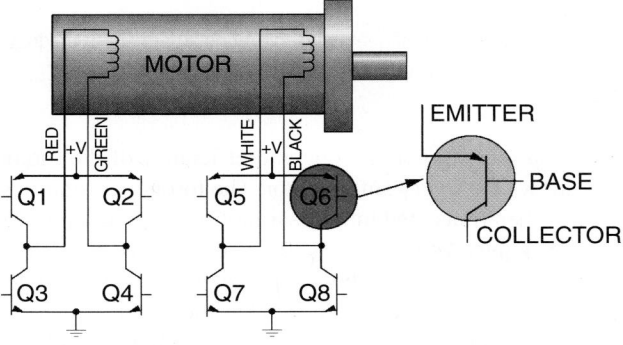

Figure 24–40 The multiple transistors that make up the step motor. *Courtesy Sporlan Valve Company*

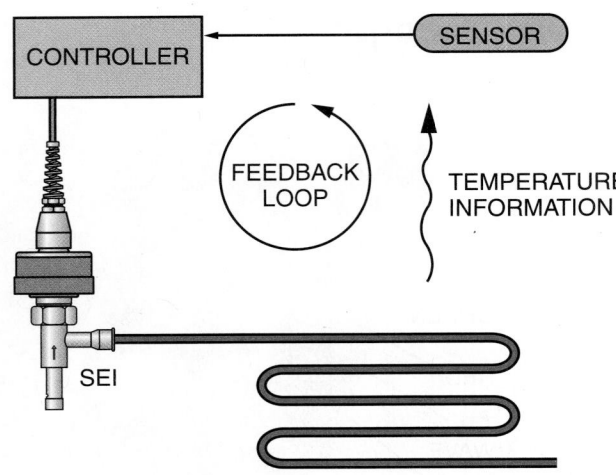

Figure 24–41 The electronic expansion valve feedback loop. *Courtesy Sporlan Valve Company*

in the feedback loop is usually a thermistor that is located on the evaporator outlet.

Often, two thermistors are used to calculate evaporator superheat. They are located on the evaporator's inlet and outlet. The difference in temperature would be the calculated evaporator superheat. With this method, there is no need for temperature/pressure tables or for a pressure transducer to sense pressure. This greatly simplifies the superheat calculation process.

24.25 ALGORITHMS AND PID CONTROLLERS

As mentioned earlier, the controls of EEVs use many different algorithms or sets of instructions along with a PID controller. The following paragraphs will explain the PID controller. Although we start with explaining proportional, integral, and derivative controls separately, in reality they work together to modify the controller's output signal to the EEV for less error.

Proportional controllers refer to a modulating controller mode. Proportional controllers generate an immediate change in the position of the expansion valve in response to changes in the evaporator superheat of the system. They generate an *analog* (modulating) output signal instead of a *binary* (on/off) output signal. The output signal of a proportional controller is always present, but its magnitude will vary. Proportional modes can only be used in closed-loop, feedback applications like that of the EEV. In proportional mode, the actual evaporator superheat will "approach" the evaporator superheat *set point* of the controller, but may never reach it in all applications. The difference between the superheat set point of the controller and the actual superheat is called *offset* or *error,* **Figure 24–42.** Proportional mode controllers change or modify the controller's output signal to the expansion valve in proportion to the size of the change in the error. It cannot sense, and has no control over, any time span that the superheat error has been in existence. For example, if the thermistor on the evaporator's outlet is very warm, meaning high superheat to

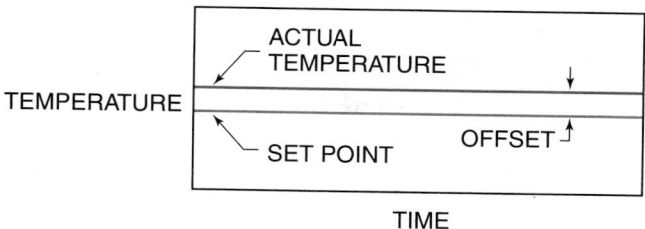

Figure 24–42   This temperature/time graph shows set point and offset or "error." *Courtesy Sporlan Valve Company*

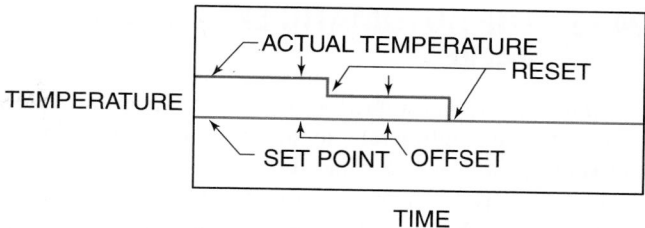

Figure 24–43   The temperature/time graph of an integral controller showing set point, offset, and reset. *Courtesy Sporlan Valve Company*

the controller, a large movement of the expansion valve in the opening direction would follow. In this example, the controller saw a large error between the evaporator superheat set point and the actual evaporator superheat, and compensated for it with a large valve movement.

A typical programmed algorithm for a proportional controller is listed here:

> If evaporator superheat is 20°F, then open the valve 175 steps.
>
> If evaporator superheat is 15°F, then open the valve 125 steps.
>
> If evaporator superheat is 10°F, then open the valve 0 steps.
>
> If evaporator superheat is 6°F, then close the valve 75 steps.
>
> If evaporator superheat is 0°F, then close the valve 1000 steps.

This algorithm will change the output (motor steps) in direct relation to the input (superheat). However, if the algorithm is used, the valve will slowly reach the controller's superheat set point, especially if the step motor has over 5000 steps. This is why a more sophisticated algorithm including integral and derivative functions must be used.

If the offset or error was a constant value as shown in **Figure 24–42**, this constant difference could simply be preprogrammed into the controller. However, offset will change over time as refrigeration system parameters change with changing heat loads. This is why some means of predicting the offset must be used to lessen its value, and let the controller reach its evaporator superheat set point value. This is where the integral controller comes into play.

Integral controller modes can be added to a proportional controller to modify the controller's output signal, which will help force the offset value error to zero. The integral mode will change the controller's output signal in proportion to the size of the error or offset and the length of time it exists. The integral mode will calculate the amount of error that exists, over a specific time interval. The integral controller actually calculates the area under the curve of the temperature/time graph, **Figure 24–43**. In mathematical terms, finding an area under a curve is often called integration. Integration, or the integral mode, of a controller will sense the average deviation of the actual evaporator superheat from the controller's superheat set point. Following is a typical integral algorithm example:

> If the average superheat for 60 seconds is 6°F high, then open the valve 50 steps.

Thus, the offset or error in superheat is constantly being changed as heat load and refrigeration system parameters change. The changing amount of offset is calculated by the control algorithm and is then added to the set point. Another name for integral control is reset control. The calculation of the time-weighted offset (error), or the area under the curve of the time/temperature graph, occurs over a period of time referred to as the reset time. The controller will constantly try to reset the measured evaporator superheat to the superheat set point of the controller.

Derivative (differential) controller modes can also be added to a proportional controller, which will ensure even closer evaporator superheat control. The terms *derivative* and *differential* are used interchangeably. Derivative algorithms estimate the slope, or rate of change, of the temperature/time curve, **Figure 24–44**. If the rate of change (slope) is great, then the controller will make larger changes. If the slope is small, the controller will make small changes. Derivative control actually calculates the instantaneous rate of change of the error with respect to time. It can tell how fast the process error is changing. The controller can then adjust the magnitude of its output signal to the EEV in proportion to the rate of change of the error in evaporator superheat of the system. This type of control reacts quickly to changes and will reduce any effect that changes in evaporator loads or system parameters have on the evaporator superheat set point. A typical derivative algorithm is shown here:

> If superheat has dropped 0.2°F in 10 seconds, then close the valve 15 steps.
>
> If superheat has dropped 2.0°F in 2 seconds, then close the valve 50 steps.

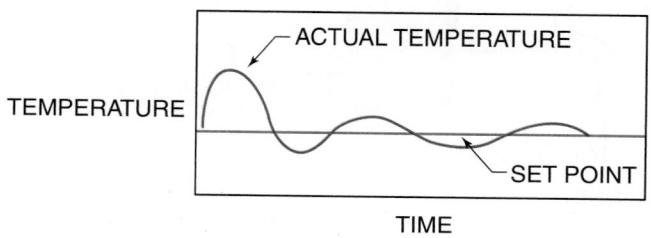

Figure 24–44   The temperature/time graph of a derivative, or differential, controller showing slopes or rates of change. *Courtesy Sporlan Valve Company*

24.26 THE AUTOMATIC EXPANSION VALVE

The *automatic expansion valve (AXV)* is an expansion device that meters the refrigerant to the evaporator by using a pressure-sensing device. This device is also a valve that changes in inside dimension in response to its sensing element. **The AXV maintains a constant pressure in the evaporator. Notice that superheat was not mentioned.** This device has a needle and seat like the TXV that is fastened to a diaphragm, **Figure 24–45.** One side of the diaphragm is common to the evaporator and the other side to the atmosphere. When the evaporator pressure drops for any reason, the valve begins to open and feed more refrigerant into the evaporator.

The AXV is built much the same as the TXV except that it does not have a sensing bulb. The body is normally made of machined brass. The adjustment of this valve is normally at the top of the valve. There may be a cap to remove, or there may be a cap to turn. This adjustment changes the spring tension that supports the atmosphere in pushing down on the diaphragm. When the tension is increased, the valve will feed more refrigerant and increase the suction pressure, **Figure 24–46.**

24.27 AUTOMATIC EXPANSION VALVE RESPONSE TO LOAD CHANGES

The AXV responds differently than the TXV to load changes. It actually acts in reverse. When a load is added to the coil, the suction pressure starts to rise. The AXV will start to throttle the refrigerant by closing enough to maintain the suction pressure at the set point. This has the effect of starving the coil slightly. A large increase in load will cause more starving. When the load is decreased and the suction pressure starts to fall, the AXV will start to open and feed more refrigerant into the coil. If the load reduces too much, liquid could actually leave the evaporator and proceed down the suction line, **Figure 24–47.**

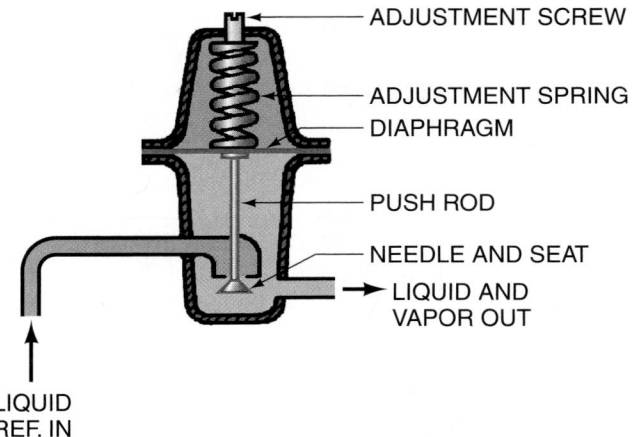

Figure 24–45 The automatic expansion valve uses the diaphragm as the sensing element and maintains a constant pressure in the evaporator but does not control superheat.

ADJUSTMENT SCREW
ADJUSTMENT SPRING
DIAPHRAGM
PUSH ROD
NEEDLE AND SEAT
LIQUID AND VAPOR OUT
LIQUID REF. IN

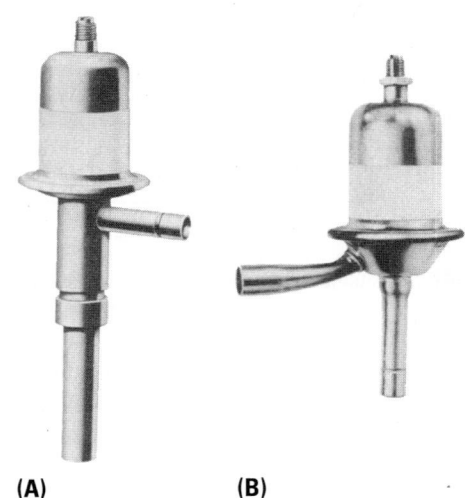

(A)   **(B)**

Figure 24–46 Automatic expansion valves **(A)–(B)**. They resemble the TXV, but they do not have the bulb for sensing temperature at the suction-line end of the evaporator. *Courtesy Singer Controls Division*

We can see from these examples that this valve responds in reverse to the load. The best application for this valve is where there is a fairly constant load. One of its best features is that it can hold a constant pressure. When this valve is applied to a water-type evaporator, freezing will not occur.

Because of the AXV's constant pressure characteristic, it cannot be used with a low-pressure motor control. When the AXV is used with a thermostat, the thermostat should be adjusted to shut the compressor off before flooding of the entire evaporator occurs. The AXV should be adjusted to maintain an evaporator pressure that corresponds to the lowest evaporator temperature desired through the entire running cycle of the compressor. AXVs are usually found on smaller equipment with somewhat constant evaporator loads, including ice cream freezers and makers and drinking fountains.

24.28 SPECIAL CONSIDERATIONS FOR THE TXV AND AXV

The TXV and the AXV both are expansion devices that allow more or less refrigerant flow, depending on the load. Both need a storage device (receiver) for refrigerant when it is not needed. The receiver is a small tank located between the condenser and the expansion device. Normally the condenser is close to the receiver. It has a *king valve* that functions as a service valve. This valve stops the refrigerant from leaving the receiver when the low side of the system is serviced. This receiver can serve both as a storage tank for different load conditions and as a tank into which the refrigerant can be pumped when servicing the system, **Figure 24–48.**

24.29 THE CAPILLARY TUBE METERING DEVICE

The *capillary tube* metering device controls refrigerant flow by pressure drop. It is a copper tube with a very small calibrated inside diameter, **Figure 24–49(A).** The diameter and

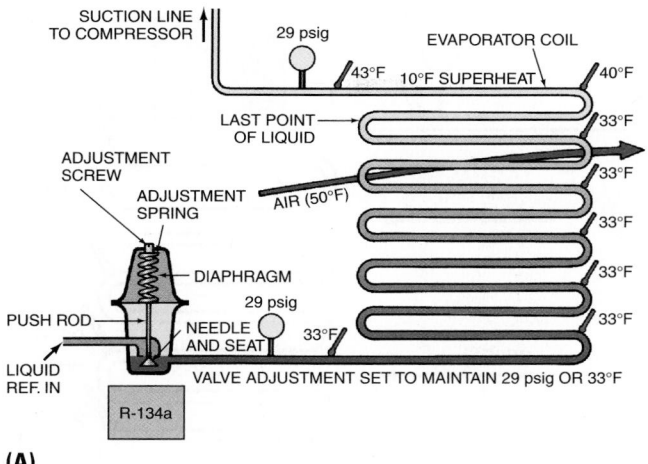

(A)

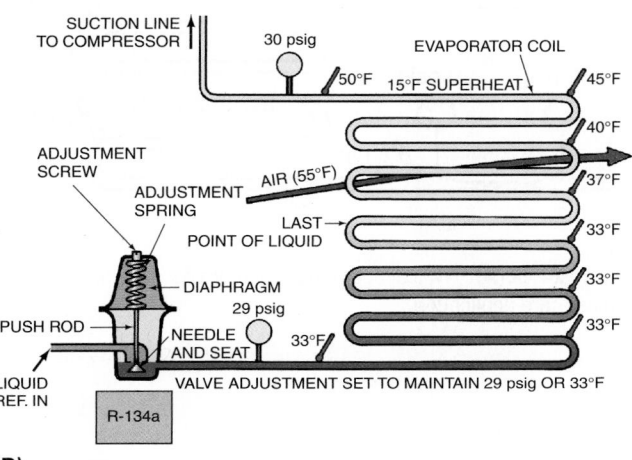

(B)

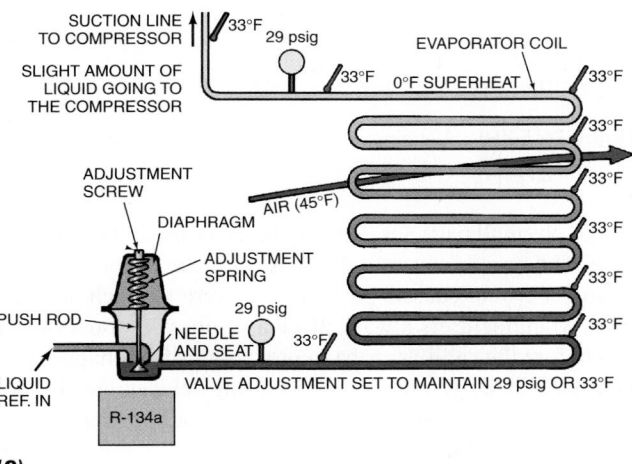

(C)

Figure 24–47 An automatic expansion valve under varying load conditions. This valve responds in reverse to a load change. **(A)** Normal operation. **(B)** When the load goes up, the valve closes down and starts to starve the coil slightly to keep the evaporator pressure from rising. **(C)** When the load goes down, the valve opens up to keep the evaporator pressure up. This valve is best applied where the load is relatively constant. When applied to a water-type evaporator, freeze protection can be a big advantage with this valve.

Figure 24–48 A refrigerant receiver. When the load increases, more refrigerant is needed and moves from the receiver into the system. *Courtesy Refrigeration Research*

the length of the tube determine how much liquid will pass through the tube at any given pressure drop, **Figure 24–49(B)**. It is much like a garden hose; the larger the hose, the more gallons of water will pass at 100 psig pressure at the hose inlet, **Figure 24–49(C)**. The water hose has pressure drop all along the length, so a longer hose will have less pressure at the outlet even though the pressure at the inlet is still 100 psig.

This length-to-pressure drop relationship is used by manufacturers to arrive at the correct pressure drop that will allow the correct amount of refrigerant to pass through the capillary tube to correctly fill the evaporator. The capillary tube can be quite long on some installations and may be wound in a coil to store the extra tubing length.

The capillary tube does not control superheat or pressure. It is a fixed-bore device with no moving parts. Because this device cannot adjust to load change, it is usually used where the load is relatively constant with no large fluctuations.

The capillary tube is an inexpensive device for the control of refrigerant and is used often in small equipment. This device does not have a valve and does not stop the liquid from moving to the low side of the system during the off cycle, so the pressures will equalize during the off cycle. This reduces the motor starting torque requirements for the compressor.

Figure 24–49(A) A capillary tube metering device. *Courtesy Parker Hannifin Corp.*

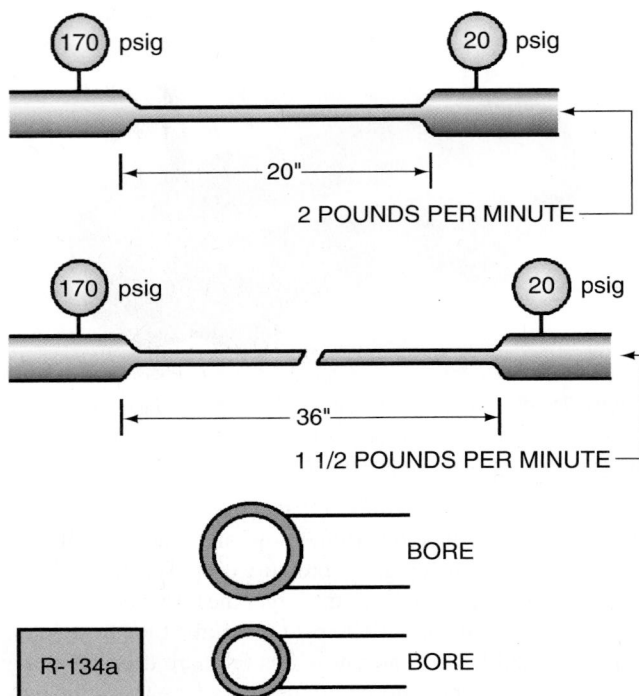

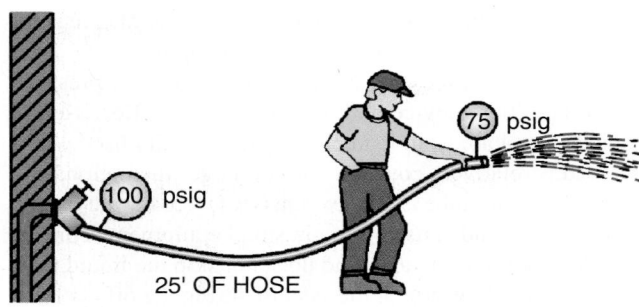

Figure 24–49(B) The length of the capillary tube as well as the bore determines the flow rate of the refrigerant, which is rated in pounds per minute.

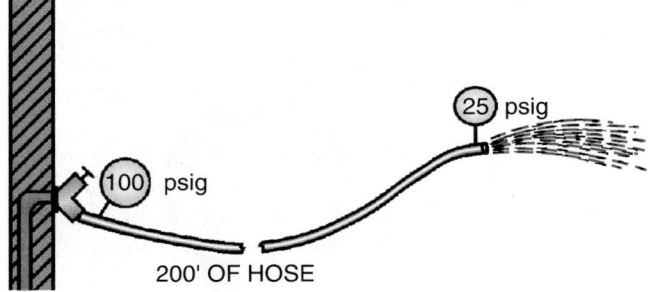

Figure 24–49(C) The longer the hose, the less water pressure at the end. A capillary tube functions the same way.

Figure 24–50(A) illustrates a capillary tube in place at the inlet of the evaporator.

The technician should become familiar with the capillary tube metering device because it is probably the most used device for metering refrigerant. It has no moving parts, so it will not wear out. About the only problems it may have

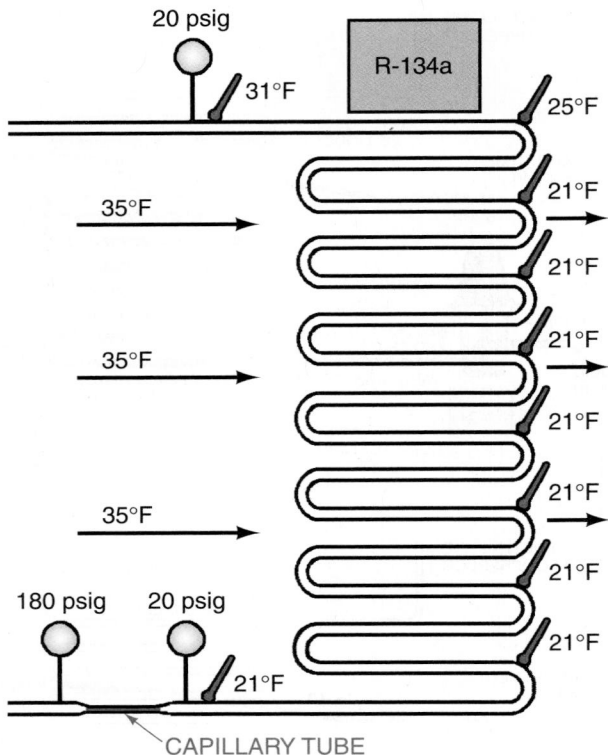

Figure 24–50(A) Typical operating conditions for a medium-temperature application.

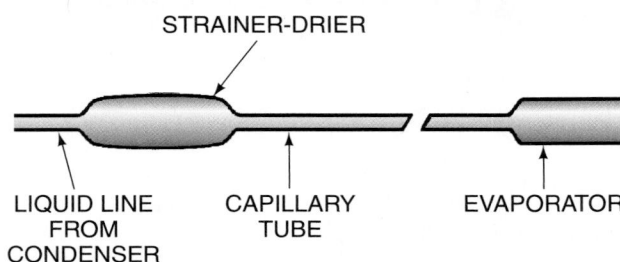

Figure 24–50(B) A strainer-drier protects the capillary tube from circulating particles.

would be small particles that may block or partially block the tube. The bore is so small that a small piece of flux, carbon, or solder would cause a problem if it were to reach the tube inlet. Manufacturers always place a strainer or strainer-drier just before the capillary tube to prevent this from happening, **Figure 24–50(B).** Capillary tube metering devices can also become oil logged. This can cause a partial or complete blockage of the capillary tube. Oil-logged capillary tubes can be caused by too much oil, or the wrong viscosity of oil, in the refrigeration system. Another cause of oil logging the capillary tube is liquid refrigerant entering the crankcase of the compressor and then flashing. This is referred to as *flooding* the crankcase when it occurs during the run cycle. It is called *migration* when it occurs during the off cycle. This flashing of refrigerant in the warm compressor's crankcase causes the oil in the crankcase to foam and be sucked into the suction valve of the compressor during the run cycle. The oil is then circulated until it eventually reaches the capillary tube.

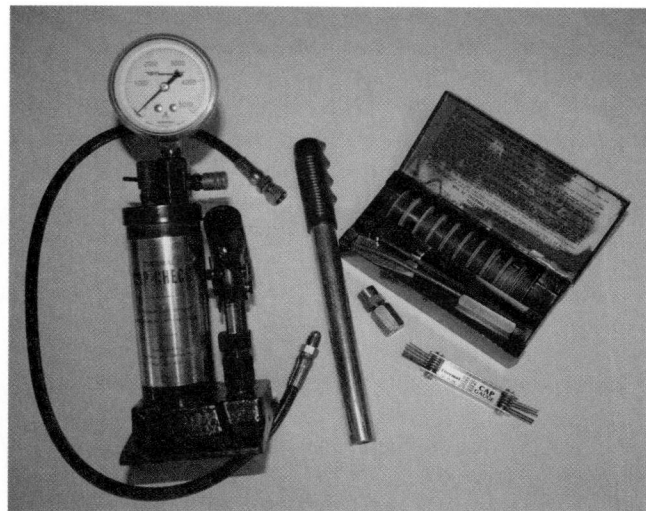

Figure 24–51 Hydraulic capillary-tube cleaning kit. *Courtesy Ferris State University. Photo by John Tomczyk*

Oil-logged capillary tubes cause sluggish operations, and systems usually run with high superheats. This often causes inefficient refrigeration systems and can damage compressors from overheating because of lack of refrigerant in circulation for cooling. Often, long running cycles will return the oil to the compressor's crankcase. If oil in the capillary tube is suspected, force the system to run for several hours. This may resolve the problem of an oil-logged capillary tube. There are hydraulic capillary tube cleaners on the market, but they should be used as a last resort, **Figure 24–51.**

24.30 OPERATING CHARGE FOR THE CAPILLARY TUBE SYSTEM

The capillary tube system requires only a small amount of refrigerant because it does not modulate (feed more or less) refrigerant according to the load. Capillary tube systems are known as critically charged systems. When the refrigerant charge is analyzed and when the unit is operating at the design conditions, a specific amount of refrigerant is in the evaporator, and a specific amount of refrigerant is in the condenser. This is the amount of refrigerant required for proper refrigeration. Any other refrigerant that is in the system is in the pipes for circulating purposes only.

The amount of refrigerant in the system is critical in capillary tube systems. It is easy for technicians to overcharge the system if they are not careful and familiar with the system. In most capillary tube systems, the charge is printed on the nameplate of the equipment. The manufacturer always recommends measuring the refrigerant into these systems either by scales or by liquid-charging cylinders. A properly charged system would maintain about 10°F superheat at the end of the coil once the system is pulled down to the desired box temperature, **Figure 24–50(A).** A characteristic of a capillary tube metering device system is to run high evaporator superheat under high evaporator heat loads. However, once the system is pulled down near the desired refrigerated box temperature, an evaporator superheat of

about 8°F to 12°F will be reached if the system is properly critically charged.

In many capillary tube systems, the capillary tube is fastened to the suction line between the condenser and the evaporator to exchange heat between the capillary tube and the suction line. To troubleshoot a possible capillary tube problem, the correct superheat reading must be taken. The capillary tube should hold the refrigerant to about 10°F superheat at the end of the evaporator. *To obtain the correct superheat reading it must be taken before the capillary tube/suction line heat exchanger,* **Figure 24–52.**

The capillary tube is very slow to respond to load changes or charge modifications. For example, if a technician were to add a small amount of refrigerant to the capillary tube system, it would take at least 15 min for the charge to adjust. The reason for this is that the refrigerant moves from one side of the system to the other through the small bore and this takes time. When you add refrigerant to the low side of the system, it moves into the compressor and is pumped into the condenser. It now must move to the evaporator before the charge is in balance. Many technicians get in a hurry and overcharge this type of system. Manufacturers do not recommend adding refrigerant to top off the charge; they recommend starting over from a deep vacuum and measuring a complete charge into the system.

Under high evaporator loads, capillary tube systems will often cause high evaporator superheats of 40°F or 50°F. This is

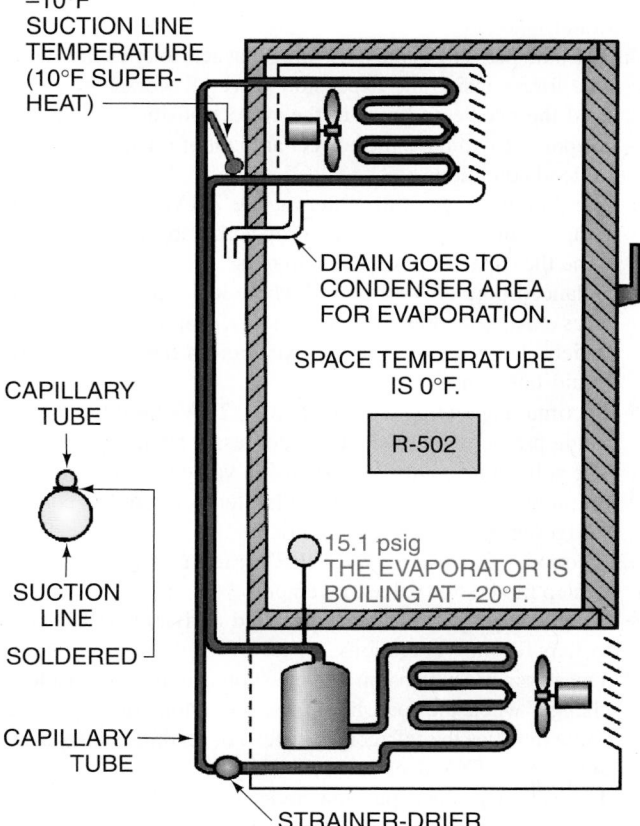

Figure 24–52 Checking the superheat before the suction–liquid heat exchange.

because the refrigerant in the evaporator will vaporize rapidly, and the last point of liquid will be toward the middle of the evaporator. This makes the system very inefficient. Many technicians have the urge to add more refrigerant to the system at these high loads because of the high superheat readings. Adding refrigerant at this point will in time only overcharge the system. Before adding refrigerant, check the superheat at the normal evaporator load and see whether it is normal. It should be from 8°F to 12°F at the normal evaporator load. If unsure, recover the refrigerant, evacuate the system, and add the specified nameplate critical charge of refrigerant.

Under low evaporator heat loads, the refrigerant in the evaporator will not vaporize as fast. Low evaporator loads are usually experienced just before the desired box temperature is reached. The last point of liquid will now be toward the end of the evaporator. The system will still have a normal superheat of about 8°F to 12°F because of its critical charge characteristics.

The capillary tube metering device is used primarily on small fractional horsepower refrigeration systems. These systems are hermetically sealed, have no bolted gasketed connections, and provide a leak-free system. They are factory assembled in a very clean environment and should run trouble free for many years.

SUMMARY

- The expansion device is one of the dividing points between the high and low sides of the system.
- The TXV valve maintains a constant *superheat* in the evaporator.
- The bulb pressure is the only force that acts to open the valve.
- The forces inside the expansion valve all work together to hold the needle and seat in the correct position so that the evaporator will have the correct amount of refrigerant under all load conditions.
- The liquid temperature entering the TXV, the evaporator temperature, and the pressure drop across the TXV all determine the capacity of the TXV in tons.
- Balanced-port TXVs are used where low ambient temperatures exist, head pressures vary widely, evaporator loads vary widely, large pressure drops exist across the TXV, or low liquid-line temperatures exist.
- Maximum operating pressure (MOP) TXVs limit the evaporating pressure to prevent the compressor from overloading.
- The solid-state controlled expansion valve uses a thermistor to monitor the evaporator outlet temperature to control refrigerant flow to the evaporator.
- Electronic expansion valves (EEVs) use either heat motors or step motors for valve movement.
- EEVs use proportional, integral, and derivative (PID) controllers for their operations.
- The automatic expansion valve responds in reverse to a load change; when the load increases, the automatic expansion valve throttles the refrigerant instead of feeding more refrigerant as the TXV does.
- The capillary tube expansion device is a fixed-bore metering device usually made of copper, with a very small inside diameter and no moving parts.
- The capillary tube system uses a very limited amount of refrigerant compared with other metering devices and is popular for small systems.

REVIEW QUESTIONS

1. Which is not a pressure or force acting on a TXV diaphragm?
   A. Head pressure
   B. Evaporator pressure
   C. Spring pressure
   D. Bulb pressure
2. The needle and seat of a TXV are normally made of _____.
3. The TXV tries to maintain a constant _____ in the evaporator.
4. The four types of charges that the TXV uses to control refrigerant flow are _____, _____, _____, and _____.
5. The bulb of the TXV is mounted on the
   A. compressor inlet.
   B. compressor outlet.
   C. evaporator inlet.
   D. evaporator outlet.
6. When do some TXVs require an external equalizer?
7. Draw a diagram showing how an external equalizer line is connected into a system.
8. The TXV responds to an increased heat load by
   A. decreasing refrigerant flow.
   B. keeping refrigerant flow constant.
   C. increasing refrigerant flow.
   D. not reacting to changing heat loads.
9. The three factors that determine the capacity of a TXV are _____, _____, and _____.
10. Explain why the entering liquid pressure has no effect on the opening force of a balanced-port TXV.
11. What is meant by a dual-port TXV?
12. Explain how a PID controller works.
13. Explain the following terms as they apply to electronic expansion valves:
    step motor
    heat motor
    offset
    error
    algorithm
    microprocessor
    feedback loop
14. What condition does the automatic expansion valve maintain in the evaporator?
15. How does the automatic expansion valve respond to a load increase?
16. What determines the amount of refrigerant that flows through a capillary tube metering device?
17. Explain what is meant by a maximum operating pressure (MOP) TXV.

UNIT 25 Special Refrigeration System Components

OBJECTIVES

After studying this unit, you should be able to

- distinguish between mechanical and electrical controls.
- explain how and why mechanical controls function.
- describe an automatic pumpdown system.
- define low ambient operation.
- describe electrical controls that apply to refrigeration.
- describe off-cycle defrost.
- describe random and planned defrost.
- explain temperature-terminated defrost.
- describe the various refrigeration accessories.
- describe the low-side components.
- describe the high-side components.

SAFETY CHECKLIST

✔ Wear goggles and gloves when attaching or removing gages to transfer refrigerant or to check pressures.

✔ Wear warm clothing when working in a walk-in cooler or freezer. A technician does not think properly when chilled.

✔ Be careful not to get your hands or clothing caught in moving parts such as pulleys, belts, or fan blades.

✔ Wear an approved back brace belt when lifting and use your legs, keeping your back straight and upright.

✔ Observe all electrical safety precautions. Be careful at all times and use common sense.

✔ Never work near live electrical circuits unless absolutely necessary. Turn the power off at the nearest disconnect, lock and tag the panel, and keep the only key in your possession.

25.1 THE FOUR BASIC COMPONENTS

The compression refrigeration cycle must have four basic components to function: the compressor, the condenser, the evaporator, and the expansion device. However, many more devices and components can enhance the performance and the reliability of the refrigeration system. Some of these protect the components, and some improve the reliability under various conditions.

Special refrigeration system components can be divided into two broad categories: controls and accessories. Control components are divided into mechanical, electrical, and electromechanical.

25.2 MECHANICAL CONTROLS

Mechanical controls generally stop, start, or modulate fluid flow and can be operated by pressure, temperature, or electricity. These controls can usually be identified because they are almost always found in the piping.

25.3 TWO-TEMPERATURE CONTROLS

Two-temperature operation is desirable when more than one evaporator is used with one compressor. This occurs if two or more evaporators are designed to operate in different temperature ranges, such as when an evaporator operating at 30°F is used on the same compressor with an evaporator operating at 20°F, **Figure 25–1**. Two-temperature application is normally accomplished with a purely mechanical valve on a temperature-controlled valve.

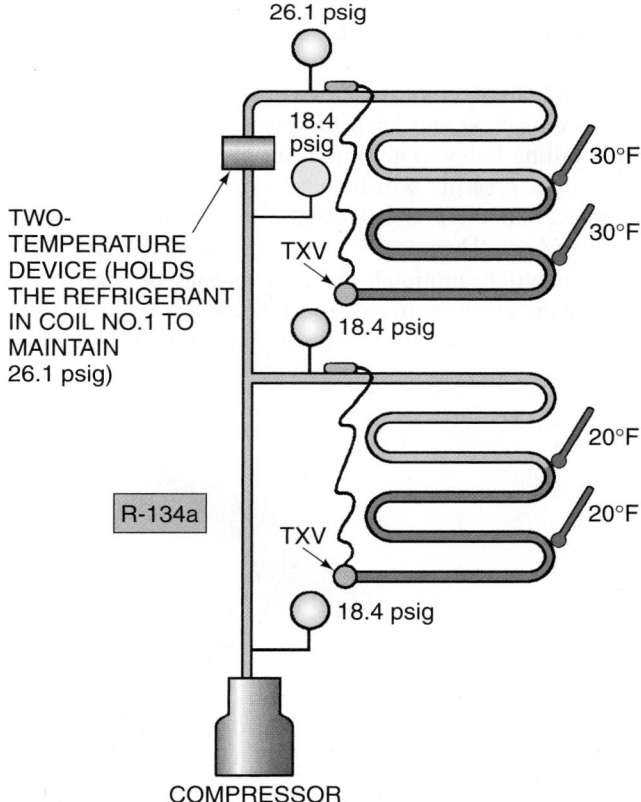

Figure 25–1 Two-temperature operation. Two evaporators are operating on one compressor. One evaporator operates at 20°F (18.4 psig), the other at 30°F (26.1 psig).

509

25.4 EVAPORATOR PRESSURE CONTROL

The *evaporator pressure-regulating valve (EPR valve)* is a mechanical control that keeps the refrigerant pressure in the evaporator from going below a predetermined point. The EPR valve is installed in the suction line at the evaporator outlet. The bellows in the EPR valve senses evaporator pressure and throttles (modulates) the suction gas to the compressor. This will then allow the evaporator pressure to go as low as the pressure setting on the valve. When the EPR valve is used with the *thermostatic expansion valve* (TXV), the system now has the characteristics of maintaining a constant superheat and keeping the pressure from going too low, **Figure 25–2.**

The EPR valve can be applied to a system that cools water. The evaporator will not go below the predetermined point, which could be freezing. For example, when the system is started, if there is a load on it, the EPR valve will be wide open. The TXV will be throttling the liquid refrigerant into the evaporator. When the evaporator pressure is to the point at which the EPR valve is set, it will begin to throttle a slow flow of refrigerant. If this setting is just above freezing, the valve will throttle off enough to keep the evaporator from freezing until the thermostat responds and shuts the system off, **Figure 25–3.** This valve is often called a "holdback valve" in the industry because it holds the refrigerant back in the evaporator, preventing the suction pressure in the evaporator from going below a set point.

The EPR valve may operate for an extended period of time while in the throttled-down position in order to satisfy the cooling load and keep the pressure in the evaporator from falling below a minimum set point. Many times, the evaporator pressure will be near the valve's set minimum pressure, but there is still some refrigeration needed because of a small heat load on the evaporator. The flow of refrigerant will be minimal. The TXV may not be able to keep a constant evaporator superheat because of the reduced mass flow rate of refrigerant. High evaporator superheat

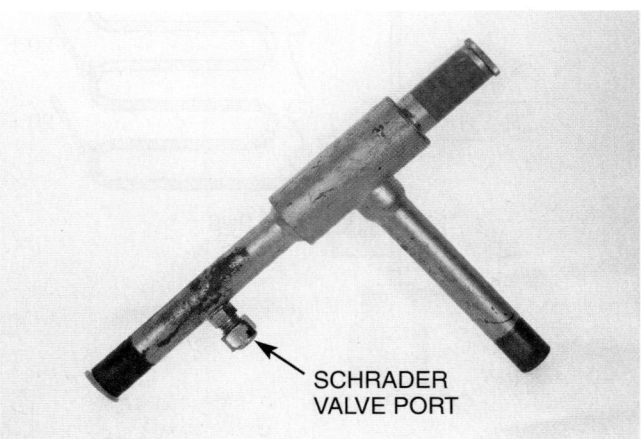

Figure 25–2 The EPR valve modulates the flow of vapor refrigerant leaving the evaporator. It limits the pressure in the evaporator and keeps it from dropping below the set point. *Courtesy Sporlan Valve Company*

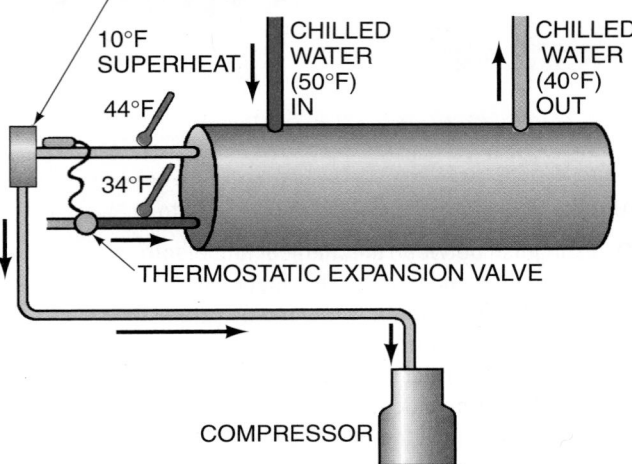

EVAPORATOR PRESSURE REGULATOR SET FOR A SUCTION PRESSURE OF NO LESS THAN 29.5 psig FOR R-134a. (THIS CORRESPONDS TO 34°F.)

Figure 25–3 The TXV keeps the evaporator at the proper level of refrigerant while the EPR valve keeps the evaporator from getting too cold. The EPR valve may be used on a water-type evaporator to keep it from operating below freezing. The thermostat should cut off the compressor soon after the EPR valve starts to throttle.

readings are often encountered at these throttled-down conditions. However, once the heat load on the evaporator increases and the EPR valve opens more to increase the mass flow rate of refrigerant, the TXV will regain control of evaporator superheat.

The two forces that control the EPR valve are the evaporator pressure, which is the valve's inlet pressure, and the valve's spring pressure. Evaporator pressure and spring pressure counteract one another. The spring pressure is a closing force, and the evaporator pressure is an opening force. The valve's outlet pressure is actually cancelled out and does not affect the valve operation because the area of the bellows is equal to the port area. The valve now has outlet pressure acting on these equal areas, but in opposite directions, and they cancel one another, **Figure 25–4(A).**

EPR valves are also referred to as *open on rise of inlet pressure (ORI)* valves, because they will throttle open on a rise in their inlet (evaporator) pressure. When the evaporator is under a high load, the EPR valve will be fully opened as if it were not there. The spring pressure can be adjusted with a screwdriver or Allen wrench, which enables the service technician to set a minimum evaporator pressure. Remember, these valves simply limit a minimum evaporator pressure. They do not hold a constant pressure or limit a maximum pressure in the evaporator.

When larger Btu capacities are called for as in supermarket installations, a larger EPR valve, or pilot-operated valve must be chosen, **Figure 25–4(B).** Some piloted EPR valves use high-pressure gas to pilot a normally open valve. These valves are held open by a spring and are forced closed by high-side gas pressure, **Figure 25–4(C).** There are actually three pressures that control the modulation of the main valve

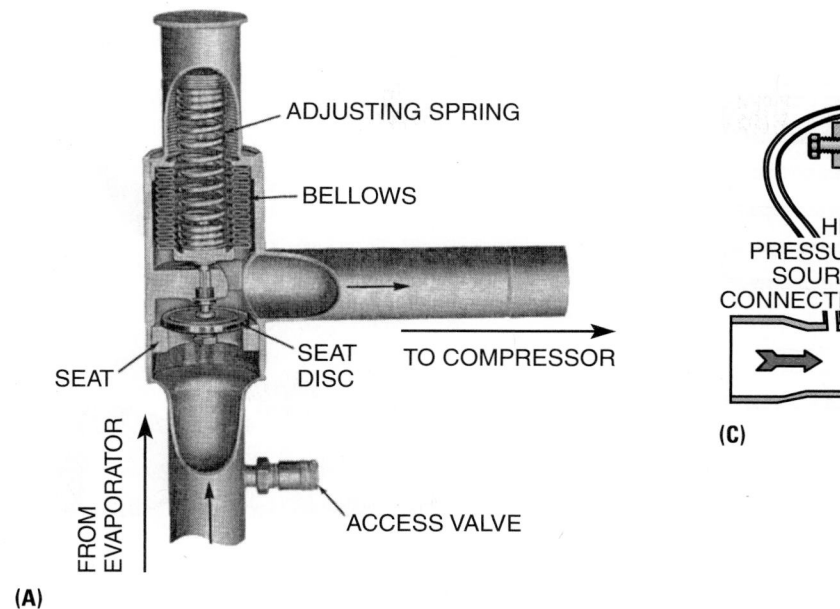

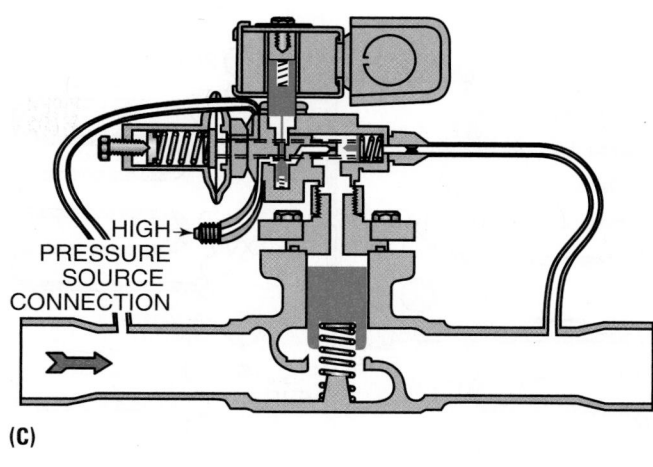

(C)

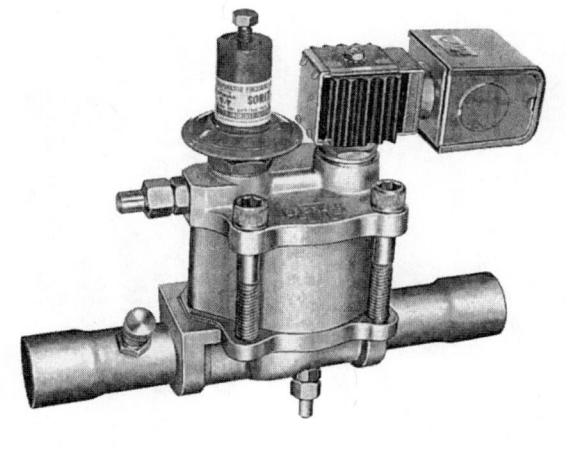

(B)

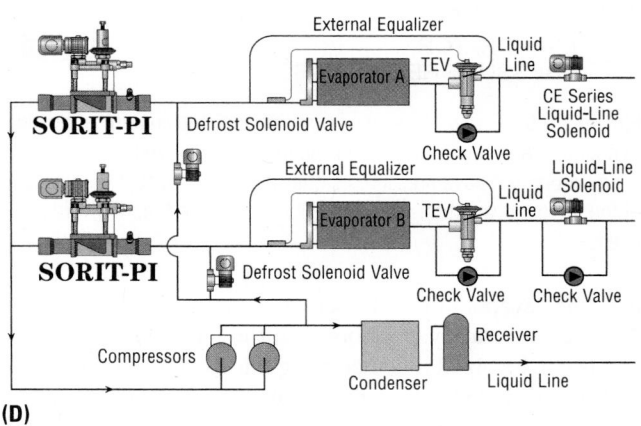

(D)

Figure 25–4 **(A)** The cutaway view of a direct-acting evaporator pressure-regulating (EPR) valve. **(B)** A pilot-operated EPR valve with a suction stop feature. **(C)** A pilot-operated EPR valve internal cutaway view showing springs, high-pressure source connection, pilot regulator, main piston, and solenoid stop feature. **(D)** A multiple-evaporator circuit showing defrost solenoid valves, EPR valves, and defrost circuit. *Courtesy Sporlan Valve Company*

piston. They include the high-pressure source, the evaporator pressure, and a spring pressure. A pilot port that is controlled by a pilot regulator valve indirectly controls the operation of the main regulating valve. These valves include a solenoid stop feature when hot gas defrost is used. This is because the hot gas is injected into the suction line between the EPR valve and the coils being defrosted. A solenoid valve in the EPR valve will be deenergized, shutting off the suction inlet pressure. This allows the internal pilot spring to force the EPR valve closed. At the same time, a hot gas defrost solenoid valve will open; the evaporator fans will stop; and the hot gas will flow in reverse through the evaporators, around the expansion valves, through a check valve, and then around a liquid-line solenoid valve. This condensed hot gas,

which is now cool liquid, will then feed into a liquid header to supply other evaporators with subcooled liquid refrigerant, **Figure 25–4(D)**.

25.5   MULTIPLE EVAPORATORS

When more than two evaporators are used with one compressor, one of them may sometimes be at a different temperature and pressure range. For example, when an evaporator that needs to operate at 15.1 psig (15°F for R-134a) is piped with another evaporator of 22 psig (25°F for R-134a) to the same compressor, an EPR valve is needed in the highest temperature evaporator's suction line. The true suction pressure for the system will be the low value of 15.1 psig, but

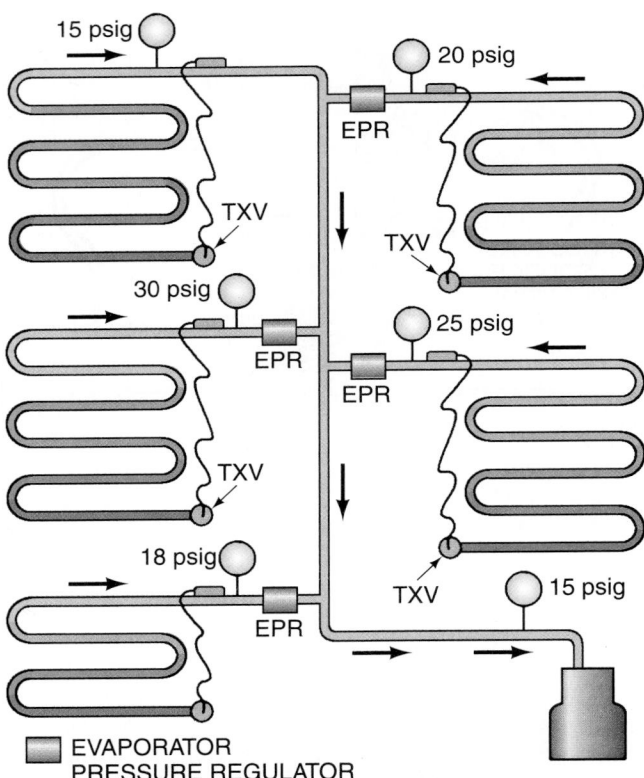

Figure 25–5   Evaporators piped together and one compressor used to maintain a suction pressure for the lowest-pressure evaporator. Evaporators that have pressures higher than the true suction pressure of the lowest evaporator all have EPR valves set to their individual needs.

the other evaporator will operate at the correct pressure of 22 psig. Several evaporators of different pressure requirements can be piped together using this method, **Figure 25–5**.

It is desirable to know the actual pressure in the evaporator because the evaporator pressure is not the same as the true suction pressure. Normally there is a gage port, known as a Schrader valve port, permanently installed in the EPR valve body, which allows the service technician to take gage readings on the evaporator side of the valve. The true suction pressure can be obtained at the compressor service valve, **Figure 25–2**.

25.6   ELECTRIC EVAPORATOR PRESSURE-REGULATING VALVE

Because of new regulations from the U.S. Food and Drug Administration *Food Code* concerning product temperature in supermarkets, a more sophisticated and accurate way of controlling the discharge air temperatures out of the evaporator section of display cases was needed. Electric evaporator pressure-regulating (EEPR) valves have been designed for very close temperature control of the discharge air in supermarket display cases, **Figure 25–6**. They are located at the evaporator outlet for single evaporator applications, or on the suction line before the common suction header in multiple evaporator applications as in parallel compressor systems

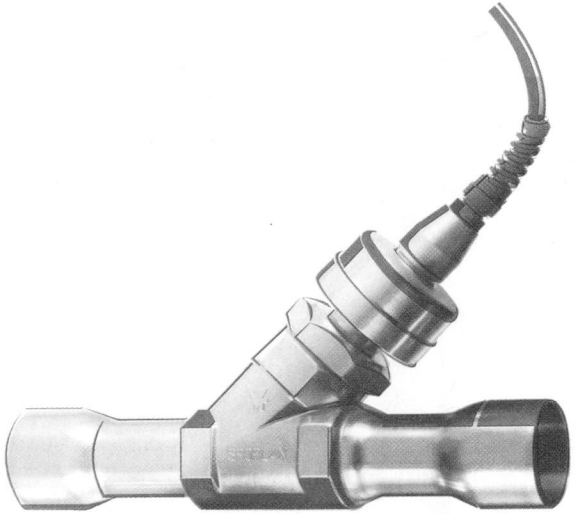

Figure 25–6   An electric evaporator pressure-regulating (EEPR) valve. *Courtesy Sporlan Valve Company*

Figure 25–7   EEPR valves in suction lines on a parallel compressor system. *Courtesy RSES Journal*

in supermarkets, **Figure 25–7**. EEPR valves and a refrigerated case discharge air sensor are wired to a microprocessor board for control purposes. The EEPR valve can move in very small increments, either opening or closing the suction line in response to the refrigerated case's discharge air temperature. Although these valves are referred to as evaporator pressure regulators, they do not serve the same purpose as the standard EPR valve mentioned earlier in this unit. A standard mechanical EPR valve prevents a minimum pressure, thus a minimum temperature, from occurring in the evaporator.

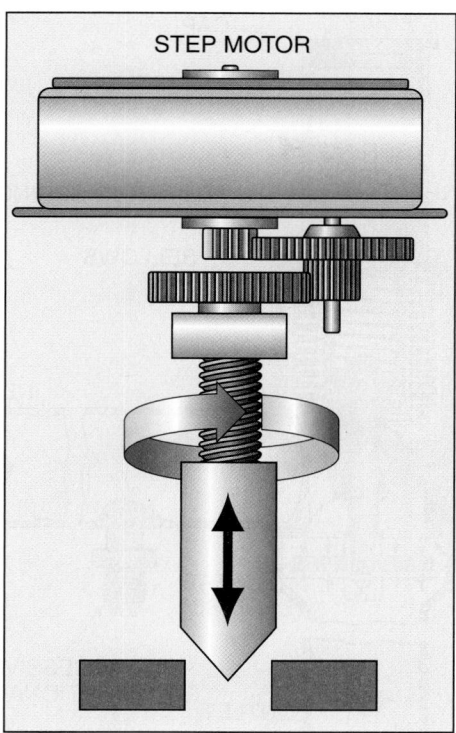

Figure 25–8 A step motor showing how the lead screw is driven through gear reduction for additional power to the valve. *Courtesy RSES Journal*

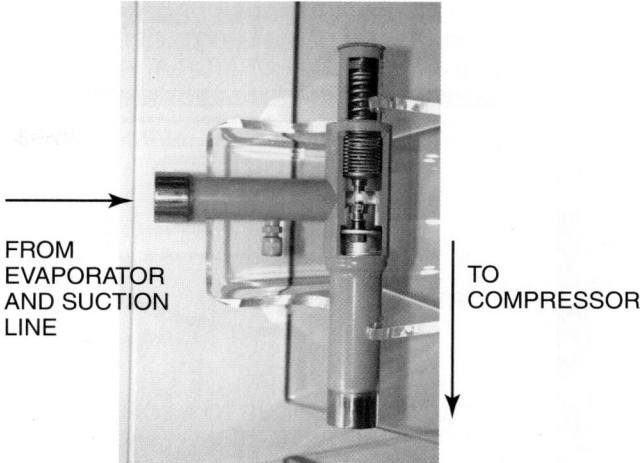

FROM EVAPORATOR AND SUCTION LINE

TO COMPRESSOR

Figure 25–9 The CPR valve to keep the compressor from running in an overloaded condition during a hot pulldown. *Courtesy Sporlan Valve Company*

EEPR valves control the discharge air temperature in the refrigerated case. They regulate the saturated pressure in the evaporator to provide the required temperature at the discharge air sensor in the refrigerated case. EEPR valves can deliver a quick temperature pulldown since they are sensing discharge air instead of evaporator pressure. EEPR valves will remain fully open until the temperature of the refrigerated case's discharge air drops to the valve's set point.

A bipolar step motor operates the EEPR valve. The step motor operates by moving the motor in discrete steps instead of a continuous rotation. The motor can step in either direction because of two different windings. A step motor may have as many as 100 electromagnets around the rotor, resulting in an angular rotation as small as 3.6° per step. The motor then drives a lead screw that converts the angular motion into straight-line or linear motion, **Figure 25–8.** The straight-line motion can have a resolution as fine as 0.0000783 in. per step. An algorithm programmed into a controller or microprocessor controls the direction and position of the step motor. Refer to Unit 24, "Expansion Devices," for a more detailed explanation of bipolar step-motor operation and control.

25.7 CRANKCASE PRESSURE REGULATOR

The *crankcase pressure-regulating valve (CPR valve)* looks much the same as the EPR valve, but it has a different function. The CPR valve is in the suction line also, but it is usually located close to the compressor rather than at the evaporator

outlet. The CPR valve sensing bellows are on the true compressor suction side of the valve and would normally have a gage port on the evaporator side of the valve, **Figure 25–9.**

This valve is used to keep the low-temperature compressor from overloading on a hot pulldown. A hot pulldown would occur (1) when the compressor has been off for a long enough time and the foodstuff has a rise in temperature, (2) on start-up with a warm box, or (3) after a defrost period. In either case the temperature in the refrigerated box or the evaporator influences the suction pressure. When the temperature is high, the suction pressure is high. When the suction pressure goes up, the density of the suction gas goes up. The compressor is a constant-volume pump and does not know when the gas it is pumping is dense enough to create a motor overload.

When a compressor is started and the refrigerated cooler is warmer than the design range of the compressor, an overload occurs. The refrigeration compressor motor can be operated at about 10% over its rated capacity without harm during this pulldown. When a motor has a full-load amperage rating of 20 A, it could run at 22 A for an extended pulldown, and there would be no harm to the motor. If the cooler temperature were to be 75°F or 80°F, the motor would be overloaded to the point that the motor overcurrent protection would shut it off. This protection is automatically reset, so the compressor would try to restart immediately, causing it to short cycle. This could continue until the manual reset motor control, the fuse, or the breaker stopped the compressor, or the system would slowly pull down during the brief running times.

The CPR valve would throttle the suction gas into the compressor to keep the compressor current at no more than the rated value, **Figure 25–10.** Before CPR valves were used, the service technician had to start the system manually by throttling the suction service valve on a hot start-up.

The CPR valve is often referred to as a *close on rise of outlet pressure (CRO)* valve. This is because it closes when its outlet (crankcase) pressure rises to a predetermined pressure. The CPR valve is controlled by crankcase (outlet) pressure

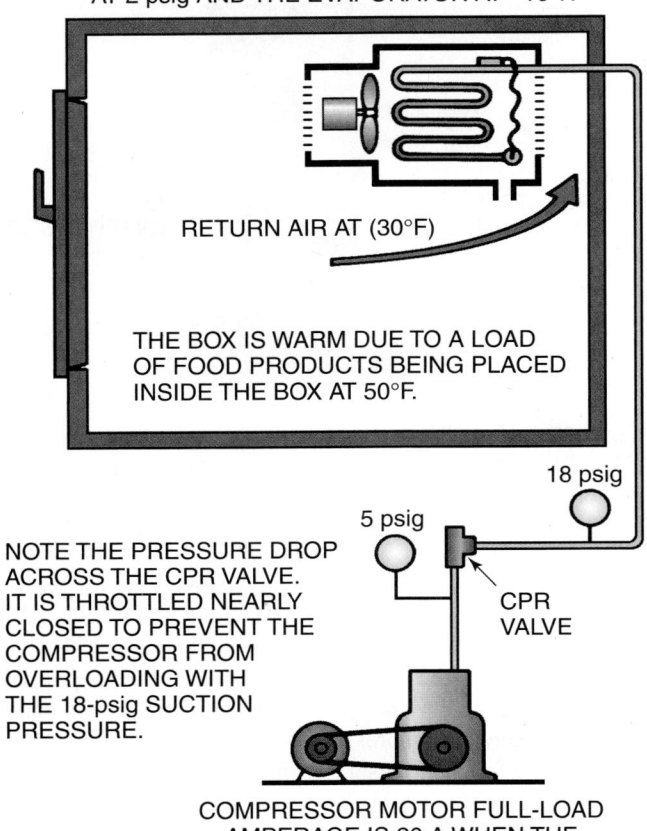

THIS IS A LOW-TEMPERATURE FREEZER WITH
THE COMPRESSOR DESIGNED TO OPERATE
AT 2 psig AND THE EVAPORATOR AT –16°F.

RETURN AIR AT (30°F)

THE BOX IS WARM DUE TO A LOAD
OF FOOD PRODUCTS BEING PLACED
INSIDE THE BOX AT 50°F.

18 psig

5 psig

NOTE THE PRESSURE DROP
ACROSS THE CPR VALVE.
IT IS THROTTLED NEARLY
CLOSED TO PREVENT THE
COMPRESSOR FROM
OVERLOADING WITH
THE 18-psig SUCTION
PRESSURE.

CPR
VALVE

COMPRESSOR MOTOR FULL-LOAD
AMPERAGE IS 20 A WHEN THE
SUCTION PRESSURE IS 5 psig.

Figure 25–10   The CPR valve throttles the vapor refrigerant entering the compressor. There will be a pressure drop across the valve when the evaporator is too warm. This valve is adjusted and set up using the compressor full-load amperage on a hot pulldown. The CPR valve can be set to throttle the suction gas to the compressor and prevent a high suction pressure on the compressor side of the valve that will not allow the compressor to overload. When the refrigerated box temperature is pulled down to the design range, the suction gas in the evaporator will be the same as the pressure at the compressor because the valve is wide open.

and its spring pressure. The crankcase pressure is a closing force, and the spring pressure is an opening force. The inlet pressure of the CPR valve is actually cancelled out because of the equal areas of the bellows and seat disc. The inlet pressure exerts an equal pressure on both the bellows and seat disc, but these pressures are in opposite directions and cancel one another out, **Figure 25–11**. The spring pressure can be adjusted with a screwdriver or an Allen wrench to allow the service technician to set the maximum crankcase pressure desired.

25.8   ADJUSTING THE CPR VALVE

The setting of the CPR valve should be accomplished on a hot pulldown or at least when the compressor has enough load that it is trying to run overloaded. This can be done by

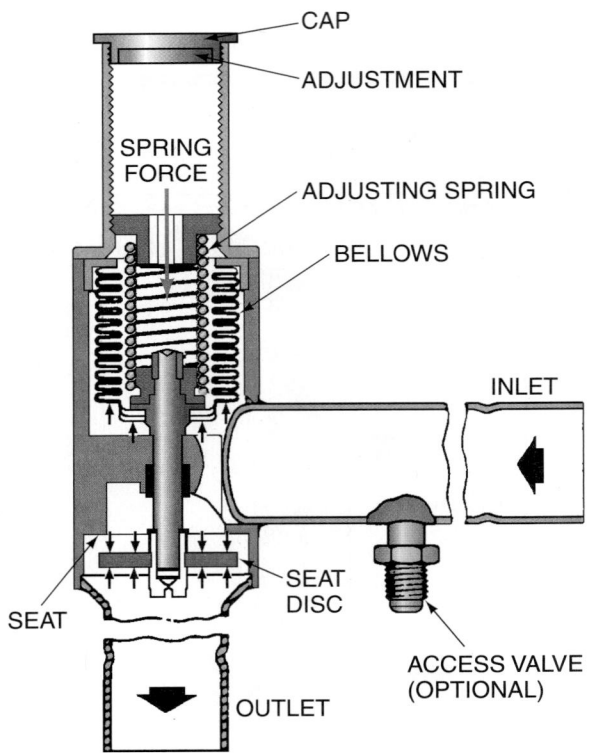

CAP

ADJUSTMENT

SPRING
FORCE

ADJUSTING SPRING

BELLOWS

INLET

SEAT
DISC

SEAT

ACCESS VALVE
(OPTIONAL)

OUTLET

Figure 25–11   The cutaway view of a crankcase pressure-regulating (CPR) valve. *Courtesy Sporlan Valve Company*

shutting off the unit until the box or cooler is warm enough to create a load on the evaporator. This loads up the compressor enough to cause it to run at a high current. Use the ammeter on the compressor while adjusting the CPR valve to the full-load amperage of the compressor. For example, if the compressor is supposed to draw 20 A, throttle the CPR valve back until the compressor amperage is 20 A and the valve is set. Full-load amperage (FLA) and run-load amperage (RLA) are one and the same when reading nameplate data on compressors.

25.9   RELIEF VALVES

Relief valves are designed to release refrigerant from a system when predetermined high pressures exist. Refrigerant relief valves come in two different types: the spring-loaded type that will reseat and the one-time type that does not close back. Spring-loaded relief valves are mainly used in today's applications because of the increased cost of refrigerants and ozone depletion scares.

The *spring-loaded* type of relief valve is normally brass with a neoprene seat. This valve is piped so that it is in the vapor space of the condenser or receiver. The relief valve must be in the vapor space, not in the liquid space, for it to relieve pressure. The object is to let vapor off to vaporize some of the remaining liquid and lower its pressure, **Figure 25–12**.

The top of the valve normally has threads, so the refrigerant can be piped to the outside of the building. SAFETY PRECAUTION: When relief valves are used to protect vessels in a fire, it is desirable for the refrigerant to be removed from

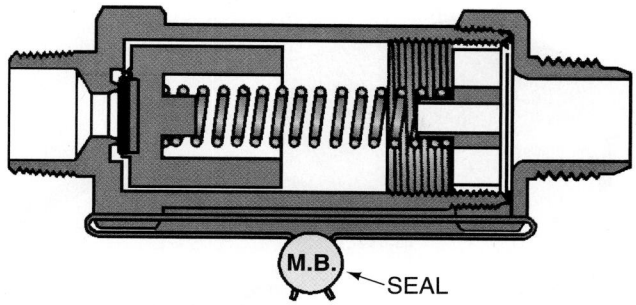

Figure 25–12 The spring-loaded type of relief valve used to protect the system from very high pressures that might occur if the condenser fan were to fail or if a water supply were to fail on a water-cooled system. This relief valve is designed to reseat after the pressure is reduced. This threads on the valve outlet enable the valve to be piped to the outside if needed. *Courtesy Superior Valve Company*

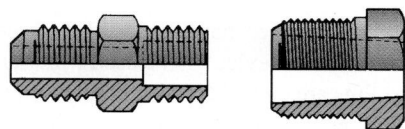

Figure 25–13 The fusible relief valve may be a low-temperature solder patch that covers a hole drilled in a fitting or pipe. Care should be taken when soldering in the vicinity of this relief device, or it will melt. *Courtesy Mueller Brass Company*

the area. Refrigerant gives off a noxious gas when it is burned. Some systems have the relief valve piped into a large, evacuated tank. The refrigerant will not be vented if the relief valve opens in this case.

The *one-time* relief valve is caused to relieve by temperature. These valves, often called fusible plugs, are designed with one of the following methods: a fitting filled with a low melting-temperature solder, a patch of copper soldered on a drilled hole in the copper line using low-temperature solder, or a fitting that has been drilled out at the end with a spot of low-temperature solder over the hole, **Figure 25–13**. Sometimes the melting temperature will be very low, about 220°F. When soldering around the compressor, make sure that the solder in the fusible plug is not melted away. This type of device will normally never relieve unless there is a fire. It can often be found on the suction side of the system close to the compressor to protect the system and the public. This will keep the compressor shell from experiencing high pressures and temperatures and rupturing during a fire. Remember, the compressor shell can have a working pressure as low as 150 psig. The plug blows at approximately 220°F.

25.10 LOW-AMBIENT CONTROLS

Low-ambient controls are important in refrigeration systems because refrigeration is needed all year. When the condenser is located outside, the head pressure will go down in the winter to the point that the expansion valve will not have enough pressure drop across it to feed refrigerant correctly. When

this happens, some method must be used to keep the head pressure up to an acceptable level. The most common methods are as follows:

1. Fan cycling using a pressure control
2. Fan speed control
3. Air volume control using shutters and fan cycling
4. Condenser flooding devices

25.11 FAN-CYCLING HEAD PRESSURE CONTROLS

Fan cycling has been used for years because it is simple. When a unit has one small fan, this is a simple and reliable method because only one fan is cycled. When there is more than one fan, the extra fans may be cycled by temperature, and the last fan can be cycled by head pressure. The control used is a pressure control that closes on a rise in pressure to start the fan and opens on a fall in pressure to stop the fan when the head pressure falls. When fan cycling is used, the fan will cycle in the winter but will run constantly in hot weather, **Figure 25–14(A)**. On multiple-fan condensing units, one or many ambient temperature controls may accomplish condenser fan cycling. These controls are usually thermal disks located somewhere on the condensing unit's frame, **Figure 25–14(B)**. As the outside ambient temperature drops, the fans begin to turn off. As the outside ambient temperature rises, the fans begin to turn on. Frequently on smaller condensing units, one fan will run continuously and the other fan will be cycled on and off with an ambient temperature controller.

These types of controls can be hard on a motor because of all the cycling. The fan motor needs to be a type that does not have a high starting current. The control must have enough contact surface area to be able to start the motor many times to be reliable. The motor current should be carefully compared with the control capabilities.

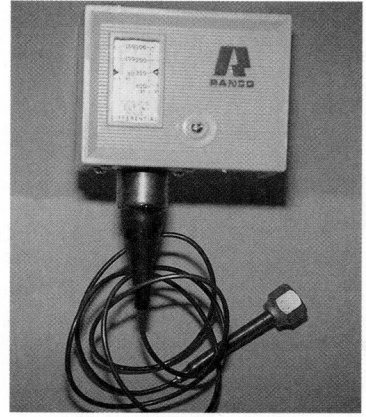

Figure 25–14(A) A pressure control with the same action as a low-pressure control; it takes a rise in head pressure to start the condenser fan. This pressure control operates at a higher pressure range than the low-pressure control normally used on the low side of the system. Fan-cycling devices will cause the head pressure to fluctuate up and down. This can affect the operation of the expansion device. *Photo by Bill Johnson*

Figure 25–14(B)   A thermal disk-type fan speed control. *Courtesy Ferris State University. Photo by John Tomczyk*

Fan cycling can vary the pressure to the expansion device a great deal. The technician must choose whether to have the control set points close together for the best expansion device performance or far apart to keep the fan from short cycling. The best application for fan-cycling devices is in the use of multiple fans, where the last fan has a whole condenser to absorb the fluctuation.

Vertical condensers may be affected by the prevailing winds. If these winds are directed into the coil, they can provide airflow similar to a fan. Sometimes a shield is installed to prevent this, **Figure 25–15.** When an air-cooled condenser is installed either on the ground or on the roof, the direction of airflow across the condenser's face should be in the prevailing wind direction. This way, when the condenser's fan is operating, it will not be bucking the prevailing winds and decreasing airflow volume (cfm) through the condenser.

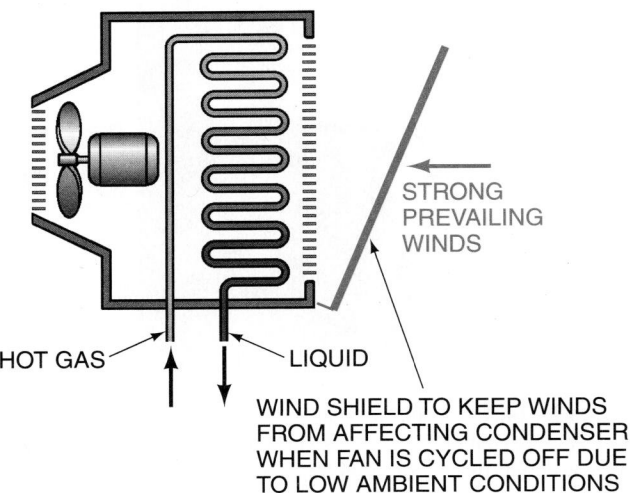

Figure 25–15   When fan-cycling controls are used, care must be taken if the condenser is horizontal. The prevailing winds can cause the head pressure to be too low even with the fan off. A shield can be used to prevent the wind from affecting the condenser.

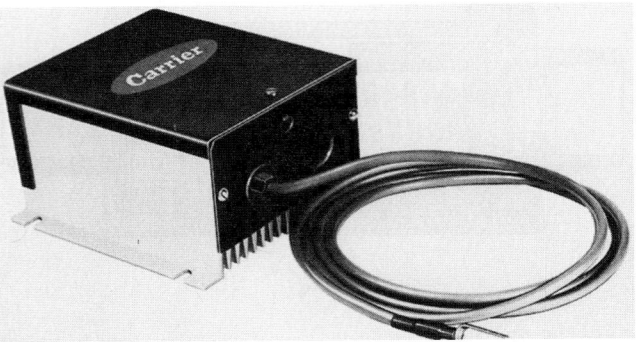

Figure 25–16   A control used to vary the fan speed on a special motor based on condenser temperature. *Courtesy Carrier Corporation*

25.12   FAN SPEED CONTROL FOR CONTROLLING HEAD PRESSURE

Fan speed control devices have been used successfully in many installations. This device can be used with multiple fans, where the first fans are cycled off by temperature, and the last fan is controlled by head pressure or condensing temperature. The device that controls this fan is normally a transducer that converts a pressure or temperature signal to an electrical signal that feeds a motor speed controller. As the temperature drops on a cool day, the motor speed is reduced. As the temperature rises, the motor speed increases. At some predetermined point the additional fans are started. On a hot day all fans will be running, with the variable-speed fan running at maximum speed. Some fan speed controls use a temperature sensor to monitor the condenser's temperature, **Figure 25–16.**

25.13   AIR VOLUME CONTROL FOR CONTROLLING HEAD PRESSURE

Air volume control using shutters is accomplished with a piston-driven damper driven with the high-pressure refrigerant. When there are multiple fans, the first fans can be cycled off using temperature, and the shutter can be used on the last fan, as in the previous two systems. This system results in a steady head pressure, as with fan speed control. The expansion valve inlet pressure does not fluctuate like it does when one fan is cycled on and off. The shutter can be located either on the fan inlet or on the fan outlet, **Figure 25–17.** Be careful that the fan motor is not overloaded with a damper.

25.14   CONDENSER FLOODING FOR CONTROLLING HEAD PRESSURE

Condenser flooding devices are used in both mild and cold weather to cause the refrigerant to move from an oversized receiver into the condenser, **Figure 25–18.** This excess refrigerant in the condenser acts like an overcharge of refrigerant and causes the head pressure to be much higher than it normally would be on a mild or cold day. The head pressure will remain the same as it would on a warm day. This method gives very steady control with no fluctuations while running. This system requires a large amount of refrigerant because in

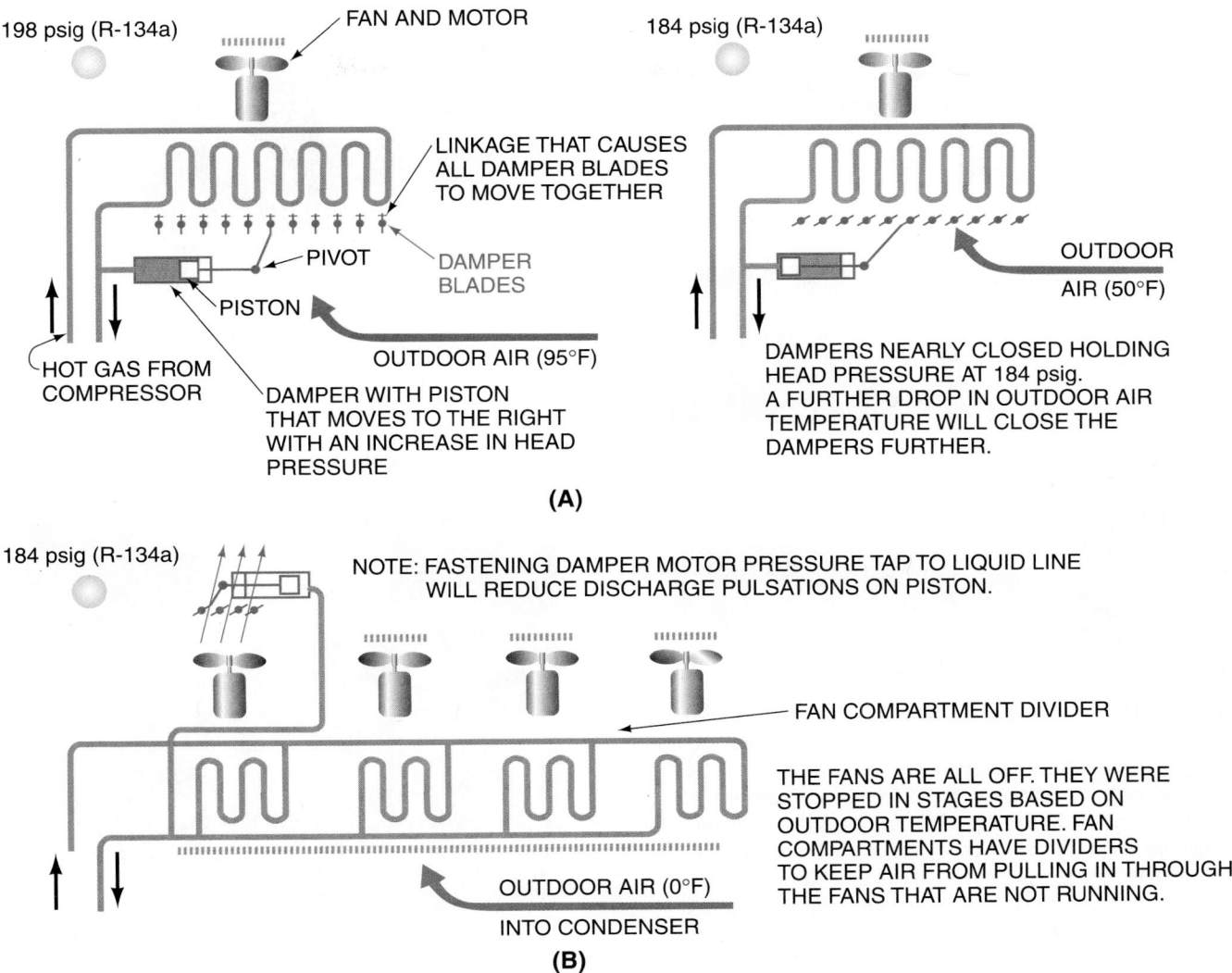

Figure 25–17   (A) Installation using dampers to vary the air volume instead of fan speed. This is accomplished with a refrigerant-operated damper that operates the shutters. This application modulates the airflow and gives steady head pressure control. **(B)** If more than one fan is used, the first fans can be cycled off based on ambient temperature. Care should be taken that the condenser fan will not be overloaded when the shutters are placed at the fan outlet.

addition to the normal operating charge, there must be enough refrigerant to flood the condenser in the winter. A large receiver is used to store the refrigerant in the summer when the valves will divert the excess liquid into the receiver.

Condenser flooding has one added benefit in the winter. Because the condenser is nearly full of refrigerant, the liquid refrigerant that is furnished to the expansion devices is well below the condensing temperature. Remember, this is called subcooling and will help improve the efficiency of the system. The liquid may be subcooled to well below freezing for systems operating in a cold climate. For a more detailed explanation of head pressure controls, refer to Section 22.18, "Head Pressure Control," in Unit 22, "Condensers."

25.15   THE SOLENOID VALVE

The *solenoid valve* is the most frequently used component to control fluid flow. This valve has a magnetic coil that when energized will lift a plunger into the coil, **Figure 25–19.** This

valve can be either normally open (NO) or normally closed (NC). The NC valve is closed until energized; then it opens. The NO valve is normally open until energized; then it closes. The plunger is attached to the valve so the plunger action moves the valve.

Solenoid valves are snap-acting valves that open and close very fast with the electrical energy to the coil. They can be used to control either liquid or vapor flow. The snap action can cause liquid hammer when installed in a liquid line, so be careful when locating the valve. Follow the manufacturer's instructions as to the location and placement of solenoid valves. Liquid hammer occurs when the fast-moving liquid is shut off abruptly by the solenoid valve, causing the liquid to stop abruptly.

The solenoid valve is responsible for stopping and starting fluid flow. Two common mistakes in installation can prevent the solenoid valve from functioning correctly: the direction in which the valve is mounted and the position in which the valve is installed. The fluid flow has to be in the correct

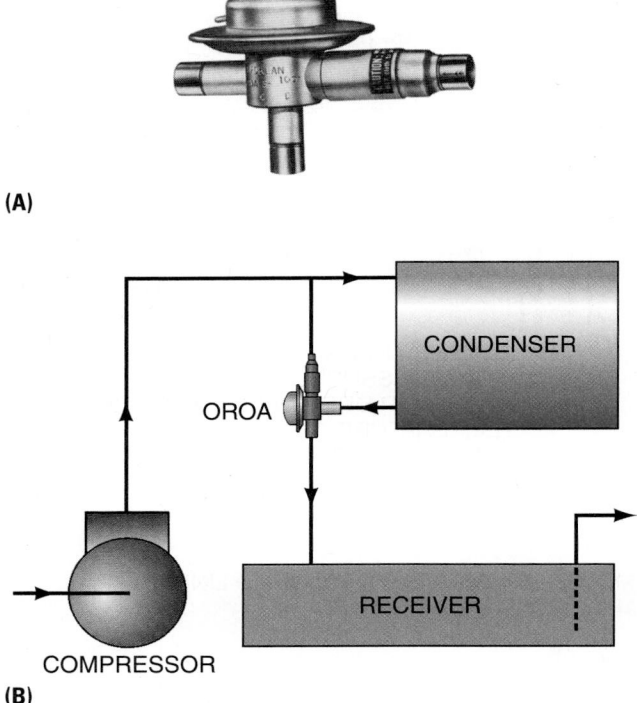

(A)

(B)

(A)

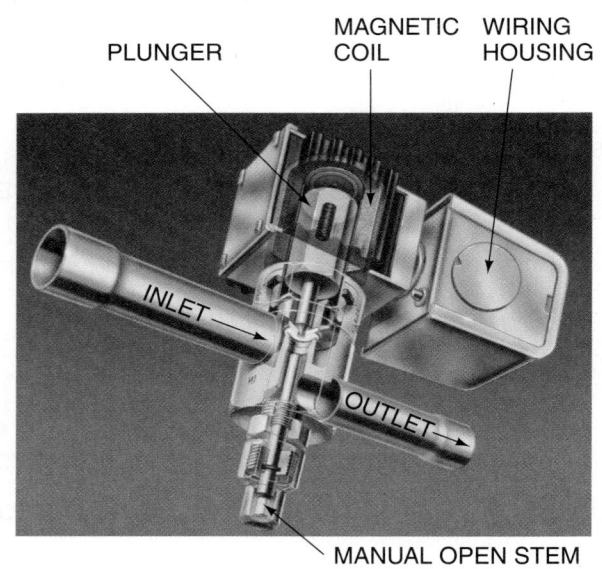

Figure 25–18 Low-ambient head pressure control accomplished with refrigerant. A large receiver stores the refrigerant when it is not needed (the summer cycle). **(A)** The valve in the system diverts the hot gas to the condenser during summer operation, and the system acts like a typical system with a large receiver. In the winter cycle, part of the gas goes to the condenser, depending on the outdoor temperature, and part of the gas goes to the top of the receiver. **(B)** The gas going to the receiver keeps the pressure up for correct expansion valve operation. The gas going to the condenser is changed to a liquid and subcooled in the condenser. It then returns to the receiver to subcool the remaining liquid. *Courtesy Sporlan Valve Company*

Figure 25–19 **(A)** Electrically operated solenoid valves. The valve is moved by the plunger attached to the seat. The plunger moves into the magnetic coil when the coil is energized to either open or close the valve. **(B)** Cutaway view of an electrically operated solenoid valve. *(A) Photo by Bill Johnson. (B) Courtesy Sporlan Division, Parker Hannifin Corp.*

direction with a solenoid valve, or the valve may not close tightly, **Figure 25–20.** The valve is mounted in the correct direction when the fluid helps to close the valve. If the high pressure is under the valve seat, the plunger may have a tendency to lift off its seat. The valve will have an arrow to indicate the direction of flow. When placing the solenoid valve in the correct direction, the position of the valve must be considered. Most solenoid valves have a heavy plunger that is lifted to open the valve. When the plunger is not energized, the weight of this plunger holds the valve on its sealing seat. If this type of valve is installed on its side or upside down, the valve will remain in the energized position when it should be closed, **Figure 25–21.**

The solenoid valve must be fastened to the refrigerant line so that it will not leak refrigerant. It can be fastened by flare, flange, or solder connections. Most valves have to be serviced at some time. The valves that are soldered in the line can be serviced easily if they can be disassembled.

A special solenoid valve known as *pilot acting* is available for larger applications for controlling fluid flow. This valve uses a very small valve seat to divert high-pressure gas that

causes a larger valve to change position. This type of solenoid valve uses this difference in pressure to cause the large movement while the solenoid's magnetic coil only has to lift a small seat. These valves are used when large vapor or liquid lines must be switched; they can have more than one inlet and outlet. Some are known as *four-way* valves, others as *three-way* valves. These have special functions. If an electrical coil had to be designed to cause the switching action of a large valve, it would be very large and draw too much current for practical use, **Figure 25–22.** Pilot action reduces the size of the electrical coil and the overall size of the valve.

Solenoid valves must be sized correctly for their particular application. For the refrigerant liquid line, they are sized according to refrigerant tonnage and a pressure drop through the valve that is acceptable. If a valve needs to be replaced, the manufacturer's data will help you choose the correct valve.

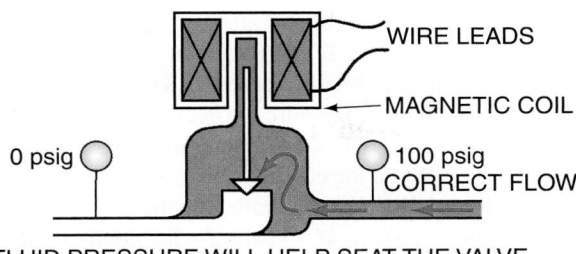

FLUID PRESSURE WILL HELP SEAT THE VALVE
WHEN THE VALVE IS INSTALLED CORRECTLY.

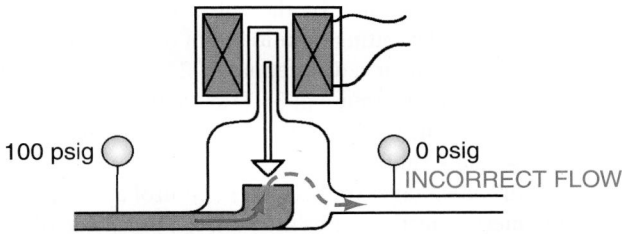

THE PRESSURE IS RAISING THE VALVE NEEDLE
UP AND LEAKING BECAUSE THE HIGH PRESSURE
IS UNDER THE SEAT.

Figure 25–20   The fluid flow in a solenoid valve must be in the correct direction, or the valve will not close tight. If the high-pressure fluid is under the seat, it will have a tendency to raise the valve off the seat. When installed correctly, the fluid helps to hold the valve on the seat.

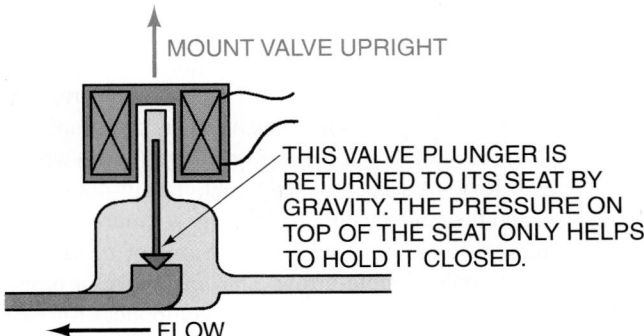

Figure 25–21   The solenoid valve installed in the correct position. Most valves use gravity to seat the valve, and the magnetic force raises the valve off its seat. If the valve is installed on its side or upside down, the valve will not seat when deenergized.

25.16 PRESSURE SWITCHES

Pressure switches are used to stop and start electrical current flow to refrigeration components. They play a very important part in the function of refrigeration equipment. The typical pressure switch can be

1. a low-pressure switch—closes on a rise in pressure.
2. a high-pressure switch—opens on a rise in pressure.
3. a low-ambient control—closes on a rise in pressure.
4. an oil safety switch—has a time delay; opens on a rise in pressure.

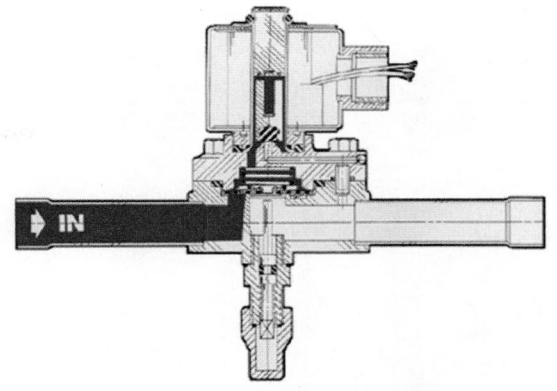

Deenergized

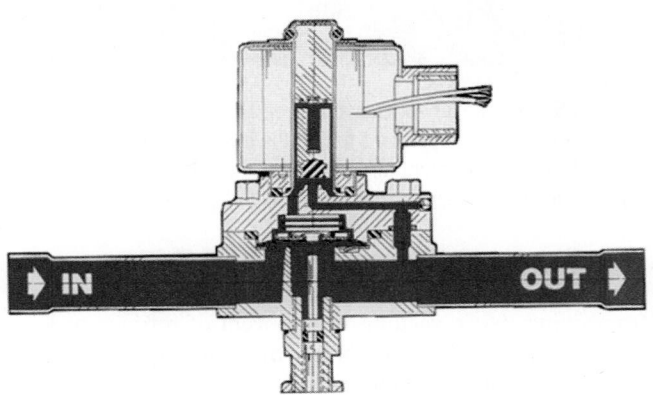

Energized

Figure 25–22   A pilot-operated solenoid valve. The magnetic coil only controls a small line that directs the high-pressure refrigerant to one end of a sliding piston. The magnetic coil only has to do a small amount of work, and the difference in pressure does the rest. *Courtesy Alco Controls Division, Emerson Electric Company*

25.17 LOW-PRESSURE SWITCH

The low-pressure switch or control is used in the following two major applications in refrigeration:

■ Low-charge protection
■ Control of space temperature

The low-pressure control can be used as a low-charge protection by setting the control to cut out at a value that is below the typical evaporator operating pressure. For example, in a medium-temperature cooler the air temperature may be expected to be no lower than 34°F. When this cooler uses R-134a for the refrigerant, the lowest pressure at which the evaporator would be expected to operate would be around 18 psig because the coil would normally operate at about 15°F colder than the air temperature in the coil (34°F − 15°F = 19°F). The refrigerant in the coil should not operate below this temperature of 19°F, which converts to 18 psig.

The low-pressure cut-out should be set below the expected operating condition of 18 psig and above the atmospheric

Figure 25–23 The low-pressure control. The control contacts make on a rise in pressure (open on a fall). If the low-side pressure goes below the control set point for any reason, the control opens the control circuit to the compressor. *Photo by Bill Johnson*

pressure of 0 psig. A setting that would cut off the compressor at 5 psig would keep the low-pressure side of the system from going into a vacuum in the event of refrigerant loss. This setting would be well below the typical operating condition. This control would normally be automatically reset. When a low-charge condition exists, the compressor would be cut off and on with this control and maintain some refrigeration. The store owner may call with a complaint that the unit is cutting off and on but not lowering the cooler temperature properly. The cut-in setting for this application should be a pressure that is below the pressure corresponding to the highest temperature the cooler is expected to experience in a typical cycle and just lower than the pressure corresponding to the thermostat cut-in point. For example, the thermostat may be set at 45°F, which would mean that the pressure in the evaporator may rise as high as 40 psig. The low-pressure control should be set to cut in at about 25 psig. The previous example of a low-pressure control set to cut out at 5 psig and to cut in at 40 psig means that the control differential is 35 psig. Remember that the compressor is not running. The wide differential helps to keep the compressor from coming back on too soon or from short cycling, **Figure 25–23.**

25.18 LOW-PRESSURE CONTROL APPLIED AS A THERMOSTAT

The low-pressure control setup described in the previous example is for low-charge protection only, to keep the system from going into a vacuum. The same control can be set to operate the compressor to maintain the space temperature in the cooler and to serve as a low-charge protection. Using the same temperatures as in the preceding paragraph, 34°F and 45°F, as the operating conditions, the low-pressure control can be set to cut out when the low pressure reaches 18 psig.

This corresponds to a coil temperature of 20°F. When the air in the cooler reaches 34°F, the coil temperature should be 19°F, with a corresponding pressure of 18 psig. This system has a room-to-coil temperature difference of 15°F (34°F room temperature − 19°F coil temperature = 15°F temperature difference). When the compressor cuts off, the air in the cooler is going to raise the temperature on the coil to 34°F and a corresponding pressure of 30 psig. As the cooler temperature goes up, it will raise the temperature of the refrigerant. When the temperature of the air increases to 45°F, the coil temperature should be 45°F and have a corresponding pressure of 40 psig. This could be the cut-in point of the low-pressure control. The settings would be cut out at 18 psig and cut in at 40 psig. This is a differential of 22 psig and would maintain a cooler temperature of 34°F to 45°F. You could say that you are using the refrigerant in the evaporator coil as the thermostat sensing fluid.

One of the advantages of this type of control arrangement is that no interconnecting wires are between the inside of the cooler and the condensing unit. If a thermostat is used to control the air temperature in the cooler, a pair of wires must be run between the condensing unit and the inside of the cooler. In some installations the condensing unit is a considerable distance from the cooler, which makes this impractical. With the temperature being controlled at the condensing unit, the owner is less likely to turn the control and cause problems.

As many low-pressure control settings exist as there are applications. Different situations call for different settings. **Figure 25–24** is a chart of settings recommended by one company.

Low-pressure controls are rated by their pressure range and the current draw of the contacts. A low-pressure control that is suitable for R-12 or R-134a may not be suitable for R-502, R-507, or R-404A because of its different pressure range. For the same application in the previous paragraph using R-502, the cut-out would be 51 psig, and the cut-in would be 88 psig. If R-404A were the refrigerant, the cut-out would be 55 psig and the cut-in would be 95 psig. A control for the correct pressure range must be chosen. Some of these controls are single-pole–double-throw. They can make or break on a rise. This control can serve as one component for two different functions.

The contact rating for a pressure control has to do with the size of electrical load the control can carry. If the pressure control is expected to start a small compressor, the inrush current should also be considered. Normally a pressure control used for refrigeration is rated so it can directly start up to a 3-hp single-phase compressor. If the compressor is any larger, or three phase, a contactor or motor starter is normally used. The pressure control then can control the contactor or motor starter coil, **Figure 25–25.**

25.19 AUTOMATIC PUMPDOWN SYSTEMS

The *automatic pumpdown system* consists of a normally closed liquid-line solenoid valve installed in the liquid line of a refrigeration system. The direction of flow of the solenoid valve should be in the direction of the evaporator. The solenoid

APPROXIMATE PRESSURE CONTROL SETTINGS
Pressure — Pounds Per Square Inch Gauge

APPLICATION	TEMP RANGE (°F)	EVAP TD (°F)	REFRIGERANT							
			22		134a		404A		507	
			Out	In	Out	In	Out	In	Out	In
Beverage Cooler										
Floral Cooler	35 to 38	15	41	66	17	33	53	82	56	86
Produce Cooler										
Smoked Meat Cooler										
Meat Reach Thru	32 to 35	15	38	62	15	30	49	77	52	81
Service Deli										
Seafood										
Multi-Deck Fresh Meat	26 to 29	15	32	54	11	25	42	68	45	72
Frozen Glass Door	-10 to 0	10	9	24	—	—	15	33	16	35
Frozen Walk-In										
Frozen Ice Cream	-30 to -20	10	0	10	—	—	4	16	4	18
Frozen Food - Open Type										

Pressure control settings assume a suction line pressure loss equivalent to 2°F

Figure 25–24 This table is to be used as a guide for setting low-pressure controls for the different applications. *Courtesy Tecumseh Products Company*

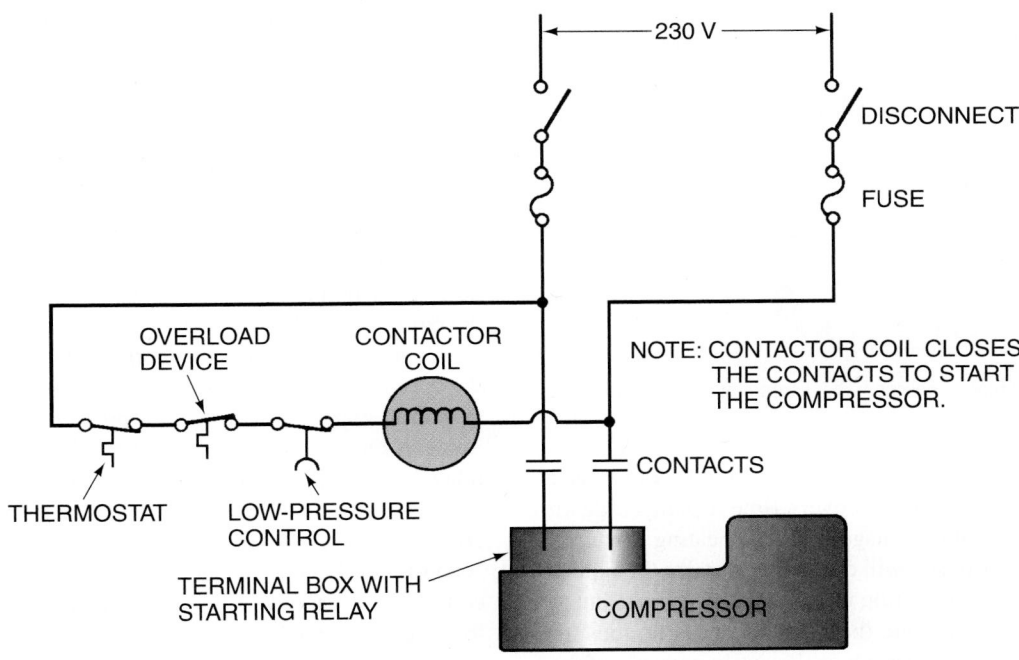

Figure 25–25 When a compressor that is larger than the electrical rating of the pressure control contacts is encountered, the compressor has to be started with a contactor. The wiring diagram shows how this is accomplished.

is a normally closed electric shutoff valve that is controlled by a thermostat. The thermostat is located somewhere in the refrigerated space. When the desired box temperature is reached in the refrigerated space, the thermostat will open and deenergize and close the liquid-line solenoid valve. The compressor will continue to run and will evacuate any refrigerant from the solenoid valve's outlet to and including the compressor. This includes part of the liquid line, evaporator, suction line, and crankcase. **Figure 25–26** is an electrical schematic diagram of an automatic pumpdown system using a liquid-line solenoid valve, thermostat, and low-pressure

control. **Figure 25–27(A)** is a system diagram showing the electrical hookup of an automatic pumpdown system.

This refrigerant will be stored in the condenser and receiver on the high side of the refrigeration system. Most of the refrigerant will be stored in the receiver. The compressor will then be shut off by the action of a low-pressure control set to open at about 5 to 10 psig. This will ensure that no refrigerant will migrate during the off cycle to the compressor. The practice years ago was to let the compressor pump down to 0 psig. This practice has proven to be hard on compressors because they reach such high compression ratios at 0-psig

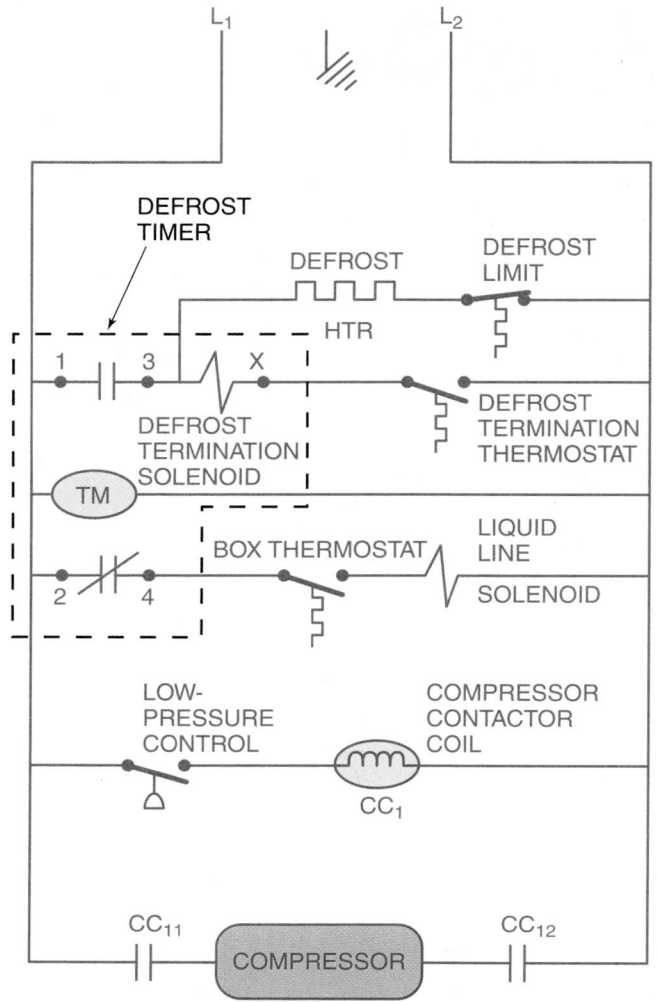

Figure 25–26 An electrical schematic diagram of an automatic pumpdown system.

pressure setting guide like the one in **Figure 25–24** will suffice. Compressor short cycling during the off cycle can also be caused by a leaky liquid-line solenoid valve or the compressor discharge valve leaking high pressure back into the compressor during the pumpdown phase. An important thing to remember when leak testing a system with automatic pumpdown is that the solenoid circuit must be energized to prevent trapping of refrigerant in the condenser and receiver.

It is a good practice to install automatic pumpdown systems on all refrigeration systems employing large amounts of refrigerant and oil. Automatic pumpdown systems are not seen on systems with small amounts of oil or refrigerant because there is not enough refrigerant to migrate to the oil and cause any damage if migration does occur during the off cycle. For example, a domestic refrigerator or freezer may have a refrigerant charge of only 8 to 16 oz of refrigerant, while the oil charge will be from 12 to 20 oz of oil. There is often more oil than refrigerant in these systems. Even if the entire refrigerant charge migrates to the crankcase during an off cycle, no real damage will be done because the refrigerant flashes off during the next on cycle. So, if the refrigerant to oil ratio is large, automatic pumpdown systems should be employed.

Figure 25–27(B) illustrates a wiring scheme for an automatic pumpdown system that will not short cycle the compressor during a pumpdown phase caused by a leaking solenoid valve or compressor discharge valve. This type of system is often referred to as a non-short-cycling or "one-time" automatic pumpdown system. Only the thermostat calling again for cooling will take the refrigeration system out of pumpdown. The diagram shows the system in a pumpdown with the compressor off.

On a call for cooling, the thermostat closes and energizes the liquid-line solenoid valve. This allows the low side of the system to experience refrigerant pressure, which allows the low-pressure switch to close, **Figure 25–27(C)**. The latching relay coil (LRC) is energized through the normally closed (NC) contacts A and B. Once the LRC is energized, it closes contacts B and C and contacts C and D, while opening contacts A and B. With the LRC energized, the compressor contactor coil (CC) is energized through contacts C and D and the low-pressure switch. This starts the compressor and then the system is in a normal cooling mode, **Figure 25–27(D)**.

When the thermostat opens, the liquid-line solenoid valve is deenergized, **Figure 25–27(E)**. This physically closes off the liquid line and initiates an automatic pumpdown. The low-pressure switch now opens from the low pressure, deenergizing the LRC coil, which opens contacts C and D and B and C. The opening of the low-pressure switch also deenergizes the compressor contactor coil (CC), which shuts off the compressor. The system is pumped down and the compressor remains off until there is another call for cooling from the thermostat. However, if the low-pressure switch closes because of a leaking component while the thermostat is still open and not calling for cooling, the compressor contactor coil (CC) will not be energized because of the normally open (NO) contacts C and D of the latching relay. This prevents the compressor from short cycling while in the pumpdown

suction pressure. Also, a refrigerant-cooled compressor would be starved of refrigerant every time it pumped down and could suffer possible damage from overheating. On a call for cooling, the thermostat will close and energize the liquid-line solenoid valve. This action will send liquid refrigerant to the expansion valve and into the empty evaporator, thus increasing evaporator pressure. Once the cut-in pressure of the low-pressure control is reached, the compressor will easily start and resume a normal refrigeration cycle. Automatic pumpdown systems do not need a larger receiver since the receiver is designed to hold the entire refrigerant charge and still have a 20% vapor head for safety.

The cut-in pressure of the low-pressure control should be set at a high enough pressure to ensure that the system will not short cycle if residual pressure does remain in the low side of the system once pumped down. Compressor short cycling can be devastating to motor windings and starting controls because of overheating. However, the cut-in pressure has to be low enough to ensure the system will cut in once the liquid-line solenoid is energized by the thermostat to start the next on cycle. These pressures are dependent on refrigerant type and desired box temperature applications. A

mode. The LRC will also not be energized because of the NO contacts B and C of the latching relay. In summary, when components leak, these latching relay contacts prevent the compressor from short cycling during pumpdown, which causes low-side pressures to increase and low-pressure switches to close prematurely.

🌐The following are three main reasons automatic pumpdown systems are employed on refrigeration systems:

■ To rid the evaporator, suction line, and crankcase of refrigerant before the off cycle and defrost cycle so migration of refrigerant to the compressor and/or compressor crankcase cannot occur

■ To prevent surges of liquid refrigerant from entering the suction port of the compressor (slugging) during start-ups

■ To rid the crankcase of refrigerant to prevent oil from foaming during start-ups and lubrication from being robbed from the compressor's mechanical parts🌐

25.20 HIGH-PRESSURE CONTROL

The high-pressure control is normally not as complicated as the low-pressure control (switch). It is used to keep the compressor from operating with a high head pressure. This

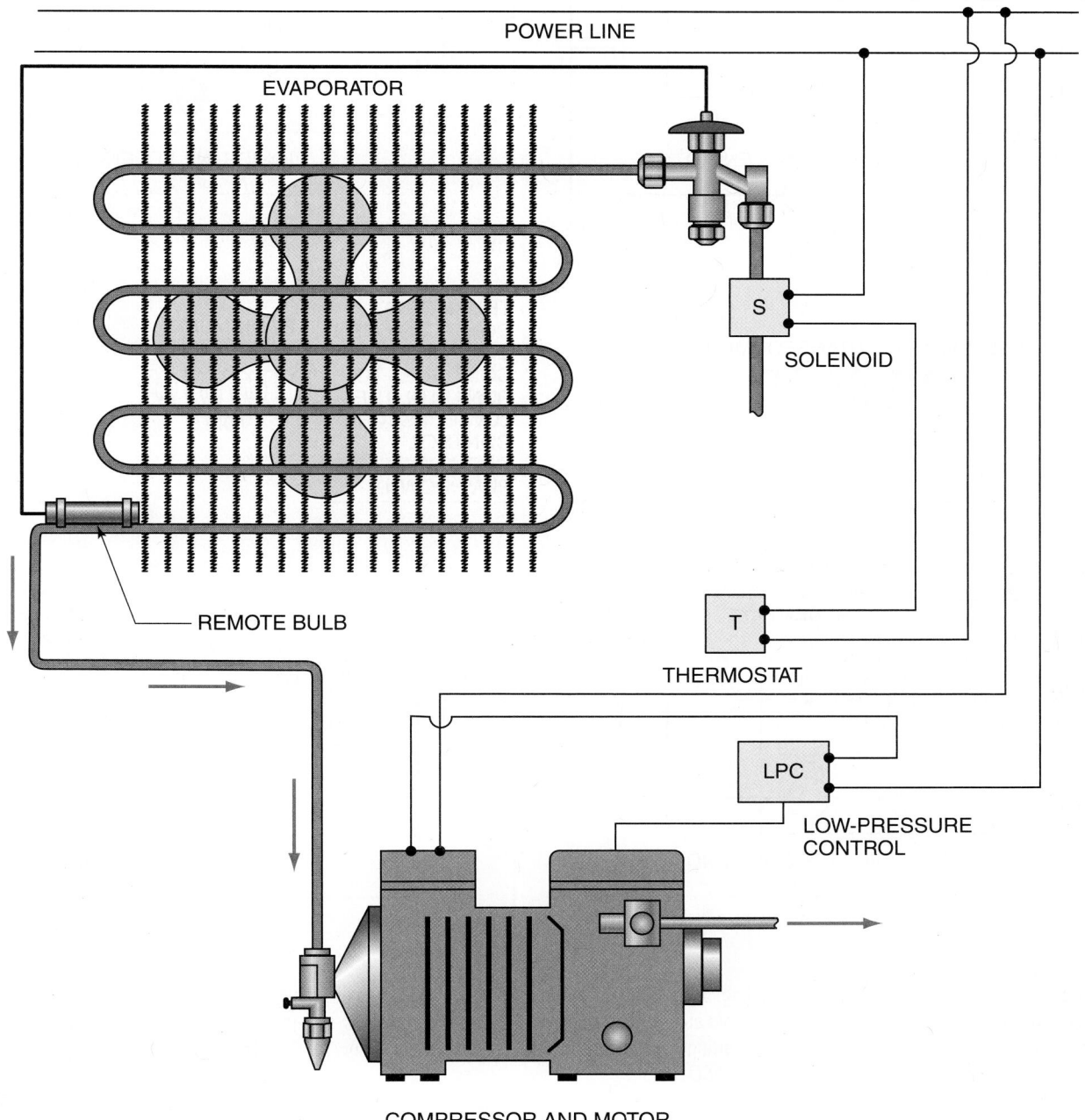

Figure 25–27   (A) The electrical and mechanical hookup of an automatic pumpdown system. *Courtesy ESCO Press*

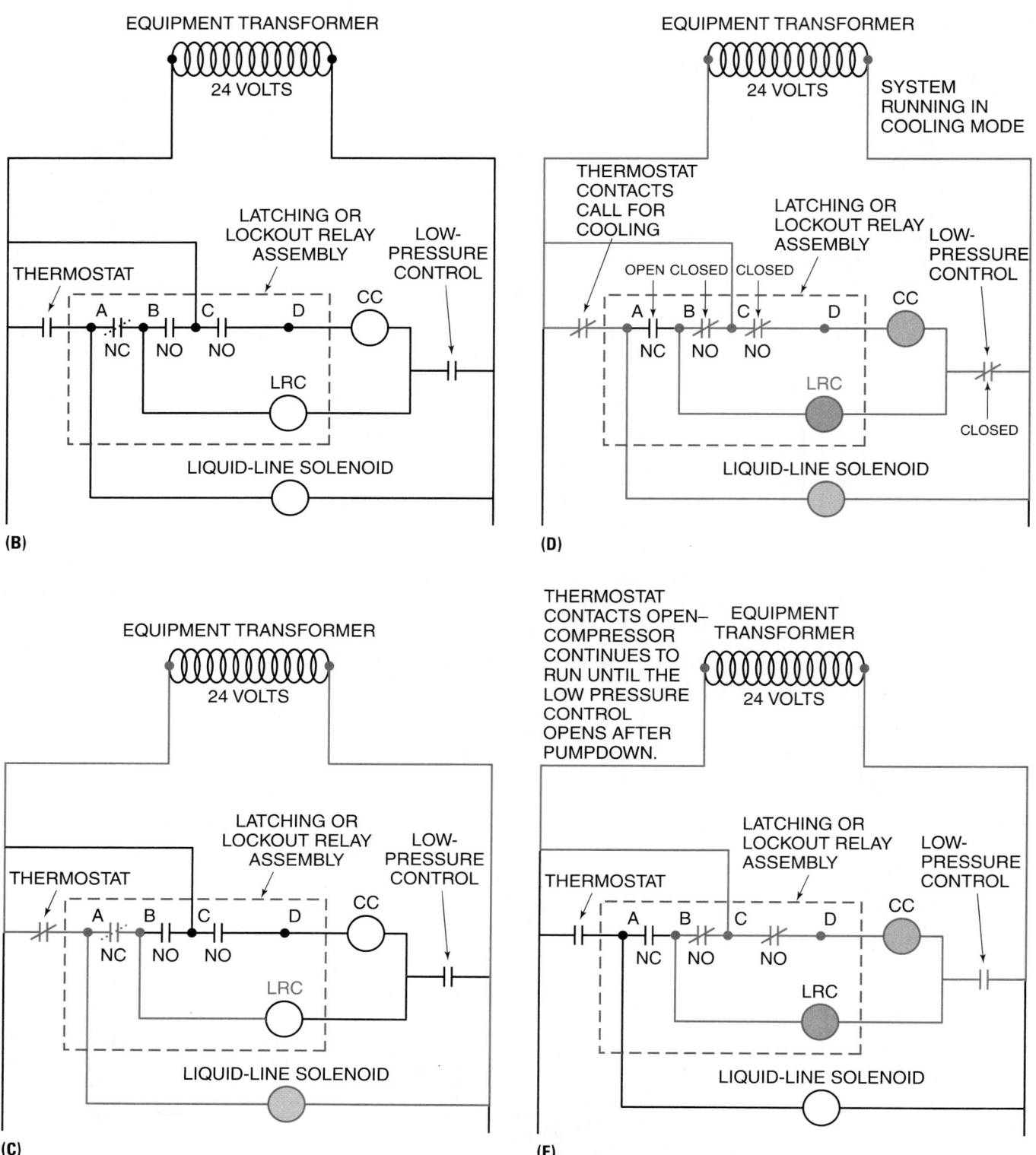

Figure 25–27 *(continued)* **(B)** A non-short-cycling, or "one-time," automatic pumpdown system using a latching or lockout relay. **(C)** The thermostat provides power to the liquid-line solenoid and refrigerant begins to flow. When the low-side pressure reaches the cut-in point on the low-pressure control, the batching relay closes, starting the compressor. **(D)** The compressor is now running. **(E)** The system will run until the low-pressure control contacts open. **(B)** *Courtesy ESCO Press*

control is necessary on water-cooled equipment because an interruption of water is more likely than an interruption of air. This control opens on a rise in pressure and should be set above the typical high pressure that the machine would

normally encounter. The high-pressure control may be either automatic or manual reset.

When an air-cooled condenser is placed outside, the condenser can be expected to operate at no more than 30°F

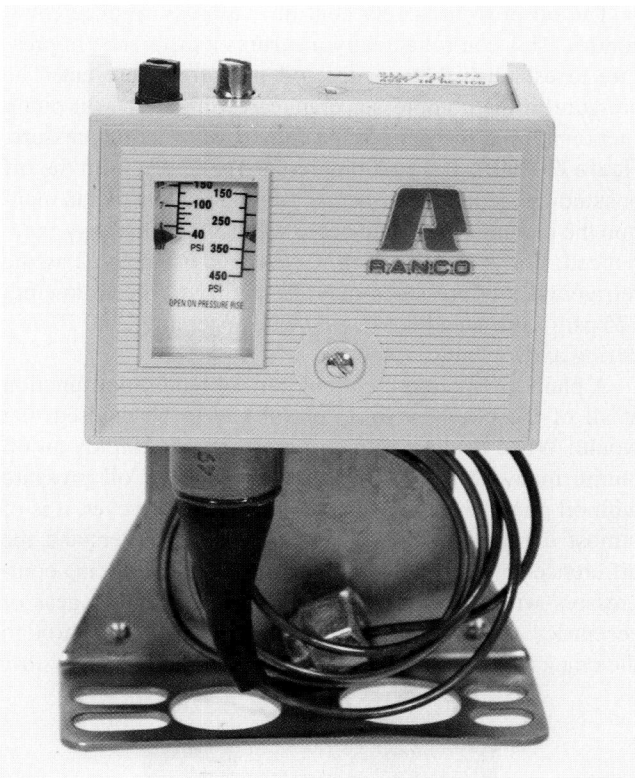

Figure 25–28 A high-pressure switch to keep the compressor from running when high system pressures occur.

warmer than the ambient air. This condition is true after the condenser has run long enough to have a coat of dirt built up on the coil. A clean condenser is important, but it is more often slightly dirty than clean. If the ambient air is 95°F, the condenser would be operating at about 170 psig if the system used R-12 (95°F + 30°F = 125°F condensing temperature, which corresponds to 169 psig). The high-pressure control should be set well above 170 psig for the R-12 system. If the control were set to cut out at 250 psig, there should be no interference with normal operation, and it still would give good protection, **Figure 25–28.** If the system incorporates R-134a as the refrigerant, the high-pressure control should be set well above 185 psig, which corresponds to a 125°F condensing temperature. Again, if the high-pressure control were set to cut out at 250 psig, it would still protect the system.

Most condensers operate at 30°F temperature difference as shipped from the factory and will continue to operate at this condition when properly maintained. Many condensers for low-temperature applications may operate at temperature differences as low as 15°F to maintain low compression ratios.

The control cut-in point must be above the pressure corresponding to the ambient temperature of 95°F. If the compressor cuts off and the outdoor fan continues to run, the temperature inside the condenser will quickly reach the ambient temperature. For example, if the ambient is 95°F, the pressure will quickly fall to 108 psig for an R-12 system. If the high-pressure control were set to cut in at 125 psig, the compressor could come back on with a safe differential of

125 psig (cut-out 250 − cut-in 125 = 125 psig differential). Some high-pressure controls have a fixed differential of 50 psig. This means they will cycle the compressor back on at 50 psig lower than the cut-out setting. For example, if a system cycles off on the high-pressure control at 250 psig, it will come back on at 200 psig (250 psig − 50 psig).

Some manufacturers specify a manual reset high-pressure control. When this control cuts out, someone must press the reset button to start the compressor. This calls attention to the fact that a problem exists. The manual reset control provides better equipment protection, but the automatic reset control may save the food by allowing the compressor to run at short intervals. An observant owner or operator should notice the short cycle of the automatic control if the compressor is near the workspace.

25.21 LOW-AMBIENT FAN CONTROL

The low-ambient fan control has the same switch action as the low-pressure control but operates at a higher pressure range. This control stops and starts the condenser fan in response to head pressure. This control must be coordinated with the high-pressure control to keep them from working against each other. The high-pressure control stops the compressor when the head pressure gets too high, and the low-ambient control starts the fan when the pressure gets to a predetermined point before the high-pressure control stops the compressor.

When a low-ambient control is used, the high-pressure cut-out should be checked to make sure that it is higher than the cut-in point of the low-ambient control. For example, if the low-ambient control is set to maintain the head pressure between 125 psig and 175 psig, a high-pressure control setting of 250 psig should not interfere with the low-ambient control setting. This can easily be verified by installing a gage and stopping the condenser fan to make sure that the high-pressure control is cutting out where it is supposed to. With the gage on the high side, the fan action can be observed also to see that the fan is cutting off and on as it should. The fan will operate all of the time on high ambient days, and the fan control will not stop the fan. The low-ambient pressure control can be identified by the terminology on the control's action. It is described as a "close on a rise" in pressure, **Figure 25–29.** This is the same terminology as used on a low-pressure control. The difference is that the low-ambient fan control has a higher operating pressure range.

25.22 OIL PRESSURE SAFETY CONTROL

Many larger compressors in the refrigeration and air-conditioning field have forced oiling systems. These compressors are usually over 5 hp. They contain an oil pump located at the end of the compressors crankshaft. The crankshaft is actually connected to the oil pump and supplies power, which turns the oil pump. Oil pumps can be of the gear or eccentric type. The oil pumps force oil through drilled holes in the crankshaft and deliver it to bearings and connecting rods. The oil then drops to the

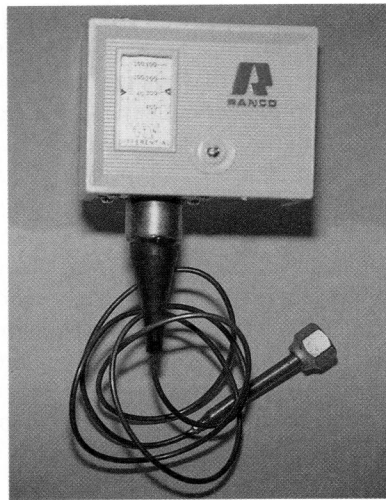

Figure 25–29 The low-ambient control used to open the contacts on a drop in head pressure to stop the condenser fan.

crankcase to be picked up again by the oil pump. Smaller compressors usually have a splash-type oiling system. These systems have an oil scoop that scoops and flings the oil throughout the crankcase, causing an oil fog as the crankshaft turns.

🌐The oil pressure safety control (switch) is used to ensure that the compressor has oil pressure when operating,🌐 **Figure 25–30(A).** This control is used on larger compressors and has a different sensing arrangement than do the high- and low-pressure controls. The high- and low-pressure controls are single-diaphragm or single-bellows controls because they are comparing atmospheric pressure with the pressures inside the system. Atmospheric pressure can be considered a constant for any particular locality because it does not vary more than a small amount.

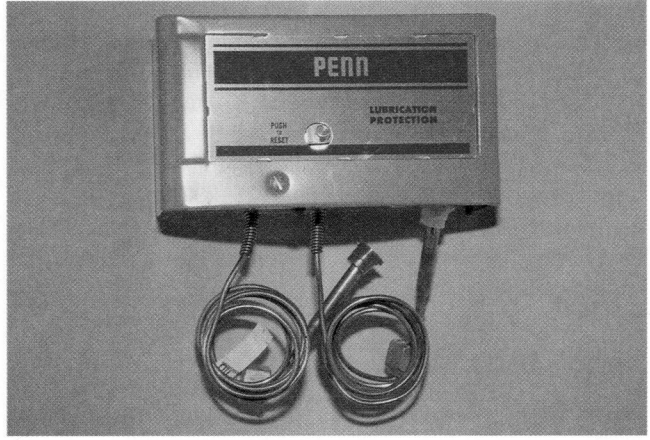

Figure 25–30(A) Oil safety controls have a bellows on each side of the control. They are opposed in their forces and measure the net oil pressure: the difference in the suction pressure (oil pump inlet) and the oil pump outlet pressure. This control has a 90-sec time delay to allow the compressor to get up to speed and establish oil pressure before it shuts down. *Photo by Bill Johnson*

The oil pressure safety control is a pressure differential control. This control actually measures a difference in pressure to establish that positive oil pressure is present. The pressure in the compressor crankcase (this is the oil pump suction inlet) is the same as the compressor suction pressure, **Figure 25–30(B).** The suction pressure will vary from the off or standing reading to the actual running reading, not to mention the reading that would occur when a low charge is experienced. For example, when a system is using R-22 as the refrigerant, the pressures may be similar to the following: 125 psig while standing, 70 psig while operating, and 20 psig during a low-charge situation.

A plain low-pressure cut-out control would not function at all of these levels, so a control had to be devised that would. When dealing with compressors that employ an oil pump, many service technicians confuse net oil pressure with oil pump discharge "outlet" pressure. However, it is of utmost importance that service technicians understand the difference between these two pressures when servicing compressors with oil pumps. The oil pump's rotating gear or eccentrics add a certain pressure to the oil pumped through the crankshaft. This pressure is considered *net oil pressure.*

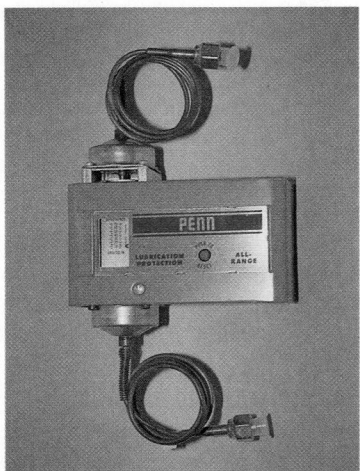

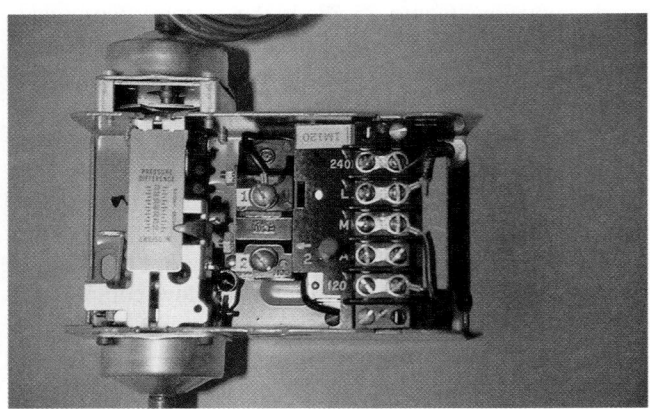

Figure 25–30(B) Two views of an oil pressure safety control. This control satisfies two requirements: how to measure net oil pressure effectively and how to get the compressor started to build oil pressure. *Photos by Bill Johnson*

Net oil pressure is *not* the pressure that can be measured at the discharge or outlet of the oil pump. The oil pump picks up oil at suction or crankcase pressure from the compressor's crankcase through a screen or filter. The oil pump discharge port's pressure includes both crankcase pressure and oil pump gear pressure it adds to the oil. This is why net oil pressure cannot be measured directly with a gage. A gage at the oil pump's discharge port would register a combination of crankcase pressure and oil pump gear pressure. The technician must subtract the crankcase pressure from the oil pump discharge port's pressure to get the net oil pressure. The following is an equation for net oil pressure:

$$\frac{\text{Oil pump discharge}}{\text{pressure}} - \frac{\text{Crankcase}}{\text{pressure}} = \frac{\text{Net oil}}{\text{pressure}}$$

Example: The oil pump discharge pressure is 80 psig. The crankcase pressure is 20 psig. What would be the net oil pressure?

Solution: Simply subtract the crankcase pressure from the oil pump discharge pressure to get net oil pressure.

$$80 \text{ psi} - 20 \text{ psi} = 60 \text{ psid net oil pressure}$$

This means the oil pump is actually putting 60 psid of pressure into the oil when delivering it into the crankshaft's drilled passages. Notice that the net oil pressure is expressed in *pounds per square inch differential (psid)* because it is a difference between two pressures.

Sometimes the compressor's crankcase may be operating in a vacuum. In this case, the crankcase pressure is negative. Remember that every 2 in. of mercury vacuum is equivalent to approximately 1 psi.

Example: What is the net oil pressure if the oil pump discharge pressure is 35 psig and the crankcase pressure is 6 in. of vacuum (-3 psi)?

Solution: Again, we must subtract the crankcase pressure from the oil pump discharge pressure to get the net oil pressure.

$$35 \text{ psi} - (-3 \text{ psi}) = 38 \text{ psid net oil pressure.}$$

This means that the oil pump is delivering 38 psi of net oil pressure through the crankshaft and bearings. NOTE: The terms *suction pressure* and *crankcase pressure* are often used interchangeably in the HVACR industry. However, on some larger refrigerant-cooled compressors there is a partition wall between the crankcase and motor barrel. During start-up periods or when system problems occur, there could be a slight difference in pressure between crankcase pressure in the crankcase and suction pressure in the motor barrel with these compressors. Because of this, it is of utmost importance to use the crankcase pressure, not suction pressure, when connecting an oil safety controller to these types of compressors.

In the basic low-pressure control, the pressure is under the diaphragm or bellows and the atmospheric pressure is on the other side. The oil pressure control uses a double bellows— one bellows opposing the other—to detect the net or actual oil pressure. The pump inlet pressure is under one bellows, and the pump outlet pressure is under the other bellows. These bellows are opposite each other either physically or by linkage.

The bellows with the most pressure is the oil pump outlet, and it overrides the bellows with the least amount of pressure. This override reads out in net pressure and is attached to a linkage that can stop the compressor when the net pressure drops for a predetermined time.

Because the control needs a differential in pressure to allow power to pass to the compressor, it must have some means to allow the compressor to get started. There is no pressure differential until the compressor starts to turn because the oil pump is attached to the compressor crankshaft. A time delay is built into the control to allow the compressor to get started and to prevent unneeded cut-outs when oil pressure may vary for only a moment. This time delay is normally about 90 to 120 sec. It is accomplished with a heater circuit and a bimetal device or electronically, **Figure 25–30(C)**.

The delay also gives the crankcase time to clear any unwanted refrigerant during periods when refrigerant migration or flooding has occurred. Furthermore, it avoids nuisance shutdowns during short fluctuations in net oil pressure on start-ups.

Net oil pressures vary from compressor to compressor. Net oil pressures usually range from 20 to 40 psi. Most oil pressure safety controllers will shut the compressor down if the net oil pressure falls below 10 psi. The following variables affect the net oil pressure:

- Compressor size
- Oil viscosity
- Oil temperature
- Bearing clearance
- Percent of refrigerant in the oil

Larger compressors need more net oil pressure because they have more surface areas to lubricate. The oil pumps must

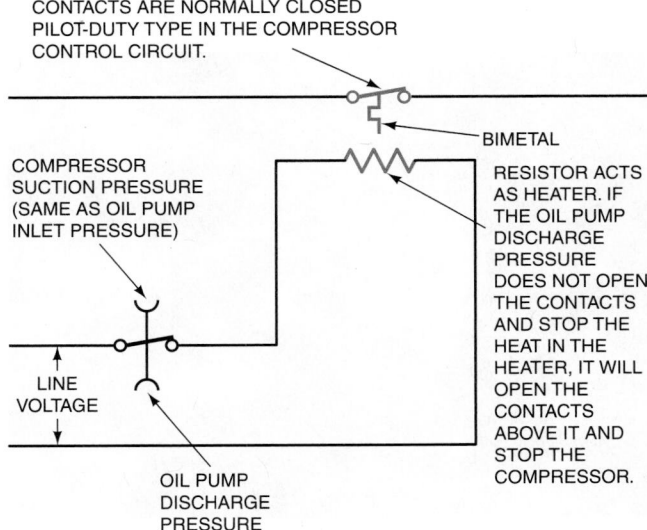

Figure 25–30(C) The oil pressure control is used for lubrication protection for the compressor. The oil pump that lubricates the compressor is driven by the compressor crankshaft. Therefore, there has to be a time delay to allow the compressor to start and build oil pressure. This time delay is normally 90 sec. The time delay is accomplished with either a heater circuit heating a bimetal or an electronic circuit.

pump and carry the oil greater distances within the larger compressor. As the oil gets hotter and its viscosity drops, the net oil pressure will also usually drop. As a compressor wears, its tolerances become greater and it becomes easier for the oil to escape through its clearances.

Transducer-type oil safety controllers use a pressure transducer, which senses a combination of oil pump discharge pressure and crankcase pressure. The pressure transducer has two separate ports to sense both crankcase pressure and oil pump discharge pressure. The subtraction or difference between these two pressures (net oil pressure) is mechanically accomplished by the pressure transducer. The pressure transducer is connected to an electronic controller by wires, **Figure 25–31(A)**. The pressure transducer is shown connected to the compressor in **Figure 25–31(B)**.

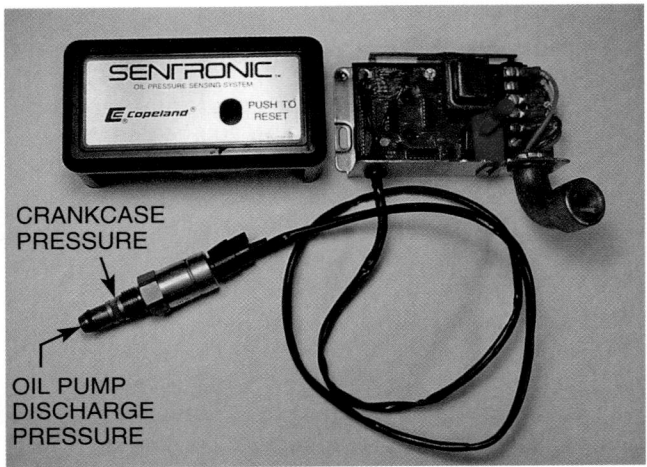

CRANKCASE PRESSURE

OIL PUMP DISCHARGE PRESSURE

(A)

(B)

ELECTRONIC OIL SAFETY CONTROL PRESSURE TRANSDUCER

Figure 25–31 **(A)** The pressure transducer wired to an electronic controller. **(B)** The pressure transducer connected to the compressor. *Courtesy Ferris State University. Photo by John Tomczyk*

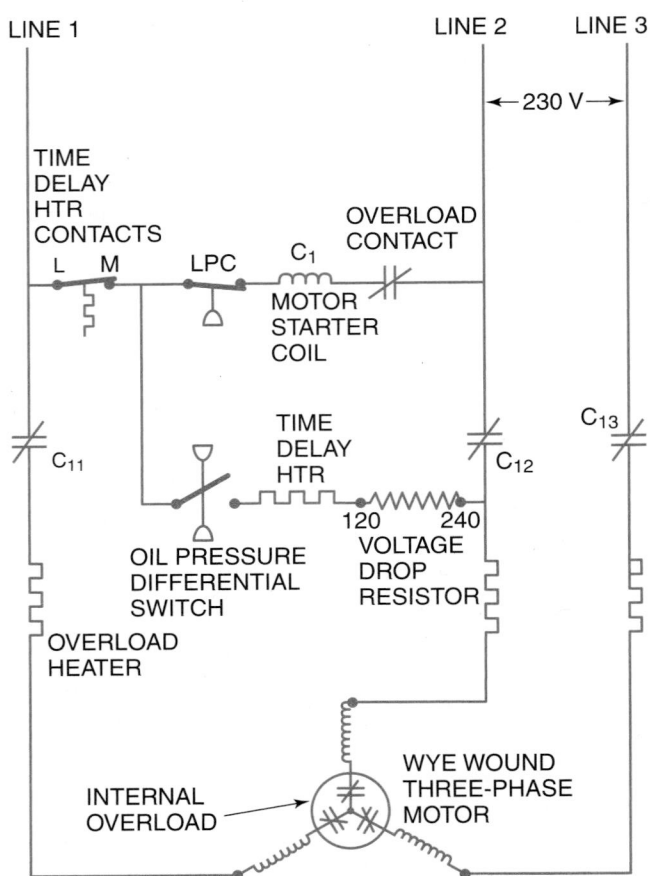

Figure 25–32 The schematic diagram of an oil safety control circuit and three-phase motor with internal overload.

It then transforms a pressure signal to an electrical signal for the electronic controller to process. The advantage of an electronic oil safety controller is that it eliminates capillary tubes. Thus, there is less chance for refrigerant leaks. Also, the electronic clock and circuitry are much more accurate and reliable. Remember, when the compressor is off, the net oil pressure is 0 psi and the differential pressure switch contacts are closed. The heater in the oil safety controller will not be energized during the off cycle because it is wired to the load side of the motor starter contacts, **Figure 25–32**. When the motor starter contacts are opened, this action takes L_2 out of the heater circuit. At start-up, when the motor starter contacts close and the compressor starts, the differential pressure switch contacts will stay closed and the heater will be energized until at least 9 psid of net oil pressure is developed on most controllers. As mentioned before, this time delay will prevent nuisance trips to the controller at compressor start-ups or when system problems like floodback and migration occur.

Internal Overloads

If a motor is equipped with both an internal inherent motor protector and an oil safety controller, the oil safety controller may trip due to a motor overheating or overloading problem on some systems. When the internal overload opens, the

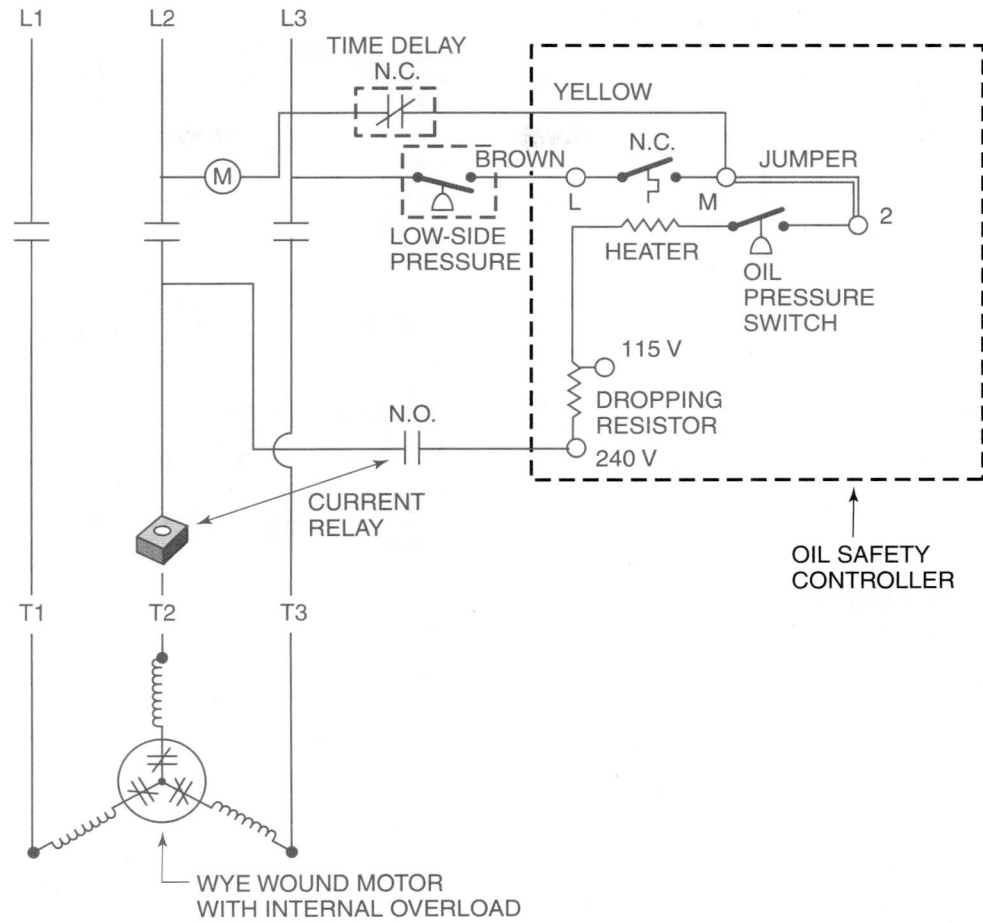

Figure 25–33(A) An oil safety circuit showing current sensing relay with motor and internal overload.

motor is shut off but the motor starter coil remains energized with contacts closed. The 2-min timing circuit would be activated due to a lack of net oil pressure. This will trip the oil safety controller in a matter of 2 min because the heater will still be energized through the load side of the motor starter, **Figure 25–32**. Remember, there is no net oil pressure when the compressor is shut off on its internal overload protector. In time, the compressor will cool off and the internal overload will reset (close). But the compressor will not restart until pushing the reset button manually resets the oil safety controller. This condition can be prevented by the use of a current sensing relay.

The use of a *current sensing relay* allows the compressor to cycle on the internal overload while not affecting the operation of the oil safety controller. The current sensing relay is wired to the load side of the contactor or motor starter. It acts like an inductive-type ammeter. One leg of the load side of the compressor motor is passed through the relay inductive coil. The inductive coil controls a normally open set of contacts in series with the time delay heater, **Figure 25–33(A)**. One leg of the compressor motor is shown passing through the inductive coil in **Figure 25–33(B)**. The contacts are closed when normal motor current is sensed. When the current drops quite low, or reduces to zero because of the motor being off, the contacts of the current relay will open and take power away from the time

CURRENT SENSING RELAY

Figure 25–33(B) A current sensing relay. *Photo by John Tomczyk*

delay heater. This prevents a tripping of the oil safety controller in case the internal overload opens (compressor off), but the motor starter's coil is still energized. The voltage-dropping resistor in **Figure 25–33(A)** makes the control more versatile for use on 230-V or 115-V systems.

It is often necessary to incorporate another field-wired relay when dealing with an electronic oil safety controller in these situations. The reason for this is that the electronic oil safety module requires power to the module on an oil safety trip for proper reset. The added relay simply provides this power to the module during a trip. However, always consult with the compressor and control manufacturer before applying or wiring any of these controls or devices. SAFETY PRECAUTION: The manufacturer's instructions should be consulted when working with any compressor that has an oil safety control.

25.23 DEFROST CYCLE

The defrost cycle in refrigeration is divided into medium-temperature and low-temperature ranges; the components that serve the defrost cycle are different.

25.24 MEDIUM-TEMPERATURE REFRIGERATION

The medium-temperature refrigeration coil normally operates below freezing and rises above freezing during the off cycle. A typical temperature range would be from 34°F to 45°F space temperature inside the cooler. The coil temperature is normally 10°F to 15°F cooler than the space temperature in the refrigerated box. This means that the coil temperatures would normally operate as low as 19°F (34°F − 15°F = 19°F). The air temperature inside the box will always rise above the freezing point during the off cycle and can be used for the defrost. This is called *off-cycle defrost* and can be either random or planned.

25.25 RANDOM OR OFF-CYCLE DEFROST

Random defrost will occur when the refrigeration system has enough reserve capacity to cool more than the load requirement. When the system has reserve capacity, it will be shut down from time to time by the thermostat, and the air in the cooler can defrost the ice from the coil. When the compressor is off, the evaporator fans will continue to run, and the air in the cooler will defrost the ice from the coil. When the refrigeration system does not have enough capacity or the refrigerated box has a constant load, there may not be enough off time to accomplish defrost. This is when it has to be planned.

25.26 PLANNED DEFROST

Planned defrost is accomplished by forcing the compressor to shut down for short periods of time so that the air in the cooler can defrost the ice from the coil. This is accomplished with a timer that can be programmed. Normally the timer stops the compressor during times that the refrigerated box is under the least amount of load. For example, a restaurant unit may defrost at 2 AM and 2 PM to avoid the rush hours, **Figure 25–34.**

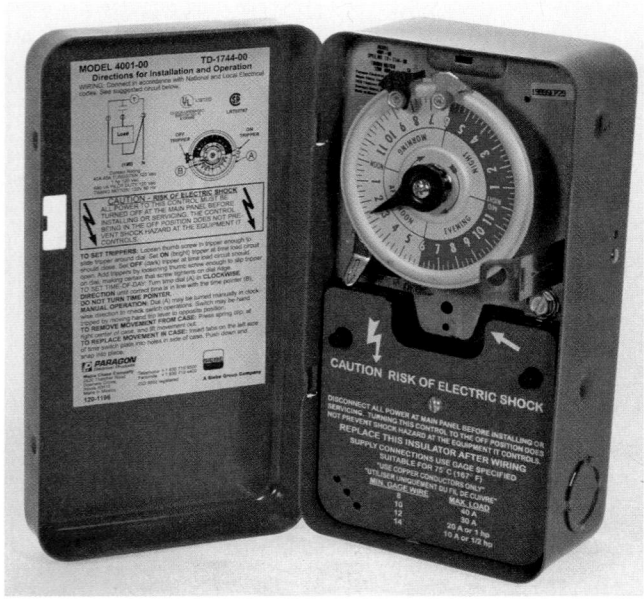

Figure 25–34 A timer to program off-cycle defrost. This is a 24-hour timer that can have several defrost times programmed for the convenience of the installation. *Photo by Bill Johnson*

25.27 LOW-TEMPERATURE EVAPORATOR DESIGN

Low-temperature evaporators all operate below freezing and must have planned defrost. Because the air inside the refrigerated box is well below freezing, heat must be added to the evaporator for defrost. This defrost is normally accomplished with *internal heat* or *external heat*.

25.28 DEFROST USING INTERNAL HEAT (HOT GAS DEFROST)

The internal heat method of defrost normally uses the hot gas from the compressor. This hot gas can be introduced into the evaporator from the compressor discharge line to the inlet of the evaporator and allowed to flow until the evaporator is defrosted, **Figure 25–35.** A portion of the energy used for hot gas defrost is available in the system. This makes it attractive from an energy-saving standpoint.

Injecting hot gas into the evaporator is rather simple if the evaporator is a single-circuit type because a T in the expansion valve outlet is all that is necessary. When a multicircuit evaporator is used, the hot gas must be injected between the expansion valve and the refrigerant distributor. This gives an equal distribution of hot gas to defrost all of the coils equally.

The defrost cycle is normally started with a timer in space temperature applications, where forced air is used to cool the product. When the defrost cycle is started, some method must be used to terminate it. Defrost can be terminated by time or temperature. The amount of time it takes to defrost the coil must be known before time alone can be used efficiently to terminate defrost. Because this time can vary from

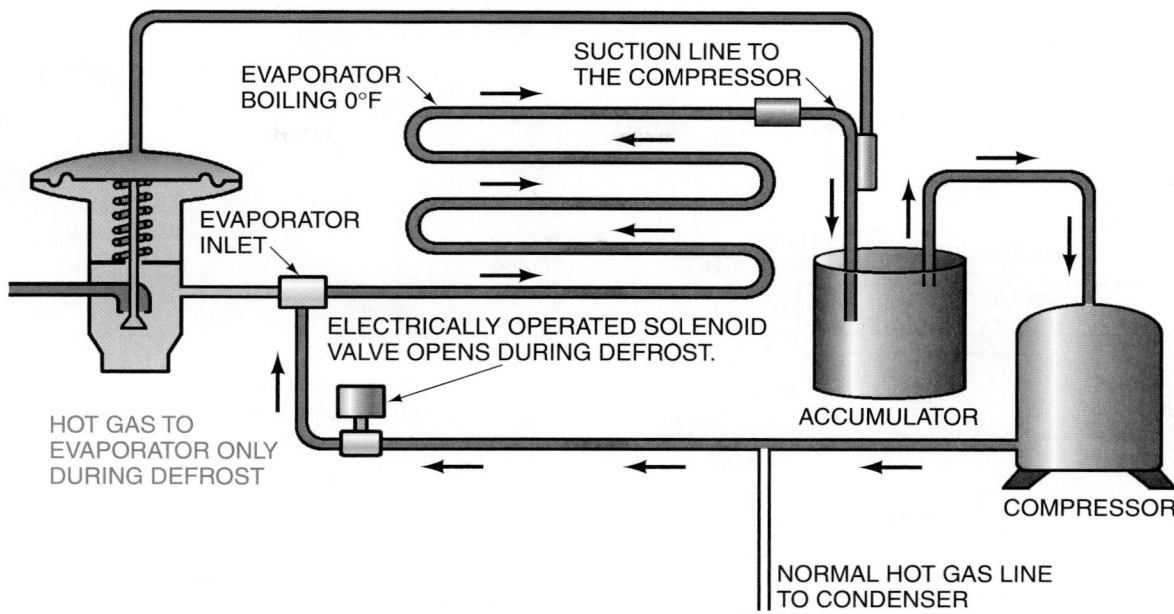

Figure 25–35 Hot gas defrost.

one situation to another, the timer could be set for too long a time, and the unit would run in defrost when it is not desirable, causing energy loss and possible product spoilage.

Defrost can be started with time and terminated with temperature. When this is done, a temperature-sensing device is used to determine that the coil is above freezing. The hot gas entering the evaporator is stopped, and the system goes back to normal operation, **Figure 25–36**.

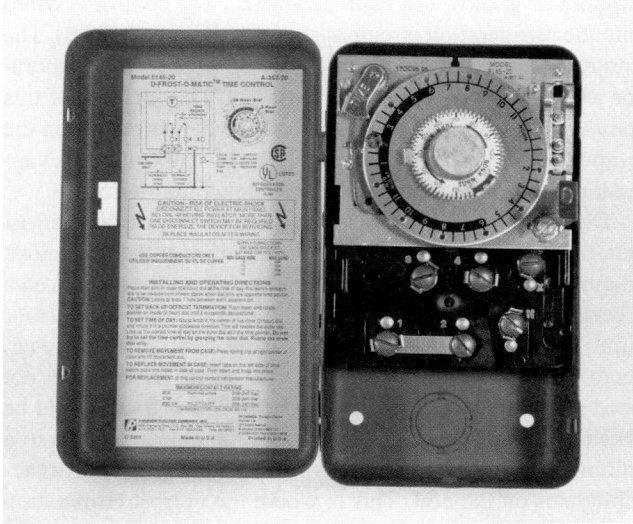

(A)

Figure 25–36 **(A)** A timer with a mechanism that can stop defrost with an electrical signal from a temperature-sensing element. When the coil is defrosted, the coil temperature will rise above freezing. There is no reason for the defrost to continue after the ice has melted. *Photo by Bill Johnson*

During the hot gas defrost cycle, several things must happen at one time. The timer is used to coordinate the following functions:

1. The hot gas solenoid must open.
2. The evaporator fans must stop, or cold air will keep defrost from occurring.
3. The compressor must continue to run.
4. A maximum defrost time must be determined and programmed into the timer in the event the defrost termination switch failed to terminate defrost.
5. Drain pan heaters may be energized.

25.29 EXTERNAL HEAT TYPE OF DEFROST

The external heat method of defrost is often accomplished with electric heating elements that are factory mounted next to the evaporator coil. This type of defrost is also a planned defrost that is controlled by a timer. The external heat method is usually not as efficient as the internal heat method because energy has to be purchased for defrost. Electric defrost may be more efficient if long runs are required for the hot gas lines. Refrigerant can condense in these long runs, causing slow defrost, and liquid refrigerant can reach the compressor, causing slugging and compressor damage. When electric defrost is used, it is more critical that the defrost be terminated at the earliest possible time. The timer controls the following events for electric defrost:

1. The evaporator fan stops on most cases.
2. The compressor stops. (There may be a pumpdown cycle to pump the refrigerant out of the evaporator to the condenser and receiver.)
3. The electric heaters are energized.

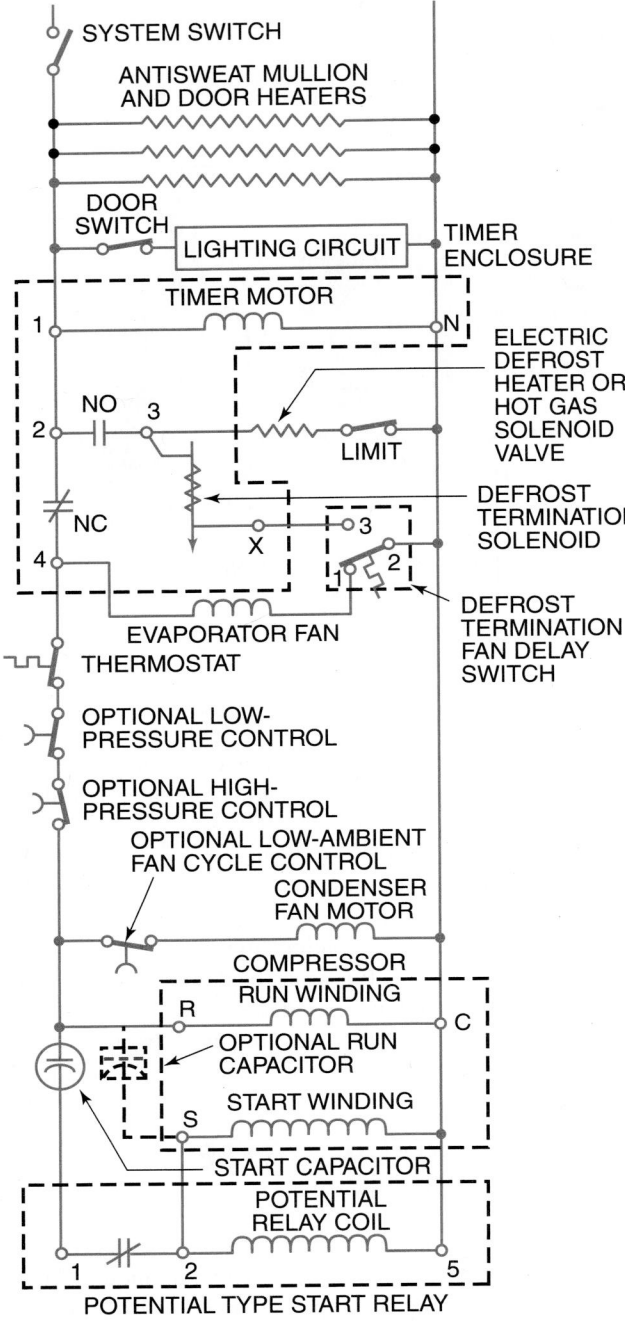

(B)

Figure 25–36 *(continued)*   **(B)** The wiring diagram of a circuit to control the defrost cycle. The events happen like this: When there is a defrost call (the timer contacts close), the solenoid valve opens, the fans stop, and the compressor continues to run and pump the hot gas into the evaporator. The coil gets warm enough to cause the thermostat to change the cold contacts to the hot contacts; the defrost will be terminated by the X terminal on the timer. When the coil cools off enough for the thermostat to change back to the cold contacts, the fan will restart. This is another method for keeping the compressor from running overloaded on a hot pulldown.

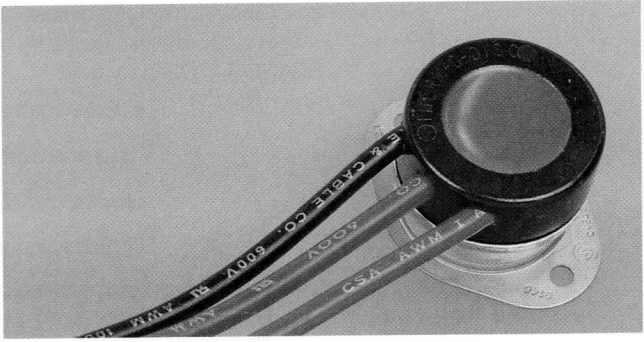

Figure 25–37   A sensor used with the timer in **Figure 25–36**. The sensor has three wires; it is a single-pole–double-throw device. It has a hot contact (made from common to the terminal that is energized on a rise in temperature) and a cold contact (made from common to the terminal that is energized when the coil is cold). This control is in either the hot or the cold mode. *Photo by Bill Johnson*

4. Drain pan heaters may be energized. NOTE: A temperature sensor may be used to terminate defrost when the coil is above freezing. A maximum defrost time should be programmed into the timer in the event the defrost termination switch failed to terminate defrost, **Figure 25–37**. This programmed defrost duration time is often referred to as the *fail safe* time.

25.30 DEFROST TERMINATION AND FAN DELAY CONTROL

The defrost termination/fan delay control is a temperature-activated, single-pole–double-throw switch controlled with a remote sensing bulb. The control shown also happens to be an adjustable type, **Figure 25–38(A)**. The control installed on an evaporator is shown in **Figure 25–38(B)**. The control is wired into the refrigeration circuit as shown in **Figure 25–36(B)**. The control's remote sensing bulb is located high on the evaporator where the frost is likely to clear last. The function of this temperature-activated switch is to terminate defrost when the evaporator coil has been defrosted, and also to delay the evaporator fans from coming on immediately after defrost.

Defrost time clocks can be programmed for certain defrost duration periods. This is a time duration set at the time clock in minute increments. For example, a defrost time clock on a freezer could be programmed to defrost every 6 hours (four times daily) and have defrost increments of 40 min. However, there will be times throughout the year where the coil does not need the entire 40 min of defrost heat. These times could be when low usage of the freezer keeps the door openings to a minimum or when the humidity is low and not much frost accumulates on the coil. This is where the defrost termination part of the control comes into play. The following paragraphs will explain the defrost termination function.

Let us assume that the system does not have a defrost termination/fan delay control. Once the normally open (NO) contacts of the defrost time control have closed and the unit is in defrost, the defrost heaters will be emitting heat and frost will be melting off the evaporator coil, **Figure 25–36(B)**.

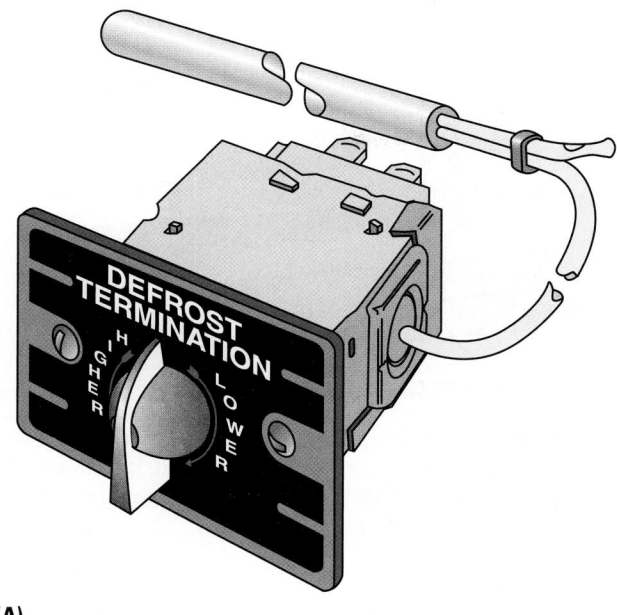

(A)

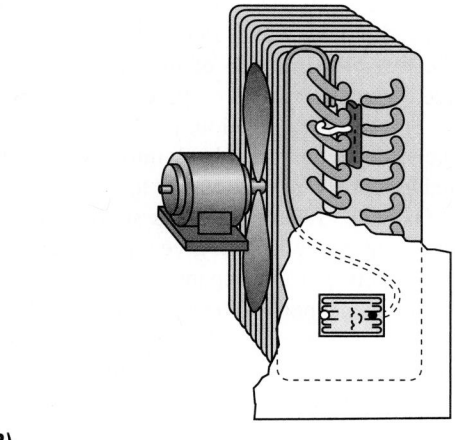

(B)

Figure 25–38 **(A)** A defrost termination–fan delay control.
(B) A defrost termination–fan delay control installed on a forced air evaporator. *Courtesy Ranco Controls Division*

If it takes only 10 min for all of the frost to leave the evaporator coil, there are still 30 min (40 min – 10 min) left in the programmed defrost duration increment. If the system has an optional heater safety switch, sometimes referred to as a defrost limit control, in series with the defrost heater as shown in **Figure 25–36(B),** this limit switch will open and take the defrost heaters out of the active circuit. However, the system's defrost timer will still have 30 min left in the defrost mode. The system will simply sit idle, and the product load will suffer in temperature because of no refrigeration for 30 min. After 30 min, the defrost timer will switch over to refrigeration and the fans will start immediately. The fans will blow the moist residual defrost heat through the refrigerated space and through the evaporator coil while the system is in refrigeration. This puts the system, thus the compressor, under an extremely high load from high suction pressures. The compressor will see high suction pressures with dense

vapors coming to its cylinders. This will cause high amperage draws and may overload the compressor to a point where its internal or external overload may open.

To prevent this long defrost period and the compressor overloading after defrost, a defrost termination/fan delay switch can be installed on the system as shown in **Figure 25–36(B).** Once the NO contacts of the defrost timer control have closed and the unit is in defrost, the defrost heaters will be emitting heat and ice will be melting off the coil. If it takes only 10 min for the ice to leave the coil, the remote bulb of the defrost termination control will sense the defrost heat and contacts between 2 and 3 will be made on the control. This will energize a defrost termination solenoid (release solenoid) in the time clock, which will mechanically put the system back into refrigeration. It does this by solenoid action and levers and will mechanically close the NC contacts and open the NO contacts of the defrost time control, **Figure 25–36(B).** This action by the defrost termination solenoid (release solenoid) prevents the system from sitting idle for 30 min in defrost with the heaters off. This action actually terminates the defrost mode and puts the refrigeration mode back into service. The fan delay part of the sequence will be explained in the following paragraph.

Now that the system is in refrigeration, the evaporator fans will be delayed from coming on. This happens until the contacts between 2 and 1 of the defrost termination/fan control close. This usually happens at about +20°F to +30°F, and is sensed and controlled by the control's remote bulb. This is an adjustable setting on most controls that lets the evaporator coil prechill itself and get rid of some of the defrost heat still in the coil. Delaying the fans prevents the suction pressure from getting too high after defrost and overloading the compressor when an automatic pumpdown cycle is not being used. It also prevents warm, moist air from being blown on the product in the refrigerated space.

25.31 REFRIGERATION ACCESSORIES

Accessories in the refrigeration cycle are devices that improve the system performance and service functions. This text will start at the condenser, where the liquid refrigerant leaves the coil, and will add various accessories as they are encountered in systems. Each system does not have all the accessories.

25.32 RECEIVERS

The *receiver* is located in the liquid line and is used to store the liquid refrigerant after it leaves the condenser. The receiver should be lower than the condenser so the refrigerant has an incentive to flow into it naturally. This is not always possible. The receiver is a tanklike device that can be either upright or horizontal, depending on the installation, **Figure 25–39.** Receivers can be quite large on systems that need to store large amounts of refrigerant. **Figure 25–40** is a photo of a large receiver that will hold several hundred pounds of refrigerant.

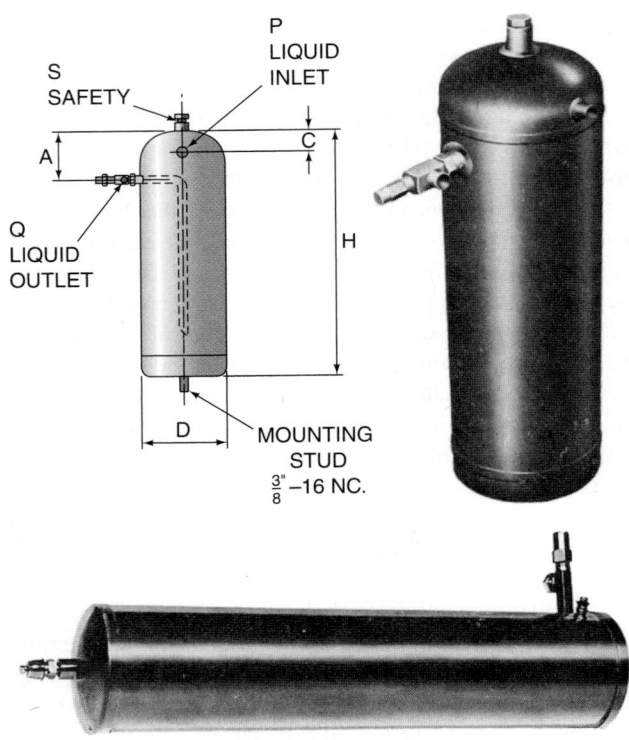

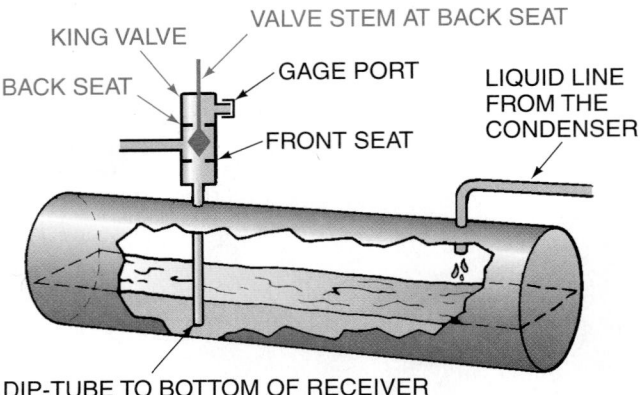

Figure 25–41 A king valve piped in the circuit. The back seat is open to the gage port when the valve is front seated. This valve has to be back seated for the gage to be removed if refrigerant is in the system.

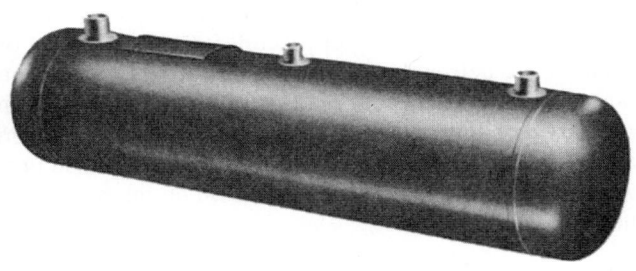

Figure 25–39 Vertical and horizontal receivers. *Courtesy Refrigeration Research*

Figure 25–40 Large receivers store the change for a condenser flooding system. It may hold more than 100 lb of refrigerant. *Courtesy Refrigeration Research*

The receiver inlet and outlet connection can be at almost any location on the outside of the tank body. On the inside of the receiver, however, the refrigerant must enter the receiver at the top in some manner. The refrigerant that is leaving the receiver must be taken from the bottom to ensure 100% liquid. This is accomplished with a dip tube if the line is at the top.

25.33 THE KING VALVE ON THE RECEIVER

The *king valve* is located in the liquid line between the receiver and the expansion valve. It is often fastened to the receiver tank at the outlet, **Figure 25–41.** The king valve is important in service work because when it is front seated, no refrigerant can leave the receiver. If the compressor is operated with this valve front seated, the refrigerant will all be pumped into the condenser and receiver. Most of it will

flow into the receiver. The receiver should be sized to hold approximately 20% more than the system's full charge of refrigerant. This leaves a vapor head for safety reasons during system pumpdowns. The other valves in the system can then be closed, and the low-pressure side of the system can be opened for service. When service is complete and the low side of the system is ready for operation, the king valve can be opened, and the system can be put back into operation.

The king valve has a pressure service port to which a gage manifold can be fastened. When the valve stem is turned away from the back seat, this port will give a pressure reading in the liquid line between the expansion valve and the compressor high-side gage port. When the valve is front seated and the system is being pumped down, this gage port is on the low-pressure side of the system common to the liquid line (this becomes the low side because the king valve is trapping the high-pressure refrigerant in the receiver). If there is any pressure in the liquid line, gage line removal may be difficult until the king valve is back seated after the repair is completed, **Figure 25–41.**

Another component that can be found in many systems is a solenoid valve. This was covered in Section 25.15, "The Solenoid Valve," earlier in this unit. It is a valve that stops and starts the liquid flow.

25.34 FILTER DRIERS

The *refrigerant filter drier* can be found at any point on the liquid line after the king valve. 🌐 The filter drier is a device that removes foreign matter from the refrigerant. This foreign matter can be dirt, flux from soldering, solder beads, filings, moisture, parts, and acid caused by moisture. Filter driers can remove construction dirt (filter only), moisture, and acid.🌐 Some sources of moisture in refrigeration systems include poor service practices, trapped air from improper evacuation techniques, leaky systems, and improper handling of polyol ester (POE) lubricants.

These filter-drying operations are accomplished with a variety of materials that are packed inside. Some of the manufacturers furnish beads of chemicals and some use a

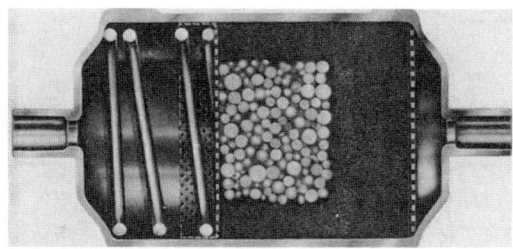

Figure 25–42(A) A filter drier. This device is for a liquid-line application and removes particles and moisture. The moisture removal is accomplished with a desiccant inside the drier shell. This desiccant material will be in bead or block form. *Courtesy Parker Hannifin Corporation*

porous block made from the drying agent. The most common agents found in the filter driers are activated alumina, molecular sieve, or silica gel. The component has a fine screen at the outlet to catch any fine particles that may be moving in the system, **Figure 25–42(A)**.

Molecular sieve desiccants are crystalline sodium alumina-silicates having cubic crystals. The crystal structure is honeycombed with uniform-sized cavities. They can selectively absorb molecules based on their charge (polarity) and size. These properties allow water to be absorbed while still permitting the larger molecules like refrigerants and their lubricants to pass by. The surface of the desiccant is coated with cations. Cations are positively charged and act like magnets that will absorb polarized molecules such as water. Molecular sieves retain the highest amount of water while keeping the concentration of water in the refrigerant low. There is a strong bond between water and the molecular sieve. System corrosion, acid formations, and freeze ups will be very low with molecular sieve filter driers. They are recommended for liquid-line filter driers because of their good water removal capabilities. Inorganic acids like hydrochloric and hydrofluoric acids are formed from the breakdown of refrigerants and water with high system temperatures. These acids can attack metal surfaces in the system and break down the crystalline structure of the molecular sieve. Organic acids then form from the breakdown of the refrigerant lubricant in the presence of the wrong desiccant and water with high system temperatures. Organic acids are sludgelike material, which can plug up metering devices. Activated alumina is more effective than the molecular sieve for removing organic acid molecules.

Activated alumina is formed from aluminum oxide. As mentioned earlier, activated alumina is more effective than the molecular sieve for removing organic acid molecules. It is not a highly crystalline structure like molecular sieves. Its pores vary in size and do not absorb molecules based on its pore size. Because of this, it can absorb much larger molecules like refrigerants and lubricants along with water. Activated alumina is even more effective in removing organic acids when it is used in the suction line of the refrigerant system. This is why some manufacturers recommend the use of a suction-line filter drier with activated alumina for acid cleanups.

Silica gel is a noncrystalline material, and its molecules are formed by bundles of polymerized silica. There is a weaker bond formed between water and the desiccant, which is why silica gels are not widely used on today's filter driers.

Filtration, another main function of a filter drier, can be accomplished with a screen. A screen will filter or catch particles that cannot fit through its wire mesh. As particles and contaminants start to cover the mesh of the screen, they act as a fine filter to remove even smaller particles. This accumulation of particles will soon cause enough pressure drop that the refrigerant will flash into a vapor. This will often cause a local cold spot on the filter drier and it may even sweat. Other filters include the following:

- Bonded desiccant cores
- Fiberglass pads
- Rigid fiberglass with a resin coating

Bonded desiccant cores have smaller openings than fiberglass pads. As refrigerant passes through, the channels trap the particles. As the channels fill, pressure drop occurs and the refrigerant will again vaporize or flash.

Fiberglass pads are less compressed than rigid fiberglass with resins. As refrigerant and particles flow through the pads, their speed decreases and they deposit themselves in the openings of the fiberglass. As the fiberglass fills with more particles, the filtration becomes finer. It soon becomes similar to a rigid fiberglass filter. A fiberglass pad can hold up to four times as many particles and contaminants as the bonded desiccant core drier can and still have the same filtration capacity.

The filter drier comes in two styles: permanent and replaceable core. Both types of driers can be used in the suction line when chosen for that application. This is discussed in more detail later.

The filter drier can be fastened to the liquid line by a flare connection in the smaller sizes up to 5/8 in. Solder connections can be used for a full range of sizes from 1/4 to 1 5/8 in. The larger solder filter driers are replaceable core types. The replaceable-core filter drier can be useful for future service, **Figure 25–42(B)**. The king valve may be front seated to pump the refrigerant into the condenser and receiver so that the drier cores may be replaced.

The filter drier can be installed anywhere in the liquid line and is found in many locations. The closer it is to the metering device, the better it cleans the refrigerant before it enters

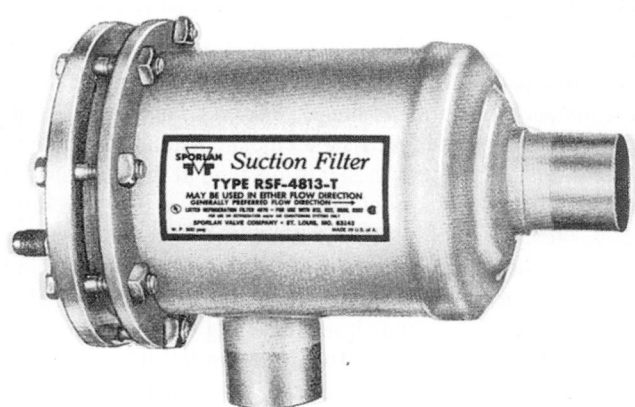

Figure 25–42(B) This filter drier casing has removable cores. It can be left in the system if needed. *Courtesy Sporlan Valve Company*

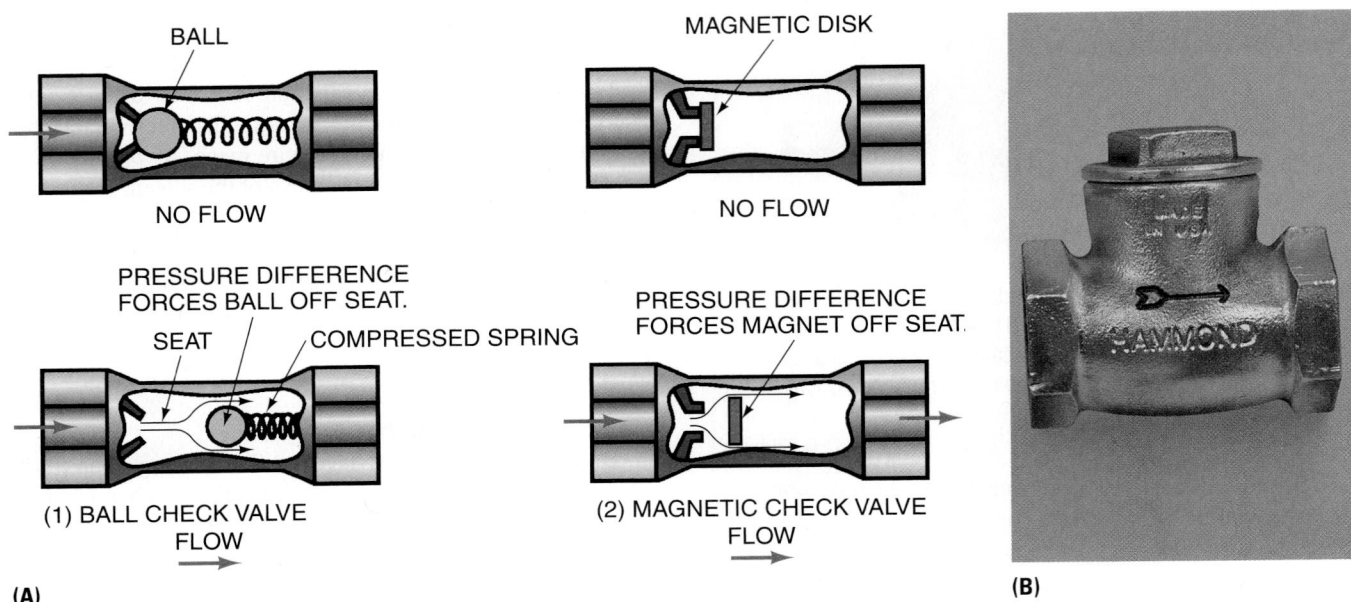

Figure 25–43 **(A)** Two types of check valves. **(B)** Photo of valve. *Photo by Bill Johnson*

the tiny orifice in the metering device. The closer it is to the king valve, the easier it is to service. Service technicians and engineers choose their own placement, based on experience and judgment.

25.35 REFRIGERANT CHECK VALVES

A *check valve* is a device that will allow fluid flow in only one direction. It is used to prevent refrigerant from backing up in a line.

Several types of mechanisms are used to allow flow in one direction. Two common types of check valves are the ball check and magnetic check, **Figure 25–43(A)**. Both of these check valves cause some pressure drop, depending on the flow rate. The technician must consult the manufacturer's catalog for flow rate and application. The check valve can be identified by an arrow on the outside that indicates the direction of flow, **Figure 25–43(B)**.

25.36 REFRIGERANT SIGHT GLASSES

The *refrigerant sight glass* is normally located anywhere liquid flow exists and anywhere it can serve a purpose. 🌎 When it is installed just prior to the expansion device, the technician can be assured that a solid column of liquid is reaching the expansion device. When it is installed at the condensing unit, it can help with troubleshooting. 🌎 More than one sight glass, one at each place, is sometimes a good investment.

Sight glasses come in two basic styles: plain glass and glass with a moisture indictor. The plain glass type is used to observe the refrigerant as it moves along the line. The sight glass with a moisture indicator in it can tell the technician what the moisture content is in the system. It has a small element in it that changes color when moisture is present. An example can be seen in **Figure 25–44**.

(A)

(B)

Figure 25–44 **(A)** A sight glass with an element that indicates the presence of moisture in the system. **(B)** A sight glass used only to view the liquid refrigerant to be sure no vapor bubbles are in the liquid line. **(A)** *Courtesy Superior Valve Company.* **(B)** *Courtesy Henry Valve Company*

25.37 LIQUID REFRIGERANT DISTRIBUTORS

The *refrigerant distributor* is used on multicircuit evaporators. It is fastened to the outlet of the expansion valve. It distributes the refrigerant to each individual evaporator circuit. This is a precision-machined device that ensures that the

Figure 25–45   A multicircuit refrigerant distributor to feed equal amounts of liquid refrigerant to the different refrigerant circuits. The combination of liquid and vapor refrigerant has a tendency to stratify with the liquid moving to the bottom. This is a precision-machined device for separating the mixture evenly. *Photo by Bill Johnson*

refrigerant is divided equally to each circuit, **Figure 25–45.** This is not a simple task because the refrigerant is not all liquid or all vapor. It is a mixture of liquid and vapor. A mixture has a tendency to stratify and feed more liquid to the bottom and the vapor to the top. Sizing a distributor to a system can be a complex decision because it must have exactly the correct pressure drop to perform correctly. The expansion valve and the distributor must be considered together because of the pressure reduction through the distributor. This sizing is determined by the manufacturer for a new installation. ♻ In the future, as refrigerants are changed from the chlorofluorocarbon refrigerants to the new refrigerants, the refrigerant distributor will need to be considered. A close relationship with the manufacturers will be necessary, or distributors may be mismatched to expansion valves. ♻

Some distributors have a side inlet that hot gas can be injected into for hot gas defrost, **Figure 25–46.**

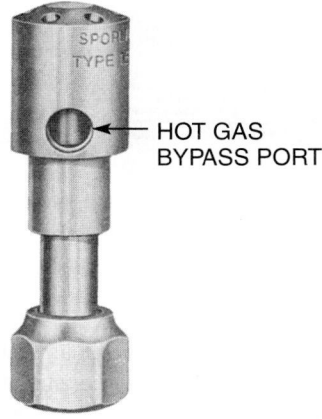

HOT GAS
BYPASS PORT

Figure 25–46   The same type of distributor as in **Figure 25–45** except that is has a side inlet for allowing hot gas to enter the evaporator evenly during defrost. *Courtesy Sporlan Valve Company*

25.38   HEAT EXCHANGERS

The *heat exchanger* is often placed in or at the suction line leaving the evaporator. This heat exchange is between the suction and the liquid line, **Figure 25–47(A),** and serves two purposes.

1. 🌐 It improves the capacity of the evaporator by subcooling the liquid refrigerant entering it, which can allow a smaller evaporator in the refrigerated box. The cooler the refrigerant entering the expansion device, the more net refrigeration effect in the evaporator. 🌐 This is not necessarily a net system efficiency improvement because the heat is given up to the suction gas and is still in the system. This heating of the suction gas expands the gas, giving it a lower density before it enters the cylinders of the compressor. This can reduce the refrigerant flow rate pumped by the compressor.

2. 🌐 It prevents liquid refrigerant from moving out of the evaporator to the suction line and into the compressor. 🌐 Most of these heat exchangers are simple and straightforward.

The heat exchanger has no electrical circuit or wires. It can be recognized by the suction line and the liquid line piped to the same device. In some small refrigeration devices, the capillary tube is soldered to the suction line. This accomplishes the same thing that a larger heat exchanger does, **Figure 25–47(B).**

The next two components that could be installed in the suction line are the EPR valve and the CPR valve. These were covered earlier in this unit as control components.

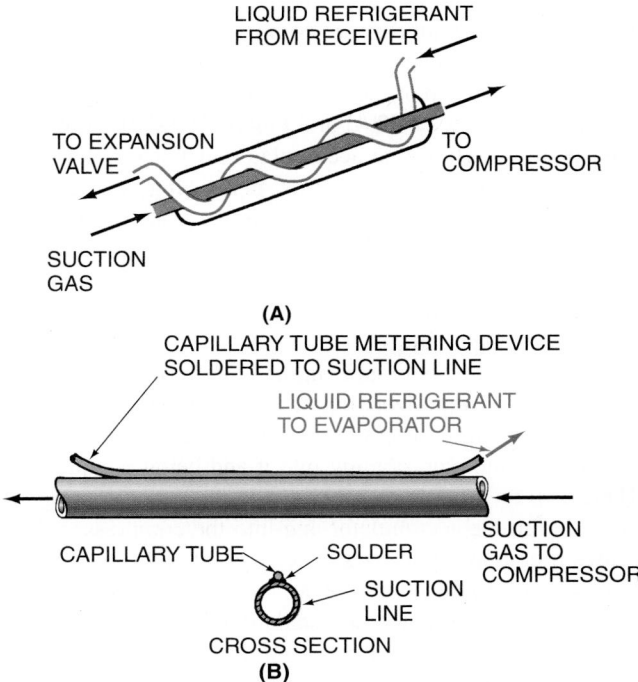

Figure 25–47   Heat exchangers. **(A)** Plain liquid-to-suction type. **(B)** Capillary tube fastened to the suction line; it accomplishes the same thing.

25.39 SUCTION-LINE ACCUMULATORS

The *suction-line accumulator* can be located in the suction line to prevent liquid refrigerant from passing into the compressor. As discussed earlier, liquid refrigerant floodback causes serious problems for a compressor. Floodback dilutes the compressor's oil with liquid refrigerant and causes foaming in the crankcase, resulting in bearing washout. Under certain conditions, usually during defrost, liquid refrigerant may leave the evaporator. The suction line is insulated and will not evaporate much refrigerant. 🌐The suction-line accumulator collects this liquid and gives it a place to boil off (evaporate) to a gas before continuing on to the compressor.🌐

As mentioned above, liquid coming into a compressor during the running cycle is called liquid floodback. The following are some causes of liquid floodback:

- Wrong TXV setting (low or no compressor superheat)
- Overcharge
- Evaporator fan not operating
- Low load on evaporator
- End of cycle (lowest load)
- Defrost clock or heater not operating, causing an iced coil
- Dirty or blocked evaporator coil
- Capillary tube overfeeding
- Capillary tube system overcharged
- Loose TXV bulb on evaporator outlet
- Oversized expansion valve
- Flooding after hot gas termination
- Heat pump changeover
- Defrost termination

When liquid is present in the bottom of the accumulator, most accumulators are designed to meter both the liquid refrigerant and the oil back to the compressor at an acceptable rate that will not damage compressor parts or cause oil foaming in the crankcase. This is done with a small metering orifice at the bottom of the outlet tube while the compressor is running, **Figure 25–48(A)** and **Figure 25–48(B)**. However, even though suction line accumulators assist in refrigerant flooding and migration problems, if flooding is severe the accumulator may also flood and cause compressor damage. The only 100% safe accumulator is one that can hold the entire system charge. Always consult with the manufacturer when sizing a suction accumulator for specific catalog numbers and tonnage ratings.

When liquid refrigerant is in the bottom of an accumulator and is being vaporized, high compressor amperage draws may occur because the compressor now sees dense saturated or near-saturated gas coming into it and being compressed. There will be little or no superheat to the vapor coming off the liquid in the accumulator and into the compressor. This causes high mass flow rates of refrigerant and can overload a compressor to the point of overheating.

Many indoor compressor installations have problems with suction accumulators sweating and dripping on floors. The only way around this problem is for the accumulator to be insulated completely and vapor sealed to prevent condensate from forming under the insulation. Because they are made of steel, accumulators can rust if exposed to moisture for any

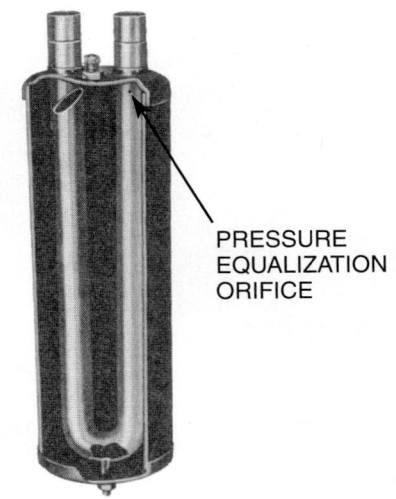

Figure 25–48(A)   A suction-line accumulator. Liquid refrigerant returning down the suction line can be trapped and allowed to boil away before entering the compressor. The oil bleed hole allows a small amount of any liquid in the accumulator to be returned to the compressor. If the liquid is refrigerant, there will not be enough of it getting back through the bleed hole to cause damage. *Courtesy AC & R Components, Inc.*

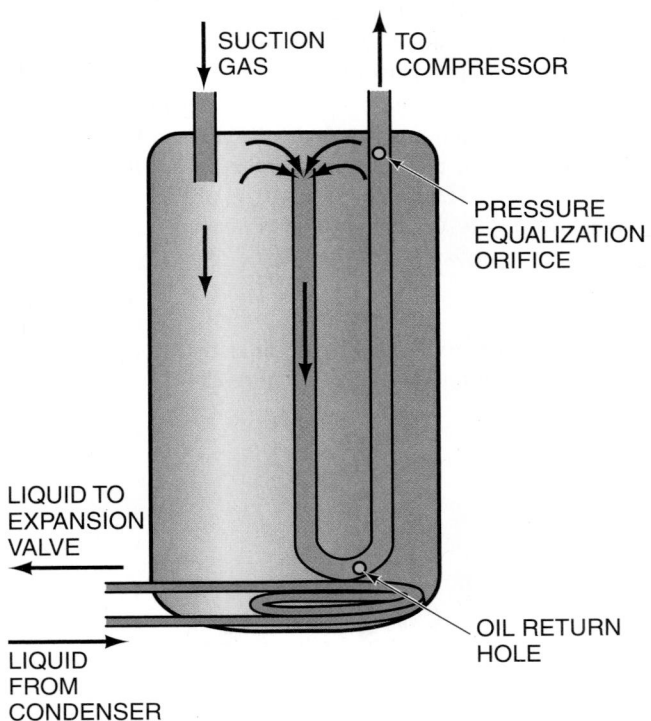

Figure 25–48(B)   A suction-line accumulator with heat exchanger.

long period of time. Rusting of an accumulator usually happens at its seams or at the piping stubs after about six to eight years of use. Although manufacturers do supply accumulators with rust preventive paints, during the welding processes these paints can be burned off, leaving exposed metals. If an accumulator is resoldered because of pinhole leaks, usually near the seams or piping stubs, the residual soldering flux should be washed away and the area cleaned with emery cloth or steel wool. The surfaces should then be covered with a silicone rubber or roofing tar for a waterproof seal. The suction lines connected to the accumulator should also be insulated to prevent sweat from forming on them and dripping onto the accumulator.

During an off cycle, the accumulator may have some liquid refrigerant move into it by gravity from the evaporator and suction line, especially when the system does not have an automatic pumpdown cycle for its off cycles. This liquid refrigerant will flow through the liquid refrigerant and oil return metering orifice and seek its own level both inside and outside the U-tube formation of the suction piping of the accumulator. This means that the U-tube of suction piping inside the accumulator will have a column of liquid refrigerant in it during the off cycle simply because of liquid inside the accumulator seeking its own level. When the compressor starts up, this column of liquid refrigerant would be sucked out of the accumulator and into the compressor if it were not for a pressure equalization orifice at the outlet of the accumulator, **Figure 25–48(C)**. This orifice will equalize pressure on both sides of the liquid column when the compressor is on or off. In other words, the liquid column will have accumulator pressure on both sides of it because of this orifice. This will prevent the column of liquid from being sucked out of the U-tube of the accumulator and possibly damaging the compressor during the running cycle. The liquid column will momentarily

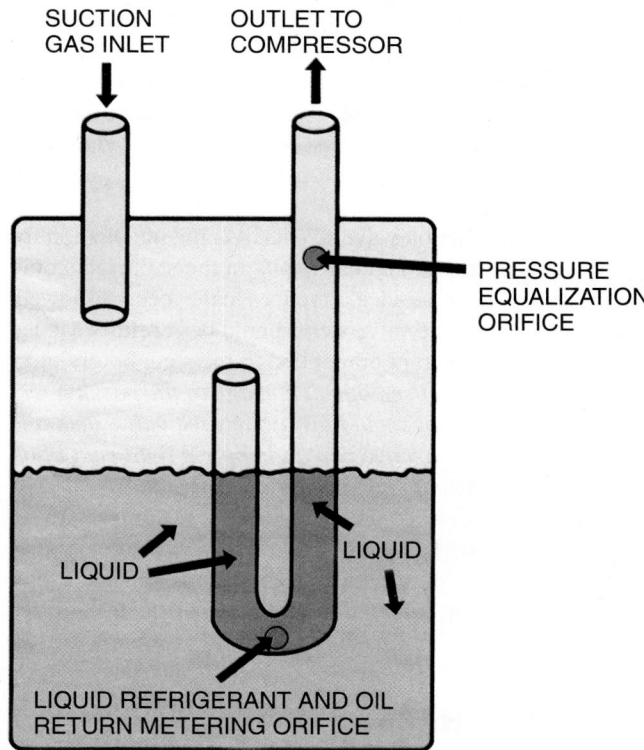

Figure 25–48(C)   A suction-line accumulator showing a pressure equalization orifice at the outlet of the accumulator.

hang in the U-tube and rapidly vaporize on a compressor start-up. This is a time where saturated gas could enter the compressor. Once this column of liquid in the accumulator is cleared or vaporized, the compressor will again see superheated gas.

Since compressors are extremely susceptible to liquid coming back to their crankcases or valves, accumulators add an extra safety precaution. Suction accumulators can assist in flooding and migration conditions. However, if flooding or migration problems are severe, suction-line accumulators have been known to flood and compressor damage still occurs. This is why the only 100% safe accumulator is one that can hold 100% of the entire system's refrigerant charge. Also, refrigerant vapor can migrate through the accumulator during the off cycle and still get to the compressor and condense. So, the accumulator is not the cure-all for refrigerant migration problems.

Accumulators should be selected with three basic considerations in mind. Consult with the manufacturer for specific catalog numbers and tonnage ratings.

1. The accumulator should have adequate liquid holding capacity. This should not be less than 50% of the entire system charge.
2. The accumulator should not add excessive pressure drop to the system.
3. Never base the accumulator on the line size of the suction line. Many times an accumulator will not have the same line size as the suction line.

The accumulator is usually located close to the compressor. It should not be insulated so that any liquid that is in it may have a chance to vaporize.

The suction-line accumulator may have a small coil in the bottom where liquid from the liquid line may be routed to boil off the refrigerant in the bottom of the accumulator, **Figure 25–48(B)**. Otherwise, part of the refrigerant charge for the system will stay in the accumulator until the heat from the vapor passing through boils it to a vapor. Actually, not much heat will exchange unless liquid is in the accumulator so it does not have much effect on the system while vapor only is in the accumulator. The liquid-line coil in the accumulator also will help subcool the liquid refrigerant before it reaches the metering device. This will increase the net refrigeration effect in the evaporator.

Suction-line accumulators are typically made of steel. They also sweat, which causes rust. Older accumulators should be observed from time to time for refrigerant leaks. Cleaning and painting the solder connection will help to prevent rust.

25.40   SUCTION-LINE FILTER DRIERS

The *suction-line filter drier* is similar to the drier in the liquid line. It is rated for suction-line use. A filter drier is placed in the suction line to protect the compressor—and it is good insurance in any installation. It is essential to install a suction-line filter drier after any failure that contaminates a system. A motor burnout in a compressor shell usually moves acid and contamination into the whole system. When a new compressor is installed, a suction-line filter drier can be installed to clean the refrigerant and oil before it reaches the compressor, **Figure 25–49(A)**. *Driers that have the capability to remove high levels of acid are typically used after motor burnout problems.*

Most suction-line filter driers have some means for checking the pressure drop across the cores. This is important because even a small pressure drop can cause the suction pressure to drop to the point that the compression ratio is increased. Compressor efficiencies will be reduced. **Figure 25–49(B)** shows the pressure fittings on a typical suction-line drier. The suction line is the most critical refrigerant line from a sizing standpoint. Any accessories such as evaporator pressure regulators, suction filters, suction accumulators, or crankcase pressure regulators must also be sized correctly when installed in the suction line. There must be the right refrigerant velocity in the suction line, whether it be a level line or a suction-line riser, to make sure there is proper oil return. Refrigerant velocity is what sweeps oil through the suction-line and back to the compressor. If the suction line is oversized, refrigerant velocity will be low and oil return will be hindered. If the suction line or its accessories are undersized, excessive pressure drop will cause lower suction pressure. This will cause higher compression ratios and lower volumetric efficiencies.

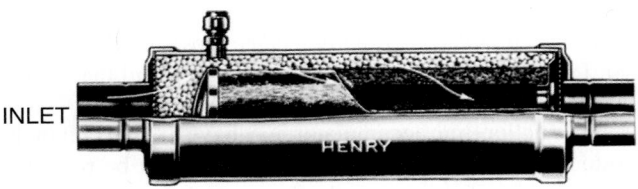

Figure 25–49(A) A suction-line filter drier. This device cleans any vapor that is moving toward the compressor. *Courtesy Henry Valve Company*

Figure 25–49(B) A suction-line filter drier with access ports for checking pressure drop. *Photo by Bill Johnson*

25.41 SUCTION SERVICE VALVES

The *suction service valve* is normally attached to the compressor. Equipment used for refrigeration installations usually has service valves. This is not always the case on air-conditioning equipment. The suction service valve can never be totally closed because of the valve's seat design. The service valve is either back seated, front seated, or midseated. A position of "just cracked" off the back seat is often used for reading system pressures and charging. An example of these functions is shown in **Figure 25–50**.

The suction service valve is used

1. as a gage port.
2. to throttle the gas flow to the compressor.
3. to valve off the compressor from the evaporator for service.
4. to charge the system with refrigerant.

The service valve consists of the valve cap, valve stem, packing gland, inlet, outlet, and valve body, **Figure 25–51**.

The *valve cap* is used as a backup to the packing gland to prevent refrigerant from leaking around the stem. It is normally made of steel and should be kept dry on the inside. An oil coating on the inside will help prevent rust. *Valve covers should always be in place except during service operations.*

The *valve stem* has a square head to accommodate the valve wrench. This stem is normally steel and should be kept rust-free. If rust does build up on the stem, a light sanding and a coat of oil will help. The valve turns in and out through the valve packing gland. Rust on the stem will destroy the packing in the gland.

The *packing gland* can be either a permanent type or an adjustable type. If the gland is not adjustable and a leak is started, it can only be stopped with the valve cap. If the gland is adjustable, the packing can be replaced. Normally it is graphite rope, **Figure 25–52**. Any time the valve stem is turned, the gland should be first loosened if it is adjustable. This will keep the valve stem from wearing the packing when the valve stem is turned.

Because the suction valve sweats, it is not uncommon for the valve stem to rust and cause pits in the stem. This only occurs when poor service practices have been followed. When this occurs, the valve stem should be carefully sanded smooth, then refrigerant oil applied to the stem before turning. *If the stem is not smoothed, damage to the packing will occur. The technician should always dry the valve stem and the inside of the protective cap and apply a light coat of oil before placing the cap on the valve stem.*

The suction service valve attaches to the compressor on one side. The refrigerant piping from the evaporator fastens to the other side of the valve. The piping can be either flanged, flared, or soldered where fastened to the service valve.

25.42 DISCHARGE SERVICE VALVES

The *discharge service valve* is the same as the suction service valve except that it is located in the discharge line. This valve can be used as a gage port and to valve off the compressor for

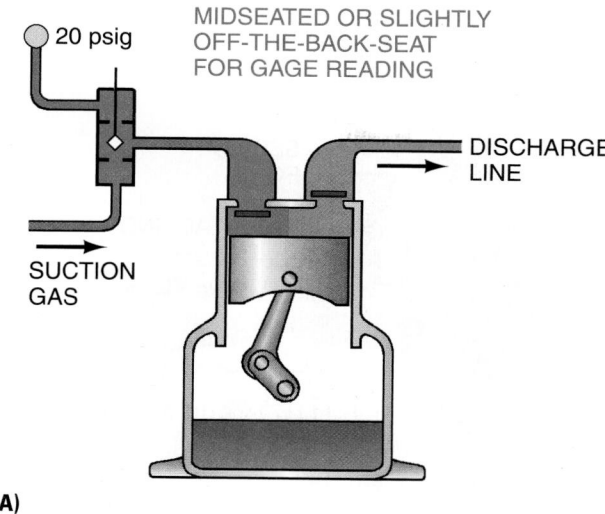

MIDSEATED OR SLIGHTLY OFF-THE-BACK-SEAT FOR GAGE READING

20 psig

DISCHARGE LINE

SUCTION GAS

(A)

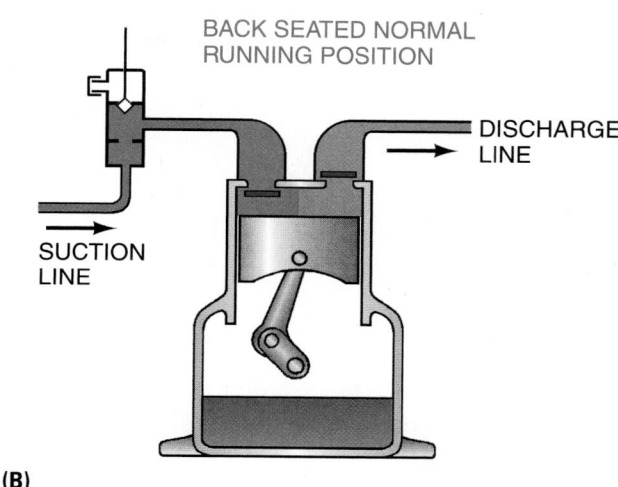

BACK SEATED NORMAL RUNNING POSITION

DISCHARGE LINE

SUCTION LINE

(B)

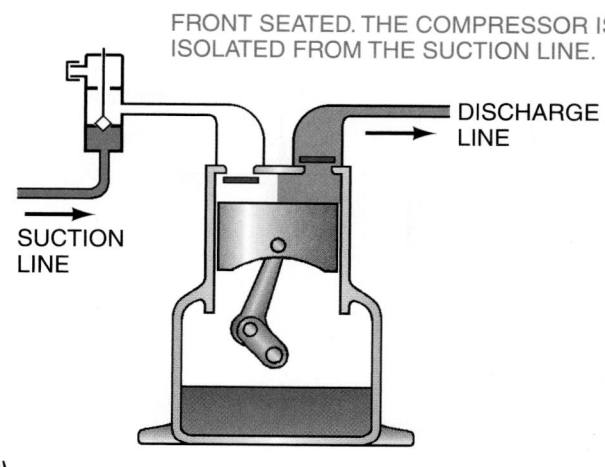

FRONT SEATED. THE COMPRESSOR IS ISOLATED FROM THE SUCTION LINE.

DISCHARGE LINE

SUCTION LINE

(C)

Figure 25–50 Three positions for the suction service valve **(A)–(C)**.

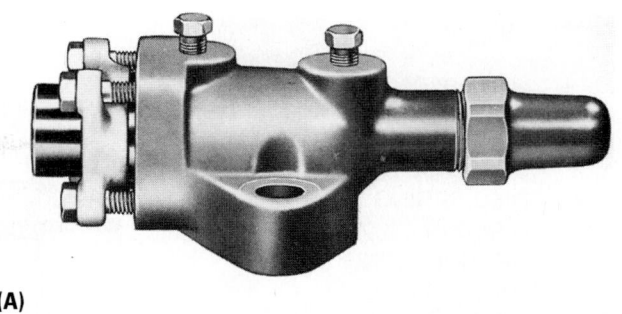

(A)

PCE NO.	DESCRIPTION	QUAN.
1	BODY	1
2	STEM	1
3	SEAT DISC	1
4	RETAINER RING	1
5	DISC SPRING	1
6	GASKET	1
7	CAPSCREW	4
8	PACKING WASHER	1
9	PACKING	1
10	PACKING GLAND	1
11	CAP	1
12	CAP GASKET	1
13	FLANGE	1
14	ADAPTER	1

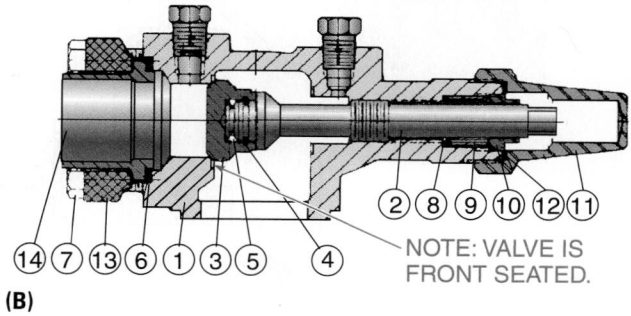

NOTE: VALVE IS FRONT SEATED.

(B)

Figure 25–51 **(A)** The suction and discharge valves are made the same except that the suction valve is often larger. **(B)** Components of a service valve. *Courtesy Henry Valve Company*

service. SAFETY PRECAUTION: The compressor cannot be operated with this valve front seated. An exception is closed-loop capacity checks that take place under experienced supervision. Extremely high pressures will result if the test equipment is not properly applied.

Compressor service valves are used for many service functions. One of the most important service functions is changing out the compressor. ♻ When both service valves are front seated, the refrigerant may be recovered from between the valves where it is trapped inside the compressor, Figure 25–53. ♻ The compressor is now isolated and can be removed. When a new one is installed, the only part of the system that must be evacuated when the new compressor is installed is the new compressor. A compressor can be changed in this fashion with no loss of refrigerant.

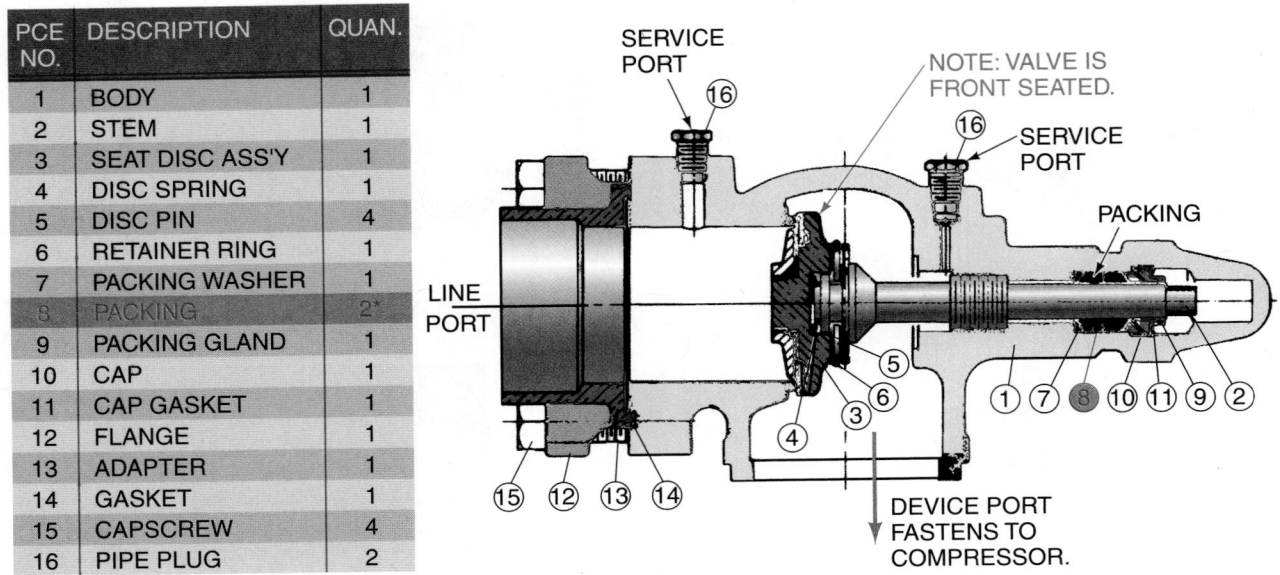

PCE NO.	DESCRIPTION	QUAN.
1	BODY	1
2	STEM	1
3	SEAT DISC ASS'Y	1
4	DISC SPRING	1
5	DISC PIN	4
6	RETAINER RING	1
7	PACKING WASHER	1
8	PACKING	2*
9	PACKING GLAND	1
10	CAP	1
11	CAP GASKET	1
12	FLANGE	1
13	ADAPTER	1
14	GASKET	1
15	CAPSCREW	4
16	PIPE PLUG	2

Figure 25–52 A service valve showing packing material. This can be replaced if it leaks. *Courtesy Henry Valve Company*

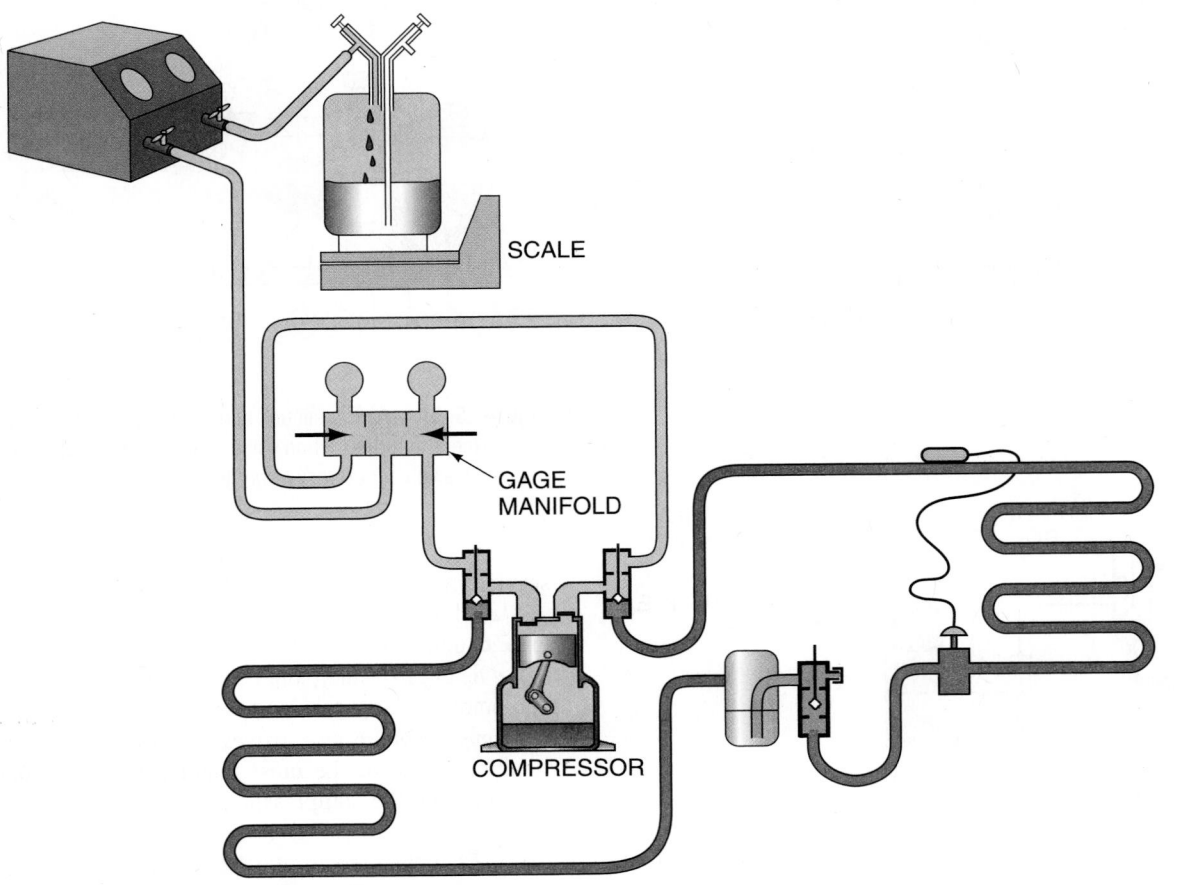

Figure 25–53 This compressor has been isolated by the service valves. The refrigerant can be recovered; then the compressor can be replaced.

25.43 REFRIGERATION LINE SERVICE VALVES

Refrigeration line service valves are normally hand-operated specialty valves used for service purposes. These valves can be used in any line that may have to be valved off for any reason. They come in two types: the diaphragm valve and the ball valve.

25.44 DIAPHRAGM VALVES

The *diaphragm valve* has the same internal flow pattern that a "globe" valve does. The fluid has to rise up and over a seat, **Figure 25–54**. There is a measurable pressure drop through this type of valve. The valve can be tightened by hand enough to hold back high pressures. This valve can be installed into the system with either a flare or soldered connection. When it is soldered, care should be taken that it is not overheated. Most of these valves have seats made of materials that would melt when the valve was being soldered into a line. The valve can be taken apart, and the internal parts can be removed for soldering purposes.

25.45 BALL VALVES

The *ball valve* is a straight-through valve with little pressure drop. This valve can also be soldered into the line, but the temperature has to be considered. All manufacturers furnish directions that show how to install their valve, **Figure 25–55.**

One of the advantages of the ball valve is that it can be opened or closed easily because it only takes a 90° turn to either open or close it. It is known as a "quick open or close" valve.

25.46 OIL SEPARATORS

🌐The *oil separator* is installed in the discharge line to separate the oil from the refrigerant and return the oil to the compressor crankcase.🌐 All reciprocating and rotary

Figure 25–54 The diaphragm-type hand valve used when servicing a system. This valve is either open or closed, unlike the suction and discharge valve with a gage port. The valve has some resistance to fluid flow, called pressure drop. It can be used anywhere a valve is needed. The larger sizes of this valve are soldered in the line. Care should be taken that the valve is not overheated when being soldered to the valve. *Courtesy Henry Valve Company*

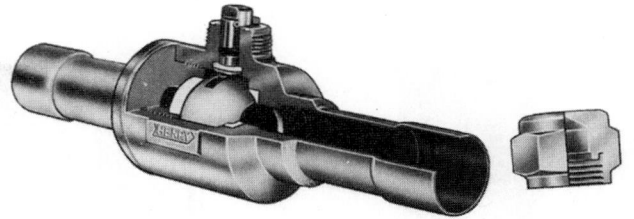

Figure 25–55 The ball valve. This valve is open straight through and creates very little pressure drop or resistance to the refrigerant flow. *Courtesy Henry Valve Company*

compressors allow a small amount of oil to pass through into the discharge line. Once the oil leaves the compressor, it would have to go through the complete system to get back to the compressor crankcase by way of the system piping and coils. The oil separator has a reservoir and float to allow this oil to shortcut and return to the crankcase, **Figure 25–56(A and B)**. Oil separators work on the principle of low refrigerant and oil velocity. As oil-laden discharge gas enters the oil separator, its velocity is immediately slowed. This slowing of velocity happens from either an abrupt change of direction, impingement screens, internal baffling, or a large opening or transitioning of the piping at the entrance of the oil separator. Oil is usually mixed with the hot discharge gas in the form of a fog. This refrigerant/oil fog runs into the internal baffling or screens, which cause the refrigerant and oil fog to change direction and slow down rapidly. This causes very fine particles of oil to collide with one another and form larger, heavier particles. Fine mesh impingement screens separate the oil and refrigerant even more, causing large oil droplets to form and drop to the bottom of the separator. The oil collects in a sump or reservoir at the bottom of the separator. A magnet is often connected to the bottom of the oil sump to collect any metallic particles.

When the oil level becomes high enough to raise a float, an oil return needle is opened, and the oil is returned to the compressor crankcase or oil reservoir through a small return line. This return line should be just above room temperature most of the time. If it is always hot, the float could be stuck open, allowing hot gas to enter the crankcase. This would also cause excessively high crankcase pressures and could overheat compressors, resulting in serious damage.

Because the oil separator is in the high side of the system and the crankcase is in the low side, a pressure difference naturally exists. This pressure difference is the driving force for the oil to travel from the separator to the crankcase when the float is in the right position. Only a small amount of oil is needed to actuate the float mechanism. This ensures that only a small amount of oil is ever absent from the compressor crankcase at any given time.

Figure 25–56(C) shows a helical oil separator. Helical oil separators are about 98% efficient in separating oil. When the oil and refrigerant fog enters the oil separator, it hits the leading edge of a helical flighting. The fog mixture is centrifugally forced along the spiral path of the helix. This causes the heavier oil particles to be forced to the perimeter of the separator's

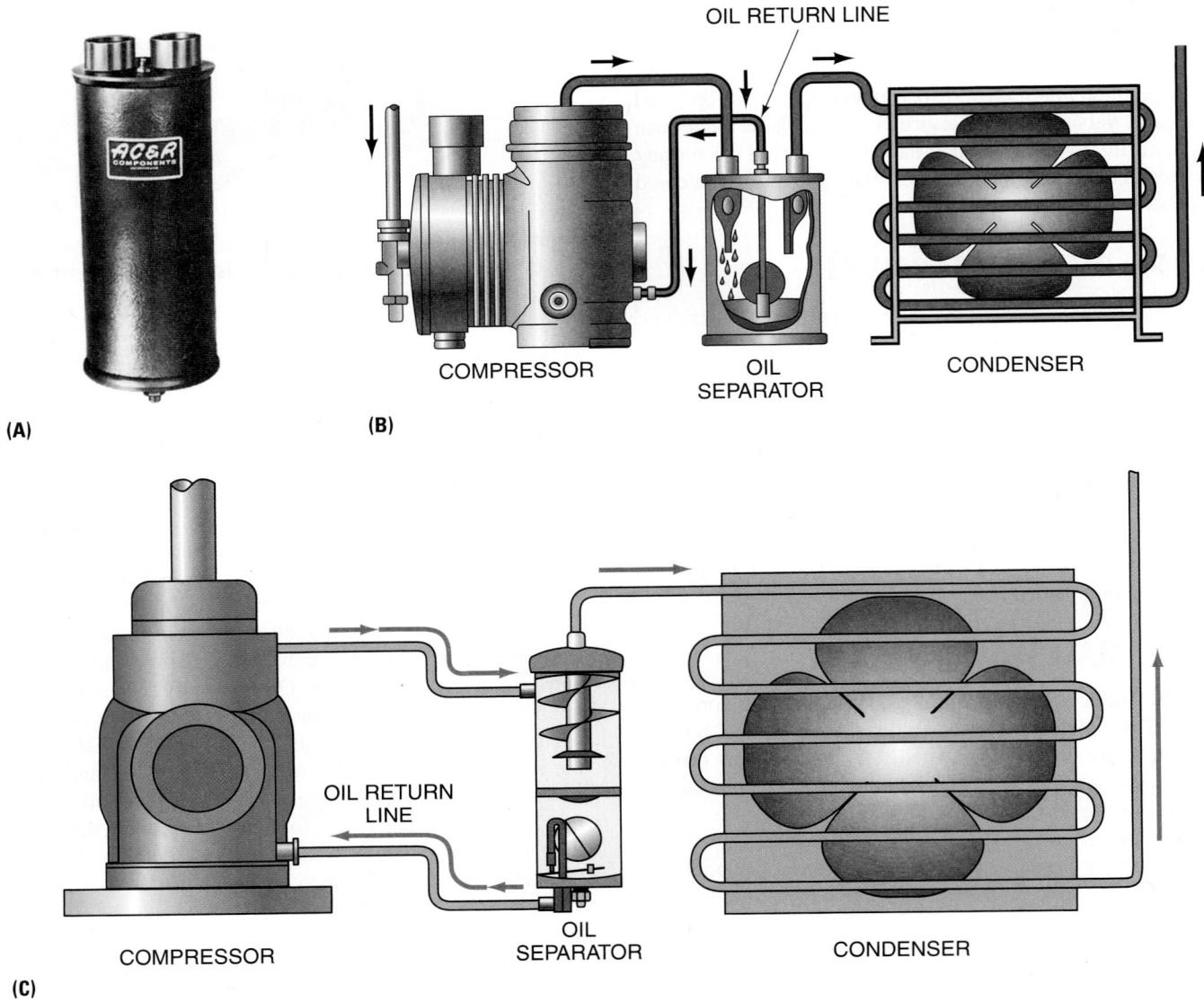

Figure 25–56 **(A)** The oil separator used in the discharge line of a compressor to return some of the oil to the compressor before it gets out into the system. **(B)** This is a float-action valve and will return liquid refrigerant as well as oil. It must be kept warm to keep refrigerant from condensing in it during the off cycle. **(C)** A helical oil separator. *Courtesy AC & R Components, Inc.*

shell. This is where impingement with a screen layer occurs. The screen layer acts as an oil stripping and draining medium. The liquid oil then flows through a baffle and into an oil collection area at the bottom of the separator, and the oil-free refrigerant gas exits the separator just below the flighting. A float-activated oil return valve allows the captured oil to return to the compressor's crankcase or oil reservoir. The oil separator should be kept warm to keep liquid refrigerant from condensing in it during the off cycle. The float does not distinguish between oil and liquid refrigerant. If liquid refrigerant were in the separator, it would return the refrigerant to the compressor crankcase. This would dilute the oil and cause marginal lubrication. If liquid refrigerant is returning to the compressor crankcase, the return line between the separator and the crankcase will be cold from vaporizing refrigerant experiencing the pressure drop. A damaged, leaking, or collapsed float in the oil separator will cause the orifice to stay closed, and excessive amounts of oil

will accumulate in the oil separator. This situation could rob the system of its valuable lubricant and cause compressor damage.

25.47 VIBRATION ELIMINATORS

Compressors produce enough vibration while running that it is often necessary to protect the tubing at the suction and discharge lines. Vibration can be eliminated on small compressors successfully with vibration loops. Large tubing cannot be routed in loops, so special vibration eliminators are often installed. These are constructed with a bellows-type lining and a flexible outer protective support, **Figure 25–57(A)**.

These devices must be installed correctly, or other problems may occur, such as extra vibration. Typically, it is recommended that they be installed close to the compressor and routed in the same direction as the compressor

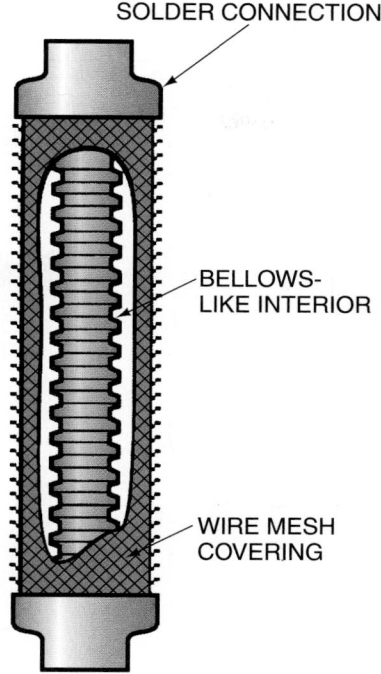

SOLDER CONNECTION

BELLOWS-LIKE INTERIOR

WIRE MESH COVERING

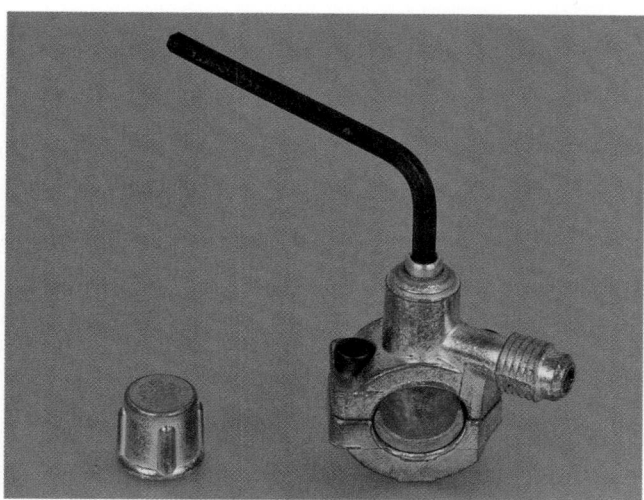

Figure 25–58   Pressure tap devices used to obtain pressure readings when there are no gage ports. *Photo by Bill Johnson*

Line-piercing valves should be installed using the manufacturer's directions, or a leak is likely to occur. It is good practice not to locate a line tap valve on the hot gas line if a gasket is involved because, in time, the heat will deteriorate the gasket and a leak will occur. If a high-side pressure reading is required, use the liquid line; it is much cooler and less likely to leak. A line tap valve with a gasket should not be left on the line as a permanent gage port.

The other type of valve that can be installed while running requires the valve to be soldered on the line with a low-temperature solder. This can only be done on a vapor line because a liquid line will not heat up as long as there is liquid in it. Manufacturers claim that there is no damage to the refrigerant in the line for this type of soldering application. This valve may be more leak free because it is soldered. After the valve is soldered on the line, a puncture is made similar to the valve previously described, **Figure 25–59.**

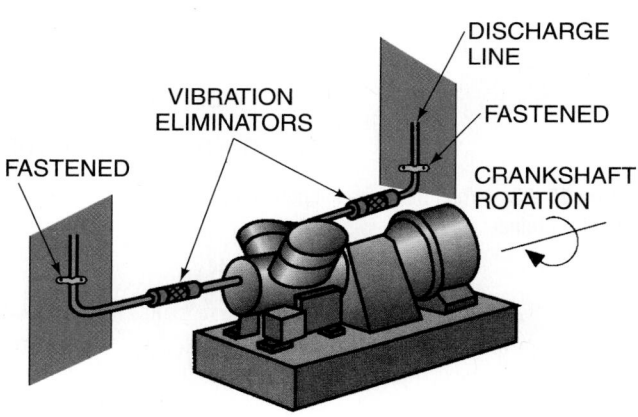

DISCHARGE LINE

VIBRATION ELIMINATORS

FASTENED

FASTENED

CRANKSHAFT ROTATION

Figure 25–57   Vibration eliminators installed on lines next to a compressor.

crankshaft, **Figure 25–57(B).** If mounted crossways of the crankshaft, excess vibration may occur. Follow the manufacturer's directions.

25.48   PRESSURE ACCESS PORTS

Pressure access ports are a method of taking pressure readings at places that do not have service ports. Several types can be used effectively. Some can be attached to a line while the unit is operating. This can be helpful when a pressure reading is needed in a hurry and the system needs to keep on running. There are two types that can be fastened while running. One type is bolted on the line and has a gasket. When this valve is bolted in place, a pointed plunger is forced through the pipe. A very small hole is pierced in the line, just enough to take pressure readings and transfer small amounts of refrigerant if needed, **Figure 25–58.**

Figure 25–59   This valve can be soldered on a vapor line while the system has pressure in it. Low-temperature solder is used. It cannot be soldered to a line with liquid in it because the liquid will not allow the line to get hot enough. *Courtesy J/B Industries*

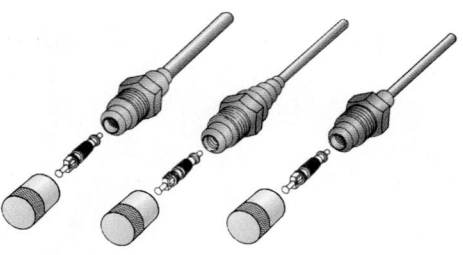

Figure 25–60 This valve port must have a hole drilled in the line and can be installed only with the system at atmospheric pressure. *Courtesy J/B Industries*

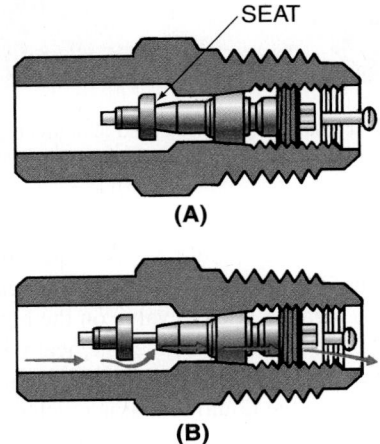

Figure 25–61 The Schrader valve is similar to the valve on a car tire. It has threads to accept the service technician's gage line threads **(A)–(B)**. *Courtesy J/B Industries*

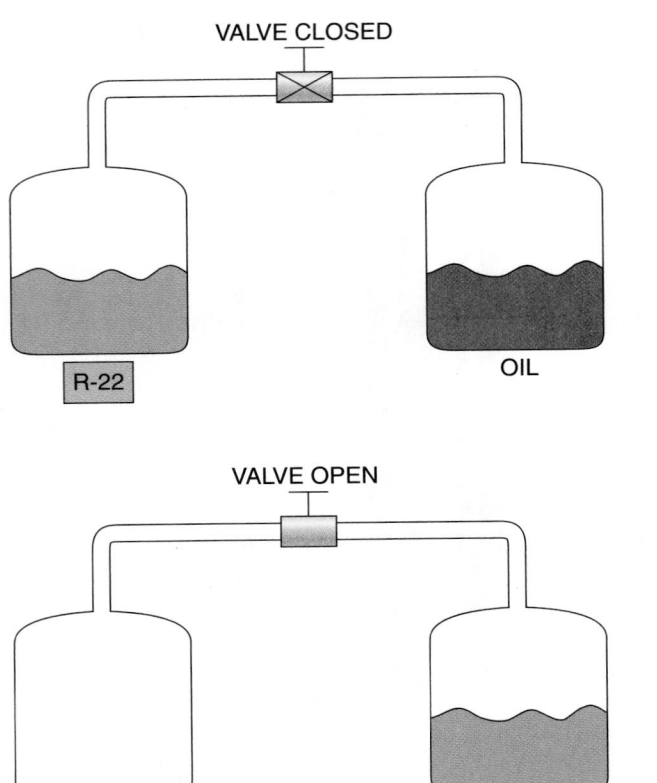

Figure 25–62 When the valve is opened, the refrigerant moves to the oil container.

Other valves used as ports must have a hole drilled in the refrigerant line. All copper filings should be removed from the line to prevent system damage. A copper tee also can be installed in the line to house the valve. This can be done without drilling. The valve stem is inserted into the line and soldered. This must be done when there is no pressure in the line, **Figure 25–60**.

Attaching a gage hose to the valves can be done in two ways. Some of the valves have handles that shut off the valve to the atmosphere. The others normally use a *Schrader* connection, which is like a tire valve on a car or bicycle except that it has threads that accept the 1/4-in. gage hose connector from a gage manifold, **Figure 25–61**.

25.49 CRANKCASE HEAT

🌎 Some refrigerants require that heat be applied to the crankcase of the compressor to keep the refrigerant from migrating to the oil during the off cycle or during any long periods of off time.🌎 Some refrigerants have an affinity for oil, almost like a magnetic attraction. ♻ The refrigerant in common use today that migrates the most is R-22. It has not been used for very many applications in refrigeration, but it is a hydrochlorofluorocarbon refrigerant, and it is being used along with other alternative refrigerants for a replacement

for many R-12 and R-502 systems because both of them have been phased out of manufacturing. *Migration* is the term used to describe refrigerant liquid or vapor entering the compressor' crankcase or suction line during the off cycle. ♻

If you take a container of R-22 and a container of refrigerant oil and pipe them together at the vapor space, the R-22 will move to the oil container in a very short time, **Figure 25–62**. This migration can be slowed by adding heat to the container of oil, **Figure 25–63**. Think of the refrigeration system as two containers attached together with piping, like the example above.

Crankcase heat is common in air-conditioning (cooling) systems where R-22 has been used for many years. The homeowner often shuts off the disconnect to the outdoor condensing unit for the winter leaving the compressor without crankcase heat. If the homeowner then starts the unit without some time for the heat to boil the refrigerant out of the oil, damage will likely occur to the compressor. On start-up, the crankcase pressure reduces as soon as the compressor starts to turn. The refrigerant will boil and turn the oil to foam. The oil and refrigerant (some of the refrigerant may be in the liquid state) will be pumped out of the compressor. Valve and bearing damage may occur and the compressor may be operated with a limited oil charge until it returns to the crankcase from the evaporator.

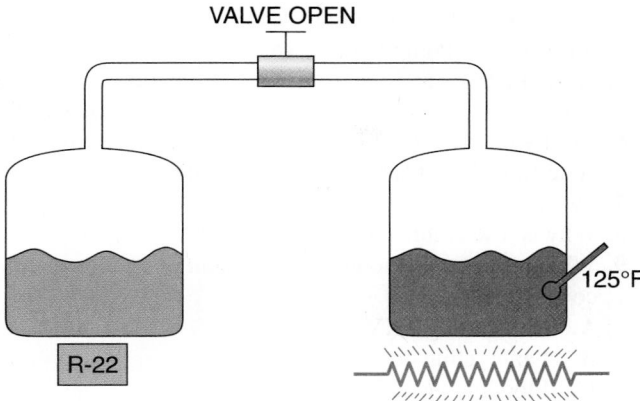

Figure 25–63 Heat keeps the refrigerant out of the oil.

Refrigeration compressors typically operate all season and will not have this seasonal shutdown, but if they are shut down for a period of time, the preceding situation will occur with R-22.

Crankcase heat can be applied in several ways. Some is applied using heaters that are inserted into the oil and some are external heaters at the base of the compressor adjacent to the oil, **Figure 25–64.** When the heater is external, the heating element must be in good contact with the compressor housing or it will overheat the element and not transfer heat into

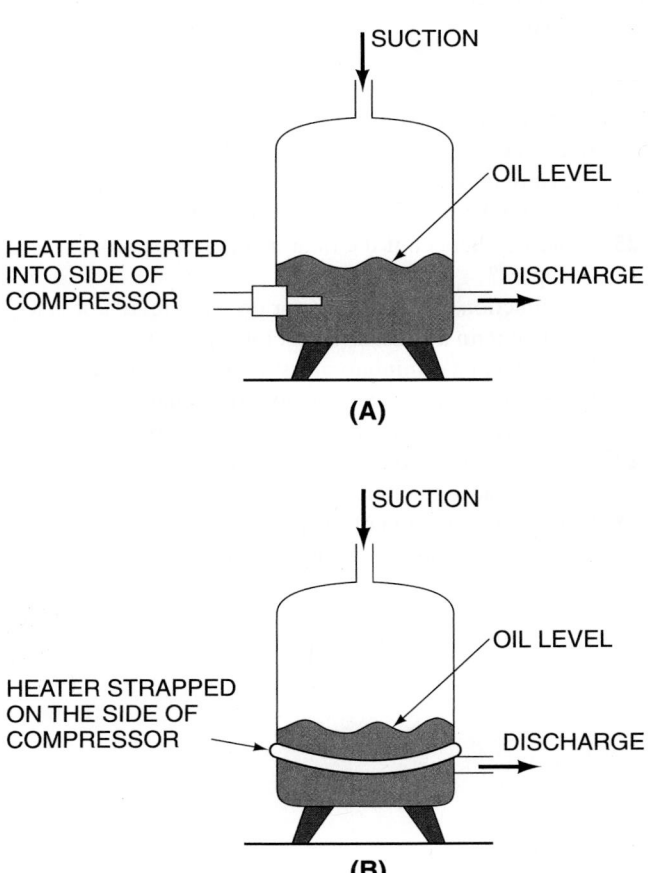

Figure 25–64 Two types of crankcase heat.

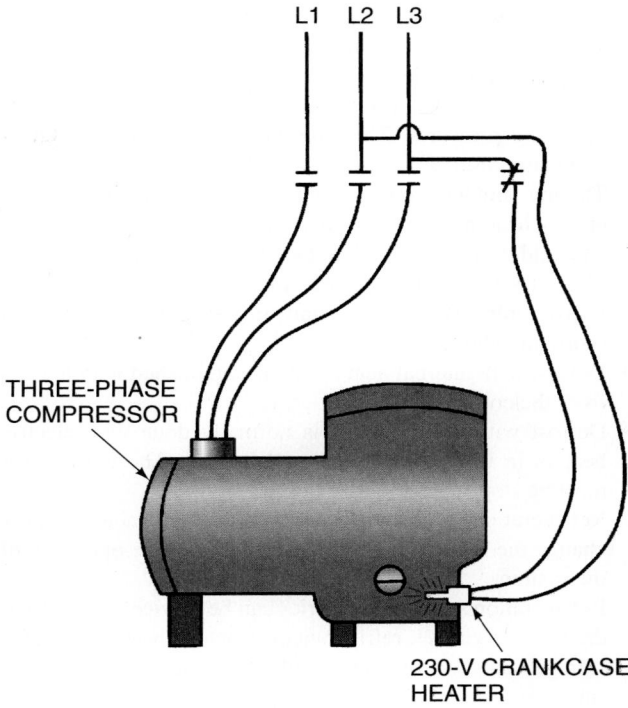

Figure 25–65 The compressor is not running, but the crankcase heater is hot. When the compressor contactor is energized, the compressor contacts will close, and the crankcase heater contacts will open, stopping the heat.

the compressor. These heaters can be installed incorrectly and left loose unless directions are followed.

Because the crankcase heat is only needed (and desired) during the off cycle, manufacturers are likely to use relays to shut the heat off during the running cycle. (Heat added during the running cycle only reduces system efficiency.) This can also be accomplished with a set of NC contacts in the compressor contactor, contacts that are closed when the contactor is deenergized, **Figure 25–65.**

The technician should be well aware of crankcase heat. If you approach a compressor and it has been off for a long time, the housing should be warm to the touch if it has crankcase heat. If the manufacturer has installed crankcase heat, do not start the compressor when it is cold, or damage may occur.

SUMMARY

- Four basic components of the compression cycle are the compressor, condenser, evaporator, and expansion device.
- Two other types of components that enhance the refrigeration cycle are control components and accessories.
- Two-temperature operation may utilize an evaporator pressure regulator.
- Relief valves prevent high pressures from occurring in the system.
- Electrical controls stop, start, or modulate electron flow for the control of motors and fluid flow.
- When energized, the solenoid valve, a valve with a magnetic coil, will open or close a valve to control fluid flow.

- Pressure switches stop and start system components.
- The low-pressure switch can be used for low-charge protection and as a thermostat.
- The high-pressure switch protects the system against high operating pressures. This control can be either manual or automatic reset.
- The low-ambient control (switch) maintains the correct operating head pressures on air-cooled equipment in both mild and cold weather by cycling the condenser fan.
- The oil safety switch ensures that the correct oil pressure is available 90 sec after start-up on larger compressors (normally above 5 hp).
- Defrost with internal heat can be accomplished with hot gas from the compressor.
- Defrost with external heat is normally done with electric heaters in the vicinity of the evaporator. The compressor must be stopped during this defrost.
- Refrigeration accessories normally *do not* automatically change the flow of refrigerant but enhance the operation of the system.
- Refrigeration system accessories can be service valves, filter driers, sight glasses, refrigerant distributors, heat exchangers, storage tanks, oil separators, vibration eliminators, or pressure taps.
- No system must have all of these components, but all systems will have some of them.
- Crankcase heat is required for many refrigeration systems, especially R-22 refrigerant systems because of refrigerant migration during downtime. Damage can occur to the compressor if it is started while excessive refrigerant is in the crankcase.

REVIEW QUESTIONS

1. Which is not a basic component of the compression refrigeration cycle?
   A. Compressor
   B. Condenser
   C. Evaporator
   D. Suction filter
2. Describe modulating fluid flow.
3. The purpose of the crankcase pressure-regulating valve is to
   A. maintain constant pressure in the crankcase.
   B. maintain constant pressure in the evaporator.
   C. prevent a compressor from going into a vacuum.
   D. prevent a compressor overload condition.
4. Describe the function of an electric evaporator pressure-regulating (EEPR) valve.
5. What are three main reasons why automatic pumpdown systems are used on refrigeration systems?
6. Briefly describe the function of a non-short-cycling, or "one-time," automatic pumpdown system.
7. Two types of relief valves are _____ and _____.
8. Why is low-ambient control necessary?
9. Which is not a method of low-ambient control on air-cooled equipment?
   A. Modulating TXV
   B. Fan-cycling controls
   C. Flooding a condenser
   D. Fan speed control
Describe the devices in Questions 10–13.
10. Low-pressure switch
11. High-pressure switch
12. Low-ambient control (switch)
13. Oil safety switch
14. What is the function of the current sensing relay in an oil safety control scheme?
15. Why is there a time delay in an oil safety control?
16. What is off-cycle defrost?
17. Two types of off-cycle defrost are _____ and _____.
18. Two methods for accomplishing defrost on low-temperature refrigeration systems are _____ and _____.

Describe how the following components enhance the refrigeration cycle:
19. Filter drier
20. Heat exchanger
21. Suction accumulator
22. Receiver
23. Pressure taps
24. Describe the function of an oil separator.
25. What will happen if the float in an oil separator collapses?
26. The function of the evaporator pressure regulator is to
   A. maintain a constant evaporator pressure.
   B. prevent a minimum pressure in the evaporator.
   C. maintain a constant evaporator temperature.
   D. prevent a maximum pressure in the evaporator.
27. What is the function of the oil return hole or orifice in a suction accumulator?
28. What is the function of the pressure equalization orifice at the outlet of an accumulator?

UNIT 26
Application of Refrigeration Systems

OBJECTIVES

After studying this unit, you should be able to

- describe the different types of display equipment.
- discuss heat reclaim.
- describe package versus remote-condensing applications.
- describe mullion heat.
- describe the various defrost methods.
- discuss walk-in refrigeration applications.
- describe basic vending machine refrigeration operation.
- explain basic refrigerated air-dry unit operation.

SAFETY CHECKLIST

✔ Wear goggles and gloves when attaching or removing gages and when using gages to transfer refrigerant or check pressures.
✔ Be careful not to get your hands or clothing caught in moving parts such as pulleys, belts, or fan blades.
✔ Observe all electrical safety precautions. Be careful at all times and use common sense.
✔ Follow manufacturers' instructions when using any cleaning or other types of chemicals. Many of these chemicals are hazardous so ensure that they are kept away from ice, drinking water, and other edible products.

26.1 APPLICATION DECISIONS

When refrigeration equipment is needed, a decision must be made as to the specific equipment to be installed. The factors that enter into this decision process are the first cost of the equipment to be installed, the conditions to be maintained, the operating cost of the equipment, and the long-term intent of the installation. When first cost alone is considered, the equipment may not perform acceptably. For example, if the wrong equipment choice is made for storing meat, the humidity may be too low in the cooler, and dehydration may occur. When too much moisture is taken out of the meat, there will be a weight loss. This weight loss is actually part of the condensate that goes down the drain during defrost.

This unit provides information regarding some of the different options to consider when equipment is purchased or installed. The service technician should know these options to help make service decisions.

26.2 REACH-IN REFRIGERATION MERCHANDISING

Retail stores use reach-in refrigeration units for merchandising their products. Customers can go from one section of the store to another and choose items to purchase. These reach-in display cases are available in high-, medium-, and low-temperature ranges. Each type of display case has open display and closed display types of cases from which to choose. These open or closed styles may be chest type, upright type with display shelves, or upright type with doors. The display boxes can be placed end to end for a continuous length of display. When this is done, the frozen food is kept together, and fresh foods (medium-temperature applications) are grouped, **Figure 26–1(A)**, **Figure 26–1(B)**, and **Figure 26–1(C)**.

All of these combinations of reach-in displays add to the customer's convenience and enhance the sales appeal of the products. The two broad categories of open and closed displays are purely for merchandising the product. A closed case is more efficient than an open case. The open case is more appealing to the customer because the food is more visible and easier to reach. Open display cases maintain conditions because the refrigerated air is heavy and settles to the bottom. These display cases can even be upright and store low-temperature products. This is accomplished by designing and

Figure 26–1(A) Display cases are used to safely store food displayed for sale. Some display cases are open to allow the customer to reach in without opening a door; some have doors. Closed cases are the most energy efficient. *Courtesy Hill Phoenix*

Figure 26–1(B)   Medium-temperature open display cases displaying an array of fresh produce. *Courtesy Meijer Corporation, Big Rapids, MI. Photo by John Tomczyk*

Figure 26–1(C)   A medium-temperature, glass-door, closed deli case. *Courtesy Meijer Corporation, Big Rapids, MI. Photo by John Tomczyk*

Figure 26–2   Open display cases accomplish even, low-temperature refrigeration because cold air can be controlled. It falls naturally. When these cases have shelves that are up high, air patterns are formed to create air curtains. These air curtains must not be disturbed by the air-conditioning system's air discharge. All open cases have explicit directions as to how to load them with load lines conspicuously marked. *Courtesy Tyler Refrigeration Corporation*

Figure 26–3   This self-contained refrigerated box has its condensing unit built in. Refrigeration tubing does not have to be run. This box rejects the heat it absorbs back into the store. It is movable because it needs only an electrical outlet and, sometimes, a drain, depending on the box. *Courtesy Hill Refrigeration*

maintaining air patterns to keep the refrigerated air from leaving the case, **Figure 26–2.** When reach-in refrigerated cases are used as storage, such as in restaurants, they do not display products, although they have the same components, and doors are used to keep the room air out.

26.3   SELF-CONTAINED REACH-IN FIXTURES

Reach-in refrigeration fixtures can be either self-contained with the condensing unit built inside the box, or they may have a remote condensing unit. Making the best decision as to which type of fixture to purchase depends on several factors.

Self-contained equipment rejects the heat at the actual case. The compressor and condenser are located at the fixture, and the condenser rejects its heat back into the store,

Figure 26–3. This is good in winter. In summer it is desirable in warmer climates to reject this heat outside. Self-contained equipment can be moved around to new locations without much difficulty. It normally plugs into either a 120-V or 230-V electrical outlet. Only the outlet may have to be moved. **Figure 26–4** shows examples of wiring diagrams of medium- and low-temperature fixtures.

When service problems arise, only one fixture is affected, and the food can be moved to another cooler. Self-contained equipment is located in the conditioned space, and the condenser is subject to any airborne particles in the

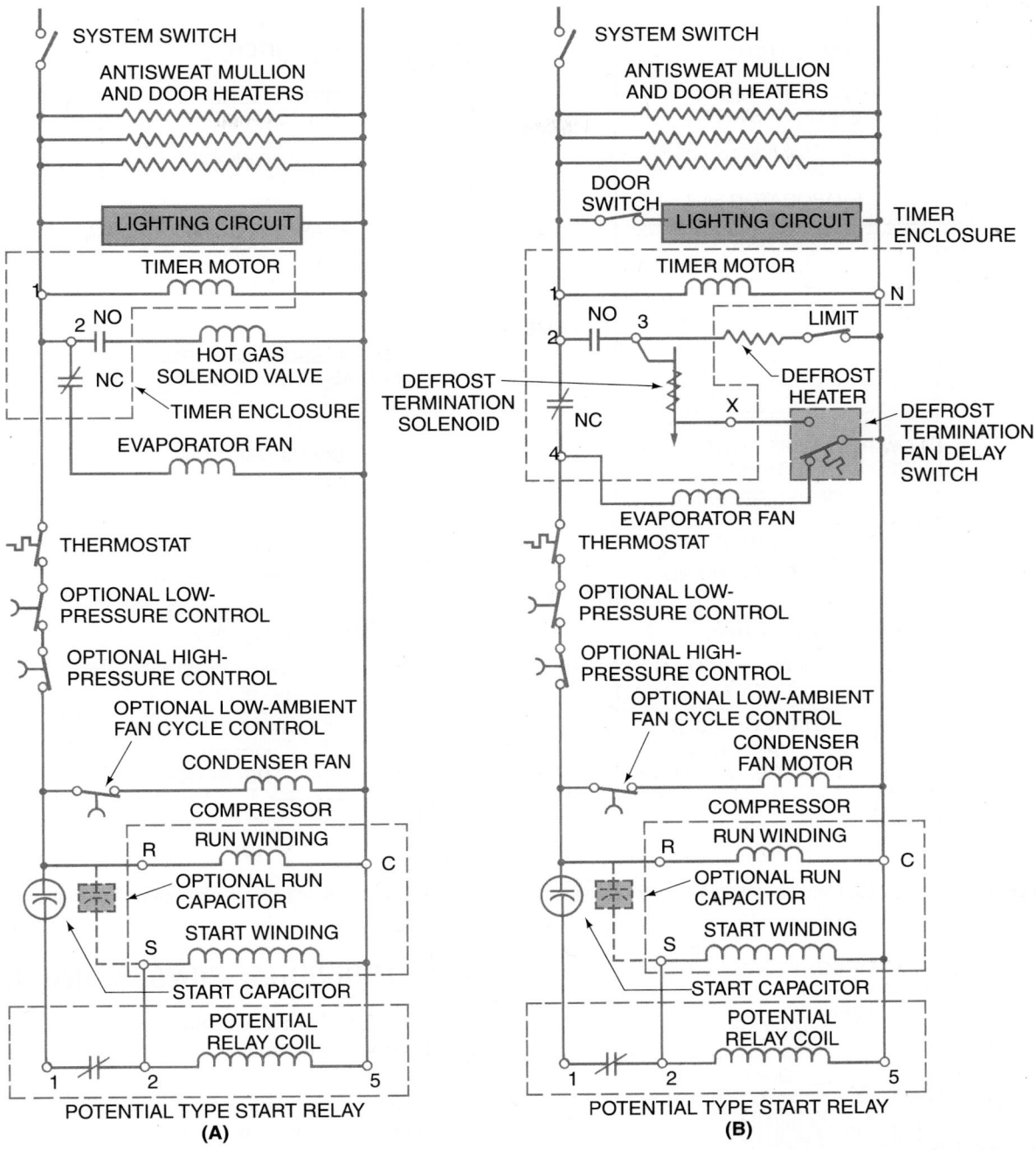

Figure 26–4 Wiring diagrams for self-contained reach-in boxes. **(A)** Low-temperature with hot gas defrost. **(B)** Low-temperature with electric defrost. **(C** and **D** on pg. 552)

conditioned space. Keeping the several condensers clean at a large installation may be difficult. For example, in a restaurant kitchen large amounts of grease will deposit on the condenser fins and coil, which will then collect dust. The combination of grease and dust is not easy to clean in the kitchen area without contaminating other areas. The versatility of self-contained equipment is good for a kitchen, but remote condensing units may be easier to maintain. The warm condensers are also a natural place for pests to locate.

All refrigeration equipment must have some means for disposing of the condensate that is gathered on the evaporator. This condensate must either be piped to a drain or be evaporated. Self-contained equipment can use the heat that the compressor gives off to evaporate this condensate, provided the condensate can be drained to the area of the condenser. This also helps to desuperheat the hot gas leaving the compressor, which will lower discharge line temperatures. To do this the condenser must be lower than the evaporator. This is not easy to do when the condenser is on top of the unit, **Figure 26–5.**

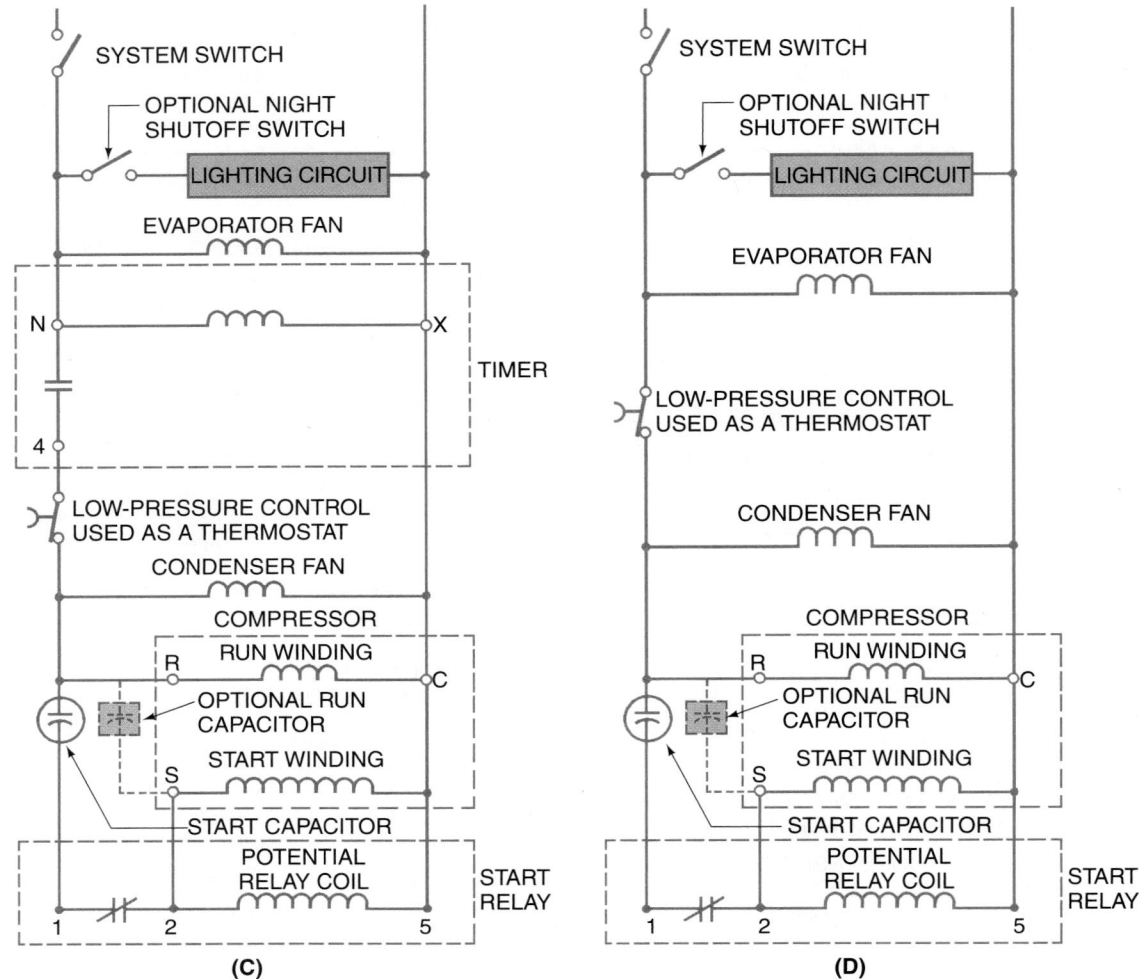

Figure 26–4 *(continued)*   **(C)** Medium-temperature with planned off-cycle defrost. **(D)** Medium- or high-temperature with random defrost.

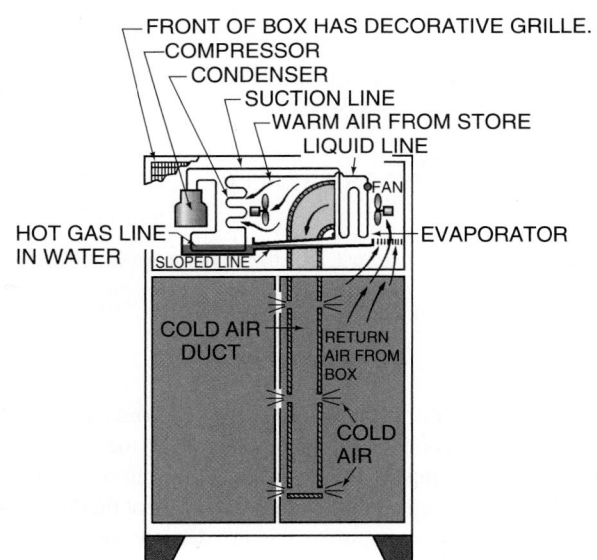

Figure 26–5   The evaporator can be high enough on a self-contained cooler so that the condensate will drain into a pan and be evaporated by heat from the hot gas line. No drain is needed in this application.

26.4   INDIVIDUAL CONDENSING UNITS

When individual condensing units are used and trouble is encountered with the compressor—or when a refrigerant leak occurs, only one system is affected. The condensing unit can be located outside, with proper weather protection, or in a common equipment room, where all equipment can be observed at a glance for routine service. The air in the room can be controlled with dampers for proper head pressure control. The equipment room can be arranged so that it recirculates store air in the winter and reuses the heat that it took out of the store through the refrigeration system. This saves buying heat from a local utility.

26.5   MULTIPLE EVAPORATORS AND SINGLE-COMPRESSOR APPLICATIONS

The method of using one compressor on several fixtures has its own advantages:

1. The compressor motors are more efficient because they are larger.

Figure 26–6 This three-way heat reclaim valve is pilot operated and solenoid activated. *Courtesy Sporlan Valve Company*

2. More heat from the equipment can be captured for use in heating the store or hot water for store use. This happens because larger compressors give off more heat.

The three-way heat reclaim valves in most heat reclaim systems are solenoid operated, **Figure 26–6.** The heat reclaim valve is a three-way valve that is pilot operated. Pilot operation means that it uses the refrigeration system's pressures to move a sliding valve when its solenoid is energized. Because the heat reclaim valve experiences very hot compressor discharge gases, it must be mounted properly and securely to counteract any expansion and contraction forces from heating and cooling effects, **Figure 26–7.** Its operation is simple. The first stage of a two-stage thermostat in the store calls for

Figure 26–7 A heat reclaim valve shown mounted securely to handle any expansion and contraction forces. *Photo by John Tomczyk*

heat. 🌐The three-way heat reclaim valve is energized and enables hot gas from the discharge header to flow to an auxiliary condenser or heat reclaim coil located in the store's ductwork, **Figure 26–8.** 🌐 If this reclaimed heat is not enough to satisfy the store's first stage of heat, the second stage of the store's thermostat will call for heat. This will initiate the store's primary heating equipment, which could be a fossil

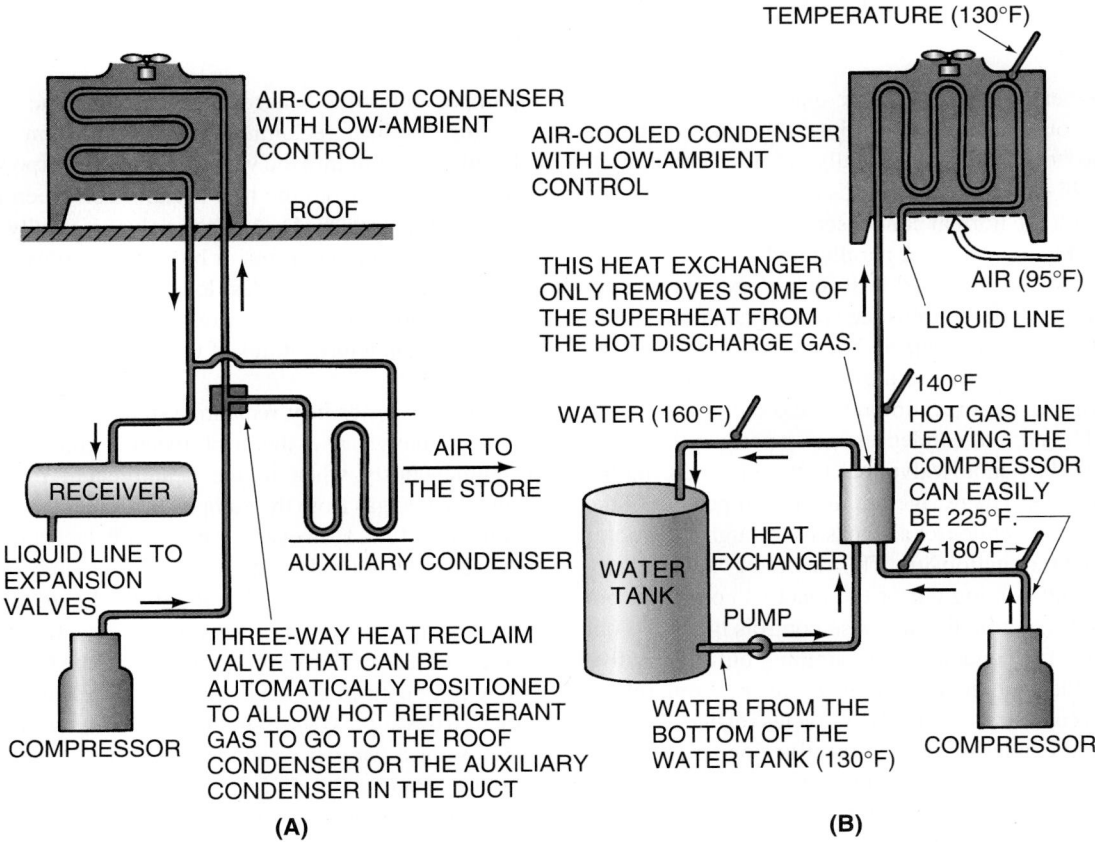

Figure 26–8 **(A)** The heat from a common discharge line can be captured and reused. This heat can be discharged from a condenser on the roof or from a coil mounted in the ductwork. **(B)** This is a hot water heat reclaim device. It uses the heat from the hot gas line to heat water.

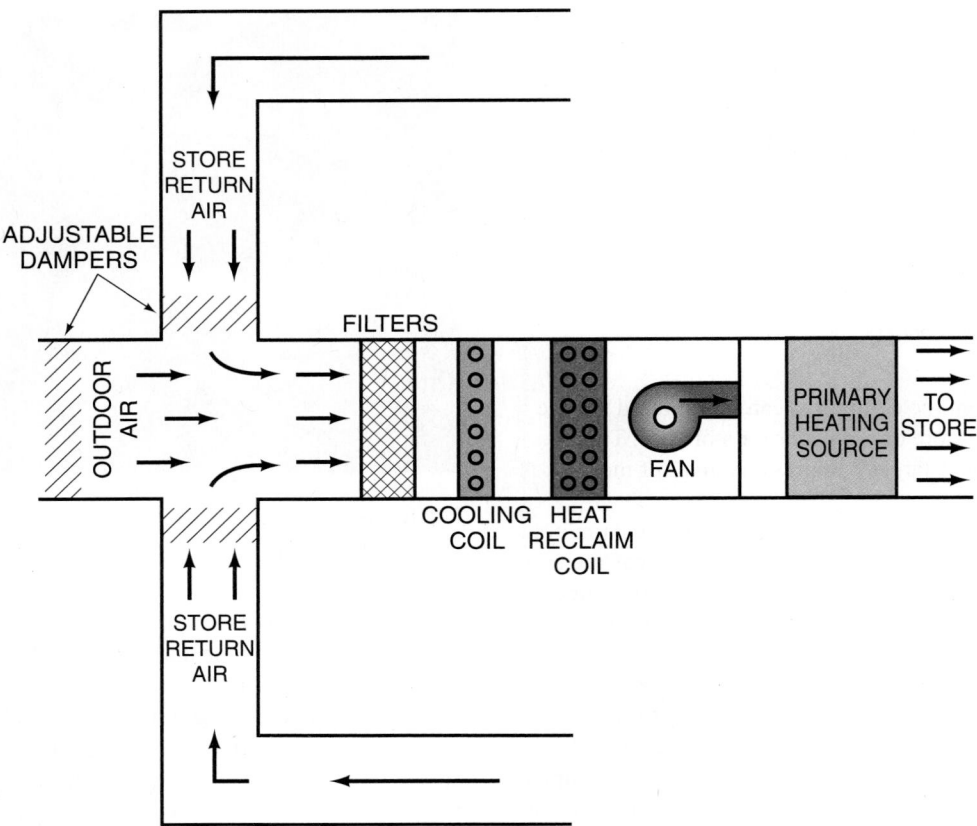

Figure 26–9 A supermarket HVACR system showing filter, cooling coil, heat reclaim coil, fan, and primary heating source in series with one another within the ductwork.

fuel furnace, heat pump, or electric heat. The heat reclaim coil, cooling coil, air filters, and primary heating source are located in the store's ductwork usually in series with one another, **Figure 26–9.** Heat reclaim coils can be used in parallel or in series with the normal condensers.

Figure 26–10 shows both a parallel and a series heat reclaim system. When a parallel heat reclaim system is used, the heat reclaim condenser must be sized to handle 100% of the refrigeration system's rejected heat when in the heat reclaim mode. This happens because the normal condenser is inactive during the heat reclaim mode and will not hold any refrigerant. The normal condenser is pumped out through a normally closed solenoid valve that is energized during the heat reclaim mode. The refrigerant that is pumped out of the normal or inactive condenser also passes through a restrictor before entering the suction line. This restrictor vaporizes any liquid refrigerant coming out of the inactive condenser before it enters the suction line and also controls the pump-out rate of the inactive condenser. By pumping out the inactive condenser in the heat reclaim mode, a proper system refrigerant charge is maintained and also prevents refrigerant from sitting idle and condensing in the inactive condenser. If the liquid were left in the inactive coil to condense, once the system switched out of the heat reclaim mode the mixing of hot gas with subcooled or saturated liquid could cause *liquid hammer.* Tremendous forces can be created, causing refrigerant lines and valves to rupture. Refrigerant must also be

pumped out of the heat reclaim coil when the system is not calling for heat reclaim. Some heat reclaim valves have bleed ports built into the valve for the sole purpose of removing refrigerant from the heat reclaim coil when it is not being used, **Figure 26–11.** A check valve is usually installed in the heat reclaim pump-out or bleed line to prevent migration of refrigerant to the coldest location when the heat reclaim coil is exposed to temperatures that are lower than the saturated suction temperatures of the refrigeration system itself, **Figure 26–10.**

When a series heat reclaim system is employed, the normal condenser and the heat reclaim condenser are used simultaneously when in the heat reclaim mode. The hot superheated gas from the compressors enters the heat reclaim condenser first. However, in series with the heat reclaim condenser is the normal condenser, and it also may receive some hot gas. Most of the condensing of the liquid takes place in the normal condenser downstream from the heat reclaim coil. Because of this series arrangement, the heat reclaim coil may not be sized to handle 100% of the heat rejected from the refrigeration system. In any heat reclaim system, the selection of a parallel or series system and the design of the piping arrangement will depend on the control scheme and the size of the heat reclaim coil being used for the particular application.

These installations are designed in two basic ways. One method uses one compressor for several cases. A 30-ton

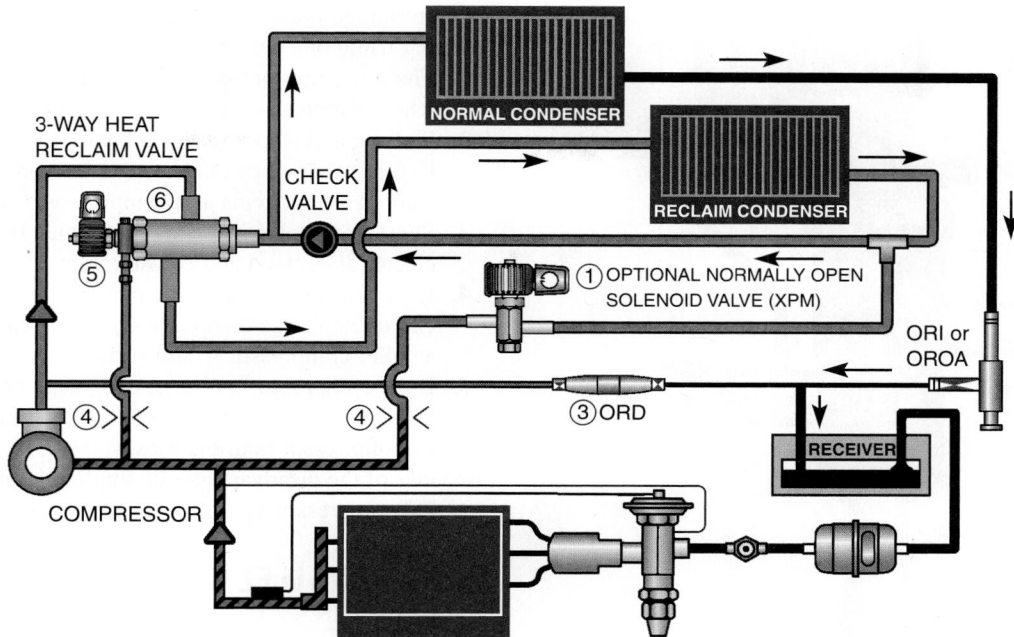

(A)

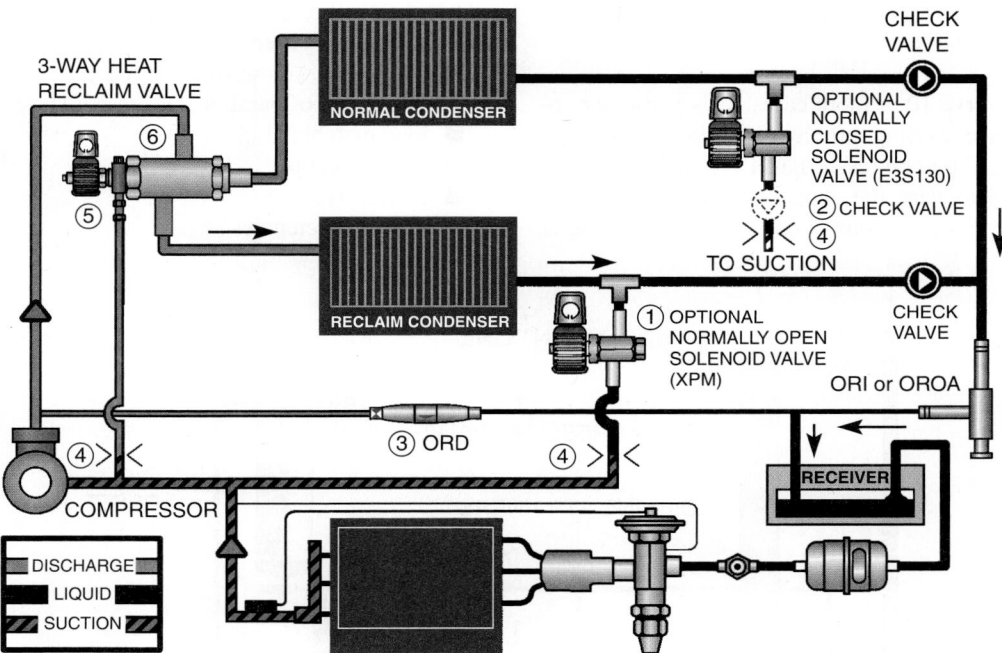

(B)

① Use optional solenoid valve and piping if pump-out is required and "C" model Heat Reclaim Valve is used; see Note 4. It is optional to omit this solenoid valve and piping on systems using "B" model Heat Reclaim Valve.

② This check valve is required if lowest operating ambient temperature is lower than evaporator temperature.

③ Not used with OROA (B, C, or D).

④ Restrictor, Part #2449-004, may be required to control pump-out rate on inactive condenser.

⑤ Pilot suction line must be open to common suction whether or not Heat Reclaim Coil is installed at time of installation and regardless of Heat Reclaim Valve model/type.

⑥ Proper support of heat reclaim valve is essential. Concentrated stresses resulting from thermal expansion or compressor vibrations can cause fatigue failure of tubing, elbows, and valve fittings. Fatigue failures can also result from vapor-propelled liquid slugging and condensation-induced shock. The use of piping brackets close to each of the 3-way valve fittings is recommended.

Figure 26–10   **(A)** Series and **(B)** parallel heat reclaim systems. *Courtesy Sporlan Valve Company*

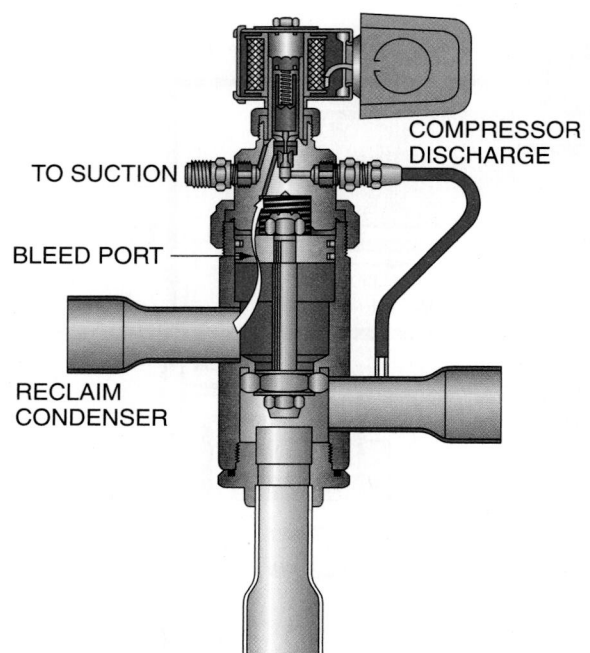

Figure 26–11 A three-way heat reclaim valve showing the bleed port for pumping down the heat reclaim coil when it is not in use. *Courtesy Sporlan Valve Company*

compressor may serve 10 or more cases. This particular application may have cycling problems at times when the load varies unless the compressor has a cylinder unloading design, a method to vary the capacity of a multicylinder compressor by stopping various cylinders from pumping at

predetermined pressures. The following are some disadvantages of using one compressor for several cases:

1. The refrigeration load cannot be closely matched even when compressor cylinder unloaders are used.
2. Starting and stopping larger compressors will draw a larger *locked-rotor amperage* (LRA) and is hard on starting components and compressor windings.
3. Short cycling larger compressors may increase both electrical demand and consumption charges.
4. Even the slightest increase in refrigeration load may call upon a large compressor to cycle on where a smaller compressor would suffice.
5. If the compressor malfunctions, many boxes will be inoperable as opposed to only one box not working if its compressor malfunctions.

Because of the disadvantages of single-compressor systems, parallel compressor systems became popular.

26.6 PARALLEL COMPRESSOR SYSTEMS

Parallel compressor systems apply two or more compressors to a common suction header, a common discharge header, and a common receiver, **Figure 26–12**. These systems are often referred to as *rack systems* because they are mounted on a steel rack, **Figure 26–13**. The compressors can be reciprocating, scroll, screw, or rotary type.

Advantages of parallel compressor systems are

- load matching.
- diversification.
- flexibility.
- higher efficiencies.

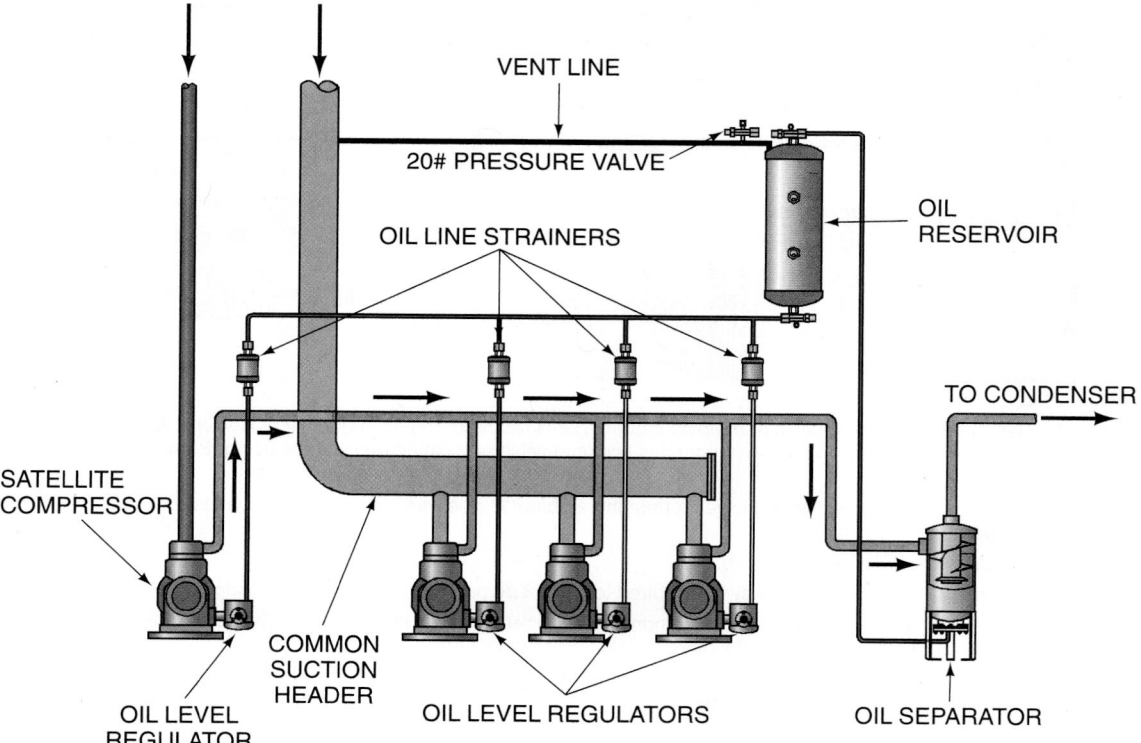

Figure 26–12 A parallel refrigeration system. *Courtesy AC & R Components, Inc.*

Figure 26–13   A parallel compressor system with load matching capabilities. The microprocessor-based electronic controller is shown as part of the package. *Courtesy Meijer Corporation, Big Rapids, MI. Photo by John Tomczyk*

- lower operating costs.
- less compressor cycling.

  Some disadvantages of parallel compressor systems are
- leaks affect the entire compressor rack.
- compressor burnouts that contaminate compressor oil will affect the entire compressor rack oiling system.

  🔘 The primary advantage of parallel compressor systems is their ability to nearly exactly match the refrigeration load, which is often referred to as *load matching.* 🔘 As the refrigeration load varies, compressors of varied capacities are turned on or off in response to the ever-changing load. The refrigeration load is sensed by a pressure transducer mounted on the common suction line or manifold, **Figure 26–14(A)**. The pressure transducer is wired to a microprocessor and an input/output control board on the compressor rack. The common suction pressure on the parallel system varies as case thermostats in the store call for cooling or as *evaporator pressure-regulating (EPR)* valves on the individual suction lines, which come into the parallel rack, throttle open and closed, according to the load of each individual line of refrigerated cases, **Figure 26–14(B)**. The pressure transducer senses this pressure increase or decrease and sends a voltage message to a computer or microprocessor-based control board located on the parallel compressor rack, **Figure 26–15(A) and (B)**. The pressure transducer changes the pressure signal to a voltage signal so that the microprocessor can process it. The pressure transducer receives its power from the same input/output board that powers the microprocessor, **Figure 26–15(B)**.

  As the refrigeration load changes, so does the common suction pressure in the suction manifold—thus the voltage coming from the pressure transducer to the microprocessor. Capacity steps that have been programmed into the microprocessor now control the parallel compressor rack cycling. Capacity steps can be changed with programming plugs or computer software and can be monitored on site or remotely

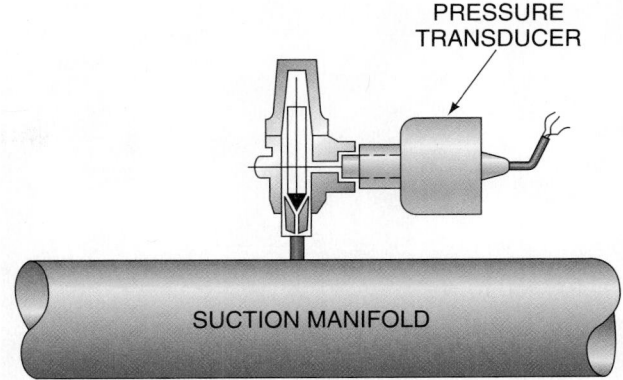

Figure 26–14(A)   A pressure transducer mounted on a common suction header. This transducer transforms a pressure signal to an electrical signal for the controller to process. This cycles the compressors to their next capacity step.

Figure 26–14(B)   Evaporator pressure-regulating (EPR) valves on suction lines of a parallel compressor rack. As the EPR valves open and close according to the load on the cases, the pressure transducer on the common suction line will sense the change in pressure. *Photo by John Tomczyk*

by a personal computer. For example, if the parallel compressor rack had four compressors of unequal sizes, and the sizes increased from compressor #1 through compressor #4, 10 capacity steps would be programmed into the microprocessor, **Figure 26–16**. As the pressure transducer on the common suction line sensed the common suction pressure increasing from the increased refrigeration loads, it would send a higher voltage signal to the microprocessor. This would trigger the capacity steps to progress from step 1. Soon the capacity of the compressor rack would match the refrigeration load. This could happen at capacity step 6 where only compressor #2 and #4 were operating. At this time, the common suction pressure would stop rising, and the parallel compressor rack would not increase or decrease in capacity steps. This same thing would happen if the load decreased, but the pressure transducer would sense a decrease in pressure.

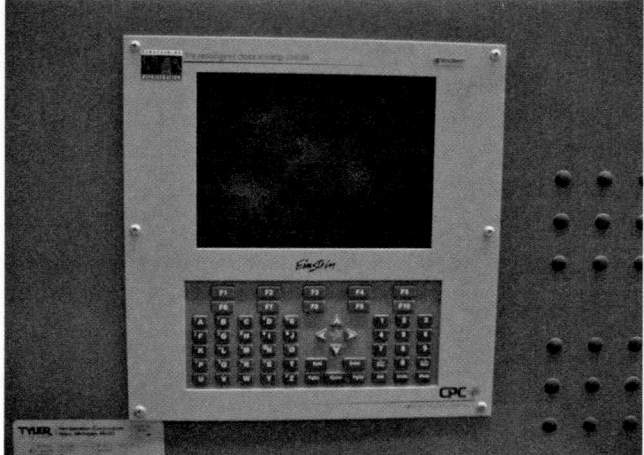

(A)

(B)

Figure 26–15 **(A)** The computer-based control for a parallel refrigeration system. **(B)** An internal view of the computer-based controller showing smaller microprocessor-based controllers. **(A)** *and* **(B)** *Courtesy Meijer Corporation, Big Rapids, MI. Photos by John Tomczyk*

	CAPACITY STEPS										
	0	1	2	3	4	5	6	7	8	9	10
COMPRESSOR #1 (10 HP)	OFF	ON	OFF	OFF	OFF	ON	OFF	OFF	ON	OFF	ON
COMPRESSOR #2 (12 HP)	OFF	OFF	ON	OFF	OFF	OFF	ON	OFF	OFF	ON	ON
COMPRESSOR #3 (15 HP)	OFF	OFF	OFF	ON	OFF	OFF	OFF	ON	ON	ON	ON
COMPRESSOR #4 (20 HP)	OFF	OFF	OFF	OFF	ON	ON	ON	ON	ON	ON	ON
	OPERATING CAPACITY IN HORSEPOWER										
	DECREASING CAPACITY										
OPERATING CAPACITY IN HORSEPOWER	INCREASING CAPACITY										
	0	10	12	15	20	30	32	35	45	47	57

Figure 26–16 Capacity steps programmed into the microprocessor-based controller for load-matching capabilities.

Figure 26–17 Line-voltage hardware, which includes circuit breakers and motor starters with current relays used for cycling the parallel compressors. *Courtesy Meijer Corporation, Big Rapids, MI. Photo by John Tomczyk*

It would now relay a decreased voltage to the microprocessor. This would cause the parallel rack's capacity steps to decrease until the load was matched with the right compressor capacity. This could happen at capacity step 3 with just compressor #3 running. **Figure 26–17** shows line-voltage hardware, which includes circuit breakers and motor starters with current-type relays, used in cycling the parallel compressors.

Parallel compressor racks usually come with compressors of unequal size and are referred to as *uneven parallel systems.* They are used for widely varying loads. However, compressors of equal sizes can be ordered and are referred to as *even parallel systems.* Even parallel systems are for refrigeration loads that are relatively constant or have known increases or decreases in capacity, as you might see in an industrial process. Compressor cycling logic for these systems may provide for even run time among all compressors, resulting in even wear on all compressors through the system's life. The parallel compressor system may have a *split suction* line. This enables the same parallel compressor rack to handle two different suction pressures with two different pressure transducers. A check valve will separate the two suction lines having different saturated suction temperatures. The split suction option enables one parallel compressor rack to handle both low- and medium-temperature applications, thus making it a more diverse refrigeration system.

To maintain the correct amount of oil in each compressor, a sophisticated oil system was designed. The oil system usually consists of four major components:

■ Oil separator
■ Oil reservoir
■ Pressure differential valve
■ Oil level regulators

A large oil separator is in the common discharge line to trap the pumped oil, **Figure 26–12.** The oil is not returned directly from the oil separator to the compressors as in conventional systems but pumped by differential pressure to an oil reservoir. Often, the oil separator and the oil reservoir are combined into one vessel, **Figure 26–18.** From the reservoir, the oil passes to an oil level regulator float valve, which is mounted on each compressor at its sightglass location, **Figure 26–19.** The oil level regulator float valve senses

Figure 26–18   A combination oil separator and oil reservoir. *Photo by John Tomczyk*

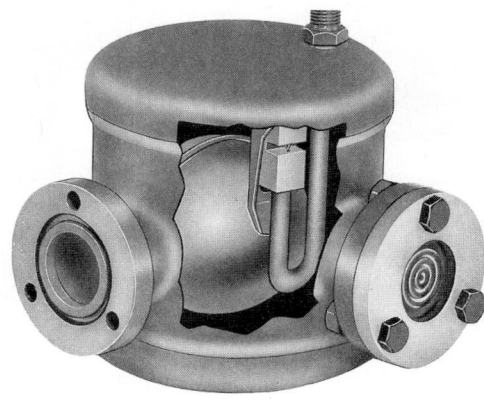

Figure 26–19   A compressor float-operated oil level regulator. *Courtesy AC & R Components, Inc.*

crankcase oil levels and opens and closes as necessary to ensure an adequate oil level in each compressor crankcase. A pressure differential valve is located on the oil reservoir. This valve usually maintains a 5- to 20-psi pressure differential between reservoir pressure and common suction pressure and is vented to the suction line, **Figure 26–12.** This valve ensures that the oil reservoir is pressurized each time an oil regulator opens. This reservoir pressure is 5 to 20 psi above common suction pressure depending on what pressure differential valve is used. The factory pressure setting is system dependent. This allows the oil level regulator orifice to be calibrated to a certain pressure difference over suction pressure. This prevents excessive inlet pressures at the oil level regulator, which may overfill the compressor crankcase with oil before the mechanical float valves can react. The pressure differential valve is shown in **Figure 26–20(B),** and an oil reservoir is shown in **Figure 26–20(C).**

Oil level regulating float valves can have fixed levels or adjustable levels and can be purely mechanical, **Figure 26–20(A),** or electromechanical. Valves that are electromechanical can use a liquid level switch, a solenoid valve, and a magnetic reed float switch to detect oil levels, **Figure 26–21(A).** These systems also offer alarm and system shutdown control if oil levels get dangerously high or low.

Some electromechanical liquid level switches detect oil level by sensing light refracting through a glass prism, **Figure 26–21(B).** An advantage is that it can be mounted on a system without disturbing the system internally.

Often, parallel compressor systems employ a *satellite* or *booster* compressor, **Figure 26–12.** Satellite compressors can be large or small and are often dedicated to the coldest evaporators. This prevents the entire parallel rack from operating at the lower suction pressures inherent to lower-temperature evaporators. This improves the efficiency of the entire parallel compressor rack by letting it run at a lower compression ratio, resulting in higher efficiency. Satellite compressors can be valved into the parallel rack. This enables the rack of compressors to assist the satellite compressor in pulling down. A check valve will then separate the parallel rack from the satellite compressor, allowing the satellite compressor to pull down further.

Satellite compressors often have variable-speed drives called *inverters* controlling their motor speeds, thus compressor capacity, and saving energy, **Figure 26–22.** The refrigeration load now can be matched even more closely by varying the frequency of an electronically altered sine wave, which affects the speed of alternating current motors. As the load decreases, the motor slows down, giving the compressor less capacity. As the load increases, the motor increases in speed, giving the compressor more capacity. Refer to Unit 17, "Types of Electric Motors," for more specific information on inverters. Cylinder unloading is another method of capacity control for satellite compressors and parallel compressors on the rack.

Another refrigeration system scheme that uses parallel compressors for capacity control and load matching also uses a *system pressure-regulating (SPR)* valve located between the condenser and the receiver, **Figure 26–23.** This

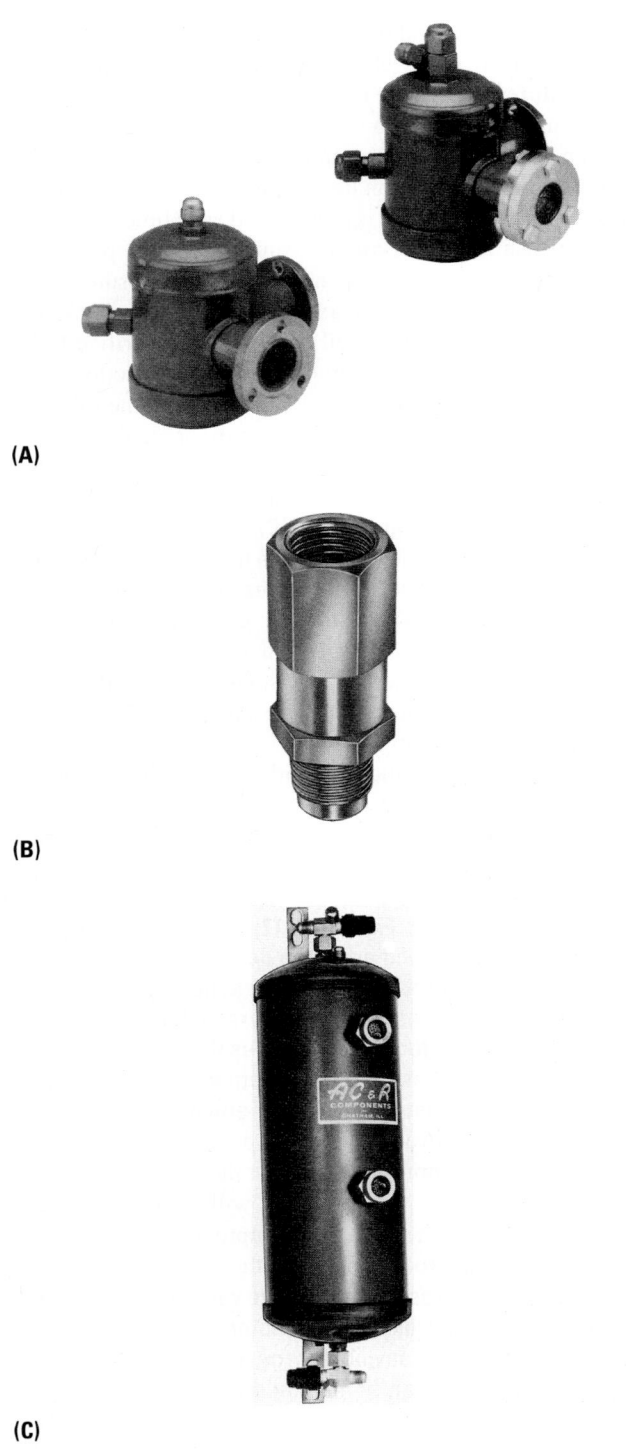

(A)

(B)

(C)

Figure 26–20 **(A)** Adjustable and fixed oil level regulators. **(B)** A differential pressure valve. **(C)** An oil reservoir. *Courtesy AC & R Components, Inc.*

(A)

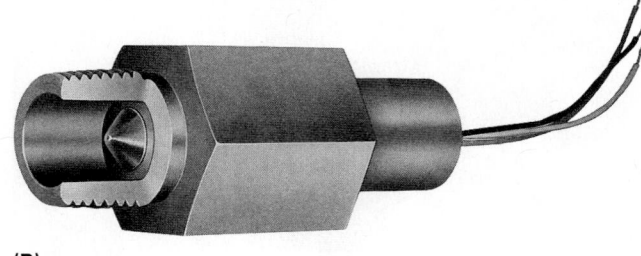

(B)

Figure 26–21 **(A)** An electromechanical oil level regulator. **(B)** A compressor oil level detector, which uses light that refracts through a glass prism to detect crankcase oil levels. *Courtesy AC & R Components, Inc.*

Figure 26–22 Inverters or variable-speed drives that change the alternating current's frequency, thus changing the satellite compressor's speed and capacity for better load matching. *Courtesy Meijer Corporation, Big Rapids, MI. Photo by John Tomczyk*

system has a receiver in parallel to the active refrigeration system, instead of a series receiver as in a conventional refrigeration system. The SPR valve has a remote, liquid-filled bulb located in the condenser's inlet air, sensing outdoor ambient temperatures. The sensor is connected to the SPR valve by a small copper tube. A more recent system design

uses sensors for determining outdoor ambient temperatures, liquid-line temperatures, and other temperatures strategic to the operation of the system. Inputs to an electronic controller operate condenser fans, a vapor relief valve out of the

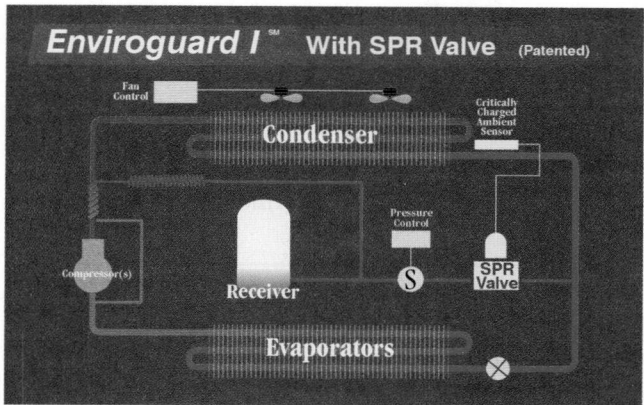

Figure 26–23 The Enviroguard Refrigerant Control System with an SPR valve. *Courtesy Tyler Refrigeration Corporation*

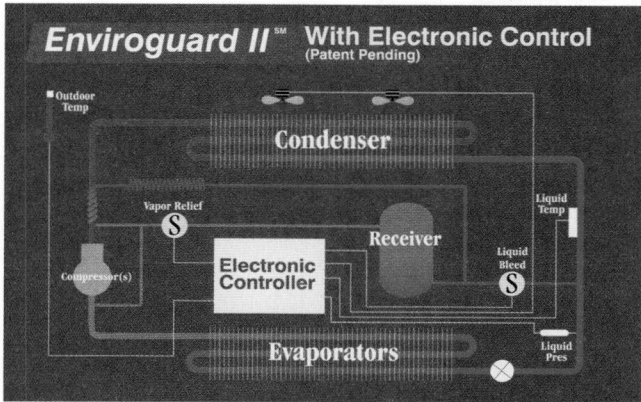

Figure 26–24 The Enviroguard II Refrigerant Control System with an electronic controller. *Courtesy Tyler Refrigeration Corporation*

receiver, liquid bleed line to receiver, and liquid pressure to the evaporator, **Figure 26–24.** A complex computer program in the electronic controller keeps the system operating at peak performance.

The SPR valve controls the amount of liquid refrigerant used in the active refrigeration system. The active system's refrigerant charge is reduced by taking the receiver out of the active refrigeration circuit and allowing liquid to return directly to the liquid manifold, which feeds the branch refrigeration circuits. With the receiver out of the active refrigeration system, a minimum receiver charge is no longer required as with a conventional system. The receiver is now simply used as a storage vessel to hold the condenser charge between summer and winter operations. On hot days, the SPR valve will bypass refrigerant to the receiver. This lowers the liquid levels in the condenser and gives it more internal volume for the hotter weather conditions. On colder days, the SPR valve throttles more closed, which bypasses less refrigerant to the receiver and keeps more in the condenser. This partially floods the condenser and keeps condensing pressures from dropping too low. The receiver has a bleed circuit that is opened to the suction manifold. This circuit bleeds refrigerant from the receiver back to the system for use,

which enables the refrigerant working charge in the system to seek a level of equilibrium relative to the ambient temperature. This means that in some conditions, the receiver is void of refrigerant and at suction pressure.

If the condenser should become damaged, fouled internally or externally, or lose a fan motor, an elevated condensing pressure will occur. This elevated pressure will cause the SPR valve to bypass refrigerant into the receiver because a design temperature difference between the ambient and the condensing temperature has been exceeded. This will prevent high compression ratios and high discharge temperatures with lower volumetric efficiencies. These occurrences often go unnoticed for long periods of time and stress compressor systems. Eventually, branch circuit evaporator temperatures will rise, because refrigerant is being bypassed out of the working system. The results will be a starved refrigeration system. Rising evaporator and product temperatures are noticed sooner, and the problem can be remedied immediately. Refrigerant leaks will also be noticed sooner, which means less refrigerant is lost to the atmosphere.

26.7 SECONDARY-FLUID REFRIGERATION SYSTEMS

With ozone depletion and global warming legislation raising the cost of refrigerants, some manufacturers are researching, and some are using, a primary (HFC or HCFC) refrigerant to cool a *secondary fluid.* The secondary fluid in this case is also considered a refrigerant. The secondary fluid or refrigerant is usually a low-temperature antifreeze solution, which stays in the liquid state and does not vaporize when it is circulated through the system with centrifugal pumps.

Some secondary fluids include *propylene glycol,* Pekasol 50, and *hydrofluoroethers (HFEs).* Even though propylene glycol is the most popular secondary fluid used today, one of the newest secondary fluids is HFEs. They have zero ozone-depletion potentials, and small global-warming potentials, and they are nontoxic. HFEs are, however, one of the most expensive secondary fluids used today. HFEs, along with other popular secondary fluids, can be used for all temperature ranges found in supermarket refrigeration from dairy to frozen food cases. This secondary fluid, which is the more abundant compound, is circulated by centrifugal pumps to the display cases in supermarkets, instead of the primary (HCFC or HFC) refrigerant. A heat exchange is then provided between the primary and secondary fluids, **Figure 26–25.** This system operates in much the same way as a direct expansion chiller but uses much less HCFC or HFC to accomplish the same task. These systems eliminate the long liquid lines and suction lines that are usually installed throughout the store. This reduces the overall system refrigerant charge and thousands of joints, which are potential leak points.

Defrost is accomplished by heating the same HFE fluid used in the secondary loop by a controlled process. It is done through a third piping loop with solenoid-controlled valving. The heat for defrost comes from the heat of compression in the discharge gases from the compressor. This heat then is

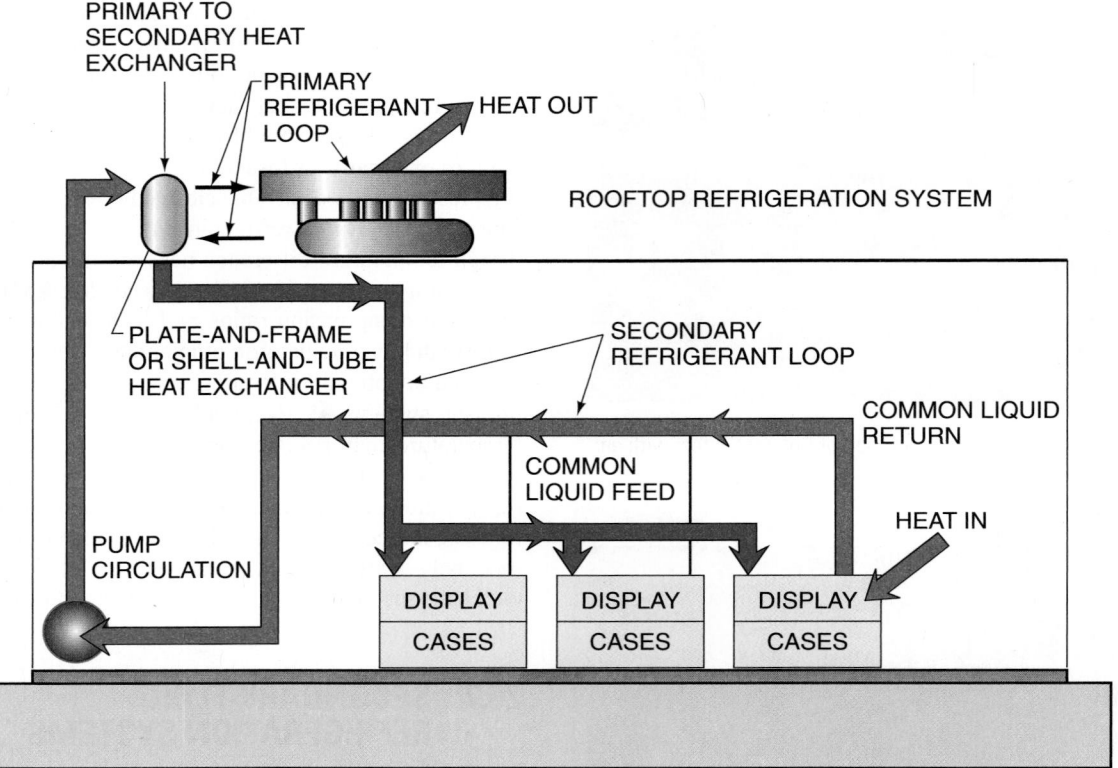

Figure 26–25   A secondary-fluid refrigeration system.

transferred to the same secondary fluid in a third piping loop for defrost purposes at the case coils.

Advantages of chilled secondary-fluid systems are

- 🌐large refrigerant reductions.
- greatly reduced leak potential.
- simpler installations.
- less critical piping.
- reduced welding and nitrogen use.
- minimized vacuum requirements to the equipment room.
- little or no oil return concerns.
- fewer expansion valve superheat adjustments.
- less compressor (total) superheat in the primary refrigerant loop because of shorter suction lines. This adds to the thermodynamic efficiency of the system.
- simpler and shorter defrosts.
- no requirement for defrost limit thermostats.
- faster defrost periods and quicker case pulldown times after defrost. This results in lower energy requirements and improved product temperatures.
- better and more stable product temperatures from higher mass flow rates in the secondary fluid.
- more efficient coil surface use, resulting in better and more stable product temperatures.
- simpler control strategies.
- higher refrigerant coil operating temperatures in secondary loop.
- lower maintenance.

- lower *total equivalent warming impact (TEWI)* than conventional refrigeration systems.
- the use of PVC piping or K-type copper tubing in the secondary loop for long line runs instead of ACR-type copper tubing used with direct-expansion refrigerants.
- only the low-pressure secondary loop is located in the store. 🌐

A few system-dependent disadvantages are

- higher cost in original equipment.
- industrial-grade solenoid valves must be used for case control in the secondary loop.
- higher electrical consumption from the extra heat-exchange energy penalty—and the added cost of secondary-fluid pumping. However, this is often offset by the higher refrigerant coil operating temperatures.
- thicker and completely sealed pipe insulation is needed in the secondary loop to prevent temperature and system loss.
- higher cost of using the low-viscosity secondary fluid that is required for low-temperature applications.

26.8   PRESSURIZED LIQUID SYSTEMS

This technology uses a small centrifugal pump to pressurize liquid entering the liquid line, **Figure 26–26(B).** A normal system is shown in **Figure 26–26(A)** for comparison purposes. The pressurized amount is equivalent to the pressure loss between the condenser outlet and the TXV inlet on receiverless systems

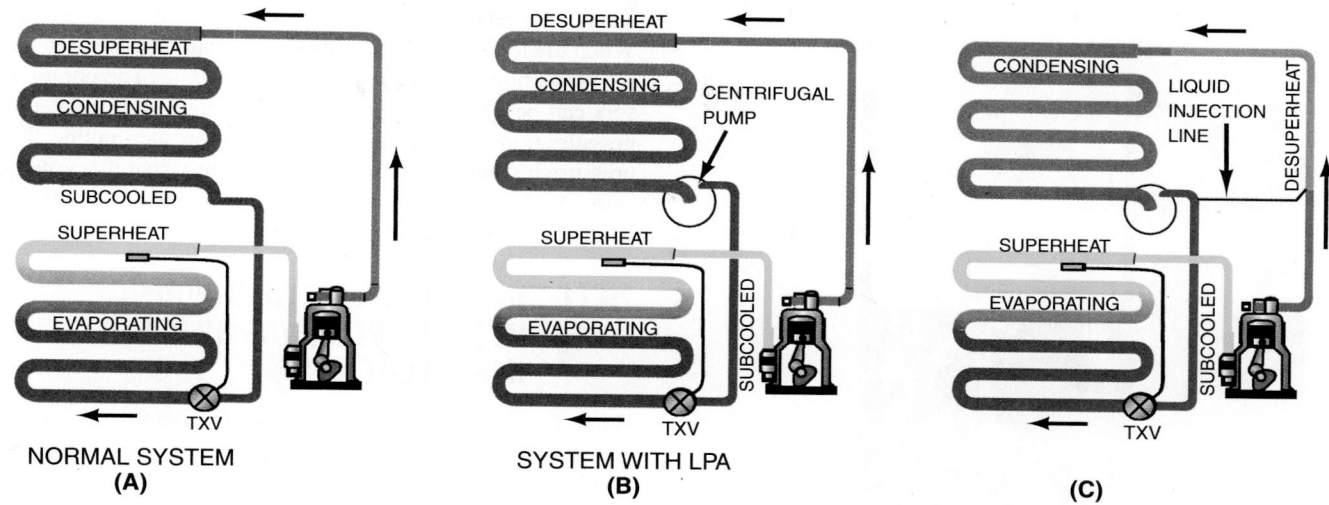

Figure 26–26   **(A)** A normal refrigeration system. **(B)** A pressurized liquid system. **(C)** A pressurized liquid system with liquid injection. *Courtesy DTE Energy Technologies*

or between the receiver and the TXV inlet on TXV/receiver systems. By increasing the pressure of liquid refrigerant, the associated saturation temperature is raised, while the actual liquid temperature remains the same. The liquid becomes subcooled and will not flash if exposed to pressure drops in the liquid line on its way to the metering device. The centrifugal pump can be sized to fit any system design.

Subcooling liquid refrigerant with this method will give the condenser more internal volume to condense refrigerant. This occurs because the condenser no longer has to use its bottom passes as a subcooling loop. This will lower the condensing temperatures and pressures, which will lower compression ratios and give efficiency boosts. In fact, because the centrifugal pump is doing the liquid subcooling, head pressures are allowed to *float* with the outside ambient temperature. Floating the head pressure is a term used for attaining the lowest possible head pressure by taking advantage of cool ambient temperatures used as the condensing medium. Very low condensing pressures can be attained with these systems. Because of this, TXV manufacturers have valves on the market that will handle a 30-psi pressure drop across them and still successfully feed the evaporator if supplied with 100% liquid. 🌐 Listed below are four advantages of a pressurized liquid system:

■ Eliminates liquid line flashing by overcoming line pressure losses.

■ Reduces energy costs because pumping liquid refrigerant is more than 40 times more efficient than using head pressure from the compressor to do the same work.

■ Increases the evaporator capacity along with the net refrigeration effect.

■ Lowers compression ratios and reduces stress on compressors. 🌐

A process of injecting liquid refrigerant into the compressor's discharge line or condenser's inlet can be used in conjunction with the pressurized liquid system. This liquid flashes to a vapor while cooling the superheated discharge gas closer to its condensing temperature. Less surface area is required for desuperheating. This leaves a more efficient condenser, which increases the overall performance of the system. The liquid comes from the same centrifugal pump, which delivers pressurized liquid to the metering device, **Figure 26–26(C)**. Although some efficiency gains are seen at low ambient temperatures, the greatest gains are realized at the higher ambient temperatures.

26.9 UNITARY STAND-ALONE REFRIGERATION SYSTEMS

Attractive, multiple compressor, stand-alone supermarket refrigeration modules that are located very close to the display cases they refrigerate are becoming increasingly popular, **Figure 26–27** and **Figure 26–28**. These modules can be cleverly hidden in the sales display area, mounted on top of display cases, or simply placed behind a wall close to the display cases. Microprocessor-based electronic controllers on

Figure 26–27   Unitary stand-alone refrigeration systems located close to display cases they are refrigerating. *Courtesy Hussmann Corporation*

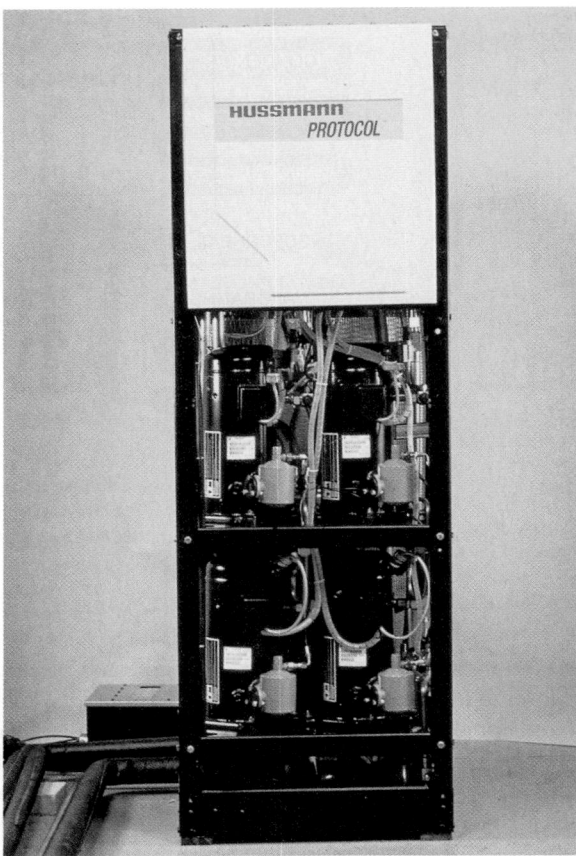

Figure 26–28 An attractive and compact commercial refrigeration module utilizing four hermetic scroll compressors. *Courtesy Hussmann Corporation*

each module manage the compressor cycling (load matching) and defrost schedules. This can be accomplished either on site or remotely by personal computers. The modules use small plate-type condensers that are cooled by a closed-loop fluid cooler. Heat reclaim can be used with these systems for both store and hot water heat. Each module can be ordered in either vertical or horizontal designs and takes up very little floor area. A remote mechanical room is not needed. 🌐 Other advantages of these modular stand-alone refrigeration systems are that they

- reduce refrigerant charge by 75% to 80%.
- use 75% less piping.
- require 75% fewer brazed joints.
- lessen the likelihood of refrigerant leaks.
- decrease or eliminate the need for EPR valves.
- lower installation costs.
- provide load matching with backup capacity protection.
- provide up to 5% energy savings as compared with conventional back-room systems.🌐

26.10 EVAPORATOR TEMPERATURE CONTROL

When multiple evaporators are used, they may not all have the same temperature rating. For example, the coldest evaporator may require a suction pressure of 28 psig. EPR valves

can be located in each of the higher temperature evaporators. When the load varies, the compressors can be cycled on and off to maintain a suction pressure of 20 psig. The compressors can be cycled using low-pressure control devices, or a pressure transducer located in the common suction line. The pressure transducer transforms a pressure signal to a voltage signal and then relays the signal to a microprocessor-based compressor controller for processing.

26.11 INTERCONNECTING PIPING IN MULTIPLE-EVAPORATOR INSTALLATIONS

When the fixtures are located in the store and the compressors are in a common equipment room, the liquid line must be piped to the fixture, and the suction line must be piped back to the compressor area. The suction line should be insulated to prevent it from picking up heat on the way back to the compressor. This can be accomplished by preplanning and by providing a pipe chase in the floor. Preferably the pipe chase should be accessible in case of a leak, **Figure 26–29.** In some installations, refrigerant lines are routed through individual plastic pipes in the floor, **Figure 26–30(A and B).** If a leak should occur, the individual lines may be replaced by pulling the old ones and replacing them with new ones. This may be more popular than the trench method in modern stores because it gives the designer more versatility for locating equipment. Under-floor refrigerant pipes are ventilated to prevent mold and mildew. **Figure 26–30(C)** shows a venting system that takes the store's ceiling air and directs it to the back of the cases, through the under-floor refrigeration piping system, and through the access pit in the mechanical room. It is hard to plan a large trench to every fixture. Long refrigerant lines can cause oil return problems. Individual runs can reduce the line length. Careful design factors must be followed for correct oil return.

Figure 26–29 A pipe chase is an effective method of running piping from the fixture to the equipment room. When a chase like this is used, the piping can be serviced if needed. *Courtesy Tyler Refrigeration Corporation*

(A)

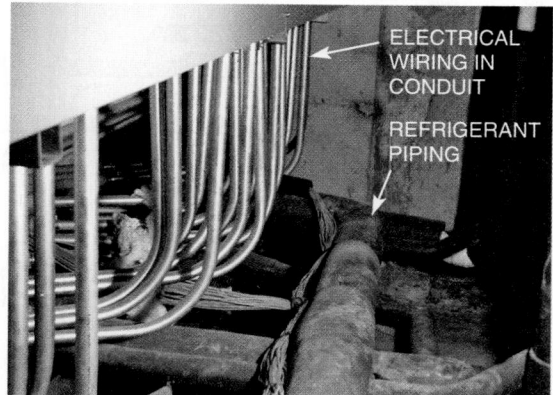

ELECTRICAL
WIRING IN
CONDUIT

REFRIGERANT
PIPING

(B)

VENTILATOR PIPING

(C)

Figure 26–30 **(A)** An access pit for refrigeration piping that is run under the floor. **(B)** Refrigeration piping and electrical wires within the access pit under the floor. **(C)** The ventilation system for under-floor refrigeration piping. **(A)–(C)** *Courtesy Meijer Corporation, Big Rapids, MI. Photos by John Tomczyk*

26.12 TEMPERATURE CONTROL OF THE FIXTURE

Control of this type of remote medium-temperature application can be accomplished without interconnecting wiring being installed between the fixture and the equipment room. A power supply for the fixture has to be located at the fixture.

Where the application is medium temperature, planned off-cycle defrost can be accomplished at the equipment room with a time clock and a liquid-line solenoid valve. The clock and solenoid valve can be located at the case, but the equipment room gives a more central location for all controls. The clock can deactivate the solenoid, closing the valve for a pre-determined time. The refrigerant will be pumped out of the individual fixture. The evaporator fan will continue to run, and the air in the fixture will defrost the coil. When the proper amount of time has passed, the solenoid-activated valve will open, and the coil will begin to operate normally. The compressor will continue to run during the defrost of the individual cases. The defrost times can be staggered by offsetting them. This is sometimes accomplished with a master time clock with many circuits. Use of electronic timing devices enables many different events to be controlled at one time.

Low-temperature installations must have a more extensive method of defrost because heat must be furnished to the coil in the fixture. This can be accomplished at the fixture with a time clock and heating elements. The power supply for the fixture is in the vicinity of the fixture. It can also be accomplished with hot gas, but a hot gas line must then be run from the compressor to the evaporator. This is a third line to be run to each case. It must also be insulated to keep the gas hot until it reaches the evaporator. The defrost for each case can be staggered with different time clock settings, **Figure 26–31.** This type of defrost is the most efficient because the heat from the defrost is coming from the other cases; however, electric defrost probably has fewer problems and is easier to troubleshoot.

These methods of defrost are typical, but by no means the only methods. Different manufacturers devise their own methods of defrost to suit their equipment.

26.13 THE EVAPORATOR AND MERCHANDISING

The evaporator is the device that absorbs the heat into the refrigeration system. Because the evaporator is located on the outside of the system, it is visible to the purchaser. Therefore, it can be built in several ways. At best, an evaporator is bulky. A certain amount of planning must be done by the manufacturers and their engineering staff to provide attractive fixtures that are also functional, **Figure 26–32.**

Customer appeal must be considered in the choice of equipment. The service technician may not understand why some equipment is installed the way it is because customer convenience or merchandising may have played a major part in the decision. Each supermarket chain is involved in staying ahead of the competition in the marketplace, and customer convenience is part of it.

Display fixtures are available as (1) chest type (open, open with refrigerated shelves, closed) and (2) upright (open with shelves, with doors).

26.14 CHEST-TYPE DISPLAY FIXTURES

Chest-type reach-in equipment can be designed with an open top or lids. Vegetables, for example, can easily be displayed in the open type. The vegetables can be stocked, rotated, and

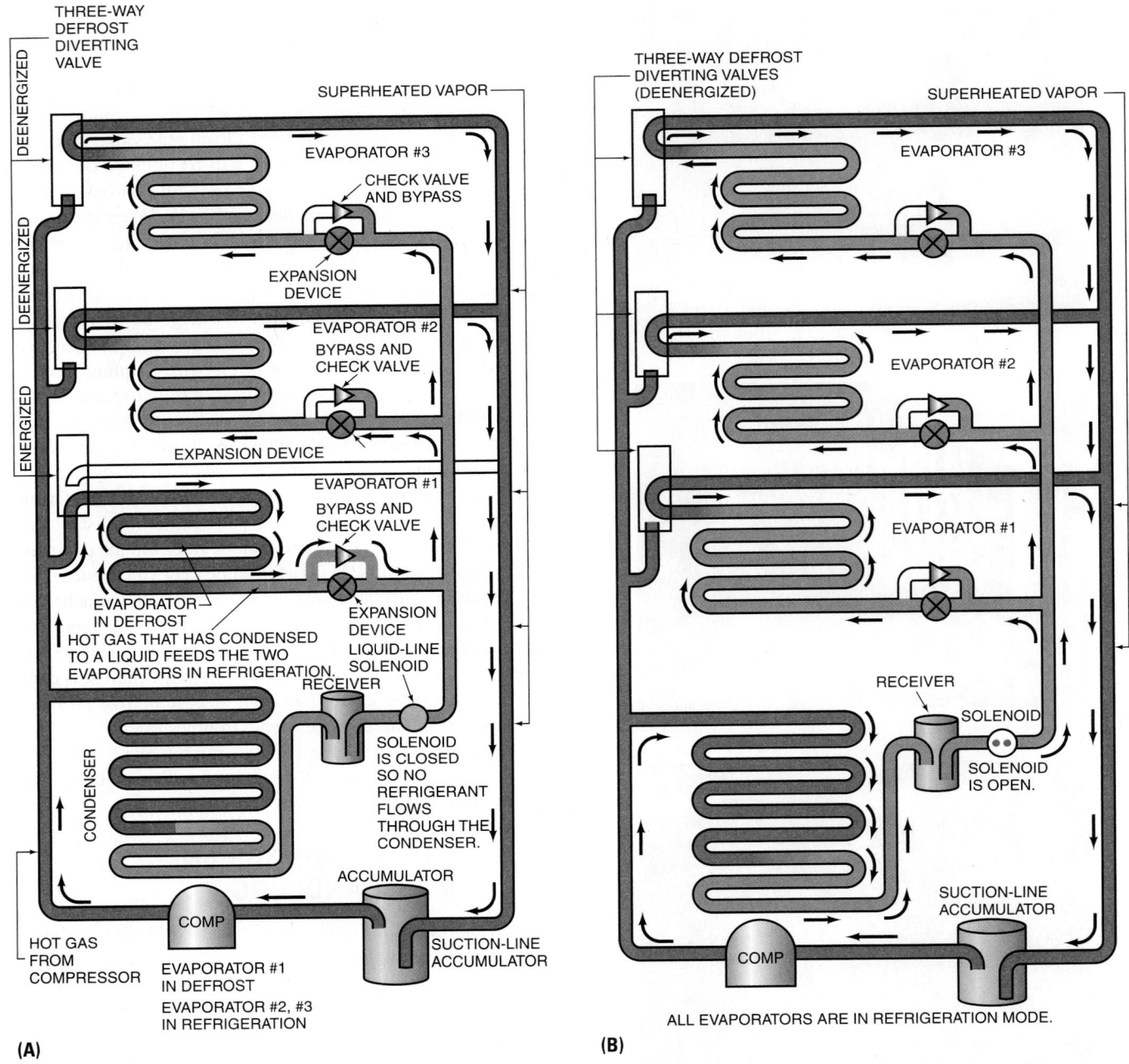

Figure 26–31 **(A)** Hot gas defrost with multiple cases. The condensed liquid from the evaporator in defrost is returned to the liquid line through the bypass and check valve around the expansion valve. The liquid then feeds the evaporators in the refrigeration mode. Three-way defrost diverting valves control the refrigerant flow in the defrost and refrigeration modes. **(B)** All three evaporators shown in the refrigeration mode.

kept damp with an automatic sprinkler system, **Figure 26–33.** The product can be covered at night with plastic lids or film. The customer can see the product because the fixture may have its own lights. Meat that is stored in the open is normally packaged in clear plastic. In this type of fixture the evaporator may be at the bottom of the box. Fans blow the cold air through grilles to give good air circulation. Service for the coil components and fan is usually through removable panels under the vegetable storage or on the front side of the fixture.

The appliance is normally placed with a wall on the back side, **Figure 26–34.**

The chest-type fixture can be furnished with the condensing unit built in or designed so that it can be located at a remote location, such as in an equipment room. If the condensing unit is furnished with the cabinet, it is usually located underneath the front and can be serviced by removing a front panel. When the condensing unit is furnished with a fixture, it is called self-contained.

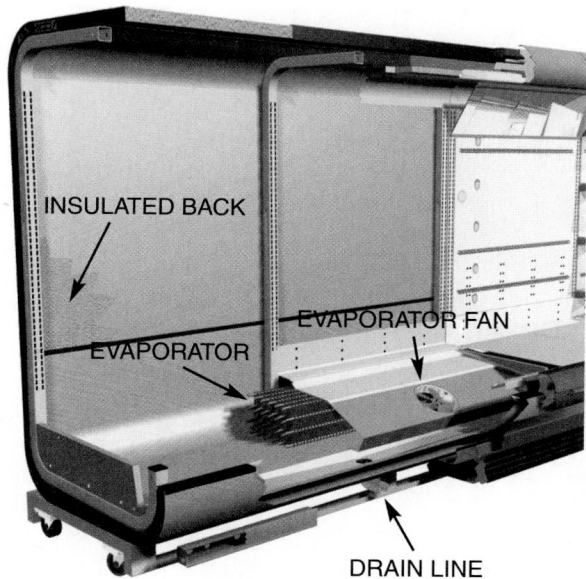

Figure 26–32 A dairy case showing the location of the evaporator. *Courtesy Hill Phoenix*

Figure 26–33 An open display case used to store fresh vegetables. The case has a drain, so the vegetables can be moistened by hand or with an automatic sprinkler. *Courtesy KES Science and Technology, Inc.*

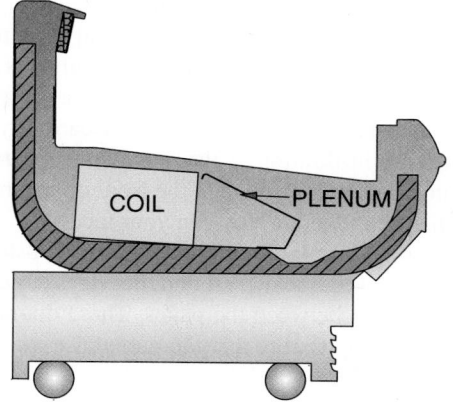

Figure 26–34 The cutaway side view of a chest-type display case. The fans can be serviced from inside the case. The case is normally located with its back to a wall or another fixture. *Courtesy Hill Phoenix*

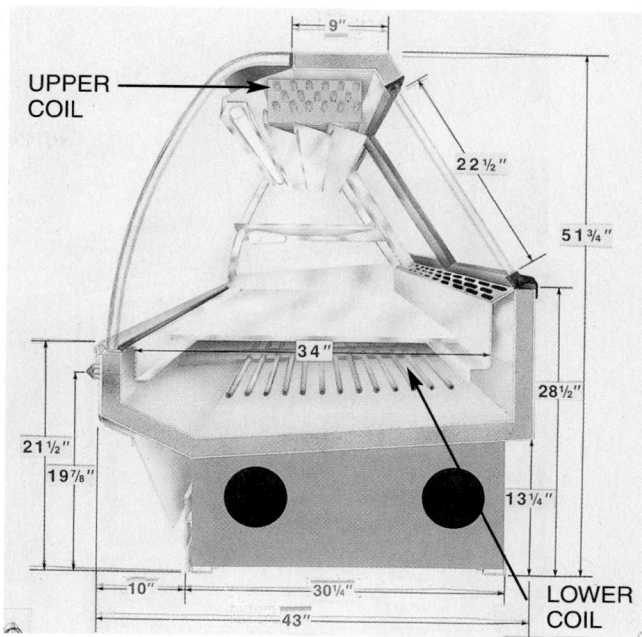

Figure 26–35 Multiple evaporators in a single case because refrigeration is needed in more than one place. *Courtesy Tyler Refrigeration Corporation*

26.15 REFRIGERATED SHELVES

In some chest-type fixtures, refrigerated shelves are located at the top. These shelves must have correct airflow around them or they must have plate-type evaporators to maintain the correct food conditions. When evaporators are at the top, they are normally piped in series with the evaporators at the bottom, **Figure 26–35.**

26.16 CLOSED-TYPE CHEST FIXTURES

The closed-type chest is normally a low-temperature fixture and can store ice cream or frozen foods. The lids may be lifted off, raised, or slid from side to side. These boxes are not as popular now as in the past because the lids are a barrier to the customer.

The upright closed display normally has doors through which the customer can look to see the product. This type of cooler may have a self-contained or remote condensing unit. It can also be piped into a system with one compressor. This display is sometimes used as one side of the wall for a walk-in cooler. The display shelves can then be loaded from inside the walk-in cooler, **Figure 26–36.** These fixtures may be found in many convenience stores.

26.17 CONTROLLING SWEATING ON THE CABINET OF FIXTURES

Display cases that have any cabinet surfaces that may operate below the dew point temperature of the room (the temperature at which moisture will form) must have some

Figure 26–36 A display case with a walk-in cooler as its back wall. The display case can be supplied from inside the cooler. *Courtesy Hill Phoenix*

means (usually small heaters) to keep this moisture from forming. This cabinet heating normally occurs around doors on closed-type equipment. The colder the refrigerated fixture, the more need there is for the protection. The heaters are made with a resistance-type wire that is run just under the surface of the cabinet. These are called *mullion* heaters. Mullion means division between panels. In refrigeration equipment, it is the panel between the doors that becomes cold. There can be a large network of these heaters in some equipment, **Figure 26–37.** Any surface that may collect moisture from sweating needs a heater. *Some of them can be thermostatically controlled or controlled by a humidistat based on the humidity in the store.* Air-conditioning (cooling) systems are used for controlling the humidity in the stores more than ever. You may have noticed how cold some supermarkets are in the shopping areas. This is designed to keep some of the load off the refrigeration equipment. Any heat and moisture that is removed with the air-conditioning system is removed at a higher efficiency level. The air-conditioning system has a much lower compression ratio than any of the refrigeration systems and will remove more moisture. Owners realize that with a low humidity in the store, they do not have to use as much mullion heat, and coils operating below freezing do not need as much defrost time.

26.18 MAINTAINING STORE AMBIENT CONDITIONS

The humidity is taken out of the store in the warm weather with the air-conditioning system and the display cases. The more humidity removed by the air-conditioning system, the less is needed to be removed by the refrigeration fixtures. This means less defrost time. Some stores keep a positive air pressure in the store with makeup air that is conditioned through the air-conditioning system. This method is different from a random method of taking in outside air when the doors are opened by customers. It is a more carefully planned approach to the infiltration of the outside air, **Figure 26–38.** You will notice this system when the front doors are opened and a slight volume of air blows in your face.

The doors on display cases are usually constructed of double-pane glass and sealed around the edges to keep moisture from entering between the panes. These doors must be rugged because any fixture that the public uses is subject to abuse, **Figure 26–39.**

The cabinets of these fixtures must be made of a strong material that is easy to clean. Stainless steel, aluminum, porcelain, and vinyl are commonly used. The most expensive cabinets are made of stainless steel, and they are the longest lasting.

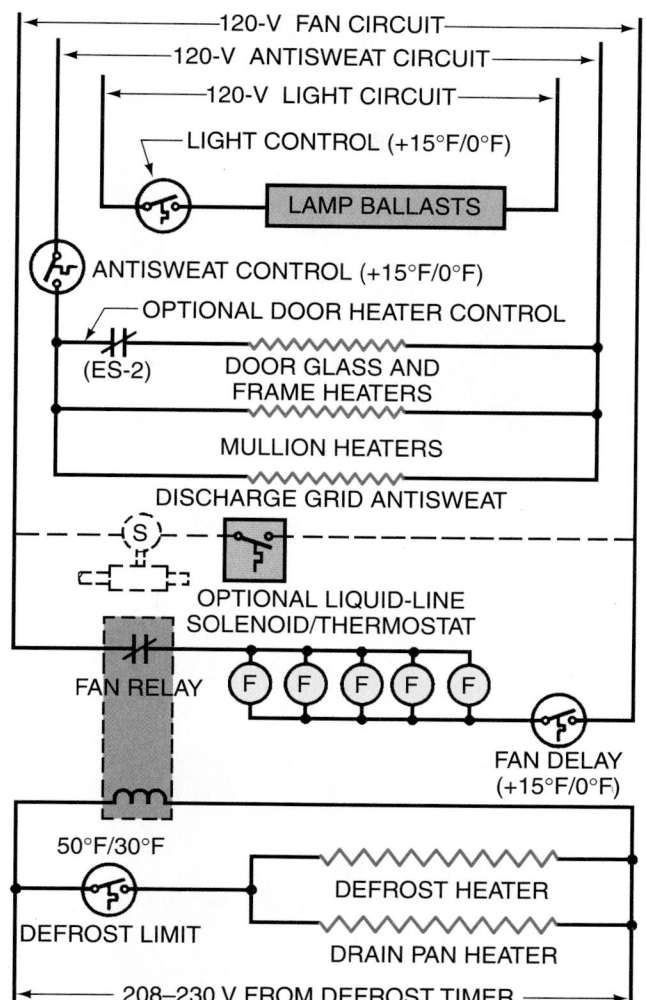

Figure 26–37   A wiring diagram of mullion heaters on a closed display case. Several circuits can be applied to the heaters. This unit has two power supplies, 120 V and 208 V or 230 V.

26.19 WALK-IN REFRIGERATION

Walk-in refrigeration equipment is either permanently erected or of the knock-down type. The permanently erected refrigerated boxes cannot be moved. Very large installations are permanent.

26.20 KNOCK-DOWN WALK-IN COOLERS

Knock-down walk-in coolers are constructed of panels from 1 in. to 4 in. thick, depending on the temperature inside the cooler. They are a sandwich type of construction with metal on each side and foam insulation between. The metal in the panels may be galvanized sheet metal or aluminum. The panels are strong enough that no internal support steel is needed for small coolers. The panels can be interlocked together. They are shipped disassembled on flats and can be assembled at the job site. This cooler can be moved from one location to another and can be reassembled, **Figure 26–40.**

Walk-in coolers come in a variety of sizes and applications. One wall of the cooler can be a display, and the shelves can be filled from inside the cooler. The coolers are normally waterproof and can be installed outdoors. The outside finish may be aluminum or galvanized sheet metal. When the panels are pulled together with their locking mechanism, they become a prefabricated structure, **Figure 26–41.**

26.21 WALK-IN COOLER DOORS

SAFETY PRECAUTION: Walk-in cooler doors are very durable and must have a safety latch on the inside to allow anyone trapped on the inside to get out, **Figure 26–42.**

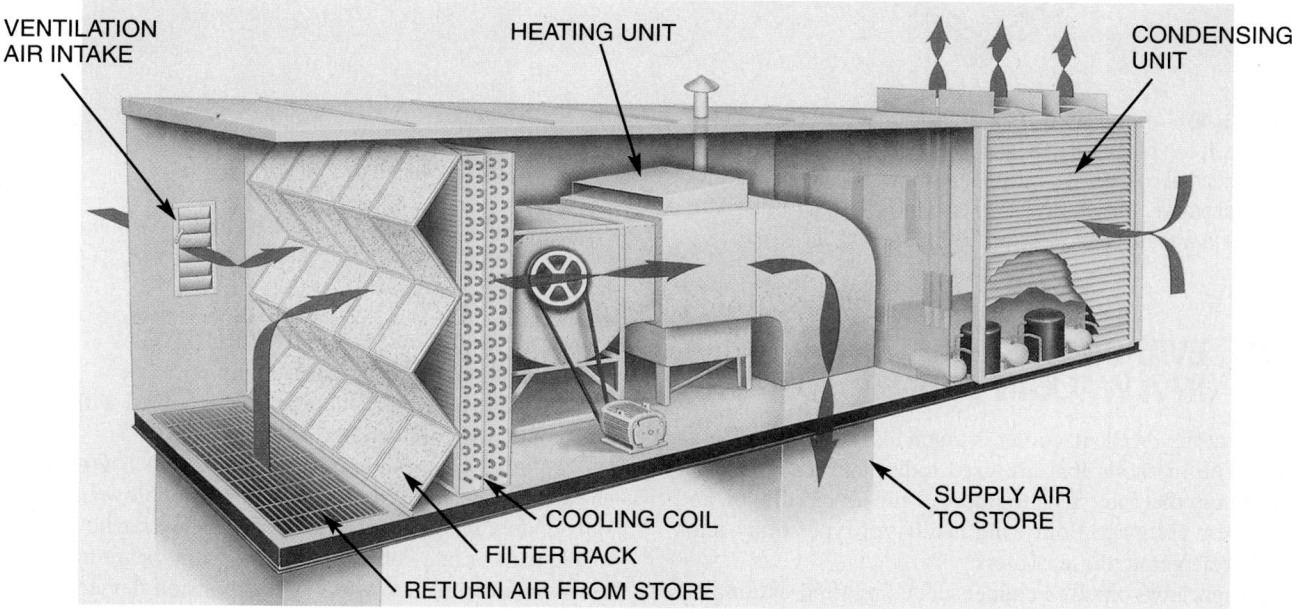

Figure 26–38   Planned infiltration known as ventilation. *Courtesy Tyler Refrigeration Corporation*

Figure 26–39 These doors are rugged and reliable. They are constructed of double-pane glass sealed on the edges to keep them from sweating between the glass. *Courtesy Hill Phoenix*

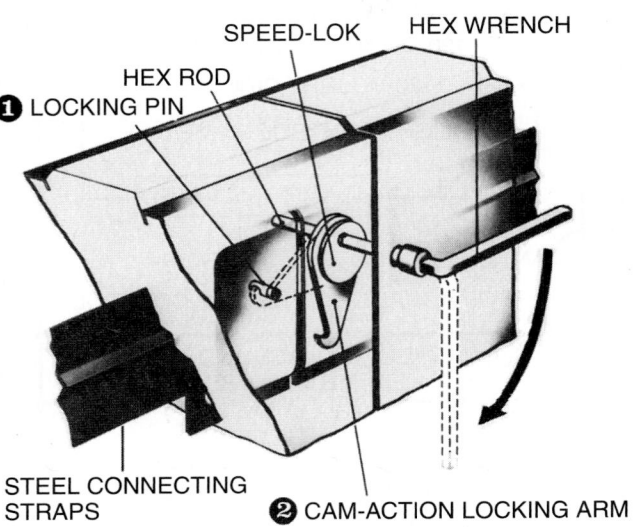

Figure 26–41 A locking mechanism for prefabricated walk-in cooler panels. They can be unlocked for moving the cooler when needed. *Courtesy Bally Case and Cooler, Inc.*

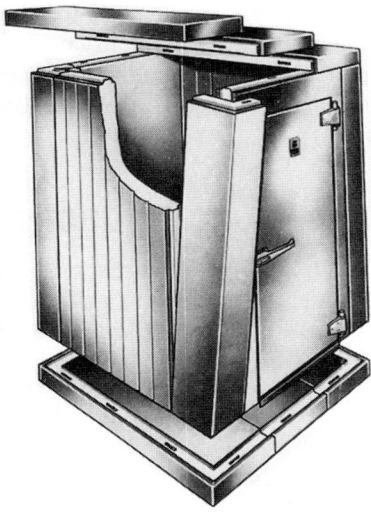

Figure 26–40 A knock-down walk-in cooler that can be assembled on the job. It can be moved at a later time if needed. The panels are foam with metal on each side, creating its own structure that needs no internal braces. This prefabricated box can be located inside or outside. It is weatherproof. *Courtesy Bally Case and Cooler, Inc.*

Figure 26–42 Walk-in cooler doors are rugged. They have a safety latch and can be opened from the inside. *Courtesy Bally Case and Cooler, Inc.*

26.22 EVAPORATOR LOCATION IN A WALK-IN COOLER

Refrigerating a walk-in cooler is much like cooling any large space. The methods that are used today take advantage of evaporators that have fans to improve the air circulation and make them compact. Following is a list of types of systems used to refrigerate these coolers:

1. Evaporators piped to condensers using field-assembled pipe

2. Evaporators with precharged piping
3. Package units, wall-hung or top-mounted with condensing units built in

The evaporators should be mounted in such a manner that the air currents blowing out of them do not blow all of the air out the door when the door is opened. They can be located on a side wall or in a corner. These evaporators are normally in aluminum cabinets in which the expansion device and electrical connections are accessed at the end panel or through

Figure 26–43 Fan coil evaporators are used in walk-in coolers. *Courtesy National Refrigeration and Air Conditioning Products*

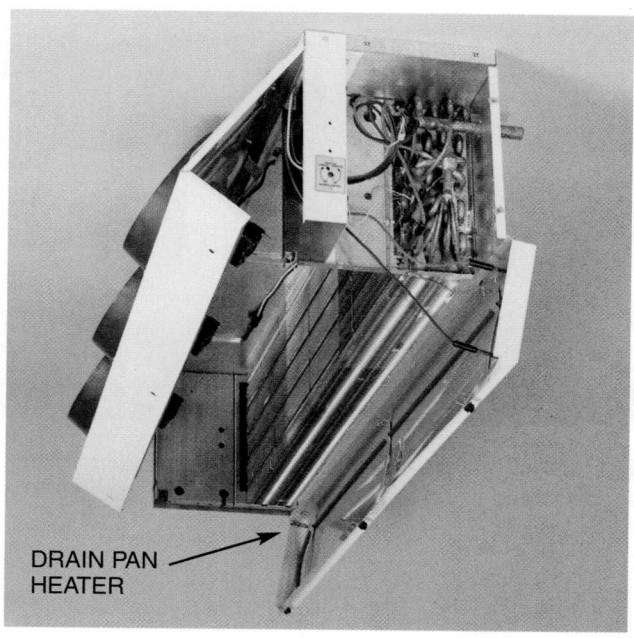

DRAIN PAN HEATER

Figure 26–44 Drain pan heaters to keep pans from freezing. *Courtesy Larkin Coils, Inc.*

the bottom of the cabinet, **Figure 26–43.** Some installations have a fan switch that shuts the fan off when the door is opened to prevent cold air from being pushed out of the cooler. This works well if the door is not propped open for long periods of time. The compressor is operating without the fans and this can cause liquid flooding to the compressor. Some coolers deenergize the liquid-line solenoid valve and pump the refrigerant from the evaporator to the condenser and receiver while the fans are off. This works also but is an added step in the control sequence.

26.23 CONDENSATE REMOVAL

The bottom of the cabinet on the evaporator contains the condensate drain pan that must be piped to the outside of the cooler. When the inside is below freezing, heat has to be provided to keep the condensate in the line from freezing, **Figure 26–44.** This heat is normally provided by an electrical resistance heater that can be field installed. The line is piped to a drain and must have a trap to prevent the atmosphere from being pulled into the cooler. The drain lines also should be sloped downward about 1/4 in. per foot for proper drainage. The line and trap must be heated if the line is run through below-freezing surroundings. These drain line heaters sometimes have their own thermostats to keep them from using energy during warm weather.

These thermostatically controlled heaters are used when the drain line is run outside. The heaters can cycle off when the outside temperature is warm. When the drain line is located inside the box, the heater is energized all of the time. Heat tape is usually used in this case.

26.24 REFRIGERATION PIPING

One of two methods may be used to install refrigeration piping. One method is to pipe the cooler in the conventional manner. The other is to use precharged tubing, called a *line set.* When the conventional method is used, the installing contractor usually furnishes the copper pipe and fittings as part of the agreement. This piping can have straight runs and factory elbows for the corners. When the piping is completed, the installing contractor has to leak check, evacuate, and charge the system. This requires the service of an experienced technician.

When precharged tubing is used, the tubing is furnished by the equipment manufacturer. It is sealed on both ends and has quick-connect fittings. This piping has no fittings for corners and must be handled with care when bends are made. The condenser has its operating charge in it, and the evaporator has its operating charge in it. The tubing has the correct operating charge for the particular length chosen for the job. This has some advantage because the installation crew does not have to balance the operating charge in the system. The system is a factory-sealed system with field connections, **Figure 26–45.** No soldering or flare connections have to be made. The system can be installed by someone with limited experience.

If, because of miscalculation, the piping is too long for the installation, you can obtain a new line set from the supplier, or you can cut the existing line set to fit. ♻ If the existing line set is to be altered, refrigerant can be recovered, the line cut, soldered, leak checked, evacuated, and charged to specifications. ♻ If the line set is too long, the extra tubing should

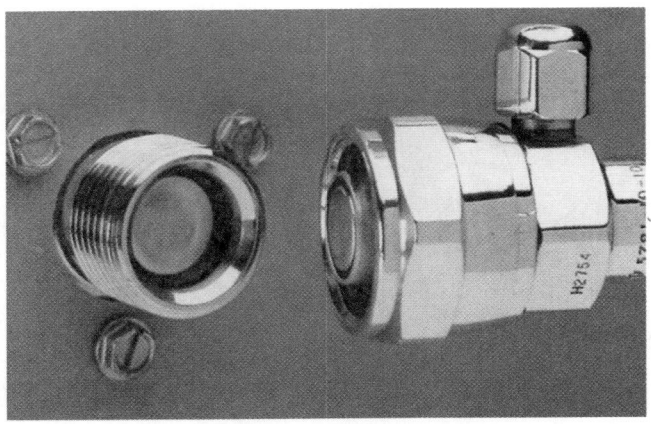

Figure 26–45 Quick-connect fittings providing the customer with a system that is factory sealed and charged. *Courtesy Aeroquip Corporation*

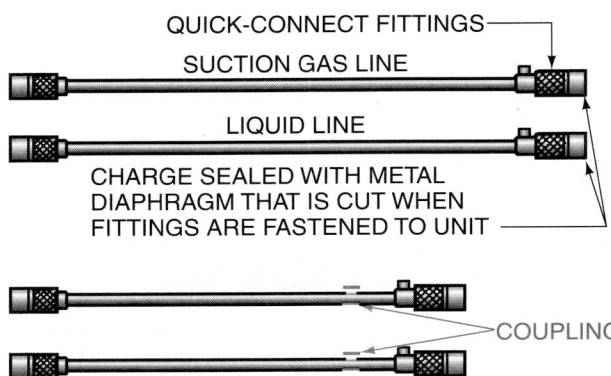

Figure 26–46 Altering quick-connect lines. NOTE: The suction line is charged with vapor. The liquid line has a liquid charge. It contains about as much liquid as may be pulled into it under a deep vacuum. ♻ Recover the charge in each tube and cut to the desired length. ♻ *Fasten together with couplings.* SAFETY PRECAUTION: Leak check, evacuate, and charge (1) the suction line with vapor, and (2) the liquid line with liquid. You now have a short line set with the correct charge. *See Unit 8, "System Evacuation," and Unit 10, "System Charging."*

be coiled and placed in a horizontal position to ensure good oil return. **Figure 26–46** shows a step-by-step illustration of how the line set can be altered.

26.25 PACKAGE REFRIGERATION FOR WALK-IN COOLERS

Wall-hung or ceiling-mount units are package-type units, **Figure 26–47.** They can be installed by personnel who do not understand refrigeration. However, proper installation techniques must be followed. The installation may be compared to installing a window air conditioner. This makes them very popular for some applications because the owner can keep the cost down. Care must be used as to where the condenser air is discharged outside. It must have room to discharge

Figure 26–47 This wall-hung "saddle unit" actually hangs on the wall of a cooler. The weight is distributed on both sides. This is a package unit that only has to be connected to the power supply to be operable. *Courtesy Bally Case and Cooler, Inc.*

without recirculating to the air inlet, or head pressure problems will occur. These units are factory assembled and require no field evacuation or charging. They come in high-, medium-, and low-temperature ranges. Only the electrical connections need to be made in the field.

26.26 VENDING MACHINE REFRIGERATION

Some of the products dispensed by a vending machine require refrigeration. These may be beverages (frozen or liquid), sandwiches, or frozen products such as ice cream, **Figure 26–48.** The refrigeration portion of the vending machine is only a small part of the mechanical and electrical operation of the machine. This refrigeration system is similar to domestic refrigeration, which is discussed later in this text. It will be a fractional horsepower hermetically sealed system. Some of the systems will be medium-temperature and some will be low-temperature applications, depending on the product dispensed.

Vending machines are self-contained, so they are plug-in devices, like a household refrigerator or freezer. They will typically operate from a 20-A electrical plug-in circuit, **Figure 26–49.** These machines are complex because the money changing devices must be built into the machine along with the system that dispenses the product. The money dispensing devices must be able to receive money, coin and paper, and dispense the correct change. This is accomplished with sophisticated electronic circuits and sensors. Often, money changers are located adjacent to the vending machines. These money changers make change for coins and frequently paper money to coins. The product handling system of a vending machine, often called the conveyor system, dispenses the correct product after the money transaction occurs. The vending machine consists of the refrigeration system, the money changing system, and the product dispensing system. Each system in the machine can be intricate. This text discusses some of the basic refrigeration systems used in vending. Further training in the other systems of the machine may be obtained in factory schools and from manufacturers' literature.

(A)

(B)

Figure 26–48 **(A)** A cold drink dispenser. **(B)** A sandwich and snack dispenser.

(C)

Figure 26–48 *(continued)* **(C)** An ice cream vending machine. *Courtesy Rowe International, Inc.*

BEVERAGE COOLING. Beverage coolers are used to cool either canned or bulk beverages. They are usually designed to dispense the product when money is fed into the machine. Medium-temperature refrigeration systems are used because the beverages must be maintained above freezing. When the beverages are in cans or bottles, the evaporators can be sized small and operate at low temperatures because evaporation of the product does not have to be considered.

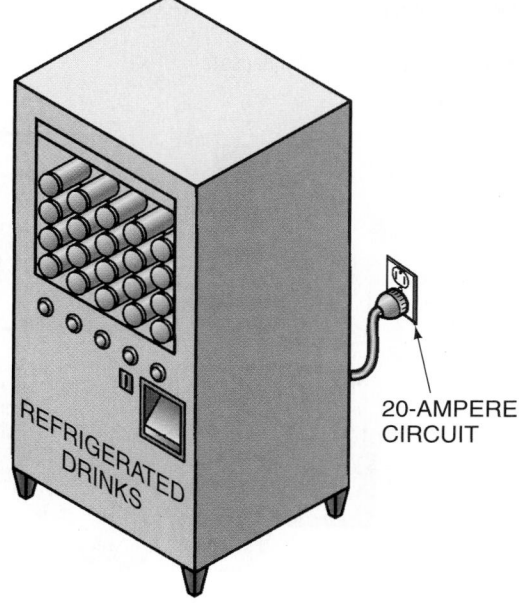

Figure 26–49 Vending machines use 20-A electrical outlets.

HOT DISCHARGE LINE ROUTED THROUGH THE
BOTTOM OF CONDENSATE PAN.
PAN SHOULD BE CHECKED AND CLEANED
DUE TO MINERAL BUILDUP AS THE
CONDENSATE EVAPORATES.

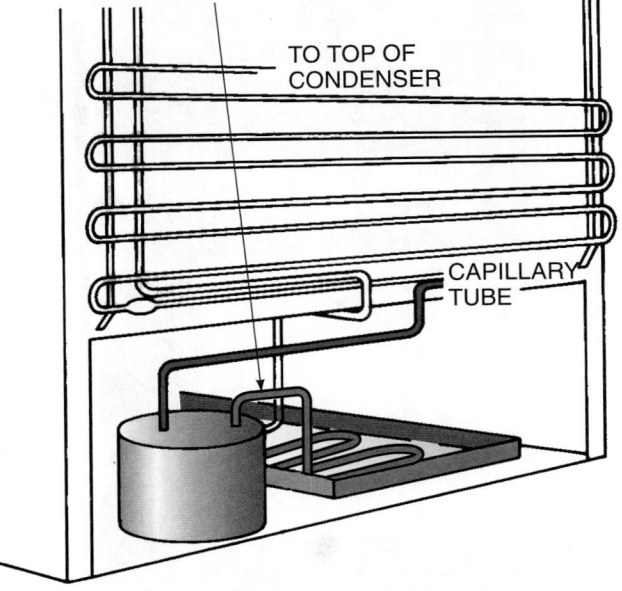

TO TOP OF
CONDENSER

CAPILLARY
TUBE

WARM AIR OVER WATER
FOR EVAPORATING CONDENSATE

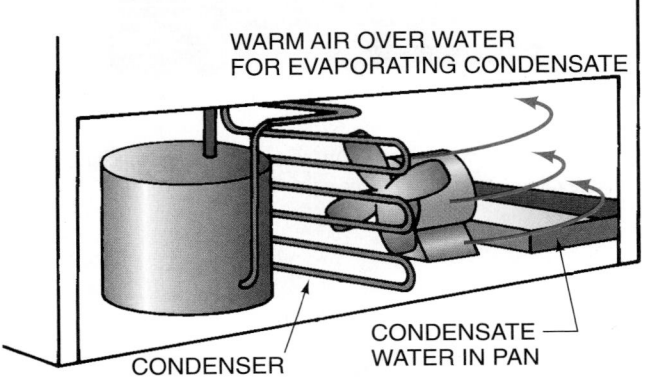

CONDENSER

CONDENSATE
WATER IN PAN

Figure 26–50   Evaporating condensate.

The evaporators in vending machines operate at below freezing so they must have a defrost cycle. Typically, the air in the vending machine is warm enough for off-cycle defrost. This may have to be timed off-cycle defrost in some cases because the machine may not have enough off time. The timed off-cycle defrost can be controlled using the electronic circuit board that is typically furnished with the vending machine. When defrost occurs, the moisture from the coil usually runs through a liquid trap to prevent atmosphere from entering the refrigerated area and then to a pan where it is evaporated using either the hot gas line or air from the condenser, **Figure 26–50**. The condensate pan can be a very dirty place where rodents and insects can be attracted because any drink can or bottle that may break will drain to the condensate pan and be evaporated, leaving behind sugar and flavoring from the drink.

In a hot climate, the drinks may be hot enough to keep the compressor running all the time if loaded off a truck in the sun to the machine. It is a good idea to precool the drinks before adding them to the machine. For example, a vending machine located close to a medium-temperature walk-in cooler is a good idea. The drinks can be brought into the area and cooled from the temperature on the back of the delivery truck to room temperature or to walk-in cooler temperature, then moved to the vending machine, **Figure 26–51**. This reduces the load on the vending machine refrigeration system and ensures cold drinks when needed. Some vending machines have a holding area in the bottom of the machine where drinks may be stored and chilled before depositing them in the dispensing racks. This again helps to ensure the customer of cold drinks when desired.

In vending machines the drinks are usually stacked on top of each other and are dispensed from the bottom. When the money is placed in the slot, the drink drops out by gravity. Change for any additional money may be dispensed also. When the machine is loaded, the drinks are loaded at the top and if they are warmer than the machine's refrigerated set point, they have time to cool before dropping out the vending slot, **Figure 26–52**. This also serves as a method for rotating

Figure 26–51   Moving hot drinks from the truck into the store and even into a walk-in cooler to precool them before placing them in the vending machine will take some of the load off the refrigeration system in the vending machine.

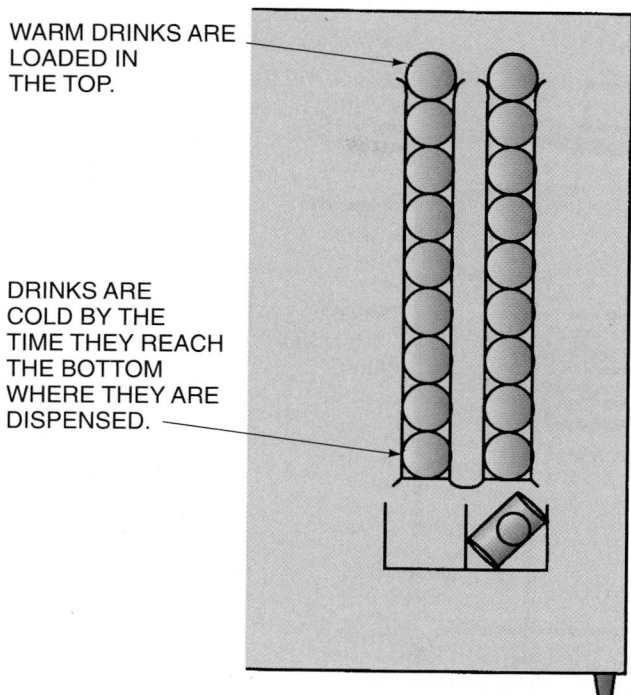

Figure 26–52 Loading a canned-drink dispensing machine.

the stock because the first cans or bottles dropped in the slot will be the first out.

Because space is important in vending machines, they use air-cooled condensers and forced-draft evaporators with capillary tube metering devices, **Figure 26–53**. The evaporator fan blows the cold air over the top of the drinks and it returns back to the evaporator at the bottom of the box. Because a

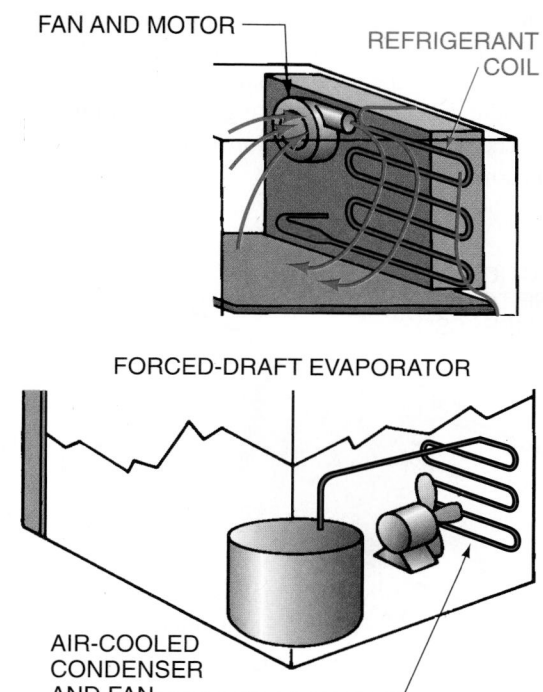

Figure 26–53 A forced-draft evaporator and condenser.

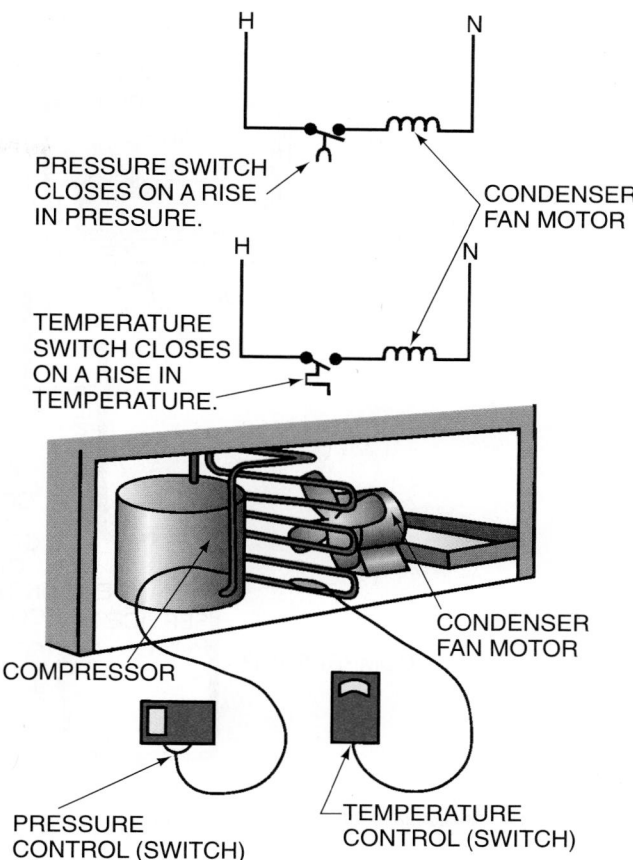

Figure 26–54 Use either a pressure control piped to the discharge line or a temperature sensor strapped on the liquid line to cycle the condenser fan motor for low-ambient control.

vending machine may be located outside or in a cool location, the condenser will typically have a low-ambient control to cycle the condenser fan to keep the head pressure high enough so the capillary tube will feed refrigerant correctly to the evaporator. This can be accomplished with either pressure control or liquid-line temperature control, **Figure 26–54**.

These systems are much like a household refrigerator using a very small hermetic compressor. Service for the refrigeration system is basically the same as with a household refrigerator. You may refer to the unit on domestic refrigerators for more information on the service of small systems.

Other beverage coolers may dispense refrigerated liquids and ice, such as soft drinks with a cup full of ice. The dispensed liquid should be refrigerated and dispensed with the ice, so the drink is ready to drink when dispensed. If the liquid is dispensed warm and into ice, it would take a few minutes to become cool. Liquids at various temperatures would also require varying amounts of ice to reach the correct temperature, and the quality of the drink would be hard to maintain.

The refrigeration system of some machines has a dual purpose: to make ice and to prechill the entering water. This is accomplished with two evaporators and one condensing unit built into the vending machine. A three-way valve in the liquid line can be used to direct the liquid to either of the evaporators, **Figure 26–55**.

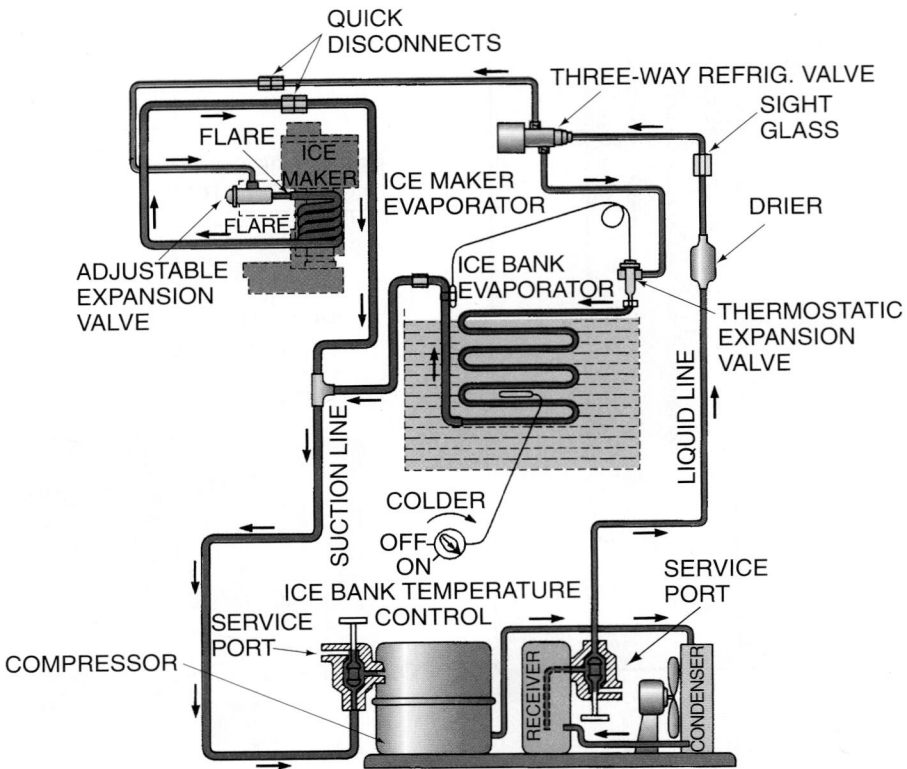

Figure 26–55 An ice maker and ice bank evaporators. *Courtesy Rowe International, Inc.*

The ice maker used in the typical vending machine may be a flake ice maker. It would have an evaporator around a chamber with an auger inside. As the ice is made, the auger scrapes it off the evaporator surface and pushes it over into a bin, **Figure 26–56.** A bin level switch will shut the compressor off or switch the compressor operation to the ice bank evaporator, depending on the demand on the machine. Because water is the big percentage of the drink's content, the water is prechilled before mixing in the consumer's container (usually a paper or plastic cup). The water may be prechilled in two ways. One is a direct heat exchange with the refrigerant, **Figure 26–57.** Another method is to use a water bath evaporator with an ice bank. The entering-water piping passes through the water bath heat exchange, which has circulating water in a tank. The evaporator is fastened to the outside wall of the tank and ice can accumulate on this evaporator plate, which gives the system reserve capacity for times of heavy use, **Figure 26–58.** When there is heavy use, the compressor can be used to make ice, which will always ensure a cold drink. The ice-making portion of the machine has first choice for use of the compressor. The ice bank has second choice.

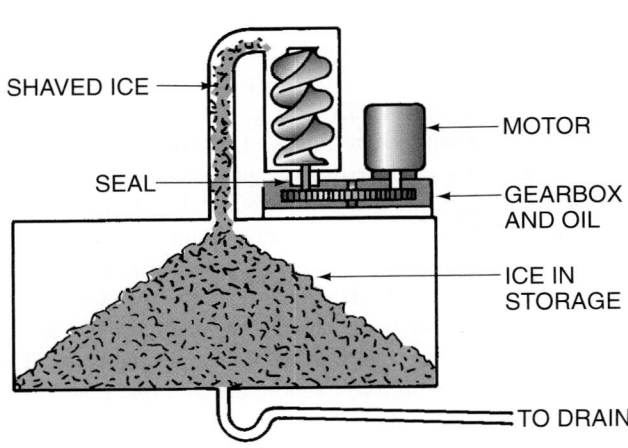

Figure 26–56 A flake ice maker.

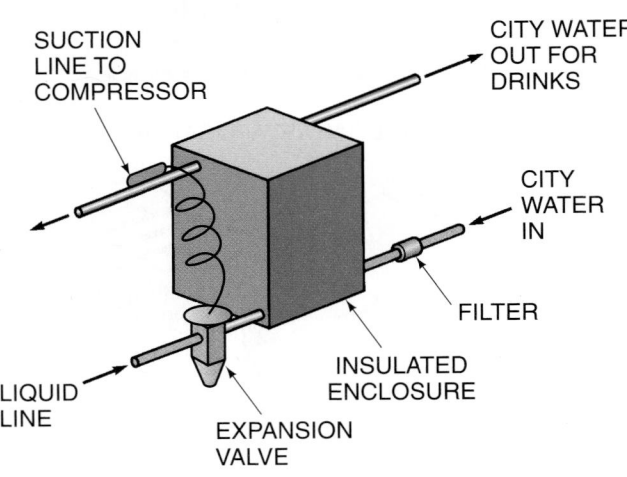

Figure 26–57 A precooling evaporator for a beverage dispenser.

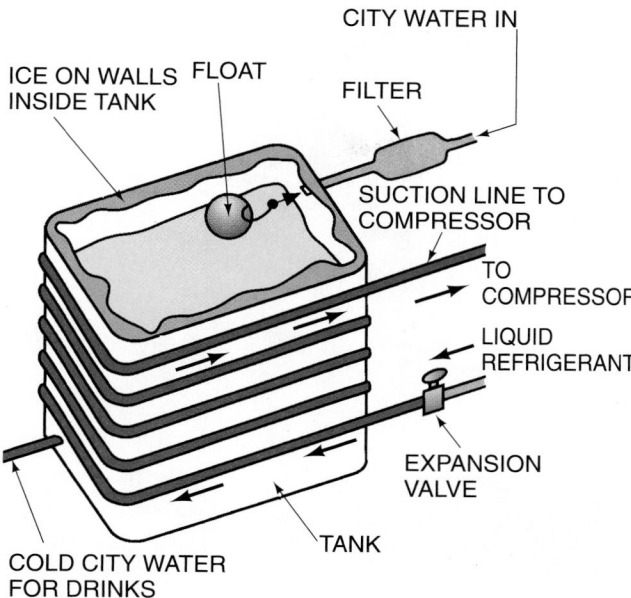

Figure 26–58 This ice bank precooler has some reserve capacity in the frozen ice.

Figure 26–59 A food vending machine with a glass front for viewing the product.

Carbonation of cold drinks is accomplished by injecting CO_2 into the drink from a small cylinder inside the vending machine. The CO_2 cylinder has a regulator to reduce the pressure of the cylinder contents as it is injected into the water while filling the cup.

VENDING MACHINES FOR SANDWICHES AND OTHER PRODUCTS. There are other products that must be refrigerated before being dispensed at medium temperature. Sandwiches and other items may be located in slots in the vending machine and dispensed like canned drinks. These machines typically have a transparent front so the customer can see the product before purchase, **Figure 26–59**.

Medium-temperature vending machines may have a totally self-contained refrigeration system that may be removed from the machine, **Figure 26–60**. This can be handy from a service standpoint. If there is a problem, the complete system can be changed out and taken to the workbench for service.

Other machines may dispense frozen products, such as ice cream. These machines must operate at about 0°F to maintain the food frozen hard and are considered low-temperature applications. These machines may have plate-type evaporators that cover the inside of the product section of the machine, **Figure 26–61**. When a large evaporator is used, automatic defrost is not practical, so defrost is handled when the machine is serviced for routine preventive maintenance.

Low-temperature machines will often have a shield inside that keeps the ambient air from loading the evaporator with moisture while the machine is being filled with product. This prevents excessive frost buildup on the evaporator.

Most vending machines have what is known as a health switch. This is a thermostat that alerts the operator that the machine is operating above the safe operating range for a

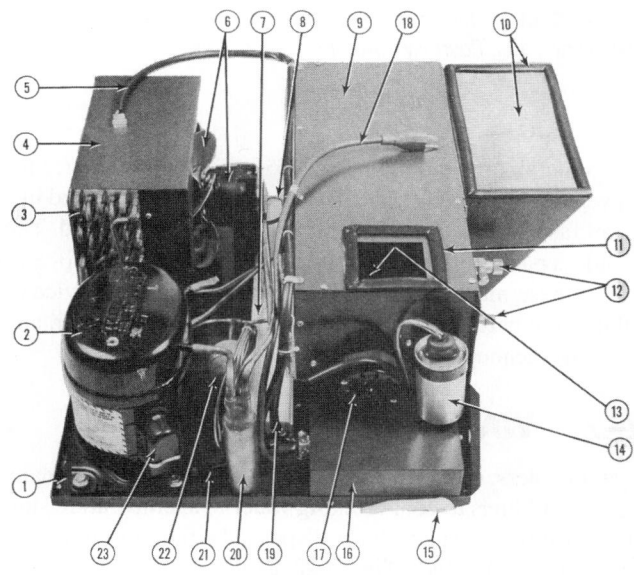

Figure 26–60 The self-contained refrigeration portion of a vending machine. This unit may be easily replaced in the field and taken to a shop for repair. *Courtesy Rowe International, Inc.*

long enough period of time that the product quality may be affected. For example, if a vending machine is dispensing sandwiches and a temperature of above 45°F is reached for a particular length of time, usually 30 min, bacteria may grow in the food products and be harmful to the health of the customers. Electronic circuits and warning lights alert the operator. The machine lighting may also shut off, alerting the public not to buy.

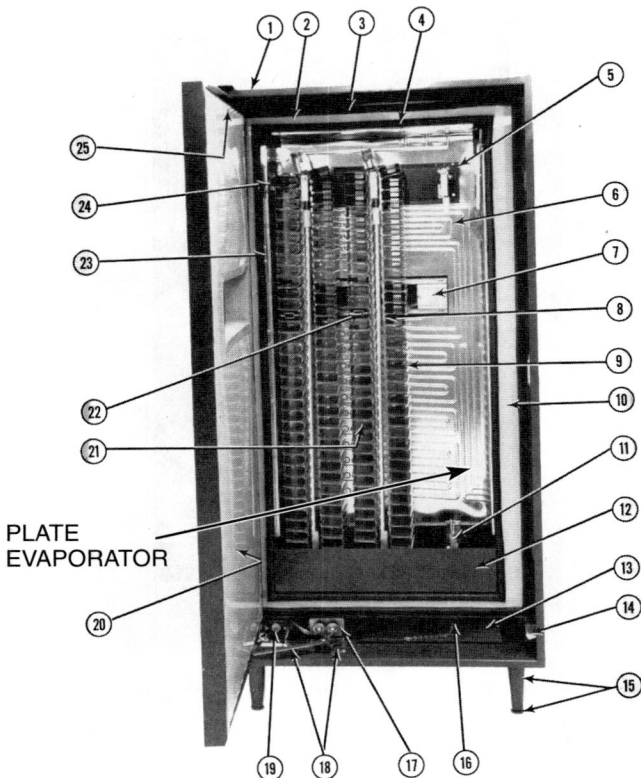

Figure 26–61 A low-temperature vending machine. Notice the plate evaporator. *Courtesy Rowe International, Inc.*

(A)

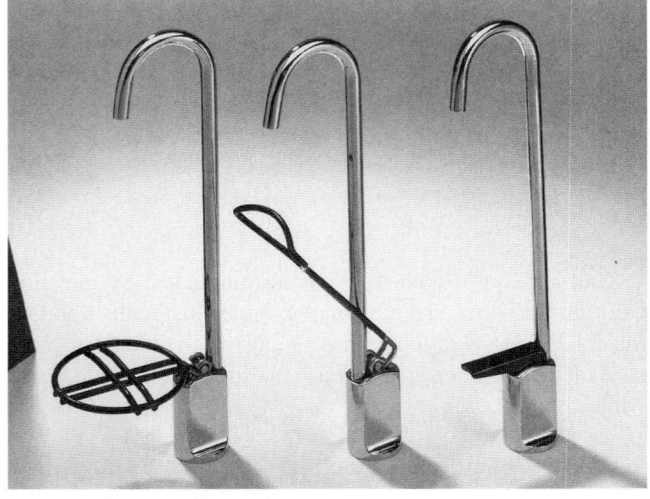

(B)

Figure 26–62 Two methods of dispensing cold water **(A)–(B)**. *Courtesy EBCO Manufacturing Company*

Sanitation of vending machines is of great importance for all vending operations. The machine must be maintained in a clean fashion or the customers' health may be affected. The health department may have some say in the vending operation if these machines are not kept clean. Regular service of all moving parts, defrost of the refrigeration evaporator, and machine cleanup must be done on a regular basis.

26.27 WATER COOLERS

Water coolers, often called water fountains, are used in many public buildings to cool drinking water to a temperature similar to ice water. The water is dispensed either through a fixture called a bubbler, **Figure 26–62(A),** or through a snout used to fill a cup, **Figure 26–62(B).** There are typically two types of package water coolers, one that has a large bottle of water located above the cooler, **Figure 26–63(A),** and one that is under regular water main pressure, **Figure 26–63(B).**

Pressure-type systems require less maintenance because they are automatic in their dispensing of water. The water may be under regular water main pressure entering the cooler and it must be reduced to a pressure that will develop an arc for people to drink from, **Figure 26–64.** The water pressure regulator for reducing the pressure is part of the system that must be serviced by the technician. The regulator typically has a screwdriver slot for adjusting the arc of the water at the bubbler, **Figure 26–65.**

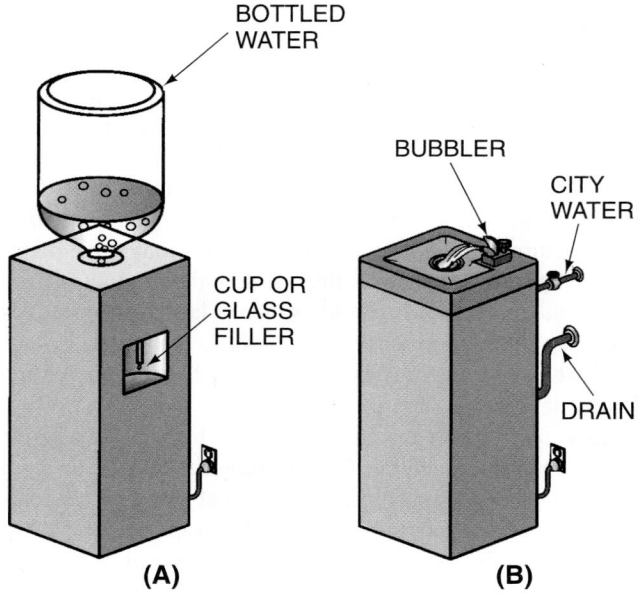

Figure 26–63 Two types of water coolers **(A)–(B)**.

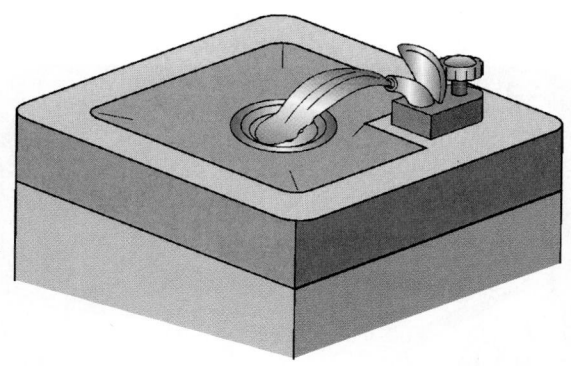

Figure 26–64   A bubbler creates an arc for drinking water.

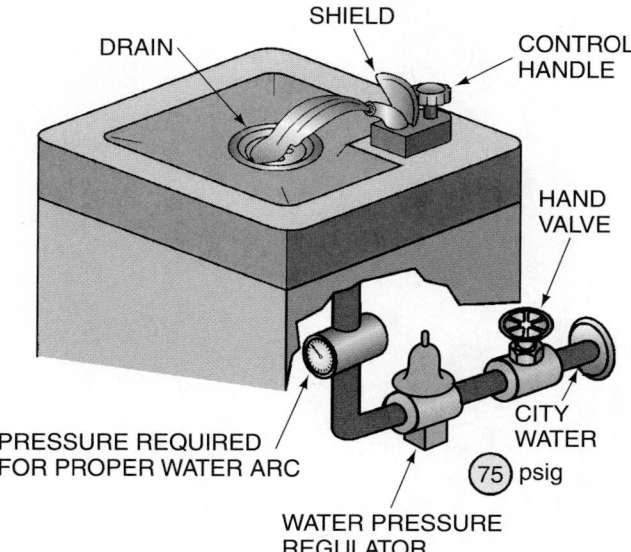

Figure 26–65   Pressure regulation is important.

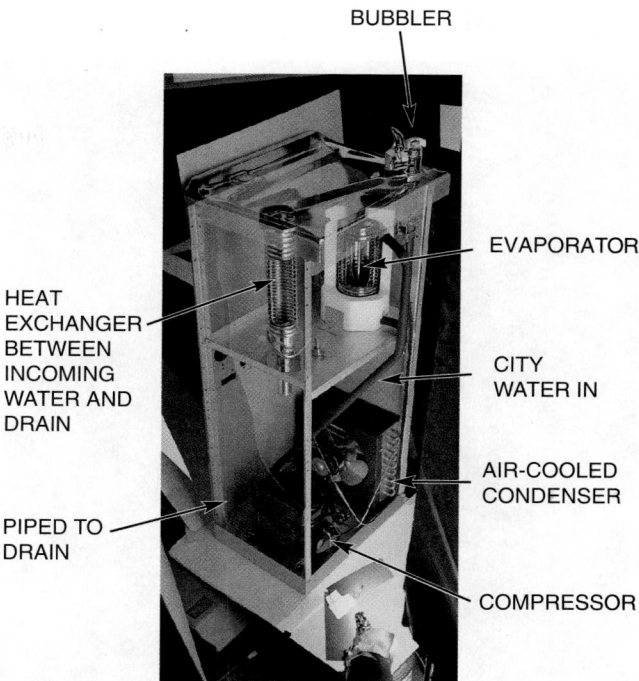

Figure 26–66   A water cooler. *Courtesy EBCO Manufacturing Company*

The refrigeration section of the appliance is a small self-contained hermetic system consisting of a compressor (fractional horsepower), a typical air-cooled condenser with a small fan, and an evaporator. The evaporator is the only special part of the refrigeration system. It is typically a small tank made of nonferrous metal such as copper, brass, or stainless steel. The refrigerant piping is wrapped around the tank. Water coolers dispense drinking water, called "potable water," and they must be sanitary, so the choice of tank is important.

It is estimated that 60% of the water dispensed in a bubbler for drinking goes down the drain. This is cold refrigerated water. Most manufacturers have some form of heat exchange to take advantage of this loss. The heat may be exchanged between the incoming water and the drain or between the liquid refrigerant line and the drain, **Figure 26–66.**

When water is dispensed from a bottle, the refrigeration system is below the bottle and enclosed in the housing, **Figure 26–67.** Water is typically dispensed by gravity into a cup when this type of cooler is used so no water is lost down the drain.

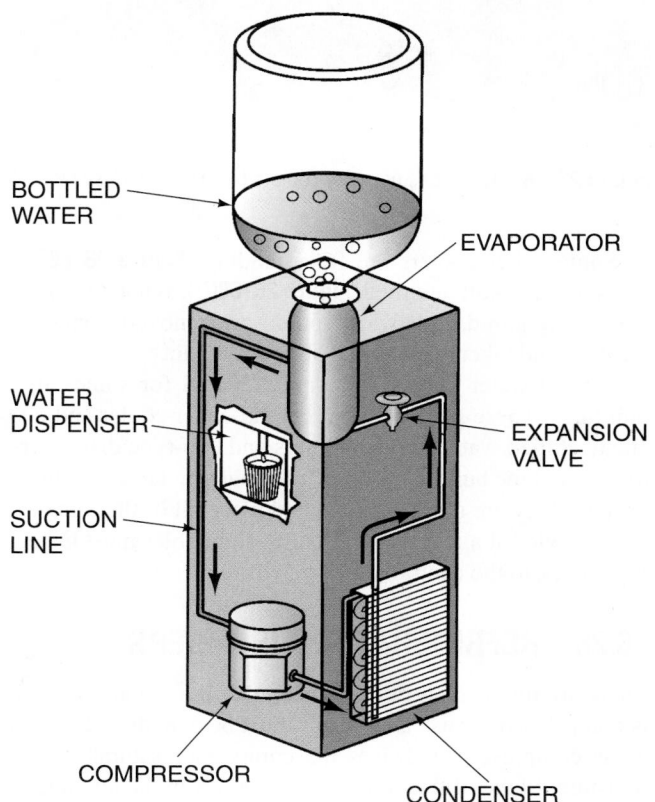

Figure 26–67   Refrigeration for a bottled water cooler.

STANDING COOLER

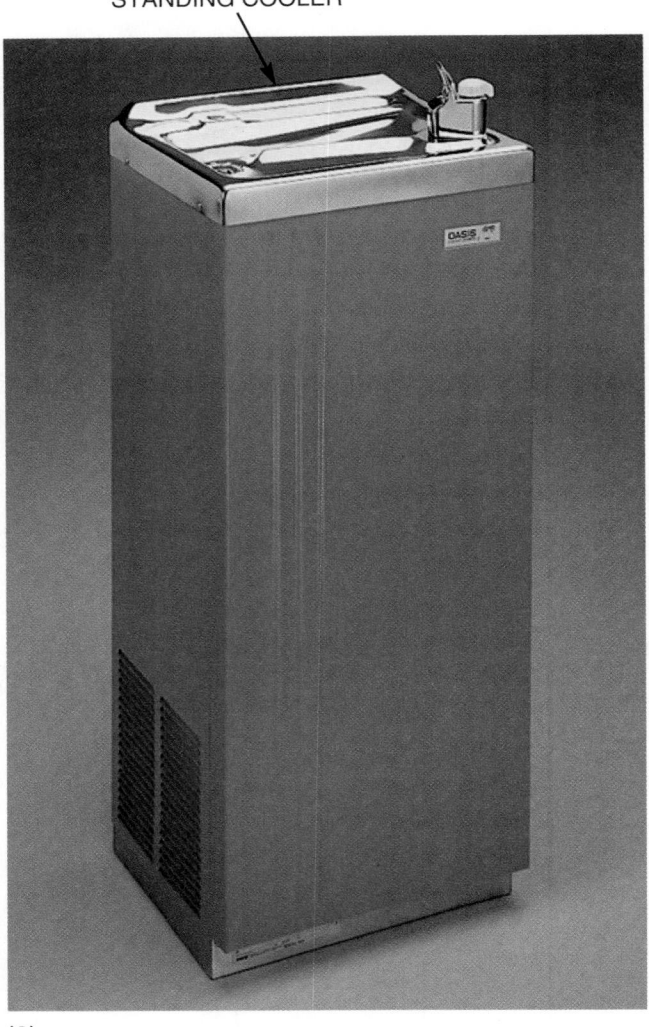

(A)

WALL-HUNG COOLER

(B)

Figure 26–68   Two types of water coolers **(A)–(B)**. *Courtesy EBCO Manufacturing Company*

Some water coolers are free standing, **Figure 26–68(A)**, and some are wall mounted, **Figure 26–68(B)**. When major repairs are required, most water coolers are removed from their location and taken to a central shop for repair.

Central water cooling systems are used for some large buildings. These systems use a central water cooler and circulate chilled water to the bubblers and cup-type dispensers throughout the building. These are popular in large buildings because they are more efficient and place all of the refrigeration service at a central location. Each bubbler must have a drain back to the central drain system.

26.28   REFRIGERATED AIR DRIERS

There are many applications where air from the atmosphere is compressed for use and the air must be dehydrated. When air is compressed, it leaves the compressor saturated with moisture. Some of this moisture condenses in the air storage tank and is exhausted through a float. The air is still very close to saturated as it leaves the storage tank. Some applications

where the air must be dry are when the air is used for special manufacturing processes, or when air is used for controls, called "pneumatic control." This air may be dehydrated using refrigeration.

Refrigerated air driers are basically refrigeration systems located in the air supply, after the storage tank. The air may be cooled in a heat exchanger, then moved to the storage tank where much of the water will separate from the air. This water may be drained using a float that rises and provides for the drainage at a predetermined level, **Figure 26–69**. The air then passes through another heat exchange where the air temperature is reduced to below the dew point temperature (the point where water will condense from the air). This heat exchanger consists of an evaporator that normally operates at a temperature just above freezing, **Figure 26–70**.

The system typically runs 100% of the time because it does not know when there is a demand for air flowing through the system. Because sometimes there is no airflow, there is virtually no load on the refrigeration system. The only time there is a load on the system is when air is passing through

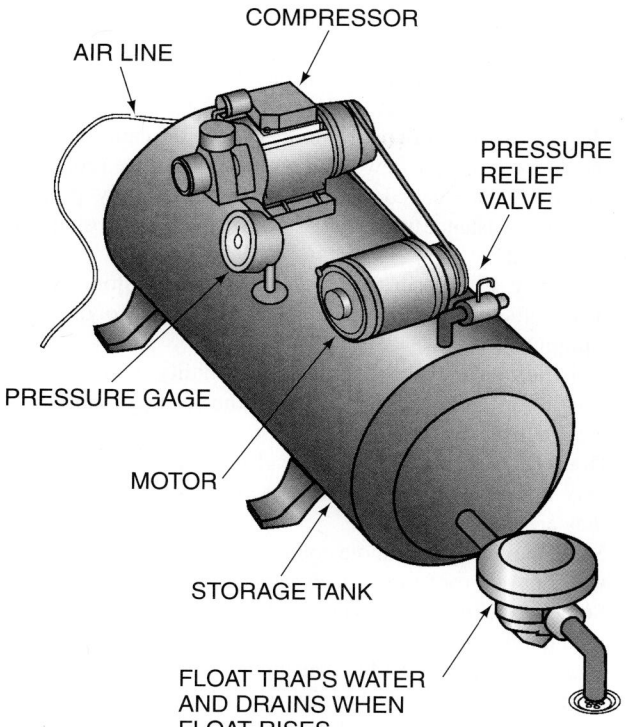

Figure 26–69 An air compressor showing a moisture-trapping system.

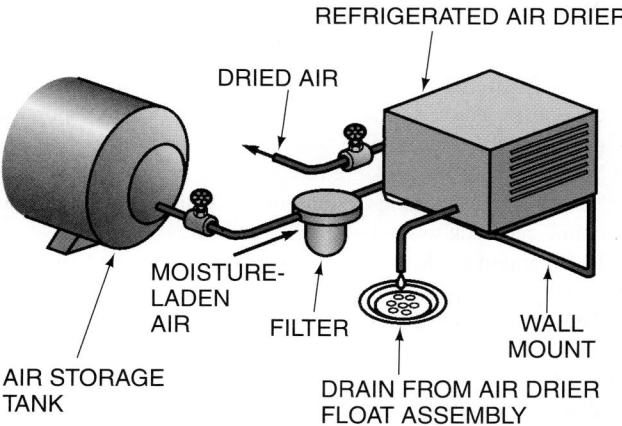

Figure 26–70 A refrigerated air drier installation.

that needs to be dried. Because the refrigeration system is running, in case there is a demand, a method must be used to apply a false load on the evaporator when there is no load. This is usually accomplished with a special refrigeration valve called a hot gas bypass valve, **Figure 26–71.** This valve is situated between the hot gas line and the inlet to the compressor or the evaporator. It monitors the pressure in the evaporator and allows hot gas to enter the suction line at any time there is no or reduced load on the evaporator. This is accomplished by monitoring the suction pressure with a pressure-sensitive valve. It is called a pressure regulator because it does not allow the pressure in the evaporator to drop below the set point. The difference in this pressure regulator and others discussed

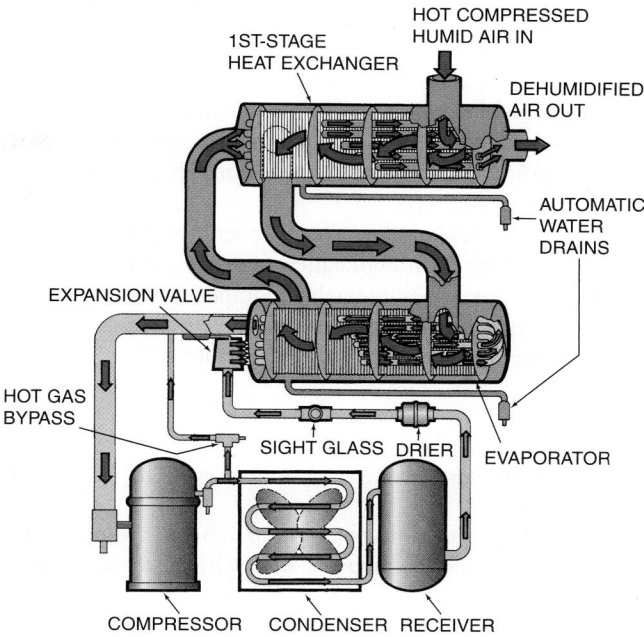

Figure 26–71 The flow schematic for a refrigerated air drier. *Courtesy Van Air Systems, Inc.*

in this text is that it is located between the high- and low-pressure sides of the system. The set point is usually just above the freezing point of water. For example, for R-22, the set point of the hot gas bypass valve may be 61.5 psig (35°F). The valve will not allow the evaporator pressure to reduce below 35°F, so the evaporator will not freeze during any no- or reduced-load situations.

Refrigerated air driers are normally self-contained refrigeration systems that may be air cooled or water cooled, **Figure 26–72.** The compressor can be either a hermetic unit

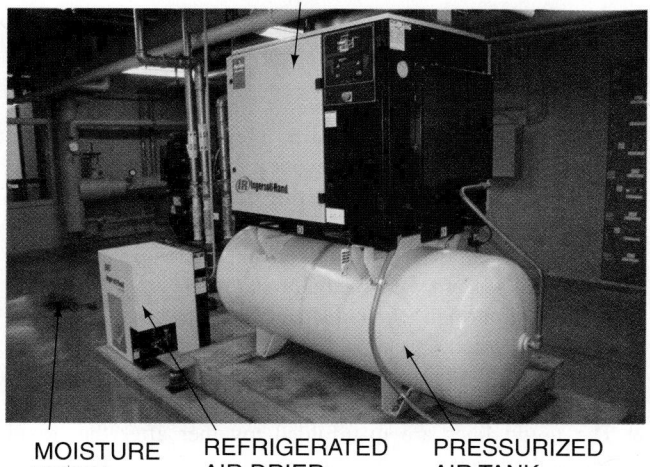

Figure 26–72 Refrigerated air drier with accompanying air compressor, controller, and pressurized air storage tank. *Courtesy Ferris State University. Photo by John Tomczyk*

for the smaller sizes or a semi-hermetic for larger sizes. The air line is piped to and through the evaporator of the air drier and then moves on to the system where it is used. The water is collected and exhausted to a convenient drain.

The evaporator may be one of several types; a pipe within a pipe or a honeycomb type are common. The compressor is usually a hermetic type.

The service technician should understand the manufacturer's intent for each application. At times, the technician must improvise when replacing components that are not available. As equipment becomes outdated, the components for exact replacement may not be available.

SUMMARY

- Product dehydration can be a factor in the choice of equipment.
- With package or self-contained equipment, the condenser rejects the heat back into the store when the equipment is located inside.
- The condensing unit in packaged equipment is normally on top or underneath the fixture.
- All fixtures have condensate that must be drained away or evaporated.
- Fixtures may also be piped to a common equipment room with either individual compressors or single-compressor units.
- When a single large compressor is used, capacity control is desirable.
- Load matching can be accomplished with parallel compressor systems. These compressors are manifolded together with a common suction and discharge line. They share one receiver.
- A pressure transducer on the common suction line signals the controller on the parallel compressor rack for capacity steps.
- A complicated oil system is used on parallel compressor systems to ensure the proper oil level in each compressor's crankcase.
- Parallel compressor systems often use satellite or booster compressors, which are usually the largest compressors on the racks.
- Secondary-fluid systems circulate an antifreeze solution to the refrigerated cases and offer many advantages over a conventional refrigeration system.
- Pressurized liquid systems offer a different way to subcool liquid refrigerant. This helps in overcoming any pressure drops in the liquid line to prevent premature flashing of the liquid before it enters the metering device.
- Unitary, stand-alone refrigeration modules offer many advantages as compared with a conventional supermarket refrigeration system with an equipment room.
- Capacity control is accomplished by cycling compressors on the multiple-compressor racks.
- Several evaporators may be piped together to a common suction line.
- The liquid lines are manifolded together after the liquid leaves the receiver.
- Each fixture has its own expansion valve.
- There must be a defrost cycle on both medium- and low-temperature fixtures.
- Heat must be added to the coil for defrost in low-temperature refrigeration systems.

- Common discharge lines concentrate the heat that was absorbed into the system.
- This heat can be used to heat the store or heat water with heat-recovery devices.
- Heaters are used around the doors to keep the cabinet around the doors above the dew point temperature of the room.
- These heaters, called mullion heaters, normally produce electric resistance heat that is sometimes thermostatically controlled.
- Humidity, usually in the summer, leaves the store by two means: the refrigeration equipment and the air-conditioning equipment.
- It is less expensive to remove humidity with the air-conditioning equipment than with the refrigeration equipment defrost cycle.
- Drain heaters normally produce electric resistance heat and may have thermostats to keep them from operating when they are not needed.
- Some coolers have remote condensing units, and some have wall-hung or roof-mount package equipment.
- Some of the remote units are field piped, and some are piped with quick-connect tubing.
- The refrigeration vending machine systems consist of the money receiving-and-changing mechanism, the food handling section (often called the conveyor), and the refrigeration system.
- Defrosted condensate flows to a sealed drain to be evaporated using moving air or the hot gas line.
- Vending refrigeration uses air-cooled condensers that have fractional horsepower compressors and often forced-draft evaporators.
- Water coolers dispense cold water, are self-contained, and use a fractional horsepower compressor and an air-cooled condenser. The evaporator is a container with the evaporator piping wrapped around it.
- Water coolers are package refrigeration machines that may be either free standing or hung on the wall. A drain must be connected back to the building drain system.
- Refrigerated air driers are refrigeration systems that are used to dehydrate air that is compressed with an air compressor.
- Refrigerated air driers typically operate all of the time, whether there is a load or not. They provide cooling to about 35°F.
- The typical refrigerated air drier uses a hermetic compressor and either an air- or water-cooled condenser.
- SAFETY PRECAUTION: There are many applications for refrigeration. There is no substitute for knowing the manufacturer's intent. Do not improvise without this knowledge.

REVIEW QUESTIONS

1. The two broad categories of display cases are _____ and _____.
2. How are conditions maintained in open display cases when there are high shelves?
3. What are mullion heaters?
4. The three temperature ranges for refrigeration systems are _____, _____, and _____.
5. The two methods for rejecting heat from refrigerated cases are _____ and _____.

6. When the compressors are located in the equipment room, the piping is routed
   A. in floor chases or in plastic pipes under the floor.
   B. in suspended ceilings.
   C. between the walls.
   D. in back of the walls.
7. When one large compressor is used, what desirable feature should it have?
   A. Discharge muffler
   B. Suction muffler
   C. Cylinder unloaders or variable-speed drive
   D. Hot gas bypass
8. Define a parallel compressor system and list five advantages of these systems.
9. Three disadvantages of parallel compressor systems are _____, _____, and _____.
10. What is the difference between an even and an uneven parallel compressor system?
11. Exactly matching the refrigeration load with the capacity of refrigeration compressors is referred to as
    A. capacity control.
    B. load shedding.
    C. load paralleling.
    D. load matching.
12. What is the function of the pressure transducer in parallel compressor systems?
13. What is the function of the microprocessor-based controller in parallel systems?
14. Explain how the oil separator, oil reservoir, pressure differential valve, and oil level regulators work together in maintaining the correct oil level in parallel compressors.
15. Large or small compressors on parallel compressor racks, which are usually dedicated to the coldest cases, are called
    A. satellite compressors.
    B. remote compressors.
    C. booster compressors.
    D. both A and C.
16. An _____ varies the frequency of an electronically altered sine wave, which affects the speed of alternating current motors.
17. Explain the function and advantages of a secondary-fluid system.
18. Explain the main idea behind pressurized liquid systems and list their advantages.
19. Three advantages of unitary, stand-alone refrigeration modules in supermarket refrigeration are _____, _____, and _____.
20. How is defrost accomplished when the equipment room is in the back and the fixtures are in the front on medium-temperature fixtures?
21. True or False: The two types of defrosts used with low-temperature fixtures are hot gas and electric defrost.
22. What special precautions should be taken with drain lines in walk-in coolers?
    A. Freeze protection
    B. Trapped to prevent air from being pulled in
    C. Sloped downward
    D. All of the above
23. What holds up the structure of a walk-in cooler?
24. What type of defrost is normally used for vending machine medium-temperature applications?
    A. Electric
    B. Hot gas
    C. Off cycle
    D. None of the above
25. What is the special valve used in a refrigerated air drier?
    A. Four-way valve
    B. Defrost valve
    C. Three-way valve
    D. Hot gas bypass valve

UNIT
27
Commercial Ice Machines

OBJECTIVES

After studying this unit, you should be able to

- describe the basic refrigeration cycle for ice flake machines.
- discuss basic troubleshooting for ice flake machines.
- state the purpose of the water fill system in an ice flake machine.
- explain the purpose of a flush cycle in an ice flake machine.
- state the purpose of a bin control in an ice flake machine.
- read and interpret ice production and performance charts for ice machines.
- describe how crescent-shaped ice is made.
- describe how cell-type ice cubes are made.
- explain the sequence of operation of an ice machine.
- describe the purpose of a harvest cycle in an ice machine.
- state the purpose of microprocessor controls in ice machines.
- explain what is meant by input/output troubleshooting for microprocessors.
- discuss the importance of water and ice quality in ice making.
- discuss the difference between cleaning and sanitizing an ice machine.
- define water filtration and treatment.

SAFETY CHECKLIST

- ✔ Follow manufacturers' instructions when using cleaning and sanitizing chemicals.
- ✔ Wear protective clothing and eyewear when handling cleaning and sanitizing chemicals.
- ✔ Be careful not to get your hands or clothes caught in any moving parts involving ice makers.
- ✔ A wet environment exists around ice machines. When servicing these machines, extra precautions must be taken to protect against electrical shock and slipping.
- ✔ Observe all electrical safety precautions. Be careful at all times and use common sense.

27.1 ICE-MAKING EQUIPMENT, PACKAGED TYPE

Ice making is accomplished in a temperature range that is somewhat different from the low-, medium-, or high-temperature refrigeration. The ice is made with evaporator temperatures between medium- and low-temperature ranges of about 10°F with the ice at 32°F. Most refrigeration applications use tube and fin evaporators and a defrost cycle to clear the ice buildup from the evaporator. Ice making is accomplished by accumulating the ice on some type of evaporator surface and then catching and saving it after a defrost cycle, commonly called the harvest cycle.

Large commercial block ice makers use cans and freeze the ice in the cans. We will discuss only small ice-making equipment, such as found in commercial kitchens, motels, and hotels. These ice makers are usually of the package type. The power is supplied by a power cord, or in some cases wired directly to the electrical circuit, called hardwired. Some ice makers may be of the split system type with the condenser on the outside.

Package ice machines store their own ice at 32°F in a bin below the ice maker. This bin is refrigerated by the melting ice in the bin (which explains the storage temperature of 32°F), so there is some melting; the hotter the day, the more melting.

A drain must be provided for the melting ice and any water that may overflow from the ice-making process. Do not confuse an ice-making machine with an ice-holding machine such as those seen on the outside of a convenience store or service station. Ice-holding machines hold ice, usually in bags, at a temperature well below freezing. This ice is made and bagged at one location, and then stored and dispensed at the retail location.

Most package ice makers are air cooled and must be located where the correct airflow across the condenser is maintained. Some package ice makers are water cooled and use wastewater systems in which the water is used to cool the condenser and then passes down the drain.

Keep the installation of a package machine simple by following these steps:
1. Set and level the machine, much like a refrigerator.
2. Supply power, plug, or hard wire.
3. Provide a water supply.
4. Provide a drain for the ice bin.

Package ice machines make two types of ice, flake ice and several forms of cube (solid) ice.

27.2 MAKING FLAKE ICE

Flake ice is normally made in a vertical refrigerated cylinder surrounded by the evaporator on the outside. Flake ice is the very thin pieces of ice often seen in restaurants and vending machine cups. Some people prefer flake ice to solid forms of

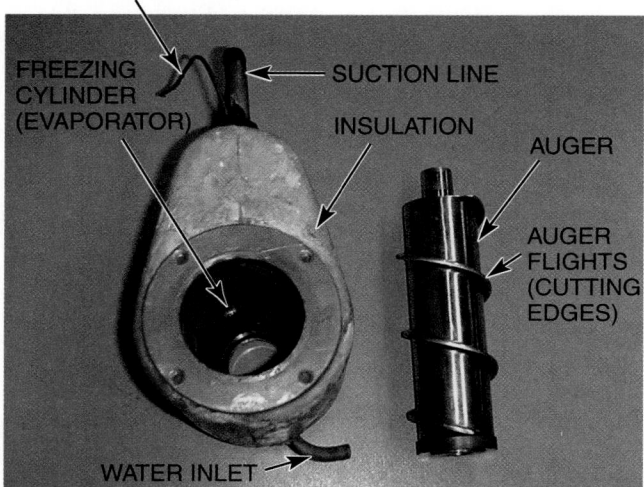

Figure 27–1   An evaporator, freezing cylinder, and auger of a flake ice machine. *Courtesy Ferris State University. Photo by John Tomczyk*

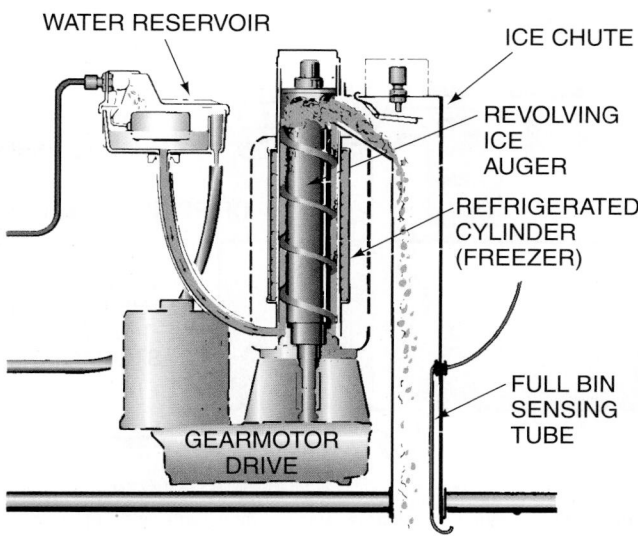

Figure 27–2   A flake ice maker, its evaporator, and its auger. The auger turns and shaves the ice off the evaporator. The tolerances are very close. *Courtesy Scotsman Ice Systems*

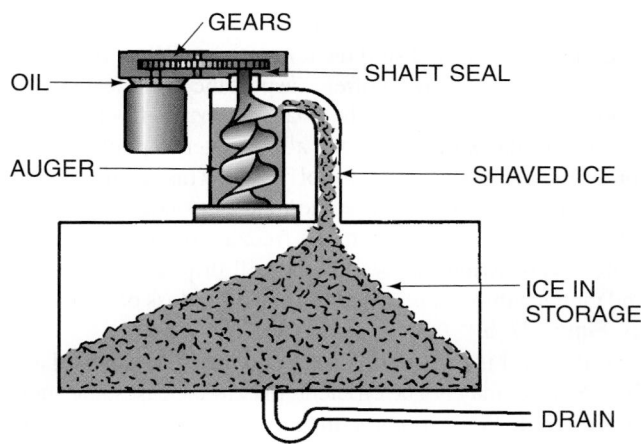

Figure 27–3   The auger is on top of the ice maker. The shaft seal prevents oil from seeping into the ice.

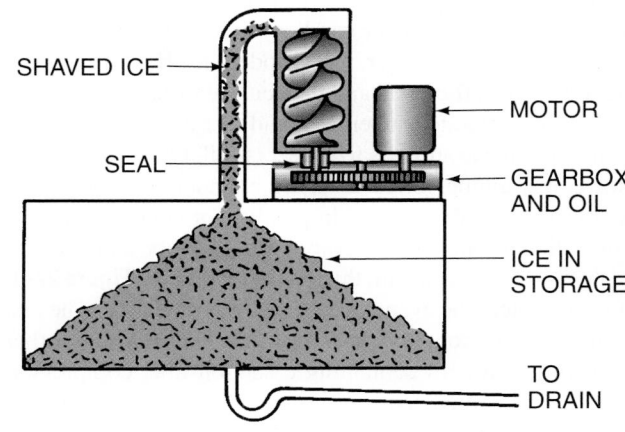

Figure 27–4   The gearbox is on the bottom of the auger. A seal leak would allow water to enter the gearbox.

ice when served in drinks. It melts fast because there is little weight with a lot of surface area to flake ice. It has a good bit of air contained in the flake. Solid forms of ice last longer because there is more weight with less surface area; remember it takes 144 Btu/lb to make or melt ice. It takes less ice by the pound to fill a cup. Solid ice costs more, and some of it is likely to be thrown out.

Flake ice is formed on the inside of the cylinder by maintaining a water level inside the cylinder. An auger is constantly turning inside the cylinder and scraping the formed ice off the evaporator surface, **Figure 27–1**. The auger forces the ice upward and out a chute on the side where it moves by gravity to the ice storage bin. The auger has cutting surfaces called *flights*. The flights cut or shave the ice off the interior of the flooded evaporator cylinder. In some modern ice flake machines, the flights on the auger add pressure to the ice toward the top of the auger. This pressurized ice can then be cut or extruded through small openings near the top of the auger to make many shapes and forms of ice. This type of ice-making machine has a constant harvest cycle, as long as the compressor and auger are operating.

A geared motor is used to turn the auger. The geared motor may be located on the top or bottom of the auger. **Figure 27–2** shows a bottom gear motor machine. The auger is in a vertical position and must have a shaft seal where the auger leaves the gearbox. If the gear motor is on top of the auger, the shaft seal prevents gear oil from seeping into the ice, **Figure 27–3**. When the gearbox is on the bottom of the auger, the shaft seal prevents water from entering the gearbox, **Figure 27–4**.

The level of the water in the evaporator is determined by a float chamber located at the top of the evaporator. This float level is critical. If the level is too low, it may impose an extra load on the gear motor. The lower suction pressure from too low a water level causes hard ice for the auger to cut. If the water level is too high in the evaporator, the auger is doing

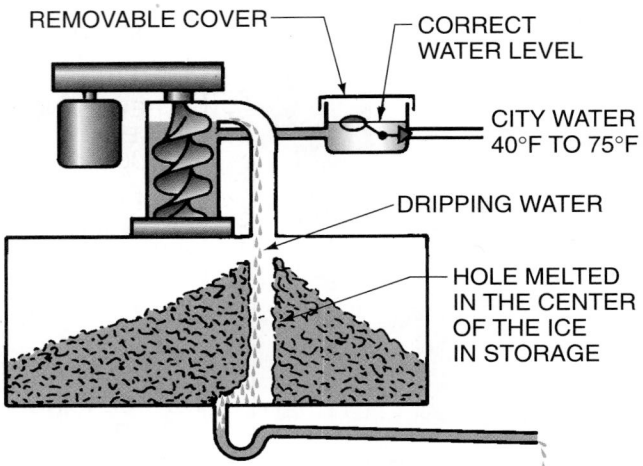

REMOVABLE COVER
CORRECT WATER LEVEL
CITY WATER 40°F TO 75°F
DRIPPING WATER
HOLE MELTED IN THE CENTER OF THE ICE IN STORAGE

Figure 27–5   When the float level is too high, water will run over and melt ice that has been made. The fresh water must be refrigerated to freezing after start-up.

more work because there is more ice to cut. The motor driving the auger may have a manual reset overload that will shut the motor off and stop the compressor. When the motor reset button must be reset often, suspect the water level as the problem. Check the seat in the float chamber. A float that will not control the water level may allow water to pour over the top of the chamber during the off cycle. This would impose an extra load on the refrigeration capacity because it must refrigerate this extra water to the freezing point to make ice. Water overflowing the evaporator will drip into the ice bin and melt ice that is already made, causing a loss of ice capacity, **Figure 27–5.**

If the level is too low, the ice quality may be too hard and the ice maker may not be efficient. In extreme cases, the auger may freeze to the evaporator. In this case the power of the motor and gears may turn the evaporator out of alignment.

Refrigeration Cycle

The refrigeration cycle for ice flakers is simple and straightforward. A TXV, AXV, or capillary tube can be used as the metering device. However, most modern ice flakers incorporate TXVs for efficiency and capacity reasons. If a TXV is used, low evaporator superheat readings of 3°F to 4°F are common to make sure the ice sheet is full. Often, the evaporator is just a tube coiled around a cylinder, and more than one expansion valve is used in parallel for better refrigerant distribution throughout the evaporator. The tube and cylinder are then both covered with thick rigid insulation, **Figure 27–1.** Once insulated, the evaporator tubes are not visible. The rest of the refrigeration system is no different from any other refrigeration system containing a suction line, compressor, water- or air-cooled condenser, liquid line, metering device, and evaporator. An EPR valve is sometimes used to prevent a minimum pressure from occurring in the evaporator. An AXV is seen when a constant evaporator pressure is needed.

Ice flake machines are usually critically charged with refrigerant. Weighing in the correct charge will let the ice machine operate to its greatest potential under all conditions. Charging by the frost line, sight glass, or pressure method may create unwanted inefficiencies because an exact critical charge of refrigerant is called for by the manufacturer. A lack of refrigerant charge, usually from a refrigerant leak, mainly shows up as an overheated compressor motor or low ice production capacity. A low-pressure control may open or the mass flow rate of refrigerant may be so low that heat cannot be removed from the evaporator, and low ice production will result.

Often, ice flake machines can be retrofitted to a parallel refrigeration system used in larger supermarket refrigeration. In this situation, the compressor and condenser from the parallel refrigeration system is used. Pressurized liquid from the liquid header of the parallel system is routed into the ice flaker. The pressurized liquid then encounters the metering device of the packaged ice flaker. The refrigeration effect takes place in the flooded evaporator freezing cylinder, and ice is cut off the freezing cylinder by the auger. An EPR valve is used on the suction line in this application to prevent too low an evaporator pressure from occurring. Liquid-line solenoid valves are also used to start and stop the liquid flow into the expansion valve. The suction gases exiting the evaporator are routed back to the suction header of the parallel compressor system. Parallel refrigeration systems are covered in detail in Unit 26, "Application of Refrigeration Systems."

Basic Troubleshooting

When water is frozen in the evaporator freezing cylinder, minerals from the water are left behind and often become attached to the interior cylinder walls and other surface areas of the ice flaker. Using an approved ice machine cleaner will dissolve these minerals. Always follow the manufacturer's instructions when using an ice machine cleaner, or serious damage can result to the ice machine. When minerals do build up on the interior of the evaporator, they add resistance to the flow of ice that is being cracked, cut, or broken away by the auger and then pushed out of the cylinder. This dirty condition causes the evaporator to run at a lower suction pressure, and thus a colder evaporator temperature. This results in the ice being very hard. Often, noises like two pieces of rigid Styrofoam being rubbed together, a crunching sound, or a loud squeal will come from the freezing cylinder when minerals are allowed to build up. A regular clicking sound usually indicates a mechanical problem, which could be in the drive train and is probably a chipped gear. A high-pitched noise is probably a worn auger drive motor. It is a good idea to start troubleshooting by simply observing and listening to the ice flaker. Ask the owner if there have been any strange or unusual noises coming from the ice machine. If the answer is yes, a cleaning may be needed. The auger and freezing cylinder may have to be inspected and removed to look at the interior of the evaporator freezing cylinder or evaporator, **Figure 27–6(A) and (B).** An ice flaker's evaporator cylinder is made of either stainless steel or a more porous material such as brass or copper. If made of copper, the cylinder must be

(A)

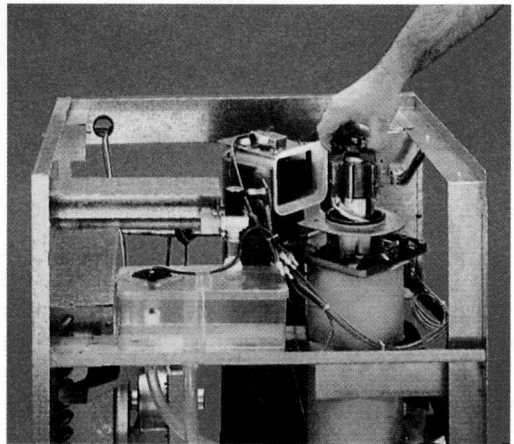

(B)

Figure 27–6 **(A)** A service technician inspecting the freezing cylinder in an ice flaker. **(B)** A service technician removing the auger from the freezing cylinder of an ice flaker. *(A)* and *(B)* *Courtesy Hoshizaki America, Inc.*

plated as required by the National Sanitation Foundation (NSF). Nickel is typically used as the plating material. Stainless steel is much more durable and resists corrosion better than copper or brass. However, copper is better for heat transfer. Manufacturers who use stainless steel for the ice-making surfaces usually have some sort of innovative design that enables the heat transfer qualities of the evaporator surface to come close to that of copper.

The auger and evaporator surface must be very clean and free from any score marks or scars. If a bearing failure has occurred, and the auger has touched the evaporator or freezing cylinder's surface, usually either the evaporator or the auger must be replaced. Sometimes both components have to be replaced. Always follow the manufacturer's instructions for disassembly. Drain the freezing cylinder and allow it to dry before inspecting it. A wet piece of equipment may

WATER RESERVOIR

INSULATED EVAPORATOR

GEAR DRIVE

STORAGE BIN

ICE CHUTE

Figure 27–7   A self-contained ice flaker. *Courtesy Scotsman Ice Systems*

look clean, but once it has had time to dry, mineral buildup will be obvious.

Dirty evaporators from mineral deposits also cause the auger motor, auger bearings, and the gear reducer to be stressed. It is these components that cause most of the service calls in ice flakers, not the refrigeration system.

Most ice flakers have a gear reduction drive that reduces the auger drive motor's speed from about 1725 rpm to 9 to 16 rpm. These drives are generally maintenance-free; however, if water has been entering the gear case, it should be disassembled and inspected. Often, there is a coupling between the auger and the output shaft of the gear reducer. This coupling may need to be lubricated on a regular basis. All ice flakers have bearings that allow the auger to rotate. The bearings can be either a sleeve bearing or a ball bearing. Bearings are a high-wear item and annual inspection is recommended. Consult the manufacturer's maintenance manual to become familiar with the ice machine before starting this procedure. **Figure 27–7** shows a self-contained ice flake machine with one side cover off. Notice the water reservoir, the gear drive, the insulated evaporator, and the ice chute that delivers ice to the storage bin. **Figure 27–8** is an exploded view of most of the mechanical parts that make up an ice flaker.

The gear motor assembly is more prone to failure than any other part in the ice flaker because of the pressures and stresses put on it. As the auger turns and extrudes the ice, the gear motor senses all of these strains. Because of this, some manufacturers use an open-type gear assembly. This means the gear assembly is open to the atmosphere. There is usually a hole in the housing of the gear assembly for hot grease to exit due to long run times with extreme stress on the drive motor, which causes high amperage draw and excessive heat. The drawback to the open-type gear drive assembly is that moisture can enter the hole in the drive housing during cool-down periods. This causes the lubricating effect of the grease in the drive housing to deteriorate. Excessive gear wear and

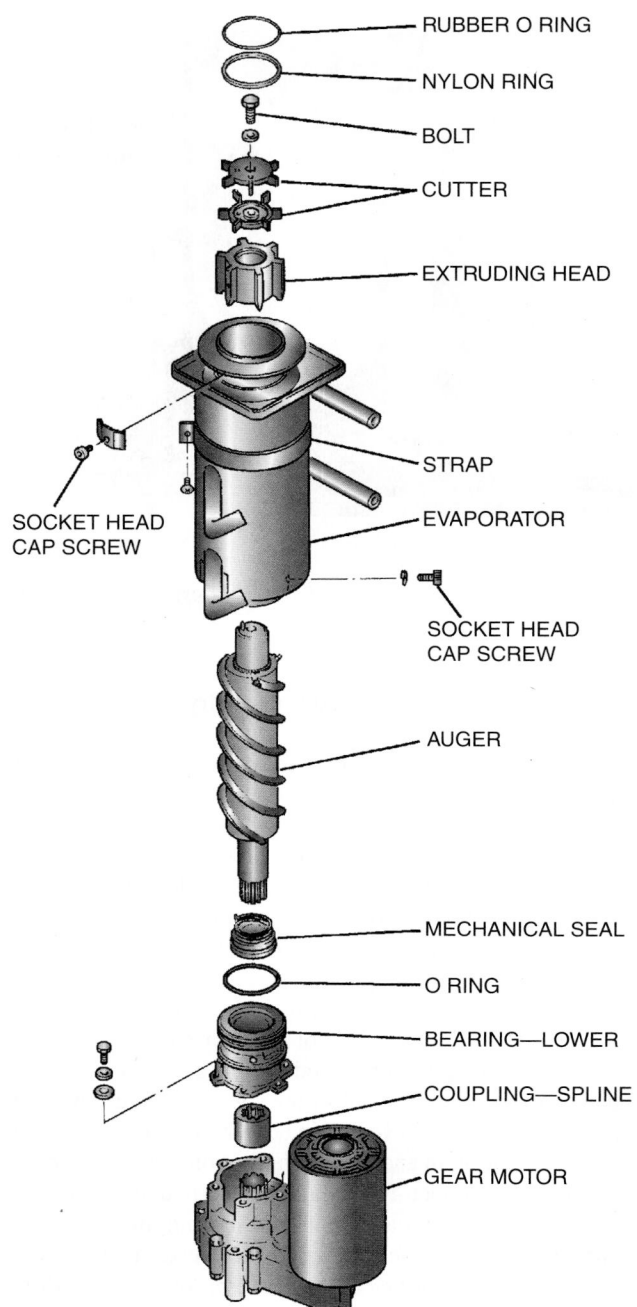

- RUBBER O RING
- NYLON RING
- BOLT
- CUTTER
- EXTRUDING HEAD
- STRAP
- EVAPORATOR
- SOCKET HEAD CAP SCREW
- SOCKET HEAD CAP SCREW
- AUGER
- MECHANICAL SEAL
- O RING
- BEARING—LOWER
- COUPLING—SPLINE
- GEAR MOTOR

Figure 27–8   An exploded view of most of the mechanical parts of an ice flaker. *Courtesy Hoshizaki America, Inc.*

Figure 27–9   A solid-state control board with light-emitting diode (LED) status indicator lights used for self-diagnostics. *Courtesy Scotsman Ice Systems*

a shorter gear life will be the outcome. Some manufacturers use gears made of a fibrous material designed to strip or break when the motor and gearbox are put under heavy loads.

It is important that the auger does not wobble within the freezing cylinder or evaporator. Sleeve-type alignment bearings made from a slippery poly material with a graphite base keep the auger from wobbling on some machines. Others use grease-lubricated steel roller bearings. Grease can leak out, causing overheating of the bearings. This usually causes auger wobble. Auger wobble causes scraping of the auger's cutting edges with the freezing cylinder's wall. Excessive wear on all surfaces, and metal shavings in the ice will result.

Manufacturers of ice flakers have thermal overloads in the motor windings or a manual-reset current-type overload in the control scheme. In fact, most modern ice machines have solid-state control boards with status indicators acting as self-diagnostic tools for service technicians. There are usually light-emitting diodes (LEDs) or indicator lights to inform the service technician whether the ice flaker is operating normally, **Figure 27–9.** Indicator lights may provide the status of the auger drive motor, auger relay, compressor relay, low- or high-pressure control, reservoir level, bin level, starting mode, condenser temperature, discharge line temperature, and more. The type of control board and its control scheme varies from model to model. When the drive motor is stressed from mineral buildup on the evaporator freezing cylinder or lack of lubrication, the extra motor torque causes higher amperage draws, which create excessive heat. The overload devices will shut the unit down, and a manual reset will often be needed. An annual cleaning with an approved ice machine cleaner along with a bearing inspection will keep the ice flake machine running longer and quieter. Remember, the buildup of mineral scale on the freezing cylinder's surface, caused by poor water quality or lack of a preventive maintenance cleaning schedule, leads to most of an ice flaker's service problems.

Water Fill System

A dual float switch and a water control valve, **Figure 27–10(A),** is one manufacturer's method to control the amount of water coming into an ice flaker's water reservoir. Another manufacturer uses a conductivity probe or water sensor located in the ice machine's water reservoir, **Figure 27–10(B).** This probe

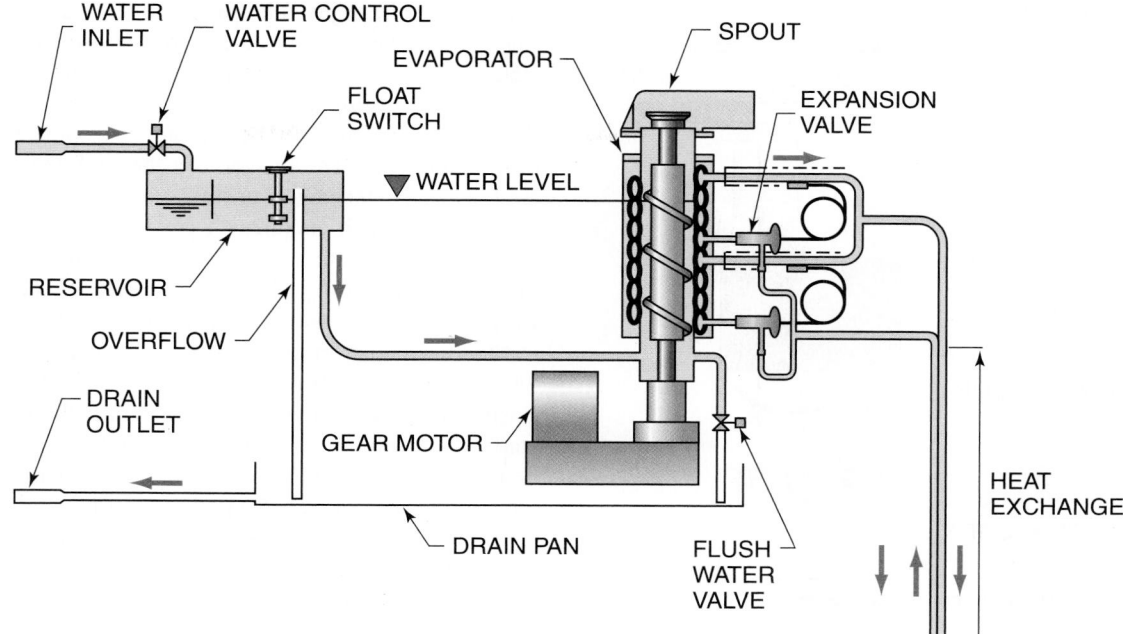

(A)

RESERVOIR

Water Sensor

- Conductivity probe located in the ice machine's water reservoir.
- Connected to the AutoSentry control board.
- Senses an electrical path from ground to the tip of the sensor.

(B)

Figure 27–10 **(A)** The water system diagram of an ice flake machine. **(B)** A conductivity probe in the water reservoir. **(A)** *Courtesy Hoshizaki America, Inc.* **(B)** *Courtesy Scotsman Ice Systems*

senses an electrical path from ground to the tip of the sensor. It is connected to an electronic control board. The sensor eliminates low- or no-water conditions and cannot be affected by adverse water conditions.

When using a dual float switch, the water inlet control valve is solenoid operated and energized through the dual float switch, **Figure 27–11.** The solid-state control board, **Figure 27–9,** monitors both of these components. The reservoir feeds water by gravity to the evaporator freezing cylinder.

The operation of the dual float switch maintains the proper level of water in the reservoir and also provides for an ice machine shutdown in case of a low-water-level situation.

The dual float switch is usually made of two reed switches and two floats. The reed switches are opened and closed by magnets inside the floats. As ice is made, the reservoir's water level drops, and the top float opens a set of latching relay contacts. Now the bottom float has control of the water control relay and the water inlet control solenoid. As the

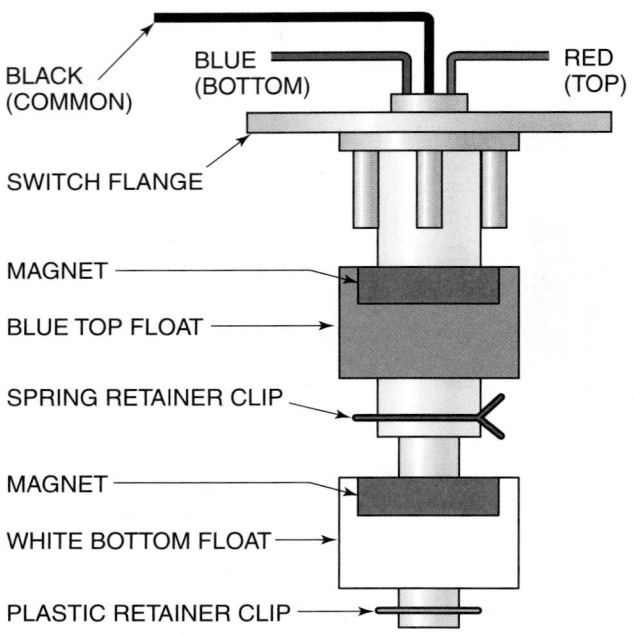

BLACK (COMMON)
BLUE (BOTTOM)
RED (TOP)

SWITCH FLANGE

MAGNET

BLUE TOP FLOAT

SPRING RETAINER CLIP

MAGNET

WHITE BOTTOM FLOAT

PLASTIC RETAINER CLIP

Figure 27–11 A dual float switch that controls reservoir water level and low-water shutdown timing. *Courtesy Hoshizaki America, Inc.*

reservoir's water level continues to drop, another set of contacts open by the action of the bottom float switch. This action deenergizes the water control relay, which also closes a circuit to the water inlet valve solenoid. At the same time, the control board energizes a low-water shutdown timer. The reservoir is now filled, and the two float switches swap jobs. The bottom switch becomes a latching relay and the top switch reenergizes the water control relay. As the top float rises, the low-water timer is shut off and the water inlet control solenoid valve is deenergized. This type of arrangement keeps a proper amount of water in the reservoir, prevents short cycling of the water inlet solenoid valve, and provides for low-water safety shutdown. **Figure 27–12** illustrates a wiring diagram of a modern ice flake machine and shows the control water valve, water control relay, control timer, and water inlet solenoid valves.

Water levels can also be sensed and controlled by a probe located in the water reservoir. The probe senses water level and relays the message to an electric water valve. The probe should be removed from the reservoir and cleaned at least twice a year.

Flush Cycles

🌐 Many ice flaker manufacturers provide for an automatic water flush cycle. When water freezes, minerals are left behind. As mentioned earlier, these minerals can build up in the evaporator freezing cylinder and cause maintenance problems and poor ice quality. A way to get rid of this unwanted buildup of minerals is to flush out the water system on a periodic basis. 🌐 One manufacturer's control scheme shuts down the refrigeration cycle for 20 min every 12 hours to allow the water system to drain. These periodic flushing periods may be adjustable for the time of day to initiate a flush cycle. The control board energizes a solenoid-operated water flush valve, which opens a drain on the bottom of the evaporator freezing cylinder, **Figure 27–10(A)**. This action drains the reservoir, freezing cylinder, and connecting piping of all impurities. The reservoir is then refilled with clean, fresh water. A mechanical or solid-state timer can initiate the flush cycle, which energizes a gravity flow flush solenoid valve. Manual flushing can also be accomplished for service purposes. The wiring diagram in **Figure 27–12** shows the flush timer, flush switch, and flush water solenoid valve. A service technician must understand the sequence of events when servicing ice flakers. It is important to remember the flush timer when diagnosing a problem if the unit will not start. Manufacturers often provide a cam wheel, which can be turned manually to advance the flush timer. This allows the unit to start if it was in a periodic flush mode.

Some models of ice flakers drain the entire water system each time the unit cycles off. This off-cycle draining can be done without the expense of a timer motor. However, when running time is heavy, flushing is not accomplished as often, and trace amounts of minerals may build up in the water system.

Bin Controls

Once the ice-holding bin is full of ice, some sort of automatic control must terminate the operation of the ice flake machine. If this is not done, flaked ice may pile up in the bin and ice chute and overflow. This overflowing ice may spill into the ice flake machine's mechanical parts and eventually spill onto the floor. Water from the melting ice may also cause damage to mechanical parts in the ice machine itself or to nearby flooring.

In the past, there have been many ways to control this overflowing ice condition. The following are some of the methods used for proper bin control:

■ Mechanical flapper-type device controlling a microswitch
■ Infrared electric eye
■ Sonar
■ Thermostats
■ Liquid-filled bulbs

Mechanical flapper-type devices are usually located in the spout. They simply sense backed-up ice by pressure and control a microswitch. The action of the microswitch shuts down the operation of the ice flake machine. When in their normal position, they supply an electronic control with a known amount of resistance. When the paddle is pushed out of its normal position by ice, the resistance value to the electronic control changes and the unit shuts down in a set amount of time.

Infrared electric eyes sense the buildup of ice by infrared technology. This control allows the ice machine to shut down before the ice chokes the spout opening. Usually two infrared sensors are mounted on the outside of the ice chute's base, and they are powered and controlled by a solid-state controller. The sensors must be kept clean to see one another. During normal operation, ice passes between two ice-level sensors

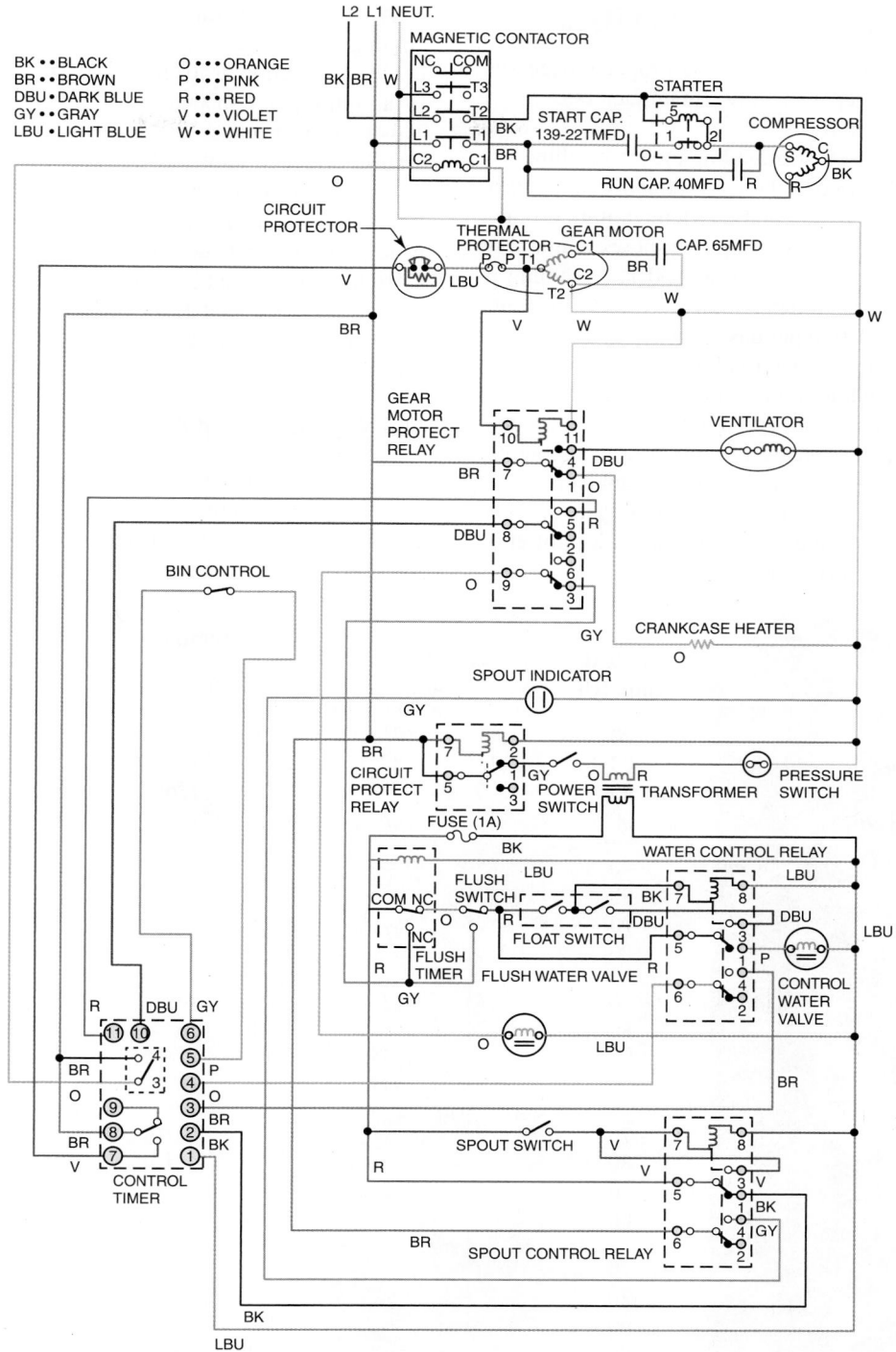

Figure 27–12 The wiring diagram of an ice flake machine. *Courtesy Hoshizaki America, Inc.*

and interrupts the infrared beam just momentarily, keeping the machine running. However, if ice builds up and blocks the path between two infrared sensors for 6 sec or longer, the machine shuts down. If the path between the infrared sensors remains clear for more than 10 sec, the ice machine will restart. In most cases, this type of bin control must be removed at least twice a year and cleaned with a soft cloth.

Sonar controls use sonic or sound waves and vibrations, which are reflected back to the control from the ice. One

disadvantage of infrared and sonic controls is condensate forming on the lenses, making these controls less accurate.

Thermostats often have liquid-filled sensing bulbs placed somewhere in the ice chute or simply at the top of the ice bin. As the ice touches the sensing bulb of the thermostat, the ice machine terminates its freezing cycle, usually within 10 sec. Most bin thermostats are adjustable and can be checked by placing a handful of ice on the remote bulb and timing the shutdown.

Ice Production and Performance Data

When flake ice machines operate at the proper refrigerant charge, ice production depends on two main factors:

■ Inlet water temperature
■ Temperature (ambient) surrounding the ice machine

Notice in **Figure 27–13** that the ice production in pounds per 24 hours decreases as the ambient temperature around the ice machine gets warmer. The top row shows four ambient temperatures ranging from 70°F to 100°F. Assuming a constant 50°F entering water temperature, at a 70°F ambient, the ice production is 2010 lb per day. However, as the ambient rises to 100°F, the ice production falls to 1570 lb per day. Italicized figures are in kilograms per day.

This loss of ice production with a hotter ambient around the ice machine happens because of more heat gain coming into the ice. For air-cooled machines, a hotter ambient causes a higher head pressure for the refrigeration system. This causes an increase in compression ratios and a loss of efficiency, and thus less ice production.

Also notice in **Figure 27–13** that as the water temperature coming into the ice machine increases from 50°F to 90°F, the ice production drops from 2010 to 1895 lb per day. This is assuming a constant 70°F ambient air temperature. This happens because it now takes more energy and refrigeration effect to cool the warmer water down to freezing temperatures. The figure also includes electric consumption, water consumption, evaporator outlet temperature, head pressure, and suction pressure as they depend on the surrounding ambient air temperature and inlet water temperature. Notice that head pressure and suction pressure both increase as the ambient temperature around the ice machine increases.

Ice Flake Size Adjustment

Some ice flaker manufacturers provide a means for the service technician to adjust the size or shape of the flake ice. This feature provides a more diverse ice flake machine to meet the needs of many customers. Flake ice is used in packing

	Ambient Temp. (F)	Water Temp (F)					
		50		70		90	
Approximate	70	2010	*(912)*	1950	*(845)*	1895	*(860)*
Ice Production per 24 hr.	80	1845	*(837)*	1795	*(814)*	1750	*(794)*
	90	1700	*(771)*	1695	*(769)*	1610	*(730)*
lbs./day (kg/day)	100	1570	*(712)*	1525	*(692)*	1410	*(640)*
Approximate Electric	70	2850	—	2850	—	2855	—
Consumption	80	2855	—	2860	—	2860	—
	90	2865	—	2865	—	2875	—
watts	100	2890	—	2890	—	2910	—
Approximate Water	70	241	*(912)*	234	*(845)*	228	*(860)*
Consumption per 24 hr.	80	222	*(837)*	216	*(814)*	210	*794)*
	90	204	*(771)*	203	*(769)*	194	*(730)*
gal./day (l/day)	100	188	*(712)*	183	*(692)*	169	*(640)*
Evaporator Outlet Temp.	70	14	*(−10)*	14	*(−10)*	14	*(−10)*
°F (°C)	80	14	*(−10)*	14	*(−10)*	14	*(−10)*
	90	14	*(−10)*	14	*(−10)*	16	*(−9)*
	100	16	*(−9)*	16	*(−9)*	16	*(−9)*
Head Pressure	70	219	*(15.4)*	219	*(15.4)*	219	*(15.4)*
	80	230	*(16.2)*	230	*(16.2)*	230	*(16.2)*
	90	241	*(16.9)*	241	*(16.9)*	241	*(16.9)*
psig (kg/sq.cmG)	100	271	*(19.0)*	271	*(19.0)*	271	*(19.0)*
Suction Pressure	70	25	*(1.8)*	25	*(1.8)*	25	*(1.8)*
psig (kg/sq.cmG)	80	26	*(1.8)*	26	*(1.8)*	25	*(1.8)*
	90	27	*(1.9)*	27	*(1.9)*	27	*(1.9)*
	100	29	*(2.0)*	29	*(2.0)*	29	*(2.0)*
Condenser Volume		214 in³					
Heat of Rejection from Condenser		16890 Btu/h (AT 90°F /WT 70°F)					
Heat of Rejection from Compressor		2860 Btu/h (AT 90°F /WT 70°F)					

Note: The data without *marks should be used for reference.

Figure 27–13   Ice production and other system performance data for an ice flake machine. *Courtesy Hoshizaki America, Inc.*

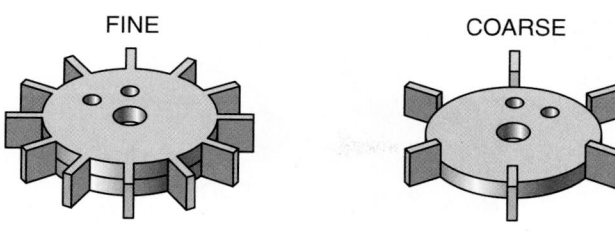

Figure 27–14   Fine and coarse ice cutting heads for a flake ice machine. *Courtesy Hoshizaki America, Inc.*

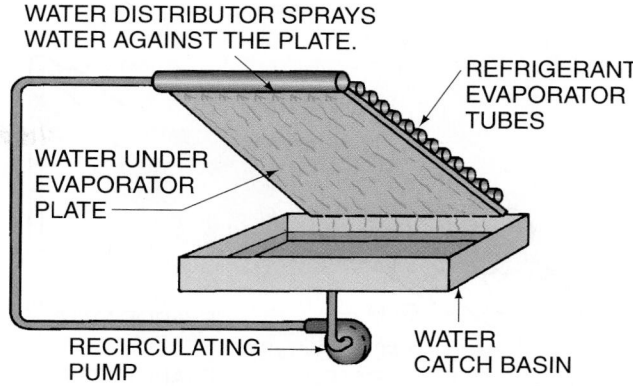

Figure 27–15   The water flows under the evaporator plate in this ice maker. This evaporator must be clean and free of mineral deposits, or the water will not follow the plate. In this case it will fall into the ice bin below and melt ice already made. Then the makeup water will have to be refrigerated, and the capacity to make ice will be reduced.

produce, fish, or poultry and in filling a salad bar or display case because the ice flakes pack tightly and keep the products cool and fresh. One method of adjustment is to rotate an ice cutter at the top of the auger. Since the pressure applied to the ice comes from the last couple of inches of the auger flights on some machines, aligning a small, medium, or large cutter opening with the pressure points on the flights of the auger will adjust ice size. Small, medium, or large flakes can come from one ice machine.

Another method is to change the cutting heads located on the upper side of the auger, **Figure 27–14.** One cutter head will produce fine flakes and another will produce coarse flakes.

The extrusion head can also be changed on some ice flakers, which allows for both cubelet and flake ice from one ice machine. The size of the cubelets or nuggets can be increased or decreased. Cubelet or nugget ice is about the size of a small marble. When using the extrusion head, the flake ice is extruded through small openings. This process squeezes out additional water, which makes the ice more dense and compact. Cubelet or nugget ice lasts much longer than flake ice when submerged in a drink. The health care industry uses cubelets or nuggets since they are much easier to chew than larger cubes, and sick and elderly people are less likely to choke when chewing them.

27.3   MAKING CUBE ICE

Several types of cube ice are made in package ice makers using several methods. Cube ice should be clear, not cloudy. Clear ice is usually preferred for beverage cooling. Ice with a high mineral content or that is aerated (has minute air bubbles inside) will appear cloudy and is not as desirable to some people because of its appearance. Some of the design features of the cube ice maker result in ice that is very clear.

Flat Ice Cut to Cubes

Flat cube ice is made on a flat evaporator to a predetermined thickness. When harvest occurs, the ice sheet is slightly defrosted and melted loose from the evaporator. The ice sheet then falls or slides to a grid of cutter wires. These wires have a very low voltage, approximately 5 V depending on the manufacturer, and will provide enough heat to cut the ice into squares or triangular shapes. The ice can be 1/4 in. to 1/2 in. thick, depending on the length of time ice is made before defrost occurs.

The evaporator may be designed for water to flow on top of the plate or under the plate by means of a water distribution

manifold. When the water flows under the plate, the surface tension of the water tends to hold the water to the plate as it flows under it, **Figure 27–15.** This evaporator must be kept clean or the water will not cling to the plate and fall into the ice bin below melting ice.

Some plate-type ice machines are designed so that the water flow is over the plate. These use a water distribution system designed for this purpose. Some of these use a small motor and cam to determine the thickness of the ice and when defrost should occur, **Figure 27–16.** With the advent of the microprocessor and more sophisticated analog and digital sensing devices to act as inputs to this microprocessor, different ways to sense ice thickness and initiate defrost have been devised. Each ice machine manufacturer has its own control scheme for making and harvesting ice.

Water level is maintained in the water reservoir for recirculation by means of a float.

Cube Ice

Many methods are used to make cube ice. Some common ones are discussed in this text. Cube ice is made by flooding water over an evaporator with cups shaped like the cube desired. When the ice reaches the desired thickness, defrost or harvest occurs. The evaporator may be horizontal or vertical.

When the evaporator is horizontal, water is sprayed up into the cups where the desired cube is formed. Some of the water falls to the catch basin and is recirculated. When defrost occurs, the water is shut off and defrost allows the cubes to fall to the first level of the catch basin. The water wand with the spray heads wipes the cubes to a chute leading to the catch basin, **Figure 27–17.**

Eyeglass-Shaped Ice Cubes

Vertical evaporators have ice cups designed to cause the ice to fall out of the cups during defrost. Water flows over these evaporators by means of gravity, like a waterfall. The evaporator piping is on the back of the evaporator plate, **Figure 27–18.** The cubes are caught during defrost in the ice bin.

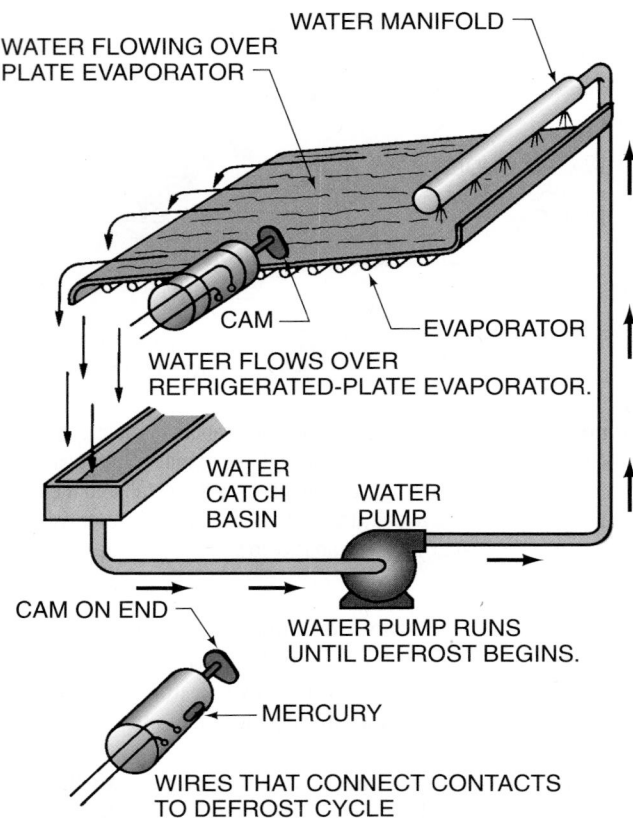

Figure 27–16 Sensors that control ice thickness in this machine. The ice thickness switch rotates when ice begins to form on the evaporator plate. The cam touches the ice during rotation (approximately 1 rpm). The switch has mercury contacts in the rear. As the cam touches the ice, it causes the mercury to roll to the back. The contacts are made, and defrost begins. Although they are no longer manufactured, a few may still be in service today.

Crescent-Shaped Ice Cubes

Making a crescent-shaped ice cube involves running water over an evaporator plate in which the copper evaporator tubing is soldered onto the plates at regular intervals. **Figure 27–19** shows two stainless-steel evaporator plates and connected tubing for making crescent-shaped ice cubes. Stainless steel is often used on evaporators because it is durable, it is very sanitary, and it resists corrosion. It is also much less porous than copper, brass, or nickel. The copper serpentine tubes soldered to the evaporator are oval to increase heat transfer because they have more surface area in contact with the stainless plates.

During the freeze cycle, water from the reservoir is circulated to the outside freezing surface. No other water is allowed to enter during the freezing cycle. Once the water reaches 32°F, ice starts to form on the plates. The coldest point on the evaporator is where the copper serpentine coil contacts the evaporator plate. This is where ice starts to form first and continues to grow outward in a crescent shape. As the ice grows, the reservoir water level falls because some of the water is freezing into ice. Once the crescent cubes are

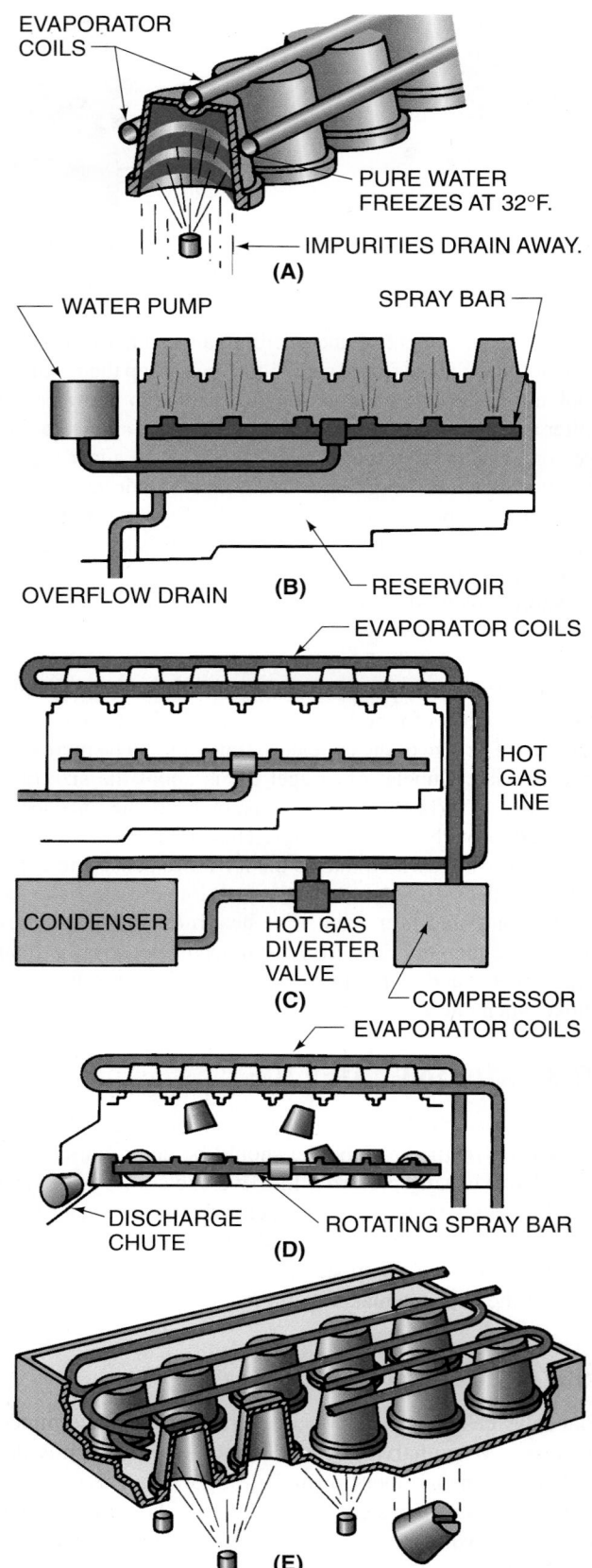

Figure 27–17 Ice cubes and the evaporator on which they are made.

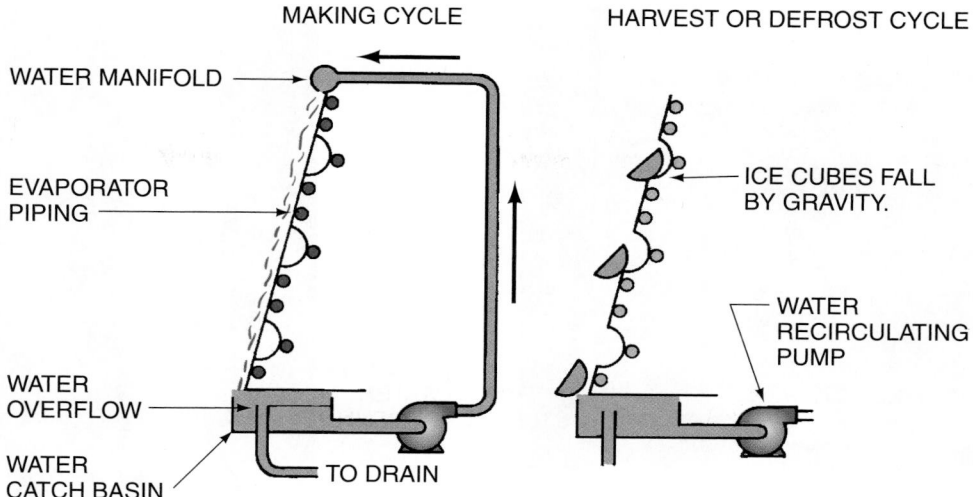

MAKING CYCLE   HARVEST OR DEFROST CYCLE

WATER MANIFOLD

EVAPORATOR
PIPING

ICE CUBES FALL
BY GRAVITY.

WATER
RECIRCULATING
PUMP

WATER
OVERFLOW

WATER
CATCH BASIN

TO DRAIN

Figure 27–18   A vertical evaporator. Water flows over the cups during the make cycle. During harvest, cubes fall out of the cups to the bin below.

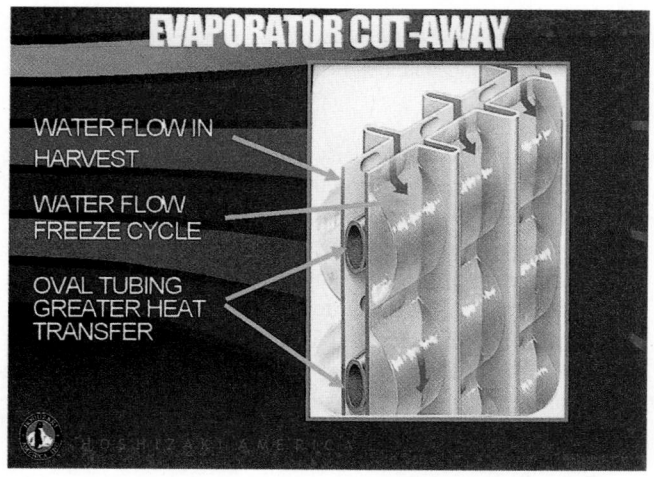

EVAPORATOR CUT-AWAY

WATER FLOW IN
HARVEST

WATER FLOW
FREEZE CYCLE

OVAL TUBING
GREATER HEAT
TRANSFER

Figure 27–19   An evaporator that produces crescent-shaped ice cubes. *Courtesy Hoshizaki America, Inc.*

Cell-Type Ice Cubes

Cell-type ice cube machines usually have vertical evaporator plates with water flowing over the individual cells, **Figure 27–20.** A plastic water curtain covers the evaporator to prevent splashing, **Figure 27–24.** A serpentine evaporator coil is soldered to the back of the evaporator plate, **Figure 27–21.** The water is distributed over the cells by a water pump through a water distributor tube shown at the top of **Figure 27–20.** The water distribution tube has evenly spaced holes for even water distribution. The water distribution tube is usually removable for cleaning and repair purposes, **Figure 27–22.** The water distribution tube carrying water from the water pump has an adjustable restrictor to let the service technician adjust the water flow rate through the tube, **Figure 27–23.** If the water flow is too slow, minerals in the water freeze, causing cloudy ice. Ideally, only water should be allowed to freeze on the evaporator plate. This

formed, a float switch in the water sump will open contacts and the harvest cycle will begin.

Hot superheated gas from the discharge of the compressor is circulated through the serpentine coil during defrost. As the heat is transferred to the stainless-steel evaporator, the cubes begin to melt away from the evaporator plate. There will be a film of water between the cube and the plate, which will cause a capillary action and attract the cubes to the plate as they slide down the plate by gravity. As the cubes slide down the plate they will contact dimples, which break the capillary attraction. This will release the cubes from the plate, and the cubes will fall into the ice bin. Water is allowed to flow during the defrost cycle, but it flows on the opposite evaporator plate. This allows the water to absorb heat from the copper coils where hot gas is being circulated and to transfer this heat to the freezing surface. This is referred to as water-assisted defrost.

Figure 27–20   The vertical evaporator of a cell-type ice machine. *Courtesy Ferris State University. Photo by John Tomczyk*

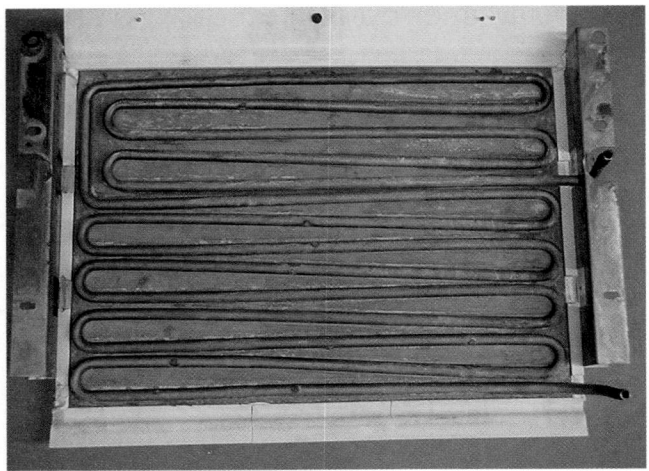

Figure 27–21   A serpentine coil soldered to the back of a cell-type ice cube evaporator. *Courtesy Ferris State University. Photo by John Tomczyk*

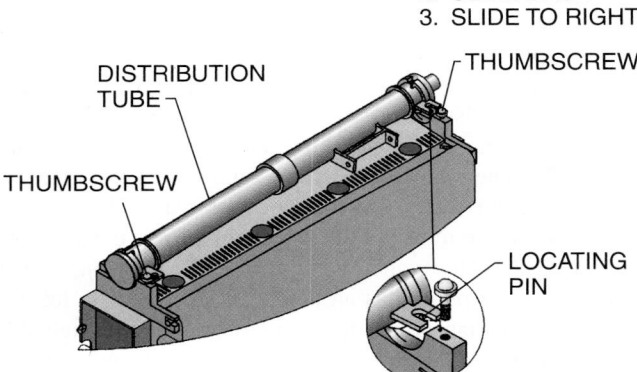

Figure 27–22   A water distribution tube that is removable for cleaning and repairs. *Courtesy Manitowoc Ice, Inc.*

Figure 27–23   A restrictor in the water line from the water pump allows for varying the water flow rate. *Courtesy Ferris State University. Photo by John Tomczyk*

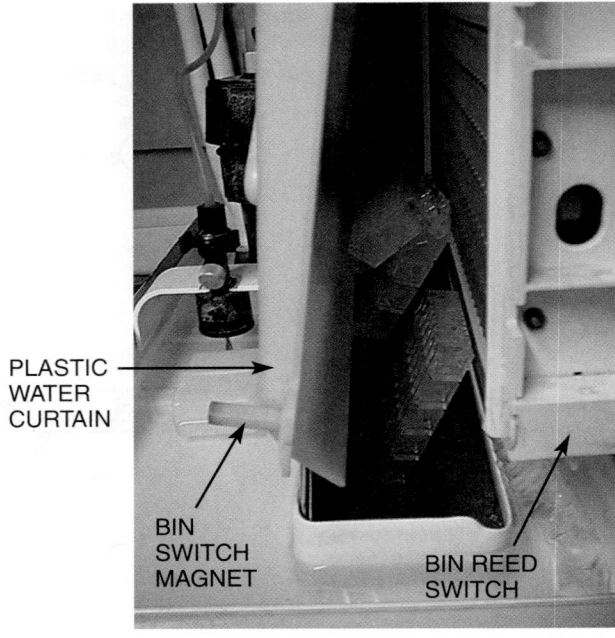

Figure 27–24   An ice slab partially breaking up as it collides with other ice in the storage bin. *Courtesy Ferris State University. Photo by John Tomczyk*

requires the right water velocity over the plates, which can be adjusted with the restrictor on the water tube. Ice will gradually form on each individual cell until it bridges with the neighboring cells. Ice bridging forms one large ice slab the size of the evaporator itself. Once the ice slab has the proper bridge thickness, a harvest or defrost cycle is initiated and the ice slab falls into the ice storage bin. Ice harvests are covered in detail later in this unit.

The proper bridge thickness is important for the individual ice cubes to be easily separated when scooped out of the ice storage bin. When a server taps the ice slab in the storage bin with an ice scoop, it should easily break up into individual cell cubes. Most of the time, when the ice slab drops off the evaporator during a harvest period, it partially breaks up as it collides with other ice in the storage bin, **Figure 27–24**. However, as the ice storage bin starts to fill, the distance the ice slab has to fall is reduced. If the bridge thickness is too great, the ice slab will not easily break up into individual cubes. A bridge thickness of 1/8 in. is not uncommon. The ice bridge thickness is adjustable through an ice thickness probe's adjustment screw, **Figure 27–25(A and B)**. **Figure 27–20** also shows the ice thickness probe and its location near the cell-type evaporator.

The centrifugal water pump, which gets its water from a water reservoir, is generally a fractional horsepower, alternating current motor, **Figure 27–26(A)**. The water pump may deliver water to more than one evaporator, **Figure 27–27**. The water reservoir's water level can be controlled by a float mechanism, **Figure 27–28**. The float controls the amount of incoming fresh water to the water reservoir. The float control is adjustable and also controls the level of water in the reservoir. As the water freezes on the evaporator's cells, the water reservoir level starts to drop. Instantly, the float or water level probe senses this drop in water level and slightly opens a valve to let more fresh water into the reservoir. This action

(A)

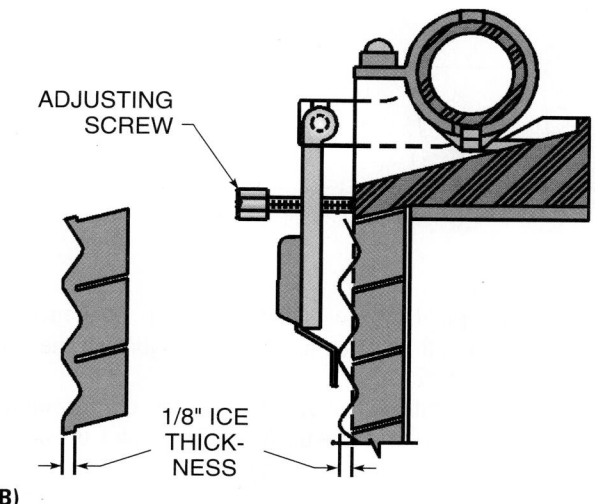

ADJUSTING
SCREW

1/8" ICE
THICK-
NESS

(B)

Figure 27–25   (A) An ice thickness probe showing a directional for changing ice bridge thickness. **(B)** An ice thickness probe and ice slab showing a 1/8-in.-thick ice bridge. **(A)** *Courtesy Ferris State University. Photo by John Tomczyk.* **(B)** *Courtesy Manitowoc Ice, Inc.*

keeps just the right level of water in the reservoir for the water pump to use. Too low a water level in the water reservoir allows the water pump to suck a mixture of water and air, which has a detrimental effect on ice quality. The water that does not freeze on the evaporator's cells will trickle into a water trough and eventually make its way into the water reservoir by gravity.

A water level probe, **Figure 27–26(A),** can also control water levels in the reservoir. It is usually nonadjustable and controls an electric, solenoid-operated water fill valve. One manufacturer uses a water level sensor that provides an electronic controller with two important pieces of information, **Figure 27–26(B).** As the machine freezes ice, the water level sensor's top photoelectric eye receives infrared light through an open control slot in the float system. When the water falls far enough, the float stem's slot is too low to allow the light to pass, and the stem breaks the top electric eye beam. This signals the controller to initiate a harvest cycle. Also, when water flows into the reservoir, the float lifts the float stem.

WATER
LEVEL
PROBE

(A)

WATER LEVEL SENSOR

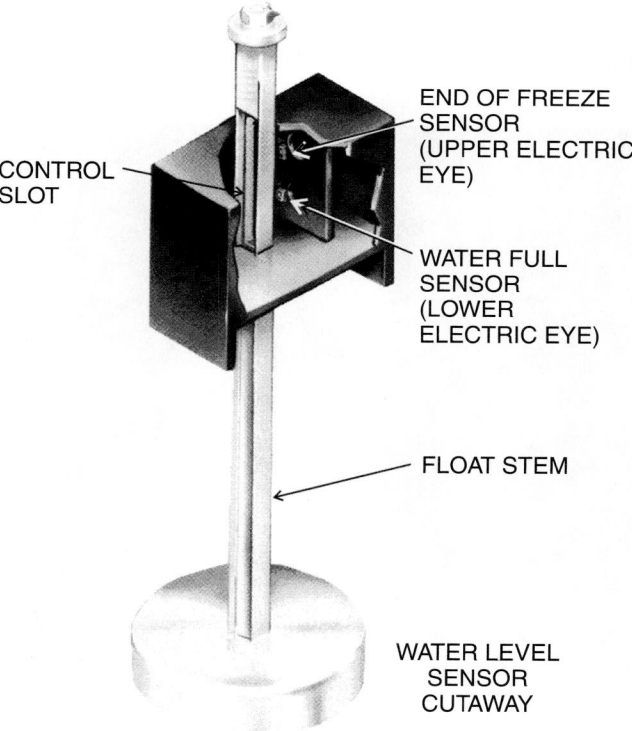

CONTROL
SLOT

END OF FREEZE
SENSOR
(UPPER ELECTRIC
EYE)

WATER FULL
SENSOR
(LOWER
ELECTRIC EYE)

FLOAT STEM

WATER LEVEL
SENSOR
CUTAWAY

(B)

Figure 27–26   (A) A fractional horsepower, alternating current, centrifugal water pump. **(B)** A water level sensor to control cube size and initiate harvest through the use of infrared technology. **(A)** *Courtesy Ferris State University. Photo by John Tomczyk.* **(B)** *Courtesy Scotsman Ice Systems*

When the bottom electric eye beam is blocked by the solid plastic body of the float stem, the controller knows that the reservoir is full. Water will continue to flow for a predetermined amount of time to overfill and rinse the reservoir.

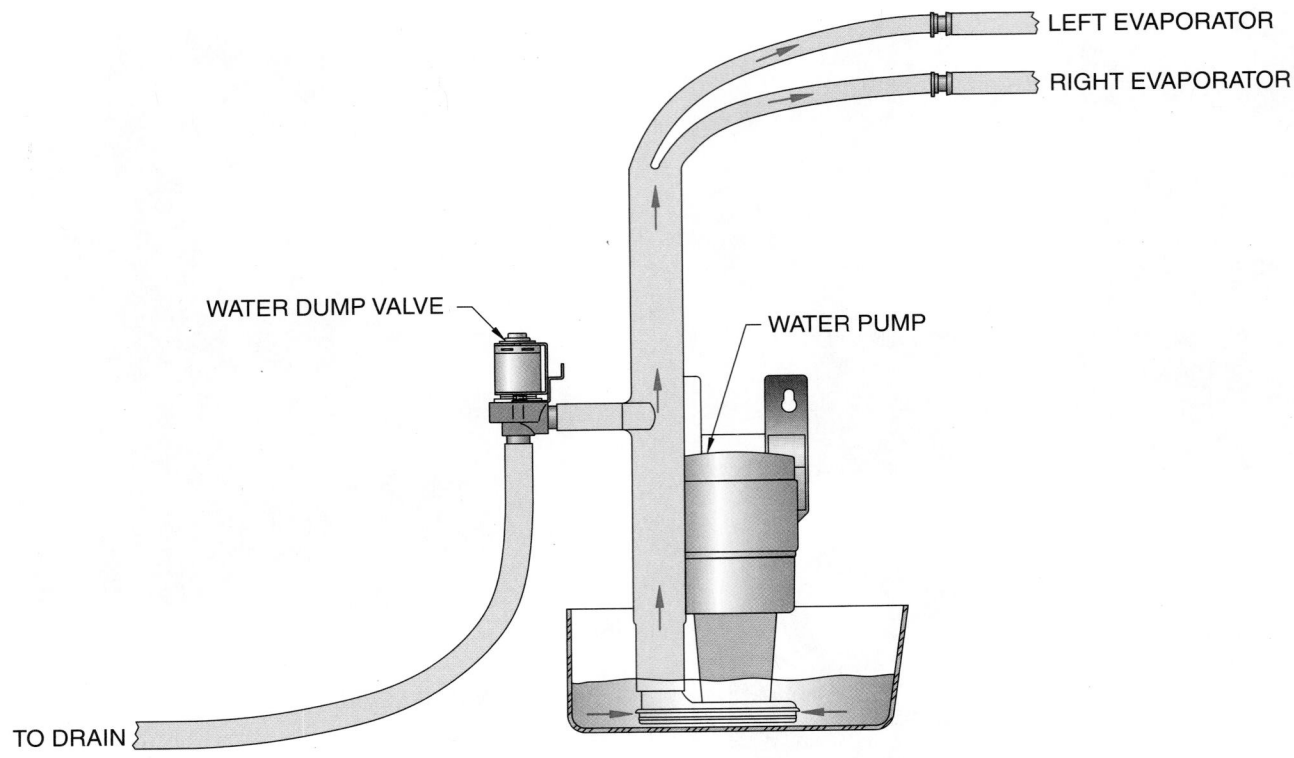

Figure 27–27 A water pump serving two evaporators. *Courtesy Manitowoc Ice, Inc.*

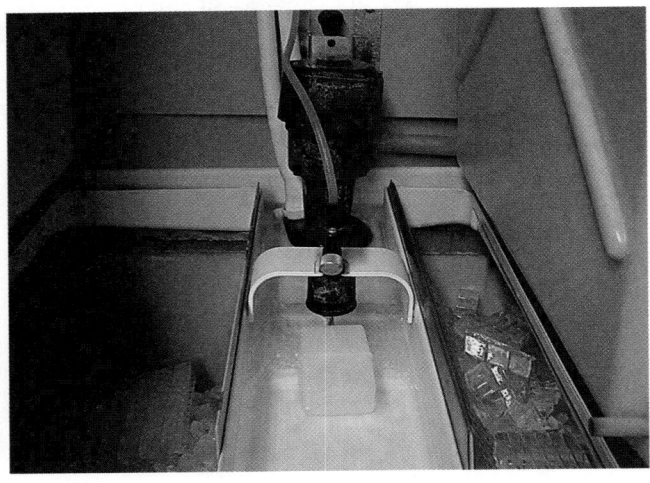

Figure 27–28 A water reservoir, float mechanism, incoming water line, and water pump. *Courtesy Ferris State University. Photo by John Tomczyk*

Sequence of Operation (Freeze and Harvest Cycle)

Figure 27–29(A) is a refrigeration system diagram of a cell-type ice maker that has two evaporators. **Figure 27–29(B)–(D)** show the water and refrigeration system, freeze cycle, and harvest cycle on one manufacturer's ice maker. **Figure 27–30** is a refrigeration system diagram of a single-evaporator system in the freeze cycle. Every manufacturer seems to have its own control schemes and sequence of operations. However, it would be impossible to cover all control schemes in this unit. A typical control scheme with its sequence of operation for a cell-type ice maker is described in the following paragraphs.

The refrigeration compressor comes on for about 30 sec before the water pump turns on. This allows the evaporator to be prechilled. Prechilling makes for a very cold evaporator when it finally experiences water flow and helps prevent slush from forming in the water sump. Slush may interfere with the water flow to the water pump and may also cause an ice dam.

The water pump comes on after about a 30-sec prechill. Water is forced through the water distribution tube and evenly distributed onto the cell-type evaporator where it begins to freeze. Either a float mechanism or an electric solenoid-operated water fill valve keeps the water in the reservoir at the proper level. Once the right thickness of ice has formed on the evaporator, water flowing over the ice contacts the ice thickness probe.

Water has to contact the ice thickness probe for at least 7 sec before an ice harvest is initiated. This keeps the harvest cycle from prematurely initiating due to splashing. Often, ice machines cannot begin harvest until their freeze cycles have been underway for a certain amount of time. A period of 6 to 7 min is not uncommon. This time period is referred to as a *freeze-lock time*. Freeze lock allows enough ice to form on the evaporator before a harvest is initiated and prevents an evaporator cookout or overheating from a premature harvest. Overheating can cause damage to the metallic plating on some evaporator surfaces.

The diagram in **Figure 27–31** shows a system during the harvest cycle. Once the ice thickness probe contacts water for

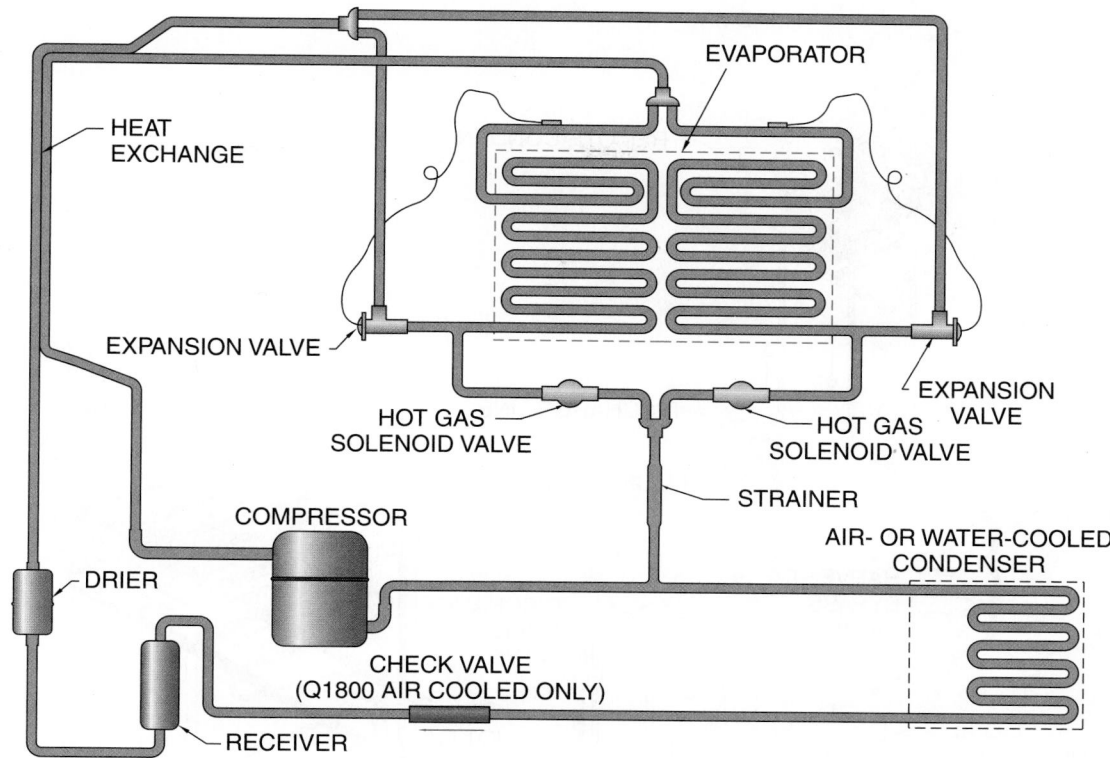

(A)

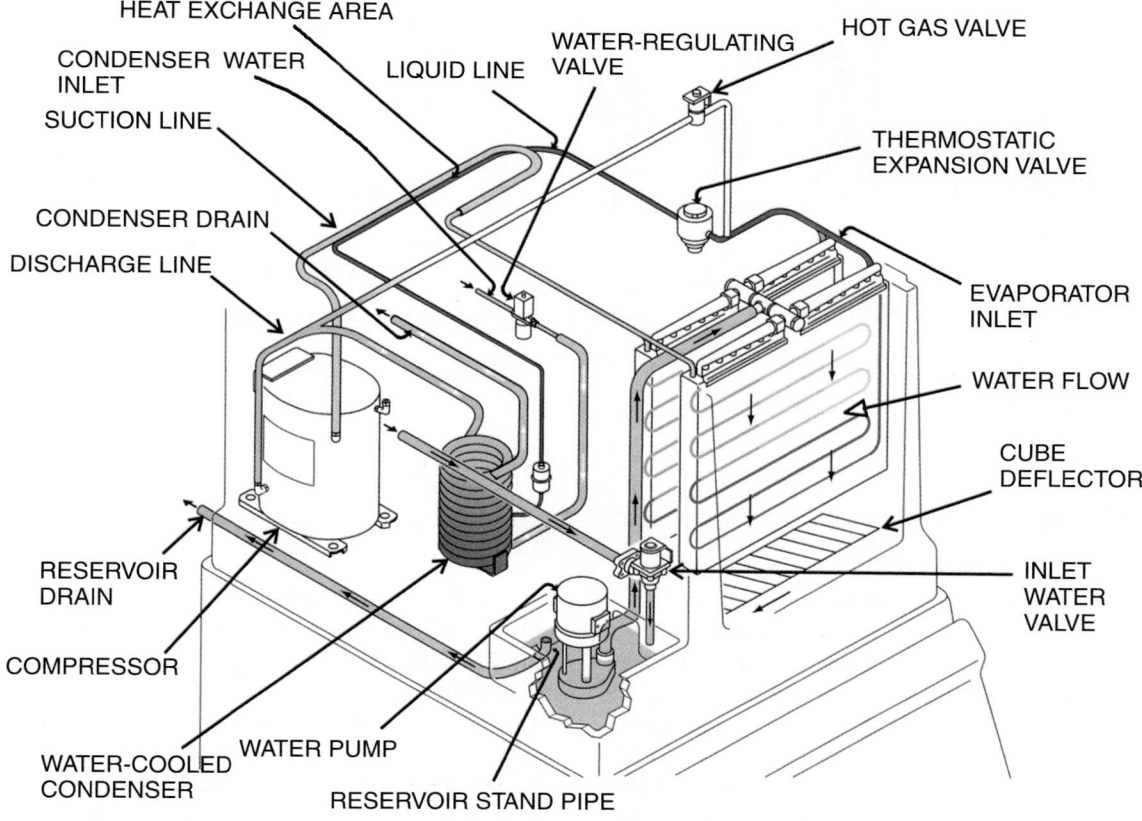

(B)

Figure 27–29 **(A)** A refrigeration system diagram of a cell-type ice maker with two evaporators. **(B)** The water and refrigeration system for a cell-type ice maker. **(A)** *Courtesy Manitowoc Ice, Inc.* **(B)–(D)** *Courtesy Scotsman Ice Systems*

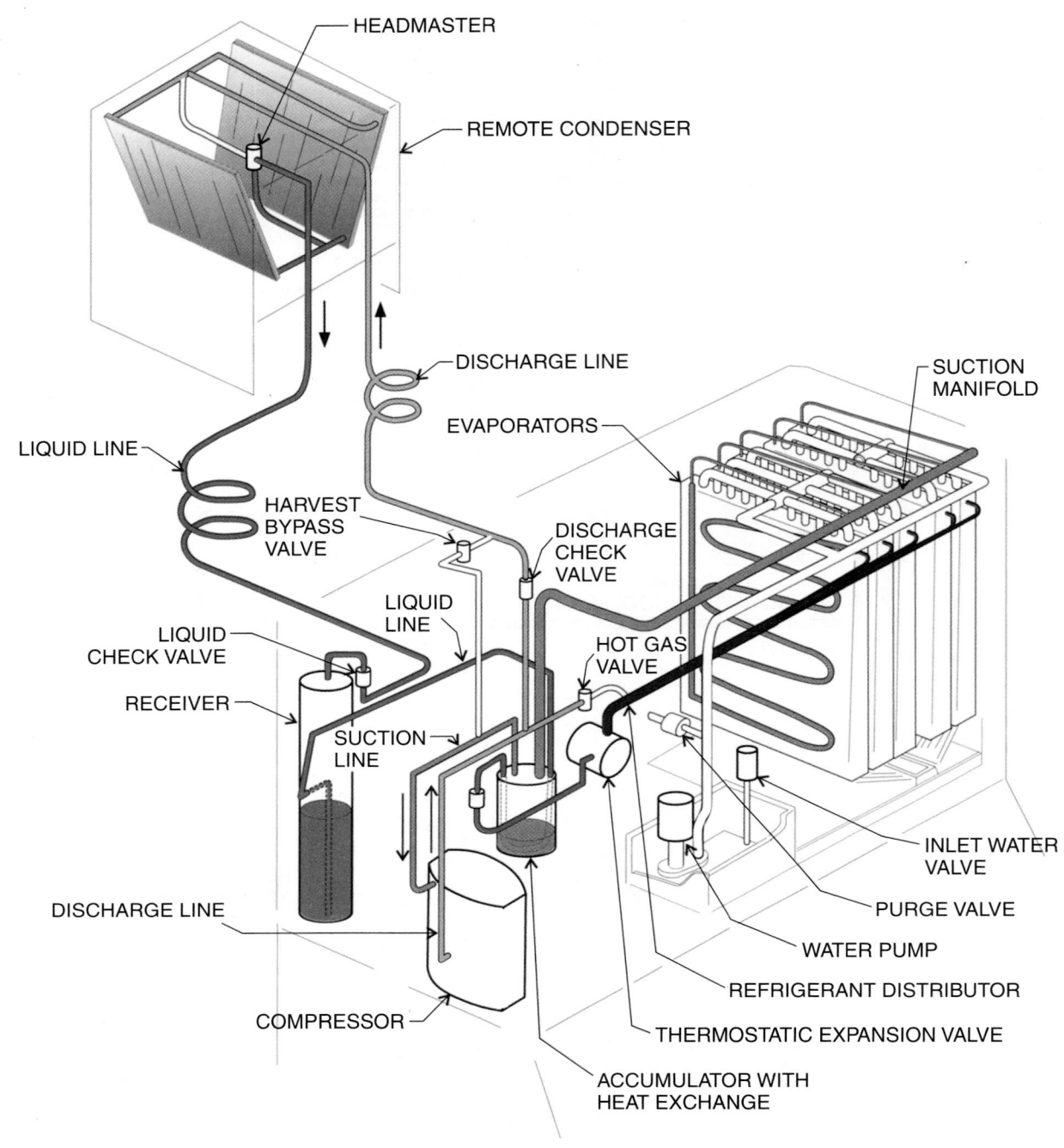

(C)

Figure 27–29 (*continued*)   **(C)** An ice machine during the freeze cycle.

more than 7 sec, the hot gas solenoid valve located on the compressor's discharge line is energized, **Figure 27–32(A).** The compressor continues to pump superheated gas through the evaporator. This hot gas warms the evaporator, causing the sheet of cubes to slide into the storage bin. One manufacturer uses a harvest bypass valve that is open for just a few seconds each harvest period, **Figure 27–32(B).** The purpose of this valve is to add the correct amount of refrigerant to the harvest circuit to promote a good release of ice. At the same time, the water pump continues to run and the electric,

solenoid-operated water dump valve, **Figure 27–33**, is energized for a specified amount of time, usually 45 sec. **Figure 27–34(A)** shows sump water being routed down a drain during the harvest cycle when the dump valve is energized. This action purges the water from the sump and gets rid of any residual minerals from the previous freeze cycle, which is very important for clear, hard ice. The water fill valve will energize for the last 15 sec of the 45-sec water sump purge time. This time may be adjustable on an electronic microprocessor-based unit. If a float mechanism is used,

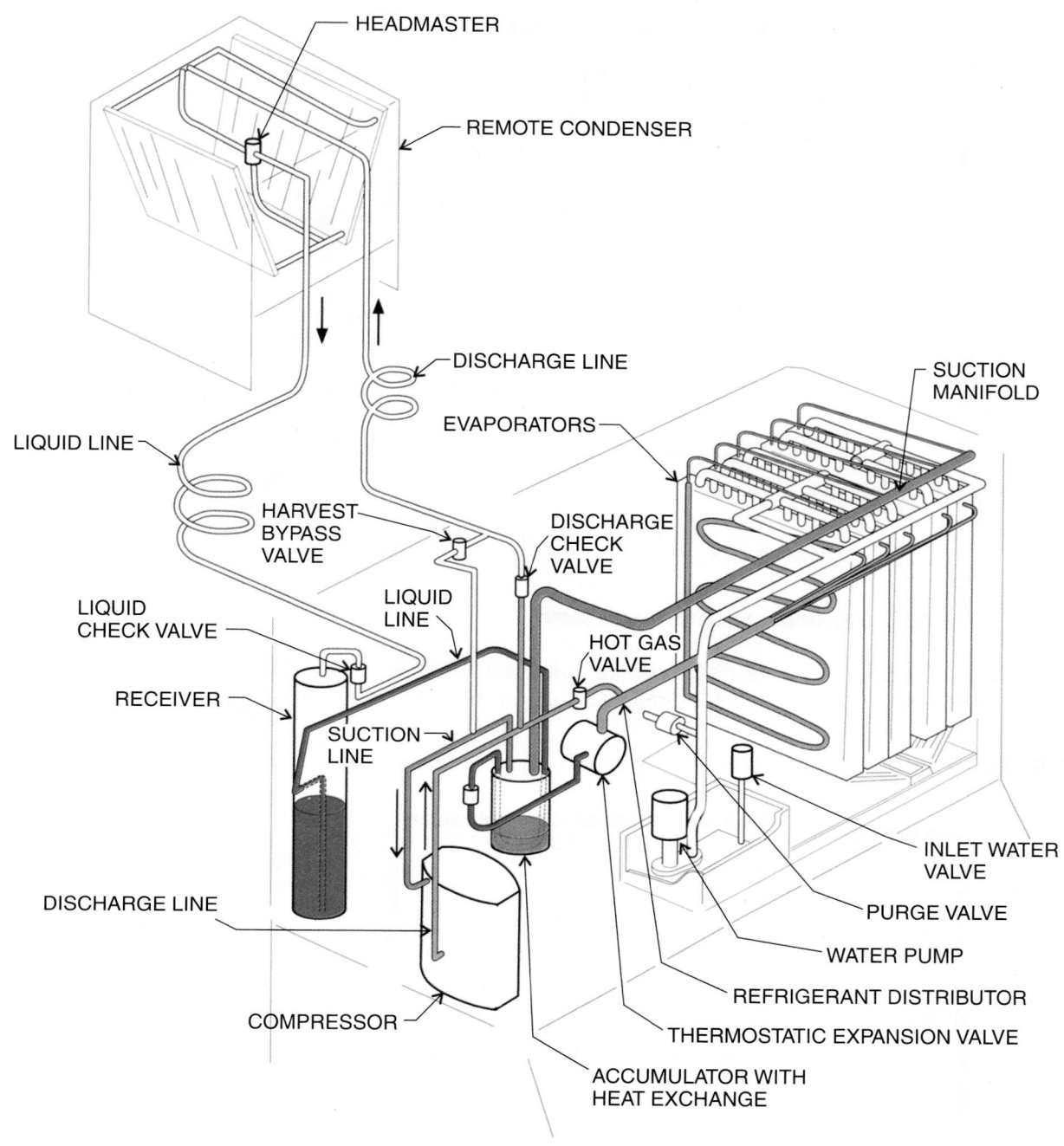

HEADMASTER

REMOTE CONDENSER

DISCHARGE LINE

SUCTION MANIFOLD

EVAPORATORS

LIQUID LINE

HARVEST BYPASS VALVE

DISCHARGE CHECK VALVE

LIQUID LINE

LIQUID CHECK VALVE

HOT GAS VALVE

RECEIVER

SUCTION LINE

DISCHARGE LINE

INLET WATER VALVE

PURGE VALVE

WATER PUMP

REFRIGERANT DISTRIBUTOR

THERMOSTATIC EXPANSION VALVE

COMPRESSOR

ACCUMULATOR WITH HEAT EXCHANGE

(D)

Figure 27–29 (*continued*)   **(D)** An ice machine during the harvest cycle.

the float valve will automatically fill the water sump to the proper level.

As the ice sheet falls off the evaporator, a water curtain swings out and opens a bin switch. This opening and closing of the bin switch ends the harvest cycle and initiates the next freeze cycle. The bin switch is a reed switch that is operated by a magnet. The magnet is attached to the water curtain, and the reed switch is attached to the evaporator mounting bracket, **Figure 27–24.** The bin switch has three functions: to terminate the harvest cycle, to return the

ice machine to the freeze cycle, or to cause an automatic shutoff.

If the ice storage bin is full at the end of a harvest cycle, the sheet of ice cubes will not let the water curtain close, **Figure 27–35.** If after a certain time period, usually 7 sec, the bin switch is still open, the ice machine will shut off automatically. It will stay off until the ice in the bin is removed and the water curtain swings back to close the bin switch. A minimum off period of 3 min is generally needed before the ice machine can automatically restart.

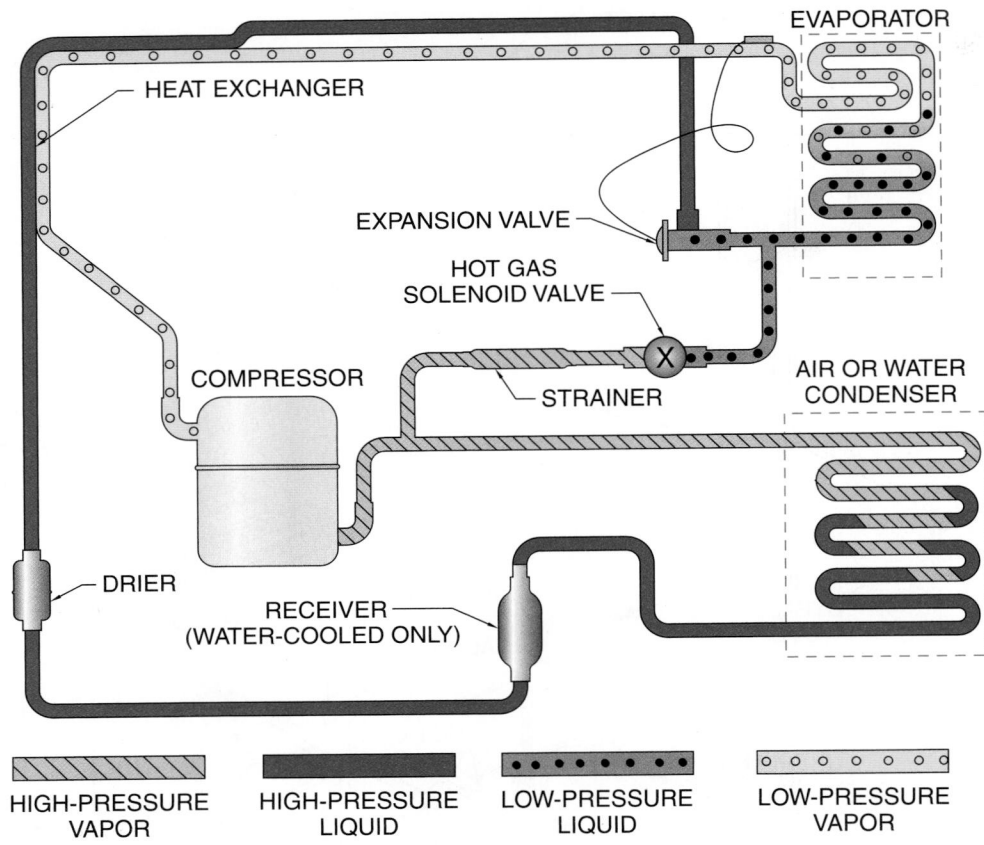

HIGH-PRESSURE VAPOR   HIGH-PRESSURE LIQUID   LOW-PRESSURE LIQUID   LOW-PRESSURE VAPOR

Figure 27–30   A refrigeration system diagram of a single-evaporator cell-type ice maker. *Courtesy Manitowoc Ice, Inc.*

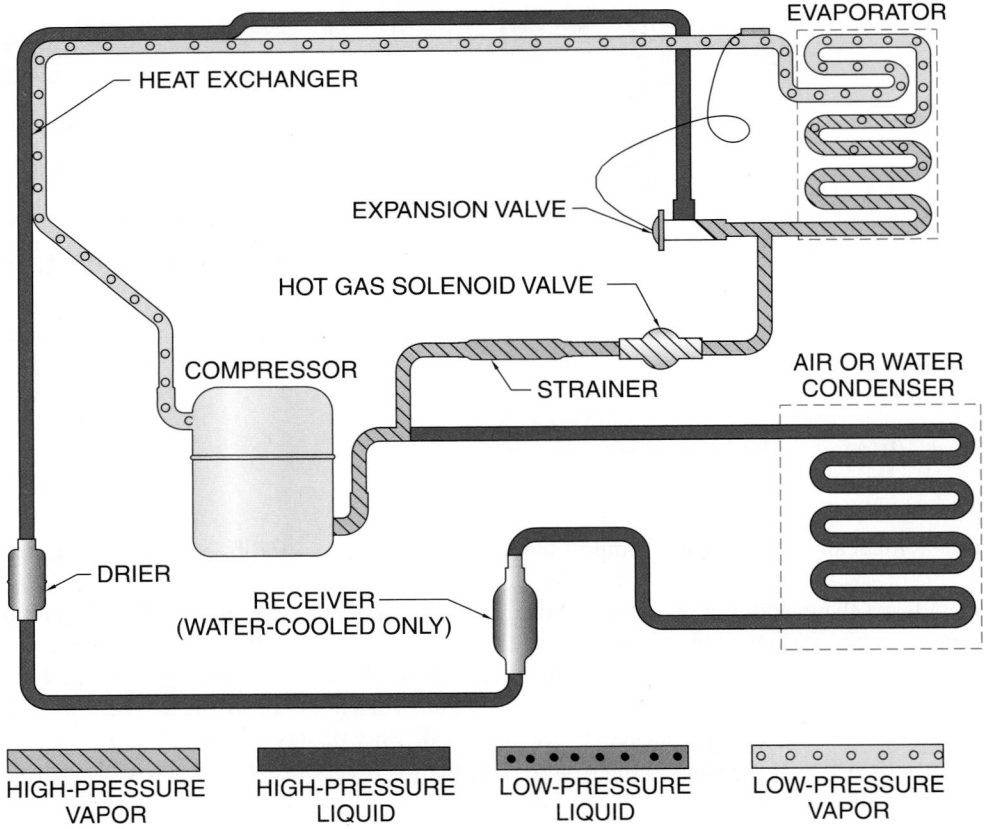

HIGH-PRESSURE VAPOR   HIGH-PRESSURE LIQUID   LOW-PRESSURE LIQUID   LOW-PRESSURE VAPOR

Figure 27–31   A system diagram during the harvest cycle with the hot gas valve energized. *Courtesy Manitowoc Ice, Inc.*

(A)

(B)

Figure 27–32 **(A)** The hot gas solenoid valve located on the compressor's discharge line. **(B)** A harvest bypass valve that is energized only for a few seconds and controls the necessary amount of refrigerant for an ice harvest. **(A)** *Courtesy Ferris State University. Photo by John Tomczyk.* **(B)** *Courtesy Scotsman Ice Systems*

On an initial start-up, or after an automatic shutoff, the water pump and the water dump valve are energized for a specific time period, usually 45 sec, **Figure 27–34(A)**. This gets rid of the old mineral-laden water. One manufacturer uses a purge valve located just above the water reservoir for controlled draining of water from the reservoir, **Figure 27–34(B)**. When it is open and the water pump is on, water flows through this valve to the ice machine's drain. This dilutes the amount of mineral scale in the reservoir. The amount of water purged is adjustable to accommodate local water conditions. ⬤The solenoid-operated hot gas valve is also energized for a specific time period during this start-up. This allows the system re-

Figure 27–33 An electric, solenoid-operated water dump valve. *Courtesy Ferris State University. Photo by John Tomczyk*

frigerant pressures to equalize between the high and low side for an easier compressor start-up. ⬤ The compressor starts after the 45-sec water purge and remains on during the freeze and harvest cycles. The compressor being on during the harvest cycle ensures quality superheated gas for melting the ice slab from the evaporator. The water fill valve is also energized at the same time the compressor is on. The water level sensor shuts off the water fill valve when the water level is satisfied. The electronic controller will not allow the water fill valve to be on longer than 6 min to prevent flooding. On air-cooled units, the condenser fan motor is energized with the compressor. **Figure 27–36** is an electrical system diagram shown in the freeze cycle.

Many manufacturers supply electrical diagrams with electrical conductor lines in bold print to illustrate which components are energized during a certain part of the electrical sequence of operation, **Figure 27–37**. This greatly assists the service technician with any systematic troubleshooting problems that may arise with the ice machine.

An energized parts chart may also be available from the ice machine manufacturer to assist the service technician in systematic troubleshooting, **Figure 27–38**. The energized parts chart tells which control board relay and contactor is energized, when, and how long. These charts, along with the electrical diagrams, can make the ice machine's sequence of operation easier to understand.

Ice Harvests

Most ice makers use a hot gas harvest or defrost to melt an ice sheet off the evaporator. **Figure 27–31** is a system diagram of an ice maker in the hot gas defrost mode. Notice that

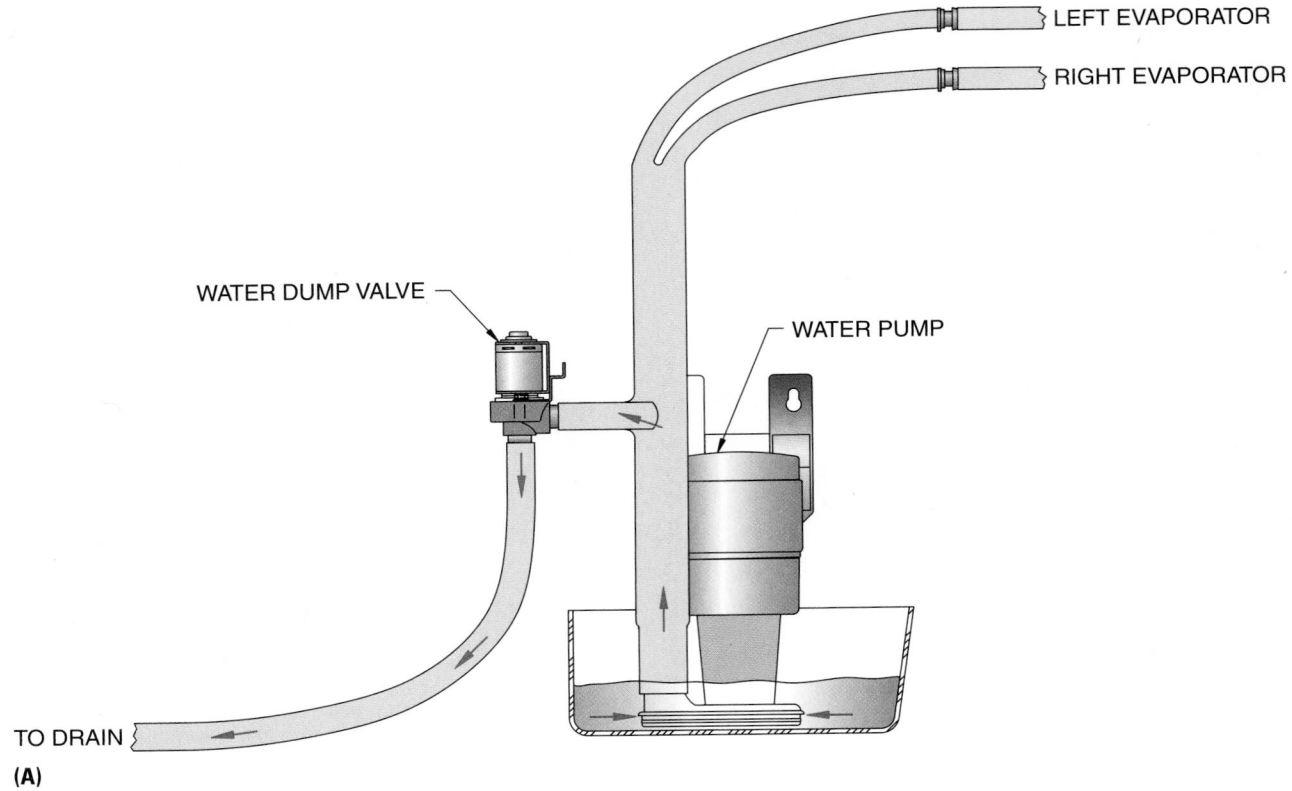

WATER DUMP VALVE

WATER PUMP

LEFT EVAPORATOR

RIGHT EVAPORATOR

TO DRAIN

(A)

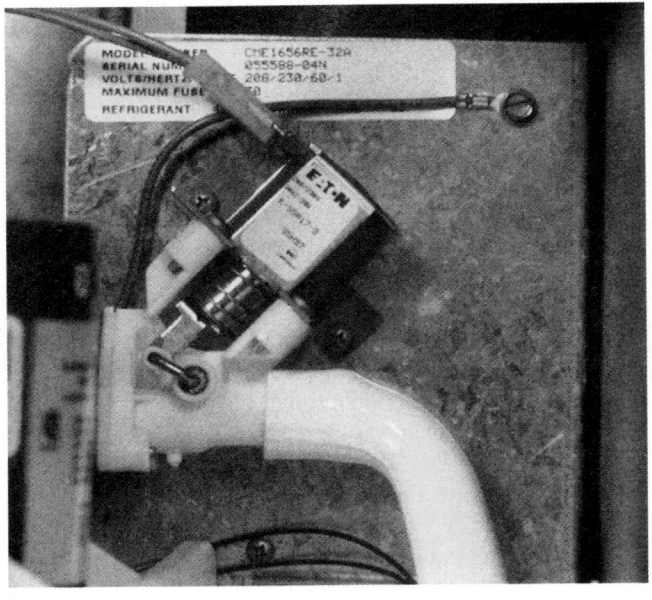

(B)

Figure 27–34 **(A)** Sump water being routed down a drain during harvest. **(B)** A purge valve located just above the water reservoir for controlled draining of water from the reservoir. **(A)** *Courtesy Manitowoc Ice, Inc.* **(B)** *Courtesy Scotsman Ice Systems*

Figure 27–35 A full ice storage bin prevents the ice slab from closing the bin switch. *Courtesy Ferris State University. Photo by John Tomczyk*

superheated gas is routed through the evaporator. As the ice sheet absorbs heat from the superheated gas, the gas becomes saturated. As more heat is absorbed from the saturated gas, it can eventually form a saturated liquid that can become a subcooled liquid. This liquid will travel to the suction line and eventually to the compressor. 🔵 Some manufacturers install a suction-line accumulator just before the compressor on the suction line in order to accumulate and vaporize this liquid before it reaches the compressor. 🔵 However, if the system has no suction-line accumulator, liquid refrigerant can reach the compressor crankcase and dilute the oil in the crankcase. Liquid refrigerant returning to the compressor while the compressor is operating is called *liquid floodback.*

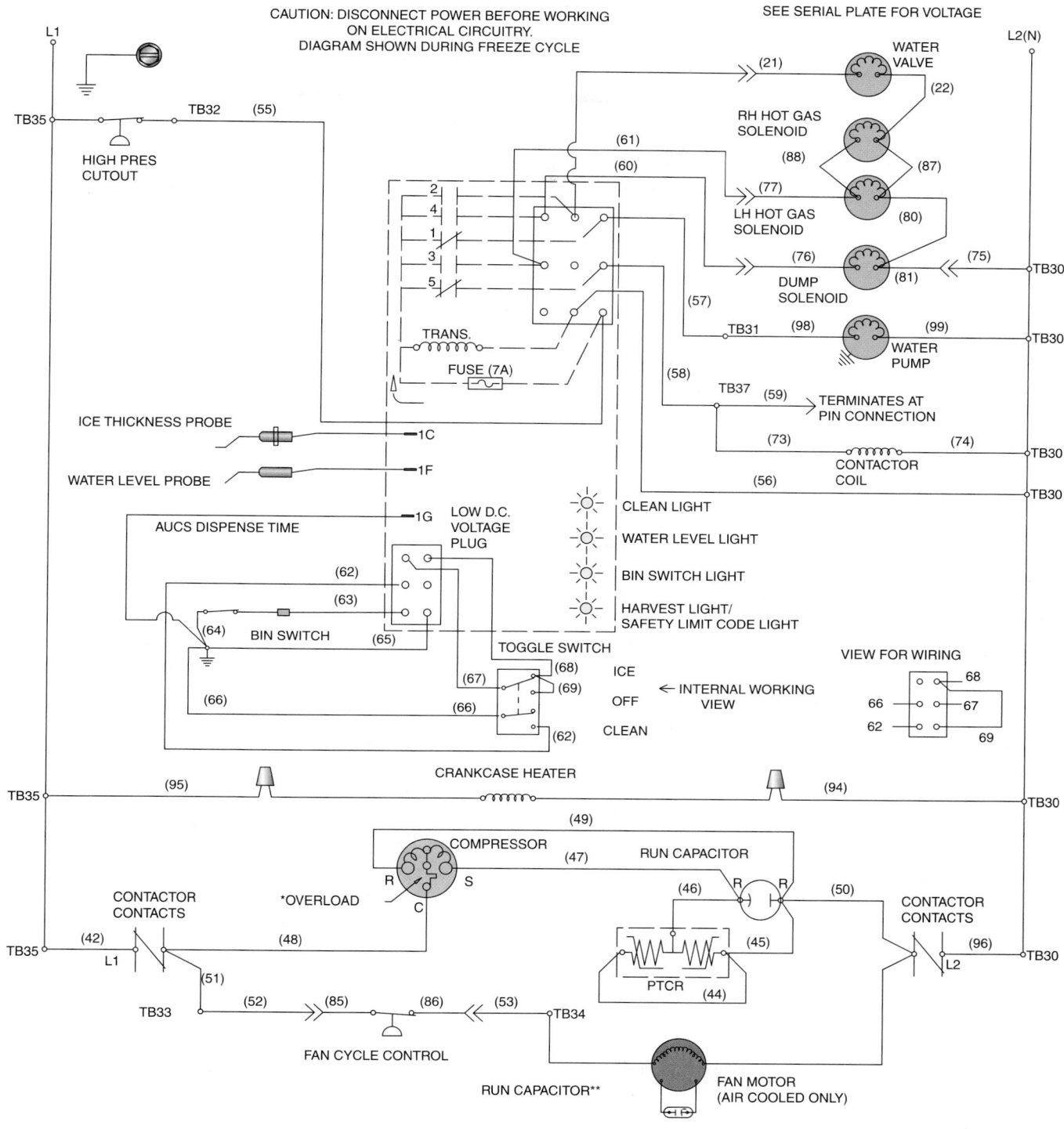

Figure 27–36 An electrical diagram of a cell-type ice maker shown in the freeze cycle. *Courtesy Manitowoc Ice, Inc.*

Liquid floodback is covered in detail in Unit 25, "Special Refrigeration System Components." Liquid floodback can cause serious compressor problems because of the diluted oil in the crankcase. Bearings and other moving parts in the crankcase can be scored from the inadequate lubrication of diluted oil. The returning liquid refrigerant eventually boils off in the compressor's hot crankcase, which causes oil foaming. A cold and sweating crankcase will result. This foaming of oil and boiling refrigerant can get sucked up into the suction valves of the compressor. Extremely high pressures can result, and valve and valve plate damage can occur. If any liquid refrigerant gets compressed, *wet compression* will result. Wet compression occurs when the compression stroke of the compressor vaporizes any liquid in its cylinder.

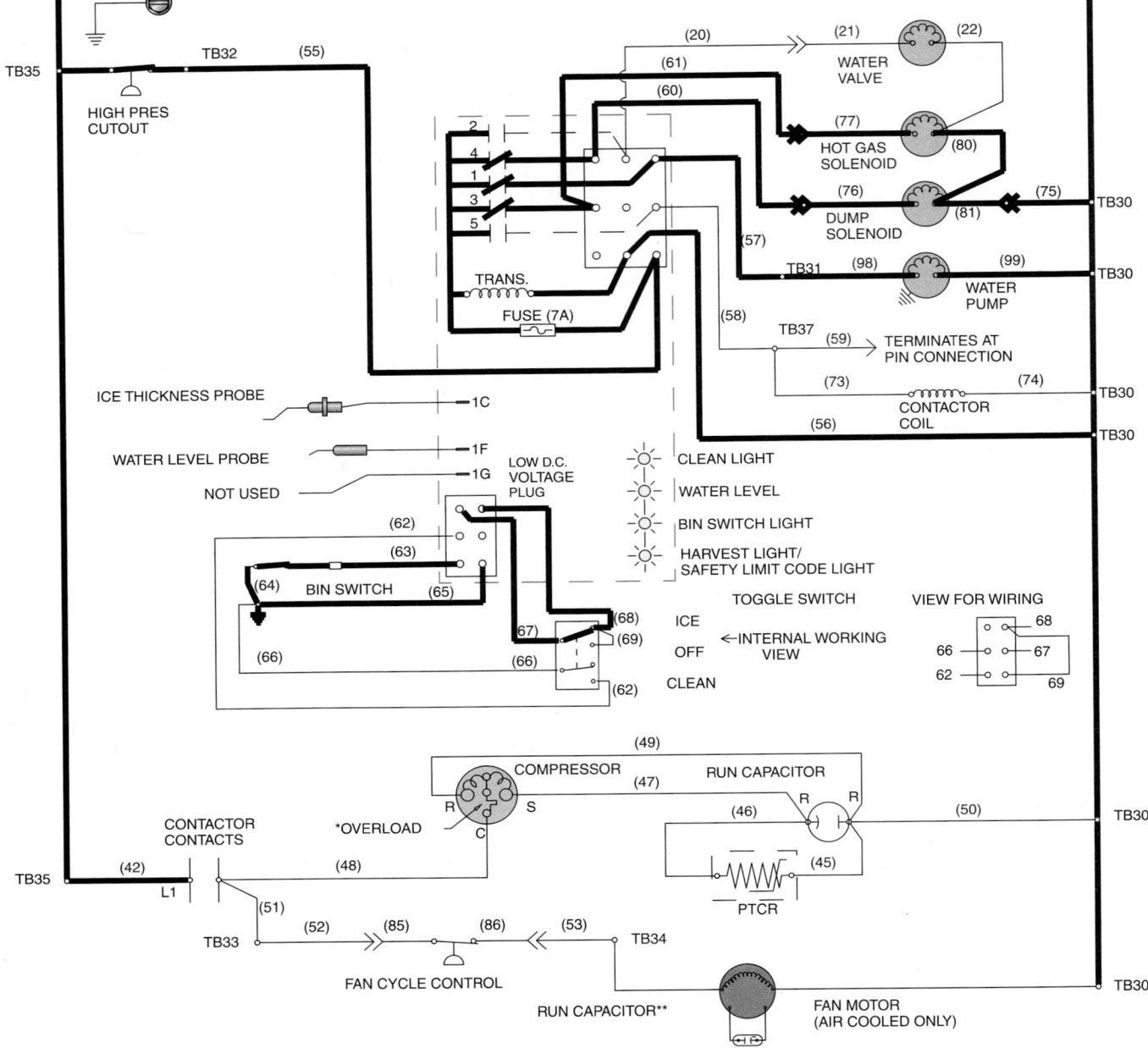

Figure 27–37 An electrical diagram with bold electric conductor lines showing energized loads. *Courtesy Manitowoc Ice, Inc.*

The vaporizing refrigerant in the cylinder absorbs a lot of the heat of compression, leaving a cool cylinder. The discharge temperature will also be much cooler.

Many manufacturers install a discharge-line thermistor, **Figure 27–39,** to sense the discharge-line temperature. If it falls below a certain temperature, for example 90°F, during a defrost period, severe liquid floodback is occurring. Since the discharge thermistor is an input to the microprocessor, the ice machine will automatically be taken out of defrost by deenergizing the hot gas solenoid. In the case of dual

evaporators, just one evaporator will be taken out of defrost. Once the discharge-line temperature increases, the evaporators will be put back into defrost. This cycling of the hot gas solenoid will repeat until all of the ice has been harvested off the evaporators.

As mentioned earlier, once the ice slab grows to a certain thickness, some means of initiating defrost will put the system into a hot gas defrost. If an ice thickness probe or feeler contacts are used, the water flowing over the ice that contacts the probe must have some electrical conductivity for defrost

	Control Board Relays					Contactor		
	1	2	3	4	5	5A	5B	
Ice Making Sequence of Operation	Water Pump	Water Fill Valve	Hot Gas Valve(s)	Water Dump Valve	Contactor Coil	Compressor	Condenser Fan Motor	**Length of Time**
Start-Up[1] 1. Water Purge	On	Off	On	On	Off	Off	Off	45 Seconds
2. Refrigeration System Start-Up	Off	On	On	Off	On	On	May cycle On/Off	5 Seconds
Freeze Sequence 3. Pre-Chill	Off	May cycle On/Off during first 45 sec.	Off	Off	On	On	May cycle On/Off	30 Seconds
4. Freeze	On	Cycles On, then Off 1 more time	Off	Off	On	On	May cycle On/Off	Until 7 sec. water contact with ice thickness probe
Harvest Sequence 5. Water Purge	On	30 sec. Off, 15 sec. On	On	On	On	On	May cycle On/Off	Factory-set at 45 Seconds
6. Harvest	Off	Off	On	Off	On	On	May cycle On/Off	Bin switch activation
7. Automatic Shut-Off	Off	Off	Off	Off	Off	Off	Off	Until bin switch re-closes

Electrical System
Energized Parts Charts
Self-Contained Air- and Water-Cooled Models

[1]Initial Start-Up or Start-Up after Automatic Shut-Off

Figure 27–38   An energized parts chart for systematic troubleshooting. *Courtesy Manitowoc Ice, Inc.*

to start. If the water is too pure because too many minerals have been removed, the conductivity will not be correct and a defrost will not be initiated. A pinch of salt may be added to the circulating water to provide conductivity to initiate a defrost when the evaporator is frozen solid. If a defrost cycle occurs after adding salt, it is an indication that the water is too pure. Special controls can be obtained from the ice maker manufacturer so that the defrost will occur under pure water conditions.

Once hot gas loosens the ice slab from the evaporator by melting some of the ice next to the evaporator surface, a suction or attraction force is formed between the ice slab and the evaporator. Air must be introduced between the ice slab and the evaporator to release this suction force. *Weep holes* between the evaporator cells allow air entering from the edges of the ice to travel along the entire ice slab and relieve the suction force. This releases the ice from the evaporator. The weep holes are shown in **Figure 27–40**.

During an ice harvest, the water pump continues to run and the electric, solenoid-operated water dump valve is energized for a specified amount of time. This action purges the water from the water sump either by gravity or water pump pressure. This removes any residual minerals from the previous freeze cycle, which is very important for clear, hard ice.

Some older ice machines do not have water dump valves. When a harvest cycle occurs, the water pump turns off. The excess water from the pumping system that is caught in the piping and water distribution header causes an overflow in the water sump. It also causes a siphon effect in the water sump from the momentum of the overflowing water. This siphon effect is planned so that some of the mineral buildup in the sump water moves down the drain. This siphon effect continues until the water sump is emptied, allowing air to break the siphon. This allows the sump to drain every cycle by gravity and prevents any mineral buildup. The float mechanism then lets fresh water into the water sump. The plumbing of the water system to allow the overflow and siphon to occur is very critical on these older machines.

Once the ice falls off the face of the evaporator, something has to sense this falling ice and terminate the harvest cycle.

Figure 27–39   A thermistor on the discharge line of an ice maker.
Courtesy Ferris State University. Photo by John Tomczyk

Figure 27–40   Weep holes between cells on the evaporator.
Courtesy Ferris State University. Photo by John Tomczyk

Most of these devices are digital inputs to a controller or microprocessor. **Figure 27–24** shows a magnetic bin reed switch that is activated by the ice curtain to tell the microprocessor that the ice has fallen from the evaporator. **Figure 27–41** shows an ice curtain switch, and **Figure 27–42** shows a paddle switch activated by a pushrod and ice curtain (the pushrod and ice curtain are not shown in this figure).

Often, a pushrod or probe controlled by a harvest motor, clutch assembly, and cam switch will push the ice off the evaporator during a harvest cycle, **Figure 27–43**. This harvest assist assembly is made up of four major parts:

- Motor
- Clutch assembly
- Probe
- Cam switch

The motor drives the harvest assist clutch assembly during the harvest cycle. The clutch assembly provides about 6 oz of force against the ice slab. It is a brass cam designed to slip as it pushes against the back of the ice slab during a harvest cycle. The upper half of the clutch slips against the lower half until the ice starts to release off the evaporator plate. The clutch then engages, which allows the probe to push the ice off the evaporator plate into the holding bin.

The probe is a stainless-steel rod, which is driven through a hole in the evaporator by the motor when the clutch engages. To ensure against the probe being frozen into the ice during the freeze cycle, it is recessed about 1/8 in. from the back of the evaporator when not in use. The probe is adjusted by loosening a nut and screwing the probe either in or out. A ball joint between the probe and the clutch allows the probe to pivot while it is being pushed against the ice.

The cam switch rides on the clutch assembly. It is a single-pole–double-throw switch. The switches operate many ice machine components. The normally closed contacts allow the water pump and water purge valve to operate during the ice-making cycle and sometimes during the first part of the harvest cycle. The normally open contacts of the switch power the harvest assist motors and hot gas solenoid during a harvest cycle. When the cam's arm starts to ride the lower side of the cam during a harvest, the ice machine starts a freeze cycle again.

Clean Cycle

Most ice machines have an ICE/OFF/CLEAN toggle switch. It is usually a double-pole–double-throw switch, which allows the owner or service technician to toggle the ice machine in either a freeze, off, or clean mode. In most cases, the clean mode energizes the water pump only, allowing the service technician or owner to add the proper chemicals for circulation through the system. Many modern ice makers have a digital switch that simply requires a touch with the fingertips.

Ice Machine Pressure Controls

Even though the microprocessor controls most of the operations of today's ice machines, there are still some safety and cycling controls remote to the microprocessor. **Figure 27–44**

(A)

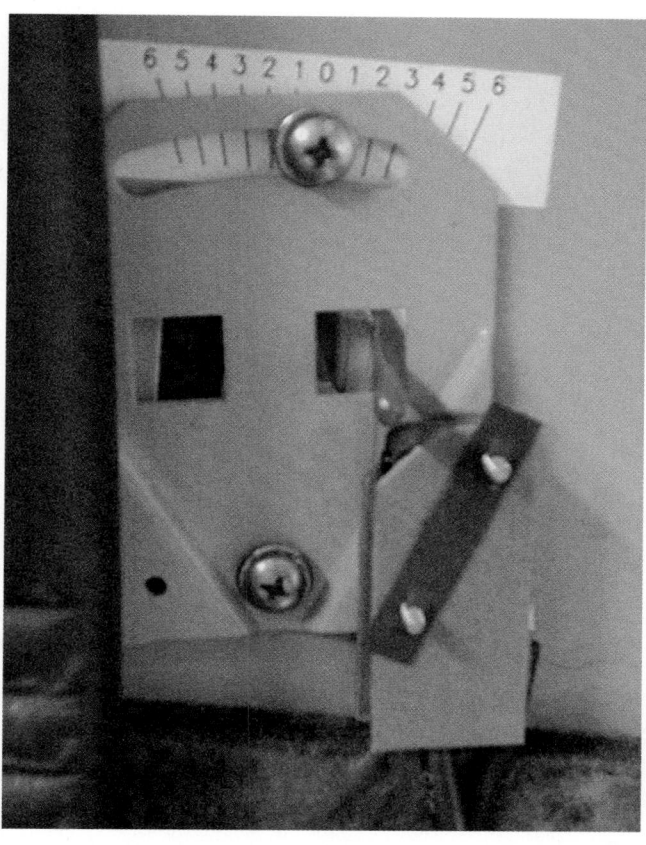

(B)

Figure 27–41 **(A)** An ice curtain switch arm for sensing fallen ice. **(B)** An ice curtain switch. **(A)** and **(B)** Courtesy Ferris State University. Photos by John Tomczyk

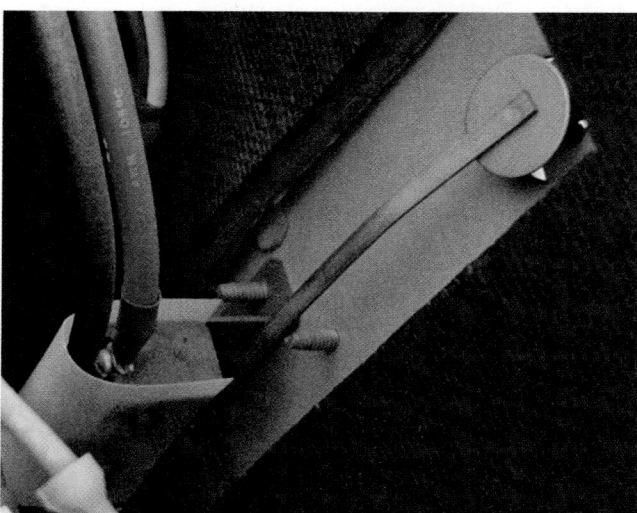

Figure 27–42 A paddle switch used to sense ice falling off an evaporator. Courtesy Ferris State University. Photo by John Tomczyk

shows a modular, nonadjustable fan-cycling control and high-pressure control. 🌐 The fan-cycling control turns the condenser fans on once the ice machine reaches a certain condensing pressure. It also cycles the fan off if the

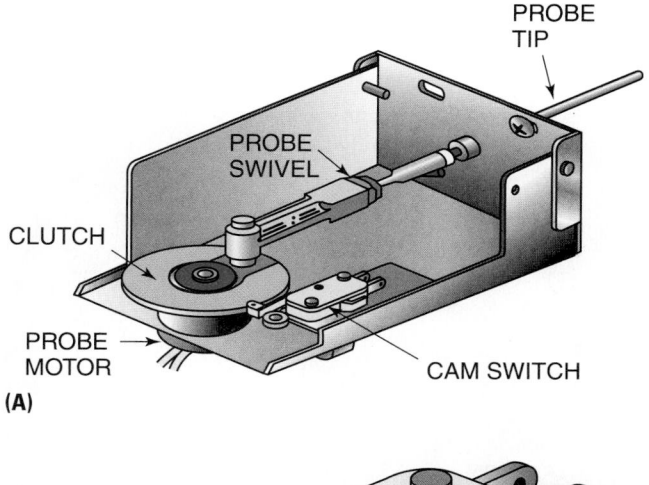

(A)

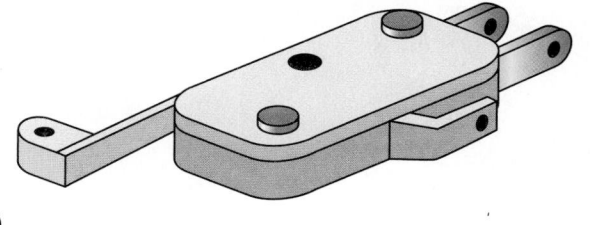

(B)

Figure 27–43 **(A)** A harvest assist clutch assembly showing motor, probe, cam switch, and frame. **(B)** Cam switch. Courtesy Ice-O-Matic

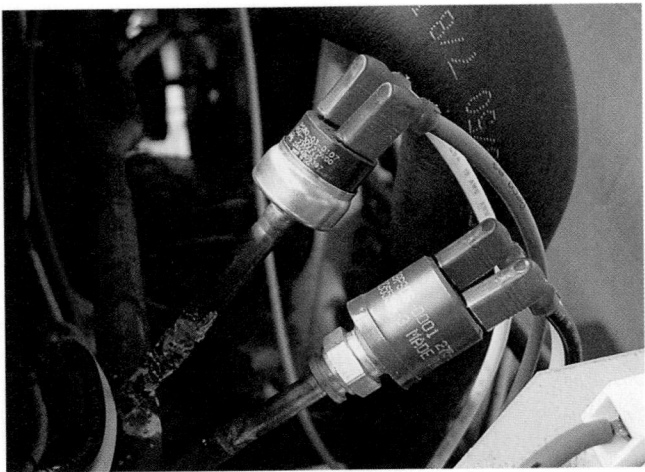

Figure 27–44 Modular, nonadjustable fan-cycling and high-pressure controls. *Courtesy Ferris State University. Photo by John Tomczyk*

condensing pressure drops below a set point. This ensures the proper pressure drop across the metering device as quickly as possible. ◯ The high-pressure control is a safety control to protect against excessively high pressures. If the high-pressure cut-out switch opens, the machine will stop immediately. It will automatically reset when the pressure falls below its cut-in point. Low-pressure controls also come in this modular style. If the low-pressure control opens, the machine will shut off. It will also automatically reset when the pressure rises above its cut-in point. Both of the control settings are dependent on what refrigerant is being used and the ice machine's application.

Figure 27–45 shows a manually resettable high-pressure control. If the machine cycles off on high head pressure, it

MANUAL RESET LEVER

Figure 27–45 A manually reset high-pressure control. *Courtesy Ferris State University. Photo by John Tomczyk*

will not continue cycling on high head pressure. A service technician must manually reset the control for the machine to operate.

Some ice machines use a reverse-acting low-pressure control to terminate the freeze cycle. As ice forms on the evaporator, the suction pressure drops from a reduced heat load. A reverse-acting low-pressure control closes its contacts at a set low pressure. This action terminates the freeze cycle and initiates a harvest cycle.

27.4 MICROPROCESSORS

Service technicians in the HVACR field will encounter the microprocessor in their daily service work, **Figure 27–46.** In fact, most manufacturers of ice machines and ice flakers incorporate microprocessors on their units. Microprocessors are simply small computers that contain stored programs or algorithms. The sequence of operation is stored in the

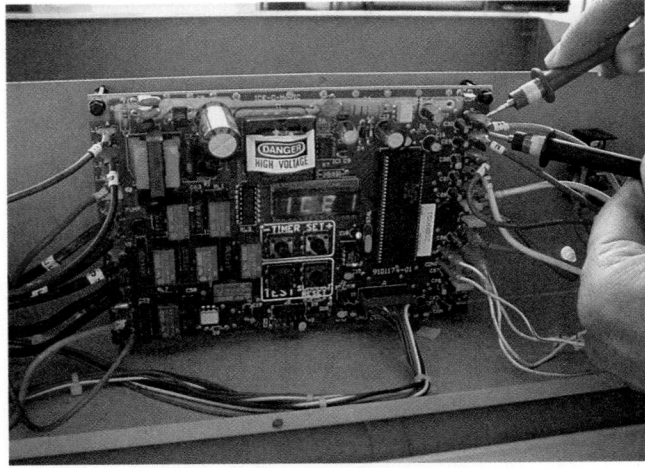

(A)

(B)

Figure 27–46 **(A)** A microprocessor used in a commercial ice machine. **(B)** Microprocessor-based controls. *(A) and (B) Courtesy Ferris State University. Photos by John Tomczyk*

software of the microprocessor. **Figure 27–47(C)** shows a small microprocessor or controller with an interfacing ice sensor and water level controller used to control a modern ice cube machine. **Figure 27–47(D)** shows a modern ice cube machine incorporating an electronic controller or microprocessor. The controller operates the ice machine according to instructions, or programs, that reside in its internal memory. Each model made by this manufacturer operates a bit differently. For maximum efficiency, individual ice machine models have different programs called *electrical erasable programmable read-only memory (EEPROM)* values. These controllers allow the ice machine to optimize efficiency of operations, recall errors that occur during running, serve self-diagnostic functions, and customize programs for individual models. Even the water fill, which takes in water in batches, is a controller function. In fact, the amount of time it takes to fill the water reservoir is measured every cycle to determine the fill time needed to maintain the correct amount of water rinse. **Figure 27–47(E)** shows a service controller that can be custom programmed for a particular model of ice machine by rotating a selector switch.

Microprocessors are often referred to as integrated circuit controllers, electronic controllers, or just controllers. 🌐HVACR equipment manufacturers are using microprocessors to make their products more reliable, less complicated, and easier to troubleshoot by including self-diagnostics. The microprocessor has cut down on the amount of hard wiring that goes into the control circuit and it has taken the place of a lot of the hard wiring that used to be inside an ice machine. 🌐 This type of technology is often referred to as *clean technology*. In fact, many years ago, there were so many wires going in and out of controls that the entire control circuitry looked like a plate of spaghetti. Now, most modern control circuits consist of a microprocessor, a few starting components like a start and run capacitor, a starting relay, a compressor contactor, and the main loads or power-consuming devices. Most service technicians are not afraid of the main loads or power-consuming devices when troubleshooting; the microprocessor is the most feared. However, the microprocessor can be the simplest component to troubleshoot with a little patience, understanding, and practice.

The microprocessor is a small computer that has a sequence of events stored in its memory. It has many sophisticated solid-state devices in its internal circuitry that are needed for its proper operation. However, a service technician does not need to understand how each solid-state device operates in order to tell if the microprocessor is good or bad. What the service technician does have to know is the microprocessor's sequence of operation, its self-diagnostic functions, and how to input/output (I/O) troubleshoot the microprocessor using its external terminals. The microprocessor's external terminals are shown at the right- and left-hand sides of **Figure 27–46(A)** where most of the wires are coming in and out of the microprocessor. These wires are the inputs and outputs of the microprocessor. Troubleshooting the external terminals of a microprocessor will be covered in detail later in this unit.

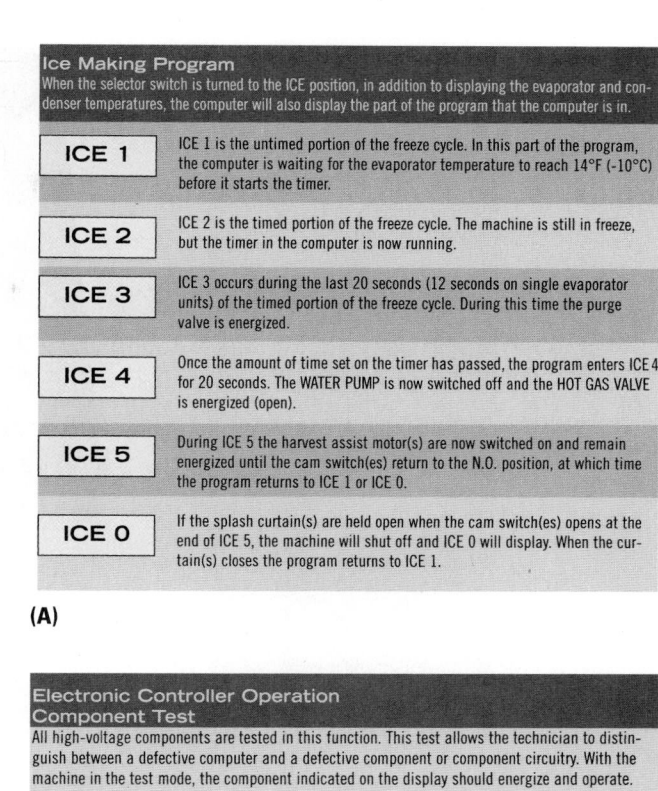

(A)

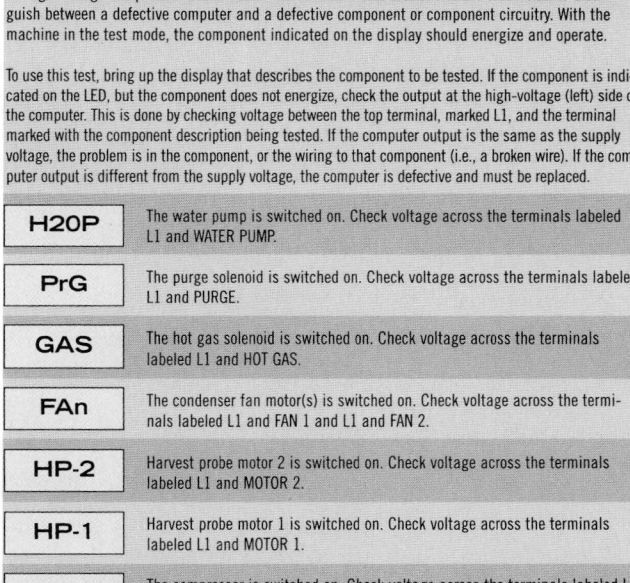

(B)

Figure 27–47   (A) A microprocessor sequence of operation. **(B)** A microprocessor component test. *(A) and (B) Courtesy Ice-O-Matic*

Microprocessor Sequence of Events and Self-Diagnostics

The sequence of events of a microprocessor is usually found in the service manual. If a service manual cannot be found on site with the piece of refrigeration equipment, the owner or manager must be contacted to see if they have filed the service manual away in some safe location. If the manual still

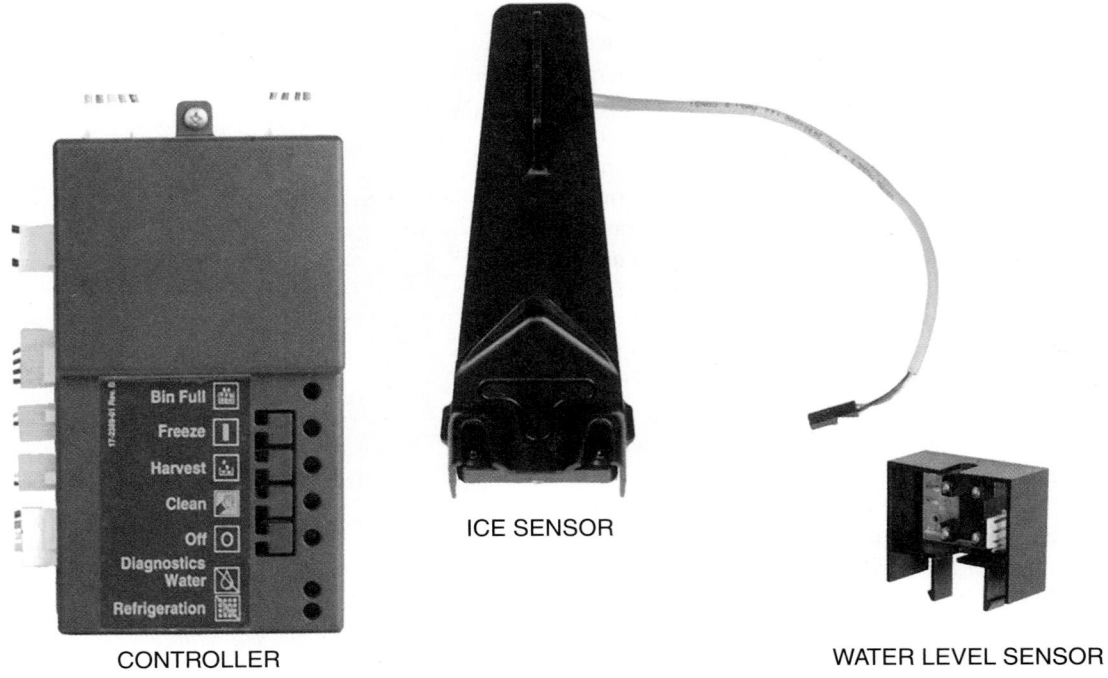

ICE SENSOR

CONTROLLER

WATER LEVEL SENSOR

(C)

WATER INLET VALVE

REFRIGERATION SERVICE ACCESS VALVES

AUTO IQ CONTROLLER

WATER PUMP

AWATER LEVEL AND CUBE-SIZE SENSOR

FLOAT BULB

EVAPORATOR

LIGHT CURTAIN BIN CONTROL AND HARVEST TERMINATION SYSTEM

CUBE DEFLECTOR

INSULATED BASE, RESERVOIR AND FREEZING COMPARTMENT

(D)

Figure 27–47 *(continued)*   **(C)** A modern controller, ice sensor, and water level sensor used on an ice cube machine. **(D)** A modern high-technology ice cube machine. **(C)–(D)** *Courtesy Scotsman Ice Systems*

Unit 27   Commercial Ice Machines   613

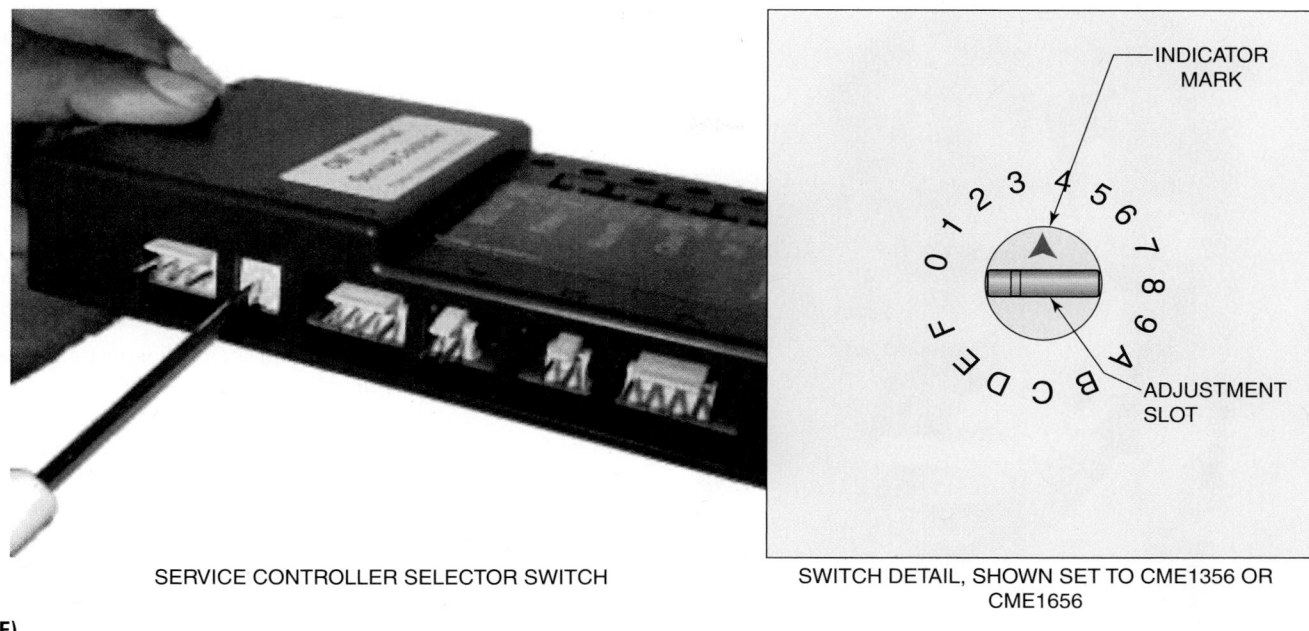

SERVICE CONTROLLER SELECTOR SWITCH

SWITCH DETAIL, SHOWN SET TO CME1356 OR CME1656

(E)

Figure 27–47 (*continued*)   **(E)** A controller that can replace many existing controllers by rotating a selector switch for a desired stored program. **(E)** *Courtesy Scotsman Ice Systems*

cannot be located, the equipment manufacturer must be contacted. Usually a model and serial number of the refrigeration machine is all that is needed for the manufacturer to locate the manual. The Internet is also a very useful tool for finding a service manual or sequence of operations. The service technician simply needs the pages from the manual that include the sequence of events of the microprocessor and how to initiate the self-diagnostics of the machine if it has any. The technician can ask the manufacturer to fax the pages that include the pertinent information for servicing and troubleshooting. With today's modern technology, this information can be in the hands of the service technician in minutes. **Figure 27–47(A and B)** shows a page from a service manual for an icemaker where the manufacturer has included the sequence of operation and the component test of the microprocessor, respectively. Simply pushing a button or moving a switch initiates the tests that the microprocessor performs. A test button is shown in **Figure 27–46(A).** Notice that the component test walks the technician through checks of the water pump, purge solenoid, hot gas solenoid, condenser fan, harvest probe motors, and compressor with a voltmeter. It even instructs the service technician on what terminal to place the voltmeter and when to check the component. This gives the service technician the knowledge of what the ice machine components are supposed to do at a certain time, temperature, or test mode.

Input/Output Troubleshooting

Even if the manufacturer of the ice machine does not include a component test mode within the programming of the microprocessor, the service technician can still troubleshoot the microprocessor. Input/output troubleshooting is actually an easy method if the technician uses some common sense. All that is needed is a voltmeter and knowledge of the sequence of events from the service manual. An ohmmeter should not be used directly on a microprocessor because of the ohmmeter's battery voltage. Many times this voltage is too high and may damage the intricate solid-state components or the magnetic memory internal to the microprocessor. However, if certain components (inputs or outputs) to the microprocessor are detached, an ohmmeter can safely be used on these detached components.

Microprocessors are fed with information from their input devices to their input terminals. Input devices can be analog, as with a thermistor (variable resistor), or digital (on/off), as with a switch. **Figure 27–48** shows a thermistor connected to the midpoint of the condenser. It is acting as an analog input device to the microprocessor for displaying the condensing temperature on an LED.

After processing the data from the inputs, an output signal is sent. The output signal can be read with a meter from the output terminals of the microprocessor. Input and output terminals of a microprocessor can be seen in **Figure 27–49.** Notice that both the input and output terminals are clearly labeled for the service technician to troubleshoot. It is from these terminals that most troubleshooting can be accomplished using a meter. Most of the time, but not always, the input signals are low-voltage (AC or DC) or resistance signals, and the output signals are of higher voltage (usually AC) going to the power-consuming devices. Always refer to the service manual for specifics. Input devices are frequently used for digital readout for a condensing temperature, an evaporating temperature, or a specific sequence

Figure 27–48 The condenser thermistor is an analog input to a microprocessor. *Courtesy Ferris State University. Photo by John Tomczyk*

mode. **Figure 27–46 (A)** shows a service technician measuring the input voltage from a microprocessor. It is important to use as short a measuring probe as possible to avoid shorts to other terminals of the microprocessor. A direct short could ruin any microprocessor. Taping the voltmeter probes halfway up will help prevent electrical shorts when measuring inputs or outputs from a microprocessor.

Microprocessor Self-Diagnostics and Error Codes

Many microprocessors come with a self-diagnostic mode. The microprocessor continually monitors all functions of the ice maker. If a malfunction should occur, the microprocessor will associate the error or malfunction with an error code in its memory. The error code will then be indicated on the LED display. **Figure 27–50** shows some error code descriptions of one manufacturer's ice machine. The service technician in **Figure 27–51** is using a voltmeter to troubleshoot the hot gas high-voltage circuit. Notice that the error code 12 (EC 12) is displayed on the LED of the microprocessor. By referring to the service manual, **Figure 27–50,** the technician finds that this error code indicates that the evaporator temperature exceeded 150°F.

Error codes are a valuable and time-saving troubleshooting tool. However not all problems can be diagnosed with error codes. Always check to see if the fuse on the microprocessor is good before condemning the processor. Most microprocessors have these fuses.

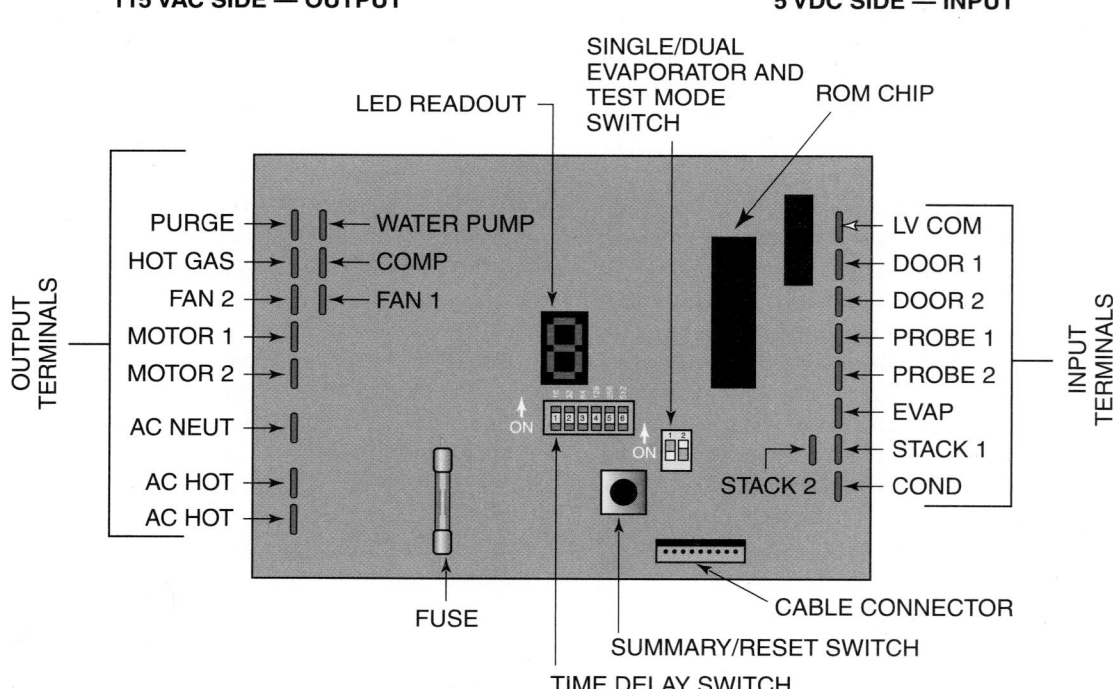

Figure 27–49 A microprocessor showing components that include the input and output terminals. *Courtesy Ice-O-Matic*

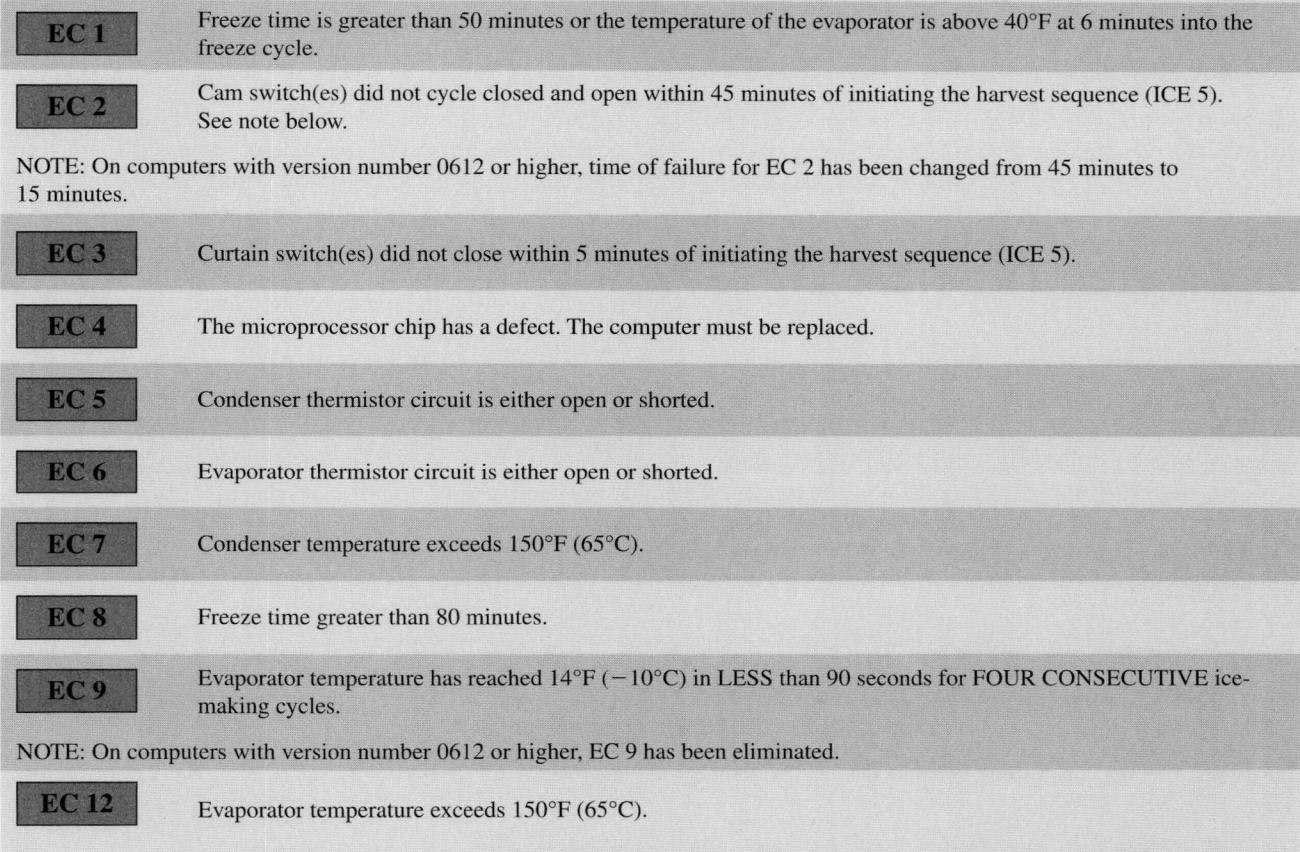

Error Codes

The computer constantly monitors the ice-making cycles to determine if everything is operating correctly. Should the computer detect a malfunction, it will record the associated error code in its memory. If the error code is a: "FATAL ERROR," the computer will shut the machine down, light the front panel "SERVICE LIGHT," and indicate the error code on the LED display. NOTE: On machines manufactured after August, 1991, error codes 2, 5, 7, and 12 will automatically make 4 attempts (one attempt every 15 minutes) to reset before the machine is shut down and the service fault light illuminates. The same error code must occur 4 consecutive times before the machine shuts down. All computers incorporating this change can be identified by a version number of 0612 or higher. See version number identification, page H-2. Error codes that do not cause a machine shut-down (EC 1 and EC 3) will ONLY display on the LED when in the "SUMMARY MODE."

Below is a description of each Error code.

EC 1   Freeze time is greater than 50 minutes or the temperature of the evaporator is above 40°F at 6 minutes into the freeze cycle.

EC 2   Cam switch(es) did not cycle closed and open within 45 minutes of initiating the harvest sequence (ICE 5). See note below.

NOTE: On computers with version number 0612 or higher, time of failure for EC 2 has been changed from 45 minutes to 15 minutes.

EC 3   Curtain switch(es) did not close within 5 minutes of initiating the harvest sequence (ICE 5).

EC 4   The microprocessor chip has a defect. The computer must be replaced.

EC 5   Condenser thermistor circuit is either open or shorted.

EC 6   Evaporator thermistor circuit is either open or shorted.

EC 7   Condenser temperature exceeds 150°F (65°C).

EC 8   Freeze time greater than 80 minutes.

EC 9   Evaporator temperature has reached 14°F (−10°C) in LESS than 90 seconds for FOUR CONSECUTIVE ice-making cycles.

NOTE: On computers with version number 0612 or higher, EC 9 has been eliminated.

EC 12   Evaporator temperature exceeds 150°F (65°C).

If a machine is shut down on a "FATAL" error code, the machine must be reset following the procedure before it can be returned to the ice making mode. Error codes will remain in the memory even after a power interruption.

Figure 27–50   Error code descriptions from a microprocessor. *Courtesy Ice-O-Matic*

Microprocessor History or Summary

Some microprocessors can record an ice machine's history. This summary of the ice machine's past operations can be retrieved from memory and reviewed at any time by using switches or buttons on the microprocessor. This history may also be erased once recorded by the service technician. Important information that happened over long periods of time such as the number of harvests, error codes stored, and average cycle times can be recalled, **Figure 27–52.**

The microprocessor is not as complicated as it may seem. All the service technician really has to know is I/O troubleshooting, the sequence of operation, and how to find and follow an instruction manual.

Ice Production and Performance Data

Ice production, cycle times, and system operating pressures for an air-cooled ice machine mainly depend on two factors:
- Temperature of the air entering the condenser
- Temperature of the water entering the ice machine

Assuming a constant water temperature entering the ice machine, as the air temperature increases through an air-cooled condenser, the ice production decreases. Notice in **Figure 27–53** that if the water temperature entering the ice machine stays constant at 50°F, the 24-hour ice production is 1880 lb at the 70°F temperature of the air entering the condenser. However, if the air entering the condenser increases in temperature to 100°F, the 24-hour ice production

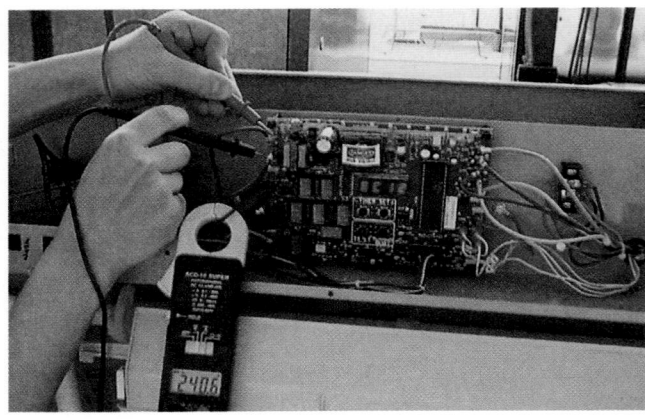

Figure 27–51 A service technician measuring output voltage from a microprocessor. *Courtesy Ferris State University. Photo by John Tomczyk*

decreases to 1550 lb. This happens because of the inefficiencies associated with a high compression ratio caused by the increase in condensing temperature with warmer air entering the condenser.

As the air entering the condenser stays constant at 70°F, the 24-hour ice production decreases from 1880 lb to 1640 lb as the water temperature entering the ice machine increases from 50°F to 90°F. This happens because it takes more refrigeration effect, and thus more time, to freeze the warmer 90°F water into ice.

Notice that the ice machine has longer freeze times as the water temperature to the ice machine and the air temperature through the air-cooled condenser increase.

For both the freeze and harvest cycles, the discharge and suction pressures increase as the air entering the air-cooled condenser increases. This happens because of higher compression ratios associated with a higher condensing temperature caused by the warmer air passing through the condenser. Knowing what the discharge and suction pressures should be for different air and water temperatures is a valuable tool for the service technician.

Summary

The machine history is recorded by the computer. This summary of past operation can be recalled and reviewed any time that the selector switch is in the ICE or OFF position. The history may only be erased when the selector switch is in the OFF position, see page H-15. The summary is not affected by power loss.

H 2 5 4	To recall the history, press the SUMMARY/RESET button momentarily. A summary of past operation will now be displayed.
–TIMER SET+ ○ ○ ○ ● **TEST SUMMARY RESET**	The drawing to the left gives an example of the first part of the summary, which is the harvest count. The H shown on the display indicates the harvest count in thousands, the number that follows (254) indicates the machine has cycled 254,000 times since last being reset.

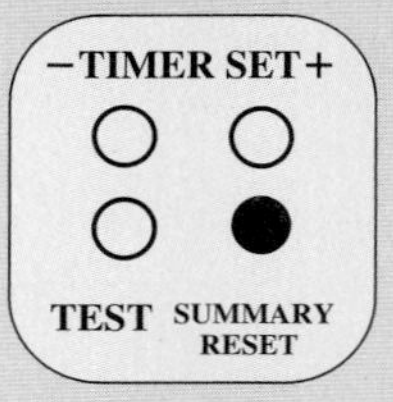

The remaining history will be displayed in the order given below.

h	The number that appears in front of h indicates the number of harvests UP TO ONE THOUSAND since last being reset.
E C	The number that appears in front of EC indicates any error code that has been stored in the memory and not yet erased.
r r r r	This tells the service tech that the computer is now going to display information or "reviews" of prior ice-making cycles. A review is the average cycle time of a group of 10 ice-making cycles.
r 1	The computer is now going to display the first review.
1 5	The number displayed after r 1 has been displayed is the average cycle time of the 10 most recent ice-making cycles. The example to the left shows a 15 minute review.
r 2	The computer is now going to display the second review. The number displayed after r 2 has been displayed is the next 10 most recent ice-making cycles.

Note: The maximum number of reviews is 10 (a total history of 100 cycles). If the computer has not operated for a minimum of 10 cycles, there will be no reviews.

Figure 27–52 A summary or history of the ice machine can be recalled by the microprocessor. *Courtesy Ice-O-Matic*

NOTE: These characteristics may vary depending on operating conditions.

Cycle Times				
Freeze Time + Harvest Time = Cycle Time				
Air Temp. Entering Condenser °F/°C	Freeze Time			Harvest Time
	Water Temperature °F/°C			
	50/10.0	70/21.1	90/32.2	
70/21.1	8.5–9.3	9.4–10.3	9.9–10.9	
80/26.7	9.0–9.9	9.8–10.8	10.5–11.5	1–2.5
90/32.2	9.6–10.5	10.4–11.5	11.1–12.2	
100/37.8	10.6–11.6	11.5–12.6	12.4–13.6	

[1]Times in minutes

24 Hour Ice Production			
Air Temp. Entering Condenser °F/°C	Water Temperature °F/°C		
	50/10.0	70/21.1	90/32.2
70/21.1	1880	1720	1640
80/26.7	1780	1650	1560
90/32.2	1690	1570	1480
100/37.8	1550	1440	1350

[1]Based on average ice slab weight of 13.0–14.12 lb
[2]Regular cube derate is 7%

Operating Pressures				
Air Temp. Entering Condenser °F/°C	Freeze Cycle		Harvest Cycle	
	Discharge Pressure PSIG	Suction Pressure PSIG	Discharge Pressure PSIG	Suction Pressure PSIG
50/10.0	220–280	40–20	155–190	60–80
70/21.1	220–280	40–20	160–190	65–80
80/26.7	230–290	42–20	160–190	65–80
90/32.2	260–320	44–22	185–205	70–90
100/37.8	300–360	46–24	210–225	75–100
110/43.3	320–400	48–26	215–240	80–100

[1]Suction pressure drops gradually throughout the freeze cycle

Figure 27–53 Cycle times, ice production, and operating pressures for an air-cooled system. *Courtesy Manitowoc Ice, Inc.*

27.5 MAKING CYLINDRICAL ICE

Cylindrical ice is made inside a tube-within-a-tube evaporator. Water flows from the center tube, and the refrigerant evaporator is in the outside tube. As the water flows through the center tube, it freezes on the outside walls of the tube. Toward the end of the cycle, the hole in the center begins to close, and the pump pressure at the inlet of the tube begins to rise. At a predetermined pressure, defrost is started. The ice shoots out the end of the tube as a long cylinder, **Figure 27–54.**

The evaporator is normally wound into a perfect circle and the ice coming from it has a slight curve. As the ice leaves the evaporator during defrost, a breaker at the end of the tube will break the ice to length. Winding the evaporator in a circle allows a large evaporator to be formed into a small machine.

27.6 WATER AND ICE QUALITY

An ice machine is an important piece of equipment to any business where it is installed and is vital to a food service customer. Most of these businesses depend heavily on ice for their applications. However, many ignore their ice machine and fail to consider that it is a self-contained "mini ice factory." An ice machine takes a raw material and manufactures a consumable product. This product is then stored and provided to the customer for whatever purpose it is needed.

Water Quality

Ice is considered a food source, and the single component material is water. The quality of the ice product depends on the quality of the water supplied to the unit. Good water quality will produce a crystal clear, hard cube that provides excellent cooling capacity and lasts a long time. Poor water quality will produce ice that can be soft and cloudy and will result in less Btu cooling capacity, ice bridging in the storage container, or both.

Water can contain many different minerals in various levels, and the quality is constantly changing. In fact, water quality varies throughout the United States and may vary across a town or even across the street. It can be improved through filtration and treatment. Water filtration and treatment is a science within itself. Technical experts know about the chemical makeup of water and how to treat specific problems. Although few service technicians have this level of expertise when it comes to resolving water quality issues, it is important to have a basic understanding of water quality in order to service an ice machine. The typical ice machine service technician is called on from time to time to recommend water filtration or treatment for ice-making equipment. Of course, the best source of information concerning filtration and treatment is a water filter manufacturer. These manufacturers will work with you and make recommendations to resolve your water quality issues. Many interesting Web sites on the Internet provide educational information on water quality and water filtration and treatment. Although it is impossible to cover every aspect of water quality, a service technician should have a basic knowledge of key water treatment terminology and how it applies to an ice machine. The

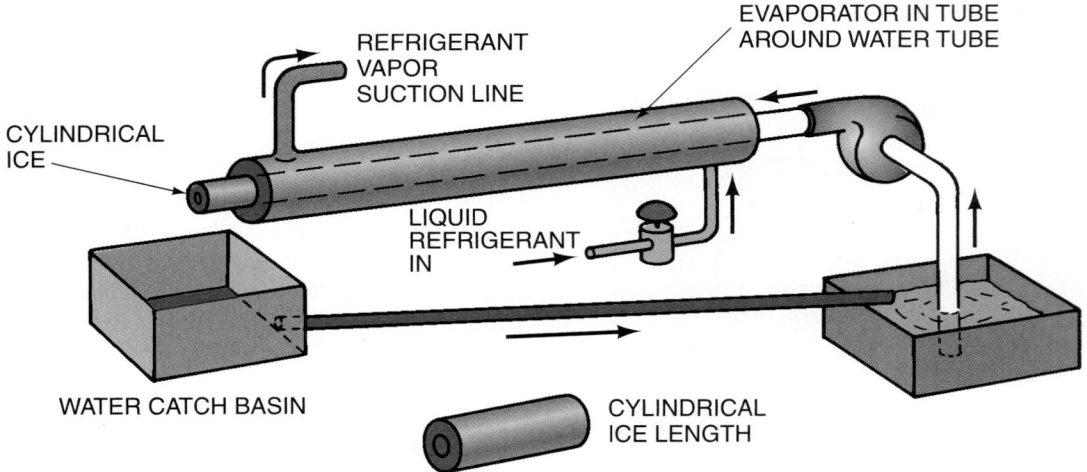

Figure 27–54 This drawing is a simulation of how cylindrical ice is made. As ice is formed on the inside of the tube, the hole in the ice becomes smaller. When a predetermined pump pressure is reached, defrost occurs, and the ice shoots out the end, as though from a gun. The ice is caught, broken to length, and then moved to a bin. In the actual machine, the evaporator would normally be wound in a coil, and the ice would have a slight curvature.

following terms are used when dealing with water quality in ice machines:

Acidic water: Water containing dissolved carbon dioxide that forms carbonic acid. Acidic water is considered aggressive to most components it comes in contact with and is represented by a low pH reading.

Alkaline water: Water that has a surplus of hydroxyl ions, which results in a pH over 7. It often contains concentrations or a mixture of soluble salts or alkali that is picked up from basic minerals in the earth such as calcium, magnesium, silica, and so on. These are scale-causing minerals.

Chloramines: Disinfectants used to treat municipal water systems against bacteria. Chloramines are a mixture of chlorine and ammonia. A higher concentration of charcoal is generally needed to reduce the amount of chloramines in the water.

Chlorine: A disinfectant used to treat municipal water systems against bacteria. Chlorine can be corrosive to metals. Charcoal filtration is generally used to reduce the amount of chlorine in the water to improve offensive chlorine taste and odor.

Filtration: Using a device installed in the water supply inlet of the ice machine to remove contaminant from the water. Coarse filtration removes the "rocks and boulders." Fine filtration removes smaller particles.

Flow rate: The volume of flow through a device measured in gallons per minute (GPM). It is important to size a filter system properly to supply the correct flow rate to the unit per the manufacturer's required specifications.

Hardness: The amount of calcium and magnesium measured in grains per gallon. See the following standards chart from the U.S. Bureau of Standards:

Water Hardness

Description	Grains Per Gallon	Parts Per Million (TDS)
Soft Water	Less than 1.0	Less than 17.1
Slightly Hard	1.0 to 3.5	17.1 to 60.0
Moderately Hard	3.5 to 7.0	60.0 to 120.0
Hard Water	7.0 to 10.5	120.0 to 180.0
Very Hard	10.5 and over	180.0 and over

Iron: A compound or bacteria found generally in well water. Iron compounds cause a rust-colored scale. Iron bacteria shows up as rust-colored slime.

pH: The term used to express the level of acid or alkaline in water or solutions. The scale for pH ranges from 0 to 14. A reading of 7 is considered a neutral pH and is the desired condition for ice machine applications. A reading of below 7 is acidic, and above 7 is basic or alkaline.

Reverse osmosis (RO): A device that forces water, under pressure, against a fine membrane to remove minute particles of contaminants.

Scale: A buildup of minerals that forms a flaky coating on the surfaces of the evaporator and water system. Scale buildup restricts water flow and reduces heat transfer in an ice machine. This buildup affects unit efficiency and usually causes service concerns.

Scale inhibitor: A material or element that slows the ability of scale to stick to a surface. Many filter manufacturers use *polyphosphate* to inhibit calcium or lime scale buildup.

Sediment: Larger particles of dirt, trash, sand, minerals, and so on. A sediment filter is generally installed prior to the filter/treatment device to remove larger contaminants. This allows for longer life of the filter/treatment device by reducing rapid plugging of the filter medium.

Total dissolved solids (TDS): The level of dissolved minerals in the water measured in parts per million (ppm). Higher concentrations of TDS can increase mineral buildup.

Treatment: Using a device that adds chemicals or minerals to the water to effect a change in the water quality. Other treatment technologies are also available in the industry. Treatment is generally required to address scale and bacteria concerns.

Turbidity: The degree of water cloudiness generally caused by high mineral saturation or the presence of air bubbles in the water.

Water softener: A device that uses salt to provide an ion exchange that reduces the hardness of water, making it less likely to produce scale.

🌐 One last term that is usually understood by the service technician but seldom considered by the customer is *preventive maintenance.* This term refers to the regularly scheduled maintenance performed on a unit including inspecting, cleaning, sanitizing, and servicing of the ice machine and external water filtration/treatment system. Preventive maintenance pays big benefits. Every equipment manufacturer provides recommendations and instructions that detail the steps necessary to perform preventive maintenance. This process, along with providing proper water filtration and treatment at initial installation, is vital for protecting the customer's investment and maintaining maximum efficiency— as well as ensuring a longer life for the ice machine.🌐 A basic understanding of these terms will help a service technician communicate any water quality concerns to a filter manufacturer and to the customer. This will definitely help in resolving ice machine water quality issues.

Ice Quality

Ice is a consumable product that serves the specific function of basic refrigeration. Most beverages taste better over ice, especially on a hot summer day. Placing some ice in your cooler keeps your picnic foods and beverages fresh and cool. Whether you crunch it, chew it, or use it as a means of cooling something, ice is an important commodity.

The amount of refrigeration effect that you get from the ice depends on the quality of the ice. Ice quality is measured in percent of hardness. Hardness is a measurement that represents the thermal cooling capacity. The higher the percent of hardness, the more cooling ability the ice possesses. The harder the ice, the denser it is and the longer it will last in a glass or a cooler. Do not confuse ice hardness with hard water. If you have hard water, you will have reduced ice hardness because of the minerals in the ice.

Ice hardness is calculated by conducting a calorimeter test. This is a specific test used by the Air Conditioning and Refrigeration Institute (ARI) to rate ice quality or cooling ability. The American National Standards Institute (ANSI) and the American Society of Heating, Refrigerating, and Air-Conditioning Engineers (ASHRAE) have standards for test procedures. ANSI/ASHRAE Standard 29-1988 lists the procedure for conducting this calorimeter test. The test requires an insulated container with a specific amount of water at a specific temperature. The ambient conditions for the test are also controlled. An exact amount of ice is stirred into the container and timed until all the ice melts. The water temperature before and after the ice is added and the actual melting time is listed on a data sheet. This information is then used to calculate the ice hardness.

Water purity definitely has an effect on ice quality. Pure ice will be crystal clear. Cloudy ice usually contains air or minerals, which will limit hardness. Ice is normally made available in blocks, cubes, chunks, flakes, and cubelets or nuggets. Customer preference and use dictates which type of ice is needed.

As block ice is frozen, pure water freezes first. Any minerals or trapped air in the cold water push toward the center of the block and finally freeze. This leaves a cloud in the center of the clear block of ice. You may have noticed this if you have ever used ice made in an ice tray in your home refrigerator. This type of ice falls in the 95% to 100% hardness range, depending on the mineral content.

Flaked ice falls in the 70% hardness range. This is due to the freezing process used in the ice flake machine. Water is frozen on a cylinder wall, broken away, and extruded into the storage bin. Generally, all the water that goes into the flaker comes out as ice. Any minerals in the water will be frozen into the ice. Because of the nature of flaked ice, more air is present. This also reduces the overall hardness and cooling effect of flaked ice.

As mentioned earlier, cubelets or nuggets are made of compressed flakes. The flakes of ice are squeezed or extruded through a smaller opening to remove more water. This extruded ice is then broken into chunks. These chunks fall in the 80% to 90% hardness range. Since cubelets or nuggets of ice are easy to chew, they are preferred by the health care industry.

Cubes fall in the 95% to 100% hardness range, depending on the evaporator style and ice-making process. Commercial ice cube machines circulate water into, across, or over an evaporator plate. As the water circulates and is cooled, pure water freezes first. The minerals tend to "wash out" during this freezing process. The result is a purer and thus harder cube of ice. The ice hardness will vary with the style of evaporator and materials used. Any restrictions to the water path on the evaporator surface can slow down the water flow and allow minerals to freeze into the cube. Such restrictions can occur when the water has to flow over a separator on the evaporator or into the cavity of a grid cell. These cubes may be cloudy and fall in the lower hardness range.

Regardless of the type of ice or its use, a cleaner ice machine and purer water will produce harder ice. Some ice machine manufacturers provide designs that include built-in flushing or purging capabilities. This helps to eliminate minerals, thus keeping the ice machine cleaner and improving ice quality. Regularly scheduled preventive maintenance checks including cleaning and sanitizing of the ice machine water system are a must. Always use the correct cleaner and follow the manufacturer's recommendations when performing the cleaning and sanitizing procedure.

Sanitizing

🌐 Sanitizing is becoming a more common practice because it helps to eliminate harmful bacteria that tend to thrive in the cool ice-making environment. 🌐 Remember that ice is consumed as a food and comes in direct contact with the food or beverage it cools. It is used to cool down beverages and ice down fruits, vegetables, fish, poultry, and meats. Because of these uses, ice is considered a food product. 🌐 This is why ice machines must be cleaned and sanitized on a regular basis. Using a sanitizer is an important step in the preventive maintenance process. An acid-based cleaner removes scale. Sanitizing kills harmful bacteria, viruses, and protozoa. 🌐

Bacteria, viruses, and protozoa can also adhere to moist areas on the inside of the evaporator compartment. These microscopic organisms can be either airborne or waterborne. Municipal water systems are relatively free of harmful waterborne organisms because chlorine has been added as a disinfectant. Water filtration or treatment, which will be covered in detail shortly, can also provide protection against bacterial contamination. However, airborne organisms can still be present. Once an organism sticks to a moist surface, it begins to grow because of the cool, damp conditions inside the evaporator compartment. The result may be a mold, algae, or slime buildup. Some of these bacteria and viruses can make people very sick. Slime is the most visible target when sanitizing an ice machine. It is usually a jellylike substance that is made up of algae, mold, and yeast spores. These spores can be either airborne or waterborne.

🌐 Lime or calcium buildup, often referred to as scale, is removed from the ice machine by using a commercial grade ice machine cleaner. 🌐 Ice machine cleaning solutions are made of a food-grade acid that is approved for use in food applications. They come in different strengths for many different applications. The weaker solutions are used for protecting different metals. Heavy scale deposits may require soaking in the cleaning solution. A stronger cleaner can be used on stainless-steel evaporators; however, a nickel-safe cleaner must be used on plated evaporators to eliminate possible flaking or peeling of the plating. Because copper and brass are more porous than stainless steel, if flaking or peeling occurs and the brass or copper is exposed, the evaporator will be more susceptible to bacterial growth. Circulating a dilute solution of acid-based cleaner through the water system will remove scale. However, it does not kill bacteria. A separate procedure called *sanitizing* must be performed to address bacteria.

Sanitizing is accomplished by using a commercial ice machine sanitizer. You cannot remove scale and sanitize with the same solution; two different solutions are needed. In fact, mixing ice machine cleaner and sanitizer together causes a strong chlorine gas that can be dangerous to breathe. The sanitizing solution must be circulated as directed by the ice machine manufacturer. Most manufacturers provide step-by-step cleaning and sanitizing instructions, which point out specific areas that must be disassembled and manually cleaned and sanitized. Sanitizing should be done at least once a month for ice machines used in the food service industry. Many service technicians mix their own sanitizing solution. This can be done by mixing 1 oz of household bleach to 2 gal of warm potable water. The water should be in the temperature range of 95°F to 115°F. This will give the solution about 200 ppm chlorine strength. However, always consult the manufacturer's directions for sanitizing the ice machine before using any sanitizing solution.

Water Filtration and Treatment

🌐 As mentioned earlier, water-related problems account for over 75% of service problems for ice makers and ice flakers. However, if cleaning and sanitizing are done on a schedule that is appropriate to your local water conditions, the ice machine will have a longer and more productive life dispensing quality ice. 🌐 Unfortunately, ice machines are often placed out of sight and seldom cleaned or sanitized until they have to be. This is where water treatment methods come into play. A good water treatment program will extend the cleaning and sanitizing intervals between service calls. But first, it is important to understand the local water conditions.

Water conditions can be broken down into the following three categories:
- Suspended solids
- Dissolved minerals and metals
- Chemicals

Sand and dirt are examples of suspended solids and are probably the easiest to remove. Mechanical filters in the water line can remove both sand and dirt.

Dissolved minerals and metals are much more expensive to capture. They require processes like reverse osmosis, distillation, or de-ionization to remove them. Reverse osmosis uses water pressure to force water through special membranes, which reject most of the dissolved minerals and organics and flush them to waste. Distillation requires water to be heated to its boiling point to create steam. The steam is then recondensed into water, leaving the minerals behind. De-ionization uses columns or beds of special filter media such as plastic beads or mineral granules to remove minerals from the water in exchange for other substances. An example of this is water softening. Water softeners are loaded up or regenerated with sodium ions from ordinary salt. These sodium ions are exchanged for calcium and magnesium, which are often referred to as the *water hardness* ions. Scale-inhibiting chemicals, which have to be food grade, can also successfully control dissolved minerals and metals.

Chemicals, on the other hand, have to be removed with carbon filtration. The most common chemicals in water are chlorine and chloramines, which are common city water disinfectants. Other types of chemicals that make their way into the water system can affect the taste, odor, and color of the water.

Figure 27–55 A triple-action filter system. *Courtesy Scotsman Ice Systems*

Water filtration and treatment can be applied to a water supply to try to treat these categories of water conditions. 🌐A common treatment for any type of ice machine is a triple-action filter system, **Figure 27–55**. These systems can filter out suspended solids, fight against scale formation by adding chemicals, and help the taste, odor, and color of water with carbon filters.🌐 Because the filters are micron filters, they do plug up and have to be changed on a regular basis. Often, a pressure gage installed on the filter will indicate a pressure drop when the filter is starting to plug up. Any time the scale inhibitor or carbon is used up is a good time to change all three cartridges. However, water treatment for a specific geographical region is never an exact science. Many times, a trial and error method is used because there is never an exact quick fix when it comes to water treatment. Always consult with your local water quality experts, ice machine distributors, or manufacturers for their professional opinions on the best water treatment for your geographical area. The water filter manufacturer can also assist you in testing the water at your site and recommending the correct water treatment combination to improve the water quality.

SAFETY PRECAUTION: These chemicals can be hazardous to humans. Always follow the manufacturer's instructions. The best time to circulate any chemicals is when there is no ice in the bin so they will not contaminate the stored ice. Because it is rare to find an ice maker empty, you may need to remove the ice from the bin and place it in a remote storage area. Then clean and rinse the ice bin according to the manufacturer's directions.

27.7 PACKAGE ICE MACHINE LOCATION

Most package ice machines are designed to be located in ambient temperatures of 40°F to 115°F. Follow the manufacturer's recommendations. If the ambient is too cold for the ice maker, freezing may occur and damage the float or other

components where water is left standing. If it is too hot, compressor failures and low ice capacities may occur. The ambient conditions under which the ice maker must operate are important.

Winter Operation

A machine located outdoors at a motel may not make ice below about 40°F ambient temperature because the thermostat in the bin is set at about 32°F. With this setting when ice touches the thermostat sensor, it will shut off the machine. The combination of 40°F outside and some ice in the bin will normally satisfy the thermostat. When the machine is located outside, the water should be shut off when the outside temperature approaches freezing. The water should be drained to prevent freezing and damage to the pump, the evaporator on some units, and the float assembly.

When machines are outside and the ambient temperature is below about 65°F, some method is normally used to prevent the head pressure from becoming so low that the expansion device will not feed enough refrigerant to the evaporator. Low-ambient control causing fan cycling is a common method used. Mixing valves, or head pressure control valves, are also used to maintain both discharge- and liquid-line pressures. These valves mix hot discharge gas with liquid coming from the receiver when the outdoor ambient drops below a certain temperature. At the same time, liquid is backed up in the condenser, causing a flooded condition. This flooding of the condenser will elevate the head pressure. Head pressure control valves are covered in detail in Unit 22, "Condensers."

The correct location of an ice machine indoors is critical because of the airflow across the condenser. Air cannot be allowed to recirculate across the condenser or high head pressures will occur. This will reduce the capacity and may in some cases harm the compressor, **Figure 27–56.**

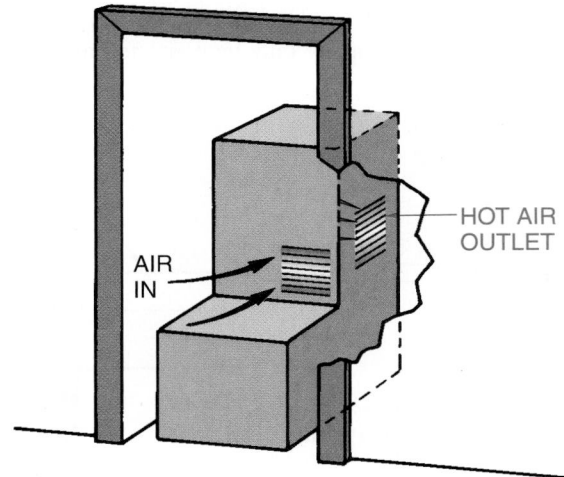

Figure 27–56 An ice machine located in an alcove with condenser air recirculating.

27.8 TROUBLESHOOTING ICE MAKERS

When an ice maker is not functioning properly, the problem can usually be found in the water circuit, the refrigeration circuit, or the electrical circuit.

Water Circuit Problems

Water level is critical in most ice makers. The manufacturer's specifications should be followed. Most manufacturers post the water level on the float chamber or at some location close to the float. Determine what the level should be and set it correctly, **Figure 27–57**.

The float valve has a soft seat, usually made of neoprene. This seat may become worn and a new one may be required. A float seat is inexpensive compared to an ice maker operating inefficiently for any period of time. When a float does not seal the incoming water, water at the supply temperature flows through the system. This water must be refrigerated to the freezing temperature and it may also melt ice that has already been made. A temporary repair may be to turn the float seat over if it can be removed from its holder, **Figure 27–58**.

The best time to determine whether or not a float is sealing correctly is when the ice maker has no ice. Remember, when there is ice in the bin, it is constantly melting. You cannot tell if the water in the drain is from melting ice or a leaking float. When there is no ice in the bin, the only water in the drain would be coming from a leaking float, **Figure 27–59**.

Water circulation over evaporators that make cube ice is accomplished by a water circulating system that is carefully designed and tested by the manufacturer. You should make every effort to understand what the manufacturer has intended before you modify anything. It is not unusual for a technician to try various methods to obtain the correct water flow over the evaporator when the manufacturer's manual would explain the correct procedure in detail. Do not overlook something as simple as a dirty evaporator. If it must be cleaned, check the manufacturer's suggestions.

The quality of the water entering the machine must be at least as good as the manufacturer recommends. Check the filter system if the water seems to have too many minerals. A water treatment company may be contacted to help you

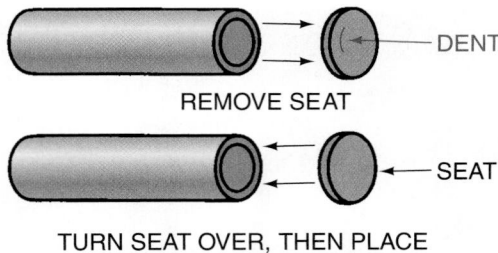

Figure 27–58   Sometimes a float seat may be reversed or turned over with the good side out for temporary repair.

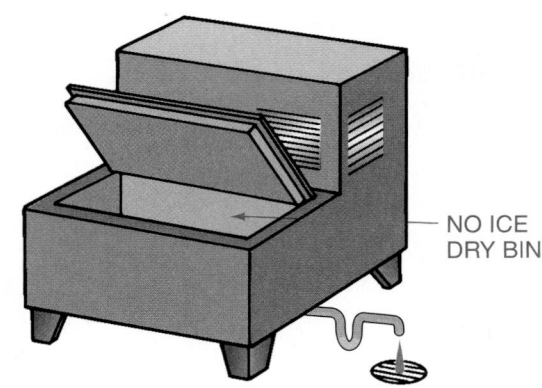

Figure 27–59   The ice machine storage bin is dry. There is no ice, but water is dripping out the drain. The float valve is leaking.

determine the water condition and the possible correction procedures.

Refrigeration Circuit Problems

Refrigeration circuit problems are either high-side or low-side pressure problems.

HIGH-SIDE PRESSURE PROBLEMS. High-side pressure problems can be caused by
■ poor air circulation over the condenser.
■ recirculated condenser air.

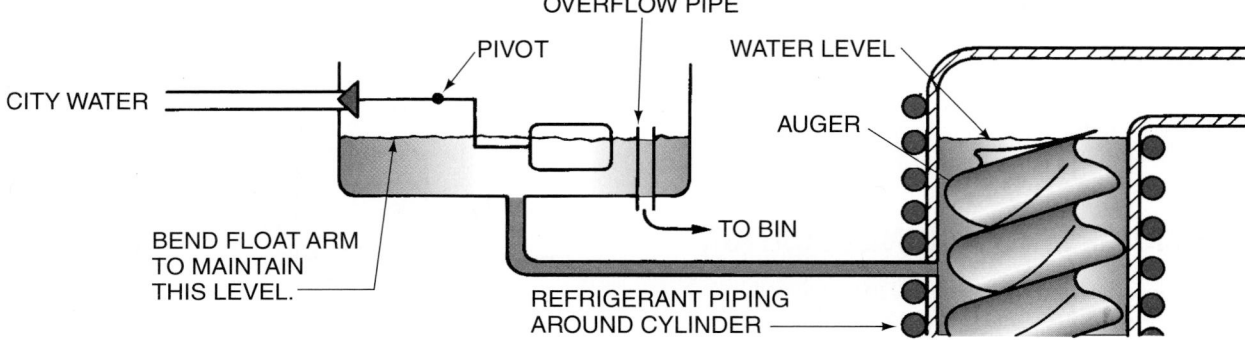

Figure 27–57   Maintain the correct float level.

- a dirty condenser.
- a defective fan motor.
- overcharge of refrigerant.
- noncondensables in the system.

Another high-side pressure problem is a result of trying to operate a system when the ambient temperature is below the design temperature for the unit. Low head pressure can cause low suction pressure, particularly in systems with capillary tube metering devices. Owners tend to try to place the machines outside and expect them to perform all year. The design conditions the manufacturer recommends should be followed.

LOW-SIDE PRESSURE PROBLEMS.   Low-side pressure problems can be caused by

- low refrigerant charge.
- incorrect water flow.
- mineral deposits on the water side of the evaporator.
- restricted liquid line or drier.
- defective metering device or thermostatic expansion valve.
- moisture in the refrigerant.
- inefficient compressor.
- blocked metering device.

Low-side problems will show a pressure that is too high or too low as indicated on the refrigeration gages. When the pressure is too low, you should suspect a starved evaporator. When the ice maker makes solid-type ice, a starved evaporator will often cause irregular ice patterns on the evaporator. The ice will become thicker where the evaporator surface is colder.

When the pressure is too high, you should suspect an expansion device that is feeding too much refrigerant, a defective TXV, or a compressor that is not pumping to capacity. When the expansion valve is overfeeding, the compressor will normally be cold from the flooding refrigerant, **Figure 27–60.**

A stuck water float can cause excess load if the entering water is warmer than normal, **Figure 27–61.** As mentioned earlier, it is hard to detect a leaking float. Look for the correct level in the float chamber. If adjusting the float does not help, the seat is probably defective.

The troubleshooting checklist in **Figure 27–62** is a general guide for the ice machine owner or service technician to

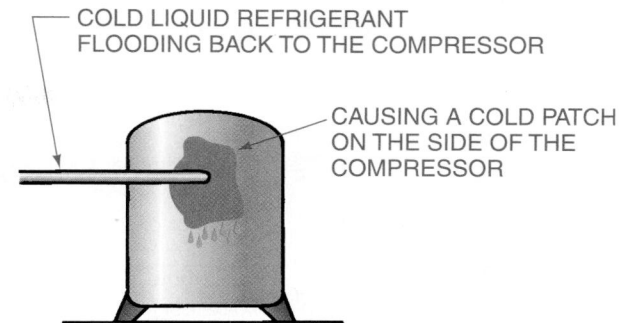

Figure 27–60   The expansion valve is allowing refrigerant to flood back to the compressor.

follow. This checklist is from a manufacturer's manual. Similar checklists can be found by contacting the manufacturer of the particular ice machine in question. As mentioned earlier, the Internet is an excellent source for quick troubleshooting information that can assist the service technician in systematically troubleshooting an ice machine.

Electrical Problems

A defrost problem may look like a low-pressure problem if the unit fails to defrost at the correct time. Heavy amounts of ice may build up. Remember, defrost is only used on machines that make cube ice. When heavy amounts of ice are found on a machine, look for the method of defrost. The machine may have a forced defrost cycle to allow you to clear the evaporator and start again. Some manufacturers may suggest that you not cause the machine to defrost or shut the machine off, that you should wait for the normal defrost cycle. This is so you can determine what type of problem caused the condition. If you arrive and find the evaporator frozen solid and force a defrost, you may not find the problem because it may not recur in the next cycle.

Line-voltage problems may be detected with a voltmeter. The unit nameplate will tell you what the operating voltage for the machine should be. Make sure the machine is within

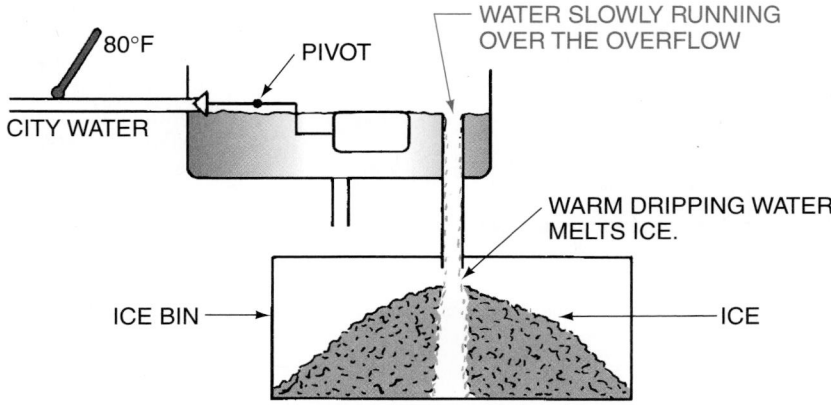

Figure 27–61   A float that does not seal will reduce the usable ice a machine can make.

Problem	Possible Cause	To Correct
Ice machine does not operate.	No electrical power to the ice machine.	Replace the fuse/reset the breaker/turn on the main switch.
	High pressure cutout tripping.	Clean dirty air filter and/or condenser coil (See page 4-1).
	ICE/OFF/CLEAN toggle switch set improperly.	Move the toggle switch to the ICE position.
	Water curtain stuck open.	Water curtain must be installed and swinging freely. See page 4-10.
	Remote receiver valve closed.	Open the valve.
Ice machine stops and can be restarted by moving the toggle switch to OFF and back to ICE.	Safety limit feature stopping the ice machine.	Refer to "Safety Limit Feature" on the next page.
Ice machine does not release ice or is slow to harvest.	Ice machine is dirty.	Clean and sanitize the ice machine. See pages 4-3 and 4-4.
	Ice machine is not level.	Level the ice machine. See page 2-6.
	Low air temperature around ice machine (air-cooled models).	Air temperature must be at least 35°F (1.6°C).
	Water regulating valve leaks in harvest mode (water-cooled models).	Refer to "Water-Cooled Condenser" on page 4-2.
Ice machine does not cycle into harvest mode.	The six-minute freeze time lock-in has not expired yet.	Wait for freeze lock-in to expire.
	Ice thickness probe is dirty.	Clean and sanitize the ice machine. See pages 4-3 and 4-4.
	Ice thickness probe wire is disconnected.	Connect the wire.
	Ice thickness probe is out of adjustment.	Adjust the ice thickness probe. See page 3-4.
	Uneven ice fill (thin at top of evaporator).	See "Shallow or Incomplete Cubes" on the next page.
Ice quality is poor (soft or not clear).	Poor incoming water quality.	Contact a qualified service company to test the quality of the incoming water and make appropriate filter recommendations.
	Water filtration is poor.	Replace the filter.
	Ice machine is dirty.	Clean and sanitize the ice machine. See pages 4-3 and 4-4.
	Water dump valve is not working.	Disassemble and clean the water dump valve. See page 4-7.
	Water softener is working improperly (if applicable).	Repair the water softener.
Ice machine produces shallow or incomplete cubes, or the ice fill pattern on the evaporator is incomplete.	Ice thickness probe is out of adjustment.	Adjust the ice thickness probe. See page 3-4.
	Water trough level is too high or too low.	Check the water level probe for damage. See page 3-4.
	Water inlet valve filter screen is dirty.	Remove the water inlet valve and clean the filter screen. See page 4-9.
	Water filtration is poor.	Replace the filter.
	Hot incoming water.	Connect the ice machine to a cold water supply. See page 2-10.
	Water inlet valve is not working.	Remove the water inlet valve and clean it. See page 4-9.
	Incorrect incoming water pressure.	Water pressure must be 20-80 psi (137.9–551.5 kPA).
	Ice machine is not level.	Level the ice machine. See page 2-6.

Figure 27–62   A basic service checklist for an ice machine. Page numbers do not refer to this text. *Courtesy Manitowoc Ice, Inc.*

Problem	Possible Cause	To Correct
Low ice capacity.	Water inlet valve filter screen is dirty.	Remove the water inlet valve and clean the filter screen. See page 4-9.
	Incoming water supply is shut off.	Open the water service valve.
	Water inlet valve is stuck open or leaking.	Remove the water inlet valve and clean it. See page 4-9.
	The condenser is dirty.	Clean the condenser. See page 4-1.
	High air temperature around ice machine (air-cooled models).	Air temperature must not exceed 110°F (43.3°C).
	Inadequate clearance around the ice machine.	Provide adequate clearance. See page 2-5.
	Objects stacked on or around ice machine, blocking air flow to condenser (air-cooled models).	Remove items blocking air flow.
	Air baffle is not installed (air-cooled models).	Install air baffle. See page 2-12.

Figure 27–62 *(continued)*

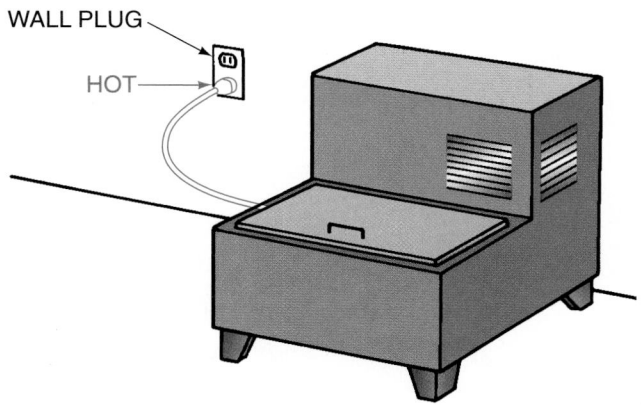

WALL PLUG

HOT

Figure 27–63 This unit has a poor connection at the wall plug.

the correct voltage range. The plug on the end of the power cord may become damaged or overheated due to a poor connection. Examine it while the machine is running by touching it. If it is hot to the touch, it is not properly conducting power to the machine, **Figure 27–63.**

When the power is wired to the junction box in the ice maker, it goes to the control circuit that operates the machine. Normally, the control voltage is the same as the line voltage. Larger machines may use relays to energize contactors for starting the compressor.

27.9 SERVICE TECHNICIAN CALLS

The following service calls show some of the most common problems experienced with ice machines.

SERVICE CALL 1

A service technician is servicing an ice flake machine. After taking the covers off the machine, the technician does not notice anything wrong. The bin is half full of ice and the machine seems to be working properly. The technician then decides to talk to the restaurant manager. The manager says that the ice flake machine makes loud screeching noises intermittently. The technician returns to the machine and does hear some small screeching noises coming from the evaporator and auger assembly. The noises get progressively louder for about 10 min and then quit. Ten minutes later, the noises start again. *The problem is mineral buildup in the evaporator freezing cylinder and auger compartment. This causes the auger to drag on the mineral deposits, making loud screeching noises like two pieces of Styrofoam being rubbed together. Sometimes the mineral deposits break loose and are delivered with the ice.*

The service technician examines the ice flakes for quality and finds that they show a lot of cloudiness. This is caused by minerals trapped in the flakes. The technician then decides that the ice flake machine needs a cleaning to rid it of the mineral deposits. The technician also decides to sanitize the machine after the cleaning to rid the machine of any harmful bacteria, viruses, and protozoa. Emptying the ice bin of all ice flakes is the first step the technician takes.

The technician follows the manufacturer's instructions in the service and maintenance manual, acquired from an Internet search, on cleaning and sanitizing the entire ice machine and bin. After restarting the ice flake machine, the technician finds that it runs quietly and that the ice has a much better quality than before the cleaning and sanitizing procedure. (Often, instructions for cleaning and sanitizing are located on a label on the ice machine itself.)

The service technician stays for about 45 min, cleaning up and talking to the restaurant manager about a preventive maintenance and water treatment program for all the ice machines in the restaurant. The technician then returns to the ice flake machine and still hears no screeching sounds. It appears that the problem has been remedied.

SERVICE CALL 2

A service technician is called to service a cell-type ice cube machine that will not harvest the ice from the evaporator. The sheet of ice grows much too thick and actually freezes the ice thickness probe into the ice sheet. *The ice thickness probe's electrodes are scaled with mineral buildup. This prevents them from completing a circuit when the thin water film next to the ice sheet contacts them. (Evaporator damage from too thick an ice sheet can occur. Also, low suction pressure from a reduced evaporator load, which may lead to compressor flooding, can occur if the problem is not fixed.)*

The service technician bypasses the freeze time lock-in feature by moving the ice/clean/off switch to *off* and back to *ice.* The technician waits for water to start running over the evaporator and then clips a jumper wire lead to the ice thickness probe and cabinet ground, **Figure 27–64.** The harvest light on the control board comes on. Eight seconds later, the ice machine goes into a harvest cycle. The technician can hear the hot gas solenoid valves energize and hot gas pass into the evaporators. The ice thickness probe

is taken off the evaporator and cleaned with the proper ice machine cleaner to get rid of any mineral deposits. While cleaning up, the technician watches the ice machine through three freeze and harvest cycles and determines that the ice machine is working as designed. After talking with the storeowner about preventive maintenance and water treatment, the technician leaves the job.

SERVICE CALL 3

A service technician is called to a restaurant to service an ice machine that makes cylindrical ice. The restaurant owner is complaining about cloudy ice that melts very fast. Customers are complaining that there is flaky sediment in the bottom of their glasses after the ice melts. *The problem is a defective water dump solenoid valve. Without dumping or purging the old water from the water reservoir every cycle, the same water is being used and the mineral content becomes very high. Excess minerals that are frozen in the ice cause it to become soft, cloudy, and dirty.*

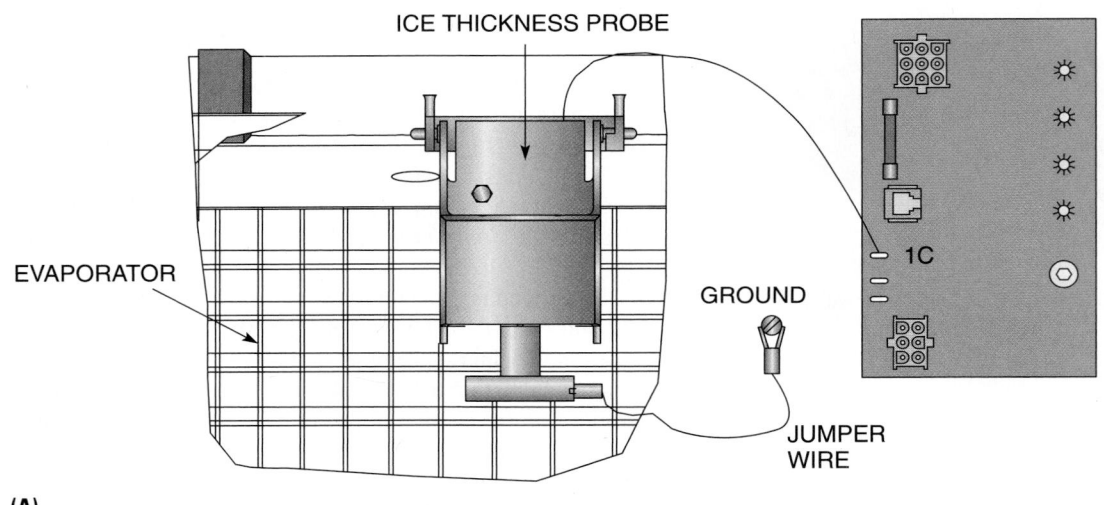

(A)

(B)

Figure 27–64 An ice thickness probe and control board. **(A)** Before jumper. **(B)** After jumper. *Courtesy Manitowoc Ice, Inc.*

The service technician inspects some ice in the ice bin and sees that it is soft, cloudy, and dirty with minerals. After observing the freeze and harvest cycles and determining that there is no purging of water during the harvest cycle, the technician turns off the ice machine and unhooks the purge solenoid from the control board. An ohmmeter is used to check resistance of the purge solenoid's coil. The resistance checks to be infinite, indicating an open solenoid coil. A new coil is put in place and the solenoid coil is hooked back up to its control board terminals. The ice machine is turned back on and a cycle is again observed. During a harvest cycle, the control board energizes the water purge solenoid and water freely flows down the drain. This action empties the water reservoir of old mineral-laden water, and fresh water now fills the reservoir. The technician catches the ice before it falls to the bin and observes it. The ice is crystal clear, hard, and clean. The service technician then talks to the restaurant owner about a preventive maintenance and water treatment program, cleans up around the ice machine, and leaves the job site.

SERVICE CALL 4

A service technician is servicing an ice maker and has determined that the ice machine will not come out of a hot gas defrost mode as quickly as it should. Too long a defrost will cook off the metal coating on the evaporator. Instead of the harvest motor's probe pushing the ice off the coil, which will terminate defrost with a curtain switch, the evaporator gets very hot from the prolonged hot gas defrost. In other words a "cookout" occurs. The ice slowly falls off the evaporator in chunks by gravity. Sometimes, one of these chunks will trip the curtain switch and bring the machine out of defrost, and other times it will not. *The problem is a defective harvest motor. Ice cannot be forced off the evaporator during harvest with a defective harvest motor.*

The service technician gets out a voltmeter and measures the output terminals of the microprocessor labeled motor 1 and motor 2 while in defrost, **Figure 27–51.** The technician measures 115 V AC at both terminals and determines that this is the correct voltage. This process tells the technician that the microprocessor is doing its job and sending an output signal to the harvest motor. The microprocessor is not to blame. The problem must lie in the harvest motor or the wires leading to the harvest motor. The service technician then checks voltage at the harvest motor itself and gets 115 V AC. This rules out the problem being in the wiring between the microprocessor and the harvest motor. The problem must lie in the harvest motor itself. (If the correct voltage is going to a power-consuming device, and it is not operating, the problem lies within the power-consuming device.)

The service technician retrieves a new harvest motor from the service van and installs it. After watching the ice machine for two complete cycles, the technician determines that everything works as it should.

SERVICE CALL 5

A service technician servicing an ice maker that has a microprocessor observes that the ice machine will not come out of ICE 1 mode. The technician consults the service manual and finds that the evaporator must reach 14°F before an internal timer will start. The technician wonders if it is the microprocessor's fault or if the problem is in one of the inputs to the microprocessor. *The problem is a defective thermistor on the evaporator. With a defective thermistor, the microprocessor does not know the evaporator's temperature and cannot advance the ice machine to the next mode (ICE 2).*

The technician reads the manual further and discovers that the microprocessor is fed an input signal from a thermistor connected to the evaporator. A thermistor is nothing but a resistor that varies its resistance as the temperature changes. This thermistor is what signals the digital LED display what mode the ice maker is in. It also tells the microprocessor when the evaporator has reached 14°F. The manufacturer has supplied a resistance versus temperature relationship table for checking thermistors. The only sure temperature that the technician can simulate accurately is 32°F. This is accomplished by placing some crushed ice in a small amount of water and stirring. The thermistor lug is then taken off the evaporator and is also electrically disconnected from the microprocessor's evaporator input terminal. The thermistor lug is given a chance to stabilize its temperature at 32°F in the ice water bath. A resistance reading taken with an ohmmeter is 5.50 thousand ohms or 5.50 Kohms. According to the temperature versus resistance chart for the thermistor, the resistance should be 9.842 Kohms. This indicates a defective thermistor and explains why the microprocessor is not going out of ICE 1 mode. A new thermistor is ordered from the manufacturer.

The technician could have measured the signal coming into the microprocessor's input terminal labeled EVAP, but that would require knowing what it was supposed to be. Most of the time these signals are a low AC or DC voltage, but they can be a resistance signal. The service technician must read the manual to find out.

SUMMARY

- Most ice flakers are critically charged refrigeration systems that come in a split or a self-contained unit.
- Ice flake machines use an auger turning inside a flooded cylinder to scrape ice off the sides of the cylinder.
- The auger has cutting surfaces called flights that do the ice cutting.
- Ice flake machines can be retrofitted to parallel refrigeration systems in a supermarket.
- When water is frozen in the evaporator or freezing cylinder, minerals from the water are left behind and often become attached to the interior cylinder walls.
- An ice machine cleaner will dissolve many minerals from the freezing section of an ice flake machine.

■ Mineral buildup on the freezing cylinder of an operating ice flake machine will cause a sound like two pieces of Styrofoam being rubbed together.

■ Ice flake machines usually have a gear reduction drive, which reduces the auger's drive motor speed.

■ Mineral buildup on a freezing cylinder of an ice flake machine will cause auger motor and auger bearings to be stressed. Gears will also be stressed.

■ The gear motor assembly is more prone to fail than any other part in the ice flake machine because of the pressures and stresses put on it.

■ Worn bearings on the auger will cause the auger to wobble and rub against the freezing cylinder walls. This often causes scrape marks on the cylinder walls and metal shavings in the ice.

■ A dual float switch and a water control valve usually control the amount of water coming into an ice flaker's water reservoir and also provide for ice machine shutdown in case of low water.

■ A water flush cycle is a way to control unwanted mineral buildup in the water reservoir. A mechanical timer or a microprocessor can initiate these cycles.

■ Ice bin controls are designed to terminate the operation of the ice machine when the bin is full of ice.

■ Ice production depends on the ice machine's water inlet temperature and the temperature around the ice machine. As the water temperature or air temperature surrounding the ice machine rises, ice production drops.

■ Some ice flake machines provide a means for the service technician or owner to adjust the size or shape of the ice flakes. Rotating the ice cutter or changing the extrusion head are ways to accomplish this.

■ Some ice machines make flat ice on a flat evaporator to a predetermined thickness. The flat ice sheet then falls or slides onto a grid of cutting wires. The wires have a low voltage to provide heat to cut the ice.

■ Making crescent-shaped ice involves running water over an evaporator plate in which copper evaporator tubing is soldered onto the plates at regular intervals. The coldest points on the evaporator are where the copper coil contacts the evaporator plate. Ice starts to form here first.

■ During a hot gas harvest, the cubes slide down the evaporator plate by gravity. Dimples in the evaporator plate will break the capillary attraction between the plate and the crescent-shaped cube. Ice then falls in a bin for storage.

■ Making cell-type ice cubes involves having a vertical evaporator plate with water flowing over the individual cells by a water pump. A serpentine evaporator coil is soldered to the back of the plate. Ice gradually forms in the cells and soon bridges with the neighboring cell.

■ Too slow a water flow will freeze minerals with the ice. A restrictor in the water line regulates the water flow rate.

■ Once a certain bridge thickness is reached, the water film on the ice will contact an adjustable ice thickness probe. A hot gas harvest is then initiated.

■ A hot gas solenoid is energized in the discharge line. The compressor keeps running and pumps superheated gas through the evaporator.

■ A hot gas defrost loosens the cubes from the evaporator plate. Either by gravity or by a pushrod, the ice comes off the evaporator plate and falls into a storage bin.

■ Weep holes between the cells on the evaporator allow air entering from the edges of the ice sheet to travel along the entire ice slab. This relieves the attractive suction forces between the evaporator plate and the ice sheet.

■ Cylindrical ice is made inside a tube-within-a-tube evaporator. As water flows through the center tube, it freezes on the outside walls of the tubes. Defrost lets the ice shoot out the end of the tube as a long cylinder.

■ Many different control schemes and controls from different ice machine manufacturers are used for water sump levels, ice thickness, harvest initiation, defrost termination, and ice storage bin controls.

■ During hot gas defrost or harvest, liquid floodback may occur to the compressor. To combat this, a thermistor is placed on the discharge line of the compressor to sense the discharge temperature. A cold discharge line during a harvest period usually means liquid floodback to the compressor.

■ Hot gas valves are often cycled by a microprocessor while in harvest in response to the discharge temperature to control liquid floodback.

■ Microprocessors are often the main controller on modern ice machines. They have stored programs or algorithms within them. This is actually a stored sequence of events in their memories.

■ Microprocessors have input and output terminals where they receive and transmit information, respectively. Inputs and outputs can be either digital devices, like switches or analog devices, like thermistors.

■ A service technician should know the microprocessor's sequence of events and how to input/output troubleshoot them. Service manuals and the Internet are often used to gain knowledge of a certain manufacturer's microprocessor sequence of events.

■ Many microprocessors have self-diagnostics to assist the service technician in systematic troubleshooting.

■ Microprocessors also have error codes, which can be displayed to assist the service technician in troubleshooting.

■ Many microprocessors can record the ice machine's history or summary of past operations. This memory can then be recalled by simply pushing buttons. Information that has been gathered over a long period of time, like the number of harvests, error codes stored, and average cycle times, can be recalled.

■ For air-cooled ice machines, ice production depends mainly on entering-water temperature and the temperature of the air entering the condenser.

■ For air-cooled ice machines, suction pressure, head pressure, electrical consumption, and other system performance factors mainly depend on entering-water temperature and the temperature of the air entering the condenser.

■ Ice quality depends on the quality of the water entering the ice machine. Good water quality will produce a crystal-clear hard cube with good cooling capacity. Poor water quality can produce soft, cloudy ice with less cooling capacity.

■ Water can contain many different minerals in various levels. Water filtration and treatment is a science in itself, but the service technician needs to know the basics of water quality to service ice makers. Important terms to know are filtration, treatment, sediment, flow rate, scale, scale inhibitor, water softener, reverse osmosis, iron, total dissolved solids, turbidity, pH, acidic water, alkaline, chlorine, chloramines, and hardness.

■ Ice quality is measured in percent of hardness. Hardness is a measurement that represents the thermal cooling capacity of the ice. The higher the percent hardness, the more cooling capacity the ice possesses.

■ The cleaner the ice machine and the purer the water, the harder the ice will be.

■ Preventive maintenance is the action of regularly scheduled maintenance performed on an ice machine. It includes inspecting, cleaning, sanitizing, and servicing the ice machine and external water filtration/treatment system.

■ Cleaning an ice machine with an acid-based cleaner simply removes scale from the mineral content in the water.

■ A sanitizing fluid that is approved for commercial ice machines kills viruses, bacteria, and protozoa.

■ If cleaning and sanitizing are done on a schedule that is appropriate for the local water conditions, the ice machine will have a longer and more productive life dispensing quality ice.

■ Water treatment programs are designed for local areas to help with certain water conditions. Water conditions can be broken down into suspended solids, dissolved minerals, and chemicals.

■ A common type of water treatment is a triple-action filter system that can filter out suspended solids, fight against scale formation by adding chemicals, and improve the taste, odor, and color of the water with carbon filters.

REVIEW QUESTIONS

1. Which type of ice maker has a continuous harvest cycle?
2. Which type of ice is made using a gear motor and auger?
   A. Cube ice
   B. Cylinder ice
   C. Flake ice
   D. Block ice
3. What would be the symptoms of a defective float on an ice maker?
4. What are the two functions of the dual float switch in an ice flake machine?
5. Which part of an ice flake machine is most prone to failure?
   A. The gear motor
   B. The float mechanism
   C. The ice chute
   D. The compressor
6. List two causes that may lead to auger wobble.
7. What is the function of a flush cycle for an ice flake machine?
8. Name and explain three types of ice bin controls in ice flake machines.
9. Ice production in an ice flake machine depends mainly on two factors. Name and briefly explain these two factors.
10. Explain the function of weep holes on the evaporator of a cell-type ice maker.
11. True or False: Stainless steel is often used on evaporators because it is durable, it is very sanitary, and it resists corrosion.

12. Explain what is meant by ice bridging in a cell-type ice maker.
13. Explain the function of the centrifugal pump in ice makers.
14. What is the function of the ice thickness probe in an ice maker?
15. Explain the function of the water dump solenoid valve in an ice maker.
16. What is the most common type of defrost during a harvest cycle for an ice maker?
    A. Electric
    B. Natural
    C. Hot gas
    D. Air
17. Explain how liquid floodback to the compressor may occur during a hot gas defrost period with ice makers.
18. Why do ice machines need to be cleaned periodically?
19. Why do ice machines need to be sanitized periodically?
20. True or False: The same solution can be used to clean and sanitize an ice machine.
21. What is a common troubleshooting technique used on microprocessors?
    A. Input/output troubleshooting
    B. Ohmmeter
    C. Ammeter
    D. Guesswork
22. One of the quickest and most thorough ways of obtaining technical service information for an ice machine is through the
    A. manufacturer.
    B. wholesale store.
    C. Internet.
    D. fax machine.
23. Most of the time, the input signals to a microprocessor are
    A. high voltage.
    B. low voltage.
    C. steady.
    D. high frequency.
24. A way in which a microprocessor communicates to a service technician about a certain malfunction in the ice machine is
    A. a summary.
    B. an error code.
    C. a history.
    D. a data link.
25. True or False: Entering-air temperature and water temperature are the main factors for determining the amount of ice produced in an air-cooled ice machine.
26. A measurement that represents the thermal cooling capacity of ice and also represents the amount of calcium and magnesium in water is
    A. hardness.
    B. turbidity.
    C. alkalinity.
    D. scale.

27. A buildup of minerals that forms a flaky coating on the surface of the evaporator and water system is called
    A. sediment.
    B. scale.
    C. hardness.
    D. turbidity.
28. A method for killing bacteria, viruses, and protozoa in ice machines is
    A. sanitizing.
    B. cleaning.
    C. rinsing.
    D. brushing.
29. True or False: Cleaning with an approved ice machine cleaner is a method of removing scale from an ice machine.

30. Water conditions can be broken down into three categories. Name these categories.
31. True or False: A common type of water filtration and treatment for ice machines is a triple-action filter system.
32. True or False: Chemicals in the water have to be removed with carbon filtration.
33. Sand and dirt in water are usually easily removed by
    A. filters.
    B. reverse osmosis.
    C. distillation.
    D. de-ionization.

UNIT 28

Special Refrigeration Applications

OBJECTIVES

After studying this unit, you should be able to

- describe methods used for refrigerating trucks.
- discuss phase-change plates used in truck refrigeration.
- describe two methods used to haul refrigerated freight using the railroad.
- discuss the basics of blast cooling for refrigeration.
- describe cascade refrigeration.
- describe basic methods used for ship refrigeration.

SAFETY CHECKLIST

✔ Ensure that nitrogen and carbon dioxide systems are off before entering an enclosed area where they have been in use. The area should also be ventilated so that oxygen is available for breathing and has not been replaced by these gases.

✔ Liquid nitrogen and liquid carbon dioxide are very cold when released to the atmosphere. Take precautions that they do not get on your skin. Wear gloves and goggles if you are working in an area where one of these gases may escape.

✔ Ammonia is toxic to humans. Wear protective clothing, goggles, gloves, and breathing apparatus when servicing ammonia refrigeration systems.

28.1 SPECIAL APPLICATION REFRIGERATION

Many special applications of refrigeration exist. Some technicians may never work with these special applications, but others will. Keep in mind the principles of heat and heat transfer as they pertain to refrigeration. These principles were covered in Unit 1.

28.2 TRANSPORT REFRIGERATION

Transport refrigeration is the process of refrigerating products while they are being transported from one place to another. This can be by truck, rail, air, or water. For example, vegetables may be harvested in California and shipped to New York City for consumption. These products may need refrigeration along the way.

Many different products must be shipped in many different ways. Some of the products are shipped in the fresh state, such as lettuce and celery, which are shipped at about 34°F. Some products are shipped in the frozen state, such as ice cream, which is shipped at about −10°F. The fresh vegetables may be shipped through climates that are very hot, such as passing through Arizona and New Mexico. They may also be shipped through very cold climates, such as in the winter when fresh vegetables are shipped through Minnesota. Different situations require different types of systems. For example, refrigeration may be needed for a load of lettuce when leaving California but when it goes to the northern states, it may need to be heated to prevent it from freezing, **Figure 28–1**.

Figure 28–1   This load of lettuce needs refrigeration when crossing the desert. It needs heat in the northern states.

Products that fall into the preceding categories are carried on trucks, trains, airplanes, and ships. The refrigeration system is responsible for getting the load to the destination in good condition.

28.3 TRUCK REFRIGERATION SYSTEMS

Several methods are used to refrigerate trucks. One of the most basic systems that is still being used is to merely pack the product in ice that is made at an ice plant. This works well for a well-insulated truck for short periods of time. The melting ice produces water that must be handled. For years, you could spot these trucks because of the trail of water left behind, **Figure 28–2.** This method can hold the load temperature to only above freezing.

Dry ice, which is solidified and compressed carbon dioxide, is also used for some short-haul loads. It is somewhat harder to hold the load temperature at the correct level using dry ice because it is so cold. Dry ice changes state from a solid to a vapor (called sublimation) at −109°F without going through the liquid state. By using these temperatures, the food products could be reduced to very low temperatures. Dehydration of food products that are not in airtight containers can become a problem when dry ice is used because the very low temperature of −109°F is a great attraction for moisture in the food.

Liquid nitrogen or liquid carbon dioxide (CO_2) may also be used for the refrigeration of products. This is accomplished by using nitrogen that has been refrigerated and condensed to a liquid and stored in a cylinder on the truck at a low pressure. The cylinder has a relief valve that releases some of the vapor to the atmosphere if the pressure in the cylinder rises above the set point of about 25 psig in the cylinder. This boils some of the remaining liquid to drop the cylinder temperature and pressure, **Figure 28–3.** Liquid is piped to a manifold where the refrigerated product is located and released to cool the air in the truck and the product. A thermostat located in the food space controls a solenoid valve in the liquid nitrogen line, stopping and starting the flow of liquid, **Figure 28–4.** Fans may be used to disperse the cold vapor evenly over the product.

Air in the atmosphere is 78% nitrogen and is the source for the nitrogen used for refrigeration. It is not toxic; it can be

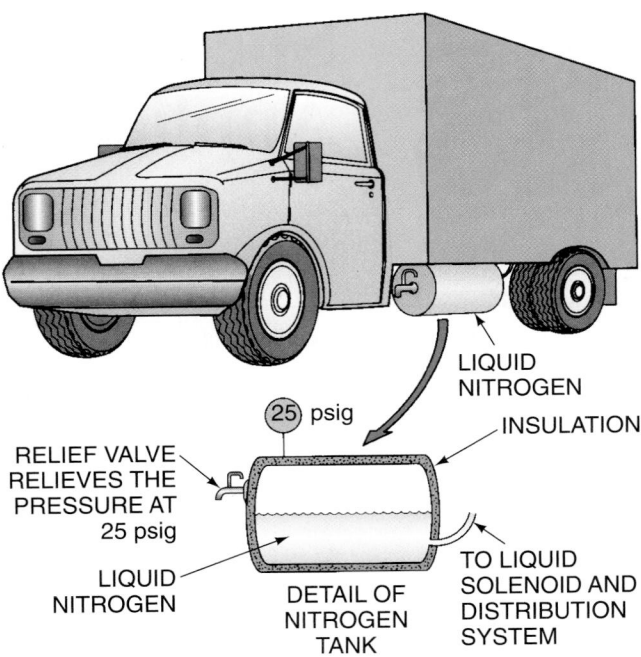

Figure 28–3 A nitrogen cylinder maintained at about 25 psig.

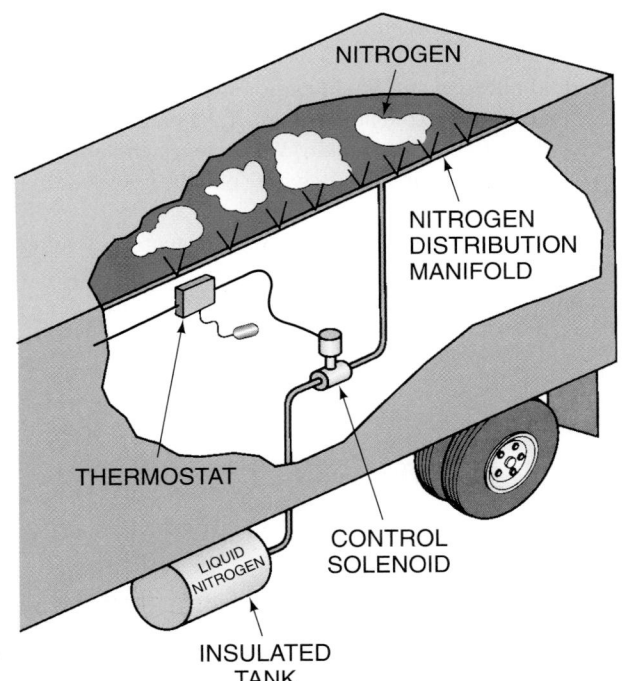

Figure 28–4 A control for a liquid nitrogen cooling system.

released to the atmosphere with confidence that no harm will occur to the environment. SAFETY PRECAUTION: Nitrogen has no oxygen in it so it is not suitable for breathing. Special safety interlocks shut the refrigeration off if the door to the refrigerated truck is opened to prevent personnel from being smothered from lack of oxygen. Nitrogen is also very cold; it evaporates at −320°F at atmospheric pressure. Any contact with the skin would cause instant freezing. Never allow liquid nitrogen to get on your skin.

Figure 28–2 When ice is used for preserving products, it melts and often leaves a trail of water on the highway or a puddle in a parking area.

Liquid CO_2 can be used in the same manner as liquid nitrogen to refrigerate products. The system would work much the same as a liquid nitrogen system; only CO_2 boils at $-109°F$.

Even though liquid nitrogen and CO_2 are both very cold, they can be used to refrigerate medium-temperature loads at $35°F$ with proper controls. These controls would include a distribution system for the refrigerant that does not impinge on the product and a space-temperature thermostat that accurately senses the refrigerated space temperature.

Both of the liquid injection methods have been used for many years as a replacement for mechanical refrigeration for permanent installations and for emergency use for refrigeration. They are probably more expensive in the long run than a permanent installation, but they also have some good points. For example, when used to refrigerate fresh foods, the presence of nitrogen or CO_2 vapor displaces any oxygen and helps preserve the food. The system is simple and easy to control because only a solenoid and a distribution manifold need to be maintained.

Many truck bodies are equipped with refrigerated plates (called cold plates) that have a phase-change solution inside called a *eutectic solution. Phase change* means to change the state of matter. This is much like the change of state from ice to water, except at a different temperature. When the change of state occurs, much more heat is absorbed per pound of material. This eutectic solution has the capability of changing the state at many different temperatures, depending on the composition of the product.

The eutectic solution in the cold plate is brine, a form of salt water using either sodium chloride, a form of table salt, or calcium chloride. These brines are very corrosive and must be handled with care.

Various strengths of solution of the brine may be used to arrive at the desirable melting temperature below freezing, called the eutectic temperature. Actually, the brine does not freeze solid; salt crystals form, and the remaining solution is liquid. Different strengths of brine will cause the crystals to form at different temperatures.

The brine solution is contained in plates that are from 1 in. to 3 in. thick. They are mounted on the wall or ceiling of the truck in such a manner that air can flow all around them, **Figure 28–5.** The room air transfers heat to the plates. The brine solution changes from a solid to a liquid while absorbing heat. To recharge the plates, they must be cooled to the point that the change of state occurs back to solid crystals. These plates will hold a refrigerated load at the correct temperature (for medium- or low-temperature application) for an all-day delivery route to be run. For example, a truck loaded with ice cream may be loaded at night for delivery the next day. The ice cream must be at its prescribed temperature, and the truck cold plates must be at the design temperature. For ice cream, this may be $-10°F$ or lower. When the truck comes back at night, it may still be $-10°F$, but most of the brine solution will be changed back to a liquid and will need regenerating to the solid crystal state. The heat absorbed during the day is in the brine solution and must be removed.

The plates are recharged in several ways. Some systems have a direct expansion coil located in the cold plate. This

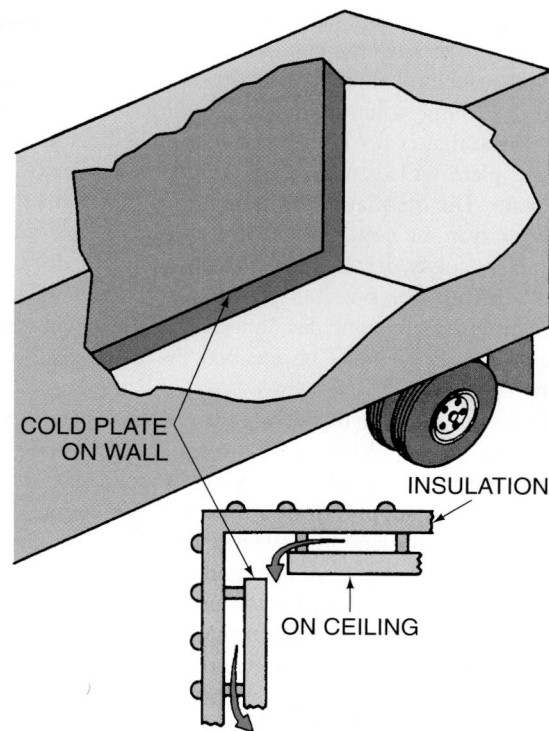

Figure 28–5   Cold plates mounted in a truck body.

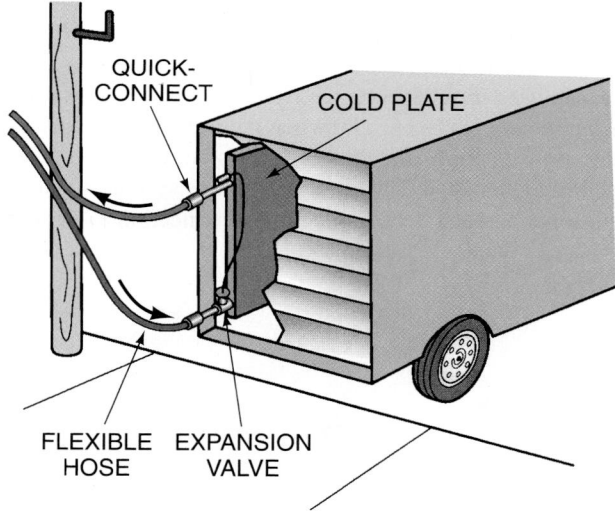

Figure 28–6   A cold plate connected to the refrigeration system at the dock.

coil is connected to the refrigeration system at the loading dock, **Figure 28–6.**

The refrigeration system may use one of the typical refrigerants, R-12, R-134a, R-500, R-404A, or R-502, or it may be R-717 ammonia or an environmentally friendly alternative refrigerant. When the truck backs up to the loading dock, quick-connect hoses are connected from the central refrigeration plant to the truck. The expansion valve may be located on the loading dock or at the actual coil in the eutectic cold plate. The central refrigeration system will reduce the pressure in the evaporator in the cold plate to the point that the brine solution will change back to solid crystals in a few hours. The truck may then be loaded with product.

Another method of recharging the cold plate is to circulate refrigerated brine solution through a coil located in the cold plate. Brine that is refrigerated at a lower temperature than the cold plate will then recharge or freeze the brine in the cold plate. The circulating brine must be much colder than the solution in the cold plate. Connecting this system to the truck also involves quick-connect fittings at the loading dock. There is always the possibility that some brine will be lost during the connecting and disconnecting of the quick-connect fittings. This brine must be cleaned up, or corrosion will occur.

Trucks may also be refrigerated using a refrigeration system on the trucks. Whatever the refrigeration source is, it must have a power supply. This power supply may be from the truck power supply, from a diesel or gas engine-driven compressor, from a land line (electric power to a building), or from an electric compressor operated by a motor generator.

Van-type delivery trucks often use a refrigeration compressor mounted under the hood that is turned by the truck engine. The refrigeration may be controlled by cycling the compressor using an electric clutch that is thermostatically controlled, **Figure 28–7**. This system works well as long as the truck engine is running. Because the truck is usually moving all day during delivery, refrigeration can easily be maintained. Some of these trucks have an auxiliary electric motor-driven compressor that may be plugged in at the loading dock while the truck engine is not running.

Trucks with small condensing units under the truck body are also used to refrigerate the load. These often are used to recharge cold plates at night when the truck is at the dock or other power supply and the cold plates hold the load during the day, **Figure 28–8**. The advantage of this system is that it is self-contained and does not require a connection to a central

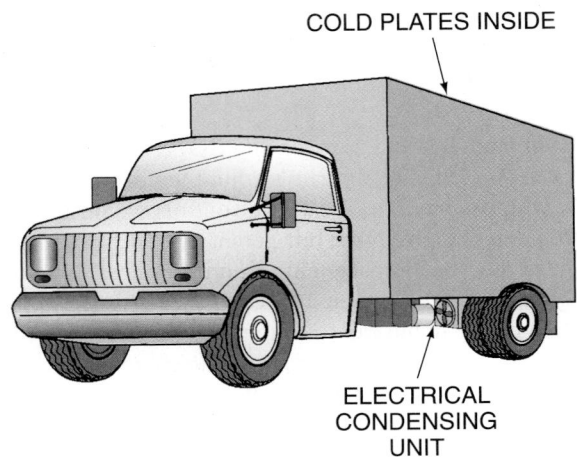

COLD PLATES INSIDE

ELECTRICAL CONDENSING UNIT

Figure 28–8   A small refrigeration system on the truck recharges the cold plates while the truck is parked at night.

refrigeration system at the loading dock. These condensing units may be powered while on the road using a small motor generator unit mounted under the chassis. In this case, there would be no need for cold plates because the generator would operate on the road and the compressor could be plugged in at night.

The motor generators have either a gas or diesel engine that turns the generator to generate 230-V, 60-cycle current to operate the compressor. Diesel is the preferred fuel because diesel engines will operate longer with less fuel and with less maintenance. Diesel engines have a greater first cost but seem to be less expensive in the long run.

A generator producing 230 V from the truck engine has been used in some cases. It must have a rather sophisticated control because the truck engine turns at so many different speeds. The stationary generator has only one speed. Different speeds would affect the power supplied to the compressor. As the engine turns faster, the voltage and the cycles per second vary; 60 cps is normal for alternating current electric motors. Electronic circuits may be used to maintain a constant voltage.

Larger trucks use either nose-mount or under-belly units that may be operated using either a diesel-engine driven compressor or a motor generator driven by a diesel engine, **Figure 28–9**. When the engine turns the compressor, it has a governor to regulate the speed of the compressor. The system may be two-speed, high and low. Control of the compressor may be accomplished using cylinder unloading on the compressor. For example, suppose the compressor is a four-cylinder compressor and has a capacity of 1 ton per cylinder. The compressor may be operated from 1 ton to 4 tons by unloading cylinders, **Figure 28–10**. There should be a minimum load of 1 ton at all times, or the diesel engine will have to be cycled off and on. It is most desirable to keep the engine running; however, controls for diesel engines have been developed that will automatically start and stop the engine, **Figure 28–11**.

The evaporator for these truck refrigeration units is typically located at the front of the truck and blows air toward the

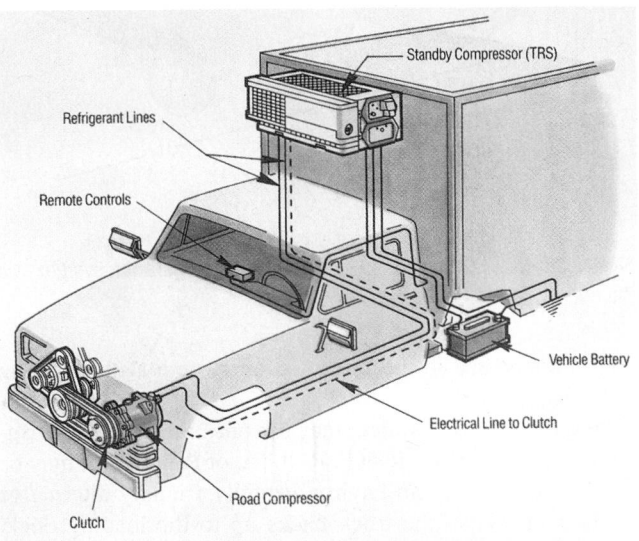

Standby Compressor (TRS)

Refrigerant Lines

Remote Controls

Vehicle Battery

Electrical Line to Clutch

Road Compressor

Clutch

Figure 28–7   This truck refrigeration is driven from the truck's engine and may be cycled by means of an electric clutch. *Reproduced courtesy of Carrier Corporation*

(A) **(B)**

Figure 28–9 **(A)** A nose-mount and **(B)** belly-mount refrigerated truck. *Reproduced courtesy of Carrier Corporation*

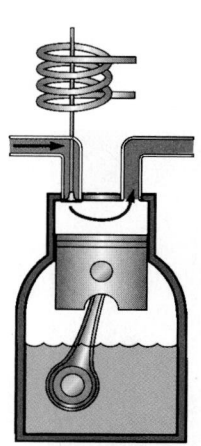 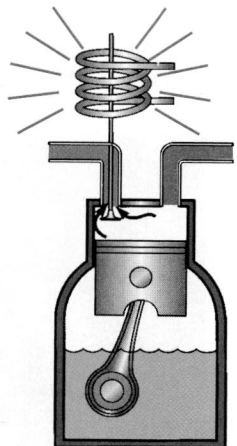

THIS CYLINDER IS PUMPING.

THIS CYLINDER IS NOT PUMPING BECAUSE THE SUCTION VALVE IS BEING HELD OPEN BY THE SOLENOID. GAS ENTERS THE CYLINDER AND IS PUSHED BACK INTO THE SUCTION LINE WITHOUT BEING COMPRESSED. COMPRESSION CAN BE CONTROLLED.

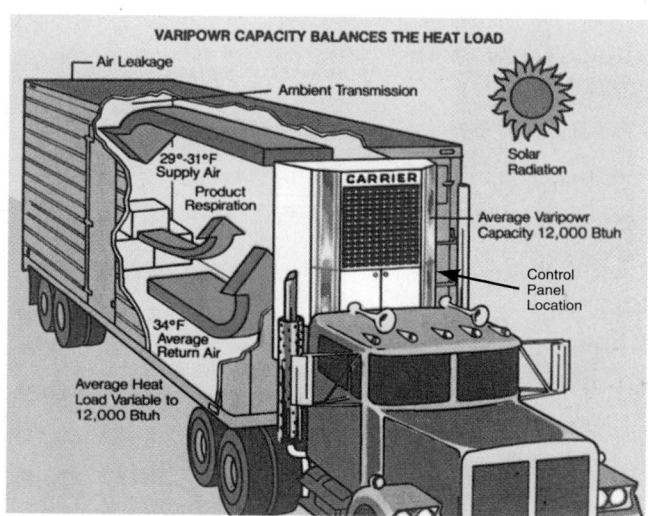

(A)

4-CYLINDER COMPRESSOR WITH 3 CYLINDERS THAT WILL UNLOAD

Figure 28–10 A capacity control for a compressor.

back of the truck, **Figure 28–12.** The fan is usually a form of centrifugal fan that distributes a high volume of air at a high velocity so the air will reach the back of the truck, **Figure 28–13.** The air is then returned to the inlet of the air handler over the product load. The fan is driven from the engine

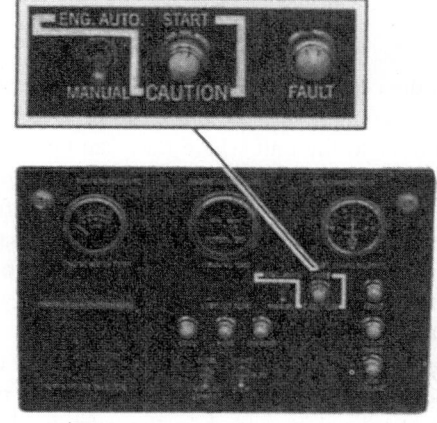

(B)

Figure 28–11 **(A)** Shows location of the panel. **(B)** An automatic control to stop and start a truck refrigeration system. *Reproduced courtesy of Carrier Corporation*

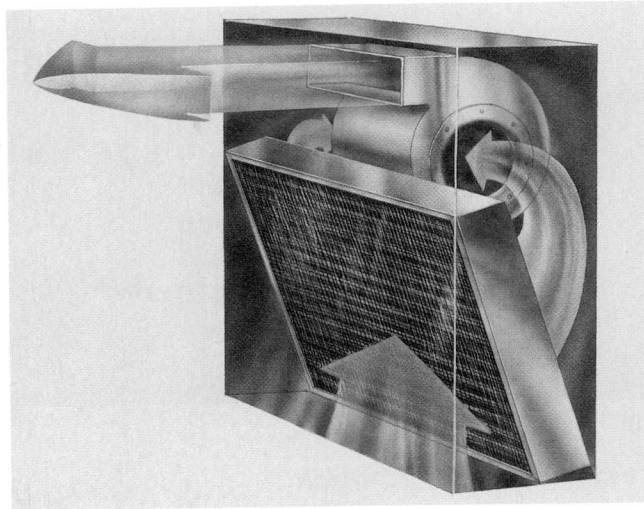

Figure 28–12 The evaporator for a nose-mount refrigeration unit.

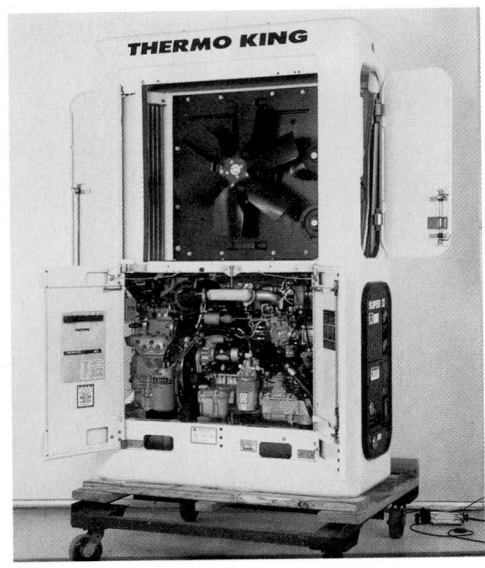

Figure 28–14 Access for service for a nose-mount unit. *Courtesy Thermo King Corp.*

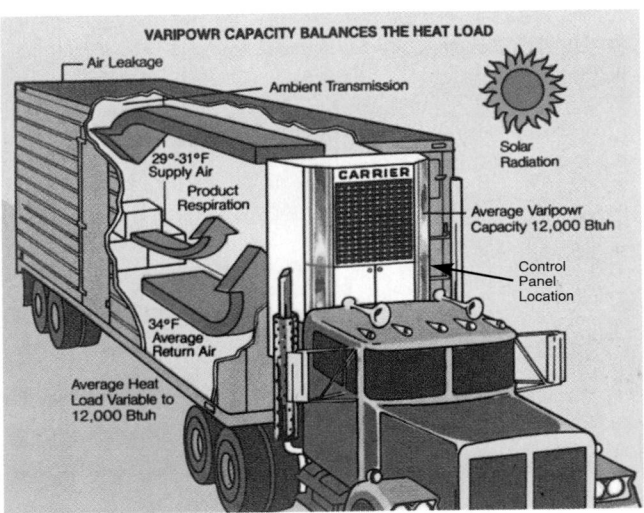

Figure 28–13 The air distribution system for a truck. *Reproduced courtesy of Carrier Corporation*

(A)

that drives the compressor and is driven with belts or gearboxes. These components are normally located just inside the access door to the unit where they can be checked and serviced easily from outside the refrigerated space, **Figure 28–14**. The evaporator fan is usually a centrifugal-type fan. The condenser fan is often driven from the same drive mechanism and is typically a prop-type fan, **Figure 28–15**.

The evaporator and condenser coils are usually made of copper with aluminum fins.

These truck systems must be designed and built to be both rugged and lightweight because each truck has a maximum load limit. Any weight added to the refrigeration unit is weight that cannot be hauled as a paying load. Many of the components are aluminum, such as the compressor and any other component that can be successfully manufactured and maintained with this lightweight material.

(B)

Figure 28–15 **(A)–(B)** Fan drive mechanisms. **(A)** *Courtesy Thermo King Corp.* **(B)** *Courtesy Carrier Corporation*

Truck refrigeration systems must be designed to haul either low- or medium-temperature loads. Part of the time, they may be hauling fresh vegetables, and part of the time the load may be frozen foods. The space temperature is controlled with a thermostat that starts out at full load and when the space temperature is within approximately 2°F of the set point, the system starts to unload the compressor. If the space temperature continues to drop, the unit will shut down.

The air distribution for truck refrigeration is designed to hold the load at the specific temperature, not to pull the load temperature down. As mentioned earlier, the food should be at the desired storage temperature before loading it onto the truck. The truck does not have much reserve capacity for pulling the load temperature down, especially if the interior of the truck warmed very much. If the refrigeration system for a medium-temperature load of food has to reduce the load temperature, the air distribution system may overcool some of the highest parts of the load because the truck body is so small that the cold air from the evaporator may impinge directly on the top product, **Figure 28–16**. If the load is frozen food and has warmed, it may be partially thawed. The truck unit may not have the capacity to refreeze the product and if it does, it may be so slow that product damage may occur. Various alarm methods are used to alert the driver of problems with the load, **Figure 28–17**. The driver should stop and check the refrigeration at the first sign of problems. It may not be easy to locate a technician when the truck is away from metropolitan areas.

Truck bodies are insulated to keep the heat out. This is normally accomplished using foam insulation sprayed onto the walls with wallboard fastened to the foam. This makes a sandwich type of wall that is very strong, **Figure 28–18**. The walls must be rugged because of the loading and unloading

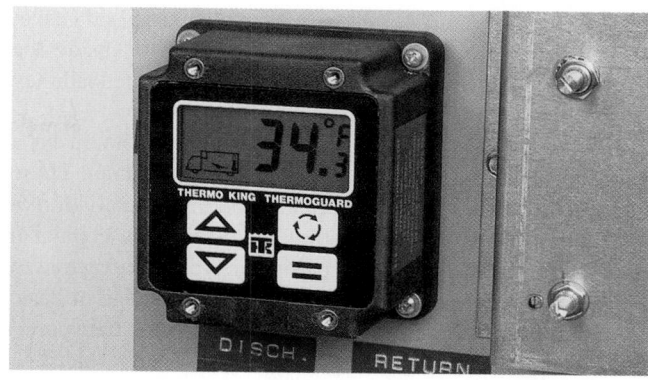

Figure 28–17 An alarm may alert the driver if the load's condition is not being maintained. *Courtesy Thermo King Corp.*

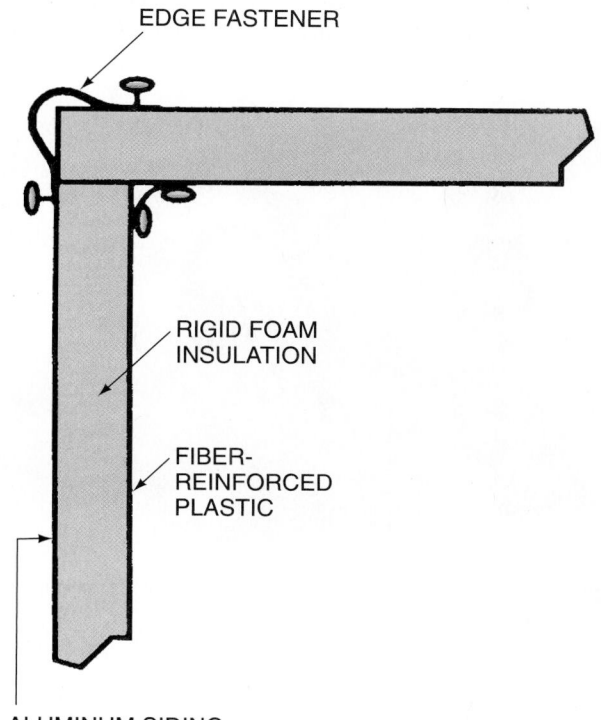

Figure 28–18 Sandwich construction for truck walls.

of the product. Aluminum- and fiber-reinforced plastic products are often used for the walls. The floor must be strong enough to support the weight of the load while traveling on a rough highway; it must also be able to support a forklift that may be used for loading.

The inside construction of a refrigerated truck must be manufactured in such a manner that it may be kept clean because food is being handled. The doors are normally in the back so that the truck may be loaded with a forklift; however, some trucks have side doors and compartments for different types of product. For example, a refrigerated truck may need to carry both low- and medium-temperature loads at the same time. This is accomplished using one refrigeration condensing unit with more than one evaporator.

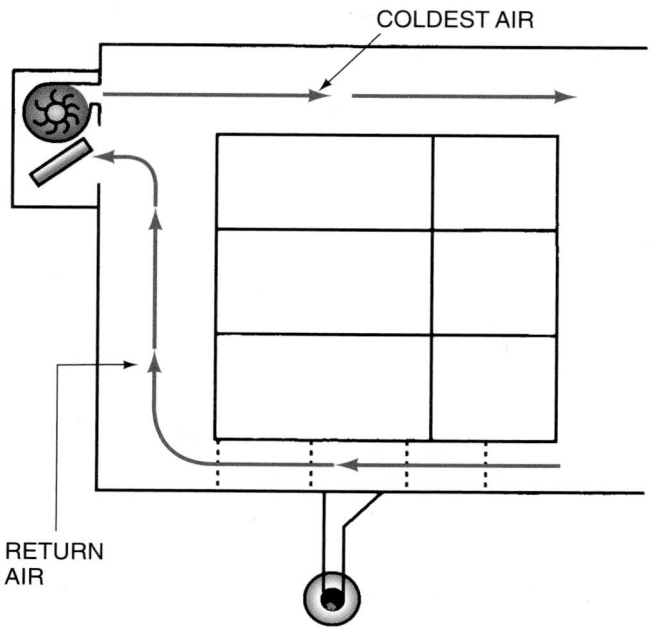

Figure 28–16 Due to air distribution, cold air may overcool food on top of the load.

The doors of a refrigerated truck must have a good gasket seal to prevent infiltration of the outside air to inside the truck. The doors must also have a secure lock or the load may be stolen.

The truck transport refrigeration technician should be a diesel technician as well as a refrigeration technician. These technicians are trained in both fields, and they usually do not work on other types of refrigeration. Many times the technician is factory trained to service a particular type of refrigeration system and engine combination. A good knowledge of basic refrigeration would be good preparation for truck transportation, and knowledge regarding diesel engines can be added.

28.4   RAILWAY REFRIGERATION

Products shipped and refrigerated by rail are mechanically refrigerated either by a self-contained unit located at one end of the refrigerated car or by an axle-driven unit. The axle unit is an older type and not being used to any extent now. The

railcar would only have refrigeration while the train is in motion using axle-driven units.

The self-contained unit is the most popular. It is generally powered by a diesel-driven motor generator, similar to truck refrigeration, **Figure 28–19**. The compressor is a standard voltage and may be powered with a power cord while the car is in a station or on a rail siding for long periods. The evaporator blower is mounted at the end of the railcar with the motor generator. All of the serviceable components can be serviced from this compartment, **Figure 28–20**.

Two-speed engines are often used when the generator has an output frequency of 60 cps at full speed and 40 cps at reduced load. The compressor electric motor must be able to operate at the different frequencies. The diesel engine speed is adjusted by the space-temperature thermostat. The diesel engine operates at a much lower fuel cost at low speed. When low speed is reached, the compressor may be unloaded using cylinder unloading. These systems have a rather wide range of capacities to suit the application. Like trucks, they are capable of reducing the space temperature to 0°F or below for holding frozen foods, or they may be operated at medium temperature for fresh foods.

Railcars have an air distribution system much like trucks and are designed to hold the refrigerated load, not reduce the temperature. They do not have much reserve capacity.

Railcars are made of steel on the outside with typically a rigid foam insulation in the walls, **Figure 28–21**. The walls are covered with various types of wallboard that can be cleaned easily. Railroad cars would ride rough, with metal wheels on metal rails, and have a lot of impact vibration when being connected and disconnected. However, they have a sophisticated suspension system for en route travel and to protect the load from the rough coupling of cars when the train is connected together. They have wide doors for forklift access. These doors must have a very good gasket to prevent air leakage. Typically, the railcar is operated under a positive air pressure and must have good door seals.

Figure 28–19 This motor generator is used to power a conventional compressor. *Courtesy Fruit Growers Express*

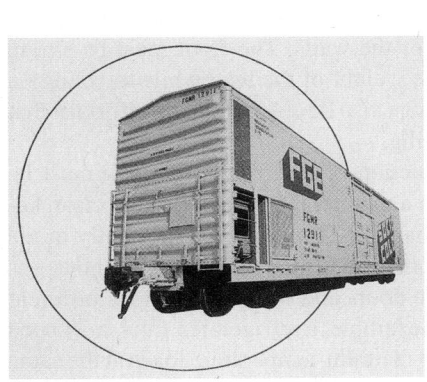

(A)

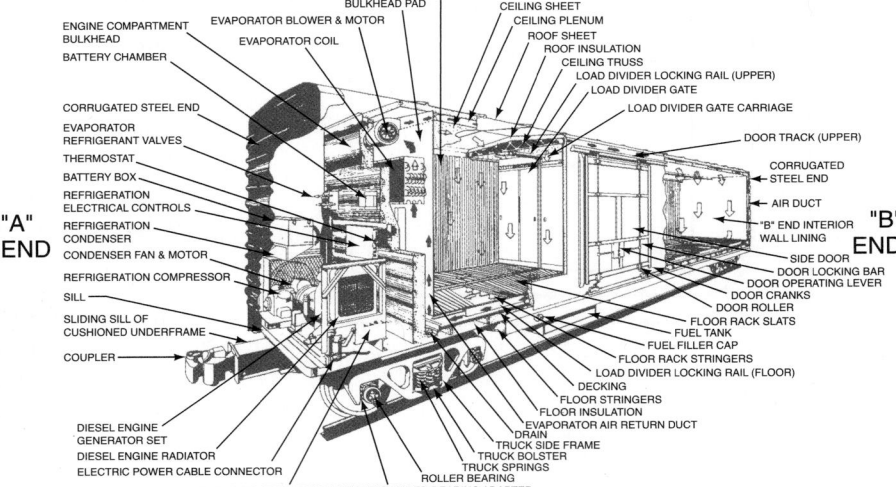

(B)

Figure 28–20 A railroad car refrigeration unit **(A)–(B)**. *Courtesy Fruit Growers Express*

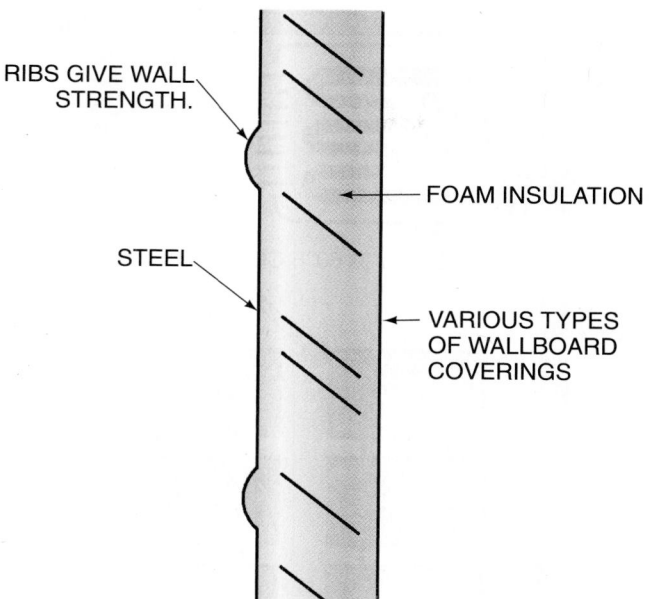

Figure 28–21   The wall construction for a railcar.

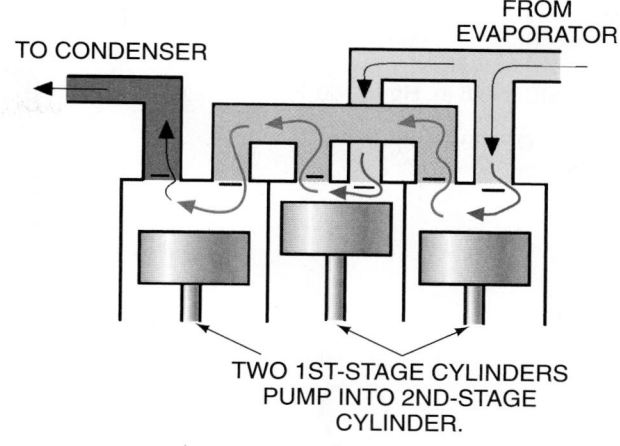

Figure 28–23   Two-stage refrigeration.

The railroad also ships refrigerated trucks in a system known as "piggyback" cars. These are cars with trucks loaded on the railcars and shipped across country by rail, **Figure 28–22**. This is popular because the refrigerated load often would have to be off-loaded to a refrigerated truck in the locality where the food is consumed or processed. With the piggyback cars, the refrigerated load arrives on the truck for final destination hauling.

28.5   EXTRA-LOW-TEMPERATURE REFRIGERATION

Extra-low-temperature refrigeration may be described as refrigeration systems that will hold the product temperature below −10°F. This text will define extra-low temperature as below −10°F. Product temperatures below −10°F are not used to any great extent for storing food; however, these temperatures are used to fast freeze foods on a commercial basis. These temperatures may be in the vicinity of −50°F.

When the product temperature needs to be reduced using temperatures lower than −50°F, a different type of system may be used. The compression ratio for the compressor becomes too great for single-stage refrigeration at extremely low evaporator temperatures. Another method, however, is to use two stages of compression. It is accomplished with a first-stage compressor (or cylinder) pumping into a second-stage compressor (or cylinder), **Figure 28–23**. The compression ratio is split over two stages of compression. If the above single-stage system were operating at a room temperature of −50°F, it would have a boiling refrigerant temperature of about −60°F, using a coil to air temperature difference of 10°F, **Figure 28–24**. This would make the evaporator pressure 6.6 in. Hg (11.48 psia) for R-404A. If the condensing temperature were to be 115°F, the head

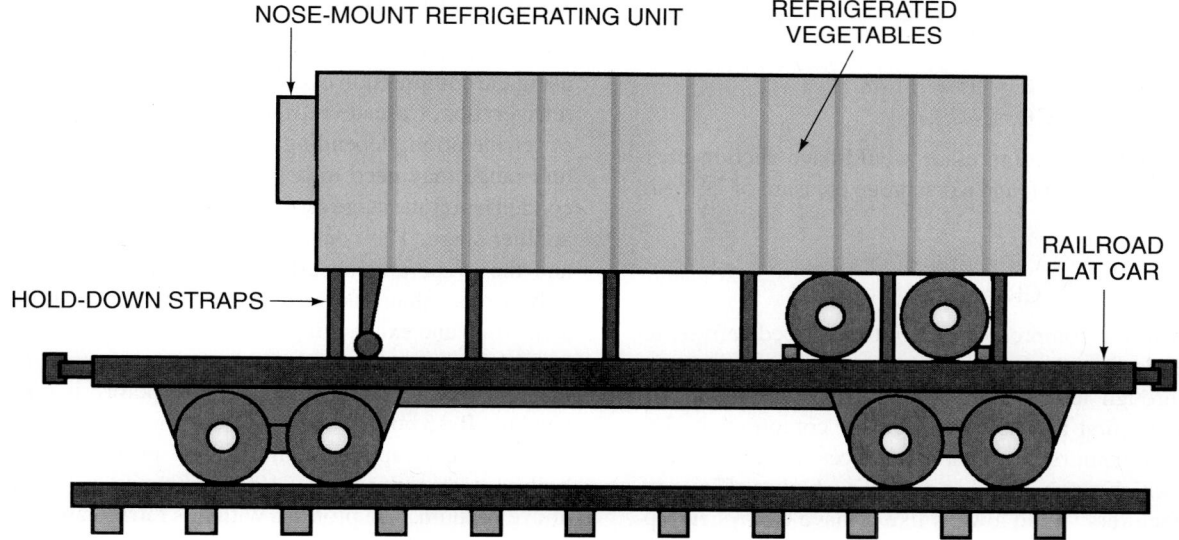

Figure 28–22   Piggyback railcar refrigeration.

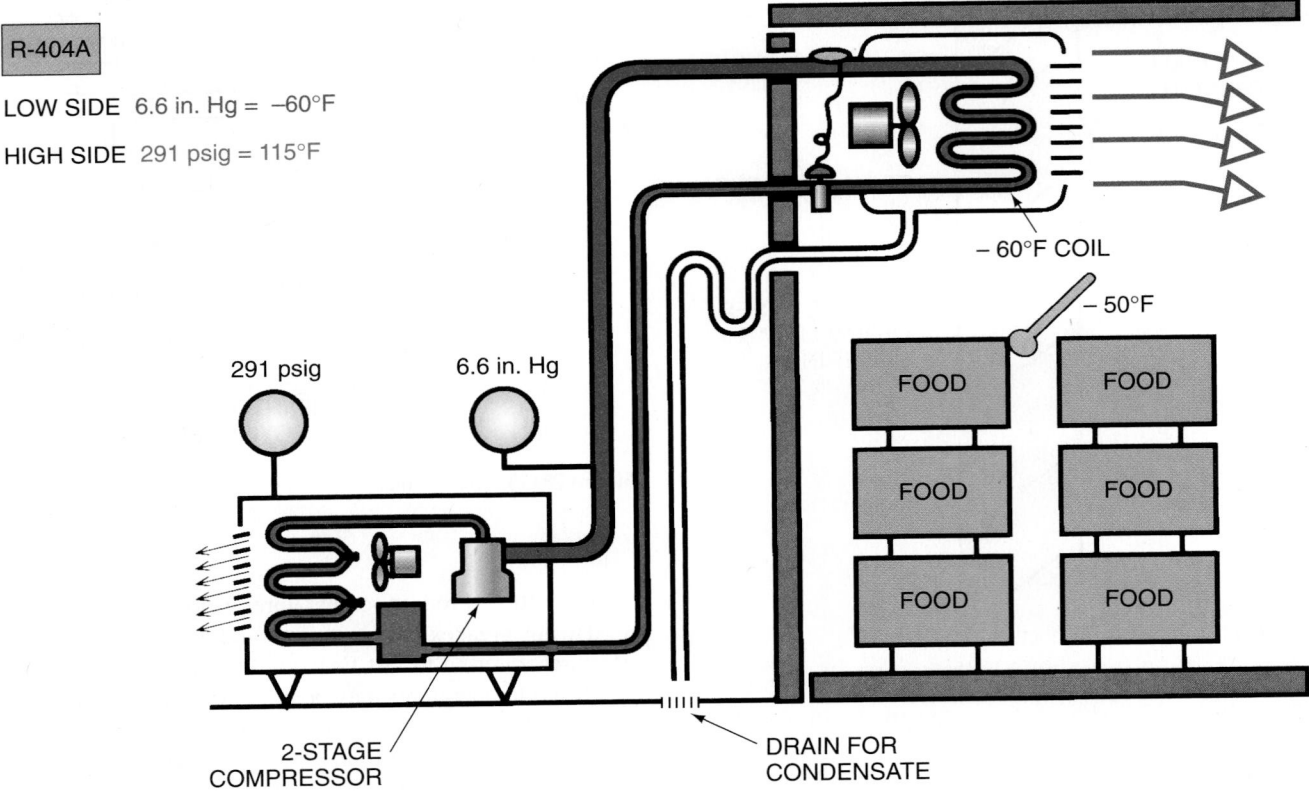

R-404A

LOW SIDE   6.6 in. Hg = −60°F

HIGH SIDE   291 psig = 115°F

291 psig

6.6 in. Hg

−60°F COIL

−50°F

FOOD   FOOD

FOOD   FOOD

FOOD   FOOD

2-STAGE
COMPRESSOR

DRAIN FOR
CONDENSATE

Figure 28–24   A two-stage refrigeration system in operation.

pressure would be 291 psig (306 psia). The compression ratio would be

$$CR = \text{Absolute discharge pressure}$$
$$\div \text{Absolute suction pressure}$$
$$CR = 306 \div 11.48$$
$$CR = 26.65:1$$

This is not a usable compression ratio. It can be reduced by using a two-stage compressor, **Figure 28–25.** The first stage of compression has a suction pressure of 11.48 psia and a head pressure of 40 psig (55 psia). The compression ratio would be

$$CR = 55 \div 11.48$$
$$CR = 4.79:1$$

The second-stage compressor would have suction pressure of 40 psig (55 psia) and a discharge pressure of 291 psig (306 psia).

$$CR = 306 \div 55$$
$$CR = 5.56:1$$

With a lower compression ratio, the trapped refrigerant gas in the clearance volume at the top of the piston's stroke will go through less reexpansion. This increases the efficiency of the first and second stages of compression. The second-stage compressor then experiences a higher suction pressure, which also will aid in giving it a higher efficiency.

Manufacturers would always like to have the system operate in a positive pressure. The system above is operating in

a vacuum, and if a leak occurs on the low-pressure side of the system, atmosphere would enter the system. This system may not be very practical, but it is very possible. The lower the compression ratio, the better the compressor efficiency. This is particularly true for the first-stage compressor. Designers will pick the best operating conditions for the system, selecting several compressor combinations until the best efficiency is found.

28.6   CASCADE SYSTEMS

Systems with operating temperatures down to about −160°F using the compression cycle may use a system called cascade refrigeration. Cascade refrigeration uses two or three stages of refrigeration, depending on how low the lower temperature range may need to be. This is accomplished when the condenser of one stage exchanges heat with the evaporator in another stage. The condenser rejects heat from the system and the evaporator absorbs heat into the system. This involves more than one refrigeration system, working at the same time and exchanging heat between them.

The first stage of the system uses a refrigerant like R-13 that boils at a very low temperature before it goes into a vacuum. R-13 boils at −114.6°F at atmospheric pressure and would have a vapor pressure of 7.58 psig at an evaporator boiling temperature of −100°F, a pressure considerably above vacuum. The problem with this refrigerant is the high-side pressure. It has a critical temperature of 83.9°F and a

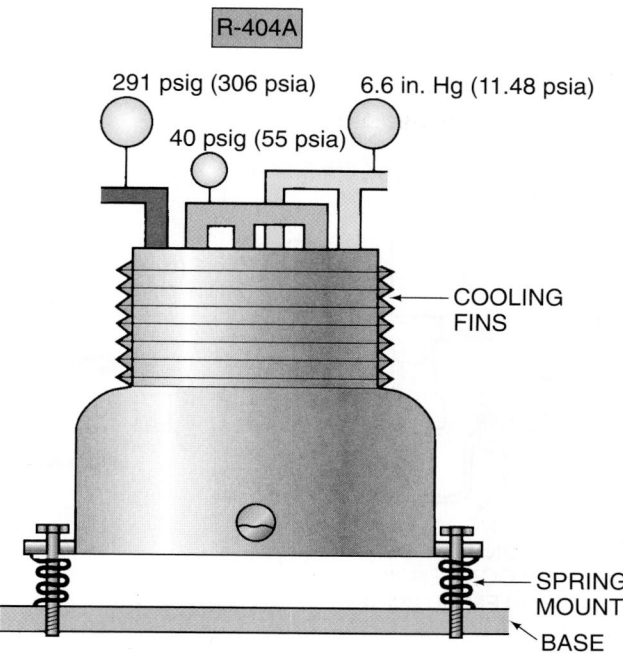

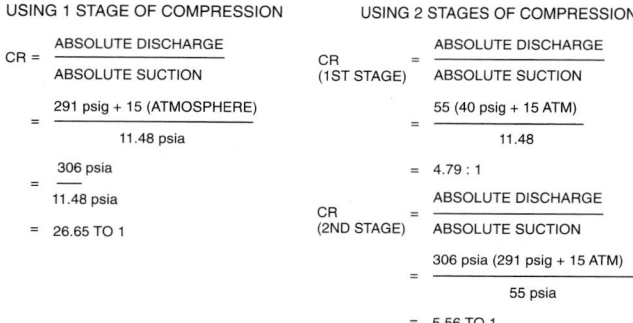

Figure 28–25 The compression ratio and two-stage refrigeration.

critical pressure of 561 psia (546 psig). The critical temperature and pressure is the highest point (temperature and pressure) at which this refrigerant will condense. For example, R-13 will not condense at pressures above 546 psig. **Figure 28–26** shows a cascade system with three stages of compression. Note the following conditions in the system:

1. The evaporator is in the first stage of refrigeration and is operating at −100°F with a suction pressure of 7.58 psig using R-13. The discharge pressure is 101.68 because condensing is occurring at −25°F. This is possible because the condenser heat is being removed with an evaporator operating at −35°F. This is the second-stage evaporator.

2. The second-stage evaporator is operating at −35°F, using R-22, and has a suction pressure of 2.6 psig.

3. The second-stage compressor has a discharge pressure of 75.5 psig because it is condensing at 44°F. The reason it is condensing this low is because the condenser is cooled by an evaporator operating at 34°F.

4. The third-stage evaporator is operating at 34°F with a suction pressure of 60.1 psig using R-22. The compressor is discharging into an air-cooled condenser with 90°F air passing over it. The condenser is operating at a 20°F temperature difference (between the air and condensing temperature) and is condensing at 110°F with a head pressure of 226.4 psig.

The compression ratios for the various stages are as follows:

1. Stage 1
   CR = Absolute discharge ÷ Absolute suction
   CR = 116.68 ÷ 22.58
   CR = 5.17:1
2. Stage 2
   CR = 90 ÷ 17.6
   CR = 5.14:1
3. Stage 3
   CR = 241.4 ÷ 75.1
   CR = 3.21:1

The designer working with these figures may try for a lower compression ratio for the first stage because this is the most critical. The compression ratio may be reduced by lowering the condensing temperature of the first-stage compressor. A different refrigerant may be used in this stage. If the evaporator temperature is dropped much lower in the second stage to lower the condensing temperature in the first stage using R-22, the second-stage evaporator will be operating in a vacuum. R-502 could be used for an evaporator in the second stage down to −49.8°F before it would operate in a vacuum. The system efficiency can also be improved by operating the condenser in the conditioned space if the system does not have too much capacity. Then room temperature air could be passed over the third-stage condenser.

Several refrigerants may be used for cascade refrigeration. The experienced designer can choose the most efficient refrigerant for the application.

28.7 QUICK FREEZING METHODS

Food that is frozen for later use must be frozen in the correct manner or it will deteriorate, the quality will be reduced, or the color may be affected. When this food is frozen, the faster the temperature is lowered, the better the quality of the food. This can be done in several ways using a refrigeration system at very low temperatures. This method is often called "quick freezing."

Generally, the meats that you buy at the grocery store and freeze in your home freezer do not have the same quality as the meats you buy already frozen. The reason for this is the home freezer does not freeze the meat fast enough. When meat is frozen slowly, ice crystals grow in the cells and often puncture them. This is why you often see a puddle of liquid when meat is thawed. This is food value and flavor escaping. When foods are frozen commercially, they are frozen quickly. This is accomplished using very cold air moving across the product at a high velocity to remove the heat quickly. When most fresh food is frozen very quickly and thawed, it should still have the quality of fresh food.

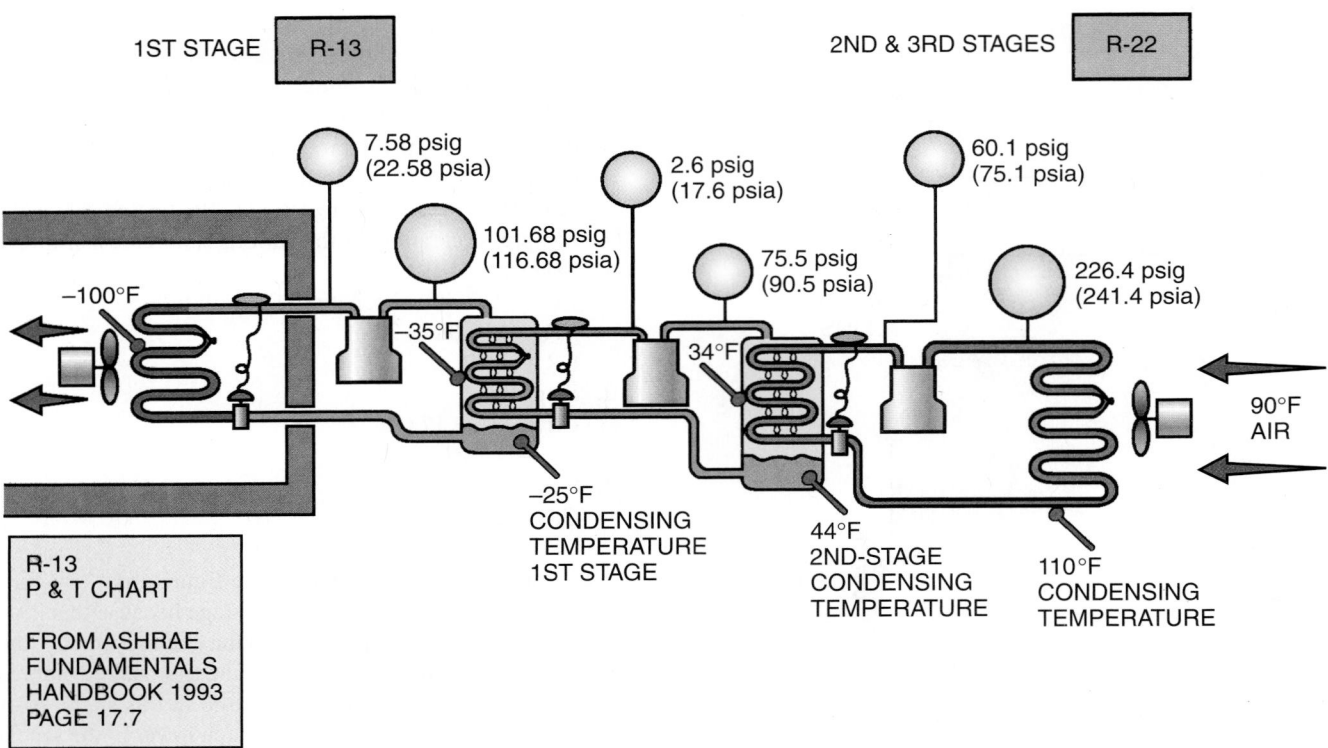

1ST STAGE R-13

2ND & 3RD STAGES R-22

7.58 psig (22.58 psia)

101.68 psig (116.68 psia)

2.6 psig (17.6 psia)

75.5 psig (90.5 psia)

60.1 psig (75.1 psia)

226.4 psig (241.4 psia)

−100°F

−35°F

34°F

90°F AIR

−25°F CONDENSING TEMPERATURE 1ST STAGE

44°F 2ND-STAGE CONDENSING TEMPERATURE

110°F CONDENSING TEMPERATURE

R-13 P & T CHART

FROM ASHRAE FUNDAMENTALS HANDBOOK 1993 PAGE 17.7

Figure 28–26 A cascade refrigeration system with three stages.

Different foods have different requirements for freezing. For example, some poultry must be frozen very quickly to preserve the color. To accomplish this, one method would be to seal the meat in an airtight plastic bag; it is then dipped in a very cold solution of brine causing the surface to freeze. The bag is then washed and the poultry removed to another area for freezing the core of the meat at a slower rate.

Foods are often frozen in bulk in packages to be thawed later for processing or cooking. When foods are frozen in this manner, they may stick together, but it makes no difference because they are going to be used in bulk.

Many foods are frozen in individual pieces. For example, shrimp may be frozen individually so they can be sold by the pound while frozen. This process may be accomplished by a process called blast freezing, which uses trays in a blast tunnel or on a conveyor system in a blast freezer room, **Figure 28–27**. The blast freeze method is accomplished at very low temperatures and a high air velocity. If the conveyor is placed in a room where the temperature is −40°F or lower, and the air is moving fast, small size food pieces may travel the length of the conveyor and leave frozen hard. When the food is placed on the conveyor in a separated fashion, it is not touching other pieces and will be frozen individually, **Figure 28–28**. This food may then be packaged and dispensed individually. Different weights of food may be placed on longer or shorter conveyor belts or the belt speed may be adjusted to the correct time for freezing the food.

Food trays and tunnels may be used for fast freezing food. The food is placed on trays with plenty of space around the food. The trays are loaded on carts and moved into a tunnel,

Figure 28–29. When the tunnel is full and a cart is pushed in, a cart is pushed out the other end with frozen food, ready to package. This method requires more labor but is less expensive to build from a first-cost standpoint than a conveyor system, which is more automated.

Larger parcels of food, such as a carcass of beef, would take longer to freeze. These would be precooled to very close to freezing and then moved to the blast freeze room for quick freezing.

This quick freezing method may be accomplished using standard refrigeration equipment. The larger refrigeration systems may use ammonia as the refrigerant with reciprocating compressors. Other systems may use R-22, or newer environmentally friendly alternative refrigerants, and screw compressors. ♻ The chlorofluorocarbon (CFC)/ozone depletion issue is causing ammonia systems to make a comeback. They are efficient and do not deplete the ozone layer. ♻

28.8 MARINE REFRIGERATION

Ship refrigeration may be used for several different applications. For example, a ship may pick up a perishable product requiring medium-temperature refrigeration (34°F) in South America and deliver it to New York City. The ship may then pick up a load of frozen food to be delivered to Spain at −15°F, a low-temperature application. The ship's refrigeration system must be able to operate at either temperature range.

Another application for ship refrigeration may be for preserving fish caught at sea. If the ship is only out to sea for a

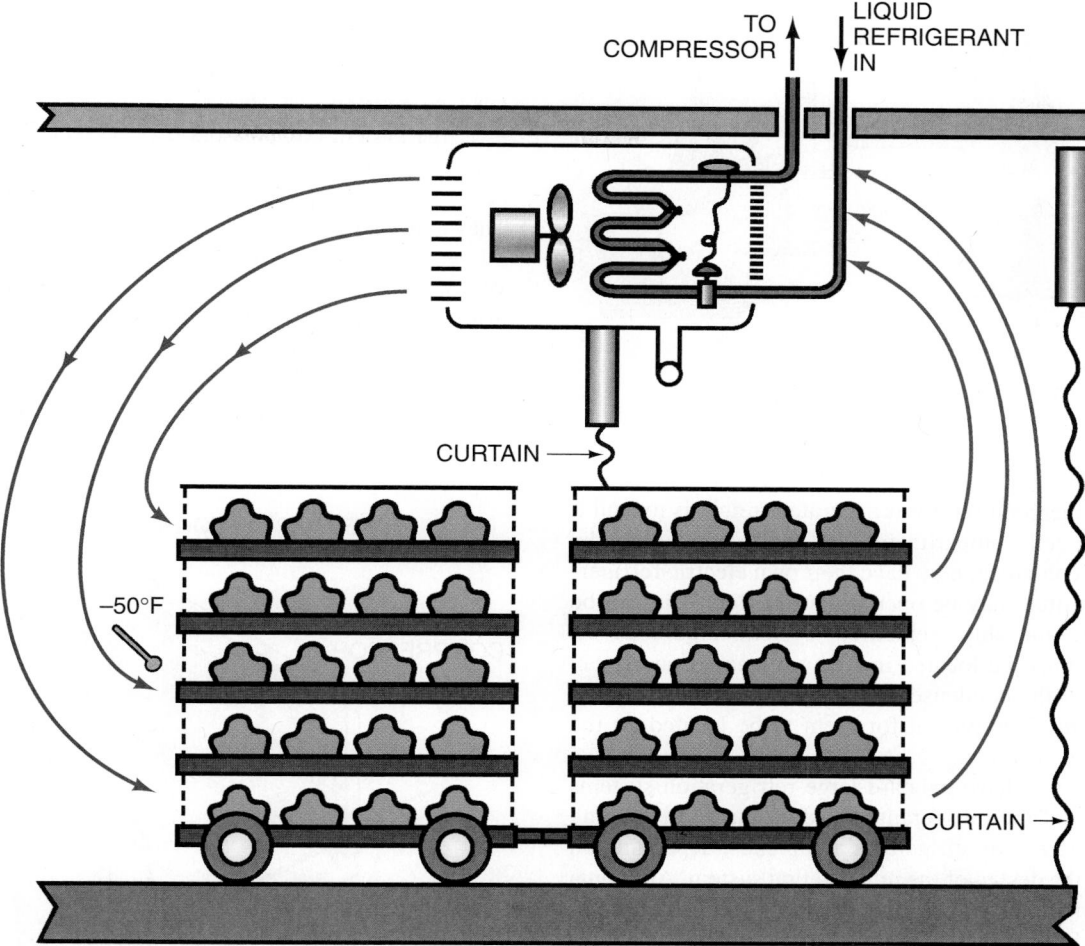

Figure 28–27   A blast freezing tunnel.

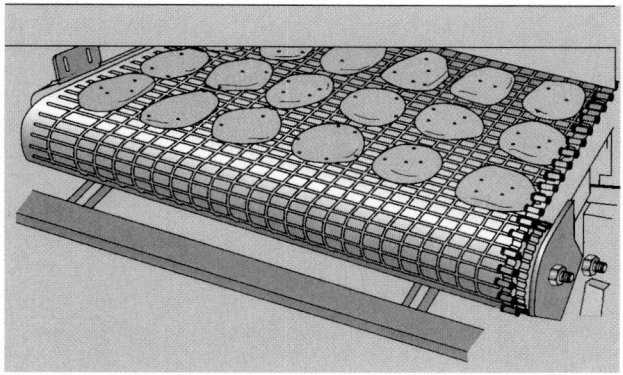

Figure 28–28   The conveyor system for a blast cooler.

THE FISH FILLETS ARE PRECOOLED TO JUST ABOVE FREEZING BEFORE ENTERING THE QUICK FREEZE PROCESS.

THE COLD AIR COMES IN CONTACT WITH ALL SIDES OF THE FISH.

Figure 28–29   Spacing of food for tunnel quick cooling.

few days, ice carried from shore is often used. This would be the case of a shrimp boat operating off the coast. Auxiliary refrigeration is also used to aid the ice in refrigerating the load, to prevent excess melting. If the ship is to be at sea for long periods, the load must be frozen for preservation. It is not unusual for a ship to stay at sea for several months gathering a load of fish. This is often accomplished with a fleet

of small boats that catch the fish and a large refrigerated processing ship called a mother ship, **Figure 28–30.** In this case, the refrigeration system may have to be very large, possibly several hundred tons. These ships must have a method for quick freezing fish to preserve the flavor and color.

Figure 28–30   The mother ship and a fleet of fishing boats.

Another method used for refrigerated freight is to haul it in containerized compartments that are loaded onto the ship. These containers may have their own electric refrigeration system that may be operated at the dock and can be plugged into the ship's electrical system, **Figure 28–31.** These units must be located in such a manner that air can circulate over the condensers, or they will have head pressure problems. It is typical for them to be located on the deck of an open ship.

The ships that have onboard large refrigeration systems may use any of the refrigerants listed in this text including the newer alternative refrigerants and blends, depending on the age and the design of the refrigeration system. Ammonia refrigeration systems are also used for ships. This refrigerant is not considered dangerous to the environment so these systems will be ongoing. It is, however, extremely toxic to humans and must be used with care.

SAFETY PRECAUTION: It is sometimes necessary for technicians to wear breathing apparatus as well as protective clothing, gloves, and goggles when working with ammonia systems.

Large system compressors may be reciprocating, screw, or centrifugal when R-11 is used. R-123 (HCFC-123) is now being used as an interim replacement for R-11 (CFC-11). They may be driven using the ship's electrical system if it has a large enough generating plant using belts or direct drive. When the system becomes large, it may be steam turbine driven using the ship's steam system.

The evaporators may be large plates in the individual refrigerated rooms or bare pipes for some systems, **Figure 28–32.** Systems using forced air are also used, **Figure 28–33.** These

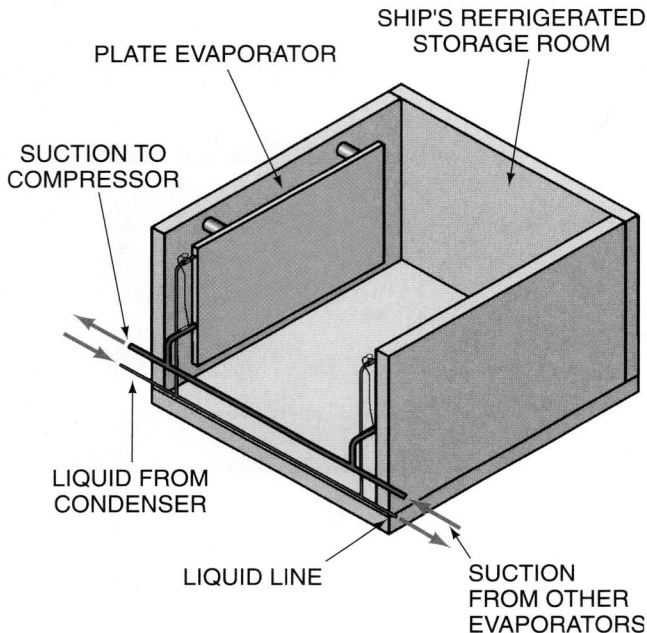

Figure 28–32   Plate evaporators for a ship.

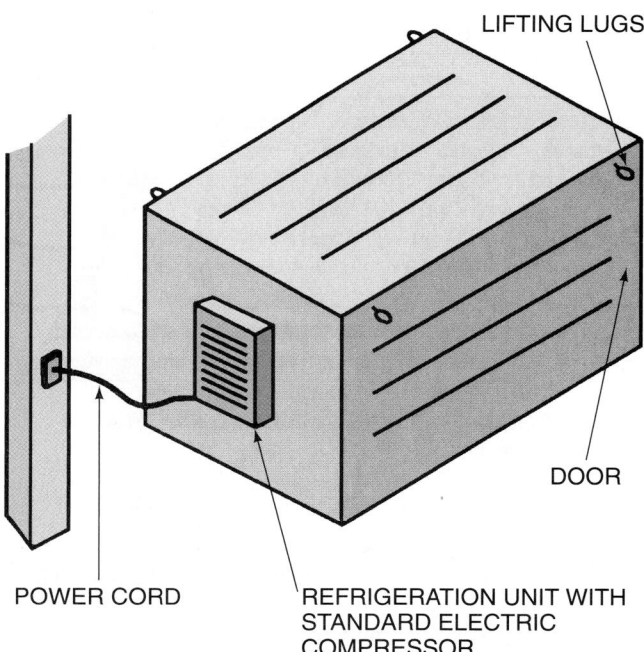

Figure 28–31   Containerized freight for ships.

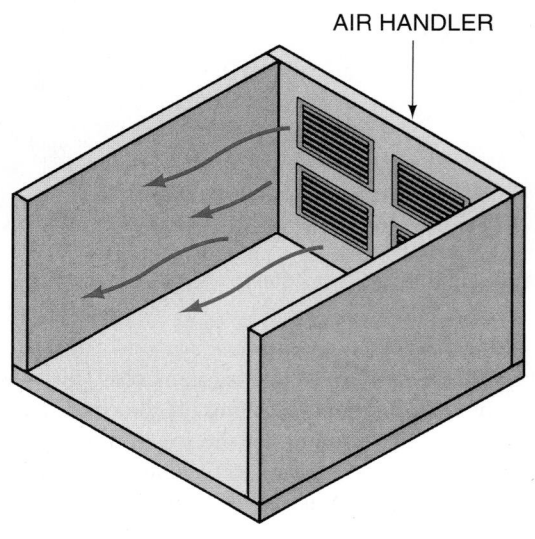

Figure 28–33   Forced-air refrigeration for a ship.

systems are direct expansion, using thermostatic expansion valves. The ship may have several refrigerated compartments so all the refrigeration is not in one large area.

These evaporators must have some means for defrost because they will normally be operating below freezing at the coils. This may be accomplished with hot gas if the evaporator is close enough to the compressor. If not, sea water may be used. If it is too cold, it may be heated and sprayed over the evaporator. This must be done without the fan running on a fan coil unit and the coil must not be refrigerating during defrost time. This can be done by stopping the fan and closing the liquid-line solenoid valve during defrost. The compressor will pump the refrigerant out of the evaporator during defrost. Defrost can be terminated using coil temperature or time. The refrigerant that is pumped out of the coil will be in the receiver during defrost, and the compressor can

continue to run to other refrigerated compartments on the ship. Defrost can be staggered so all coils will not call for defrost at the same time.

The condensers will use sea water, which is readily available everywhere the ship goes. Sea water contains debris and must be filtered with filters that are easy to access and clean. This is commonly done with valves that allow the pumps to reverse the flow through the filters and piping to backwash the system, **Figure 28–34.** The refrigerant condensers for the typical refrigerants, R-11, R-12, R-22, R-502, and the newer alternative refrigerants and blends, must be designed and built for easy cleaning because they will become dirty. These condensers have removable water box covers, called marine water boxes, where the tubes can be cleaned using a brush, **Figure 28–35.** The tubes for these condensers are made of cupronickel (copper and nickel), a tough metal that does not

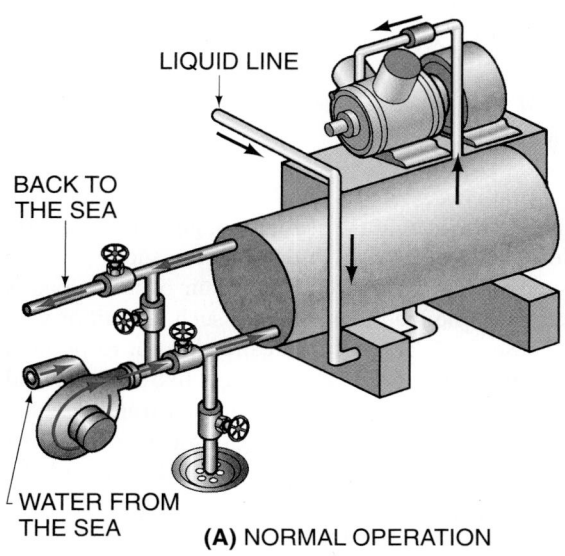

LIQUID LINE

BACK TO THE SEA

WATER FROM THE SEA

(A) NORMAL OPERATION

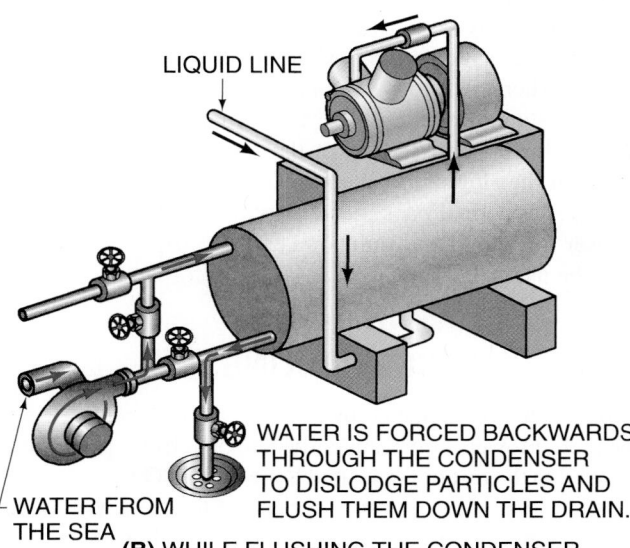

LIQUID LINE

WATER IS FORCED BACKWARDS THROUGH THE CONDENSER TO DISLODGE PARTICLES AND FLUSH THEM DOWN THE DRAIN.

WATER FROM THE SEA

(B) WHILE FLUSHING THE CONDENSER

Figure 28–34 Flushing the ship's seawater-cooled condensers **(A)–(B).**

(A)

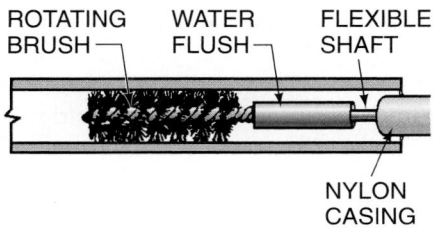

ROTATING BRUSH   WATER FLUSH   FLEXIBLE SHAFT

NYLON CASING

(B)

Figure 28–35   (A) Cleaning a condenser using a brush. **(B)** A cutaway of a tube showing a rotating brush.

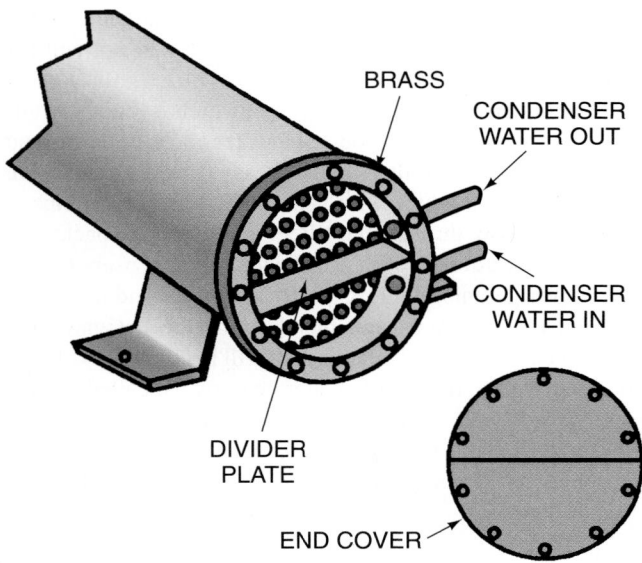

Figure 28–36 Tubes and water boxes for a condenser.

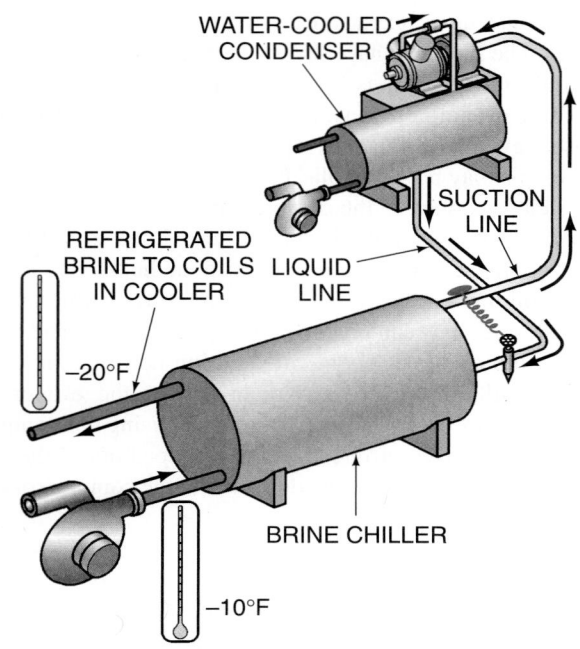

Figure 28–37 The brine circulation system for a central refrigeration system.

corrode easily. The tube sheet that holds the tubes is typically made of steel, and the water box is made of brass where possible, **Figure 28–36.**

Reserve capacity refrigeration and a spare parts inventory for making repairs at sea are recommended by the American Bureau of Shipping. They recommend that two compressor systems be used, either one of which should have enough capacity to maintain the load, running continuously while in tropical waters. This is 100% standby capacity of the compressor, which is the most likely part of the system to cause a problem. These systems can become quite large. There must be someone aboard with experience who is familiar with the system and can make any necessary repairs. Repairs at sea are not uncommon. It can take days to reach a port with a repair facility.

The individual evaporators can be located at great distances from the ship's condensers and compressors and long refrigerant lines will have to be run. This brings on the problems of returning lubricating oil that leaves the compressor. Oil separators may be used at the compressor to keep most of the oil in the compressor, but some will get into the system and will have to be returned.

Circulating brine systems have some advantages over individual evaporators in the refrigerated spaces. The primary refrigeration components and the primary refrigerant can all be contained in the equipment room with a central compressor or compressors. The brine may be chilled and circulated to the various locations for desired applications, **Figure 28–37.**

As with truck refrigeration, cargo ship refrigeration is designed to hold the refrigerated load temperature. It usually does not have much reserve capacity for reducing the load temperature. The ship storage rooms are not designed and built for the type of air circulation necessary to reduce the temperature of the load even if reserve capacity is available.

When the ship is loaded, every square foot of storage counts. The ship must be designed to accept the load and store it for the trip even if the seas are rough. The placement of the load will have much to do with the success of the refrigerated load. Correct circulation of air over the load is important when the load is fresh food, such as fruit or vegetables. These foods give off heat called heat of respiration as they are stored, **Figure 28–38.** This is not like storing a load of frozen food that only has to be protected from heat that may enter from the outside, like a walk-in cooler. It is heat generated by the load and is calculated in the size of the refrigeration equipment.

All ships must have refrigeration for the ship storage of foods for the crew and passengers. Refrigeration for the ship stores may be drawn from the main refrigeration system as long as it is running when the ship hauls refrigerated cargo. When the ship is not hauling refrigerated cargo, auxiliary systems may be used or the ship may have a separate system for the crew. Many of the same appliances may be used on the ship as in a supermarket. Some of these units may be individual package units with the compressor condenser built into the unit, or central refrigeration rooms may be used.

28.9 AIR CARGO HAULING

Products are often hauled by air to hasten the product trip to market. Flowers are frequently shipped from places like Hawaii to the rest of the United States by air. Onboard refrigeration is out of the question because of the weight of the refrigeration systems. When the product needs to be cooled during flight, ice or dry ice is used.

Specially designed containers are built to slide into the cargo compartment of the plane, **Figure 28–39.**

Thermal Properties of Food

	Heat of Respiration, Btu/ton d						
Commodity	32°F	41°F	50°F	59°F	68°F	77°F	Reference
Beans, Lima Unshelled	2306-6628	4323-7925	–	22,046-27,449	29,250-39,480	–	Lutz and Hardenburg (1968) Tewfik and Scott (1954)
Shelled	3890-7709	6412-13,436	–	–	46,577-59,509	–	Lutz and Hardenburg (1968) Tewfik and Scott (1954)
Beans, Snap	∗b	7529-7709	12,032-12,824	18,731-20,533	26,044-28,673	–	Ryall and Lipton (1972) Watada and Morris (1966)
Beets, Red, Roots	1189-1585	2017-2089	2594-2990	3711-5115	–	–	Ryall and Lipton (1972)
Sweet	901-1189	2089-3098	–	5512-9907	6196-7025	–	Lutz and Hardenburg (1968), Micke et al. (1965), Gerhardt et al. (1942)
Corn, Sweet with Husk, Texas	9366	17,111	24,676	35,878	63,543	89,695	Scholz et al. (1963)
Cucumbers, Calif.	∗b	∗b	5079-6376	5295-7313	6844-10,591	–	Eaks and Morris (1956)

Figure 28–38 The heat of respiration for some fresh vegetables. *Reprinted with permission of the American Society of Heating, Refrigerating, and Air-Conditioning Engineers from the 1994 ASHRAE Handbook—Refrigeration*

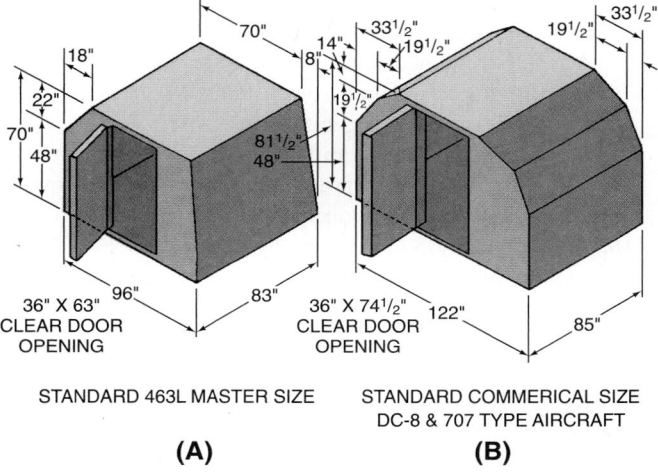

STANDARD 463L MASTER SIZE **(A)**

STANDARD COMMERICAL SIZE DC-8 & 707 TYPE AIRCRAFT **(B)**

Figure 28–39 Containers for shipping refrigerated products for airlines **(A)–(B)**. *Reprinted with permission of the American Society of Heating, Refrigerating, and Air-Conditioning Engineers from the 1998 ASHRAE Handbook—Refrigeration*

SUMMARY

- Transport refrigeration is the process of transporting various refrigerated products by truck, rail, air, or water.
- Vegetables shipped from California to New York will need refrigeration while crossing the hot desert and may even need heat when traveling in the cold northern states.
- Some products are shipped at medium temperatures and some at low temperatures in the frozen state.
- Dry ice, made from compressed carbon dioxide (CO_2), may be used to refrigerate loads. It changes state from a solid to a vapor without going through the liquid state, called sublimation, at −109°F. It is particularly good for low-temperature loads.
- Liquid nitrogen or CO_2 may also be used for refrigeration. The liquid is stored in low-pressure cylinders with a relief valve system that maintains the cylinder pressure at about 25 psig by relieving excess pressure from the cylinder. Nitrogen evaporates at −320°F and CO_2 at −109°F.

- These systems are simple, with only a thermostat controlling a solenoid valve. A distribution system distributes the cold nitrogen vapor.
- SAFETY PRECAUTION: Nitrogen and CO_2 must be handled with care because they are so cold. They can freeze flesh immediately on contact. Be sure to wear protective clothing, gloves, and goggles when working with either of these systems. Both of these vapors also displace oxygen so the system cannot be operating while people are in the refrigerated space. Special controls shut the system off if the door is opened to the refrigerated space.
- Trucks are often equipped with refrigerated plates that contain a solution called a "eutectic solution" that goes through a phase change when cooled and warmed. The solution is typically a brine solution of either sodium chloride (table salt) or calcium chloride.
- These brine solutions are corrosive to ferrous metals.
- The brine solution does not actually turn solid; it turns to crystals when frozen and the crystals turn to liquid when warmed.
- Trucks may also be refrigerated using mechanical refrigeration located on the truck.
- When at the dock, the compressor may be plugged into a power supply.
- Larger trucks use either nose-mount or under-belly refrigeration units.
- The evaporator and fan are located in the front of the truck on a typical unit.
- Truck systems are not designed to have the capacity to refrigerate a load that is not down to temperature. They are meant to hold the load temperature while in transit.
- Truck bodies are insulated and the doors have gaskets to prevent infiltration.
- Two-speed engines and compressors that unload are used for capacity control.

- Extra-low-temperature refrigerated applications at temperatures below $-10°F$ often use the compression cycle by using either two-stage compressors or cascade refrigeration systems.
- Cascade refrigeration systems use more than one stage of refrigeration to reach very low temperatures.
- Fresh food that is frozen quickly retains its fresh color and taste. The quicker the food is frozen, the better these qualities will be maintained.
- When meats are frozen slowly, the ice crystals form slowly and puncture the cell walls of the meat.
- Food may be frozen quickly by dipping it in a cold solution, such as brine (salt water), or it may be frozen quickly by using blast freezers, very cold air blowing directly on the food. The cold air and the velocity help to remove heat from the food quickly.
- Conventional refrigeration systems may be used for blast freezing.
- Ship refrigeration may have several applications, from medium temperature to low temperature.
- A fish processing ship must have the capability of quick freezing and holding the fish after they are processed.
- Refrigerated cargo ships may have large refrigerated holds where plate-type or forced-draft evaporators are used to maintain the temperature.
- The condensers for ship refrigeration usually use sea water and are made of cupronickel.
- Another method for shipping refrigerated cargo is to use self-contained air-cooled refrigeration on containerized freight.
- Ships must also have refrigeration for the ship's stores.
- Many products are shipped by air to speed them to market (e.g., flowers). They are typically kept cool using ice or dry ice.

REVIEW QUESTIONS

1. List the evaporating temperatures for liquid nitrogen and CO_2. _____ and _____
2. What is the working pressure of a typical liquid nitrogen storage tank, and how is this pressure maintained?
3. True or False: Solutions used in eutectic plate refrigeration are brine, calcium chloride, and sodium chloride solutions.
4. Describe phase-change refrigeration in the eutectic cold plates.
5. What is the advantage of cold plate refrigeration?
6. What is the method of refrigerating cold plates?
7. Two locations for truck refrigeration systems are _____ and _____.
8. Which are power supplies used for truck refrigeration?
   A. Gas-engine driven compression
   B. Electric compressor operated with a motor generator
   C. Diesel compressor
   D. All of the above
9. What type of reserve capacity does truck refrigeration have?
10. Why is truck refrigeration designed to operate on low- and medium-temperature refrigeration?
11. How is the air distributed over the cargo in truck refrigeration?
12. Name two types of engines used for truck refrigeration systems.
13. How are railcars refrigerated?
14. Truck trailers with their own refrigeration systems, which are transported on railway cars, are called
    A. self-contained refrigeration systems.
    B. piggyback refrigeration systems.
    C. double-layer refrigeration systems.
    D. dependent refrigeration systems.
15. Why is two-stage compression popular for extra-low-temperature refrigeration systems?
16. An R-404A system is operating with a $-50°F$ evaporator, and the condenser is operating at $115°F$. What is the compression ratio of this system?
    A. 16.9 to 1
    B. 17.8 to 1
    C. 19.1 to 1
    D. 21.33 to 1
17. Describe cascade refrigeration and tell why more than one refrigerant is used.
18. Why is a circulating brine system more attractive for ship refrigeration than individual evaporators?
19. Two methods used to power refrigeration compressors on a ship are _____ and _____.

Troubleshooting and Typical Operating Conditions for Commercial Refrigeration

OBJECTIVES

After studying this unit, you should be able to

- list the typical operating temperatures and pressures for the low-pressure side of a refrigeration system for high, medium, and low temperatures.
- list the typical operating pressures and temperatures for the high-pressure side of a system.
- state how different refrigerants compare on the high-pressure and low-pressure sides of the system.
- diagnose an inefficient evaporator.
- diagnose an inefficient condenser.
- diagnose an inefficient compressor.

SAFETY CHECKLIST

✔ Wear goggles and gloves when attaching and removing gages and when transferring refrigerant or checking pressures.

✔ Wear warm clothing when working in a cooler or freezer. A cold technician does not make good decisions.

✔ Observe all electrical safety precautions. Be careful at all times and use common sense.

✔ Disconnect electrical power whenever possible when working on a system. Lock and tag the panel and keep the key on your person.

✔ Wear rubber gloves when using chemicals to clean an evaporator or condenser. Use only approved chemicals. This is particularly important when cleaning an evaporator in a food storage unit.

✔ Do not attempt to perform a compressor vacuum or closed-loop test without the close supervision of an instructor or other experienced person.

29.1 ORGANIZED TROUBLESHOOTING

To begin troubleshooting any area of a refrigeration system, you need some idea of what the typical conditions should be. In commercial refrigeration we may be dealing with air temperatures inside a refrigerated box or outside at the condenser. We may need to know the current draw of the compressor or fan motors. The pressures inside the system may be important for troubleshooting and diagnoses. There are many conditions both outside and within the system that are important to know and understand.

When a piece of equipment has been running well for some period of time without noticeable problems, normally only one problem will trigger any sequence of events that may be encountered. For instance, it is common for the technician to approach a piece of equipment and think that two or three components have failed at one time. The chance of two parts failing at once is remote, unless one part causes the other to fail. These failures can almost always be traced to one original cause.

Knowing how the equipment is supposed to be functioning helps. You should know how it is supposed to sound, where it should be cool or hot, and when a particular fan is supposed to be operating. Knowing the correct operating pressures for a typical system can help you get started. **Before you do anything, look over the whole system for obvious problems.**

These troubleshooting procedures are divided into high-, medium-, and low-temperature ranges with some typical pressures for each. Each system has its own temperature range, depending on what it is supposed to be refrigerating. The table in **Figure 29–1** shows the temperature ranges for some refrigerated storage applications. This is only a partial table from the ASHRAE Fundamental Guide. Refer to this guide for a complete table of all foods. Notice that there is a different temperature requirement for almost every food under more than one situation. There may be a different temperature requirement for long-term storage than for short-term storage. The temperatures and pressures on the low-pressure side of the system are a result of the product load. **An evaporator acts the same whether there is a single compressor close to the evaporator or a large compressor in the equipment room.** The examples given relate to evaporators. The function of the compressor is to lower the pressure in the evaporator to the correct boiling point to achieve a desired condition in the evaporator.

The service technician is going to be faced with many numbers as the conditions are discussed for various applications. In the past, basically one refrigerant was used for medium- and high-temperature applications, and one was used for low-temperature applications. For routine service calls, the technician needed to think only of pressures that were common to these two refrigerants for the specific application. Several new refrigerants recently have entered the equation, which complicates the amount of numbers. The technician should concentrate only on the temperatures that are suggested for the particular applications. The temperatures can readily be converted to the correct pressures that

COMMODITY	STORAGE TEMP. °F	RELATIVE HUMIDITY %	APPROXIMATE STORAGE LIFE
Carrots			
Prepackaged	32	80–90	3–4 weeks
Topped	32	90–95	4–5 months
Celery	31–32	90–95	2–4 months
Cucumbers	45–50	90–95	10–14 days
Dairy products			
Cheese	30–45	65–70	
Butter	32–40	80–85	2 months
Butter	0 to −10	80–85	1 year
Cream (sweetened)	−15	—	several months
Ice cream	−15	—	several months
Milk, fluid whole			
Pasteurized Grade A	33	—	7 days
Condensed, sweetened	40	—	several months
Evaporated	Room Temp	—	1 year, plus
Milk, dried			
Whole milk	45–55	low	few months
Non-fat	45–55	low	several months
Eggs			
Shell	29–31	80–85	6–9 months
Shell, farm cooler	50–55	70–75	
Frozen, whole	0 or below	—	1 year, plus
Frozen, yolk	0 or below	—	1 year, plus
Frozen, white	0 or below	—	1 year, plus
Oranges	32–34	85–90	8–12 weeks
Potatoes			
Early crop	50–55	85–90	—
Late crop	38–50	85–90	—

(A)

COIL-TO-AIR TEMPERATURE DIFFERENCES TO MAINTAIN PROPER BOX HUMIDITY

TEMPERATURE RANGE	DESIRED RELATIVE HUMIDITY	TD (REFRIGERANT TO AIR)
25° F to 45° F	90%	8° F to 12° F
25° F to 45° F	85%	10° F to 14° F
25° F to 45° F	80%	12° F to 16° F
25° F to 45° F	75%	16° F to 22° F
10° F and Below	—	15° F or less

(B)

Figure 29–1 **(A)** This chart shows some storage requirements for typical products. **(B)** A coil-to-air temperature differences chart for selecting desired relative humidity. *Reprinted with permission of the American Society of Heating, Refrigerating, and Air-Conditioning Engineers, Inc.*

should be displayed on the gages. The text makes many references to R-134a and R-404A because many refrigeration systems are using these environmentally friendly, alternative refrigerants. R-134a is the refrigerant replacement for R-12, and R-404A is the refrigerant replacement for R-502 in many refrigeration applications. However, R-134a and R-404A are *not* drop-in replacements for any refrigerant. The text does have some system examples with R-12 and R-502 because of older systems still using these refrigerants. Always consult

with the refrigeration system's manufacturer before retrofitting any system's refrigerant. You may convert to any other refrigerant using a pressure-temperature chart.

29.2 ▶ TROUBLESHOOTING HIGH-TEMPERATURE APPLICATIONS

High-temperature refrigerated box temperatures start at about 45°F and go up to about 60°F. Normally a product temperature will be at one end of the range or the other depending on the product (flowers may be at 60°F and candy at 45°F). The intent is to provide the low- and high-temperature and pressure conditions that may be encountered for high-temperature refrigeration systems. The coil temperature is normally 10°F to 20°F cooler than the box temperature. This difference is referred to as temperature difference (TD). This means that the coil will normally be operating at 22 psig (45°F − 20°F = 25°F coil temperature) at the lowest temperature for an R-134a system. The pressure would be 57 psig (60°F coil temperature) at the end of the cycle when the compressor is off for the highest temperature for a high-temperature application. **Figure 29–2** is a chart of some typical operating conditions for high-, medium-, and low-temperature applications. These figures are for R-134a and R-404A for their typical applications. If any other refrigerant or application is used, the temperature readings and pressure readings may be converted to the different refrigerant.

When the technician suspects problems and installs the gages for pressure readings, the following readings should be found on a 20°F TD high-temperature system using R-134a:

1. With the compressor running just before the cut-out point, a reading below 22 psig would be considered too low.
2. With the compressor running just after it started up with a normal box temperature, 60°F (60°F − 20°F = 40°F coil temperature), a reading above 35.1 psig would be considered too high.
3. When the compressor is off and the box temperature is up to the highest point, the coil pressure would correspond to the air temperature in the box.

If a 60°F box is the application, the suction pressure could be 57 psig. This would be just before the compressor starts in a normal operation.

These pressures are fairly far apart but serve as reference points for most high-temperature applications. The preceding illustration is for a 20°F TD. There will be times when a 10°F TD is considered normal for a particular application. The best procedure for the technician is to consult the chart in **Figure 29–1(A)** for the correct application and determine what the pressures should be by converting the box temperatures to pressure. This chart shows what the storage temperatures are.

The correct box humidity conditions may also be found by studying **Figure 29–1(B)**. The box humidity conditions will determine the coil-to-air temperature difference for a particular application. For example, you may want to store

This table is intended to show the typical operating pressures for commercial refrigeration systems. Column 1 is with the compressor off and the box temperature just at the cut-in point. Column 2 is just after the compressor comes on. Column 3 is just before the compressor cuts off. This is not intended to be the only operating pressures, but the upper and lower limits as applied to high-, medium-, and low-temperature typical applications. A coil-to-inlet air temperature of 20°F (TD) was used as an average. Many systems will use 10°F or 20°F TD.

R-134a			
	Column 1 Compressor Off	Column 2 Compressor On	Column 3 Compressor On
HIGH-TEMPERATURE			
Box Temperature	60°F	60°F	45°F
Coil Temperature	60°F	40°F	25°F
Temperature Difference	0°F	20°F	20°F
Suction Pressure	57 psig	35.1 psig	22 psig
MED-TEMPERATURE			
Box Temperature	45°F	45°F	30°F
Coil Temperature	45°F	25°F	10°F
Temperature Difference	0°F	20°F	20°F
Suction Pressure	40 psig	22 psig	11.9 psig
LOW-TEMPERATURE			
Box Temperature	5°F	5°F	−20°F
Coil Temperature	5°F	−15°F	−40°F
Temperature Difference	0°F	2 °F	20°F
Suction Pressure	9.1 psig	0 psig	14.7 in. Hg
R-404A			
LOW-TEMPERATURE			
Box Temperature	5°F	5°F	−20°F
Coil Temperature	5°F	−15°F	−40°F
Temperature Difference	0°F	20°F	20°F
Suction Pressure	38 psig	20.5 psig	4.8 psig
R-404A			

ICE MAKING
Ice begins to form at 20°F coil temperature.
Suction pressure 56#
End of cycle, ice about to harvest at 5°F coil temperature.
Suction pressure 38#

When a service technician installs the gage manifold, the readings for a typical installation should not exceed or go below these readings. For instance, if the gages were applied to a medium-temperature vegetable box, the highest suction pressure that should be encountered on a typical box should be 22 psig with the compressor running and a low of 11.9 psig at the end of the cycle. Any readings above or below these will be out of range for a box that has a 20°F temperature difference between the air temperature entering the coil and the coil's refrigerant boiling temperature.

Figure 29–2(A)   These charts can be used for typical operating temperatures and pressures for refrigerating systems using forced-air evaporators.

TEMPERATURE °F	REFRIGERANT 12	22	134a	502	R-404A	R-410A	TEMPERATURE AND PRESSURE RANGES FOR THE LOW-PRESSURE SIDE OF THE SYSTEM
−60	19.0	12.0		7.2	6.6	0.3	
−55	17.3	9.2		3.8	3.1	2.6	
−50	15.4	6.2		0.2	0.8	5.0	
−45	13.3	2.7		1.9	2.5	7.8	
−40	11.0	0.5	14.7	4.1	4.8	9.8	
−35	8.4	2.6	12.4	6.5	7.4	14.2	
−30	5.5	4.9	9.7	9.2	10.2	17.9	
−25	2.3	7.4	6.8	12.1	13.3	21.9	← LOW TEMPERATURE
−20	0.6	10.1	3.6	15.3	16.7	26.4	
−18	1.3	11.3	2.2	16.7	18.7	28.2	
−16	2.0	12.5	0.7	18.1	19.6	30.2	
−14	2.8	13.8	0.3	19.5	21.1	32.2	
−12	3.6	15.1	1.2	21.0	22.7	34.3	
−10	4.5	16.5	2.0	22.6	24.3	36.4	
−8	5.4	17.9	2.8	24.2	26.0	38.7	
−6	6.3	19.3	3.7	25.8	27.8	40.9	
−4	7.2	20.8	4.6	27.5	30.0	42.3	
−2	8.2	22.4	5.5	29.3	31.4	45.8	
0	9.2	24.0	6.5	31.1	33.3	48.3	
1	9.7	24.8	7.0	32.0	34.3	49.6	
2	10.2	25.6	7.5	32.9	35.3	50.9	
3	10.7	26.4	8.0	33.9	36.4	52.3	
4	11.2	27.3	8.6	34.9	37.4	53.6	
5	11.8	28.2	9.1	35.8	38.4	55.0	
6	12.3	29.1	9.7	36.8	39.5	56.4	
7	12.9	30.0	10.2	37.9	40.6	57.8	
8	13.5	30.9	10.8	38.9	41.7	59.3	
9	14.0	31.8	11.4	39.9	42.8	60.7	
10	14.6	32.8	11.9	41.0	43.9	62.2	
11	15.2	33.7	12.5	42.1	45.0	63.7	
12	15.8	34.7	13.2	43.2	46.2	65.3	
13	16.4	35.7	13.8	44.3	47.4	66.8	
14	17.1	36.7	14.4	45.4	48.6	68.4	
15	17.7	37.7	15.1	46.5	49.8	70.0	
16	18.4	38.7	15.7	47.7	51.0	71.6	
17	19.0	39.8	16.4	48.8	52.3	73.2	← MEDIUM TEMPERATURE
18	19.7	40.8	17.1	50.0	53.5	75.0	
19	20.4	41.9	17.7	51.2	54.8	76.7	
20	21.0	43.0	18.4	52.4	56.1	78.4	
21	21.7	44.1	19.2	53.7	57.4	80.1	
22	22.4	45.3	19.9	54.9	58.8	81.9	
23	23.2	46.4	20.6	56.2	60.1	83.7	
24	23.9	47.6	21.4	57.5	61.5	85.5	
25	24.6	48.8	22.0	58.8	62.9	87.3	
26	25.4	49.9	22.9	60.1	64.3	90.2	
27	26.1	51.2	23.7	61.5	65.8	91.1	
28	26.9	52.4	24.5	62.8	67.2	93.0	
29	27.7	53.6	25.3	64.2	68.7	95.0	
30	28.4	54.9	26.1	65.6	70.2	97.0	
31	29.2	56.2	26.9	67.0	71.7	99.0	← HIGH TEMPERATURE
32	30.1	57.5	27.8	68.4	73.2	101.0	
33	30.9	58.8	28.7	69.9	74.8	103.1	
34	31.7	60.1	29.5	71.3	76.4	105.1	
35	32.6	61.5	30.4	72.8	78.0	107.3	
36	33.4	62.8	31.3	74.3	79.6	108.4	
37	34.3	64.2	32.2	75.8	81.2	111.6	
38	35.2	65.6	33.2	77.4	82.9	113.8	
39	36.1	67.1	34.1	79.0	84.6	116.9	
40	37.0	68.5	35.1	80.5	86.3	118.3	
41	37.9	70.0	36.0	82.1	88.0	120.5	

Vacuum—Red Figures
Gage Pressure—Bold Figures

Figure 29–2(B) Typical low, medium, and high temperature and pressure ranges for the low-pressure side of the refrigeration system.

cucumbers in a cooler. The chart shows that cucumbers should be stored at a temperature of 45°F to 50°F and a humidity of 90% to 95% to prevent dehydration. Examining the second part of the chart indicates that to maintain these conditions the coil-to-air temperature difference should be 8°F to 12°F; probably 10°F should be used.

The pressures from one piece of equipment to another may vary slightly. *Do not forget that any time the compressor is running, the coil will be 10°F to 20°F cooler than the fixture air temperature.*

29.3 ▶ TROUBLESHOOTING MEDIUM-TEMPERATURE APPLICATIONS

Medium-temperature refrigerated box temperatures range from about 30°F to 45°F in the refrigerated space. Many products will not freeze when the box temperature is as low as 30°F. Using the same methods that were used in the high-temperature examples to find the low- and high-pressure readings that would be considered normal, we find the following. The lowest pressures that would normally be encountered would be at the lowest box temperatures of 30°F. If the coil were to be boiling the refrigerant at 20°F below the inlet air temperature, the refrigerant would be boiling at 10°F (30°F − 20°F = 10°F). This corresponds to a pressure for R-134a of 11.9 psig. If the service technician encounters pressures below 11.9 psig for an R-134a medium-temperature system, there is most likely a problem. The highest pressures that would normally be encountered while the system is running would be at the boiling temperature just after the compressor starts at 45°F. This is 45°F − 20°F = 25°F, or a corresponding pressure of 22 psig, **Figure 29–2**. The pressure in the low side while the compressor is off would correspond to the air temperature inside the cooler. When 45°F is considered the highest temperature before the compressor comes back on, the pressure would be 40 psig. Consult the chart in **Figure 29–1** for any specific application because each is different. The chart is used as a guideline for the product type. The equipment operating conditions may vary to some extent from one manufacturer to another.

29.4 ▶ TROUBLESHOOTING LOW-TEMPERATURE APPLICATIONS

Low-temperature applications start at freezing and go down in temperature. There is not much application at just below freezing. The first usable application is the making of ice, which normally occurs from 20°F evaporator temperature to about 5°F evaporator temperature. Ice making has many different variables. These variables are associated with the type of ice being made. Flake ice is a continuous process so the pressures are about the same regardless of the manufacturer. The manufacturer of the specific piece of equipment should be consulted for exact temperatures and pressures for flake ice machines.

Normally for cube ice makers you will find that a suction pressure of 36 psig (20°F) at the beginning of the cycle will begin to make ice using R-404A. The ice will normally be harvested before the suction pressure reaches the pressure corresponding to 5°F or 38 psig, **Figure 29–2**. If you discover that the suction pressure would not go below 56 psig for an R-404A machine, suspect a problem. If the suction pressure goes below 38 psig, suspect a problem. These pressures are for the normal running cycle.

The low-temperature application for frozen foods is generally considered to start at 5°F space temperature and goes down from there. Foods require different temperatures to be frozen hard. Some are hard at 5°F, others at −10°F. The designer or operator will use the most economical design or operation. For instance, ice cream may be frozen hard at −10°F, but it may be maintained at −30°F without harm. The economics of cooling it to a cooler temperature may not be wise. As the temperature goes down, the compressor has less capacity because the suction gas becomes thinner. The compressor may not even cut off at the very low temperatures if the box thermostat is set this low.

Using the same guidelines as in high- and medium-temperature applications, we find the highest normal refrigerant boiling temperature for low-temperature refrigeration applications to be 5°F − 20°F = −15°F for R-134a. The refrigerant boils at 20°F below the space temperature because the compressor is running and creating a suction pressure of 0 psig. The lowest suction pressure may be anything down to the evaporator pressure corresponding to −20°F space temperature. This would be −20°F minus 20°F = −40°F or 14.7 in. Hg vacuum for an R-134a system, **Figure 29–2(B)**. With the compressor off just at the time the thermostat would call for cooling for the highest temperature application, the evaporator would be 5°F, the same as the space temperature or 9.1 psig.

A different refrigerant for low-temperature applications may be used to keep the pressures positive, above atmospheric. If we used R-404A for this application, we would find a new set of figures. 5°F corresponds to 38 psig when the space temperature is 5°F with the compressor off. When the space temperature is 5°F and the compressor is running, the refrigerant will be boiling at about 20°F below the space temperature or 5°F − 20°F = −15°F. This −15°F corresponds to a pressure of 20.3 psig. The lowest temperature normally encountered in low-temperature refrigeration is a space temperature of about −20°F. This would make the suction pressure about −20°F minus 20°F = −40°F or 4.8 psig, **Figure 29–2(B)**. We can easily see that by using R-404A in this application the system is going to operate well above a vacuum. R-404A boils at −50°F at 0 psig, so the cooler would get down to −30°F air temperature before the suction would get down to atmospheric pressure of 0 psig with a 20°F TD.

Temperatures below −20°F space or product temperature are obtainable and used.

The chart in **Figure 29–2(A)** can be used as a guideline for typical low-side operating conditions for the aforementioned systems. Its intent is to illustrate the high and low temperatures and pressures for each application. The machine you may be working on should fall within these limits.

29.5 TYPICAL AIR-COOLED CONDENSER OPERATING CONDITIONS

Air-cooled condensers have an operating temperature range from cold to hot as they are exposed to outside temperatures. The equipment may even be located inside a conditioned building. The condenser has to maintain a pressure that will create enough pressure drop across the expansion device for it to feed correctly. This pressure drop across the expansion device will push enough refrigerant through the device for

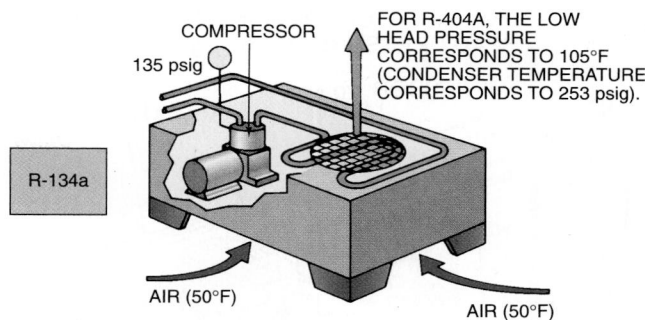

Figure 29–3 Refrigeration system shows the relationship of the air-cooled condenser to the entering air temperature on a cool day. This unit has head pressure control for mild weather. The outside air entering the condenser is 50°F. This would normally create a head pressure of about 85 psig. This unit has a fan cycle device to hold the head pressure up to 125 psig minimum.

proper evaporator refrigerant levels in the evaporator. However, some modern thermostatic expansion valves (TXVs) can operate with much less pressure drop across them as long as 100% liquid is supplied to them. The balanced-port TXV is noted for its low-pressure drop performance.

Expansion valves for R-134a are normally sized at a 75- to 100-psig pressure drop. For example, when a valve manufacturer indicates that a valve has a capacity of 1 ton at an evaporator temperature of 20°F, this is for a pressure drop of 80 psig. The same valve will have less capacity if the pressure drop goes to 60 psig.

Because any piece of equipment located outside must have head pressure control, the minimum values at which the head pressure control is set would be the minimum expected head pressure. For R-134a this is normally about 135 psig for air-cooled equipment, **Figure 29–3**; this corresponds to 105°F condensing temperature. R-404A has a head pressure of 253 psig at 105°F. When the discharge pressure is less than 135 psig for an R-134a machine, the condenser is operating at the lowest pressure to maintain 80 psig drop across the expansion device, and a lower pressure may not be feeding the expansion valve correctly. An exception to this is when the manufacturer *floats* the head pressure with the ambient temperature. Head pressure controls can still be used, but they are usually set at much lower settings to take advantage of the cooler ambient temperatures. Floating head pressures will increase efficiencies by enabling the compressor to run at lower compression ratios. Review 22.21, "Floating Head Pressures," for more in-depth information on floating the head pressure.

29.6 CALCULATING THE CORRECT HEAD PRESSURE FOR AIR-COOLED EQUIPMENT

The maximum normal high-side pressure would correspond to the maximum ambient temperature in which the condenser is operated. Most air-cooled condensers will condense the

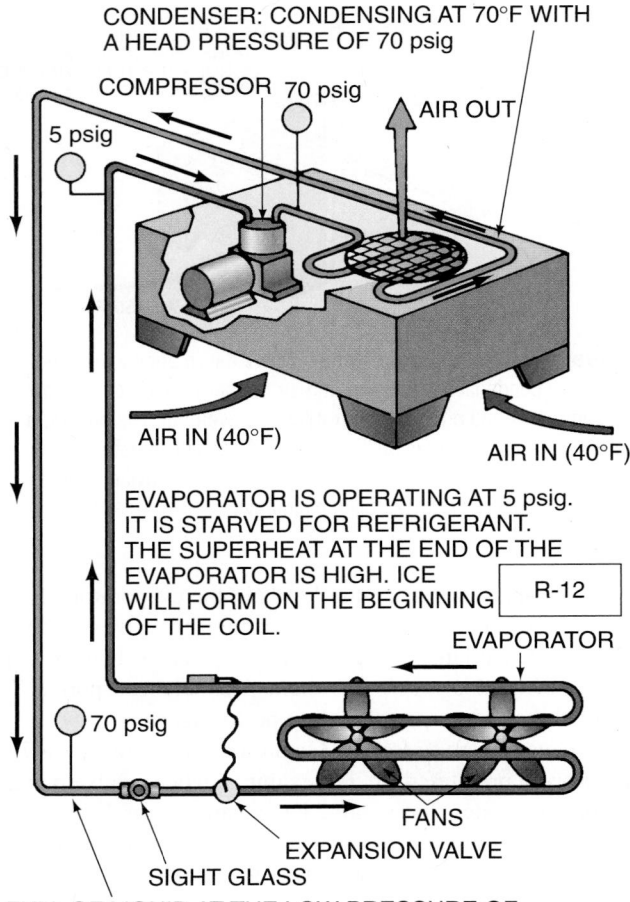

CONDENSER: CONDENSING AT 70°F WITH A HEAD PRESSURE OF 70 psig

COMPRESSOR 70 psig

AIR OUT

5 psig

AIR IN (40°F)   AIR IN (40°F)

EVAPORATOR IS OPERATING AT 5 psig. IT IS STARVED FOR REFRIGERANT. THE SUPERHEAT AT THE END OF THE EVAPORATOR IS HIGH. ICE WILL FORM ON THE BEGINNING OF THE COIL.

R-12

EVAPORATOR

70 psig

FANS

EXPANSION VALVE

SIGHT GLASS

FULL OF LIQUID AT THE LOW PRESSURE OF 70 psig. THE PRESSURE WILL NOT FEED ENOUGH REFRIGERANT THROUGH THE EXPANSION VALVE.

MEDIUM-TEMPERATURE EVAPORATOR SHOULD BE OPERATING AT 21 psig (20°F BOILING TEMPERATURE).

Figure 29–4 A refrigeration system operating at an ambient temperature that caused the head pressure to be too low for normal operation. The expansion valve is not feeding correctly because there is not enough head pressure to push the liquid refrigerant through the valve.

refrigerant at a temperature of about 30°F above the ambient temperature when the ambient is above 70°F. **Figure 29–4** presents an example of a refrigeration system operating under low ambient conditions without head pressure control. When the ambient temperature drops below 70°F, the relationship changes to a lower value.

If the condenser is a higher-efficiency condenser or oversized, the condenser may condense at 25°F above the ambient. For energy-efficiency purposes, many manufacturers of high-efficiency systems are manufacturing condensers with improved heat transfer materials and surfaces. Condenser sizes are also increasing to keep head pressures lower and compression ratios lower for better efficiencies. This will cause some condensing temperatures to be only 10°F to 15°F above the ambient.

Using the 30°F figure, we see that if a unit were located outside and the temperature were 95°F, the condensing

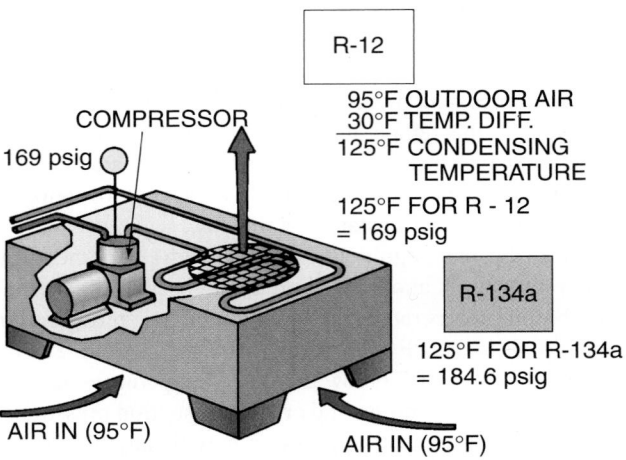

R-12

COMPRESSOR

169 psig

95°F OUTDOOR AIR 30°F TEMP. DIFF. 125°F CONDENSING TEMPERATURE

125°F FOR R - 12 = 169 psig

R-134a

125°F FOR R-134a = 184.6 psig

AIR IN (95°F)   AIR IN (95°F)

STANDARD-EFFICIENCY AIR-COOLED CONDENSER CONDITIONS WITH A TYPICAL LIGHT FILM OF AIRBORNE DIRT CONDENSING AT 30°F ABOVE THE AMBIENT ENTERING-AIR TEMPERATURE.

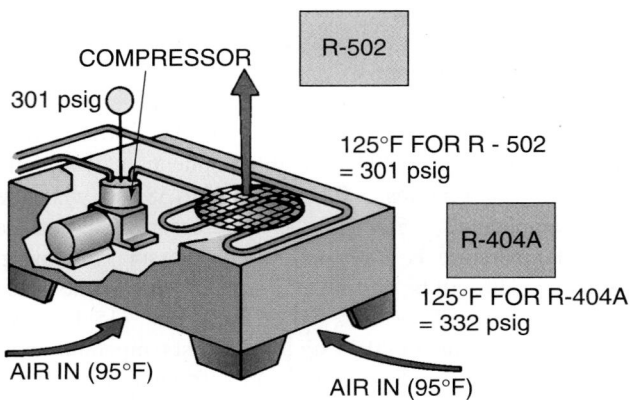

COMPRESSOR

R-502

301 psig

125°F FOR R - 502 = 301 psig

R-404A

125°F FOR R-404A = 332 psig

AIR IN (95°F)   AIR IN (95°F)

Figure 29–5 An air-cooled refrigeration system with a condenser outside. Typical pressures are shown.

temperature would be 95°F + 30°F = 125°F. This corresponds to 169 psig for R-12 and 301 psig for R-502, **Figure 29–5.** These were the two most common refrigerants for commercial refrigeration; however, R-134a, R-404A, R-22, and replacement refrigerant blends are now being used because of the ozone depletion and global-warming problems. The pressure for R-134a would be 185 psig, for R-404A it would be 332 psig, and for R-22 it would be 278 psig at 125°F condensing temperature.

29.7 TYPICAL OPERATING CONDITIONS FOR WATER-COOLED EQUIPMENT

Water-cooled condensers are used in many systems in two ways. Some are wastewater, and some reuse the same water by extracting the heat with a cooling tower. Typically a water-cooled condenser that uses fresh water, such as city water or well water, uses about 1.5 gal of water per minute per ton while a system using a cooling tower circulates about 3 gpm/ton. These two applications have different operating conditions and will be discussed separately.

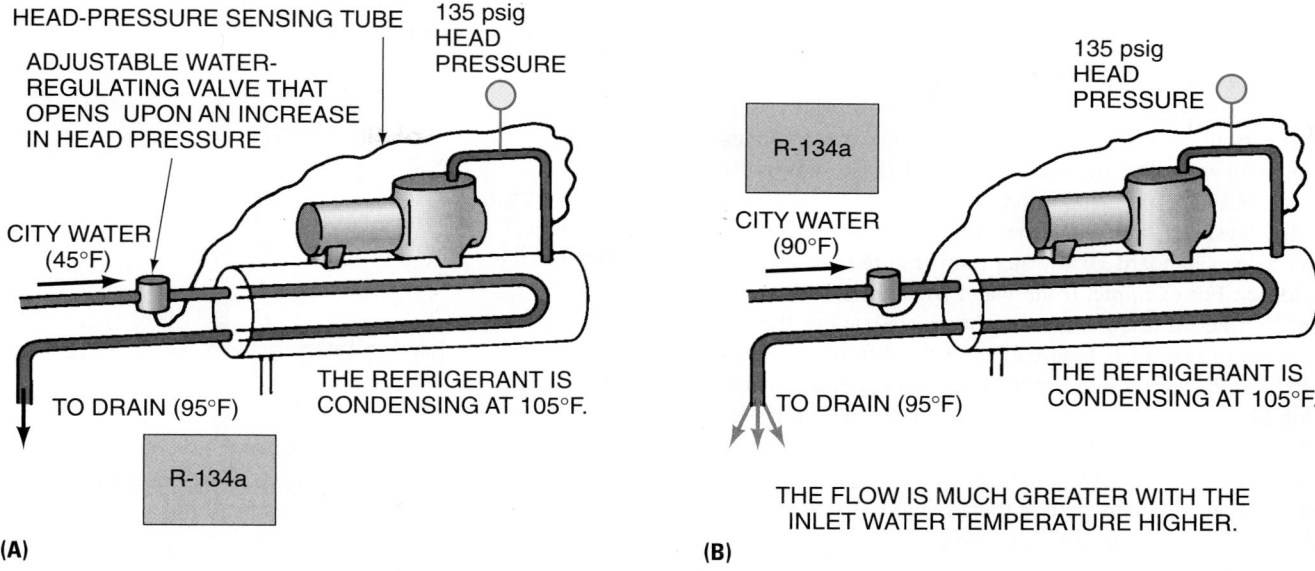

Figure 29–6 **(A)** A water-cooled wastewater refrigeration system. The water flow is adjusted by the water-regulating valve, which keeps too much water from flowing, maintains a constant head pressure, and shuts the water off at the end of the cycle. **(B)** Water flow is greater with a 90°F inlet water temperature.

29.8 TYPICAL OPERATING CONDITIONS FOR WASTEWATER CONDENSER SYSTEMS

Wastewater systems use the same condensers as the cooling tower applications, but the water is wasted down the drain. Normally a water-regulating valve is used to regulate the water flow for economy and to regulate the head pressure. The condensers are either cleanable or a coil type that cannot be cleaned mechanically. With either type of condenser, it is advantageous to know from the outside how the condenser is performing on the inside.

When a water-regulating valve is used to control the water flow, the water flow will be greater if the condenser is not performing correctly. When the head pressure goes up, the water will start to flow faster to compensate. This will take place until the capacity of the valve opening is reached. Then the head pressure will increase with maximum water flow. Sometimes the inlet water is colder, such as in winter. It would not be unusual for the inlet water to be 45°F in the winter. 90°F may be the high value if the water travels through a hot ceiling in the summer, **Figure 29–6.** Because of the variable flow and variable temperature of water flowing through a wastewater system during different seasons of the year, there is no refrigerant-to-water temperature relationship or rule of thumb for the technician to go by for these systems. The service technician must simply rely on the water-regulating valve to throttle more or less water at different temperatures through the condenser to maintain a constant head pressure. For more detailed information on condensers and water-regulating valves, refer to Unit 22, "Condensers."

If the condensing temperature were much above 105°F, you would suspect a dirty condenser. If the gages indicated 207 psig (140°F condensing temperature) and the leaving-water temperature were to be 95°F, the difference in

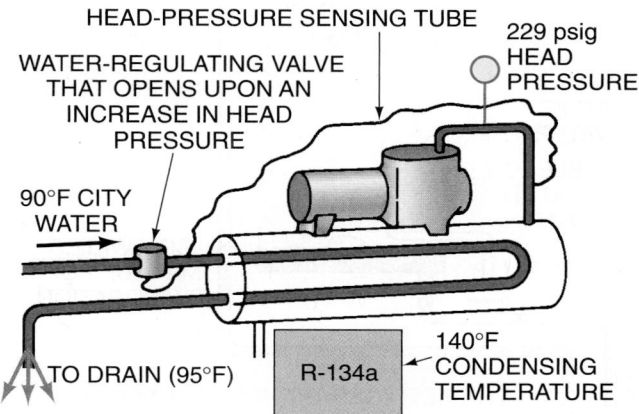

Figure 29–7 A water-cooled refrigeration system has dirty condenser tubes, and the water is not taking the heat out of the system.

temperature of the condensing refrigerant is 140°F − 95°F = 45°F. This indicates that the condenser is not removing the heat; the coil is dirty, **Figure 29–7.** It can be cleaned chemically or with brushes if it is a cleanable condenser. Whenever there is a noticeable increase in water flow, a low-temperature differential between water in and out of the condenser, or a high head-pressure situation, a dirty condenser should be suspected.

29.9 TYPICAL OPERATING CONDITIONS FOR RECIRCULATED WATER SYSTEMS

Water-cooled condensers that use a cooling tower to remove the heat from the water normally do not use water-regulating valves to control the water flow. It is normally a constant volume of water that is pumped by a pump. The volume is customarily designed into the system in the beginning and can be verified by checking the pressure drop across the water circuit at the condenser inlet to outlet. There has to be some pressure drop for there to be water flow. The original specifications of the system should include the engineer's intent with regard to the water flow, but these may not be obtainable on an old installation.

Most systems that reuse the same water with a cooling tower have a standard 10°F water temperature rise across the condenser. For example, if the water from the tower were to be 85°F entering the condenser, the water leaving the condenser should be 95°F, **Figure 29–8.** If the difference were to be 15°F or 20°F, you might think that the condenser is doing its job of removing the heat from the refrigerant, but the water flow is insufficient, **Figure 29–9.** If the water temperature entering the condenser were to be 85°F and the leaving water were to be 90°F, it may be that there is too much water flow. If the head pressure is not high, the condenser is removing the heat, and there is too much water flow. If the head pressure is high and there is the right amount of water flow, the condenser is dirty. See **Figure 29–10** for an example of a dirty

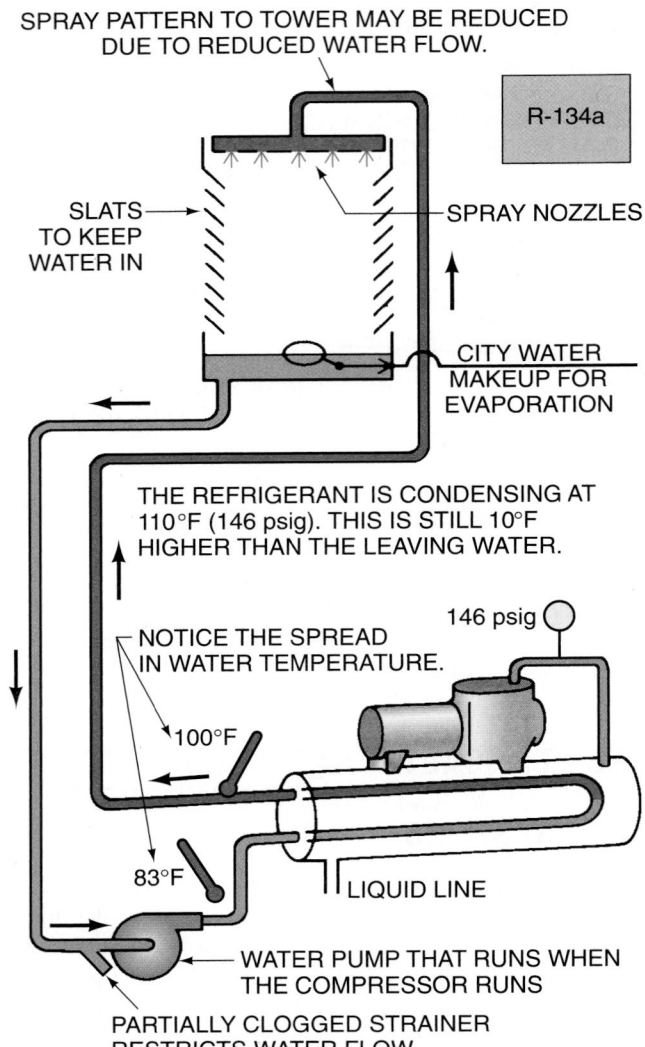

SPRAY PATTERN TO TOWER MAY BE REDUCED DUE TO REDUCED WATER FLOW.

R-134a

SLATS TO KEEP WATER IN

SPRAY NOZZLES

CITY WATER MAKEUP FOR EVAPORATION

THE REFRIGERANT IS CONDENSING AT 110°F (146 psig). THIS IS STILL 10°F HIGHER THAN THE LEAVING WATER.

146 psig

NOTICE THE SPREAD IN WATER TEMPERATURE.

100°F

83°F

LIQUID LINE

WATER PUMP THAT RUNS WHEN THE COMPRESSOR RUNS

PARTIALLY CLOGGED STRAINER RESTRICTS WATER FLOW.

Figure 29–9 This system has too much temperature rise, indicating there is not enough water flow. The water strainers may be stopped up. The condensing temperature is higher than normal, which causes the head pressure to rise. A decrease in water flow may be detected by water pressure drop if it is known what it is supposed to be.

condenser. **Figure 29–11** shows a 10°F water temperature rise across the condenser and related condensing pressures for a 105°F condensing temperature when R-12, R-502, R-134a, R-22, and R-404A are used as the refrigerants.

In this type of system the condenser is getting its inlet water from the cooling tower. A cooling tower has a heat exchange relationship with the ambient air. Usually the cooling tower is located outside and may be natural draft or forced draft (where a fan forces the air through the tower). Either tower will be able to supply water temperature according to the humidity or moisture content in the outside air. The cooling tower can normally cool the water to within 7°F of the wet-bulb temperature of the outside air, **Figure 29–12.** If the outside wet-bulb temperature (taken with a psychrometer) is 78°F, the leaving water will be about 85°F if the tower is performing correctly. Wet-bulb temperature relates to the moisture content in air and is discussed in more detail in Unit 35, "Comfort and Psychrometrics."

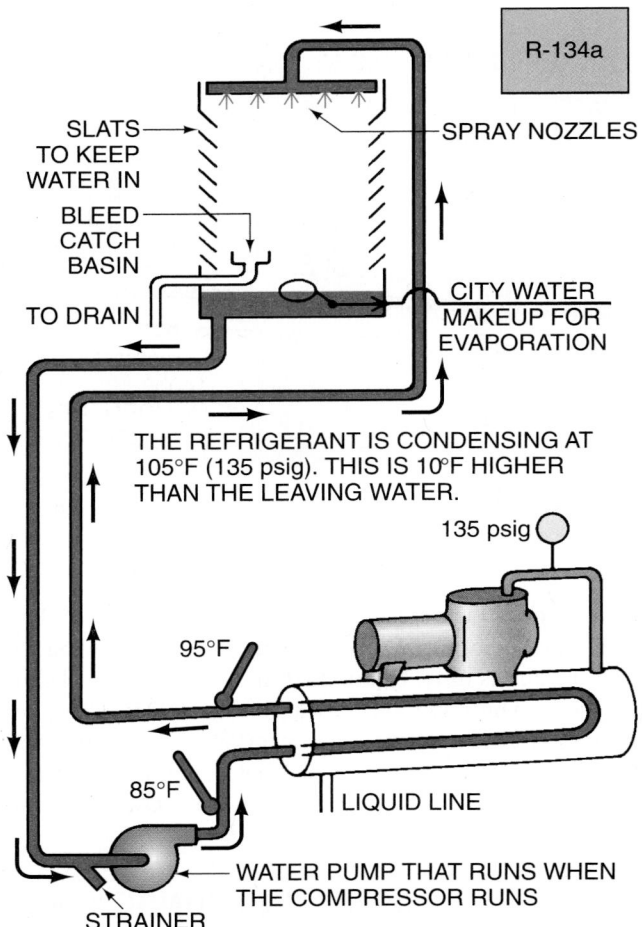

R-134a

SLATS TO KEEP WATER IN

SPRAY NOZZLES

BLEED CATCH BASIN

TO DRAIN

CITY WATER MAKEUP FOR EVAPORATION

THE REFRIGERANT IS CONDENSING AT 105°F (135 psig). THIS IS 10°F HIGHER THAN THE LEAVING WATER.

135 psig

95°F

85°F

LIQUID LINE

WATER PUMP THAT RUNS WHEN THE COMPRESSOR RUNS

STRAINER

Figure 29–8 A water-cooled refrigeration system reusing the water after the heat is rejected to the atmosphere. There is a constant bleed of water to keep the system from overconcentrating with the minerals left behind when the water is evaporated. The temperature difference between the incoming water and the outgoing water is 10°F. This is the typical temperature rise across a cooling tower system.

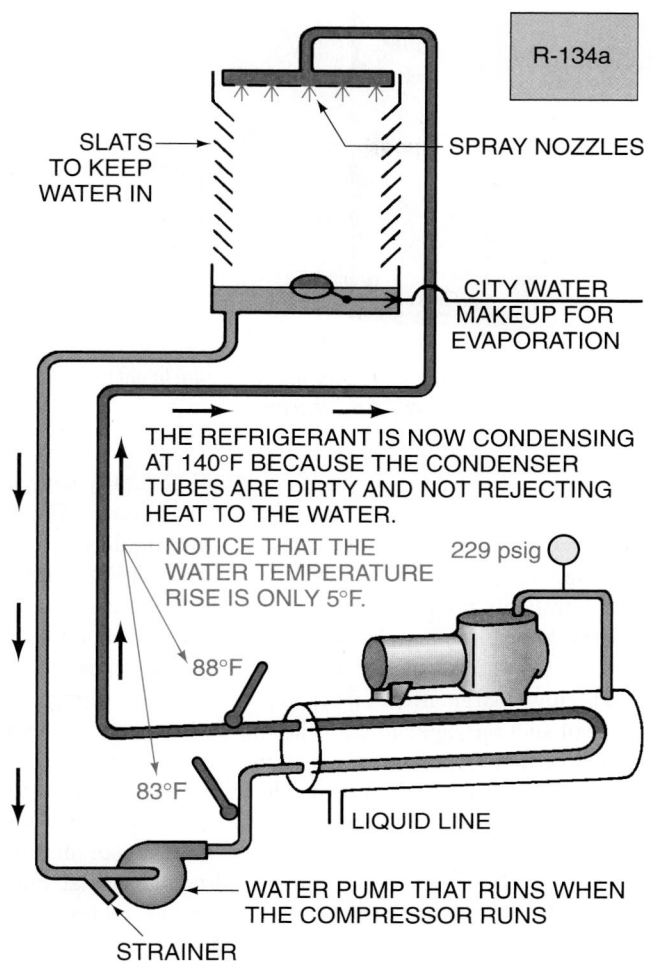

R-134a

SLATS TO KEEP WATER IN

SPRAY NOZZLES

CITY WATER MAKEUP FOR EVAPORATION

THE REFRIGERANT IS NOW CONDENSING AT 140°F BECAUSE THE CONDENSER TUBES ARE DIRTY AND NOT REJECTING HEAT TO THE WATER.

NOTICE THAT THE WATER TEMPERATURE RISE IS ONLY 5°F.

229 psig

88°F

83°F

LIQUID LINE

WATER PUMP THAT RUNS WHEN THE COMPRESSOR RUNS

STRAINER

Figure 29–10   This system has dirty condenser tubes.

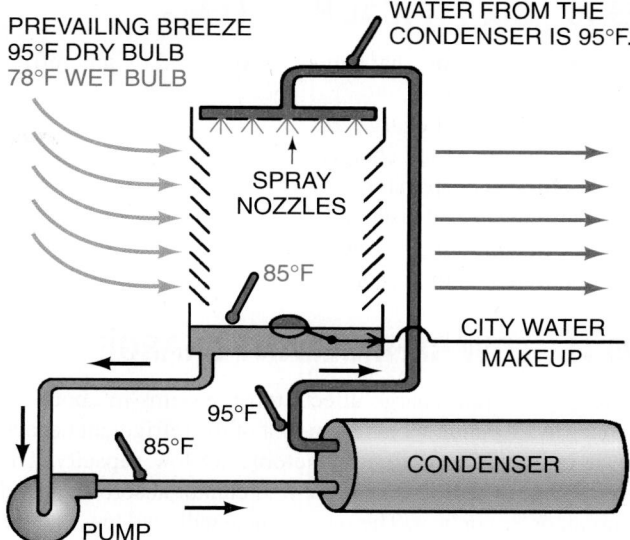

PREVAILING BREEZE 95°F DRY BULB 78°F WET BULB

WATER FROM THE CONDENSER IS 95°F.

SPRAY NOZZLES

85°F

CITY WATER MAKEUP

95°F

85°F

CONDENSER

PUMP

COOLED WATER IN BASIN OF TOWER IS 85°F. IT CAN NORMALLY BE COOLED TO WITHIN 7°F OF THE OUTDOOR WET-BULB TEMPERATURE DUE TO EVAPORATION.
NOTICE THAT THE FINAL WATER TEMPERATURE IS MUCH COOLER THAN THE OUTDOOR DRY-BULB TEMPERATURE.

Figure 29–12   This cooling tower has a temperature relationship with the air that is cooling the water. Most cooling towers can cool the water that goes back to the condenser to within 7°F of the wet-bulb temperature of the ambient air. For example, if the wet-bulb temperature is 78°F, the tower should be able to cool the water to 85°F. If it does not, a tower problem should be suspected.

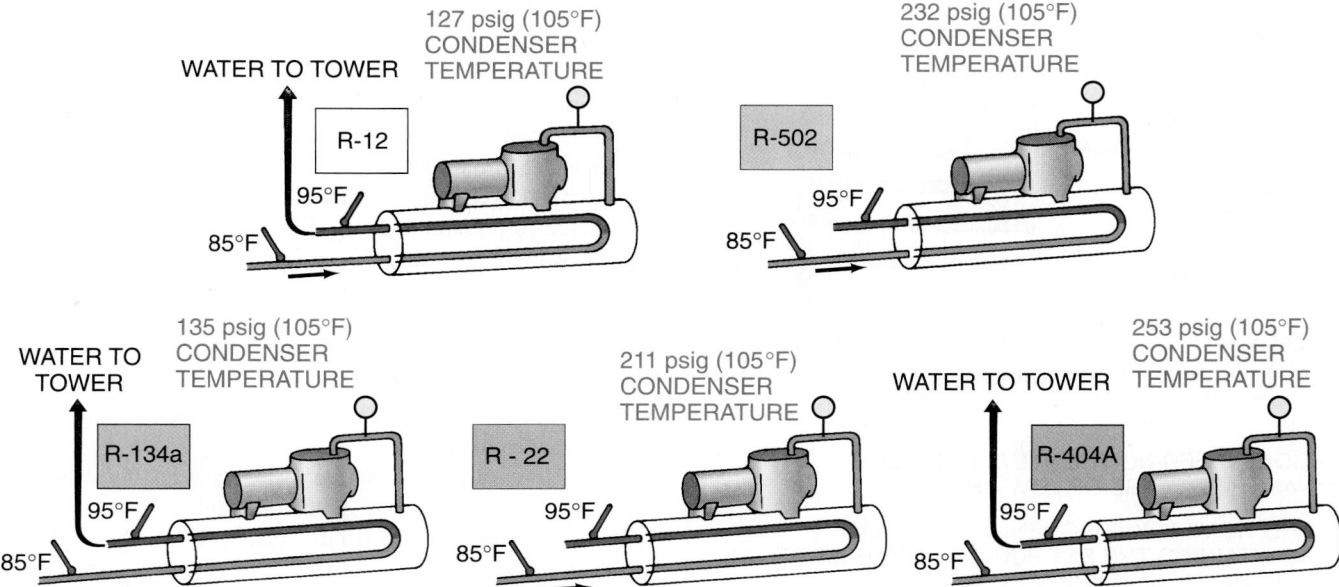

WATER TO TOWER

127 psig (105°F) CONDENSER TEMPERATURE

R-12

95°F

85°F

232 psig (105°F) CONDENSER TEMPERATURE

R-502

95°F

85°F

WATER TO TOWER

135 psig (105°F) CONDENSER TEMPERATURE

R-134a

95°F

85°F

211 psig (105°F) CONDENSER TEMPERATURE

R - 22

95°F

85°F

WATER TO TOWER

253 psig (105°F) CONDENSER TEMPERATURE

R-404A

95°F

85°F

Figure 29–11   A cooling tower system showing a 10°F water temperature rise across the condenser and related condensing pressures for a 105°F condensing temperature when R-12, R-502, R-134a, R-22, and R-404A are used as the refrigerants.

29.10 SIX TYPICAL PROBLEMS

Six typical problems that can be encountered by any refrigeration system are

1. low refrigerant charge.
2. excess refrigerant charge.
3. inefficient evaporator.
4. inefficient condenser.
5. restriction in the refrigerant circuit.
6. inefficient compressor.

29.11 LOW REFRIGERANT CHARGE

A low refrigerant charge affects most systems in about the same way, depending on the amount of the refrigerant needed to be correct. The normal symptoms are low capacity. The system has a starved evaporator and cannot absorb the rated amount of Btu or heat. The suction gage will read low, and the discharge gage will read low. The exception to this is a system with an automatic expansion valve, **Figure 29–13**. It will be discussed later in this unit.

If the system has a sight glass, it will have bubbles in it that look like air, but which are actually vapor refrigerant. **Figure 29–14** is a typical liquid-line sight glass. Remember, a sight glass that is full of vapor or liquid may look the same. If there is only vapor in the glass, a slight film of oil may be present. This is a good indicator of vapor only.

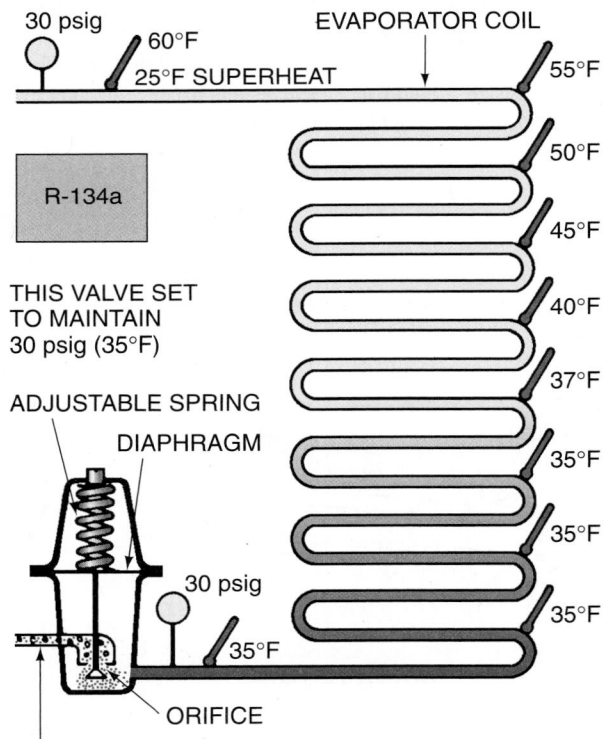

Figure 29–13 Low-refrigerant charge system characteristics when the metering device is an automatic expansion valve.

30 psig
60°F
25°F SUPERHEAT
EVAPORATOR COIL
55°F
50°F
45°F
R-134a
40°F
37°F
THIS VALVE SET
TO MAINTAIN
30 psig (35°F)
35°F
ADJUSTABLE SPRING
DIAPHRAGM
35°F
30 psig
35°F
35°F
ORIFICE

LIQUID FEEDING THE VALVE HAS SOME VAPOR BECAUSE OF LOW CHARGE.

THE VALVE IS WIDE OPEN AND IS STILL MAINTAINING THE SET POINT FOR PRESSURE. COIL HAS A HIGH SUPERHEAT.

Figure 29–14 A sight glass to indicate when pure liquid is in the liquid line. *Courtesy Henry Valve Company*

When a system has a sight glass, it will generally have a TXV or automatic expansion valve and a receiver. These valves will hiss when a partial vapor–partial liquid mixture is going through the valve. If the system has a capillary tube, it will probably not have a sight glass. The technician needs to know how the system feels to the touch at different points to determine the gas charge level without using gages.

The low charge affects the compressor by not supplying the cool suction vapor to cool the motor. Low charges also cause low suction pressures, which increase the compression ratio. A higher compression ratio causes inefficiencies along with higher discharge temperatures. Most compressors are suction cooled, so the result is a hot compressor motor. It may even be off due to the motor-winding thermostat, **Figure 29–15**. If the compressor is air cooled, the suction line coming back to the compressor will not need to be as cool as a suction-cooled compressor. It may be warm by comparison, **Figure 29–16**.

Low compressor amperage will occur with an undercharge of refrigerant because the higher superheat coming back to the compressor inlet will cause the inlet vapors to expand, which decreases their density. Low-density vapors entering the compressor will mean low refrigerant flow through the compressor. This will cause low amperage (amp) draw, because the compressor does not have to work as hard compressing the low-density vapors. This low refrigerant flow may also cause refrigerant-cooled compressors to overheat.

Because of the low refrigerant flow rate through the compressor and system that occurs with an undercharge of refrigerant, the 100% saturated-vapor point in the condenser will be very low in the condenser, causing low condenser subcooling. The condenser will not receive enough refrigerant vapors to condense it to a liquid and feed the receiver (if the system has one). Condenser subcooling is a good indication of how much refrigerant charge is in the system. Low condenser subcooling may mean a low charge, while high condenser subcooling may mean an overcharge.

This is not true for capillary tube systems because they usually have no receiver. A capillary tube system can run high subcooling simply from a restriction in the capillary tube or liquid line. The excess refrigerant will accumulate in the condenser, causing high subcooling and high head pressures. If a TXV/receiver system is restricted in the liquid line, some refrigerant will accumulate in the condenser but

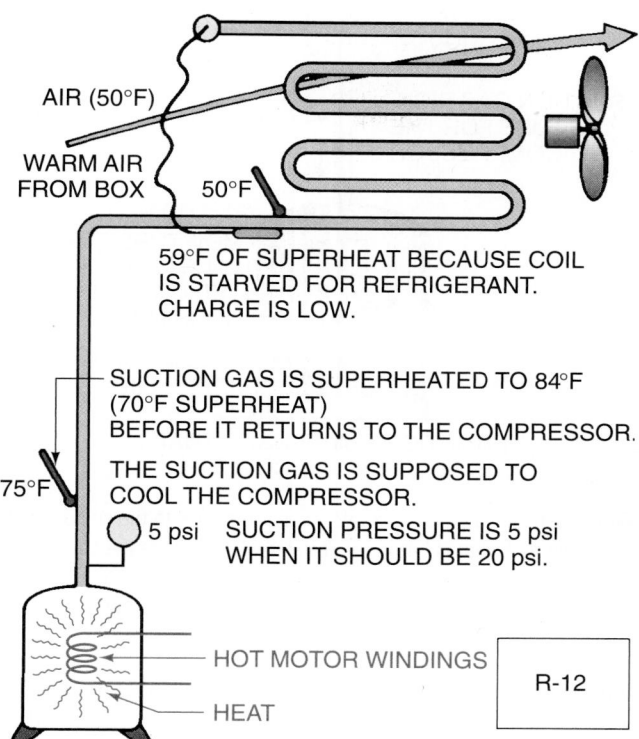

AIR (50°F)

WARM AIR FROM BOX   50°F

59°F OF SUPERHEAT BECAUSE COIL IS STARVED FOR REFRIGERANT. CHARGE IS LOW.

SUCTION GAS IS SUPERHEATED TO 84°F (70°F SUPERHEAT) BEFORE IT RETURNS TO THE COMPRESSOR.

THE SUCTION GAS IS SUPPOSED TO COOL THE COMPRESSOR.

75°F

5 psi   SUCTION PRESSURE IS 5 psi WHEN IT SHOULD BE 20 psi.

HOT MOTOR WINDINGS

HEAT

R-12

Figure 29–15 A system with a suction-cooled hermetic compressor. The compressor is hot enough to cause it to cut off because of motor temperature. It is hard to cool off a hot hermetic compressor from the outside because the motor is suspended inside a vapor atmosphere.

Figure 29–16 An air-cooled compressor. The compressor motor will not normally get hot as a result of a low refrigerant charge, but the discharge gas leaving the compressor may get too warm because the refrigerant entering the compressor is too warm. Most compressor manufacturers require the suction gas temperature to be not more than 65°F for continuous operation. *Courtesy Tyler Refrigeration Company*

most will collect in the receiver. Remember, a receiver is designed to handle the system's entire refrigerant charge and still have a 25% vapor head for safety. This will cause low condenser subcooling and a low head pressure.

The symptoms of a low refrigerant charge are summarized below:

■ Low system capacity
■ Starved evaporator (high superheat)
■ Low suction pressure
■ Low discharge pressure
■ High compression ratio
■ Bubbling sight glass
■ Low compressor amperage (amp) draws
■ Low condenser subcooling

29.12 REFRIGERANT OVERCHARGE

A refrigerant overcharge also acts much the same way from system to system. The discharge pressure is high, and the suction pressure may be high. The automatic expansion valve system will not have a high suction pressure because it maintains a constant suction pressure, **Figure 29–17**. The TXV system may have a slightly higher suction pressure if the head pressure is excessively high because the system capacity may be down, **Figure 29–18**.

The capillary tube system will have a high suction pressure because the amount of refrigerant flowing through it depends on the difference in pressure across it. The more head pressure, the more liquid it will pass. The capillary tube will allow enough refrigerant to pass so that it will allow liquid into the compressor. When the compressor is sweating down the side or all over, it is a sign of liquid refrigerant in the compressor, **Figure 29–19** and **Figure 29–20**.

Critically charging a capillary tube system may help the severity of these flooding problems. Often, in high head-pressure situations in which the system is critically charged

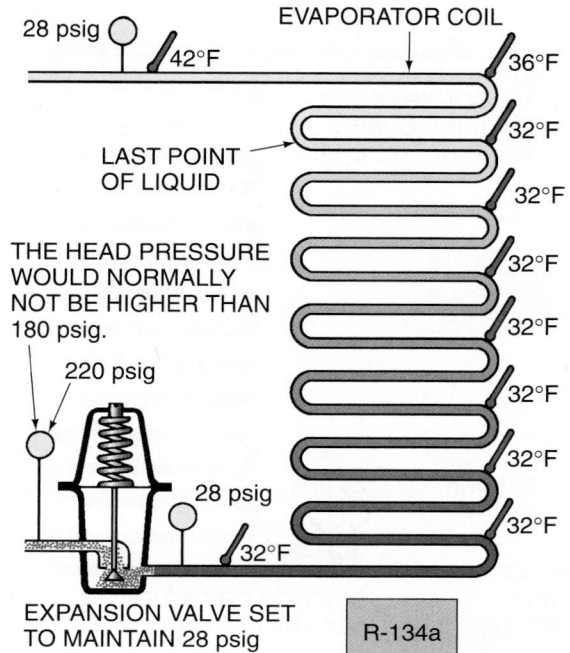

28 psig   42°F   EVAPORATOR COIL   36°F

32°F

LAST POINT OF LIQUID   32°F

32°F

THE HEAD PRESSURE WOULD NORMALLY NOT BE HIGHER THAN 180 psig.   32°F

220 psig   32°F

28 psig   32°F

32°F

32°F

EXPANSION VALVE SET TO MAINTAIN 28 psig   R-134a

Figure 29–17 System has an automatic expansion valve for a metering device. This device maintains a constant suction pressure. The head pressure is higher than normal, but the suction pressure remains the same.

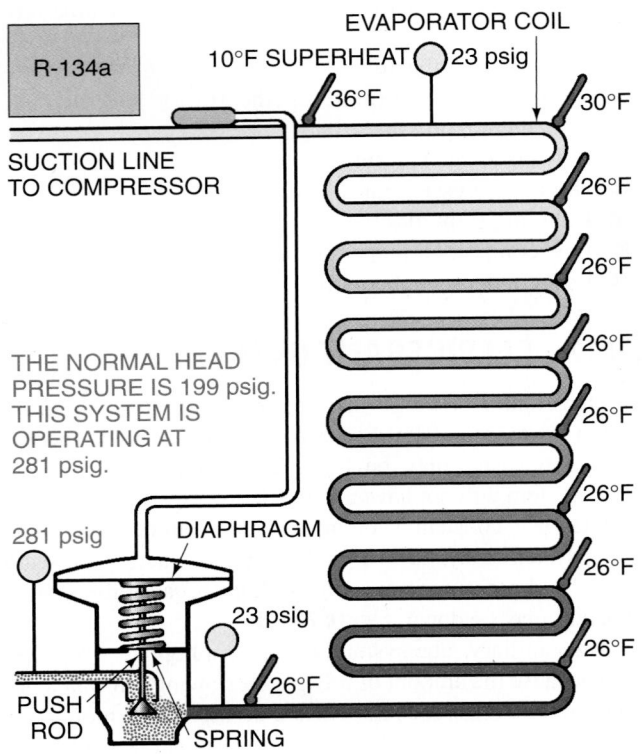

Figure 29–18 A TXV system using R-134a, which has a higher-than-normal head pressure.

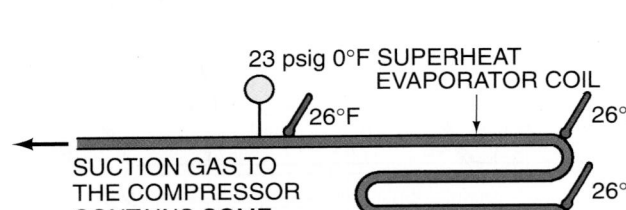

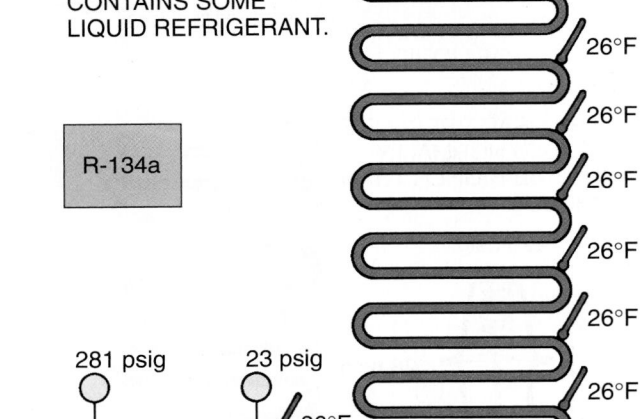

Figure 29–19 The capillary tube system using R-134a has an overcharge of refrigerant, and the head pressure is higher than normal. This has a tendency to push more refrigerant than normal through the metering device.

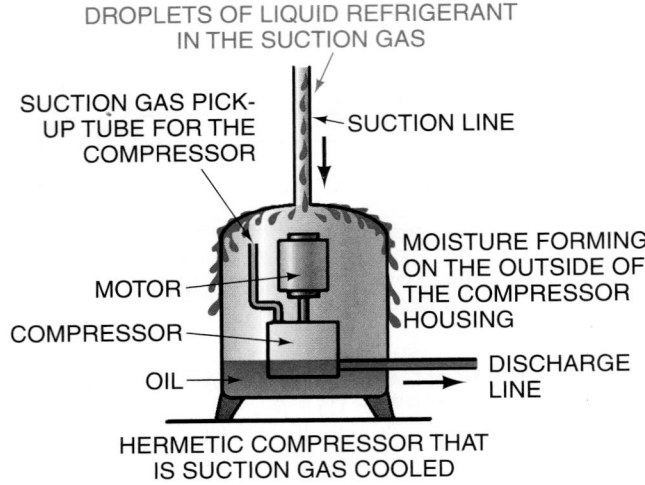

Figure 29–20 When liquid refrigerant gets back to the compressor, the latent heat that is left in the refrigerant will cause the compressor to sweat more than normal. When vapor only gets back to the compressor, the vapor changes in temperature quickly, and the compressor does not sweat a lot.

with refrigerant, there may not be enough refrigerant in the system to adversely affect the compressor when flooding occurs. The manufacturer will determine the amount of critical charge and stamp it on the unit's nameplate. If it is not stamped on the unit, contact the manufacturer to determine the correct charge.

The compressor should only have vapor entering it. Vapor will rise in temperature as soon as it touches the compressor shell. When liquid is present, it will not rise in temperature and will cool the compressor shell. **Liquid refrigerant still has its latent heat absorption capability and will absorb a great amount of heat without changing temperature. A vapor absorbs only sensible heat and will change in temperature quickly, Figure 29–20.** Another reason that liquid may get back to the compressor is poor heat exchange in the evaporator in a capillary tube system. If liquid is getting back to the compressor, the evaporator heat exchange should be checked before removing refrigerant.

As mentioned earlier, when a system is overcharged with refrigerant, the discharge pressure and suction pressure will be higher. The condenser will be flooded with subcooled liquid, causing the discharge pressure to be excessively high. Remember, any liquid in the condenser with temperatures lower than the condensing temperature is considered to be subcooled. Excessive condenser subcooling will cause a high compression ratio and low efficiencies. Because of this, low refrigerant flow rates and capacities will be experienced. Forced-air condensers should have at least 6 to 8 degrees of liquid subcooling; however, subcooling amounts depend on system piping configurations and liquid-line static and friction pressure drops.

If the system has a TXV for a metering device, the TXV will try to maintain evaporator superheat even with an excessive overcharge. The TXV may overfeed slightly during its opening strokes, but it should stabilize if still in its pressure operating range.

The symptoms of a refrigerant overcharge are summarized below:

- High discharge pressures
- High suction pressures
- High compression ratios
- High condenser subcooling
- Normal evaporator superheats with a TXV; low superheats with a capillary tube

29.13   INEFFICIENT EVAPORATOR

An inefficient evaporator does not absorb the heat into the system and will have a low suction pressure. The suction line may be sweating or frosting back to the compressor. An inefficient evaporator can be caused by a dirty coil, a fan running too slowly, an expansion valve flooding or starving the coil, recirculated air, ice buildup, or product interference causing blocked airflow, **Figure 29–21.**

Whenever the evaporator sees reduced airflow across its face, there is a reduced heat load on the coil. This reduced airflow and heat load will cause the refrigerant in the evaporator coil to remain a liquid and not vaporize. This causes low suction pressure because there is no vapor pressure being generated from vaporizing refrigerant. This liquid refrigerant will travel past the evaporator coil and eventually reach the compressor and cause compressor damage from flooding and slugging. Often, the system will have no evaporator and compressor superheat. The refrigerant in the suction line is what causes the suction line and compressor to get cold and sweat or frost. If liquid refrigerant enters the crankcase, it will mix with the oil and often vaporize. This vaporization in the crankcase will cause a cold and sweaty crankcase. The suction valves may see liquid refrigerant or a dense vapor entering—depending on whether the compressor is air or refrigerant cooled. In either case, the amp draw of the compressor will be high.

If a thermostatic expansion valve is starving the evaporator coil, the evaporator will also be very inefficient. This situation will cause low suction pressure with very high superheats. The low suction pressure will cause a low evaporator temperature—causing the suction line and compressor to frost and sweat also. Even though the suction pressure and evaporator temperatures are low, the refrigerant flow rate through the evaporator is also very low. This causes low capacities and a very starved and inefficient evaporator.

The reduced heat load on the evaporator causes the condenser to see a reduced heat load, and have, therefore, less heat to reject. This causes the head pressure to be low.

All of these can be checked with an evaporator performance check. This check can be performed by using a superheat check to make sure that the evaporator has the correct amount of refrigerant, **Figure 29–22.** The heat exchange surface should be clean. The fans should be blowing enough air and not recirculating it from the discharge to the inlet of the coil.

The refrigerant boiling temperature should not be more than 20°F colder than the entering air on an air evaporator coil, **Figure 29–2(A).** A water coil should have no more than a

Canned goods and other nonrefrigerated items are great for presenting imposing piles and eye-catching displays. Don't try the same gimmicks with perishables. The case will fail to refrigerate any of the merchandise when the air ducts are blocked.

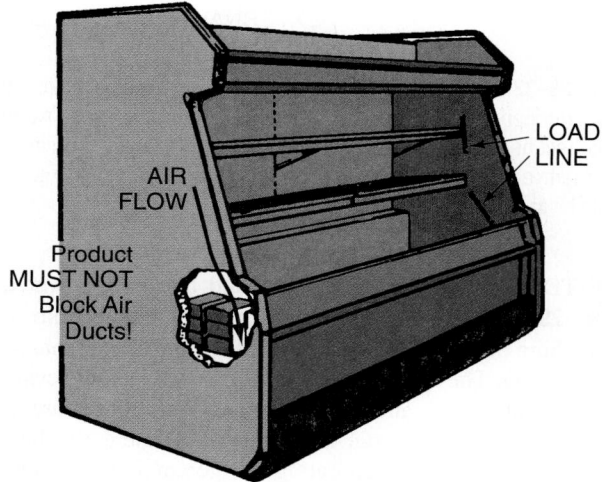

Low-temperature multishelf cases are particularly sensitive to air pattern changes as well as extremes of humidity and temperature in the store.

Jumble displays may have some sales benefits, but they really foul up the protective layer of cold air in the case. Observe the LOAD LINE stickers!

Figure 29–21   An evaporator that cannot absorb the required Btu because of product interference. There is a load line on the inside where the product is stored. *Courtesy Tyler Refrigeration Company*

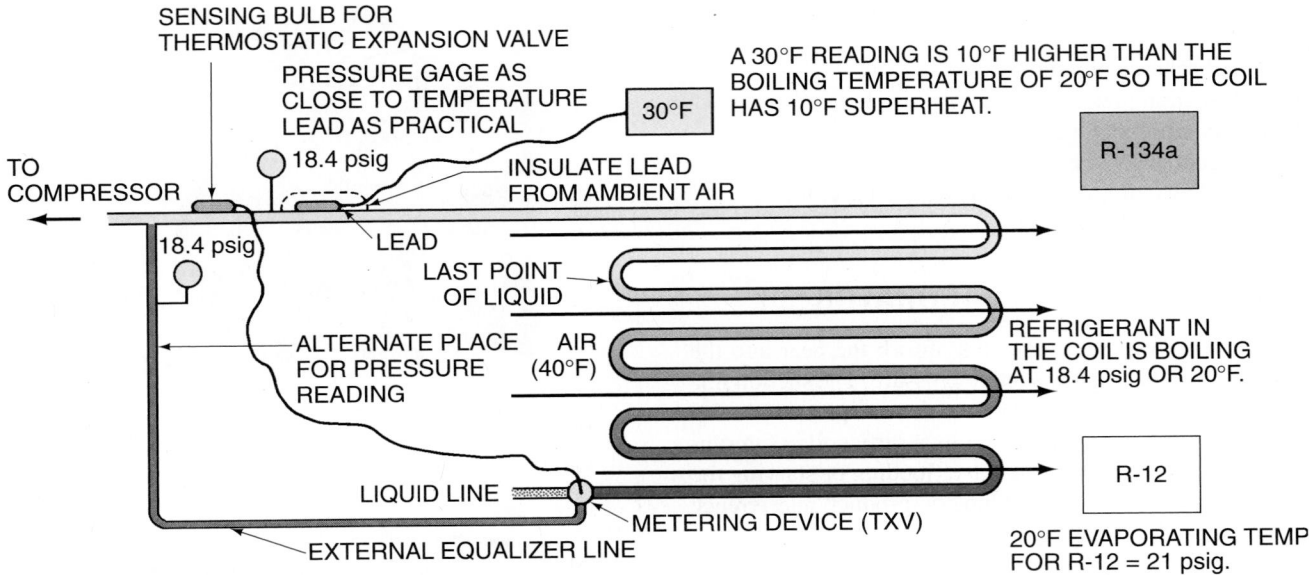

Figure 29–22 Analyzing a coil for efficiency. This requires temperature and pressure checks at the outlet of the evaporator. When a coil has the correct refrigerant level, checked by superheat and the correct air-to-refrigerant heat exchange, the coil will absorb the correct amount of heat. The correct heat exchange is taking place if the refrigerant is boiling at 10°F to 20°F cooler than the entering air. Note that this occurs when the cooler temperature is within the design range, for example, 30°F to 45°F for medium temperature. Every circuit on coils with multiple circuits should be checked for even distribution.

10°F TD in the boiling refrigerant and the leaving water, **Figure 29–23.** When the boiling refrigerant relationship to the medium being cooled starts increasing, the heat exchange is decreasing. The temperature difference between the evaporating temperature and the leaving-water temperature is referred to as the *approach.* The more efficient the heat exchange between the evaporating refrigerant and the water, the smaller the approach will be.

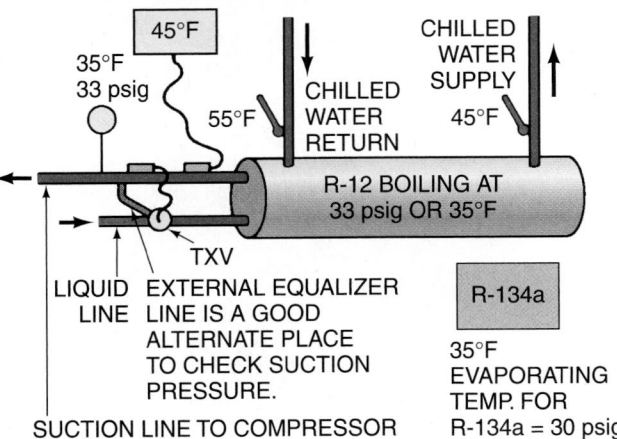

Figure 29–23 Water coils can be analyzed in much the same way as air coils. The refrigerant temperature and pressure at the end of the evaporator will indicate the refrigerant level. The refrigerant boiling temperature should not be more than 10°F cooler than the leaving water.

The symptoms of an inefficient evaporator are summarized below:

- Low suction pressure
- Sweaty or frosted suction line to compressor
- Low head pressure
- Low superheat
- Cold compressor crankcase
- High compressor amp draw

29.14   INEFFICIENT CONDENSER

An inefficient condenser acts the same whether it is water cooled or air cooled. If the condenser cannot remove the heat from the refrigerant, the head pressure will go up. The condenser does three things and has to be able to do them correctly, or excessive pressures will occur.

1. Desuperheat the hot gas from the compressor. This gas may be 200°F or hotter on a hot day on an air-cooled system. Desuperheating is accomplished in the beginning of the coil.
2. Condense the refrigerant. This is done in the middle of the coil, the only place that the coil temperature will correspond to the head pressure. You could check the temperature against the head pressure if a correct temperature reading can be taken, but the fins are usually in the way.
3. Subcool the refrigerant before it leaves the coil. This subcooling is cooling the refrigerant to a point below the actal condensing temperature. A subcooling of 5°F to 20°F is typical. The subcooling can be checked just like the superheat, only the temperature is checked at the liquid line and compared with the high-side pressure converted to condensing temperature, **Figure 29–24.**

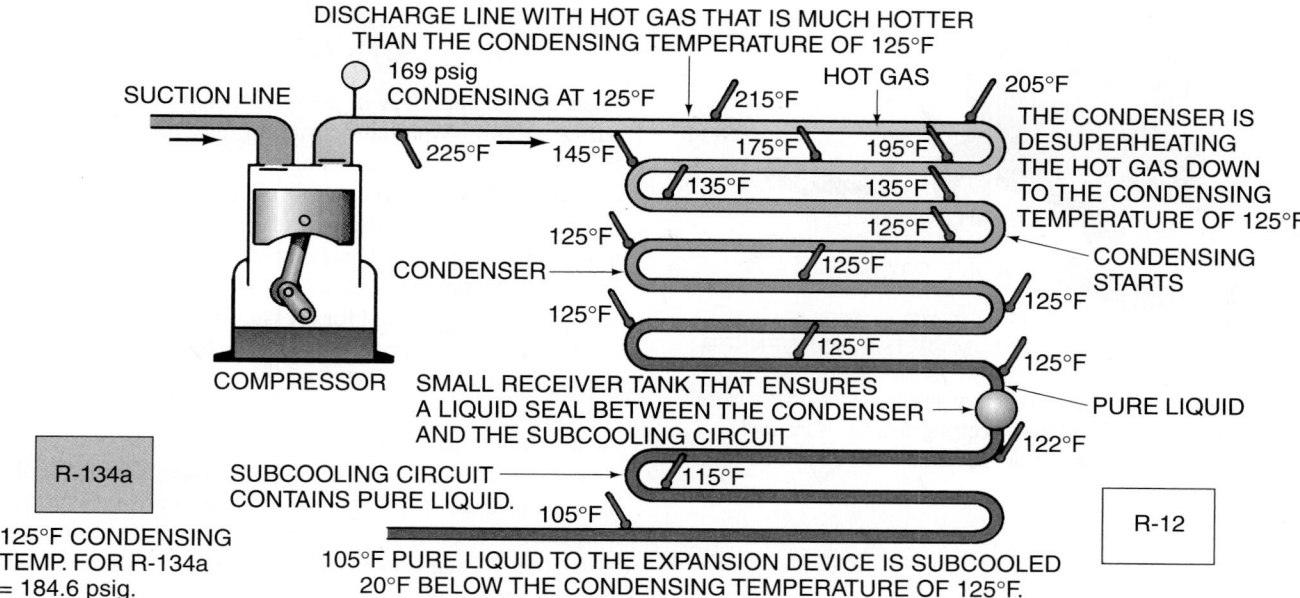

DISCHARGE LINE WITH HOT GAS THAT IS MUCH HOTTER
THAN THE CONDENSING TEMPERATURE OF 125°F

SUCTION LINE

169 psig
CONDENSING AT 125°F

HOT GAS

215°F   205°F

THE CONDENSER IS
DESUPERHEATING
THE HOT GAS DOWN
TO THE CONDENSING
TEMPERATURE OF 125°F.

225°F   145°F   175°F   195°F

135°F   135°F

125°F   125°F

CONDENSER   125°F

125°F   125°F

125°F   125°F

125°F

COMPRESSOR

SMALL RECEIVER TANK THAT ENSURES
A LIQUID SEAL BETWEEN THE CONDENSER
AND THE SUBCOOLING CIRCUIT

CONDENSING
STARTS

PURE LIQUID

122°F

R-134a

SUBCOOLING CIRCUIT
CONTAINS PURE LIQUID.   115°F

105°F

R-12

125°F CONDENSING
TEMP. FOR R-134a
= 184.6 psig.

105°F PURE LIQUID TO THE EXPANSION DEVICE IS SUBCOOLED
20°F BELOW THE CONDENSING TEMPERATURE OF 125°F.

Figure 29–24   Checking the subcooling on a condenser. The condenser does three jobs: (1) it desuperheats the hot gas in the first part of the condenser, (2) it condenses the vapor refrigerant to a liquid in the middle of the condenser, and (3) it subcools the refrigerant at the end of the condenser. The condensing temperature corresponds to the head pressure. Subcooling is the temperature of the liquid line subtracted from the condensing temperature. A typical condenser can subcool the liquid refrigerant to 5°F to 20°F cooler than the condensing temperature.

The condenser must have the correct amount of cooling medium (air or water). This medium must **not** be recirculated (mixed with the incoming medium) without being cooled. Ensure that air-cooled equipment is not located so that the air leaving the condenser circulates back into the inlet. This air is hot and will cause the head pressure to go up in proportion to the amount of recirculation, **Figure 29–25**. An air-cooled condenser should not be located down low, close to the roof, even though the air comes in the side. The temperature is higher at the roof level than it is a few inches higher. A clearance of about 18 in. will give better condenser performance, **Figure 29–26**. An air-cooled condenser that has a vertical coil may be influenced by prevailing winds. If the fan is trying to discharge its air into a 20-mph wind, it may not move the correct amount of air, and a high head pressure may occur, **Figure 29–27**.

Dirty condenser coils often result in inefficient condensers. Once the condenser coil gets dirty, the condenser has a hard time rejecting heat. Since heat from the evaporator, suction, compressor's motor, and compression is rejected in the condenser, the condenser coil must be kept clean with the proper amount of airflow through it. If the condenser cannot reject heat fast enough, it will accumulate heat until its temperature and pressure is high enough to reject the heat to the surrounding ambient. The system is now operating at an elevated condensing temperature and pressure and causing the unwanted inefficiencies that result from high compression ratios.

Now that the system is running at elevated head pressures with unwanted inefficiencies, the suction pressure will be a bit high. Higher-than-normal suction pressures are caused by low refrigerant flow rates. Low refrigerant flow rates are caused by the low volumetric efficiencies resulting from the high condensing pressures. The evaporator may not be able to

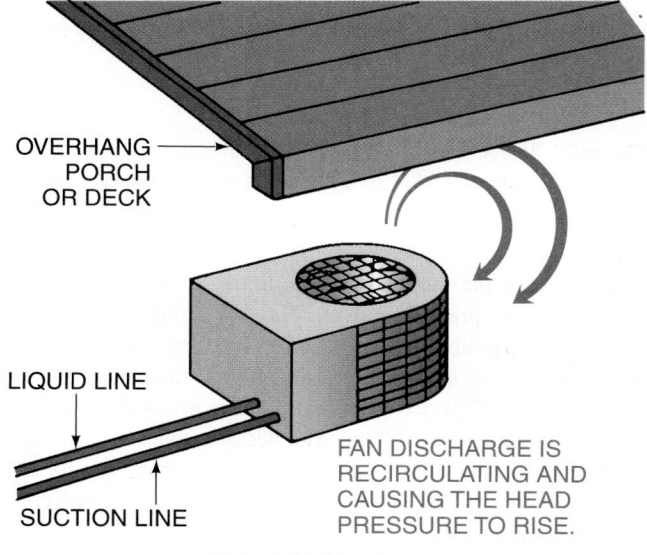

OVERHANG
PORCH
OR DECK

LIQUID LINE

SUCTION LINE

FAN DISCHARGE IS
RECIRCULATING AND
CAUSING THE HEAD
PRESSURE TO RISE.

THE CONDENSER SHOULD NOT
HAVE ANY AIR OBSTRUCTIONS.

Figure 29–25   The air-cooled condenser is located too close to an obstacle, and the hot air leaving the condenser is recirculated back into the inlet of the coil.

keep up with the heat load, which creates higher-than-normal suction pressures.

High head pressures cause high compression ratios and low volumetric efficiencies. High compression ratios will cause a greater pressure range in which the suction vapors can be compressed, requiring more work for the compressor and increasing the amp draw.

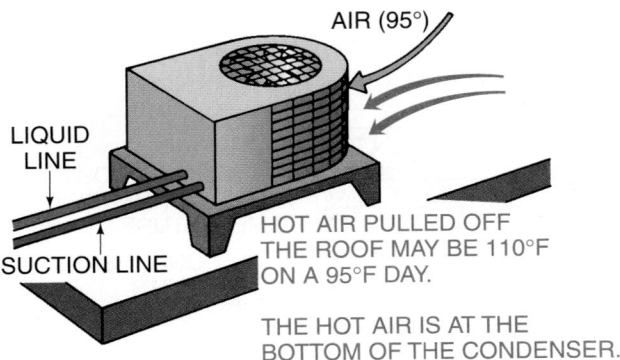

Figure 29–26 An air-cooled condenser should not be located close to a roof. The temperature of the air coming directly off the roof is much warmer than the ambient air because the roof acts like a solar collector.

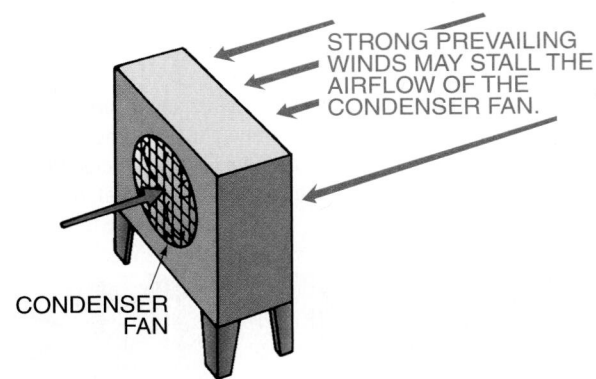

Figure 29–27 A condenser that is located in such a manner that it is discharging its air into a strong prevailing wind may not get enough air across the coil.

If the head pressure gets too high, the TXV may be operating out of its pressure ranges. The TXV may let too much refrigerant into the evaporator during its opening stroke from the higher head pressures, causing higher-than-normal suction pressures. This action may also cause lower-than-normal evaporator superheats. Otherwise, because the TXV is maintaining superheat and doing its job, there may be normal evaporator pressures, depending on the severity of the condenser's condition.

What liquid subcooling is formed in the condenser will be at an elevated temperature and will reject heat to the ambient faster. Because of this faster heat rejection, the liquid in the condenser will cool faster and have a greater temperature difference as compared with the condensing temperature. Therefore, the condenser subcooling may be a bit high when the condensing temperature is high.

The symptoms of an inefficient condenser are summarized below:

- High head pressure
- Normal to slightly high suction pressures
- Normal to slightly low evaporator superheats
- High compression ratio
- High compressor amp draws
- Normal-to-high condenser subcooling

29.15 REFRIGERANT FLOW RESTRICTIONS

Restrictions that occur in the refrigeration circuit are either partial or full. A partial restriction may be in the vapor or the liquid line. A restriction always causes pressure drop at the point of the restriction. Different conditions will occur, depending on where the restriction is. Pressure drop can always be detected with gages, but the gages cannot always indicate the correct location of the restriction. Gage ports may need to be installed for pressure testing.

If a restriction occurs due to something outside the system, it is usually physical damage, such as flattened or bent tubing. These can be hard to find if they are in hidden places such as under the insulation or behind a fixture.

If a partial restriction occurs in a liquid line, it will be evident because the refrigerant will have a pressure drop and will start to act like an expansion device at the point of the restriction. **When there is a pressure drop in a liquid line, there is always a temperature change.** A temperature check on each side of a restriction will locate the place. Sometimes when the drop is across a drier, the temperature difference from one side to the other may not be enough to feel with bare hands, but a thermometer will detect it, **Figure 29–28.**

If a system has been running for a long time and a restriction occurs, physical damage may have occurred. If the restriction occurs soon after start, a filter or drier may be plugging up. When this occurs in the liquid line drier, bubbles will appear in the sight glass when the drier is located before the sight glass, **Figure 29–29.**

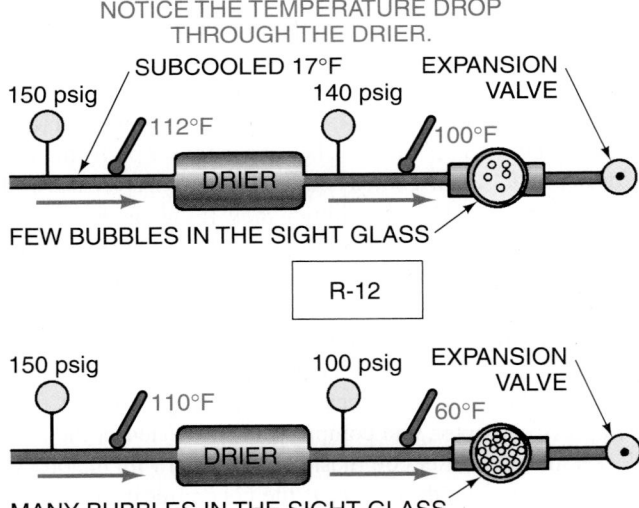

Figure 29–28 The driers each have a restriction. One of them is very slight. Where there is pressure drop in a liquid line, there is temperature drop. If the temperature drop is very slight, it can be detected with a thermometer. Sometimes gages are not easy to install on each side of a drier to check for pressure drop. If possible, use the same gage when checking the pressure drop through a drier to avoid inaccuracies.

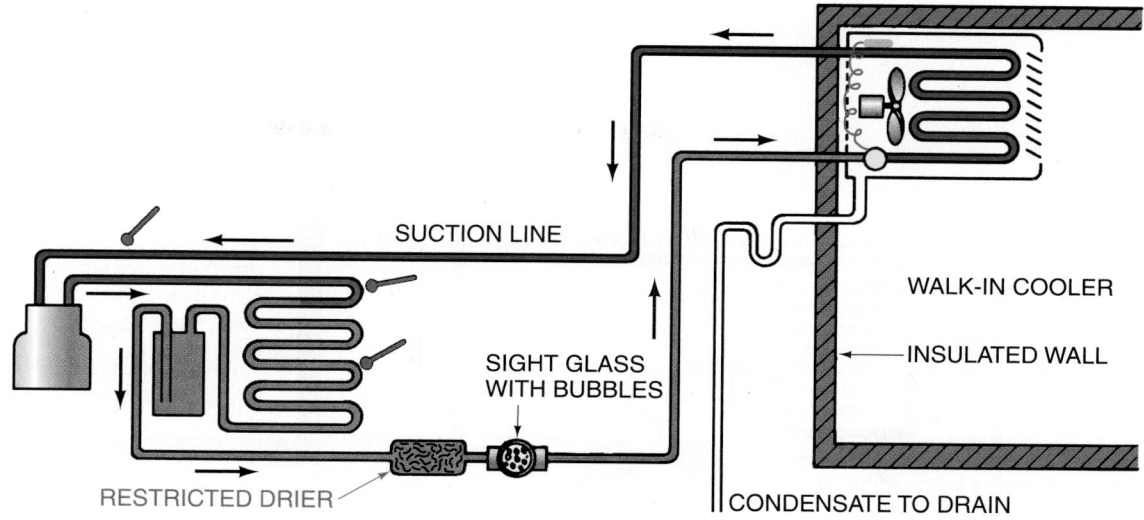

SUCTION LINE

WALK-IN COOLER

INSULATED WALL

SIGHT GLASS
WITH BUBBLES

RESTRICTED DRIER

CONDENSATE TO DRAIN

Figure 29–29 The restriction in a system may occur shortly after start-up. This indicates that solid contaminants from installing the system must be in the drier. If the restriction occurs after the system has been running for a long time, the restriction may be physical damage, such as a bent pipe. Normally, loose contamination will make its way through the system to the drier in a matter of hours.

Another occurrence that may create a partial restriction could be valves that do not open all the way. Normally the TXV either works or it does not. It will function correctly, but if it loses its charge in the thermal bulb, it will close and cause a full restriction, **Figure 29–30.**

There is a strainer at the inlet to most expansion devices to trap particles, and it may stop up slowly. If the device is a valve that can be removed, it can be inspected and cleaned if necessary. If it is a capillary tube, it will be soldered into the line and not easy to inspect, **Figure 29–31.**

Water circulating in any system that operates below freezing will freeze at the first cold place it passes through. This would be in the expansion device. One drop of water can stop a refrigeration system. Sometimes a piece of equipment that has just been serviced will show signs of moisture on the first hot day, **Figure 29–32.** This is because the drier in the liquid line will have more capacity to hold moisture when it is cool. The first hot day, the drier may turn a drop or two of water loose, and it will freeze in the expansion device, **Figure 29–33.** When you suspect this, apply heat to thaw the ice to a liquid. SAFETY PRECAUTION: Care must be used when applying heat. A hot wet cloth is a good source. If applying a hot cloth to the metering device causes the system to start functioning properly, the problem is free water in the system. ♻ Recover the refrigerant, change the drier, and evacuate the system. Recharge with recycled or virgin refrigerant. ♻

THE BULB HAS
LOST ITS CHARGE.   0 psig

SUCTION LINE
TO COMPRESSOR

EVAPORATOR
15 in.   COIL
VACUUM

THE SPRING PRESSURE
CLOSES THE VALVE.

DIAPHRAGM

SPRING

PUSH
ROD   15 in.
VACUUM

LIQUID REFRIGERANT
FROM RECEIVER

Figure 29–30 The charge in the bulb of a TXV is the only force that opens the valve, so when it loses its pressure the valve closes. The system can go into a vacuum if there is no low-pressure control to stop the compressor. A partial restriction can occur if the bulb loses part of the charge or if the inlet strainer stops up.

STRAINER

Figure 29–31 A capillary tube metering device with the strainer at the inlet. This strainer is soldered into the line and is not easy to service. *Courtesy Parker Hannifin Corporation*

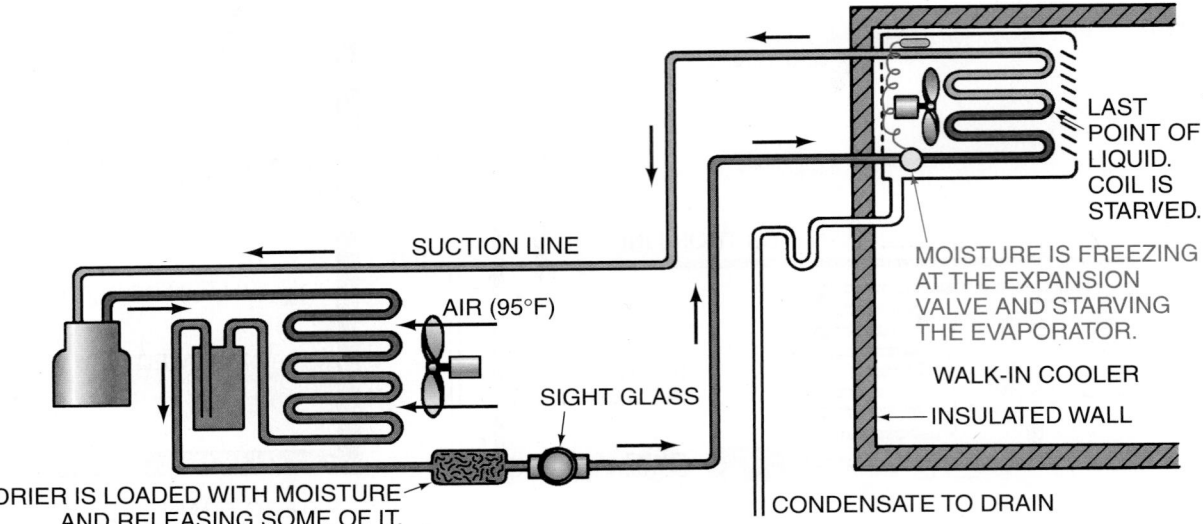

Figure 29–32 This system has had some moisture in it. The drier has all the moisture it can hold at mild temperatures. When the weather gets warm, the drier cannot hold all of the moisture and turns some of it loose. The moisture will freeze at the expansion device where the first refrigeration is experienced if the system operates below freezing.

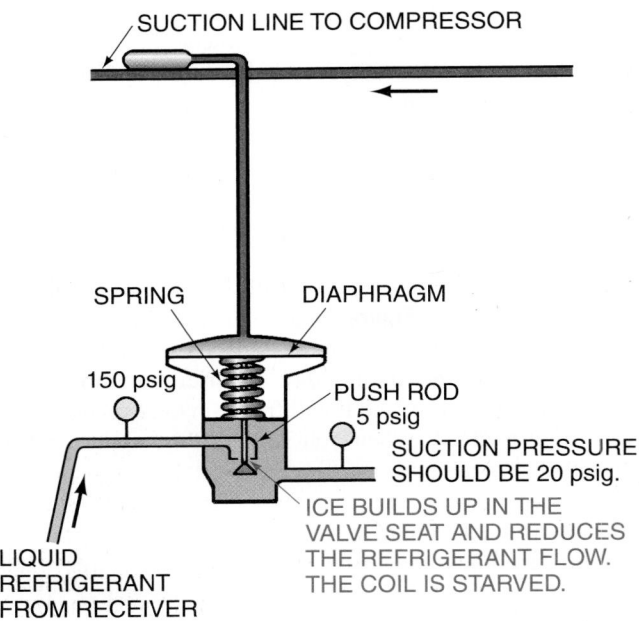

Figure 29–33 If it is suspected that moisture is frozen in the metering device, mild heat can be added. Heat from a hot wet cloth will normally thaw the moisture out, and the system will start refrigerating again.

Other components that may cause restrictions are automatic valves in the lines, such as the liquid-line solenoid, crankcase pressure regulator, or the evaporator pressure regulator. These valves may easily be checked with gages applied to both sides of them where pressure taps are provided, **Figure 29–34**. Restrictions in the liquid line usually occur after the receiver and will cause the evaporator to be starved of refrigerant, resulting in a low suction pressure. High superheats will also be experienced because of the starved evaporator. Because the evaporator is starved of refrigerant,

there will not be much heat transferred to it from its surroundings. From the highly superheated vapors coming into it, the compressor will also be starved of refrigerant and will run hot and have a low amp draw. Highly superheated vapors mean low-density vapors for the compressor to compress.

Because the evaporator and compressor are starved of refrigerant, the condenser will also be starved. The condenser will not have a lot of heat to reject because the evaporator did not absorb much heat. This will cause a low condensing (head) pressure.

If the restriction is before the sight glass, the sight glass may bubble, depending on the severity of the restriction.

The symptoms of a restricted liquid line after the receiver are summarized below:

- Low suction pressures
- High superheats
- Overheated compressor
- Low amp draw
- Low head pressure
- Bubbles in the sight glass
- Temperature drop in the liquid line across the restriction

29.16 INEFFICIENT COMPRESSOR

Inefficient compressor operation can be one of the most difficult problems to find. When a compressor will not run, it is evident where the problem is. Motor troubleshooting procedures are covered in a separate unit. When a compressor is pumping at slightly less than capacity, it is hard to determine the problem. It helps at this point to remember that a compressor is a vapor pump. It should be able to create a pressure from the low side of the system to the high side under design conditions.

If a reciprocating compressor is inefficient because its valves are bad, the suction pressure will be high and the head

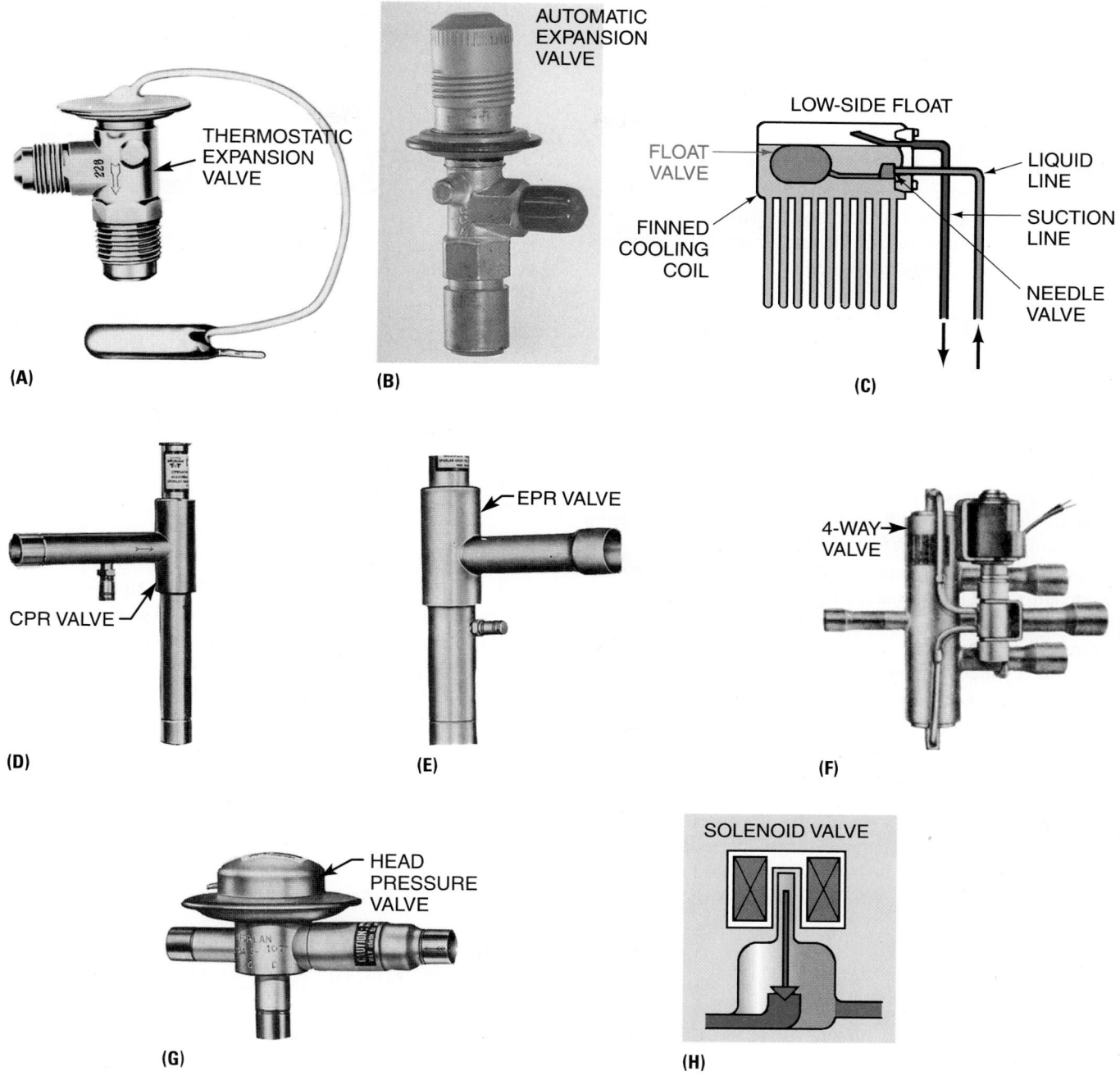

Figure 29–34   Valve components that can close are all subject to closing and causing a restriction **(A)–(H)**. **(A)**, **(B)**, and **(D)** *Courtesy Singer Controls Division.* **(E)** and **(G)** *Courtesy Sporlan Valve Company.* **(F)** *Courtesy ALCO Controls Division, Emerson Electric Company*

pressure will be low. There is no other situation that will give a system both low head and high suction pressures at the same time, other than worn piston rings causing blow-by of gases around the rings.

If a discharge valve is not seating properly, a low head pressure will result. Refrigerant vapor will be forced out of the cylinder and into the discharge line during the upstroke of the compressor. On the downstroke, this same refrigerant that is in the discharge line will be drawn back into the cylinder because of the discharge valve not seating properly. This short cycling of refrigerant will cause high

discharge temperatures out of the compressor and low head (condensing) pressures.

Refrigerant vapors will be drawn from the suction line into the compressor cylinder during the downstroke of the compressor. However, if the suction valve is not seating properly, this same refrigerant may leak back into the suction line during the upstroke, reducing the refrigerant flow and causing higher-than-normal suction pressures.

Bad valves will cause a low mass flow rate of refrigerant through the system, which will usually cause higher-than-normal superheats on the low side of the system. System capacities will be low—with higher-than-normal box or conditioned space temperatures. Because of the reduced refrigerant flow rates, compressor amp draws will be lower.

The symptoms of an inefficient compressor due to bad valves are summarized below:

■ Low head pressure
■ High suction pressure
■ Higher-than-normal superheats
■ Lower-than-normal compressor amp draws

The following methods are all used by service technicians to discover the compressor problems.

29.17 COMPRESSOR VACUUM TEST

The compressor vacuum test is usually performed on a test bench with the compressor out of the system. This test may be performed in the system when the system has service valves. NOTE: Care should be taken not to pull air into the system while in a vacuum. Make sure that there is no residual refrigerant mixed in the oil to affect the outcome of the test. Do this by running the system for 30 minutes or leaving the crankcase heater on for several hours.

All reciprocating compressors should immediately go into a vacuum if the suction line is valved off when the compressor is running. This test proves that the suction valves are seating correctly on at least one cylinder. This test is *not satisfactory* on a multicylinder compressor. If one cylinder will pump correctly, a vacuum will be pulled. A reciprocating compressor should pull 26 in. to 28 in. Hg vacuum with the atmosphere as the discharge pressure, **Figure 29–35**. The compressor should pull about 24 in. Hg vacuum, against 100 psig discharge pressure, **Figure 29–36**. When the compressor has pumped a differential pressure and is stopped, the pressures should not equalize. For example, a compressor has been operated until the suction pressure is 24 in. Hg vacuum and the head pressure is 100 psig, then it is stopped. These pressures should stay the

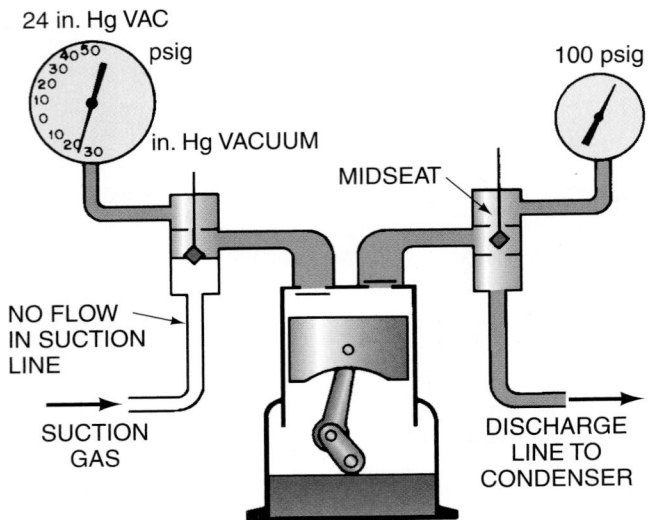

Figure 29–36 Perform a compressor check by pulling a vacuum while pumping against a head pressure. A reciprocating compressor normally can pull 24 in. Hg vacuum against 100 psig head pressure.

same while the compressor is off. When refrigerant is used for this pumping test, the 100 psig will drop some because of the condensing refrigerant. Nitrogen is a better choice to pump in the test because the pressure will not drop. SAFETY PRECAUTION: Care should be taken when operating a hermetic compressor in a deep vacuum (below 1000 microns) because the motor is subject to damage. This low vacuum cannot be obtained with a reciprocating compressor. Also, most compressor motors are cooled with suction gas and will get hot if operated for any length of time performing these tests. This vacuum test should not take more than 3 to 5 min. The motor will not overheat in this period of time. The test should be performed only by experienced technicians.

29.18 CLOSED-LOOP COMPESSOR RUNNING BENCH TEST

Doing a running bench test on the compressor can be accomplished by connecting a line from the discharge to the suction of the compressor and operating the compressor in a closed loop. A difference in pressure can be obtained with a valve arrangement or gage manifold. This will prove the compressor will pump. NOTE: When the compressor is hermetic, it should operate at close to full-load current in the closed loop when design pressures are duplicated. Nitrogen or refrigerant can be used as the gas to compress. Typical operating pressures will have to be duplicated for the compressor to operate at near the full-load current rating. For example, for a medium-temperature compressor, a suction pressure of 20 psig and a head pressure of 170 psig will duplicate a typical condition for R-12 on a hot day. The compressor should operate at near to nameplate full-load current when the design voltage is supplied to the motor.

The following is a step-by-step procedure for performing this test. Use this procedure with the information in

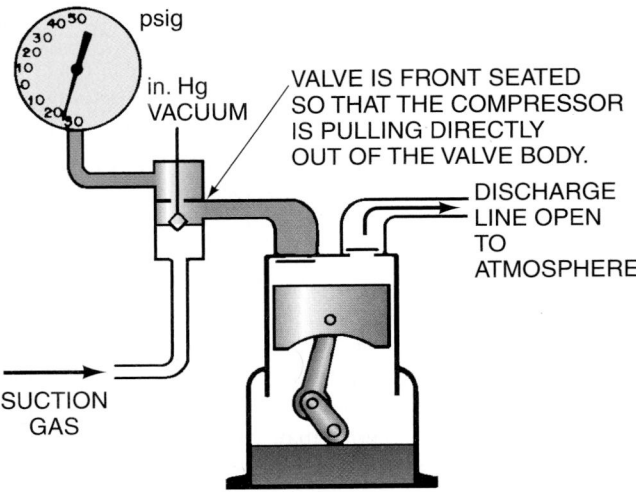

Figure 29–35 A compressor pulling a vacuum with the atmosphere as the head pressure. Most reciprocating compressors can pull 26 in. to 28 in. Hg vacuum with the atmosphere as the discharge pressure.

Figure 29–37. SAFETY PRECAUTION: This test should be performed only under the close supervision of an instructor or by a qualified person and on equipment under 3 hp. Safety goggles must be worn.

Steps for performing a closed-loop test with the compressor out of the system:

1. Use the gage manifold and fasten the suction line to the low-pressure gage line in such a manner that the compressor is pumping only from the gage line.
2. Fasten the discharge gage line to the discharge valve port in such a manner that the compressor is pumping only into the gage manifold.
3. Plug the center line of the gage manifold.

THE TYPICAL SUCTION AND DISCHARGE PRESSURE FOR THIS MEDIUM-TEMPERATURE R-12 SYSTEM

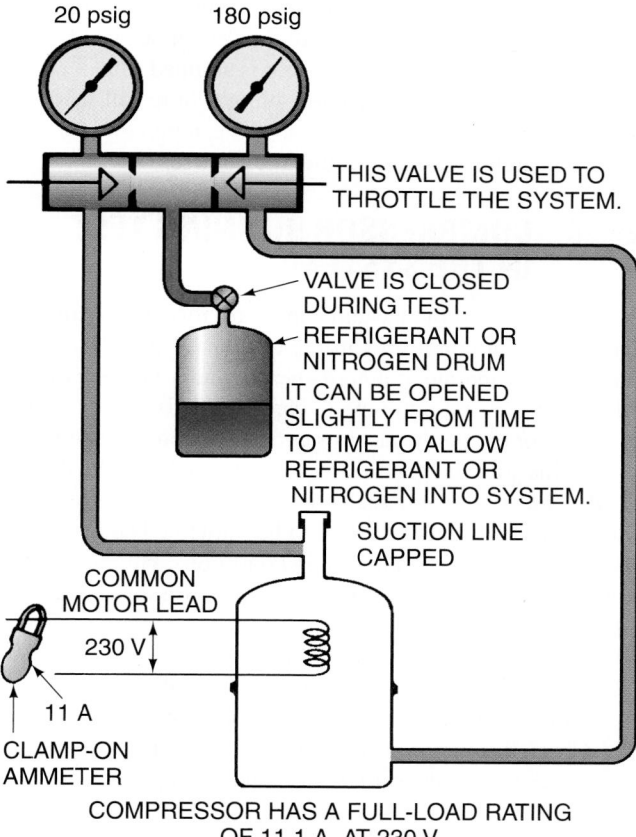

Figure 29–37 Checking a compressor's pumping capacity by using a closed loop. The test is accomplished by routing the discharge gas back into the suction port with a piping loop. The discharge gage manifold valve is gradually throttled toward closed (do not entirely close). It can never be closed, or tremendous pressure will occur. When the design suction and discharge pressures are reached, the compressor should be pulling close to nameplate full-load amperage. *A suction-cooled compressor cannot be run for long in this manner or the motor will get hot.* The refrigerant that is characteristic to the compressor (or nitrogen) can be used to circulate for pumping. Nitrogen will not produce the correct amperage, but it will be close enough. SAFETY PRECAUTION: This test should only be performed by experienced technicians.

4. SAFETY PRECAUTION: Open both gage manifold valves wide open, counterclockwise.
5. SAFETY PRECAUTION: Start the compressor; keep your hand on the off switch.
6. The compressor should now be pumping out the discharge line and back into the suction line. The discharge gage manifold valve may be slowly throttled (*do not close entirely*) toward closed until the discharge pressure rises to the design level and the suction pressure drops to the design level.
7. When the desired pressures are reached, the amperage reading on the compressor motor should compare closely to the full-load amperage of the compressor. If the correct pressures cannot be obtained with the amount of gas in the compressor, a small amount of gas may be added to the loop system by attaching the center line to a nitrogen cylinder and slightly opening the cylinder valve.
8. The amperage may vary slightly from full load because of the input voltage. For example, a voltage above the nameplate will cause an amperage below full load and vice versa.

Figure 29–38 illustrates a situation in which a technician accidentally completely closed the valve.

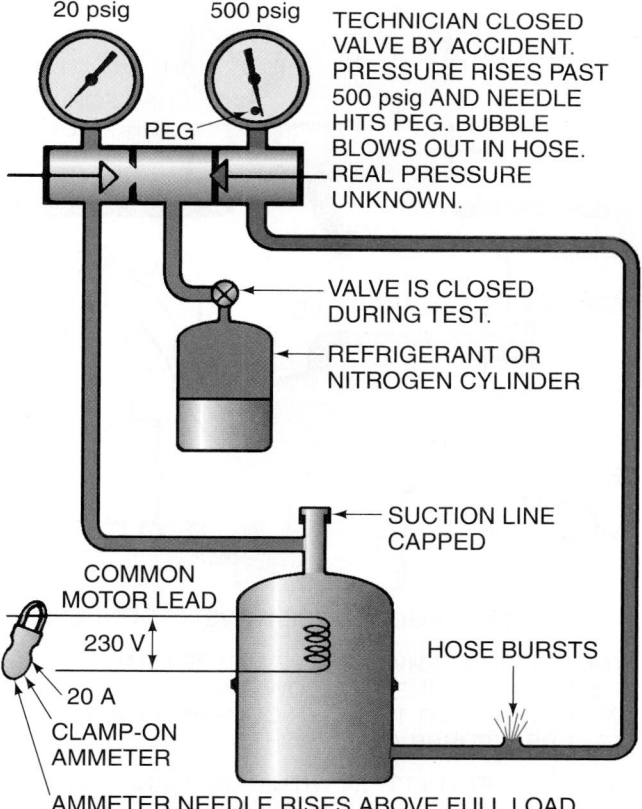

Figure 29–38 SAFETY PRECAUTION: What happens if a compressor is started up in a closed loop with no place for the discharge gas to go (discharge gage manifold closed by accident)? One cylinder full of gas could be enough to build tremendous pressures.

29.19 CLOSED-LOOP COMPRESSOR RUNNING FIELD TEST

When a compressor has suction and discharge service valves, this test can be performed in place in the system using a gage manifold as the loop, **Figure 29–39**. The compressor is started with the compressor service valves turned all the way to the front seat and the gage manifold valves open. The center line on the gage manifold must be plugged. SAFETY PRECAUTION: This can be a dangerous start-up and should be performed only under supervision of an experienced person. The compressor has to have a place to pump the discharge gas because reciprocating compressors are positive displacement pumps. When the compressor cylinder is full of gas, it is going to pump it somewhere or stall. In this test the gas goes around through the gage manifold and back into the suction. Should the compressor be started before the gage manifold is open for the escape route through the gage manifold, tremendous pressures will result before you can stop the compressor. It takes only one cylinder full of vapor to fill the gage manifold to more than capacity. This test should only be performed on compressors under 3 hp. Safety goggles must be worn.

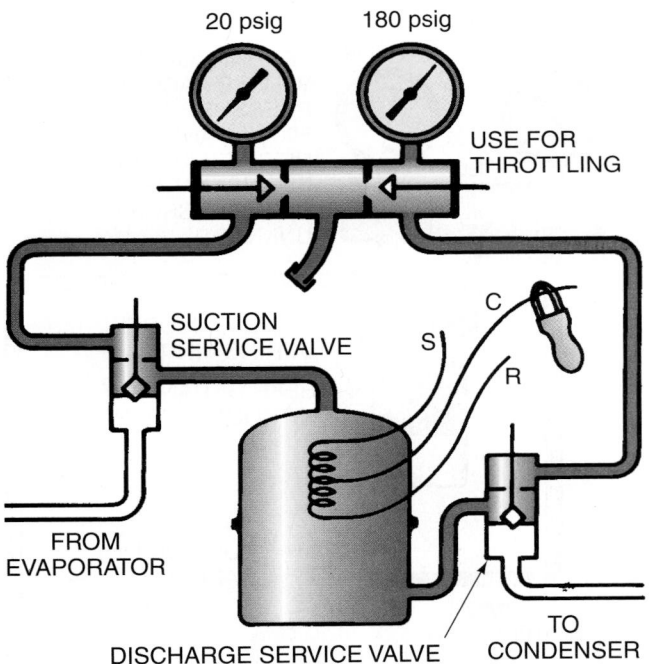

THE SUCTION SERVICE VALVE MAY BE CRACKED FROM TIME TO TIME TO ALLOW SMALL AMOUNTS OF REFRIGERANT TO ENTER COMPRESSOR UNTIL TEST PRESSURES ARE REACHED.

NOTE: START THIS TEST WITH GAGE MANIFOLD VALVES WIDE OPEN. THEN USE THE HIGH-SIDE VALVE FOR THROTTLING.

Figure 29–39   A compressor test being performed in a system using a gage manifold as a closed loop. Notice that this test can be performed in the system because the compressor has service valves. Do not perform this test without experienced supervision.

Steps for performing the closed-loop compressor test with the compressor in the system:

1. Turn the unit OFF and fasten the suction line of the gage manifold to the suction service valve. Plug the center line of the gage manifold.
2. Turn the suction service valve stem to the front seat.
3. Fasten the discharge line to the discharge valve and turn the discharge service valve stem to the front seat.
4. SAFETY PRECAUTION: Open both gage manifold valves all the way, counterclockwise.
5. SAFETY PRECAUTION: Start the compressor and keep your hand on the switch.
6. The compressor should now be pumping out through the discharge port, through the gage manifold, and back into the suction port.
7. The discharge gage manifold valve may be throttled toward the seat to restrict the flow of refrigerant and create a differential in pressure. Throttle (*do not entirely close*) the valve until the design head and suction pressure for the system is attained (the compressor should then be pulling near to full-load amperage). As in the bench test the amperage may vary slightly because of the line voltage.

29.20 COMPRESSOR RUNNING TEST IN THE SYSTEM

A running test in the system can be performed by creating typical design conditions in the system. Typically a compressor will operate at a *high suction* and a *low head pressure* when it is not pumping to capacity. This will cause the compressor to operate at a low current. When the technician gets to the job, the conditions are not usually at the design level. The fixture is usually not refrigerating correctly—this is what instigated the call to begin with. The technican may not be able to create design conditions, but the following approach should be tried if the compressor capacity is suspected:

1. Install the high- and low-side gages.
2. Make sure that the charge is correct (not over or under), using manufacturer's recommendations.
3. Check the compressor current and compare to full load.
4. Block the condenser airflow and build up the head pressure.

If the compressor will not pump the head pressure up to the equivalent of a 95°F day (95°F + 30°F = 125°F condensing temperature or 170 psig for R-12, 185 psig for R-134a, 278 psig for R-22, 301 psig for R-502, and 332 psig for R-404A) and draw close to nameplate full-load current, the compressor is not pumping. When the compressor is a sealed hermetic type, it may whistle when it is shut down. This whistle is evidence of an internal leak from the high side to the low side.

If the compressor has service valves a closed-loop test can be performed on the compressor using the methods we have explained, while it is in the system. NOTE: Make sure no air is drawn into the system while it is in a vacuum. SAFETY PRE-CAUTION: The compressor temperature should be monitored

PREVENTIVE MAINTENANCE FOR REFRIGERATION

PACKAGED EQUIPMENT. Packaged equipment is built and designed for minimum maintenance because the owner may be the person that takes care of it until a breakdown occurs. Most of the fan motors are permanently lubricated and will run until they quit, at which time they are replaced with new ones.

The owners should be educated to keep the condensers clean and not to stack inventory so close as to block the condenser airflow. When the unit is a reach-in cooler, the owner should be cautioned to follow the manufacturer's directions in loading the box. The load line on the inside should be observed for proper air distribution.

The owner or manager should examine each refrigerated unit frequently and be aware of any peculiar noises or actions from the box. Each box should have a thermometer that should be monitored each day by management. Any rise in temperature that does not reverse itself should alert the manager that a problem is occurring. This can prevent unnecessary loss of perishable foods.

The electrical service for all package equipment should be visually inspected for frayed wires and overheating. Power cord connections may become loose and start to build up heat. If the end that plugs into the wall receptacle becomes hot, the machine should be shut down until the problem is found and corrected. SAFETY PRECAUTION: If only the plug is replaced, the wall receptacle may still be a problem. The whole connection, the wall receptacle and the cord plug, should be inspected and repaired.

Ice machines require special attention if they are to be reliable. Management should know what the quality of the ice should be when the machine is operating correctly. When the quality begins to deteriorate, the reason should be found. This can be as simple as looking for the correct water flow on a cube maker or the correct ice-cutting pattern on another type of cube maker. The ice falling into the bin of a flake maker should have the correct quality, neither soft and mushy nor brittle and hard. The level of the ice in the bin the morning after the machine has been left to run all night can reveal if the machine is making enough ice. For example, if it is observed that the bin is full every morning and the bin thermostat is satisfied but then one morning the bin is only half full, trouble can be suspected. Keep a close watch on the machine.

Drains and drain lines for all refrigeration equipment should be maintained and kept clean and free. This can normally be accomplished when the unit is cleaned and sanitized on the inside. When the cleaning water is flushed down the drain, the speed that the drain moves the water should indicate if the drain is partially plugged or draining freely.

SPLIT-SYSTEM REFRIGERATION EQUIPMENT, EVAPORATOR SECTION, CONDENSER SECTION, AND INTERCONNECTING PIPING. In refrigeration split systems the evaporator section is located in one place and the con-

densing unit in another. The interconnecting piping makes the system complete. The evaporator section contains the evaporator, metering device, motor, defrost heaters, drain pan, and drain line. Refrigeration evaporators do not need cleaning often but need to be inspected and cleaned. The technician cannot always tell when a coil is dirty by looking at the evaporator. Grease or dirt may be in the core of the coil. Routine cleaning once a year for the evaporator will usually keep the coil clean. SAFETY PRECAUTION: Use only approved cleaning compounds where food is present. Turn off the power before cleaning any system. Cover the fan motors and all electrical connections when cleaning to prevent water and detergent from getting in them.

The motors in the evaporator unit are normally sealed motors and permanently lubricated. If not, they should be lubricated at recommended intervals. These are normally marked on the motor. Observe the fan blade for alignment and look for bearing wear. This may be found by lifting the motor shaft. Most small motors have considerable end play. Do not mistake this for bearing wear. You must lift the shaft to discover bearing wear. The fan blade may also gather weight from dirt and become out of balance.

All wiring in the evaporator section should be visually inspected. If it is cracked or frayed, shut off the power and replace it. Do not forget the defrost heaters and any heater tapes that may be used to keep the drain line warm during cold weather.

The complete case should be cleaned and sanitized at regular intervals. You may need to be the judge of these intervals. Do not let the unit become dirty, including the floor and storage racks in walk-in coolers. SAFETY PRECAUTION: Keep all ice off the floor of walk-in coolers, or an accident may occur.

The condensing unit may be located inside an equipment room in the store or it may be outside. When the condenser is located inside an equipment room, it must have proper ventilation. Most equipment rooms have automatic exhaust systems that turn fans on when the equipment room reaches a certain temperature. In large equipment rooms, the temperature will stay warm enough for the exhaust fan to run all the time.

Condensers in equipment rooms become dirty like they do outdoors and must be cleaned. Most system condensers should be cleaned at least once a year to ensure the best efficiency and to prevent problems. The condensers may be cleaned with an approved cleaner for condensers and then washed. SAFETY PRECAUTION: Turn off the power before cleaning any system. All motors and wiring must be covered to prevent them from getting wet. Watch the equipment room floor. If oil is present and water gets on it, it will become slippery.

The technician can find many future problems by close inspection of the equipment room. Leaks may be found by observing oil spots on piping. Touch testing the various

(continued)

components will often tell the technician that a compressor is operating too hot or too cold. Tape-on-type temperature indicators may be fastened to discharge lines that indicate the highest temperature the discharge line has ever reached. For example, most manufacturers would consider that oil inside the compressor will begin to break down at discharge temperatures above 250°F. A tape-on temperature indicator will tell if high temperatures are occurring while no one is around, such as a defrost problem in the middle of the night. This can lead to a search for a problem that is not apparent.

The crankcase of a compressor should not be cold to the touch below the oil level. This is a sure sign that liquid refrigerant is in the refrigerant oil on some systems, not necessarily in great amounts, but enough to cause diluted refrigerant oil. Diluted refrigerant oil causes marginal lubrication and bearing wear.

Fan motors may be inspected for bearing wear. Again, do not mistake end play for bearing wear. Some fan motors require lubrication at the correct intervals.

A general cleaning of the equipment room at regular intervals will ensure the technician good working conditions when a failure occurs.

The interconnecting piping should be inspected for loose insulation and oil spots (indicating a leak) and to ensure the pipe is secure. Some piping is in trenches in the floor. A trench that is full of water because of a plugged floor drain will hurt capacity of the equipment because the insulation value of the suction line insulation is not as great when wet. A heat exchange between the liquid lines and suction lines or the ground and suction lines will occur. The pipe trenches should be kept as clean as practical.

at any time these tests are being conducted. If the compressor gets too warm, it should be stopped and allowed to cool. If the compressor has internal motor-temperature safety controls, do not operate it when the control is trying to stop the compressor.

HVAC GOLDEN RULES

When making a service call to a business:
- Never park your truck or van in a space reserved for customers.
- Look professional and be professional.
- Before starting troubleshooting procedures, get all the information you can regarding the problem.
- Be extremely careful not to scratch tile floors or to soil carpeting with your tools or by moving equipment.
- Be sure to practice good sanitary and hygiene habits when working in a food preparation area.
- Keep your tools and equipment out of the customers' and employees' way if the equipment you are servicing is located in a normal traffic pattern.
- Be prepared with the correct tools and ensure that they are in good condition.
- Always clean up after you have finished. Try to provide a little extra service by cleaning filters, oiling motors, or providing some other service that will impress the customer.
- Always discuss the results of your service call with the owner or representative of the company. Try to persuade the owner to call if there are any questions as a result of the service call.

29.21 SERVICE TECHNICIAN CALLS

In addition to the six typical problems encountered in the refrigeration system, many more problems are not so typical. The following service situations will help you understand troubleshooting. Most of these service situations have already been described, although not as an actual troubleshooting

procedure. Sometimes the symptoms do not describe the problem, and a wrong diagnosis is made. Do not draw any conclusion until the whole system has been examined. Become a system doctor. Examine the system and say, "This needs further examination," and then do it.

Refrigeration systems often cool large amounts of food so they are slow to respond. The temperature may drop very slowly on a hot or warm pulldown.

SERVICE CALL 1

A customer calls and complains that a medium-temperature walk-in cooler with a remote condensing unit has a compressor that is short cycling and not cooling correctly. *The evaporator has two fans, and one is burned. The unit is short cycling on the low-pressure control because there is not enough load on the coil.*

On the way to the job the technician goes over the possibilities of the problem. This is where it helps to have some familiarity with the job. The technician remembers that the unit has a low-pressure control, a high-pressure control, a thermostat, an overload, and an oil safety control. Defrost is accomplished with an off cycle using the refrigerated space air with the fans running. Process of elimination helps make the decision that the thermostat and the oil safety control are not at fault. The thermostat has a 10°F differential, and the cooler temperature should not vary 10°F in a short cycle. The oil safety control is manual reset and will not short cycle. This narrows the possibilities down to the motor protection device and the high- or low-pressure controls.

On arrival, the technician looks over the whole system before doing anything and notices that one evaporator fan is not running, and the coil is iced so the suction is too low, **Figure 29–40**. This causes the low-pressure control to shut off the compressor. When the examination is complete, the fan motor is changed, and the system is put back in operation. The compressor is shut off, and the fans allowed to run long enough to defrost the coil. The temperature

NOTICE THAT THE HEAD PRESSURE WOULD
NORMALLY BE 185 psig (95°F + 30°F = 125°F)
BUT IT IS 135 psig BECAUSE OF THE REDUCED
LOAD OF ONE EVAPORATOR FAN.

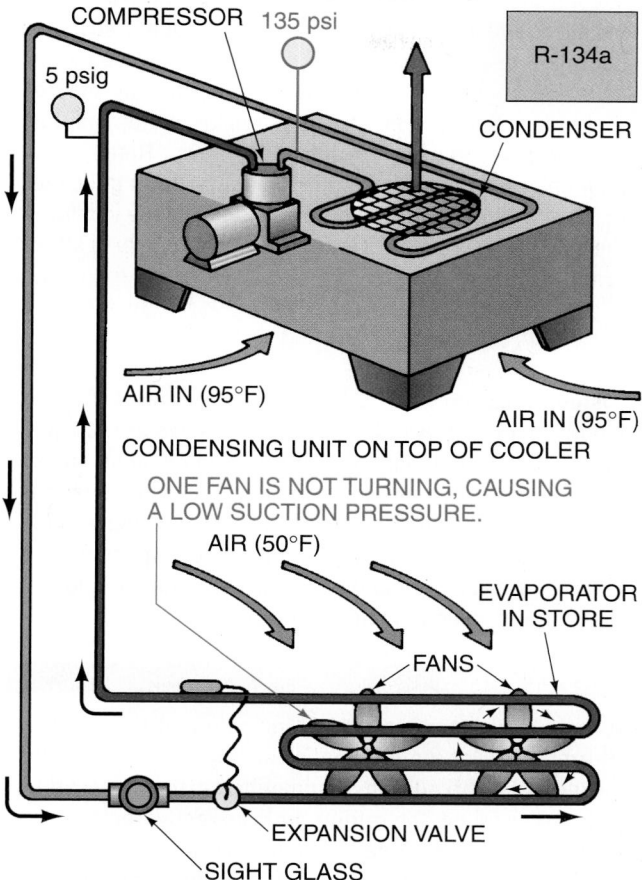

MEDIUM-TEMPERATURE EVAPORATOR SHOULD BE
OPERATING AT 18.4 psig (20°F BOILING
TEMPERATURE) BUT IT IS OPERATING ALL OF THE
TIME AND THE COOLER TEMPERATURE IS 50°F.

Figure 29–40   Symptoms of a medium-temperature system with
one of the two evaporator fans burned out. This system has a TXV.

inside the cooler is 50°F, and because of the amount of food
it will take a long while for the cooler to pull down to the cut-
out point of 35°F. The technician cautions the owner to
watch the thermometer in the cooler to make sure that
the temperature is going down. A call later in the day will
confirm that the unit is working properly.

SERVICE CALL 2

A medium-temperature reach-in cooler is not cooling, and
the compressor is cutting on and off. In this system the
condensing unit is on top. *The unit has a TXV and a low
charge. Two tubes rubbed together and created a leak.*

The technician arrives at the job and finds the unit is
short cycling on the low-pressure control. The sight glass
has bubbles in it, indicating a low charge, **Figure 29–41.**
The technician discovers that the small tube leading to the
low-pressure control has rubbed a hole in the suction line.

NOTICE THAT THE HEAD PRESSURE IS
DOWN DUE TO REDUCED LOAD.

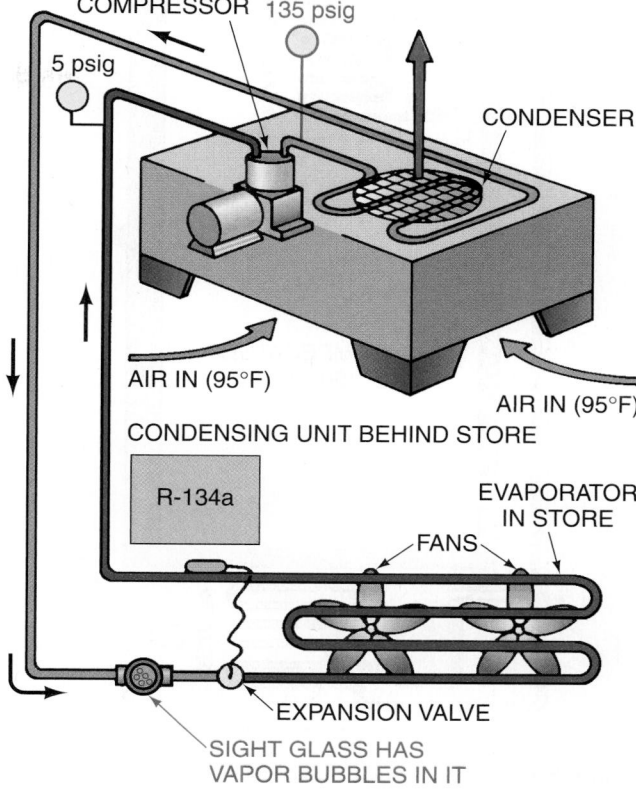

MEDIUM-TEMPERATURE EVAPORATOR SHOULD BE
OPERATING AT 18.4 psig (20°F BOILING
TEMPERATURE) BUT IT IS OPERATING AT 5 psig
BECAUSE OF A LOW CHARGE CAUSED BY A LEAK.

Figure 29–41   Symptoms of a medium-temperature system
operating in a low-charge situation. This system has a TXV.

There is oil around the point of the leak. The system is
pumped down by closing the king valve, and the leak is
repaired. The system is pressurized with nitrogen and leak
checked. The system is then evacuated to the proper
micron level. Once evacuation is complete and the vacuum
pump has been isolated from the system, the king valve can
be opened.

After the repair a leak check is performed. The system is
then started and charged to the correct charge. A call later
in the day verifies that the system is functioning correctly.

SERVICE CALL 3

A factory cafeteria manager reports that the reach-in
cooler in the lunchroom is not cooling and is running all the
time. This is a medium-temperature cooler with the condensing unit at the bottom of the fixture. It has not had a service call in a long time. *The evaporator is dirty and icing over,*
Figure 29–42. *The unit has a TXV and there is no off time for
defrost because the box does not get cool enough.*

The technician arrives and sees that the system is iced.
The compressor is sweating down the side; liquid is slowly

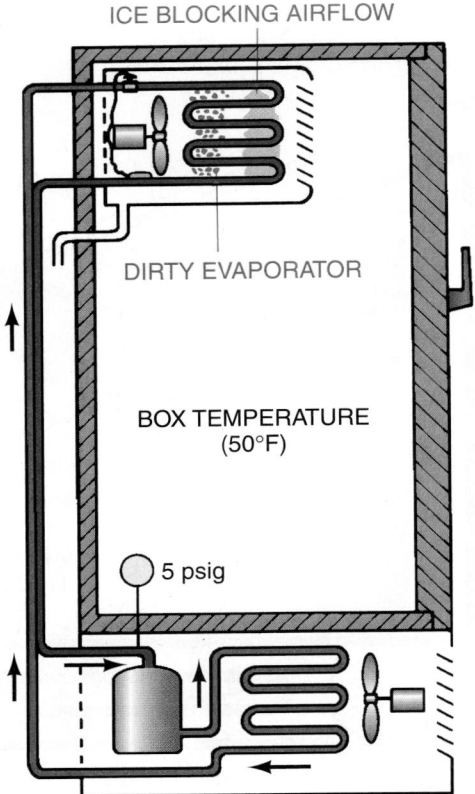

ICE BLOCKING AIRFLOW

DIRTY EVAPORATOR

BOX TEMPERATURE
(50°F)

5 psig

Figure 29–42 Symptoms of a medium-temperature cooler with a dirty evaporator coil. This system has a TXV.

coming back to the compressor—not enough to cause a noise from slugging. The TXV should maintain a constant superheat, but it may lose control if the pressure drops too low. The first thing that has to be done is to defrost the evaporator. This is done by stopping the compressor and using a heat gun (like a high-powered hair drier). The evaporator has a lot of dirt on it. SAFETY PRECAUTION: The evaporator is cleaned with a coil cleaner that is approved for food-handling areas. When the evaporator is cleaned, the system is started. The unit is now operating with a full sight glass, and the suction line leaving the evaporator is cold. From this point it will take time for the unit to pull the cooler temperature down. The service technician leaves and calls later in the day to confirm that the repair has solved the problem.

SERVICE CALL 4

A convenience store manager reports that the reach-in cooler that stores the dairy products is not cooling properly. *The temperature is 55°F. It had been cooling well until early this morning. There is a leak due to a stress crack (caused by age and vibration) in the suction line near the compressor. The compressor is vibrating because the customer has moved the unit, and the condensing unit has fallen down in the frame. This system has a capillary tube metering device,* **Figure 29–43.**

The technician's examination discloses that the compressor is vibrating because the condensing unit is not set-

ting straight in the frame. While securing the condensing unit, the technician notices an oil spot on the bottom of the suction line and that the compressor shell is hot. Gages are installed, and it is discovered that the suction pressure is operating in a vacuum. NOTE: Care must be used when installing gages if a vacuum is suspected or air may be drawn into the system. The compressor is stopped, and the low-side pressure is allowed to rise. There is not enough pressure in the system to accomplish a good leak test, so refrigerant is added. A leak check in the vicinity of the suction line reveals a leak. This appears to be a stress crack due to the vibration. ♻ The refrigerant is recovered from the system to repair the leak. Air must have been pulled into the system while it was operating in a vacuum, so the refrigerant charge must be recovered. ♻ **This system does not have a low-pressure control.** A short length of pipe is installed where the stress crack was found. A new liquid-line drier is installed because the old one may not have any capacity left. The system is leak checked, triple evacuated, and charged, using a measured charge that the manufacturer recommended. The technician calls the manager later in the day and learns that the unit is functioning correctly.

SERVICE CALL 5

A restaurant manager calls to indicate that the reach-in freezer used for ice storage is running all the time. This system was worked on by a competitor in the early spring. A leak was found, the refrigerant was recovered, and the leak was repaired. A deep vacuum was then pulled and the unit was recharged. Hot weather is here, and the unit is running all the time. *This system has a capillary tube metering device, and the system has an overcharge of refrigerant. The other service technician did not measure the charge into the system,* **Figure 29–44.**

The service technician examines the fixture and notices that the compressor is sweating down the side. The condenser feels hot for the first few rows and then warm. This appears to be an overcharge of refrigerant. The condenser should be warm near the bottom where the condensing is occurring. The evaporator fan is running, and the evaporator looks clean, so the evaporator must be doing its job. Gages are installed, and the head pressure is 400 psig with an outside temperature of 95°F; the refrigerant is R-502. The head pressure should be no more than 301 psig on a 95°F day (95°F + 30°F = 125°F condensing temperature or 301 psig). For refrigerant R-404A, a condensing temperature of 125°F would give a pressure of 332 psig. This system calls for a measured charge of 2 lb 8 oz. ♻Two approaches can be taken: (1) alter the existing charge; or (2) recover the charge and measure a new charge into the unit while the unit is in a deep vacuum. It is a time-consuming process to recover the charge and evacuate the system.♻

The technician chooses to alter the existing charge. This is a plain capillary tube system with a heat exchanger (the capillary tube is soldered to the suction line after it leaves the evaporator). A thermometer lead is fastened to the

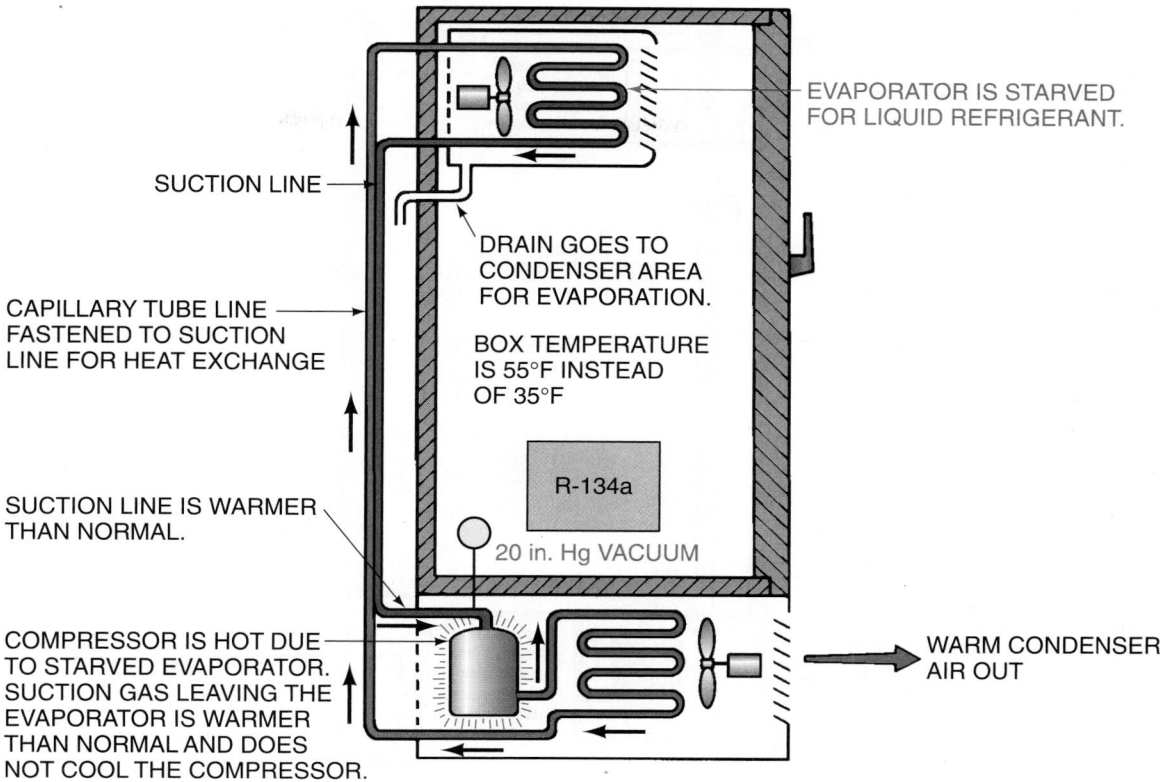

SUCTION LINE

CAPILLARY TUBE LINE
FASTENED TO SUCTION
LINE FOR HEAT EXCHANGE

SUCTION LINE IS WARMER
THAN NORMAL.

COMPRESSOR IS HOT DUE
TO STARVED EVAPORATOR.
SUCTION GAS LEAVING THE
EVAPORATOR IS WARMER
THAN NORMAL AND DOES
NOT COOL THE COMPRESSOR.

EVAPORATOR IS STARVED
FOR LIQUID REFRIGERANT.

DRAIN GOES TO
CONDENSER AREA
FOR EVAPORATION.

BOX TEMPERATURE
IS 55°F INSTEAD
OF 35°F

R-134a

20 in. Hg VACUUM

WARM CONDENSER
AIR OUT

Figure 29–43 Symptoms of a medium-temperature system with a low charge. This system has a capillary tube.

suction line **after the evaporator** but **before the heat exchanger.** The suction pressure is checked for the boiling point of the refrigerant and compared with the suction line temperature. The superheat is 0°F with the existing charge of refrigerant. ♻ Refrigerant is removed to an approved recovery cylinder until the superheat is 5°F at this point. ♻ NOTE: The heat exchanger will allow a lower superheat than normal. When the system charge is balanced, the technician leaves the job and will call the manager later in the day for a report on how the system is functioning.

SERVICE CALL 6

An office manager calls and reports that the reach-in medium-temperature beverage cooler in the employee cafeteria is running all the time. The suction line is covered with frost back to the compressor. *A small liquid-line drier is stopped up with sludge from a compressor changeout after a motor burnout. A suction-line drier should have been installed,* **Figure 29–45.**

The technician looks the system over closely and sees that the suction line is frosting and that the compressor has a frost patch on it. A first glance indicates that the fan at the evaporator is off and that the coil is dirty. Further examination, however, shows that the liquid line is frosting starting at the outlet of the drier. This means the drier is partially stopped up. The pressure drop across the drier makes the drier act like an expansion device. This effectively means there are two expansion devices in series because

the drier is feeding the capillary tube. ♻ The refrigerant charge must be recovered, and the drier replaced. ♻ After the refrigerant has been recovered, the drier is sweated into the liquid line. There are no service valves. While the system is open, the technician solders a suction-line drier in the suction line close to the compressor. A deep vacuum is then pulled and the system is recharged.

The reason the unit was frosting instead of sweating is that the evaporator was starved for liquid refrigerant. The suction pressure went down to a point below freezing. The unit's capacity was reduced to the point where it was running constantly and had no defrost. The frost will become more dense, blocking the air through the coil and acting as an insulator. This will cause the coil to get even colder with more frost. This condition will continue with the frost line moving on to the compressor. The ice or frost acts as an insulator and an air blockage. Air has to circulate across the coil for the unit to produce at its rated capacity.

SERVICE CALL 7

A restaurant owner calls indicating that the reach-in freezer for ice cream is running but the temperature is rising. *The system has a capillary tube metering device and the evaporator fan is defective. The customer hears the compressor running and thinks the whole unit is running,* **Figure 29–46.**

The technician has never been to this installation and has to examine the system thoroughly. It is discovered that the

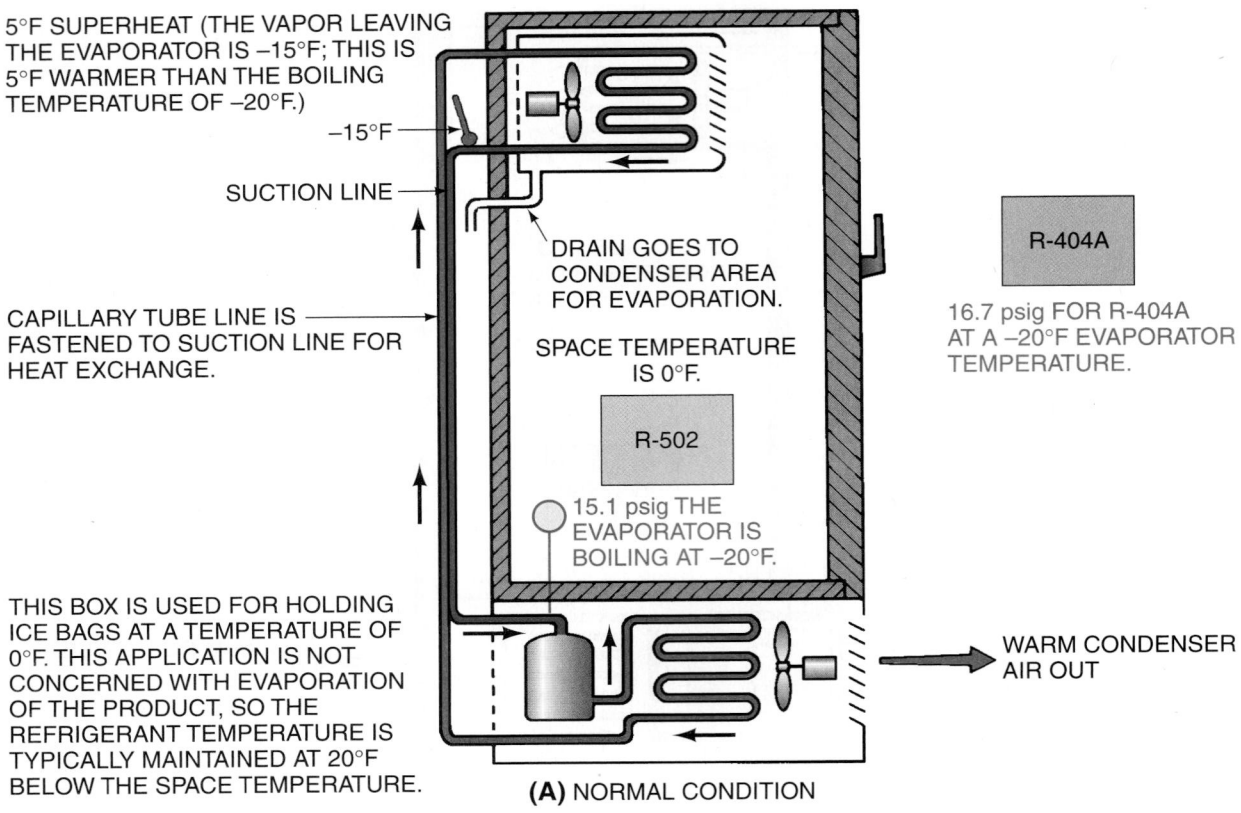

5°F SUPERHEAT (THE VAPOR LEAVING THE EVAPORATOR IS −15°F; THIS IS 5°F WARMER THAN THE BOILING TEMPERATURE OF −20°F.)

−15°F

SUCTION LINE

CAPILLARY TUBE LINE IS FASTENED TO SUCTION LINE FOR HEAT EXCHANGE.

THIS BOX IS USED FOR HOLDING ICE BAGS AT A TEMPERATURE OF 0°F. THIS APPLICATION IS NOT CONCERNED WITH EVAPORATION OF THE PRODUCT, SO THE REFRIGERANT TEMPERATURE IS TYPICALLY MAINTAINED AT 20°F BELOW THE SPACE TEMPERATURE.

DRAIN GOES TO CONDENSER AREA FOR EVAPORATION.

SPACE TEMPERATURE IS 0°F.

R-502

15.1 psig THE EVAPORATOR IS BOILING AT −20°F.

R-404A

16.7 psig FOR R-404A AT A −20°F EVAPORATOR TEMPERATURE.

WARM CONDENSER AIR OUT

(A) NORMAL CONDITION

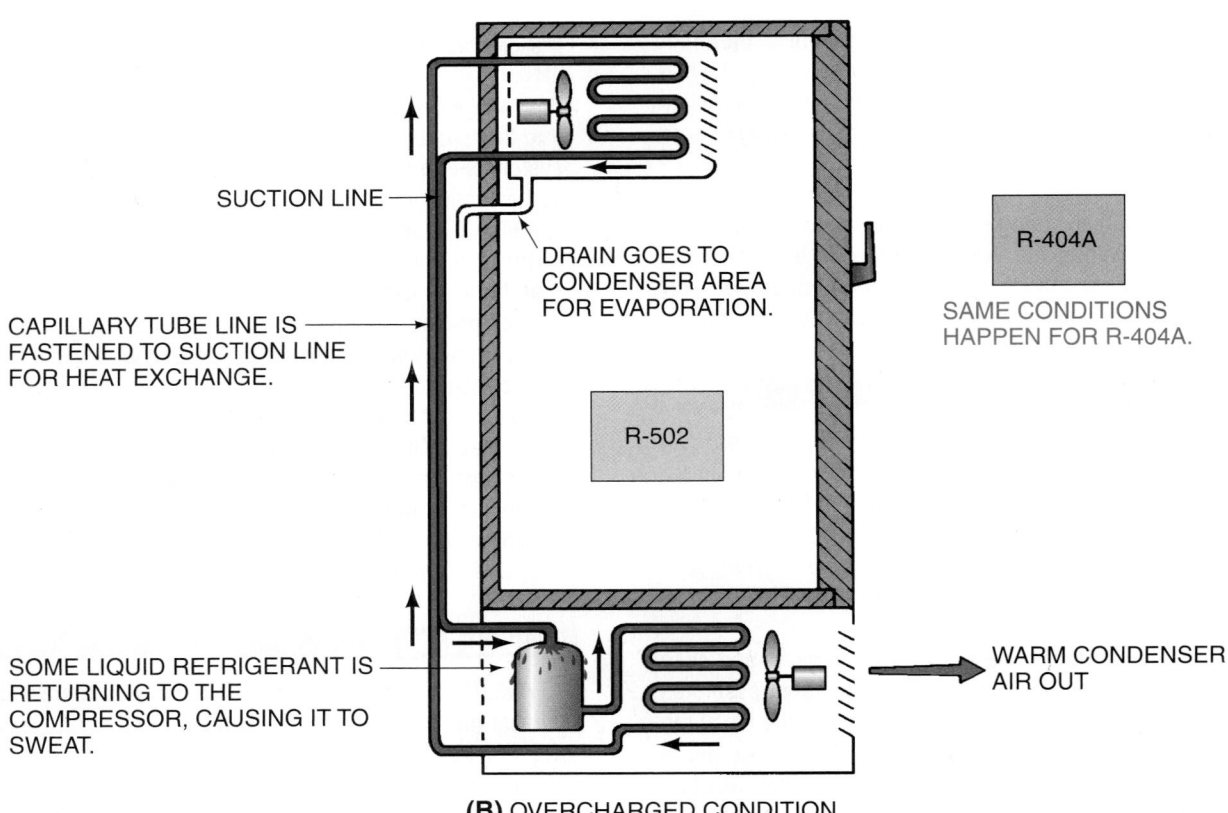

SUCTION LINE

CAPILLARY TUBE LINE IS FASTENED TO SUCTION LINE FOR HEAT EXCHANGE.

DRAIN GOES TO CONDENSER AREA FOR EVAPORATION.

R-502

R-404A

SAME CONDITIONS HAPPEN FOR R-404A.

SOME LIQUID REFRIGERANT IS RETURNING TO THE COMPRESSOR, CAUSING IT TO SWEAT.

WARM CONDENSER AIR OUT

(B) OVERCHARGED CONDITION

Figure 29–44 **(A)** A reach-in freezer in normal conditions. **(B)** Symptoms of a reach-in freezer with an overcharge of refrigerant. This system has a capillary tube.

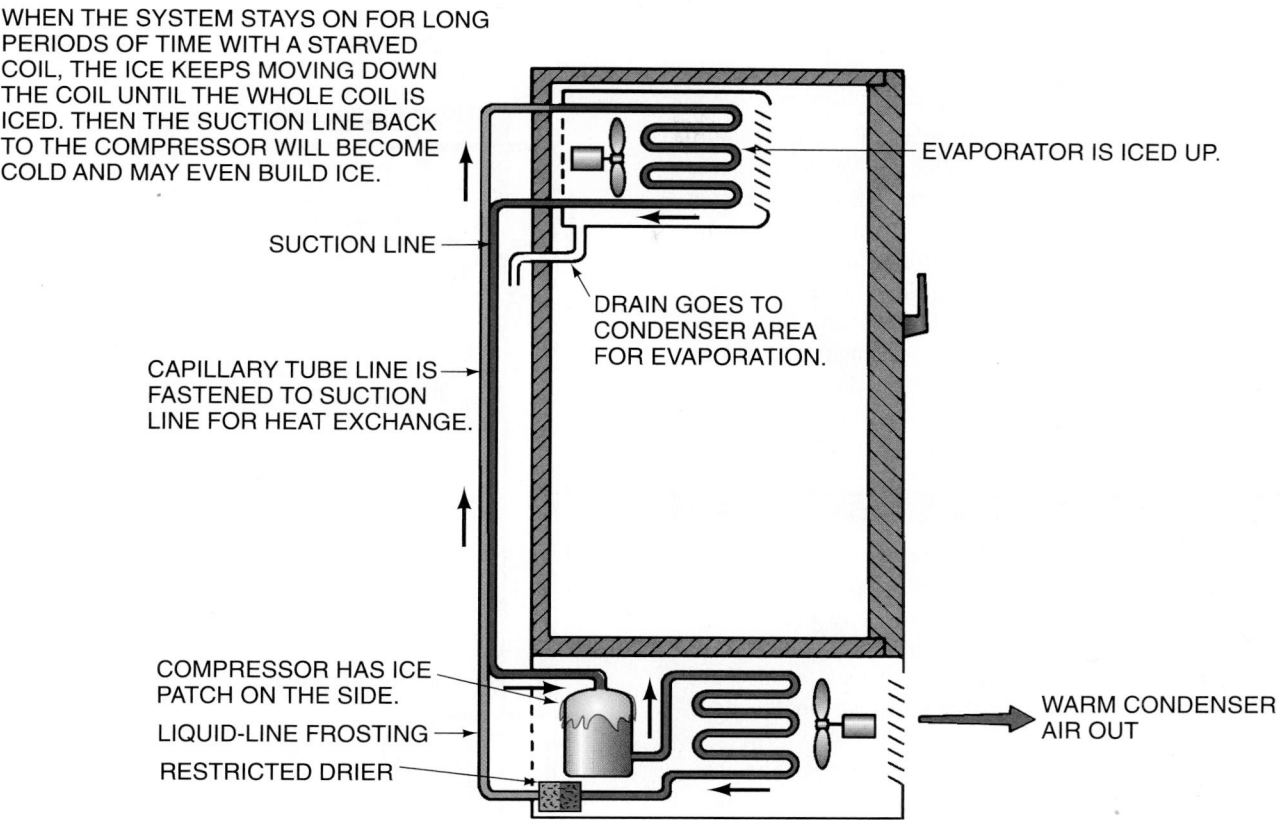

WHEN THE SYSTEM STAYS ON FOR LONG PERIODS OF TIME WITH A STARVED COIL, THE ICE KEEPS MOVING DOWN THE COIL UNTIL THE WHOLE COIL IS ICED. THEN THE SUCTION LINE BACK TO THE COMPRESSOR WILL BECOME COLD AND MAY EVEN BUILD ICE.

SUCTION LINE

CAPILLARY TUBE LINE IS FASTENED TO SUCTION LINE FOR HEAT EXCHANGE.

EVAPORATOR IS ICED UP.

DRAIN GOES TO CONDENSER AREA FOR EVAPORATION.

COMPRESSOR HAS ICE PATCH ON THE SIDE.

LIQUID-LINE FROSTING

RESTRICTED DRIER

WARM CONDENSER AIR OUT

Figure 29–45   Symptoms of a medium-temperature reach-in cooler with a partially stopped-up liquid-line drier.

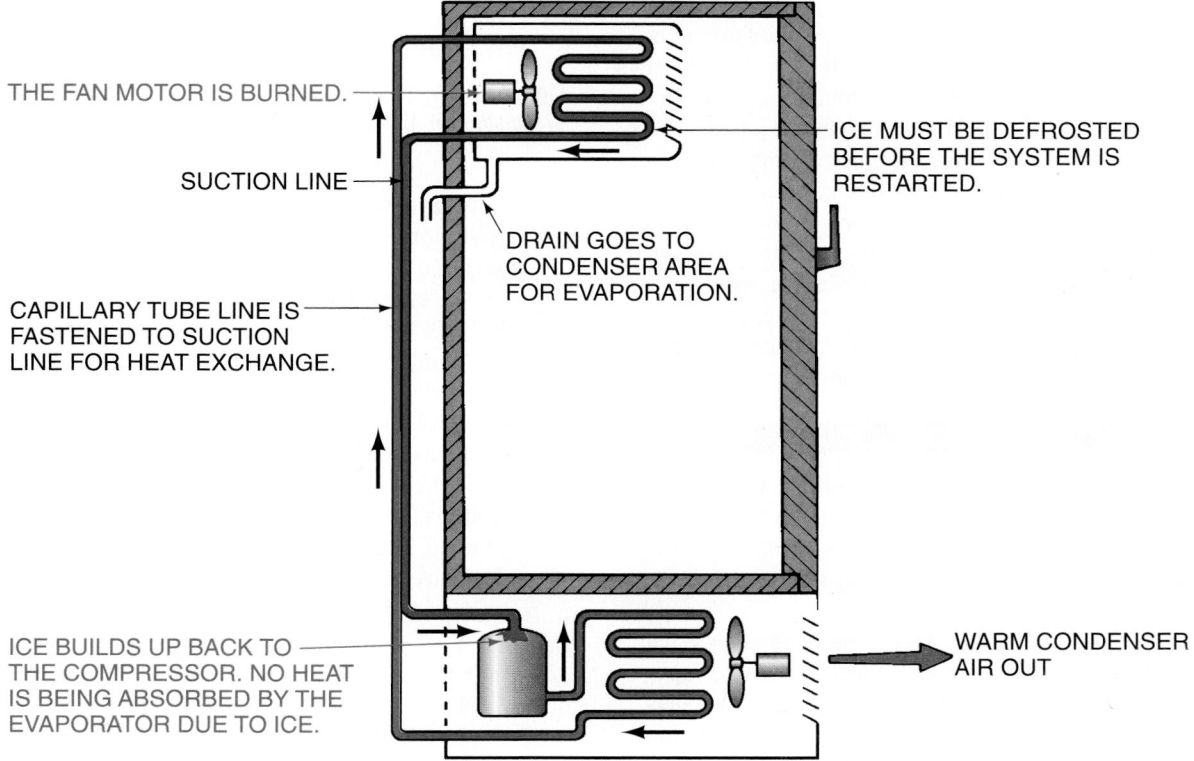

THE FAN MOTOR IS BURNED.

SUCTION LINE

CAPILLARY TUBE LINE IS FASTENED TO SUCTION LINE FOR HEAT EXCHANGE.

ICE MUST BE DEFROSTED BEFORE THE SYSTEM IS RESTARTED.

DRAIN GOES TO CONDENSER AREA FOR EVAPORATION.

ICE BUILDS UP BACK TO THE COMPRESSOR. NO HEAT IS BEING ABSORBED BY THE EVAPORATOR DUE TO ICE.

WARM CONDENSER AIR OUT

Figure 29–46   Symptoms of a reach-in freezer with a defective evaporator fan motor. This system has a capillary tube.

frost on the suction line goes all the way to and down the side of the compressor. The first thought is that the system is not going through the proper defrost cycle. The coil is iced and has to be defrosted before anything can be done. The defrost cycle clears the coil of ice. When defrost is over, the fan does not restart as it should. The technician takes off the panel to the fan compartment and checks the voltage. There is voltage, but the fan will not run. When the motor is checked for continuity through the windings, it is discovered that the fan motor winding is open. The fan motor is changed, and the system is started again. It will take several hours for the fixture to pull back down to the normal running temperature of −10°F. A call later in the day verifies that the freezer temperature is going down.

SERVICE CALL 8

A call is received indicating that the customer's reach-in dairy case used for milk storage is running constantly, and the temperature is 48°F. This unit has not had a service call in 10 years of service. *The evaporator is dirty. This unit has an automatic expansion valve. These coils never have filters, and years of dust will accumulate on the coils,* **Figure 29–47.**

The technician finds the suction line very cold. The compressor is sweating. This is evidence of an overcharge, or that the refrigerant is not boiling to a vapor in the evaporator. A close examination of the evaporator indicates that it is not exchanging heat with the air in the cooler. SAFETY PRECAUTION: The evaporator is cleaned with a special detergent approved for evaporator cleaning and for use in food-handling areas. Areas that have dairy products are particularly difficult because dairy products absorb odors easily. The system is started; the sweat gradually moves off the compressor, and the suction line feels normally cool. The technician leaves the job and will call back later to see how the unit is performing.

An automatic expansion valve maintains a constant pressure. It responds in reverse to a load change. If a load of additional product is added to the cooler, the rise in suction pressure will cause the automatic expansion valve to throttle back and slightly starve the evaporator. If the load is reduced, such as with a dirty coil, the valve will overfeed to keep the refrigerant pressure up.

SERVICE CALL 9

A golf club restaurant manager calls and says that a small beverage cooler is not cooling the drinks to the correct temperature. *A small leak at the flare nut on the outlet to the automatic expansion valve has caused a partial loss of refrigerant,* **Figure 29–48.**

The technician arrives and hears the expansion valve hissing. This means the expansion valve is passing vapor along with the liquid it is supposed to pass. The sight glass shows some bubbles. Gages are installed; the suction pressure is normal, and the head pressure is low. The suction pressure is 17.5 psig; this corresponds to 15°F boiling temperature for R-12. 15°F for R-134a is 15.1 psig.

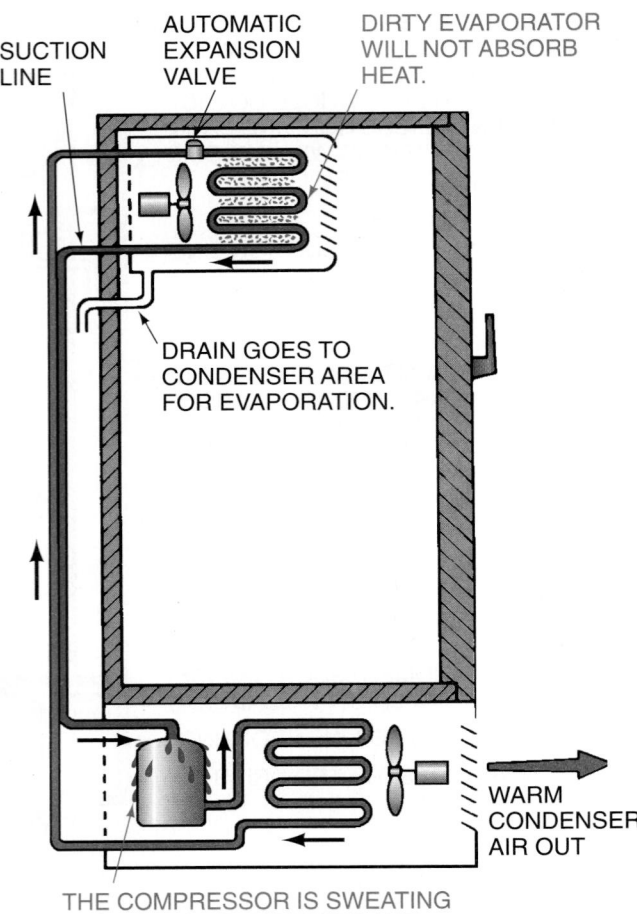

Figure 29–47 Symptoms of a reach-in dairy case with a dirty evaporator coil. This system has an automatic expansion valve.

The boiling refrigerant normally is about 20°F cooler than the beverages to be cooled. The liquid beverage in the cooler is 50°F and should be 35°F. The head pressure should be about 126 psig, the pressure corresponding to 105°F. 105°F for R-134a is 135 psig. The ambient is 75°F, and the condensing temperature should be 105°F (75°F + 30°F = 105°F). The head pressure is 100 psig. All of these signs point to an undercharge. The technician turns the cooler off and allows the low-side pressure to rise so that there will be a better chance to detect a leak. A leak is found at the flare nut leaving the expansion valve. The nut is tightened, but it does not stop the refrigerant from leaking. The flare connection must be defective.

The charge is recovered, and the flare nut is removed. There is a crack in the tubing at the base of the flare nut. The flare is repaired, and the system is leak checked and evacuated. A new charge of refrigerant is measured into the system from a vacuum. This gives the most accurate operating charge. The system is started. The technician will call back later to ensure that the unit is operating properly.

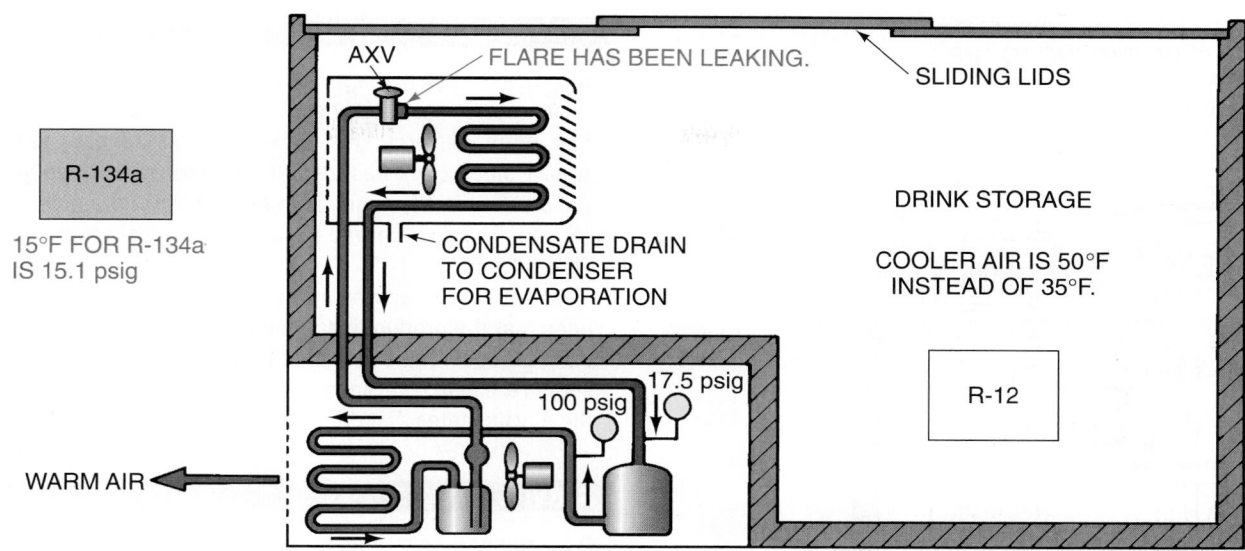

Figure 29–48   Symptoms of a reach-in beverage cooler with a low charge. This system has an automatic expansion valve.

SERVICE CALL 10

A restaurant manager reports that the compressor in the pie case is cutting off and on and sounds like it is straining while it is running, **Figure 29–49**. This unit was charged

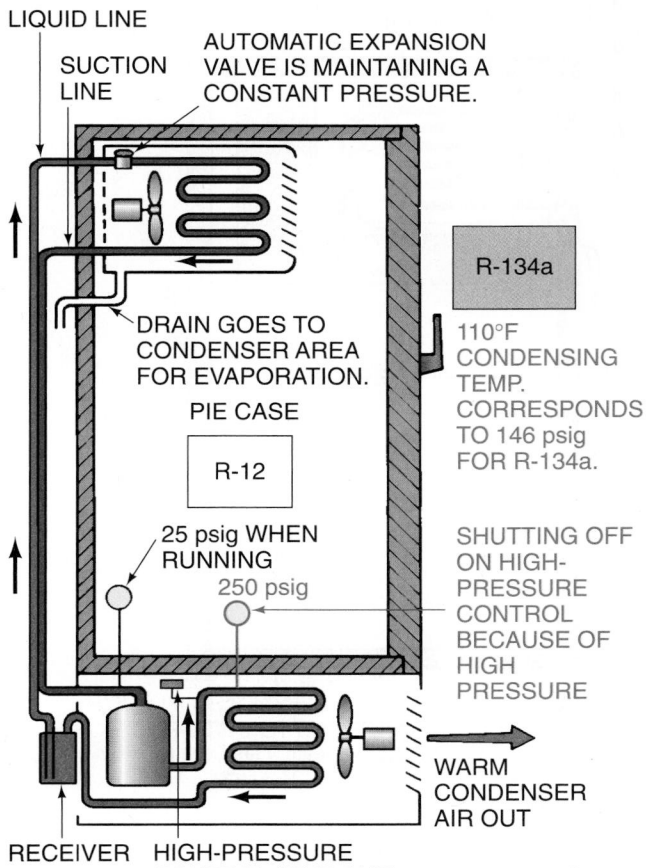

Figure 29–49   Symptoms of a reach-in pie case with an overcharge of refrigerant. This system has an automatic expansion valve.

after a leak and then shut down and put in storage. It has not been operated in several months. *The unit has an automatic expansion valve and has an overcharge of refrigerant. When the unit was charged it was in the back room in the winter, and it was cold in the room. The unit was charged to a full sight glass with a cold condenser.*

The service technician remembers that the unit was started and charged in a cold ambient and assumes that it may have an overcharge of refrigerant. When the technician arrives, the unit is started in the location where it is going to stay; the ambient is warm. Gages are installed, and the head pressure is 250 psig. The head pressure would be 281 psig for R-134a at the overcharged condition. The pressure should be 136 psig at the highest because the ambient temperature is 80°F. This should create a condensing temperature of no more than 110°F (80°F + 30°F = 110°F) or 136 psig for R-12. The compressor is cutting off because of high pressure at 260 psig. A 110°F condensing temperature for R-134a corresponds to 146 psig. ♻ Refrigerant is removed to an approved recovery cylinder from the machine until the head pressure is down to 136 psig, the correct head pressure for the ambient temperature. ♻ The sight glass is still full. A call back to the owner later in the day verifies that the system is working correctly.

SERVICE CALL 11

A store manager calls to say that a reach-in cooler is rising in temperature, and the unit is running all the time. It is a medium-temperature cooler and should be operating between 35°F and 45°F. It has an automatic expansion valve. The unit has not had a service call in 3 years. *The evaporator fan motor is not running. This system has off-cycle defrost,* **Figure 29–50**.

The service technician arrives at the site and examines the system. The compressor is on top of the fixture and is

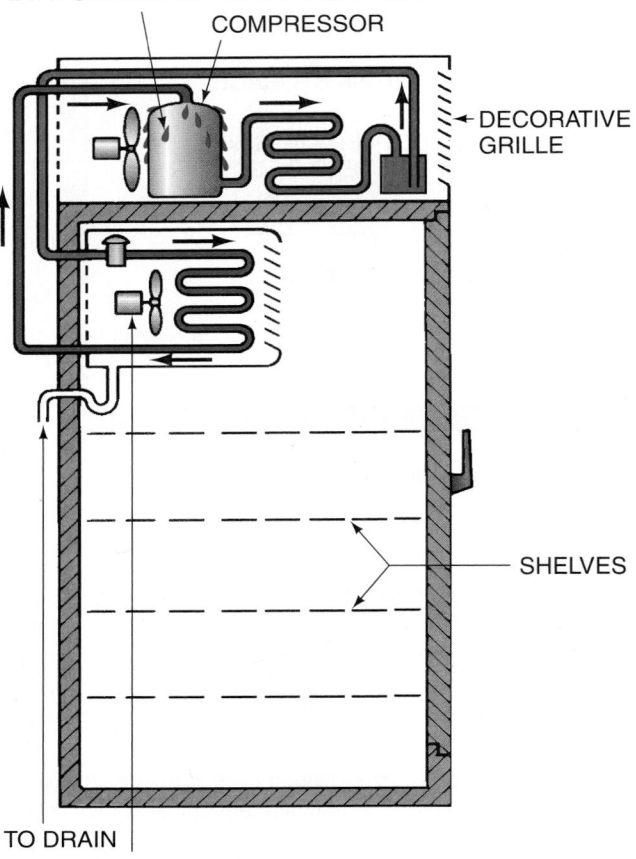

THE COMPRESSOR IS SWEATING.
LIQUID IS RETURNING TO THE
COMPRESSOR THAT IS NOT
EVAPORATED IN THE EVAPORATOR.

COMPRESSOR

DECORATIVE GRILLE

SHELVES

TO DRAIN

THE EVAPORATOR FAN IS STALLED
AND NO HEAT IS BEING ABSORBED
BY THE EVAPORATOR. THE
AUTOMATIC EXPANSION VALVE
OPENS WIDE TO KEEP THE SUCTION
PRESSURE UP.

Figure 29–50 Symptoms of a reach-in cooler with a defective fan motor. This system has an automatic expansion valve.

sweating down the side. Because this uses an automatic expansion valve, the sweating is a sign that the evaporator is not boiling the refrigerant to a vapor. The evaporator could be dirty, or the fan may not be moving enough air. After removing the fan panel, the technician sees that the evaporator fan motor is not running. The blades are hard to turn, indicating that bearings are tight. The bearings are lubricated; the fan is started and seems to run like it should. This is a nonstandard motor and cannot be purchased locally. The system is left in running condition until a new fan motor can be obtained.

The technician returns in a week with the correct motor and exchanges it for the old one. The system has been working correctly all week. The old motor would fail again because the bearings are scored, which is the reason for exchanging a working motor for a new one.

A store assistant manager reports that a reach-in medium-temperature cooler is rising in temperature. *This unit has been performing satisfactorily for several years. A new stock clerk has loaded the product too high, and the product is interfering with the airflow of the evaporator fan. The system has an automatic expansion valve,* **Figure 29–51**.

The technician arrives at the store and immediately notices that the product is too high in the product area. The extra product is moved to another cooler. The stock clerk is shown the load level lines in the cooler. A call back later in the day indicates that the cooler is now working correctly.

A customer reports that the condensing unit on a medium-temperature walk-in cooler is running all the time, and the cooler is rising in temperature. *This is the first hot day*

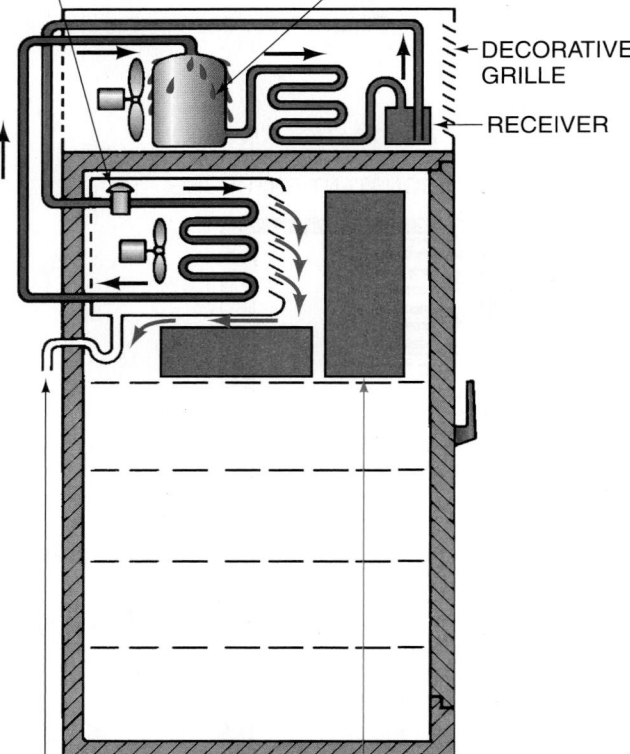

AUTOMATIC EXPANSION
VALVE OPENS MORE
DURING REDUCED LOAD.

THE COMPRESSOR
IS SWEATING DUE
TO REDUCED LOAD.

DECORATIVE GRILLE

RECEIVER

TO DRAIN

THE FOOD IS STACKED IN SUCH
A MANNER AS TO INTERFERE
WITH AIRFLOW. THE AIR IS
TURNING BACK AND
RECIRCULATING.

Figure 29–51 Symptoms of a reach-in medium-temperature cooler where the product is interfering with the air pattern. This system has an automatic expansion valve.

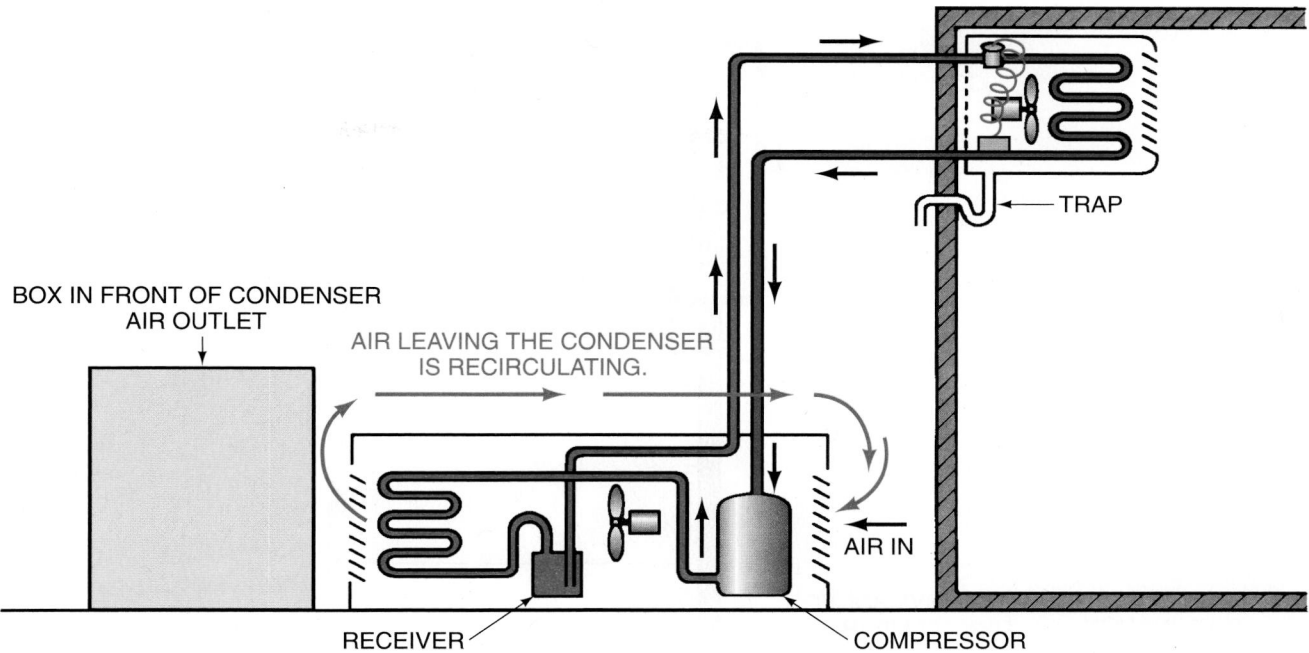

BOX IN FRONT OF CONDENSER
AIR OUTLET

AIR LEAVING THE CONDENSER
IS RECIRCULATING.

TRAP

AIR IN

RECEIVER   COMPRESSOR

Figure 29–52   Symptoms of an air-cooled condenser with the hot discharge air recirculating back into the condenser inlet.

of the season, and some boxes have been stacked too close to the condenser outlet. The hot air is leaving the condenser and recirculating back into the fan inlet, **Figure 29–52.**

The service technician arrives and looks over the system. The temperature is 52°F inside the cooler. The liquid-line temperature is hot instead of warm. This is a TXV system, and the sight glass is clear. The unit appears to have a full charge with a high head pressure. Gages are installed, and the head pressure is 207 psig. The system is using R-12 as the refrigerant. If the system had R-134a as the refrigerant, the high head pressure would be 229 psig. The head pressure should be no more than 158 psig, which is a condensing temperature of 120°F; it is 90°F (90°F + 30°F = 120°F or 158 psig). For R-134a, a condensing pressure of 171 psig corresponds to 120°F condensing temperature. Further examination indicates that air is leaving the condenser and recirculating back to the condenser inlet. Several boxes have been stored in front of the condenser. These are moved, and the head pressure drops to 158 psig. The reason there has not been a complaint up to now is that the air has been cold enough in the previous mild weather to keep the head pressure down even with the recirculation problem. A call back later in the day verifies that the cooler temperature is back to normal.

SERVICE CALL 14

A store frozen food manager calls to say that the temperature is going up in the low-temperature walk-in freezer. *The defrost time clock motor is burned, and the system will not go into defrost,* **Figure 29–53.**

THE SUCTION PRESSURE FOR R-502 SHOULD BE ABOUT 18 psig (–16°F) FOR A 0°F FREEZER TEMPERATURE. THIS FREEZER IS WARMER THAN 0°F AND THE SUCTION PRESSURE IS LOW.

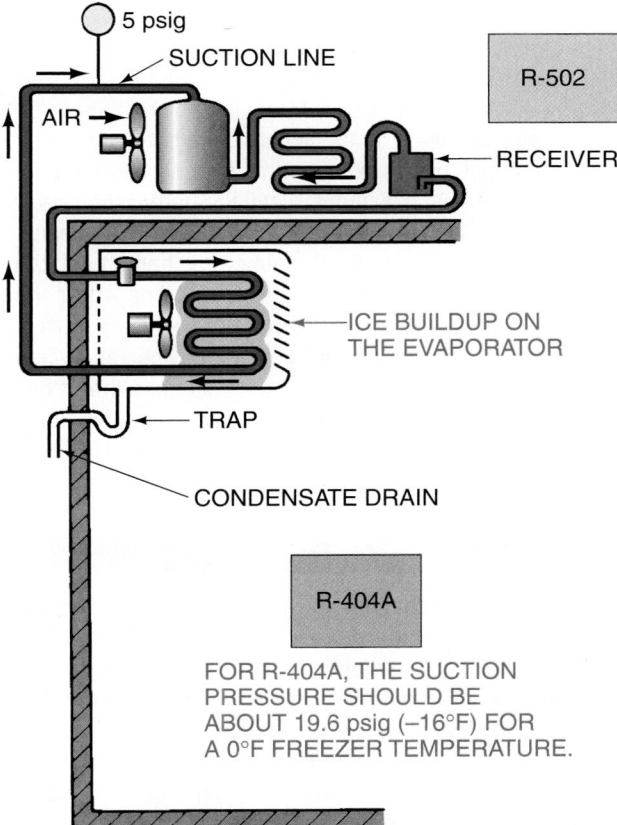

5 psig

SUCTION LINE

AIR

R-502

RECEIVER

ICE BUILDUP ON
THE EVAPORATOR

TRAP

CONDENSATE DRAIN

R-404A

FOR R-404A, THE SUCTION PRESSURE SHOULD BE ABOUT 19.6 psig (–16°F) FOR A 0°F FREEZER TEMPERATURE.

Figure 29–53   Symptoms of a defrost problem with a defective time clock.

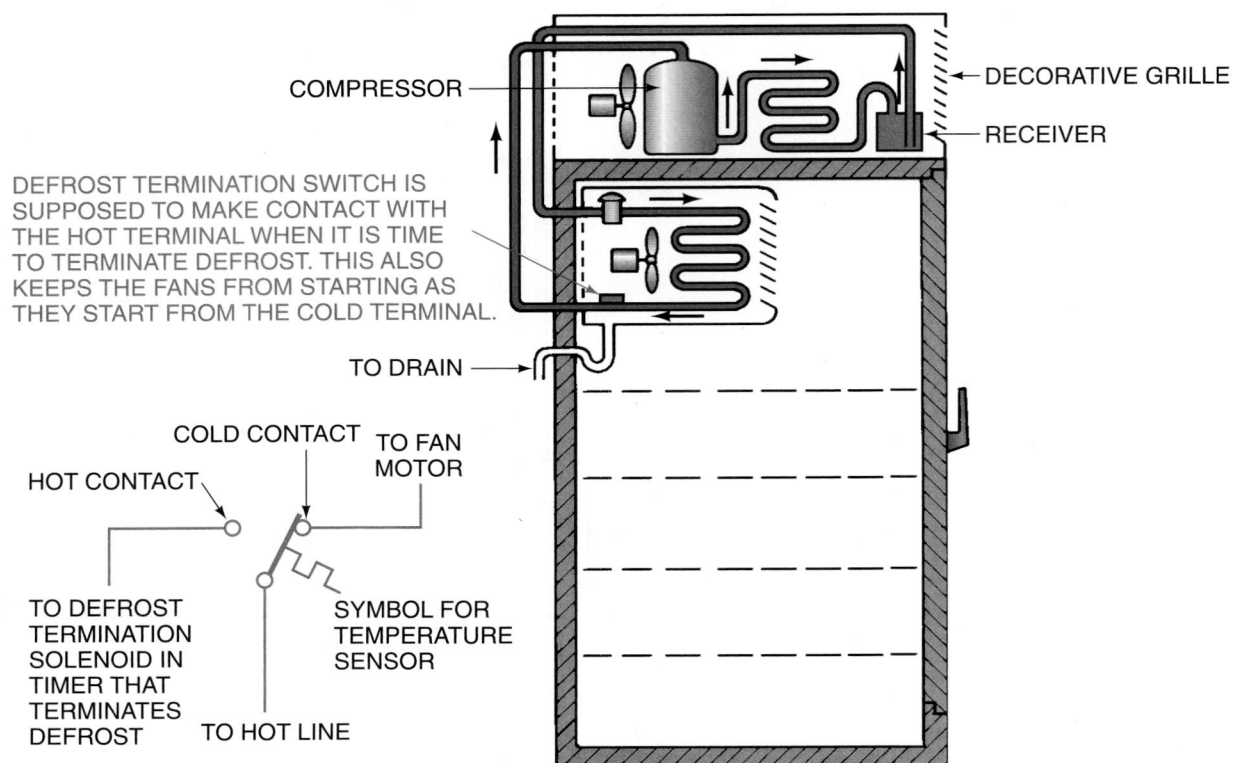

COMPRESSOR

DECORATIVE GRILLE

RECEIVER

DEFROST TERMINATION SWITCH IS
SUPPOSED TO MAKE CONTACT WITH
THE HOT TERMINAL WHEN IT IS TIME
TO TERMINATE DEFROST. THIS ALSO
KEEPS THE FANS FROM STARTING AS
THEY START FROM THE COLD TERMINAL.

TO DRAIN

COLD CONTACT   TO FAN
MOTOR

HOT CONTACT

TO DEFROST
TERMINATION
SOLENOID IN
TIMER THAT
TERMINATES
DEFROST

TO HOT LINE

SYMBOL FOR
TEMPERATURE
SENSOR

Figure 29–54   Symptoms of a defective defrost termination thermostat (stuck with cold contacts closed) in a low-temperature freezer.

The service technician arrives, examines the system, and sees that the evaporator is coated with thick ice. It is evident that it has not been defrosting. The first thing to do is to force a defrost. The technician goes to the timer and when the clock dial is examined, it is found that the time indicator says 4:00 AM, but it is 2:00 PM. Either the power has been off, or the timer motor is not advancing the timer. The technician advances the timer by hand until the defrost cycle starts and then marks the time. After the defrost is terminated by the temperature-sensing device, the system goes into normal operation. The technician gives the clock about one-half hour and sees that it has not moved. A new time clock is installed. The customer is cautioned to look out for heavy ice buildup. A call the next day verifies that the system is working correctly.

SERVICE CALL 15

A customer states that the defrost seems to be lasting too long in the low-temperature walk-in cooler. It used to defrost, and the fans would start back up in about 10 min. It is now taking 30 min. *The defrost termination switch is not terminating defrost with the temperature setting. This is an electric defrost system. The system is staying in defrost until the time override in the timer takes it out of defrost. This also causes the compressor to operate at too high a current because the fans should not start until the coil cools to below 30°F. With the defrost termination*

switch stuck in the cold position, the fans will come on when the timer terminates defrost. The compressor will be overloaded by the heat left in the coil from the defrost heaters, **Figure 29–54.**

The technician arrives at the job and examines the system. The coil is free of ice so defrost has been working. The cooler is cold, −5°F. The technician advances the time clock to the point that defrost starts. The termination setting on the defrost timer is 30 min. This may be a little long unless the cooler is only defrosted once a day. The timer settings are to defrost twice a day, at 2:00 PM and 2:00 AM. The coils are defrosted in approximately 5 min, but the defrost continues. The technician allows defrost to continue for 15 min to allow the coil to get to maximum temperature, so the defrost termination switch will have a chance to close. After 15 min the system is still in defrost, so the timer is advanced to the end of the timed cycle. When the system goes out of defrost, the fans start. They should not start until the coil gets down to below 30°F. The defrost termination switch is removed. It has three terminals, and the wires are marked for easy replacement of a new control. A new control is acquired and installed. The system is started. The owner called the next day, reported a normal defrost, and that the system was acting normally. The old defrost termination switch is checked with an ohmmeter after having been room temperature for several hours, and it is found that the cold contacts are still closed. The control is definitely defective.

The frozen food manager in a supermarket calls indicating that the walls between the doors are sweating on a reach-in freezer. *This has never happened. The freezer is about 15 years old. The mullion heater in the wall of the cooler that keeps the panel above the dew point temperature of the room is not functioning,* **Figure 29–55.**

The technician knows what the problem is before going to the site. After arriving, the technician examines the fixture for the reason that the mullion heater is not getting hot. If the heater is burned out, the panel needs to be removed. The circuit is traced to the back of the cooler, and the wires to the heater are found. An ohm check proves the heater still has continuity. A voltage check shows there is no voltage going to the heater. After further tracing by the technician, a loose connection is located in a junction box. The wires are connected properly, and a current check is performed to prove the heater is working.

A restaurant maintenance person calls to say that the medium-temperature walk-in cooler is off. The breaker was off, and it was reset, but it tripped again. *The compressor motor is burned,* **Figure 29–56.**

The technician arrives at the job and examines the system. The food is warming up, and the cooler is up to 55°F. Before resetting the breaker, the technician uses an ohm-meter and finds the compressor has an open circuit through the run winding. Nothing can be done except to change the compressor. A refrigerant line is opened slightly to determine the extent of the burn. The refrigerant has a high acid odor, indicating a bad motor burn. ♻ The technician uses a recovery/recycle unit to capture, clean, and recycle the refrigerant in this cooler. ♻

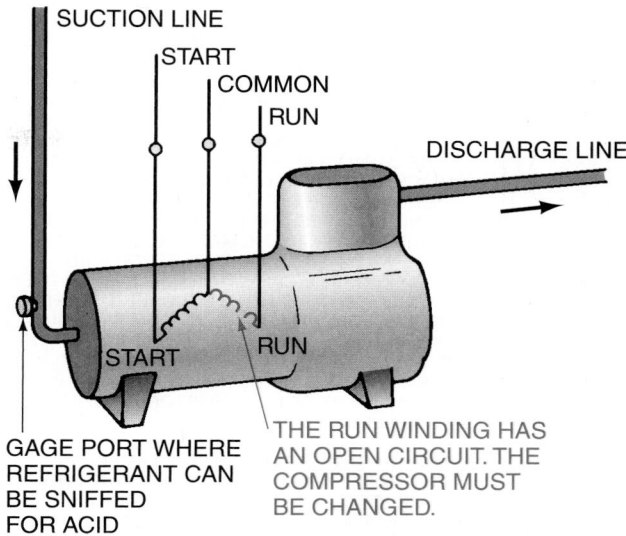

GAGE PORT WHERE REFRIGERANT CAN BE SNIFFED FOR ACID

THE RUN WINDING HAS AN OPEN CIRCUIT. THE COMPRESSOR MUST BE CHANGED.

Figure 29–56   A system with a burned compressor motor.

It will be several hours before the system can be put back into service, so the food is moved to another cooler. The technician goes to the supply house and gets the required materials, including a suction-line filter drier with high acid-removing qualities. This drier has removable cores for each changeout. The compressor is changed, and the suction-line drier is installed. The system is purged with dry nitrogen to push as much of the free contaminants as possible out of the system. The system is leak checked, and a deep vacuum is pulled. The system is then charged and started. After the compressor runs for an hour, an acid test is taken on the crankcase oil. It shows a slight acid count. The unit is run for another 4 hours, and an acid check shows even less acid. The system is pumped down, and the cores are changed in the suction-line drier. The liquid-line drier is changed, and the system is allowed to operate overnight.

The technician returns the next day and takes another acid check. It shows no sign of acid. The system is left to run in this condition with the drier cores still in place in case some acid becomes loose at a later date. *The motor burn is attributed to a random motor failure.*

A medium-temperature reach-in cooler is not cooling properly. The temperature is 55°F inside, and it should be no higher than 45°F. The compressor sounds like it is trying to start but then cuts off. *The starting capacitor is defective.*

The technician arrives at the job in time to hear the compressor try to start. Several things can keep the compressor from starting. It is best to give the compressor the benefit of the doubt and assume it is good. Check the starting components first and then check the compressor. The starting capacitor is removed from the circuit for

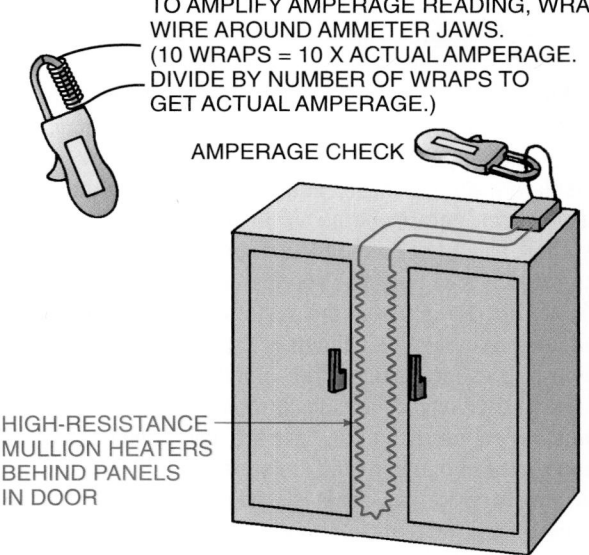

TO AMPLIFY AMPERAGE READING, WRAP WIRE AROUND AMMETER JAWS. (10 WRAPS = 10 X ACTUAL AMPERAGE. DIVIDE BY NUMBER OF WRAPS TO GET ACTUAL AMPERAGE.)

AMPERAGE CHECK

HIGH-RESISTANCE MULLION HEATERS BEHIND PANELS IN DOOR

Figure 29–55   A mullion heater on a low-temperature freezer.

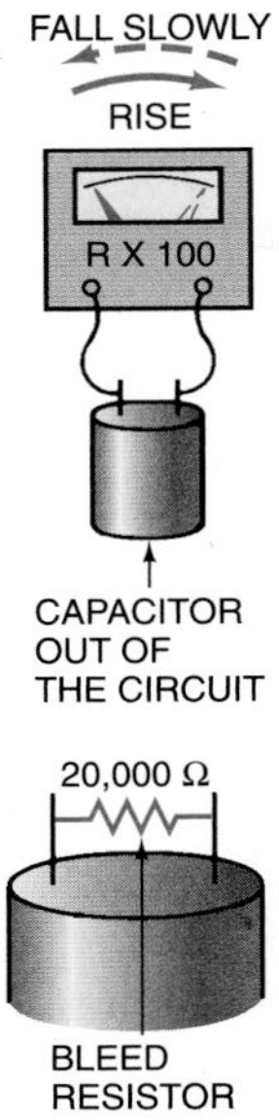

GOOD CAPACITOR

THE CAPACITOR HAS BEEN CHECKED TO SEE THAT IT IS NOT CHARGED BY PLACING A 20,000-OHM 5-WATT RESISTOR FROM TERMINAL TO TERMINAL FOR AT LEAST 5 SECONDS.

FALL SLOWLY

RISE

R X 100

1. PLACE THE LEADS OF AN OHMMETER ONTO THE CAPACITOR TERMINALS. NOW, SWITCH EACH OHMMETER LEAD TO THE OPPOSITE TERMINALS OF THE CAPACITOR. REPEAT SEVERAL TIMES. THE NEEDLE SHOULD RISE THEN FALL BACK. THE MAGNITUDE OF NEEDLE RISE WILL DEPEND ON THE MICROFARAD RATING OF THE CAPACITOR. THE HIGHER THE MICROFARAD RATING, THE HIGHER THE NEEDLE RISE BEFORE FALLING BACK.

CAPACITOR OUT OF THE CIRCUIT

2. IF THERE IS A RESISTOR BETWEEN THE TERMINALS THE NEEDLE WILL FALL BACK TO THE VALUE OF THE RESISTOR (KNOWN AS A BLEED RESISTOR).

20,000 Ω

3. TO PERFORM THE TEST AGAIN, THE LEADS MUST BE REVERSED. THE METER'S BATTERY IS DIRECT CURRENT. YOU WILL NOTICE AN EVEN FASTER RISE IN THE NEEDLE THE SECOND TIME BECAUSE THE CAPACITOR HAS THE METER'S BATTERY CHARGE.

BLEED RESISTOR

DEFECTIVE CAPACITOR

BLEED THE CAPACITOR AS DESCRIBED ABOVE.

1. TOUCH THE METER LEADS TO THE CAPACITOR TERMINALS. IF IT IS DEFECTIVE, IT WILL NOT RISE ANY HIGHER THAN THE VALUE OF THE BLEED RESISTOR. IF THE CAPACITOR DOES NOT HAVE A BLEED RESISTOR, THE NEEDLE WILL NOT RISE AT ALL. THIS SHOWS AN "OPEN" CAPACITOR. ELECTRONIC CAPACITOR TESTERS ARE A QUICK WAY TO TEST FOR AN OPEN OR A WEAK CAPACITOR. SIMPLY PLACE THE LEADS OF THE CAPACITOR TESTER ON THE CAPACITOR BEING TESTED AND PUSH A BUTTON. A MODERN DIGITAL CAPACITOR TESTER WILL INDICATE THE CAPACITANCE OF THE CAPACITOR. IT WILL ALSO TELL IF THE CAPACITOR IS OPEN.

Figure 29–57 Symptoms of a system with a defective starting capacitor.

checking, **Figure 29–57**. The capacitor has discoloration around the vent at the top, and the vent is pushed upward. These are signs the capacitor is defective. After bleeding the capacitor with a 20,000-Ω, 5-W resistor, an ohm test shows the capacitor open. The capacitor is replaced with a similar capacitor, and the compressor is started. The compressor is allowed to run for several minutes to allow the suction gas to cool the motor; then it is stopped and restarted to make sure that it is operating correctly. A call back the next day indicates that the compressor is still stopping and starting correctly.

SERVICE CALL 19

The compressor in the low-temperature walk-in freezer of a supermarket is not starting. The box temperature is 0°F, and it normally operates at −10°F. *The compressor is locked. This is a multiple-evaporator installation with four evaporators piped into one suction line. An expansion valve on one of the evaporators has been allowing a small amount of liquid refrigerant to get back to the compressor. This has caused marginal lubrication, and*

the compressor has scored bearings that are bad enough to lock the compressor.

The service technician arrives and examines the system. The compressor is a three-phase compressor and does not have a starting relay or starting capacitor. It is important that the compressor be started within 5 hours, or the food must be moved. The technician turns off the power to the compressor starter. The leads are then removed from the load side of the starter. The motor is checked for continuity with an ohmmeter, **Figure 29–58(A)**. The motor windings appear to be normal. NOTE: Some compressor manufacturers furnish data that tell what the motor-winding resistances should be. The meter is then turned to the voltage selection for a 230-V circuit. The starter is then energized to test the load side of the starter to make sure that each of the three phases has the correct power. This is a no-load test. The leads have to be disconnected from the motor terminals and the voltage applied to the leads while the motor is trying to start before the technician knows for sure there is a full 230 V under the starting load, **Figure 29–58(B)**.

When the load is connected, the technician connects the motor leads back up and gets two more meters for checking voltage. This allows the voltage to be checked from phase 1 to phase 2, phase 1 to phase 3, and phase 2 to phase 3 at the same time while the motor is being started. When the power is applied to the motor, there is 230 V from each phase to the other, and the motor will not start, **Figure 29–58(C)**. **This is conclusive. The motor is locked.** The motor can be reversed (by changing any two leads, such as L1 and L2) and it may start, but it is not likely that it will run for long.

The technician now has 4 hours to either change the compressor or move the food products. A call to the local compressor supplier shows that a compressor is in stock in town. The technician calls the shop for help. Another technician is dispatched to pick up the new compressor, and a crane is ordered to set the new compressor in and take the old one out. The compressor is 30 hp and weighs about 800 lb.

When the crane gets to the job, the old compressor is disconnected and ready to set out. This is accomplished by front seating the suction and discharge service valves to isolate the compressor and then recovering the compressor's refrigerant charge. When the new compressor is installed, all that will need to be done after evacuating the compressor is to open the service valves and start it. The original charge is still in the system. By the time the old compressor is removed, the new compressor arrives. The new compressor is set in place, and the service valves are connected. While a vacuum is being pulled, the motor leads are connected. In an hour the new compressor has been evacuated and is ready to start.

When the new compressor is started, it is noticed that the suction line is frosting on the side of the compressor. Liquid is getting back to the compressor. A thermometer lead is fastened to each of the evaporators at the evaporator

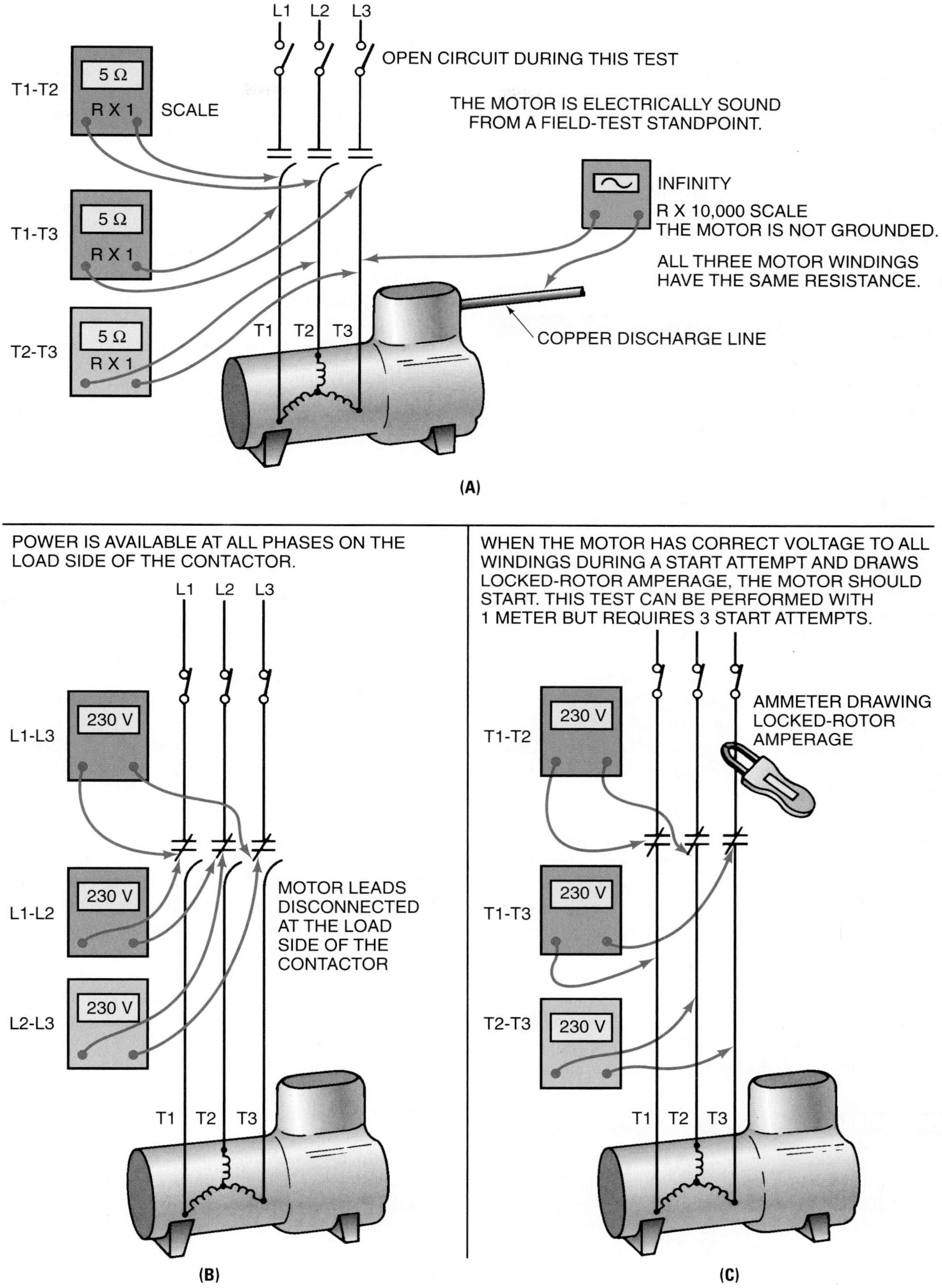

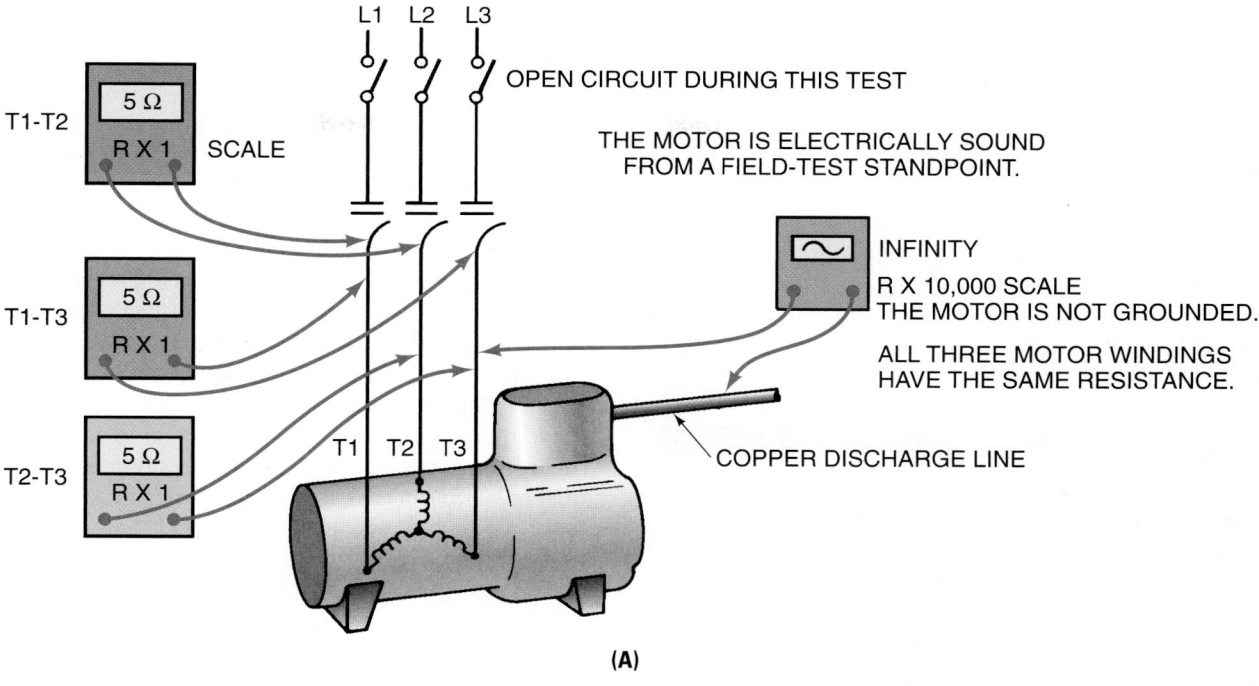

L1 L2 L3

T1-T2   5 Ω   R X 1   SCALE

OPEN CIRCUIT DURING THIS TEST

THE MOTOR IS ELECTRICALLY SOUND
FROM A FIELD-TEST STANDPOINT.

T1-T3   5 Ω   R X 1

INFINITY

R X 10,000 SCALE
THE MOTOR IS NOT GROUNDED.

ALL THREE MOTOR WINDINGS
HAVE THE SAME RESISTANCE.

T2-T3   5 Ω   R X 1

T1 T2 T3

COPPER DISCHARGE LINE

(A)

POWER IS AVAILABLE AT ALL PHASES ON THE
LOAD SIDE OF THE CONTACTOR.

L1 L2 L3

L1-L3   230 V

L1-L2   230 V

MOTOR LEADS
DISCONNECTED
AT THE LOAD
SIDE OF THE
CONTACTOR

L2-L3   230 V

T1 T2 T3

(B)

WHEN THE MOTOR HAS CORRECT VOLTAGE TO ALL
WINDINGS DURING A START ATTEMPT AND DRAWS
LOCKED-ROTOR AMPERAGE, THE MOTOR SHOULD
START. THIS TEST CAN BE PERFORMED WITH
1 METER BUT REQUIRES 3 START ATTEMPTS.

T1-T2   230 V

AMMETER DRAWING
LOCKED-ROTOR
AMPERAGE

T1-T3   230 V

T2-T3   230 V

T1 T2 T3

(C)

Figure 29–58   Symptoms of a system with a three-phase compressor that is locked.

outlet. It is discovered that one of the evaporator suction lines is much colder than the others. The expansion valve bulb is examined and is found to be loose and not sensing the suction-line temperature. Someone has taken the screws out of the mounting strap.

The expansion valve bulb is then secured to the suction line, and the system is allowed to operate. The frost line moves back from the side of the compressor, and the system begins to function normally.

The technician stops by the store the next day and determines that the frost line to the compressor is correct, not frosting the compressor.

SERVICE CALL 20

A restaurant owner with a cube ice-making machine calls and indicates that the evaporator is frozen solid and no ice is dropping into the bin. *The problem is the defrost probes are dirty. These probes sense the ice thickness by making contact with the ice and creating a circuit by conducting through the ice to start defrost when the ice reaches a predetermined thickness. The probe is monitored by an electronic circuit to determine when the correct thickness has been reached by the conductivity of the ice.*

The technician arrives and examines the entire system carefully. This technician has worked on this type of ice maker before and knows that the manufacturer recommends that the machine not be shut off at this time until the problem is found. If the technician were to shut the machine off and defrost the evaporator, the condition may not happen again and the trouble might not be found. The technician knows this machine is older and suspects dirty defrost sensor probes. The manufacturer has furnished the technician with a special jumper cord with a fixed resistor of the value of the resistance through ice. This jumper can be used to jump the ice probes and determine if the sensor is defective. The technician jumps the probe connections and the unit starts a defrost cycle. This is a sure sign that the sensor is dirty where it contacts the ice.

The technician waits until the unit completes the defrost cycle and the ice drops to the storage bin, then shuts the machine off and turns the power off. The probe is removed and cleaned. It is reinstalled and the machine power is restored. The machine is started. About 25 min later, the machine starts another defrost cycle on its own. The technician leaves the job, telling the owner to watch the machine for the rest of the day for correct operation.

SERVICE CALL 21

The owner of a restaurant calls and states that one of the ice machines is not dropping new ice to the bin. This is a new installation in a new restaurant. *The problem is the water has been filtered to the point that so many minerals have been removed that the water does not have enough conductivity for the electronic circuit to determine when the ice is touching the probes.*

The technician arrives and surveys the situation. The evaporator is frozen solid. This is a new machine, with the best water filtration system. There is a chance the filtration is so effective that the conductivity of the water has been reduced to a point where the defrost sensor probe will not sense the conductivity. The technician remembers the last service school she attended where these symptoms were described and goes to the shelf for a box of salt. A pinch of salt (about 1/8 teaspoon) is placed in the water sump where the pump will pick it up and circulate the slightly salted water over the probes. Within a very few minutes, the machine starts to defrost.

The technician knows the owner cannot put salt in the filtered water every day, so the manual is consulted. A resistor is available from the manufacturer that can be placed across certain terminals on the circuit board and allow the probe and electronic circuit board to sense the conductivity of water that has been filtered. Normal defrost can then be expected.

The technician obtains the resistor from the supplier and installs it across the appropriate terminals. The machine is started; the technician watches it through two defrost cycles and is then satisfied the machine will operate satisfactorily.

SERVICE CALL 22

A restaurant owner calls to request a checkup for the cube-type ice maker in the kitchen. This maker has been in the kitchen for several years and the owner cannot remember when the last service was performed. *There is no particular problem.*

The technician arrives and examines the machine thoroughly. The condenser appears to be dirty, and the evaporator section needs cleaning. The machine is full of ice. The technician decides to remove the ice and clean the bin at the same time.

The machine is turned off, and the power supply is shut off, locked, and tagged. The technician starts transferring the ice to trays to be moved to the walk-in cooler. As the technician gets toward the bottom of the bin, it is found to be dirty. So cleaning the bin is a good idea.

The panels to the condenser are removed. A vacuum cleaner is used to remove lint and light dust. The condenser fan motor and terminal box are covered with a plastic bag to prevent moisture from entering. An approved coil cleaner is applied to the condenser and allowed to set on the coil so it can soak into the dirty surface. While the condenser coil cleaner is working, the bin is cleaned and sanitized. When the coil cleaner has been on the condenser for about 30 min, the technician takes a water hose with a nozzle to flush the coil cleaner off the condenser and down a nearby drain. The restaurant owner is surprised at the dirt washed out of the condenser.

The water is allowed to drain off the machine and is mopped from the floor around the machine. The plastic is removed from the motor and the unit is dried with towels.

This unit has a clean cycle for cleaning the evaporator. The clean cycle is a cycle where only the water pump circulates and chemicals may be circulated with the water for the prescribed period of time. The technician pours the recommended amount of chemicals in the sump and starts the pump. The chemicals circulate for the prescribed time and the technician dumps the chemicals down the drain and flushes fresh water through the system.

The evaporator section is then sanitized using the clean cycle and the water dumped. A hose is used to wash the bin area one more time to be sure no chemicals are still present and the machine is started.

The technician allows the machine to make two harvests of ice and dumps them down the drain. This ensures there are no residual chemicals and that the machine is making proper ice.

The machine is then put back into service.

The following service calls do not include solutions. The solutions can be found in the *Instructor's Guide*.

SERVICE CALL 23

A store manager reports that the food is thawing out in the low-temperature walk-in freezer. This freezer lost its charge in the winter, the leak was repaired, and the unit recharged. The service technician remembers that this unit was serviced by a new technician when the loss-of-charge incident happened. The technician suspects that the unit may be off because of the manual reset high-pressure control. This is the only control on the unit that will keep it off, unless there were a power problem. The technician arrives at the job and finds that the unit is off because of the high-pressure control. Before resetting the control, the technician installs a gage manifold so that the pressures can be observed. When the high-pressure control is reset, the compressor starts, but the head pressure rises to 375 psig and shuts the compressor off. The system uses R-502. The ambient temperature is 90°F, so the head pressure should not be more than the pressure corresponding to 120°F. This was arrived at by adding 30°F to the ambient temperature: 30°F + 90°F = 120°F. The head pressure corresponding to 120°F is 282 psig. It is obvious the head pressure is too high. The condenser seems clean enough, and air is not recirculating back to the condenser inlet.

What is the likely problem and recommended solution?

SERVICE CALL 24

A customer calls to indicate that the medium-temperature walk-in cooler unit that was just installed is running all the time and not cooling the box.

The technician goes to the job and looks it over. This is a new installation of a walk-in medium-temperature cooler and was just started up yesterday. The hermetic

compressor is sweating down the side of the housing. The evaporator fan is operating, and the box temperature is not down to the cut-off point, so there is enough load to boil the refrigerant to a vapor.

What is the likely problem and the recommended solution?

SERVICE CALL 25

A customer reports that the low-temperature reach-in cooler that had a burned compressor last week is not cooling properly. The food is beginning to thaw.

The technician arrives and examines the whole system. The evaporator seems to be starving for refrigerant. When the technician approaches the condensing unit in the back of the store, it is noticed that the liquid line is sweating where it leaves the drier. A suction-line drier was not installed at the time of the motor burn.

What is the likely problem and the recommended solution?

SERVICE CALL 26

A customer calls on the first hot day in the spring and says the walk-in freezer is rising in temperature. The food will soon start to thaw if something is not done.

The service technician arrives, examines the whole system, and notices that the liquid line is hot, not warm. This is a TXV system and has a sight glass. It is full, indicating a full charge of refrigerant. The compressor is cutting off because of high pressure and then restarting periodically. This system uses R-502, and the head of the compressor is painted purple to signify the refrigerant type. Gages are installed, and the head pressure is starting out at 345 psig and rising to the cut-out point or 400 psig.

What is the likely problem and the recommended solution?

SERVICE CALL 27

A restaurant manager indicates that the compressor on a water-cooled unit is cutting off from time to time, *and the walk-in cooler is losing temperature. This is a medium-temperature cooler, using R-12.*

The service technician arrives to find a wastewater condenser (the water is regulated by a water-regulating valve to control the head pressure and then goes down the drain). The water is coming out of the condenser and going down the drain at a rapid rate. The liquid line is hot, not warm. The water is not taking the heat out of the refrigerant like it should. Gages are installed at the compressor, and the head pressure is 200 psig. The compressor is cutting off and on because of the high-pressure control. The head pressure should be about 125 psig with a condensing temperature of 105°F.

What is the likely problem and the recommended solution?

SERVICE CALL 28

A customer calls and says the temperature is rising in a frozen food walk-in freezer. Although it tries, the compressor will not start.

The service technician listens to the compressor try to start and hears the overload cut it off, making a clicking sound. The compressor is hot, too hot to start many more times before it will overheat to the point that it will be off for a long time. This is because every start the motor goes through makes it hotter. The technician needs to make sure that it starts and stays on the line the next time it is started. This is a single-phase compressor and has a starting relay and starting capacitor. The problem can be in several places: (1) the starting capacitor, (2) the starting relay, or (3) the compressor. The technician uses an ohmmeter to check for continuity in the compressor, from the common terminal to the run terminal, and from the common terminal to the start terminal. The compressor has continuity and should start, provided it is not locked. An ohm check of the starting capacitor shows that it will charge and discharge. (This test is accomplished by first shorting the capacitor terminals with a 20,000-Ω, 5-W resistor and then using the R × 100 scale on the ohmmeter.) The capacitor should cause the meter to rise and then fall. Reversing the leads will cause it to rise and fall again. NOTE: This check does not indicate the actual capacity of a capacitor. It indicates that the capacitor will charge and discharge.

The starting relay is checked for continuity from terminal 1 to terminal 2. This is the circuit the starting capacitor is energized through. The circuit shows no resistance; the contacts are good.

What are the likely problems and the recommended solutions?

SERVICE CALL 29

A motel maintenance person calls indicating that the ice maker on the second floor of the motel is not making ice. This ice maker is a flake ice machine. It sits outside and the temperature is hot.

The technician arrives and notices right away there is no ice in the bin. The compressor is running and hot air is coming from the condenser. Therefore, it must be refrigerating. The suction line leaving the evaporator is cold. Gages are fastened to the compressor. The suction pressure is high, but so is the discharge pressure. The machine uses R-502 and the suction pressure is 55 psig (30°F boiling temperature). No ice can be made under these conditions. The discharge pressure is 341 psig (135°F condensing temperature). It is a hot day, 97°F, but not hot enough to cause a head pressure this high. The technician wonders if the unit is overcharged or if the condenser is dirty. This could cause high head pressure, which would then cause high suction pressure and low or no ice production.

After sitting down and thinking for a minute, the technician thinks of the water circuit, then feels the water line. It is cooler than hand temperature (which is about 90°F). This is not the answer. The technician then looks at the float chamber and discovers that water is running into the chamber.

What is the likely problem and the recommended solution?

SUMMARY

- The technician should know how the equipment should sound, where it should be cool, and where it should be warm.
- All evaporators that cool air have a relationship with the air they are cooling. The coil will generally boil the refrigerant between 10°F and 20°F colder than the air entering the evaporator.
- High-temperature evaporators normally operate between liquid-refrigerant boiling temperatures of 25°F and 40°F.
- Medium-temperature evaporators normally operate between liquid-refrigerant boiling temperatures of 10°F and 25°F.
- Low-temperature food storage systems normally operate between liquid-refrigerant boiling temperatures of −15°F down to about −49°F.
- There is a relationship between entering air and condensing refrigerant. Usually refrigerant should not condense at more than 30°F higher than the entering air, and high-efficiency condensers at no more than 25°F higher.
- The automatic expansion valve maintains a constant low-side pressure. For this reason the low-side pressure does not go down during low charge operation as with a capillary tube and a TXV.
- An overcharge of refrigerant always causes an increase in head pressure.
- An increase in head pressure causes an increase in suction pressure when the TXV or the capillary tube is the metering device.
- An increase in head pressure causes more refrigerant to flow through a capillary tube. Liquid refrigerant may flood into the compressor with an overcharge.
- The automatic expansion valve responds in reverse to load changes. When the load is increased, the valve throttles back and will slightly starve the evaporator. When the load is decreased excessively, the valve is subject to overfeed the coil and may allow small amounts of liquid refrigerant to enter the compressor.
- When an inefficient evaporator is encountered, the refrigerant boiling temperature will be too low for the entering-air temperature.
- When an inefficient condenser is encountered, the condensing refrigerant temperature will be too high for the heat rejection medium.
- The best check for an inefficient compressor is to see if it will pump to capacity.
- Design load conditions can almost always be duplicated by building the head pressure up and duplicating the design suction pressure.

REVIEW QUESTIONS

1. What is the first thing a technician should know before beginning troubleshooting?
2. What is the coil-to-air temperature relationship for a refrigeration system designed for minimum food dehydration?
   A. 12 to 15°F
   B. 10 to 15°F
   C. 8 to 20°F
   D. 8 to 12°F
3. When food dehydration is not a factor, the coil-to-air temperature relationship is _____ to _____ °F.
4. How is minimum food dehydration accomplished?
5. Why is minimum dehydration not used on every job?
6. What is the coil-to-air temperature relationship for an average coil where dehydration is not a factor?
7. At what evaporator temperature does an ice maker usually begin to make ice?
   A. 35°F
   B. 32°F
   C. 25°F
   D. 20°F
8. What is the evaporator coil-to-air relationship when the compressor is not running?
9. What is meant by the term *approach* when dealing with an evaporator chilling water?
10. How can an evaporator be tested for efficiency?
11. How can a condenser be tested for efficiency?
12. How can a hermetic compressor be tested for efficiency?
13. A TXV system responds to an overcharge of refrigerant by
    A. low suction pressure.
    B. high head pressure.
    C. a flooded compressor.
    D. both A and B.
14. An automatic expansion valve system responds to an overcharge of refrigerant by
    A. low suction pressure.
    B. high head pressure.
    C. a flooded compressor.
    D. both B and C.
15. A capillary tube system responds to an overcharge of refrigerant by
    A. high suction pressure.
    B. high head pressure.
    C. a flooded compressor.
    D. all of the above.

How do the following systems respond to an undercharge?
16. TXV
17. Automatic expansion valve
18. Capillary tube

DIAGNOSTIC CHART FOR COMMERCIAL REFRIGERATION

Commercial refrigeration equipment that controls the storage temperatures for perishable products must be kept operating properly to prevent spoilage. The technician must restore the system back to good working order in the least amount of time. Orderly troubleshooting is the most effective way to arrive at sound conclusions in a minimum of time. Following are some recommendations to follow for commercial refrigeration systems:

1. Listen to the customer's description of the system's symptoms. Most customers of commercial refrigeration equipment monitor the system and the temperatures carefully. Question the customer about sounds, cycle times, and temperatures on normal days and ask what is different now.

2. Observation may be one of the most important steps you can follow. Inspect the complete system before making any adjustments or repairs. Check the condenser for dirt, leaves, or trash in the coil or fan. Look for blocked airflow. Inspect the wiring for discoloration of connections. Look for loose or missing insulation on the suction line or kinks or flattened tubing. Check the evaporator for air blockage due to frost, ice, or dirt. Make sure the fan is circulating air.

3. Once the customer has been consulted and the system has been observed, proceed and determine what is functioning and what is not functioning. The following chart will help to determine possible causes of the problem.

DIAGNOSTIC CHART FOR COMMERCIAL REFRIGERATION (Continued)

Problem	Possible Cause	Possible Repair	Heading Number
Compressor will not start or attempt to start, makes no sound	Open disconnect switch	Close disconnect switch.	15.1, 15.2, 15.7
	Open fuse or breaker	Replace fuse or reset breaker and determine why it opened.	15.1, 15.2, 15.7
	Tripped overload	Reset and determine why it tripped.	19.6
	Pressure control stuck in open position	Reset or replace and determine why it opened.	25.16, 25.20
	Open internal or external overload	Allow to cool down.	
	Open coil on motor starter	Check coil and replace if needed.	
	Faulty wiring	Repair or replace faulty wiring connectors.	20.13
Compressor will not start, hums and trips on overload	Incorrect wiring, **Single Phase**	Check common, run, and start connections.	17.17, 17.18, 17.19, 20.6, 20.7, 20.8, and 20.9
	High discharge pressure	Overcharge of refrigerant, not enough condenser cooling, closed discharge service valve.	29.5, 29.6, 29.7 29.8, and 29.9
	Incorrect wiring, **Three Phase**	Check all three phases, phase-to-phase.	17.7, 17.16
	Start relay wired wrong or defective, **Single Phase**	Rewire relay if needed, or replace if defective.	17.18, 17.19, 17.20
	Start or run capacitor defective	Replace capacitor.	17.12, 17.13, 17.14, 20.11
	Compressor winding open or shorted	Replace compressor.	17.20
	Internal compressor problems, stuck or tight	Replace compressor.	Unit 29 Service Call 19
Compressor starts but stays in start, **Single Phase**	Wired incorrectly	Check wiring as compared with diagram.	17.18, 17.19
	Low-voltage supply	Correct low voltage.	17.17
	Defective start relay	Replace relay.	17.18, 17.19
	Tight or binding compressor	Replace compressor.	Unit 29 Service Call 19
Compressor starts and runs for a short time and then shuts off due to overload	Defective overload	Check actual load on overload and replace if it trips below specifications.	19.6
	Excess current in overload circuit	Check circuit for added load, such as fans or pumps.	19.6
	Low voltage	Determine reason and correct.	17.7
	Unbalanced voltage, **Three Phase**	Correct unbalance, possible redistribution of loads.	48.46
	Defective run capacitor	Replace.	Unit 29 Service Call 19
	Excessive load on compressor A. HIGH SUCTION PRESSURE: defrost heater on all the time	Check and repair defrost control circuit.	29.1, 29.2, 29.3, 29.4
	B. Hot food placed in cooler	Operator education.	
	C. HIGH DISCHARGE PRESSURE: Recirculating condenser air	Move object causing recirculation.	29.6, 29.14
	Reduced airflow	Remove blockage, make sure all fans are on.	29.6, 29.14
	Dirty condenser	Clean condenser.	29.6
	Compressor windings shorted	Replace compressor.	20.6, 20.7, 20.8 20.9, 20.10, 20.11

(continued)

Problem	Possible Cause	Possible Repair	Heading Number
Compressor starts but short cycles	Overload protector shutting compressor off	See above category.	
	Thermostat	Differential too close, thermostat setting; bulb in cold air stream, move location.	29.1, 29.2, 29.3, 29.4
	High-pressure condition	Check airflow or water flow.	29.6, 29.14
		Air blockage or recirculating, correct.	29.6, 29.7, 29.8, 29.9
		Overcharge, remove refrigerant.	9.15
		Air in system, remove.	29.6
	Low-pressure condition with pumpdown control		25.19
	A. Solenoid valve leaking	Replace solenoid.	25.15
	B. Compressor discharge valve leaking during off cycle (will also leak while running)	Repair valve or replace compressor.	29.17, 29.19
	With or without pumpdown control		
	A. Low refrigerant charge	Repair leak and add refrigerant.	29.11
	B. Restriction in system such as expansion valve, drier, or crimped pipe	Repair or replace.	29.15
Compressor runs continuously or space temperature is too high	Low refrigerant charge	Repair leak, add refrigerant.	29.15
	Operating control (thermostat or low-pressure control) contacts stuck closed	Replace control.	
	Excessive load		
	A. Unit undersized	A. Reduce load or replace unit.	29.1, 29.2, 29.3, 29.5
	B. Door of cooler open too often	B. Train operator.	29.13
	C. Product too hot when placed in cooler	C. Precool food in air-conditioned space.	26.26
	D. Poor door gaskets or infiltration	D. Seal air leaks.	26.16
	E. Cooler setting in too hot a location	E. Move or reduce heat.	
	Evaporator coil not defrosting	Check defrost method and repair.	25.23, 25.24, 25.25, 25.26, 25.27, 25.28 25.29, 25.30
	Restriction in refrigerant circuit, drier, metering device, crimped pipe	Find and repair.	29.15
	Poor condenser performance		
	A. Restricted airflow	Restore airflow.	29.14
	B. Recirculated air	Move objects causing recirculation.	29.14
	C. Dirty condenser	Clean condenser.	29.14

SECTION 6

Air Conditioning (Heating and Humidification)

UNIT 30 Electric Heat

OBJECTIVES

After studying this unit, you should be able to

- ■ discuss the efficiency and relative operating costs of electric heat.
- ■ list types of electric heaters and state their uses.
- ■ describe how sequencers operate in electric forced-air furnaces.
- ■ trace the circuitry in a diagram of an electric forced-air furnace.
- ■ perform basic tests in troubleshooting electrical problems in an electric forced-air furnace.
- ■ describe typical preventive maintenance procedures used in electric heating units and systems.

SAFETY CHECKLIST

- ✔ Be careful that nails or other objects driven into or mounted on radiant heating panels do not damage the electrical circuits.
- ✔ Any heater designed to have air forced across the element should not be operated without the fan.
- ✔ Always observe all electrical safety precautions.

30.1 INTRODUCTION

Electric heat is produced by converting electrical energy to heat. This is done by placing a known resistance of a particular material in an electrical circuit. The resistance has relatively few free electrons and does not conduct electricity easily. The resistance to electron flow produces heat at the point of resistance. One type of material used for this resistance is a special wire called *nichrome,* for nickel chromium.

Electric heat is efficient but is more expensive to operate compared with many sources of heat. It is efficient because very little electrical energy is lost from the meter to the heating element. It is expensive because it takes large amounts of electrical energy to produce the heat, and the cost of electrical energy in most areas of the country can be expensive compared to fossil fuels (coal, oil, and gas).

The purchase price of electrical heating systems is usually less than other systems. The installation and maintenance is also usually less expensive. This makes electric heating systems attractive to many purchasers.

When electric heat is used as the primary heat source, a high value of insulation is normally used throughout the structure to lower the amount of heat required.

This unit briefly describes several types of electric heating devices. Emphasis, however, is placed on central forced-air electric heat because it is frequently serviced by technicians in this industry in some areas of the country.

30.2 PORTABLE ELECTRIC HEATING DEVICES

Portable or small space heaters are sold in many retail stores and by many industrial distributors and manufacturers, **Figure 30–1.** Some have glowing coils (due to the resistance of the wire to electron flow). These transfer heat by radiation (infrared rays) to the solid objects in front of the heater. The radiant heat travels in a straight line and is absorbed by solid objects that warm the space around them. Radiant heat also provides heating comfort to individuals. The heat concentration decreases by the square of the distance and is soon dissipated into the space, **Figure 1–13.** Quartz and glass panel heaters are also used to heat small spaces by radiation, **Figure 30–2.**

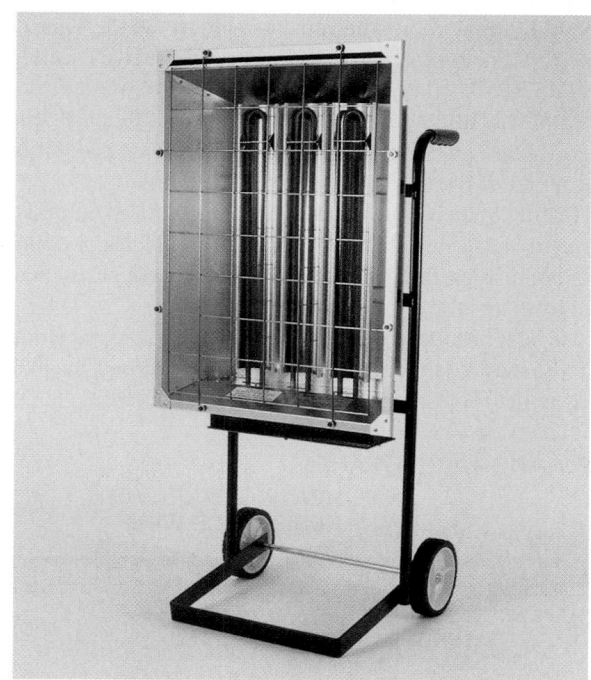

Figure 30–1 A portable electric space heater. *Courtesy Fostoria Industries, Inc.*

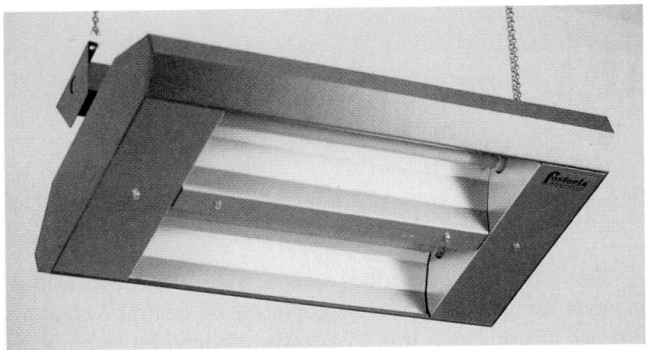

Figure 30–2   A quartz heater. *Courtesy Fostoria Industries, Inc.*

Other space heaters use fans to move air over the heating elements and into the space. This is called forced-convection heat because the heat-laden air is moved mechanically. The units may be designed to move enough air across the heating elements so that they do not glow.

Radiant spot heating can be effectively used at doorways, warehouses, work areas, and even outdoors. The effect is much like a sunlamp pointing at the heated area. Distance has a great bearing on the effectiveness of radiant heating.

30.3   RADIANT HEATING PANELS

Radiant electric heating panels are used in residential and light commercial buildings. The panel is often made of gypsum board with wire heating circuits running throughout. The panels are usually installed in ceilings and controlled with individual room thermostats, **Figure 30–3.** This provides room or zone control. SAFETY PRECAUTION: Be careful that nails or other objects driven into or mounted on the panels do not damage the electrical circuits. Be particularly careful when installing a cooling system when radiant heat is provided if the outlets penetrate the ceiling.

Heating panels must have good insulation behind (or above) to keep heat from escaping. The electrical connections are also on the back, so the junctions are easily accessible from an attic.

The heat produced is even, is easy to control, and tends to keep the mass of the room warm by its radiation. This makes items in the room pleasingly warm to the touch.

INSULATION ABOVE PANEL

HEATING PANEL WITH RESISTANCE WIRE

LINE-VOLTAGE THERMOSTAT

Figure 30–3   A radiant ceiling heating panel.

30.4   ELECTRIC BASEBOARD HEATING

Baseboard heaters are popular convection heaters used for whole house, spot, or individual room heating, **Figure 30–4.** They are economical to install and can be controlled by individual room thermostats. These thermostats are normally line-voltage thermostats. Unused rooms can be closed off

(A)

(B)

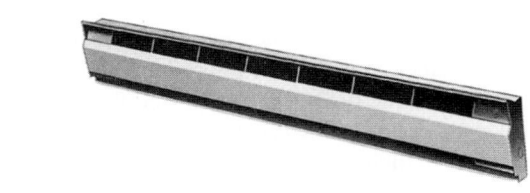

(C)

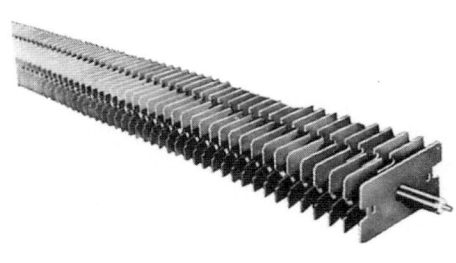

(D)

Figure 30–4   **(A)** A wall thermostat. **(B)** A thermostat built into a baseboard unit. **(C)** An electric baseboard heater. **(D)** A finned element used in an electric element used in an electric baseboard heater. **(A)** *Photo by Bill Johnson.* **(B)–(D)** *Courtesy W. W. Grainger, Inc.*

and the heat turned down, which makes baseboard heaters economical in certain applications.

Baseboard units are mounted on the wall just above the floor or carpet. Outside walls are usually used. The heater is a natural-draft unit. Air enters near the bottom, passes over the electric element, is heated, and rises in the room. As the air cools, it settles, setting up a natural-convection air current.

Baseboard heat is easy to control, safe, quiet, and evenly distributed throughout the house.

30.5   UNIT HEATERS

Unit heaters are usually suspended from the ceiling and use a fan to force the air across the elements into the space to be heated, **Figure 30–5.** These heaters are controlled with line-voltage thermostats. SAFETY PRECAUTION: Any heater designed to have air forced across the element should not be operated without the fan operating.

30.6   ELECTRIC HYDRONIC BOILERS

The electric hydronic (hot water) boiler system is used for some residential and light commercial applications. Except for the control arrangement and safety devices, the boiler is somewhat like an electric domestic water heater. It is also connected to a closed loop of piping and requires a pump to move water through the loop consisting of the boiler and the terminal heating units, **Figure 30–6.**

The electric boiler is small and compact for easy location and installation. It is relatively easy to troubleshoot and repair.

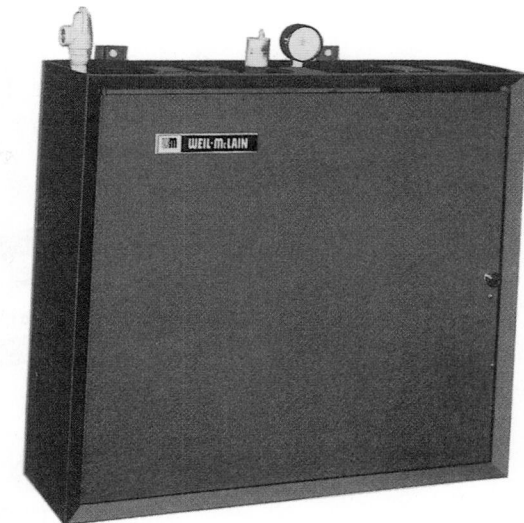

Figure 30–6   An electric hydronic boiler. *Courtesy Wiel-McLain/A United Dominion Company*

Any boiler handling water is subject to all the problems of a water-circulating system. For instance, if the boiler shuts down and the room temperature drops below freezing, pipes will burst, and the boiler will be damaged; water treatment against scale and corrosion is also needed in a boiler installation.

The boiler itself is efficient because it converts virtually all of its electrical energy input into heat energy and transfers it to the water. When the boiler is located in the conditioned space, any heat loss through the boiler's walls or fittings is not lost from the heat structure.

This system is quiet and reliable, but cooling and humidification systems cannot easily be added.

30.7   CENTRAL FORCED-AIR ELECTRIC FURNACES

Central forced-air furnaces are used with ductwork to distribute the heated air to rooms or spaces away from the furnace. The heating elements are factory installed in the furnace unit with the air-handling equipment (fan), or they can be purchased as duct heaters and installed within the ductwork, **Figure 30–7** and **Figure 30–8.**

Central heaters are usually controlled by a single thermostat, resulting in one control point—no individual room or

Figure 30–5   An electric unit heater. *Reznor-Thomas & Belts Corp., HVAC Div.*

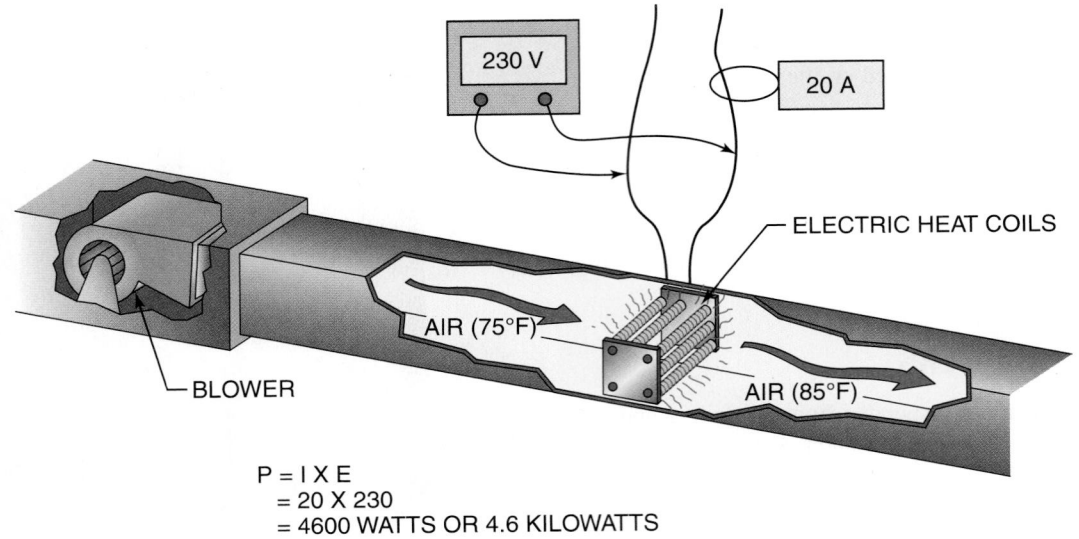

$$P = I \times E$$
$$= 20 \times 230$$
$$= 4600 \text{ WATTS OR } 4.6 \text{ KILOWATTS}$$

Figure 30–7 This electric duct heater raises the temperature of the air in the duct.

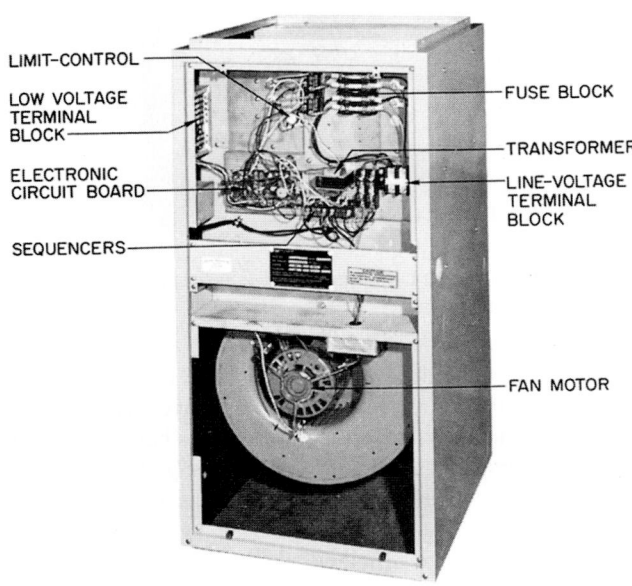

Figure 30–8 A central forced-air electric furnace with multiple sequencers. *Courtesy The Williamson Company*

zone control exists. An advantage with duct heaters is that temperature in individual rooms or zones can be controlled with multiple heaters placed in the ductwork system. However, an interlock system must be incorporated. The interlock system will not allow power to reach the heating element unless the fan is operating, which keeps the heating elements from overheating and possibly causing a fire.

The heating elements are made of nichrome resistance wire mounted on ceramic or mica insulation. They are enclosed in the ductwork or the furnace housing in an electric furnace.

Air conditioning (cooling and humidification) can usually be added because of the air-handling feature.

SAFETY PRECAUTION: Be careful when servicing these systems because many exposed electrical connections are behind the inspection panels. Electric shock can be fatal.

30.8 AUTOMATIC CONTROLS FOR FORCED-AIR ELECTRIC FURNACES

Automatic controls are used to maintain temperature at desired levels in given spaces and to protect the equipment and occupant. Three common controls used in electric heat applications are *thermostats, sequencers* (discussed later in this unit), and *contactors (or relays)*. **Figure 30–9** shows a typical thermostat.

30.9 THE LOW-VOLTAGE THERMOSTAT

The low-voltage thermostat is used for sequencers and contactors because it is compact, very responsive, safe, and easy to install and troubleshoot. The low-voltage wiring may be installed without an electrical license in many localities.

Figure 30–9 A low-voltage thermostat. *Photo by Bill Johnson*

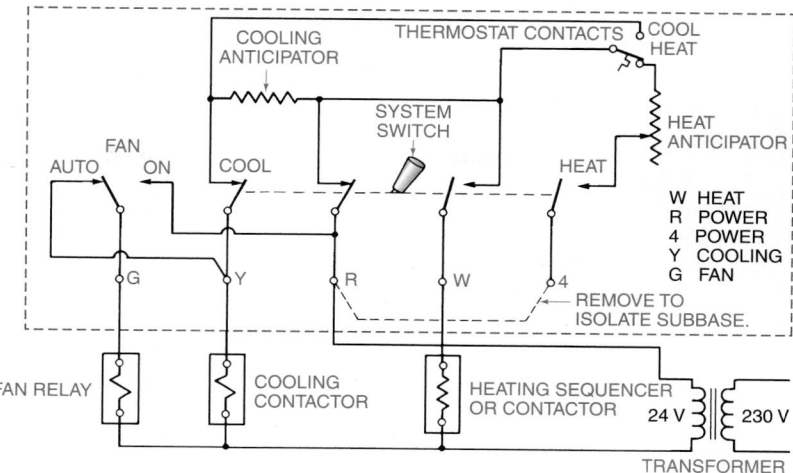

Figure 30–10 A wiring diagram of a low-voltage thermostat. Some manufacturers use different terminal designations.

Figure 30–10 shows a wiring diagram of a low-voltage thermostat typical of those used with an electric forced-air heating furnace.

The thermostat has an isolated subbase that allows two power supplies to be run to the thermostat. This isolated subbase may be needed when air conditioning is added after the furnace is installed. The furnace will have its low-voltage power supply, and the air-conditioning unit may have its own low-voltage power supply. If one power supply is used for both, a jumper is required from the R terminal to the 4 terminal. This is shown in **Figure 30–10** by a dotted line. The R terminal is the cooling terminal, and the 4 terminal is assigned to heating if the subbase has two power supplies.

The *heat anticipator* used in a thermostat with electric heat must be set at the time of the installation. The setting is determined by adding all the current draw in the 24-V heating circuit that passes through the thermostat control and heat anticipator. When this current is determined, it is then set with the indicator on the heat anticipator. For example, if the current passing through terminal 4 is 0.75 A, this number is used to set the heat anticipator in the low-voltage thermostat subbase. An ammeter with a ten-wrap multiplier may be used to determine the amperage. This is explained in **Figure 12–52.**

Most sequencers will have the ampere load of the low-voltage circuit printed on the sequencer. If there is more than one, they can be added. For example, for three sequencers, each with a heater load of 0.3 A, the heat anticipator is set on 0.3 + 0.3 + 0.3 = 0.9 A.

30.10 CONTROLLING MULTIPLE STAGES

Most electric heating furnaces have several heating elements activated in stages to avoid putting a high-power load in service all at once. Some furnaces may have as many as six heaters to be connected to the electrical load at the proper time. The *sequencer* is used to do this. It uses low-voltage control power to start and stop the electric heaters. A sequencer can be described as a heat motor type of device. It uses a bimetal strip with a low-voltage wire wrapped around

it. When the thermostat calls for heat, the low-voltage wire heats the strip and warps it out of shape in a known direction for a known distance. As it warps or bends, it closes electrical contacts to the electric heat circuit. This bending takes time, and each set of contacts closes quietly and in a certain order with a time delay between the closings. When the heat requirement has been met, the thermostat opens the low-voltage circuit, and the steps are reversed.

See **Figure 30–11** for a diagram of a package sequencer. This sequencer can start or stop three stages of strip heat. Some sequencers have five circuits, **Figure 30–12**: three heat circuits, a fan circuit, and a circuit to pass low-voltage power to another sequencer for three more stages of heat. There are five sets of contacts, and none is in the same circuit.

Another sequencer design, called an *individual sequencer,* has only a single circuit that can be used for starting and stopping an electric heat element, but it could have two other circuits: one to energize another sequencer through a set of low-voltage contacts and one for starting the fan motor. Several stages of electric heat may be controlled with several of these sequencers, one for each heat strip.

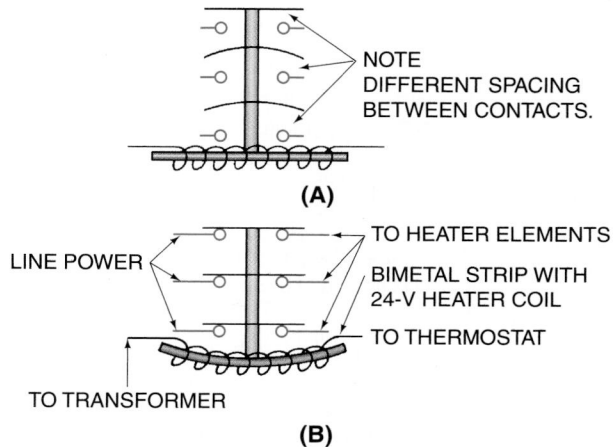

Figure 30–11 A sequencer with three contacts. **(A)** OFF position. **(B)** ON position.

Figure 30–12 A multiple-type sequencer. *Photo by Bill Johnson*

30.11 WIRING DIAGRAMS

Individual manufacturers vary in how they illustrate electrical circuits and components. Some use pictorial, some schematic, and others use both types of diagrams. The *pictorial* type shows the location of each component as it actually appears to the person installing or servicing the equipment. When the panel door is opened, the components inside the control box are in the same location as on the pictorial diagram.

The *schematic wiring diagram* (sometimes called the *line wiring diagram* or *ladder* type) shows the current path to the components. Schematic diagrams help the technician to understand and follow the intent of the design engineer.

Figure 30–13 contains a legend of electrical symbols in an electric heating circuit. Such a legend is vital in following a wiring diagram. Many symbols are standardized throughout the industry, even throughout the world.

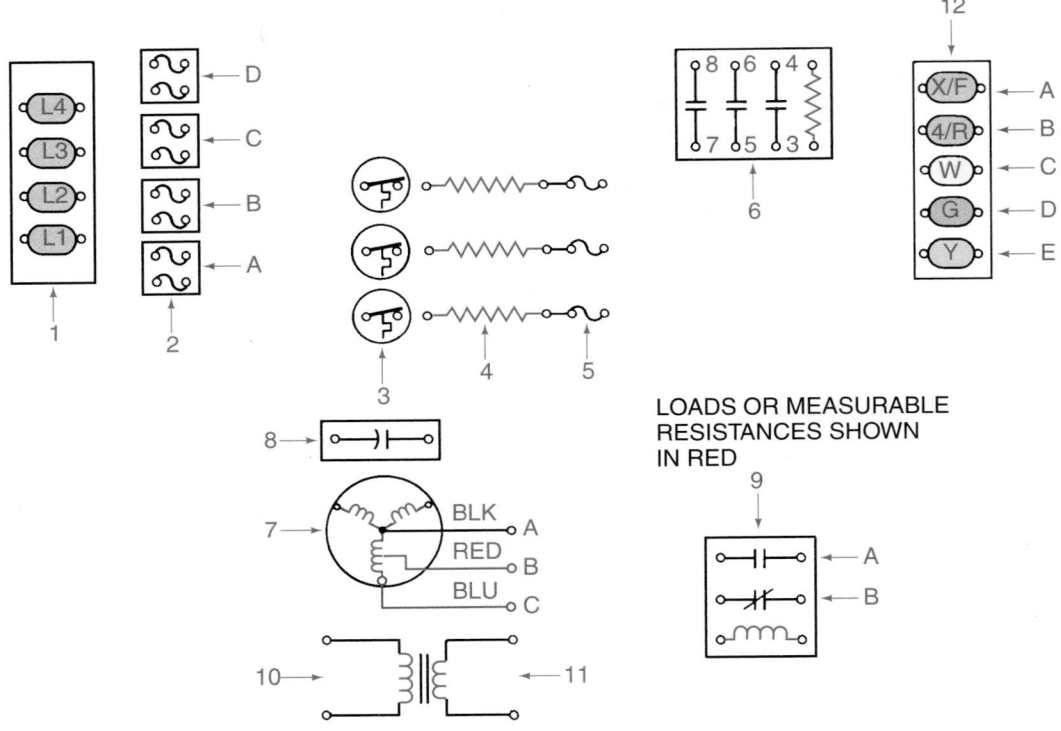

1. LINE-VOLTAGE TERMINAL BLOCK
   FROM DISCONNECT
2. FUSE BLOCK
   A. L1 FUSE BLOCK
   B. L2 FUSE BLOCK
   C. L3 FUSE BLOCK
   D. L4 FUSE BLOCK
3. AUTOMATIC RESET LIMIT SWITCH
4. ELECTRIC HEAT ELEMENT
5. FUSIBLE LINK
6. SEQUENCER
7. FAN MOTOR
   A. HIGH-SPEED MOTOR LEAD
   B. MEDIUM-SPEED MOTOR LEAD
   C. LOW-SPEED MOTOR LEAD

8. FAN MOTOR CAPACITOR
9. FAN RELAY (DOUBLE-POLE, SINGLE-THROW)
   A. NORMALLY OPEN CONTACTS
   B. NORMALLY CLOSED CONTACTS
10. PRIMARY SIDE OF TRANSFORMER (LINE VOLTAGE)
11. SECONDARY SIDE OF TRANSFORMER
    (LOW VOLTAGE—24 V)
12. LOW-VOLTAGE TERMINAL BLOCK
    A. COMMON TERMINAL FROM TRANSFORMER
    B. TERMINAL FOR HOT LEAD FROM
       TRANSFORMER
    C. HEATING
    D. FAN RELAY FOR COOLING
    E. COOLING

Figure 30–13 Components of a pictorial diagram.

30.12 CONTROL CIRCUITS FOR FORCED-AIR ELECTRIC FURNACES

The low-voltage control circuit safely and effectively controls the heating elements that do the work, and they pull the most current. The circuit contains devices for safety and for control. For example, the limit switch, a safety device, shuts off the unit if high temperature occurs; the room thermostat, control, or operating device stops and starts the heat based on room temperature.

Safety and operating devices can consume power and do work or pass power to a power-consuming device. For example, the sequencer heater coil that operates the bimetal and a magnetic solenoid that moves the armature in a contactor are power-consuming devices in the low-voltage circuit. The contacts of a contactor or a limit control pass power to the power-consuming devices and are in the high-voltage circuit. **Power-passing devices are wired in series with the power-consuming devices. Power-consuming devices are wired in parallel with each other.**

Figure 30–14 is a diagram of a low-voltage control circuit. By tracing this circuit, you can see that when the thermostat contacts are closed, a circuit is completed. The transformer is the power source. The common leg of power in the diagram is blue and furnishes power to the top side of the sequencer coil. The red leg of power furnished power to the thermostat and on to the bottom of the sequencer coil. When the room cools down to the set point of the thermostat, the thermostat contacts close.

This in turn activates the contacts in the sequencer, which activate the heating elements in the high-voltage or line-voltage circuit. The line to the G terminal is used for the cooling circuit and has no other purpose. The heating elements and the fan circuit have been omitted to simplify the circuit.

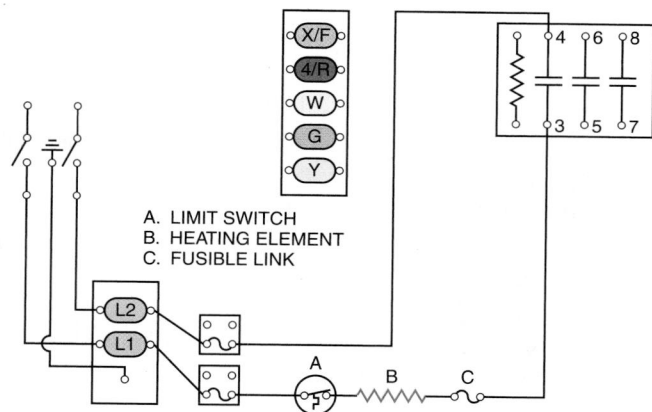

A. LIMIT SWITCH
B. HEATING ELEMENT
C. FUSIBLE LINK

Figure 30–15 A single-heating-element circuit with sequencer (high voltage).

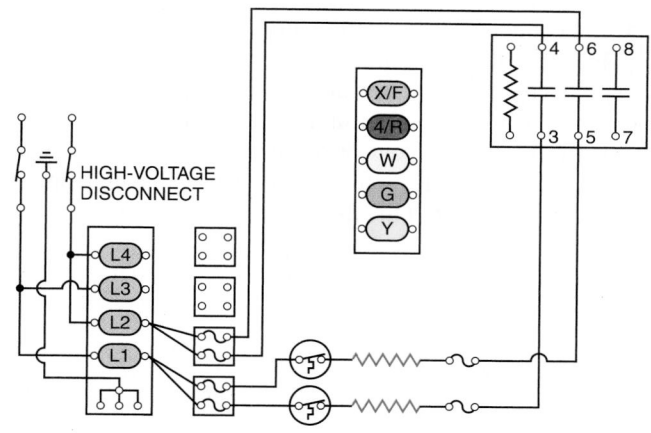

Figure 30–16 Double-heating-element circuits.

Single-heating-element control is illustrated in **Figure 30–15.** The electrical-current from the L1 fuse block goes directly to the limit switch. This is a temperature-actuated switch (A) that opens under excessive heat and provides protection to the furnace. It is usually an automatic reset switch that closes when the temperature cools. It is wired to the heating element (B) and to a fusible link (C) that provides additional protection from overheating of the elements. Under higher temperatures this link melts and must be replaced. The link is wired to terminal 3 of the sequencer with the circuit completed through L2 when the sequencer contact is closed.

Figure 30–16 presents a diagram of a furnace with two heating elements. **Figure 30–17** is an example of one with three elements. Each of these examples is using the same package sequencer as shown in **Figure 30–12.**

30.13 FAN MOTOR CIRCUITS

The fan motor, a power-consuming device, that forces the air over the electric heat elements, must be started and stopped at the correct time. **Figure 30–18** is an example of the fan

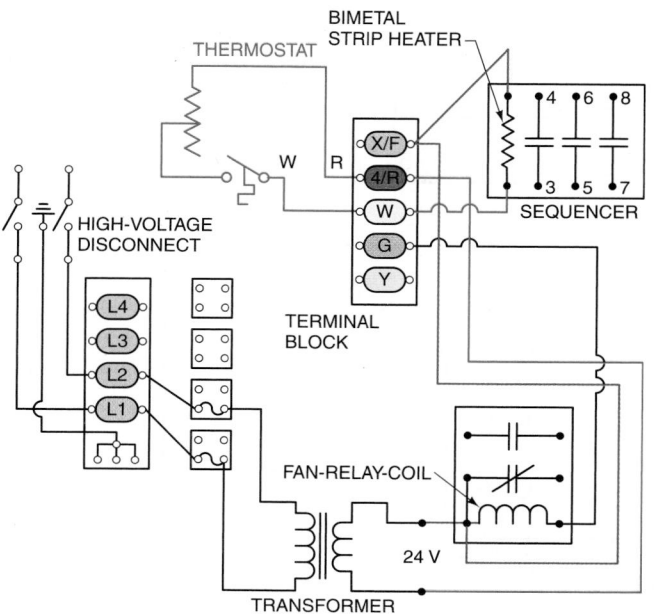

Figure 30–14 A diagram of a low-voltage control circuit.

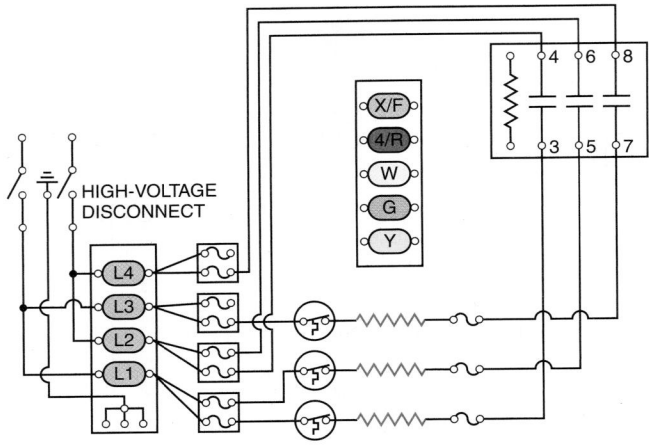

Figure 30–17   Triple-heating-element circuits.

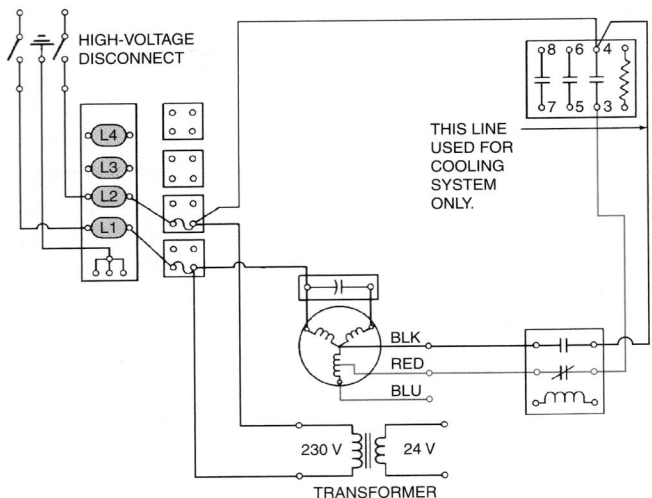

Figure 30–18   The fan wiring circuit.

wiring circuit. NOTE: It must run before the furnace gets too hot and continue to run until the furnace cools down.

Note that the L1 terminal is wired directly to the fan motor and that the L2 terminal is wired directly to terminal 4 on the sequencer. From terminal 4 a circuit is made to the normally open (NO) contact on the fan relay. This circuit could have been made directly to L2. This is the high-speed fan circuit used to start the fan in the cooling mode.

Power is passed through terminal 4 to terminal 3 when the sequencer is energized long enough for the bimetal to bend and close the contacts. The power then passes to the normally closed (NC) terminal on the fan relay and on to the slow-speed winding required for heating. **Figure 30–19** combines **Figure 30–17** and **Figure 30–18** into a line diagram.

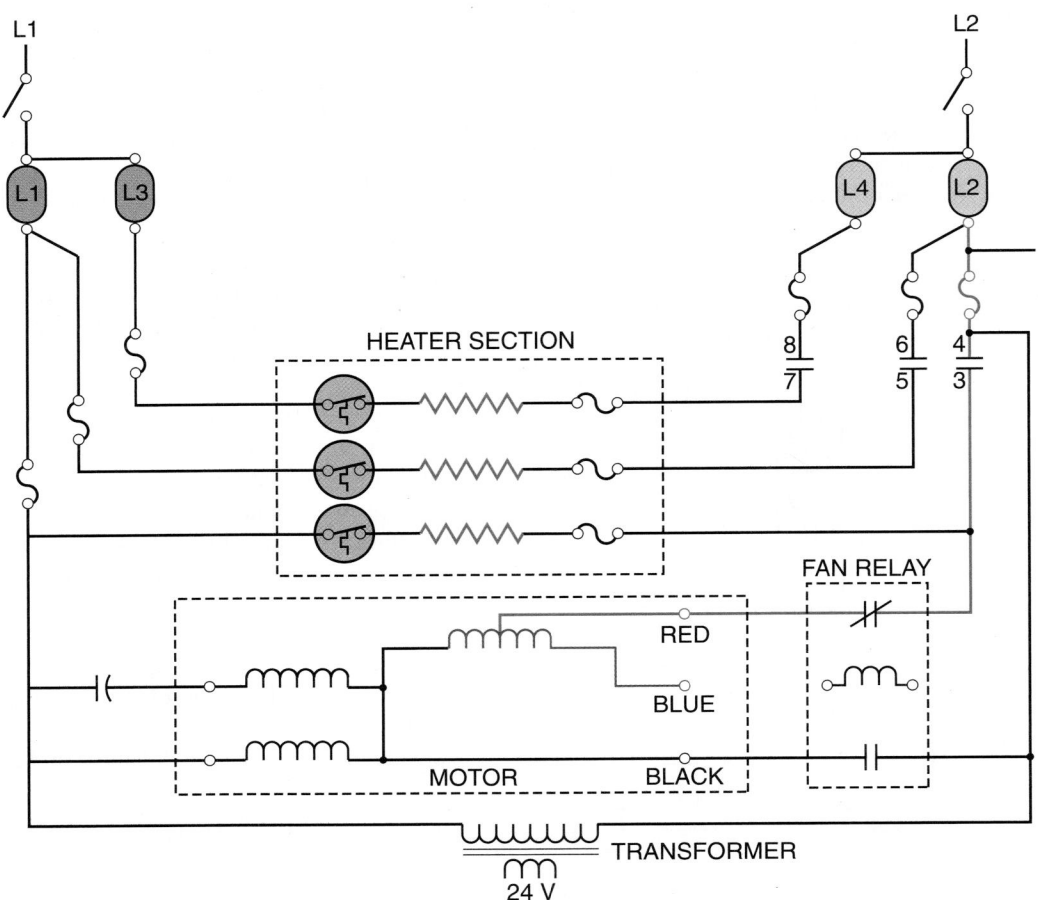

Figure 30–19   A simplified line drawing combining components in **Figure 30–17** and **Figure 30–18**.

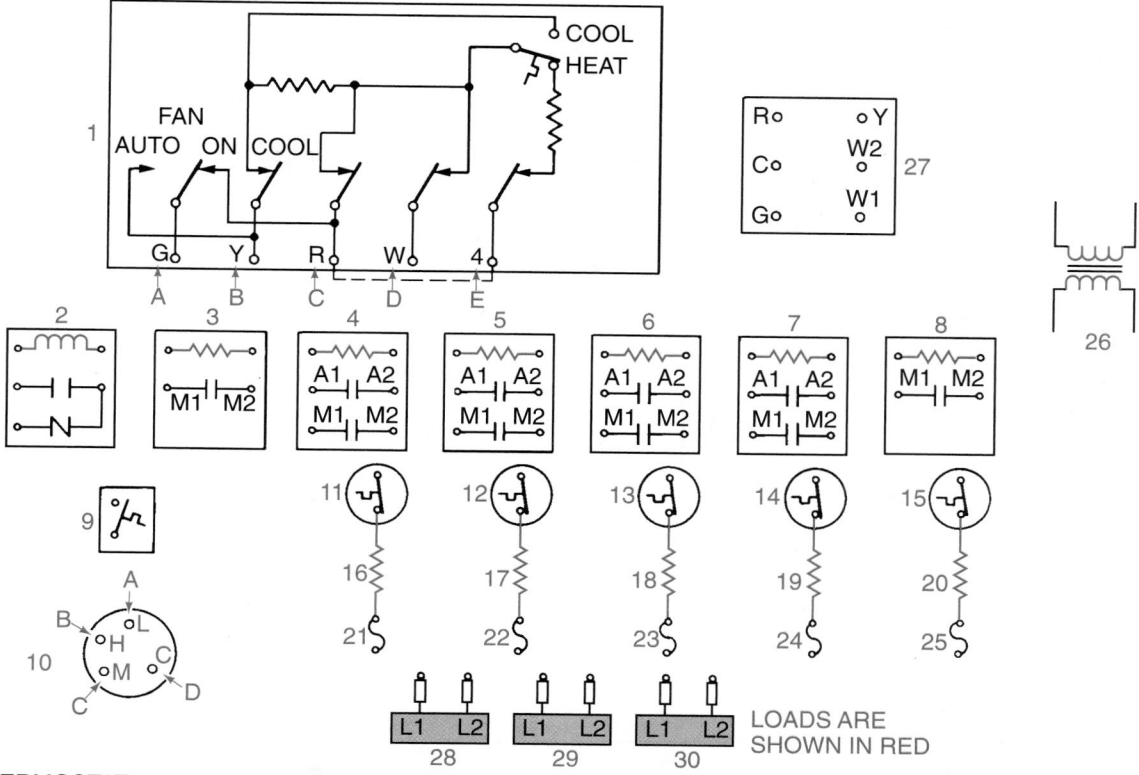

1. THERMOSTAT
   A. G TERMINAL (TO FAN RELAY)
   B. Y TERMINAL (TO COOLING CONTACTOR)
   C. R TERMINAL (FROM TRANSFORMER SECONDARY)
   D. W TERMINAL (TO HEATING SEQUENCER)
   E. 4 TERMINAL (RED OR JUMPED)
2. FAN RELAY (FOR COOLING)
3. TIMED FAN CONTROL
4. HEAT SEQUENCER I
5. HEAT SEQUENCER II
6. HEAT SEQUENCER III
7. HEAT SEQUENCER IV
8. HEAT SEQUENCER V
   (A1 AND A2 TERMINALS FOR 24 VOLTS ONLY. THESE
   ARE FOR AUXILIARY SEQUENCER HEATER STRIPS
   THAT ENERGIZE NEXT SEQUENCER. M1 AND M2
   ARE TERMINALS FOR HEATING ELEMENTS.)

9. TEMPERATURE-ACTUATED FAN CONTROL
10. FAN MOTOR
    A. LOW SPEED
    B. HIGH SPEED
    C. MEDIUM SPEED
    D. COMMON
11-15. TEMPERATURE-ACTUATED LIMIT SWITCHES
       (AUTOMATIC RESET)
16-20. ELECTRIC HEATING ELEMENTS
21-25. FUSIBLE LINKS
   26. PRIMARY SIDE OF TRANSFORMER
   27. LOW-VOLTAGE TERMINAL BOARD
28-30. LINE TERMINAL BLOCKS

Figure 30–20   The electrical legend for **Figure 30–21**.

Remember that the sequencer contacts and the relay contacts *pass* power and do not consume power. They are wired in series.

Figure 30–20 is an example legend of terminal designations used on electric forced-air furnaces that use multiple sequencers instead of one package sequencer.

Figure 30–21 is a wiring diagram for a forced-air furnace with individual sequencers. Note the low-voltage circuit wiring sequence:

1. Power is wired directly from the transformer to terminal C and on to all power-consuming devices. No power-passing devices are in this circuit. This can be called the common circuit.

2. Power is wired directly from the other side of the transformer to terminal R and then on to the R terminal on the room thermostat. The room thermostat passes (or distributes) power to the respective power-consuming device circuits on a call for HEAT, COOL, or FAN ON.

3. On a call for heat, the thermostat energizes the W terminal and passes power to the timed fan control and the first sequencer. The first sequencer passes power to the next sequencer heater coil and then to the next.

See **Figure 30–21** for the high-voltage sequence:

1. Although there are six fuses, they are labeled either L1 or L2, meaning that there are only two lines supplying power to the unit. The six fuses break the power-consuming heaters and fan motor down into smaller amperage increments.

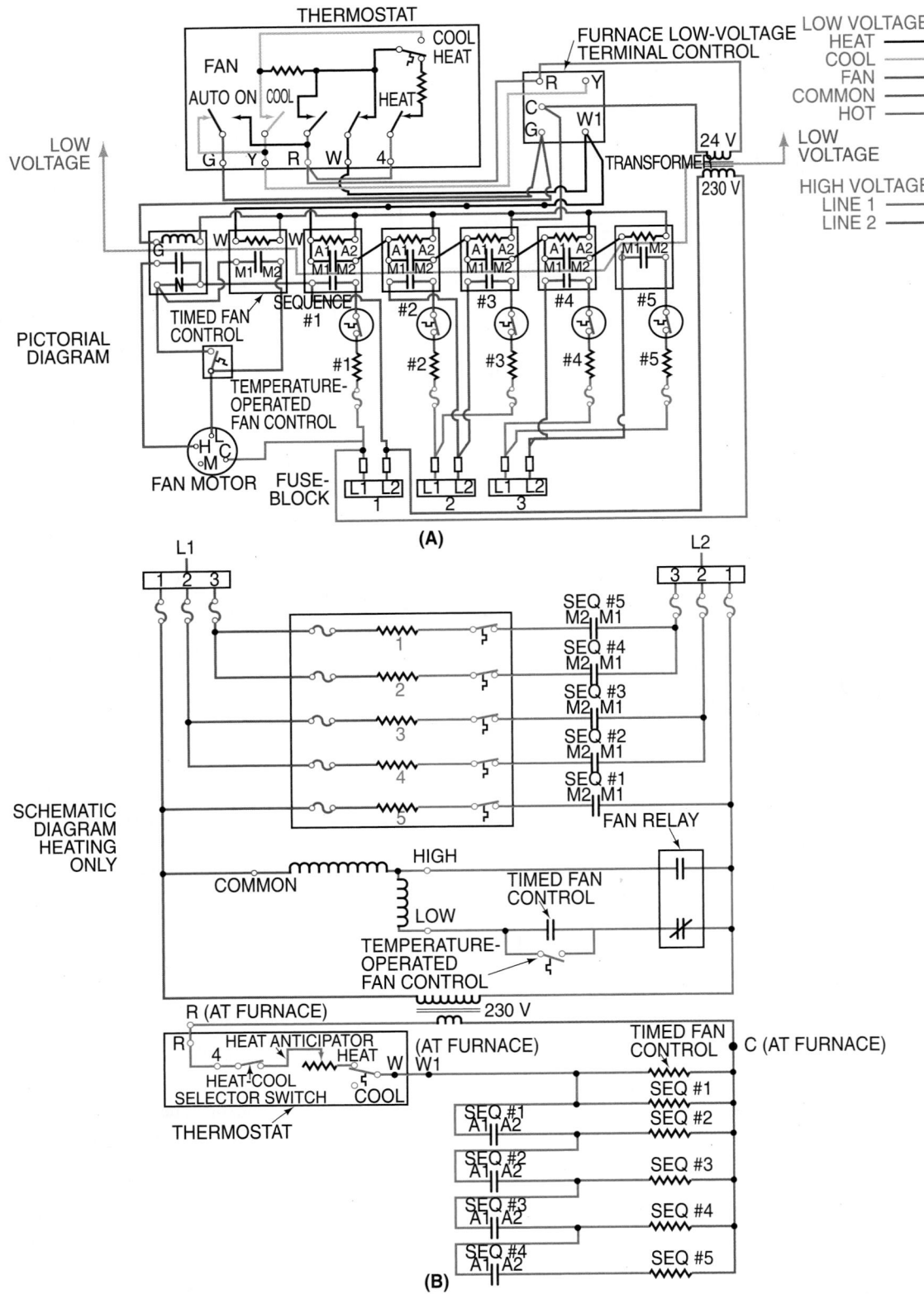

Figure 30–21 **(A)** The wiring diagram for an electric furnace. **(B)** The line diagram.

2. On a call for heat, the contacts in sequencer 1 close and pass power from L2 to the limit switch and on to heater 1. At the same time, the timed fan control is energized and its contacts close, starting the fan in low speed.

3. When the sequencer contacts close for the high-voltage load, the low-voltage contacts (A1 to A2) also close and pass power in the low-voltage circuit to the next sequencer's operating coil, which sets up the same sequence for the second heater. This continues down the line until all sequencer contacts are closed and all the heat is on.

NOTE: This sequence may take 20 sec or more per sequencer, and there are five heat sequencers for the heat to be completely on. When the room thermostat is satisfied, it only takes about 20 sec to turn off the heat because the power to all sequencers is interrupted by the room thermostat. The temperature-operated fan switch takes command of the fan at this time and keeps the fan on until the heaters cool.

The fan circuit is a circuit by itself. The fan will start in low speed if the fan relay is not energized. If the fan is running in slow speed and someone turns the fan switch to ON, the fan will switch to high speed. It cannot run in both speeds at the same time due to the relay with NO and NC contacts.

30.14 CONTACTORS FOR CONTROLLING ELECTRIC FURNACES

Using contactors (or relays) to control electric heat means that all of the heat will be started at one time unless separate time delay relays are used. Contactors are magnetic and make noise. They snap in and may hum. Contactors are normally used on commercial systems where they may not be directly attached to the ductwork. This prevents the noise from traveling through the system, **Figure 30–22**. The contactor magnetic coils are 24 V and 230 V. The 24-V coils are controlled by the room thermostat, and they energize the 230-V coil. There are four contactors and their coils in this diagram for the electric heat and one for the fan. The 24-V coils are energized and pull in together.

30.15 AIRFLOW IN ELECTRIC FURNACES

The airflow in cubic feet per minute (cfm) may be verified in an electric heating system using the following formula and example. This formula is often called the sensible heat formula and is restated as it is seen in many other text calculations as

$$Q_s = 1.08 \times \text{cfm} \times \text{TD}$$
$$Q_s = \text{Sensible heat in Btu/h}$$
$$1.08 = \text{A constant}$$
$$\text{cfm} = \text{Cubic feet per minute}$$
$$\text{TD} = \text{Temperature difference across furnace}$$

The formula is restated and solved for cfm:

$$\text{cfm} = \frac{Q_s}{1.08 \times \text{TD}}$$

The total heat (in watts) added to the airstream may be calculated by taking the total amperage to the electric furnace and multiplying it by the applied voltage. For example, a furnace ampere draw is 85 A. At the same time the applied voltage is 208 V, and the temperature rise across the furnace is 50°F.

$$\text{Watts} = \text{Amperes} \times \text{Volts}$$
$$\text{Watts} = 85 \times 208$$
$$\text{Watts} = 17,680$$

NOTE: This calculation includes the fan motor, which is treated like a resistance load. This will give a slight error on the motor calculation. The motor would actually consume about 200 W less power if measured with a wattmeter. Because this is a fraction of a percent error, it is ignored.

The watts must be converted to Btu/h by multiplying it by 3.413. (There are 3.413 Btu heat energy in each watt.)

$$\text{Btu/h} = \text{Watts} \times 3.413$$
$$\text{Btu/h} = 17,680 \times 3.413$$
$$\text{Btu/h} = 60,341.8 \ \textbf{(Figure 30–23)}$$

We can now use the sensible heat formula to find the cfm.

$$\text{cfm} = \frac{Q_s}{1.08 \times \text{TD}}$$

$$\text{cfm} = \frac{60,341.8}{1.08 \times 50}$$

$$\text{cfm} = 1117.4$$

The technician may use the electric heat section of a system to establish the correct cfm for the cooling system. Control the fan speed by operating the fan in the FAN ON position. This will ensure that the fan is running at the same speed for cooling. Reduced airflow causes real problems with cooling systems and is covered in a later unit.

HVAC GOLDEN RULES

When making a service call to a residence:
- Try to park your vehicle so that you do not block the customer's driveway.
- Look professional and be professional. Clean clothes and shoes must be maintained. Wear coveralls for under-the-house work and remove them before entering the house.
- Ask the customer about the problem. The customer can often help you solve the problem and save you time.
- Check humidifiers where applicable. Annual service is highly recommended for health reasons.

Added Value to the Customer

Here are some simple, inexpensive procedures that may be included with the basic service call.
- Repair or replace all frayed wires.
- Clean or replace dirty filters.
- Lubricate fan bearings when needed.
- Fasten all cabinet doors securely, and replace any missing screws.

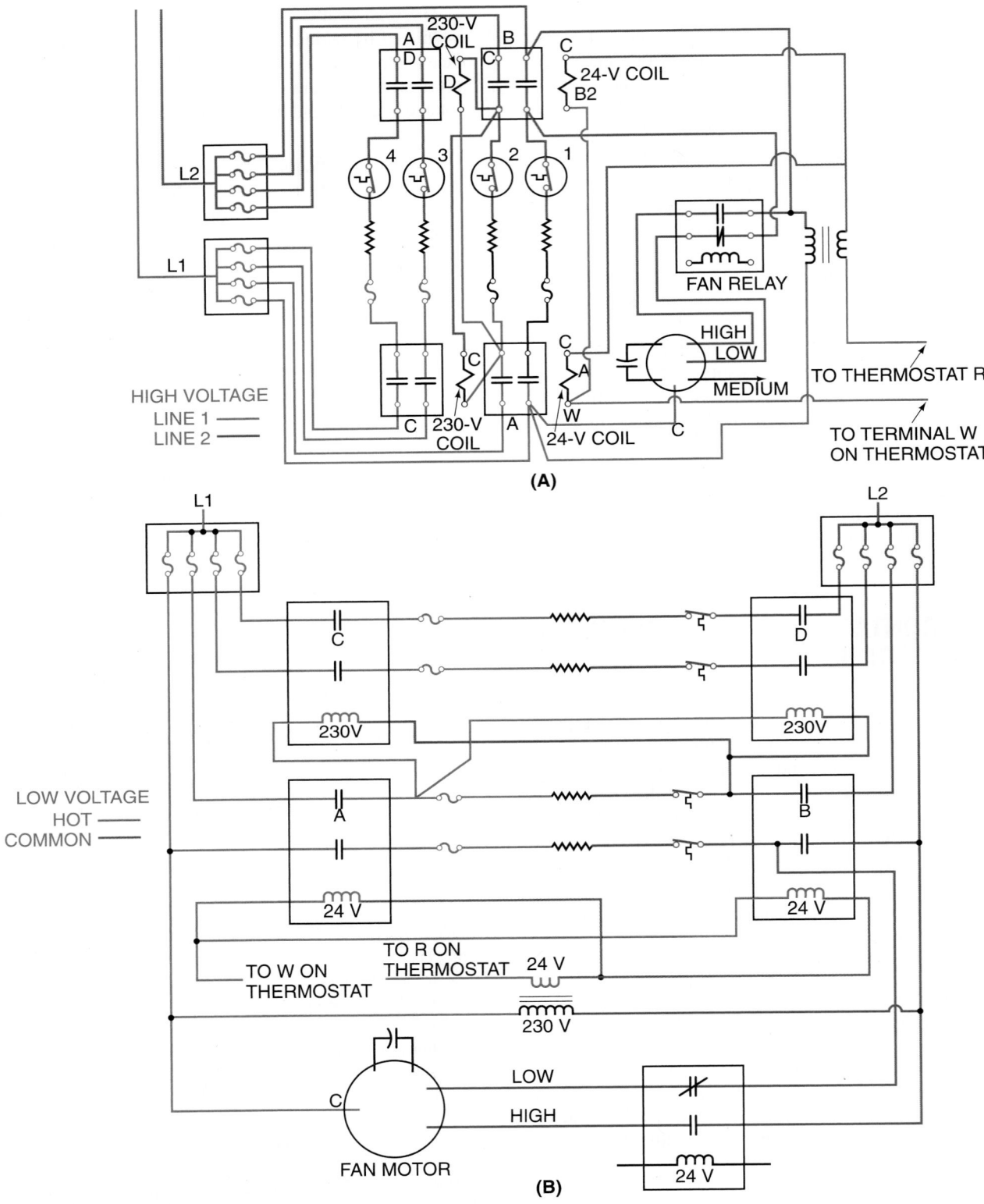

Figure 30–22 **(A)** The wiring diagram for an electric furnace using contractors. **(B)** A line or ladder diagram.

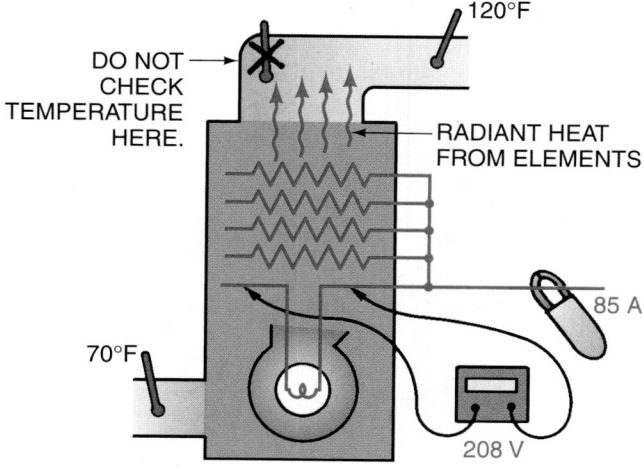

WATTS = AMPERES X VOLTS
= 85 X 208
= 17,680 W

Btu/h = WATTS X 3.413
= 17,680 X 3.413
= 60,341.8 Btu/h

Figure 30–23 An electric furnace airflow calculation.

30.16 SERVICE TECHNICIAN CALLS

SERVICE CALL 1

A customer with a 20-kW electric furnace calls. *There is no heat. The customer says that the fan comes on, but the heat does not. This is a system that has individual sequencers with a fan-starting sequencer. The first-stage sequencer has a burned-out coil and will not close its contacts. The first stage starts the rest of the heat, so there is none.*

The technician is familiar with this system and had started it after the initial installation. The technician realizes that because the fan starts, the 24-V power supply is working. On arriving at the job, the thermostat is set to call for heat. The technician goes under the house with a spare sequencer, a volt-ohmmeter, and an ammeter.

When approaching the electric furnace, the fan can be heard running. The panel covering the electric heat elements and sequencers is removed. The ammeter is used to check to see whether any of the heaters are using current; none are. SAFETY PRECAUTION: The technician observes electrical safety precautions. The voltage is checked at the coils of all sequencers. See **Figure 30–21** for a diagram. It is found that 24 V is present at the timed fan control and the first-stage sequencer, but the contacts are not closed. The electrical disconnect is opened, and the continuity is checked across the first-stage sequencer coil. The circuit is open. When the sequencer is changed and power is turned on, the electric heat comes on.

SERVICE CALL 2

A residential customer calls. *There is no heat, but the customer smells smoke. The company dispatcher advises that the customer should turn the system off until the technician arrives. The fan motor has an open circuit and will not run. The smoke smell is coming from the heating unit cycling on the limit control.*

The technician arrives and goes to the room thermostat and turns the system to heat. This is a system that does not have cooling, or the thermostat would be turned to FAN ON to see whether the fan will start. If it did not start, this would help to solve the problem. When the technician goes to the electric furnace in the hall closet, it is noticed that a smoke smell is present. When an ammeter is used to check the current at the electric heater, it is found that it is pulling 40 A, but the fan is not running. SAFETY PRECAUTION: The technician observes electrical safety precautions.

The technician looks at the wiring diagram and discovers that the fan should start with the first-stage heat from a set of contacts on the sequencer. The voltage is checked to the fan motor; it is getting voltage but is not running. The unit power is turned off, locked, and tagged, and a continuity check of the motor proves that the motor winding is open. The technician changes the motor, and the system operates normally.

SERVICE CALL 3

The service technician is on a routine service contract call and inspection. *The terminals on the electric heat units have been hot. The insulation on the wire is burned at the terminals.*

When removing the panels on the electric heat panel, the technician sees the burned insulation on the wires and shows this to the store manager. It is suggested that these wires and connectors be replaced. The store manager asks what the consequences of waiting for them to fail would be. The technician explains that they may fail on a cold weekend and allow the building water to freeze. If the overhead sprinkler freezes and thaws, all of the merchandise will get wet; it is not worth the chance. SAFETY PRECAUTION: The technician turns off the power and completes the job, using the correct wire size, high-temperature insulation, and connectors.

SERVICE CALL 4

A customer at a retail store calls and reports that their electric heat system is not performing. The room thermostat is set at 72°F and the store temperature is 60°F. The technician arrives and talks to the owner to find out what he knows. The owner tells the technician that he changed the thermostat out for a newer model yesterday and the system has not worked correctly since. He explains that he changed the thermostat wire for wire.

The technician goes to the electric furnace in a closet and starts to check around. The system has five individual sequencers, **Figure 30–21.** He then checks all of the circuits and only three have amperage. While he is looking at his ammeter, the #3 sequencer opens and that circuit stops heating.

The technician then goes to the room thermostat and removes the cover. He watches the mercury bulbs for a minute and notices that the bulbs are turning counterclockwise, or fixing to shut the heat off. He is puzzled because with it this cool in the store, the bulbs should be all of the way clockwise and calling for heat. Then he looks at the heat anticipator setting and it is set at 0.3 A. With all of these sequencers, it should draw more current than that.

The technician goes to the furnace and finds that each sequencer coil draws 0.2 A. The heat anticipator should be set at 1 A, (0.2 × 5 = 1). He goes back to the room thermostat and sets the anticipator and the thermostat remains steady. He then goes to the furnace and checks the amperage at each individual heater coil and they are all heating. The technician explains to the owner that he did a good job of changing the thermostat, but that he left out the step about the heat anticipator. The heating system is now operating correctly.

SERVICE CALL 5

A customer calls at the end of the heating season and explains that the fan on his electric furnace is starting and stopping from time to time. He has the thermostat set to OFF so there should be no activity.

The technician arrives and goes to the furnace that is located in a closet in the basement. When the technician touches the cabinet of the furnace, he can feel that it is warm; it shouldn't be—it should be room temperature. He removes the cabinet cover, takes his ammeter, checks the current at each of the three heating elements, and discovers that one of the elements is drawing 8 A. There should be no current draw to this heater. The technician is really puzzled for a few minutes. He turns off the power and removes that heating element and finds that one of the elements has been burned in two and that one of the coils is touching the frame. The element is grounded and drawing current, **Figure 30–24.** (NOTE that the sequencer has no control over this grounded element. The fuse did not open because the current flow was not enough—but there is enough current flow to create heat and activate the temperature control for the fan. The running fan will cool off the element quickly and turn off, only to turn on again.) Thank goodness the furnace frame has a good ground connection for carrying the current to ground instead of through the technician. SAFETY PRECAUTION: It is a good electrical safety practice to check from the frame of any appliance to ground to make sure there is no danger. Do this *before* touching the appliance! This is particularly important when you are lying on the ground under a house checking equipment.

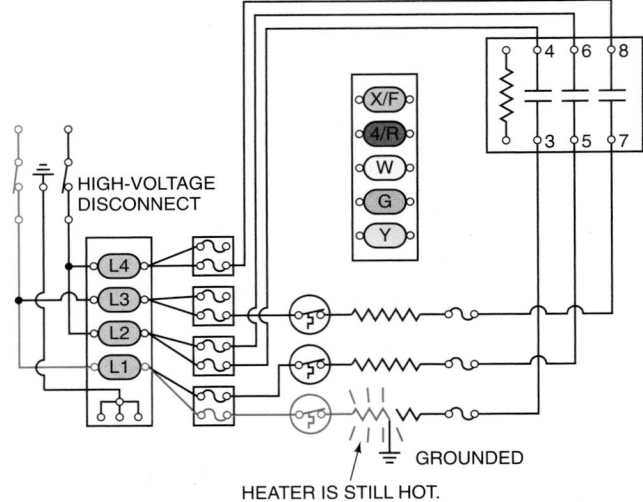

Figure 30–24 One side of this heating element is grounded to the frame of the furnace and is heating all of the time. Technicians should always check an appliance to ground for safety. This furnace is grounded and safe. If the ground were not correct, the technician could be injured.

The technician shows the element to the homeowner and leaves for the supply house to get a replacement element.

When the new element is installed, the unit operates correctly.

The following service call does not include the solution. The solution can be found in the *Instructor's Guide.*

SERVICE CALL 6

A small business calls. Their heating system is not putting out enough heat. This is the first day of very cold weather, and the space temperature is only getting up to 65°F with the thermostat set at 72°F. The system is a package air-conditioning unit with 30 kW (six stages of 5 kW each) of strip heat located on the roof. The fuse links in two of the heaters are open due to previously dirty air filters.

The technician arrives, goes to the room thermostat, checks the setting of the thermostat, and finds that the space temperature is 5°F lower than the setting. The technician goes to the roof with a volt-ohmmeter and an ammeter. SAFETY PRECAUTION: After removing the panels on the side of the unit where the strip heat is located, the amperage is carefully checked. Two of the six stages are not pulling any current. It looks like a sequencer problem at first. A voltage check of the individual sequencer coils shows that all of the sequencers should have their contacts closed; there is 24 V at each coil. A voltage check at each heater terminal shows that all stages have voltage but are not pulling any current.
What is the problem and the recommended solution?

PREVENTIVE MAINTENANCE

Maintenance of electrical heating equipment may involve small appliances or central heating systems. Electric heat appliances, units, or systems all consume large amounts of power. This power is routed to the units by way of wires, wall plugs (appliance heaters), and connectors. These wires, plugs, and connectors carry the load and must be maintained.

ELECTRICAL HEATING APPLIANCES.   The power cord at the plug-in connector is usually the first place that trouble may occur on a heating appliance. The heater should be taken out of service at the first sign of excess heat at the wall plug. It may become warm to the touch, but if it becomes too hot to hold comfortably, shut it off and check the plug and the wall receptacle. Fire may start in the vicinity of the plug if this is not corrected. It is a situation that will become worse every time the heater is operated.

An electric appliance heater should be observed at all times to ensure safe operation. Do not locate the heater too close to combustible materials. Only operate the heater in the upright position.

CENTRAL ELECTRICAL HEATING SYSTEMS.   The maintenance for large electrical heating systems may include either panel type, baseboard type, or forced air. The panel and baseboard systems do not require much maintenance when properly installed. Baseboard heat has more air circulation over the elements than panel type, so the units must be kept dust free. Some areas will have more dust and require cleaning more often. Both panel and baseboard type systems use line-voltage thermostats. These thermostats contain the only moving part and will normally be the first to fail. These thermostats will collect some dust and must be cleaned inside at some time. Be sure to turn the power off before servicing.

Central forced-air systems have a fan to move the air and consequently require filter maintenance. Filters should be changed or cleaned on a regular basis, depending on the rate of dust accumulation.

Some fan bearings require lubrication at regular intervals. Motor and fan shafts with sleeve bearings must be kept in good working order because these bearings will not take much abuse. Lift the motor or fan shaft and look for movement. Do not confuse shaft end-play with bearing wear. Many motors have up to 1/8-in. end-play.

Some systems may have belt drives and require belt tension adjustment or belt changing. Frayed or broken belts must be replaced to ensure safe and trouble-free operation. Belts should be checked at least once a year. It is very dangerous to operate electric heating systems with little or no airflow.

Filters should be checked every 30 days of operation unless it is determined that longer periods may be allowed. Use the filters that fit the holder and be sure to follow the direction arrows on the filter for the correct airflow direction. When the filter system becomes clogged on a forced-air system, the airflow is reduced and causes overheating of the elements. The automatic reset limit switches should be the first to trip and then reset. If the condition continues, these switches may fail, and the fuse link in the heater may open the circuit. When this happens, the link must be replaced by a qualified technician. This requires a service call that could have been prevented.

The electric heating elements are usually controlled by low-voltage thermostats that open or close the contacts on sequencers or contactors. The contacts in the sequencers or contactors carry large electrical loads, which will probably produce the most stress. They will probably be the first component to fail in a properly maintained system. You may visually check the contacts on a contactor, but the contacts on sequencers are concealed. When the contacts on a contactor become pitted, change the contacts on the contactor. The condition will only become worse and could then cause other problems, namely, terminal and wire damage to the circuit.

The only method you can use to check a sequencer is to visually check the wires for discoloration and check the current flow through each circuit with the sequencer energized. If power will not pass through the sequencer circuit, the sequencer may be defective. Do not forget that the heater must have a path through which electrical energy can flow. Many sequencers have been declared defective only to find later that the fuse links in the heater or the heater itself had an open circuit.

Line-voltage electrical connections in the heater junction box should be examined closely. If there is any wire discoloration or connectors that appear to have been hot, these components must be changed. If the wire is discolored, and it is long enough, it may be cut back until the conductor is its normal color.

Fuse holders may lose their tension and not hold the fuses tightly. This is caused from heating and will be apparent because the fuses will blow when no overload exists and for no other apparent reason. These fuses and holders must be changed for the unit to perform safely and correctly.

SUMMARY

- Electric heat is a convenient way to heat individual rooms and small spaces.
- Central electric heating systems are usually less expensive to install and maintain than other types.
- Operational costs may be higher than with other types of fuels.
- The low-voltage thermostat, sequencer, and relays are control mechanisms used in central electric forced-air systems.
- The sequencer is used to activate the heating elements in stages. This avoids putting heavy kilowatt loads in service at one time. If this were done, it could cause fluctuations in power, resulting in voltage drop, flickering lights, and other disturbances.
- Sequencers have bimetal strips with 24-V heaters. These heaters cause the bimetal strip to bend, closing contacts and activating the high-voltage heating elements.
- The fan in a central system must operate while the heating elements are on. The systems must be wired to ensure that this occurs.
- Systems are protected with limit switches and fusible links (temperature controlled).
- Preventive maintenance inspections should be made periodically on electric heating units and systems.

REVIEW QUESTIONS

1. The type of material often used for the resistance wire in electric heat is _____.
2. Portable or small space heaters with glowing coils transfer heat by _____.
3. Baseboard heaters transfer heat by _____.
4. A disadvantage of a central forced-air electric furnace is that
   A. there is no zone control for individual rooms.
   B. a thermostat must be installed in each room.
   C. only line-voltage thermostats can be used.
   D. it is difficult to add air conditioning (cooling).
5. Three common controls used in central electric heat applications are thermostats, contactors (or relays), and
   A. capacitors.
   B. sequencers.
   C. cool anticipators.
6. The low-voltage thermostat is used for central electric heat because it is compact, it is responsive, and
   A. it provides fuse protection.
   B. it acts as a limit switch.
   C. it is safe.
7. The heat anticipator setting is determined by
   A. adding all the current draw in the 24-V circuit that passes through the thermostat control bulb and heat anticipator.
   B. using the amperage of the fan motor.
   C. using the total of the high-voltage amperage in the circuit.
   D. using the voltage at the limit control.
8. The sequencer uses _____ (high or low) voltage control power to start and stop the electric heaters.
9. True or False: A package sequencer can start or stop only one stage of strip heat.
10. Two types of wiring diagrams are the pictorial and _____.
11. The limit switch shuts off the unit if
    A. high temperature occurs.
    B. the furnace is unable to maintain the correct heat.
    C. the heat anticipator does not function properly.
    D. the cool anticipator does not function properly.
12. The _____ wire from the transformer furnishes power to the thermostat.
    A. red
    B. blue
    C. orange
    D. yellow
13. State the formula used for determining the cubic feet per minute (cfm) airflow across the furnace heating elements.
14. Power-consuming devices are wired in _____ (series or parallel).
15. Power-passing devices are wired in _____ (series or parallel).

DIAGNOSTIC CHART FOR ELECTRIC HEAT

Electric heat may involve either boiler systems or forced-air systems. The discussion in this text has been primarily for forced-air systems as they are the most common. These forced-air systems may be found as the primary heating system or as the secondary heating system when used with heat pumps. The technician should always listen to the customer for symptoms. The customer can often lead you to the cause of the problem with a simple description of how the system is performing.

Problem	Possible Cause	Possible Repair	Heading Number
No heat—thermostat calling for heat	Open disconnect switch	Close disconnect switch.	30.11
	Open fuse or breaker	Replace fuse or reset breaker and determine why it opened.	30.11
	High-temperature fuse link open circuit	Tighten loose connection at fuse link causing heat.	30.12
	Faulty high-voltage wiring or connections	Repair or replace faulty wiring or connections.	30.12, 30.13
	Control-voltage power supply off	Check control-voltage fuses and safety devices.	30.12
	Faulty control-voltage wiring or connections	Repair or replace faulty wiring or connections.	30.12
	Heating element burned, open circuit	Replace heating element—check airflow.	30.12, 30.15
Insufficient heat	Portion of heaters or limits open circuit	See above.	30.10, 30.12
	Low voltage	Correct voltage.	30.12, 30.13

UNIT 31 Gas Heat

OBJECTIVES

After studying this unit, you should be able to

- describe each of the major components of a gas furnace.
- list two fuels burned in gas furnaces and describe characteristics of each.
- discuss a multipoise furnace and its safety devices.
- discuss flame rollout switches, auxiliary limit switches, and draft safeguard switches.
- discuss gas pressure measurement in inches of water column and describe how a manometer is used to make this measurement.
- discuss gas combustion.
- describe a solenoid, diaphragm, and heat motor gas valve.
- list the functions of an automatic combination gas valve.
- describe the function of a servo-operated gas pressure regulator.
- discuss the meaning of a redundant gas valve.
- discuss different gas burners and heat exchangers.
- describe the difference between induced-draft and forced-draft systems.
- describe and discuss different ways of controlling the warm air fan.
- state the function of an off-delay timing device for a warm air fan control.
- describe the standing pilot, intermittent pilot, direct-spark, and hot surface ignition systems.
- list three flame-proving devices and describe the operation of each.
- discuss reasons and the systems used for the delay in starting and stopping the furnace fan.
- state the purpose of a limit switch.
- describe flue-gas venting systems.
- discuss flame rectification and how it pertains to local and remote flame sensing.
- apply flame rectification troubleshooting and maintenance procedures.
- discuss the components of a high-efficiency gas furnace.
- describe direct-vented, non-direct-vented, and positive pressure systems.
- explain dew point temperature as it applies to a high-efficiency condensing furnace.
- discuss excess air, dilution air, combustion air, primary air, and secondary air.
- describe the condensate disposal system of a high-efficiency condensing gas furnace.
- identify furnace efficiency ratings.
- discuss electronic ignition modules and integrated furnace controllers.
- describe a two-stage furnace, a modulating gas furnace, and a variable output thermostat.
- explain gas piping as it applies to gas furnaces.
- interpret gas furnace wiring diagrams and troubleshoot flowcharts or guides.
- compare the designs of a high-efficiency gas furnace and a conventional furnace.
- describe procedures for taking flue-gas carbon dioxide and temperature readings.
- describe typical preventive maintenance procedures.

SAFETY CHECKLIST

- ✔ Fuel gases are dangerous in poorly vented areas. They can replace oxygen in the air and cause suffocation. They are also explosive. If a gas fuel leak is suspected, prevent ignition by preventing any open flame or spark from occurring. SAFETY PRECAUTION: A spark from a flashlight switch can ignite gas. Vent all enclosed areas adequately before working in them.
- ✔ Always shut off the gas supply when installing or servicing a gas furnace.
- ✔ A gas furnace that is not functioning properly can produce carbon monoxide, a poisonous gas. It is absolutely necessary that the production of carbon monoxide be avoided.
- ✔ Yellow tips on the flame indicate an air-starved flame emitting poisonous carbon monoxide.
- ✔ It is essential that the furnace be vented properly so that the flue gases will all be dissipated into the atmosphere.
- ✔ When taking flue-gas samples do not touch the hot vent pipe.
- ✔ A furnace with a defective heat exchanger (one with a hole or crack) must not be allowed to operate because the flue gases can mix with the air being distributed throughout the building.
- ✔ Wear goggles when using compressed air for cleaning parts.
- ✔ The limit switch must operate correctly to keep the furnace from overheating due to a restriction in the airflow or other furnace malfunction.
- ✔ Proper replacement air must be available in the furnace area.
- ✔ All new gas piping assemblies should be tested for leaks.

✔ Gas piping systems should be purged only in well-ventilated areas.

✔ Follow all safety procedures when troubleshooting electrical systems. Electric shock can be fatal.

✔ Turn off the power before attempting to replace any electrical component. Lock and tag the disconnect panel. There should be only one key. Keep this key on your person while making any repairs.

✔ Perform a gas leak check after completing an installation or repair.

31.1   INTRODUCTION TO GAS-FIRED, FORCED-HOT-AIR FURNACES

Gas-fired, forced-hot-air furnaces have a heat-producing system and a heated air distribution system. The heat-producing system includes the manifold and controls, burners, the heat exchanger, and the venting system. The heated air distribution system consists of the blower that moves the air throughout the ductwork and the ductwork assembly. See **Figure 31–1** for a photo of a modern high-efficiency condensing gas furnace.

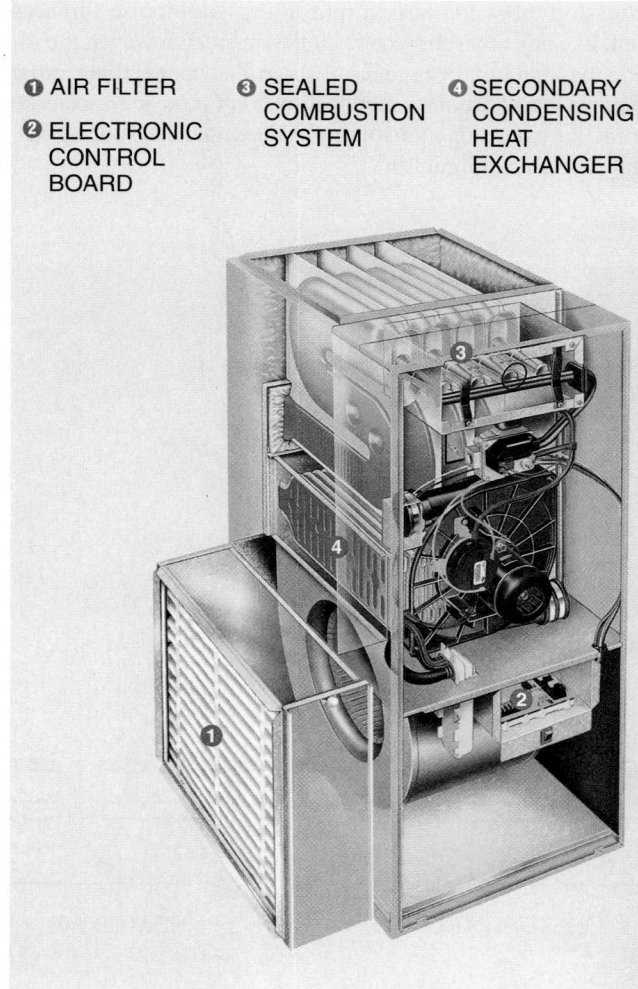

- ❶ AIR FILTER
- ❷ ELECTRONIC CONTROL BOARD
- ❸ SEALED COMBUSTION SYSTEM
- ❹ SECONDARY CONDENSING HEAT EXCHANGER

Figure 31–1   A modern high-efficiency condensing gas furnace. *Courtesy Bryant Heating and Cooling Systems*

The manifold and controls meter the gas to the burners where the gas is burned, which creates flue gases in the heat exchanger and heats the air in and surrounding the heat exchanger. The venting system allows the flue gases to be exhausted into the atmosphere. The blower distributes the heated air through ductwork to the areas where the heat is wanted.

31.2   TYPES OF FURNACES

Upflow

The *upflow* furnace stands vertically and needs headroom. It is designed for the first floor installation with the ductwork in the attic or for basement installation with the ductwork between or under the first floor joists. The furnace takes in cool air from the rear, bottom, or sides near the bottom. It discharges hot air out the top, **Figure 31–2.**

"Low-boy"

The *low-boy* furnace is approximately 4 ft high. It is used primarily in basement installations with low headroom where the ductwork is located under the first floor. Air intake and discharge are both at the top, **Figure 31–3.**

The low profile of the low-boy is accomplished by placing the blower behind the furnace in a separate cabinet rather than in line with the heat exchanger.

Downflow

The *downflow* furnace, sometimes referred to as a *counterflow* furnace, looks like the upflow furnace. The ductwork may be in a concrete slab floor or in a crawl space under the house. The air intake is at the top, and the discharge is at the bottom, **Figure 31–4.**

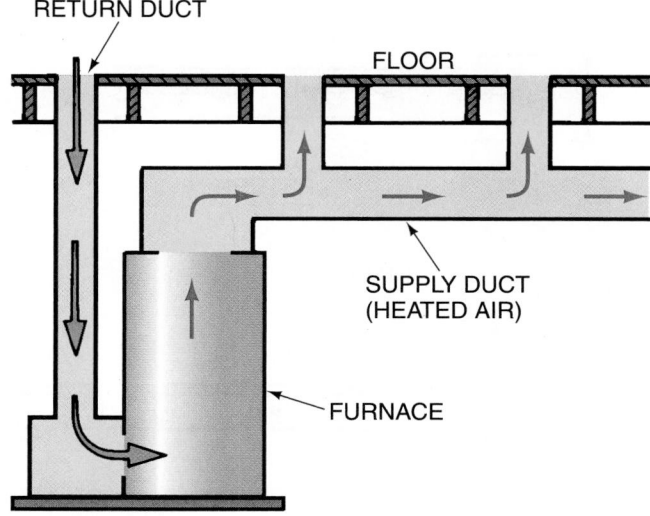

Figure 31–2   The airflow for an upflow gas furnace.

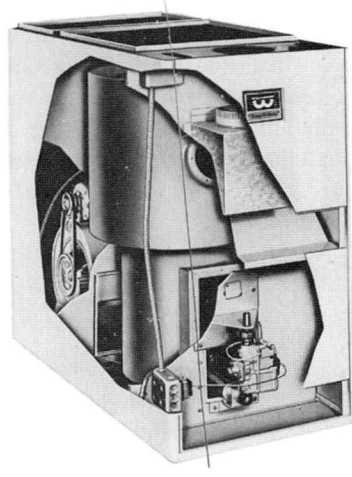

(A)

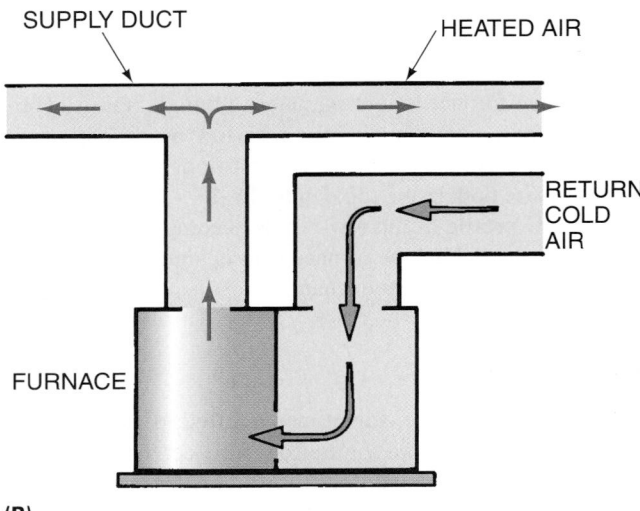

(B)

Figure 31–3   **(A)** A low-boy gas furnace. **(B)** The airflow for a low-boy furnace. **(A)** *Courtesy The Williamson Company*

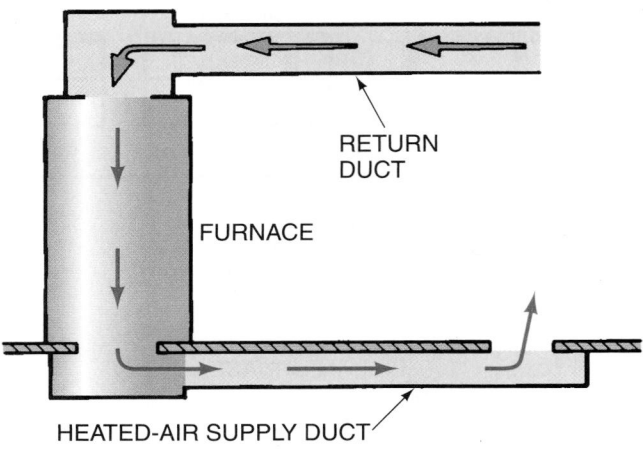

Figure 31–4   The airflow for a downflow, or counterflow, gas furnace.

Horizontal

The *horizontal* furnace is positioned on its side. It is installed in crawl spaces, installed in attics, or suspended from floor joists in basements. In these installations it takes no floor space. The air intake is at one end; the discharge is at the other, **Figure 31–5.**

Multipoise or Multipositional

A multipoise furnace can be installed in any position. This adds versatility to the furnace, especially for the installation contractor. Multipoise furnaces can be upflow, downflow, horizontal right, or horizontal left air discharge, **Figure 31–6.** Many times, these furnaces come from the factory configured for the more common upflow applications but can be changed with very little modification in the field. Often, the combustion air pipe and exhaust pipe can be attached to either side of the furnace. Condensate drain lines can also be attached to either side of the furnace as long as the one not in use is capped. **Figure 31–7(A)** shows a high-efficiency condensing furnace with two combustion air and condensate drain line connection options. Many multipoise furnaces come with vent and drain extension tubes for ease in retrofitting. Multipoise furnaces look like any other furnace from the outside; however, the inside may have extra condensate drain line connections, more than one combustion air connection, vent pipe or condensate drain line extensions, and plugs for lines not in use during certain position configurations.

(A)

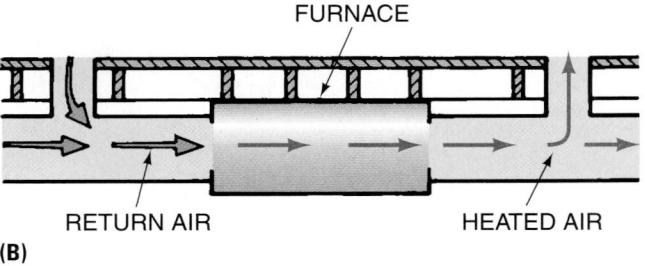

(B)

Figure 31–5   **(A)** A horizontal gas furnace with air conditioning. **(B)** The airflow for a horizontal gas furnace. **(A)** *Courtesy BPD Company*

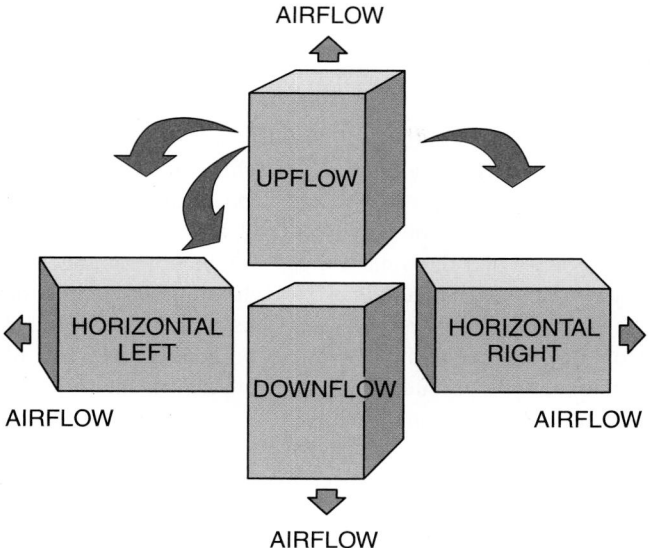

AIRFLOW

UPFLOW

HORIZONTAL
LEFT

DOWNFLOW

HORIZONTAL
RIGHT

AIRFLOW

AIRFLOW

AIRFLOW

Figure 31–6 Four different positions for a multipoise furnace.

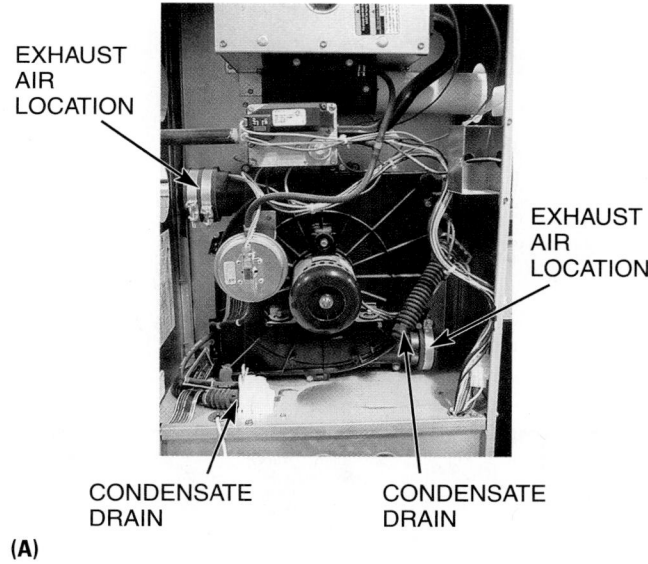

EXHAUST
AIR
LOCATION

EXHAUST
AIR
LOCATION

CONDENSATE
DRAIN

CONDENSATE
DRAIN

(A)

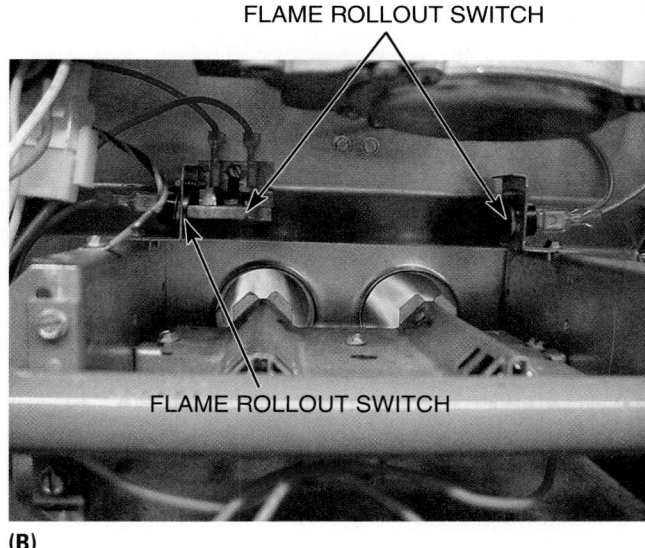

FLAME ROLLOUT SWITCH

FLAME ROLLOUT SWITCH

(B)

Figure 31–7   (A) A multipositional furnace showing two exhaust air location options and two condensate drain location options. **(B)** A multipoise furnace exhibiting multiple flame rollout switches. **(A)** and **(B)** Courtesy Ferris State University. Photos by John Tomczyk

Because multipoise furnaces can be placed in four different configurations, multiple safety controls must be designed into the furnace. **Figure 31–7(B)** shows multiple bimetallic flame rollout or overtemperature limits very near to one another. This protects the furnace no matter what configuration it is in.

Normally closed (NC) flame rollout switches are usually located near the top of the burner box, where the flame is most likely to roll out. When a furnace is installed horizontally, additional rollout switches are wired in series to provide extra flame rollout protection. Control schemes on some furnaces allow for the combustion blower to operate continuously when a rollout switch has opened. Flame rollout switches may have to be manually reset by pushing a button on top of the switch.

SAFETY PRECAUTION: Never use an automatic reset flame rollout switch to replace a damaged or inoperative manual reset flame rollout switch.

Auxiliary limit switches are used for counterflow and horizontal furnaces. They are usually bimetallic, similar to flame rollout switches, but are mounted on the warm air blower housing, **Figure 31–7(C)**. They monitor room return air temperature and interrupt burner operation when the temperature gets too hot. This protects the furnace's filter from exceeding its maximum allowable temperature. Usually the combustion blower or warm air blower will continue to run if any auxiliary limit has opened. Most of these controls are automatically reset.

Vent limit or draft safeguard switches are NC switches usually mounted on the draft hood if the furnace is equipped with one. They are typically bimetallic switches similar to flame rollout and auxiliary limit switches. They monitor the draft hood temperature and open when there is any spillage of flue gas, **Figure 31–7(D)**.

31.3   GAS FUELS

Natural gas, manufactured gas, and liquefied petroleum (LP) are commonly used in gas furnaces.

Natural Gas

Natural gas has been forming along with oil for millions of years from dead plants and animals. This organic material accumulated for years and years and gradually was washed or deposited into hollow spots in the earth, **Figure 31–8.** This material accumulated to great depths, causing tremendous

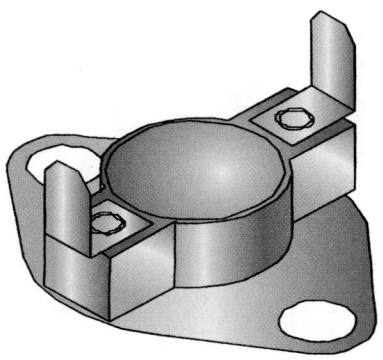

(C)

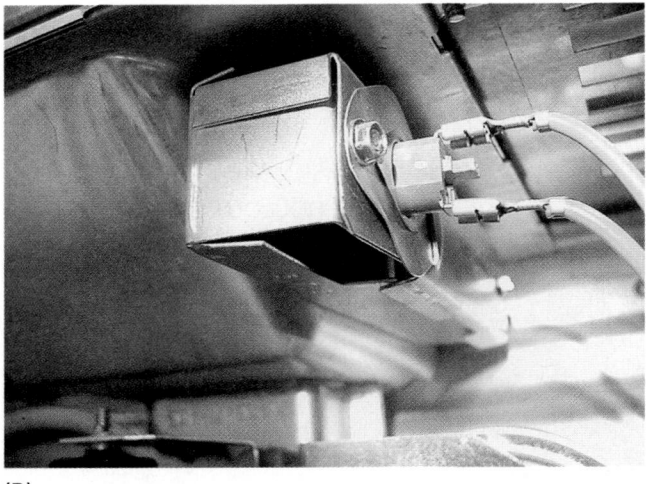

(D)

Figure 31–7 (*continued*)   **(C)** An auxiliary limit switch for horizontal and counterflow furnaces. They are located on the warm air blower housing and monitor return air temperature. **(D)** The draft safeguard switch that opens the electrical circuit when there is any spillage of flue gas. **(D)** *Courtesy Ferris State University. Photo by John Tomczyk*

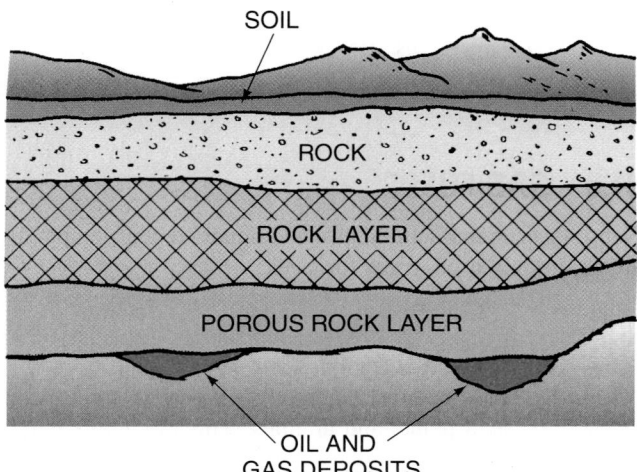

Figure 31–8   Gas and oil deposits deep in the earth.

pressure and high temperature from its own weight. This caused a chemical reaction, changing this organic material to oil and gas, which accumulated in pockets and porous rocks deep in the earth.

Natural gas is composed of 90% to 95% methane and other hydrocarbons almost all of which are combustible, which makes this gas efficient and clean burning. Natural gas has an average specific gravity of 0.60, and dry air has a specific gravity of 1.0, so natural gas weighs 60% as much as dry air. Therefore, natural gas rises when discharged into the air. The characteristics of natural gas will vary somewhat from one location to another. When burned with air, 1 ft³ of natural gas will produce about 1050 Btu of heat energy. Contact the utility company providing the gas in your area for its specific characteristics.

Gas from these pockets deep in the earth (more generally called wells) may contain moisture and other gases that must be removed before distributing the gas.

Natural gas by itself has neither odor nor color and is not poisonous. SAFETY PRECAUTION: But it is dangerous because it can displace oxygen in the air, causing suffocation, and when it accumulates, it can explode.

Sulfur compounds, called *odorants*, which have a garlic smell, are added to the gas to make leak detection easier. No odor is left, however, when the gas has been burned.

Liquefied Petroleum

Liquefied petroleum (LP) is liquefied propane, butane, or a combination of propane and butane. When in the vapor state these gases may be burned as a mixture of one or both of these gases with air. The gas is liquefied by keeping it under pressure until ready to be used. A regulator at the tank reduces the pressure as the gas leaves the tank. As the vapor is drawn from the tank, it is replaced by a slight boiling or vaporizing of the remaining liquid. LP gas is obtained from natural gas or as a by-product of the oil-refining process. The boiling point of propane at atmospheric pressure is −44°F, which makes it feasible to store in tanks for use in low temperatures during northern winters.

In the liquid state, butane has a high boiling point. At any temperature below 31°F the tank would be in a negative pressure, so the tank must be buried or have a small heater supplied to keep the fuel pressure high enough to move the gas to the main burner. For these reasons, this fuel is not very popular. It has a very high heat content but is hard to work with because of the pressures.

When burned, 1 ft³ of *propane* produces 2500 Btu of heat. However, it requires 24 ft³ of air to support this combustion. The specific gravity of propane gas is 1.52, which means that it is 1 1/2 times heavier than air, so it sinks when released in air. SAFETY PRECAUTION: This is dangerous because it will replace oxygen and cause suffocation; it will also collect in low places to make pockets of highly explosive gas.

Butane produces approximately 3200 Btu/ft³ of gas when burned. LP gas contains more atoms of hydrogen and carbon

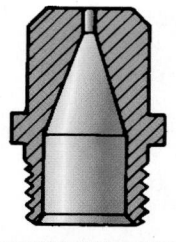 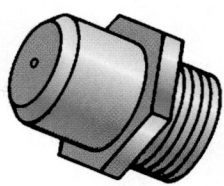

CROSS SECTION

Figure 31–9 A spud showing the orifice through which the gas enters the burner.

than natural gas does, which causes the Btu output to be greater. Occasionally, butane and propane are mixed with air to alter its characteristics. It is then called propane or butane air gas. The mixture is accomplished with blending equipment under pressure. This reduces the Btu/ft³ and the specific gravity. The characteristics may be changed with air to closely duplicate natural gas. This allows it to be substituted and burned in natural gas burners without adjustment for standby purposes. LP gas is used primarily where natural gas is not available.

The specific gravity of a gas is important because it affects the gas flow through the piping and through the orifice of the furnace. The orifice is a small hole in a fitting (called a *spud*) through which the gas must flow to the burners, **Figure 31–9**. The gas flow rate also depends on the pressure and on the size of the orifice. At the same pressure more light gas than heavy gas will flow through a given orifice.

SAFETY PRECAUTION: LP gas alone must not be used in a furnace that is set up for natural gas because the orifice will be too large. This results in overfiring and soot buildup.

When the proper mixture of LP gas and air is used, it will operate satisfactorily in natural gas furnaces because the proper mixture will match the orifice size, and the burning rate will be the same.

Manifold Pressures

The manifold pressure at the furnace should be set according to the manufacturer's specifications because they relate to the characteristics of the gas to be burned. The manifold gas pressure is much lower than 1 psi and is expressed in inches of water column (in. WC). *One inch of water column* pressure is the pressure required to push a column of water up 1 in. One psi of pressure will support a column of water 27.7 in. high or 1 psi = 27.7 in. WC.

SOME COMMON MANIFOLD PRESSURES

Natural gas and propane/air mixture	3 to 3 1/2 in. WC
Manufactured gas (below 800 Btu quality)	2 1/2 in. WC
LP gas	11 in. WC

A *water manometer* is used to measure gas pressure. It is a tube of glass or plastic formed into a U. The tube is

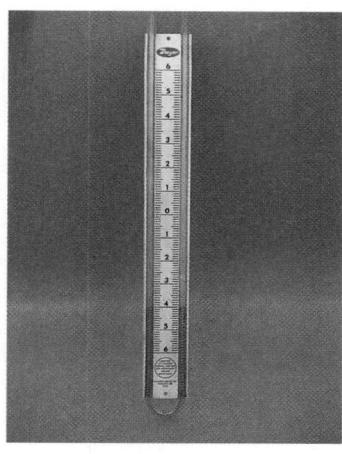

Figure 31–10 A manometer used for measuring pressure in inches of water column. *Photo by Bill Johnson*

about half full of plain water, **Figure 31–10**. The standard manometer used in domestic and light commercial installations is graduated in. WC. The pressure is determined by the difference in levels in the two columns. Gas pressure is piped to one side of the tube, and the other side is open to the atmosphere to allow the water to rise, **Figure 31–11**. The instrument often has a sliding scale, so it can be adjusted for easier reading. Some common pressure conversions follow:

PRESSURES PER SQUARE INCH

1 lb/in² = 27.71 in. WC
1 oz/in² = 1.732 in. WC
2.02 oz/in² = 3.5 in. WC (standard pressure for natural gas)
6.35 oz/in² = 11 in. WC (standard pressure for LP gas)

A digital manometer is often used to measure gas pressures, **Figure 31–12**. It can measure both positive and negative pressures in inches of water column. One advantage of a digital manometer is that there is no fluid to spill. Another advantage is that the pressure reading appears without the service technician having to measure a difference in height of water column. A disadvantage is that the digital manometer requires batteries.

SAFETY PRECAUTION: Turn the gas off while connecting any manometer.

31.4 GAS COMBUSTION

To properly install and service gas heating systems, the heating technician must know the fundamentals of combustion. These systems must be installed to operate safely and efficiently.

Combustion needs fuel, oxygen, and heat. A reaction between the fuel, oxygen, and heat is known as *rapid oxidation* or the process of burning.

The fuel in this case is the natural gas (methane), propane, or butane; oxygen comes from the air, and the heat comes from a pilot flame or other type of ignition system. The fuels contain hydrocarbons that unite with the oxygen when heated

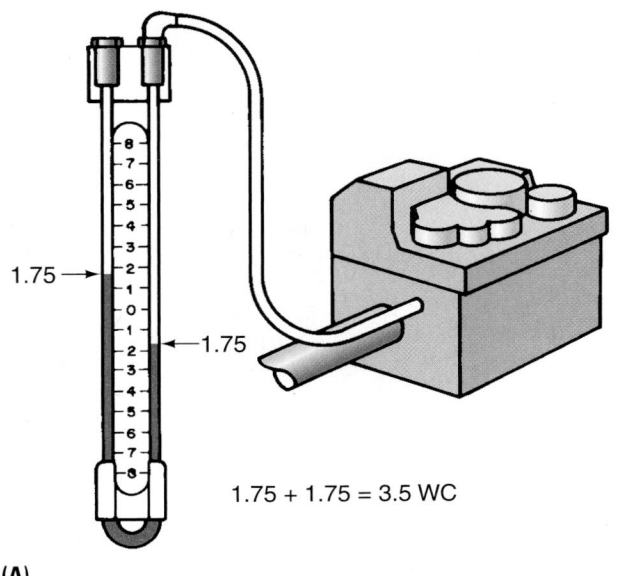

1.75 + 1.75 = 3.5 WC

(A)

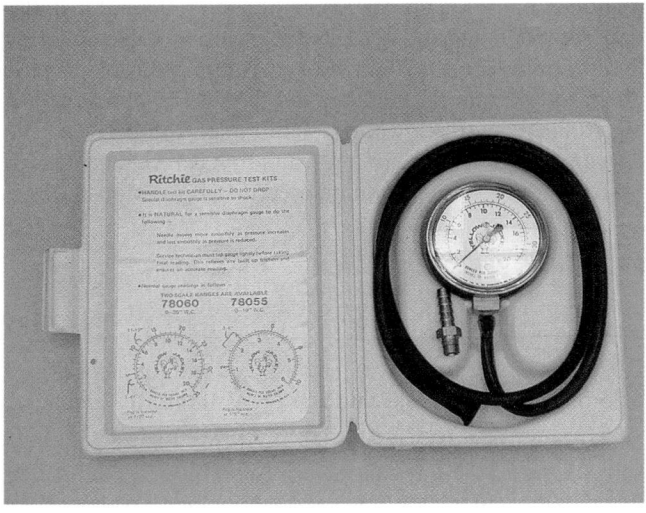

(B)

Figure 31–11 (A) A manometer connected to a gas valve for checking the manifold pressure. (B) A gage calibrated in inches of water column. *(B) Photo by Bill Johnson*

Figure 31–12 A digital manometer being used to measure gas pressure in inches of water column. *Courtesy Ferris State University. Photo by John Tomczyk*

By-products of poor combustion are carbon monoxide—a poisonous gas—soot, and minute amounts of other products. It should be obvious that carbon monoxide production must be avoided, as it is a colorless, odorless poisonous gas.

Soot lowers furnace efficiency because it collects on the heat exchanger and acts as an insulator, so it also must be kept at a minimum.

Most gas furnaces and appliances use an atmospheric burner because the gas and air mixtures are at atmospheric pressure during burning. The gas is metered to the burner through the orifice. The velocity of the gas pulls in the primary air around the orifice, **Figure 31–13(A).**

The burner tube diameter is reduced where the gas is passing through it to induce the air. This burner diameter reduction is called the *venturi,* **Figure 31–13(A).** The gas-air mixture moves on through the venturi into the mixing tube where the gas is forced by its own velocity through the burner ports or slots where it is ignited as it leaves the port, **Figure 31–13(B).**

When the gas is ignited at the port, secondary air is drawn in around the burner ports to support combustion. The flame should be a well-defined blue with slightly orange, *not yellow,* tips, **Figure 31–13(C).** SAFETY PRECAUTION: Yellow tips indicate an air-starved flame emitting poisonous carbon monoxide. Orange streaks in the flame are not to be confused with yellow. Orange streaks are dust particles burning.

Other than pressure to the gas orifice, the primary air is the only adjustment that can be made to the flame. Modern furnaces have very little adjustability; this is an intentional design to maintain minimum standards of flame quality. The only thing that will change the furnace combustion characteristics to any extent, except gas pressure, is dirt or lint drawn in with the primary air. If the flame begins to burn *yellow,* primary air restriction should be suspected.

For a burner to operate efficiently, the gas flow rate must be correct, and the proper quantity of air must be supplied. Modern furnaces use a primary and a secondary air supply, and the

to the ignition temperature. The air contains approximately 21% oxygen. Enough air must be supplied to furnish the proper amount of oxygen for the combustion process. The ignition temperature for natural gas is 1100°F to 1200°F. The pilot flame or other type of ignition must provide this heat, which is the minimum temperature for burning. When the burning process occurs, the chemical formula is

$$CH_4 \text{ (methane)} + 2O_2 \text{ (oxygen)} \rightarrow CO_2 \text{ (carbon dioxide)} + 2H_2O \text{ (water vapor)} + heat$$

This formula represents perfect combustion. This process produces carbon dioxide, water vapor, and heat. Although perfect combustion seldom occurs, slight variations create no danger. SAFETY PRECAUTION: The technician must see to it that the combustion is as close to perfect as possible.

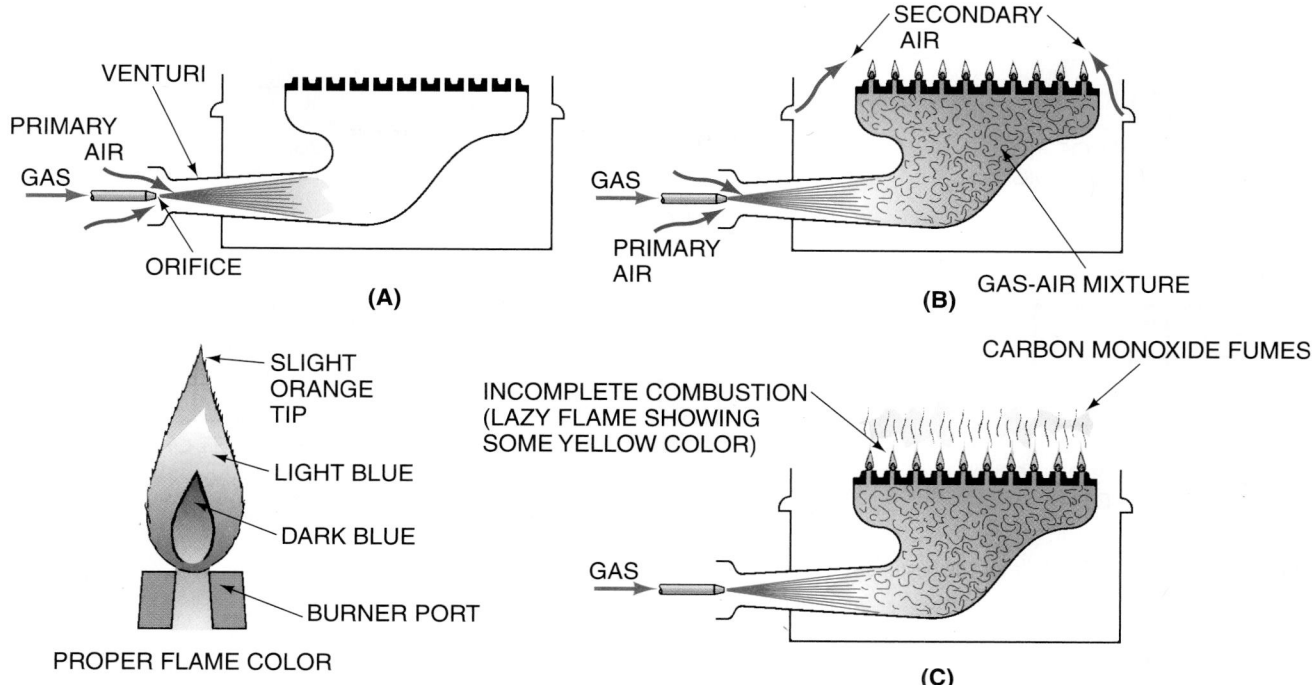

Figure 31–13 **(A)** Primary air is induced into the air shutter by the velocity of the gas stream from the orifice. **(B)** Ignition of the gas is on top of the burner. **(C)** Incomplete combustion yields yellow "lazy" flame. Any orange color indicates dust particles drawn in with primary air.

gas supply or flow rate is determined by the orifice size and the gas pressure. A normal pressure at the manifold for natural gas is 3 1/2 in. WC. It is important, however, to use the manufacturer's specifications when adjusting the gas pressure.

A gas-air mixture that is too lean (not enough gas) or too rich (too much gas) will not burn. If the mixture for natural gas contains 0% to 4% natural gas, it will not burn. If the mixture contains from 4% to 15% gas, it will burn. SAFETY PRECAUTION: However, if the mixture is allowed to accumulate, it will explode when ignited. If the gas in the mixture is 15% to 100%, the mixture will not burn or explode. The burning mixtures are known as the *limits of flammability* and are different for different gases.

The rate at which the flame travels through the gas and air mixture is known as the *ignition* or *flame velocity* and is determined by the type of gas and the air-to-gas ratio. Natural gas (CH_4) has a lower burning rate than butane or propane because it has fewer hydrogen atoms per molecule. Butane (C_4H_{10}) has 10 hydrogen atoms, whereas natural gas has 4. The speed is also increased in higher gas-air mixtures within the limits of flammability; therefore, the burning speed can be changed by adjusting the air flow.

Perfect combustion requires two parts oxygen to one part methane. The atmosphere consists approximately of one-fifth oxygen and four-fifths nitrogen with very small quantities of other gases. Approximately 10 ft^3 of air is necessary to obtain 2 ft^3 of oxygen to mix with 1 ft^3 of methane. This produces 1050 Btu and approximately 11 ft^3 of flue gases, **Figure 31–14.**

The air and gas never mix completely in the mixing tube chamber. Extra primary air is always supplied so that the methane will find enough oxygen to burn completely. This excess air is normally supplied at 50% over what would be needed if it were mixed thoroughly. This means that to burn 1 ft^3 of methane, 15 ft^3 of air is supplied. There will be 16 ft^3 of flue gases to be vented from the combustion area, **Figure 31–14.** Additional air also will be added at the draft hood. This is discussed later.

Flue gases contain approximately 1 ft^3 of oxygen (O_2), 12 ft^3 of nitrogen (N_2), 1 ft^3 of carbon dioxide (CO_2), and 2 ft^3 of water vapor (H_2O). SAFETY PRECAUTION: It is essential that the furnace be vented properly so that these gases will all be dissipated into the atmosphere.

Cooling the flame will cause inefficient combustion. This happens when the flame strikes the sides of the combustion chamber (due to burner misalignment) and is called *flame impingement.* When the flame strikes the cooler metal of the chamber, the temperature of that part of the flame is lowered below the ignition temperature. This results in poor combustion and produces carbon monoxide and soot.

The percentages and quantities indicated above are for the burning of natural gas only. These figures vary considerably when propane, butane, propane-air, or butane-air are used. Always use the manufacturer's specifications for each furnace when making adjustments. SAFETY PRECAUTION: While taking flue-gas samples, be careful not to touch the hot vent pipe.

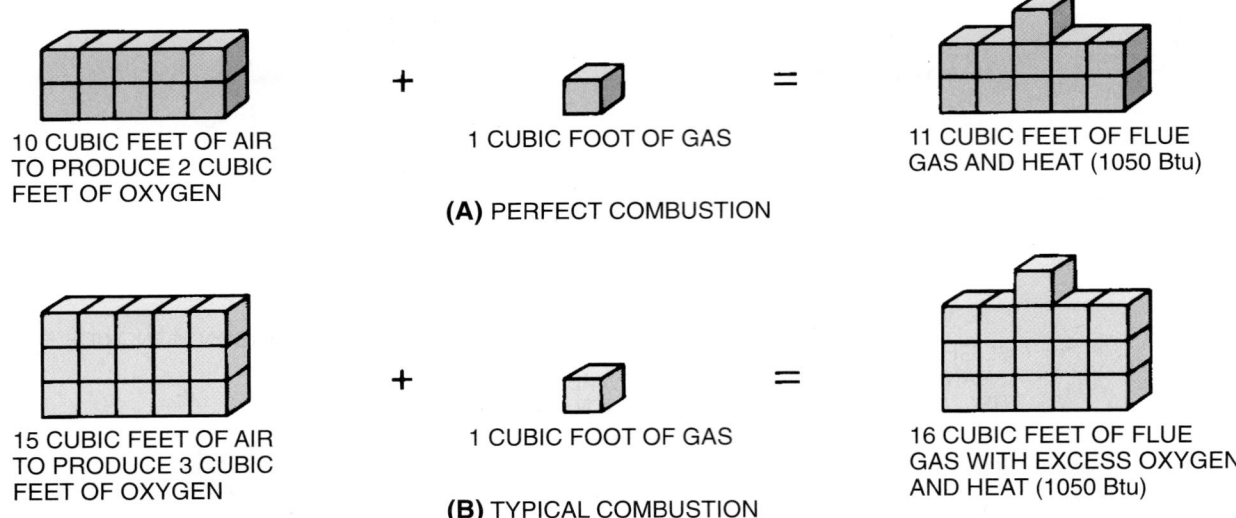

10 CUBIC FEET OF AIR TO PRODUCE 2 CUBIC FEET OF OXYGEN + **1 CUBIC FOOT OF GAS** = **11 CUBIC FEET OF FLUE GAS AND HEAT (1050 Btu)**

(A) PERFECT COMBUSTION

15 CUBIC FEET OF AIR TO PRODUCE 3 CUBIC FEET OF OXYGEN + **1 CUBIC FOOT OF GAS** = **16 CUBIC FEET OF FLUE GAS WITH EXCESS OXYGEN AND HEAT (1050 Btu)**

(B) TYPICAL COMBUSTION

Figure 31–14 The quantity of air for combustion.

31.5 GAS REGULATORS

Natural gas pressure in the supply line does not remain constant and is always at a much higher pressure than required at the manifold. The *gas regulator* drops the pressure to the proper level (in. WC) and maintains this constant pressure at the outlet where the gas is fed to the gas valve. Many regulators can be adjusted over a pressure range of several inches of water column, **Figure 31–15**. The pressure is increased when the adjusting screw is turned clockwise; it is decreased when the screw is turned counterclockwise. Some regulators have limited adjustment capabilities; others have no adjustment. Such regulators are either fixed permanently or sealed so that an adjustment cannot be made in the field. The natural gas utility company should be contacted to determine the uproper setting of the regulator. In most modern furnaces, the regulator is built into the gas valve, and the manifold pressure is factory set at the most common pressure: 3 1/2 in. WC for natural gas.

LP gas regulators are located at the supply tank. These regulators are furnished by the gas supplier. Check with the supplier to determine the proper pressure at the outlet from the regulator. The outlet pressure range is normally 10 to 11 in. WC. The gas distributor would normally adjust this regulator. In some localities a higher pressure is supplied by the distributor. The installer then provides a regulator at the appliance and sets the water column pressure to the manufacturer's specifications, usually 11 in. WC. LP gas installations normally *do not* use gas valves with built-in regulators.

Combination gas valves have built-in regulators in their bodies. Manufacturers usually will provide some sort of retrofit kit for converting from natural gas to LP gas. Many times a higher tension spring will have to be installed in the regulator at the gas valve to hold the valve open all the way. Some gas valves incorporate a push-pull knob. Pushing the knob down into the regulator will open the regulator all the way, which allows the regulator at the tank or the regulator at the side of the house to take control and regulate at 10 1/2–11 in. WC. Gas valves sold as LP valves already may have had this done to their regulators in some fashion. Always consult the manufacturer's specifications in setting regulators for any type of gas.

Converting from natural gas to LP gas involves many steps that may differ from manufacturer to manufacturer because of design differences of gas valves. For example, converting from natural gas to LP gas will include changing the regulator pressure from 3.5 in. WC by some method, installing smaller main burner orifices, and installing a new pilot orifice. This process also may include changing the ignition/control module to a 100% shutoff module because LP gas is heavier than air.

SAFETY PRECAUTION: Only experienced technicians should adjust gas pressures.

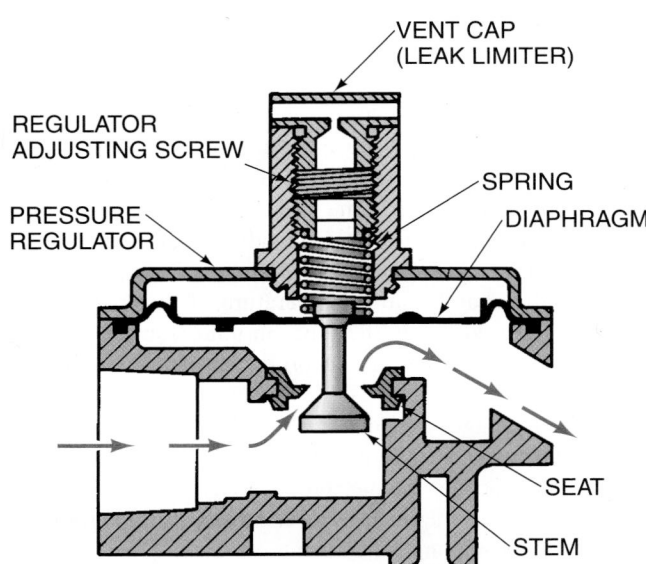

REGULATOR ADJUSTING SCREW

PRESSURE REGULATOR

VENT CAP (LEAK LIMITER)

SPRING

DIAPHRAGM

SEAT

STEM

Figure 31–15 The diagram of a standard gas pressure regulator.

31.6 GAS VALVE

From the regulator the gas is piped to the *gas valve* at the manifold, **Figure 31–16**. Several types of gas valves exist. Many are combined with pilot valves and called *combination* gas valves, **Figure 31–17**. We will first consider the gas valve separately and then in combination. Valves are generally classified as solenoid, diaphragm, and heat motor.

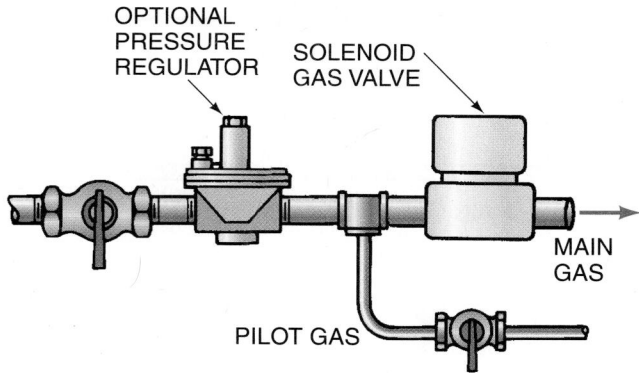

Figure 31–16 Natural gas installation where the gas passes through a separate regulator to the gas valve and then to the manifold.

(A)

(B)

Figure 31–17 Two gas valves with pressure regulator combined **(A)–(B)**. *(A) Courtesy Honeywell, Inc., Residential Division. (B) Courtesy Robertshaw Controls Company*

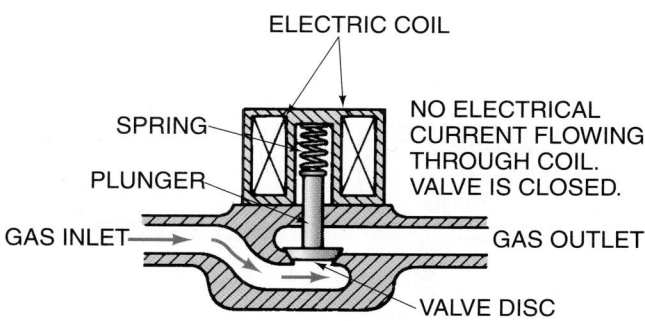

Figure 31–18 A solenoid gas valve in its normally closed position.

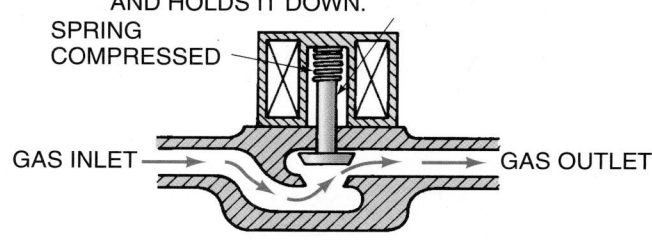

Figure 31–19 A solenoid valve in the open position.

31.7 SOLENOID VALVE

The gas-type *solenoid* valve is an NC valve, **Figure 31–18**. The plunger in the solenoid is attached to the valve or is in the valve. When an electric current is applied to the coil, the plunger is pulled into the coil. This opens the valve. The plunger is spring loaded so that when the current is turned off the spring forces the plunger to its NC position, shutting off the gas, **Figure 31–19**.

31.8 DIAPHRAGM VALVE

The *diaphragm* valve uses gas pressure on one side of the diaphragm to open the valve. When there is gas pressure above the diaphragm and atmospheric pressure below it, the diaphragm will be pushed down, and the main valve port will be closed, **Figure 31–20**. When the gas is removed from above the diaphragm, the pressure from below will push the diaphragm up and open the main valve port, **Figure 31–20**. This is done by a very small valve, called a *pilot-operated* valve because of its small size. It has two ports—one open while the other is closed. When the port to the upper chamber is closed and not allowing gas into the chamber above the diaphragm, the port to the atmosphere is opened. The gas already in this chamber is vented or bled to the pilot where it is burned. The valve controlling the gas into this upper chamber is operated electrically by a small magnetic coil, **Figure 31–21**.

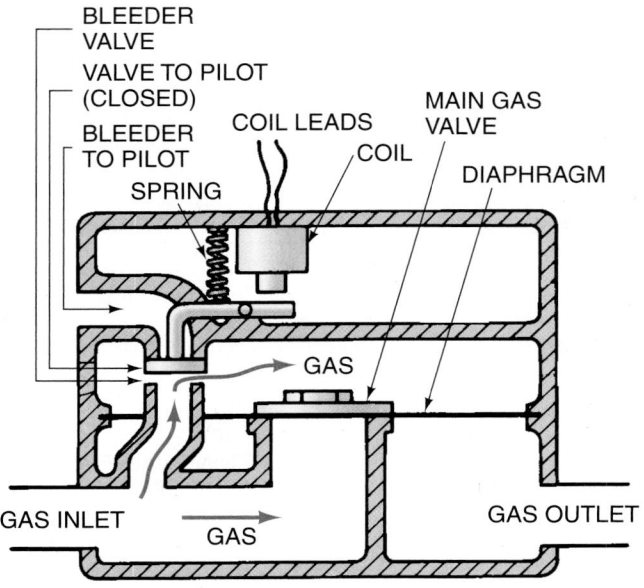

Figure 31–20 An electrically operated magnetic diaphragm valve.

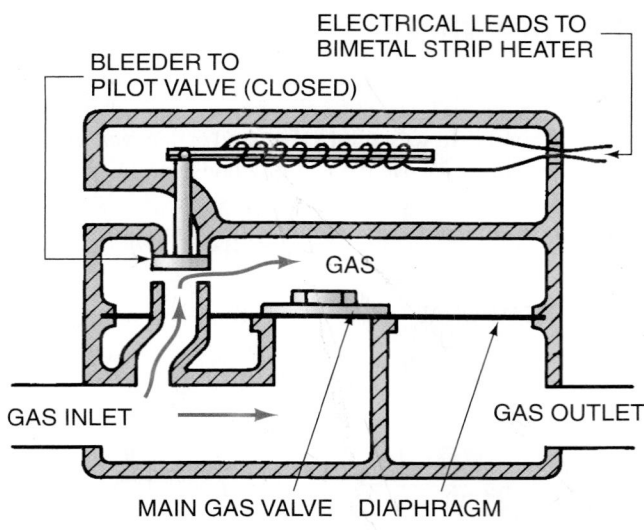

Figure 31–22 The thermally operated diaphragm gas valve.

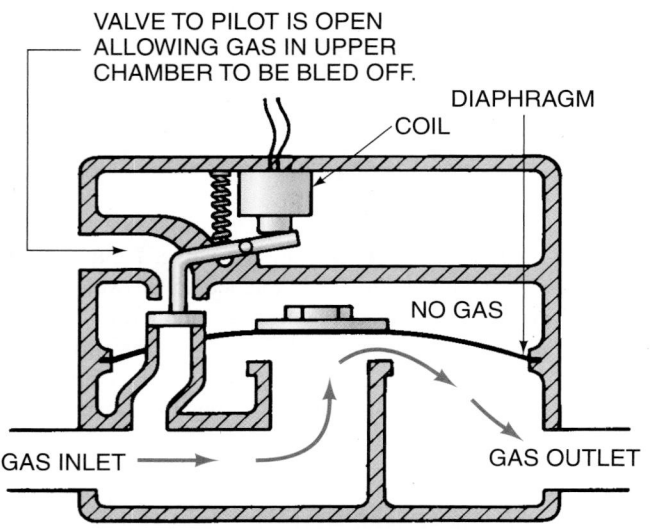

Figure 31–21 When an electric current is applied to the coil, the valve to the upper chamber is closed as the lever is attracted to the coil. The gas in the upper chamber bleeds off to the pilot, reducing the pressure in this chamber. The gas pressure from below the diaphragm pushes the valve open.

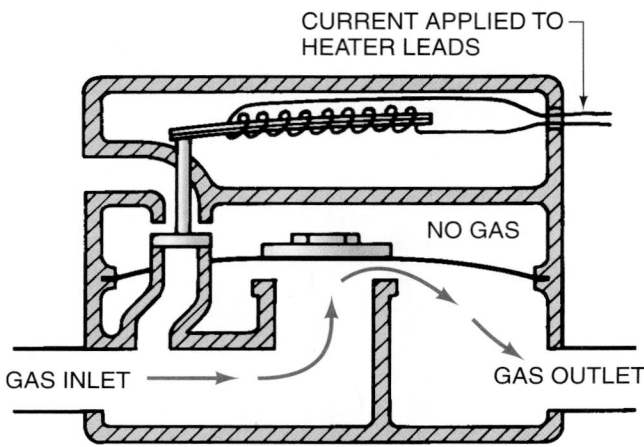

Figure 31–23 When an electric current is applied to the leads of the bimetal strip heater, the bimetal warps, closing the valve to the upper chamber, and opening the valve to bleed the gas from the upper chamber. The gas pressure is then greater below the diaphragm, pushing the valve open.

When the thermostat calls for heat in the thermally operated valve, a bimetal strip is heated, which causes it to warp. A small heater is attached to the strip, or a resistance wire is wound around it, **Figure 31–22.** When the strip warps, it closes the valve to the upper chamber and opens the bleed valve. The gas in the upper chamber is bled to the pilot where it is burned, reducing the pressure above the diaphragm. The gas pressure below the diaphragm pushes the valve open, **Figure 31–23.**

31.9 HEAT MOTOR–CONTROLLED VALVE

In a heat motor–controlled valve an electric heating element or resistance wire is wound around a rod attached to the valve, **Figure 31–24.** When the thermostat calls for heat, this heating coil or wire is energized and produces heat, which expands, or elongates, the rod. When elongated, the rod opens the valve, allowing the gas to flow. As long as heat is applied to the rod, the valve remains open. When the heating coil is deenergized by the thermostat, the rod contracts. A spring will close the valve.

It takes time for the rod to elongate and then contract. This varies with the particular model but the average time is 20 sec to open the valve and 40 sec to close it.

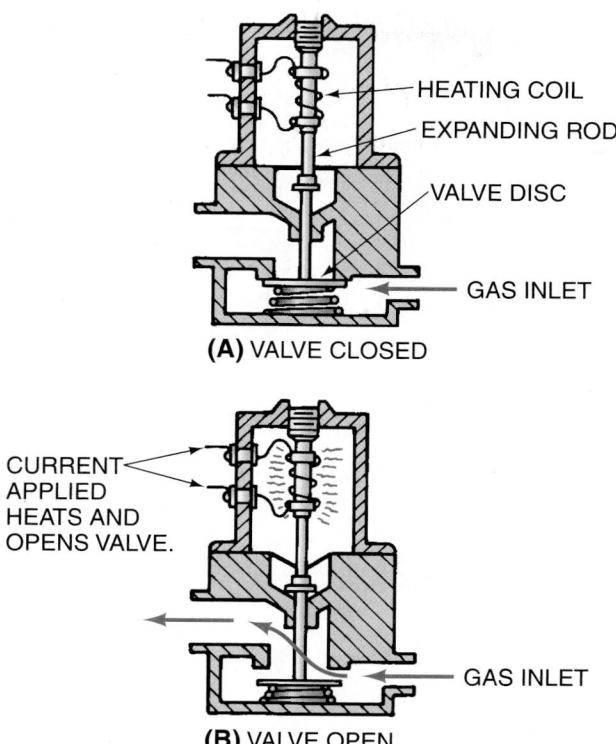

Figure 31–24 A heat motor–operated valve, **(A)** closed and **(B)** open.

Figure 31–25 An automatic combination gas valve. *Courtesy Honeywell, Inc., Residential Division*

SAFETY PRECAUTION: Be careful while working with heat motor gas valves because of the time delay. Because there is no audible click, you cannot determine the valve's position. Gas may be escaping without your knowledge.

31.10 AUTOMATIC COMBINATION GAS VALVE

Many modern furnaces designed for residential and light commercial installations use an automatic combination gas valve (ACGV), **Figure 31–25**. These valves incorporate a manual control, the gas supply for the pilot, the adjustment and safety shutoff features for the pilot, the pressure regulator, and the controls to operate the main gas valve. They often have dual shutoff seats for extra safety protection. This is also called the redundant gas valve. These valves also combine the features described earlier relating to the control and safety shutoff of the gas.

Modern combination gas valves also may include programmed safe lighting features, servo pressure regulation, choices of different valve operators, and installation aids.

Standing Pilot Automatic Gas Valves

The internals of a typical combination gas valve for a standing pilot system are shown in **Figure 31–26**. A standing pilot system has a pilot burner that is lit all the time. The operation of the valve is as follows.

The gas inlet is on the left of the valve. A screen or filter is usually installed at the factory to keep the valve free from dirt and miscellaneous debris. The gas then encounters the safety shutoff valve. Pushing the red reset button can manually open this valve. In fact, this is how the pilot burner is lit. Pushing the red reset button allows gas to flow to the pilot gas outlet. The pilot gas burner is then manually lit. The pilot flame engulfs the thermocouple, which starts generating a direct-current (DC) voltage. The voltage generation of a single standard thermocouple is usually 24 to 30 millivolts DC. A direct current is then generated through the thermocouple and the power unit. Thermocouples will be covered in more detail later in this unit.

The power unit consists of a low-resistance coil and an iron core. When the power unit is energized, it has enough power to hold the safety shutoff valve open against its own spring pressure. If the pilot flame is ever lost, the power unit will be deenergized because the thermocouple will quickly stop generating a millivoltage as it cools. This action will stop all gas flow.

The power unit does not have enough power to pull the assembly open once closed because it is simply a holding coil. Once closed, pushing the red reset button on the combination gas valve will manually reset the unit. The red reset button can only be pushed down when it is in the pilot position. **Figure 31–27** shows how a safety shutoff mechanism assembly works along with its shutoff circuit.

Once the gas passes the safety shutoff valve, it comes to the first automatic valve. The first automatic valve is controlled by the first automatic valve solenoid coil. The closing

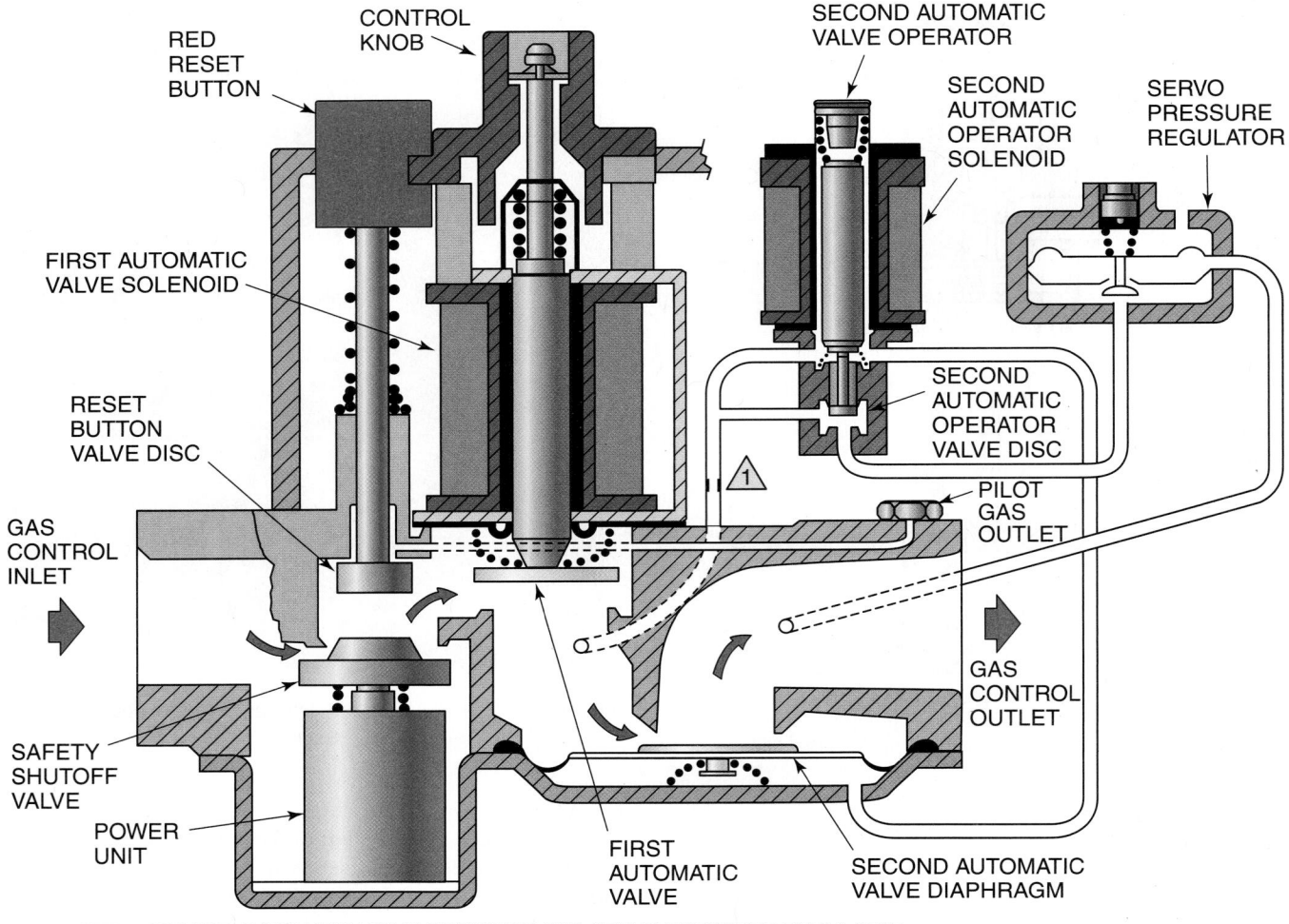

NOTE: SECOND AUTOMATIC VALVE OPERATOR AND SERVO PRESSURE REGULATOR SHOWN OUTSIDE GAS CONTROL FOR EASE IN TRACING GAS FLOW.

△1 SLOW-OPENING GAS CONTROL HAS A GAS FLOW RESTRICTOR IN THIS PASSAGE.

Figure 31–26 A standing pilot combination gas valve with servo pressure regulation. *Courtesy Honeywell, Inc., Residential Division*

SAFETY SHUTOFF

GAS COCK KNOB

TAPERED PLUG GAS COCK ⑤

SAFETY SHUTOFF VALVE DISC ④

CONTROL INLET

① PILOTSTAT POWER UNIT

② LOADING (DROP OUT) SPRING

③ ROCKER ARM

SAFETY SHUTOFF CIRCUIT

PILOTSTAT POWER UNIT

THERMOCOUPLE

(A)

(B)

Figure 31–27 **(A)** A safety shutoff mechanism. **(B)** A safety shutoff circuit. *Courtesy Honeywell, Inc., Residential Division*

of the room thermostat energizes its solenoid. The valve is spring loaded and opens against flowing gas pressure. When the solenoid is deenergized, the spring closes the valve. This is one of the valve's built-in safeties.

Once the room thermostat calls for heat, the first automatic valve is opened. The closing of the room thermostat also energizes the second automatic valve solenoid, which raises the second automatic operator valve disc. This in turn controls the position of the second automatic valve diaphragm. The second automatic valve diaphragm is a servo-operated valve, which means that its position is controlled by the pressure of gas coming to a chamber beneath its diaphragm. This modulates the valve diaphragm between the open and closed position. Putting more pressure under the valve diaphragm modulates the valve closed, and less pressure modulates it more in the open position.

The servo pressure regulator, an outlet or working pressure regulator, is an integral part of the second automatic gas valve. During the furnace's running cycle, the servo pressure regulator closely monitors the outlet or working pressure of the gas valve, even when inlet pressures and flow rates vary widely. This is quite different from a standard gas pressure regulator, which monitors the gas valve's inlet pressure, **Figure 31–28**. Any gas outlet pressure change is instantly sensed and reflected back to the diaphragm in the servo pressure regulator. The servo pressure regulator then repositions its diaphragm and disc to change the flow rate through the servo pressure regulator valve. This changes the pressure under the second automatic valve diaphragm and thus the position of the diaphragm.

If the working pressure or outlet pressure of the gas valve rises, the servo pressure regulator diaphragm moves slightly higher. This allows less gas to bleed out of the servo pressure regulator to the gas valve outlet. However, it increases the pressure under the second automatic valve diaphragm, causing the valve to close. This action reduces gas valve outlet pressure.

If the working pressure or outlet pressure of the gas valve drops, the servo pressure regulator diaphragm moves

slightly lower. This allows more gas to bleed out of the servo pressure regulator to the gas valve outlet. This decreases the pressure under the second automatic valve diaphragm, causing the valve to open. This action increases gas valve outlet pressure.

The servo pressure regulation system is a self-balancing system that works off gas valve outlet or working pressure and is independent of flow rate. It can operate over a wide range of gas flow rates. The gas valve's outlet pressure is adjusted at the top of the servo pressure regulator's body. Putting more or less tension on a spring will change the outlet or working gas valve pressure.

Intermittent Pilot Automatic Gas Valves

In an intermittent pilot (IP) automatic gas valve system, the pilot is lit every time there is a call for heat by the thermostat. Once the pilot is lit and proved, it lights the main burner. When the call for heat ends, the pilot is extinguished and does not relight until the next call for heat. In other words, there is automatic and independent control of the gas flow for both the pilot and the main burner.

Figure 31–29 shows the internals of a typical combination gas valve for an IP system. Note that there is no power unit or reset button. In fact, there is no thermocouple circuit either. There are simply two automatic valves. The first valve has an electric solenoid operator, and the second valve is servo operated. The passage for the pilot gas is located between these two main valves.

The first automatic valve solenoid is energized by an electronic module or integrated furnace controller (IFC) when there is a call for heat from the thermostat. This opens the passage for pilot gas to flow. The electronic module or IFC has about 90 sec to light the pilot gas. Four electronic modules and an IFC are shown in **Figure 31–30**. Once the pilot is lit and proved, usually through flame rectification, the ignition module or IFC energizes the second automatic valve solenoid. Flame rectification will be covered in detail later in this unit.

Because the second automatic valve is servo operated, the second main valve will open and the main burner will be lit. Its servo operator will operate the same way as described earlier for the standing pilot automatic gas valve. The main burner and the pilot burner will remain lit until the room thermostat opens and the call for heat ends. Both automatic valves will close once the call for heat ends. This action will shut off the main burner and pilot burner until the room thermostat initiates another call for heat.

Direct Burner Automatic Gas Valves

In a direct burner automatic gas valve system, the electronic module or IFC lights the main burner directly without a pilot flame. A spark, a hot surface igniter, or a glow coil usually accomplishes ignition. These types of ignition will be covered in detail later in this unit.

Figure 31–31 shows the internals of a typical combination gas valve for a direct burner system. Notice that there is no

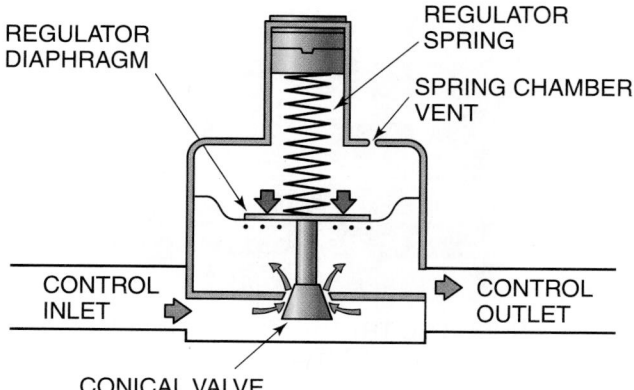

STANDARD PRESSURE REGULATOR

REGULATOR DIAPHRAGM

REGULATOR SPRING

SPRING CHAMBER VENT

CONTROL INLET

CONTROL OUTLET

CONICAL VALVE

Figure 31–28 A standard pressure regulator that monitors gas inlet pressure. *Courtesy Honeywell, Inc., Residential Division*

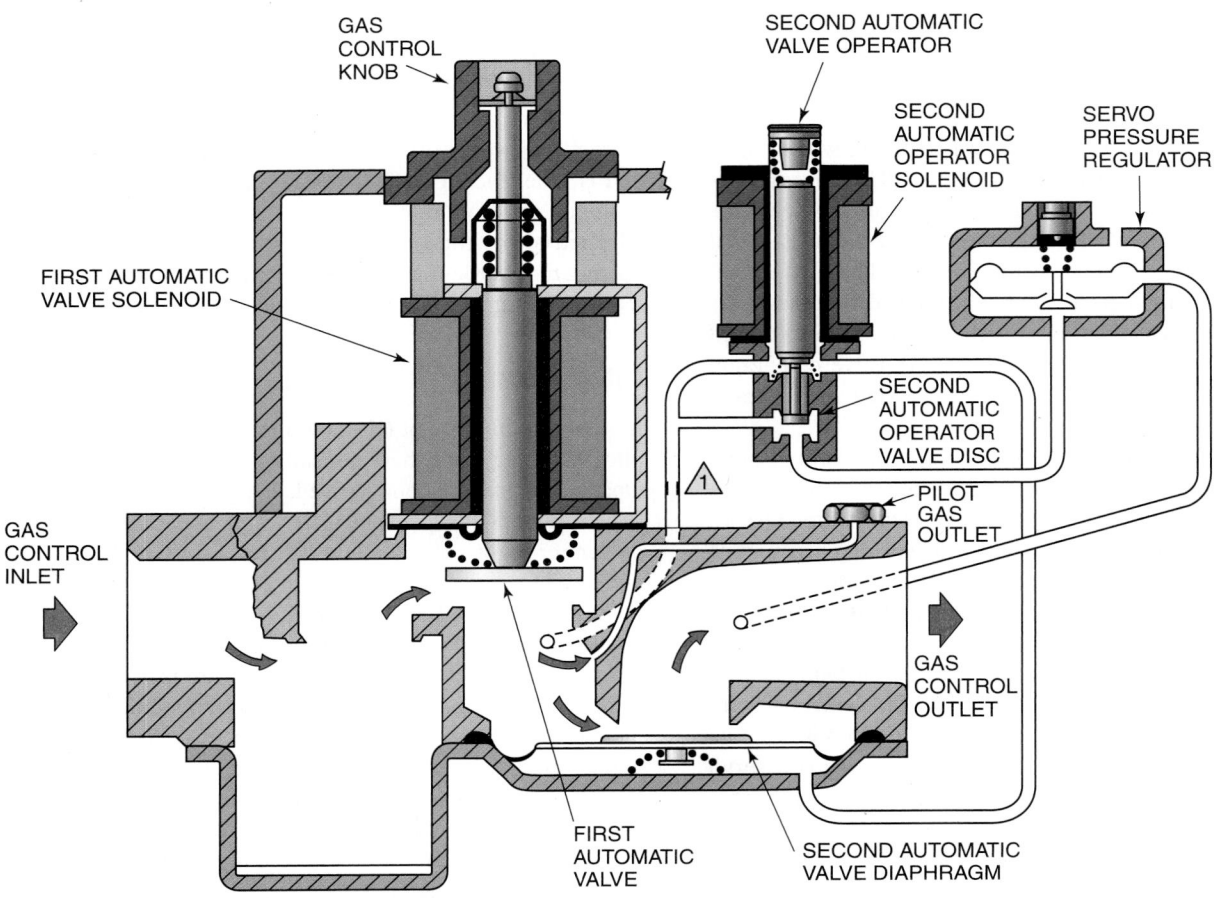

INTERMITTENT PILOT VALVE

NOTE: SECOND AUTOMATIC VALVE OPERATOR AND SERVO PRESSURE REGULATOR SHOWN OUTSIDE GAS CONTROL FOR EASE IN TRACING GAS FLOW.

⚠ SLOW-OPENING GAS CONTROL HAS A GAS FLOW RESTRICTOR IN THIS PASSAGE.

Figure 31–29 An intermittent pilot combination gas valve with servo pressure regulation. *Courtesy Honeywell, Inc., Residential Division*

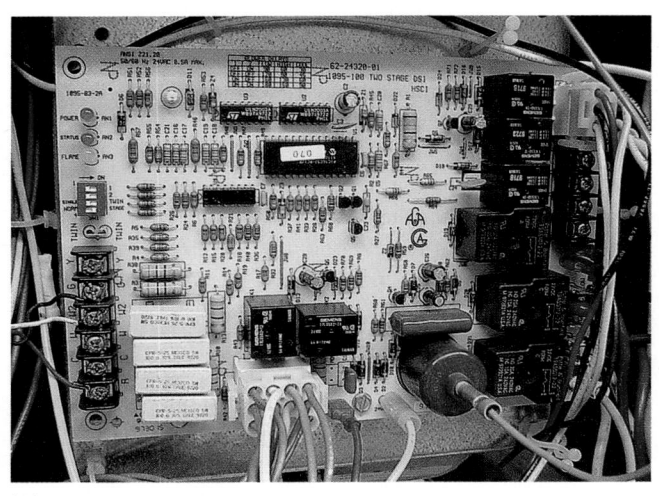

(A)

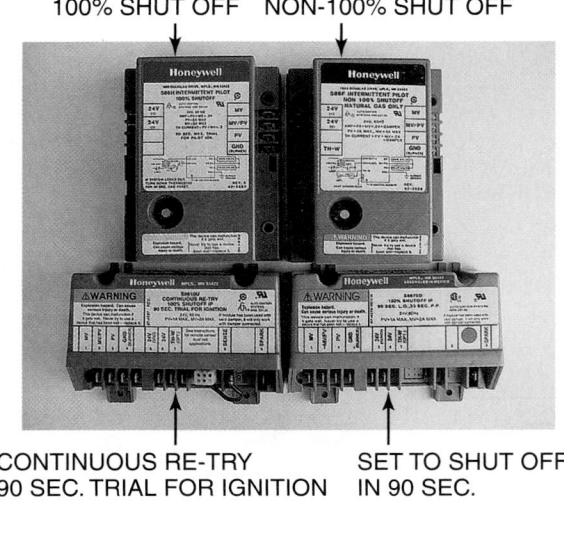

(B)

Figure 31–30 **(A)** An integrated furnace controller (IFC). **(B)** Four electronic modules for gas furnaces. *(A) and (B) Courtesy Ferris State University. Photos by John Tomczyk*

DIRECT BURNER IGNITION VALVE

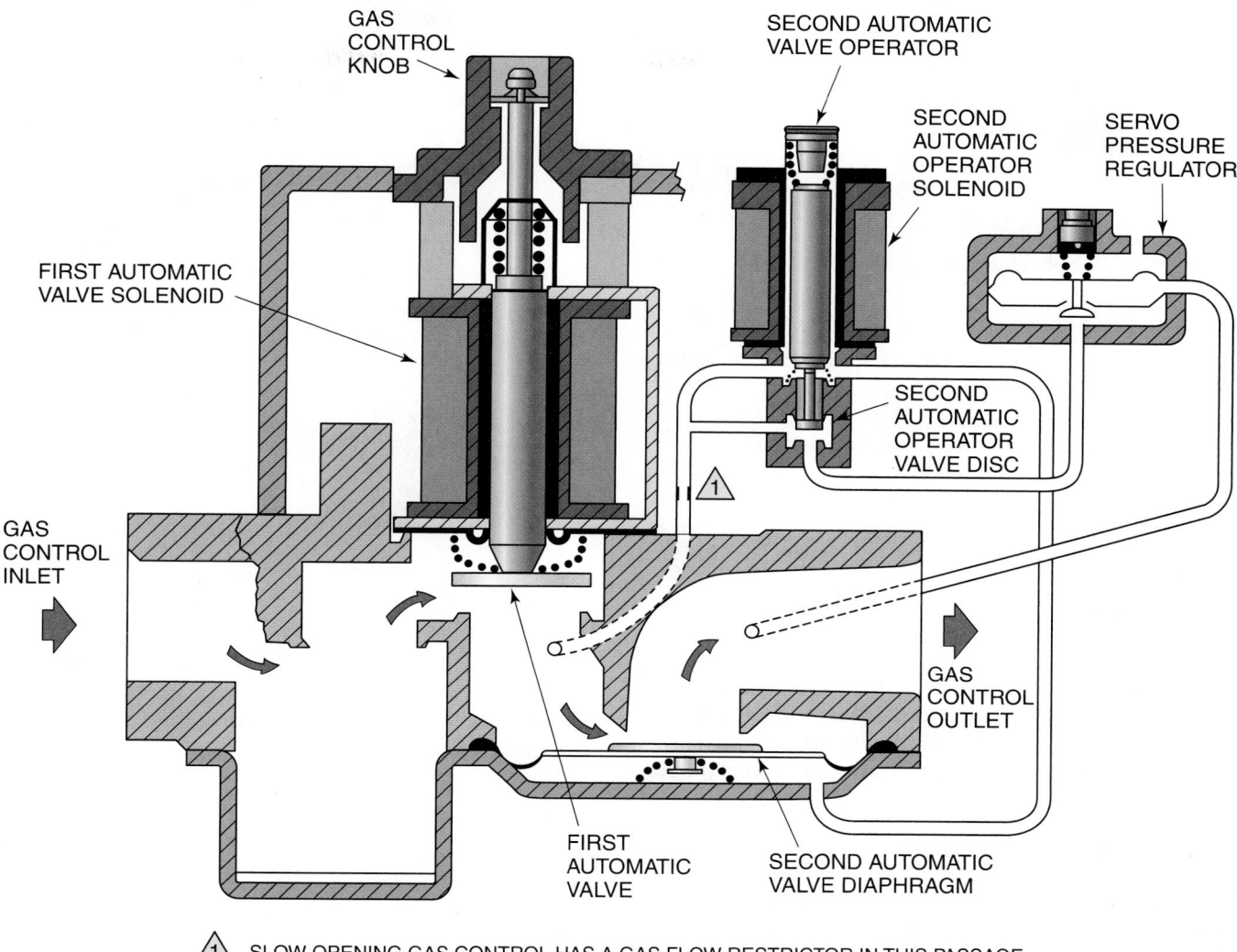

① SLOW-OPENING GAS CONTROL HAS A GAS FLOW RESTRICTOR IN THIS PASSAGE.

Figure 31–31   A direct burner combination gas valve with servo pressure regulation. *Courtesy Honeywell, Inc., Residential Division*

power unit or reset button. There are two automatic valves, and one is servo operated. The servo-operated valve will operate the same way as described for both the standing pilot and intermittent pilot combination gas valves.

In this system, when the room thermostat calls for heat, an electronic module or IFC energizes both first and second automatic valve solenoids at the same time along with the ignition source. Remember, there is no pilot in a direct burner system. Gas flows to the main burner and must be ignited within a short period of time, usually 4 to 12 sec. The reason for the short time period is safety because of the large flow of gas to the main burners compared to the small flow of gas to a pilot burner. Once the main burner is lit and proved, usually through flame rectification, the furnace will run until the room thermostat opens and the call for heat ends. If the main burner flame is not proved, gas flow will be shut off to the main burner through the electronic module or IFC.

Most modern combination gas valves have a built-in slow-opening feature. The second automatic valve opens slowly, letting the flame grow slowly in the main burner. This eliminates noise and concussion in the heat exchanger by giving the main burner a slow and controlled ignition. Slow opening may be accomplished through a restriction in the passage to the second automatic valve operator. This restriction controls and limits the rate at which gas pressure builds in the servo pressure regulator system. The restriction is shown in **Figure 31–26, Figure 31–29,** and **Figure 31–31.**

All three types of gas valves—standing pilot, intermittent pilot, and direct burner—are classified as redundant gas valves. A redundant gas valve has two or three valve operators physically in series but wired in parallel with one another. This built-in safety feature allows any operator (pilot or main) to block gas from getting to the main burner. **Figure 31–26, Figure 31–29,** and **Figure 31–31** are all examples of redundant

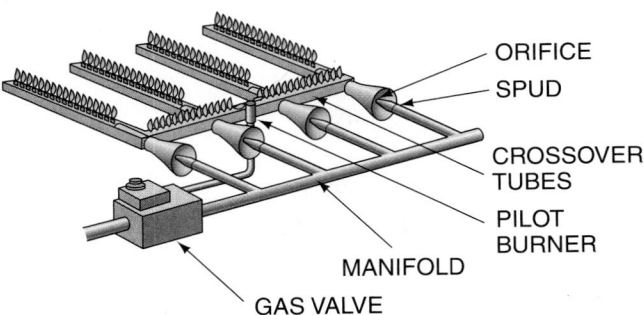

Figure 31–32   The gas valve feeds gas to the manifold.

gas valves. Most gas valves manufactured today are redundant gas valves.

SAFETY PRECAUTION: Most combination gas valves look alike. Their model number may be the only way to differentiate them. The correct combination gas valve must be used in each application, or serious injury can occur. Always contact the furnace or gas valve manufacturer if you are unsure of what gas valve to use.

31.11   MANIFOLD

The *manifold* in the gas furnace is a pipe through which the gas flows to the burners and on which the burners are mounted. The gas orifices are drilled into a spud, which is threaded into the manifold, and direct the gas into the venturi in the burner. The manifold is attached to the outlet of the gas valve, **Figure 31–32.**

31.12   ORIFICE

The *orifice* is a precisely sized hole through which the gas flows from the manifold to the burners. The orifice is located in the spud, **Figure 31–9.** The spud is screwed into the manifold. The orifice allows the correct amount of gas into the burner.

31.13   BURNERS

Gas combustion takes place at the burners. Combustion uses *primary* and *secondary air.* Primary air enters the burner from near the orifice, **Figure 31–13.** The gas leaves the orifice with enough velocity to create a low-pressure area around it. The primary air is forced into this low-pressure area and enters the burner with the gas. The procedure for adjusting the amount of primary air entering the burner is explained later in this unit.

The primary air is not sufficient for proper combustion. Additional air, called secondary air, is available in the combustion area. Secondary air is vented into this area through ventilated panels in the surface. Both primary and secondary air must be available in the correct quantities for proper combustion. The gas is ignited at the burner by the pilot flame.

The drilled-port burner is generally made of cast iron with the ports drilled. The slotted-port burner is similar to the drilled port except that the ports are slots, **Figure 31–33** and **Figure 31–34.** The ribbon burner produces a solid flame down

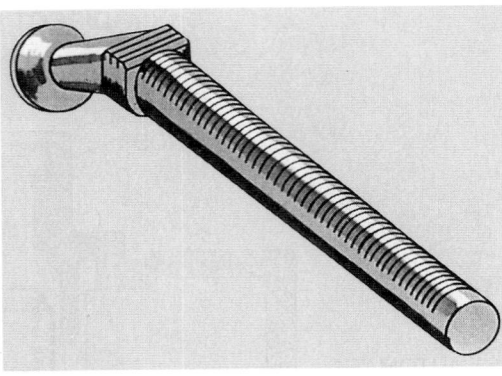

Figure 31–33   A cast-iron burner with slotted ports. *Reproduced courtesy of Carrier Corporation*

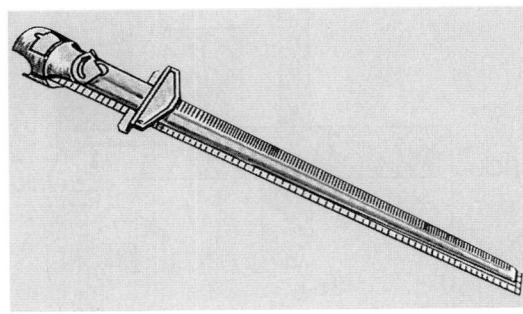

Figure 31–34   A stamped-steel slotted burner. *Reproduced courtesy of Carrier Corporation*

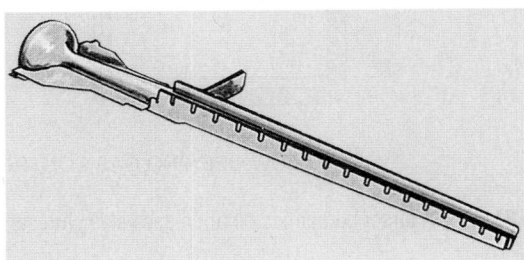

Figure 31–35   A ribbon burner. *Reproduced courtesy of Carrier Corporation*

the top of the burner, **Figure 31–35.** The single-port burner is the simplest and has, as the name implies, one port. This is often called the *inshot* or *upshot* burner, **Figure 31–36.** Most inshot burners are made of aluminized steel that has crossover porting for good flame retention, proper flame carryover to each main burner, and excellent flame stability. Inshot burners have no primary or secondary air adjustments. The burners simply slip over the orifice spuds and are screwed in place. **Figure 31–37** shows both inshot and upshot burners. Inshot burners are generally used with induced-draft systems, which pull or suck combustion gases through the heat exchanger. All of these burners are known as *atmospheric* burners because the air for the burning process is at atmospheric pressure. Most modern gas furnaces use either induced- or forced-draft systems. Induced-draft systems

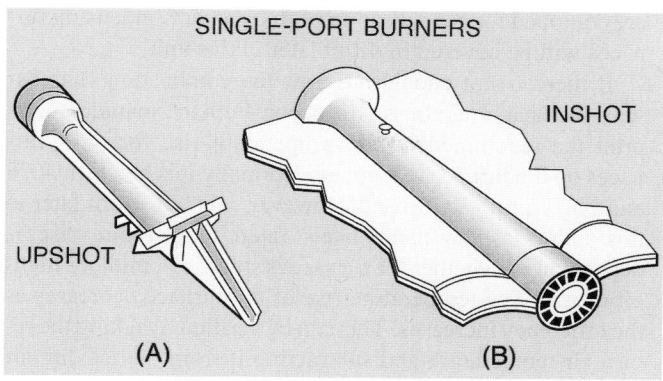

Figure 31–36   An inshot burner. *Reproduced courtesy of Carrier Corporation*

(A)

(B)

Figure 31–37   (A) Upshot burners. **(B)** Inshot burners. *(A) and (B) Courtesy Ferris State University. Photos by John Tomczyk*

have a combustion blower motor located on the outlet of the heat exchanger. They pull or suck combustion gases through the heat exchanger, usually causing a slight negative pressure in the heat exchanger itself. Forced-draft systems have the combustion blower motor on the inlet of the heat exchanger.

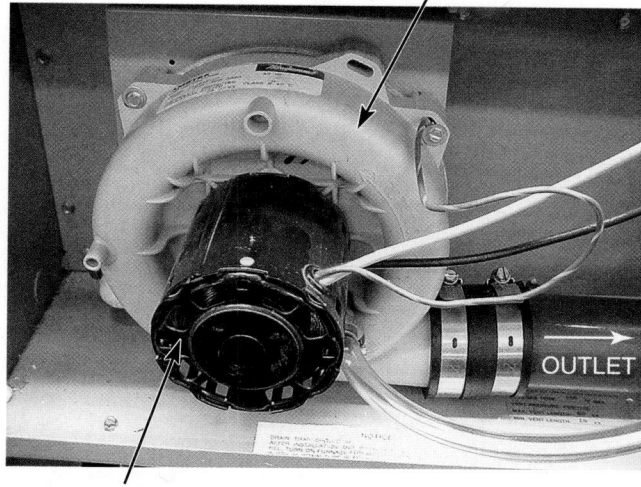

Figure 31–38   A combustion blower assembly. *Courtesy Ferris State University. Photo by John Tomczyk*

They push or blow combustion gases through the heat exchanger and cause a positive pressure in the heat exchanger. **Figure 31–38** shows a combustion blower motor assembly. Both induced- and forced-draft systems will be covered in detail later in this unit.

31.14   HEAT EXCHANGERS

Many burners are located at the bottom of the heat exchanger, **Figure 31–39**. However, modern high-efficiency gas furnaces have their burners located at the top of the

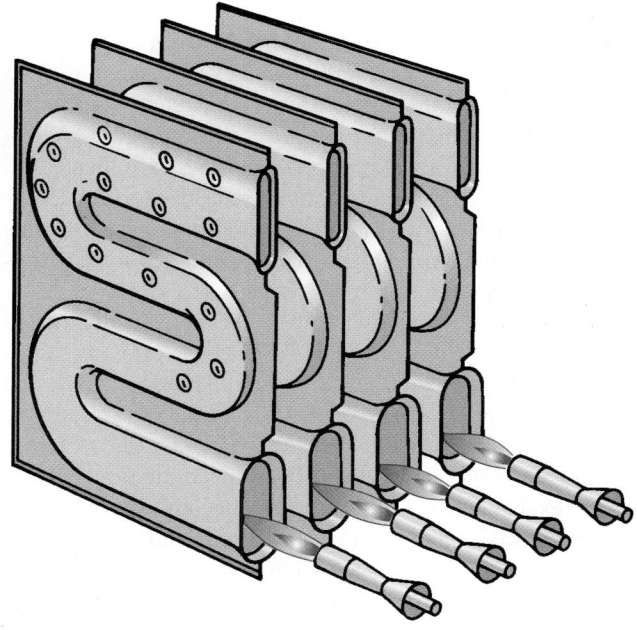

Figure 31–39   A heat exchanger with four sections.

❶ AIR FILTER

❷ ELECTRONIC
CONTROL
BOARD

❸ SEALED
COMBUSTION
SYSTEM

❹ SECONDARY
CONDENSING
HEAT
EXCHANGER

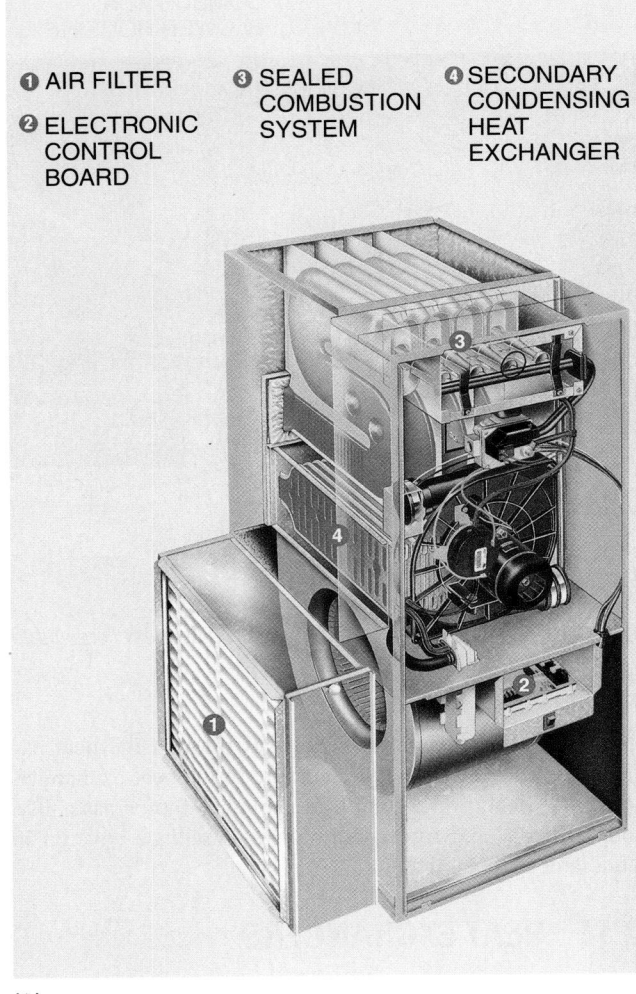

(A)

Figure 31–40 A   (A) A high-efficiency furnace with burners on top of the heat exchanger. *Courtesy Bryant Heating and Cooling Systems*

furnace in a sealed combustion chamber, **Figure 31–40(A).** The combustion gases are pulled or sucked through the heat exchanger. The combustion gases soon reach dew point, and liquid condensate forms and drains by gravity to a condensate sump or drain trap. The burners are located on top to allow the draining of condensate by gravity. High-efficiency furnaces will be covered in detail later in this unit. The *heat exchanger* is divided into sections with a burner in each section.

Heat exchangers are made of sheet steel or aluminized steel and are designed to provide rapid transfer of heat from the hot combustion products through the steel to the air that will be distributed to the space to be heated.

Heat exchangers must have the correct airflow across them or problems will occur. If too much air flows across a heat exchanger, the flue gas will become too cool and the products of combustion may condense and run down the flue pipe. These products are slightly acid and will deteriorate the flue system if not properly removed. However, high-efficiency condensing gas furnaces and their venting systems

are equipped for condensation of flue gases. Condensing furnaces will be covered in detail later in this unit.

If there is not enough airflow, the combustion chamber will overheat, and stress will occur. Furnace manufacturers print the recommended air temperature rise for most furnaces on the furnace nameplate. Normally it is between 40°F and 70°F for a gas furnace. However, you will learn later in this unit that the nameplate's rated temperature rise is dependent on whether the furnace is standard-, mid-, or high-efficiency. The temperature rise of the furnace decreases as the efficiency increases. This can be verified by taking the return air temperature and subtracting it from the leaving-air temperature, **Figure 31–40(B).** Be sure to use the correct temperature probe locations. Radiant heat can affect the temperature readings.

The following formula may be used to calculate the exact airflow across a gas furnace:

$$cfm = \frac{Q_s}{1.08 \times TD}$$

where Q_s = Sensible heat in Btu/h
$\quad$ cfm = Cubic feet of air per minute
$\quad$ 1.08 = Constant used to change cfm to pounds of air per minute for the formula
$\quad$ TD = Temperature difference between supply and return air

This requires some additional calculations because gas furnaces are rated by the input. If a gas furnace is 80% efficient during steady-state operation, you may multiply the input times the efficiency to find the output. For example, if a gas furnace has an input rating of 80,000 Btu/h and a temperature rise of 55°F, how much air is moving across the heat exchanger in cfm? The first thing we must do is find out what the actual heat output to the airstream is: 80,000 × 0.80 (80% estimated furnace efficiency) = 64,000 Btu/h. Use this formula:

$$cfm = \frac{Q_s}{1.08 \times TD}$$

$$cfm = \frac{64,000}{1.08 \times 55}$$

$$cfm = 1077.4$$

There is some room for error in the calculation unless you take several temperature readings at both the supply and return duct and average them. The more readings you take, the more accurate your answer. Also, the more accurate your thermometer is, the more accurate the results. Glass stem thermometers graduated in 1/4 degree increments are preferred. There is also the heat added by the fan motor. This will be about 300 W for a furnace this size. If you will multiply 300 W × 3.413 Btu/W, you will find some extra heat added to the air. 300 × 3.413 = 1023.9 Btu. For an even more accurate reading, you may take a watt reading of the motor or multiply the amperage reading times the applied voltage for a watt reading approximation.

Watts = Amperage × Applied voltage

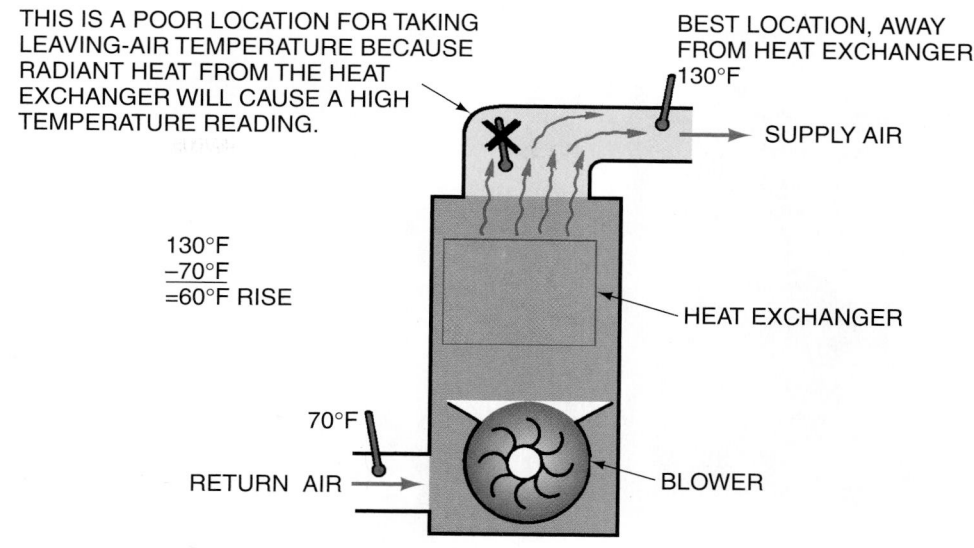

Figure 31–40 B *(continued)*   **(B)** Checking the air temperature rise across a gas furnace.

Poor combustion can corrode the heat exchanger, so good combustion is preferred. The steel in the exchangers may be coated or bonded with aluminum, glass, or ceramic material. These materials are more corrosion resistant, and the unit will last longer. Some exchangers are made of stainless steel, which is more expensive but resists corrosion extremely well. SAFETY PRECAUTION: The exchanger must not be corroded or pitted enough to leak since one of its functions is to keep combustion gases separated from the air to be heated and circulated throughout the building.

Modern heat exchangers come in many sizes, shapes, and materials. **Figure 31–41** shows a heat exchanger with both L-shaped and S-shaped tubes made of aluminized steel. The L tubes carry the hot combustion gases to a transition box. The S tubes then carry the combustion gases from the transition box to a collection box. The S tubes are smaller than the

L tubes. There is actually a serpentine path for combustion gases to travel. The collection box directs the combustion gases to an induced-draft centrifugal fan and then to the furnace's vent. The transition box is usually made of a highly corrosion-resistant stainless steel because it may see some condensation. The tubes are often swaged to the connection and transition boxes. The collection and transition box flanges are crimped for a tight fit.

Condensing furnaces may have three heat exchangers. The third or tertiary heat exchanger is typically made of stainless steel or aluminum to resist corrosion from the slightly acidic condensate, **Figure 31–42.** It may also have extended surfaces or fins for better heat conduction to the heating air. This condensing heat exchanger is designed to intentionally handle condensate and eventually drain it away from the furnace. One side of the heat exchanger has a plastic condensate drain

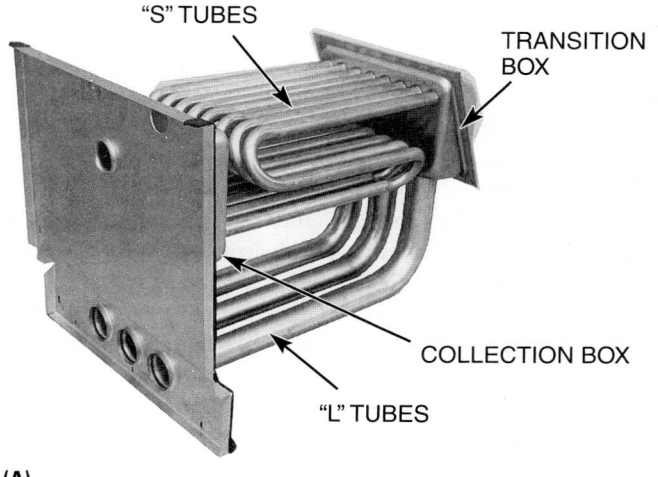

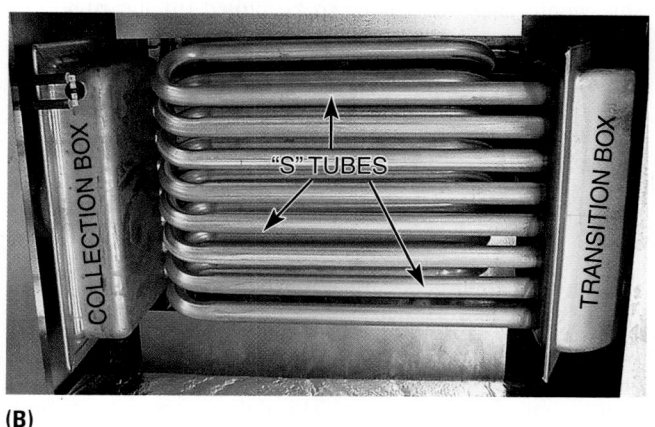

Figure 31–41   **(A)** A modern heat exchanger showing L and S tubes, a transition box, and a collection box. **(B)** A close-up of S tubes, a transition box, and a collection box. *Courtesy Rheem Manufacturing Company*

Figure 31–42 A stainless-steel condensing heat exchanger with aluminum fins. Notice the plastic condensate drain header on the left end. *Courtesy Ferris State University. Photo by John Tomczyk*

header. Some high-efficiency condensing furnaces have only two heat exchangers. If this is the case, the second heat exchanger is the condensing heat exchanger. **Figure 31–43** shows a high-efficiency condensing gas furnace with only two heat exchangers. The second heat exchanger resists corrosion from the slightly acidic condensate with a polypropylene plastic laminate coating.

Because of the serpentine path of combustion gases in most modern heat exchangers, they do not lend themselves well to the ribbon or slotted types of burners found in most traditional standard- and mid-efficiency furnaces, **Figure 31–34** and **Figure 31–35.** One of the main reasons for this is the height of the flame in the ribbon or slotted-port burner. Flame impingement would occur on the tubelike passages of these modern heat exchanger surfaces if ribbon or slotted-port burners were used. Also, if ribbon or slotted burners were used in modern tube-style heat exchangers, the secondary combustion air would flow at right angles to the burner flame. This would cause the flame to be pulled off the ribbon or slotted-port burner. It is for these reasons that the inshot burner is used in tube-style heat exchangers. The inshot burner, sometimes called a *jet burner,* fires a flame straight out its end, making it ideal for modern tube-style serpentine heat exchangers, **Figure 31–37(B).**

31.15 FAN SWITCH

The fan switch automatically turns the blower on and off. The blower circulates the heated air to the conditioned space. The switch can be temperature controlled or on a time delay. In either instance, the heat exchanger is given a chance to heat up before the fan is turned on. This delay keeps cold air from being circulated through the duct system before the heat exchanger gets hot at the beginning of the cycle. The heat exchanger must be hot in a conventional furnace because the

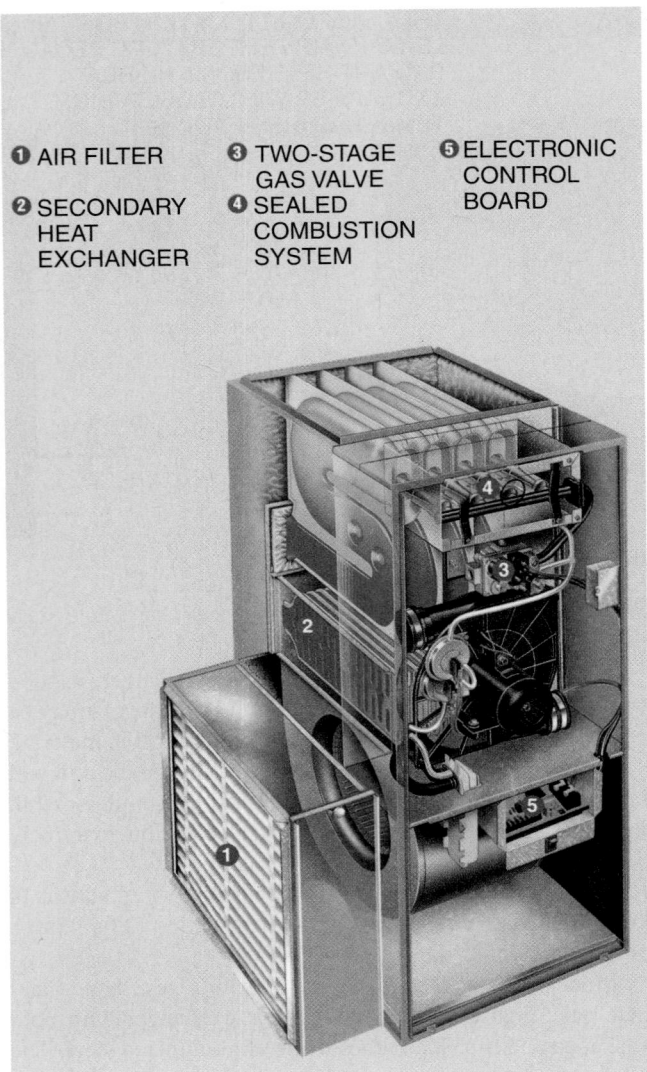

① AIR FILTER
② SECONDARY HEAT EXCHANGER
③ TWO-STAGE GAS VALVE
④ SEALED COMBUSTION SYSTEM
⑤ ELECTRONIC CONTROL BOARD

Figure 31–43 A modern, high-efficiency condensing gas furnace with a plastic polypropylene laminate coating on the second heat exchanger. *Courtesy Bryant Heating and Cooling Systems*

heat provides a good draft for properly venting the combustion gases.

There is also a delay in shutting off the fan. This allows the heat exchanger time to cool off and dissipate the furnace heat at the end of the cycle.

The temperature-sensing element of the switch, usually a bimetal helix, is located in the airstream near the heat exchanger, **Figure 31–44.** When the furnace comes on, the air is heated, which expands the bimetal and closes the contacts, thus activating the blower motor. This is called a *temperature on–temperature off* fan switch.

The fan switch could activate the blower with a time delay and shut it off with a temperature-sensing device. When the thermostat calls for heat and the furnace starts, a small resistance heating device is activated. This heats a bimetal strip that will close electrical contacts to the blower when heated. This provides a time delay and allows the furnace to heat the

(A)

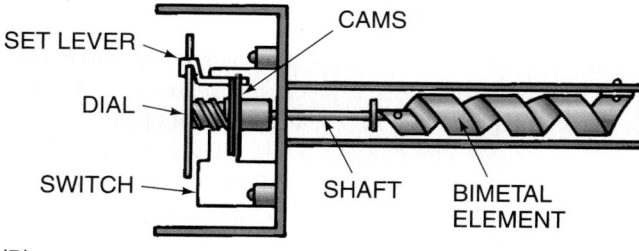

SET LEVER

CAMS

DIAL

SWITCH

SHAFT

BIMETAL
ELEMENT

(B)

HELICAL BIMETAL

(C)

Figure 31–44 **(A)** A temperature on–temperature off fan switch.
(B) Cutaway. **(C)** Helical bimetallic fan and limit controller located in
close proximity to the heat exchanger. **(A)** *Photo by Bill Johnson.*
(C) *Courtesy Ferris State University. Photo by John Tomczyk*

air before the blower comes on. The bimetal helix in the
airstream keeps the contacts closed even after the room thermostat is satisfied, and the burner flame goes out. When the
furnace shuts down, the bimetal cools at the same time as the
heat exchanger and turns off the blower. This fan switch provides a positive starting of the fan with temperature stopping
it and is called *time on–temperature off*.

A third type, called a *time on–time off* switch, uses a
small heating device, such as the one used in the time
on–temperature off switch. The difference is that this switch
is not mounted in such a way that the heat exchanger heat

will influence it. The time delay is designed into the switch
and is not adjustable. Most models of the other two switches
are adjustable.

Many modern gas furnaces use electronic modules or
IFCs to control the fan or blower's off-delay timing. **Figure 31–45(A)** shows an electronic module that uses dual in-
line pair (DIP) switches to control the fan's off-delay time.
The on-delay timing is for 30 sec and is nonadjustable. The
service technician cannot control when the fan will come on
once the burners light. This will always be within 30 sec for
the heating cycle. However, the technician does have adjustments for the fan or blower's off-delay timing for the fan to
shut off once the burners shut off. The four configurations
for the DIP switches in **Figure 31–45(B)** let the fan have a
60-, 100-, 140-, or 180-sec on-time period after the burners
shut off to cool down the heat exchanger. This electronic
module also controls the self-diagnostic functions of the gas
furnace.

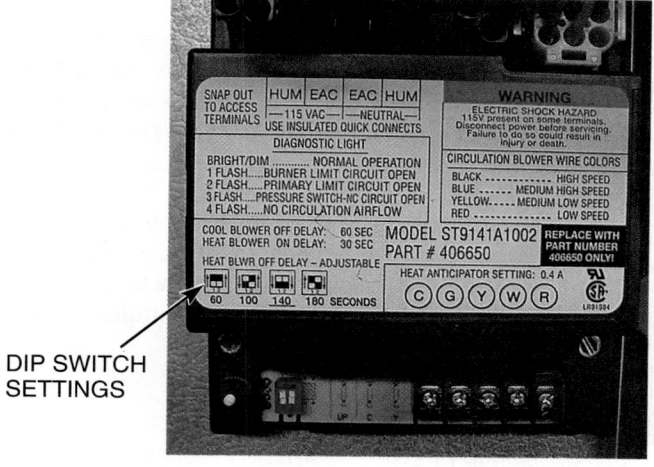

DIP SWITCH
SETTINGS

(A)

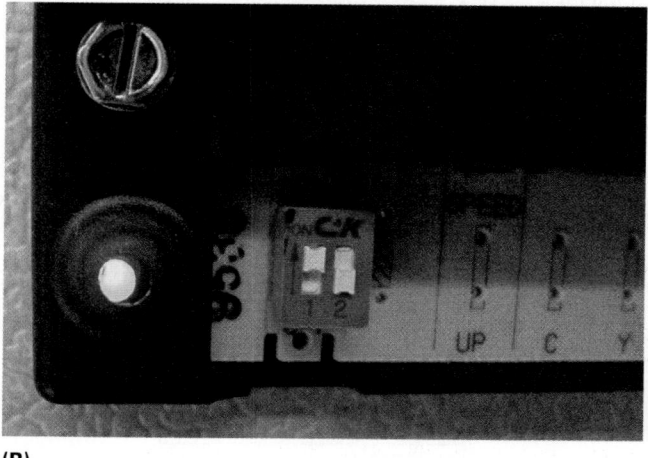

(B)

Figure 31–45 **(A)** An electronic module to control the blower's
off-delay timing. **(B)** The dual in-line pair switches for controlling
blower off-delay timing. **(A)** *and* **(B)** *Courtesy Ferris State University.*
Photos by John Tomczyk

IFCs often have nonadjustable blower fixed timing periods for both on and off delays, which are programmed into their software, **Figure 31–30(A)**. Because there are many manufacturers of gas furnaces, the service technician must understand the furnace's sequence of operation, including the blower's on and off times and delays.

31.16 LIMIT SWITCH

The *limit switch* is a safety device. If the fan does not come on or if there is another problem causing the heat exchanger to overheat, the limit switch will open its contacts, which closes the gas valve. SAFETY PRECAUTION: Almost any circumstance causing a restriction in the airflow to the conditioned space can make the furnace overheat, for example, dirty filters, a blocked duct, dampers closed, fan malfunctioning, or a loose or broken fan belt. The furnace may also be overfired due to an improper setting or malfunctioning of the gas valve. It is extremely important that the limit switch operate as it is designed to do.

The limit switch has a heat-sensing element. When the furnace overheats, this element opens contacts, thus directly or indirectly closing the main gas valve. This switch can be combined with a fan switch, **Figure 31–46(A)**. Different state codes have different high-limit cut-out requirements, often around 200°F to 250°F.

The combination fan and high-limit control in **Figure 31–46(B)** can be a high-voltage, low-voltage, or combination high- and low-voltage control. The actions of both the fan and the high-limit contacts are controlled by the same helical bimetal positioned next to the heat exchanger, **Figure 31–44(C)**. The combination control can be easily retrofitted for different voltages in the field. If the furnace's circuit calls for both the high-limit and fan control contacts to be high voltage, a jumper tab must remain in place in the fan/limit switch control, **Figure 31–46(B)**. However, if the furnace's circuit calls for a high-voltage fan control and a low-voltage limit control, the jumper tab must be removed to separate the different voltages. This makes the control more versatile for circuit wiring purposes. SAFETY PRECAUTION: Always read the manufacturer's literature for the contact's voltage rating before wiring or retrofitting any furnace control. Failure to do so may lead to a fire or explosion.

Many modern gas furnaces use snap-action, bimetallic, high-limit controls located in close proximity to the heat exchanger, **Figure 31–47**. These disks are set to snap open at a certain temperature dictated by the furnace manufacturer. Most high-limit controls will automatically reset back to the closed position when cooled to a certain temperature. However, some must be manually reset.

Bimetallic, snap-action disks are also used as flame roll-out safety controls, but, they usually have to be manually reset. An example is shown in **Figure 31–48**. Notice that it is wired in series with the high-limit control. If either of these controls overheats and opens its contacts, the circuit to the main gas valve will be interrupted. The stem coming out of the disk must be pushed in once the control has cooled down. This resets the control. Manual reset requires a service call by a technician to troubleshoot why the furnace has flame rollout problems.

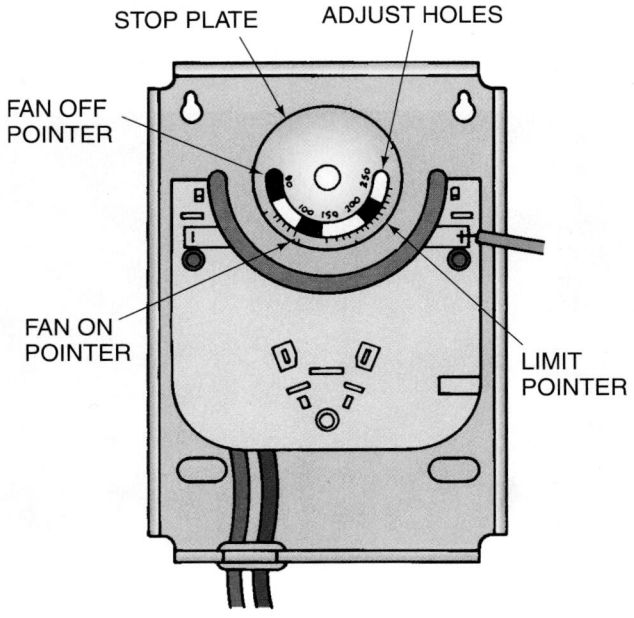

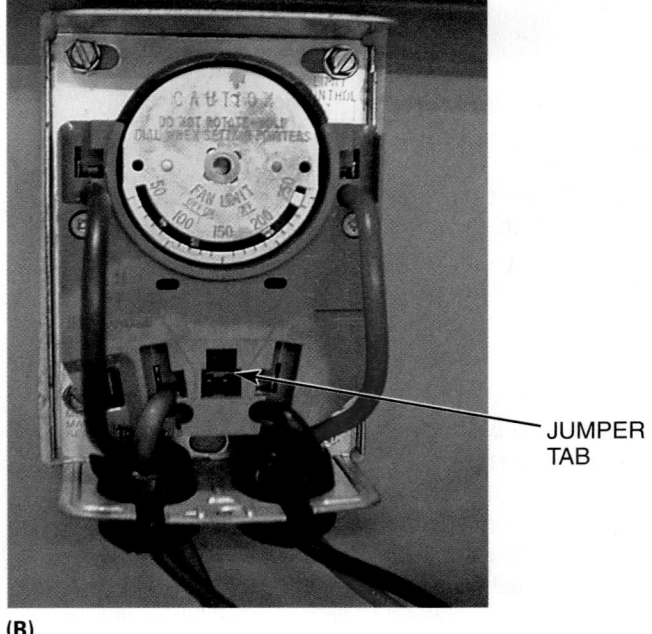

(A)   **(B)**

Figure 31–46 **(A)** A fan on-off and limit control in one switch. **(B)** A combination fan and limit control. *Courtesy Ferris State University. Photo by John Tomczyk*

(A) **(B)**

Figure 31–47 **(A)** A snap-action, bimetallic, high-limit control. The control is nonadjustable. **(B)** A snap-action, bimetallic, high-limit control positioned close to the heat exchanger. **(A)** and **(B)** Courtesy Ferris State University. Photos by John Tomczyk

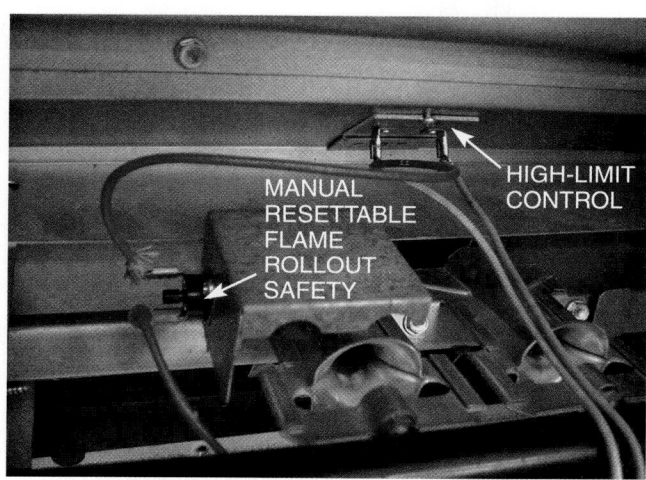

(A) **(B)**

Figure 31–48 **(A)** A manually reset flame rollout safety control. **(B)** A manual resettable flame rollout safety control. **(A)** and **(B)** Courtesy Ferris State University. Photos by John Tomczyk

31.17 PILOTS

Pilot flames are used to ignite the gas at the burner on most conventional gas furnaces. Pilot burners can be aerated or nonaerated. In the aerated pilot the air is mixed with the gas before it enters the pilot burner, **Figure 31–49.** The air openings, however, often clog and require periodic cleaning if there is dust or lint in the air. Nonaerated pilots use only secondary air at the point where combustion occurs. Little maintenance is needed with these, so most furnaces are equipped with nonaerated pilots, **Figure 31–50.**

The pilot is actually a small burner, **Figure 31–51.** It has an orifice, similar to the main burner, through which the gas passes. If the pilot goes out or does not perform properly, a safety device will stop the gas flow.

Standing pilots burn continuously; other pilots are ignited by an electric spark or other ignition device when the thermostat calls for heat. In furnaces without pilots, the ignition system ignites the gas at the burner. In furnaces with pilots, the pilot must be ignited and burning, and this must be proved before the gas valve to the main burner will open.

The pilot burner must direct the flame so there will be ignition at the main burners, **Figure 31–52.** The pilot flame also provides heat for the safety device that shuts off the gas flow if the pilot flame goes out.

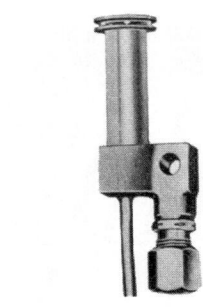

(A)

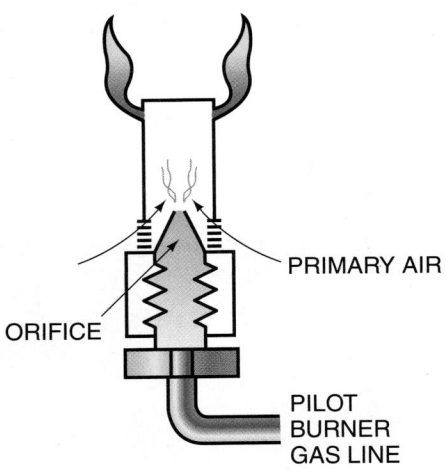

PRIMARY AIR

ORIFICE

PILOT
BURNER
GAS LINE

(B)

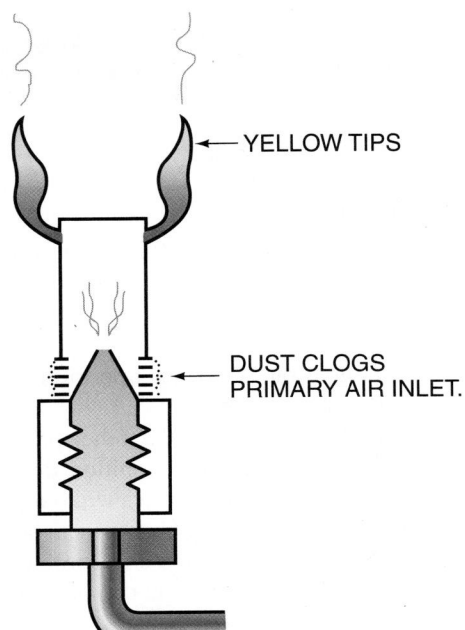

YELLOW TIPS

DUST CLOGS
PRIMARY AIR INLET.

(C)

Figure 31–49 **(A)** The aerated pilot. **(B)** Cutaway. **(C)** Clogged, and starved for air. *Courtesy Robertshaw Controls Company*

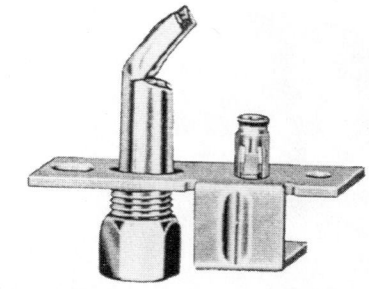

(A)

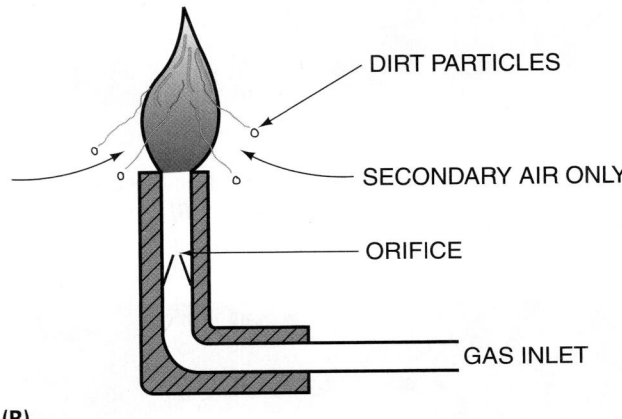

DIRT PARTICLES

SECONDARY AIR ONLY

ORIFICE

GAS INLET

(B)

Figure 31–50 **(A)** A nonaerated pilot. **(B)** Cutaway. Notice how dust particles burn away. *Courtesy Robertshaw Controls Company*

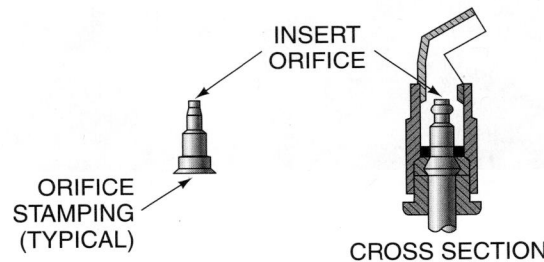

INSERT
ORIFICE

ORIFICE
STAMPING
(TYPICAL)

CROSS SECTION

Figure 31–51 A nonaerated pilot burner showing an orifice.

31.18 SAFETY DEVICES AT THE STANDING PILOT

Three main types of safety devices, called *flame-proving* devices, keep the gas from flowing through the main valve if the pilot flame goes out: the thermocouple or thermopile, the bimetallic strip, and the liquid-filled remote bulb.

Thermocouples and Thermopiles

The *thermocouple* consists of two dissimilar metals welded together at one end, **Figure 31–53,** called the "hot junction." When this junction is heated, it generates a small voltage (approximately 15 mV with load; 30 mV without load) across

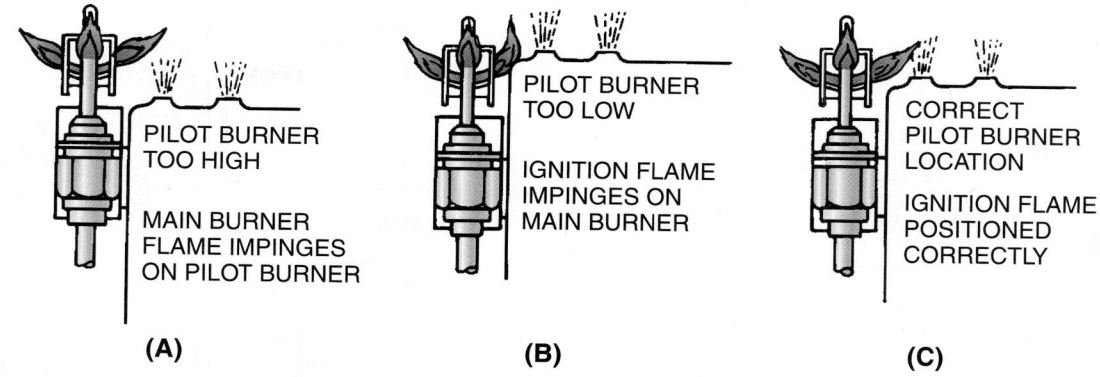

Figure 31–52 The pilot flame must be directed at the burners and adjusted to the proper height **(A)–(C)**.

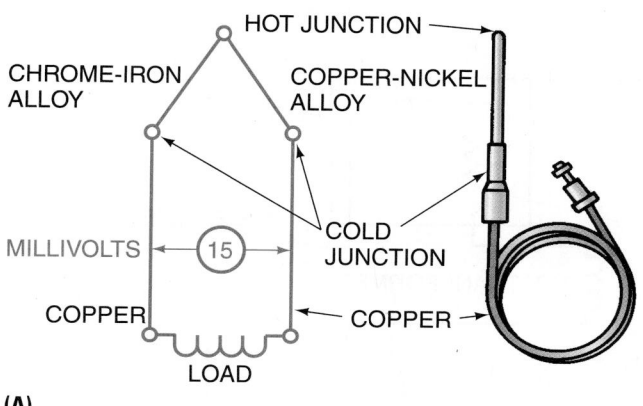

(A)

the two wires or metals at the other end. The other end is called the "cold junction." The thermocouple is connected to a shutoff valve, **Figure 31–54**. As long as the electrical current in the thermocouple energizes a coil, the gas can flow. If the flame goes out, the thermocouple will cool off in about 30 to

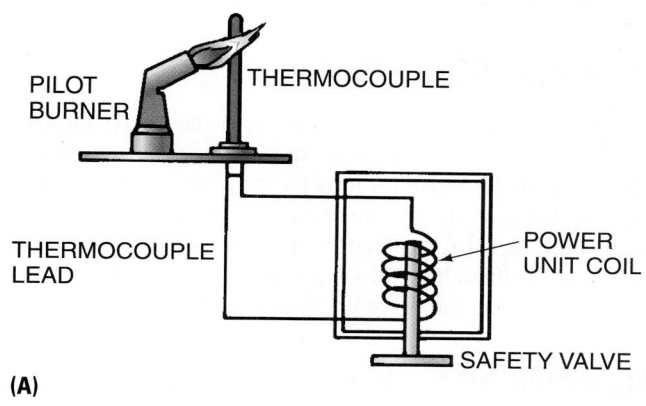

(A)

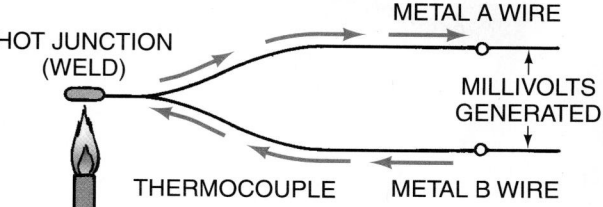

WHEN HEATED, ELECTRONS FLOW IN ONE DIRECTION IN ONE TYPE OF METAL (A) AND IN THE OPPOSITE DIRECTION IN THE OTHER TYPE (B).

(C)

Figure 31–53 The thermocouple **(A)–(C)**. *Courtesy Robertshaw Controls Company*

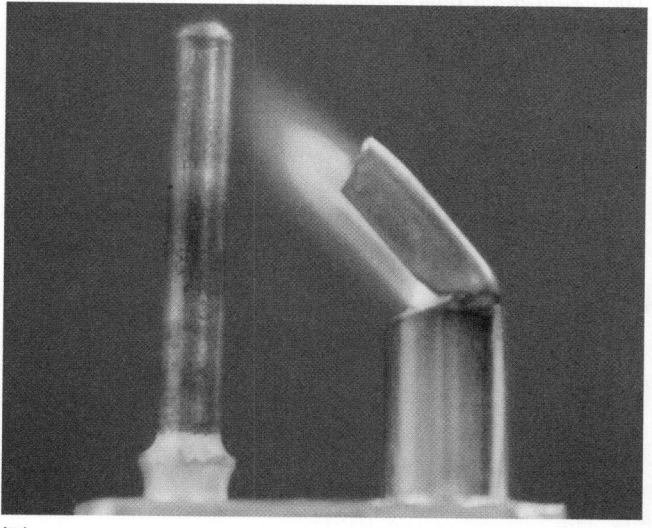

(B)

Figure 31–54 The thermocouple generates electrical current when heated by the pilot flame. This induces a magnetic field in the coil of the safety valve holding it open. If flame is not present, the coil will deactivate, closing the valve, and the gas will not flow. *Photo by Bill Johnson*

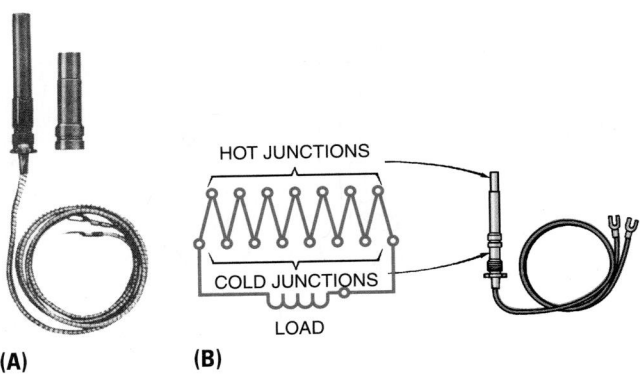

Figure 31–55 **(A)** A thermopile consists of a series of thermocouples in one **(B)** housing. **(A)** *Photo Courtesy Honeywell*

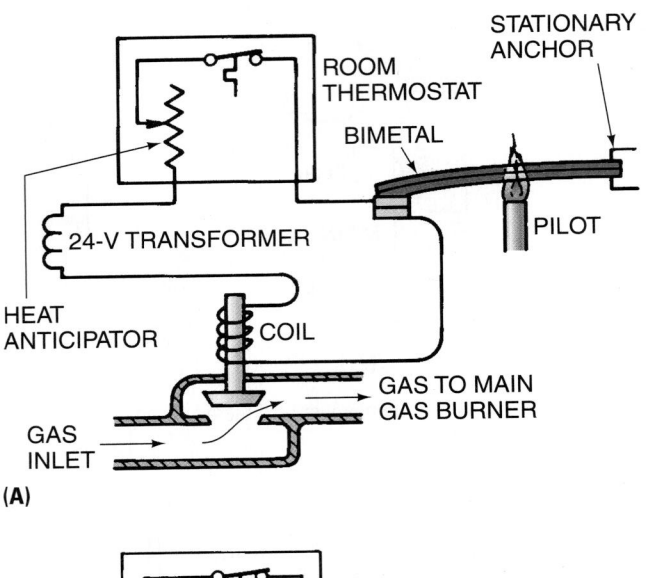

(A)

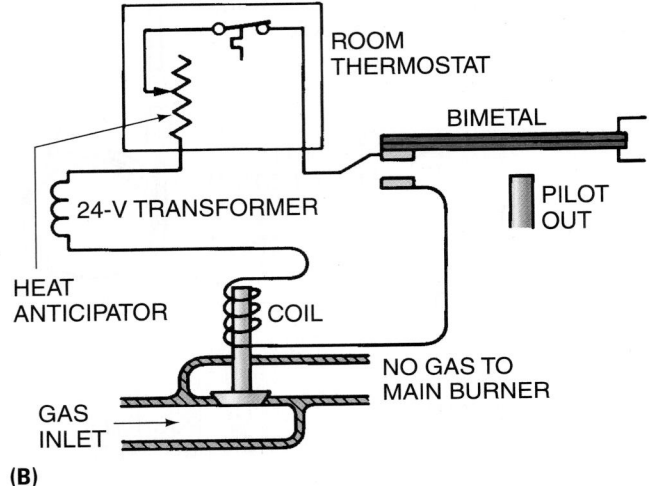

(B)

120 sec, and no current will flow and the gas valve will close. A *thermopile* consists of several thermocouples wired in series to increase the voltage. If a thermopile is used, it performs the same function as the thermocouple, **Figure 31–55**.

Bimetallic Safety Device

In the *bimetallic* safety device the pilot heats the bimetal strip, which closes electrical contacts wired to the gas safety valve, **Figure 31–56(A)**. As long as the pilot flame heats the bimetal, the gas safety valve remains energized and gas will flow when called for. When the pilot goes out, the bimetal strip cools within about 30 sec and straightens, opening contacts and causing the valve to close, **Figure 31–56(B)**.

Liquid-Filled Remote Bulb

The *liquid-filled remote bulb* includes a diaphragm, a tube, and a bulb, all filled with a liquid, usually mercury. The remote bulb is positioned to be heated by the pilot flame, **Figure 31–57**. The pilot flame heats the liquid at the remote bulb. The liquid expands, causing the diaphragm to expand, which closes contacts wired to the gas safety valve. As long as the pilot flame is on, the liquid is heated and the valve is open, allowing gas to flow. If the pilot flame goes out, the liquid cools in about 30 sec and contracts, opening the electrical contacts and closing the gas safety valve.

31.19 IGNITION SYSTEMS

Ignition systems can ignite either a pilot or the main burner. They can be divided into three categories:

■ Intermittent pilot (IP)
■ Direct-spark ignition (DSI)
■ Hot surface ignition (HSI)

Intermittent Pilot Ignition

In the intermittent pilot or spark-to-pilot type of gas ignition systems, a spark from the electronic module ignites the pilot, which ignites the main gas burners, **Figure 31–58**. The pilot

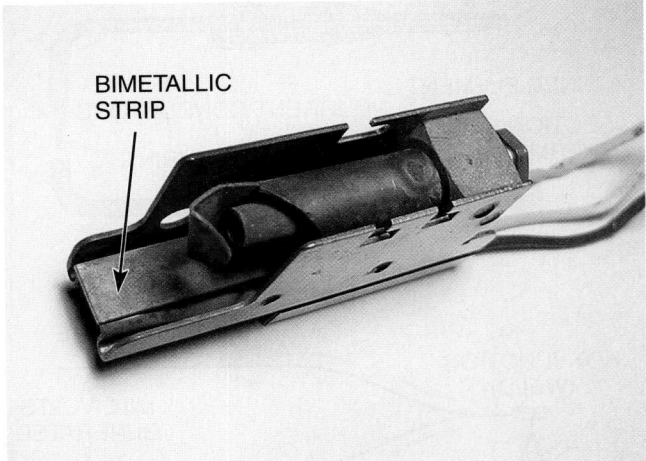

(C)

Figure 31–56 **(A)** When the pilot is lit, the bimetal warps, causing the contacts to close. The coil is energized, pulling the plunger into the coil opening the valve. This allows the gas to flow to the main gas valve. **(B)** When the pilot is out, the bimetal straightens, opening the contacts. The safety valve closes. No gas flows to the furnace burners. **(C)** A bimetallic safety device assembly. *Courtesy Ferris State University. Photo by John Tomczyk*

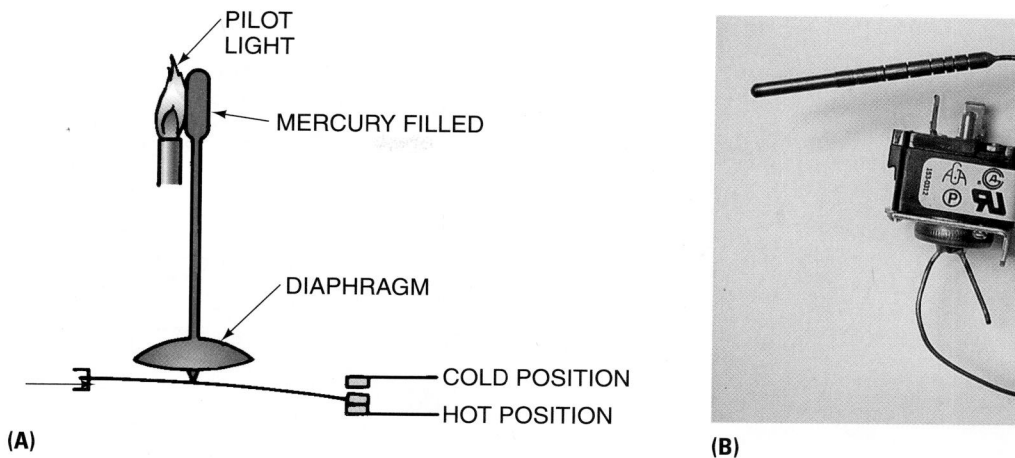

Figure 31–57   **(A)** A liquid-filled remote bulb. **(B)** A liquid-filled remote bulb assembly. *Courtesy Ferris State University. Photo by John Tomczyk*

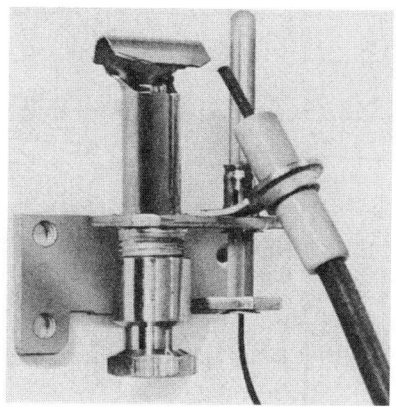

Figure 31–58   A spark-to-pilot ignition system. *Courtesy Robertshaw Controls Company*

burns only when the thermostat calls for heat. This system is popular because fuel is not wasted with the pilot burning when not needed. Two types of control schemes are used with this system. One is used a lot with natural gas and is not considered a 100% shutoff system. If the pilot does not ignite, the pilot valve will remain open, the spark will continue, and the main gas valve will not open until the pilot is lit and proved. The other type is used with LP gas and some natural gas applications and is a 100% shutoff system. If there is no pilot ignition, the pilot gas valve will close and may go into safety lockout after approximately 90 sec, and the spark will stop. Safety lockout systems will be covered in detail later in this unit. This system must be manually reset, usually at the thermostat or power switch.

The 100% shutoff is necessary for any gas fuel heavier than air, because it will accumulate in low places. Even the small amount of gas from a pilot light could be dangerous over time. Natural gas rises up through the flue system and is not dangerous in small quantities, such as the volume from a pilot light. Even though natural gas is lighter than air, some codes call for 100% shutoff. All technicians that service fuel gas products should be totally aware of the local codes. Both

100% shutoff and less than 100% shutoff systems will be covered in detail later in this unit.

When there is a call for heat in the natural gas system, contacts will close in the thermostat, providing 24 V to the ignition module which will send power to the pilot igniter and to the pilot valve coil. The coil opens the pilot valve, and the spark ignites the pilot.

Once the pilot flame is lit, it must be proved. Flame rectification is the fastest, safest, and most reliable method for proving a flame.

In the *flame rectification* system, the pilot flame changes the normal alternating current to direct current. The electronic components in the system will energize and open the main gas valve only with a direct current measured in microamps. It is important that the pilot flame quality be correct to ensure proper operation. Consequently, the main gas valve will open only when pilot flame is present. Flame rectification will be covered in detail later in this unit.

The spark is intermittent and arcs approximately 100 times per minute. It must be a high-quality arc, or the pilot will not ignite. The arc comes from the control module and is very high voltage. The voltage can reach 10,000 V in some systems but is usually very low amperage. A direct path to earth ground must be provided because the arc actually arcs to ground. A ground strap near the pilot assembly or the pilot hood often acts as ground, **Figure 31–59.**

Direct-Spark Ignition (DSI)

Many modern furnaces are designed with a spark ignition direct to the main burner or a ground strap, **Figure 31–59.** No pilot is used in this system. Components in the system are the igniter/sensor assembly and the DSI module. The sensor rod verifies that the furnace has fired and sends a microamp signal through flame rectification to the DSI module confirming this. The furnace will then continue to operate. This system goes into a "safety lockout" if the flame is not established within the "trial for ignition" period (approximately 4 to 11 sec). Gas is being furnished to the main burner, so there

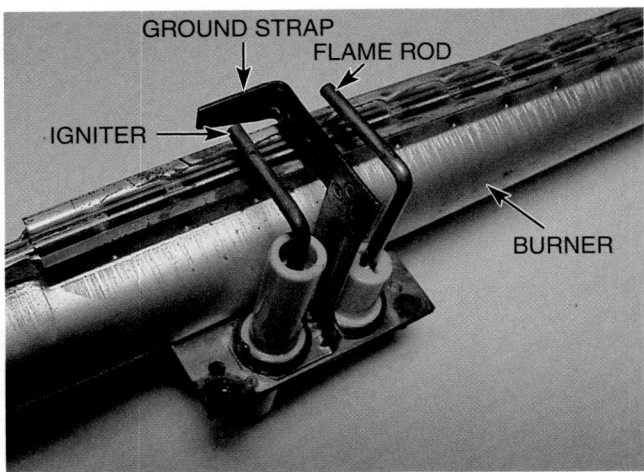

Figure 31–59 A direct-spark ignition assembly near the gas burner. *Courtesy Ferris State University. Photo by John Tomczyk*

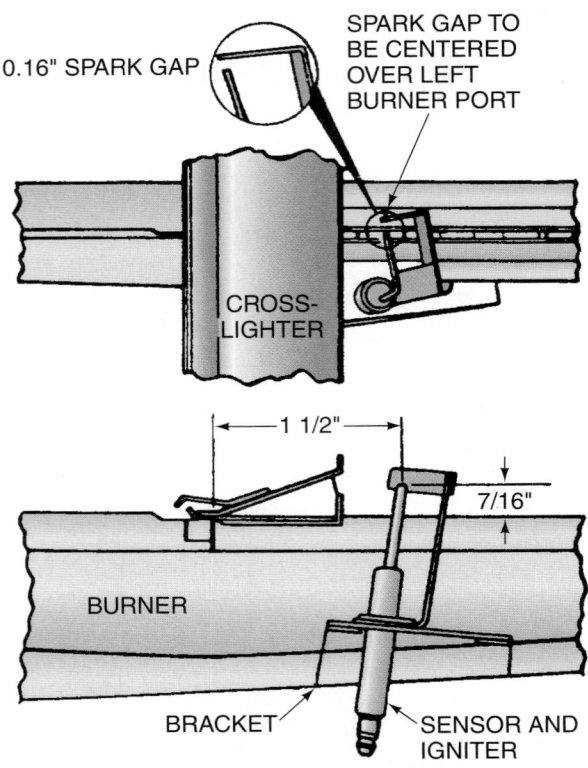

Figure 31–60 The spark gap and igniter position for a DSI system. *Courtesy Heil-Quaker Corporation*

cannot be much time delay compared with a 90-sec trial for ignition for IP systems. The system can then only be reset by turning the power off to the system control and waiting 1 min before reapplying the power. This is a "typical" system. The technician should follow the wiring diagram and manufacturer's instructions for the specific furnace being installed or serviced.

Most ignition problems are caused from improperly adjusted spark gap, igniter positioning, and bad grounding, **Figure 31–60.** The igniter is centered over the left port. Most manufacturers also provide specific troubleshooting instructions. Once the main burner flame is proved by the flame rectification system, the sparking will stop. If the sparking does not stop once the main burner is lit, the electronic DSI module could be defective. The continual sparking may not be harmful to the system, but it is noisy and often can be heard in the living or working area. Read and follow the manufacturer's instructions to repair the system.

Hot Surface Ignition

The hot surface ignition (HSI) system uses a special product called silicon carbide that offers a high resistance to electrical current flow but is very tough and will not burn up, like a glow coil. This substance is placed in the gas stream and is allowed to get very hot before the gas is allowed to impinge on the glowing hot surface. Immediate ignition should occur when the gas valve opens.

The hot surface igniter is usually operated off 120 V, the line voltage to the furnace, and draws considerable current when energized. It is energized for only a short period of time during start-up of the furnace so this current is not present for more than a very few minutes per day. **Figure 31–61(A,B,C)** shows the hot surface igniter.

Even though the HSI system has been used successfully for many years, some problems do occur. The igniter is very brittle and breaks very easily. If you bump it with a screwdriver, it will break.

If repeated failures occur with an HSI system, look for the following:

- Higher-than-normal applied voltage. Voltage that is in excess of 125 V can shorten the igniter's life.
- Accumulation of drywall dust, fiberglass insulation, or sealant residue.
- Delayed ignition can stress the igniter due to the small explosion.
- Overfiring condition.
- Furnace short cycling as with a dirty filter may cause the furnace to cycle on high limit.

The hot surface igniter may be used for lighting the pilot or for direct ignition of the main furnace burner. When it is used as the direct igniter for the burner, it will have very little time, usually 4 to 11 sec, to ignite the burner; then a safety lockout will occur to prevent too much gas from escaping.

Newer hot surface igniters are made of a different material that does not break as easily as older ones. Care should still be taken when working around hot surface igniters because of their fragile nature.

Some HSI systems operate on 24 V. These systems are usually hot surface to pilot, meaning that the 24-V hot surface igniter lights a pilot flame, **Figure 31–61(D)** and **Figure 31–61(E)**. The pilot flame has been the most reliable ignition source for lighting the main burner for many decades. Also, by using a low-voltage hot surface igniter, there is less chance for high-voltage electrical shock.

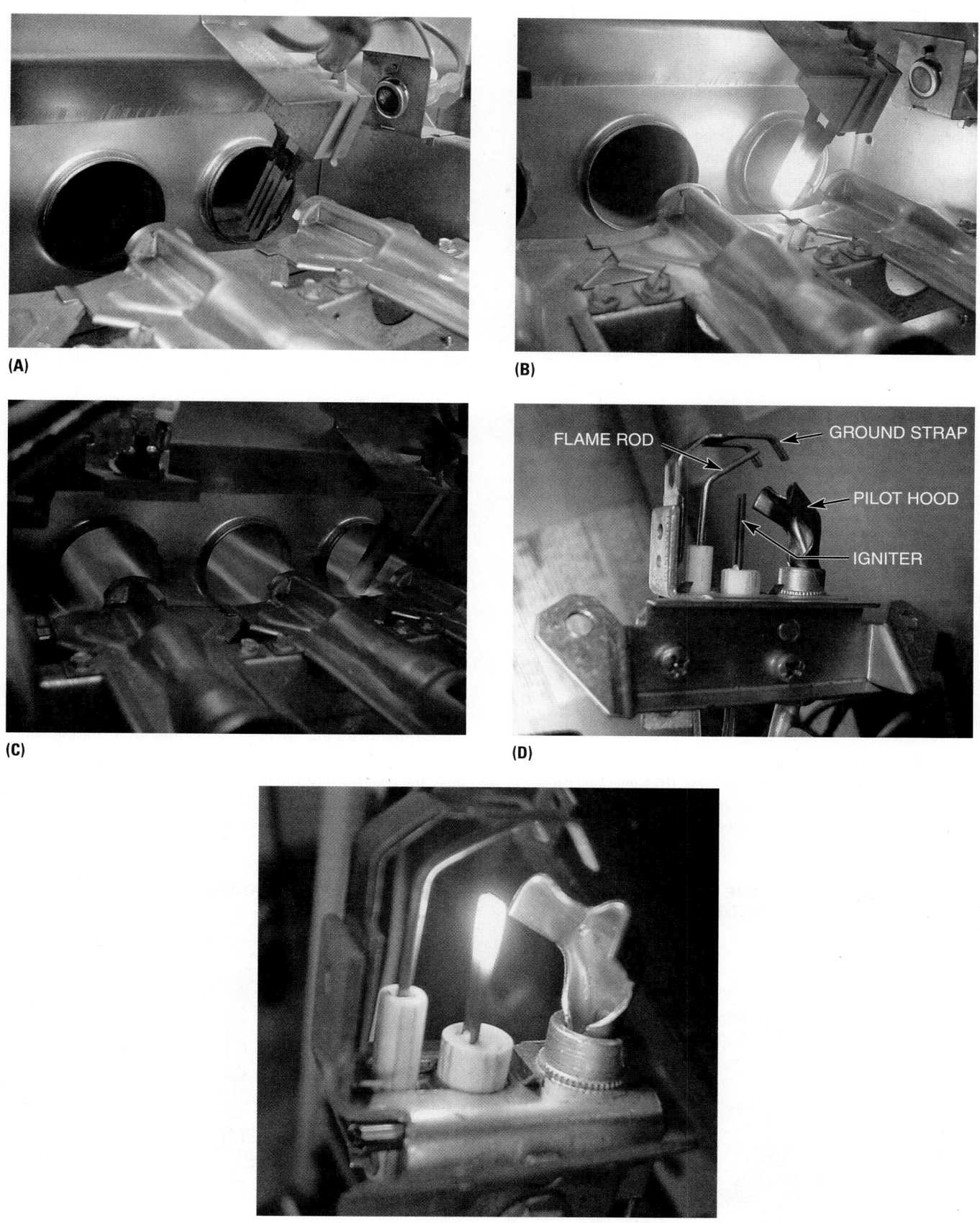

Figure 31–61 **(A)** A hot surface igniter deenergized. **(B)** The hot surface igniter system energized before main gas is allowed to flow. **(C)** The hot surface igniter system deenergized during the furnace heating cycle. **(D)** A 24-V hot surface igniter that lights a pilot flame. **(E)** The 24-V hot surface igniter energized. *(A)–(E) Courtesy Ferris State University. Photos by John Tomczyk*

31.20 FLAME RECTIFICATION

In the flame rectification system, the pilot flame, or main flame, can conduct electricity because it contains ionized combustion gases, which are made up of positively and negatively charged particles. The flame is located between two electrodes of different sizes. The electrodes are fed with an alternating-current (AC) signal from the furnace's electronic module.

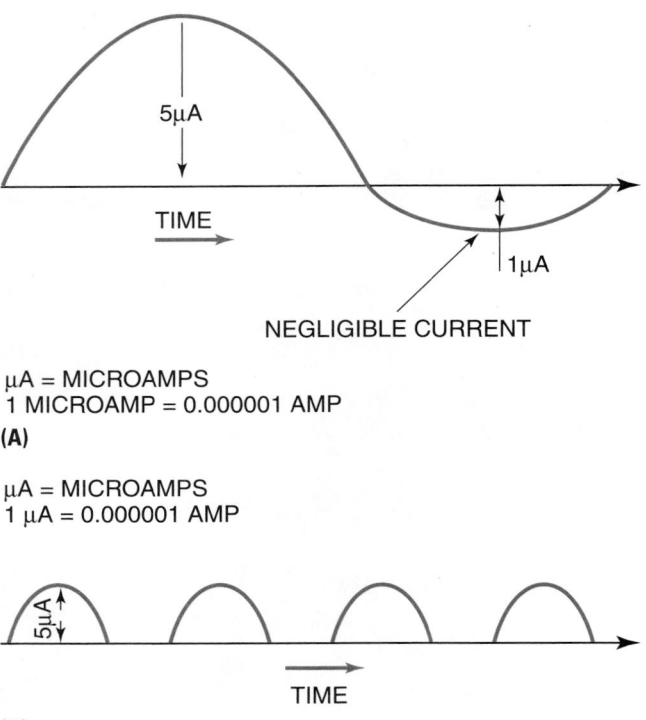

μA = MICROAMPS
1 MICROAMP = 0.000001 AMP
(A)

μA = MICROAMPS
1 μA = 0.000001 AMP

(B)

Figure 31–62 **(A)** The different sizes of the electrodes cause current to flow better in one direction than the other. **(B)** A pulsating direct-current (DC) signal that can be measured in microamps with a microammeter in series with one of the electrodes.

Because of the different size of the electrodes, current will flow better in one direction than in the other, **Figure 31–62(A)**. The flame actually acts as a switch. If there is no flame, the switch is open, and electricity will not flow. When the flame is present, the switch is closed and will conduct electricity.

The two electrodes in a spark-to-pilot (intermittent pilot) ignition system are the pilot hood (earth ground) and a flame rod or flame sensor, **Figure 31–63**. In a DSI system, the two electrodes can be the main burners (earth ground) and a flame rod or sensor. These types of systems are often referred to as dual-rod, or remote, sensing systems because one rod is for flame sensing and the other is for the ignition process. Often, the flame rod or sensor can be part of the ignition system and is referred to as a combination sensor and igniter, **Figure 31–60**. This type of system is referred to as a single-rod, or local, sensing system because it has a combination sensor and igniter in one rod.

As alternating current switches the polarity of the electrodes 60 times each second, the positive ions and negative electrons move at various speeds to the different size electrodes. This causes a rectified AC signal, which looks much like a pulsating DC signal, because the positively charged particles (ions) are relatively larger, heavier, and slower than the negatively charged particles (electrons), **Figure 31–62(B)**. In fact, the electrons are about 100,000 times lighter than the ions. This DC signal is what the electronics of the furnace recognize. The DC signal is proof that a flame exists, rather than a short resulting from humidity or direct contact between the electrodes. Humidity or direct contact between the electrodes would cause an AC current to flow, and the furnace's electronic module would not recognize the AC signal. The module only recognizes the DC signal. An AC signal would then prevent the main gas valve from opening in a spark-to-pilot system. It would shut down the main gas in a DSI system once the trial for ignition timing has elapsed and may put the furnace in a lockout mode. This DC signal can be measured with a microammeter in series with one of the electrode leads.

**FLAME RECTIFICATION
PILOT AND PROBE**

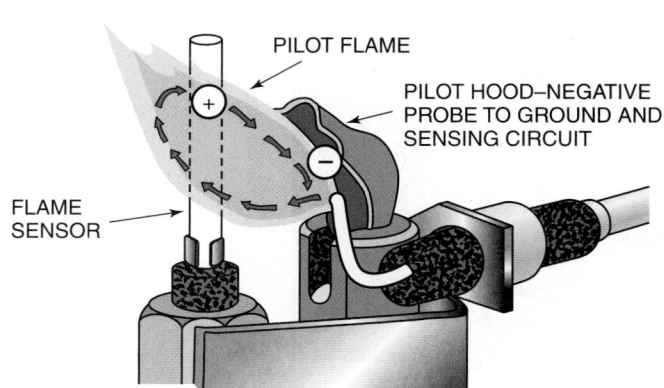

RECTIFICATION CIRCUIT

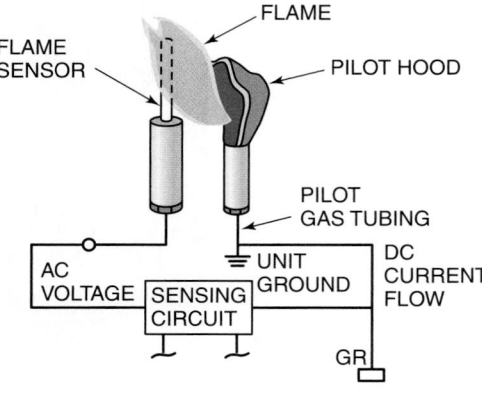

Figure 31–63 An intermittent pilot dual-rod system showing the pilot hood and the flame sensor rod as the two electrodes for flame rectification. Dual-rod systems are often referred to as remote sensing systems. *Courtesy Johnson Controls*

This signal usually ranges between 1 and 25 microamps. The magnitude of the microamp signal may depend on the quality, size, and stability of the flame and on the electronic module and electrode design. Always consult with the furnace manufacturer for specific information on flame rectification measurements.

Single-rod, or local, sensing consists of an igniter and a sensor all in one rod. It is often referred to as a one-rod pilot. Single-rod flame rectification systems use a single rod to accomplish both ignition and sensing. There is only one wire, which is the large ignition wire, running from the pilot assembly to the ignition module, **Figure 31–64.** This high-voltage lead does both the sparking and the sensing. The same rod used for ignition is used for flame rectification. The ignition pulse is accomplished on one half cycle and the sensing pulse current is accomplished on the other half cycle, **Figure 31–65.** Signal analyzers within the electronic module can differentiate between the two signals.

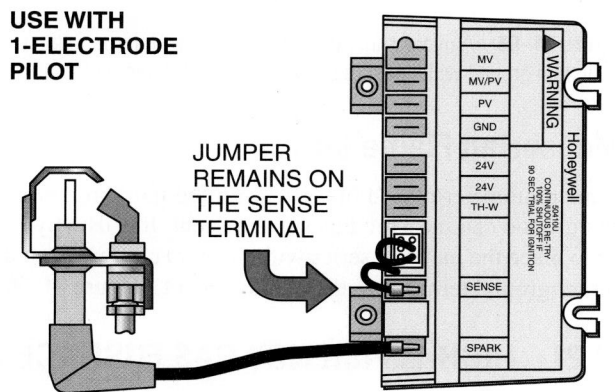

Figure 31–64 A single-rod, or local, sensing system connected to a furnace electronic module. *Courtesy Honeywell*

SPARKING — SENSING

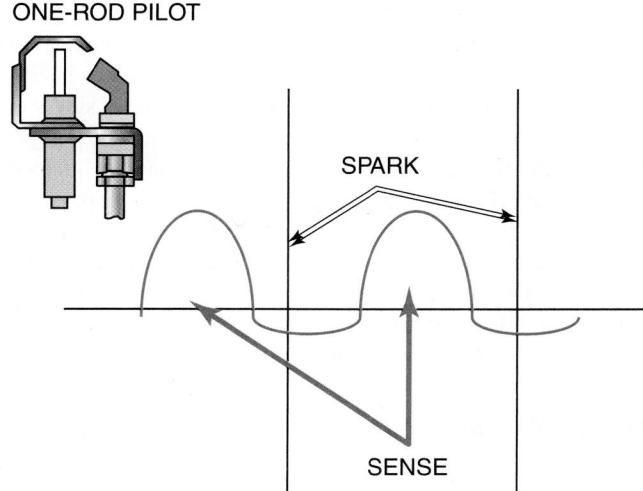

Figure 31–65 Single-rod, or local, flame rectification sensing system senses flame current during one half cycle and the ignition pulse during the other half cycle. *Courtesy Honeywell*

TWO-ROD PILOT

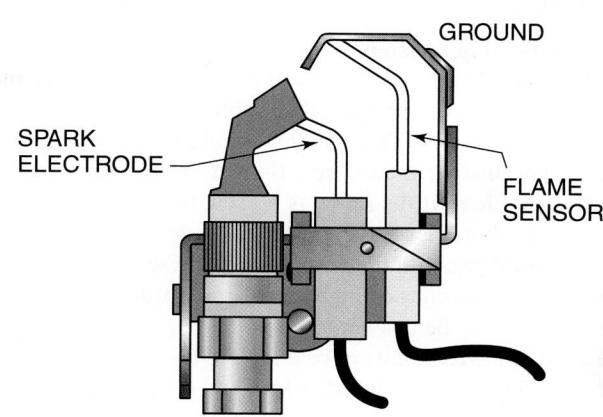

Figure 31–66 A dual-rod, or remote, flame rectification sensing system showing a separate igniter spark electrode and flame sensing rod. *Courtesy Honeywell*

Dual-rod, or remote, sensing uses a separate igniter and flame-sensing rod or electrode, **Figure 31–66.** The sparking is accomplished with one rod and the sensing is done with the other rod. There are two wires running from the pilot assembly to the module. The spark electrode wire goes to the spark terminal on the module. The flame rod or sensing wire goes to the sense terminal. A jumper wire will have to be cut off on some models, **Figure 31–67.** The flame sensor works with the grounded burner hood and the ground strap to achieve flame rectification.

The examples in the preceding two paragraphs are only a few of the furnace modules seen in the field today. Many newer furnaces have much more sophisticated IFCs. However, the principles of flame rectification still hold true for these modern furnaces.

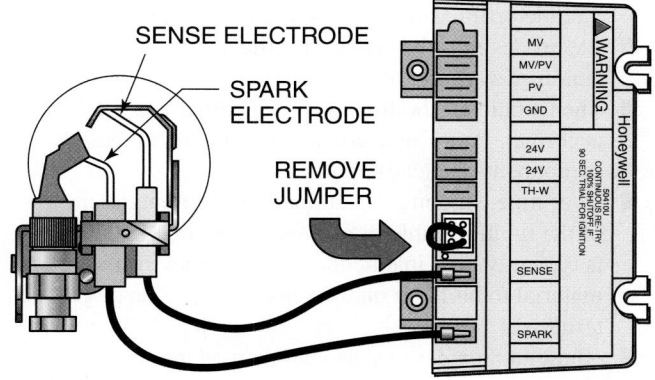

Figure 31–67 A dual-rod, or remote, flame rectification sensing system showing two separate wires going to the electronic furnace module. *Courtesy Honeywell*

Troubleshooting Flame Rectification Systems

Troubleshooting a flame rectification system must be handled in steps. During a normal call for heat, if the pilot flame or main flame does not light, the problem is not in the flame rectification system. The problem would be in the ignition system because there must be a flame to have a flame rectification problem. If the flame is present but goes out due to the system shutting off the gas valve, there could be one of several problems with the flame-sensing system:

- Flame electrodes could be dirty or corroded.
- A wire may be disconnected.
- The flame rod may not be engulfed by the flame.
- The flame signal amplifier (module or IFC) may have failed.
- There may be a poor ground wire connection.

The service technician must understand the ignition control sequence to see if the ignition source is working. The system may have a spark or hot surface ignition element. It is of utmost importance to know the sequence of events of the ignition module or IFC. If the ignition sequence seems to be working fine, but the pilot or main burner does not ignite, it is not a flame rectification problem. It could be one of the following:

- Misadjusted pilot burner
- Out-of-place ignition source
- Low or no gas pressure
- Lack of power to the system
- Insufficient spark or insufficient power to the hot surface element
- Broken or disconnected wires or cables
- Broken electrode or HSI element

Flame Rectification Maintenance

Several factors can affect the very small flame current measured in DC microamps by the service technician. Remember, 1 microamp equals 0.000001 amp. The following are some basic maintenance factors:

- Wires should be replaced every 3 to 5 years.
- Wiring between the flame electrodes and ignition module or IFC should be as short as possible.
- Electrodes must be replaced if there are signs of bending, warping, or deterioration.
- The technician should make sure burner gas pressure is correct. Poor flame contact with the metal burner components can cause failure.
- Over time, silicone contamination from outgassing of the ignition cable or ordinary room dust, which is typically high in silicone, can cause an insulating material to build up on the sensing electrode or ground terminal.
- Aluminum oxides can also build up on the sensing electrode's surface. The source of the aluminum can be the sensing electrode itself. Most flame rods are made of a Kanthol element, which contains aluminum and gives the rod the ability to withstand high temperatures.

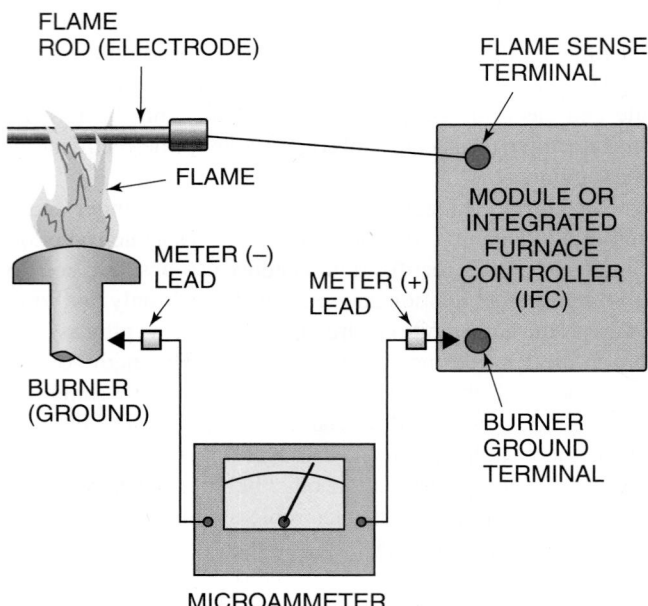

Figure 31–68 A microammeter placed in series with the flame rod or system ground will measure DC microamps of flame rectification.

Measuring Flame Current

A microammeter placed in series with the flame rod or system ground can measure this small current. It is usually easier to place the meter in series with the wire coming from the burner ground terminal to the module or IFC, **Figure 31–68**.

31.21   HIGH-EFFICIENCY GAS FURNACES

Besides for their high efficiencies, high-efficiency gas furnaces have gained popularity due primarily to the lack of sidewall venting material for mid-efficiency furnaces. Stack temperatures are the limiting factor for sidewall venting. Sidewall venting has been in demand by both builders and contractors for many years. It involves inexpensive materials like polyvinyl chloride (PVC) venting material with quiet, energy-efficient operations due to their sealed combustion characteristics. A sealed combustion chamber is shown in **Figure 31–69**. High-efficiency furnaces that are direct sidewall vented with sealed combustion chambers also offer lower gas bills. There is less outdoor air infiltration into the indoor heated space because the combustion chamber never sees conditioned room air. This prevents the heated, conditioned room air from being used as combustion air and pulling the conditioned room into a slight vacuum for outdoor infiltration to take place. Depressurization is the term for rooms that have lost pressure and are in a slight vacuum.

High-Efficiency Gas Furnace Anatomy

Unlike conventional and mid-efficiency gas furnaces, high-efficiency gas furnaces have some unique characteristics:

- A "recuperative" or "secondary" (and sometimes "tertiary") heat exchanger with connecting piping for flue-gas passages from the primary heat exchanger

Figure 31–69   A sealed combustion chamber on a high-efficiency furnace. *Courtesy Ferris State University. Photo by John Tomczyk*

- A flue-gas condensate disposal system, which includes drain, taps, drain lines, and drain traps, **Figure 31–70(A)**
- A venting system that is either induced or forced power drafted, **Figure 31–70(B)**

A direct-vented high-efficiency gas furnace has a sealed combustion chamber. This means that the combustion air is brought in from the outside through an air pipe, which is usually made of PVC plastic, **Figure 31–71**. The combustion chamber never sees conditioned room air. Since conditioned room air is not used for combustion, it prevents the conditioned room from being depressurized or pulled into a slight vacuum. This cuts down on the amount of infiltration entering the room, and less cold, infiltrated air has to be reheated.

Direct venting also minimizes the interaction of the furnace with other combustion appliances. Depressurization of a building or home may lead to combustion products coming in from the vent systems of water heaters, clothes driers, and exhaust products from attached garages. Even toxic fumes and particulates from a connecting or attached garage or building can infiltrate a conditioned space when depressurized. These are reasons why direct venting a furnace is an added safety feature.

A non-direct-vented high-efficiency furnace uses indoor conditioned air for combustion. A room can become depressurized using a non-direct-venting system.

Both direct-vented and non-direct-vented systems are positive pressure systems and can be vertically or sidewall

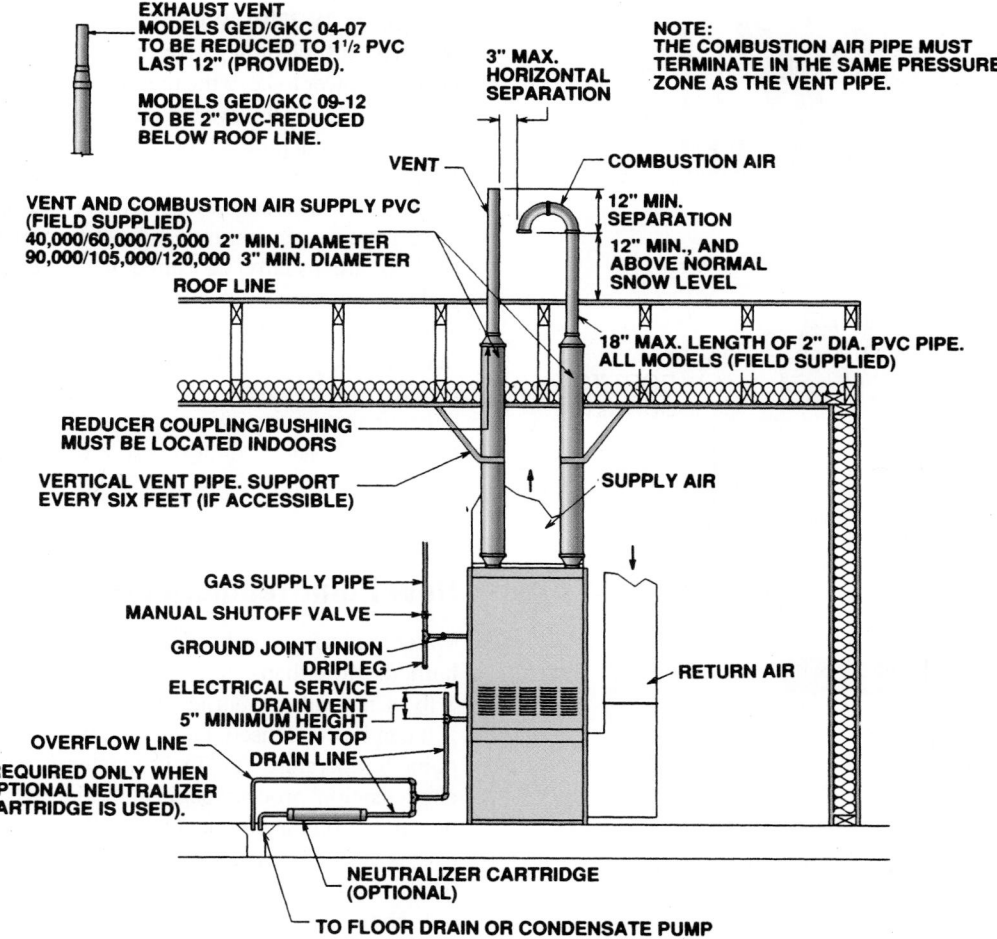

(A)

Figure 31–70   **(A)** A flue-gas condensate disposal system for a high-efficiency condensing furnace. **(A)** *Courtesy Rheem Manufacturing Company*

COMBUSTION BLOWER MOTOR

PRESSURE SWITCH AND DIAPHRAGM FOR PROVING DRAFT

BLOWER HOUSING

(B)

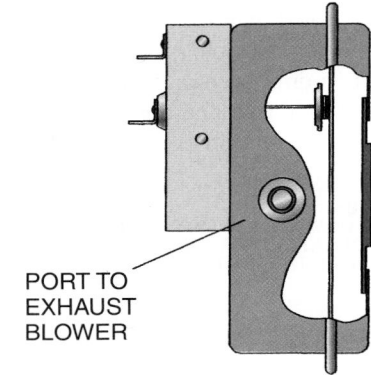

PORT TO EXHAUST BLOWER

(C)

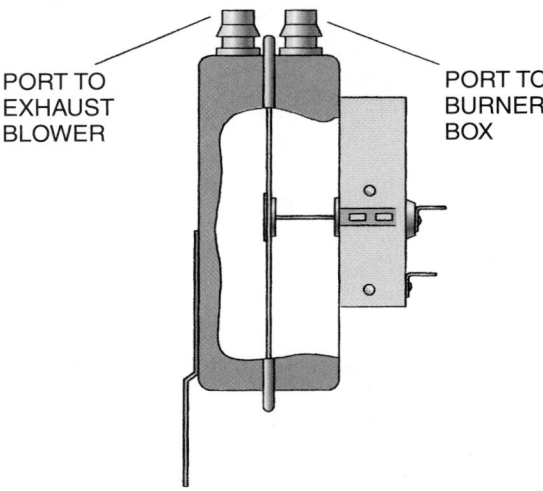

PORT TO EXHAUST BLOWER

PORT TO BURNER BOX

(D)

Figure 31–70 (*continued*)   **(B)** A combustion blower motor assembly with a pressure switch and diaphragm used for safely venting a furnace. **(C)** A single-port pressure switch. **(D)** A dual-port, or differential, pressure switch. **(B)** *Courtesy Ferris State University. Photo by John Tomczyk.* **(C)** *and* **(D)** *Courtesy International Comfort Products Corporation LLC*

vented. A positive pressure system's flue pipe pressure is positive the entire distance to its terminal end. However, a system can be direct vented and still not be a positive pressure system.

Some direct-vented designs need stronger flue fan or combustion blower systems to overcome the additional pressure drop created by the sealed combustion chamber and added combustion air pipe. In fact, the combustion blower must overcome three system pressure drops:

- The heat exchangers' system pressure drops
- The vent system (vent and combustion air pipe) pressure drop
- Vent termination wind resistance pressure

Combustion blower motors use an air-proving switch or pressure switch to prove that the mechanical draft has been established before allowing the main burner to light, **Figure 31–70(B)**. The pressure switches are connected to the combustion blower housing or heat exchanger (burner box) by rubber tubing. These pressure switches have diaphragms that are sensitive to pressure differentials. Some furnaces use a single-port pressure switch that senses the negative pressure inside the heat exchanger caused by an induced-draft combustion blower. Others use a dual-port, or differential, pressure switch that senses the difference in pressure between the combustion blower and the burner box, **Figure 31–70(C)** and **Figure 31–70(D)**. When the venting system is operating properly, sufficient negative pressure is created by the combustion blower to keep the pressure switch closed and the furnace operating. However, if the heat exchanger has a leak or the furnace vent pipe is blocked, the pressure switch will open from insufficient negative pressure. This will interrupt the operation of the main burners. Different furnace manufacturers use different pressure settings for their pressure switches. The settings are usually measured in inches of water column and are sometimes indicated on the pressure switch housing. The pressures can be measured using a digital or liquid-filled manometer. SAFETY PRECAUTION: Never replace a pressure switch with another pressure switch without knowing its operating pressures. Always consult with the furnace manufacturer or use a factory-authorized substitute pressure switch.

Dew Point Temperature

Water vapor is a combustion by-product when natural gas is burned. More than 2 lb of water vapor is produced for every 1 lb of natural gas burned. This water vapor contains heat. If it can be condensed, latent heat will be given off. The dew point temperature (DPT) is the temperature at which the condensation process begins. In flue gas, the DPT varies depending on the composition of flue gas and the amount of excess air (combustion air and dilution air).

For every 1 lb of water vapor that is condensed into liquid, about 970 Btu are given off. This is the latent heat of condensation of water vapor at atmospheric pressure. This latent heat adds significantly to the efficiency of the condensing furnace. The condensate can be somewhat acidic

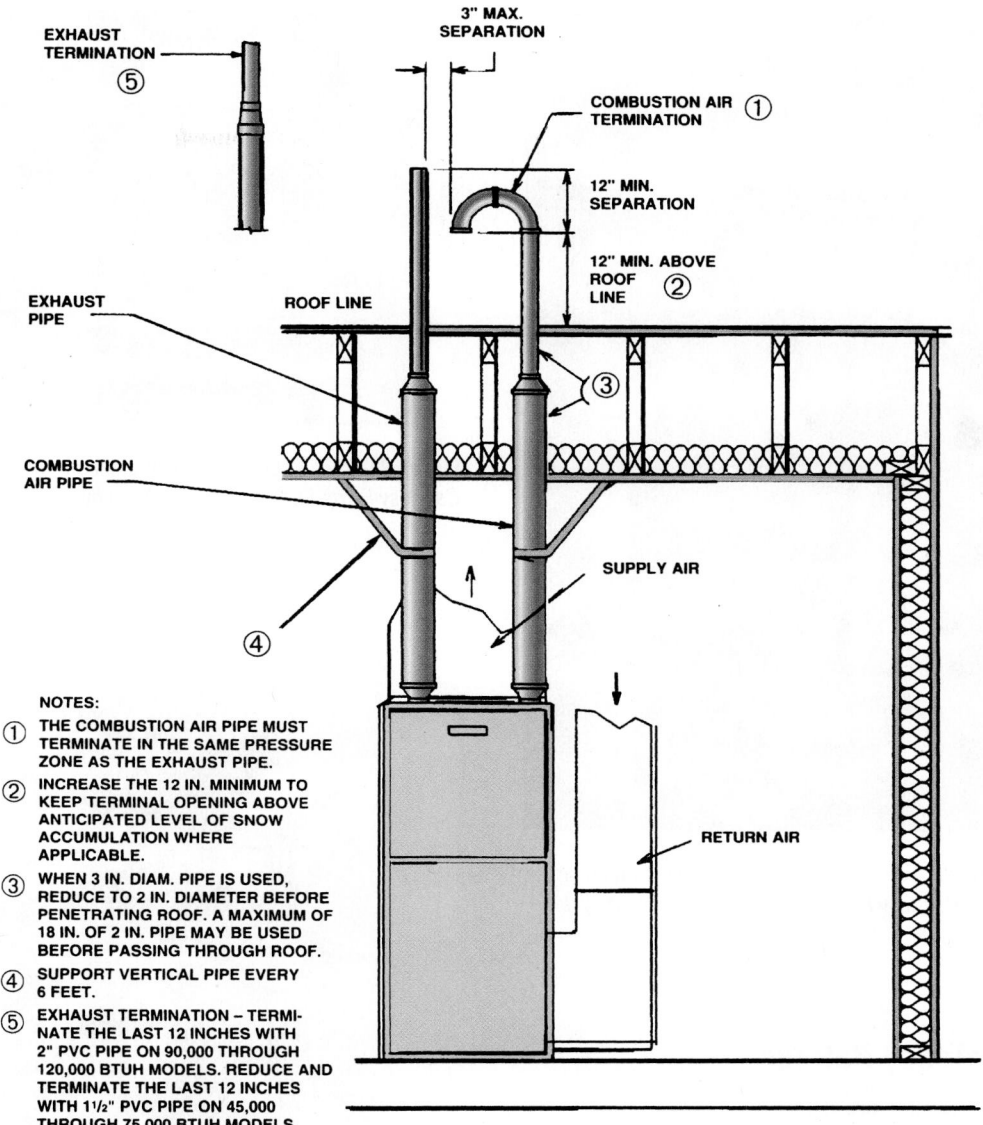

NOTES:

① THE COMBUSTION AIR PIPE MUST TERMINATE IN THE SAME PRESSURE ZONE AS THE EXHAUST PIPE.

② INCREASE THE 12 IN. MINIMUM TO KEEP TERMINAL OPENING ABOVE ANTICIPATED LEVEL OF SNOW ACCUMULATION WHERE APPLICABLE.

③ WHEN 3 IN. DIAM. PIPE IS USED, REDUCE TO 2 IN. DIAMETER BEFORE PENETRATING ROOF. A MAXIMUM OF 18 IN. OF 2 IN. PIPE MAY BE USED BEFORE PASSING THROUGH ROOF.

④ SUPPORT VERTICAL PIPE EVERY 6 FEET.

⑤ EXHAUST TERMINATION – TERMINATE THE LAST 12 INCHES WITH 2" PVC PIPE ON 90,000 THROUGH 120,000 BTUH MODELS. REDUCE AND TERMINATE THE LAST 12 INCHES WITH 1½" PVC PIPE ON 45,000 THROUGH 75,000 BTUH MODELS.

Figure 31–71   A direct-vented high-efficiency gas furnace. *Courtesy Rheem Manufacturing Company*

and corrosive. The flue gas is carried away with PVC piping, and the liquid condensate can be piped down an appropriate drain, **Figure 31–70(A)**.

Excess Air

Excess air consists of combustion air and dilution air. Combustion air is primary and/or secondary air. Primary air enters before combustion takes place. It mixes with raw gas in the burner before a flame has been established. Secondary air happens after combustion and supports combustion. Dilution air is excess air after combustion and usually enters at the end of the heat exchanger. It is brought in by the draft hood of the furnace, **Figure 31–72**.

As excess air decreases, the DPT increases. The DPT for undiluted flue gas is around 140°F. In diluted flue gas, the temperature may drop to about 105°F. This is the main reason high-efficiency condensing furnaces require very little excess

air for combustion. High-efficiency condensing furnaces operate below the DPT of the combustion gases in order for condensation to happen. The flue gas is intentionally cooled below the DPT to promote condensation.

Gas furnaces with high-efficiency ratings have been developed and are being installed in many homes and businesses. The U.S. Federal Trade Commission requires manufacturers to provide an annual fuel utilization efficiency rating (AFUE). This rating allows the consumer to compare furnace performances before buying. Annual efficiency ratings have been increased in some instances from 65% to 97% or higher. A part of this increase in efficiency is accomplished by keeping excessive heat from being vented to the atmosphere. Stack temperatures in conventional furnaces are kept high to provide a good draft for proper venting. They also prevent condensation and therefore corrosion in the vent. All forced-air gas furnaces have efficiency ratings.

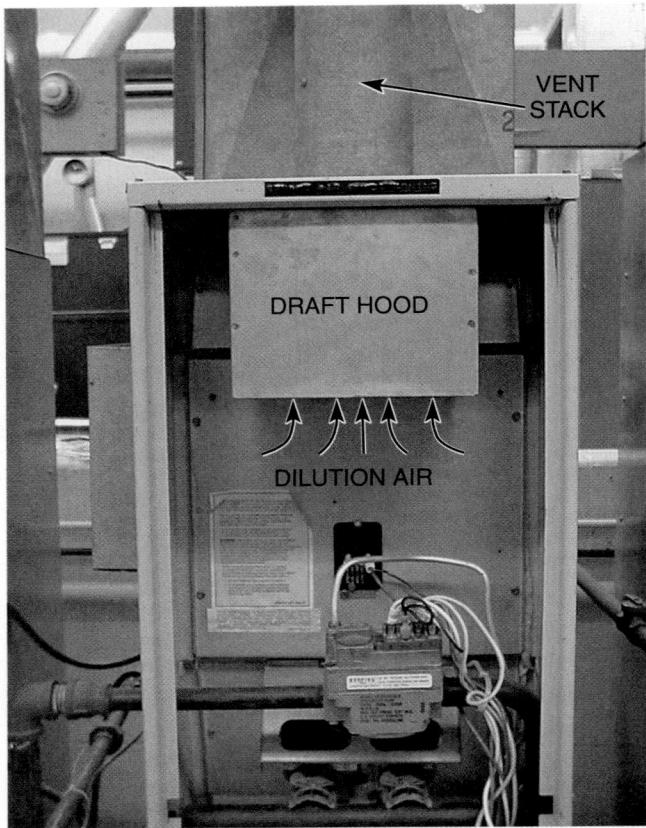

Figure 31–72 A standard-efficiency furnace showing dilution air entering a draft hood. *Courtesy Ferris State University. Photo by John Tomczyk*

Figure 31–73 A conventional, or standard-efficiency, furnace. *Courtesy Ferris State University. Photo by John Tomczyk*

Figure 31–74 A mid-efficiency furnace. *Courtesy Bryant Heating and Cooling Systems*

Furnace Efficiency Ratings

Furnace efficiency ratings are determined by the amount of heat that is transferred to the heated medium, which is usually water or air. The following factors determine the efficiency of the furnace:

- Type of draft (natural atmospheric injection, induced, or forced)
- Amount of excess air used in the combustion chamber
- Delta-T or temperature difference of the air or water entering versus leaving the heating medium side of the heat exchanger
- Flue stack temperature

The following are furnace classifications and approximate AFUE ratings:

Conventional, or Standard-Efficiency, Furnace (78% to 80% AFUE), Figure 31–73

- Atmospheric injection or draft hood
- 40% to 50% excess air usage
- Delta-T of (70°–100°)F (lowest cfm airflow of furnaces)
- Stack temperatures of (350°–450°)F
- Noncondensing
- One heat exchanger

Mid-Efficiency Furnace (78% to 83% AFUE), Figure 31–74

- Usually induced or forced draft
- No draft hoods
- 20% to 30% excess air
- Delta-T of (45°–75°)F (more cfm airflow than conventional units)
- Stack temperatures of (275°–300°)F
- Noncondensing
- One heat exchanger

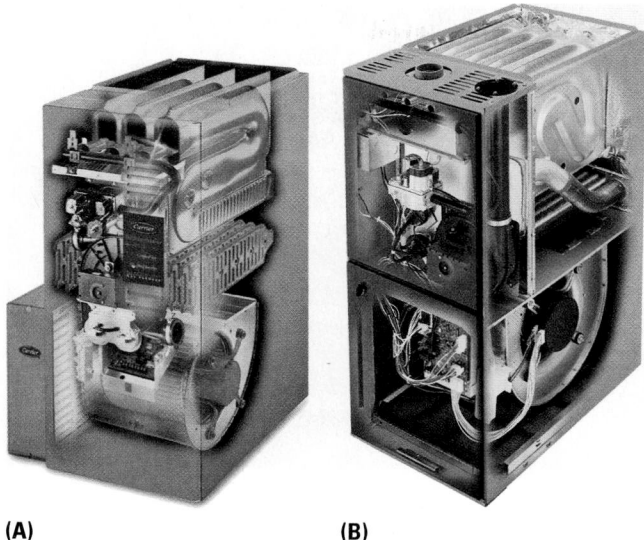

(A) **(B)**

Figure 31–75 Two high-efficiency gas furnaces. *(A) Courtesy Carrier. (B) Courtesy Trane*

High-Efficiency Furnace (87% to 97% AFUE), Figure 31–75

- Usually induced or forced draft
- No draft hoods
- 10% excess air
- Delta-T of (35°–65°)F (highest cfm of all furnaces)
- Stack temperatures of (110°–120°)F
- Two or three heat exchangers
- Condensing furnace (reclaims the latent heat of condensation by condensing flue-gas water vapors to liquid water)
- Use PVC pipe to avoid corrosion

As mentioned earlier, as excess air decreases, the DPT increases. This is the main reason high-efficiency condensing gas furnaces require very little excess air in the combustion processes. High-efficiency condensing furnaces operate below the DPT of the combustion gases in order for condensation to happen. The reason for the low delta-T for high- efficiency furnaces is the increased amount of air (cfm) through their heat exchangers. This keeps the air in contact with the heat exchanger for a shorter period of time.

31.22 ELECTRONIC IGNITION MODULES AND INTEGRATED FURNACE CONTROLLERS

Electronic ignition modules and IFCs control the ignition and sequence of operation of most modern gas furnaces. **Figure 31–76** shows examples of electronic ignition modules and an IFC. Electronic modules usually indicate on their faceplate what basic sequence of events or operations they include within their electronics. Some of their most common control sequences and control definitions follow.

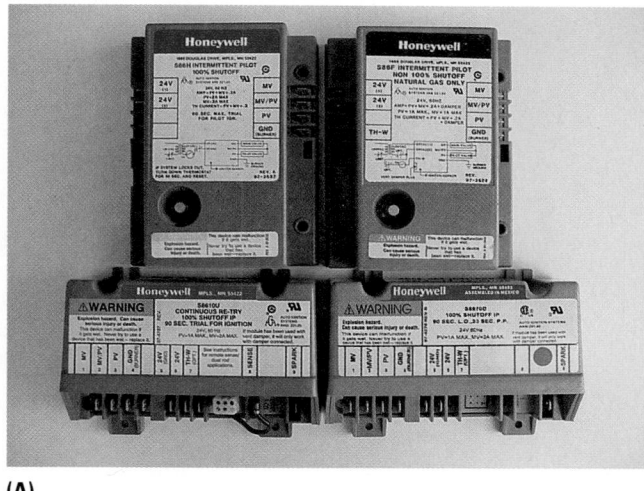

(A)

(B)

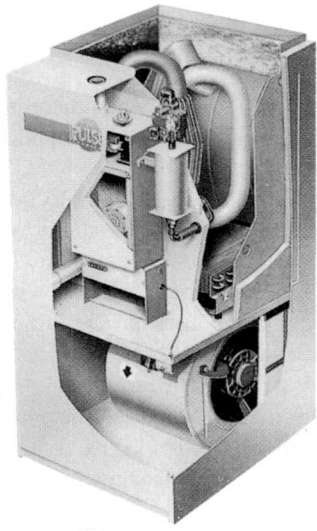

(C)

Figure 31–76 **(A)** Four electronic modules for gas furnaces. **(B)** An integrated furnace controller (IFC). **(C)** A high-efficiency pulse furnace. *(A) and (B) Courtesy Ferris State University. Photos by John Tomczyk. (C) Courtesy Lennox International*

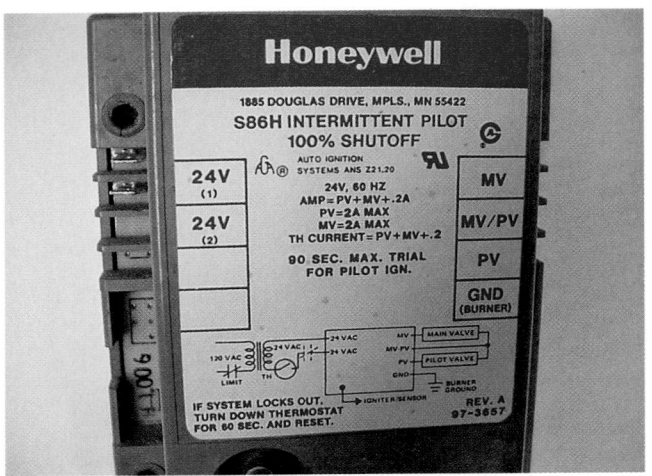

Figure 31–77 A 100% shutoff electronic ignition module. *Courtesy Ferris State University. Photo by John Tomczyk*

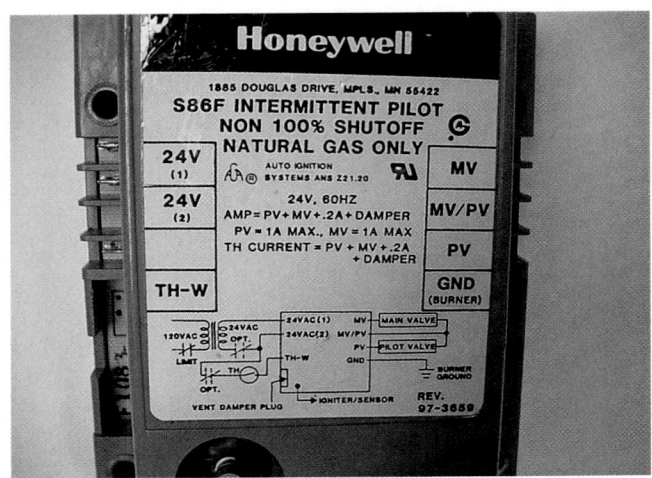

Figure 31–78 A non-100% shutoff ignition module. *Courtesy Ferris State University. Photo by John Tomczyk*

100% Shutoff System

A system is 100% shutoff when there is a failure in the flame-proving device system and both the pilot valve and main gas valve are shut down (closed). In the past, these systems had to be used on liquid petroleum gas systems since LP gas is heavier than air and will not vent naturally up the flue. It was thought that letting the pilot gas bleed could create a dangerous situation. This is no longer true. Today, there are LP gas systems that use non-100% shutoff systems in a safe manner by having short shutdown or soft lockout times between ignition retries to let any residual gas dissipate. Lockout timing will be covered in later paragraphs. A 100% shutoff module is shown in **Figure 31–77**. Notice that it is designed for an IP system and usually has a 90-sec maximum trial for pilot ignition before going into a shutoff.

Non-100% Shutoff System

A system is non-100% shutoff when there is a failure in the flame-proving device system and the main gas valve shuts down but the pilot valve continues to bleed gas. These systems are used where there is a need to continually try to relight the pilot to avoid nuisance shutdowns. Many commercial rooftop heating units are non-100% shutdown, which prevents nuisance shutdowns on windy days. Many newer ignition control modules have ignition retry sequences good for both natural and LP gas. Many modules have soft shutdown periods of 5 min or more to allow any accumulation of fuels to dissipate. The system must have a pilot mechanism to be classified as either 100% or non-100% shutoff. A non-100% shutoff module is shown in **Figure 31–78**.

Continuous Retry with 100% Shutoff

A module with this control scheme is shown in **Figure 31–79**. It has 100% shutoff with a 90-sec trial for ignition. However, it will continuously try to relight the pilot after the shutoff period. Once in the shutoff period, there will not be any bleeding of pilot gas.

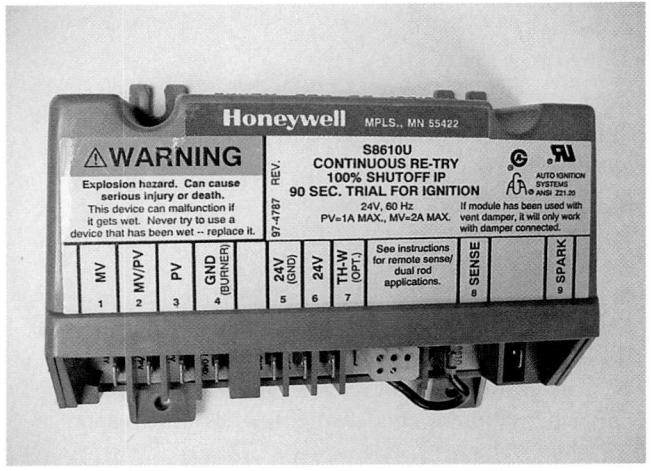

Figure 31–79 A continuous retry 100% shutoff electronic ignition module. *Courtesy Ferris State University. Photo by John Tomczyk*

Lockout

A lockout is a time period built into the module that allows for a certain time to light or relight the pilot or main burner. If this time period is exceeded, the module will go into either a soft lockout or a hard lockout.

Soft Lockout

A soft lockout is a time period built into the module that allows for a certain time to light or relight the pilot or main burner. If this time period is exceeded, the module will go into a semi-shutdown for a certain time period, but will eventually keep trying to relight the system. This system may go into a hard lockout once several soft lockouts have been attempted and still no flame has been established. Soft lockouts can range from 5 min to 1 hour. Soft lockouts allow time to pass. Hopefully, a change in the local ambient or outside environment will allow the unit to light normally. They also allow for any power interrupts to clear themselves without going into a hard lockout.

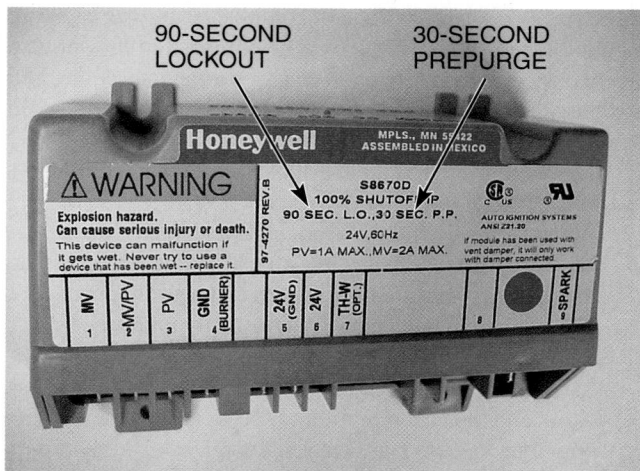

Figure 31–80 A 100% shutoff module with a 90-sec lockout and a 30-sec prepurge. *Courtesy Ferris State University. Photo by John Tomczyk*

Hard Lockout

A hard lockout is a time period built into the module that allows for a certain time to light or relight the pilot or main burner. If this time period is exceeded, the module will go into a hard lockout or shutdown. To reset the module after this shutdown, power has to be interrupted to the module and then turned back on. This usually takes a service call and costs the building owner money.

Prepurge

The module in **Figure 31–80** is 100% shutoff for an IP system. There is a 90-sec lockout (LO) with a 30-sec prepurge (PP). Before each heating cycle, the PP control scheme allows the combustion blower motor to run for 30 sec to clear or PP the heat exchanger of any unwanted flue gases, household fumes, or dust that may have accumulated during the last heating cycle.

Interpurge

Interpurge allows the combustion blower to operate for a certain time period between ignition tries. Interpurge is generally used when the furnace does not light or does not sense a flame. The heat exchanger must be purged of residual unburned gas or combustion by-products from the previous unsuccessful ignition attempt.

Postpurge

A postpurge allows the combustion blower motor to operate for a certain time period after each heating cycle. This control strategy, along with a PP, is double insurance that the heating cycle will start with only air in the heat exchanger, allowing a quick and safe light-off. Both PP and postpurge control strategies can be used alone or with one another.

The IFC provides all of the ignition, safety, and sequence of operation or control schemes for the furnace. These controllers are used by many furnace manufacturers because of their safe, reliable operations and versatility. The controller carefully monitors all thermostat, temperature, resistance, and current inputs to ensure stable and safe operation of the combustion blower, gas valve, ignition, flame rectification system, and warm air blower. They often come with many DIP switches that allow the controller to be programmed for different timing functions and control scheme changes. **Figure 31–81(A)** shows a section of an IFC with DIP switches and status lights. DIP switches 1 and 2 are for setting the warm air blower's off-delay timing in seconds, **Figure 31–81(B)**. DIP switch 3 is for furnace twinning applications. DIP switch 4 is not used in this application.

Twinning involves the operation of two furnaces side by side, connected by a common ducting system. High-voltage power is supplied by a common source. Low-voltage power supplied off a transformer is controlled by one common thermostat for both furnaces. Status lights will usually blink a certain number of times to tell the service technician that the furnace is in the twinning mode. Twinning is used when a heating capacity greater than that of the largest capacity

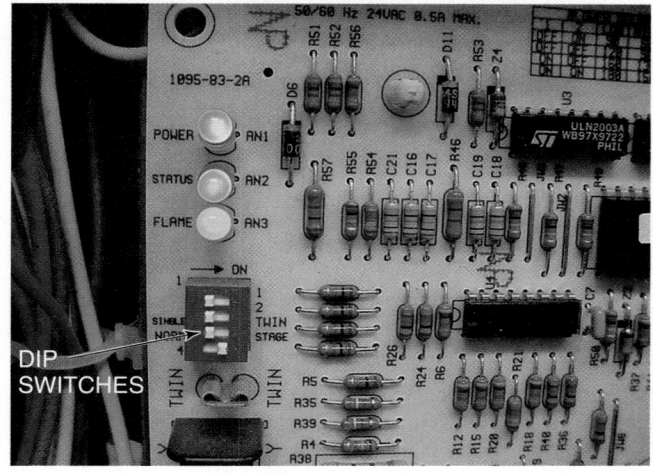

(A)

BLOWER OFF TIMES (SECONDS)				
LOW FIRE	HIGH FIRE	COOLING	SWITCH 1	SWITCH 2
90	60	30	OFF	ON
120	90	45	OFF	OFF
160	130	60	ON	OFF
180	150	90	ON	ON

(B)

Figure 31–81 **(A)** An integrated furnace controller showing DIP switches and status lights. **(B)** DIP switches for setting the warm air blower off-delay timing. **(A)** *Courtesy Ferris State University. Photo by John Tomczyk.* **(B)** *Courtesy Rheem Manufacturing Company*

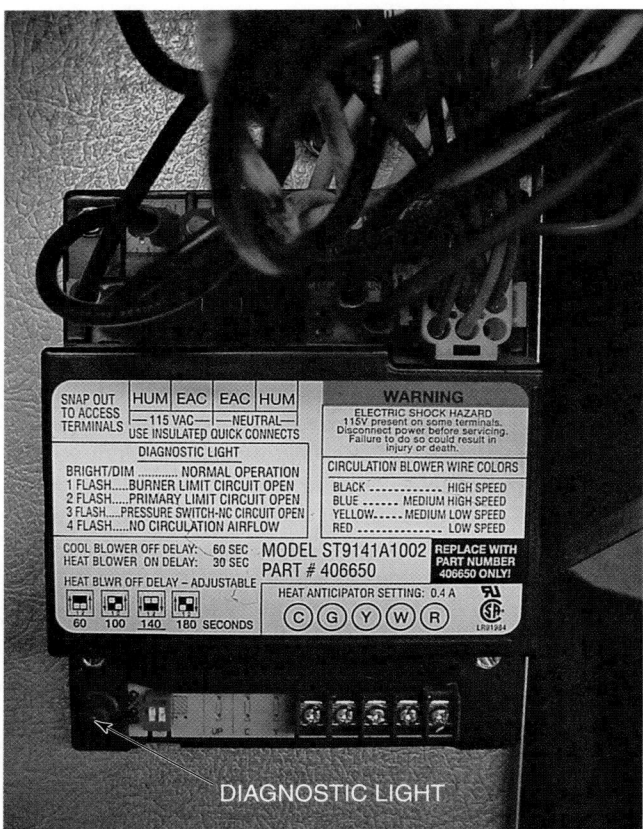

Figure 31–82 An integrated furnace controller (IFC) with self-diagnostic lights and flashing fault codes. *Courtesy Ferris State University. Photo by John Tomczyk*

furnace manufactured is needed. Twinning is also used when the airflow of one furnace, for air-conditioning purposes, is greater than a single furnace can provide.

The status lights monitor incoming power, furnace setup modes, diagnostic status for troubleshooting, and the flame rectification signal. The status check can be a self-diagnostic flashing light with a code to help the service technician systematically troubleshoot the furnace. **Figure 31–82** shows a furnace controller with a self-diagnostic light and flashing codes for some common furnace faults. The diagnostic light is located on the bottom left corner of the controller. Care must be taken when troubleshooting an IFC because static electricity can damage the sensitive hardware and software. Always make sure you are touching a well-grounded surface before attempting to troubleshoot an IFC board.

The small voltage of an ohmmeter can also damage an IFC board. Never put an ohmmeter on an IFC board unless instructed to do so by the furnace manufacturer's troubleshooting procedures. Always disconnect the wires of a power-consuming device or sensor from the IFC before using an ohmmeter in troubleshooting these devices.

Following is a sequence of operation or control scheme for a residential direct burner ignition, natural gas furnace employing an IFC and a single-stage thermostat. Either a single- or dual-stage thermostat can be used on this system with the proper settings. This is a two-stage furnace, meaning the gas

valve has two stages of heat accomplished by gas pressure regulation and solenoid valves. The combustion blower also has two speeds. The sequence includes a PP, interpurge, and postpurge with a soft lockout. The control board or IFC used on this furnace is shown in **Figure 31–76(B)**.

1. The W terminal on the thermostat is connected to the W2 terminal on the IFC board.
2. When there is a call for heat, the R and W2 contacts close and the IFC runs a self-check routine to verify that the draft-proving pressure switch contacts are open. The limit switch contacts are constantly monitored.
3. The induced-draft motor starts on high for a few seconds, to ensure that the low-pressure switch contacts close, and then changes to low speed. After a 30-sec prepurge, the spark indicator is energized and the low stage of the gas valve opens, lighting the burners.
4. After the gas valve opens, the remote flame sensor must prove ignition for 1 sec using the process of flame rectification. If the burners do not light, the system goes through another ignition sequence. It does this up to four times (two tries on low fire and two tries on high fire).
5. The main blower starts on low speed 30 sec after the burners light. The furnace will operate on low fire for 12 min. Then, if the thermostat is not satisfied, it will shift to high fire, causing the draft inducer to go to heat speed. (If a two-stage thermostat is used, the second stage of the thermostat, W2, will initiate high fire.) The draft inducer will continue running for a 5-sec (low speed) or 10-sec (high speed) postpurge.

The main blower will run for 90 sec on high speed or 120 sec on low speed. (This timing is field adjustable for 60, 90, 120, or 150 sec.)

The following is the sequence if the system does not light or does not sense flame:

1. If a flame is not sensed within 8 sec after the gas valve opens, the valve closes and the igniter is deenergized. The induced-draft motor will run for 60 sec on low, stop momentarily, and then restart. The ignition process will go through one more try on low fire. If this fails, there will be two tries on high fire with a 30-sec interpurge between trials. If there is no ignition after the second trial on high fire, the furnace will go into soft lockout for 1 hour.
2. The sequence will repeat after a 1-hour delay and continue repeating until ignition is successful or the call for heat is terminated.
3. To reset the lockout, make and break power either at the thermostat or at the unit disconnect switch for 5 to 10 sec. The furnace will then go through another set of trials for ignition.

Electronic Ignition Module Wiring and Terminal Naming

Most electronic modules have terminal markings with abbreviations, **Figure 31–76(A)**. Some electronic modules include the wiring diagram on their faceplate, **Figure 31–77** and **Figure 31–78**. This can assist the service technician in wiring and troubleshooting the furnace's electrical and mechanical system. The following are important terminal marking abbreviations and terminal naming:

MV	Main valve (gas valve)
PV	Pilot valve
MV/PV	Common terminal for main and pilot valves
GND	Burner ground
24V	24-V source out of the transformer
24V(GND)	Common or ground out of the 24-V transformer
TH-W	Thermostat lead to the module
IGNITER/SENSOR	High-voltage igniter and flame rectification rod or sensor for local sensing (single-rod systems)
SENSE	Flame rectification rod or sensor for remote sensing (dual-rod systems)
SPARK	High-voltage igniter

Smart Valve

Smart valve is a gas valve and an electronic control module in one package. The electronic module and gas valve are in one enclosure, **Figure 31–83.** The smart valve combines the features of an IP system and an HSI system in one. The smart valve's logic lights the pilot with a 24-V hot surface igniter system, **Figure 31–84.** The pilot then lights the main burners like any other intermittent ignition system. The smart valve system is a continuous trial for pilot for both natural and LP gas. The pilot flame is probably one of the most reliable ignition sources in the industry. Using a 24-V HSI system instead of a high-voltage spark to light the pilot simplifies the electronics, because there is no need for a noisy spark generator or high-voltage ignition transformer. It allows for a quiet ignition system.

The 24-V hot surface igniter is made of a material different from the conventional silicone carbide line-voltage igniters. It is designed to be stronger and less brittle. In fact, it is well protected by the pilot burner and grounding strap. The ground strap is there to provide a ground electrode for the flame rectification system. The other electrode is the flame rod, **Figure 31–84.**

Once the flame is proved, the 24-V hot surface igniter is deenergized and the electronic fan timer within the module is energized. The electronic fan timer tells when to start timing the fan-on delay and the fan-off delay.

The smart valve system is easy to troubleshoot because of simplified wiring and modular electrical connections. A voltmeter can be used at the valve and electrical plug connections, **Figure 31–85. Figure 31–86(A)** is a sequence of operation for a smart valve control. **Figure 31–86(B)** is a systematic troubleshooting chart. SAFETY PRECAUTION: Because there are different models and makes of gas valves, electronic modules, and IFCs, always consult with the gas-valve and electronic-module or IFC manufacturer for specific information on systematic troubleshooting and sequence of operation to avoid personal injury.

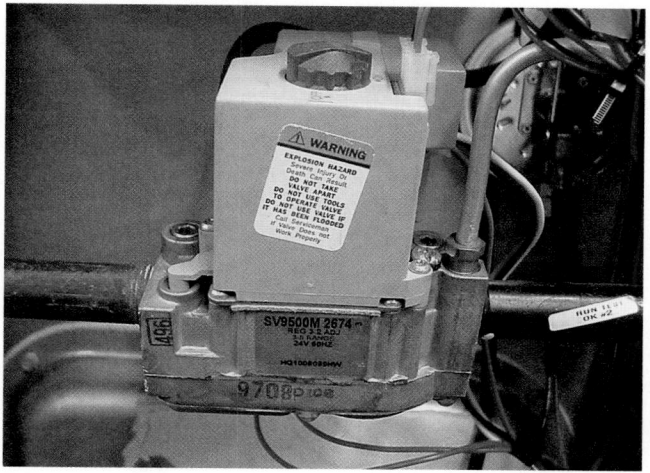

(A)

(B)

Figure 31–83 **(A)** A smart valve gas control system. **(B)** Internal view of a smart valve showing electronic module and gas valve. **(A)** and **(B)** Courtesy Ferris State University. Photos by John Tomczyk

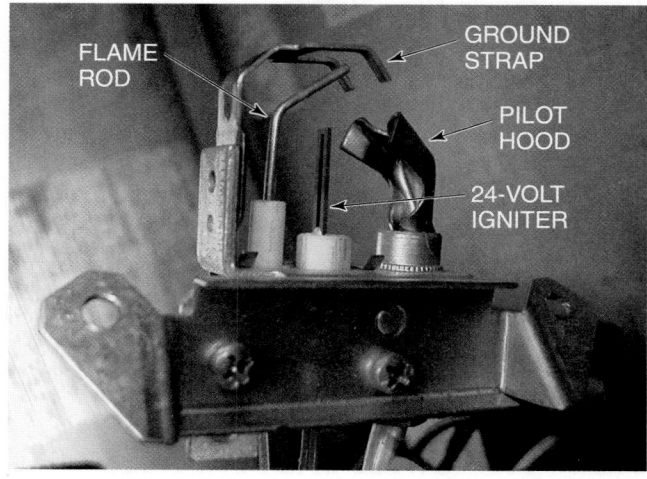

Figure 31–84 A 24-V hot surface igniter that lights a pilot flame. Courtesy Ferris State University. Photo by John Tomczyk

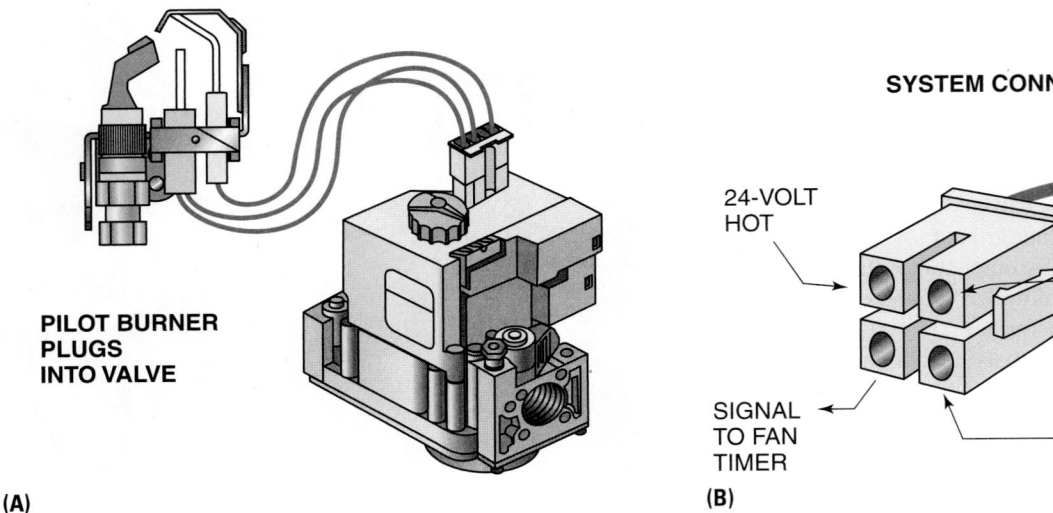

(A)

**PILOT BURNER
PLUGS
INTO VALVE**

SYSTEM CONNECTIONS

24-VOLT
HOT

24-VOLT
COMMON

SIGNAL
TO FAN
TIMER

THERMOSTAT OR
PRESSURE SWITCH

(B)

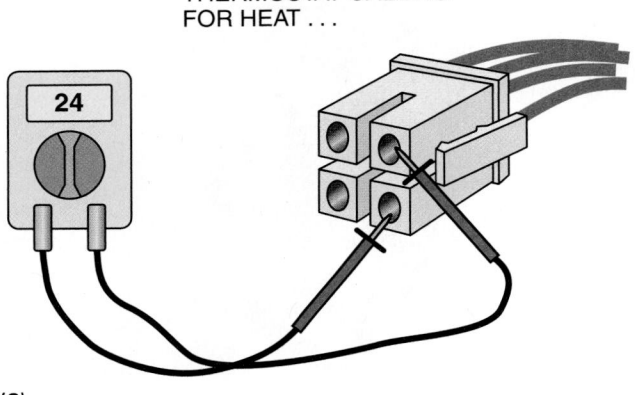

CHECK POWER TO VALVE

THERMOSTAT CALLING
FOR HEAT . . .

24

(C)

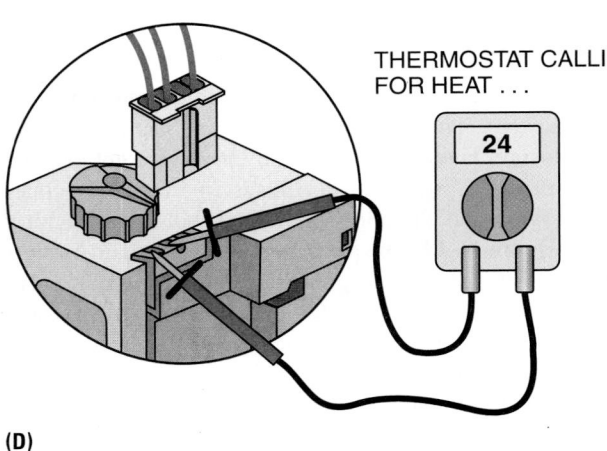

CHECK VOLTAGE TO IGNITER

THERMOSTAT CALLING
FOR HEAT . . .

24

(D)

Figure 31–85 The smart valve's modular electrical connections **(A)–(B)**. A voltmeter being used to troubleshoot the smart valve system **(C)–(D)**. *(A)–(D) Courtesy Honeywell*

Pulse Furnace

The pulse furnace, **Figure 31–76(C),** ignites minute quantities of gas 60 to 70 times per second in a closed combustion chamber. The process begins when small amounts of fuel gas and air enter the combustion chamber. This mixture is ignited with a spark igniter. This ignition forces the combustion materials down a tailpipe to an exhaust decoupler. The pulse is reflected back to the combustion chamber, igniting another gas and air mixture, and the process is repeated. Once this process is started, the spark igniter can be turned off because the pulsing will continue on its own 60 to 70 times per second. This furnace also uses heat exchangers to absorb much of the heat produced. Air is circulated over the exchanger and then to the space to be heated.

31.23 TWO-STAGE GAS FURNACES

Modern two-stage gas furnaces use a two-stage gas valve and a two-speed combustion blower motor with two pressure switches to prove the draft. An IFC usually controls modern two-stage furnaces. **Figure 31–76(B)** shows a modern IFC for a two-stage gas furnace. **Figure 31–87** shows the two-stage gas valve, two-speed combustion blower motor, and dual pressure switches to prove the draft. One pressure switch is a low-pressure switch and the other is a high-pressure switch to prove a low and high speed on the combustion blower motor.

The first stage of heating puts out a manifold pressure of 1.75 in. WC by energizing the first solenoid valve in the gas valve. The first stage of heating usually operates at about 50% to 70% of the total furnace heating output. The second stage of heating puts out a manifold pressure of 3.5 in. WC by energizing the second solenoid valve in the gas valve. This is 100% of the total furnace output. Two-stage gas furnaces control the temperature in the conditioned space much closer to the set point of the thermostat. Because of this, room temperature swings are smaller and comfort levels are increased. A sequence of operation for a two-stage furnace is given in Section 31.22.

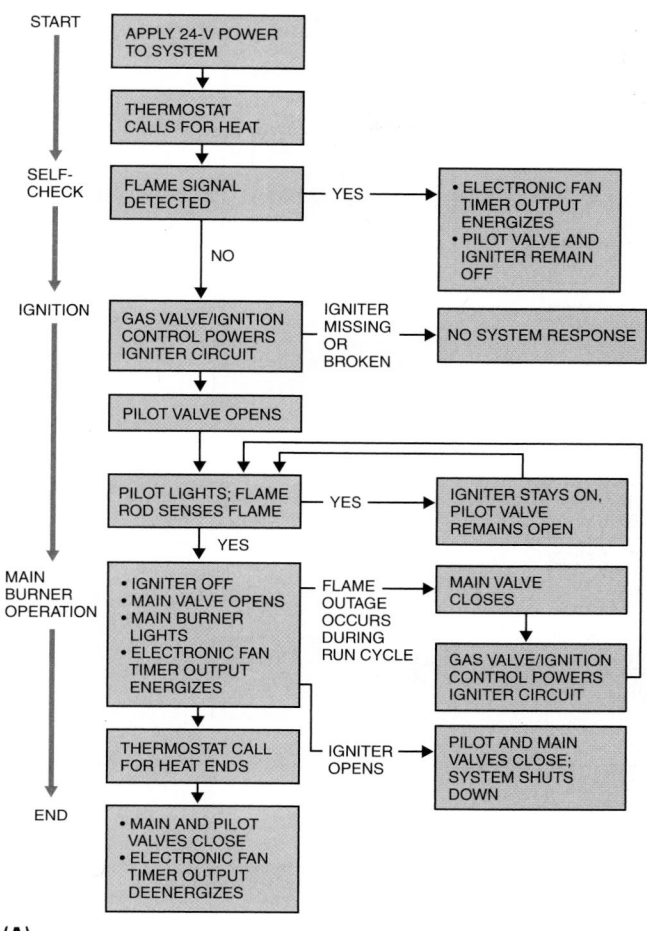

Figure 31–86 (A) The smart valve sequence of operation.
(A) Courtesy Honeywell

31.24 MODULATING GAS FURNACES

The industry has always needed a furnace that follows the heat loss of a structure. For example, a house that requires a 100,000-Btu/h furnace on the coldest day of winter, which may be 0°F, may require only 60,000 Btu/h of heat when it is 20°F. The furnace begins to cycle off and then back on as the weather warms. The warmer the weather, the more cycle times per day the furnace will experience. These cycle times cause the furnace to be less efficient. It is estimated that an oversized furnace may have a seasonal efficiency as low as 50%, while having a steady-state efficiency of 80%. When this cycling can be reduced or eliminated, the efficiency of the furnace becomes greater. 🌎 A modern modulating gas furnace can follow the heat loss of a structure and have longer run times, thus less cycling. This makes it a very efficient furnace. 🌎

Modern modulating gas furnaces use a modulating gas valve instead of a staged gas valve. They also come with a variable-speed warm air blower to vary the speed and amount of warm air. Two-stage combustion blower motors are standard on most modulating gas furnaces. An advanced intelligent IFC monitors and controls the furnace operations to vary the heating input and warm airflow.

Modern modulating gas furnaces have a variable operating solenoid or servo valve within the gas valve that is fed with a varying or proportional milliamp signal from the IFC. This servo valve acts as a capacity controller. It varies the input rate of the furnace proportionally to the signal of the controller. This allows the furnace to operate over a heating output range between 40% and 100%. The controller gets its proportional signal from a variable output programmable thermostat, which will be covered in detail later in this unit.

Modulating gas furnaces may have supply and return air sensors installed in the supply and return air system. These sensors are usually thermistors that send an analog signal to the IFC. They can sense the temperature of the return and discharge air and relay this information to the furnace controller. The controller can then calculate the temperature rise of the furnace and proportionally adjust the modulating gas valve to maintain a comfortable discharge air temperature. The IFC monitors the thermostat inputs and all other temperature inputs from the furnace and provides stable operation of the variable input gas valve, variable warm air blower motor, and two-speed inducer motor. The modulating gas furnace also has the ability to adjust itself to changes in gas heating values and air density.

The variable output programmable thermostat is not a conventional on-off thermostat. It sends a varying or proportional signal to the furnace controller. It can sense the exact heating or cooling requirements of the conditioned space by a "fuzzy logic" control routine resident in its software. These routines calculate the room's load by evaluating recent room conditions from a sensor within the thermostat or remote to the thermostat, and reactions to supplying heating and cooling. The thermostat and IFC then calculate a heating or cooling load factor. This load factor is used to determine when, for how long, and at what firing rate and airflow the furnace should be activated.

This control routine optimizes furnace efficiency and reduces temperature swing in the controlled space. Modulating furnaces will have longer run times, allowing greater comfort levels in the conditioned space. Temperature swings in the conditioned space will not vary by more than one-half degree of the thermostat set point. **Figure 31–88** shows the internals of a variable output programmable thermostat used in a modulating furnace.

A variable-speed warm air blower provides varied airflow through the furnace's ductwork to meet the conditioned space demands. A brushless, permanent-magnet, internally commutated motor (ICM) with variable speed capabilities can deliver airflows as low as 300 cfm. These furnaces orchestrate the use of the variable-speed warm air blower, variable input gas valve, supply and return air sensors, advanced IFC, and variable input thermostat to achieve a targeted temperature rise. For a detailed review of ICMs, refer to Unit 17, "Types of Electric Motors."

31.25 VENTING

Conventional gas furnaces use a hot flue gas and natural convection to vent the products of combustion (flue gases). The hot gas is vented quickly, primarily to prevent cooling of

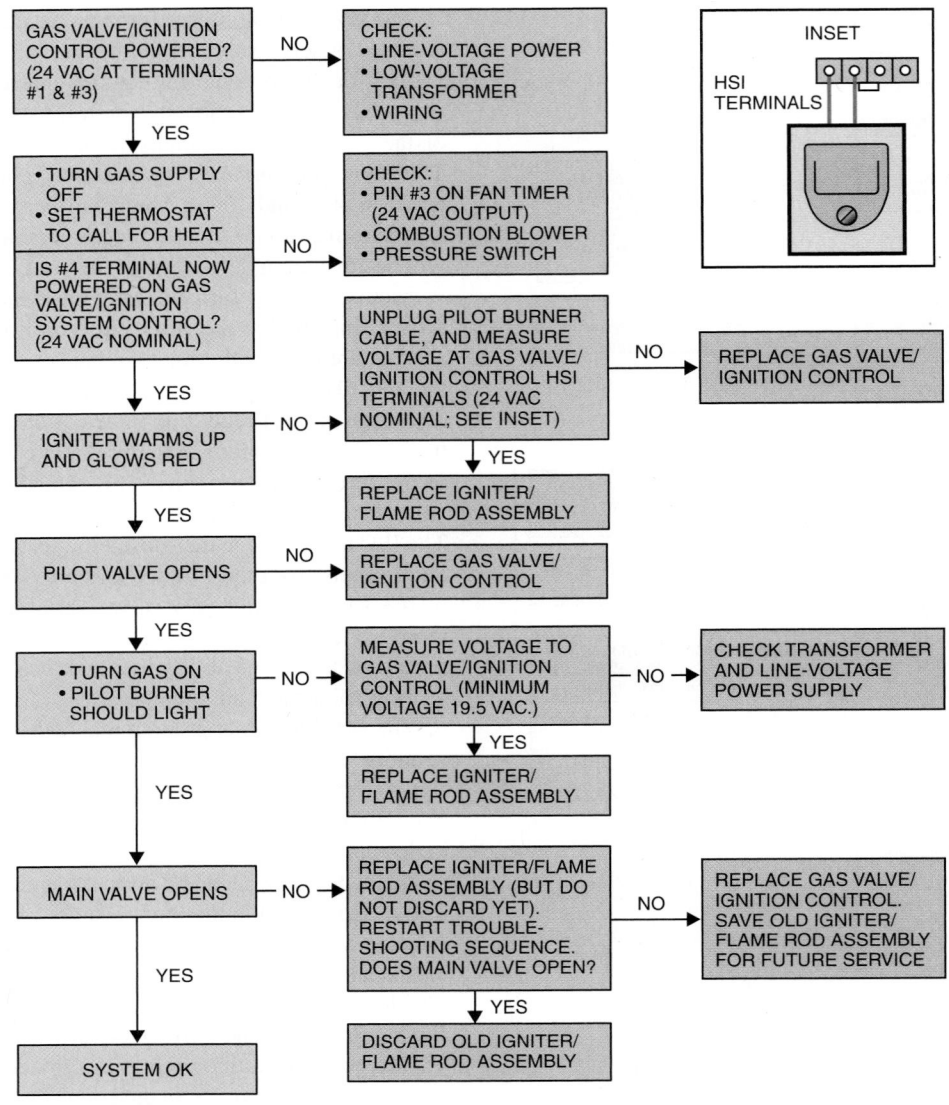

Figure 31–86 *(continued)* **(B)** A smart valve troubleshooting chart. *(B) Courtesy International Comfort Products*

the flue gas, which produces condensation and other corrosive actions. However, these furnaces lose some efficiency because considerable heat is lost up the flue.

High-efficiency furnaces (90% and higher) recirculate the flue gases through a special extra heat exchanger to keep more of the heat available for space heating in the building. The gases are then pushed out the flue by a small fan. This causes condensation and thus more corrosion in the flue. Plastic pipe is used because it is not damaged by the corrosive materials. These furnaces may have efficiencies of up to 97%. SAFETY PRECAUTION: Regardless of the type, a gas furnace must be properly vented. The venting must provide a safe and effective means of moving the flue gases to the outside air. In a conventional furnace, it is important for the flue gases to be vented as quickly as possible. Conventional furnaces are equipped with a draft hood that blends some room air with the rising flue gases, **Figure 31–89(A)**. The products of combustion enter the draft hood and are mixed with air from the area around the furnace. Approximately

100% additional air (called *dilution* air) enters the draft hood at a lower temperature than the flue gases. The heated gases rise rapidly and create a draft, bringing the dilution air in to mix and move up the vent.

All furnaces must have the correct amount of excess air. Excess air consists of combustion air and dilution air. Combustion air consists of both primary and secondary air. This amount of air for gas-burning appliances is determined by the appliance size and is mandated by local codes or the National Fuel Gas code, which is considered the standard by many states. Obviously, the larger the furnace, the more excess air is needed. One rule of thumb is that a grill with a free area of 1 in^2 per 1000 Btu/h must be provided below the furnace (if the structure is over a crawl space) and above the furnace (to the attic space) to allow for adequate air for combustion. For a 100,000-Btu/h furnace, this would be 100 in^2 of free area. This grill must be larger than 10 in. × 10 in. = 100 in^2 because of the space the louvres would take up. (Approximately 30% of the grill area is louvres.) The grill would

Figure 31–87 A two-stage gas furnace showing the two-stage gas valve, the two-stage combustion blower motor, and dual pressure controls. *Courtesy Ferris State University. Photo by John Tomczyk*

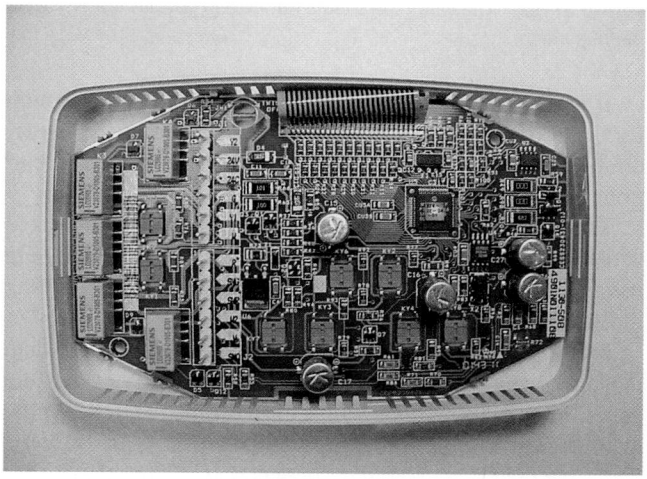

Figure 31–88 Internal view of a variable output programmable thermostat for a modulating furnace. *Courtesy Ferris State University. Photo by John Tomczyk*

have to be $100/0.70 = 142.9$ in^2 or 12 in. × 12 in. The reason for the vent above is for dilution air. The vent below is for combustion air. This example of figuring the ventilation air for this situation is only one possibility.

In the past, excess air was merely extracted from the volume of air in the structure. Structures today are built with less

infiltration air entering around the windows and doors, so makeup air for the furnace must be considered. Even old structures are being tightened by the use of storm windows and weather stripping around the doors, so there is less infiltration. A furnace installed in an older structure, which has been tightened, may have venting problems evidenced by flue gases spilling out of the draft diverter rather than exiting out the flue.

If wind conditions produce a downdraft, the opening in the draft hood provides a place for the gases and air to go, diverting it away from the pilot and main burner flame. This *draft diverter* helps reduce the chance of the pilot flame being blown out and the main burner flame being altered, **Figure 31–89(B).**

When high-efficiency furnaces were introduced, venting categories were created. Following are the four venting categories listed by the American National Standards Institute (ANSI). They are based on ANSI Standard Z21.47A-1990

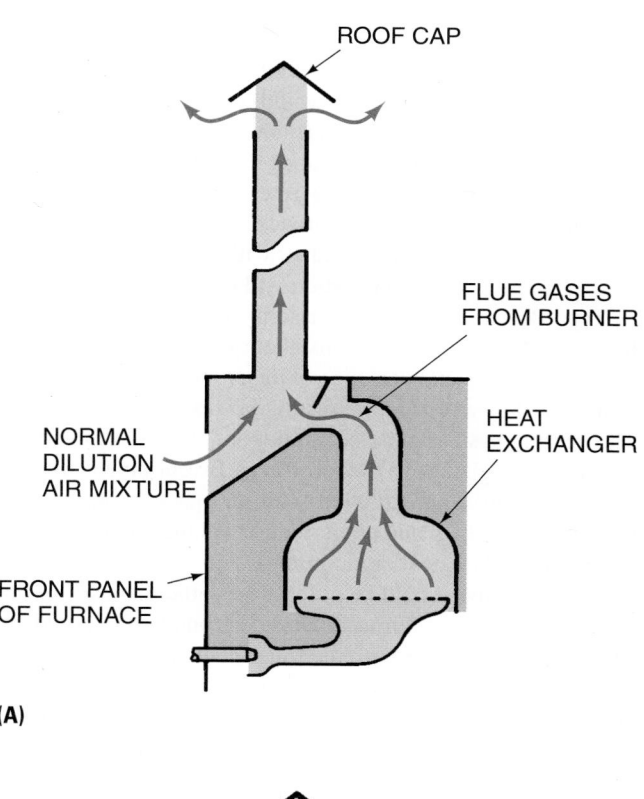

(A)

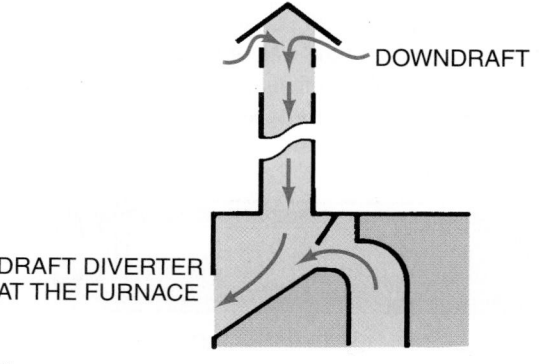

(B)

Figure 31–89 Draft hood operation.

and set venting standards for temperature and pressure requirements for the proper venting of gas furnaces.

■ Category I—Furnace has a nonpositive vent pressure and operates with a vent gas temperature at least 140°F above its dew point (conventional, or standard-efficiency, furnace).

■ Category II—Furnace has a nonpositive vent pressure and operates with a vent gas temperature less than 140°F above its dew point (no longer manufactured because of condensate problems at start-up).

■ Category III—Furnace has a positive vent pressure and operates with a vent gas temperature at least 140°F above its dew point (mid-efficiency furnace).

■ Category IV—Furnace has a positive vent pressure and operates with a vent gas temperature less than 140°F above its dew point (high-efficiency condensing furnace). SAFETY PRECAUTION: Make sure replacement air is available in the furnace area. Remember it takes 10 ft^3 of air for each 1 ft^3 of natural gas to support combustion. To this is added 5 ft^3 more to ensure enough oxygen. Another 15 ft^3 is added at the draft hood, making a total of 30 ft^3 of air for each 1 ft^3 of gas. A 100,000-Btu/h furnace would require 2857 ft^3/h of fresh air (100,000 ÷ 1050 Btu/h × 30 ft^3 = 2857 ft^3/h). All of this air must be replaced in the furnace area and also must be vented as flue gases rise up the vent. If the air is not replaced in the furnace area, it will become a negative pressure area, and air will be pulled down the flue. Products of combustion will then fill the area.

Type B vent or approved masonry materials are required for conventional gas furnace installations. *Type B* venting systems consist of metal vent pipe of the proper thickness, which is approved by a recognized testing laboratory, **Figure 31–90.** The venting system must be continuous from the furnace to the proper height above the roof. Type B vent pipe is usually of a double-wall construction with air space between. The inner wall may be made of aluminum and the outer wall may be constructed of steel or aluminum.

The vent pipe should be at least the same diameter as the vent opening at the furnace. The horizontal run should be as short as possible. Horizontal runs should always be sloped upward as the vent pipe leaves the furnace. A slope of 1/4 in. per foot of run is the minimum recommended, **Figure 31–91.** Use as few elbows as possible. Long runs lower the temperature of the flue gases before they reach the vertical vent or chimney and they reduce the draft. Do not insert the vent connector beyond the inside of the wall of the chimney and make sure that there are no obstructions in the chimney when the gases are vented into a masonry-type chimney, **Figure 31–92.**

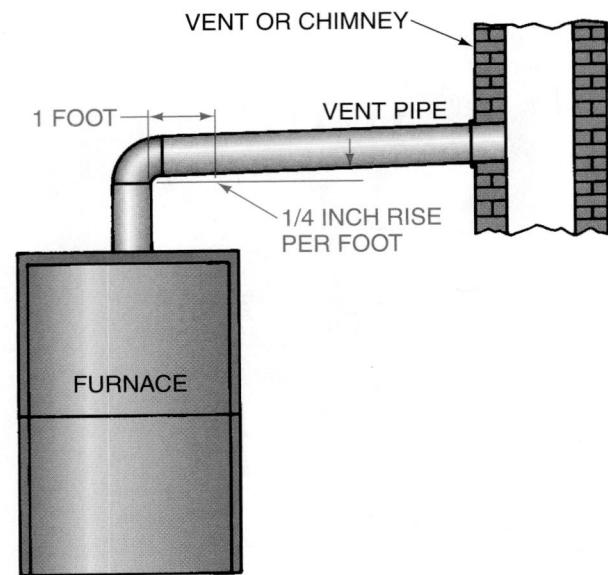

Figure 31–91 The minimum rise should be 1/4 in. per foot of horizontal run.

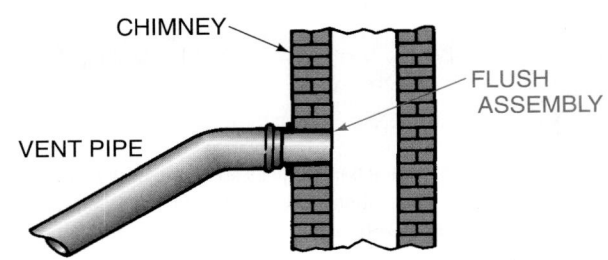

Figure 31–92 The vent pipe should not extend beyond the inside of the flue lining.

If two or more gas appliances are vented into a common flue, the common flue size should be calculated based on recommendations from the National Fire Protection Association (NFPA) or local codes.

The top of each gas furnace vent should have an approved vent cap to prevent the entrance of rain and debris and to help prevent wind from blowing the gases back into the building through the draft hood, **Figure 31–93.**

Metal vents reach operating temperatures quicker than masonry vents do. Masonry chimneys tend to cool the flue gases. They must be lined with a glazed-type tile or vent pipe installed in the chimney. If an unlined chimney is used, the corrosive materials from the flue gas will destroy the mortar joints. Condensation occurs at start-up and short running

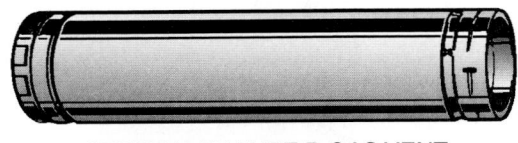

SECTION OF TYPE B GAS VENT

Figure 31–90 A section of type B gas vent.

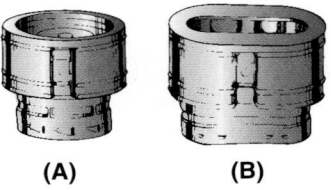

(A)   (B)

Figure 31–93 Gas vent caps **(A)–(B).**

times in mild weather. It takes a long running cycle to heat a heavy chimney. The warmer the gases, the faster they rise and the less damage there is from condensation and other corrosive actions. Vertical vents can be masonry with lining and glazing or prefabricated metal chimneys approved by a recognized testing laboratory. When an unlined chimney must be used, a special corrosion-resistant flexible liner often can be installed. It is worked down into the chimney and connected at the bottom to the furnace vent. This provides a corrosion-proof vent system.

When the furnace is off, heated air can leave the structure through the draft hood. Automatic vent dampers that close when the furnace is off and open when the furnace is started can be placed in the vent to prevent air loss.

Several types of dampers are available. One type uses a helical bimetal with a linkage to the damper. When the bimetal expands, it turns the shaft fastened to the helical strip and opens the damper, **Figure 31–94.** Another type uses damper blades constructed of bimetal. When the system is off, the damper is closed. When the vent is heated by the flue gases, the bimetal action on the damper blades opens them, **Figure 31–95.**

In high-efficiency systems a small combustion blower is installed in the vent system near the heat exchanger. This blower mainly operates when the furnace is on so that heated room air is not being vented 24 hours a day. Heated room air will only be vented if the furnace is non–direct vented. This happens because non-direct-vented furnaces get their combustion air from the conditioned space. However, if the furnace is direct vented, it will have a sealed combustion chamber and will get its combustion air from the outside through an air pipe usually made of PVC plastic, **Figure 31–71.** There is no possible way for conditioned room air to be vented because the combustion chamber never sees conditioned room air. Since conditioned room air is not used for combustion, it prevents the conditioned room from being depressurized, or being pulled into a slight vacuum. This will cut down on

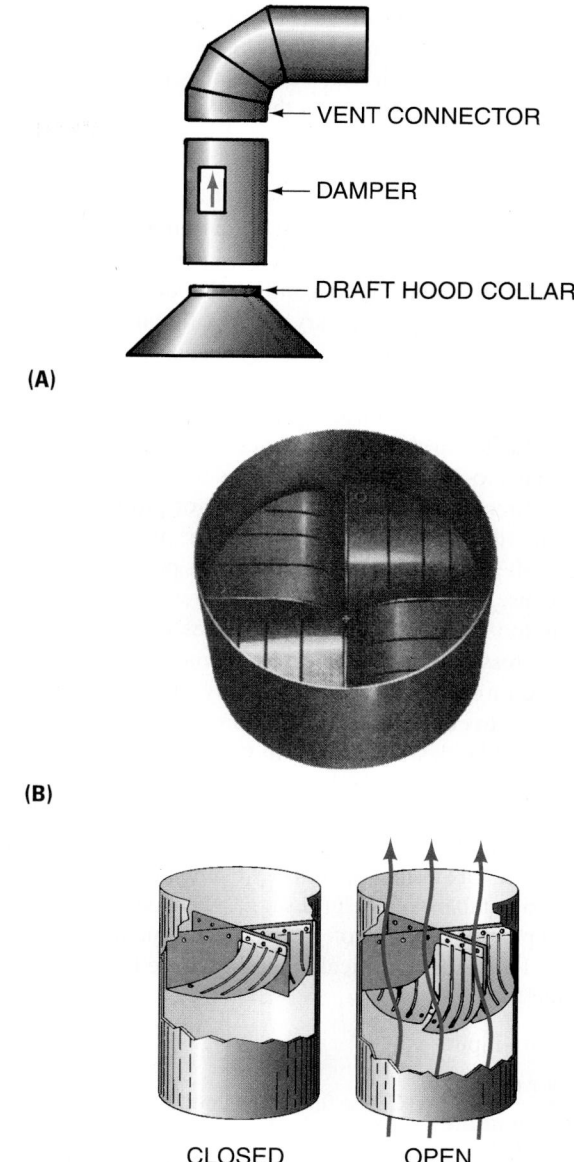

(A)

(B)

CLOSED   OPEN

(C)

Figure 31–95 An automatic vent damper with bimetal blades **(A)–(B). (C)** Cutaway.

the amount of infiltration entering the room, and less cold infiltrated air will have to be reheated. Direct venting also minimizes the interaction of the furnace with other combustion appliances.

As mentioned earlier in this unit, combustion blowers can be forced- or induced-draft systems. An induced-draft system has the combustion blower at the outlet of the heat exchangers, causing a negative pressure that actually sucks the combustion gases out of the heat exchanger. A forced-draft system has the combustion blower before the heat exchangers and causes a positive pressure that blows the combustion gases through the heat exchanger. This blower will vent gases at a lower temperature than is possible with natural draft, allowing the heat exchanger to absorb more of the heat from the burner for distribution to the conditioned space,

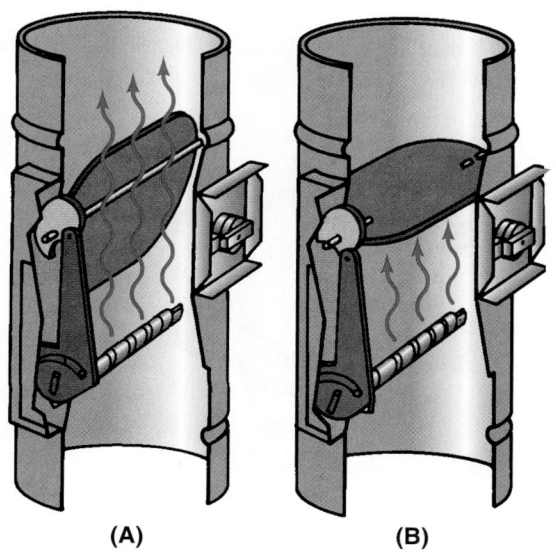

(A)   **(B)**

Figure 31–94 A rotating helix and vent damper operated with a bimetal helix **(A)–(B).**

Figure 31–75. When combustion blowers are used for furnace venting, the room thermostat normally energizes the combustion blower. Blower air can be established by airflow switches or a centrifugal switch on the motor.

31.26 GAS PIPING

Installing or replacing gas piping can be an important part of a technician's job. The first thing a technician should do is to become familiar with the national and local codes governing gas piping. Local codes may vary from national codes. A technician should also be familiar with the characteristics of natural gas and LP gas in the particular area in which he or she works. Pipe sizing and furnace Btu ratings will vary from area to area due to varying gas characteristics.

The piping should be kept as simple as possible with the pipe run as direct as possible. Distributing or piping gas is similar to distributing electricity. The piping must be large enough, and there should be as few fittings as possible in the system because each of these fittings creates a resistance. Pipe that is too small with other resistances in the system will cause a pressure drop, and the proper amount of gas will not reach the furnace. The specific gravity of the gas must also be taken into consideration when designing the piping. Systems should be designed for a maximum pressure drop of 0.35 in. WC. In designing the piping system the amount of gas to be consumed by the furnace must be determined. The gas company should be contacted to determine the heating value of the gas in that area. They can also furnish you with pipe sizing tables and helpful suggestions. Most natural gas will supply 1050 Btu/ft^3 of gas. To determine the gas to be consumed in 1 h for a typical natural gas, use the following formula:

$$\frac{\text{Furnace Btu input}}{\substack{\text{per hour rating} \\ \hline \text{Gas heating rating} \\ \text{in Btu/ft}^3}} = \text{Cubic feet of gas needed per hour}$$

Suppose the Btu rating of the furnace were 100,000 and the gas heating rating were 1050. Then

$$\frac{100,000}{1050} = 95.2 \text{ ft}^3 \text{ of natural gas needed per hour}$$

Tables give the size (diameter) of pipe needed for the length of the pipe to provide the proper amount of gas needed for this furnace. The designer of the piping system would have to know whether natural gas or LP gas were going to be used because the pipe sizing would be different due to the differences in specific gravities of the gases.

Steel or wrought iron pipe should be used. Aluminum or copper tubing may be used, but each requires treatment to prevent corrosion. They are used only in special circumstances.

SAFETY PRECAUTION: Ensure that all piping is free of burrs and that threads are not damaged, **Figure 31–96.** All scale, dirt, or other loose material should be cleaned from pipe threads and from the inside of the pipe. Any loose particles can move through the pipe to the gas valve and keep it from closing properly. It may also stop up the small pilot light orifice.

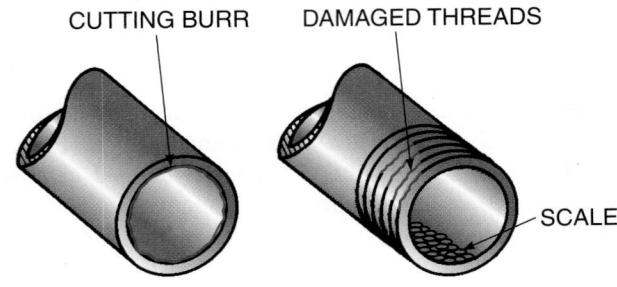

Figure 31–96 Ensure that threads are not damaged. Deburr pipe. Clean all scale, dirt, and other loose material from pipe threads and inside of pipe.

When assembling threaded pipe and fittings, use a joint compound, commonly called *pipe dope*. Do not apply this compound on the first two threads at the end of the pipe because it could get into the pipe and plug the orifice or prevent the gas valve from closing, **Figure 31–97.** Also be careful when using Teflon tape to seal the pipe threads because it can also get into the pipe and cause similar problems.

At the furnace the piping should provide for a drip trap, a shutoff valve, and a union, **Figure 31–98.** A manual shutoff valve within 2 ft of the furnace is required by most local

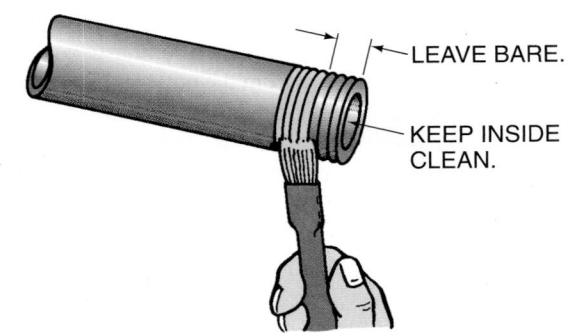

Figure 31–97 Pipe dope should be applied to male threads, but do not apply to the last two threads at the end of the pipe.

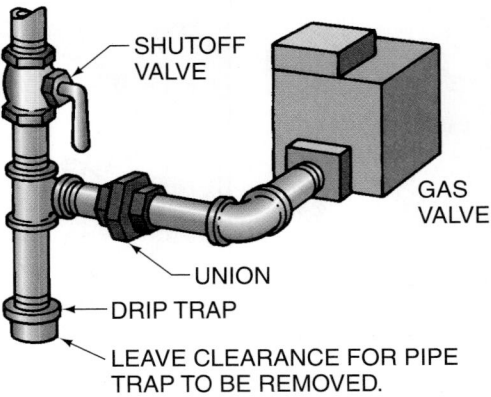

Figure 31–98 A shutoff valve, drip trap, and union should be installed ahead of the gas valve.

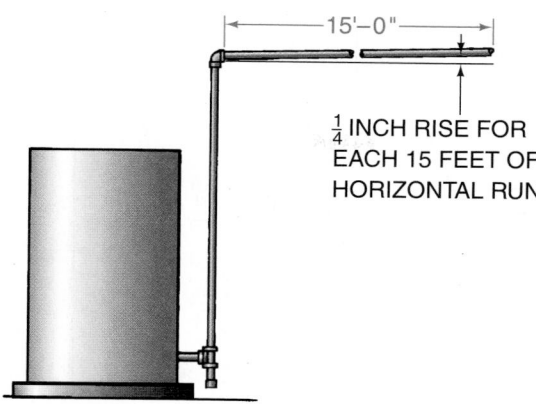

Figure 31–99   Piping from the furnace should have a rise of 1/4 in. for each 15 ft of horizontal run.

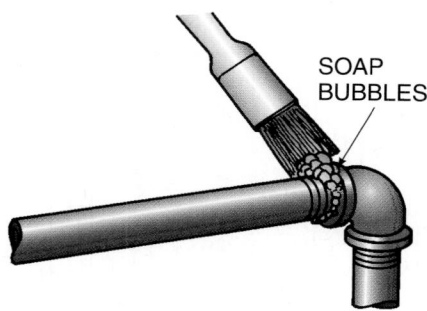

Figure 31–101   Apply liquid soap to joints. Bubbling indicates a leak.

codes. The drip trap is installed to catch dirt, scale, or condensate (moisture) from the supply line. A union between the tee and the gas valve allows the gas valve or the entire furnace to be removed without disassembling other piping. Piping should be installed with a pitch of 1/4 in. for every 15 ft of run in the direction of flow. This will prevent trapping of the moisture that could block the gas flow, **Figure 31–99**. Small amounts of moisture will move to the drip leg and slowly evaporate. Use pipe hooks or straps to support the piping adequately, **Figure 31–100**.

SAFETY PRECAUTION: When completed, test the piping assembly for leaks. There are several methods; one is to use the gas in the system. Do not use other gases, and especially *do not* use oxygen. Turn off the manual shutoff to the furnace. Make sure all joints are secure. Turn the gas on in the system. Watch the gas meter dial to see if it moves, indicating that gas is passing through the system. A check for 5 to 10 min will indicate a large leak if the meter dial continues to show gas flow. An overnight standing check is better. If the dial moves, indicating a leak, check each joint with soap and water, **Figure 31–101**. A leak will make the soap bubble. When you find the leak and repair it, repeat the same procedure to ensure that the leak has been repaired and that there are no other leaks.

Leaks can also be determined with a manometer. Install the manometer in the system to measure the gas pressure. Turn the gas on and read the pressure in in. WC on the instrument. Now turn the gas off. There should be no loss of pressure. Some technicians allow an overnight standing period with no pressure drop as their standard. Use soap and water to locate the leak as before.

NOTE: If a standing pilot system that is not 100% shut off is being checked, the pilot light valve *must be closed* because it will indicate a leak.

A high-pressure test is more efficient. For a high-pressure leak check, use air pressure from a bicycle tire pump. Ten pounds of pressure for 10 min with no leakdown will prove there are no leaks. SAFETY PRECAUTION: Do not let this pressure reach the automatic gas valve. The manual valve must be closed during the high-pressure test.

If there are no gas leaks, the system must be *purged,* that is, the air must be bled off to rid the system of air or other gases. SAFETY PRECAUTION: Purging should take place in a well-ventilated area. It can often be done by disconnecting the pilot tubing or loosening the union in the piping between the manual shutoff and the gas valve. Do not purge where the gas will collect in the combustion area. After the system has been purged, allow it to set for at least 15 min to allow any accumulated gas to dissipate. If you are concerned about gas collecting in the area, wait for a longer period. Moving the air with a hand-operated fan can speed this up. When it is evident that no gas has accumulated, the pilot can be lit. SAFETY PRECAUTION: Be extremely careful because gas is explosive. The standing leak test must be performed and all conditions satisfied before a gas system is put in operation. Be aware of and practice any local code requirements for gas piping.

31.27   GAS FURNACE WIRING DIAGRAMS AND TROUBLESHOOTING FLOWCHARTS

A wiring diagram in both schematic and pictorial form is shown in **Figure 31–102(A)**. The diagram is for a modern two-stage gas furnace with a direct-spark ignition system and

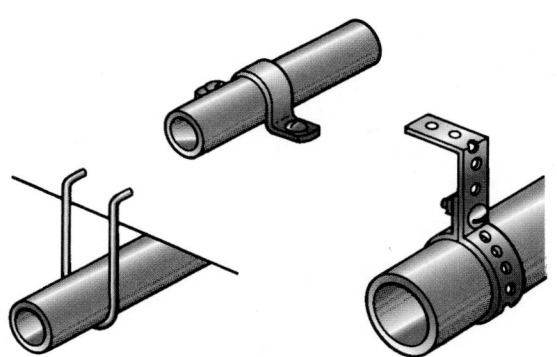

Figure 31–100   Piping should be supported adequately with pipe hooks or straps.

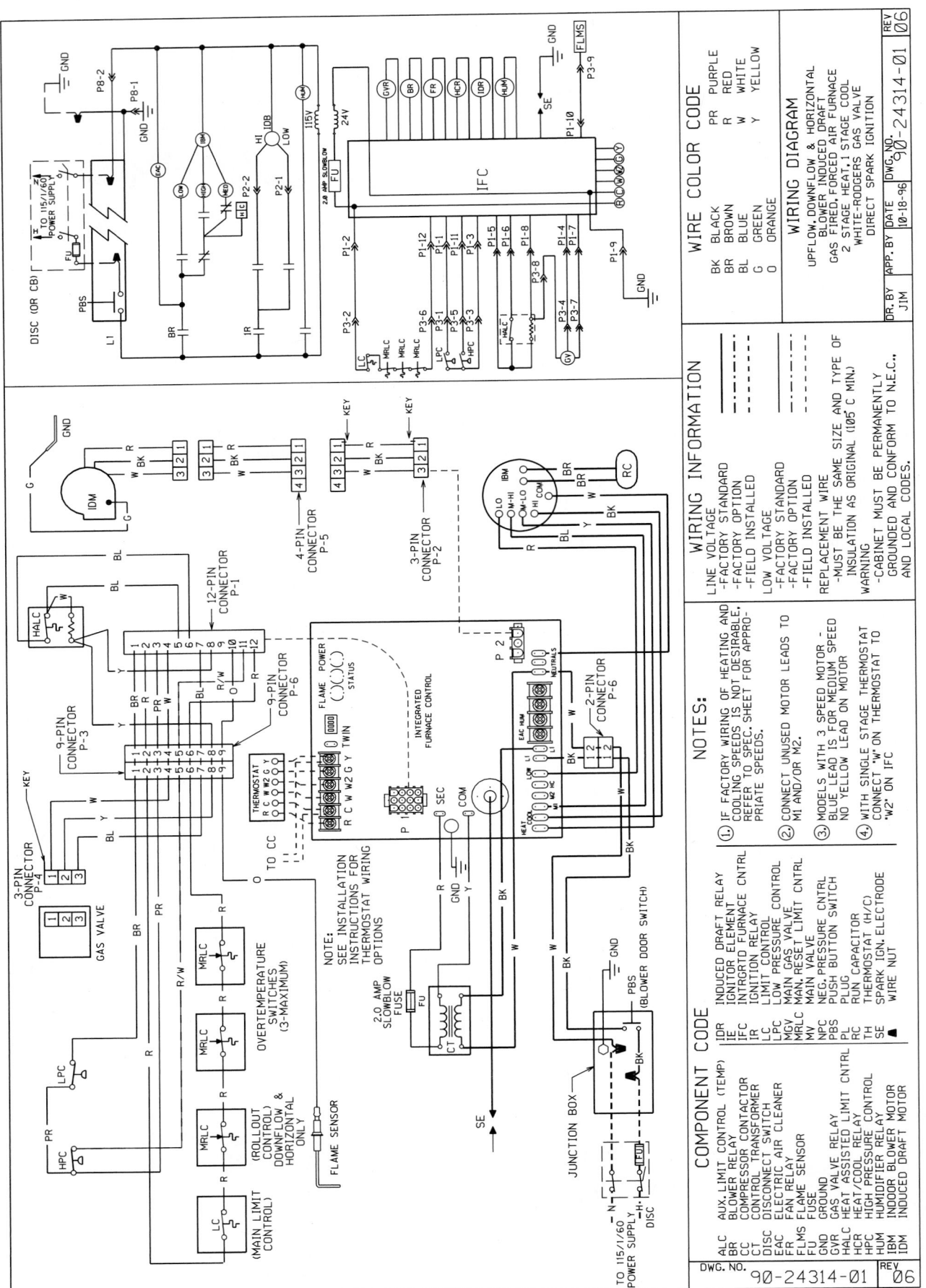

Figure 31-102(A) The schematic wiring diagram of a gas furnace. **(A)** *Courtesy Rheem Manufacturing*

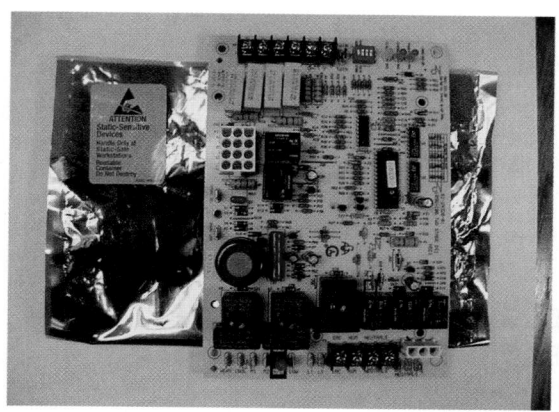

Figure 31–102(B)   An IFC board with a reusable static shielding bag. *Courtesy Ferris State University. Photo by John Tomczyk*

an integrated furnace controller. Notice that the IFC is shown as an empty box in the schematic diagram. However, the pictorial diagram shows the IFC wiring terminals. It is beyond the scope of this text to explain the internal operations of the IFC. Although service technicians do not troubleshoot the circuits within the IFC board, they do troubleshoot circuits coming to and from the IFC board by following the input/output troubleshooting procedure. Troubleshooting guides can be followed as the technician understands the IFC's sequence of operation.

The IFC is a complicated microprocessor or small computer and must be handled with care. A small static discharge can damage the IFC. Service technicians must make sure they are physically grounded before attempting to troubleshoot the IFC. To prevent electrical static discharge from their body, some technicians even wear a grounding strap on their wrist, which is securely wired to ground, when they are replacing an IFC board.

An IFC board is shipped in a reusable static shielding bag, **Figure 31–102(B).** The service technician must be careful not to expose the IFC board to a static shock once it is removed from the static shielding bag. As mentioned before, the service technician must be securely grounded before handling the IFC board.

The first step in systematically troubleshooting a furnace that has an IFC is to obtain the electrical diagram of the furnace and the sequence of operation of the IFC. Then the service technician must obtain a systematic troubleshooting guide or flowchart like the one in **Figure 31–103.** The sequence of operation for the two-stage gas furnace in the electrical diagrams in **Figure 31–102(A)** is listed in Section 31.22. This sequence of operation, the electrical diagrams, and the systematic troubleshooting guide are the technician's most important tools for furnace troubleshooting.

When voltage-troubleshooting the IFC board with a voltmeter, make sure the voltmeter leads do not touch unintended parts of the board. This can lead to short circuits and damage to the IFC board. The voltmeter leads should be kept as short as possible. Applying some electrical tape on the bottom part of the voltmeter leads can shorten their conducting length.

Figure 31–104(A) is another example of both pictorial (connect) and schematic (ladder) diagrams for a modern gas furnace incorporating a furnace and fan control module. **Figure 31–104(B)** is a flowchart for systematically troubleshooting a gas furnace that has an IFC and DSI system. **Figure 31–105(A)** is an example of a connect and ladder diagram for a gas furnace having both an IFC and a 115-V hot surface ignition system. **Figure 31–105(B)** is a flowchart for systematically troubleshooting a different furnace that has an IFC and a spark ignition system.

31.28   TROUBLESHOOTING THE SAFETY PILOT-PROVING DEVICE— THE THERMOCOUPLE

The thermocouple generates a small electrical current when one end is heated. When the pilot flame is lit and heating the thermocouple, it generates a current that energizes a coil holding a safety valve open. This valve is manually opened when lighting the pilot; the coil only holds it open. If the pilot flame goes out, the current would no longer be generated and the safety valve would close, shutting off the supply of gas.

To light the pilot, turn the gas valve control to PILOT position. Depress the control knob and light the pilot. Hold the knob down for approximately 45 sec, release, and turn the valve to the ON position. The thermocouple may be defective if the pilot goes out when the knob is released. **Figure 31–106** describes the no-load test for a thermocouple.

It is important to remember that the thermocouple only generates enough current to *hold* the coil armature in. The armature is pushed in when the valve handle is depressed.

To check under-load conditions, unscrew the thermocouple from the gas valve. Insert the thermocouple testing adapter into the gas valve. The testing adapter allows you to take voltage readings while the thermocouple is operating. Screw the thermocouple into the top of this adapter, **Figure 31–107.** Relight the pilot and check the voltage produced with a millivoltmeter, using the terminals on the adapter. The pilot flame must cover the entire top of the thermocouple rod. If the thermocouple produces at least 9 mV under load (while connected to its coil), it is good. Otherwise replace it. The manufacturer's specifications must be checked to determine the acceptable voltage for holding the different valve coils. If the thermocouple functions properly but the flame will not continue to burn when the knob is released, the coil in the safety valve must be replaced. This is normally done by replacing the entire gas valve. Thermopiles can generate much more voltage than thermocouples, sometimes in the range of 500 mV (0.5 V) to 750 mV (0.750 V). When more power is required to operate the circuit, the thermopile is a good choice. **Figure 31–108** is a systematic troubleshooting guide for a standing pilot gas furnace system.

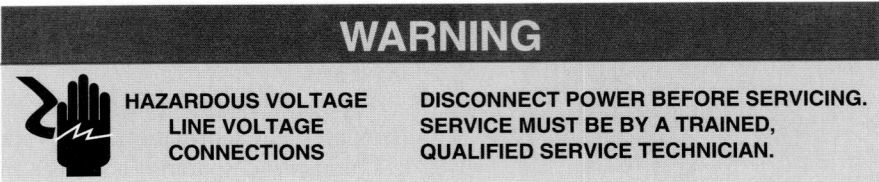

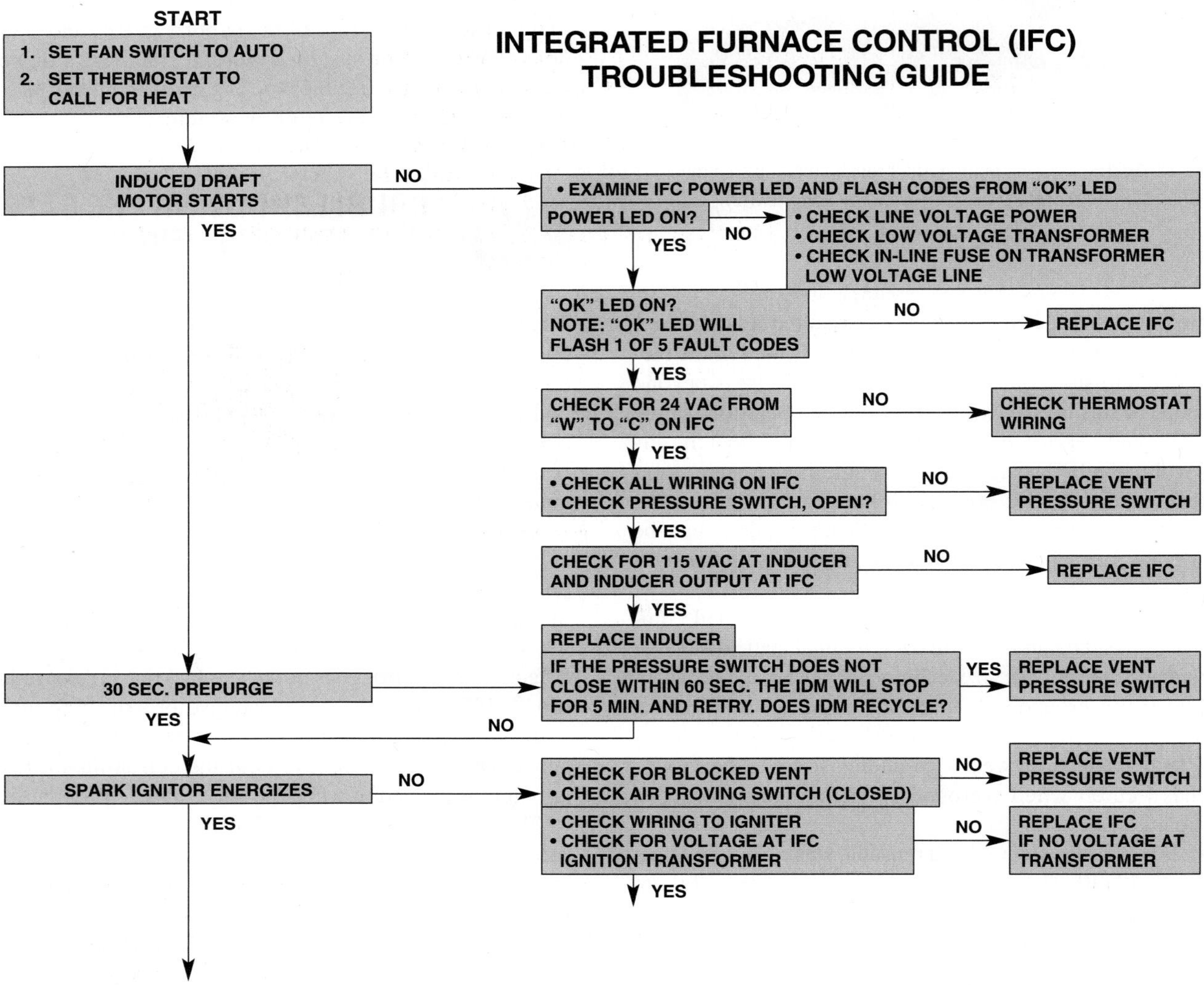

Figure 31–103 A troubleshooting guide for a spark ignition system with an integrated furnace controller (IFC). *Courtesy Rheem Manufacturing Company*

31.29 ▶ TROUBLESHOOTING SPARK IGNITION AND INTERMITTENT PILOT SYSTEMS

Most spark ignition or intermittent pilot-light assemblies have internal circuits or printed circuit boards. The technician can troubleshoot only the circuit to the board, not circuits within the board. If the trouble is in the circuit board, the board must be changed. **Figure 31–109** shows how one manufacturer accomplishes spark ignition. There are two-diagrams; one is a pictorial and the other is a line type.

Notice that the manufacturer has placed the terminal board on the bottom of the pictorial and on the side of the line diagram. Components are not placed in the same position from one diagram to another.

The spark ignition board in the previous example has terminals provided for adding other components, such as an electronic air cleaner and a vent damper shutoff motor. This is handy because the installing technician only has to follow the directions to wire these components when they are used in conjunction with this furnace. The electronic air cleaner is properly interlocked to operate only when the fan is running. This is not

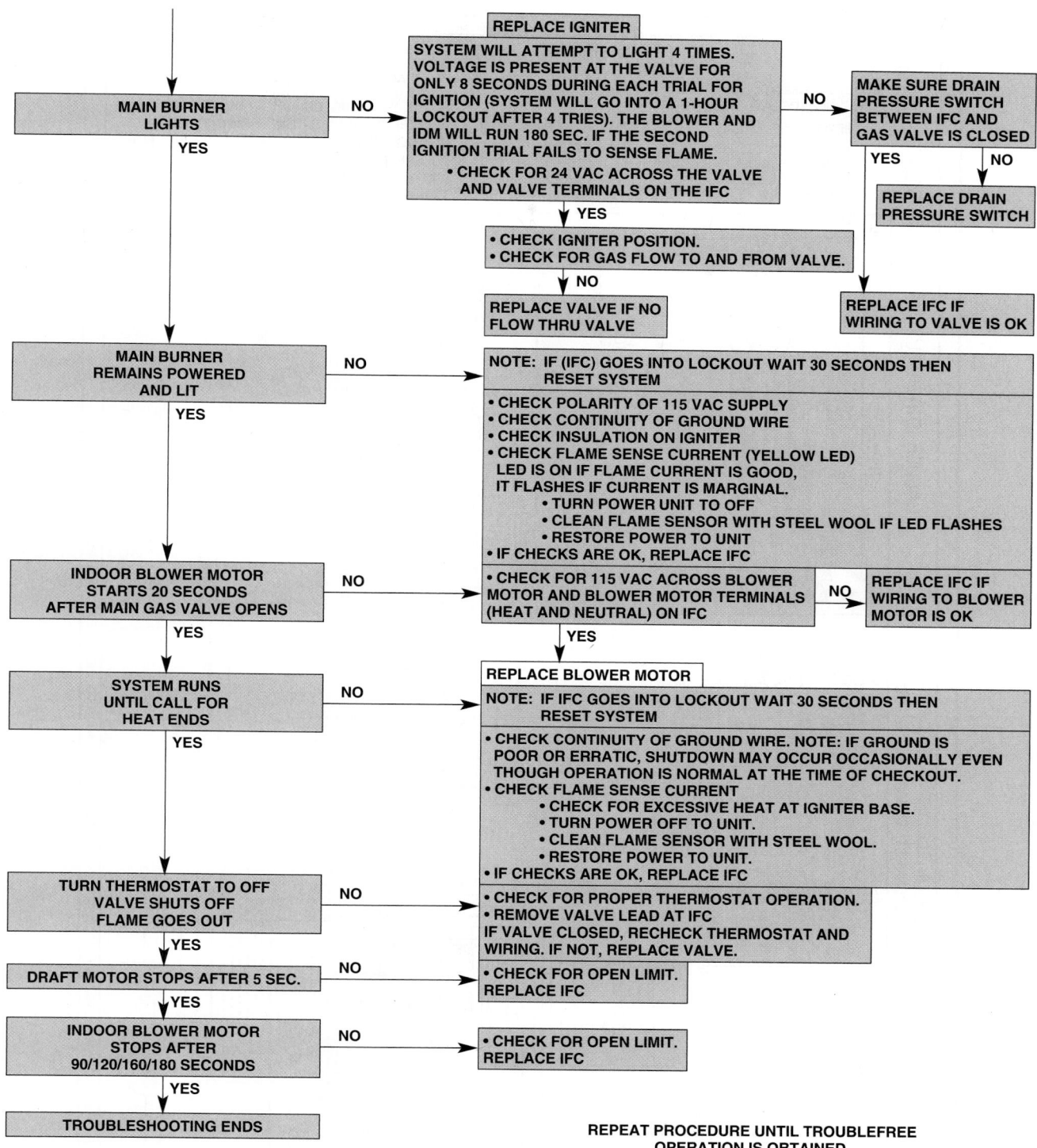

REPLACE IGNITER

SYSTEM WILL ATTEMPT TO LIGHT 4 TIMES. VOLTAGE IS PRESENT AT THE VALVE FOR ONLY 8 SECONDS DURING EACH TRIAL FOR IGNITION (SYSTEM WILL GO INTO A 1-HOUR LOCKOUT AFTER 4 TRIES). THE BLOWER AND IDM WILL RUN 180 SEC. IF THE SECOND IGNITION TRIAL FAILS TO SENSE FLAME.

• CHECK FOR 24 VAC ACROSS THE VALVE AND VALVE TERMINALS ON THE IFC

MAIN BURNER LIGHTS — NO →   YES

MAKE SURE DRAIN PRESSURE SWITCH BETWEEN IFC AND GAS VALVE IS CLOSED — NO →
YES   NO

REPLACE DRAIN PRESSURE SWITCH

• CHECK IGNITER POSITION.
• CHECK FOR GAS FLOW TO AND FROM VALVE.
NO

REPLACE VALVE IF NO FLOW THRU VALVE

REPLACE IFC IF WIRING TO VALVE IS OK

MAIN BURNER REMAINS POWERED AND LIT — NO →   YES

NOTE: IF (IFC) GOES INTO LOCKOUT WAIT 30 SECONDS THEN RESET SYSTEM

• CHECK POLARITY OF 115 VAC SUPPLY
• CHECK CONTINUITY OF GROUND WIRE
• CHECK INSULATION ON IGNITER
• CHECK FLAME SENSE CURRENT (YELLOW LED)
 LED IS ON IF FLAME CURRENT IS GOOD,
 IT FLASHES IF CURRENT IS MARGINAL.
  • TURN POWER UNIT TO OFF
  • CLEAN FLAME SENSOR WITH STEEL WOOL IF LED FLASHES
  • RESTORE POWER TO UNIT
• IF CHECKS ARE OK, REPLACE IFC

INDOOR BLOWER MOTOR STARTS 20 SECONDS AFTER MAIN GAS VALVE OPENS — NO →   YES

• CHECK FOR 115 VAC ACROSS BLOWER MOTOR AND BLOWER MOTOR TERMINALS (HEAT AND NEUTRAL) ON IFC — NO → REPLACE IFC IF WIRING TO BLOWER MOTOR IS OK
YES

REPLACE BLOWER MOTOR

SYSTEM RUNS UNTIL CALL FOR HEAT ENDS — NO →   YES

NOTE: IF IFC GOES INTO LOCKOUT WAIT 30 SECONDS THEN RESET SYSTEM

• CHECK CONTINUITY OF GROUND WIRE. NOTE: IF GROUND IS
 POOR OR ERRATIC, SHUTDOWN MAY OCCUR OCCASIONALLY EVEN
 THOUGH OPERATION IS NORMAL AT THE TIME OF CHECKOUT.
• CHECK FLAME SENSE CURRENT
  • CHECK FOR EXCESSIVE HEAT AT IGNITER BASE.
  • TURN POWER OFF TO UNIT.
  • CLEAN FLAME SENSOR WITH STEEL WOOL.
  • RESTORE POWER TO UNIT.
• IF CHECKS ARE OK, REPLACE IFC

TURN THERMOSTAT TO OFF VALVE SHUTS OFF FLAME GOES OUT — NO →   YES

• CHECK FOR PROPER THERMOSTAT OPERATION.
• REMOVE VALVE LEAD AT IFC
IF VALVE CLOSED, RECHECK THERMOSTAT AND WIRING. IF NOT, REPLACE VALVE.

DRAFT MOTOR STOPS AFTER 5 SEC. — NO →   YES

• CHECK FOR OPEN LIMIT.
REPLACE IFC

INDOOR BLOWER MOTOR STOPS AFTER 90/120/160/180 SECONDS — NO →   YES

• CHECK FOR OPEN LIMIT.
REPLACE IFC

TROUBLESHOOTING ENDS

REPEAT PROCEDURE UNTIL TROUBLEFREE OPERATION IS OBTAINED.

NOTE:
STATIC DISCHARGE CAN DAMAGE INTEGRATED FURNACE CONTROL (IFC)
* "OK" LED BLINKS TO INDICATE EXTERNAL FAULTS:

(1) BLINK FOLLOWED BY A 2 SEC. PAUSE – 1 HOUR LOCKOUT
(2) BLINKS FOLLOWED BY A 2 SEC. PAUSE – PRESSURE SWITCH IS OPEN
(3) BLINKS FOLLOWED BY A 2 SEC. PAUSE – LIMIT SWITCH IS OPEN
(4) BLINKS FOLLOWED BY A 2 SEC. PAUSE – PRESSURE SWITCH CLOSED
(5) TWINNING FAULT

Figure 31–103   *(continued)*

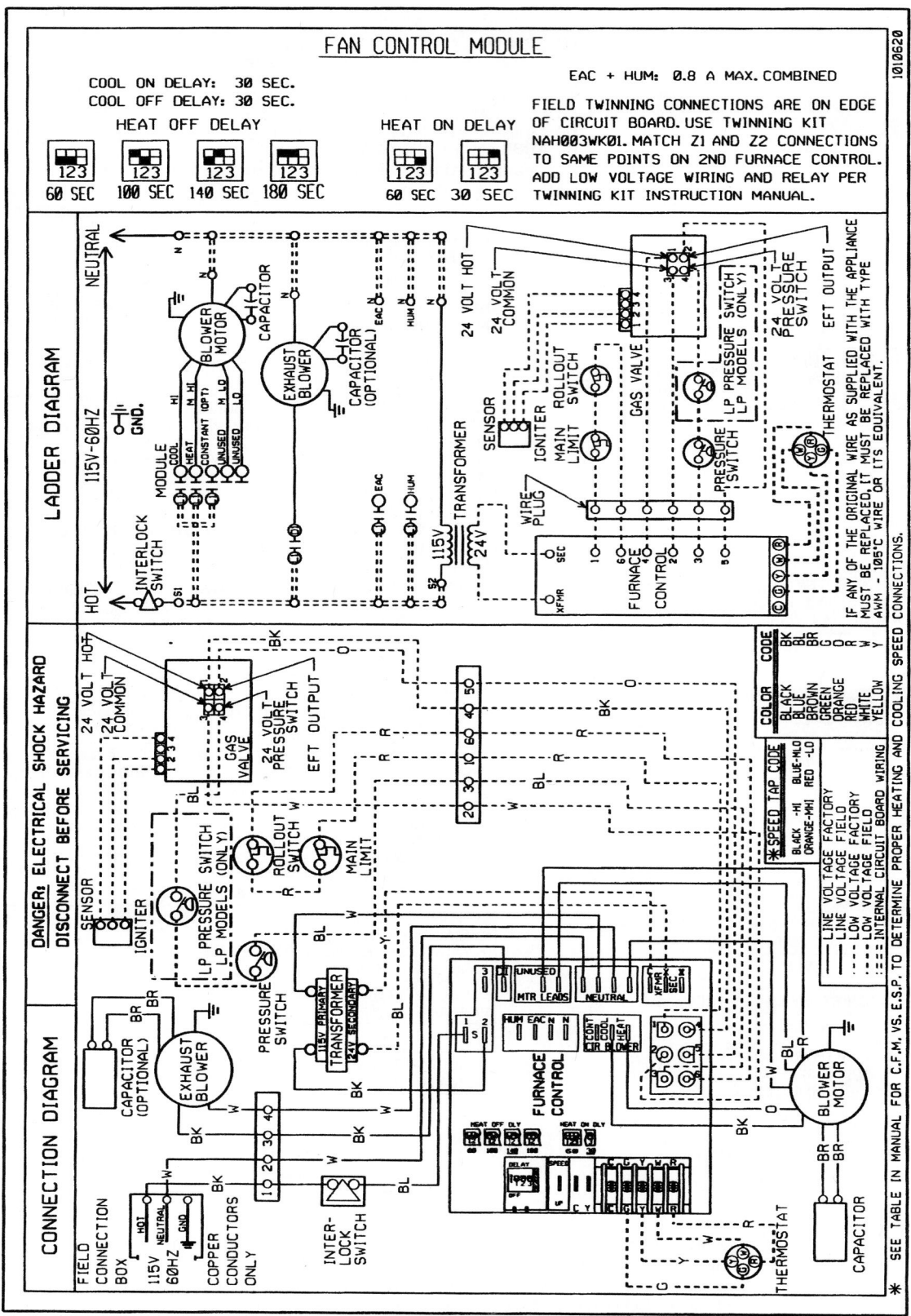

Figure 31–104(A)   An electrical diagram for a modern gas furnace that incorporates a furnace and fan control module. *Courtesy International Comfort Products Corporation*

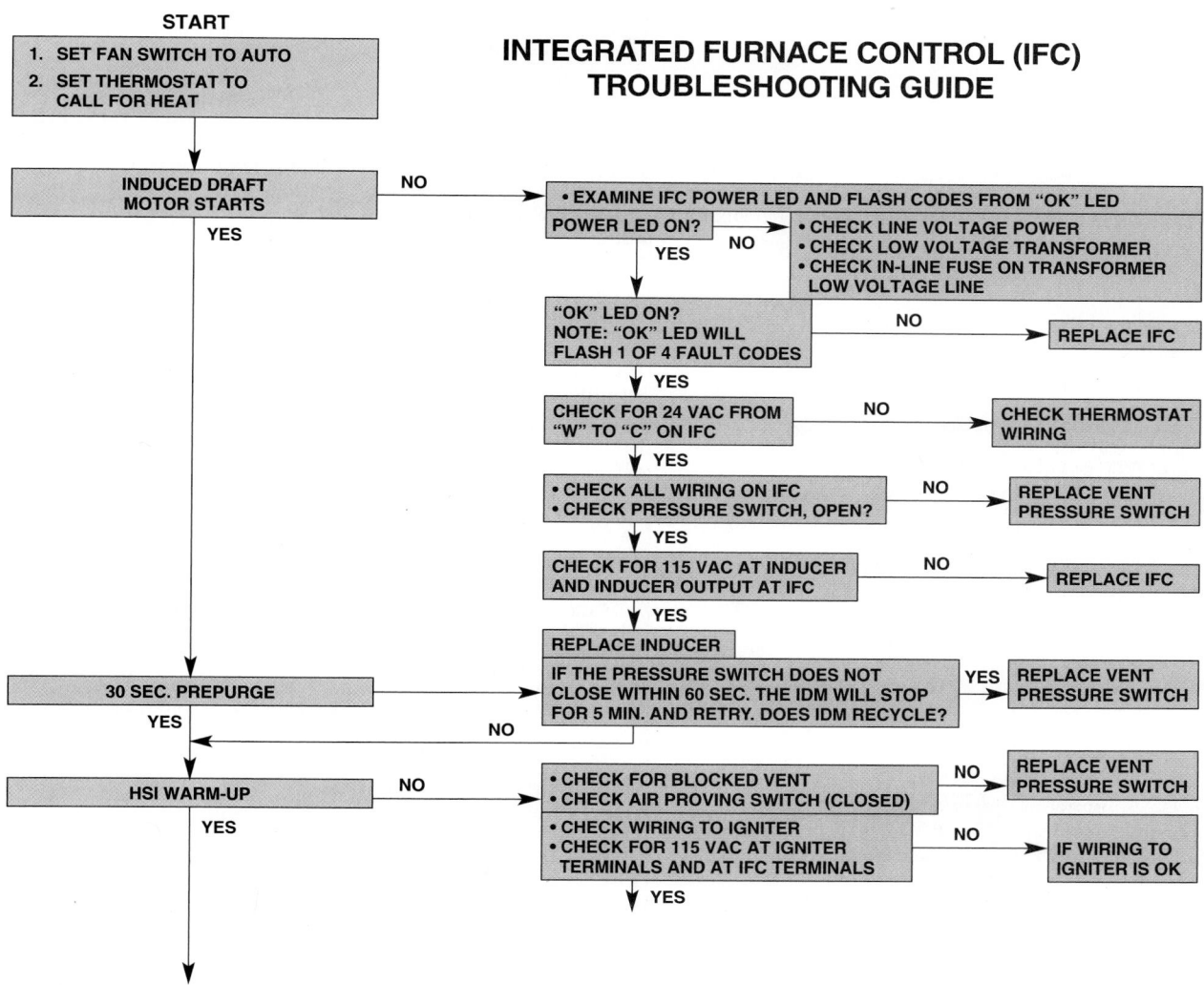

WARNING

| HAZARDOUS VOLTAGE LINE VOLTAGE CONNECTIONS | DISCONNECT POWER BEFORE SERVICING. SERVICE MUST BE BY A TRAINED, QUALIFIED SERVICE TECHNICIAN. |

INTEGRATED FURNACE CONTROL (IFC) TROUBLESHOOTING GUIDE

START

1. SET FAN SWITCH TO AUTO
2. SET THERMOSTAT TO CALL FOR HEAT

INDUCED DRAFT MOTOR STARTS — NO →
- EXAMINE IFC POWER LED AND FLASH CODES FROM "OK" LED

YES

POWER LED ON? — NO →
- CHECK LINE VOLTAGE POWER
- CHECK LOW VOLTAGE TRANSFORMER
- CHECK IN-LINE FUSE ON TRANSFORMER LOW VOLTAGE LINE

YES

"OK" LED ON?
NOTE: "OK" LED WILL FLASH 1 OF 4 FAULT CODES — NO → **REPLACE IFC**

YES

CHECK FOR 24 VAC FROM "W" TO "C" ON IFC — NO → **CHECK THERMOSTAT WIRING**

YES

- CHECK ALL WIRING ON IFC
- CHECK PRESSURE SWITCH, OPEN? — NO → **REPLACE VENT PRESSURE SWITCH**

YES

CHECK FOR 115 VAC AT INDUCER AND INDUCER OUTPUT AT IFC — NO → **REPLACE IFC**

YES

REPLACE INDUCER

30 SEC. PREPURGE →
IF THE PRESSURE SWITCH DOES NOT CLOSE WITHIN 60 SEC. THE IDM WILL STOP FOR 5 MIN. AND RETRY. DOES IDM RECYCLE? — YES → **REPLACE VENT PRESSURE SWITCH**

YES
NO

HSI WARM-UP — NO →
- CHECK FOR BLOCKED VENT
- CHECK AIR PROVING SWITCH (CLOSED) — NO → **REPLACE VENT PRESSURE SWITCH**

- CHECK WIRING TO IGNITER
- CHECK FOR 115 VAC AT IGNITER TERMINALS AND AT IFC TERMINALS — NO → **IF WIRING TO IGNITER IS OK**

YES

YES

FLOW CHART CONTINUED ON NEXT PAGE

Figure 31–104(B)   A guide for systematically troubleshooting a spark ignition system containing an integrated furnace controller (IFC). *Courtesy Rheem Manufacturing Company*

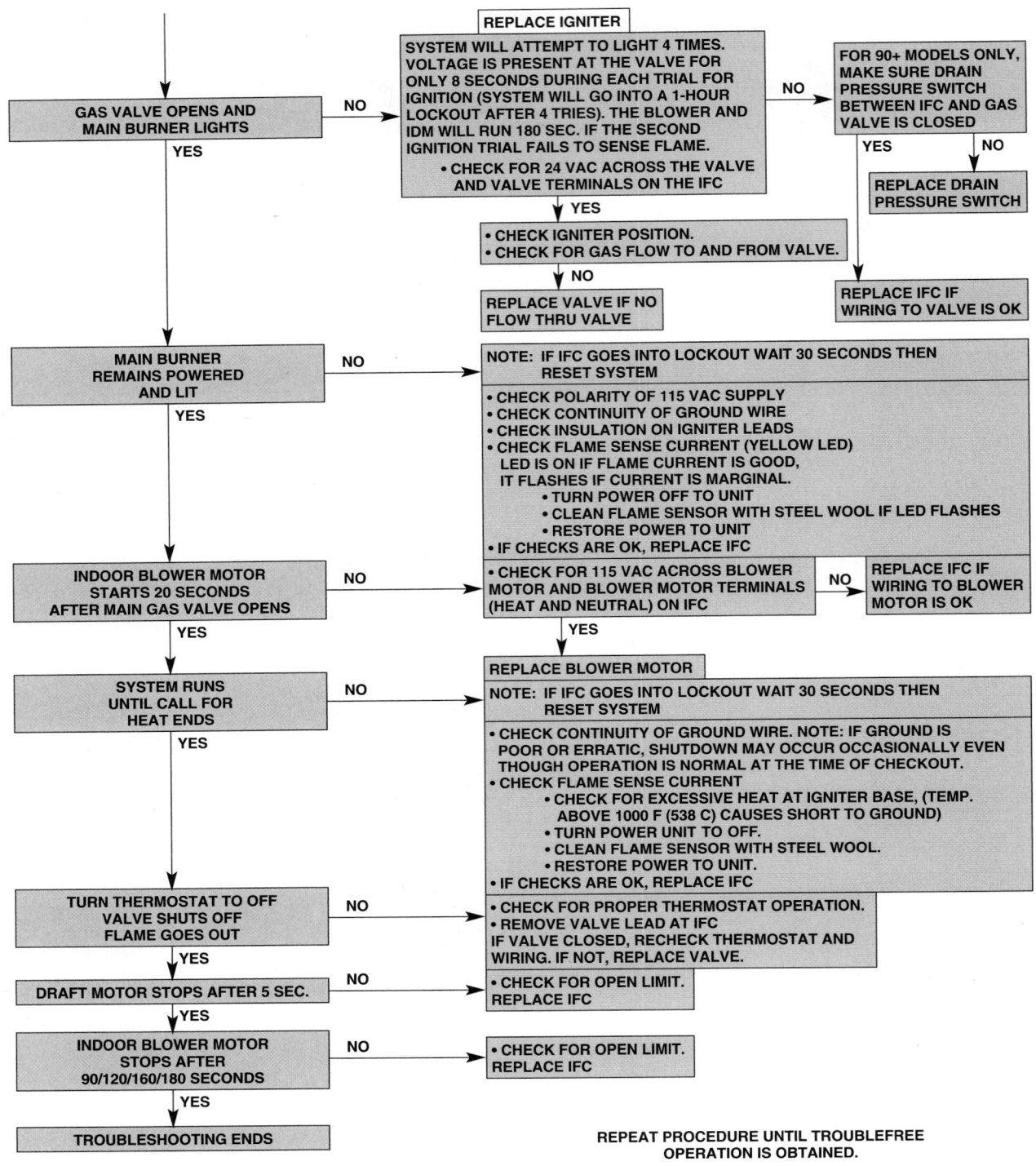

REPLACE IGNITER

SYSTEM WILL ATTEMPT TO LIGHT 4 TIMES. VOLTAGE IS PRESENT AT THE VALVE FOR ONLY 8 SECONDS DURING EACH TRIAL FOR IGNITION (SYSTEM WILL GO INTO A 1-HOUR LOCKOUT AFTER 4 TRIES). THE BLOWER AND IDM WILL RUN 180 SEC. IF THE SECOND IGNITION TRIAL FAILS TO SENSE FLAME.
• CHECK FOR 24 VAC ACROSS THE VALVE AND VALVE TERMINALS ON THE IFC

FOR 90+ MODELS ONLY, MAKE SURE DRAIN PRESSURE SWITCH BETWEEN IFC AND GAS VALVE IS CLOSED

REPLACE DRAIN PRESSURE SWITCH

GAS VALVE OPENS AND MAIN BURNER LIGHTS

• CHECK IGNITER POSITION.
• CHECK FOR GAS FLOW TO AND FROM VALVE.

REPLACE VALVE IF NO FLOW THRU VALVE

REPLACE IFC IF WIRING TO VALVE IS OK

MAIN BURNER REMAINS POWERED AND LIT

NOTE: IF IFC GOES INTO LOCKOUT WAIT 30 SECONDS THEN RESET SYSTEM
• CHECK POLARITY OF 115 VAC SUPPLY
• CHECK CONTINUITY OF GROUND WIRE
• CHECK INSULATION ON IGNITER LEADS
• CHECK FLAME SENSE CURRENT (YELLOW LED) LED IS ON IF FLAME CURRENT IS GOOD, IT FLASHES IF CURRENT IS MARGINAL.
    • TURN POWER OFF TO UNIT
    • CLEAN FLAME SENSOR WITH STEEL WOOL IF LED FLASHES
    • RESTORE POWER TO UNIT
• IF CHECKS ARE OK, REPLACE IFC

INDOOR BLOWER MOTOR STARTS 20 SECONDS AFTER MAIN GAS VALVE OPENS

• CHECK FOR 115 VAC ACROSS BLOWER MOTOR AND BLOWER MOTOR TERMINALS (HEAT AND NEUTRAL) ON IFC

REPLACE IFC IF WIRING TO BLOWER MOTOR IS OK

SYSTEM RUNS UNTIL CALL FOR HEAT ENDS

REPLACE BLOWER MOTOR

NOTE: IF IFC GOES INTO LOCKOUT WAIT 30 SECONDS THEN RESET SYSTEM
• CHECK CONTINUITY OF GROUND WIRE. NOTE: IF GROUND IS POOR OR ERRATIC, SHUTDOWN MAY OCCUR OCCASIONALLY EVEN THOUGH OPERATION IS NORMAL AT THE TIME OF CHECKOUT.
• CHECK FLAME SENSE CURRENT
    • CHECK FOR EXCESSIVE HEAT AT IGNITER BASE, (TEMP. ABOVE 1000 F (538 C) CAUSES SHORT TO GROUND)
    • TURN POWER UNIT TO OFF.
    • CLEAN FLAME SENSOR WITH STEEL WOOL.
    • RESTORE POWER TO UNIT.
• IF CHECKS ARE OK, REPLACE IFC

TURN THERMOSTAT TO OFF VALVE SHUTS OFF FLAME GOES OUT

• CHECK FOR PROPER THERMOSTAT OPERATION.
• REMOVE VALVE LEAD AT IFC
IF VALVE CLOSED, RECHECK THERMOSTAT AND WIRING. IF NOT, REPLACE VALVE.

DRAFT MOTOR STOPS AFTER 5 SEC.

• CHECK FOR OPEN LIMIT.
REPLACE IFC

INDOOR BLOWER MOTOR STOPS AFTER 90/120/160/180 SECONDS

• CHECK FOR OPEN LIMIT.
REPLACE IFC

TROUBLESHOOTING ENDS

REPEAT PROCEDURE UNTIL TROUBLEFREE OPERATION IS OBTAINED.

NOTE:
STATIC DISCHARGE CAN DAMAGE INTEGRATED FURNACE CONTROL (IFC)
* "OK" LED BLINKS TO INDICATE EXTERNAL FAULTS:
(1) BLINK FOLLOWED BY A 2 SEC. PAUSE – 1 HOUR LOCKOUT
(2) BLINKS FOLLOWED BY A 2 SEC. PAUSE – PRESSURE SWITCH IS OPEN
(3) BLINKS FOLLOWED BY A 2 SEC. PAUSE – LIMIT SWITCH IS OPEN
(4) BLINKS FOLLOWED BY A 2 SEC. PAUSE – PRESSURE SWITCH CLOSED

Figure 31–104(B)   *(continued)*

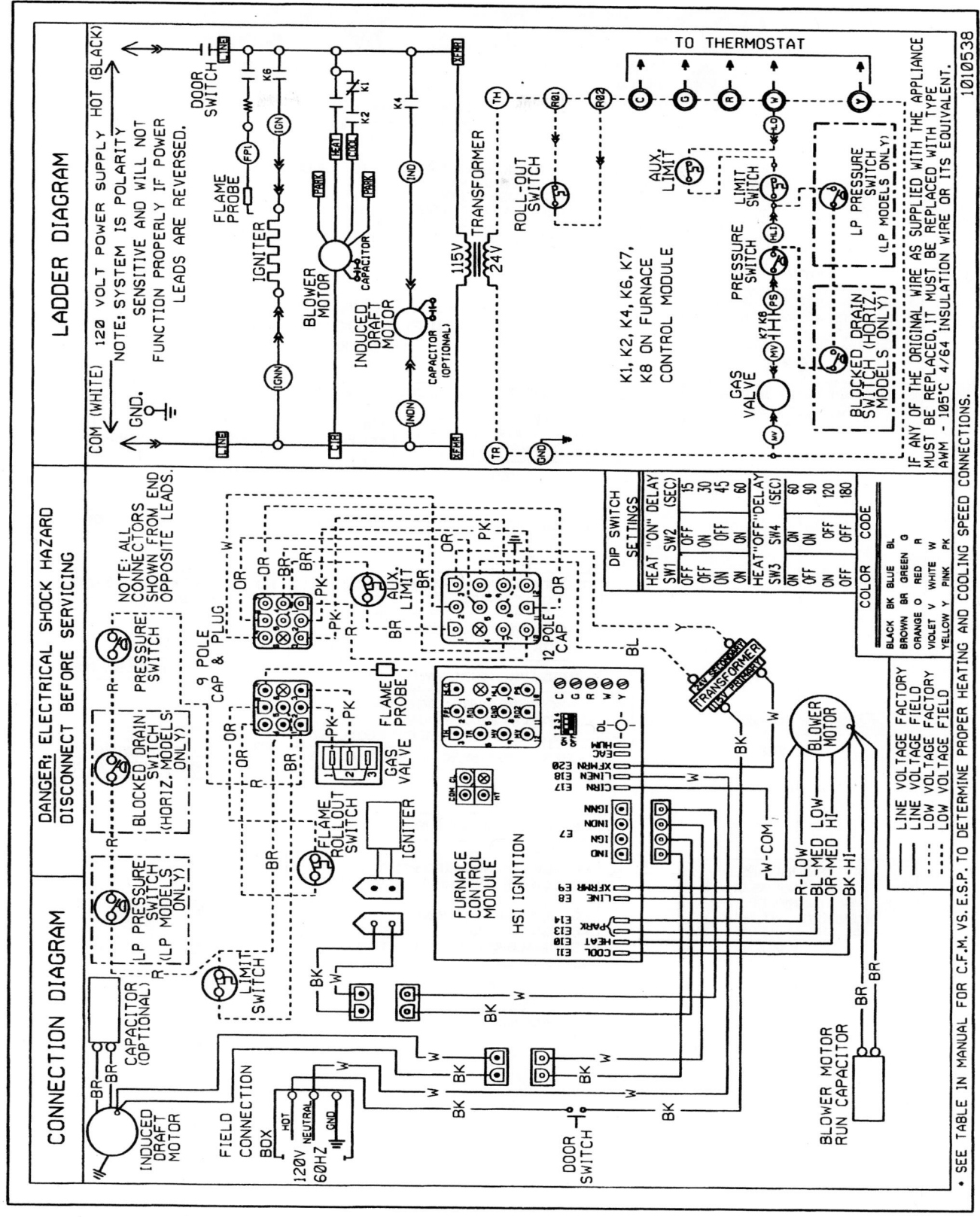

Figure 31–105(A) A gas furnace connect and ladder diagram for an IFC and a 115-V hot surface ignition system. *Courtesy International Comfort Products Corporation*

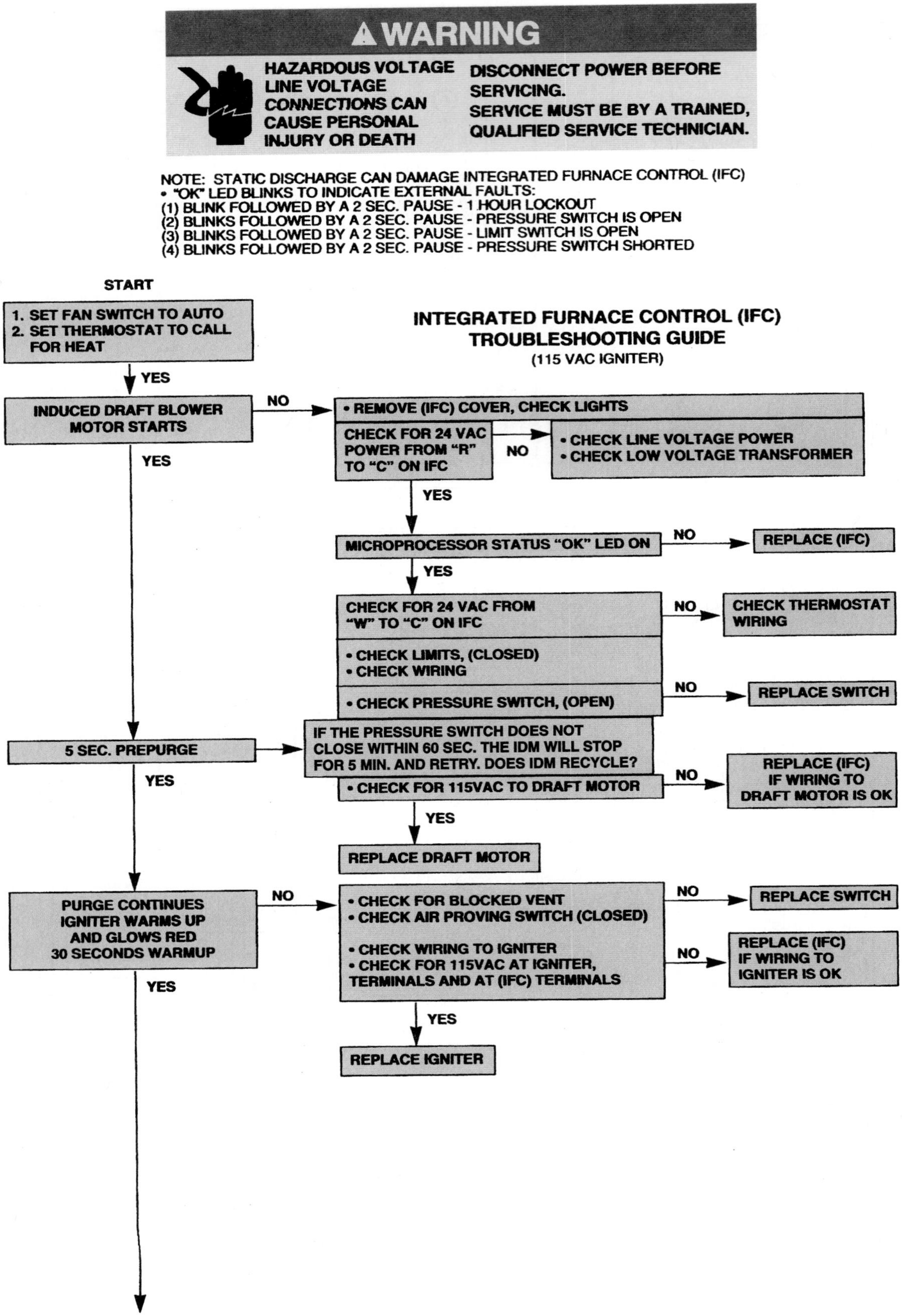

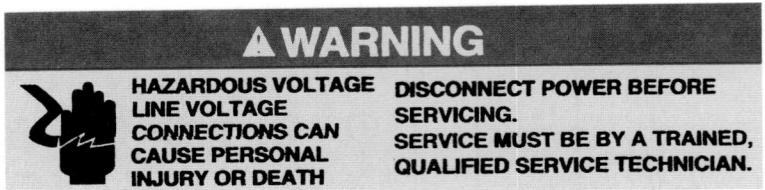

WARNING
HAZARDOUS VOLTAGE LINE VOLTAGE CONNECTIONS CAN CAUSE PERSONAL INJURY OR DEATH
DISCONNECT POWER BEFORE SERVICING.
SERVICE MUST BE BY A TRAINED, QUALIFIED SERVICE TECHNICIAN.
NOTE: STATIC DISCHARGE CAN DAMAGE INTEGRATED FURNACE CONTROL (IFC)
• "OK" LED BLINKS TO INDICATE EXTERNAL FAULTS:
(1) BLINK FOLLOWED BY A 2 SEC. PAUSE - 1 HOUR LOCKOUT
(2) BLINKS FOLLOWED BY A 2 SEC. PAUSE - PRESSURE SWITCH IS OPEN
(3) BLINKS FOLLOWED BY A 2 SEC. PAUSE - LIMIT SWITCH IS OPEN
(4) BLINKS FOLLOWED BY A 2 SEC. PAUSE - PRESSURE SWITCH SHORTED
START
1. SET FAN SWITCH TO AUTO
2. SET THERMOSTAT TO CALL FOR HEAT
INTEGRATED FURNACE CONTROL (IFC) TROUBLESHOOTING GUIDE
(115 VAC IGNITER)
YES
INDUCED DRAFT BLOWER MOTOR STARTS
NO
• REMOVE (IFC) COVER, CHECK LIGHTS
CHECK FOR 24 VAC POWER FROM "R" TO "C" ON IFC
NO
• CHECK LINE VOLTAGE POWER
• CHECK LOW VOLTAGE TRANSFORMER
YES
YES
MICROPROCESSOR STATUS "OK" LED ON
NO
REPLACE (IFC)
YES
CHECK FOR 24 VAC FROM "W" TO "C" ON IFC
NO
CHECK THERMOSTAT WIRING
• CHECK LIMITS, (CLOSED)
• CHECK WIRING
• CHECK PRESSURE SWITCH, (OPEN)
NO
REPLACE SWITCH
5 SEC. PREPURGE
IF THE PRESSURE SWITCH DOES NOT CLOSE WITHIN 60 SEC. THE IDM WILL STOP FOR 5 MIN. AND RETRY. DOES IDM RECYCLE?
• CHECK FOR 115VAC TO DRAFT MOTOR
NO
REPLACE (IFC) IF WIRING TO DRAFT MOTOR IS OK
YES
YES
REPLACE DRAFT MOTOR
PURGE CONTINUES IGNITER WARMS UP AND GLOWS RED 30 SECONDS WARMUP
NO
• CHECK FOR BLOCKED VENT
• CHECK AIR PROVING SWITCH (CLOSED)
NO
REPLACE SWITCH
• CHECK WIRING TO IGNITER
• CHECK FOR 115VAC AT IGNITER, TERMINALS AND AT (IFC) TERMINALS
NO
REPLACE (IFC) IF WIRING TO IGNITER IS OK
YES
REPLACE IGNITER
YES

Figure 31–105(B) The troubleshooting guide for a 115-V hot surface ignition system with an integrated furnace controller. *Courtesy Rheem Manufacturing Company*

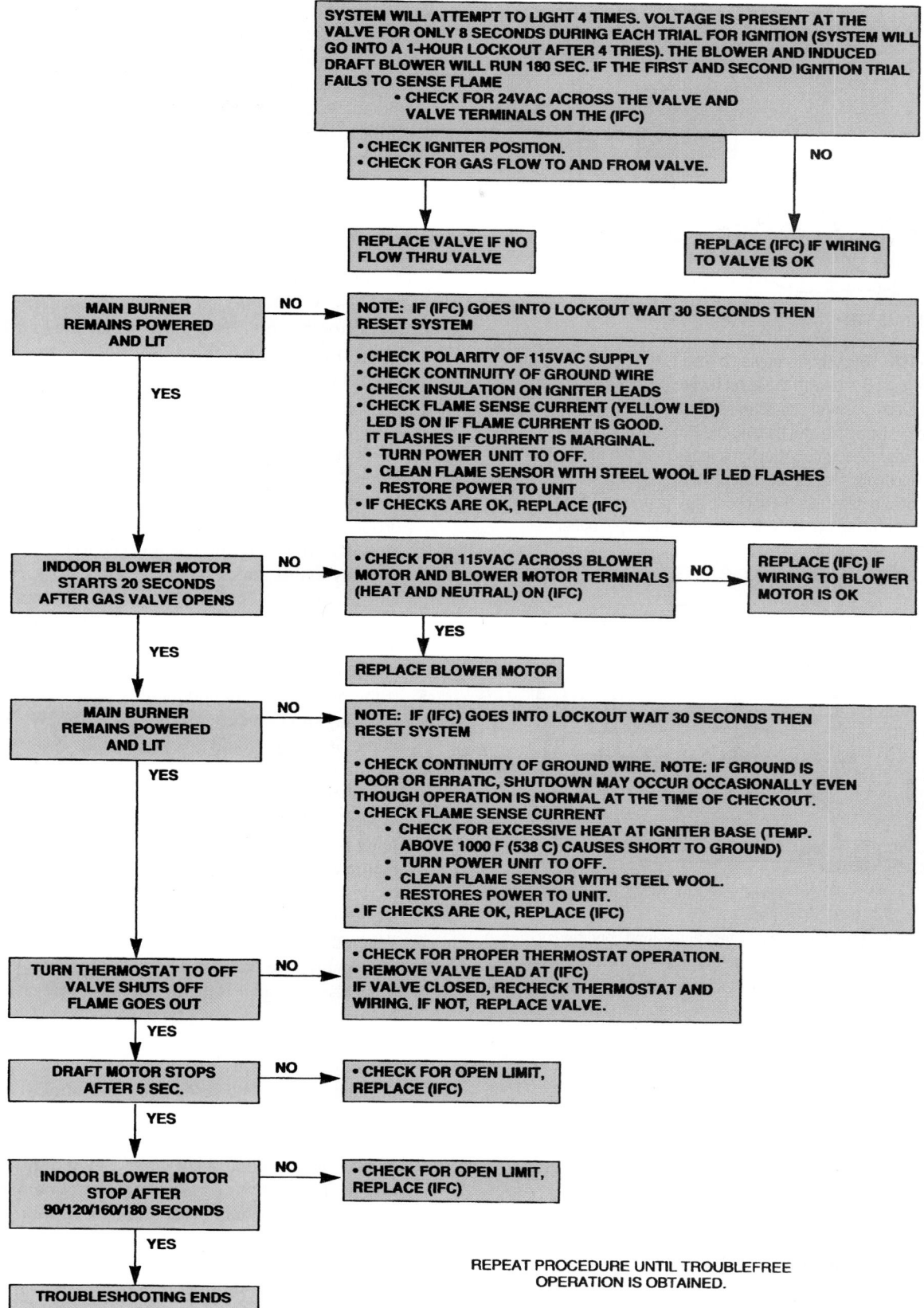

SYSTEM WILL ATTEMPT TO LIGHT 4 TIMES. VOLTAGE IS PRESENT AT THE VALVE FOR ONLY 8 SECONDS DURING EACH TRIAL FOR IGNITION (SYSTEM WILL GO INTO A 1-HOUR LOCKOUT AFTER 4 TRIES). THE BLOWER AND INDUCED DRAFT BLOWER WILL RUN 180 SEC. IF THE FIRST AND SECOND IGNITION TRIAL FAILS TO SENSE FLAME
• CHECK FOR 24VAC ACROSS THE VALVE AND VALVE TERMINALS ON THE (IFC)

• CHECK IGNITER POSITION.
• CHECK FOR GAS FLOW TO AND FROM VALVE.

NO

REPLACE VALVE IF NO FLOW THRU VALVE

REPLACE (IFC) IF WIRING TO VALVE IS OK

MAIN BURNER REMAINS POWERED AND LIT

NO

NOTE: IF (IFC) GOES INTO LOCKOUT WAIT 30 SECONDS THEN RESET SYSTEM

• CHECK POLARITY OF 115VAC SUPPLY
• CHECK CONTINUITY OF GROUND WIRE
• CHECK INSULATION ON IGNITER LEADS
• CHECK FLAME SENSE CURRENT (YELLOW LED)
  LED IS ON IF FLAME CURRENT IS GOOD.
  IT FLASHES IF CURRENT IS MARGINAL.
  • TURN POWER UNIT TO OFF.
  • CLEAN FLAME SENSOR WITH STEEL WOOL IF LED FLASHES
  • RESTORE POWER TO UNIT
• IF CHECKS ARE OK, REPLACE (IFC)

YES

INDOOR BLOWER MOTOR STARTS 20 SECONDS AFTER GAS VALVE OPENS

NO

• CHECK FOR 115VAC ACROSS BLOWER MOTOR AND BLOWER MOTOR TERMINALS (HEAT AND NEUTRAL) ON (IFC)

NO

REPLACE (IFC) IF WIRING TO BLOWER MOTOR IS OK

YES

REPLACE BLOWER MOTOR

YES

MAIN BURNER REMAINS POWERED AND LIT

NO

NOTE: IF (IFC) GOES INTO LOCKOUT WAIT 30 SECONDS THEN RESET SYSTEM

• CHECK CONTINUITY OF GROUND WIRE. NOTE: IF GROUND IS POOR OR ERRATIC, SHUTDOWN MAY OCCUR OCCASIONALLY EVEN THOUGH OPERATION IS NORMAL AT THE TIME OF CHECKOUT.
• CHECK FLAME SENSE CURRENT
  • CHECK FOR EXCESSIVE HEAT AT IGNITER BASE (TEMP. ABOVE 1000 F (538 C) CAUSES SHORT TO GROUND)
  • TURN POWER UNIT TO OFF.
  • CLEAN FLAME SENSOR WITH STEEL WOOL.
  • RESTORES POWER TO UNIT.
• IF CHECKS ARE OK, REPLACE (IFC)

YES

TURN THERMOSTAT TO OFF VALVE SHUTS OFF FLAME GOES OUT

NO

• CHECK FOR PROPER THERMOSTAT OPERATION.
• REMOVE VALVE LEAD AT (IFC)
IF VALVE CLOSED, RECHECK THERMOSTAT AND WIRING. IF NOT, REPLACE VALVE.

YES

DRAFT MOTOR STOPS AFTER 5 SEC.

NO

• CHECK FOR OPEN LIMIT, REPLACE (IFC)

YES

INDOOR BLOWER MOTOR STOP AFTER 90/120/160/180 SECONDS

NO

• CHECK FOR OPEN LIMIT, REPLACE (IFC)

YES

TROUBLESHOOTING ENDS

REPEAT PROCEDURE UNTIL TROUBLEFREE OPERATION IS OBTAINED.

Figure 31–105(B)   *(continued)*

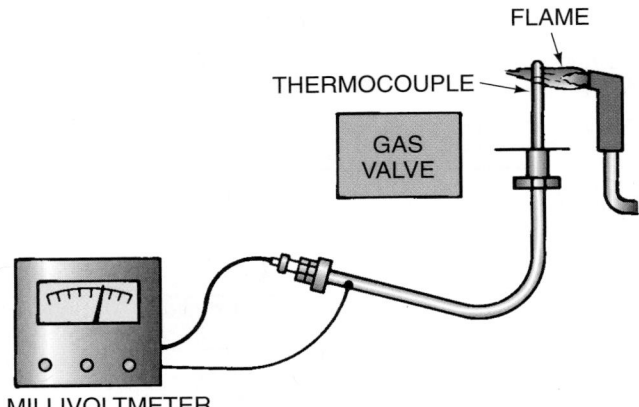

Figure 31–106 The thermocouple no-load test. Operate the pilot flame for at least 5 min with the main burner off. (Hold the gas cock knob in the PILOT position and depress.) Disconnect the thermocouple from the gas valve while still holding the knob down to maintain the flame on the thermocouple. Attach the leads from the millivoltmeter to the thermocouple. Use direct current voltage. Any reading below 20 mV indicates a defective thermocouple or poor pilot flame. If the pilot flame is adjusted correctly, test the thermocouple under load conditions.

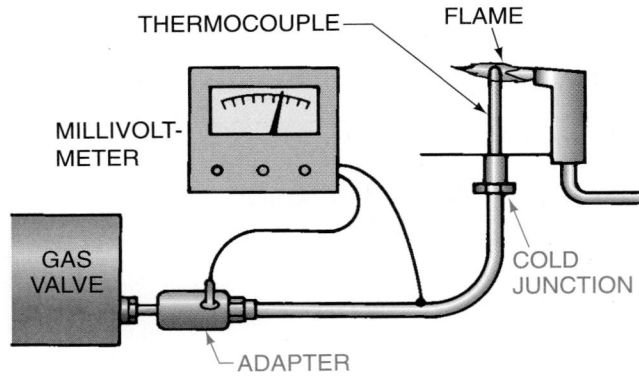

Figure 31–107 The thermocouple test under load conditions. Disconnect the thermocouple from the gas valve. Screw the thermocouple test adapter into the gas valve. Screw the thermocouple into the test adapter. Light the pilot and main burner and allow them to operate 5 min. Attach one lead from the millivoltmeter to either connecting post of the adapter and the other lead to the thermocouple tubing. Any reading under 9 mV would indicate a defective thermocouple or insufficient pilot flame. Adjust the pilot flame. Replace the thermocouple if necessary. Ensure that the pilot, pilot shield, and thermocouple are positioned correctly. Too much heat at the cold junction would cause a satisfactory voltage under no-load conditions but an unsatisfactory voltage under a load condition.

always easy to wire into the circuit to accomplish the proper sequence, particularly with two-speed fan applications.

The following reference points can be used for this circuit board:

1. SAFETY PRECAUTION: When the front panel is removed, everything stops until the switch is blocked closed. This should only be done by a qualified service technician.

2. The fan motor is started through the 2A contacts and in single speed.

3. The 2A contacts are held open during the off cycle by energizing the 2A coil. This coil is deenergized during the on cycle.

4. If the control transformer is not functioning, the fan will run all the time. If the fan runs constantly but nothing else operates, suspect a bad control transformer.

5. The fan starts and stops through the time delay in the heating cycle.

The following description of the electrical circuit can be used for an orderly troubleshooting procedure. See **Figure 31–109** for the voltages in the pictorial and line diagrams.

1. Line power should be established at the primary of the control transformer.

2. 24 V should be detected from the C terminal (common to all power-consuming devices) to the LIM-1 terminal. See the meter lead on the 24-V meter labeled 1.

3. Leaving the probe on the C terminal, move the other probe to positions 2, 3, and 4. 24 V should be detected at each terminal, or there is no need to proceed.

4. The key to this circuit is to have 24 V between the C terminal and the GAS-1 terminal. At this time the pilot should be trying to light. The spark at the pilot should be noticeable. It makes a ticking sound with about one-half second between sounds.

5. If the pilot light is lit and the ticking has stopped, the 6H relay should have changed over. The NC contacts in the 6H relay should be open and the NO contacts should be closed. This means that you should have 24 V between the C terminal, which is the same as the GAS-2 terminal, and the GAS-3 terminal. The 6H relay is actually a bimetallic pilot-proving relay, **Figure 31–56(C),** and heat from the pilot changes its position. If it will not pass power, it must be replaced.

6. When the circuit board will pass power to the GAS-3 terminal (probe position 5) and the burner will not light, the gas valve is the problem. Before changing it, be sure that its valve handle is turned in the correct direction and the interconnecting wire is good. The procedure of leaving one probe on the C terminal and moving the other probe can be done quickly when a furnace has a circuit board with terminals. This makes troubleshooting easier.

7. If the R terminal has 24 V and the W terminal does not, the thermostat and the interconnecting wiring are the problem. You can place a jumper from R to W for a moment; if the spark starts to arc, the problem is the wiring or the thermostat.

8. The fan is started through the circuit board and is part of the sequence of operation also. The fan is started when the 2A relay is deenergized through the time-delay circuit. This is accomplished when the 6H relay in the circuit board changes position. When 24 V is detected from terminal C to terminal GAS-3, the

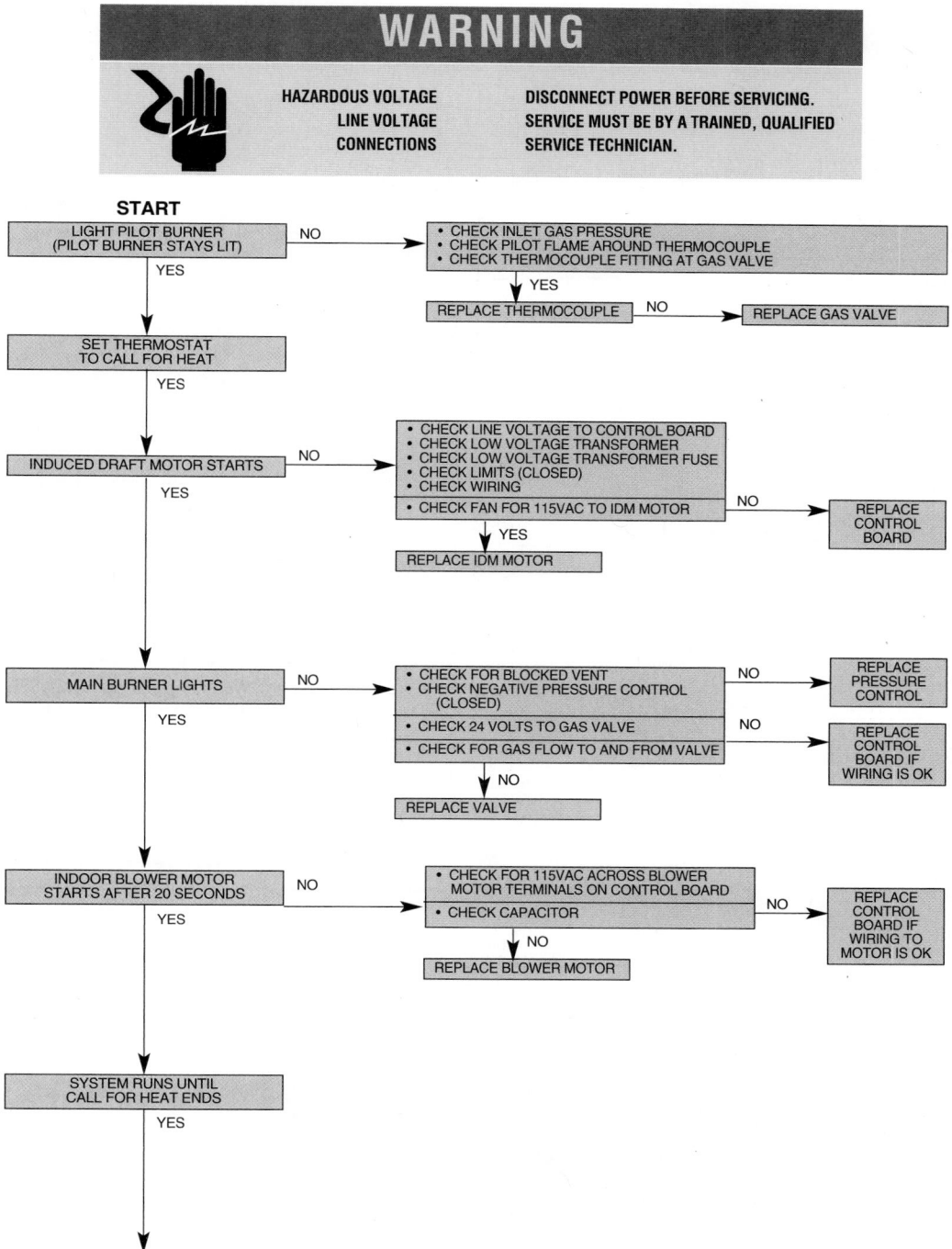

WARNING

HAZARDOUS VOLTAGE
LINE VOLTAGE
CONNECTIONS

DISCONNECT POWER BEFORE SERVICING.
SERVICE MUST BE BY A TRAINED, QUALIFIED
SERVICE TECHNICIAN.

START

LIGHT PILOT BURNER
(PILOT BURNER STAYS LIT) — NO →
- CHECK INLET GAS PRESSURE
- CHECK PILOT FLAME AROUND THERMOCOUPLE
- CHECK THERMOCOUPLE FITTING AT GAS VALVE

YES ↓

REPLACE THERMOCOUPLE — NO → REPLACE GAS VALVE

YES ↓

SET THERMOSTAT
TO CALL FOR HEAT

YES ↓

INDUCED DRAFT MOTOR STARTS — NO →
- CHECK LINE VOLTAGE TO CONTROL BOARD
- CHECK LOW VOLTAGE TRANSFORMER
- CHECK LOW VOLTAGE TRANSFORMER FUSE
- CHECK LIMITS (CLOSED)
- CHECK WIRING
- CHECK FAN FOR 115VAC TO IDM MOTOR — NO → REPLACE CONTROL BOARD

YES ↓

REPLACE IDM MOTOR

YES ↓

MAIN BURNER LIGHTS — NO →
- CHECK FOR BLOCKED VENT
- CHECK NEGATIVE PRESSURE CONTROL (CLOSED) — NO → REPLACE PRESSURE CONTROL
- CHECK 24 VOLTS TO GAS VALVE
- CHECK FOR GAS FLOW TO AND FROM VALVE — NO → REPLACE CONTROL BOARD IF WIRING IS OK

NO ↓

REPLACE VALVE

YES ↓

INDOOR BLOWER MOTOR
STARTS AFTER 20 SECONDS — NO →
- CHECK FOR 115VAC ACROSS BLOWER MOTOR TERMINALS ON CONTROL BOARD
- CHECK CAPACITOR — NO → REPLACE CONTROL BOARD IF WIRING TO MOTOR IS OK

NO ↓

REPLACE BLOWER MOTOR

YES ↓

SYSTEM RUNS UNTIL
CALL FOR HEAT ENDS

YES ↓

Figure 31–108 A guide for systematically troubleshooting a standing pilot furnace. *Courtesy Rheem Manufacturing Company*

time-delay circuit starts timing. When the time is up, the contacts open and deenergize the fan relay coil; the contacts close, and the fan starts.

9. The voltmeter can be relocated to the COM terminal and the LO terminal. When relay 2A is passing power, 115 V should be detected here. If there is power to the fan motor and it will not start, open up the fan section and check the motor for continuity. Check the

capacitor. See if the motor is hot. If the motor cannot be started with a good capacitor, replace the motor. **Figure 31–110** shows a flowchart for systematically troubleshooting an intermittent pilot module. SAFETY PRECAUTION: Not all ignition modules are alike. The service technician must consult with the ignition module manufacturer for specific systematic troubleshooting charts or procedures.

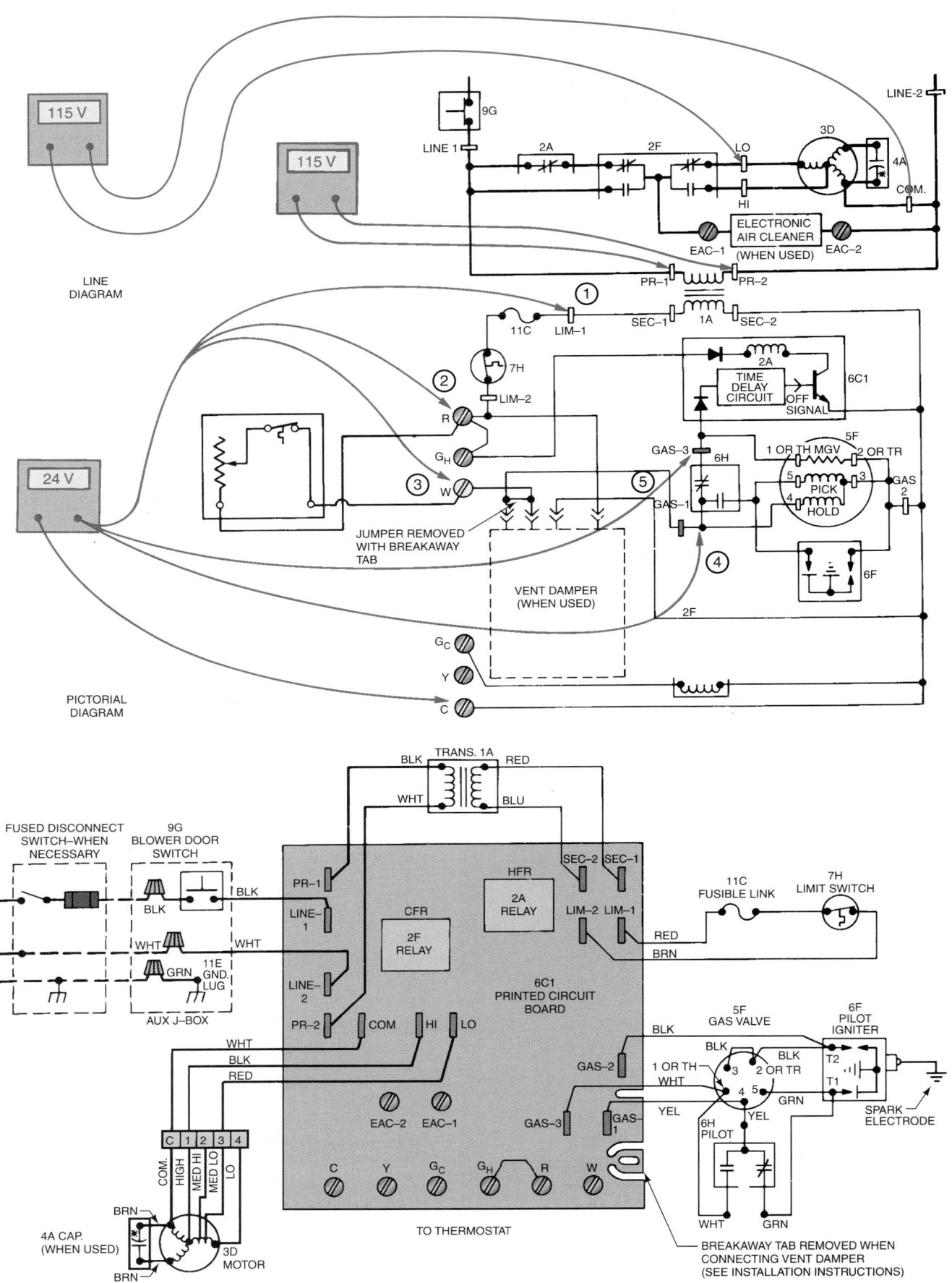

Figure 31–109 Note the voltages indicated. *Courtesy BDP Company*

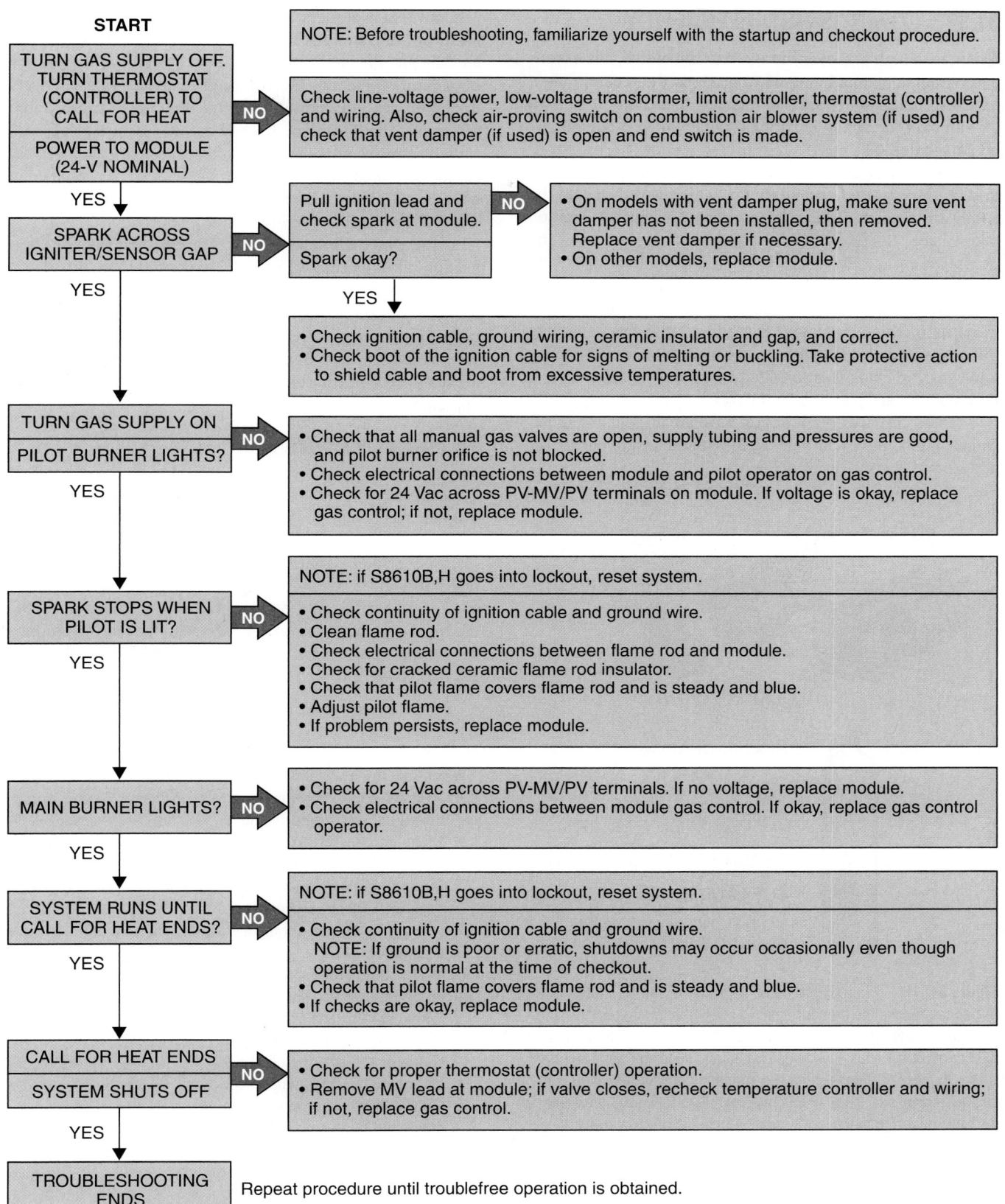

Figure 31–110   A chart for systematically troubleshooting an intermittent pilot module. *Courtesy Honeywell*

31.30 COMBUSTION EFFICIENCY

🌐With the high and ever-increasing cost of fuel, it is essential that adjustments be made to produce the most efficient but safe combustion possible. It is also necessary to ensure that an overcorrection is not made that would produce carbon monoxide.🌐

Atmospheric gas burners use primary air, which is sucked in with the gas, and secondary air, which is pulled into the combustion area by the draft produced when the ignition takes place. SAFETY PRECAUTION: Incomplete combustion produces carbon monoxide. Enough secondary air must be supplied so that carbon monoxide will not be produced.

🌐Primary air intake can be adjusted with shutters near the orifice. This adjustment is made so that the flame has the characteristics specified by the manufacturer. Generally, however, the flame is blue with a small orange tip when it is burning efficiently, **Figure 31–111.**🌐

Modern furnaces often do not have an air adjustment because of the types of burners used. With the inshot burner, air adjustment is not as critical.

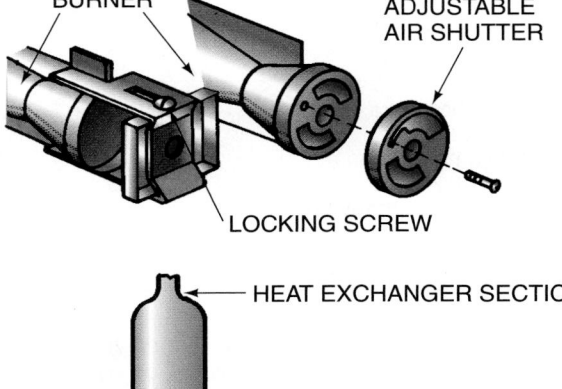

BURNER
ADJUSTABLE AIR SHUTTER
LOCKING SCREW

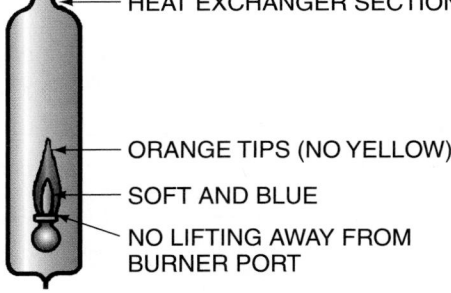

HEAT EXCHANGER SECTION

ORANGE TIPS (NO YELLOW)
SOFT AND BLUE
NO LIFTING AWAY FROM BURNER PORT

Figure 31–111 Main burner flame adjustment. The air shutter is turned to produce the proper flame. To check the flame, turn the furnace on at the thermostat. Wait a few minutes to ensure that any dust or other particles have been burned and no longer have an effect on the flame. The flames should be stable, quiet, soft, and blue with slightly orange tips. They should not be yellow. They should extend directly upward from the burner ports without curling downward, floating, or lifting off the ports, and they should not touch the side of the heat exchanger.

Conversion burners are used to convert coal or oil furnaces or boilers to gas fired. They normally have secondary air adjustments, and these must be adjusted in the field for maximum efficiency. The carbon dioxide test is used to get the necessary information from the flue gases to make the secondary air adjustment. Carbon dioxide in the flue gases is actually a measurement of the excess secondary air supplied. As the secondary air supplied is decreased, the percentage of carbon dioxide in the flue gases increases, and the chance of producing carbon monoxide becomes greater.

If the exact amount of air is supplied and perfect combustion takes place, a specific amount of carbon dioxide by volume will be present in the flue gases. This is called the ultimate carbon dioxide content. The following table indicates the ultimate percentage of carbon dioxide content for various gases (these figures are for perfect combustion and are not used as the carbon dioxide reading for a gas-burning appliance):

FOR PERFECT COMBUSTION

Natural gas	11.7–12.2
Butane gas	14.0
Propane gas	13.7

SAFETY PRECAUTION: A gas furnace is never adjusted to produce the ultimate carbon dioxide because there is too much danger of producing carbon monoxide. Therefore, the accepted practice is to adjust the secondary air to produce excess air—resulting in less carbon dioxide by volume in the flue gases.

HVAC GOLDEN RULES

When making a service call to a residence:
■ Always carry the proper tools and avoid extra trips through the customer's house.
■ Make and keep firm appointments. Call ahead if you are delayed. The customer's time is valuable also.
■ Keep customers informed if you must leave the job for parts or other reasons. Customers should not be upset when you inform them of your return schedule if it is reasonable.

Here are some simple, inexpensive procedures that may be included in the basic service call:
■ Leave the furnace and area clean.
■ If standing pilot is used, perform a service check.
■ Clean or replace dirty filters.
■ Lubricate bearings when needed.
■ Check the burner compartment to ensure burners are clean and burning correctly.
■ Replace all panel covers.

PREVENTIVE MAINTENANCE

Gas equipment preventive maintenance consists of servicing the air-side components: filter fan, belt and drive, and the burner section. Air-side components are the same for gas as for electric forced-air heat, so only the burner section will be discussed here.

The burner section consists of the actual burner, the heat exchanger, and the venting system for the products of combustion.

The burner section of the modern gas furnace does not need adjustment from season to season. It will burn efficiently and be reliable if it is kept clean. Rust, dust, or scale must be kept from accumulating on top of the burner where the secondary air supports combustion or in the actual burner tube where primary air is induced into the burner. When a burner is located in a clean atmosphere (no heavy particles in the air as in a manufacturing area) most of the dust particles in the air will burn and exit through the venting system. The burner may not need cleaning for many years. A vacuum cleaner with a crevice tool may be used to remove small amounts of scale or rust deposits.

The burner and manifold may be removed for more extensive cleaning inside the burner tube where deposits may be located. Each burner may be removed and tapped lightly to break the scale loose. An air hose should be used to blow down through the burner ports while the burner is out of the system. The particles will blow back out the primary air shutter and out of the system. SAFETY PRECAUTION: Wear goggles when using compressed air for cleaning. This cleaning function cannot be accomplished while the burners are in the furnace.

Burner alignment should be checked to make sure that no flame impingement to the heat exchanger is occurring. The burners should set straight and feel secure while in place. Large deposits of rust scale may cause flame impingement and must be removed.

A combustion analysis is not necessary as routine maintenance on the modern gas furnace. The primary air is sometimes the only adjustment on new furnaces. Many times, even primary air cannot be adjusted because of inshot burners being used. All gas burners should burn with a clear flame, with only orange streaks in the flame. **There should never be any yellow or yellow tips on the flame. If yellow tips cannot be adjusted from the flame, the burner contains dirt or trash, and it should be removed for cleaning.**

Observe the burner for correct ignition as it is lit. Often, the crossover tubes that carry the burner flame from one burner to the next get out of alignment and the burner lights with an irregular pattern. Sometimes it will "puff" when slow ignition of one of the burners occurs.

The pilot light flame should be observed for proper characteristics. It should not have any yellow tips and it should not impinge on the burner. It should impinge on the thermocouple only. If the pilot flame characteristics are not correct, the pilot assembly may have to be removed and cleaned. Compressed air usually will clear any obstructions from the pilot parts. SAFETY PRECAUTION: Do not use a needle to make the orifice holes larger. If they are too small, obtain the correct size from a supply house.

The venting system should be examined for obstructions. Birds may nest in the flue pipe, or the vent cap may become damaged. Visual inspection is the first procedure. Then, with the furnace operating, the technician should strike a match in the vicinity of the draft diverter and look for the flame to pull into the diverter. This is a sign that the furnace is drafting correctly.

Observe the burner flame with the furnace fan running. A flame that blows with the fan running indicates a crack in the heat exchanger. This would warrant further examination of the heat exchanger. It is difficult to find a small crack in a heat exchanger and one may only be seen with the heat exchanger out of the furnace housing. Extensive tests have been developed for proving whether or not a gas furnace heat exchanger has a crack. These tests may be beyond what a typical technician would use because the recommended tests require expensive equipment and experience.

High-efficiency condensing furnaces require maintenance of the forced-draft blower motor if it is designed to be lubricated. The condensate drain system should be checked to be sure it is free of obstructions and that it drains correctly. Manufacturers of condensing furnaces sometimes locate the condensing portion in the return air section where all air must move through the coil. This is a finned coil and will collect dirt if the filters are not cleaned. Routine inspection of this condensing coil for obstruction is a good idea.

All installations should be checked by setting the room thermostat to call for heating and by making sure the proper sequence of events occurs. For example, when checking a standard furnace, the pilot may be lit all the time and the burner should ignite. Then the fan should start. Some furnaces may have intermittent pilot lights. The pilot should light first, the burner next, and then the fan should start. With a condensing furnace, the power vent motor may start first, the pilot light next, then the burner, and the fan motor last. Always know the sequence of operation for the furnace.

Some technicians disconnect the circulating fan motor and allow the furnace to operate until the high-limit control shuts it off. This proves the safety device will shut off the furnace. Technicians will often perform a combustion analysis test to check furnace efficiency and then make air shutter adjustment. The adjustment should be made to produce from 65% to 80% of the ultimate carbon dioxide. This would produce 8% to 9.5% carbon dioxide in the flue. It is important that manufacturers' specifications be followed when making any adjustments.

(continued)

The flue-gas temperature should also be taken if the percent of combustion efficiency is to be determined. The percent combustion efficiency is an index of the useful heat obtained. It is not possible to obtain 100% efficiency because heated air does leave the building through the flue except on direct-vented furnaces.

The carbon dioxide and flue-gas temperatures are taken in the flue between the draft diverter and the furnace. A sampling tube can be inserted through the draft diverter projecting into the inlet side of a gas-burning furnace, **Figure 31–112.** An oxygen and carbon dioxide analyzer and a stack thermometer are used for making these tests, **Figure 31–113.** You

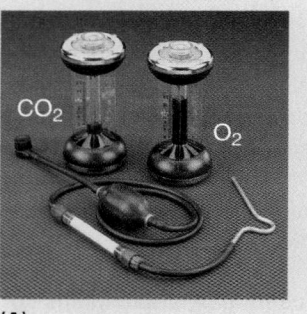

(A) **(B)**

Figure 31–113 **(A)** An oxygen and carbon dioxide analyzer. **(B)** A stack thermometer. *(A) and (B) Photos by Bill Johnson*

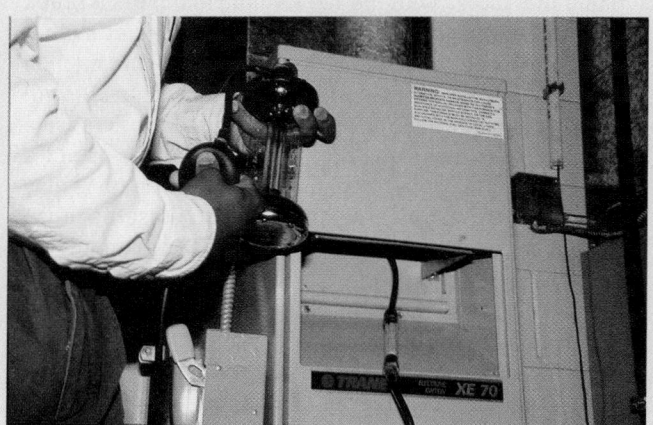

Figure 31–112 A sampling tube in the draft diverter. *Photo by Bill Johnson*

can refer to a chart provided by the manufacturer of the test equipment to determine the best efficiency. Average gas-burning furnaces manufactured before 1982 should produce 75% to 80% efficiency. Modern digital furnace efficiency testing equipment is available and provides good accuracy.

Test equipment is also available for checking the carbon monoxide content in the flue gases. Carbon monoxide–free combustion is defined as less than 0.04% carbon monoxide in an air-free sample of the flue gas. Carbon monoxide can be produced by flame impingement on a cool surface and by insufficient primary air (yellow flame) as well as insufficient secondary air.

Other adjustments that can be checked and compared with manufacturers' specifications are suction at the draft hood, draft at the chimney, gas pressure, and gas input.

31.31 SERVICE TECHNICIAN CALLS

SERVICE CALL 1

The manager of a retail store calls. *There is no heat. This is an upflow gas furnace with a standing pilot and air conditioning. The low-voltage transformer is burned out because the gas valve coil is shorted. This causes excess current for the transformer. The furnace is located in the stockroom.*

The technician arrives at the job, goes to the room thermostat, and places the fan switch to ON to see whether the indoor fan will start. It will not, so the technician suspects that the low-voltage power supply is not working. The thermostat is set to call for heat. The technician goes to the stockroom where the furnace is located. The voltage is checked at the transformer secondary, and it indicates 0 V. The power is turned off and the ohmmeter is used to check continuity of the transformer. The secondary (low-voltage) coil has an open circuit, so the transformer is changed. Before connecting the secondary wires, the continuity of the low-voltage circuit is checked. There is only 2 Ω of resistance in the gas valve coil. This

is so low that it will cause a high current flow in the secondary circuit. (The resistance should be at least 20 Ω.) The technician goes to the truck and checks the continuity of another gas valve and finds it to be 50 Ω, so the gas valve is also changed.

The system is started with the new gas valve and transformer. A current check of the gas valve shows that it is only pulling 0.5 A, which is not overloaded. If the technician had just installed the transformer and turned the power on, it is likely that another transformer failure would have occurred quickly.

The technician changes the air filters and calls the store manager back to the stockroom for a conference before leaving. The store manager is informed that the boxes of inventory must be kept away from the furnace because they may present a fire hazard.

SERVICE CALL 2

A customer in a residence reports that *fumes can be smelled and are probably coming from the furnace in the hall closet. The furnace has not been operated for the past two weeks because the weather has been mild. Because the fumes might be harmful, the dispatcher tells the customer to shut the furnace off until a technician arrives. The flue is stopped up with a shingle laid on top of the brick chimney by a roofer who had been making repairs.*

The technician arrives and starts the furnace. A match is held at the draft diverter to see whether the flue has a negative pressure. The flue-gas fumes are not rising up the flue for some reason; the match flame blows away from the flue pipe and draft diverter. The technician turns off the burner and examines it and the heat exchanger area with a flashlight. There is no soot that would indicate the burner had been burning incorrectly. The technician goes to the roof to check the flue and notices a shingle on top of the chimney. The shingle is lifted off, and heat then rises out of the chimney.

The technician goes back to the basement and starts the furnace again. A match is held at the draft diverter, and the flame is drawn toward the flue. The furnace is now operating correctly. The furnace filter is changed, and the fan motor is oiled before the technician leaves the job.

SERVICE CALL 3

A customer requests that an efficiency check be performed on a furnace. *This customer thinks that the gas bill is too high for the conditions and wants the system completely checked out.*

The technician arrives and meets the customer, who wants to watch the complete procedure of the service call. The technician will run an efficiency test on the furnace burners at the beginning and then at the end, so the customer can see the difference.

The furnace is in the basement and easily accessible. The technician turns the thermostat to 10°F above the room temperature setting to ensure that the furnace will not shut off in the middle of the test. The technician then goes to the truck and gets the flue-gas analyzer kit. SAFETY PRECAUTION: Be careful not to touch the stack; the stack temperature is taken at the heat exchanger side of the draft diverter. A sample of the flue gas is drawn into the sample chamber after the flue-gas temperature has stopped rising and the indoor fan has been running for about 5 min. This ensures that the furnace is up to maximum temperature.

The flue-gas reading shows that the furnace is operating at 80% efficiency. This is normal for a standard-efficiency-type furnace. After the test is completed and recorded, the technician shuts off the gas to the burner and allows the fan to run to cool down the furnace.

When the furnace has cooled down enough to allow the burners to be handled, the technician takes them out. The burners are easily removed in this furnace by removing the burner shield and pushing the burners forward one at a time while raising up the back. They will then clear the gas manifold and can be removed.

There is a small amount of rust, which is normal for a furnace in a basement. The draft diverter and the flue pipe are removed so that the technician and customer can see the top of the heat exchanger. All is normal—no rust or scale.

With a vacuum cleaner the technician removes the small amount of dirt and loose rust from the heat exchanger and burner area. The burners are taken outside where they are blown out with a compressed air tank at the truck. SAFETY PRECAUTION: The technician wears eye protection. After the cleaning is complete, the technician assembles the furnace and tells the customer that the furnace was in good condition and that no difference in efficiency will be seen. Modern gas burners do not stop up as badly as the older ones, and the air adjustments will not allow the burner to get out of adjustment more than 2% or 3% at the most.

The fan motor is oiled, and the filter is changed. The system is started and allowed to get up to normal operating temperature. While the furnace is heating to that temperature, the technician checks the fan current and finds it to be running at full load. The fan is doing all the work that it can. The efficiency check is run again, and the furnace is still operating at about 80%. It is now clean, and the customer has peace of mind. The thermostat is set before leaving.

SERVICE CALL 4

A customer indicates that *the pilot light will not stay lit. The pilot light goes out after it has been lit a few minutes. The heat exchanger has a hole in it very close to the pilot light. The pilot light will light, but when the fan starts it blows out the pilot light.*

The technician arrives and checks the room thermostat to see that it is set above the room temperature. The technician then goes to the basement. The standing pilot light is not lit, so the technician lights it and holds the button down until the thermocouple will keep the pilot light lit. When the gas valve is turned to ON, the main burner lights. Everything is normal until the fan starts, then the pilot light flame starts to wave around and after a short period it goes out. The thermocouple cools, and the gas to the main burner shuts off. The technician shows the customer the hole in the heat exchanger. The gas valve is turned off, and it is explained to the customer that the furnace cannot be operated in this condition because of the potential danger of gas fumes.

The technician explains to the customer that this furnace is 18 years old, and the customer really should consider getting a new one. A heat exchanger can be changed in this furnace, but it requires considerable labor plus the price of the heat exchanger. The customer decides to replace the furnace.

SERVICE CALL 5

A residential customer calls. *There is no heat. The main gas valve coil is shorted, and the low-voltage transformer is burned. This is an electronic intermittent ignition system,*

Figure 31–109. This symptom can lead the technician to believe that the printed circuit board is defective.

The technician arrives and notices the indoor fan is running. The technician cannot stop the fan from running by turning the room thermostat to OFF. (The low-voltage circuit holds the fan motor off in this system.) A low-voltage problem is suspected. The technician goes to the furnace and checks the output of the low-voltage transformer; the meter reads 0 V at the transformer. The transformer is burned. SAFETY PRECAUTION: The technician turns off the power and replaces the transformer. When the power is restored, the electronic ignition circuit lights the pilot light. Everything seems normal, but the transformer becomes extra hot and smells like it is burning.

The technician turns off the power and changes the electronic circuit board and 1-A fuse. When power is restored, the pilot lights and everything is normal for a moment, but the transformer is still getting hot.

The technician turns off the power and fastens a clamp-on ammeter in the low-voltage system, using 10 wraps of wire to amplify the ampere reading, **Figure 12–52**. The amperage is normal. When power is restored, the pilot lights again and the amperage is normal (about 0.5 A). When the pilot proves and it is time for the main gas valve to be energized, the current goes up to 3 A and the transformer begins to heat up.

The technician turns off the power and uses the ohmmeter to check the resistance of all power-consuming components in the circuit. When the main gas valve coil in the gas valve is checked, it shows 2 Ω of resistance.

The gas valve is replaced. Power is restored. The pilot light is lit and proved. The main gas valve opens and the burner ignites. The system operates normally.

The technician turns off the power and changes the circuit board back to the original board and then restores the power. The furnace goes through a normal start-up.

SERVICE CALL 6

A residential customer indicates that the furnace is heating the house. However, it is blowing cold air just before the fan goes off at the end of each heating cycle. *The warm air fan's off-delay timing is set too long. The furnace has an IFC with programmable DIP switches for the fan's off-delay timing, Figure 31–45(A and B). The DIP switches have to be repositioned so the fan's programmed off-delay timing is shortened. This will shut the fan off sooner after each heating cycle and prevent the cooler air from being blown into the conditioned space.*

The service technician arrives at the location and takes the front panel off the furnace. Some duct tape is then put on the furnace's door switch so the furnace keeps running while the front panel is removed. The thermostat is then turned 10°F higher than the room temperature to initiate a heating cycle. Once the furnace has run for 5 min, the thermostat is turned down. As usual, the flame goes out but the warm air fan remains running. The technician

times the fan's run time with a wristwatch once the flame goes out and finds that the off-delay timing for the fan is 180 sec. The technician then checks the DIP switches on the IFC and finds that switch 1 is in the up position and switch 2 is in the down position, **Figure 31–45(B)**. According to the faceplate of the IFC, the DIP switches in this position will give a 180-sec off-delay time for the warm air fan. The service technician then changes the switches so that both switch 1 and switch 2 are in the up position, **Figure 31–45(A)**. This will give the warm air fan a 60-sec off-delay time.

The technician then cycles the furnace through a complete cycle and checks the fan's off-delay timing again. The timing is now 60 sec as planned. The technician notices that the air coming from the furnace just before it shuts off is much warmer than it was with the longer off-delay timing. The technician explains to the homeowner what has been adjusted and how it pertains to the warm air fan's off-delay timing, thanks the customer, and sets the thermostat to the proper temperature before leaving the house.

SERVICE CALL 7

A residential customer says that the furnace is not heating the house. No airflow is coming from any of the warm air ducts. *The combustion blower motor is burned out and the draft pressure switch is not closing, preventing the burners from firing, Figure 31–70(B).*

Before taking the front cover off the furnace, the service technician notices a self-diagnostic light flashing through a clear plastic inspection hole on the front panel of the furnace, **Figure 31–114**. A technician must always look into this inspection hole before removing the front cover. Many times, taking the front cover off the furnace will take power away from the IFC board and reset the self-diagnostic flashing lights. The technician makes note of the flashing intervals. The front cover is then taken off the furnace, which interrupts power to the IFC and resets the self-diagnostics. The furnace door switch is taped closed with duct tape to keep the furnace running while the door is off. The diagnostic light is blinking in intervals of three flashes, indicating that the draft pressure switch is open, **Figure 31–82** and **Figure 31–45(B)**. The service technician turns power off to the furnace to make sure the IFC is reset. Power is then turned back on. The combustion blower motor never comes on, which prevents the draft pressure switch from closing. The technician then shuts off power to the furnace and unplugs the wire leads going to the combustion blower motor from the IFC. An ohmmeter reading of infinity across the combustion blower motor leads proves to the technician that the combustion blower motor is open. A new combustion blower motor is installed and plugged into the IFC. Care must be taken in making sure the technician is grounded so as not to cause a static shock, which can easily damage the delicate IFC.

Figure 31–114   A furnace with a clear plastic inspection hole on the front panel for observing the self-diagnostic flashing light. *Courtesy Ferris State University. Photo by John Tomczyk*

Power is turned back on to the furnace. The combustion blower motor is operating, the pressure switch has closed, and the hot surface igniter is glowing red. The burners light in about 5 sec. The technician politely explains to the customer what was wrong with the furnace and that the problem has been remedied. The technician then makes sure the furnace successfully operates through three complete heating cycles.

SERVICE CALL 8

A residential customer complains that there is no heat in their residence, but the warm air blower runs continuously. *The burner high-limit control has opened due to a dirty return air filter.*

The service technician notices that the self-diagnostic status light is blinking with only one flash, **Figure 31–81(A)**. The self-diagnostic code on the faceplate of the IFC indicates one flash is for an open burner limit switch. Because the warm air fan is running continuously, the technician pulls out the furnace filter. The filter is completely plugged. No or low airflow will heat up the heat exchanger until the burner's high-limit control opens, **Figure 31–47(A)**. The technician then interrupts power to reset the IFC. Power is then resumed and the furnace runs fine. A new filter is

installed and the furnace continues to run correctly. The technician politely advises the customer to replace the furnace filter at least monthly, and sometimes even more often, to avoid having this problem again.

The following service calls do not include solutions. The solutions can be found in the *Instructor's Guide*.

SERVICE CALL 9

A residential customer calls and reports that *the furnace stopped in the middle of the night. The furnace is old and has a thermocouple for the safety pilot.*

The technician arrives and goes directly to the thermostat. The thermostat is set correctly for heating. The thermometer in the house shows a full 10°F below the thermostat setting. This unit has air conditioning, so the technician turns the fan switch to ON to see if the indoor fan will start; it does.

The furnace is in an upstairs closet. The technician sees that the pilot light is not burning. This system has 100% shutoff, so there is no gas to the pilot unless the thermocouple holds the pilot-valve solenoid open. The technician positions the main gas valve to the PILOT position and presses the red button to allow gas to the pilot light. The pilot light burns when a lit match on an extender is placed next to it. The technician then holds the red button down for 30 sec and slowly releases it. The pilot light goes out. **What is the likely problem and the recommended solution?**

SERVICE CALL 10

A residential customer calls. *The furnace is not heating the house. The furnace is located in the basement and is very hot, but no heat is moving into the house. The dispatcher tells the customer to turn the furnace off until a service technician can get to the job.*

The technician has some idea of what the problem is from the symptoms. The technician knows what kind of furnace it is from previous service calls, so some parts are brought along. When the technician arrives, the room thermostat is set to call for heat. From the service request it is obvious that the low-voltage circuit is working because the customer says the burner will come on. The technician then goes to the basement and hears the burner operating. When the furnace has had enough time to get warm and the fan has not started, the technician takes the front control panel off the furnace and notices that the temperature-operated fan switch dial (circular dial type, **Figure 31–46**) has rotated as if it were sensing heat. SAFETY PRECAUTION: The technician carefully checks the voltage entering and leaving the fan switch. **What is the likely problem and the recommended solution?**

SERVICE CALL 11

A residential customer calls. *There is no heat. The furnace is in the basement, and there is a sound like a clock ticking.*

The service technician arrives and goes to the basement. The furnace door is removed. The technician hears the arcing sound. The shield in front of the burner is removed. The technician sees the arc. A match on an extender is lit and placed near the pilot to see whether there is gas at the pilot and to see whether it will light. It does. The arcing stops, as it should. The burner lights after the proper time delay.

What is the likely problem and the recommended solution?

SUMMARY

- Forced-hot-air furnaces are normally classified as upflow, downflow (counterflow), horizontal, low-boy, or multipoise.
- Multipoise furnaces can be installed in any position, which adds versatility. Multiple safety controls like bimetallic flame rollout or overtemperature limits must be designed into the furnace for safety.
- Auxiliary limit switches, which sense room return air temperature, interrupt the burner operation when the temperature gets too hot.
- Vent limit or draft safeguard switches are normally closed switches and are usually mounted on the draft hood or near the exhaust vent pipe exit. They monitor the draft hood temperature and open when there is any spillage of flue gas.
- Furnace components consist of the cabinet, gas valve, manifold, pilot, burners, heat exchangers, blower, electrical components, and venting system.
- Gas fuels are natural gas, manufactured gas (used in few applications), and LP gas (propane, butane, or a mixture of the two).
- Inches of water column is the term used when determining or setting gas fuel pressures.
- A water manometer or digital manometer measures gas pressure in inches of water column.
- For combustion to occur there must be fuel, oxygen, and heat. The fuel is the gas, the oxygen comes from the air, and the heat comes from the pilot flame or other igniter.
- Gas burners use primary and secondary air. Excess air is always supplied to ensure as complete combustion as possible.
- Gas valves control the gas flowing to the burners. The valves are controlled automatically and allow gas to flow only when the pilot is lit or when the ignition device is operable.
- Some common gas valves are classified as solenoid, diaphragm, or heat motor valves.
- A servo pressure regulator is an outlet or working pressure regulator. It monitors gas outlet pressure when gas flow rates and inlet pressures vary widely.
- A redundant gas valve has two or three valve operators physically in series with one another but wired in parallel.
- Ignition at the main burners is caused by heat from the pilot or from an electric spark. There are standing pilots that burn

continuously, and intermittent pilots that are ignited by a spark when the thermostat calls for heat. There is also a direct-spark or direct burner ignition, in which the spark or hot surface igniter ignites the gas at the burners.

- The thermocouple, the bimetal, and the liquid-filled remote bulb are three types of safety devices (flame-proving devices) to ensure that gas does not flow unless the pilot is lit.
- The manifold is a pipe through which the gas flows to the burners and on which the burners are mounted.
- The orifice is a precisely sized hole in a spud through which the gas flows from the manifold to the burners.
- The burners have holes or various designs of slots through which the gas flows. The gas burns immediately on the outside of the burners at the top or end depending on the type.
- Inshot burners are usually made of aluminized steel and have crossover porting for good flame retention, proper flame carryover to each main burner, and excellent flame stability.
- Modern gas furnaces use either induced- or forced-draft systems. Induced-draft systems have a combustion blower motor located on the outlet of the heat exchanger. They pull or suck combustion gases through the heat exchanger, usually causing a slight negative pressure in the heat exchanger itself.
- Forced-draft systems have a combustion blower motor on the inlet of the heat exchanger. They push or blow combustion gases through the heat exchanger and cause a positive pressure in the heat exchanger.
- The gas burns in an opening in some heat exchangers. Air passing over the heat exchanger is heated and circulated to the conditioned space.
- A blower circulates this heated air. The blower is turned on and off by a fan switch, which is controlled either by time or by temperature.
- Modern heat exchangers come in many sizes, shapes, and materials. Many are L-shaped and S-shaped and made of aluminized steel. There is actually a serpentine path for combustion gases to travel. Serpentine heat exchangers use inshot burners instead of ribbon or slotted-port burners.
- Modern condensing furnaces may have two or even three heat exchangers. The last heat exchanger is usually made of stainless steel or aluminum to resist corrosion. It is designed to intentionally handle condensate and eventually drain it away from the furnace.
- Many modern gas furnaces use electronic modules or integrated furnace controllers (IFCs) to control operations. These controllers often have dual in-line pair (DIP) switches, which can be switched to reprogram the furnace to make it more versatile.
- The limit switch is a safety device. If the fan does not operate or if the furnace overheats for another reason, the limit switch causes the gas valve to close.
- High-limit switches come in a variety of styles such as bimetallic snap action, helical bimetal, or liquid filled. They come as either manual or automatic resettable.
- In a flame rectification system, the pilot flame, or main flame, can conduct electricity because it contains ionized combustion gases that are made up of positively and negatively charged particles. The flame is located between two electrodes of different sizes. The electrodes are fed with an alternating current from the furnace's electronic module or

controller. Current will flow in one direction more than the other because of the different size electrodes. The flame acts as a switch. When the flame is present, the switch is closed and will conduct electricity.

- Flame rectification systems can be classified as either single rod (local sensing) or dual rod (remote sensing). Single-rod systems consist of an igniter and a sensor all in one rod. Dual-rod systems use a separate igniter and flame-sensing rod.
- A direct-vented high-efficiency gas furnace has a sealed combustion chamber. This means that the combustion air is brought in from the outside through an air pipe, which is usually made of PVC plastics. A non-direct-vented high-efficiency gas furnace uses indoor air for combustion.
- A positive pressure system can be vertically or sidewall vented. Its flue pipe pressure is positive the entire distance to its terminal end.
- The dew point temperature is the temperature at which the condensation process begins in a condensing furnace. The DPT varies depending on the composition of the flue gas and the amount of excess air. As excess air decreases, the DPT increases.
- Excess air consists of combustion air and dilution air. Combustion air is primary air and/or secondary air. Primary air enters the burner before combustion takes place. Secondary air happens after combustion and supports combustion.
- Dilution air is excess air after combustion and usually enters at the end of the heat exchanger.
- Furnace efficiency ratings are determined by the amount of heat that is transferred to the heated medium. Factors that determine the efficiency of a furnace are the type of draft, the amount of excess air, the temperature difference of the air or water entering versus leaving the heating medium side of the heat exchanger, and the flue stack temperature.
- Electronic ignition modules come in 100% shutoff, non-100% shutoff, continuous retry with 100% shutoff, and many other custom control schemes that are manufacturer dependent. The combustion blower can prepurge, interpurge, or postpurge the heat exchanger depending on the modules' control scheme.
- Twinning involves the operation of two side-by-side furnaces connected by a common ducting system. Twinning is often used when the required heating capacity is greater than the largest-capacity furnace manufactured. It is also used when the airflow of one furnace, for air-conditioning purposes, is greater than a single furnace can provide.
- Modern two-stage gas furnaces use a two-stage gas valve and a two-speed combustion blower motor with two pressure switches to prove the draft.
- Modern modulating gas furnaces follow the heat loss of a structure. They use a modulating gas valve instead of a staged valve. They also come with variable-speed warm air blowers to vary the speed and amount of warm air. Variable output thermostats that send a proportional signal instead of an on-off signal to the furnace controller often control modulating furnaces.
- One of the first steps in systematically troubleshooting a furnace with an integrated furnace controller is to obtain the electrical diagram of the furnace and sequence of operation of the IFC.

- **SAFETY PRECAUTION:** Venting systems must provide a safe and effective means of moving the flue gases to the outside atmosphere. Flue gases are mixed with other air through the draft hood. Venting may be by natural draft or by forced draft. Flue gases are corrosive.
- Gas piping should be kept simple with as few turns and fittings as possible. It is important to use the correct pipe size.
- **SAFETY PRECAUTION:** All piping systems should be tested carefully for leaks.
- The combustion efficiency of furnaces should be checked and adjustments made when needed.
- Preventive maintenance calls should be made periodically on gas furnaces.

REVIEW QUESTIONS

1. The four types of gas furnace air flow patterns are _____, _____, _____, and _____.
2. Describe the function of a multipoise or multipositional furnace.
3. Describe the function of a draft safeguard switch.
4. Where are auxiliary limit switches used in heating applications?
5. The specific gravity of natural gas is
   A. 0.08.
   B. 1.00.
   C. 0.42.
   D. 0.60.
6. What is a water and digital manometer?
7. The pressure at the manifold for natural gas is typically
   A. 10 psig.
   B. 8 in. WC.
   C. 5.5 psia.
   D. 3.5 in. WC.
8. What is the approximate oxygen content of air?
9. The typical manifold pressure for propane gas is _____ in. WC.
10. Why is excess air supplied to all gas-burning appliance burners?
11. The purpose of the gas regulator at a gas burning appliance is to
    A. increase the pressure to the burner.
    B. decrease the pressure to the burner.
    C. filter out any water vapor.
    D. cause the flame to burn yellow.
12. True or False: All gas valves snap open and closed.
13. Which of the following features does an automatic combination gas valve have?
    A. Pressure regulator
    B. Pilot safety shutoff
    C. Redundant shutoff feature
    D. All of the above
14. What is the function of the servo pressure regulator in a modern combination gas valve?
15. What is a redundant gas valve?
16. Briefly describe an intermittent pilot gas valve system.
17. Briefly describe a direct burner gas valve system.

18. Where are inshot burners used in heating applications?
19. Describe the difference between a forced-draft system and an induced-draft system.
20. What is an integrated furnace controller (IFC)?
21. What is meant by the warm air blower's off-delay timing?
22. What are dual in-line pair (DIP) switches?
23. What is a flame rollout safety switch used for in heating applications?
24. Describe the two types of pilot lights.
25. True or False: A thermocouple develops direct current and voltage.
26. Describe how a thermocouple flame-proving system functions.
27. Describe how a bimetallic flame-proving system functions.
28. Describe how a liquid-filled flame-proving system functions.
29. Explain why the preceding flame-proving systems are called safety devices.
30. Describe a flame rectification flame-proving system.
31. What is a hot surface ignition system?
32. Describe the difference between a single-rod (local sensing) and a dual-rod (remote sensing) system as they pertain to flame rectification.
33. Flame current is measured in _____.
34. What is a direct-vented furnace?
35. What is a non-direct-vented furnace?
36. What is meant by a furnace that has a positive pressure venting system?
37. Define the dew point temperature as it applies to a condensing furnace.
38. What is excess air?
39. What is dilution air?

40. What are four factors that determine the efficiency of a furnace?
41. What is a 100% shutoff system?
42. What is the difference between a hard and a soft lockout?
43. What is meant by prepurging a heat exchanger?
44. When would a combustion blower ever interpurge a furnace's heat exchange?
45. What is a smart valve system?
46. An orifice is a drilled hole in a _____.
47. A _____ distributes gas to the various burners.
48. True or False: An orifice measures the quantity of gas to the burner as well as directs the gas stream.
49. Name four types of gas burners.
50. The _____ _____ transfers heat from the hot gas to the airstream to heat the structure.
51. Name two types of fan switch.
52. When the limit switch circuit opens, it opens the circuit to the _____ _____.
53. Describe the function of the draft diverter.
54. Describe how a two-stage furnace operates.
55. Describe how a modulating furnace operates.
56. What is a variable output programmable thermostat?
57. Why is a metal vent often preferred rather than a masonry chimney?
58. Describe how a vent damper shutoff functions.
59. Describe why you should never use a flame to leak check a gas system.
60. What is the first step a service technician should take when systematically troubleshooting a furnace that contains an IFC?
61. A _____ _____ sample is taken from the flue gas to check the efficiency of a gas furnace.

UNIT 32 Oil Heat

OBJECTIVES

After studying this unit, you should be able to

- describe how fuel oil and air are prepared and mixed in the oil burner unit for combustion.
- list products produced as a result of combustion of the fuel oil.
- list the components of gun-type oil burners.
- describe basic service procedures for oil burner components.
- sketch wiring diagrams of the oil burner primary control system and the fan circuit.
- state tests used to determine oil burner efficiency.
- explain corrective actions that may be taken to improve burner efficiency, as indicated from the results of each test.
- describe preventive maintenance procedures.

SAFETY CHECKLIST

- ✔ Do not reset any primary control too many times because unburned oil may accumulate after each reset. If a puddle of oil is ignited, it will burn intensely. If this should happen, stop the burner motor but allow the furnace fan to run. The air shutter should be shut off to reduce the air to the burner. Notify the fire department. Do not try to open the inspection door to put the fire out; let it burn itself out with reduced air.
- ✔ Do not start a burner if the heat exchanger is cracked or otherwise defective. The flue gases from combustion should never mix with the circulating air.
- ✔ Do not start the burner with the fuel pump bypass plug in place unless the return line is in place. Without this line, there is no place for the excess oil to go except to possibly rupture the shaft seal.
- ✔ When conducting flue-gas efficiency tests, avoid burns by not touching the hot flue-gas pipe.
- ✔ The ignition is produced by an arc of 10,000 V or more. Keep your distance from this arc.
- ✔ Observe all electrical safety precautions.

32.1 INTRODUCTION TO OIL-FIRED FORCED-WARM AIR FURNACES

The oil-fired forced-warm-air furnace has two main systems: a heat-producing system and a heat-distributing system. The *heat-producing system* consists of the oil burner, fuel supply components, combustion chamber, and heat exchanger. The

Figure 32–1 An oil-fired furnace. *Courtesy Williamson-Thermoflo*

heat-distributing system is composed of the blower fan, which moves the air through the ductwork, and other related components, **Figure 32–1**.

Following is a brief description of how such a furnace operates. When the thermostat calls for heat, the ignition system is powered and the oil burner motor is started, which turns the fuel pump and combustion air blower wheel. The oil-air mixture is sprayed into the combustion area, where it is ignited. After the mixture has heated the combustion chamber and the heat exchanger is up to temperature, the heating air fan comes on and distributes the heated air through the ductwork to the space to be heated. When this space has reached a predetermined temperature, the thermostat will cause the burner motor to shut down. The heating air fan will continue to run until the heat exchanger cools to a set temperature.

32.2 PHYSICAL CHARACTERISTICS

The physical appearance and characteristics of forced-air furnaces vary to some extent. The *low-boy* is often used when there is not much headroom, **Figure 32–2**. Low-boys may have a cooling coil on top to provide air conditioning.

An *upflow* furnace is a vertical furnace in which the air is taken in at the bottom and is forced across the heat exchanger

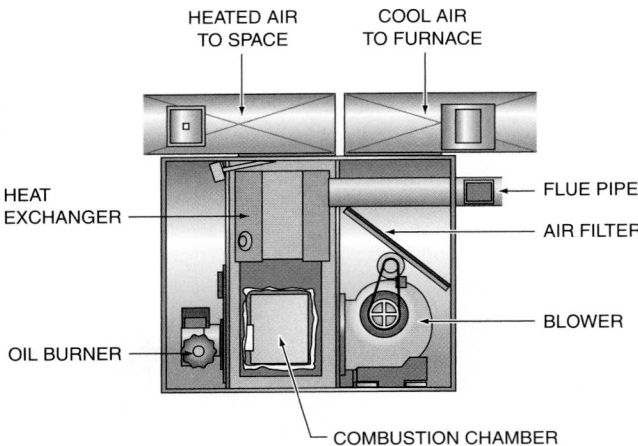

Figure 32–2 A low-boy oil-fired furnace. *Courtesy Thermo Pride*

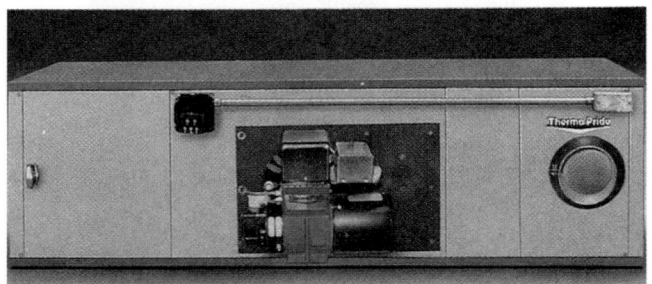

Figure 32–3 A horizontal, oil-fired forced-air furnace. *Courtesy Williamson-Thermoflo*

and out the top. This furnace is installed with the ductwork above it, such as in a hall closet, or it can be installed in a basement with the ductwork above it under the floor.

A *downflow* furnace looks similar to an upflow except that the air is drawn in from the top and forced out the bottom. In the downflow configuration, the supply ductwork that carries the heated air to the occupied space is located just below or in the floor below the furnace. The cooler air coming from the occupied space to the appliance enters at the top of the furnace.

A *horizontal* furnace is usually installed in a crawl space under a house or in an attic, **Figure 32–3**. These units are available with right-to-left or left-to-right airflow.

Oil-fired forced-air heating is a popular method of heating residences and light commercial buildings. Millions of furnaces of this type are being used in the United States at the present time. It is essential that the air-conditioning and heating technician understand proper installation and servicing techniques involved with oil burners. Safety is always a concern of technicians but with the high cost of energy, efficiency in the combustion process is also extremely important.

32.3 FUEL OILS

Fuel oil as we know it is a fossil fuel, as is the natural gas that was discussed in Unit 31. Fuel oil is delivered to the customer in liquid form and stored in tanks, either aboveground

Figure 32–4 A local oil dealer delivers the fuel oil in small trucks to aboveground or underground storage tanks.

or underground, **Figure 32–4**. The oil used in residential and most commercial systems in the United States is basically diesel fuel that has been dyed red for tax purposes. (Taxes are higher when diesel fuel is used for transportation purposes. By dyeing the fuel, authorities can identify the cheaper, lower-taxed product.)

Fuel oil is derived from crude oil by a process of distillation called *cracking*. There are six grades of fuel oil and a numbering system, 1 through 6, is used to identify each grade. The lower-numbered oils are called *light oils* because they weigh less per gallon than the higher-numbered oils. The reason for this is the carbon content. Oils are hydrocarbons made up of carbon and hydrogen, and, since carbon is heavier than hydrogen, it can be concluded that the higher-numbered oils contain more carbon than the lower-numbered oils. The higher-numbered fuel oils also have higher heat contents than the lower-numbered oils do. The lightest of the fuel oils, No. 1, is most commonly known as kerosene.

Since fuel oil is made up of carbon and hydrogen, it is classified as a hydrocarbon. If you recall from earlier units, other pure hydrocarbons such as methane, ethane, propane, and butane are very good refrigerants but are typically not used because of their high flammability. Fuel oil No. 2 is also known as pentane, which is a hydrocarbon molecule made up of 5 carbon atoms surrounded by 12 hydrogen atoms. **Figure 32–5** shows the molecular structure of some very common hydrocarbons. Methane, **Figure 32–5(A)**, is the base molecule for refrigerants such as R-22 and R-12. Ethane, **Figure 32–5(B)**, is the base molecule for refrigerants such as R-123 and R-134a. **Figure 32–5(E)** represents the pentane molecule.

The most common fuel oil in use is No. 2, as it is used in most residences and commercial installations. The heavier

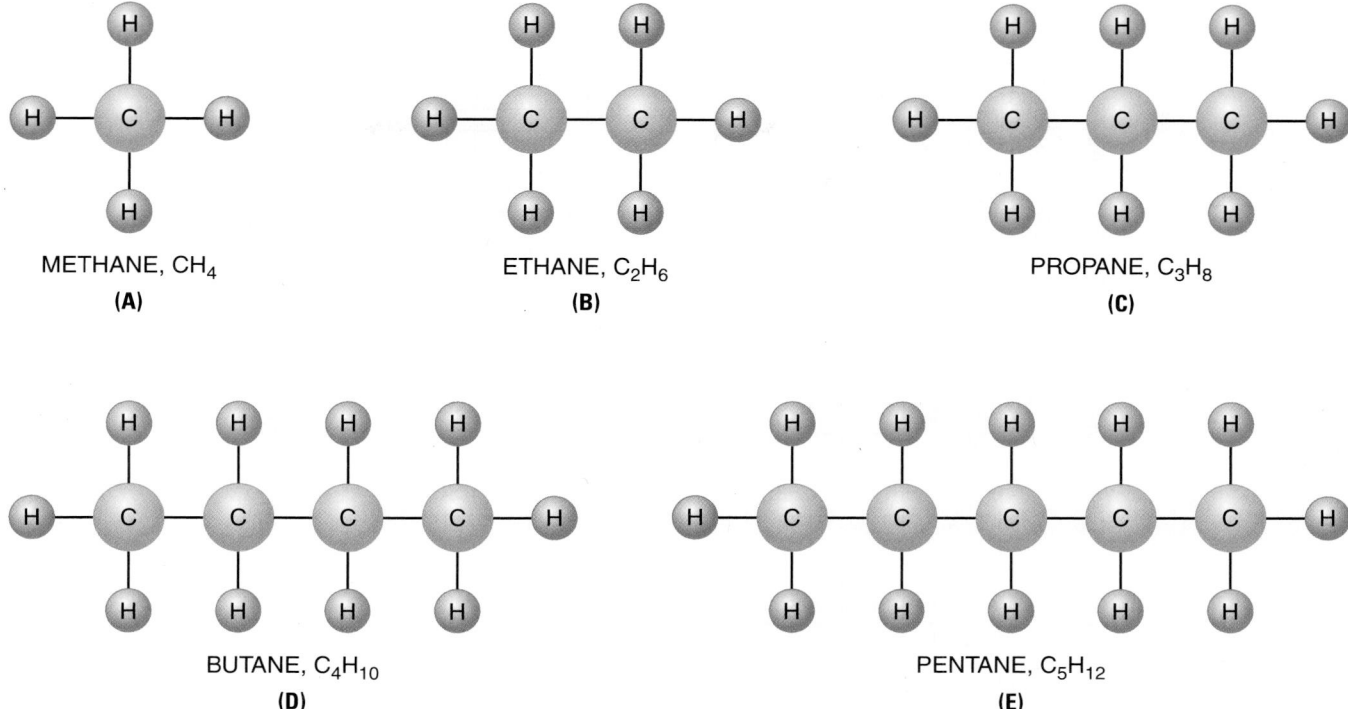

METHANE, CH$_4$
(A)

ETHANE, C$_2$H$_6$
(B)

PROPANE, C$_3$H$_8$
(C)

BUTANE, C$_4$H$_{10}$
(D)

PENTANE, C$_5$H$_{12}$
(E)

Figure 32–5 Chemical composition of some common hydrocarbons.

oils are used in large commercial and industrial installations. No. 2 fuel oil has a heat content of 137,000 to 141,800 Btu/gal and weighs 6.960 to 7.296 lb/gal. This text will use the average of 140,000 Btu/gal, which is an industry standard. No. 6 fuel oil has a heat content of about 153,000 Btu/gal and weighs 8.053 to 8.448 lb/gal. We will use No. 2 fuel oil in all of our examples because it is the most common.

The American Society for Testing and Materials (ASTM) publishes standards for a number of industries, including the oil heat industry. It has established minimum standards for the production and refinement of oil used for heating purposes. The fuel oil standards set acceptable ranges for fuel oil characteristics that include

- Flash point
- Ignition point
- Viscosity
- Carbon residue
- Water and sediment content
- Pour point
- Ash content
- Distillation quality

Flash Point

The flash point has to do with the maximum safe storage and handling temperature for the fuel. The *flash point* is the lowest temperature at which vapors in the air above the fuel oil ignite for a short period of time when exposed to a flame. The flash is caused by vapors rising from the liquid fuel oil. The lighter the oil, the lower the flash point. The minimum flash point for No. 2 oil is 100°F.

Ignition Point

The *ignition point* is a few degrees higher than the flash point. The flame will continue to burn because vapors will continue to rise from the liquid.

Viscosity

Viscosity is the thickness of the oil under normal temperatures. This thickness can be compared with other oils and fluids. The heavier oils are thicker at the same temperatures. Viscosity is expressed in Saybolt Seconds Universal (SSU). This describes how much oil will drip through a calibrated hole at a certain temperature. As the temperature becomes colder, the same oil will have a reduced flow through the calibrated hole. **Figure 32–6** shows the viscosity of various oils. Warmer oil has a lower viscosity and is therefore desireable for ensuring better flow through the oil lines as well as through the system filters, pumps, and nozzles.

Carbon Residue

Carbon residue is the amount of carbon left in a sample of oil after it is changed from a liquid to a vapor by being boiled in an oxygen-free atmosphere. Properly burned oil does not have any appreciable carbon residue.

Water and Sediment Content

Ideally, the process of oil refining and delivering should occur in a manner that will ensure that no water or sediment is present in the oil once it arrives at its final destination.

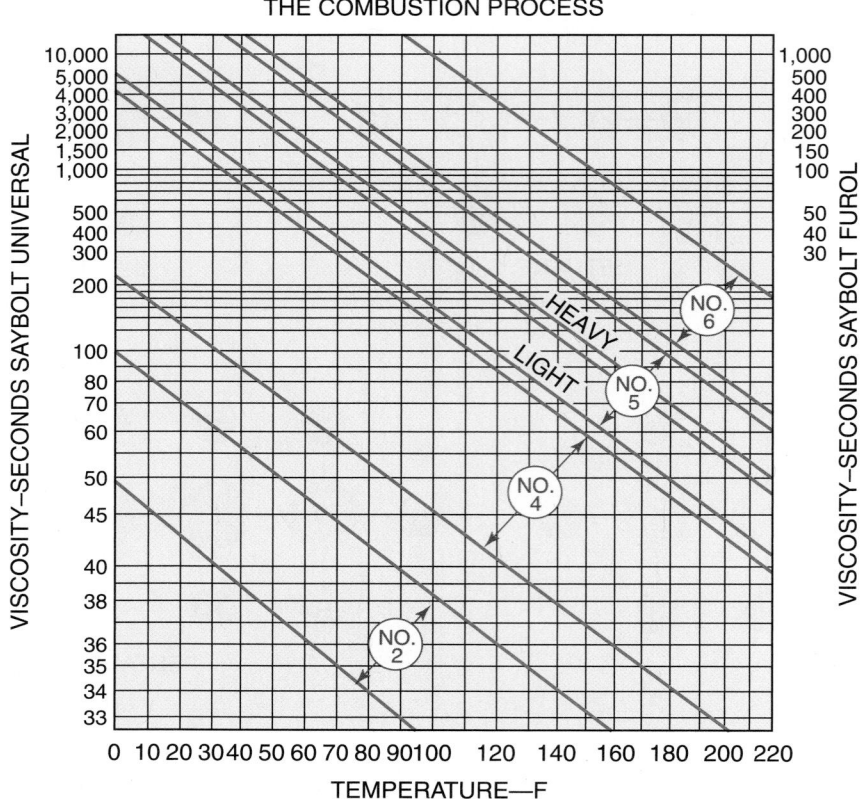

Figure 32–6 This graph shows the viscosities of the different grades of fuel oil. *Courtesy American Society of Heating, Refrigerating, and Air-Conditioning Engineers (ASHRAE)*

In the real world, however, this is impossible to expect. At the very least, a strong effort should be made to ensure that the water, sediment, and other contaminant levels are as low as possible under the given set of conditions.

Sediment can form as a result of rust forming on the internal pipe and tank surfaces, which may be made of iron or steel. In addition, sludge can form in the tank when the water, in the form of condensation, reacts with the fuel oil. Higher condensation concentrations are present when the oil tank is not filled to capacity and the bare steel walls are exposed to the moisture/humidity in the air above the oil level.

Sludge and sediment can lead to clogged oil lines, nozzles, and filters. Oil impurities and fuel-line restrictions can lead to poor flame characteristics and inefficient burning of the fuel. Following proper tank installation and maintenance procedures will help minimize fuel-related problems that occur after product delivery.

Pour Point

Pour point is the lowest temperature at which the fuel can be stored and handled. No. 2 fuel oil is one of the lower pour-point fuels. The oil does not freeze, but there are waxes in the oil that become thick so that it will not flow correctly. Combustion will stop, and a no-heat call will result. No. 2 fuel oil can be successfully stored and used down to about 20°F. If the storage temperature is expected to be colder than 20°F,

blending the fuel oil with 25% kerosene, known as fuel oil No. 1, will lower the pour point of the fuel. Other commercially available pour-point suppressants are available as well. Storing the oil in a basement or underground tank will often keep the oil warm enough for adequate flow. The heavier oils may have to be heated to facilitate flow when low temperatures are encountered.

Ash Content

The *ash content* indicates the amount of noncombustible materials contained in the fuel oil. These materials pass through the flame without burning and are contaminants. They can also be abrasive and wear down burner components. The refinery is responsible for keeping the ash content within the required tolerance. Many states have inspection departments that regularly check oil that is distributed in the state.

Distillation Quality

Oil must be turned into a vapor before it can be burned. The *distillation quality* describes the ability of the oil to be vaporized. The lighter oils turn into a vapor easier than the heavier oils.

However, there are many reasons to use the heavier oils. They are less expensive and have more heat content per gallon. Heavier oils are less volatile and therefore safer in some situations, for example, on board a ship. These oils do not have storage codes as strict as the lighter oils. Many large buildings and commercial establishments must store the oil on the premises. Heavy oil is safer because it does not ignite as easily.

Fuel oil is a combination of liquid hydrocarbons containing hydrogen and carbon in chemical combination. Some of these hydrocarbons in the fuel oil are light, and others are heavy. Combustion occurs when the hydrocarbons unite rapidly with oxygen (O_2). Heat, carbon dioxide (CO_2), and water vapor are produced when the combustion occurs.

32.4   OIL STORAGE

Once oil is delivered to the customer, it must be stored until it is needed. For this reason, heating with fuel oil requires the installation of a storage tank on the property. Fuel oil storage tanks are available in a variety of shapes and sizes and can be installed either indoors or outside. Outdoor oil tanks can be installed either aboveground or underground. Whether an aboveground or an underground tank should be used is determined mainly by the ground conditions at a particular location.

When installing an oil tank, be sure to follow all local codes as well as those outlined by the International Code Council (ICC) and the National Fire Protection Association (NFPA). Be sure to comply with the required clearances that must be present between oil storage tanks and property lines, electric meters, combustion sources, open flames, windows, doors, air intakes, and other relevant property elements.

Underground Tanks

Tanks that are to be installed underground must be protected against corrosion. The two most popular types of tanks for in-ground residential fuel storage applications are the STI-P3 tanks and fiberglass tanks. Fiberglass tanks, **Figure 32–7,** are

Figure 32–8   An STI-P3 steel oil storage tank. *Courtesy Oilheat Associates. Photo by John Levey*

not subject to corrosion, but steel tanks are. What follows is a brief description of the steel tank and the methods by which this tank type is protected from corrosion.

The acronym STI-P3 stands for "Steel Tank Institute—Protected 3 Ways." These steel tanks, **Figure 32–8,** are manufactured with an external corrosion-resistant coating that prevents the rapid deterioration of the base metal. In addition, these tanks also have dielectric bushings and sacrificial anodes. Dielectric bushings provide a noncorrosive barrier at the point where dissimilar metals are joined. It is typically at this point in any piping circuit that oxidation and corrosion take place. The sacrificial anodes are rods typically manufactured from either zinc or magnesium. These rods are secured to the base metal of the tank and serve to protect the tank from corrosion. This is possible because the anode material oxidizes and corrodes before the base metal and, since the anode rods are in contact with the steel tank, the steel will not corrode. Once the anodes have completely corroded, the steel tank will then begin to corrode. Therefore, it is beneficial to replace the anodes prior to this point. Field technicians can perform tests to determine the condition of the anodes, even though the tank is buried in the ground. Replacing the anodes, however, involves exposing the top of the tank.

Aboveground Tanks

For many years, aboveground oil storage tanks were constructed of either 12- or 14-gauge steel. Over the past decade, technological advances have paved the way for a new generation of tanks that are made of steel, fiberglass, or a combination of steel and plastic. Modern, residential aboveground tanks are typically fabricated from 12-gauge steel and hold between 275 and 330 gallons of oil. Smaller tanks are available and some applications call for the twinning, or joining, of smaller tanks to create larger storage capacity. The capacities of residential fiberglass tanks typically range from 240 to 300 gallons.

Figure 32–7   An underground fiberglass oil storage tank. *Courtesy Oilheat Associates. Photo by John Levey*

Figure 32–9   A polyethylene/steel oil storage tank. *Photo by Eugene Silberstein*

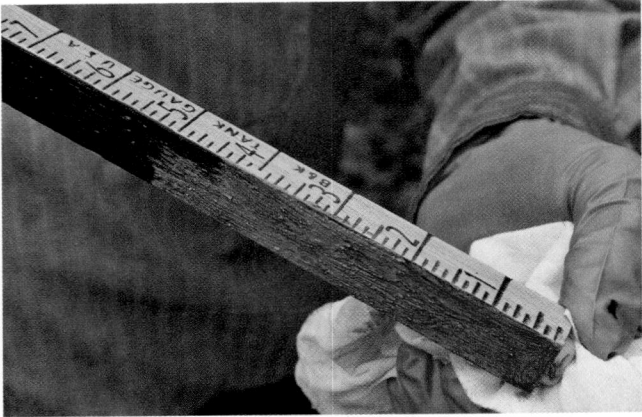

Figure 32–10   Water-sensing paste changes color when water is present in an oil tank. *Courtesy Oilheat Associates. Photo by John Levey*

Oil Tank Maintenance

For the most part, oil storage tanks do not require much maintenance. What follows in this section are some key points and suggestions that should be followed to help maintain both the integrity of the tank itself and the valuable oil contained within it.

- Aboveground tanks should painted periodically and inspected on a regular basis. Deficiencies should be addressed promptly to reduce the possibility of premature tank failure.
- STI-P3 tanks should be checked periodically to ensure that the protective anodes are still in good working order. Remember that these anodes are *designed* to deteriorate in order to protect the integrity of the tank itself.
- Tank piping should be inspected regularly for imperfections and leaks. Always check for loose or missing vent and fill caps.
- Tanks should be checked for water at least once a year. If water is present in the tank, the source of the water should be identified and the situation should be corrected. Tanks are tested for water content by placing a water-sensing paste on the oil tank stick. When dry, the paste is green. In the presence of water, the paste changes color, **Figure 32–10.**

32.5   FUEL OIL SUPPLY SYSTEMS

In order to get the fuel oil from the storage location to the oil burner itself, oil lines must be run between these two components. There are two common piping configurations that can be used: the one-pipe system and the two-pipe system. In a one-pipe system, **Figure 32–11,** there is only one pipe run between the oil storage tank and the oil burner. As oil is needed, it flows from the tank and through the oil line to the burner—where it is ultimately atomized, mixed with air, and ignited. Since only one pipe is being used, there is a lower potential for oil line leaks.

In a two-pipe setup, **Figure 32–12,** one pipe is the supply and the other acts as the return. Oil flows at maximum pump

Polyethylene/steel tanks are constructed as a tank within a tank, **Figure 32–9.** The inner oil-holding tank is made of polyethylene and the outer galvanized-steel tank is intended to protect the inner tank. The outer tank provides secondary containment in the event that the inner tank develops a leak.

Indoor Oil Storage

When oil tanks are located outside, the temperature of the fuel oil will change as the outside ambient temperature changes. This can lead to oil flow issues as well as condensation problems. When the option presents itself, indoor oil storage is desired. By storing the oil inside, there will be less temperature fluctuation of the fuel and there will also be less moisture and condensation on the interior surfaces of the tank. Condensation in the tank will increase the rate of sludge formation, leading to tank corrosion. In addition, warmer fuel oil has better flow characteristics and helps the oil burner provide more even combustion and more even heating.

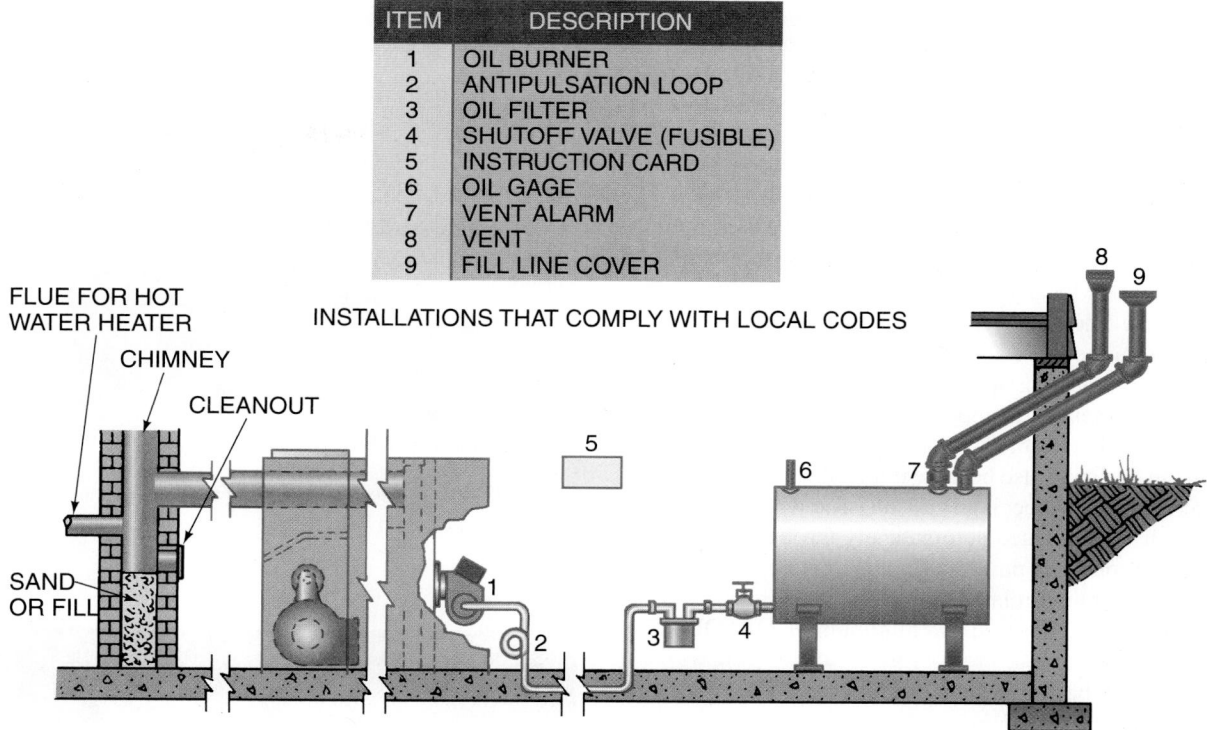

ITEM	DESCRIPTION
1	OIL BURNER
2	ANTIPULSATION LOOP
3	OIL FILTER
4	SHUTOFF VALVE (FUSIBLE)
5	INSTRUCTION CARD
6	OIL GAGE
7	VENT ALARM
8	VENT
9	FILL LINE COVER

INSTALLATIONS THAT COMPLY WITH LOCAL CODES

FLUE FOR HOT WATER HEATER
CHIMNEY
CLEANOUT
SAND OR FILL

Figure 32–11 A one-pipe system connecting the oil tank to the burner.

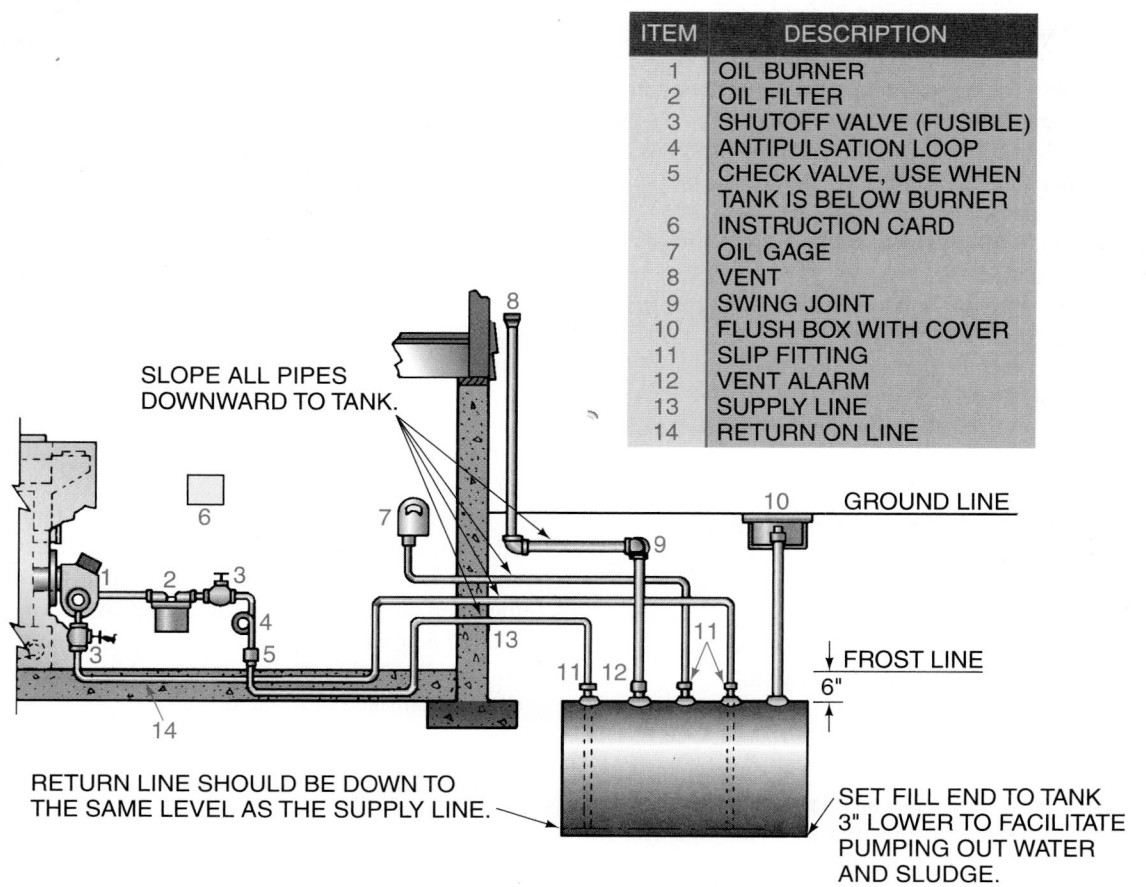

ITEM	DESCRIPTION
1	OIL BURNER
2	OIL FILTER
3	SHUTOFF VALVE (FUSIBLE)
4	ANTIPULSATION LOOP
5	CHECK VALVE, USE WHEN TANK IS BELOW BURNER
6	INSTRUCTION CARD
7	OIL GAGE
8	VENT
9	SWING JOINT
10	FLUSH BOX WITH COVER
11	SLIP FITTING
12	VENT ALARM
13	SUPPLY LINE
14	RETURN ON LINE

SLOPE ALL PIPES DOWNWARD TO TANK.

GROUND LINE

FROST LINE
6"

RETURN LINE SHOULD BE DOWN TO THE SAME LEVEL AS THE SUPPLY LINE.

SET FILL END TO TANK 3" LOWER TO FACILITATE PUMPING OUT WATER AND SLUDGE.

Figure 32–12 A two-pipe system connecting the oil tank to the burner.

capacity from the tank to the burner—and whatever oil the system needs for combustion, it takes. Any oil that is not needed is pumped back to the tank via the return line.

The location of the fuel storage tank in relation to the burner determines, for the most part, whether a one-pipe or a two-pipe system will be used. If the oil tank is located above the oil burner, a one-pipe system may be used. The oil will then flow by gravity from the tank and through a fuel filter, **Figure 32–11**, and then on to the burner itself. When the oil storage tank is located below the oil burner, a two-pipe system is sometimes used. In the two-pipe arrangement, oil is pumped from the tank.

Whenever the heating appliance is supplied oil by means of a one-pipe system, it is important that all of the air in the connecting fuel line is bled out to ensure proper fuel flow. Bleeding the line must also be done if the oil tank has been allowed to run dry. Air bleeding is not an issue with a two-pipe system as air is eliminated automatically through the return pipe. Here are some key points to keep in mind regarding fuel oil lines and other tank-related piping:

- Oil lines should be sized at a minimum of 3/8" OD copper lines.
- Oil lines can be constructed of pipe made of copper, wrought iron, steel, or brass; the piping should be connected with malleable fittings. Never use cast-iron (steam) fittings.
- Never use PVC pipe for oil lines.
- Use only flare connections to join copper oil lines, **Figure 32–13**. DO NOT USE COMPRESSION FITTINGS.
- When needed, use only a nonhardening, oil-resistant pipe joint compound.
- When installing oil lines, make the runs as short as possible, using as few fittings as possible. Avoid kinking the lines.
- Be sure that all oil lines are secured and otherwise protected from damage.
- When oil lines are run outside, make certain that they are well insulated.
- Underground oil lines should be installed with secondary containment, **Figure 32–13**.
- There should be a shutoff valve on the suction line at the oil burner.
- Fill and vent pipes should be pitched toward the tank opening to prevent the formation of oil traps.
- A vent alarm, **Figure 32–14**, should be installed in all tanks at the vent pipe. While the oil tank is being filled, the vent alarm will whistle. As the tank nears capacity, the whistling sound will stop, alerting the oil delivery person that the tank is nearly full.

Oil Deaerators

The oil deaerator, **Figure 32–15**, is an oil piping component that allows a one-pipe system to function as a two-pipe system. This is accomplished by connecting two pipes between the deaerator and the fuel pump on the oil burner, but only one pipe between the deaerator and the oil storage tank. This helps

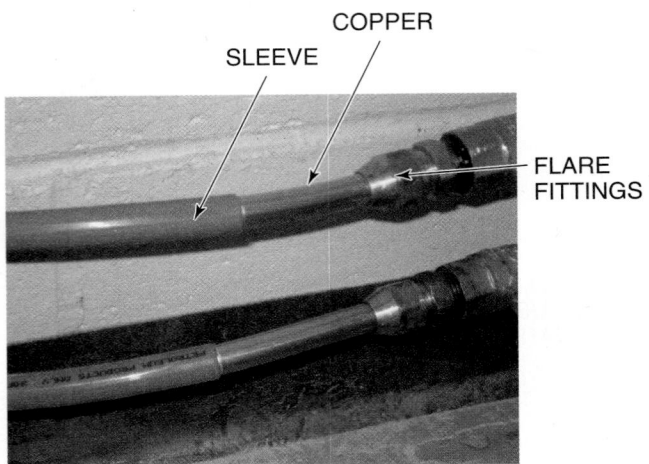

Figure 32–13   A sleeve over oil lines and flare connections. *Photo by Eugene Silberstein*

Figure 32–14   A vent alarm. *Photo by Eugene Silberstein*

eliminate the air problems that are typically found on one-pipe oil supply systems.

In operation, the fuel travels from the oil tank, through the oil filter, and into the deaerator. It is in the deaerator that any air that is mixed with the oil is separated from the fuel. The fuel then passes on to the fuel pump. At the fuel pump, the desired amount of oil will be pushed through the burner for combustion, while the remaining oil is pumped back to the deaerator instead of to the oil storage tank, **Figure 32–16**.

Auxiliary Fuel Supply Systems

In some installations, oil burner units will be located above the fuel oil supply tank at a height that will be beyond the lift capabilities of the burner pump. This occurs primarily in commercial applications. Heights exceeding 15 ft from the

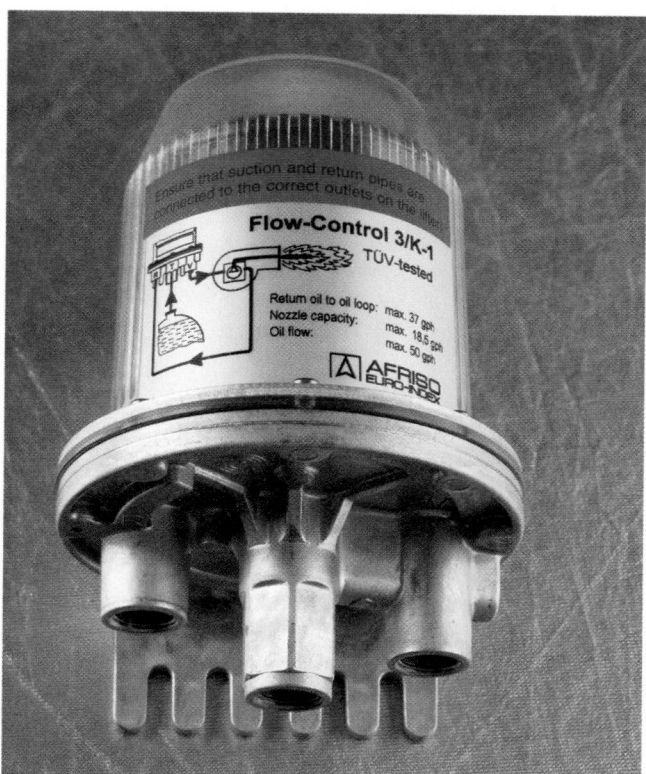

Figure 32–15 An oil deaerator.

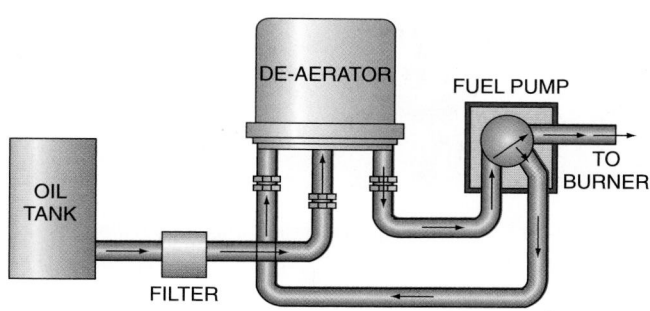

Figure 32–16 Piping diagram of the oil deaerator.

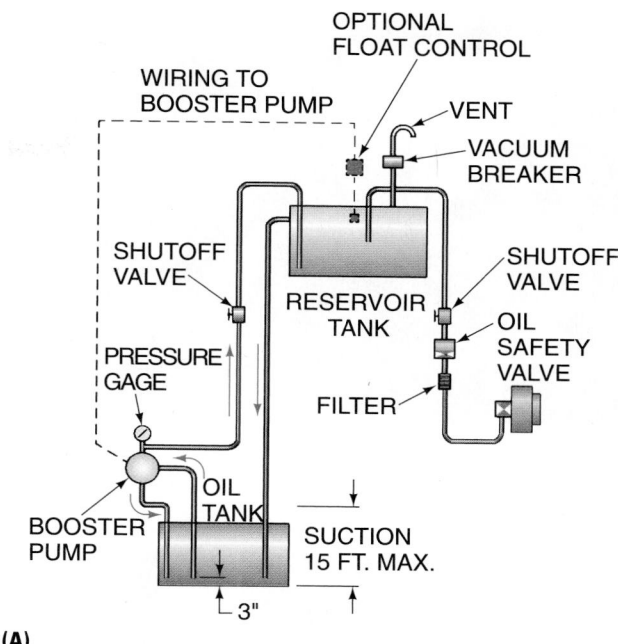

(A)

(B)

Figure 32–17 **(A)** A diagram of a system using an auxiliary booster pump and reservoir tank. **(B)** A booster pump. *Courtesy Suntec Industries Incorported, Rockford, Illinois*

bottom of the tank are beyond the capabilities of even the two-stage systems. In such installations an auxiliary fuel system must be used to get the fuel supply to the burner.

In this instance, a fuel oil booster pump is used to pump the oil to an accumulator or reservoir tank, **Figure 32–17**. This booster pump is wired separately from the burner unit and pumps the fuel oil through a separate piping system. The accumulator tank is kept full. Check valves are used to maintain the prime of the booster pump when it is not operating. An adjustable pressure regulator valve is a part of the fuel oil booster pump, and this regulator must be set to maintain a fuel oil pressure of 5 psi measured at the accumulator tank. Failure to properly adjust this regulator could result in seal damage to the fuel unit.

Fuel Line Filters

A filter should be located between the tank and the pump to remove any fine solid impurities from the fuel oil before it reaches the pump, **Figure 32–18**. Filters may also be located in the fuel line at the pump outlet to further reduce the impurities that may reach tiny oil passages in the nozzle, **Figure 32–19**. There are different types of filter media available. **Figure 32–20** shows a cutaway of one type of oil filter.

Figure 32–18 An oil filter.

Figure 32–20 Cutaway of an oil filter. *Courtesy Oilheat Associates. Photo by John Levey*

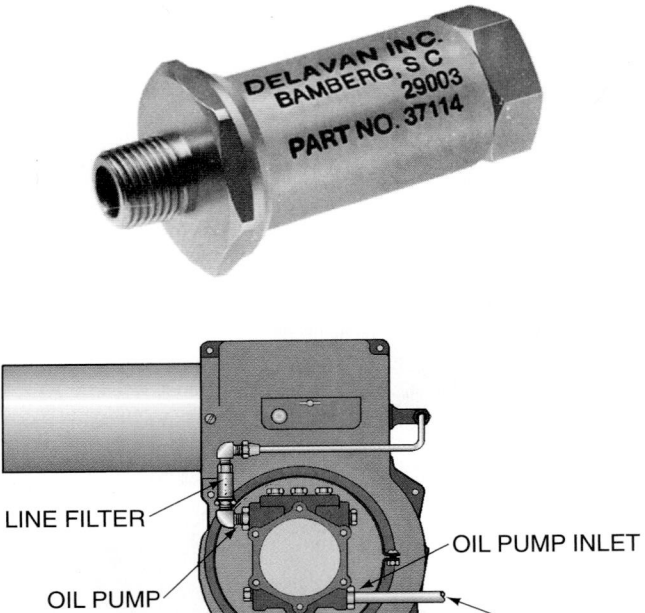

Figure 32–19 A fuel-line filter at the pump outlet. *Courtesy Delavan Corporation*

Figure 32–21 Oil filter media and gaskets. *Courtesy Oilheat Associates. Photo by John Levey*

NOTE: Be sure to replace the gaskets on the fuel filter when changing the filter cartridge, **Figure 32–21.**

In many cases, systems are equipped with dual in-line oil filtration systems, **Figure 32–22.** This helps ensure that the oil that reaches the appliance is as clean as possible. Make certain that all filters are installed with the arrow on the device pointing in the direction of fuel flow, **Figure 32–23.** Sometimes, a sludge/water isolator is installed in the fuel line as well. This device is located before the filter and is nothing more than an empty canister, **Figure 32–24,** that has a drain plug at the bottom. The fuel oil enters from the side and leaves from the top. Since water and sediment are heavier than the fuel oil, these substances fall to the bottom of the canister. Periodically, the drain plug is removed and the water and sediment are removed.

Figure 32–22 Dual in-line oil filtration. *Courtesy Oilheat Associates. Photo by John Levey*

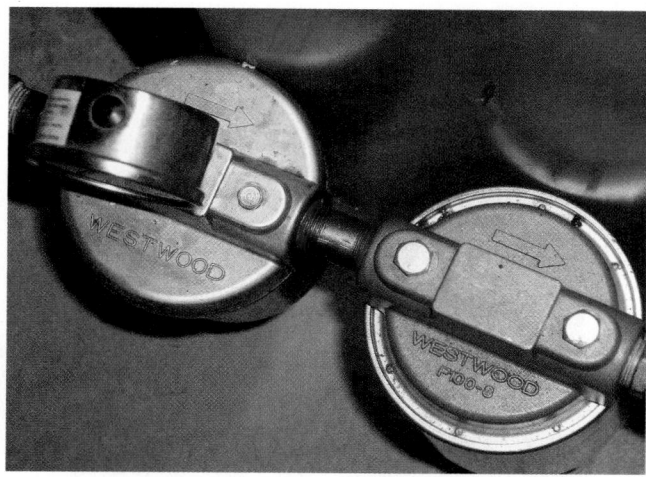

Figure 32–23   Make certain that the arrows on oil filters point in the direction of oil flow. *Courtesy Oilheat Associates. Photo by John Levey*

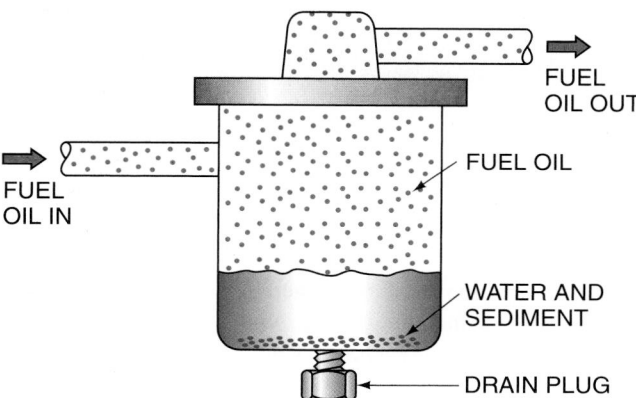

Figure 32–24   Diagram of a sludge/water isolator.

Oil Safety Valves (OSV)

Another system component commonly found in the piping arrangement of an oil-fired heating appliance is the oil safety valve. The oil safety valve (OSV) is a device that acts to stop oil flow in the event of a leak in the suction line. The valve itself is located in the suction line, which is the line that carries oil from the tank to the oil burner. When the pump on the oil burner operates, a vacuum is created in the oil line, causing oil to flow from the tank to the pump. The OSV senses this vacuum and opens to allow oil flow. In the event of a suction-line leak, the vacuum will be lost and the valve will close.

32.6   COMBUSTION

Combustion is the process of burning the fuel to generate heat. The combustion process is the same regardless of the fuel being burned. In order for combustion to take place, there must be ample supplies of fuel, heat, and oxygen. If any of these three components is missing, the combustion process cannot occur. If, during the combustion process, the

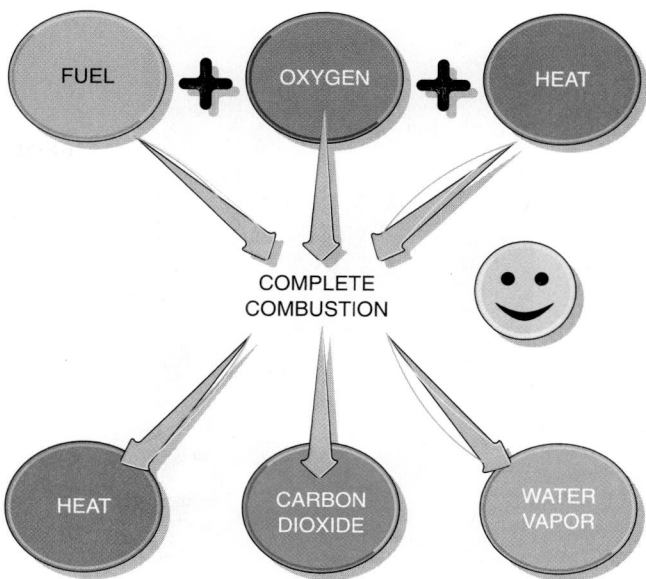

Figure 32–25   Diagram representing complete combustion.

supply of one or more of these three components is depleted, the combustion process will stop.

During the combustion process, the oxygen reacts with the fuel oil to produce heat, carbon dioxide, and water vapor, **Figure 32–25**. If only these three by-products are produced, perfect combustion will have occurred. Perfect combustion exists only in theory—as achieving this delicate balance is not possible in the real world. When combustion is not complete, carbon monoxide, soot or free carbon, and other undesirable substances form in addition to the heat, carbon dioxide, and water vapor, **Figure 32–26**. Carbon monoxide is

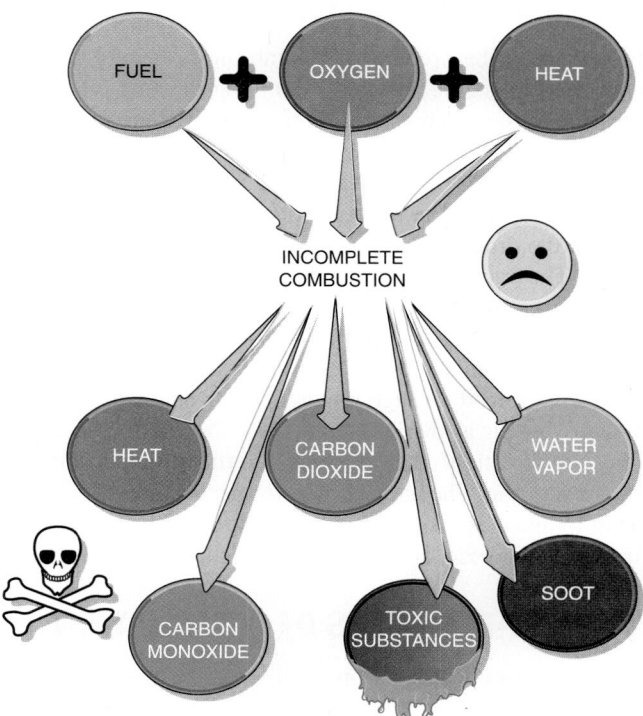

Figure 32–26   Diagram representing incomplete combustion.

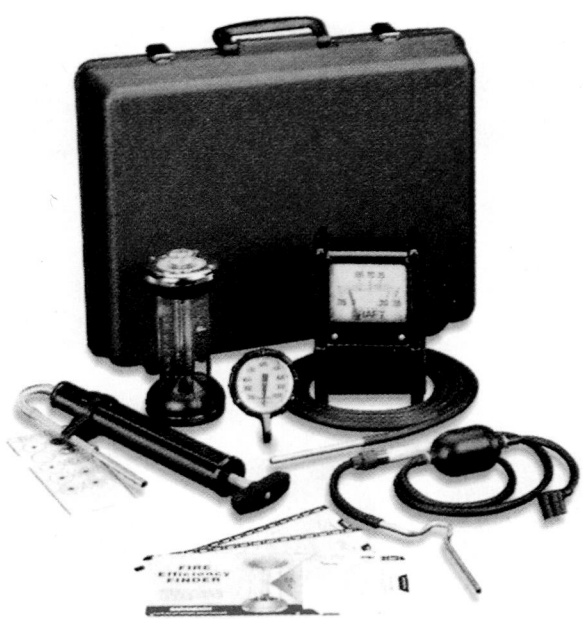

Figure 32–27 A combustion test kit. *Courtesy Bacharach*

Figure 32–28 Atomized fuel oil droplets and air leaving the burner nozzle. *Courtesy Delavan Corporation*

a colorless and odorless gas. Carbon monoxide, even in small quantities, is highly toxic and dangerous. Fortunately, with the proper use of combustion analysis equipment, **Figure 32–27,** the level of carbon monoxide can be measured. In addition, combustion testing provides important data to aid in the troubleshooting process.

32.7 PREPARATION OF FUEL OIL FOR COMBUSTION

Fuel oil must be prepared for combustion. It must first be converted to a gaseous state by forcing the fuel oil, under pressure, through a nozzle. The nozzle breaks the fuel oil up to form tiny droplets. This process is called *atomization,* **Figure 32–28.** The oil droplets are then mixed with air, which contains oxygen. The lighter hydrocarbons form gas covers around the droplets. Heat is introduced at this point with a spark. The vapor ignites (combustion takes place), and the temperature rises, causing the droplets to vaporize and burn.

High-pressure gun-type oil burners are generally used to achieve this combustion. Oil is fed under pressure to a nozzle. Air is forced through a tube that surrounds this nozzle. Usually the air is swirled in one direction and the oil in the opposite direction. The oil forms into tiny droplets combining with the air. The ignition transformer or ignitor furnishes a high-voltage spark between two electrodes located near the front of the nozzle, and combustion occurs.

32.8 BY-PRODUCTS OF COMBUSTION

The correct ratio of fuel oil to air must be maintained for efficient combustion. The fuel to be burned is made of carbon and hydrogen. When 1 lb of fuel oil is to be burned, it must

be mixed with about 3 lb of oxygen for complete combustion to take place. Since air is made up of approximately 79% nitrogen and 21% oxygen, we would need 14.4 lb of air to give us 3 lb of oxygen (3 lb of oxygen/0.21 lb of oxygen per lb of air = 14.4 lb of air). Since 1 lb of air takes up 13.33 ft³ at standard atmospheric conditions, we would need about 192 ft³ (13.33 ft³ of air per pound × 14.4 lb of air = 192 ft³) of air to burn a single pound of fuel oil. This is the correct amount of air needed for perfect and complete combustion and, as with fuel gases, it would be dangerous to set up a burner for exactly the correct amount of air. If the air were restricted for some reason, incomplete combustion would occur. Practical combustion requires excess air to ensure that there is enough oxygen to come in contact with all of the carbon and hydrogen particles for complete combustion. The air supplied to an oil burner is normally in excess of the exact amount needed. Most oil burners are set up to burn with approximately 50% excess air. When 1 lb of oil with 50% excess air is burned, 21.6 lb or 288 ft³ of air must be furnished with the 1 lb of oil (21.6 lb × 13.33 ft³/lb of standard air = 287.9 ft³ of air). One gallon of No. 2 fuel oil weighs approximately 7 lb, so 1 lb of oil is one-seventh of a gallon; this is 0.88 of a pint, or nearly the contents of a soda can. **Figure 32–28** shows a picture of the process of atomizing oil to mix with air.

When the correct amounts of oil and air are mixed and the mixture is ignited, heat is given off by the appliance for use in heating. The typical oil-burning appliance uses nearly

75% of the heat energy, with approximately 25% escaping with the flue gases. When a gallon of fuel oil contains 140,000 Btu of heat, 105,000 Btu are used for the heating and the remaining 35,000 Btu push the by-products of combustion up the flue ($140,000 \times 0.75 = 105,000$ and $140,000 \times 0.25 = 35,000$).

The by-products of combustion result from the chemical reaction in which the stored energy in the fuel is released as heat. The nitrogen in the air passes through the combustion process and is just heated. Since air contains about 79% nitrogen, 227.5 ft^3 of heated nitrogen rising up the flue ($288 \times 0.79 = 227.5$ ft^3 of nitrogen).

The excess air contains oxygen that also passes through the combustion process and is not combined with the carbon and hydrogen particles of the fuel oil and burned. Of the 288 ft^3 of air, the combustion process uses only 192 ft^3, so 96 ft^3 pass through and are only heated. This air contains 21%, or 20 ft^3, of oxygen ($96 \times 0.21 = 20$ ft^3).

The other by-products of combustion are CO_2 and water vapor. The flue gases contain 27 ft^3 of CO_2 and more than a pound of water vapor per pound of fuel. The flue gases must remain hot enough to prevent the water vapor from condensing and hot enough to rise up the flue to the outside.

32.9   GUN-TYPE OIL BURNERS

The term *gun-type* describes how the oil and air are prepared for burning. Oil and air are forced into the burner head for mixing and ignition. Modern oil burners are referred as high-pressure burners, **Figure 32–29,** because the oil is delivered through the burner at a pressure of about 100 psi. Older oil burners, classified as low-pressure burners, deliver oil at about 10 psi. Although there are still many low-pressure burners in service, new oil burners are all of

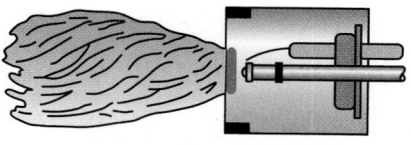
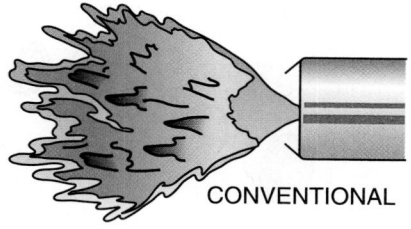

Figure 32–30   Diagram that compares the flames created by conventional and flame-retention head oil burners.

the high-pressure variety. Many newer oil burners utilize a flame-retention design.

The design of flame-retention-type burners is a radical improvement over the conventional burner design that was utilized a few decades ago. Benefits of flame-retention burners include the following:

- Faster motor speeds (3450 rpm versus 1725 rpm)
- Higher static pressure
- Improved air-oil mixing
- Lower emissions
- Higher efficiency
- Cleaner combustion (less soot formation)
- Lower system maintenance requirements
- Production of a hotter flame (300°F–500°F hotter)

As part of the improved burner design, the air-oil mixture is swirled at high pressure to produce a tightly formed flame. **Figure 32–30** compares a conventional burner flame and a flame-retention burner flame. The flame created by a conventional oil burner is erratic, so there are large temperature fluctuations in the flame and it is common for unburned fuel to remain behind in the combustion area. The flame-retention flame is tighter and has a more uniform temperature. Since the flame is tight and uniform, there is substantially less unburned fuel left behind.

The main parts of a gun-type oil burner are the burner motor, blower or fan wheel, pump, nozzle, air tube, electrodes, transformer/igniter, and primary controls, **Figure 32–31.**

Burner Motor

The oil burner *motor,* **Figure 32–32,** provides power for both the fan and the fuel pump. A flexible coupling, **Figure 32–33,** is used to connect the shaft of the motor to the shaft of the pump, **Figure 32–34.** The motor speeds may be either 1750 or 3450 rpm. The fuel pump should always match the motor rpm. Most flame-retention burners have motors that operate at 3450 rpm, while most conventional burners are equipped with motors that operate at 1725 rpm.

Figure 32–29   High-pressure oil burner. *Courtesy R. W. Beckett Corp.*

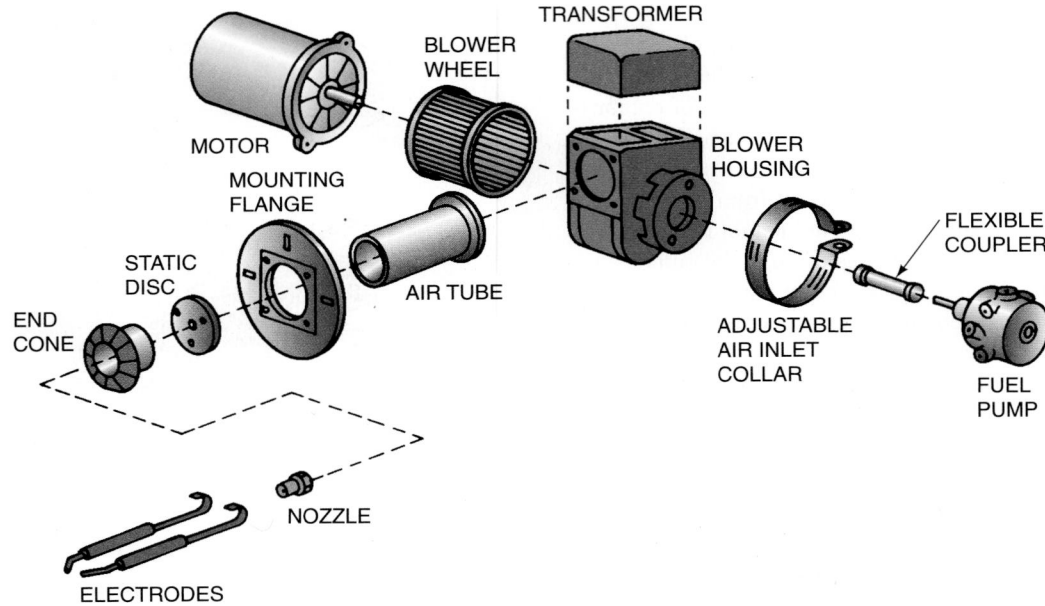

Figure 32–31   Exploded view of an oil burner assembly.

Figure 32–32   An oil burner motor. *Courtesy R. W. Beckett Corp.*

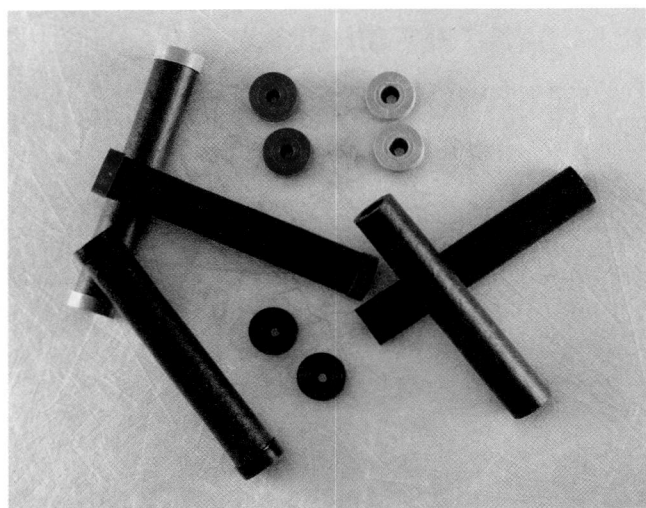

Figure 32–33   Flexible couplers.

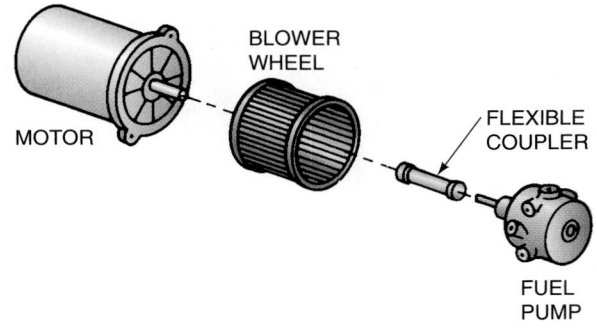

Figure 32–34   Oil burner motor, pump, connector, and blower assembly.

The most common type of motor found on modern oil burners is the permanent split-capacitor (PSC) motor. The PSC motor is manufactured and designed to operate with a start winding, a run winding, and a run capacitor. By design, both the start and run windings are energized whenever the motor is operating. These motors have relatively low starting torque but, since they are operating under a relatively low load, they are perfect for this application. The main benefits of this type of motor are that they have excellent running

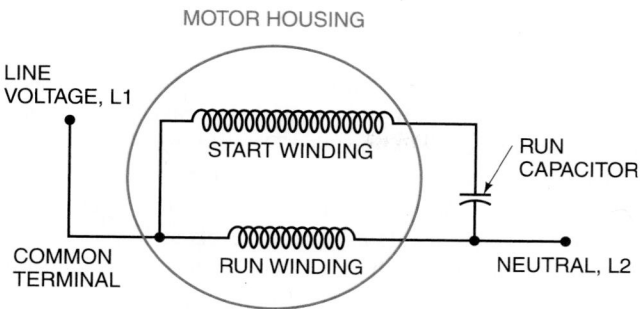

Figure 32–35   Simplified wiring diagram of a single-speed permanent split-capacitor (PSC) motor.

efficiency and have no external or mechanical control relays or contacts to wear out. The wiring diagram for a simple, single-speed PSC motor is shown in **Figure 32–35**. Refer to Unit 17, "Types of Electric Motors," for more information regarding electric motors.

Burner Fan or Blower

The burner *fan* or blower is a squirrel-cage type with adjustable air inlet openings in a collar attached to the blower housing. This provides a means for regulating the volume of air being drawn into the blower. The fan forces air through the air tube to the combustion chamber where it is mixed with the atomized fuel oil to provide the necessary oxygen to support the combustion, **Figure 32–36**.

Fuel Oil Pumps

Just like the burner fan or blower, the fuel pump, **Figure 32–37**, is connected to the shaft of the burner motor and turns at the same speed as the motor itself. The pump performs three major functions in the system. First, the pump moves the fuel oil from the storage tank to the burner for ignition. Second, the pump acts to filter or pulverize particulate matter that may

have found its way past the filter. Third, the pump regulates the pressure at which the oil is fed through the burner into the combustion area. Oil pumps are typically set at the factory to deliver oil at a pressure of 100 psig. This is the same pressure at which oil nozzles are rated.

There are two main types of oil pumps used on gun-type burners. These two types are the single-stage pump and the two-stage pump, **Figure 32–38**. Fuel pumps are manufactured with gear sets that facilitate the movement of oil through the system, **Figure 32–39**. Fuel pumps that have one set of gears to pull oil from the storage tank are referred to as one-stage, or single-stage, pumps, **Figure 32–40**. Single-stage pumps

Figure 32–37   An oil pump. *Courtesy R. W. Beckett Corp.*

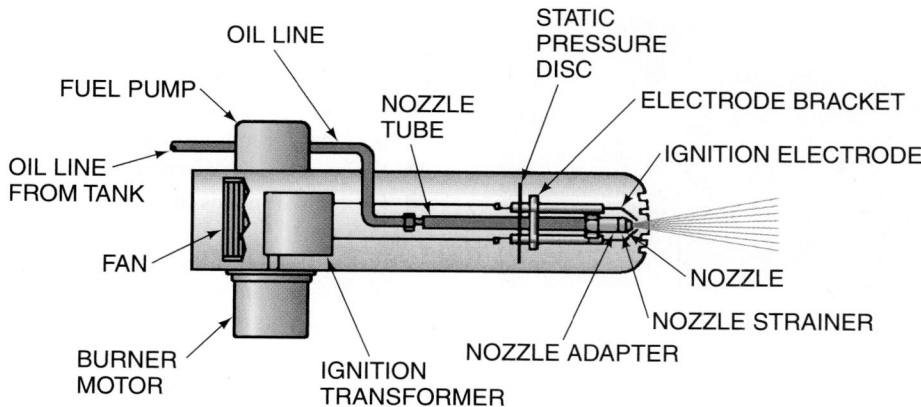

Figure 32–36   Top view diagram of a typical gun-type oil burner. *Courtesy Honeywell*

(A)

(B)

Figure 32–38 **(A)** A single-stage fuel oil pump. **(B)** A two-stage fuel oil pump. **(A)** *Courtesy Webster Electric Company.* **(B)** *Courtesy Suntec Industries Incorporated, Rockford, Illinois*

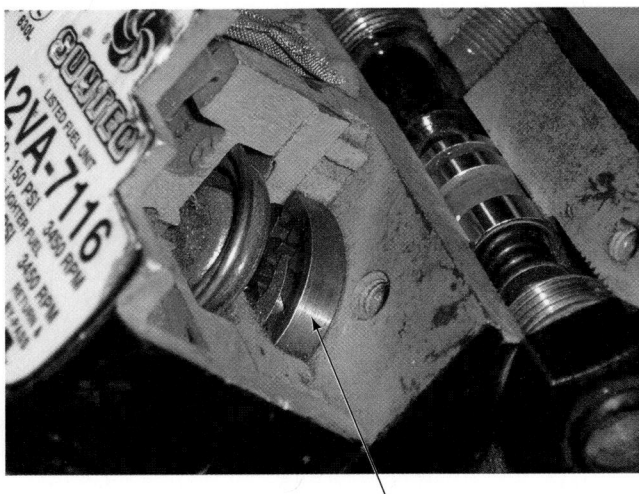

GEAR SET

Figure 32–39 Gear set on an oil pump. *Courtesy Oilheat Associates. Photo by John Levey*

can be used in conjunction with one-pipe or two-pipe systems as long as the vacuum necessary for operation does not exceed 12 in. Two-stage pumps, **Figure 32–41,** have two sets of gears: one set to pull oil from the oil storage tank and one set to push oil onto the nozzle. The additional gear set is called the suction pump gear and allows for a greater vacuum to lift the oil further than a single-stage gear set can. Two-stage pumps can operate at vacuum levels as high as 17 in. Two-stage pumps are highly recommended for high-lift applications. Refer to the next section for more information regarding pump pressures and vacuums.

Oil pumps are equipped with a bypass plug, **Figure 32–42.** This plug will be either left in place or removed, depending on the type of piping configuration being used. For a one-pipe installation, **Figure 32–11,** the bypass plug needs to be removed. This is so that any unused oil can be recirculated back to the inlet of the pump. When the piping configuration is a two-pipe setup, **Figure 32–12,** or if the system is piped with a deaerator, the bypass plug is left in place.

Built into each pump is a pressure-regulating valve, **Figure 32–40.** The pump provides excessive pressure. The pressure-regulating valve can be set so that the fuel oil being delivered to the nozzle is under a set pressure. Typically the nozzle pressure is 100 psig. There are occasions where higher nozzle pressures are used. Higher nozzle pressures will yield more oil flow for the same size nozzle. **Figure 32–43** is a chart that shows nozzle capacities at various pressures. As mentioned previously, oil is also delivered in greater quantities than the nozzle can handle. This excess oil is diverted back to the inlet side of the pump in one-pipe systems and back to the storage tank in two-pipe systems.

NOTE: Fuel pumps are equipped with screens, **Figure 32–44,** at the inlet to prevent particulate matter from entering the pump. Before replacing what is thought to be a defective pump, be sure to check this screen. If the screen is indeed clogged, be sure to replace it—as well as the gasket, **Figure 32–45.**

PUMP PRESSURES. Oil pumps generate a vacuum during the process of pulling oil from the storage tank. This vacuum is measured in inches of mercury. Field technicians can estimate the expected vacuum by using the following guidelines:

- Estimate 1 inch of vacuum for each foot of vertical lift between the oil tank and the oil burner.
- Estimate 1 inch of vacuum for every 10 feet of horizontal run between the oil tank and the oil burner.
- Estimate 1 inch of vacuum for each oil filter.
- Estimate 3 inches of vacuum for an oil safety valve (OSV).

So, a system that has the oil burner 5 ft above the tank at a distance of 40 ft will have an approximate vacuum at the oil burner of 9 in. of mercury, not including the filters and other components. Adding an oil filter and an OSV valve will increase the expected vacuum to about 13 in. of mercury.

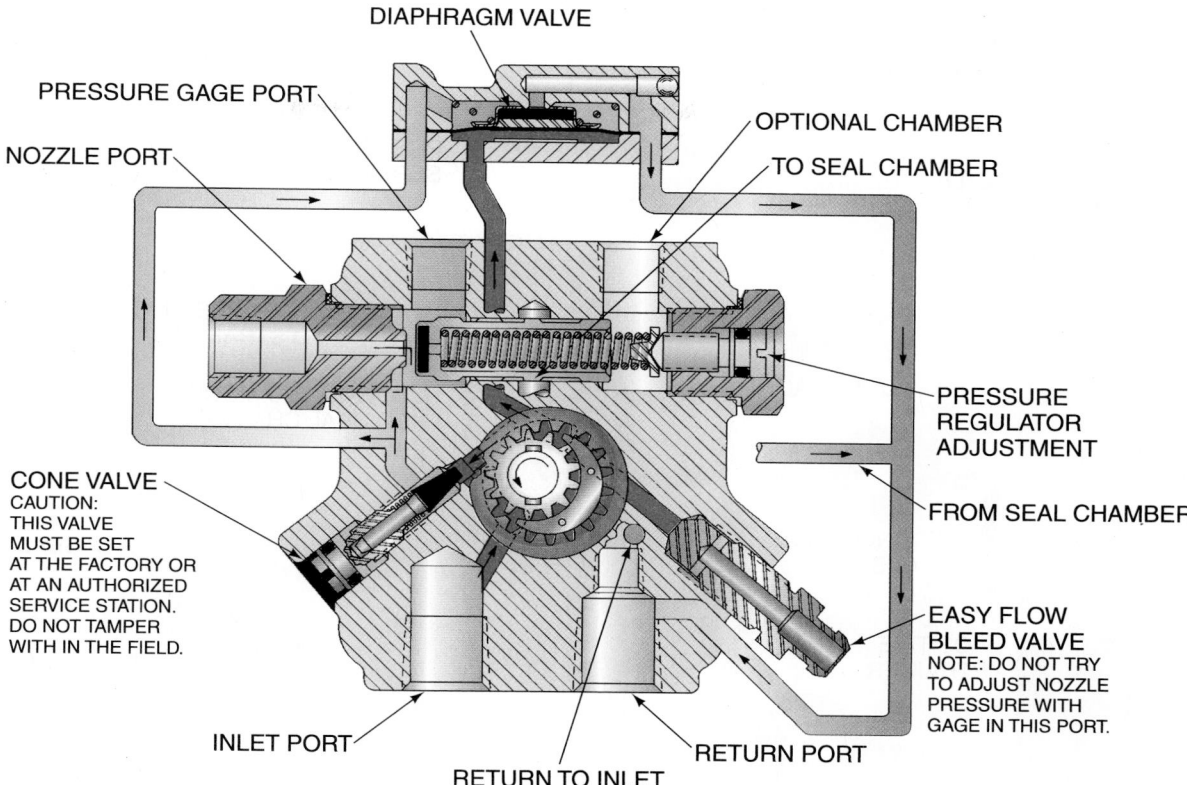

Figure 32–40   A single-stage fuel oil pump showing the flow of oil. *Courtesy Suntec Industries Incorporated, Rockford, Illinois*

Excessive vacuum at the inlet of the pump can cause cavitation (oil changing to a vapor) and oil foaming. This will have a negative effect on system operation and efficiency. During normal system operation when the pump is operating within its design vacuum range, there should be no bubbling or foaming of the oil, **Figure 32–46(A).** As the vacuum level in the oil supply line increases, the amount of foaming at the inlet of the pump increases and the amount of oil that is being pumped decreases, **Figure 32–46(B).** At extremely high vacuums, the oil is nearly completely foam upon entering the pump, **Figure 32–46(C).** NOTE: If there is an exceptionally high vacuum at the inlet of the pump, the technician should check for crimped or kinked supply oil lines or a dirty oil filter.

Nozzle

The *nozzle* prepares the fuel oil for combustion by atomizing, metering, and patterning the fuel oil. The process of atomization is best described as when the fuel oil is broken down into tiny droplets. To give some idea of how small these droplets are, one gallon of fuel oil, when atomized, creates approximately 55,000,000,000 (55 billion) droplets of oil. By atomizing the oil, the surface area of the oil is increased, leading to its rapid vaporization and ignition. It should be noted that fuel oil will not burn in the liquid state, so atomization is necessary in order for the combustion process to occur. The

smallest of the atomized droplets will be ignited first. The larger droplets (there are more of these) provide more heat transfer to the heat exchanger when they are ignited. The atomization of the fuel oil is a complex process. The straight lateral movement of the flowing fuel oil must be changed to a circular motion. This is accomplished in the swirl chamber of the nozzle. **Figure 32–47** shows a cutaway view of a nozzle—as well as a diagram showing some of the parts, including the swirl chamber, that make up this precision element of the oil burner.

The orifice, or bore size, of the nozzle is designed to allow a certain amount of fuel through at a given pressure to produce the amount of heat desired in Btu/h. Each nozzle is marked as to the amount of fuel it will deliver. A nozzle marked 1.00 will deliver 1 gallon of fuel oil per hour (gph) if the input pressure is 100 psi. With No. 2 fuel oil at a temperature of 60°F, approximately 140,000 Btu would be produced each hour. A 0.8 nozzle would deliver 0.8 gph, and so on. By increasing the pressure of the fuel oil, the quantity of oil that the nozzle allows to flow through it will increase as well. For example, if you refer to **Figure 32–43** you can see that a nozzle rated to deliver 0.75 gallons of fuel per hour at 100 psig will deliver about 0.92 gallons per hour if the oil pressure is increased to 150 psig.

The nozzle must be designed so that the spray will ignite smoothly and provide a steady, quiet fire that will burn cleanly and efficiently. It must provide a uniform spray

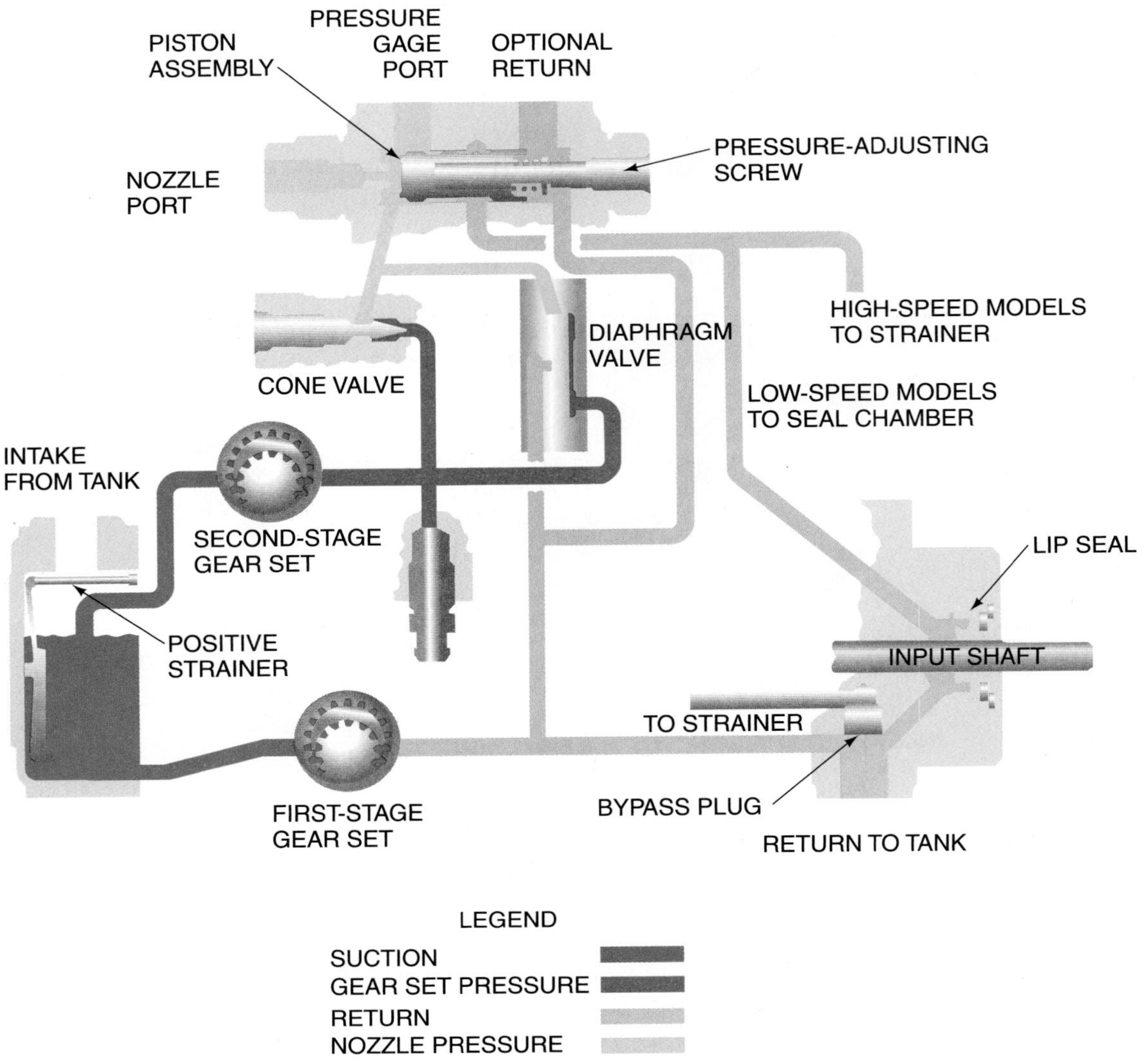

PISTON ASSEMBLY

PRESSURE GAGE PORT

OPTIONAL RETURN

NOZZLE PORT

PRESSURE-ADJUSTING SCREW

HIGH-SPEED MODELS TO STRAINER

CONE VALVE

DIAPHRAGM VALVE

LOW-SPEED MODELS TO SEAL CHAMBER

INTAKE FROM TANK

SECOND-STAGE GEAR SET

LIP SEAL

POSITIVE STRAINER

INPUT SHAFT

TO STRAINER

FIRST-STAGE GEAR SET

BYPASS PLUG

RETURN TO TANK

LEGEND

SUCTION
GEAR SET PRESSURE
RETURN
NOZZLE PRESSURE

Figure 32–41   A two-stage fuel oil pump. *Courtesy Suntec Industries Incorporated, Rockford, Illinois*

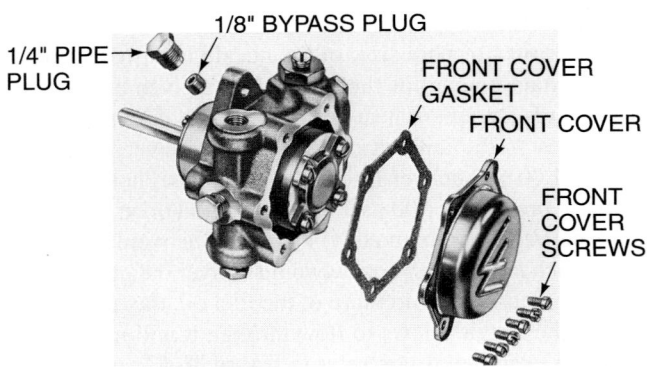

1/4" PIPE PLUG

1/8" BYPASS PLUG

FRONT COVER GASKET

FRONT COVER

FRONT COVER SCREWS

Figure 32–42   The bypass plug location. *Courtesy Webster Electric Company*

pattern and angle that is best suited to the requirements of the specific burner. There are three basic spray patterns—hollow, semihollow, and solid—with angles from 30° to 90°, **Figure 32–48**.

Hollow cone nozzles generally produce a more stable spray angle and pattern than do solid cone nozzles of the same flow rate. These nozzles are often used where the flow rate is under 1 gph.

Solid cone nozzles distribute droplets fairly evenly throughout the pattern. These nozzles are often used in larger burners.

Semihollow cone nozzles are often used in place of the hollow or solid cone nozzles. The higher flow rates tend to produce a more solid spray pattern, and the lower flow rates tend to produce a more hollow spray pattern.

Nozzle manufacturers often offer nozzles with hybrid spray patterns that are a combination of the three just mentioned. **Figure 32–49** shows some different spray patterns that are available.

The angle of the spray should be selected by evaluating the shape of the chamber in which combustion takes place. If the combustion chamber is round or square, a nozzle with a larger spray angle is desired, **Figure 32–50**. If the combustion

NOZZLE CAPACITIES U.S. Gallons per Hour No. 2 Fuel Oil								
Rate gph @ 100 psi	Operating Pressure: pounds per square inch							
	125	140	150	175	200	250	275	300
.40	.45	.47	.49	.53	.56	.63	.66	.69
.50	.56	.59	.61	.66	.71	.79	.83	.87
.60	.67	.71	.74	.79	.85	.95	1.00	1.04
.65	.73	.77	.80	.86	.92	1.03	1.08	1.13
.75	.84	.89	.92	.99	1.06	1.19	1.24	1.30
.85	.95	1.01	1.04	1.13	1.20	1.34	1.41	1.47
.90	1.01	1.07	1.10	1.19	1.27	1.42	1.49	1.56
1.00	1.12	1.18	1.23	1.32	1.41	1.58	1.66	1.73
1.10	1.23	1.30	1.35	1.46	1.56	1.74	1.82	1.91
1.20	1.34	1.42	1.47	1.59	1.70	1.90	1.99	2.08
1.25	1.39	1.48	1.53	1.65	1.77	1.98	2.07	2.17
1.35	1.51	1.60	1.65	1.79	1.91	2.14	2.24	2.34
1.50	1.68	1.77	1.84	1.98	2.12	2.37	2.49	2.60
1.65	1.84	1.95	2.02	2.18	2.33	2.61	2.73	2.86
1.75	1.96	2.07	2.14	2.32	2.48	2.77	2.90	3.03
2.00	2.24	2.37	2.45	2.65	2.83	3.16	3.32	3.46
2.25	2.52	2.66	2.76	2.98	3.18	3.56	3.73	3.90
2.50	2.80	2.96	3.06	3.31	3.54	3.95	4.15	4.33
2.75	3.07	3.25	3.37	3.64	3.90	4.35	4.56	4.76
3.00	3.35	3.55	3.67	3.97	4.24	4.75	4.97	5.20
3.25	3.63	3.85	3.98	4.30	4.60	5.14	5.39	5.63
3.50	3.91	4.14	4.29	4.63	4.95	5.53	5.80	6.06
3.75	4.19	4.44	4.59	4.96	5.30	5.93	6.22	6.50
4.00	4.47	4.73	4.90	5.29	5.66	6.32	6.63	6.93
4.50	5.40	5.32	5.51	5.95	6.36	7.11	7.46	7.79
5.00	5.59	5.92	6.12	6.61	7.07	7.91	8.29	8.66
5.50	6.15	6.51	6.74	7.27	7.78	8.70	9.12	9.53
6.00	6.71	7.10	7.35	7.94	8.49	9.49	9.95	10.39
6.50	7.26	7.69	7.96	8.60	9.19	10.28	10.78	11.26
7.00	7.82	8.28	8.57	9.25	9.90	11.07	11.61	12.12
7.50	8.38	8.87	9.19	9.91	10.61	11.86	12.44	12.99
8.00	8.94	9.47	9.80	10.58	11.31	12.65	13.27	13.86
8.50	9.50	10.06	10.41	11.27	12.02	13.44	14.10	14.72
9.00	10.06	10.65	11.02	11.91	12.73	14.23	14.93	15.59
9.50	10.60	11.24	11.64	12.60	13.44	15.02	15.75	16.45
10.00	11.18	11.83	12.25	13.23	14.14	15.81	16.58	17.32
10.50	11.74	12.42	12.86	13.89	14.85	16.60	17.41	18.19
11.00	12.30	13.02	13.47	14.55	15.56	17.39	18.24	19.05
12.00	13.42	14.20	14.70	15.88	16.97	18.97	19.90	20.79

Figure 32–43   Nozzle flow rates at different pressures.

chamber is deep, a nozzle with a smaller spray angle is desired, **Figure 32–51.**

The high-pressure oil burner nozzle is a precision device that must be handled carefully. SAFETY PRECAUTION: Do not attempt to clean these nozzles. Metal brushes or other cleaning devices will distort or otherwise damage the precision-machined surfaces. When the nozzle is not performing properly, replace it. Use a nozzle changer designed specifically for this purpose to remove and install a nozzle, **Figure 32–52.**

Certain conditions can cause the nozzle to drip after the burner has been shut off. This can cause a rumble at the end

Figure 32–44   Cutaway of an oil pump showing the screen. *Courtesy Oilheat Associates. Photo by John Levey*

(A)

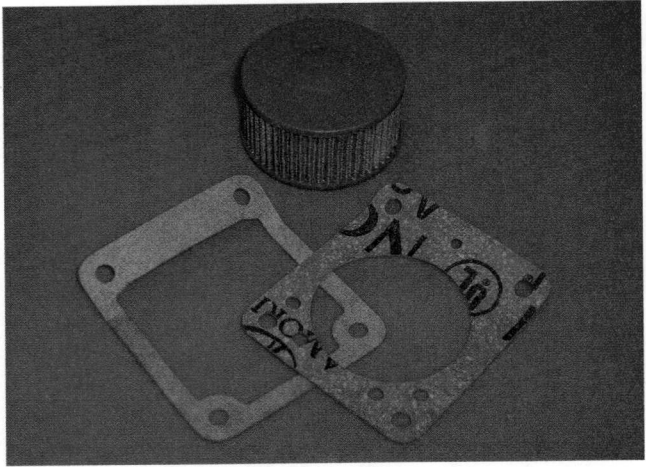

Figure 32–45   Oil pump screen and gaskets. *Courtesy Oilheat Associates. Photo by John Levey*

(B)

of the cycle. This is actually an ignition of the dripped oil. Components within the oil pump should eliminate this, but they sometimes fail to work correctly.

To avoid an afterfire drip at the nozzle, a solenoid-type cut-off valve can be installed. This valve reduces smoking after the burner is shut off, **Figure 32–53.**

Air Tube

Air is blown into the combustion chamber through the *air tube,* **Figure 32–54.** Within this tube the straight or lateral movement of the air is changed to a circular motion. This circular motion is opposite that of the circular motion of the fuel oil pattern leaving the nozzle. The air moving in a circular motion in one direction mixes with the fuel oil moving in a circular motion in the opposite direction in the combustion chamber. The air tube of some burners contains a stationary

(C)

Figure 32–46   **(A)** No foaming at normal vacuum level. **(B)** Foaming begins to occur as vacuum level increases. **(C)** High vacuum levels at the inlet of the oil pump cause excessive oil foaming. *Courtesy Oilheat Associates. Photos by John Levey*

(A)

Figure 32–47 An oil burner nozzle. *Courtesy Delavan Corporation*

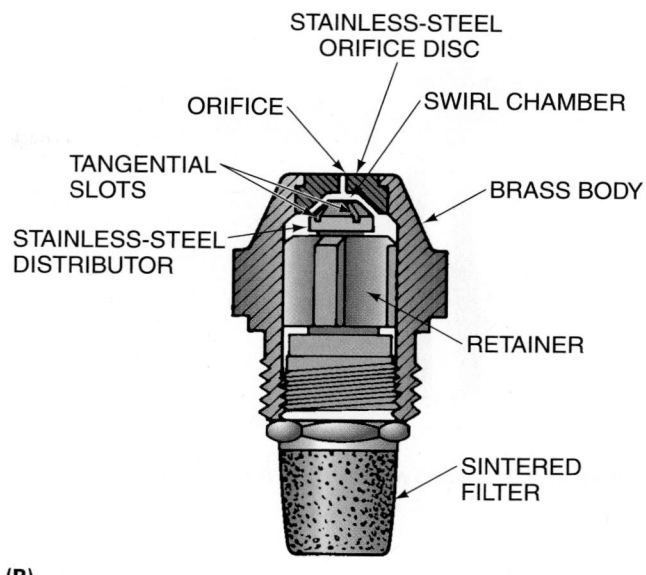

STAINLESS-STEEL
ORIFICE DISC

ORIFICE — SWIRL CHAMBER

TANGENTIAL
SLOTS

STAINLESS-STEEL
DISTRIBUTOR

BRASS BODY

RETAINER

SINTERED
FILTER

(B)

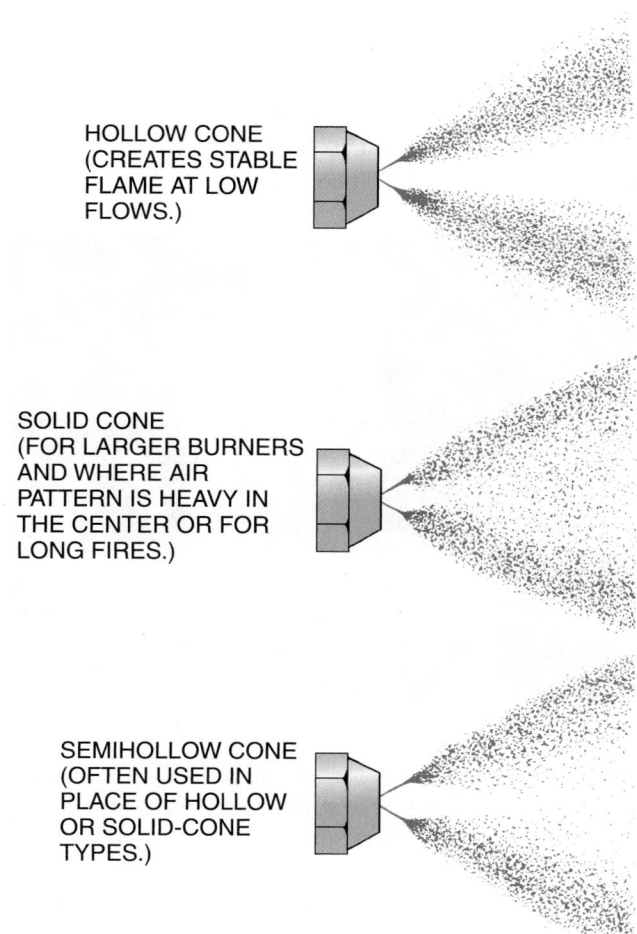

HOLLOW CONE
(CREATES STABLE
FLAME AT LOW
FLOWS.)

SOLID CONE
(FOR LARGER BURNERS
AND WHERE AIR
PATTERN IS HEAVY IN
THE CENTER OR FOR
LONG FIRES.)

SEMIHOLLOW CONE
(OFTEN USED IN
PLACE OF HOLLOW
OR SOLID-CONE
TYPES.)

Figure 32–48 Nozzle spray patterns. *Courtesy Delavan Corporation*

SPRAY PATTERNS

B   H   ES   SS   P

Hago Manufacturing Co., Inc.
1120 Globe Avenue   Mountainside, NJ USA 07092
Tel (908) 232-8687   Fax (908) 232-7246
E-Mail: sales@hagonozzles.com
www.hagonozzles.com

Figure 32–49 Additional spray patterns. *Courtesy Delavan Corporation*

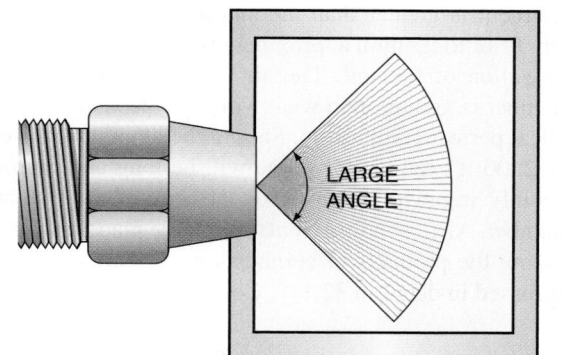

LARGE
ANGLE

Figure 32–50 Large spray angle used for round, square, and more shallow combustion chambers.

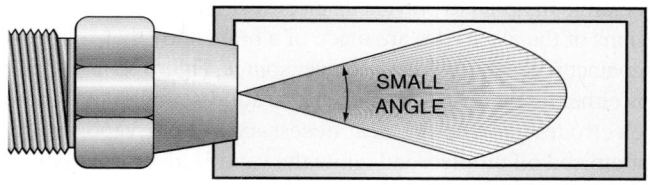

SMALL
ANGLE

Figure 32–51 Small spray angle used on deeper combustion chambers.

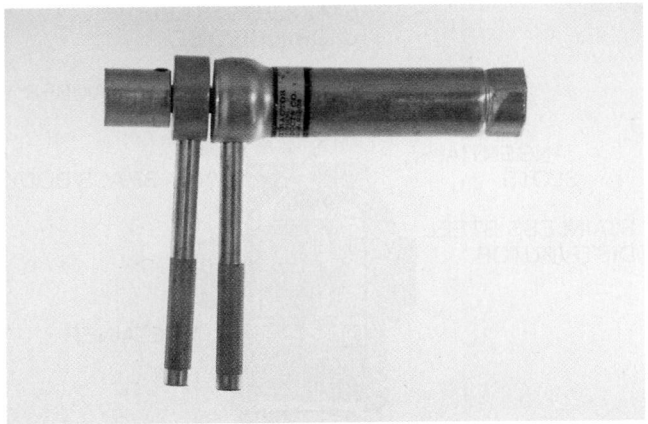

Figure 32–52   A nozzle wrench. *Photo by Bill Johnson*

disc, **Figure 32–55,** referred to as an end cone. This end cone increases the static air pressure and reduces the air volume. The increase in static pressure causes an increase in air velocity to create a better air-oil mixture. The circular motion is achieved by vanes located in the air tube after the nozzle. The head of the air tube chokes down the air and the fuel oil mixture, causing a higher velocity **Figure 32–56.** These end cones are often referred to as flame-retention rings, as they are designed to retain the flame on the end of the air tube. This creates more turbulence within the air-oil mixture, resulting in greater burning efficiency. It also provides for more complete burning once the atomized air-oil mixture is ignited. It is in the air tube that the atomized oil from the nozzle mixes with the air. The nozzle, therefore, is positioned inside the air tube, **Figure 32–57.**

The amount of air required for the fuel oil preignition treatment is greater than the amount required for combustion. Prior to ignition approximately 2000 ft^3 of air is needed per gallon of fuel oil. The air blown into the combustion chamber is greater than what would be needed for the theoretical perfect combustion. Should the amount of air exceed the 2000 ft^3, the burner flame will be long and narrow and possibly impinge on or strike the rear of the combustion chamber. Airflow adjustments should be made only after analyzing the products of combustion, the flue gases. This is discussed in detail in 32.17, "Combustion Efficiency."

Electrodes

Two electrodes are positioned close to the orifice of the nozzle, **Figure 32–58.** The electrodes are metal rods insulated with ceramic material to prevent an electrical ground. The rear portions of the electrodes are made of a brass alloy that comes in contact with a high-voltage power source, **Figure 32–59,** which is either an ignition transformer or a solid-state igniter. These electrodes provide the heat necessary to both vaporize the atomized oil droplets and ignite the vaporized air-oil mixture. A high-voltage spark is generated between the electrodes to facilitate these processes. The exact position of the electrode tips varies from manufacturer to manufacturer, so be sure to

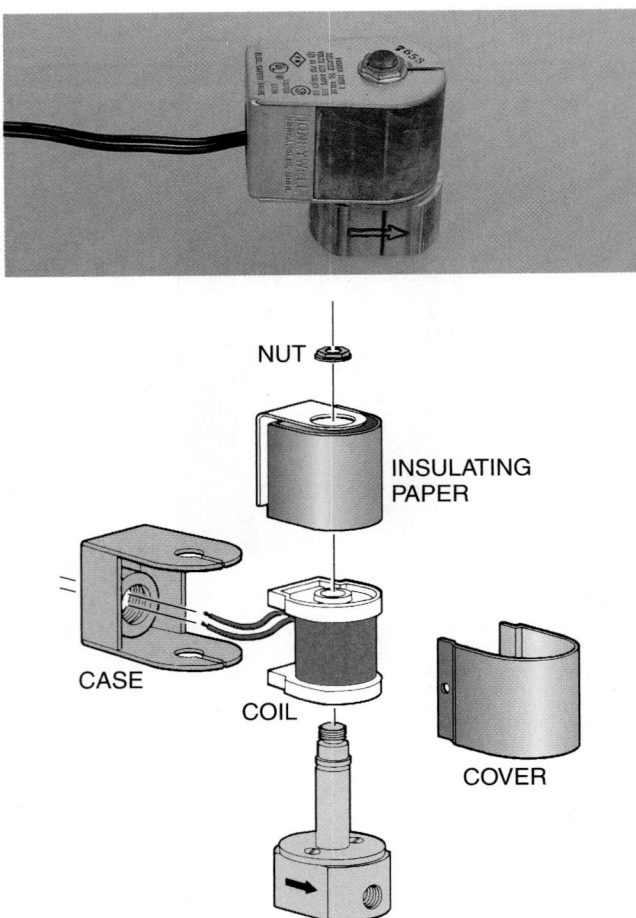

(A)

(B)

Figure 32–53   (A) Solenoid valve provides positive fuel oil cut-off to help prevent afterdrip on the burner. **(B)** Solenoid installed on an oil burner. **(A)** *Photo by Bill Johnson.* **(B)** *Courtesy Oilheat Associates. Photo by John Levey*

check the literature that accompanies the oil burner for specific information regarding adjusting the electrodes. Refer to Section 32.16, "Service Procedures," for more information

AIR TUBE

Figure 32–54   Air tube on an oil burner. *Courtesy Oilheat Associates. Photo by John Levey*

Figure 32–55   End cone on a flame-retention oil burner. *Courtesy Oilheat Associates. Photo by John Levey*

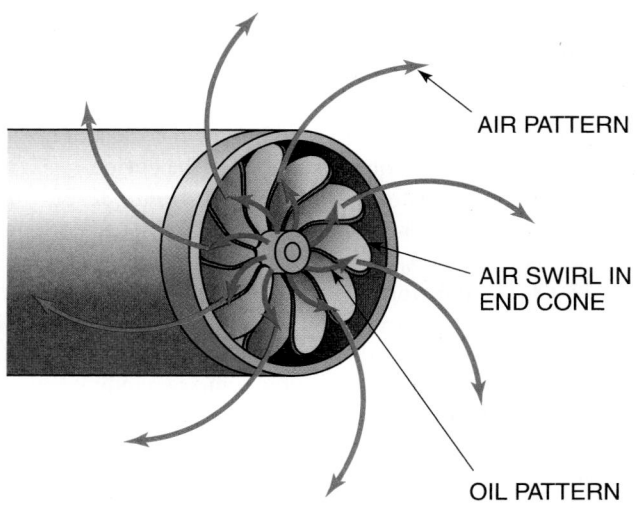

AIR PATTERN

AIR SWIRL IN
END CONE

OIL PATTERN

Figure 32–56   The end cone swirls the air into the swirling oil to achieve the best air-oil mixture.

Figure 32–57   Hollow air tube showing placement of nozzle and electrodes in the tube. Hollow tube is for illustrative purposes only. *Courtesy Oilheat Associates. Photo by John Levey*

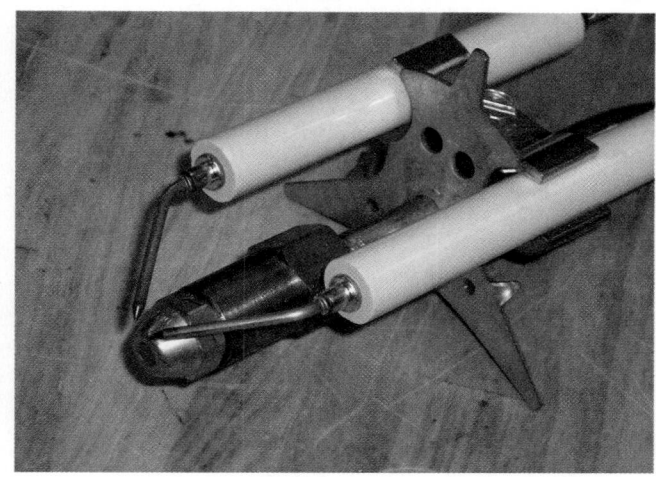

Figure 32–58   Electrode location with respect to the nozzle. *Courtesy Oilheat Associates. Photo by John Levey*

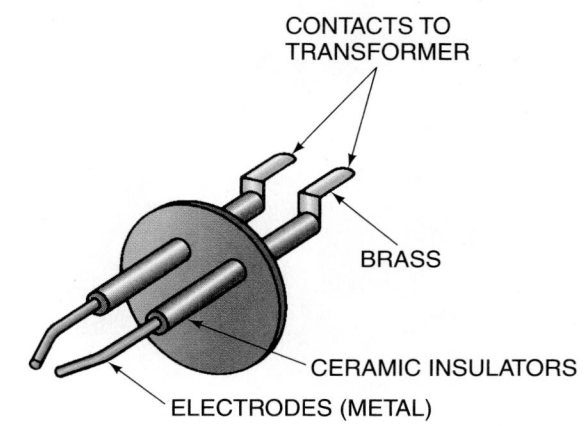

CONTACTS TO
TRANSFORMER

BRASS

CERAMIC INSULATORS

ELECTRODES (METAL)

Figure 32–59   An electrode assembly.

regarding electrode positioning. **Figure 32–57** shows the nozzle assembly and the electrodes in position within the air tube.

The design of the ignition system determines the length of time that the electrodes will provide a spark. The two most commonly encountered ignition strategies are called *interrupted ignition* and *intermittent ignition*. Interrupted ignition employs an electrode spark for a brief period at the beginning of the burner's run cycle. This spark is intended to facilitate the lighting of the burner; then, once the burner flame has been established, the spark stops. On systems with intermittent ignition, the spark is generated continuously during the burner's run cycle. Intermittent ignition used to be referred to as constant ignition, but it isn't any more. Of the two types of ignition, the interrupted type is a better option for a number of reasons. Here are a few:

- Because the spark is only generated for a short period at the beginning of the cycle, transformer/igniter life is extended.
- The elimination of the constant spark reduces burner operation noise.
- Combustion problems become more evident.
- The useful life of the electrodes is lengthened because sparks are generated for shorter periods of time.

Ignition Transformer

Some oil burners use a *step-up transformer* to provide high voltage to the electrodes, which produce the spark for ignition. This step-up ignition transformer is located on top of the oil burner, **Figure 32–60**. Step-up transformers are constructed with more turns in the secondary winding than in the primary, **Figure 32–61**. Notice that the size of the secondary winding is substantially larger than the primary winding. The ignition transformer is supplied with 115 volts at its primary winding; the voltage output of the ignition transformer, at the secondary winding, is between 10,000 volts and 14,000 volts.

IGNITION TRANSFORMER

Figure 32–60 Ignition transformer on an oil burner. *Courtesy Amercian Oil Burner Corporation*

PRIMARY WINDING   SECONDARY WINDING

Figure 32–61 Clear transformer shows the size difference between the primary and the secondary windings on the ignition transformer. *Courtesy Oilheat Associates. Photo by John Levey*

SPRINGS

Figure 32–62 Springs on the secondary (high-voltage) winding of the ignition transformer. *Photo by Bill Johnson*

The voltage at the secondary winding of the transformer is directly related to the voltage supplied to the device.

The ignition transformer has springs, **Figure 32–62,** that allow the secondary winding of the transformer to come in contact with the back, or flat portion, of the electrodes, **Figure 32–63**. The transformer is hinged, so the springs and the back end of the electrode rods can be accessed. Ignition transformers are not intended to be serviced in the field. When there is voltage present at the primary winding and no voltage present at the secondary winding, the ignition transformer needs to be replaced.

Service technicians can test the operation of the ignition transformer by using a tool designed for such purposes, **Figure 32–64**. The ignition transformer tester is connected to the primary winding of the ignition transformer. The display on

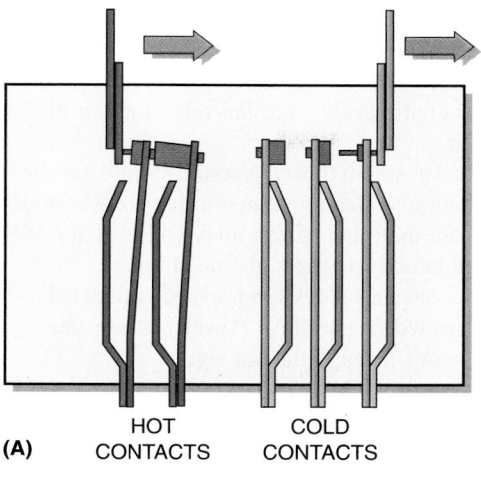

(A)

HOT
CONTACTS

COLD
CONTACTS

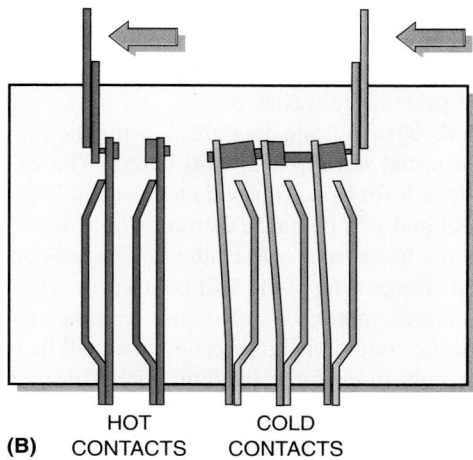

(B)

HOT
CONTACTS

COLD
CONTACTS

Figure 32–70 **(A)** The hot contacts are closed and the cold contacts are open. High temperature is being sensed in the flue pipe. **(B)** The hot contacts are open and the cold contacts are closed. Low temperature is being sensed in the flue pipe.

Figure 32–71 A cad cell relay. *Photo by Bill Johnson*

Figure 32–72 A cad cell. *Photo by Bill Johnson*

FACE OF THE CAD CELL

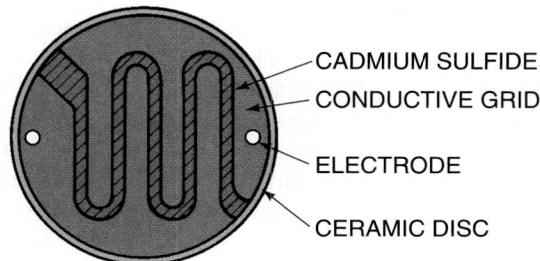

CADMIUM SULFIDE

CONDUCTIVE GRID

ELECTRODE

CERAMIC DISC

THE CAD CELL IS SEALED IN GLASS.

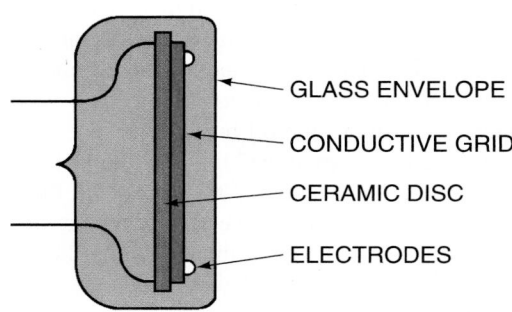

GLASS ENVELOPE

CONDUCTIVE GRID

CERAMIC DISC

ELECTRODES

Figure 32–73 A diagram of the face and side of a cad cell.

CAD CELL LOCATION IS A CRITICAL
FACTOR IN PERFORMANCE.

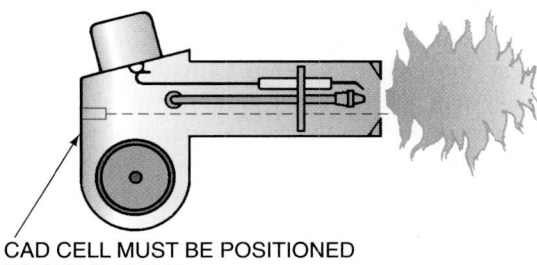

CAD CELL MUST BE POSITIONED
TO SIGHT FLAME.

1. CELL REQUIRES A DIRECT VIEW OF FLAME.

2. ADEQUATE LIGHT FROM THE FLAME MUST REACH THE CELL TO LOWER ITS RESISTANCE SUFFICIENTLY.

3. CELL MUST BE PROTECTED FROM EXTERNAL LIGHT.

4. AMBIENT TEMPERATURE MUST BE UNDER 140°F.

5. LOCATION MUST PROVIDE ADEQUATE CLEARANCE. METAL SURFACES MUST NOT AFFECT CELL BY MOVEMENT, SHIELDING, OR RADIATION.

Figure 32–74 Typical cad cell location in an oil burner.

32.10 OIL FURNACE WIRING DIAGRAMS

Figure 32–75 is a wiring diagram for a typical forced-warm-air oil furnace. This diagram is designed to be used with an air-conditioning system as well. It includes wiring for the fan circuit for the air distribution, the oil burner primary control circuit, and the 24-V control circuit. Note that this diagram illustrates the circuitry for a multispeed blower motor.

Figure 32–76 illustrates the wiring for the blower fan motor operation only. The fan relay [A] protects the blower motor from current flowing in the high- and low-speed circuits at the same time. If the normally open (NO) contacts on the high-speed circuit close, the NC contacts on the low-speed circuit will open. The NO fan limit switch [B] is operated by a temperature-sensing bimetal device. When the temperature at the heat exchanger reaches approximately 140°F, these contacts close, providing current to the low-speed terminal on the fan motor. The high-speed circuit can be activated only from the thermostat. The high speed on the fan motor is *not* used for heating. It will operate only when the thermostat is set to FAN ON manually or for cooling (air conditioning).

The fan circuit for heating would then be from the power source (hot leg) through the temperature-actuated fan limit switch, to the NC fan relay, to the low-speed fan motor terminal, and back through the neutral wire.

Figure 32–77 shows the oil burner wiring diagram. The burner is wired through an NC limit switch [A]. This limit switch is temperature actuated to protect the furnace and the building. If there is an excessive temperature buildup from the furnace overheating, this switch will open, causing the burner to shut down. This overheating could result from the burner being overfired, from a fan problem, from airflow restriction, or from similar problems. The wiring can then be followed from the limit switch to the primary control [B].

The black wire and orange wire at the primary pass the same current, but the orange is wired through the 24-V primary high-voltage contacts [C]. When the thermostat calls for heat, these contacts close, providing current to the ignition system, burner motor, and fuel valve, if one is used. The return is through the neutral (white wire) to the power source. The transformer in the primary control reduces the 115-V line voltage to 24 V for the thermostat control circuit.

If there is a problem in the start-up of the burner, the safety device (stack switch or cad cell) will shut down the burner. **Figure 32–78** illustrates the wiring for the 24-V thermostat control circuit. The letter designations on the diagram indicate what the wire colors would normally be: R (red), "hot" leg from the transformer; W (white), heat; Y (yellow), cooling; G (green), manual fan relay.

It will occasionally be necessary to convert a system from one designed for heating only, to one that will also accommodate air conditioning. A heating-cooling system requires a larger 24-V control transformer than one normally supplied with a heating-only system. To make this conversion, a 40-VA transformer is required in addition to the one supplied in the normal primary control circuit. When adding a cooling package, the larger 40-VA transformer is required because of the added cooling relay and fan-relay load in the 24-V control circuit.

The thermostat in this conversion should also have an isolating subbase so that one transformer is not connected electrically with the other. A fan relay [A] is also needed in the fan circuit to activate the high-speed fan.

Power-consuming devices must be connected to both legs (i.e., connected in parallel). Power-passing devices will be wired in series through the hot leg.

32.11 STACK SWITCH SAFETY CONTROL

The bimetal element of the stack switch is positioned in the flue pipe. When the bimetal element is heated, proving that there is ignition, it expands and pushes the drive shaft in the direction of the arrow, **Figure 32–79**. This closes the hot contacts and opens the cold contacts.

Figure 32–80 is a wiring diagram showing the current flow during the initial start-up of the oil burner. The 24-V room thermostat calls for heat. This will energize the 1K coil, closing the 1K1 and 1K2 contacts. Current will flow through the safety switch heater and cold contacts. The hot contacts remain open. The closing of the 1K1 contacts provides current to the oil burner motor, oil valve, and ignition transformer. Under normal conditions ignition and heat will be produced from the combustion of the air-oil mixture. This will provide heat to the stack switch in the flue pipe, causing the hot contacts to close and the cold contacts to open. The current flow then is shown in **Figure 32–81**. The circuit is completed through the 1K2 contact, the hot contacts to the 1K coil. The 1K1 contact remains closed, and the furnace continues to run as in a normal safe start-up. The safety switch heater is no longer in the circuit.

If there is no ignition and consequently no heat in the stack, the safety switch heater remains in the circuit. In approximately 90 sec the safety switch heater opens the safety switch, which causes the burner to shut down. To start the cycle again, depress the manual reset button. Allow approximately 2 min for the safety switch heater to cool, before attempting to restart.

32.12 CAD CELL SAFETY CONTROL

In a standard modern primary circuit, the cad cell discussed earlier may be coupled with a triac, which is a form of a solid-state device designed to pass current when the cad cell circuit resistance is high (no flame). When the burner fires, the cad cell senses light and its resistance drops. This causes the triac to open, stopping current flow through it. When the burner is not firing, the resistance of the cad cell is high and the triac is essentially closed, allowing current to pass through the triac.

The circuit in **Figure 32–82** shows how the triac and the cad cell are wired in the primary control. The red lines indicate the electrical paths through which electric current is

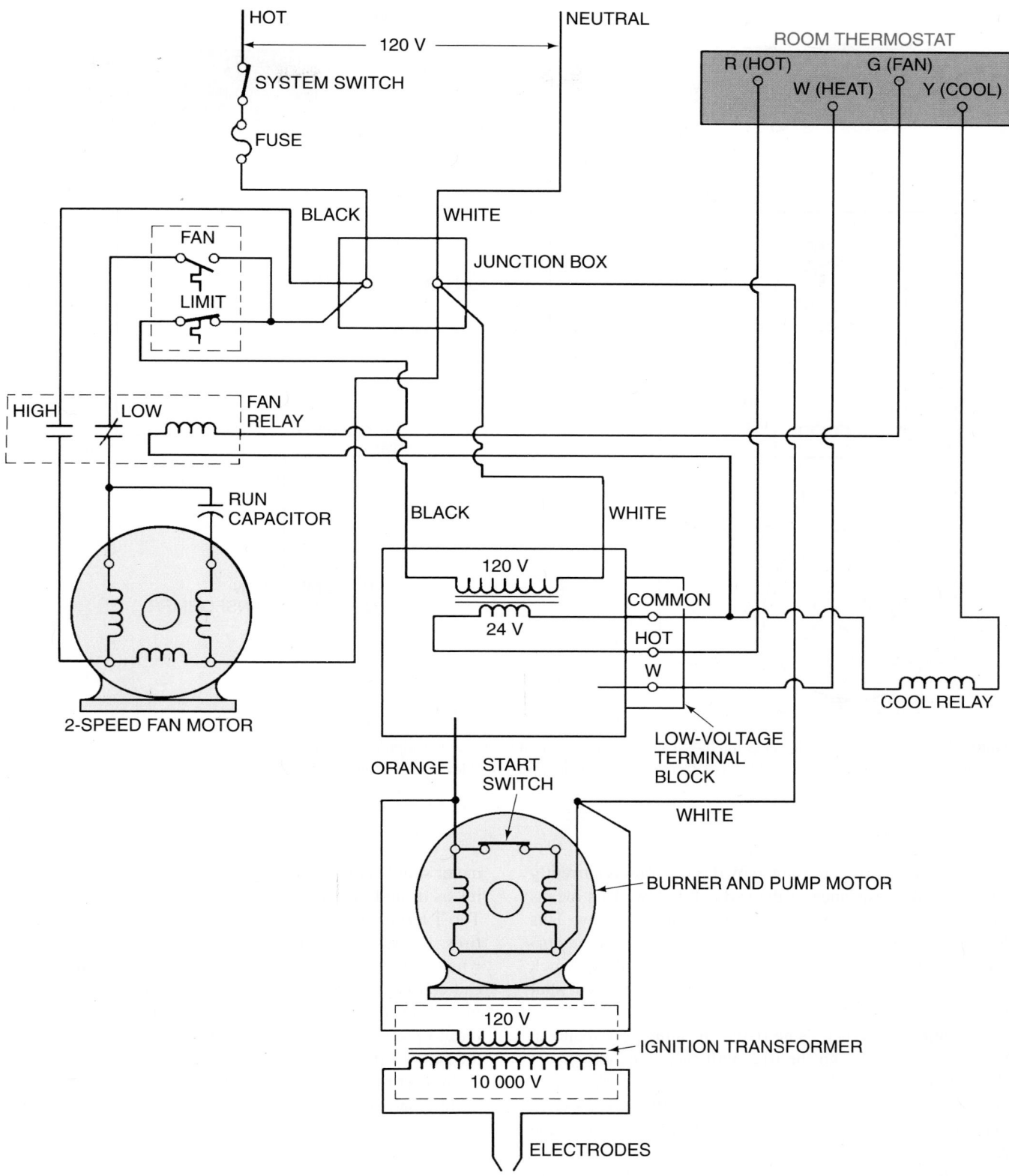

Figure 32–75 The primary control has a hot wire (black) and a neutral wire (white) feeding into it. When there is a call for heat from the thermostat, the orange wire is energized. This starts the burner motor and the sparking at the electrodes.

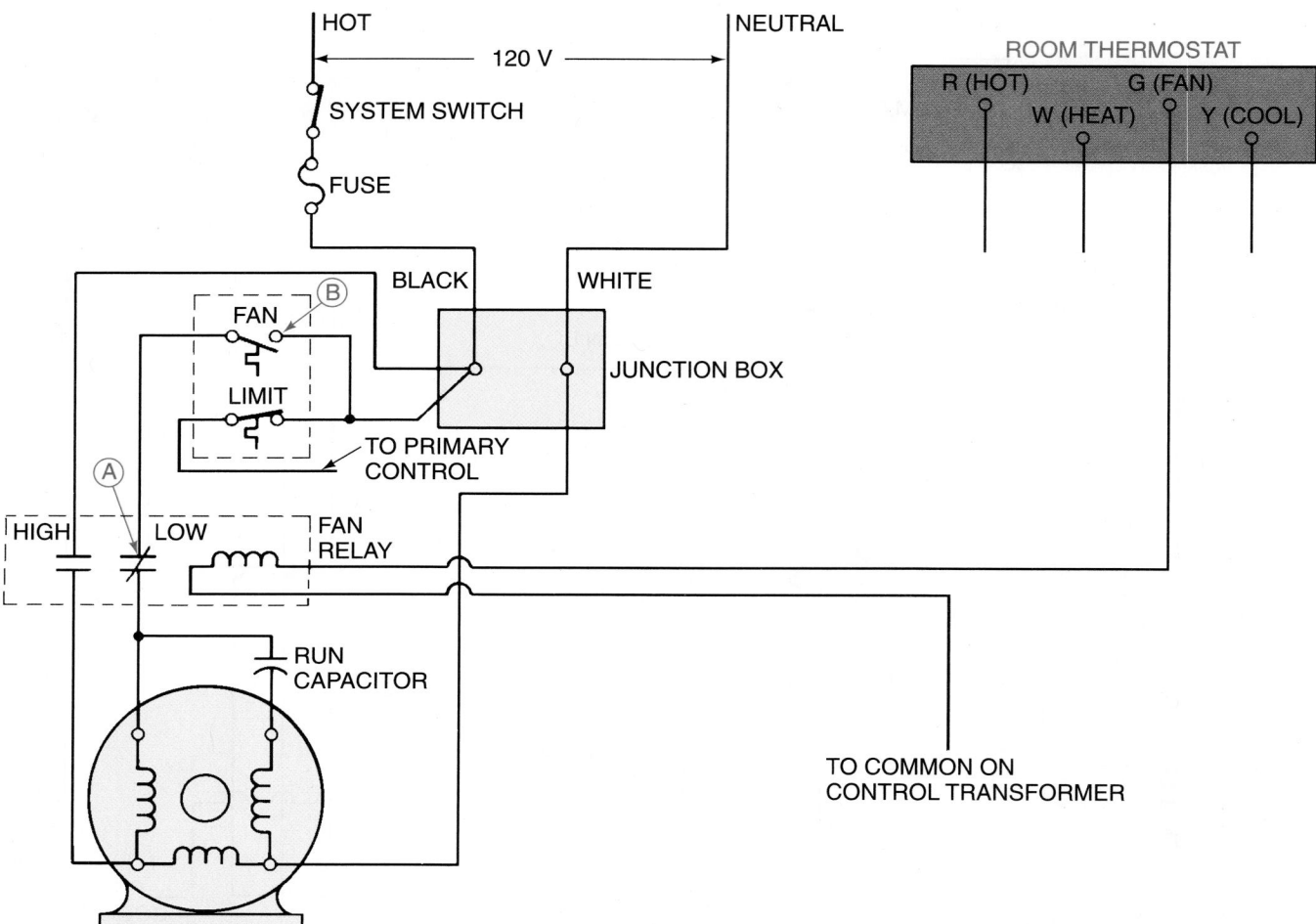

Figure 32–76 The furnace blower or fan runs using the thermally (temperature) activated fan switch in the heating cycle. If the coil on the fan relay were to be energized while in the HEAT mode, the fan would switch from low speed to high speed.

flowing. The burner represented by this diagram is currently firing and the resistance of the cad cell is low. This means that the resistance between the two F terminals is low. This low resistance causes the triac to stop passing electric current through the safety switch heater. If the safety switch heater remains in the circuit too long, the burner will shut down.

In **Figure 32–83** there is no flame and a high resistance in the cad cell. By design the triac will then conduct current that will pass through the safety switch heater. If this current continues through the heater, it will open the safety switch and cause the burner to shut down. To recycle, press the reset button.

The technician must be able to check the cad cell. It connects to the primary control on the outside at terminals F and F. To check the cad cell for its resistance reading, disconnect the two wires at the F and F terminal block; they are yellow. With the furnace off, it should be dark inside the combustion chamber. Set the ohmmeter to check out 1000 Ω. Calibrate the meter to 0 Ω with the meter leads touching each other. Now place the meter leads on the two yellow wires. The

meter should read a very high resistance, maybe even infinity, as if the leads were held apart in air. Now jump out the T to T terminals and the furnace should fire. As the furnace fires, the resistance will become less on the two yellow wires. It should be somewhere around 600 to 1000 Ω. The cad cell is good if this test proves out. The furnace should run for a short while and then lock out. This is a safety feature to prevent someone from running the furnace without the cad cell protection.

In most modern installations the cad cell is preferred over the thermal stack switch. It acts faster, has no mechanical moving parts, and consequently is considered more dependable.

SAFETY PRECAUTION: Care should be taken not to reset any primary control too many times because unburned oil may accumulate after each reset. If a puddle of oil is ignited, it will burn intensely. If this should happen, the technician should stop the burner motor but should allow the furnace fan to run. The air shutter should be shut off to reduce the air to the burner. Notify the fire department. Do *not* try to open the inspection door to put the fire out. Let the fire burn itself out with reduced air.

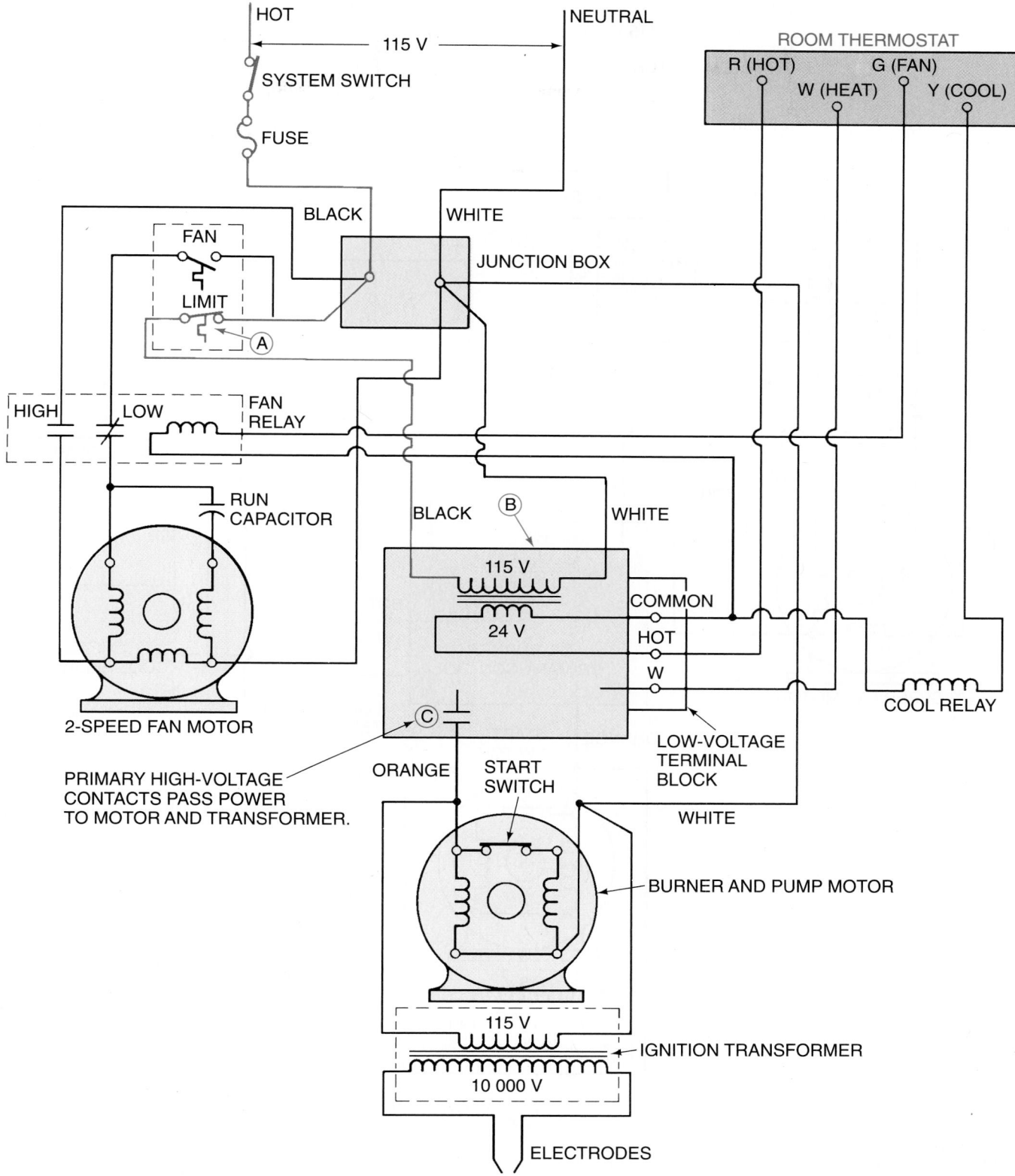

Figure 32–77   The power to operate the burner motor and transformer passes through the NC limit switch to the primary. On a call for heat, the thermostat energizes an internal 24-V relay in the primary control. If the primary safety circuit is satisfied, power will pass through the orange wire to the burner motor and ignition transformer.

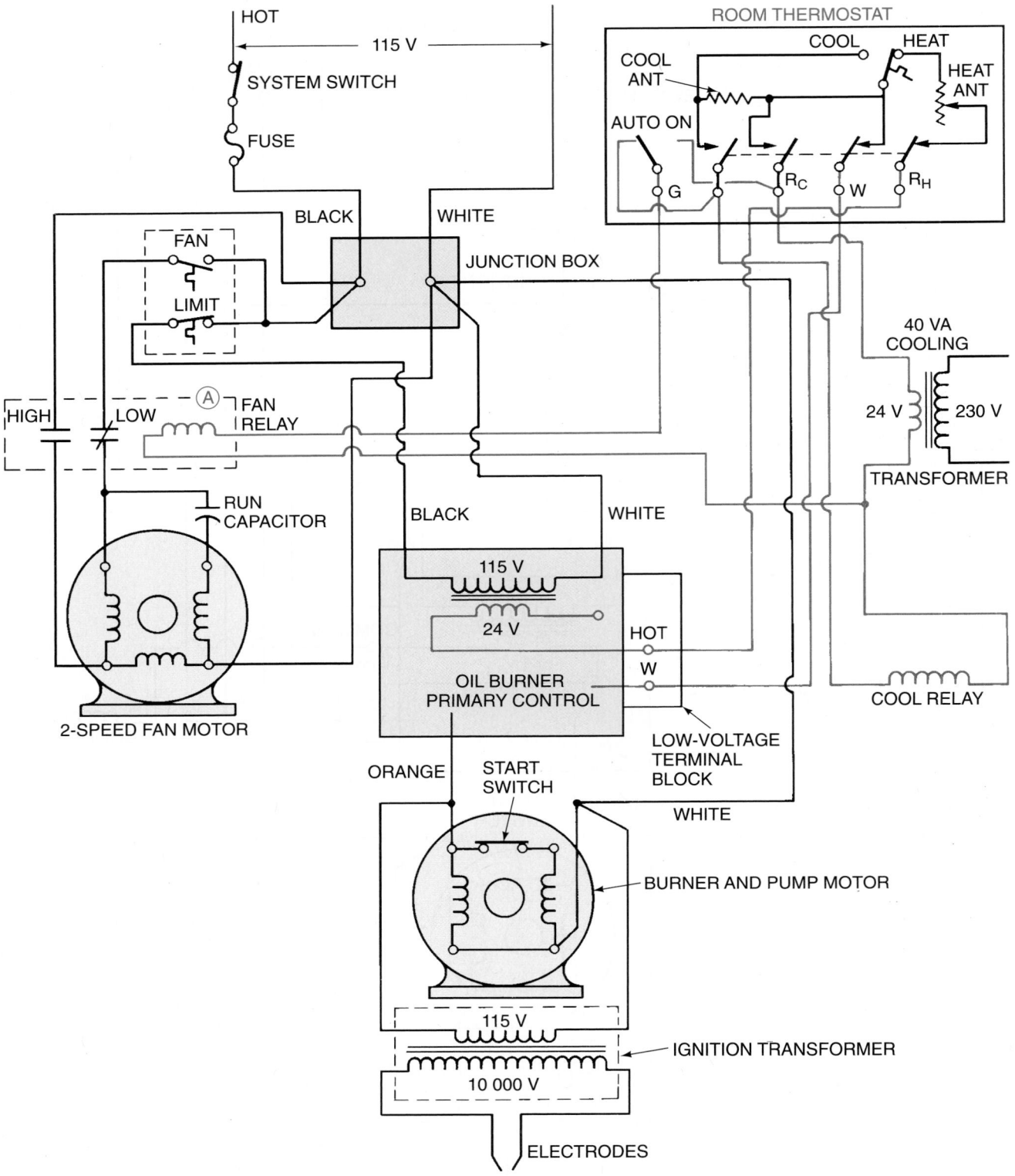

Figure 32–78 The common wire in the primary transformer does not extend into the field circuit. The room thermostat has a split subbase with the hot wire from two different transformers feeding it. These two hot circuits do not come in contact with each other because of the split, or isolating, subbase.

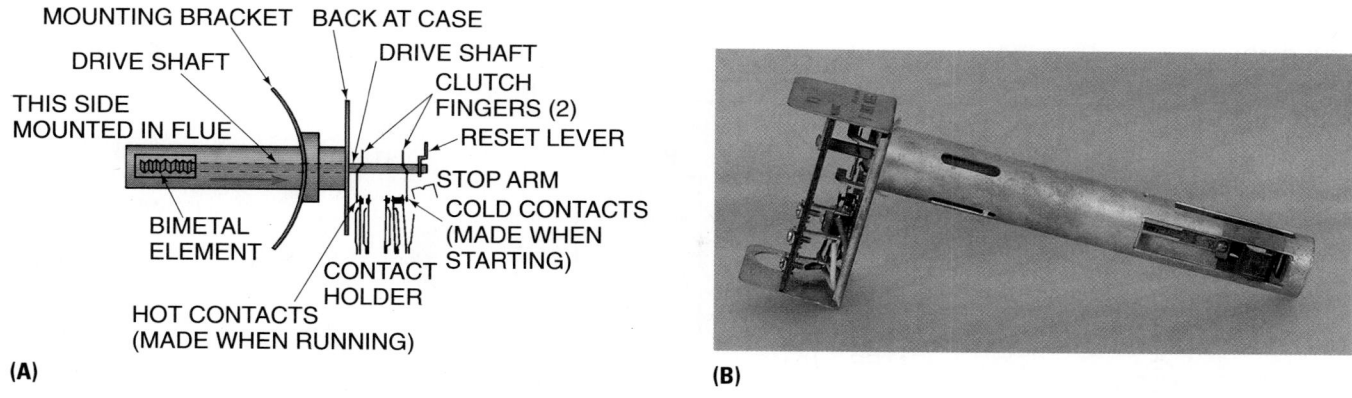

Figure 32–79 **(A)** An illustration of a stack switch or relay. **(B)** A side view of an actual stack switch. **(A)** *Courtesy Honeywell.* **(B)** *Photo by Bill Johnson*

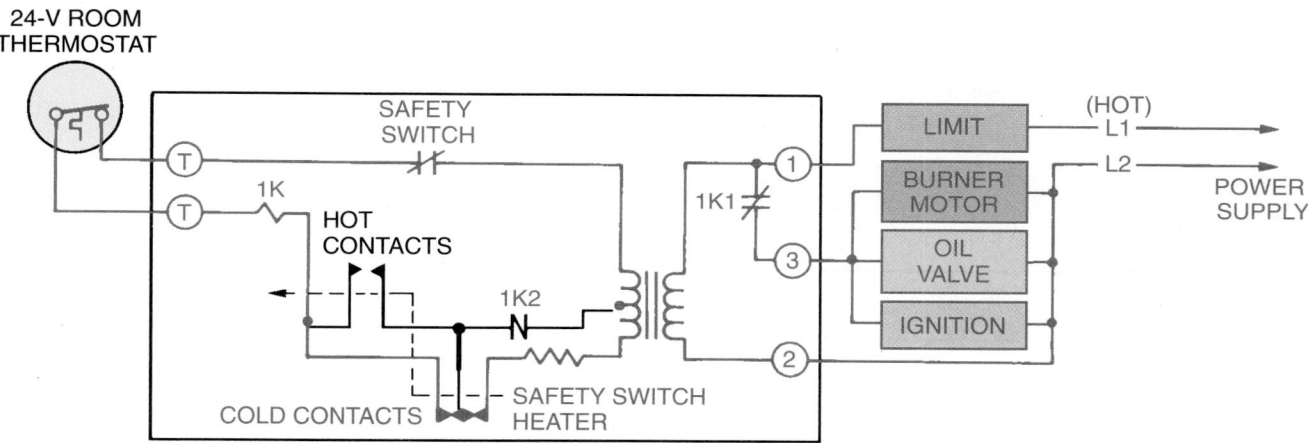

Figure 32–80 The stack switch circuit when the burner first starts. Note that current flows through the safety switch heater.

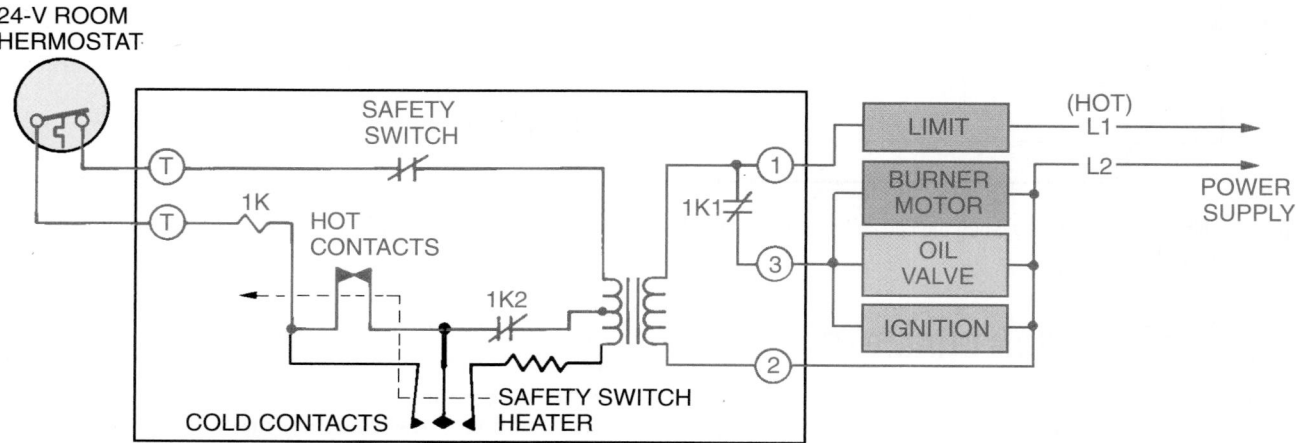

Figure 32–81 Typical stack switch circuitry with flame on, hot contacts closed, and cold contacts open. The safety switch heater is not in the circuit.

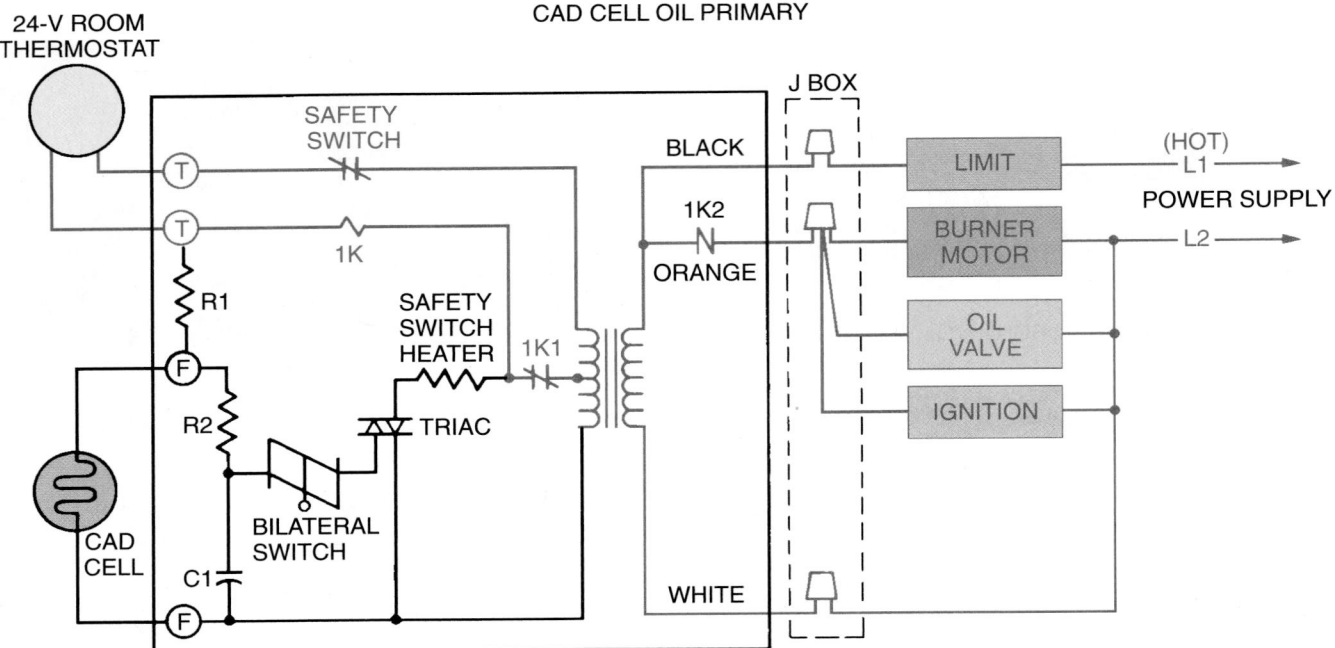

Figure 32–82 Cad cell circuitry when the cad cell "sights" flame. The safety switch heater is not in the circuit.

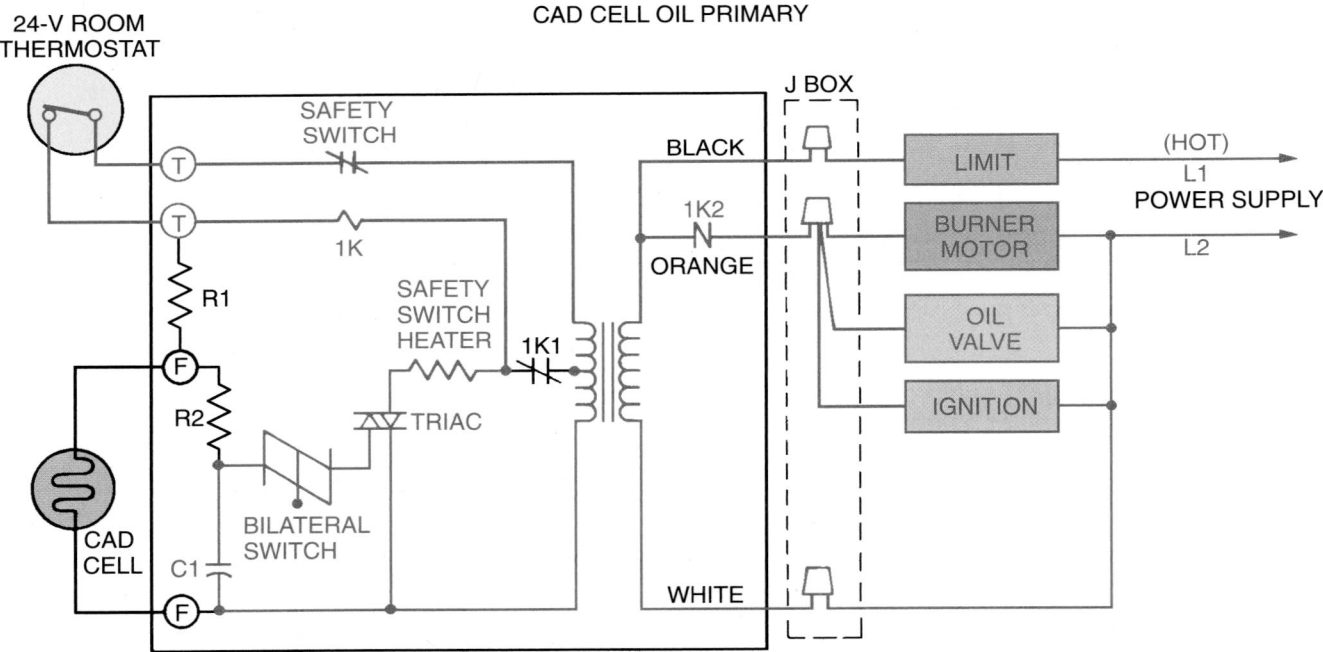

Figure 32–83 Cad cell circuitry when the cad cell does not "sight" flame. The safety switch heater will heat up and open the safety switch, shutting the burner down.

32.13 COMBUSTION CHAMBER

The atomized oil and air mixture is blown from the air tube into the combustion chamber where it is ignited by a spark from the electrodes. The atomized oil must be burned in suspension; that is, it must be burned while in the air in the combustion chamber. If the flame hits the wall of the combustion chamber before it is totally ignited, the cooler wall will cause the oil vapor to condense, and efficient combustion will not occur. Combustion chambers should be designed to avoid this condition. These chambers may be built of steel or a refractory material. A silicon refractory is used in many modern furnaces, **Figure 32–84. Figure 32–85** shows a flame in a combustion chamber.

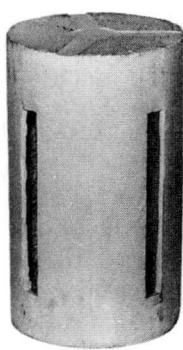

Figure 32–84 An aluminum silicon refractory combustion chamber. *Courtesy Ducane Corporation*

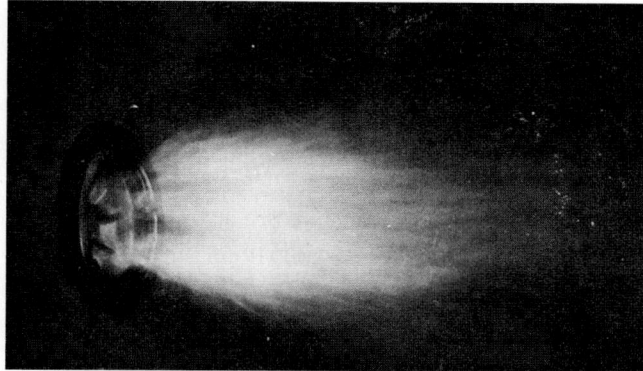

Figure 32–85 Inside a combustion chamber. Note the retention ring with flame locked to it. *Courtesy R. W. Beckett Corp.*

The combustion chamber is supposed to be matched to the burner flame at the factory. It would be a rare case where the flame and combustion chamber were not a fit from the beginning. It is not unusual for a technician to change oil burner nozzles to what may be on the service truck at the time of the call and get the flame and combustion chamber out of step with each other. This could be likely for after-hours calls for heat. On older systems, the technician may encounter several replacement nozzles lying close to the furnace. It is always best to follow the manufacturer's recommendations regarding nozzle size.

Recall from our earlier discussion on nozzles that the spray angle of the nozzle is, at least in part, determined by the shape and configuration of the combustion chamber. Nozzles with smaller spray angles are intended for use in conjunction with deeper combustion chambers, while nozzles with large spray angles are intended for use with square or round combustion chambers. Oil burner manufacturers often have original equipment manufacturer (OEM) guides that provide information regarding the original nozzle size that is intended for use on a specific furnace using a specific model of oil burner. This information is extremely valuable to the service technician.

The combustion chamber is merely a firebox, or a box to contain the fire. They sometimes develop cracks. There is

no connection between the combustion chamber and the recirculating air, so the combustion chamber oftentimes may be repaired. The repair kit may consist of a wet blanket material that can be installed through the hole when the burner is removed. The wet blanket is placed over the crack and the furnace is reassembled and fired. The blanket will become hard and is fireproof. The other option is to replace the combustion chamber.

32.14 HEAT EXCHANGER

A heat exchanger in a forced-air system takes heat caused by the combustion in the furnace and transfers it to the air that is circulated to heat the building, **Figure 32–86.** The heat exchanger is made of material that will cause a rapid transfer of heat. Modern exchangers, particularly in residential furnaces, are made of sheet steel, which is frequently coated with special substances to resist corrosion. Acids produced by combustion cause corrosion.

The heat exchanger is also designed to keep flue gases and other combustion materials separate from the air circulated through the building. SAFETY PRECAUTION: The flue gases from combustion should never mix with the circulating air. When a technician discovers a crack in a heat exchanger there is the temptation to repair it by welding it closed. Many state codes prohibit this type of repair.

Heat exchangers and combustion chambers are often inspected by the technician. A mirror made of chrome-plated steel is an ideal tool for this job, **Figure 32–87.** New products on the market allow for technicians to inspect heat exchangers with more accuracy and reliability than ever before. These products include a small, live video camera on the end

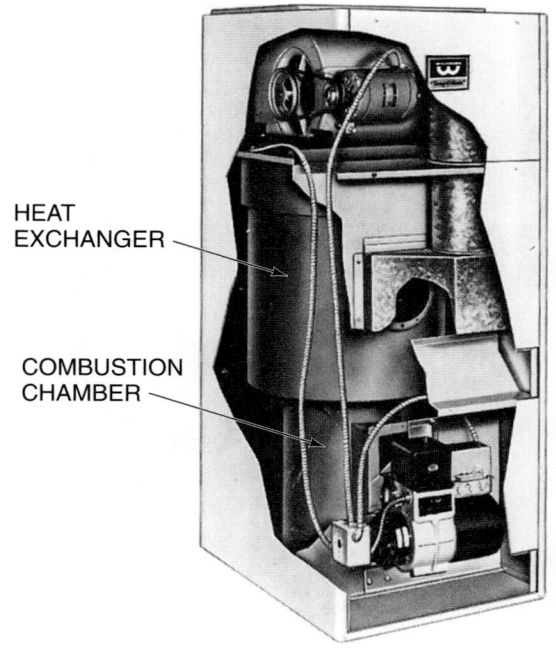

HEAT EXCHANGER

COMBUSTION CHAMBER

Figure 32–86 A heat exchanger in an oil-fired forced-air furnace. *Courtesy The Williamson-Thermoflo Company*

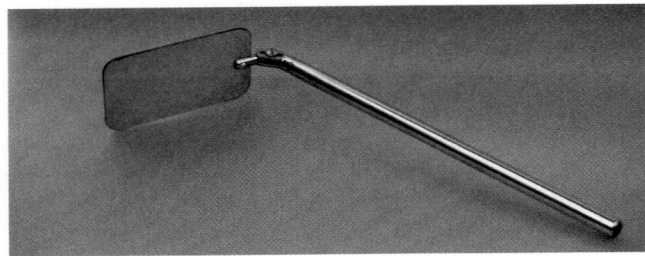

Figure 32–87   A mirror used by the heating technician to inspect the heat exchanger, the inside of the combustion chamber, and the flame. The handles on some models telescope, and the mirror angle is adjustable. *Photo by Bill Johnson*

of a long, snakelike, flexible rod that can be inserted into the furnace. The tip of the rod contains a light so the technician can inspect the heat exchanger by monitoring the screen on the device, **Figure 32–88.**

One way a technician can get a sense if there is a crack in the heat exchanger is to insert the probe from the combustion analyzer into the supply air duct far from the furnace location. If there is a measurable quantity of carbon monoxide in the supply air stream, a leak in the heat exchanger should be suspected. Damage to the heat exchanger can be the result of, among other things, insufficient airflow through the appliance, or an oversized furnace that cycles on and off too often.

Oil furnaces should have the correct airflow across the heat exchanger. An air temperature rise check may be performed if the installation is new or if it is suspected that the airflow is not correct. Often, the quantity of airflow must be known for air-conditioning equipment. The unit

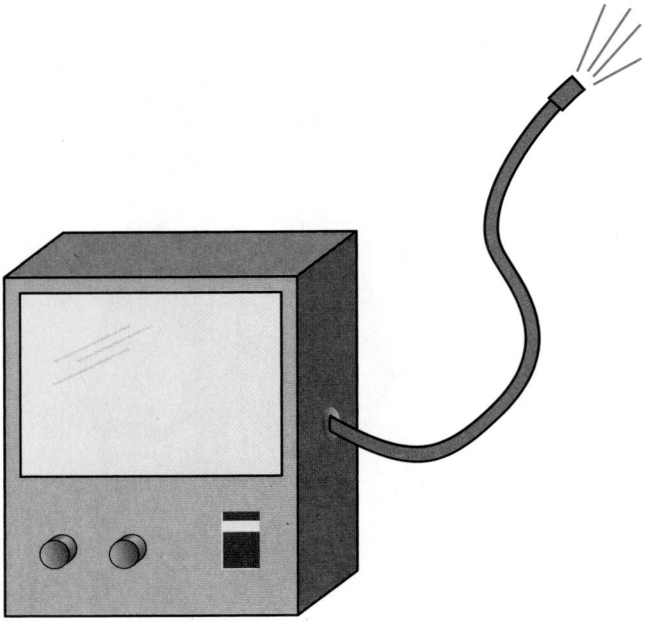

Figure 32–88   Small video cameras can be used to inspect heat exchangers and other furnace components.

capacity may be found on the unit nameplate or calculated from the oil burner nozzle size. For example, the unit may have a 0.75-gph nozzle and a 65°F temperature rise. What should the cfm be?

Since one gallon of No. 2 fuel oil can produce about 140,000 Btu of heat energy per hour, a 0.75-gph nozzle will produce approximately 105,000 Btu per hour as shown here:

$$140,000 \text{ Btu/gal} \times 0.75 \text{ gal/h} = 105,000 \text{ Btu/h}$$

The 105,000 Btu/h is the input of the furnace. There will be a certain percentage of this heat that is wasted up the chimney. If we assume that the heating appliance is 80% efficient, we will have a heat output of

$$105,000 \text{ Btu/h} \times 0.80 = 84,000 \text{ Btu/h}$$

This result represents the amount of heat that is being added to the air in the structure. In order to calculate the air volume (in cubic feet per minute), that is required to transfer this amount of heat, we use the following formula:

$$Q_S = \text{cfm} \times 1.08 \times \Delta T$$

Where Q_S = The quantity of heat in Btu/h
          cfm = The air volume in cubic feet per minute
          1.08 = A constant
          ΔT = The temperature rise across the furnace

In this example, the Q_S is 84,000 Btu/h and the ΔT is 65°F. Plugging these values into the formula gives us

$$84,000 \text{ Btu/h} = \text{cfm} \times 1.08 \times 65$$
$$84,000 \text{ Btu/h} = \text{cfm} \times 70.2$$
$$\text{cfm} = 84,000 \div 70.2$$
$$\text{cfm} = 1196.6$$

So, the required airflow is about 1200 cfm.

32.15   CONDENSING OIL FURNACE

Some manufacturers have worked to design more efficient oil furnaces since energy prices have increased and since the need for fossil-fuel conservation has become so evident. The *condensing oil furnace* is one of the product designs for increased efficiency.

The condensing oil furnace, like the more conventional furnace, has two systems: the combustion or heat-producing system and the heated air circulation system. The combustion system includes the burner and its related components: the combustion chamber, as many as three heat exchangers, and a vent fan and pipe. The heated air circulation system includes the blower fan, housing, motor, plenum, and duct system.

The following describes the operation of a condensing oil furnace manufactured by Yukon Energy Corporation, **Figure 32–89.** The burner forces air and fuel oil into the combustion chamber where they are mixed and ignited. Much of this combustion heat is transferred to the main heat exchanger, which surrounds the combustion chamber. Combustion gases still containing heat are forced through a second heat exchanger where additional heat is transferred. These gases still containing some heat are forced through a third heat

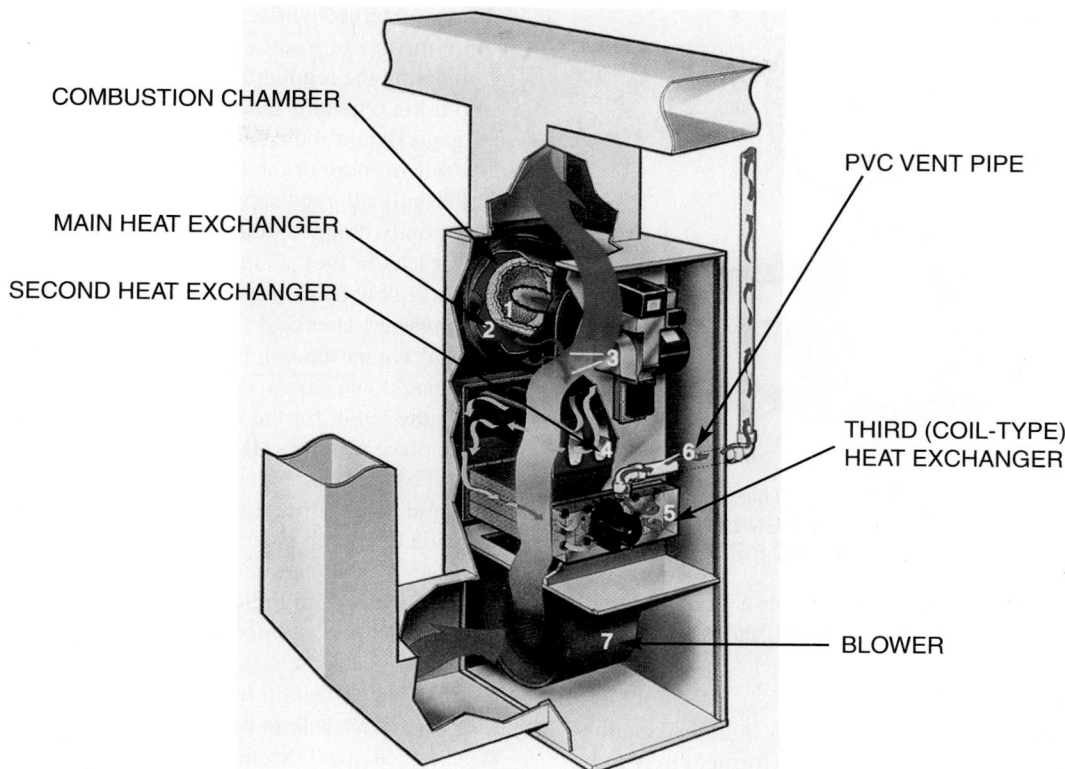

COMBUSTION CHAMBER

MAIN HEAT EXCHANGER

SECOND HEAT EXCHANGER

PVC VENT PIPE

THIRD (COIL-TYPE) HEAT EXCHANGER

BLOWER

Figure 32–89 A condensing oil furnace. *Courtesy Yukon Energy Corporation*

exchanger where nearly all the remaining heat is removed. The third heat exchanger is a coil-type where the temperature is reduced below the dew point, low enough for a change of state to occur. This results in the moisture condensing from the flue gases. This is a latent-heat high-efficiency exchange. About 1000 Btu are transferred to the airstream for each pound of moisture condensed (1 lb water = about 1 pint). This can result in considerable savings. The remaining exhaust gases are vented to the outside. These gases are normally forced out by a blower through polyvinyl chloride (PVC) pipe. The condensing coil must be made of stainless steel to resist the acid in the flue gases. A drain is located at the condensing coil to remove the condensate to a container or to a suitable area or drain outside.

In the heated air circulation system the blower draws air through the return air duct and moves it across the condensing heat exchanger where this air is preheated by removing the sensible and latent heat from the coil. The air then passes around the second heat exchanger where additional heat is added.

The circulating air then moves over the main heat exchanger, after which it is circulated to the rooms to be heated through the duct system. This leaving air is approximately 60°F warmer than when it entered the furnace through the return air duct.

These furnaces have a 90+ AFUE. They also can be installed without a chimney as they are vented through a PVC pipe.

32.16 SERVICE PROCEDURES
Pumps

The performance of the fuel oil system from the tank to the nozzle in residential and commercial units without booster pumps can be determined with a vacuum gage and a pressure gage. Before the vacuum and pressure checks are made, however, the following should be determined:

1. The tank has sufficient fuel oil.
2. The tank shutoff valve is open.
3. The tank location is noted (above or below burner).
4. The type of system is noted (one or two pipe).

Connect the vacuum and pressure gages to the fuel oil unit as shown in **Figure 32–90**. If the tank is above the burner level, the supply system may be either a one- or two-pipe system. If the tank is above the burner and with the unit in operation, the vacuum gage should read 0 in. Hg. If the vacuum gage indicates a vacuum, one or more of the following may be the problem:

■ The fuel line may be kinked.
■ The line filter may be clogged or blocked.
■ The tank shutoff valve may be partially closed.

If the supply tank is below the burner level, the vacuum gage should indicate a reading. Generally, the gage should read 1 in. Hg for every foot of vertical lift and 1 in. Hg for every 10 ft of horizontal run. The lift and horizontal run combination should not total more than 17 in. Hg. (Remember,

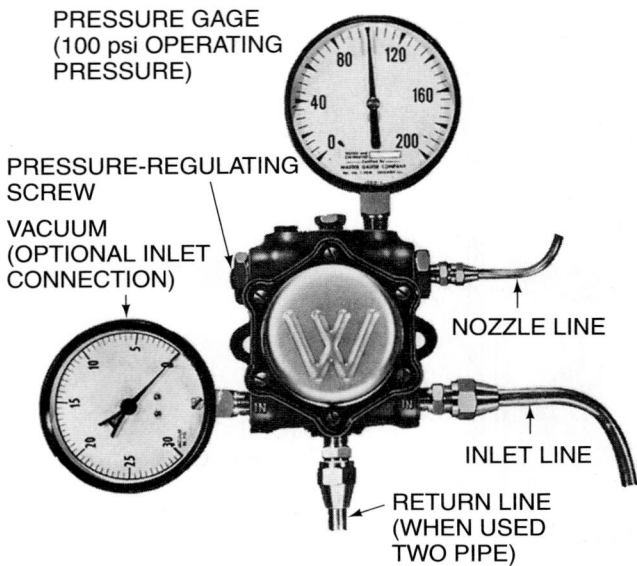

PRESSURE GAGE
(100 psi OPERATING
PRESSURE)

PRESSURE-REGULATING
SCREW

VACUUM
(OPTIONAL INLET
CONNECTION)

NOZZLE LINE

INLET LINE

RETURN LINE
(WHEN USED
TWO PIPE)

Figure 32–90 A pump with a pressure gage and a vacuum gage in place of tests and servicing. *Courtesy Webster Electric Company*

the lift capability of a dual-stage pump is approximately 15 ft.) Vacuum readings in excess of this formula may indicate a tubing size too small for the run. This may cause the oil to separate into a partial vapor and become milky.

The pressure gage indicates the performance of the pump and its capability to supply a steady even pressure to the nozzle. The pressure regulator should be adjusted to 100 psi for typical systems. See **Figure 32–91** for an illustration of a pressure-adjusting valve. To adjust the regulator, with the pressure gage in place, turn the adjusting screw until the gage indicates 100 psi. With the pump shown in **Figure 32–91**, the valve-screw cover screw should be removed and the valve screw turned with a 1/8-in. Allen wrench. Some pumps require the use of a screwdriver instead of an Allen wrench for adjusting the regulator.

With the regulator set and the unit in operation, the pressure gage should indicate a steady reading. If the gage pulsates, one or more of the following may be the problem:

- Partially clogged supply filter element
- Partially clogged unit filter or screen
- Air leak in fuel oil supply line
- Air leak in fuel oil pump cover

A pressure check of the valve differential is necessary when oil enters the combustion chamber after the unit has been shut down. To make this check, insert the pressure gage in the outlet for the nozzle line, **Figure 32–92**. With the gage in place and the unit running, record the pressure. With the unit shutdown, the pressure gage should drop about 15 psi and hold. If the gage drops to 0 in. psi, the fuel oil cutoff inside the pump is defective, and the pump may need to be replaced or an external solenoid installed, **Figure 32–53**. If the pressure gage indicates a hold of 15 psi but there is still an indication of improper fuel oil shutoff, the problem may be

- Possible air leak in fuel unit (pump)
- Possible air leak in fuel oil supply system
- Clogged nozzle strainer

Burner Motor

If the burner motor does not operate:

1. Press the reset button if there is one. Newer oil burners are equipped with permanent split-capacitor motors that do not have reset buttons—they may just reset automatically from the internal overload.

NOTE: Reset may not function if motor is too hot. Wait for motor to cool down.

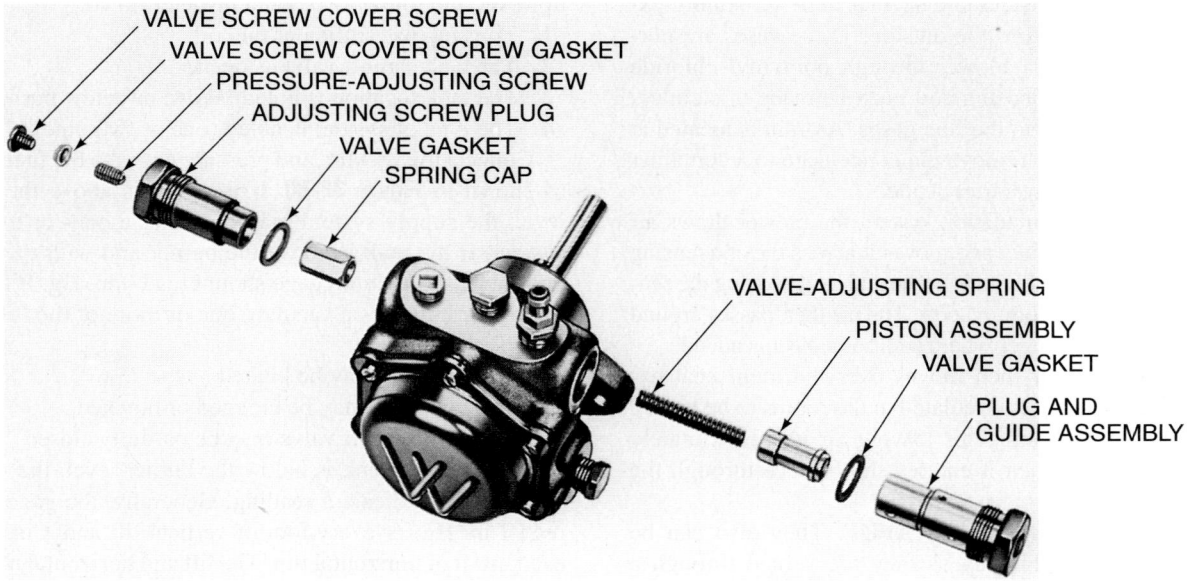

VALVE SCREW COVER SCREW
VALVE SCREW COVER SCREW GASKET
PRESSURE-ADJUSTING SCREW
ADJUSTING SCREW PLUG
VALVE GASKET
SPRING CAP

VALVE-ADJUSTING SPRING
PISTON ASSEMBLY
VALVE GASKET
PLUG AND
GUIDE ASSEMBLY

Figure 32–91 A fuel pump illustrating a pressure-adjusting valve. *Courtesy Webster Electric Company*

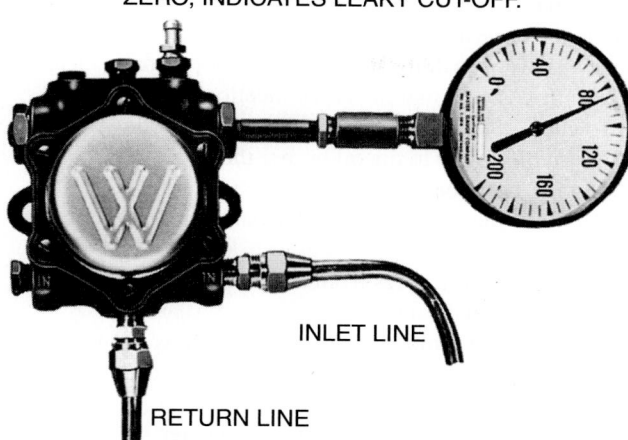

75–90 psi CUT-OFF POINT SHOULD HOLD. IF PRESSURE DROPS BACK TO ZERO, INDICATES LEAKY CUT-OFF.

INLET LINE

RETURN LINE

Figure 32–92   A pressure gage in place to check valve differential. *Courtesy Webster Electric Company*

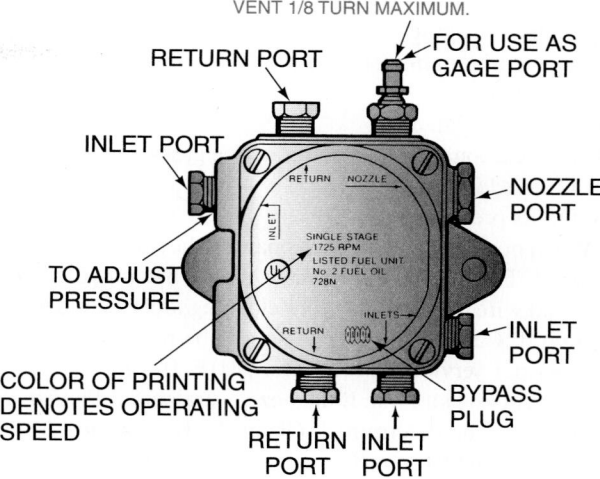

TO BLEED OR VENT PUMP
ATTACH 1/4" ID PLASTIC TUBE.
USE 3/8" WRENCH TO OPEN
VENT 1/8 TURN MAXIMUM.

RETURN PORT

FOR USE AS GAGE PORT

INLET PORT

NOZZLE PORT

TO ADJUST PRESSURE

INLET PORT

COLOR OF PRINTING DENOTES OPERATING SPEED

BYPASS PLUG

RETURN PORT   INLET PORT

Figure 32–93   A diagram of a fuel oil pump. *Courtesy Webster Electric Company*

2. Check the electrical power to the motor. This can be done with a voltmeter across the orange and white or black and white leads to the motor. If there is power to the motor and there is nothing binding it physically and it still does not operate, it should be replaced.

Bleeding a One-Pipe System

When starting an oil burner with a one-pipe system for the first time or whenever the fuel-line filter or pump is serviced, the system piping will have air in it. The system must be bled to remove this air. The system should also be bled if the supply tank is allowed to become empty.

To bleed the system, 1/4-in. flexible transparent tubing should be placed on the vent or bleed port, and the free end of the tube placed in a container, **Figure 32–93.** Turn the bleed port counterclockwise. Usually a turn from 1/8 to 1/4 is sufficient. The burner should then be started, allowing the fuel oil to flow into the container. Continue until the fuel flow is steady and there is no further evidence of air in the system; then shut off the valve.

Converting a One-Pipe System to a Two-Pipe System

Most residential burners are shipped to be used with a one-pipe system. To convert these units to a two-pipe system, do the following:
1. Shut down all electrical power to the unit.
2. Close the fuel oil supply valve at the tank.
3. Remove the inlet port plug from the unit, **Figure 32–94.**
4. Insert the bypass plug shipped with the unit into the deep seat of the inlet port with an Allen wrench.
5. Replace the inlet port plug and install the flare fitting and copper line returning to the tank.

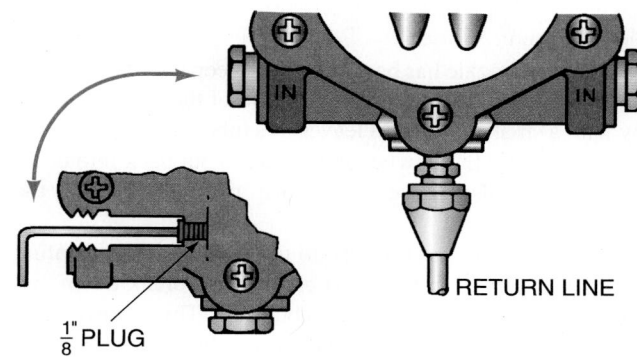

RETURN LINE

1/8" PLUG

Figure 32–94   Installation of the bypass plug for conversion to a two-pipe system. *Courtesy Webster Electric Company*

NOTE: Make certain that the specific manufacturer's instructions are followed when converting a one-pipe system to a two-pipe system. **The procedure just outlined is for Webster pumps only.** Pumps manufactured by Danfoss, for example, do not use a bypass. In addition, pumps manufactured by Suntec require that the bypass plug be installed in the return port.

SAFETY PRECAUTION: Do not start the pump with the bypass plug in place unless the return line to the tank is in place. Without this line, there is no place for the excess oil to go except to possibly rupture the shaft seal.

Nozzles

Nozzle problems may be discovered by observing the condition of the flame in the combustion chamber, taking readings on a pressure gage, and analyzing the flue gases. (Flue-gas analysis is discussed in 32.17, "Combustion Efficiency.")

The following are common conditions that may relate to the nozzle:

- Pulsating pressure gage
- Flame changing in size and shape
- Flame impinging (striking) on sides of combustion chamber
- Sparks in flame
- Low carbon dioxide reading in flue gases (less than 8%)
- Delayed ignition
- Odors present

When nozzle problems are apparent, the nozzle should be replaced. The nozzle is a delicate, finely machined component, and wire brushes and other cleaning tools should not be used on it. Generally, a nozzle should be replaced annually as a normal servicing procedure. NOTE: Do *not* attempt to clean the nozzle strainer. If the nozzle strainer is clogged, the cause should be determined. Clean the fuel oil nozzle line and the fuel unit (pump) strainer. Change the supply filter element. There may also be water in the supply tank, and the fuel oil may otherwise be contaminated.

Carbon formation at the nozzle or fuel oil burning at the nozzle may be caused by bent or distorted nozzle features. SAFETY PRECAUTION: Nozzles should be removed and replaced with a special nozzle wrench. Never use adjustable pliers or a pipe wrench.

When the nozzle has been removed, keep the oil from running from the tube by plugging the end of the nozzle assembly with a small cork. If oil leaves this tube, air will enter and cause an erratic flame when the burner is put back into operation. It may also cause an afterdrip or afterfire when the burner is shut down.

It is important that nozzles do not overheat. Overheating causes the oil within the swirl chamber to break down and cause a varnish-like substance to build up. The following are major causes of overheating and varnish buildup:

- Fire burning too close to the nozzle
- Firebox too small
- Nozzle too far forward
- Inadequate air-handling components on burner
- Burner overfired

If the fire burns too close to the nozzle, the blast tube opening may be narrowed. This will increase the air velocity as it leaves the tube and move the flame away from the nozzle. Increasing the pressure of the pump will increase the velocity of the oil droplets leaving the nozzle and will help alleviate this problem.

A nozzle that is too far forward will clog from reflected heat from the firebox from the gum formation caused by the cracking (overheated) oil. The nozzle may be moved back into the air tube by using a short adapter or by shortening the oil line.

SAFETY PRECAUTION: Do not allow an afterdrip or afterfire to go unchecked. Any leakage or afterfire will result in carbon formation and clogging of the nozzle. Check for an oil leak, air in the line, or a defective cut-off valve if there is one. An air bubble behind the nozzle will not always be pushed out during burning. It may expand when the pump is shut off and cause an afterdrip.

Ignition System

Of all phases of oil burner servicing, ignition problems are among the easiest to recognize and solve. Much ignition service consists of making the proper gap adjustment, cleaning the electrodes, and making all connections secure.

To check the ignition transformer, do the following:

1. Turn off power to the oil burner unit.
2. Swing back the ignition transformer.
3. Shut off the fuel supply or disconnect the burner motor lead.
4. Disconnect the wires from the primary winding of the ignition transformer.
5. Connect the ignition transformer tester as shown in **Figure 32–64.**
6. Turn the ignition transformer tester on.
7. Observe the display on the tester to determine whether the transformer is good. SAFETY PRECAUTION: Follow the manufacturer's instructions and keep your distance from the 10,000-V leads.

To check the electrodes, do the following:

1. Ensure that the three spark gap settings of the electrodes are set properly. Check **Figure 32–95** for the settings of the gap, height above the center of the nozzle, and distance of the electrode tips forward from the nozzle center.
2. Make sure that the tips of the electrodes are in back of the oil spray. This can be checked by using a flame mirror, **Figure 32–87.** If this is not possible, remove the electrode assembly and spray the fuel oil into an open

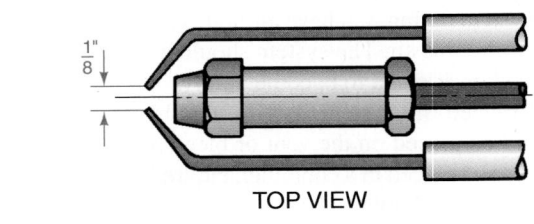

TOP VIEW

ELECTRODES

MINIMUM $\frac{1}{4}$" TO NOZZLE

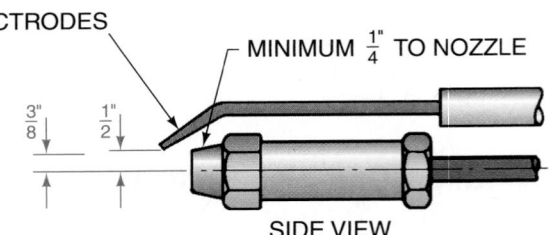

SIDE VIEW

HEIGHT ADJUSTMENT

$\frac{1}{2}$" RESIDENTIAL INSTALLATION

$\frac{3}{8}$" COMMERCIAL INSTALLATION

POSITION OF ELECTRODES IN FRONT OF NOZZLE IS DETERMINED BY SPRAY ANGLE OF NOZZLE.

Figure 32–95 Electrode adjustments. Electrodes cannot be closer than 1/4 in. to any metal part.

container, making sure that the electrodes are not in the fuel oil path. NOTE: Oil burner manufacturers have special gauges that help technicians properly position the electrode tips with respect to the nozzle orifice and to each other. When they are available, it is advisable to use these gauges. These gauges can be easily obtained by placing a phone call to the burner manufacturer.

ELECTRODE INSULATORS. Wipe the insulators clean with a cloth moistened with a solvent. If they are cracked or remain discolored after a good cleaning, they are filled with carbon throughout their porous surfaces and should be replaced.

32.17   COMBUSTION EFFICIENCY

Until a few years ago the heating technician was primarily concerned with ensuring that the heating equipment operated cleanly and safely. Technicians made adjustments by using their eyes and ears. High costs of fuel, however, now make it necessary for the technician to make adjustments using test equipment to ensure efficient combustion.

Fuels consist mainly of hydrocarbons in various amounts. In the combustion process, new compounds are formed and heat is released. The following are simplified formulas showing what happens during a perfect combustion process:

$$C + O_2 \rightarrow CO_2 + \text{heat}$$

$$H_2 + \tfrac{1}{2}O_2 \rightarrow H_2O + \text{heat}$$

During a normal combustion process, carbon monoxide, soot, smoke, and other impurities are produced along with heat. Excess air is supplied to ensure that the oil has enough oxygen for complete combustion, but even then the air and fuel may not be mixed perfectly. The technician must make adjustments so that near perfect combustion takes place, producing the most heat and reducing the quantity of unwanted impurities. On the other hand, too much excess air will absorb heat, which will be lost in the stack (flue gas) and reduce efficiency. Air contains only about 21% oxygen, so the remaining air does not contribute to the heating process.

The following tests can be made for proper combustion:

1. Draft
2. Smoke
3. Net temperature (flue stack)
4. Carbon dioxide

Technicians develop their own procedures in combustion testing. Making an adjustment to help correct one problem will often correct or help to correct others. Compromises have to be made also. When correcting one problem, another may be created. In some instances problems may be caused by the furnace design.

Some technicians use individual testing devices for each test, **Figure 32–96.** Others use electronic combustion analyzers with digital readouts, **Figure 32–97.**

To make these tests with the individual testing devices, a hole must be drilled or punched in the flue pipe 12 in. from

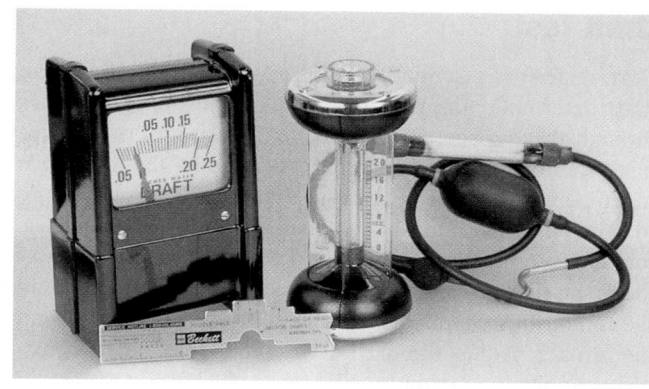

(A)

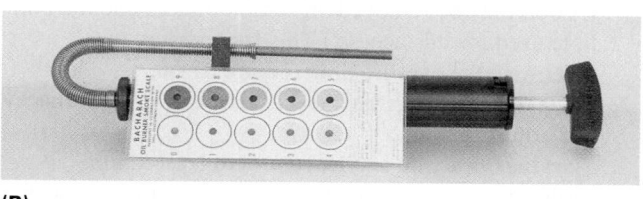

(B)

Figure 32–96   Combustion efficiency testing equipment. **(A)** A draft gage, nozzle gage, and CO_2 tester. **(B)** A smoke tester. *Photos by Bill Johnson*

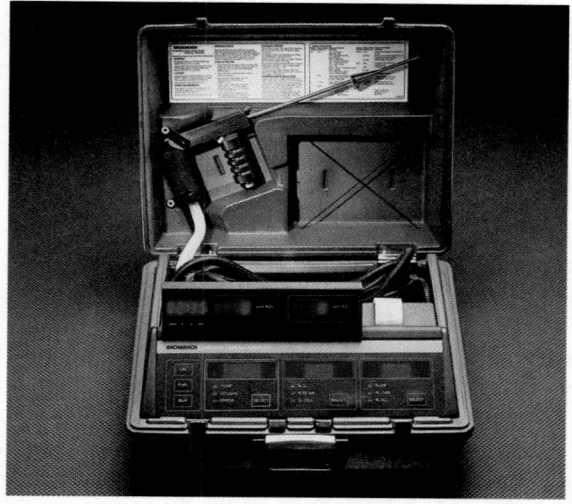

Figure 32–97   A combustion efficiency analyzer. *Courtesy Bacharach, Inc., Pittsburgh, PA USA*

the furnace breaching, on the furnace side of the draft regulator, and at least 6 in. away from it. The hole should be of the correct size so that the stem or sampling tube of the instrument can be inserted into it.

SAFETY PRECAUTION: The manufacturer's instructions furnished with the instruments should be followed carefully. Procedures indicated here are very general and should not take the place of the manufacturer's instructions.

Draft Test

Correct draft is essential for efficient burner operation. The draft determines the rate at which combustion gases pass through the furnace, and it governs the amount of air supplied for combustion. The draft is created by the difference in temperatures of the hot flue gases and is negative pressure in relation to the atmosphere. Excessive draft can increase the stack temperature and reduce the amount of carbon dioxide in the flue gases. Insufficient draft may cause pressure in the combustion chamber, resulting in smoke and odor around the furnace. Adjust the draft before you make other adjustments to obtain maximum efficiency.

To make the test, do the following:

1. Drill a hole into the combustion area for the draft tube. (A bolt may be removed on some furnaces and the bolt hole used for this access.) This is necessary to determine the overfire draft, **Figure 32–98.**
2. Place the draft gage on a level surface near the furnace and adjust to 0 in.
3. Turn the burner on and let run for at least 5 min.
4. Insert the draft tube into the combustion area to check the overfire draft, **Figure 32–98.**
5. Insert the draft tube into the flue pipe to check the flue draft.

The overfire draft should be set to the manufacturer's specifications, typically −0.02 in. WC. The flue draft should be adjusted with the draft regulator to maintain the proper overfire draft. Most residential oil burners require a flue draft of −0.04 in. to −0.06 in. WC to maintain the proper overfire draft. These drafts are updrafts and are negative in relationship to atmospheric pressure. Longer flue passages require a higher flue draft than shorter flue passages.

Smoke Test

Excessive smoke is evidence of incomplete combustion. This incomplete combustion can result in a fuel waste of up to 15%. A 5% fuel waste is not unusual. Excessive smoke also results in a soot buildup on the heat exchanger and other heat-absorbing areas of the furnace. Soot is an insulator. This results in less heat being absorbed by the heat exchanger and

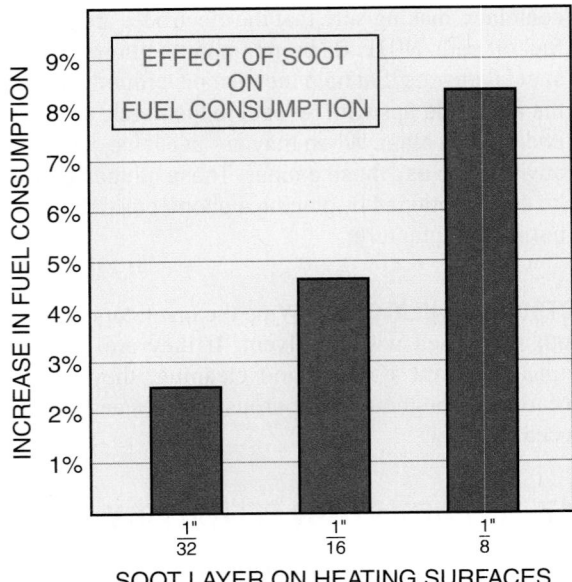

Figure 32–99 The effect of soot on fuel consumption. *Courtesy Bacharach, Inc., Pittsburgh, PA USA*

increased heat loss to the flue. A 1/16-in. layer of soot can cause a 4.5% increase in fuel consumption, **Figure 32–99.**

The smoke test is accomplished by drawing a prescribed number of cubic inches of smoke-laden flue products through a specific area of filter paper. The residue on this filter paper is then compared with a scale furnished with the testing device. The degree of sooting can be read off the scale. A smoke tester such as the one illustrated in **Figure 32–100** may be used.

To make the smoke test, **Figure 32–101,** do the following:

1. Turn on the burner and let run for at least 5 min or until the stack thermometer stops rising.
2. Ensure that filter paper has been inserted in tester.
3. Insert sampling tube of test instrument into hole in flue.
4. Pull the tester handle the number of times indicated by the manufacturer's instructions.
5. Remove the filter paper and compare with the scale furnished with the instrument.

Excessive smoke can be caused by

- improper fan collar setting (burner air adjustment).
- improper draft adjustment (draft regulator may be required or need adjustment).
- poor fuel supply (pressure).
- oil pump not functioning properly.
- defective or incorrect type of nozzle.

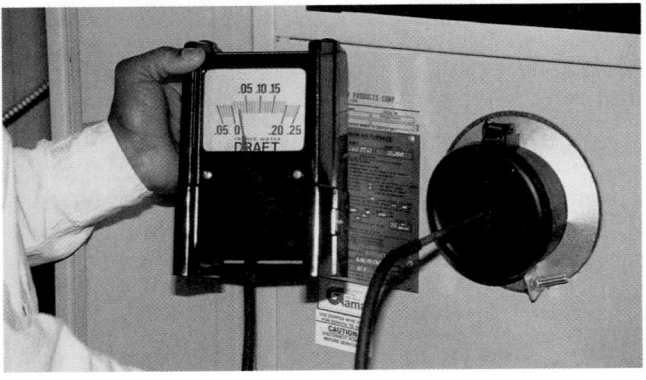

Figure 32–98 Checking overfire draft. *Photo by Bill Johnson*

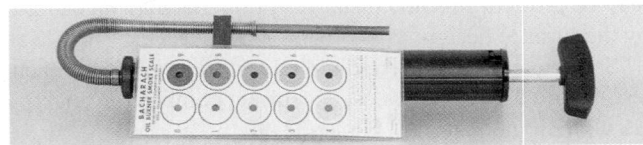

Figure 32–100 A smoke tester. *Photo by Bill Johnson*

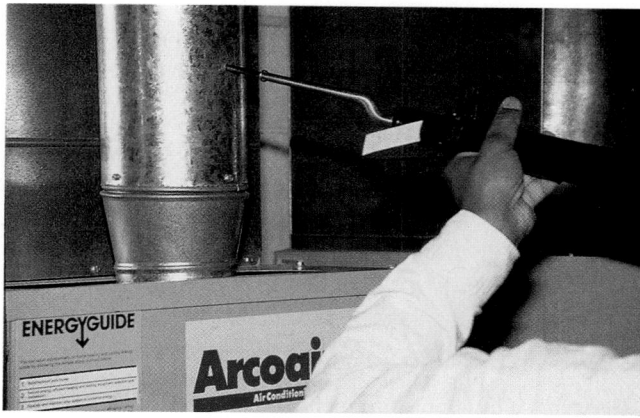

Figure 32–101 Making a smoke test. *Photo by Bill Johnson*

- excessive air leaks in furnace (air diluting flame).
- improper fuel-to-air ratio.
- defective firebox.
- improper burner air-handling parts.

Net Stack Temperature

The net stack temperature is important because an abnormally high temperature is an indication that the furnace may not be operating as efficiently as possible. The net stack temperature is determined by subtracting the air temperature around the furnace from the measured stack or flue temperature. For instance, if the flue temperature reading were 650°F and the basement air temperature where the furnace was located were 60°F, the net stack temperature would be 650°F − 60°F = 590°F. The manufacturer's specifications would be consulted to determine normal net stack temperatures for particular furnaces.

To determine the stack temperature, **Figure 32–102,** do the following:

1. Insert the thermometer stem into the hole in the flue.
2. Turn on the burner and allow to run for at least 5 min or until the stack thermometer stops rising in temperature.
3. Subtract the basement or ambient temperature from the stack temperature reading.

Figure 32–102 Making a stack gas temperature test. *Photo by Bill Johnson*

4. Compare with manufacturer's specifications.

A high stack temperature may be caused by

- excessive draft through the combustion chamber.
- dirty or soot-covered heat exchanger.
- lack of baffling.
- incorrect or defective combustion chamber.
- overfiring (check nozzle size and pressure).

Carbon Dioxide Test

The carbon dioxide test is an important combustion efficiency test. A high carbon dioxide reading is good. If the test reading is low, it indicates that the fuel oil has not burned efficiently or completely. The reading should be considered with all of the test readings. Under most normal conditions a carbon dioxide reading greater than 10% should be obtained. If problems exist that would be very difficult to correct but the furnace was considered safe to operate and had a net stack temperature of 400°F or less, a carbon dioxide reading of 8% could be acceptable. However, if the net stack temperature is over 500°F, a reading of at least 9% carbon dioxide should be obtained. SAFETY PRECAUTION: A CO_2 reading over 12.5% can be dangerous.

To make the carbon dioxide test, **Figure 32–103,** do the following:

1. Turn on the burner and operate for at least 5 min.
2. Insert thermometer and wait for the temperature to stop rising.
3. Insert the sampling tube into the hole previously made in the flue pipe.
4. Remove a test sampling, using the procedure provided by the manufacturer of the test instrument.
5. Mix the fluid in the test instrument with the sample gases from the flue according to instructions.
6. Read the percent carbon dioxide from the scale on the instrument.

A low percent carbon dioxide reading may be caused by one or more of the following:

- High draft or draft regulator not working properly
- Excess combustion air

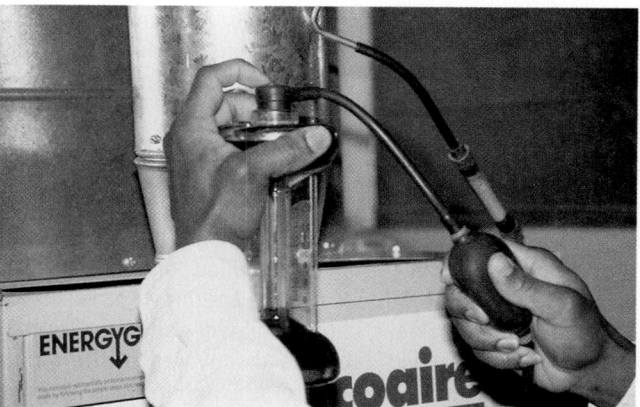

Figure 32–103 Making a carbon dioxide test. *Photo by Bill Johnson*

- Air leakage into combustion chamber
- Poor oil atomization
- Worn, clogged, or incorrect nozzle
- Oil-pressure regulator set incorrectly

Electronic Combustion Analyzers

The technician may have an electronic combustion analyzer. This equipment provides information similar to that provided by the test instruments already described. These instruments may indicate carbon monoxide and oxygen concentrations rather than carbon dioxide percentages, but the technician can determine acceptable levels and necessary corrections to be made on the furnace from the manufacturer's instructions.

Generally, the tests are made from samplings from the same position in the flue, as with the other instruments. These electronic instruments are easier to use and save time. The readings are provided by a digital readout system that is convenient to use, **Figure 32–97.**

HVAC GOLDEN RULES

When making a service call to a residence, do the following:
- Ask the customer about the problem. The customer can often help you solve the problem and save you time.
- Check the humidifiers where applicable. Annual service is highly recommended for health reasons.
- Be prepared and do not let any fuel oil escape on the ground or floor around the furnace. The customer will notice the odor.

Added Value to the Customer

Here are some simple, inexpensive procedures that may be included in the basic service call.
- Check flame characteristics; a smoke test is good insurance and may lead to extra paying service work.
- Replace all panels with the correct fasteners.
- Inspect heat exchanger for soot buildup.

PREVENTIVE MAINTENANCE

Oil heat, like gas heat, has air-side and burner-side components. The air-side has been discussed in the electric heat unit. Only the burner section will be discussed here.

Oil heat requires more consistent regular service than any heat system discussed in this book. The reason is that the unit uses fuel oil that must be properly metered and burned for the best efficiency. The burning efficiency is accomplished with the proper fuel supply and properly adjusted and maintained burner section. This includes the combustion chamber, the heat exchanger, and the flue system. They must all be working perfectly for proper burning efficiency. Improper burning efficiency will cause soot to form, which will slow down the heat exchange between the combustion products and the room air. This condition will deteriorate and cause the combustion efficiency to worsen. It will pay the customer to have the system serviced each year before the season starts, using the appropriate sections of the following procedures. This may look like a long service call for each furnace every year, but a competent service technician should be able to accomplish the maintenance portions in an hour for a typical furnace. Corrective actions are not considered as part of the maintenance. Discussion of the preventive maintenance procedures will start at the tank and move to the flue.

The tank will either be aboveground or underground and can be either above or below the burner. Underground tanks are exposed to the various soils that may corrode the tank. Leaks in the tank may allow oil to seep out or water to seep in. Oil leaks to the soil may not be discovered, but water leaks into the oil will cause fuel burning problems. They will either be discovered at the time of routine preventive maintenance or when an emergency service call is made.

Every year, the service technician should check the tank for water accumulation in the bottom of the tank. A commercial paste that changes color when in contact with water may be spread on the oil-tank measuring stick and inserted into the tank. If there is water in the bottom of the tank when the stick is pulled out of the oil, the paste will show the level of the water. If the tank is leaking, it should be replaced. Often, an aboveground tank may be substituted.

The oil lines should be inspected for bends, dents, and rust where they leave the ground and enter the house. A small pinhole in a supply line will cause air to be pulled into the system if the tank is below the burner. This will cause an erratic flame at the burner. The burner may also have an afterburn when the unit is stopped. This is caused by the expanding air in the system. Any flame problems, problems priming the pump, or afterburn problems should prompt you to check the supply line for leaks.

A vacuum test may be performed by disconnecting the oil supply line from the tank and letting the system pump the oil supply line into a vacuum and shutting the pump off. The line should stay in a vacuum if it is leak free. NOTE: Oil and air may leak backwards through the pump. This may be prevented by disconnecting the oil return line and holding your finger over it once the pump is stopped. A pressure test on the oil supply line will also prove whether there is a leak in the supply line. The line must be disconnected and plugged on each end. Apply about 10 psig of pressure (the filter housing is the weakest point and its working pressure is approximately 10 psig) to the line and it should hold indefinitely. No pressure drop in 15 min should satisfy you that no leak is present. If there is a pressure drop, the leak must be found and repaired. If the line is full of oil, you should see it seep out. Do not forget to check the filter housing gasket.

The gun burner should be removed from the unit and the combustion chamber inspected to be sure the refractory is in good condition. Look for signs of soot, and vacuum out as needed. While the burner is out, change the oil nozzle; be sure to replace the old one with the correct nozzle and use only a nozzle wrench. Check the burner head for overheating and cracks. Examine the static tube for signs of oil or soot, and clean as needed. Set the electrodes using a gage, and examine the insulators for cracks and soot deposits. Clean as needed and change if cracked. Examine the flexible coupling for signs of wear. Make sure the cad cell is clean and aligned correctly and that the bracket is tight. Lubricate the burner motor. Replace the gun burner back into the furnace. Be sure to install the gasket and tighten all bolts.

Change the in-line oil filter. Be sure to fill the cartridge with fresh oil. Change the gasket and tighten it correctly. This will help ensure a good start.

Start the furnace and while it is heating up to operating temperature, prepare to perform a combustion analysis. Check the draft. If the draft is not correct (see text for where to check and approximate draft readings), clean the flue and heat exchanger if needed.

Insert the thermometer in the flue and when it reaches the correct temperature, perform a smoke test. If the fire is smoking, adjust the air until the correct smoke spot is obtained. If you suspect any problem, check the oil pressure to be sure that you have 100 psig at the nozzle. When the smoke is correct, perform a combustion analysis and adjust to the correct carbon dioxide reading.

When this is accomplished, the furnace should operate correctly for the next year.

32.18   SERVICE TECHNICIAN CALLS

SERVICE CALL 1

A new customer calls requesting *a complete checkup of the oil furnace, including an efficiency test. This customer will stay with the service technician and watch the complete procedure.*

The technician arrives and explains to the customer that the first thing to do is run the furnace and perform an efficiency test. By running a test before and after adjustment, the technician can report results. The thermostat is set to about 10°F above the room temperature to allow time for it to warm up while the technician sets up. This will also ensure that the furnace will not shut off during the test. The technician gets the proper tools and goes to the basement where the furnace is located. The customer is already there and asks whether the technician would mind an observer. The technician explains that a good technician should not mind being watched.

The technician inserts the stack thermometer in the flue and observes the temperature; it is no longer rising. The indoor fan is operating. A sample of the combustion gas is taken and checked for efficiency. A smoke test is also performed. The unit is operating with a slight amount of smoke, and the test shows the efficiency to be 65%. This is about 5% to 10% lower than normal.

The technician then removes a low-voltage wire from the primary, which shuts off the oil burner and allows the fan to continue to run and cool the furnace. The technician then removes the burner nozzle assembly and replaces the nozzle with an exact replacement. Before returning the nozzle assembly to its place, the technician also sets the electrode spacing and then changes the oil filter, changes the air filter, and oils the furnace motor. The furnace is ready to start again.

The technician starts the furnace and allows it to heat up. When the stack temperature has stopped rising, the technician pulls another sample of flue gas. The furnace is now operating at about 67% efficiency and still a little smoky in the smoke test. The technician checks the furnace nameplate and notices that this furnace needs a 0.75-gph nozzle but has a 1-gph nozzle. It is common (but poor practice) for some technicians to use what they have, even if it is not exactly correct.

This technician removes the nozzle just installed as an exact replacement and then gets more serious about the furnace because it has been mishandled. The technician installs the correct nozzle, starts the unit, and checks the oil pressure. The pressure has been reduced to 75 psig to correct for the oversized nozzle. The technician changes the pressure to 100 psig (the correct pressure), standard for most residential gun-type burners.

The air to the burner is adjusted; meanwhile the stack temperature is reading 660°F (this is a net temperature of 600°F because the room temperature is 60°F). The technician runs another smoke test, which now shows minimum smoke. The efficiency is now 73%. This is much better.

The customer has been kept informed all through the process and is surprised to learn that oil burner service is so exact. The technician sets the room thermostat back to normal and leaves a satisfied customer.

SERVICE CALL 2

A residential customer calls and reports that the *oil furnace in a basement under the family room is making a noise when it shuts off. The oil pump is not shutting off the oil fast enough.*

The technician arrives, goes to the room thermostat, and turns it up above the room temperature to keep the furnace running. This furnace has been serviced in the last 60 days, so a nozzle and oil filter change is not needed. The technician goes to the furnace and removes a low-voltage wire from the primary control to stop the burner. The fire does not extinguish immediately but shuts down slowly with a rumble.

The technician installs a solenoid in the small oil line that runs from the pump to the nozzle and wires the solenoid coil in parallel with the burner motor so that it will be energized only when the burner is operating. The technician disconnects the line where it goes into the burner housing and places the end into a bottle to catch any oil that may escape. The burner is then turned on for a few seconds. This clears any air that may be trapped in the solenoid out of the line leading to the nozzle.

The technician reconnects the line to the housing and starts the burner. After it has been running for a few minutes, the technician shuts down the burner. It has a normal shutdown. The start-up and shutdown is repeated several times. The furnace operates correctly. The technician then sets the room thermostat to the correct setting before leaving.

SERVICE CALL 3

A customer from the duplex apartment calls. There is no heat. The customer is out of fuel. The customer had fuel delivered last week, but the driver filled the wrong tank. This system has an underground tank and it is sometimes hard to get the oil pump to prime (pull fuel to the pump). The technician is misled for some time, thinking that there is fuel and that the pump is defective.

The technician arrives and goes to the room thermostat. It is set at 10°F above the room temperature. The technician goes to the furnace, which is in the garage at the end of the apartment. The furnace is off because the primary control has tripped. The technician takes a flashlight and examines the combustion chamber for any oil buildup that may have accumulated if the customer has been resetting the primary. SAFETY PRECAUTION: If a customer had repeatedly reset the primary trying to get the furnace to fire and the technician then starts the furnace, this excess oil would be dangerous. There is no excess oil in the combustion chamber, so the reset button is pushed. The burner does not fire.

The technician suspects that the electrodes are not firing correctly or that the pump is not pumping. The first thing to do is to fasten a piece of flexible tubing to the bleed port side of the oil pump. The hose is placed in a bottle to catch any fuel that may escape, then the bleed port is opened and the primary control is reset. This is a two-pipe system and when the correct oil quality is found at the bleed port the oil supply is verified. The burner and pump motor start. No oil is coming out of the tubing, so the pump must be defective. Before changing the pump, the technician decides to check to make sure there is oil in the line. The line entering the pump is removed. There is

no oil in the line. The technician opens the filter housing and finds very little oil in the filter. The technician now decides to check the tank and borrows the stick the customer uses to check the oil level. The stick is pushed through the fill hole in the tank. There is no oil.

The technician tells the customer, who calls the oil company. The driver that delivered the oil is close by and dispatched to the job. The driver points to the tank that was filled, the wrong one, and then fills the correct tank.

The technician starts the furnace and bleeds the pump until a full line of liquid oil flows, then closes the bleed port. The burner ignites and goes through a normal cycle. The furnace is not under contract for maintenance, so the technician suggests to the occupant that a complete service call with nozzle change, electrode adjustment, and filter change be done. The customer agrees. The technician completes the service and turns the thermostat to the normal setting before leaving.

The following service calls do not include solutions. The solutions may be found in the *Instructor's Guide*.

SERVICE CALL 4

A customer calls indicating a smell of smoke when the furnace starts up. This customer does not have the furnace serviced each year.

The technician goes to the furnace in the basement, and the customer follows. The technician turns the system switch off and goes back upstairs and sets the thermostat 5°F above the room temperature so the furnace can be started from the basement. When the technician gets back to the basement, the furnace is started and a puff of smoke is observed. The technician inserts a draft gage in the burner door port; the draft is +0.01 in. WC positive pressure.

What is the likely problem and the recommended solution?

SERVICE CALL 5

A retail store manager calls stating that there is no heat in the small retail store.

The technician arrives, goes to the room thermostat, and discovers it set at 10°F higher than the room temperature. The thermostat is calling for heat. The technician goes to the furnace in the basement and discovers it needs resetting for it to run. The technician examines the combustion chamber with a flashlight and finds no oil accumulation. The primary control is then reset. The burner motor starts, and the fuel ignites. It runs for 90 sec and shuts down.

What is the likely problem and the recommended solution?

SERVICE CALL 6

A customer reports that there is no heat and the furnace will not start when the reset button is pushed.

The technician arrives and checks to see that the thermostat is calling for heat. The set point is much higher than the room temperature. The technician goes to the garage where the furnace is located and presses the reset; nothing happens. The primary is carefully checked to see whether there is power to the primary control. It shows 120 V. Next the circuit leaving the primary, the orange wire, is checked. (The technician realizes that power comes into the primary on the white and black wires, the white being neutral, and leaves on the orange wire.) There is power on the white to black but none on white to orange, **Figure 32–78.**

What is the likely problem and the recommended solution?

SUMMARY

- No. 2 grade fuel oil is most commonly used in heating residences and light commercial buildings.
- Fuel oil is composed primarily of hydrogen and carbon in chemical combination.
- Gun-type oil burner parts are the burner motor, blower or fan wheel, pump, nozzle, air tube, electrodes, ignition system, and primary controls.
- The pump can be single or dual stage.
- Air is blown through the air tube into the combustion chamber and mixed with the atomized fuel oil.
- The ignition system provides the high voltage, producing the spark across the electrodes.
- Ignition can be interrupted or intermittent.
- Interrupted ignition systems provide a spark only at the beginning of the cycle.
- Intermittent ignition provides a continuous spark.
- Fuel storage and supply systems can be one-pipe or two-pipe design.
- An auxiliary or booster supply system must be used where the burner is more than 15 ft above the storage tank.
- The atomized oil and air mixture is ignited in the combustion chamber.
- The heat exchanger takes heat caused by the combustion and transfers it to the air that is circulated to heat the building.
- The performance of the pump can be checked with a vacuum gage and a pressure gage.
- One-pipe systems must be bled before the burner is started for the first time or whenever the fuel supply lines have been opened.
- To convert a pump from a one-pipe to a two-pipe system, insert a bypass plug in the return or inlet port.
- Nozzles should not be cleaned or unplugged. They should be replaced.
- Electrodes should be clean, connections should all be secure, and the spark gap should be adjusted accurately.

- Combustion efficiency tests should be made when servicing oil burners, and corrective action taken to obtain maximum efficiency.
- Preventive maintenance procedures should be performed annually.

REVIEW QUESTIONS

1. How many grades of fuel oil are normally considered as heating oils?
2. What two elements make up fuel oil?
3. The most common fuel oil used for residential and light commercial is
   A. No. 1.
   B. No. 2.
   C. No. 3.
   D. No. 4.
4. How many pounds of oxygen are required to burn one pound of No. 2 fuel oil?
5. Explain the concept of excess air when it is used to describe the combustion process.
6. Before fuel oil can be burned, it must be
   A. dripped.
   B. decompressed.
   C. vitalized.
   D. atomized.
7. What products of combustion are produced with ideal combustion when fuel oil is burned?
8. When incomplete combustion occurs, what additional by-products of combustion are created?
9. The typical nozzle pressure for residential and light commercial gun-type oil burners is
   A. 100 psig.
   B. 175 psia.
   C. 200 psig.
   D. 250 psia.
10. Two types of oil burner pumps are _____ stage and _____ stage.
11. Describe the three functions of the oil burner nozzle.
12. Describe three common spray patterns for an oil burner nozzle.
13. The purpose of the electrodes of an oil burner is to
   A. create a blue flame.
   B. establish best efficiency.
   C. put out the flame electronically.
   D. ignite the fuel oil at start-up.
14. What is the purpose of the ignition transformer?
15. Explain the differences between an ignition transformer and a solid-state igniter.
16. List the advantages a solid-state igniter has over an ignition transformer.
17. Which of the following is the typical output voltage of the ignition transformer?
   A. 24 V
   B. 200,000 V
   C. 115 V
   D. 10,000 V

18. What are the two functions of the primary control on an oil burner?
19. List the two types of primary controls that are commonly found on oil-fired heating equipment.
20. Describe the operation of a stack relay.
21. Explain the difference between the hot contacts and the cold contacts on a stack relay.
22. The purpose of the cad cell in an oil burner control circuit is to
    A. make sure that the burner ignites by sensing the flame.
    B. turn the burner off when the room temperature is satisfied.
    C. start the burner when the thermostat calls for heat.
    D. prevent the furnace from overheating.
23. Explain why it is important to keep the cad cell clean at all times.
24. A one-pipe system may be used when the oil tank is higher or lower than the oil pump?
25. Describe where a two-pipe system is used.
26. When is an auxiliary oil pump used?
27. The oil flame must not hit the combustion chamber because it will
    A. overheat.
    B. cool off and create soot and smoke.
    C. use too much fuel oil.
    D. prevent the furnace from starting up.
28. Describe the purpose of the heat exchanger.
29. When servicing an oil burner, a service technician must have a _____ gage and a _____ gage.
30. Describe the function of the flame-retention ring on a modern oil burner.

DIAGNOSTIC CHART FOR OIL HEAT

The oil-burning appliances discussed here are the forced-air type. (Many of the same rules apply to oil boilers as far as their burner operation is concerned.) These oil-burning systems may be found in homes and businesses as the primary heating systems. Two basic types of oil safety controls are the cad cell and the stack switch. Both are discussed in the text. Only the basic procedures are discussed here.

For more specific information about any particular step of the procedure, refer to the particular oil heat unit as it is discussed in the text or consult the manufacturer. Always listen to customers; their input can help you to locate the problem. Remember, they are present most of the time and often pay attention to how the equipment functions.

Problem	Possible Cause	Possible Repair	Heading Number
Furnace will not start—no heat	Open disconnect switch	Close disconnect switch.	32.10
	Open fuse or breaker	Replace fuse or reset breaker and determine why it opened.	32.10
	Faulty wiring	Repair or replace faulty wiring or connections.	32.10
	Defective low-voltage transformer	Replace transformer and look for possible overload condition.	32.10
	Primary safety control off—needs reset	Check for oil accumulation in combustion chamber; if none, reset control and observe fire and flame characteristics.	32.10, 32.12 32.16
	Tripped burner motor reset	Press reset button, check amperage; if too much, check motor or pump for binding.	32.16
Furnace starts after reset but shuts off after 90 sec	Cad cell may be out of alignment or dirty	Check cad cell for alignment and smoke on lens—if smoke deposits on lens adjust burner for correct fire.	32.12
	Defective cad cell	Replace cad cell and reset.	32.12
	Defective primary control	Change primary control.	32.9
Furnace starts but no ignition occurs	No fuel	Fill fuel tank.	32.4
	Electrodes out of alignment	Align electrodes.	32.16
	Defective ignition transformer	Replace transformer.	32.16
	Defective oil pump	Replace oil pump.	32.16
	Restricted fuel filter	Replace filter.	32.16
	Defective coupling between pump and motor	Replace coupling.	32.9

Problem	Possible Cause	Possible Repair	Heading Number
Burner runs and ignition occurs, but fan does not start	Faulty wiring or connections in fan circuit	Repair or replace faulty wiring or connectors.	32.10
	Defective fan switch	Replace fan switch.	32.10
	Defective fan motor	Replace fan motor.	32.10
Burner has delayed ignition, makes noise on start-up	Electrode out of alignment	Align electrodes.	32.16
	Clogged nozzle	Change nozzle and line filter.	32.16
	Transformer weak	Change transformer.	32.16
	Too much or too little air	Adjust air.	32.16
	Nozzle position in burner	Adjust nozzle position.	32.16
Burner makes noise on shutdown	Fuel cut-off at fuel pump	Check fuel cut-off using gages. If not correct, change fuel pump or add solenoid in oil supply line to burner.	32.16
High stack temperature	Too much air to burner	Correct air by adjustment.	32.17
	Too much draft	Adjust or add draft regulator.	32.17
	Overfired	Change to correct nozzle size with correct oil pressure.	32.17
Smoking	Too little air	Correct air by adjustment.	32.17
	Not enough draft	Clean flue and clean furnace heat exchanger on oil side.	32.17
Smoke in conditioned space	Cracked heat exchanger	Replace heat exchanger.	32.2
	Smoke puffing out around burner or inspection door and pulling in through fan compartment	See preceding smoking problem.	
Oil smell	Oil leak or spill during service	Clean up oil spill or leak.	32.16

OBJECTIVES

After studying this unit, you should be able to

- describe a basic hydronic heating system.
- describe reasons for a hydronic system to have more than one zone.
- list four heat sources commonly used in hydronic heating systems.
- explain the difference between a wet-base and a dry-base boiler.
- state the reason a boiler is constructed in sections or tubes.
- discuss the reasons why air should be eliminated from hydronic heating systems.
- explain the effects air has on a cast-iron or steel boiler.
- describe the function of the air cushion or expansion tank.
- explain the operation of circulator pumps as they apply to hydronic heating systems.
- describe the importance of the "point of no pressure change."
- describe the purpose of limit controls and low-water cut-off devices.
- state the purpose of a pressure relief valve.
- state the purpose of a zone valve.
- list the various types of zone valves that are available.
- explain how "outdoor reset" can be used to increase system efficiency.
- sketch a series loop hydronic heating system.
- sketch a one-pipe hydronic heating system.
- explain the function of the diverter tee.
- explain the differences between a two-pipe direct-return hydronic heating system and a two-pipe reverse-return hydronic heating system.
- explain the application that requires the use of a balancing valve.
- list the benefits of primary–secondary pumping.
- describe the operation and function of mixing valves.
- describe the differences between radiant and conventional hydronic heating systems.
- list three common types of radiant heating system installations.
- describe a tankless domestic hot water heater used with a hydronic space-heating system.
- list preventive maintenance procedures for hydronic heating systems.

SAFETY CHECKLIST

✔ When checking a motor pump coupling, turn off the power and lock and tag the disconnect before removing the cover to the coupling.
✔ If you are checking the electrical service to a pump that has a contactor, shut off the power and remove the cover to the contactor. Do not touch any electrical connections with meter leads until you have placed one voltmeter lead on a ground source such as a conduit. Then touch the other meter lead to every electrical connection in the contactor box. The reason for this is that some pumps are interlocked electrically with other disconnects, and a circuit from another source may be in the box.
✔ Ensure that all controls are working properly in a hot water system because an overheated boiler has great explosion potential.

33.1 INTRODUCTION TO HYDRONIC HEATING

Hydronic heating systems are systems in which water or steam carries the heat through pipes to the areas to be heated. This text will cover only hot water systems because they are generally used in residential and light commercial installations.

In hydronic systems, water is heated and circulated through pipes to a heat transfer component called a *terminal unit,* such as a radiator or finned-tube baseboard unit. Here heat is given off to the air in the room. The cooler water is then returned to the heat source to be reheated. There is no forced moving air in these systems in most residential installations, and if properly installed there will be no hot or cold spots in the conditioned space. Hot water stays in the tubing and heating units even when the boiler is not running, so there are no sensations of rapid cooling or heating, which might occur with forced-warm-air systems. Air conditioning cannot readily be added to a structure that has hydronic heat. Usually, the hydronic system will not accommodate air conditioning, and a separate duct system must be added. Then the structure has two complete systems.

These systems are designed to include more than one zone when necessary. If the system is heating a small home or area, it may have one zone. If the house is a long ranch type or multilevel, there may be several zones. Separate zones are often installed in bedrooms so that these temperatures can be kept lower than those in the rest of the house, **Figure 33–1.**

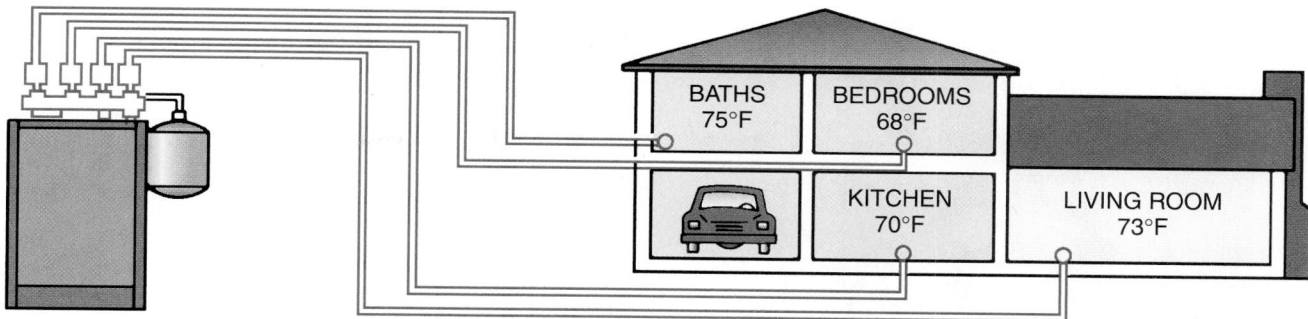

Figure 33–1 A four-zone hydronic heating system.

The water is heated using an oil, gas, or electrical heat source. These burners or heating elements are similar to those discussed in other units in this section of the book. Sensing elements start and stop the heat source according to the water temperature. The water is circulated with a centrifugal pump. A thermostatically controlled zone control valve allows the heated water into the zone needing the heat. Most residential installations use finned-tube baseboard units to transfer heat from the water to the air.

The design process of sizing of the system will not be covered in detail in this unit, but it is necessary that the boiler, piping, and terminal units be the proper size. All components including the pump and valves should also be sized properly for the correct water flow.

33.2   THE HEAT SOURCE

A major part of the hydronic heating system is the component that heats the water. The piping circuits and strategies that follow later on in this unit, however, are for the most part independent of the heat source—because the strategies indicate what is done with the water once it is heated. The heat used in hydronic systems can be generated by a number of different methods including the burning of fossil fuels, the collection of solar energy, the use of electric heaters, and the use of heat pump technology.

For the most part, the heated water used in hydronic systems comes from a boiler, but geothermal heat pump systems are becoming more and more popular for low-temperature (radiant) applications. Boilers can be used for both high- and low-temperature applications, but accommodations must be made to keep the temperature of the water that returns to the boiler hot enough to prevent the flue gases from condensing—if in fact the boiler is not intended to be used for flue-gas condensation.

The Boiler

A *boiler* is, in its simplest form, an appliance that heats water, using oil, gas, or electricity as the heat source. Some larger commercial boilers use a combination of two fuels. When one fuel is more readily available than another, it can be used, or the boiler can easily be changed over or converted to use the other fuel. **Figure 33–2** shows a gas-fired boiler, **Figure 33–3** shows an oil-fired boiler, and **Figure 33–4** shows an electric boiler.

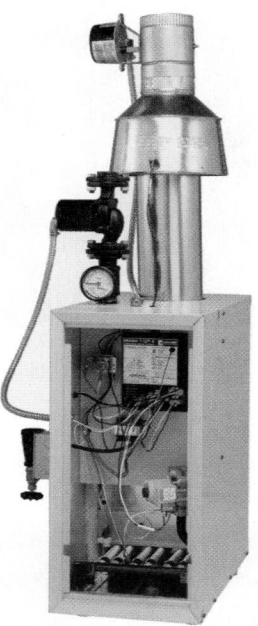

Figure 33–2 A gas-fired boiler. *Courtesy Weil-McLain Corporation*

Boilers can be used to supply water at various temperatures to the areas being heated, but the most common temperature to which water is heated is about 180°F. Newer technology, however, provides for a wide range of attainable boiler water temperatures that are based on the heating requirements of the space and the outside ambient temperature. Refer to the "Outdoor Reset" section in this unit for more on this very important topic. Hydronic systems can be designed to operate with water temperatures ranging from 90°F to over 200°F.

The temperature of the water returning to the boiler affects the operation of the boiler as well as process of removing flue gases from the structure. As the temperature of the return water to the boiler decreases, the temperature of the flue gases leaving the combustion chamber will also decrease. This is because the temperature of the heat exchanger will be lower and more heat will be transferred from the hot flue gases to the cooler heat exchanger. If the temperature of the flue gases is permitted to drop below about 130°F, the water vapor in the flue gases may begin to condense back into a liquid. Most boilers, classified as *conventional*

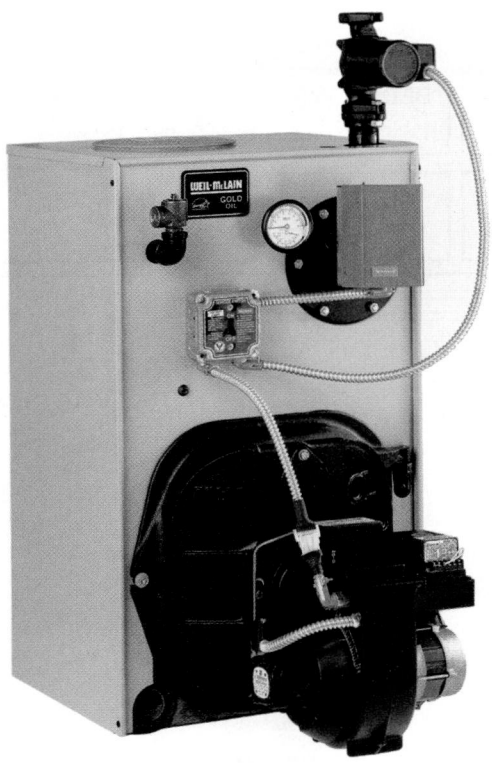

Figure 33–3 An oil-fired boiler. *Courtesy Weil-McLain Corporation*

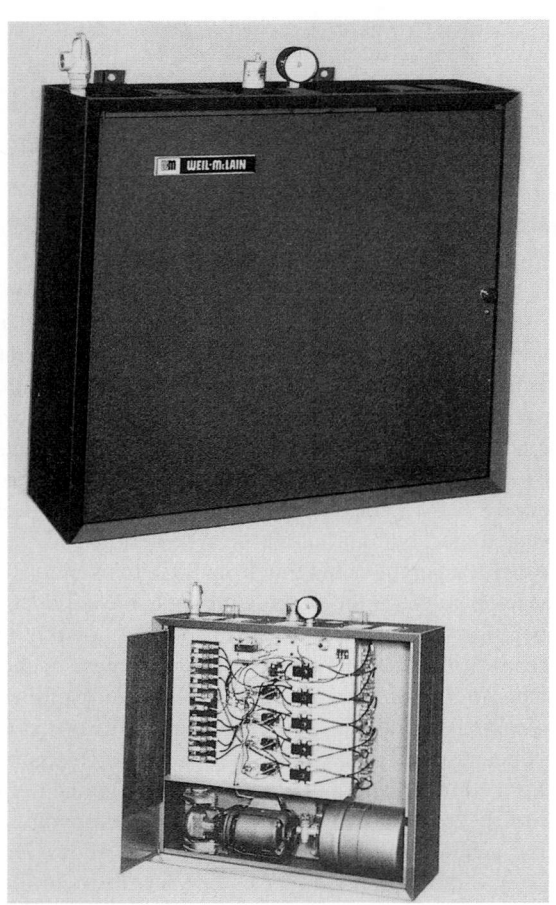

Figure 33–4 An electric boiler. *Courtesy Weil-McLain Corporation*

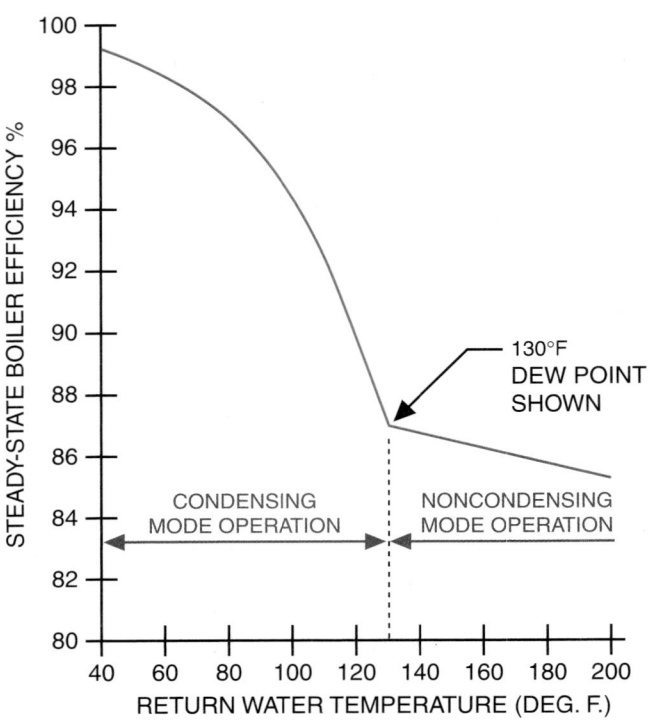

Figure 33–5 Efficiency of a boiler in both condensing and noncondensing modes.

boilers, are designed to operate with the flue-gas temperature above the 130°F dew point temperature to prevent the condensing of the flue gases. These condensing flue gases are highly corrosive and can damage the chimney and the heating equipment. Noncondensing boilers are designed with efficiencies up to about 86%. **Figure 33–5** shows the relationship between boiler return-water temperature, system efficiency, and dew point temperature for the flue gases.

In order to increase the efficiency of boilers, additional and/or more efficient heat exchangers are incorporated into the systems. These improved heat transfer surfaces remove more heat from the flue gases, thereby lowering the temperature of the flue gases below the dew point temperature. Condensing flue gases result. But, since these systems are designed to condense the flue gases, equipment design engineers are able to plan for and design around the condensing-flue-gas issue. These systems are referred to as *condensing boilers,* **Figure 33–6.** Condensing boilers are classified as 90+ because their efficiencies are over 90%. Condensing boilers are available in both gas- and oil-fired models.

CAST-IRON BOILERS. The most commonly encountered boiler in residential and light commercial applications is the cast-iron boiler. It is often made up of individual sections, **Figure 33–7.** Two classifications of boiler section assemblies are the wet-base type and the dry-base type. **Figure 33–7(A)** shows a section of a dry-base boiler. The term dry-base means that the area under the combustion chamber is dry, as there is no water there. **Figure 33–7(B)** shows a section of a wet-base boiler. In a wet-base boiler setup, the water being heated is

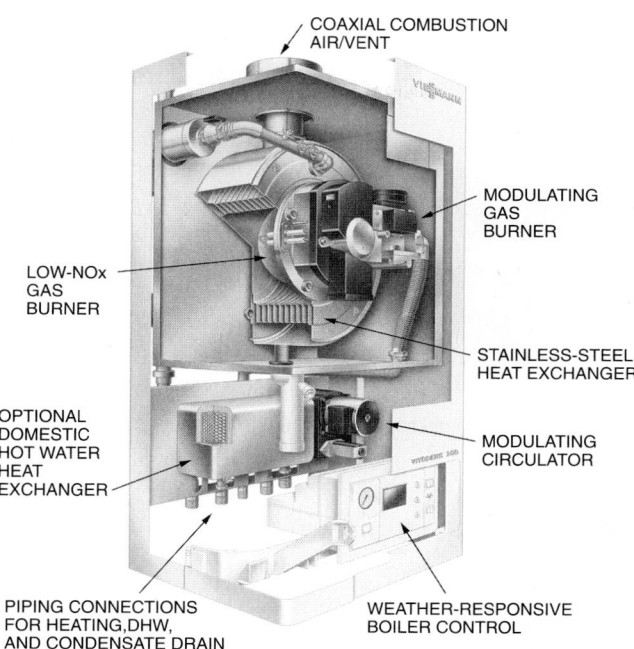

COAXIAL COMBUSTION AIR/VENT

MODULATING GAS BURNER

LOW-NOx GAS BURNER

STAINLESS-STEEL HEAT EXCHANGER

OPTIONAL DOMESTIC HOT WATER HEAT EXCHANGER

MODULATING CIRCULATOR

PIPING CONNECTIONS FOR HEATING, DHW, AND CONDENSATE DRAIN

WEATHER-RESPONSIVE BOILER CONTROL

Figure 33–6 Example of a wall-hung condensing boiler. *Courtesy Veissmann Manufacturing*

(A)

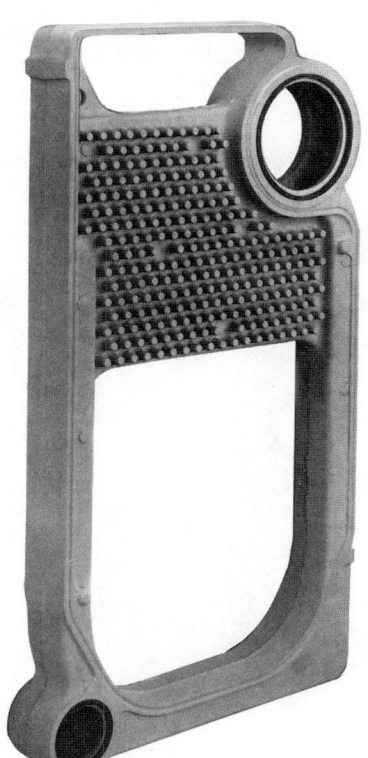

(B)

Figure 33–7 **(A)** Cast-iron dry-base boiler section. **(B)** Wet-base boiler section. *(A)* and *(B)* *Courtesy Weil-McLain Corporation*

located both above and below the combustion area. Multiple sections are bolted together to form the complete heat exchange surface between the burning fuel and the water being heated, **Figure 33–8.** The more sections a boiler has, the higher its capacity will be. The following are some key characteristics associated with the cast-iron boiler:

- They are very heavy appliances.
- They hold between 15 and 30 gallons of water.
- They are classified as high-mass boilers.
- They take a long time to heat up.
- They hold onto heat for a long period of time.
- They have longer run times and longer off cycles.
- They are used on closed-loop systems.
- Air in the system can result in boiler corrosion.

STEEL BOILERS. Within steel boilers, **Figure 33–9,** the water being heated surrounds a bundle of steel tubes through which the combustion gases pass on their way to the chimney. These tubes have a series of baffles located in them to increase the heat transfer surface area and to slow the speed of the flue gases as they pass through the tubes. Without the baffles, the hot flue gases would flow out of the appliance very quickly at a hotter temperature. By slowing the rate of flue-gas flow, more heat can be extracted from the gases, increasing the efficiency of the boiler.

COPPER WATER-TUBE BOILERS. The heat exchanger in a copper-tube boiler is constructed as a series of finned copper tubes, **Figure 33–10,** that are heated by the flue gases in the combustion chamber of the boiler, **Figure 33–11.** The heat exchangers on copper-tube boilers have much less surface area than cast-iron or steel boilers do—but, since copper is such a good conductor of heat, it is still able to heat water

quickly and efficiently. Copper-tube boilers are classified as low-mass boilers—given their light weight and the fact that they contain only a few gallons of water as compared with the 15 to 30 gallons that are found in the high-mass boilers made of cast iron or steel.

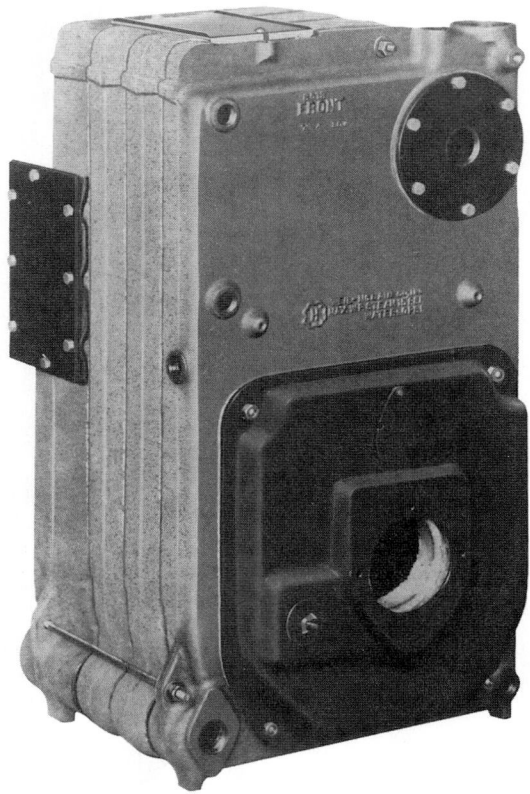

Figure 33–8 Boiler block assembly. *Courtesy Weil-McLain Corporation*

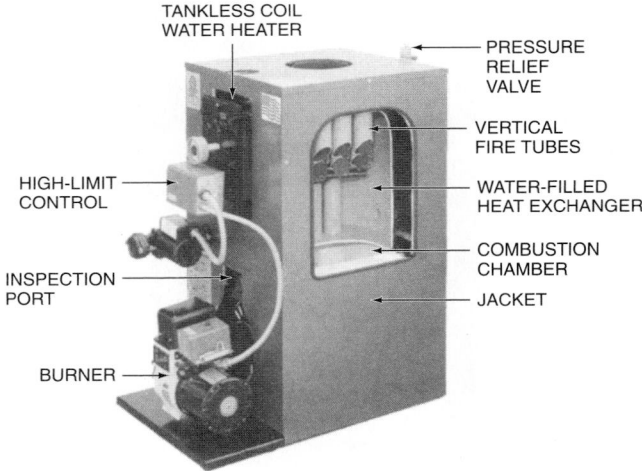

Figure 33–9 Cutaway view of a vertical fire-tube steel boiler. *Courtesy Columbia Boiler Company*

The Geothermal Heat Pump

One heat source that is becoming popular for use in conjunction with hydronic heating systems is the geothermal heat pump. To provide heat to the occupied space, the geothermal heat pump transfers heat from the earth to a water/water-antifreeze mixture. This heat-laden water mixture then transfers its heat to the space to be heated. Geothermal heat pump systems can heat system water to a temperature as high

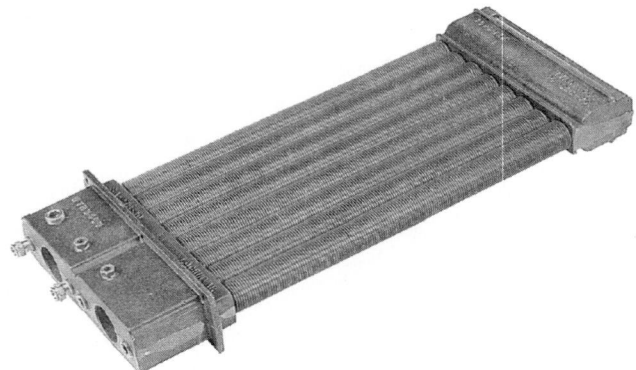

Figure 33–10 Finned copper-tube heat exchanger assembly.

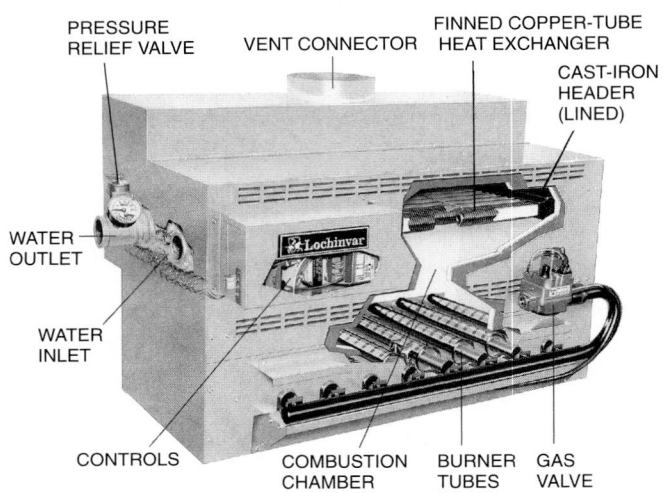

Figure 33–11 Cutaway view of a copper water-tube boiler. *Courtesy Lochinvar Corporation*

as 130°F, so it is a good option for low-temperature, radiant, hydronic heating systems. Refer to Unit 44, "Geothermal Heat Pumps," for more information on this topic. Radiant heating systems will be discussed later on in this unit.

33.3 THE BASIC HYDRONIC SYSTEM

Although there are many individual components that make up a hydronic heating system, we will begin our discussion of hydronic systems by examining two system devices to see how they function in a completely sealed water loop. This will enable us to gain a basic understanding of some of the properties of water and of the function of the circulator pump in the hydronic system. As the unit progresses, more components will be discussed to help broaden your understanding of these popular systems.

Expansion Tank

Hot water hydronic systems are closed-loop systems and are, ideally, air free. As water is heated, it expands—and if the system is totally air free, the excess pressure in the system will cause the pressure relief valve to open each and every time the water is heated. To prevent this from happening, an extra volume or space is provided to accommodate the extra volume of water that is created when the water is heated. The expansion tank, **Figure 33–12,** provides this extra volume.

There are two types of expansion tanks. One type is called the compression tank or standard expansion tank, **Figure 33–13.** This tank is nothing more than a steel tank located above the boiler. Upon installation, the tank is filled with air at an atmospheric pressure of 0 psi. As water is added to the boiler, the air that is being displaced is pushed up into the expansion tank, causing the pressure in the tank to increase, **Figure 33–14.** This creates a pressurized air cushion within the shell of the tank. As the water in the system is heated, the volume of water in the system increases and further compresses the air in the tank. Over time, however, this tank can become completely filled with water, thereby removing the air cushion. This situation can be easily identified because the pressure relief valve will continuously open and close as the water in the system is heated.

A more common type of expansion tank is the diaphragm-type expansion tank. This tank has a rubber, semipermeable membrane within the shell of the tank, **Figure 33–15.** One side of the tank contains pressurized air, whereas the other side of the tank is open to the water circuit. The air portion of the tank is pressurized at the factory, but it is a good idea to check the pressure with a tire gage prior to installing the tank in the system. For residential applications, these tanks often come precharged with 12 psi of air. The tank pressure is noted on the nameplate of the tank itself, **Figure 33–16.**

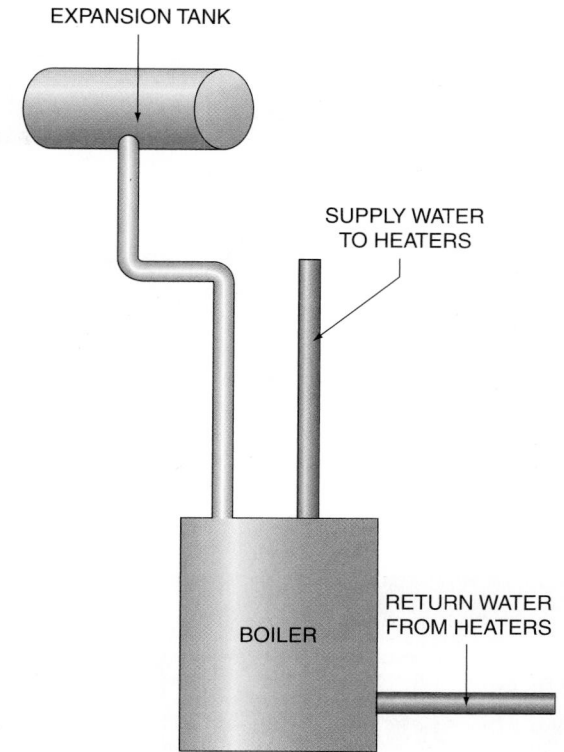

Figure 33–13 A steel expansion tank.

Figure 33–12 An expansion tank. *Photo by Eugene Silberstein*

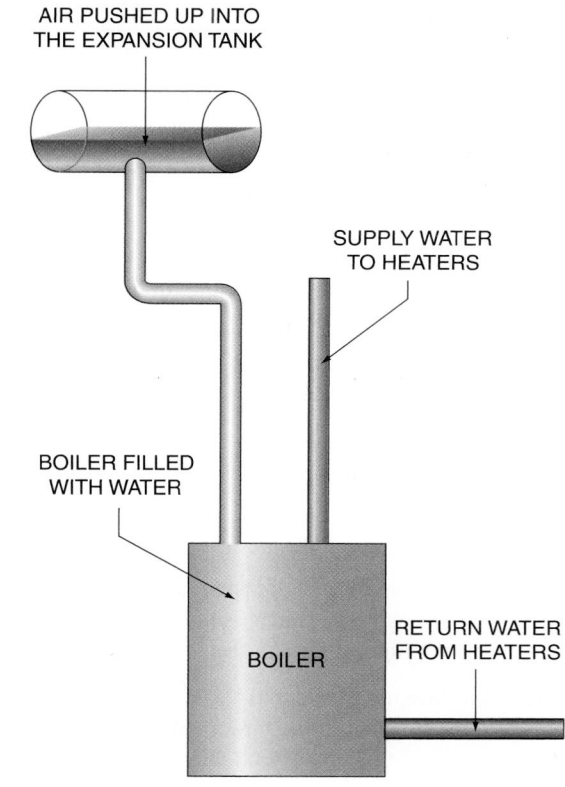

Figure 33–14 Pressure in the tank increases as water is added to the system.

Figure 33–15 Cutaway of a diaphragm-type expansion tank. *Photo by Eugene Silberstein*

Figure 33–16 Expansion tank data tag. *Photo by Eugene Silberstein*

To determine the actual pressure required for the expansion tank, it is necessary to measure the vertical distance (in feet) between the highest pipe in the system and the inlet of the expansion tank. We can then use the following formula:

$$P_{tank} = (H \div 2.31) + 5$$

Where P_{tank} = The pressure in the expansion tank in psi

    H = The vertical distance between the highest pipe in the system and the inlet of the expansion tank

    2.31 = The amount of vertical lift (head) we get for each pound of pressure

    5 = The additional pressure added to the system to ensure that there is a positive pressure at the highest pipe in the system

Consider the following residential application:
- The boiler is located in the basement.
- The opening of the expansion tank is 4 feet from the basement ceiling.
- The home has two floors, each with 9-foot ceilings.
- The second floor has baseboard heating and the highest pipe is 2 feet from the floor.

From this we can conclude that the "H" measurement is 15 feet (9 ft + 2 ft + 4 ft) and that therefore the required tank pressure is

$$P_{tank} = (H \div 2.31) + 5$$
$$P_{tank} = (15 \div 2.31) + 5$$
$$P_{tank} = 6.49 + 5$$
$$P_{tank} = 11.49 \text{ psi}$$

Since most residential applications will provide similar results, it is easy to see why the tanks are supplied by the manufacturer with a pressure of 12 psi.

Circulator/Centrifugal Pumps

Centrifugal pumps, also called *circulators*, **Figure 33–17**, force the hot water from the heat source through the piping to the heat transfer units and back to the boiler. These pumps use centrifugal force to circulate the water through the system. Centrifugal force is generated whenever an object is rotated around a central axis. The object or matter being rotated tends to fly away from the center due to its velocity. This force increases proportionately with the speed of the rotation, **Figure 33–18.**

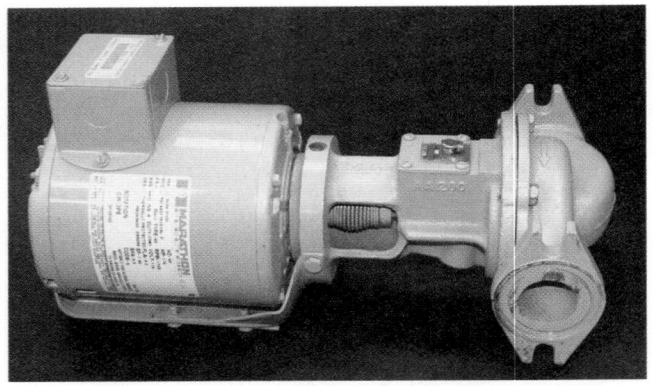

Figure 33–17 A centrifugal pump. *Courtesy Ferris State University. Photo by John Tomczyk*

Figure 33–18 When the impeller is rotated, it "throws" the water away from the center of the pump and out through the opening.

The *impeller* is that part of the pump that spins and forces water through the system. The proper direction of rotation of the impeller is essential. The vanes or blades in the impeller must "slap" and then throw the water, **Figure 33–19.** Impellers used on circulating pumps in hot water heating systems usually have sides enclosing the vanes. Such types are called *closed* impellers. Many pumps, **Figure 33–20,** used in closed systems where some makeup water is used (which will cause corrosion) are called *bronze-fitted* pumps. They generally have a cast-iron body with the impeller and other moving parts made of bronze or nonferrous metals. Others have stainless-steel or all bronze parts.

Centrifugal water pumps are not positive displacement pumps as are most compressors. The term *circulating pump* is used because the centrifugal pumps used in small systems do not add much pressure to the water from the inlet to the outlet. It is important to understand how a centrifugal pump responds to pressure and load changes.

The centrifugal pump is responsible for creating a pressure difference between the water at its inlet and the water at the outlet of the pump. It is this pressure differential that makes the water flow through the piping circuit. However, before water can flow through the circuit, the pressure difference generated by the pump must overcome the resistance of the piping circuit itself. Consider a pump that is operating in a system that offers no resistance to flow. If this pump were

supplied with an unlimited amount of water to pump—and this water were immediately pumped to atmospheric pressure once it left the pump, it is safe to say that the amount of water moved by the pump would be the pump's maximum capacity. As resistance to flow is added, the volume of water moved by the pump will decrease, reducing the pump's capacity. A point will be reached where the amount of water pumped will reach zero because the resistance of the piping arrangement is too much for the pump to overcome.

The following example will demonstrate how a centrifugal pump performance curve is developed. This is a progressive example. See **Figure 33–21(A)–(J).**

A. A pump is connected to a reservoir that maintains the water level just above the pump inlet. The pump meets no resistance to the flow, and maximum flow exists. The water flow is 80 gpm.

B. A pipe is extended 10 ft high (known as feet of head, or pressure), and the pump begins to meet some pumping resistance. The water flow reduces to 75 gpm.

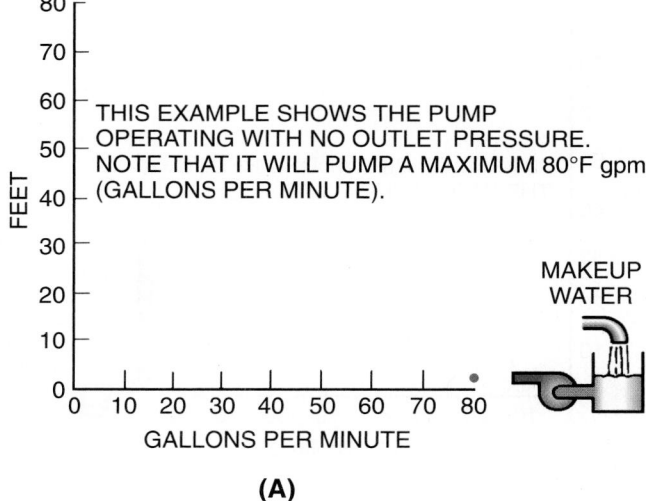
(A)

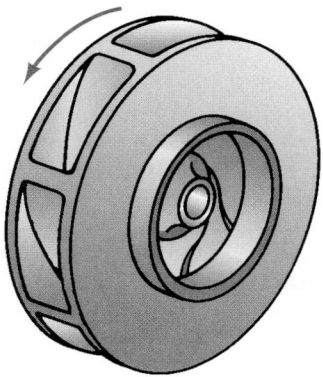

Figure 33–19 A pump impeller. Note the direction of the rotation as indicated by the arrow. This is an example of a closed impeller.

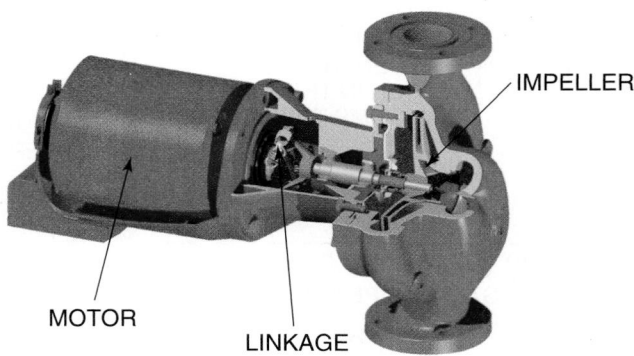

Figure 33–20 Cutaway of a centrifugal pump. *Courtesy Bell and Gossett*

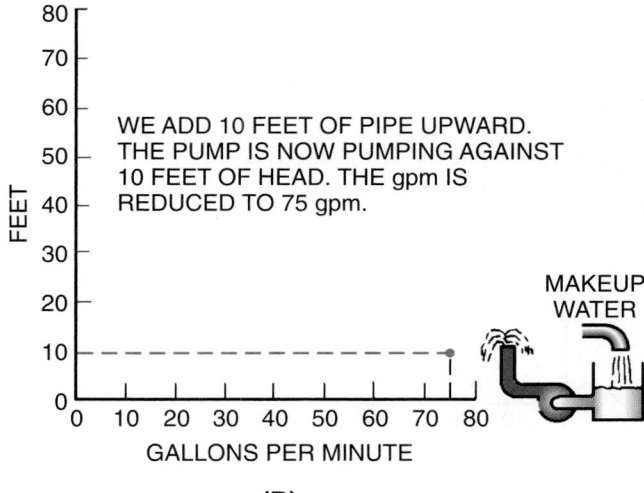
(B)

Figure 33–21 Developing a centrifugal pump performance curve **(A)–(J).**

C. The pipe is extended to 20 ft, and the flow reduces to 68 gpm.

D. When 30 ft of pipe is added, the water flow reduces to 63 gpm.

E. The pipe is extended to 40 ft. The pump is meeting more and more resistance to flow. The water flow is now 58 gpm.

F. At 50 ft, the flow slows to 48 gpm.

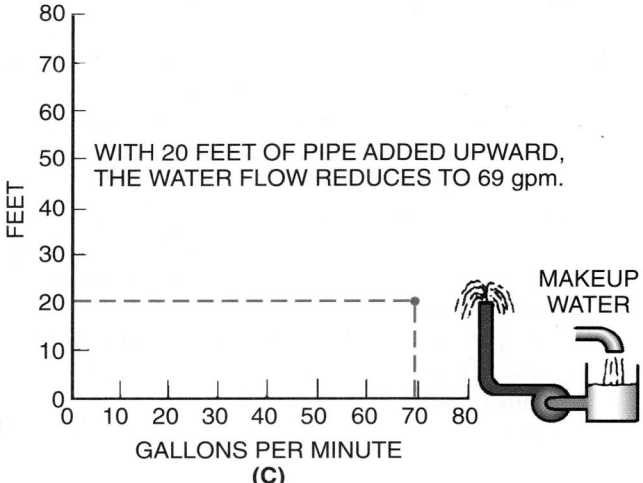

(C)

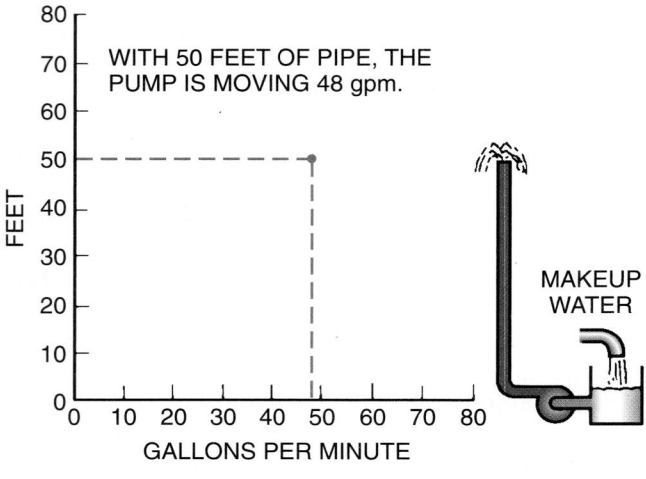

(F)

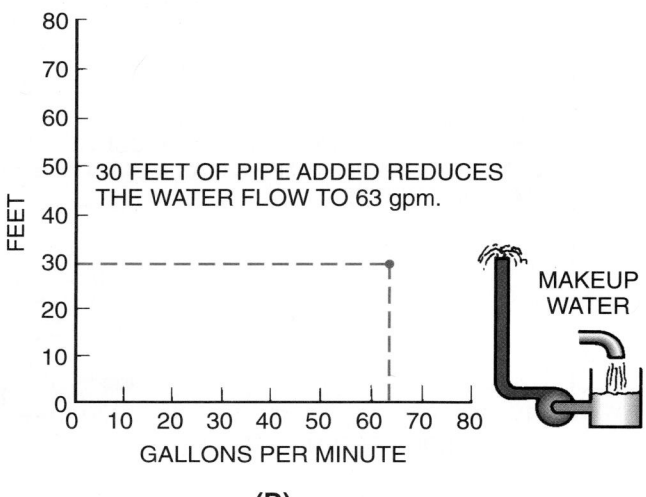

(D)

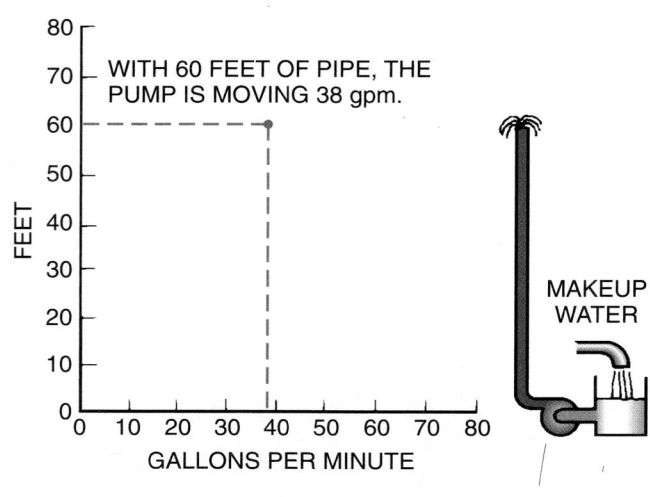

(G)

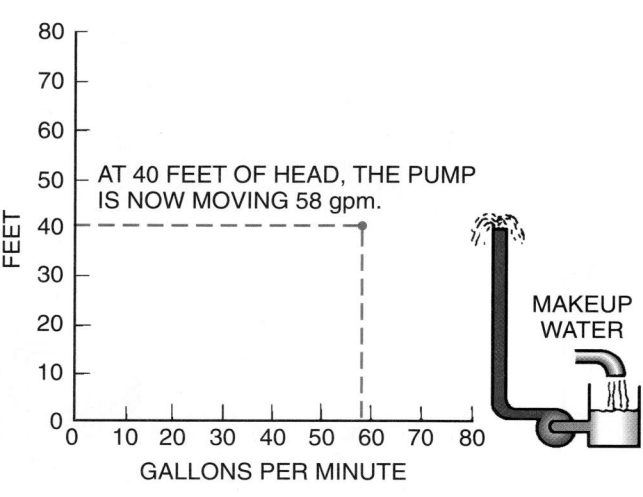

(E)

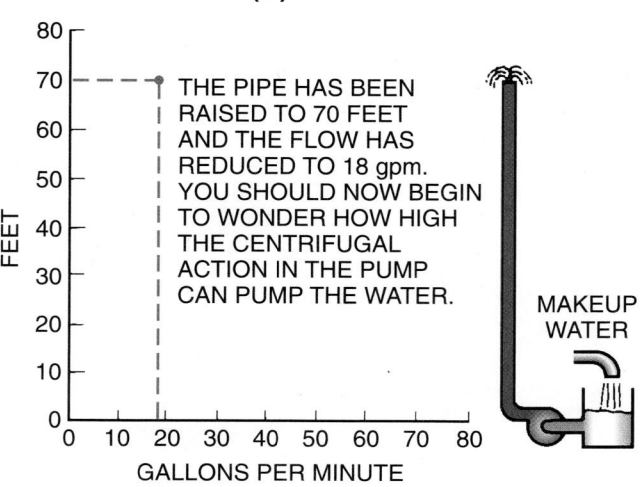

(H)

Figure 33–21 *(continued)*

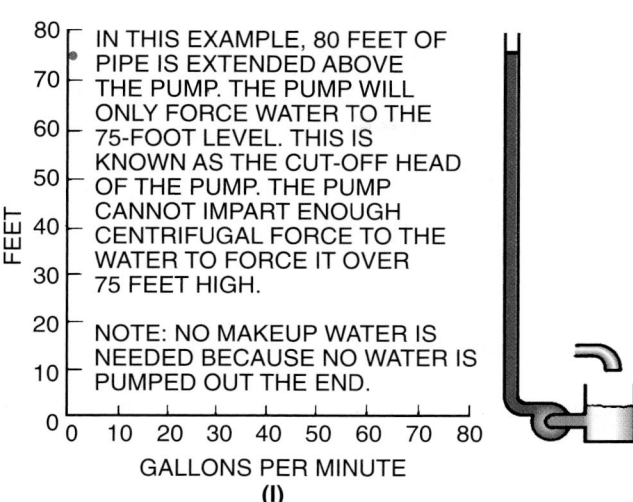

IN THIS EXAMPLE, 80 FEET OF PIPE IS EXTENDED ABOVE THE PUMP. THE PUMP WILL ONLY FORCE WATER TO THE 75-FOOT LEVEL. THIS IS KNOWN AS THE CUT-OFF HEAD OF THE PUMP. THE PUMP CANNOT IMPART ENOUGH CENTRIFUGAL FORCE TO THE WATER TO FORCE IT OVER 75 FEET HIGH.

NOTE: NO MAKEUP WATER IS NEEDED BECAUSE NO WATER IS PUMPED OUT THE END.

GALLONS PER MINUTE

(I)

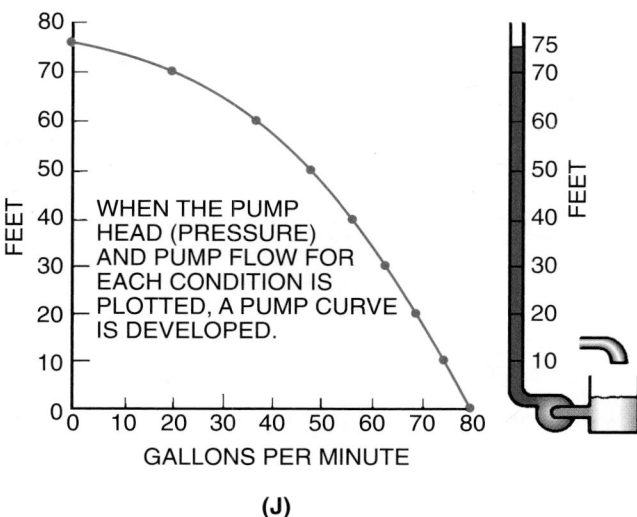

WHEN THE PUMP HEAD (PRESSURE) AND PUMP FLOW FOR EACH CONDITION IS PLOTTED, A PUMP CURVE IS DEVELOPED.

GALLONS PER MINUTE

(J)

Figure 33–21   *(continued)*

G. At 60 ft, the flow slows to 38 gpm.

H. At 70 ft, the water flow has been reduced to 18 gpm. The pump has almost reached its pumping limit. The pump can impart only a fixed amount of centrifugal force.

I. When 80 ft of pipe is added to the pump outlet, the water will rise to the 75-ft level. If you could look down into the pipe, you would be able to see the water gently moving up and down. The pump has reached its pumping pressure head capacity. This is known as the shutoff point of the pump.

J. This is a manufacturer's pump curve and the preceding examples show how the manufacturer arrives at the pump curve.

It would be awkward for the manufacturer to add vertical pipe to all pump sizes for testing purposes so they use valves to simulate vertical head. Before we can use valves, we must be able to convert liquid water vertical head to gage readings. Gages read in psig. When a gage reads 1 psig, it is the same pressure that a column of water 2.309 ft (27.7 in.) high exerts at the bottom, **Figure 33–22.** This would be the same as saying that a column of water 1 ft high (feet of head) is equal to 0.433 psig because 1 psig divided by 2.3093 ft = 0.433. Some examples of water columns and pressure are shown in **Figure 33–23.**

A pump manufacturer may use gages to establish pump head (pressure) by placing a gage at the inlet and a gage at the outlet and reading the difference. This is known as psi difference. Psi difference is then converted to feet of head, **Figure 33–24.**

The power consumed by the motor driving a centrifugal pump is in proportion to the quantity of water the pump circulates. For example, in the first pump example in **Figure 33–21(A),** the pump would require more horsepower than in example (I) where the pump was pumping against the pump cut-off pressure. When a valve is installed in a piping circuit and closed, it would seem that the power consumption would

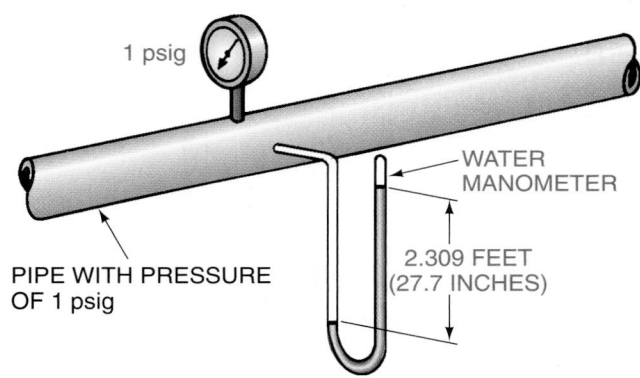

1 psig

WATER MANOMETER

PIPE WITH PRESSURE OF 1 psig

2.309 FEET (27.7 INCHES)

Figure 33–22   A gage pressure reading of 1 psi is equal to 2.309 feet of water.

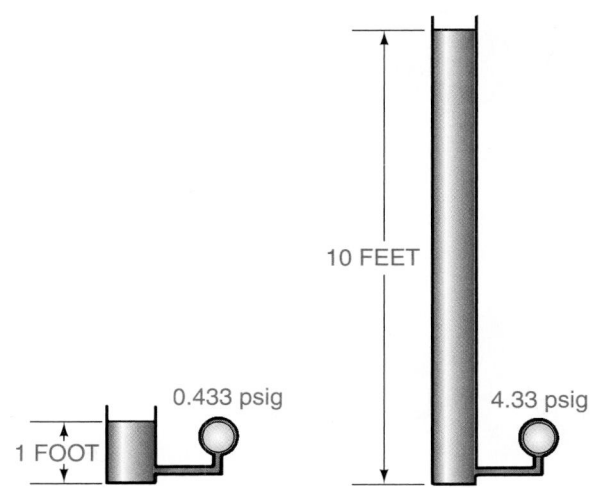

10 FEET

0.433 psig

4.33 psig

1 FOOT

Figure 33–23   This figure shows that different heights of water columns exert different pressures. Each time you increase the water column one foot, the pressure rises 0.433 psi.

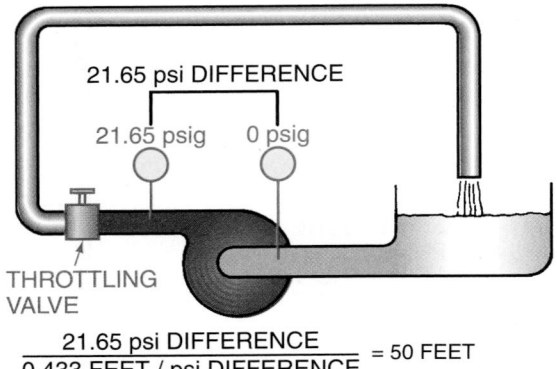

21.65 psi DIFFERENCE

21.65 psig   0 psig

THROTTLING
VALVE

$$\frac{21.65 \text{ psi DIFFERENCE}}{0.433 \text{ FEET / psi DIFFERENCE}} = 50 \text{ FEET}$$

Figure 33–24 This figure illustrates how a pump manufacturer may test a pump for pumping head. More elaborate systems that measure gallons per minute at the same time are used to establish a pump curve.

rise because of an increase in pump head, but it does not. As the discharge valve is closed, the water begins to recirculate in the pump, and the power consumption reduces. This can be demonstrated by shutting the valve at the outlet of a centrifugal pump and monitoring the amperage, **Figure 33–25**.

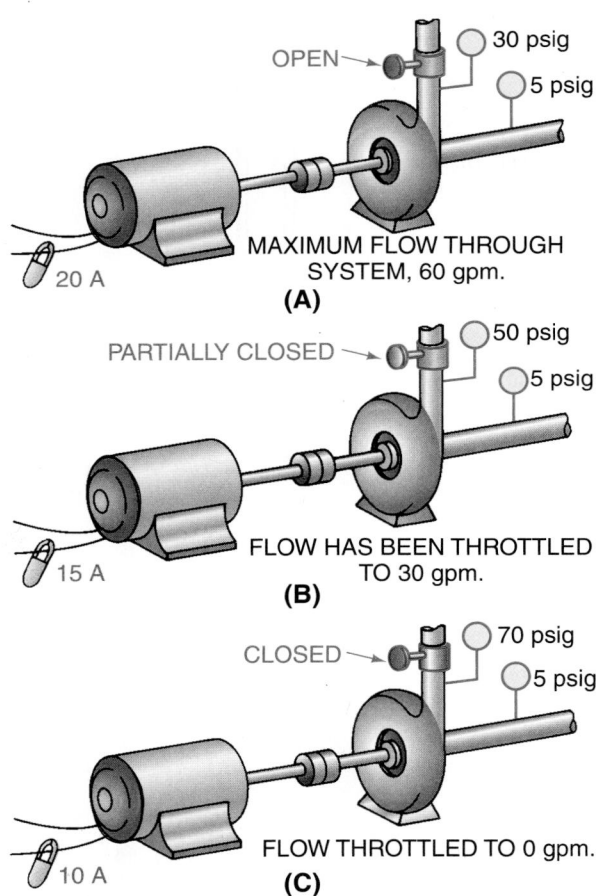

OPEN → 30 psig
5 psig
20 A
MAXIMUM FLOW THROUGH
SYSTEM, 60 gpm.
(A)

PARTIALLY CLOSED → 50 psig
5 psig
15 A
FLOW HAS BEEN THROTTLED
TO 30 gpm.
(B)

CLOSED → 70 psig
5 psig
10 A
FLOW THROTTLED TO 0 gpm.
(C)

Figure 33–25 **(A)** Maximum flow. **(B)** As the flow is reduced by throttling the outlet valve, the power to drive the motor reduces the current required to turn the pump impeller in the water. **(C)** Throttled to 0 gpm.

33.4 THE POINT OF NO PRESSURE CHANGE

Probably one of the most important and busy locations in the hydronic hot water system is the "point of no pressure change." As its name implies, this is the point in the system where the pressure, no matter what the system is doing, will remain the same. This provides a great reference point for system evaluation and also provides a location for multiple system-component connections.

Consider a piping arrangement that contains only the two components discussed so far in this unit: the expansion tank and the circulator pump, **Figure 33–26**. Notice that, in the figure, the pump is discharging in a direction pointing toward the expansion tank. Assume that the pressure in the expansion tank prior to installation was 10 psi and that the pump generates a pressure difference of 15 psig between its inlet and outlet. When the piping loop is filled, the pressure will be 10 psi, as determined by the pressure in the expansion tank. Since the loop is completely filled and our expansion tank has a 10-psi cushion of air inside it, what will happen when the pump is turned on? Will the pump push water into the expansion tank? Will the pump pull water out of the expansion tank?

The answer is that water will neither be pushed into nor pulled out of the expansion tank and the pressure in the circuit at the inlet of the expansion tank will remain unchanged at 10 psi. Water cannot be transferred from the loop to the expansion tank because there would be nothing to replace the water that leaves the loop. Also, water cannot be relocated from the expansion tank to the loop because there is no room in the loop to accept any additional water. So, if water cannot enter or leave the expansion tank, the pressure in the loop at the inlet of the expansion tank will remain constant. Note: This will change slightly once we actually heat the water, as the water will expand and some will be pushed into the expansion tank when this heating occurs.

So, if the pump creates a pressure difference of 15 psi and the pressure at the outlet of the pump cannot change, the pressure at the inlet of the pump will have to drop to account

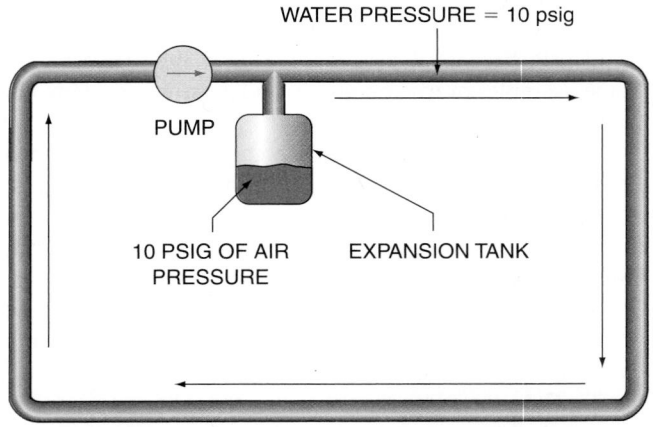

WATER PRESSURE = 10 psig

PUMP

10 PSIG OF AIR
PRESSURE   EXPANSION TANK

Figure 33–26 Simple loop with only an expansion tank and a circulator.

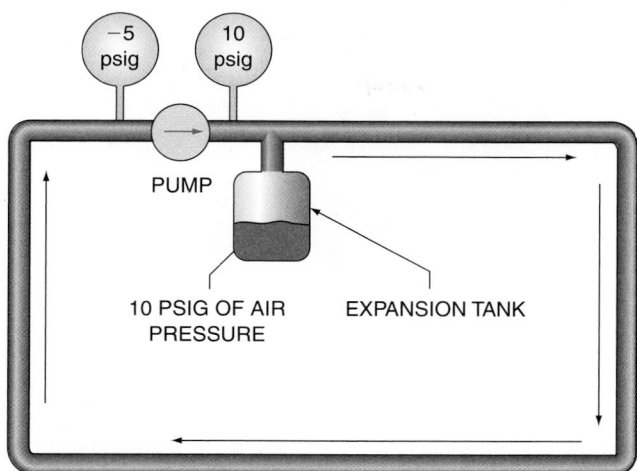

Figure 33–27 Pumping toward the expansion tank can result in a vacuum in the piping circuit.

for the 15-psi drop across the pump, **Figure 33–27**. This will create a vacuum in the piping circuit. If a leak should develop (at the pump's mounting flange, for example), air will be pulled into the system, creating system problems. Therefore, the operating pressure range for this system will be between −5 psi and +10 psi.

By moving the pump to the other side of the expansion tank, we now eliminate the vacuum situation. The pressure at the inlet of the circulator will be 10 psi and the pressure at the outlet of the pump will be 25 psig (10 psi + 15 psi), **Figure 33–28**. Relocating the pump in this manner minimizes the possibility of having the system pull into a vacuum and also reduces the effects of air in the system. We will be addressing air in the next few sections, as it is an important aspect of hot water systems—but one issue needs to be mentioned here. At higher system pressures, air bubbles are smaller and are more likely to remain in solution. Air bubbles that remain in solution with the water are far less likely to have a large negative effect on system operation and performance.

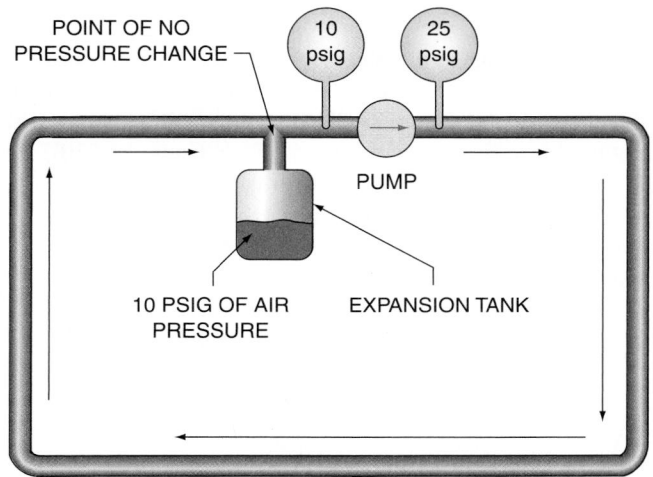

Figure 33–28 Point of no pressure change.

As we move through this unit, the importance of the "point of no pressure change" will become more and more evident. This point in the system is the desired connection point for the following system components:

- Inlet of the circulator pump
- Inlet of the expansion tank
- Air separator
- Air vent
- Outlet of the pressure-reducing (water-regulating) valve

33.5 OTHER HYDRONIC SYSTEM COMPONENTS

The expansion tank and the circulator are the two components needed to make up the most basic of hydronic systems. A system such as this, however, would not function effectively because a number of important operational and safety issues would not be accounted for. In addition to simply moving the water and allowing for expansion, we must also do the following:

- Separate air from the water
- Remove the air from the system
- Provide for makeup water to the system
- Provide safety controls in the event of system overpressurization
- Provide for desired flow control
- Provide for desired temperature control

The system components discussed in the next sections allow the system to operate as needed and desired—while simultaneously ensuring that it operates in a safe and efficient manner.

Air Separator and Air Scoop

When working with hot water systems, it is important to remove any air that may be trapped in the piping circuit, the boiler, or any other system component. Air, as mentioned earlier in the unit, can lead to excessive corrosion in cast-iron and steel boilers. As the temperature of the water in the system increases, the rate of corrosion increases. Corrosion takes time to occur, so the effects of air on the structural integrity of the boiler may not be readily seen. However, there are more significant and immediate problems associated with air in the hydronic hot water system. Since air is lighter than water, air pockets will form at the highest points in the system. These trapped air pockets prevent water flow and can result in insufficient heating. In addition, trapped air can lead to noisy pipes and noisy system operation.

The air separator, **Figure 33–29**, is the hydronic system component that separates air from the water as it flows through the system. It is important to note that the air separator does not remove air from the system—it simply separates the air from the water. Another system component, the air vent, is responsible for venting, or removing the air from the system. Air separators are capable of removing small "microbubbles" from the system. The new generation of air separators operates on the concept of "collision and adhesion." As water passes through the separator, it comes into contact with a mesh screen or similar material, **Figure 33–30**.

Figure 33–29 An air separator. *Courtesy Bell and Gossett*

Figure 33–30 Wire screen in the air separator. *Photo by Eugene Silberstein*

The air bubbles collide with the screen and adhere to it. As more and more air bubbles adhere to the mesh, the bubbles get larger, break loose, and travel up into the air vent, where they are vented from the system.

Other air separation devices, known as air scoops, **Figure 33–31,** are installed on horizontal runs of straight pipe, which cause the air in the pipe to separate from the water and rise to the top of the pipe. Using a baffle, the air scoops then

Figure 33–31 A cast-iron air scoop. *Courtesy Watts Regulator, Co.*

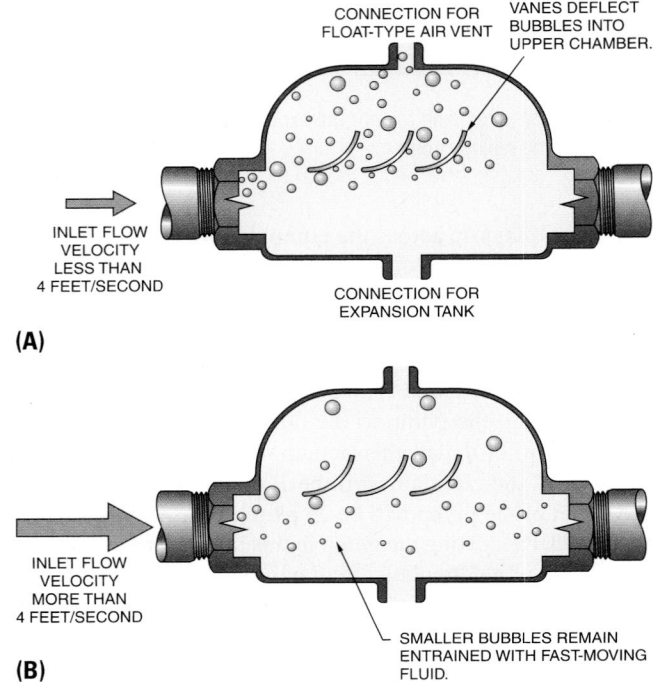

(A)

(B)

Figure 33–32 **(A)** When the flow velocity is lower than 4 feet per second, the vanes will deflect the air bubbles into the upper chamber. **(B)** If the velocity of the water is more than 4 feet per second, the bubbles remain trapped in the water and cannot be scooped by the baffle.

scoop the air out of the pipe, **Figure 33–32.** In order to work properly, the velocity or speed of the water must be lower than 4 feet per second, **Figure 33–32(A),** or the smaller bubbles will not be able to rise to the top of the device, **Figure 33–32(B).** At the top of the device, the trapped air is directed to the air vent for removal from the system. Air scoops are directional, so be sure to install them in the system with the arrow pointing in the direction of water flow.

Air Vent

Once the air has been separated from the circulating water, the air vent is responsible for removing it from the system. Air vents can be manually operated devices or automatic devices. Air vents are typically located at high points in the hydronic system as well as directly above air separators and air scoops, as these are devices that accumulate air that is separated from the water in the system. Manual air vents are the simplest of the air vents found the hydronic system, **Figure 33–33.** They are often referred to as "coin-operated" air vents, or "bleeders." As the name implies, these vents are opened and closed manually as the need arises. Automatic air vents, **Figure 33–34,** rely on an internal disc that swells when water is sensed. In operation, if there is no air at the vent location, the disc will be wet and swollen. This will seal off the air vent port. As air accumulates, the disc will dry out and shrink, allowing the air to pass through to the vent port. As the air is removed and replaced with water, the disc will swell, closing off the port. Since it takes time to fully close, there is often a small amount of water loss when the device vents.

Another type of automatic air vent utilizes a float mechanism, **Figure 33–35.** The position of the float is determined by the amount of air in the device. When there is no air in the air vent, the float is at the top of the vent and the vent port is closed, **Figure 33–36(A).** As air accumulates in the vent, the float drops, opening the vent port and removing the air from the system, **Figure 33–36(B).**

Figure 33–35 A float-type air vent. *Courtesy Maid O' Mist*

Figure 33–33 A "coin-operated" air vent. *Courtesy Bell and Gossett*

Figure 33–34 An automatic air vent. *Courtesy Bell and Gossett*

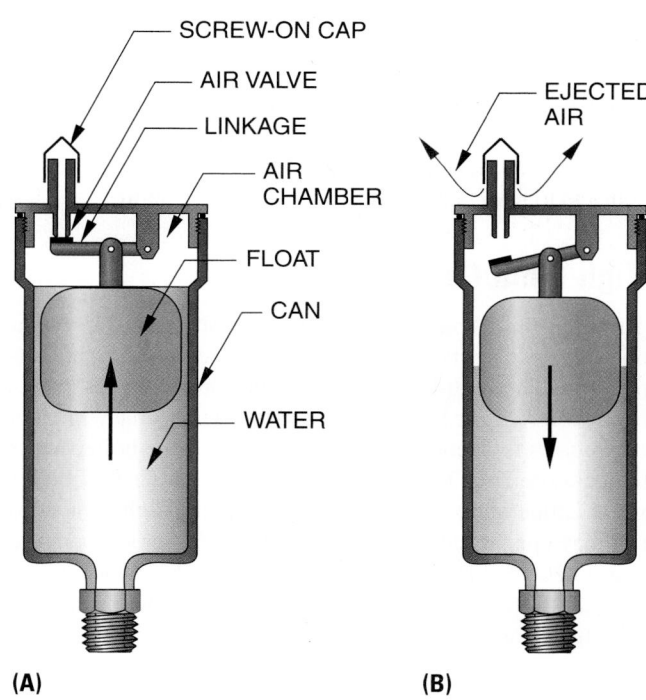

(A) **(B)**

Figure 33–36 **(A)** The float closes the valve when there is no air in the air vent. **(B)** When air is present, the float falls and the valve opens to vent the air.

Temperature-Limiting Control (Aquastat)

The temperature-limiting control, more commonly known as the aquastat, is a temperature-controlled switch that is intended to maintain the temperature of the water in a hydronic heating system, **Figure 33–37.** If a heating system is to maintain a temperature of 180°F with a 10-degree differential, the boiler will cycle off when the water is heated to 180°F. When the water cools to 170°F (180°F − 10°F), the boiler cycles back on. The water temperature will, therefore, fluctuate between 170°F and 180°F. The aquastat is commonly equipped

Figure 33–37 A temperature-limiting control, or aquastat. This is the system component responsible for maintaining the boiler water at the desired temperature. *Photo by Bill Johnson*

Figure 33–38 A pressure-reducing valve. *Courtesy Bell and Gossett*

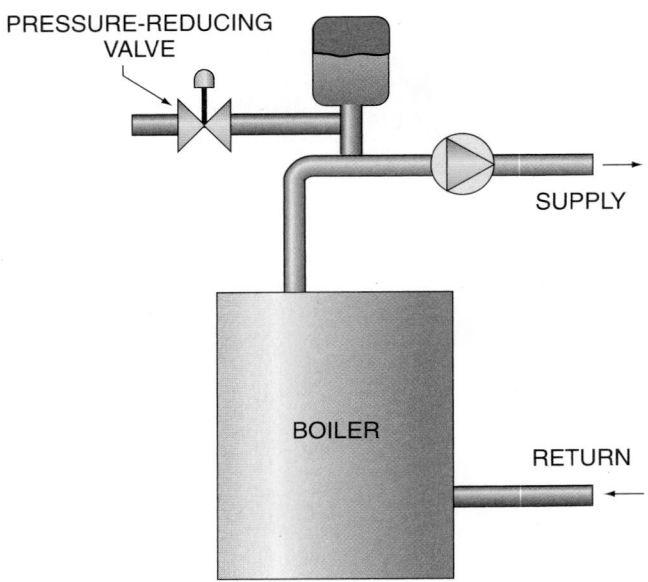

Figure 33–39 Proper location for the pressure-reducing valve.

with a remote sensing bulb that is positioned in the boiler to provide accurate control of the boiler water temperature. The aquastat is an operational control that opens and closes its contacts to maintain the desired water temperature.

High-Limit Control

The high-limit control, like the aquastat, is a device that opens and closes its contacts in response to boiler water temperature. The high-limit control is not, however, an operational control. The high-limit is a safety device that will open its contacts to deenergize the heat source if the boiler water temperature rises too high. If the maximum desired boiler water temperature is 180°F, for example, the high-limit control may open its contacts when the boiler water temperature rises to 200°F. Since 200°F boiler water is above the desired level, at this temperature a system problem is likely and the high-limit control will respond accordingly.

Water-Regulating Valve (Pressure-Reducing Valve)

Water heating systems should have an automatic method of adding water back into the system if water is lost due to leaks. The source for this water may have a pressure too great for the system. A *water-regulating valve*, **Figure 33–38,** is installed in the water makeup line leading to the boiler and is set to maintain the pressure on its leaving side (entering the boiler) at less than the relief valve on the boiler. These low-pressure water boilers have a working pressure of 30 psig.

The pressure-reducing valve should be connected to the point of no pressure change to ensure that the system is filled to the desired pressure, **Figure 33–39.** Since the pressure of

the system will change at every other point in the system as the system cycles on and off, the system fill pressure will also change unless the point of no pressure change is the system connection point.

Pressure Relief Valve

The American Society of Mechanical Engineers (ASME) Boiler and Pressure Vessel Codes require that each hot water heating boiler have at least one officially rated *pressure relief valve* set to relieve at or below the maximum allowable working pressure of the low-pressure boiler. This would be 30 psig. This valve discharges excessive water when pressure is created by expansion. It also releases excessive pressure if there is a runaway overfiring emergency, **Figure 33–40.**

Figure 33–40 A safety relief valve. *Courtesy Bell and Gossett*

Low-Water Cut-Off

The low-water cut-off valve, **Figure 33–41,** is responsible for deenergizing the system in the event the level of water in the system drops below the desired point. In some states, low-water cut-offs are required by law, but even in states where they are not required, it is good field practice to equip all systems with one. This is especially true on radiant heating systems, as the tubing is usually below the level of the heat source and often buried in concrete. A leak in the tubing circuit can cause the system to drain itself.

Figure 33–41 A low-water cut-off. *Photo by Eugene Silberstein*

Zone Valves

Zone control valves, or simply zone valves, are thermostatically controlled valves that control water flow to the various zones in the hydronic heating system. Some zone valves are equipped with gear-motor actuators, whereas others have heat-motor actuators, **Figure 33–42.** When a particular zone

(A)

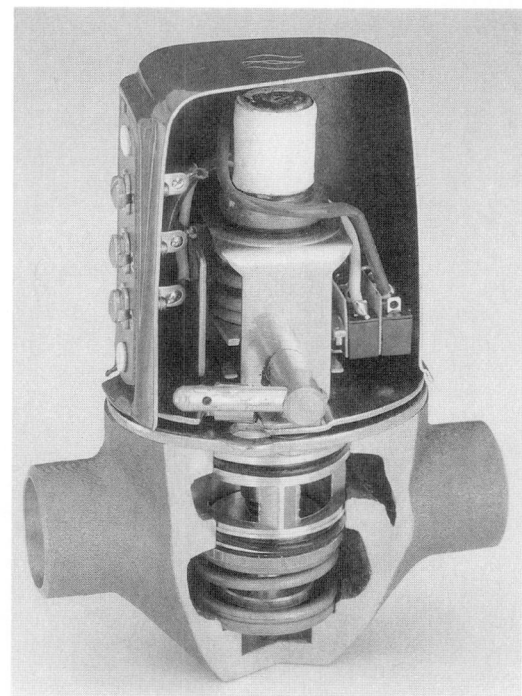

(B)

Figure 33–42 Two examples of hydronic zone valves. **(A)** Zone valve with a gear-motor actuator. **(B)** Zone valve with a heat-motor actuator. *Courtesy Sparco, Inc.*

calls for heat, the zone valve is energized and it opens. On the gear-motor variety of zone valve, when potential is applied to the valve, a motor-driven gear assembly slowly opens the valve. When potential is applied to the heat-motor variety of zone valve, electrical current generates heat in the valve and a bimetal strip warps to open the valve.

Zone valves are often equipped with a setting that allows the valve to be opened manually. This position can be used to help troubleshoot the heating system and, in the event of zone-valve malfunction, provide temporary heat in the conditioned space.

Although there are many different system control strategies and options, here is one application that would employ zone valves. When an individual zone calls for heat, the zone valve for that zone opens. When the zone valve has completely opened, an end switch on the valve closes to start the circulator pump. This will circulate heated water from the boiler to the occupied space until the thermostat is satisfied. When the space reaches the desired temperature, the zone valve closes and the circulator is deenergized. **Figure 33–43** shows two zones being supplied hot water by a single circulator. When either zone is calling for heat, the circulator is energized.

Figure 33–43 Gear-motor zone valves on a multiple-zone loop. *Photo by Eugene Silberstein*

Zone valves are available in both two-port and three-port varieties. The piping circuit in **Figure 33–44** shows an application that uses two-port zone valves. If only one of the three heating zones are calling for heat, the other two supply circuits will be closed off and the pressure in the supply header will likely increase. Refer to the section called "Pressure Differential Bypass Valve," for a discussion of valves that help alleviate the problem associated with increased system pressure due to closed zone valves.

When three-port zone valves are used, the water flow rate remains constant because water flows through all valves, whether or not there is a call for heat, as long as at least one zone is calling for heat. **Figure 33–45** shows a piping circuit that uses three-port zone valves. It can be seen that Zone 2 is the only zone calling for heat. When hot water enters zone valve #2, it leaves and flows through the terminal or heating units in that zone before returning to the boiler. When hot water flows through zone valve #1 and zone valve #3, it can be seen that, when the water leaves the valves, it is directed right back to the boiler without flowing through the terminal units in those zones.

Balancing Valves

When designing a hot water heating system, consideration must be given to the flow rate of the water and the friction in the system. Friction is caused by the resistance of the water flowing through the piping, valves, and fittings in the system. The flow rate is the number of gallons of water flowing each minute through the system. A system is considered to be in balance when the resistance to the water flow is the same in each flow path. A means for balancing the system should be provided in all installations. One method is to install a *balancing valve,* **Figure 33–46,** in each heating circuit branch. This valve is adjustable.

This valve has pressure taps on either side of the valve where a very accurate pressure gage can be fastened to check the pressure drop across the valve. The gage is calibrated in in. WC. The pressure drop then can be plotted on a chart for the specific valve to show the gallons of water per minute (gpm), and then compared with the system specifications. The valve then can be adjusted to the correct flow rate. NOTE: The valve adjustment is generally determined by the system designer, and the installer sets them accordingly. Be sure to read the manufacturer's instructions for the correct procedure for setting the valve.

Pressure Differential Bypass Valve

On systems that use two-port zone valves, excessive pressure in the supply header may result when multiple zone valves close. The piping circuit may become excessively noisy as a result of the increased water velocity and water banging within the piping circuit. In order to help alleviate this situation, the pressure differential bypass valve, **Figure 33–47,** is often used. The bypass valve is located between the supply-and-return headers of the system, **Figure 33–48**. When all, or most, of the zone valves are open, the bypass valve is in the closed position

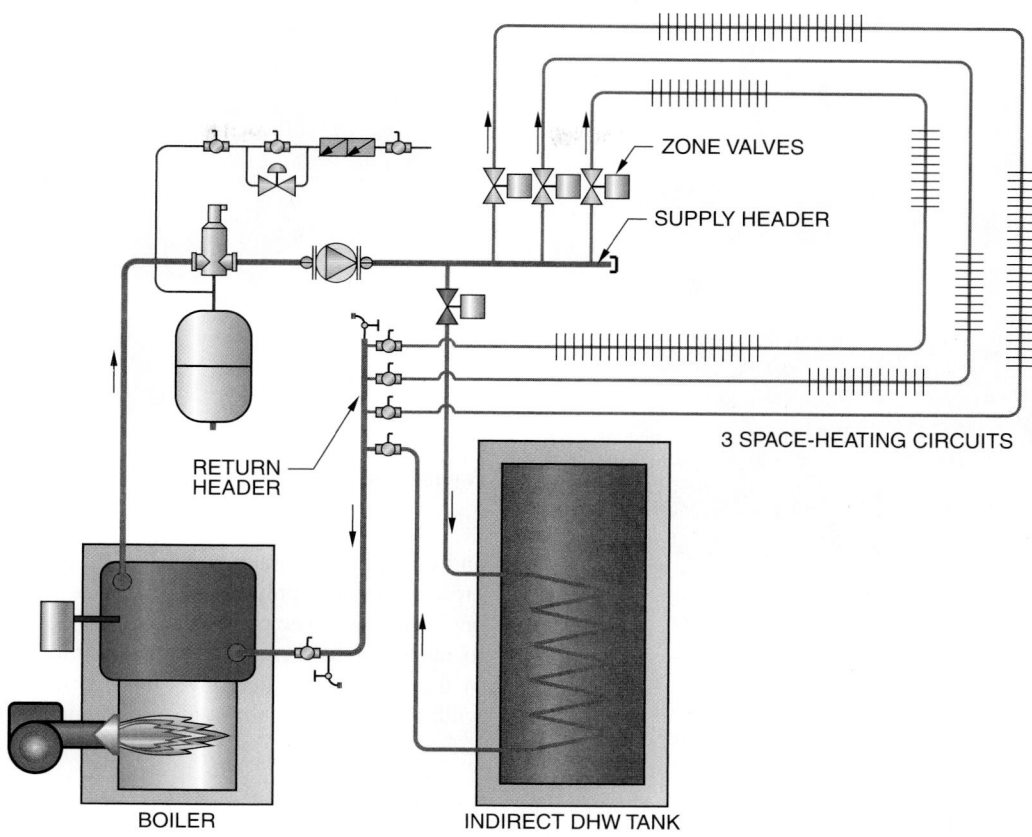

Figure 33–44   Hydronic piping circuit using two-port zone valves.

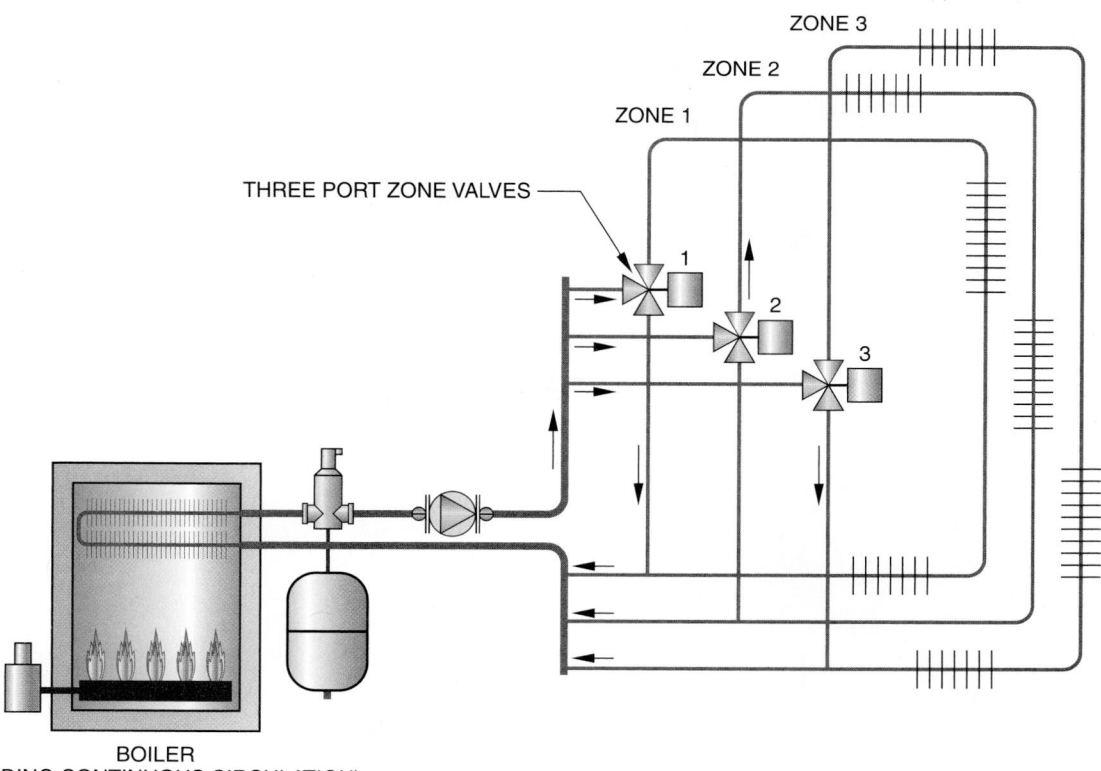

Figure 33–45   Hydronic piping circuit using three-port zone valves.

Figure 33–46 A balancing valve. *Photo by Bill Johnson*

Figure 33–47 A pressure differential bypass valve.

and the system operates normally. As zone valves close, the pressure in the supply header increases and the bypass valve begins to open. The partially open valve allows the circulator to move a fixed amount of water (although only a portion of the circulated water flows through the heating zones)—while the remaining hot water recirculates back to the boiler.

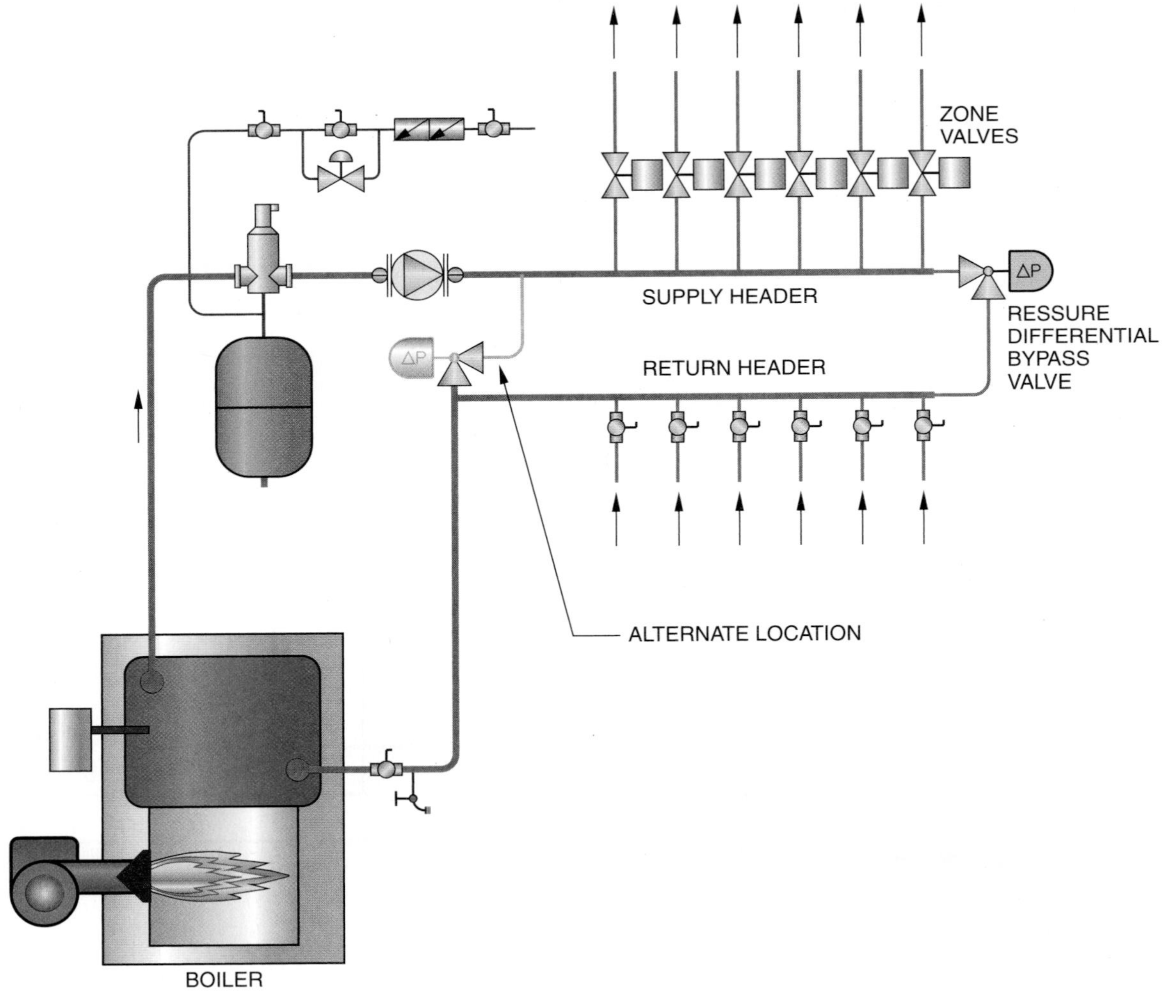

Figure 33–48 Location of the pressure differential bypass valve on systems with zone valves.

Flow Control Valves

The flow control valve, **Figure 33–49,** is actually a flow check valve that prevents hot water in the system from flowing through a heating loop when no flow through that particular loop is desired. This is often referred to as *ghost flow.* When the system uses zone valves to provide a positive shutoff between the heat source and the heating loops, flow control valves are not needed. However, when the system utilizes only circulators, flow control valves are desirable. Some circulator manufacturers are now installing flow control valves right on the circulators themselves to ease the installation process, **Figure 33–50.**

Outdoor Reset Control

The outdoor reset control is the device that senses the outside ambient temperature and adjusts the boiler water temperature accordingly. In theory, as the outside ambient temperature increases, the occupied space will require less heating, so cooler water can be used. Conversely, when the outside ambient temperature drops, hotter water will be needed to heat the space. When cooler water can be used, energy is saved.

The reset control measures two temperatures: the outside ambient temperature and the temperature of the water in the boiler. At the time of initial installation, the control must be set up manually to select which set of "commands" the control will obey. **Figure 33–51** shows a set of control options for a particular reset control. In this case, a reset control setting of 1.0 will cause the boiler to maintain a water temperature of 160°F if the outside ambient temperature is 20°F. The same control setting will cause the boiler water temperature to drop to 140°F if the outside ambient temperature rises to 40°F.

Thermostatic Radiator Valves

On hydronic systems with multiple zones, one common method for controlling the temperature in each individual zone is to use thermostatic radiator valves, **Figure 33–52.** These valves can be used on many different types of hydronic piping circuits—with the exception of the series loop system, which will be discussed later on in this unit. It should be noted that when the thermostatic radiator valve is installed, it should be protected from the heat that will rise

Figure 33–49   A flow control valve. *Courtesy Bell and Gossett*

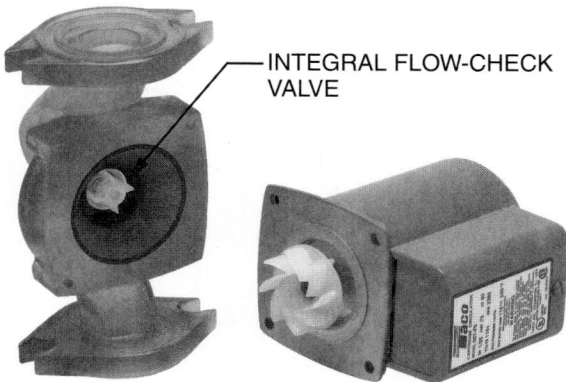

INTEGRAL FLOW-CHECK VALVE

Figure 33–50   Example of a circulator pump with an integral flow-check valve. *Courtesy Taco, Inc.*

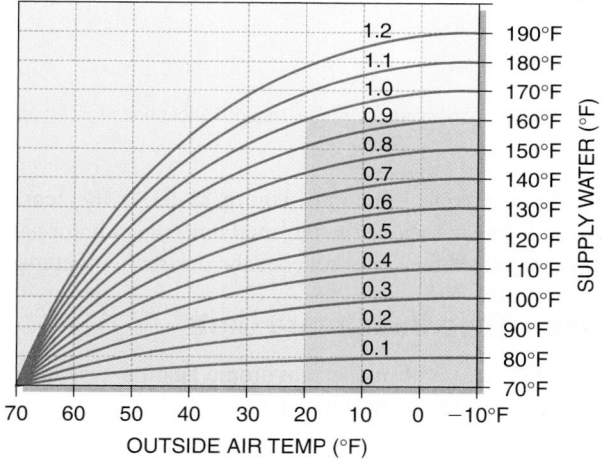

Figure 33–51   Residential outdoor reset control curve.

Figure 33–52   Thermostatic radiator valve. *Courtesy Danfoss, Inc.*

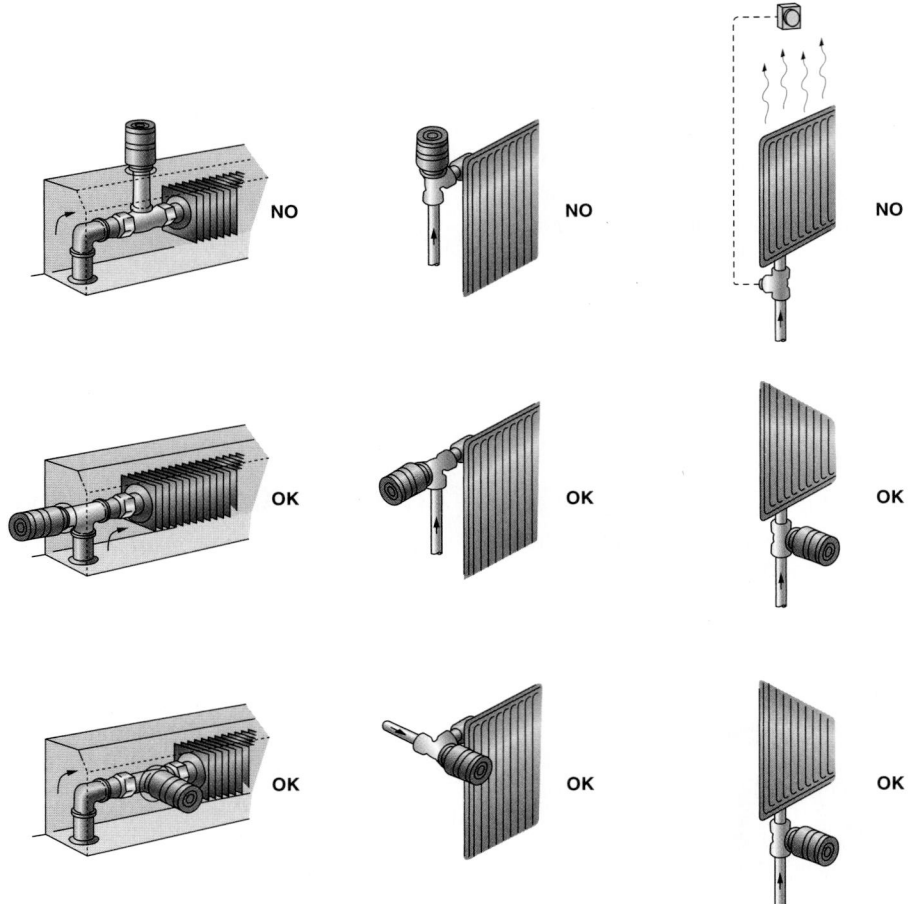

Figure 33–53   Proper and improper mounting locations for the operator on the thermostatic radiator valve.

from the terminal heater unit, **Figure 33–53.** If the sensing element is located above the terminal unit, the radiator valve will close before the space reaches the desired temperature.

Finned-Tube Baseboard Units

Most residences with modern hydronic heating systems use finned-tube baseboard heat transfer for terminal heating units, **Figure 33–54.** Air enters the bottom of these units and passes over the hot fins. Heat is given off to the cooler air, causing it to rise by convection. The heated air leaves the unit through the damper area. The damper can be adjusted to regulate the heat flow. Baseboard units are generally available in lengths from 2 ft to 8 ft and are relatively easy to install following a manufacturer's instructions. **Figure 33–55** shows a radiator and fan coil, which are other types of terminal units.

Finned-tube heating units are rated in Btuh/ft of pipe in two different flow rates, approximately –500 lb/h and 2000 lb/h. Five hundred lb/h is the equivalent to 1 gal of water flow per minute, and 4 gal/min equals 2000 lb/h.

$$1 \text{ gal/min} = 8.33 \text{ lb/min}$$
$$8.33 \text{ lb/min} \times 60 \text{ min/h} = 499.8 \text{ lb/h (Rounded to 500 lb/h)}$$

$$8.33 \text{ lb/gal} \times 4 \text{ gal/min} = 33.32 \text{ lb/min}$$
$$33.32 \text{ lb/min} \times 60 \text{ min/h} = 1999.2 \text{ lb/h}$$
$$\text{(Rounded to 2000 lb/h)}$$

1. RETURN TUBING
2. FINNED TUBING
3. SUPPORT BRACKET
4. FRONT COVER
5. DAMPER

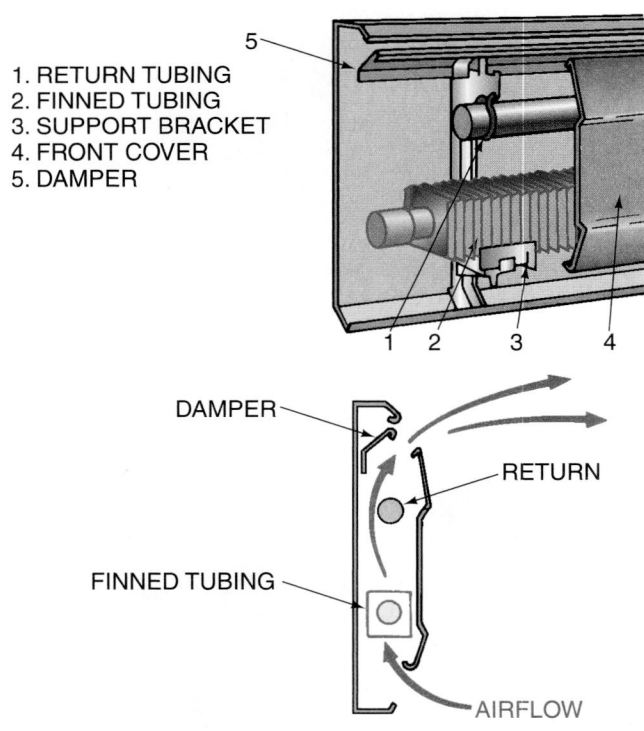

Figure 33–54   A finned-tube baseboard unit.

(A)

(B)

Figure 33–55 Terminal hydronic heating units. **(A)** Water in the radiator radiates heat out into the room. **(B)** The fan coil unit is similar to baseboard units but has a blower that circulates air over the hot surface of the terminal unit. **(A)** *Photo by Bill Johnson.* **(B)** *Courtesy Ferris State University. Photo by John Tomczyk*

The manufacturer will rate the finned-tube radiation at so many Btuh/ft at different temperatures, **Figure 33–56.**

These same convectors are available in 3/4-in. ID tubing. This will give you some idea of how the convectors are chosen. If a room needs 16,000 Btu/h and the flow rate has been determined to be 1 gal/min and the system's average water temperature is 180°F, the convector length would be 27.6 ft.

16,000 Btu/h ÷ 580 Btu/ft = 27.6 ft

Other types of terminal heating units are selected from manufacturers' literature in much the same manner.

When installing baseboard terminal units, provide for expansion due to heat. The hot water passing through the unit makes it expand. Expansion occurs toward both ends of the unit so the ends should not be restricted. It is a good practice to install expansion joints in longer units, **Figure 33–57.** If one expansion joint is used, place it in the center. If two are used,

RATINGS: Fine/Line 15 Series
(Hot water ratings, BTU/HR per linear ft. with 65°F (18.3°C) entering air)

ELEMENT	WATER FLOW	PRESSURE DROP *	140°F†	150°F	160°F	170°F	180°F	190°F	200°F	210°F	215°F	220°F
No. 15-75E Baseboard with 3/4" element	1 GPM	47	290	350	420	480	550	620	680	750	780	820
	4 GPM	525	310	370	440	510	580	660	720	790	820	870
No. 15-50 Baseboard with 1/2" element	1 GPM	260	310	370	430	490	550	610	680	740	770	800
	4 GPM	2880	330	390	450	520	580	640	720	780	810	850

* Millinches per foot.

† Ratings at 140°F determined by multiplying 150°F rating by the I=B=R conversion multiplier of .84.

NOTE: Ratings are for element installed with damper open, with expansion cradles. Ratings are based on active finned length (5" to 6" less than overall length) and include 15% heating effect factor. Use 4 gpm ratings only when flow is known to be equal to or greater than 4 gpm; otherwise, 1 gpm ratings must be used.

Figure 33–56 Example of a thermal rating table for a finned-tube baseboard. *Courtesy Slant/Fin Corp.*

Figure 33–57   An expansion joint. *Courtesy Edwards Engineering Corporation*

TUBE SIZE		RECOMMENDED MINIMUM HOLE (INCHES)
NOMINAL (INCHES)	O. D. (INCHES)	
$\frac{1}{2}$	$\frac{5}{8}$	1
$\frac{3}{4}$	$\frac{7}{8}$	$1\frac{1}{4}$

Figure 33–58   This table indicates the diameter of the holes required for each pipe size to allow for expansion.

AVERAGE WATER TEMPERATURE °F	MAXIMUM LENGTH OF STRAIGHT RUN (FEET)
220	26
210	28
200	30
190	33
180	35
170	39
160	42
150	47

Figure 33–59   The maximum lengths of a baseboard unit that can be installed when the temperature is raised from 70°F to the temperature indicated in the table.

space them evenly at intervals one third of the length of the unit. When piping drops below the floor level, the vertical risers should each be at least 1 ft long. The table in **Figure 33–58** shows the diameter of the holes needed through the floor for the various pipe sizes. The table in **Figure 33–59** lists the maximum lengths that can be installed when the water temperature is raised from 70°F to the temperature shown.

33.6   HIGH-TEMPERATURE HYDRONIC PIPING SYSTEMS

Up to this point in the unit, we have discussed the components that are used to make up hydronic heating systems. Some of these components are connected together and make up what is known as the "near boiler piping." A common near-boiler-piping configuration is shown in **Figure 33–60**. It is now that we will start putting these pieces together to form a number of different types of heating systems. As you will see, each of these systems will have advantages and disadvantages. It is important to evaluate each project separately as the priorities and concerns of the equipment owner

and operator will undoubtedly vary from one job to the next. The systems that will be discussed in the following sections are basic high-temperature applications, but variations or combinations of these configurations are possible depending on the design requirements for the system.

The Series Loop System

The series loop system is the hydronic system that is found most often, primarily because of the low installation costs involved in executing this piping configuration. Similar to a series electric circuit, all of the terminal units are piped in series with each other so that the outlet of one heat emitter is the inlet of the next, **Figure 33–61**. The main drawback of the series loop system is that it is not possible to have individual temperature control for each area that is being heated. This is because there is only one path for the hot water to take and stopping the water flow through one terminal unit will stop water flow in the entire system. Another drawback is that the first terminal units being supplied will be warmer than the later ones. This is because, as the water travels through the loop and heat is transferred from the hot water to the occupied space, the temperature of the water in the loop will drop.

Consider the series loop system in **Figure 33–62**. This system has three terminal units that are designed to provide 20,000 Btu/h, 30,000 Btu/h, and 50,000 Btu/h, respectively. We will also conclude that the water supplied by the boiler is 180°F and that the temperature of the water returning to the boiler is 160°F. From this information, we can determine the amount of water flow by using the following formula:

$$\text{Water flow (gpm)} = \frac{Q_T}{500 \times \Delta T}$$

$$\text{Water flow (gpm)} = [(20{,}000 \text{ Btu/h} + 30{,}000 \text{ Btu/h} + 50{,}0000 \text{ Btu/h})] \div [500 \times (180 - 160)]$$

$$\text{Water flow (gpm)} = 100{,}000 \div 10{,}000$$

$$\text{Water flow (gpm)} = 10 \text{ gpm}$$

We now know that there will be 10 gpm of water flow through the piping circuit when the circulator is operating. We can determine the temperature of the water at the outlet of the first terminal unit by using the following variation on the previous formula:

$$\Delta T = Q_1 \div (500 \times \text{gpm})$$
$$\Delta T = 20{,}000 \div (500 \times 10 \text{ gpm})$$
$$\Delta T = 20{,}000 \div (500 \times 10 \text{ gpm})$$
$$\Delta T = 4°F$$

So, if the temperature of the water entering the first terminal unit is 180°F, the temperature of the water at the outlet of the first terminal unit will be 4 degrees less than that, or 176°F, **Figure 33–63**. The temperature of the water at the outlet of the second terminal units can be calculated as follows:

$$\Delta T = Q_2 \div (500 \times \text{gpm})$$
$$\Delta T = 30{,}000 \div (500 \times 10 \text{ gpm})$$
$$\Delta T = 30{,}000 \div (500 \times 10 \text{ gpm})$$
$$\Delta T = 6°F$$

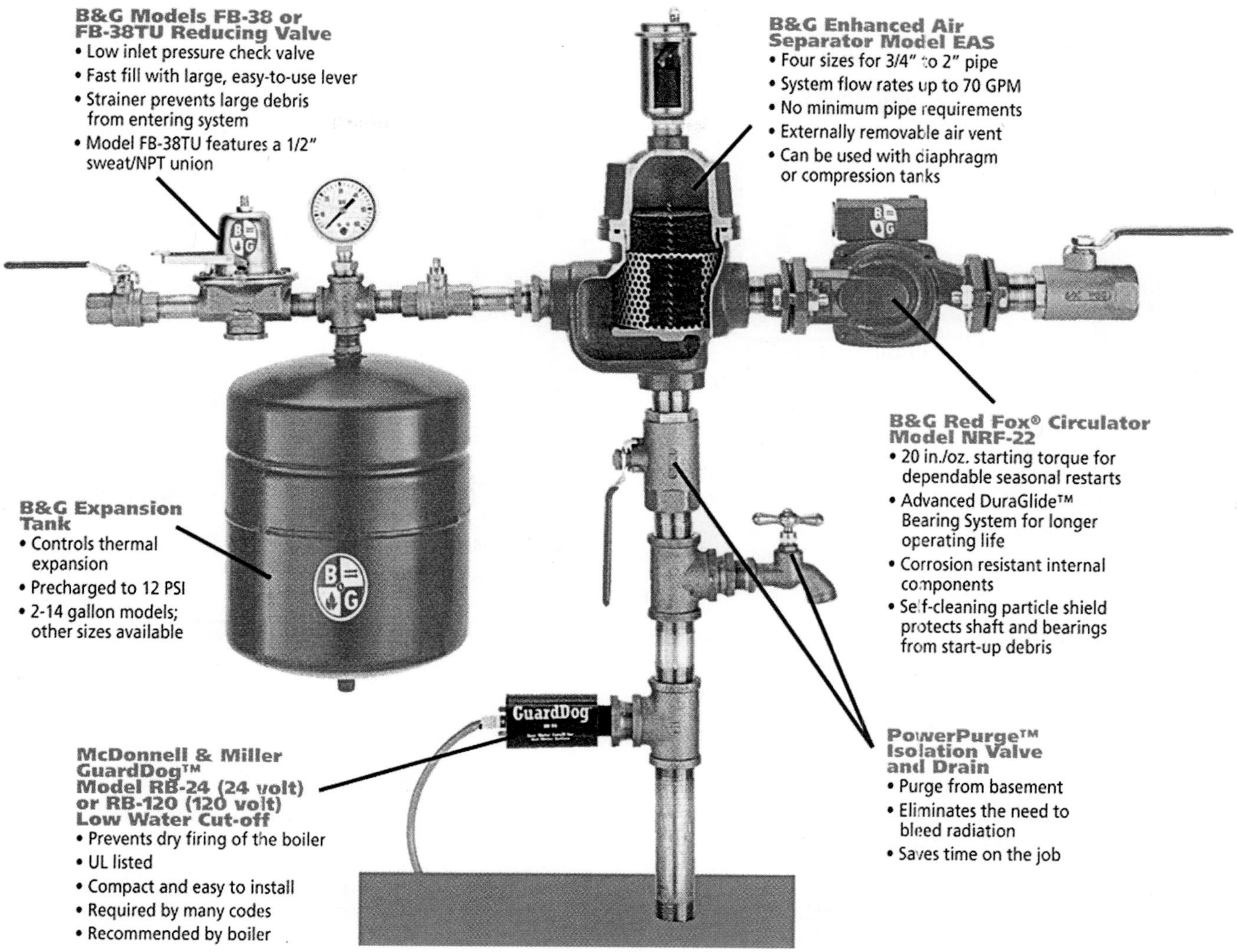

B&G Models FB-38 or FB-38TU Reducing Valve
- Low inlet pressure check valve
- Fast fill with large, easy-to-use lever
- Strainer prevents large debris from entering system
- Model FB-38TU features a 1/2" sweat/NPT union

B&G Enhanced Air Separator Model EAS
- Four sizes for 3/4" to 2" pipe
- System flow rates up to 70 GPM
- No minimum pipe requirements
- Externally removable air vent
- Can be used with diaphragm or compression tanks

B&G Red Fox® Circulator Model NRF-22
- 20 in./oz. starting torque for dependable seasonal restarts
- Advanced DuraGlide™ Bearing System for longer operating life
- Corrosion resistant internal components
- Self-cleaning particle shield protects shaft and bearings from start-up debris

B&G Expansion Tank
- Controls thermal expansion
- Precharged to 12 PSI
- 2-14 gallon models; other sizes available

McDonnell & Miller GuardDog™ Model RB-24 (24 volt) or RB-120 (120 volt) Low Water Cut-off
- Prevents dry firing of the boiler
- UL listed
- Compact and easy to install
- Required by many codes
- Recommended by boiler

PowerPurge™ Isolation Valve and Drain
- Purge from basement
- Eliminates the need to bleed radiation
- Saves time on the job

Figure 33–60   Hot water supply manifold. *Courtesy Bell and Gossett*

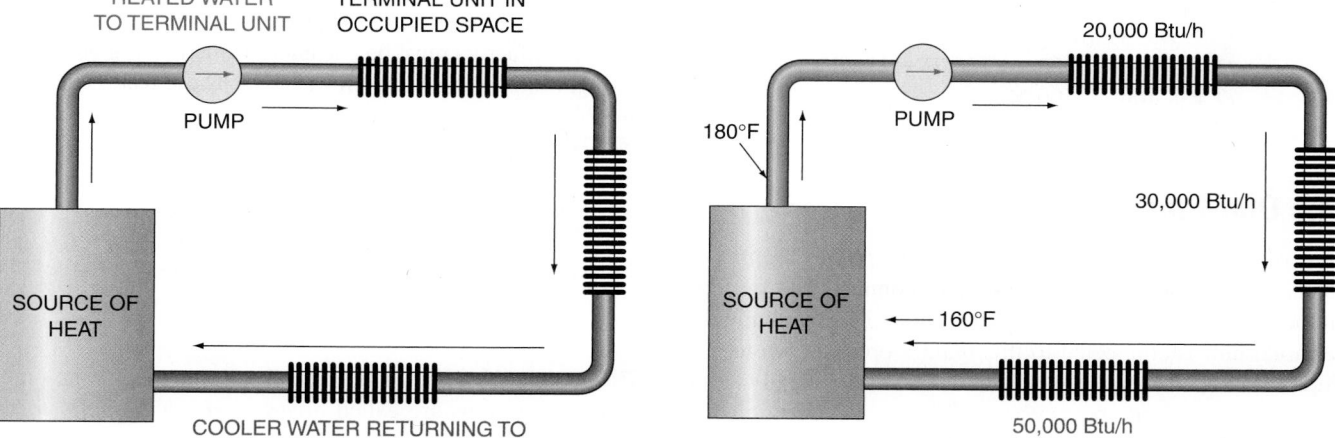

Figure 33–61   Series loop piping circuit.

HEATED WATER TO TERMINAL UNIT   TERMINAL UNIT IN OCCUPIED SPACE

PUMP

SOURCE OF HEAT

COOLER WATER RETURNING TO THE HEAT SOURCE

Figure 33–62   A series loop piping system with a 100,000 Btu/h output and a ΔT of 20°F across the boiler.

20,000 Btu/h

180°F   PUMP

30,000 Btu/h

SOURCE OF HEAT

160°F

50,000 Btu/h

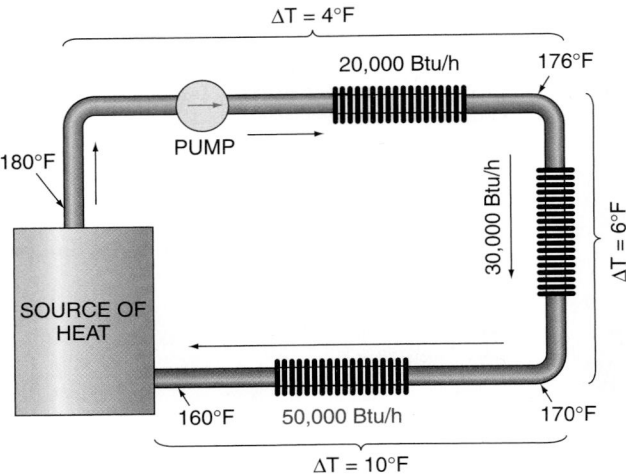

Figure 33–63 The water temperature at the outlet of each terminal unit. The ΔT is determined by taking both the Btu output of each terminal unit and the water flow through each terminal unit into account.

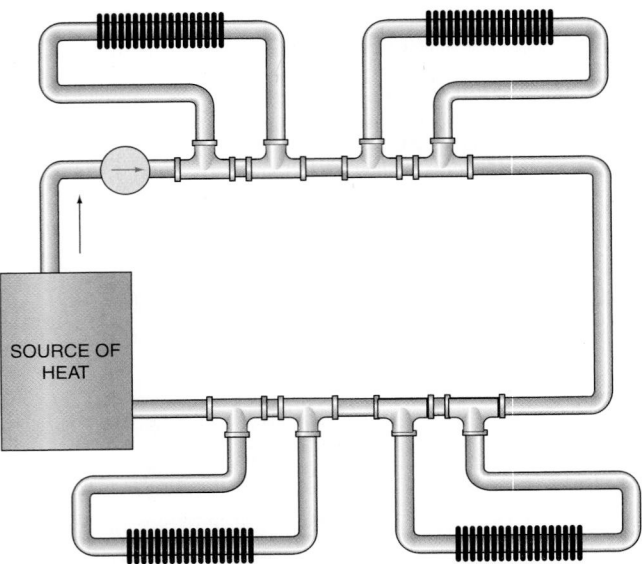

Figure 33–64 Simplified one-pipe hydronic system.

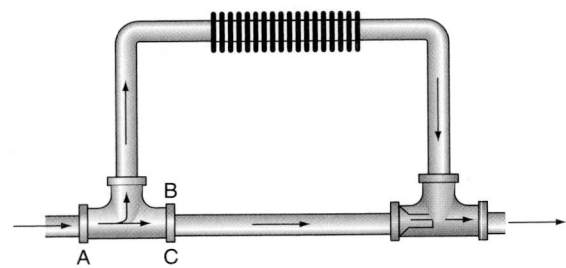

Figure 33–65 Common piping between the tees. In this application, the diverter tee is located at the return side of the terminal unit.

Since the temperature of the water entering the second terminal unit is 176°F, the temperature at the outlet of the second terminal unit is 170°F (176°F − 6°F), **Figure 33–63.** Finally, let's calculate the temperature at the outlet of the third terminal:

$$\Delta T = Q_3 \div (500 \times gpm)$$
$$\Delta T = 50{,}000 \div (500 \times 10 \text{ gpm})$$
$$\Delta T = 50{,}000 \div (500 \times 10 \text{ gpm})$$
$$\Delta T = 10°F$$

Since the temperature of the water entering the third terminal unit is 170°F, the temperature at the outlet of the third terminal unit is 160°F (170°F − 10°F), **Figure 33–63.**

Using the formula a little differently, we can determine the output of a particular terminal unit if we know the temperature difference across it. Consider a terminal unit that has 8 gpm of hot water flowing through it. The temperature at the inlet of the terminal unit is 180°F and the temperature at the outlet of the terminal unit is 178°F. The capacity of the terminal unit is as follows:

$$Q = \text{Water flow (gpm)} \times 500 \times \Delta T$$
$$Q = 8 \text{ gpm} \times 500 \times (180°F − 178°F)$$
$$Q = 8 \text{ gpm} \times 500 \times 2$$
$$Q = 8{,}000 \text{ Btu/h}$$

The One-Pipe System

In a one-pipe hydronic system, one main piping loop extends around the occupied space and connects the outlet of the boiler back to the return of the boiler. Each individual terminal unit is connected to this main loop with two tees, **Figure 33–64,** one or more of which may be specially designed for use on one-pipe systems. The proper operation of a one-pipe hydronic system relies on the proper ratios of resistance between the terminal unit branch and the resistance to flow in the section of pipe between the two tees. Consider

the setup in **Figure 33–65.** If there is a water flow of 4 gpm entering the tee at point A, 4 gpm must leave the tee. How much water will flow through the terminal unit and how much water will bypass the terminal unit depends on the resistances of each branch. If the resistance of the terminal unit branch is three times the resistance of the bypass branch, the flow through the bypass branch will be three times the flow through the terminal unit, **Figure 33–66.** For this reason, a number of factors must be considered when laying out, evaluating, or installing a one-pipe system. These factors include the following:

- The length of the terminal unit branch circuit
- The distance between the tees
- The size of the piping in the branch circuit
- The size of the piping between the tees
- The location of the terminal unit with respect to the main loop

THE DIVERTER TEE. The special tees used in one-pipe hydronic systems are called *diverter tees* or *Monoflo tees*, **Figure 33–67.** Diverter tees are designed to increase the resistance in the main loop pipe section between the two tees. By increasing the resistance in the main loop, more water will be

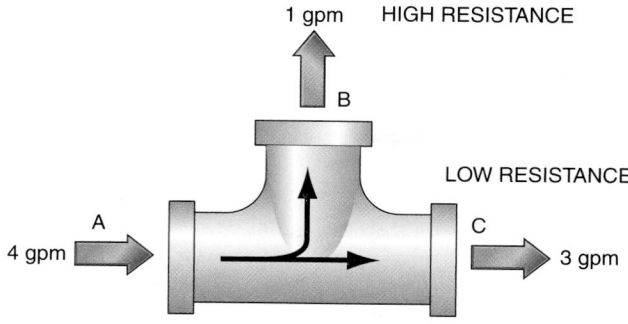

Figure 33–66 The high resistance at point B lowers the flow through that branch, while the lower resistance at point C results in greater flow through that branch.

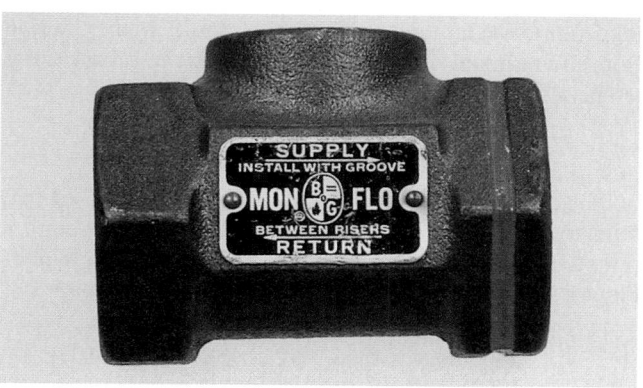

Figure 33–67 Diverter tee for use on one-pipe systems.

directed through the terminal heating units. The diverter tee is constructed with an interior cone, **Figure 33–68,** that reduces the diameter of the pipe, thereby increasing resistance to water flow. When working with diverter tees, keep the following in mind:

- If the terminal unit is located above the hot water main and the length of the terminal unit branch is not excessive, one diverter tee should be used on the return side of the terminal unit, **Figure 33–65.**
- If the terminal unit is located above the hot water main and the length of the terminal unit branch is very long, two diverter tees should be used—one on the supply side of the terminal unit and one on the return side of the terminal unit, **Figure 33–69.**
- If the terminal unit is located below the hot water main, two diverter tees should be used, **Figure 33–70.**
- If the terminal unit is located above the main hot water loop, the tees should be no closer than 6 or 12 inches depending on the manufacturer of the diverter tee.
- If the terminal unit is located below the main hot water loop, the tees should be as far apart as the ends of the terminal unit. If the terminal unit is 2 ft long, the tees should be 2 ft apart.
- If there are multiple terminal units, and some are above and some below the main hot water loop, it is recommended that the tees be staggered, **Figure 33–71.**

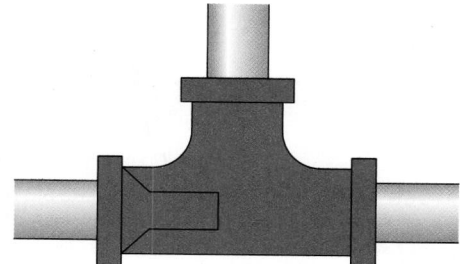

Figure 33–68 Cross-sectional view of a diverter tee.

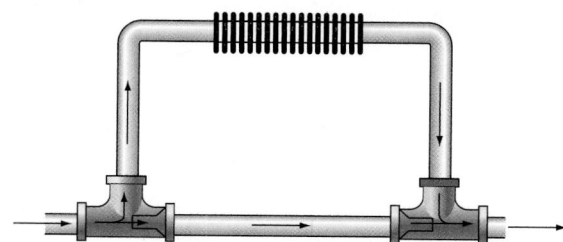

Figure 33–69 Two diverter tees are recommended if the radiator is located above the main loop and there is significant resistance in that branch.

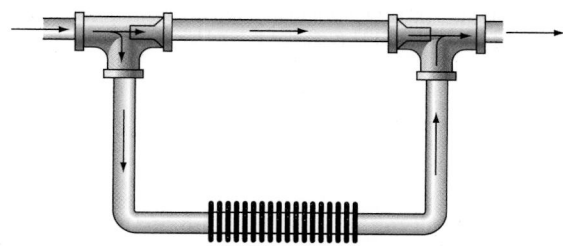

Figure 33–70 Two diverter tees are recommended if the radiator is located below the main loop.

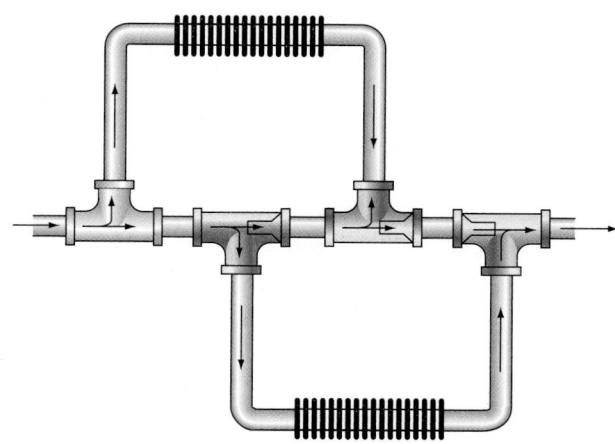

Figure 33–71 Alternate the diverter tees if there are terminal units both above and below the main loop.

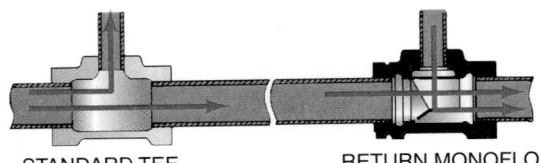

STANDARD TEE   RETURN MONOFLO

FOR RADIATORS ABOVE THE MAIN—NORMAL RESISTANCE
FOR MOST INSTALLATIONS WHERE RADIATORS ARE ABOVE
THE MAIN, ONLY ONE MONOFLO FITTING NEED BE USED FOR
EACH RADIATOR.

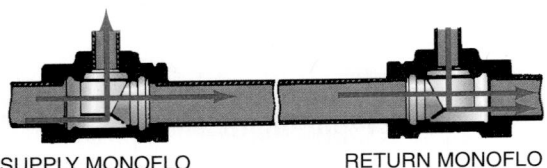

SUPPLY MONOFLO   RETURN MONOFLO

FOR RADIATORS ABOVE THE MAIN—HIGH RESISTANCE
WHERE CHARACTERISTICS OF THE INSTALLATION ARE SUCH
THAT RESISTANCE TO CIRCULATION IS HIGH, TWO FITTINGS
WILL SUPPLY THE DIVERSION CAPACITY NECESSARY.

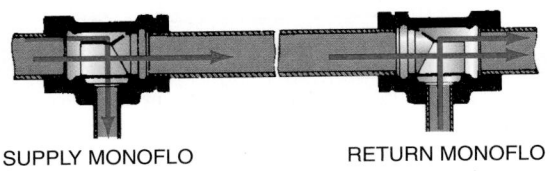

SUPPLY MONOFLO   RETURN MONOFLO

FOR RADIATORS BELOW THE MAIN
RADIATORS BELOW THE MAIN REQUIRE THE USE OF BOTH A
SUPPLY AND RETURN MONOFLOW FITTING. (AN EXCEPTION IS
A 3/4" MAIN, WHICH USES A SINGLE RETURN FITTING.)

Figure 33–72 Partial water flow is diverted through the tee to and from the terminal unit. *Courtesy ITT Fluid Handling Division*

■ Always check the installation literature that comes with the diverter tees to ensure that they are installed correctly and that they are pointing in the right direction.

The above information is summarized in **Figure 33–72**.

THE HEAT FORMULA AND THE ONE-PIPE HYDRONIC SYSTEM.
Consider a portion of the one-pipe system shown in **Figure 33–73**. Let's assume we have a one-pipe hydronic system that has a total heating load of 100,000 Btu/h. If there is a ΔT from boiler outlet to boiler return of 20°F, the water flow through the main hot water loop will be as follows:

$$\text{Water flow (gpm)} = \frac{Q_T}{500 \times \Delta T}$$

$$\text{Water flow (gpm)} = 100,000 \div (500 \times 20)$$

$$\text{Water flow (gpm)} = 100,000 \div 10,000$$

$$\text{Water flow (gpm)} = 10 \text{ gpm}$$

If the first terminal unit is to provide 20,000 Btu of heating, this will require that we have 2 gpm (20,000 ÷ 10,000) of hot water flowing through the terminal unit. The selected diverter tee should be capable of providing the desired flow.

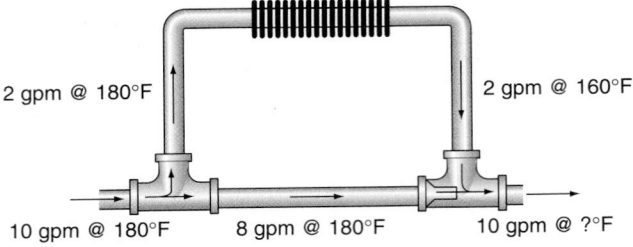

2 gpm @ 180°F   2 gpm @ 160°F

10 gpm @ 180°F   8 gpm @ 180°F   10 gpm @ ?°F

Figure 33–73 The flow rate through this terminal unit is 2 gpm.

Since 2 gpm of water will flow through the terminal unit, 8 gpm will continue to flow through the main hot water loop. At the second tee, there will be 8 gpm of 180°F water entering from the main loop and 2 gpm of 160°F water entering from the terminal unit. We can determine the mixed water temperature by using the following formula:

$$(\text{Flow 1} \times \text{Temp. 1}) + (\text{Flow 2} \times \text{Temp. 2})$$
$$= (\text{Flow 3} \times \text{Temp. 3})$$

In the above formula, Flow 1, Flow 2, Temp. 1, and Temp. 2 refer to water entering the tee, while Flow 3 and Temp. 3 refer to the water leaving the tee. Flow 3 must be the sum of Flow 1 and Flow 2. Here's how the numbers work out:

$$(\text{Flow 1} \times \text{Temp. 1}) + (\text{Flow 2} \times \text{Temp. 2})$$
$$= (\text{Flow 3} \times \text{Temp. 3})$$
$$(8 \text{ gpm} \times 180°F) + (2 \text{ gpm} \times 160°F)$$
$$= (10 \text{ gpm} \times \text{Temp. 3})$$
$$1440 + 320 = 10 \text{ gpm} \times \text{Temp. 3}$$
$$1760 = 10 \text{ gpm} \times \text{Temp. 3}$$
$$\text{Temp. 3} = 1760 \div 10$$
$$\text{Temp. 3} = 176°F$$

From this we can conclude that the temperature of the water entering the second terminal unit will be 176°F, **Figure 33–74**. A sample one-pipe system using thermostatic radiator valves is shown in **Figure 33–75**.

The Two-Pipe Direct-Return System

Two-pipe hydronic systems use two pipes. One pipe is used to carry water to the terminal units, while the other pipe is used to carry water from the terminal units back to the

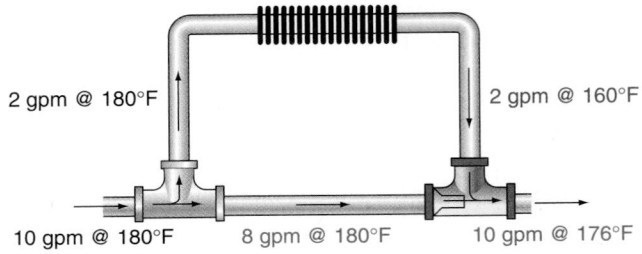

2 gpm @ 180°F   2 gpm @ 160°F

10 gpm @ 180°F   8 gpm @ 180°F   10 gpm @ 176°F

Figure 33–74 The mixed water temperature at the outlet of the terminal unit is 176°F.

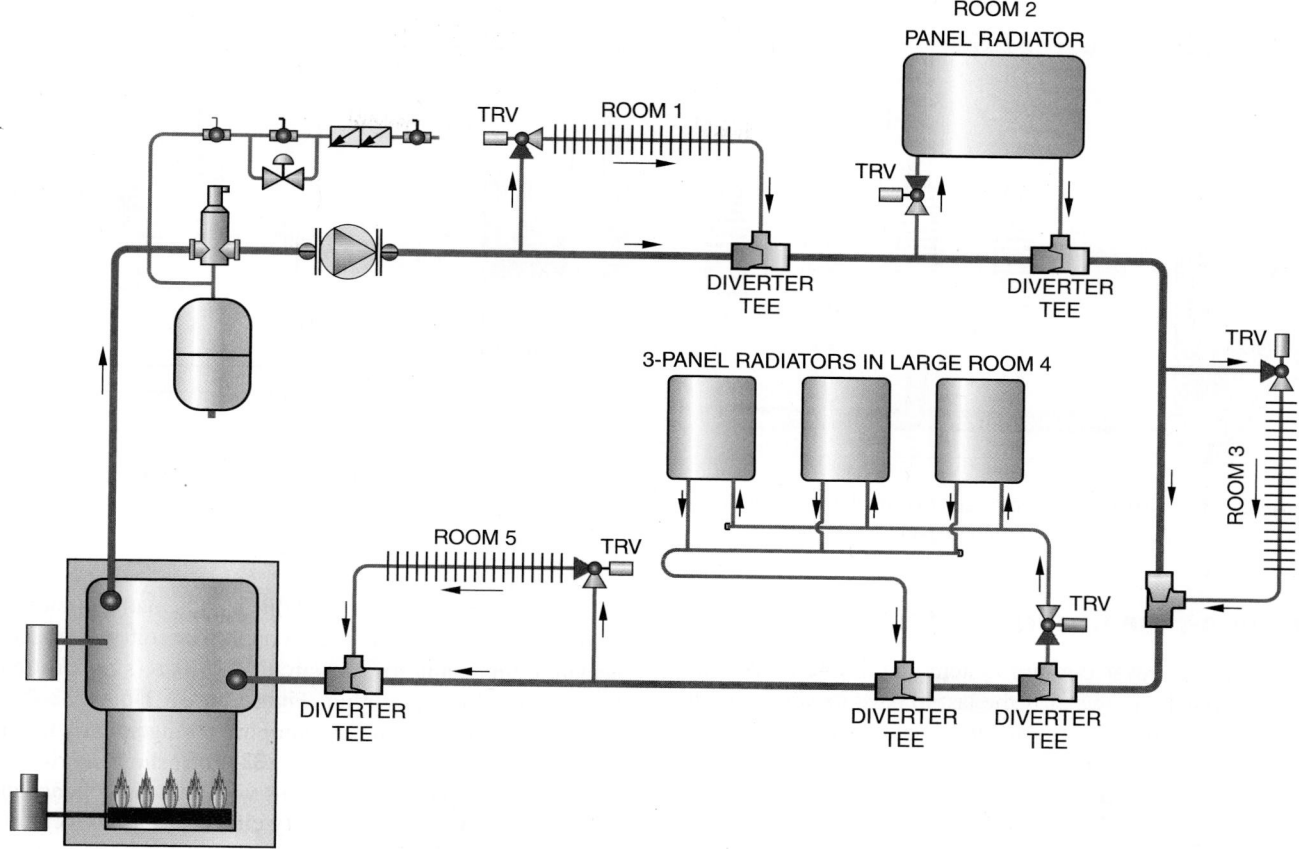

Figure 33–75 One-pipe hydronic system utilizing thermostatic radiator valves.

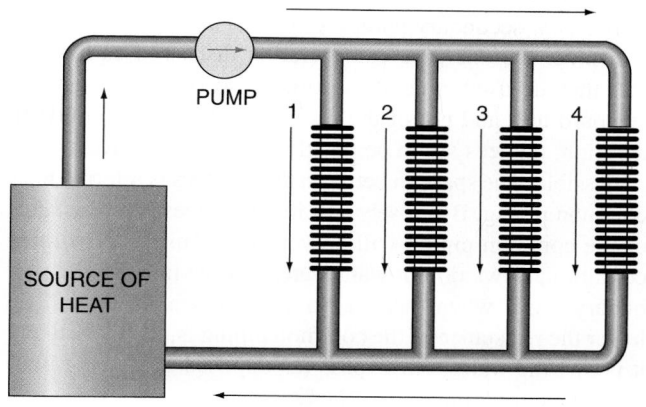

Figure 33–76 Simplified two-pipe direct-return piping arrangement.

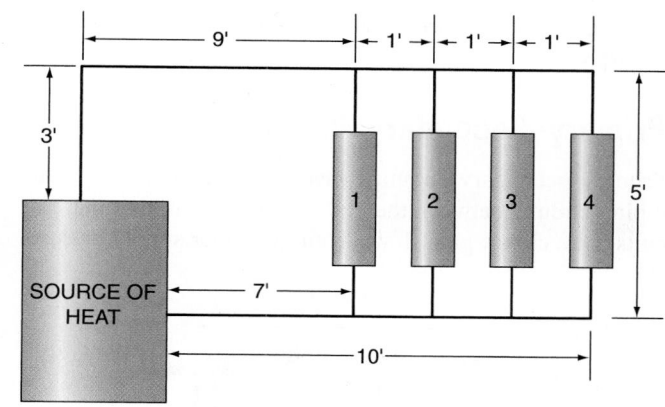

Figure 33–77 Water flowing through terminal unit 1 must travel 24 feet, while water flowing through terminal unit 4 must travel 30 feet.

boiler, **Figure 33–76.** In a two-pipe direct-return system, the terminal unit closest to the boiler will have the shortest piping run, while the terminal unit farthest from the boiler will have the longest piping run. This means that the resistance of the closer terminal unit circuits is lower than that of the terminal units located far from the boiler. Consider the piping circuit shown in **Figure 33–77.** This circuit is a two-pipe direct-return system that has four terminal units numbered 1 through 4. For water to flow from the boiler

through terminal unit 1 and back to the boiler, the water must travel 24 feet (3 + 9 + 5 + 7). For water to make a complete path through terminal unit 4, the water must travel 30 feet.

Because the piping circuit through the more distant terminal unit location is longer, less water will flow to this location. In order to even out the water flow, balancing valves, **Figure 33–46,** are used to increase the resistance of the shorter piping loops, **Figure 33–78.**

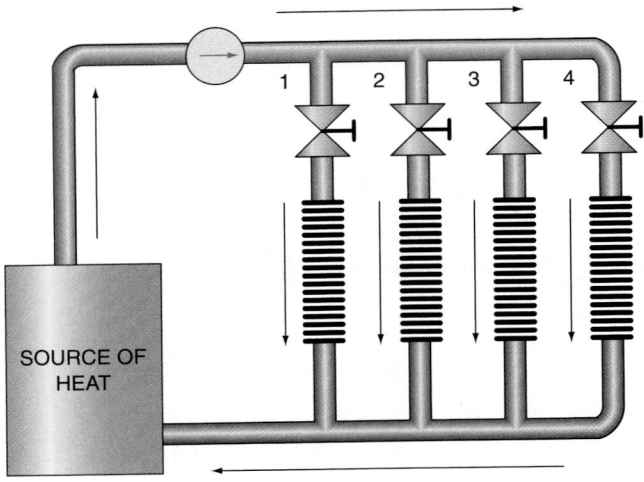

Figure 33–78 Balancing valves are used to even the resistance in each branch of a two-pipe direct-return system.

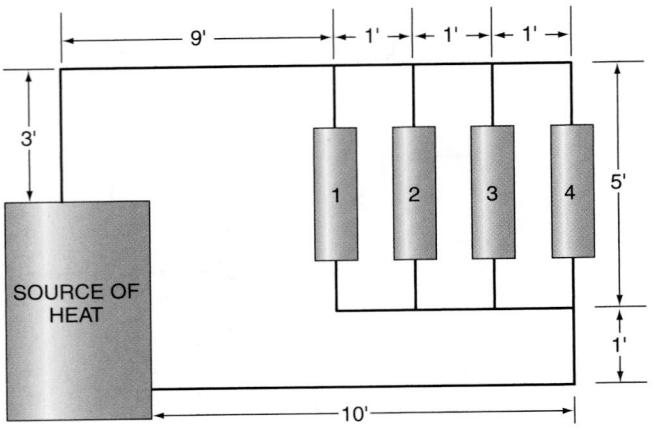

Figure 33–80 The length of the piping through any one branch is exactly the same as the length through any other branch.

The Two-Pipe Reverse-Return System

In the two-pipe reverse-return system, the first terminal unit that is supplied with water is the last to have its water return to the boiler. In other words, the terminal unit with the shortest supply pipe will have the longest return pipe and vice versa, **Figure 33–79.** By configuring the piping in this manner, the distance traveled by the water will be the same for all terminal units in the piping arrangements. It is left as an exercise for the reader to confirm that water must flow 31 feet to get from the boiler through any of the terminal units and back to the boiler in the piping arrangement shown in **Figure 33–80.** Two-pipe reverse-return systems do not require the use of balancing valves.

Primary–Secondary Pumping

Primary–secondary pumping involves at least two separate piping circuits between the boiler and the terminal-unit circuits. One circuit path flows from the boiler supply back to

the boiler return. Flow through this path is made possible by using a circulator pump. The secondary circuit(s) is/are connected to this main/primary circuit and share(s) a portion of the piping with that circuit, **Figure 33–81. Figure 33–81(A)** shows a primary–secondary pumping arrangement with one secondary circuit, whereas **Figure 33–81(B)** shows a primary–secondary pumping arrangement with three secondary circuits. Notice that each secondary circuit has its own circulator pump that will push water into the individual secondary circuits when the corresponding pump is energized. The industry term that is used to describe any system that uses a circulator to push water into a secondary heating loop (as in the case of primary–secondary pumping) is *injection.*

Unlike the one-pipe configuration discussed earlier, the tees that are used to create a primary–secondary system are standard tees and not of the diverter or Monoflo variety. In addition, the tees are to be positioned as close to each other as possible. The space in between the two tees is referred to as common piping. By closely spacing these tees, the resistance of the common piping will be very low. This is a desirable condition, as we do not want water to flow through the secondary loops when there is no call for heat in them. The lower the resistance of the common piping is, the more likely it is that the water will bypass (as opposed to flowing into) the secondary loops, **Figure 33–82.**

CIRCULATOR PUMPS IN PRIMARY–SECONDARY SYSTEMS.

It can be seen in **Figure 33–81(B)** that primary–secondary pumping involves the use of multiple circulator pumps. One circulator pump is located in the primary loop and is typically the largest pump in the system. It is this pump that is responsible for making the heated water available to the secondary loops. When no remote areas or zones are calling for heat, all system pumps are off but, depending on system design, the boiler may be cycling on and off to maintain the water in the boiler at the desired operating temperature. When an individual zone calls for heat, both the secondary pump for that loop and the primary pump will cycle on. The primary pump will

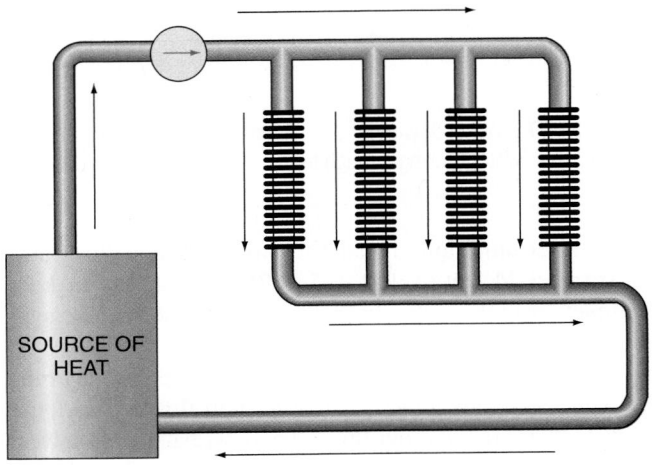

Figure 33–79 A two-pipe reverse-return piping arrangement.

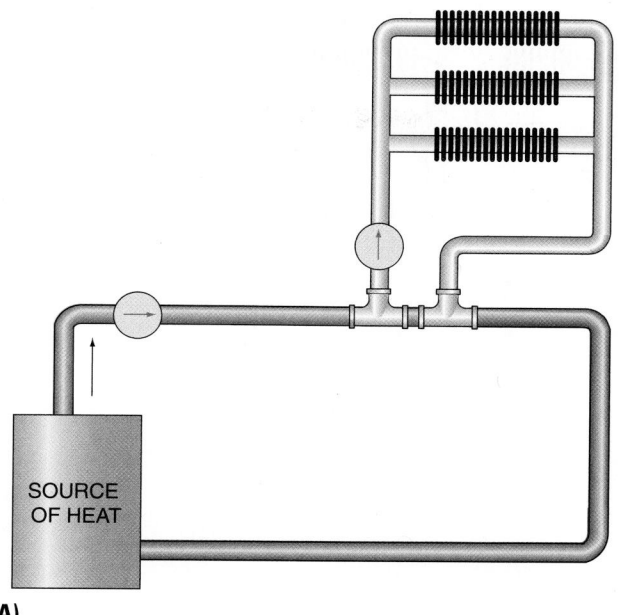

(A)

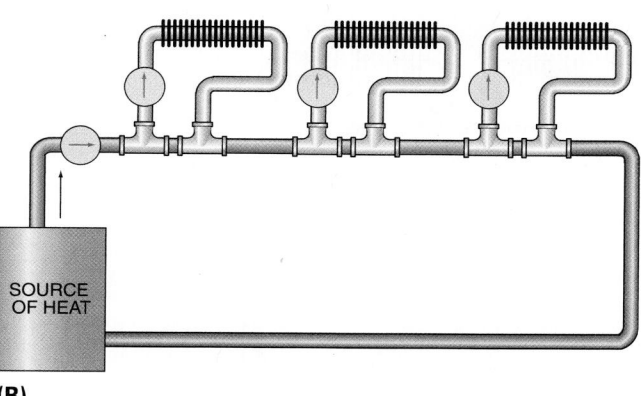

(B)

Figure 33–81 **(A)** Primary–secondary piping arrangement where one secondary loop is feeding three terminal units. **(B)** Primary–secondary piping arrangement where three secondary loops are each supplying hot water to one terminal unit.

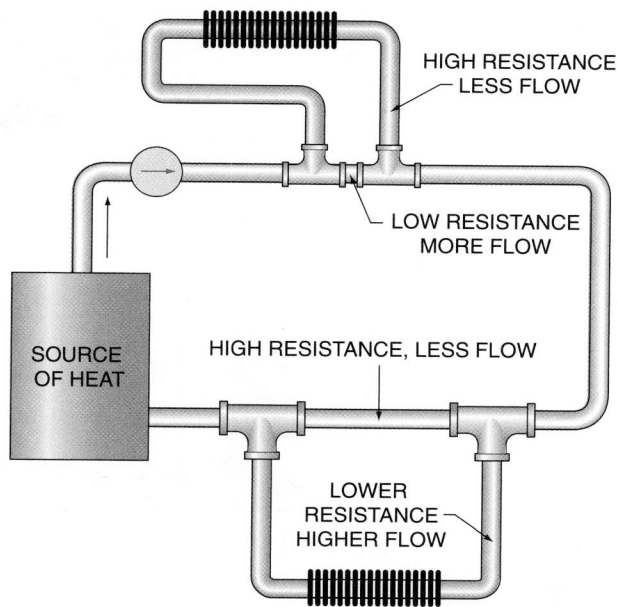

Figure 33–82 More resistance between the tees results in more water flow through the terminal unit connected to the tees. Less resistance between the tees results in less water flow through the terminal unit.

bring the heated water to the common piping section and the secondary pump will push heated water into the secondary loop to satisfy the heating requirements of the space.

Consider the primary–secondary setup in **Figure 33–83.** The system has a capacity of 200,000 Btu/h and the boiler operates with a ΔT of 20°F. There are three secondary loops with the following heating capacities:

■ Zone 1: 120,000 Btu/h
■ Zone 2: 50,000 Btu/h
■ Zone 3: 30,000 Btu/h

The total water flow that should be made available by the system is as follows:

$$\text{Water flow (gpm)} = \frac{Q_T}{500 \times \Delta T}$$

Water flow (gpm) = 200,000 ÷ (500 × 20)

Water flow (gpm) = 200,000 ÷ 10,000

Water flow (gpm) = 20 gpm

If all three secondary loops are calling for heat, the water temperature at the outlet of each secondary loop is as shown **Figure 33–84.** Notice that the temperature of the water at the outlet of the third secondary loop is the temperature of the boiler supply water (180°F) less the 20°F that was indicated as the ΔT of the boiler. Let's now assume that the thermostat for the first secondary loop is satisfied. In that case, the corresponding secondary pump will cycle off and there will be no water flow into that loop. When the first loop is no longer active, water at a temperature of 180°F is fed to the second loop. The temperature of the water at the outlet of the second loop, **Figure 33–85,** is calculated as follows:

(Flow 1 × Temp. 1) + (Flow 2 × Temp. 2)
     = (Flow 3 × Temp. 3)
(15 gpm × 180°F) + (5 gpm × 160°F)
     = (20 gpm × Temp. 3)
      2700 + 800 = 20 gpm × Temp. 3
           3500 = 20 gpm × Temp. 3
        Temp. 3 = 3500 ÷ 20
        Temp. 3 = 175°F

The water temperature at the outlet of the third secondary loop will be as follows:

(Flow 1 × Temp. 1) + (Flow 2 × Temp. 2)
     = (Flow 3 × Temp. 3)
(17 gpm × 175°F) + (3 gpm × 155°F)
     = (20 gpm × Temp. 3)
      2975 + 465 = 20 gpm × Temp. 3
           3440 = 20 gpm × Temp. 3
        Temp. 3 = 3440 ÷ 20
        Temp. 3 = 172°F

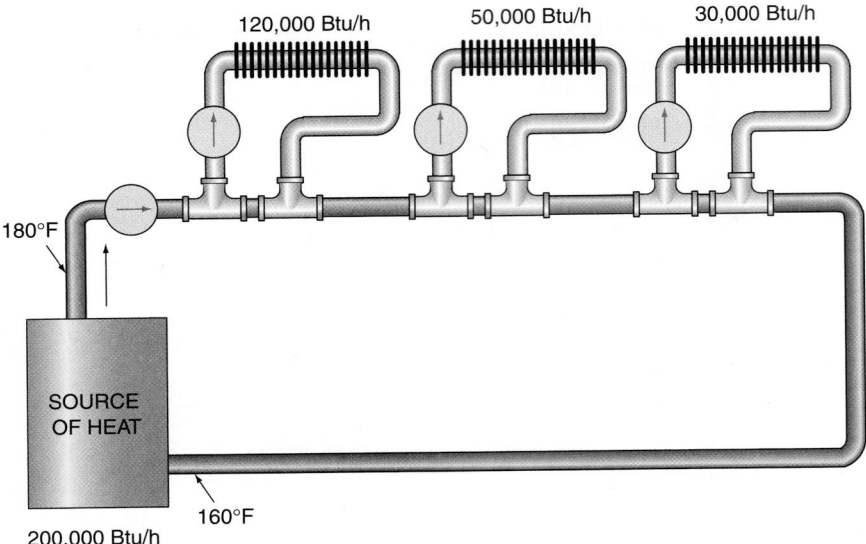

Figure 33–83   Sample one-pipe system with three terminal units providing 200,000 Btu/h of heating to the occupied space.

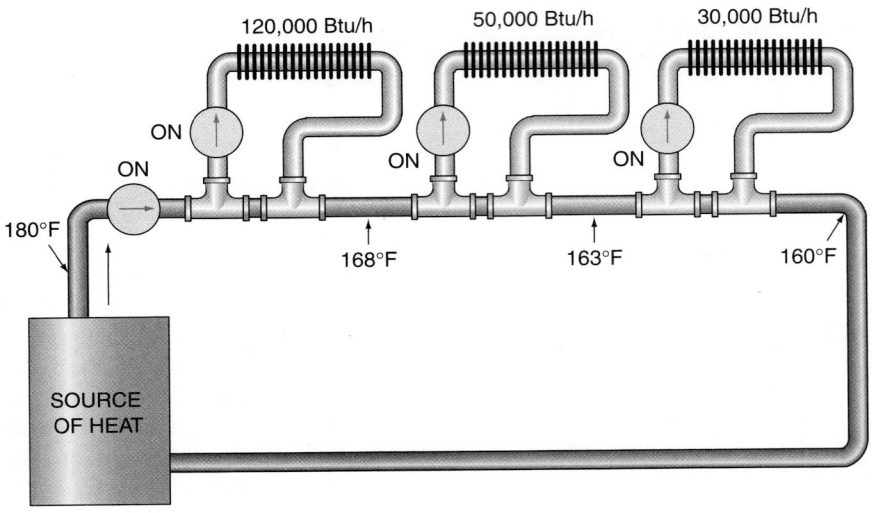

Figure 33–84   Water temperatures at various points in the system when all three zones are calling for heat.

The temperature of the water returning to the boiler will be 172°F, **Figure 33–85**. Notice that the water left the boiler at 180°F and returned at 172°F. This represents a ΔT of only 8°F across the boiler, when we originally anticipated a ΔT of 20°F. This is because the first secondary loop is not calling for heat and this loop represents 60% (0.6) of the total system capacity (120,000 Btu ÷ 200,000). Since 60% of our 20°F ΔT is 12 degrees, we only see the portion of the system (20°F − 12°F = 8°F) that is actively being heated. Notice that when all three secondary loops are actively heating, the return temperature at the boiler reflects the ΔT of 20°F.

MIXING VALVES IN PRIMARY–SECONDARY SYSTEMS.
Mixing valves are used to combine two water streams of different temperatures to produce water flow that is at a temperature between the two entering temperatures. Common types of mixing valves include three-way valves, **Figure 33–86**, and four-way valves, **Figure 33–87**. Three-way valves have two inlet ports and one outlet port, whereas four-way valves have two inlet ports and two outlet ports.

Mixing valves can be manually set, **Figure 33–86**, or they can be thermostatically controlled, **Figure 33–88**. A manually set mixing valve, is set to the desired position on the basis of the percentages of water flow desired from each inlet. At either extreme, the water at the outlet of the valve will be 100% from the hot water inlet or 100% from the cold water inlet. At any point in between, the outlet water will be a mixture of the two, **Figure 33–89**. Four-way mixing valves allow some of the water leaving the boiler to return to the boiler to keep the return-water temperature high enough to prevent flue-gas condensation, **Figure 33–90**.

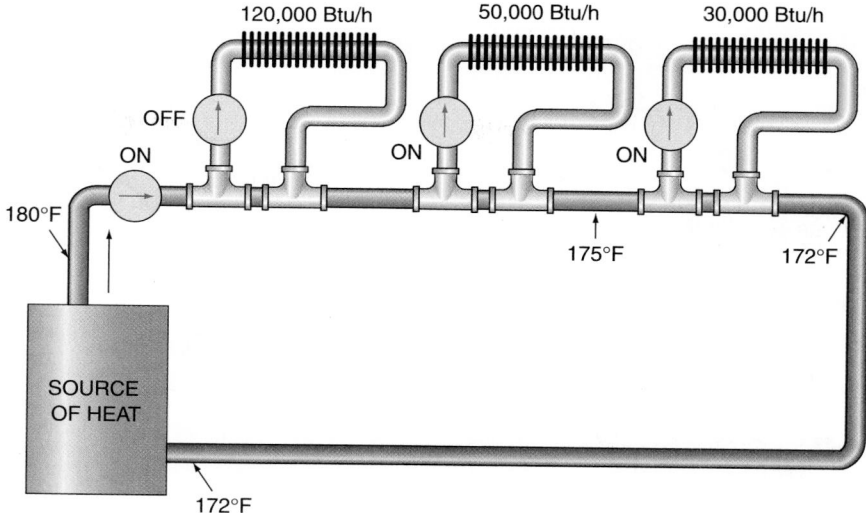

Figure 33–85 Water temperatures at various points in the system when only the second and third zones are calling for heat.

Figure 33–86 A three-way mixing valve with a manual adjustment knob. *Courtesy Danfoss, Inc.*

Figure 33–87 Four-way mixing valves. *Courtesy Tekkmar Control Systems*

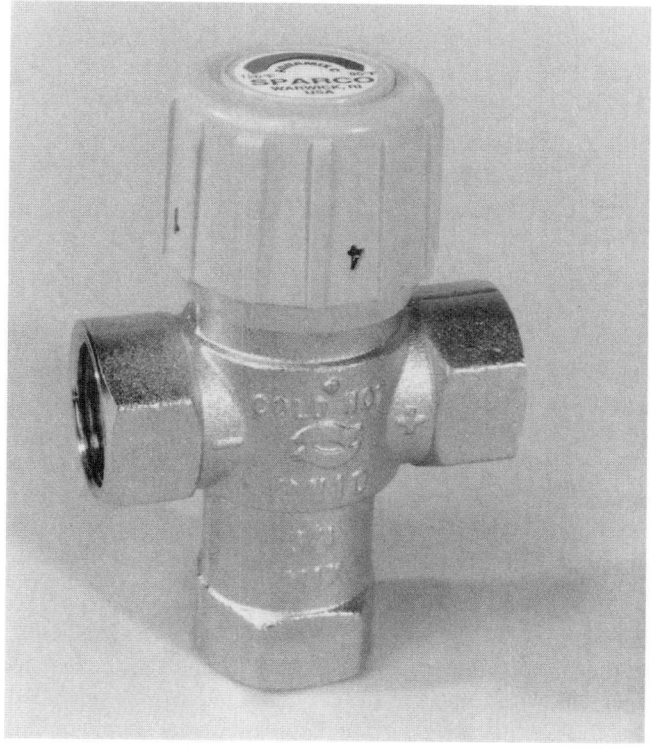

Figure 33–88 Three-way thermostatic mixing valve. *Courtesy Sparco, Inc.*

One of the many benefits of primary–secondary pumping is that each individual loop can be used to supply water at different temperatures to the terminal units in that loop. For example, one portion of the structure may have baseboard heating elements installed, while another portion of the structure may have radiant heat. The area with baseboard heating elements will require higher-temperature water than will the area with radiant heat. By using primary–secondary

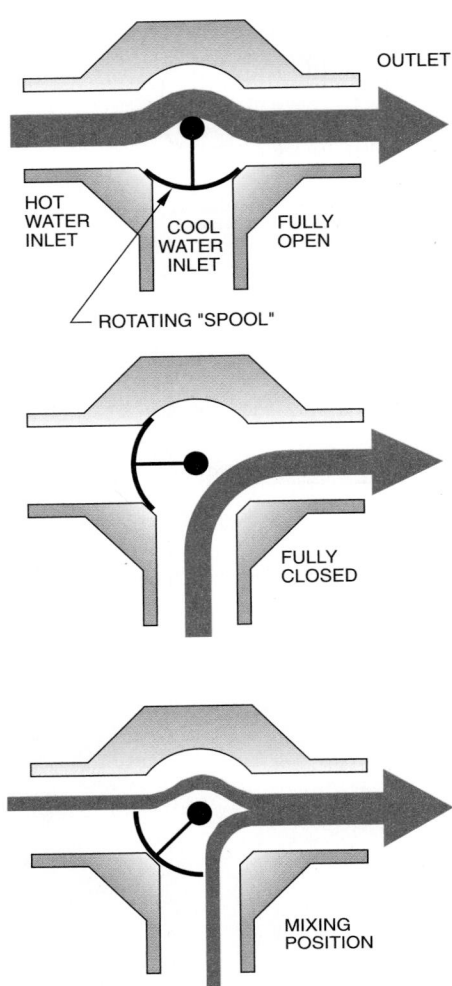

Figure 33–89 Cutaway illustration showing internal construction of a three-way valve.

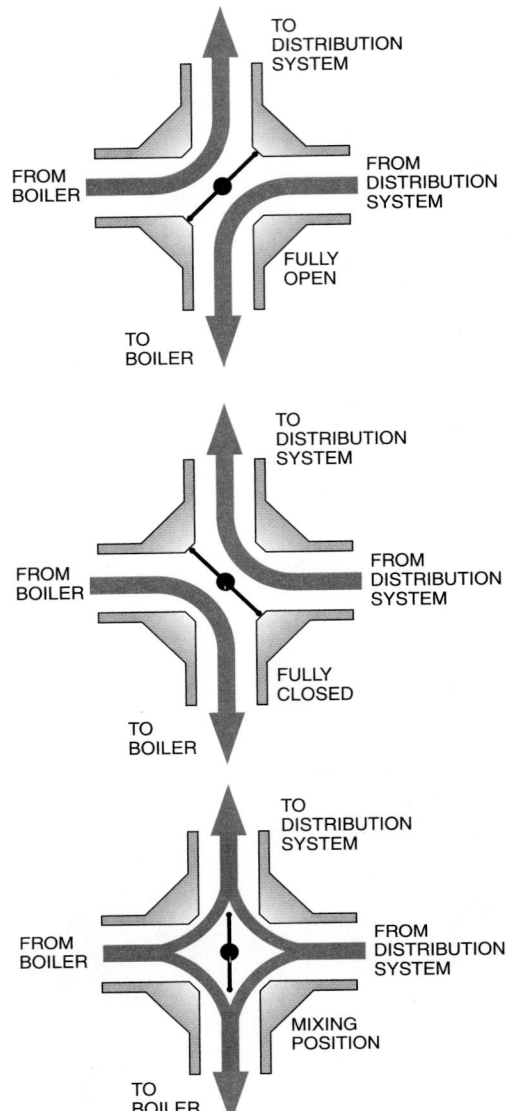

Figure 33–90 Cutaway illustration showing internal construction of a four-way valve.

pumping in conjunction with a mixing valve, both conditions can be met. Multitemperature combination systems will be discussed later on in this unit.

EXPANSION TANKS IN PRIMARY–SECONDARY SYSTEMS.

Earlier in this unit, we discussed the location of the expansion tank with respect to the circulator pump. We concluded that the best place for the circulator was on the supply piping coming from the boiler, pumping away from the expansion tank. This was referred to as the "point of no pressure change." Since primary–secondary systems have additional circulator pumps, it may seem logical that there should be expansion tanks in each of the secondary loops to provide for the expansion of heated water in these locations. The truth of the matter is that the common piping between the primary and secondary loops serves as the expansion tank for that loop. As water is heated, it will move from the secondary loop into the primary loop and then into the expansion tank located in the primary loop.

Since we want to pump away from the point of no pressure change, the circulators in the secondary loops should be

located on the pipe closest to the primary pump—and should be directed to pump into the secondary circuit and away from the common piping.

Regardless of the type of piping layout, the estimator or engineer will pay close attention to the sizing of the pipe for any system. Pipe sizing and pump size determines how much water will flow in each circuit. If the pipe is too large, it will be expensive and it may not meet the low-limit flow requirements for convectors. For example, the minimum design flow rates for different diameter convector pipes are:

1/2 in. ID	0.3 gpm
3/4 in. ID	0.5 gpm
1 in. ID	0.9 gpm
1 1/4 in. ID	1.6 gpm

If the flow rates below these are used at the convectors, the heat exchange between the water and the air will be less. The velocity of the water in the convector helps the heat exchange. Flow that is too slow is called *laminar flow*.

If a system is designed with the flow too fast, the pipe will be too small and the velocity noise of the water will be objectionable. The designer tries to reach a balance between oversized and undersized pipe. The following table lists some water velocities in ft per minute (fpm) recommended for hydronic piping systems.

Pump discharge	8 to 12 fpm
Pump suction	4 to 7 fpm
Drain lines	4 to 7 fpm
Header	5 to 15 fpm
Riser	3 to 10 fpm
General service	5 to 10 fpm
City water lines	3 to 7 fpm

33.7 RADIANT, LOW-TEMPERATURE HYDRONIC PIPING SYSTEMS

The concept of radiant heat differs a great deal from that of convective heating as it has been discussed thus far in this unit. Radiant heating systems rely on heating the shell of the structure as opposed to heating the air in the structure. The purpose of convective heating is to determine the rate at which heat is lost from the room by conduction, convection, and radiation at the desired conditions. Radiant heating involves regulating the rate at which the human body loses heat.

How the Body Functions as a Radiator

Under normal conditions, the human body produces about 500 Btu/h. The body requires about 100 Btu/h to remain alive. The rest of this heat must be rejected, making the body act as a radiator. By controlling the rate at which we give up heat, we can control our comfort level. If we give up heat too fast, we will feel cold. If we do not give up heat fast enough, we will feel warm and uncomfortable. Typically, a room at a temperature of 68°F allows us to shed at an acceptable rate the extra 400 Btu/h that our body generates. For this reason, it is desirable that the occupied space be about 68°F—with the exception of the areas close to the ceiling and floor. Since our heads are not close to the ceiling, there is no need to keep that area heated, so it can be somewhat cooler than 68°F. In addition, we lose a lot of heat from our feet, so the floor should be somewhat warmer than 68°F. **Figure 33–91** shows

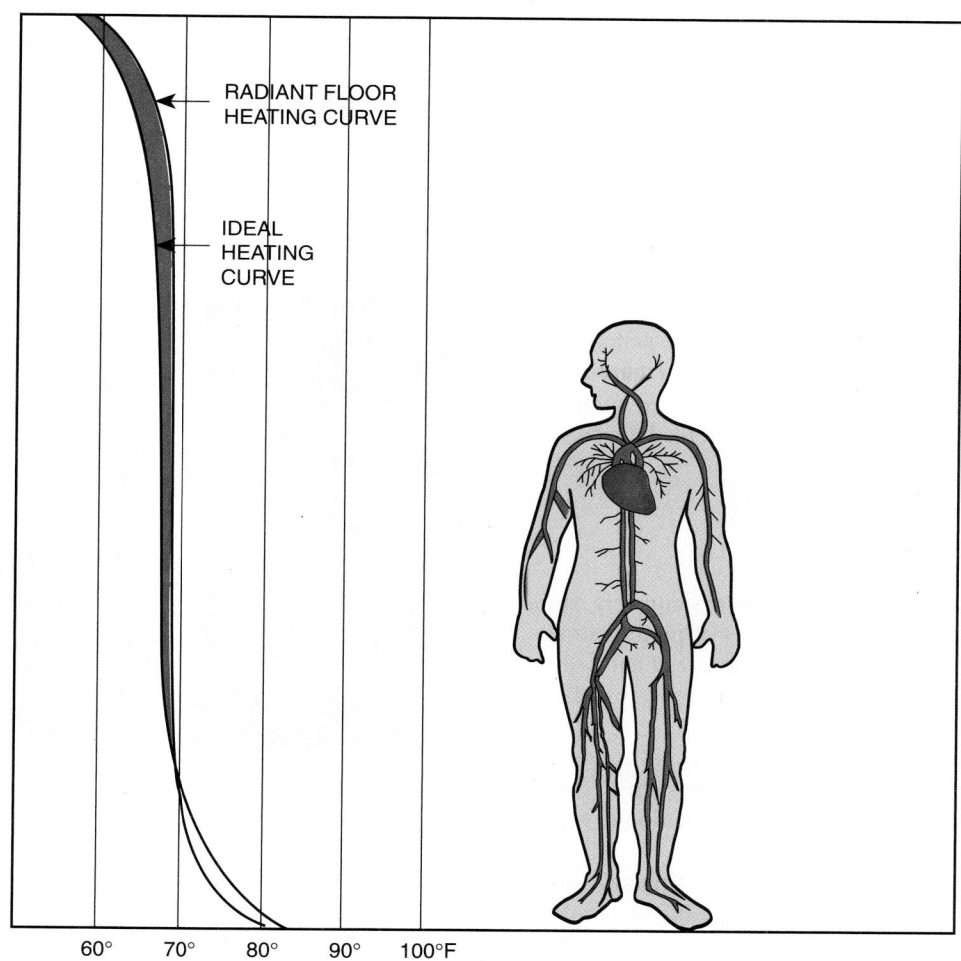

Figure 33–91   Ideal comfort chart compared with that of a radiant heating system. *Courtesy Uponor*

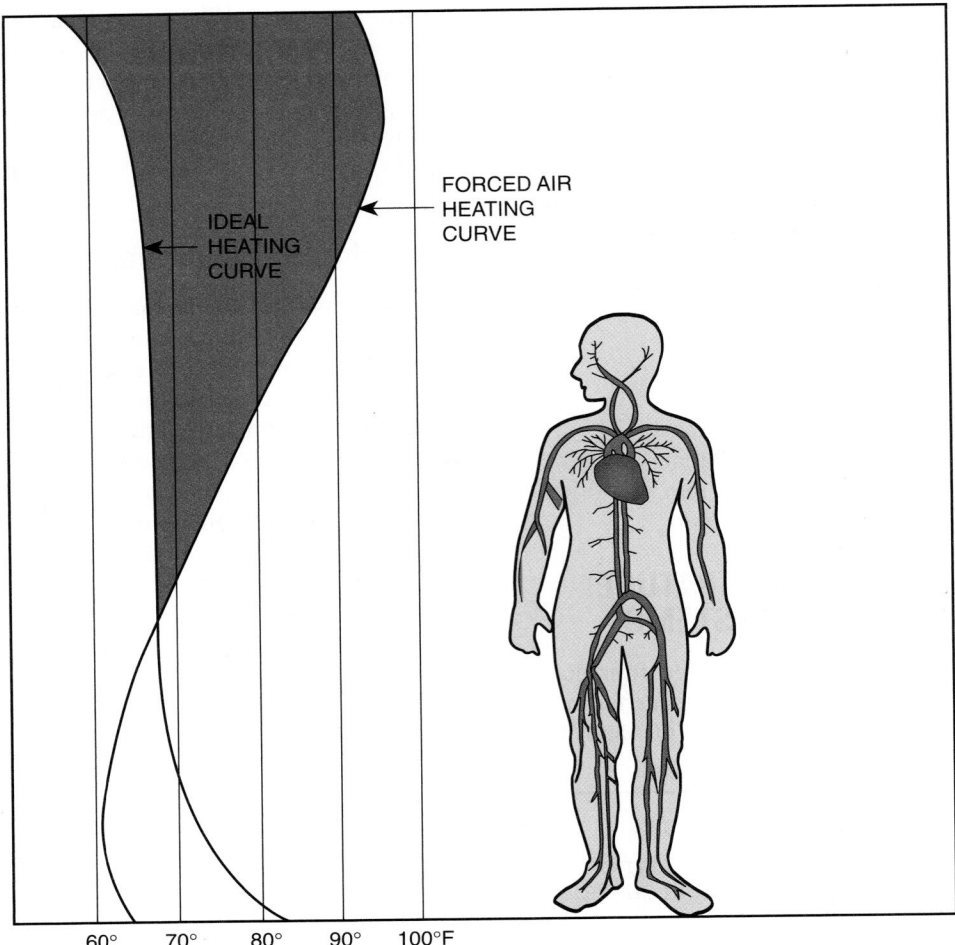

Figure 33–92   Ideal comfort chart compared with that of a forced-air heating system. *Courtesy Uponor*

room temperature changes as we go from the floor to the ceiling. This represents an ideal temperature pattern, one which radiant heating systems closely follow. **Figure 33–92,** shows the same ideal temperature pattern as compared with the temperature pattern created by forced-air heating systems.

The Radiant System

Like hot water hydronic systems, radiant systems rely on the circulation of hot water through a series of piping circuits. There are three main differences between radiant heating systems and conventional hot water hydronic systems. These differences are as follows:

■ Radiant heating systems are nearly invisible as compared with conventional hydronic systems. The piping arrangements are concealed in the floors, walls, and sometimes ceilings of the structure.

■ The piping material used on radiant heating systems is different. Instead of using copper piping materials, radiant heating systems often use *polyethylene tubing,* more commonly known as *PEX tubing,* **Figure 33–93.** PEX is installed using a minimum of pipe fittings, **Figure 33–94.**

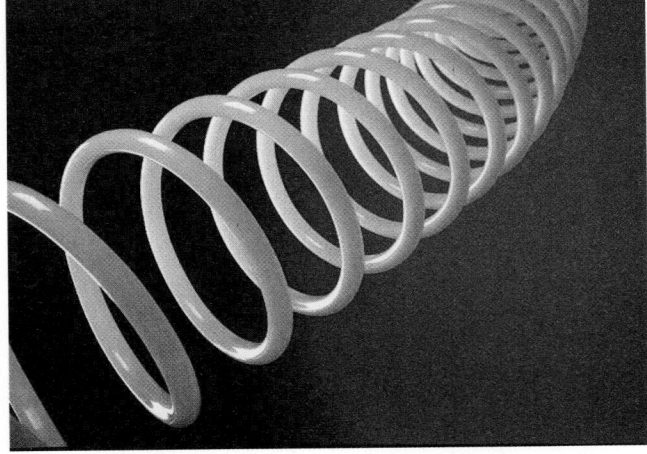

Figure 33–93   PEX tubing. *Courtesy Uponor*

■ The temperature of the water in radiant heating systems is considerably lower than the temperature of the water utilized by conventional hydronic systems. Typical hydronic systems use water as hot as 180°F, while radiant systems use water at an average temperature of about 110°F.

(A)

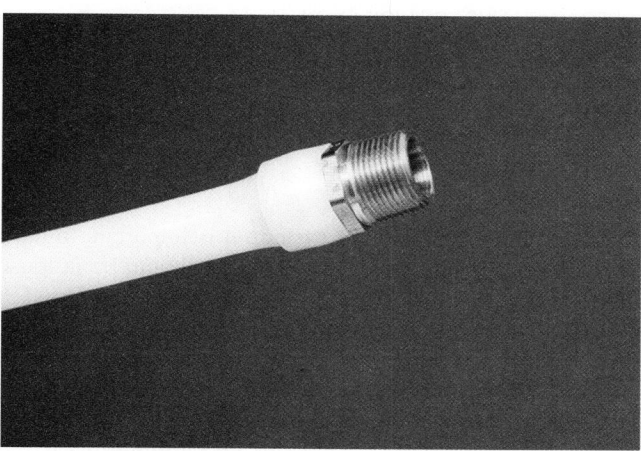

(B)

(C)

Figure 33–94 Fittings used to connect PEX tubing **(A)–(C)**.
Courtesy Uponor

Heat Sources for Radiant Heating Systems

As mentioned at the beginning of this unit, the heat source for hydronic systems is quite often a boiler. It was also mentioned that, in the case of low-temperature systems such as radiant applications, geothermal heat pumps are becoming more and more popular. Here are some quick notes and suggestions regarding the heat sources used for radiant heating systems:

- When using a geothermal heat pump system for radiant heating, be sure to use a buffer tank. This will prevent water flow restrictions on the condenser water side, which can lead to high-pressure problems in the refrigerant circuit.
- The buffer tank serves as a storage tank for the heated water; the heat pump will cycle on and off to maintain the desired water temperature in the tank.
- Larger temperature differential settings on the buffer tank increases the run time of the heat pump, but it also increases the off time of the system—which thereby increases the efficiency of the system.
- When using a copper-tube, low-mass boiler for radiant applications, it is a good idea to use a buffer tank as well. This will prevent short cycling of the boiler, **Figure 33–95.**
- If direct piping is being used, **Figure 33–96,** be sure that the boiler is designed to handle the potentially low temperature of the water returning to the appliance.
- When the boiler return-water temperature may be a problem, be sure to provide a means for keeping the return-water temperature high enough to prevent flue-gas condensation, **Figure 33–97.**

Radiant Heating Piping

Given the versatility of radiant heating systems, the possibilities are boundless when it comes to installation variations and custom applications. The most common types of radiant heating system applications are the following:

- Slab on grade
- Thin slab
- Dry or concrete-free applications

SLAB ON GRADE. The slab-on-grade application is very popular—especially for new construction applications and projects. Since the concrete for these applications has not been poured, it is a relatively simple task to position the PEX tubing loops right in the concrete slab. On such applications, the tubing material is secured to the steel reinforcement mesh in the floor using straps or clips specially designed for that use, **Figure 33–98.** The spacing between the PEX tubes is determined by the depth of the tubes below the concrete surface, the size of the tubes used, the type of floor being installed, and the location of the tubes with respect to the outside walls of the structure. For higher system efficiency, insulation should be placed below the concrete slab as well as below a vapor barrier to prevent the ground from robbing the slab of its heat.

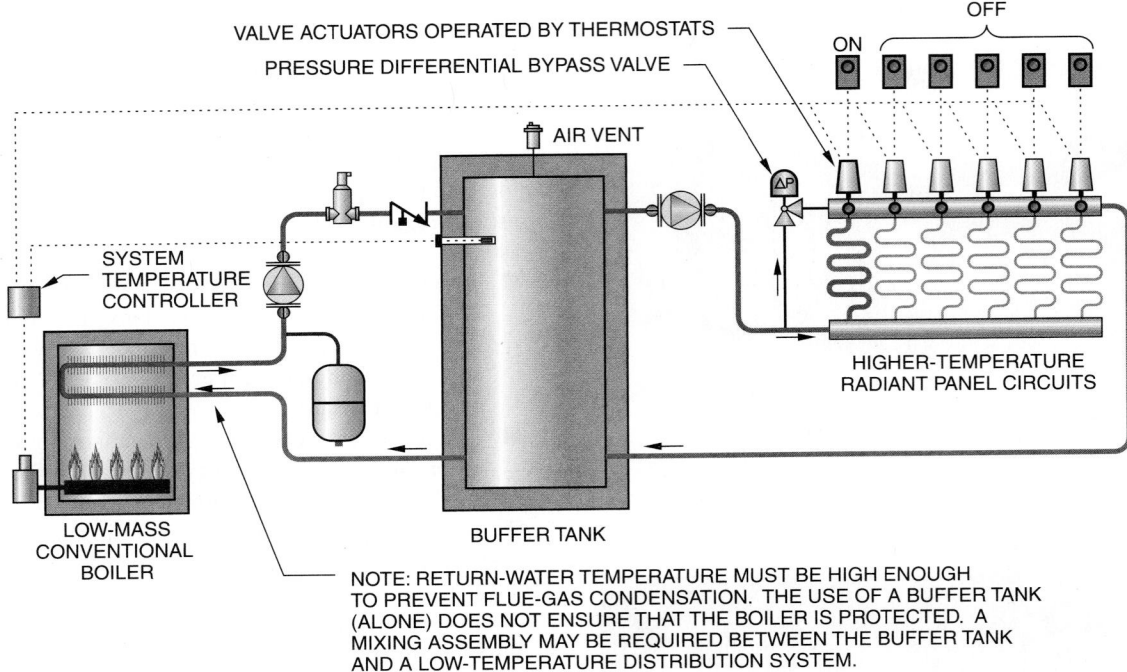

Figure 33–95   A buffer tank used in conjunction with a low-mass boiler.

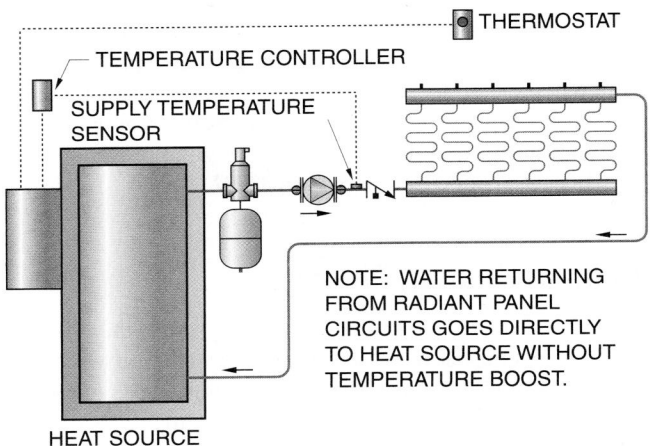

Figure 33–96   Direct piping from the heat source to the heating circuits. There is no mixing taking place in this piping arrangement.

THIN SLAB.   When there is an existing floor in place, thin-slab installations are common. The PEX tubing is stapled in place to the top of the frame floor and concrete is then poured over the existing floor. The thickness of the concrete can range from 1.5 in. to 2 in. and should completely cover the PEX tubing with at least ¾ in. of poured material, **Figure 33–99.** To ensure that the existing floor will be able to handle the weight of the concrete, be sure to consult an engineer or architect when installing a thin-slab system.

DRY APPLICATIONS.   Another option for installing radiant systems is to staple the tubing to the bottom of the flooring material. This is referred to as a staple-up job, **Figure 33–100.**

When installing this type of system, it is important to have the staples closely spaced, as the tubing should be in loose contact with the floor material for effective heat transfer. Insulation should be positioned under the stapled-up tubing as illustrated in the figure.

Another option for a concrete-free radiant heating system installation is the use of aluminum "wings" or fins, **Figure 33–101,** that can be attached to PEX tubing that is fed through the floor joists (as opposed to being stapled up to the underside of the floor). These fins are installed in sets of two and are designed to completely surround the PEX tubing, **Figure 33–102.** Installation is relatively simple, given that holes are predrilled in the material. The completed installation is illustrated in **Figure 33–103.**

Piping Arrangements

There are many different configurations of piping that can be utilized on radiant heating systems. What follows is intended to provide merely a small sampling of the options available—as attempting to discuss all installation and piping schemes here would be beyond the scope of this text.

DIRECT PIPING.   By far, the simplest configuration of piping in a radiant system is referred to as direct piping. In a direct-piping scheme, the water supplied to the boiler is pumped into the radiant heating circuits, **Figure 33–96.** This type of piping configuration is more system friendly when the heat source is a buffer tank as opposed to a boiler. If the heat source is a boiler and the desired water temperature at the radiant heating circuits is low, there is a great chance that the flue gases in the boiler will condense, because the

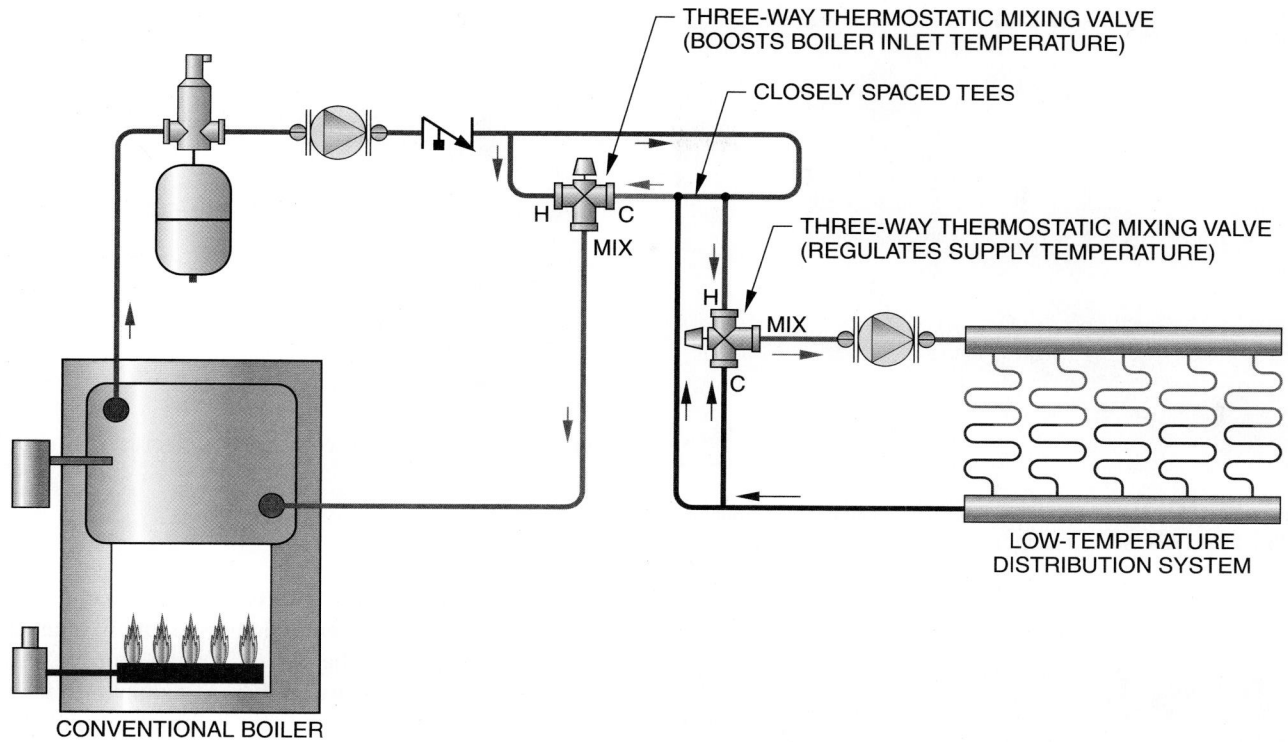

Figure 33–97 Boiler piping allows for control of boiler return-water temperature.

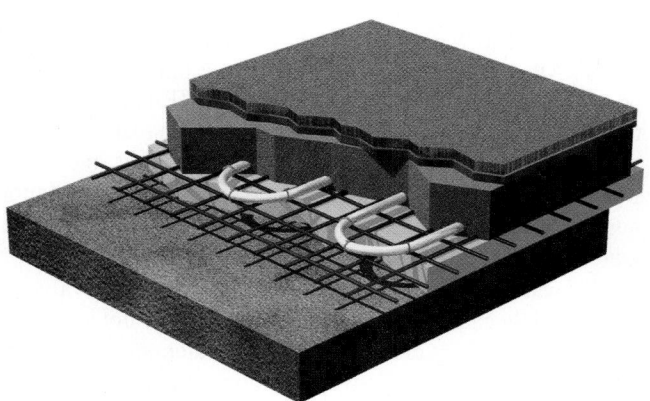

Figure 33–98 Cutaway of a slab-on-grade radiant piping layout.
Courtesy Uponor

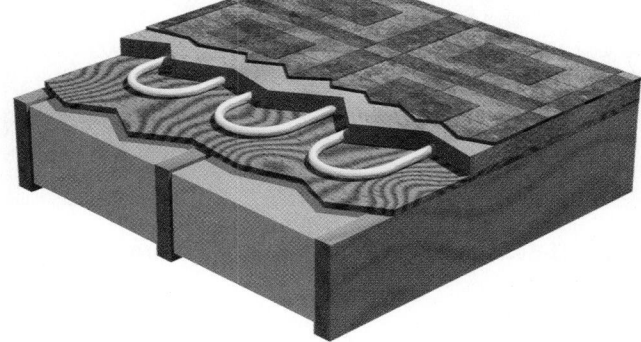

Figure 33–99 Cutaway of a thin-slab radiant piping layout.
Courtesy Uponor

return-water temperature will be low as well. If a buffer tank is used, the water that returns to the boiler comes from the buffer tank, not the outlet of the radiant heating circuits.

MANUALLY SET MIXING VALVES. As mentioned earlier in the radiant section, manually set mixing valves do not respond to changes in water temperature or to changing water flow rates. For this reason, using these valves is not the best option, as they do not prevent low-temperature water from returning to the boiler, **Figure 33–104.**

Figure 33–100 Cutaway of a staple-up radiant piping layout.
Courtesy Uponor

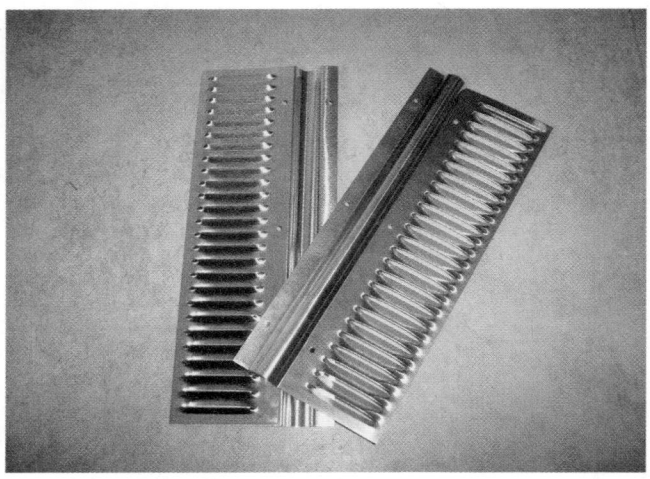

Figure 33–101   Aluminum "fins" used to increase the heat transfer surface of the PEX tubing in a dry, concrete-free radiant application.

PREDRILLED HOLES

Figure 33–102   Predrilled holes ease the installation process.

Figure 33–103   Completed installation.

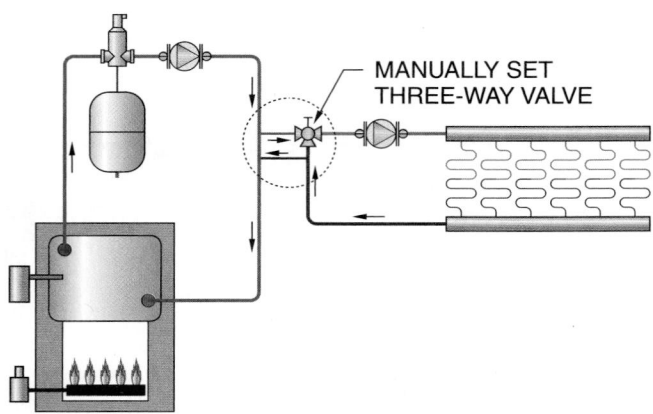

MANUALLY SET
THREE-WAY VALVE

Figure 33–104   An example of a hydronic system using a manually set three-way valve.

THERMOSTATIC MIXING VALVES. Thermostatic mixing valves will adjust their internal settings to supply water at the desired temperature to the radiant heating loops. Consider the system shown in **Figure 33–105**. As a point of reference, let's assume that the entire system is completely filled with 180°F water and the desired water temperature in the radiant loops is 110°F. The radiant loops operate with a ΔT of 20°F. Since the water temperature at the inlet of the radiant loops is 70°F higher (180°F − 110°F) than desired, the hot water inlet port of the thermostatic mixing valve is completely closed. The 180°F water enters the radiant loops and leaves at 160°F. All of the 160°F water flows back into the cold inlet port of the mixing valve, **Figure 33–106.** NOTE: No water from the radiant loop can return to the boiler because no water has entered the radiant loop from the hot water inlet port. Because the radiant loops are completely filled, we cannot add water without removing water, and we cannot remove water without adding water.

The water enters the cold inlet port of the mixing valve at 160°F and, because the water is still too hot, the hot water inlet port still remains closed. The 160°F water enters the radiant loops and the water leaves the loops at a temperature of 140°F. This process continues until the water leaving the radiant loops is at a temperature of 100°F. Since the water enters the cold water inlet port of the mixing valve at 100°F, and since the desired water temperature is 110°F, the hot water inlet port will open slightly to mix some 180°F water with the 100°F water to provide 110°F water at the outlet of the mixing valve, **Figure 33–107.** Because some water enters the mixing valve through the hot water inlet port, the same amount of water will return to the boiler from the outlet of the radiant heating circuits.

Another application for thermostatic mixing valves is to ensure that the temperature of the return water to the boiler is high enough to prevent flue-gas condensation. The system in **Figure 33–108** utilizes a mixing valve to regulate water supplied to the heating circuits and also uses another mixing valve to regulate the temperature of the water returning to the boiler.

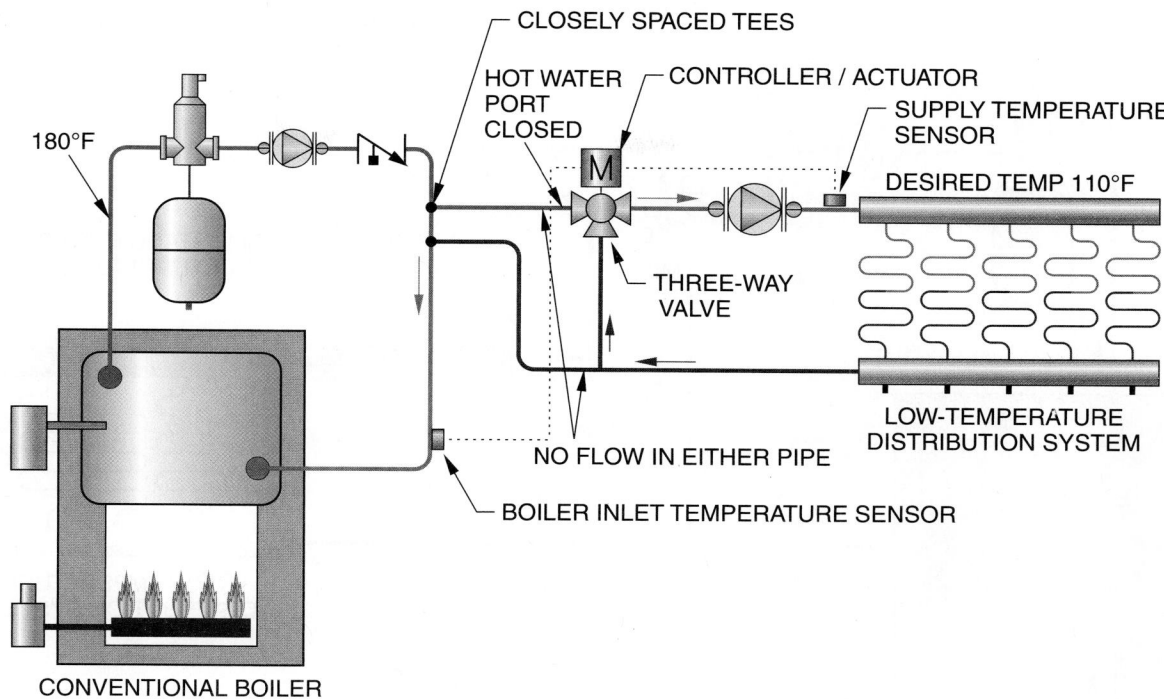

Figure 33–105   Low-temperature hydronic system using a three-way thermostatic mixing valve. Since the water in the heating loop is warmer than desired, the hot water inlet port on the mixing valve is in the closed position. Under these conditions, no water can enter or leave the low-temperature heating circuit.

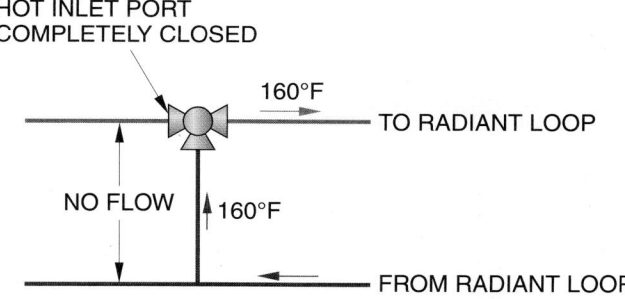

Figure 33–106   Water that is too hot for the heating loop will continue to circulate without mixing until the temperature at the outlet of the heating loop (cold water inlet) falls below the desired water temperature.

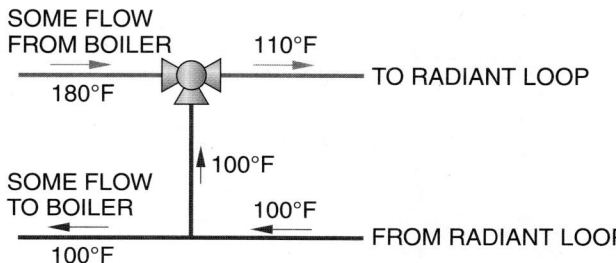

Figure 33–107   Since the water at the outlet of the heating loop is at a temperature that is lower than desired, the hot water inlet port will open slightly. This will allow hot water to enter the mixing valve—and also allow cool water from the heating loop to leave the loop and return to the boiler.

33.8   COMBINATION (HIGH- AND LOW-TEMPERATURE) PIPING SYSTEMS

In a case where a structure has both high- and low-temperature heating circuits, the same boiler can serve both applications by using a primary–secondary pumping arrangement, **Figure 33–109.** In the figure we can see that water is being supplied to the first two secondary loops at a temperature close to 180°F. The water in the main loop after these two secondary loops will be about 165°F. This water can then return to the boiler, while a portion of the water will mix with the water in the radiant loop to maintain the desired water temperature in that circuit. The temperature of the water returning to the boiler will be about 160°F.

33.9   TANKLESS DOMESTIC HOT WATER HEATERS

Most hot water heating boilers (oil and gas fired) can be furnished with a domestic hot water heater consisting of a coil inserted into the boiler. The domestic hot water is contained within the coil and heated by the boiler quickly, which eliminates the need for a storage tank. This system can meet most domestic hot water needs without using a separate tank, **Figure 33–110.** This is generally an efficient way to produce hot water.

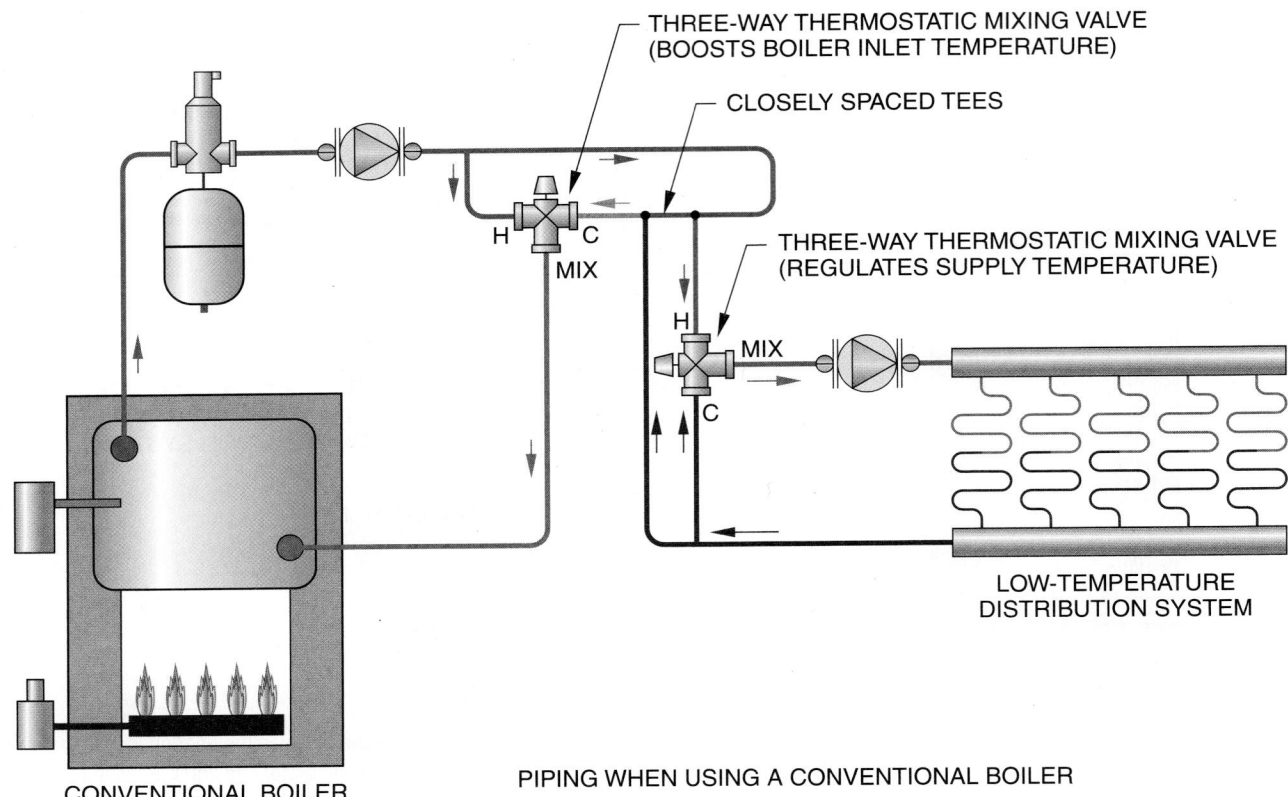

Figure 33–108 This system uses two thermostatic mixing valves: one to control the temperature of the heating loop, and one to control the temperature of the water returning to the boiler.

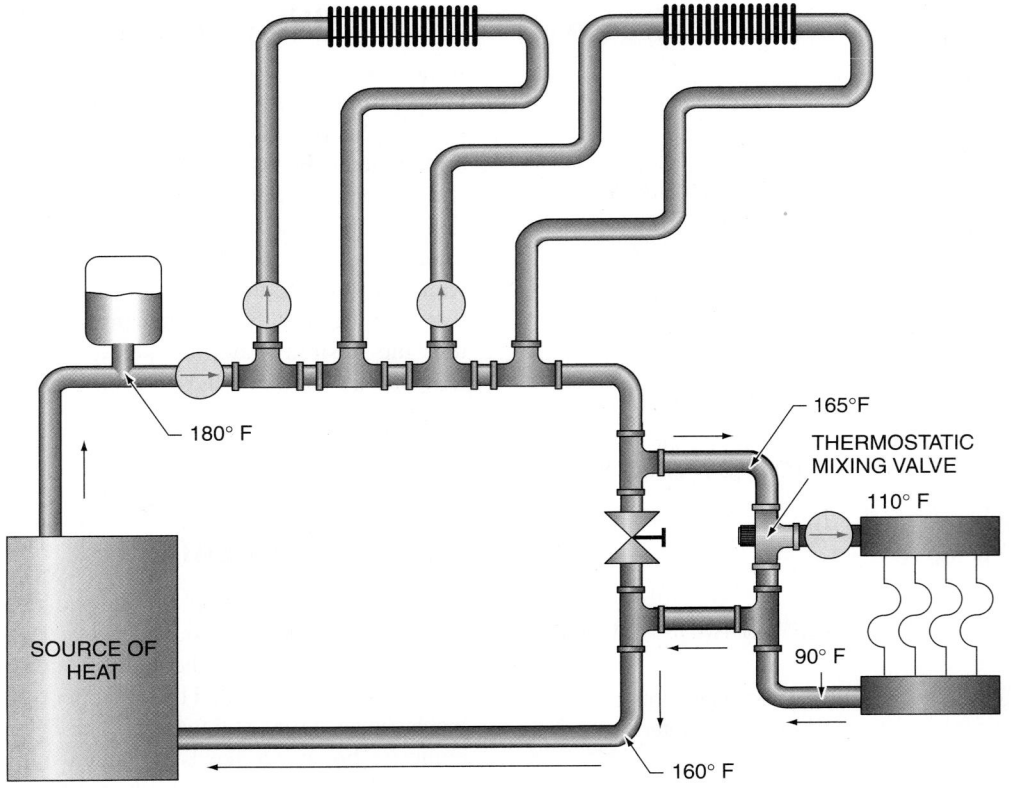

Figure 33–109 A primary–secondary piping arrangement that incorporates both high- and low-temperature heating circuits.

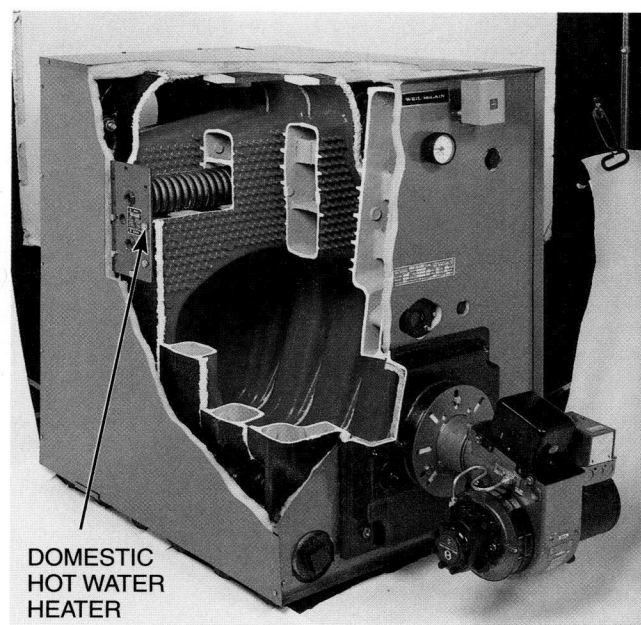

DOMESTIC
HOT WATER
HEATER

Figure 33–110   A tankless domestic water heater installed in a boiler. *Courtesy Weil McLain Corporation*

PREVENTIVE MAINTENANCE

The air side of hydronic systems can be either natural draft or forced draft. If the system is natural draft, the convectors will be next to the floor and will be subjected to dust on the bottom side where the air enters. There are no filters on these natural-draft systems and consequently there is no filter maintenance. The coil becomes the filter, but the air moves so slowly that not much dust is picked up. The cover should be removed after several years and the coil vacuumed clean.

If the system is forced air, routine filter maintenance will be required as described in Unit 30. Coils may become dirty after many years of good filter maintenance because all particles of dust are not filtered and some will deposit on the coil. When the coil becomes dirty, capacity will be reduced. Most coils are oversized and the capacity reduction may not be noticed for many years.

Cleaning a coil is accomplished with approved liquid detergents mixed with water and applied to the coil. If the coil is in an equipment room, it may be cleaned in place. If the coil is in the attic or above a ceiling, it may have to be removed from the air handler for cleaning.

The fan blades may also have dust deposits built up on them and need cleaning. The fan wheel may be cleaned in the fan if the motor is not in the way. This may be accomplished by scraping the dust buildup with a screwdriver blade, using a vacuum cleaner to remove it. If the fan wheel is very dirty, it may be best to remove the fan section and clean it outdoors. NOTE: Do not allow water to enter the bearings or motor. Plastic bags may be used to protect them. If the bearings get soaked, purge them with oil or grease to remove the water. If

the motor gets wet, set it in the sun and use an ohmmeter to verify that the motor is not grounded. Setting the motor on an operating air-conditioning condensing unit fan outlet will quickly dry it as the warm, dry air passes over it.

The water side of a hydronic system requires some maintenance for the system to give long, trouble-free performance. Fresh water contains oxygen, which causes corrosion in the pipes. Corrosion is a form of rust. The proper procedure for the start-up of a hydronic system calls for the system to be washed inside with an approved solution to remove any oils and construction debris. This may or may not have been performed on any system you may encounter. When the system is cleaned initially, it is supposed to be filled with fresh water and chemicals that will treat the initial charge of water from corrosion. Again, this may or may not have been done. If the system is several years old and no water treatment has been used, the system will surely need maintenance. It should be drained and flushed. Water treatment should be added when the system is filled with fresh water. Be sure to follow the water treatment company's instructions.

Systems with leaks have a continuous water makeup of fresh water, causing continuing corrosion because the fresh water contains oxygen. Water leaks must be repaired. It is common for water from leaks in coils to be carried away in drain lines and not be discovered. When a system has an automatic air bleed, air will not build up.

Water pumps must be maintained. They have lubrication ports for oil or grease. Keep the oil reservoir at the correct

(continued)

level on the pump with the correct oil. The motor will not have an oil reservoir but will require a certain amount of oil. Do not exceed this recommended amount.

When a motor has grease fittings, proper procedures must be used when applying grease. You cannot continue to pump grease into the bearings without some relief. The seal will rupture. NOTE: Unscrew the relief fitting on the bottom of the bearing and allow grease to escape through the relief hole. Grease may be added until fresh grease is forced out the relief opening.

Pumps also have couplings that should be inspected. SAFETY PRECAUTION: Turn off the power and lock and tag the disconnect before removing the cover to the coupling. Check the coupling filings caused by wear that may be in the housing. Check for loose bolts and play in the coupling

if it is a rubber type. Play in the coupling is natural for the spring type.

The electrical service should be checked if the pump has a contactor. SAFETY PRECAUTION: Shut off the power and remove the cover to the contactor. Do not touch any electrical connections with meter leads until you have placed one voltmeter lead on a ground source, a conduit pipe will do, and touch the other meter lead to every electrical connection in the contactor box. The reason for this is some pumps are interlocked electrically with other disconnects, and there may be a circuit from another source in the box. Use all precautions.

When contactor contacts are found burned, replace them; if wire is found to be frayed or burned, repair or otherwise correct the situation.

33.10   SERVICE TECHNICIAN CALLS

SERVICE CALL 1

A motel manager calls indicating that there is no heat at the motel. *The problem is on the top two floors of a four-story building. The system has just been started for the first time this season, and it has air in the circulating hot water. The automatic vent valves have rust and scale in them due to lack of water treatment and proper maintenance.*

The technician arrives and goes to the boiler room. The boiler is hot, so the heat source is working. The water pump is running, because the technician can hear water circulating. The technician goes to the top floor to one of the room units and removes the cover; the coil is at room temperature. The technician listens closely to the coil and can hear nothing circulating. There may be air in the system, and the air vents, sometimes called bleed ports, must be located. The technician knows they must be at the high point in the system and normally would be on the top of the water risers from the basement.

The technician goes to the basement where the water lines start up through the building and finds a reference point. The technician then goes to the top floor and finds the top portions of these pipes. The pipes rise through the building next to the elevator shaft. The technician goes to the top floor to the approximate spot and finds a service panel in the hall in the ceiling. Using a ladder, the automatic bleed port for the water supply is found over the panel. This is the pump discharge line. See **Figure 33–34** for an example of an automatic air vent or bleed port.

The technician removes the rubber line from the top of the automatic vent. The rubber line carries any water that

bleeds to a drain. There is a valve stem in the automatic vent (as in an automobile tire). The technician presses the stem, but nothing escapes. There is a hand valve under the automatic vent, so the valve is closed and the automatic vent is removed. The technician then carefully opens the hand valve and air begins to flow out. Air is allowed to bleed out until water starts to run. The water is very dirty. SAFETY PRECAUTION: Care should be used around hot water.

The technician then bleeds the return pipe in the same manner until all the air is out, and water only is at the top of both pipes. The technician then goes to the heating coil where the cover was previously removed. The coil now has hot water circulating.

The technician then takes the motel manager to the basement and drains some water from the system and shows the manager the dirty water. The water should be clear or slightly colored with water treatment. The technician suggests that the manager call a water treatment company that specializes in boiler water treatment, or the system will have some major troubles in the future.

The technician then replaces the two automatic vents with new ones and opens the hand valves so that they can operate correctly.

SERVICE CALL 2

A customer in a small building calls. *There is no heat in one section of the building although the system is heating well in most of the building. The building has four hot water circulating pumps, and one of them is locked up—the bearings are seized. These pumps are each 230 V, three-phase, and 2 hp.*

The technician arrives at the building and consults the customer. They go to the part of the building where there is no heat. The thermostat is calling for heat. They go to the basement where the boiler and pumps are located. They can tell from examination that the number 3 pump, which serves the part of the building that is cool, is not turning. The technician carefully touches the pump motor; it is cool and has not been trying to run. Either the motor or the pump has problems.

The technician chooses to check the voltage to the motor first and then goes to the disconnect switch on the wall and measures 230 V across phases 1 to 2, 2 to 3, and 1 to 3. This is a 230-V three-phase motor with a motor starter. The motor overload protection is in the starter, and it is tripped.

The technician still does not know whether the motor or the pump has problems. All three windings have equal resistance, and there is no ground circuit in the motor. The power is turned on, an ammeter is clamped on one of the motor leads, and the overload reset is pressed. The technician can hear the motor try to start, and it is pulling locked-rotor current. The disconnect is pulled to stop the power from continuing to the motor. It is evident that either the pump or the motor is stuck.

The technician turns off the power and locks it out and then returns to the pump. It is on the floor next to the others. The guard over the pump shaft coupling is removed. The technician tries to turn the pump over by hand—remember, the power is off. The pump is very hard to turn. The question now is whether it is the pump or the motor that is tight.

The technician disassembles the pump coupling and tries to turn the motor by hand. It is free. The technician then tries the pump; it is too tight. The pump bearings must be defective. The technician gets permission from the owner to disassemble the pump. The technician shuts off the valves at the pump inlet and outlet, removes the bolts around the pump housing, then removes the impeller and housing from the main pump body. The technician takes the pump impeller housing and bearings to the shop and replaces them, installs a new shaft seal, and returns to the job.

While the technician is assembling the pump, the customer asks why the whole pump was not taken instead of just the impeller and housing assembly. The technician shows the customer where the pump housing is fastened to the floor with bolts and dowel pins. The relationship of the pump shaft and motor shaft has been maintained by not removing the complete pump. They will not have to be aligned after they are reassembled. The pump is manufactured to be rebuilt in place to avoid this.

After completing the pump assembly, the technician turns it over by hand, assembles the pump coupling, and affirms that all fasteners are right. Then the pump is started. It runs and the amperage is normal. The technician turns off the power, locks it out, and assembles the pump coupling guard. The power is turned on and the technician leaves.

SERVICE CALL 3

An apartment house manager reports that *there is no heat in one of the apartments. This apartment house has 25 fan coil units with a zone control valve on each and a central boiler. One of the zone control valves is defective.*

The technician arrives and goes to the apartment with no heat. The room thermostat is set above the room setting. The technician goes to the fan coil unit located in the hallway ceiling and opens the fan coil compartment door. It drops down on hinges. The technician stands on a short ladder to reach the controls. There is no heat in the coil, and there is no water flowing.

This unit has a zone control valve with a small heat motor. The technician checks for power to the valve's coil; it has 24 V, as it should. See **Figure 33–42** for an example of this type of valve. This valve has a manual open feature, so the valve is opened by hand. The technician can hear the water start to flow and can feel the coil get hot. The valve heat motor or valve assembly must be changed. The technician chooses to change the assembly because the valve and its valve seat are old.

The technician obtains a valve assembly from the stock of parts at the apartment and proceeds to change it. *The technician shuts off the power and the water to the valve and water coil. A plastic drop cloth is then spread to catch any water that may fall. Then a bucket is placed under the valve before the assembly is removed.* The old valve is removed, and the new valve installed. The electrical connections are made, and the system valves are opened allowing water back into the coil.

The technician turns on the power and can see the valve begin to move after a few seconds (this is a heat motor valve, and it responds slowly). The technician can now feel heat in the coil. The compartment door is closed. The plastic drop cloth is removed, the room thermostat is set, and the technician leaves.

The following service calls do not include solutions. The solutions can be found in the *Instructor's Guide.*

SERVICE CALL 4

A customer calls and indicates that one of the pumps beside the boiler is making a noise. This is a large home with three pumps, one for each zone.

The technician arrives and goes to the basement, hears the coupling, and notices there are metal filings around the pump shaft.

What is the problem and the recommended solution?

SERVICE CALL 5

A homeowner calls and states that *there is water on the floor around the boiler in the basement. The relief valve is relieving water.*

The technician goes to the basement and sees the boiler relief valve seeping. The boiler gage reads 30 psig, which is the rating of the relief valve. The system normally operates at 20 psig. The technician looks at the flame in the burner section and sees the burner burning at a low fire. A voltmeter check shows there is no voltage to the gas valve operating coil.
What is the problem and the recommended solution?

- Popular radiant heating system installation types include slab on grade, thin slab, and dry/staple-up.
- Tankless domestic hot water heaters are commonly used with hot water heating systems. The water is contained within a coil and heated by the boiler.
- Hydronic systems should be serviced with proper preventive maintenance procedures annually.

SUMMARY

- Hydronic heating systems use hot water or steam to carry heat to the areas to be heated.
- A boiler is an appliance that heats water. In gas and oil systems the part of the boiler containing the water is constructed in sections or tubes.
- Cast-iron and steel boilers are classified as high-mass boilers.
- Copper-tube boilers are classified as low-mass boilers.
- A limit control is used to shut the heat source off if the water temperature gets too high.
- A low-water cut-off shuts down the heat source if the water level gets too low.
- A pressure relief valve is required to discharge water when excessive pressure is created by expansion due to overheating.
- An air cushion tank or expansion tank is necessary to provide space for trapped air and to allow for expansion of the heated water.
- Zone control valves are thermostatically operated and control the flow of hot water into individual zones.
- A centrifugal pump circulates the water through the system.
- The "point of no pressure change" is the point in a hydronic system where the water pressure does not change; it is located at the inlet of the expansion tank.
- Finned-tube baseboard terminal units are commonly used in residential hydronic heating systems.
- A balancing valve is used to equalize the flow rate in different heating circuits.
- Outdoor reset control is used to change the temperature of the boiler water in response to changes in the outside ambient temperature.
- The series loop system is the most economical hydronic system to install but does not allow for individual terminal-unit temperature control.
- One-pipe systems and two-pipe reverse-return piping systems are most commonly used in residential and light commercial installations.
- A one-pipe system requires a specially designed tee to divert some water to the baseboard unit while allowing the rest to flow through the main pipe.
- Primary–secondary pumping allows for multiple-temperature operation for multizone systems.
- Mixing valves are used to achieve different water temperatures at different points in a hydronic system.
- Radiant heating systems use lower water temperatures to provide heating.

REVIEW QUESTIONS

1. A hydronic heating system uses _____ to move heat around in the system.
2. Describe a *zone* in a hydronic heating system.
3. Name four sources of heat commonly used for hydronic heating systems.
4. Give three reasons that air is detrimental to a hydronic heating system.
5. Explain the difference between a wet-base and a dry-base boiler.
6. Where is air vented from a hydronic heating system?
7. The "point of no pressure change" is located at the
   A. outlet of the circulator pump.
   B. inlet of the expansion tank.
   C. inlet of the pressure-reducing valve.
   D. outlet of the pressure relief valve.
8. The purpose of a limit control in a boiler is to
   A. vent the air.
   B. cycle the boiler on and off to maintain the desired water temperature in the boiler.
   C. shut the boiler off before it freezes.
   D. fill the system with water.
9. The device that relieves the pressure in a boiler when it is above the design working pressure is the _____ valve.
10. Describe the function of a low-water cut-off.
11. The relief valve setting for a typical low-pressure boiler is _____ psig.
12. Describe how an air cushion tank functions in a hydronic heating system.
13. True or False: City water does not contain air.
14. State two ways that air can get into a hydronic heating system.
15. The most common type of pump used for hydronic heating systems is the
    A. reciprocating.
    B. pulsating.
    C. scroll.
    D. centrifugal.
16. Another term for pump pressure is feet of _____.
17. Two types of zone control valves are the _____ and the _____.
18. Why must expansion be considered when sizing finned-tube radiation?
19. How is expansion in hot water pipes handled?
20. Sketch the following systems: one pipe, two-pipe direct return, and two-pipe reverse return.

21. Describe the advantage of a two-pipe reverse-return system.

22. The pressure differential bypass valve is most likely found on systems equipped with
 A. two-port zone valves.
 B. diverter tees.
 C. three-port zone valves.
 D. manually adjustable mixing valves.

23. What is the heat output, in Btu/h, of a terminal unit with a ΔT of 10°F and a water flow rate of 5 gpm?

24. In a primary–secondary pumping arrangement, the largest circulator pump will be located
 A. in the primary loop.
 B. in the zone with the largest terminal unit.
 C. in the zone with the smallest terminal unit.
 D. anywhere in the system because all of the pumps are the same size.

25. The temperature of the water at the outlet of a three-way mixing valve will most likely be
 A. equal to the temperature of the water at the hot water inlet.
 B. lower than the temperature of the water at the cold water inlet.
 C. higher than the temperature of the water at the hot water inlet.
 D. higher than the temperature of the water at the cold water inlet.

26. A tankless water-heating system gets its heat from the system's _____.

UNIT
34
Indoor Air Quality

SAFETY CHECKLIST

✔ Do not disturb materials containing asbestos.
✔ Use all electrical safety precautions when servicing or troubleshooting electrical or electronic circuits. Use common sense.
✔ When replacing or servicing line-voltage components, ensure that power is off, that the panel is locked and tagged, and that you have the only key.

34.1 INTRODUCTION

In the last several years, evidence has suggested that the air in homes and other buildings is becoming increasingly polluted. Some evidence indicates that indoor air may be more seriously polluted than the outdoor air in some areas. Other research indicates that many people spend as much as 90% of their time indoors. Some people, such as the young, elderly, and chronically ill, may spend more time indoors than others, and they may be the most susceptible to the effects of indoor pollution. These people face a serious risk from the cumulative effects of this indoor air pollution. ASHRAE Standard 62-2001, "Ventilation for Acceptable Indoor Air Quality," is the standard generally used by engineers and technicians in this industry to determine acceptable indoor air quality.

34.2 SOURCES OF INDOOR AIR POLLUTION

Following are some potential sources of indoor air pollution:

Moisture	Carbon monoxide
Mold	Carpets
Pressed wood furniture	Pressed wood subflooring
Humidifiers	Drapes
Moth repellents	Fireplaces
Dry-cleaned goods	Household chemicals
House dust mites	Asbestos floor tiles
Personal care products	Pressed wood cabinets
Air freshener	Unvented gas stoves
Stored fuels	Asbestos pipe wrap
Car exhaust	Radon
Paint supplies	Unvented clothes drier
Paneling	Pesticides
Wood stoves	Stored hobby products
Tobacco smoke	Lead-based paints

Figure 34–1 illustrates sources and causes of indoor pollution. In older homes, outdoor air may enter through openings, joints, and around windows and doors. Most modern homes and office buildings, however, are designed and constructed to keep the outside air out in order to maximize heating and cooling efficiencies. This may result in an accumulation of pollution inside. Often a means of *mechanical ventilation* may be required.

Inadequate ventilation can increase the effects of indoor air pollution. Outdoor air in sufficient quantities must be brought inside to mix with and dilute the pollutant emissions, and the indoor air must be vented to the outside to carry out pollutants. Without this ventilation, the pollutants may accumulate to the point where they can cause health and comfort problems. Mechanical ventilation can include exhaust fans in a single room, such as in a bathroom or kitchen, or an air-handling system to intermittently or continuously remove indoor air and replace it with filtered and conditioned air from outside.

Results of inadequate ventilation may include moisture condensation on windows or walls, smelly or stuffy air, dirty central heating and cooling equipment, and areas in which books, shoes, or other items become moldy.

34.3 CONTROLLING INDOOR AIR CONTAMINATION

The following methods are used to control indoor air contamination: eliminate the source of the contamination, including moisture; provide adequate ventilation; and provide a means for cleaning the air.

1. CHIMNEY
- MOISTURE INTRUSION (IMPROPER FLASHING OR POOR MASONRY)
- DOWN-DRAFTING OF COMBUSTION GASES
- IMPROPER COMBUSTION

22. PAINTS/STAINS
- VOCs
- LEAD (PAINT PRE-1978)
- CONTAIN PRESERVATIVES AND FUNGICIDES
- FLAKING/AEROSOLIZATION

21. ATTIC
- ASBESTOS
- FIBERGLASS
- VERMICULITE
- MOLD
- BACTERIA
- DUST/DEBRIS

20. PEOPLE
- PERFUME/COLOGNE
- BODY ODOR/SMOKING
- SKIN FRAGMENTS
- TRANSMITTABLE DISEASE (TUBERCULOSIS/COMMON COLD/ETC.)

19. DUCTWORK/VENTILATION
- IMPROPER EXCHANGE RATE
- FIBERGLASS INSULATION
- CONTAMINATED DUCTWORK (DUST/POLLEN/FUNGI/BACTERIA)
- TEMPERATURE EXTREMES
- IMPROPER FILTRATION

18. CARPETING/FLOORING
- POTENTIAL MICROBIAL & CHEMICAL RESERVOIR
- VOCs
- ASBESTOS FLOOR TILES
- IMPROPERLY SEALED CONCRETE
- NON-HEPA VACUUMING

17. MOISTURE INTRUSION
- LEAKS
- WICKING OF MOISTURE (SLAB/WALLS/ROOF)
- MOLD, MVOCs & MYCOTOXINS
- BACTERIA & ENDOTOXINS

16. PLANTS
- POLLEN/ALLERGENS
- FUNGI
- INSECTS
- INSECTICIDES/ FUNGICIDES/ HERBICIDES
- STANDING WATER

2. STORAGE/CLEANING ROOM
- CLEANING PRODUCTS (VOCs)
- PESTICIDES
- HERBICIDES
- FUNGICIDES
- AIR FRESHENERS
- STANDING WATER

3. INFANT/CHILD ROOM
- HUMIDIFIER CAUSING EXCESS HUMIDITY & MICROBIAL GROWTH
- HUMIDIFIER FEVER
- PET ALLERGENS
- DUST MITES IN BEDDING

15. FURNACE ROOM
- IMPROPER COMBUSTION (PARTICULATES/GASES)
- IMPROPER VENTILATION
- BYPASSING PROPER FILTRATION
- ASBESTOS INSULATION
- FIBERGLASS INSULATION
- *LEGIONELLA* FROM IMPROPERLY HEATED WATER TANK

14. GARAGE
- EXHAUST INFILTRATION
- VOCs FROM CLEANING PRODUCTS
- STORED PESTICIDES/ HERBICIDES
- GASES FROM STORED FUEL
- GARBAGE CONTAINERS

4. STRUCTURAL AREAS
- ICE DAMS
- GUTTER OBSTRUCTION
- EXTERIOR CRACKS
- POOR CONSTRUCTION
- IMPROPERLY SEALED CONCRETE
- IMPROPER EXTERIOR GRADING

13. BASEMENT
- EXHAUST/COMBUSTION FUMES
- CARBON MONOXIDE
- MOISTURE INTRUSION
- CHEMICAL STORAGE/ SPILLS
- RADON RISK
- PARTICULATES/ WOODWORKING

5. HOME OFFICE
- VOCs (FURNISHING)
- COMPUTERS (VOCs/OZONE/ PARTICULATES)
- ELECTROMAGNETIC FIELDS/PARTICLES
- IMPROPER VENTILATION

6. BATHROOM
- AIR FRESHENERS (VOCs)
- CLEANING PRODUCTS & PERSONAL CARE ITEMS (VOCs)
- SEWER GAS
- WATER LEAKS (MOLD/ BACTERIA/VIRUSES)

7. KITCHEN
- COOKING ODORS/OVEN FUMES
- TOBACCO SMOKE
- WATER LEAKS (MOLD/BACTERIA)
- ALLERGENS (COCKROACH/ MOUSE/RAT)

8. INTERIOR WATER SOURCES
- DISHWASHER
- CLOTHES WASHER
- TOILET
- BATHTUB/SHOWER
- WATER HEATER
- WATER PURIFIER

9. BIRDS
- ODORS
- ALLERGENS
- HISTOPLASMOSIS
- CRYPTOCOCCOSIS
- PSITTICOSIS

10. PESTICIDES/HERBICIDES
- IMPROPER STORAGE
- EXTERIOR & INTERIOR APPLICATIONS
- INHALATION & TRACKING OF CHEMICALS ONTO FLOORING MATERIALS

11. HOME MAINTENANCE
- NON-HEPA VACUUM
- VOCs FROM CLEANING ITEMS
- MACHINERY EXHAUST
- PESTICIDES
- HERBICIDES
- FUNGICIDES

12. TRASH CONTAINERS
- ODORS
- BACTERIA
- FUNGI
- RODENT/INSECT/AVIAN ALLERGENS
- PARTICULATES

Figure 34–1   Sources and causes of indoor pollution. *Courtesy Aerotech Laboratories, Inc., 800.651.4802. www.aerotechpk.com*

34.4   COMMON POLLUTANTS

The following is a description of some of the more common pollutants.

RADON. *Radon* is a colorless, odorless, and radioactive gas. The main health risk of breathing air polluted with radon is lung cancer. Research also has indicated that drinking water with high radon levels may also cause health risks; these are believed to be much lower than breathing air with radon gas. The most common source of radon is uranium in the soil or rock on which homes are built. As uranium breaks down, it releases radon gas, which can enter buildings through cracks in concrete floors and walls, floor drains, and sumps. Testing equipment should be approved by the Environmental Protection Agency (EPA) or be state certified. Some types of contamination can be determined easily by purchasing a radon test kit. After the exposure to the test materials has occurred, the test canister can be sent to a laboratory where an analysis will be made. If it is determined that the radon level, whether in the air or water, is above the acceptable limits, corrective action should be taken to eliminate the source. Procedures for the reduction of the radon gas should be performed by contractors or individuals who have satisfactorily completed the EPA Radon Contractor Proficiency Test or who are otherwise certified to do this.

ENVIRONMENTAL TOBACCO SMOKE.   *Environmental tobacco smoke* (ETS) is the mixture of smoke that comes from the burning end of a cigarette, pipe, or cigar, and smoke exhaled by the smoker. It is a mixture of more than 4000 compounds, more than 40 of which are known to cause cancer in humans or animals and many of which are strong irritants. ETS is often referred to as "secondhand smoke." To reduce the exposure of environmental tobacco smoke, smoking should not be permitted indoors. Separating smokers and nonsmokers in different rooms in the same house will help reduce but will not eliminate exposure to ETS.

BIOLOGICAL CONTAMINANTS.   Many sources of *biological contaminants* exist. Central air-handling systems can grow mold, mildew, and other contaminants. Viruses can be transmitted by people and animals. House dust mites, the source of a biological allergen, are found frequently in carpeting and bedding and possibly other fabrics. Many allergens can become airborne when dried. Water-damaged materials and wet surfaces provide opportunities for molds, mildew, and bacteria to grow.

Dust mites (Dermatophagoides farinae) are microscopic spiderlike insects usually found indoors. In warm, humid conditions they thrive and produce waste pellets. Their bodies disintegrate when they die, producing very small particulates that mix with household dust. These dust mites and their remains are often thought to be a primary irritant to people sensitive to dust irritants. Dust mites may be found in carpeting, bedding, and other fabrics. Carpeting should be vacuumed often, but vacuum cleaners stir up the dust in the carpeting and often raise dust levels in the air. Vacuum cleaners with high-efficiency particulate arrestor (HEPA) filters or central vacuuming systems that discharge air outside the building can be used to control most of this dust. Replacing carpeting with hard surface floors would be helpful. Smaller area rugs can be used if cleaned regularly.

Bedding should be washed often with hot water. Mattresses and pillows can be covered with plastic so the dust is more easily removed.

Molds are simple organisms commonly present both outdoors and indoors. Molds found outside the house break down many organic substances necessary to plant, animal, and human life. Mold, a fungus, develops and releases spores into the air. Many of the spores found indoors come from outside. However, quantities of mold growing indoors can produce many spores that are harmful to humans. Mold is usually found in areas where there is moisture or high humidity. **Figure 34–2** shows one kind of mold as seen under a

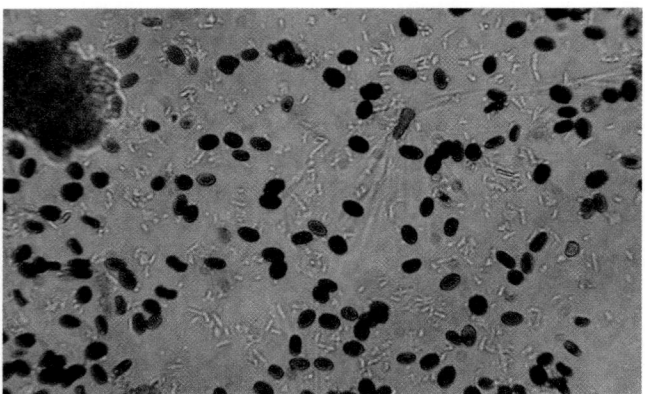

Figure 34–2   Photo of *Stachybotrys* mold under a microscope. *Courtesy Aerotech Laboratories, Inc. 800.651.4802. www.aerotechpk.com*

microscope. Estimates indicate that there may be over 100,000 different molds.

To grow, molds need moisture and food, which includes paper, textiles, dirt, leaves, drywall, or wood. These molds digest this material and gradually destroy it. As long as there is moisture and food, mold will not go away. **Figure 34–3 (A and B)** show examples of extreme cases of mold found

(A)

(B)

Figure 34–3   **(A)** A mold, *Stachybotrys,* colonized on a ceiling and ceiling cavity. **(B)** Mixture of *Cladosporium, Aspergillus,* and *Penicillium* mold on a painted drywall surface. *(A)* and *(B)* *Courtesy Aerotech Laboratories, Inc. 800.651.4802. www.aerotechpk.com*

indoors. Molds can grow on the surface of damp inorganic matter such as glass and concrete when there is a thin invisible layer of organic material that is food for the mold. Left unchecked, mold can continue to grow and cause health problems for sensitive people. Mold can usually be seen, or it may be smelled. It has a musty odor. It may grow in wall cavities, above ceilings, under carpets, and behind wallpaper. Because air-conditioning and heating systems are responsible for controlling humidity inside structures, the technicians in the HVACR field work with these systems and therefore it is wise for them to be able to recognize mold. There are many different types of mold and each has a different look and description. See the appendix for photos of the different molds and a description of each. A moisture meter may be used to help detect moisture, **Figure 34–4.**

Molds consist of multicelled filaments called hyphae. These filaments can accumulate into balls or mats and may penetrate the food material, making removal and cleanup difficult.

Molds reproduce through the production of spores. Spores are released into the air after they are produced and dried. They move with the air currents, land, germinate, and grow. Spores can survive for long periods of time in hot and cold dry environments. They require only moisture and food to germinate and grow. Spores contain allergens, which are substances that have the potential to cause allergic reactions in humans. Individuals may ingest, inhale, or be exposed to spores through skin contact. A susceptible person may experience a temporary effect from this exposure or in some cases a long-term effect.

Molds can produce a variety of health problems in humans. These problems are a result of the type and quantity of the mold as well as the sensitivity of the individual. Molds produce volatile organic compounds (VOCs). Exposure to a large quantity of these VOCs can irritate mucous membranes and produce headaches, inability to concentrate, and dizziness.

Some molds may have the potential for producing *mycotoxins*. As the name implies, these molds may have a toxic effect on humans. However, it is believed that certain environmental conditions must be present for a mold to produce

these mycotoxins. As of this time, these environmental conditions are not completely understood.

Sick building syndrome (SBS) is a term given to situations when persons who spend considerable time in a particular building experience health problems. These illnesses may be caused by poor ventilation or biological contaminants such as molds or bacteria.

Mold in buildings has become a significant issue in the United States. As of this time, a bill has been introduced in the House of Representatives and hearings are being held. This bill, the United States Toxic Mold Safety and Protection Act of 2005, is designed to protect home buyers and consumers from moving into mold-infested homes and to provide legal protection to home buyers and renters who are exposed to dangerous levels of indoor mold. This bill is expected to establish national standards, including insurance provisions, health benefits, and education programs.

There are many legal ramifications related to mold. Contractors are being sued. Insurance companies may or may not cover expenses and liability costs when claims involving mold are presented. It has been suggested by some sources that HVACR and other contractors obtain legal advice to ascertain where they stand on this issue. Contractors should meet with their insurance representatives to ensure they have the coverage they want.

MOLD REMEDIATION.   The following general recommendations regarding mold remediation relate primarily to small areas of mold growth. Larger areas should be cleaned and/or materials should be removed and disposed of by contractors certified to do so.

Removing the source of the moisture should be the first step in mold cleanup, and it should be done quickly. All leaks should be repaired, and cleanup of all water and damp areas should take place. The humidity in the air may need to be decreased.

SAFETY PRECAUTION: Technicians involved with removing mold should wear outer clothing designed for this purpose, including gloves, hat, and respirator or an approved alternative.

As molds can penetrate many surfaces they are feeding on, merely cleaning the surface may not be enough. Using a household bleach or chemicals available for the purpose may be helpful in eradicating mold growing on a hard surface. Carpets with mold growing within the fabric or underneath must be removed, cleaned, and dried. The floor beneath the carpet must be cleaned, ensuring that all mold is removed and any moisture problems are resolved. Wood, wallboard, or wallpaper that has been heavily contaminated with mold may need to be removed and replaced. This should be done by contractors and technicians certified to do so. Dead molds can still cause problems to human health, so complete removal is necessary.

Air-conditioning (cooling) equipment should not be oversized. In some instances when a house is being constructed, owners want the equipment to be oversized as they may have planned an addition to the building. However, oversized air-conditioning equipment may not run for long enough periods to remove sufficient humidity. When the cooling requirement is met with oversized equipment, the unit shuts

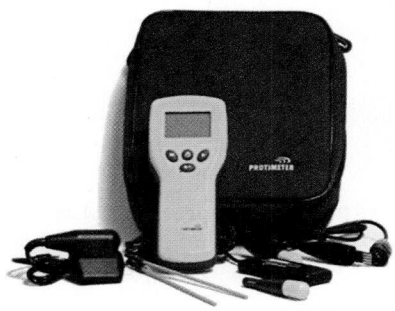

Figure 34–4   A digital moisture meter. This meter has a pointed probe that can be inserted into carpet or wood to measure the moisture level. *Courtesy Aerotech Laboratories, Inc. 800.651.4802. www.aerotechpk.com*

down and excess humidity may still be in the air. Dehumidifiers may be needed.

The ground around the foundation should slope away from the building to ensure that water does not collect around or under the house.

Air-conditioning condensate pans should be cleaned regularly and the drain lines kept free so the condensate will drain from the equipment properly. Humidification equipment should also be cleaned regularly. If mold is or has been present within a house the HVACR duct may need to be inspected. If mold is found the duct should be cleaned or replaced by qualified technicians. Clothes driers and stoves should be vented to the outside. Exhaust fans should be used and vented to the outside when bathing or showering, when cooking, or when running the dishwasher.

An architect or engineer may be consulted to determine whether whole house ventilation is needed.

POLLUTANTS FROM COMBUSTION PRODUCTS.
Wood stoves, wood-burning fireplaces, vented and unvented gas-burning fireplaces, gas stoves, and unvented kerosene and gas space heaters may be the source of pollutants from combustion products. The pollutants released may be carbon monoxide, nitrogen dioxide, sulfur dioxide, and particles that may be suspended in the air. The combustion gases may come from improperly sealed stove doors, from chimneys and flues that are improperly installed or maintained, or from cracked furnace heat exchangers.

Combustion also produces water vapor. Although not a pollutant in itself, water vapor can result in high humidity and wet surfaces. This can encourage the growth of biological pollutants such as those mentioned previously.

Carbon monoxide (CO) is a very poisonous gas that is lighter than air. It is produced by incomplete combustion in appliances that use the combustion process. This incomplete combustion may be a result of lack of oxygen or the incomplete mixing of oxygen and fuel. Excess combustion air or an insufficient fuel supply can also result in the production of CO. When a flame temperature is reduced below 1128°F, CO will be produced.

CO is very dangerous because it is colorless, odorless, tasteless, and nonirritating. It can only be measured with a testing device, **Figure 34–5**. However, alarms can be purchased that if properly installed will respond when CO levels become dangerous. CO combines with blood hemoglobin in the bloodstream, gradually replacing the oxygen. It is cumulative to the point that when there is too little oxygen in the bloodstream, it will no longer support life.

This chemical especially affects fetuses, infants, and people with anemia or a history of heart disease. Low levels of exposure can cause fatigue and increase chest pain in people with chronic heart disease. Higher levels may cause headaches, dizziness, and weakness in healthy people as well as sleepiness, nausea, vomiting, confusion, disorientation, loss of consciousness, and death. Carbon monoxide in the ambient air is measured by the number of CO molecules present in a million molecules of air. This is referred to as parts per million (ppm). See **Figure 34–6** for concentrations and symptoms

Figure 34–5   A testing device that will measure ppm of carbon monoxide. *Courtesy Bacharach, Inc., Bacharach Institute of Technical Training*

CO CONCENTRATIONS & SYMPTOMS DEVELOPED	
Concentrations of CO in the air	Inhalation time and toxic symptoms developed.
9 ppm (0.0009%)	The maximum allowable concentration for short-term exposure in a living area, according to ASHRAE.
50 ppm (0.0050%)	The maximum allowable concentration for continuous exposure in any 8-hour period, according to federal law.
200 ppm (0.02%)	Slight headache, tiredness, dizziness, and nausea after 2–3 hours.
400 ppm (0.04%)	Frontal headaches within 1–2 hours, life-threatening after 3 hours, also maximum parts per million in flue gas (on an air-free basis), according to EPA and AGA.
800 ppm (0.08%)	Dizziness, nausea, and convulsions within 45 minutes. Unconsciousness within 2 hours. Death within 2–3 hours.
1,600 ppm (0.16%)	Headache, dizziness, and nausea within 20 minutes. Death within 1 hour.
3,200 ppm (0.32%)	Headache, dizziness, and nausea within 5–10 minutes. Death within 30 minutes.
6,400 ppm (0.64%)	Headache, dizziness, and nausea within 1–2 minutes. Death within 10–15 minutes.
12,800 ppm (1.28%)	Death within 1–3 minutes.

Figure 34–6   This chart indicates carbon monoxide concentrations and symptoms that may develop. *Courtesy Bacharach, Inc., Bacharach Institute of Technical Training*

developed. Carbon monoxide monitors or alarms should be installed in all areas of a building where a combustion process is being used and in other parts of a building where CO may collect.

Auto exhaust is considered to be the leading cause of carbon monoxide alarm responses. When individuals let their cars warm up or otherwise let them run in a garage that is attached to the house or below the house, the CO may be trapped and infiltrate to the living quarters, setting off the alarm and producing a dangerous condition.

It is estimated that unvented, poorly installed, and improperly maintained oil and gas appliances are the second most frequent causes of CO alarms responding.

The third most frequent cause of CO alarm responses is believed to be the backdrafting of natural-vented combustion appliances. These backdrafts may be caused by blockages in the vent such as a bird nest or foreign debris or by pressure differences. As the exhaust systems in the bathrooms, kitchen, attic, and other areas of the house exhaust air, the house will have a negative pressure with respect to the outside air. This tends to push or backdraft vented air back into the house. **Figure 34–7** shows air being exhausted from within the house and other air returning through the venting or chimneys to make up for the exhausted air. This backdrafting air may contain carbon monoxide.

Combustion appliances must be properly installed and adjusted, and a proper supply of fresh air that includes oxygen must be available. In all vented appliances a proper flue passageway must be available to vent the flue gases to the atmosphere.

Nitrogen dioxide causes irritation of the respiratory tract and can cause shortness of breath.

Sulfur dioxide can cause eye, nose, and respiratory tract irritation and breathing problems.

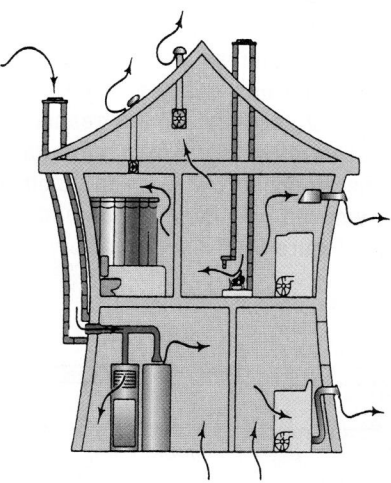

Figure 34–7   Air exhausted from a house causes a negative pressure within the house. This interferes with the venting of combustion appliances, causing a backdraft that forces some of the flue gases back down the chimneys and vents into the living areas. *Courtesy Bacharach, Inc., Bacharach Institute of Technical Training*

Particles from combustion suspended in the air can cause eye, nose, throat, and lung irritation. Certain chemicals attached to these particles can cause lung cancer. All of these risks depend on the length of exposure, the size of the particles, and the chemical makeup. Wood smoke contains both polycyclic organic matter (POM) and nonpolycyclic organic matter, which are considered to be health hazards. Both are by-products of wood combustion.

When a space heater is used in one room, doors should be opened to allow the air to move to other rooms. Stove hoods and fans should be used over cooking stoves.

It is important to keep wood stove emissions to a minimum and to make sure that stove doors are tight fitting. All joints in stove pipe and/or fittings should be tight or sealed. EPA-certified wood stoves should be used whenever possible. Central air-handling systems, including furnaces, flues, and chimneys, need to be inspected regularly. Cracked heat exchangers should be replaced before using. Cracked or damaged flues and chimneys should be repaired. Ductwork in central air-handling systems may need to be cleaned. See Section 34.8, "Duct Cleaning."

HOUSEHOLD PRODUCTS CONTAINING ORGANIC CHEMICALS.
Paints, varnishes, cleaning products, disinfecting and hobby products, and various fuels contain *organic chemicals.* The compounds from these products are released primarily when they are being used but may be released to some extent if the products are not stored properly.

These household products should be properly used and stored. Follow manufacturers' recommendations when using these materials. Use proper disposal procedures to throw away partially full containers of these materials when not needed. If the containers are stored, ensure that they are sealed.

FORMALDEHYDE.
Formaldehyde is a chemical used in the manufacture of some building materials and certain household products. It is also contained in combustion gases and may exist in the gases from unvented fuel-burning appliances such as unvented gas stoves or kerosene space heaters. Pressed wood products, which have adhesives containing urea-formaldehyde (UF) resins, include particleboard, which may be used as subflooring and shelving, hardwood plywood paneling, medium-density fiberboard, cabinets, and some furniture tops. Pressed wood, such as softwood plywood and flake board, are produced for exterior use and contain phenolformaldehyde (PF) resin. Formaldehyde is present in both resins, but the PF resin emits at a lower rate.

PESTICIDES.
Pesticides used in and around a residence may be used to control common insects, termites, rodents, or fungi. Microbes used in disinfectants are considered a pesticide. Research has indicated that a high percentage of people's exposure to pesticides occurs indoors.

Use pesticides only as instructed by the manufacturer. When storing containers, ensure that they are sealed properly.

ASBESTOS.
Asbestos is found to a greater extent in older homes and buildings. The use of asbestos now is very limited.

It may be found in pipe and furnace insulation materials, shingles, textured paints, and floor tiles. Concentrations of airborne asbestos can be found when asbestos-containing materials have been disturbed by cutting, sanding, or removing activities. The most dangerous asbestos fibers are too small to be seen, can accumulate in the lungs, and cause lung cancer and other diseases. SAFETY PRECAUTION: If there is any possibility that deteriorating, damaged, or disturbed materials contain asbestos, these materials should be examined and tested by a person certified to do so. If it is found that materials do contain asbestos, they must be removed by contractors licensed and equipped to do so. No exposure to others is allowed.

LEAD. Humans may be exposed to *lead* through air, drinking water, food, contaminated soil, deteriorating paint, and dust. Old lead-based paint is probably the greatest source of exposure. Lead particles can result from lead dust from outdoor sources and indoor activities using lead solder. SAFETY PRECAUTION: Deteriorating, damaged, or disturbed materials that contain lead, particularly lead paint, should be removed by contractors licensed and equipped to do so. No exposure to others is allowed.

34.5 CONTAMINATION SOURCE DETECTION AND ELIMINATION

The existence of some types of contamination can be determined easily and steps can be taken to eliminate the source.

Many types of indoor air-quality test and monitoring instruments are available. **Figure 34–8** is an example of these. Monitoring instruments can measure such materials as carbon dioxide, carbon monoxide, hydrogen sulfide, sulfur dioxide, chlorine, nitrogen dioxide, nitric oxide, hydrogen cyanide, ammonia, ethylene oxide, oxygen, hydrogen, hydrogen chloride, ozone, and others.

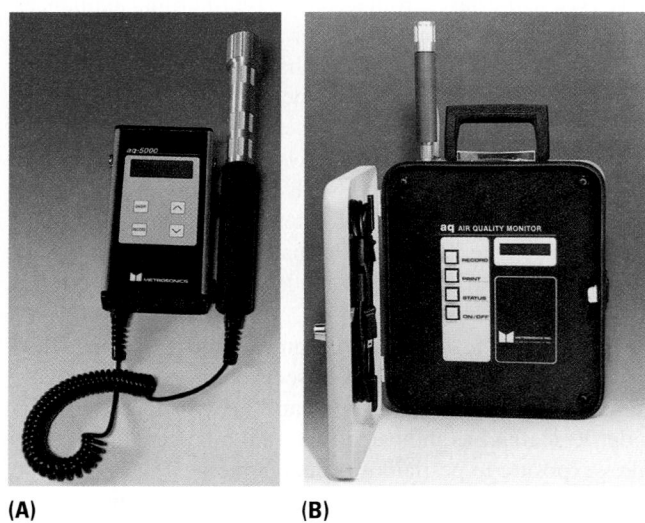

(A)   **(B)**

Figure 34–8 **(A)** A handheld indoor air-quality monitor that monitors and records carbon dioxide, temperature, humidity, and gas levels. **(B)** An indoor air-quality monitor with sensors that monitors and records many gases. *Courtesy Metrosonics, Inc.*

34.6 VENTILATION

Ventilation is the process of supplying and removing air by natural or mechanical means to and from any space. Indoor air quality can be improved when proper ventilation procedures are followed. These procedures remove some of the air with pollutants and dilute the remaining pollutants with acceptable outside air. Ventilation air is part of the supplied air from outdoors plus any recirculated air that has been cleaned or treated.

Ventilation air should be supplied to the occupant's breathing space. Outside air and/or cleaned filtered air does not benefit building occupants if it is short circuited from the supply diffuser to the return grille.

Ventilation has an adverse effect on heating and cooling efficiencies as the air that is brought in from the outside must be heated or cooled to maintain the comfort temperature. See Unit 35, "Comfort and Psychrometrics," for calculations on how much heating or cooling capacity must be added to account for outdoor air from ventilation. ASHRAE Standard 62-2001 prescribes ventilation and other measures to ensure acceptable air quality.

Commercial buildings are often designed with the appropriate mechanical ventilation. This ventilation should consider the number of occupants and any special requirements arising from the activities taking place in the building. However, few homes have heating, cooling, or ventilation systems that mechanically bring outside air into the home. Therefore, opening windows and doors or operating window or attic fans becomes necessary. Bathroom or kitchen fans may be used to exhaust indoor air outside. Replacement air may infiltrate into the house, or a door or window may be opened for short periods of time.

ENERGY RECOVERY VENTILATORS. 🌐One design of mechanical equipment that has been developed to provide more dilution consists of two insulated ducts run to an outside wall. One duct provides for fresh outside air to enter the building, and the other exhausts stale polluted air to the outside. One type utilizes a desiccant-coated heat transfer disc that rotates between the two airstreams. In winter this disc recovers heat and moisture from the exhaust air, transferring it to the incoming fresh air. In summer, heat and moisture are removed from incoming air and transferred to the exhaust air. **Figure 34–9** illustrates how this equipment can be installed.

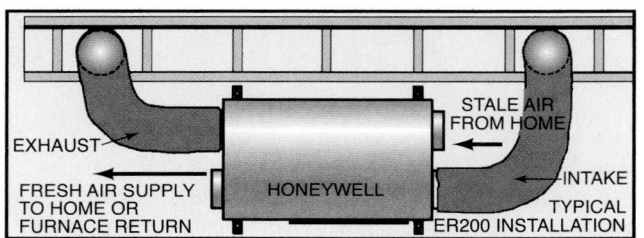

Figure 34–9 🌐An air dilution system with energy recovery.🌐 *Courtesy Honeywell, Inc.*

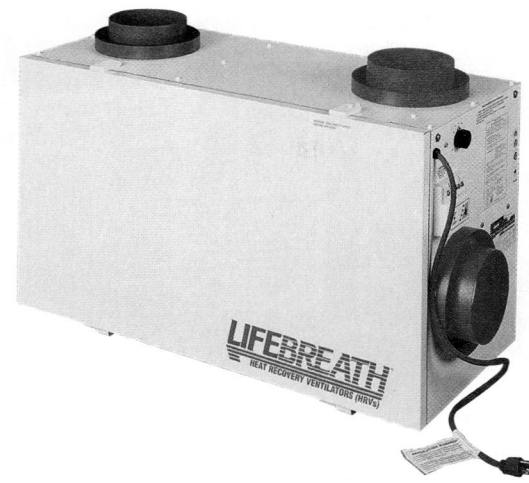

Figure 34–10 🌐 An energy recovery ventilator.🌐 *Courtesy Nutech Energy Systems, Inc. (Life Breath)*

Other designs use different types of heat transfer systems. **Figure 34–10** is a photo of an energy recovery ventilator (ERV). In most units, indoor humidity levels are prevented from becoming too moist or too dry. Some units, heat recovery ventilators (HRVs) provide heat-only recovery. These are used primarily in northern climates where cooling is not a major factor. This controlled type of ventilation may be designed to reduce pollen, dust, and other household pollutants, providing a healthier indoor environment.🌐

SECURITY.   Security is a relatively new topic for HVACR technicians and others as a result of events that have occurred during the past few years, including war, terrorism, accidents, and natural disasters. Building owners, managers, and technicians should meet and assess all risks and dangers that could possibly be present. A plan should be developed for each of these potential events. Universal plans can seldom be used because of the differences in building architecture, HVACR systems, and how buildings are occupied and used.

Outdoor air intakes for buildings should be located away from public areas when possible. An elevated position or a roof location would be the best in most instances. These locations may also produce cleaner air in normal situations. If an event does occur that results in an air pollutant, it is recommended by some that ventilation systems not be shut down until the source of the pollutant is determined. This source should be determined as quickly as possible. If it is from outside, the ventilation system would be shut down. If the intrusion is from the inside, the building will need all the outside ventilation it can get to dilute the pollutant. If it is decided to shut down the ventilation, the outside louvres should be closed and sealed if they are movable.

Buildings should use filtering systems with the highest physically and economically possible minimum efficiency reporting value (MERV) ratings. MERV ratings range from 1 to 20, with the upper levels providing the most filtering. Filters should be chosen not only by their capacity to arrest particles but also by the amount of pressure drop they would cause. The filter retainer frame should be inspected regularly

to ensure that it has not come loose or deteriorated. Clips should hold the frame tight. For the best filtration, gaskets should be used. Considerable amounts of air can be passed around the filter frames if they are not installed and maintained properly. It would take a HEPA filter to arrest anthrax because its spores are in the 1-micron size. There will be more on filters in the next section.

A qualified and knowledgeable technician can be very valuable when preparing a plan and maintaining building equipment. A technician can acquire the Testing, Adjusting and Balancing Bureau (TABB) certification developed by the National Energy Management Institute (NEMI).

34.7   AIR CLEANING

Air cleaning is another method along with source control and ventilation to help provide acceptable indoor quality. 🌐 Four general types of air cleaners exist: mechanical filters, electronic air cleaners, ion generators, and ultraviolet C light. Some systems are considered "hybrid" devices because they may contain two or more particle-removing devices. For instance, a mechanical filter may be combined with an electrostatic precipitator or an ion generator. Air cleaners may contain absorbent or adsorbent materials to remove gaseous materials. *Absorption* is the process of one substance being absorbed by another. For instance, carbon dioxide and carbon monoxide can be absorbed in either solid or liquid-spray filters. *Adsorption* is the process of a thin film of the liquid or gas adhering to the surface of a solid substance. Charcoal is used extensively as an adsorber.🌐

HIGH-LOFT POLYESTER FILTER MEDIA.   This medium is spray bonded and can be purchased in bulk, in media rolls, or in cut pads, **Figure 34–11**.

FIBERGLASS THROWAWAY FILTERS.   This filter medium is constructed of continuous glass fibers bonded together with a thermosetting agent. It is coated with a dust-holding adhesive

Figure 34–11   Filter media in rolls and pads. *Courtesy Aerostar Filtration Group*

Figure 34–12 Fiberglass throwaway filters. *Courtesy Aerostar Filtration Group*

Figure 34–14 A pleated filter. These filters have more filter surface area per front area of the filter. *Courtesy Aerostar Filtration Group*

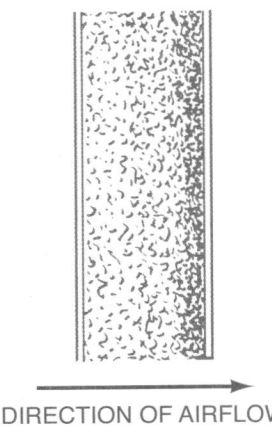

DIRECTION OF AIRFLOW

Figure 34–13 A section of fiberglass medium that shows that the medium becomes denser as the air passes through it. This helps the filter to have more capacity to filter particles because the larger particles are filtered first in the less dense parts of the filter.

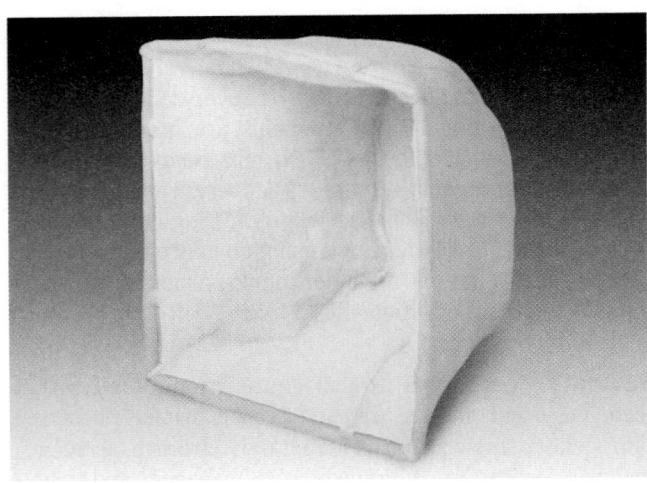

Figure 34–15 A cube filter. This filter has a lot of surface area and requires a deep filter case. *Courtesy Aerostar Filtration Group*

that helps prevent particulates from migrating downstream, **Figure 34–12.** Filter material is designed to get progressively denser as the air passes through it, **Figure 34–13.**

PLEATED FILTERS. These filters can accumulate heavier, more restrictive particles at the bottom of the pleats, leaving the sides open longer for effective filtration, **Figure 34–14.**

CUBE FILTERS. These filters have a very high dust-holding capacity. This medium has three filter layers heat sealed together. The first layer acts as a high-porosity impingement filter, the second acts as a high-density interceptor, and the third acts as a durable strainer, **Figure 34–15.**

POCKET FILTERS. Vertical separators for each pocket channel air throughout the medium to prevent turbulence and allow even contaminant loading, **Figure 34–16.**

Figure 34–16 A pocket filter. *Courtesy Aerostar Filtration Group*

Figure 34–17 ⊗ A HEPA filter.⊗ *Courtesy Aerostar Filtration Group*

HEPA FILTERS. ⊗ This type of filter is used where the highest degree of filtration is desired or required. Many dust particles are in the 20–30 micron (μm) range. Other particles are much smaller and require a much more efficient filter. HEPA filters can filter to a very high efficiency—down to a 0.3-μm sized particle or lower. These minute particles can contain mold, dust mites, bacteria, pollen, and other allergens that if breathed in may remain in the lungs.

Figure 34–17 is an example of a HEPA filter. This particular filter can remove from the air 99.99% of particulates having a minimum size of 0.3 μm. Many HEPA filters remove 99.97% of particulates having a minimum size of 0.3 μm. The symbol μm stands for micrometer, often called a micron. For comparison, a human hair ranges from 70 to 100 μm in diameter, and an object 300 μm in diameter would fit easily through the eye of a sewing needle.⊗

SAFETY PRECAUTION (FILTER APPLICATION WARNING): The technician would naturally want to use the most effective filtration possible for any application. Caution must be used when changing from one filter type to another. The filter enclosure or rack for a particular application has a fixed dimension at the time of installation. This dimension is often small and therefore a low-density filter is chosen at the time of installation. If a technician installs a high-density filter in this application, it is very likely that the airflow will be reduced as the same filter area is used with a denser filter. If the airflow is reduced too much, problems with the heat exchanger will occur. In summer, it will cause coil freezing problems. In winter, it may cause overheating problems with the heating system. The technician must be aware of airflow requirements. The filter rack may need to be expanded. Some of the filters increase the filter area for the same filter size by means of the pleated filters mentioned earlier. Filters are rated in cfm at a prescribed pressure drop expressed in inches of water column. Be sure to match the new filter with the old filter airflow, or expand the filter rack.

ELECTROSTATIC PRECIPITATORS OR ELECTRONIC AIR CLEANERS. ⊗ Several types of *electrostatic precipitators* exist. These cleaners are designed to be mounted (1) at the furnace, (2) within the throwaway filter-type frame, or (3) within duct systems. These cleaners also may be stand-alone portable systems, **Figure 34–18**. These systems generally have a prefilter section that filters out the larger airborne particles and an ionizing or charging section. Some also have a charcoal section.

The prefilter traps larger particles and airborne contaminants. In the ionizing or charging section, particles are charged with a positive charge. These particles then pass through a series of negatively and positively charged plates. The charged contaminants are repelled by the positive plates and attracted to the negative plates. The air then passes through a charcoal filter, on those systems that have them, to remove many of the odors and gases by adsorption.⊗

ACTIVATED CHARCOAL AIR PURIFIER. ⊗ *Activated charcoal,* also called *activated carbon,* is manufactured from coal or coconut shells into pellets. This material is treated during manufacturing to increase the internal surface area of the pellets. Activated charcoal is especially good when used to adsorb solvents, other organic materials, and odors. The purifier, **Figure 34–19** is mounted in the return air duct of a heating and air-conditioning system. The velocity of the air within the duct forces the gaseous material to impinge on the activated charcoal, which is held in tubes suspended in the airstream. The gaseous material is attracted down into the carbon particle, and eventually it condenses into a liquid particle and is held in place. These gases and/or odors are held in the carbon media until the carbon is replaced. Some typical gases that can be removed by this type of purifier are:

- compounds, such as alcohols, aldehydes, and acids from household products, construction materials, carpeting, furniture, or other household items.
- chlorinated hydrocarbons.
- inorganic compounds, such as halogens, phosgene, and sulfur compounds.
- a variety of odors from cooking, waste storage, pets, and humans.⊗

Ion Generators

⊗ *Ion generators* charge particles in a room. The charged particles are then attracted to walls, floors, draperies, and other objects in the room. The ion generator may have a collector that collects these charged particles. Those generators that do not have a collector may cause soiling of walls and the other surfaces previously mentioned.⊗

Ultraviolet Light

⊗ Ultraviolet (UV) light has been studied for use in providing cleaner indoor air. UV light represents the frequency of light between 200 and 400 nanometers (nm). Within this range there are three frequency bands:

UV-A (long wave) 315 to 400 nm
UV-B (midrange) 280 to 315 nm
UV-C (low range) 200 to 280 nm

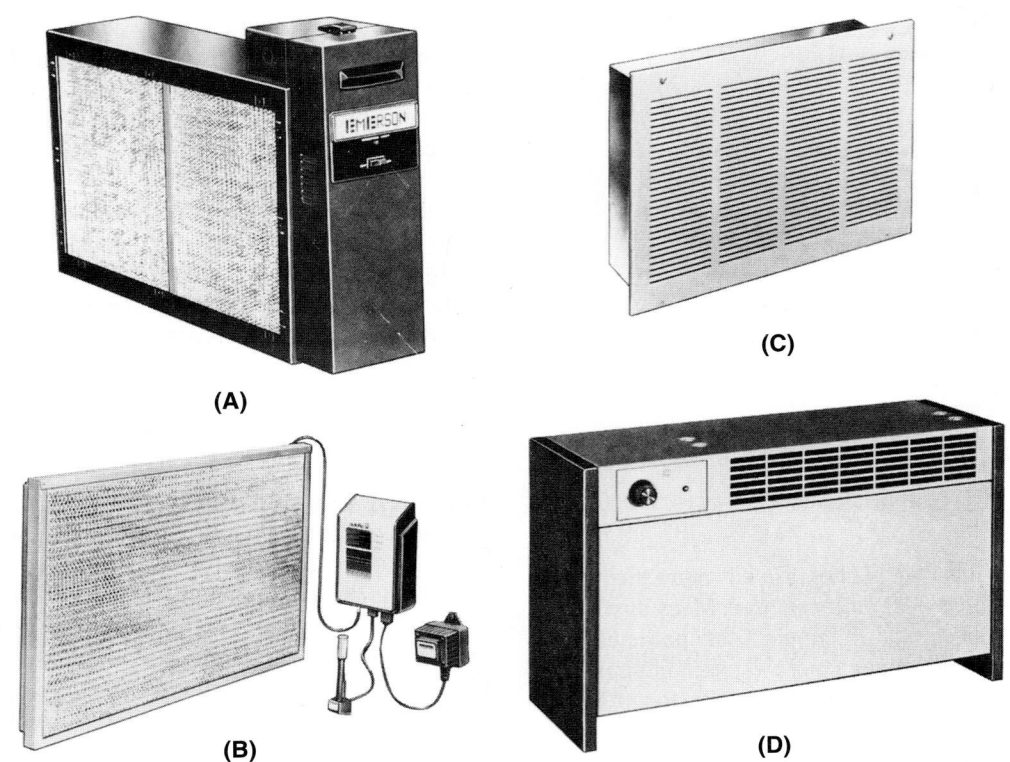

(A)

(C)

(B)

(D)

Figure 34–18 🌎 Electrostatic precipitators. 🌎 **(A)** Furnace mount. **(B)** Throwaway-filter frame mount. **(C)** Single-intake return system. **(D)** Stand-alone portable system. *Courtesy W. W. Grainger, Inc.*

Figure 34–19 🌎 An activated carbon adsorber removes many gaseous contaminants found in the home. 🌎 *Courtesy General Filters, Inc.*

The UV-C (germicidal) range has been studied and researched for use in destroying microbes that may cause unhealthy air in homes, offices, or other buildings, including hospitals and schools.

UV-C light can penetrate the wall of a microbe and damage its DNA. It may kill smaller microbes while rendering larger microbes harmless so that they cannot reproduce.

This light can be particularly effective when used near cooling coils in HVACR systems and when used near higher efficiency air filters. A moist cooling coil can attract dust and provide a place for microbes such as mold to live. Air filters collect dust and microbes. When placed properly in the duct system, UV-C light can kill the living microbes.

Generally these lights should be installed where there is low airflow velocity. However, they vary in size and design just as HVACR systems vary in size and design. It is important to follow the manufacturer's instructions. 🌎

SAFETY PRECAUTIONS: Although the UV light may not be seen, do not look at it when it is activated. The unit should be placed where accidental exposure to technicians or building occupants will not occur.

Do not turn the light on before installing it.

Place warning signs for future technicians and building occupants.

Check the duct for possible light leaks, and seal them if any are found.

Turn off the light before performing any maintenance or service on the HVACR system.

Replace the light element annually.

34.8 DUCT CLEANING

Dirty air ducts may be a possible source of pollutants that are present in homes. Homeowners should consider having the air ducts cleaned if:

■ there is visible mold growth present on hard surface (sheet metal) ducts or on other components of the heating and cooling system.

- ducts are infested with vermin (rodents or insects).
- ducts are clogged with excessive amounts of dust and debris and/or particles are actually released into the home from supply registers.

Duct cleaning generally refers to the cleaning of the interior of ductwork and the cleaning of various heating and cooling system components of forced-air systems. This includes the supply and return air ducts and registers, grilles and diffusers, heat exchangers, heating and cooling coils, condensate drain pans, fan motor and fan housing, any humidification components within the system, and the air handling unit housing. If the duct system is to be cleaned, all of the listed components should be cleaned. Failure to clean all units may result in recontamination of the entire system. If there is one inlet to the return duct system, having the filter placed there would help keep the return duct system clean.

SAFETY PRECAUTION: The entire system should be inspected before cleaning to make sure that no asbestos-containing materials are in the heating and cooling system. If asbestos is found, specialized procedures are required. The asbestos should not be disturbed or removed except by specially trained and equipped contractors.

The vacuum system used should exhaust particles outdoors or high-efficiency particle air-vacuuming equipment should be used if the exhaust is inside the building. All carpeting and household furnishings should be protected.

Equipment with controlled brushing of duct surfaces and contact vacuuming cleaning should be used to dislodge and contain dust and other particles, **Figure 34–20.** Soft-bristled brushes should be used for fiberglass duct board and sheet metal ducts lined internally with fiberglass.

Figure 34–20 🌐 Duct cleaning equipment. 🌐 *Courtesy Atlantic Engineering*

When finished cleaning, all access holes and/or access joints should be resealed and reinsulated when appropriate so that they are airtight. If moisture and/or mold exist, the source of this problem should be determined, and the problem should be corrected, or it will reoccur.

Duct cleaning technicians should be familiar with and follow the National Air Duct Cleaning Association's standards, and if the ducts are constructed of flex duct, duct board, or lined with fiberglass, guidelines of the North American Insulation Manufacturers Association (NAIMA) should be followed.

34.9 AIR HUMIDIFICATION

In fall and winter, homes often are dry because cold air from the outside infiltrates the conditioned space. The infiltration air in the home is artificially dried out when it is heated because it expands, spreading out the moisture. The amount of moisture in the air is measured or stated by a term called *relative humidity.* It is the percentage of moisture in the air compared to the capacity of the air to hold moisture. In other words, if the relative humidity is 50%, each cubic foot of air is holding one-half the moisture it is capable of holding. The relative humidity of the air decreases as the temperature increases, because air with higher temperatures can hold more moisture. When a cubic foot of 20°F outside air at 50% relative humidity is heated to room temperature (75°F), the relative humidity of that air drops.

For comfort, the dried-out air should have its moisture replenished. The recommended relative humidity for a home is between 40% and 60%. When the relative humidity varies above these limits, studies have shown that bacteria, viruses, fungi, and other organisms become more active. In conditioned spaces with lower relative humidity, the dry warm air draws moisture from everything in the conditioned space, including carpets, furniture, woodwork, plants, and people. Furniture joints loosen, nasal and throat passages dry out, and skin becomes dry. Dry air causes more energy consumption than necessary because the air gets moisture from the human body through evaporation from the skin. The person then feels cold and sets the thermostat a few degrees higher to become comfortable. With more humidity in the air, a person is more comfortable at a lower temperature.

Static electricity is also much greater in dry air. A person also may receive a small electrical shock when touching something after having walked across the room.

Years ago people placed pans of water on radiators or on stoves. They even boiled water on the stove to make moisture available to the air. The water evaporated into the air and raised the relative humidity. Although this may still be done in some homes, efficient and effective equipment called humidifiers produce this moisture and make it available to the air by evaporation. The evaporation process is speeded up by using power or heat or by passing air over large areas of water. The area of the water can be increased by spreading it over pads or by atomizing it.

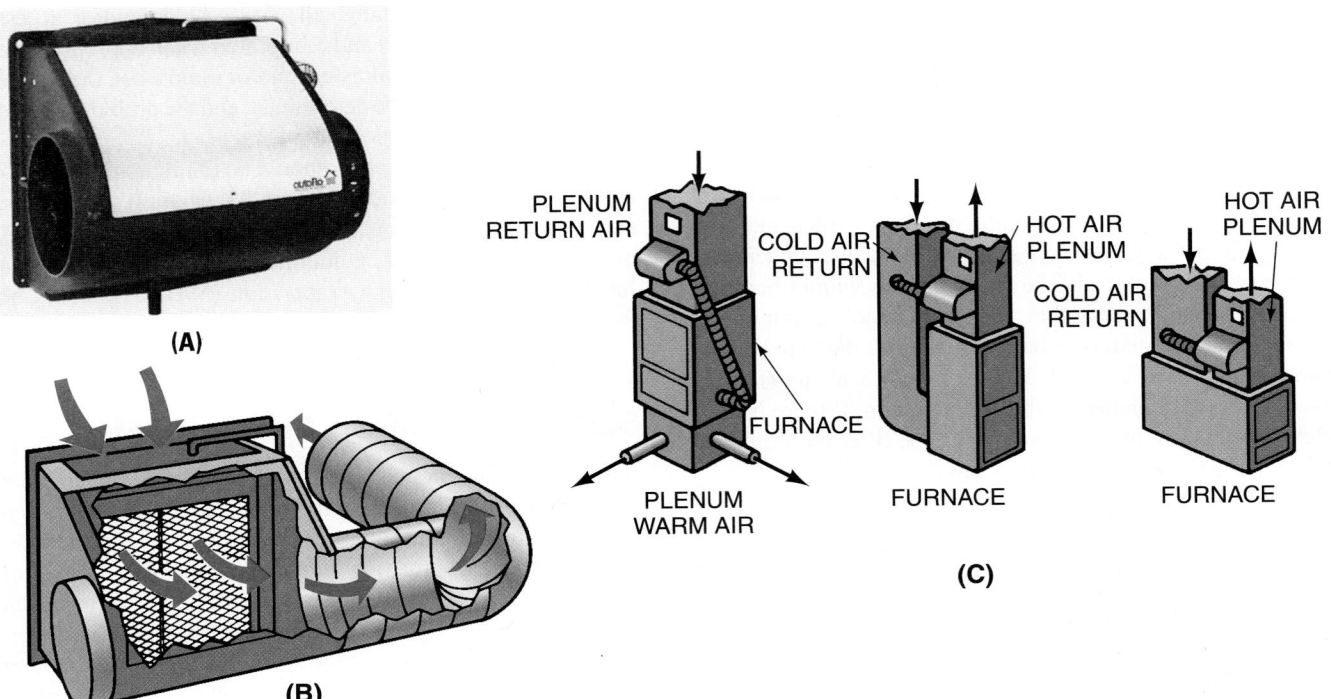

(A)

(B)

(C)

Figure 34–21   **(A)** A bypass humidifier. **(B)** Cutaway showing airflow from plenum through media to the return. **(C)** Typical installations. *Courtesy AutoFlo Company*

HUMIDIFIERS. *Evaporative humidifiers* work on the principle of providing moisture on a surface called a media and exposing it to the dry air. This is normally done by forcing the air through or around the media and picking up the moisture from the media as a vapor or a gas. Several types of evaporative humidifiers exist.

The *bypass humidifier* relies on the difference in pressure between the supply (warm) side of the furnace and the return (cool) side. It may be mounted on either the supply plenum or duct or the cold air return plenum or duct. Piping must be run from the plenum or duct where it is mounted to the other plenum or duct. If mounted in the supply duct, it must be piped to the cold air return, **Figure 34–21** and **Figure 34–22**. The difference in pressure between the two plenums draws some heated air through the humidifier to the return duct and is distributed throughout the house.

The *plenum-mount humidifier* is mounted in the supply plenum or the return air plenum. The furnace fan forces heated air through the media where it picks up moisture. The air and moisture are then distributed throughout the conditioned space, **Figure 34–23**.

The *under-duct humidifier* is mounted on the underside of the supply duct so that the media is extending into the heated airflow where moisture is picked up in the airstream. **Figure 34–24** illustrates an under-duct humidifier.

HUMIDIFIER MEDIA. Humidifiers are available in several designs with various kinds of media. **Figure 34–24** is a photo of a type using disc screens mounted at an angle. These discs are mounted on a rotating shaft, causing the slanted discs to pick up moisture from the reservoir. The moisture is then

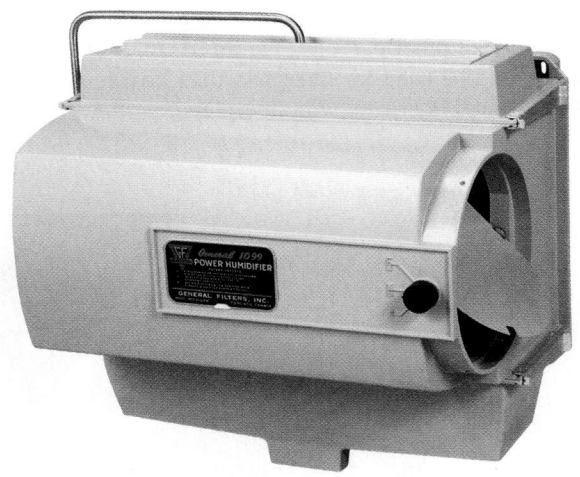

Figure 34–22   A bypass humidifier. *Courtesy General Filters, Inc.*

evaporated into the moving airstream. The discs are separated to prevent electrolysis, which causes the minerals in the water to form on the media. The wobble from the discs mounted at an angle washes the minerals off and into the reservoir. The minerals then can be drained from the bottom of the reservoir.

Figure 34–25 illustrates a type of media in a drum design. A motor turns the drum, which picks up moisture from the reservoir. The moisture then is evaporated from the drum into the moving airstream. The drums can be screen or sponge types.

A plate- or pad-type media is shown in **Figure 34–23**. The plates form a wick that absorbs water from the reservoir. The

Figure 34–23   A plenum-mounted humidifier with a plate-type medium that absorbs water from the reservoir and evaporates it into the air in the plenum. *Courtesy AutoFlo Company*

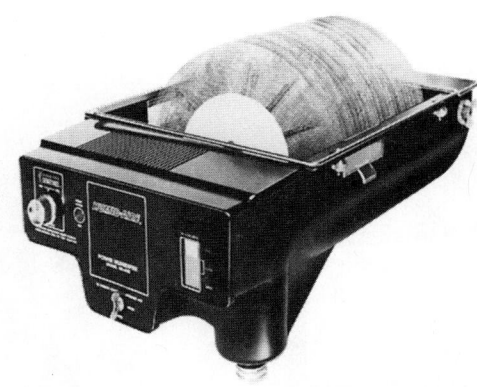

Figure 34–24   An under-duct humidifier using disc screens as the medium. *Courtesy HumidAire Division*

Figure 34–25   An under-duct humidifier using drum-style media. *Courtesy Herrmidifier Company, Inc.*

airstream in the duct or plenum causes the water to evaporate from the wicks or plates.

Figure 34–26 shows an electrically heated water humidifier. In electric furnace and heat pump installations, the

Figure 34–26   A humidifier with electric heating elements. *Courtesy AutoFlo Company*

temperature in the duct is not as high as in other types of hot-air furnaces. Media evaporation is not as easy with lower temperatures. The electrically heated humidifier heats the water with an electric element, causing it to evaporate and be carried into the conditioned space by the airstream in the duct.

An *infrared humidifier* is shown in **Figure 34–27,** which is mounted in the duct and has infrared lamps with reflectors to reflect the infrared energy onto the water. The water thus evaporates rapidly into the duct airstream and is carried throughout the conditioned space. This action is similar to the sun's rays shining on a lake and evaporating the water into the air. Both the electrical and the infrared types of humidifier use electrical energy.

Humidifiers are often controlled by a *humidistat,* **Figure 34–28.** The humidistat controls the motor and the heating elements in the humidifier. The humidistat has a moisture-sensitive element, often made of hair or nylon ribbon. This material is wound around two or more bobbins and shrinks or expands, depending on the humidity. Dry air causes the element to shrink, which activates a snap-action switch and starts the humidifier. Many other devices are used, including solid-state electronic components that vary in resistance with the humidity.

Figure 34–27   An infrared humidifier. *Courtesy HumidAire Division*

(A)

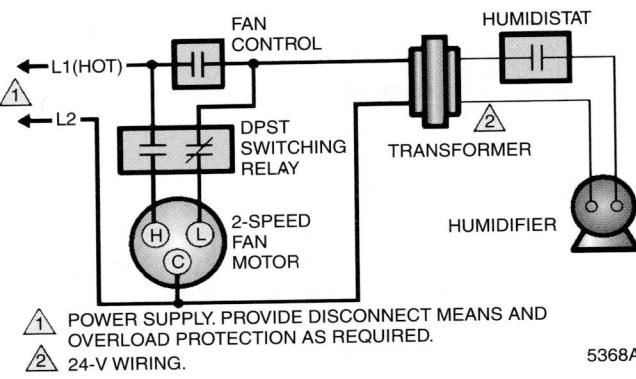

POWER SUPPLY. PROVIDE DISCONNECT MEANS AND OVERLOAD PROTECTION AS REQUIRED.

24-V WIRING.   5368A

(B)

Figure 34–28 **(A)** A humidistat. **(B)** A wiring diagram with humidistat wired in. *Courtesy of Honeywell, Inc., Residential Division*

ATOMIZING HUMIDIFIERS. *Atomizing humidifiers* discharge tiny water droplets (mist) into the air, which evaporate very rapidly into the duct airstream or directly into the conditioned space. These humidifiers can be *spray-nozzle* or *centrifugal* types, but they should not be used with hard water because it contains minerals (lime, iron, etc.) that leave the water vapor as dust and will be distributed throughout the house or building. Eight to ten grains of water hardness is the maximum recommended for atomizing humidifiers.

The spray-nozzle type sprays water through a metered bore of a nozzle into the duct airstream where it is distributed to the occupied space. Another type sprays the water onto an evaporative media where it is absorbed by the airstream as a vapor. They can be mounted in the plenum, under the duct, or on the side of the duct. It is generally recommended that atomizing humidifiers be mounted on the hot air or supply side of the furnace. **Figure 34–29** and **Figure 34–30** illustrate two types.

The *centrifugal atomizing humidifier* uses an impeller or slinger to throw the water and break it into particles that are evaporated in the airstream, **Figure 34–31**. SAFETY PRECAUTION: Atomizing humidifiers should operate only when the furnace is operating, or moisture will accumulate and cause corrosion, mildew, and a major problem with moisture where the humidifier is located.

Figure 34–29 A combination spray-nozzle and evaporative pad humidifier. *Courtesy Aqua-Mist, Inc.*

Figure 34–30 An atomizing humidifier. *Courtesy AutoFlo Company*

Figure 34–31 A centrifugal humidifier. *Courtesy Herrmidifier Company, Inc.*

Some models operate with a thermostat that controls a solenoid valve turning the unit on and off. The furnace must be on and heating before this type will operate. Others, wired in parallel with the blower motor, operate when the blower motor operates. Most are also controlled with a humidistat.

SELF-CONTAINED HUMIDIFIERS. Many residences and light-commercial buildings do not have heating equipment with ductwork through which the heated air is distributed. Hydronic heating systems, electric baseboards, or unit heaters, for example, do not use ductwork.

To provide humidification where these systems are used, *self-contained humidifiers* may be installed. These generally use the same processes as those used with forced-air furnaces. They may use the evaporative, atomizing, or infrared processes. These units may include an electric heating device to heat the water, or the water may be distributed over an evaporative media. A fan must be incorporated in the unit to distribute the moisture throughout the room or area. **Figure 34–32** illustrates a drum type. A design using steam is shown in **Figure 34–33**. In this system the electrodes heat the water, converting it to steam. The steam passes through a hose to a stainless steel duct. Steam humidification is also used in large industrial applications where steam boilers are available. The steam is distributed through a duct system or directly into the air.

PNEUMATIC ATOMIZING SYSTEMS. *Pneumatic atomizing* systems use air pressure to break up the water into a mist of tiny droplets and disperse them. SAFETY PRECAUTION: These systems as well as other atomizing systems should be applied only where the atmosphere does not have to be kept clean or where the water has a very low mineral content because the minerals in the water are also dispersed in the mist throughout the air. The minerals fall out and accumulate on surfaces in the area. These are often used in manufacturing areas, such as textile mills.

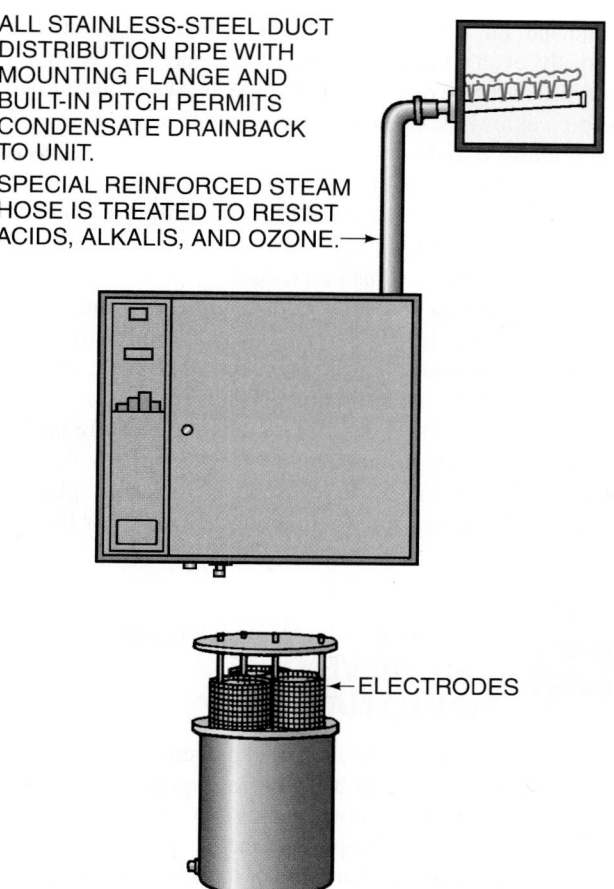

ALL STAINLESS-STEEL DUCT DISTRIBUTION PIPE WITH MOUNTING FLANGE AND BUILT-IN PITCH PERMITS CONDENSATE DRAINBACK TO UNIT.

SPECIAL REINFORCED STEAM HOSE IS TREATED TO RESIST ACIDS, ALKALIS, AND OZONE. →

← ELECTRODES

Figure 34–33 A self-contained steam humidifier.

34.10 SIZING HUMIDIFIERS

The proper size of humidifier should be installed. This text emphasizes installation and service, so the details for determining the size or capacity of humidifiers will not be covered. However, the technician should be aware of some general factors involved in the sizing process:

1. The number of cubic feet of space to be humidified. This is determined by taking the number of heated square feet of the house and multiplying it by the ceiling height. A 1500-ft^2 house with an 8-ft ceiling height would have 1500 ft$^2 \times$ 8 ft = 12,000 ft^3.
2. The construction of the building. This includes quality of insulation, storm windows, fireplaces, building "tightness," and so on.
3. The amount of air change per hour and the approximate lowest outdoor temperature.
4. The level of relative humidity desired.

34.11 INSTALLATION

The most important factor regarding installation of humidifiers is to follow the manufacturer's instructions. Evaporative humidifiers often are operated independent of the furnace. It is normally recommended that they be controlled by a

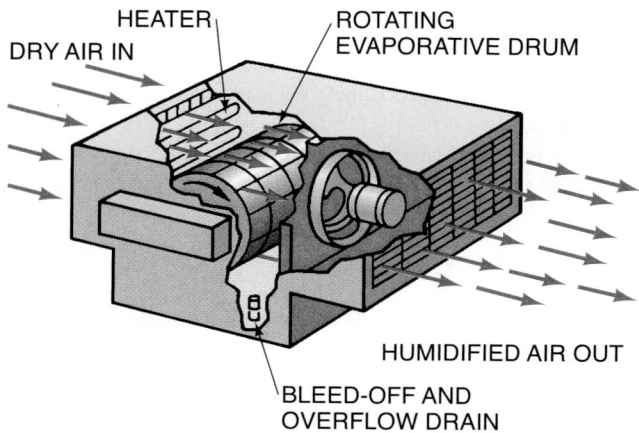

HEATER
DRY AIR IN
ROTATING EVAPORATIVE DRUM
HUMIDIFIED AIR OUT
BLEED-OFF AND OVERFLOW DRAIN

Figure 34–32 A drum-type self-contained humidifier.

humidistat, but it does no real harm for them to operate continuously, even when the furnace is not operating. Atomizing humidifiers, however, should not operate when the furnace and blower are not operating. Moisture will accumulate in the duct if allowed to do so.

Particular attention should be given to clearances within the duct or plenum. The humidifier should not exhaust directly onto air-conditioning coils, air filters, electronic air cleaners, blowers, or turns in the duct.

If mounting on a supply duct, choose one that serves the largest space in the house. The humid air will spread throughout the house, but the process will be more efficient when given the best distribution possible.

Plan the installation carefully, including locating the humidifier, as already discussed, and providing the wiring and plumbing (with drain). A licensed electrician or plumber must provide the service where required by code or law.

34.12 ▶ SERVICE, TROUBLESHOOTING, AND PREVENTIVE MAINTENANCE

Proper service, troubleshooting, and preventive maintenance play a big part in keeping humidifying equipment operating efficiently. Cleaning the components that are in contact with the water is the most important factor. The frequency of cleaning depends on the hardness of the water: the harder the water, the more minerals in the water. In evaporative systems, these minerals collect on the media, on other moving parts, and in the reservoir. In addition, mold, algae, bacteria, and virus growth can cause problems, even to the extent of blocking the output of the humidifier. Algaecides can be used to help neutralize algae growth. The reservoir should be drained regularly if possible, and components, particularly the media, should be cleaned periodically.

Indoor air quality can be adversely affected by the humidifier because water is used, and any mold, mildew, or algae can be distributed by the moving airstream. Mold, mildew, or algae may cause allergic reactions or upper respiratory problems with the occupants. One of the maintenance factors that must be attended to on a regular basis is the cleanliness of the water reservoir and evaporative surface when this type of humidifier is used. These surfaces must be cleaned and sanitized on a regular basis to prevent mold and algae from accumulating. Some manufacturers have used UV lights to kill these growing molds and algae and to prevent them from contaminating the area. Follow the manufacturer's recommendations for cleaning a particular unit. If the humidifier is an atomizing type, look for mold or mildew in the duct downstream from the nozzle.

HUMIDIFIER NOT RUNNING. When the humidifier does not run, the problem is usually electrical, or a component is bound tight or locked due to a mineral buildup. A locked condition may cause a thermal overload protector to open. Using typical troubleshooting techniques, check overload protection, circuit breakers, humidistat, and low-voltage controls if there are any. Check the motor to see whether it is burned out. Clean all components and disinfect if appropriate.

EXCESSIVE DUST. If excessive dust is caused by the humidifier, the dust will be white due to mineral buildup on the media. Clean or replace the media. If excessive dust occurs in an atomizing humidifier, the wrong equipment has been installed.

WATER OVERFLOW. Water overflow indicates a defective float valve assembly. It may need cleaning, adjusting, or replacing.

MOISTURE IN OR AROUND DUCTS. Moisture in ducts is found primarily in atomizing humidifiers. **Remember, this equipment should operate only when the furnace operates.** Check the control to see whether it operates at other times, such as COOL or FAN ON modes. A restricted airflow may also cause this problem.

LOW OR HIGH LEVELS OF HUMIDITY. If the humidity level is too high or too low, check the calibration of the humidistat by using a sling psychrometer. If it is out of calibration, it may be possible to adjust it. The humidistat may need to be relocated if it is too close to a window or door. Ensure that the humidifier is clean and operating properly.

SUMMARY

- Many modern homes and office buildings are being designed and constructed to keep the outside air out in order to maximize heating and cooling efficiencies. This may result in the accumulation of air pollutants inside.
- Eliminating the source of indoor pollution should be the first step in improving indoor air quality.
- Mold growth indoors has become a major problem.
- Indoor air quality can be improved through the use of ventilation, which removes some air pollutants and dilutes those remaining.
- Air cleaning through the use of filtering, absorption, and adsorption is another means for improving indoor air quality.
- Ducts may be cleaned if there is significant evidence that it should be done.
- In cool or cold weather, humidifiers may be needed for comfort and to protect furniture and other household materials.

REVIEW QUESTIONS

1. List five different sources of indoor air pollution.
2. ASHRAE Standard _____, "Ventilation for Acceptable Indoor Air Quality," is the standard generally used to determine acceptable indoor air quality.
3. Radon is a _____, _____, and _____ gas.

4. Three different biological contaminants are _____, _____, and _____.

5. What are two substances molds need in order to grow?

6. Three methods to control indoor air contamination are _____, _____, and _____.

7. If radon gas is suspected, what should be done?

8. How does ventilation improve air quality?

9. Three general types of air cleaners are _____, _____, and _____.

10. True or False: Carbon monoxide is produced by mixing vapors from nitrogen dioxide and sulfur dioxide.

11. Describe the difference between absorption and adsorption.

12. True or False: Fiberglass filter media are generally coated with a special nondrying, nontoxic adhesive on each fiber.

13. An electrostatic precipitator is the same as
    A. an extended surface air filter.
    B. a fiberglass media filter.
    C. an electronic air cleaner.
    D. a steel washable air filter.

14. The purifier ingredient in an activated charcoal air purifier may be
    A. activated carbon.
    B. coconut shell pellets.
    C. manufactured from coal.
    D. any of the above.

15. Ion generators _____ particles in an area.
    A. disintegrate
    B. charge
    C. wash
    D. remove odors from

16. The relative humidity in heated homes is generally the _____ in the winter season.
    A. highest
    B. lowest

17. The recommended relative humidity for a home is between
    A. 50% and 70%.
    B. 20% and 40%.
    C. 30% and 50%.
    D. 40% and 60%.

18. True or False: Some humidifiers have their own water-heating devices.

19. Static electricity is much greater in _____ air.
    A. dry
    B. moist

20. The plenum-mounted humidifier is installed in the _____ plenum.
    A. supply
    B. return
    C. supply or return

21. Describe the difference between an evaporative and an atomizing humidifier.

22. Why is it essential for the furnace to be operating when an atomizing humidifier is being used?

DIAGNOSTIC CHART FOR FILTRATION AND HUMIDICATION SYSTEMS

All forced-air systems have a filtration system of some sort. These vary from the basic air filter to the more complex electronic air filter. Most of the problems for basic air filters are reduced airflow due to lack of maintenance. Problems for electronic air-filter systems are more complex.

Humidification systems all contain water, and water can cause problems if the system is not properly cared for. The following chart lists some of the common problems of filtration and humidification systems.

Problem	Possible Cause	Possible Repair	Heading Number
Filters			
Media type			
Restricted	Filters are not changed often enough	Change on a more regular schedule.	34.7
Dust entering conditioned space, possibly causing dirty evaporator coil and fan blades	Filter media too coarse	Change to finer filter media.	34.7
	Air bypassing filter	Install filters of the correct size.	
Electronic Filter			
No power to filter	Filter interlock with fan	Establish correct fan interlock circuit.	34.7
Power to filter but filter not operating	Filter grounded in element	Clean filter.	
	Contacts in filter element not making contact	Clean contacts and reassemble.	34.7
	Defective power assembly	Change power assembly.	34.7
Filter makes cracking sound too often	High dust content in air	Change to finer prefilter.	

(continued)

DIAGNOSTIC CHART FOR FILTRATION AND HUMIDICATION SYSTEMS (Continued)

Problem	Possible Cause	Possible Repair	Heading Number

Humidifiers
Unit not operating—
  depends on electricity
	Faulty electrical circuits	Change fuse or reset breaker.
		Repair faulty electrical circuits or
		  connections.
			34.12

	Humidistat not calling for
	  humidity
		Check humidistat calibration; adjust or
		  change humidistat.	

	Interlock with fan circuit	Determine what type of interlock and correct.	
Unit operating, but humidity is low	No water supply	Reestablish water supply.	34.12
	Defective float	Change float.	

	Evaporation media saturated
	  with minerals and will not
	  absorb water
		Change media.	

	Power assembly that turns
	  evaporation media not turning
		Make sure that power is available; if it is,
		  change unit.	

SECTION 7

Air Conditioning (Cooling)

UNIT
35 Comfort and Psychrometrics

35.1 COMFORT

Comfort describes a delicate balance of pleasant feeling in the body produced by its surroundings. A comfortable atmosphere describes our surroundings when we are not aware of discomfort. Providing a comfortable atmosphere for people becomes the job of the heating and air-conditioning profession. Comfort involves four things: (1) temperature, (2) humidity, (3) air movement, and (4) air cleanliness.

The human body has a sophisticated control system for both protection and comfort. The human body can move from a warm house to 0°F outside, and it starts to compensate for the surroundings. It can move from a cool house to 95°F outside, and it will start to adjust to keep the body comfortable and from overheating. Body adjustments are accomplished by the circulatory and respiratory systems. When the body gets too warm, the vessels next to the skin dilate to get the blood closer to the surrounding air in an effort to increase the heat exchange with the air. This is why you turn red when you become hot. If this does not cool the body, it will perspire. When this perspiration is evaporated, it takes heat from the body and cools it. Excess perspiration in hot weather explains why you must drink more fluids.

35.2 FOOD ENERGY AND THE BODY

The human body may be compared to a coal hot water boiler. The coal is burned in the boiler to create heat. Heat is energy. Food to the human body is like coal to a hot water boiler. The coal in a boiler is converted to heat for space heating. Some heat goes up the flue, some escapes to the surroundings, and some is carried away in the ashes. If fuel is added to the fire

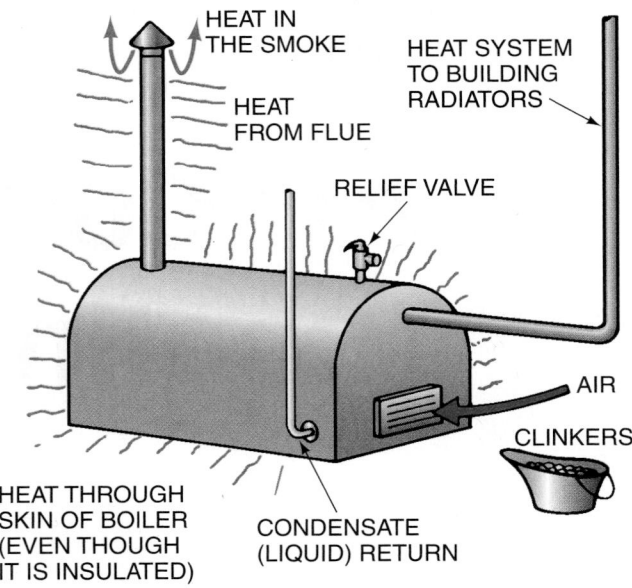

Figure 35–1 The boiler can be compared to the human body in that it uses fuel for energy.

and the heat cannot be dissipated, the boiler will overheat, **Figure 35–1.**

The body uses food to produce energy. Some energy is stored as fatty tissue, some leaves as waste, some leaves as heat, and some is used as energy to keep the body functioning. If the body needs to dissipate some of its heat to the surroundings and cannot, it will overheat. The average body temperature is 98.6°F, but this is the core temperature of the body, not the surface temperature. The surface temperature of the body, or the temperature of our skin, is considerably lower than 98.6°F and varies according to the temperature of its surroundings. So, if someone is standing outside in 100°F air, the body will not be able to reject heat to its surroundings, **Figure 35–2.** Instead, the body will actually absorb heat from its surroundings. When such is the case, the body's internal temperature-control system will cause perspiration to form on the surface of the skin in an attempt to cool the body. As the surrounding temperature drops, the body will reject heat naturally, as a warmer substance transfers heat to cooler ones, **Figure 35–3.** We are comfortable when our body is transferring heat to its surroundings at the correct rate, but certain conditions must be met for this comfortable, or balanced, condition to exist.

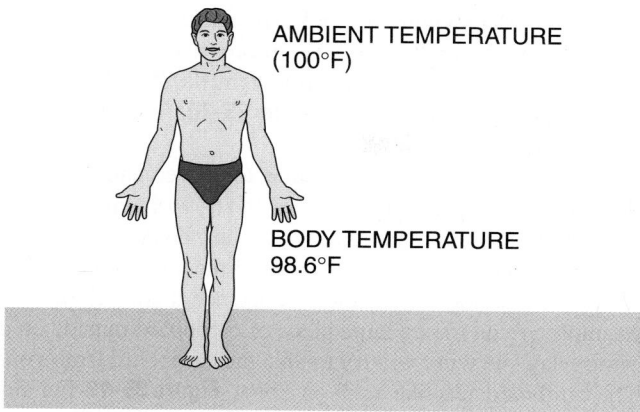

Figure 35–2   The human body must give off some of its generated heat to the surroundings or it will overheat.

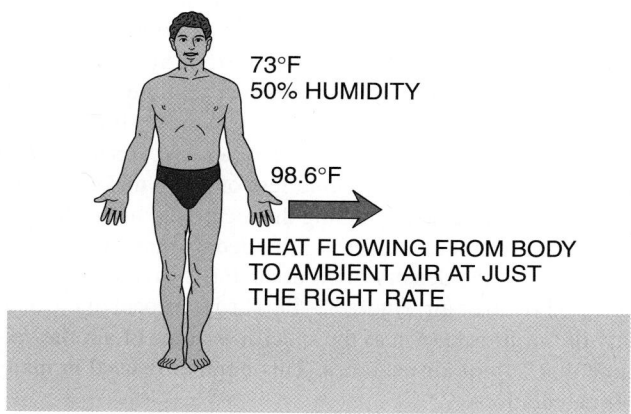

Figure 35–3   Normal (average) body temperature is 98.6°F.

35.3   HEAT TRANSFER TO AND FROM THE BODY

The body gives off and absorbs heat by the three methods of heat transfer: conduction, convection, and radiation, **Figure 35–4**. The evaporation of perspiration could be considered a fourth way, **Figure 35–5**. When the surroundings are at a particular comfort condition, the body is giving up heat at a steady rate that is comfortable. The surroundings must be cooler than the body for the body to be comfortable. Typically, when the body is at rest (sitting) and in surroundings of 75°F and 50% humidity with a slight air movement, the body is close to being comfortable. In cooler weather a different set of conditions applies (e.g., we wear more clothing). The following statements can be used as guidelines for comfort.

1. In winter:
   A. Lower temperature can be offset with higher humidity.
   B. The lower the humidity is, the higher the temperature must be.
   C. Air movement is more noticeable.
2. In summer:
   A. When the humidity is high, air movement helps.

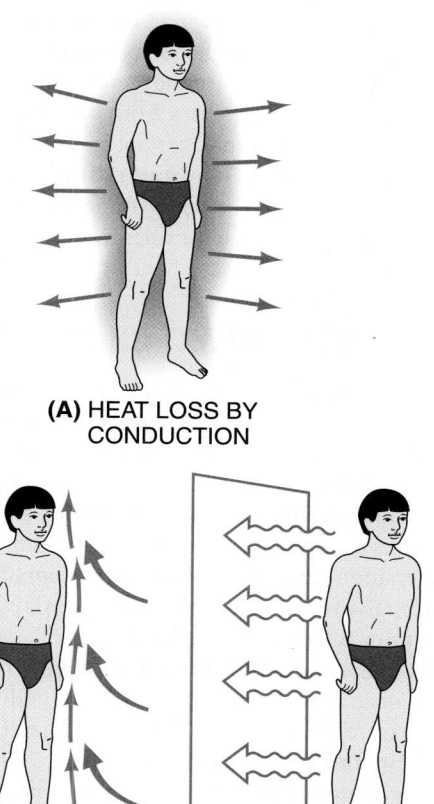

(A) HEAT LOSS BY CONDUCTION

(B) HEAT LOSS BY CONVECTION

(C) HEAT LOSS BY RADIATION

Figure 35–4   The three direct ways the human body gives off heat. **(A)** Conduction. **(B)** Convection. **(C)** Radiation.

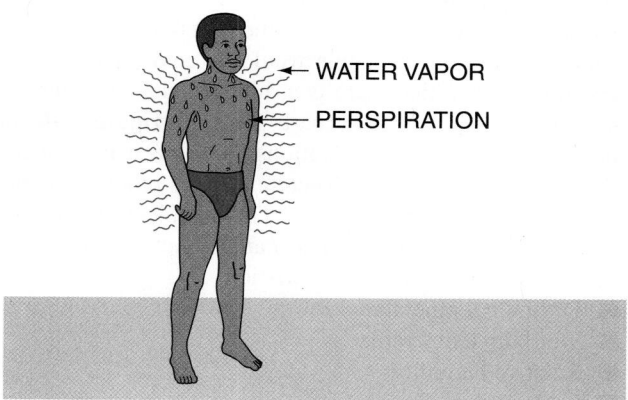

WATER VAPOR

PERSPIRATION

Figure 35–5   A fourth way the body gives off heat—the evaporation of perspiration.

   B. Higher temperatures can be offset with lower humidity.
3. The comfort conditions in winter and in summer are different.
4. Styles of clothes in different parts of the country make a slight difference in the conditioned space-temperature requirements for comfort. For example,

in Maine, the styles would be warmer in the winter than in Georgia, so the inside temperature of a home or office will not have the same comfort level.

5. Body metabolism varies from person to person. Women, for example, are not as warm-bodied as men. The circulatory system generally does not work in older people as well as in younger people. Women seem to expect a closer tolerance for temperature control.

35.4 THE COMFORT CHART

The chart in **Figure 35–6,** often called a *generalized comfort chart,* can be used to compare one situation with another. **Figure 35–6(A)** represents the comfort chart for the summer months. In the center of this chart there is a bold rectangular square that encompasses the temperature and humidity combinations at which most people will be comfortable. The air temperatures are located along the bottom of the chart, while the (relative) humidity percentages are shown by the curved lines as indicated in **Figure 35–7.**

From the comfort chart for the summer months, it can be determined that the average individual will be comfortable if the surrounding air is 75°F with a relative humidity of 50%, **Figure 35–8.** This is because the 75°F line and the 50% relative-humidity line cross at a point that is located within the "comfort" square. It can also be determined that the average person will *not* be comfortable if the surrounding air is 85°F and the relative humidity is 50%, **Figure 35–9.** This is because these two lines intersect at a point that is outside the comfort region. The same concept applies to the comfort chart for the winter months, as shown in **Figure 35–6(B).**

A technician may use charts like those shown in **Figure 35–6** to plot and compare comfort conditions in an occupied space. In general, the closer the plot falls to the middle of the chart, the more people would be comfortable at that condition. Notice that there is a different chart for summer and winter conditions. The charts shown in **Figure 35–8** and **Figure 35–9** show only the relationship between air temperature and relative humidity. There are many other properties of air that affect comfort levels. The complete graph is referred to as a *psychrometric chart* and compiles information about the following air characteristics:

■ Dry-bulb temperature
■ Wet-bulb temperature
■ Relative humidity
■ Specific volume
■ Moisture content
■ Heat content

Each of these items will be discussed in more detail in this unit. Section 35.13, "Plotting on the Psychrometric Chart," will show you how to plot these conditions on the chart.

35.5 PSYCHROMETRICS

The study of air and its properties is called *psychrometrics.* When we move through a room, we are not aware of the air inside the room, but the air has weight and occupies space

like water in a swimming pool. The water in a swimming pool is denser than the air in the room; it weighs more per unit of volume. Perfectly dry air weighs 0.075 lb/ft^3 at standard atmospheric pressure, **Figure 35–10.** The density of water is 62.4 lb/ft^3.

Air, like water, offers resistance to movement. To prove this, take a large piece of cardboard and try to swing it around with the flat side moving through the air, **Figure 35–11.** It is hard to do because of the resistance of the air. The larger the area of the cardboard, the more resistance there will be. For example, if you took a large piece of cardboard outside on a windy day, the wind will try to take the cardboard from you. The cardboard acts like a sail on a boat, **Figure 35–12.** For another example, **Figure 35–13,** invert an empty glass and push it down in water. The air in the glass resists the water going up into the glass.

The weight of air in a room can be calculated by multiplying the room volume by the weight of a cubic foot of air. In a room 10 ft × 10 ft × 10 ft, the volume is 1000 ft^3, so the room air weighs 1000 ft^3 × 0.075 lb/ft^3 = 75 lb, **Figure 35–14.** It is important to remember that the 0.075 lb/ft^3 that is used in this example assumes that the air is *dry*. Deviations from these conditions will have an effect on the weight of the air. The number of cubic feet of air to make a pound of air can be obtained by taking the reciprocal of the density. The reciprocal of a number is 1 divided by that number. The reciprocal of the density of air at 70°F is 1 divided by 0.075 or 13.33 ft^3/lb of air, **Figure 35–15.** The reciprocal of the density of the air is known as the specific volume of air, the volume that 1 lb of air occupies. This number is used in many air calculations.

35.6 MOISTURE IN AIR

Air consists of approximately 78% nitrogen, 21% oxygen, and 1% other gases, **Figure 35–16.** But air is not totally dry. Surface water and rain keep moisture in the atmosphere everywhere at all times, even in a desert. Two-thirds of the earth's surface is covered with water. Water, in the form of low-pressure vapor, is suspended in the air and is called *humidity,* **Figure 35–17.** There are two types of humidity that are studied: absolute humidity and relative humidity.

35.7 ABSOLUTE AND RELATIVE HUMIDITY

The moisture content in air (humidity) is measured by weight and expressed in pounds or grains (7000 gr/lb) per pound of air. Air can hold very little water vapor. Each pound of air at sea level (standard atmospheric pressure) at a temperature of 70°F can hold 110.5 grains of moisture, or 0.01578 pounds of water. The amount of water vapor that the air sample can hold will change as the air conditions change. The actual amount of water vapor that is contained in an air sample is referred to as the *absolute* humidity, and is expressed in grains per pound. For example, a particular air sample that contains 10 grains of moisture per pound of air has an absolute humidity of 10 grains per pound. Absolute humidity is

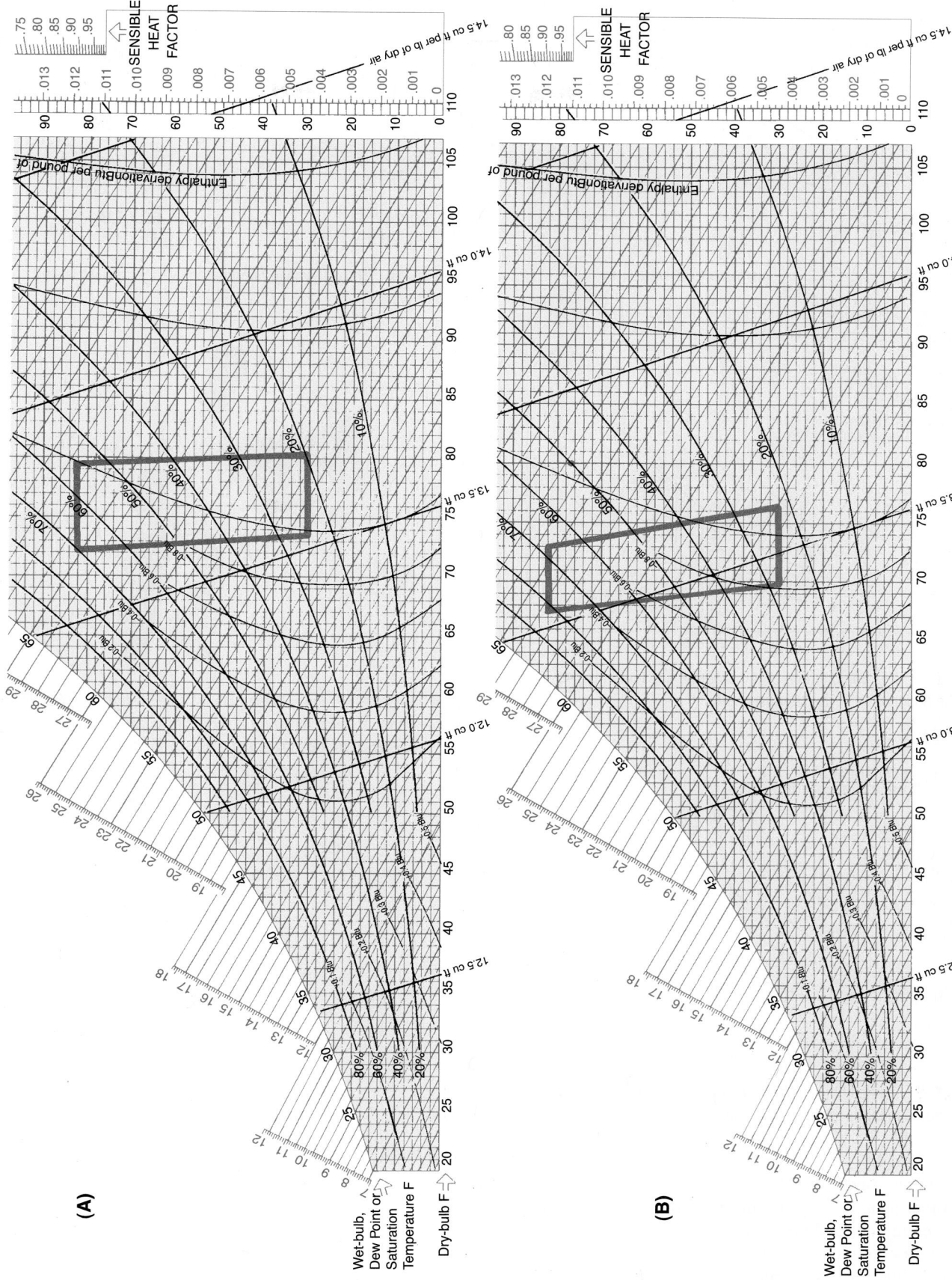

Figure 35–6 Comfort charts for **(A)** summer and **(B)** winter. *Adapted from Carrier Corporation Psychrometric Chart*

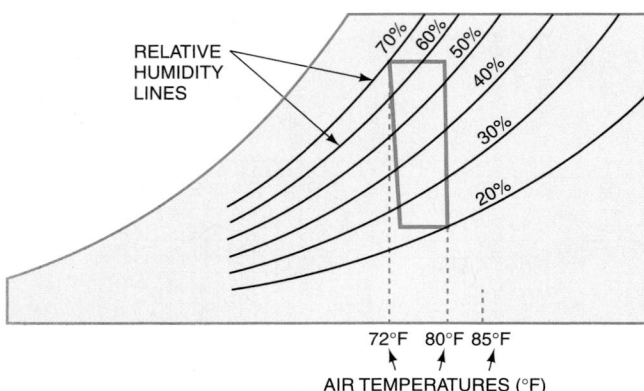

Figure 35–7   Combinations of temperature and humidity make up the boundaries of the comfort charts.

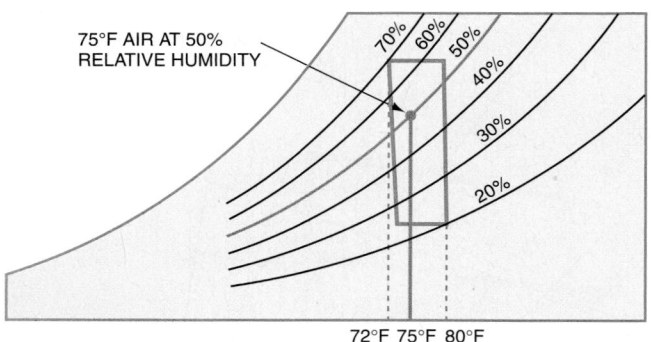

Figure 35–8   This combination of temperature and humidity lies within the boundaries of the comfort chart.

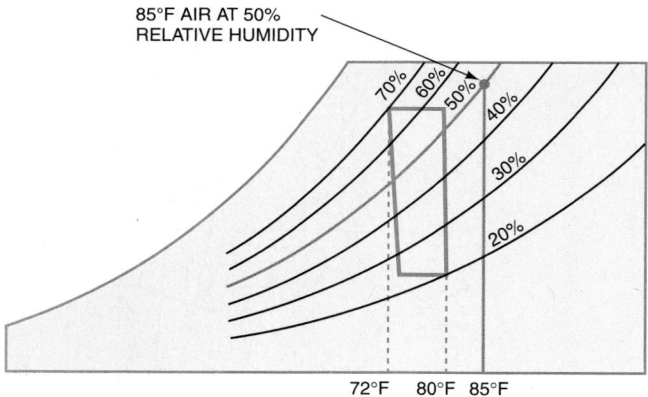

Figure 35–9   This combination of temperature and humidity lies outside the boundaries of the comfort chart.

not commonly used for field calculations. Relative humidity, on the other hand, is very useful and is often used to determine air conditions and the operating effectiveness and efficiency of air-conditioning equipment.

Relative humidity is the relationship between the weight of water vapor in a pound of air compared with the weight of water vapor that a pound of air could hold if it were 100%

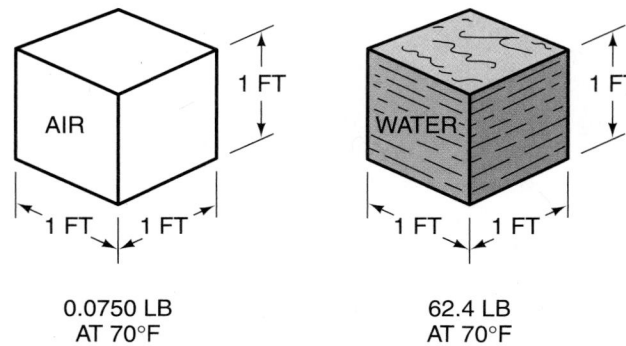

Figure 35–10   The difference between the weight of air and the weight of water.

Figure 35–11   This man is having a hard time swinging a piece of cardboard around because the air around him is taking up space and causing resistance to the movement of the cardboard.

Figure 35–12   This man is walking in a breeze. He is pushed along by the breeze against the cardboard because the air has weight and takes up space.

saturated, **Figure 35–18. Figure 35–18(A)** shows that 100% saturated air at 70°F holds 110.5 grains of moisture per pound, and **Figure 35–18(B)** shows that 50% saturated air will hold 55.25 grains at the same temperature. The air in

Figure 35–13   This illustration shows another way to prove that air takes up space.

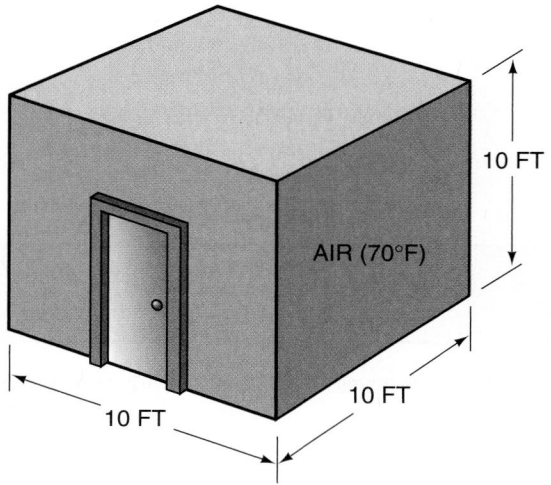

Figure 35–14   This room is 10 ft × 10 ft × 10 ft. The volume of the room is 1000 ft³ of air that weighs 0.075 lb/ft³. 1000 ft³ × 0.075 lb = 75 lb of air in the room.

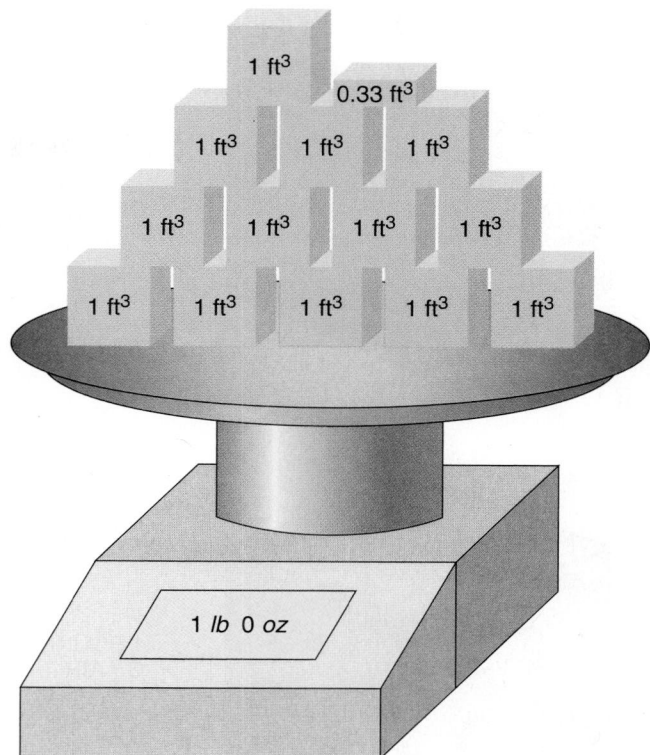

Figure 35–15   At standard atmospheric pressure, it will take 13.33 cubic feet of air to make up one pound of air.

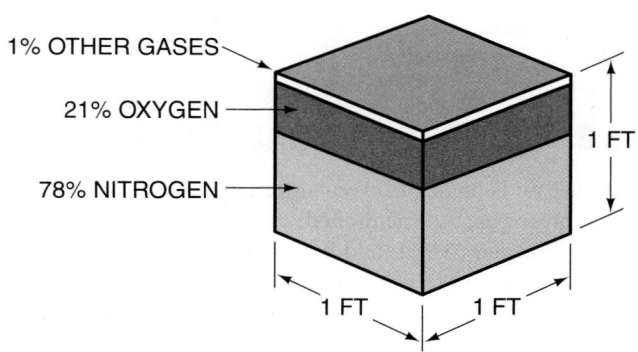

Figure 35–16   Substances that make up air.

Figure 35–18(A) has a relative humidity of 100%, whereas the air sample in **Figure 35–18(B)** has a relative humidity of only 50%.

Large buildings should have a set of specifications that tell the technician what the design conditions are for the building occupied space. The system should be able to maintain these conditions when the system and controls are set and operating correctly. When there are complaints in local areas, the technician may be called on to see whether these spaces are within the correct temperature and humidity ranges. Often, the only way to settle a dispute is to plot the air conditions on the comfort chart or compare the readings with the building's specifications. The technician must be prepared to do some accurate testing and record the results to see what the problem is.

Although meteorologists talk about relative humidity, they do not like to use it for comparing conditions because as air temperature changes, it holds different amounts of moisture per pound of air. Meteorologists use dew point temperature for this comparison, which will be discussed in more detail later in this unit.

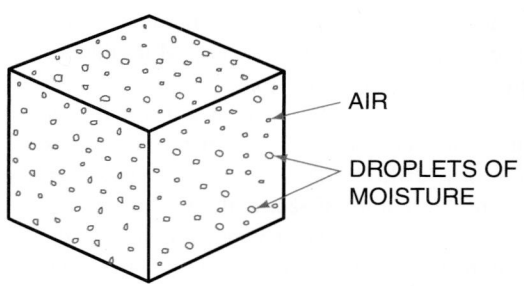

Figure 35–17   Moisture suspended in air.

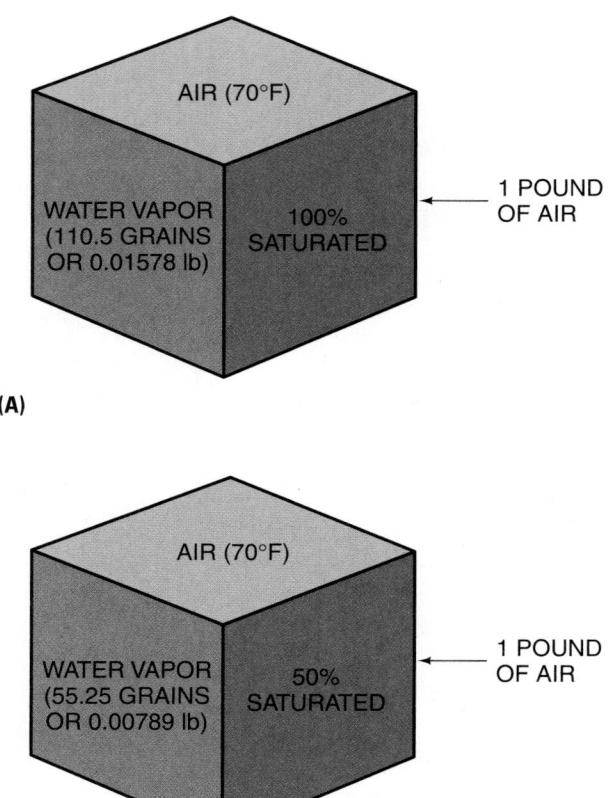

(A)

(B)

Figure 35–18 **(A)** This air sample is 100% saturated, so the relative humidity is 100%. **(B)** This air sample is 50% saturated, so the relative humidity is 50%.

35.8 SUPERHEATED GASES IN AIR

Because air is made of several different gases, it is not a pure element or gas. As mentioned, air is made up of nitrogen (78%), oxygen (21%), and approximately 1% other gases, **Figure 35–16.** These gases in the air are highly superheated. Nitrogen, for instance, boils at −319°F, and oxygen boils at −297°F at atmospheric pressure, **Figure 35–19.** Hence, nitrogen and oxygen in the atmosphere are *superheated gases*— they are superheated several hundred degrees above absolute zero (0° Rankine). Each gas exerts pressure according to Dalton's Law of Partial Pressures. This law states that each gas in a mixture of gases acts independently of the other gases and the total pressure of a gas mixture is the sum of the pressures of each gas in the mixture. More than one gas can occupy a space at the same time.

Water vapor suspended in air is a gas that exerts its own pressure and occupies space with the other gases. Water at 70°F in a dish in the atmosphere exerts a pressure of 0.7392 in. Hg, **Figure 35–20.** If the water vapor pressure in the air is less than the water vapor pressure in the dish, the water in the dish will evaporate slowly to the lower pressure area of the water vapor in the air. For example, the room may be at a temperature of 70°F with a humidity of 30%.

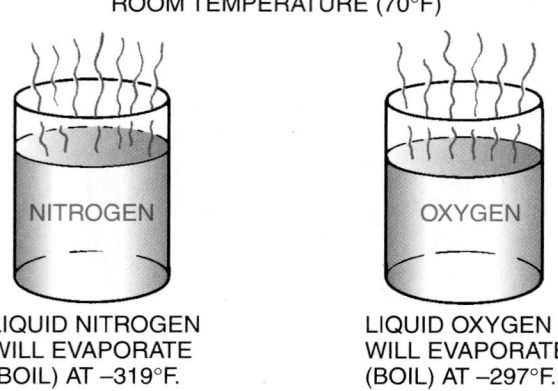

Figure 35–19 If you place a container of liquid nitrogen and a container of liquid oxygen in a room, the nitrogen and oxygen will start to evaporate or boil.

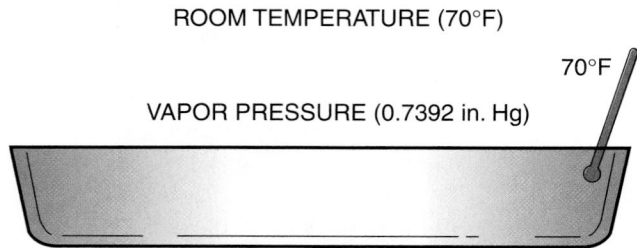

Figure 35–20 Vapor pressure of water at 70°F in an open dish in a room.

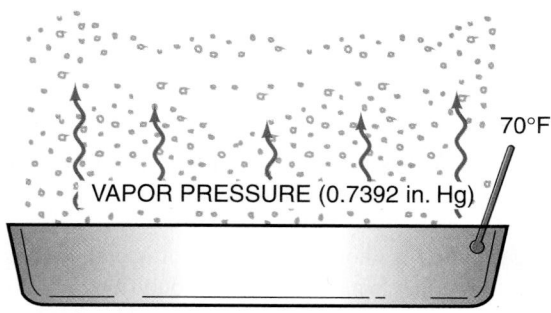

Figure 35–21 Moisture suspended in the air has a pressure controlled by the humidity in the room.

The vapor pressure for the moisture suspended in the air is 0.101 psia × 2.036 = 0.206 in. Hg, **Figure 35–21.** Vapor pressure for moisture in air can be found in some psychrometric charts and in saturated water tables. When reverse pressures occur, the action of the water vapor reverses. For example, if the water vapor pressure in the dish is less than the pressure of the vapor in the air, water from the air will condense into the water in the dish, **Figure 35–22.**

When water vapor is suspended in the air, the air is sometimes called "wet air." If the air has a large amount of moisture, the moisture can be seen (e.g., fog or a cloud).

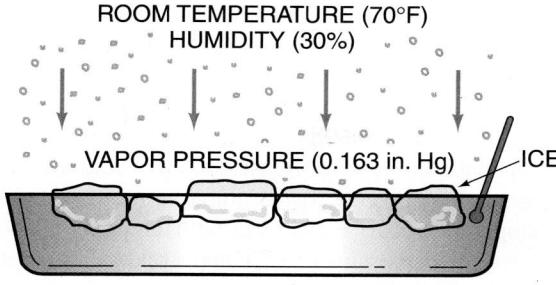

Figure 35–22 The moisture in the dish has ice in it, which lowers the vapor pressure to 0.163 (0.08 psia × 2.036 = 0.163 in. Hg). The room temperature is still 70°F with a humidity of 30%, which has a vapor pressure of 0.206 in. Hg.

Actually, the air is not wet because the moisture is suspended in the air. This could more accurately be called a nitrogen, oxygen, and water vapor mixture.

35.9   DRY-BULB AND WET-BULB TEMPERATURES

Up to this point in the unit, we have made several references to the temperature of air. When doing so, we have been referring to the level of heat intensity that is obtained when a standard thermometer is used. In actuality, there are two types of temperature readings that are very useful to the field technician. These are the *dry-bulb* temperature and the *wet-bulb* temperature.

The moisture content of air can be checked by using a combination of dry-bulb and wet-bulb temperatures. *Dry-bulb* temperature is the sensible-heat level of air and is taken with an ordinary thermometer. Wet-bulb temperature is taken with a thermometer with a wick on the end that is soaked with distilled water, **Figure 35–23**. Both the wet-bulb

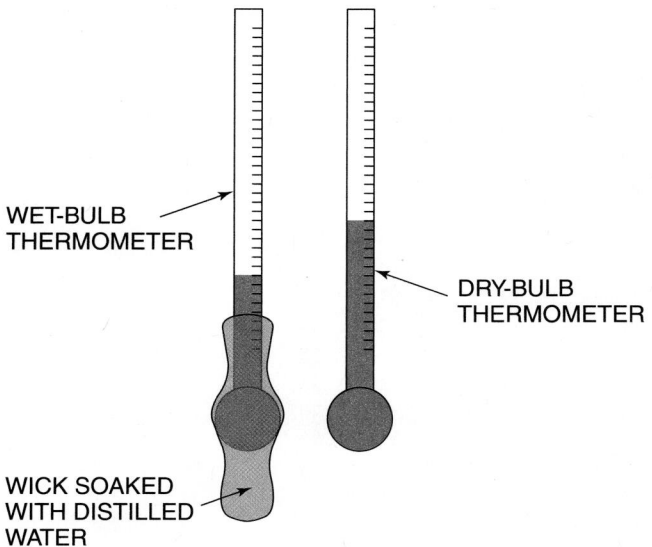

Figure 35–23 Wet- and dry-bulb thermometer bulbs.

WET-BULB THERMOMETER

DRY-BULB THERMOMETER

WICK SOAKED WITH DISTILLED WATER

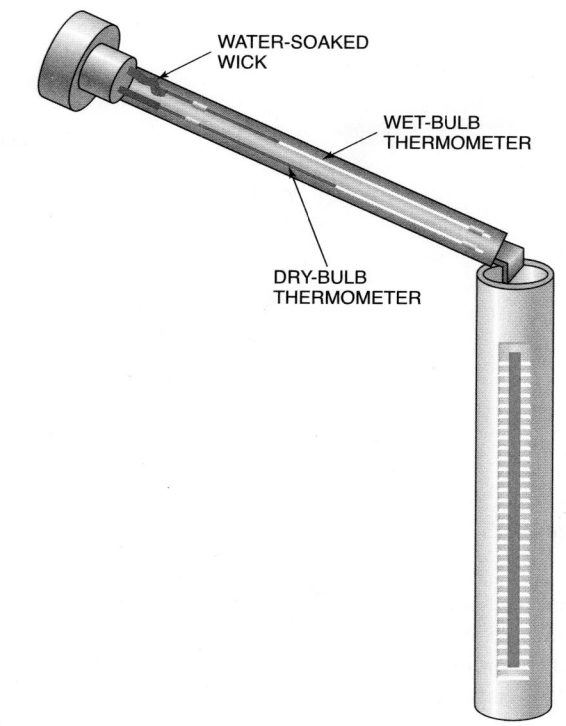

WATER-SOAKED WICK

WET-BULB THERMOMETER

DRY-BULB THERMOMETER

(A)

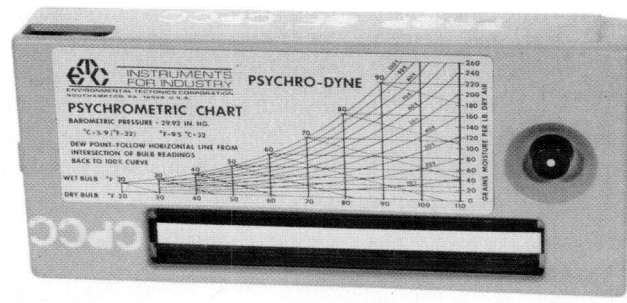

(B)

(C)

Figure 35–24 **(A)** Sling psychrometer. **(B)** An electric psychrometer with a small fan. **(C)** An electronic psychrometer. *Photos **(B)** and **(C)** by Bill Johnson*

and dry-bulb temperature readings are obtained at the same time by using an instrument referred to as a sling psychrometer, **Figure 35–24**. The reading from a wet-bulb thermometer takes into account the moisture content of the air. It reflects the total heat content of air. To take accurate wet- and dry-bulb temperature readings, the sling psychrometer is spun in the air, **Figure 35–25**. The wet-bulb thermometer will get cooler than the dry-bulb thermometer because of the evaporation of the distilled water. The rate at which water will

Figure 35–25   Technician spinning a sling psychrometer.

water is used because some water has undesirable mineral deposits. Some minerals will change the boiling temperature.

The difference between the dry-bulb reading and the wet-bulb reading is called the *wet-bulb depression*. **Figure 35–26** shows a wet-bulb depression chart. As the amount of moisture suspended in the air decreases, the wet-bulb depression increases and vice versa. For example, a room with a dry-bulb temperature of 76°F and a wet-bulb temperature of 64°F has a wet-bulb depression of 12°F and a relative humidity of 52%. If the 76°F dry-bulb temperature is maintained and moisture is added to the room so that the wet-bulb temperature rises to 74°F, the relative humidity increases to 91% and the new wet-bulb depression is 2°F. If the wet-bulb depression is allowed to go to 0°F (e.g., 76°F dry bulb and 76°F wet bulb), the relative humidity will be 100%. The air is holding all of the moisture it can—it is *saturated* with moisture.

35.10   DEW POINT TEMPERATURE

The *dew point temperature* is the temperature at which moisture begins to condense out of the air. For example, if you were to set a glass of warm water in a room with a temperature of 75°F and 50% relative humidity, the water in the glass would evaporate slowly to the room. If you gradually cool the glass with ice, when the glass surface temperature becomes 55.5°F, water will begin to form on the surface of the glass, **Figure 35–27**. Moisture from the room will also collect in the water in the glass and the level will begin to rise. This temperature at which water forms is called the dew point temperature of the air. Air can be dehumidified by passing it over a surface that is below the dew point temperature of the air; moisture will collect on the cold surface, for example, an

evaporate from the wick on the wet-bulb thermometer is determined by the moisture content of the surrounding air. If the surrounding air is very dry, the moisture will evaporate quickly, causing the wet-bulb temperature to drop lower. If the surrounding air is very wet (high relative humidity) the rate of evaporation will be very low and the wet-bulb temperature reading will be closer to the dry-bulb temperature reading. The closer the wet-bulb and dry-bulb temperatures are to each other, the higher the relative humidity. For example, if the wet-bulb and dry-bulb temperature readings are both 78°F, the relative humidity will be 100%. The wet-bulb temperature can never be higher than the dry-bulb temperature. If it ever indicates higher, there is a mistake. Distilled

DB TEMP.	WB DEPRESSION																														
---	1	2	3	4	5	6	7	8	9	10	11	12	13	14	15	16	17	18	19	20	21	22	23	24	25	26	27	28	29	30
32	90	79	69	60	50	41	31	22	13	4																				
36	91	82	73	65	56	48	39	31	23	14	6																			
40	92	84	76	68	61	53	46	38	31	23	16	9	2																	
44	93	85	78	71	64	57	51	44	37	31	24	18	12	5																
48	93	87	80	73	67	60	54	48	42	36	34	25	19	14	8															
52	94	88	81	75	69	63	58	52	46	41	36	30	25	20	15	10	6	0												
56	94	88	82	77	71	66	61	55	50	45	40	35	34	26	24	17	12	8	4											
60	94	89	84	78	73	68	63	58	53	49	44	40	35	31	27	22	18	14	6	2										
64	95	90	85	79	75	70	66	61	56	52	48	43	39	35	34	27	23	20	16	12	9									
68	95	90	85	81	76	72	67	63	59	55	51	47	43	39	35	31	28	24	21	17	14									
72	95	91	86	82	78	73	69	65	61	57	53	49	46	42	39	35	32	28	25	22	19									
76	96	91	87	83	78	74	70	67	63	59	55	52	48	45	42	38	35	32	29	26	23									
80	96	91	87	83	79	76	72	68	64	61	57	54	54	47	44	41	38	35	32	29	27	24	21	18	16	13	11	8	6	1
84	96	92	88	84	80	77	73	70	66	63	59	56	53	50	47	44	41	38	35	32	30	27	25	22	20	17	15	12	10	8
88	96	92	88	85	81	78	74	71	57	64	61	58	55	52	49	46	43	41	38	35	33	30	28	25	23	21	18	16	14	12
92	96	92	89	85	82	78	75	72	69	65	62	59	57	54	51	48	45	43	40	38	35	33	30	28	26	24	22	19	17	15
96	96	93	89	86	82	79	76	73	70	67	74	61	58	55	53	50	47	45	42	40	37	35	33	31	29	26	24	22	20	18
100	96	93	90	86	83	80	77	74	71	68	65	62	59	57	54	52	49	47	44	42	40	37	35	33	31	29	27	25	23	21
104	97	93	90	87	84	80	77	74	72	69	66	63	61	58	56	53	51	48	46	44	41	39	37	35	33	31	29	27	25	24
108	97	93	90	87	84	81	78	75	72	70	67	64	62	59	57	54	52	50	47	45	43	41	39	37	35	33	31	29	28	26

Figure 35–26   A wet-bulb depression chart.

Figure 35–27   The glass was gradually cooled until beads of water began to form on the outside of the glass.

air-conditioning coil, **Figure 35–28.** The condensed moisture is drained, and this is the moisture that you see running out of the condensate line of an air conditioner, **Figure 35–29.**

35.11   ENTHALPY

As air is heated or cooled, the heat content of the air changes. When cooled, the air's heat content decreases, and when heated, the air's heat content increases. Enthalpy of air is the measure of heat content in Btu per pound of air. Given the conditions of an air sample, the heat content of the sample can be determined by using the psychrometric chart, which will be discussed in the next section. If the conditions of the air in the occupied space are known and we can measure them after the air passes through the heating or cooling equipment, we can easily determine how much heat has been either added or removed from the air as it passes through the heat-transfer surface. This information, together with some other basic calculations, will enable the service technician to verify system capacity and operating efficiency. As we will soon see, enthalpy and wet-bulb temperatures are closely

related. So, by measuring the wet-bulb temperature of the air before and after it moves through a heat exchanger, we can get a good idea of how much heat is being transferred—as long as we know how much air is moving through the system, **Figure 35–30.** This will give a fairly accurate account of the performance of the heat exchanger.

Enthalpy is the total heat content of air and is made up of sensible- and latent-heat components. Just to recap, sensible heat is the term used to describe a heat transfer that can be measured with a thermometer, while a latent-heat transfer is one that represents a change in state of a substance with no change in temperature.

35.12   THE PSYCHROMETRIC CHART

All of the properties of air that have been discussed in this unit are compiled on the psychrometric chart shown in, **Figure 35–31(A).** The bottom scales on the chart represent the

Figure 35–29   The water dripping from the back of this window air conditioner is moisture that was collected from the room onto the cooling coil of the unit.

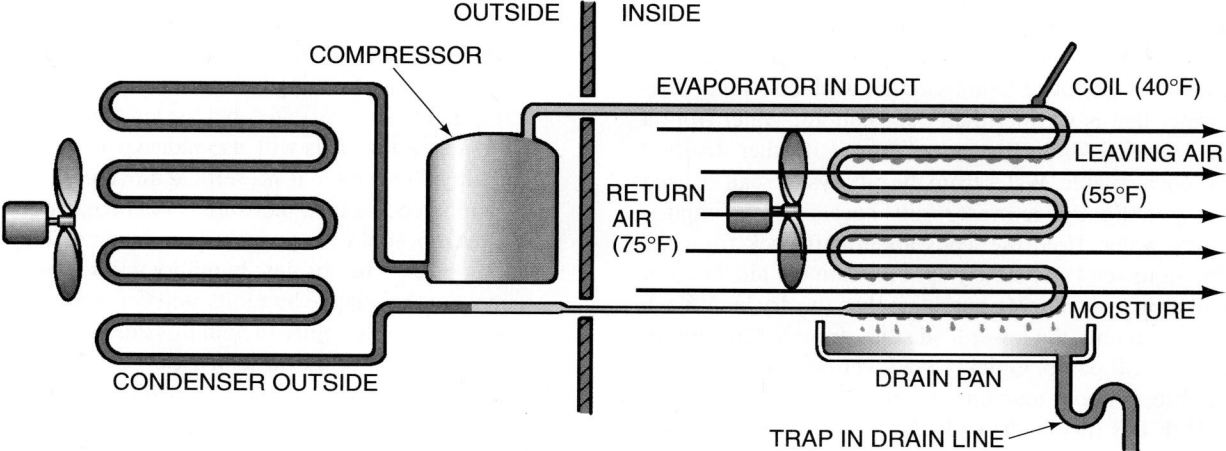

Figure 35–28   The cold surface of the evaporator coil causes moisture to condense from the air as the air passes over the surface of the coil.

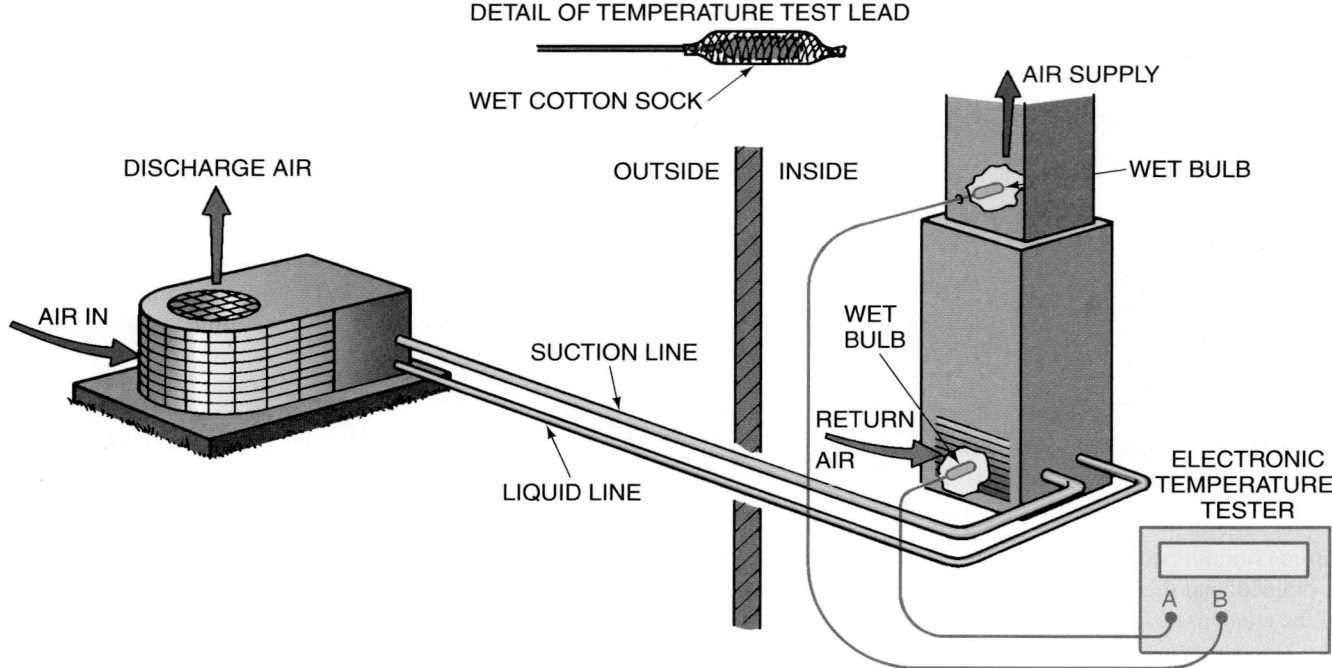

Figure 35–30   A wet-bulb reading can be taken on each side of a heat exchanger.

sensible-heat portion, while the scales on the vertical right-hand side of the chart represent the latent-heat portion. Since enthalpy is the total heat, the enthalpy scale, which is located on the curved top portion of the chart, is a combination of the latent-heat (vertical) and sensible-heat (horizontal) properties of the air. The chart looks complicated, but a clear plastic straightedge and a pencil will help you understand it. See **Figure 35–31(B)** through **Figure 35–31(G)** for some examples of plottings of the different conditions on a psychrometric chart.

If you know any two conditions previously mentioned, you can easily determine any of the other conditions. The easiest conditions to determine from room air are the wet-bulb and the dry-bulb temperatures. For example, you can take an electronic thermometer and make a wet-bulb thermometer if you do not have one. Take two leads and tape them together with one lead about 2 in. below the other one, **Figure 35–32.** A simple wick can be made from a piece of white cotton. Make sure that it does not have any perspiration on it. Wet the lower bulb (with the wick on it) with distilled water that is warmer than the room air. Water from a clean condensate drain line may be used if other distilled water is not available. Water from the city system can be used but may give slightly incorrect results because of the impurities in the water. Hold the leads about 3 ft back from the element on the end and slowly spin them in the air. The wet lead will drop to a colder temperature than the dry lead. Keep spinning them until the lower lead stops dropping in temperature but is still damp. Quickly read wet-bulb and dry-bulb temperatures without touching the bulbs. Suppose the reading is 75°F DB (dry bulb) and 62.5°F WB (wet bulb). Put your pencil point at this place on the psychrometric chart, **Figure 35–33,** and make a dot. Draw a light circle around it so

you can find the dot again. The following information can be concluded from this plot:

1. Dry-bulb temperature   **75°F**
2. Wet-bulb temperature   **62.5°F**
3. Dew point temperature   **55.5°F**
4. Total heat content of 1 lb of air   **28.2 Btu**
5. Moisture content of 1 lb of air   **65 gr**
6. Relative humidity   **50%**
7. Specific volume of air   **13.7 ft³/lb**

35.13 PLOTTING ON THE PSYCHROMETRIC CHART

The condition of air can be plotted on the psychrometric chart as it is being conditioned. The following examples show how different applications of air conditioning are plotted:

■ Air is heated. Movement through the heating equipment can be followed as a sensible-heat direction on the chart, **Figure 35–34(A)**. It can be seen from this diagram that the arrow is moving in a horizontal direction from left to right. This indicates that the temperature of the air is increasing. This will also indicate an increase in enthalpy. Since there is no vertical movement on the system's process line, moisture is not being added or removed from the air, so there is no latent activity taking place. The absolute humidity will remain the same, but the relative humidity will decrease. This is because the air's ability to hold moisture increases as the temperature of the air increases.

■ Air is cooled. There is no moisture removal. This shows a sensible-heat direction on the chart, **Figure 35–34(B)**. Once again there is no latent-heat activity in this process. The temperature of the air is dropping, so the

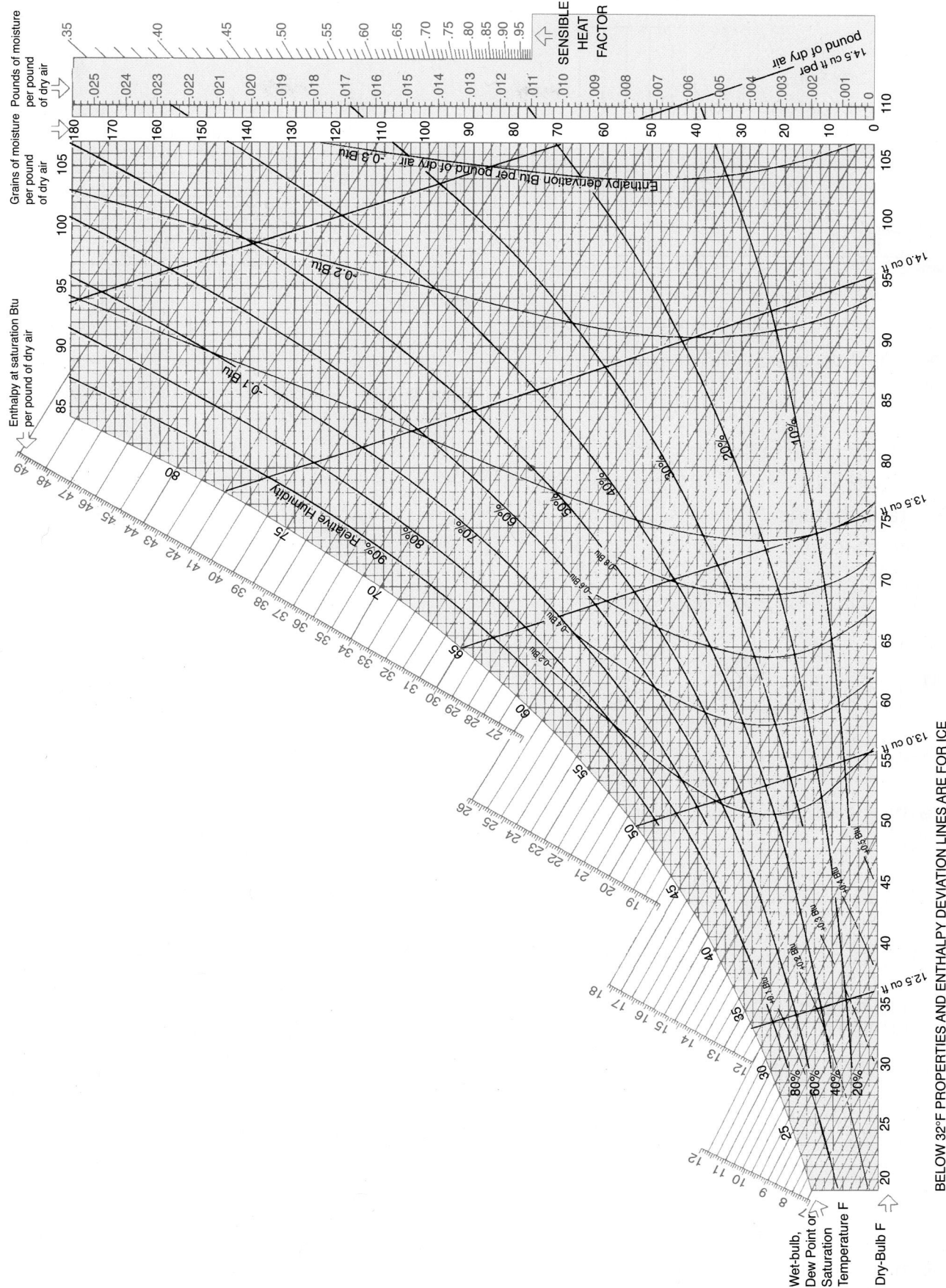

BELOW 32°F PROPERTIES AND ENTHALPY DEVIATION LINES ARE FOR ICE

Figure 35–31(A)   A psychrometric chart. *Reproduced courtesy of Carrier Corporation*

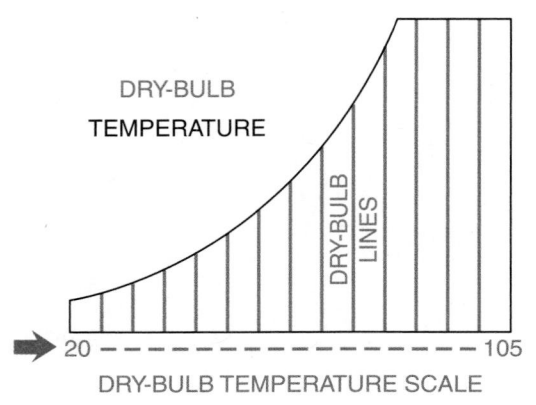

Figure 35–31(B)   A skeleton chart showing the dry-bulb temperature.

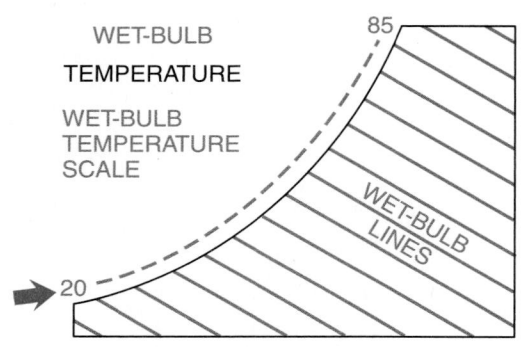

Figure 35–31(C)   A skeleton chart showing the wet-bulb temperature.

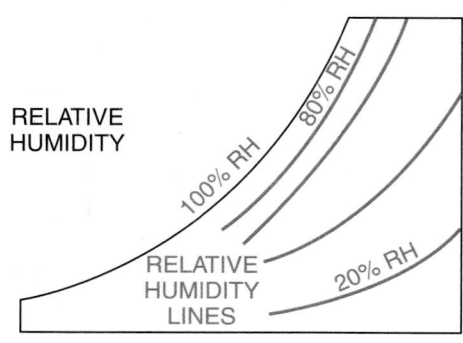

Figure 35–31(D)   A skeleton chart showing the relative humidity lines.

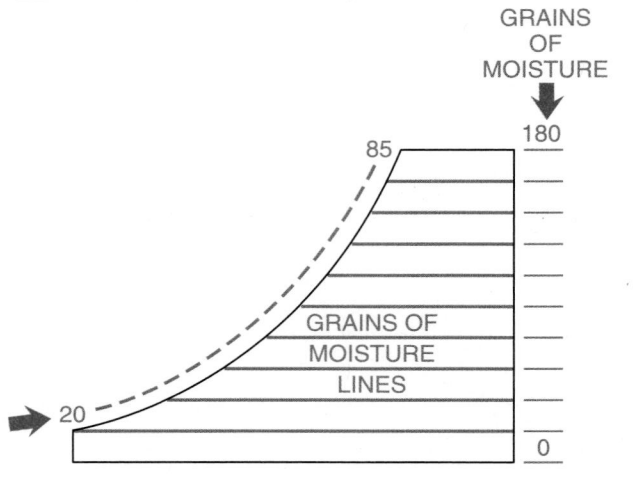

Figure 35–31(E)   A skeleton chart showing the moisture content of air expressed in grains per pound of air.

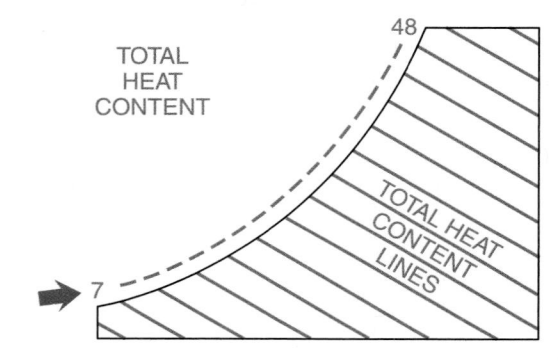

Figure 35–31(F)   A skeleton chart showing the total heat content of the air in Btu/lb. These lines are almost parallel to the wet-bulb lines.

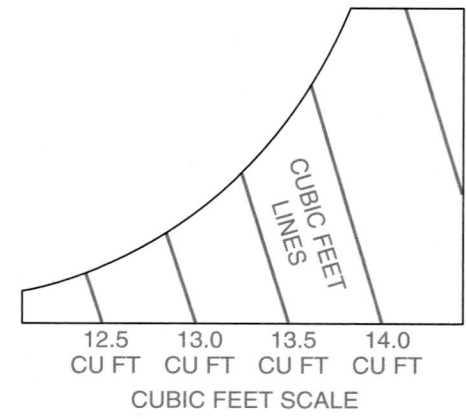

Figure 35–31(G)   A skeleton chart showing the specific volume of air at different conditions.

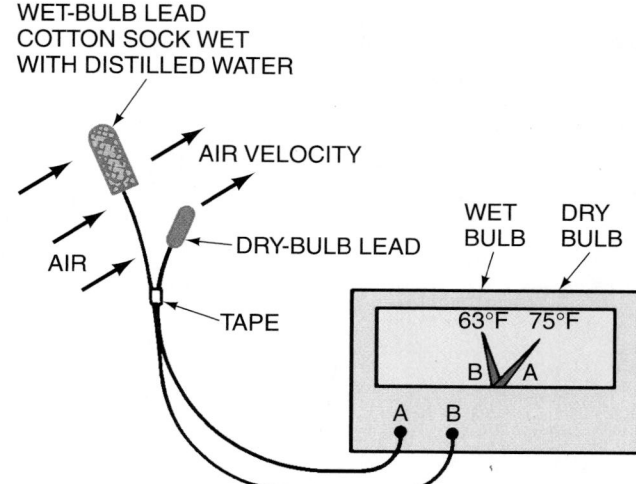

WET-BULB LEAD
COTTON SOCK WET
WITH DISTILLED WATER

AIR VELOCITY

DRY-BULB LEAD

AIR

TAPE

WET
BULB   DRY
BULB

63°F   75°F

B   A

A   B

Figure 35–32   How to make a simple sling psychrometer.

enthalpy of the air is dropping. In addition, the relative humidity of the air is increasing because, as air is cooled, its ability to hold moisture decreases.

■ Air is humidified. No heat is added or removed. An increase in moisture content and dew point temperature shows, **Figure 35–34(C)**. This is not typical.

■ Air is dehumidified. No heat is added or removed. A decrease in moisture content and dew point temperature shows, **Figure 35–34(D)**. This is not typical.

■ Air is cooled and humidified using an evaporative cooler. These are popular in hot, dry climates, **Figure 35–34(E)**.

When air enters air-conditioning equipment, it can be plotted on a chart. **Figure 35–35** shows a chart indicating that from the reference point in the middle, the air may be conditioned to heat, cool, humidify, or dehumidify. Some apparatus will both add heat and moisture, or cool and remove moisture. The following examples will show what happens in the most common heating and cooling systems.

■ The most common winter application is to heat and humidify. This will show both a rise in temperature and an increase in moisture and dew point temperature, **Figure 35–36**.

■ The most common summer application is to cool and dehumidify air. A decrease in temperature, moisture content, and dew point will occur, **Figure 35–37**.

It is important to notice that any change in heat content or moisture content of air will cause a change in the wet-bulb reading and, therefore, a change in the total heat content.

35.14 FRESH AIR, INFILTRATION, AND VENTILATION

The air that surrounds us has to be maintained at the correct conditions for us to be comfortable. The air in our homes is treated by heating it, cooling it, dehumidifying it, humidifying it, and cleaning it so that our bodies will give off the

correct amount of heat for comfort. A small amount of air is induced from the outside into the conditioner to keep the air from becoming oxygen starved and stagnant. This is called fresh air intake or *ventilation*. If a system has no ventilation, it is relying on air infiltrating the structure around doors and windows.

⊕ Modern energy-efficient homes can be built so tightly that infiltration does not provide enough fresh air. Recent studies indicate that indoor pollution in homes and buildings has increased as a result of an increase in energy conservation in heating and air-conditioning systems and the structures they reside in. People are more energy conscious now than in the past, and modern buildings are constructed to allow much less air infiltration from the outside. ⊕ A typical homeowner may take the following measures to prevent outside air from entering the structure:

■ Install storm windows
■ Install storm doors
■ Caulk around windows and doors
■ Install dampers on exhaust fans and dryer vents

All of these will reduce the amount of outside air that leaks into the structure and improve energy costs. This may not be all good because of the following indoor pollution sources:

■ Chemicals in new carpets, drapes, and upholstered furniture
■ Cooking odors
■ Vapors from cleaning chemicals
■ Bathroom odors
■ Vapors from freshly painted rooms
■ Vapors from aerosol cans, hair sprays, and room deodorizers
■ Vapors from particle-building-board epoxy resins
■ Pets and their upkeep
■ Radon gas leaking into the structure from the soil

These indoor pollutants may be diluted with outdoor air in the form of ventilation. Infiltration is the term used for random air that leaks into a structure. Ventilation is planned, fresh air added to the structure. When air is introduced into the system before the heating or air-conditioning system, it is called ventilation. This may be accomplished with a duct from the outside to the return air side of the equipment, **Figure 35–38**.

There is some discussion as to how much air should be introduced, but it is generally agreed that at least a 0.25 air change per hour for the entire structure is desirable. This means that 25% of the indoor air is pushed out by inducing air into the system. For example, suppose a 2000-ft² home with an 8-ft ceiling needs ventilation because the home is very tight. How many cubic feet of air per minute must be introduced to change 25% of the air per hour?

$$2000 \text{ ft}^2 \times 8\text{-ft ceiling} \times 0.25 = 4000 \text{ ft}^3/\text{h}$$

$$\frac{4000 \text{ ft}^3/\text{h}}{60 \text{ min/h}} = 67 \text{ cfm}$$

This adds a considerable load to the equipment. For example, suppose the house is located in Atlanta, Georgia, where the outdoor design temperature is 17°F in the winter. (See **Figure 36–13**, Design Dry-Bulb column 99%.) If the

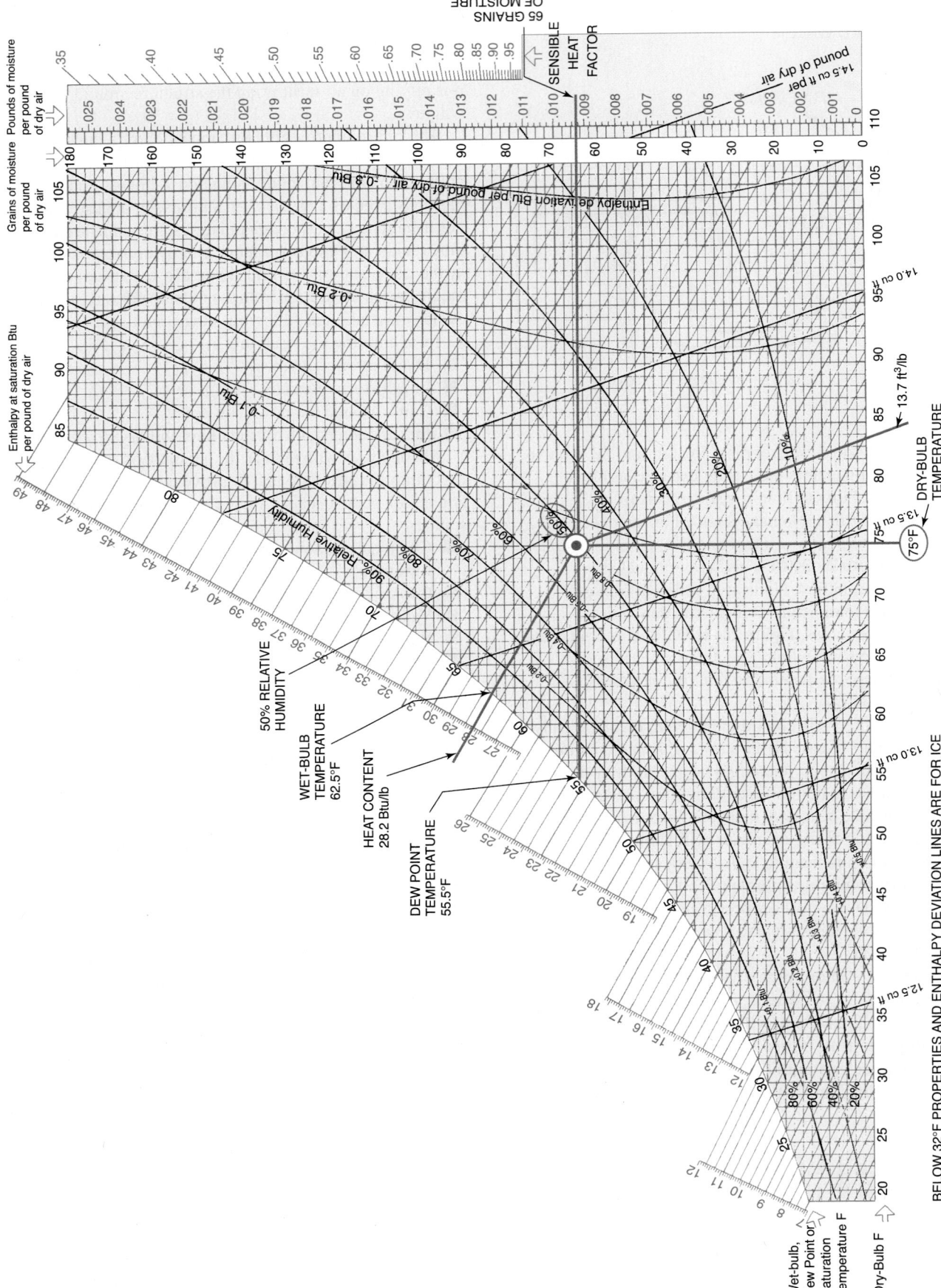

Figure 35–33 Psychrometric chart plotting example. *Adapted from Carrier Corporation Psychrometric Chart*

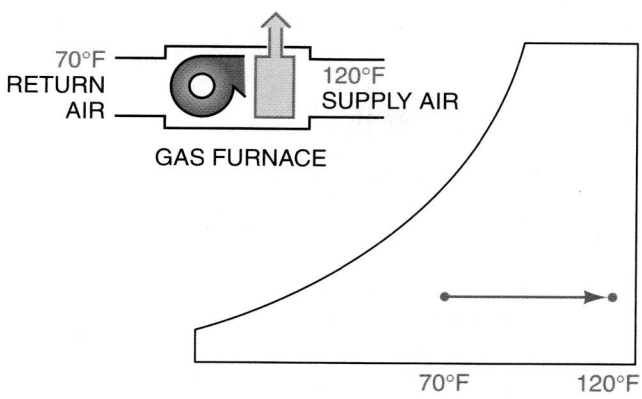

Figure 35–34(A)   Air passing through a sensible-heat exchange furnace.

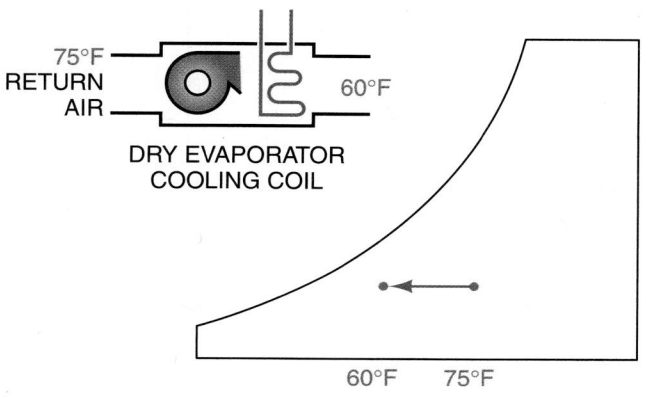

Figure 35–34(B)   Air is cooled with a dry evaporator coil operating above the dew point temperature of the air. No moisture is removed. This is not a typical situation.

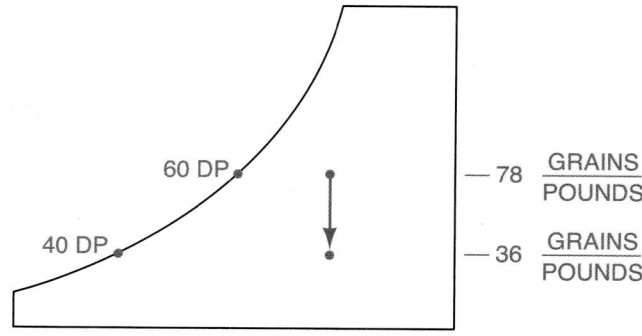

Figure 35–34(D)   Moisture is removed from the air. This is not a typical application and is shown as an example only.

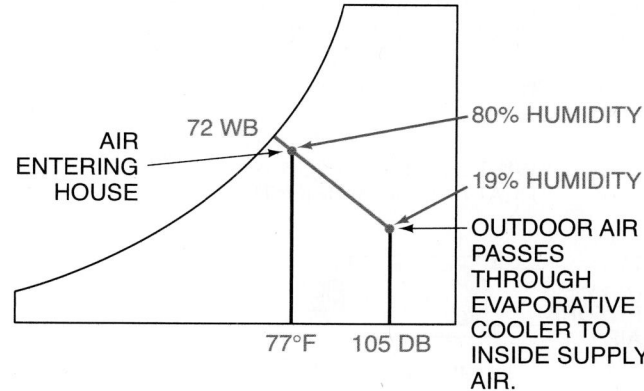

Figure 35–34(E)   Hot, dry air passes through the water circuit of the evaporative cooler. Heat is given up to the cooler water, and the air entering the house is cooled and humidified; water evaporates to cool the air.

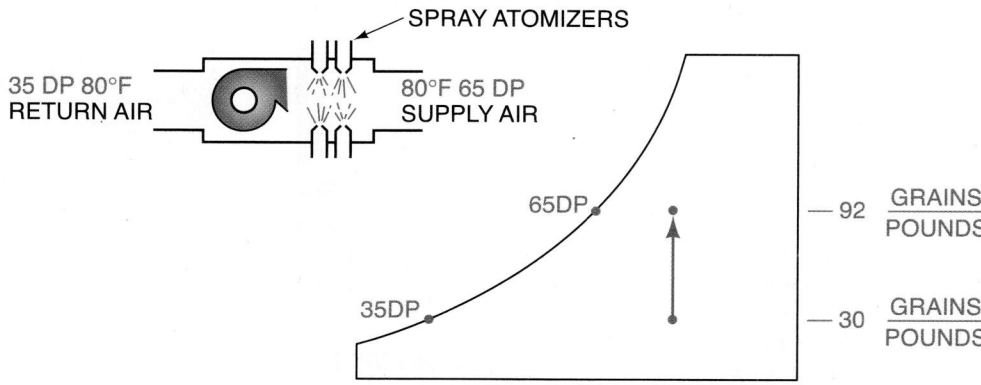

Figure 35–34(C)   Spray atomizers are used to add moisture to the air. The dew point temperature and moisture content both increase.

home is to be maintained at 70°F indoors when the outdoor temperature is 17°F, there is a 53°F temperature difference (TD). The load on the heating equipment due to ventilation alone will be:

$$Q_s = 1.08 \times cfm \times TD$$
$$Q_s = 1.08 \times 67 \times 53$$
$$Q_s = 3835 \text{ Btu/h}$$

This formula is explained in detail in Unit 30, "Electric Heat."

Outside temperature at 17°F is only for 1% of the year. The fresh air will be warmer than 17°F the other 99% of the year. This problem depicts the worst possible case. Many system designers will use the 97.5% column for design of the system. Smaller equipment may be selected because of the

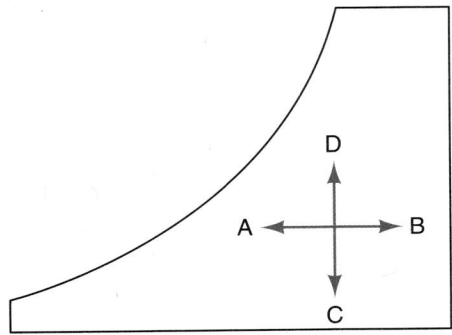

WHEN AIR IS CONDITIONED AND THE
PLOT MOVES IN THE DIRECTION OF:

(A) SENSIBLE HEAT IS REMOVED.
(B) SENSIBLE HEAT IS ADDED.
(C) LATENT HEAT IS REMOVED, MOISTURE REMOVED.
(D) LATENT HEAT IS ADDED, MOISTURE ADDED.

Figure 35–35   A summation of sensible and latent heat.

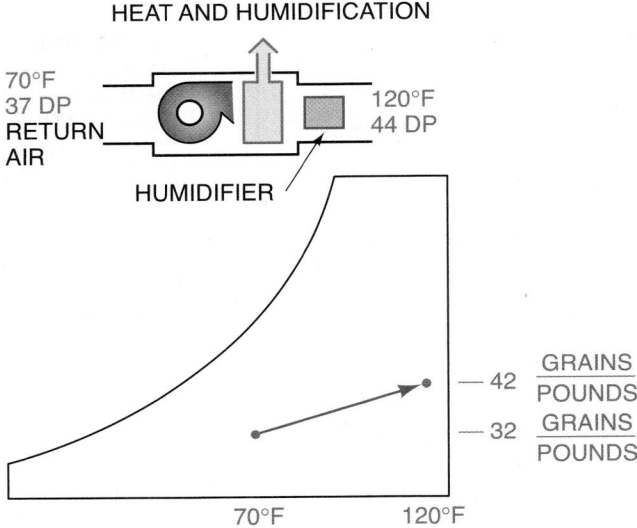

Figure 35–36   Sensible heat raises the temperature of the air from 70°F to 120°F. Moisture is evaporated, and latent heat is added to the air.

warmer design temperatures. This may be taking a chance in the event of a cold winter.

The summertime calculation is made in much the same manner, except a different formula is used. Sensible and latent heat must be considered. In the summer, the design temperatures are 95°F dry-bulb and 74°F wet-bulb. The total heat formula may be used for this calculation.

$$Q_t = 4.5 \times \text{cfm} \times \Delta h$$

where Q_t = total heat
    4.5 = a constant used to change pounds of air to cfm
    Δh = difference in total heat (enthalpy) indoors and outdoors

To solve the problem of fresh air, plot the indoor air and the outdoor air on the psychrometric chart.

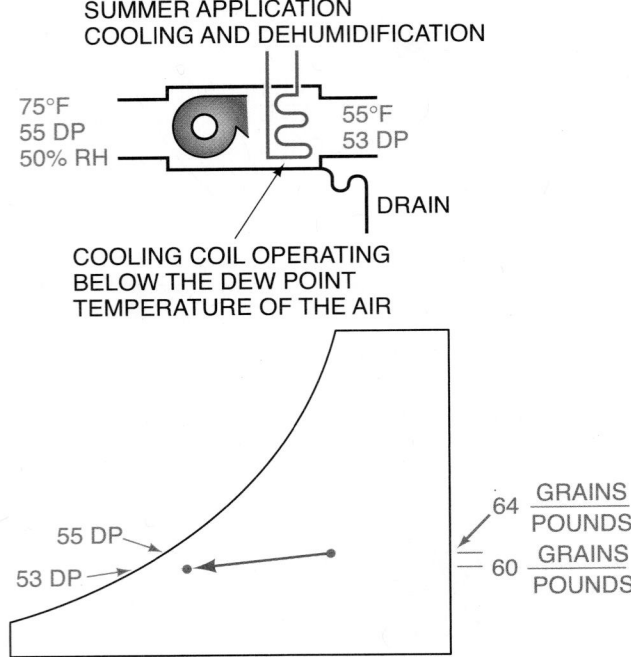

Figure 35–37   Removal of sensible heat cools the air. Removal of latent heat removes moisture from the air.

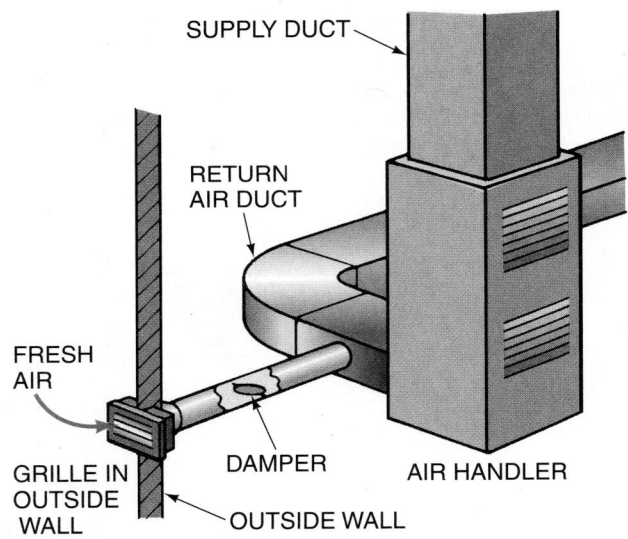

Figure 35–38   Fresh air is drawn into the return air duct to improve air quality inside the house.

Outdoors: 95°F dry-bulb and 74°F wet-bulb = 37.68 Btu/lb
Indoors: 75°F dry-bulb and 50% relative humidity indoors = 28.2 Btu/lb, **Figure 35–39**.

Total heat difference = 37.68 − 28.2 = 9.48 Btu/lb

$Q_t = 4.5 \times \text{cfm} \times$ total heat difference
$Q_t = 4.5 \times 67 \times 9.48$
$Q_t = 2858$ Btu/h total heat added due to ventilation

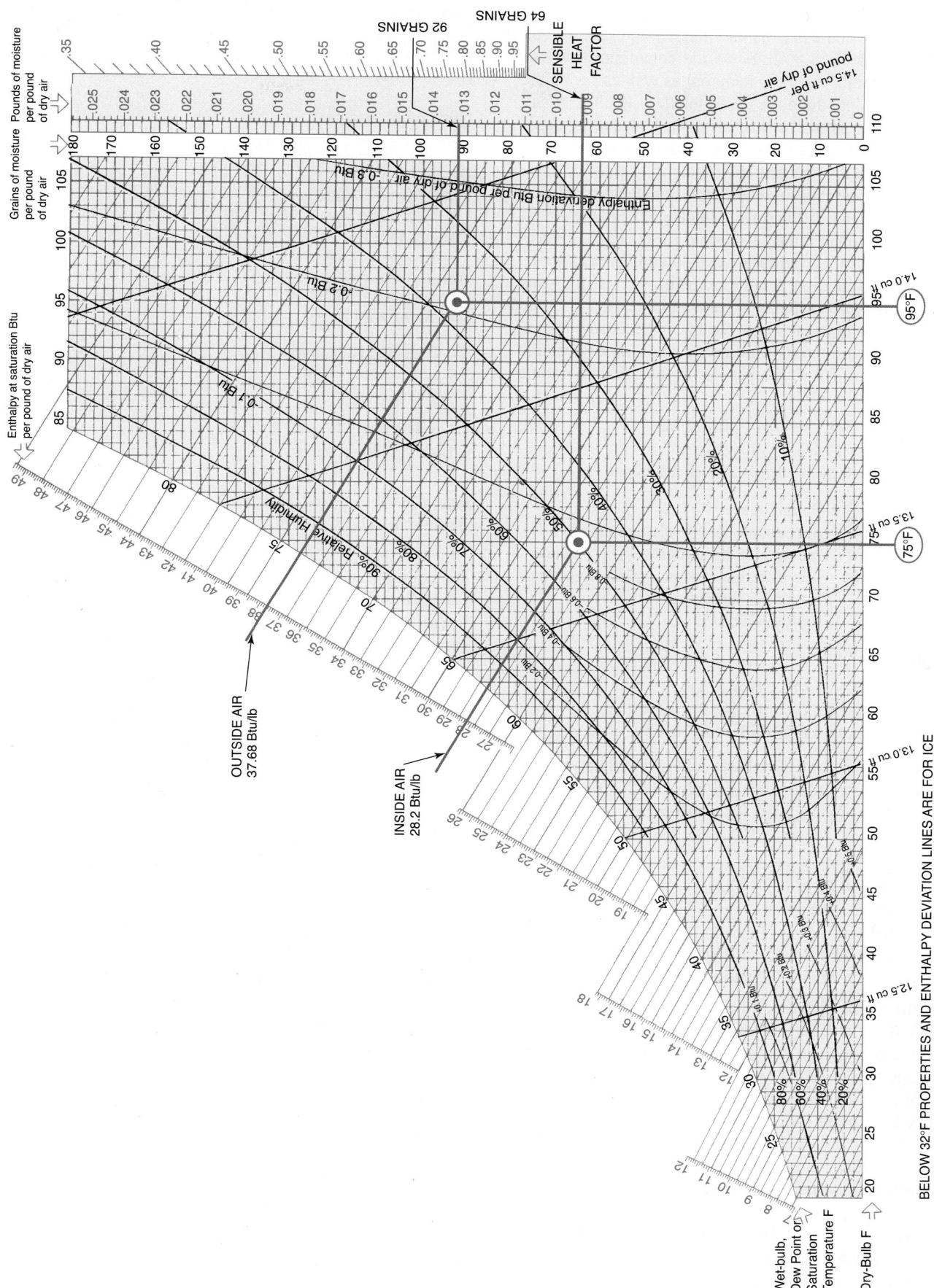

Figure 35–39 Graphic representation of ventilation example.

NOTE: The 4.5 that is used in the total heat calculation is only an estimate and assumes that the average specific volume of the air is 13.33 ft³/lb. Since the conversion involves changing time units (hours to minutes) as well as taking the specific volume into account, the 60 min/h ÷ 13.33 ft³/lb calculation will yield a result of 4.5. However, if the average specific volume differs greatly from the 13.33 value, there will be small calculation errors down the road.

System designers would require this total heat calculation to be broken down into the sensible-heat gain and the latent-heat gain. Equipment must be selected for the correct sensible- and latent-heat capacities or space humidity will not be correct. This calculation is done by using the sensible-heat and the latent-heat formulas as separate calculations. You will not arrive at the exact same total heat, because you cannot see the lines on the psychrometric chart closely enough for total accuracy. (In addition, refer to the calculation note above.)

$$Q_s = 1.08 \times \text{cfm} \times \text{TD}$$
plus
$$Q_l = 0.68 \times \text{cfm} \times \text{grains difference}$$

where Q_l = latent heat
$\quad$ 0.68 = a constant used to change cfm to pounds of air and grains per pound
grains difference = the difference in the grains per pound of air for the indoor air and the outdoor air

Example:

$$Q_s = 1.08 \times \text{cfm} \times \text{TD}$$
$$Q_s = 1.08 \times 67 \times 19$$
$$Q_s = 1375 \text{ Btu/h sensible heat}$$

From the psychrometric chart points plotted earlier, you will find that the outdoor air contains 92.8 gr/lb and the indoor air contains 64 gr/lb for a difference of 28.8 gr/lb of air.

$$Q_l = 0.68 \times \text{cfm} \times \text{grains difference}$$
$$Q_l = 0.68 \times 67 \times 28.8$$
$$Q_l = 1312$$
$$\text{total heat} = Q_s + Q_l$$
$$\text{total heat} = 1375 + 1312$$
$$\text{total heat} = 2687 \text{ Btu/h}$$

Notice the small difference between the 2858 Btu/h and the 2687 Btu/h results. This difference is not large enough to cause a problem in the evaluation of the total system performance.

Office buildings have the same pollution problems, except they have more sources of pollution because there are more people and different and varied activities. Office buildings have a tendency to be remodeled more often leading to more construction-type pollution. Copying machines that use liquid copy methods give off vapors. Also, there are more people per square foot in office buildings.

The national code requirements for fresh air in buildings where the public works is based on the number of people in

RECOMMENDED OUTDOOR AIR VENTILATING RATES Abbreviated from ASHRAE Standard 62-1989	
	cfm or see footnote
Dining Room	20
Bars & cocktail lounges	30
Hotel conference rooms	20
Office spaces	20
Office conference rooms	35
Retail stores	.02 to .03(a)
Beauty shops	25
Ballrooms & discos	25
Spectator areas	15
Theater auditoriums	15
Transportation waiting rooms	15
Classrooms	15
Hospital patient rooms	25
Residences	.35(b)
Smoking lounges	60

a   cfm per square feet of space
b   air changes per hour

Figure 35–40   Fresh air requirements for some typical applications. *Reprinted with permission from ASHRAE, Inc.—Standard 62-1989*

the building and the type of use. Different buildings may have different indoor pollution rates because of the activity in the building. For example, buildings that have a lot of copying machines or blueprint machines may have more indoor pollution than ones with only electric processors, such as computers. Department stores have different requirements than restaurants do. **Figure 35–40** shows some of the fresh air requirements for different applications recommended by ASHRAE. These figures are constantly changing as new research is done. These numbers are for comparison only.

The design engineer is responsible for choosing the correct fresh air makeup for a building, and the service technician is responsible for regulating the airflow in the field. The technician may be given a set of specifications for a building and told to regulate the outdoor air dampers to induce the correct amount of outdoor air into the building.

The mixed air condition may be determined by calculation for the purpose of setting the outside air dampers. The condition may be plotted on the psychrometric chart for ease of understanding. The dampers may then be adjusted for the correct conditions. For example, suppose a building with a return air condition of 73°F dry-bulb and 60°F wet-bulb is taking in and mixing air from the outside of 90°F dry-bulb and 75°F wet-bulb. When these two conditions are plotted on the psychrometric chart and a line is drawn between them, the condition of the mixture air will fall on the line, **Figure 35–41**. If the mixture were half outdoor air and half indoor air, the condition of the mixture air would be midway between points A and B.

The point on the line between A and B may be calculated for any air mixture if the percentage of either air, outdoor or indoor, is known. For example, when the point for 25% outdoor

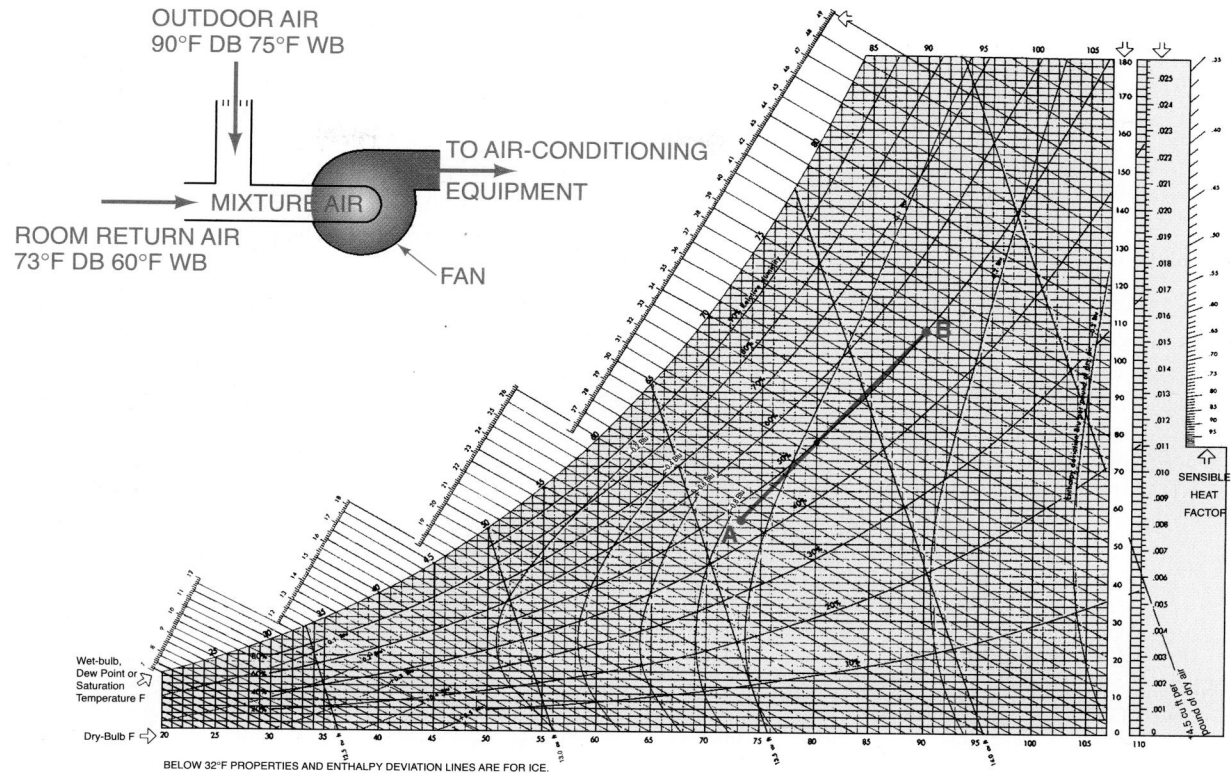

OUTDOOR AIR
90°F DB 75°F WB

TO AIR-CONDITIONING
EQUIPMENT

MIXTURE AIR

ROOM RETURN AIR
73°F DB 60°F WB

FAN

Wet-bulb,
Dew Point or
Saturation
Temperature F

Dry-Bulb F

BELOW 32°F PROPERTIES AND ENTHALPY DEVIATION LINES ARE FOR ICE.

Figure 35–41   Point A is the condition of the indoor air, and point B is the condition of the outdoor air. When the two are mixed, the mixed air condition will fall somewhere on the line that connects point A to point B. *Adapted from Carrier Corporation Psychrometric Chart*

air and 75% indoor air mixture was calculated, the following calculation was used to find the dry-bulb point on the line. When the dry-bulb point is known and it falls on the line, you have a coordinate to find any other information for this air condition.

25% outdoor air: 0.25 × 90°F = 22.50
75% indoor air: 0.75 × 72°F = 54
The mixture (dry-bulb) = 76.5

When you look on the chart in **Figure 35–42,** you will find the point (C). Note, the point always falls closest to the air temperature with the most percentage of air.

An actual problem may be something like the following:

Building Air-Conditioning Specifications

■ Total building air = 50,000 cfm
■ Recirculated air = 37,500 cfm, 75% return air
■ Makeup air = 12,500 cfm, 25% return air
■ Indoor design temperature = 75°F dry-bulb, 50% relative humidity (62.5 wet-bulb)
■ Outdoor design temperature (Atlanta, Georgia) = 94°F dry-bulb, 74°F wet-bulb

A technician arrives on the job to check the outdoor air percentage to find that the outdoor temperature is not at the design temperature. It is 93°F dry-bulb/75°F wet-bulb and the indoor conditions are 73°F dry-bulb/59°F wet-bulb. What should the mixture air temperature be?

A calculation for the mixture air is made. The actual conditions are plotted in **Figure 35–43**.

93°F × 0.25 = 23.25
73°F × 0.75 = 54.75
The mixture (dry-bulb) = 78°F

This calculation tells you that 25% of the mixed air dry-bulb temperature comes from the outside air, which is at 95°F. That temperature representation is 23.25°F, and 75% of the representation is from the 73°F inside air dry-bulb temperature and is 54.75°F. Add them together and you get the mixed air dry-bulb temperature of 78°F (23.25 + 54.75 = 78). The technician can use dry-bulb temperatures alone to calculate air mixture percentages.

The technician would now check the mixture air temperature. If it is too high, the outdoor air dampers would be slightly closed until the mixture air temperature stabilizes at the correct mixture temperature. If the mixture temperature is too low, the outdoor dampers may be opened slightly until the mixture air temperature is stable and correct. NOTE: The outdoor and indoor conditions should be rechecked after adjustment because of weather changes and indoor condition changes. These calculations and proper damper settings also depend on correct instrumentation while checking the air temperatures.

A technician may also use the psychrometric chart for field checking the capacity of a piece of air-conditioning equipment. For example, suppose the capacity of a 5-ton

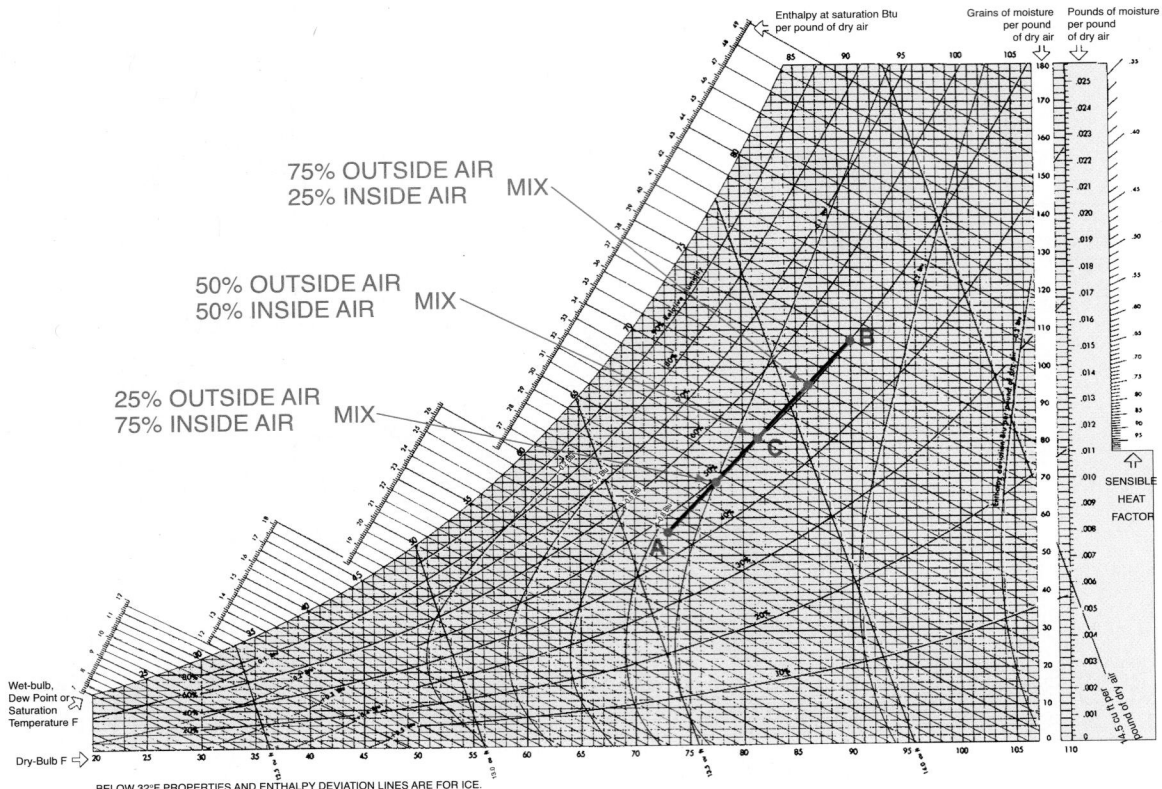

Figure 35–42   Mixture of different percentages of outside and inside air. *Adapted from Carrier Corporation Psychrometric Chart*

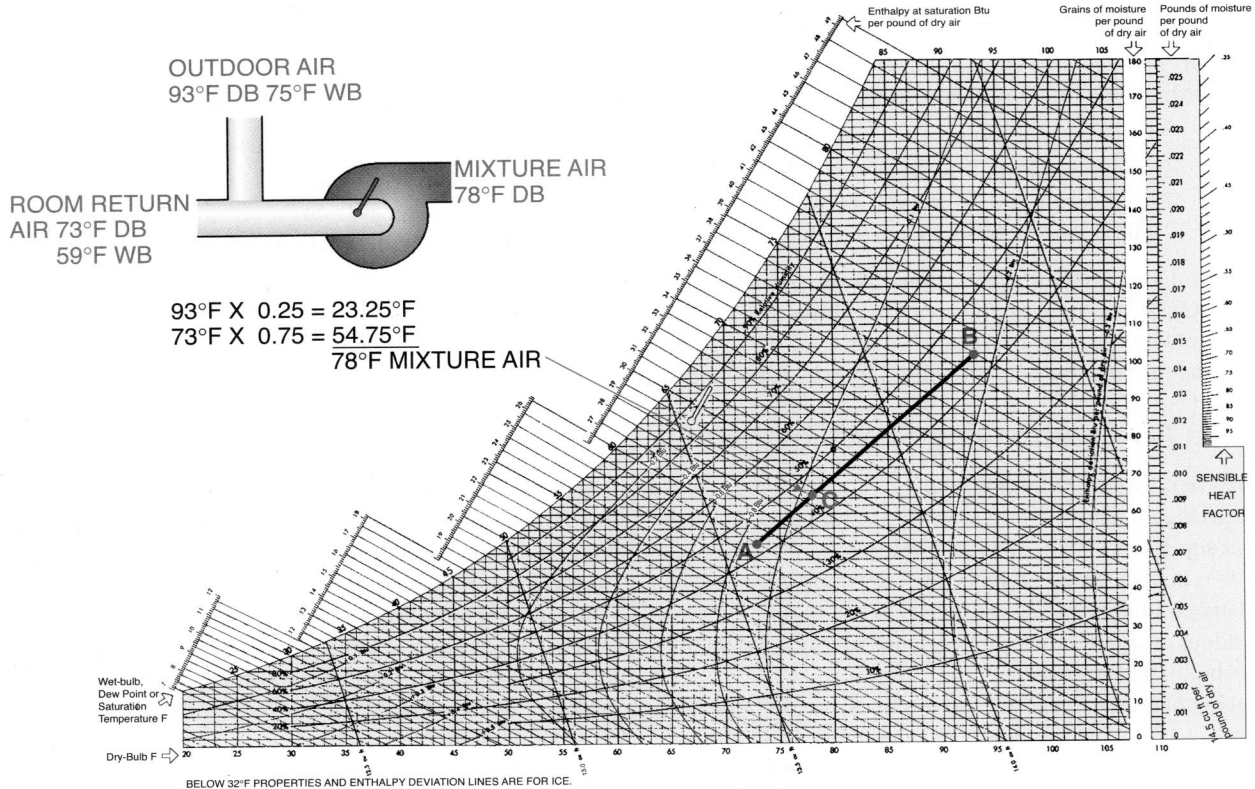

Figure 35–43   This practical problem of finding the mixed air condition is shown plotted on the chart. Point C is the mixed air point. *Adapted from Carrier Corporation Psychrometric Chart*

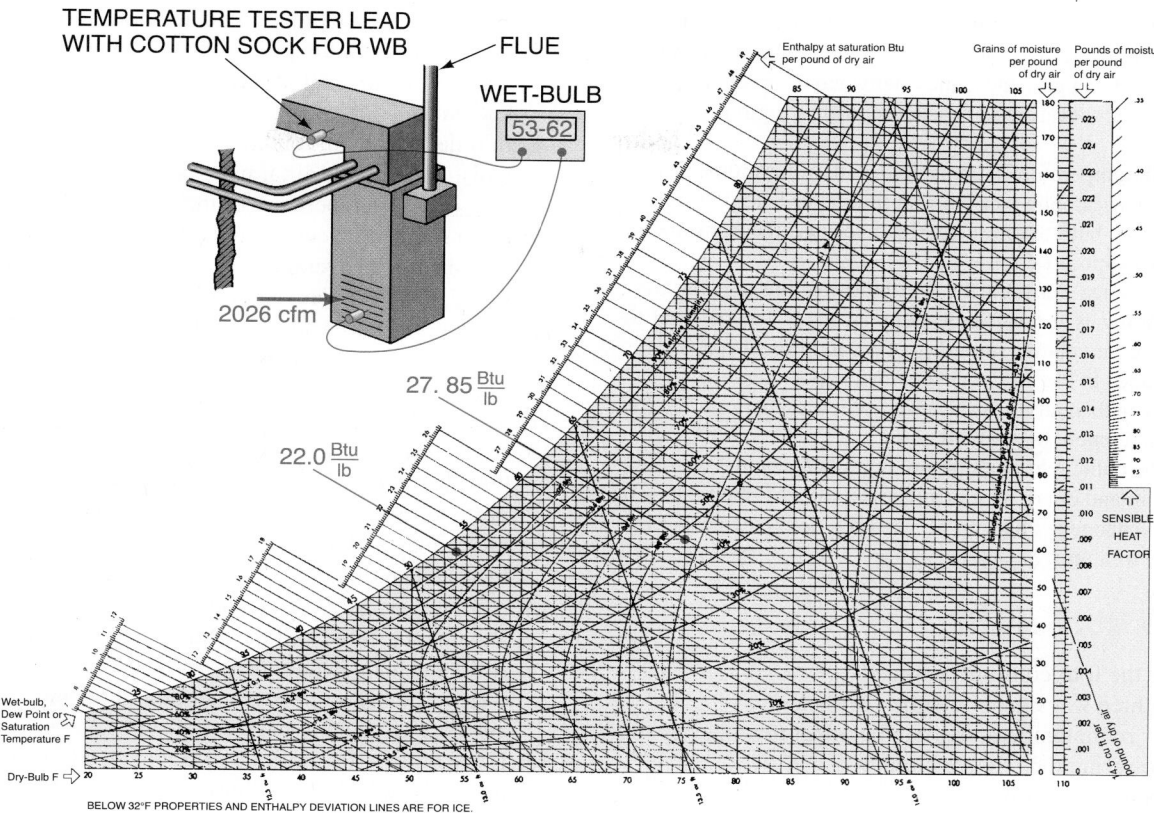

TEMPERATURE TESTER LEAD
WITH COTTON SOCK FOR WB

FLUE

WET-BULB

53-62

2026 cfm

27. 85 $\frac{Btu}{lb}$

22.0 $\frac{Btu}{lb}$

Enthalpy at saturation Btu
per pound of dry air

Grains of moisture
per pound
of dry air

Pounds of moisture
per pound of dry air

SENSIBLE
HEAT
FACTOR

Wet-bulb,
Dew Point or
Saturation
Temperature F

Dry-Bulb F

BELOW 32°F PROPERTIES AND ENTHALPY DEVIATION LINES ARE FOR ICE.

Figure 35–44   Capacity check for cooling mode. *Adapted from Carrier Corporation Psychrometric Chart*

(60,000 Btu/h) unit on a shoe store is in question and the unit uses gas as the heat source. The heat source is used to determine the cfm of the equipment.

A technician may want to perform the cfm part of this problem first thing in the morning because it will involve operating the heating system long enough to arrive at the cfm. The unit heat is turned on and an accurate temperature rise across the gas heat exchanger is taken at 69.5°F. NOTE: The technician was careful not to let radiant heat from the heat exchanger influence the thermometer lead.

The unit nameplate shows the unit to have 187,500 Btu/h input. The output would be 80% of the input, or 187,500 Btu/h × 0.80 = 150,000 Btu/h output to the airstream. The fan motor would add about 600 W × 3.431 Btu/W = 2048 Btu for a total of 152,048 Btu/h. NOTE: The fan is operated using the FAN ON position on the room thermostat. This ensures the same fan speed in heating and cooling.

Using the sensible-heat formula, the technician can find the cfm.

$$Q_s = 1.08 \times cfm \times TD$$

Solved for cfm:

$$cfm = \frac{Q_s}{1.08 \times TD}$$

$$cfm = \frac{152,048}{1.08 \times 69.5}$$

$$cfm = 2026 \text{ ft}^3/\text{min}$$

The technician then would turn the unit to cooling and check the wet-bulb temperature in the entering and leaving airstream. Suppose that the entering wet-bulb is 62°F and the leaving wet-bulb is 53°F, **Figure 35–44**.

Total heat at 62°F wet-bulb is 27.85
Total heat at 53°F wet-bulb is 22.00
Total heat difference = 5.85 Btu/lb of air

Using the total heat formula:

$$Q_t = 4.5 \times cfm \times \text{total heat difference}$$
$$Q_t = 4.5 \times 2026 \times 5.85$$
$$Q_t = 53,334 \text{ Btu/h total heat}$$

The unit has a capacity of 60,000 Btu/h so it is operating very close to capacity. This is about as close as a technician can expect to obtain.

SUMMARY

■ *Comfort* describes the delicate balance of feeling in relationship to our surroundings.
■ The body stores energy, wastes it, consumes it in work, or gives the heat off to the surroundings.
■ For the body to be comfortable, it has to be warmer than the surroundings, so it can give up excess heat to the surroundings.
■ Air contains 78% nitrogen, 21% oxygen, 1% other gases, and suspended water vapor.
■ The specific volume of air is the reciprocal of the density: 1/0.075 = 13.33 ft³/lb.

- The moisture content of air can vary the transfer of heat from the human body; therefore, different temperatures and moisture content can give the same relative comfort level.
- Dry-bulb temperature is registered with a regular thermometer.
- Wet-bulb temperature is registered with a thermometer that has a wet wick. The wet-bulb thermometer lead gets colder than the dry-bulb thermometer lead because the moisture on the wick evaporates.
- The difference between the wet-bulb reading and the dry-bulb reading is the wet-bulb depression. It can be used to determine the relative humidity of a conditioned space.
- Water vapor in the air creates its own vapor pressure.
- The wet-bulb reading on a psychrometric chart shows the total heat content of a pound of air.
- When the cubic feet of air per minute is known, the wet-bulb reading in and out of an air exchanger can give the total heat being exchanged. This can be used for field calculating the capacity of a unit.

REVIEW QUESTIONS

1. Name the four comfort factors.
2. State three ways the body gives off heat.
3. Lower room temperatures can be offset in winter by
   A. lowering the relative humidity.
   B. being very still.
   C. activity.
   D. raising the relative humidity.
4. Perspiration cools the body by _____.
5. Relative humidity is measured using a
   A. thermotyporiter.
   B. sling psychrometer.
   C. dry-bulb thermometer.
   D. volt-ohmmeter.
6. Name the two unknowns that are easiest to obtain for making plots on the psychrometric chart.
7. Air that contains all of the moisture that it can hold is known as _____.
8. In order for an air-conditioning coil to remove moisture from the air, it must be below the _____ _____ temperature of the air.
9. For a room to be comfortable, the following conditions are considered average: _____ °F dry-bulb and _____ % relative humidity.
10. With a dry-bulb reading of 70°F and a wet-bulb reading of 61°F, the dew point temperature is _____ °F.
11. Relative humidity in question 10 is _____ %.
12. A gas furnace has an output of 60,000 Btu/h. The return air temperature is 72°F, and the temperature out of the furnace is 130°F. Disregarding the heat the fan motor may add to the airstream, how much air is the furnace handling?
13. A house has 3500 ft² of floor space and a 9-ft ceiling. What is the cubic volume of the house?
14. If the fresh air requirements for the preceding house were to be 0.4 air changes per hour, how much fresh air must be taken in per minute?
15. If the outside fresh-air conditions were to be 93°F dry-bulb and 74°F wet-bulb, what would the total heat content of the air be?
16. If the air leaving the cooling coil were to be 55°F dry-bulb and 53°F wet-bulb, what would the total heat content of the air be?
17. Using the answers from questions 14, 15, and 16, what would be the total heat gain due to the fresh air for the house in question 13?
18. Describe why fresh air is important for a house.

UNIT 36
Refrigeration Applied to Air Conditioning

OBJECTIVES

After studying this unit, you should be able to

■ explain three ways in which heat transfers into a structure.

■ state two ways that air is conditioned for cooling.

■ explain refrigeration as applied to air conditioning.

■ describe an air-conditioning evaporator.

■ describe three types of air-conditioning compressors.

■ describe an air-conditioning condenser.

■ describe an air-conditioning metering device.

■ list different types of evaporator coils.

■ identify different types of condensers.

■ explain how "high efficiency" is accomplished.

■ describe package air-conditioning equipment.

■ describe split-system equipment.

SAFETY CHECKLIST

✔ Use caution when loading or unloading equipment from the dock or truck. Wear an approved back brace belt when lifting and use your legs, keeping your back straight and upright. It may be necessary to use a small crane. Trucks with lift gates may be used to lower the equipment to the ground.

✔ When installing equipment in attics and under buildings be alert for stinging insects such as spiders. Avoid being scratched or cut by exposed nails and other sharp objects in attics and in crawl spaces under buildings.

✔ Wear goggles and gloves when connecting charged line sets or gages because escaping refrigerant can cause serious frostbite.

36.1 REFRIGERATION

Air conditioning (cooling) is refrigeration applied to keeping the space temperature of a building cool during the hot summer months. The air-conditioning system (refrigeration) removes the heat that leaks into the structure from the outside and deposits it outside the structure where it came from. Some people living in the warmer climate sections of the country may never have air conditioning, but at times, they are probably uncomfortable. When the nights are warm (above 75°F) and the humidity is high, it is hard to be comfortable enough to rest well.

The basics of refrigeration and its components were discussed in Unit 3. Some of the material is covered again here so that it will be readily available. As you study this unit, you will find that many of the components used for refrigeration applied to air conditioning are different from the components applied to commercial refrigeration.

36.2 STRUCTURAL HEAT GAIN

Heat leaks into a structure by conduction, infiltration, and radiation (the sun's rays or solar load). The summer *solar load* on a structure is greater on the east and west sides because the sun shines for longer periods of time on these parts of the structure, **Figure 36–1.** If a building has an attic, the air space can be ventilated to help relieve the solar load on the ceiling, **Figure 36–2(A and B).** There are two kinds of attic

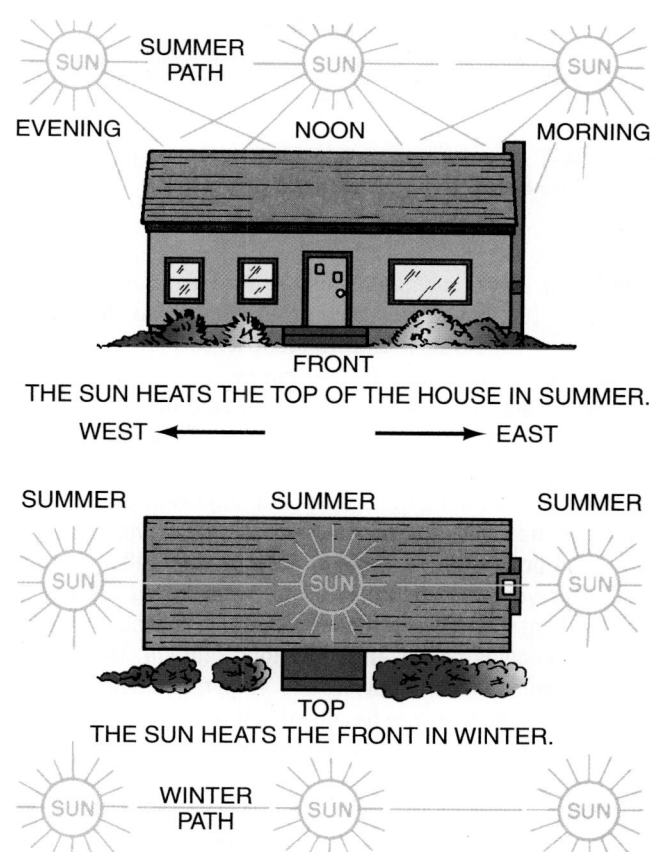

SUMMER PATH

EVENING   NOON   MORNING

FRONT

THE SUN HEATS THE TOP OF THE HOUSE IN SUMMER.

WEST ⟵   ⟶ EAST

SUMMER   SUMMER   SUMMER

TOP

THE SUN HEATS THE FRONT IN WINTER.

WINTER PATH

Figure 36–1   The solar load on a home.

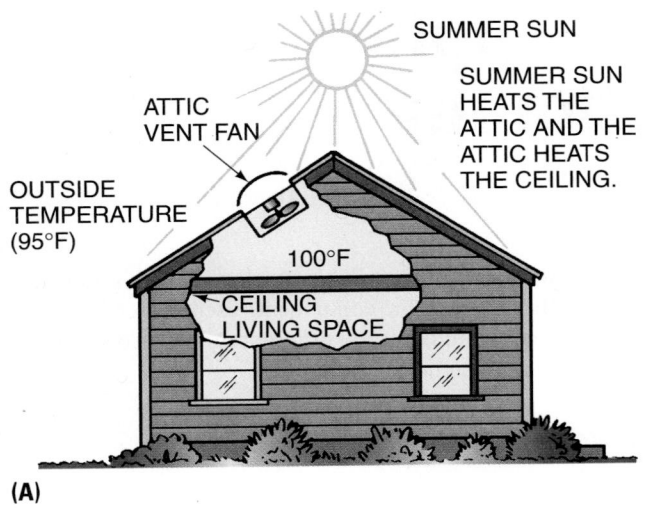

(A)

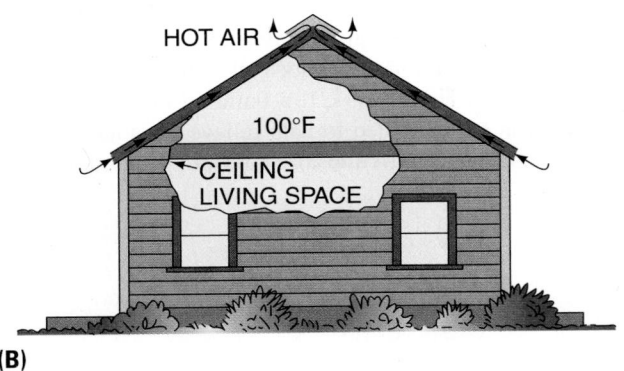

(B)

Figure 36–2 A ventilated attic helps keep the solar heat from the ceiling of the house.

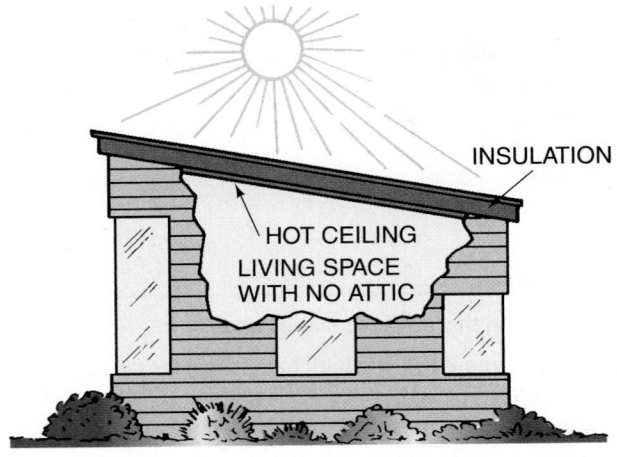

Figure 36–3 This house has no attic. The sun shines directly on the ceiling of the living space.

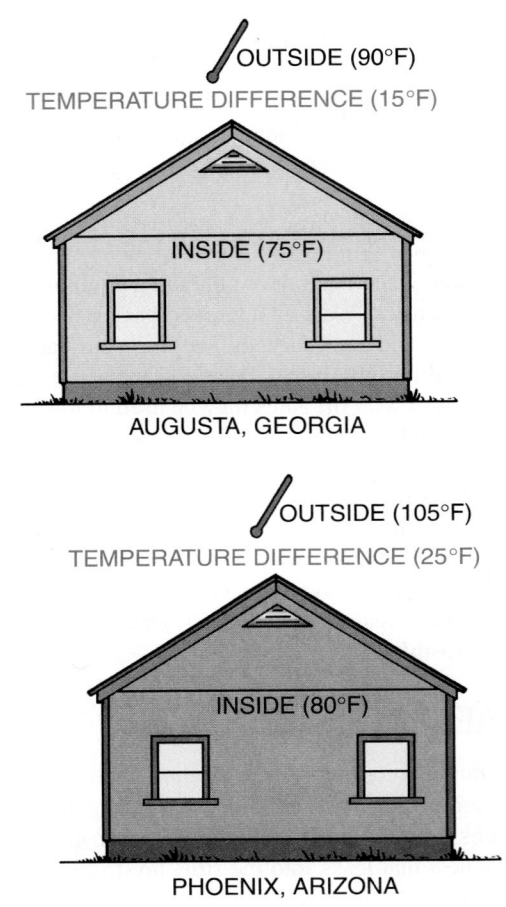

Figure 36–4 The difference between the inside and outside temperatures of a home in Augusta, Georgia, and a home in Phoenix, Arizona.

ventilation, power and natural. **Figure 36–2 (A)** shows a power ventilator. It is a thermostatically controlled fan that starts operating when the attic temperature reaches a predetermined temperature, typically 85°F. This fan must have a high-limit control to shut the fan off in case of an attic fire. The power ventilator would just feed air to a fire. 🌐**Figure 36–2(B)** shows a natural-draft ventilation system that is used more extensively. It is called a ridge vent. The ridge vent is cut along the top of the entire roof ridge. It has a vent that typically runs along the bottom of the roof overhang and is called a soffit vent. This vent system has air circulation along the entire underside of the roof surface. Both systems are said to keep the roof shingles cooler and to prolong roof life. The power vent consumes some power and requires service when broken.🌐 If the structure has no attic, it is at the mercy of the sun unless it is well insulated, **Figure 36–3.**

Conduction heat enters through walls, windows, and doors. The rate depends on the temperature difference between the inside and the outside of the house, **Figure 36–4.**

Some of the warm air that gets into the structure is infiltrated through the cracks around the windows and doors. Air also leaks in when the doors are opened for persons entering and leaving the building. This infiltrated air has different characteristics in different parts of the country. Using the example in **Figure 36–4,** the typical design condition in Phoenix is 105°F dry-bulb and 71°F wet-bulb. In Augusta, the air may be 90°F dry-bulb and 73°F wet-bulb. When the air leaks into the structure in Phoenix, it is cooled to the space temperature. This air contains a certain amount of humidity for each cubic foot that leaks in. In Augusta, there will normally be more humidity with the infiltration than in Phoenix.

The humidity difference in the two different locations must be accounted for in the conditioning equipment choice. Equipment selected for parts of the country that are humid will have more capacity to remove moisture from the conditioned space. For this reason, used equipment cannot just be readily moved from one locality to another.

36.3 EVAPORATIVE COOLING

🌐Air has been conditioned in more than one way to achieve comfort. In the climates where the humidity is low, a device called an *evaporative cooler*, **Figure 36–5,** has been used for years. This device uses fiber mounted in a frame with water slowly running down the fiber as the cooling media. Fresh air is drawn through the water-soaked fiber and cooled by evaporation to a point close to the wet-bulb temperature of the air. The air entering the structure is very humid but cooler than the dry-bulb temperature. For example, in Phoenix, Arizona, the design dry-bulb temperature in summer is 105°F. At the same time the dry-bulb is 105°F, the wet-bulb temperature may be 70°F. An evaporative cooler may lower the air temperature entering the room to 80°F dry-bulb. 80°F air is cool compared with 105°F, even if the humidity is high.🌐

These units use 100% outdoor air. There must be an outlet through which the air that enters the structure can leave the structure. The conditioned air will have a tendency to move toward the outlet used to exhaust the air from the structure. Open a window on the far side of the structure from where the air enters to allow air out, otherwise, the structure will pressurize with air that is being pushed in.

36.4 REFRIGERATED COOLING OR AIR CONDITIONING

Refrigerated air conditioning is similar to commercial refrigeration because the same components are used to cool the air: (1) the evaporator, (2) the compressor, (3) the condenser, and (4) the metering device. These components are assembled in several ways to accomplish the same goal, refrigerated air to cool space. Review Unit 3, "Refrigeration and Refrigerants," if you are not familiar with the basics of refrigeration.

Package Air Conditioning

The four components are assembled into two basic types of equipment for air-conditioning purposes: package equipment and split-system equipment. With *package equipment* all of the components are built into one cabinet. It is also called *self-contained* equipment, **Figure 36–6.** Air is ducted to and from the equipment. Package equipment may be located beside the structure or on top of it. In some instances the heating equipment is built into the same cabinet.

Split-System Air Conditioning

In *split-system* air conditioning the condenser is located outside, remote from the evaporator, and uses interconnecting refrigerant lines. The evaporator may be located in the attic, a crawl space, or a closet for upflow or downflow applications. The fan to blow the air across the evaporator may be included in the heating equipment, or a separate fan may be used for the air-conditioning system, **Figure 36–7.**

36.5 THE EVAPORATOR

The *evaporator* is the component that absorbs heat into the refrigeration system. It is a refrigeration coil made of aluminum or copper with aluminum fins on either type attached to the coil to give it more surface area for better heat exchange. The

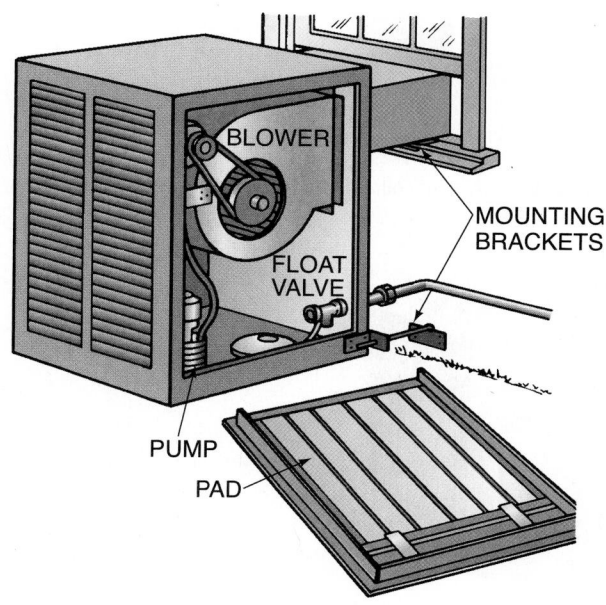

Figure 36–5 🌐An evaporative cooler.🌐

Figure 36–6 🌐A package air conditioner.🌐 *Courtesy Heil-Quaker Corporation*

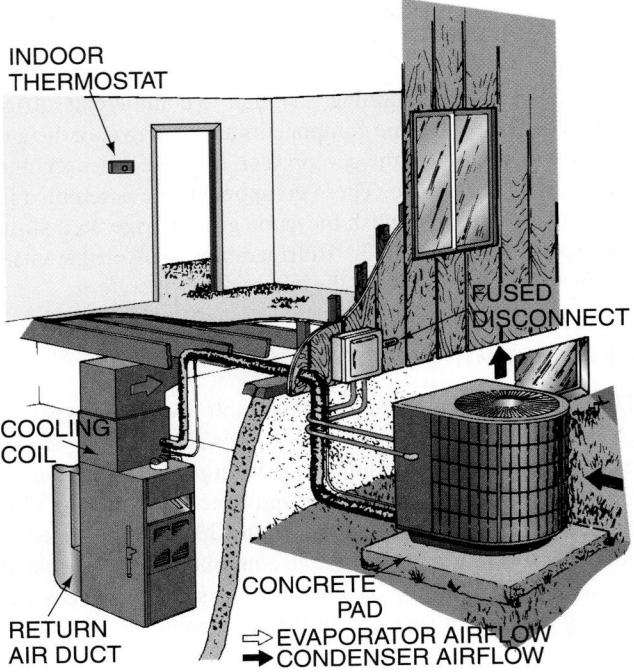

Figure 36–7   A split air-conditioning system. *Courtesy Climate Control*

evaporator coil has several designs for airflow through the coil and draining the condensate water from the coil, depending on the installation. The different designs are known as the A coil, the *slant* coil, and the *H* coil.

The A Coil

The *A coil* is used for upflow, downflow, and horizontal flow applications. It consists of two coils with their circuits side by side and spread apart at the bottom in the shape of the letter A, **Figure 36–8.** When used for upflow or downflow, the

condensate pan is at the bottom of the A pattern. When used for horizontal flow, a pan is placed at the bottom of the coil, and the coil is turned on its side. The airflow through an A coil is through the core of the coil. It cannot be from side to side with the two coils in the series. When horizontal airflow is needed, slant or H coils may be more desirable.

The Slant Coil

The *slant coil* is a one-piece coil mounted in the duct on an angle (usually 60°) or slant to give the coil more surface area. The slant of the coil causes the condensate water to drain to the condensate pan located at the bottom of the slant. The coil can be used for upflow, downflow, or horizontal flow when designed for these applications, **Figure 36–9.**

The H Coil

The *H coil* is normally applied to horizontal applications, although it can be adapted to vertical applications by using special drain pan configurations. The drain is normally at the bottom of the H pattern, **Figure 36–10.**

Coil Circuits

All of the aforementioned coils may have more than one circuit for the refrigerant. As was indicated in Unit 21, "Evaporators and the Refrigeration System," when a coil becomes

Figure 36–9   A slant coil. *Courtesy BDP Company*

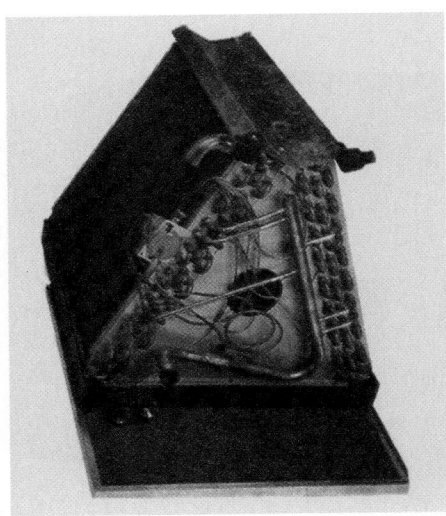

Figure 36–8   An A coil. Notice that two coils are standing at an angle and touching at the top—shaped like an A.

Figure 36–10   An H coil. *Courtesy BDP Company*

Figure 36–11 A multicircuit coil. Notice that the expansion valve has four outlet tubes and that the suction line has four connections. There are four completely different circuits. *Courtesy Sporlan Valve Company*

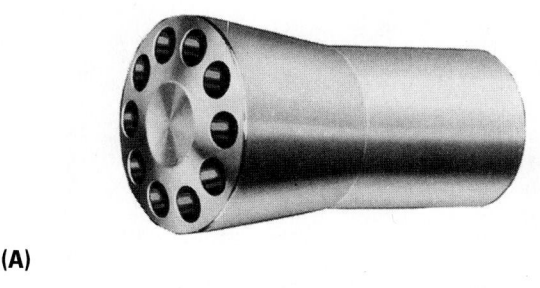

(A)

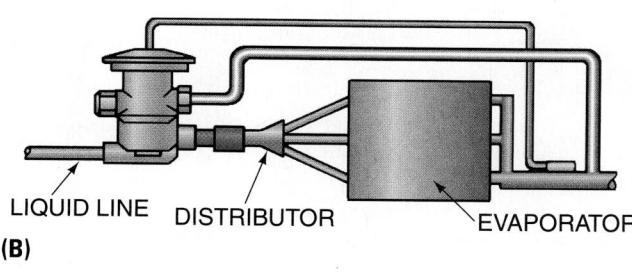

LIQUID LINE DISTRIBUTOR EVAPORATOR

(B)

Figure 36–12 A refrigerant distributor. It is responsible for equally distributing refrigerant to each of the three circuits in this coil. *Courtesy Sporlan Valve Company*

too long and excessive pressure drop occurs, it is advisable to have more than one coil in parallel, **Figure 36–11**. The coil may have as many circuits as necessary to do the job. However, when more than one circuit is used, a distributor must be used to distribute the correct amount of refrigerant to the individual circuits, **Figure 36–12**.

36.6 THE FUNCTION OF THE EVAPORATOR

The evaporator is a heat exchanger that takes the heat from the room air and transfers it to the refrigerant. Two kinds of heat must be transferred: sensible heat and latent heat. *Sensible heat* lowers the air temperature, and *latent heat* changes the water vapor in the air to condensate. The condensate collects on the coil and runs through the drain pan to a trap

(to stop air from pulling into the drain), and then it is normally piped to a drain.

Typically, room air may be 75°F dry-bulb and have a humidity of 50%, which is 62.5°F wet-bulb. The coil generally operates at a refrigerant temperature of 40°F to remove the required amount of sensible heat (to lower the air temperature) and latent heat (to remove the correct amount of moisture). The air leaving the coil is approximately 55°F dry-bulb with a humidity of about 95%, which is 54°F wet-bulb. Notice the high humidity leaving the coil. This is because the air has been cooled. When it mixes with room air, it is heated and expands where it can absorb moisture from the room air. The result is dehumidification. These conditions are average for a climate with high humidity. If the humidity is very high, such as in coastal locations, the coil temperature may be a little lower to remove more humidity. If the system is in a locality where the humidity is very low, such as in desert areas, the coil temperature may be a little higher than 40°F. The coil temperature is controlled with the airflow across the evaporator coil. More airflow will cause higher coil temperatures, and less airflow will cause lower coil temperatures. The same equipment can be sold in all parts of the country, and the airflow can be varied to accomplish the proper evaporator temperature for the proper humidity. Fin spacing on the evaporator may also affect the leaving air's humidity levels. Closer fin spacing will take out more humidity from the air.

36.7 DESIGN CONDITIONS

A house built in Augusta, Georgia, has less sensible-heat load and more latent-heat load than one constructed in Phoenix, Arizona. The designer or engineer must be familiar with local design practices, **Figure 36–13**. The airflow across the same coil in the two different parts of the country may be varied to accomplish different air conditions.

People who choose equipment for conditioning homes and buildings either use the chart shown in **Figure 36–13** or some other chart that has been derived from this chart. The estimator may either be a practical-design person or an engineer and will want the equipment to be sized for the best efficiency for the owner. Using as a basis for comparison the summer conditions of the two cities Augusta, Georgia, and Phoenix, Arizona, let's explore the different possibilities. There are three columns for the design dry-bulb (DB) and wet-bulb (WB) conditions. The 1% column for Augusta has 97 DB/77 WB listed. This means that on average, from records taken over time, Augusta conditions will be at 97 DB/77 WB 1% of the time during the summer. The 2.5% column shows 95/76 and the 5% column shows 93/76. The person selecting the equipment may use any of these figures.

The equipment will perform at its best efficiency when running at full load. Stopping and starting accounts for much of the inefficiency of equipment. Therefore, many designers try to select equipment to run as much as possible. If the equipment is selected for the 1% conditions, it will only run at full load 1% of the time. The rest of the

CLIMATIC CONDITIONS FOR THE UNITED STATES[a]															
					Winter[d]		Summer[e]								
Col. 1	Col. 2		Col. 3		Col. 4	Col. 5		Col. 6			Col. 7		Col. 8		
State and Station	Lati-tude[b]		Longi-tude[b]		Eleva-tion[c]	Design Dry-Bulb		Design Dry-Bulb and Mean Coincident Wet-Bulb			Mean Daily Range	Design Wet-Bulb		
	°	'	°	'	Ft	99%	97.5%	1%	2.5%	5%		1%	2.5%	5%
ARIZONA														
Douglas AP	31	3	109	3	4098	27	31	98/63	95/63	93/63	31	70	69	68
Flagstaff AP	35	1	111	4	6973	−2	4	84/55	82/55	80/54	31	61	60	59
Fort Huachuca AP (S)	31	3	110	2	4664	24	28	95/62	92/62	90/62	27	69	68	67
Kingman AP	35	2	114	0	3446	18	25	103/65	100/64	97/64	30	70	69	69
Nogales	31	2	111	0	3800	28	32	99/64	96/64	94/64	31	71	70	69
Phoenix AP (S)	33	3	112	0	1117	31	34	109/71	107/71	105/71	27	76	75	75
Prescott AP	34	4	112	3	5014	4	9	96/61	94/60	92/60	30	66	65	64
Tucson AP (S)	32	1	111	0	2584	28	32	104/66	102/66	100/66	26	72	71	71
Winslow AP	35	0	110	4	4880	5	10	97/61	95/60	93/60	32	66	65	64
Yuma AP	32	4	114	4	199	36	39	111/72	109/72	107/71	27	79	78	77
CALIFORNIA														
Bakersfield AP	35	2	119	0	495	30	32	104/70	101/69	98/68	32	73	71	70
Barstow AP	34	5	116	5	2142	26	29	106/68	104/68	102/67	37	73	71	70
Blythe AP	33	4	114	3	390	30	33	112/71	110/71	108/70	28	75	75	74
Burbank AP	34	1	118	2	699	37	39	95/68	91/68	88/67	25	71	70	69
Chico	39	5	121	5	205	28	30	103/69	101/68	98/67	36	71	70	68
Los Angeles AP (S)	34	0	118	2	99	41	43	83/68	80/68	77/67	15	70	69	68
Los Angeles CO (S)	34	0	118	1	312	37	40	93/70	89/70	86/69	20	72	71	70
Merced-Castle AFB	37	2	120	3	178	29	31	102/70	99/69	96/68	36	72	71	70
Modesto	37	4	121	0	91	28	30	101/69	98/68	95/67	36	71	70	69
Monterey	36	4	121	5	38	35	38	75/63	71/61	68/61	20	64	62	61
Napa	38	2	122	2	16	30	32	100/69	96/68	92/67	30	71	69	68
Needles AP	34	5	114	4	913	30	33	112/71	110/71	108/70	27	75	75	74
Oakland AP	37	4	122	1	3	34	36	85/64	80/63	75/62	19	66	64	63
Oceanside	33	1	117	2	30	41	43	83/68	80/68	77/67	13	70	69	68
Ontario	34	0	117	36	995	31	33	102/70	99/69	96/67	36	74	72	71
GEORGIA														
Albany, Turner AFB	31	3	84	1	224	25	29	97/77	95/76	93/76	20	80	79	78
Americus	32	0	84	2	476	21	25	97/77	94/76	92/75	20	79	78	77
Athens	34	0	83	2	700	18	22	94/74	92/74	90/74	21	78	77	76
Atlanta AP (S)	33	4	84	3	1005	17	22	94/74	92/74	90/73	19	77	76	75
Augusta AP	33	2	82	0	143	20	23	97/77	95/76	93/76	19	80	79	78
Brunswick	31	1	81	3	14	29	32	92/78	89/78	87/78	18	80	79	79
Columbus, Lawson AFB	32	3	85	0	242	21	24	95/76	93/76	91/75	21	79	78	77
Dalton	34	5	85	0	720	17	22	94/76	93/76	91/76	22	79	78	77
Dublin	32	3	83	0	215	21	25	96/77	93/76	91/75	20	79	78	77
Gainesville	34	2	83	5	1254	16	21	93/74	91/74	89/73	21	77	76	75
NEW YORK														
Albany AP (S)	42	5	73	5	277	−6	−1	91/73	88/72	85/70	23	75	74	72
Albany CO	42	5	73	5	19	−4	1	91/73	88/72	85/70	20	75	74	72
Auburn	43	0	76	3	715	−3	2	90/73	87/71	84/70	22	75	73	72
Batavia	43	0	78	1	900	1	5	90/72	87/71	84/70	22	75	73	72
Binghamton AP	42	1	76	0	1590	−2	1	86/71	83/69	81/68	20	73	72	70
Buffalo AP	43	0	78	4	705r	2	6	88/71	85/70	83/69	21	74	73	72
Cortland	42	4	76	1	1129	−5	0	88/71	85/71	82/70	23	74	73	71
Dunkirk	42	3	79	2	590	4	9	88/73	85/72	83/71	18	75	74	72
Elmira AP	42	1	76	5	860	−4	1	89/71	86/71	83/70	24	74	73	71
Geneva (S)	42	5	77	0	590	−3	2	90/73	87/71	84/70	22	75	73	72

[a] Table 1 was prepared by ASHRAE Technical Committee 4.2, Weather Data, from data compiled from official weather stations where hourly weather observations are made by trained observers, See also Ref 1, 2, 3, 5 and 6.

[b] Latitude, for use in calculating solar loads, and longitude are given to the nearest 10 minutes. For example, the latitude and longitude for Anniston, Alabama are given as 33 34 and 85 55 respectively, or 33° 40, and 85° 50.

[c] Elevations are ground elevations for each station. Temperature readings are generally made at an elevation of 5 ft above ground, except for locations marked r, indicating roof exposure of thermometer.

[d] Percentage of winter design data shows the percent of the 3-month period, December through February.

[e] Percentage of summer design data shows the percent of 4-month period, June through September.

Figure 36–13   This excerpt from a table shows different design conditions for various parts of the United States. *Reprinted with permission from American Society of Heating, Refrigerating, and Air-Conditioning Engineers*

time, it will be stopping and starting. If the equipment is selected for 5% conditions, it will be stopping and starting at conditions that fall below the 5% conditions. The problem with selecting equipment for 5% is that when the 1% conditions occur, the equipment will run all of the time and will not always maintain the conditions at a comfortable level for the occupants. This would also mean that the equipment would have no reserve capacity to lower the structure temperature in case of equipment downtime. For example, if the equipment is shut off for any reason for a period of time and the space conditions rise, it may not be able to reduce the space conditions until an overnight run time has occurred. Suppose that a storm causes a power failure for several hours during the day—the unit will probably not catch up until the middle of the night when the unit can run at reduced load.

In order to alleviate difficulties such as this, many designers compromise by selecting equipment for the 2.5% conditions.

36.8   EVAPORATOR APPLICATION

The evaporator may be installed in the airstream in several different ways. It may have a coil case that encloses the coil, **Figure 36–14**, or it may be located in the ductwork. The coil will normally operate below the dew point temperature. The coil enclosure should be insulated to keep it from absorbing heat from the surroundings. An insulated coil cabinet will not sweat on the outside, **Figure 36–15**. Many evaporators are built into the air handler by the manufacturer, **Figure 36–16**.

Figure 36–14   An evaporator with a coil case. *Courtesy BDP Company*

36.9   THE COMPRESSOR

The following types of compressors are used in air-conditioning systems: reciprocating, rotary, scroll, centrifugal, and screw. These are of the same design as similar types used in commercial refrigeration. The centrifugal and screw compressors, described briefly in Unit 23, "Compressors," are used primarily in large commercial and industrial applications and will be discussed further later in this text.

Compressors are vapor pumps that pump heat-laden refrigerant vapor from the low-pressure side of the system, the evaporator side, to the high-pressure, or condenser, side. To do this they compress the vapor from the low side, increasing the pressure and raising the temperature.

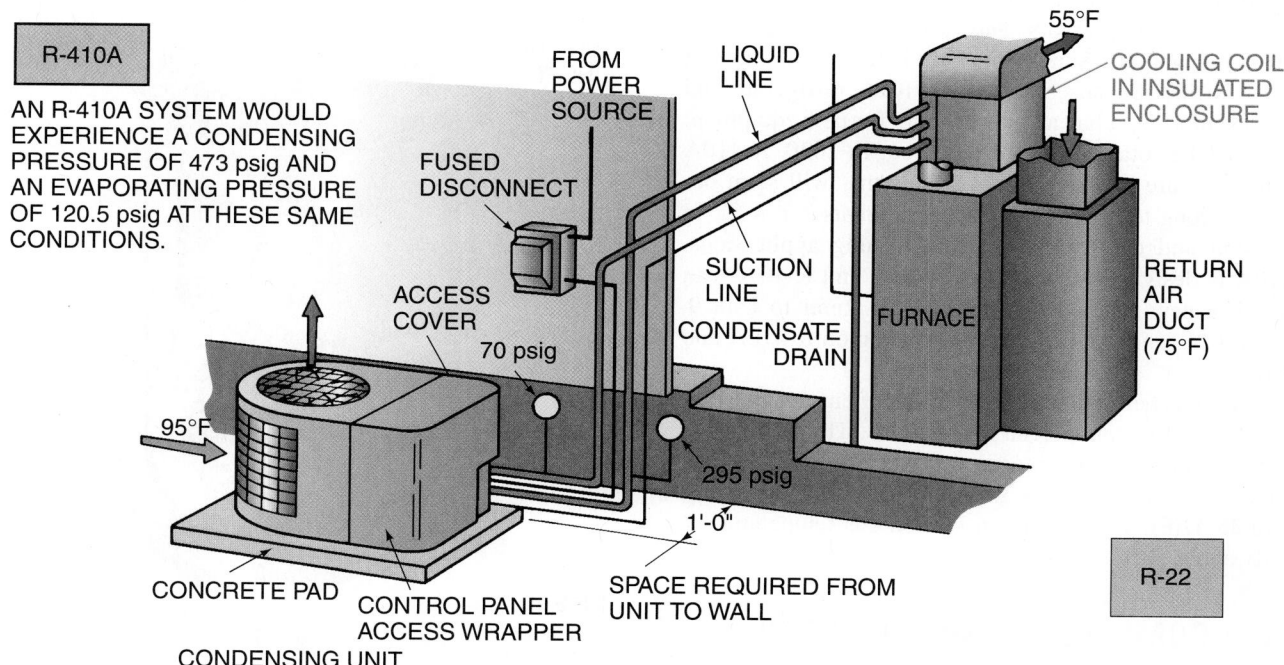

Figure 36–15   The operating conditions of an evaporator and a condenser. Notice that the cooling coil is in an enclosure to prevent sweating on the outside.

Figure 36–16 An evaporator mounted in the air handler. *Courtesy Carrier Corporation*

36.10 THE RECIPROCATING COMPRESSOR

The reciprocating compressor used in residential and light commercial air-conditioning applications is similar to those discussed in Unit 23.

These compressors may be either the fully hermetic or the serviceable hermetic type, **Figure 36–17.** Modern residential systems with reciprocating compressors usually are equipped with the fully hermetic type. These are suction gas cooled, **Figure 36–18.** These compressors are positive-displacement compressors and use R-22, R-410A, or R-407C refrigerants. Some units built before 1970 may use R-12 or R-500. Because R-22 is an HCFC refrigerant and contains chlorine, it has a phaseout date for new equipment in 2010 and a total production phaseout in 2020. R-410A and R-407C are two refrigerant blends that will soon become the long-term replacement refrigerants for R-22 in residential and commercial air-conditioning applications. R-410A is designed for new applications and R-407C can be used in new and retrofit applications. Refer to Unit 9, Section 9.11 for more detailed information on R-410A and R-407C.

Serviceable hermetic compressors are often used in larger systems found in commercial applications. These compressors may be either gas or air cooled. If suction gas cooled, the suction line is piped to the motor end of the compressor, **Figure 36–17(B).** The maximum suction gas temperature is usually about 70°F.

36.11 COMPRESSOR SPEEDS (RPM)

Modern air-conditioning compressors used in the small and medium size ranges must turn standard motor speeds of 3450 rpm or 1750 rpm. Early compressors were 1750 rpm,

(A)

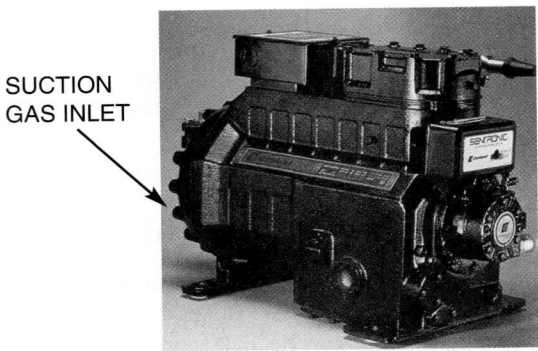

SUCTION GAS INLET

(B)

Figure 36–17 **(A)** A hermetic compressor. **(B)** A serviceable hermetic compressor. **(A)** *Courtesy Bristol Compressor Company.* **(B)** *Courtesy Copeland Corporation*

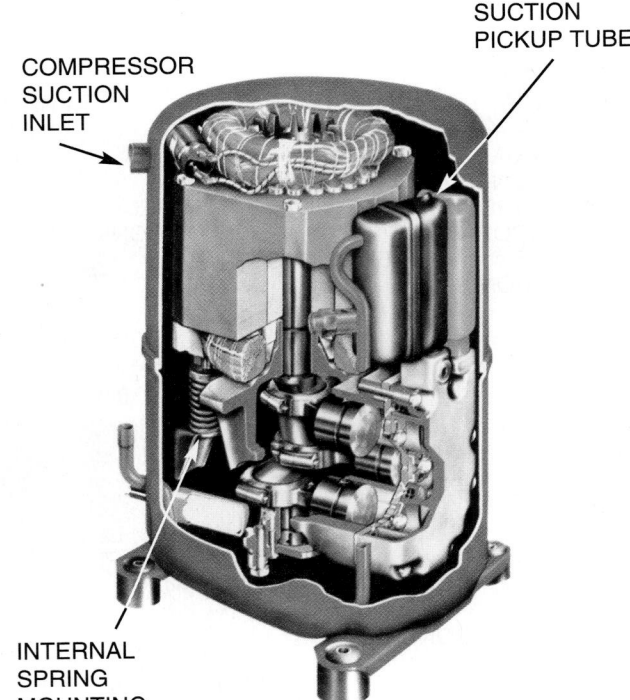

COMPRESSOR SUCTION INLET

SUCTION PICKUP TUBE

INTERNAL SPRING MOUNTING

Figure 36–18 A suction gas–cooled compressor. *Courtesy Tecumseh Products Company*

used R-12, and were large and heavy. Present compressors use the faster motor and the more efficient refrigerants like R-22, R-410A, and R-407C and other environmentally friendly alternative refrigerant blends so equipment can be smaller and lighter.

36.12 COOLING THE COMPRESSOR AND MOTOR

All compressors must be cooled because some of the mechanical energy in the compressor converts to heat energy. Without cooling, the compressor motor would burn and the oil would become hot enough to break down and form carbon. Hermetic compressors have always been considered suction gas cooled. The large ones are directly suction gas cooled. Small compressors up to 7 1/2 horsepower may not be directly cooled by the suction gas. Some compressors, such as the rotary and scroll compressors, are cooled by the discharge gas leaving the compressor. This discharge gas temperature is influenced by the suction gas temperature. If the correct superheat is not maintained at the compressor, the compressor will overheat because the discharge gas temperature will rise. Because of this, the compressor motor temperature is still controlled by the suction gas temperature.

The newer compressor motors are able to run at higher temperatures because of better materials in the motor windings and higher-quality lubricating oils. These are materials that will still insulate the winding wires and not break down while running under high heat conditions and oils that do not break down at reasonably high temperatures.

NOTE: The technician should be aware of the discharge-cooled compressor and determine that the compressor is not running too hot. The case of the compressor is hotter than a suction-cooled compressor. A check of the discharge line temperature would be the sure test. Most manufacturers agree that the discharge line leaving the compressor should not exceed 225°F.

Some serviceable hermetic compressors are air cooled, **Figure 36–19**. The suction line enters the compressor at the side of the cylinder rather than near the motor. These compressors have ribs that help dissipate the heat into the air passing over it. It is essential that these compressors be located in a moving airstream.

In some compressors used on water-cooled equipment, the inlet water line is wrapped around the compressor motor body, **Figure 36–20**.

36.13 COMPRESSOR MOUNTINGS

Welded hermetic reciprocating compressors all have rubber mounting feet on the outside, and the compressor is mounted on springs inside the shell, **Figure 36–18**. Older compressors were mounted on springs outside the shell, and the compressor was pressed into the shell. New compressors have a vapor space between the motor and the shell, so the

Figure 36–19 An air-cooled compressor. Notice that the suction line enters at the side of the cylinder. *Courtesy Copeland Corporation*

Figure 36–20 This compressor is similar to that in **Figure 36–19**, but it is applied to a water-cooled condenser. *Photo by Bill Johnson*

motor-temperature sensor must be on the inside to sense the motor temperature quickly.

The suction gas dumps out into the shell, usually in the vicinity of the motor. Some compressors dump the suction gas directly into the rotor. The turning rotor tends to dissipate any liquid drops in the return suction gas. The suction pickup tube for the compressor, which is inside the shell, is normally located in a high position so that liquid refrigerant or foaming oil cannot enter the compressor cylinders, **Figure 36–18**.

The compressor for air-conditioning equipment is normally located outside with the condenser. These hermetic compressors cannot be field serviced and often are not factory serviced. When one becomes defective, the manufacturer may authorize the technician to discard it or may require that it be returned to the factory to determine what made it fail.

36.14 REBUILDING THE HERMETIC COMPRESSOR

Some manufacturers are remanufacturing hermetic compressors by opening the shell and repairing the compressor inside. The manufacturer will be the one to decide whether this is economical or not. The larger the compressor, the more advantage there would be to repairing it.

The standard serviceable hermetic compressor is cast iron, and the manufacturer will want this compressor returned for remanufacturing. Small cast-iron compressors are not used widely in small air-conditioning equipment because of the initial cost.

36.15 THE ROTARY COMPRESSOR

The rotary compressor is small and light. It is sometimes cooled with compressor discharge gas, which makes the compressor appear to be running too hot. A warning may be posted in the compressor compartment that the housing will appear to be too hot. The compressor and motor are pressed into the shell of a rotary compressor, and no vapor space is between the compressor and the shell. This also reduces the size of the rotary compressor as compared with a reciprocating compressor.

The rotary compressor is more efficient than the reciprocating compressor and is used in small- to medium-sized systems, **Figure 36–21**. These compressors are manufactured in two basic design types: the stationary vane and the rotary vane.

Stationary Vane Rotary Compressor

The components for this compressor are the housing, a blade or vane, a shaft with an off-center (eccentric) rotor, and a discharge valve. The shaft turns the off-center rotor so that it "rolls" around the cylinder, **Figure 36–22**. The blade or vane keeps the intake and compression chambers of the cylinder separate. The tolerances between the rotor and cylinder must be very close, and the vane must be machined so that gas does not escape from the discharge side to the intake. As the rotor turns, the vane slides in and out to remain tight against the rotor. The valve at the discharge keeps the compressed gas from leaking back into the chamber and into the suction side during the off cycle.

As the shaft turns, the rotor rolls around the cylinder, allowing suction gas to enter through the intake and compresses the gas on the compression side. This is a continuous process as long as the compressor is running. The results are similar to those with a reciprocating compressor. Low-pressure suction gas enters the cylinder, is compressed (which also causes its temperature to rise), and leaves through the discharge opening.

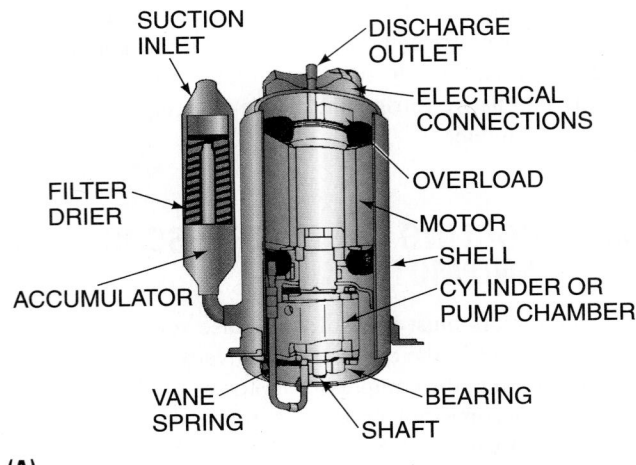

(A)

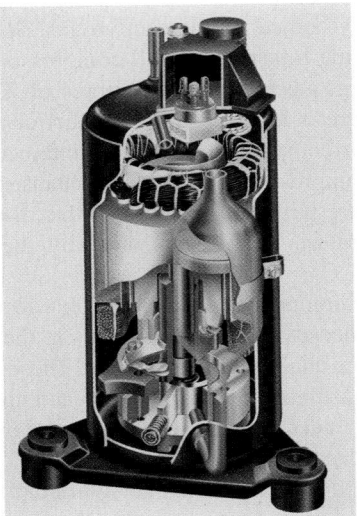

(B)

Figure 36–21 A rotary compressor **(A)–(B)**. *Courtesy Motors and Armatures, Inc.*

Rotary Vane Rotary Compressor

The rotary vane compressor has a rotor fitted to the center of the shaft. The rotor and the shaft have the same center. The rotor has two or more vanes that slide in and out and trap and compress the gas, **Figure 36–23**. The shaft and rotor are positioned off center in the cylinder. As a vane passes by the suction intake opening, low-pressure gas follows it. This gas continues to enter and is trapped by the next vane, which compresses it and pushes it out the discharge opening. Notice that the intake opening is much larger than the discharge opening. This is to allow the low-pressure gas to enter more readily. The intake and compression is a continuous process as long as the compressor is running.

All the refrigerant that enters the intake port is discharged through the exhaust or discharge port. There is no clearance volume as in the reciprocating compressor. This is the primary reason why the rotary compressor is highly efficient.

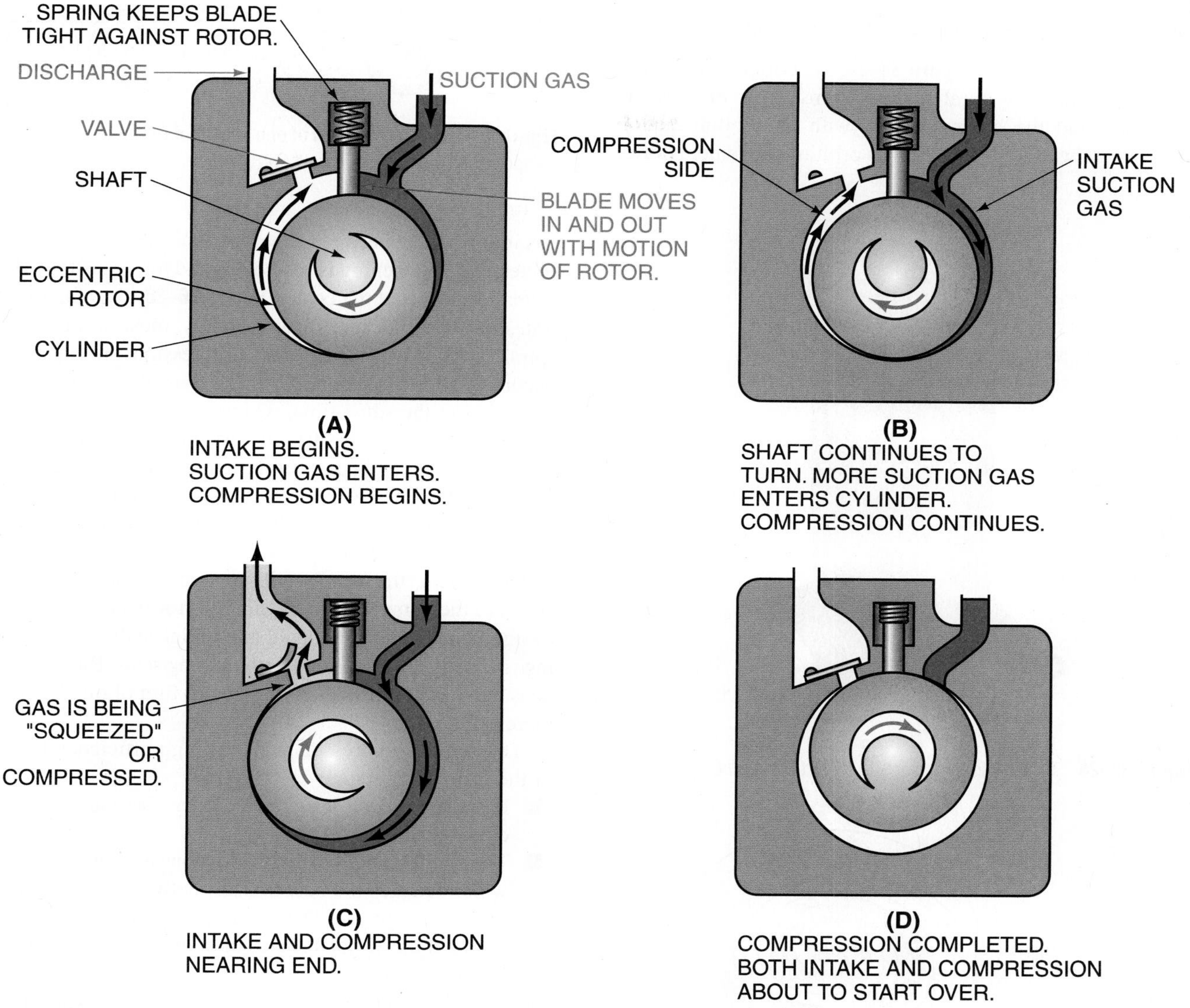

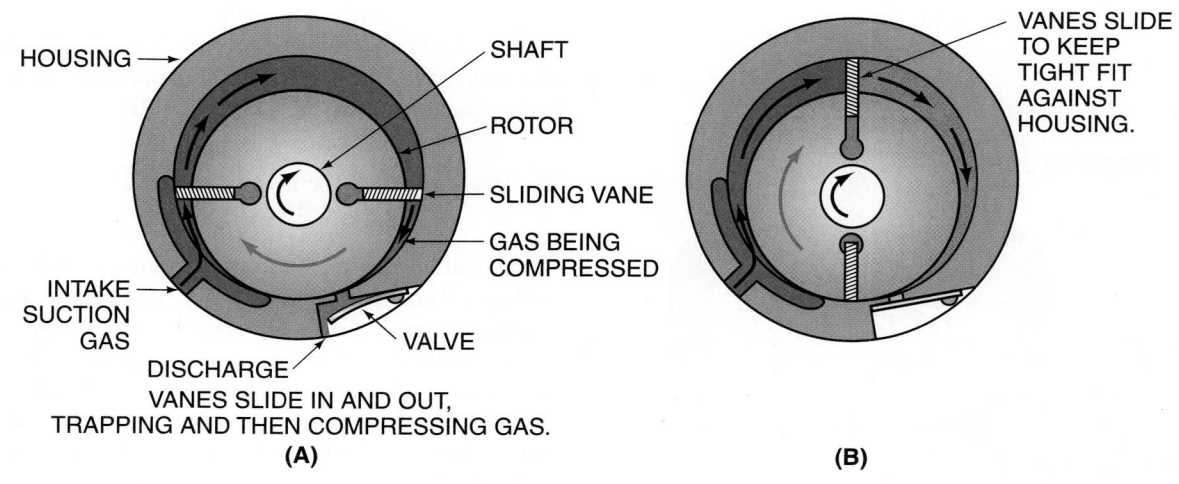

SPRING KEEPS BLADE
TIGHT AGAINST ROTOR.

DISCHARGE

VALVE

SHAFT

ECCENTRIC
ROTOR

CYLINDER

SUCTION GAS

BLADE MOVES
IN AND OUT
WITH MOTION
OF ROTOR.

(A)
INTAKE BEGINS.
SUCTION GAS ENTERS.
COMPRESSION BEGINS.

COMPRESSION
SIDE

INTAKE
SUCTION
GAS

(B)
SHAFT CONTINUES TO
TURN. MORE SUCTION GAS
ENTERS CYLINDER.
COMPRESSION CONTINUES.

GAS IS BEING
"SQUEEZED"
OR
COMPRESSED.

(C)
INTAKE AND COMPRESSION
NEARING END.

(D)
COMPRESSION COMPLETED.
BOTH INTAKE AND COMPRESSION
ABOUT TO START OVER.

Figure 36–22 Operation of a stationary vane rotary compressor.

HOUSING

SHAFT

ROTOR

SLIDING VANE

GAS BEING
COMPRESSED

INTAKE
SUCTION
GAS

VALVE

DISCHARGE
VANES SLIDE IN AND OUT,
TRAPPING AND THEN COMPRESSING GAS.
(A)

VANES SLIDE
TO KEEP
TIGHT FIT
AGAINST
HOUSING.

(B)

Figure 36–23 Operation of a rotary vane rotary compressor.

36.16 THE SCROLL COMPRESSOR

In a scroll compressor, **Figure 36–24,** the compression takes place between two spiral-shaped forms. One of these is stationary, and the other operates with an orbiting action within the other, **Figure 36–25.** This orbiting movement draws

Figure 36–24   A scroll compressor. *Courtesy Copeland Corporation*

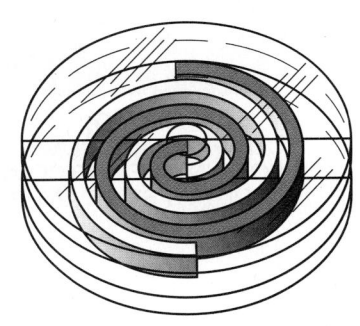

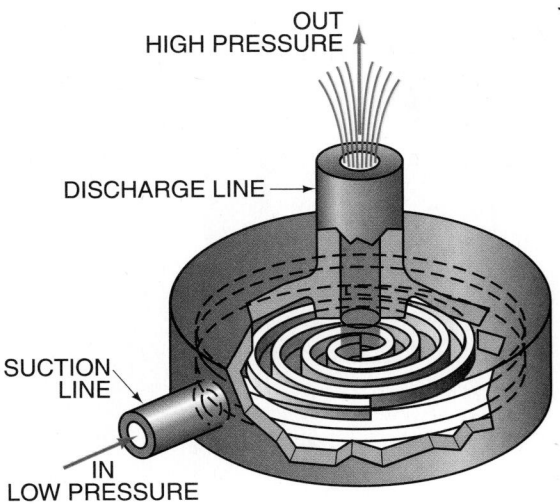

Figure 36–25   Orbiting action between the scrolls in a scroll compressor.

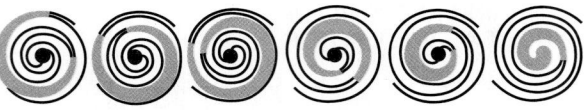

Figure 36–26   Compression of one pocket of refrigerant in a scroll compressor. *Courtesy Copeland Corporation*

gas into a pocket between the two spirals. As this action continues, the gas opening is sealed off, and the gas is forced into a smaller pocket at the center, **Figure 36–26.** This figure illustrates only one pocket of gas. Actually, most of the scroll or spiral is filled with gas, and compression or "squeezing" occurs in all the pockets. Several stages of compression are occurring at the same time, which acts to balance the compression action. The scroll compressor is quiet compared with reciprocating compressors of the same size.

The contact between the spirals sealing off the pockets is achieved using centrifugal force. This minimizes gas leakage. However, should liquid refrigerant or debris be present in the system, the scrolls will separate without damage or stress to the compressor. The high performance of the scroll design depends on good axial and radial sealing. Axial sealing is usually accomplished with floating seals. Radial sealing is usually accomplished by only a thin film of oil. Therefore, there is little wear.

The scroll compressor provides high efficiency because of the following factors:

- It requires no valves; consequently, there are no valve compression losses.
- The suction and discharge locations are separate, which reduces the heat transfer between the suction and discharge gas. There is less temperature difference between the various stages of compression occurring in the compressor.
- There is no clearance volume providing reexpansion gas as in a reciprocating compressor.

36.17 THE CONDENSER

The condenser for air-conditioning equipment is the component that rejects the heat from the system. Most equipment is air cooled and rejects heat to the air. The coils are copper or aluminum, and both types have aluminum fins to add to the heat exchange surface area, **Figure 36–27.**

36.18 SIDE-AIR-DISCHARGE CONDENSING UNITS

Air-cooled condensers all discharge hot air loaded with the heat absorbed from within the structure. Early condensers discharged the air out the side and were called side discharge, **Figure 36–28.** The advantage of this equipment is that the fan and motor are under the top panel, **Figure 36–29.** However, any noise generated inside the cabinet is discharged into the leaving airstream and may be clearly heard in a neighbor's yard. NOTE: The heat from the condenser coil can be hot enough to kill plants that it blows on. These condensers are still being used.

Figure 36–27   Condensers have either copper or aluminum tubes with aluminum fins. *Courtesy Carrier Corporation*

36.19 TOP-AIR-DISCHARGE CONDENSERS

The modern trend in residential equipment is for the condenser to be a top-discharge type, **Figure 36–30.** In this type of unit, the hot air and noise are discharged from the top of the unit into the air. This is advantageous as far as air and noise are concerned, but the fan and motor are on top of the equipment, and rain, snow, and leaves can fall directly into the unit, so the fan motor should be protected with a rain shield, **Figure 36–31.** The fan motor bearings are in a vertical position, which means there is more thrust on the end of the bearing, so the bearing needs a thrust surface for this type of application, **Figure 36–32.**

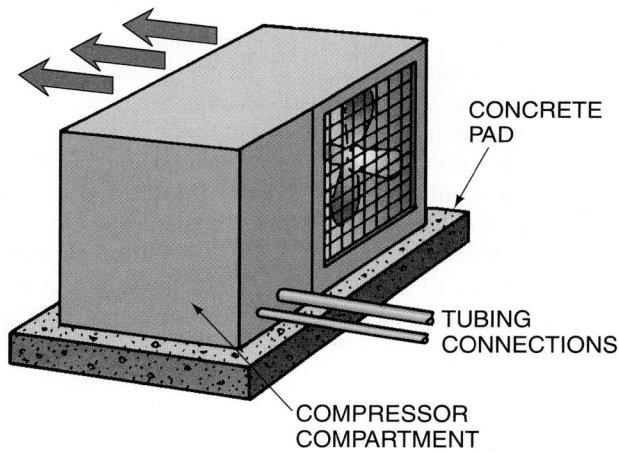

Figure 36–28   This condensing unit discharges hot air out of the side.

Figure 36–30   Equipment with air discharged out of the top of the cabinet. *Courtesy York Corporation*

Figure 36–29   The fan motor and all components are located under the top panel. *Courtesy Carrier Corporation*

Figure 36–31   In the top-air-discharge units, the fan is usually on the top.

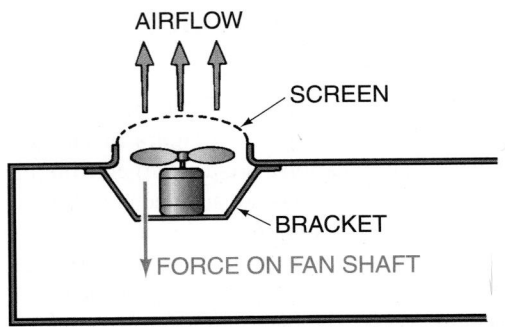

Figure 36–32 Top-air-discharge units have an additional load placed on the bearings. The fan is trying to fly down while it is running and pushing air out of the top. During the starting and stopping of the fan, the bottom bearing has the fan blade and shaft resting on it. This bearing must have a thrust surface.

36.20 CONDENSER COIL DESIGN

Some coil surfaces are positioned vertically, and grass and dirt can easily get into the bottom of the coil. The coils must be clean for the condenser to operate efficiently. 🌐Some equipment uses the bottom few rows for a subcooling circuit to lower the condensed refrigerant temperature below the condensing temperature. It is common for the liquid line to be 10°F to 15°F cooler than the condensing temperature. Each degree of subcooling will add approximately 1% to the efficiency of the system. If the subcooling circuit is dirty, it could affect the capacity by 10% to 15%.🌐 This could mean the difference in a piece of equipment being able to cool or not being able to cool a structure if the equipment were sized too close to the design cooling load.

Some manufacturers use horizontal or slant-type condensers to position the coil off the ground. These condensers are less likely to pick up leaves and grass at the ground level, **Figure 36–33**.

36.21 HIGH-EFFICIENCY CONDENSERS

🌐Modern times and the federal government have demanded that air-conditioning equipment become more efficient. Probably the best way to improve efficiency is to lower

the head pressure so that the compressor is not working as hard because it now has a lower compression ratio. More surface area in the condenser reduces the compressor head pressure even in the hottest weather. This means lower compressor current and less power consumed for the same amount of air conditioning. Some manufacturers with oversized condensers use two-speed condenser fans—one speed for mild weather, one for hot. Without two-speed fans the condensers would be too efficient in mild weather, causing the head pressure to be too low, starving the expansion device and resulting in less capacity.🌐

36.22 CABINET DESIGN

The condenser cabinet is usually located outside, so it needs weatherproofing. Most cabinets are galvanized and painted to give them more years of life without rusting. Some cabinets are made of aluminum, which is lighter but may not last as long in a salty environment, such as coastal areas.

Most small equipment is assembled with self-tapping sheet metal screws. These screws are held by a drill screw holder during manufacturing and are threaded into the cabinet when turned with an electric drill. See **Figure 36–34** for an example of a portable electric drill and screw holders. These screws should be made of weather-resistant material that will last for years out in the weather. The weather may be salt air as in the coastal areas. In these locations, stainless steel is a good choice of metal for sheet metal screws. When equipment that is assembled with drill screws is installed in the field, all of the screws should be fastened back into the cabinet tightly, or the unit may rattle. After being threaded many times, the screw holes may become oversized; if so, use the next 1/2 size screw to tighten the cabinet panels.

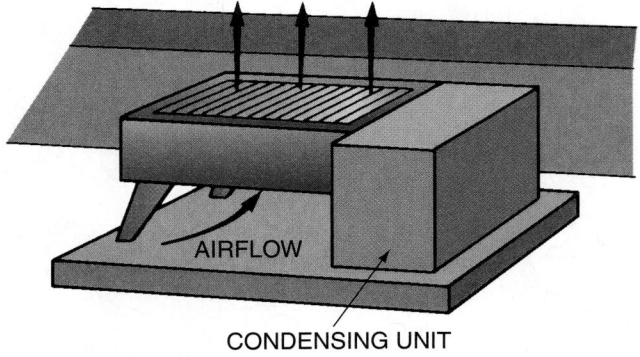

Figure 36–33 Some condensers are designed so that the coil is off the ground.

Figure 36–34 A portable electric drill and screw holders. *Photo by Bill Johnson*

36.23   EXPANSION DEVICES

The expansion device meters the refrigerant to the evaporator. The TXV and the fixed-bore metering device (either a capillary tube or an orifice) are the types most often used.

The TXVs are the same types as those described in Unit 24, "Expansion Devices," except that they have a different temperature range. Air-conditioning expansion devices are in the high-temperature range. TXVs are more efficient than fixed-bore devices because they allow the evaporator to reach peak performance faster. They allow more refrigerant into the evaporator coil during a hot pulldown—when the conditioned space is allowed to get too warm before the unit is started.

When a TXV is used, the refrigerant pressures do not equalize during the off cycle unless the valve is made to equalize them with a planned bleed port. When the pressures do not equalize during the off cycle, a high-starting torque motor must be used for the compressor. This means that the compressor must have a start capacitor and starting relay for start-up after the system has been off. Some valve manufacturers have a bleed port that always allows a small amount of refrigerant to bleed through. During the off cycle this valve will allow pressures to equalize, and a compressor with a low-starting-torque motor can be used. Low-starting-torque motors are less expensive and have fewer parts that may malfunction.

36.24   AIR-SIDE COMPONENTS

The air side of an air-conditioning system consists of the supply air and the return air systems. The airflow in an air-conditioning system is normally 400 cfm/ton in the average humidity climates. For other climates, a different airflow may be used. In humid coastal areas 350 cfm/ton may be used, whereas 450 cfm/ton may be used in desert areas. The air leaving the air handler in a typical application could be expected to be about 55°F in the average system. The ductwork carrying this air must be insulated when it runs through an unconditioned space, or it will sweat and gain unwanted heat from the surroundings. The insulation should have a vapor barrier to keep the moisture from penetrating the insulation and collecting on the metal duct. All connections of the insulation must be fastened tight, and a vapor barrier must be used at all seams.

The return air will normally be about 75°F. If the return air duct is run through the unconditioned space, it may not need insulation. If the unconditioned space is a crawl space or basement, the temperature may be 75°F, and no heat will exchange. Even if a small amount of heat did exchange, a cost evaluation should be made to see whether it is more economical to insulate the return duct or allow the small heat exchange. If the duct is run through a hot attic, the duct must be insulated.

Cool air will distribute better from a supply register (diffuser) located high in the room because it will fall after leaving the register. The final distribution point is where the cool air is mixed with the room air to arrive at a comfort condition. **Figure 36–35** shows some examples of air-conditioning diffusers.

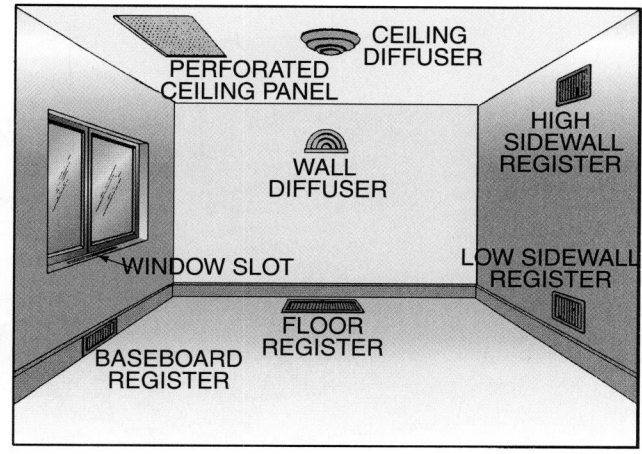

Figure 36–35   The final task of the air-conditioning system is to get the refrigerated air properly distributed in the conditioned space.

36.25   INSTALLATION PROCEDURES

Package air-conditioning systems were described earlier as systems in which the whole air conditioner is built into one cabinet. Actually, a window unit is a package air conditioner designed to blow freely into the conditioned space. Most package air conditioners described in this text have two fan motors, one for the evaporator and one for the condenser, **Figure 36–36**. A window unit, however, uses one fan motor with a double shaft that drives both fans. The indoor fan is not designed to push air down a duct.

The package air conditioner has the advantage that all equipment is located outside the structure, so all service can be performed on the outside. ⊛This equipment is more efficient than the split system because it is completely factory assembled and charged with refrigerant. The refrigerant lines are short, so there is less efficiency loss in the refrigeration lines.⊛

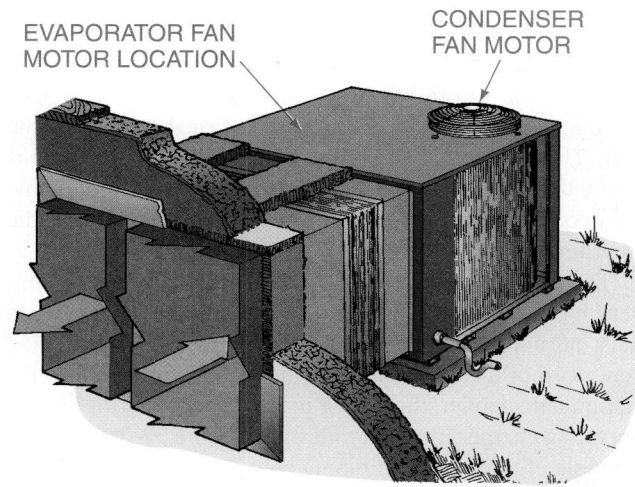

Figure 36–36   A package air conditioner with two fan motors—one for the evaporator coil and one for the condenser. *Courtesy Climate Control*

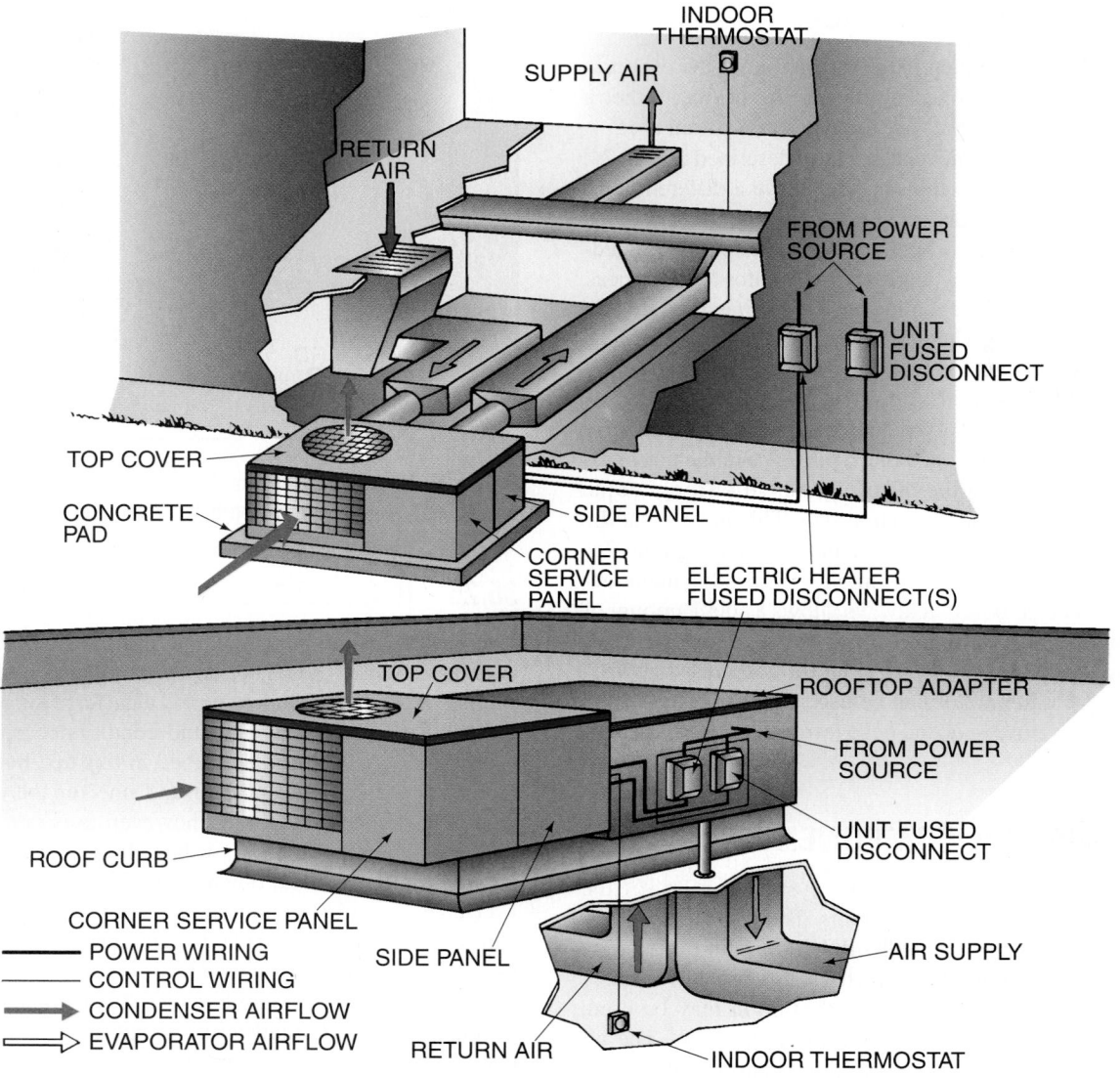

Figure 36–37   Package unit installations.

The installation of a package system consists of mounting the unit on a firm foundation, fastening the package unit to the ductwork, and connecting the electrical service to the unit. These units can be installed easily in some locations where the ductwork can be readily attached to the unit. Some common installation types are on the rooftop, beside the structure, and in the eaves of structures, **Figure 36–37**.

The split air-conditioning system is used when the condensing unit must be located away from the evaporator. There must be interconnecting piping furnished by the installing contractor or the equipment supplier. When furnished by the equipment supplier, the tubing may be charged with its own operating charge or with dry nitrogen. **Figure 36–38** shows an example of three types of tubing connections.

The condensing unit should be located as close as possible to the evaporator to keep the interconnecting tubing short. The tubing consists of a cool gas line and a liquid line.

The cool gas line is insulated and is the larger of the two lines. The insulation keeps unwanted heat from conducting into the line and keeps the line from sweating and dripping. **Figure 36–39** is an example of a precharged line set.

A typical installation with pressures and temperatures is shown in **Figure 36–40**. This illustration gives some guidelines about the operating characteristics of a typical system. SAFETY PRECAUTION: Installation of equipment may require the technician to unload the equipment from a truck or trailer and move it to various locations. The condensing unit may be located on the other side of the structure, far away from the driveway, or it may even be located on a rooftop. Proper care should be taken in handling the equipment. Small cranes are often used for lifting. Lift gate trucks may be used to set the equipment down to the ground.

When installing the equipment, care should be taken while working under structures and in attics. Spiders and other

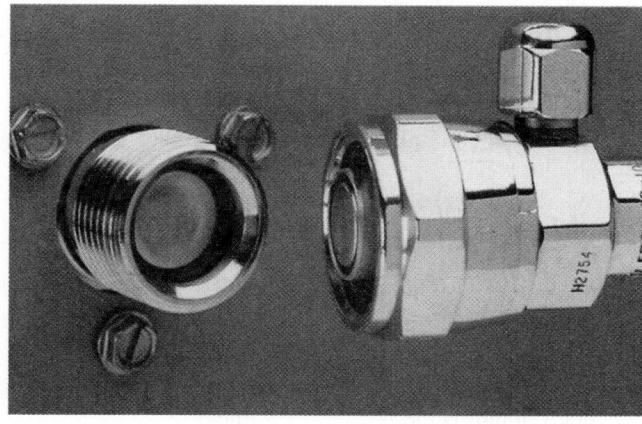

(A)

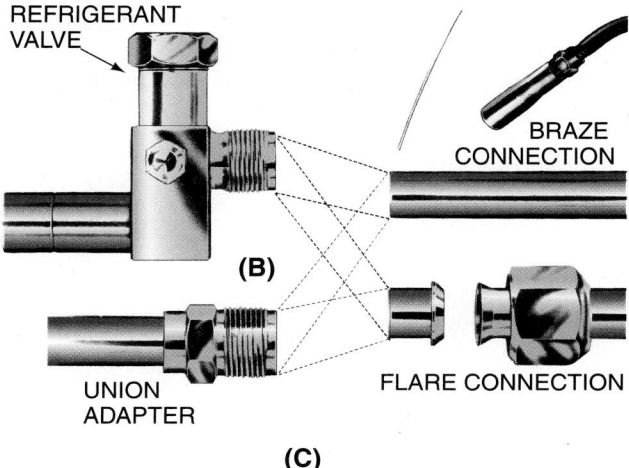

REFRIGERANT VALVE

BRAZE CONNECTION

(B)

UNION ADAPTER

FLARE CONNECTION

(C)

Figure 36–38   Three types of tubing connectors. **(A)** Quick-connect. **(B)** Solder type. **(C)** Compression type. *Courtesy Aeroquip Corporation*

Figure 36–39   A suction (cool gas line) line set. *Photo by Bill Johnson*

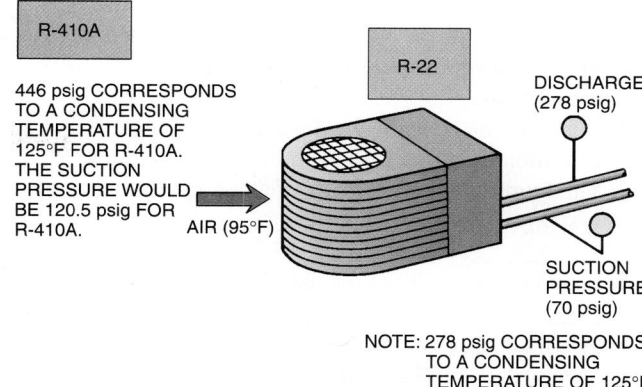

R-410A

446 psig CORRESPONDS TO A CONDENSING TEMPERATURE OF 125°F FOR R-410A. THE SUCTION PRESSURE WOULD BE 120.5 psig FOR R-410A.

AIR (95°F)

R-22

DISCHARGE (278 psig)

SUCTION PRESSURE (70 psig)

NOTE: 278 psig CORRESPONDS TO A CONDENSING TEMPERATURE OF 125°F FOR R-22.

95°F AMBIENT TEMP.
+ 30°F DIFFERENCE IN AIR TEMPERATURE AND CONDENSING TEMPERATURE ON STANDARD-GRADE EQUIPMENT.
‾‾‾‾‾‾
= 125°F CONDENSING TEMPERATURE.

Figure 36–40   An installation showing the pressures and temperatures for a system in the humid southern part of the United States.

types of stinging insects are often found in these places, and sharp objects such as nails are often left uncovered. When working in an attic, be careful not to step through a ceiling.

Care should be taken while handling the line sets during connection if they have refrigerant in them. Liquid R-22 boils at −41°F and R-410A boils at −60°F in the atmosphere and can inflict serious frostbite to the hands and eyes. Goggles and gloves must be worn when connecting line sets.

SUMMARY

- Evaporative cooling may be used in areas where the temperature is high and the humidity is low.
- Refrigerated air conditioning cools the air and removes moisture.
- Refrigerated air conditioning is used in hot temperatures with high or low humidity.
- Evaporators are made in three types: the A coil, slant coil, and H coil.
- Air-conditioning evaporators operate at about 40°F and remove sensible heat and latent heat.
- Removal of sensible heat lowers the air temperature; removal of latent heat removes moisture.
- The compressor is the positive displacement pump that pumps the heat-laden vapor from the evaporator to the condenser.
- Condensers are located outside to reject the heat to the outside.
- High efficiency in a condenser is achieved by increasing the condenser surface area. Two-speed fans may be used, one speed for mild weather and the other for hot weather.
- Package air conditioners are installed through the roof or through the wall at the end of the structure, wherever the duct can be fastened.
- The package unit is charged at the factory and is factory assembled. Under similar conditions, it is more efficient than the split system.

REVIEW QUESTIONS

1. True or False: Evaporative air conditioning will work in any climate.
2. What are the advantages of using refrigerated air conditioning rather than evaporative conditioning?
3. The four major components of an air-conditioning system are the _____, _____, _____, and _____.
4. Most compressors are cooled using the _____ _____.
5. What advantage does a thermostatic expansion valve have over a capillary tube?
6. What advantage does a capillary tube have over a thermostatic expansion valve?
7. The _____ expansion device ordinarily must use a compressor with a high starting torque.
8. The most popular refrigerant used in the past in residential air conditioning is R-_____.
9. What are two higher-efficiency refrigerant blends that can be used as long-term replacements for R-22 in residential and light commercial air-conditioning applications?
10. What refrigerant blend can be used in retrofit applications as a long-term replacement for R-22 in residential and light commercial air-conditioning applications?
11. What are the two types of heat that are removed by the evaporator coil in the air-conditioning process?
12. The design boiling point for most evaporator coils in residential air conditioning is _____ °F.
13. What are the three most common types of compressors used in air conditioning?
14. The large interconnecting line between the indoor and outdoor coil is the _____ line.
15. The small interconnecting line between the indoor and outdoor coil is the _____ line.
16. The outside design temperatures for summer in Tucson, Arizona, for 2.5% of the time are _____ °F dry-bulb and _____ °F wet-bulb.
17. The dew point temperature tells you how much _____ is in the air.

OBJECTIVES

After studying this unit, you should be able to

■ describe characteristics of the propeller and the centrifugal blowers.
■ take basic air pressure measurements.
■ measure air quantities.
■ list the different types of air-measuring devices.
■ describe the common types of motors and drives.
■ describe duct systems.
■ explain what constitutes good airflow through a duct system.
■ describe a return air system.
■ plot airflow conditions on the air friction chart.

SAFETY CHECKLIST

✔ Use a grounded, double-insulated, or cordless portable electric drill when installing or drilling into metal duct.
✔ Sheet metal can cause serious cuts. Wear gloves when installing or working with sheet metal.
✔ Wear goggles whenever working around duct that has been opened with the fan on.

37.1 CONDITIONING EQUIPMENT

As indicated previously, in many climates air has to be conditioned for us to be comfortable. One way to condition air is to use a fan to move the air over the conditioning equipment. This equipment may consist of a cooling coil, a heating device, a device to add humidity, or a device to clean the air. The forced-air system uses the same room air over and over again. Air from the room enters the system, is conditioned, and is returned to the room. Fresh air enters the structure either by infiltration around the windows and doors or by ventilation from a fresh air inlet connected to the outside, **Figure 37–1.** Blower-assisted airflow is referred to as forced convection.

The forced-air system is different from a natural-draft system, where the air passes naturally over the conditioning equipment. Baseboard hot water heat is an example of natural-draft heat. The warmer water in the pipe heats the air in the vicinity of the pipe. The warmer air expands and rises. New air from the floor at a cooler temperature takes the place of the heated air, **Figure 37–2.** There is very little concern for the amount of air moving in a natural convection system.

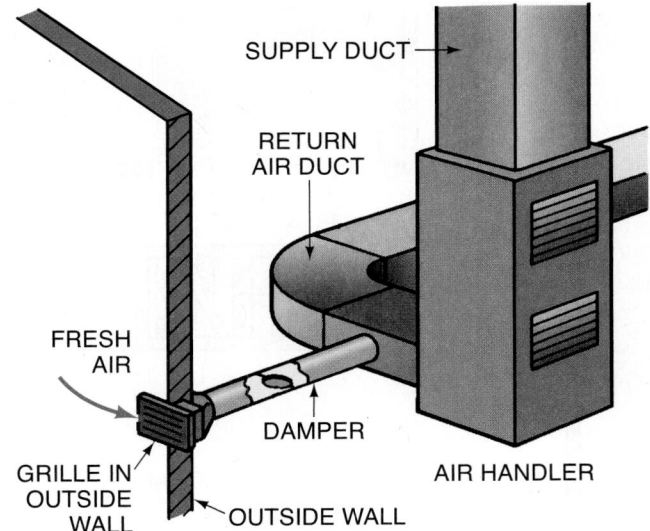

Figure 37–1 Ventilation using fresh air.

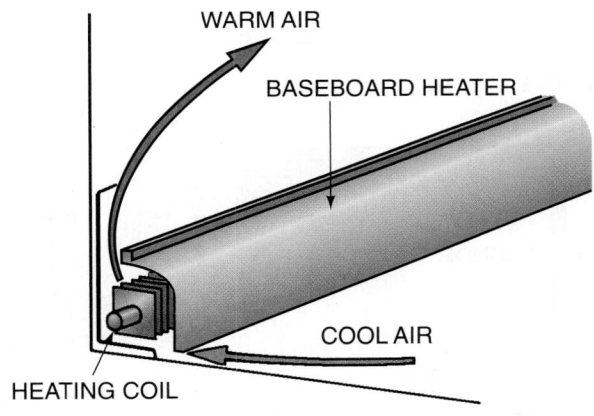

Figure 37–2 Air near the heated baseboard will become heated; it will expand and then rise.

37.2 CORRECT AIR QUANTITY

🌐 The object of the forced-air system is to *deliver the correct quantity of conditioned air to the occupied space.* When this occurs, the air mixes with the room air and creates a comfortable atmosphere in that space. Delivering the correct quantity of air to the space also increases the efficiency of the system. 🌐 Different spaces have different air quantity requirements; the same structure may have several different cooling requirements. For example, a house has rooms of different sizes with different requirements. A bedroom requires less heat and cooling than does a large living room. Different amounts of air need to be delivered to these rooms to maintain comfortable conditions, **Figure 37–3** and **Figure 37–4**.

Another example is a small office building with a high cooling requirement in the lobby and a low requirement in the individual offices. A factor that is taken into account is, among other things, the windows in each particular room or area in the structure. The number, type, and direction of the

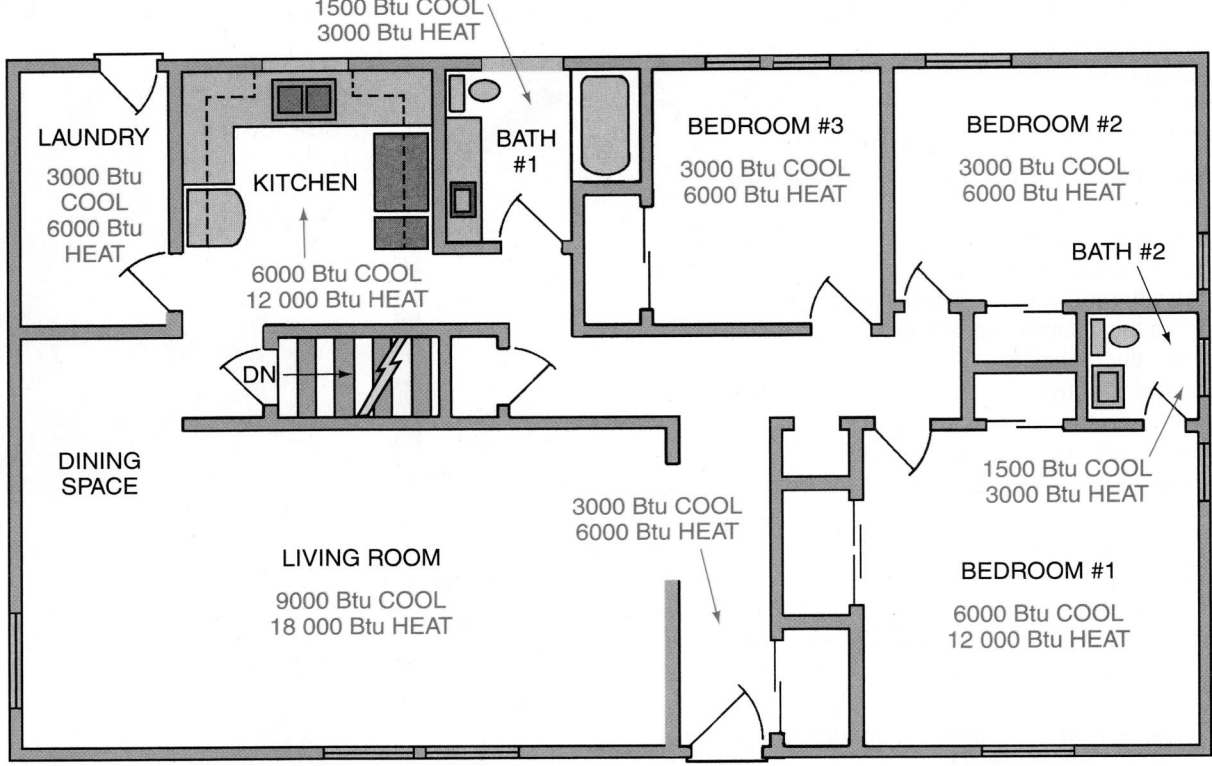

Figure 37–3 This floor plan has the heating and cooling requirements for each room indicated.

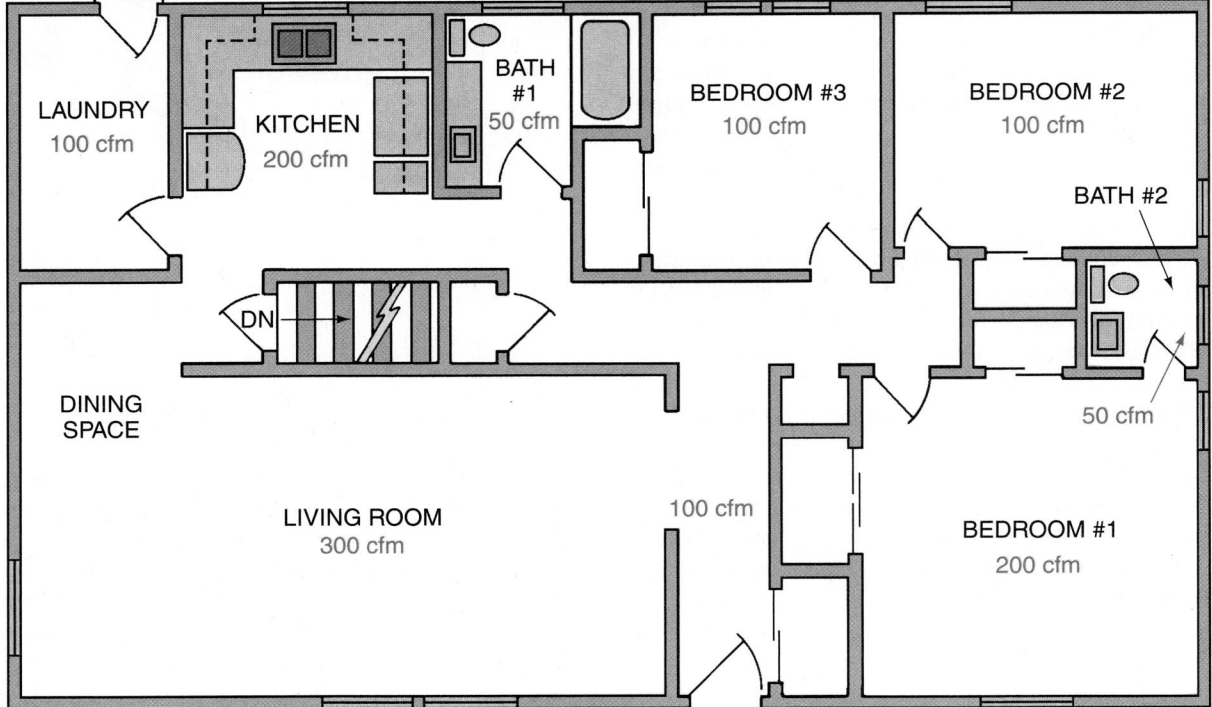

Figure 37–4 This is the same floor plan as that shown in **Figure 37–3**. The quantity of air, in cfm, to be delivered to each room is indicated.

windows in the structure will have a significant effect on the amount of conditioned air that is required for each individual area or room. The correct amount of air must be delivered to each part of the building so that one area will not be over-cooled while other areas are cooled correctly.

37.3   THE FORCED-AIR SYSTEM

The components that make up the *forced-air system* are the *blower* (fan), the *air supply system,* the *return air system,* and the *grilles* and *registers* where the circulated air enters the room and returns to the conditioning equipment. See **Figure 37–5** for an example of duct fittings. When these components are correctly chosen, they work together as a system with the following characteristics:

1. No air movement will be felt in the conditioned space that would normally be occupied.
2. No air noise will be noticed in the conditioned space.
3. No temperature swings will be felt by the occupants.
4. The occupants will not be aware that the system is on or off unless it stops for a long time and the temperature changes.

The lack of awareness of the conditioning system is important.

37.4   THE BLOWER

The *blower* or fan provides the pressure difference to force the air into the duct system, through the heat transfer surfaces in the conditioning equipment, through the grilles and registers, and into the room. Air has weight and has a resistance to movement. This means that it takes energy to move the air to the conditioned space. If the blower on a 3-ton air-conditioning system is required to move 1200 cubic feet of

air through the system per minute (400 cfm/ton), this is about 90 pounds of air moving through the system every minute (1200 ft³/min ÷ 13.33 ft³/lb = 90.02 lb/min). Ninety pounds of air moving through the system every minute equates to 5400 pounds of air every hour, or 129,600 pounds of air a day. In order to move the required air quantities, the blower must overcome the resistance and friction created by other components located in the air distribution system: the return duct, the supply duct, the air filter, the cooling coil, the heat exchanger, the supply registers, and the return air grilles, **Figure 37–6.** The blower motor consumes power to facilitate the moving of the correct air volumes through the system.

The pressure in a duct system for a residence or a small office building is too small to be measured in psi. It is measured in a unit of pressure that is still force per unit of area but in a smaller graduation. The pressure in ductwork is measured in *inches of water column* (in. WC). A pressure of 1 in. WC is the pressure necessary to raise a column of water 1 in. Air pressure in a duct system is measured with a *water manometer,* which uses colored water that rises up a tube. **Figure 37–7** shows a water manometer that is included for more accuracy at very low pressures. **Figure 37–8** shows some other instruments that may be used to measure very low air pressures; they are graduated in inches of water even though they may not contain water.

The atmosphere exerts a pressure of 14.696 psi at standard atmospheric conditions. If you recall from earlier units, 14.696 psi is also used to convert absolute pressure (psia) to gage pressure (psig). Atmospheric pressure will support a column of water 34 ft high. Since 14.696 psi will support 34 ft, one psi will support 2.31 ft (34 ft ÷ 14.696 = 2.31). A distance of 2.31 ft is equal to 27.7 in. (2.31 ft × 12 in./ft = 27.7 in.), **Figure 37–9.** The average

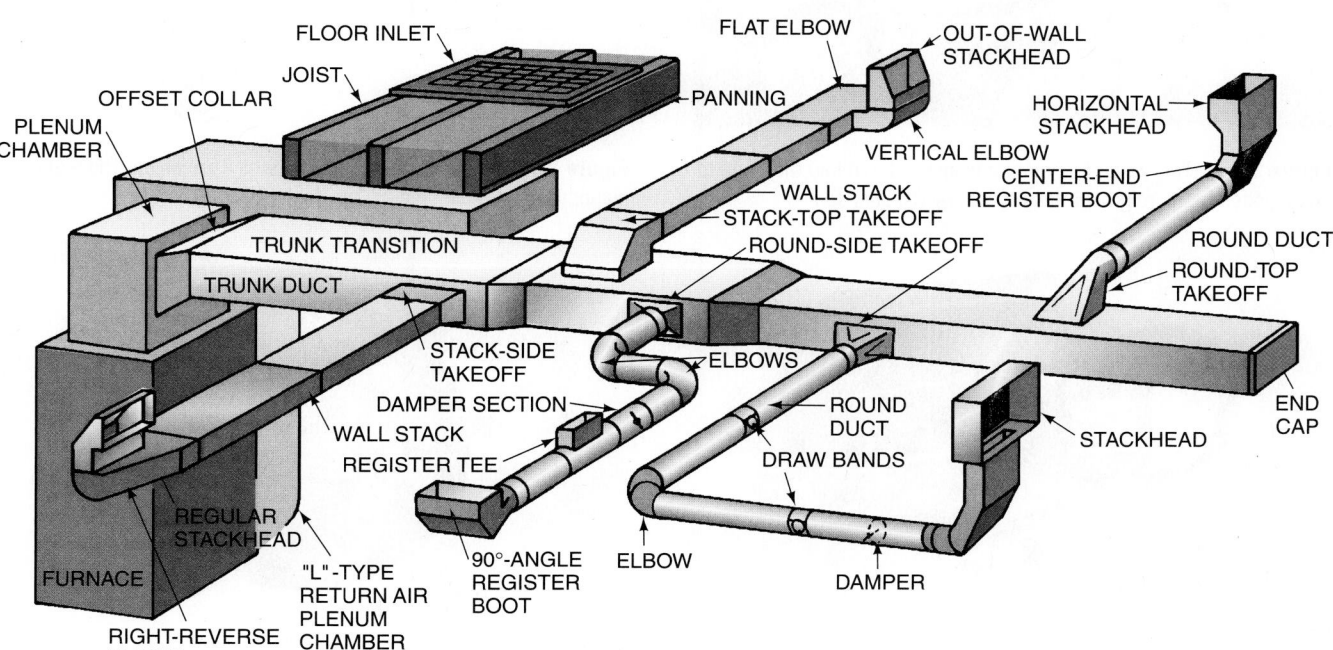

Figure 37–5 Duct fittings.

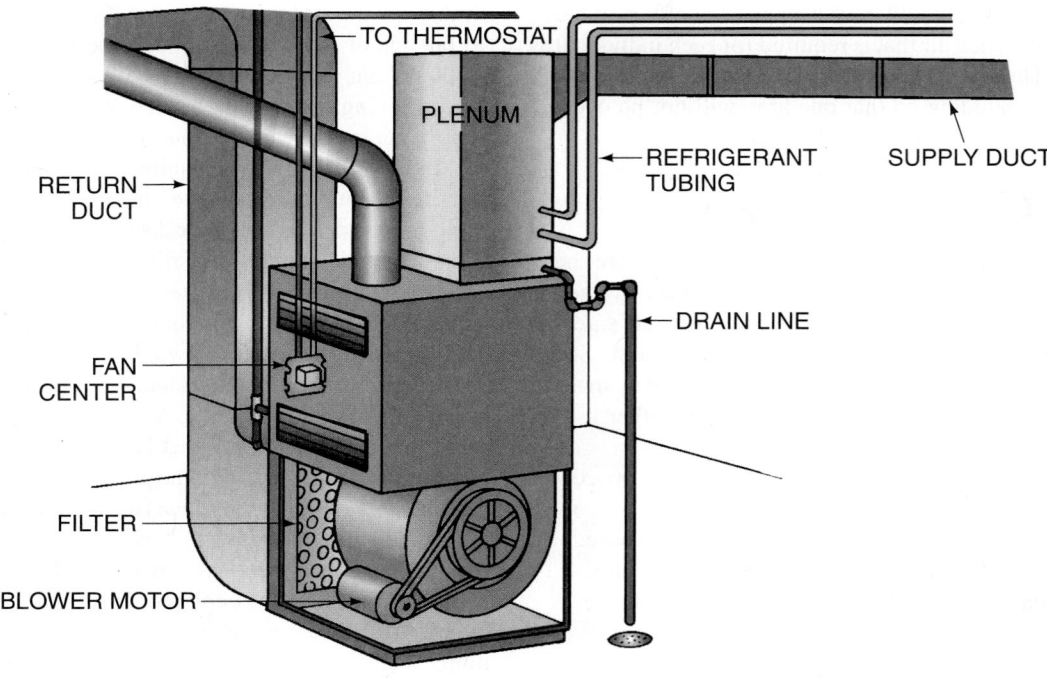

Figure 37–6 A fan and motor for moving air.

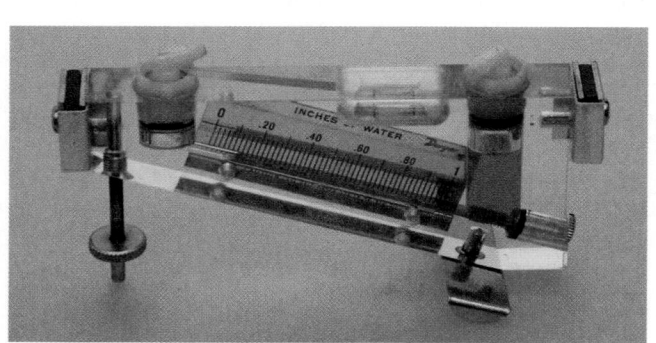

Figure 37–7 This water manometer is inclined to allow the scale to be extended for more graduations. *Photo by Bill Johnson*

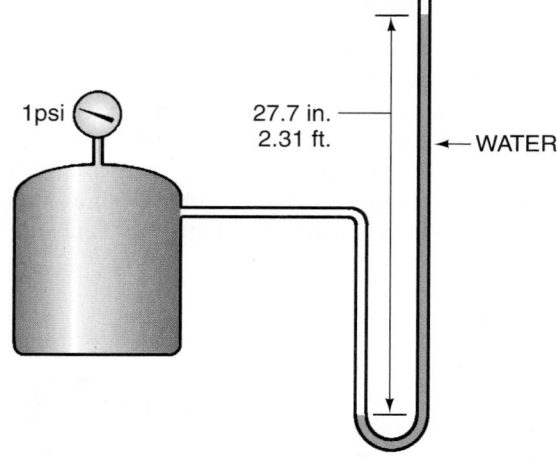

Figure 37–9 A vessel with a pressure of 1 psi inside. The water manometer has a column 27.7 in. high (2.31 ft).

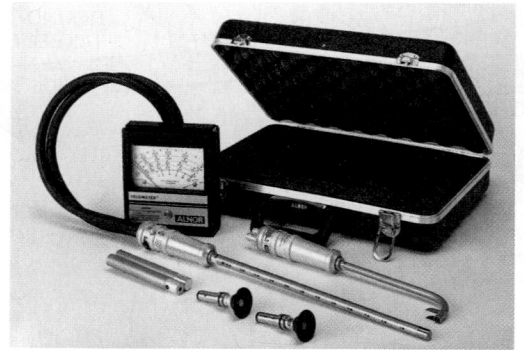

(A)

(B)

Figure 37–8 Other instruments used to measure air pressures. *(A) Courtesy Alnor Instrument Company. (B) Photo by Bill Johnson*

duct system will not exceed a pressure of 1 in. WC. A pressure of 0.05 psig will support a column of water 1.39 in. high ($27.7 \times 0.05 = 1.39$ in.). Airflow pressure in ductwork is measured in some very low figures.

37.5 SYSTEM PRESSURES

A duct system is pressurized by three pressures—*static pressure*, *velocity pressure*, and *total pressure*. Static pressure + velocity pressure = total pressure. Static pressure is the same as the pressure on an enclosed vessel and can be thought of as the pressure of the air pushing against the walls of the duct sections. This is like the pressure of the refrigerant in a cylinder that is pushing outward. **Figure 37–10** shows a manometer for measuring static pressure. Notice the position of the sensing tube. The probe has a very small hole in the end, so the air rushing by the probe opening will not cause incorrect readings.

The air in a duct system is moving along the duct, parallel to the walls of the duct section, and therefore has velocity. The velocity and weight of the air create velocity pressure. **Figure 37–11** shows a manometer for measuring velocity

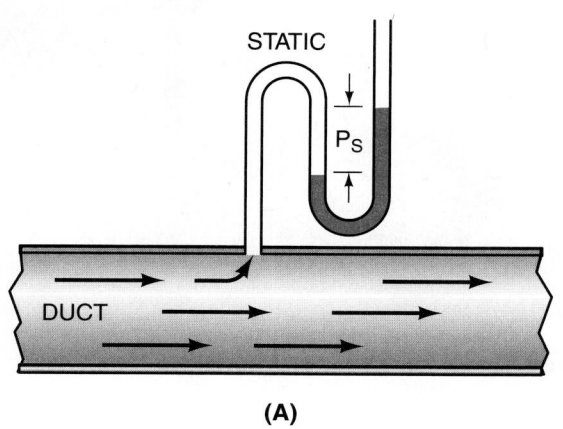

(A)

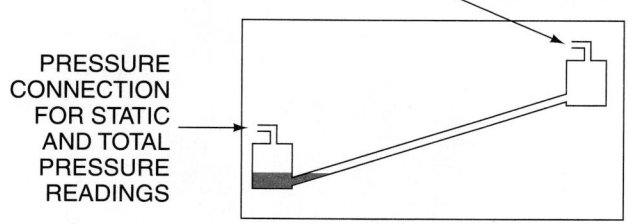

NOTE: THIS IS AN INCLINED MANOMETER. THE RISE IN PRESSURE IS AMPLIFIED BECAUSE OF THE INCLINE. IF THE INCLINE IS ON A GRADE OF 1 TO 3, THE FLUID WILL MOVE UP THE INCLINE 3" FOR EACH ACTUAL 1" OF WATER COLUMN PRESSURE.

(B)

Figure 37–10 **(A)** A manometer connected to measure the static pressure. **(B)** An inclined manometer.

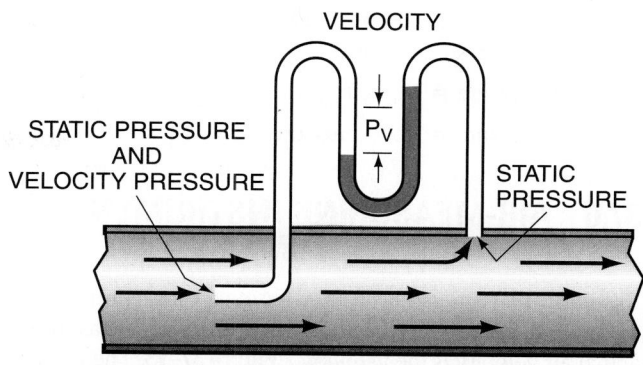

THE STATIC PRESSURE PROBE CANCELS THE STATIC PRESSURES AT THE VELOCITY PRESSURE END AND INDICATES TRUE VELOCITY PRESSURE.

Figure 37–11 A manometer connected to measure the velocity pressure of the air moving in the duct.

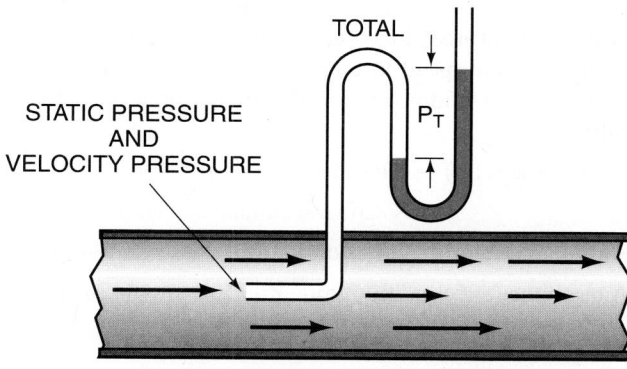

Figure 37–12 A manometer connected to measure the total air pressure in the duct.

pressure in an air duct. Notice the position of the sensing tube. The air velocity goes straight into the tube inlet, which registers both velocity and static pressure. The probe and manometer arrangement reads the velocity pressure by canceling the static pressure with the second probe. The static pressures balance each other and the velocity pressure is the difference.

The total pressure of a duct can be measured with a manometer applied a little differently, **Figure 37–12**. Notice that the velocity component or probe of the manometer is positioned so that the air is directed into the end of the tube. This will register static and velocity pressures. The main difference between the manometer connections in **Figure 37–11** and **Figure 37–12** is that there is no second connection to cancel out the static pressure from the measurement. In **Figure 37–11**, we have velocity pressure and static pressure pushing on the left side of the fluid in the manometer and only the static pressure pushing on the right side of the fluid in the manometer. This gives us the following duct pressure reading:

Duct pressure = Velocity pressure + Static pressure

− Static pressure

Duct pressure = Velocity pressure

With the connection as shown in **Figure 37–12,** we get the following duct pressure reading:

Duct pressure = Velocity pressure + Static pressure

Duct pressure = Total pressure

37.6 AIR-MEASURING INSTRUMENTS FOR DUCT SYSTEMS

The water manometer has been mentioned as an air pressure–measuring instrument. An instrument used to measure the actual air velocity is the *velometer,* **Figure 37–13.** The velometer actually measures how fast air is moving past a particular point in an air duct. Always use tools according to the manufacturer's instructions and recommendations. Once the average air speed, in ft/min, has been determined using an instrument similar to that shown in **Figure 37–13,** the air volume that is moving in the duct can be easily determined. The only other piece of information that is needed is the cross-sectional area of the duct section, in ft^2. For example, if the average air speed through a 9" × 8" rectangular duct section is 700 ft/min, we can determine the air volume as follows:

Air volume (cfm) = Air velocity (ft/min)
$$\times \text{ Cross-sectional area (ft}^2)$$

Air volume (cfm) = 700 ft/min × [(9" × 8") ÷ 144 in^2/ft^2]

Air volume (cfm) = 700 ft/min × (72 in^2 ÷ 144 in^2/ft^2)

Air volume = 350 cfm

(A)

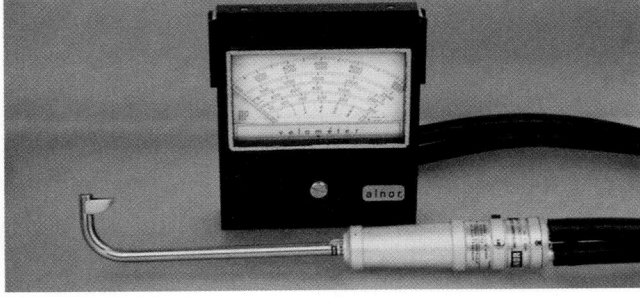

(B)

Figure 37–13 Two types of velometers. *Photos by Bill Johnson*

These calculations and concepts are covered in more detail in Section 37.31, "Measuring Air Movement for Balancing."

New-generation velometers have incorporated a number of additional functions into their design, **Figure 37–14.** This instrument, for example, has the capability to measure the temperature and relative humidity of the air as it passes through the impeller of the unit. This is accomplished by positioning sensors in the air stream, **Figure 37–15.** Other instruments have a modified keypad so that the service technician can enter the duct measurements into the device. This enables the instrument to display the actual cfm of the air, saving the technician time on the job and reducing the number of manual calculations that need to be performed.

A special device called a *pitot tube* was developed many years ago and is used with special manometers for

Figure 37–14 A new-generation digital velometer. *Photo by Eugene Silberstein*

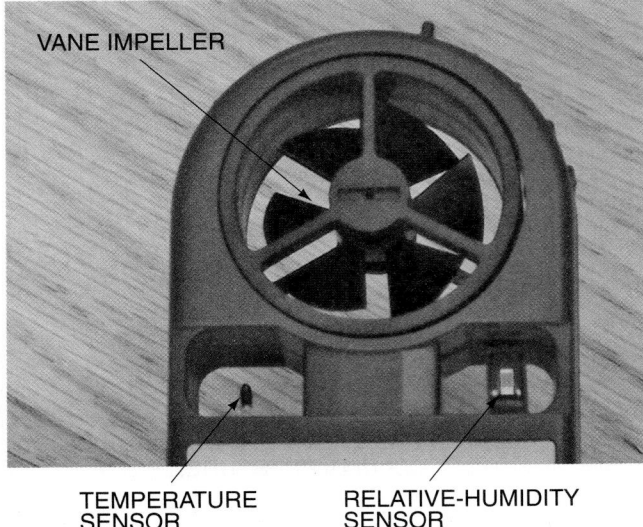

VANE IMPELLER

TEMPERATURE SENSOR

RELATIVE-HUMIDITY SENSOR

Figure 37–15 Temperature and relative-humidity sensors. *Photo by Eugene Silberstein*

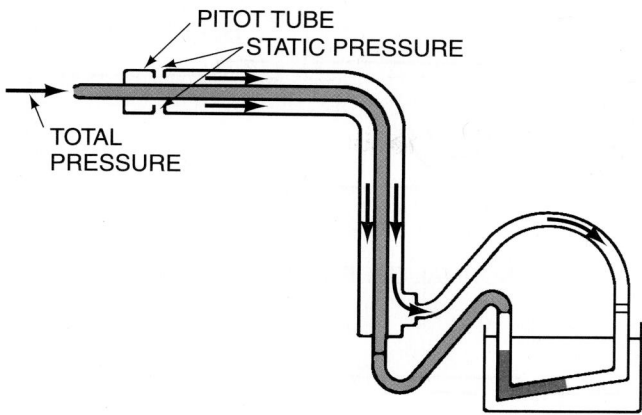

Figure 37–16 A pitot tube set up to measure velocity pressure.

checking duct air pressure at most pressure levels, **Figure 37–16**. The pitot tube is constructed as two concentric tubes and is bent into a 90-degree angle. The open end of the bent tube is inserted into the duct section with the open end directed into the air stream. This way, the air flowing in the duct will flow into the tube. The outer tube has small holes in it that are positioned perpendicular to airflow. The airflow into the center tube will provide the total pressure, while the pressure that is indicated by the perpendicular holes will give the static pressure. **Figure 37–17** shows such a setup where PT is the total pressure, PV is the velocity pressure, and PS is the static pressure. The velocity pressure reading is obtained by subtracting the static pressure from the total pressure. This is accomplished at the manometer.

37.7 TYPES OF FANS

The blower, or fan as it is sometimes called, can be described as a device that produces airflow or movement. Several different types of blowers produce this movement, but all can be described as non-positive-displacement air movers. Remember, the compressor (with the exception of the centrifugal compressor) is a positive-displacement pump. When the cylinder is full of refrigerant (or air for an air compressor), the compressor is going to empty that cylinder or break something. The fan is not positive displacement—it cannot build the kind of pressure that a compressor can. The fan has other characteristics that have to be dealt with.

The two fans that we discuss in this text are the *propeller fan* and the *forward curve centrifugal fan,* also called the *squirrel cage fan,* or *blower wheel.*

The propeller fan is used in exhaust-fan and condenser-fan applications. It will handle large volumes of air at low-pressure differentials. The propeller can be cast iron, aluminum, or stamped steel and is set into a housing called a *venturi* to encourage airflow in a straight line from one side of the fan to the other, **Figure 37–18**. The propeller fan makes more noise than the centrifugal fan so it is normally used where noise is not a factor.

The squirrel cage or centrifugal fan has characteristics that make it desirable for ductwork. It builds more pressure from the inlet to the outlet and moves more air against more pressure. This fan has a forward curved blade and a cut-off to shear the air spinning around the fan wheel. This air is thrown by centrifugal action to the outer perimeter of the fan wheel. Some of it would keep going around with the fan wheel if it were not for the shear that cuts off the air and sends it out the fan outlet, **Figure 37–19**. The centrifugal fan is very quiet when properly applied. It meets all requirements of duct systems up to very large systems that are considered high-pressure systems. High-pressure systems have pressures of 1 in. WC and more. Different types of fans, some of them similar to the forward curve centrifugal fan, are used in larger systems. The rotation of a centrifugal fan is critical to its capability to move air. The rotation can be determined by looking at the fan from the side and finding the fan outlet. The fan wheel must turn so that the top of the fan wheel is rotating toward the fan outlet, **Figure 37–19**.

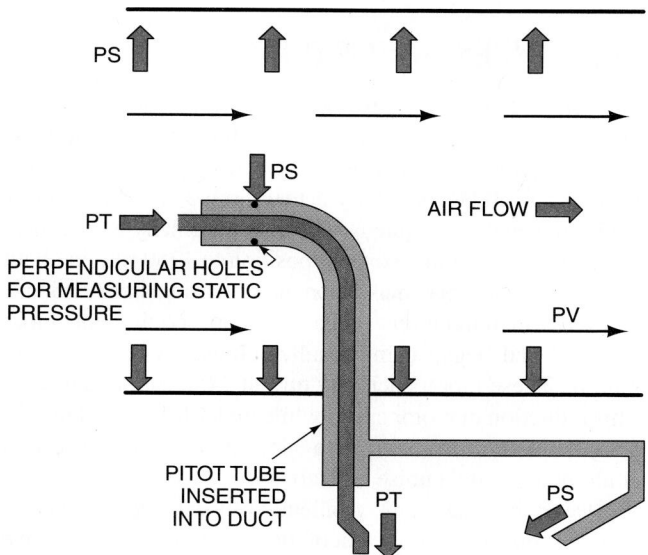

Figure 37–17 Pressures exerted on the pitot tube.

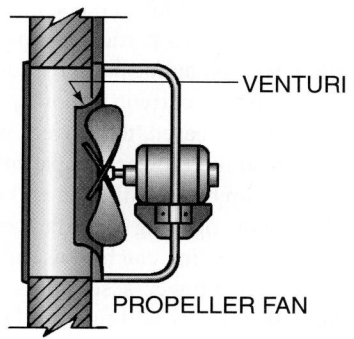

Figure 37–18 A propeller-type fan.

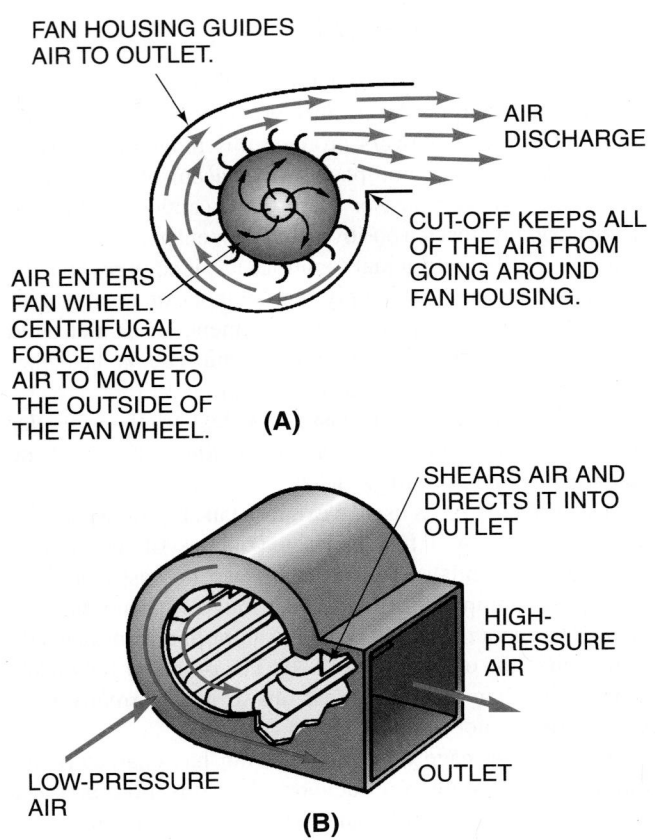

FAN HOUSING GUIDES AIR TO OUTLET.

AIR DISCHARGE

CUT-OFF KEEPS ALL OF THE AIR FROM GOING AROUND FAN HOUSING.

AIR ENTERS FAN WHEEL. CENTRIFUGAL FORCE CAUSES AIR TO MOVE TO THE OUTSIDE OF THE FAN WHEEL.

(A)

SHEARS AIR AND DIRECTS IT INTO OUTLET

HIGH-PRESSURE AIR

OUTLET

LOW-PRESSURE AIR

(B)

Figure 37–19 A centrifugal fan.

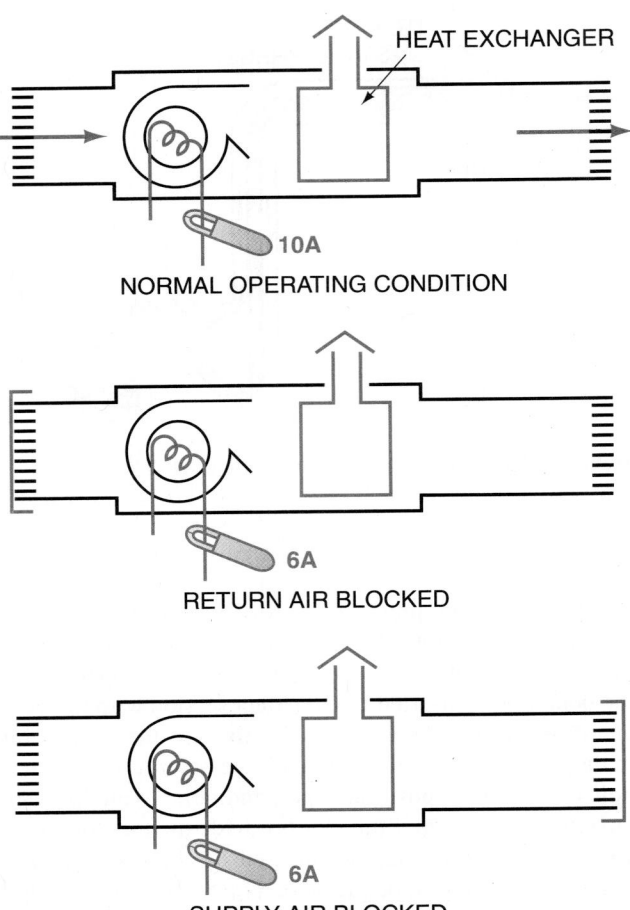

HEAT EXCHANGER

10A

NORMAL OPERATING CONDITION

6A

RETURN AIR BLOCKED

6A

SUPPLY AIR BLOCKED

Figure 37–20 Airflow situations.

One characteristic that makes troubleshooting the centrifugal fan easier is the volume of air-to-horsepower requirement. This fan uses energy at the rate at which it moves air through the ductwork. The current draw of the fan motor is in proportion to the pounds of air it moves or pumps. For example, a fan motor that pulls full-load amperage at the rated fan capacity will pull less than full-load amperage at any value less than the fan capacity. The fan is supposed to pull 10 A while moving only this amount of air. The weight of this volume of air can be calculated by dividing (3000 cfm ÷ 13.35 ft³/lb = 224.7 lb/min). If the fan inlet is blocked, the suction side of the fan will be starved for air, and the current will go down. If the discharge side of the fan is blocked, the pressure will go up in the discharge of the fan, and the current will go down because the fan is not handling as many pounds of air, **Figure 37–20**. The air is merely spinning around in the fan and housing and not being forced into the ductwork. This particular type of fan can be checked for airflow with an ammeter when you make simple field measurements. If the current is down, the airflow is down. If the airflow is increased, the current will go up. For example, if the door on the blower compartment is opened, the fan current will go up because the fan will have access to more air through the large opening of the fan compartment. In addition, the amperage reading can be a useful tool for properly setting belt-driven blower assemblies. If the driven pulley is too small, the blower will turn too fast and the amperage draw of the motor will increase. The same condition will be present if the drive pulley is too large. This is a relatively common situation given that the drive pulley is often a variable-pitch component. Some technicians will use the variable-pitch feature on the drive pulley to adjust the tension on the belt. This is not desirable as doing so will alter the rate at which air is moving through the system and will also affect the amperage draw of the motor.

37.8 TYPES OF FAN DRIVES

The centrifugal blower must be turned by a motor. Two drive mechanisms are used: belt drive or direct drive. The belt-drive blower was used exclusively for many years. The motors were usually 1800 rpm. They actually run at 1750 rpm under load and operate very quietly. The motor normally has a capacitor and will go from a stopped position to 1750 rpm in about 1 sec. The motor may make more noise starting than running.

Later, manufacturers began making equipment more compact and began using smaller blowers with 3600-rpm motors. These motors actually turn at 3450 rpm under load. This reduction in motor speed while under load is called *slip*. They had to turn from 0 rpm to 3450 rpm in about 1 sec, and could make quite a noise on start-up.

Sleeve bearings and resilient (rubber) mountings are used to keep bearing noise out of the blower section. Belt-drive blowers have two bearings on the fan shaft and two bearings on the motor. Sometimes these bearings are permanently lubricated by the manufacturer, so a technician cannot oil them.

The drive pulley on the motor, the driven pulley on the fan shaft, and the belt must be maintained. This motor and blower combination has many uses because the pulleys can be adjusted or changed to change fan speeds, **Figure 37–21**.

Recently, most manufacturers have been using a direct-drive blower for cooling equipment up to 5 tons. The motor is mounted on the blower housing, usually with rubber mounts, with the motor shaft extending into the fan wheel. The motor is a permanent split-capacitor (PSC) motor that starts up very slowly, taking several seconds to get up to speed, **Figure 37–22**. It is very quiet and does not have a belt and pulleys to wear out or to adjust. Shaded-pole motors are used on some direct-drive blowers; however, they are not as efficient as PSC motors.

With PSC motors the fan wheel bearing is located in the motor, which reduces the bearing surfaces from four to two. The bearings may be permanently lubricated at the factory. The front bearing in the fan wheel may be hard to lubricate if a special oil port is not furnished. No belts or pulleys exist to maintain or adjust. The fan turns at the same speed as the motor, so multispeed

Figure 37–22 A fan motor that is normally energized with a relay. *Courtesy Universal Electric Company*

motors are common. The air volume may be adjusted with the different fan motor speeds instead of a pulley. The motor may have up to four different speeds that can be changed by switching wires at the motor terminal box. Common speeds are from about 1500 rpm down to about 800 rpm. The motor can be operated at a faster speed in the summer (as two-speed motors) for more airflow for cooling, **Figure 37–23**.

37.9 THE SUPPLY DUCT SYSTEM

The supply duct system distributes air to the terminal units, registers, grilles, or diffusers in the conditioned space. Starting at the fan outlet, the duct can be fastened to the blower or blower housing directly or have a fireproof vibration eliminator between the blower and the ductwork. This vibration eliminator is often referred to as a canvas collar, although these are very rarely made of canvas anymore. Flexible rubber or a similar material is more commonly used. The vibration eliminator is recommended on all installations but is not always used. If the blower is quiet, it may not be necessary, **Figure 37–24**.

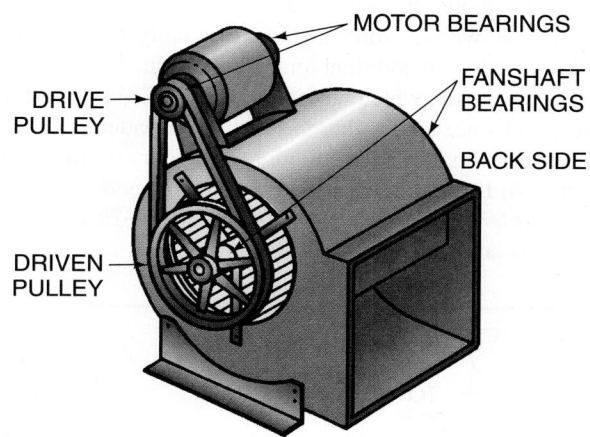

Figure 37–21 This blower is driven with a motor using a belt and two pulleys to transfer the motor energy to the fan wheel.

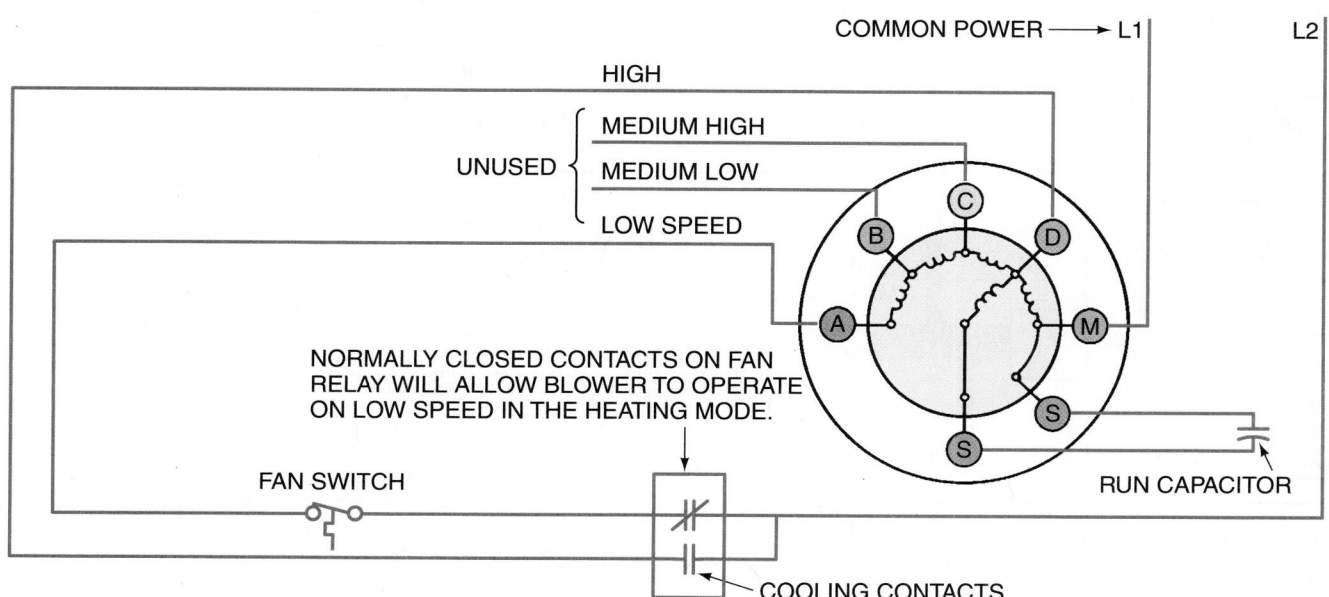

Figure 37–23 The wiring diagram of a multiple-speed motor.

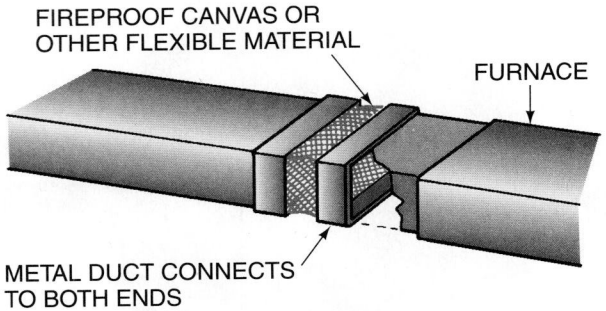

FIREPROOF CANVAS OR
OTHER FLEXIBLE MATERIAL

FURNACE

METAL DUCT CONNECTS
TO BOTH ENDS

Figure 37–24   A vibration eliminator.

Since the equipment casing and ductwork are both made of metal, any sounds or vibrations are easily transmitted through the duct system. Given the relatively low cost of the vibration eliminator material, it is wise to incorporate this component on all air distribution systems.

The supply duct system must be designed to allow the air to move toward the conditioned space as freely as possible. The duct system must be sized properly: neither undersized nor oversized. Oversized duct systems cost more to fabricate and install and will also result in the slowing of the air as it flows through the duct. If you recall from earlier in this unit, we determined that the volume of airflow through a duct section was calculated by multiplying the velocity of the air by the cross-sectional area of the duct. Consider an air duct that is designed to move 350 cfm and has a cross-sectional area of one-half of a square foot (72 in^2). The velocity of the air

is determined as follows:

$$\text{Air volume (cfm)} = \text{Air velocity (ft/min)}$$
$$\times \text{Cross-sectional area (ft}^2\text{)}$$
$$350 \text{ cfm} = \text{Air velocity (ft/min)} \times 0.5 \text{ ft}^2$$
$$\text{Air velocity (ft/min)} = 350 \text{ cfm} \div 0.5 \text{ ft}^2$$
$$\text{Air velocity (ft/min)} = 700 \text{ ft/min}$$

Let's now assume that the ductwork was oversized by 50% for this application. The desired airflow is still 350 cfm, but the cross-sectional area of the duct is now 0.75 ft^2. The velocity of the air is now as follows:

$$\text{Air volume (cfm)} = \text{Air velocity (ft/min)}$$
$$\times \text{Cross-sectional area (ft}^2\text{)}$$
$$350 \text{ cfm} = \text{Air velocity (ft/min)} \times 0.75 \text{ ft}^2$$
$$\text{Air velocity (ft/min)} = 350 \text{ cfm} \div 0.75 \text{ ft}^2$$
$$\text{Air velocity (ft/min)} = 467 \text{ ft/min}$$

By reducing the velocity of the air, the path the air takes as it leaves the supply register will also be changed. The air will not come out with the force that was originally planned for at the design stage. In addition, undersizing a duct will result in excessive velocity, which will increase the noise levels of the system and, once again, alter the desired air patterns.

Duct systems typically follow one of four common configurations. Duct systems can be *plenum, extended plenum, reducing plenum,* or *perimeter loop,* **Figure 37–25.** Each system has its advantages and disadvantages.

(A) PLENUM OR RADIAL DUCT SYSTEM

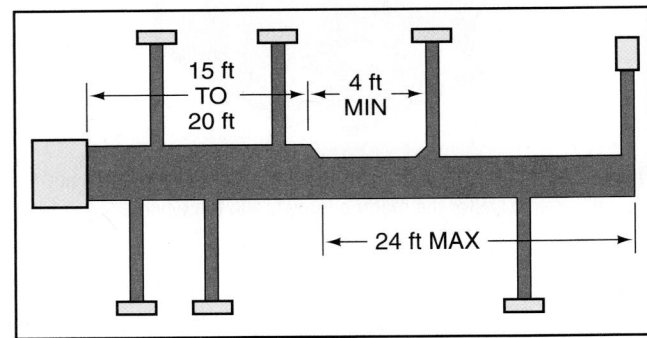

(C) REDUCING EXTENDED PLENUM SYSTEM

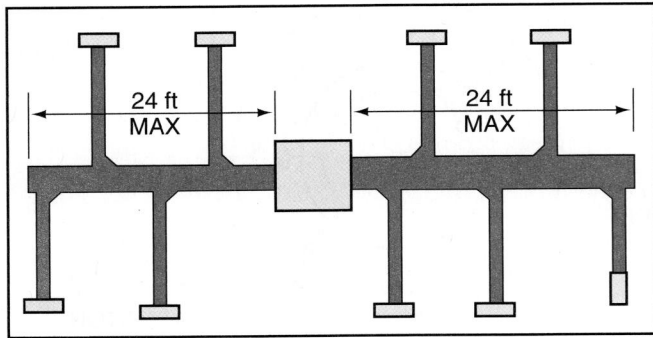

(B) EXTENDED PLENUM SYSTEM

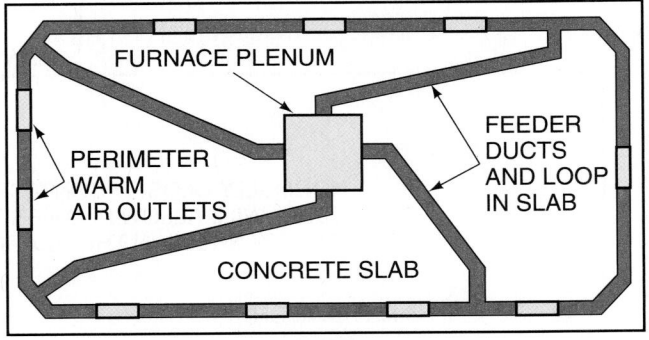

(D) PERIMETER LOOP SYSTEM WITH FEEDER
AND LOOP DUCTS IN CONCRETE SLAB

Figure 37–25   **(A)** A plenum system. **(B)** An extended plenum system. **(C)** A reducing extended plenum system. **(D)** A perimeter loop system.

37.10   THE PLENUM SYSTEM

The plenum system has an individual supply system that makes it well suited for a job in which the room outlets are all close to the unit. This system is economical from a first-cost standpoint and can be installed easily by an installer with a minimum of training and experience. The supply diffusers (where the air is diffused and blown into the room) are normally located on the inside walls and are used for heating systems with very warm or hot air as the heating source. Plenum systems work better on fossil-fuel (coal, oil, or gas) systems than with heat pumps because the leaving-air temperatures are much warmer in fossil-fuel systems. They are often applied to inside walls for short runs, for example, in apartment houses.

When the supply diffusers are located on the inside walls, a warmer air is more desirable. The supply air temperature on a heat pump without strip heat is rarely more than 100°F, whereas on a fossil-fuel system it could easily reach 130°F.

The return air system can be a single return located at the air handler, which makes materials economical. The single-return system will be discussed in more detail later.

Figure 37–26 shows an example of a plenum system with registers in the ceiling.

37.11   THE EXTENDED PLENUM SYSTEM

The extended plenum system can be applied to a long structure such as the ranch-style house. This system takes the plenum closer to the farthest point. The extended plenum is called the *trunk duct* and can be round, square, or rectangular, **Figure 37–27**. The system uses small ducts called *branches* to complete the connection to the terminal units. These small ducts also can be round, square, or rectangular. In small sizes they are usually round because it is less expensive to manufacture and assemble round duct. An average home probably has 6-in. round duct for the branches. The main drawback of the extended plenum system is that the velocity of the air is lower as the air reaches the last take-offs on the extended plenum. Consider the portion of the air distribution system shown in **Figure 37–28**. There is 1000 cfm of air moving through the duct at a velocity of 700 ft/min. It can be determined that the cross-sectional area of the duct is 1.43 square feet (1000 cfm ÷ 700 ft/min). After the first

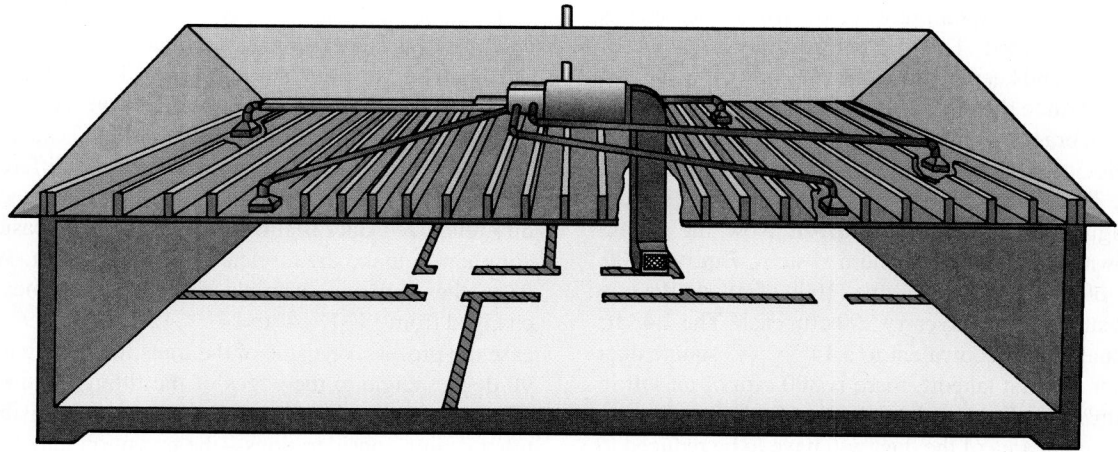

Figure 37–26   Using a plenum system.

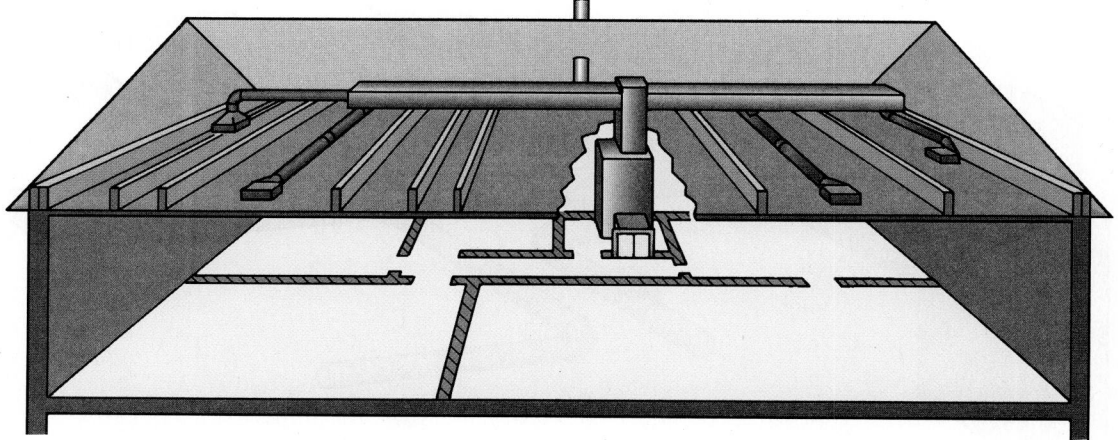

Figure 37–27   An extended plenum system.

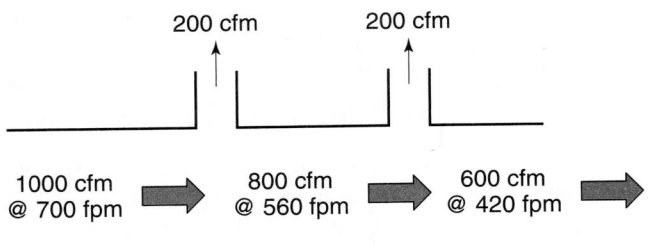

Figure 37–28 In an extended plenum system, the velocity of the air decreases as the end of the trunk line is reached.

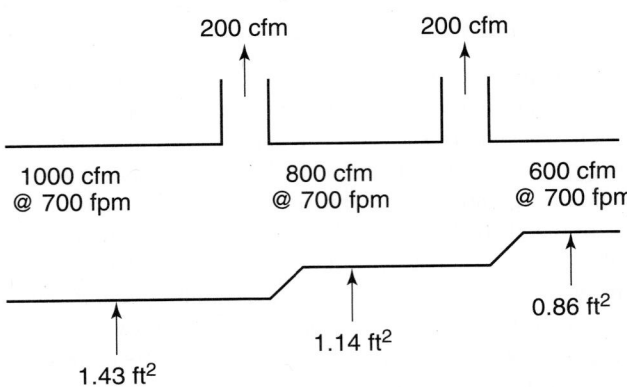

Figure 37–30 The velocity of the air that reaches the end of the trunk line can be maintained by reducing the size of the trunk line.

takeoff, where 200 cfm has been directed to an area in the conditioned space, there is 800 cfm of air remaining. Because this duct system is an extended plenum system, the duct size remains the same. This means that the velocity of the air after the first 200-cfm takeoff will be only 560 ft/min (800 cfm ÷ 1.43 ft^2). After the second 200-cfm takeoff, the air velocity will be only 420 ft/min.

37.12 THE REDUCING PLENUM SYSTEM

The reducing plenum system reduces the trunk duct size as branch ducts are added. This system has the advantage of saving materials and keeping the same pressure from one end of the duct system to the other, when properly sized. This ensures that each branch duct has approximately the same pressure and velocity pushing air into its takeoff from the trunk duct, **Figure 37–29**. Let's take a look at the same duct system that was originally shown in **Figure 37–28**. Now, the system will be shown as a reducing plenum system, **Figure 37–30**. The initial volume of air is the same (1000 cfm) and the two takeoffs are still designed to carry 200 cfm each. The 1.43 ft^2 cross-sectional area is equivalent to a 14" × 14" square duct section. After the first takeoff, there is 800 cfm of air left in the main trunk line. To keep the velocity of the air the same, the cross-sectional area of the duct will have to be reduced to 1.14 ft^2 (800 cfm ÷ 700 ft/min). This is the equivalent to a

rectangular duct section that is about 14" × 12". After the second takeoff, there is 600 cfm of air left in the main trunk line. To keep the velocity of the air the same, the cross-sectional area of the duct will have to be reduced to 0.86 ft^2 (600 cfm ÷ 700 ft/min). This last duct section is equivalent to a rectangular duct section that is 14" × 9".

It can be seen that manufacturing a 14" × 9" duct section will use less material than fabricating a 14" × 14" duct section. The same holds true for the 14" × 12" section. The main downside to this type of system, however, is the need for transition duct sections, **Figure 37–31(A)**. These duct fittings are used to connect duct sections that have different dimensions. Most of the time, when the system is designed, the two different-sized duct sections will have one measurement in common to make the fabrication process easier. We can see from the previous example that the duct measurements changed from 14" × 14" to 14" × 12" to 14" × 9". To further ease the process, one side of the transition fitting is often at a 90-degree angle to the edges of the fitting, **Figure 37–31(B)**. When the air distribution system is to remain visible—or for other design considerations—the transition may have a symmetrical layout, **Figure 37–31(C)**.

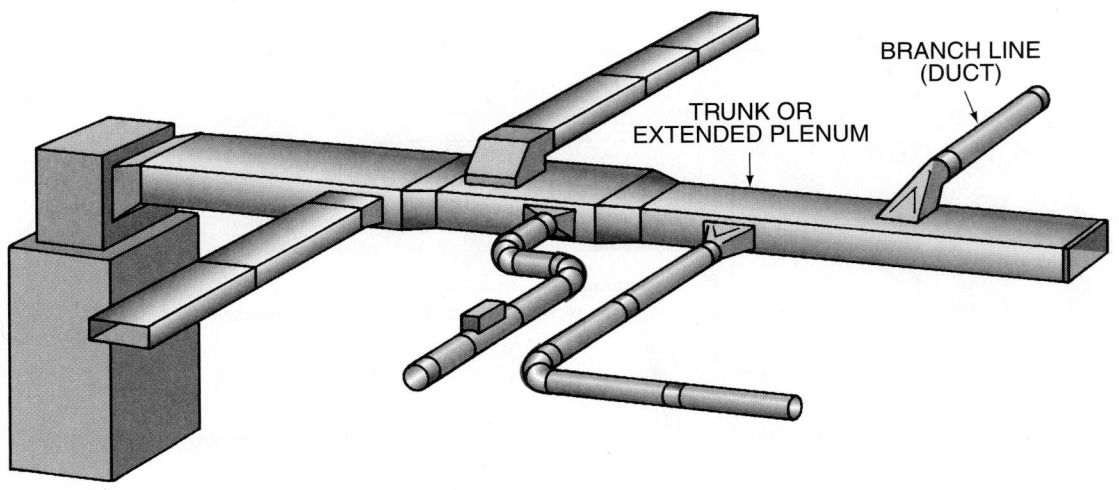

Figure 37–29 A reducing plenum system. *Courtesy Climate Control*

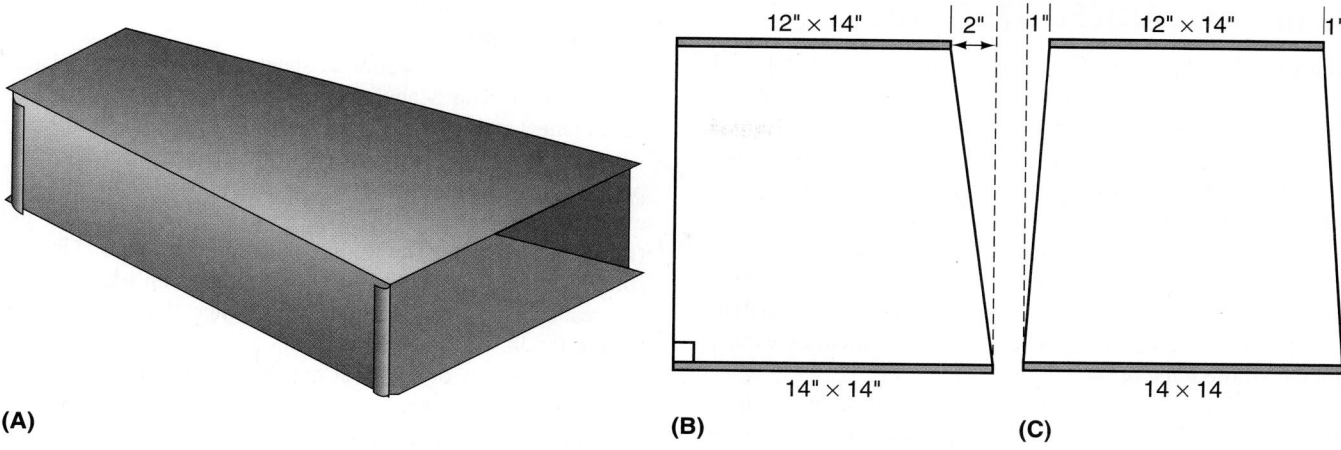

Figure 37–31   **(A)** A transition fitting. **(B)** Common transition fitting layout. **(C)** Symmetrical transition fitting.

37.13   THE PERIMETER LOOP SYSTEM

The perimeter loop duct system is particularly well suited for installation in a concrete floor in a colder climate. The loop can be run under the slab close to the outer walls with the outlets next to the wall. Warm air is in the whole loop when the furnace fan is running, and this keeps the slab at a more even temperature. The loop has a constant pressure around the system and provides the same pressure to all outlets, **Figure 37–32**.

37.14   DUCT SYSTEM STANDARDS

All localities should have some minimum standard for air-conditioning and heating systems. The regulating body might use a state standard, or it might have a local standard. These are called codes. Many state and local standards adopt other known standards, such as Building Officials and

Code Administrators International (BOCA). All technicians should become familiar with the local code requirements and follow them. These code requirements are not standard across the nation.

37.15   DUCT MATERIALS

The ductwork for carrying the air from the fan to the conditioned space can be made of different materials. For many years galvanized sheet metal was used exclusively, but it is expensive to manufacture and assemble at the job. Galvanized metal is by far the most durable material. It can be used in walls where easy access for servicing is not available. Aluminum, fiberglass ductboard, spiral metal duct, and flexible duct have all been used successfully. SAFETY PRECAUTION: The duct material must meet the local codes for fire protection.

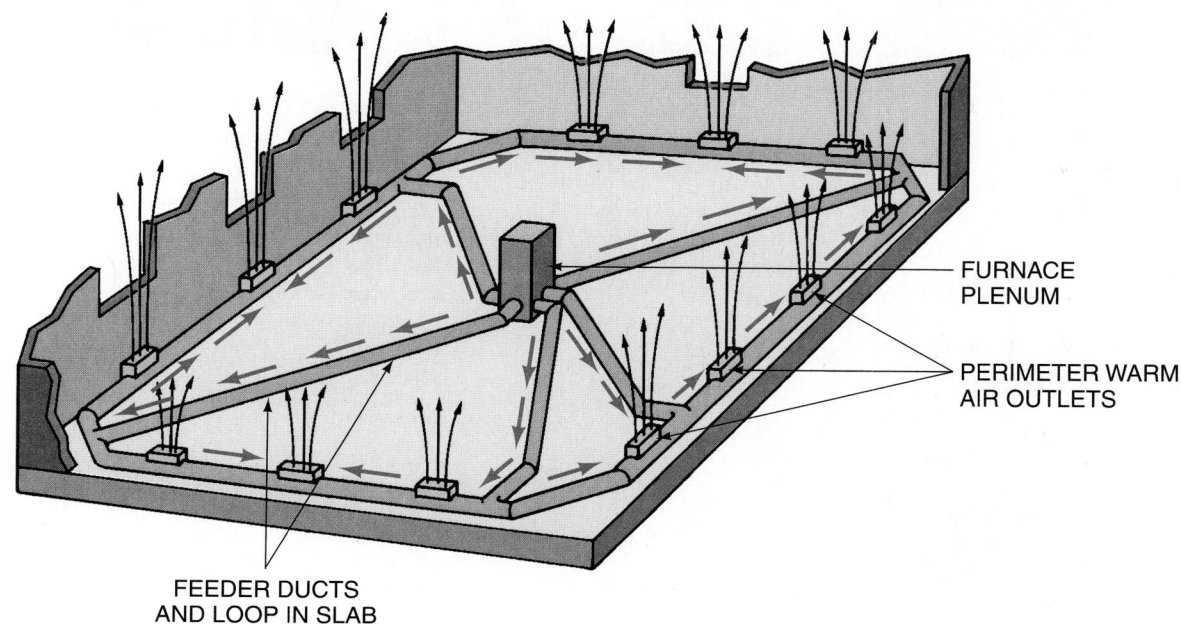

Figure 37–32   A perimeter loop system.

37.16 GALVANIZED-STEEL DUCT

Many air distribution systems are fabricated from a material known as galvanized-steel sheet metal. Galvanized steel is steel sheet metal that has been treated with zinc to form a coating on the surface of the metal. This zinc coating will ultimately form zinc carbonate on the surface of the metal, which acts to prevent corrosion of the steel underneath. The galvanizing process takes place at about 860°F. The material is easily identified by the crystal-like patterning on the surface of the metal that is often referred to as spangle. Galvanized sheet metal comes in several different thicknesses that are called the *gage* of the metal.

As the gage number gets larger, the thickness of the sheet metal gets thinner. For example, the thickness of 30-gage galvanized sheet metal is 0.0157 inches, while the thickness of 22-gage galvanized sheet metal is 0.0336 inches. It can be seen that 22-gage galvanized sheet metal is about twice as thick as 30 gage. **Figure 37-33** shows the thicknesses of some commonly used gages of steel, galvanized-steel, and aluminum sheet metal.

The thickness of a section of sheet metal is directly related to its weight. The Manufacturer's Standard Gage for Sheet Metal (sheet steel) is 41.82 lb/ft². This is a constant factor and, when multiplied by the thickness of the steel sheet metal, will provide the weight of one square foot of the material. For example, the thickness of 24-gage sheet steel is 0.0239 in. (from **Figure 37-33**). The weight of 1 ft² of this material is 1 lb (0.0239 inches × 41.82 lb/ft²). It is left as an exercise for the reader to confirm that 1 ft² of 30-gage steel sheet metal weighs 0.5 lb.

The thickness of the duct can be less when the dimensions of the ductwork are small. When the ductwork is larger, it must be more rigid, or it will swell and make noises when the fan starts or stops. **Figure 37-34** shows a table to be used as a guideline for choosing duct metal thickness. Quite often the duct manufacturer will cross-break or make a slight bend from corner to corner on large fittings to make the duct more rigid.

Metal duct is normally furnished in lengths of 4 ft and can be round, square, or rectangular. Smaller round duct can be purchased in lengths up to 10 ft. Duct lengths can be fastened together with special fasteners (called S fasteners or slip fasteners) and drive cleats if the duct is square or rectangular, or fastened with self-tapping sheet metal screws if the duct is round. These fasteners make a secure connection that is almost airtight at the low pressures at which the duct is normally operated. See **Figure 37-35** for an example of these fasteners. If there is any question that the air might leak out, special tape or sealant can be applied to the connections. **Figure 37-36** shows a duct that is fastened together with self-tapping sheet metal screws. **Figure 37-37** shows a connection that has been taped after being fastened.

SHEET THICKNESS IN INCHES			
GAGE	STEEL	GALVANIZED STEEL	ALUMINUM
22	0.0299	0.0336	0.0253
24	0.0239	0.0276	0.0201
26	0.0179	0.0217	0.0159
28	0.0149	0.0187	0.0126
30	0.012	0.0157	0.0100

Figure 37-33 Common thicknesses for steel, galvanized-steel, and aluminum sheet metal.

GAGES OF METAL DUCTS AND PLENUMS USED FOR COMFORT HEATING OR COOLING FOR A SINGLE DWELLING UNIT				
	COMFORT HEATING OR COOLING			Comfort Heating Only
	Galvanized Steel			
	Nominal Thickness (In Inches)	Equivalent Galvanized Sheet Gage No.	Approximate Aluminum B & S Gage	Minimum Weight Tin-Plate Pounds Per Base Box
Round Ducts and Enclosed Rectangular Ducts				
14" or less	0.016	30	26	135
Over 14"	0.019	28	24	—
Exposed Rectangular Ducts				
14" or less	0.019	28	24	—
Over 14"	0.022	26	23	—

Figure 37-34 A table of recommended metal thickness for different sizes of duct.

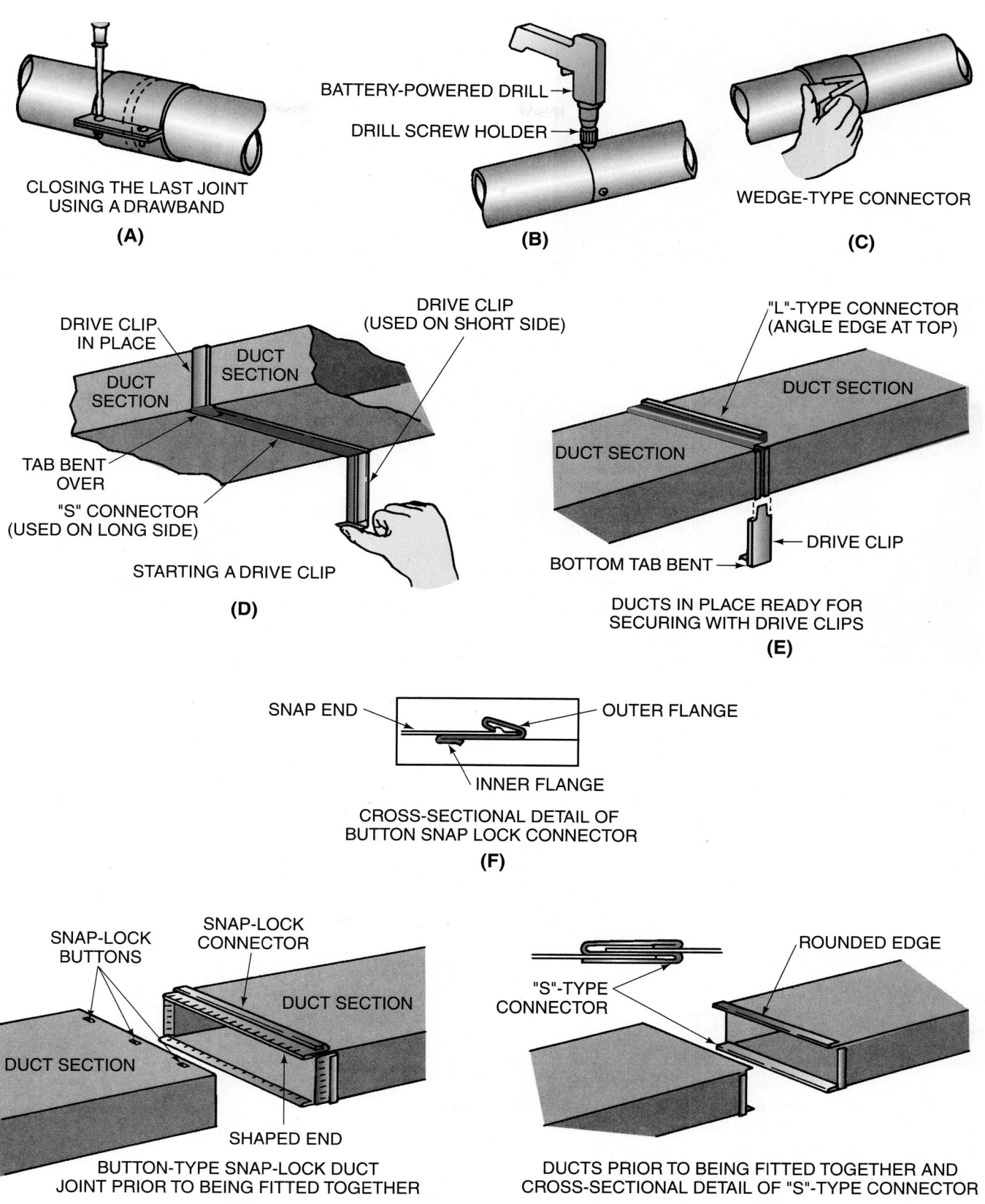

Figure 37–35 Fasteners for square and round duct for low-pressure systems only **(A)–(H)**.

Figure 37–36 Fastening duct using a portable electric drill and sheet metal screws. *Photos by Bill Johnson*

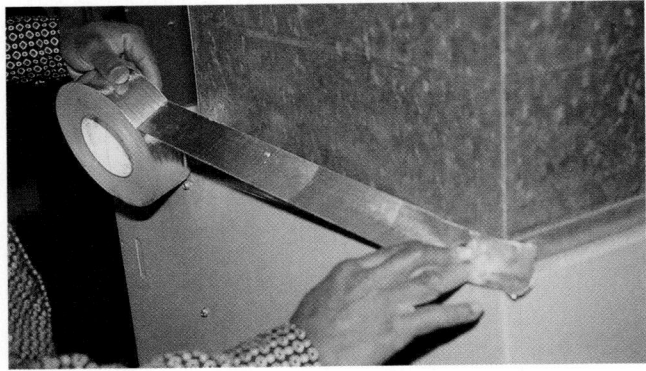

Figure 37–37 A taped duct connection. *Photo by Bill Johnson*

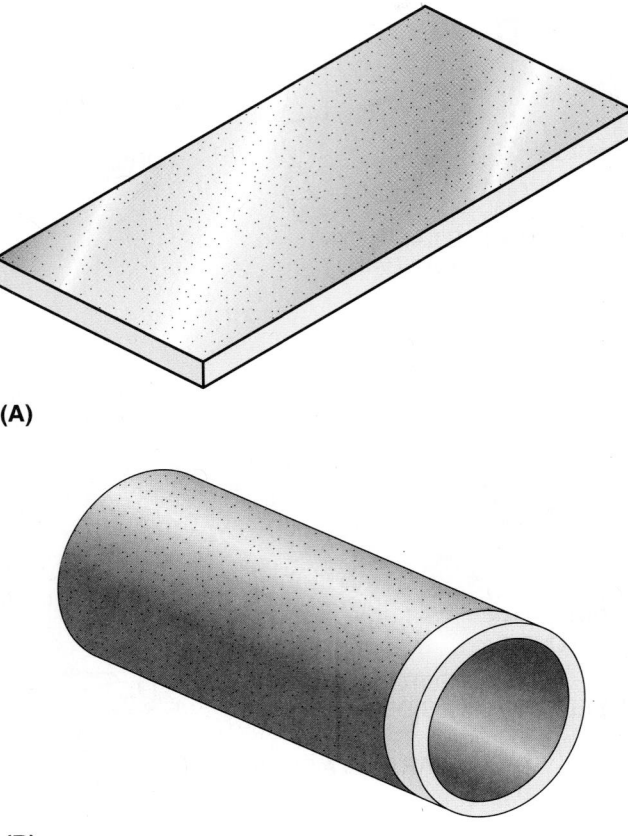

(A)

(B)

Figure 37–38 **(A)** Rigid fiberglass ductboard. **(B)** Prefabricated rigid round fiberglass duct sections.

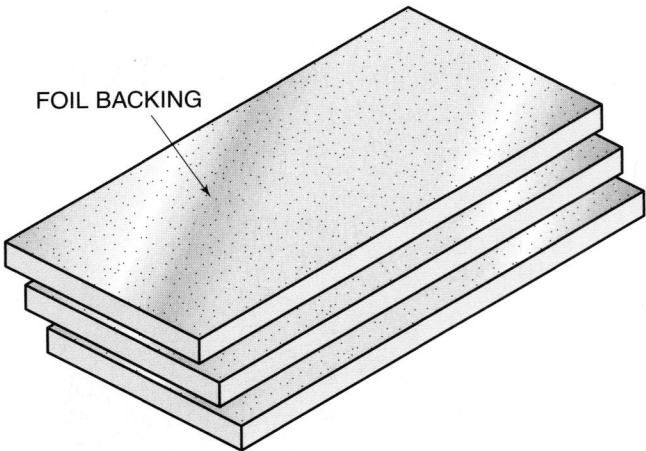

FOIL BACKING

Figure 37–39 Foil backing on the fiberboard material.

A special product called mastic also can be applied to duct connections to make them airtight. Mastic is a very thick puttylike substance, which is applied with a stiff bristle paintbrush. It dries to a very hard, but slightly flexible, sealant. Most mastics are water soluble, so they are easy to clean up after use.

Aluminum duct follows the same guidelines as galvanized duct. The cost of aluminum prevents this duct from being used for many applications.

37.17 FIBERGLASS DUCT

Fiberglass duct is furnished in two styles: flat sheets for fabrication and round prefabricated duct, **Figure 37–38**. Fiberglass duct is normally 1 or 1 1/2 in. thick with an aluminum foil backing, **Figure 37–39**. The foil backing has a fiber reinforcement to make it strong. When the duct is fabricated, the fiberglass is cut to form the edges, and the reinforced foil backing is left intact to support the connection and provide an air barrier, **Figure 37–40**. These duct systems are easily transported and assembled in the field. Fiberglass duct acts as a sound dampener because of the rough inside lining. These systems are very quiet.

Figure 37–41 shows a layout of a fiberglass ductboard system. Main trunk lines, offsets, and takeoffs can be fabricated on the job. Special round takeoffs are available that can be inserted into the main trunk line and connected to prefabricated round duct sections for supplying air to the remote

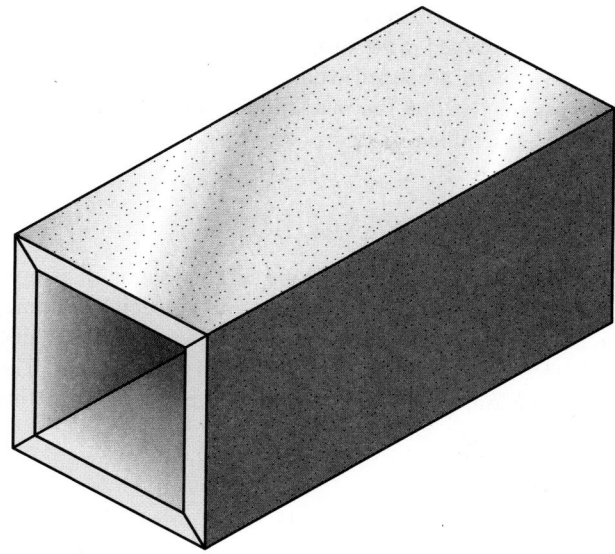

Figure 37–40 Foil backing supports the connection and provides a vapor barrier.

locations. One of the main benefits of using the fiberboard duct system is that the sections can be fabricated or modified on the job. In the event that a section has not been fabricated properly, the job can continue because the piece can be remade relatively quickly.

The ductboard can be made into duct in several different ways. Special knives that cut the board in such a manner as to produce overlapping connections can be used in the field by placing the ductboard on any flat surface and using a straightedge to guide the knife, **Figure 37–42.** When fabricating a duct section, the fiberboard is positioned on the work surface with the fiberglass facing up and the foil backing against the work surface, **Figure 37–43(A).** Using the desired knife, appropriate cuts are made in the material, **Figure 37–43(B).** Notice that only the fiberglass is cut away and that the foil backing is left intact. Once all of the cuts have been made, the duct section is folded, **Figure 37–43(C).** After it has been folded, there will be a flap of foil backing that will serve as on overlap for completing the connection, **Figure 37–43(D).** To finish the section, the flap is stapled and taped to seal the section, **Figure 37–43(E).** Special ductboard machines can be used to fabricate the duct in the field, at the job site, or in the shop to be transported to the job. An operator has to be able to set up the machine for different sizes of duct and fittings. When the duct is made in the shop, it can be cut and left flat and stored in the original boxes. This makes transportation easy. The pieces can be marked for easy assembly at the job site.

The machines or the knives cut away the fiberglass to allow the duct to be folded with the foil on the back side. When two pieces are fastened together, an overlap of foil is left so that one piece can be stapled to the other. A special

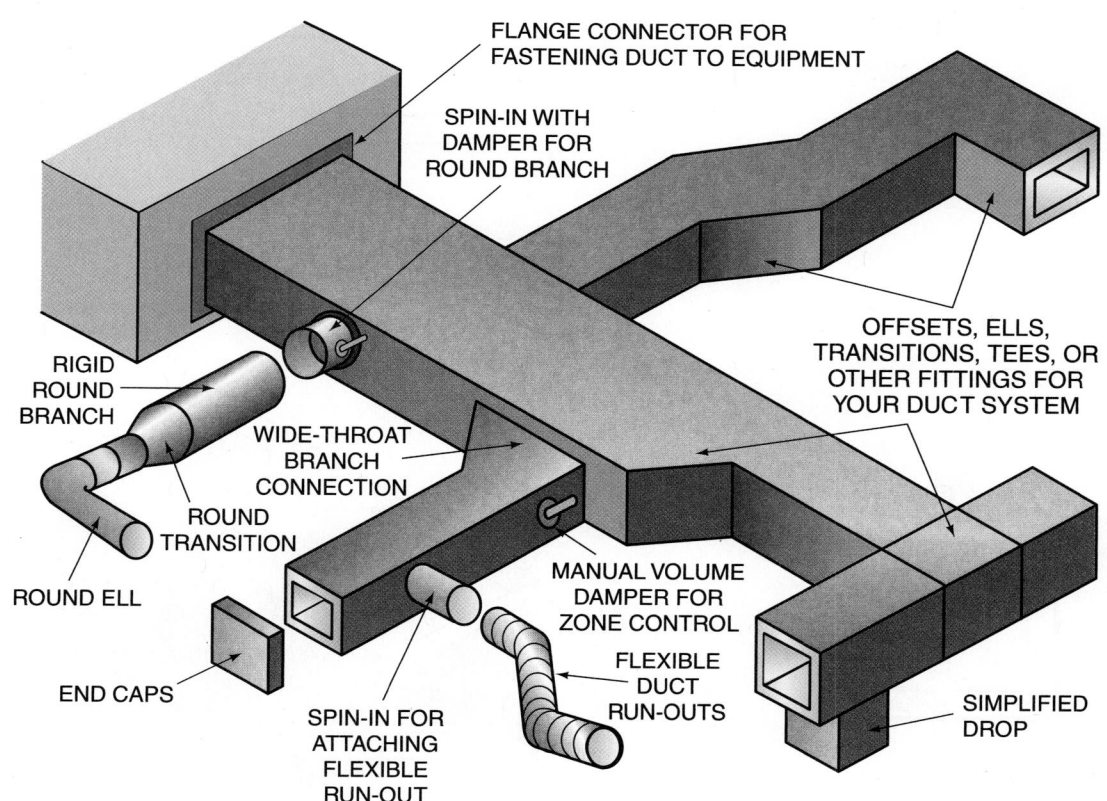

Figure 37–41 Fiberboard duct system layout.

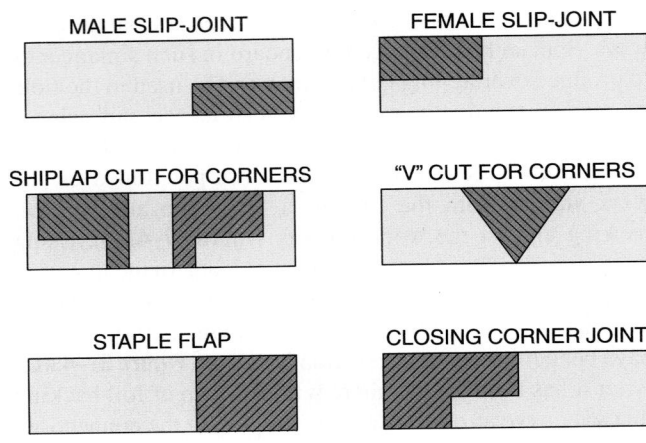

MALE SLIP-JOINT

FEMALE SLIP-JOINT

SHIPLAP CUT FOR CORNERS

"V" CUT FOR CORNERS

STAPLE FLAP

CLOSING CORNER JOINT

Figure 37–42 Special knives cut various shapes into the fiberboard panels for easy duct fabrication and assembly.

staple is used that turns out and up on the ends. Then the connection is taped with special tape. The tape should be pressed on with a special iron. One of the advantages of fiberglass duct is that the insulation is already on the duct when it is assembled.

37.18 SPIRAL METAL DUCT

Spiral metal duct is used more on large systems. It is normally manufactured at the job site with a special machine. The duct comes in rolls of flat narrow metal. The machine winds the metal off the spool and folds a seam in it. The length of a run of duct can be made very long with this method.

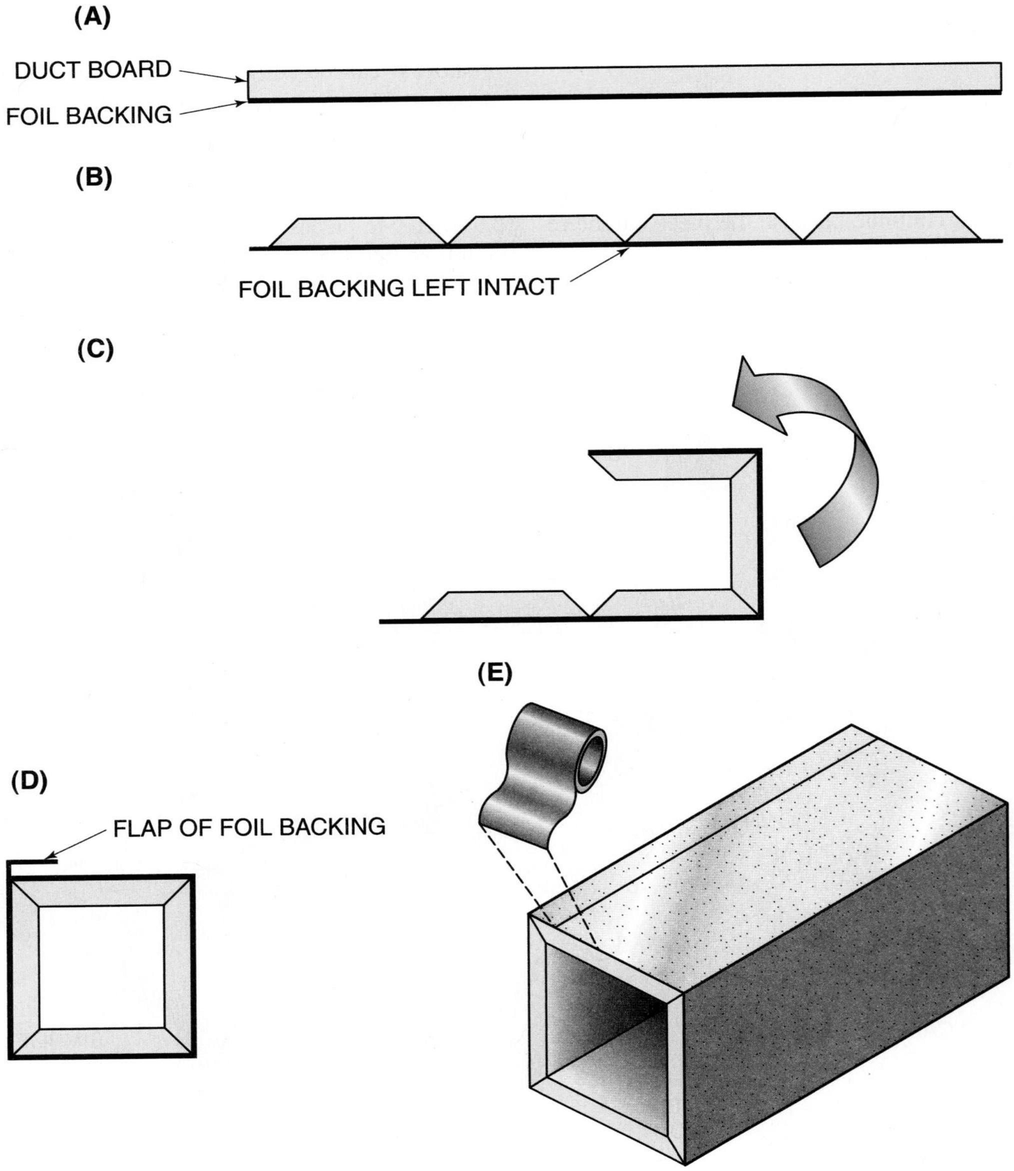

(A)

DUCT BOARD

FOIL BACKING

(B)

FOIL BACKING LEFT INTACT

(C)

(E)

(D)

FLAP OF FOIL BACKING

Figure 37–43 Fabricating a fiberboard duct section.

37.19　FLEXIBLE DUCT

Flexible round duct comes in sizes up to about 24 in. in diameter. Some of it has a reinforced aluminum foil backing and comes in a short box with the duct material compressed. Without the insulation on it, the duct looks like a coil spring with a flexible foil backing. Flexible duct installation is an easy installation because the runs can be aligned easily unlike rigid duct, which must be aligned exactly.

Some flexible duct comes with vinyl or foil backing and insulation on it, in lengths of 25 ft to a box, **Figure 37–44**. It is compressed into the box. Flexible duct is easy to route around corners. Keep the duct as short as practical and do not allow tight turns that may cause the duct to collapse. This duct has more friction loss inside it than metal duct does, but it also serves as a sound attenuator to reduce blower noise down the duct. For best airflow, flexible duct should be stretched as tightly as is practical.

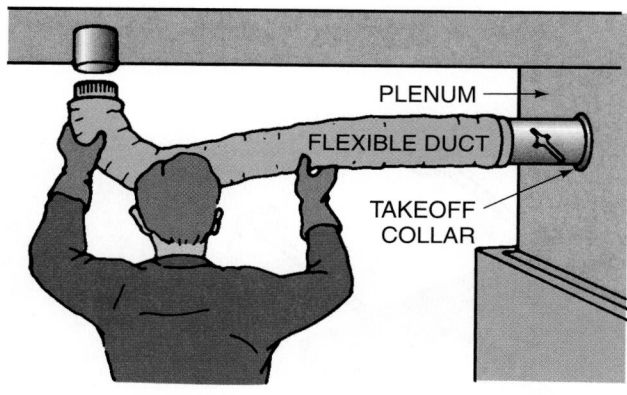

Figure 37–44　Flexible duct.

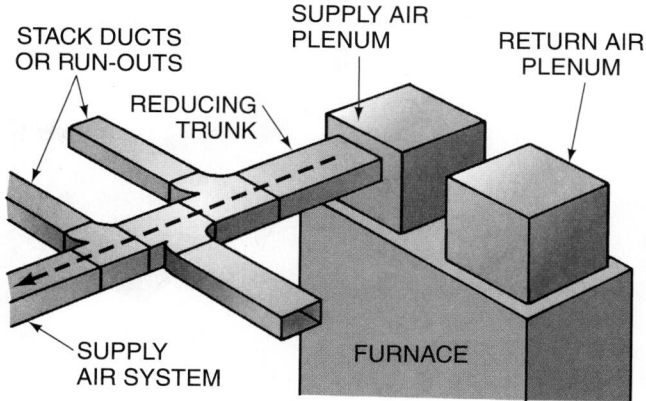

Figure 37–45　Square or rectangular duct.

37.20　COMBINATION DUCT SYSTEMS

Duct systems can be combined in various ways. For example:

1. All square or rectangular metal duct, the trunk line, and the branches are the same shape, **Figure 37–45**.
2. Metal trunk lines with round metal branch duct, **Figure 37–46**.
3. Metal trunk lines with rigid round fiberglass branch duct, **Figure 37–47**.
4. Metal trunk lines with flexible branch duct, **Figure 37–48**.
5. Ductboard trunk lines with rigid round fiberglass branches, **Figure 37–49**.
6. Ductboard trunk lines with round metal branches, **Figure 37–50**.
7. Ductboard trunk lines with flexible branches, **Figure 37–51**.
8. All round metal duct with round metal branch ducts, **Figure 37–52**.
9. All round metal trunk lines with flexible branch ducts, **Figure 37–53**.

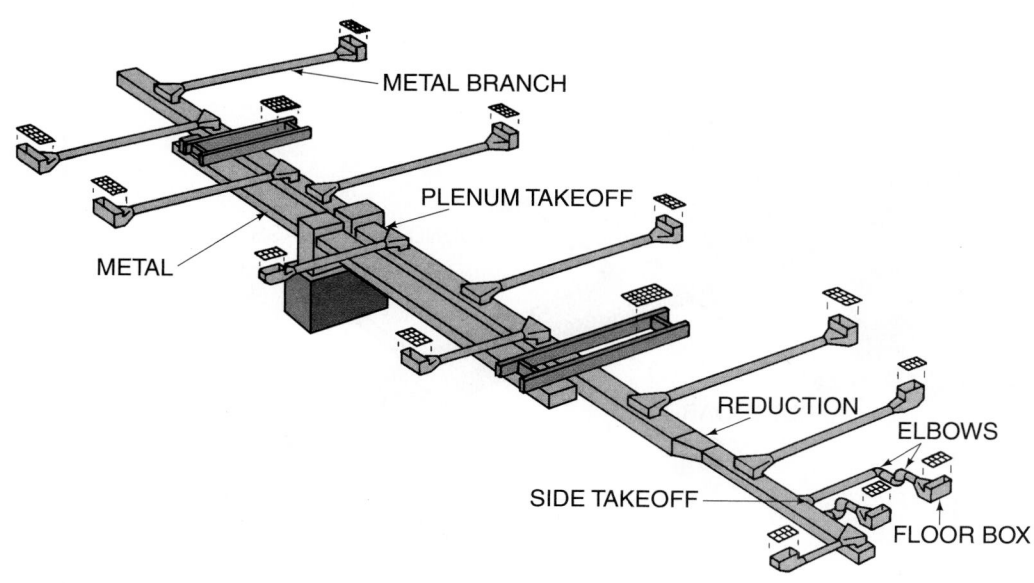

Figure 37–46　Rectangular metal trunk duct with round metal branch ducts.

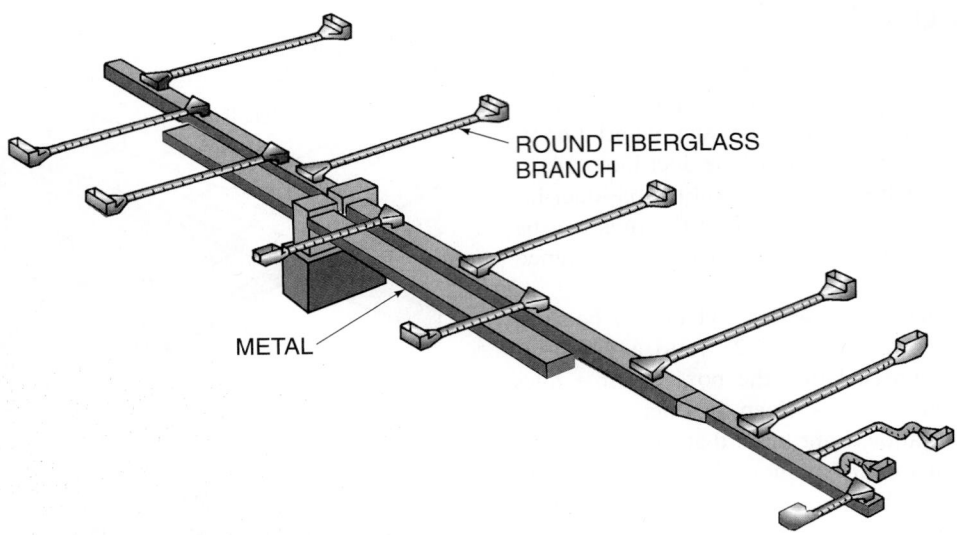

Figure 37–47   Metal trunk duct with round fiberglass branch ducts.

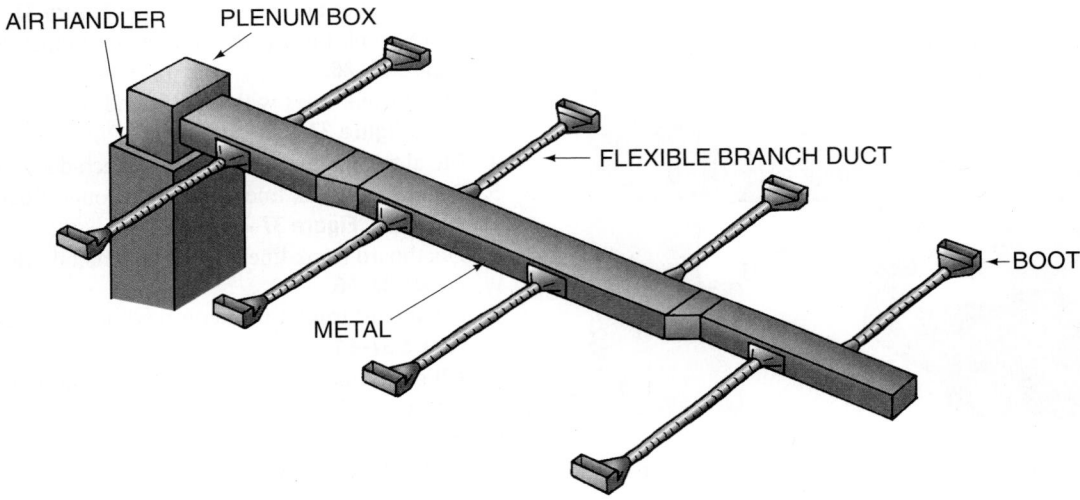

Figure 37–48   Metal trunk duct with flexible branch ducts.

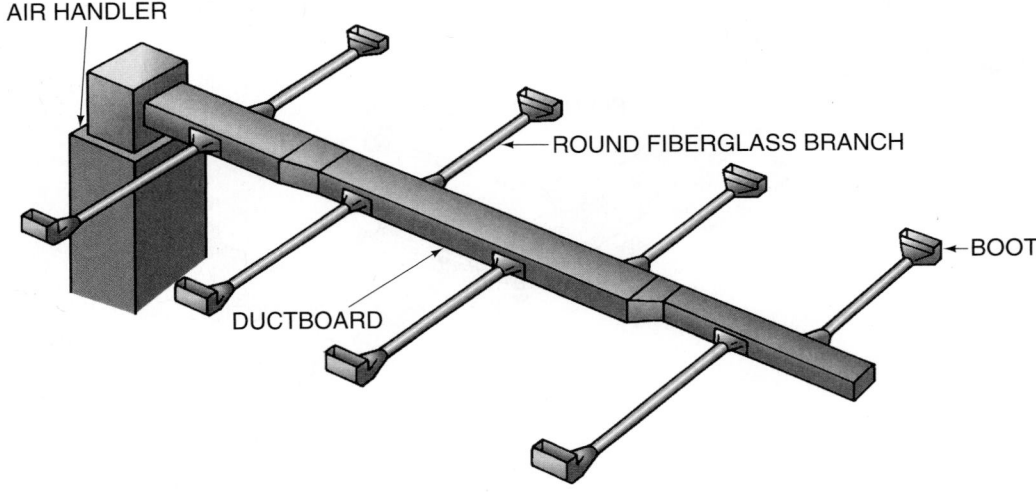

Figure 37–49   Fiberglass duct with round fiberglass ducts.

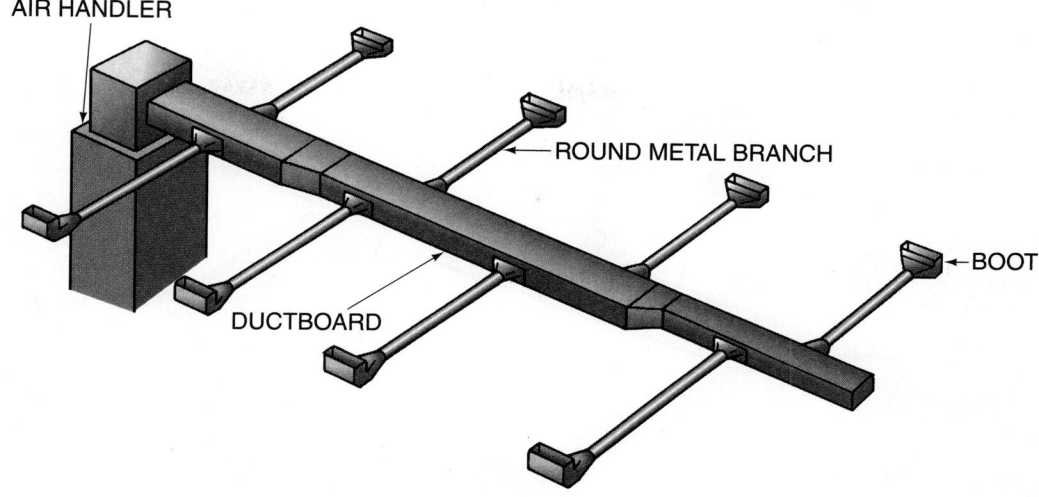

Figure 37–50 Fiberglass ductboard trunk and round metal branches.

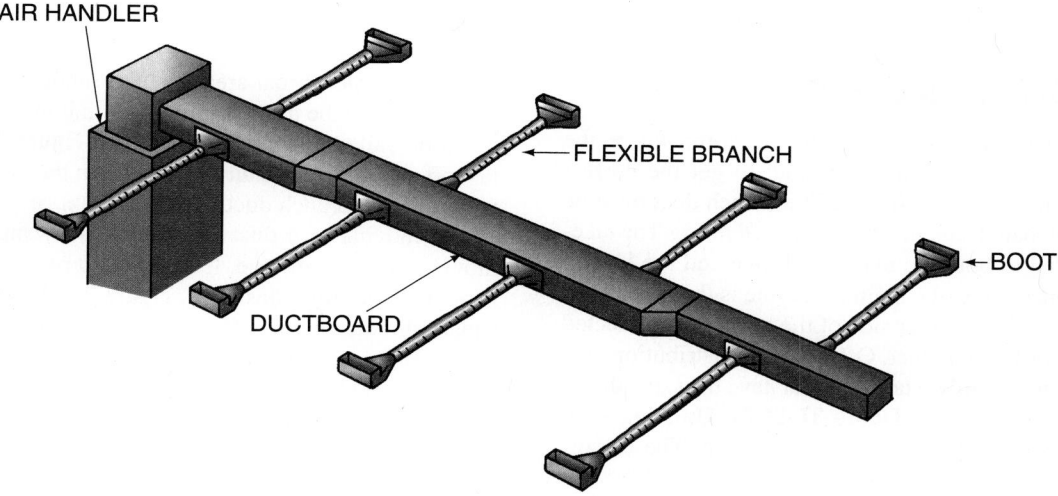

Figure 37–51 Fiberglass ductboard trunk and flexible branch ducts.

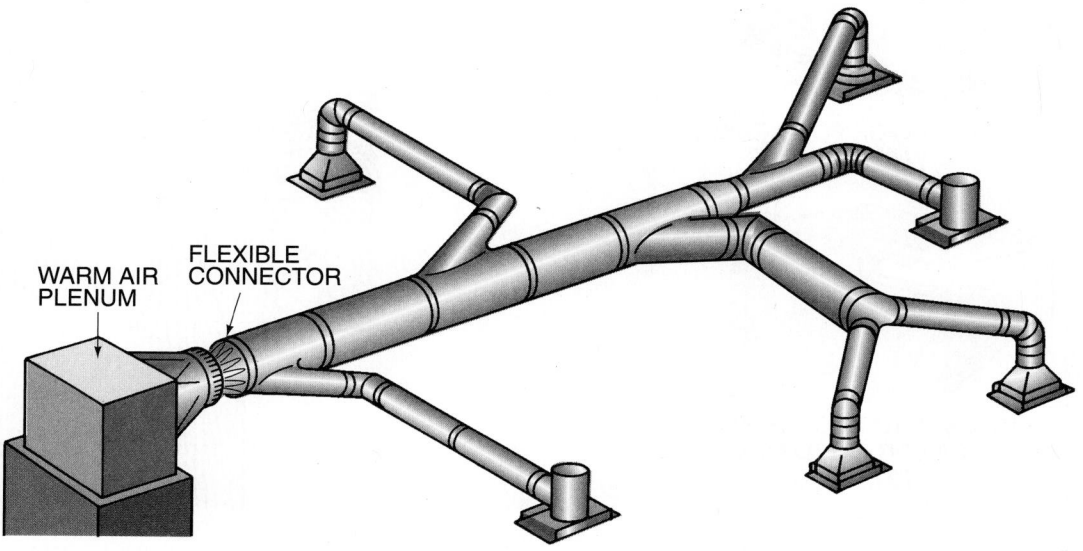

Figure 37–52 Round metal duct.

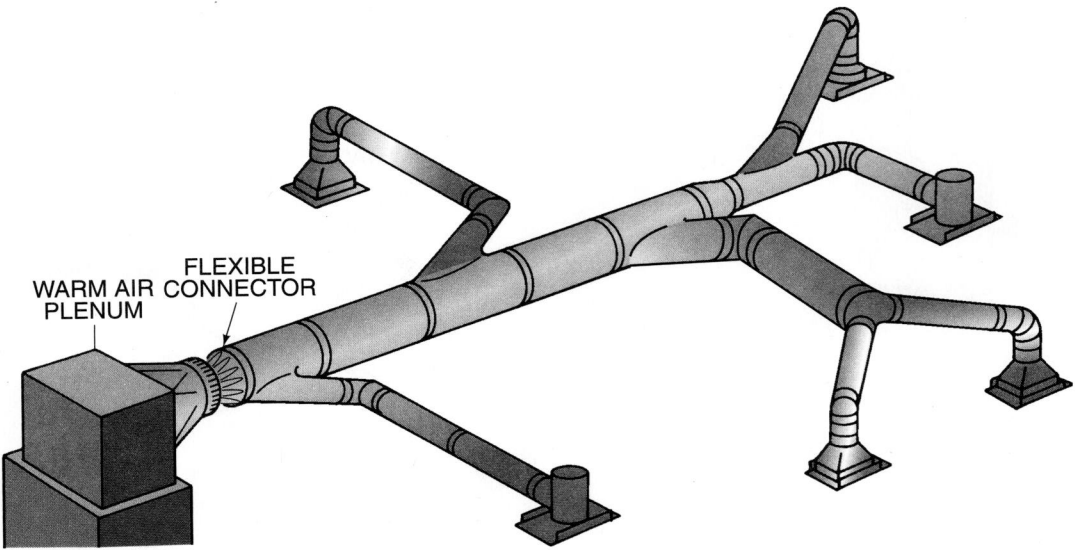

Figure 37–53 All round metal trunk line with flexible branch lines.

37.21 DUCT AIR MOVEMENT

Special attention should be given to the point where the branch duct leaves the main trunk duct to get the correct amount of air into the branch duct. The branch duct must be fastened to the main trunk line with a *takeoff fitting*. The takeoff is connected to both the main trunk line and the branch duct. The connection at the main trunk line is the throat portion of the takeoff. The other side of the takeoff is connected to the runout, or branch, duct. Often, an air distribution system installation will utilize takeoffs that have the same throat and runout measurements, **Figure 37–54(A)**. Utilizing such takeoffs may very well not be the best option. The takeoff

that has a larger throat area than the runout duct will allow the air to leave the trunk duct with a minimum of effort. This could be called a streamlined takeoff, **Figure 37–54(B)**. The takeoff encourages the air moving down the duct to enter the takeoff to the branch duct.

Air moving in a duct has *inertia*—it wants to continue moving in a straight line. If air has to turn a corner, the turn should be carefully designed, **Figure 37–55**. For example, a square-throated elbow offers more resistance to airflow than a round-throated elbow does. If the duct is rectangular or square, turning vanes will improve the airflow around a corner, **Figure 37–56**.

37.22 BALANCING DAMPERS

A well-designed system will have *balancing dampers* in the branch ducts to balance the air in the various parts of the system. Balancing the air with dampers enables the technician to direct the correct volume of air to the correct run of duct for

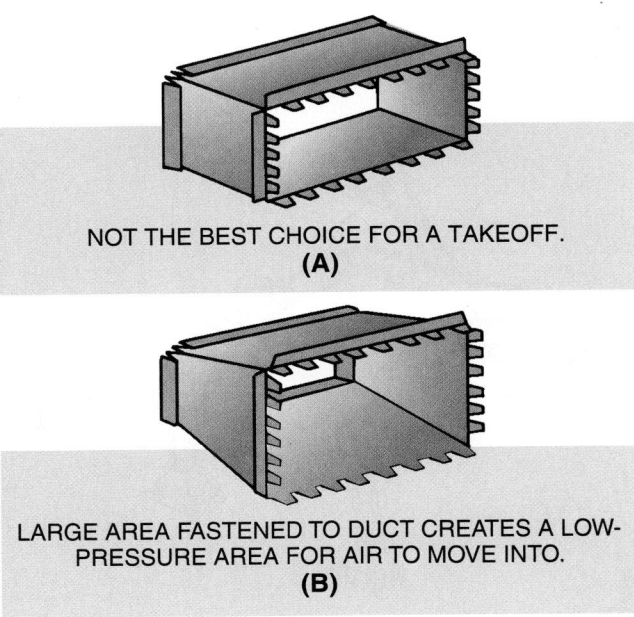

NOT THE BEST CHOICE FOR A TAKEOFF.
(A)

LARGE AREA FASTENED TO DUCT CREATES A LOW-PRESSURE AREA FOR AIR TO MOVE INTO.
(B)

Figure 37–54 **(A)** A standard takeoff fitting. **(B)** A streamlined takeoff fitting.

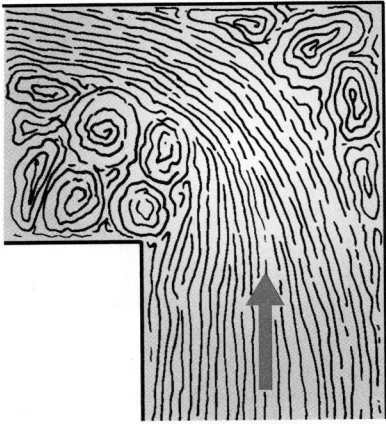

Figure 37–55 This illustration shows what happens when air tries to go around a corner.

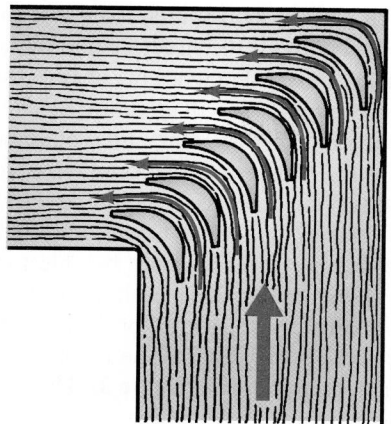

Figure 37–56   A square elbow with turning vanes.

helps ensure that the space is heated and/or cooled evenly. Once the system is properly balanced, the balancing dampers should never have to be touched again. Another type of damper that is found in an air distribution system is the zone damper. Here are some characteristics that will differentiate between the two types:

- ■ Balancing dampers are set to the desired position and locked in place at the time of initial system start-up or shortly thereafter.
- ■ Zone dampers are opened and closed in response to changes in room temperature.
- ■ Balancing dampers are manually opened and closed.
- ■ Zone dampers open and close automatically in response to temperature changes. These dampers can be pneumatically (air) or electrically controlled.
- ■ Balancing dampers are used to set airflow at some point between minimum and maximum on the basis of the size of the duct and the velocity of the air moving through the duct.
- ■ Zone dampers will move from the fully open to the fully closed position depending on whether or not the temperature requirements for the zone are satisfied.

🌐 The zoning process utilizes zone dampers, **Figure 37–58**, and enables a single forced-air heating and/or cooling system to maintain different temperatures in different rooms or areas of the conditioned space. By zoning a system, efficiency is increased.🌐 Consider, for example, a single central air-conditioning system that is being used to cool a two-story house. Typically, the second floor of the structure will be considerably warmer than the lower level. This temperature difference can be as high as 10 degrees. If the thermostat is located on the lower level, the upper floor will never get cool. If the thermostat is located upstairs, the lower level will be too cool. The same condition can surface in some ranch-style homes. In this case, one side of the house may be warmer than the other. Depending on the location of the thermostat in this situation, one side of the

better room temperature control. The dampers should be located as close as practical to the trunk line, with the damper handles uncovered if the duct is insulated. The place to balance the air is near the trunk, so if there is any air velocity noise it will be absorbed in the branch duct before it enters the room. A damper consists of a piece of metal shaped like the inside of the duct with a handle protruding through the side of the duct to the outside. The handle allows the damper to be turned at an angle to the airstream to slow the air down, **Figure 37–57**. Damper handles are directional. When the handle is crosswise of the duct, the damper is closed. When the handle is running with the duct, the damper is open. When a damper is closed, it will restrict the air to only about 15%.

37.23   ZONING

As we mentioned in the previous section, balancing dampers are used to ensure that the proper amount of air is delivered to each area in the conditioned space. Balancing the system

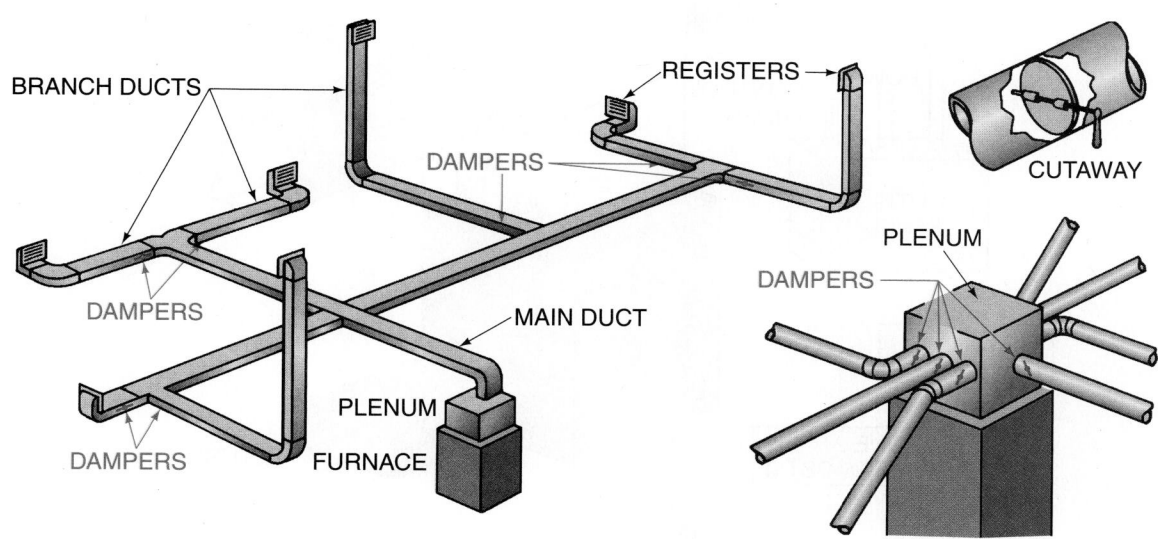

Figure 37–57   Balancing dampers.

home will be comfortable, while the other will be either too hot or too cold.

Zoning an air-conditioning system involves installing a thermostat and a zone damper for each room or zone in the structure. These thermostats and dampers are wired to a central control panel that will control the operation of the system on the basis of the actual conditions in the space. The complexity of these systems varies greatly depending on what features and tasks the system is able to perform. A very basic zoning system is shown in **Figure 37–58**. This system has two separate zones. Each zone has its own zone damper

Figure 37–58 A zone damper.

and thermostat. On a system similar to the one shown in this figure, it is easy to divide the structure into two separate areas, as the main plenum branches off in two distinct directions. For this example, we will assume that each of the two zones have the same airflow, cfm, and requirements. When configured in this manner (as two otherwise equal zones), the ductwork serving each zone should be oversized about 50%. If the total system were to provide 2000 cfm, this would mean that each trunk should be able to carry 1000 cfm of air. For zoning purposes, each trunk should be sized to carry approximately 1500 cfm, or 75%, of the total airflow requirement for the system. The wiring/control system for the zoned system shown in **Figure 37–59** is comprised of a central control panel, two damper motors, and two thermostats. One thermostat is centrally located in each of the two zones.

Of course there are more intricate and complicated zoning systems available. Typically, the two-zone system just discussed can identify zones in one of the following ways:

■ Zone 1: upstairs; Zone 2: downstairs
■ Zone 1: living room and kitchen; Zone 2: bedrooms and bathrooms

As homes get larger, the zoning requirements for these structures change as well. For example, some higher-end homes have the master bedroom suite as its own zone. The possibility for zoning is limited only by the system designer's imagination. It is not unreasonable to have a home that has every room as a separate zone. Obviously, systems like this are very costly to design and install and more complicated to troubleshoot.

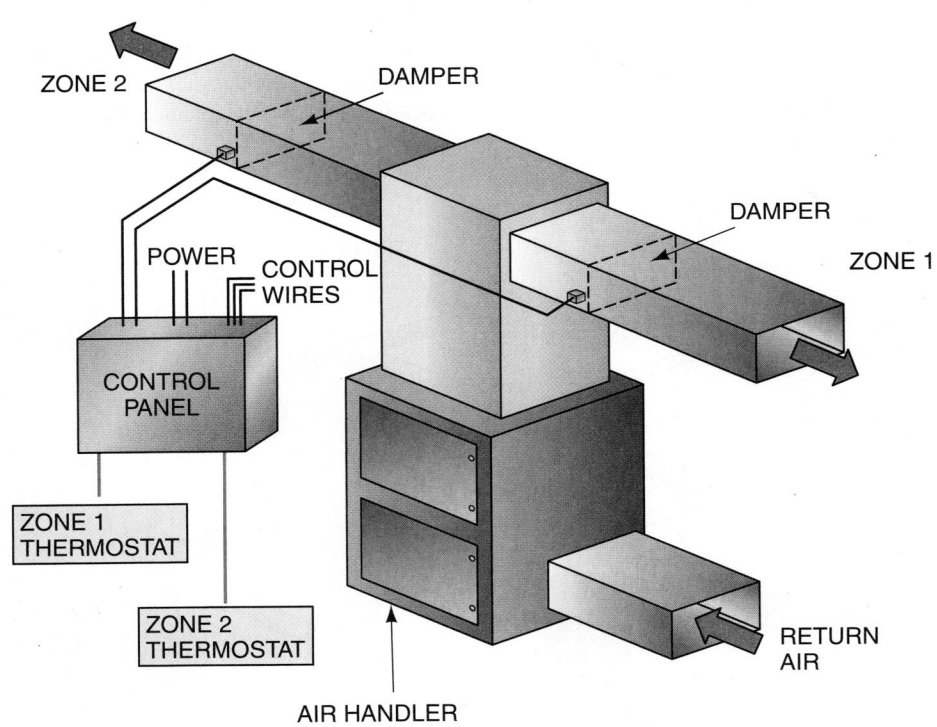

Figure 37–59 A common two-zone system layout.

37.24 ZONING WITH A SINGLE-SPEED BLOWER MOTOR

The opening and closing of the zone dampers will have an effect on the velocity and volume of the air moving through the system. This is not something that should be looked upon lightly, as a reduction in airflow through a cooling coil, for example, can cause the coil to freeze, further reducing airflow and causing a major reduction in cooling. On a two-zone system, as in the situation discussed in the previous section, oversizing the duct system by approximately 50% will alleviate this problem for the most part. Let's take a closer look. If the system were designed to move 2000 cfm at a velocity of 700 ft/min, the supply plenum would have to have a cross-sectional area of 2.86 ft² (2000 cfm ÷ 700 ft/min). If the air is to be divided between the two zones as shown in **Figure 37–58,** each zone will have to carry 1000 cfm at the same 700 ft/min. This means that the cross-sectional area of the duct is going to be 1.43 ft². By oversizing this zone duct by 50%, the cross-sectional area will be increased to 2.15 ft². This duct will still be carrying the 1000 cfm, so the velocity of the air will be reduced to 465 ft/min (1000 cfm ÷ 2.15 ft²). This will be the velocity when both zones are calling for cooling, **Figure 37–60(A).** When only one zone is calling for cooling, the entire 2000 cfm will flow through the duct feeding

that zone. The speed of the air flowing through that duct will be 930 ft/min (2000 cfm ÷ 2.15 ft²), **Figure 37–60(B).**

On systems with three or more zones, oversizing the duct to each zone by 50% is not practical. In such instances, oversizing the ducts by 10% to 15% will often suffice. In addition to this minimal oversizing of the duct, a bypass duct should be installed in the system, **Figure 37–61.** The bypass duct allows the blower to move the required amount of air even if a number of zones are not calling for conditioning. The bypass duct should be equipped with a damper so that the airflow can be adjusted. This damper can be manually or automatically controlled. When the damper is a manual device, it should be set for the "worst-case scenario," which is when only the smallest zone is calling for conditioning.

Consider the system shown in **Figure 37–62(A).** This is an air-conditioning system with four zones. In this case, all four zones are calling for cooling, so there is no problem with excess or unwanted air volume. In **Figure 37–62(B),** only the smallest zone is calling for cooling, so all zone dampers are closed with the exception of the one serving Zone 4. The blower will still deliver 2000 cfm, but only 400 cfm are needed. This is the worst-case scenario, as only the smallest zone is calling for cooling. The bypass duct should be sized to handle enough air for this condition, which, in this case, is 1600 cfm (2000 cfm − 400 cfm). When the damper in the bypass zone is automatically controlled, a pressure switch is often used to open and close the damper. The pressure switch senses the pressure in the duct and, as the pressure in the duct rises, causes the bypass damper to open.

When using the bypass-duct method, it is important to sense the temperature of the air as it moves through the system. By feeding conditioned air back through the system, the air can be overheated in the heating mode or overcooled in the cooling mode. This can result in damage to the equipment. For example, if the return air is too cool, it may very well cause the evaporator coil to freeze. This can lead to liquid floodback to the compressor, resulting in major system damage. The sensor should send a signal to the central control panel to cycle the compressor off when such conditions are present. Different manufacturers tackle this situation differently, so be sure to follow the recommendations and instructions provided with each specific zoning product.

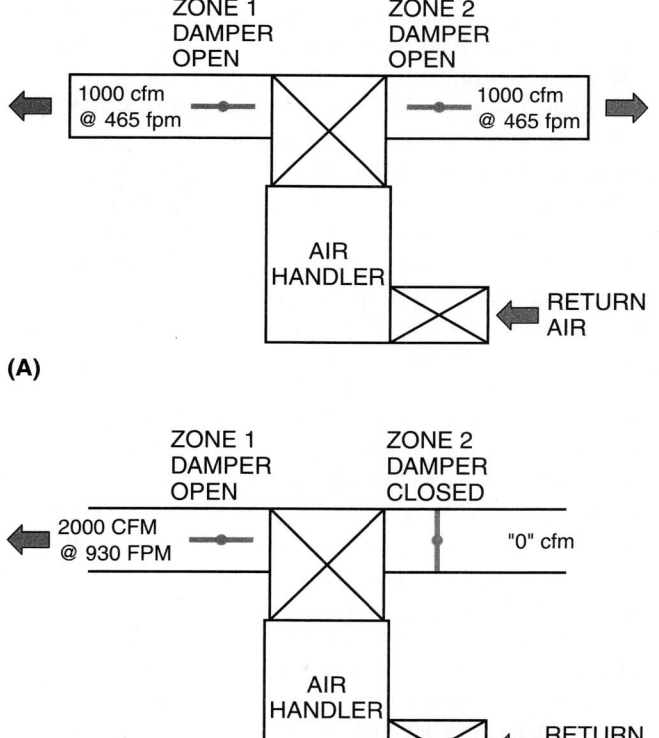

Figure 37–60 **(A)** Both zones are calling for cooling so each zone is being supplied with 1000 cfm of cooled air. **(B)** Only Zone 1 is calling for cooling, so this zone is receiving all of the air.

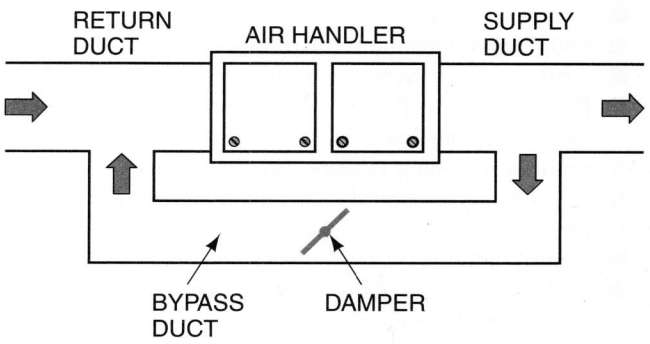

Figure 37–61 A bypass duct.

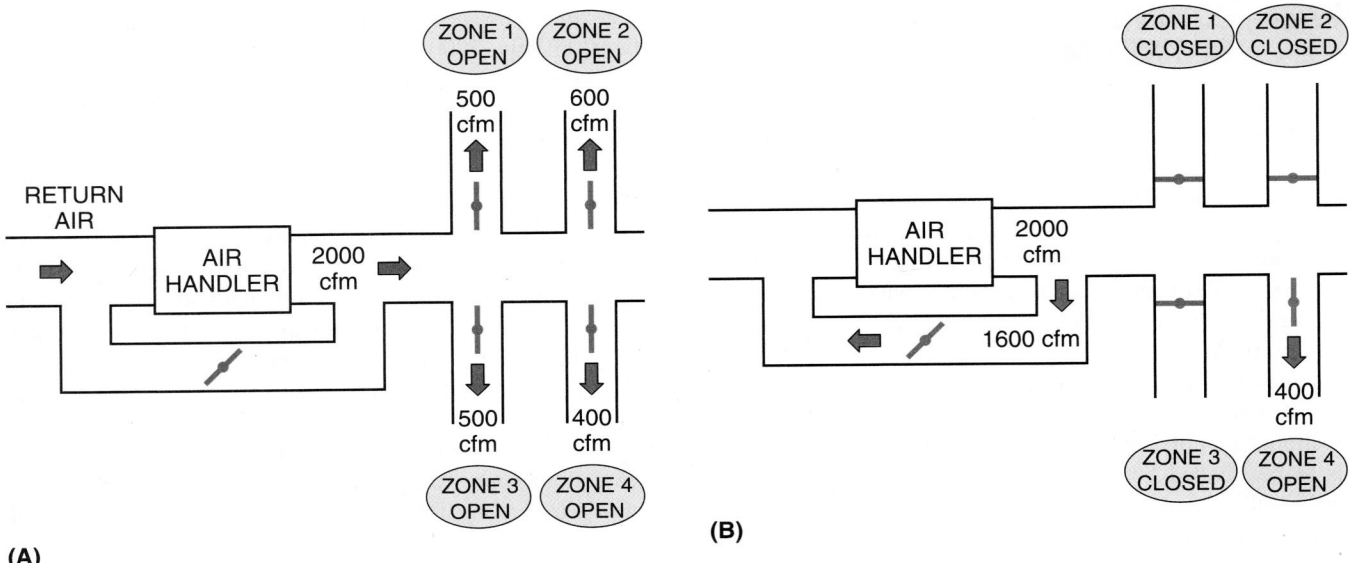

Figure 37–62 **(A)** All four zones are calling for system operation. **(B)** This is the worst-case scenario as only the smallest zone is calling for system operation. The bypass damper must be able to handle enough air for this situation.

One alternative to using the bypass duct is to incorporate a dump zone into the system. Instead of recirculating the air back through the equipment, the excess air is directed to an area in the structure that does not have critical temperature requirements. Such areas include basements, common hallways, or similar spaces. Since this may be objectionable to the occupants of the structure, options should be discussed prior to the design and installation stages of the job.

37.25 ZONING WITH A MULTISPEED COMPRESSOR AND VARIABLE-SPEED BLOWER

Some of the most desirable zoning systems utilize multi-speed compressors and variable-speed blower motors. These systems use computer-controlled panels that store information regarding the air requirements for each zone. Some of these zoning systems use the following pieces of data:

■ Return air temperature
■ Supply air temperature
■ Airflow requirements for each zone
■ Percentage of airflow for each zone as a percentage of total system airflow

More advanced zoning systems have the ability to

■ provide simultaneous heating and cooling by zone.
■ provide continuous fan operation options by zone.
■ control individual zone temperature.
■ set heating or cooling priority by zone.

Consider the following system setup:

■ Zone 1—200 cfm—den and bedroom 4
■ Zone 2—450 cfm—bedrooms 2 and 3

■ Zone 3—500 cfm—master bedroom suite
■ Zone 4—450 cfm—living room and dining room area

The total cfm requirement for this home is 1600 cfm, so the percentage of air going to each zone is as follows:

■ Zone 1 → (200 cfm ÷ 1600 cfm) × 100% = 13%
■ Zone 2 → (450 cfm ÷ 1600 cfm) × 100% = 28%
■ Zone 3 → (500 cfm ÷ 1600 cfm) × 100% = 31%
■ Zone 4 → (450 cfm ÷ 1600 cfm) × 100% = 28%

Notice that the total air volume is equal to 100%. These air volume percentages are rounded off and will get programmed into the central control panel. The average air volumes for Zones 1 through 4 will then be estimated as 10%, 30%, 30%, and 30% respectively. On the basis of this information and those zones that are calling for conditioning, the compressor and blower speeds will be selected automatically by the system. For example, if Zones 3 and 4 are calling for cooling, the blower will operate at a speed that will deliver 960 cfm of air (60% of 1600 cfm). The compressor speed will be selected on the basis of this airflow as well as the data gathered from the return and/or supply air sensors. If another zone now calls for cooling, the blower will ramp up its speed to deliver more air volume.

If the system is equipped to provide both heating and cooling, the zoning equipment may have the capability of selecting a priority mode of operation. For example, if a heating priority is programmed into the system and there are simultaneous calls for heating and cooling, the system will satisfy the heating demand first and then switch over to satisfy the cooling demand. Similarly, if there is a cooling priority programmed into the system, the system will satisfy the cooling demand first and then switch over to the heating mode for those zones that require it. There may also be a priority referred to as *zone priority*. Zone priority gives preference to whatever mode of operation is called for by the priority zone. For example, some homeowners set the master bedroom suite as the priority zone. So, if that zone is calling for cooling and the rest of the home is calling for heating, the cooling demand will be satisfied first.

37.26   ADDING ZONING TO AN EXISTING SYSTEM

Although it is easier to zone a system when it is initially designed and installed, retrofitting existing systems has become a very popular option. With the new 13 SEER regulations in place as of January of 2006, more and more attention is being put on energy efficiency and conservation. Since many of the 13 SEER units incorporate variable-speed blowers, adding some sort of zoning equipment seems to be a logical step toward increasing the operating efficiency of systems. Adding zones to an existing system has become an easier task as well. Compact, easy-to-install dampers are readily available in both round, square, and rectangular configurations to fit into existing duct sections. The manufacturers of these products provide templates along with their products to further ease the installation process. **Figure 37–63(A)–(D)** shows the installation of round and square zone dampers.

37.27   DUCT INSULATION

When ductwork passes through an unconditioned space, heat transfer may take place between the air in the duct and the air in the unconditioned space. If the heat exchange adds or removes very much heat from the conditioned air, insulation should be applied to the ductwork. A 15°F temperature difference from inside the duct to outside the duct is considered the maximum difference allowed before insulation is necessary.

The insulation is built into a fiberglass duct by the manufacturer. Metal duct can be insulated in two ways: on the outside or on the inside. When applied to the outside, the insulation is usually a foil- or vinyl-backed fiberglass. It comes in several thicknesses, with 2-in. thickness the most common. The backing creates a moisture vapor barrier. This is important where the duct may operate below the dew point temperature of the surroundings, and moisture would form on the duct. The insulation is joined by lapping it and stapling it. It is then taped to prevent moisture from entering the seams. External insulation can be added after the duct has been installed if the duct has enough clearance all around.

When applied to the inside of the duct, the insulation is either glued or fastened to tabs mounted on the duct by spot

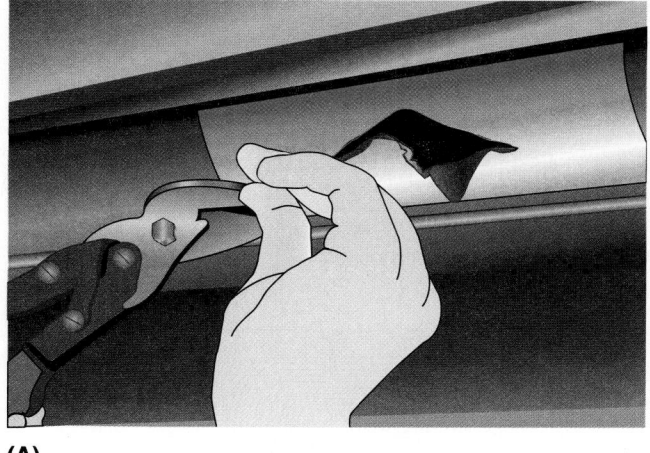

(A)

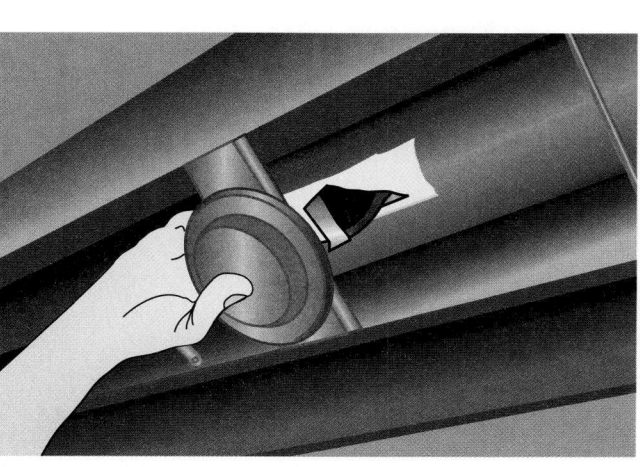

(B)

(C)

(D)

Figure 37–63   Installing a zone damper in a round duct **(A)–(B)**. Installing a zone damper in a square duct **(C)–(D)**.

weld or glue. This insulation must be applied when the duct is being manufactured.

37.28   BLENDING THE CONDITIONED AIR WITH ROOM AIR

When the air reaches the conditioned space, it must be properly distributed into the room so that the room will be comfortable without anyone being aware that a conditioning system is operating. This means that the final components in the system must place the air in the proper area of the conditioned space for proper air blending. Following are guidelines that can be used for room air distribution.

1. When possible, air should be directed on the walls. They are the load where the heat exchange occurs. For example, in winter air can be directed on the outside walls to cancel the load (cold wall) and keep the wall warmer. This will keep the wall from absorbing heat from the room air. In summer the same distribution will work; it will keep the wall cool and keep room air from absorbing heat from the wall. The *diffuser* spreads the air to the desired air pattern, **Figure 37–64.**
2. Warm air for heating distributes better from the floor because warm air tends to rise, **Figure 37–65.**
3. Cool air distributes better from the ceiling because cool air tends to fall, **Figure 37–66.**
4. The most modern concept for both heating and cooling is to place the diffusers next to the outside walls to accomplish this load-canceling effect, **Figure 37–67.**
5. The amount of throw (how far the air from the diffuser will blow into the room from the diffuser) depends on the air pressure behind the diffuser and the style of the diffuser blades. Air pressure in the duct behind the diffuser creates the velocity of the air leaving the diffuser.

Figure 37–68 shows some air registers and diffusers. The various types can be used for low-side wall, high-side wall, floor, ceiling, or baseboard.

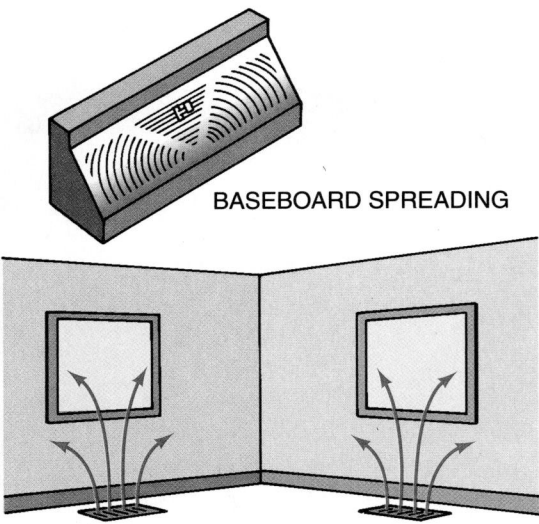

Figure 37–64   Diffusers spread and distribute the air.

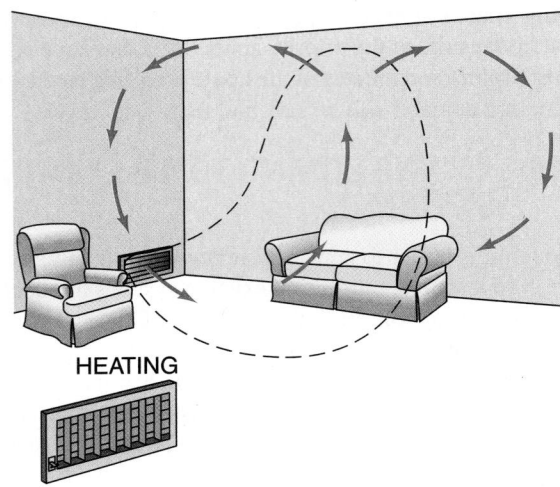

Figure 37–65   Warm air distribution.

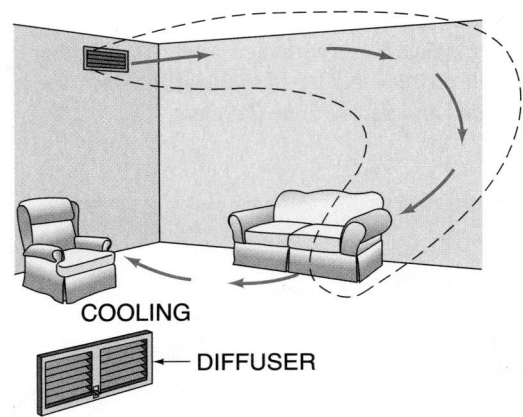

Figure 37–66   Cool air distribution.

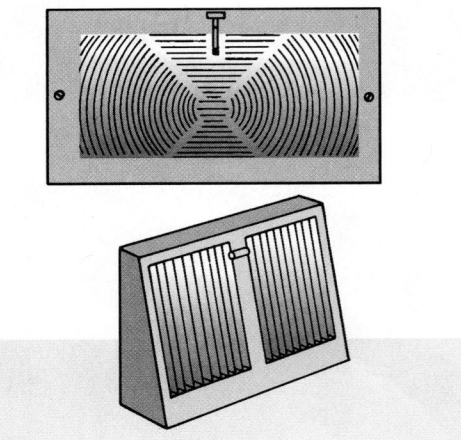

Figure 37–67   Diffusers used on the outside wall.

37.29   THE RETURN AIR DUCT SYSTEM

The return air duct is constructed in much the same manner as the supply duct except that some installations are built with central returns instead of individual room returns. Individual

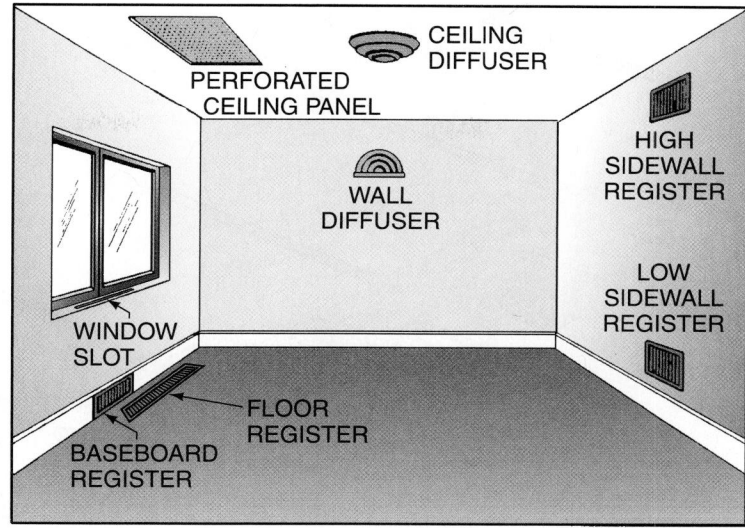

	FLOOR	BASEBOARD	LOW SIDEWALL	HIGH SIDEWALL	CEILING
COOLING PERFORMANCE	Excellent	Excellent if used with perimeter systems	Excellent if designed to discharge upward	Good	Good
HEATING PERFORMANCE	Excellent	Excellent if used with perimeter systems	Excellent if used with perimeter systems	Fair–should not be used to heat slab houses in northern climates	Good–should not be used to heat slab houses in northern climates
INTERFERENCE WITH DECOR	Easily concealed because it fits flush with the floor and can be painted to match	Not quite so easy to conceal because it projects from the baseboard	Hard to conceal because it is usually in a flat wall	Impossible to conceal because it is above furniture and in a flat wall	Impossible to conceal but special decorative types are available
INTERFERENCE WITH FURNITURE PLACEMENT	No interference–located at outside wall under a window	No interference–located at outside wall under a window	Can interfere because air discharge is not vertical	No interference	No interference
INTERFERENCE WITH FULL-LENGTH DRAPES	No interference–located 6 or 7 inches from the wall	When drapes are closed they will cover the outlet	When located under a window, drapes will close over it	No interference	No interference
INTERFERENCE WITH WALL-TO-WALL CARPETING	Carpeting must be cut	Carpeting must be notched	No interference	No interference	No interference
OUTLET COST	Low	Medium	Low to medium, depending on the type selected	Low	Low to high—wide variety of types are available
INSTALLATION COST	Low because the sill need not be cut	Low when fed from below–sill need not be cut	Medium–requires wall stack and cutting of plates	Low on furred-ceiling system; high when using under-floor system	High because attic ducts require insulation

Figure 37–68   Air registers and diffusers.

return air systems have a return air grille in each room that has a supply diffuser (with the exception of restrooms and kitchens). Individual return systems provide the best return air, but they are expensive. The return air duct is normally sized at least slightly larger than the supply duct, so there is less resistance to the airflow in the return system than in the supply system. There will be more details when we discuss duct sizing. See **Figure 37–69** for an example of a system with individual room returns.

The central return system is usually satisfactory for a one-level residence. Larger return air grilles are located so that air from common rooms can easily move back to the common returns. For air to return to central returns, there must be a path, such as doors with grille work, open doorways, and undercut doors in common hallways. These open areas in the doors can interfere with the privacy that some people desire.

In a structure with more than one floor level, install a return at each level. Remember, cold air moves downward and warm air moves upward naturally without encouragement. **Figure 37–70** illustrates air stratification in a two-level house.

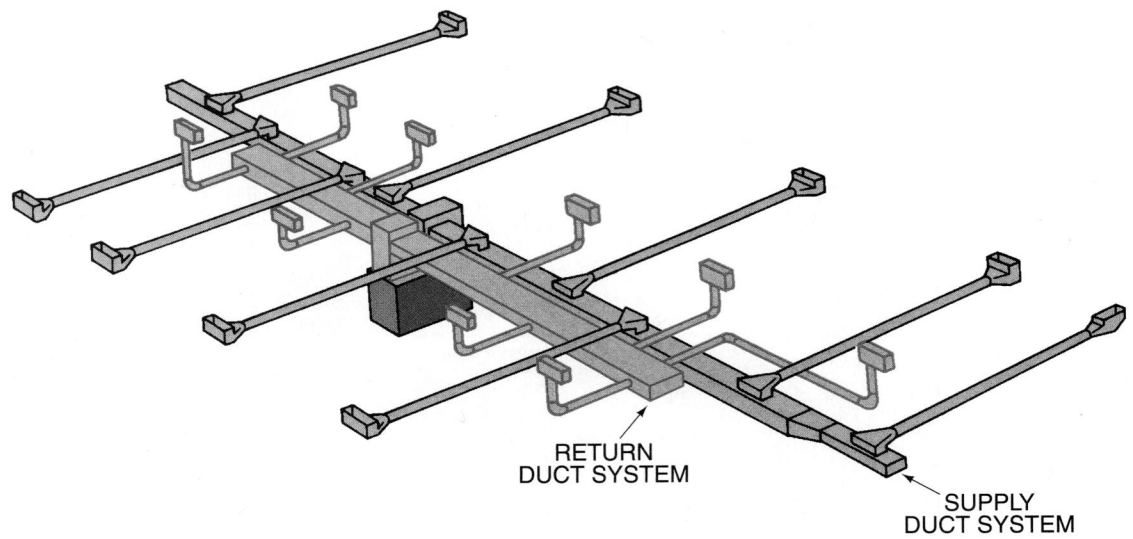

Figure 37–69   A duct plan of individual room return air inlets. *Reproduced courtesy of Carrier Corporation*

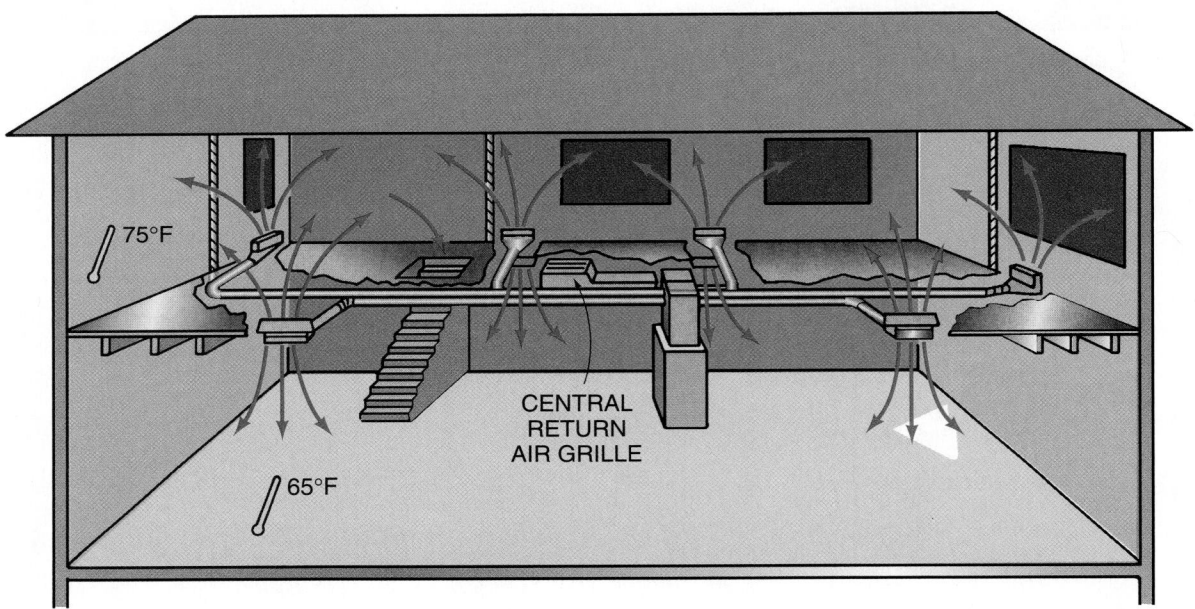

Figure 37–70   Air stratifies even when distributed because warm air rises and cold air falls.

A properly constructed central return air system helps to eliminate fan noise in the conditioned space. The return air plenum should not be located on the furnace because fan running noise will be noticeable several feet away. The return air grille should be around an elbow from the furnace. If this cannot be done, the return air plenum can be insulated on the inside to help deaden the fan noise.

Return air grilles are normally large and meant to be decorative. They do not have another function unless they house a filter. They are usually made of stamped metal or have a metal frame with grille work, **Figure 37–71.**

37.30   SIZING DUCT FOR MOVING AIR

To move air takes energy because (1) air has weight, (2) the air tumbles down the duct, rubbing against itself and the ductwork, and (3) fittings create resistance to the airflow.

Friction loss in ductwork is due to the actual rubbing action of the air against the side of the duct and the turbulence of the air rubbing against itself while moving down the duct.

Friction due to air rubbing the walls of the ductwork cannot be eliminated but can be minimized with good design practices. Proper duct sizing for the amount of airflow helps

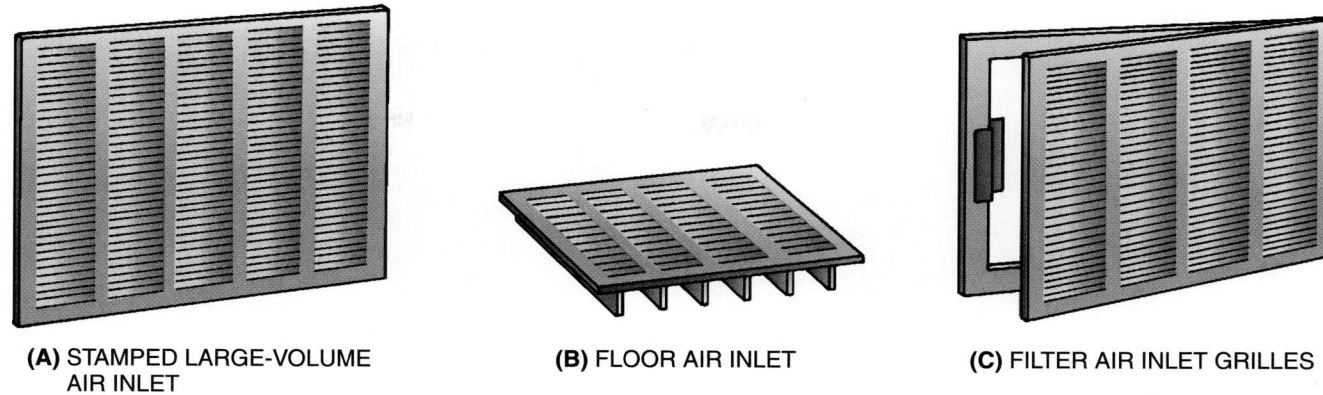

(A) STAMPED LARGE-VOLUME AIR INLET

(B) FLOOR AIR INLET

(C) FILTER AIR INLET GRILLES

Figure 37–71   Return air grilles.

maximize system performance. The smoother the duct surface is, the less friction there is. The slower the air is moving, the less friction there will be. It is beyond the scope of this text to go into details of duct design. However, the following information can be used as basic guidelines for a typical residential installation.

Each foot of duct offers a known resistance to airflow. This is called *friction loss.* It can be determined from tables

and special slide calculators designed for this purpose. The following example is used to explain friction loss in a duct system, **Figure 37–72.**

1. Ranch-style home requires 3 tons of cooling.
2. Cooling provided by a 3-ton cooling coil in the ductwork.
3. The heat and fan are provided by a 100,000-Btu/h furnace input, 80,000-Btu/h output.

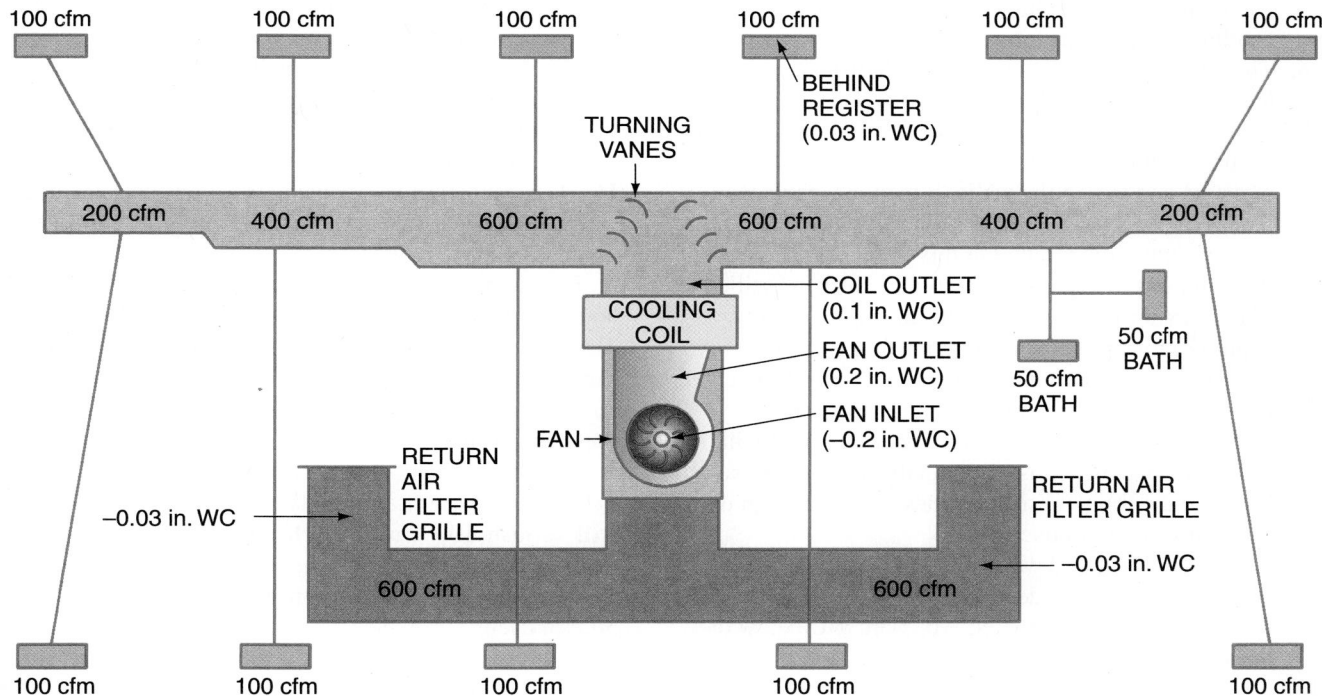

SYSTEM CAPACITY 3 TONS
cfm REQUIREMENT 400 cfm PER TON 400 X 3 = 1200 cfm
FAN STATIC PRESSURE (0.4 in. WC)

SUPPLY DUCT STATIC PRESSURE (0.2 in. WC)
RETURN DUCT STATIC PRESSURE (–0.2 in. WC)

Figure 37–72   A duct system.

SIZE	Blower Motor HP	Speed	External Static Pressure in. W.C.							
			0.1	0.2	0.3	0.4	0.5	0.6	0.7	0.8
048100	1/2 PSC	High	1750	1750	1720	1685	1610	1530	1430	—
		Med-High	1360	1370	1370	1360	1340	1315	—	—
		Med-Low	1090	1120	1140	1130	1100	—	—	—
		Low	930	960	980	980	965	945	—	—

Figure 37–73 A manufacturer's table for furnace airflow characteristics. *Courtesy BDP Company*

4. The fan has a capacity of 1360 cfm of air while operating against 0.40 in. WC static pressure with the system fan operating at medium-high speed. The system needs only 1200 cfm of air in the cooling mode. The system fan will easily be able to achieve this with a small amount of reserve capacity using a 1/2-hp motor, **Figure 37–73**. The cooling mode usually requires more air than the heating mode. As a rule of thumb, cooling normally requires 400 cfm of air per ton; 3 tons × 400 = 1200 cfm.

5. The system has 11 outlets, each requiring 100 cfm, in the main part of the house and 2 outlets, each requiring 50 cfm, located in the bathrooms. Most of these outlets are on the exterior walls of the house and distribute the conditioned air on the outside walls.

6. The return air is taken into the system from a common hallway, one return at each end of the hall.

7. While reviewing this system, think of the entire house as the system. The supply air must leave the supply registers and sweep the walls. It then makes its way across the rooms to the door adjacent to the hall. The air is at room temperature at this time and goes under the hall door to make its way to the return air grille.

8. The return air grille is where the duct system starts. A slight negative pressure (in relation to the room pressure) at the grille gives the air the incentive to enter the system. The filters are located in the return air grilles. The pressure on the fan side of the filter will be −0.03 in. WC, which is less than the pressure in the room, so the room pressure pushes the air through the filter into the return duct.

9. As the air proceeds down the duct toward the fan, the pressure continues to decrease. The lowest pressure in the system is in the fan inlet, −0.20 in. WC below the room pressure.

10. The air is forced through the fan, and the pressure increases. The greatest pressure in the system is at the fan outlet, 0.20 in. WC above the room pressure. The pressure difference in the inlet and the outlet of the fan is 0.40 in. WC.

11. The air is then pushed through the heat exchanger in the furnace where it drops to a new pressure that is not useful to the service technician.

12. The air then moves through the cooling coil where it enters the supply duct system at a pressure of 0.10 in. WC.

13. The air will take a slight pressure drop as it goes around the corner of the tee that splits the duct into two reducing plenums, one for each end of the house. This tee in the duct has turning vanes to help reduce the pressure drop as the air goes around the corner.

14. The first section of each reducing trunk has to handle an equal amount of air, 600 cfm each. Two branch ducts are supplied in the first trunk run, each with an air quantity of 100 cfm. This reduces the capacity of the trunk to 400 cfm on each side. A smaller trunk can be used at this point, and materials can be saved, while maintaining the correct air velocity.

15. The duct is reduced to a smaller size to handle 400 cfm on each side. Because another 200 cfm of air is distributed to the conditioned space, another reduction can be made.

16. The last part of the reducing trunk on each side of the system needs to handle only 200 cfm for each side of the system.

This supply duct system will distribute the air for this house with minimal noise and maximum comfort. The pressure in the duct will be about the same all along the duct because the air was distributed off the trunk line, and the duct size was reduced to keep the pressure inside the duct at the prescribed value.

At each branch, duct dampers should be installed to balance the system air supply to each room. This system will furnish 100 cfm to each outlet, but if a room does not need that much air, the dampers can be adjusted. The branch to each bathroom will need to be adjusted to 50 cfm each.

The return air system is the same size on each side of the system. It returns 600 cfm per side with the filters located in the return air grilles in the halls. The furnace fan is located far enough from the grilles so that it will not be heard.

Complete books have been written on duct sizing. Manufacturers' representatives may also help you with specific applications. They also offer schools in duct sizing that use their methods and techniques.

37.31 MEASURING AIR MOVEMENT FOR BALANCING

Air balancing is sometimes accomplished by measuring the air leaving each supply register. When one outlet has too much air, the damper in that run is throttled to slow down the air. This, of course, redistributes air to the other outlets and will increase their flow.

The air quantity of an individual duct can be measured in the field to some degree of accuracy by using instruments to determine the velocity of the air in the duct. A velometer can be inserted into the duct to do this. The velocity must be measured in a cross section of the duct, and an average of the readings taken is used for the calculation. This is called *traversing* the duct. For example, if the air in a 1-ft² duct (12 in. × 12 in. = 144 in²) is traveling at a velocity of 1 ft/min, the volume of air passing a point in the duct is 1 cfm. If the velocity is 100 ft/min, the volume of air passing the same point is 100 cfm. The cross-sectional area of the duct is multiplied by the average velocity of the air to determine the volume of the moving air, **Figure 37–74. Figure 37–75** shows the patterns used while traversing the duct.

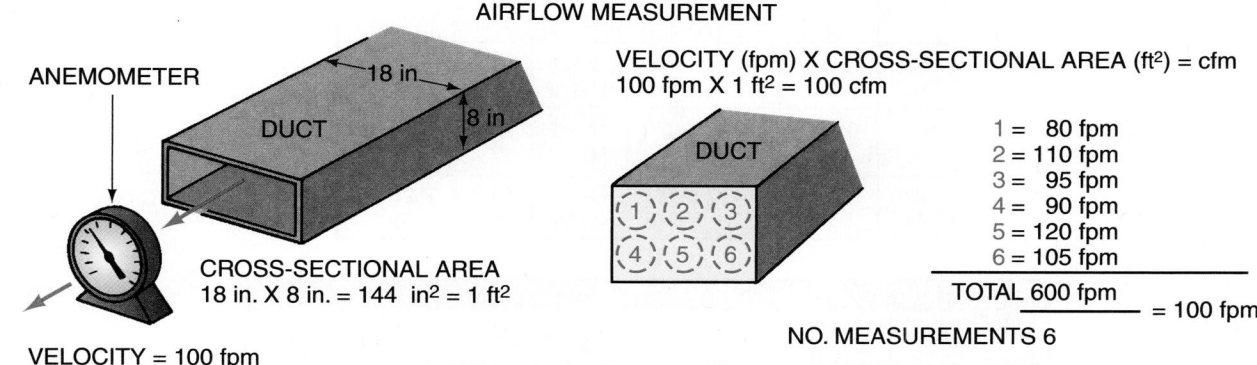

Figure 37–74 This cross section of duct with airflow shows how to measure duct area and air velocity.

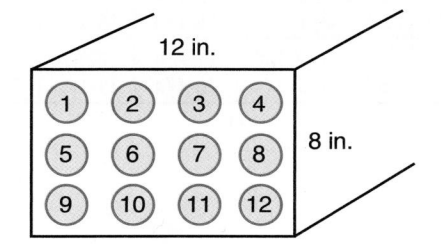

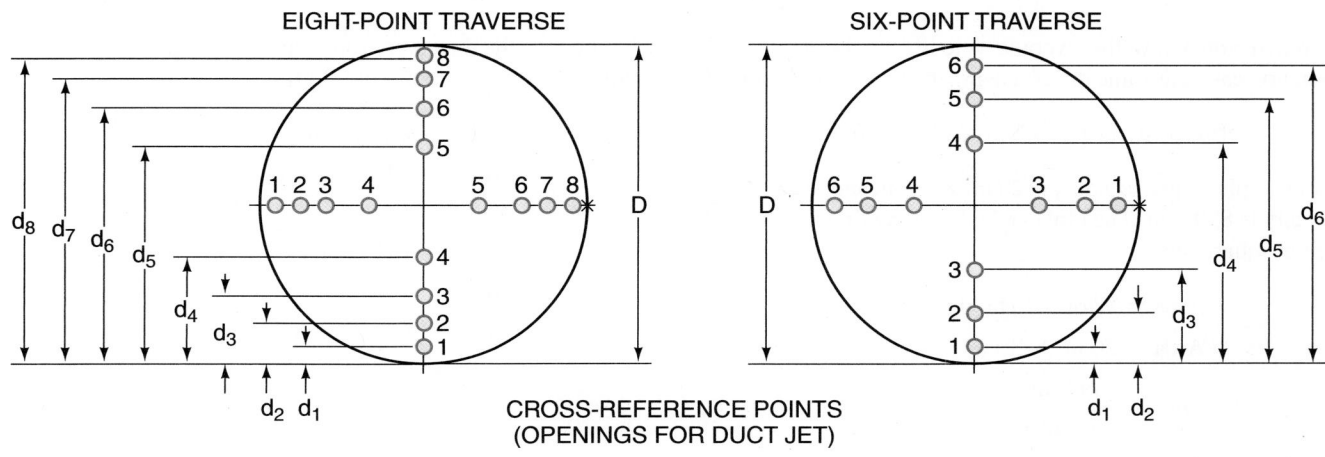

(A)

Figure 37–75 This figure shows the pattern for square, rectangular, and round duct that should be used for traversing the duct for the most accurate average reading **(A)–(D).**

TRAVERSE METHOD ↓	PROBE IMMERSION IN DUCT DIAMETERS									
	d_1	d_2	d_3	d_4	d_5	d_6	d_7	d_8	d_9	d_{10}
6 POINT	0.043	0.147	0.296	0.704	0.853	0.957	–	–	–	–
8 POINT	0.032	0.105	0.194	0.323	0.677	0.806	0.895	0.968	–	–
10 POINT	0.025	0.082	0.146	0.226	0.342	0.658	0.774	0.854	0.918	0.975

(B)

DUCT. DIA. (IN.)	PROBE IMMERSION FOR 6-PT. TRAVERSE					
	d_1	d_2	d_3	d_4	d_5	d_6
10	3/3	1-1/2	3	7	8-1/2	9-5/8
12	1/2	1-3/4	3-1/2	8-1/2	10-1/4	11-1/2
14	5/8	2	4-1/8	9-7/8	12	13-3/8
16	3/4	2-3/8	4-3/4	11-1/4	13-5/8	15-1/4
18	3/4	2-5/8	5-3/8	12-5/8	15-3/8	17-1/4
20	7/8	3	6	14	17	19-1/8
22	1	3-1/4	6-1/2	15-1/2	18-3/4	21
24	1	3-1/2	7-1/8	16-7/8	20-1/2	23

(C)

DUCT. DIA. (IN.)	PROBE IMMERSION FOR 8-PT. TRAVERSE							
	d_1	d_2	d_3	d_4	d_5	d_6	d_7	d_8
10	5/16	1	2	3-1/4	6-3/4	8	9	9-5/8
12	3/8	1-1/4	2-3/8	3-7/8	8-1/8	9-5/8	10-3/4	11-1/2
14	7/16	1-1/2	2-3/4	4-1/2	9-1/2	11-1/4	12-1/2	13-1/2
16	1/2	1-5/8	3-1/8	5-1/8	10-7/8	12-7/8	14-3/8	15-1/2
18	9/16	1-7/8	3-1/2	5-7/8	12-1/4	13-1/2	16-1/8	17-1/2
20	5/8	2-1/8	3-7/8	6-1/2	18-1/2	16-1/8	17-7/8	19-3/8
22	11/16	2-3/8	4-1/4	7-1/8	14-7/8	17-3/4	19-3/4	21-1/4
24	3/4	2-1/2	4-5/8	7-3/4	16-1/4	19-1/2	21-1/2	23-1/4

(D)

Figure 37–75 *(continued)*

When you know the average velocity of the air in any duct, you can determine the cfm using the following formula:

$$\text{cfm} = \text{Area (in ft}^2) \times \text{Velocity (in fpm)}$$

For example, suppose a duct is 20 in. × 30 in. and the average velocity is 850 fpm. The cfm can be found by first finding the area in square feet:

$$\text{Area} = \text{Width} \times \text{Height}$$
$$\text{Area} = 20 \text{ in.} \times 30 \text{ in.}$$
$$\text{Area} = \frac{600 \text{ in}^2}{144 \text{ in}^2/\text{ft}^2}$$
$$\text{Area} = 4.2 \text{ ft}^2$$
$$\text{cfm} = \text{Area} \times \text{Velocity}$$
$$\text{cfm} = 4.2 \text{ ft}^2 \times 850 \text{ fpm}$$
$$\text{cfm} = 3570 \text{ ft}^3 \text{ of air per minute}$$

Suppose the duct were round and had a diameter of 12 in. with an average velocity of 900 fpm.

$$\text{Area} = \pi \times \text{Radius squared}$$
$$\text{Area} = 3.14 \times 6 \times 6$$
$$\text{Area} = \frac{113 \text{ in}^2}{144 \text{ in}^2/\text{ft}^2}$$
$$\text{Area} = 0.78 \text{ ft}^2$$
$$\text{cfm} = \text{Area} \times \text{Velocity}$$
$$\text{cfm} = 0.78 \text{ ft}^2 \times 900$$
$$\text{cfm} = 702 \text{ ft}^3 \text{ of air per minute}$$

Special techniques and good instrumentation must be used to find the correct average velocity in a duct. For example, readings should not be taken within 10 duct diameters of the nearest fitting. This means that long straight runs must be used for taking accurate readings. This is not always possible because of the fittings and takeoffs in a typical system, particularly residential systems.

The proper duct traverse must be used and it is different for round duct versus square or rectangular duct. **Figure 37–75** shows the pattern for square, rectangular, and round duct that should be used for traversing the duct for the most accurate average reading. See Section 37.33 for some practical tips for measuring airflow.

37.32 THE AIR FRICTION CHART

The previous system can be plotted on the friction chart in **Figure 37–76.** This chart is for volumes of air up to 2000 cfm. Using the 400 cfm/ton mentioned earlier for air conditioning,

we can use this chart for systems having up to 5 tons of cooling. **Figure 37–77** shows a chart for larger systems, up to 100,000 cfm.

The friction chart has cubic feet on the left, and round-pipe sizes angle from left to right toward the top of the page. These duct sizes are rated in round-pipe sizes on the chart and can be converted to square or rectangular duct by using the table in **Figure 37–78.** The round-pipe sizes are for air with a density of 0.075 lb/ft^3 using galvanized pipe. Other charts are available for ductboard and flexible duct. The air velocity in the pipe is shown on the diagonal lines that run from left to right toward the bottom of the page. The friction loss in inches of water column per 100 ft of duct is shown along the top and bottom of the chart. For example, a run of pipe

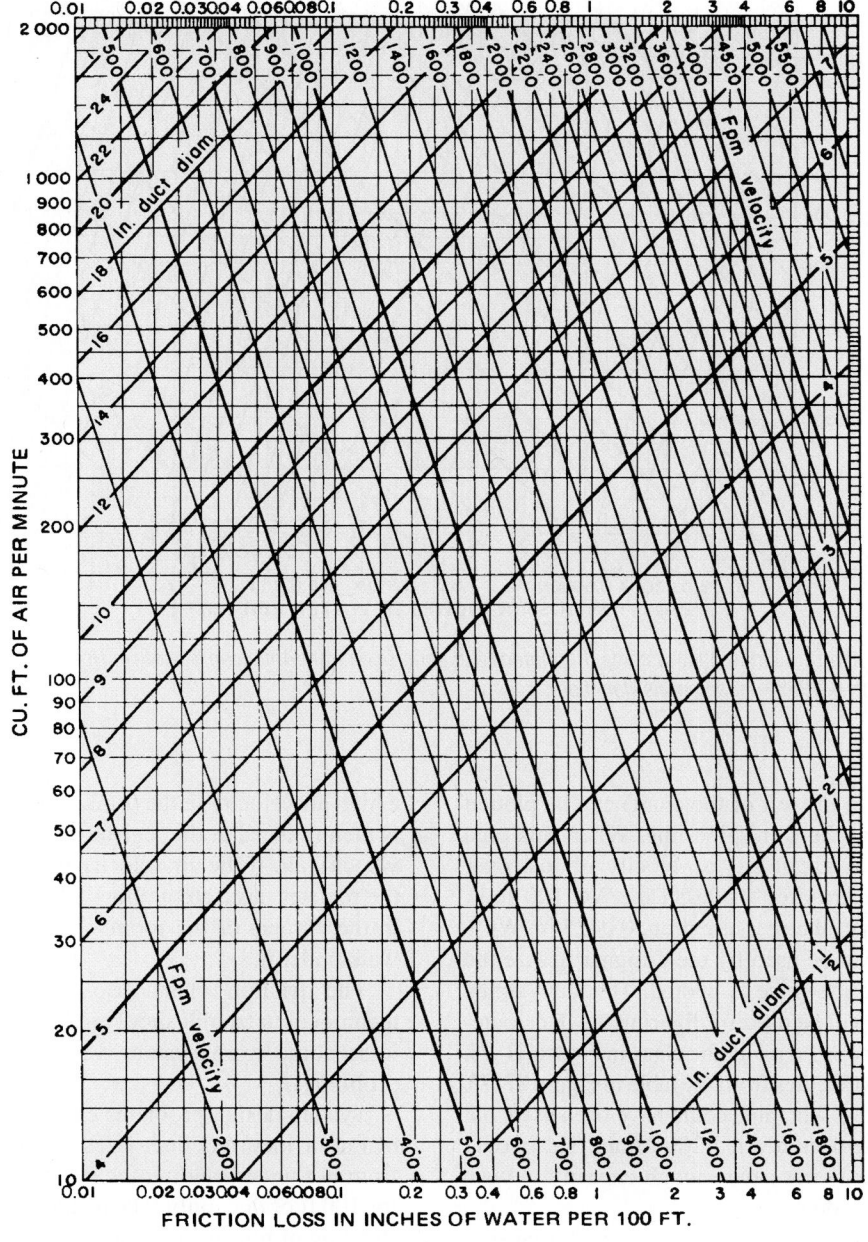

Figure 37–76 An air friction chart. *Used with permission from ASHRAE, Inc.*

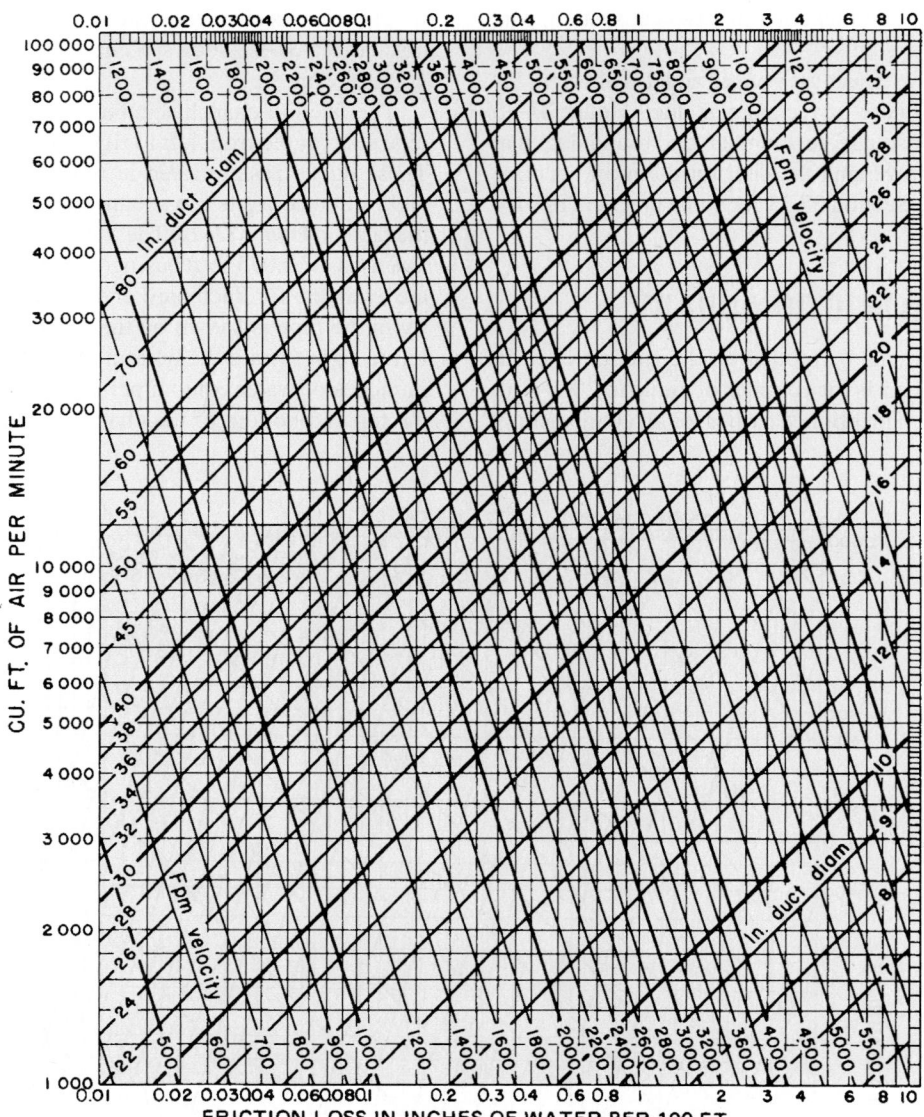

Figure 37–77   A friction chart for larger duct systems. *Reprinted with permission from American Society of Heating, Refrigerating, and Air-Conditioning Engineers, Inc.—1985 Fundamentals Handbook*

that carries 100 cfm (on the left of the chart) can be plotted over the intersection of the 6-in. pipe line. When this pipe is carrying 100 cfm of air, it has a velocity of just over 500 ft/min and a pressure drop of 0.085 in. WC/100 ft. A 50-ft run would have half the pressure drop, 0.0425 in. WC.

The friction chart can be used by the designer to size the duct system before the job price is quoted. This duct sizing provides the sizes that will be used for figuring the duct materials. The duct should be sized using the chart in **Figure 37–76,** and the recommended velocities from the table in **Figure 37–79.** In the previous example a 4-in. pipe could have been used, but the velocity would have been nearly 1200 ft/min. This would be noisy, and the fan may not have enough capacity to push sufficient air through the duct.

High-velocity systems have been designed and used successfully in small applications. Such systems normally have

a high air velocity in the trunk and branch ducts; the velocity is then reduced at the register to avoid drafts from the high-velocity air. If the air velocity is not reduced at the register, the register has a streamlined effect and is normally located in the corners of the room where someone is not likely to walk under it.

The friction chart also can be used by the field technician to troubleshoot airflow problems. Airflow problems come from system design, installation, and owner problems.

System design problems can be a result of a poor choice of fan or incorrect duct sizes. Design problems may result from not understanding the fact that the air friction chart is for 100 ft of duct only. The technician should have an understanding of how to size duct to determine whether or not the duct is sized correctly.

Lgth Adj.[b]	6	7	8	9	10	11	12	13	14	15	16	17	18	19	20	22	24	26	28	30	Lgth Adj.[b]
6	6.6																				6
7	7.1	7.7																			7
8	7.6	8.2	8.7																		8
9	8.0	8.7	9.3	9.8																	9
10	8.4	9.1	9.8	10.4	10.9																10
11	8.8	9.5	10.2	10.9	11.5	12.0															11
12	9.1	9.9	10.7	11.3	12.0	12.6	13.1														12
13	9.5	10.3	11.1	11.8	12.4	13.1	13.7	14.2													13
14	9.8	10.7	11.5	12.2	12.9	13.5	14.2	14.7	15.3												14
15	10.1	11.0	11.8	12.6	13.3	14.0	14.6	15.3	15.8	16.4											15
16	10.4	11.3	12.2	13.0	13.7	14.4	15.1	15.7	16.4	16.9	17.5										16
17	10.7	11.6	12.5	13.4	14.1	14.9	15.6	16.2	16.8	17.4	18.0	18.6									17
18	11.0	11.9	12.9	13.7	14.5	15.3	16.0	16.7	17.3	17.9	18.5	19.1	19.7								18
19	11.2	12.2	13.2	14.1	14.9	15.7	16.4	17.1	17.8	18.4	19.0	19.6	20.2	20.8							19
20	11.5	12.5	13.5	14.4	15.2	16.0	16.8	17.5	18.2	18.9	19.5	20.1	20.7	21.3	21.9						20
22	12.0	13.0	14.1	15.0	15.9	16.8	17.6	18.3	19.1	19.8	20.4	21.1	21.7	22.3	22.9	24.0					22
24	12.4	13.5	14.6	15.6	16.5	17.4	18.3	19.1	19.9	20.6	21.3	22.0	22.7	23.3	23.9	25.1	26.2				24
26	12.8	14.0	15.1	16.2	17.1	18.1	19.0	19.8	20.6	21.4	22.1	22.9	23.5	24.2	24.9	26.1	27.3	28.4			26
28	13.2	14.5	15.6	16.7	17.7	18.7	19.6	20.5	21.3	22.1	22.9	23.7	24.4	25.1	25.8	27.1	28.3	29.5	30.6		28
30	13.6	14.9	16.1	17.2	18.3	19.3	20.2	21.1	22.0	22.9	23.7	24.4	25.2	25.9	26.6	28.0	29.3	30.5	31.7	32.8	30
32	14.0	15.3	16.5	17.7	18.8	19.8	20.8	21.8	22.7	23.5	24.4	25.2	26.0	26.7	27.5	28.9	30.2	31.5	32.7	33.9	32
34	14.4	15.7	17.0	18.2	19.3	20.4	21.4	22.4	23.3	24.2	25.1	25.9	26.7	27.5	28.3	29.7	31.0	32.4	33.7	34.9	34
36	14.7	16.1	17.4	18.6	19.8	20.9	21.9	22.9	23.9	24.8	25.7	26.6	27.4	28.2	29.0	30.5	32.0	33.3	34.6	35.9	36
38	15.0	16.5	17.8	19.0	20.2	21.4	22.4	23.5	24.5	25.4	26.4	27.2	28.1	28.9	29.8	31.3	32.8	34.2	35.6	36.8	38
40	15.3	16.8	18.2	19.5	20.7	21.8	22.9	24.0	25.0	26.0	27.0	27.9	28.8	29.6	30.5	32.1	33.6	35.1	36.4	37.8	40
42	15.6	17.1	18.5	19.9	21.1	22.3	23.4	24.5	25.6	26.6	27.6	28.5	29.4	30.3	31.2	32.8	34.4	35.9	37.3	38.7	42
44	15.9	17.5	18.9	20.3	21.5	22.7	23.9	25.0	26.1	27.1	28.1	29.1	30.0	30.9	31.8	33.5	35.1	36.7	38.1	39.5	44
46	16.2	17.8	19.3	20.6	21.9	23.2	24.4	25.5	26.6	27.7	28.7	29.7	30.6	31.6	32.5	34.2	35.9	37.4	38.9	40.4	46
48	16.5	18.1	19.6	21.0	22.3	23.6	24.8	26.0	27.1	28.2	29.2	30.2	31.2	32.2	33.1	34.9	36.6	38.2	39.7	41.2	48
50	16.8	18.4	19.9	21.4	22.7	24.0	25.2	26.4	27.6	28.7	29.8	30.8	31.8	32.8	33.7	35.5	37.2	38.9	40.5	42.0	50
52	17.1	18.7	20.2	21.7	23.1	24.4	25.7	26.9	28.0	29.2	30.3	31.3	32.3	33.3	34.3	36.2	37.9	39.6	41.2	42.8	52
54	17.3	19.0	20.6	22.0	23.5	24.8	26.1	27.3	28.5	29.7	30.8	31.8	32.9	33.9	34.9	36.8	38.6	40.3	41.9	43.5	54
56	17.6	19.3	20.9	22.4	23.8	25.2	26.5	27.7	28.9	30.1	31.2	32.3	33.4	34.4	35.4	37.4	39.2	41.0	42.7	44.3	56
58	17.8	19.5	21.2	22.7	24.2	25.5	26.9	28.2	29.4	30.6	31.7	32.8	33.9	35.0	36.0	38.0	39.8	41.6	43.3	45.0	58
60	18.1	19.8	21.5	23.0	24.5	25.9	27.3	28.6	29.8	31.0	32.2	33.3	34.4	35.5	36.5	38.5	40.4	42.3	44.0	45.7	60
62		20.1	21.7	23.3	24.8	26.3	27.6	28.9	30.2	31.5	32.6	33.8	34.9	36.0	37.1	39.1	41.0	42.9	44.7	46.4	62
64		20.3	22.0	23.6	25.1	26.6	28.0	29.3	30.6	31.9	33.1	34.3	35.4	36.5	37.6	39.6	41.6	43.5	45.3	47.1	64
66		20.6	22.3	23.9	25.5	26.9	28.4	29.7	31.0	32.3	33.5	34.7	35.9	37.0	38.1	40.2	42.2	44.1	46.0	47.7	66
68		20.8	22.6	24.2	25.8	27.3	28.7	30.1	31.4	32.7	33.9	35.2	36.3	37.5	38.6	40.7	42.8	44.7	46.6	48.4	68
70		21.1	22.8	24.5	26.1	27.6	29.1	30.4	31.8	33.1	34.4	35.6	36.8	37.9	39.1	41.2	43.3	45.3	47.2	49.0	70
72			23.1	24.8	26.4	27.9	29.4	30.8	32.2	33.5	34.8	36.0	37.2	38.4	39.5	41.7	43.8	45.8	47.8	49.6	72
74			23.3	25.1	26.7	28.2	29.7	31.2	32.5	33.9	35.2	36.4	37.7	38.8	40.0	42.2	44.4	46.4	48.4	50.3	74
76			23.6	25.3	27.0	28.5	30.0	31.5	32.9	34.3	35.6	36.8	38.1	39.3	40.5	42.7	44.9	47.0	48.9	50.9	76
78			23.8	25.6	27.3	28.8	30.4	31.8	33.3	34.6	36.0	37.2	38.5	39.7	40.9	43.2	45.4	47.5	49.5	51.4	78
80			24.1	25.8	27.5	29.1	30.7	32.2	33.6	35.0	36.3	37.6	38.9	40.2	41.4	43.7	45.9	48.0	50.1	52.0	80
82				26.1	27.8	29.4	31.0	32.5	34.0	35.4	36.7	38.0	39.3	40.6	41.8	44.1	46.4	48.5	50.6	52.6	82
84				26.4	28.1	29.7	31.3	32.8	34.3	35.7	37.1	38.4	39.7	41.0	42.2	44.6	46.9	49.0	51.1	53.2	84
86				26.6	28.3	30.0	31.6	33.1	34.6	36.1	37.4	38.8	40.1	41.4	42.6	45.0	47.3	49.6	51.7	53.7	86
88				26.9	28.6	30.3	31.9	33.4	34.9	36.4	37.8	39.2	40.5	41.8	43.1	45.5	47.8	50.0	52.2	54.3	88
90				27.1	28.9	30.6	32.2	33.8	35.3	36.7	38.2	39.5	40.9	42.2	43.5	45.9	48.3	50.5	52.7	54.8	90
92					29.1	30.8	32.5	34.1	35.6	37.1	38.5	39.9	41.3	42.6	43.9	46.4	48.7	51.0	53.2	55.3	92
96					29.6	31.4	33.0	34.7	36.2	37.7	39.2	40.6	42.0	43.3	44.7	47.2	49.6	52.0	54.2	56.4	96

Figure 37–78 A chart used to convert from round to square or rectangular duct. *Reprinted with permission from American Society of Heating, Refrigerating, and Air-Conditioning Engineers, Inc.—1985 Fundamentals Handbook*

To size duct properly the technician must understand the calculation of pressure drop in fittings and pipe. The pressure drop in duct is calculated and read on the friction chart in feet of duct, from the air handler to the terminal outlet. This is a straightforward measurement. The fittings are calculated in a different manner. Pressure drop across different types of fittings in in. WC has been determined by laboratory experiment and is known. This pressure drop has been converted to equivalent feet of duct for the convenience of the designer and technician. The technician must know what type of fittings are used in the system, and then a chart may be consulted for the equivalent feet of duct, **Figure 37–80**. The equivalent feet of duct for all of the fittings for a particular run must be added together and then added to the actual duct length for the technician to determine the proper duct size for a particular run.

When a duct run is under 100 ft, a corrected friction factor must be used or the duct will not have enough

Structure	Supply Outlet	Return Openings	Main Supply	Branch Supply	Main Return	Branch Return
Residential	500–750	500	1,000	600	800	600
Apartments, Hotel Bedrooms, Hospital Bedrooms	500–750	500	1,200	800	1,000	800
Private Offices, Churches, Libraries, Schools	500–1,000	600	1,500	1,200	1,200	1,000
General Offices, Deluxe Restaurants, Deluxe Stores, Banks	1,200–1,500	700	1,700	1,600	1,500	1,200
Average Stores, Cafeterias	1,500	800	2,000	1,600	1,500	1,200

Figure 37–79   A chart of recommended velocities for different duct designs.

resistance to airflow and too much flow will occur. The corrected friction factor may be determined by multiplying the design friction loss in the duct system times 100 (the friction loss used on the chart) and dividing by the actual length of run. For example, a system may have a duct run handling 300 cfm of air to an inside room. This duct is very close to the main duct and air handler. There are 60 effective feet, including the friction loss in the fittings and the duct length, **Figure 37–81.**

The fan may be capable of moving the correct airflow in the supply duct with a friction loss of 0.36 in. WC. The cooling coil may have a loss of 0.24 and the supply registers a loss of 0.03. When the coil and register loss are subtracted from the friction loss for the supply duct, the duct must be designed for a loss of 0.09 in. WC (0.36 − 0.27 = 0.09).

The following describes a mistake the designer made and how the technician discovered it. The designer chose 9-in. duct from the friction chart at the junction of 0.09 in. WC and

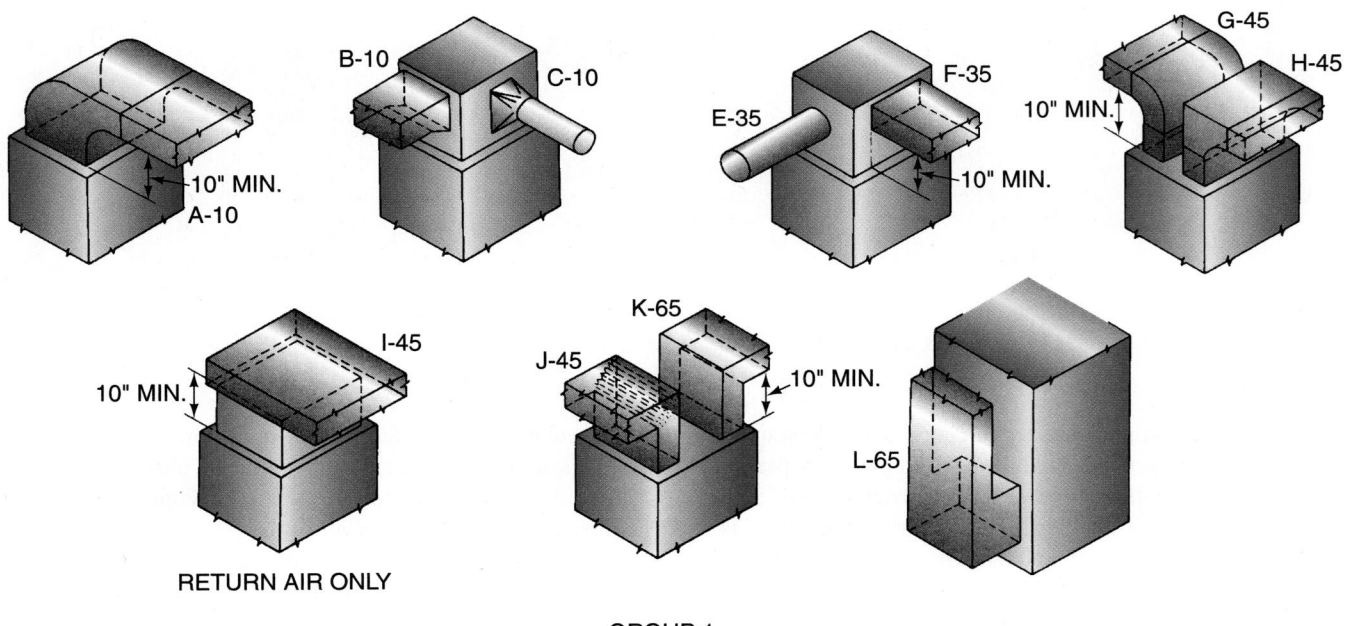

Figure 37–80   Various duct fittings and their equivalent feet. *Used with permission from ASHRAE, Inc.*

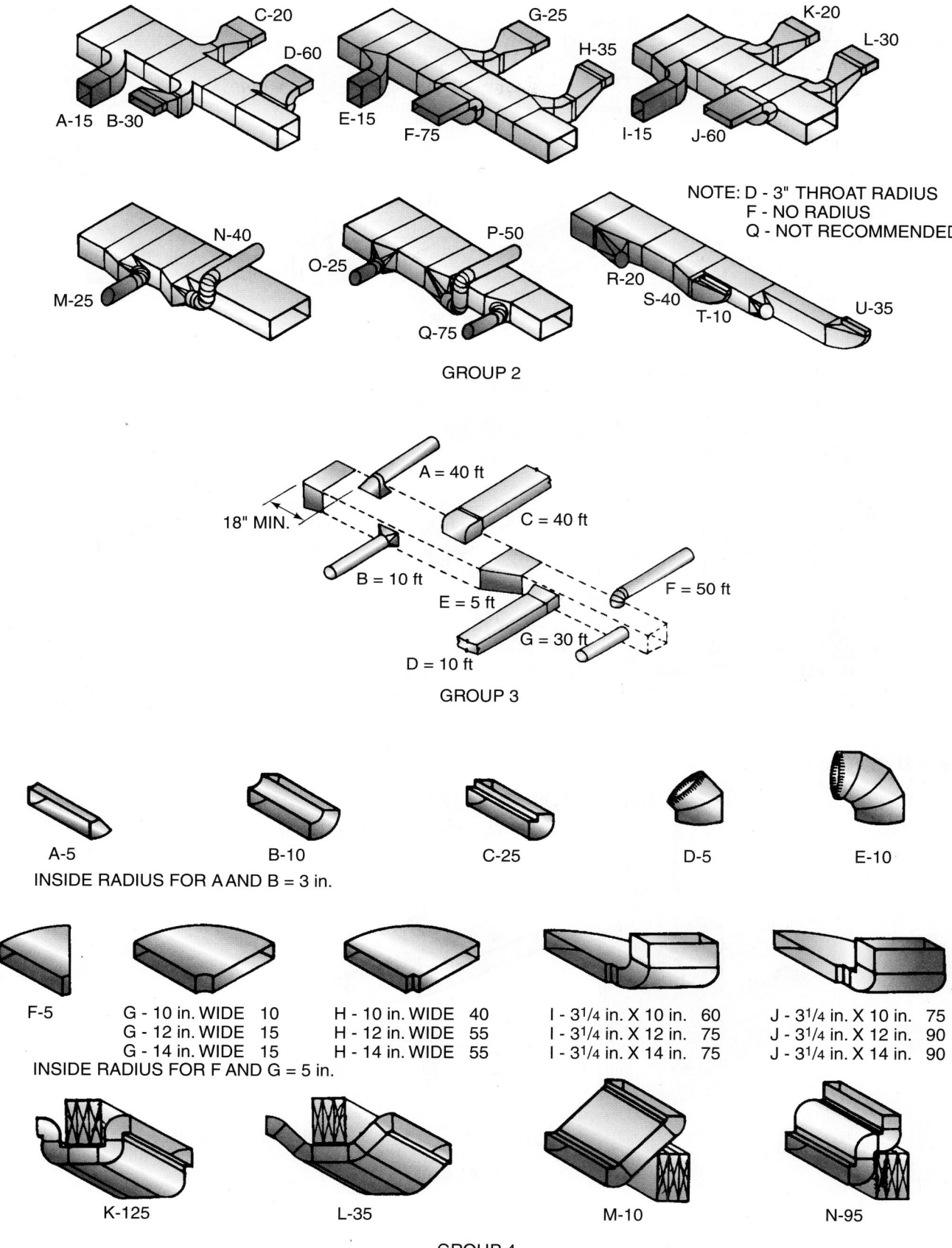

NOTE: D - 3" THROAT RADIUS
F - NO RADIUS
Q - NOT RECOMMENDED

GROUP 2

18" MIN.

GROUP 3

A-5   B-10   C-25   D-5   E-10

INSIDE RADIUS FOR A AND B = 3 in.

F-5

G - 10 in. WIDE	10
G - 12 in. WIDE	15
G - 14 in. WIDE	15

H - 10 in. WIDE	40
H - 12 in. WIDE	55
H - 14 in. WIDE	55

I - 3¹/4 in. X 10 in.	60
I - 3¹/4 in. X 12 in.	75
I - 3¹/4 in. X 14 in.	75

J - 3¹/4 in. X 10 in.	75
J - 3¹/4 in. X 12 in.	90
J - 3¹/4 in. X 14 in.	90

INSIDE RADIUS FOR F AND G = 5 in.

K-125   L-35   M-10   N-95

GROUP 4

Figure 37–80 *(continued)*

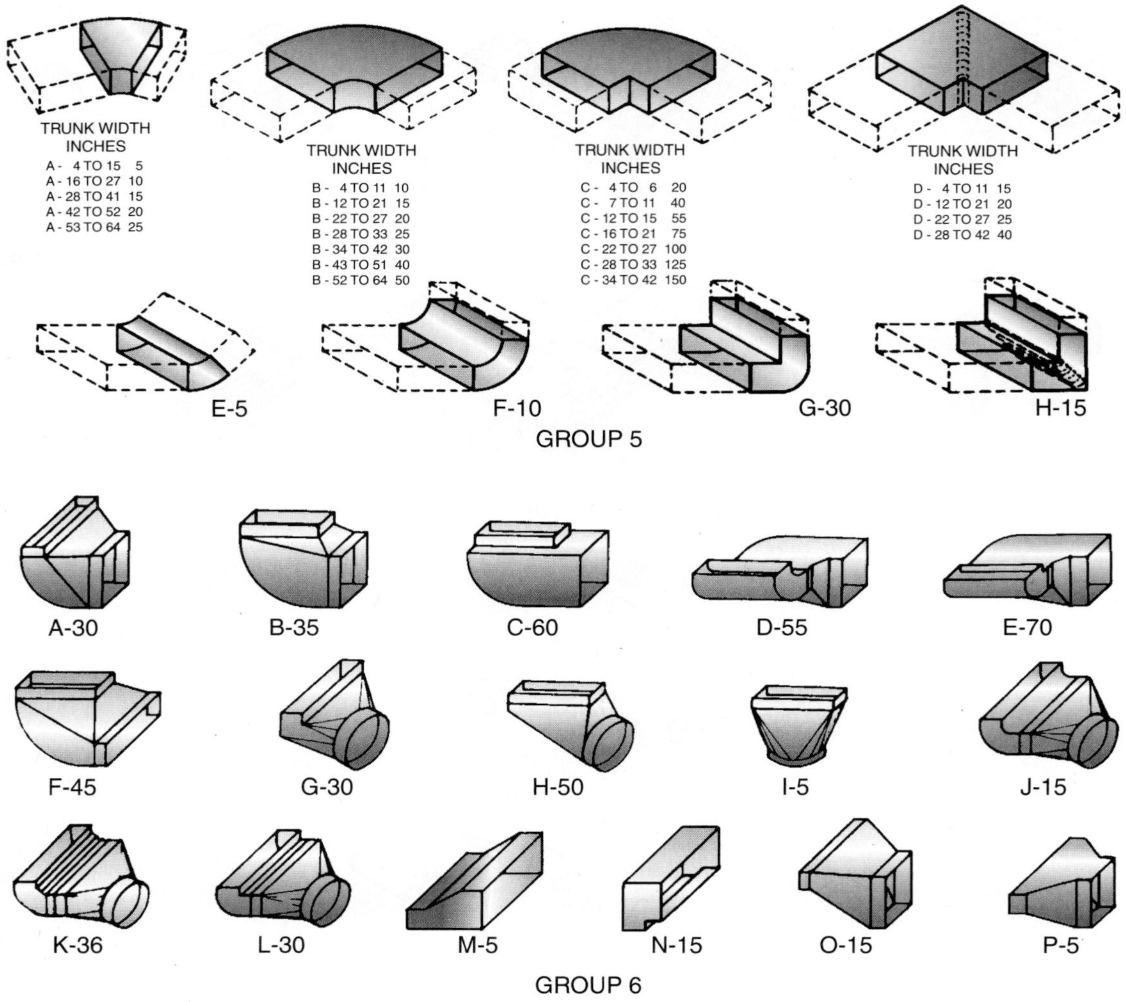

TRUNK WIDTH
INCHES
A - 4 TO 15 5
A - 16 TO 27 10
A - 28 TO 41 15
A - 42 TO 52 20
A - 53 TO 64 25

TRUNK WIDTH
INCHES
B - 4 TO 11 10
B - 12 TO 21 15
B - 22 TO 27 20
B - 28 TO 33 25
B - 34 TO 42 30
B - 43 TO 51 40
B - 52 TO 64 50

TRUNK WIDTH
INCHES
C - 4 TO 6 20
C - 7 TO 11 40
C - 12 TO 15 55
C - 16 TO 21 75
C - 22 TO 27 100
C - 28 TO 33 125
C - 34 TO 42 150

TRUNK WIDTH
INCHES
D - 4 TO 11 15
D - 12 TO 21 20
D - 22 TO 27 25
D - 28 TO 42 40

E-5   F-10   G-30   H-15

GROUP 5

A-30   B-35   C-60   D-55   E-70

F-45   G-30   H-50   I-5   J-15

K-36   L-30   M-5   N-15   O-15   P-5

GROUP 6

Figure 37–80 *(continued)*

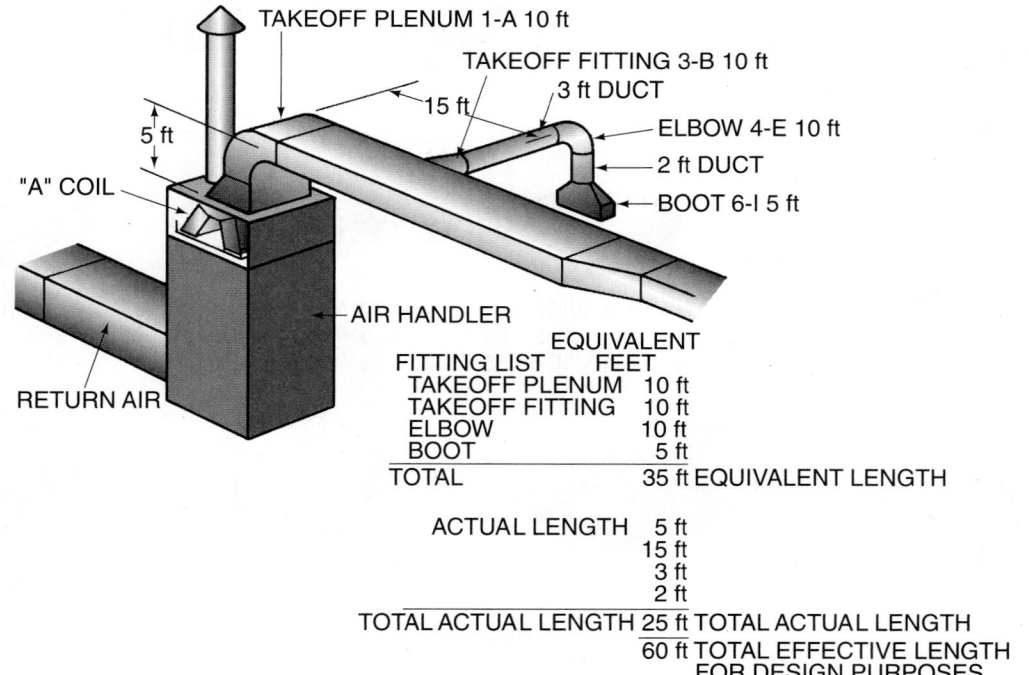

TAKEOFF PLENUM 1-A 10 ft

TAKEOFF FITTING 3-B 10 ft

3 ft DUCT

15 ft

5 ft

"A" COIL

ELBOW 4-E 10 ft

2 ft DUCT

BOOT 6-I 5 ft

AIR HANDLER

RETURN AIR

	EQUIVALENT
FITTING LIST	FEET
TAKEOFF PLENUM	10 ft
TAKEOFF FITTING	10 ft
ELBOW	10 ft
BOOT	5 ft
TOTAL	35 ft EQUIVALENT LENGTH

ACTUAL LENGTH	5 ft
	15 ft
	3 ft
	2 ft
TOTAL ACTUAL LENGTH	25 ft TOTAL ACTUAL LENGTH
	60 ft TOTAL EFFECTIVE LENGTH FOR DESIGN PURPOSES

Figure 37–81 This duct is run less than 100 feet from the air handler.

300 cfm without making a correction for the fact that the duct is only 60 ft long. The technician corrected the static pressure loss of 60 ft of duct using the formula:

$$\text{Adjusted static} = \frac{\text{Design static} \times 100}{\text{Total equivalent length of duct}}$$

$$\text{Adjusted static} = \frac{0.09 \times 100}{60}$$

$$\text{Adjusted static} = 0.15$$

By using this adjusted static pressure, the technician referred to the friction chart and plotted at 300 cfm, then moved to the right to 0.15 adjusted static. The duct size now became 8-in. duct. This is not a significant savings in materials, but it does reduce the airflow to the correct amount. An alternative is to use a throttling damper where the branch line leaves the trunk duct. **The key to this problem is correcting the static pressure to an adjusted value.** This causes the chart to undersize the duct for the purpose of reducing the airflow to this run because it is less than 100-ft long. Otherwise, the air will have so little resistance that too much air will flow from this run, starving other runs. One run that is oversized will not make too much difference, but a whole system sized using the same technique will have problems.

This problem illustrates a run that is less than 100 ft. The same technique will apply to a run that is more than 100 ft, except that the run will be undersized and starved for air. For example, suppose that another run off the same system requires 300 cfm of air and is a total of 170 ft (fittings and actual duct length), **Figure 37–82.** When the designer sized the duct for 300 cfm at a static pressure loss of 0.09, a duct size of 9 in. round was chosen. The technician finds the airflow is too low and checks the figures using the formula:

$$\text{Adjusted static} = \frac{\text{Design static} \times 100}{\text{Total equivalent length of duct}}$$

$$\text{Adjusted static} = \frac{0.09 \times 100}{170}$$

$$\text{Adjusted static} = 0.053$$

When the duct is sized using the adjusted static, the adjustment causes the chart to oversize the duct, and it moves the correct amount of air. In this case, the technician enters the chart at 300 cfm, moves to the right to 0.053, and finds the correct duct size to be 10 in. instead of 9 in. This is the reason this system is starved for air. All duct is first sized in round, then converted to square or rectangular. NOTE: All runs of duct are measured from air handler to the end of the run for sizing.

As shown from the previous examples, when some duct is undersized and some oversized the result will probably produce problems. The technician should pay close attention to any job with airflow problems and try to find the original duct sizing calculations. This would enable the technician to fully analyze a problem duct installation. Any equipment supplier will help you choose the correct fan and equipment. They may even help with the duct design.

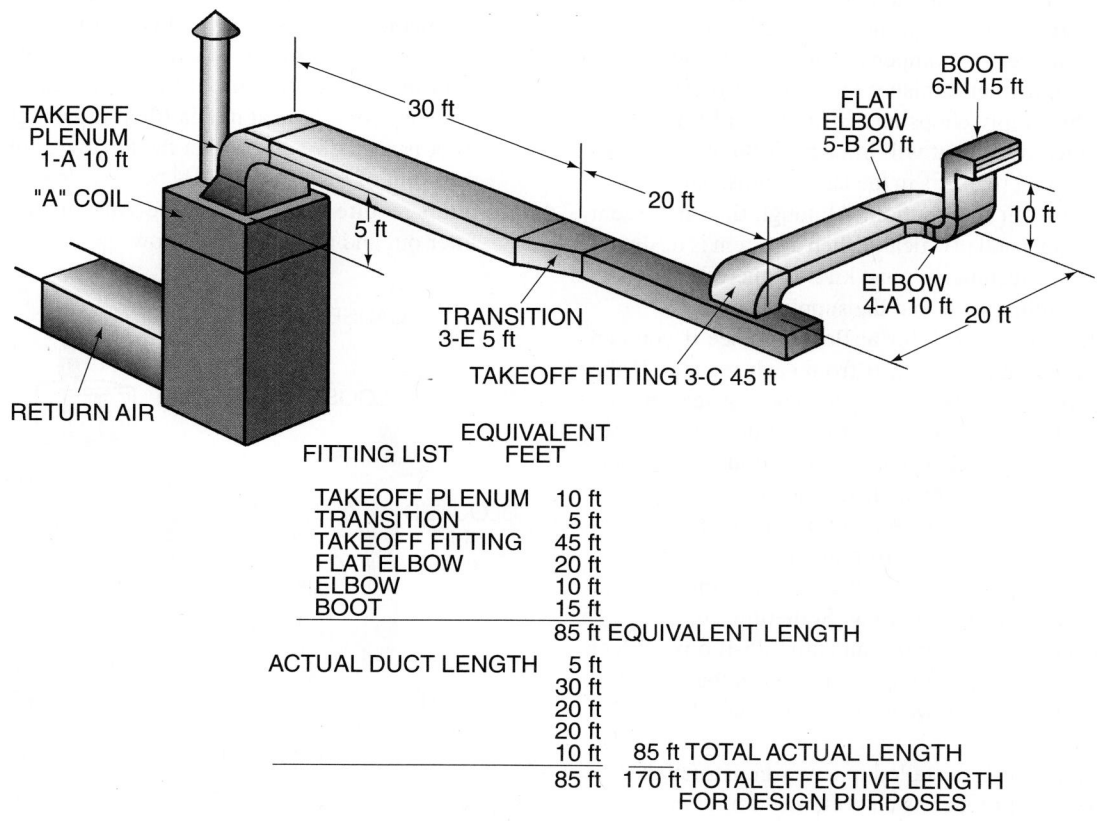

Figure 37–82   This duct is run more than 100 feet from the air handler.

37.33 PRACTICAL TROUBLESHOOTING TECHNIQUES

Most duct systems are either residential or commercial installations. Residential systems will be discussed first. Generally, technicians service residential systems that have less than 5 tons of cooling and 200,000 Btu/h of heating. These systems are often installed from experience rather than from the use of tedious design techniques. Many contractors know how much air should be delivered to a particular room in a home; unless they have a design department, the duct systems will be similar in configuration. Therefore, the use of instruments may not be necessary for locating and repairing airflow problems. As explained earlier, for instruments to work correctly, measurements must be made in long runs of duct. Residential systems are usually short runs. Most testing of residential systems is performed at the registers and grilles because of the lack of long duct runs. Commercial systems are much different and will require the use of instruments.

37.34 RESIDENTIAL DUCT SYSTEM PROBLEMS

Many contractors size the branch duct system in residences using two sizes of duct for the economy of the installation. The duct to the rooms is often sized as 6-in. or 8-in. round (which may be converted to square or rectangular duct) with dampers used in each run to adjust the airflow. A 6-in. round duct will move 100 cfm of air with a friction loss of about 0.08 in. WC pressure drop per 100 ft. An 8-in. round duct will move 200 cfm of air with a friction loss of about 0.075 in. WC. Many contractors divide the house into 100- and 200-cfm zones and use dampers to compensate for any airflow unbalance. The trunk duct that furnishes the air to the branch ducts is often sized by simply choosing a size that will furnish the number of branch ducts that will branch off the main. There will be several outlet registers in the large rooms, but that will also provide more air distribution. Although this may seem careless, it is simple and practical. When a system is designed in this manner, the technician must know if there is adequate airflow at the terminal units and the supply registers.

The simple way to check for airflow is to see if you can feel the velocity at about 2 to 3 ft from each register. If the registers are floor registers and it is the cooling season, you can walk to each register, one at a time, while holding your hand down at your side. This can often be done while touring the home with the homeowner, who may not even be aware of what you are doing. When you find one that has a very low flow, follow the duct from the trunk to the register. The following are some problems that you may encounter:

- Flex duct runs may be too long. Sometimes the installing contractor will use an entire 25-ft box of duct for a 15-ft run and just leave the excess in the attic. Some installers do not want to cut and haul off the extra duct.
- When the system is installed in a new construction home, construction debris can often be found in the duct. The register holes should be covered when floor registers are used, because it is convenient to sweep scraps into the holes in the floor.
- Runs may have been disconnected. Duct systems that are run where people, such as cable TV or telephone repair technicians, crawl over or step on them can become disconnected or crushed.
- It is common for metal duct that is run under a concrete slab to get stepped on and collapse just before the concrete is poured. There will be practically no airflow at the register end of the duct. There is no simple remedy for this.
- The damper on one or more of the branch ducts may be closed. Look for the damper handle that is supposed to be installed where you can see it. Even if the duct is insulated on the outside, the handle should protrude out of the insulation. If the damper handle is under the insulation, it can often be found by feel.
- Flexible duct will collapse inside if it is installed in too short a turn. It may not be easy to see from the outside.
- Installers have placed pieces of insulation in the duct with the intention of getting back to them, but the duct was connected before the pieces were removed.
- When the duct is lined on the inside, it is hard to determine where the obstruction may be. The sure sign of this obstruction is reduced airflow at the outlet registers for that trunk run.
- Look for furniture that is placed over registers. In some cases homeowners will block registers and grilles because they feel a draft. Floor registers or grilles can be covered with area rugs, **Figure 37–83**.
- The wrong filter media may have been chosen. Many companies advertise better filters for residential systems. A better filter will have a tighter weave and offer more air restriction. If the filter rack was designed for a typical media filter and a high-density filter is used, it is likely that the airflow is reduced to the point of low flow. This is because the high-density filters require more area for the same airflow. Take the filter out and recheck the airflow.

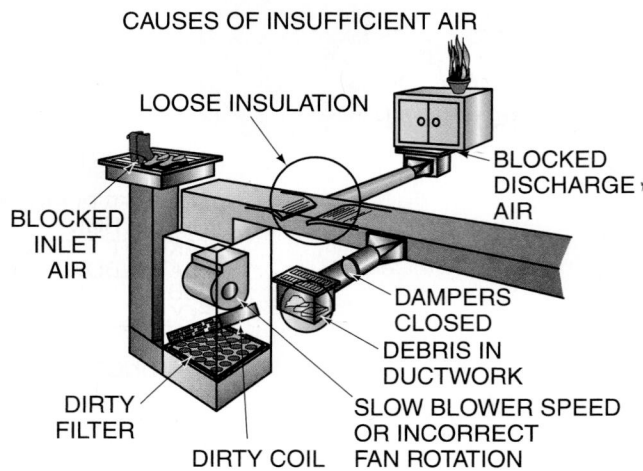

CAUSES OF INSUFFICIENT AIR

LOOSE INSULATION

BLOCKED DISCHARGE AIR

BLOCKED INLET AIR

DAMPERS CLOSED

DEBRIS IN DUCTWORK

DIRTY FILTER

SLOW BLOWER SPEED OR INCORRECT FAN ROTATION

DIRTY COIL

Figure 37–83 Obstructed airflow. *Reproduced courtesy of Carrier Corporation*

After checking each individual register for noticeable airflow, look for signs of airflow into the return air grilles. If the return air grille is oversized, you may not notice airflow. A piece of paper placed close to the grille will be sucked onto the grille surface with only a slight amount of airflow velocity. Oversized return air grilles are good, because they reduce the air noise.

To verify airflow for floor and sidewall registers, tie a piece of fine sewing thread or ribbon about 12 in. long to the end of a stick or piece of wire. Hold it over the airflow. The thread or ribbon will show the air pattern by waving in the air. If there is no airflow, it will not wave in the air. This does not work as well with ceiling registers because the airflow is usually down, just as the thread or ribbon would normally hang.

When the technician has air velocity instruments, readings can be taken at the various registers. The industry standard for the maximum recommended air velocity for a register in a home is 750 fpm, **Figure 37–79.** You should be able to feel this airflow.

The airflow from a register is not the same across the entire surface. It depends on how the register is fastened to the ductwork. Although it may seem that the static pressure in a duct would provide even pressure to all surfaces of the register, there are other considerations. For example, when a register is mounted on the duct itself, the air turning into the register can create a low-pressure area at one end of the register, **Figure 37–84.** This situation may be remedied by the use of turning vanes at the register.

Manufacturers of registers and grilles provide tables that show what the airflow through each device is at a particular velocity. The problem is that the velocity is often not a constant across the entire device.

If you think you have an overall low airflow problem because there does not seem to be enough air at all of the registers, do a total system examination starting with the filter. Many times the air handler is shipped with a filter in the cabinet. The installer who installs a return air filter grille at the common return may forget to remove the filter shipped with the unit. If this common return filter is faithfully changed, the filter at the air handler will still eventually become obstructed. An overall airflow reduction would show

up in the summer cooling application as low suction pressure and in the winter as a high temperature rise across the furnace or heating system. Less airflow can cause the furnace to trip off because of high heat exchanger temperatures. The technician can look at most furnaces and check the recommended air temperature rise. If the rise is above the maximum, there is not enough air. For example, if a gas furnace manufacturer recommends an air temperature rise of between 45°F and 65°F and the rise is 80°F, there is not enough airflow to satisfy the furnace manufacturer's specifications. It is easy for the technician to check for temperature rise with a temperature check at the supply and return registers.

An amperage reading of the fan motor may provide a clue to overall reduced airflow. If the fan is supposed to have a draw of 7 A and it is only 4 A, suspect an obstruction or even an undersized duct. When a new system is started up for the first time and the airflow is established, the technician should record the fan motor amperage on the unit to use later for comparison testing.

When a system has one or two common return air grilles, the average air velocity can be taken at the grilles and calculated to find the overall airflow. The free or usable area of the grille is first measured, then the average velocity across the grille is multiplied times the area (cubic feet per minute = velocity in feet per minute × area in square feet). The real area of the grille is not an exact measurement of size because of the obstruction of the fins. The real area is about 70% of the measured area in square feet. For example, suppose that the average velocity of a return air grille is 350 fpm and the measured size of the grille opening is 14 in. × 20 in. The real area would be 14 × 20/144 in² per square foot (14 × 20/144 = 1.944 ft²). Since only 70% of the grille area is usable because of the fins, the actual usable area is 1.944 × 0.7 or 1.44 ft². The volume of air would be cfm = velocity × area, 350 × 1.44 = 504 cfm. This is approximately the amount of airflow required for a 1 1/2-ton system for cooling. If the return air grille is too small, it will normally make a whistling sound from the velocity of air passing through it.

Other methods may be used for determining the flow for a residential system. These are discussed in the individual units that apply to gas, oil, or electric heat. These methods determine the airflow by measuring the energy input to the airstream and calculating the airflow.

No matter what method you use, the accuracy will probably be within + or −20%. This is about as close as you can get for residential systems, assuming they are not leaking air at the duct connections. NOTE: All air measurements for residential systems should be performed at maximum airflow. Any system that is both heating and cooling and that has a multiple-speed motor may have a different airflow for heating versus cooling. Switch the room thermostat to FAN ON for all testing. This will be the maximum airflow.

If you discover there is not enough air for the total system and there are no physical barriers, the next step is to see if the air volume can be increased. If the system has a belt-drive fan, check the amperage as compared with the motor full-load amperage. You can close the pulley down a few turns and see if the amperage and airflow rise to normal.

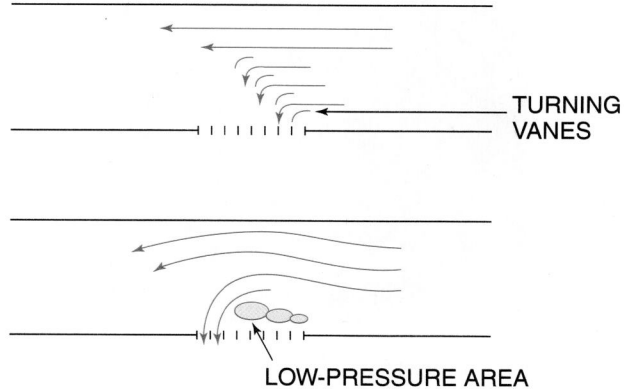

TURNING VANES

LOW-PRESSURE AREA

Figure 37–84   This register is installed just after a reducing transition and will need turning vanes to turn the air into the register.

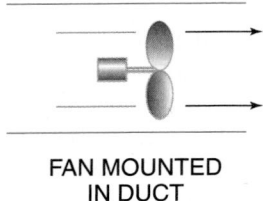

**FAN MOUNTED
IN DUCT**

Figure 37–85 This in-line duct fan will help increase airflow in a duct that is short of airflow.

When a reduced airflow is discovered on a long run of duct to a room and there are no physical barriers, a small fan can be installed in the duct to move more air to that outlet, **Figure 37–85.** This will make up for slight design deficiencies.

37.35 COMMERCIAL DUCT SYSTEMS

Commercial duct systems vary from small buildings to large buildings. Commercial duct systems almost always have a blueprint of the duct layout along with a set of specifications that tell how much air should be flowing at any particular part of the system. This makes it easier to get the proper instruments and discover problems.

When a large multistory building is erected, a certified testing and balancing company should be at the building systems start-up to verify that the airflow is correct before the building is turned over to the owner-operator. This company will verify the total building airflow, including the flow in all of the trunk ducts, in all of the branch ducts, and at the terminal units. These airflows are supposed to compare with the building specifications. If not, notes should be made of the discrepancies. The field technician or building management should take over from there.

An advantage of commercial high-rise buildings is that there is usually one duct system for each floor and they are usually duplicates. This means learning one system for an entire building.

Troubleshooting airflow problems in large systems involves reading the prints to determine what the airflow is supposed to be under the conditions you are trying to check. If the system is a variable volume system, the airflow may be reduced because of the thermostat setting. If it is a constant flow system, the flow should be known. In either case, when the correct air volume is known, measure the airflow with quality instruments. This is often the calculation discussed previously, cfm = area × velocity.

There should be test ports in the duct where the testing and balancing company did their test work. If not, you will have to install your own. Be sure to use the correct drill size and wear safety goggles because the airflow in the duct may blow chips into your eyes. Testing will involve taking traverse readings of the duct, determining the average velocity of the air, and then using the formula to determine the cfm of air. Commercial duct systems often have *splitter dampers* or *diversion dampers* that divert or divide the airflow in ductwork. These dampers can be adjusted to cause more or less

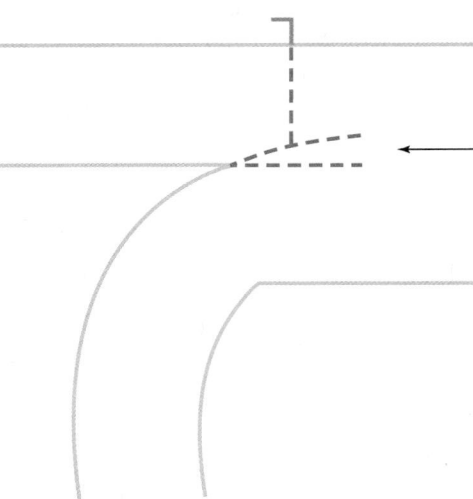

Figure 37–86 This splitter, or diverter, damper can help control how much airflow is in the two different ducts.

air to flow into branch ducts, **Figure 37–86.** The building blueprint should tell you the correct cfm in each duct run. If there are no prints for the building, you may be able to get a copy of the original drawings from the installing contractor, the engineering firm, or the building architect. Be sure to plug the test ports when you are finished. Tapered rubber stoppers are commonly used for this.

Flow hoods are available for measuring the airflow at terminal units. These flow hoods fit over the register, sealing off the air, **Figure 37–87.** Because they read out directly in cfm, they can be compared with the specifications.

Figure 37–87 This hood fits directly over ceiling registers up to about 2 ft × 2 ft. The hood forces the air from the register through the meter and displays the airflow in cfm. *Courtesy Alnor Instrument Company*

Total airflow of commercial systems can be verified by checking the main duct for the correct airflow. If it is not correct, the technician should check for any problem that could reduce the airflow. These are much the same as with residential systems. Filters should be checked first. Dampers may be blocking airflow if they are not positioned correctly. All large fans are powered by three-phase power, and fan motors should be checked for proper rotation. It is not uncommon for a fan to be running backward because some technician reversed two wires while servicing the electrical system. A squirrel cage fan will run as well backward as forward, but it will only discharge about two-thirds of the correct volume. This may not be noticed until peak season on some systems. SAFETY PRECAUTION: The technician is responsible for obtaining air pressure and velocity readings in ductwork. This involves drilling and punching holes in metal duct. Use a grounded drill cord or a cordless drill and be careful with drill bits and the rotating drill. Sheet metal can cause serious cuts, and air blowing from the duct can blow chips in your eyes. Protective eye covering is necessary.

SUMMARY

■ Air is passed through conditioning equipment and then circulated into the room to condition it.

■ Infiltration is air leaking into a structure.

■ Ventilation is air being induced into the conditioning equipment and conditioned before it is allowed to enter the conditioned space.

■ The duct system distributes air to the conditioned space. It consists of the blower (or fan), the supply duct, and the return duct.

■ The blower or fan uses energy to move the air.

■ The propeller type is used to move a lot of air against a small pressure. It can make noise.

■ The centrifugal type is used to move large amounts of air in ductwork, which offers resistance to the movement of air.

■ A fan is not a positive-displacement device.

■ Small centrifugal fans use energy in proportion to the amount of air they move.

■ Duct systems, both supply and return, are large pipes or tunnels that the air flows through.

■ 1 in. WC is the amount of pressure needed to raise a column of water 1 in.

■ The atmospheric pressure of 14.696 psia will support a column of water 34 ft high.

■ 1 psi will support a column of water 27.7 in. or 2.31 ft high.

■ Static pressure is the pressure pushing outward on the duct.

■ Velocity pressure is moving pressure created by the velocity of the air in the duct.

■ Total pressure is the velocity pressure plus the static pressure.

■ The pitot tube is a probe device used to measure the air pressures.

■ Air velocity (fpm) in a duct can be multiplied by the cross-sectional area (ft²) of the duct to obtain the amount of air passing that particular point in cubic feet per minute.

■ Typical supply duct systems used are plenum, extended plenum, reducing plenum, and perimeter loop.

■ Branch ducts should always have balancing dampers to balance the air to the individual areas.

■ Zoning allows a single air-conditioning system to provide individual room temperature control.

■ When a system is zoned, each zone has its own zone damper and space thermostat.

■ Zoned systems can provide simultaneous heating and cooling to different zones.

■ When the air is distributed in the conditioned space, it is common practice to distribute the air on the outside wall to cancel the load.

■ The amount of throw tells how far the air from a diffuser will reach into the conditioned space.

■ Each foot of duct, supply or return, has a friction loss that can be plotted on a friction chart for round duct.

■ Round duct sizes can be converted to square or rectangular equivalents for sizing and friction readings.

REVIEW QUESTIONS

1. Name five changes that are made to air to condition it.
2. The two ways that fresh air enters a structure are _____ and _____.
3. Which of the following types of fan is used to move large amounts of air against a low static pressure drop?
   A. Vane axial
   B. Centrifugal
   C. Large
   D. Propeller
4. True or False: Turbulence in airflow causes friction drop.
5. Name two types of blower drives.
6. Pressure in ductwork is expressed in which of the following?
   A. Inches of water column
   B. Psig
   C. Psia
   D. Inches of Hg
7. Name two types of blower drives.
8. What is a common instrument used to measure pressure in ductwork?
9. The three pressures created by moving air in ductwork are _____ pressure, _____ pressure, and _____ pressure.
10. Name the two types of return air systems in residential installations.
11. What is the component that distributes air in the conditioned space?
12. Warm air heats a room better if distributed
    A. low in the room.
    B. high in the room.
    C. behind the furniture.
    D. under the curtains.
13. Name four materials used to manufacture duct.
14. The name of the fitting where the branch line leaves the duct is the _____ _____.

15. True or False: Air for cooling distributes better from down next to the floor.

16. Why are dampers recommended in all branch line ducts?

17. Zoning an air-conditioning or heating system allows for
  A. temperature control of individual rooms or areas in the space.
  B. simultaneous heating and cooling of individual rooms in a structure.
  C. Neither A nor B is a characteristic of a zoned system.
  D. Both A and B are characteristics of a zoned system.

18. Multiple-zone systems often use a _____ or a _____ _____ to alleviate the problem of excess air when only some zones are calling for conditioning.

19. True or False: Zoning an air-conditioning system can only be done when the system is initially designed and installed.

20. The duct sizes on the air friction chart are expressed in which of the following?
  A. Square
  B. Round
  C. Oval
  D. Rectangular

21. The average velocity of a 16-in. round duct is 800 fpm; the airflow in the duct is _____ cfm.

22. If a rectangular duct system (12 × 26 in.) has an average velocity of 700 fpm, what is the airflow?

23. The equivalent feet of pressure drop for a group 3-G branch line takeoff is
  A. 30 ft.
  B. 20 ft.
  C. 10 ft.
  D. 5 ft.

UNIT 38 Installation

OBJECTIVES

After completing this unit, you should be able to

- list three crafts involved in air-conditioning installation.
- identify types of duct system installations.
- describe the installation of metal duct.
- describe the installation of ductboard systems.
- describe the installation of flexible duct.
- recognize good installation practices for package air-conditioning equipment.
- discuss different connections for package air-conditioning equipment.
- describe the split air-conditioning system installation.
- recognize correct refrigeration piping practices.
- state start-up procedures for air-conditioning equipment.

SAFETY CHECKLIST

- ✔ Be careful when handling sharp tools, fasteners, and metal duct. Wear gloves whenever practicable.
- ✔ Use only grounded, double-insulated, or cordless electrical tools.
- ✔ Wear gloves, goggles, and clothing that will protect your skin when working with or around fiberglass insulation.
- ✔ Extreme care should be used when working with any electrical circuit. All safety rules must be adhered to. Particular care should be taken while working with any primary power supply to a building because the fuse that protects that circuit may be on the power pole outside. Keep the power off whenever possible when installing or servicing equipment. Lock and tag the panel. There should be only one key. Keep this on your person at all times.
- ✔ Although low-voltage circuits are generally harmless, you can experience shock if you are wet and touch live wires. Use precautions for all electrical circuits.
- ✔ Before starting a system for the first time, examine all electrical connections and moving parts. Connections, pulleys, or other moving parts may be loose.
- ✔ Follow all manufacturer's instructions before and during initial start-up.

38.1 INTRODUCTION TO EQUIPMENT INSTALLATION

Installing air-conditioning equipment requires three crafts: duct, electrical, and mechanical, which includes refrigeration. Some contractors use separate crews to carry out the different tasks. Others may perform the duties of two of the crafts within their own company and subcontract the third to a more specialized contractor. Some small contractors do all three jobs with a few highly skilled people. The three job disciplines are often licensed at local and state levels; they may be licensed by different departments.

Included in the following are duct installation procedures for (1) all-metal square or rectangular, (2) all-metal round, (3) fiberglass ductboard, and (4) flexible duct. NOTE: Local codes must be followed while performing all work.

38.2 SQUARE AND RECTANGULAR DUCT

Square metal duct is fabricated in a sheet metal shop by qualified sheet metal layout and fabrication personnel. The duct is then moved to the job site and assembled to make a system. Because the all-metal duct system is rigid, all dimensions have to be precise, or the job will not go together. The duct must sometimes rise over objects or go beneath them and still measure out correctly so the takeoff will reach the correct branch location. The branch duct must be the correct dimension for it to reach the terminal point, the boot for the room register. **Figure 38–1** illustrates an example of a duct system layout.

Square duct or rectangular duct is assembled with S fasteners and drive clips, which make a duct connection nearly airtight. If further sealing is needed, the connection can be caulked and taped. A special product called mastic may be applied to the duct joints with a stiff paintbrush. It dries hard but is slightly pliable. It becomes an airtight connection for the air pressures within the duct. While the duct is being assembled, it has to be fastened to the structure for support. This can be accomplished in several ways. It may be laid flat in an attic installation or be hung by hanger straps. The duct should be supported so that it will be steady when the fan starts and will not transmit noise to the structure. The rush of air (which has weight) down the duct will move the duct if it is not fastened. Vibration eliminators next to the fan section will prevent the transmission of fan vibrations down the duct. See **Figure 38–2** for a flexible duct connector that can be installed between the fan and the metal duct. These are always recommended, but not always used. **Figure 38–3** illustrates an example of a flexible connector that can be installed at the end of each run used to dampen sound.

Metal duct can be purchased in the popular sizes from some supply houses for small and medium systems. This makes metal duct systems available to the small contractor who may not have a sheet metal shop. The assortment of

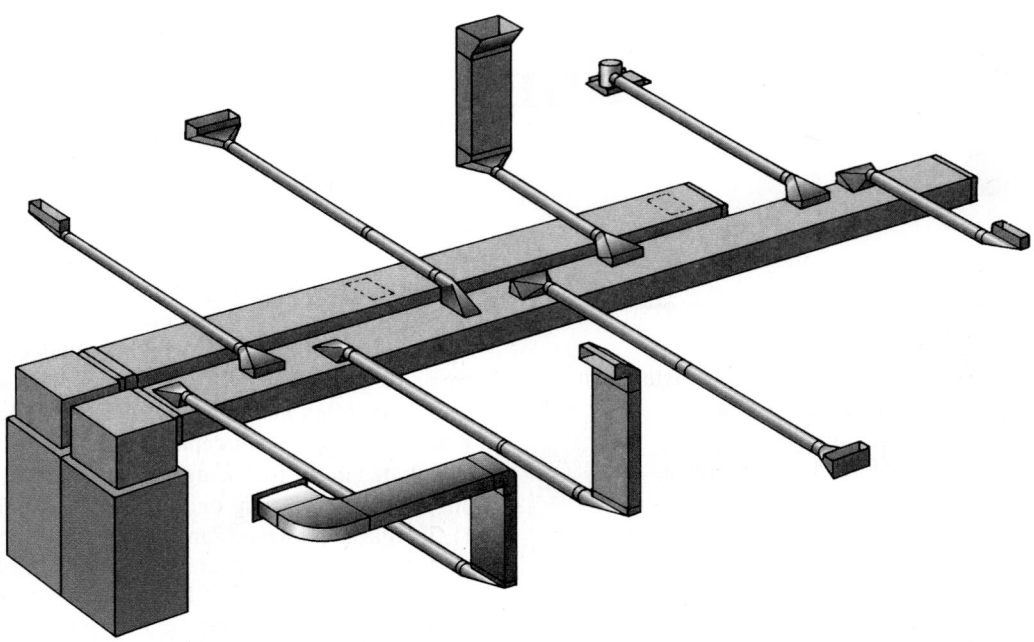

Figure 38–1   A duct system.

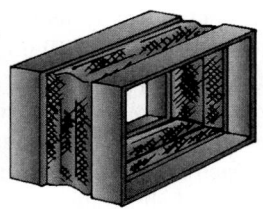

Figure 38–2   A flexible duct connector.

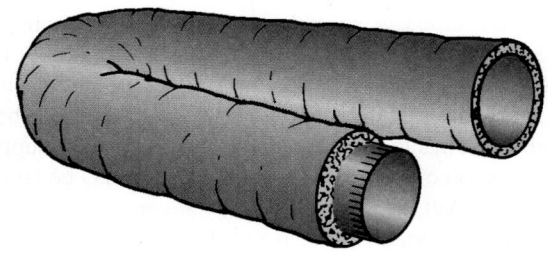

Figure 38–3   A round flexible duct connector.

standard duct sizes can be assembled with an assortment of standard fittings to build a system that appears to be custom made for the job, **Figure 38–4.**

38.3   ROUND METAL DUCT SYSTEMS

Round metal duct systems are easy to install and are available from some supply houses in standard sizes for small and medium systems. These systems use reducing fittings from a main trunk line and may be assembled in the field. This type of system must be fastened together at all connections.

Self-tapping sheet metal screws are popular fasteners. The screws are held by a magnetized screw holder while an electric drill turns and starts the screw, **Figure 38–5.** Each connection should have a minimum of three screws spaced evenly to keep the duct steady. A good installer can fasten the connections as fast as the screws can be placed in the screw holder, **Figure 38–6.** A reversible, cordless, variable-speed drill is the ideal tool to use.

Round metal duct takes more clearance space than square or rectangular duct. It must be supported and mounted at the correct intervals to keep it straight. Exposed, round metal duct does not look as good as square duct and rectangular duct, so it is often used in places that are out of sight. SAFETY PRECAUTION: Be careful when handling the sharp tools and fasteners. Use grounded, double-insulated, or cordless electrical tools.

38.4   INSULATION FOR METAL DUCT

Insulation for metal duct can be applied to the inside or the outside of the duct. When insulation is applied to the inside, it is usually done in the fabrication shop. The insulation can be fastened with tabs, glue, or both. The tabs are fastened to the inside of the duct and have a shaft that looks like a nail protruding from the duct wall. The liner and a washer are pushed over the tab shaft. The tabs have a base to which the shaft is fastened, **Figure 38–7.** This base is fastened to the inside of the duct by glue or spot welding. Spot welding is the most permanent method but is difficult to do and is expensive. An electrical spot welder must be used to weld the tab.

The liner can be glued to the duct, but it may come loose and block airflow, perhaps years from the installation date. This is difficult to find and repair, particularly if the duct is in

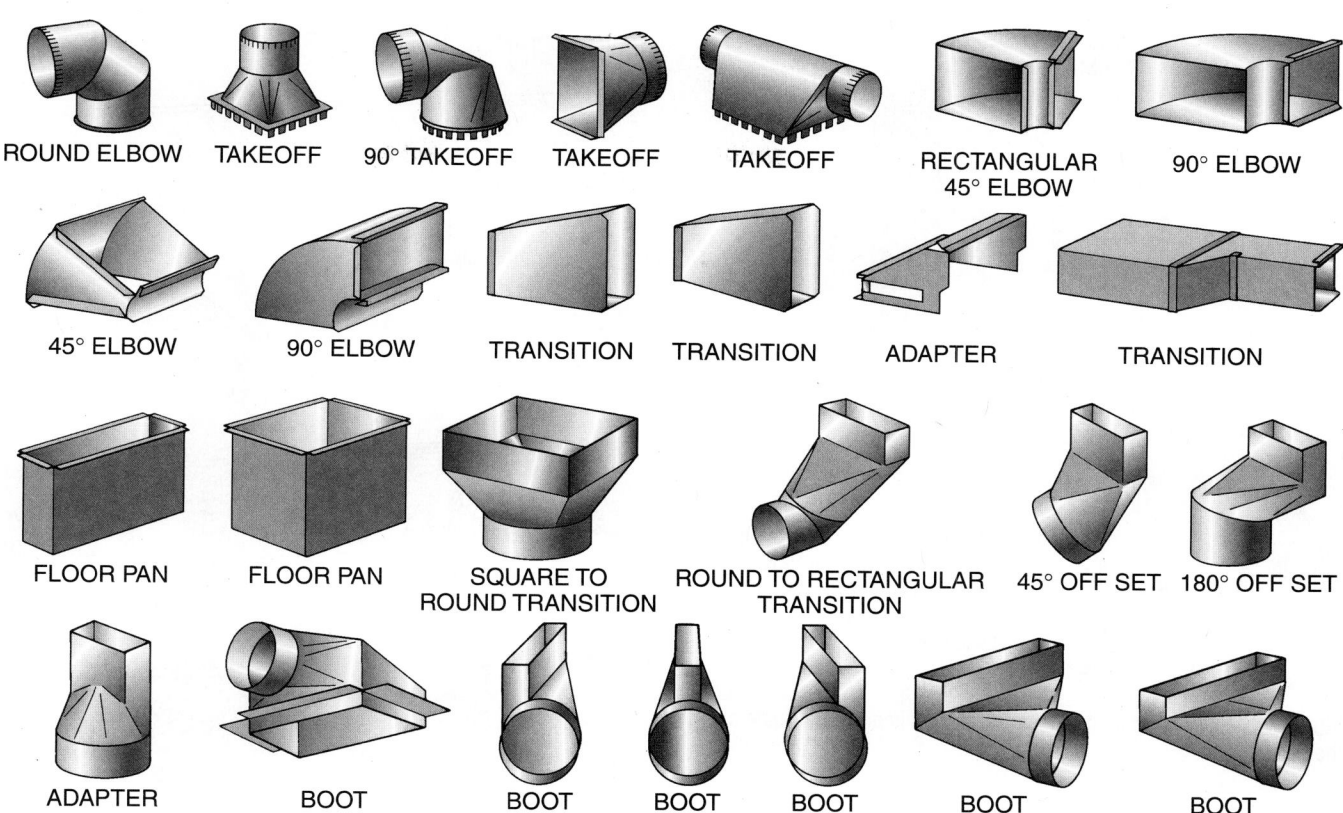

ROUND ELBOW TAKEOFF 90° TAKEOFF TAKEOFF TAKEOFF RECTANGULAR 45° ELBOW 90° ELBOW

45° ELBOW 90° ELBOW TRANSITION TRANSITION ADAPTER TRANSITION

FLOOR PAN FLOOR PAN SQUARE TO ROUND TRANSITION ROUND TO RECTANGULAR TRANSITION 45° OFF SET 180° OFF SET

ADAPTER BOOT BOOT BOOT BOOT BOOT BOOT

Figure 38–4 Standard duct fittings.

Figure 38–5 A battery-powered hand drill. *Photo by Bill Johnson*

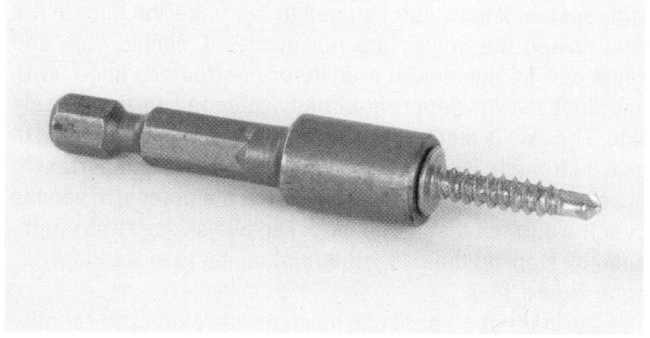

Figure 38–6 A self-tapping sheet metal screw with a magnetic screw holder.

a wall or framed in by the building structure. The glue *must* be applied correctly and even then may not hold forever. Using tabs with glue is a more permanent method.

The liner is normally fiberglass and is coated on the air side to keep the airflow from eroding the fibers. An antimicrobial coating is often added to the duct liner to reduce the potential of microbial growth on the insulation. Some dust will accumulate on the duct liner over time. This dust can support mold growth should the liner become damp for any reason.

Many pounds of air pass through the duct each season. An average air-conditioning system handles 400 cfm/ton of air. A 3-ton system handles 1200 cfm or 72,000 cfh (1200 cfm × 60 min/h = 72,000 cfh). Then 72,000 cfh ÷ 13.35 ft³/lb = 5393 lb of air per hour. This is more than 2 1/2 tons of air per hour. SAFETY PRECAUTION: Fiberglass insulation may irritate the skin. Gloves and goggles should be worn while handling it.🌐

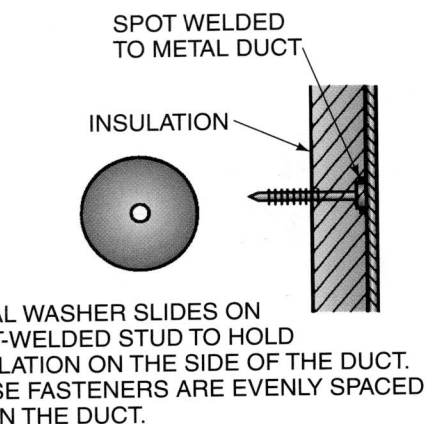

Figure 38–7   A tab to hold the fiberglass duct liner to the inside of the duct.

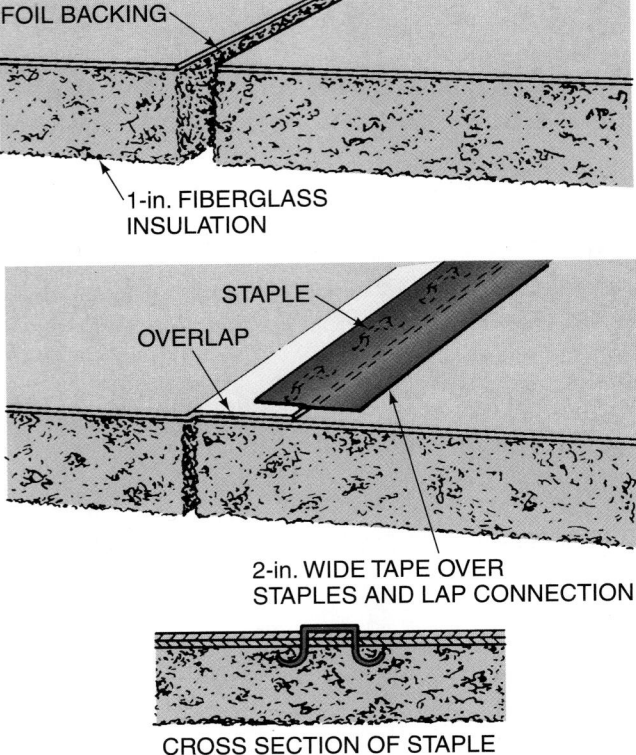

Figure 38–8   The outer skin is lapped over, and the backing of one piece is stapled to the backing of the other piece.

38.5   DUCTBOARD SYSTEMS

🌎 Fiberglass ductboard is popular among many contractors because it requires little special training to construct a system. Special knives can be used to fabricate the duct in the field. When the knives are not available, simple cuts and joints can be made with a utility or contractor's knife. With this duct the insulation is already attached to the outside skin. The skin is made of foil with a fiber running through it to give it strength. When assembling this duct, it is important to cut some of the insulation away so that the outer skin can lap over the surface that it is joining. The two skins are strengthened by stapling them together and taping to make them airtight, **Figure 38–8.**

Fiberglass ductboard can be made into almost any configuration that metal duct can be made into. It is lightweight and easy to transport because the duct can be cut, laid out flat, and assembled at the job site. Metal duct fittings, on the other hand, are large and take up a lot of space in a truck. The original shipping boxes can be used to transport the ductboard and to keep it dry.

Round fiberglass duct is as easy to install as the ductboard because it can also be cut with a knife.

Fiberglass duct must be supported like metal duct to keep it straight. The weight of the ductboard itself will cause it to sag over long spans. A broad type of hanger that will not cut the ductboard cover is necessary.

Fiberglass duct deadens sound because the inside of the duct has a coating (like the coating in metal duct liner that keeps the duct fibers from eroding) that helps to deaden any air or fan noise that may be transmitted into the duct. The duct itself is not rigid and does not carry sound. 🌎

38.6   FLEXIBLE DUCT

Flexible duct has a flexible liner and may have a fiberglass outer jacket for insulation if needed, for example, where a heat exchange takes place between the air in the duct and the space where the duct is routed. The outer jacket is held by a moisture-resistant cover made of fiber-reinforced foil or vinyl. The duct may be used for the supply or return and should be run in a direct path to keep bends from closing the duct. Sharp bends can greatly reduce the airflow and should be avoided.

When used in the supply system, flexible duct may be used to connect the main trunk to the boot at the room diffuser. The boot is the fitting that goes through the floor. In this case, it has a round connection on one end for the flexible duct and a rectangular connection on the other end where the floor register fastens to it. The duct flexibility makes it valuable as a connector for metal duct systems. The metal duct can be installed close to the boot and the flexible duct can be used to make the final connection. Long runs of flexible duct are not recommended unless the friction loss is taken into account.

Flexible duct must be properly supported. A band 1 in. or more in width is the best method to keep the duct from collapsing and reducing the inside dimension. Some flexible duct has built-in eyelet holes for hanging the duct. Tight turns should be avoided to keep the duct from collapsing on the inside and reducing the inside dimension. Flexible duct should be stretched to a comfortable length to keep the liner from closing and creating friction loss.

Flexible duct used at the end of a metal duct run will help reduce any noise that may be traveling through the duct. This can help reduce the noise in the conditioned space if the fan or heater is noisy.

38.7 ELECTRICAL INSTALLATION

The air-conditioning technician should be familiar with certain guidelines regarding electrical installation to make sure that the unit has the correct power supply and that the power supply is safe for the equipment, the service technician, and the owner. The control voltage for the space thermostat is often installed by the air-conditioning contractor even if the line-voltage power supply is installed by a licensed electrical technician.

The power supply must include the correct voltage and wiring practices, including wire size. The law requires the manufacturer to provide a nameplate with each electrical device that gives the voltage requirements and the current the unit will draw, **Figure 38–9**. The applied voltage (the voltage that the unit will actually be using) should be within ±10% of the rated voltage of the unit. That is, if the rated voltage of a unit is 230 V, the maximum operating voltage at which the unit should be allowed to operate would be 253 V (230 × 1.10 = 253). The minimum voltage would be 207 V (230 × .90 = 207). Some motors are rated at 208/230 V. This is a versatile rating because the unit will operate at ±208/230 V. The lower limit would be 187 V (208 × .90 = 187) and the upper limit would be 253 V (230 × 1.10 = 253). NOTE: If the unit operates for a long time beyond these limits, the motors and controls will be damaged.

When package equipment is installed, there is one power supply for the unit. If the system is split, there are two power supplies, one for the inside unit and one for the outside unit. Both power supplies will go back to a main panel, but there will be a separate fuse or breaker at the main panel, **Figure 38–10**. SAFETY PRECAUTION: A disconnect or cut-off switch should be within 25 ft of each unit and within sight of the unit. If the disconnect is around a corner, someone may turn on the unit while the technician is working on the electrical system. SAFETY PRECAUTION: Care should be used while working with any electrical circuit. All safety rules *must* be adhered to. Particular care should be taken while working in the primary power supply to any building because the fuse that protects that circuit may be on the power pole outside. A screwdriver touched across phases may throw off pieces of hot metal before the outside fuse responds.

Wire-sizing tables specify the wire size that each component will need. The *National Electrical Code (NEC)* provides installation standards, including workmanship, wire sizes, methods of routing wires, and types of enclosures for wiring and disconnects. Use it for all electrical wiring installation unless a local code prevails.

The control-voltage wiring in air-conditioning and heating equipment is the line voltage reduced through a step-down transformer. It is installed with color-coded or numbered wires so that the circuit can be followed through the various components. For example, the 230-V power supply is often at the air handler, which may be in a closet, an attic, a basement, or a crawl space. The interconnecting wiring may have to leave the air handler and go to the room thermostat and the condensing unit at the back of the house. The air handler may be used as the junction for these connections, **Figure 38–11**.

The control wire is a light-duty wire because it carries a low voltage and current. The standard wire size is 18 gage. A standard air-conditioning cable has four wires each of 18 gage in the same plastic-coated sheath and is called 18–4 (18 gage, 4 wires). Some wires may have eight conductors and are called 18–8, **Figure 38–12**. Red, white, yellow, and green wires are in the cable. The cable or sheath can be installed by the air-conditioning contractor in most areas because it is low voltage. An electrical license may be required in some areas. SAFETY PRECAUTION: Although low-voltage circuits are generally harmless, you can be hurt if you are wet and touch live wires. Use precautions for all electrical circuits.

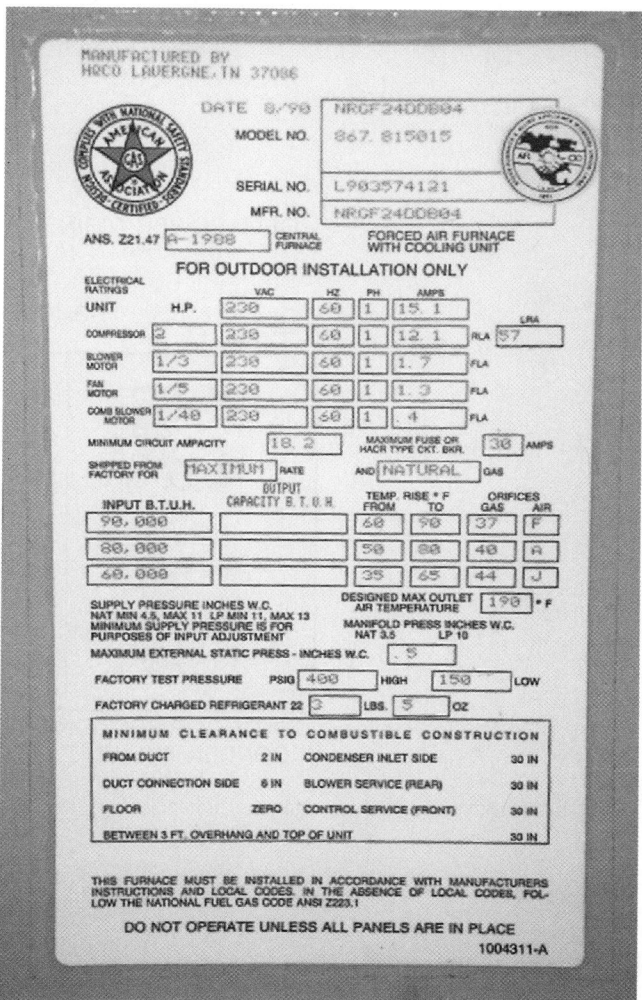

Figure 38–9   An air-conditioner unit nameplate. *Photo by Bill Johnson*

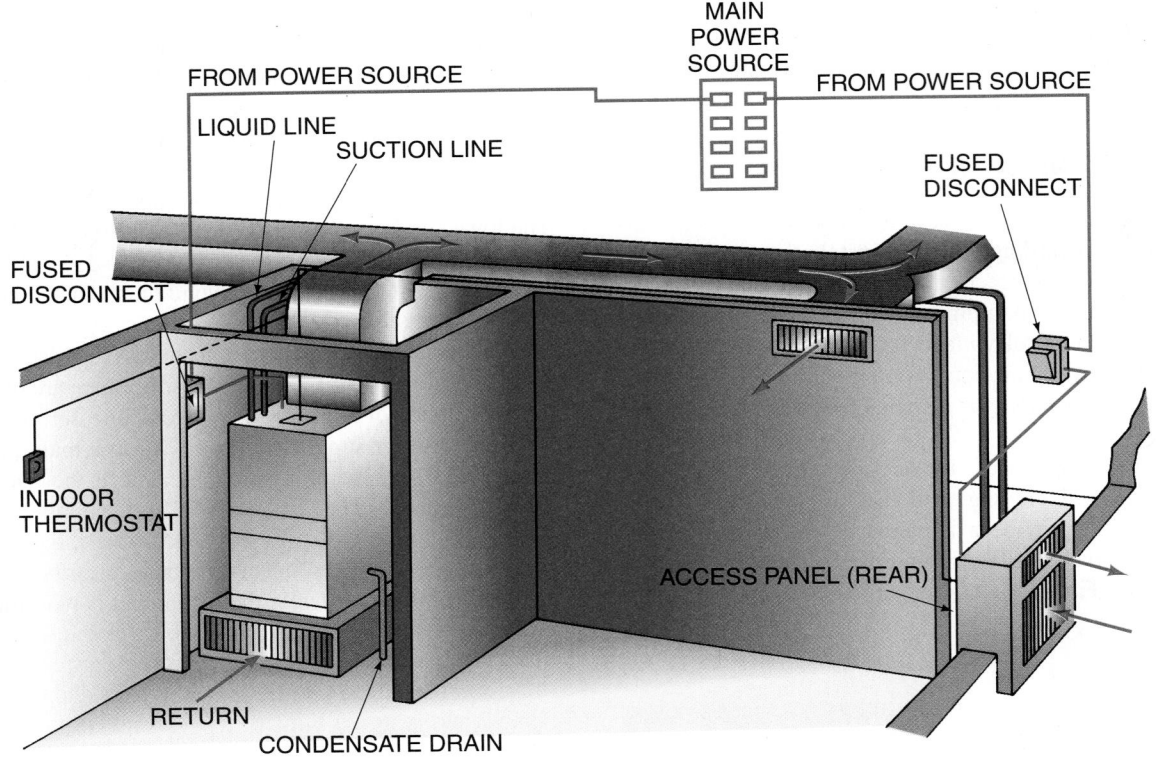

Figure 38–10 Wiring connecting indoor and outdoor units to the main power service.

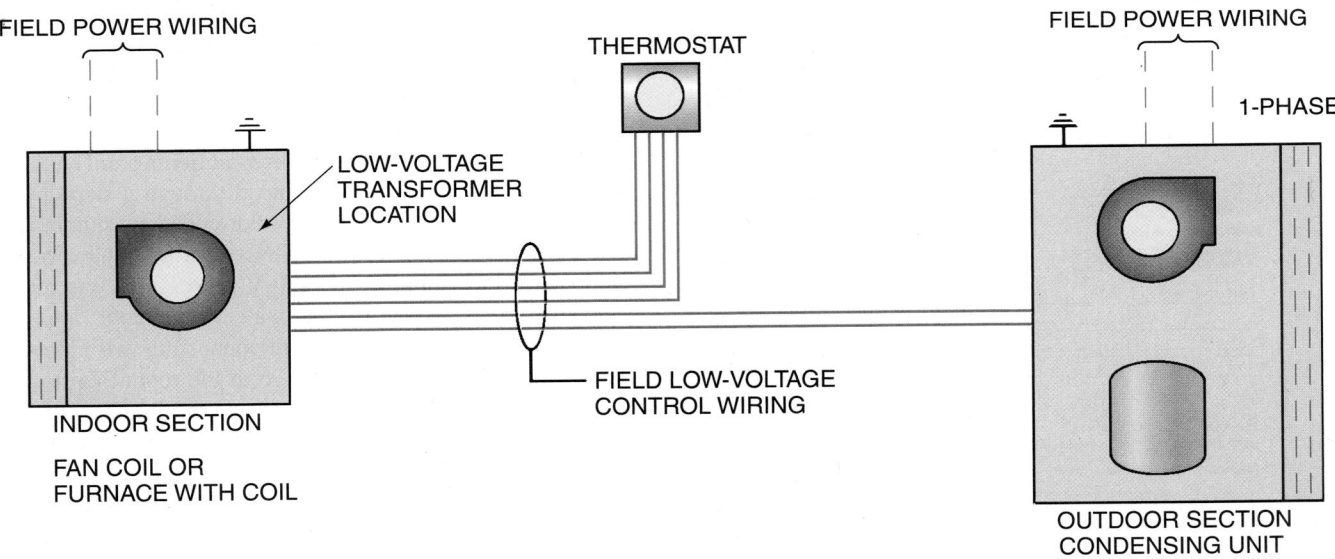

Figure 38–11 This pictorial diagram shows the relative position of the wiring as it is routed to the room thermostat, the air handler, and the condensing unit.

38.8 INSTALLING THE REFRIGERATION SYSTEM

The mechanical or refrigeration part of the air-conditioning installation is included in either a package or split system.

Package Systems

Package or self-contained equipment is equipment in which all components are in one cabinet or housing. The window air-conditioner is a small package unit. Larger

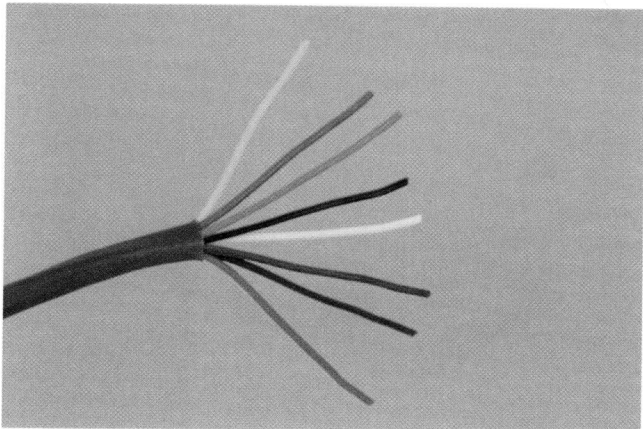

Figure 38–12 This is called 18–8 thermostat wire. Its size is 18 gage and there are 8 wires in the cable. *Photo by Bill Johnson*

package systems may have 100 tons of air-conditioning capacity.

The package unit is available in several different configurations for different applications, some of which are described here.

1. An air-to-air application is similar to a window air conditioner except that it has two motors. This unit will be discussed because it is the most common. The term *air-to-air* is used because the refrigeration unit absorbs its heat from the air and rejects it into the air, **Figure 38–13.**

2. **Figure 38–14** shows an air-to-water unit. This system absorbs heat out of the conditioned space air and rejects the heat into water. The water is wasted or passed through a cooling tower to reject the heat to the atmosphere. This system is sometimes called a water-cooled package unit.

3. Water-to-water describes the equipment in **Figure 38–15.** It has two water heat exchangers and is used in large

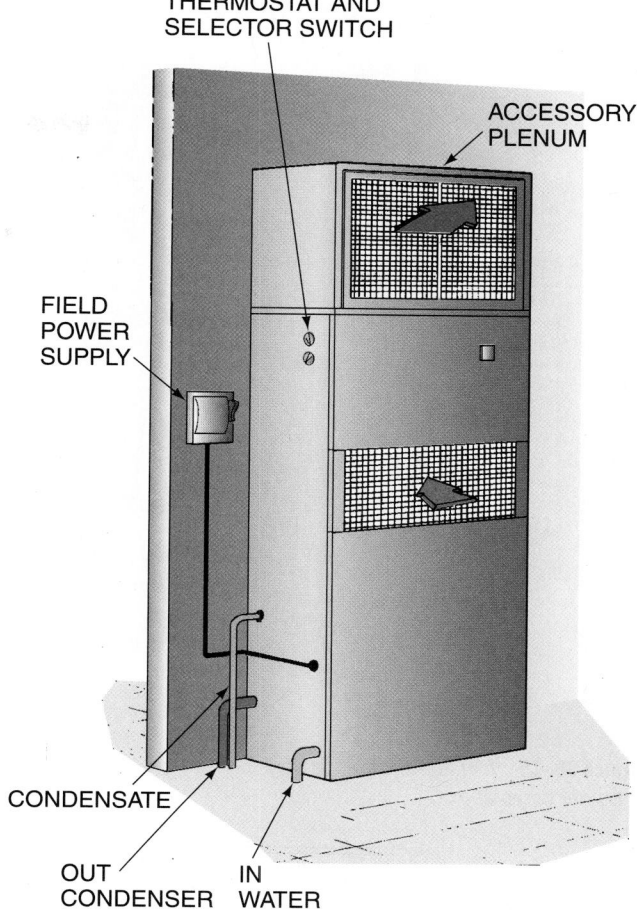

Figure 38–14 An air-to-water package unit. *Reproduced courtesy of Carrier Corporation*

commercial systems. The water is cooled and then circulated through the building to absorb the structure's heat, **Figure 38–16.** This system uses two pumps and two water circuits in addition to the fans to circulate the air in the conditioned space. To properly maintain this system, you need to be able to service pumps and water circuits.

4. Water-to-air describes the equipment in **Figure 38–17.** The unit absorbs heat from the water circuit and rejects the heat directly into the atmosphere. This equipment is used for larger commercial systems.

Air-cooled air-conditioning systems are by far the most common systems in residential and light commercial installations. The air-to-air system installation requires that the unit be set on a firm foundation. The unit may be furnished with a roof curb for rooftop installations. The roof curb will raise the unit off the roof and provide waterproof duct connections to the conditioned space below. When a unit is to be placed on a roof in new construction, the roof curb can be shipped separately, and a roofer can install it, **Figure 38–18.** The air-conditioning contractor can then set the package unit on the roof curb for a watertight installation. The foundation for another type of installation may be located beside

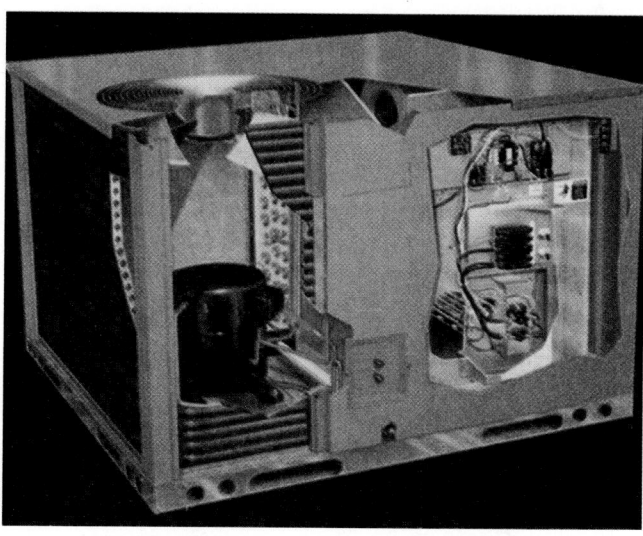

Figure 38–13 An air-to-air unit. *Courtesy Heil-Quaker Corporation*

Figure 38–15 A water-to-water package unit. *Reproduced courtesy of Carrier Corporation*

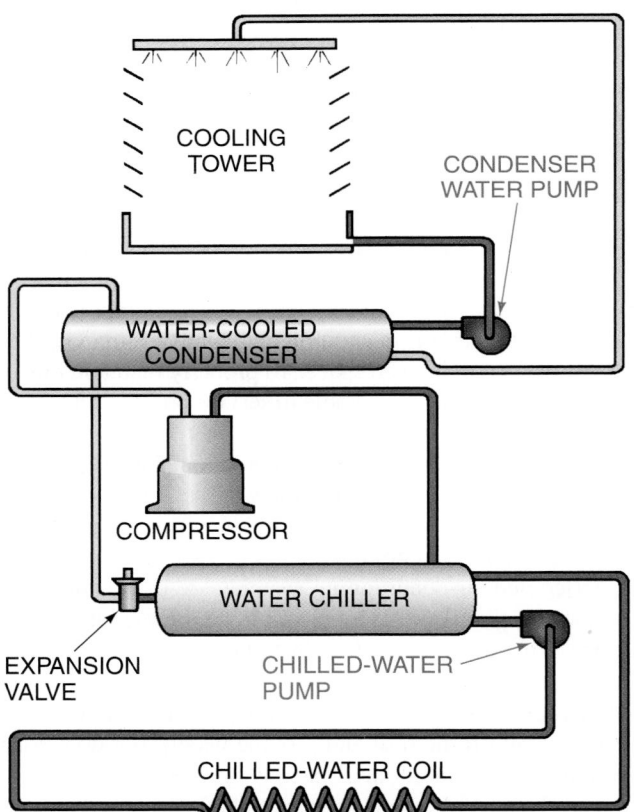

Figure 38–16 The total water-to-water system must have two pumps to move the water in addition to the fans that move the air across the coils.

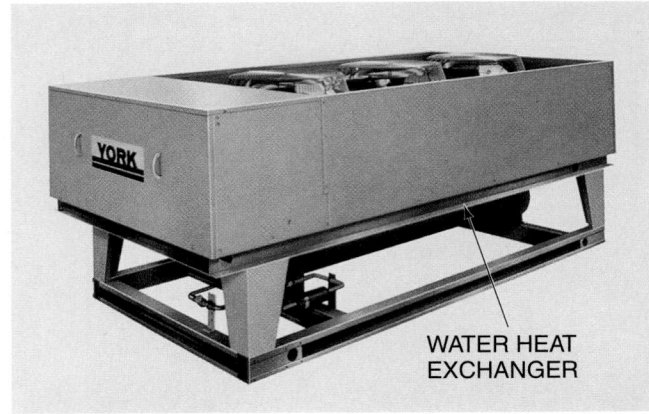

Figure 38–17 A water-to-air package unit. *Courtesy York International Corporation*

the conditioned space and may be a high-impact plastic pad, a concrete pad, or a metal frame, **Figure 38–19.** The unit should be placed where the water level cannot reach the unit.

Vibration Isolation

The foundation of the unit should be placed in such a manner that the unit vibration is not transmitted into the building. The unit may need vibration isolation from the building structure. Two common methods for preventing vibration are rubber and cork pads and spring isolators. Rubber and cork pads are the simplest and least expensive method for simple installations, **Figure 38–20.** The pads come in sheets that can be cut to the desired size. They are placed under the unit at the point where the unit rests on its foundation. If there are raised areas on the bottom of the unit for this contact, the pads may be placed there.

Spring isolators may also be used and need to be chosen correctly for the particular weight of the unit. The springs will compress and lose their effectiveness under too heavy a load. They also may be too stiff for a light load if not chosen correctly.

Duct Connections for Package Equipment

The ductwork entering the structure may be furnished by the manufacturer or made by the contractor. Manufacturers furnish a variety of connections for rooftop installations but do not normally have many factory-made fittings for going through a wall. See **Figure 38–21** for a rooftop installation.

Package equipment comes in two different duct connection configurations relative to the placement of the return and supply ducts, either side by side or over and under. In the *side-by-side* pattern the return duct connection is beside the supply connection, **Figure 38–22.** The connections are almost the same size. There will be a large difference in the connection

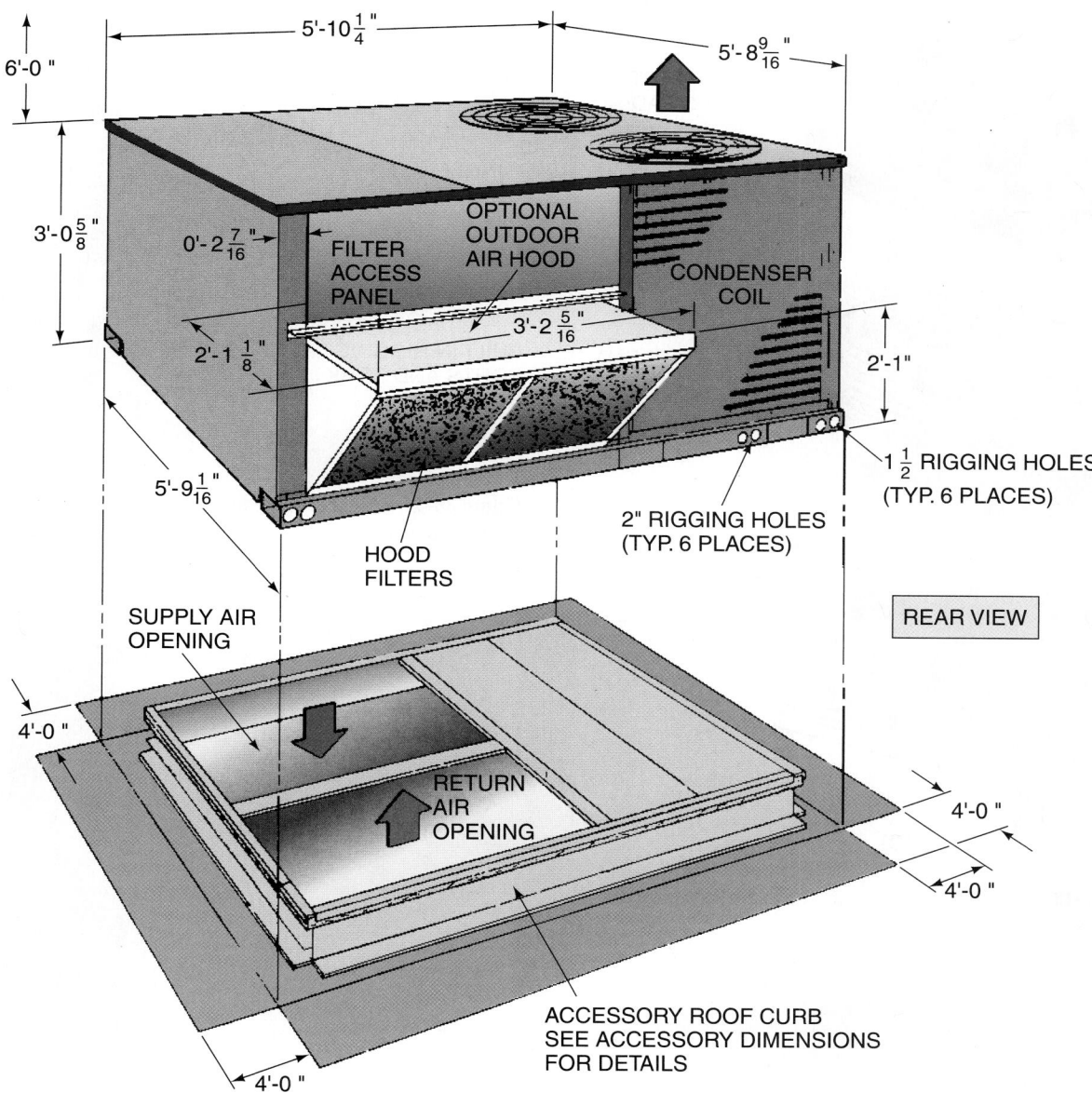

Figure 38–18 This unit will sit on the roof curb. *Courtesy Heil Heating and Cooling Products*

sizes on an over-and-under unit. If the unit is installed in a crawl space, the return duct must run on the correct side of the supply duct if the crawl space is low. It is difficult to cross a return duct over or under a supply duct if there is not enough room. The side-by-side duct design makes it easier to connect the duct to the unit because of the sizes of the connections on the unit. They are closer to square in dimension than the connections on an over-and-under duct connection. If the duct has to rise or fall going into the crawl space, a standard duct size can be used.

The *over-and-under* duct connection is more difficult to fasten ductwork to because the connections are more oblong, **Figure 38–23**. They are wide because they extend from one side of the unit to the other and shallow because one is on top of the other. A duct transition fitting is almost always necessary to connect this unit to the ductwork. The transition is normally from wide shallow duct to almost square. When over-and-under systems are used, the duct system may be designed to be shallow and wide to keep the duct transition from being so complicated.

The duct connection must be watertight and insulated, **Figure 38–24**. Some contractors cover the duct with a weatherproof hood.

At some time the package equipment will need replacing. The duct system usually outlasts two or three units. Manufacturers may change their equipment style from over and under to side by side. This complicates the new choice of equipment because an over-and-under unit may not be as easy to find as a side-by-side unit, and the duct already exists for an over-and-under connection, **Figure 38–25**.

Package air conditioners have no field run refrigerant piping. The refrigerant piping is assembled within the unit at the

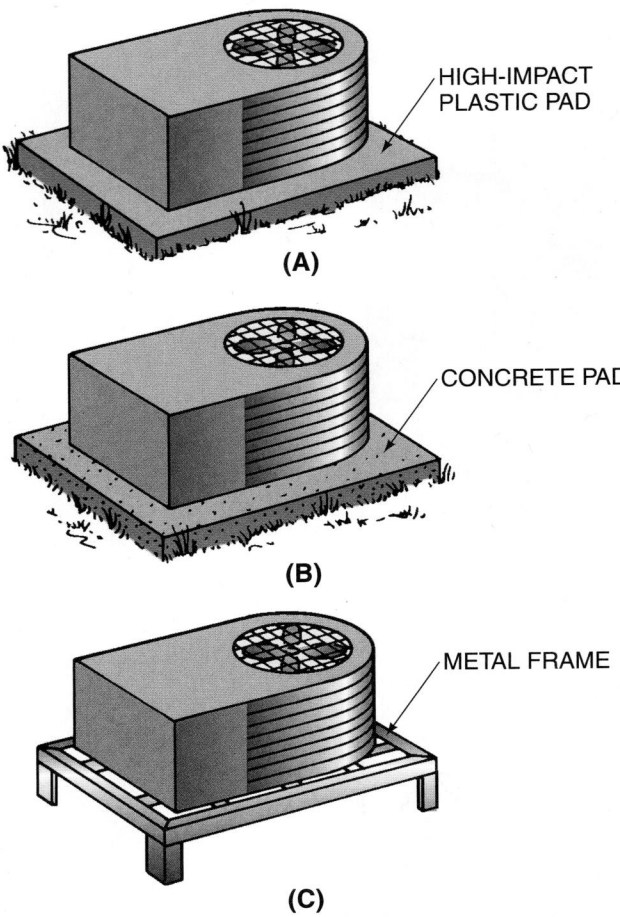

Figure 38–19 Various types of unit pads or foundations **(A)–(C)**.

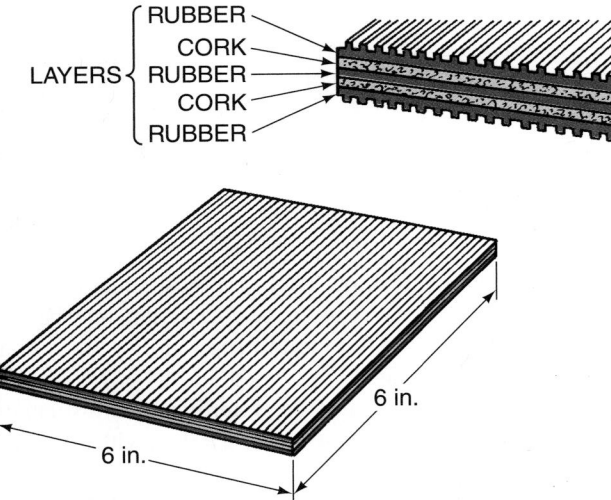

Figure 38–20 Rubber and cork pads may be placed under the unit to reduce vibration.

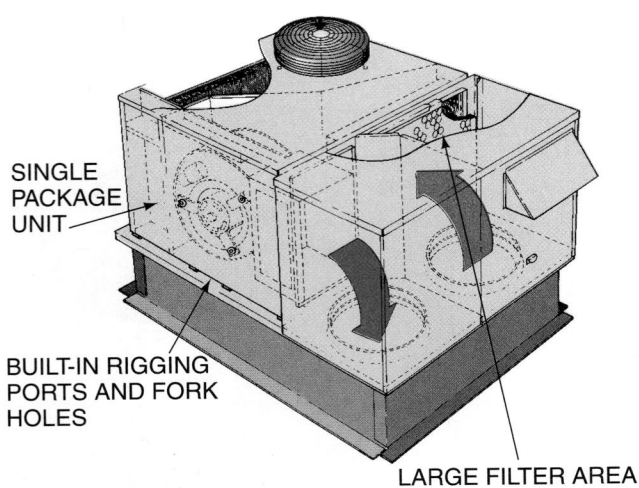

Figure 38–21 Duct connector for a rooftop installation. *Courtesy Climate Control*

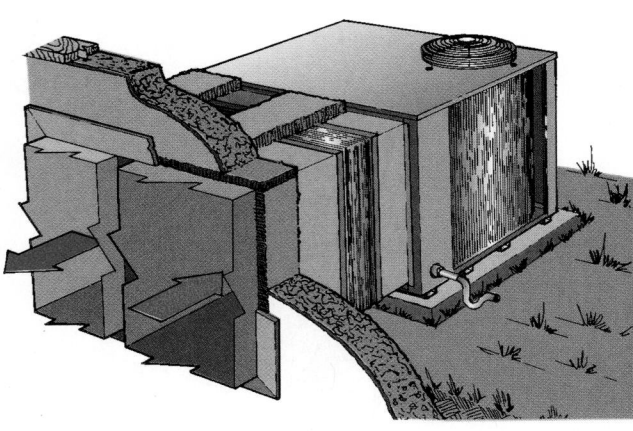

Figure 38–22 This air-to-air package unit has side-by-side duct connections. *Courtesy Climate Control*

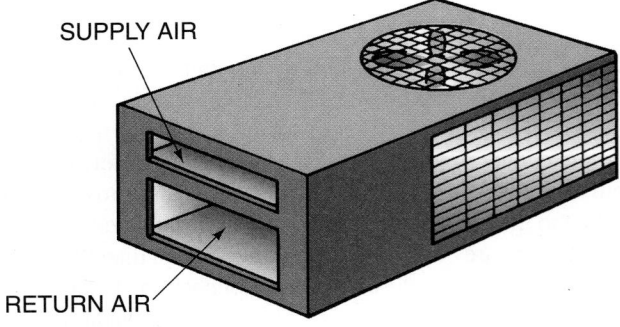

Figure 38–23 Over-and-under duct connections.

factory. The refrigerant charge is included in the price of the equipment. The equipment is ready to start except for electrical and duct connections. One precaution must be observed, however. Most of these systems use refrigerants that have an affinity (attraction) for the oil in the system. The

refrigerant in the unit will all be in the compressor if it is not forced out by a crankcase heater. NOTE: All manufacturers that use crankcase heaters on their compressors supply a warning to leave the crankcase heater energized for some time, perhaps as long as 12 hours, before starting the

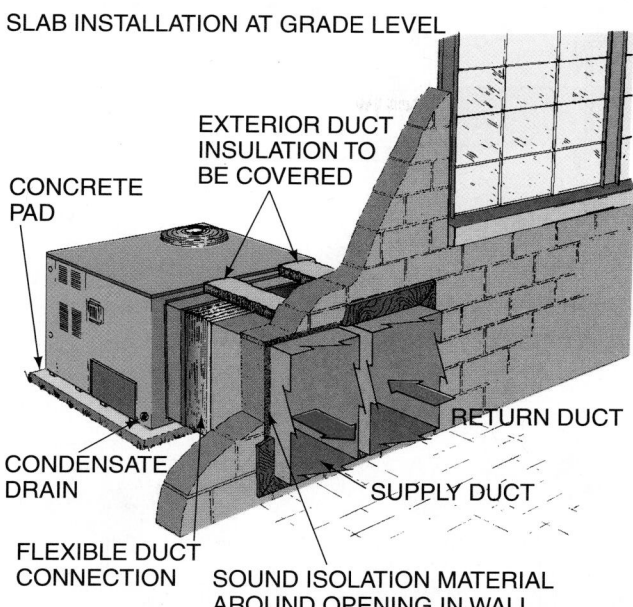

SLAB INSTALLATION AT GRADE LEVEL

EXTERIOR DUCT
INSULATION TO
BE COVERED

CONCRETE
PAD

RETURN DUCT

CONDENSATE
DRAIN

SUPPLY DUCT

FLEXIBLE DUCT
CONNECTION

SOUND ISOLATION MATERIAL
AROUND OPENING IN WALL

Figure 38–24 An air-to-air package unit installed through a wall.
Courtesy Climate Control

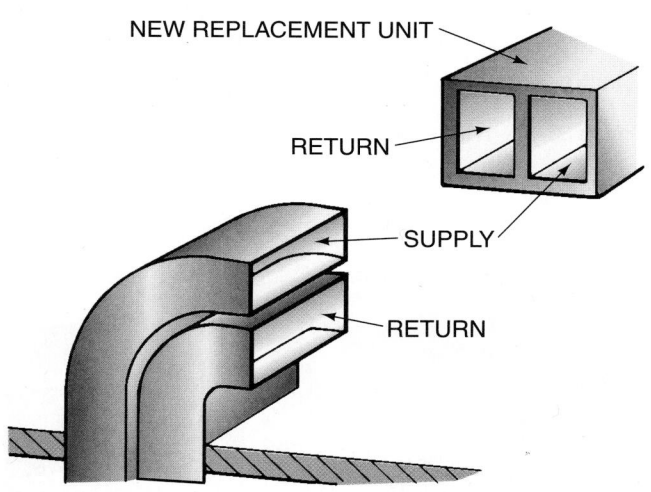

NEW REPLACEMENT UNIT

RETURN

SUPPLY

RETURN

Figure 38–25 An air-cooled package unit with over-and-under duct connections. The manufacturer has changed the design to a side-by-side duct connection, and the installing contractor has a problem.

compressor. The installing contractor must coordinate the electrical connection and start-up times closely because the unit should not be started immediately when the power is connected.

38.9 INSTALLING SPLIT-SYSTEM AIR CONDITIONERS

The evaporator is normally located close to the fan section regardless of whether the fan is in a furnace or in a special air handler. The air handler (fan section) and coil must be located on a solid base or suspended from a strong support.

Upflow and downflow equipment often has a base that may be fireproof and rigid for the air handler to rest on. Some vertical-mount air-handler installations have a wall-hanging support for the unit.

When the unit is installed horizontally, it may rest on the ceiling joists in an attic, **Figure 38–26,** or on a foundation of blocks or concrete in a crawl space. The air handler may be hung from the floor or ceiling joists in different ways, **Figure 38–27.** If the air handler is hung from above, vibration isolators are often located under it to keep fan noise or vibration from being transmitted into the structure. **Figure 38–28** is a trapeze hanger using vibration isolation pads.

The air handler (fan section) should always be installed so that it will be easily accessible for future service. The air handler will always contain the blower and sometimes the controls and heat exchanger.

Most manufacturers have designed their air handlers or furnaces so that, when installed vertically, they are totally accessible from the front. This works well for closet installations where there is insufficient room between the closet walls and the sidewalls of the furnace or air handler.

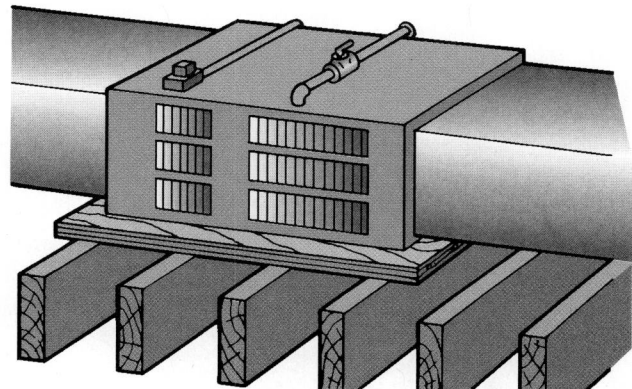

Figure 38–26 A furnace located in an attic crawl space. See the manufacturer's instructions for suggestions. It may require a fireproofed base.

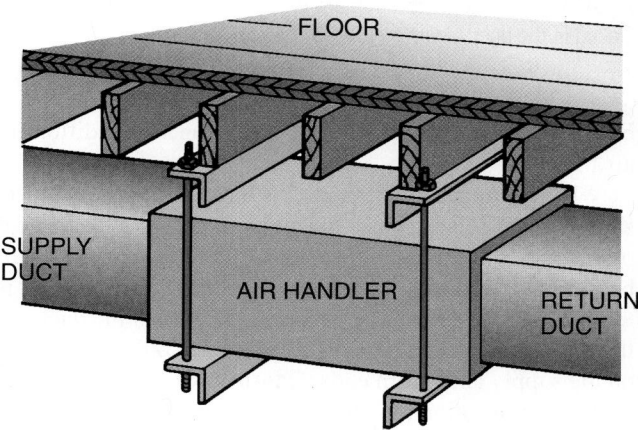

FLOOR

SUPPLY
DUCT

AIR HANDLER

RETURN
DUCT

Figure 38–27 The air handler is hung from above, and the ductwork is connected and hung at the same height.

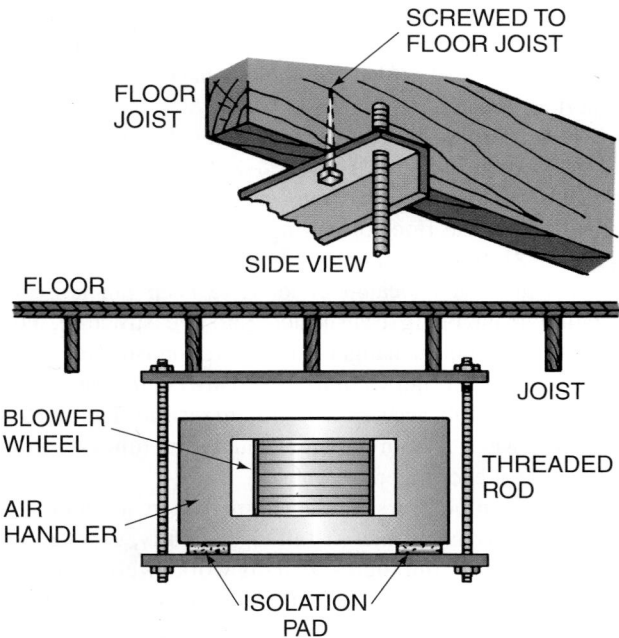

Figure 38–28 A trapeze hanger with vibration isolation pads.

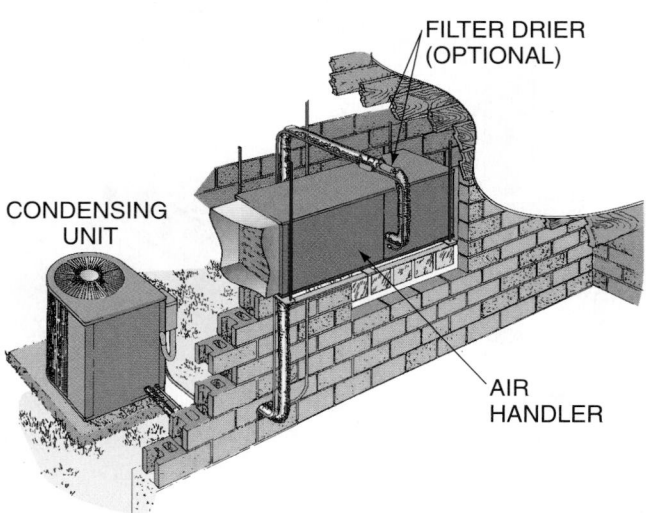

Figure 38–29 An electric furnace with an air-conditioning coil placed in the duct. *Courtesy Climate Control*

Figure 38–29 is an electric furnace used as an air-conditioning air handler.

When a furnace or air handler is located in a crawl space, side access is important, so the air handler should be located off the ground next to the floor joists. The ductwork also will be installed off the ground, **Figure 38–30**. If the air handler is top access, it will have to be located lower than the duct and the duct must make a transition downward to the air handler on the supply and return ends, **Figure 38–31**.

The best location for a top-access air handler is in an attic crawl space where the air handler can be set on the joists. The technician can work on the unit from above, **Figure 38–32**.

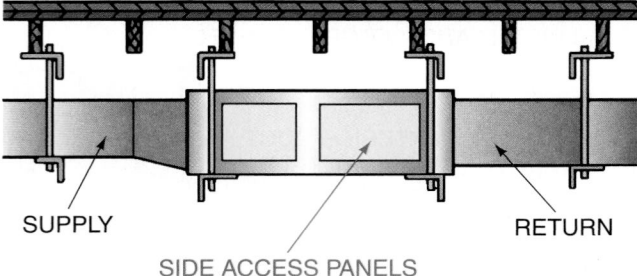

Figure 38–30 A side-access air handler in a crawl space.

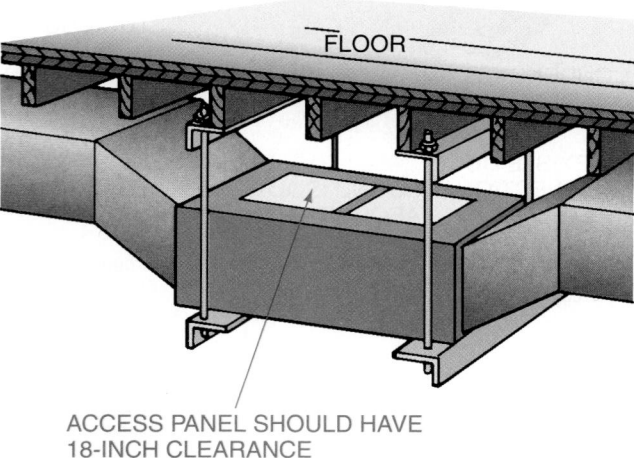

Figure 38–31 A top-access air handler.

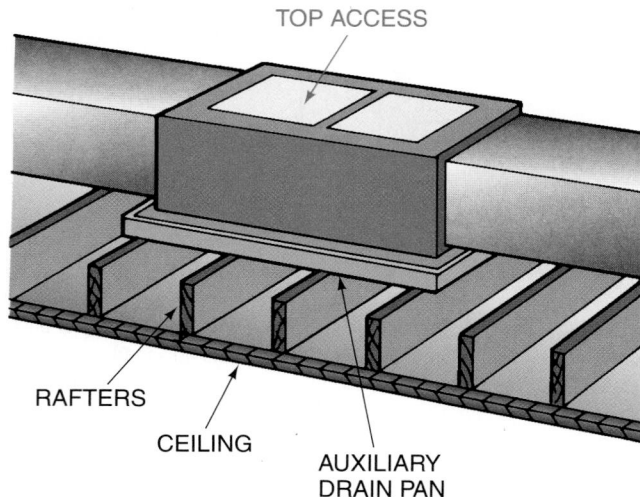

Figure 38–32 An evaporator installed in an attic crawl space.

Condensate Drain Piping

When the evaporator is installed, provisions must be made for the condensate that will be collected in the air-conditioning cycle. An air conditioner in a climate with average humidity will collect about 3 pints (pt) of condensate per hour of operation for each ton of air conditioning. A 3-ton system would condense about 9 pt per hour of operation. This is more than a gallon of condensate per hour or more than 24 gal in a 24-hour operating period. This can add up to a great deal of water over a period of time. If the unit is near a drain that is below the drain pan, simply pipe the condensate to the drain, **Figure 38–33**. A trap in the drain line will hold some water and keep air from pulling into the unit from the termination point of the drain. The drain may terminate in an area where foreign particles may be pulled into the drain pan. The trap will prevent this, **Figure 38–34**. If no drain is close to the unit, the

condensate must be drained or pumped to another location, **Figure 38–35**.

Some locations call for the condensate to be piped to a dry well. A dry well is a hole in the ground filled with stones and gravel. The condensate is drained into the well and absorbed into the ground, **Figure 38–36**. For this to be successful, the soil must be able to absorb the amount of water that the unit will collect.

When the evaporator and drain are located above the conditioned space, an auxiliary drain pan under the unit is recommended and often required, **Figure 38–37**. Airborne particles, such as dust and pollen, can get into the drains. Algae will also grow in the water in the lines, traps, and pans and may eventually plug the drain. If the drain system is plugged, the auxiliary drain pan will catch the overflow and keep the water from damaging whatever is below it. This auxiliary drain should be piped to a conspicuous place. The owner should be warned that if the water ever comes from this drain line, a service call is necessary. Some contractors pipe this drain to the end of the house and out next to a driveway or patio so that if water were to ever drain from this point it would be readily noticed, **Figure 38–38**.

38.10 THE SPLIT-SYSTEM CONDENSING UNIT

The condensing unit location is removed from the evaporator. The following must be considered carefully when a condensing unit is being placed:

1. Proper air circulation
2. Convenience for piping and electrical service

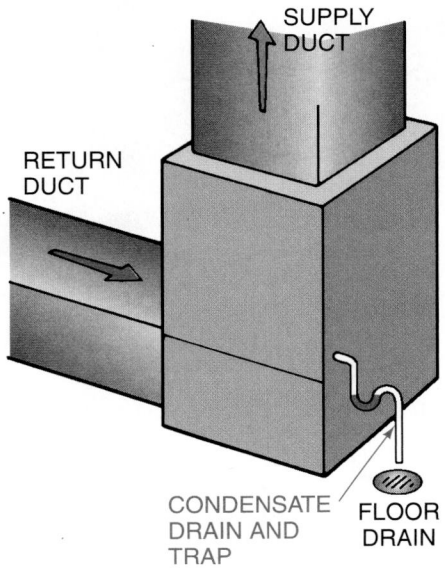

Figure 38–33 Condensate piped to a drain below the evaporator drain pan.

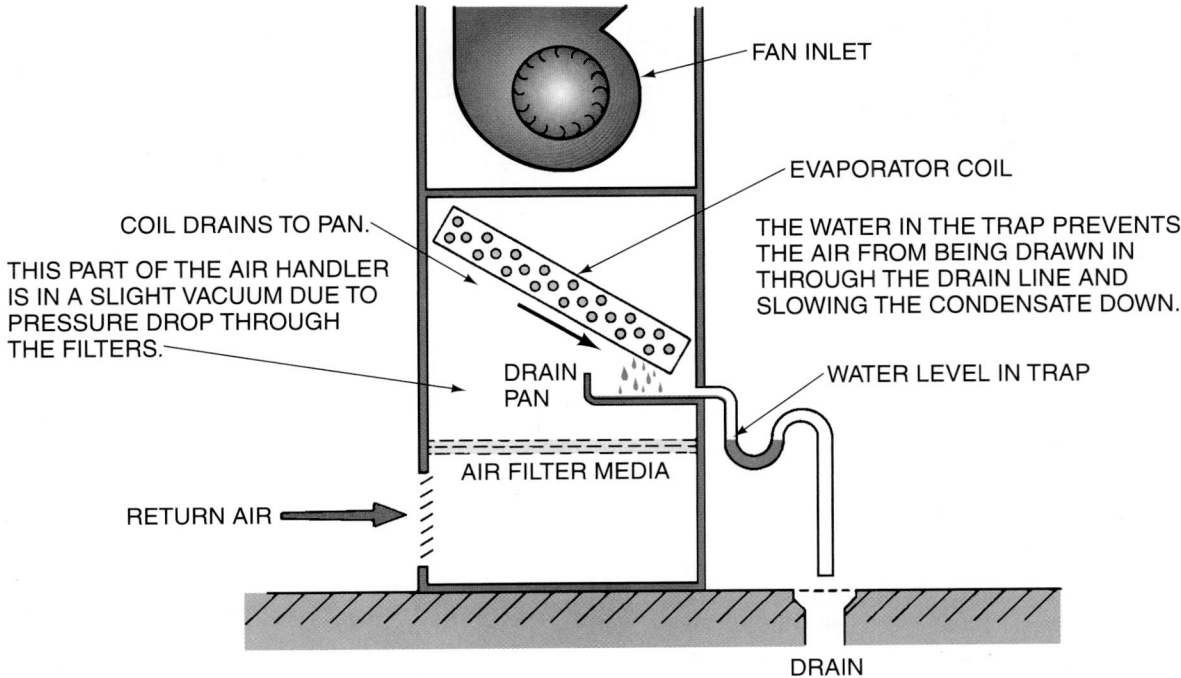

Figure 38–34 Cutaway of a drain trap.

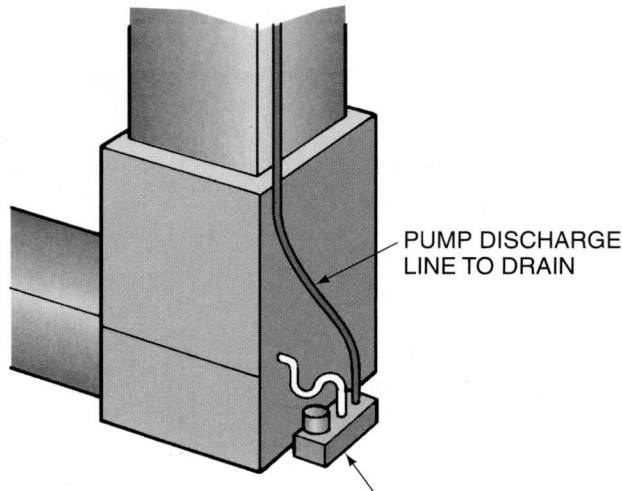

PUMP DISCHARGE LINE TO DRAIN

CONDENSATE PUMP HAS A FLOAT FOR TURNING THE PUMP ON. SOME PUMPS HAVE A SECOND FLOAT AND SWITCH THAT STOPS THE UNIT IF THE FIRST FLOAT FAILS.

Figure 38–35 The drain in this installation is above the drain connection on the evaporator, and the condensate must be pumped to a drain at a higher level.

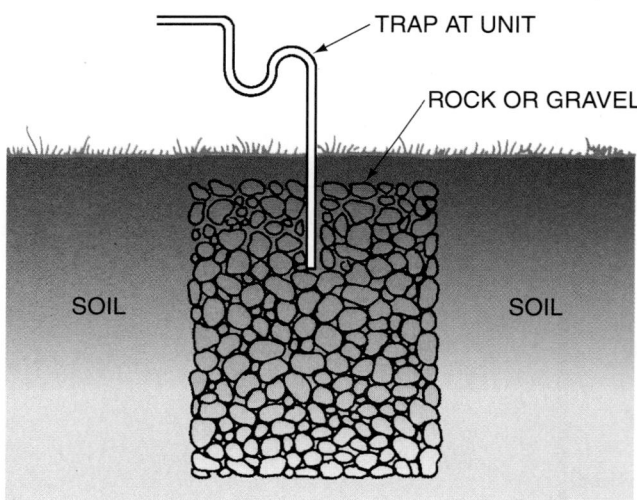

TRAP AT UNIT

ROCK OR GRAVEL

SOIL   SOIL

Figure 38–36 A dry well for condensate.

3. Future service
4. Natural water and roof drainage
5. Solar influence
6. Appearance

Air Circulation and Installation

The unit must have adequate air circulation. The air discharge from the condenser may be from the side or from the top. Discharged air must not hit an object and circulate back through the condenser. The air leaving the condenser is warm or hot and will create high head pressure and poor operating efficiency if it passes through the condenser again. Follow

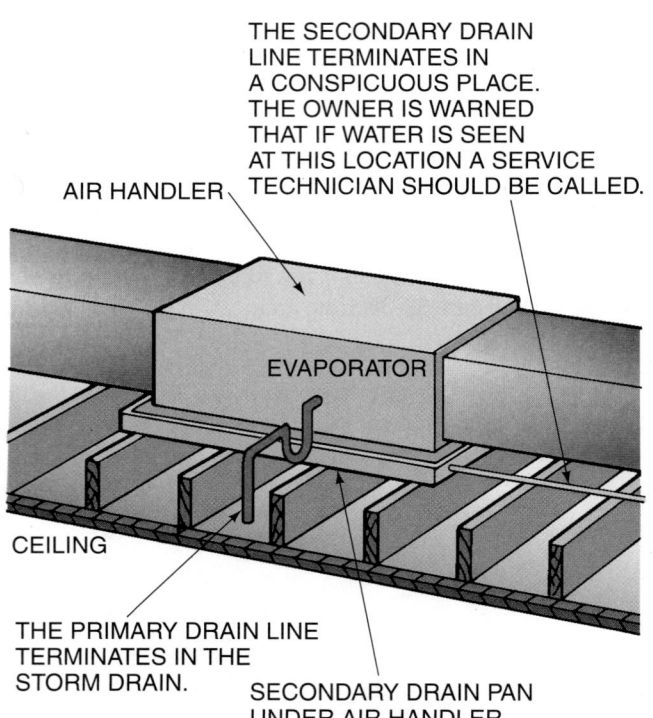

THE SECONDARY DRAIN LINE TERMINATES IN A CONSPICUOUS PLACE. THE OWNER IS WARNED THAT IF WATER IS SEEN AT THIS LOCATION A SERVICE TECHNICIAN SHOULD BE CALLED.

AIR HANDLER

EVAPORATOR

CEILING

THE PRIMARY DRAIN LINE TERMINATES IN THE STORM DRAIN.

SECONDARY DRAIN PAN UNDER AIR HANDLER

Figure 38–37 An auxiliary drain-pan installation.

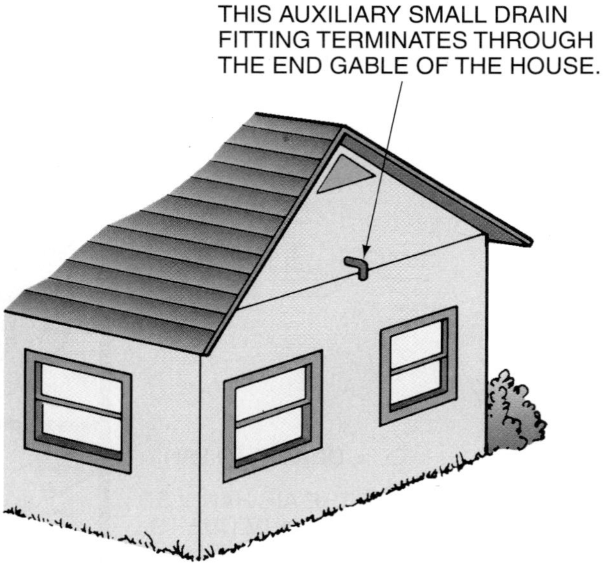

THIS AUXILIARY SMALL DRAIN FITTING TERMINATES THROUGH THE END GABLE OF THE HOUSE.

Figure 38–38 An auxiliary drain piped to the end of the house. If water is noticed at this location, the owner should alert the service technician.

the manufacturer's literature for minimum clearances for the unit, **Figure 38–39.**

Electrical and Piping Considerations

The refrigerant piping and the electrical service must be connected to the condensing unit. Piping is discussed later in this unit, but, for now, realize that the piping will be

Figure 38–39   The condensing unit for a split system located so that it has adequate airflow and service room. *Courtesy Carrier Corporation*

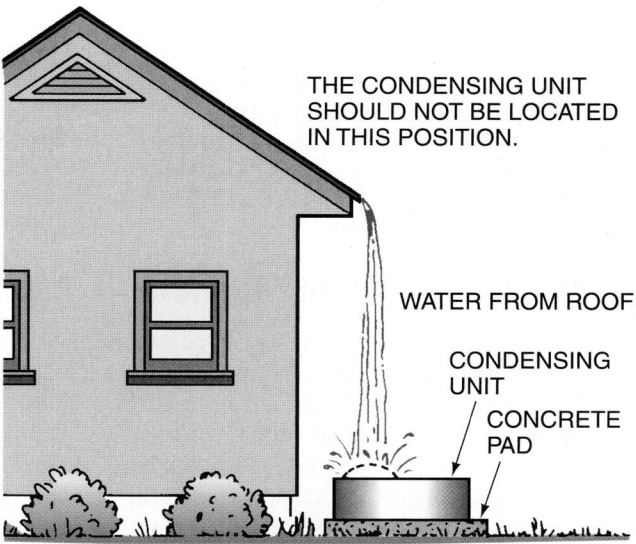

THE CONDENSING UNIT SHOULD NOT BE LOCATED IN THIS POSITION.

WATER FROM ROOF

CONDENSING UNIT

CONCRETE PAD

Figure 38–40   This unit is located improperly and allows the roof drain to pour down into the top of the unit.

routed between the evaporator and the condenser. When the condenser is located next to a house, the piping must be routed between the house and the unit, usually behind or beside the house with the piping routed next to the ground. If the unit is placed too far from the house, the piping and electrical service become natural obstructions between the house and the unit. Children may jump on the piping and electrical conduit. If the unit is placed too close to the house, it may be difficult to remove the service panels. Before placing the unit, study the electrical and the refrigeration line connections and make them as short as practical, leaving adequate room for service.

Service Accessibility

Unit placement may determine whether a service technician gives good service or barely adequate service. The technician must be able to see what is being worked on. A unit is often placed so that the technician can touch a particular component but is unable to see it, or, by shifting positions, can see the component but not touch it. The technician should be able to both see and touch the work at the same time.

Water Drainage from Natural Sources

The natural drainage of the ground water and roof water should be considered in unit placement. SAFETY PRECAUTION: The unit should not be located in a low place where ground water will rise in the unit. If this happens, the controls may short to ground, and the wiring will be ruined. All units should be placed on a base pad of some sort. Concrete and high-impact plastic are commonly used. Metal frames can be used to raise the unit when needed.

The roof drainage on a structure may run off in gutters. The condensing unit should not be located where the gutter drain or roof drainage will pour down onto the unit. Condensing units are made to withstand rain water but not large volumes of drainage. If the unit is a top-discharge unit, the drainage from above will be directly into the fan motor, **Figure 38–40.**

Solar Influence

If possible, put the condensing unit on the shady side of the house because the sun shining on the panels and coils will lower the efficiency. However, the difference is not crucial, and it would not pay to pipe the refrigerant tubing and electrical lines long distances just to keep the sun from shining on the unit.

Shade helps cool the unit, but it may also cause problems. Some trees have small leaves, sap, berries, or flowers that may harm the finish of the unit. Pine needles may fall into the unit, and the pine pitch that falls on the unit's cabinet may harm the finish and outweigh the benefit of placing the unit in the shade of many types of trees.

Placing a Condensing Unit for Best Appearance

The condensing unit should be located where it will not be noticeable or will not make objectionable noise. When located on the side of a house, the unit may be hidden from the street with a low shrub. If the unit is a side-discharge type, the fan discharge must be away from the shrub or the shrub may not live. If the unit is top discharge, the shrub is unaffected but the unit's sound will rise. This noise may be objectionable in a bedroom located above the unit.

Locating the condensing unit at the back of the house may place the unit closer to the evaporator and would mean shorter piping, but the back of the house is the usual location of patios and porches, and the homeowner, while sitting outside, may not want to hear the unit. In such a case, a side location at the end of the house where there are no bedrooms may be the best choice.

Each location has its considerations. The salesperson and the technician should consult with the owner about locating the various components. The salesperson should be familiar with all local code requirements. A floor plan of the structure may help. Some companies use large graph paper to draw a rough floor plan to scale when estimating the job. The equipment can be located on the rough floor plan to help solve these problems. The floor plan can be shown to the homeowners to help them understand the contractor's suggestions.

38.11 INSTALLING REFRIGERANT PIPING

The refrigerant piping is always a big consideration when installing a split-system air conditioner. The choice of the piping system may make a difference in the start-up time for the system. The piping should always be kept as short as practical. For an air-conditioning installation three methods are used to connect the evaporator and the condensing unit on almost all equipment under 5 tons (refrigeration systems for commercial refrigeration are not the same): (1) contractor-furnished piping, (2) flare or compression fittings, with the manufacturer furnishing the tubing (called a line set), and (3) precharged tubing with sealed quick-connect fittings, called a precharged line set. **In all of these systems the operating charge for the system is shipped in the equipment.**

The Refrigerant Charge

Regardless of who furnishes the connecting tubing, the complete operating charge for the system is normally furnished by the manufacturer. The charge is shipped in the condensing unit. The manufacturer furnishes enough charge to operate the unit with a predetermined line length, typically 30 ft. The manufacturer holds and stores the refrigerant charge in the condensing unit with service valves if the tubing uses flare or compression fittings. When quick-connect line sets are used, the correct operating charge for the line set is included in the actual lines. See **Figure 38–41** for an example of both service valves and quick-connect fittings.

When service valves are used, the piping is fastened to the service valves by flare or compression fittings. Some manufacturers provide the option of soldering to the service valve connection. The piping is always hard-drawn or soft copper. In installations where the piping is exposed, straight pipe may look better. Hard-drawn tubing may be used for this with factory elbows used for the turns. When the piping is not exposed, soft copper is easily formed around corners where a long radius will be satisfactory.

The Line Set

When the manufacturer furnishes the tubing, it is called a *line set*. ⓢ The suction line is insulated, ⓢ and the tubing may be charged with nitrogen, contained with rubber plugs in the tube ends. When the rubber plug is removed, the nitrogen rushes out with a loud hiss. This indicates that the tubing is not leaking.

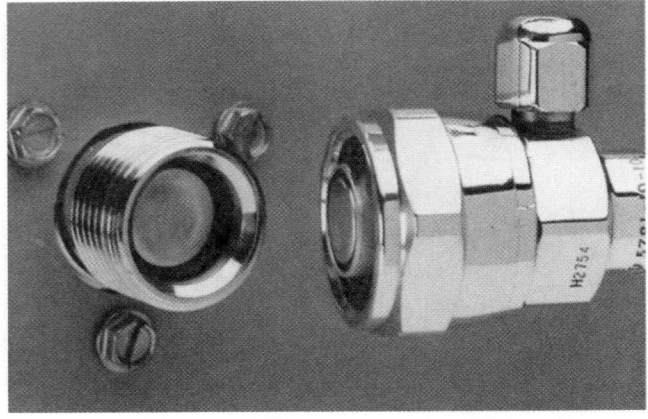

(A)

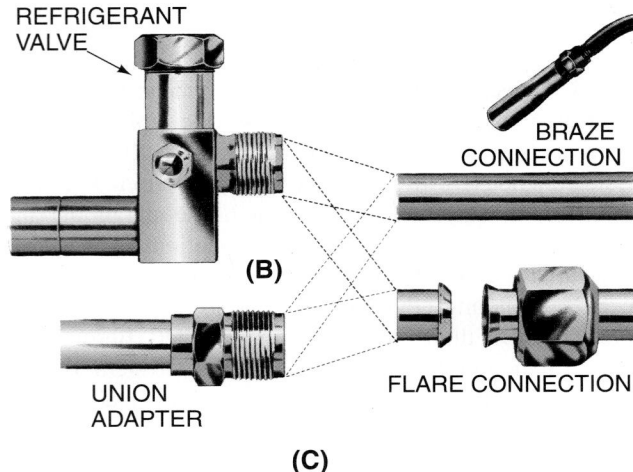

Figure 38–41 Two basic piping connections for split-system air conditioning **(A)–(B)**. *Courtesy Aeroquip Corporation*

When the tubing is used, uncoil it from the end of the coil. Place one end of the coil on the ground and roll it while keeping your foot on the tubing end on the ground, **Figure 38–42.** NOTE: Be careful not to kink the tubing when going around corners, or it may collapse. Because of the insulation on the pipe, you may not even see the kink.

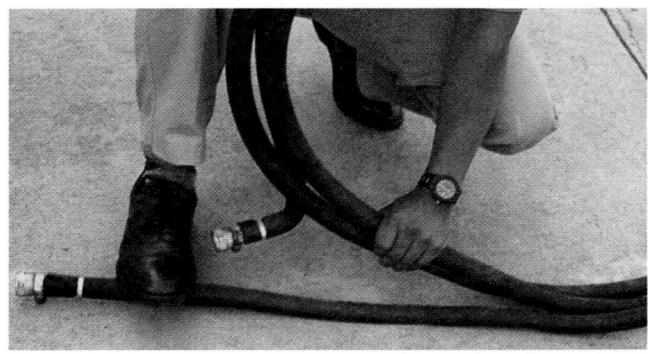

Figure 38–42 Uncoiling tubing. *Photo by Bill Johnson*

Tubing Leak Test and Evacuation

When the piping is routed and in place, the following procedure is commonly used to make the final connections:

1. The tubing is fastened at the evaporator end. Many tubing sets offer either compression fittings or a braze connection. Flaring also may be an option. Brazed connections are the connection of choice when done correctly. When the tubing has flare nuts, a drop of oil applied to the back of the flare will help prevent the flare nut from turning the tubing when the flare connection is tightened. The suction line may be as large as 1 1/8-in. OD tubing. Large tubing connections will have to be made very tight. Two adjustable wrenches are recommended. The liquid line will not be any smaller than 1/4-in. OD tubing, and it may be as large as 1/2-in. OD on larger systems.
2. The smaller (liquid) line is fastened to the condensing unit service valve.
3. The larger (suction) line is fastened.
4. The service valves are still closed. Allow a charge of a small amount of R-22 into the line set and evaporator through the Schrader valve ports for leak-checking purposes. Add nitrogen to the test pressure.
5. After the leak check, vent the small amount of R-22 and nitrogen to the atmosphere. Connect the vacuum pump to the line set and evaporator portion of the system and evacuate it to a low vacuum, **Figure 38–43.** ♻ EPA rules allow this process using R-22 and nitrogen under de minimis loss of refrigerant. ♻
6. When the evacuation is complete, open the valves at the condensing unit. The complete charge should be stored in the unit. Check the manufacturer's installation manual for the correct charge for the length of line set. Follow their instructions for the correct charge if needed.

Line sets come in standard lengths of 10, 20, 30, 40, or 50 ft. If lengths other than these are needed, the manufacturer should be consulted. Some manufacturers have a maximum allowable line length, normally in the vicinity of 50 ft. If the line length has to be changed from the standard length, the manufacturer will also recommend how to adjust the unit charge for the new line length. Most units are shipped with a charge for 30 ft of line. If the line is shortened, refrigerant must be recovered. If the line is longer than 30 ft, refrigerant must be added.

Altered Line-Set Lengths

When line sets must be altered, they may be treated as a self-contained system of their own. The following procedures should be followed:

1. Alter the line sets as needed for the proper length. The nitrogen charge will escape during this alteration. **Do not remove the rubber plugs.**
2. When the alterations are complete, the lines may be pressured to about 25 psig to leak check any connections you have made. The rubber plugs should stay in place. If you desire to pressure test with higher pressures, hook up the evaporator and condenser ends and you may safely pressurize to 150 psig with R-22 and nitrogen. The line's service port is common to the line side of the system. If the valves are not opened at this time, the line set and evaporator may be thought of as a sealed system of their own for pressure testing and evacuation, **Figure 38–44.**
3. After the leak test has been completed, evacuate the line set (and evaporator if connected). When all connections are made and checked, the system valves may be opened and the system started.

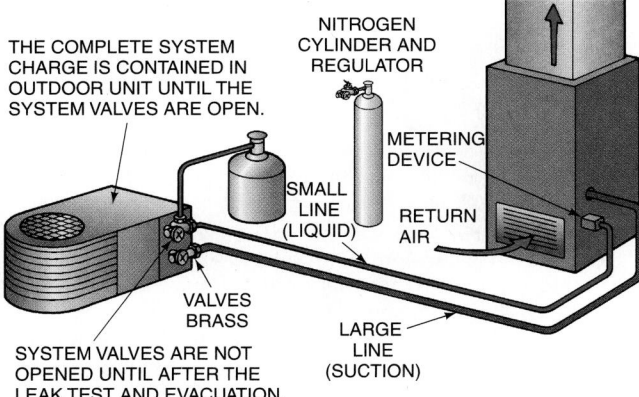

Figure 38–43　After all line set connections are made tight, the line set is leak checked using refrigerant and nitrogen. The line set is then evacuated, and the system valves are opened.

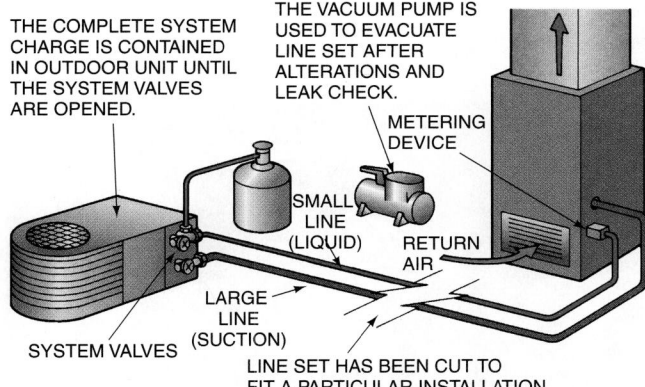

Figure 38–44　Altering the line set. After the line set has been altered, it is connected as in **Figure 38–43** and charged with nitrogen and a trace gas of R-22 for leak-checking purposes. It is checked for leaks. R-22 trace gas and nitrogen are released to the atmosphere. The line set and evaporator are then evacuated. After evacuation the system valves are opened, and the system is ready to run.

Precharged Line Sets (Quick-Connect Line Sets)

Precharged line sets with quick-connect fittings are shipped in most of the standard lengths. The difference in line sets and precharged line sets is that the correct refrigerant operating charge is shipped in the precharged line set. Refrigerant does not have to be added or taken out unless the line set is altered. The following procedures are recommended for connecting precharged line sets with quick-connect fittings.

1. Roll the tubing out straight.
2. Determine the routing of the tubing from the evaporator to the condenser and put it in place.
3. Remove any protective plastic caps on the evaporator fittings. Place a drop of refrigerant oil (this is sometimes furnished with the tubing set) on the neoprene O rings of each line fitting. The O ring is used to prevent refrigerant from leaking out while the fitting is being connected, and it serves no purpose after the connection is made tight. It may tear while making the connection if it is not lubricated.
4. Start the threaded fitting and tighten hand tight. Making sure that several threads can be tightened by hand will ensure that the fitting is not cross-threaded.
5. When the fitting is tightened hand tight, finish tightening the connection with a wrench. You may hear a slight escaping of vapor while making the fitting tight; this purges the fitting of any air that may have been in the fitting. The O ring should be seating at this time. Once you have started tightening the fitting, do not stop until you are finished. Two adjustable wrenches are recommended, **Figure 38–45.** If you stop in the middle, some of the system charge may be lost.
6. Tighten all fittings as indicated.
7. After all connections have been tightened, leak check all connections that you have made.

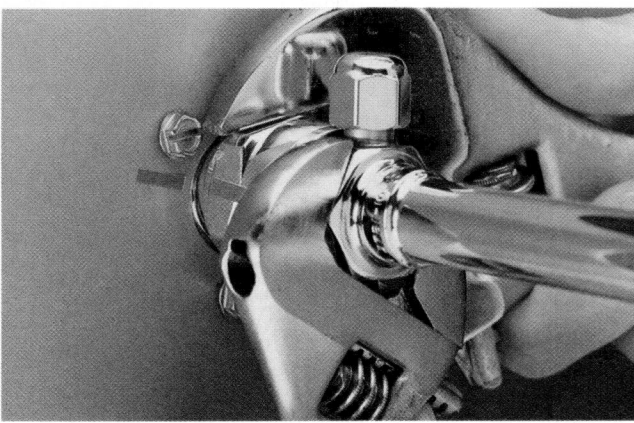

Figure 38–45 Using two wrenches to tighten fittings—one to hold the nut with the service tap on it and one to turn the movable nut next to the cabinet. *Courtesy Aeroquip Corporation*

Altered Precharged Line Sets

When the lengths of quick-connect line sets are altered, the charge will have to be altered. The line set may be treated as a self-contained system for alterations. The following is the recommended procedure.

1. ♻ Before connecting the line set, recover the refrigerant from the line set. Do not just cut it because the liquid and suction lines contain some liquid refrigerant. Recovering the refrigerant is required. ♻
2. Cut the line set and alter the length as needed.
3. Pressure test the line set.
4. Evacuate the line set to a low vacuum.
5. Valve off the vacuum pump and pressurize it to about 10 psig (using refrigerant that is the same as the refrigerant in the system) on just the line set and connect it to the system using the procedures already given.
6. Read the manufacturer's recommendation as to the amount of charge for the new line lengths and add this much refrigerant to the system. If no recommendations are given, see **Figure 38–46** for a table of liquid-line capacities. The liquid line should contain the most refrigerant, and the proper charge should be added to the liquid line to make up for what was lost when the line was cut. No extra refrigerant charge needs to be added for the suction line if it is filled with vapor before assembly.

Piping Advice

The piping practices given for air-conditioning equipment are typical of most manufacturers' recommendations. **The manufacturer's recommendation should always be followed.**

LIQUID-LINE DIAMETER INCHES	OUNCES OF R-22 PER FOOT OF LENGTH OF LIQUID LINE
$\frac{3}{8}$	0.58
$\frac{5}{16}$	0.36
$\frac{1}{4}$	0.21

If 3 feet of liquid line must be added to a ($\frac{3}{8}$-in.) 30-foot line, an additional 1.74 ounces must be added to the system upon starting the unit. (3 × 0.58 = 1.74).

If the unit is shipped with a precharged line set, the complete charge for that particular line set is contained in the lines. If this set is altered, the complete liquid-line length must be used. For example, if a 50-foot set ($\frac{1}{4}$ in.) is cut to 25 feet of length, the charge is exhausted from the lines, and 5.25 ounces must be added to the charge contained in the condenser when started. (25 × 0.21 = 5.25).

Figure 38–46 A table of liquid-line capacities.

Each manufacturer ships installation and start-up literature inside the shipping crate. If it is not there, request it from the manufacturer.

38.12 EQUIPMENT START-UP

The final step in the installation is starting the equipment. The manufacturer will furnish start-up instructions. After the unit or units are in place, are leak checked, have the correct factory charge furnished by the manufacturer, and are wired for line and control voltage, follow these guidelines:

1. The line voltage must be connected to the unit disconnect panel. This should be done by a licensed electrician. Disconnect a low-voltage wire, such as the Y wire at the condensing unit, to prevent the compressor from starting. Turn on the line voltage. The line voltage allows the crankcase heater to heat the compressor crankcase. NOTE: This applies to any unit with crankcase heaters. Heat must be applied to the compressor crankcase for the amount of time the manufacturer recommends (usually not more than 24 h). Heating the crankcase boils any refrigerant out of the crankcase before the compressor is started. If the compressor is started with liquid in the crankcase, some of the liquid will reach the compressor cylinders and may cause damage. The oil also will foam and provide only marginal lubrication until the system has run for some time.

2. It is normally a good idea to plan on energizing the crankcase heater in the afternoon and starting the unit the next day. Before you leave, make sure that the crankcase heater is hot.

3. If the system has service valves, open them. NOTE: Some units have valves that *do not* have back seats. Do not try to back seat these valves that do not have back seats, or you may damage the valve. Open the valve until resistance is felt, then stop, **Figure 38–47.**

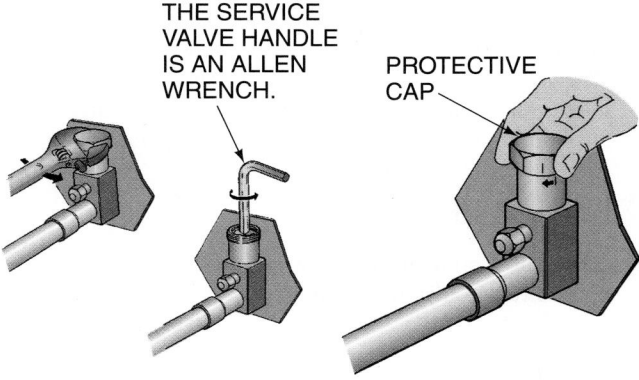

THE SERVICE VALVE HANDLE IS AN ALLEN WRENCH.

PROTECTIVE CAP

Figure 38–47 Service valves furnished with equipment may not have a "back seat." Do not try to turn the valve system all of the way out. *Courtesy Aeroquip Corporation*

4. Check the line voltage at the installation site and make sure that it is within the recommended limits.

5. Check all electrical connections including those made at the factory and ensure that they are tight and secure. SAFETY PRECAUTION: Turn the power off when checking electrical connections.

6. Set the fan switch on the room thermostat to FAN ON and check the indoor fan for proper operation, rotation, and current draw. You should feel air at all registers, normally about 2 to 3 ft above a floor register. Make sure that no air blockages exist at the supply and return openings.

7. Turn the fan switch to FAN AUTO and with the HEAT-OFF-COOL selector switch to the off position, replace the Y wire at the condensing unit. SAFETY PRECAUTION: The power at the unit is turned off and the panel is locked and tagged while making this connection.

8. Place your ammeter on the common wire to the compressor, have someone move the HEAT-OFF-COOL selector switch to COOL, and slide the temperature setting to call for cooling. The compressor should start. NOTE: Some manufacturer's literature will recommend that you have a set of gages on the system at this time. Be careful of the line length on the gage you install on the high-pressure side of the system. If the system has a critical charge and a 6-ft gage line is installed, the line will fill up with liquid refrigerant and will alter the charge, possibly enough to affect performance. A short gage line is recommended, **Figure 38–48.** Leak check the gage port when the gages are removed. Replace the gage port cover.

If the manufacturer recommends that gages be installed for start-up, install them before you start the system. If they do not recommend them, follow these recommendations:

9. When gages are not installed, certain signs can indicate correct performance. The suction line coming back to the compressor should be cool, although the "coolness" can vary. Two things will cause the suction-line temperature to vary and still be correct: the metering device and the ambient temperature. Most modern systems are fixed-bore metering devices (capillary tube or orifice). When the outside temperature is down to 75°F or 80°F, for example, the suction line will not be as cool as on a hot day because the condenser becomes more efficient and liquid refrigerant is retained in the condenser, partially starving the evaporator. If the day is cool, 65°F or 70°F, some of the air to the condenser may be blocked, which causes the head pressure to rise and the suction line to become cooler. The ambient temperature in the conditioned space causes the evaporator to have a large load, which makes the suction line not as cool until the inside temperature is reduced to near the design temperature, about 75°F.

The amperage of the compressor is a good indicator of system performance. If the outside temperature is hot and there is a good load on the evaporator, the compressor will be

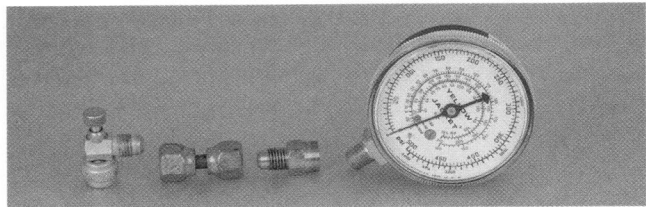

(A)

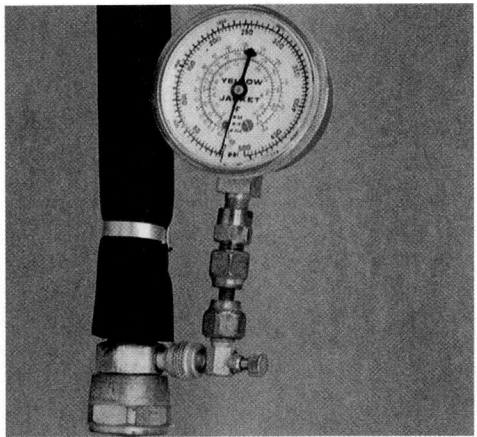

(B)

Figure 38–48 A short gage line for the high-pressure side of the system **(A)–(B)**. *Photos by Bill Johnson*

pumping at near capacity. The motor current will be near nameplate amperage. It is rare for the compressor amperage to be more than the nameplate rating due to only the system load. It is also normal for the compressor amperage to be slightly below the nameplate rating.

When you are satisfied that the unit is running satisfactorily, inspect the installation. Check the following:

🌐 1. All air registers are open.
 2. There are no air restrictions.
 3. The duct is hung correctly, and all connections are taped.
 4. All panels are in place with all screws.
 5. The customer knows how to operate the system.
 6. All warranty information is completed.
 7. The customer has the operation manual.
 8. The customer knows how to contact you.🌐

SAFETY PRECAUTION: Before starting the system examine all electrical connections and moving parts. The equipment may have faults and defects that can be harmful (e.g., a loose pulley may fly off when the motor starts to turn). Remember that vessels and hoses are under pressure. Always be mindful of potential electrical shock.

SUMMARY

■ Duct systems are normally constructed of square, rectangular, or round metal; ductboard; or flexible material.
■ Square and rectangular metal duct systems are assembled with S fasteners and drive clips.

■ The first fitting in a duct system may be a vibration eliminator to keep any fan noise or vibration from being transmitted into the duct.
■ Insulation may be applied to the inside or outside of any metal duct system that may exchange heat with the ambient.
■ When the duct is insulated on the outside, a vapor barrier must be used to keep moisture from forming on the duct surface if the duct surface is below the dew point temperature of the ambient.
■ Flexible duct is a flexible liner that may have a cover of fiberglass that is held in place with vinyl or reinforced foil.
■ The electrical installation includes choosing the correct enclosures, wire sizes, and fuses or breakers.
■ The electrical contractor will normally install the line-voltage wiring, and the air-conditioning contractor will usually install the low-voltage control wiring. Local codes should be consulted before any wiring is done.
■ The low-voltage control wiring is normally color coded.
■ Air-conditioning equipment is manufactured in package systems and split systems.
■ Air-to-air package equipment installation consists of placing the unit on a foundation, connecting the ductwork, and connecting the electrical service and control wiring.
■ The duct connections in package equipment may be made through a roof or through a wall at the end of a structure.
■ Roof installations have waterproof roof curbs and factory-made duct systems.
■ Isolation pads or springs placed under the equipment prevent equipment noise from traveling into the structure.
■ Two refrigerant lines connect the evaporator and the condensing unit; the large line is the insulated suction line, and the small line is the liquid line.
■ The air handler and condenser should be installed so that they are accessible for service.
■ A condensate drain provision must be made for the evaporator.
■ A secondary drain pan should be provided if the evaporator is located above the conditioned space.
■ The condensing unit should not be located where its noise will be bothersome.
■ Line sets come in standard lengths of 10, 20, 30, 40, and 50 ft. The system charge is normally for 30 ft when the whole charge is stored in the condensing unit.
■ Line sets may be altered in length. The refrigerant charge must be adjusted when the lines are altered.
■ The start-up procedure for the equipment is in the manufacturer's literature. Before start-up, check the electrical connections, the fans, the airflow, and the refrigerant charge.

REVIEW QUESTIONS

1. The three crafts normally associated with an air-conditioned installation are _____, _____, and _____.
2. True or False: The flexible duct system is the most economical.
3. Name the two fasteners that typically fasten square or rectangular connections.
4. What fastener is normally used to fasten round duct connections?

5. True or False: Insulation on duct does not prevent duct from sweating.
6. When insulation on the inside of the duct comes loose it causes
   A. the system to shut off.
   B. high head pressure.
   C. increased airflow.
   D. reduced airflow.
7. Most duct insulation is made of _____.
8. What is the purpose of the flexible duct connector?
9. What happens if flexible duct is turned too sharply around a corner?
10. Why must flexible duct be stretched slightly for straight runs?
11. Explain how flexible duct is supported when installed.
12. The two duct connection combinations for a package unit are _____ and _____.
13. What are the differences in a package system and a split system?
14. True or False: A package unit comes from the factory with the charge inside.
15. Name three criteria of a good location for an air-cooled condensing unit.

UNIT 39 Controls

OBJECTIVES

After studying this unit, you should be able to

- describe the control sequence for an air-conditioning system.
- explain the function of the 24-V control voltage.
- describe the space thermostat.
- describe the compressor contactor.
- explain the operation of the high- and low-pressure controls.
- discuss the function of the overloads and motor-winding thermostat.
- discuss the winding thermostat and the internal relief valve.
- identify operating and safety controls.
- compare modern and older control concepts.
- describe how crankcase heat is applied in some modern equipment.

39.1 CONTROLS FOR AIR CONDITIONING

Equipment control for maintaining correct air conditions involves the control of three components: the indoor fan, the compressor, and the outdoor fan. These components are used in air-cooled air-conditioning equipment. Water-cooled equipment for small air-conditioning applications is seldom used. Some of these controls were covered in earlier units. They are discussed again here for the convenience of the reader.

Manufacturers are working harder to create better efficiencies with their equipment. The controls can be part of the efficiency improvements. For example, the indoor fan may be allowed to run after the outdoor unit is shut off. The cold coil can then absorb a few Btu of heat energy from the air. The outdoor fan may be controlled at lower speeds in mild weather for better head pressure control.

The indoor fan, compressor, and outdoor fan must be started and stopped automatically at the correct times. The normal sequence of operation is as follows:

1. The indoor fan must operate when the compressor operates. The exception to this is where the fan may be allowed to run for a few minutes to warm up the cold coil and continue removing heat from the circulating air.

2. The outdoor fan must operate when the compressor operates (except for units that may have a fan cycle device which allows the fan to be cycled off for short periods of time during cool weather for head pressure control). Some outdoor fan motors are multiple-speed or variable-speed motors. Multiple-speed motors can

be controlled by a bimetallic disc located on the condensing unit. This disc senses outdoor ambient temperatures. Variable-speed motors vary their rpms with the assistance of a controller to compensate for changing outdoor ambient temperatures.

3. The indoor fan may have a continuous-operation switch (FAN ON) at the thermostat. In this position the indoor fan will run continuously, and the compressor and outdoor fan will cycle on and off on demand.

Operating and safety controls are the two main types of functional controls. A room thermostat, for example, is an operating control used to sense the space temperature to stop and start the compression system (compressor and outdoor fan), **Figure 39–1**. The high-pressure control is a safety control that keeps the unit from operating when the head pressure is too high, **Figure 39–2(A)**. However, most high-pressure

Figure 39–1 A room thermostat for controlling space temperature. *Courtesy Honeywell*

Figure 39–2(A) The commercial type of high-pressure control used in earlier residential equipment. *Photo by Bill Johnson*

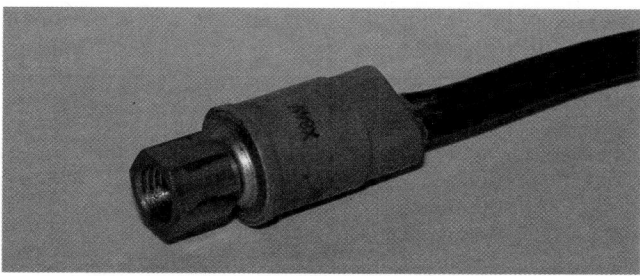

Figure 39–2(B)   The permanently mounted, modular type of high-pressure control used in air-conditioning applications. This control is factory set and nonadjustable. *Photo by Bill Johnson*

controls for air-conditioning systems are a modular type. They are permanently installed on the discharge line of the air-conditioning system. These newer high-pressure controls are factory set and nonadjustable, **Figure 39–2(B).** Both of these controls have something in common: neither consumes power. They pass power to other devices, such as the compressor's magnetic contactor.

39.2   PRIME MOVERS—COMPRESSORS AND FANS

The prime movers of the system are the fans (indoor and outdoor) and the compressor. These devices consume most of the power in the air-conditioning process and are operated with high voltage, usually 230 V in a residence. The controls are operated with 24 V for safety and convenience. These low-voltage controls can be enhanced by means of electronic circuit boards that allow an infinite number of control sequences. The price and reliability have given the manufacturers a great deal of room to figure out control sequences for convenience, comfort, and efficiency. Many units today have small microprocessors (computers) built into the circuit boards. The voltage reduction is accomplished by a transformer located in the condensing unit or air handler, **Figure 39–3.**

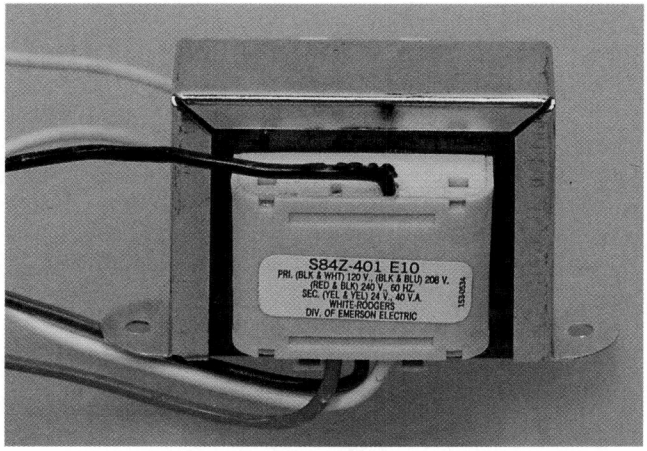

Figure 39–3   A low-voltage transformer. *Photo by Bill Johnson*

39.3   LOW-VOLTAGE CONTROLS

The low-voltage circuit operates the various power-consuming devices in the control system, which start and stop the compressor and fan motors. These low-voltage devices do not use much current. The control transformer is usually a 40-VA transformer. This means that the highest amperage the control transformer will be able to produce on the secondary side is 1.7 A. This was arrived at by dividing the VA rating of 40 by the rated voltage (40 VA ÷ 24 V = 1.666 A rounded to 1.7). If the circuit current is greater than 1.666 A, the voltage will begin to drop, and the transformer will get hot. In this situation a larger VA transformer is needed.

The contactor that actually starts and stops the compressor is considered one of the controls. It consumes power in its 24-V magnetic holding coil, **Figure 39–4.** The contacts of the contactor pass power when the magnetic coil is energized, **Figure 39–5.** The contact circuit to the compressor is 230 V.

Figure 39–4   A contactor. *Photo by Bill Johnson*

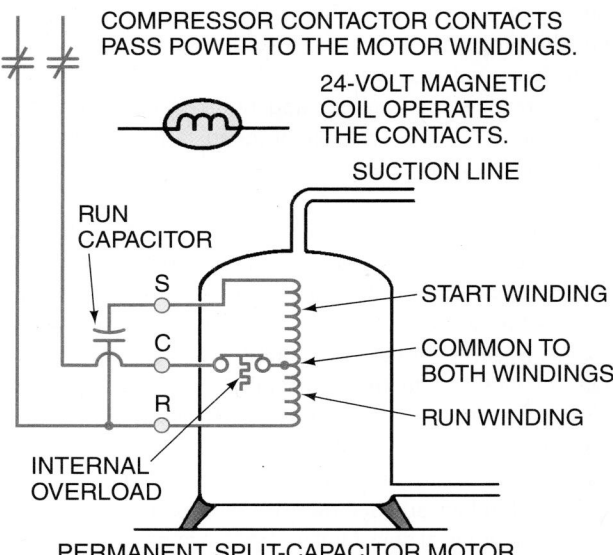

COMPRESSOR CONTACTOR CONTACTS
PASS POWER TO THE MOTOR WINDINGS.

24-VOLT MAGNETIC
COIL OPERATES
THE CONTACTS.

SUCTION LINE

RUN
CAPACITOR

S — START WINDING

C — COMMON TO BOTH WINDINGS

R — RUN WINDING

INTERNAL
OVERLOAD

PERMANENT SPLIT-CAPACITOR MOTOR

Figure 39–5   A diagram of the circuit that passes power to the motor windings.

39.4 SOME HISTORY OF RESIDENTIAL CENTRAL AIR CONDITIONING

Residential central air conditioning became popular in the late 1950s, and its popularity has continued to grow. This popularity has caused air-conditioning equipment to become competitive in price. In the warm climates in the United States, central air conditioning is desirable in all new construction because of the resale value of the structure, even if the owner or builder does not desire air conditioning. Lending institutions may even require that a new home have air conditioning before they will finance the loan. The reason is that the lender may have to resell the house, and central air conditioning is a strong selling feature, which may not be easy to add in the future.

The first central air-conditioning systems installed in residences were small commercial systems applied to homes. They were all water cooled and very efficient and reliable. Air-cooled equipment became popular later. Air-cooled systems do not pump and handle water; therefore, freezing water and mineral deposits are not problems.

39.5 ECONOMICS OF EQUIPMENT DESIGN

The first air-cooled systems were heavy duty, bulky, and hard to handle. Slow-speed hermetic compressors (1800 rpm) or belt-drive open compressors were used with R-12 as the refrigerant. The residential- and light-commercial-equipment market began to grow to the point that price-conscious buyers began to look for less expensive equipment. Speculating home builders began to build large developments and wanted to save on the air-conditioning equipment. Manufacturers were thus forced to find more economical ways to build equipment. A more efficient refrigerant was also needed. When R-22 became available, the evaporator and the condenser tubing could be made smaller, making equipment more compact and lighter, which made storage and shipping much easier. In addition, the suction and liquid lines that connect the condensing unit and the evaporator are one size smaller and reduced the price of the installation. The physical size of the installation was scaled down, and the manufacturing and installation costs were reduced.

The compressors were then manufactured to turn at a higher speed, so a small compressor could pump more refrigerant and do the job of a larger compressor. Present compressors turn at 3600 rpm and are much smaller than earlier models. Because of the faster compressors and the more efficient R-22 refrigerant, along with the newer R-22 alternative refrigerants and blends such as R-410A and R-407C, the equipment of today is smaller and more efficient than the earlier equipment. Every manufacturer constantly seeks to make equipment lighter, smaller, more efficient, and less expensive.

🌎 Every central split-cooling system manufactured in the United States today must have a Seasonal Energy Efficiency Ratio (SEER) of at least 13. This energy requirement was mandated by federal law as of January 23, 2006. In addition,

with the phaseout of R-22 just around the corner, manufacturers of HVACR equipment are looking for energy-efficient methods to apply to their equipment to meet these new energy requirements. The timeline for R-22 is as follows:

- 2010—R-22 use is banned at the original equipment manufacturer (OEM) level with a 75% reduction on HCFC production from baseline year 1989.
- 2015—there is a 90% reduction on HCFC production from baseline year 1989.
- 2020—there is a total ban on R-22 production. 🌎

Equipment covered in this federal mandate includes the following:

- Unitary equipment ranging from 1 1/2 to 5 tons
- Split/packaged air conditioners and heat pumps

Equipment not covered includes the following:

- Commercial equipment greater than 6 tons
- Space-constrained units smaller than 3 tons (room air conditioners)
- Water-source units

🌎 SEER is calculated on the basis of the total amount of cooling (in Btu) the system will provide over the entire season, divided by the total number watt-hours it will consume. Higher SEERs reflect a more efficient cooling system. This federal mandate will impact 95% of the unitary market in the United States, which is about 8 million units at the time of this writing. Because of this new federal mandate of 13 SEER, most air-conditioning and heat pump manufacturers are looking for more efficient evaporator and condenser designs, more efficient compressors and fan motors, and more sophisticated control systems to meet this energy efficiency requirement. Refer to Unit 21, "Evaporators and the Refrigeration System," and Unit 22, "Condensers," for more detailed information on evaporator and condenser designs. 🌎

Air-conditioning equipment manufactured to compete economically with more expensive equipment will have only the essential controls, yet may be very reliable. The following sections describe in detail typical controls for an air-cooled system built in the mid-1960s when central air-conditioning systems were being installed in large numbers for the first time. Many of these systems are still in use today. The detailed description is important because many of these controls are no longer used in recently manufactured equipment.

39.6 OPERATING CONTROLS FOR OLDER AIR-COOLED SYSTEMS

The room thermostat senses and controls the space temperature. The thermostat is not a power-consuming device, but it passes power to the power-consuming devices. It is sensitive to temperature changes because of the bimetal element or thermistor under the cover. Early thermostats were larger than modern ones.

The fan relay starts and stops the indoor fan. The relay is a power-consuming device that receives power through the room thermostat to energize the magnetic coil and to close the contacts that start the fan.

The compressor contactor starts and stops the compressor and outdoor fan, which are normally wired in parallel,

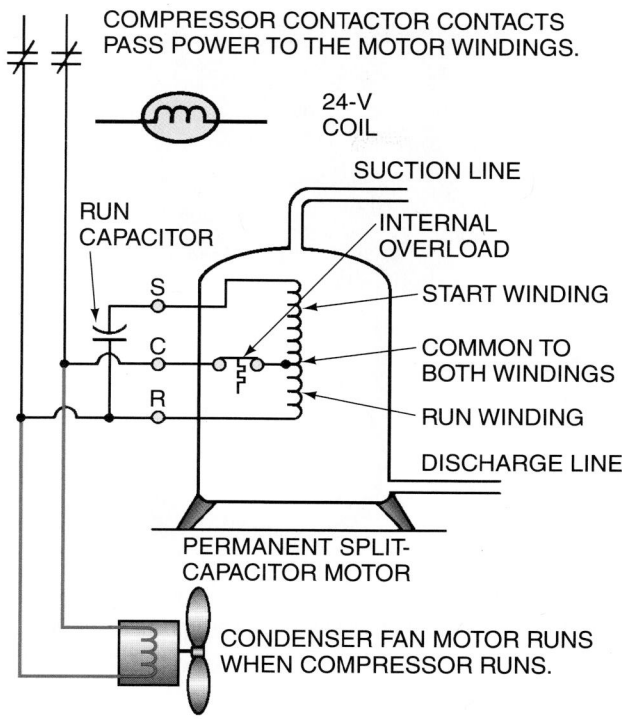

COMPRESSOR CONTACTOR CONTACTS PASS POWER TO THE MOTOR WINDINGS.

24-V COIL

SUCTION LINE

RUN CAPACITOR

INTERNAL OVERLOAD

START WINDING

S
C
R

COMMON TO BOTH WINDINGS

RUN WINDING

DISCHARGE LINE

PERMANENT SPLIT-CAPACITOR MOTOR

CONDENSER FAN MOTOR RUNS WHEN COMPRESSOR RUNS.

Figure 39–6   The compressor and outdoor fan wired in parallel.

Figure 39–6. This control is a power-consuming device because of the magnetic coil. When the coil is energized, it closes the contacts and starts the compressor and the outdoor fan motor, **Figure 39–7.**

The compressor starting and running circuits are actually not controls in the strictest sense, but most service technicians treat them as such. The compressors that were used for many years had capacitor-start, capacitor-run motors. These motors have a high starting torque and were used because thermostatic expansion valves were used as metering devices. They do not equalize the high- and low-side pressures during the off cycle, and a high-torque compressor motor is required. The following components were part of the starting

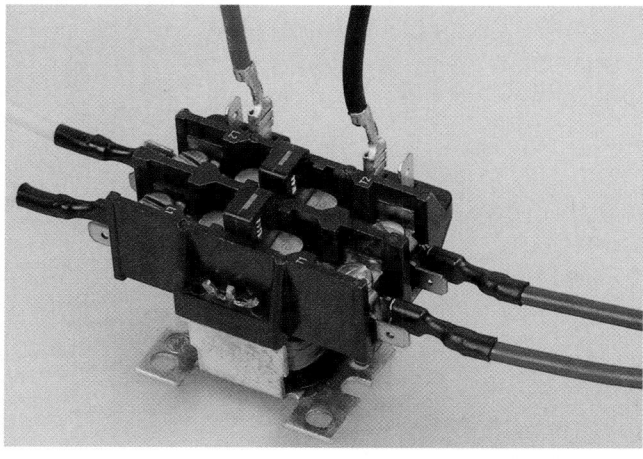

Figure 39–7   A typical compressor contactor. *Photo by Bill Johnson*

system: (1) a potential starting relay, (2) a start capacitor, and (3) a run capacitor, **Figure 39–8.** See **Figure 39–9** for an example of the starting circuit diagram.

The motor starting circuit is not easy to troubleshoot but can be simplified by realizing that the "run" and "start" terminals on the compressor are on the same electrical line. It makes a circuit from run to start just long enough to start

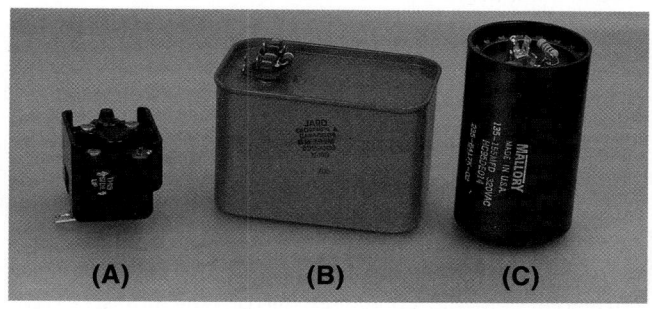

(A)   **(B)**   **(C)**

Figure 39–8   The components to start and run a capacitor-start, capacitor-run motor. **(A)** Potential relay. **(B)** Run capacitor. **(C)** Start capacitor. *Photo by Bill Johnson*

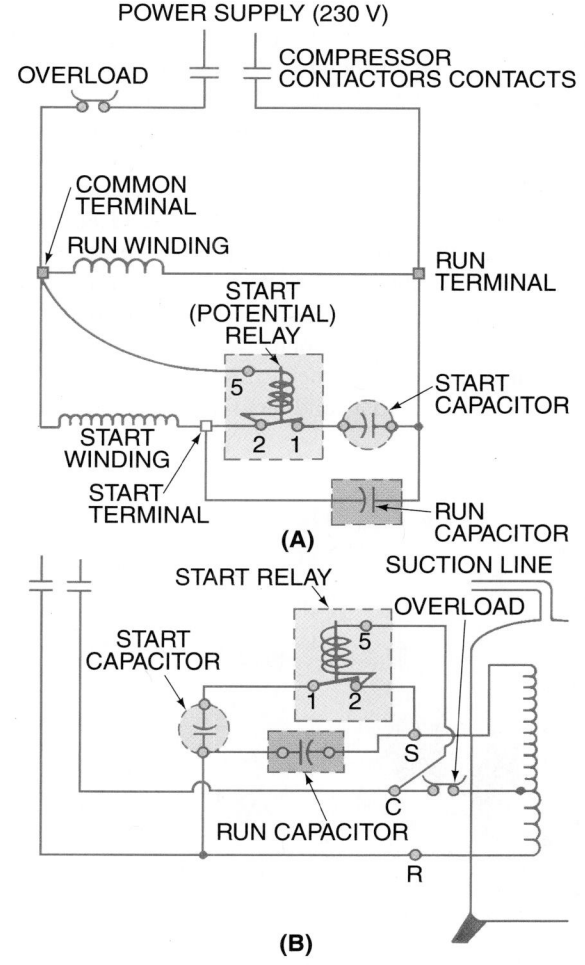

Figure 39–9   The starting circuit for a capacitor-start, capacitor-run motor. **(A)** Schematic. **(B)** Pictorial.

the compressor. The technician should isolate the compressor from the control circuit by disconnecting the compressor terminals (common, run, and start) and insulating them. Then the compressor can be checked electrically and the control circuit can be checked to determine which may have a problem. Many technicians use a test cord to operate the compressor so they can isolate it from the system's control circuit. If the compressor will start with the test cord and not the compressor starting circuit, the problem is the start circuit.

39.7 SAFETY CONTROLS FOR OLDER AIR-COOLED SYSTEMS

The high-pressure control stops the compressor when a high-pressure condition exists, for example, if the condenser fan stops when it should be running. The compressor has no way of knowing that the fan has stopped, and it keeps pumping refrigerant into the condenser. NOTE: The pressures can quickly rise to a dangerous level, so the compressor must be stopped. **Figure 39–10** illustrates a commercial high-pressure control.

For protection, the low-pressure control stops the compressor from pulling the suction pressure below a predetermined point. If the system loses all or part of its refrigerant charge, the low side of the system may pump into a vacuum without a low-pressure control. The low-pressure control can be set to shut off the compressor before a vacuum is reached. **Figure 39–11** is a photo of a commercial low-pressure control. If the leak is on the low side of the system and the compressor pumps into a vacuum, air will be drawn into the system. Then there would be two problems: a leak and air in the system. The low-pressure control also provides freeze protection. When the air across the indoor coil is reduced, the suction pressure will be reduced to the point that a freeze condition may result. The condensate on the evaporator coil may turn to ice and block the airflow even more. The evaporator coil may turn to a solid block of ice. Reduced airflow

Figure 39–11   The type of low-pressure control used on commercial equipment and on early residential equipment. *Photo by Bill Johnson*

may also be caused from dirty filters, closed air registers, a blocked return air grille, or a dirty evaporator coil.

The early systems had overload protection for the common and the run circuit of the compressor. This overload protection was usually a heat-sensitive bimetal device with a current rating. If the motor current went above the rating on the overload device, the bimetal strip was heated and opened the circuit to the compressor contactor coil to stop the compressor before damage occurred, **Figure 39–12**. This overload protection was typically located in the motor terminal box outside the refrigerant atmosphere.

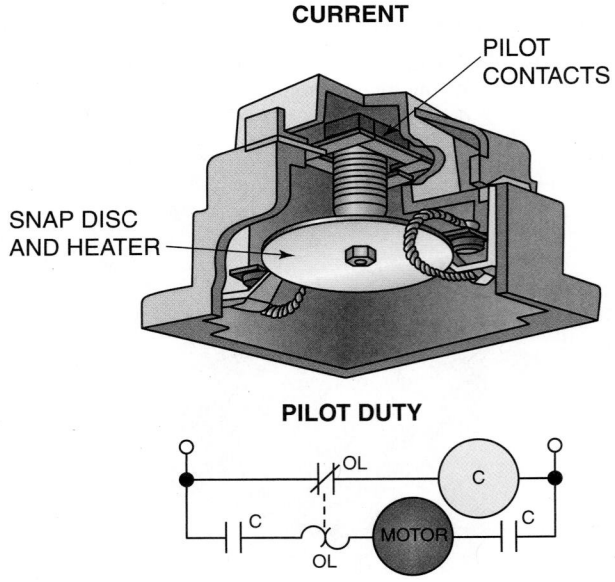

Figure 39–12   This overload device is a heat-sensitive bimetal strip. *Reproduced courtesy of Carrier Corporation*

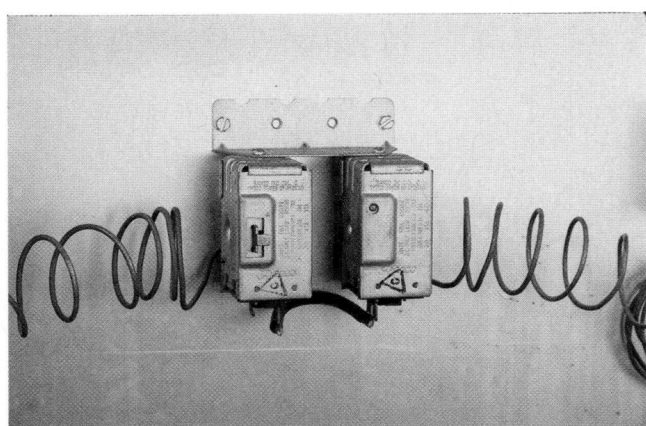

Figure 39–10   This commercial high-pressure control is much larger than the residential high-pressure controls that are currently used. *Photo by Bill Johnson*

Internal motor protection senses the actual motor temperature and stops the motor when it is too hot. Compressor internal protection may be the type that breaks the line power inside the compressor, or it may be the pilot type that interrupts the circuit to the contactor coil. This motor protection is located inside the compressor inside its own sealed container.

Short-cycle protection was used to prevent the compressor from short cycling when a safety or operating control would open the circuit and then close the circuit in a short cycle. The overload is an example. Some types of overloads are current sensitive. These controls will open the circuit when the current is too high but will make back the instant it is reduced. If a compressor had a bad starting relay and stopped due to high current, it would try to start again immediately after stopping if no short-cycle control existed, **Figure 39–13**. This same control protects the compressor from short cycling if the homeowner turns the thermostat up and down several times while adjusting it.

39.8 OPERATING CONTROLS FOR MODERN EQUIPMENT

The room thermostat is now much smaller, **Figure 39–14**. Some room thermostats are electronic and use a thermistor as the sensing element, **Figure 39–15**. These thermostats may have programs for night and day work schedule adjustments for both the heating and the cooling settings. They can be

Figure 39–13(B) A solid-state anti-short-cycle timer. *Courtesy Ferris State University. Photo by John Tomczyk*

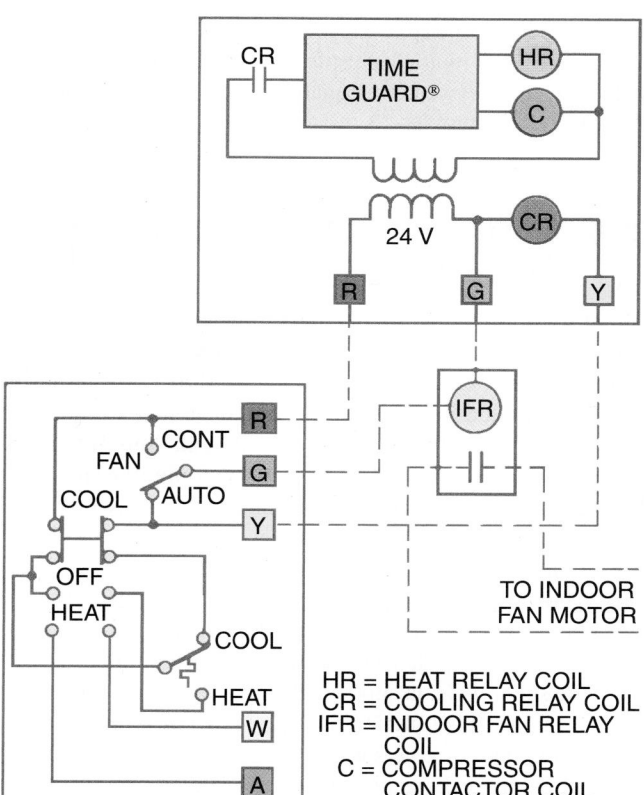

Figure 39–13(A) A timer to keep the compressor from short cycling.

HR = HEAT RELAY COIL
CR = COOLING RELAY COIL
IFR = INDOOR FAN RELAY COIL
C = COMPRESSOR CONTACTOR COIL

Figure 39–14 This is a typical room thermostat. It is a combination heating and cooling thermostat with a bimetal element for sensing room temperature. *Photo by Bill Johnson*

thought of as power consuming, because they must have a circuit with a timer. They also pass power to various components (the fan relay and compressor contactor). The circuit that passes power is the contact circuit.

The fan relay for the modern equipment may be smaller than the older versions, but it performs the same function, **Figure 39–16**.

The modern compressor contactor has one big difference on some manufacturers' equipment. Some contactors have only one set of contacts, **Figure 39–17**. Older contactors always had two sets of contacts that interrupted the power to the common and run circuits on the compressor, both line 1

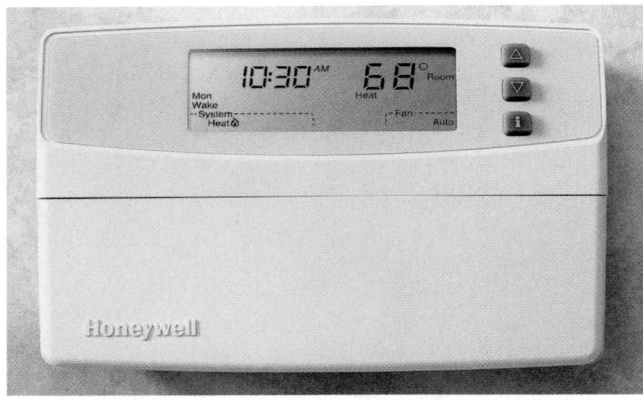

Figure 39–15 An electronic thermostat with a thermistor for the sensing element. *Courtesy Honeywell*

Figure 39–16 A fan relay. *Photo by Bill Johnson*

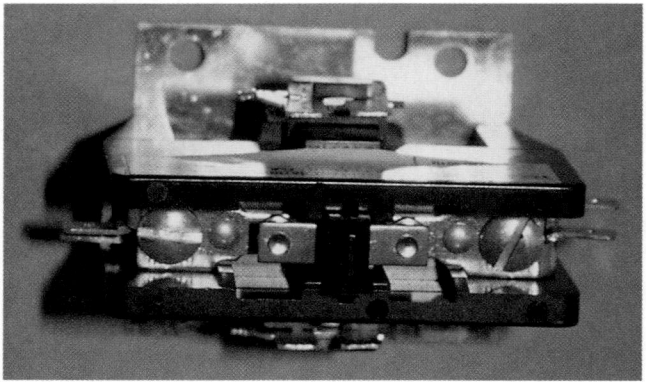

Figure 39–17 A compressor contactor with one set of contacts. *Photo by Bill Johnson*

and line 2. The modern contactor may interrupt the power to only one circuit. NOTE: The circuit with continuous power may feed power through a run capacitor to provide a small amount of current through the compressor windings to keep the compressor warm during the off cycle. This is a substitute for crankcase heat. If a contactor with two contacts is substituted, there will not be heat for the crankcase.

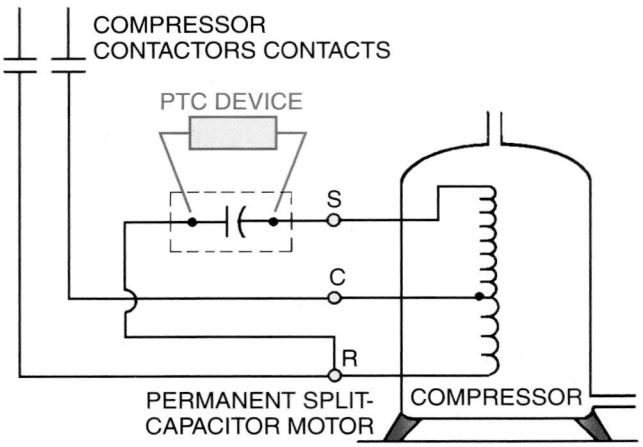

Figure 39–18 This wiring diagram shows how a PSC motor is wired when a PTC device is used as a start assist for the compressor.

The starting circuit does not have as many components. If a retired service technician suddenly opened the panel of a modern piece of equipment, the conclusion would quickly be drawn that the equipment could not work correctly. It would seem unprotected because not as many components are there as before. The components used to start and run modern equipment are a run capacitor and possibly a start-assist unit such as a *positive temperature coefficient (PTC) start device.* The PTC device has no moving parts as a start relay does. It makes a circuit from the start terminal to the run terminal for a short time to get the compressor started. These devices are used on permanent split-capacitor (PSC) motors as a start assist. The PSC motor has little starting torque and is used with metering devices that equalize system pressures during the off cycle, **Figure 39–18.**

39.9 SAFETY CONTROLS FOR MODERN EQUIPMENT

Modern equipment may not have any visible safety controls. It may have only a motor-temperature control, which senses the temperature of the motor windings. This control may be mounted in the windings and have no external terminals. A note on the compressor may read "this compressor is internally protected with a winding thermostat," **Figure 39–19.** The intent of the manufacturer is to let this control function cut off the compressor during the following situations:

1. When a system is operating with a low refrigerant charge, the motor windings will overheat, and the motor-winding thermostat will stop the compressor. The suction gas is supposed to cool the compressor windings. It is functioning as low-charge protection. When the motor gets warm or hot, the heat in the mass of the motor prevents the control from closing its contacts until the motor cools. This is a built-in short-cycle protection.

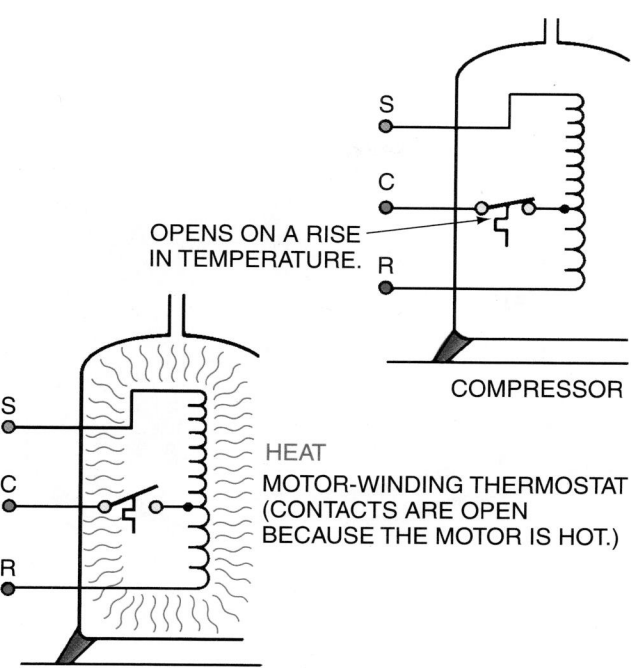

Figure 39–19 A motor-winding thermostat.

2. When the condenser is dirty, the compressor head pressure rises. This makes the current increase, and the motor gets hot and cuts off because of the motor temperature cut-out. The motor-winding thermostat is functioning as a high-pressure control in this situation. It still has the same short-cycle protection because of the mass of the motor.

3. In most compressors there is a pressure relief valve that will relieve hot gas from the compressor discharge into the motor-winding thermostat if a great pressure differential is experienced, **Figure 39–20**. Suppose that the condenser fan motor burned and stopped. High

head pressure would occur immediately. When these pressures exceed the relief valve setting, the valve relieves and blows hot gas on the winding thermostat to cut the compressor off. Internal relief valve settings may have a differential of 450 psig. This means the head pressure has to be 450 psig higher than the suction pressure for the relief valve to open. If a condenser fan stopped, the head pressure could go to 540 psig, and the suction pressure could go to 90 psig before the internal relief valve opened (540 − 90 = 450 psig). NOTE: A reading of the high-pressure gage is not enough to determine that the relief valve should function. The winding thermostat in this application serves as the high-pressure cut-out when fast action is needed.

4. The run capacitor may become defective, and the compressor will pull too much current trying to start. The motor will heat up and cut out on the winding thermostat. The winding thermostat is functioning as an overload in this situation.

5. Someone may fool with the space temperature thermostat by turning it up and down several times in quick succession. The compressor will try to start each time, and the motor will get hot enough to cause the motor-winding thermostat to cut it off. The motor-winding thermostat is preventing the compressor from short cycling in this situation. Often, a time delay within the room thermostat is used to help solve this shorting-cycling problem.

SAFETY PRECAUTION: When an overcharge of refrigerant is great enough to cause the compressor windings to be cool, the winding thermostat may not give adequate protection, **Figure 39–21**.

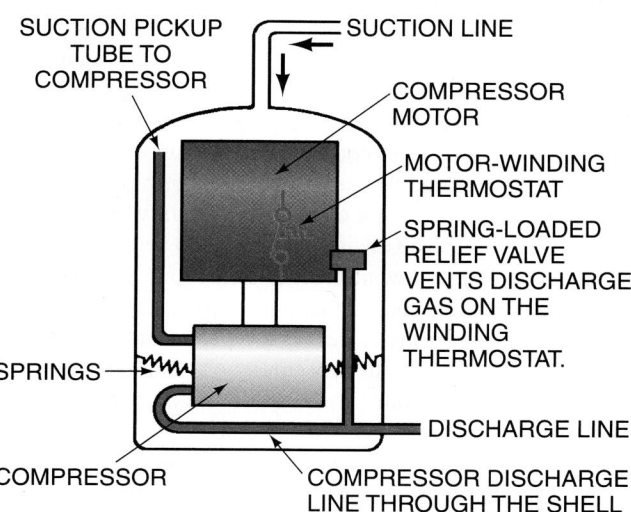

Figure 39–20 The internal relief valve in a welded hermetic compressor.

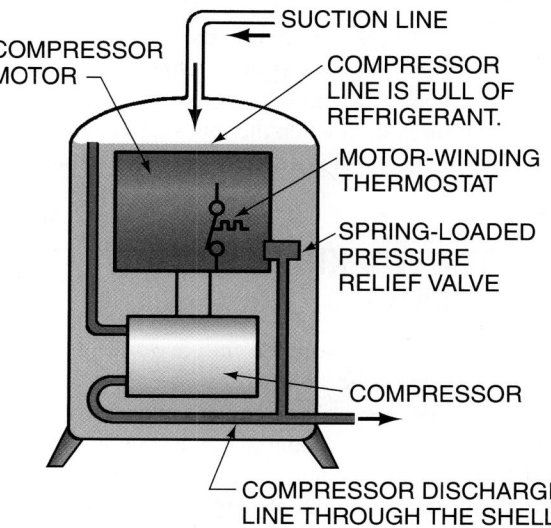

Figure 39–21 This is an example of one time when the relief valve and motor temperature combination may not give high-pressure protection to stop the compressor.

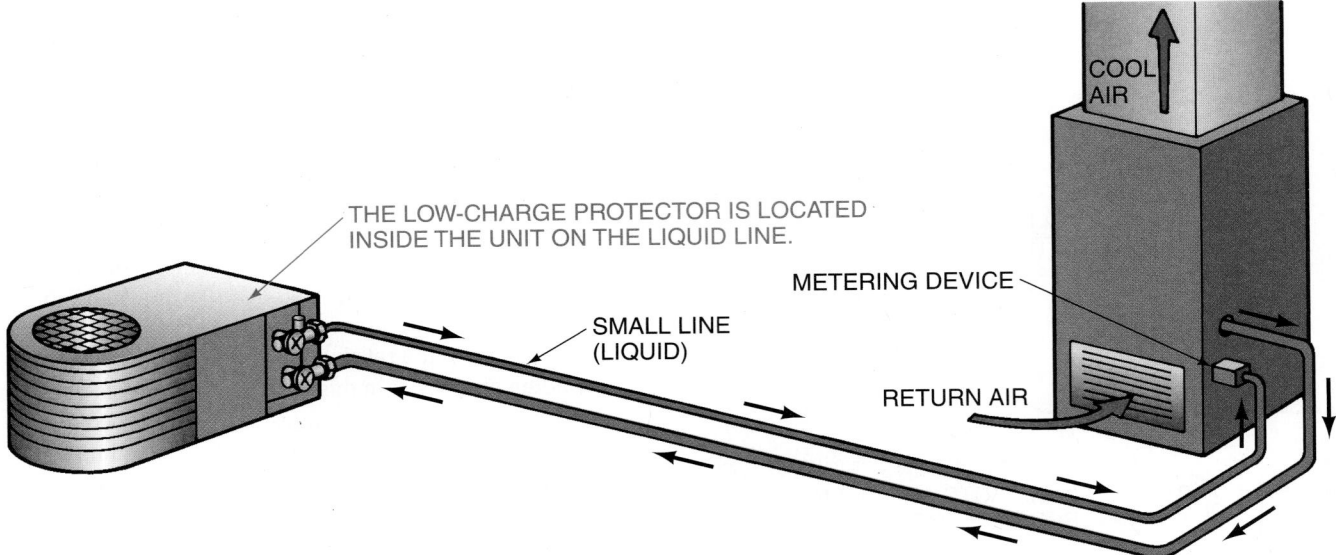

THE LOW-CHARGE PROTECTOR IS LOCATED INSIDE THE UNIT ON THE LIQUID LINE.

COOL AIR

METERING DEVICE

SMALL LINE (LIQUID)

RETURN AIR

Figure 39–22 This loss-of-charge protector is located in the liquid line and will stop the compressor if the pressure goes as low as 5 psig.

If the motor should burn while the compressor is full of liquid, the motor-winding thermostat may not function because of the amount of liquid refrigerant in which it is submerged. The liquid refrigerant may become very hot with electric current applied, and high pressures may occur. The pressure may even get high enough to split the compressor's shell. The only protection the compressor has is the internal relief valve, and it will serve no purpose in this situation.

Some modern systems have a loss-of-charge protection in the form of a low-pressure cut-out control located on the liquid line. It may have a setting as low as 5 psig and will stop the system only if there is complete loss of charge, **Figure 39–22**. This may not seem like much protection, but it will shut the compressor off when a complete loss of charge occurs. The compressor will not try to restart until the charge is put back into the system. One thing to keep in mind for this type of system is when a low-side leak occurs, air will enter the system when the low side goes into a vacuum because this control is on the high-pressure side of the system.

39.10 THE WORKING CONTROL PACKAGE

All controls must be assembled into a working assembly. Manufacturers must be competitive, so they are always trying to provide a simple effective control package that will protect the equipment and give it a long life span. **Figure 39–23(A)** illustrates some typical diagrams that apply to simple residential equipment. **Figure 39–23(B)** shows a modern rooftop cooling unit and control package incorporating four parallel scroll compressors.

Figure 39–24 summarizes the controls used on earlier equipment and newer equipment. The technician should be

familiar with both because the older controls may still be in use on older equipment.

In some modern high-efficiency equipment, part of the efficiency package is the method of controlling the equipment. Many modern air-conditioning units have electronic controllers on board that have terminals for both analog and digital sensor inputs. These same electronic controllers will also have terminals for many outputs. Short-cycle timers can also be built into the electronics within the controller. **Figure 39–25(A)–(C)** is an example of a modern electronic controller used on air-conditioning equipment.

39.11 ELECTRONIC CONTROLS AND AIR-CONDITIONING EQUIPMENT

An electronic circuit board can monitor high voltage, whereas electromechanical controls cannot easily do this. Power companies may supply voltage that is higher than the equipment specification. For example, when a system is rated as 230 V, the equipment may be operated at ±10% of this voltage. If the voltage is too low, the motor will pull more than normal current, and the overload device (or winding thermostat) should stop the motor. If the motor operates at a slightly high-voltage condition, the overload devices will not help because the motor current will be below normal. The electronic circuit board may have a voltage monitor that will stop the compressor at a high voltage even if the current is low. The electronic circuit board can also react quickly to low voltage (known as "brownout"). It can monitor and cut off the compressor before an overload has time to heat up and react.

Each manufacturer who uses a circuit board has a recommended checkout procedure. For practical troubleshooting, remember that a circuit board usually looks like a single control on the wiring diagram. The control circuit goes into and comes out of the board. However, sometimes the

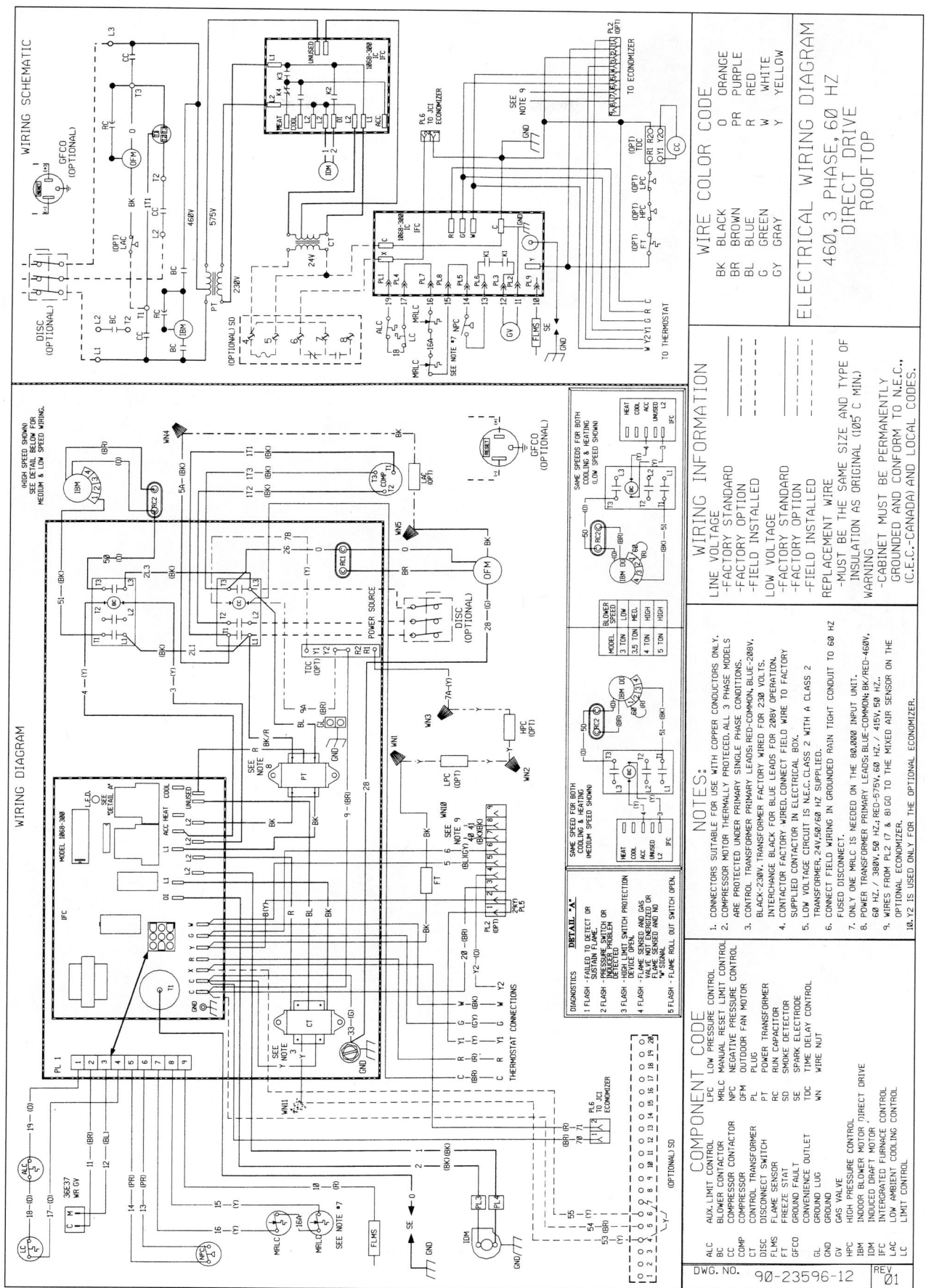

Figure 39–23   (A) Wiring diagram and wiring schematic for a small, commercial, rooftop air-conditioning system.

(B)

Figure 39–23 *(continued)* **(B)** A modern rooftop air-conditioning unit with four parallel scroll compressors. *Photo by John Tomcyzk*

OLDER EQUIPMENT	NEWER EQUIPMENT
Room thermostat	Room thermostat
Fan relay	Fan relay
Compressor contactor	Compressor contactor
Winding thermostat (maybe)	Winding thermostat
Run capacitor	Run capacitor
Internal relief valve (maybe)	Internal relief valve
Low-pressure control	No charge protection
High-pressure control	Winding thermostat
Short-cycle protection	Winding thermostat
Overloads (usually two)	Winding thermostat
Crankcase heater	Through the capacitor

Figure 39–24 Controls used on older and newer equipment.

(B)

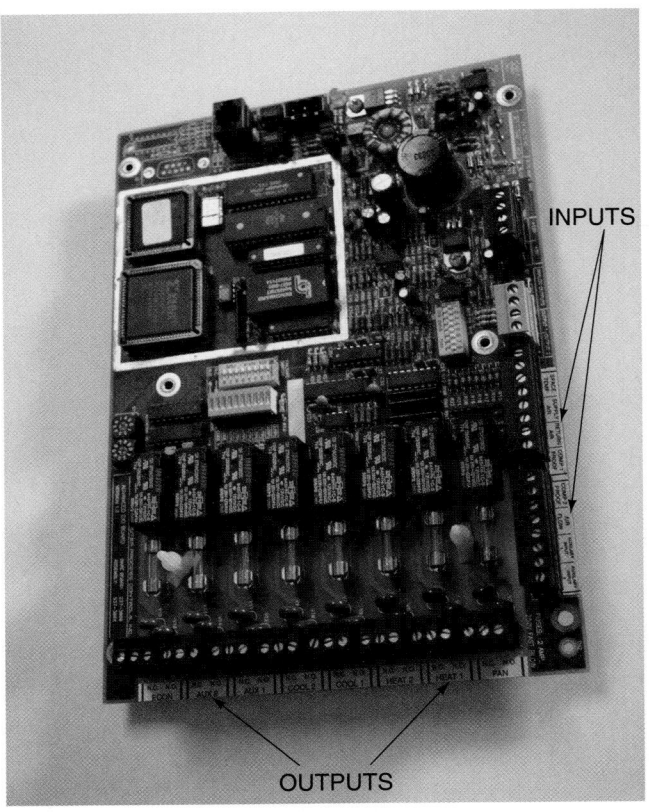

(A)

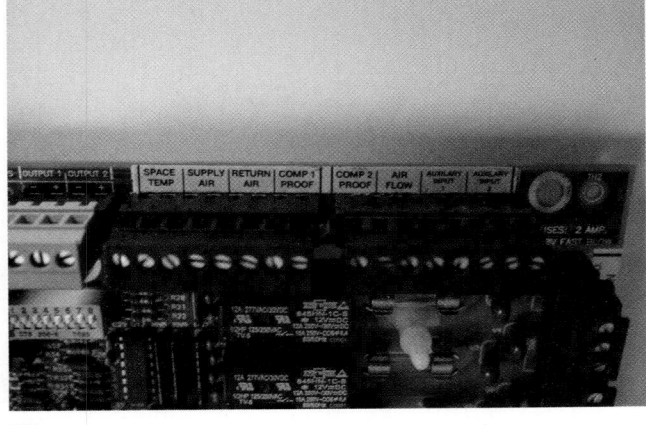

(C)

Figure 39–25 **(A)** Modern solid-state electronic board showing input and output terminals. **(B)** Output terminals, fuses, and relays on the same board. **(C)** Input terminals on the same board. *Courtesy Ferris State University. Photos by John Tomczyk*

circuits on the board may be checked by using a jumper wire from one circuit to another to determine whether the board is defective.

Always consult with the manufacturer or read the unit's troubleshooting manual before using a jumper wire to troubleshoot a circuit board. Some manufacturers may specify that a certain resistance wire or resistor be used in place of the jumper wire to avoid any shorts within the circuit board. Many manufacturers will give detailed, step-by-step instructions on how to systematically troubleshoot their circuit boards. Also, many modern circuit boards are self-diagnostic and have *light-emitting diodes (LEDs)* that spell out the problem.

SUMMARY

- The control of air-conditioning equipment consists of controlling the various components in order to start and stop them in the correct sequence to maintain space temperature for comfort.
- The three main components controlled are the indoor fan, the compressor, and the outdoor fan.
- The control circuit uses low voltage (24 V) for safety.
- Low voltage is obtained from high voltage by a transformer located in the condensing unit or in the air handler.
- Some electronic thermostats use a thermistor as the sensing device and may incorporate other features, such as temperature setback, for night and day.
- Air-cooled equipment is used more widely today than water-cooled equipment.
- Controls are either operating controls or safety controls.
- Some controls, such as contactors and electronic thermostats, have more than one circuit.
- The compressor overload protectors are used to keep the compressor from drawing too much current and over-working.
- The motor-winding thermostat shuts off the compressor when the compressor motor winding gets too hot.
- The compressor internal relief valve vents high-pressure gas from the high-pressure side to the low-pressure side of the compressor internally. It can be used as a high-pressure indicator by directing this high-pressure, high-temperature gas onto the winding thermostat.
- Modern equipment may have only a few controls and components. It is common for the unit to have a single-pole contactor and a run capacitor as the only visible components.
- Electronic circuit boards are furnished with some equipment. The circuit boards may include short-cycle protection, as well as low- and high-voltage protection.

REVIEW QUESTIONS

1. Name two types of controls.
2. Some controls _____ power, and some controls _____ power.
3. What type of control is the room thermostat?
   What protection do the controls in questions 4–8 offer?
4. High-pressure control.
5. Low-pressure control.
6. Winding thermostat.
7. Internal relief valve.
8. Electronic circuit board.
9. Why can you not substitute a two-pole contactor for a single-pole contactor on some units?
10. The standard low-voltage control voltage for residential air conditioning is
    A. 24 V.
    B. 48 V.
    C. 12 V.
    D. 16 V.
11. What changes have manufacturers made in residential air-conditioning equipment?
12. The _____ starts the compressor in a residential air conditioner.
13. What component starts the indoor fan motor in residential air conditioners?
    A. Fan relay
    B. Thermistor
    C. Bimetallic element
    D. Motor starter
14. What component starts the outdoor fan motor in residential air conditioners?
    A. Fan-relay
    B. Compressor motor contactor
    C. Bimetallic element
    D. Sequencer

UNIT 40 Typical Operating Conditions

OBJECTIVES

After studying this unit, you should be able to

- explain what conditions will vary the evaporator pressures and temperatures.
- define how the various conditions in the evaporator and ambient air affect condenser performance.
- state the relationship of the evaporator to the rest of the system.
- describe the relationship of the condenser to the total system performance.
- 🌐compare high-efficiency equipment and standard-efficiency equipment.🌐
- establish reference points when working on unfamiliar equipment to know what the typical conditions should be.
- describe how humidity affects equipment suction and discharge pressure.
- 🌐explain three methods that manufacturers use to make air-conditioning equipment more efficient.🌐

SAFETY CHECKLIST

✔ The technician must use caution while observing equipment to obtain operating conditions. Electrical, pressure, and temperature readings must be taken while the technician observes the equipment, and many times the readings must be taken while the equipment is in operation.

✔ Wear goggles and gloves when attaching gages to a system.

✔ Observe all electrical safety precautions. Be careful at all times and use common sense.

Air-conditioning technicians must be able to evaluate both mechanical and electrical systems. The mechanical operating conditions are determined or evaluated with gages and thermometers; electrical conditions are determined with electrical instruments.

40.1 MECHANICAL OPERATING CONDITIONS

Air-conditioning equipment is designed to operate at its rated capacity and efficiency at one set of design conditions. This design condition is generally considered to be at an outside temperature of 95°F and at an inside temperature of 80°F

with a humidity of 50%. This rating is established by the Air Conditioning and Refrigeration Institute (ARI). Equipment must have a rating as a standard from which to work. Equipment is also rated so that the buyer will have a common basis with which to compare one piece of equipment with another. All equipment in the ARI directory is rated under the same conditions—80°F dry-bulb and 50% humidity indoors, and 95°F dry-bulb outdoors. When an estimator or buyer finds that a piece of equipment is rated at 3 tons, it will perform at a 3-ton capacity or 36,000 Btu/h under the stated conditions. When the conditions are different, the equipment will perform differently. Most occupants will not be comfortable at 80°F with a relative humidity of 50%. They will normally operate their system at about 75°F, and the relative humidity in the conditioned space will be close to 50%. The equipment will not have quite the capacity at 75°F that it had at 80°F. If the designer wants the system to have a capacity of 3 tons at 75°F and 50% relative humidity, the manufacturer's literature may be consulted to make the change in equipment choice. **This 75°F, 50% humidity condition will be used as the design condition for this unit because it is a common operating condition of the equipment in the field.** ARI rates condensers with 95°F air passing over them. A new standard-efficiency condenser will condense refrigerant at about 125°F with 95°F air passing over it. As the condenser ages, dirt accumulates on the outdoor coil, and the efficiency decreases. The refrigerant will then condense at a higher temperature. This higher temperature can easily approach 130°F. This is the value often found in the field. This unit uses a 125°F condensing temperature, assuming that equipment is properly maintained in the field. It is not easy to see the change in load conditions on your gages and instruments used in the field. An increase in humidity is not followed with a proportional rise in suction pressure and amperage. Perhaps the most noticeable aspect of an increase in humidity is an increase in the condensate accumulated in the condensate drain system.

40.2 RELATIVE HUMIDITY AND THE LOAD

The inside relative humidity adds a significant load to the evaporator coil and has to be considered as part of the load. When conditions vary from the design conditions, the equipment will vary in capacity. The pressures and temperatures will also change.

40.3 SYSTEM COMPONENT RELATIONSHIPS UNDER LOAD CHANGES

If the outside temperature increases from 95°F to 100°F, the equipment will be operating at a higher head pressure and will not have as much capacity. The capacity also varies when the space temperature goes up or down or when the humidity varies. There is a relationship among the various components in the system. The evaporator absorbs heat. When anything happens to increase the amount of heat that is absorbed into the system, the system pressures will rise. The condenser rejects heat. If anything happens to prevent the condenser from rejecting heat from the system, the system pressures will rise. The compressor pumps heat-laden vapor. Vapor at different pressure levels and saturation points (in reference to the amount of superheat) will hold different amounts of heat and will require different energy inputs.

40.4 EVAPORATOR OPERATING CONDITIONS

The evaporator normally will operate at a 40°F boiling temperature when operating at the 75°F 50% humidity condition. This will cause the suction pressure to be 70 psig for R-22. (The actual pressure corresponding to 40°F is 68.5 psig; for our purposes we will round this off to 70 psig.) If R-410A is used as the refrigerant, the suction pressure would be 118 psig for a 40°F evaporating temperature. This example is at design conditions and a steady-state load. At this condition the evaporator is boiling the refrigerant exactly as fast as the expansion device is metering it into the evaporator. As an example, suppose that the evaporator has a return air

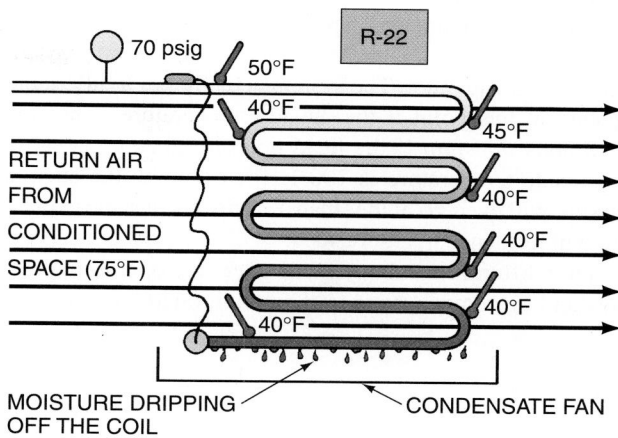

Figure 40–1   An evaporator operating at typical conditions. The refrigerant is boiling at 40°F in the coil. This corresponds to 68.5 psig, which is typically rounded off to 70 psig for R-22 and 118 psig for R-410A.

temperature of 75°F and the air has a relative humidity of 50%. The liquid refrigerant goes nearly to the end of the coil, and the coil has a superheat of 10°F. This coil is operating as intended at this typical condition, **Figure 40–1.**

Late in the day, after the sun has been shining on the house, the heat load inside becomes greater. A new condition is going to be established. The example evaporator in **Figure 40–2** has a fixed-bore metering device that will feed only a certain amount of liquid refrigerant. The space temperature in the

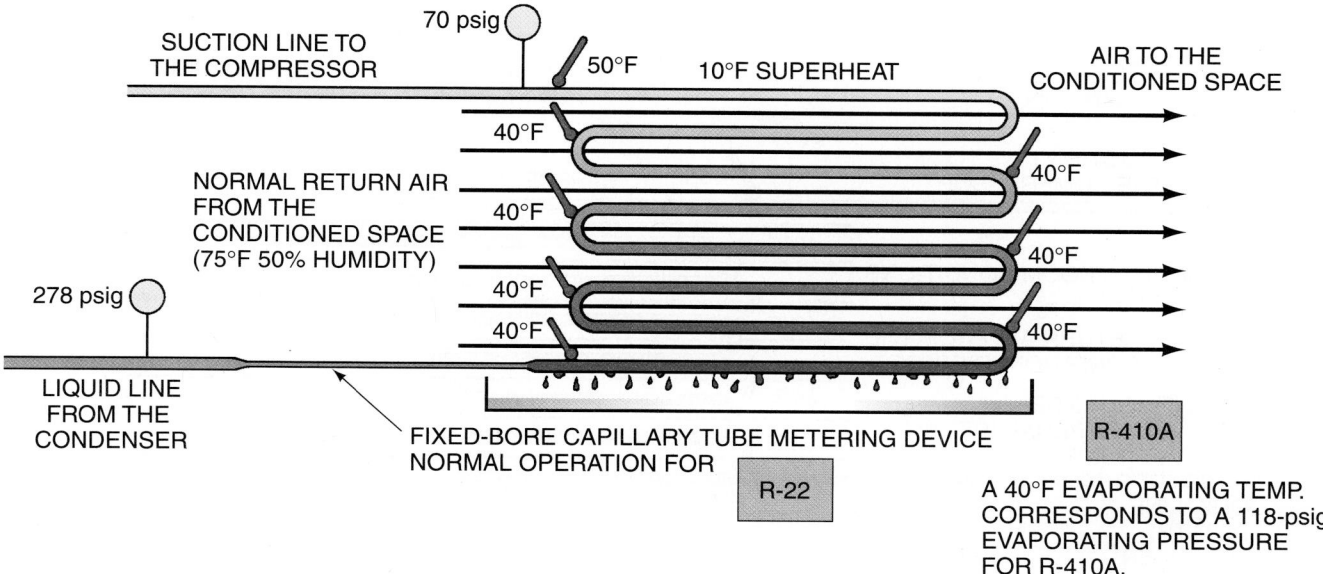

Figure 40–2   This fixed-bore metering device will not vary the amount of refrigerant feeding the evaporator as much as the thermostatic expansion valve. It is operating at an efficient state of 10°F superheat.

house has climbed to 77°F and is causing the liquid refrigerant in the evaporator to boil faster, **Figure 40–3.** This causes the suction pressure and the superheat to go up slightly. The new suction pressure is 73 psig, and the new superheat is 13°F. This is well within the range of typical operating conditions for an evaporator. The system actually has a little more capacity at this point if the outside temperature is not too much above design. If the head pressure goes up because the outside temperature is 100°F, for example, the suction pressure will even go higher than 73 psig, because head pressure will influence it, **Figure 40–4.**

Many different conditions will affect the operating pressures and air temperatures to the conditioned space. There can actually be as many different pressures as there are different inside and outside temperature and humidity variations. This can be confusing, particularly to the new service technician. However, the service technician can use some common conditions in troubleshooting. This is necessary because the technician rarely has a chance to work on a piece of equipment when the conditions are perfect. Most of the time when the technician is assigned the job, the system has been off for some time or not operating correctly for long enough that the conditioned space temperature and humidity are higher than normal, **Figure 40–5.** After all, a rise in space temperature is what often prompts a customer to call for service.

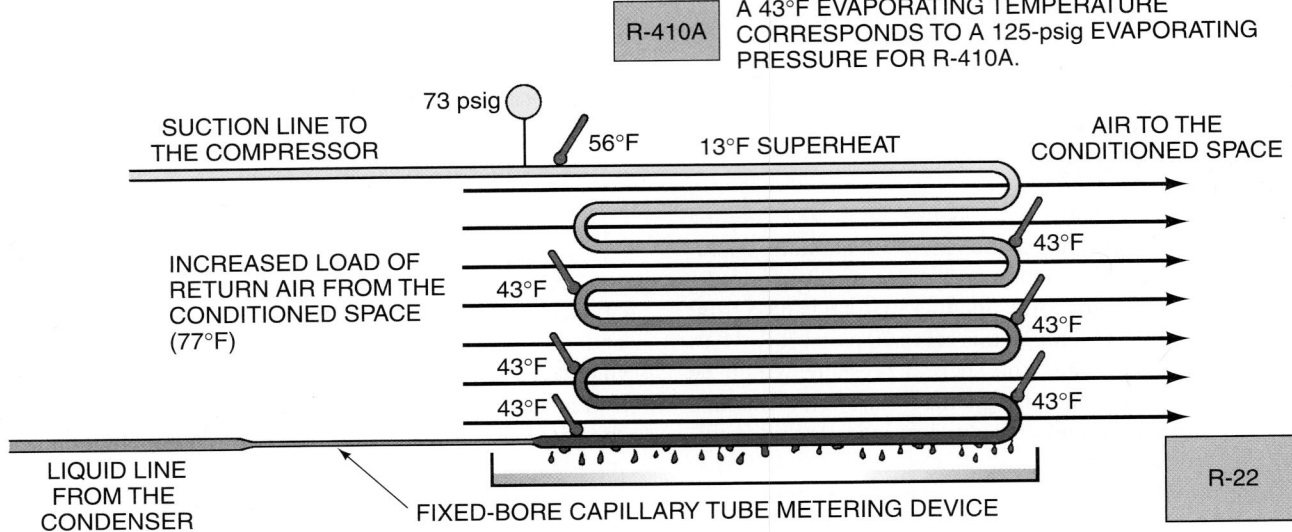

Figure 40–3   This is a fixed-bore metering device when an increase in load has caused the suction pressure to rise.

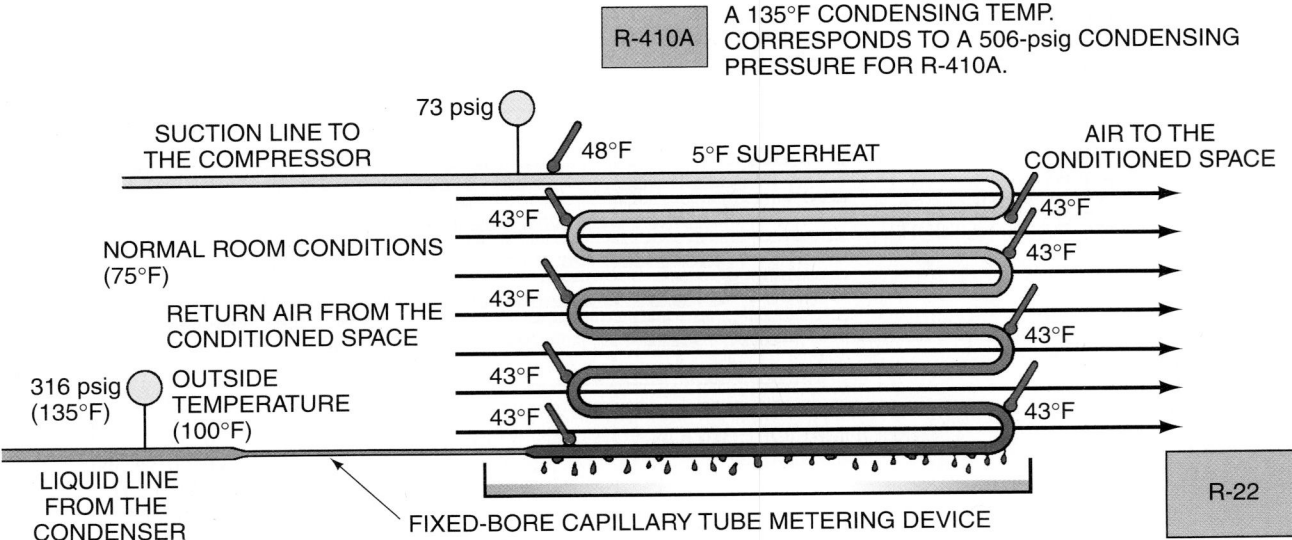

Figure 40–4   A system with a high head pressure. The increase in head pressure has increased the flow of refrigerant through the fixed-bore metering device, and the superheat is decreased. This system has a 135°F condensing temperature.

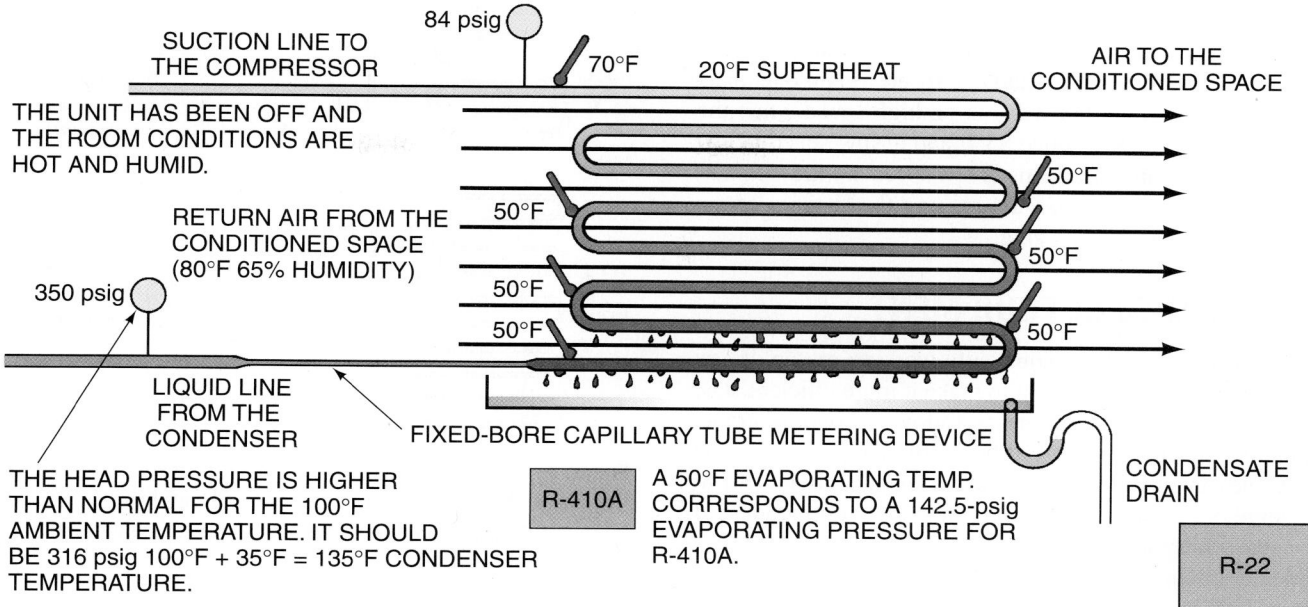

SUCTION LINE TO
THE COMPRESSOR

84 psig

70°F

20°F SUPERHEAT

AIR TO THE
CONDITIONED SPACE

THE UNIT HAS BEEN OFF AND
THE ROOM CONDITIONS ARE
HOT AND HUMID.

50°F

50°F

RETURN AIR FROM THE
CONDITIONED SPACE
(80°F 65% HUMIDITY)

50°F

50°F

50°F

50°F

350 psig

50°F

50°F

LIQUID LINE
FROM THE
CONDENSER

FIXED-BORE CAPILLARY TUBE METERING DEVICE

CONDENSATE
DRAIN

THE HEAD PRESSURE IS HIGHER
THAN NORMAL FOR THE 100°F
AMBIENT TEMPERATURE. IT SHOULD
BE 316 psig 100°F + 35°F = 135°F CONDENSER
TEMPERATURE.

R-410A

A 50°F EVAPORATING TEMP.
CORRESPONDS TO A 142.5-psig
EVAPORATING PRESSURE FOR
R-410A.

R-22

Figure 40–5   This system has been off long enough that the temperature and the humidity inside the conditioned space have gone up. Notice the excess moisture forming on the coil and going down the drain.

40.5   HIGH EVAPORATOR LOAD AND A COOL CONDENSER

High temperature in the conditioned space is not the only thing that will cause the system to have different pressures and capacity. The reverse can happen if the inside temperature is warm and the outside temperature is cooler than normal, **Figure 40–6.** For example, before going to work, a couple may turn the air conditioner off to save electricity. They may not get home until after dark to turn the air conditioner back on. By this time it may be 75°F outside but still be 80°F inside the structure. The air passing over the condenser is now cooler than the air passing over the evaporator.

The evaporator may also have a large humidity load. The condenser becomes so efficient that it will hold some of the charge because it starts to condense refrigerant in the first part of the condenser. Thus, more of the refrigerant charge is in the condenser tubes, and this will slightly starve the evaporator. The system may not have enough capacity to cool the home for several hours. The condenser may hold back enough refrigerant to cause the evaporator to operate below freezing and freeze up before it can cool the house and satisfy the thermostat. This condition has been improved by some manufacturers by means of a variable-speed fan operated through an inverter or by two-speed fan operation for the condenser fan. The fan is controlled by a

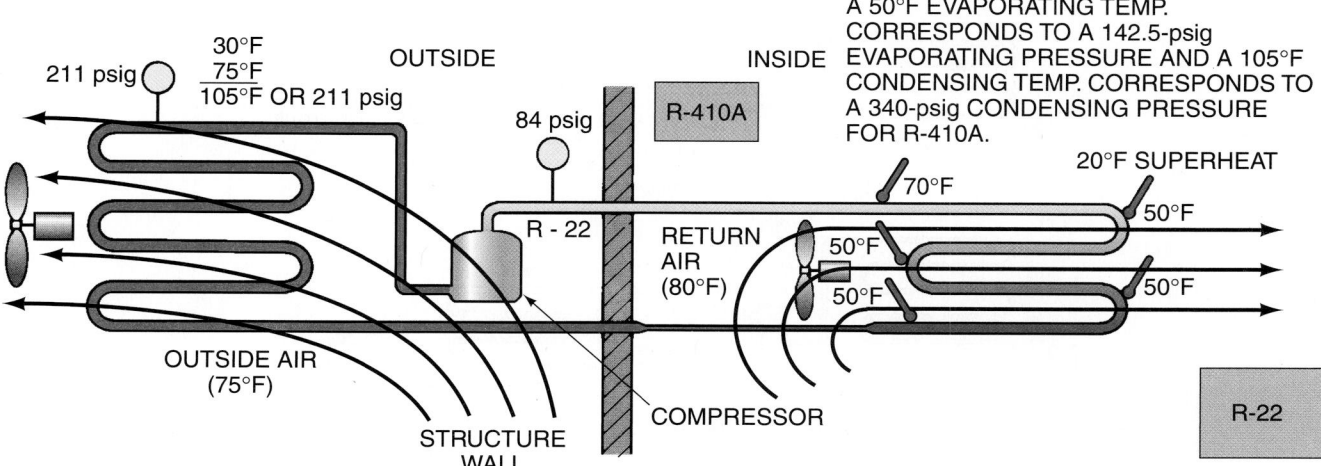

211 psig

30°F
75°F
105°F OR 211 psig

OUTSIDE

INSIDE

A 50°F EVAPORATING TEMP.
CORRESPONDS TO A 142.5-psig
EVAPORATING PRESSURE AND A 105°F
CONDENSING TEMP. CORRESPONDS TO
A 340-psig CONDENSING PRESSURE
FOR R-410A.

R-410A

84 psig

R - 22

20°F SUPERHEAT

70°F

50°F

RETURN
AIR
(80°F)

50°F

50°F

50°F

OUTSIDE AIR
(75°F)

COMPRESSOR

R-22

STRUCTURE
WALL

Figure 40–6   A system operating when the outside ambient air is cooler than the inside space temperature air. As the return air cools down, the suction pressure and boiling temperature will decrease.

modulating controller and fan motor or a single-pole–double-throw thermostat that operates the fan at high speeds (when the outdoor temperature is high) and low speeds (in mild temperatures). Typically, the fan will run on high speed when the outdoor temperature is above 85°F and at low speed below 85°F. When the fan is operating at low speed in mild weather, the head pressure is increased, reducing the chances of the evaporator condensate freezing.

40.6 GRADES OF EQUIPMENT

Manufacturers have been constantly working on the design of air-conditioning equipment to make it more efficient. The three grades of equipment are as follows: economy grade, standard-efficiency grade, and high-efficiency grade. Some companies manufacture all three grades and offer them to the supplier. Some manufacturers offer only one grade and may take offense at someone calling their equipment the lower grade. Economy grade and standard-efficiency grade are about equal in efficiency, but the materials they are made of and their appearances are different. 🌐 High-efficiency equipment may be much more efficient and will not have the same operating characteristics, 🌐 **Figure 40–7**. A condenser normally will condense the refrigerant at a temperature of about 30°F to 35°F higher than the ambient temperature. For example, when the outside temperature is 95°F, the average condenser will condense the refrigerant at 125°F to 130°F, and the head pressure for these condensing temperatures would be 278 to 297 psig for R-22 or 446 to 475 psig for R-410A. 🌐 High-efficiency air-conditioning equipment may have a much lower operating head pressure. The high efficiency is gained by using a larger condenser surface or by using more modern alloys with more extended surfaces or fins. The condensing temperature may be as low as 20°F greater than the outdoor ambient temperature. This would bring the head pressure down to a temperature corresponding to 115°F or 243 psig for R-22 and 390 psig for R-410A. The compressor will

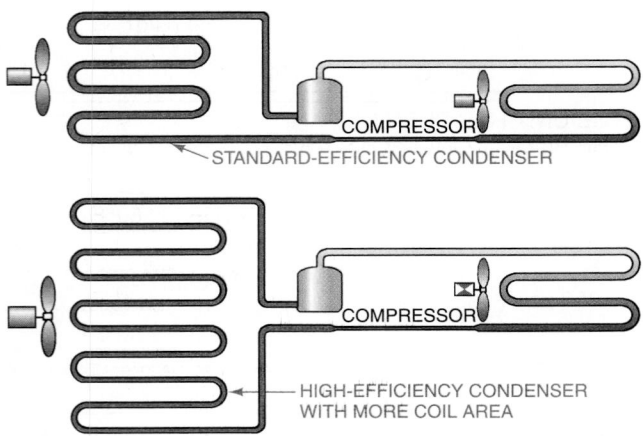

Figure 40–7 A standard-efficiency condenser and a high-efficiency condenser. The high-efficiency condenser will have more surface area and may be made of a different metal or alloy with more extended surfaces or fins.

not use as much power at this condition because of the lower compression ratio, **Figure 40–8**. When large condensers are used, the head pressure is reduced. Lower power requirements are a result. More efficient condensers require some means of head pressure control, such as that mentioned previously. 🌐

This discussion about condensers and their typical operating conditions holds true when the air-conditioning system is operating at a typical load. However, when high evaporator heat loads are experienced, the temperature difference between the condenser and the ambient will be much higher. The reason for this is simple. As the evaporator absorbs heat, it must reject that heat in the condenser. As more heat is absorbed in the evaporator, more heat must be rejected in the condenser. However, for a fixed-sized condenser and a relatively constant outside ambient, this amount of heat

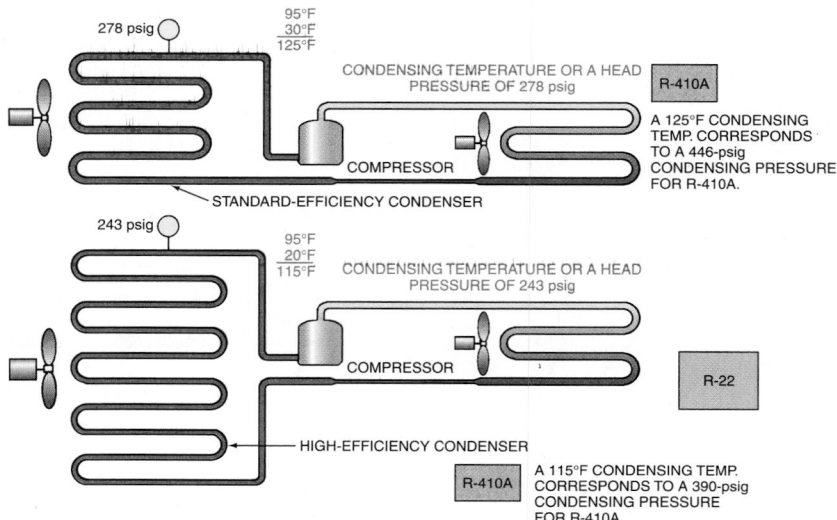

Figure 40–8 A standard-efficiency unit and a high-efficiency unit.

rejection cannot be accomplished at a 20°F temperature difference for a high-efficiency condenser or at a 30°F temperature difference for a standard condenser. Because of this, excess heat starts to accumulate in the condenser, and the condensing temperature increases. However, after the condensing temperature increases, a greater temperature difference between the ambient and the condensing temperature exists. This temperature difference increases the heat transfer between the condenser and the ambient, and the condenser will be able to reject the higher evaporator loads. However, it will be rejecting this higher heat load at a higher condensing temperature and higher compression ratio.

The same is true when low evaporator loads are experienced. However, the condenser will operate at a lower temperature difference between itself and the ambient because of the reduced load.

40.7 DOCUMENTATION WITH THE UNIT

The technician needs to know what the typical operating pressures should be at different conditions. Some manufacturers furnish a chart with the unit to tell what the suction and discharge pressures should be at different conditions, **Figure 40–9** and **Figure 40–10**. Some publish a bulletin that lists all of their equipment along with the typical operating pressures and temperatures. Others furnish this information with the unit in the installation and start-up manual, **Figure 40–11**. The homeowner may have this booklet, which may be helpful to the technician.

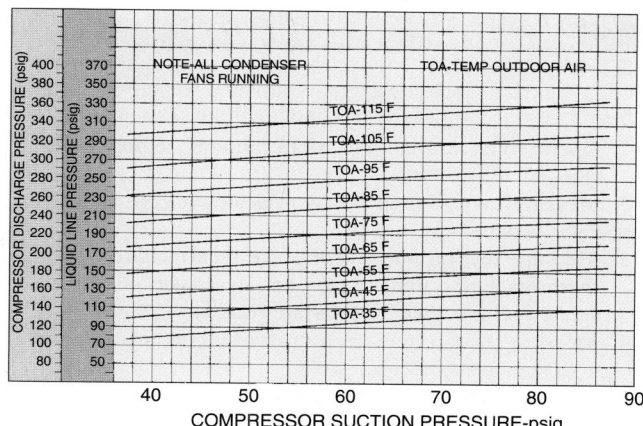

Figure 40–10 🌐 A page from a manufacturer's bulletin explaining how to charge one model of equipment. 🌐 *Reproduced courtesy of Carrier Corporation*

Three things must be considered when the manufacturer publishes typical operating conditions:
1. The load on the outdoor coil, which is influenced by the outdoor temperature and the evaporator heat loading.
2. The sensible-heat load on the indoor coil, which is influenced by the indoor dry-bulb temperature.
3. The latent-heat load on the indoor coil, which is influenced by the humidity. The humidity is determined by taking the wet-bulb temperature of the indoor air.

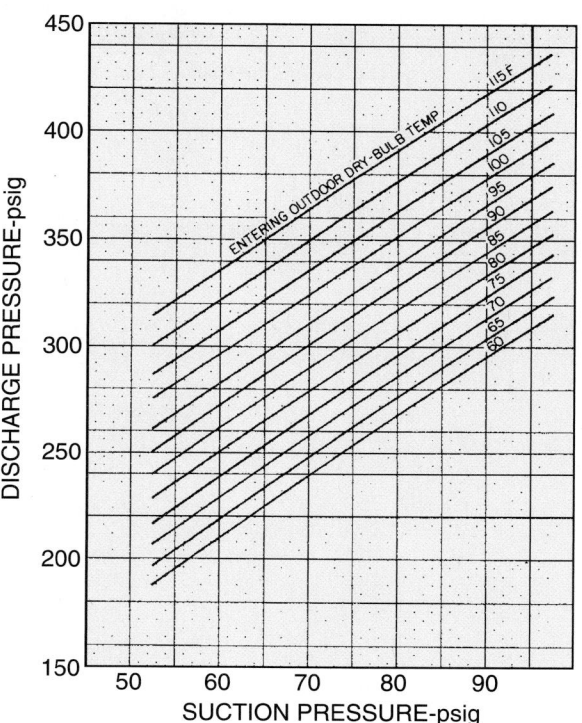

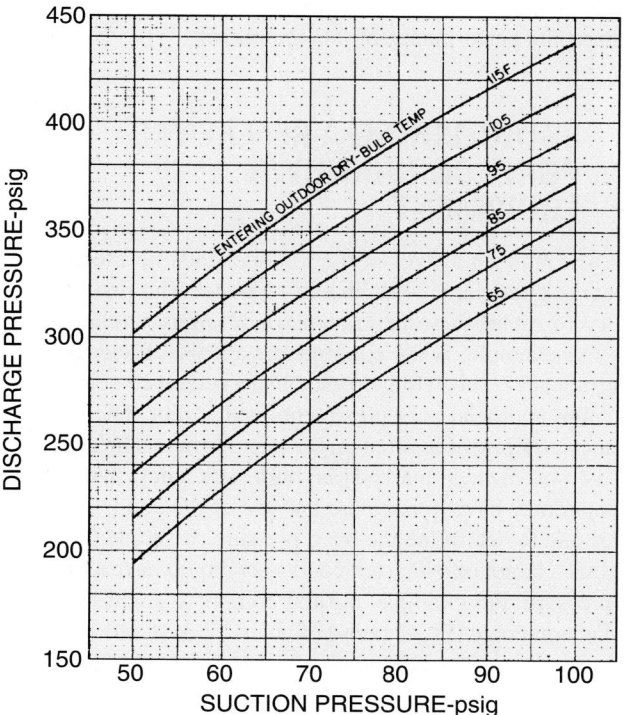

Figure 40–9 🌐 Charts are furnished by some manufacturers to check unit performance. 🌐 *Reproduced courtesy of Carrier Corporation*

REFRIGERANT CHARGING

SAFETY PRECAUTION: to prevent personal injury, wear safety glasses and gloves when handling refrigerant. Do not overcharge system. This can cause compressor flooding.

1. Operate unit a minimum of 15 minutes before checking charge.
2. Measure suction pressure by attaching a gage to suction valve service port.
3. Measure suction line temperature by attaching a service thermometer to unit suction line near suction valve. Insulate thermometer for accurate readings.
4. Measure outdoor coil inlet air dry-bulb temperature with a second thermometer.
5. Measure indoor coil inlet air wet-bulb temperature with a sling psychrometer.
6. Refer to table. Find air temperature entering outdoor coil and wet-bulb temperature entering indoor coil. At this intersection note the superheat.
7. If unit has higher suction line temperature than charted temperature, add refrigerant until charted temperature is reached.
8. If unit has lower suction line temperature than charted temperature, bleed refrigerant until charted temperature is reached.
9. If air temperature entering outdoor coil or pressure at suction valve changes, charge to new suction line temperature indicated on chart.
10. This procedure is valid, independent of indoor air quantity.

SUPERHEAT CHARGING TABLE
(SUPERHEAT ENTERING SUCTION SERVICE VALVE)

Outdoor Temp (°F)	INDOOR COIL ENTERING AIR °F WB													
	50	52	54	56	58	60	62	64	66	68	70	72	74	76
55	9	12	14	17	20	23	26	29	32	35	37	40	42	45
60	7	10	12	15	18	21	24	27	30	33	35	38	40	43
65	—	6	10	13	16	19	21	24	27	30	33	36	38	41
70	—	—	7	10	13	16	19	21	24	27	30	33	36	39
75	—	—	—	6	9	12	15	18	21	24	28	31	34	37
80	—	—	—	—	5	8	12	15	18	21	25	28	31	35
85	—	—	—	—	—	—	8	11	15	19	22	26	30	33
90	—	—	—	—	—	—	5	9	13	16	20	24	27	31
95	—	—	—	—	—	—	—	6	10	14	18	22	25	29
100	—	—	—	—	—	—	—	—	8	12	15	20	23	27
105	—	—	—	—	—	—	—	—	5	9	13	17	22	26
110	—	—	—	—	—	—	—	—	—	6	11	15	20	25
115	—	—	—	—	—	—	—	—	—	—	8	14	18	23

Figure 40–11 🌐A chart furnished by a manufacturer in the installation and start-up literature.🌐 *Reproduced courtesy of Carrier Corporation*

The manufacturer would require the technician to record these temperatures and plot them against a graph or table to determine the performance of the equipment.

40.8 ESTABLISHING A REFERENCE POINT ON UNKNOWN EQUIPMENT

When a technician arrives at the job and finds no literature and cannot obtain any, what should be done? The first thing is to try to establish some known condition as a reference point. For example, is the equipment standard efficiency or high efficiency? This will help establish a reference point for the suction and head pressures mentioned earlier. The high-efficiency equipment is often larger than normal. The equipment is not always marked as high efficiency, and the technician may have to compare the size of the condenser to another one to determine the head pressure. It should be obvious that a larger or oversized condenser would have a lower head pressure. For example, a 3-ton compressor will have a full-load amperage (FLA) rating of about 17 A at 230 V. The amperage rating of the compressor may help to determine the rating of the equipment, **Figure 40–12**. Although a 3-ton high-efficiency piece of equipment will have a lesser amperage rating than a standard piece of equipment, the ratings will be

APPROXIMATE FULL-LOAD AMPERAGE VALUES FOR ALTERNATING-CURRENT MOTORS

Motor	Single Phase		3-Phase Squirrel Cage Induction		
HP	115 V	230 V	230 V	460 V	575 V
$\frac{1}{6}$	4.4	2.2			
$\frac{1}{4}$	5.8	2.9			
$\frac{1}{3}$	7.2	3.6			
$\frac{1}{2}$	9.8	4.9	2	1.0	0.8
$\frac{3}{4}$	13.8	6.9	2.8	1.4	1.1
1	16	8	3.6	1.8	1.4
$1\frac{1}{2}$	20	10	5.2	2.6	2.1
2	24	12	6.8	3.4	2.7
3	34	17	9.6	4.8	3.9
5	56	28	15.2	7.6	6.1
$7\frac{1}{2}$			22	11.0	9.0
10			28	14.0	11.0

Does not include shaded pole.

Figure 40–12 A table of current ratings for different-sized motors at different voltages. *Courtesy BDP Company*

close enough to compare and to determine the capacity of the equipment. If the condenser is very large for an amperage rating that should be 3 tons, the equipment is probably high efficiency, and **the head pressure will not be as high as in a standard piece of equipment.**

40.9 METERING DEVICES FOR HIGH-EFFICIENCY EQUIPMENT

High-efficiency air-conditioning equipment often uses a thermostatic expansion valve rather than a fixed-bore metering device because a certain amount of efficiency may be gained. The evaporator may also be larger than normal. An oversized evaporator will add to the efficiency of the system. It is more difficult to make a determination regarding the evaporator size because, being enclosed in a casing or the ductwork, it is not as easy to see.

The operating conditions of a system can vary so much that all possibilities cannot be covered; however, a few general statements committed to memory may help. Standard-efficiency equipment and high-efficiency equipment will follow the general conditions listed in the following subsections.

Operating Conditions Near Design Space Conditions for Standard-Efficiency Equipment

1. The suction temperature is 40°F (70 psig for R-22 and 118 psig for R-410A), **Figure 40–13.**
2. The head pressure should correspond to a temperature of no more than 35°F above the outside ambient temperature. This would be 297 psig for R-22 and 475 psig for R-410A when the outside temperature is 95°F (95°F + 35°F = 130°F condensing temperature).

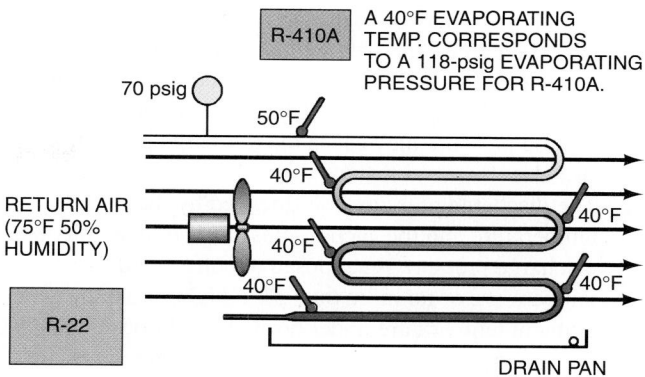

Figure 40–13 An evaporator operating at close-to-design conditions.

A typical condition would be condensing at 30°F above the ambient, or 95°F + 30°F = 125°F, **Figure 40–14.** The head pressure would be 278 psig for R-22 and 446 psig for R-410A.

Space Temperature Higher Than Normal for Standard-Efficiency Equipment

1. The suction pressure will be higher than normal. Normally, the refrigerant boiling temperature is about 35°F cooler than the entering-air temperature. (Recall the relationship of the evaporator to the entering air temperature described in the refrigeration section.) When the conditions are normal, the refrigerant boiling temperature would be 40°F when the return air temperature is 75°F. This is true when the humidity is normal. When the space temperature is higher than normal because the equipment has been off for a long time, the return air temperature may go up to 85°F, and

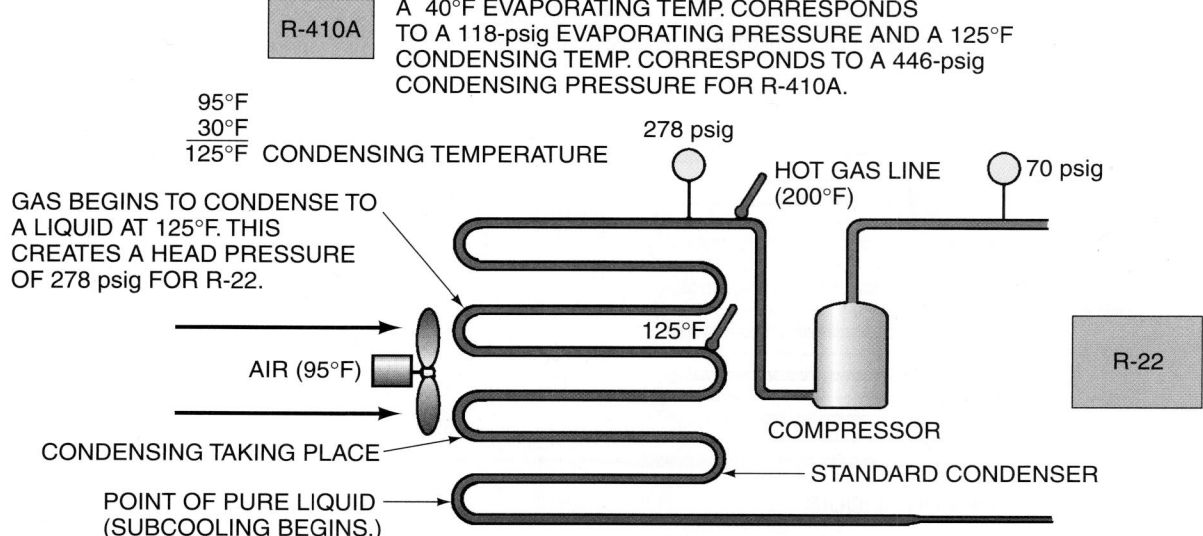

Figure 40–14 The normal conditions of a standard condenser operating on a 95°F day.

the humidity may be high also. The suction temperature may then go up to 50°F, with a corresponding pressure of 84 to 93 psig for R-22 and 142 to 156 psig for R-410A, **Figure 40–15**. This higher-than-normal suction pressure may cause the discharge pressure to rise also.

2. The discharge pressure is influenced by the outside temperature and the suction pressure. For example, the discharge pressure is supposed to correspond to a temperature of no more than 30°F higher than the ambient temperature under normal conditions. When the suction pressure is 80 psig, the discharge pressure is going to rise accordingly. It may move up to a new condensing temperature 10°F higher than normal. This would mean that the discharge pressure for R-22 could be 317 psig (95°F + 30°F + 10°F = 135°F) or 506 psig

for R-410A while the suction pressure is high, **Figure 40–16**. When the unit begins to reduce the space temperature and humidity, the evaporator pressure will begin to reduce. The load on the evaporator is reduced. The head pressure will come down also.

Operating Conditions Near Design Conditions for High-Efficiency Equipment

1. The evaporator design temperature may in some cases operate at a slightly higher pressure and temperature on high-efficiency equipment because the evaporator is larger. The boiling refrigerant temperature may be 45°F with the larger evaporator at design conditions, and this would create a suction pressure of 76 psig for R-22 and 130 psig for R-410A and be normal, **Figure 40–17**.

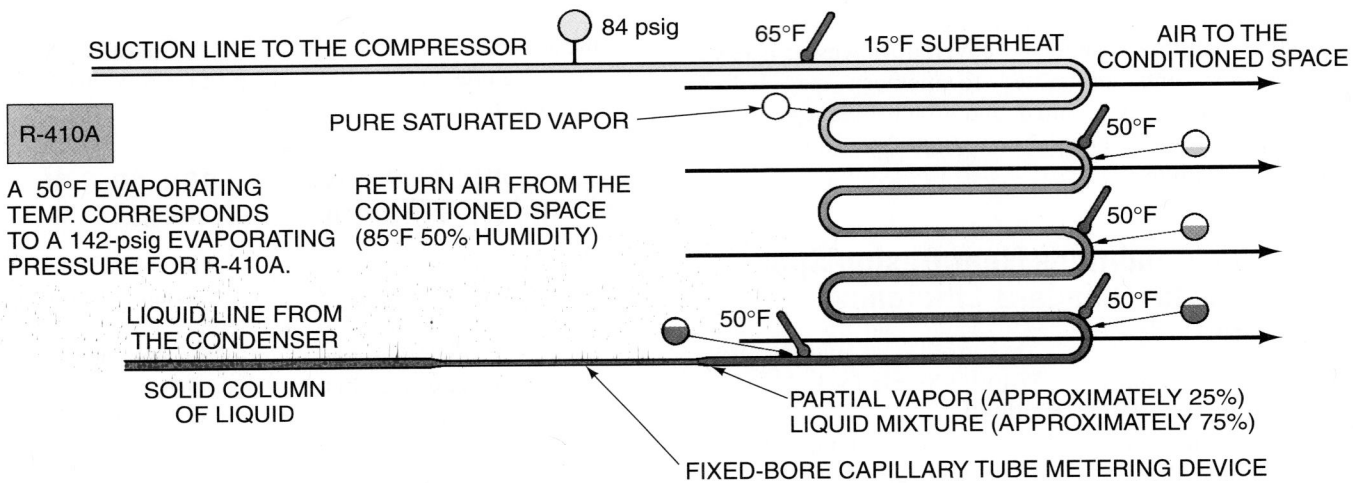

Figure 40–15   Pressures and temperatures as they may occur with an evaporator when the space temperature and humidity are above design conditions.

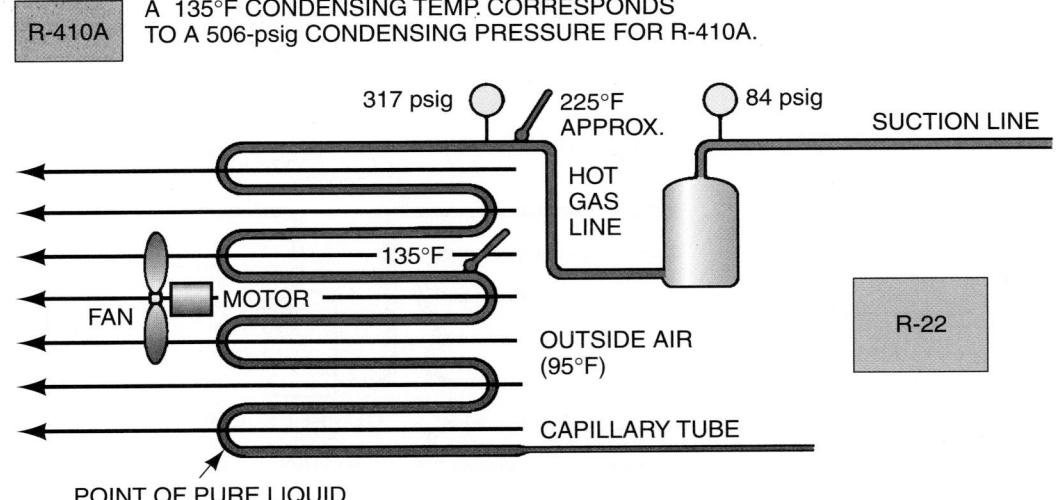

Figure 40–16   The condenser is operating at the design condition as far as the outdoor ambient air is concerned, but the pressure is high because the evaporator is under a load that is above design.

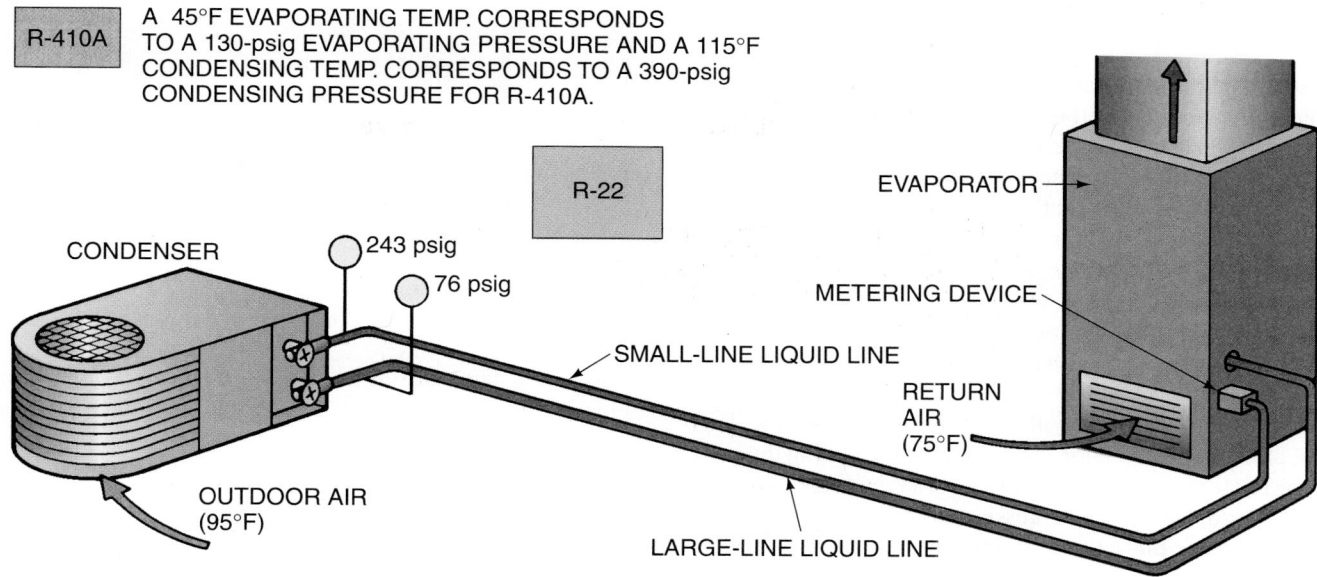

Figure 40–17 🌐 High-efficiency system operating conditions—45°F evaporator temperature and 115°F condensing temperature. 🌐

2. The refrigerant may condense at a temperature as low as 20°F more than the outside ambient temperature. For a 95°F day with R-22, the head pressure may be as low as 243 psig and as low as 390 psig for R-410A, **Figure 40–17.** If the condensing temperature were as high as 30°F above the ambient temperature, you should suspect a problem. For example, the head pressure should not be more than 277 psig on a 95°F day for R-22 and 446 psig for R-410A.

Other-Than-Design Conditions for High-Efficiency Equipment

1. When the unit has been off long enough for the load to build up, the space temperature and humidity are above design conditions, and the high-efficiency system

pressures will be higher than normal, as they would in the standard-efficiency system. With standard equipment, when the return air temperature is 75°F and the humidity is approximately 50%, the refrigerant boils at about 40°F. This is a temperature difference of 35°F. A high-efficiency evaporator is larger, and the refrigerant may boil at a temperature difference of 30°F, or around 55°F when the space temperature is 85°F, **Figure 40–18.** The exact boiling temperature relationship depends on the manufacturer's design and how much coil surface area was selected.

2. The high-efficiency condenser, like the evaporator, will operate at a higher pressure when the load is increased. The head pressure will not be as high as it would with a standard-efficiency condenser because the condenser has extra surface area.

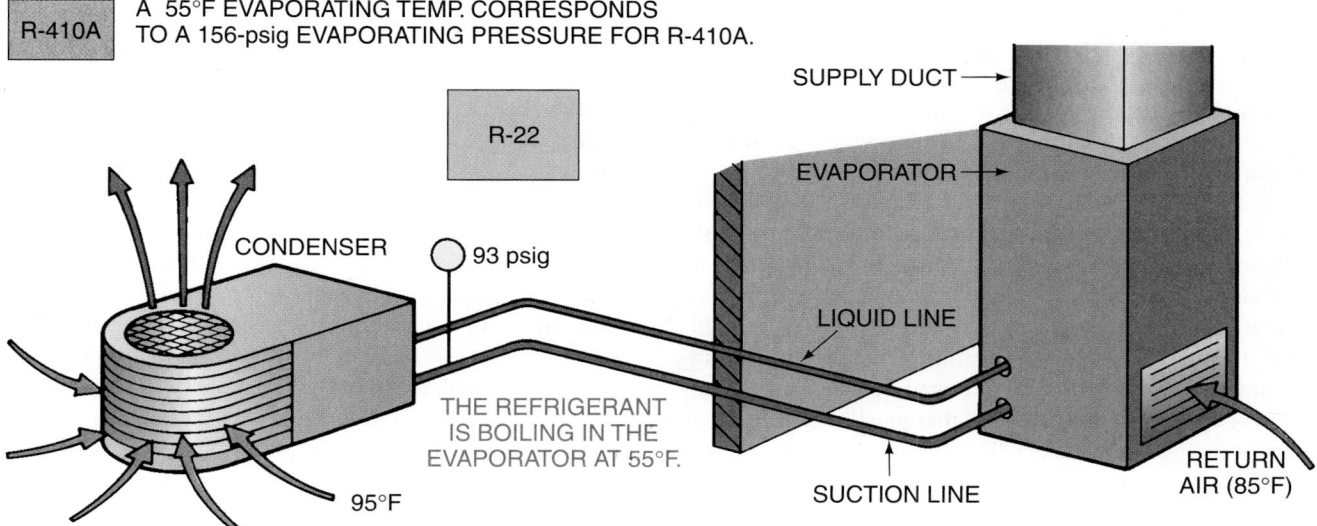

Figure 40–18 An evaporator in a high-efficiency system with a load above design conditions.

The capacity of high-efficiency systems is not up to the rated capacity when the outdoor temperature is much below design. The earlier example of the family that shut off the air-conditioning system before going to work and then turned it back on when they came home from work will be much worse with a high-efficiency system. The fact that the condenser became too efficient at night when the air was cooler with a standard condenser will be much more evident with the larger high-efficiency condenser. Most manufacturers produce variable-speed or two-speed condenser fans so that lower fan speeds can be used in mild weather to help compensate for this temperature difference. The service technician who tries to analyze a component of high-efficiency equipment on a mild day will find the head pressure low. This will also cause the suction pressure to be low. Using the coil-to-air relationships for the condenser and the evaporator will help determine the correct pressures and temperatures.

Remember these two statements:

1. The evaporator absorbs heat, which is related to its operating pressures and temperatures.
2. The condenser rejects heat and has a predictable relationship to the evaporator load and outdoor ambient temperature.

40.10 EQUIPMENT EFFICIENCY RATING

Manufacturers have a method of rating equipment so that the designer and the owner can tell high efficiency from low efficiency at a glance. This rating was originally called the EER rating. EER stands for energy efficiency ratio and is actually the output in Btu/h divided by the input in watts of power used to produce the output. For example, a system may have an output of 36,000 Btu/h with an input of 4000 W.

$$\frac{36,000 \text{ Btu/h}}{4000 \text{ W}} = \text{an EER of } 9$$

The larger the EER rating, the more efficient the equipment. For example, suppose that the 36,000 Btu/h air conditioner only required 3600 W input.

$$\frac{36,000 \text{ Btu/h}}{3600 \text{ Btu/h}} = \text{an EER of } 10$$

The customer is getting the same capacity using less power. The equipment is more efficient. **The larger the EER rating, the more efficient the equipment.**

The EER rating is a steady-state rating and does not account for the time the unit operates before reaching peak efficiency. This operating time has an unknown efficiency. It also does not account for shutting the system down at the end of the cycle (when the thermostat satisfies), leaving a cold coil in the duct. The cold coil continues to absorb heat from the surroundings, not the conditioned space. Refrigerant pressures equalize to the cold coil that must be pumped out at the beginning of the next cycle. This accounts for some of the inefficiency at the beginning of the cycle. Some manufacturers have controls to keep the fan running at the end of the cycle to take advantage of this cold-coil heat exchange.

MODEL	CAPACITY (Btu/x)	SEER
A	24,000	9.00
B	24,000	10.00
C	24,000	10.50
D	24,000	11.00
E	24,000	11.50
F	24,000	12.00

Figure 40–19 Examples of SEER ratings.

The indoor fan will run for several minutes after the compressor shuts off. This is accomplished with the electronic circuit board that controls the fan operation.

The picture is not complete using the EER rating system, so a rating of seasonal efficiency has been developed, called seasonal energy efficiency ratio (SEER). This rating is tested and verified by a rating agency and includes the start-up and shutdown cycles. ⊕ The government has been encouraging industry to go to equipment ratings of 13 SEER. In fact, 13-SEER equipment is now mandatory. Equipment manufactured with an efficiency rating of 12 SEER is common. ⊕ This equipment has a higher first cost and not everyone will make this investment, so lower-efficiency equipment is still popular. Speculative home builders have a tendency to install the lower price and lower-efficiency equipment for quicker sale of the home. The rating agency is the ARI. Ratings of manufacturers' equipment are published by ARI, and these manufacturers list the ratings in their catalogs. A typical rating may look like **Figure 40–19.**

40.11 TYPICAL ELECTRICAL OPERATING CONDITIONS

The electrical operating conditions are measured with a volt-ohmmeter and an ammeter. Three major power-consuming devices may have to be analyzed from time to time: the indoor fan motor, the outdoor fan motor, and the compressor. The control circuit is considered a separate function.

The starting point for considering electrical operating conditions is to know what the system supply voltage is supposed to be. For residential units, 230 V single-phase is the typical voltage. Light-commercial equipment will nearly always use 208 V or 230 V single-phase or three-phase. Single- and three-phase power may be obtained from a three-phase power supply. The equipment rating may be 208/230 V. The reason for the two different ratings is that 208 V is the supply voltage that some power companies provide, and 230 V is the supply voltage provided by other power companies. Some light-commercial equipment may have a supply voltage of 460 V three-phase if the equipment is installed at a large commercial installation. For example, an office may have a 3- or 4-ton air-conditioning unit that operates separately from the main central system. If the supply voltage is 460 V three-phase, the small unit may operate from the same power supply. When 208/230-V equipment is used at a commercial installation, the compressor may be three-phase, and the fan motors single-phase. The number of phases that the power company furnishes makes a difference in the method of starting the compressor. Single-phase compressors may have a start assist, such as the positive temperature coefficient device or a start relay and start capacitor. Three-phase compressors will have no start-assist accessories.

40.12   MATCHING THE UNIT TO THE CORRECT POWER SUPPLY

The typical operating voltages for any air-conditioning system must be within the manufacturer's specifications. This is ±10% of the rated voltage. For the 208/230-V motor the minimum allowable operating voltage would be 208 × 0.90 = 187.2 V. The maximum allowable operating voltage would be 230 × 1.10 = 253 V. Notice that the calculations used the 208 V for the base for figuring the low voltage and the 230-V rating for calculating the highest voltage. This is because this application is 208/230 V. If the motor is rated at 208 V or 230 V alone, that value (208 V or 230 V) is used for evaluating the voltage. The equipment may be started under some conditions that are beyond the rated conditions. The technician must use some judgment. For example, if 180 V is measured, the motor should not be started because the voltage will drop further with the current draw of the motor. If the voltage reads 260 V, the motor may be started because the voltage may drop slightly when the motor is started. If the voltage drops to within the limits, the motor is allowed to run.

40.13   STARTING THE EQUIPMENT WITH THE CORRECT DATA

When the correct rated voltage is known and the minimum and maximum voltages are determined, the equipment may be started if the voltages are within the limits. The three motors—indoor fan, outdoor fan, and the compressor—can be checked for the correct current draw.

The indoor fan is building the air pressure to move the air through the ductwork, filters, and grilles to the conditioned space. By law, the voltage characteristics must be printed on the motor in such a manner that they will not come off. This information may be printed on the motor, but the motor might be mounted so that the data cannot be easily seen. In some cases the motor may be inside the squirrel cage blower. If so, removing the motor is the only way of determining the fan current. When the supplier can be easily contacted, you may obtain the information there. If the motor electrical characteristics cannot be obtained from the nameplate, you might be able to get them from the unit nameplate. However, the fan motor may have been changed to a larger motor for more fan capacity.

40.14   FINDING A POINT OF REFERENCE FOR AN UNKNOWN MOTOR RATING

When a motor is mounted so that the electrical characteristics cannot be determined, you must improvise. We know that air-conditioning systems normally move about 400 cfm of air per ton. This can help you determine the amperage of the indoor fan motor—you can compare the fan amperage on an unknown system with the amperage of a known system. All you need is the approximate system capacity. As discussed earlier, you can find this by comparing compressor amperages of the unit in question and a known unit. For example, the compressor amperage of a 3-ton system is about 17 A when operating on 230 V. If you notice that the amperage of the compressor in the system you are checking is 17 A, you can assume that the system is close to 3 tons. The indoor fan motor for a 3-ton system should be about 1/3 hp for a typical duct system. The fan motor amperage for a 1/3-hp permanent split-capacitor motor is 3.6 A at 230 V. If the fan in question were pulling 5 A, suspect a problem.

Fan motors may be shipped in a warm-air furnace and may have been changed if air conditioning was added at a later date. In this case the furnace nameplate may not give the correct fan motor data. The condensing unit will have a nameplate for the condenser fan motor. This motor should be sized fairly close to its actual load and should pull close to nameplate amperage.

40.15   DETERMINING THE COMPRESSOR RUNNING AMPERAGE

The compressor current draw may not be as easy to determine as the fan motor current draw because compressor manufacturers do not all stamp the compressor run-load amperage (RLA) rating on the compressor nameplate. Because so many different compressor sizes exist, it is hard to state the correct full-load amperage. For example, motors normally come in the following increments: 1, 1 1/2, 2, 3, and 5 hp. A unit rated at 34,000 Btu/h is called a 3-ton unit, although it actually takes 36,000 Btu/h to be a 3-ton unit. Ratings that are not completely accurate are known as *nominal ratings*. They are rounded off to the closest rating. A typical 3-ton air-conditioning unit would have a 3-hp compressor motor. A unit with a rating of 34,000 Btu/h does not need a full 3-hp motor, but it is supplied with one because there is no standard horsepower motor to meet its needs. If the motor amperage for a 3-hp motor were stamped on the compressor, it could cause confusion because the motor may never operate at that amperage. A unit nameplate lists electrical information, **Figure 40–20.** NOTE: The manufacturer may stamp the compressor RLA on the unit nameplate. This amperage should not be exceeded.

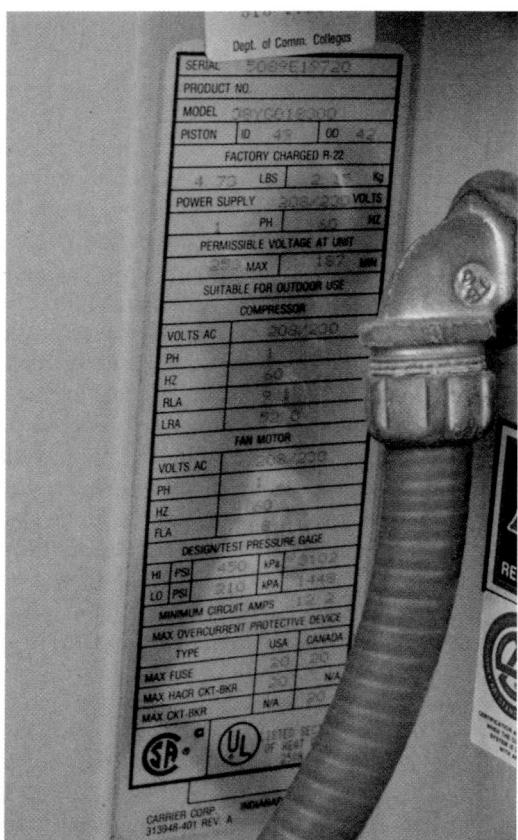

Figure 40–20   A condensing unit nameplate. *Photo by Bill Johnson*

40.16   COMPRESSORS OPERATING AT FULL-LOAD CURRENT

It is rare for a compressor motor to operate at its RLA rating. If design or above-design conditions were in effect, the compressor would operate at close to full load. When the unit is operating at a condition greater than design, such as when the unit has been off for some time in very hot weather, it might appear that the compressor is operating at more than RLA. However, usually other conditions keep the compressor from drawing too much current. The compressor is pumping vapor, and vapor is very light. It takes a substantial increase in pressure difference to create a significantly greater workload.

40.17   HIGH VOLTAGE, THE COMPRESSOR, AND CURRENT DRAW

A motor operating at a voltage higher than the voltage rating of the motor is a condition that will prevent the motor from drawing too much current. A motor rated at 208/230 V has an amperage rating at some value between the amperage at 208 and 230 V. Therefore, if the voltage is 230 V, the amperage may be lower than the nameplate amperage even during

overload. The compressor motor may be larger than needed most of the operating time and may not reach its rated horsepower until it gets to the maximum design rating for the system. It may be designed to operate at 105°F or 115°F outdoor ambient for very hot regions, but when the unit is rated at 3 tons, it would be rated down at the higher temperatures. The unit nameplate may contain the compressor amperage at the highest operating condition for which the unit is rated, 115°F ambient temperature.

40.18   CURRENT DRAW AND THE TWO-SPEED COMPRESSOR

Some air-conditioning manufacturers use two-speed compressors to achieve better seasonal efficiencies. These compressors may use a motor capable of operating as a two-pole or a four-pole motor. A four-pole motor runs at 1800 rpm, and a two-pole motor runs at 3600 rpm. A motor control circuit will slow the motor to reduced speed for mild weather.

Variable speed may be obtained with electronic circuits. The system's efficiency is greatly improved because there is no need to stop the compressor and restart it at the beginning of the next cycle. However, this means more factors for a technician to establish typical running conditions under other-than-design conditions. The equipment should perform just as typical high-efficiency equipment would when operating at design conditions.

SAFETY PRECAUTION: The technician must use caution while observing equipment to obtain the operating conditions. Electrical, pressure, and temperature readings must be taken while the technician observes the equipment. Many times the readings must be made while the equipment is in operation.

SUMMARY

- The inside conditions that vary the load on the system are the space temperature and humidity.
- When the discharge pressure rises, the system capacity decreases.
- High-efficiency systems often use larger or oversized evaporators, and the refrigerant will boil at a temperature of about 45°F at typical operating conditions. This is a temperature difference of 30°F compared with the return air.
- High-efficiency equipment has a lower operating head pressure than standard-efficiency equipment partly because the condenser is larger. It has a different relationship to the outside ambient temperature, normally 20°F to 25°F.
- The first thing that the technician needs to know for electrical troubleshooting is what the operating voltage for the unit is supposed to be.
- Equipment manufacturers require that operating voltages of ±10% of the rated voltage of the equipment be maintained.
- The unit may have the compressor's RLA printed on it. It should not be exceeded.
- Some manufacturers are now using two-speed motors and variable-speed motors for variable capacity.

REVIEW QUESTIONS

1. The standard design conditions for air-conditioning systems established by the Air Conditioning and Refrigeration Institute (ARI) are outside temperature _____°F, inside temperature _____°F, with a humidity of _____%.

2. The typical temperature relationship between a standard-efficiency air-cooled condenser and the ambient temperature is
   A. 50°F.
   B. 40°F.
   C. 30°F.
   D. 20°F.

3. The evaporator will normally operate at a _____°F boiling temperature when operating at a 75°F indoor temperature with a 50% humidity condition.
   A. 40
   B. 50
   C. 60
   D. 70

4. A typical temperature relationship between a high-efficiency condenser and the ambient temperature is
   A. 10°F.
   B. 20°F.
   C. 30°F.
   D. 40°F.

5. How is condenser high efficiency obtained?

6. High-efficiency air-conditioning systems may use a _____ rather than a fixed-bore metering device.

7. What does the head pressure do if the suction pressure rises?

8. What will cause the suction pressure to rise?

9. The evaporator design temperature may in some cases operate at a slightly _____ (higher or lower) pressure and temperature on high-efficiency equipment because the evaporator is larger.

10. The seasonal energy efficiency ratio (SEER) includes the energy used in the _____ and _____ cycles.

11. The electrical operating conditions are measured with a (an) _____ and a (an) _____.

12. The three major power-consuming devices on an air-conditioning system that may have to be analyzed are the _____, _____, and the _____.

13. Air-conditioning systems normally move about _____ cfm of air per ton.

14. The compressor amperage of a 3-ton system operating on 230 V is approximately _____A.

15. Many ratings for components of an air-conditioning system are not completely accurate due to many different operating conditions and are called _____ ratings.
   A. nominal
   B. SEER
   C. high-efficiency
   D. standard-efficiency

UNIT 41 Troubleshooting

OBJECTIVES

After studying this unit, you should be able to

■ select the correct instruments for checking an air-conditioning unit with a mechanical problem.
■ calculate the correct operating suction pressures for both standard- and high-efficiency air-conditioning equipment under various conditions.
■ calculate the standard operating discharge pressures at various ambient conditions.
■ select the correct instruments for troubleshooting electrical problems in an air-conditioning system.
■ check the line- and low-voltage power supplies.
■ troubleshoot basic electrical problems in an air-conditioning system.
■ use an ohmmeter to check the various components of the electrical system.

SAFETY CHECKLIST

✔ Be extremely careful when installing or removing gages. Escaping refrigerant can injure your skin and eyes. Wear gloves and goggles. If possible turn off the system to install a gage on the high-pressure side. Most air-conditioning systems are designed so that the pressures will equalize when turned off.
✔ ♻ Wear goggles and gloves when transferring refrigerant from a cylinder to a system or when recovering refrigerant from a system. Do not recover refrigerant into a disposable cylinder. Use only tanks or cylinders approved by the Department of Transportation (DOT). ♻
✔ Be very careful when working around a compressor discharge line. The temperature may be as high as 220°F.
✔ All safety practices must be observed when troubleshooting electrical systems. Often, readings must be taken while the power is on. Only let the contact tips on the insulated meter leads touch the hot terminals. Never use a screwdriver or other tools around terminals that are energized. Never use tools in a hot electrical panel.
✔ Turn the power off whenever possible when working on a system. Lock and tag the disconnect panel and keep the only key on your person.
✔ Do not stand in water when making any electrical measurements.
✔ Short capacitor terminals with a 20,000-Ω resistor before checking with an ohmmeter.

✔ Students and other inexperienced persons should perform troubleshooting tasks only under the supervision of an experienced person.

41.1 INTRODUCTION

Troubleshooting air-conditioning equipment involves troubleshooting mechanical and electrical systems, and they may have symptoms that overlap. For example, if an evaporator fan motor capacitor fails, the motor will slow down and begin to get hot. It may even get hot enough for the internal overload protector to stop it. While it is running slowly, the suction pressure will go down and give symptoms of a restriction or low charge. If the technician diagnoses the problem based on suction pressure readings only, a wrong decision may be made.

41.2 MECHANICAL TROUBLESHOOTING

Gages and temperature-testing equipment are used when performing mechanical troubleshooting. The gages used are those on the gage manifold as shown in **Figure 41–1**. The suction or low-side gage is on the left side of the manifold, and the discharge or the high-side gage is on the right side. The most common refrigerants used in air conditioning are R-22 and R-410A. An R-22 temperature chart is printed on many gages

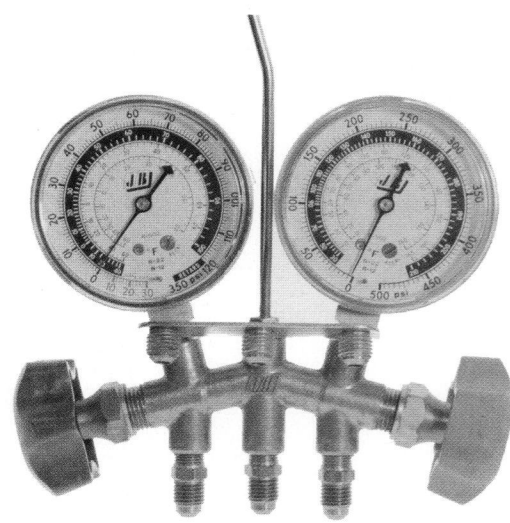

Figure 41–1 A low-side gage on the left and a high-side gage on the right. *Photo by Bill Johnson*

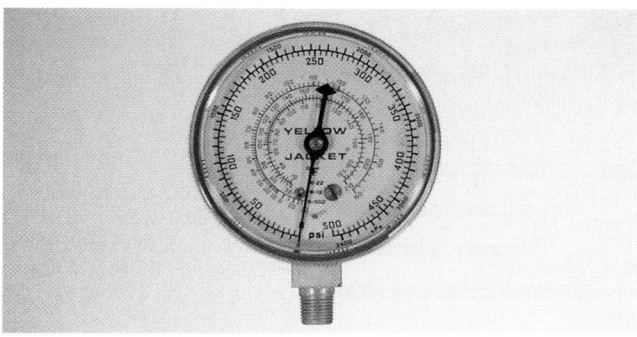

(A)

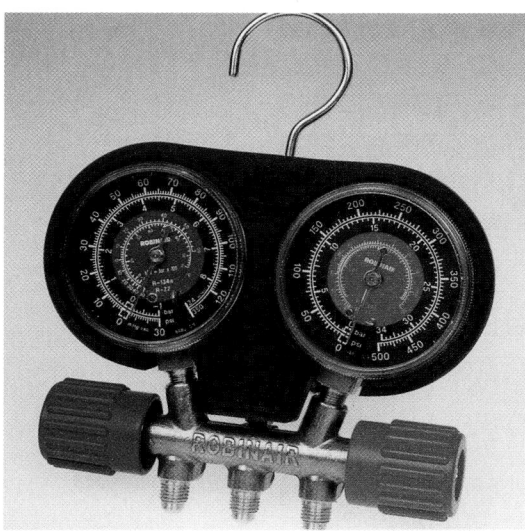

(B)

Figure 41–2 **(A)** Many gages have common refrigerant temperature relationships for R-12, R-22, and R-502 printed on them. **(B)** Other gages are universal with no printed temperature relationships, or they may be for specific refrigerants. *(A) Photo by Bill Johnson.* *(B) Courtesy of Robinair Division, SPX Corporation*

for determining the saturation temperature for the low- and high-pressure sides of the system. Because these same gages are often used for refrigeration, a temperature scale for R-12 and R-502 is printed on most gages also, **Figure 41–2(A)**. The R-12 scale is not used for residential air conditioning unless the equipment is old enough that R-12 is used. ♻ The common refrigerant for automobile air conditioning starting in 1992 is R-134a. ♻ Older automobiles normally would use R-12.

Many newer gages are more universal and may not have the temperature relationship printed on them, or they may have the temperature relationship printed on them for a specific refrigerant, **Figure 41–2(B)**.

Many manufacturers use R-410A as their refrigerant. This refrigerant has a much higher working pressure than R-22. The gage ports are the same size, ¼ in.; either gage manifold will attach to the R-410A gage manifold, but gages rated for R-22 will be overpressurized if attached to an R-410A system. If R-22 gages are used on an R-410A system, they will read at the highest point on the gage. Gages are most

accurate in the middle of their scale range. The technician should be sure what the refrigerant is and apply the correct gage manifold and hoses to the correct refrigerant.

41.3 APPROACH TEMPERATURE AND TEMPERATURE DIFFERENCE

Many technicians use the terms *approach temperature* or *temperature split* to describe the difference in temperature between heat exchange mediums. For example, the difference in temperature between the indoor coil and the return air for a central air-conditioning system has been mentioned to be 35°F for standard air-conditioning equipment, **Figure 41–3**. The condenser also has an approach, or split, temperature, **Figure 41–4**. These temperatures are used by the technician to determine when a heat exchanger is functioning correctly. For example, when an air-cooled condenser approach temperature begins to increase, it is a sign that the condenser is not condensing efficiently. This tells the technician that the coil is dirty or that there is not enough air passing over the coil.

Temperature difference (TD) is another term that is used by technicians but it has a different meaning. Temperature difference applies to the inlet versus the outlet temperature of the same medium. For example, the air entering an evaporator may be 75°F and the leaving-air temperature may be 55°F. The temperature difference is 20°F. This only applies to the air temperature difference; in the approach temperature example mentioned above, the approach, or split, temperature was between the refrigerant and the air temperature.

41.4 GAGE MANIFOLD USAGE

The gage manifold displays the low- and high-side pressures while the unit is operating. These pressures can be converted to the saturation temperatures for the evaporating

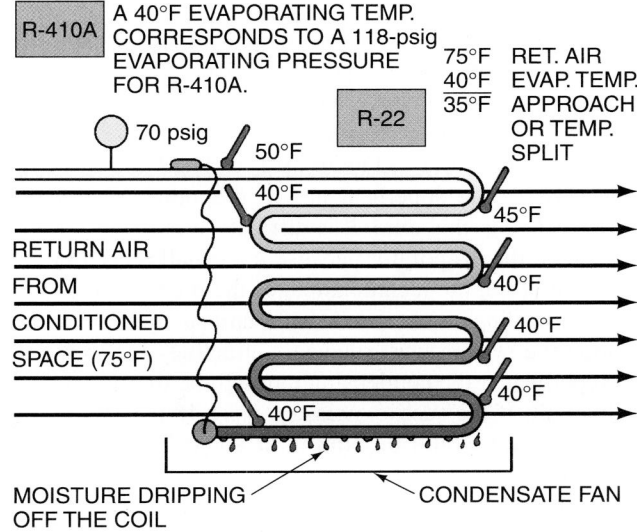

Figure 41–3 The difference between the entering-air temperature and the refrigerant boiling temperature is known as the "approach temperature," or "temperature split."

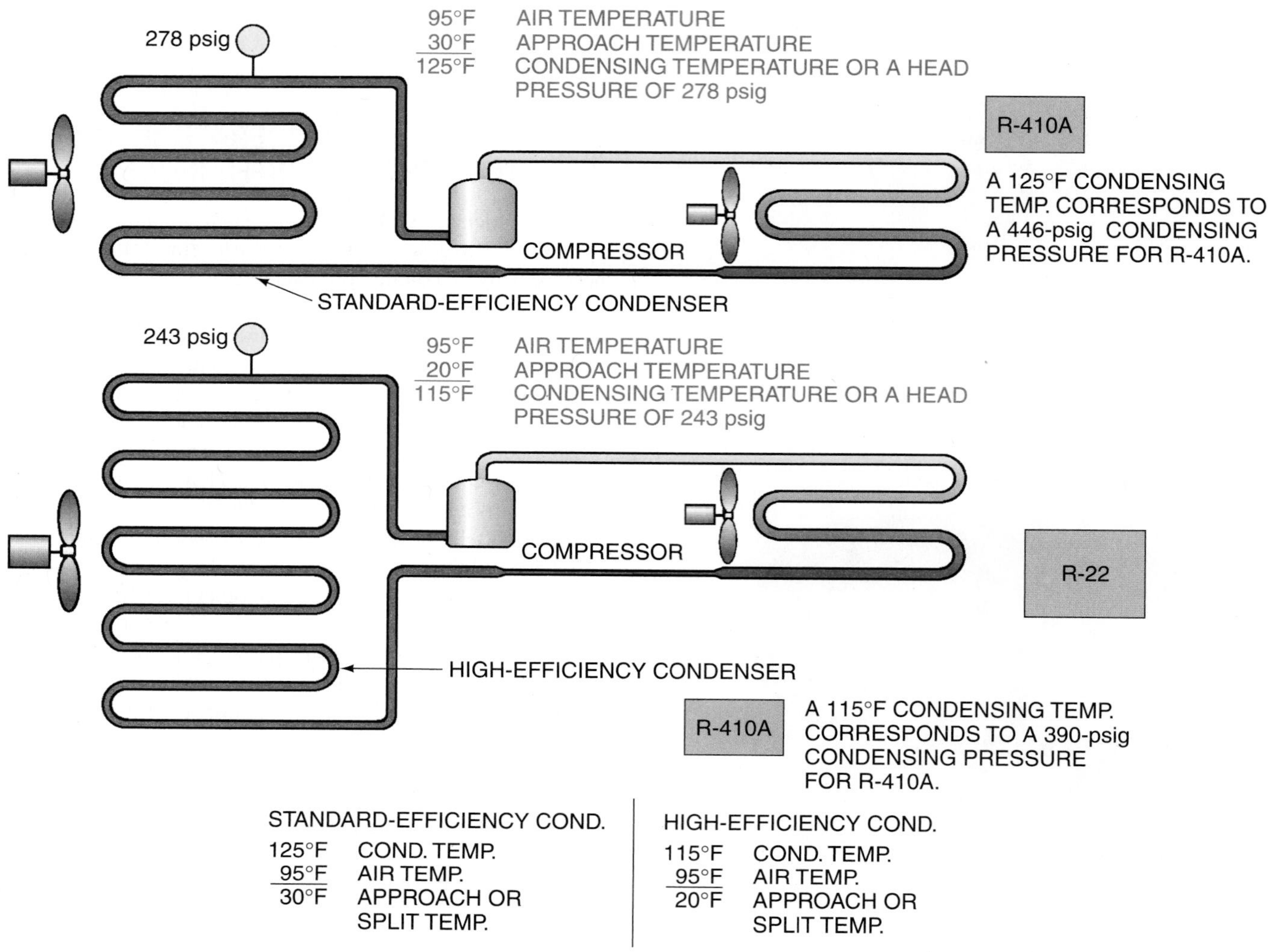

278 psig

95°F   AIR TEMPERATURE
30°F   APPROACH TEMPERATURE
125°F CONDENSING TEMPERATURE OR A HEAD
       PRESSURE OF 278 psig

R-410A

A 125°F CONDENSING
TEMP. CORRESPONDS TO
A 446-psig CONDENSING
PRESSURE FOR R-410A.

COMPRESSOR

STANDARD-EFFICIENCY CONDENSER

243 psig

95°F   AIR TEMPERATURE
20°F   APPROACH TEMPERATURE
115°F CONDENSING TEMPERATURE OR A HEAD
       PRESSURE OF 243 psig

COMPRESSOR

R-22

HIGH-EFFICIENCY CONDENSER

R-410A

A 115°F CONDENSING TEMP.
CORRESPONDS TO A 390-psig
CONDENSING PRESSURE
FOR R-410A.

STANDARD-EFFICIENCY COND.		HIGH-EFFICIENCY COND.	
125°F	COND. TEMP.	115°F	COND. TEMP.
95°F	AIR TEMP.	95°F	AIR TEMP.
30°F	APPROACH OR SPLIT TEMP.	20°F	APPROACH OR SPLIT TEMP.

Figure 41–4   The "approach temperature," or "temperature split," works the same for the condenser.

(boiling) refrigerant and the condensing refrigerant by using the pressure and temperature relationship. The low-side pressure can be converted to the boiling temperature. If the boiling temperature for a system should be close to 40°F, it can be converted to 70 psig for R-22 and 118 psig for R-410A. The superheat at the evaporator should be close to 10°F at this time. It is difficult to read the suction pressure at the evaporator because it normally has no gage port. Therefore, the technician takes the pressure and temperature readings at the condensing unit suction line to determine the system performance. Guidelines for checking the superheat at the condensing unit will be discussed later in this unit. If the suction pressure were 48 psig for R-22, the refrigerant would be boiling at about 24°F, which is cold enough to freeze the condensate on the evaporator coil and too low for continuous operation. The probable causes of low boiling temperature are low charge or restricted airflow, **Figure 41–5**.

SAFETY PRECAUTION: You must be extremely careful while installing manifold gages. High-pressure refrigerant will injure your skin and eyes. Wearing goggles and gloves is

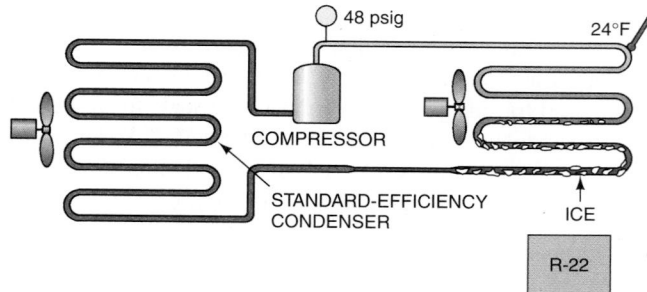

Figure 41–5   This coil was operated below freezing until the condensate on the coil froze.

necessary. The danger from attaching the high-pressure gages can be reduced by shutting off the unit and allowing the pressures to equalize to a lower pressure. Liquid R-22 can cause serious frostbite because it boils at −42°F at atmospheric pressure, while R-410A boils at −60°F.

The high-side gage may be used to convert pressures to condensing temperatures. For example, if the high-side gage reads 278 psig for R-22 and the outside ambient temperature is 80°F, the head pressure may seem too high. The gage manifold chart shows that the refrigerant is condensing at 125°F. However, the refrigerant should not condense at a temperature more than 30°F higher than the ambient temperature. The ambient temperature is 80°F + 30°F = 110°F, so the condensing temperature is actually 15°F too high. Probable causes are a dirty condenser, an overcharge of refrigerant, or noncondensables in the system.

The gage manifold is used whenever the pressures need to be known. Two types of pressure connections are used with air-conditioning equipment: the Schrader valve and the service valve, **Figure 41–6.** The *Schrader valve* is a pressure connection only. The *service valve* can be used to isolate the system for service. It may have a Schrader connection for the gage port and a service valve for isolation, **Figure 41–7.**

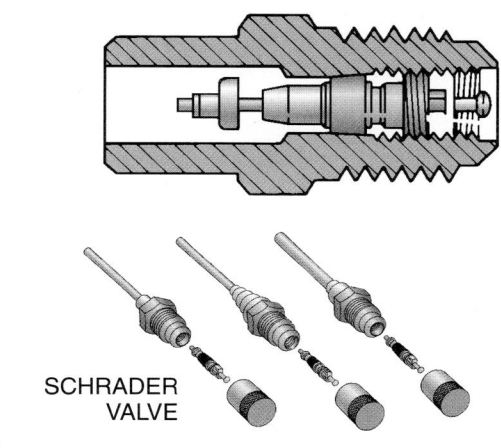

SCHRADER
VALVE

(A)

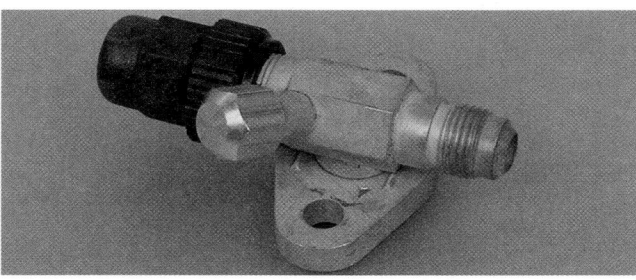

(B)

Figure 41–6   Two valves commonly used when taking pressure readings on modern air-conditioning equipment. **(A)** A Schrader valve. **(B)** A service valve. **(A)** *Courtesy JB Industries.* **(B)** *Photo by Bill Johnson*

Figure 41–7   This service valve has a Schrader port instead of a back seat like a refrigeration service valve. *Photo by Bill Johnson*

41.5   WHEN TO CONNECT THE GAGES

When servicing small systems, the technician should not connect a gage manifold every time the system is serviced. A small amount of refrigerant escapes each time the gage is connected. Some residential and small commercial systems have a critical refrigerant charge. When the high-side line is connected, high-pressure refrigerant will condense in the gage line. The refrigerant will escape when the gage line is disconnected from the Schrader connector. A gage line full of liquid refrigerant lost while checking pressures may be enough to affect the system charge. A short gage line connector for the high side will help prevent refrigerant loss, **Figure 41–8.** This can be used only to check pressure. You cannot use it to transfer refrigerant out of the system because it is not a manifold.

Another method of connecting the gage manifold to the Schrader valve service port is with a small hand valve, which has a depressor for depressing the stem in the Schrader valve, **Figure 41–9.** The hand valve can be used to keep the refrigerant in the system when disconnecting the gages. Follow these steps: (1) Turn the stem on the valve out; this allows the Schrader valve to close. (2) Make sure that a plug is in the center line of the gage manifold. (3) Open the gage manifold valves. This will cause the gage lines to equalize to the low side of the system. The liquid refrigerant will move from the high-pressure gage line to the low side of the system. The only refrigerant that will be lost is an insignificant amount of vapor in the three gage lines. This vapor is at the suction pressure while the system is running and should be about 70 psig for R-22. NOTE: This procedure should only be done with a clean and purged gage manifold, **Figure 41–10.**

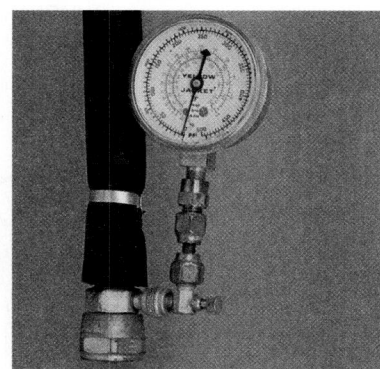

Figure 41–8   This short service connection is used on the high-side gage port to keep too much refrigerant from condensing in the gage line. *Photo by Bill Johnson*

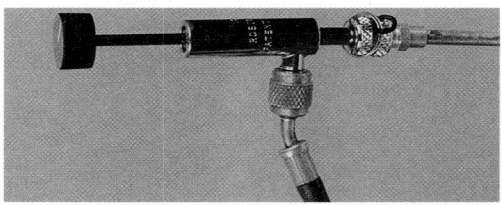

Figure 41–9   A hand valve used to press the Schrader valve stem and control the pressure. It can be used to back the valve stem out. *Photo by Bill Johnson*

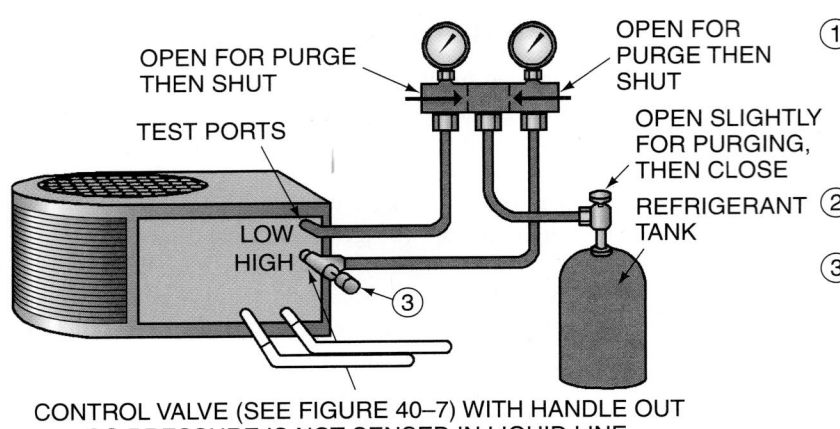

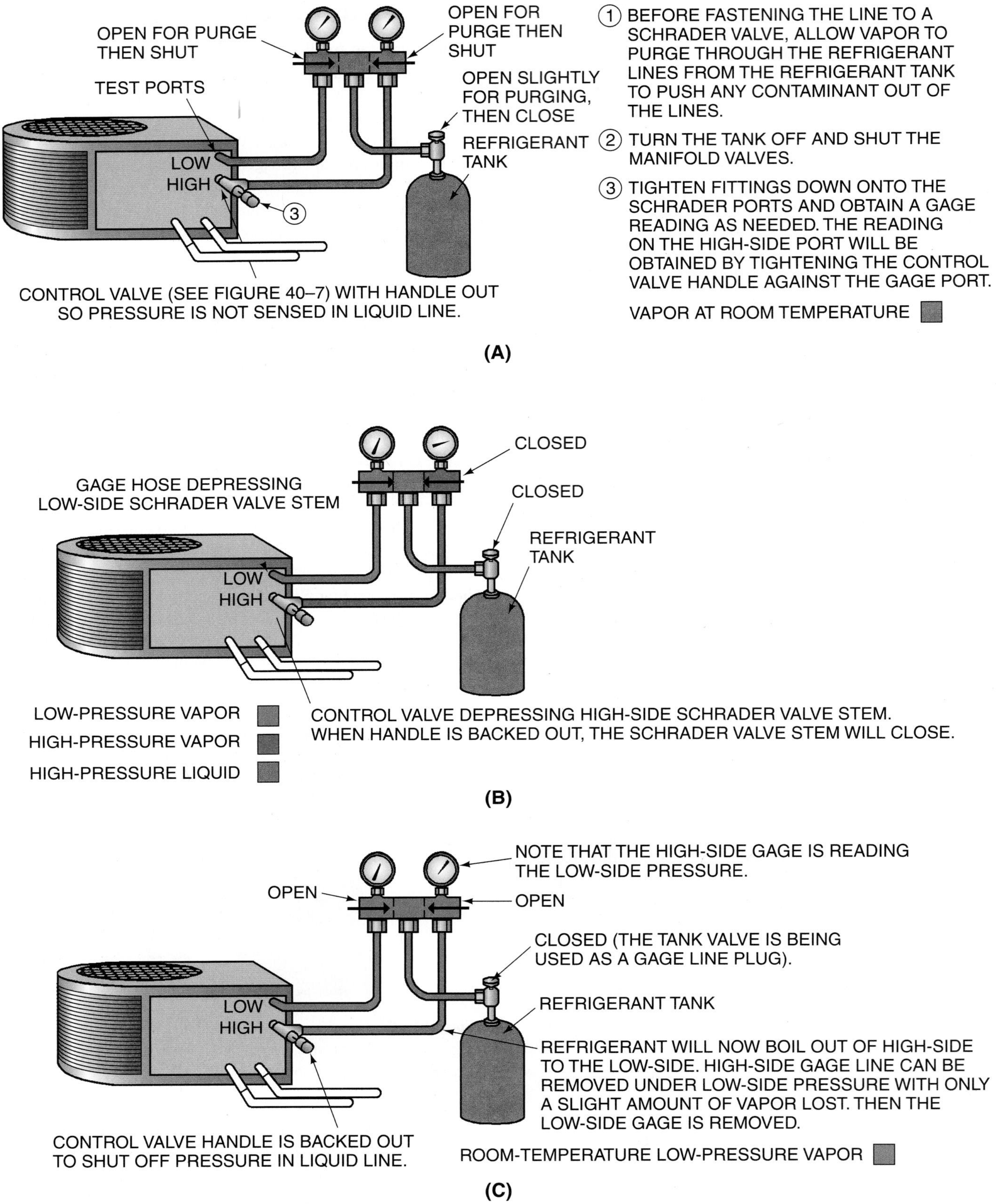

OPEN FOR PURGE THEN SHUT

TEST PORTS

OPEN FOR PURGE THEN SHUT

OPEN SLIGHTLY FOR PURGING, THEN CLOSE

REFRIGERANT TANK

LOW

HIGH

③

CONTROL VALVE (SEE FIGURE 40–7) WITH HANDLE OUT SO PRESSURE IS NOT SENSED IN LIQUID LINE.

① BEFORE FASTENING THE LINE TO A SCHRADER VALVE, ALLOW VAPOR TO PURGE THROUGH THE REFRIGERANT LINES FROM THE REFRIGERANT TANK TO PUSH ANY CONTAMINANT OUT OF THE LINES.

② TURN THE TANK OFF AND SHUT THE MANIFOLD VALVES.

③ TIGHTEN FITTINGS DOWN ONTO THE SCHRADER PORTS AND OBTAIN A GAGE READING AS NEEDED. THE READING ON THE HIGH-SIDE PORT WILL BE OBTAINED BY TIGHTENING THE CONTROL VALVE HANDLE AGAINST THE GAGE PORT.

VAPOR AT ROOM TEMPERATURE ▓

(A)

CLOSED

CLOSED

GAGE HOSE DEPRESSING LOW-SIDE SCHRADER VALVE STEM

REFRIGERANT TANK

LOW

HIGH

LOW-PRESSURE VAPOR ▓

HIGH-PRESSURE VAPOR ▓

HIGH-PRESSURE LIQUID ▓

CONTROL VALVE DEPRESSING HIGH-SIDE SCHRADER VALVE STEM. WHEN HANDLE IS BACKED OUT, THE SCHRADER VALVE STEM WILL CLOSE.

(B)

NOTE THAT THE HIGH-SIDE GAGE IS READING THE LOW-SIDE PRESSURE.

OPEN

OPEN

CLOSED (THE TANK VALVE IS BEING USED AS A GAGE LINE PLUG).

LOW

HIGH

REFRIGERANT TANK

REFRIGERANT WILL NOW BOIL OUT OF HIGH-SIDE TO THE LOW-SIDE. HIGH-SIDE GAGE LINE CAN BE REMOVED UNDER LOW-SIDE PRESSURE WITH ONLY A SLIGHT AMOUNT OF VAPOR LOST. THEN THE LOW-SIDE GAGE IS REMOVED.

CONTROL VALVE HANDLE IS BACKED OUT TO SHUT OFF PRESSURE IN LIQUID LINE.

ROOM-TEMPERATURE LOW-PRESSURE VAPOR ▓

(C)

Figure 41–10 A method of pumping refrigerant condensed in the high-pressure line over into the suction line. **(A)** Purging a manifold. **(B)** Taking pressure readings. **(C)** Removing liquid refrigerant from the high-side gage line.

41.6 LOW-SIDE GAGE READINGS

When using the gage manifold on the low side of the system, you can compare the actual evaporating pressure to the normal evaporating pressure. This verifies that the refrigerant is boiling at the correct temperature for the low side of the system at some load condition. High-efficiency systems often have oversized evaporators. This makes the suction pressure slightly higher than normal. A standard-efficiency system usually has a refrigerant boiling temperature of about 35°F cooler than the entering-air temperature at the standard operating condition of 75°F return air with a 50% humidity.

If the space temperature is 85°F and the humidity is 70%, the evaporator has an oversized load. It is absorbing an extra heat load, both sensible heat and latent heat, from the moisture in the air. You need to wait a sufficient time for the system to reduce the load before you can determine whether the equipment is functioning correctly. Gage readings at this time will not reveal the kind of information that will verify the system performance unless a manufacturer's performance chart is available.

It is good practice to start the unit and check for problems. If none are apparent, let the unit run for the rest of the day and come back the next morning before evaluating system performance. You may notice that a larger-than-normal volume of condensate is pouring from the drain line. The moisture content in the form of condensate must be removed before the actual air temperature will begin to drop significantly.

41.7 HIGH-SIDE GAGE READINGS

Gage readings obtained on the high-pressure side of the system are used to check the relationship of the condensing refrigerant to the ambient air temperature. Standard air-conditioning equipment condenses the refrigerant at no more than 30°F higher than the ambient temperature. For a 95°F entering air temperature, the head pressure should correspond to 95°F + 30°F = 125°F for a corresponding head pressure of 278 psig for R-22, 169 psig for R-12, and 445 psig for R-410A. If the head pressure shows that the condensing temperature is higher than this, something is wrong.

When checking the condenser air temperature, be sure to check the actual temperature. Do not take the weather report as the ambient temperature. Air-conditioning equipment located on a black roof has solar influence from the air being pulled across the roof, **Figure 41–11**. If the condenser is located close to any obstacles, such as below a sundeck, air may be circulated back through the condenser and cause the head pressure to be higher than normal, **Figure 41–12**.

High-efficiency condensers perform the same as standard-efficiency condensers except that they operate at lower pressures and condensing temperatures. High-efficiency condensers normally condense the refrigerant at a temperature as low as 20°F higher than the ambient temperature. On a 95°F day the head pressure corresponds to a temperature of 95°F + 20°F = 115°F, which is a pressure of 243 psig for R-22 and 390 psig for R-410A.

41.8 TEMPERATURE READINGS

Temperature readings also can be useful. The electronic thermometer performs very well as a temperature-reading instrument, **Figure 41–13**. It has small temperature leads that respond quickly to temperature changes. The leads can be attached easily to the refrigerant piping with electrical tape and insulated from the ambient temperature with short pieces of foam line insulation, **Figure 41–14**.

It is important that the lead be insulated from the ambient if the ambient temperature is different from the line temperature. This temperature lead is better and easier to use than the glass thermometers used in the past. It was almost impossible to get a true line reading by strapping a glass thermometer to a copper line, **Figure 41–15**.

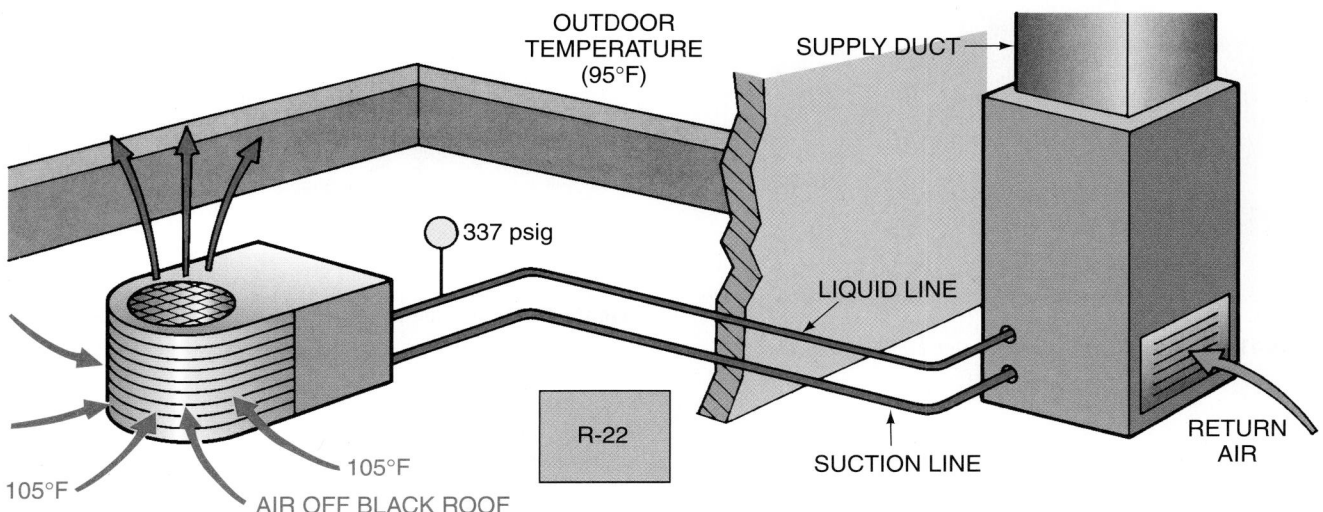

Figure 41–11 A condenser located low on a roof. Hot air from the roof level enters it. A better installation would be to have the condenser mounted up about 20 in. so that air could at least mix with the roof ambient air. Ambient air is 95°F and air off the roof is 105°F.

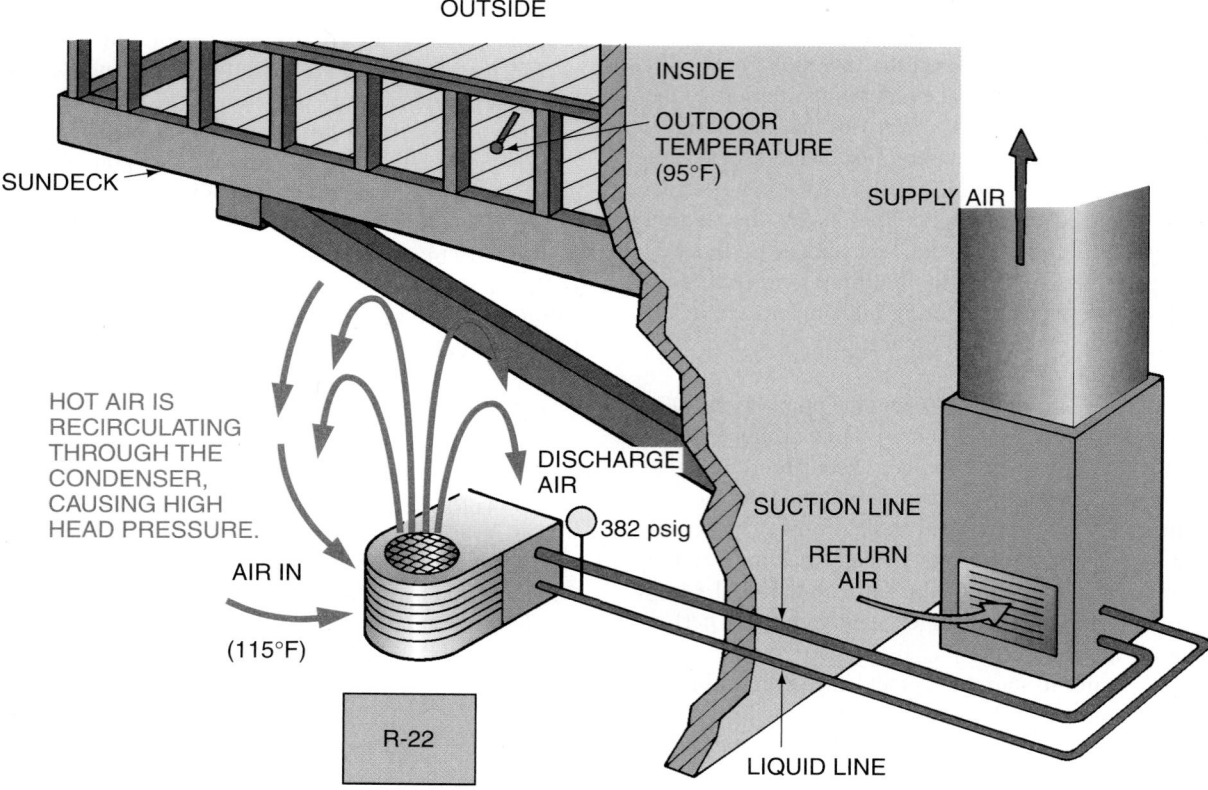

Figure 41–12   A condenser installed so that the outlet air is hitting a barrier in front of it and recirculating to the condenser inlet.

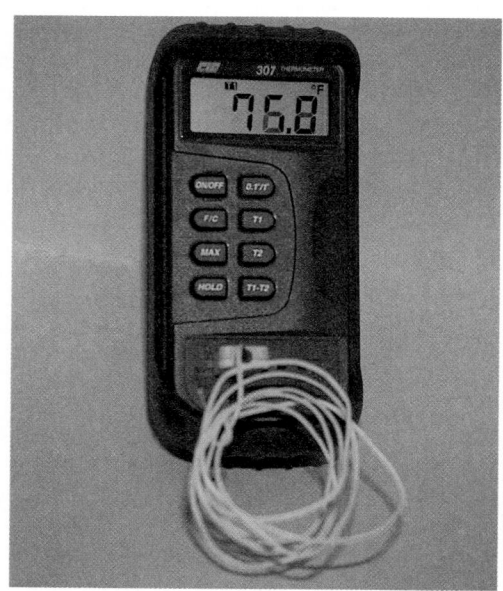

Figure 41–13   An electronic thermometer. *Photo by Bill Johnson*

Figure 41–14   A temperature lead attached to a refrigerant line in the correct manner. It must also be insulated. *Photo by Bill Johnson*

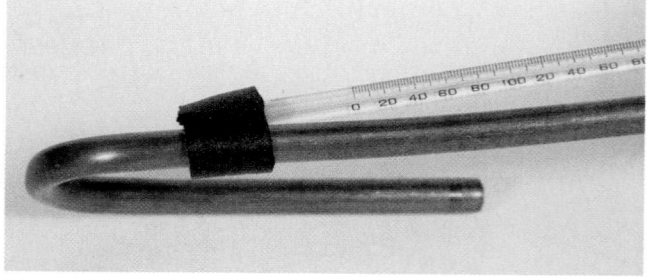

Figure 41–15   A glass thermometer strapped to a refrigerant line. *Photo by Bill Johnson*

Temperatures vary from system to system. The technician must be prepared to accurately record these temperatures to evaluate the various types of equipment. Some technicians record temperature readings of various equipment under different conditions for future reference. The common temperatures used would be inlet air wet- and dry-bulb, outdoor air dry-bulb, and the suction-line temperature. Sometimes, the compressor discharge temperature needs to be known. A thermometer with a range from $-50°F$ to $+250°F$ is a common instrument to be used for all of these tests.

Inlet Air Temperatures

It may be necessary to know the inlet air temperature to the evaporator for a complete analysis of a system. A wet-bulb reading for determining the humidity may be necessary. Such a reading can be obtained by using one of the temperature leads with a cotton sock that is saturated with pure water. A wet-bulb and a dry-bulb reading may be obtained by placing a dry-bulb temperature lead next to a wet-bulb temperature lead in the return airstream, **Figure 41–16.** The velocity of the return air will be enough to accomplish the evaporation for the wet-bulb reading.

Evaporator Outlet Temperatures

The evaporator outlet air temperature is seldom important. It may be obtained, however, in the same manner as the inlet air temperature. The outlet air dry-bulb temperature will normally be about 20°F less than the inlet air temperature. The temperature drop across an evaporator coil is about 20°F when it is operating at typical operating conditions of 75°F and 50% relative humidity return air. If the conditioned space temperature is high with a high humidity, the temperature drop across the same coil will be much less because of the latent-heat load of the moisture in the air.

If a wet-bulb reading is taken, there will be approximately a 10°F wet-bulb drop from the inlet to the outlet during standard operating conditions. The outlet humidity will be almost 90%. This air is going to mix with the room air and will soon drop in humidity because it will expand to the room air temperature. It is very humid because it is contracted or shrunk from the cooling process. SAFETY PRECAUTION: The temperature lead must not be allowed to touch the moving fan while taking air temperature readings.

Suction-Line Temperatures

The temperature of the suction line returning to the compressor and the suction pressure will help the technician understand the characteristics of the suction gas. The suction gas may be part liquid if the return air filters are stopped up or if the evaporator coil is dirty, **Figure 41–17.** The suction gas may have a high superheat if the unit has a low charge or if there is a refrigerant restriction, **Figure 41–18.** The combination of suction-line temperature and pressure will help the service technician decide whether the system has a low charge or a stopped-up air filter in the air handler. For example, if the suction pressure is too low and the suction line is warm, the system has a starved evaporator, **Figure 41–19.** If the suction-line temperature is cold and the pressure indicates that the refrigerant is boiling at a low temperature, the coil is not absorbing heat as it should. The coil may be dirty, or the airflow may be insufficient, **Figure 41–20.** The cold suction line indicates that the unit has enough charge because the evaporator must be full for the refrigerant to get back to the suction line, **Figure 41–21.**

Discharge-Line Temperatures

The temperature of the discharge line may tell the technician that something is wrong inside the compressor. If there is an internal refrigerant leak from the high-pressure side to the low-pressure side, the discharge gas temperature will go up,

Figure 41–16 This electronic thermometer temperature lead has a damp cotton cloth wrapped around it to convert it to a wet-bulb lead. *Photo by Bill Johnson*

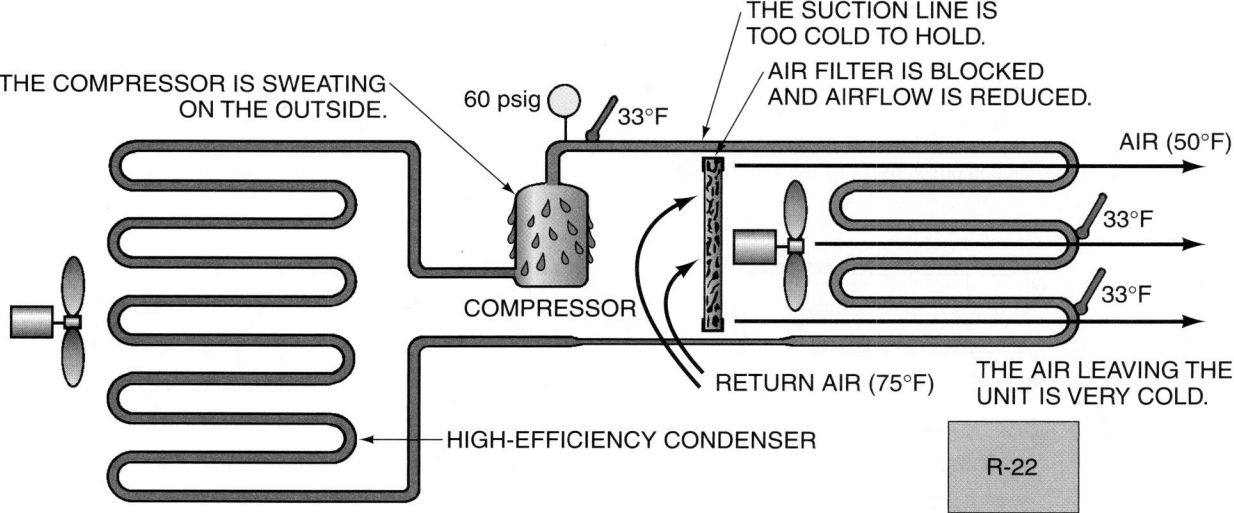

Figure 41–17 An evaporator flooded with refrigerant.

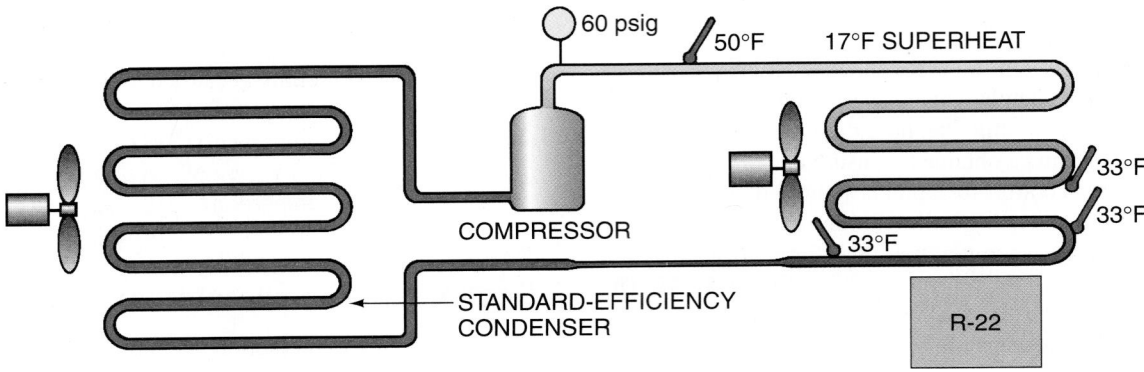

Figure 41–18   The evaporator is starved for refrigerant because of a low refrigerant charge.

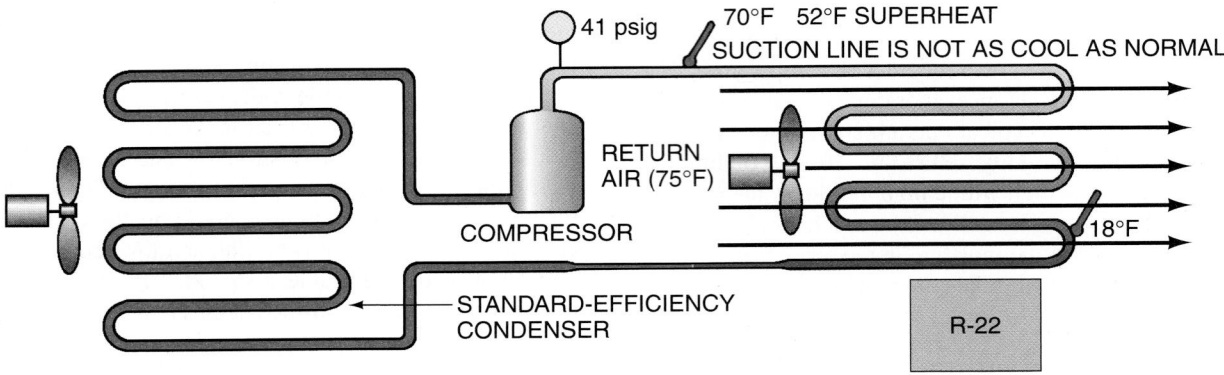

Figure 41–19   If the suction pressure is too low, and the suction line is not as cool as normal, the evaporator is starved for refrigerant. The unit refrigerant charge may be low.

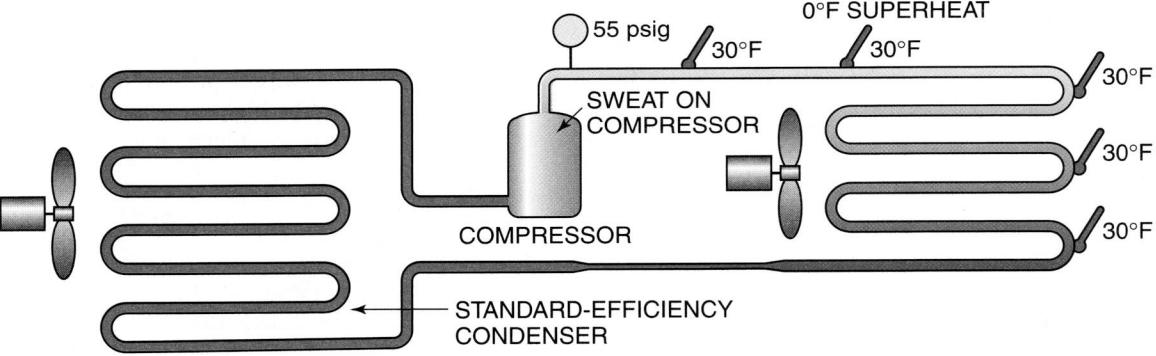

Figure 41–20   When the suction pressure is too low and the superheat is low, the unit is not boiling the refrigerant in the evaporator. The coil is flooded with refrigerant.

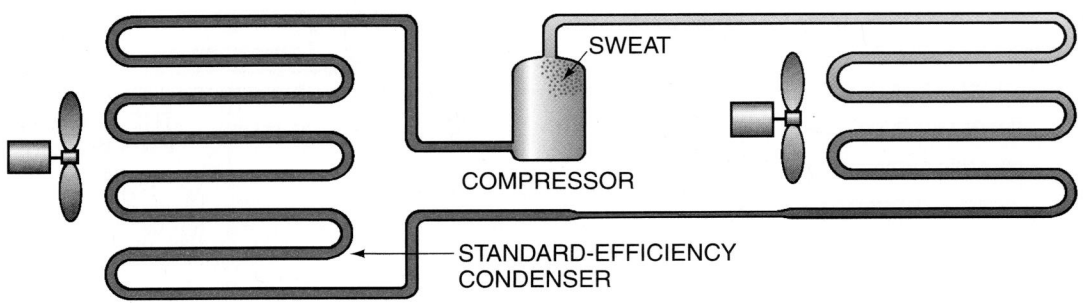

Figure 41–21   The evaporator is full of refrigerant.

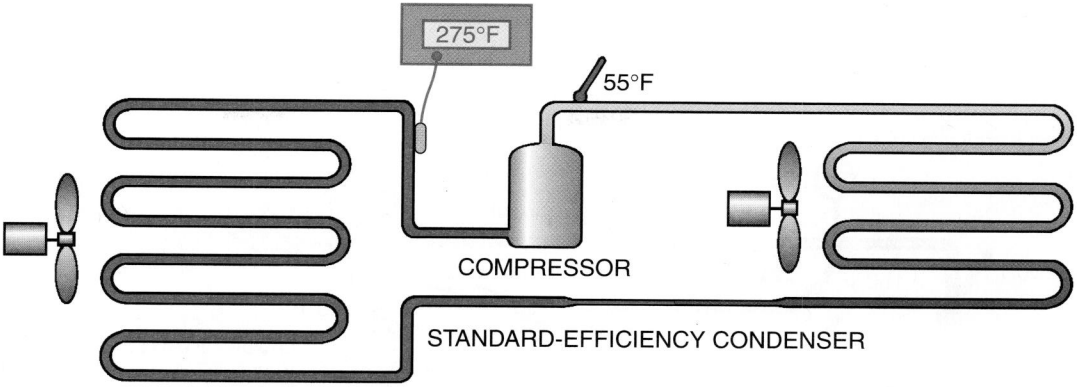

Figure 41–22 A thermometer is attached to the discharge line on this compressor. If the compressor has an internal leak, the hot discharge gas will circulate back though the compressor. The discharge gas will be abnormally hot. When a compressor is cooled by suction gas, a high superheat will cause the compressor discharge gas to be extra hot.

Figure 41–22. Normally the discharge-line temperature at the compressor would not exceed 220°F for an air-conditioning application even in very hot weather. When a high discharge-line temperature is discovered, the probable cause is an internal leak. The technician can prove this by building up the head pressure as high as 300 psig for R-22 and then shutting off the unit. If there is an internal leak, this pressure difference between the high and the low sides can often be heard (as a whistle) equalizing through the compressor. If the suction line at the compressor shell starts to warm up immediately, the heat is coming from the discharge of the compressor.

High discharge temperatures also can be created by high compression ratios. High compression ratios can be the result of low suction pressures, high condensing pressures, or a combination of both. High discharge temperatures also can be caused from high superheats coming back to the compressor, which is the result of an undercharge of refrigerant, restricted filter drier, or metering device starving the system.

SAFETY PRECAUTION: The discharge line of a compressor may be as hot as 220°F under normal conditions, so be careful while attaching a temperature lead to this line.

Liquid-Line Temperatures

Liquid-line temperature may be used to check the subcooling efficiency of a condenser. Most condensers will subcool the refrigerant to between 10°F and 20°F below the condensing temperature of the refrigerant. If the condensing temperature is 125°F on a 95°F day, the liquid line leaving the condenser may be 105°F to 115°F when the system is operating normally. If there is a slight low charge, there will not be as much subcooling, and the system efficiency therefore will not be as good. The condenser performs three functions: (1) removes the superheat from the discharge gas, (2) condenses the refrigerant to a liquid, and (3) subcools the liquid refrigerant below the condensing temperature. All three of these functions must be successfully accomplished for the condenser to operate at its rated capacity.

It is a good idea for a new technician to take the time to completely check out a working system operating at the correct pressures. Apply the temperature probes and gages to all points to actually verify the readings. This will provide reference points to remember, **Figure 41–23.**

41.9 CHARGING PROCEDURES IN THE FIELD

While establishing field charging procedures the technician should keep in mind what the designers of the equipment intended and how the equipment should perform. The charge consists of the correct amount of refrigerant in the evaporator, the liquid line, the discharge line between the compressor and the condenser, and the suction line. The discharge line and the suction line do not hold as much refrigerant as the liquid line because the refrigerant is in the vapor state. Actually, the liquid line is the only interconnecting line that contains much refrigerant. When a system is operating correctly, under design conditions, a prescribed amount of refrigerant should be in the condenser, the evaporator, and the liquid line.

Understanding the following statements is important for the technician in the field:

1. The amount of refrigerant in the evaporator can be measured by the superheat method.
2. The amount of refrigerant in the condenser can be measured by the subcooling method.
3. The amount of refrigerant in the liquid line may be determined by measuring the length and calculating the refrigerant charge. However, in field service work, if the evaporator is performing correctly, the liquid line has the correct charge.

When the preceding statements are understood, the technician can check for the correct refrigerant level to determine the correct charge.

A field charging procedure may be used to check the charge of most typical systems. The technician sometimes needs typical reference points to add small amounts of vapor refrigerant for adjusting the amount of refrigerant in equipment that has no charging directions. NOTE: Remember, if the refrigerant is a zeotropic blend and may fractionate

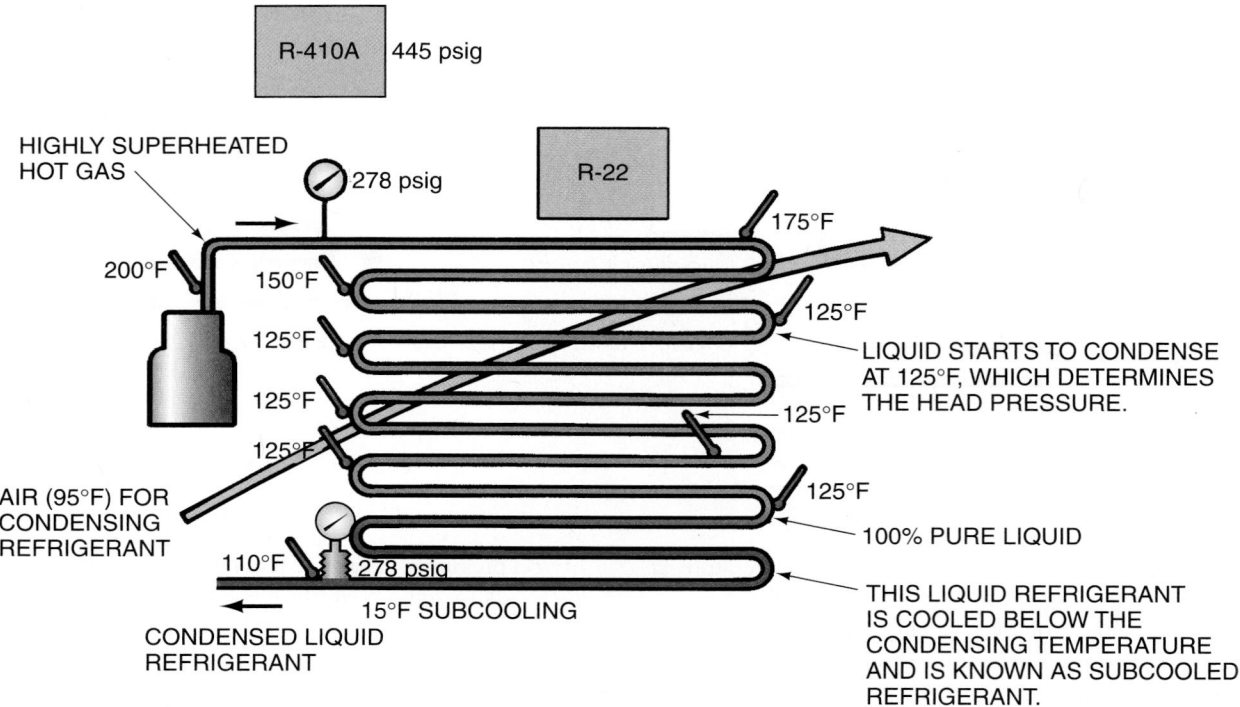

Figure 41–23 This condenser is operating at normal conditions. A technician can use these readings as points of reference for a condenser operating at full-load normal conditions.

when vapor charged, the refrigerant must be throttled in the low side of the system as a liquid to avoid fractionation.

Often the technician arrives and finds the system charge needs adjusting. This can occur due to an over- or under-charge from the factory or from a previous technician's work. It can also occur from system leaks. *System leaks should always be repaired to prevent further loss of charge to the atmosphere.* The technician must establish charging procedures to use in the field for all types of equipment. These procedures will help get the system back on line under emergency situations. Following are some methods used for different types of equipment.

Fixed-Bore Metering Devices— Capillary Tube and Orifice Type

Fixed-bore metering devices like the capillary tube do not throttle the refrigerant as the *thermostatic expansion valve* (TXV) does. They allow refrigerant flow based on the difference in the inlet and the outlet pressures. The one time when the system can be checked for the correct charge and everything will read normal is at the typical operating condition of 75°F and 50% humidity return air and 95°F outside ambient air. If other conditions exist, different pressures and different superheat readings will occur. The condition that most affects the readings is the outside ambient temperature. When it is lower than normal, the condenser will become more efficient and will condense the refrigerant sooner in the coil. This will have the effect of partially starving the evaporator for refrigerant. Refrigerant in the condenser that should be in the evaporator starves the evaporator.

When you need to check the system for correct charge or to add refrigerant, the best method is to follow the manufacturer's instructions. If they are not available, the typical operating condition may be simulated by reducing the airflow across the condenser to cause the head pressure to rise. On a 95°F day the highest condenser head pressure is usually 278 psig for R-22 (95°F ambient + 30°F added condensing temperature difference = 125°F condensing temperature or 278 psig). The pressure for R-410A would be 445 psig. Because the high pressure pushes the refrigerant through the metering device, when the head pressure is up to the high-normal end of the operating conditions there is no refrigerant held back in the condenser.

When the condenser is pushing the refrigerant through the metering device at the correct rate, the remainder of the charge must be in the evaporator. A superheat check at the evaporator is not always easy with a split air-conditioning system, so a superheat check at the condensing unit for a split system can be made. The suction line from the evaporator to the condensing unit may be long or short. Let us use two different lengths: up to 30 ft and from 30 to 50 ft for a test comparison. When the system is correctly charged, the superheat should be 10°F to 15°F at the condensing unit with a line length of 10 to 30 ft. The superheat should be 15°F to 18°F when the line is 30 to 50 ft long. Both of these conditions are with a head pressure of 278 psig for R-22 and 445 psig for R-410A ± 10 psig. At these conditions the actual superheat at the evaporator will be close to the correct superheat of 10°F. When using this method, be sure that you allow enough time for the system to settle down after adding refrigerant, before you draw any conclusions, **Figure 41–24.**

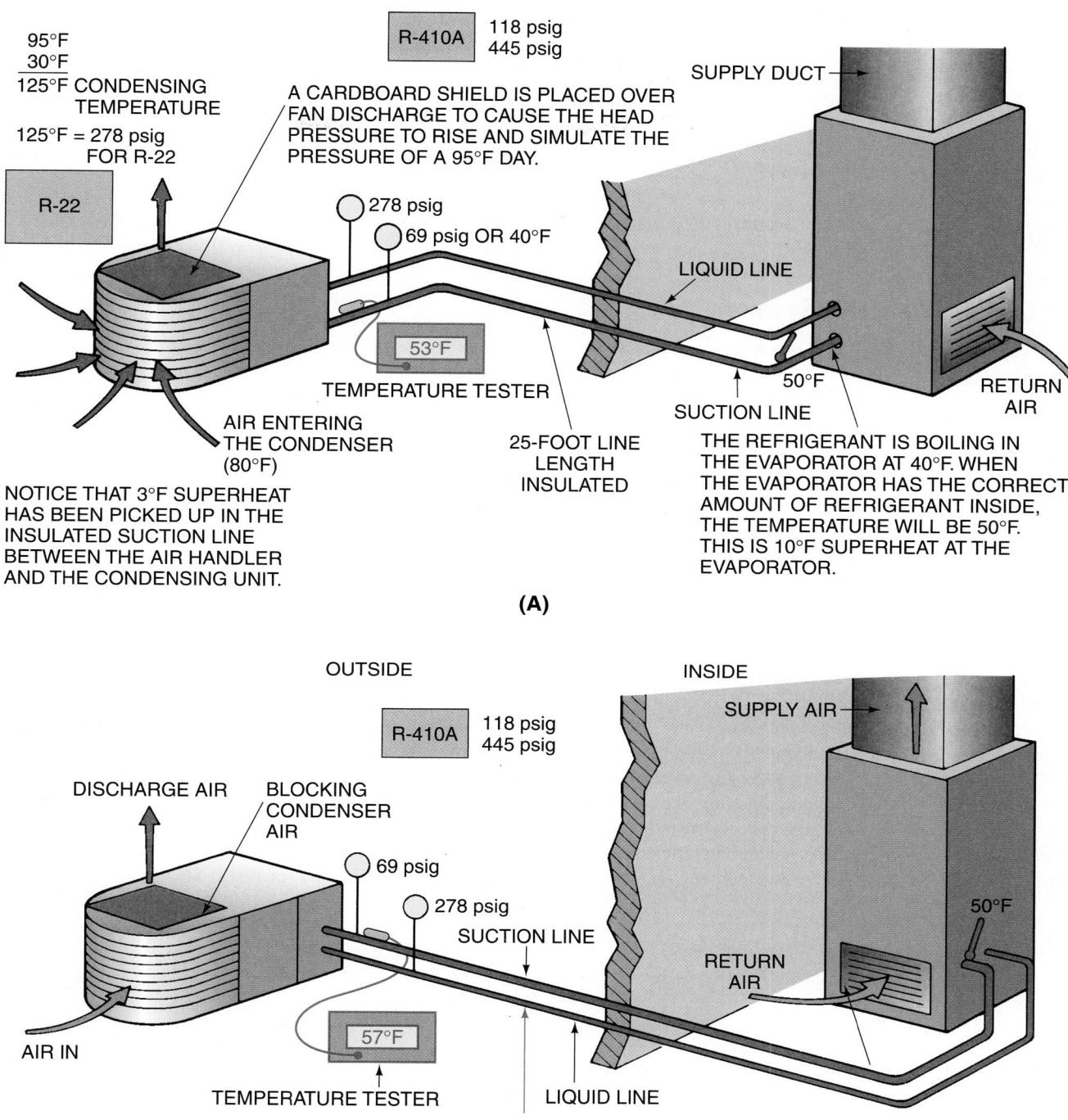

95°F
30°F
125°F CONDENSING
TEMPERATURE

125°F = 278 psig
FOR R-22

R-22

R-410A 118 psig
445 psig

A CARDBOARD SHIELD IS PLACED OVER
FAN DISCHARGE TO CAUSE THE HEAD
PRESSURE TO RISE AND SIMULATE THE
PRESSURE OF A 95°F DAY.

278 psig
69 psig OR 40°F

SUPPLY DUCT

LIQUID LINE

53°F
TEMPERATURE TESTER

AIR ENTERING
THE CONDENSER
(80°F)

25-FOOT LINE
LENGTH
INSULATED

50°F
SUCTION LINE

RETURN
AIR

NOTICE THAT 3°F SUPERHEAT
HAS BEEN PICKED UP IN THE
INSULATED SUCTION LINE
BETWEEN THE AIR HANDLER
AND THE CONDENSING UNIT.

THE REFRIGERANT IS BOILING IN
THE EVAPORATOR AT 40°F. WHEN
THE EVAPORATOR HAS THE CORRECT
AMOUNT OF REFRIGERANT INSIDE,
THE TEMPERATURE WILL BE 50°F.
THIS IS 10°F SUPERHEAT AT THE
EVAPORATOR.

(A)

OUTSIDE

INSIDE

R-410A 118 psig
445 psig

SUPPLY AIR

DISCHARGE AIR   BLOCKING
CONDENSER
AIR

69 psig

278 psig
SUCTION LINE

57°F
TEMPERATURE TESTER

AIR IN

LIQUID LINE

45-FOOT LINE SET

RETURN
AIR

50°F

R-22

NOTICE THAT IN THIS EXAMPLE 7°F SUPERHEAT WAS GAINED
BETWEEN THE AIR HANDLER AND THE CONDENSING UNIT.

(B)

Figure 41–24 **(A)** This system has a 25-ft line set and is charged for 10 to 15 degrees of superheat. **(B)** This line set is 45 ft and is charged for 15 to 18 degrees of superheat.

Superheat taken at the compressor will vary when typical operating conditions are not met. Factors that cause this variation are outside ambient temperatures and inside dry- and wet-bulb temperatures. An air-conditioning system may be operating normally at 30°F superheat at the compressor if the evaporator is seeing a high latent- and sensible-heat load. On the other hand, superheats of 8°F to 15°F may be normal if outside ambient temperatures are high. This is because refrigerant is being forced through the capillary tube at a higher rate. The manufacturer's superheat charging information should be used if available. It will take into account all of the varying conditions that affect superheat at the compressor.

Orifice-type metering devices have a tendency to hunt while reading steady-rate operation. This hunting can be observed by watching the suction pressure rise and fall accompanied by the superheat as the suction-line temperature changes. When this occurs, the technician will have to use averaging to arrive at the proper superheat for the coil. For example, if it is varying between 6°F and 14°F, the average value of 10°F superheat may be used.

Field Charging the TXV System

The TXV system can be charged in much the same way as the fixed-bore system, with some modifications. The condenser on a TXV system will also hold refrigerant back in mild ambient conditions. This system always has a refrigerant reservoir or receiver to store refrigerant and will not be affected as much as the capillary tube by lower ambient conditions. To check the charge, restrict the airflow across the condenser until the head pressure simulates a 95°F ambient, 278 psig head pressure for R-22 or 445 psig for R-410A. Using the superheat method will not work for this valve because if there is an overcharge, the superheat will remain the same. Superheats of 15°F to 18°F are not unusual when measured at the condensing unit for TXVs. If the sight glass is full, the unit has at least enough refrigerant, but it may have an overcharge. If the unit does not have a sight glass, a measure of the subcooling of the condenser may tell you what you want to know. For example, a typical subcooling circuit will subcool the liquid refrigerant from 10°F to 20°F cooler than the condensing temperature. A temperature lead attached to the liquid line should read 115°F to 105°F, or 10°F to 20°F cooler than the condensing temperature of 125°F, **Figure 41–25**. If the subcooling temperature is 20°F to 25°F cooler than the condensing temperature, the unit

has an overcharge of refrigerant, and the bottom of the condenser is acting as a large subcooling surface.

The charging procedures just described will also work for high-efficiency equipment. The head pressure does not need to be operated quite as high. A head pressure of 250 psig will be sufficient for an R-22 system, and 390 psig for R-410A, when charging.

SAFETY PRECAUTION: Refrigerant in cylinders and in the system is under great pressure (as high as 300 psi for R-22 and 500 psig for R-410A). ♻ Use proper safety precautions when transferring refrigerant, and be careful not to overfill tanks or cylinders when recovering refrigerant. Do not use disposable cylinders by refilling them. Use only DOT-approved tanks or cylinders. ♻ The high-side pressure may be reduced for attaching and removing gage lines by shutting off the unit and allowing the unit pressures to equalize.

41.10 ► ELECTRICAL TROUBLESHOOTING

Electrical troubleshooting is often required at the same time as mechanical troubleshooting. The volt-ohm-milliammeter (VOM) and the clamp-on ammeter are the primary instruments used, **Figure 41–26**.

You need to know what the readings should be to know whether the existing readings on a particular unit are correct. This is often not easy to determine because the desired reading may not be furnished. **Figure 41–27** shows typical horsepower-to-amperage ratings. It is a valuable tool for determining the correct amperage for a particular motor.

For a residence or small commercial building, one main power panel will normally serve the building. This panel is divided into many circuits. For a split-system type of cooling system, separate breakers (or fuses) are usually in the main

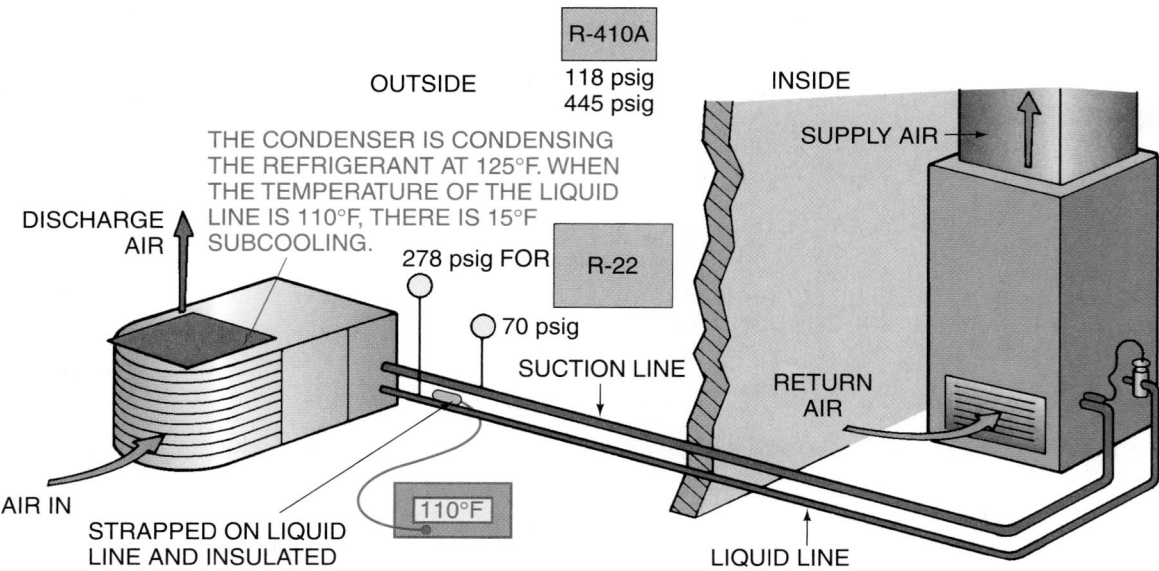

Figure 41–25 A unit with a TXV cannot be charged using the superheat method because the TXV controls superheat. The head pressure is raised to simulate a 95°F day, and the temperature of the liquid line is checked for the subcooling level. A typical system will have 10°F to 20°F of subcooling when the condenser contains the correct charge.

 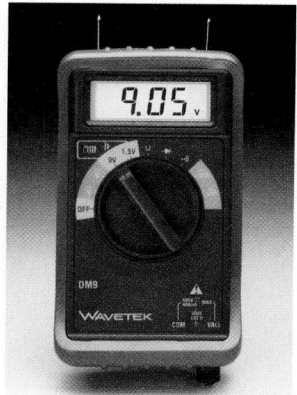

(A) **(B)**

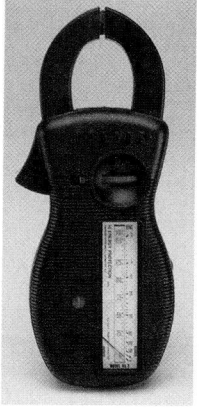 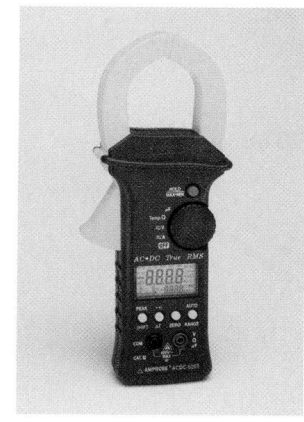

(C)

Figure 41–26 Instruments used to troubleshoot the electrical part of an air-conditioning system. **(A)** Analog VOM. **(B)** Digital VOM. **(C)** Clamp-on ammeter. *(A) and (B) Courtesy Wavetek. (C) Courtesy Amprobe Instrument Division of Core Industries, Inc.*

Motor	Single Phase		3-Phase Squirrel Cage Induction		
HP	120V	230V	230V	460V	575V
$\frac{1}{6}$	4.4	2.2			
$\frac{1}{4}$	5.8	2.9			
$\frac{1}{3}$	7.2	3.6			
$\frac{1}{2}$	9.8	4.9	2	1.0	0.8
$\frac{3}{4}$	13.8	6.9	2.8	1.4	1.1
1	16	8	3.6	1.8	1.4
$1\frac{1}{2}$	20	10	5.2	2.6	2.1
2	24	12	6.8	3.4	2.7
3	34	17	9.6	4.8	3.9
5	56	28	15.2	7.6	6.1
$7\frac{1}{2}$			22	11.0	9.0
10			28	14.0	11.0

APPROXIMATE FULL-LOAD AMPERAGE VALUES FOR ALTERNATING CURRENT MOTORS

Does not include shaded pole.

Figure 41–27 This chart shows approximate full-load amperage values. *Courtesy BDP Company*

SAFETY PRECAUTION: All safety practices must be observed while troubleshooting electrical systems. Many times the system must be inspected while power is on. Only let the insulated meter lead contact probes touch the hot terminals. Special care should be taken while troubleshooting the main power supply because the fuses may be correctly sized large enough to allow great amounts of current to flow before they blow (e.g., when a screwdriver slips in the panel and shorts across hot terminals). *Never* use a screwdriver in a hot panel.

If the power supply voltages are correct, move on to the various components. The path to the load may be the next item to check. If you are trying to get the compressor to run, remember that the compressor motor is operated by the compressor contactor. Is the contactor energized? Are the contacts closed? See **Figure 41–30** for a diagram. Note that in the diagram the only thing that will keep the contactor coil from being energized is the thermostat, the path, or the low-pressure control. If the outdoor fan is operating and the compressor is not, the contactor is energized because it also starts the fan. If the fan is running and the compressor is not, either the path (wiring or terminals) is not making good contact, or the compressor internal overload protector is open.

41.11 COMPRESSOR OVERLOAD PROBLEMS

When the compressor overload protector is open, touch the motor housing to see whether it is hot. If you cannot hold your hand on the compressor shell, the motor is too hot. Ask yourself these questions: Can the charge be low (this compressor is suction gas cooled)? Can the start-assist circuit not be working and the compressor not starting?

panel for the air handler or furnace for the indoor unit and for the outdoor unit. For a package or self-contained system, usually one breaker (or fuse) serves the unit, **Figure 41–28**. The power supply voltage is stepped down by the control transformer to the control voltage of 24 V.

Begin any electrical troubleshooting by verifying that the power supply is energized and that the voltage is correct. One way to do this is to go to the room thermostat and see whether the indoor fan will start with the FAN ON switch. See **Figure 41–29** for a wiring diagram of a typical split-system air conditioner.

The air handler or furnace, where the low-voltage transformer is located, is frequently under the house or in the attic. This quick check with the FAN ON switch can save you a trip under the house or to the attic. If the fan will start with the fan relay, several things are apparent: (1) the indoor fan operates, (2) there is control voltage, and (3) there is line voltage to the unit because the fan will run. When taking a service call over the phone, ask the homeowner whether the indoor fan will run. If it does not, take a transformer. This could save a trip to the supply house or the shop.

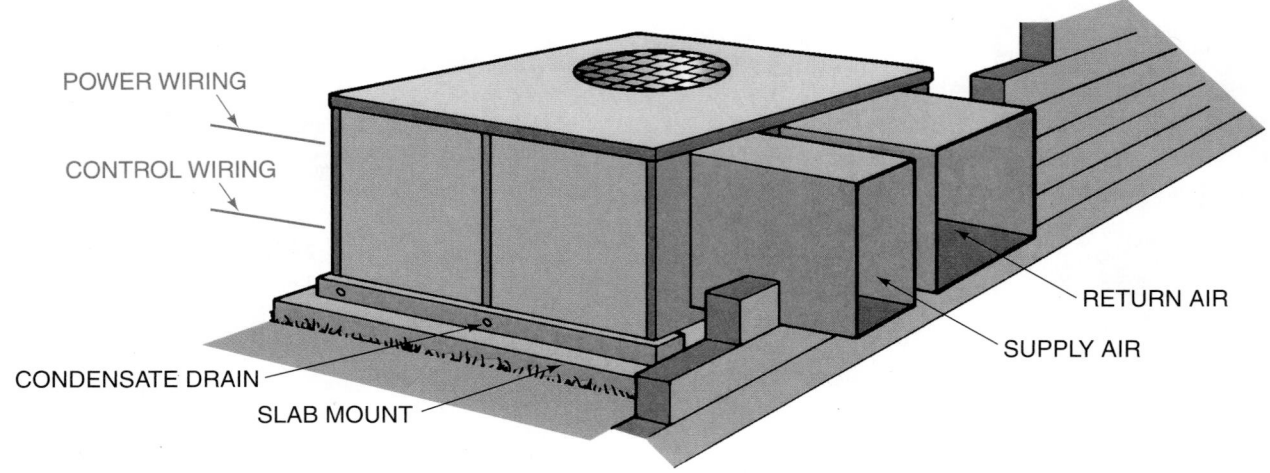

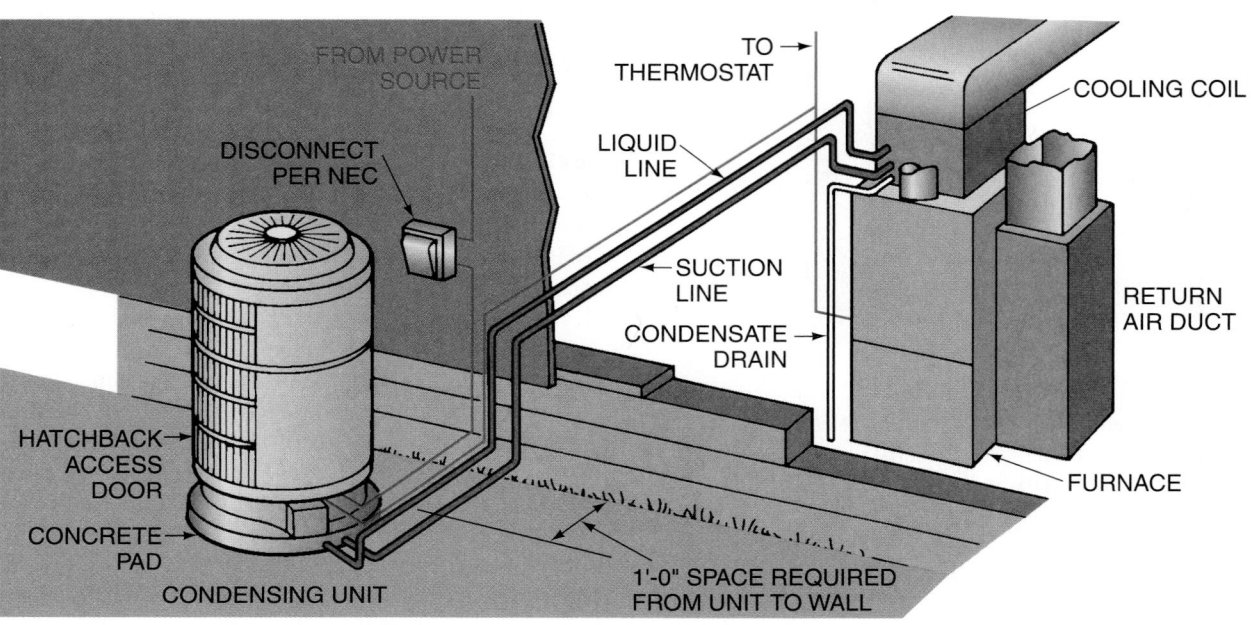

Figure 41–28   A typical package and split-system installation.

Allow the compressor to cool before restarting it. It is best that the unit be fixed so that it will not come back on for several hours. The best way to do this is to remove a low-voltage wire. If you pull the disconnect switch and come back the next day, the refrigerant charge may have migrated to the crankcase because there was no crankcase heat. If you want to start the unit within the hour rather than waiting, pull the disconnect switch and run a small amount of water through a hose over the compressor. SAFETY PRECAUTION: The standing water poses a potential electrical hazard. Be sure that all electrical components are protected with plastic or other waterproof covering. Do not come in contact with the water or electrical current when working around live electricity. It will take about 30 min to cool. Have the gages on the unit and a cylinder of refrigerant connected because when the compressor is started up by closing the disconnect, it may need refrigerant. If the system has a low charge and you have to get set up to charge after starting the system, the compressor may cut off again from overheating before you have a chance to get the gages connected.

41.12 COMPRESSOR ELECTRICAL CHECKUP

Manufacturers report that a fairly high percentage of compressors that are returned to them in warranty are actually not defective but were reported by the technician as defective. The technician should use great care when condemning a compressor, as it costs a lot of money in labor and materials to reclaim the refrigerant, ship in a new compressor, and ship the defective compressor back to the manufacturer for in-warranty work. When the compressor is out of warranty, the owner must bear that burden. The technician must be correct with the diagnosis.

When a compressor will not start, the problem can be either electrical or mechanical. The electrical portion of the problem is rather easy to prove. When the technician can

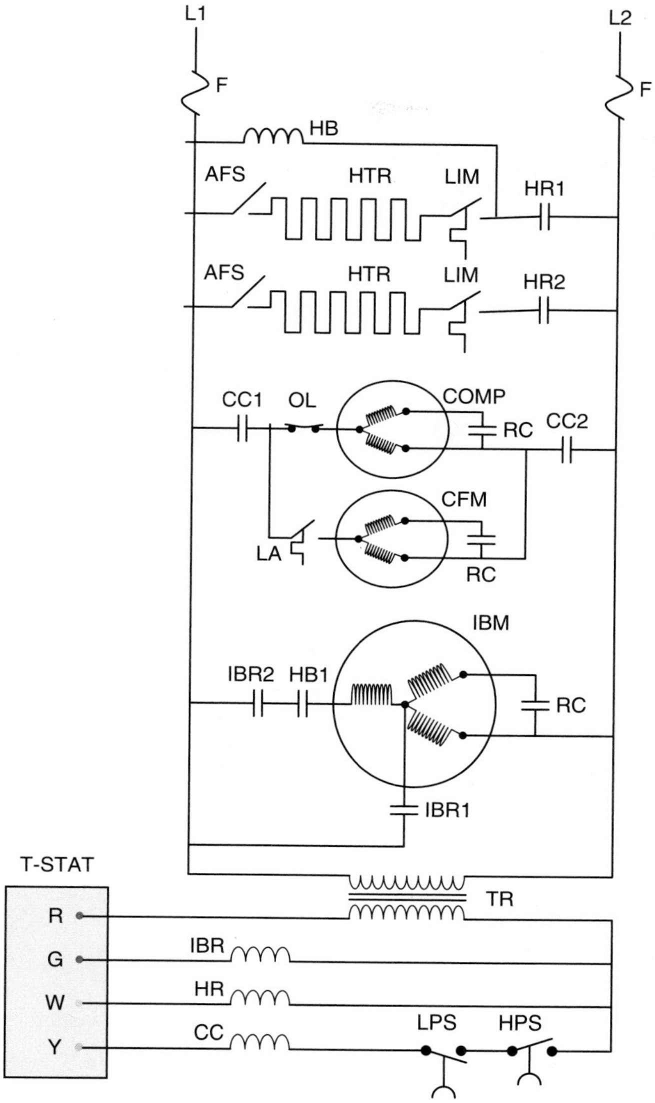

AFS	Airflow Switch
CC	Compressor Contactor (Located Outdoors)
CC1, CC2	Compressor Contactor Contacts (Located Outdoors)
CFM	Condenser Fan Motor (Located Outdoors)
COMP	Compressor (Located outdoors)
F	Fuse
G	Thermostat Fan Terminal
HB	Blower Relay Coil (Heating)
HB1	Blower Relay Contacts (Heating, Located Indoors)
HPS	High-Pressure Switch
HR	Heating Relay/Sequencer (Located Indoors)
HR1, HR2	Heating Relay Contacts
HTR	Heater (Located Indoors)
IBM	Indoor Blower Motor (Located Indoors)
IBR	Indoor Blower Relay Coil (Located Indoors)
IBR1, IBR2	Indoor Blower Relay Contacts
L1, L2	Incoming Power Lines
LA	Low Ambient Control
LIM	High-Limit Switch (Located Outdoors)
LPS	Low-Pressure Switch (Located Outdoors)
OL	Compressor Overload (Located Outdoors)
R	Thermostat Power Terminal
RC	Run Capacitor
T-STAT	Space Thermostat (Located in the Conditioned Space)
TR	Control Transformer (Located Indoors)
W	Thermostat Heating Terminal
Y	Thermostat Cooling Terminal

Figure 41–29   The wiring diagram of a split-system summer air conditioner. *Courtesy Climate Control*

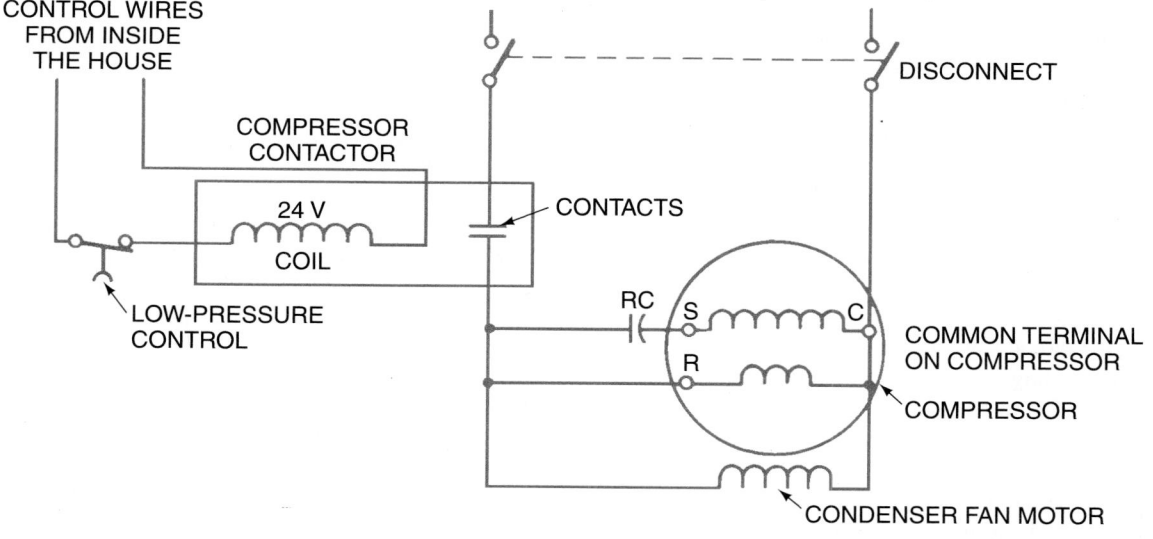

Figure 41–30   The wiring diagram of basic components in a control-and-compressor circuit.

prove the compressor is electrically sound and it will not start, the only other conclusion would be mechanical.

The steps for checking a single-phase compressor electrically are as follows:

1. Check the compressor from winding to ground using either an ohmmeter or a megohmmeter. A megohmmeter will detect much smaller ground circuits and is used on larger equipment exclusively. If there is a circuit showing to ground, remove all wires from the compressor terminals and check just the terminals. If the ground is no longer there, the ground is somewhere in the circuit wiring, not the compressor. If there is still a ground indicated, clean the compressor terminal block if you think there is a possibility that current can leak to ground due to dirt. If there is still a ground, condemn the compressor as grounded.
   SAFETY PRECAUTION: If a compressor has a large amount of liquid refrigerant in the compressor housing, such as when the crankcase heater has been off for a long period, a circuit to ground may be detected. If the compressor crankcase heat has been off and the compressor housing is cool, it is advisable to let it heat for about 24 hours and check it again. The reason for this condition is probably that the refrigerant has floated some system contamination that is normally dormant up into the motor windings. After the 24-hour heating period, the system should operate correctly. Also, dirt around the motor terminals can cause a slight circuit. If there is a question, clean them with a degreasing solvent and try again.

2. Check the start winding for correct resistance from common to start.

3. Check the run winding for correct resistance from common to run. If the correct resistances are not readily available, make sure that the start winding and run winding have different resistances. The start winding should have much more resistance than the run winding. Either winding can have some shorted turns in the motor. **Figure 41–31** shows a start winding with *shorted*, sometimes called *shunted*, winding. **Figure 41–32** shows a start winding that has an open circuit.

4. Check for continuity from the run to the common terminal. If this circuit is open, the winding thermostat may be open. If the compressor is hot to the touch, you may want to let it cool for several hours or cool it with water, **Figure 41–33.**
   SAFETY PRECAUTION: Make sure all power is off and do not stand in water when electrically checking any circuit.

5. Check the voltage from common to run and from common to start using two different voltmeters while starting the motor. This voltage must be within ±10% of the motor's ratings. For example, many motors are rated at 208/230. The minimum voltage would be 187.2 V (208 × 0.90 = 187.2) and the maximum voltage would be 253 V (230 × 1.10 = 253). This voltage must be verified as close to the compressor terminals as possible, but do not use the terminals themselves. You can follow

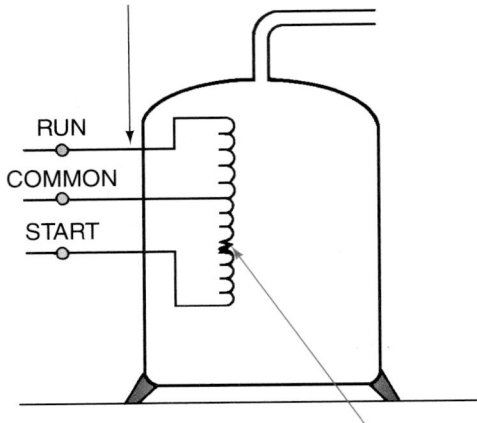

THE MOTOR TERMINALS ARE INSULATED WHERE THEY PASS THROUGH THE COMPRESSOR HOUSING.

RUN
COMMON
START

SOME OF THE WINDINGS ARE TOUCHING EACH OTHER AND REDUCING THE RESISTANCE IN THE START WINDING.

Figure 41–31 A compressor with a shorted, or shunted, winding.

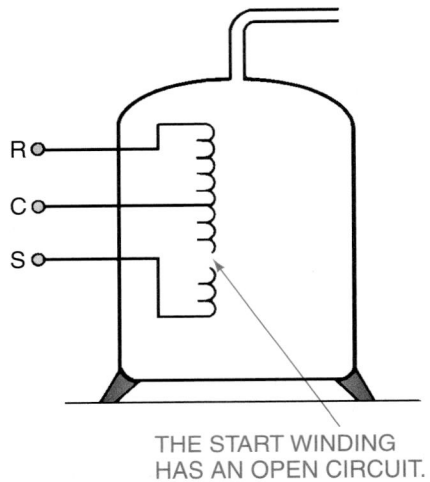

R
C
S

THE START WINDING HAS AN OPEN CIRCUIT.

Figure 41–32 A compressor with an open start winding.

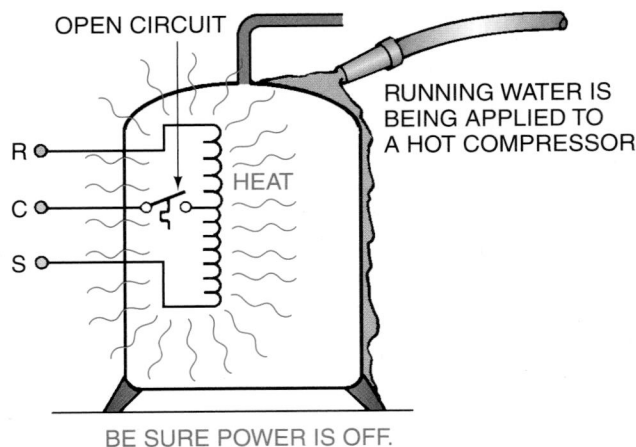

OPEN CIRCUIT

RUNNING WATER IS BEING APPLIED TO A HOT COMPRESSOR.

R
C
S

HEAT

BE SURE POWER IS OFF.

Figure 41–33 Cooling a compressor with water.

the wiring back to the first junction, such as the load side of the compressor contactor. It is not unusual for a loose connection to create a low line-voltage problem at the compressor. If there is any discoloration at the motor terminals, suspect a loose connection. SAFETY PRECAUTION: Never start a compressor with the motor terminal cover box removed. If a motor terminal were to blow out while starting, there is the potential for a high-velocity blowtorch type of fire. The electrical arc may ignite the escaping oil with the refrigerant, propelling it outward. The box and cover are built to enclose this type of malfunction.

The steps to check a three-phase compressor electrically are as follows:

1. Check the compressor from all windings to ground using either an ohmmeter or a megohmmeter as described previously in step 1. Pay attention to the Safety Precautions.

2. Check each set of windings from winding to winding, T1 to T2, T1 to T3, and T2 to T3 on a three-lead motor. Some larger motors may be dual motors in the same housing; check both motors. There should be the same resistance across each motor winding on a three-phase motor. There should be no circuit from motor to motor in a dual motor application.

When you have completed these tests, and the motor passes them, you have done all you can in the field to prove that the motor is sound. There is one more thing you may do for single-phase motors. You may use a test cord that manually takes the place of the starting circuit components to start the compressor. If the compressor will not start after being declared sound and with a test cord, you can condemn the motor with confidence.

Mechanically checking the compressor comes next. When you have full voltage to the compressor terminals and it will not start, you should expect that the compressor is stuck. To accomplish this, you should have two voltmeters for single-phase compressors and check from common to run and common to start at the same time. For a three-phase compressor you should have three voltmeters and check all three windings at the same time. With three-phase compressors you may try reversing any two of the main leads to the contactor and start it again. If the motor has two contactors, it is a part winding start compressor, and it has two motors. Be sure to reverse the leads that feed both contactors. This will reverse the motor. If it will start backward, it may run long enough for you to make a plan to change the compressor at a more convenient time, but you should not rely on it running normally for an extended period. Internally the compressor may be binding, and the bind will reoccur. Changing the compressor is the most permanent solution.

The most difficult problem to troubleshoot is to check the capacity of a compressor. The complaint will probably be that the compressor runs all the time but the space temperature is not cool enough. This will be the same for cooling systems and refrigeration systems. Many of the compressors will have two, three, or four cylinders, and one or more of them may not be functioning. The smaller the compressor, the more difficult

this is to prove. For any compressor that does not have the capacity to unload, or to run at reduced load or speed, follow these guidelines:

1. Try to get the system to operate as close to design conditions as possible. The amperage is the key to this procedure. If the amperage is considerably lower than the RLA and the voltage is correct, the compressor is not pumping to capacity.

2. If you have gages on the system, the suction pressure may be a little high with the discharge pressure a little low. This is a sign the compressor is not doing maximum work. **Figure 41–34** shows an example of a residential system that is not operating at the proper gage readings. Notice that the outdoor temperature is hot and the indoor unit is operating at inside design temperature. This system will not remove humidity because the coil temperature is well above the dew point temperature of the indoor air. The indoor air may be cool, but it will feel warm because of the humidity, and the compressor will run all the time. This could very well be first thing in the morning after the unit has run all night because as the day goes by, the indoor air temperature will rise.

3. This is probably the only time that the suction pressure rises, the head pressure drops, and the amperage drops.

4. You may want to see if you can raise the head pressure by partially blocking the condenser airflow. Usually you cannot get the head pressure to rise to the head pressure of a design day; if you do, the suction pressure will rise accordingly.

The other mechanical problem that can occur with a compressor is an internal leak, such as when a discharge valve reed becomes split or a head gasket leaks through. To determine what the problem is, you should install gages and an ammeter. The unit may appear to be running correctly, but the compressor will be hot and will likely shut off from time to time from high temperature. Try to raise the head pressure to about 300 psig for R-22 and 475 psig for R-410A and then shut the system off. Be prepared to listen closely to the compressor discharge line. You may use a long screwdriver or stick with one end placed on the discharge line and the other end to your ear. This rudimentary "stethoscope" works well for listening to internal sounds. SAFETY PRECAUTION: Never use an ohmmeter to check a live circuit.

41.13 TROUBLESHOOTING THE CIRCUIT ELECTRICAL PROTECTORS—FUSES AND BREAKERS

One service call that must be treated cautiously is when a circuit protector, such as a fuse or breaker, opens the circuit. The compressor and fan motors have protection that will normally guard them from minor problems. The breaker or fuse

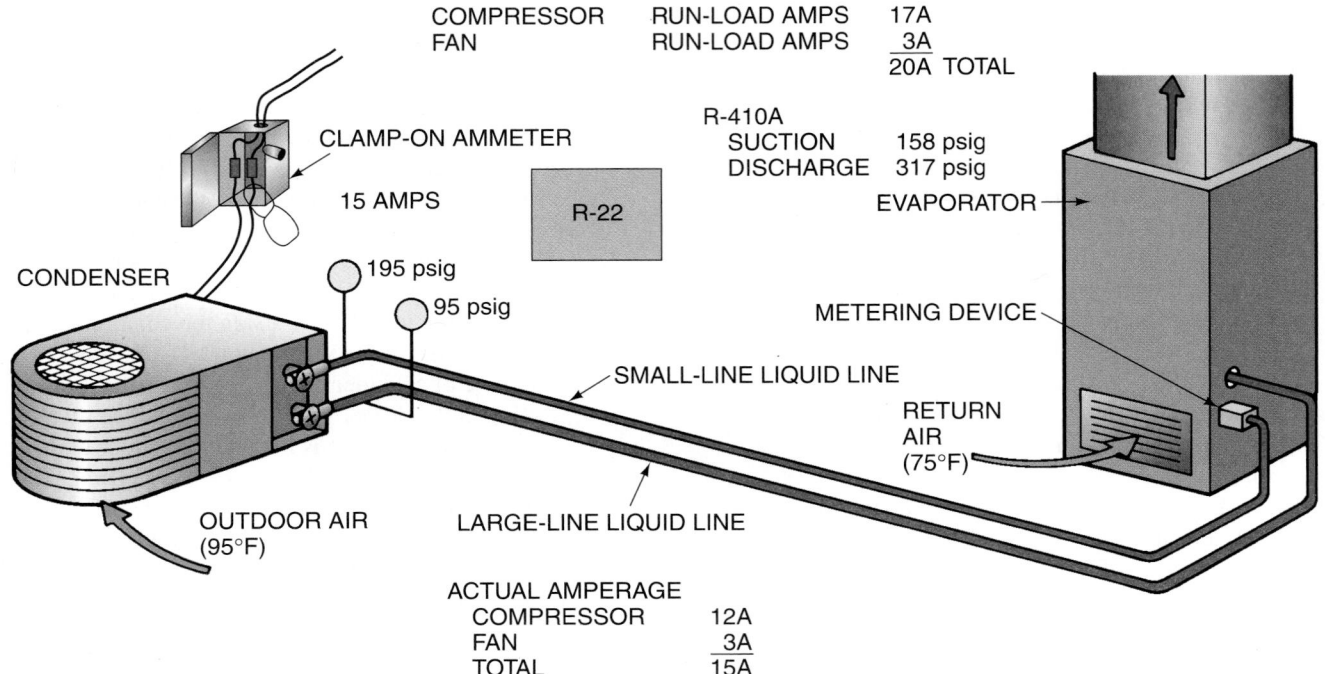

Figure 41–34 This system compressor is not functioning up to capacity. The head pressure is low, the suction pressure is high, and the amperage is low. The compressor is defective.

is for large current surges in the circuit. When one is tripped, do not simply reset it. Perform a resistance check of the compressor section, including the fan motor. SAFETY PRECAUTION: The compressor may be grounded (may have a circuit to the case of the compressor), and it could be dangerous to try to start it. **Be sure to isolate the compressor circuit before condemning the compressor.** Take the motor leads off the compressor to check for a ground circuit in the compressor, **Figure 41–35.**

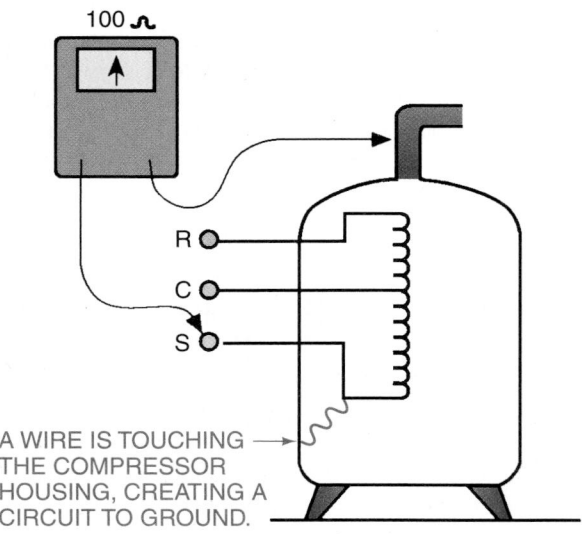

Figure 41–35 A compressor with winding shorted to the casing.

HVAC GOLDEN RULES

- Always carry the proper tools and avoid extra trips through the customer's house.
- Make and keep firm appointments. Call ahead if you are delayed. The customer's time is valuable, too.
- If possible, do not block the customer's driveway.

Added Value to the Customer

Here are some simple, inexpensive procedures that may be included in the basic service call.

- Touch-test the suction and liquid lines to determine possible over- or undercharge.
- Clean or change filters as needed.
- Lubricate all bearings as needed.
- Replace all panels with the correct fasteners.
- Inspect all contactors and wiring. This may lead to more paying service.
- Make sure that the condensate line is draining properly. Clean if needed. Algaecide tablets may be necessary.
- Check evaporator and condenser coils for blockage or dirt.

41.14   SERVICE TECHNICIAN CALLS

Troubleshooting can take many forms and cover many situations. The following actual troubleshooting situations will help you understand what the service technician does while solving actual problems.

PREVENTIVE MAINTENANCE

Air-conditioning equipment preventive maintenance involves the indoor airside, the outdoor airside (air cooled and water cooled), and electrical circuits.

The indoor airside maintenance is much the same for air conditioning as for electric heat, where motor and filter maintenance is involved. The only difference is that the evaporator coil operates below the dew point temperature of the air and is wet. It will become a superfilter for any dust particles that may pass through the filter or leak in around loose panel compartment doors. Many air handlers have draw-through coil-fan combinations. The air passes through the coil before the fan. When the fan blades become dirty and loaded with dirt, it is a sure sign that the coil is dirty. The dirt has to pass through the coil first, and it is wet much of the time. When the fan is dirty, the coil must be cleaned.

The coil may be cleaned in place in the unit by two methods. One is by the use of a special detergent manufactured to work while the coil is wet. This detergent is sprayed on the coil and into the core of the coil with a hand-pump-type sprayer, similar to a garden sprayer. When the unit is started, the condensate will carry the dirt down the coil and down the condensate drain line. This type of cleaner is for light-duty cleaning. Care must be taken that the condensate drain line does not become clogged with the dirt from the coil and pan.

The coil may also be cleaned by shutting the unit down and applying a more powerful detergent to the coil, forcing it into the coil core. After the detergent has had time to work, the coil is then sprayed with a water source, such as a water hose. Care must be taken not to force too much water into the coil, or the drain pan will overflow. The water must not be applied faster than the drain can accept it.

Special pressure cleaners that have a nozzle pressure of 500 to 1000 psig may be used. These units have a low water flow, about 5 gal/min, and will not overflow an adequate drain system. The high nozzle pressure enables the water to clean the core of the coil.

It is always best to "backwash" a coil when cleaning with water. The water is forced through the coil in the opposite direction to the airflow for backwashing. The reason for this is most of the dirt will accumulate on the inlet of the coil and be progressively less as the air moves through the coil. If the coil is not backwashed, you may only drive dirt accumulation to the center of the coil. SAFETY PRECAUTION: Never use hot water or a steam cleaner on refrigeration equipment if refrigerant IS in the unit because pressure will rise high enough to burst the unit at the weakest point. This may be the compressor shell. A refrigerant system must be open to the atmosphere before hot water or steam is used for cleaning.

In some cases, the coil cannot be cleaned in the unit and must be removed for cleaning. If the unit has service valves, the refrigerant may be pumped into the condenser-receiver and the coil removed from the unit. The coil will not have any refrigerant in it and may be cleaned with approved detergent and hot water or with a steam cleaner. Approved cleaners may be purchased at any air-conditioning supply house. SAFETY PRECAUTION: Follow the directions. Make sure no water can enter the coil through the piping connections.

The outdoor unit may be either air cooled or water cooled. Air-cooled units have fan motors that must be lubricated. Some motors require lubrication only after several years of operation. Then a recommended amount of approved oil is added to the oil cup. Some motors require more frequent lubrication.

The fan blades should be checked to make sure that they are secure on the shaft. At the same time, the rain shield on top of the fan motor should be checked if there is one. It should not allow water to enter the motor on upflow units where the motor is out in the open.

Coils may become dirty from pulling dust through the coils. It is hard to look at any coil and determine how much dirt is in the core. The coil should be cleaned at the first sign of dirt buildup or high operating head pressure.

Condenser coils may be easier to clean because they are on the outside. SAFETY PRECAUTION: Make sure that all power is turned off, locked, and tagged where someone cannot accidentally turn it on. The fan motor must be covered and care should be taken that water does not enter the controls. They may not all be in the control cabinet. Apply an approved detergent on the coil using a hand-pump sprayer. Soak the coil to the middle. Let the detergent soak for 15 to 30 min, then backwash the coil using a garden hose or spray cleaner.

Water-cooled equipment must be maintained like air-cooled equipment. Two things are done as part of the maintenance program to minimize the mineral concentrations in the water. One is to make sure the cooling tower has an adequate water bleed system. This is a measured amount of water that goes down the drain. This water is added back to the system with the supply water makeup line. Owners may suggest that the bleed system be shut off because all they see is water going down the drain. They may not realize that the purpose of this is to dilute the minerals in the water.

The other maintenance procedure is to use the correct water treatment. A water treatment specialist should supervise the setup of the treatment procedures. This will ensure the best quality water in the system.

Electrical preventive maintenance consists of examining any contactors and relays for frayed wire and pitted contacts. When these occur, they should be taken care of before you leave the job. There can be no prediction as to how long they will last. Single-phase compressors will just stop running if a wire burns in two or a contact burns away. A three-phase compressor will try to run on two phases and this is very hard on the motor. Wires should be examined to make sure they are not rubbing on the frame and wearing the insulation off the wire. Wires resting on copper lines may rub a hole in the line, resulting in a refrigerant leak.

SERVICE CALL 1

A residential customer calls to report that the central air conditioner at a residence is not cooling enough and runs continuously. *The problem is low R-410A refrigerant charge.*

The technician arrives and finds the unit running. The temperature indicator on the room thermostat shows the thermostat is set at 72°F, and the thermometer on the thermostat indicates that the space temperature is 80°F. The air feels very humid. The technician notices that the indoor fan motor is running. The velocity of the air coming out of the registers seems adequate, so the filters are not stopped up.

The technician goes to the condensing unit and hears the fan running. The air coming out of the fan is not warm. This indicates that the compressor is not running. The door to the compressor compartment is removed, and it is noticed that the compressor is hot to the touch. This is an indication that the compressor has been trying to run. The gages are installed on the service ports, and they both read the same because the system has been off. The gages read 235 psig, which corresponds to 80°F. The residence is 80°F inside, and the ambient is 90°F, so it can be assumed that the unit has some liquid refrigerant in the system because the pressure corresponds so closely to the chart. If the system pressure were 170 psig, it would be obvious that little liquid refrigerant is left in the system. A large leak should be suspected.

The technician decides that the unit must have a low charge and that the compressor is off because of the internal overload protector. The technician pulls the electrical disconnect and takes a resistance reading across the compressor terminals with the motor leads removed. The meter shows an open circuit from common to run and common to start. It shows a measurable resistance between run and start. This indicates that the motor-winding thermostat must be open. This indication is verified by the hot compressor.

All electrical components and terminals are covered with plastic. The electrical disconnect is locked out. A water hose is connected, and a small amount of water is allowed to run over the top of the compressor shell to cool it. A cylinder of refrigerant is connected to the gage manifold so that when it is time to start the compressor, the technician will be ready to add refrigerant and keep the compressor running. The gage lines are purged of any air that may be in them. While the compressor is cooling, the technician changes the air filters and lubricates the condenser and the indoor fan motors. After about 30 min the compressor seems cool. The water hose is removed, and the water around the unit is allowed a few minutes to run off. The technician is careful not to stand in the water. When the disconnect switch is closed, the compressor starts. The suction pressure drops to 75 psig. The normal suction pressure is 118 psig for the system because the refrigerant is R-410A. Refrigerant is added to the system to bring the charge up to normal. The space temperature is

80°F, so the suction pressure will be higher than normal until the space temperature becomes normal. The technician must have a reference point to get the correct charge. The following reference points are used in this situation because no factory chart is available.

1. The outside temperature is 90°F; the normal operating head pressure should correspond to a temperature of 90°F + 30°F = 120°F or 417 psig. It should not exceed this when the space temperature is down to a normal 75°F. This is a standard-efficiency unit. The technician restricts the airflow to the condenser and causes the head pressure to rise to 440 psig as refrigerant is added.
2. The suction pressure should correspond to a temperature of about 35°F cooler than the space temperature of 80°F, or 45°F, which corresponds to a pressure of 128 psig for R-410A.
3. The system has a capillary tube metering device, so some conclusions can be drawn from the temperature of the refrigerant coming back to the compressor. A thermometer lead is attached to the suction line at the condensing unit. As refrigerant is added in the liquid state, the technician notices that the refrigerant returning from the evaporator is getting cooler. The evaporator is about 30 ft from the condensing unit, and some heat will be absorbed into the suction line returning to the condensing unit. The technician uses a guideline of 15°F of superheat with a suction line this length. This assumes that the refrigerant leaving the evaporator has about 10°F of superheat and that another 5°F of superheat is absorbed along the line. When these conditions are reached, the charge is very close to correct and no more refrigerant is added.

A leak check is performed, and a flare nut is found to be leaking. It is tightened, and the leak is stopped. The technician loads the truck and leaves. A call later in the day shows that the system is working correctly.

SERVICE CALL 2

A residential customer calls the air-conditioning service company. The air-conditioning unit has been cooling correctly until afternoon, when it quit cooling. This is a residential unit that has been in operation for several years. *The problem is a complete loss of charge.*

The technician arrives at the residence and finds the thermostat to be set at 75°F and the space temperature to be 80°F. The air coming out of the registers feels the same as the return air temperature. There is plenty of air velocity at the registers, so the filters appear to be clean.

The technician goes to the back of the house and finds that the fan and compressor in the condensing unit are not running. The breaker at the electrical box is in the ON position, so power must be available. A voltage check shows that the voltage is 235 V. A look at the wiring diagram shows that there should be 24 V between the C and Y terminals to energize the contactor. The voltage actually reads 25 V, slightly above normal, but so is the line voltage of 235 V. The conclusion is that the thermostat is calling for cooling, but the contactor is not energized. The only safety control in the contactor coil circuit is the low-pressure control, so the unit must be out of refrigerant.

Gages are fastened to the gage ports, and the pressure is 0 psig. The technician connects a cylinder of R-22 and starts adding a trace amount of refrigerant for the purpose of leak checking. When the pressure is up to about 2 psig, the refrigerant is stopped. Nitrogen is added to push the pressure up to 50 psig and gas can be heard leaking from the vicinity of the compressor suction line. A hole is found in the suction line where the cabinet had been rubbing against it. This accounts for the fact that the unit worked well up to a point and quit working almost immediately.

The trace refrigerant (R-22) and nitrogen is allowed to escape to the atmosphere from the system, and the hole is patched with silver solder. A liquid-line drier is installed in the liquid line, and a triple evacuation is performed to remove any contaminants that may have been pulled into the system. The system is charged and started. The technician follows the manufacturer's charging chart to ensure the correct charge.

SERVICE CALL 3

A commercial customer calls and reports that the air-conditioning unit at a small office building is not cooling. The unit was operating and cooling yesterday afternoon when the office closed. *The problem is that someone turned the thermostat down to 55°F late yesterday afternoon, and the condensate on the evaporator froze solid overnight trying to pull the space temperature down to 55°F.*

The technician arrives at the site and notices that the thermostat is set at 55°F. An inquiry shows that one of the employees was too warm in an office at the end of the building and turned down the thermostat to cool that office. The technician notices no air is coming out of the registers. The air handler is located in a closet at the front of the building. Examination shows that the fan is running, and the suction line is frozen solid at the air handler.

The technician stops the compressor and leaves the evaporator fan operating by turning the heat-cool selector switch to OFF and the fan switch to ON. It is going to take a long time to thaw the evaporator, probably an hour. The technician leaves the following directions with the office manager.

1. Let the fan run until air comes out of the registers, which will verify that airflow has started through the ice.
2. When air is felt at the registers, wait 30 min to ensure that all ice is melted and turn the thermostat back to COOL to start the compressor.

The technician then looks in the ceiling to check the air damper on the duct run serving the back office of the employee who was not cool enough and turned down the thermostat. The damper is nearly closed; it probably was

brushed against by a telephone technician who had been working in the ceiling. The damper is reopened.

A call later in the day indicates that the system is working again.

SERVICE CALL 4

A residential customer reports that the air-conditioning unit is not cooling correctly. *The problem is a dirty condenser.* The unit is in a residence and next to the side yard. The homeowner mows the grass, and the lawnmower throws grass on the condenser. This is the first hot day of summer, and the unit had been cooling the house until the weather became hot.

The technician arrives at the job and notices that the thermostat is set for 75°F but the space temperature is 80°F. Plenty of slightly cool air is coming out of the registers. The technician goes to the side of the house where the unit is located. The suction line feels cool, but the liquid line is very hot. An examination of the condenser coil shows that the coil is clogged with grass and is dirty. SAFETY PRECAUTION: The technician shuts off the unit with the breaker at the unit and locks it out, then takes enough panels off to be able to spray coil cleaner on the coil. A high-detergent coil cleaner is applied to the coil and allowed to stand for about 15 min to soak into the coil dirt. While this is occurring, the motors are oiled and the filters are changed. SAFETY PRECAUTION: To keep the condenser fan motor from getting wet when the coil cleaner is washed off, it is covered with a plastic bag.

A water hose with a nozzle to concentrate the water stream is used to wash the coil in the opposite direction of the airflow. One washing is not enough, so the coil cleaner is applied to the coil again and allowed to set for another 15 min. The coil is washed again and is now clean.

The unit is assembled and started. The suction line is cool, and the liquid line is warm to the touch. The technician decides not to put gages on the system and leaves. A call back later in the day to the homeowner indicates that the system is operating correctly.

SERVICE CALL 5

A homeowner reports that the air-conditioning unit is not cooling. This is a residential high-efficiency unit. *The problem is that the TXV is defective.*

The service technician arrives and finds the house warm. The thermostat is set at 74°F, and the house temperature is 82°F. The indoor fan is running, and plenty of air is coming out of the registers.

The technician goes to the condensing unit and finds that the fan and compressor are running, but they quickly stop. The suction line is not cool. Gages are attached to the gage ports. The low-side pressure is 25 psig, and the head pressure is 170 psig; this corresponds closely to the ambient air temperature. The unit is off because of the low-pressure control. When the pressure rises, the compressor restarts but stops in about 15 sec. The liquid-line sight glass is full of refrigerant. The technician concludes that a restriction exists on the low side of the system. The TXV is a good place to start when there is an almost complete blockage. ♻ The system does not have service valves, so the charge has to be removed and recovered in an approved cylinder. ♻

A new TXV is soldered into the system. It takes an hour to complete the TXV change. When the valve is changed, the system is leak checked and evacuated to a deep vacuum. A charge is measured into the system, and the system is started. It is evident from the beginning that the valve change has repaired the unit.

SERVICE CALL 6

An office manager calls to report that the air-conditioning unit in a small office building is not cooling. The unit was cooling correctly yesterday afternoon when the office closed. *The problem is that a night electrical storm tripped a breaker on the air-conditioner air handler.* The power supply is at the air handler.

The technician arrives and goes to the space thermostat. It is set at 73°F, and the temperature is 78°F. There is no air coming out of the registers. The fan switch is turned to ON, and the fan still does not start. The technician checks the low-voltage power supply, which is in the attic at the air handler, and finds the tripped breaker. SAFETY PRECAUTION: Before resetting it, the technician decides to check the unit electrically.

A resistance check of the fan circuit and the low-voltage control transformer proves there is a measurable resistance. The circuit seems to be safe. The circuit breaker is reset and stays in. The thermostat is set at COOL, and the system starts. A call later shows that the system is operating properly.

SERVICE CALL 7

A residential customer calls. The unit is not running. The homeowner found the breaker at the condensing unit tripped and reset it several times, but it did not stay set. *The problem is that the compressor motor winding is grounded to the compressor shell and tripping the breaker.* SAFETY PRECAUTION: The homeowner should be warned to reset a breaker no more than once.

The technician arrives and goes straight to the condensing unit with electrical test equipment. The breaker is in the tripped position and is moved to the OFF position. SAFETY PRECAUTION: The voltage is checked at the load side of the breaker to ensure that there is no voltage. The breaker has been reset several times and is not to be trusted. When it is determined that the power is definitely off, the ohmmeter is connected to the load side of the breaker. The ohmmeter reads infinity, meaning no circuit. The compressor contactor is not energized, so the fan and compressor are not included in the reading. The technician pushes in the armature of the compressor contactor to make the con-

tacts close. The ohmmeter now reads 0, or no resistance, indicating that a short exists. It cannot be determined whether the fan or the compressor is the problem, so the compressor wires are disconnected from the bottom of the contactor. The short does not exist when the compressor is disconnected. This verifies that the short is in the compressor circuit, possibly the wiring. The meter is moved to the compressor terminal box and the wiring is disconnected. The ohmmeter is attached to the motor terminals at the compressor. The short is still there, so the compressor motor is condemned.

The technician must return the next day to change the compressor. The technician disconnects the control wiring and insulates the disconnected compressor wiring so that power can be restored. This must be done to keep crankcase heat on the unit until the following day. If it is not done, most of the refrigerant charge will be in the compressor crankcase and too much oil will move with the refrigerant when it is recovered.

The technician performs one more task before leaving the job. The Schrader valve fitting at the compressor is slightly depressed to determine the smell of the refrigerant. The refrigerant has a strong acid odor. A suction-line filter drier with high acid removal capacity needs to be installed along with the normal liquid-line drier. ♻ The technician returns on the following day, recovers the refrigerant from the system, and then changes the compressor, adding the suction- and liquid-line driers. ♻ The unit is leak checked, evacuated, and the charge measured into the system with a set of accurate scales. The system is started, and the technician asks the customer to leave the air conditioner running, even though the weather is mild, to keep refrigerant circulating through the driers to clean it in case any acid is left in the system.

The technician returns on the fourth day and measures the pressure drop across the suction-line drier to make sure that it is not restricted with acid from the burned-out compressor. It is well within the manufacturer's specification.

SERVICE CALL 8

A residential customer calls. The customer can hear the indoor fan motor running for a short time; then it stops. This happens repeatedly. *The problem is that the indoor fan motor is cycling on and off on its internal thermal overload protector because the fan capacitor is defective.* The fan will start and run slowly then stop.

The technician knows from the work order to go straight to the indoor fan section. It is in the crawl space under the house, so electrical instruments are carried on the first trip. The fan has been off long enough to cause the suction pressure to go so low that the evaporator coil is frozen. The breaker is turned off during the check. The technician suspects the fan capacitor or the bearings, so a fan capacitor is carried along. After bleeding the charge from the fan capacitor, it is checked with the ohmmeter to see whether it will charge and discharge. SAFETY PRE-CAUTION: Ensure that no electrical charge is in the capacitor before checking it with an ohmmeter. A 20,000-Ω resistor should be used between the two terminals to bleed off any charge. The capacitor will not charge and discharge, so it is changed.

The technician oils the fan motor and starts it from under the house with the breaker. The fan motor is drawing the correct amperage. The coil is still frozen, and the technician must set the space thermostat to operate only the fan until the homeowner feels air coming out of the registers. Then only the fan should be operated for one-half hour (to melt the rest of the ice from the coil) before the compressor is started.

The following service calls do not include solutions. The solutions can be found in the *Instructor's Guide*.

SERVICE CALL 9

A commercial customer reports that an air-conditioning unit in a small office building is not cooling on the first warm day.

The technician first goes to the thermostat and finds that it is set at 75°F, and the space temperature is 82°F. The air feels very humid. No air is coming from the registers. The air handler is in the attic. An examination of the condensing unit shows that the suction line has ice all the way back to the compressor. The compressor is still running.

What is the likely problem and the recommended solution?

SERVICE CALL 10

A residential customer calls. The customer reports the unit in their residence was worked on in the early spring under the service contract. The earlier work order shows that refrigerant was added by a technician new to the company.

The technician finds that the thermostat is set at 73°F, and the house is 77°F. The unit is running, and cold air is coming out of the registers. Everything seems normal until the condensing unit is examined. The suction line is cold and sweating, but the liquid line is only warm.

Gages are fastened to the gage ports, and the suction pressure is 85 psig, far from the correct pressure of 74 psig (77°F − 35°F = 42°F, which corresponds to about 74 psig). The head pressure is supposed to be 243 psig to correspond to a condensing temperature of 115°F (85°F + 30°F = 115°F or 243 psig). The head pressure is 350 psig. The unit shuts off after about 10 min running time.

What is the likely problem and the recommended solution?

SERVICE CALL 11

A residential customer calls to report that the air-conditioning unit is not cooling enough to reduce the space temperature to the thermostat setting. It is running continuously on this first hot day.

The technician arrives and finds the thermostat set at 73°F; the house temperature is 78°F. Air is coming out of the registers, so the filters must be clean. The air is cool but not as cold as it should be. The air temperature should be about 55°F, and it is 63°F.

At the condensing unit the technician finds that the suction line is cool but not cold. The liquid line seems extra cool. It is 90°F outside, and the condensing unit should be condensing at about 120°F. If the unit had 15°F of subcooling, the liquid line should be warmer than hand temperature, yet it is not.

Gages are fastened to the service ports to check the suction pressure. The suction pressure is 95 psig, and the discharge pressure is 225 psig. The airflow is restricted to the condenser, and the head pressure gradually climbs to 250 psig; the suction pressure goes up to 110 psig. The compressor should have a current draw of 27 A, but it only draws 15 A.

What is the likely problem and the recommended solution?

SERVICE CALL 12

A commercial customer calls to indicate that the air-conditioning unit in a small office building was running but suddenly shut off.

The technician arrives, goes to the space thermostat, and finds it set at 74°F; the space temperature is 77°F. The fan is not running. The fan switch is moved to ON, but the fan motor does not start. It is decided that the control voltage power supply should be checked first. The power supply is in the roof condensing unit on this particular installation. SAFETY PRECAUTION: The ladder is placed against the building away from the power line entrance. Electrical test instruments are taken to the roof along with tools to remove the panels. The breakers are checked and seem to be in the correct ON position. The panel is removed where the low-voltage terminal block is mounted. A voltmeter check shows there is line voltage but no control voltage.

What is the likely problem and the recommended solution?

SAFETY PRECAUTION: All troubleshooting by students or inexperienced people should be performed under the supervision of an experienced person. If you do not know whether a situation is safe, consider it unsafe.

SUMMARY

- The superheat for an operating evaporator coil is used to prove coil performance.
- Standard air-conditioning conditions for rating equipment are 80°F return air with a humidity of 50% when the outside temperature is 95°F.
- The typical customer operates the equipment at 75°F return air with a humidity of 50%. This is the condition that the normal pressures and temperatures in this unit are based on.

- A high-efficiency evaporator normally has a boiling temperature of 45°F.
- Gages fastened to the high side of the system are used to check the head pressure. The head pressure is a result of the refrigerant condensing temperature.
- The condensing refrigerant temperature has a relationship to the medium to which it is giving up heat.
- A standard-efficiency unit normally condenses the refrigerant at no more than 30°F higher than the air to which the heat is rejected.
- A high-efficiency condensing unit normally condenses the refrigerant at a temperature as low as 20°F warmer than the air used as a condensing medium.
- *Approach temperature,* or *temperature split,* are terms used to describe the temperature difference between two different heat exchange mediums, for example, the return air temperature and coil boiling temperature of a refrigerant evaporator.
- *Temperature difference* is a term used to describe the difference between the inlet and outlet temperature of a heat exchanger, for example, the inlet air and the outlet air of an evaporator.
- The electronic thermometer may be used to check wet-bulb (using a wet wick on one bulb) and dry-bulb temperatures.
- The condenser has three functions: to take the superheat out of the discharge gas, to condense the hot gas to a liquid, and to subcool the refrigerant.
- A typical condenser may subcool the refrigerant 10°F to 20°F lower than the condensing temperature.
- Two types of metering devices are normally used on air-conditioning equipment: the fixed bore (orifice or capillary tube) and the TXV.
- The fixed-bore metering device uses the pressure difference between the inlet and outlet of the device for refrigerant flow. It does not vary in size.
- The TXV modulates or throttles the refrigerant to maintain a constant superheat.
- To correctly charge an air-conditioning unit, the technician must follow the manufacturer's recommendations.
- The TXV system normally has a sight glass in the liquid line to aid in charging. A subcooling temperature check may be used when no sight glass is available.
- SAFETY PRECAUTION: Before checking a unit electrically, the proper voltage and the current draw of the unit should be known.
- The main electrical power panel may be divided into many circuits. The air-conditioning system is normally on two separate circuits for a split system and one circuit for a package system.
- When a hot compressor is started, assume that the system is low in refrigerant. A cylinder of refrigerant should be connected, so refrigerant may be added before the compressor shuts off again.

REVIEW QUESTIONS

1. Before installing a gage on the high-pressure side of a system, you should do which of the following?
   - A. Turn on the system.
   - B. Turn off the system.
   - C. Adjust the system to a neutral position.
   - D. Have a cylinder approved by the DOT nearby.

2. Capacitors should be shorted with a _____ before checking with an ohmmeter.
3. The high- and low-side pressures on an operating air-conditioning system can be converted to the _____ and _____ temperatures by using a temperature/pressure relationship chart.
4. Why do most air-conditioning systems not need a defrost system?
5. A typical standard-efficiency evaporator under standard conditions will operate at a temperature of
   - A. 30°F.
   - B. 40°F.
   - C. 50°F.
   - D. 60°F.
6. The typical temperature difference between the entering air and the boiling refrigerant on a standard air-conditioning evaporator is _____°F.
7. When the ambient outside air temperature is 95°F and the unit is a high-efficiency unit, the lowest condensing refrigerant temperature when the unit is operating properly is
   - A. 110°F.
   - B. 115°F.
   - C. 120°F.
   - D. 125°F.
8. When the outside ambient air temperature is 90°F, the temperature of the condensing refrigerant in a standard-efficiency unit should be approximately
   - A. 120°F.
   - B. 130°F.
   - C. 140°F.
   - D. 150°F.
9. If a condensing pressure is 260 psig for R-22, the condensing temperature will be
   - A. 110°F.
   - B. 120°F.
   - C. 130°F.
   - D. 140°F.
10. How is high efficiency accomplished in an air-conditioning condensing unit?
11. True or False: When troubleshooting a small system, the first step is to attach the high- and low-pressure gages.
12. The suction gas may have a high superheat if the unit has a _____ charge.
13. If the suction pressure is low and the suction line is warm, the system has a _____ evaporator.
    - A. full
    - B. starved
14. What electrical test instrument is used to measure continuity in a compressor winding?
15. What electrical test instrument is used to measure the current draw of a compressor?
16. Most condensers will subcool the refrigerant to between _____°F below the condensing temperature of the refrigerant.
    - A. 0 and 5
    - B. 5 and 10
    - C. 10 and 20
    - D. 20 and 30

17. What are two types of fixed-bore metering devices?

18. If no crankcase heat is at the compressor, what may happen to the refrigerant charge?

19. What is the purpose of the internal overload protector in the compressor motor?

20. If a condenser circuit breaker is tripped, you should
   A. reset it.
   B. check the suction pressure.
   C. check the high-side pressure.
   D. perform a resistance check of the compressor section.

DIAGNOSTIC CHART FOR COOLING AIR-CONDITIONING SYSTEMS

Many homes and businesses have air conditioning (cooling). These systems operate in the spring, summer, and fall for the purpose of comfort for the occupants. These systems are often connected to the heating system. Air conditioning may also be a part of a heat pump system. The air-conditioning system controls will be discussed here. Always listen to the owner to help to identify the potential problem.

Problem	Possible Cause	Possible Repair	Heading Number
No cooling, outdoor unit not running, indoor fan running	Open outdoor disconnect switch	Close disconnect switch.	41.13
	Open fuse or breaker	Replace fuse, reset breaker, and determine problem.	41.13
	Faulty wiring	Repair or replace faulty wiring or connections.	41.9, 41.13
No cooling; indoor fan and outdoor unit will not run	Low-voltage control problem A. Thermostat	Repair loose connections or replace thermostat and/or subbase.	41.10
	B. Interconnecting, wiring, or connections	Repair or replace wiring or connections.	41.10
	C. Transformer	Replace if defective. SAFETY PRECAUTION: Look out for too much current draw due to ground circuit or shorted coil.	41.10
No cooling; indoor and outdoor fan running, but compressor not running	Tripped compressor internal overload A. Low line voltage	Correct low voltage, call power company, repair loose connections.	41.10
	B. High head pressure		41.7
	1. Dirty outdoor (condenser)	Clean condenser.	41.7
	2. Condenser fan not running all the time	Check condenser fan motor and capacitor.	41.7
	3. Condenser air recirculating	Correct recirculation problem.	41.7
	4. Overcharge of refrigerant	Correct charge.	41.9
	C. Low charge, motor not being properly cooled	Correct charge; if due to leak, repair leak.	41.9
Indoor coil freezing	Restricted airflow	Change filters.	41.6
		Open all supply register dampers.	41.6
		Clear return air blockage.	41.6
		Clean fan blades.	41.6
		Speed fan up to higher speed.	41.6
	Low charge	Adjust unit charge—repair leak if refrigerant has been lost.	41.9
	Metering device	Change or clean metering device.	41.9
	Restricted filter drier	Change filter drier.	41.9
	Operating unit during low ambient conditions without proper head pressure control	Add proper low-ambient head pressure control.	22.18

SECTION 8

All-Weather Systems

UNIT 42

Electric, Gas, and Oil Heat with Electric Air Conditioning

OBJECTIVES

After studying this unit, you should be able to

- ■ describe year-round air conditioning.
- ■ discuss the three typical year-round air-conditioning systems.
- ■ list the five ways to condition air.
- ■ determine airflow for a cooling system.
- ■ describe why a heating system normally uses less air than a cooling system.
- ■ explain two methods used to vary the airflow in the heating season from that in the cooling season.
- ■ describe two types of control-voltage power supplies used in add-on air conditioning.
- ■ describe add-on air conditioning.
- ■ explain package all-weather systems.

SAFETY CHECKLIST

- ✔ Care should be used around any power supply because electrical shock hazard is possible.
- ✔ If you suspect a gas leak, ensure that ventilation around a gas furnace is adequate before troubleshooting.

42.1 COMFORT ALL YEAR

Year-round air conditioning describes a system that conditions the living space for heating and cooling throughout the year. This is done in several ways. The most common are electric air conditioning with electric resistance heat, electric air conditioning with gas heat, and electric air conditioning with oil heat. This unit describes how these systems work together. Each system has been covered individually in other units. Another common method is the heat pump, which is discussed in Unit 43, "Air Source Heat Pumps," and Unit 44, "Geothermal Heat Pumps." A less frequently used method is gas heat with gas air conditioning. This is a special system and will not be covered in this text.

42.2 FIVE PROCESSES FOR CONDITIONING AIR

Air is *conditioned* when it is heated, cooled, humidified, dehumidified, or cleaned. This unit describes how air is heated and cooled with the same system. The systems discussed in this unit are called *forced-air* systems. Air is distributed through ductwork to the conditioned space. The fan is normally a component of the heating system and provides the

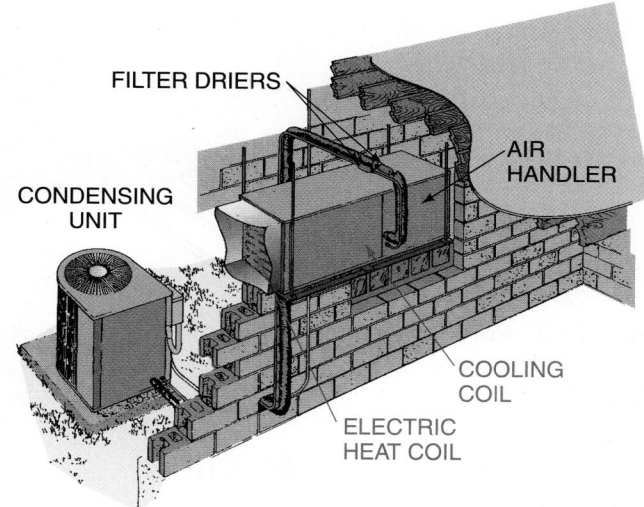

Figure 42–1 This is a typical electric furnace that has a cooling coil placed in the duct. The fan in the furnace is used to circulate air for both cooling and heating. *Courtesy Climate Control*

force to move the air through the duct system. A typical system may have an electric, gas, or oil furnace with an evaporator in the airstream for cooling. **Figure 42–1** shows an air-conditioning system with electric heat. The evaporator, located in the indoor airstream, is used in conjunction with a condensing unit outside the conditioned space. Interconnecting piping connects the evaporator to the condensing unit, **Figure 42–2**.

42.3 ADD-ON AIR CONDITIONING

Many systems are installed in stages. The furnace may be installed when the structure is built. The air conditioning may be added at the same time or later (*add-on air conditioning*). When the heating system is installed first, some considerations must be made for air conditioning if it is to be added later.

Air-conditioning systems must have the correct air circulation. Typically, they require an airflow of 400 cfm/ton, which is more than needed for an average forced-air heating system. A 3-ton cooling system, for example, requires 1200 cfm. The furnace on an existing heating system where 3 tons of cooling are added must be able to furnish the required amount of air, and the ductwork must be sized for the airflow.

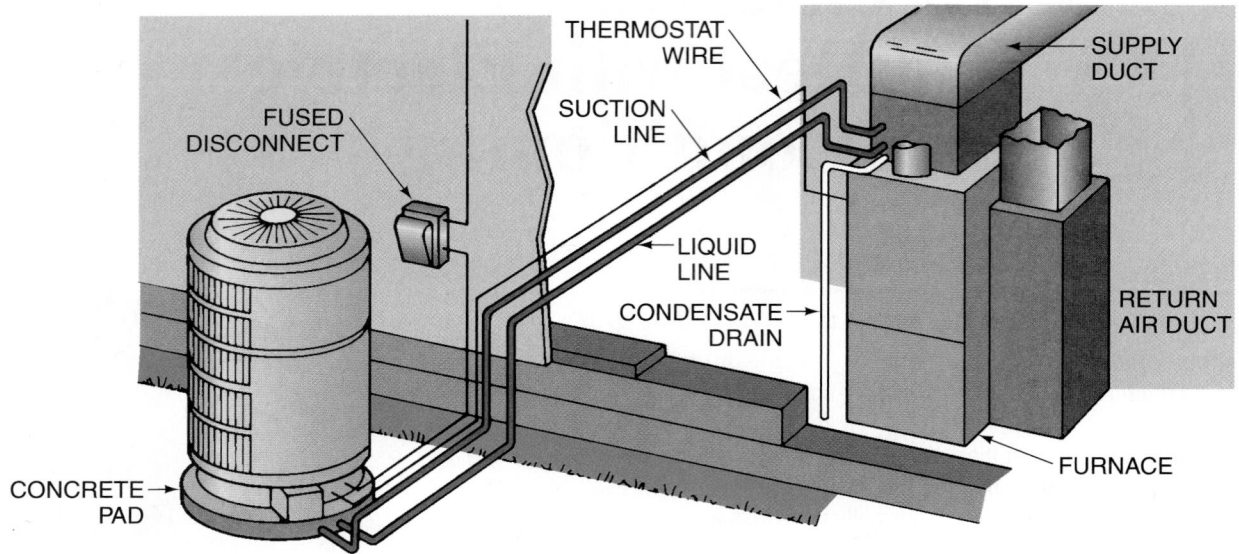

Figure 42–2   A complete installation with the piping shown. The refrigerant to and from the evaporator is piped to the condensing unit on the outside of the structure.

42.4   INSULATION FOR EXISTING DUCTWORK

It is popular in warm climates to install heating systems in the crawl space below the structure. The heat from the warm air duct rises and warms the floor, **Figure 42–3**. Some people believe that heat lost under the house is not all lost because some of it rises through the floor. This makes a significant difference only when the space under the house is sealed, and duct systems are not insulated. If air conditioning (cooling) is added to these systems, the duct will sweat. The system would be more efficient if the duct were insulated from the beginning.

Therefore, if the ductwork for a system with air conditioning is installed outside the conditioned space, it must be insulated or moisture ("sweat") from the ambient air will form on it during the cooling cycle. The insulation also helps prevent heat exchange between the air in the duct and the ambient air.

A typical system installed for only heating may have an undersized duct system and a blower that is too small if air conditioning is added. For example, a customer may decide to add 3 tons of air conditioning, which requires 1200 cfm of air, to an existing furnace system. A typical gas furnace must have a minimum temperature rise of 45°F for proper venting. This is a minimum furnace size of 58,320 Btu/h for the airflow of 1200 cfm. The smallest standard furnace has 50,000 Btu/h output.

Q = sensible heat in Btu/h
cfm = cubic feet of air per min
1.08 = constant used to change cfm to pounds of air per minute for the formula
TD = temperature difference between supply and return air

$$Q = 1.08 \times \text{cfm} \times \text{TD}$$
$$Q = 1.08 \times 1200 \times 45$$
$$Q = 58{,}320 \text{ Btu/h output}$$

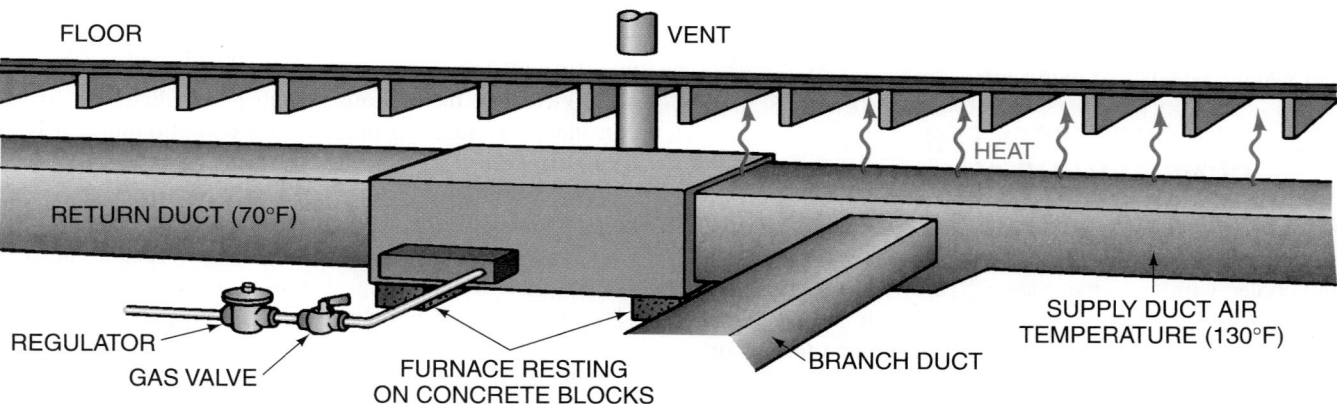

Figure 42–3   Heat rising off the duct in a crawl space under a house.

Using a 60,000 Btu/h output

$$\frac{60,000}{0.8} = 75,000 \text{ Btu/h input}$$

or a 75,000-Btu/h input gas furnace for a standard 80% efficiency furnace.

In a warm climate location, the furnace requirements may only be 60,000 Btu/h. This may be the furnace already installed at the job site. Adding a 3-ton unit requiring 1200 cfm may cause airflow problems if the duct system is sized for the airflow for heating only. This may be a 70°F rise, or

$$\text{cfm} = \frac{Q}{1.08 \times \text{TD}} = \frac{60,000}{1.08 \times 70} = 794 \text{ (round to 800)}$$

The furnace fan may not be capable of moving 1200 cfm of air, so some installation technicians will install a larger fan motor (for a belt-drive system) or a larger fan wheel and motor (for a direct-drive system). This is not what the furnace manufacturer intended and may cause venting problems. If the technician succeeds in moving 1200 cfm through duct sized for 800 cfm, the air will make noise because of the increased velocity. The alternative would be to reduce the air-conditioning load, increase the furnace size, or install a system producing two different fan volumes, one for summer and one for winter. Two-speed fan operation is discussed later in this unit.

🌎 The air-conditioning load may be reduced by attic ventilation, insulation, storm windows, shades, or awnings. For example, if the air-conditioning load in the preceding problem could be reduced to 2 tons, the airflow could be reduced to 800 cfm, and the current furnace would be adequate for the airflow. The duct would probably be adequate also. The customer would save money on the installation, and the technician would gain a satisfied customer. 🌎

Computer load-calculation programs can be used to work "what if" situations for a typical structure. For example, when a computer load calculation is entered, storm windows may be added with only a few keystrokes to determine the Btu/h savings. Then attic ventilation may be tried. Different combinations may be recalculated until the most economical combination of reductions is found. The customer benefits from monthly power reduction in addition to a less expensive air-conditioning installation.

42.5   EVALUATION OF AN EXISTING DUCT SYSTEM

When an air-conditioning system is to be added after the original furnace installation, the following considerations are the most important: the airflow, the ductwork, and the registers and grilles.

Air distribution was discussed in detail in Unit 37, "Air Distribution and Balance." A heating system that is already installed needs to be evaluated to see whether air distribution changes need to be made before air conditioning is installed. The blower size and the motor horsepower may be guides as to the amount of air the fan section is capable of moving. The manufacturer of the air handler or furnace is the best source

of this information. However, it is not always readily available. **Figure 42–4** is a chart used by estimators and service technicians to help determine what the blower capabilities of a typical system may be. The blower wheel dimensions are an important factor.

The duct system may be evaluated with an evaluation chart also, **Figure 42–4**. NOTE: This chart should not be used when designing a system from the beginning. It should be used only as a reference when estimating or troubleshooting. The estimator or service technician who suspects an airflow problem may consult the chart and compare the duct system to the chart. If the system does not meet the chart's minimum requirements, further investigation should be conducted.

For example, the system described previously (a system with a 75,000-Btu/h furnace, 80% efficiency, an output of 60,000 Btu/h, and an airflow of 800 cfm) can be evaluated using the top table in **Figure 42–4**. The output capacity falls between 60,000 and 70,000 Btu/h. The airflow requirement is 700 cfm for this range of furnace. If 2 tons of air conditioning are added, the required airflow is 800 cfm (from the bottom table in **Figure 42–4**). Looking to the right, the blower horsepower and blower wheel size can be found. The existing furnace may be compared to the data from the chart. The fan motor may be too small but may be changed to the larger one required for air conditioning if the fan wheel dimensions are correct. Various combinations of duct size may be found. A survey of the existing "heat only" duct system may show that the return air is undersized, and an extra return duct may be required. Further to the right you will find the required supply duct information. It can be compared to the system and changed to suit the application. This type of chart is valuable for job surveys.

The *grilles* and *registers* used for the final air distribution are important also. **Figure 42–4** will help you choose a return air grille of the correct size. It is important that the air volume meet the minimum before the air-conditioning system is added. The registers are responsible for distributing an air pattern into the conditioned space. If the original purpose of the installation was heating only, the registers may not provide the correct air pattern. This is particularly true for floor baseboard registers. They may be designed to keep the air down low on the floor if they are for heating only. Heated air will rise and warm the room. These registers will not work well in the air-conditioning season because the cool air will stay on the floor, **Figure 42–5**. Contact a supplier for the correct supply register replacements.

NOTE: When a distribution system does not meet the minimum standards of the charts discussed previously, it should be changed. If it is not changed, there will probably not be enough air for the air-conditioning system, and the air might not be properly distributed. The changes may involve adding new duct or return air runs or changing the blower or motor. The installation may require an airflow that is too great for the existing furnace because it will not allow enough air to be circulated. Because the furnace heat exchanger is not large enough, a new furnace may be the only solution. The air distribution registers are usually easy to change to accommodate air conditioning.

QUICK-SIZING TABLE for HEATING (FURNACE) ONLY DUCT SYSTEM
(Without a Cooling Coil in Place)

OUTPUT CAPACITY (SEE NOTES)	MIN. AIR FLOW REQ'D.	SUPPLY DUCT OR EXTENDED PLENUM @ 800 FPM	MIN. SQ. INCH NEEDED FOR SPEC. CFM (TOTAL AREA OF ALL SUPPLY DUCT)	MIN. NUMBER SUPPLY RUNS @ 600 FPM			MINIMUM SIZE	
				5" RUNS 80 CFM	6" RUNS 115 CFM	7" RUNS 155 CFM	RETURN DUCT FURNACE OR AIR HANDLER @ 800 FPM	RETURN AIR GRILLE (OR EQUIVALENT) @ FREE VELOCITY OF 500 FPM
45,000 to 55,000	500 CFM	14" x 8" or 12" round	100	7	5	4	14" x 8" or 12" round	12" x 12"
60,000 to 70,000	700 CFM	18" x 8" or 14" round	140	10	6	5	18" x 8" or 14" round	24" x 10"
75,000 to 85,000	800 CFM	22" x 8" or 14" round	170	10	7	5	22" x 8" or 14" round	24" x 12"
95,000 to 105,000	900 CFM	24" x 8" or 15" round	190	12	8	6	24" x 8" or 14" round	24" x 12"
105,000 to 115,000	1100 CFM	22" x 10" or 16" round	220	—	10	7	22" x 10" or 16" round	30" x 12"
125,000 to 150,000	1400 CFM	24" x 12" or 18" round	280	—	12	9	24" x 12" or 18" round	30" x 14"
155,000 to 160,000	1600 CFM	1-35" x 10" or 20" round or 2-22" x 8"	360	—	14	10	32" x 10" or 20" round	30" x 18"

Notes:
1. BTUH with **maximum** temperature rise.
2. Gas furnaces are rated in input capacity. Rated output capacity is 80% of input.
3. Oil and electric furnace are rated in output capacity.

QUICK-SIZING TABLE for HEATING and COOLING DUCT SYSTEM
Air-conditioning systems should never be sized on the basis of floor area only, but knowledge of the approximate floor area (sq. ft.) that can be cooled with a ton of air conditioning will be of invaluable assistance to you in avoiding serious mathematical errors.

SIZE OF O.D. UNIT	NORMAL AIR FLOW REQ'D @ 400 CFM PER TON	FURNACE		SUPPLY DUCT OR EXTENDED PLENUM @ 800 FPM	MIN. NUMBER SUPPLY RUNS @ 600 FPM				MIN. RETURN DUCT SIZE AT FURNACE OR AIR HANDLER @ 800 FPM	MIN. RETURN AIR GRILLE SIZE (OR EQUIVALENT) @ FACE VELOCITY OF 500 FPM
		BLOWER MOTOR H.P.	BLOWER WHEEL DIA. X WIDTH		5" RUNS 80 CFM	6" RUNS 115 CFM	7" RUNS 155 CFM	3½" X 14" 170 CFM		
1½ ton 18,000 BTUH	600 CFM	1/4 H.P.	9" X 8" 10" X 8"	16" X 8" or 12" round	8	5	4	4	16" X 8" or 12" round	24" X 8"
2 ton 24,000 BTUH	800 CFM	1/4 H.P.	9" X 9" 10" X 8"	22" X 8" or 14" round	10	7	5	5	22" X 8" or 14" round	22" X 12"
2½ ton 30,000 BTUH	1000 CFM	1/3 H.P.	10" X 8" 10" X 10" 12" X 9"	20" X 10" or 18" round	13	9	7	6	20" X 10" or 16" round	30" X 12"
3 ton 36,000 BTUH	1200 CFM	1/3 H.P.	10" X 8" 10" X 10" 12" X 9"	24" X 10" or 18" round	—	11	8	7	24" X 10" or 18" round	30" X 12"
3½ ton 42,000 BTUH	1400 CFM	1/2 H.P. 3/4 H.P.	10" X 8" 10" X 10" 12" X 9" 12" X 10"	24" X 12" or 18" round	—	12	9	8	24" X 12" or 18" round	30" X 14"
4 ton 48,000 BTUH	1600 CFM	1/2 H.P. 3/4 H.P.	10" X 10" 12" X 9" 12" X 10" 12" X 12"	32" X 10" or 20" round	—	14	11	10	28" X 12" or 20" round	30" X 18"

Figure 42–4   Duct estimating tables. These charts show many system characteristics, including the return air characteristics of typical systems used for heating only and for heating and cooling. *Copyright Trane Company 2000*

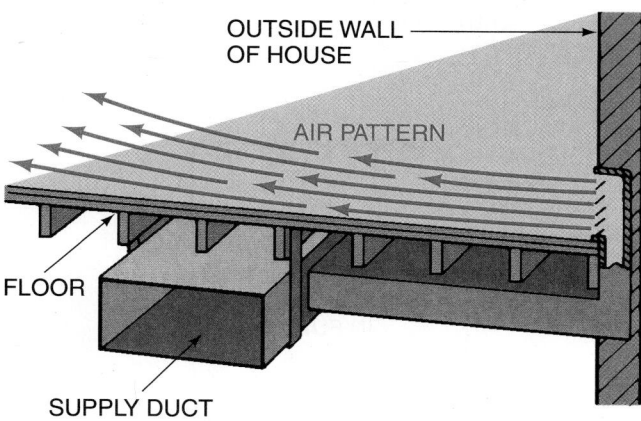

Figure 42–5   Low sidewall registers, not adequate for cooling.

42.6   COOLING VERSUS HEATING AIR QUANTITY

It is usually desirable to have less airflow in the winter, so the air that is entering the room through the registers will be hot. If the airflow of 400 cfm/ton of air conditioning is used in the winter, the air may create drafts of slightly warm air instead of hot air. Correct airflow is necessary for gas and oil furnaces for correct venting. Too much air will cause products of combustion to be too cool and condensation may occur. Changing the air volume is accomplished with dampers or a multispeed fan.

Dampers are sometimes installed by the original contractor. They have summer and winter positions. Someone must change the damper position for each change of season. Because this is often overlooked, some contractors install a low-voltage end switch that will not allow the cooling contactor to be energized unless the damper is in the summer position, **Figure 42–6.**

Multispeeds from the fan may be achieved by adjusting the fan motor pulley or using multispeed motors. When the pulley

must be adjusted to change fan speeds for a new season, a service technician must visit the location twice a year. The pulley may be adjusted at the spring start-up for air conditioning and at the fall furnace checkout. The filters may be changed and the motors oiled at the same time. This is a routine service call for some installations each year. Changing the pulley is not a job that homeowners should perform unless they have experience with this type of service. Multispeed motors have a winding for each motor speed, **Figure 42–7.**

42.7   CONTROL WIRING FOR COOLING AND HEATING

Another consideration for year-round air conditioning is the control wiring. The control system must be capable of operating heating and air-conditioning equipment at the proper times. NOTE: The heating must not be operated at the same time as the cooling. The thermostat is the control that accomplishes this. **Figure 42–8** shows an example of a typical HEAT-COOL thermostat with manual changeover from season to season. **Figure 42–9** is a HEAT-COOL thermostat with automatic changeover.

42.8   TWO LOW-VOLTAGE POWER SUPPLIES

The control circuit may have more than one low-voltage power supply, which can cause much confusion for the technician trying to troubleshoot a problem. If the heating system was installed first and the air conditioning added later, there may be two power supplies. The furnace must have a low-voltage transformer to operate in the heating mode. When the air conditioning was added, a transformer may have been furnished with it because the air-conditioning manufacturer did not know if the low-voltage power supply furnished with the furnace was adequate. Air-conditioning systems usually require a 40-VA transformer to supply enough current to energize the compressor contactor coil and the fan relay coil.

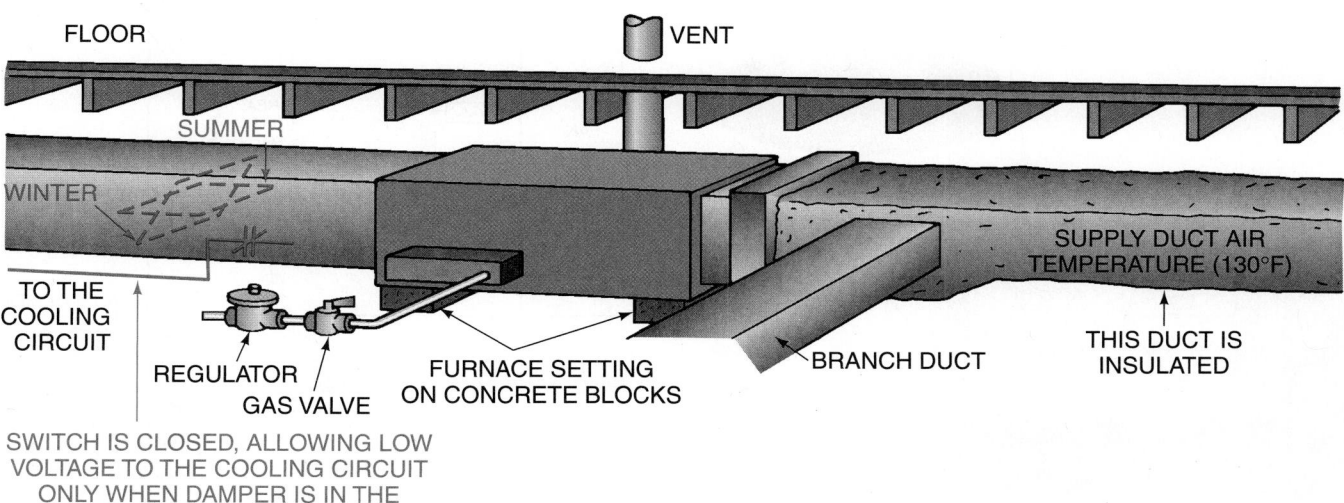

Figure 42–6   This system has a damper to allow more airflow in the summer than in the winter.

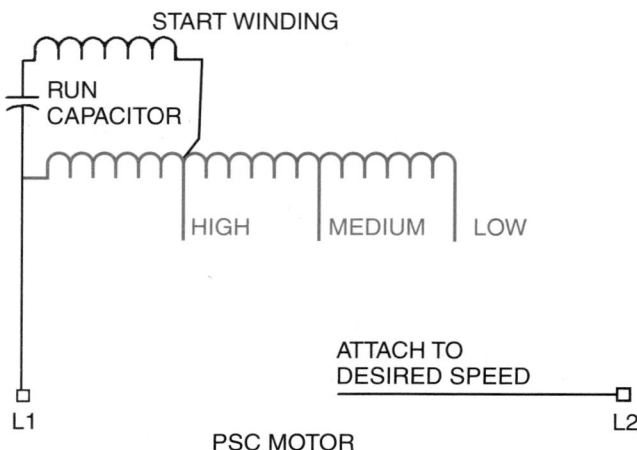

FAN RELAY: WHEN ENERGIZED, SUCH AS IN COOLING, THE FAN CANNOT RUN IN THE LOW-SPEED MODE. WHEN DEENERGIZED, THE FAN CAN START IN THE LOW-SPEED MODE THROUGH THE CONTACTS IN THE HEAT-OPERATED FAN SWITCH.

Figure 42–7 Wiring diagram for a three-speed fan motor.

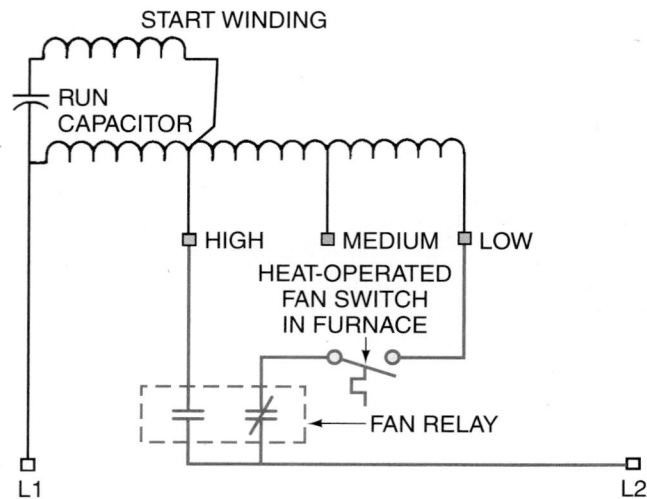

IF THE FAN SWITCH AT THE THERMOSTAT IS ENERGIZED WHILE THE FURNACE IS HEATING, THE FAN WILL MERELY SWITCH FROM LOW TO HIGH. THIS RELAY PROTECTS THE MOTOR FROM TRYING TO OPERATE AT 2 SPEEDS AT ONCE.

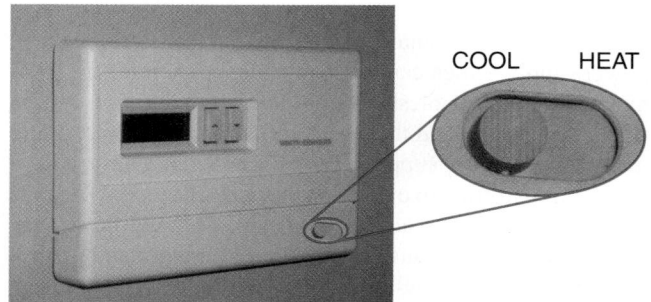

Figure 42–8 A typical HEAT-COOL thermostat. *Photo by Eugene Silberstein*

Most basic furnaces can operate on a 25-VA transformer and may be furnished with one.

When two power supplies are used, a special arrangement must be made in the thermostat and subbase. In one arrangement the two circuits are separated; in the other the two transformers are wired in parallel by phasing them. SAFETY PRECAUTION: If the transformers are to be separated, two hot wires are wired into the subbase. The two circuits must be kept apart, or damage may occur, **Figure 42–10.** Care should be used around any power supply because electrical shock hazard is possible. Notice that one leg of power from the heating transformer is in the cooling circuit. No problem should occur unless the other leg is involved.

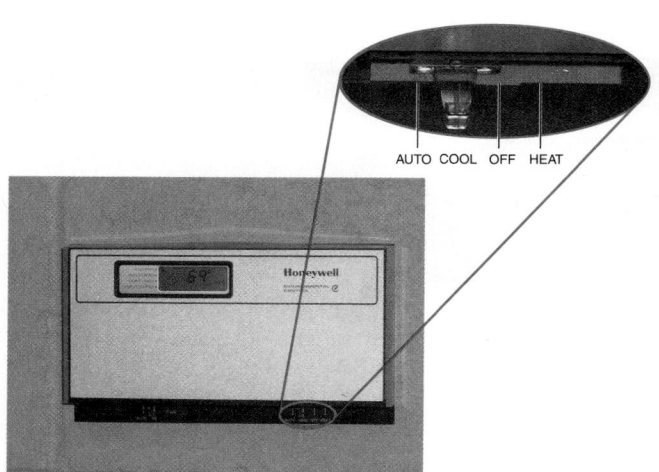

Figure 42–9 A thermostat with automatic changeover from cooling to heating. *Photo by Eugene Silberstein*

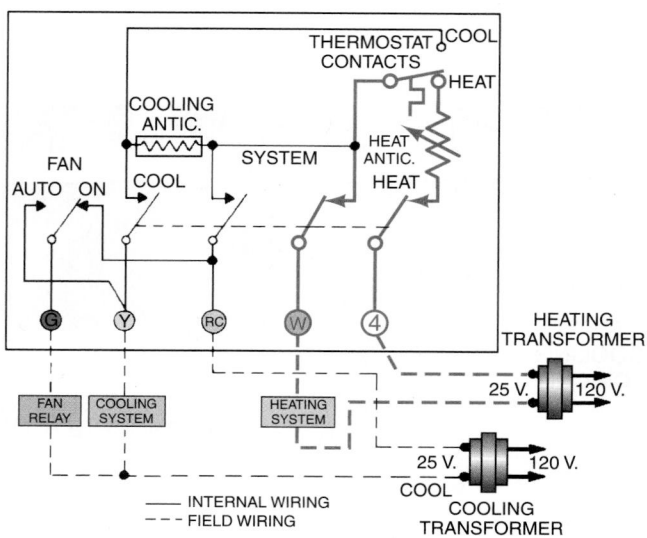

Figure 42–10 The wiring of an integrated subbase thermostat. *Courtesy White-Rodgers Division, Emerson Electric Co.*

42.9   PHASING TWO LOW-VOLTAGE TRANSFORMERS

NOTE: When the two transformers are wired in parallel, they must be kept in phase or the transformers will have opposing phases, **Figure 42–11**.

The preferred method for determining whether or not they are in phase is to connect a voltmeter to the two hot terminals. In the diagram in **Figure 42–12** they are terminals RC and 4. High voltage, about 50 V, indicates the transformers are out of phase. The meter will read 0 V from hot lead to hot lead if the transformers are in phase. If out of phase, it can be corrected with the primary or secondary wiring, **Figure 42–11** and **Figure 42–12**.

42.10   ADDING A FAN RELAY

When air conditioning is added to an existing system, the furnace may not have a fan relay to start the fan. Most fossil-fuel furnaces (oil or gas) will start the fan with a thermal type of fan switch in the heating mode and will have no provision to start the fan in the cooling mode. Some electric furnaces have a fan relay already built in. When there is none, a separate fan relay must be added when the air-conditioning system is installed. This relay is sometimes furnished in a package with the control transformer called a *transformer relay package*, or *fan center*, **Figure 42–13**. This fan relay is controlled by the 'G' terminal on the thermostat. SAFETY PRECAUTION: The fan relay is part of the low-voltage and high-voltage circuits. Use proper caution whenever you work with electricity.

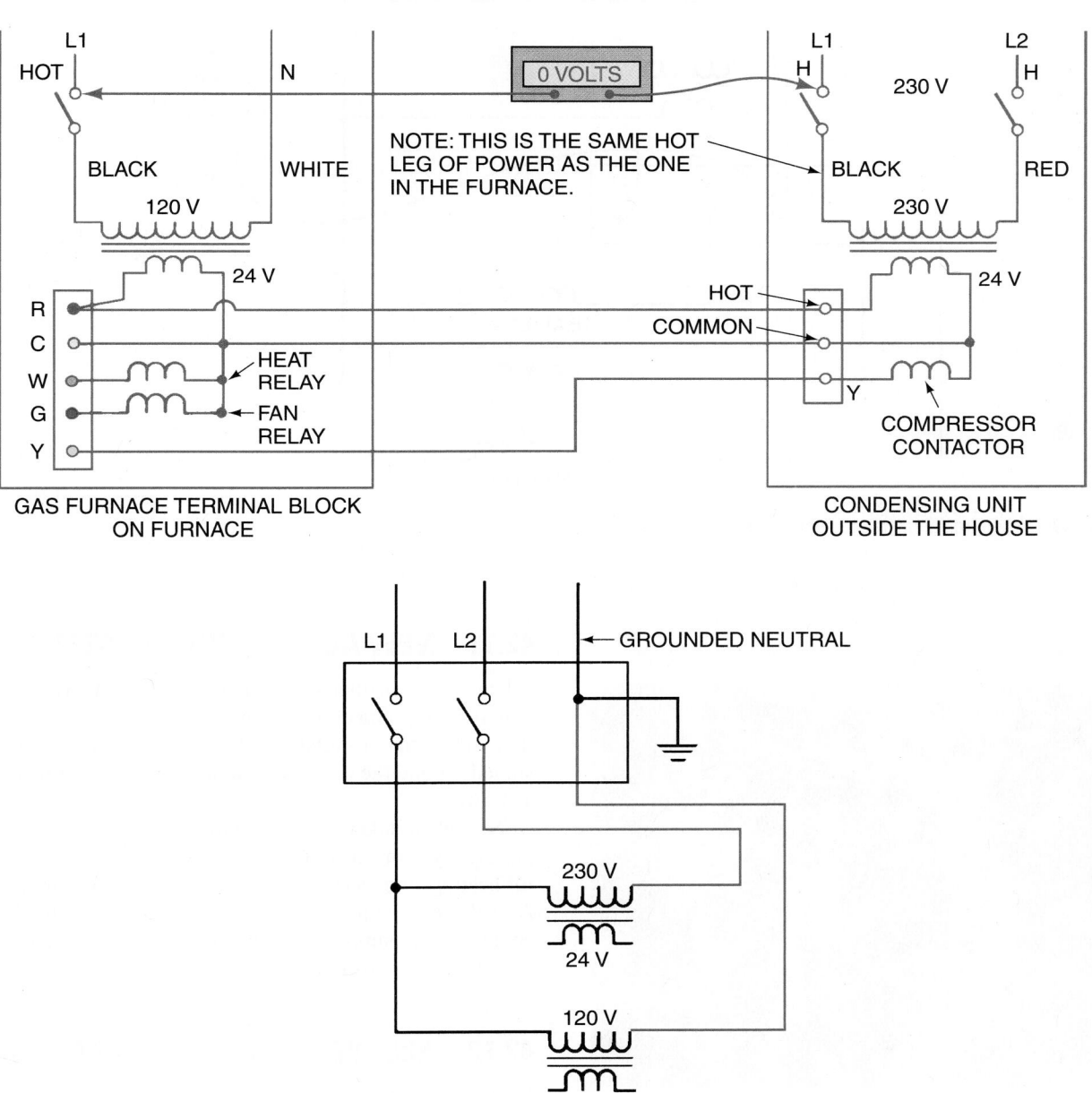

Figure 42–11   Transformers wired in parallel.

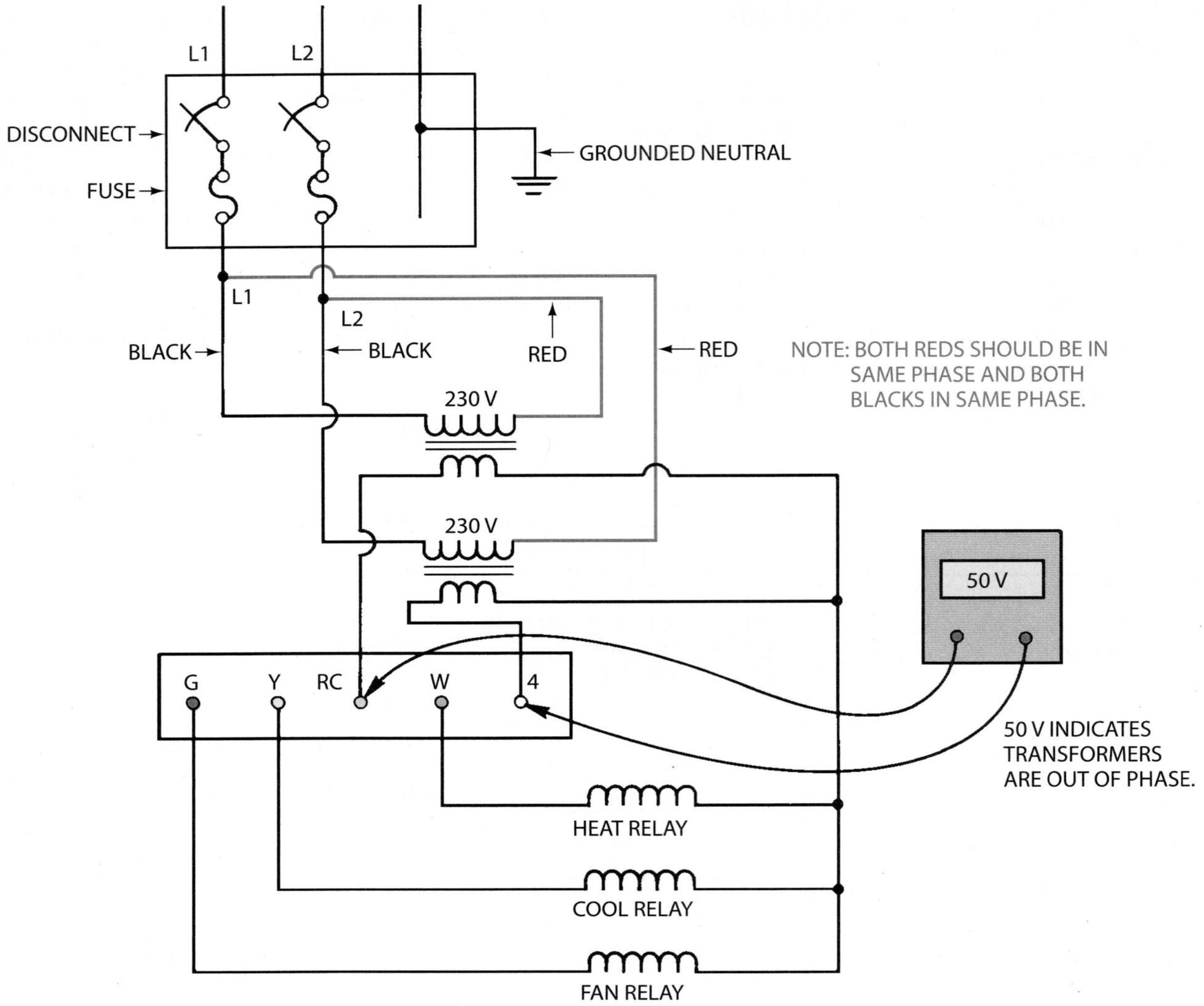

Figure 42–12   When transformers are out of phase, high voltage will occur.

Figure 42–13   A transformer and fan relay package.
Photo by Bill Johnson

42.11   NEW ALL-WEATHER SYSTEMS

When an all-weather system is installed as an original system, the foregoing considerations can be correctly designed into the initial installation. The ductwork is always designed around the cooling system because it requires more airflow.

New all-weather system installations are split systems or package systems. If split, a gas, oil, or electric furnace is used for heating, and electric air conditioning is used for cooling. A package system will be gas or electric heat with electric cooling. Heat pumps also come in package systems; they are covered in the next unit.

42.12   ALL-WEATHER SPLIT SYSTEMS

Split systems are installed in the same way as a "furnace only" system with the duct sized to handle the air for the cooling. Remember, a cooling system must have more cubic feet per minute of air than a heating system alone. Some consideration may be given to the air distribution at the grilles for proper air distribution for cooling. Most furnace manufacturers have matching cooling coil packages that will fit their furnaces. Using the matching coil package for a furnace is a good idea because the coil will come in its own insulated enclosure and will properly fit the furnace for the best airflow through the coil. Using the matching coil package may cause a hardship at installation time because the existing ductwork will need to be cut back for the coil package to be fitted.

42.13　PACKAGE OR SELF-CONTAINED ALL-WEATHER SYSTEMS

Package systems are not normally made for oil heating with air conditioning. These systems are gas and electric or electric and electric, **Figure 42–14,** and are installed like electric air-conditioning package systems. A review of Unit 38, "Installation," will show how this is done. The units that use gas must have a gas line installed. The flue installed with a gas package unit is a fixed component and part of the unit. These units are sometimes called *gas packs,* **Figure 42–15.**

42.14　WIRING THE ALL-WEATHER SYSTEM

The wiring of all-weather systems is similar to that of air-conditioning systems except for the extra power that may be required to operate the electric heat, **Figure 42–16.** The control wiring is much the same as a gas furnace and electric air conditioning except that the wiring is all done between the thermostat and the package unit. The wires that normally run to a remote furnace are not needed.

(A)

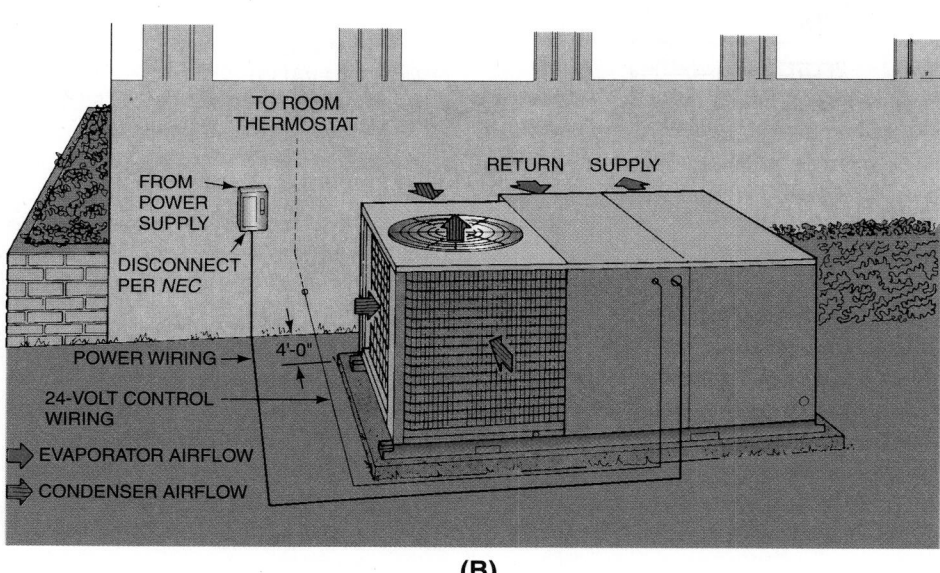

TO ROOM THERMOSTAT

FROM POWER SUPPLY

DISCONNECT PER *NEC*

RETURN　SUPPLY

POWER WIRING

4'-0"

24-VOLT CONTROL WIRING

EVAPORATOR AIRFLOW

CONDENSER AIRFLOW

(B)

Figure 42–14　These two package units look alike. **(A)** Gas used as the heat source and electricity used for cooling. **(B)** Electric heat used with electric air conditioning. **(A)** *Courtesy Climate Control.* **(B)** *Reproduced courtesy of Carrier Corporation*

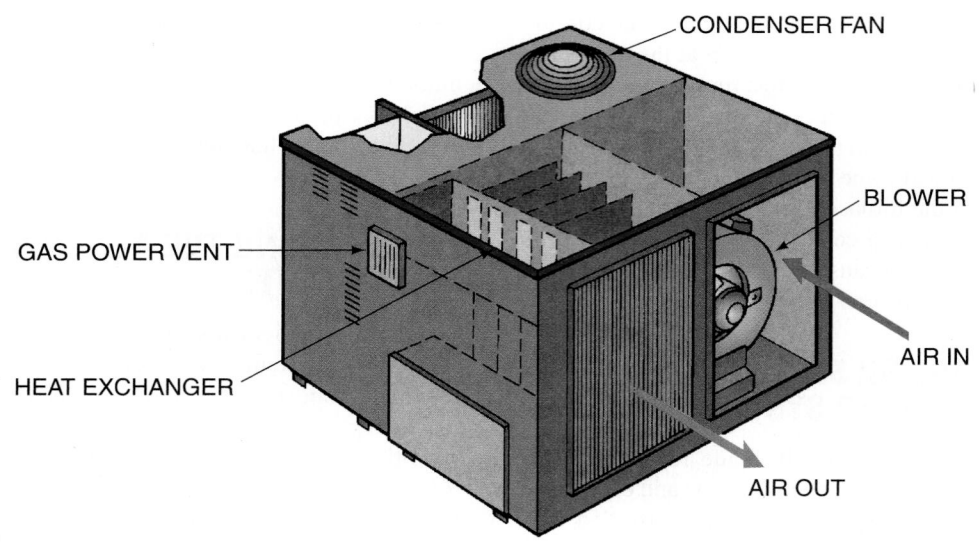

Figure 42–15   The flue vent on a gas package unit. *Courtesy Climate Control*

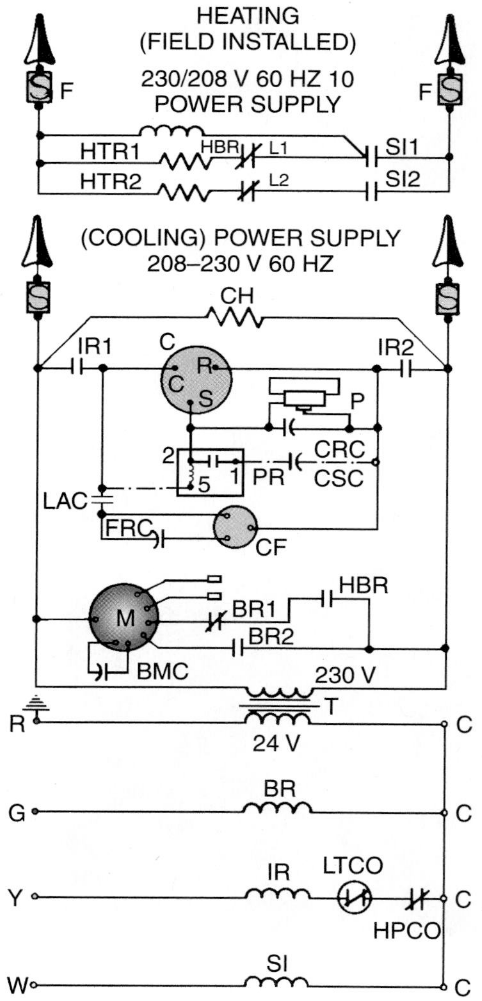

COMPONENT PART IDENTIFICATION

BMC	BLOWER MOTOR CAPACITOR
BR	BLOWER RELAY
C	COMPRESSOR
CF	CONDENSER FAN MOTOR
CH	CRANKCASE HEATER
CRC	COMPRESSOR RUN CAPACITOR
CS	EXHAUSTER CENTRIFUGAL SWITCH
CSC	COMPRESSOR START CAPACITOR
EM	EXHAUSTER MOTOR
F	FUSE
FC	FAN CONTROL KLIXON
FRC	FAN RUN CAPACITOR
FT	FAN TIMER
GV	GAS VALVE
HBR	BLOWER RELAY (HEATING)
HPCO	HIGH-PRESSURE CUT-OUT
HR	HEAT RELAY
HTR	ELECTRIC HEATER
IR	COMPRESSOR CONTACTOR
L	LIMIT
LAC	LOW-AMBIENT CONTROL (0°F) (ALL RD/RG-D UNITS & ALL R-H RG-H3 PHASE UNITS)
LTCO	LOW-TEMPERATURE CUT-OUT (50°F) (ALL R-H/RG-H 1 PHASE UNITS)
M	BLOWER MOTOR
P	PTC STANDARD START ASSIST
PPK	POST PURGE KLIXON
PR	POTENTIAL RELAY
PR+	
CDC	OPTIONAL START ASSIST (REPLACES "P")
SI	SEQUENCER
T	TRANSFORMER

Figure 42–16   The wiring diagram for an electric heat and electric air-conditioning package unit. *Courtesy Climate Control*

42.15 SERVICING THE ALL-WEATHER SYSTEM

Package equipment installations have the advantage that the whole system is outside the house. SAFETY PRECAUTION: Any gas hazard is virtually eliminated because gas leaks are dissipated outside. A technician can service the unit without crawling to a furnace in an attic or under a house. All of the control wiring is at the unit and is easily accessible. **Figure 42–17** shows an example of a typical installation.

When major repairs are needed on the compressor or expansion device, for example, the repair is simplified because they are all together and accessible. The proper panels may be removed to service the various components.

When troubleshooting an all-weather system, the technician usually is called to deal with a heating problem in winter or a cooling problem in summer. Seldom does a heating problem carry over to the cooling season, or a cooling problem to the heating season, because the controls are separated sufficiently. Dirty air filters are one type of problem that may not show up at the end of the cooling or heating season but will cause problems at the peak of next season. For example, when a homeowner does not change the air filters at the end of the heating season, the cooling will normally work fine until the first hot day that the unit has long running times. When the running times are short, the unit may run below freezing due to the dirty filters but will cut off before the coil freezes up. When the running time is longer, the coil may freeze solid before the room thermostat shuts off the unit.

The reverse is true with filters when they are not changed at the end of the cooling season. The cooling unit may not run long enough to freeze solid before the thermostat shuts it off. When the heating season starts, the furnace will appear to be working fine when the weather is mild. SAFETY PRECAUTION: When the weather gets cold enough to cause the furnace to run for long periods, it may overheat—with the high-limit control shutting off. If a technician is on a cooling-related service call at the end of the cooling season, it is important to start up the heating system to ensure that it is functioning properly.

The other components of year-round air conditioning, such as the humidifier and the air cleaner, are discussed in Section 6 in this text.

SUMMARY

- Summer air conditioning is sometimes added to an existing heating system. This is called add-on cooling.
- Different air volumes are sometimes desirable for the heating and cooling seasons. The air in the heating season is warmer at the terminal units when the air volume is reduced in the heating season.
- Different air volumes are accomplished with dampers and variable fan speeds.
- The control circuit may have two transformers—one furnished with the furnace and one with the air-conditioning unit.
- When two power supplies (transformers) exist, the two may be kept separated in the thermostat, or they may be wired in parallel and in phase.
- Package all-weather systems normally consist of gas heat and electric air conditioning or electric heat and electric air conditioning.

REVIEW QUESTIONS

1. Air is conditioned when it is _____, _____, _____, _____, or _____.
2. A typical all-weather system may have a (an) _____, _____, or _____ furnace with an evaporator in the airstream for cooling.
3. Typically, air-conditioning systems require an airflow of _____ cfm/ton.
4. If add-on air conditioning is installed and the duct is outside the conditioned space, it should be _____, or moisture will form on it during the cooling cycle.
5. What are some of the things that must be considered with add-on air conditioning?
6. A heating system requires _____ a cooling system.
   A. less airflow than
   B. greater airflow than
   C. about the same airflow as
7. The add-on air-conditioning load may be reduced by
   A. add-on insulation.
   B. storm windows.
   C. shades or awnings.
   D. all of the above.
8. List two ways in which multispeeds may be achieved with the airflow fan.
9. What test instrument is used to determine whether or not two low-voltage transformers are in phase?

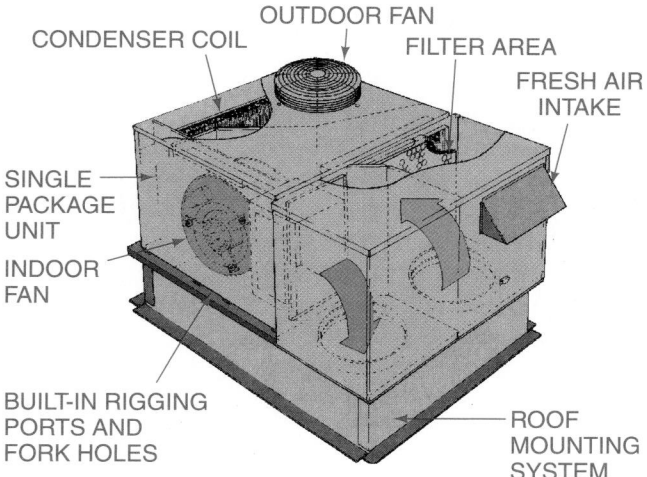

Figure 42–17   A typical package unit installation with a roof mounting system. *Courtesy Climate Control*

CONDENSER COIL
OUTDOOR FAN
FILTER AREA
FRESH AIR INTAKE
SINGLE PACKAGE UNIT
INDOOR FAN
BUILT-IN RIGGING PORTS AND FORK HOLES
ROOF MOUNTING SYSTEM

10. Too much duct airflow in the heating season may cause the products of combustion to be too _____, and condensation may occur.
    - A. warm
    - B. cool
    - C. dry

11. Correct duct airflow in the heating season for gas and oil furnaces is necessary to provide proper
    - A. venting.
    - B. absorption.
    - C. electrical phasing.
    - D. flame retention.

12. Most air-conditioning systems require a _____ transformer to supply enough current to energize the compressor contactor and the fan relay.
    - A. 208-V
    - B. 230-V
    - C. 30-VA
    - D. 40-VA

13. When two transformers in a heating and an add-on air-conditioning system are wired in parallel, they must be kept
    - A. in phase.
    - B. out of phase.
    - C. in relay.
    - D. in a circuit with the same amperage.

14. In a new all-weather system, the ductwork is designed around the
    - A. cooling system.
    - B. heating system.

15. What is one advantage of a package all-weather system?

Air Source Heat Pumps

OBJECTIVES

After studying this unit, you should be able to

- describe a reverse-cycle heat pump.
- list the components of a reverse-cycle heat pump.
- explain a four-way valve.
- state the various heat sources for heat pumps.
- compare electric heat to heat with a heat pump.
- state how heat pump efficiency is rated.
- determine by the line temperatures whether a heat pump is in cooling or heating.
- discuss the terminology of heat pump components.
- define coefficient of performance.
- explain auxiliary heat.
- describe the control sequence on an air-to-air heat pump.
- describe techniques being used to improve the efficiency of heat pump systems.
- discuss recommended preventive maintenance procedures for heat pump systems.

SAFETY CHECKLIST

✔ When installing insulated ductwork or otherwise working around fiberglass, wear gloves, goggles, and clothing that will cover your skin. If fiberglass particles are in the air, wear a mask that covers your nose and mouth. To avoid skin irritation, do not wear short-sleeve shirts or short pants when working with fiberglass insulation.

✔ Be careful not to cut your hands when working with metal duct, panels, and fasteners.

✔ Exercise all precautions when installing units on rooftops or in other hazardous locations.

✔ Observe all safety precautions when troubleshooting the electrical components of a heat pump. When units are located in crawl spaces and you are in contact with the ground, be careful not to establish a path for current to flow through your body to ground.

✔ Troubleshoot with the electrical power on only when it is absolutely necessary. At all other times ensure that the power is off. When the power is off, lock and tag the disconnect panel. There should be only one key. Keep it with you so that another person will not be able to turn the power on while you are working on the equipment.

✔ Be extra cautious if conditions are wet when you are troubleshooting an outside unit.

✔ When connecting gage lines or transferring refrigerant, wear gloves, goggles, and clothing to protect your skin, because refrigerant can freeze your skin and eye tissue. High-pressure refrigerant can pierce your skin and blow particles into your eyes.

✔ The hot gas line can cause serious burns and should be avoided.

✔ If water is used to melt the ice from the coil of the outdoor unit, turn the power off and protect all electrical components. If troubleshooting later with the power on, do not stand in this water.

✔ Use common sense and caution at all times while installing, maintaining, or troubleshooting any piece of equipment. As a technician you are constantly exposed to *potential danger.* Refrigerant, electrical shock hazard, rotating equipment, hot metal, sharp metal, and lifting heavy objects are among the most common hazards you will face.

43.1 REVERSE-CYCLE REFRIGERATION

Heat pumps, like refrigerators, are refrigeration machines. Refrigeration involves the removal of heat from a place where it is not wanted and depositing it in a place where it makes little or no difference. The heat can actually be deposited as heat reclaim in a place where it is wanted. This is the difference between a heat pump and a cooling-only air-conditioning system. The air conditioner can pump heat only one way. The heat pump is a refrigeration system that can pump heat two ways. Since it has the ability to pump heat into as well as out of a structure, the heat pump system can provide both heating and cooling.

All compression-cycle refrigeration systems are heat pumps in that they pump heat-laden vapor. The evaporator of a heat pump absorbs heat into the refrigeration system, and the condenser rejects the heat. The compressor pumps the heat-laden vapor. The metering device controls the refrigerant flow. These four components—the evaporator, the condenser, the compressor, and the metering device—are essential to compression-cycle refrigeration equipment. The same components are in a heat pump system along with the *four-way valve,* which is used to switch the unit between the heating and cooling modes of operation. By changing the mode of operation of a heat pump system, the functions of the indoor and outdoor coils change as well. In the cooling mode, the indoor coil acts as the evaporator and the outdoor coil functions as the condenser. In the heating mode, the indoor coil functions as the condenser and the outdoor coil operates as the evaporator.

43.2 HEAT SOURCES FOR WINTER

The cooling system in a typical residence absorbs heat into the refrigeration system through the evaporator and rejects this heat to the outside of the house through the condenser. The house might be 75°F inside while the outside temperature might be 95°F or higher. The air conditioner pumps heat from a low temperature inside the house to a higher temperature outside the house; that is, it pumps heat up the temperature scale. A freezer in a supermarket takes the heat out of ice cream at 0°F to cool it to −10°F so that it will be frozen hard.

The heat removed can be felt at the condenser as hot air. This example shows that there is usable heat in a substance even at 0°F, **Figure 43–1.** There is heat in any substance until it is cooled down to −460°F. Review Unit 1, "Heat and Pressure," for examples of heat level.

If heat can be removed from 0°F ice cream, it can be removed from 0°F outside air. The typical heat pump does just that. It removes heat from the outside air in the winter and deposits it in the conditioned space to heat the house. (Actually, about 85% of the usable heat is still in the air at 0°F.) Hence, it is called an *air-to-air* heat pump, **Figure 43–2.** In summer the heat pump acts like a conventional air conditioner and removes heat from the house and deposits it outside. From the outside an air-to-air heat pump looks like a central cooling air conditioner, **Figure 43–3.** The term "air to air" indicates the source of heat while the system is operating in the heating

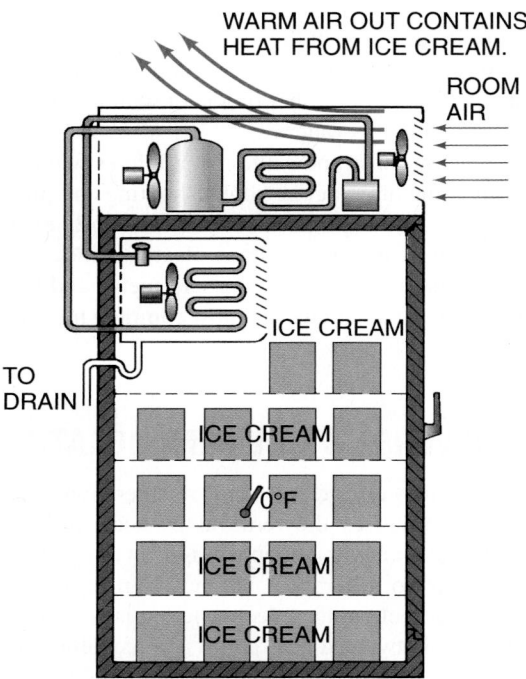

Figure 43–1   This low-temperature refrigerated box is removing heat from ice cream at 0°F. Part of the heat coming out of the back of the box is coming from the ice cream.

Figure 43–3   The outdoor unit of an air-to-air heat pump. It absorbs heat from the outside air for use inside the structure. *Reproduced courtesy of Carrier Corporation*

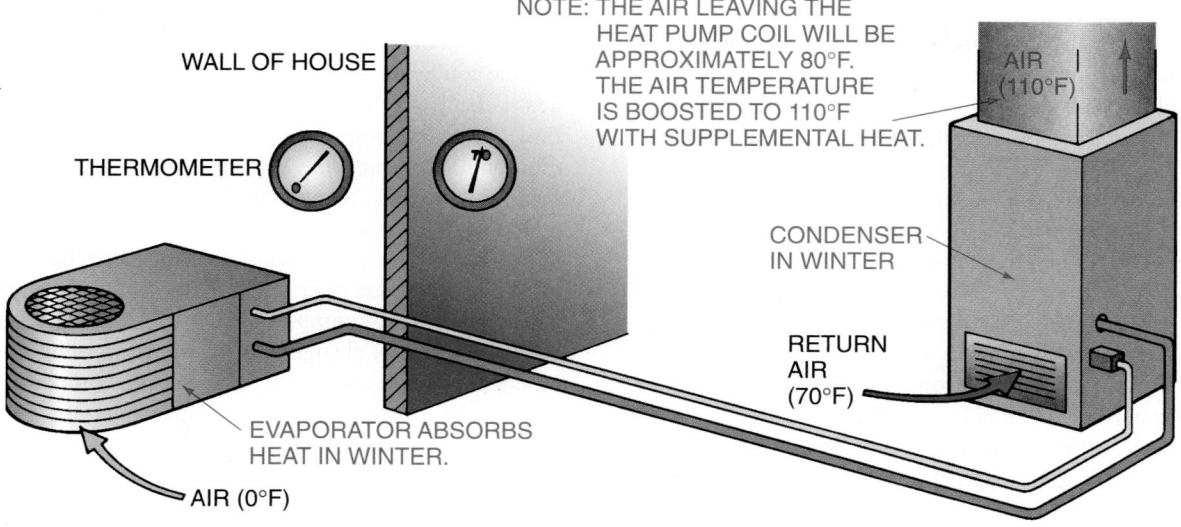

Figure 43–2   An air-to-air heat pump removing heat from 0°F air and depositing it in a structure for winter heat.

mode and the medium that is ultimately being treated. The first "air" represents the heat source, while the second "air" represents the medium being heated. For example, an "air-to-water" heat pump uses "air" to heat "water." Water-source heat pumps use water as the heat source in the heating mode and can be "water to water" or "water to air." The water-to-air heat pump, therefore, uses water as the heat source to heat air. The water source can be, for example, lakes or wells. Water-source heat pumps are discussed in Unit 44.

43.3 THE FOUR-WAY REVERSING VALVE

The refrigeration principles that a heat pump uses are the same as those stated previously. However, a new component is added to allow the refrigeration equipment to pump heat in either direction. The air-to-air heat pump in **Figure 43–4**

shows the heat pump moving heat from inside the conditioned space in summer to the outside. Then in winter the heat is moved from the outside to the inside, **Figure 43–5**. This change of direction is accomplished with a special component called a *four-way reversing valve*, **Figure 43–6**. This valve can best be described in the following way. The heat absorbed into the refrigeration system is pumped through the system with the compressor. The heat is contained and concentrated in the discharge gas. The four-way valve diverts the discharge gas and the heat in the proper direction to either heat or cool the conditioned space. This valve is controlled by the space temperature thermostat, which positions it to either HEAT or COOL. The four-way reversing valve is a four-port valve that has a slide mechanism in it, **Figure 43–7**. The position of the slide is determined by a solenoid that is energized in one mode of operation and not the other. The

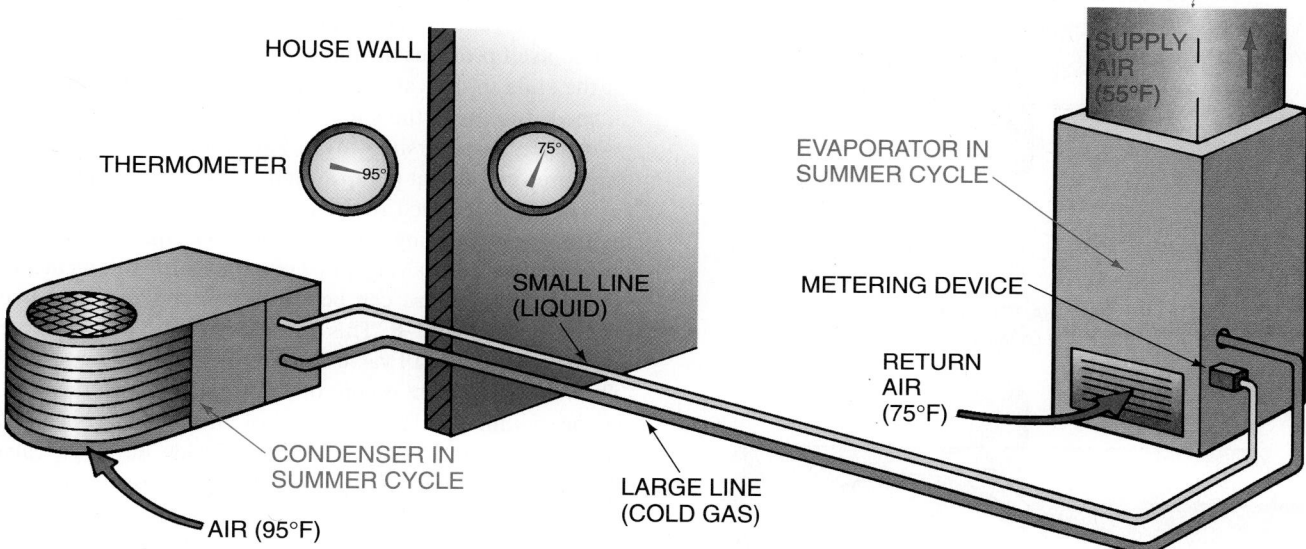

Figure 43–4 An air-to-air heat pump moving heat from the inside of a structure to the outside.

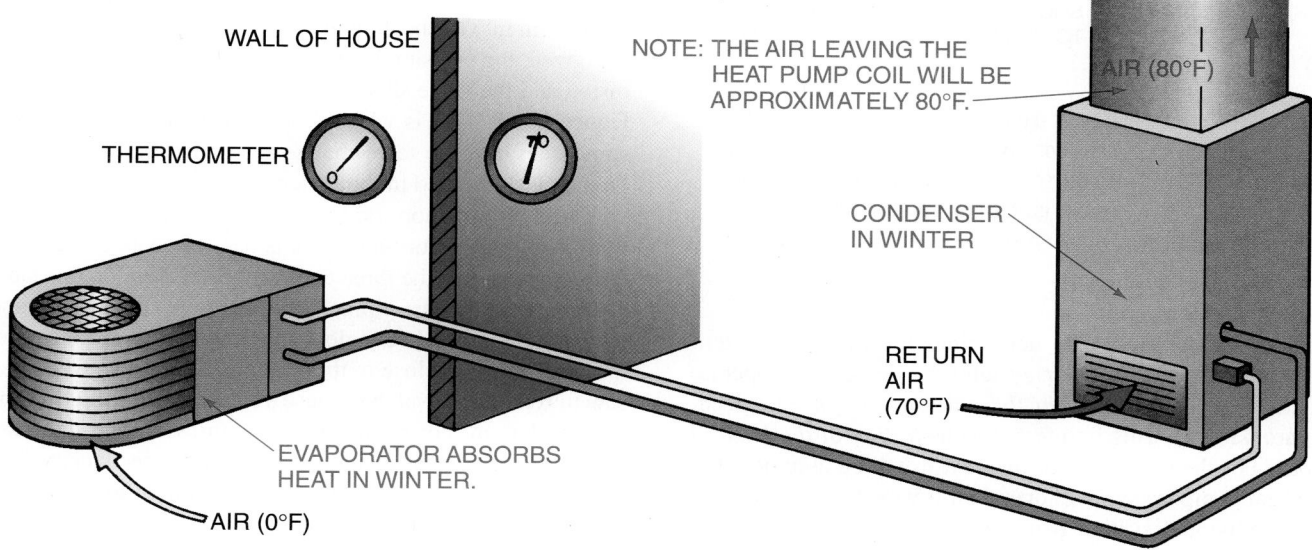

Figure 43–5 In the winter, the heat pump moves heat into the structure.

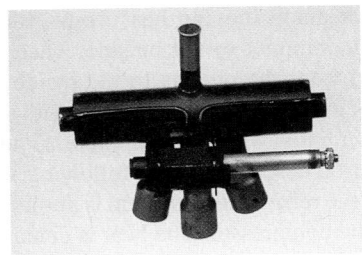

Figure 43–6   A four-way valve. *Photo by Bill Johnson*

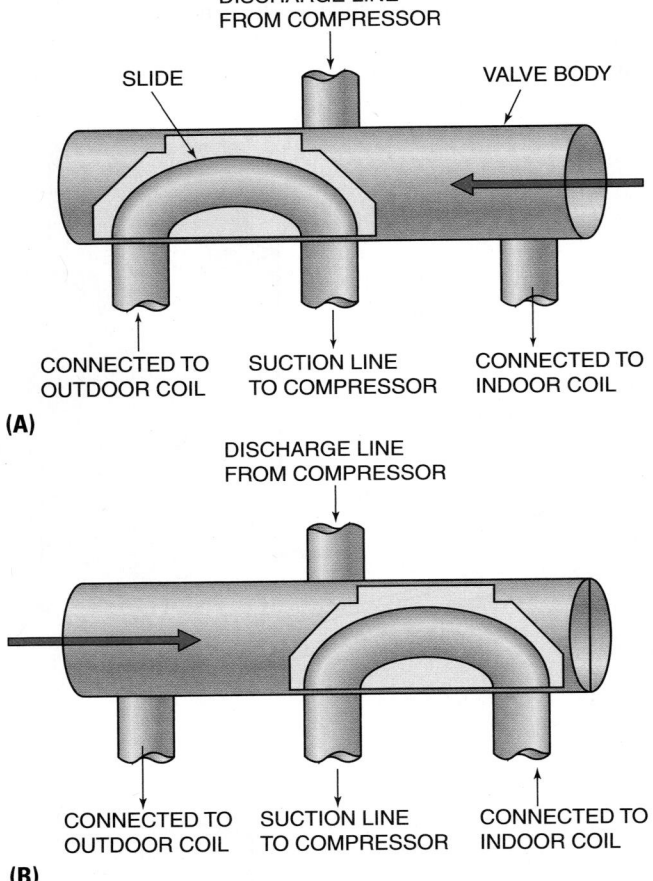

Figure 43–7   Internal slide in the four-way reversing valve. The solenoid determines the position of the slide. Most systems operate in the heating mode when the solenoid is deenergized. **(A)** Position of the slide when the solenoid is deenergized. **(B)** Position of the slide when the solenoid is energized.

position of the slide also determines the mode of system operation. Most heat pump systems are designed to operate in the heating mode when the solenoid is not energized, **Figure 43–7(A).** This is referred to as *failing in the heating mode.* For the remainder of this text it will be assumed that the heat pump system will operate this way. Be sure to check each individual system being worked on, as this may not always be the situation. **Figure 43–7(B)** shows the position of

the slide when the solenoid is energized and the system is operating in the cooling mode.

Typical reversing valves have one isolated port on one side and the remaining three ports on the other. The isolated port is where the hot gas from the compressor enters the valve. **Figure 43–7** indicates the piping connections on the valve. Depending on the position of the slide, the hot gas will be directed to either the outdoor coil or the indoor coil. In either case, the coil that accepts the hot gas from the compressor is functioning as the condenser coil. Note that in **Figure 43–7(A)** the hot gas enters the valve at the top and is directed to the right-hand port, which is connected to the indoor coil. Since the indoor coil is receiving the hot gas from the compressor, the indoor coil is the condenser and the system is operating in the heating mode. Compare this with **Figure 43–7(B)**, where the hot gas from the compressor is being directed to the outdoor coil. Under these conditions, the system is operating in the cooling mode.

Smaller heat pump systems utilize smaller, direct-acting four-way reversing valves. *Direct acting* implies that the solenoid itself provides the force needed to change the position of the slide in the valve. On larger systems, additional help is needed to move the slide. These valves are referred to as pilot-operated reversing valves, **Figure 43–8,** and they use the pressures in the heat pump system to help move the slide.

The pilot-operated four-way valve is actually two valves built into one housing. The control wires control a small solenoid valve that has very small piping that runs along the outside of the valve, as can be seen in the bottom right-hand corner of **Figure 43–8.** This valve is called a pilot-operated valve because a small valve controls the flow in a large valve. The valve movement is controlled by pressure difference on a piston within the large portion of the valve. This piston slides back and forth on a platform inside the large valve. When the solenoid is deenergized, high pressure from the compressor discharge pushes through the tubes on the pilot valve as shown in **Figure 43–9.** This exerts a pressure on the right side of the slide, pushing it to the left. The refrigerant on the left side of the slide is pushed back through the pilot valve and into the compressor's suction line. In this position, the hot gas from the compressor is directed to the indoor coil, putting the system into the heating mode. When the solenoid valve is energized, the pressure is then applied to the other end of the piston and the valve slides to the right for the cooling mode, **Figure 43–10.** This is why if you start up a heat pump, you sometimes hear the valve change position as the pressure difference is developed in the system.

There will be more on checking the four-way valve later in this unit. The four-way valve is hard to change in case of failure because of the three connections that are on one side. These are the largest piping connections on the system and they are often piped very close to each other. You cannot get a tubing cutter in close to the valve. You should never use a saw to remove the valve because you will leave filings in the system that are not good for the compressor, and there are very close tolerances inside the valve slide mechanism. The high-temperature brazing alloy is hard to melt loose using a torch with a single tip. The connections would have to be melted loose one at a time by applying heat to both sides of

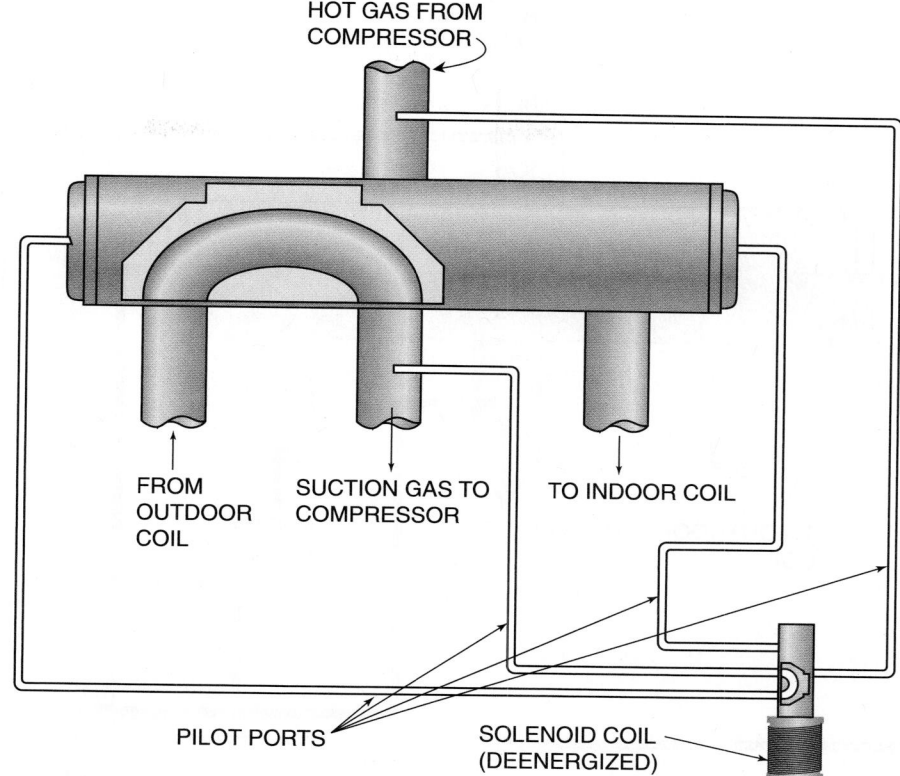

Figure 43–8 A pilot-operated four-way reversing valve. Notice the pilot valve in the bottom right-hand corner of the figure.

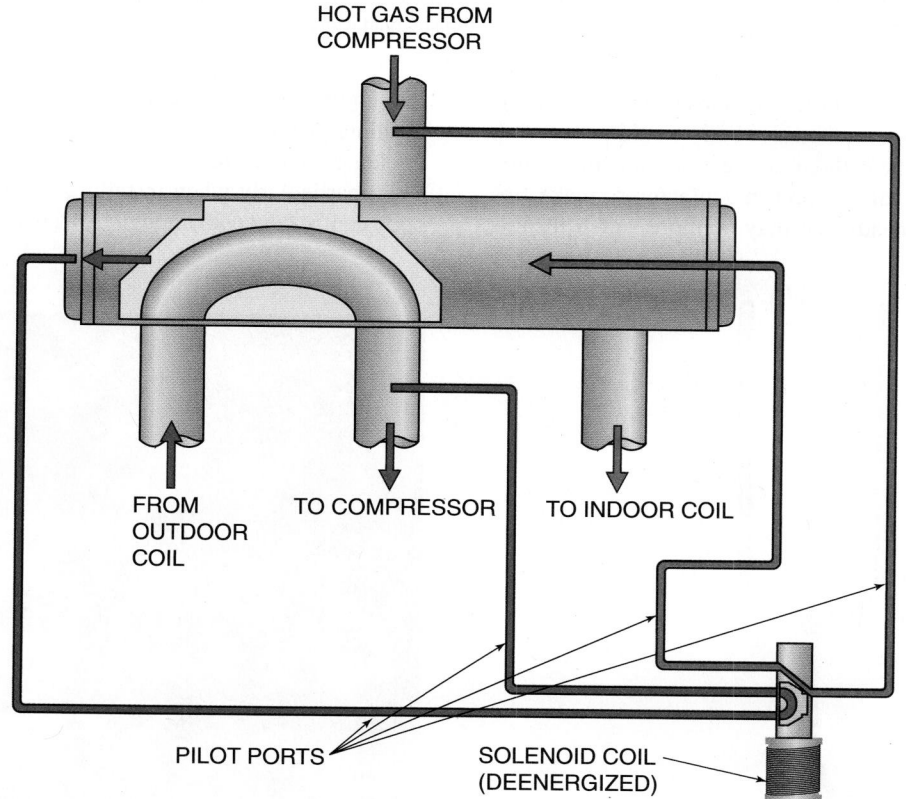

Figure 43–9 In the heating mode, the high-pressure hot gas from the compressor is directed to push the main valve's slide to the left. This directs the hot gas from the compressor to the indoor coil to provide space heating. Notice that the solenoid is deenergized.

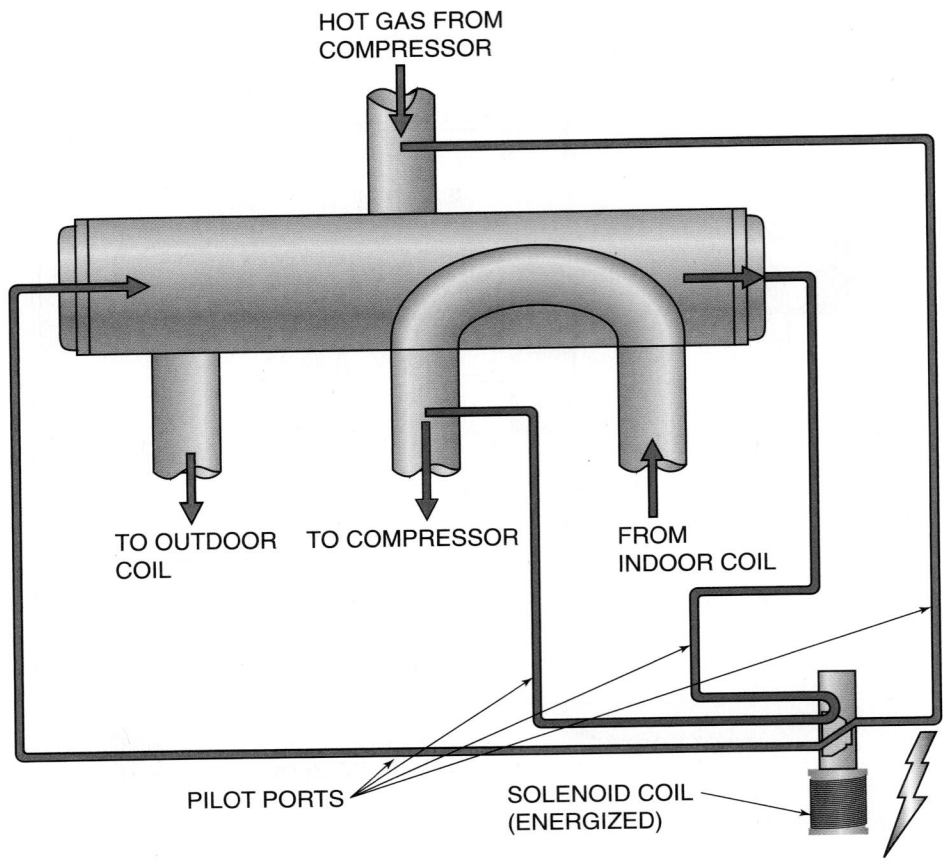

HOT GAS FROM
COMPRESSOR

TO OUTDOOR
COIL

TO COMPRESSOR

FROM
INDOOR COIL

PILOT PORTS

SOLENOID COIL
(ENERGIZED)

Figure 43–10 In the cooling mode, the high-pressure hot gas from the compressor is directed to push the main valve's slide to the right. This directs the hot gas from the compressor to the outdoor coil. Notice that the solenoid is energized.

the pipe to get it hot enough to turn loose. By the time the third one was melted loose, the valve body would be very hot. It is even harder to heat three connections together while assembling the new valve. Torch manufacturers make tips with multiple flame heads that may be more suited for this job, **Figure 43–11.** Remember, inside the valve body there is a sliding piston. It slides on a nylon pad that has a low melting temperature. The valve body must be kept as cool as possible, particularly when installing the new valve. The valve body can be kept cool enough by a heat sink paste or by a wet

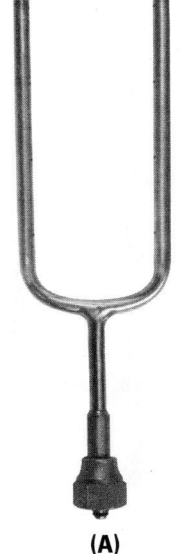

(A)

(B)

Figure 43–11 **(A)** Specialized torch tip designed for removing reversing valves. **(B)** Using the torch tip. *Photos courtesy Uniweld Products, Inc.*

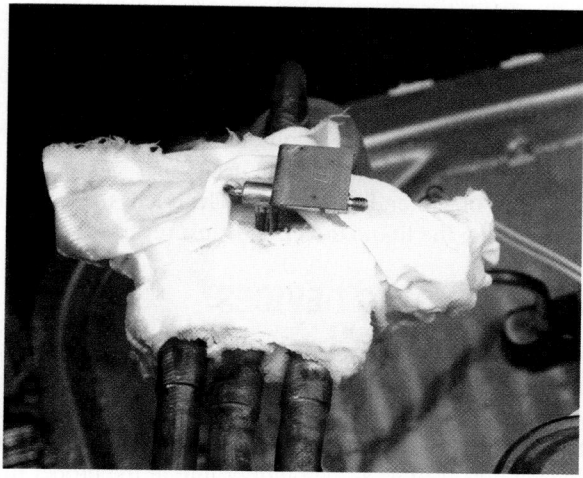

Figure 43–12 This valve body has a wet rag wrapped around it and water is dripped on the rag to keep the valve body cool. *Photo by Bill Johnson*

rag wrapped around the valve body while it is being brazed in, **Figure 43–12.** Water can actually be dripped on the rag while the connection is being heated if you have help. If the wet rag option is used, make certain that no water is allowed to enter the piping, as this will pose problems later on. Avoid pointing the torch directly at the valve, as damage to the component can result.

When removing a defective four-way valve, it is often better to cut the piping back far enough with a tubing cutter and just have additional pipe stubs brazed in the system, **Figure 43–13.** You can get a clean installation this way and could actually braze the stubs into the four-way valve in the vise where you can control the heat much better. If the old valve is removed by using one of the specialty torch tips

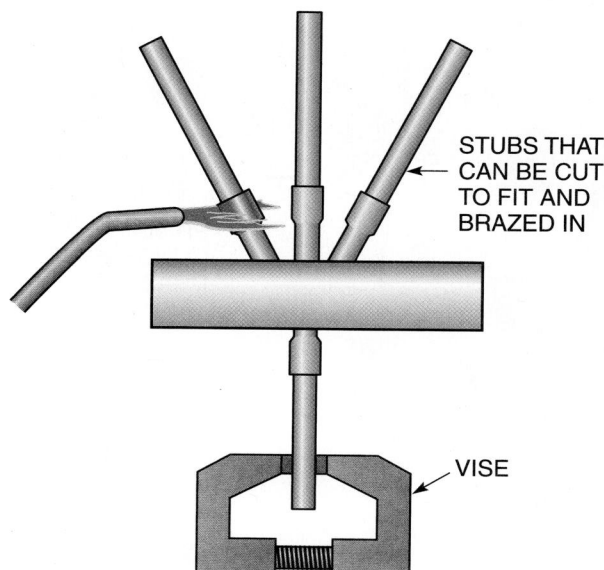

Figure 43–13 The valve and stubs are held in a vise and the stubs are brazed in. Remember, the valve body must be kept cool during this process. The valve and stubs can be installed in the unit by cutting the stubs to fit.

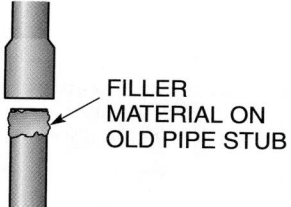

Figure 43–14 This stub has been pulled from a connection on a valve and has residue of silver brazing stuck to the stub. This braze material is very hard to clean off the stub before the stub is inserted into a new fitting.

without cutting the valve stubs back, cleaning up the old stubs that will be put back into the valve will be an important task. All of the old braze filler material must be removed before the old stub will insert back into the valve, **Figure 43–14.** This silver braze filler material is hard to remove. It must be filed off and the pipe must have something in it to prevent the filings from entering the piping. When the piping is still in the system, it is difficult to clean up these stubs correctly.

43.4 THE AIR-TO-AIR HEAT PUMP

The air-to-air heat pump resembles the central air-conditioning system. It has indoor and outdoor system components. When discussing typical air-conditioning systems, these components are often called the *evaporator* (indoor unit) and the *condenser* (outdoor unit). This terminology will work for air conditioning but not for a heat pump.

The system's coils have new names when applied to a heat pump. The coil that serves the inside of the house is called the *indoor coil*. The unit outside the house is called the outdoor unit and contains the *outdoor coil*. The reason is that the indoor coil is a condenser in the heating mode and an evaporator in the cooling mode. The outdoor coil is a condenser in the cooling mode and an evaporator in the heating mode. This is all determined by which way the hot gas is flowing. In winter the hot gas is flowing toward the indoor unit and will give up heat to the conditioned space. The heat must come from the outdoor unit, which is the evaporator. See **Figure 43–15** for an example of the direction of the gas flow. The system mode can be determined easily by gently touching the insulated gas line to the indoor unit. SAFETY PRECAUTION: If it is hot (it can be 200°F, so be careful), the unit is in the heating mode.

Like cooling equipment, heat pumps are manufactured in split systems and package systems.

43.5 REFRIGERANT LINE IDENTIFICATION

When an air-to-air heat pump is a split system, the same lines are connected between the indoor unit and the outdoor unit except that they have a new name. The large line is called a *gas line* because it always carries gas. In previous units it was called a cold gas line or a suction line. In a heat pump it is

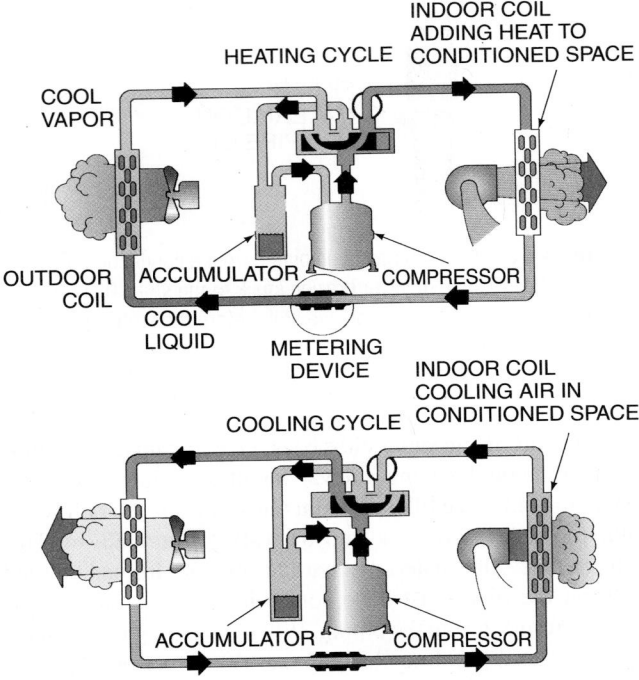

Figure 43–15 The heat pump refrigeration cycle shows the direction of the refrigerant gas flow. *Reproduced courtesy of Carrier Corporation*

always a gas line because it is a suction line or cold gas line in summer and a hot gas line in winter, **Figure 43–16.**

The small line is a *liquid line* in summer and winter, so it keeps the same name. Some changes do occur in the liquid line between summer operation and winter operation. The liquid flows toward the inside unit in summer and toward the outside unit in winter, **Figure 43–17.** The line is the same size as for cooling; the liquid direction is just reversed.

43.6 METERING DEVICES

Because the direction of the liquid flow is reversed from one season to another, some of the refrigeration components are slightly different. The metering devices used on heat pumps are different because there must be a metering device at the indoor unit as well as at the outdoor unit at the proper time. For example, when the unit is in the cooling mode, the metering device is at the indoor unit. When the system changes over to heating, a metering device must then meter refrigerant to the outdoor unit. This is accomplished in various ways with several combinations of metering devices.

43.7 THERMOSTATIC EXPANSION VALVES

The *thermostatic expansion valve* (TXV) was the first metering device in common use with heat pumps. Because this device will only allow liquid to flow in one direction, it had

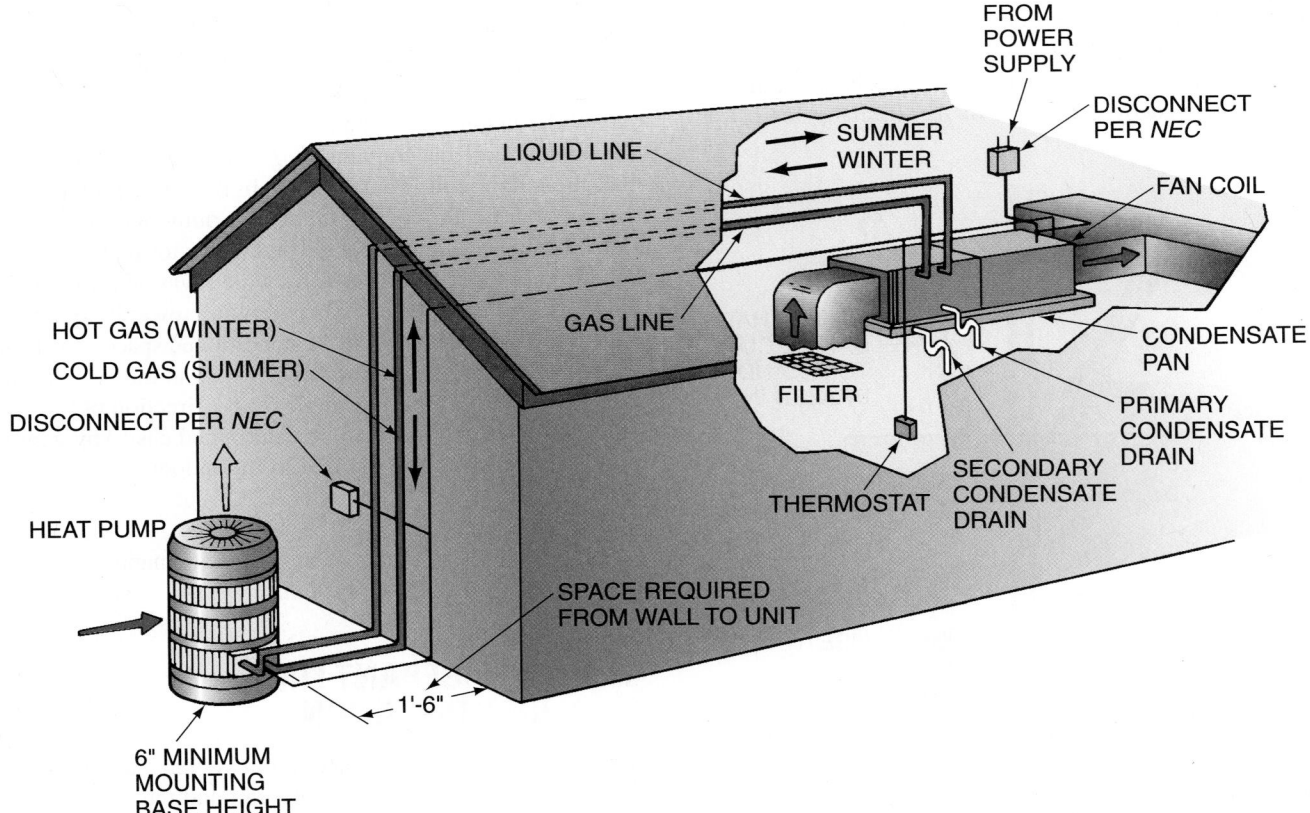

Figure 43–16 This split-system heat pump shows the interconnecting refrigerant lines.

SUMMER OPERATION		
REFRIGERANT LINE	**REFRIGERANT STATE**	**DIRECTION OF FLOW**
Larger refrigerant line	Cold gas	Indoors → Outdoors
Smaller refrigerant line	warm liquid	Outdoors → Indoors

WINTER OPERATION		
REFRIGERANT LINE	**REFRIGERANT STATE**	**DIRECTION OF FLOW**
Larger refrigerant line	Hot gas	Outdoors → Indoors
Smaller refrigerant line	Warm liquid	Indoors → Outdoors

(A)

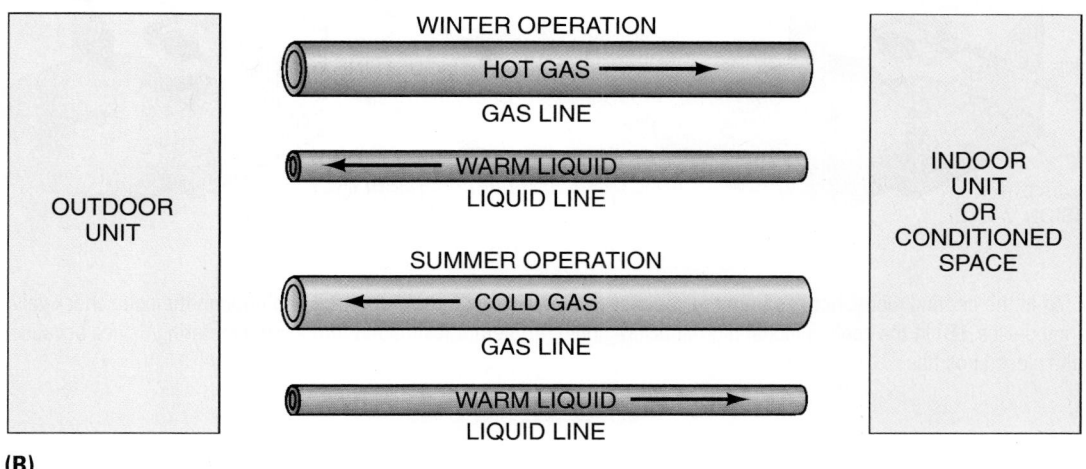

(B)

Figure 43–17   (A) Table identifying the refrigerant lines for summer and winter operation as well as the direction of refrigerant flow. **(B)** The small line is the liquid line for both summer and winter operation.

to have a check valve piped in parallel with it to allow flow around it in the other mode of operation. For example, in the cooling mode the valve needs to meter the flow with the liquid refrigerant moving toward the indoor unit. When the system reverses to the heating mode, the indoor unit becomes a condenser, and the liquid needs to be able to move freely toward the outdoor unit. The liquid flows through the check valve in winter and is metered through the TXV in the summer cycle, **Figure 43–18.**

NOTE: The TXV applied to a heat pump must be chosen carefully. The sensing bulb for the valve is on the gas line, which may become too hot for a typical TXV bulb. It is not unusual for the hot gas line to reach temperatures of 200°F. If the sensing bulb for a typical TXV is exposed to these temperatures, it is subject to rupture, **Figure 43–19.**

A vapor-charged TXV, one with a limited charge in the bulb, is usually used for this application. Remember from Unit 24, "Expansion Devices," that a limited charge bulb does not continue to build pressure with added heat. It builds only a certain amount of pressure; then additional heat causes no appreciable increase in pressure.

The use of thermostatic expansion valves, along with filter driers and check valves, adds to the complexity of the refrigerant piping circuit. These additional components,

piping, and pipe connections increase the possibility of system malfunction and refrigerant leak. For these reasons, manufacturers offer *bidirectional thermostatic expansion valves*. These valves have a check valve built into the body of the TXV. The valve looks and operates in a manner similar to that of other TXVs, with the exception of the internal check valve.

In operation, when refrigerant flow is desired through the expansion valve, the internal check valve will be in the closed position and the refrigerant is directed through the needle and seat of the expansion valve. When refrigerant flows through the valve in the opposite direction, the check valve opens and the refrigerant is able to bypass the needle and seat in the valve.

43.8   THE CAPILLARY TUBE

The capillary tube metering device is also used on heat pumps. This device allows refrigerant flow in either direction, but it is not normally used in this manner because capillary tubes of different sizes are required in summer and winter. The capillary tube may be used in several ways. One is to use a check valve to reverse the flow. When this is done, two capillary tubes are usually used, **Figure 43–20.**

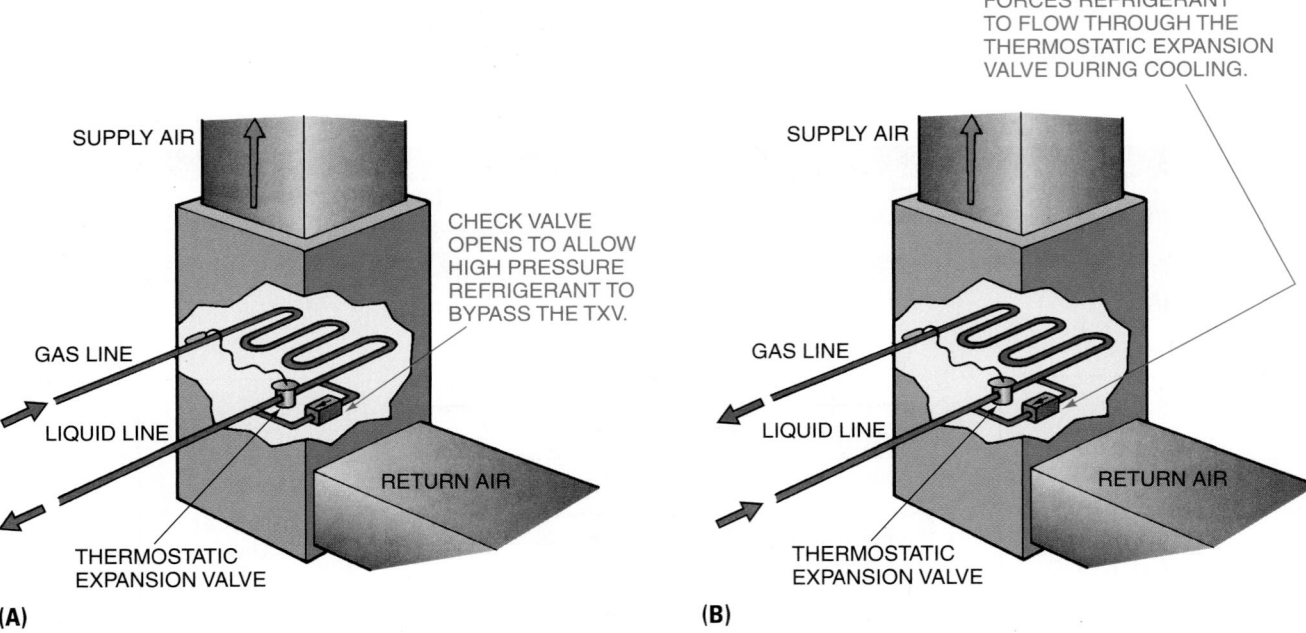

SUPPLY AIR

CHECK VALVE
OPENS TO ALLOW
HIGH PRESSURE
REFRIGERANT TO
BYPASS THE TXV.

GAS LINE

LIQUID LINE

RETURN AIR

THERMOSTATIC
EXPANSION VALVE

(A)

CLOSED CHECK VALVE
FORCES REFRIGERANT
TO FLOW THROUGH THE
THERMOSTATIC EXPANSION
VALVE DURING COOLING.

SUPPLY AIR

GAS LINE

LIQUID LINE

RETURN AIR

THERMOSTATIC
EXPANSION VALVE

(B)

Figure 43–18   **(A)** In the heating mode, hot gas from the compressor flows through the indoor coil and through the open check valve to bypass the indoor metering device. **(B)** In the cooling mode, high-pressure liquid refrigerant is directed through the metering device because the check valve will be in the closed position.

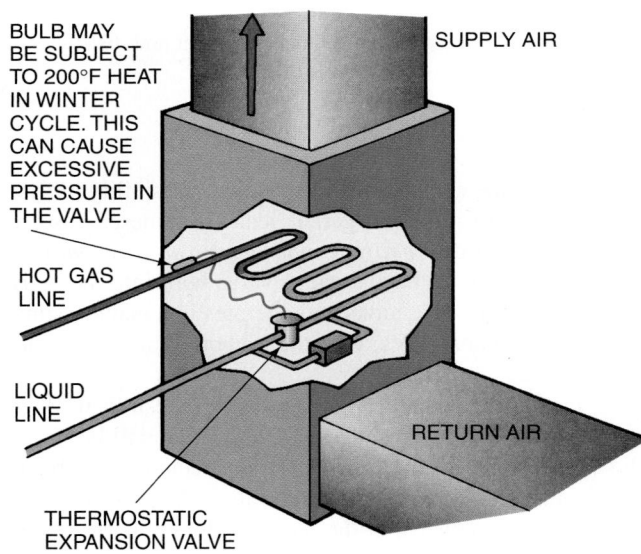

BULB MAY
BE SUBJECT
TO 200°F HEAT
IN WINTER
CYCLE. THIS
CAN CAUSE
EXCESSIVE
PRESSURE IN
THE VALVE.

SUPPLY AIR

HOT GAS
LINE

LIQUID
LINE

RETURN AIR

THERMOSTATIC
EXPANSION VALVE

Figure 43–19   This illustration shows that care must be taken in the choice of TXVs. If an ordinary TXV sensing bulb were to be mounted on this gas line, it would get hot enough to rupture the valve diaphragm.

43.9   COMBINATIONS OF METERING DEVICES

Sometimes a combination of two metering devices is used. One popular combination is to use the capillary tube for the indoor metering device for summer operation, with a check valve piped parallel to allow flow in the other direction for winter operation. Then the outside unit may have a TXV with a check valve. This system is efficient because it uses the capillary tube in the summer mode only. It is an efficient metering device for summer operation because the load conditions are nearly constant. The TXV is used in the winter on the outdoor coil. This is efficient for the winter cycle because winter conditions are not constant. The TXV will reach maximum efficiency sooner than the capillary tube in the same application because it can open its metering port when needed. This allows the evaporator to become full of refrigerant earlier in the running cycle. This dual system uses each device at its best, **Figure 43–21.**

43.10   ELECTRONIC EXPANSION VALVES

The electronic expansion valve was discussed in the unit on refrigeration and is sometimes applied to heat pumps. If the heat pump is a close-coupled unit with the indoor coil close to the outdoor unit, such as a package system (discussed in Unit 24), a single valve can be used. The reason the valve is best applied to a close-coupled unit is that the valve can meter in both directions, and if the liquid line were long it would need to be insulated. The valve will meter in either direction and maintain the correct superheat at the compressor's common suction line in the heating and cooling modes. The sensing element may be located on the common suction line just before the compressor to maintain this correct refrigerant control.

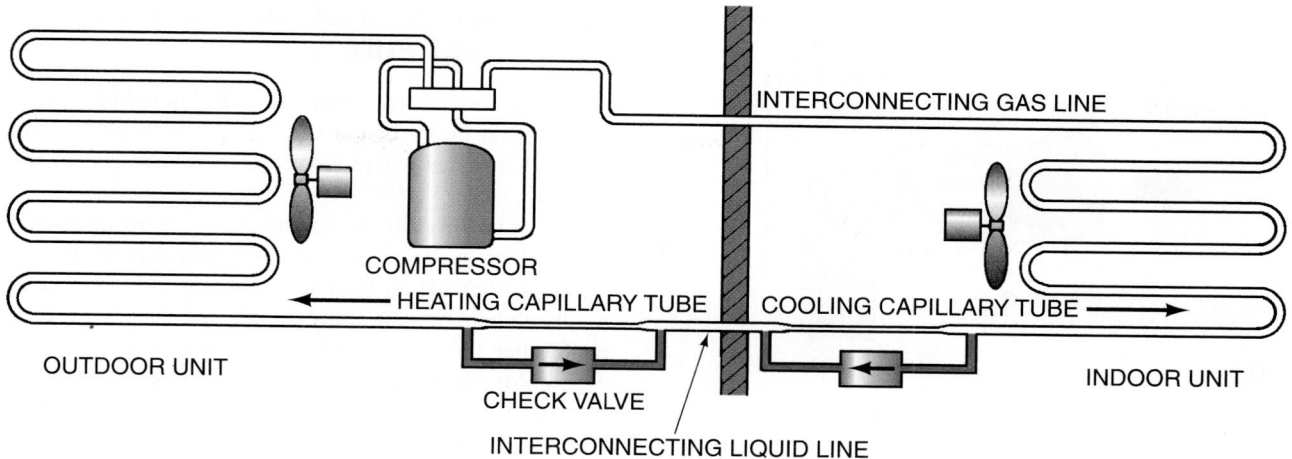

Figure 43-20   Capillary tube metering devices used on a heat pump. Notice that check valves are used and there are two different capillary tubes.

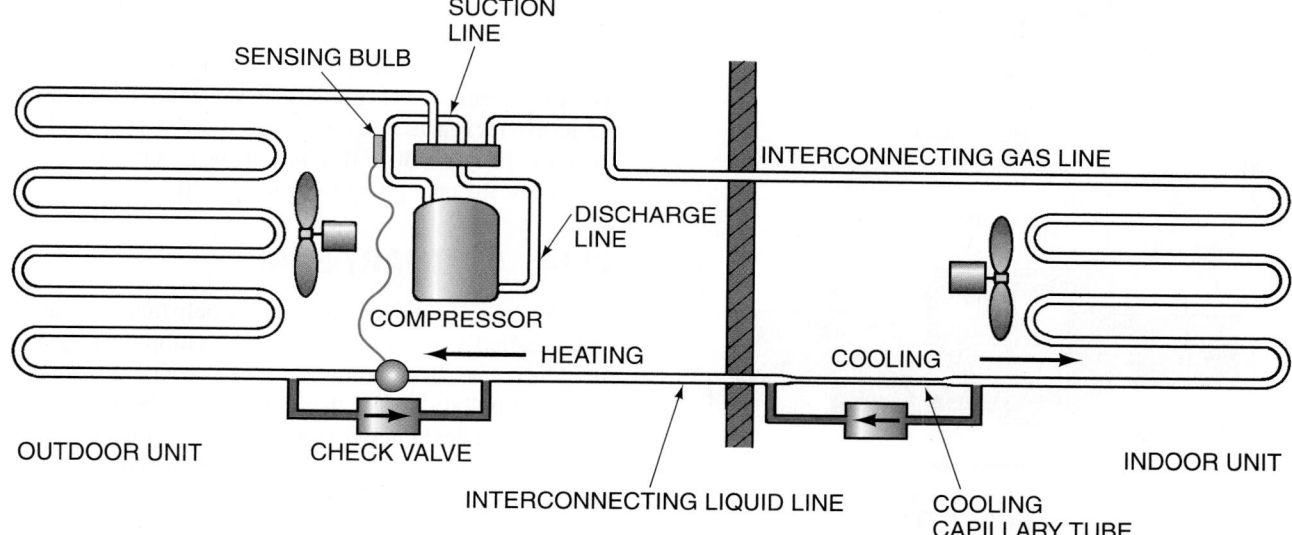

Figure 43-21   This unit uses a capillary tube for the indoor cycle and a TXV for the outdoor unit. The sensing element for this TXV is mounted on the compressor permanent suction line.

43.11   ORIFICE METERING DEVICES

Another common metering device used by some manufacturers is a combination flow device and check valve. This device allows full flow in one direction and restricted flow in the other direction, **Figure 43-22(A)**. Two of these devices are necessary with a split system—one at the indoor coil and one at the outdoor coil. The metering device at the indoor coil has a larger bore than the one at the outdoor coil because the two coils have different flow characteristics. The summer cycle uses more refrigerant in normal operation. When this device is used, a biflow filter drier is normally used in the liquid line when field repairs are done.

All of these components may be found on water-to-air heat pumps or air-to-air heat pumps. The components that control and operate the systems are much the same.

43.12   LIQUID-LINE ACCESSORIES

Liquid-line filter driers must be used in conjunction with all of the metering devices just described. Liquid-line filter driers help prevent particulate matter that may be in the refrigerant piping circuit from clogging or blocking the metering device. When standard liquid-line filter driers are installed in a system with check valves to control the flow through the metering devices, each drier is installed in series with one check valve and one expansion device. The flow direction is the same as with the metering device. If the same filter drier were installed in the common liquid line, it would filter in one direction, and the particles would wash out in the other direction, **Figure 43-23**.

In the heating mode, the high-pressure refrigerant leaves the indoor coil and bypasses the cooling capillary tube by flowing through the indoor check valve and filter drier,

IN COOLING

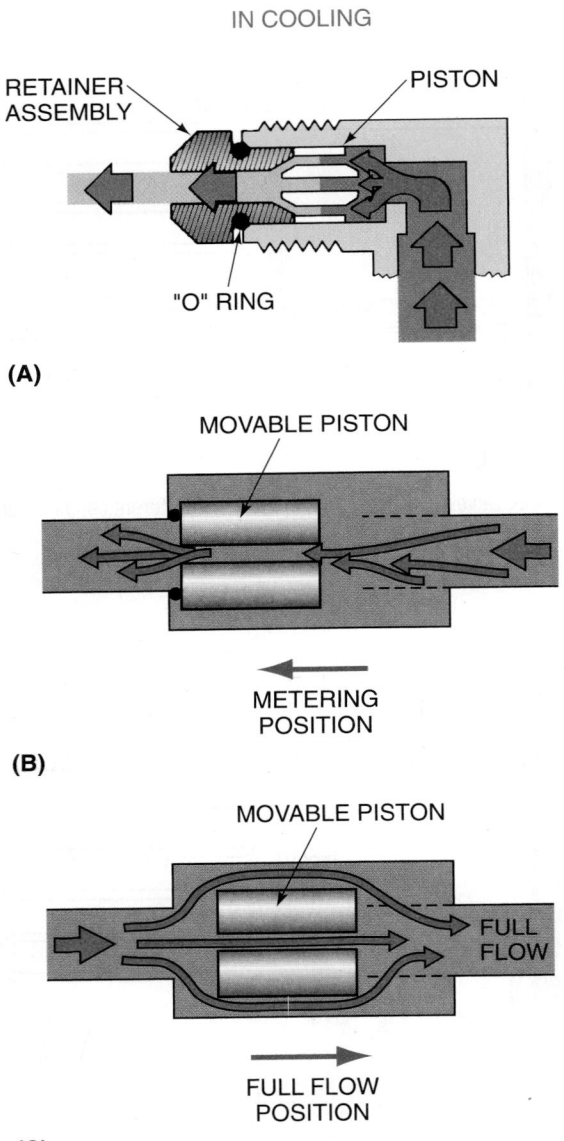

(A)

(B)

METERING
POSITION

(C)

FULL FLOW
POSITION

Figure 43–22 **(A)** This piston works like the ones illustrated in **(B)** and **(C)**. **(B)** When the piston is positioned to the left, it only allows flow through the orifice, creating a metering device. **(C)** When the flow is reversed during the next season, the piston slides to the right and full flow is allowed. A heat pump may have two of these: one at the indoor unit for cooling and one at the outdoor coil for heating. These are very small devices and may be in a flare fitting. **(A)** *Reproduced courtesy of Carrier Corporation*

Figure 43–24. Once at the outdoor unit, the refrigerant is directed through the outdoor metering device, as the outdoor check valve is in the closed position. In the cooling mode, the outdoor heating capillary tube is bypassed and the refrigerant is directed through the cooling metering device located at the indoor coil, **Figure 43–25.**

Special *biflow* filter driers are manufactured to allow flow in either direction. They are actually two driers in one shell with check valves inside the drier shell to cause the liquid to flow in the proper direction at the proper time, **Figure 43–26.**

43.13   APPLICATION OF THE AIR-TO-AIR HEAT PUMP

Air-to-air systems are normally installed in milder climates—in the "heat pump belt," which is basically those parts of the United States where winter temperatures can be as low as 10°F.

The reason for this geographical line is the characteristic of the air-to-air heat pump. It absorbs heat from the outside air; as the outside air temperature drops, it is more difficult to absorb heat from it. For example, the evaporator must be cooler than the outside air for the air to transfer heat into the evaporator. Normally, in cold weather, the heat pump evaporator will be about 20°F to 25°F cooler than the air from which it is absorbing heat. We will use 25°F temperature difference (TD) in this text as the example, **Figure 43–27.** On a 10°F day the heat pump evaporator will be boiling (evaporating) the liquid refrigerant at 25°F lower than the 10°F air. This means that the boiling refrigerant temperature will be −15°F. As the evaporator temperature goes down, the compressor loses capacity. The compressor is a fixed-size pump. It will have more capacity on a 30°F day than on a 10°F day. **The heat pump loses capacity as the capacity need of the structure increases.** On a 10°F day the structure needs more capacity than on a 30°F day, but the heat pump's capacity is less. Thus, the heat pump must have help.

43.14   AUXILIARY HEAT

In an air-to-air heat pump system, the help that the heat pump gets is called *auxiliary heat.* The heat pump itself is the primary heat, and the auxiliary heat may be electric, oil, or gas. Electric auxiliary heat is the most popular because it is easier to adapt a heat pump to an electric system.

The structure that a heat pump is intended to heat has a different requirement for every outside temperature level. For example, a house might require 30,000 Btu of heat during the day when the outside temperature is 30°F. As the outside temperature drops, the structure requires more heat. It might need 60,000 Btu of heat in the middle of the night when the temperature has dropped to 0°F. The heat pump could have a capacity of 30,000 Btu/h at 30°F and 20,000 Btu/h at 0°F. The difference of 40,000 Btu/h must be made up with auxiliary heat.

43.15   BALANCE POINT

The *balance point* occurs when the heat pump can pump in exactly as much heat as the structure is leaking out. At this point the heat pump will completely heat the structure by running continuously. Above this point the heat pump will cycle off and on. Below this point the heat pump will run continuously but will not be able to maintain the desired temperature. Although the auxiliary heat portion of a heat pump system will be operational at some temperature above the balance point, it is impossible for a heat pump system to satisfy the heating requirements of a structure without auxiliary heat if the temperature falls below the balance point.

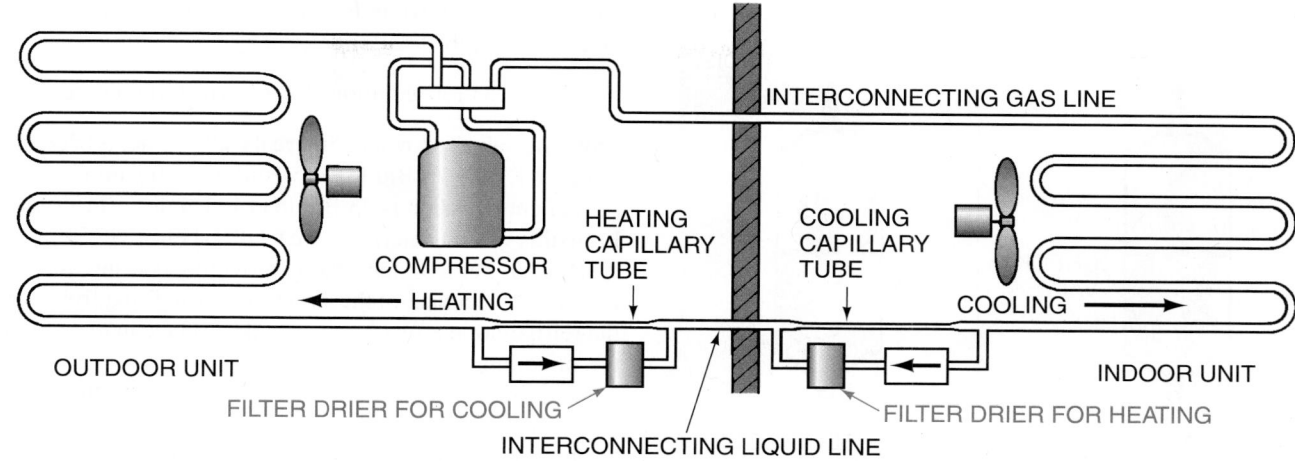

Figure 43–23   Placing a liquid-line filter drier in the heat pump refrigerant piping.

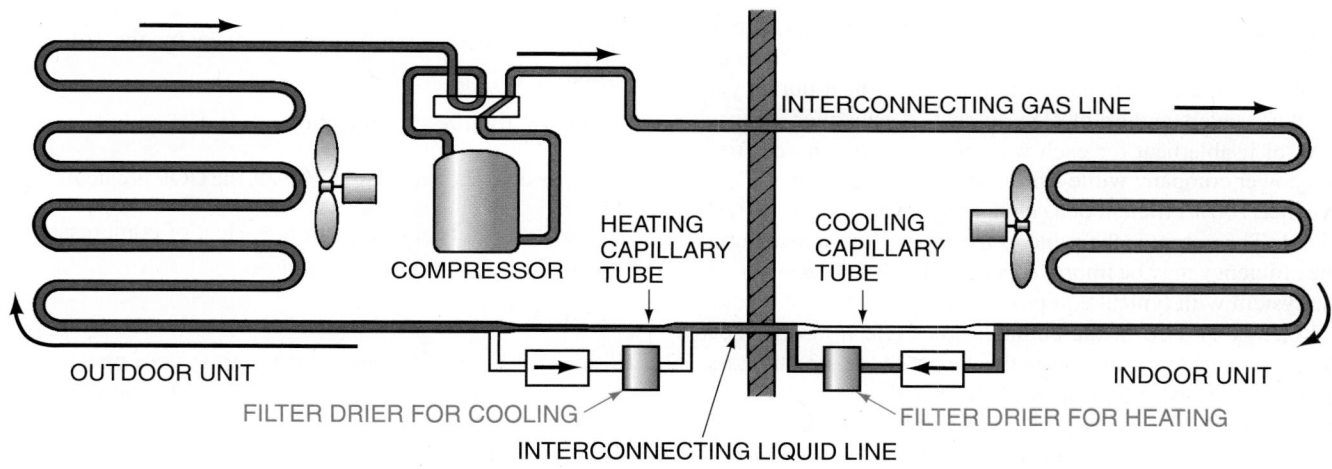

Figure 43–24   In the heating mode, the cooling metering device is bypassed.

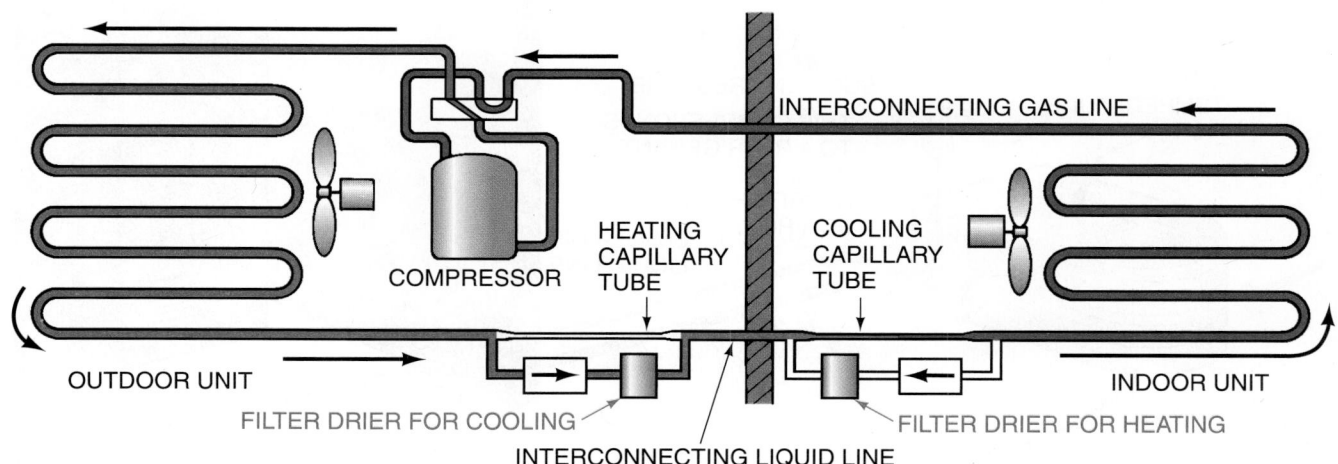

Figure 43–25   In the cooling mode, the heating metering device is bypassed.

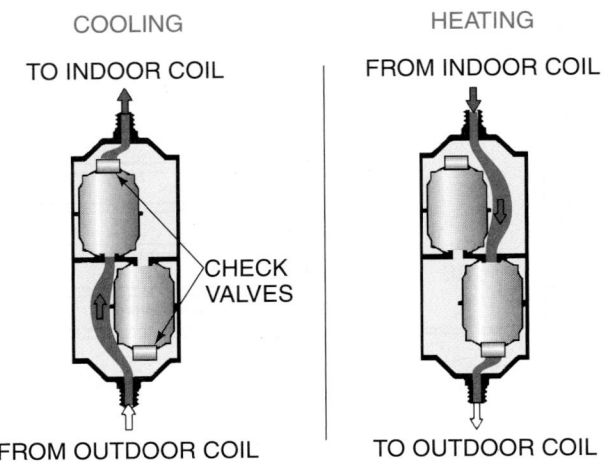

COOLING

TO INDOOR COIL

HEATING

FROM INDOOR COIL

CHECK VALVES

FROM OUTDOOR COIL

TO OUTDOOR COIL

Figure 43–26   A biflow drier. *Reproduced courtesy of Carrier Corporation*

43.16   COEFFICIENT OF PERFORMANCE

If the heat pump will not heat the structure all winter, why is it so popular? This can be answered in one word, *efficiency*. To understand the efficiency of an air-to-air heat pump, you need an understanding of electric heat. A customer receives 1 W of usable heat for each watt of energy purchased from the power company while using electric resistance heat. This is called 100% efficient or a *coefficient of performance (COP)* of 1:1. The output is the same as the input. With a heat pump the efficiency may be improved as much as 3.5:1 for an air-to-air system with typical equipment. When the 1 W of electrical energy is used in the compression cycle to absorb heat from the outside air and pump this heat into the structure, the unit could furnish 3.5 W of usable heat. Thus its COP is 3.5:1, or it can be thought of as 350% efficient. See **Figure 43–28** for a manufacturer's rating table for an air-to-air heat pump.

To provide some insight into why the COP of a heat pump system is so high, it is necessary to take a look at the pressure enthalpy chart shown in **Figure 43–29**. For a system that provides cooling, the COP is defined as

COP = Net refrigeration effect ÷ Heat of compression

From the values shown in **Figure 43–29**, the net refrigeration effect (NRE) is 75 Btu/lb (115 Btu/lb − 40 Btu/lb) and the heat of compression is 25 Btu/lb (140 Btu/lb − 115 Btu/lb). From this we can determine the COP as being 3 (75 Btu/lb ÷ 25 Btu/lb). For a system that is providing cooling, the NRE is what the occupant of the structure is benefiting from, since the refrigeration effect is what is taking place in the evaporator, or cooling, coil.

On a heat pump system that is operating in the heating mode, it is not the NRE that is providing the heat. Instead, the total heat of rejection (THOR) is what is conditioning the space. A portion of the THOR is the heat that is generated during the compression process. So, the compressor itself is contributing in two ways: it is providing the pressure difference to facilitate refrigerant flow through the system and it is also generating heat that is directly transferred to the medium being heated. The COP for a heat pump system operating in the heating mode is given by the following formula:

COP = Total heat of rejection ÷ Heat of compression

For the system shown in **Figure 43–29**, the COP is calculated as

COP = Total heat of rejection ÷ Heat of compression

COP = 100 Btu/lb ÷ 25 Btu/lb

COP = 4

Notice that the COP of the heat pump system operating in the heating mode is higher than the COP of the same heat pump system operating in the cooling mode. This example is greatly simplified and intended to illustrate the contribution of the compressor heat to the efficiency of the heat pump system.

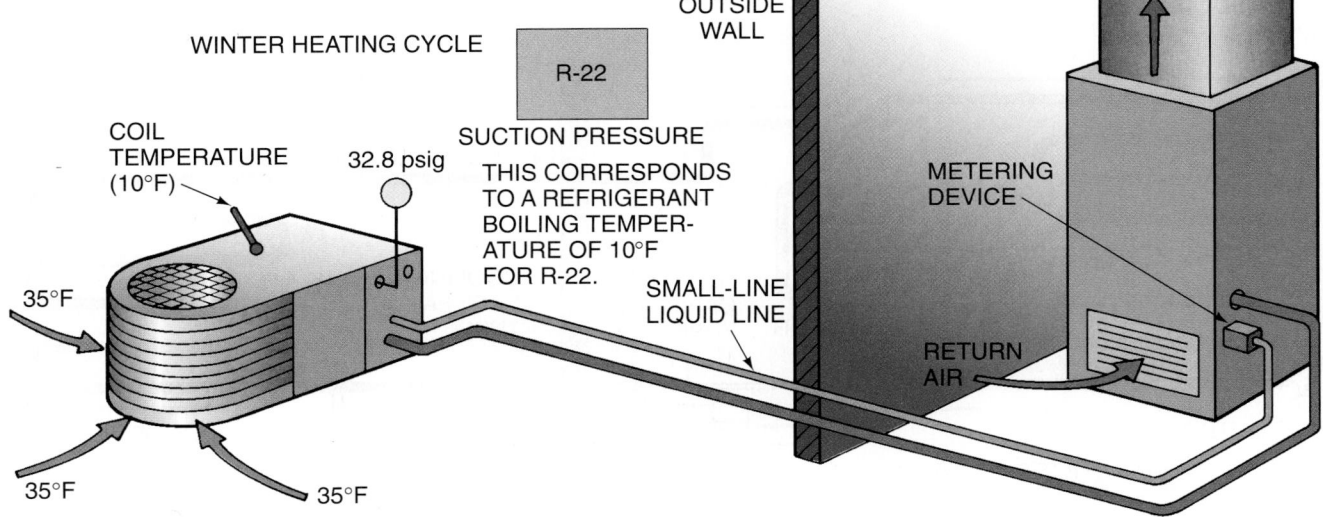

WINTER HEATING CYCLE

R-22

OUTSIDE WALL

COIL TEMPERATURE (10°F)

32.8 psig

SUCTION PRESSURE

THIS CORRESPONDS TO A REFRIGERANT BOILING TEMPERATURE OF 10°F FOR R-22.

METERING DEVICE

SMALL-LINE LIQUID LINE

35°F

35°F   35°F

RETURN AIR

Figure 43–27   A heat pump outdoor coil operating in the winter cycle.

OUTDOOR UNIT WITH HEAT PUMP COILS							
	TXC030D4	TXC031C4	TXC031D4	TXC035C4	TXC035D4	TXC036C4	TXC036D4
EXPANSION TYPE	CHG TO 59	CHG TO 59	CHG TO 59	CHG TO 59	CHG TO 59	CHG TO 59	CHG TO 59
RATINGS (COOLING) ①							
BTUH (TOTAL)	29000	29200	29200	29400	29400	29800	29800
BTUH (SENSIBLE)	22100	22200	22200	23200	23200	23900	23900
INDOOR AIRFLOW (CFM)	1000	1000	1000	1100	1100	1115	1115
SYSTEM POWER (KW)	2.70	2.70	2.70	2.75	2.75	2.77	2.77
EER / SEER (BTU/WATT-HR.)	10.75/11.75	10.80/12.00	10.80/12.00	10.70/11.75	10.70/11.75	10.75/11.75	10.75/11.75
RATINGS (HEATING) ①							
(HIGH TEMP.) BTUH	28200	28400	28400	28400	28400	28600	28600
SYSTEM POWER (KW)	2.77	2.65	2.65	2.74	2.74	2.56	2.56
COP	2.98	3.14	3.14	3.04	3.04	3.28	3.28
HSPF (BTU/WATT-HR.)	7.75	8.05	8.05	7.80	7.80	8.30	8.30

(A)

OUTDOOR UNIT WITH AIR HANDLERS							
	4TEE3F31A	4TEE3F37A	4TEE3F40A	4TEP3F24A	4TEP3F30A	4TEP3F36A	4TEP3F42A
EXPANSION TYPE	TXV-NB	TXV-NB	TXV-NB	TXV-NB	TXV-NB	TXV-NB	TXV-NB
RATINGS (COOLING) ①							
BTUH (TOTAL)	29600	30000	31400	29000	29200	29600	30200
BTUH (SENSIBLE)	23400	23200	24300	21400	22000	23700	24200
INDOOR AIRFLOW (CFM)	1060	1000	1010	900	940	1110	1125
SYSTEM POWER (KW)	2.54	2.48	2.50	2.62	2.62	2.73	2.73
EER / SEER (BTU/WATT-HR.)	11.65/14.00	12.10/14.25	12.55/14.75	11.05/13.00	11.15/13.00	10.85/12.50	11.05/13.00
RATINGS (HEATING) ①							
(HIGH TEMP.) BTUH	28400	27600	28000	28000	28000	28600	28800
SYSTEM POWER (KW)	2.49	2.42	2.23	2.72	2.66	2.60	2.54
COP	3.34	3.34	3.68	3.02	3.08	3.22	3.32
HSPF (BTU/WATT-HR.)	8.45	8.50	9.00	7.80	8.00	8.20	8.40

(B)

Figure 43–28 Rating tables of air-to-air heat pumps. *Courtesy Trane Corporation*

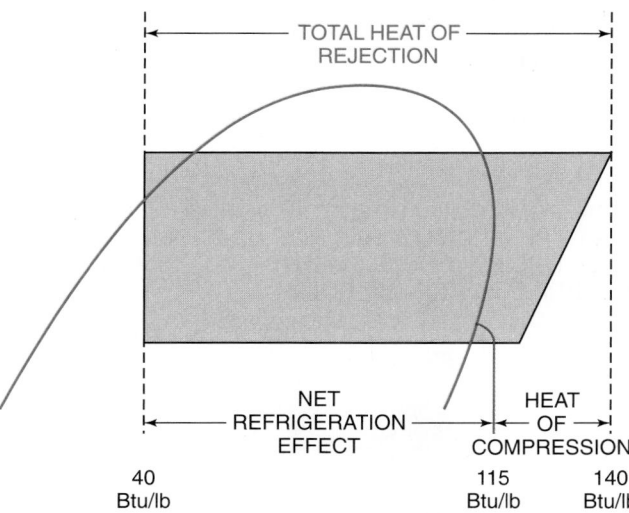

Figure 43–29 Pressure enthalpy chart showing the relationship between the total heat of rejection (THOR), net refrigeration effect (NRE), and the heat of compression (HOC).

A high COP occurs only during higher outdoor winter temperatures. As the temperature falls, the COP also falls. A typical air-to-air heat pump will have a COP of 1.5:1 at 0°F, so it is still economical to operate the compression system at these temperatures along with the auxiliary heating system. Some manufacturers have controls to shut off the compressor at

low temperatures. When the temperature rises, the compressor will come back on. Some temperatures that manufacturers use to stop the compressor are 0°F to 10°F. The long running times of the compressor will not hurt it or wear it out to any extent, so some manufacturers do not shut off the compressor at all. They would rather have the compressor run all the time than to restart it with a cold crankcase when the temperature rises.

The *water-to-air* heat pump might not need auxiliary heat because the heat source (the water) can be a constant temperature all winter. The heat loss (the heat requirement) and the heat gain (the cooling requirement) of the structure might almost be equal. Consequently, these heat pumps have COP ratings as high as 4:1. Just remember that the higher COP ratings of water-to-air heat pumps are due not to the components of the refrigeration equipment but to the temperatures of the heat source. Earth and lakes have a more constant temperature than air. Such pumps are very efficient for winter heating because of the COP and for summer cooling because they operate at water-cooled air-conditioning temperatures and head pressures.

43.17 SPLIT-SYSTEM AIR-TO-AIR HEAT PUMP

Like cooling air-conditioning equipment, air-to-air heat pumps come in two styles: split systems and package (self-contained) systems.

The split-system air-to-air heat pump resembles the split-system cooling air-conditioning system. The components look exactly alike. An expert normally cannot tell whether the equipment is air-conditioning or heat pump equipment from the outside.

43.18 THE INDOOR UNIT

The indoor unit of an air-to-air system may be an electric furnace with a heat pump indoor coil where the summer air-conditioning evaporator would be placed, **Figure 43–30.** The outdoor unit may resemble a cooling condensing unit.

The indoor unit is the part of the system that circulates the air for the structure. It contains the fan and coil. The airflow pattern may be upflow, downflow, or horizontal to serve different applications. Some manufacturers have cleverly designed their units so that one unit may be adapted to all of these flow patterns. This is done by correctly placing the pan for catching and containing the summer condensate, **Figure 43–31.**

Coil placement in the airstream is important in the indoor unit. The electric auxiliary heat and the primary heat (the heat pump) may need to operate at the same time in weather below the balance point of the house. **The refrigerant coil must be located in the airstream before the auxiliary heating coil.** Otherwise, heat from the auxiliary unit will pass through the refrigerant coil when both are operating in winter. If the

Figure 43–30 An air-to-air heat pump indoor unit. It is an electric furnace with a heat pump coil in it. Notice that the air flows through the heat pump coil and then through the heating elements. *Photo by Bill Johnson*

HEATING AND COOLING UNIT

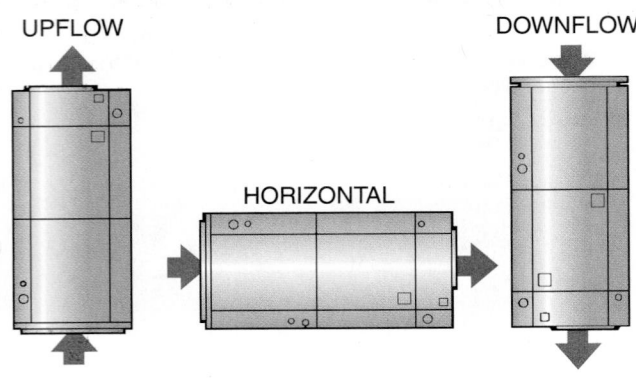

Figure 43–31 This heat pump indoor unit can be positioned either in the vertical (upflow, downflow) or horizontal position by placing the condensate pan under the coil in the respective position. *Reproduced courtesy of Carrier Corporation*

auxiliary heat is operating and is located before the heat pump coil, the head pressure will be too high and could rupture the coil or burn the compressor motor. Remember, the coil is operating as a *condenser* in the heating mode. It rejects heat from the refrigeration system, so any heat added to it will cause the head pressure to rise, **Figure 43–32.**

The indoor unit may be a gas or oil furnace. If this is the case, the indoor coil must be located in the outlet airstream of the furnace. When the auxiliary heat is gas or oil, the heat pump does not run while the auxiliary heat is operating. If it is not located after the furnace heat exchanger it would sweat in summer. The indoor coil would be operating as an evaporator in the summer, and the outlet air temperature could be lower than the dew point temperature of the air

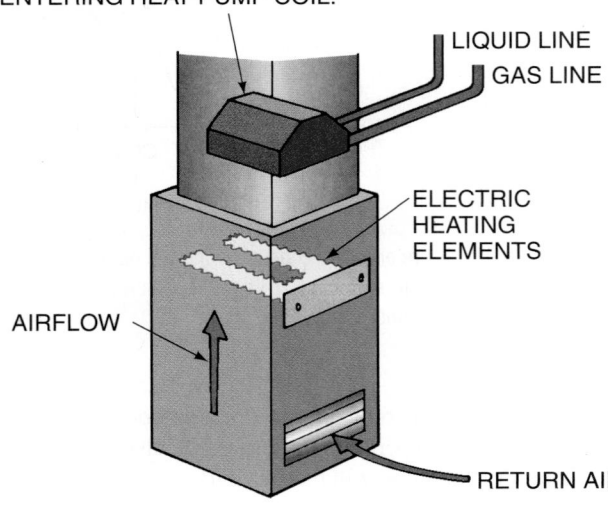

Figure 43–32 This illustration shows a heat pump coil installed after electric heating elements (oil and gas are the same).

THE HEAT PUMP COIL IS INSTALLED IN THE
HOT AIR STREAM OF THE GAS FURNACE. THIS
SYSTEM HAS A CONTROL ARRANGEMENT
THAT WILL NOT ALLOW THE FURNACE TO
RUN AT THE SAME TIME AS THE HEAT PUMP.
THIS CONTROL ARRANGEMENT CAN BE USED
FOR THE ELECTRIC FURNACE IN FIGURE 43-32.

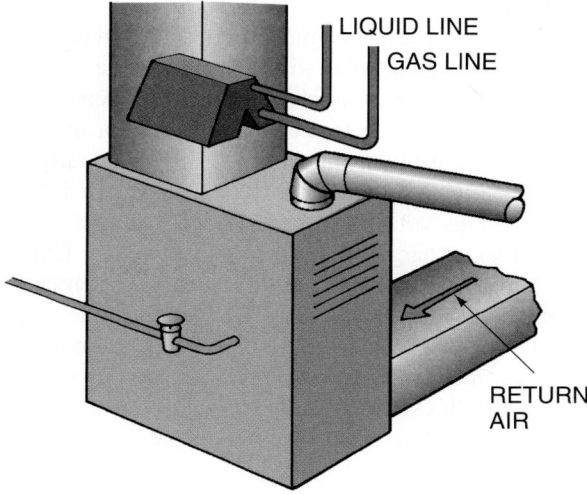

LIQUID LINE
GAS LINE
RETURN
AIR

Figure 43–33   A gas furnace with a heat pump indoor coil instead
of an air-conditioner evaporator.

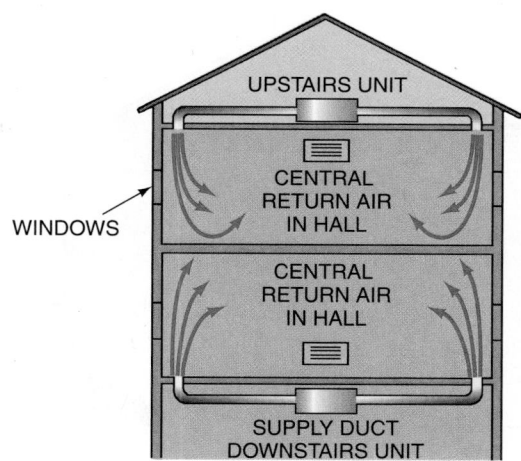

UPSTAIRS UNIT
WINDOWS
CENTRAL
RETURN AIR
IN HALL
CENTRAL
RETURN AIR
IN HALL
SUPPLY DUCT
DOWNSTAIRS UNIT

Figure 43–34   This heat pump installation shows a good method of
distributing the warm air in the heating mode.

surrounding the heat exchanger. Most local codes will not
allow a gas or oil heat exchanger to be located in a cold
airstream for this reason. When the auxiliary heat is gas or
oil, special control arrangements must be made so that the
heat pump will not operate while the gas or oil heat is operating. When a heat pump is added onto an electric furnace
and the coil must be located after the heat strips, follow the
same rules as for gas or oil. See **Figure 43–33** for an example of this installation.

43.19   AIR TEMPERATURE OF THE CONDITIONED AIR

The heat pump indoor unit is installed in much the same way
as a split-system cooling unit. The air distribution system
must be designed more precisely due to the air temperatures
leaving the air handler during the heating mode. The air temperature is not as hot as with gas and oil systems that are normally the heating system with electric cooling. The heat
pump usually has leaving-air temperatures of 100°F or lower
when just the heat pump is operating. If the airflow is restricted, the air temperature will go up slightly but the unit
COP will not be as good. The efficiency will be reduced.
Most heat pumps require a minimum of 400 cfm/ton of capacity in the cooling mode. This will equate to approximately 400 cfm/ton in heating.

When 100°F air is distributed in the conditioned space, it
must be distributed carefully or drafts will occur. Normally,
the air distribution system is on the outside walls. The air registers are either in the ceiling or the floor, depending on the
structure. It is common in two-story houses for the first-story

registers to be in the floor with the air handler in the crawl
space below the house. The upstairs unit is commonly located
in the attic crawl space with the registers in the ceiling near
the outside walls. The outside walls are where the heat leaks
out of a house in the winter and leaks into the house in the
summer. If these walls are slightly heated in the winter with
the airstream, less heat is taken out of the air in the conditioned space. The room air stays warmer, **Figure 43–34**. 100°F
air mixed with room air does not feel like 130°F air mixed
with room air from a gas or oil furnace. It may feel like a draft.

When air is pushed into the room from the air register at
100°F, the air that is around the register is induced and mixed
with the register air. This air may be at a temperature of 75°F.
These two airs will mix resulting in a mixed air temperature
of about 85°F. If this mixed air at 85°F blows on a person
with a skin temperature of about 91°F, it will seem like the
room is cool, not warm. Many people have called heat pumps
cold heat for this reason. Proper air distribution can prevent
complaints.

When 100°F air is distributed from the inside walls, such
as high sidewall registers, the system may not heat satisfactorily. The air will mix with the room air and might feel
drafty. The outside walls will also be lower in temperature
than when the air is directed on them, and a *cold wall* effect
could be noticed in cold weather, **Figure 43–35**. This is generally considered a poor application for a heat pump.

43.20   THE OUTDOOR UNIT INSTALLATION

The outdoor installation for a heat pump is much like a central air-conditioning system from an airflow standpoint. The
unit must have a good air circulation around it, and the discharge air must not be allowed to recirculate.

Some more serious aspects should be considered. The direction of the prevailing wind in the winter could lower the
heat pump performance. If the unit is located in a prevailing
north wind or a prevailing wind from a lake, the performance
may not be up to standard. A prevailing north wind might

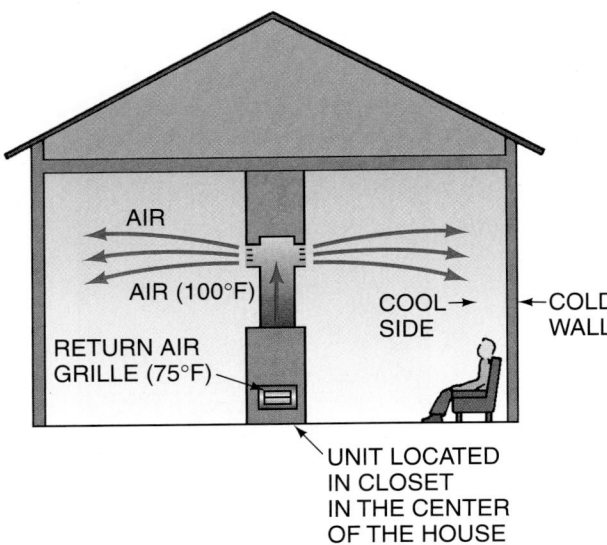

Figure 43–35 This air distribution system with the air being distributed from the inside using high sidewall registers shows the heat pump's 100°F air mixing with the room air of 75°F.

cause the evaporator to operate at a lower-than-normal temperature. A wind blowing inland off a lake will be very humid and might cause freezing problems in the winter.

The outdoor unit must *not* be located where roof water will pour into it. The outdoor unit will be operating at below freezing much of the time, and any moisture or water that is not in the air itself should be kept from the unit's coil. If not, excess freezing will occur.

The outdoor unit is an evaporator in winter and will attract moisture from the outside air. If the coil is operating below freezing, the moisture will freeze on the coil. If the coil is above freezing, the moisture will run off the coil as it does in an air-conditioning evaporator. This moisture must have a place to go. If the unit is in a yard, the moisture will soak into the ground. SAFETY PRECAUTION: If the unit is on a porch or walk, the moisture could freeze and create slippery conditions, **Figure 43–36.**

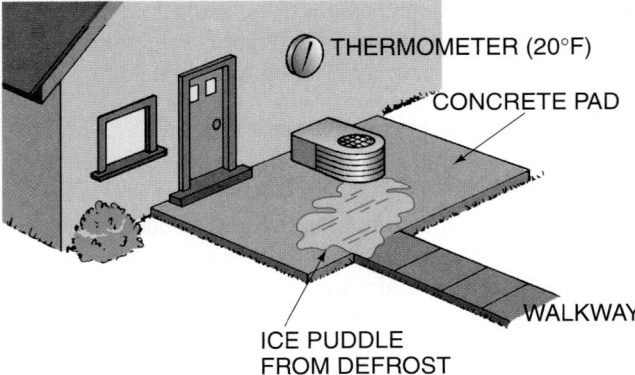

Figure 43–36 This heat pump was placed in the wrong location. The water that forms on the outside coil will run onto the walkway next to the coil and freeze.

The outdoor unit is designed with drain holes or pans in the bottom of the unit to allow free movement of the water away from the coil. If they are inadequate, the coil will become a solid block of ice in cold weather. When the coil is frozen solid, it is a poor heat exchanger with the outside air, and the COP will be reduced. Defrosting methods are discussed in Section 43.23, "The Defrost Cycle."

Manufacturers have recommended installation procedures for their particular units that will show the installer how to locate the unit for the best efficiency and water drainage.

The refrigerant lines that connect the indoor unit to the outdoor unit are much the same as for air conditioning. They come in line sets with the large line insulated. Quick-connect fittings with precharged lines or flare connectors with nitrogen-charged lines are typical. See Unit 38, "Installation," for line installation. The only difference that should be considered is the large line, the gas line. In winter it may be 200°F and should be treated as a hot line in which heat must be contained. Therefore, many manufacturers use a thicker insulation on the gas line for heat pumps. The gas line should not be located next to any object that will be affected by its warm outside temperature. If the line set must be run underground, it must be kept dry because moisture will transfer heat to the ground. Underground piping should be routed through plastic sleeves for waterproofing and water should not be allowed to flood the sleeves. SAFETY PRECAUTION: Installing a heat pump involves observing the same safety precautions as installing a "cooling only" air-conditioning system. The ductwork must be insulated, which requires working with fiberglass. Be careful when working with metal duct and fasteners and especially when installing units on rooftops and other hazardous locations.

43.21 PACKAGE AIR-TO-AIR HEAT PUMPS

Package air-to-air heat pumps are much like package air conditioners. They look alike and are installed in the same way. Therefore, give the same considerations to the prevailing wind and water conditions. The heat pump outdoor coil must have drainage in summer and winter. The package heat pump has all of the components in one housing and is easy to service, **Figure 43–37.** They have optional electric heat compartments that usually accept different electric heat sizes from 5000 W (5 kW) to 25,000 W (25 kW).

The metering devices used for package air-to-air heat pumps are much like those for split systems. Some manufacturers use a common metering device for the indoor and outdoor coils because the two are so close together. When one metering device is used, it must be able to meter both ways at a different rate in each direction because a different amount of refrigerant is used in the summer evaporator (indoor coil) than in the winter evaporator (outdoor coil).

Package air-to-air heat pumps must be installed correctly. For example, the duct must extend to the unit. Because the unit contains the outdoor coil also, the duct will need to extend all the way to an outside wall of the structure for a

Figure 43–37   A package heat pump. All components are contained in one housing. *Reproduced courtesy of Carrier Corporation*

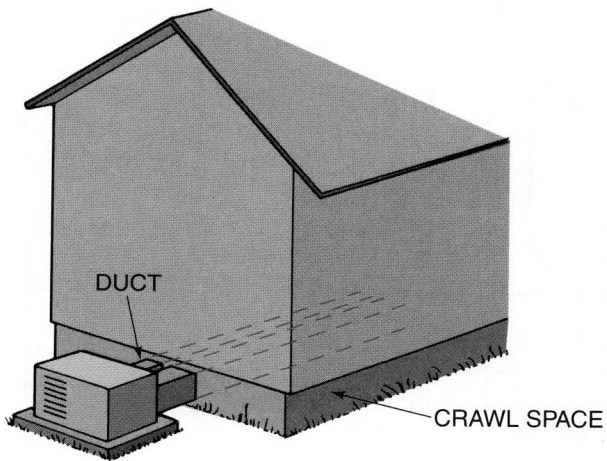

Figure 43–38   The self-contained unit contains the outdoor coil and the indoor coil. The ductwork must be routed to the unit.

crawl space installation, **Figure 43–38.** Supply and return ducts must be insulated to prevent heat exchange between the duct and the ambient air.

Package air-to-air heat pumps have the same advantage from a service standpoint as package cooling air conditioners or package gas-burning equipment. All of the controls and components may be serviced from the outside. They also have the advantage of being factory assembled and charged. They are leak checked and may even be operated before leaving the factory. This reduces the likelihood of having a noisy fan motor or other defect.

43.22   CONTROLS FOR THE AIR-TO-AIR HEAT PUMP

The air-to-air heat pump is different from any other combination of heating and cooling equipment. The following items must be controlled at the same time for the heat pump to be efficient: space temperature, defrost cycle, indoor fan, the compressor, the outdoor fan, auxiliary heat, and emergency

heat. Each manufacturer has its own method of controlling the heating and cooling sequence.

The space temperature for an air-to-air heat pump is not controlled in the same way as a typical heating and cooling system. There are actually two complete heating systems and one cooling system. The two heating systems are the refrigerated heating cycle from the heat pump and the auxiliary heat from the supplemental heating system—electric, oil, or gas. The auxiliary heating system may be operated as a system by itself if the heat pump fails. When the auxiliary heat becomes the primary heating system because of heat pump failure, it is called *emergency heat* and is normally operated only long enough to get the heat pump repaired. The reason is that the COP of the auxiliary heating system is not as good as the COP of the heat pump.

The space-temperature thermostat is the key to controlling the system. It is normally a two-stage heating and two-stage cooling type of thermostat used exclusively for heat pump applications. Other variations in the thermostat concern the number of stages of heating or cooling. **Figure 43–39** shows a

(A)

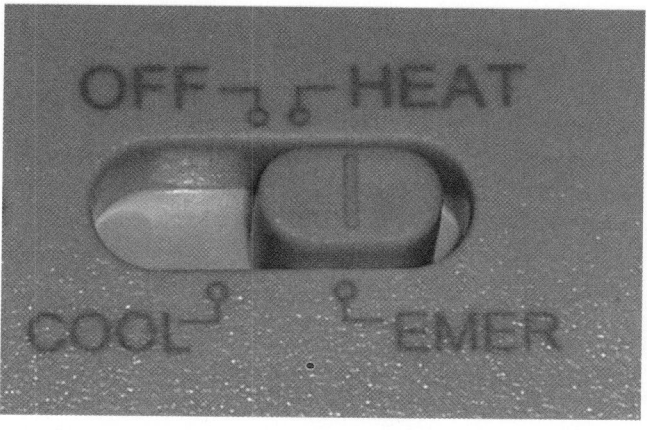

(B)

Figure 43–39   **(A)** A heat pump thermostat. **(B)** Notice the EMERGENCY HEAT option on the thermostat. *Photos by Eugene Silberstein*

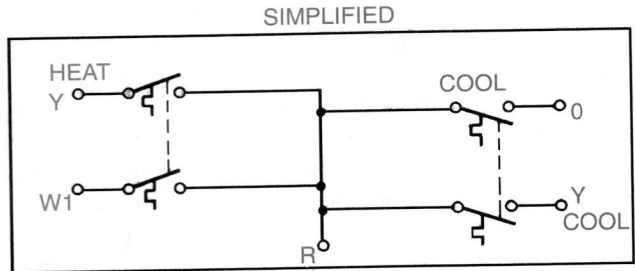

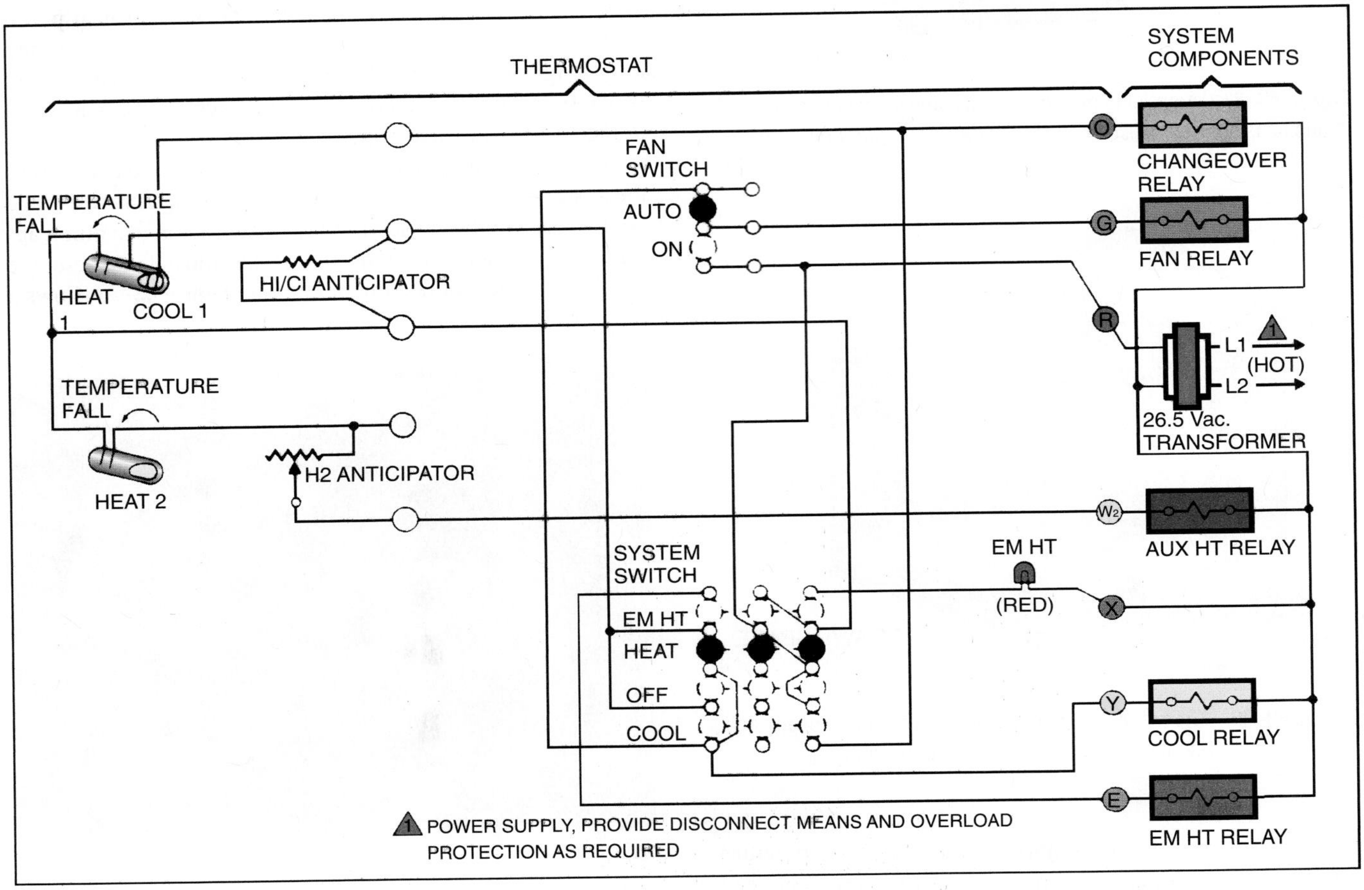

SIMPLIFIED

Figure 43–40   A heat pump thermostat with two-stage heating and two-stage cooling. *Courtesy Honeywell Corporation*

typical heat pump thermostat. Heat pump thermostats are easily identified because they have an EMERGENCY HEAT position on the selector switch, **Figure 43–39(B).**

The wiring diagram in **Figure 43–40** represents the control wiring of a typical heat pump system.

The following is the sequence of cooling and heating control for an automatic changeover thermostat—one that automatically changes from cooling to heating and back. NOTE: This control sequence is only one of many. Each manufacturer has its own. The temperature-sensing element is a bimetal that controls mercury bulb contacts. The auxiliary heat is electric. The thermostat fan switch is in AUTO.

Cooling Cycle Control

When the first-stage bulb of the thermostat closes its contacts (on a rise in space temperature), the four-way reversing valve solenoid coil is energized, **Figure 43–41.**

When the space temperature rises about 1°F, the second-stage cooling contacts of the thermostat close. The second-stage contacts energize the compressor contactor coil and the indoor fan relay coil to start the compressor, outdoor fan, and the indoor fan, **Figure 43–42.** When the compressor starts the second stage of operation, it diverts the hot gas from the compressor to the outdoor coil, and the system is in the cooling cycle.

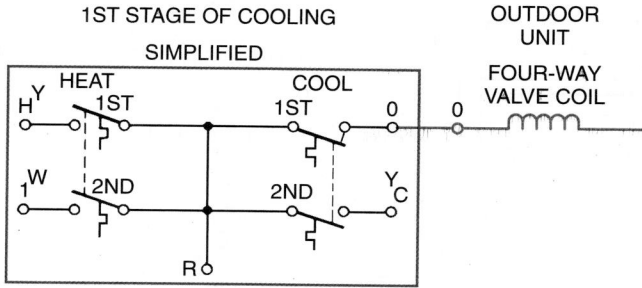

Figure 43–41   The first stage of the cooling cycle in which the four-way valve magnetic holding coil is energized.

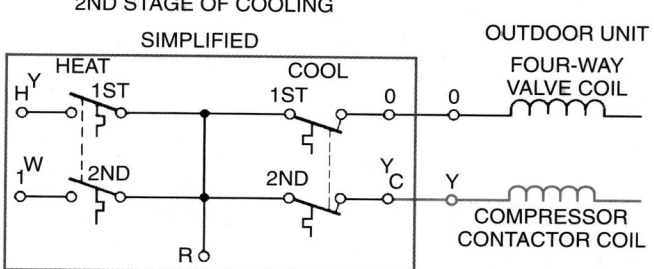

Figure 43–42   This diagram shows what happens with a rise in temperature. The first stage of the thermostat is already closed; then the second stage closes and starts the compressor. The compressor was the last on and will be the first off.

When the space temperature begins to fall and is satisfied (because the compressor is running and removing heat), the first full contacts to open are the second-stage contacts. The compressor (and outdoor fan) and the indoor fan stop. The first-stage contacts remain closed, and the four-way valve remains energized. The system pressure will now equalize through the metering device for ease of compressor starting, but the four-way valve will not change position.

When the space temperature rises again, the compressor (and outdoor fan) and indoor fan will start again. If the outdoor temperature is getting cooler (e.g., in autumn), the space temperature will continue to fall, and the first-stage contacts will open. This deenergizes the four-way reversing valve solenoid coil, and the unit can change over to heat when it starts up. **The four-way valve is a pilot-operated valve. It will not change over until the compressor starts and a pressure differential is built up.**

Space Heating Control

When the space temperature continues to fall due to outdoor conditions, the first-stage heating contacts close. This starts the compressor (and outdoor fan) and the indoor fan. The four-way valve is not energized, so the compressor hot gas is directed to the indoor coil. The unit is now in the heating mode, **Figure 43–43**.

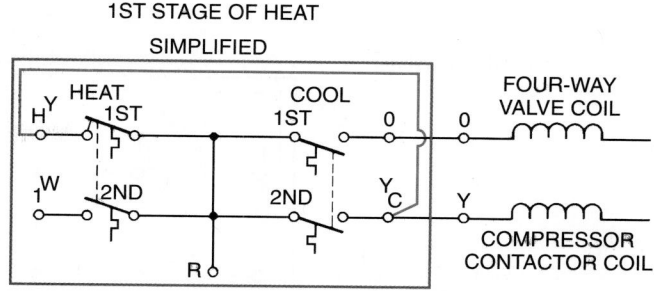

Figure 43–43   The space temperature drops to below the cooling set points for the first stage of heating; the compressor starts. This time, the four-way valve magnetic coil is deenergized, and the unit will be in the heating mode.

When the space temperature warms, the first-stage heating contacts open, and the compressor (and outdoor fan) and indoor fan stop.

If the outdoor temperature is cold, below the balance point of the structure, the space temperature will continue to fall because the heat pump alone will not heat the structure. The second-stage heating contacts will close and start the auxiliary electric heat to help the compressor. The second-stage heat was the last component to be energized, so it will be the first component off. This is the key to how a heat pump can get the best efficiency from the refrigerated cycle. It will operate continuously because it was the first on, and the second-stage auxiliary heat will stop and start to assist the compressor, **Figure 43–44**. There are as many methods of controlling the foregoing cooling and heating sequence as there are manufacturers. One common variation for controlling the heating system is to use outdoor thermostats to control the auxiliary heat, **Figure 43–45**. This keeps all of the auxiliary heat from coming on with the second stage of heating. It will come on based on outdoor temperature by the use of outdoor thermostats. These thermostats only allow the auxiliary heat on when needed, using the structure balance point as a guideline.

There are actually several balance points for any structure when a sophisticated heat pump system is installed. The first one was mentioned previously for the heat pump alone. Several stages of electric strip heat may be applied using balance points. It is vital that the strip heat not be operated any more than necessary. Later in this unit, it will be explained why.

Using the earlier example, suppose that at 30°F the thermal balance point for the heat pump is reached. There will be 30 kW of total electric strip heat available in increments as needed. Some strip heat should be available to start just before the first balance point, let's say at 30°F. When the outdoor temperature reaches 30°F, 10 kW of electric heat will be available because it is set up to start and stay on for a period of time if the space temperature drops to the second stage of heat set point, 68°F, if the first stage of the thermostat is set at 70°F. The condition that must be met for the first stage is that the indoor thermostat must call for heat, for

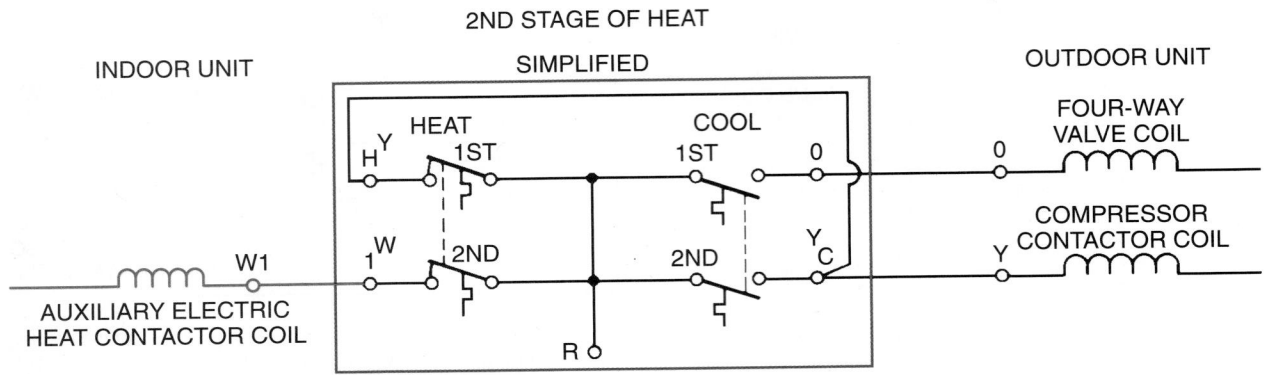

Figure 43–44   When the outside temperature continues to fall, it will pass the balance point of the heat pump. When the space temperature drops approximately 1.5°F, the second-stage contacts of the thermostat will close, and the auxiliary heat contactors will start the auxiliary heat.

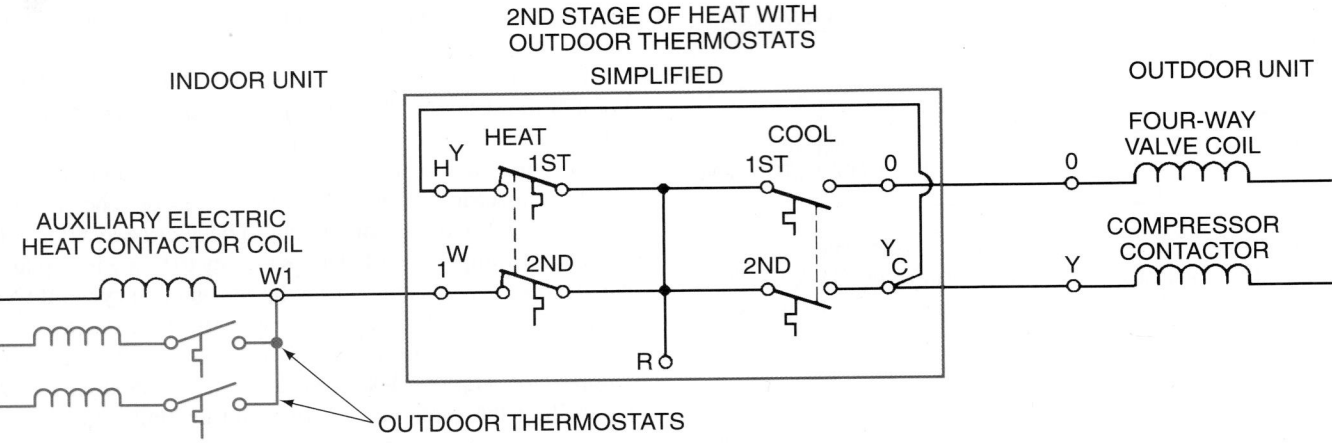

Figure 43–45   A wiring diagram with outdoor thermostats for controlling the electric auxiliary heat.

example, 70°F space temperature. When the temperature drops to 68°F, the second stage will start. It is the last on and will be the first off. This gives the owner some reserve heat capacity.

If the outside temperature drops to 20°F and a new balance point is reached, it takes all of the output of the heat pump and the 10 kW strip heat to heat the house. If the temperature drops any lower there will not be enough heat.

The third set of strip heat may be set to come on at 15°F, the lowest expected outdoor temperature. The outdoor thermostats are low-voltage thermostats. They are mounted with the sensing bulbs outside to monitor the outdoor temperature, to anticipate needs, and to conserve energy.

One more aspect of this type of system should be discussed. There should be an emergency heat setup that allows the owner to start all of the strip heat in the event of a heat pump failure. This requires an emergency heat relay that shuts off the heat pump and makes all strip heat available to the owner and thermostatically controlled. This is accomplished with 24-V low-voltage circuit wiring, **Figure 43–46**.

There are two theories about how to apply strip heat. One is that there should be enough strip heat to totally heat the structure in the event of heat pump failure; the other is that

there should only be enough strip heat to help the heat pump during low temperatures. This is the designer's choice and should be discussed with the owner before installation. The installation of strip heat is fairly inexpensive at the time of the initial installation.

When strip heat is energized and operates along with the heat pump it is called auxiliary heat. When strip heat is used instead of the heat pump, in case of heat pump failure, it is called emergency heat.

Many manufacturers use a signal light to alert the owner that the auxiliary heat is energized. This may be in the form of a blue light at the thermostat. If the owner were to notice the blue light on in mild weather, it might be assumed that the heat pump itself is not working properly because the strip auxiliary heat is helping it more than usual. This should prompt a call for service. When the system is switched to emergency heat, a red light is often switched on as well as the blue light. This should remind the owner to get the system serviced soon. Often it is not practical to service the heat pump because of bad outdoor conditions, and it may be several days before service is available.

The technician should be familiar with four wiring sequences. The next four figures show how they work. These

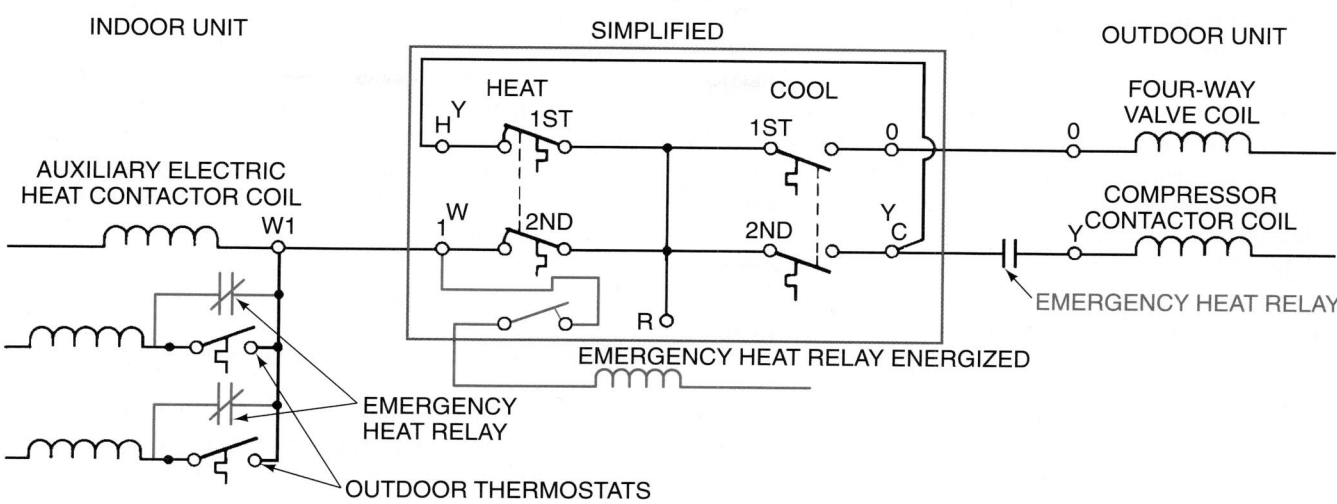

Figure 43–46 The wiring diagram of a space-temperature thermostat set to the emergency heat mode.

illustrations are from a generic HEAT-COOL thermostat that may be used for several manufacturers' equipment. Following the arrows, notice there is a change of direction in the way the power feeds some of these devices from one mode to the other, **Figure 43–47**, while the unit is in the cooling mode. Notice what devices are energized. The

four-way valve is energized by the thermostat selector switch when it is placed in the cool mode. The other devices that must be energized are the compressor, the outdoor fan, and the indoor fan. These components must operate in a coordinated fashion; they can be checked from the thermostat subbase by removing the thermostat. The terminals

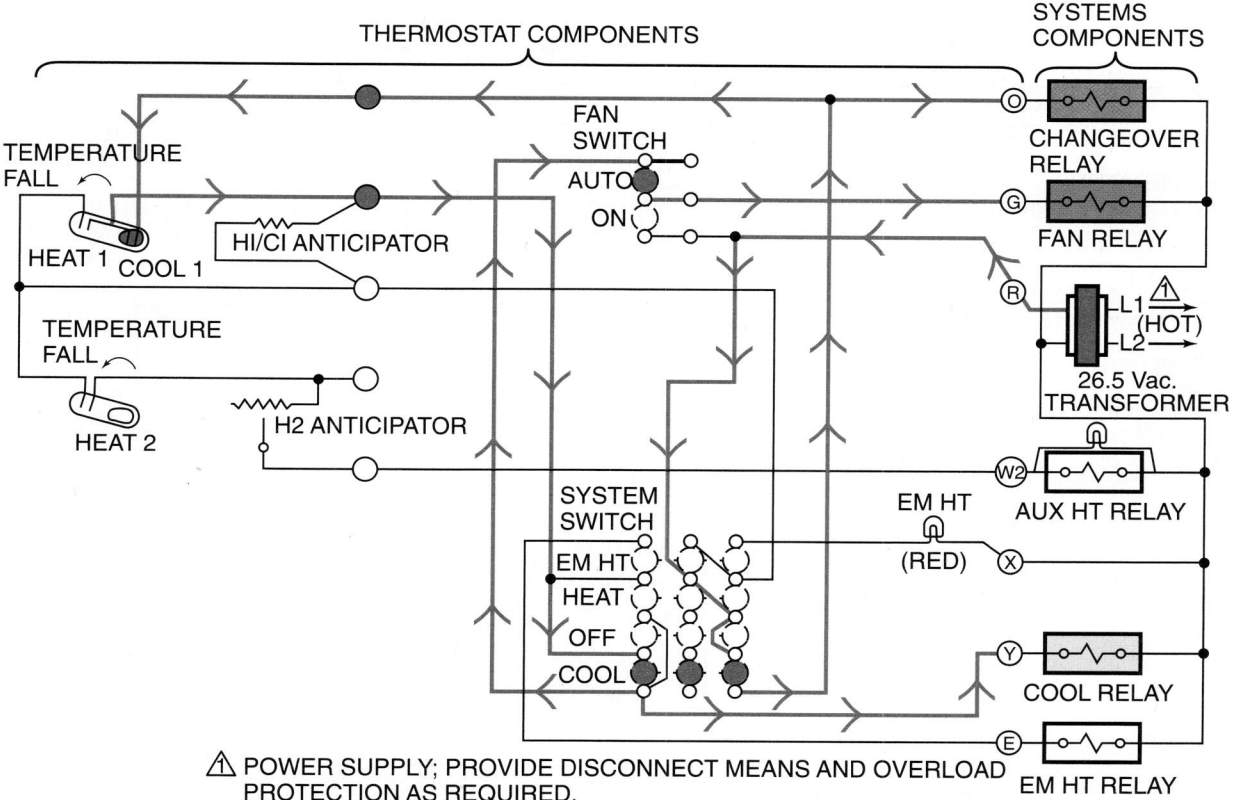

Figure 43–47 A generic HEAT-COOL heat pump thermostat in the cooling mode.

shown with letter designations are on the thermostat sub-base. These may be jumped one at a time to see if the component will start. For example, if the technician jumps from R, which is the hot terminal, to G, the indoor fan will start up if the relay and the wiring are good. The technician can also shut off all power and use an ohmmeter between each terminal to the R terminal to establish that the circuit is good. Remember, there must be a circuit consisting of the conductors and the measurable resistance for current to flow.

Figure 43–48 shows the first stage of heat. The four-way valve is not energized, so the system will start up in the heat mode with the compressor and outdoor fan starting from the cool relay. The indoor fan is running.

Figure 43–49 shows the system operating in the heat mode with the first stage of electric heat operating from the second stage of the thermostat. Notice that the blue light on the thermostat is energized to show the homeowner that the strip heat is operating. The outdoor temperature should be below the heat pump balance point of the house for this heat to be operating. If it continues to operate in mild weather, either the homeowner is turning the thermostat up to make it call or the heat pump system itself is not functioning correctly. Owners should be advised not to turn the heat down when they leave the house. When they return and turn the thermostat up to the required temperature, the strip heat will start up and the power bill will be higher. It is best to set the temperature and leave it alone.

Figure 43–50 shows the system operating in the emergency heat mode. Notice there is no circuit to start the indoor fan. It must be started from an electric heat sequencer (see Unit 30) or from a thermal fan control (see Unit 30). The emergency heat relay is energized. It will provide a circuit to jump out all outdoor thermostats and make all electric heat controllable through the second stage of the indoor thermostat. Both the blue auxiliary heat light and the red emergency heat light will remind the owner that excess energy is being used.

Heat Anticipator

Conventional heat pump thermostats have heat anticipators just as gas and oil heat systems do. They can be more complicated with heat pumps because there are often two of them. The thermostat in **Figure 43–40** has two anticipators. The one in the first stage is preset and not adjustable. The one in the second stage is adjustable and is coordinated with the auxiliary heat relay. A study of Unit 15, "Troubleshooting Basic Controls," would help to answer any heat anticipator questions. When electric heat is involved, there can be many control variables that require understanding the heat anticipator because often the auxiliary heat has several controls that must be taken into consideration while working with heat anticipation. Unit 30 contains a lengthy discussion on electric heat that could also apply to auxiliary heat for a heat pump system.

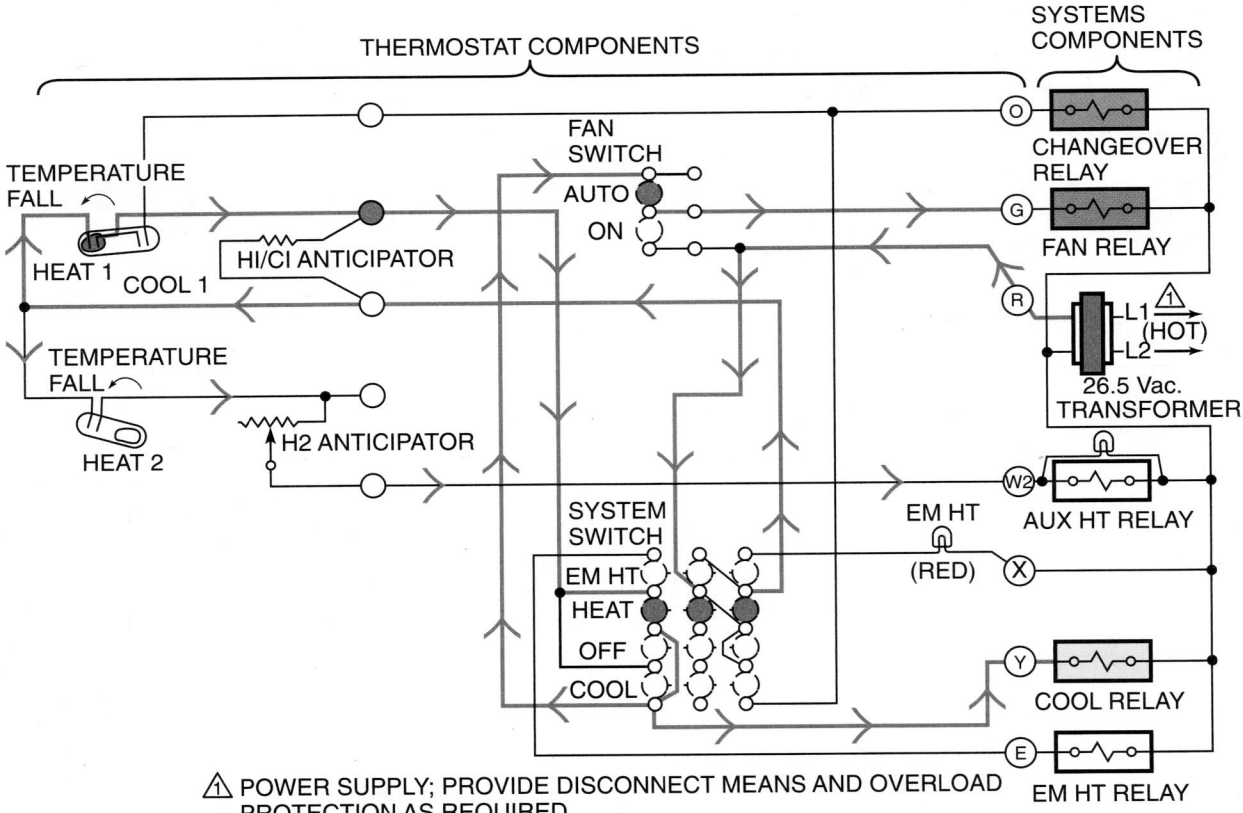

⚠ POWER SUPPLY; PROVIDE DISCONNECT MEANS AND OVERLOAD PROTECTION AS REQUIRED.

Figure 43–48 A generic HEAT-COOL heat pump thermostat in the first-stage heat mode with only the heat pump operating.

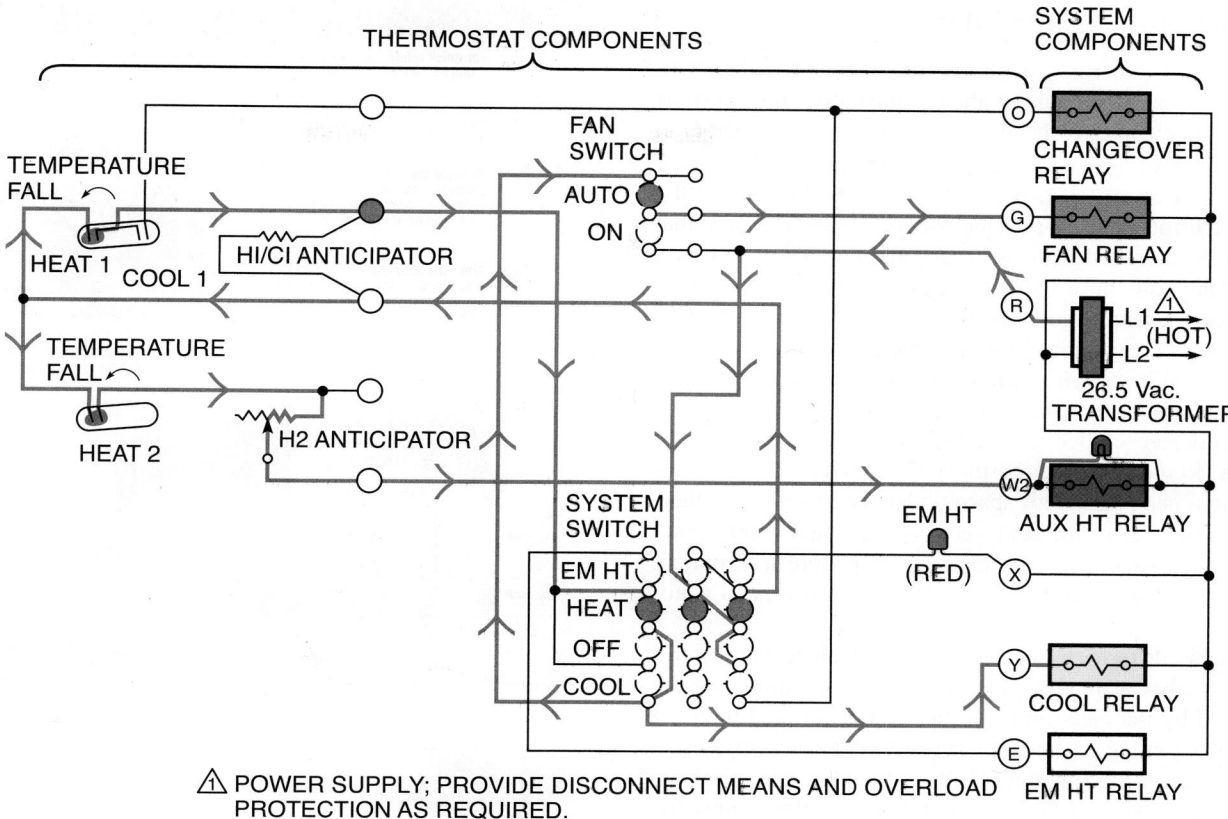

Figure 43–49 A generic HEAT-COOL heat pump thermostat operating with both the heat pump and the auxiliary heat working.

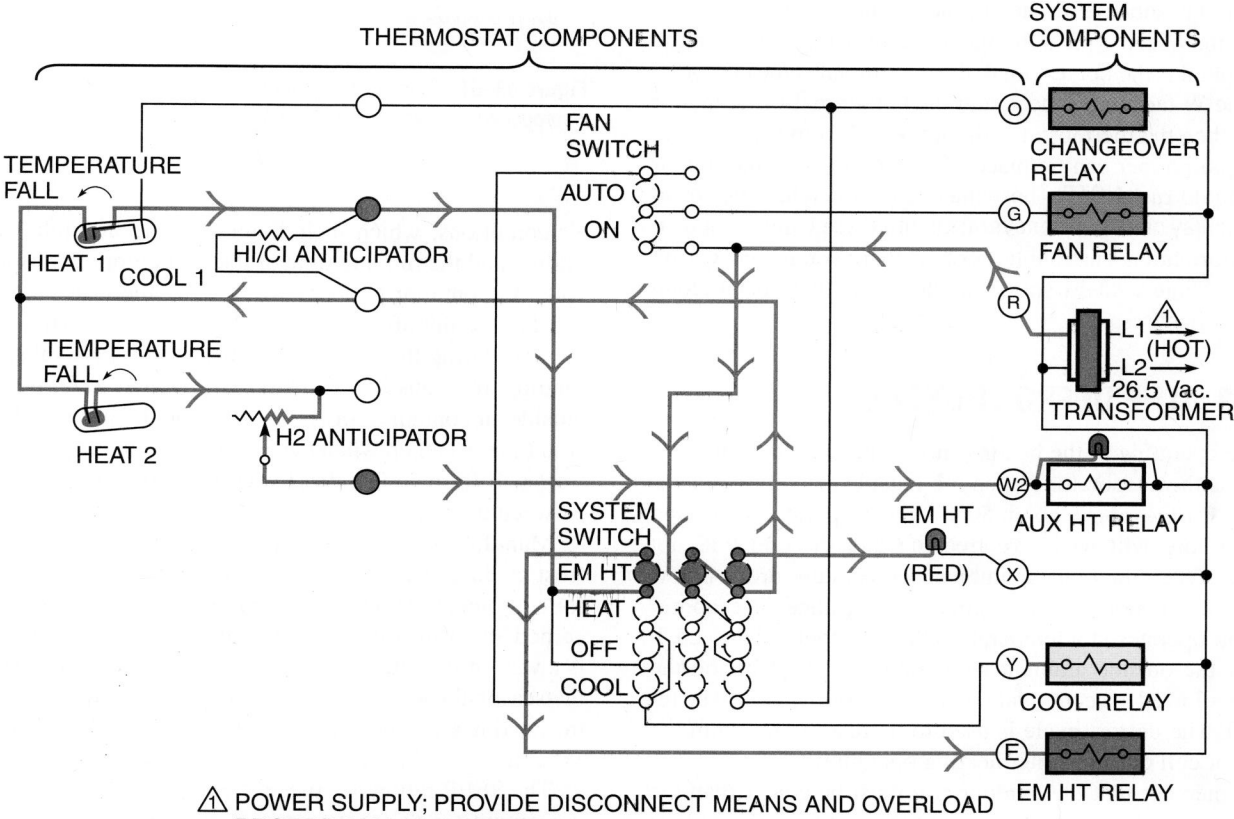

Figure 43–50 A generic HEAT-COOL heat pump thermostat operating in the emergency heat mode.

Electronic Thermostats

Most modern thermostats have thermistors for the heat-sensing element. Remember, the thermistor is a resistor-type device that changes resistance with a change in temperature. There are several reasons why the thermistor is being used. For one, mercury (Hg), which was used exclusively in the earlier thermostats, is a poisonous environmental hazard that should be disposed of as a hazardous material. There is a very small amount of mercury in each bulb of a thermostat. A two-stage heat, two-stage cooling thermostat will often have four of these bulbs. This can amount to a lot of mercury over time. When Hg is used, the temperature-sensing device is a bimetal strip or coil. The mass of the bimetal that was used to detect temperature caused rather large temperature swings. Remember, the thermostat's sensing element is located in a nearly dead air space under the cover of the thermostat. The thermistor has a very low mass and will respond much faster to temperature changes. For more information about electronic thermostats, see Unit 16, "Advanced Automatic Controls."

One big difference in a system with an electronic control is the number of features the thermostat may have that are provided by the electronics in the system coupled with the electronics in the thermostat. These have to do with timing functions, sensing functions, and error codes, which are explained in detail in Unit 16. Otherwise, troubleshooting the electronic thermostat is much like troubleshooting a mercury bulb thermostat. If it has error codes, you follow them. Otherwise, treat it like a switch and jump from terminal to terminal in the subbase to determine if the thermostat has a problem. With a typical thermostat terminal designation, the R terminal is the hot terminal, the Y terminal is the compressor, the W terminal is heat, and the G terminal is the indoor fan. When the thermostat is disconnected from the subbase, a simple jumper can be placed from R to G and the indoor fan should run. NOTE: There may be a time delay built in, so the fan may not start for several minutes. See Units 15 and 16 for information on troubleshooting thermostats and wiring circuits. **Figure 43–51** shows an electronic thermostat along with its features.

43.23 THE DEFROST CYCLE

While operating in the heating mode, the outdoor coil functions as the evaporator. When the outside ambient temperature is warm—above about 50°F—the evaporator saturation temperature will be above freezing and the coil will not freeze. When the outside ambient temperature drops below 50°F, the outdoor coil can begin to freeze, since the evaporator coil operates at a temperature that is about 20°F to 25°F below the outside ambient temperature. If the frost on the coil is allowed to accumulate, system performance will be reduced. The defrost cycle is used to defrost the ice from the outdoor coil during winter heating operation.

Remember that the outdoor coil must be colder than the outside air if the coil is going to absorb heat from the outside air, **Figure 43–27**. The need for defrost varies with the outdoor

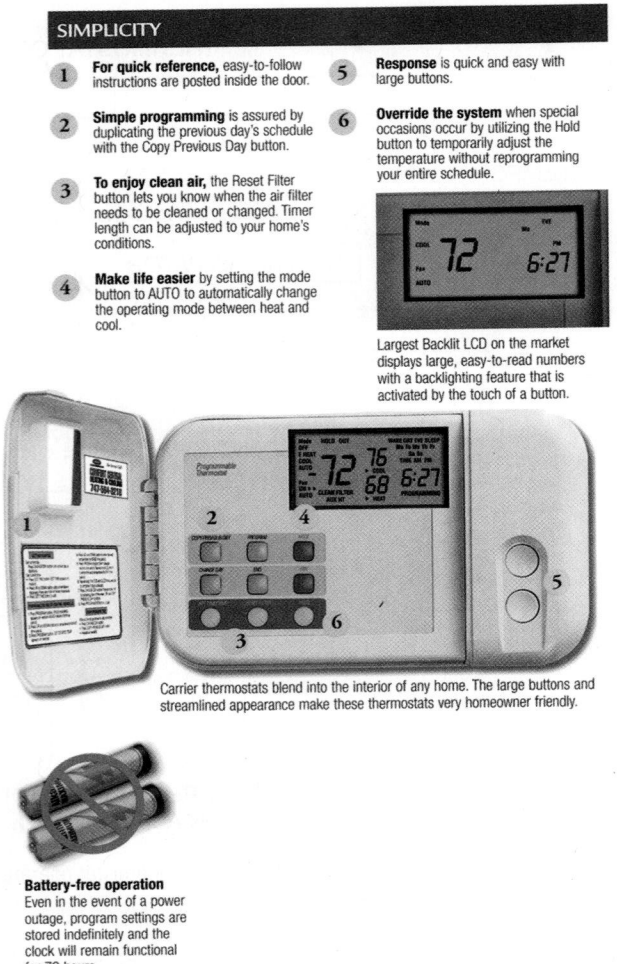

Figure 43–51 This electronic thermostat has many features.
Reproduced courtesy of Carrier Corporation

air conditions, which include temperature and relative humidity, and the running time of the heat pump. For example, when the outdoor temperature is 45°F, the heat pump will satisfy and shut off from time to time. The frost will melt off the coil during this off time. As the temperature falls, more running time causes more frost to form on the coil. When the outside air contains more moisture, more frost forms. During cold rainy weather, when the air temperature is 35°F to 45°F, frost will form so fast that the coil will be covered with ice between defrost cycles.

Manufacturers have been working on managing the defrost cycle as part of their energy efficiency improvements for equipment. At one time, they did not need to be too concerned about the efficiency of equipment. Today, the public is aware of the numbers that spell out an efficient system—the seasonal energy efficiency ratio (SEER) for cooling and the heating seasonal performance factor (HSPF) for heating. What happens during defrost affects the HSPF.

The SEER rating of a piece of equipment is the total cooling output for the system for an estimated typical cooling season divided by the actual power input for the entire season in watt-hours. This factor started with the energy efficiency ratio (EER) rating where the steady-state output was divided by the power required to obtain the output. This was expanded to a seasonal ratio because the EER did not take into consideration the effects of starting and stopping the system, or different weather conditions, such as mild spring and fall seasons. Remember, a system does not get up to maximum efficiency for about 15 min after start-up. When it is shut off, a cold cooling coil is still available to absorb more heat. Many manufacturers have a time-delay start on the indoor fan to let the coil become cool before air is passed through it. They may also have a time delay incorporated into the control scheme to keep the indoor blower operating at the end of the cooling cycle. This will allow the cooling coil to remove more heat from the airstream at the end of the cycle. All of this is available because of electronic control systems.

The HSPF rating describes a seasonal performance for a particular piece of equipment. In the United States, we have cold dry areas, moderate humid areas, and everything in between these extremes. The HSPF rating for each piece of equipment can be used to calculate the cost of operation in six different zones in the United States. These zones were decided on by the Department of Energy (DOE). This gives the potential owner or the calculating contractor a method to predict the cost of operation for a piece of equipment and compare it to other equipment for both summer and winter operation. The defrost cycle figures into this cost analysis.

The SEER and the HSPF are a result of federal energy policies and are used by the Air Conditioning and Refrigeration Institute (ARI) to rate equipment. The ARI is a national trade association composed of the majority of the equipment manufacturers in the field. They maintain a test laboratory for verifying equipment standards and publish the SEER and HSPF for equipment. The higher the number, either SEER or HSPF, the more Btu output you get for the same input. For more about the ARI, go to www.ari.org.

How Is Defrost Accomplished?

In order for the heat pump system to operate in the defrost mode, a number of things must take place:

■ The reversing valve switches the system into the cooling mode. This involves energizing the reversing valve coil for systems that fail in the heating mode—or deenergizing the reversing valve coil for heat pump systems that fail in the cooling mode. The outdoor fan motor is functioning as the condenser coil during defrost.

■ The outdoor fan motor is deenergized. With the fan not operating, more heat will be concentrated in the coil, allowing the frost to melt much faster than if the fan were operating.

■ The indoor blower will remain on in most cases. Since the indoor coil is functioning as the evaporator in the defrost mode, the coil can freeze if the indoor blower is not operating.

■ Some auxiliary heat is energized during defrost. This will help temper the cooler air being introduced to the occupied space during defrost. During defrost, the temperature of the occupied space should not be increased. If the temperature of the space is increased, the thermostat's set point may be reached and the heating cycle will end. If this happens, the defrost cycle will end as well and the ice removal process will stop.

When the heat pump system is in defrost with the auxiliary heat energized, the system is cooling and heating at the same time and is not efficient—so the defrost time must be held to a minimum. It must not operate except when needed. *Demand defrost* means defrosting only when needed. Combinations of time, temperature, and pressure drop across the outdoor coil are used to determine when defrost should be started and stopped.

The factors that vary the operating cost of a piece of equipment are the number of defrosts per season, length of defrost, and whether the manufacturer tempers the air to the conditioned space during defrost with auxiliary heat. Remember, the unit will be in the cooling mode and blowing cold air during defrost; some owners will object to that. If the manufacturer can get the equipment rated without the auxiliary heat, the rating will be better.

Initiating the Defrost Cycle

Starting the defrost cycle with the correct frost buildup is desirable. Manufacturers design the systems to start defrost as close as possible to when the coil builds frost that affects performance. Some manufacturers use time and temperature to start defrost. This is called *time and temperature initiated* and is performed with a timer and temperature-sensing device. The timer closes a set of contacts for 10 to 20 sec for a trial defrost every 30, 60, or 90 min. The timer contacts are in series with the temperature sensor contacts so that both must be made at the same time. This means that two conditions must be met before defrost can start, **Figure 43–52.** In operation, every time the allotted time delay has elapsed, the timer will attempt to bring the heat pump system into defrost. If the temperature sensor determines that there is ice on the coil, defrost will start. The sensing device contacts will close if the coil temperature is as low as 25°F, and defrost will start. **Figure 43–53** shows a typical timer and a sensor. Typically, the timer runs any time the compressor runs, even in the cooling mode.

Another common method of starting defrost uses an air-pressure switch that measures the air pressure drop across the outdoor coil. When the unit begins to accumulate ice, a pressure drop occurs and the air switch contacts close. This can be wired in conjunction with the timer and temperature sensor to ensure that an actual ice buildup is on the coil. The combination of time and temperature only ensures that time has passed and the coil is cold enough to actually accumulate ice. It does not actually sense an ice thickness. It can be more efficient to defrost when an actual buildup of ice exists, **Figure 43–54.**

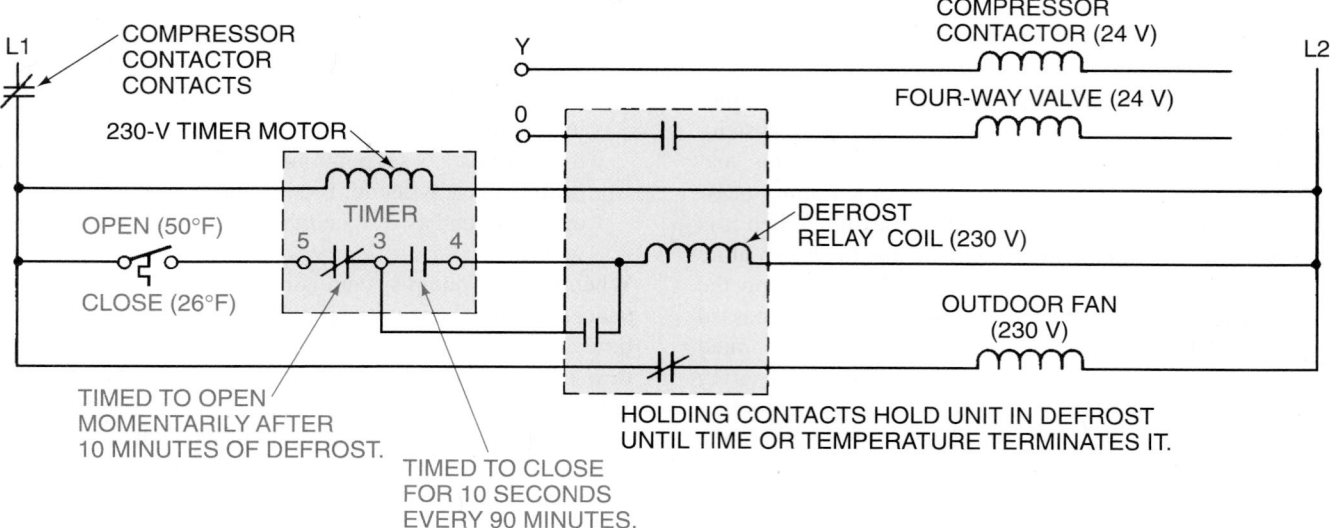

Figure 43–52   A wiring diagram of conditions to be met for defrost to start.

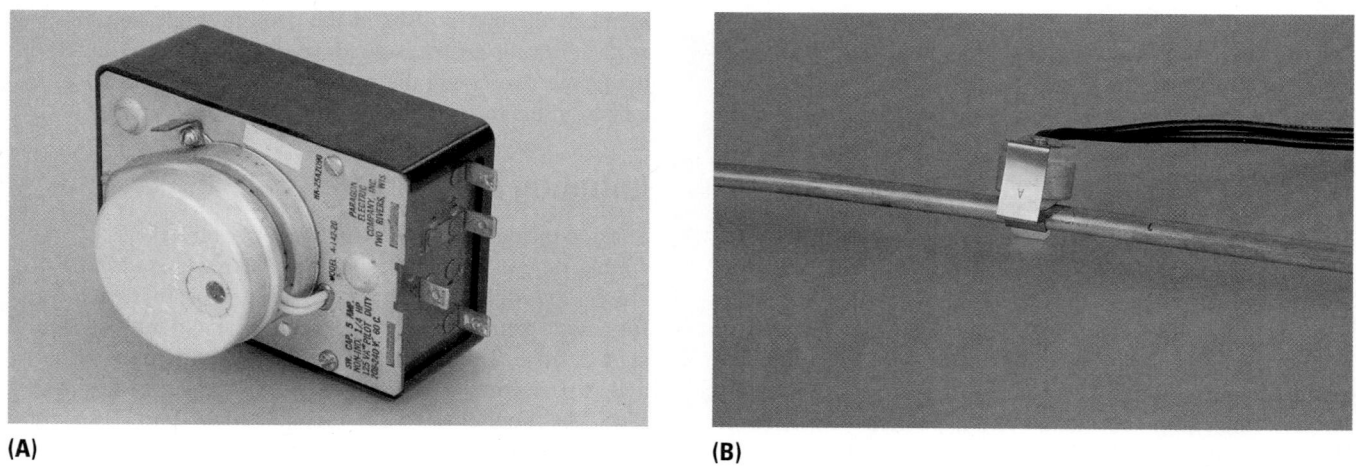

(A)   **(B)**

Figure 43–53   **(A)** Timer. **(B)** Sensor. *Photos by Bill Johnson*

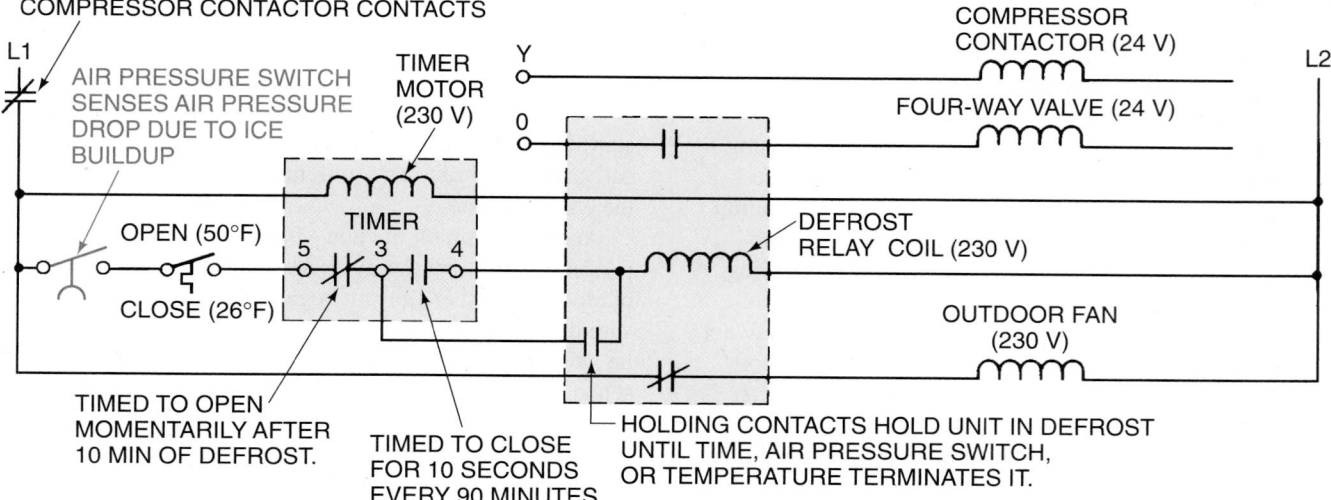

Figure 43–54   A wiring diagram taking ice buildup into consideration.

Terminating the Defrost Cycle

Terminating the defrost cycle at the correct time is just as important as starting defrost only when needed. This is done in several ways. *Time and temperature* was the terminology used to start defrost. *Time or temperature terminated* is the terminology used to terminate one type of defrost cycle. The difference is that two conditions must be satisfied to start defrost, and either one of two conditions can terminate defrost. After defrost has started, time or temperature can terminate it. For example, if the defrost temperature sensor warms up to the point that it is obvious that ice is no longer on the coil and its contacts break, defrost will stop. This temperature termination is normally 50°F at the location of the temperature sensor. If the contacts do not open because it is too cold outside, the timer will terminate defrost within a predetermined time period. Normally, the maximum time the timer will allow for defrost is 10 min, **Figure 43–55.**

Electronic Control of Defrost

Electronic circuit boards and their controls can help the manufacturer more closely control the defrost cycle. They have the ability to accurately control time and temperature with electronic timers and thermistors.

The electronic timer and the thermistor may be used the same as the mechanical timer and line thermostat shown in **Figure 43–53,** but with much more accuracy. The manufacturer may use "time and temperature" initiated and "time or temperature" terminated, just like the combination timer and thermostat. Some systems may use the difference in the entering air temperature and the coil temperature to arrive at a temperature split where defrost is desired. For example, suppose the outdoor temperature is 35°F. The coil will operate at about 20°F to 25°F colder than the outdoor coil. Using 20°F temperature split as an example, suppose the coil begins to develop ice on the surface. We have seen from previous examples that the pressure will drop across the coil, but the temperature will also drop inside the coil as ice develops. The manufacturer may decide that it is time for a defrost when the coil temperature has dropped to 8°F below the normal operating temperature for that condition, **Figure 43–56.** If the time portion of the defrost initiation has been satisfied, usually 90, 60, or 30 min, the unit will start to defrost and will be terminated by the coil temperature. All of this is accomplished with the electronic circuit board and two thermistors.

Manufacturers of heat pumps may have other combinations of conditions to operate the defrost cycle that include air pressure drop across the coil. The manufacturers look for something economical, reliable, and effective for making their HSPF number higher for a better rating. **Figure 43–57** is an example of an electronic circuit board from a typical heat pump system. Many variations of this type of circuit board are designed by and for different manufacturers.

43.24 INDOOR FAN MOTOR CONTROL

Starting the indoor fan motor of a heat pump differs from other heating systems. In other heating systems, the indoor fan is started with a temperature-operated fan switch. With a heat pump the fan must start at the beginning of the cycle, which is controlled by the thermostat. The indoor fan is started with the fan terminal, sometimes labeled G. The compressor operates whenever the unit is calling for cooling or heating, so the G or fan terminal is energized whenever there is a call for cooling or heating.

43.25 AUXILIARY HEAT

During cold weather the air-to-air heat pump must have auxiliary heat. This is usually accomplished with an electric furnace with a heat pump coil for cooling and heating. Electric heat is often started with a sequencer with a time-start and

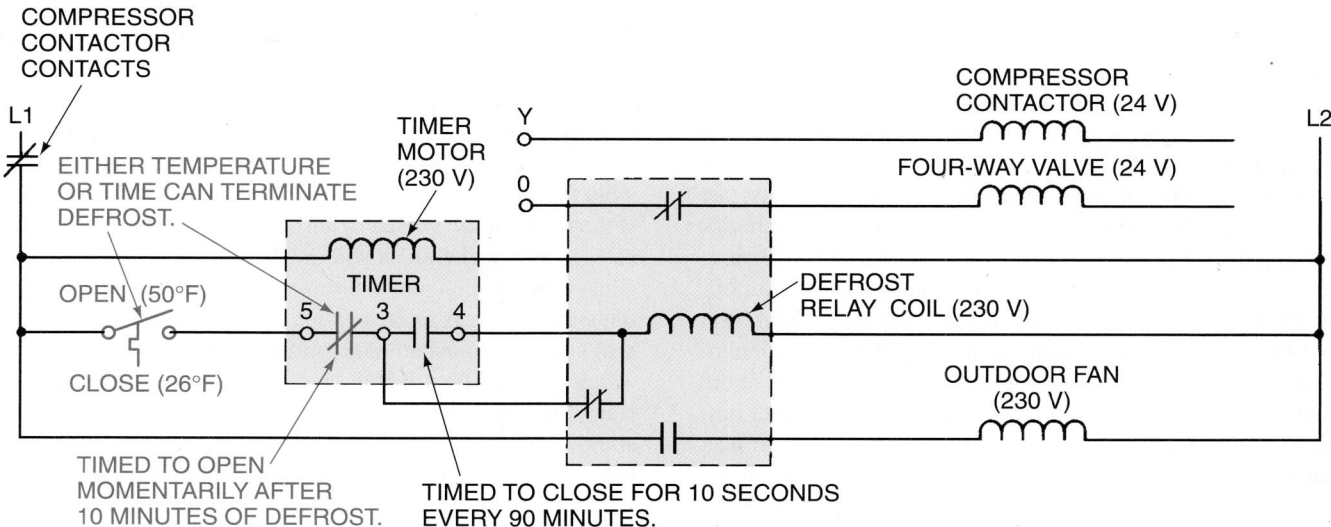

Figure 43–55 A wiring diagram for a time- or temperature-terminated method of defrost.

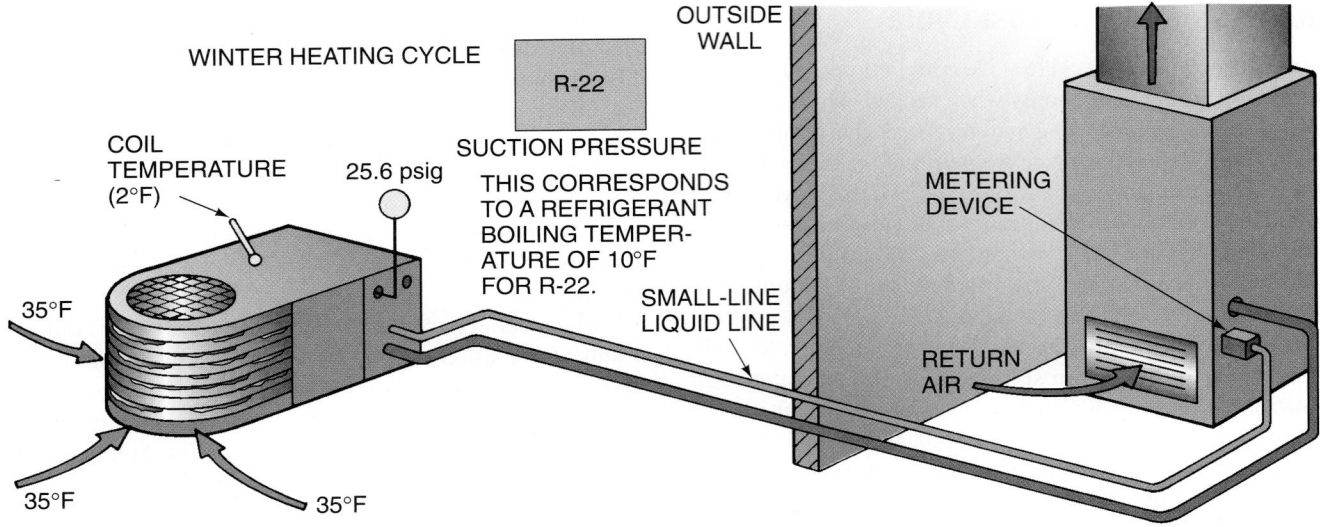

Figure 43–56 This system has ice accumulating on the outdoor coil and is in need of defrost. The manufacturer is using the difference between the outdoor air entering the coil and the actual coil temperature, measured with a thermistor mounted on the coil.

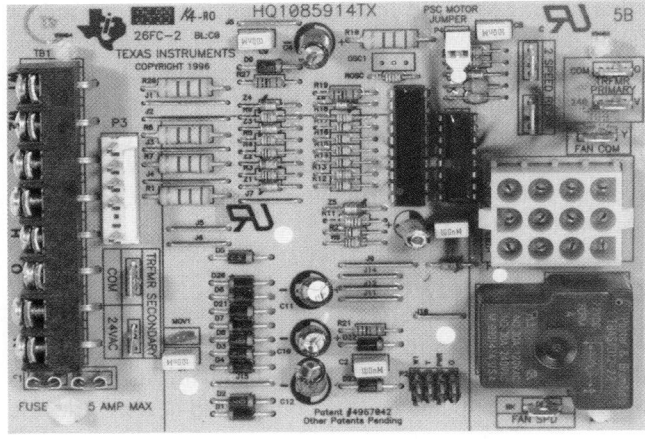

Figure 43–57 This is a typical heat pump circuit board used in modern heat pumps. *Photo by Bill Johnson*

time-stop feature for the electric heating element. This means that the electric heat element is energized for a time period if the unit is shut off with the thermostat on-off switch. When the thermostat is allowed to operate the system normally, the heat pump will be the first component to come on and the last to shut off. The indoor fan will start with the compressor through the fan relay circuit. The electric strip heat will operate as the last component to come on and the first to shut off. When the owner decides to shut off the heat at the thermostat and the strip heat is energized, the heating elements will continue to heat until the sequencers stop them. A temperature switch is often used to sense this heat and keep the fan running until the heaters cool. The electric furnaces used for heat pumps are like the ones described in Unit 30, "Electric Heat."

Electric strip heaters are used for three purposes in the heat pump system. They provide

- supplementary heat when the heating system is operating below the balance point of the system.
- emergency heat when the reverse-cycle heat pump system is not functioning.
- air tempering when the system is operating in the defrost mode.

Depending on the function these electric strip heaters are intended to serve, the number of heaters to be energized will change. When acting to provide supplementary heat, the number of heaters energized is determined by the outside ambient temperature. When the temperature is colder outside, more heaters can be energized. When the outside temperature is warmer, fewer, if any, heaters need to be brought into operation.

When providing emergency heat, all of the available strip heaters are energized. This is because the main/primary heat source is not operating and the strip heaters are attempting to satisfy the heating requirements of the space by themselves.

When used to temper the air during defrost, it is very important that too many strip heaters are not energized. As mentioned earlier in this unit, the heaters used to temper the air during defrost should not raise the temperature of the occupied space. Consider a heat pump system that just went into defrost. The space temperature is 70°F and the thermostat is set to 71°F. If the heaters temper the air too much, it is possible that the temperature of the space may increase by 1°F. If the temperature sensed by the thermostat reaches 71°F, the call for heat will end, as will the defrost cycle. If the defrost ends, any unmelted ice will remain on the outdoor coil when the system returns to the heating mode. If this continues, ice will continue to accumulate on the outdoor coil and system efficiency will be affected. So, to prevent this from occurring, the amount of heat added to the air during defrost should not exceed 80% of the sensible cooling capacity of the system. Consider the following example.

A heat pump system has a sensible cooling capacity of 60,000 Btu/h. During defrost, the system switches over to the cooling mode, so the air introduced to the occupied space will be cooled. We want to temper this air with enough heat to

Unit 43   Air Source Heat Pumps   1101

offset 80% of the cooling capacity—which, in this case, is 48,000 Btu/h (60,000 Btu/h × 0.80). The conversion factor between Btu and watts is 3.413 Btu/W, so we will need to energize no more than 14,064 W (48,000 Btu/h ÷ 3.413 Btu/W) of electric heat. If the heat pump system is equipped with six 3.5-kW electric strip heaters (3500 W each), we can have up to four heaters energized during defrost. These four heaters will provide 14 kW of heat (3.5 kW × 4) to the air passing over the strip heaters. Since the 14 kW (14,000 watts) is less than the maximum calculated value, the space will not be heated above its present temperature during defrost. This will allow the system to complete its defrost cycle as needed.

43.26 SERVICING THE AIR-TO-AIR HEAT PUMP

Servicing the air-to-air heat pump is much like servicing a refrigeration system. During the cooling season, the unit is operating as a high-temperature refrigeration system, and during the heating season it is operating as a low-temperature system with planned defrost. The servicing of the system is divided into electrical and mechanical servicing.

Servicing the electrical system is much like servicing any electrical system that is used to operate and control refrigeration equipment. The components that the manufacturer furnishes are built to last for many years of normal service. However, parts can also fail due to manufacturing faults and misuse. One significant point about a heat pump is that it has much more running time than a cooling-only unit. A typical cooling unit in a residence in Atlanta, Georgia, will run about 120 days per year. A heat pump may well run 250 days, more than twice the time. In winter the heat pump may run for days and not stop when the temperature is below the balance point of the structure.

SAFETY PRECAUTION: The same safety precautions must be observed when troubleshooting electrical components of a heat pump as were observed when servicing an air-conditioning system. When the units are located in crawl spaces and you are in contact with the ground, be careful not to establish a path for current to flow through your body to ground.

43.27 TROUBLESHOOTING THE ELECTRICAL SYSTEM

Troubleshooting the electrical system of an air-to-air heat pump is similar to troubleshooting a cooling air-conditioning system. Recall that there must be a power supply, a complete path for the electrical current to take, and a load before electricity will flow. The power supply provides the electrical energy to the unit, the complete path is the wiring, and the closed switches or electrical contacts and the loads are the electrical components that perform the work. Every circuit in the heat pump wiring can be reduced to these items. It is important that you think in terms of power, conductor, switch, and load, **Figure 43–58.**

It is usually easy to find electrical problems if some component will not function. For example, when a compressor

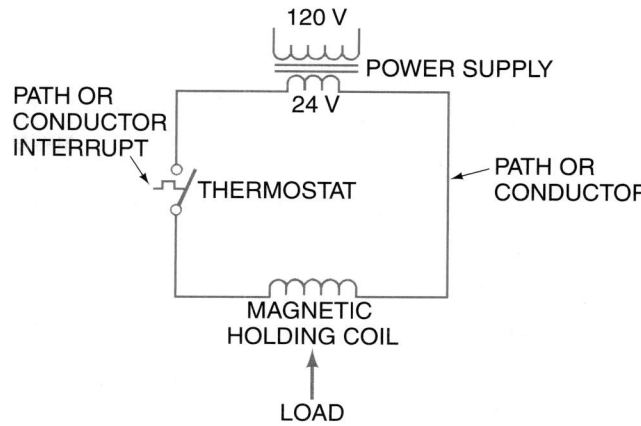

Figure 43–58   This is the basic example of a power supply, a path or conductor, and a load for consuming power.

contactor coil is open, it is evident that the problem is in the contactor. A voltage reading can be taken on each terminal of the contactor coil. If there is voltage and it will not energize the contacts, a resistance check of the coil will show whether there is a circuit through it. A correctly operating circuit has the correct resistance. No resistance or too little resistance will create too much current flow. Infinite resistance results in no current flow as there is no path. It sometimes helps to be able to compare a suspected bad component (e.g., one with too little resistance) to a component of correct resistance, **Figure 43–59.**

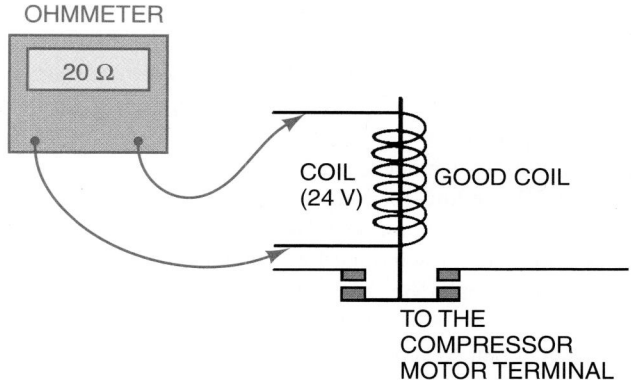

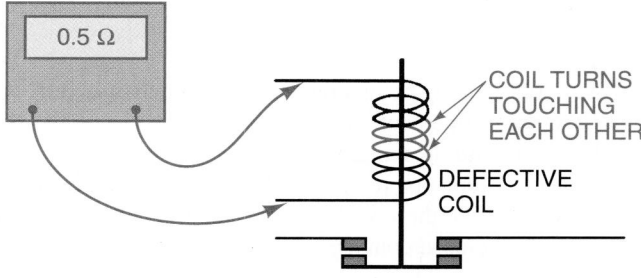

Figure 43–59   Magnetic holding coils used for the contactor. One of them has a measurable resistance of 0.5 Ω. The other one, the correct one, has a measurable resistance of 20 Ω. The one with the lower resistance is shorted. It will draw too much current and place a load on the control transformer. If the load is too much, it will overload the transformer.

Electrical malfunctions that cause the technician real problems are the intermittent ones, such as a unit that does not always defrost so that ice builds up on the outdoor coil. The heat pump can be difficult to troubleshoot in winter because of the weather. SAFETY PRECAUTION: If it is wet outside the emergency heat feature of the heat pump should be used until the weather improves. It is hard enough to find electrical problems in cold weather; the moisture makes it dangerous.

Typical Electrical Problems

1. Indoor fan motor and outdoor fan motor
    A. Open winding: motor will not start; may draw locked rotor amperage; may have infinite resistance across one winding, normal resistance across the other winding; motor will overheat.
    B. Shorted winding: motor may or may not start; motor will draw higher amperage.
    C. Shorted winding to ground: motor may or may not start; circuit breaker may trip or fuse may blow; motor may draw excessive current; there will be measurable resistance between the motor terminals and ground.
    D. Open run capacitor: motor will not start; motor will draw locked rotor amperage; motor will overheat.
    E. Open start capacitor: motor will likely not start; motor will draw high amperage; motor will overheat if unable to start.
    F. Shorted run capacitor: motor may or may not start; motor will draw higher amperage; motor will run slowly if it does start.
    G. Shorted start capacitor: motor will likely not start; if motor starts it will turn slowly and draw high amperage.
2. Compressor contactor, fan relays
    A. Coil winding open: device will not close NO contacts; will not open NC contacts.
    B. Shorted coil winding: low-voltage control circuit protection device may trip; control transformer may overload; device will not close NO contacts; will not open NC contacts.
    C. Pitted contacts: reduced voltage to the load being controlled; there will be a measurable voltage across closed contacts.
3. Defrost relay
    A. Coil winding open: contacts on the relay will not close or open automatically.
    B. Coil winding shorted: circuit overload.
    C. Pitted contacts: reduced voltage to the load being controlled; there will be a measurable voltage across closed contacts.
4. Compressor
    A. Open compressor motor winding: will not start; may draw locked-rotor amperage if only one winding is open.
    B. Shorted compressor motor winding: circuit breaker trips when compressor is energized.

C. Shorted winding to ground: compressor may operate normally if the ground is on the neutral side of the power supply; circuit breaker trips when compressor is energized if ground is on the line side of the power supply.
    D. Open internal overload: no current draw; no compressor operation; compressor may be hot to the touch; there may be infinite resistance readings between all combinations of the compressor motor terminals.
5. Reversing valve solenoid
    A. Open coil: system will remain in the mode in which the system fails (usually heating); no defrost cycle.
    B. Shorted coil: fuse protection trips when system changes over.
6. Electric strip heaters
    A. Open heater: may or may not be noticeable, depending on how many heaters are affected; there may be a reduced air-tempering effect, a reduced heating capacity (second stage and emergency heat modes), and/or no current draw through the heater circuit.
    B. Open safety switch: no current flow through the heater.
    C. Open fan-proving device: no voltage supplied to the coil of the heater control relays or sequencer; no current flow through the heaters.
7. General wiring problems
    A. Loose electrical connections: burned connections; localized heat; intermittent system problems.
    B. Short circuit: tripping breakers or blowing fuses.

43.28 ▶ TROUBLESHOOTING MECHANICAL PROBLEMS

Mechanical problems can be hard to find in a heat pump, particularly in winter operation. Summer operation of a heat pump is similar to that of a conventional cooling unit. The pressures and temperatures are the same, and the weather is not as difficult to work in. Mechanical problems are solved with gage manifolds, wet- and dry-bulb thermometers, and air-measuring instruments.

Some Typical Mechanical Problems

1. Indoor unit
    A. Air filter dirty (winter); high head pressure; low COP.
    B. Dirty coil (winter); high head pressure; low COP.
    C. Air filter dirty (summer); low suction pressure.
    D. Dirty coil (summer); low suction pressure.
    E. Refrigerant restriction (summer); low suction.
    F. Refrigerant restriction (winter); high head pressure.
    G. Defective TXV on the indoor coil (summer); low suction pressure.
    H. Defective TXV on the outdoor coil (winter); low suction.
    I. Leaking check valve (summer); flooded indoor coil.

2. Outdoor unit

    A. Dirty coil (winter); low suction pressure.

    B. Dirty coil (summer); high head pressure.

    C. Four-way valve will not shift; stays in same mode.

    D. Four-way valve halfway; pressures will equalize while running and appear as a compressor that will not pump.

    E. Compressor efficiency down; capacity down, suction and discharge pressures too close together. Compressor will not pump pressure differential.

    F. Defective TXVs (winter); low suction pressure.

    G. Leaking check valve (winter); flooded coil.

The technician should be aware of three mechanical problems: (1) four-way valve leaking through, (2) compressor not pumping to capacity, and (3) charging the heat pump. These particular problems are different enough from commercial refrigeration that they need individual attention.

SAFETY PRECAUTION: Be aware of the high-pressure refrigerant in the system while connecting gage lines. High-pressure refrigerant can pierce your skin and blow particles in your eyes; use eye protection. Liquid refrigerant can freeze your skin; wear gloves. The hot gas line can cause serious burns and should be avoided.

43.29 ▶ TROUBLESHOOTING THE FOUR-WAY VALVE

The four-way valve can cause a technician a lot of confusion when troubleshooting. Several things can cause the four-way valve to malfunction:

1. The valve may be stuck in heating or cooling.

2. The coil may be defective.

3. The valve may have an internal leak.

The sign of a stuck valve may be that it will not change from heating to cooling and back. When the valve is energized in the cooling mode and is trying to operate in the heating position, the first thing the technician should do is make sure the coil is energized. The technician may do this by using a voltmeter at the coil lead wires. Another procedure is to hold a screwdriver close to the coil to see if there is a magnetic field when the valve is supposed to be energized. If the valve is getting power and will not change over while the unit is running and there is a pressure difference from the high to the low side, the valve may be stuck. Remember, the valve will not change over unless there is a pressure difference between the high- and the low-pressure sides of the system because it is a pilot operated valve. The technician may also feel the valve coil to see if it is hot. That is a sign that it is energized.

If the valve is truly stuck, the technician may take a soft face hammer or some other soft surface instrument and tap the valve on one end and then the other to see if it will break loose. Something inside may be binding it. If it breaks loose, the technician should force the valve to change positions from heat to cool several times to see if it is free to

operate. If so, it may be assumed that the problem has been resolved. If it reoccurs, the technician should change the valve.

The valve coil can have an open circuit, which will cause it to operate only in the deenergized mode. Since the coil is removable, it can be changed. It is always desirable to change the coil if that will repair the valve, rather than to change the complete valve.

The four-way valve leaking through can easily be confused with a compressor that is not pumping to capacity. The suction pressure on a system with a leaking reversing valve will be higher than normal. The capacity of the system will not be up to normal in summer or winter cycles. This is the same symptom as small amounts of hot gas leaking from the high-pressure side of the system to the low-pressure side. When the gas is pumped around and around, work is accomplished in the compression process, but usable refrigeration is not available. When you suspect that a four-way valve is leaking from the high-pressure line to the low-pressure line, use a good-quality thermometer to check the temperature of the low-side line, the suction line from the evaporator (the indoor coil in summer or the outdoor coil in winter), and the permanent suction line between the four-way valve and the compressor. The temperature difference should not be more than about 3°F. NOTE: Take these special precautions when recording temperatures: (1) Take the temperatures at least 5 in. from the valve body (to keep valve body temperature from affecting the reading). (2) Insulate the temperature lead that is fastened tightly to the refrigerant line, **Figure 43–60.**

Replacing the Four-Way Reversing Valve

If it is determined that the four-way reversing valve needs to be replaced, a few things should be kept in mind to ensure that the system will operate as intended after the repair is made. Make certain that

- the new valve is thoroughly inspected before installation. Check the valve for scratches, dents, and other imperfections on the bodies of the main and pilot valves. Dents can prevent the internal slide from moving within the valve. If possible, inspect the valve while you are still at the parts depot to prevent a second trip if the valve is damaged.

- the new valve is not dropped or mishandled.

- excessive heat is not used when installing the valve.

- the new valve is piped so that the valve will fail in the same mode of operation as the original.

- the new valve is installed horizontally with the pilot tubes on top. This will prevent oil from entering the pilot ports and causing the valve to malfunction. Always follow the manufacturer's guidelines when replacing the valve.

- an exact part replacement is selected whenever possible. If the ports on the new valve are too small, excessive pressure drops will be present across the valve, compromising system performance. If the new valve is too large, the switchover time will be increased, resulting in sluggish system operation.

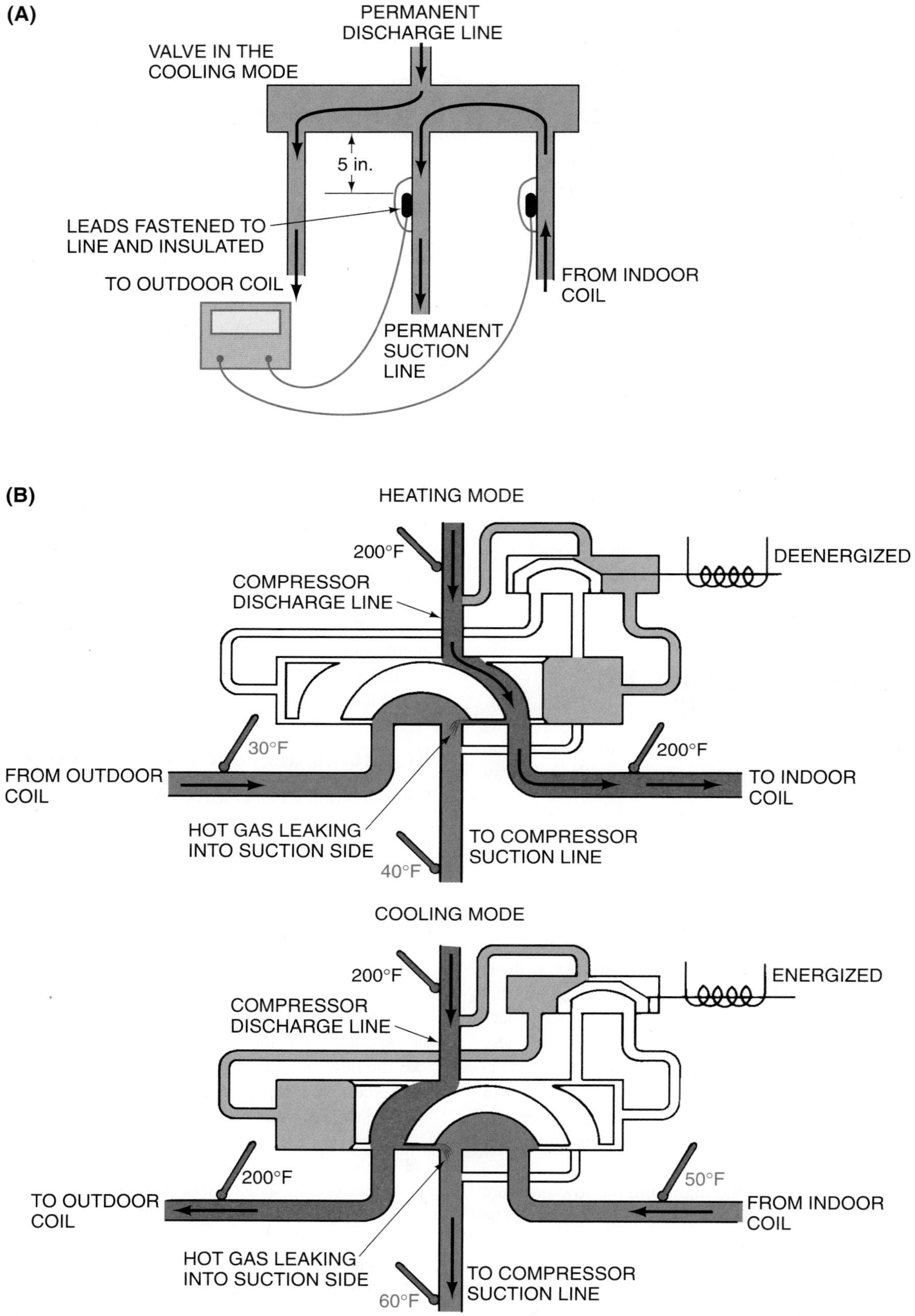

(A)

PERMANENT
DISCHARGE LINE

VALVE IN THE
COOLING MODE

5 in.

LEADS FASTENED TO
LINE AND INSULATED

TO OUTDOOR COIL

FROM INDOOR
COIL

PERMANENT
SUCTION
LINE

(B)

HEATING MODE

200°F

DEENERGIZED

COMPRESSOR
DISCHARGE LINE

30°F

200°F

FROM OUTDOOR
COIL

TO INDOOR
COIL

HOT GAS LEAKING
INTO SUCTION SIDE

TO COMPRESSOR
SUCTION LINE

40°F

COOLING MODE

200°F

ENERGIZED

COMPRESSOR
DISCHARGE LINE

200°F

50°F

TO OUTDOOR
COIL

FROM INDOOR
COIL

HOT GAS LEAKING
INTO SUCTION SIDE

TO COMPRESSOR
SUCTION LINE

60°F

Figure 43–60 **(A)** A method for checking performance of a four-way valve using a temperature comparison. **(B)** A line diagram showing a defective valve in cooling and heating.

43.30 ▶ TROUBLESHOOTING THE COMPRESSOR

Checking a compressor in a heat pump for pumping to capacity is much like checking a refrigeration compressor except that there are normally no service valves to work with. Some manufacturers furnish a chart with the unit to show what the compressor characteristic should be under different operating conditions, **Figure 43–61**. If such a chart is available, use it.

HEATING CYCLE CHECK CHART (R-22)

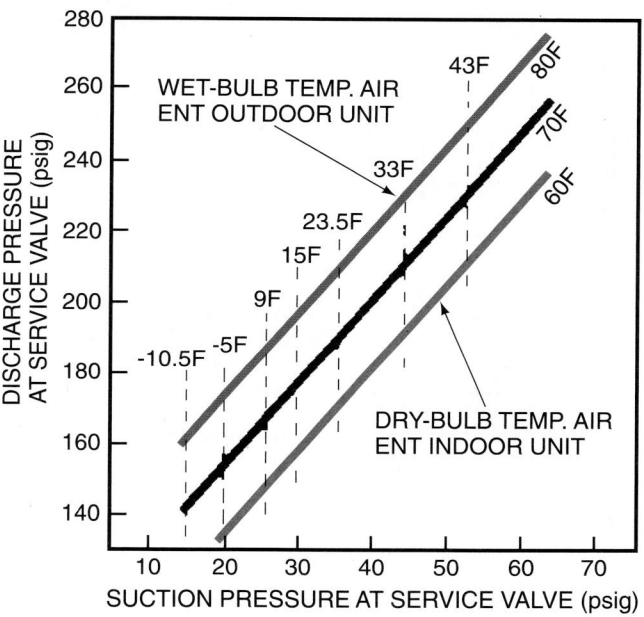

COOLING CYCLE CHARGING CHART (R-22)

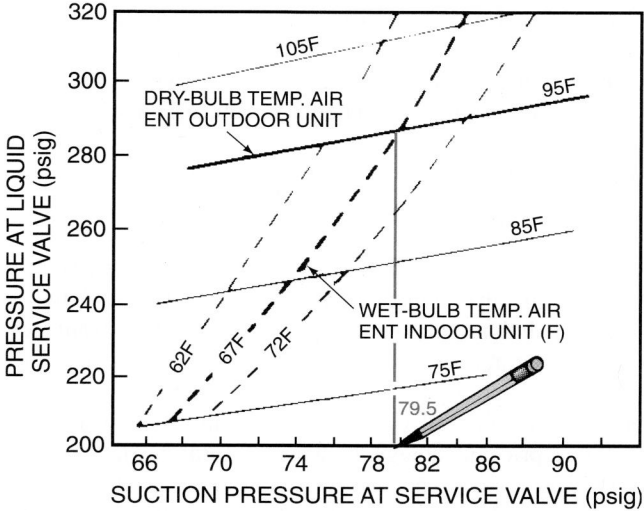

Figure 43–61   The performance chart for a 36,000-Btu/h heat pump system. *Reproduced courtesy of Carrier Corporation*

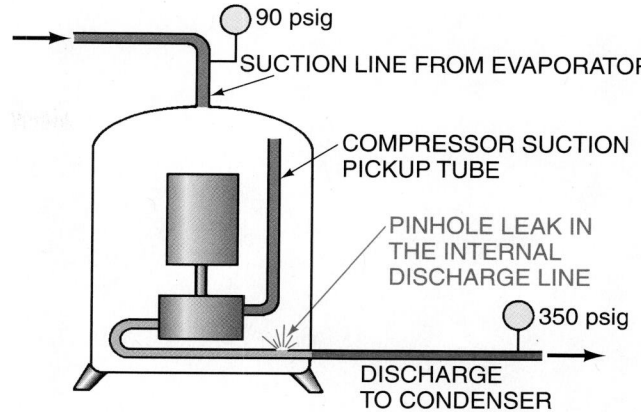

Figure 43–62   This compressor has vapor leaking from the high side to the low side.

The following is a reliable test, using field working conditions, which will tell whether the compressor is pumping at near capacity. Any large inefficiencies will appear.

1. Whether summer or winter, operate the unit in the cooling mode. In winter the four-way valve may be switched by energizing it with a jumper or deenergizing it by disconnecting a wire to switch the unit to the summer mode. This will allow the auxiliary heat to heat the structure and keep it from getting too cold.
2. Block the condenser airflow until the head pressure is 275 psig (this simulates a 95°F day) and the suction pressure is about 70 psig.
3. The compressor amperage should be at close to full load. The combination of discharge pressure, suction pressure, and amperage should reveal whether the compressor is working at near full load. If the suction pressure is high and the discharge pressure is low, the amperage will be low. This indicates that the compressor is not pumping to capacity. Sometimes you can hear a whistle when the compressor is shut off under these conditions. The suction line may also get warm immediately after shutting off the unit under these conditions. Such symptoms indicate that the compressor is leaking from the high side to the low side internally, **Figure 43–62**.

NOTE: If there is any doubt as to where the leak through may be, perform a temperature check on the four-way valve.

43.31 CHECKING THE CHARGE

Most heat pumps have a critical refrigerant charge. The tolerance could be as close as ±1/2 oz of refrigerant. **Therefore do not install a standard gage manifold each time you suspect a problem.** See **Figure 43–63** for an example of a very short coupled system that may be used on the high-side line that will not alter the operating charge. When a heat pump has a partial charge, it is obvious that the refrigerant leaked

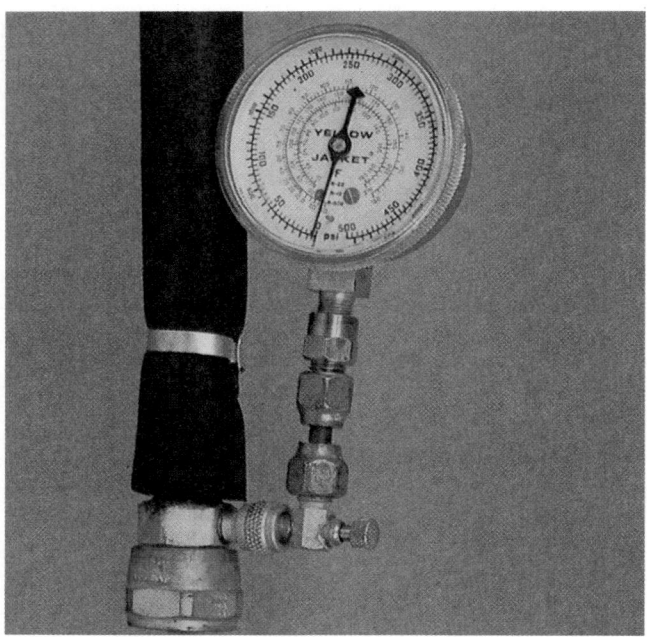

Figure 43–63 A high-pressure gage for checking the head pressure on a heat pump with a critical charge. *Photo by Bill Johnson*

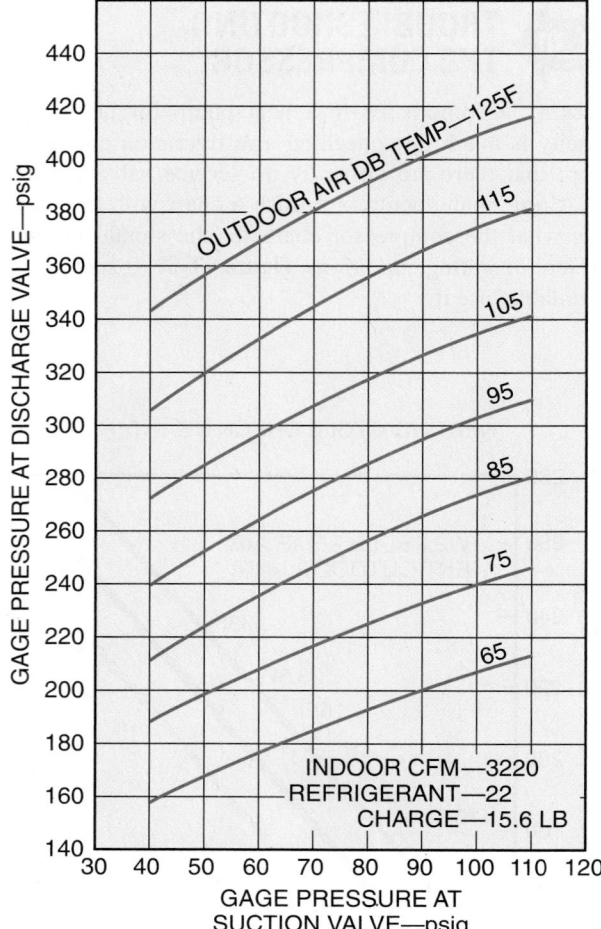

Figure 43–64 A heat pump performance chart.

out of the system. Leak check the system and repair the leak before charging.

When a partial charge is found in a heat pump, some manufacturers recommend that the system be evacuated to a deep vacuum and recharged by measuring the charge into the system. Some manufacturers furnish a charging procedure to allow you to add a partial charge. It is always best to follow the manufacturer's recommendation. **Figure 43–64** is a sample performance chart for a heat pump.

If the manufacturer's recommendations are not available, the following method may be used to partially charge a system. This is specifically for systems with *fixed-bore metering devices*, not TXVs.

1. Start the unit in the cooling mode. If in the winter mode, leave the electric heat where it can heat the structure. Block the condenser and cause the head pressure to rise to 275 psig. This will simulate a 95°F day operation. If the head pressure will not go up, refrigerant may have to be added.

2. Fasten a thermometer to the gas line—it will be cold while in the cooling mode. The suction pressure should be about 70 psig when fully charged but may be lower if there is a reduced latent-heat load because of low humidity. If the system is split with a short gas line (10 to 30 ft), charge until the system has a 10°F to 15°F superheat at the line entering the outdoor unit, **Figure 43–65**. If the line is long (30 to 50 ft), charge until the superheat is 12°F to 18°F.

This charging procedure is close to correct for a typical heat pump with no liquid-line–gas-line heat exchange (some manufacturers have liquid-to-gas heat exchange to improve their particular performance). This heat exchange is normally inside the outdoor unit and may not affect these procedures.

NOTE: When working with the charge on any heat pump that has a suction-line accumulator, and this includes most heat pumps, part of the charge can be stored in the accumulator and will boil out later. You can heat the accumulator by running water over it to drive the refrigerant out if you are in doubt. You may also give the liquid time to boil out on its own. Often, the accumulator will frost or sweat at a particular level if liquid refrigerant is contained in it, **Figure 43–66**.

43.32 SPECIAL APPLICATIONS FOR HEAT PUMPS

The use of oil or gas furnaces for auxiliary heat is a special application. Several manufacturers have systems designed to accomplish this. This type of system has several advantages. Usually, natural gas is less expensive to use as the auxiliary fuel than electricity is at a 1:1 COP. Fuel oil may be considered where a fuel oil system is already installed, where fuel oil has a price advantage, or where natural gas is not available. Both systems have similar control arrangements. The heat pump indoor coil can be used in conjunction with an oil or gas furnace, but the coil must be downstream of the oil or gas heat exchanger. The air must flow through the oil or gas heat exchanger first and then through the heat pump indoor

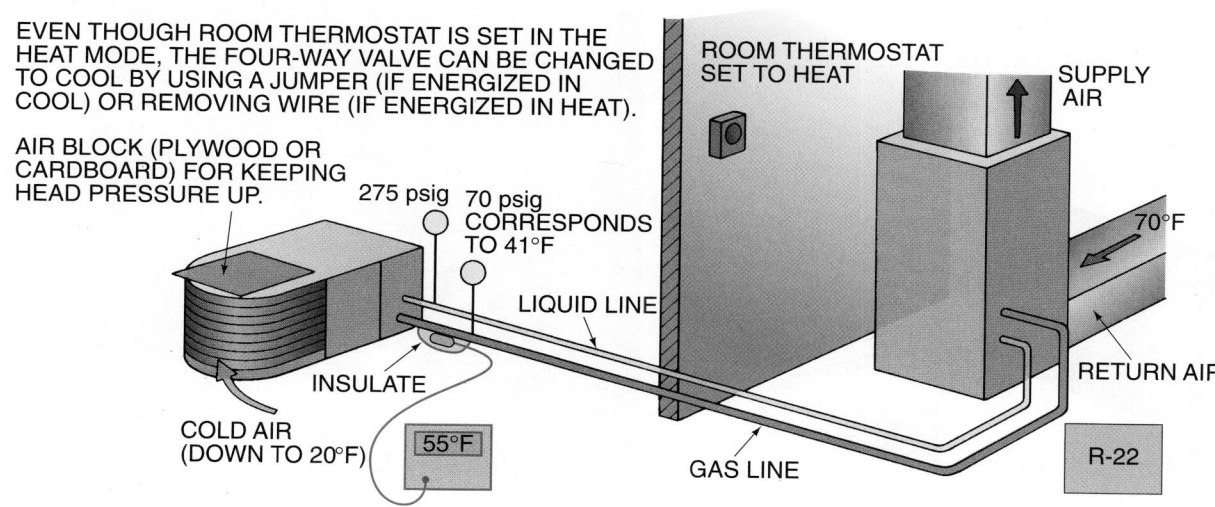

Figure 43–65   This illustration shows how a heat pump with a fixed-bore metering device may be charged when the manufacturer's chart or information is not available. See text for procedure.

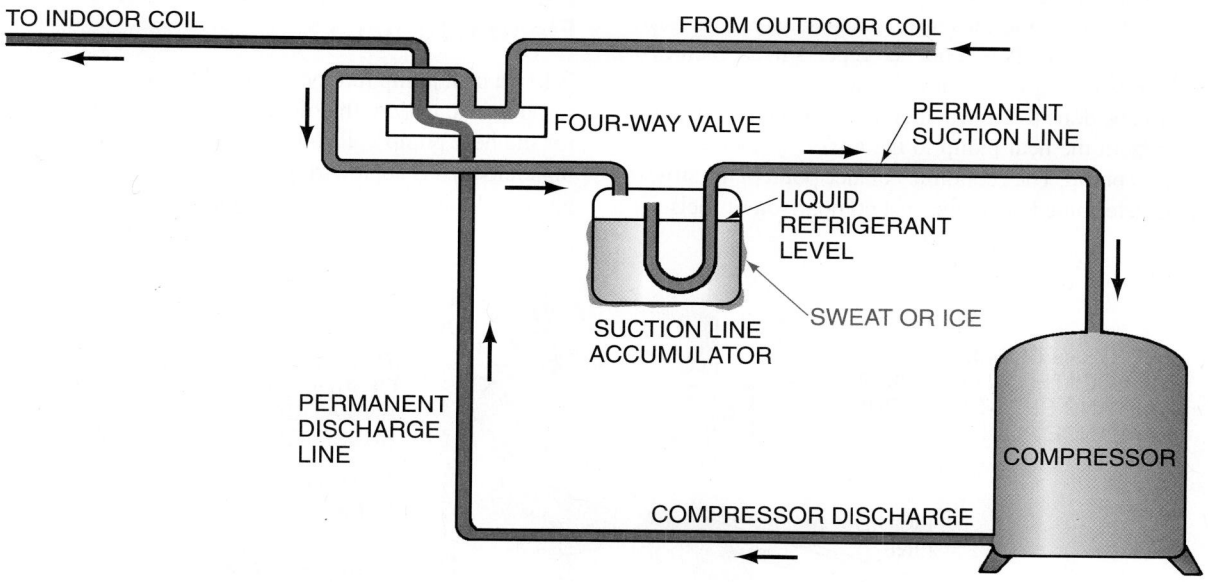

Figure 43–66   A suction-line accumulator that is sweating.

coil. SAFETY PRECAUTION: This means that the gas or oil furnace must not operate at the same time that the heat pump is operating or high head pressures will result, **Figure 43–67.**

The control function can be accomplished with an outdoor thermostat set at the balance point of the structure to change the call for heating to the oil or gas furnace and to shut off the heat pump. This allows the heat pump to operate down to the balance point when the oil or gas furnace will take over. When defrost occurs, the oil or gas furnace will come on during defrost because the same controls are used as for a standard heat pump, and a call for auxiliary heat to warm the cool

air during defrost is still part of the control sequence. SAFETY PRECAUTION: A high-pressure control can prevent the compressor from overloading if a defrost does not terminate due to a defective defrost control.

Maximum Heat Pump Running Time

Other control modifications allow the heat pump to run below the balance point. Such arrangements have more relays and controls. The heat pump is designed to operate whenever it can. This is accomplished by using the second-stage contacts

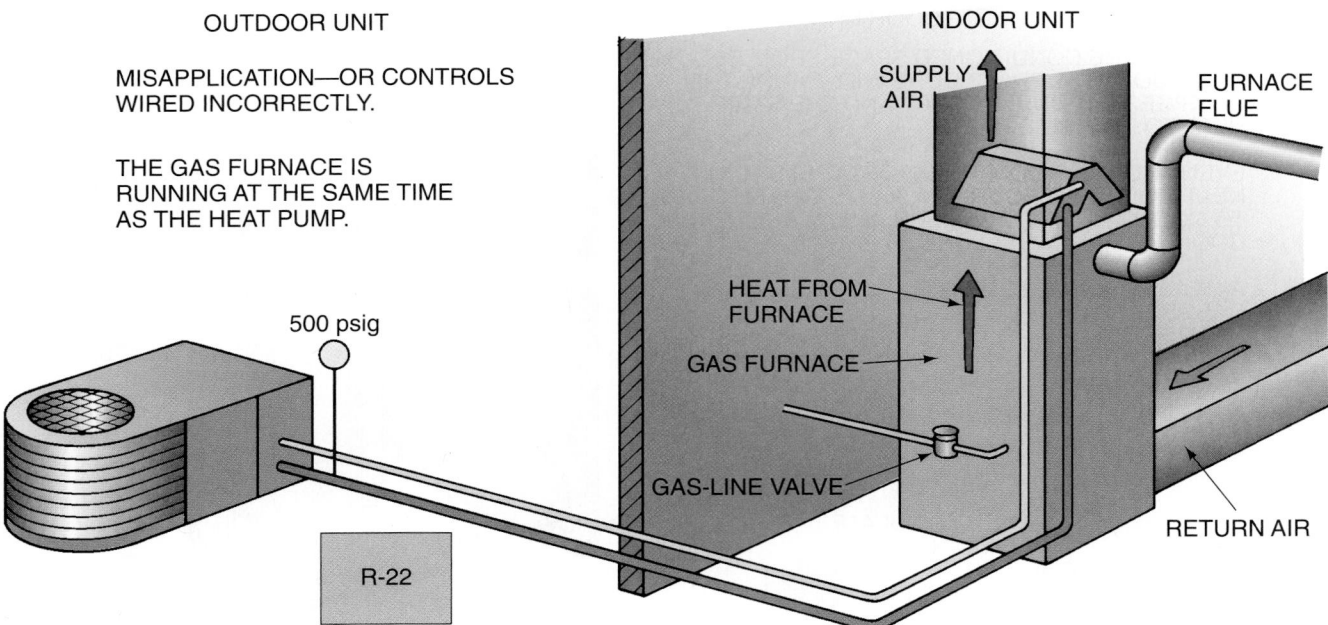

Figure 43–67   The operating conditions inside a heat pump coil while a gas furnace is operating at the same time as the heat pump.

to start the oil or gas heat. This sequence stops the heat pump until the second-stage thermostat contacts satisfy. The heat pump then starts again. This sequence repeats itself each cycle. Before considering this control sequence, a study of the system should be done to see whether it is more economical to switch over from the heat pump to the auxiliary source or to restart the heat pump. The economic balance point of the structure may be determined using the cost comparison of fuels.

Heat Pump Added On to Existing Electric Furnace

When a heat pump indoor coil is added to an existing electric furnace installation, the older furnace might not be equipped for the heat pump coil to be located before the electric heating elements. If so, a wiring configuration similar to an oil or gas installation may be used, **Figure 43–68.**

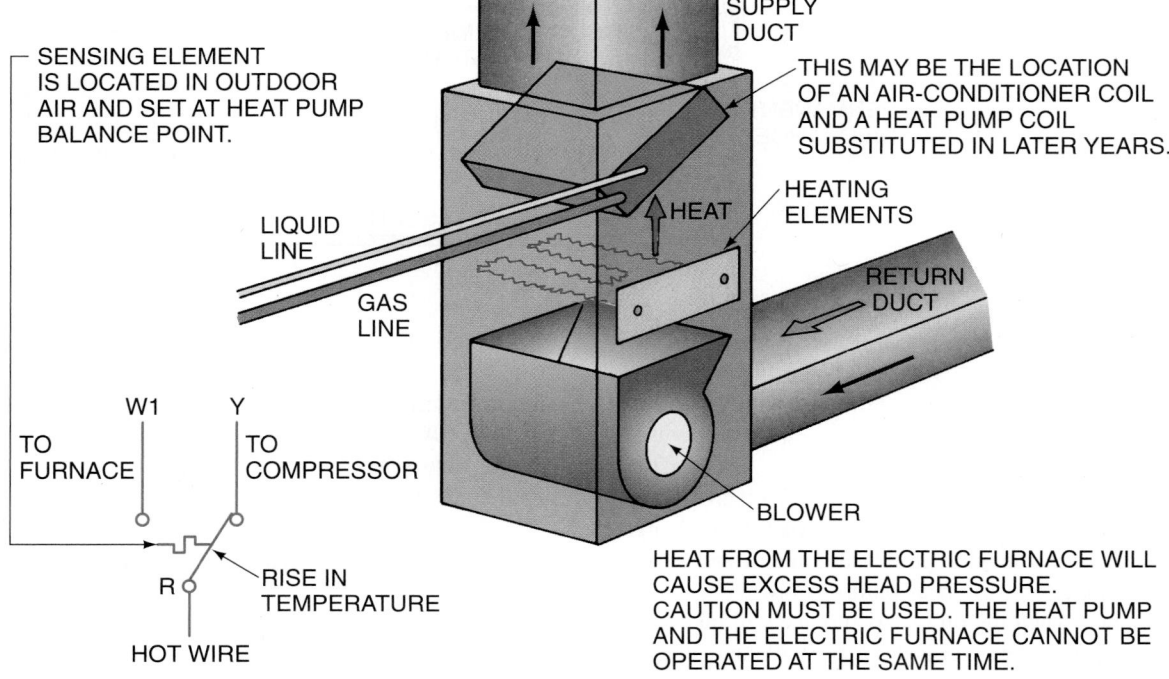

Figure 43–68   The electric furnace heat coils are below the heat pump coil and will cause a high head pressure if used incorrectly with a heat pump.

Many manufacturers have different methods of building a heat pump for their own performance characteristics. These methods involve many different piping configurations with special heat exchangers. This text is intended to explain the typical heat pump. If a different type of system is encountered, the manufacturer should be consulted.

43.33 HEAT PUMPS USING SCROLL COMPRESSORS

Manufacturers are continually working to make heating and cooling equipment more efficient and less expensive to purchase. Some of the new techniques involve the use of improved compressors.

Compressor pumping efficiency has been improved with the *scroll compressor*, **Figure 43–69**. This compressor is ideally suited for heat pump application because of its pumping characteristics. The compression takes place between two spiral-shaped forms, **Figure 43–70**. The scroll compressor does not lose as much capacity as the reciprocating compressor at the higher head pressures of summer or the lower suction pressures of winter operation. This is because the scroll compressor does not have the top-of-the-stroke clearance volume loss that a reciprocating compressor has. The gas trapped in the clearance volume of a reciprocating compressor must reexpand before the cylinder starts to fill. This reduces the efficiency. See Unit 23, "Compressors."

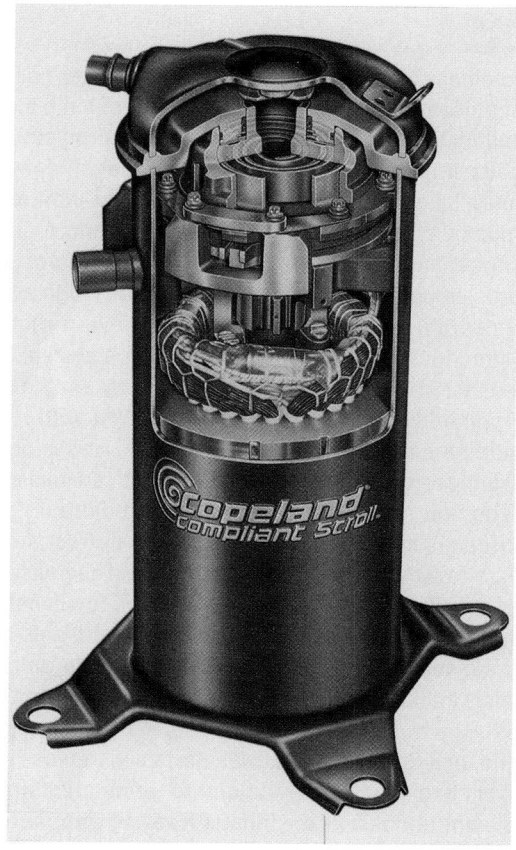

Figure 43–69 A scroll compressor. *Courtesy Copeland Corporation*

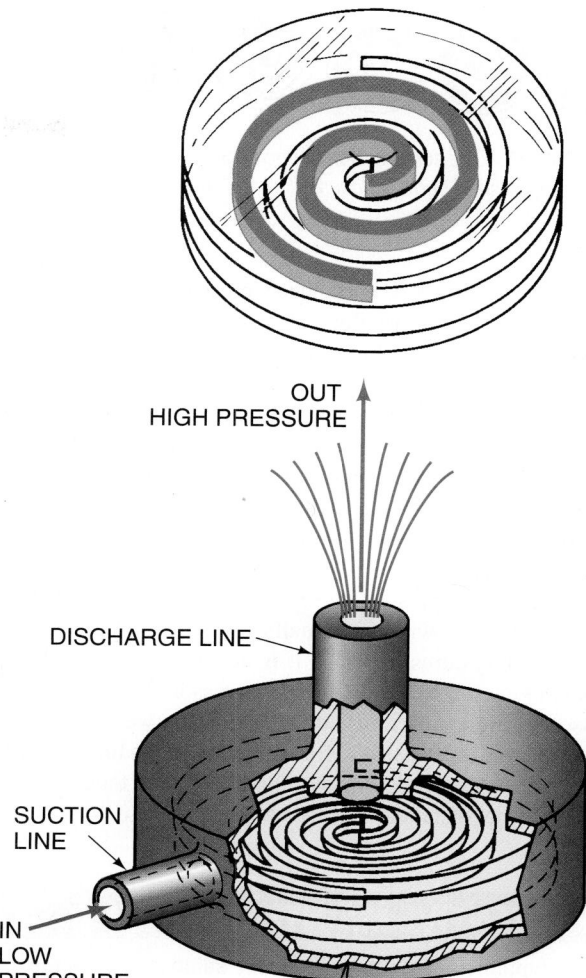

OUT
HIGH PRESSURE

DISCHARGE LINE

SUCTION
LINE

IN
LOW
PRESSURE

Figure 43–70 The orbiting action between the scrolls in a scroll compressor provides compression.

The scroll compressor is about 15% more efficient than the reciprocating compressor. It has many of the same pumping characteristics as the rotary compressor. The scroll compressor does not require crankcase heat because it is not as sensitive to liquid as the reciprocating compressor. This results in a saving in power.

Scroll compressor pressures are about the same as reciprocating compressors, so the technician should not experience any difference in gage readings. They are discharge gas cooled so the technician will notice that the compressor shell is much hotter than a suction-cooled compressor. The scroll compressor runs with less vibration and a lower noise level. Several stages of compression are occurring at the same time in the scroll compressor. Each pocket of gas represents a stage of compression. **Figure 43–71** illustrates one pocket.

The scroll compressor has a check valve in the discharge leaving the compressor to prevent pressures from equalizing back through the compressor during the off cycle. When the compressor is shut off, the gas trapped between the check valve and the compressor scroll will back up through the scroll, and the compressor will make a strange sound at the

Figure 43–71 The compression of one pocket of refrigerant in a scroll compressor. *Courtesy Copeland Corporation*

moment it is shut off until these pressures equalize. This sound resembles the rustling of a handful of marbles. The high-side pressure will then equalize through the metering device to the low side, like a reciprocating compressor system.

The equalizing of pressure between the check valve and the compressor scroll allows the compressor to start up without hard-start assistance because there is no pressure differential for the first starting revolutions of the compressor. This results in fewer start components to purchase and fewer components to cause trouble.

The scroll compressor does not normally require a suction-line accumulator, which is usually included in reciprocating compressor systems, because it is not as sensitive to liquid floodback as the reciprocating compressor. The scroll compressor is more efficient, stronger, lighter weight, and requires fewer accessories for good performance. Unlike reciprocating compressors, scroll compressors tend to "wear in" with time, instead of wear out.

43.34 HEAT PUMP SYSTEMS WITH VARIABLE-SPEED MOTORS

Use of variable-speed motors for the compressor and both fan motors is another method to improve efficiency in heat pump systems. This technology allows for a different selection of heat pump capacity. Previously, heat pump system capacities have been determined by the cooling capacity of the heat pump. A typical structure in the heat pump belt requires about 2 to 2 1/2 times as much heat as cooling. For example, a home in Atlanta, Georgia, may require 30,000 Btu/h for cooling. It would typically require 60,000 to 75,000 Btu/h for heating. When the heat pump is selected for 30,000 Btu/h, it requires considerable auxiliary heat for winter conditions below the balance point of the structure. If the heat pump is selected for the winter capacity need, it would be oversized in the summer. Poor performances and high humidity would occur, along with low efficiency. The variable-speed heat pump closes this gap because the system may be selected at closer to the heating requirement at full load and will run at part load and reduced power in warmer months. Less auxiliary heat is required, so year-round efficiency is achieved.

Variable speed is accomplished with electronics and electronically controlled motors. Each manufacturer has its own method of control and checkout procedure. Repair of a unit with variable-speed motors and electronic controls should not be attempted without the manufacturer's checklist.

PREVENTIVE MAINTENANCE

The typical heat pump requires the same maintenance as an air-cooled air-conditioning system with electric heat. The air side of the system should always handle the maximum air for which it is designed. This is important because some homeowners shut registers off in unused rooms. This should never be done with a heat pump. The indoor system is the condenser in the winter cycle. Any reduction in airflow will cause high head pressure and a low COP. This will cause higher-than-normal power bills with less cooling. Filters should be maintained more closely in heat pump systems for the same reason. Reduced airflow cannot be allowed.

The outdoor unit should be examined for a dirty coil. The insulation on the outdoor gas line should extend all the way to the unit or heat will be lost to the outdoor air. The gas line temperature may be as high as 220°F and will transfer a lot of heat to the cold outside air. This is heat that is paid for and lost.

The contactor should be examined at least once a year for pitting and loose connections. This contactor functions about twice as many times each year as an air-conditioning system contactor. Frayed wires should be repaired or replaced. Wires or capillary tubes rubbing on the cabinet or refrigerant lines will soon cause problems and they should be adjusted or isolated.

The fan motor may require lubrication. Follow the manufacturer's directions.

The customer should be quizzed about the winter performance of a heat pump to include the amount of ice buildup on the outdoor unit. If excess ice seems to be a problem, defrost problems may be causing the unit to stay in defrost. Defrost may be simulated even in hot weather with most heat pumps by disconnecting the outdoor fan and running the unit in heating mode. With the outdoor fan motor disconnected, the unit will not overload because of high evaporator temperatures. Most manufacturers have a recommended procedure for checking defrost. If the system has the same timer shown in **Figure 43–53,** you may run the unit until the coil and sensor are cold and covered with frost. This should be below the make point of 26°F. At this point, carefully jump terminal 3 to 4 on the timer with an insulated jumper. The unit should immediately go into defrost if the sensor circuit is made. Should it not go into defrost at this time, wait a few more minutes and try again. The sensor is large and may take a few minutes to make. If it will not go into defrost, leave the jumper on the 3 to 4 terminals and jump the sensor. If the unit immediately goes into defrost, the sensor is either defective or not in good contact with the liquid line.

After the unit goes into defrost normally, remove the jumper from 3 to 4 and allow the unit to terminate defrost. If the ambient temperature is above freezing and there is no ice accumulation, this should take about a minute. If it takes a full 10 min, the defrost control contacts may be permanently closed. They may be checked by allowing the control to rise to ambient temperature above 26°F and check with an ohmmeter to see whether the contacts remain closed. They should be open when the sensor is above 50°F. One sign of permanently closed contacts is if the unit will go into defrost every time the timer calls for it to do so, whether the sensor is cold or not, and stay in defrost the full 10 min. This is hard on the compressor if the unit does not have a high-pressure control, because high pressure will occur at this time.

When performing maintenance on systems that have electronic circuit boards, the technician will need to work with the timing circuits. If the technician has to wait for each sequence to time out, the service will take too long. Most manufacturers have a method of advancing the timing circuits to a "quick time" that will speed up the service call and any checking procedures being performed.

The last thing the technician should do is make sure that all fasteners are in place and tight. It is common for the cabinet screws to become loose. These should be tight to prevent rattling of the cabinet. If the screw holes are oversized because of service, go to a screw one size larger. It is good practice to have a supply of rust-resistant, one-half oversize screws. Stainless screws cost more, but the heads do not rust off them with time.

HVAC GOLDEN RULES

When making a service call to a residence:

- Look professional and be professional. Clean clothes and shoes should be worn.
- Have a neat haircut, trim facial hairs, and practice good oral hygiene by brushing teeth daily. Showering and daily use of underarm deodorants are strongly recommended.
- Wear coveralls for work under the house and remove them before entering the house.
- Ask the customer about the problem. The customer can often help you solve the problem and save you time.

Added Value to the Customer

Here are some simple, inexpensive procedures that may be included in the basic service call.

- Check line temperatures by touch in summer or winter cycle for the correct temperature.
- Clean or replace air filters as needed. This is very important for heat pumps.
- Make sure that all air registers are open—again very important for heat pumps.
- Replace all panels with the correct fasteners.
- Check indoor and outdoor coils for dirt or debris.
- Ask the owners whether they have noticed any ice buildup in winter operation. This may lead to a defrost checkout for the system.
- Check the gas line for proper insulation. Heat pumps may lose capacity due to loss of line insulation.

43.35 SERVICE TECHNICIAN CALLS

When reading through these service calls, keep in mind that three things are necessary for an electrical circuit to be complete and current to flow: a power supply (line voltage), a path (wire), and a load (a measurable resistance). The mechanical part of the system has four components that work together: The evaporator absorbs heat, the condenser rejects heat, the compressor pumps the heat-laden vapor, and the metering device meters the refrigerant.

SERVICE CALL 1

A homeowner calls reporting that the air-conditioning system is heating, not cooling. This is the first time the unit has been operated in the cooling mode this season. *The problem is that the magnetic coil winding on the four-way valve is open and will not switch the valve over into the cooling mode.*

The service technician turns the space thermostat to the cooling mode and starts the system. This is a split system with the air handler in a crawl space under the house. SAFETY PRECAUTION: The technician goes to the outdoor unit and carefully touches the gas line (the large line); it is hot, meaning the unit is in the heating mode.

The panel is removed, and the voltage is checked at the four-way valve holding coil. A voltage of 24 V is present; the coil should be energized. The unit is shut off, and one side of the four-way valve coil is disconnected. A continuity check shows that the coil is open. The coil is replaced and the unit is started. It now operates correctly in the cooling mode.

SERVICE CALL 2

A store owner calls stating that the air-conditioning system in the small store will not start. This is a split-system heat pump with the outdoor coil on the roof. *The problem is that the four-way valve 24-V holding coil is burned and has overloaded the control transformer.* This has blown the fuse in the 24-V control circuit.

The technician begins by trying to start the unit at the space temperature thermostat. The unit will not start. When the FAN ON switch is turned to FAN ON, the fan will not start. A control voltage problem is suspected. The indoor unit is in a utility closet downstairs. SAFETY PRECAUTION: The power is turned off and the door to the indoor unit is removed. The power is restored, and no power is at the low-voltage transformer secondary. The low-voltage fuse is checked and found to be blown. A new fuse is in a box in the utility room. Before installing the new fuse, the technician turns off the power and then replaces the fuse.

One lead is then removed from the control transformer. An ohmmeter is fastened to the two leads leaving the transformer and the resistance is 0.5 Ω. This seems very low and will likely blow another fuse if power is restored. Some 24-V component must be burned and have a shorted coil. The meter should indicate a 20-Ω resistance. This will increase the current flow enough to blow the fuse. A look at the unit diagram will show which components are 24 V. One of them must be burned. It is decided to check the components in the indoor unit first and save a trip onto the roof, if possible. Each component is checked, one at a time, by removing one lead and applying the ohmmeter. There are three sequencers for the electric heat, and one fan relay. There is no problem here. SAFETY PRECAUTION: The technician goes to the roof and turns off the power. The compressor contactor is checked. It shows a resistance of 26 Ω. The four-way valve holding coil shows a resistance of 0.5 Ω. It is replaced with a new coil. The new coil has a resistance of 20 Ω. The problem has been found. The unit is put in operation with an ammeter attached to the lead leaving the control transformer to check for excessive current. An amperage multiplier made of thermostat wire wrapped 10 times around the ammeter lead is used. The current is 1 A, which is normal.

SERVICE CALL 3

A customer reports that the heat pump has an ice bank built up on the outdoor coil. The customer is advised to turn the thermostat to the emergency heat mode until a technician can get there. *The problem is that the defrost relay coil winding is open, and the unit will not go into defrost.*

The weather is below freezing, so when the technician arrives, there is still an ice bank on the outdoor coil. SAFETY PRECAUTION: With the unit off, the technician applies a jumper wire across terminals 3 and 4 on the defrost timer, **Figure 43–72**. The unit is then started. This should force the unit into defrost if the timer contacts are not closing. This does not put the unit into defrost, so the jumper is left in place, and the defrost temperature sensor is jumped. This satisfies the two conditions that must be satisfied to start defrost, but nothing happens. SAFETY PRECAUTION: The technician turns off the unit, turns off the power, locks and tags the panel, removes the jumpers, and checks the defrost relay. The coil is open. The relay is changed for a new one, and the unit is started again. When the jumper is again applied to terminals 3 and 4, the unit goes into defrost. Remember that two conditions, time and temperature, must be satisfied before defrost can start. The temperature condition is satisfied because the coil is frozen. When the connection from terminals 3 to 4 is made, defrost is started. This defrost would have occurred normally when the time clock advanced to a trial defrost. By jumping from 3 to 4, the technician did not have to wait for the timer. The unit is allowed to go through defrost, and the coil is still iced. The technician decides to

shut off the unit and use artificial means to defrost the coil. SAFETY PRECAUTION: The power is turned off. The panel is locked and tagged. Electrical components are protected with plastic. A water hose is pulled over to the unit and city water is used to melt the ice. City water is about 45°F. Care is taken that water is not directed on to the electrical components. When the ice is melted, the unit is put back in operation. A call back the next morning to the customer verifies that the unit is not icing any more.

SERVICE CALL 4

A homeowner reports that the heat pump unit is running continuously and not heating as well as it used to. *The problem is that the outdoor fan is not running because the defrost relay has a set of burned contacts—the ones that furnish power to the outdoor fan.* These contacts are used to stop the outdoor fan in the defrost cycle. They should also start the fan at the end of the cycle.

The technician finds that the outdoor fan is not running during the normal running cycle. The gas line is warm, and the coil is iced up, so the unit is in the heating cycle. The fan is not running. The unit is shut off, and the panel to the fan control compartment is removed. A voltmeter is fastened to the fan motor leads, and the unit is started. No voltage is going to the fan motor. The wiring diagram shows that the power supply to the fan goes through the defrost relay. SAFETY PRECAUTION: The power is checked entering and leaving the defrost relay. Power is going in but not coming out. The defrost relay is changed, and the unit starts and runs as it should.

SERVICE CALL 5

A residential customer calls. The heat pump serving the residence is blowing cold air for short periods of time. This is a split-system heat pump with the indoor unit in the crawl space under the house. *The problem is that the contacts in the defrost relay that bring on the auxiliary heat during defrost are open and will not allow the heat to come on during defrost.*

The technician is familiar with the particular heat pump and believes that the problem is in the defrost relay or the first-stage heat sequencer. The first thing to do is to see whether the sequencer is working correctly. The space-temperature thermostat is turned up until the second-stage contacts are closed. The technician goes under the house with an ammeter and verifies that the auxiliary heat will operate with the space thermostat. Checking the defrost relay is the next step. This can be done by falsely energizing the relay, by waiting for a defrost cycle, or by simulating a defrost cycle. The technician decides to simulate a defrost cycle in the following manner. SAFETY PRECAUTION: The power is turned off. To prevent the outdoor fan from running, one of the motor leads is removed. Then the unit is restarted. This causes the outdoor coil to form ice very quickly and make the defrost temperature thermostat contacts close. When the coil has ice on it for 5 min, the technician jumps the contacts on the defrost timer from 3 to 4, and the unit goes into defrost, **Figure 43–72.** The technician goes under the house and checks the electric strip heat; it is not operating. The problem must be the defrost relay. By the time the technician gets back to the outdoor unit, the unit is out of defrost. It is decided that the defrost cycle is functioning as it should, so the technician carefully energizes the defrost relay with a jumper wire. The 24-V power supply feeds terminal 4 and then should go out on terminal 6 to the auxiliary heat sequencer. These contacts are not closing. The defrost relay is changed, and the new relay is energized. The technician goes under the house and verifies that the strip heat is working. A call back later verifies that the unit is operating properly.

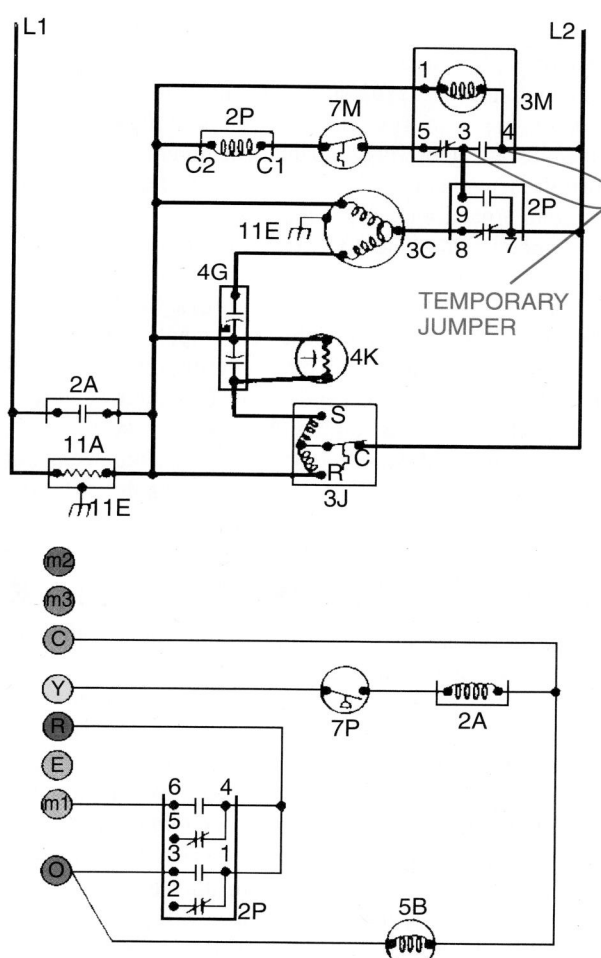

Figure 43–72   With the unit off, the technician applies a jumper wire from terminals 3 to 4 to cause a defrost. This is to keep from waiting for the timer. This is a high-voltage circuit, and this exercise should not be performed unless the technician is very experienced.
Reproduced courtesy of Carrier Corporation

SERVICE CALL 6

A customer calls and says that a large amount of ice is built up on the outdoor unit of the heat pump. *The problem is that the changeover contacts in the defrost relay are open.* The system is not reversing in the defrost cycle. When this happens, the 10-min maximum time causes the unit to stay in heating without defrosting the outdoor coil, because the timer is still functioning.

The technician goes to the outdoor unit first because it is obvious there is a defrost problem. The coil looks like a large bank of ice. The technician removes the control box panel and jumps the timer contacts from 3 to 4; the unit fan stops, but the cycle does not reverse as it should. See **Figure 43–72** for the wiring diagram. A voltage check at the four-way valve coil shows that no voltage is getting to the valve. The technician then jumps from the R terminal to the O terminal, and the four-way valve changes over to cooling, so the jumper is removed. The valve must not be getting voltage through the defrost relay. The defrost relay contacts are jumped from 1 to 3, and the valve reverses. The defrost relay is changed, and another defrost is simulated by jumping the 3 to 4 contacts. The defrost thermostat contacts are still closed, and the unit goes into defrost. The unit has so much ice accumulation that the technician uses water to defrost the coil and get the unit back to a normal running condition. A call the next day indicates that the unit is defrosting correctly.

SERVICE CALL 7

A shop owner calls reporting that the heat pump in their small shop is running constantly. The unit is blowing cold air for short periods every now and then. The power bill was excessive last month, and it was not a cold month. This is a split system with 20-kW strip heat for quick recovery, so the customer can turn the thermostat back to 60°F on the weekend and have quick recovery on Monday morning. The outdoor unit is in the rear of the shop. *The problem is that the contact in the defrost relay that changes the unit to cooling in the defrost mode is stuck closed.* The unit is running in the cooling mode, and the strip heat is heating the structure. It must also make up for the cooling effect of the cooling mode.

The technician sets the thermostat to the first stage of heating. The air at the registers is cool, not warm. The unit should have warm air in the first stage of heat (85°F to 100°F), depending on the outside air temperature. The indoor coil is in a closet in the back of the shop. The gas line is not hot but cold. This indicates that the unit is in the cooling mode, but a voltage check should be performed. See **Figure 43–73** for a wiring diagram. The technician removes the control voltage panel cover and finds 24 V at the C to O terminals. This means the four-way valve coil is energized, and the unit should be in the cooling mode. Now, the technician must decide whether the 24 V is coming from the space thermostat or the defrost relay. The field wire from the space thermostat is removed, and the

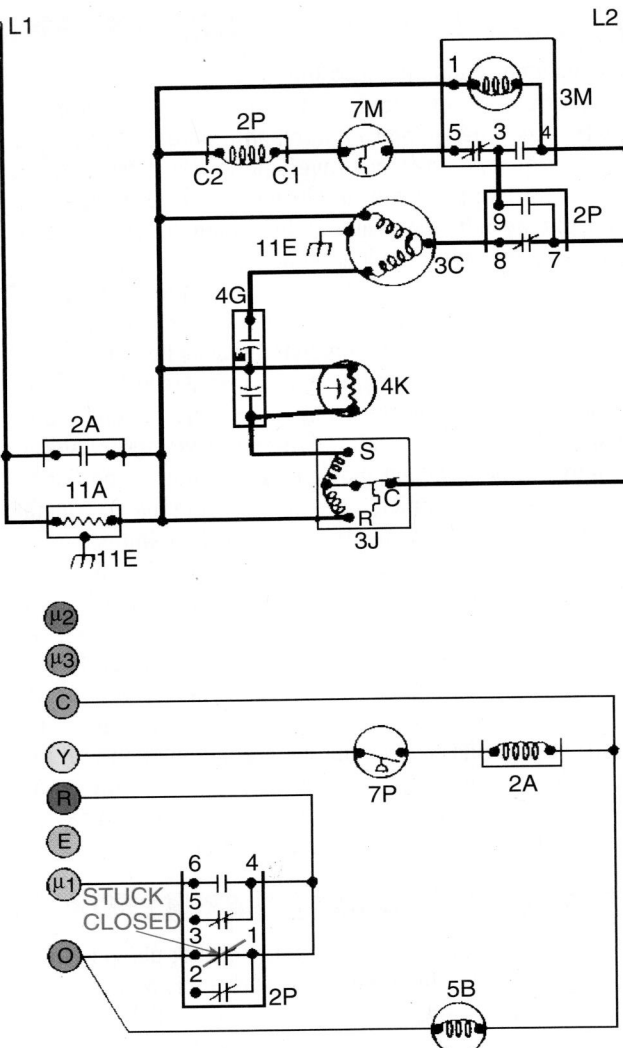

Figure 43–73 The contacts that change the unit to cooling for defrost are stuck closed. *Reproduced courtesy of Carrier Corporation*

is a unit that has a timer that can be advanced by hand. The timer and the defrost sensor are built into one control. It is obvious that the sensor is cold enough for the contacts to make, so the dial of the timer is advanced all the way around. The unit does not go into defrost, so it must be the contacts in the timer sensor that are faulty. A new timer is installed and advanced by hand until the contacts make. The unit goes into defrost normally. One defrost cycle does not melt enough ice from the coil to leave the heat pump on its own. The technician forces the unit through three more defrosts before the ice is melted to a normal level. If a water hose had been available, it would have been used to melt the ice to save time.

SERVICE CALL 9

While on a routine service call at an apartment house complex, the technician notices a unit go into defrost when there is no ice on the coil. The unit stays in defrost for a long time, probably the full 10 min allowable. This is a split-system heat pump and should have defrosted for only 2 or 3 min with a minimum of ice buildup. The compressor was loud, as if it were pumping against a high head pressure. *The problem is that the defrost thermostat contacts are shut.* The unit is going into defrost every 90 min and staying for the full 10 min. The timer is terminating the defrost cycle after 10 min. Toward the end of the cycle, the head pressure is very high.

The technician shuts off the unit, removes the panel to the control box, removes one lead of the defrost thermostat, and takes a continuity check across the thermostat. The circuit should be open if the coil temperature is above 50°F but the contacts are closed. (This control should close at about 25°F and open at about 50°F. Allow plenty of time for this control to function because it is large, and it takes a while for the unit's line to exchange heat from and to the control.) The technician changes the defrost thermostat, and the unit is put back into operation.

SERVICE CALL 10

A residential customer reports that no heat is in their home. It smells like something is hot or burning. SAFETY PRECAUTION: The customer is told to turn off the unit at the indoor breaker to stop all power to the unit. *The problem is that the indoor fan motor has an open circuit, and the electric strip heat is coming on from time to time and then shutting off because of excessive temperature.*

The technician goes to the space thermostat and sets it to OFF. The electrical breakers are then energized. The thermostat is turned to FAN ON. The fan relay can be heard energizing, but the fan will not start. The indoor unit is under the house in the crawl space, so the technician takes tools and electrical test equipment to the unit. SAFETY PRECAUTION: The breaker under the house is turned off, and the door to the fan compartment is opened. The fan motor is checked and found to have an open winding. It is

valve stays energized. The voltage must be coming from the defrost relay. The wire is replaced on the 0 terminal, and the wire on terminal 3 on the defrost relay is carefully removed. The unit changes to heating. The defrost relay is changed, and the unit operates correctly. A call back later finds the unit to be operating as expected.

SERVICE CALL 8

A store manager calls stating that the heat pump heating a small retail store has large amounts of ice built up on the outdoor unit. The unit is located outside in back of the store. *The problem is that the defrost thermostat contacts are not closing when the timer advances to the defrost cycle.* This is a combination control—the timer and the sensor built into the same control.

The technician goes to the outdoor unit and finds a large bank of ice on the coil. The control panel is removed. This replaced with a new one, and the unit is started. The technician uses an ammeter to check all three stages of electric heat for current draw. One stage draws no current. SAFETY PRECAUTION: The breaker is again turned off, and the electric heat is checked with an ohmmeter. The circuit in one unit is open. This happened because the fan was not running, and the unit overheated. A new fuse link is installed, and the heat is started again. This time it is drawing current. The technician puts the unit panels back together and leaves the job.

SERVICE CALL 11

A homeowner reports that the residential heat pump smells hot and not much air is coming out of the registers. SAFETY PRECAUTION: The customer is told to shut off the unit at the breaker for the indoor unit. *The problem is that the indoor fan capacitor is defective, and the fan motor is not running up to speed.* The fan motor is cutting off from an internal overload periodically.

The technician sets the thermostat to OFF and turns on the breaker. The FAN ON switch at the thermostat is set to ON. The fan relay can be heard energizing, and the fan starts. There does not seem to be enough air coming out of the registers. The fan motor is not running smoothly. The technician goes to the attic crawl space where the indoor unit is located, taking tools and electrical test instruments. SAFETY PRECAUTION: The breaker next to the indoor unit is shut off, and the panel to the control box is removed. The breaker is then turned on. It is noticed that the fan is not turning up to speed. The breaker is turned off again. The fan capacitor is removed, and a new one is installed. The breaker is energized again, and the fan comes up to speed. A current check of the fan motor shows the fan is operating correctly. SAFETY PRECAUTION: The power is shut off, locked, and tagged, and the motor is oiled. The technician then replaces all panels and turns on the power.

SERVICE CALL 12

A homeowner calls indicating that the unit in their residence is blowing cool air from time to time. One of the air registers is close to their television chair, and the cool air bothers them. *The problem is that the breaker serving the outdoor unit is tripped, apparently for no reason.* The control voltage is supplied at the indoor unit so the electric auxiliary heat is still operable.

The technician turns the space temperature thermostat to the first stage of heat and notices that the outdoor unit does not start. When the thermostat is turned to a higher setting, the auxiliary heat comes on, and heat comes out of the registers. With the thermostat in the first-stage setting, only the outdoor unit should be operating. The auxiliary heat should not be operating. In the second stage, the heat pump and the auxiliary heat should be operating. The technician goes to the outdoor unit and finds that the breaker is tripped. SAFETY PRECAUTION: There is no way

of knowing how long the breaker has been off, but it is not wise to start the unit because there has been no crankcase heat on the compressor. The breaker is switched to the off position. The panel is locked and tagged. The technician uses an ohmmeter and checks the resistance across the contactor load-side circuit to make sure there is a measurable resistance. It is normal, about 2 Ω (there is a compressor motor, fan motor, and defrost timer motor to read resistance through). The contactor load-side terminals are checked to ground to see whether any of the power-consuming devices are grounded. The meter reads infinity, the highest resistance. It is assumed that someone may have turned the space-temperature thermostat off and back on before the system pressures were equalized and the compressor motor would not start. This means that the compressor tried to start in a locked-rotor condition, and the breaker tripped before the motor overload protector turned off the compressor. A sudden, brief power outage then back on will cause the same thing.

The technician does not reset the breaker before going to the indoor thermostat, setting it to the emergency heat mode. This prevents the compressor from starting with a cold crankcase. The technician then tells the homeowner to switch back to normal heat mode in 10 hours. This gives the crankcase heater time to warm the compressor. The homeowner is instructed to see whether the outdoor unit starts correctly or if the breaker trips. If the breaker trips, the thermostat should be set back to emergency heat, the breaker reset, and the technician should be called. If the technician must go back, the compressor could have a starting problem, and it may not start the next time. The customer does not call back, so the unit must be operating correctly.

SERVICE CALL 13

A customer says that the heat pump is running constantly in the heating mode in mild weather. It did not do this last year; it cut off and on in the same kind of weather. *The problem is that the check valve in the liquid line (parallel to the TXV) is stuck open and not forcing liquid refrigerant through the TXV.* This is overfeeding the evaporator and causing the suction pressure to be too high, hurting the efficiency of the heat pump. For example, if it is 40°F outside, the evaporator should be operating at about 15°F for the outside air to give up heat to the evaporator. If the evaporator is operating at 30°F because it is overfeeding, it will not exchange as much heat as it should. The capacity will be down.

The technician turns the room thermostat to the first stage of heating and goes to the outdoor unit at the back of the house. The unit is running. The gas line temperature is warm but not as hot as it should be. The compressor compartment door is removed so the compressor can be observed. It is sweating all over from liquid refrigerant returning to the compressor. This means that the TXV is not controlling the liquid refrigerant flow to the evaporator. A gage manifold is fastened to the high and low sides of the system. The suction pressure is higher than normal, and the head pressure is slightly lower than normal. The head pressure is down slightly because some of the charge that would normally be in the condenser is in the evaporator. The thermometer lead is attached to the suction line just before the four-way valve, and the superheat is found to be 0°F. The outdoor coil is flooded. Not enough refrigerant is flooding to the compressor to make a noise because the liquid is not reaching the cylinders. It is boiling away in the compressor crankcase. The technician tries adjusting the TXV and it makes no difference. The TXV is open and not closing or else the check valve is open and bypassing the TXV. The technician goes to the supply house and gets a TXV and a check valve. ♻The unit charge is removed and recovered into an approved recovery cylinder. ♻ The check valve is removed first and tested. The technician can blow nitrogen vapor through the valve in either direction, so the check valve is defective. It is changed, and the TXV is left alone. The unit is pressured, leak checked, evacuated, and charged. When the unit is started in the heating mode, it operates correctly. The check valve was the problem. The superheat is checked to be sure that it is normal (8°F to 12°F) and the unit panels are assembled. A call later verifies that the unit is cycling as it should.

SERVICE CALL 14

A residential customer reports that the heat pump serving the home is not performing to capacity. The power bills are much higher than last year's bills although the heating season is much the same. The customer would like a performance check on the heat pump. *The problem is that the heat pump has a two-cylinder compressor, and one of the suction valves has been damaged in the compressor due to liquid floodback on start-up.* The owner had shut the disconnect off at the outdoor unit last summer and left it off all summer. When the unit was started in the fall, the compressor had enough liquid in the crankcase to damage one of the suction valves. If the crankcase heat had been allowed to operate for 8 to 10 hours before starting the unit, this would not have happened. Note that if one cylinder of a two-cylinder compressor is not pumping, the compressor will pump at half capacity.

The technician fastens gages to the system and starts it. The manufacturer's performance chart is used to compare pressures. The unit is running a high suction pressure and a low discharge pressure. The technician switches the unit to cooling and blocks the outdoor coil airflow to build up head pressure. The head pressure goes to 200 psi and the suction to 100 psi. The compressor current is below normal. The compressor is not pumping to capacity; it is defective. A temperature check of the gas lines at the four-way valve proves the valve is not bypassing gas.

The technician goes to a supply house and gets a new compressor. The old compressor is not in warranty, so there is no compressor exchange. ♻The refrigerant is recovered and the compressor is changed. ♻ A new liquid-line filter drier is installed. The system is pressured, leak checked, and evacuated before charging. The charge is measured into the system, and the system is started. The system is operating correctly. A call to the customer later in the day proves that the system is functioning correctly. The outdoor unit is satisfying the room thermostat and shutting the unit off from time to time because the weather is above the balance point of the house.

SERVICE CALL 15

A customer calls the heating and cooling service company indicating that the heat pump serving the residence must not be operating efficiently because the power bill has been abnormally high for the month of January compared with last year's bill. Both years have had about the same weather pattern. *The problem is that the four-way valve in the heat pump is stuck in mid-position and will not change all the way to heat.* There is no apparent reason for the sticking valve; it must be a factory defect.

The technician turns the room thermostat to the first stage of heating and goes to the outdoor unit. The gas line leaving the unit is warm, not hot as it should be. Gages are installed on the system. The head pressure is low, and the suction pressure is high. The compressor amperage is low. All of these are signs of a compressor that is not pumping up to capacity. The outside temperature is 40°F, and the coil does not have an ice buildup, so the unit should be working at a high capacity. The technician suspects the compressor, so a compressor performance test is performed. A jumper wire is installed between the R and O terminals to energize the four-way valve coil and change the unit over to the cooling mode. The thermostat is left in the first-stage heat mode. If the house begins to cool off, the second stage of heat will come on and keep the house from getting too cool. A plastic bag is wrapped around the condenser coil to slow the air entering the coil. The head pressure begins to climb, but the suction pressure rises with it. The compressor will not build a sufficient differential between the high and low sides. Generally, the unit should be able to build 300 psig of head pressure with 70 to 80 psig suction. This unit has a 120-psig suction at 250-psig head pressure. Before the compressor is condemned, the technician decides to check the four-way valve operation. An electronic thermometer is fastened with the two leads on the cold gas line entering and leaving the four-way valve. A difference in temperature should not be more than 3°F between these two lines, but it is 15°F. ♻The four-way valve is leaking hot gas into the suction gas. The four-way valve is changed after the refrigerant is recovered. ♻ The unit is started after the pressure leak test, evacuation, and charging.

The unit now performs to capacity with normal suction and head pressures.

SAFETY PRECAUTION: Use common sense and caution while troubleshooting any piece of equipment. As a technician you are constantly exposed to *potential danger.* High-pressure refrigerant, electrical shock hazard, rotating equipment shafts, hot metal, sharp metal, and lifting heavy objects are among the most common hazards you will face. Experience and listening to experienced people will help to minimize the dangers.

SERVICE CALL 16

An apartment house complex with 100 package heat pumps is under contract with a local company for routine service. While on a routine visit, the technician notices that a compressor is sweating all over in the heating cycle. *The problem is that one of the newer technicians serviced this unit during the last days of the cooling season and added refrigerant to the system.* It is a capillary tube metering device system. The unit now has an overcharge.

The technician knows that a compressor should never sweat all over, so a gage manifold is installed to check the pressures. A charging chart is furnished with this unit to use when adjusting the charge. When the gage readings are compared with the chart readings, both the suction and the discharge pressures are too high. This symptom calls for reducing the charge. ♻Refrigerant is allowed to flow into an approved empty cylinder until the pressures on the chart compare to the pressure in the machine.♻ An observant technician has saved a compressor.

The following service calls do not include solutions. The solutions may be found in the *Instructor's Guide*.

SERVICE CALL 17

A beauty shop owner calls and says the unit serving the beauty shop is blowing cool air from time to time.

This is a split system with the outdoor unit on the roof. The technician turns the thermostat to the first stage of heat and cool air comes out of the air registers. The technician goes to the outdoor unit—the outdoor fan is running, but the compressor is not. SAFETY PRECAUTION: The power is turned off. The panel is locked and tagged before the technician removes the cover to the compressor compartment. An ohm check shows that there is an open circuit between the R and C terminal and between the S and C terminal, but there is continuity between R and S. The compressor body temperature is warm, not hot. **What is the likely problem and the recommended solution?**

SERVICE CALL 18

A residential customer calls and says the heat pump outdoor unit is off because of a tripped breaker. The customer reset it three times, but it would not stay reset.

The technician goes to the outdoor unit first and sees that the breaker is in the tripped position. It is decided that before resetting it a motor test should be performed. SAFETY PRECAUTION: A volt reading is taken at the compressor contactor line side and it is found that 50 V is present. This must be caused by the repeated resetting of the breaker. Voltage is leaking through the breaker. The main heating breaker is turned off, and the 50 V disappears. SAFETY PRECAUTION: If the technician had taken a continuity reading first, the meter would have been damaged. With the main breaker off, a resistance check on the load side of the compressor contactor shows a reading of 0 Ω to ground (the compressor suction line is a convenient ground). **What is the likely problem and the recommended solution?**

SERVICE CALL 19

An owner of a two-story house calls indicating that the unit serving the upstairs of the house is blowing cold air part of the time and not heating properly.

The technician turns the space-temperature thermostat to the first stage of heating and notices that cool air is coming out of the air registers. The outdoor unit can be seen from a window and the fan is running. The technician goes to the outdoor unit and finds that the gas line leaving the unit is barely warm, not hot. The compressor is either not running, or it is running and not working. The compressor access panel is removed, and the compressor can be touched and seen. It is hot to the touch. **What is the likely problem and the recommended solution?**

SERVICE CALL 20

A dress shop manager calls to report that the heat pump serving the dress shop will not cool properly. This is the first day of operation for the system this summer. The compressor was changed about two weeks ago after a bad motor burn. The system has not operated much during the two-week period because the weather has been mild.

The technician turns the thermostat to the cooling mode and starts the unit. The compressor is on the roof, and the indoor unit is in a closet. The technician goes to the indoor unit and feels the gas line; it is hot. The unit is in the heating mode with the thermostat set in the cooling mode. The technician goes to the roof and removes the low-voltage control panel. In this unit the four-way valve is energized in the cooling mode, so it should be energized at this time. A voltmeter check shows that power is at the C to O terminals (a common terminal and the four-way valve coil terminal). The technician removes the wire from the O terminal and hears the pilot solenoid valve change (with a click). Every time the wire is touched to the terminal, the valve can be heard to position. This means that the pilot valve is moving.

What is the likely problem and the recommended solution?

SERVICE CALL 21

A retail store owner reports that the heat pump serving the store is blowing cold air from time to time and not heating as it normally does. It runs constantly. The unit has a TXV.

The technician turns the thermostat to the first stage of the heating mode. The air coming out of the air registers is cool; it feels like the return air. The outdoor unit is on the roof, and the indoor unit is in an attic crawl space. The technician goes to the roof; the thermostat is set with the first stage calling for heat. The outdoor unit should be on. The outdoor unit is running, but the gas line is cool. A gage manifold is installed on the high and low sides of the system. The low side is operating in a vacuum, and the high-side pressure is 122 psig, which corresponds to the inside temperature of 70°F. This indicates that liquid refrigerant is in the condenser (the indoor coil). If the condenser pressure were 60 psig, well below the 122 psig corresponding to 70°F, it would indicate that the unit is low in refrigerant charge.

What is the likely problem and the recommended solution?

SUMMARY

- All compression-cycle refrigeration systems are similar to heat pumps; however, heat pumps have the capability of pumping heat either way.
- A new component, the four-way reversing valve, enables the heat pump to reverse the refrigeration cycle and reject heat in either direction.
- *Water to air* is the term used to describe heat pumps that absorb heat from water and transfer it to air.
- Commercial buildings may use water to absorb heat from one part of a building and reject the same heat to another part of the same building.
- Air-to-air heat pumps are similar to summer air-conditioning equipment.
- There are two styles of equipment: split systems and package (self-contained) systems.
- New names are applied to the heat pump components because of the reverse-cycle operation.
- The terms "indoor coil" and "outdoor coil" are also used with package heat pumps to avoid confusion.
- The refrigerant lines that connect the indoor coil with the outdoor coil are the gas line (hot gas in winter and cold gas in summer) and the liquid line.
- The liquid line is always the liquid line; the flow reverses from season to season.

- Several metering devices are used with heat pumps. The first was the TXV. There may be two of them.
- Capillary tube metering devices are common. There may be two of them, one for cooling and one for heating operations.
- A fixed-bore metering device that will allow full flow in one direction and restricted flow in the other direction is used by many manufacturers.
- The electronic expansion valve is used by some manufacturers with close-coupled equipment because it will meter in both directions and maintain the correct superheat.
- When standard filter driers are used with a heat pump installation, they must be placed in the circuit with the check valve to ensure correct flow. It must have a check valve piped in series with it, or when the flow reverses, it will backwash and the particles will be pushed back into the system.
- The air-to-air heat pump loses capacity as the outside temperature goes down.
- At the balance point the heat pump alone will run constantly and just heat the structure.
- Auxiliary heat is the heat that a heat pump uses as a supplement.
- Auxiliary heat is normally electric resistance heat. Oil and gas may be used in some installations.
- When the auxiliary heat is used as the only heat source, such as when the heat pump fails, it is called emergency heat.
- Emergency heat is controlled with a switch in the room thermostat.
- Coefficient of performance (COP) is determined from the heat pump's heating output divided by the input. A COP of 3:1 is common with air-to-air heat pumps at the 47°F outdoor temperature level. A COP of 1.5:1 is common at 0°F.
- The efficiency of a heat pump is a result of its being used to capture heat from the outdoors and to pump that heat indoors.
- A heat pump is installed in much the same manner as a cooling air conditioner.
- The terminal air must be distributed correctly because it is not as hot as the air in oil or gas installations. Heat pump air normally is not over 100°F with only the heat pump operating.
- SAFETY PRECAUTION: Provision for water drainage at the outdoor unit must be provided in winter.
- The indoor fan must run when the compressor is operating, and the compressor operates in both summer and winter modes.
- The four-way valve determines whether the unit is in the heating or cooling mode.
- The technician can tell which mode the unit is in by the temperature of the gas line. It can get very hot.
- The space-temperature thermostat controls the direction of the hot gas by controlling the position of the four-way valve.
- Because the heat pump evaporator operates below freezing in the winter, frost and ice will build up on the outdoor coil.
- When the system is in defrost, it is in the cooling mode with the outdoor fan off to aid in the buildup of heat.
- Defrost is normally instigated (started) by **time and temperature** and terminated by **time or temperature.**
- The refrigerant charge is normally critical with a heat pump.
- The use of scroll compressors increases the efficiency of heat pump systems.

REVIEW QUESTIONS

1. How does a heat pump resemble a refrigeration system?
2. Name the three common sources of heat for heat pumps.
3. What large line connects the indoor unit to the outdoor unit?
   A. A discharge line
   B. A liquid line
   C. An equalizer line
   D. A gas line
4. Why is the indoor unit not called an evaporator?
5. When is the outdoor unit called an evaporator?
   A. Heating season
   B. Cooling season
   C. Defrost periods
   D. Spring and fall seasons
6. The indoor coil is called a condenser during the _____ season.
7. Why is it important to have drainage for the outdoor unit?
8. Which metering device is most efficient in heat pumps?
   A. A thermostatic expansion valve
   B. An automatic expansion valve
   C. A capillary tube
   D. An orifice or restrictor
9. Where is the only permanent suction line on a heat pump?
10. What type of drier may be used in the liquid line of a heat pump?
11. Where must a suction-line drier be placed in a heat pump after a motor burnout?
12. What controls the heat pump to determine whether it is in the heating cycle or the cooling cycle?
13. Can a heat pump switch from heating to cooling and from cooling to heating automatically?
14. What must be done when frost and ice build up on the outdoor coil of a heat pump?
15. True or False: All heat pumps are practical anywhere in the United States. Explain your answer.
16. Why are scroll compressors more efficient than reciprocating compressors?
17. Describe why variable-speed compressor motors are more efficient than single-speed motors in heat pump systems.

DIAGNOSTIC CHART FOR HEAT PUMPS IN THE HEATING MODE

The heat pump is the most complex of the heating and air-conditioning systems. These are found in many homes and businesses in the south and southwest parts of the United States. This climate is best suited for heat pumps. These systems operate in both heating and cooling modes. The heating mode is discussed here.

Problem	Possible Cause	Possible Repair	Heading Number
Cooling Season **Heating Season**	See summer air conditioning (Unit 42)		
No heat—outdoor unit will not run—indoor fan runs	Open outdoor disconnect	Close disconnect.	41.10
	Open fuse or breaker	Replace fuse, reset breaker, and determine problem.	41.10
	Faulty wiring	Repair or replace faulty wiring or connection.	41.10
No heat—indoor fan or outdoor unit will not run	Low-voltage control problem A. Thermostat	Repair loose connections or replace thermostat and/or subbase.	41.10
	B. Interconnecting—wiring or connections	Repair or replace wiring or connections.	41.10
	C. Transformer	Replace if defective. SAFETY PRECAUTION: Look out for too much current draw due to ground circuit or shorted coil.	41.10

(continued)

DIAGNOSTIC CHART FOR HEAT PUMPS IN THE HEATING MODE (Continued)

Problem	Possible Cause	Possible Repair	Heading Number

No heating; indoor and outdoor fans running, but compressor not running	Compressor overload tripped
	A. Low line voltage
	
	B. High head pressure
	  1. Dirty indoor (condenser) coil
	  2. Indoor fan not running all the time
	  3. Overcharge of refrigerant
	  4. Restricted indoor airflow
		Correct low voltage, call power company, or correct loose connections.
		
		Clean indoor coil.
		Check indoor fan motor and capacitor.
		Correct charge.
		Change filters.
		Open all supply registers.
		Clear return air blockage.
			41.10, 41.11, 43.27
			
			43.28
			43.28
			43.31
			43.28
			43.28
			43.28

Outdoor coil (evaporator) freezes and ice will not melt	Defrost control sequence not operating Low charge, not enough refrigerant to perform adequate defrost	Follow manufacturer's directions and correct defrost sequence. Correct charge and run unit through enough defrost cycles to clear ice off; then allow unit to run normally.	43.23 43.30
Unit will not change from cooling to heating or heating to cooling	Four-way valve not changing over A. Defective defrost relay or circuit board B. Four-way valve stuck C. Thermostat not changing to heat in subbase	Replace relay or circuit board. Change four-way valve. Repair or replace subbase.	43.29 43.29 43.22
Excessive power bill	Compressor not running, heating off or operating on auxiliary heat Suction pressure too high and head pressure too low for conditions	Repair compressor circuit or replace compressor. Perform four-way valve temperature check. If valve is defective, change valve. If valve is good, change compressor.	43.30 43.29

UNIT 44

Geothermal Heat Pumps

OBJECTIVES

After studying this unit, you should be able to

- describe an open- and closed-loop geothermal heat pump system.
- explain how water quality affects an open-loop geothermal heat pump.
- describe different ground-loop configurations for closed-loop geothermal heat pump systems.
- explain the advantages and disadvantages of series- and parallel-flow configurations in geothermal heat pump systems.
- explain the different system fluids and heat exchanger materials.
- describe different geothermal well types and water sources for heat pumps.
- explain some of the most common service problems with geothermal heat pump systems.
- list and explain the governing formulas that calculate the amount of heat rejected or absorbed by the water side of a geothermal heat pump.
- describe a waterless, earth-coupled, closed loop, geothermal heat pump system.

SAFETY CHECKLIST

- ✔ Always wear protective clothing and eye shields when working with geothermal heat pumps. They contain high-pressure refrigerants, which can freeze your skin and cause serious injury.
- ✔ When working around geothermal heat pumps, antifreeze solution or water leaking from the system can cause an electrical hazard. When troubleshooting with the electrical power on, extra care must be taken to prevent electrical shocks.
- ✔ Closed-loop geothermal heat pumps also contain a solution in the ground loop that can be harmful if not handled properly. Both protective clothing and eye protection must be worn when contacting this fluid. Also, some antifreeze solutions used in closed-loop systems can be flammable, so never expose them to extreme heat or an open flame.

44.1 REVERSE-CYCLE REFRIGERATION

Geothermal heat pumps are refrigeration machines. They are very similar to air source heat pumps in that they can remove heat from one place and transfer it to another. However, geothermal heat pumps use the earth, or water in the earth, for their heat source and heat sink. Energy is transferred daily to and from the earth by the sun's radiation, rain, and wind. Each year, more than 6000 times the amount of energy currently used by humans is striking the earth from the sun. In fact, less than 4% of the stored energy in the earth's crust comes from its hot molten center. Heat pumps use the energy stored in the earth's crust for heating. Also, because the earth's crust temperature is cooler than the air just above it in the summer months, summer air-conditioning heat loads can be transferred to the earth.

Geothermal heat pumps can pump heat in two ways. Because of this, they are normally used for space conditioning: heating and cooling. The same four basic components of heat pumps mentioned in Unit 43, "Air Source Heat Pumps," operate and control geothermal heat pumps. They are the compressor, condenser, evaporator, and metering device. The *four-way valve* also controls the direction of heat flow in geothermal heat pumps. For a review of basic heat pump theory and operation, review Unit 43.

44.2 GEOTHERMAL HEAT PUMP CLASSIFICATIONS

Geothermal heat pumps are classified as either *open-loop* or *closed-loop* systems. Open-loop, or *water-source,* systems use water from the earth as the heat transfer medium and then expel the water back to the earth in some manner. This usually involves a well, lake, or pond. Open-loop systems need a large volume of clean water to operate properly. This same water supply also can be used for drinking and cooking.

Closed-loop, or *earth-coupled,* systems reuse the same heat transfer fluid, which is circulated in buried plastic pipes within the earth or within a lake or pond. Closed-loop or earth-coupled systems are used where the water is rich in minerals, where local codes prohibit open-loop systems, or where not enough water exists to support an open-loop well water system. Both open-loop and closed-loop systems will be covered in detail in the following paragraphs.

Whether the geothermal heat pump system is an open-loop, closed-loop, or earth-coupled system, water is still the source of heat when the system is operating in the heating mode. For this reason, these heat pumps are referred to as water-source heat pumps. Water-source heat pumps can be classified as either water to air or water to water. Water-to-air heat pumps are used to heat air in the occupied space, while water-to-water heat pumps are used to heat water, which would be appropriate for an application such as a radiant heating system.

44.3 OPEN-LOOP SYSTEMS

Open-loop, water-source heat pump systems involve the transfer of heat between a water source and the air or water being circulated to a conditioned space. Remember, open-loop systems use water from the earth as the heat transfer medium and then expel the water back to the earth. During the heating mode, heat is being transferred from the water source to the conditioned space. During the cooling mode, the heat removed from the conditioned space is deposited into the water. **Figure 44–1** illustrates both a heating and a cooling application of an open-loop, water-to-air heat pump. Defrost systems are not needed in geothermal heat pump systems.

In the *heating mode,* water is supplied from a water source to a coiled, coaxial heat exchanger, **Figure 44–2,** by a circulating pump, **Figure 44–1(A).** This heat exchange takes place between the water and the refrigerant. The coaxial heat exchanger carries refrigerant in the outer section and the water flows in the inner tube, **Figure 44–2(A).** Notice that the inner tube is ribbed to increase both the surface area and the heat transfer rate between the fluids, **Figure 44–2(B).** The refrigerant side of the heat exchanger is the refrigeration system's evaporator. Heat is absorbed from the water into the vaporizing refrigerant. The refrigerant vapor then travels through the reversing valve to the compressor, where it is compressed. The heat-laden hot gas from the compressor then travels to the condenser. The condenser is a refrigerant-to-air, finned-tube heat exchanger located in the ductwork. It is often referred to as the *air coil.* Heat is then rejected to the air as the refrigerant condenses. A fan delivers the heated air to the conditioned space. The condensed liquid then travels through the expansion valve and vaporizes in the evaporator, absorbing heat from the water source. The process is then repeated. High-resistance, electric strip heaters can be used for auxiliary and/or emergency heat when the heat pump needs assistance.

In the *cooling mode,* the water loop acts as the condensing medium for the refrigerant, **Figure 44–1(B).** Discharge gas from the compressor travels through the reversing valve to the outer portion of the coaxial (tube-within-a-tube) heat exchanger of the water loop. The refrigerant side of the heat exchanger is the refrigerant system's condenser. A refrigerant-to-water heat exchange takes place. The water loop absorbs heat from the refrigerant and condenses the refrigerant. Subcooled liquid refrigerant then travels to the expansion valve and on to the *air coil* where it evaporates and absorbs heat from the air. The air is cooled and dehumidified. A fan delivers this air to the conditioned space. The heat-laden, superheated, refrigerant gas then travels from the air coil and through the reversing valve to the compressor, where the refrigerant is compressed. The process is then repeated.

For smaller residences, a single water-source heat pump is normally used, but larger houses may require more than one. For commercial applications, multiple heat pumps are combined into a system with a common water-piping loop. This loop provides a means for transferring the rejected heat from one heat pump in the cooling mode to another heat pump, which is in the heating mode. Commercial applications may also be connected to a well, a pond, a lake, or an earth-coupled closed-loop system. These systems may also contain a *boiler* and *cooling tower* to maintain desired loop temperatures, **Figure 44–3.**

44.4 WATER QUALITY

Water quality is the most important factor to consider when dealing with open-loop systems that rely on well water. Listed here are three of the most important questions involving water quality that technicians and designers have to look at before choosing an open-loop, well-water system for heating and cooling applications.

- Will the well deliver enough water in gallons per minute (gpm) to the heat pump?
- What is the temperature of the well water?
- Is the well water clean and low in minerals?

HEATING MODE

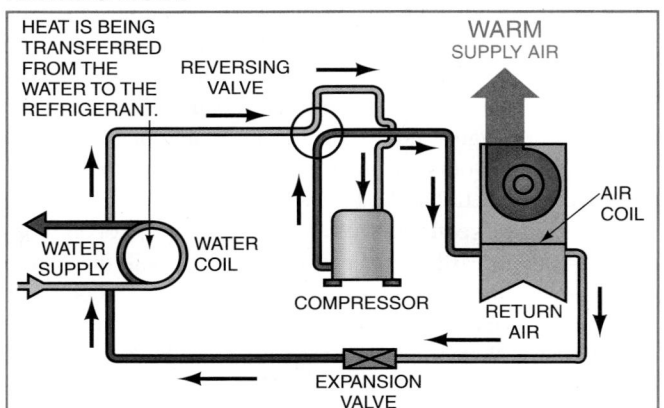

IN THE HEATING MODE, HOT REFRIGERANT FLOWS THROUGH THE AIR COIL, SUPPLYING WARM AIR TO THE CONDITIONED SPACE.

(A)

COOLING MODE

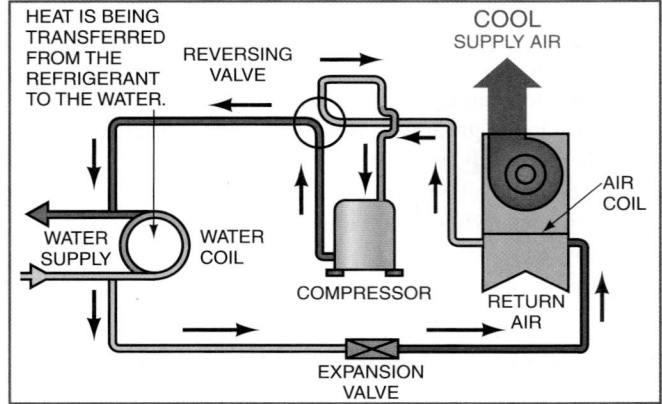

IN THE COOLING MODE, COLD REFRIGERANT FLOWS THROUGH THE AIR COIL, SUPPLYING COOL AIR TO THE CONDITIONED SPACE.

(B)

Figure 44–1   An open-loop, water-source heat pump in both **(A)** heating and **(B)** cooling mode. *Courtesy Mammoth Corporation*

(A)

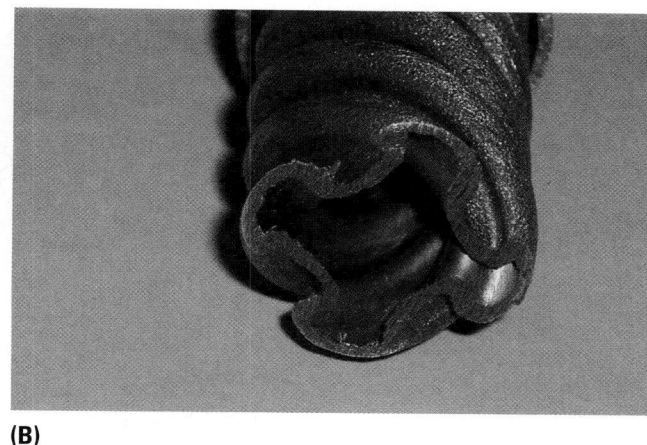

(B)

Figure 44–2   Cutaway of a coaxial (tube-within-a-tube) heat exchanger. **(A)** Refrigerant flows in the outer tube, while water flows in the inner tube. **(B)** Notice the shape of the inner tube. This increases the surface area and increases the rate of heat transfer. *Photos by Eugene Silberstein*

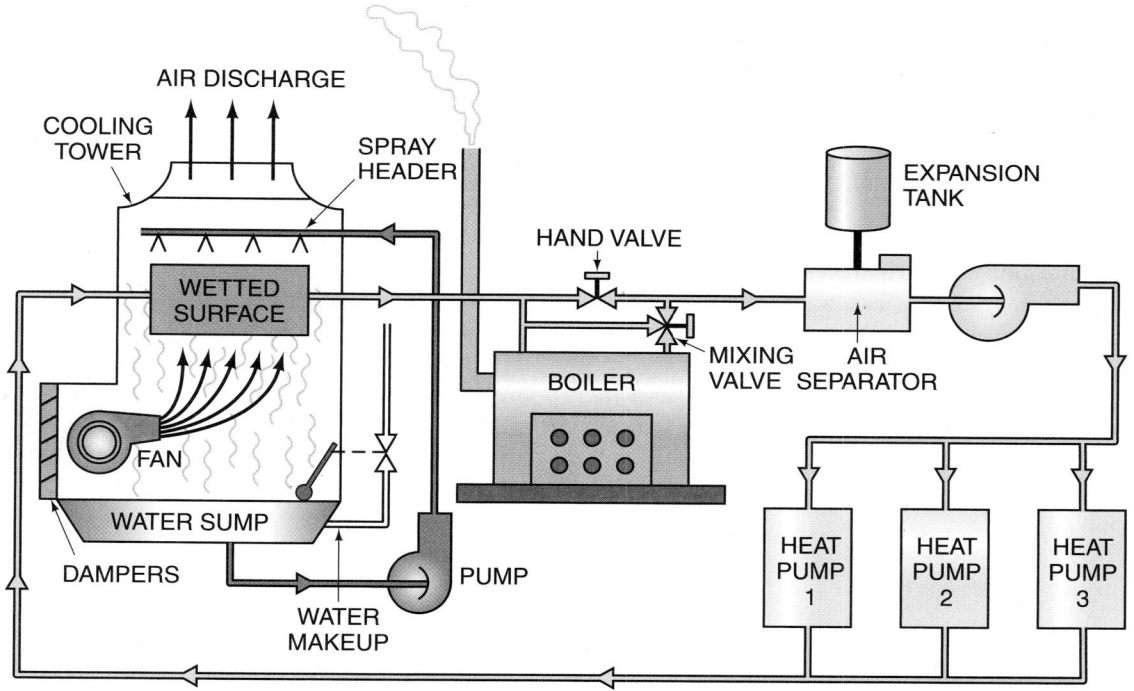

Figure 44–3   An open-loop, water-source heat pump with boiler and cooling tower to maintain the loop temperature.

Enough water flow in gpm has to be available to properly feed the system and handle the required capacities for heating and cooling. Remember, this well water is the heat transfer medium between the earth and the refrigerant in the heat pump. Consult with the manufacturer of the heat pump for specific information on water flow rates in gpm for their heat pumps. The local well driller can determine whether enough water is in the well to properly operate the heat pump.

Well-water temperature is another important factor. Water temperature is one of the factors that determines the heat

pump's capability to transfer heat energy. This is because the temperature difference between the water and the refrigerant is the driving potential for heat transfer. The larger the temperature difference, the greater the heat transfer. **Figure 44–4** illustrates water temperatures in wells ranging from 50 to 150 ft in depth within the United States.

Cleanliness of the well water is another major factor to consider when dealing with an open-loop, well-water system. The water cannot be contaminated with sand, dirt, clay, or any solid mineral particulate. They all may cause a gradual wearing down of the coaxial heat exchanger in which the

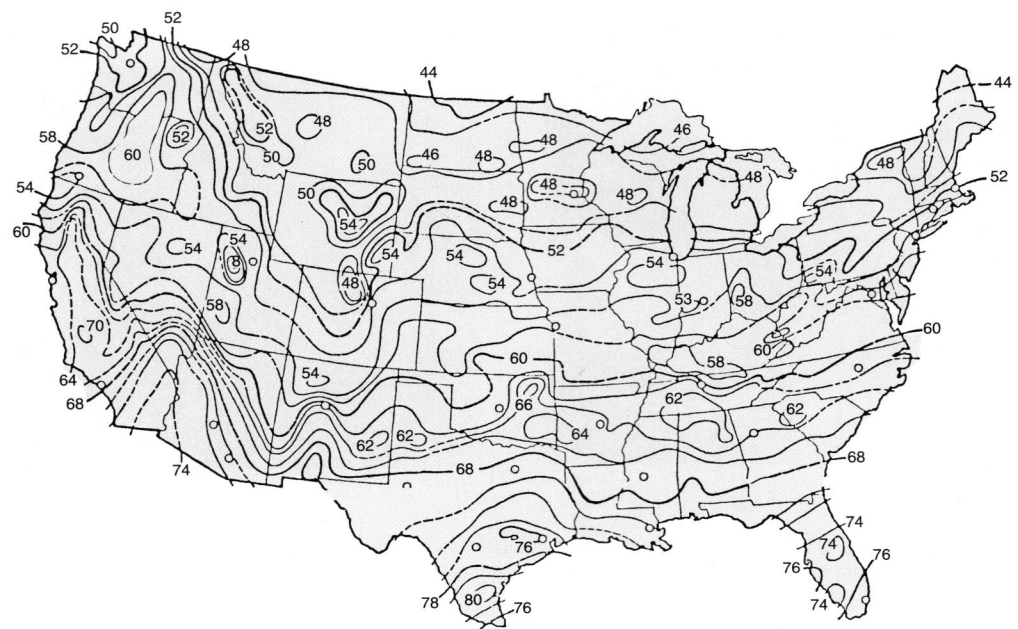

Figure 44–4 Temperature (°F) in wells ranging from 50 ft to 150 ft deep. *Courtesy Mammoth Corporation*

water flows with a significant velocity. Minerals eventually will foul the walls of the water side of the coaxial heat exchanger.

Heat exchangers of water-source heat pumps allow water to flow in one direction and refrigerant to flow in the other. This is called *counterflowing.* The water and refrigerant never come in physical contact with one another. These heat exchangers are the coiled, coaxial, tube-within-a-tube type. Water flows through the inner tube, and refrigerant flows in the outer jacket. The outer jacket is made of steel. The heat exchanger and a cross section of the heat exchanger are

shown in **Figure 44–5**. Notice that the inner tubes have extended surfaces called *leads,* which increase the surface area for better heat transfer. These leads also cause turbulence in both the water and refrigerant loops, which increases the heat transfer between the water and the refrigerant. The temperature differential (delta-T) between the entering and leaving water usually ranges from 7°F to 10°F for open-loop systems. This means that if water is entering the heat exchanger at 55°F and leaving at 47°F for a heating application, the delta-T would be 8°F. This is an acceptable delta-T because it falls between the 7°F to 10°F

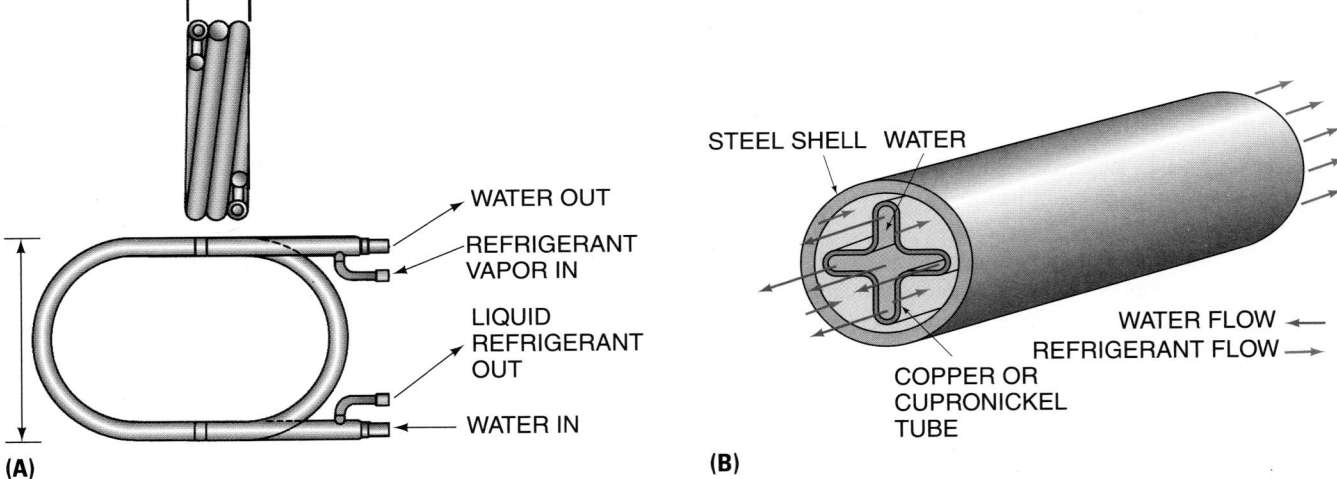

(A) WATER OUT
REFRIGERANT VAPOR IN
LIQUID REFRIGERANT OUT
WATER IN

(B) STEEL SHELL WATER
WATER FLOW
REFRIGERANT FLOW
COPPER OR CUPRONICKEL TUBE

Figure 44–5 **(A)** A coiled coaxial, tube-within-a-tube heat exchanger. The tubes are sealed in such a manner that the inside tube is separate from the outside tube. **(B)** The cross section of a counterflow, coaxial heat exchanger showing leads or extended surfaces for better heat transfer. *(A) Courtesy Noranda Metal Industries, Inc.*

MAMMOTH WATER COIL SELECTION GUIDE

POTENTIAL PROBLEM	USE COPPER COIL	USE CUPRONICKEL COIL
SCALING- Calcium and magnesium salts (hardness)	Less than 350 ppm (25 grain/gallon)	More than 350 ppm (up to sea water)
Iron oxide	Low	High
CORROSION-* pH	7 to 9	5-7 and 8-10
Hydrogen sulfide	Less than 10 ppm	10-50 ppm
Carbon dioxide	Less than 50 ppm	50 to 75 ppm
Dissolved oxygen	Only with pressurized water tank	All systems
Chloride	Less than 300 ppm	300 to 600 ppm
Total dissolved solids	Less than 1000 ppm	1000 to 1500 ppm
BIOLOGICAL GROWTH- Iron bacteria	Low	High
SUSPENDED SOLIDS-	Low	High

*Important—If the concentration of these corrosives exceeds the maximum tabulated in the cupronickel column, then the potential for serious corrosion problems exists. Water treatment may be required.

Figure 44-6 A selection guide for determining whether a copper or a cupronickel heat exchanger should be used. *Courtesy Mammoth Corporation*

range. However, always consult with the manufacturer of the heat pump for specific temperature differences; temperature differences may vary between manufacturers and geographical locations.

Most coaxial heat exchangers have inner tubes made of copper, which handle the water circuit. However, an alloy of copper is also available. This alloy is called *cupronickel*. Heat exchangers made of cupronickel have a higher corrosion resistance for acid cleaning. They also have a higher resistance to abrasion than copper heat exchangers. Because of this, the heat exchangers will have a longer service life. However, cupronickel heat exchangers will foul just as fast as a copper heat exchanger. **Figure 44-6** can be used as a guideline for determining whether a cupronickel heat exchanger should be used in an open-loop heat pump system.

44.5   CLOSED-LOOP SYSTEMS

In closed-loop heat pump systems, many yards of plastic pipe are buried in the earth. The pipe can be placed in either a horizontal or vertical configuration depending on how much land is available and what the soil configuration consists of. These loops of piping are called either *ground loops* or *water loops,* **Figure 44-7** through **Figure 44-13.**

A completely sealed and pressurized loop of water, or water and antifreeze solution, is circulated through the buried pipe in the ground by a low-wattage centrifugal pump. Heat is transferred from the ground, through the plastic pipe, to the liquid in the ground loop in the winter or heating mode. Heat is rejected away from the circulating fluid,

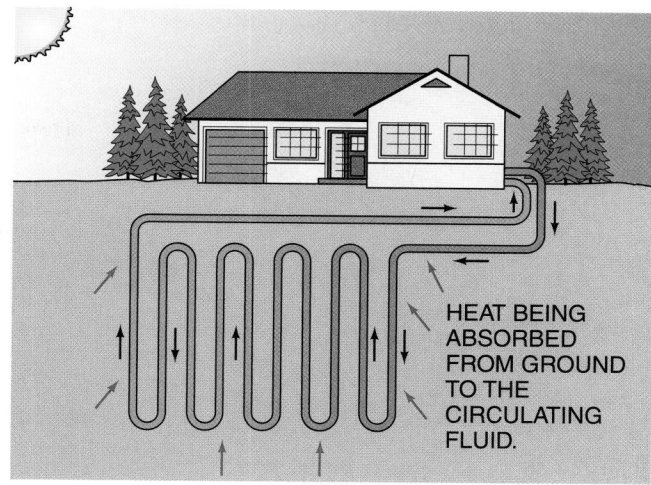

Figure 44-7 A ground loop showing a series-vertical configuration in the heating mode.

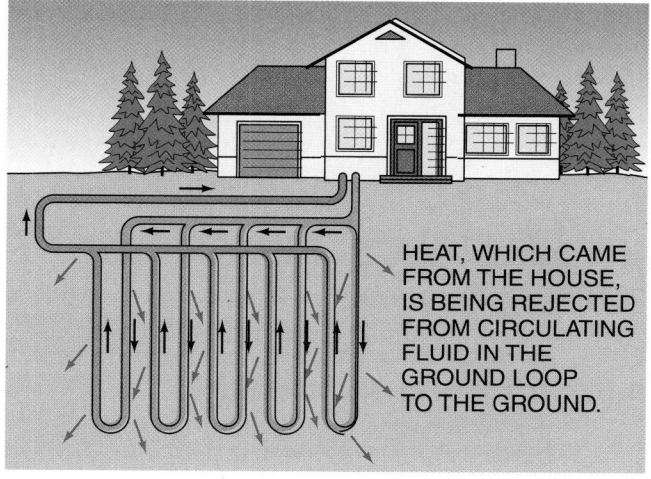

Figure 44-8 A ground loop showing a parallel-vertical configuration in the cooling mode.

Figure 44-9 A single-layer, horizontal ground loop in the heating mode.

Figure 44–10 A two-layer, horizontal ground loop in the cooling mode.

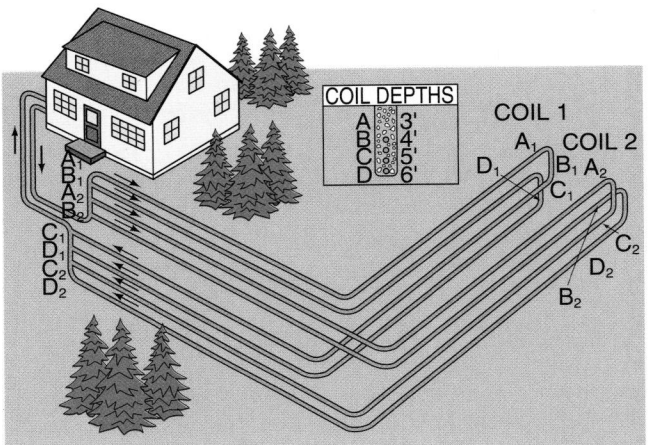

Figure 44–11 A four-pipe, horizontal ground loop.

Figure 44–12 A slinky ground loop.

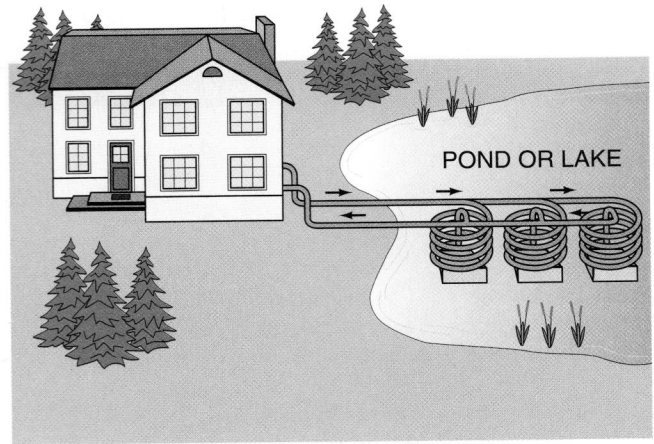

Figure 44–13 A pond or lake loop.

through the plastic pipe, and into the ground during the summer or cooling mode.

This circulating fluid in the plastic pipes or ground loop exchanges its energy with a *refrigerant loop.* The antifreeze-to-refrigerant heat exchange occurs within the heat pump's antifreeze-to-refrigerant heat exchanger, which is contained within the cabinet of the heat pump. This heat exchanger is a coiled, coaxial, tube-within-a-tube heat exchanger. No fouling of this heat exchanger occurs because of the closed-loop system using the same treated water in its loop. *This is probably one of the main advantages a closed-loop system has over an open-loop system. Water-quality problems, which often foul the water side of coaxial heat exchangers, are almost nonexistent with closed-loop systems.*

An *air loop* is used to distribute heated or cooled air to the building. This is accomplished through the use of a finned-coil, air-to-refrigerant heat exchanger located in the duct-work. A *squirrel cage blower* is used to move the air through the air distribution system. **Figure 44–14** and **Figure 44–15** illustrate the ground, refrigerant, and air loops in the heating and cooling modes respectively.

A fourth loop, or *domestic hot water loop,* is often used for heating domestic hot water from the hot discharge gas coming from the heat pump's refrigerant compressor. Because the discharge of the compressor is the hottest part of the refrigeration system, heat can be transferred easily from the hot refrigerant to the cool domestic water. However, a separate heat exchanger is again needed. This heat exchanger is also a coiled, coaxial, tube-within-a-tube heat exchanger. The domestic hot water is pressurized and circulated through the loop by a circulating pump. The refrigerant loop, which is in the outer shell of the heat exchanger, gives up heat to the cooler domestic water, which is in the inner tube of the same heat exchanger. The hot refrigerant and cool water counter-flow one another, meaning they flow in opposite directions through the heat exchanger. This helps desuperheat the refrigerant and at the same time heat domestic water. More modern systems have heat exchangers that not only desuperheat the refrigerant, but also condense it for increased heat transfer. These systems can often supply 100% of heated domestic hot water demands. **Figure 44–16** illustrates a domestic hot water heat-exchanger loop.

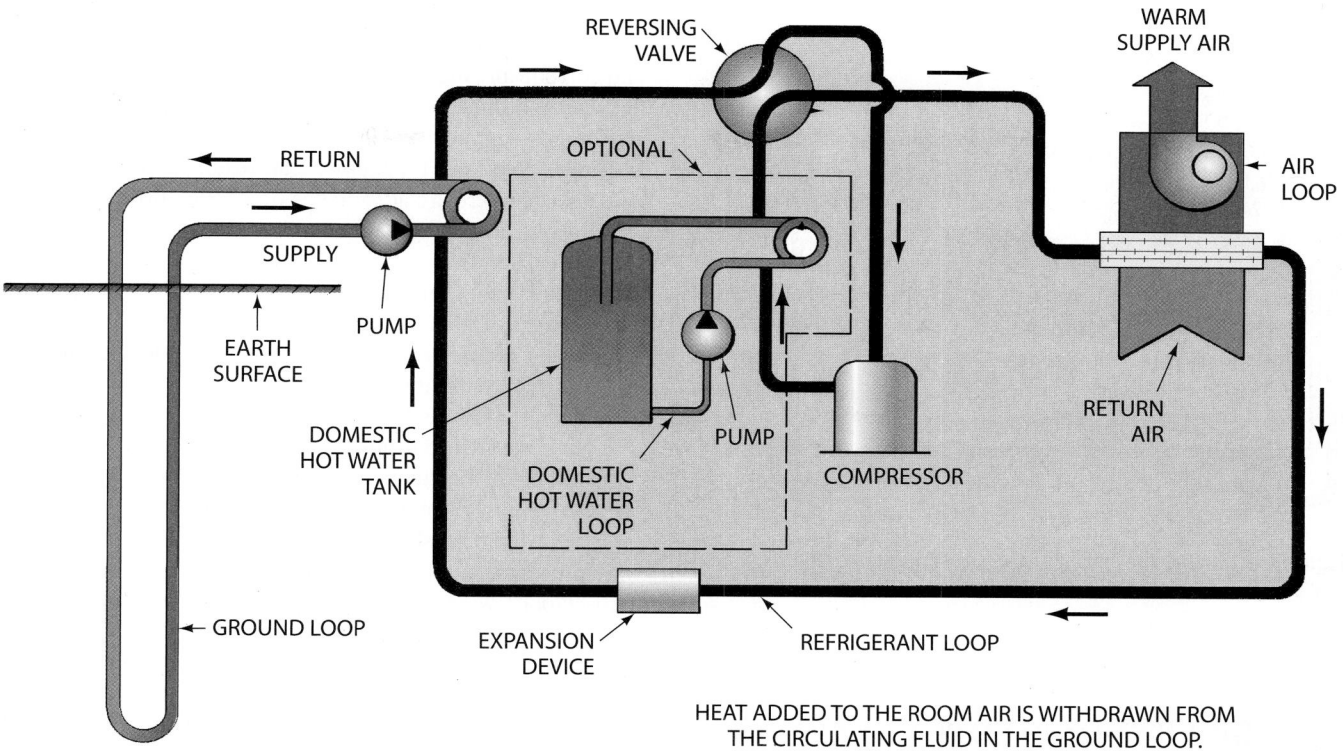

HEAT ADDED TO THE ROOM AIR IS WITHDRAWN FROM
THE CIRCULATING FLUID IN THE GROUND LOOP.

Figure 44–14   A closed-loop, water-source heat pump in the cooling mode. *Courtesy Oklahoma State University*

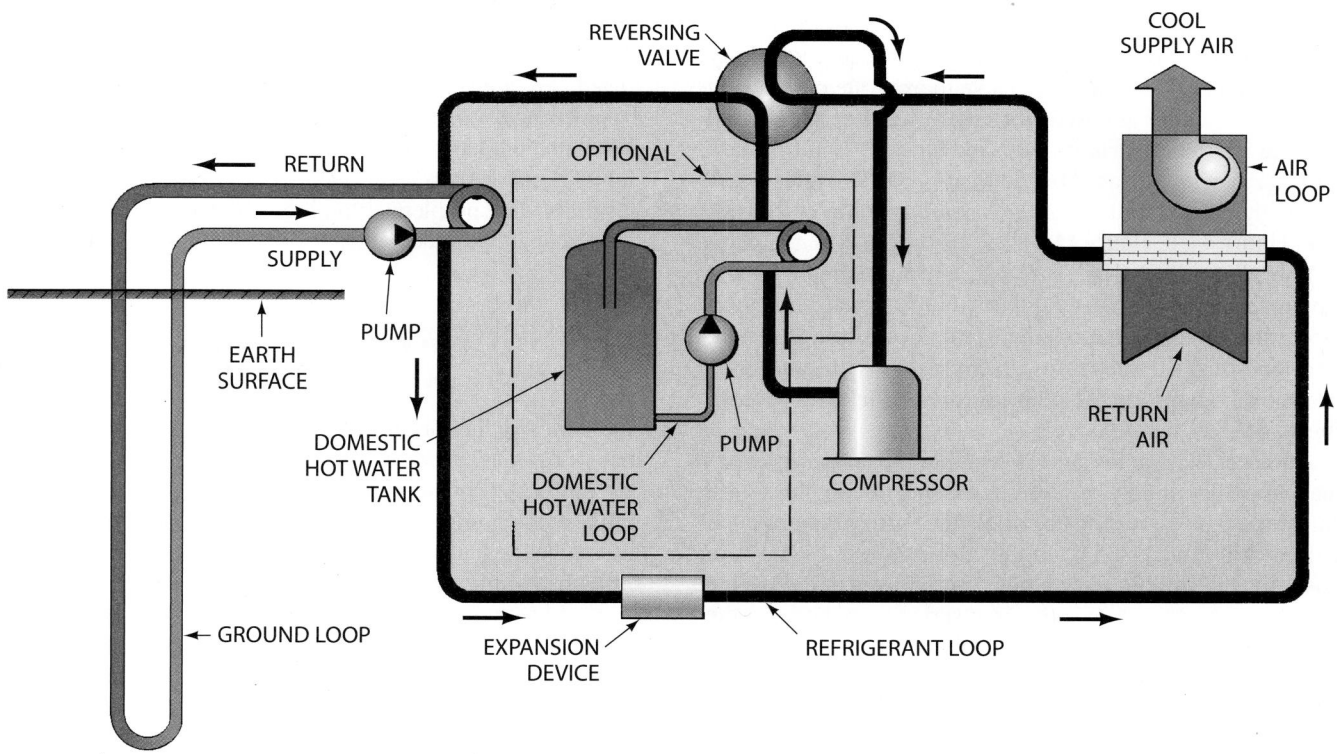

Figure 44–15   A closed-loop, water-source heat pump in the heating mode. *Courtesy Oklahoma State University*

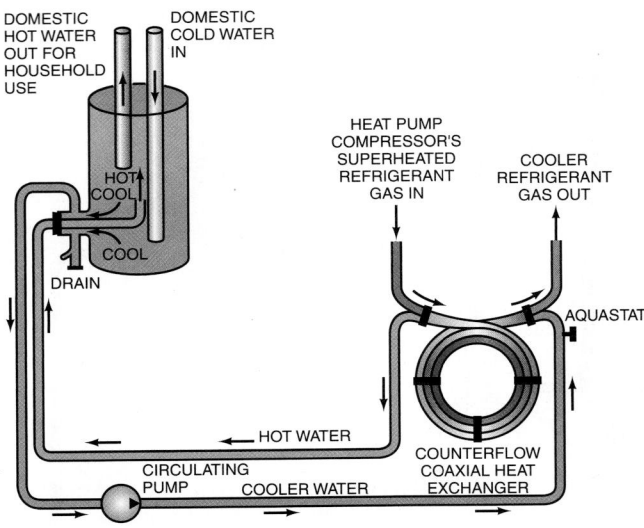

Figure 44–16 A domestic hot water loop showing a counterflow, coaxial heat exchanger. The water is heated by the refrigerant loop.

44.6 GROUND-LOOP CONFIGURATIONS AND FLOWS

Ground loops can either be a vertical, horizontal, slinky, or pond/lake type. Loops can also have series or parallel fluid flows, which will be covered later in this section. **Figure 44–7** through **Figure 44–13** illustrate these loop configurations. Loop configuration choices will depend on how much land is available. The choice also depends on the type of soil found on the land and the contour of the land. Both of these factors affect excavation and drilling costs because very hilly or rocky land can become expensive to excavate or drill.

Vertical systems are used when there is a shortage of land or space restrictions, **Figure 44–7** and **Figure 44–8**. If the soil is rocky, a rock bit is used on the rotary drill. If land is available without hard rock, a *horizontal loop* should be considered, **Figure 44–9**, **Figure 44–10**, and **Figure 44–11**.

The *slinky loop* is a flattened, circular coil of plastic pipe resembling a Slinky®, **Figure 44–12**. Because of their configuration, slinky loops can reduce the trench length from one-third to two-thirds as compared with other system loops. Loops can also be installed in lakes or ponds, **Figure 44–13**.

Another decision to make is whether the ground heat exchanger or loop should be *series* or *parallel flow*. **Figure 44–17** illustrates both series and parallel flow paths in horizontal and vertical ground loops. In the *series flow*, only one path exists for the fluid to flow along. Air trapped in the plastic pipe is much easier to remove in a series loop, because the fluid's path is very well defined. This trapped air can be removed by power flushing. Power flushing involves connecting the heat pump and ground loop to a purging unit or pump stand that has a built-in, high-volume, high-velocity, and high-head circulating pump. The pump stand or purging unit also has a built-in air separator. The air then can be purged

from the system, **Figure 44–18**. It is important that all the air is removed before the system is started, because air can erode and corrode any metal components in the water loop. Air can also cause a fluid flow blockage in the loop. Power flushing also removes any debris that may damage circulating pump bearings. A system breakdown can result in either of these cases. Listed here are advantages of a series-flow system:
- Ease of removing trapped air
- Simplified flow path
- Higher heat transfer per foot of pipe

Listed here are some disadvantages of a series-flow system:
- Larger-diameter plastic pipe is needed, meaning more antifreeze solution
- Increases installation costs
- Higher pressure drops

Parallel-flow systems can use smaller-diameter pipe because of their lower pressure drops. This smaller-diameter pipe has a lower cost associated with it. If air problems exist in the water loop, it is very difficult to remove the air with high-velocity water flushing in parallel-flow systems because of the path choices. Excess air can also cause blockage in the water flow in parallel systems. To make sure that equal pressure drops are in each parallel path, pipe lengths must be fairly equal in length. If not, the parallel path with the least resistance, or shortest length, will be fed with the most fluid flow. The path with the most resistance will be partially starved of fluid. Heat pump capacity will be seriously affected with unequal flow paths. Large-diameter headers are used at the inlet and outlets of the water loops to ensure that each loop experiences the same pressure for equal fluid flow. Listed here are advantages of a parallel-flow system:
- Smaller-diameter and lower-cost piping
- Lower installation and labor costs
- Less antifreeze required

Listed here are some disadvantages of parallel fluid flow in ground loops:
- Much harder to remove air in the ground loops because of parallel paths
- Balancing problems if piping is of unequal length or unequal resistance

44.7 SYSTEM MATERIALS AND HEAT EXCHANGE FLUIDS

The buried piping or underground heat exchanger is usually made of *polyethylene* or *polybutylene*. These two materials can be bonded or joined through a heating process, which will give the piping joints a very long life. The antifreeze solutions inside the buried piping are used to prevent freezing of the heat pump heat exchanger and for good heat transfer. A lot of heat pumps in the southern states use only pure water as their heat exchange fluid for the ground loop, because there is no threat of freezing the loop in the winter. Three choices for antifreeze solutions are:
- *Salts*—Calcium chloride and sodium chloride
- *Glycols*—Ethylene glycol and propylene glycol
- *Alcohols*—Methyl, isopropyl, and ethyl

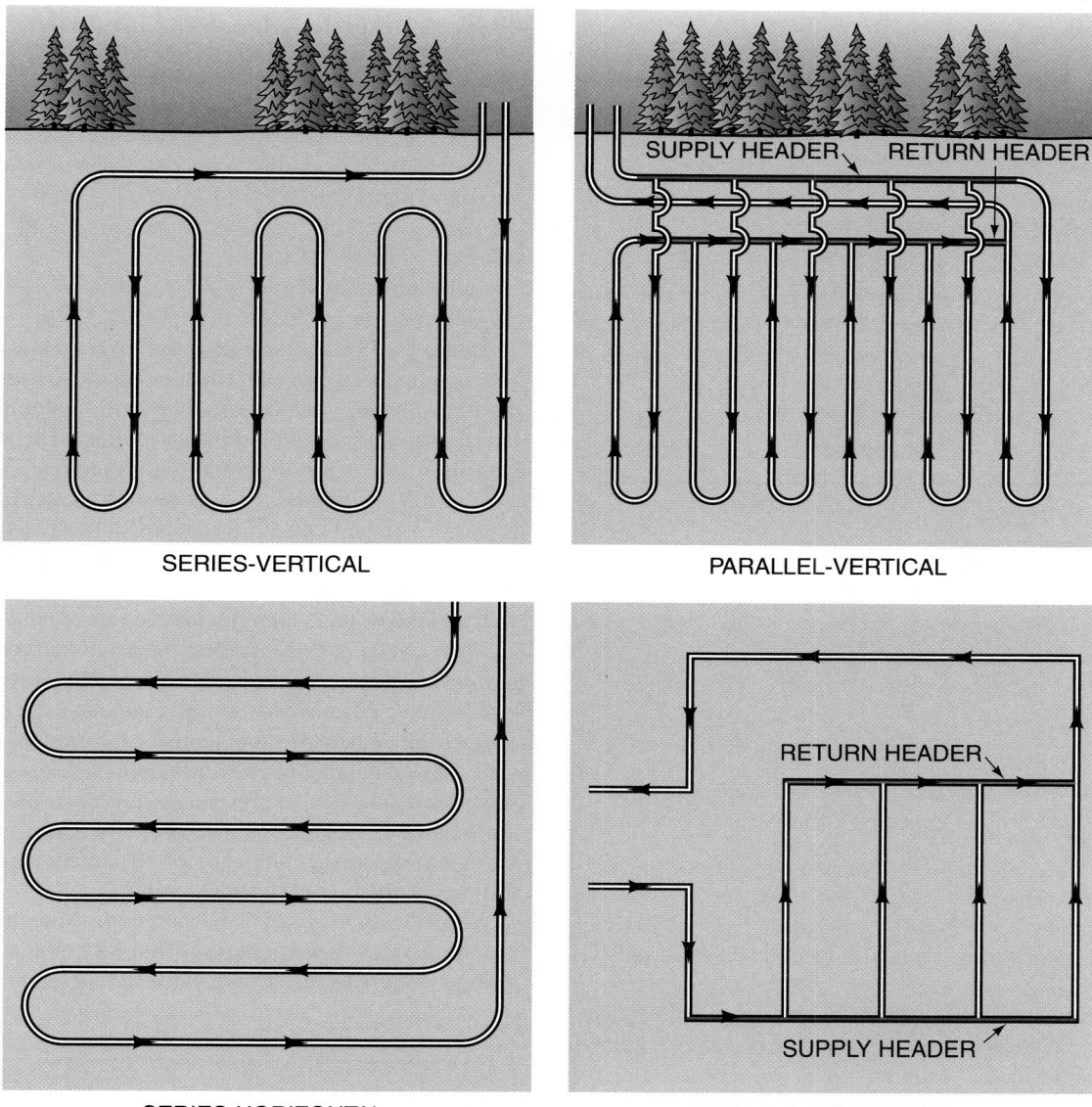

SERIES-VERTICAL

SUPPLY HEADER   RETURN HEADER

PARALLEL-VERTICAL

SERIES-HORIZONTAL

RETURN HEADER

SUPPLY HEADER

PARALLEL-HORIZONTAL

Figure 44–17   Different flow paths in ground loops.

For any fluid to be used, it must
- last a long time without breaking down.
- transfer heat readily.
- have a low cost.
- be environmentally safe.
- be noncorrosive.
- be nontoxic.

Salts are safe, are nontoxic, and transfer heat well. Salts also have low costs, are environmentally safe, and will not break down in time. Salts are, however, corrosive when in contact with some metals, especially when mixed with air. Because of this, designers must select the proper metals for system use. Installers and service technicians must also make sure that the systems remain air free if salts are to be used successfully in ground loops.

Glycols are safe, most of the time are noncorrosive, transfer heat fairly well, and also have low costs. Glycols do have a track record of turning to gel at really low temperatures, which decreases their heat transfer capabilities and labors the circulating pump responsible for their flows. Glycols have a shorter life than salts.

Alcohols can burn and are combustible if they are mixed with air. This is their major disadvantage. Alcohols are somewhat toxic, are relatively noncorrosive, transfer heat fairly well, and are relatively inexpensive to purchase. If handled properly and not mixed with air, alcohols can be a safe alternative to salts and glycols.

All system components must be carefully chosen when used in contact with either salts, glycols, or alcohols. These components include the coaxial heat exchangers,

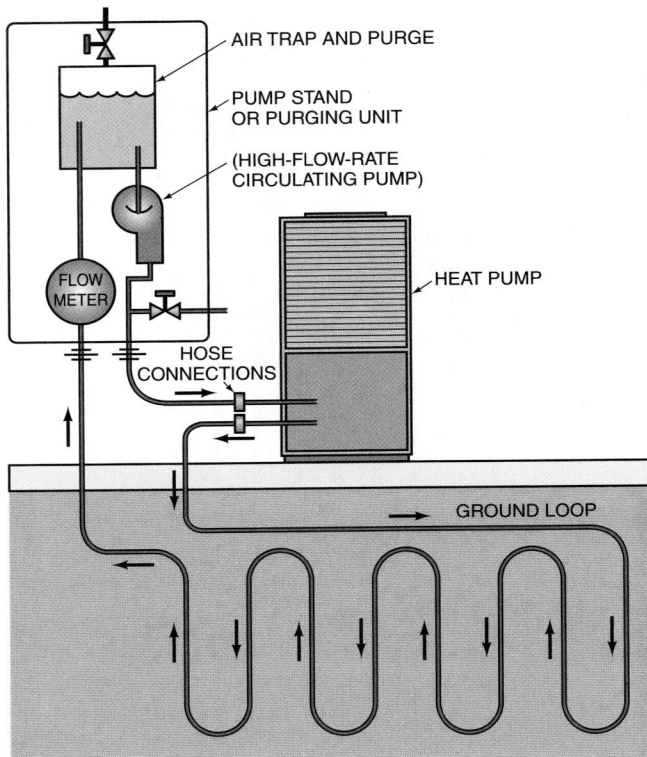

Figure 44–18   A schematic of a purging unit hooked up to a heat pump for power flushing air and debris from the ground loop.

circulating pumps and flanges, and of course all plastic piping.

Historically, the refrigerant used in water-source heat pumps was R-22. However, with the Clean Air Act giving R-22 a phaseout date of manufacture because of its chlorine content—and ozone-depleting potential—chlorine-free environmentally friendly alternative refrigerants will soon be used in place of R-22. R-410A is a leading alternative refrigerant to R-22 in air-conditioning and heat pump systems. However, other alternative refrigerants and refrigerant blends have entered the market and may also be substitutes for R-22. Refer to Unit 3, "Refrigeration and Refrigerants," for more detailed information on alternative refrigerants and refrigerant blends.

Beginning in the year 2010, there can be no importing and no production of HCFC-22 (R-22), except for use in equipment manufactured before January 1, 2010. Beginning in the year 2015, there can be no production and no importing of any HCFC refrigerant, except for use as a refrigerant in equipment manufactured before January 1, 2020. In the year 2020, there will be a total production and importing ban on HCFC-22. Beginning in the year 2030, there will be a total production and importing ban on all HCFC refrigerants. However, existing systems with HCFC refrigerants will still be in use. Also, recovered and reclaimed HCFC refrigerants will still be legal to use.

44.8   GEOTHERMAL WELLS AND WATER SOURCES

Water sources for open-loop systems may be an existing well or a new well. A well pump delivers the water from the well to the heat pump. Some popular well categories are the following:

- Drilled well
- Return well
- Geothermal well
- Dry well

Figure 44–19 shows a *drilled well*. Notice that an electric water pump, complete with electrical lines, is encased underground within the well casing. This type of pump is often referred to as a submersible well-water pump. The water pump gets its water from an underground aquifer. Other water sources may be a pond, lake, or swimming pool. The water is then discharged into a lake, stream, or marsh. **Figure 44–20** illustrates an open-loop, geothermal heat pump system in which a water pump delivers water to the heat pump from a well and the water is then discharged to a pond.

Most wells for geothermal heat pump systems are *grouted.* Grouting is a procedure in which a cement-like material is injected between the well casing and the hole drilled for the well. When the grout hardens, it forms a seal and prevents any contamination between other water sources in the same local area. It also prevents rain or other miscellaneous surface water from seeping into the well and polluting it. Grouting also makes the entire well structure sturdier and can prevent rusting of the well casing.

Return wells are used to discharge the water back into the ground after it has experienced the heat pump's heat exchanger, **Figure 44–21.** Return wells should be spaced at a far

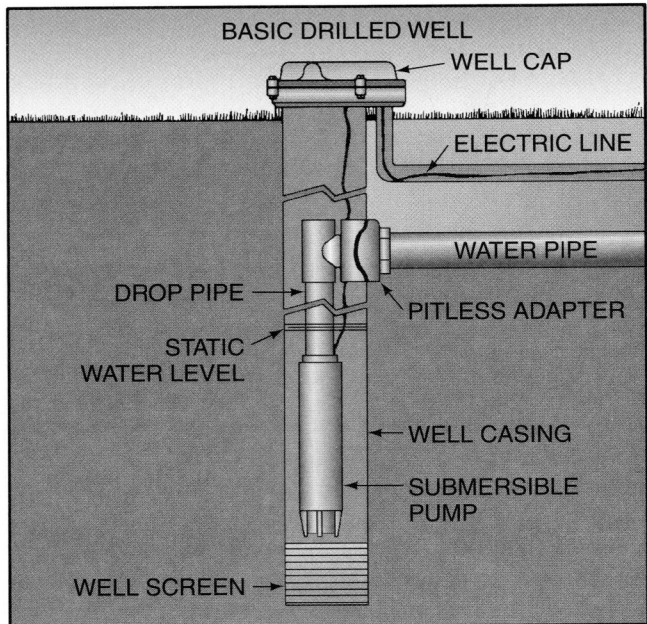

Figure 44–19   A basic drilled well. *Courtesy Mammoth Corporation*

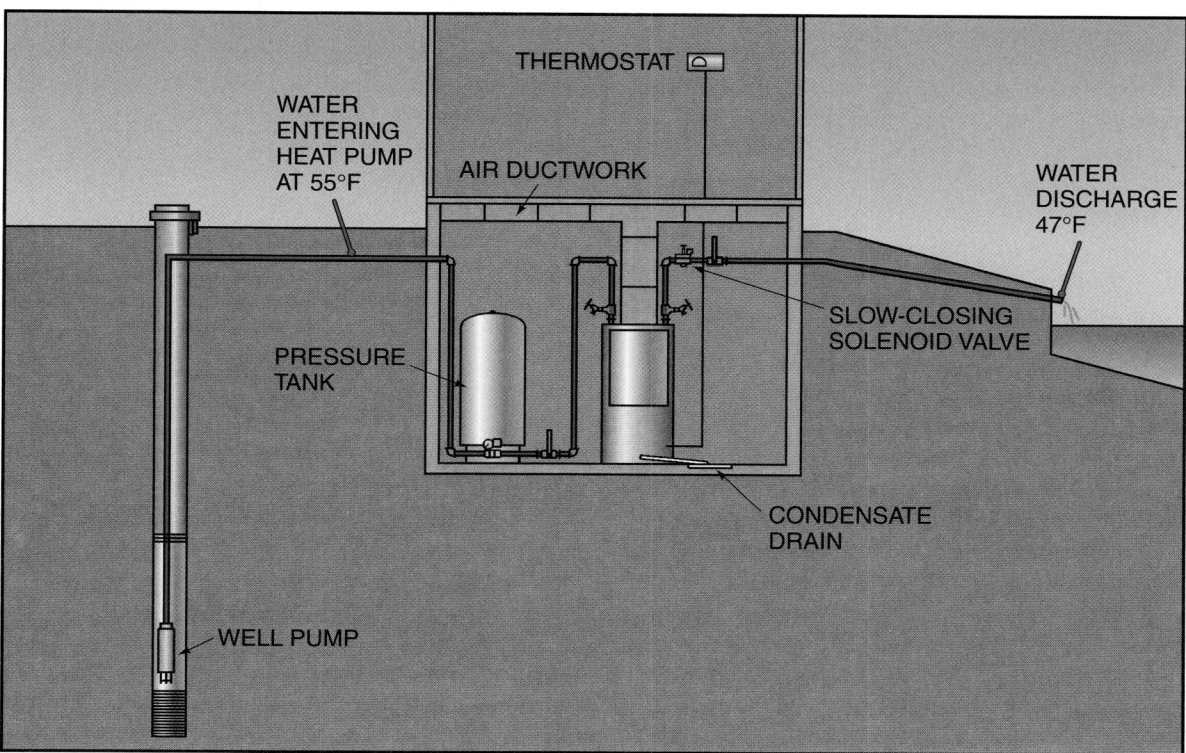

Figure 44–20 Lake, stream, pond, or marsh discharge for an open-loop geothermal heat pump in the heating mode. *Courtesy Mammoth Corporation*

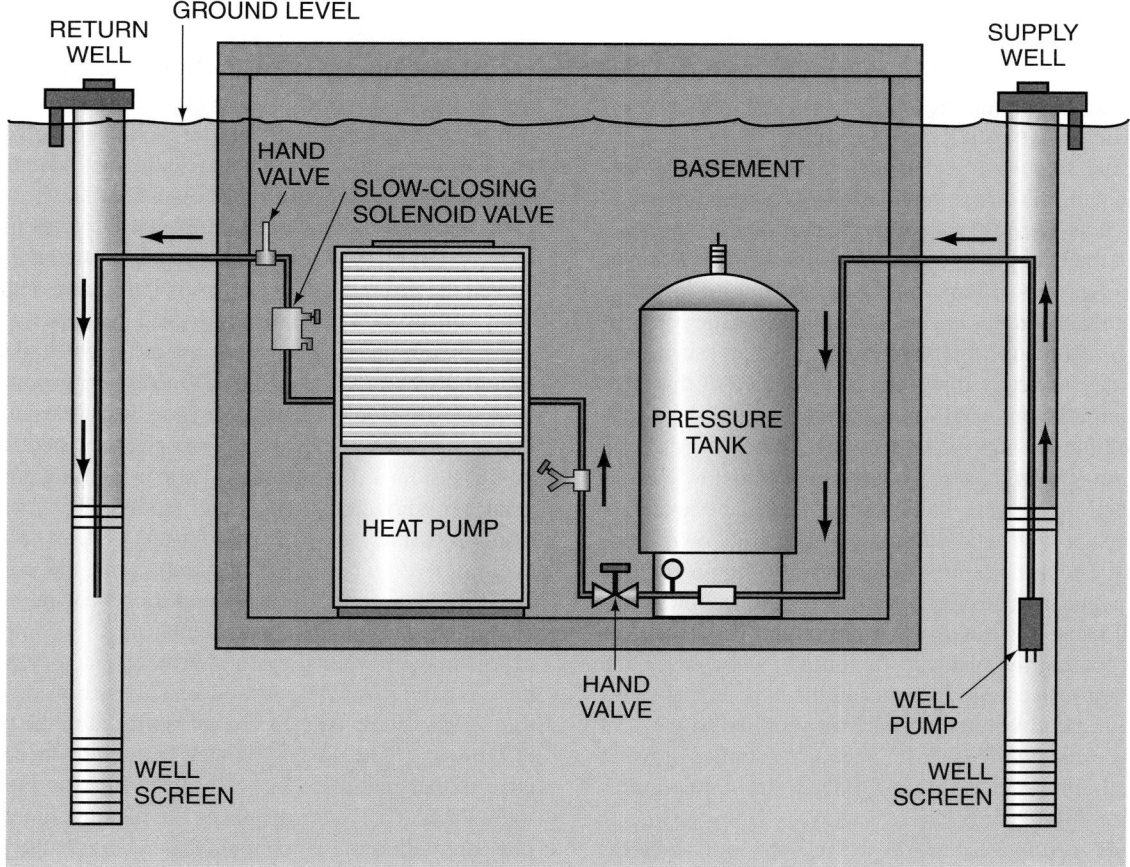

Figure 44–21 A return well system.

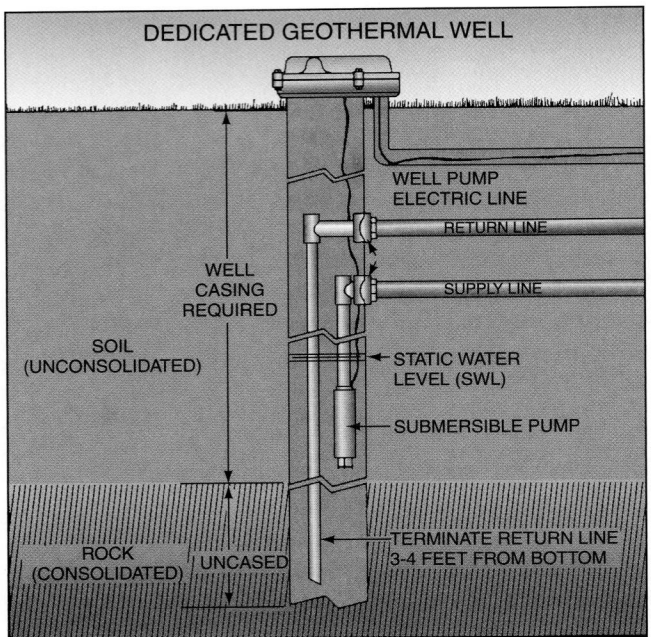

Figure 44–22 A dedicated geothermal well. *Courtesy Mammoth Corporation*

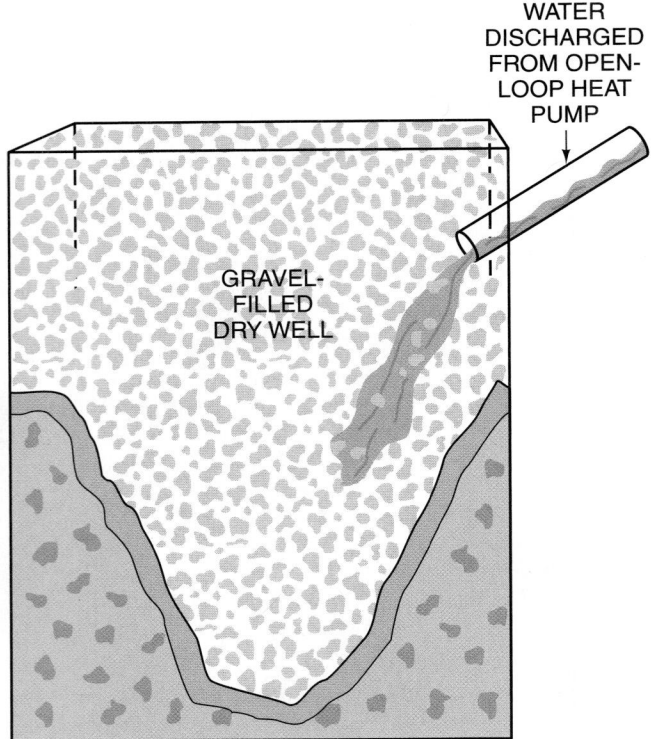

Figure 44–23 A dry well.

enough distance from the supply well to prevent early mixing of the supply and return water. The return well must be at least the size of the supply well to handle the water flow. If early mixing of supply and return water occurs, the supply water's temperature could be increased or decreased depending on the season. This could seriously affect the capacity of the heat pump. Supply and return wells should have at least 100 ft of spacing between them. Notice in **Figure 44–22** that a slow-closing *solenoid valve* is installed in the return line. The slow closing of this solenoid valve prevents a pressure pulse, known as *water hammer,* from occurring each time it opens when the heat pump cycles on. These solenoid valves are almost always positioned in the return line to keep the coaxial heat exchanger in the heat pump at the same pressure as the *pressure tank* during the off cycle. This pressurization helps keep any minerals dissolved in the water so that they will not precipitate out and foul the heat exchanger. Minerals are more soluble in water as the pressure increases. Also, the return or drop pipe in the return well must end below the static water level in the well. Static water level is the level water will rise in a well when it naturally seeks its own level. This keeps the return pipe free of air and helps to prevent the growth of algae and bacteria within the return line. This growth will cause restricted water flows and decreased capacities by coating the inside of the line. In some cases, the return well may discharge its water in the supply well. This ensures that a good volume of water is supplied to the heat pump's heat exchanger. However, adequate supplies of water must be available to ensure that supply water temperatures are not seriously affected.

Dedicated geothermal wells are closed-loop systems that draw water from the top of the water column, circulate it through the heat pump where it either absorbs or rejects heat

energy, and then return the water at a different temperature to the bottom of the water column, **Figure 44–22.** The water discharged to the bottom of the well will be at its normal temperature before it is drawn to the top of the well. Dedicated geothermal wells are used where there is not enough water in the underground aquifer to use other standard well systems.

A *dry well* is used for the discharge water in an open-loop system. Dry wells are nothing more than a large reservoir in the ground filled with gravel and sand, **Figure 44–23.** They are usually used in sandy soil conditions. The used water is filtered as it seeps through the gravel and sand and it eventually makes its way back to the underground aquifer.

A pressure tank is used on most well systems in houses today. It is also used extensively on open-loop geothermal heat pump systems. The pressure tank is nothing more than a small pressurized tank for water storage. The main purpose of the pressure tank is to prevent the well pump from short cycling. Even when the well pump is off, the pressure tank can supply water when the open-loop heat pump calls for heating or cooling. Pressure tanks have a compressed air charge in them when they come from the factory. Some tanks have a fitting for the addition or subtraction of air to customize a system. This air charge is placed in the tank on top of the water. The air is separated from the water in the tank by a rubber bladder. The well pump fills the pressure tank with water. The water displaces the bladder, creating higher pressure within the air charge. This increases the pressure in the pressure tank. When an operating maximum pressure is reached, the well pump automatically will shut off by a

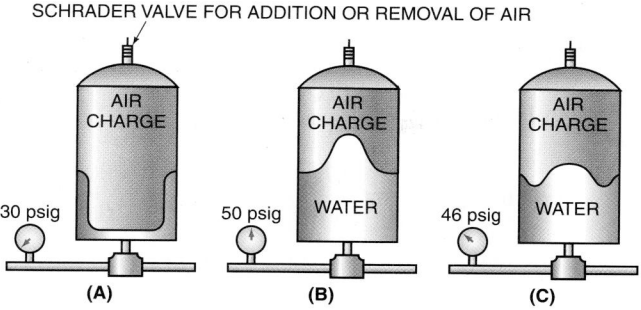

(A) FACTORY AIR CHARGE.

(B) WELL PUMP HAS PRESSURIZED AIR CHARGE AND WATER PRESSURE IS 50 psig. THE WELL PUMP WILL NOW SHUT OFF BECAUSE PRESSURE SWITCH HAS OPENED ON A RISE IN PRESSURE.

(C) WHEN WATER IS USED BY THE HEAT PUMP, THE PRESSURE IN THE AIR CHARGE PUSHES WATER INTO THE SYSTEM. THE PUMP STAYS OFF. THE WELL PUMP ONLY COMES ON WHEN PRESSURE SWITCH CLOSES ON A DROP IN PRESSURE.

Figure 44–24 The operation of a well system's pressure tank.

pressure switch. The well pump will remain off until the pressure in the tank reaches a minimum pressure. The pump is then cycled on, and the process is repeated. The pressure tank should be sized so that the well pump does not start any more often than one time every 10 min. **Figure 44–24** illustrates the operation of a pressure tank.

44.9 WATER-TO-WATER HEAT PUMPS

Up to this point in this unit, our discussion has primarily been geared toward the water-to-air heat pump system. Now we will examine another heat pump configuration that is gaining popularity at a surprisingly fast rate: the water-to-water heat pump, **Figure 44–25**. In the past, the applications for water-

to-water systems were rather limited given the fact that the attainable water temperatures were rather low for many heating applications. For an R-22 heat pump system, water can be heated to a temperature as high as 130°F; for an R-410A system, the water temperature will not likely be higher than 120°F. Given the increase in popularity of radiant heating systems that rely on lower-temperature water than do conventional baseboard heating systems, the water-to-water heat pump has become a viable option for providing hot water at a reasonable cost.

Water-to-water heat pumps utilize two coaxial heat exchangers, as compared with only one on the water-to-air configuration, **Figure 44–26**. These water-to-water systems can be configured as either open-loop or closed-loop systems. When operating in the heating mode, the coaxial heat exchanger that is linked to the ground loops is functioning as the evaporator, while the heat exchanger that is heating the water is operating as the condenser. In order to ensure continued satisfactory operation of the system, the water flow through the condenser must be unrestricted. Restricting the water flow through the condenser will cause the head pressure to rise, possibly causing the system to shut off on safety. This could pose a problem, especially since the heat pump system is often used to control more than one heating circuit. When only one or two zones are calling for heat, a high head-pressure situation may arise. For this reason, it is very common to see a buffer tank installed on the condenser water side of the water-to-water heat pump system, **Figure 44–27**.

Figure 44–25 A water-to-water heat pump installed as part of a hydronic heating system.

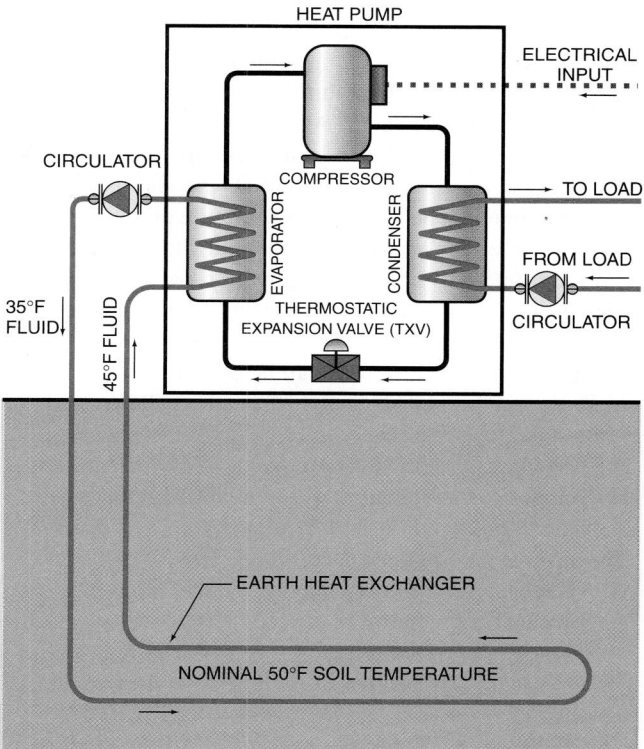

Figure 44–26 Heat exchanger configuration on a water-to-water heat pump system.

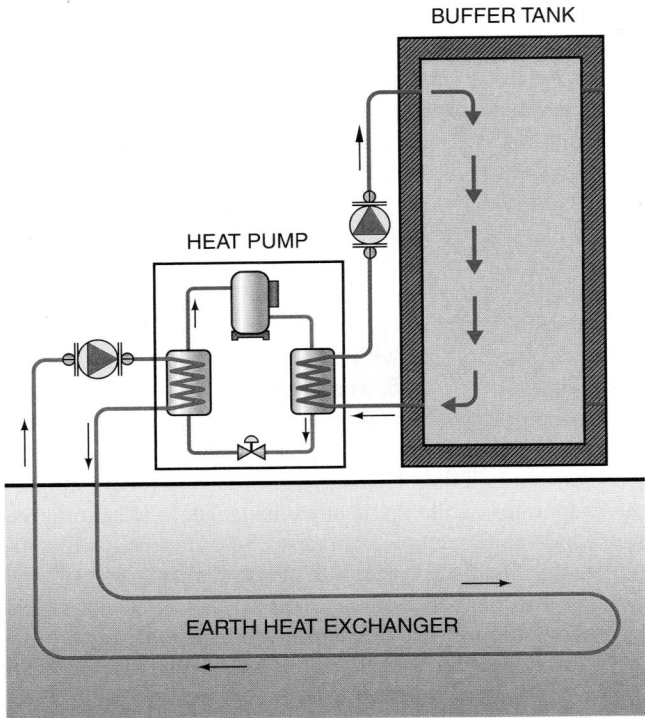

Figure 44–27 Buffer tank location on a water-to-water heat pump system.

The buffer tank is nothing more than a large tank that both receives water from the condenser and supplies water to the condenser. The heat pump will cycle on and off on the basis of the temperature of the water in the tank. The cut-in temperature is the temperature at which the heat pump will cycle on; the cut-out temperature is the temperature at which the heat pump will cycle off. As with other controls, the difference between the cut-in and cut-out temperature is the differential. The smaller the differential, the shorter the heat pump run time will be—but the temperature of the water will vary very little. The larger the differential, the longer the heat pump run time will be. The longer run-time option is more efficient because more power is consumed when the system first starts up.

The size of the buffer tank can be determined by crunching some numbers. The factors that affect the size of the buffer tank are as follows:

■ The capacity of the heat pump, $Q_{\text{heat pump}}$
■ The desired temperature differential (ΔT) of the tank water
■ The heat load on the space, Q_{space}
■ The desired run time of the heat pump, T

The formula for the buffer tank volume is

$$\text{Volume}_{\text{buffer tank}} = \frac{T(Q_{\text{heat pump}} - Q_{\text{space}})}{\Delta T \times 500}$$

Let's assume that we have a heat pump with a 24,000-Btu capacity and that we want the temperature of the water in the buffer tank to range from 90°F to 105°F. There is also a 10,000-Btu load on the space and we want the heat pump to

run for 10 minutes. From this information, we get the following:

$$\text{Volume}_{\text{buffer tank}} = \frac{T(Q_{\text{heat pump}} - Q_{\text{space}})}{\Delta T \times 500}$$

$$\text{Volume}_{\text{buffer tank}} = \frac{10 \times (24,000 - 10,000)}{(105°F - 95°F) \times 500}$$

$$\text{Volume}_{\text{buffer tank}} = 140,000 \div 5,000$$

$$\text{Volume}_{\text{buffer tank}} = 28 \text{ Gallons}$$

By using a tank that is larger than calculated, the run time of the heat pump can be increased.

The buffer tank functions as the water supply tank for the radiant heating system. There is one pump that moves the water from the condenser to the tank and from the tank to the condenser. The second pump moves water from the buffer tank to the hot water heating circuits and from the heating circuits back to the tank, **Figure 44–28.** As long as the heating zones are calling for heat, the pump that carries the heated water into the heating zones will be operating. This does not mean, however, that the heat pump is operating, as the heat pump will cycle on and off to maintain the desired water temperature in the buffer tank. As more and more zones are calling for heating, the temperature of the water returning to the buffer tank will be lower, so the heat pump will cycle on for a longer period of time to maintain temperature.

44.10 ▶ TROUBLESHOOTING

Troubleshooting geothermal heat pumps is much like servicing the air-to-air heat pumps covered in Unit 43. However, one of the minor differences is how to gain access to pressures and temperatures in the water or ground loop. **Figure 44–29(A)** shows pressure and temperature port locations and mountings on the water or ground loop. **Figure 44–29(B)** illustrates a pressure and temperature port adapter used in performance testing the heat pump. The temperature probe is used for measuring the temperature difference between the inlet and outlet of the water's coaxial heat exchanger. The pressure gage is used for determining the flow rate through the heat exchanger by acquiring the pressure drop across the heat exchanger. A curve or chart of flow in gpm versus pressure drop has to be used in conjunction with the pressure measurements. The refrigerant system is accessed just like any other heat pump system. Troubleshooting is very similar to troubleshooting any other refrigeration or air-conditioning system. Even the electrical system is very similar to other refrigeration and air-conditioning systems. Because of this, this troubleshooting section will cover mainly the water loop or ground loop of geothermal heat pump systems. If a review of the refrigeration system (refrigerant loop) or airflow system (air loop) is needed, review Unit 36, "Refrigeration Applied to Air Conditioning," Unit 37, "Air Distribution and Balance," Unit 41, "Troubleshooting," and Unit 43, "Air Source Heat Pumps."

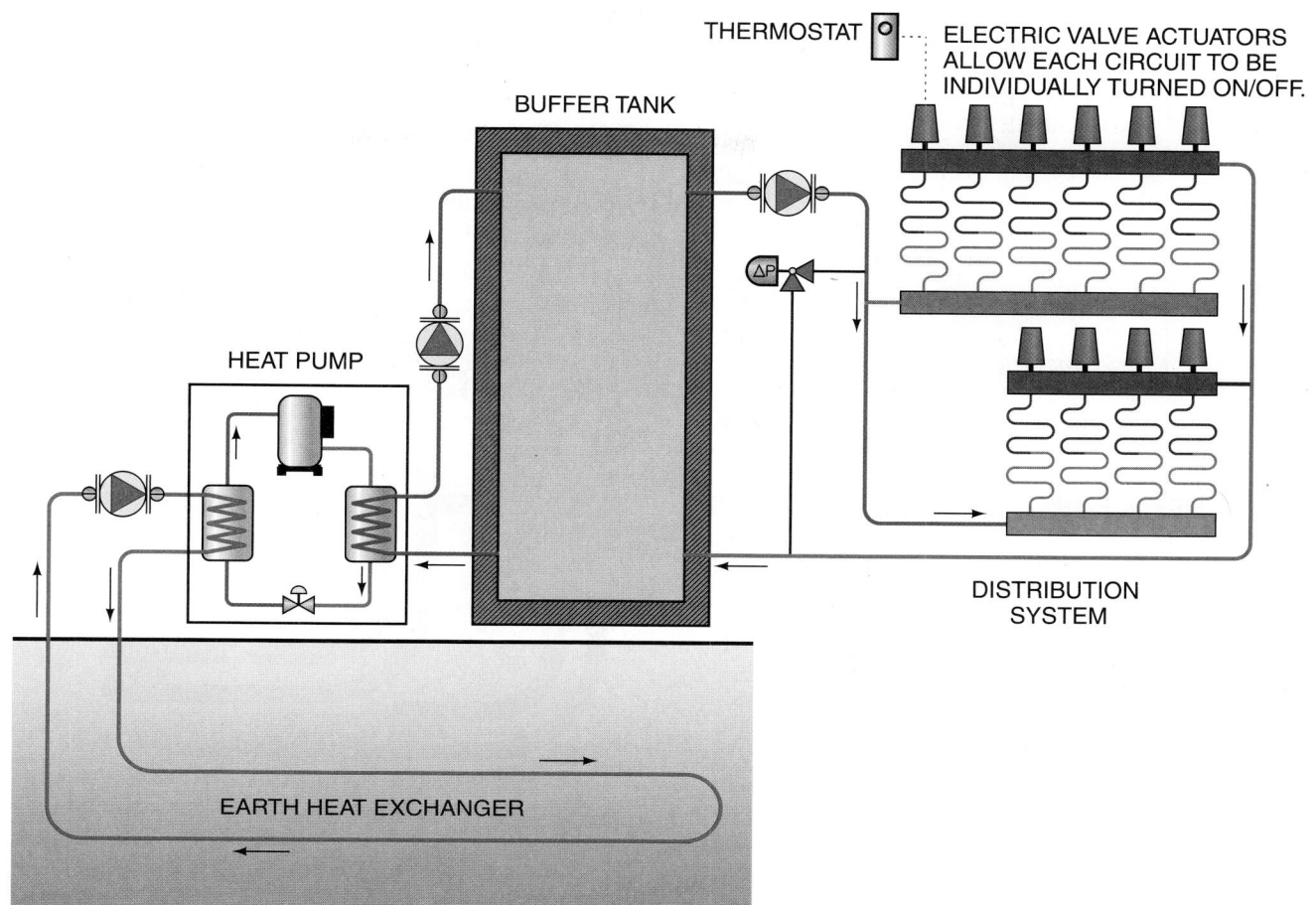

Figure 44–28   Piping connections between (1) the heat pump and the buffer tank and (2) the buffer tank and the heating circuits.

The *ground loop,* which is sometimes referred to as the water loop, consists of the antifreeze solution, which is contained in the plastic pipes buried within the earth or lake in closed-loop systems. In an open-loop system, the water comes from a well and is discarded back to the earth in some way. In either case, a water-to-refrigerant heat exchange takes place in the coaxial heat exchanger within the geothermal heat pump.

The *governing formula* that determines the amount of heat absorbed in the heating mode or the amount of heat rejected in the cooling mode is

Heat quantity in Btu/h = (gpm) × (Temp. Diff.) × (500)

gpm = Gallons per minute of antifreeze circulated.

Temp. Diff. = Difference in temperature between water in and out of the heat exchanger.

500 = Accounts for specific heat of fluid circulated, the weight of one gallon of water (8.33 pounds), and includes a time factor to convert from gpm to Btu/h. 500 is used for pure water. However, this number can vary depending on the specific heat of the antifreeze solution. 485 is usually used for an antifreeze solution.

The two variables in the preceding equation are the gpm of antifreeze solution and the temperature difference of the antifreeze solution in and out of the coaxial, water-to-refrigerant heat exchanger. If either of these two variables change, the heating and/or cooling capacity of the heat pump will be affected. Flow rates of antifreeze in the ground loop can be measured by flowmeters installed in the pipes or by the pressure drop across the coaxial heat exchanger. If the pressure drop technique is used, a curve of pressure drop versus flow rate in gpm must be used.

For example, a low gpm flow rate of antifreeze can be caused from a bad circulating pump, air restriction in the piping, contamination or a kinked pipe in the closed loop, or a low water supply pressure in the open-loop system. If the heat pump is in the *heating mode,* the refrigerant side of the coaxial heat exchanger is the evaporator, and low suction pressure will be noticed in the refrigeration loop. This causes low evaporating temperatures. Because of the reduced flow rates in the water side of the coaxial heat exchanger, the water will stay in contact longer with a colder coil. This will increase the temperature difference within the water. So, the service technician will notice a lower suction pressure and a higher temperature difference between water in and water out of the heat exchanger.

If the heat pump is in the *cooling mode,* the refrigerant side of the coaxial heat exchanger is now the condenser, and the ground loop or water loop is the condensing medium. A

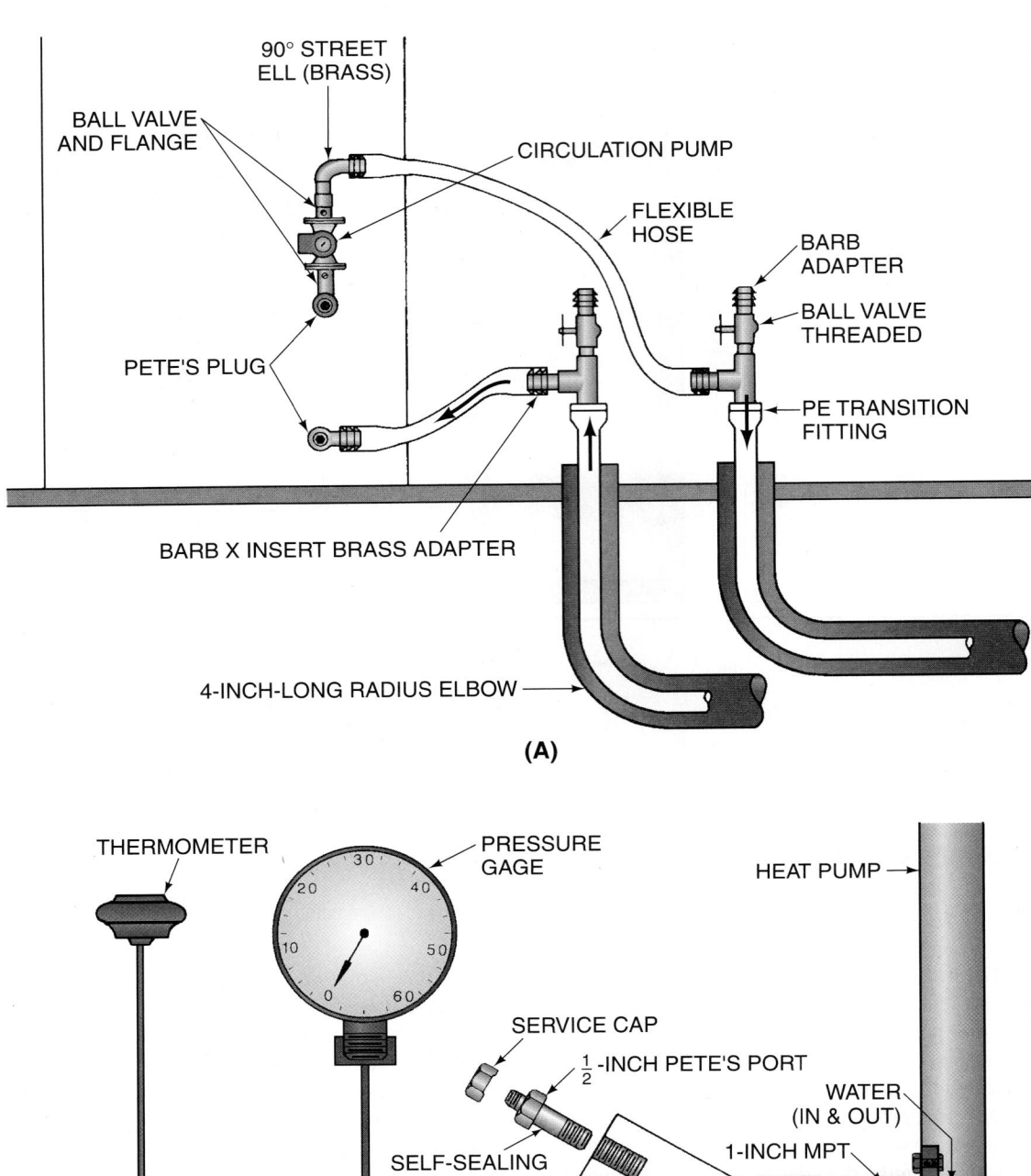

Figure 44–29 **(A)** Hookup from ground loop to heat pump. **(B)** Pressure- and temperature-sensing adapters for ground-loop access. *Courtesy Oklahoma State University*

reduced flow of water will cause high head pressures in the refrigerant loop. An increase in the temperature difference of the water in and out of the heat exchanger will also be seen. This is because the water stays in contact with a hotter coil for a longer time.

If minerals in the water of an open-loop system foul and restrict the coaxial heat exchanger's surface on the water side, poor heat transfer will take place between the water and the refrigerant. This is because the scale will act as an insulator. This will cause a low temperature difference of water in and out of the heat exchanger in both the heating and cooling mode. This results in a high head pressure in the cooling mode and low suction pressure in the heating mode. Closed-loop systems hardly ever have this problem because of the same, treated water/antifreeze solution that is circulated throughout the loop.

If for some reason the loop is designed too short on a closed-loop system, low incoming water or antifreeze solution temperatures coming into the heat pump will be experienced in the heating mode. This will cause low suction pressures in the refrigerant loop. In the cooling mode, high incoming water or antifreeze solution temperatures coming into the heat pump will be experienced, which will cause high head pressures. These same symptoms will occur if the loop is in the wrong soil type or if poor thermal contact exists between the soil and the water loop.

44.11 WATERLESS, EARTH-COUPLED, CLOSED-LOOP GEOTHERMAL HEAT PUMP SYSTEMS

As a review, the traditional, closed-loop geothermal heat pump system consists of an antifreeze or water solution circulated through plastic pipes buried in the ground. This solution, which is circulated by a small liquid centrifugal pump, acts as a heat transfer fluid. The plastic pipes containing the heat transfer fluid are often referred to as the *ground loop* or *water loop,* **Figure 44–7.**

In the heating mode, heat is transferred from the ground, through the plastic pipe, to the ground loop. The circulated heat transfer fluid within the plastic pipes then exchanges its heat energy with a *refrigerant loop.* This heat exchange occurs within the heat pump's antifreeze-to-refrigerant heat exchanger contained within the cabinet of the heat pump. The antifreeze-to-refrigerant heat exchanger is usually a coiled, coaxial, tube-within-a-tube heat exchanger, **Figure 44–5. Figure 44–14** illustrates the entire traditional, closed-loop geothermal heat pump system. These systems are often referred to as a *two-step* or *indirect* heat exchange with the earth because the antifreeze or water solution is a secondary refrigerant. The heat added to the conditioned space is withdrawn from the circulating fluid in the ground loop by the refrigerant in the refrigerant loop. Various ground-loop configurations for traditional, closed-loop geothermal heat pump systems are illustrated in **Figure 44–7** through **Figure 44–13.**

Waterless Systems

The waterless, earth-coupled, closed-loop geothermal heat pump system consists of small 3/8- to 7/16-in. polyethylene plastic–coated copper pipes buried in the ground with refrigerant flowing through them, **Figure 44–30.** The copper pipes act as the evaporator where the refrigerant experiences a phase change from liquid to vapor. This is often referred to as the *waterless earth loop, phase-change loop,* or *refrigerant loop.* For clarification purposes, this loop will be referred to as the waterless earth loop for the rest of this unit. In the heating mode, heat is absorbed from the earth "directly" to the vaporizing refrigerant. The refrigerant vapor then travels through the reversing valve to the compressor, where it is compressed and superheated to a higher temperature. The superheated, heat-laden hot gas from the compressor then

Figure 44–30 Polyethylene plastic–coated pipe being readied to be buried in a trench. *Courtesy CoEnergies LLC, Traverse City, Michigan*

(A)

SUPPLY DUCTWORK

LIQUID LINE
GAS LINE

FINNED-
TUBE
AIR COIL

RETURN
AIR

(B)

Figure 44–31 **(A)** A finned-tube air coil heat exchanger. **(B)** A finned-tube air coil heat exchanger located in the ductwork of the conditioned space or building. **(A)** *Courtesy Carrier Corporation*

travels to the condenser. The condenser is a refrigerant-to-air, finned-tube heat exchanger located in the ductwork of the conditioned space or building, **Figure 44–31.** It is often referred to as the *air coil.* Heat is then rejected to the air as the refrigerant condenses at a higher temperature than the air. A

HEATING-MODE SCHEMATIC

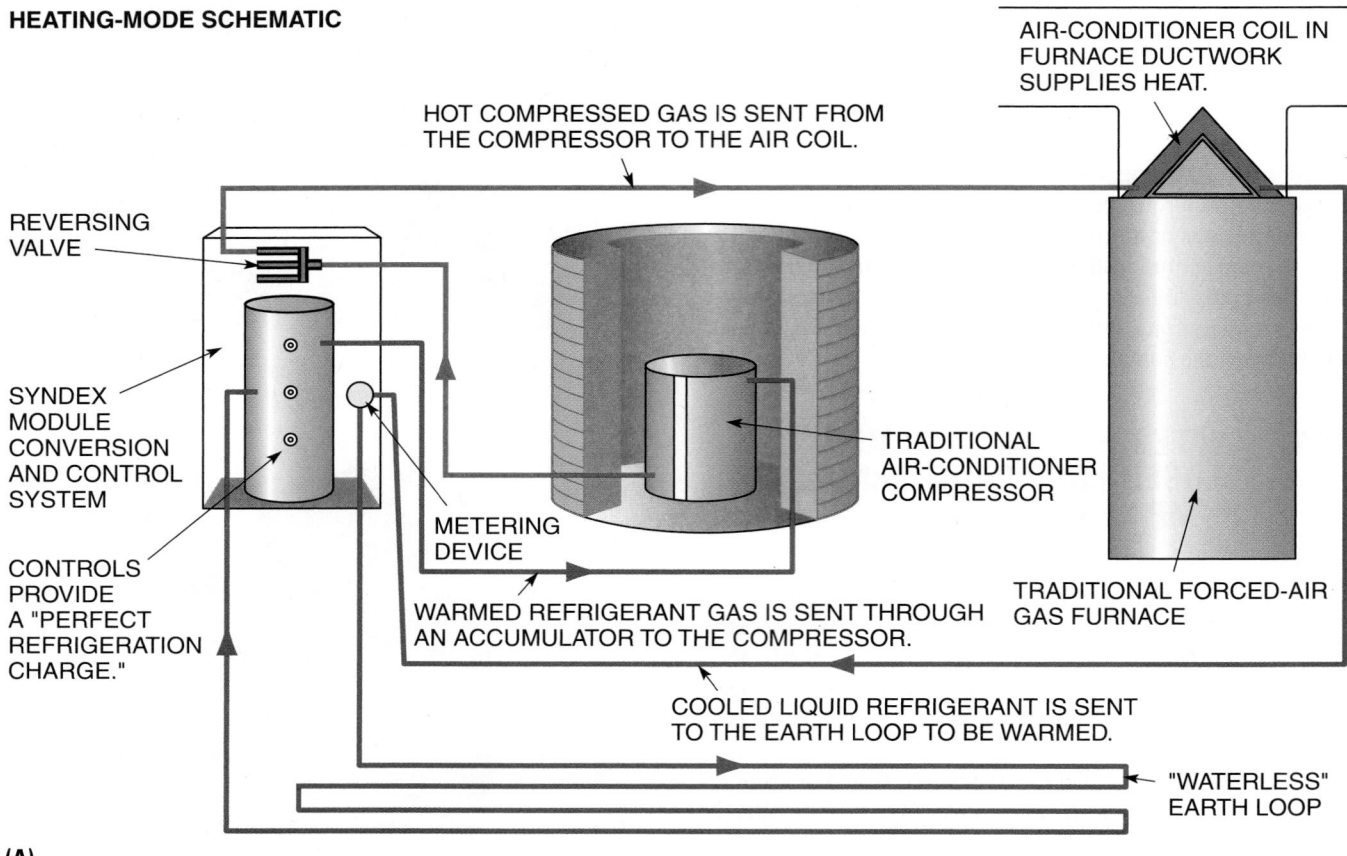

Figure 44–32   (A) A heating-mode schematic.

fan delivers the heated air to the conditioned space. The condensed liquid then travels through the expansion valve and vaporizes in the evaporator or waterless earth loop buried in the ground. The process is then repeated. Both heating and cooling schematics of a waterless, earth-coupled, closed-loop geothermal heat pump system are shown in **Figure 44–32.**

There is no water or antifreeze solution acting as a heat transfer fluid in these systems. The waterless earth loop or evaporator, which is buried in the earth, extracts heat directly from the earth. This is often referred to as a *direct* or *one-step* heat exchange with the earth. This eliminates the coaxial heat exchanger, the liquid centrifugal pump, antifreeze or water heat transfer fluids, and the need for system flushing common to traditional closed-loop, indirect geothermal heat pump systems. Because of this, the thermal efficiency of the heat pump is improved.

The waterless earth loop phase changes liquid refrigerant to vapor and takes advantage of the *latent heat of vaporization* of the refrigerant being used. The waterless earth loop also absorbs heat directly from the earth, which allows for higher efficiencies. The buried waterless earth loop is not utilized in the cooling mode, and the conventional air conditioner's outside condenser takes its place.

The waterless, earth-coupled, ground-loop heat pump system can be retrofitted to a common home furnace gas-heating

and conventional air-conditioning system, **Figure 44–33.** These systems are easily retrofitted to the average home. Once retrofitted, the home's conventional air-conditioning outside condenser coil and fan is not used for the heating mode. However, the condenser coil will be used during cooling, as before the conversion, but will operate with a perfect or critical refrigerant charge because of a refrigerant management system which will be mentioned shortly.

Home energy costs for heating and cooling can be reduced as much as 50% when retrofitted to an earth-coupled system. This is especially true when fossil fuels are used exclusively. Because the system uses a precise refrigerant management module that will eliminate the common problems of an improperly charged air-conditioning system, the cooling system's performance will be greatly improved. The improved heating and cooling efficiencies result in overall energy savings for almost every geographical region of North America.

The refrigerant management module is installed between the outside air conditioner and the home's furnace. It manages the refrigerant circulated in the buried waterless earth loop, **Figure 44–33.** This buried waterless earth loop, combined with a conventional furnace, air-conditioning compressor, and inside air coil, provides geothermal heating in the winter months and cooling in the summer months. The

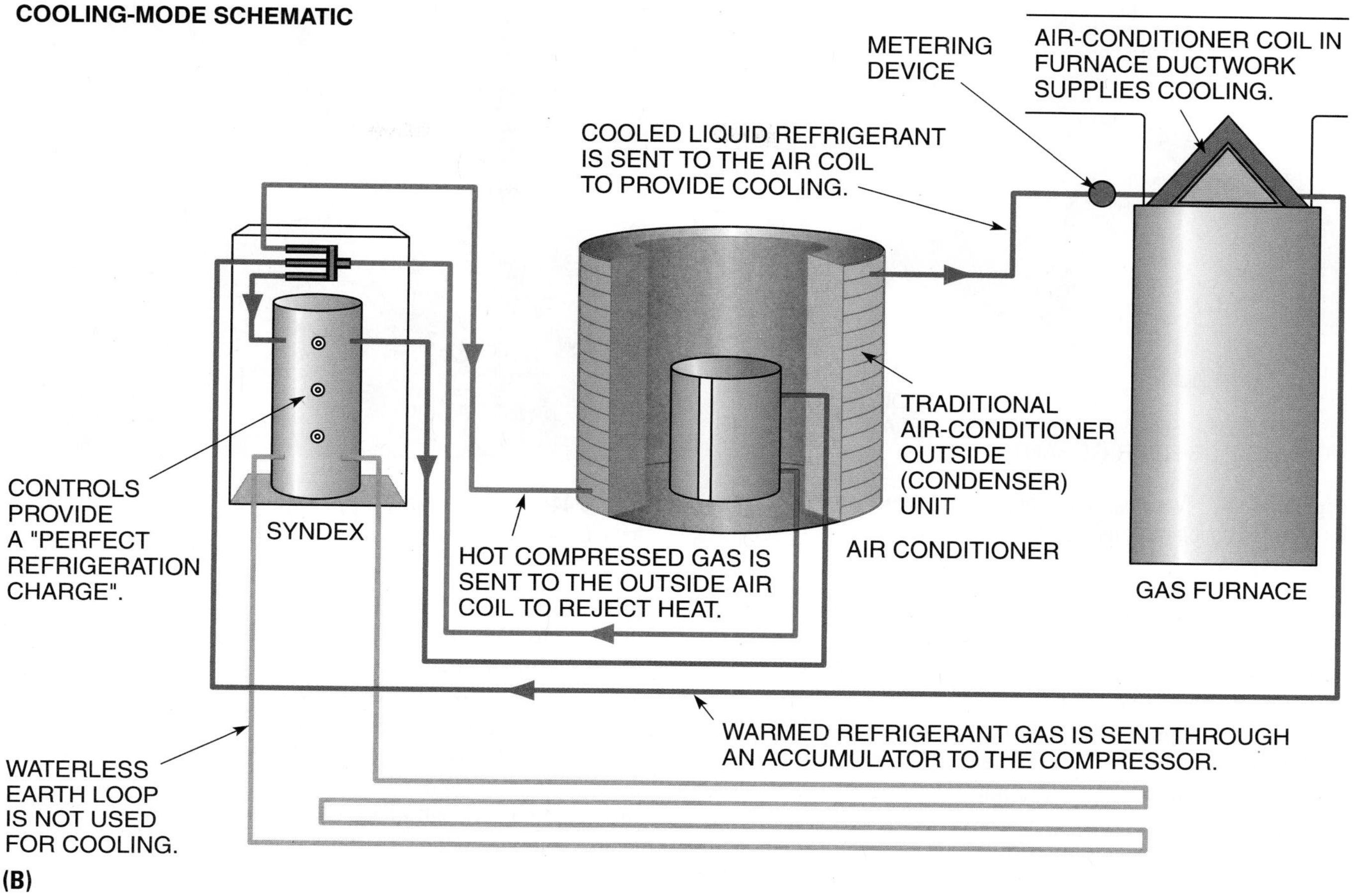

COOLING-MODE SCHEMATIC

METERING DEVICE

AIR-CONDITIONER COIL IN FURNACE DUCTWORK SUPPLIES COOLING.

COOLED LIQUID REFRIGERANT IS SENT TO THE AIR COIL TO PROVIDE COOLING.

CONTROLS PROVIDE A "PERFECT REFRIGERATION CHARGE".

SYNDEX

TRADITIONAL AIR-CONDITIONER OUTSIDE (CONDENSER) UNIT

HOT COMPRESSED GAS IS SENT TO THE OUTSIDE AIR COIL TO REJECT HEAT.

AIR CONDITIONER

GAS FURNACE

WARMED REFRIGERANT GAS IS SENT THROUGH AN ACCUMULATOR TO THE COMPRESSOR.

WATERLESS EARTH LOOP IS NOT USED FOR COOLING.

(B)

Figure 44–32 *(continued)* **(B)** A cooling-mode schematic. **(A)** *and* **(B)** *Courtesy CoEnergies LLC, Traverse City, Michigan*

REFRIGERANT MANAGEMENT MODULE

Figure 44–33 A retrofitted system showing the refrigerant management module. *Courtesy CoEnergies LLC, Traverse City, Michigan*

system responds automatically to the thermostat inside the home.

The home's conventional furnace can be used as second-stage or emergency heating with the waterless, earth-coupled, closed-loop geothermal heat pump providing first-stage heat.

The furnace will not operate at the same time as the heat pump system because once second-stage heat is turned on, the heat pump will be locked out until the thermostat is satisfied.

Installation and Refrigerant-Loop Piping

Installation costs for the retrofit are generally less than the installation of a comparable traditional, stand-alone, earth-coupled geothermal heating and cooling system. By using the one-step or direct heat exchange process between the phase-changing refrigerant and the earth, the home-owner's existing conventional air-conditioning compressor acts as the refrigerant pump and heat generator. All other heat energy to and from the system is directly transferred heat energy.

As mentioned earlier, the waterless earth loop buried in the ground consists of 3/8- to 7/16-in. copper pipe covered with polyethylene plastic. The polyethylene plastic pipe has two small vents or grooves that allow the pipe to act as a vented double-wall heat exchanger, **Figure 44–34.** Any refrigerant or oil leaks in the buried copper piping can vent to the grooves in the plastic coating, which will be vented to the surface. This surface venting allows for ease of leak detection and protects the ground from refrigerant and oil

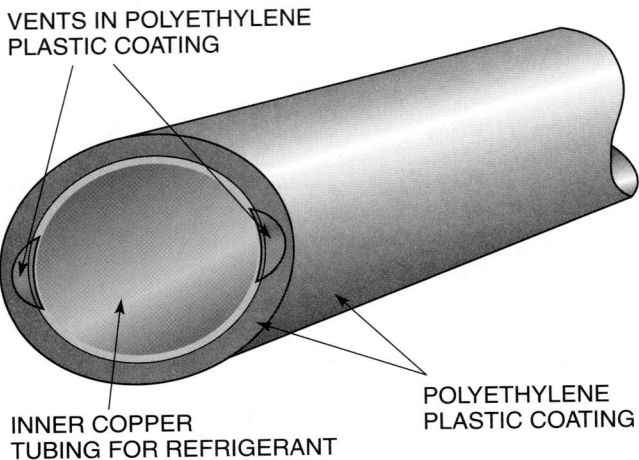

VENTS IN POLYETHYLENE
PLASTIC COATING

POLYETHYLENE
PLASTIC COATING

INNER COPPER
TUBING FOR REFRIGERANT

Figure 44–34   Two small vents, or grooves, allowing the pipe to act as a vented double-wall heat exchanger.

Figure 44–35   A trencher digging the trench for the waterless earth loop. *Courtesy CoEnergies LLC, Traverse City, Michigan*

contamination, making it a more environmentally conscious system. Each copper pipe loop is approximately 70 to 100 ft long and is buried in trenches 3 to 4 ft deep. A trencher is used to dig the trench for the polyethylene plastic–coated copper pipe, **Figure 44–35,** and the piping is then covered with sand for good thermal efficiency. The trenches are then backfilled and restored to their natural grade. These shallow trenches are generally quicker to install than traditional closed-loop, indirect, geothermal heat pump systems containing antifreeze or water in their ground loops. Each buried waterless earth loop is a one-piece circuit without any buried joints. The number of loops depends on the capacity of the system and the soil composition.

Each waterless earth loop circuit's supply and return ends of pipe are brazed to a copper header that is contained in an easily accessible header box, **Figure 44–36(A–D).** In the unlikely event that a leak should occur, the header is easily accessible through the header box. The individual refrigerant

loops in separate trenches all meet at the header box or header pit. The circuits are soldered together at the header pit. Remember, there are no joints or connections anywhere below the ground. The two supply and return lines are then insulated and routed to the air conditioner, **Figure 44–37.** The refrigerant lines of the existing conventional air conditioner are tapped into and diverted to the header pit, where the refrigerant can be circulated through the waterless earth loop buried in the ground.

44.12   SERVICE TECHNICIAN CALLS

The following service calls will deal with both open- and closed-loop systems. They are some of the most common problems experienced with geothermal heat pump systems. Notice that a problem with one loop will usually affect the performance of the other loops in the heat pump.

(A) The entire assembly is pressure tested to ensure that there are no leaks.

(B)

Figure 44–36   **(A)** A copper header to which the waterless earth supply and return ends are brazed. **(B)** A copper header being brazed to supply and return loops by a technician.

(C)

(D)

Figure 44–36 *(continued)*   **(C)** Copper headers showing supply and return ends of piping. **(D)** A header box being positioned over a copper header. *(A)–(D) Courtesy CoEnergies LLC, Traverse City, Michigan*

LINES ARE INTERRUPTED AND REROUTED TO THE LOOP FIELD. IN HEATING MODE, THE REVERSING VALVE ROUTES THE REFRIGERANT TO THE LOOP FIELD. IN COOLING MODE, THE REFRIGERANT IS ROUTED TO THE EXISTING AIR COIL ASSEMBLY.

Figure 44–37   Insulated supply and return lines routed to the retrofitted air conditioner. *Courtesy CoEnergies LLC, Traverse City, Michigan*

SERVICE CALL 1

A rural homeowner reports that his heat pump seems to be running much too often. The homeowner also has noticed that his monthly electric bills have risen steadily in the past two months. It is the heating season, and the heat pump is an open-loop system that gets its water from a drilled well located near the house. The heat pump's refrigerant loop contains R-22. *The problem is that the water side of the coaxial heat exchanger has been fouled with mineral deposits. This will not allow a good heat exchange between the water loop and the refrigerant loop.*

A service technician turns the space thermostat to a high setting in an attempt to keep the unit running while troubleshooting. The technician installs gages on the refrigerant loop and notices that the suction pressure is running at 35 psig. The normal suction pressure for that particular system should be about 55 psig for this geographical region with an entering-water temperature of 47°F. The technician then measures the temperature difference between the inlet and outlet of the water through the heat exchanger. The technician finds that there is only a 4°F temperature drop across the heat exchanger. The normal temperature drop for this geographical region is usually 7°F to 10°F. A combination of the low suction pressure and a low temperature drop in the water indicates that a reduced heat transfer is taking place between the refrigerant loop and the open water loop. This could be caused from minerals in the water, which have built up and fouled the heat exchanger on the water side. This explains the high monthly electric bills. Because a good heat exchange could not take place between the water and the refrigerant, the heat pump's capacity was reduced. This forced the system to rely on electric strip heaters to meet heating capacity demands of the home. The technician then tells the homeowner that the heat exchanger will have to be chemically cleaned and maybe replaced.

SERVICE CALL 2

Homeowners are complaining that their heat pump keeps shutting down automatically. Power has to be manually switched off and then back on to reset the electrical lock-out relay. The heat pump is a closed-loop system with two circulating pumps hooked in parallel in the water loop. It is the heating season, and the heat pump's refrigeration circuit runs on R-22. *The problem is that one of the two parallel pumps in the water loop is not operating, causing low flow rates.*

The service technician turns the space thermostat to a high setting in an effort to keep the heat pump running. The heat pump runs for about 5 min and then shuts down on its lock-out relay. After turning the power off, the technician uses an ohmmeter and determines that the heat pump's low-pressure control (LPC) is opening, which causes the lock-out relay to hold the circuit open. The technician lets the heat pump stay off for about 20 min and analyzes the situation. The technician then quickly restores power to the heat pump. This resets the lock-out relay, and the heat pump runs again.

The technician then measures the temperature difference between the water in and out of the refrigerant-to-water heat exchanger. The temperature drop is 9°F instead of the normal 5°F for this closed-loop system. This indicates that the antifreeze solution in the ground loop is staying in contact with the heat exchanger longer because of the slower flow rate, causing an increased temperature drop in the solution. A lower suction pressure on the refrigerant loop is noticed after gages are installed.

The service technician then examines the circulating pumps and determines that one of the two parallel pumps is not running. The antifreeze flowmeter indicates that there is only 7 gpm of fluid being circulated instead of the normal 14 gpm. This explains the increase in temperature difference of the antifreeze solution through the heat exchanger. The technician then takes a voltmeter to the motor terminals of the circulating pump that is not operating. The correct line voltage of 115 V is measured at the terminals. This means that the motor is defective. The technician then shuts down the heat pump and replaces the circulating pump's motor with a new one. The power is turned back on, and the system runs normally.

SERVICE CALL 3

A store owner is complaining of hardly any cooling and a reduced airflow coming from the registers in the building. The owner also has to manually reset the heat pump's electronic lock-out relay by turning the power off and then back on about eight times daily. It is the cooling season.

The problem is that the heat pump's air loop air filter is plugged with lint and dust. This prevents the correct air volume from entering the air coil.

A service technician arrives, installs gages on the unit, and notices that the suction pressure is 35 psig. This is a very low suction pressure for an R-22 system. The normal suction pressure should be about 60 to 65 psig for that geographical region. The technician also checks out the airflow problem and agrees with the store owner that something is restricting the airflow. The technician takes a current reading of the air coil's fan motor and finds it to be 3 A. The nameplate says the motor should be pulling around 9 A. This low amperage indicates that the fan is not moving enough air, and the fan motor is only partially loaded. The technician then pulls out the air filter located in the return air just before the air coil and notices that it is almost completely filled with lint and dust. However, even with the air filter pulled, a restricted airflow problem still exists, low suction pressure still exists, and the fan motor continues to pull low current. The technician then decides to have a look at the air coil. The power to the unit is shut off, and the plenum to the heat pump is taken off. The air coil is completely covered with a blanket of ice and frost. The technician then melts the iced coil with a large blow drier while the power is turned off.

The technician then explains to the building owner that a dirty air filter has caused a restricted airflow to the air coil. This restriction in the airflow has caused a low suction pressure because of the reduced heat load from the building entering the air coil. The low suction pressure made the refrigerant flowing through the air coil's tubes colder. This finally froze the coil solid. The low suction pressure was tripping the LPC, and the unit was being locked out by its lock-out relay. This is why the store owner would have to reset the unit about eight times daily. The restricted airflow also unloaded the fan motor causing it to draw low current. The technician then explains to the store owner the importance of keeping the air coil filters clean and also tells the building owner that it is not wise to reset the heat pump any more than once before calling a service technician. After putting the plenum back on the unit and installing a clean air filter, the technician starts the heat pump. The proper airflow has been established, and the suction pressure is normal at 65 psig. The fan motor is also drawing normal current.

SERVICE CALL 4

A homeowner is complaining about his heat pump running continuously in the heating mode. Another complaint is that his winter electric bills are unusually high. When the homeowner enters the basement where the heat pump is located, a crackling sound can be heard in the water loop as it enters the circulating pump. The ground loop is a parallel-vertical bore, **Figure 44–8**. *The problem is that a flange leak at the inlet of the circulating pump has let air enter the water or ground loop and has blocked one of its* parallel paths. This actually shortens the loop and causes colder return-water temperatures—which decreases the heat pump capacity.

A service technician arrives at the scene and carefully listens to the crackling sound within the water in the ground loop as it flows through the circulating pump. The technician also notices that the electric strip heaters are on. This is what has been causing the high electric bills. The technician then carefully questions the homeowner about what type of system configuration the water loop consists of. The homeowner tells the service technician that the water loop configuration is a parallel-vertical bore system. The service technician then measures the temperature difference between the water in and the water out of the coaxial heat exchanger. The measured temperature difference is only 2°F. The temperature difference should be 4°F to 5°F for that geographical region. This indicates to the service technician that the water is coming back to the heat exchanger too cold. One loop of the parallel system is being blocked by air, causing the entire loop to be shortened. This severely decreases the capacity of the heat pump and explains why the electric strip heaters are coming on, causing high electric bills. This would also explain the low temperature difference across the heat exchanger and the crackling sounds at the circulating pump.

The technician determines that the ground loop needs to be power flushed to rid it of air. The system's circulating pump is valved off, and a new flange is installed on the circulating pump to prevent any more air from entering the system. A portable pump cart or purging unit is brought in and hooked up to the ground loop and heat pump. The purging unit power flushes the loop and heat exchangers. The air that was trapped in the ground loop is purged from the top of the pump cart, **Figure 44–18**. The system is then hooked back up to the normal circulating pump and put back into operation.

After about 2 hours of run time, the technician measures the temperature difference across the coaxial heat exchanger. The temperature drop is a normal 5°F. No crackling sound can be heard in the water circuit. The house is up to temperature without the electric strip heaters operating. The technician is satisfied that the heat pump is operating normally.

SERVICE CALL 5

A homeowner is complaining about high electric bills in the winter. The residence has a geothermal heat pump that has a closed-loop, horizontal, two-layer ground loop, **Figure 44–10**. The system uses R-22 as a refrigerant. The owner of the house uses a setback (night/day) thermostat, which is kept at 72°F during the day and 65°F at night. *The problem is with the setback thermostat. At 8 AM the thermostat switches its set point from 65°F to 72°F. However, the microprocessor (smart board) on the heat pump is programmed to kick on the electric strip heat whenever more than a 2°F differential exists between the set point*

on the thermostat and the actual room temperature. In this case, there would be a 7°F temperature differential at 8 AM every morning. A technician is called after the homeowner has received the third high electric bill.

A service technician arrives at the residence at 10 AM and immediately enters the basement where the heat pump is housed. The service technician installs gages on the high and low sides of the system and temperature probes on the inlet and outlet of the coaxial heat exchanger. While the heat pump is running, it is determined that the temperature difference across the heat exchanger is 5°F. This is the recommended temperature difference for a closed-loop system in that geographical region. The suction pressure also reads normal at 45 psig for an entering-water temperature of 33°F. The technician then checks the flow rate of the antifreeze solution at the flowmeter and determines it to be normal. Just as the technician tells the homeowner that everything looks good, the unit shuts down for a normal off cycle. The technician then asks the homeowner where the thermostat is located so the unit can be cycled back on and more tests can be made. As the technician approaches the thermostat, it is found to be a setback thermostat. The homeowner is questioned as to the setback function on the thermostat with this heat pump. The homeowner explains how it is set back from 72°F to 65°F in the evening. The technician then explains how the heat pump's electric strip heaters will kick on if a 2°F differential is reached between the set point of the thermostat and the room temperature. The technician also explains to the homeowner that this is happening at 8 AM every morning and running up his electric bill. The technician then tells the homeowner that it is best not to use the setback function with this style of heat pump. The homeowner agrees and thanks the technician for the advice. The technician reprograms the thermostat to a normal operation, makes sure the heat pump is operating correctly, and then leaves.

SUMMARY

- Energy is transferred daily to and from the earth by solar radiation, rainfall, and wind.
- Geothermal heat pumps use the earth, or water in the earth, for their heat source and heat sink.
- Because the earth's underground temperature in the summer is cooler than the outside air, heat loads from summer air conditioning can be rejected underground more efficiently.
- Geothermal heat pumps are very similar to air source heat pumps in that they both use reverse-cycle refrigeration.
- Geothermal heat pumps are classified as either open- or closed-loop systems.
- Water quality is one of the most important considerations in the design of an open-loop geothermal heat pump system.
- Open-loop systems usually use well water as their heat source and heat sink. The water is then returned to the earth

in some way. Heat exchanger fouling can be a problem if water quality is poor.

■ Water sources for open-loop systems may be an existing well or a new well. A well pump delivers the water from the well to the heat pump. Wells have many categories.

■ Pressure tanks are used in conjunction with wells in open-loop systems.

■ Closed-loop heat pump systems recirculate the same antifreeze fluid in a closed loop. This eliminates water-quality problems.

■ Closed-loop or earth-coupled systems are used where there is insufficient water quality or quantity or where local codes prohibit open-loop systems.

■ The ground loop for closed-loop systems can either be a vertical, horizontal, slinky, or pond/lake type. Loops can also have series or parallel fluid flows.

■ The buried piping or underground heat exchanger is usually either polyethylene or polybutylene pipe. Both of these materials can be heat welded by a heat fusion process.

■ The antifreeze solutions inside the buried piping are used to prevent freezing of the heat pump heat exchanger and for heat transfer purposes. Some southern climates do not need an antifreeze additive.

■ The water-to-water heat pump is becoming popular for use in radiant heating systems.

■ Water-to-water heat pump systems often use a buffer tank to store the heated water until it is needed by the heating circuits.

■ Water-to-water heat pump systems often have three circulator pumps: one on the ground-water circuit, one between the heat pump and the buffer tank, and one between the buffer tank and the heating circuits.

■ Waterless heat pump systems utilize buried refrigerant lines instead of buried water lines.

■ Waterless heat pump systems transfer heat into and out of the refrigerant by using the ground as the heat source in the winter and as the heat sink in the summer.

REVIEW QUESTIONS

1. Geothermal heat pumps, or water-source heat pumps, are classified as either _____ loop or _____ loop systems.
2. An important factor (or factors) involving water quality with regard to open-loop heat pump systems is (are)
   A. water temperature.
   B. water cleanliness.
   C. water volume in gpm.
   D. all of the above.
3. Explain the *ground loop* or *water loop* of a closed-loop, geothermal heat pump system.
4. Explain the *refrigerant loop* of a closed-loop, geothermal heat pump system.

5. The _____ loop is used to distribute heated or cooled air to the building of a geothermal heat pump system.
6. True or False: Domestic hot water can be heated with a geothermal heat pump system.
7. Two materials used in the buried piping of the ground loop of a closed-loop, geothermal heat pump system are _____ and _____.
8. Antifreeze solutions that can be used in the ground loop of a closed-loop geothermal heat pump system can be
   A. alcohol.
   B. salts.
   C. glycols.
   D. all of the above.
9. What is the main difference between a copper and a cupronickel coaxial heat exchanger, and where is each used?
10. A _____ well draws water from the top of the water column, circulates it through the heat pump, and then returns the water to the bottom of the water column.
11. Which of the following is a process in which a cement-like material is injected between the well casing and the hole drilled for the well?
    A. Grouting
    B. Cementing
    C. Sealing
    D. Separating
12. The function of the pressure tank in an open-loop, geothermal heat pump well system is to prevent the well pump from _____.
13. If 14 gpm of water is flowing through an open-loop heat pump with a temperature difference of 7°F, how much heat in Btu/h is being absorbed or rejected in the heat pump? Show all work and units.
14. Explain the operation/purpose of the buffer tank on a water-to-water heat pump system.
15. How will oversizing the buffer tank on a water-to-water heat pump system affect the run time of the system?
16. Explain why a waterless, earth-coupled, closed-loop geothermal heat pump system is more efficient than a conventional closed-loop geothermal heat pump system.
17. Explain how refrigerant and oil leakage are found in a waterless, earth-coupled, closed-loop geothermal heat pump system.
18. What kind of material is used in the waterless ground loop in a waterless, earth-coupled, closed-loop geothermal heat pump system?

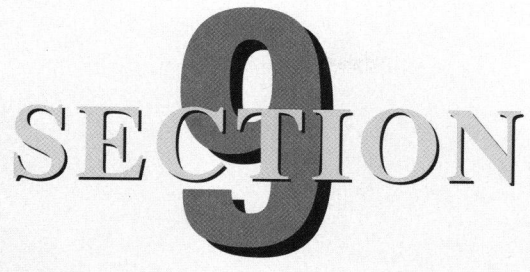

SECTION 9

Domestic Appliances

Section 9 can be found on the back of book CD.

Chilled-Water Air-Conditioning Systems

OBJECTIVES

After studying this unit, you should be able to

- list different types of chilled-water air-conditioning systems.
- describe how chilled-water air-conditioning systems operate.
- state the types of compressors often used with high-pressure refrigerant water chillers.
- describe the operation of a centrifugal compressor in a high-pressure chiller.
- explain the difference between direct-expansion and flooded chiller evaporators.
- explain what is meant by approach temperature in a water-cooled condenser.
- state two types of condensers used in chilled-water systems.
- explain subcooling.
- list the types of metering devices used in high-pressure chillers.
- list the types of refrigerants typically used in low-pressure chillers.
- state the type of compressor used in low-pressure chiller systems.
- describe the metering devices used in low-pressure chiller systems.
- explain the purge system used on a low-pressure chiller condenser.
- describe the absorption cooling system process.
- state the refrigerant generally used in large absorption chillers.
- state the compound normally used in salt solutions in large absorption chillers.
- state the type of electric motors typically used on chiller air-conditioning systems.
- discuss the various start mechanisms for these motors.
- describe a load-limiting device on a chiller motor.
- discuss various motor overload protection devices and systems.

SAFETY CHECKLIST

- ✔ Wear gloves and use caution when working around hot steam pipes and other heated components.
- ✔ Use caution when working around high-pressure systems. Do not attempt to loosen fittings or connections when system is pressurized. Follow recommended procedures when using nitrogen.
- ✔ Follow all recommended safety procedures when working around electrical circuits.
- ✔ Lock and tag disconnect boxes or panels when power has been turned off to work on an electrical system. There should be one key for this lock, and it should be in the technician's possession.
- ✔ Never start a motor with the door to the starter components open.
- ✔ Check pressures regularly as indicated by the manufacturer of the system.

Chilled-water systems are used for larger applications for central air conditioning because of the ease with which chilled water can be circulated in the system. If refrigerant were piped to all floors of a multistory building, there would be too many possibilities for leaks to occur, in addition to the expense of the refrigerant to charge the system.

The design temperature for boiling refrigerant in a coil used for cooling air is 40°F. If water can be cooled to approximately the same temperature, it can also be used to cool or condition air, **Figure 48–1**. This is the logic used for circulating chilled-water systems. Water is cooled to about 45°F and circulated throughout the building to air-heat exchange coils that absorb heat from the building air. When

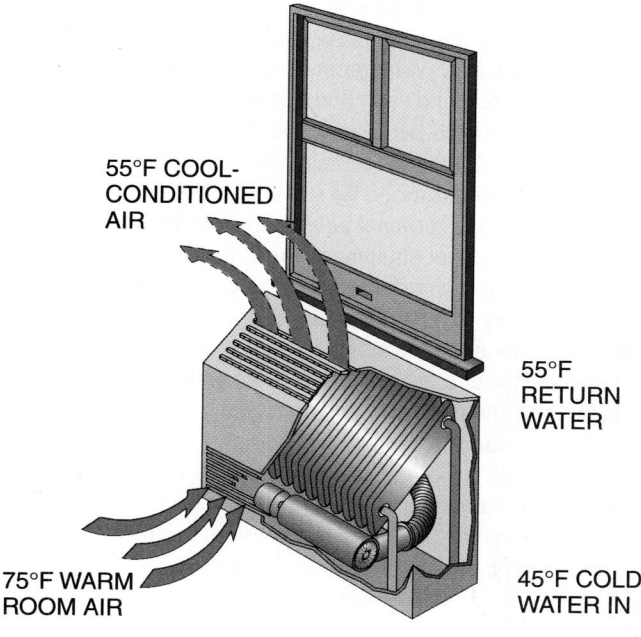

55°F COOL-CONDITIONED AIR

55°F RETURN WATER

75°F WARM ROOM AIR

45°F COLD WATER IN

Figure 48–1 Water circulating in a fan coil unit for cooling a room.

COOLING
TOWER

COOLING
COIL

55°F

45°F

CHILLED-WATER
EVAPORATOR

COMPRESSOR

Figure 48–2   Water is circulated in the building's piping to the fan coil unit. The water is considered a secondary refrigerant.

water is used for circulation in a building, the water is called a secondary refrigerant. It is much less expensive to circulate than refrigerant, **Figure 48–2.**

The chilled water circulated through the building is typically 45°F. ⬢ Lower water temperatures may be used to make the building side of the chilled-water system more efficient. ⬢ Many systems use chilled water at 42°F. With the development of electronic controls, some manufacturers are actually able to furnish 34°F chilled water. ⬢ The colder the circulating water, the smaller the terminal equipment can be. The coils and fans are the terminal equipment used. With colder water, the interconnecting piping for the entire system can be smaller. There would be savings on piping supplies, fittings, and insulation for the entire system. Labor is saved by installing smaller piping circuits. Electrical savings would come from smaller fans and pumps, which consume less power. ⬢

All installations must deal with variable loads. The load on the equipment is not constant. In a residential system, it is typical to just turn the unit off when the desired conditions are met. In large installations, it is not wise to simply cycle the compressor or chiller. Much of the wear occurs when the chiller is started. The bearings are not as well lubricated at start-up as they are when the compressor is running. The motor is stressed when the compressor is started up in most installations. Manufacturers of large equipment have developed many capacity-control methods for their equipment.

In most comfort-cooling installations, the load varies during the season as well as during the day. We saw in Unit 36 that when a system is planned for very high-temperature design conditions the equipment will only run at full load about 3% of the time. The other 97% of the time it will run at reduced load. Often, the equipment will need to operate at 40% or 50% of its capacity. There are times that it will need to run at 15% or 20% of its capacity in mild weather.

Different types of chillers will use different types of reduced capacity, such as variable-speed, or unloading, compressors. This is typically called unloading the compressor. These will be discussed in sections of this unit that cover each type of chiller. The manufacturers usually sell their chillers by rating the capacity at full load. Their unit efficiency is usually stated at full load. With energy costs going up, manufacturers are trying harder to get their equipment to operate efficiently at part load. In the past, the equipment did not typically operate really efficiently at part load. Many technologies are being used to ensure the best efficiencies at all load conditions for the equipment.

48.1   CHILLERS

A chiller refrigerates circulating water. As the water passes through the evaporator section of the machine, the temperature of the water is lowered. It is then circulated throughout the building where it picks up heat. The typical design temperatures for a circulating chilled-water system are 45°F water furnished to the building and 55°F water returned to the chiller from the building. The heat from the building adds 10°F to the water that returns to the chiller. Here the heat is removed and the water is recirculated.

Cooling buildings is not the only application for water chillers. This unit, however, will mostly cover comfort-cooling chillers—as a thorough discussion of all chiller applications would be beyond the scope of this text.

Many manufacturers use chillers for process cooling. For example, some chillers are designed to circulate glycol (antifreeze) at temperatures well below freezing. An example of a process-cooling application would be in the plastic-molding manufacturing business. When milk bottles are manufactured by the process of injecting hot plastic into a mold, the mold must be cooled to solidify the plastic. A water valve opens at the correct time, the plastic milk bottle is water-cooled slightly until solid, and then it is ejected from the mold. Then the flow of chilled water is stopped and the mold is ready to make another bottle. The chilled water helps this process happen very quickly.

Textile mills use huge amounts of chilled water to maintain the conditions that allow the thread in the mill to move fast. If the temperature and humidity are held at the correct conditions, maximum production can be accomplished. In the years before chilled water was used for maintaining optimal conditions, mills could only operate at full capacity about three months of the year. With chilled water, the conditions can be maintained 24 hours a day for 365 days a year. Many mills run 24 hours a day year-round, except during one week of maintenance. This is not an example of comfort cooling; the conditions are maintained for production.

There are two basic categories of chillers: the compression cycle chiller and the absorption chiller. The compression type of chiller uses a compressor to provide the pressure differences inside the chiller to boil and condense refrigerant. The absorption chiller uses a salt solution and water to accomplish the same results. These chillers are very different and are discussed separately.

48.2 COMPRESSION CYCLE CHILLERS

The compression cycle chiller has the same four basic components as a typical air conditioner: a *compressor*, an *evaporator*, a *condenser*, and a *metering device*. However, these components are generally larger to be able to handle more refrigerant and they may use a different refrigerant.

The heart of the compression cycle refrigeration system is the compressor. As mentioned in Unit 3, "Refrigeration and Refrigerants," there are several types of these compressors. The compressors common to water chillers are the *reciprocating, scroll, screw,* and *centrifugal.* Photos of these compressors can be found in Unit 23, "Compressors." The compressor can be thought of as a *vapor pump.* The technician should think of the compressor as a component in the line that lowers the evaporator pressure to the desired boiling point of the refrigerant. Typically this is about 38°F for a chiller. It then builds the pressure in the condenser to the point that vapor will condense to a liquid for reuse in the evaporator. The typical condensing temperature is 105°F. The technician can use these temperatures to determine whether a typical chiller is operating within the design parameters. The compressor must be of a design that will pump and compress the vapor to meet the needs of a particular installation.

Compression cycle chillers may be classified as either *high-pressure* or *low-pressure systems.* Following is a discussion of high-pressure refrigerant water chillers.

48.3 RECIPROCATING COMPRESSOR CHILLERS

Large reciprocating compressors used for water chillers operate the same as those for any other reciprocating compressor application with a few exceptions. A review of Unit 23 will help you to understand how compressors function. These compressors range in size from about 1/2 hp to approximately 150 hp, depending on the application. Most manufacturers have stopped using one large compressor for a large reciprocating chiller and have started using multiple smaller compressors. They are positive displacement compressors and cannot pump liquid refrigerant without risk of damage to the compressor.

Several refrigerants have been used for reciprocating compressor chillers; R-500, R-502, R-12, R-134a, and R-22 are the most common. Because of the ozone depletion and global warming scares, environmentally friendly alternative refrigerants like R-134a and certain refrigerant blends are now being used.

The large reciprocating compressor will have many cylinders to produce the pumping capacity needed to move large amounts of refrigerant. Some of these compressors have as many as 12 cylinders. This becomes a machine with many moving parts and much internal friction. If one cylinder of the compressor fails, the whole system is off the line. With multiple compressors, if one compressor fails, the others can carry the load. Multiple compressors give some backup from total failure. For this reason and because of capacity control, many manufacturers use multiple compressors of a smaller size.

All large chillers must have some means for controlling the capacity or the compressor will cycle on and off. This is not satisfactory because most compressor wear occurs during start-up before oil pressure is established. A better design approach is to keep the compressor on the line and operate it at reduced capacity. Reduced capacity operation also smooths out temperature fluctuations that occur from shutting off the compressor and waiting for the water to warm up to bring the compressor back on.

48.4 CYLINDER UNLOADING AND VARIABLE-FREQUENCY DRIVES

Reduced capacity for a reciprocating compressor is accomplished by *cylinder unloading* and *variable-frequency drives* (VFDs). For example, suppose a 100-ton compressor with eight cylinders is used for the chiller for a large office building and the chiller has 12.5 tons of capacity per cylinder. When all eight cylinders are pumping, the compressor has a capacity of 100 tons ($8 \times 12.5 = 100$). As the cylinders are unloaded, the capacity is reduced. For example, the cylinders may unload in pairs, which would be 25 tons per unloading step. The compressor may have four steps of unloading—so the compressor has four different capacities: 100 tons (eight cylinders pumping), 75 tons (six cylinders pumping), 50 tons (four cylinders pumping), and 25 tons (two cylinders pumping). In the morning when the system first starts, the building may only need 25 tons of cooling. As the temperature outside rises, the chiller may need more capacity and the compressor will automatically load two more cylinders for 50 tons of capacity. As the temperature rises, the compressor can load up to 100% capacity or 100 tons. If the building stays open at night, for example, a hotel, the compressor will start to unload as the outside temperature cools. It will unload down to 25 tons; then if this is too much capacity, the chiller will shut off. When the chiller is restarted, it will start up at the reduced capacity, lowering the starting current. A compressor cannot be unloaded to 0 pumping capacity or it would not move any refrigerant through the system to return oil that is in the system. Usually compressors will unload down to 25% to 50% of their full-load pumping capacity.

Another big advantage of cylinder unloading is that the power to operate the compressor is reduced as the capacity is reduced. The reduction in power consumption is not in direct proportion with a compressor capacity, but the power consumption is greatly reduced at part load. In addition to reducing the workload by unloading the cylinders, the compression ratio is reduced; when a cylinder is unloaded, the suction pressure rises slightly, and the head pressure is reduced slightly.

Cylinder unloading is accomplished in several ways; *blocked suction* and *suction-valve-lift unloading* are the most common.

Blocked Suction

Blocked suction is accomplished by placing a solenoid valve in the suction passage to the cylinder being unloaded, **Figure 48–3**. If the refrigerant gas cannot reach the cylinder, no gas is pumped. If a compressor has four cylinders and the suction gas is blocked to one of the cylinders, the capacity of the compressor is reduced by 25%, and the compressor then pumps at 75% capacity. The power consumption also goes down approximately 25%. Power consumption is related to the amperage draw of the compressor. Amperage is typically measured in the field by using a clamp-on ammeter. When a compressor is running at half capacity, the amperage will be about half full-load amperage. The power consumption of the compressor is actually measured in watts. Using amperage as a measure of compressor capacity is close enough for field troubleshooting.

Suction-Valve-Lift Unloading

If the suction valve is lifted off the seat of a cylinder while the compressor is pumping, this cylinder will quit pumping. Gas that enters the cylinder will be pushed back out into the suction side of the system on the upstroke. There is no resistance to pumping the refrigerant back into the suction side of the system, so it requires no energy. The power consumption will be reduced as in the example of the blocked suction. One of the advantages of lifting the suction valve is that the gas that enters the cylinder will contain oil, and good cylinder lubrication will occur even while the cylinder is not pumping.

Compressor unloading could be accomplished by letting hot gas back into the cylinder, but this is not practical because power reduction will not occur. When the gas has been pumped from the low-pressure side to the high-pressure side of the system, the work has been accomplished. Also the cylinder would become overheated by compressing the hot gas again.

Except for cylinder unloading, the reciprocating chiller compressor is the same as smaller compressors. Most compressors over 5 hp have pressure-lubricating systems. The pressure for lubricating the compressor is an oil pump that is typically mounted on the end of the compressor shaft and driven by the shaft, **Figure 48–4**. The oil pump picks up oil in the sump at evaporator pressure and delivers the oil to the bearings at about 30 to 60 psig greater than suction pressure. This is called net oil pressure, **Figure 48–5**. These compressors also have an oil safety shutdown in case of oil pressure failure, which has a time delay of about 90 sec to allow the compressor to get started and to establish oil pressure before it shuts the compressor off.

VFDs electronically alter the frequency of alternating-current (AC) sine waves. This frequency variation is done through a device called an inverter. Today, the compressor motor can operate with a range of speeds that will vary its capacity without shutting off. ● The speeds are used in conjunction with electronic expansion valves that also have an almost unlimited ability to vary the refrigerant flow to the evaporator. The condenser water pump and the chilled-water

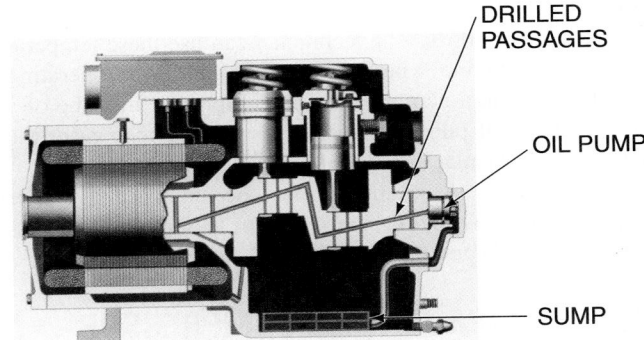

Figure 48–4 The pressure lubrication system for a reciprocating compressor. Notice that the oil pump is on the end of the shaft. The oil is picked up from the sump and pumped through the drilled passages to all moving parts. *Courtesy Trane Company*

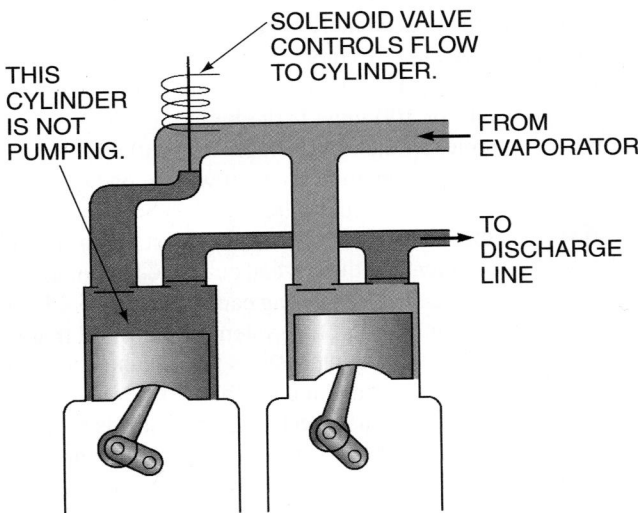

Figure 48–3 Blocking the flow to one of the cylinders in a reciprocating compressor will unload that cylinder's load from the system.

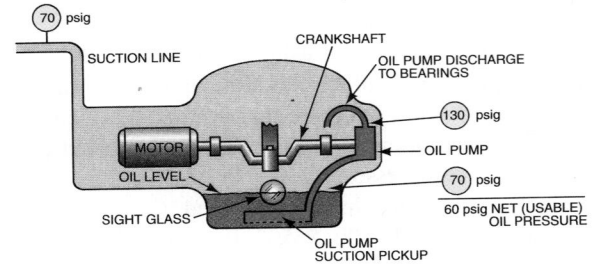

Figure 48–5 The oil pump's inlet pressure is the suction pressure of the crankcase. The oil pump then builds pressure to the pump discharge pressure.

pumps may also have a VFD drive to vary the water flow. The system flow rates must be synchronized by using a computer and taking into account the desired system conditions. This must be done to prevent problems and maintain minimum and maximum flow rates for all devices that are coordinated together.🌐 See Unit 17 for details.

48.5   SCROLL COMPRESSOR CHILLERS

The scroll compressor is a positive displacement compressor. The scroll compressors applied to chillers are larger than those scroll compressors described earlier in this text. These compressors are in the 10- to 25-ton range but operate the same as the smaller compressors. See **Figure 48–6** for an illustration of a scroll compressor. They are welded hermetic compressors. When these compressors are used in chillers, the capacity control of the chiller is maintained by cycling the compressors off and on in increments of 10 and 25 tons. For example, a 25-ton chiller would have a 10- and a 15-ton compressor and would have capacity control of 10 and 15 tons. A 60-ton chiller may have four 15-ton compressors and be able to operate at 100% (60 tons), 75% (45 tons), 50% (30 tons), and 25% (15 tons).

Some of the advantages of the scroll compressor include the following:
1. Efficiency
2. Quietness
3. Fewer moving parts

4. Size and weight
5. Can pump small amounts of liquid refrigerant without compressor damage

The scroll compressor offers little resistance to refrigerant flow from the high side of the system to the low side during the off cycle. A check valve is provided to prevent backward flow when the system is shut down, **Figure 48–6**.

Lubrication of the scroll compressor is provided by an oil pump at the bottom of the crankshaft. The oil pump picks up oil and lubricates all moving parts on the shaft.

48.6   ROTARY SCREW COMPRESSOR CHILLERS

Most of the major manufacturers are building rotary screw compressors for larger-capacity chillers. This is larger-capacity equipment using high-pressure refrigerants. The rotary screw compressor is capable of handling large volumes of refrigerant with few moving parts, **Figure 48–7**. This type of compressor is a positive displacement compressor that has the characteristic of being able to handle some liquid refrigerant without compressor damage. This is unlike the reciprocating compressor that cannot handle liquid refrigerant. Rotary screw compressors are manufactured in sizes from about 50 to 700 tons of capacity. These compressors are reliable and trouble-free.

Manufacturers build both *semi-hermetic rotary screw compressors* and *open-drive compressors*. The open-drive models are direct drive and must have a shaft seal to contain the refrigerant where the shaft penetrates the compressor shell. It is common practice to start up open-drive compressors regularly during the off-season to lubricate the shaft seal, which may become dry due to lack of rotation.

Capacity control may be accomplished with a rotary screw compressor by means of a *slide valve* that blocks the suction gas before it enters the rotary screws in the compressor, or by means of compressor speed control, VFD. This slide valve is typically operated by differential pressure in the system. 🌐 The slide valve may be moved to the completely

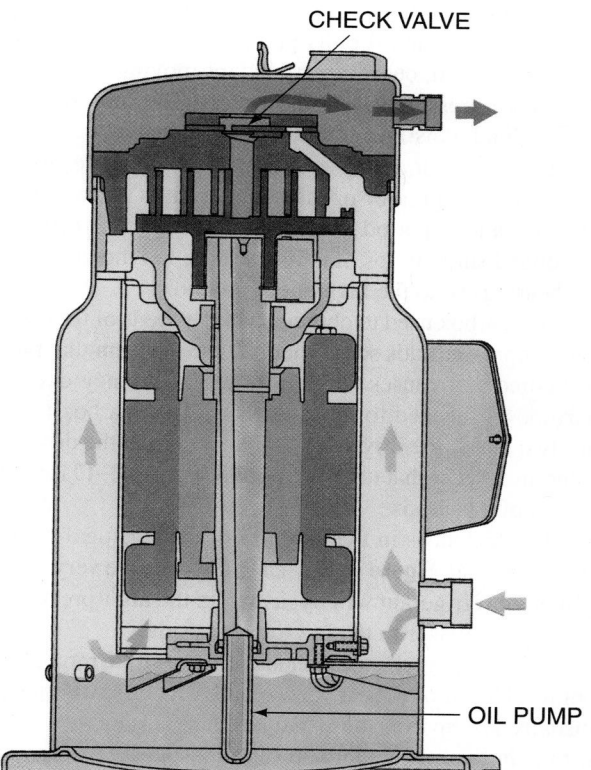

Figure 48–6   The check valve in a scroll compressor prevents discharge pressure from flowing back through the compressor when the compressor is shut down. *Courtesy Trane Company*

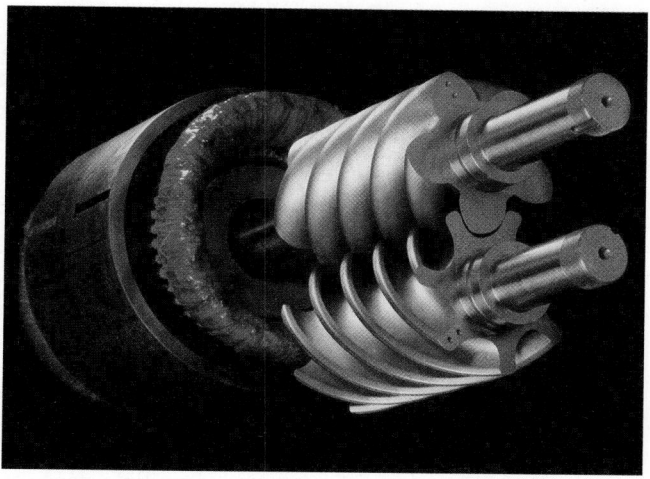

Figure 48–7   Rotary screw compressor package chiller. *Reproduced courtesy of Trane Company*

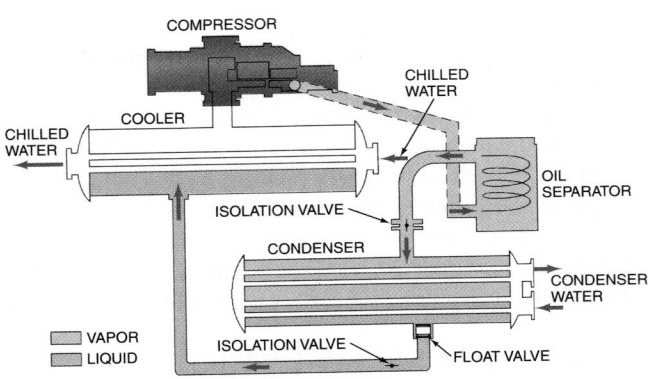

Figure 48–8   An oil separator for a rotary screw compressor.
Reproduced courtesy of Carrier Corporation

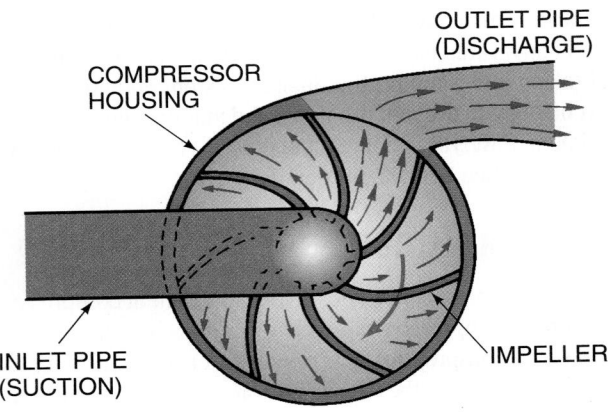

THE TURNING IMPELLER IMPARTS CENTRIFUGAL
FORCE ON THE REFRIGERANT, FORCING THE
REFRIGERANT TO THE OUTSIDE OF THE IMPELLER.
THE COMPRESSOR HOUSING TRAPS THE
REFRIGERANT AND FORCES IT TO EXIT INTO THE
DISCHARGE LINE. THE REFRIGERANT MOVING TO
THE OUTSIDE CREATES A LOW PRESSURE IN THE
CENTER OF THE IMPELLER WHERE THE INLET IS
CONNECTED.

Figure 48–9   Centrifugal action is used to compress refrigerant.

unloaded position before shutdown so that on start-up, the compressor is unloaded, reducing the inrush current on start-up. Most of these compressors can function from about 10% load to 100% load with sliding graduations because of the nature of the slide valve unloader. 🌐 This is as opposed to the step capacity control of a reciprocating compressor, which unloads one or two cylinders at a time.

The nature of the screw compressor is to pump a great deal of oil while compressing refrigerant so these compressors typically have an oil separator to return as much oil to the compressor reservoir as possible, **Figure 48–8**. The oil is moved to the rotating parts of the compressor by means of pressure differential within the compressor instead of an oil pump. Oil is also accumulated in the oil reservoirs and moved to the parts that need lubrication by means of gravity. The rotating screws are close together but do not touch. The gap between the screws is sealed by oil that is pumped into the rotary screws as they turn. This oil is separated from the hot gas in the discharge line and returned to the oil sump through an oil cooler.

This oil separation is necessary because if too much oil reaches the system, a poor heat exchange will occur in the evaporator and loss of capacity will occur. The natural place to separate the oil from the refrigerant is in the discharge line. Different manufacturers use different methods, but all of them have some means of oil separation.

48.7   CENTRIFUGAL COMPRESSOR CHILLERS (HIGH PRESSURE)

The centrifugal compressor uses only the centrifugal force applied to the refrigerant to move the refrigerant from the low- to the high-pressure side of the system, **Figure 48–9**. It is like a large fan that creates a pressure difference from one side of the compressor to the other. It does not have a great deal of force, but some companies manufacture centrifugal compressors for use in high-pressure systems. Centrifugal compressors can handle a large volume of refrigerant. To provide the pressure difference from the evaporator to the condenser for high-pressure systems the compressor is turned very fast by means of a gearbox or multiple stages of compression. Typical motor speeds used for direct-drive reciprocating, scroll, and rotary screw compressors are near 3600 rpm. When faster compressor speeds are needed for

centrifugal systems, a gearbox is used to speed the compressor up to higher rpm levels. Speeds of about 30,000 rpm are used for some single-stage centrifugal compressors.

If the head pressure becomes too high or the evaporator pressure too low, the compressor cannot overcome the pressure difference, and it quits pumping. The motor and compressor still turn, but refrigerant stops moving from the low- to the high-pressure side of the system. The compressor may make a loud whistling sound. This is called a "surge." Even though this is a loud noise, it will normally not cause damage to the compressor or motor unless it is allowed to continue for a long period of time. If damage does occur due to prolonged surge times, it is likely to be the thrust surface of the bearings or to the high-speed gearbox.

The gearbox used to obtain the high speeds of the centrifugal compressor adds some friction to the system that must be overcome and causes a slight loss of efficiency due to the horsepower needed to turn the gears. The gearbox typically has two gears; the drive gear is the larger and the driven gear is the small gear that turns the fastest. **Figure 48–10** is an illustration of a gearbox.

The clearances in the impeller of a centrifugal compressor are critical, **Figure 48–11**. The parts must be very close together, or refrigerant will bypass from the high-pressure side of the compressor to the low-pressure side.

Lubrication is accomplished with a separate motor and oil pump. This motor is a three-phase fractional horsepower (usually 1/4 hp to 3/4 hp) and is located inside the oil sump. It runs in the refrigerant and oil atmosphere. A three-phase motor is used so there will be no motor internal start switch as in a single-phase motor. This motor allows the manufacturer to start the oil pump before the centrifugal motor and establishes lubrication before the compressor and gearbox start to turn. It also allows the oil pump to run during coast

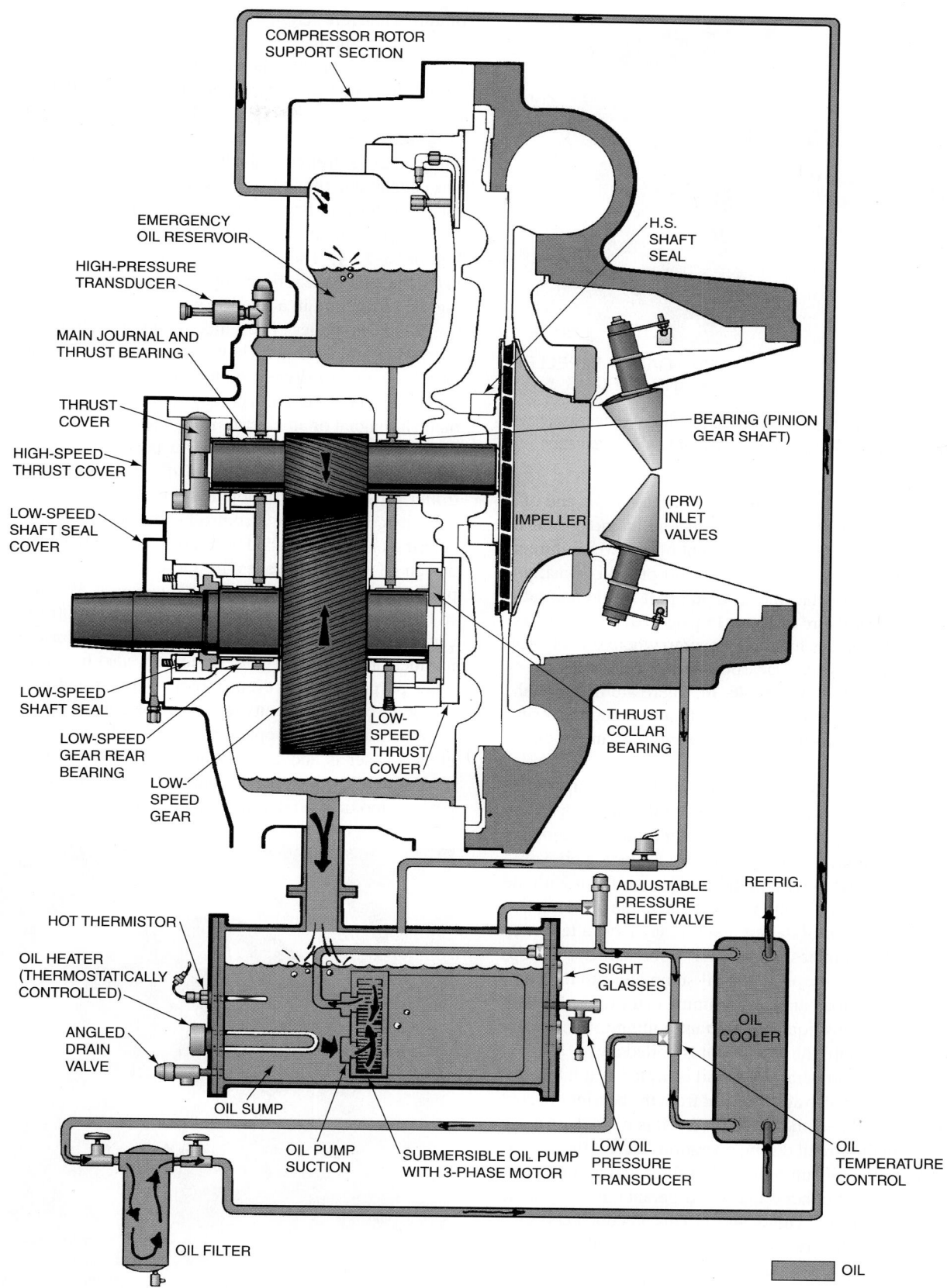

Figure 48–10 The gear box for a high-speed compressor. *Courtesy York International*

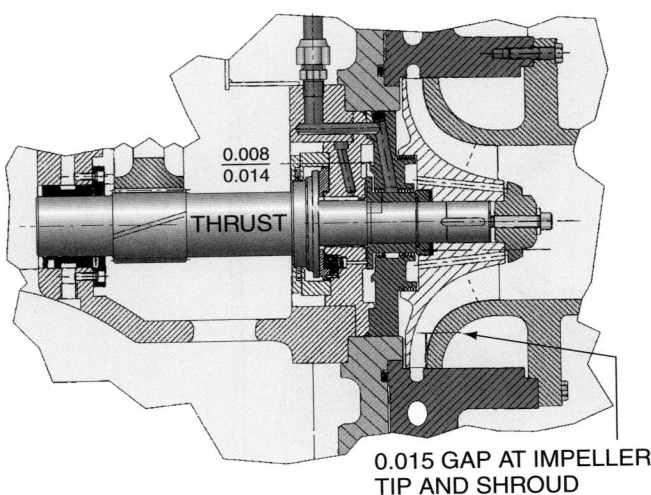

Figure 48–11 The parts in a high-speed centrifugal compressor have a very close tolerance. *Reproduced courtesy of Carrier Corporation*

down of the centrifugal impeller and gearbox at the end of a cycle when the machine is shut off. There is enough oil in reservoirs for a coast down in the event of a power failure to prevent bearing and gear failure. The oil pump is usually a gear-type positive displacement oil pump. It would normally furnish a net oil pressure of about 15 psig to the bearings and with some compressors to an upper sump where oil is fed by gravity to the bearings and gearbox during a coast down because of power failure, **Figure 48–10.** There are many methods of routing the oil, but the oil must have enough pressure to reach the farthest bearings with enough pressure to circulate the oil around the bearings.

The lubrication system has a heater in the oil sump to prevent liquid refrigerant from migrating to the oil sump when the system is off, **Figure 48–10.** The heater typically maintains the oil sump at about 140°F. Without this heater the oil would soon become saturated with liquid refrigerant. Even with the heater operating, some liquid will migrate to the oil sump and when the compressor is started, the oil may have a tendency to foam for a few minutes immediately after start-up. The operator should keep an eye on the oil sump and run the compressor at reduced load until any foaming in the oil is reduced. Foaming oil contains liquid refrigerant boiling off and is not a good lubricant until the refrigerant is boiled away.

When a chiller is operating, the oil is heated as it lubricates the moving parts. It also absorbs heat from the bearings to cool them. The oil will become overheated if it is not cooled so an oil cooler is in the circuit during operation. The oil typically is pumped from the oil sump through a filter and an oil cooler heat exchanger with either water or refrigerant removing some of the heat from the oil, **Figure 48–10.** This oil cooler cools the oil from a sump temperature of 140°F–160°F to about 120°F. It will then pick up heat again while lubricating.

The lubrication system for a centrifugal chiller is a sealed system, and oil is not intended to be mixed with the refrigerant and separated for recovery like the reciprocating, scroll, or rotary screw systems. If the oil gets into the refrigerant, it is difficult to remove because only vapor should leave the evap-

orator. When refrigerant in the evaporator becomes oil logged or saturated, the heat exchange becomes less efficient. Oil-logged refrigerant must be distilled to remove the oil. This is a long, time-consuming process. The intent of the manufacturer is to keep the two separated.

🌐Capacity control is accomplished by means of guide vanes at the entrance to the impeller eye. These guide vanes are usually pie-shaped devices located in a circle and can rotate to allow full flow or reduced flow down to about 15% or 20%, depending on the clearances in the vane mechanism, **Figure 48–12.**🌐 The guide vanes also serve another purpose in that they help the refrigerant enter the impeller eye by starting the refrigerant in a rotation pattern that matches the rotation of the impeller. The guide vanes are often called the prerotation guide vanes. They are controlled using electronic or pneumatic (air-driven) motors to vary the capacity of the centrifugal compressor. The controller sends either a pneumatic air signal or an electronic signal to the inlet guide vane operating motor, which is either a pneumatic or electronic type motor. When the building load starts to satisfy, the controller tells the vane motor to start closing the vanes. The capacity control motor is controlled by the temperature of the entering or leaving chilled water, depending on the engineer's design. If the guide vanes are closed, the compressor only pumps at 15% to 20% of its rated capacity. When open all of the way, the compressor pumps 100%. A 1000-ton chiller can operate from 150–200 tons to 1000 tons of capacity in any number of steps. The guide vanes can also be used for two other purposes—to prevent motor overload and to start the motor at reduced capacity to reduce the current on start-up.

When a chiller is operating at a condition in which the chilled water is above the design temperature, for example, when a building is hot on Monday morning start-up, the compressor motor would run at an overloaded condition. The

Figure 48–12 The prerotation vanes start the refrigerant rotating and also serve as capacity control for a centrifugal compressor. *Courtesy York International*

(A)

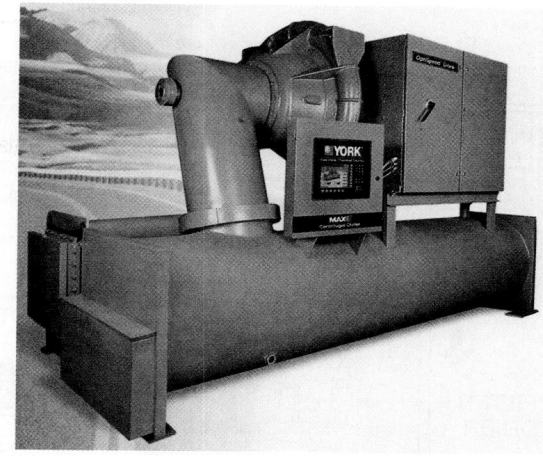

(B)

Figure 48–13 **(A)** A hermetically sealed centrifugal chiller. **(B)** An open-drive centrifugal chiller. *(A) Reproduced courtesy of Carrier Corporation. (B) Courtesy York International*

return chilled water may be 75°F instead of the design temperature of 55°F. The chiller control panel will have a control known as a *load limiter* that will sense the motor amperage and partially close the guide vanes to limit the compressor amperage to the full-load value. When the controls have established that the compressor motor is up to speed, the control system will take over and open the vanes to the correct size for the application. When the compressor is started, the prerotation guide vanes are closed and do not open until the motor is up to speed. Therefore, the compressor starts up unloaded. This reduces the power required to start the compressor.

The gearbox and impeller have bearings to support the weight of the turning shaft. These bearings are typically a soft material known as babbitt. The babbitt is backed and supported by a steel sleeve. With a single-stage compressor, there must also be a thrust bearing to counter the thrust of the refrigerant entering the compressor impeller. This thrust bearing counters the sideways movement of the shaft.

The motor rotation for many centrifugal chillers is critical. If the motor is started in the wrong direction, some chillers can be damaged, as discussed later in this unit.

Some of these centrifugal compressors are hermetically sealed and some have open drives, **Figure 48–13**. The hermetically sealed compressors have refrigerant-cooled motors. **Figure 48–14** shows a liquid-cooled motor. Liquid refrigerant is allowed to flow around the motor housing. Systems with open-drive motors use air in the equipment room to cool the motor. The heat from a large motor is considerable and must be exhausted from the equipment room.

48.8   EVAPORATORS FOR HIGH-PRESSURE CHILLERS

The *evaporator* is the component that absorbs heat into the system. Liquid refrigerant is boiled to a vapor by the circulating water.

The heat exchange surface is typically copper for water chillers. Other materials may be used if corrosive fluids are

circulated for manufacturing processes. In typical smaller air-conditioning systems, air is on one side of the heat exchange process and liquid or vapor refrigerant on the other. The rate of heat exchange between air and vapor refrigerant is fair. The heat exchange between air and liquid refrigerant is better and the best heat exchange is between water and liquid refrigerant. This is where the water chiller gets part of its versatility. The heat exchange surface can be small and still produce the desired results.

The evaporators used for high-pressure chillers are either direct-expansion evaporators or flooded evaporators.

48.9   DIRECT-EXPANSION EVAPORATORS

Direct-expansion evaporators are also known as dry-type evaporators, meaning they have an established superheat at the evaporator outlet. They normally use thermostatic expansion devices for metering the refrigerant. Direct-expansion evaporators are used for smaller chillers, up to approximately 150 tons on older types and to about 100 tons for more modern chillers. Direct-expansion chillers introduce the refrigerant into the end of the chiller barrel with the water being introduced into the side of the shell, **Figure 48–15**. The water is on the outside of the tubes, and baffles cause the water to be in contact with as many tubes as possible for the best heat exchange. The problem with this arrangement is if the water side of the circuit ever gets fouled or dirty, the only way it can be cleaned is with the use of chemicals because the chillers cannot be cleaned with brushes. These are different from chillers in which the water circulates in the tubes and the refrigerant circulates around the tubes—as with flooded chillers.

48.10   FLOODED EVAPORATOR CHILLERS

Flooded chillers introduce the refrigerant at the bottom of the chiller barrel and the water circulates through the tubes. There are some advantages to this in that the tubes may be totally submerged under the refrigerant for the best heat exchange, **Figure 48–16**. The tubes may also be physically cleaned using

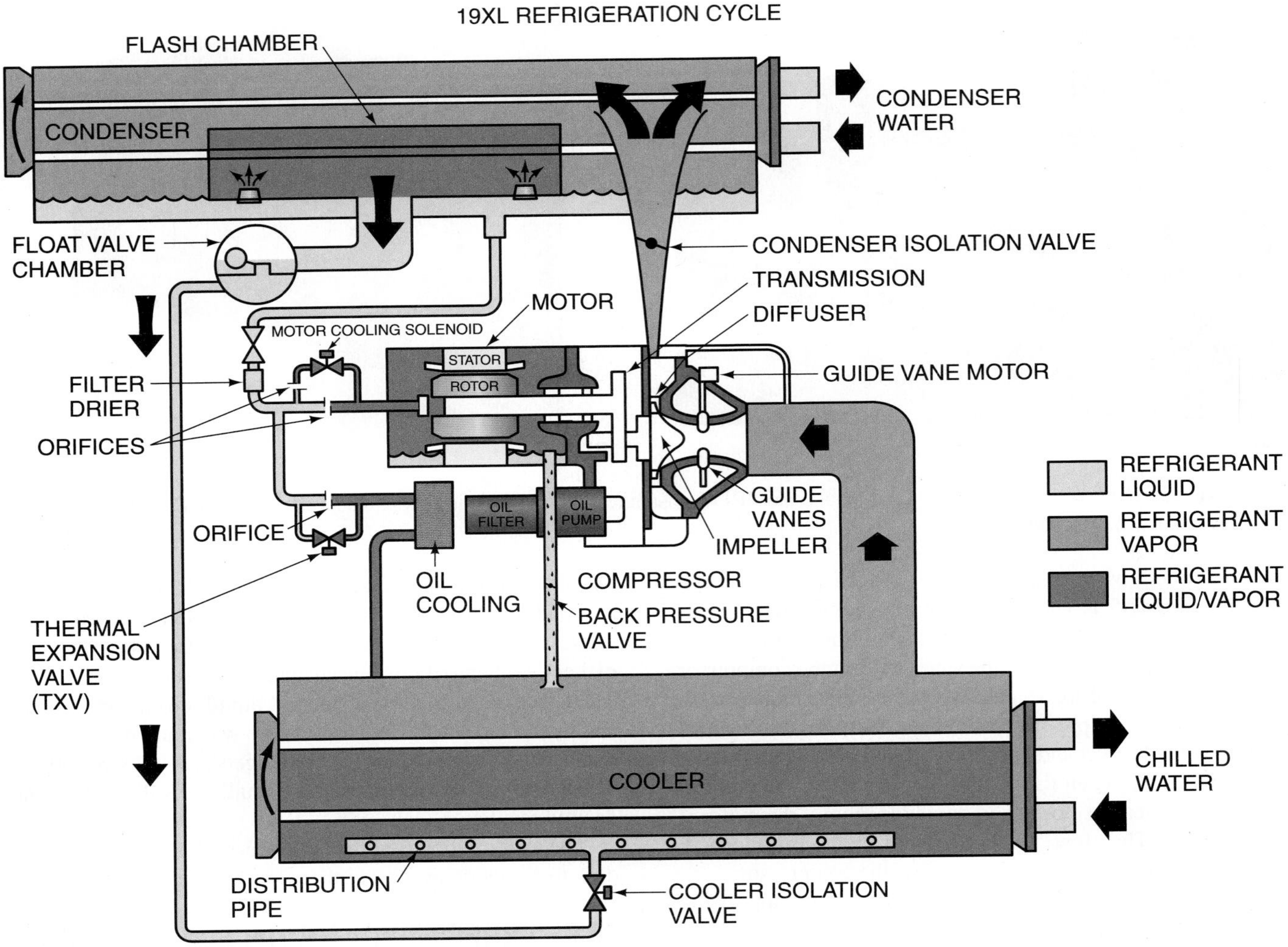

19XL REFRIGERATION CYCLE

FLASH CHAMBER

CONDENSER

CONDENSER WATER

FLOAT VALVE CHAMBER

CONDENSER ISOLATION VALVE

TRANSMISSION

MOTOR

DIFFUSER

MOTOR COOLING SOLENOID

FILTER DRIER

STATOR

ROTOR

GUIDE VANE MOTOR

ORIFICES

ORIFICE

OIL FILTER

OIL PUMP

GUIDE VANES

IMPELLER

REFRIGERANT LIQUID

REFRIGERANT VAPOR

REFRIGERANT LIQUID/VAPOR

THERMAL EXPANSION VALVE (TXV)

OIL COOLING

COMPRESSOR BACK PRESSURE VALVE

COOLER

CHILLED WATER

DISTRIBUTION PIPE

COOLER ISOLATION VALVE

Figure 48–14 A refrigerant-cooled hermetic centrifugal compressor motor. *Reproduced courtesy of Carrier Corporation*

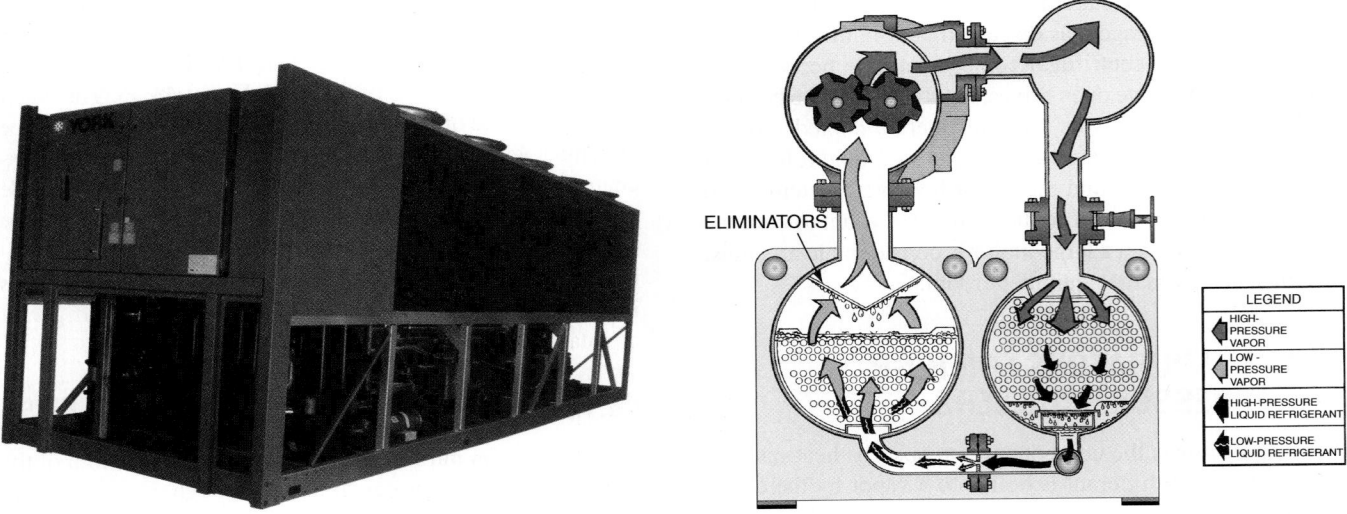

ELIMINATORS

LEGEND	
	HIGH-PRESSURE VAPOR
	LOW-PRESSURE VAPOR
	HIGH-PRESSURE LIQUID REFRIGERANT
	LOW-PRESSURE LIQUID REFRIGERANT

Figure 48–15 A direct-expansion chiller with an expansion device on the end of the chiller. *Courtesy York International*

Figure 48–16 Water tubes submerged under the liquid refrigerant. *Courtesy York International*

Figure 48–17 🌐 Cleaning the tubes in a condenser by using a power brush and flushing with water.🌐 *Courtesy Goodway Tools Corporation*

several different methods. **Figure 48–17** shows tubes being cleaned with a brush. The water side of an evaporator should not become dirty unless the chiller is applied to an open process, such as a manufacturing operation. Most chiller systems use a closed water circuit that should remain clean unless there are many leaks and water must be continually added.

Flooded chillers use much more refrigerant charge than direct-expansion chillers so leak monitoring must be part of regular maintenance with high-pressure systems.

When the water is introduced into the end of the chiller, the water is contained in water boxes. When these water boxes have removable covers they are known as marine water boxes, **Figure 48–18.** When the piping is attached to the cover, the piping must be removed to remove the water box cover and

Figure 48–18 Marine water boxes allow the technician to clean the tubes by removing the water-box cover; piping does not have to be disconnected. *Courtesy Trane Company*

Figure 48–19 Standard water-box covers with piping connections. *Courtesy York International*

these are known as standard covers, **Figure 48–19.** The water box also is used to direct the water for different applications. For example, the water may need to pass through the chiller one, two, three, or four times for different applications. The water box directs the water by means of partitions. This enables the manufacturer to use one chiller with combinations of water boxes and partitions for different applications.

When water passes through a one-pass chiller, it is only in contact with the refrigerant for a short period of time. Using two, three, or four passes keeps the water in contact longer but also creates more pressure drop and more pumping horsepower, **Figure 48–20.**

The design water temperatures are typically 55°F inlet water and 45°F outlet water for a two-pass chiller. The refrigerant

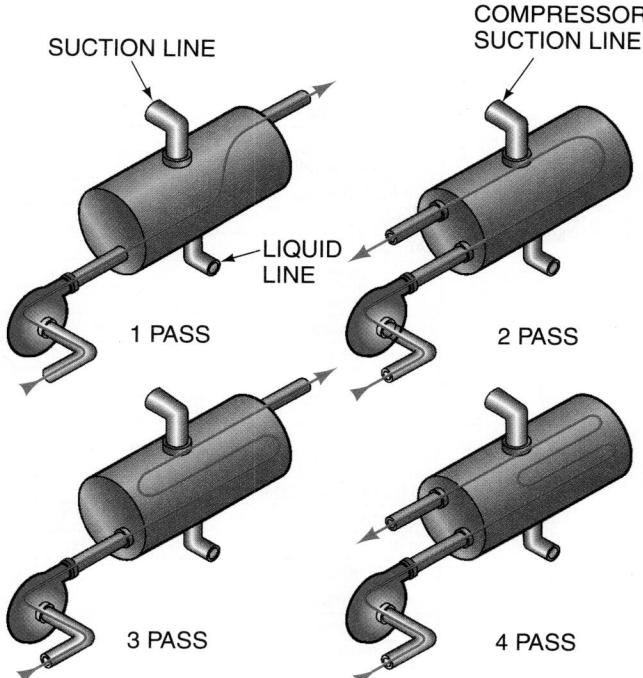

Figure 48–20 Heat exchangers for evaporators or condensers that are one-, two-, three-, and four-pass shells.

is absorbing heat from the water so it is typically about 7°F cooler than the leaving water. This is called the approach temperature, **Figure 48–21. The approach temperature is very important to the technician for troubleshooting chiller performance.** If the chiller tubes become dirty, the approach becomes greater because the heat exchange becomes poor. The approach temperature is different for chillers with a different number of passes. Direct-expansion chillers have only one pass of water (with baffles to sweep the tubes), and the approach temperature may be about 8°F. Chillers with three-pass evaporators will have an approach temperature of about 5°F, and chillers with four passes may have an approach temperature of 3°F or 4°F. The longer the refrigerant is in contact with the water, the closer the approach temperature will be.

When a chiller is first started, a record should be started known as an operating performance log. This log of machine performance should be recorded when the chiller is at full-load conditions. Manufacturers rather than contractors often choose to start up larger chillers because of the cost of warranty. They will always make entries in an operating log and keep the log on file for future reference. **Figure 48–22** is an example of a manufacturer's log sheet.

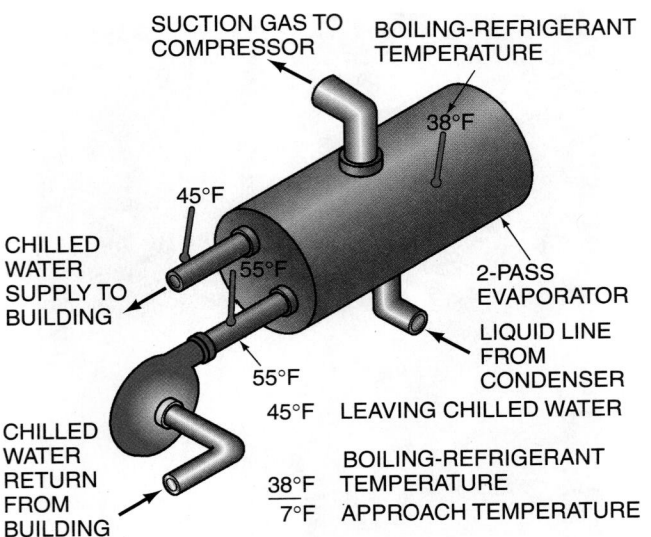

Figure 48–21   Approach temperature for a two-pass chiller.

YORK		**ROTARY SCREW** **LIQUID CHILLER LOG SHEET** **125 - 400 TONS**					**CHILLER LOCATION**				
							SYSTEM NO.				

Date												
Time												
Hour Meter Reading												
O.A. Temperature D.B./W.B.			/	/	/	/	/	/	/	/	/	/
Compressor	Oil Level											
	Oil Pressure											
	Oil Temperature											
	Suction Temperature											
	Discharge Temperature											
	Filter PSID											
	Slide Valve Position %											
	Oil Added (gallons)											
Motor	Volts											
	Amps											
Cooler	Refrig.	Suction Pressure										
	Liquid	Inlet Temperature										
		Inlet Pressure										
		Outlet Temperature										
		Outlet Pressure										
		Flow Rate — GPM										
Condenser	Refrig.	Discharge Pressure										
		Corresponding Temperature										
		High Pressure Liquid Temperature										
		System Air — Degrees										
	Water	Inlet Temperature										
		Inlet Pressure										
		Outlet Temperature										
		Outlet Pressure										
		Flow Rate — GPM										

FORM 160.47-F6

Figure 48–22   A manufacturer's operation log sheet. *Courtesy York International*

Flooded chillers generally have a means of measuring the actual refrigerant temperature in the evaporator either with a thermometer well or a direct readout on the control panel. Direct-expansion chillers typically have a pressure gage on the low-pressure side of the system and the pressure must be converted to temperature for the evaporating refrigerant temperature.

When water is boiled in an open pot and the heat is turned up to the point that the water is boiling vigorously, water will splash out of the pot. Refrigerant in a flooded chiller acts much the same way; the compressor removes vapor from the chiller cavity, which will accelerate the process due to vapor velocity. Liquid eliminators are often placed above the tubes to prevent liquid from carrying over into the compressor suction intake at full load, **Figure 48–16**. The suction gas is often saturated as it enters the suction intake of a compressor on a flooded chiller. This is different from a direct-expansion chiller, which will always have some superheat leaving the evaporator.

🌐The heat exchange surface of both types of tube (direct expansion and flooded) may be improved by machining fins on the outside of the tube. Direct-expansion evaporator tubes may have inserts that cause the refrigerant to sweep the sides of the tube, or the inside of the tube may be rifled like the bore of a rifle, **Figure 48–23**.🌐 These processes cause slight pressure drops, but the results are worth the pressure drop caused.

Evaporators are constructed in a shell with end sheets to hold the tubes. Typically the tubes are made leak free in the tube sheets with a process called rolling. The tube sheet has grooves cut in the hole through which the tube fits. After the tube is inserted in the tube sheet hole, it is expanded with a roller into the grooves for a leak-free connection, **Figure 48–24**. Some chiller evaporators may have silver-soldered tubes at the tube sheet. These tubes would be hard to repair by replacement compared with the ones that are rolled in place.

The evaporator has a working pressure for the refrigerant circuit and for the water circuit. High-pressure chillers must be able to accommodate the refrigerant used, generally R-22, or R134a. Other newer, environmentally friendly alternative refrigerants for newer models may also be used. The water side of the chiller may have one of two working pressures, 150 or 300 psig. The 300-psig chillers are used when the chiller is located on the bottom floor of a tall building where

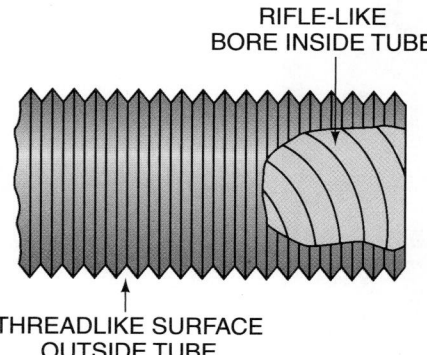

Figure 48–23 🌐Fins on the outside of a chiller tube and a rifle-like bore on the inside are used to expand the surface area and increase the heat exchange per length of tube.🌐

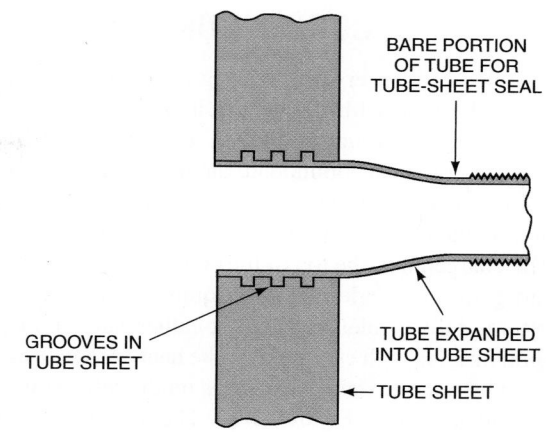

Figure 48–24   The tubes are rolled into the tube-sheet grooves to make a leak-free connection between the refrigerant and water.

the water column creates a high pressure from the standing column of water. Unit 37, "Air Distribution and Balance," indicates that 1 psi of pressure will support a column of water 2.31 ft high. A column of water 1 ft high will exert a pressure of 0.433 psig (1 psi/2.31 ft = 0.4329 or 0.433 psi/ft). A chiller with a 300-psig working pressure can have only 693 ft of water standing above it, which would create a pressure of 300 psig (300 psi/0.433 ft/psi = 692.8 ft). Because a multistory building usually has about 12 ft per story, the chiller could serve only a building that is 58 stories high (692.8/12 = 57.5 stories). Multistory buildings can locate the chillers on upper floors. Service becomes more difficult, but not many other options exist.

The evaporator is normally manufactured, then leak tested. After it is proven to be leak free, it is then insulated. Many larger chillers are insulated with rubber-type foam insulation that is glued directly to the shell. Smaller chillers may be insulated by being placed in a skin of sheet metal and foam insulation applied between the chiller shell and the skin. If any repair is necessary, the skin will have to be removed and the foam cut away. The skin can later be replaced and foam insulation added back to insulate the chiller barrel.

Some evaporators may be located outside, as in the case of an air-cooled package chiller, so freeze protection for the water in the shell is necessary for these chillers. This can be accomplished by adding resistance heaters to the chiller barrel under the insulation. These heaters may be wired into the chiller control circuit and energized by a thermostat if the ambient temperature approaches freezing and water is not being circulated.

48.11   CONDENSERS FOR HIGH-PRESSURE CHILLERS

The condenser is the component in the system that transfers heat out of the system. The condenser for high-pressure chillers may be either water cooled or air cooled. Both will take the heat from the system and transfer the heat out of the system. Usually the heat is transferred to the atmosphere. In some manufacturing processes, the heat may be recovered and used for other applications.

48.12 WATER-COOLED CONDENSERS

Water-cooled condensers used for high-pressure chillers are shell-and-tube type with the water circulating in the tubes and the refrigerant around the tubes. The shell must have a working pressure to accommodate the refrigerant used and a water-side working pressure of 150 to 300 psig, like the evaporators mentioned earlier. The hot discharge gas is normally discharged into the top of the condenser, **Figure 48–25**. The refrigerant is condensed to a liquid and drips down to the bottom of the condenser where it gathers and drains into the liquid line. As with evaporators, the heat exchange can be improved by use of extended surfaces on the refrigerant side of the condenser tube, **Figure 48–23**. The inside of the condenser tubes may also be grooved, as with the evaporator tubes, for improved water-side heat exchange.

Condensers must also have a method for the water to be piped into the shell. There are two basic types of connections to the water box that is attached to the condenser shell. In a standard water box the piping is attached to the removable water box and some of the piping must be removed before the inspection cover can be removed on the piping end. A marine water box has a removable cover for easy access to the piping end of the machine. Tube access is more important for the condenser than the evaporator in most installations because

the open cooling tower provides greater potential for getting the tubes dirty.

Some comfort-cooling installations have both cooling and heating needs at the same time. While the outside of the building may a have heating need in the winter, the inside or core of the building may be generating heat from lighting and equipment that needs to be removed. These applications may have what is known as double-bundled condensers. There are two condensers in one shell. When the heat can be used for the building perimeter, the warm condenser water can be circulated in the heating circuit. When the heat needs to be transferred to the outside, the other condenser may be used and the heat transferred to the cooling tower. With this type of installation, the temperature of the water from the condenser may not be warm enough, so the water temperature may be increased with a boiler. The net result is that the water is preheated to the boiler with the heat from the building interior.

48.13 CONDENSER SUBCOOLING

The design condensing temperature for a water-cooled condenser is around 105°F on a day when 85°F water is supplied to the condenser and the chiller is operating at full load. The saturated liquid refrigerant would be 105°F. Many condensers have subcooling circuits that reduce the liquid temperature to

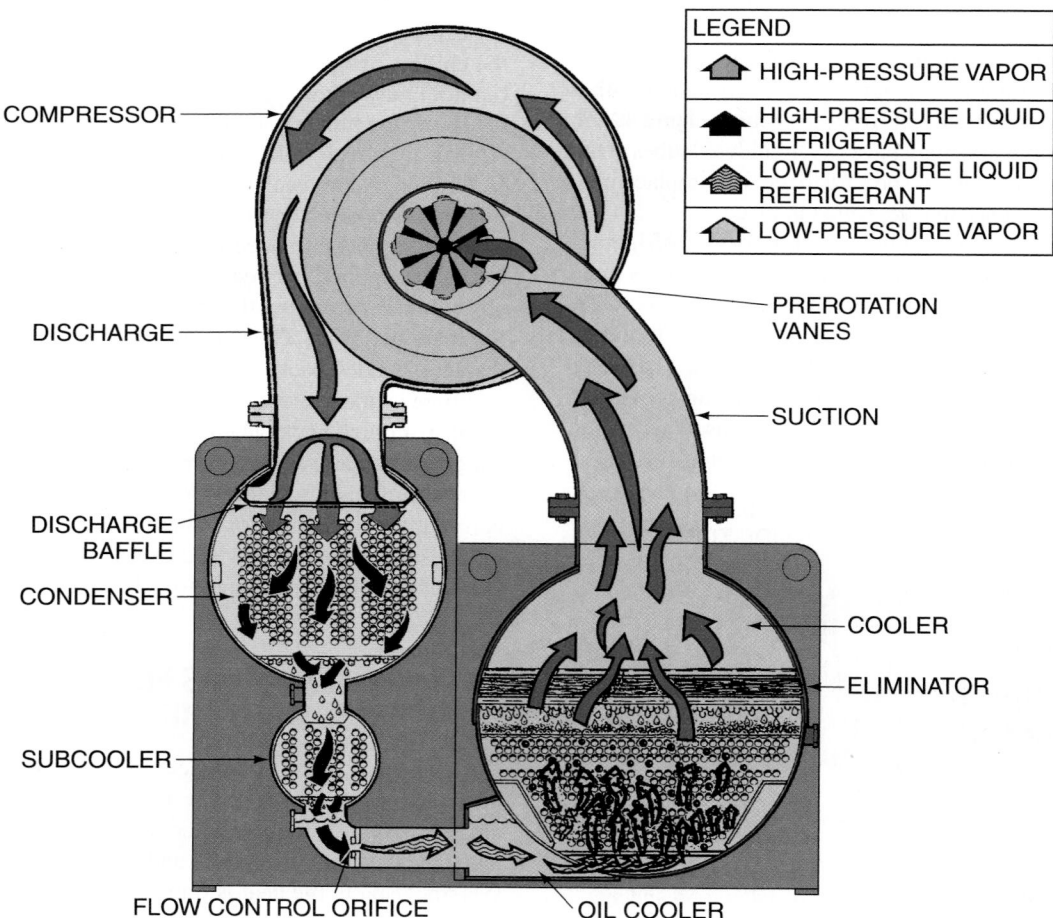

LEGEND	
⬆	HIGH-PRESSURE VAPOR
⬆	HIGH-PRESSURE LIQUID REFRIGERANT
⬆	LOW-PRESSURE LIQUID REFRIGERANT
⬆	LOW-PRESSURE VAPOR

COMPRESSOR

DISCHARGE

DISCHARGE BAFFLE

CONDENSER

SUBCOOLER

FLOW CONTROL ORIFICE

PREROTATION VANES

SUCTION

COOLER

ELIMINATOR

OIL COOLER

Figure 48–25 The subcooling circuit improves the efficiency of the chiller. *Courtesy York International*

below the 105°F. If the entering water at 85°F is allowed to exchange heat from the saturated refrigerant, the liquid-line temperature can be reduced. 🌐 If it can be reduced to 95°F, a considerable amount of capacity can be gained for the evaporator. This subcooling can increase the machine capacity about 1%/°F of subcooling. For a 300-ton chiller, 10°F subcooling would increase the capacity about 30 tons with very little energy use. 🌐 The only expense is the extra circuit in the condenser. To subcool refrigerant, it must be isolated from the condensing process with a separate circuit, **Figure 48–25**.

The condenser, like the evaporator, has an approach relationship between the refrigerant condensing temperature and the leaving-water temperature. Most water-cooled condensers are two-pass condensers and are designed for 85°F entering water and 95°F leaving water with a condensing temperature of about 105°F, **Figure 48–26**. This is an approach temperature of 10°F and is important to the technician. Other configurations of condenser may be either one-, three-, or four-pass with different approach temperatures. As with evaporators, the longer the refrigerant is in contact with the water, the closer the approach temperature will be. The original operating log will reveal what the approach temperature was in the beginning. It should be the same for similar conditions at a later date, provided the tubes are clean.

Because the condenser rejects heat from the system, its capacity to reject heat determines the head pressure for the refrigeration process. There must be some pressure in the condenser to push the refrigerant through the expansion device. The head pressure must be controlled. Head pressure may become too low, and problems may occur at start-up and in very cold weather when there is still a call for air conditioning.

When there is a call to start up a water-cooled chiller, and the water in the cooling tower is cold, it may cause the head pressure to become so low that the suction pressure will be

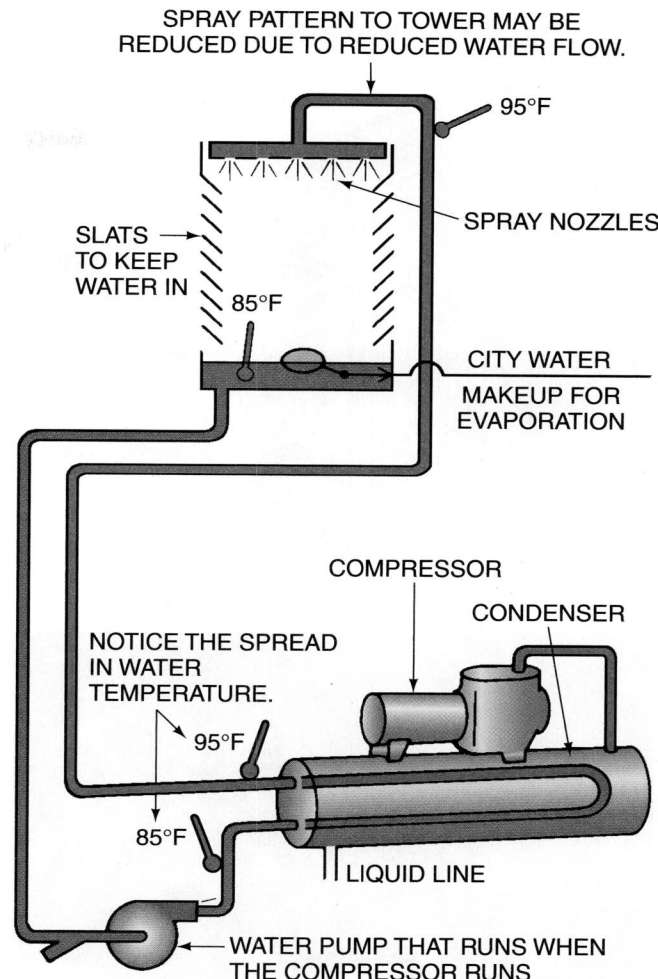

Figure 48–27   Heat moves from the refrigerant to the water and then into the atmosphere from the cooling tower.

low enough to trip either the low-pressure control (direct-expansion chillers) or the evaporator freeze control (flooded chillers). Continued operation will also cause oil migration in many machines. A bypass valve is typically located in the condenser circuit to bypass water during start-up to prevent nuisance shutdowns due to cold condenser water. This same bypass may be used to bypass water during normal operation. This is discussed in more detail in Unit 49, "Cooling Towers and Pumps."

When water-cooled condensers are used, the heat from the refrigerant is transferred into the water circulating in the condenser circuit. The heat must now be transferred to the atmosphere in a typical installation. This is done by means of a cooling tower, **Figure 48–27**. These are discussed in Unit 49.

48.14   AIR-COOLED CONDENSERS

Air-cooled condensers are generally constructed of copper tubes with the refrigerant circulating inside and with aluminum fins on the outside to provide a larger heat exchange surface, **Figure 48–28**. These air-cooled heat exchange surfaces

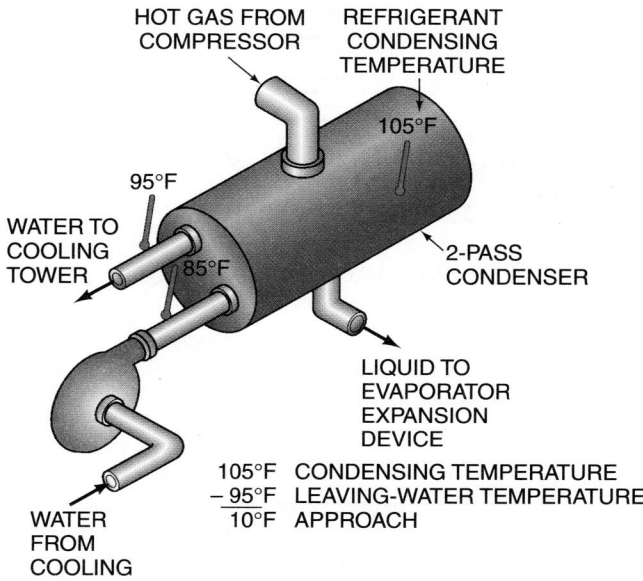

Figure 48–26   The approach temperature for a two-pass condenser.

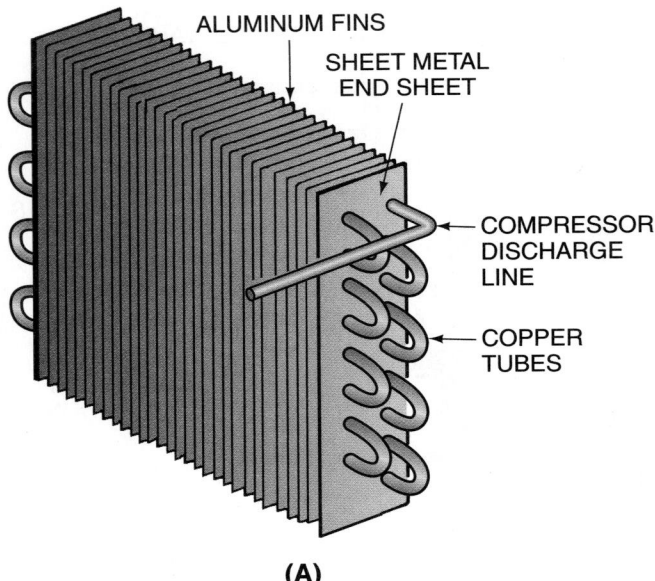

(A)

(B)

Figure 48–28 **(A)** An air-cooled condenser coil. **(B)** An air-cooled chiller. The condenser coils are at the top. **(B)** *Courtesy York International*

have been used for many years with great success. Some manufacturers furnish copper finned tubes or special coatings for the aluminum fins for locations where salt air may corrode the aluminum.

These coils are positioned differently by the manufacturer for their own design purposes. Some coils are positioned in a horizontal and some in a vertical mode.

Multiple fans are used for many of these condensers. The fans may be cycled on and off for head pressure control. Various types of head pressure control for air-cooled condensers are discussed in Unit 25, "Special Refrigeration System Components."

Air-cooled condensers range in size from very small to several hundred tons. Air-cooled condensers eliminate the need for using water towers and the problems that occur when using water. Systems using air-cooled condensers do not operate at the low condensing temperatures and head

pressures found with water-cooled condensers. Water-cooled condensers typically condense at about 105°F. Air-cooled condensers condense at 20°F to 30°F higher than the entering-air temperature depending on the condenser surface area per ton of condenser. For example, an R-22 system with a water-cooled condenser would have a head pressure about 211 psig (105°F compares to 211 psig). An air-cooled condenser on a 95°F day would have a head pressure of from 243 psig (95°F + 20°F = 115°F) to 278 psig (95°F + 30°F = 125°F). This is a considerable difference in head pressure and will increase the operating cost of the chiller. Air-cooled chillers are popular because there is less maintenance.

48.15 SUBCOOLING CIRCUIT

Subcooling in an air-cooled condenser is just as important as in a water-cooled unit. It gives the unit more capacity by about 1%/°F of subcooling. Subcooling is accomplished in an air-cooled condenser by means of a small reservoir in the condenser to separate the subcooling circuit from the main condenser, **Figure 48–29.** A condenser operating on a 95°F day can expect to achieve 10°F to 15°F of subcooling.

Many units use the bottom of the condenser for the subcooling process by flooding the bottom of the condenser with refrigerant. This would be like an overcharge of refrigerant, except that the condenser has extra tubes to hold the extra charge. The technician must be aware of this subcooling circuit and charge the unit to the correct subcooling level, or the unit will not operate at rated capacity. The manufacturers that use a reservoir use a sight glass in the reservoir and charge is added until there is a level of refrigerant in the

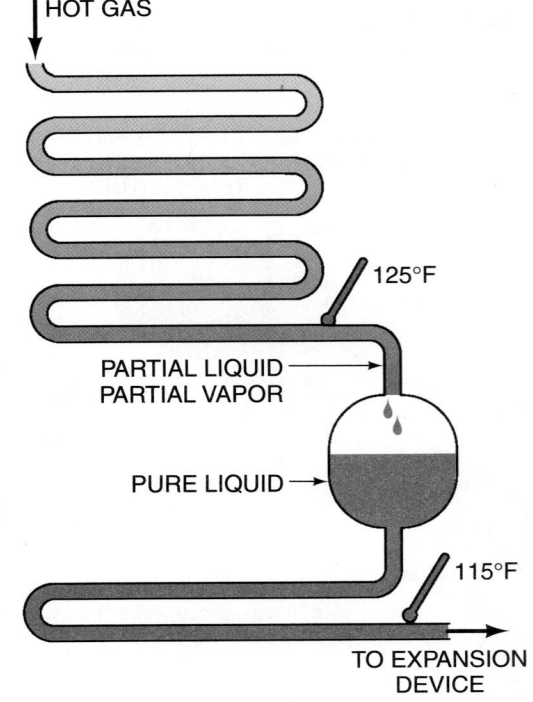

Figure 48–29 The subcooling circuit in an air-cooled condenser.

reservoir. Some manufacturers use sight glasses in the condenser end sheet to show the level of charge in the condenser. Otherwise, the technician must use a temperature lead attached to the liquid line to determine the subcooling temperature. This is accomplished by converting head pressure to condensing temperature and adding refrigerant until the liquid line is lower in temperature than the condensing temperature. For example, the condensing temperature may be 125°F. To arrive at a subcooling of 15°F, the technician would charge the unit until the liquid-line temperature is 110°F. This is a much longer process than using a sight glass and there is much more room for technician error.

48.16 METERING DEVICES FOR HIGH-PRESSURE CHILLERS

The metering device holds liquid refrigerant back and meters the liquid into the evaporator at the correct rate. Four types of metering devices may be used for large chillers: thermostatic expansion valve, orifice, high- and low-side floats, and electronic expansion valve.

48.17 THERMOSTATIC EXPANSION VALVE

The *thermostatic expansion valve (TXV)* was discussed in detail in Unit 24, "Expansion Devices." The TXVs used with chillers are the same type, only larger. The TXV maintains a constant superheat at the end of the evaporator. Usually this is 8°F to 12°F of superheat. The more superheat, the more vapor there will be in the evaporator and less heat exchange will occur. The more liquid the evaporator has, the better the heat exchange, so TXV valves are not used except on smaller chillers, up to about 150 tons.

48.18 ORIFICE

The orifice metering device is a fixed-bore metering device that is merely a restriction in the liquid line between the condenser and the evaporator, **Figure 48–30**. These are trouble free, because they have no moving parts.

The flow of refrigerant through an orifice is a constant at a given pressure drop. When there is a greater pressure difference, more flow will occur. The flow at greater loads is accomplished because of a higher head pressure at greater loads. When the head pressure is allowed to rise, more flow will occur. Any chiller with an orifice for a metering device has a critical charge. The unit must not be overcharged or liquid refrigerant will enter the compressor inlet. This could be detrimental to a reciprocating compressor so orifices may not be used where these compressors are used. Small amounts of liquid refrigerant, often in the form of wet vapor, will not harm the scroll, rotary screw, or most centrifugal compressors.

48.19 FLOAT-TYPE METERING DEVICES

There are two types of float-type metering devices: the *low-side float* and the *high-side float*. The low-side float rises and throttles back the refrigerant flow when the liquid refrigerant level rises in the low side of the chiller, **Figure 48–31**. It is located at the correct level to produce the level of refrigerant required for the chiller.

The high-side float is located in the liquid line entering the evaporator. When the level of refrigerant is greater in the liquid line, it is because the evaporator needs liquid refrigerant and the float rises, allowing liquid refrigerant to enter the evaporator, **Figure 48–32**.

Floats can be a problem if they rub the side of the float chamber causing resistance or if a hole rubs in the float. If a hole develops in the float, it will sink. In the case of the low-side float, it will open and allow full flow of refrigerant to the evaporator and no refrigerant will be held back to the liquid line and in the bottom of the condenser. If a hole wears in the high-side float, it will sink and block liquid to the evaporator.

The refrigerant charge is critical for both the low- and high-side floats. An overcharge for a low-side float will back refrigerant up into the liquid line and condenser. An overcharge of refrigerant will flood the evaporator when a high-side float is used.

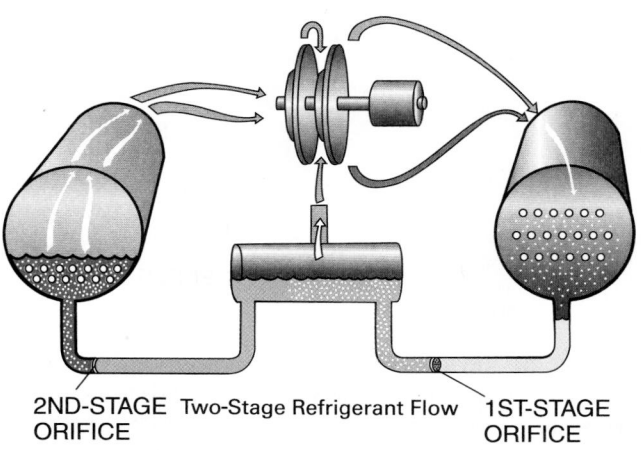

2ND-STAGE ORIFICE Two-Stage Refrigerant Flow **1ST-STAGE ORIFICE**

Figure 48–30 An orifice metering device. *Courtesy Trane Company*

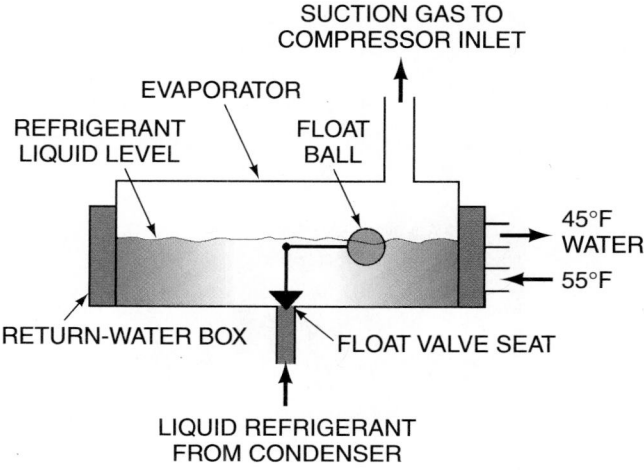

Figure 48–31 A low-side float.

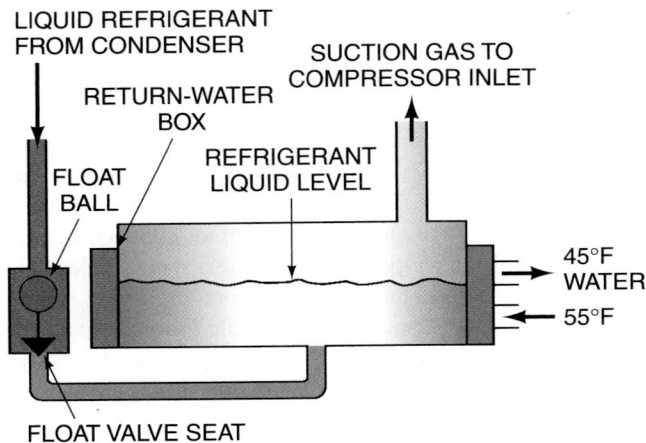

Figure 48–32 A high-side float.

48.20 ELECTRONIC EXPANSION VALVE

The electronic expansion valve for large chillers is similar to the one explained in Unit 24. These valves operate with a thermistor to monitor the refrigerant temperature. The liquid does not reach the sensing element; an electronic circuit checks the actual temperature. This is much like a thermometer except it does not read out in temperature. A pressure transducer may be used to give the electronic circuit a pressure reading that can be converted to temperature. With the pressure reading from the evaporator converted to temperature and with the actual suction-line temperature, the real superheat can be determined and close control can be achieved.

The electronic expansion valve looks much like a solenoid valve in the line, **Figure 48–33.** The thermistor sensor can be inserted in the evaporator piping.

🌐 An advantage of the electronic expansion valve is the electronic control circuit versatility. The valve can be used for a wider variation in loads than a typical TXV. It can also be

used to allow more refrigerant to flow during low head pressure conditions during low-ambient operation. 🌐 This is particularly helpful for air-cooled units when the weather is cold. The large liquid handling capacity of the electronic expansion valve can allow maximum liquid refrigerant flow.

These electronic expansion valves are often called step-control valves because there can be so many increments of modulated flow with their controls. **Figure 48–34** is a photo of a step-control expansion valve with an illustration showing that the motor turns a shaft with a needle and seat to modulate flow through the valve seat. **Figure 48–35** shows

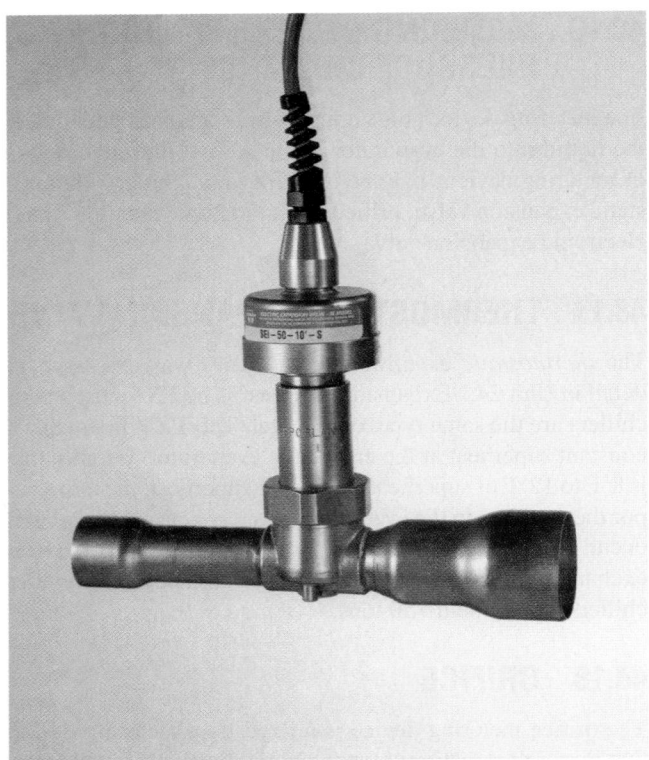

(A)

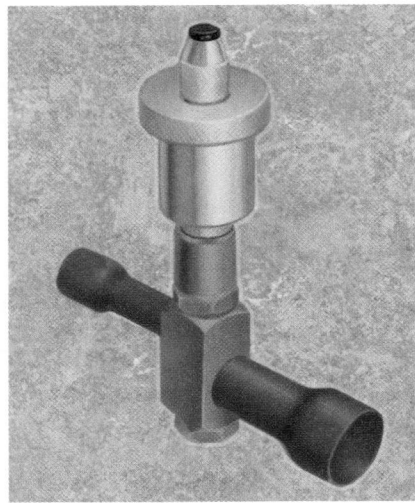

Figure 48–33 An electronic expansion valve. *Courtesy Trane Company*

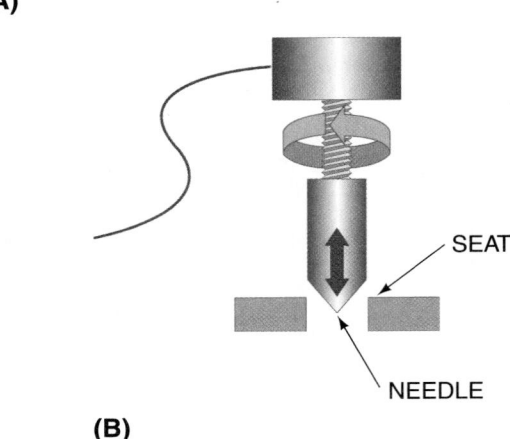

(B)

Figure 48–34 **(A)** A step-motor-control expansion valve for large systems. **(B)** This illustration shows how the needle is threaded into the seat by means of the motor. This device can provide an infinite number of steps of control. *(A and B) Courtesy Sporlan Division, Parker Hannifin Corp.*

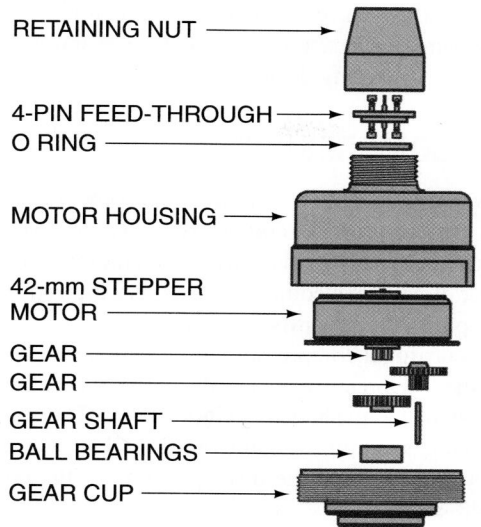

Figure 48–35 This is an exploded view of the gears and motor that accomplish step control. *Courtesy Sporlan Division, Parker Hannifin Corp.*

an exploded view of the valve with all of its intricate parts. **Figure 48–36** is a wiring diagram showing how a valve may be wired into the circuit. Notice that there are two sensors (thermistors): one on the inlet of the coil and one on the outlet. There is no pressure connection for checking superheat. Temperature difference is used. There is also an optional pumpdown feature. The valve can be closed when the system thermostat is satisfied and the refrigerant can be pumped into the condenser. The valve would be reopened at the beginning of the next cycle to allow refrigerant to flow naturally during the ON cycle. The pumpdown cycle

is used for any system that might suffer if liquid refrigerant can be pumped into the compressor at the beginning of a new cycle because the refrigerant has migrated to the evaporator during the OFF cycle.

This ability to allow maximum flow on a cold day can give the condenser much more capacity to pass on to the system, for example, if it is 40°F outside and the building is calling for cooling. The condenser can condense the refrigerant at about 60°F to 70°F. This is a head pressure of 102 to 121 psig for R-22. R-12 and R-500 would have head pressures corresponding to these temperatures. This is not enough head pressure for a typical TXV. But an electronic expansion valve (because of its flow capability) can use this low pressure to feed the evaporator. If the refrigerant is then subcooled 10°F more, the evaporator can be furnished 50°F to 60°F liquid refrigerant. 🌎 The compressor is only required to pump from an evaporator pressure of approximately 65.6 psig (corresponding to 38°F evaporator boiling temperature) to 121 psig maximum head pressure.🌎 This is a big difference from a typical summer condition where the evaporator suction pressure may be 65.6 psig and the head pressure 278 psig (corresponding to a condensing temperature of 125°F) for R-22.

Each manufacturer has different design features, but all of them attempt to build a reliable, low-cost chiller that will last a long time.

48.21 LOW-PRESSURE CHILLERS

Low-pressure chillers typically have used R-11, R-113, or R-123 for their refrigerants. Most small low-pressure chillers in the range up to approximately 150 tons have used R-113, and chillers above this range, R-11. Newer low-pressure chillers

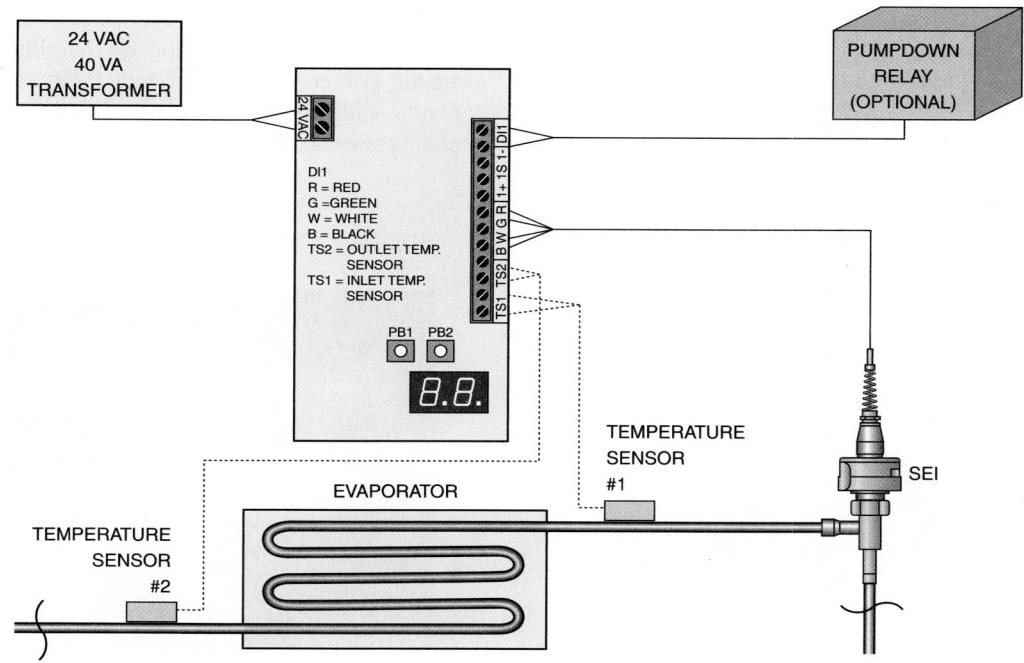

Figure 48–36 This wiring and piping diagram shows two temperature sensors and the wiring diagram that would be used for a step-control expansion valve.

and those that have been converted are using R-123. Some use R-114 for special applications. ♻ As of July 1, 1992, it became unlawful to vent CFC and HCFC refrigerants into the atmosphere. As of November 15, 1995, it became unlawful to vent HFC refrigerants and any other alternative refrigerant into the atmosphere. Also, on January 1, 1996, the phaseout of CFC refrigerants began. This phaseout made it unlawful to manufacture any CFC refrigerant. R-11, R-113, and R-114 are all low-pressure, CFC refrigerants. R-123 is an HCFC refrigerant. ♻ **Figure 48–37** shows a temperature/pressure chart for the above low-pressure refrigerants.

VAPOR PRESSURES				
TEMP °F	113	11	114	123
−150.0				
−140.0				
−130.0				
−120.0				
−110.0			29.7	
−100.0			29.5	
−90.0		29.7	29.3	
−80.0		29.6	29.0	29.7
−70.0		29.4	28.6	29.6
−60.0		29.2	28.0	29.4
−50.0	29.6	28.9	27.1	29.1
−40.0	29.5	28.4	26.1	28.8
−35.0	29.4	28.1	25.4	28.6
−30.0	29.3	27.8	24.7	28.3
−25.0	29.2	27.4	23.8	28.1
−20.0	29.0	27.0	22.9	27.7
−15.0	28.8	26.6	21.8	27.4
−10.0	28.7	26.0	20.6	26.9
−5.0	28.4	25.4	19.3	26.4
0.0	28.2	24.7	17.8	25.9
5.0	27.9	23.9	16.2	25.2
10.0	27.5	23.1	14.4	24.5
15.0	27.2	22.1	12.4	23.7
20.0	26.7	21.1	10.2	22.9
25.0	26.3	19.9	7.8	21.9
30.0	25.7	18.6	5.1	20.8
35.0	25.1	17.1	2.2	19.5
40.0	24.4	15.6	0.4	18.2
45.0	23.7	13.8	2.1	16.7
50.0	22.9	12.0	3.9	15.0
55.0	21.9	9.9	5.9	13.2
60.0	20.9	7.7	8.0	11.2
65.0	19.8	5.2	10.3	9.1
70.0	18.6	2.6	12.7	6.7
75.0	17.3	0.1	15.3	4.1
80.0	15.8	1.6	18.2	1.3
85.0	14.2	3.3	21.2	0.9
90.0	12.5	5.0	24.4	2.5
95.0	10.6	6.9	27.8	4.2
100.0	8.6	8.9	31.4	6.1
105.0	6.4	11.1	35.3	8.1
110.0	4.0	13.4	39.4	10.3
115.0	1.4	15.9	43.8	12.6
120.0	0.7	18.5	48.4	15.1
125.0	2.1	21.3	53.3	17.8
130.0	3.7	24.3	58.4	20.6
135.0	5.3	27.4	63.9	23.6
140.0	7.1	30.8	69.6	26.8
145.0	9.0	34.3	75.6	30.2
150.0	11.1	38.1	82.0	33.9

Figure 48–37   A temperature/pressure chart that includes low-pressure refrigerants.

These chillers have all of the same components as the high-pressure chillers: a compressor, an evaporator, a condenser, and a metering device.

48.22 COMPRESSORS

All low-pressure chillers as well as some high-pressure chillers use centrifugal compressors. These low-pressure centrifugal compressors also turn from a motors synchronous speed for direct-drive compressors to very high speed (about 30,000 rpm) with gear-drive machines.

The centrifugal compressor has an impeller that turns and creates the low-pressure area in the center where the suction line is fastened to the housing, **Figure 48–38**. The compressed refrigerant gas is trapped in the outer shell known as the volute and guided down the discharge line to the condenser, **Figure 48–39**.

Centrifugal compressors can be manufactured to produce more pressure by building more than one compressor and operating them in series, called multistage, or operating one stage of compression at a high speed. Actually with multistage compressors, the discharge from one compressor becomes the suction for the next compressor, **Figure 48–40**. It is common for compressors for air-conditioning applications to have up to three stages of compression. **Figure 48–38** shows a modern three-stage compressor that can be used for either R-11 or R-123. One of the advantages of multistage operation is that the compressor can be operated at slow speed, typically the speed of the motor, about 3600 rpm.

🌐Refrigerant vapor can also be drawn off the liquid leaving the condenser to subcool the liquid entering the evaporator metering device.🌐 The refrigerant vapor is drawn off using the intermediate compressor stages so the liquid temperature corresponds to the pressure of the compressor intermediate suction pressure; it is less than the pressure corresponding to the condensing temperature. 🌐The subcooling process gives the evaporator more capacity for its respective size at very little additional pumping cost. This process is called an economizer cycle, **Figure 48–40.**🌐

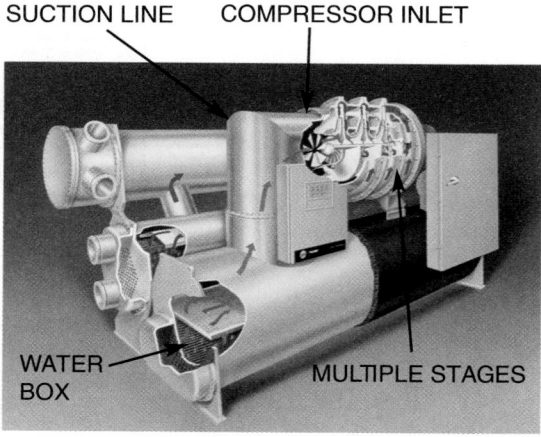

SUCTION LINE   COMPRESSOR INLET

WATER BOX   MULTIPLE STAGES

Figure 48–38   A multiple-stage compressor and chiller. *Courtesy Trane Company*

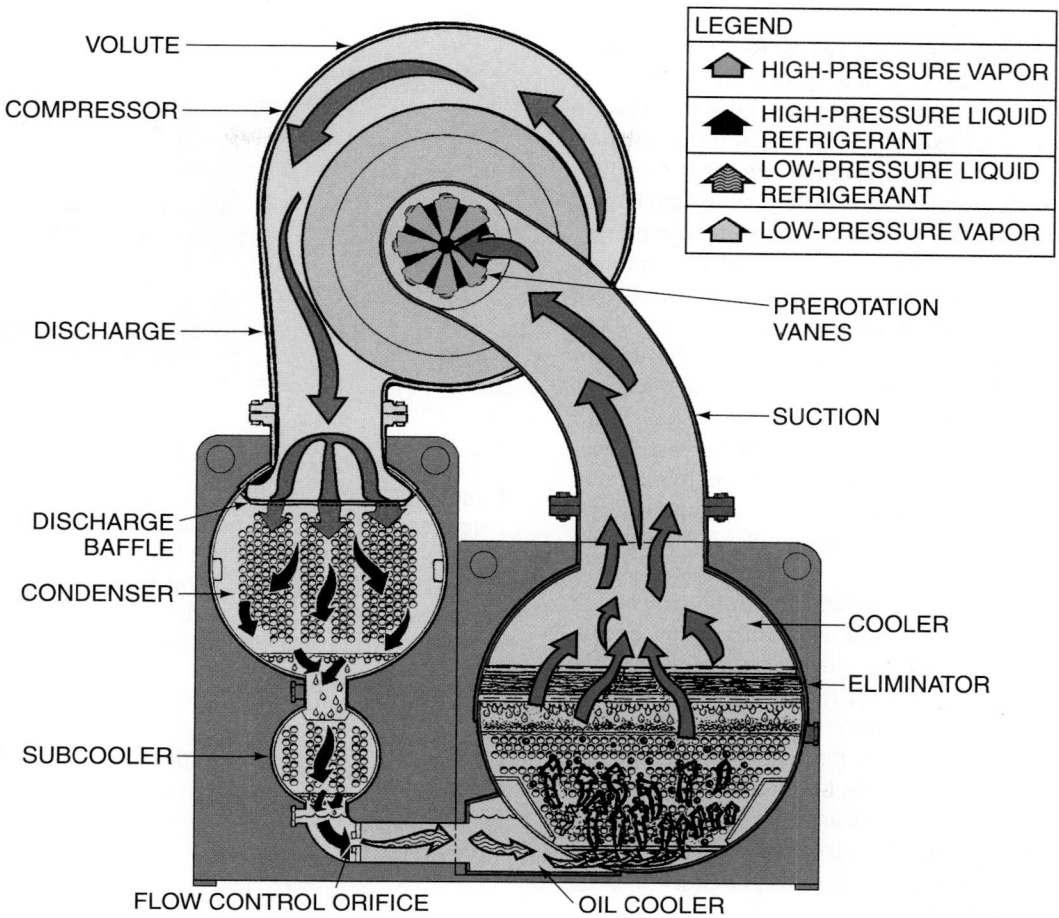

LEGEND

⬆	HIGH-PRESSURE VAPOR
⬆	HIGH-PRESSURE LIQUID REFRIGERANT
⬆	LOW-PRESSURE LIQUID REFRIGERANT
⬆	LOW-PRESSURE VAPOR

Figure 48–39   The hot gas enters the top of the condenser through the discharge line. *Courtesy Trane Company*

THREE-STAGE OPERATING CYCLE

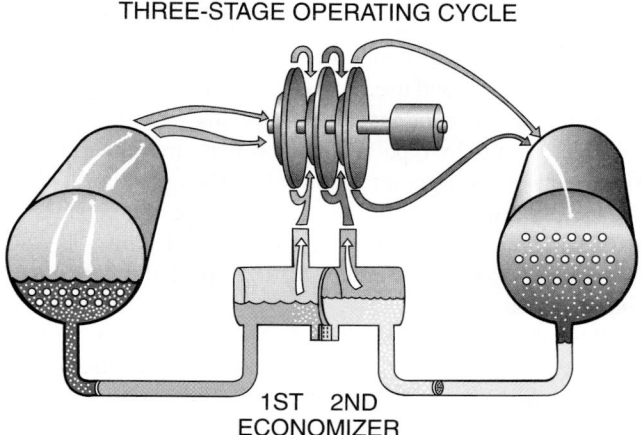

Figure 48–40   A three-stage centrifugal chiller with two economizer cycles. *Courtesy Trane Company*

Single-stage compression can be accomplished using a single compressor that turns at a high speed, up to about 30,000 rpm. This is done by means of a 3600-rpm motor and a gearbox to step up the speed of the compressor, **Figure 48–10.** The advantage of this compressor is size and weight.

The amount of compression needed depends on the refrigerant used. For example, most centrifugal compressors used for chillers use two-pass evaporators and two-pass condensers. If the evaporator has a 7°F approach temperature, the evaporator would boil the refrigerant at 40°F for a chiller with 47°F leaving water. NOTE: We are using 47°F water to get an evaporator temperature of 40°F for use on a typical temperature/pressure chart. If the condenser has a 10°F approach temperature and is receiving 85°F water from the tower and has 95°F leaving water, the condenser should be condensing the refrigerant at about 105°F, **Figure 48–26.** The compressor must overcome the following pressures to meet the requirement for this system:

Refrigerant Type	Evap Press 40°F	Cond Press 105°F
R-113	24.4 in. Hg Vac	6.4 in. Hg Vac
R-11	15.6 in. Hg Vac	11.1 psig
R-123	18.1 in. Hg Vac	8.1 psig
R-114	0.44 psig	35.3 psig
R-500	46 psig	152.2 psig
R-502	80.5 psig	231.7 psig
R-12	37 psig	126.4 psig
R-134a	35 psig	134.9 psig
R-22	68.5 psig	210.8 psig

Notice that the compressor has only to raise the pressure from 24.4 in. Hg Vac to 6.4 in. Hg Vac for R-113; this is only 18 in. Hg Vac (or $18/2.036 = 8.84$ psig). The compressor that has R-502 in the system would have to raise the pressure from 80.6 psig to 231.7 psig or 151.1 psig. A different compressor would have to be used than the one for R-113. The compressors do not raise the pressure the same amount nor do they operate in the same pressure ranges. One compressor operates in a vacuum on both the high- and low-pressure sides of the system and the other operates under a discharge pressure of 231.7 psig. The physical thickness of the shell of the two compressors can be very different. The above comparison will give you some idea of the different applications for centrifugals.

Many of the features and operation of low-pressure centrifugal chillers are the same or similar to those found on or with high-pressure units. **Figure 48–41** shows a cutaway of a high-pressure chiller shell but looks similar to most low-pressure chillers.

The evaporator on a low-pressure system has a working pressure for the refrigerant circuit and for the water circuit. Low-pressure chillers must be able to accommodate the refrigerant used, typically R-113, R-11, or R-123. Because these are low-pressure chillers, the refrigerant side of the shell does not have to be as strong as the high-pressure chillers. These can be much lighter weight because of the pressures. The low-pressure chiller shell may have a refrigerant working pressure as low as 15 psig. The refrigerant safety device for the system is called a rupture disc; its relief pressure is 15 psig, and it is located in the low-pressure side of the system,

Figure 48–42 A rupture disc mounted on the suction line of the compressor. These rupture discs are typically set at 15 psig. The disc is on the suction side so it is under low pressure during operation. If the shell were to be subject to fire or if a technician tried to overpressure the shell for leak-testing purposes, the disc would blow out and protect the shell from splitting. *Courtesy York International*

on the evaporator section or the suction line, **Figure 48–42.** This device prevents excess pressure on the shell. If a fire occurs or a technician uses too much pressure, the shell is safe from rupture.

48.23 CONDENSERS FOR LOW-PRESSURE CHILLERS

The condenser is the component that rejects heat from the system. The condensers for low-pressure chillers are water cooled. As with high-pressure systems, the heat is usually transferred to the atmosphere. In some manufacturing processes, the heat may be recovered and used for other applications.

Water-cooled condensers used for low-pressure chillers are shell-and-tube type with the water circulating in the tubes and the refrigerant around the tubes. The shell must have a working pressure to accommodate the refrigerant used and a water-side working pressure of 150 to 300 psig. The hot discharge gas enters at the top of the condenser, **Figure 48–39.**

Because the condenser rejects heat from the system, its capacity to reject heat determines the head pressure for the refrigeration process. The condenser is usually located above the evaporator with low-pressure chillers. The compressor lifts the refrigerant to the condenser level in the vapor state. When it is condensed to a liquid, it flows by gravity with very little pressure difference to the evaporator through the metering device. The head pressure must be controlled so that it does not drop too low below the design pressure drop.

Water-cooled condensers may also have a subcooling circuit. The purpose of the subcooling circuit is to lower the liquid refrigerant temperature to a level below the condensing

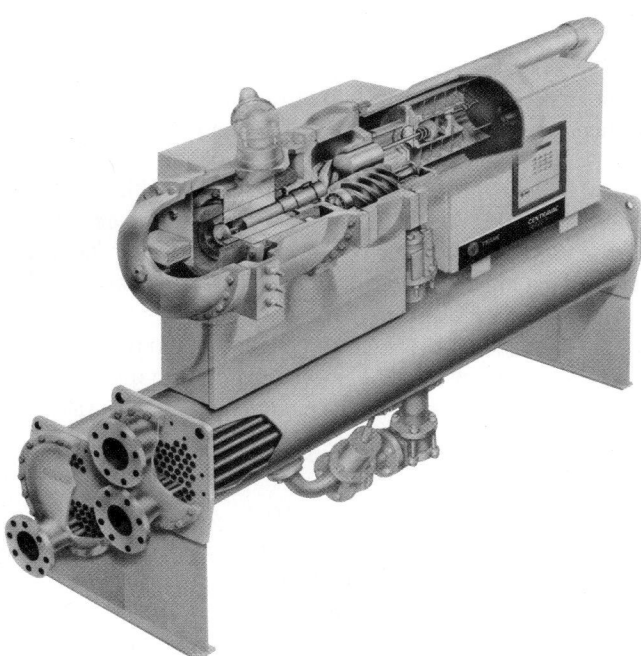

Figure 48–41 A cutaway of a high-pressure chiller shell. Low-pressure shells are much like this. *Reproduced courtesy of Trane Company*

temperature. It is accomplished by means of a separate chamber in the bottom of the condenser where the entering condenser water (at about 85°F) can remove heat from the condensed liquid refrigerant at about 105°F. This gives the system more efficiency, **Figure 48–25.**

48.24 METERING DEVICES FOR LOW-PRESSURE CHILLERS

The metering device holds liquid refrigerant back and meters the liquid into the evaporator at the correct rate. Two types of metering devices are typically used for low-pressure chillers, the orifice and the high- or low-side float. These were discussed under high-pressure chillers.

48.25 PURGE UNITS

When a centrifugal uses a low-pressure refrigerant, the low-pressure side of the system is always in a vacuum when in operation. If the system uses R-113, the complete system is in a vacuum. This means that any time the machine has a leak, it is in a vacuum, and air will enter the machine, **Figure 48–43.** Air may cause several problems. It contains oxygen and moisture and will mix with the refrigerant, which will cause a mild acid that will eventually attack the motor windings and cause damage, or motor burn. Air will move to the condenser while the machine is running, but it will not condense. It will cause the head pressure to rise. If much air enters the system,

the head pressure will rise to the point that the machine will either cause a "surge" or shut down on high head pressure.

When the air moves to the condenser, it will collect at the top and take up condenser space. This air can be collected and removed with a purge system. The purge is a separate device that often has a compressor to collect a sample of whatever is in the top of the condenser for the purpose of trying to condense it in a separate condenser. If it can condense the product from the top of the condenser, it is refrigerant and it will return it to the system at the evaporator level, **Figure 48–44.** If it cannot condense the sample, the purge pressure will rise and a relief valve will allow this sample to be exhausted to the atmosphere.

Several types of purge systems exist, but they all perform the same function, some more efficiently than others. Some purge systems use the difference in pressure from the high-pressure side of the system to the low-pressure side of the system to create the same pressure difference that a purge compressor will create—they use a condenser to condense any refrigerant from the sample.

♻ Older purges were not efficient, and a high percentage of the sample that was relieved to the atmosphere contained refrigerant. This is considered poor practice today because of environmental concerns and the price of lost refrigerant. Modern-day purges must be efficient and allow only a small part of 1% of loss during the relieving process. ♻ **Figure 48–45** shows one of the modern purges that releases only a very small amount of refrigerant when relieving the purge pressure. The vapor is relieved to the atmosphere only when air is present. A leak-free system will contain no air and the purge system will not have to function.

Some manufacturers relieve the purge pressure by means of a pressure switch and a solenoid valve. A counter tells the operator when the purge functions. The operator can use this as a guide for determining when a system has a leak and air is entering. When too many purge reliefs occur, the system is considered to have a leak.

48.26 ABSORPTION AIR-CONDITIONING CHILLERS

Absorption refrigeration is a process that is considerably different from the compression refrigeration process discussed in the preceding paragraphs. The absorption process uses heat as the driving force instead of a compressor. 🌐When heat is plentiful or economical, or when it is a by-product of some other process, absorption cooling can be attractive. For example, in a manufacturing process where steam is used, it is often used at 100 psig and must be condensed before it can be reintroduced to the boiler.🌐 Condensing steam in the absorption system process is a natural choice for chilling water for air conditioning. 🌐In this case, chilled water for air conditioning can be furnished at an economical cost, basically for the cost of operating all related pumps because no compressor is involved.🌐 In other cases, natural gas is an inexpensive fuel in the summer. The gas company is looking for ways to market gas in the summer because their winter

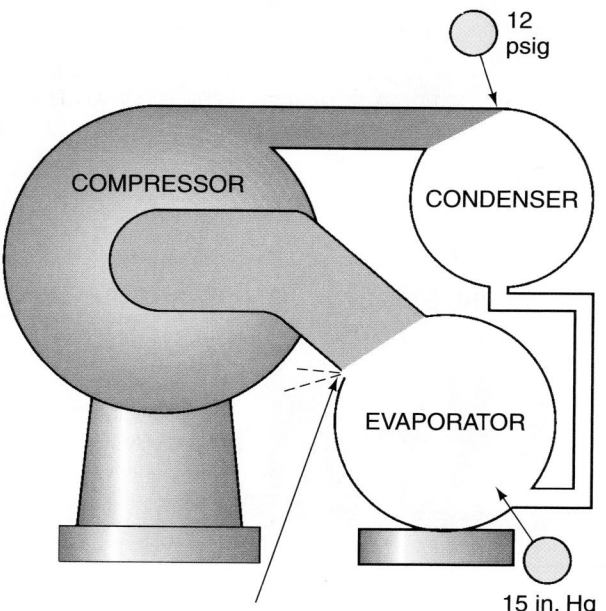

LEAK AT GASKET. AIR ENTERS BECAUSE EVAPORATOR IS IN A VACUUM. THE AIR WILL BE PUMPED TO THE CONDENSER BY THE COMPRESSOR, WHERE IT WILL STAY BECAUSE IT CANNOT CONDENSE.

Figure 48–43 When a low-pressure chiller leaks while in a vacuum, atmosphere enters the chiller shell.

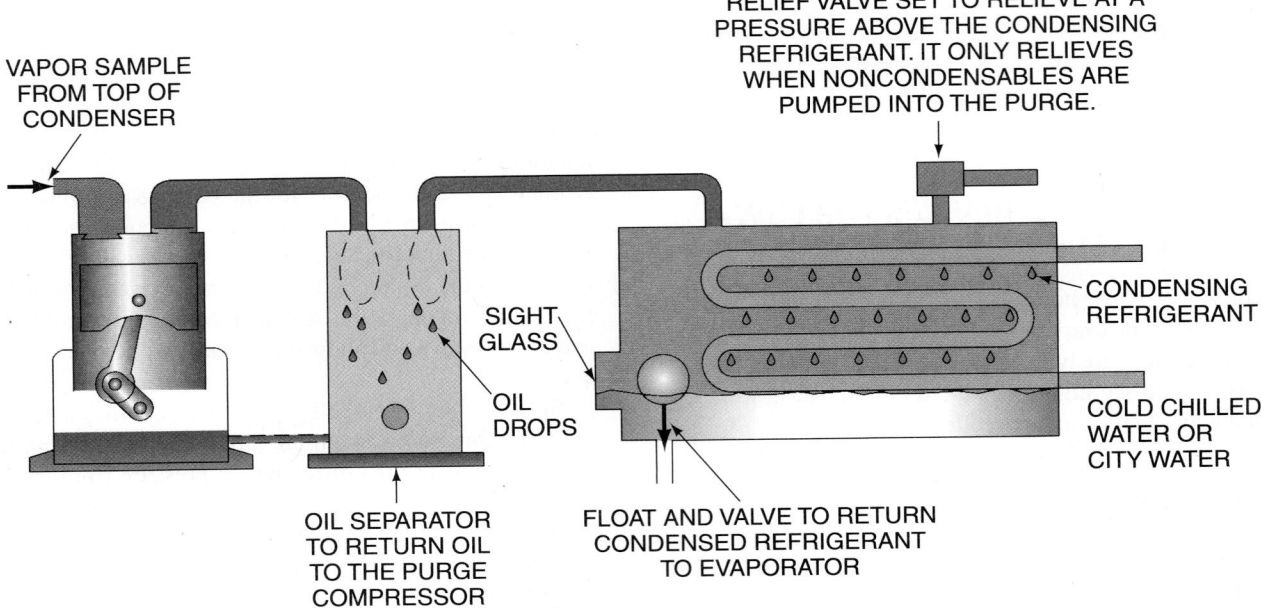

VAPOR SAMPLE
FROM TOP OF
CONDENSER

RELIEF VALVE SET TO RELIEVE AT A
PRESSURE ABOVE THE CONDENSING
REFRIGERANT. IT ONLY RELIEVES
WHEN NONCONDENSABLES ARE
PUMPED INTO THE PURGE.

SIGHT
GLASS

OIL
DROPS

CONDENSING
REFRIGERANT

COLD CHILLED
WATER OR
CITY WATER

OIL SEPARATOR
TO RETURN OIL
TO THE PURGE
COMPRESSOR

FLOAT AND VALVE TO RETURN
CONDENSED REFRIGERANT
TO EVAPORATOR

Figure 48–44 A simplified purge unit in operation.

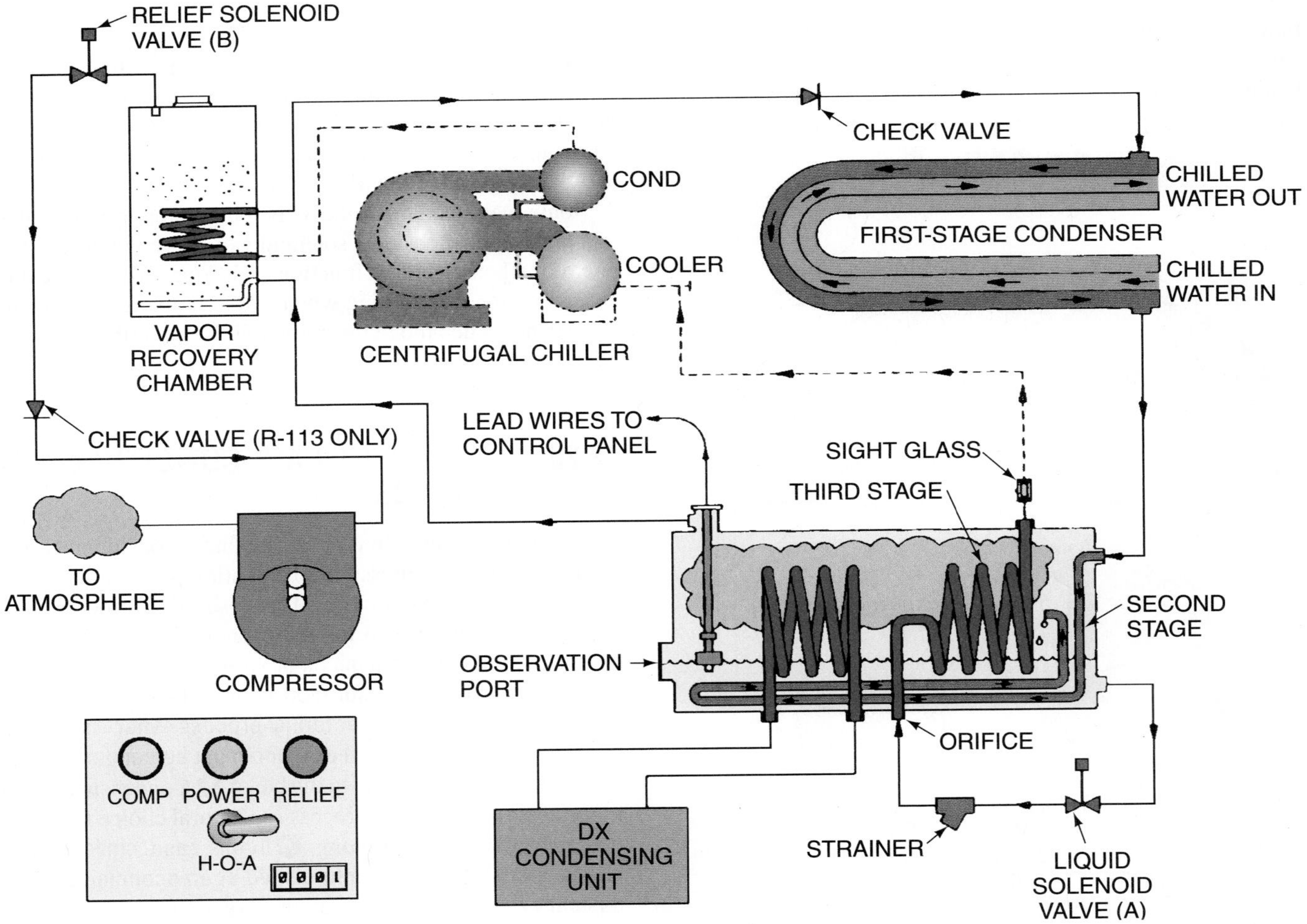

RELIEF SOLENOID
VALVE (B)

CHECK VALVE

COND

COOLER

CHILLED
WATER OUT

FIRST-STAGE CONDENSER

CHILLED
WATER IN

VAPOR
RECOVERY
CHAMBER

CENTRIFUGAL CHILLER

CHECK VALVE (R-113 ONLY)

LEAD WIRES TO
CONTROL PANEL

SIGHT GLASS

THIRD STAGE

TO
ATMOSPHERE

SECOND
STAGE

COMPRESSOR

OBSERVATION
PORT

ORIFICE

COMP POWER RELIEF

H-O-A

DX
CONDENSING
UNIT

STRAINER

LIQUID
SOLENOID
VALVE (A)

Figure 48–45 A modern purge with features that reduce the amount of refrigerant that escapes with the noncondensable gases. *Courtesy Carolina Products, Inc.*

quota may depend on how much gas they can sell in the summer. Some gas companies may offer incentives to install gas appliances that consume fuel in the off-peak summer months. ♻ Absorption refrigeration equipment also operates in a vacuum and contains non-CFC refrigerants. ♻

The absorption system equipment looks much like a boiler except it has chilled-water piping and condenser-water piping routed to it in addition to the piping for steam or hot water. Oil or gas burners are part of the system if it is a direct-fired chiller. **Figure 48–46** shows several absorption systems by different manufacturers. The small machines are considered package

machines, meaning they are completely assembled and tested at the factory and shipped as one component. They are all built in one shell or a series of chambers so the range of sizes of equipment may be limited because it must be moved as a single component. The machines are rated in tons and range in size from approximately 100 to 1700 tons. A 1700-ton system is extremely large. The large tonnage machines may be manufactured in sections for ease of rigging and moving into machine rooms, **Figure 48–47.** These sections are typically assembled in the factory for proper alignment except for the welds that must be accomplished in the field.

The absorption process is very different in some respects but also similar to the compression cycle refrigeration process in others. It would be helpful to understand the compression cycle before trying to understand the absorption cycle. The boiling temperature for the refrigerant in compression cycle chillers is controlled by controlling the pressure above the boiling liquid. This boiling pressure is controlled in the compression cycle by the compressor. **Figure 48–48** shows a refrigeration system using water as the

(A)

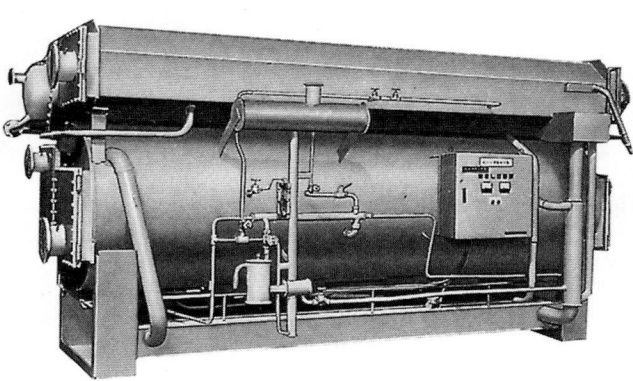

(B)

(C)

Figure 48–46   Absorption chillers. **(A)** *Courtesy Trane Company.* **(B)** *Reproduced courtesy of Carrier Corporation.* **(C)** *Courtesy York International*

Figure 48–47   This absorption chiller is assembled after it is placed in the equipment room. *Reproduced courtesy of Carrier Corporation*

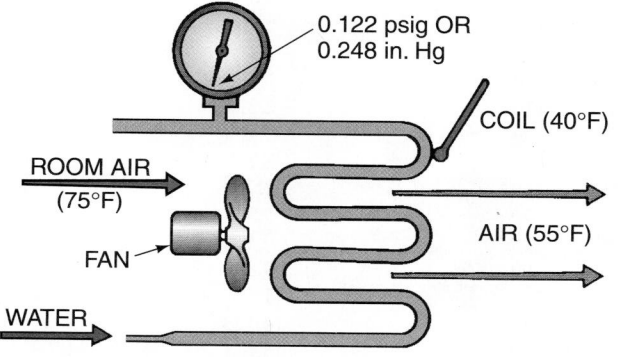

Figure 48–48   Water can be used as a refrigerant when the pressure is lowered enough.

refrigerant. The water must be boiled at 0.248 in. Hg absolute or 0.122 psia to boil the water at 40°F. **Figure 48–49** shows a limited temperature/pressure chart for water. This refrigeration cycle would be workable using a compressor except that the volume of vapor rising from the boiling water is excessive. The compressor would have to remove 2444 ft³ of water vapor for every pound of water boiled at 40°F. **Figure 48–50** shows a limited specific-volume chart for water at different temperatures. A compressor is impractical for a refrigeration system using water as the refrigerant.

The absorption machine uses water for the refrigerant but does not use a compressor to create the pressure difference. It uses the fact that certain salt solutions have enough attraction for water that they may be used to create the pressure difference. You may have noticed that a grain of salt on a hot humid day will attract moisture to the point that it will become wet, **Figure 48–51**. This same concept is used in an absorption refrigeration system to reduce the pressure of the water to a point where it will boil at a low temperature. A type of salt solution called lithium-bromide (Li-Br) is used as the absorbent (attractant) for the water. Li-Br is a solution in the liquid form as used in absorption cooling systems. The liquid solution of Li-Br is diluted with distilled water. It

TEMPERATURE °C	TEMPERATURE °F	SPECIFIC VOLUME OF WATER VAPOR ft³/lb	ABSOLUTE PRESSURE lb/in.²	ABSOLUTE PRESSURE kPa	ABSOLUTE PRESSURE in. Hg
−12.2	10	9054	0.031	0.214	0.063
−6.7	20	5657	0.050	0.345	0.103
−1.1	30	3606	0.081	0.558	0.165
0.0	32	3302	0.089	0.613	0.180
1.1	34	3059	0.096	0.661	0.195
2.2	36	2837	0.104	0.717	0.212
3.3	38	2632	0.112	0.772	0.229
4.4	40	2444	0.122	0.841	0.248
5.6	42	2270	0.131	0.903	0.268
6.7	44	2111	0.142	0.978	0.289
7.8	46	1964	0.153	1.054	0.312
8.9	48	1828	0.165	1.137	0.336
10.0	50	1702	0.178	1.266	0.362
15.6	60	1206	0.256	1.764	0.522
21.1	70	867	0.363	2.501	0.739
26.7	80	633	0.507	3.493	1.032
32.2	90	468	0.698	4.809	1.422
37.8	100	350	0.950	6.546	1.933
43.3	110	265	1.275	8.785	2.597
48.9	120	203	1.693	11.665	3.448
54.4	130	157	2.224	15.323	4.527
60.0	140	123	2.890	19.912	5.881
65.6	150	97	3.719	25.624	7.573
71.1	160	77	4.742	32.672	9.656
76.7	170	62	5.994	41.299	12.203
82.2	180	50	7.512	51.758	15.295
87.8	190	41	9.340	64.353	19.017
93.3	200	34	11.526	79.414	23.468
98.9	210	28	14.123	97.307	28.754
100.0	212	27	14.696	101.255	29.921

Figure 48–50   A specific-volume chart for water.

TEMPERATURE °F	ABSOLUTE PRESSURE lb/in.²	ABSOLUTE PRESSURE in. Hg
10	0.031	0.063
20	0.050	0.103
30	0.081	0.165
32	0.089	0.180
34	0.096	0.195
36	0.104	0.212
38	0.112	0.229
40	0.122	0.248
42	0.131	0.268
44	0.142	0.289
46	0.153	0.312
48	0.165	0.336
50	0.178	0.362
60	0.256	0.522
70	0.363	0.739
80	0.507	1.032
90	0.698	1.422
100	0.950	1.933
110	1.275	2.597
120	1.693	3.448
130	2.224	4.527
140	2.890	5.881
150	3.719	7.573
160	4.742	9.656
170	5.994	12.203
180	7.512	15.295
190	9.340	19.017
200	11.526	23.468
210	14.123	28.754
212	14.696	29.921

Figure 48–49   A temperature/pressure chart for water.

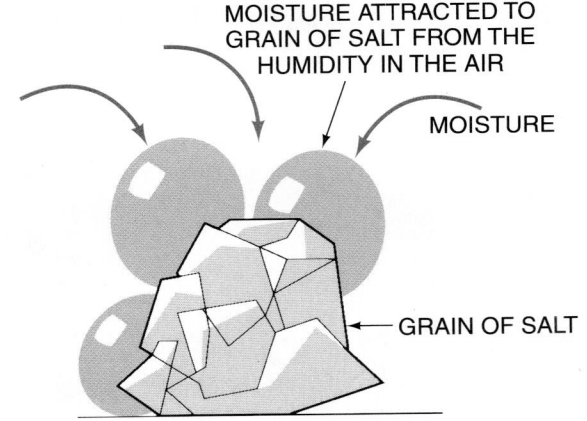

Figure 48–51   A grain of salt in a humid climate will attract water.

is actually a mixture of about 60% Li-Br and 40% water. The term absorption means to attract moisture.

A simplified absorption system would look like the one in **Figure 48–52**. This is not a complete cycle but is simplified

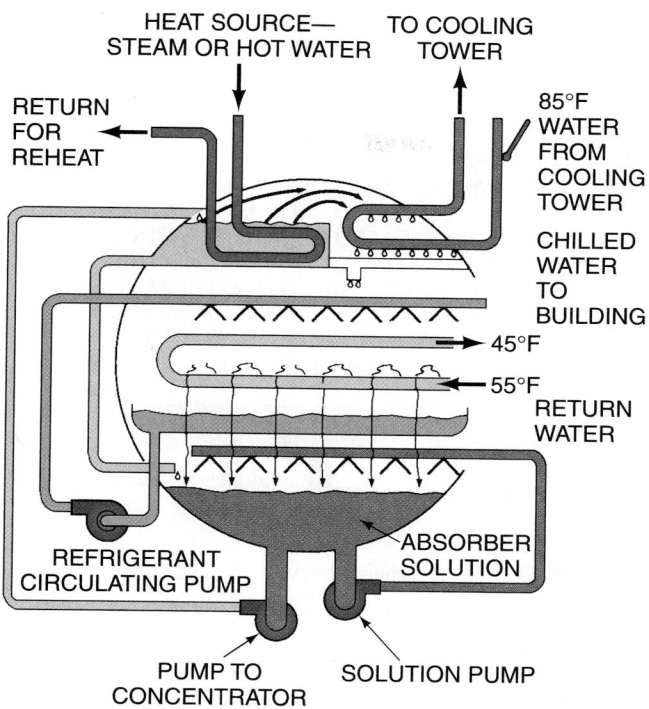

HEAT SOURCE—
STEAM OR HOT WATER
TO COOLING
TOWER

RETURN
FOR
REHEAT

85°F
WATER
FROM
COOLING
TOWER

CHILLED
WATER
TO
BUILDING

45°F

55°F
RETURN
WATER

REFRIGERANT
CIRCULATING PUMP

ABSORBER
SOLUTION

PUMP TO
CONCENTRATOR

SOLUTION PUMP

Figure 48–52 A simplified absorption refrigeration cycle.

for ease of understanding. The absorption cycle can be followed in this progressive description.

1. **Figure 48–53** shows the equivalent of the evaporator system. Refrigerant (water) is metered into the evaporator section through a restriction (orifice). It is warm until it passes through the orifice where it flashes to a low-pressure area (about 0.248 in. Hg absolute or 0.122 psia). The reduction in pressure also reduces the temperature of the water. The water drops to the pan below the evaporator tube bundle. A refrigerant-

circulating pump circulates the water through spray heads to be sprayed over the evaporator tube bundle. This wets the tube bundle through which the circulating water from the system passes. The heat from the system water evaporates the refrigerant. Water is constantly being evaporated and must be made up through the orifice at the top.

2. **Figure 48–54** shows the absorber section, which would be the equivalent of the suction side of the compressor in a compression cycle system. The salt solution (Li-Br) spray is a very low-pressure attraction for the evaporated water vapor so it readily absorbs into the solution. The solution is recirculated through the spray heads to give the solution more surface area for attracting the water. As the solution absorbs the water, it becomes diluted with the water. If this water is not removed, the solution will become so diluted that it will no longer have any attraction and the process will stop, so another pump constantly removes some of the solution and pumps it to the next step called the concentrator. This solution that is pumped to the concentrator is called the weak solution because it contains water absorbed from the evaporator.

3. **Figure 48–55** shows the concentrator and condenser segment. The dilute solution is pumped to the concentrator, where it is boiled. This boiling action changes the water to a vapor; it leaves the solution and is attracted to the condenser coils. The water is condensed to a liquid where it gathers and is metered to the evaporator section through the orifice. The heat source for boiling the water is either steam or hot water in this example. Direct-fired machines are also available and are discussed later. The concentrated solution is drained back to the absorber area for circulation by the absorber pump.

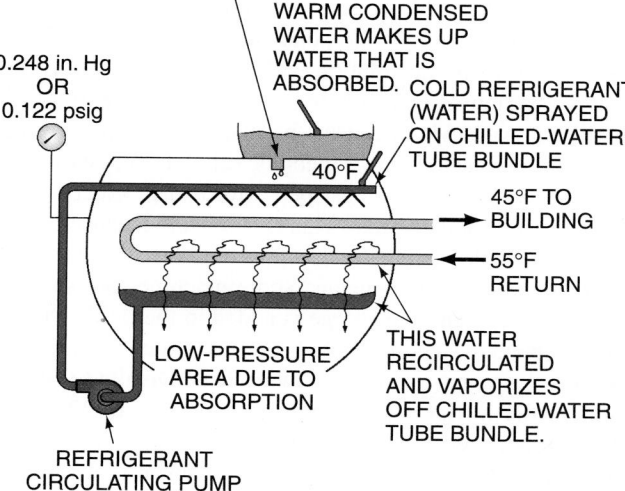

ORIFICE WARM WATER FLASHES
TO COLD WATER WHEN IT ENTERS THE
LOW-PRESSURE SIDE OF THE MACHINE.

WARM CONDENSED
WATER MAKES UP
WATER THAT IS
ABSORBED.

0.248 in. Hg
OR
0.122 psig

COLD REFRIGERANT
(WATER) SPRAYED
ON CHILLED-WATER
TUBE BUNDLE

40°F

45°F TO
BUILDING

55°F
RETURN

LOW-PRESSURE
AREA DUE TO
ABSORPTION

THIS WATER
RECIRCULATED
AND VAPORIZES
OFF CHILLED-WATER
TUBE BUNDLE.

REFRIGERANT
CIRCULATING PUMP

Figure 48–53 The evaporator section of an absorption machine.

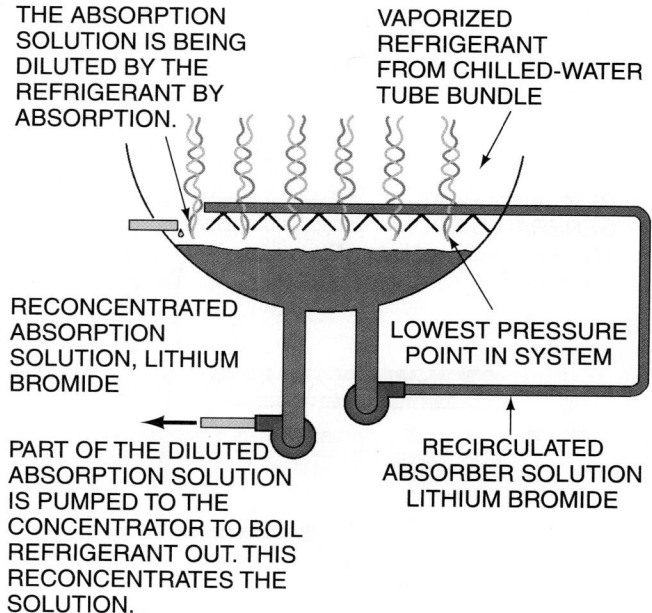

THE ABSORPTION
SOLUTION IS BEING
DILUTED BY THE
REFRIGERANT BY
ABSORPTION.

VAPORIZED
REFRIGERANT
FROM CHILLED-WATER
TUBE BUNDLE

RECONCENTRATED
ABSORPTION
SOLUTION, LITHIUM
BROMIDE

LOWEST PRESSURE
POINT IN SYSTEM

PART OF THE DILUTED
ABSORPTION SOLUTION
IS PUMPED TO THE
CONCENTRATOR TO BOIL
REFRIGERANT OUT. THIS
RECONCENTRATES THE
SOLUTION.

RECIRCULATED
ABSORBER SOLUTION
LITHIUM BROMIDE

Figure 48–54 The absorber section of an absorption machine.

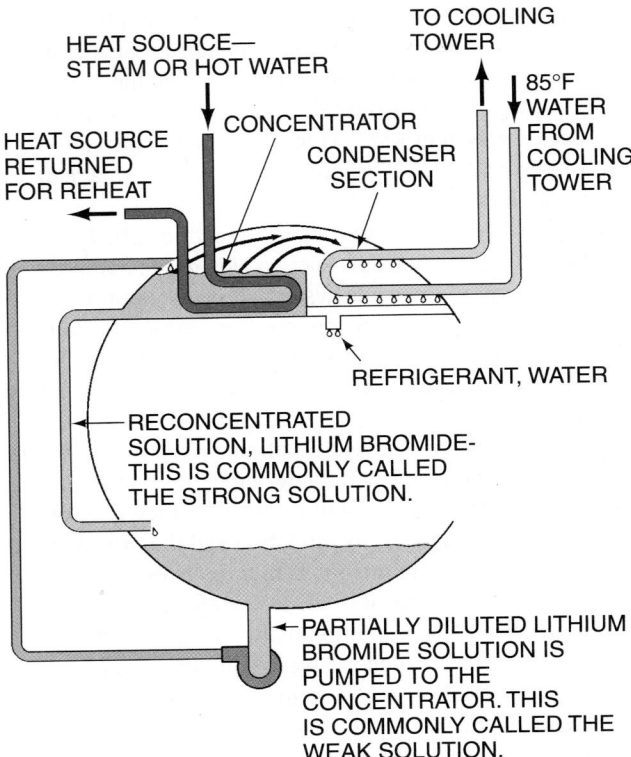

Figure 48–55 The concentrator and condenser section of an absorption machine.

As you can see from this description, the absorption process is not complicated. The only moving parts are the pump motors and impellers.

Some other features are added to the cycle to make it more efficient. One of these is shown in **Figure 48–56**. It is a heat exchange between the cooling tower water and the absorber solution in the absorber. This heat exchange removes heat

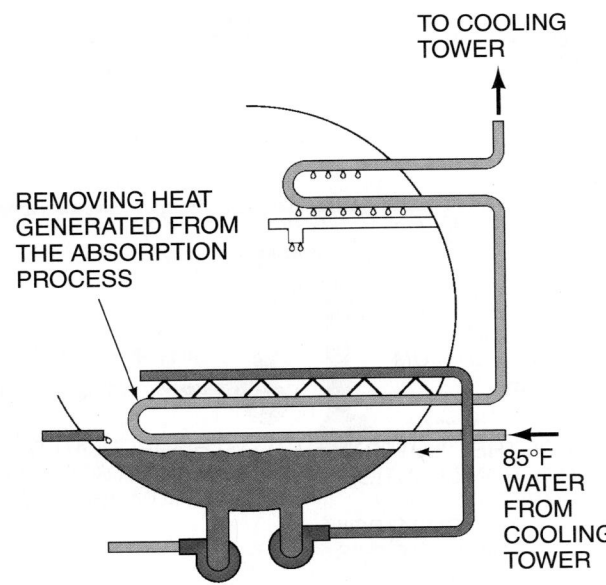

Figure 48–56 🌐 A heat exchange in the absorber to increase efficiency.🌐

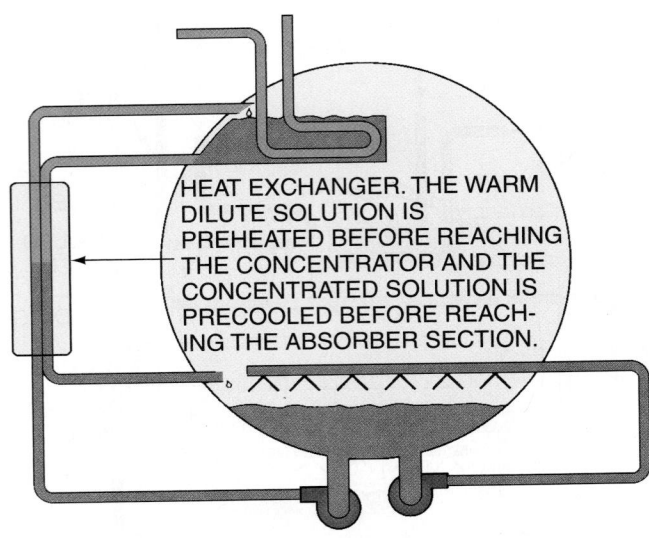

Figure 48–57 A heat exchange between the solutions.

that is generated when the water vapor is absorbed into the absorber solution.

Figure 48–57 is a heat exchanger between the dilute and the concentrated solution. This heat exchange serves two purposes; it preheats the dilute solution before it enters the concentrator and precools the concentrated solution before it enters the absorber section. It is much like the heat exchange between the suction and liquid line on a household refrigerator. Without this heat exchange, the machine would be less efficient.

Some manufacturers have developed a two-stage absorption machine that uses a higher pressure steam or hot water, **Figure 48–58**. The steam pressure may be 115 psig or hot water at 370°F. These machines are more efficient than the single-stage machine mentioned earlier.

48.27 SOLUTION STRENGTH

The concentration strength of the solutions determines the ability of the machine to perform. The wider the spread between the dilute and the strong solution, the more capacity the machine has to lower the pressure to absorb the water from the evaporator. If the concentrated solution becomes too concentrated, the solution will become rock salt. If this occurs, the flow of concentrated solution may be partially or totally blocked, which causes the machine to stop cooling. The solution is said to have "become solid" or "crystallized," and service procedures must be performed in order to dissolve the hard salt crystals. This is a delicate balance of water and Li-Br. Adjusting the strengths of these solutions is the job of the start-up technician. Some machines are shipped with no charge and the charge is added in the field. The Li-Br is shipped in steel drums to be added to the machine. The estimated amount of Li-Br is added first. Distilled water is used as the refrigerant charge and added next. When the charge is adjusted, the technician calls it the trim. You charge a compression cycle system and trim an absorption system.

Typically, a technician trims the machine with the approximate correct amount of Li-Br and water, then starts the

TWO-STAGE STEAM-FIRED ABSORPTION UNIT

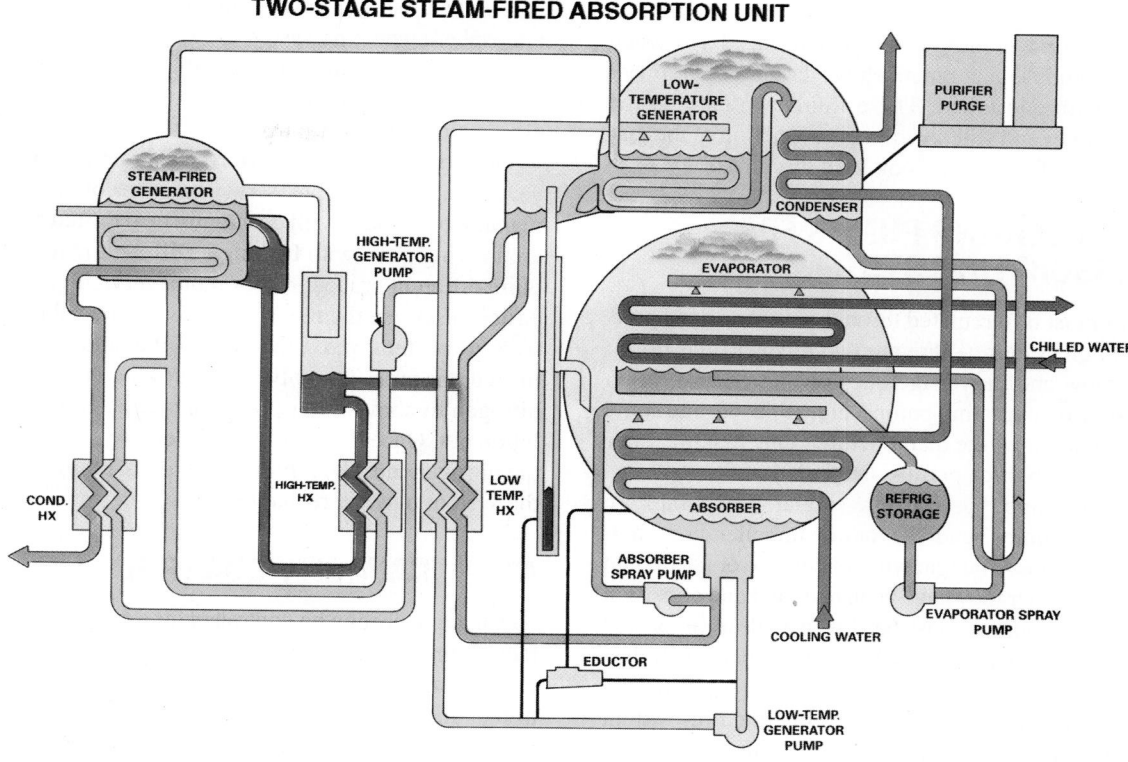

Figure 48–58   A two-stage absorption machine. *Courtesy Trane Company*

machine. The machine is then gradually run up to full load. Full load may be determined by the temperature drop across the evaporator and full steam pressure, which is normally 12 to 14 psig (hot water or direct-fire equipment would be similar). When full load is obtained, the technician pulls a sample of dilute and strong solution and measures the specific gravity using a hydrometer, **Figure 48–59**. Specific

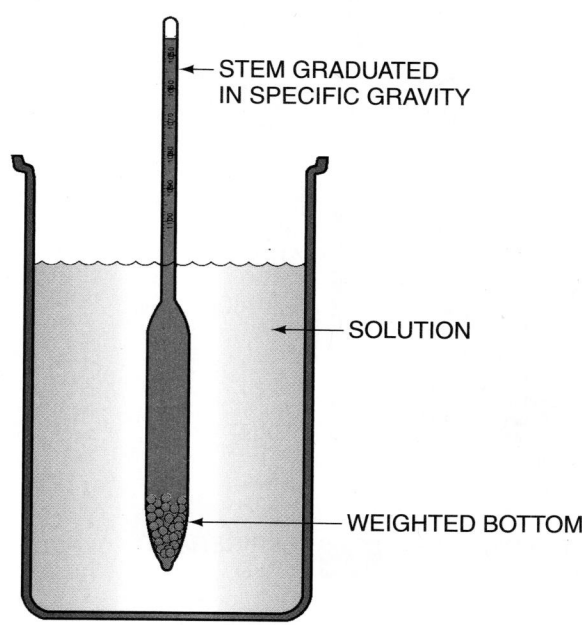

Figure 48–59   A hydrometer for measuring the specific gravity of absorption solutions.

gravity is the weight of a substance compared with the weight of an equal volume of water. The technician would know what these specific gravity readings should be from the manufacturer's literature or a Li-Br temperature/pressure chart. Water is either added or removed to obtain the correct specific gravity.

This may be a relatively long process in some cases because the technician cannot always obtain full-load operation at start-up. The machine may have to be trimmed to an approximate trim and finished later when full load can be achieved. Some manufacturers furnish a trim chart for a partial load.

48.28   SOLUTIONS INSIDE THE ABSORPTION SYSTEM

The solutions inside the absorption system are corrosive. Rust is oxidation and occurs where corrosive materials and oxygen are present. The absorption machine is manufactured with materials such as steel and contains a salt water solution that will corrode if air is present. It is next to impossible to manufacture and keep a system from ever letting air with oxygen be exposed to the working parts.

These solutions must be circulated through the system and the system must be kept as clean as possible. Some of the passages that the solution must circulate through are very small, such as the spray heads in the absorber. Manufacturers use different methods of removing solid materials that may clog any small passages. Filters will stop solid particles, and magnetic attraction devices will attract steel particles that may be in circulation. Even though the liquid circuits have filtration, the solutions in an absorption machine become discolored. The first opinion a technician may have when looking at the fluids is that the machine should be scrapped, and a new machine installed. These fluids can take on the color of rusty water and still not affect the function of the machine. It is normal for them to look very discolored.

48.29   CIRCULATING PUMPS FOR ABSORPTION SYSTEMS

The solutions must be circulated throughout the various parts of the absorption system. There are two distinct fluid flows: the solution flow and the refrigerant flow. Some absorption machines use two circulating pumps, **Figure 48–60,** and some use three, depending on the manufacturer. One manufacturer uses one motor with three pumps on the end of the shaft.

Whatever the pump type, they are similar. The pumps are centrifugal-type pumps, and the pump impeller and shaft must be made of a material that will not corrode or they could never be serviced after they have been operated and corrosion has occurred. The pumps must be driven with a motor and special care must be taken not to let the atmosphere enter during the pumping process. Because of this, it is typical for the motor to be hermetically sealed and to operate only within the system atmosphere. The pump is in the actual solution to

be pumped, but the motor windings are sealed in their own atmosphere where no system solution can enter.

These motors must be cooled so cold refrigerant water from the evaporator or normal supply water is used in a closed circuit for this purpose.

Motors must also be serviced over a period of time. Many years may pass before servicing is recommended by the manufacturer, but it will eventually be required because there are moving parts with bearings. Different manufacturers require different procedures for motor servicing, but it is much easier to service the motor and drive if the solution does not have to be removed. However, if the solution must be removed, the process involves pressurizing the system with dry nitrogen to above atmospheric pressure and pushing the solution out. Once this is accomplished and service completed, the nitrogen must be removed, which is a long process. Then the system must be recharged.

48.30   CAPACITY CONTROL

The capacity can be controlled for a typical absorption system by throttling the supply of heat to the concentrator. A typical system that operates on steam uses 12 to 14 psig of steam at full load and 6 psig at half capacity. The steam can be controlled with a modulating valve, **Figure 48–61.** Hot water

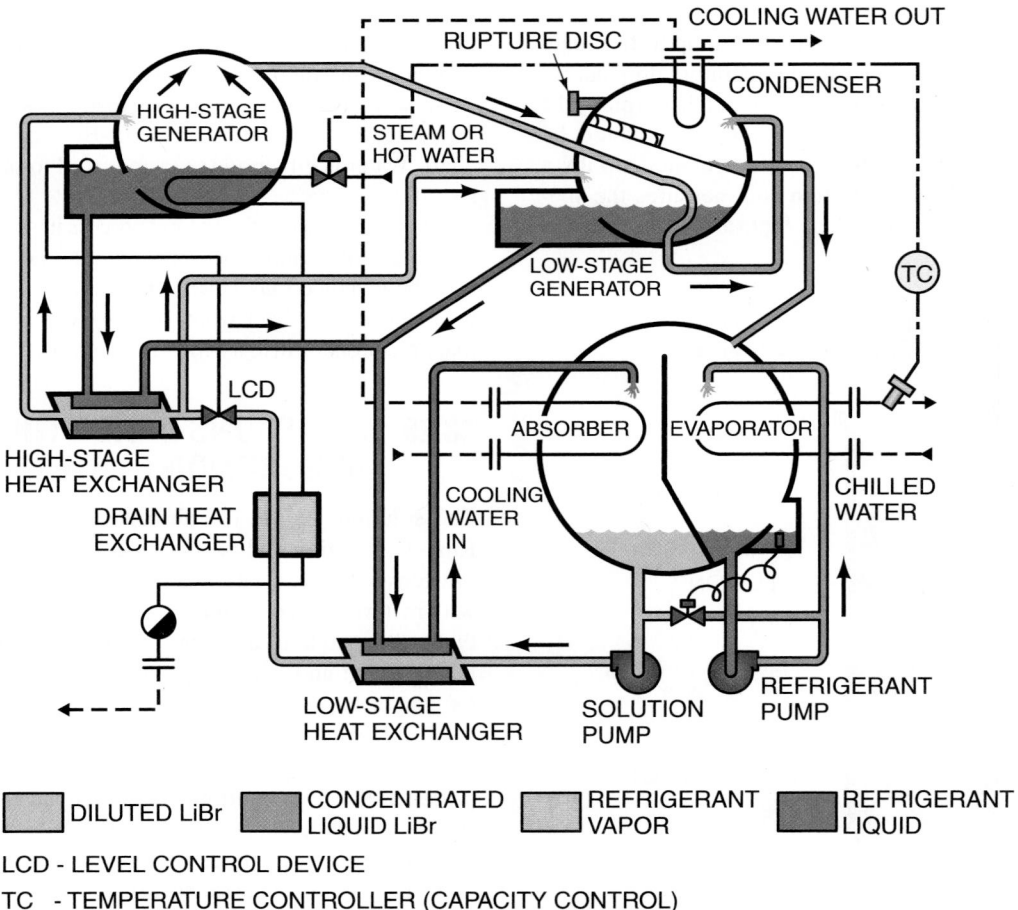

Figure 48–60   A two-pump absorption machine application. *Reproduced courtesy of Carrier Corporation*

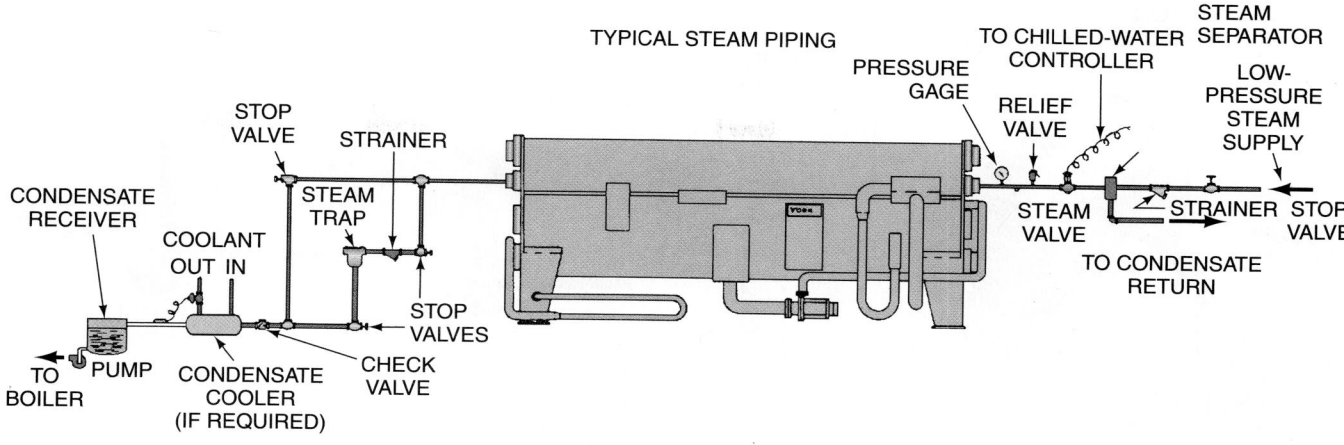

Figure 48–61 Capacity control using a modulating steam valve. *Courtesy York International*

and direct-fired machines can be controlled in a similar manner. Other manufacturers may have other controls for capacity control. These may control the internal fluid flow in the machine. For example, the flow of weak (dilute) solution to the concentrator is practical because at reduced load, less solution needs to be concentrated.

48.31 CRYSTALLIZATION

The use of a salt solution for absorption cooling creates the possibility of the solution becoming too concentrated and actually turning back to rock salt. This is called *crystallization* and may occur if the machine is operated under the wrong conditions. For example, with some systems if the cooling tower water is allowed to become too cold while operating at full load, the condenser will become too efficient and remove too much water from the concentrate. This will result in a strong solution that has too little water. When this solution passes through the heat exchanger, it will turn to crystals and restrict the flow of the solution. If this is not corrected, a complete blockage will occur and the machine will stop cooling. Because this is a difficult problem to overcome when it happens, manufacturers have developed various methods to prevent this condition. One manufacturer uses pressure drop in the strong solution across the heat exchanger as a key to the problem. The action taken may be to open a valve between the refrigerant circuit and the absorber fluid circuit to make the weak solution very weak for long enough to relieve the problem. When the situation is corrected, the valve is closed and the system resumes normal operation. Another manufacturer may shut the machine down for a dilution cycle when overconcentration occurs.

Crystallization can occur for several reasons. Cold condenser water is one cause, and a shutdown of the machine due to power failure while operating at full load is another. An orderly shutdown calls for the solution pumps to operate for several minutes after shutdown to dilute the strong solution. With a power failure, the shutdown is not orderly and crystallization can occur.

Because the machine operates in a vacuum, atmospheric air can be pulled into the machine at any point where a leak occurs. Air in the system can also cause crystallization.

48.32 PURGE SYSTEM

The *purge* system removes noncondensables from the absorption machine during the operating cycle.

All absorption systems operate in a vacuum. If there is any source for a leak, the atmosphere will enter. A soft drink bottle full of air in a 500-ton machine will affect the capacity. The air will expand greatly when pulled into a vacuum. These machines must be kept absolutely leak free. All piping of the solutions have factory-welded connections wherever possible. When the typical packaged absorption machine is assembled at the factory, it is put through the most rigid of leak tests, called the mass spectrum analysis. For this test, the machine is surrounded with an envelope of helium. The system is pulled into a deep vacuum and the exhaust of the vacuum pump is analyzed with a mass spectrum analyzer for any helium, **Figure 48–62**. Helium is a gas with very small molecules that will leak in at any leak source.

Even with these rigid leak-check and welding procedures, the system is subject to leaks developing during shipment to the job site. Leaks may also develop after years of operation and maintenance. When a leak occurs, the noncondensables must be removed.

Absorption systems also generate small amounts of noncondensables while in normal operation. A by-product of the internal parts causes hydrogen gas and some other noncondensables to form inside during normal operation. These are held to a minimum with the use of additives, but they still occur and it is the responsibility of the purge and the operator to keep the machine free of these noncondensables.

Two kinds of purge systems are used on typical absorption cooling equipment: the nonmotorized purge and the motorized purge. The nonmotorized purge uses the system pumps to create a flow of noncondensable products to a chamber where they are collected and then bled off by the

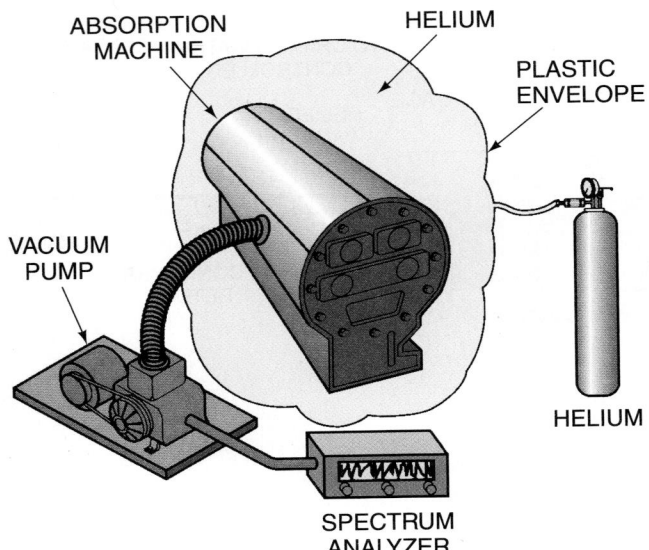

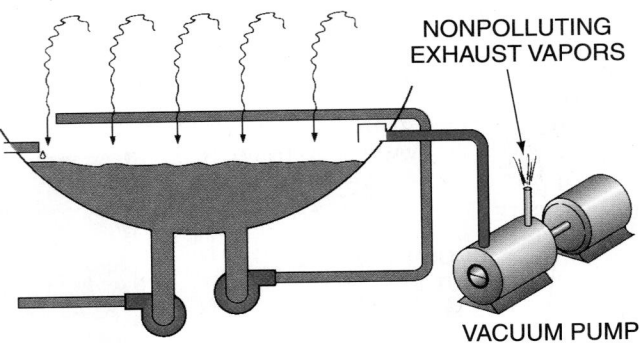

Figure 48–64 The vapors purged from an absorption machine are nonpolluting.

Figure 48–62 Leak checking an absorption machine in the factory. A vacuum is pulled on the machine that is enclosed in a helium atmosphere. The vacuum pump's exhaust is checked for helium with a spectrum analyzer. Helium has very small molecules and will easily leak through a very small hole.

machine operator, **Figure 48–63**. This purge operates well while the machine is in operation but may not be of much use if the machine were to need service and is pressured to atmosphere using nitrogen. This machine manufacturer offers an optional motorized purge for removal of large amounts of noncondensables.

The motorized purge is essentially a two-stage vacuum pump that removes a sample of whatever gas is in the absorber and pumps it to the atmosphere. These vapors are not harmful to the atmosphere. Because the absorber operates in such a low vacuum, this gas sample would contain only non-condensables or water vapor, **Figure 48–64**. The motorized purge requires some maintenance because of the corrosive nature of the vapors that will be pulled through the vacuum

pump. The vacuum pump oil must be changed on a regular basis to prevent vacuum pump failure. Vacuum pumps are expensive and must be properly maintained.

48.33 ABSORPTION SYSTEM HEAT EXCHANGERS

The absorption machine has heat exchangers as do the compression cycle chillers. There are actually more heat exchangers—the evaporator, the absorber, the concentrator, the condenser, and the first-stage heat exchanger—for two-stage systems. They are similar to compression cycle heat exchangers but have some differences. They also contain tubes. These tubes are either copper or cupronickel (a tube made of copper and nickel).

The chilled-water heat exchange tubes remove heat from the building water and add it to the refrigerant (water). The chilled-water circuit is usually a closed circuit, so the tubes rarely need maintenance except for water treatment. A flange at the end of the tube bundle allows access. These machines have either marine or standard water boxes for this access. This section of the absorption system must have some means of preventing heat from the equipment room from transferring

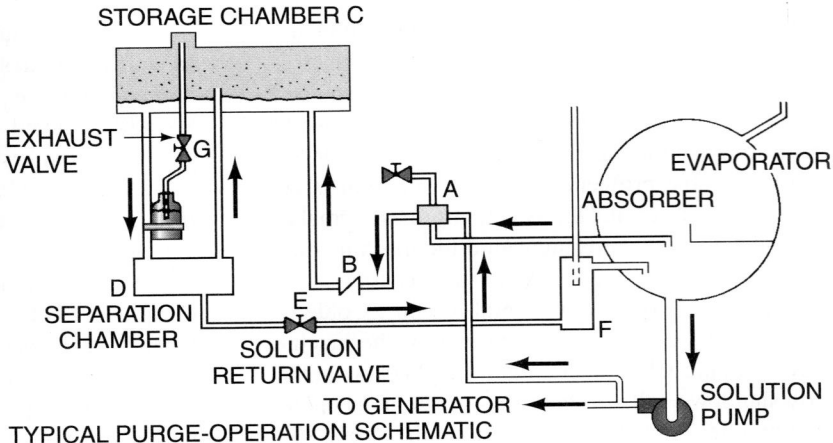

Figure 48–63 A collection chamber for noncondensable gasses. *Reproduced courtesy of Carrier Corporation*

into the cold machine parts. Some refrigeration systems are insulated in this section and some have a double-wall construction to prevent heat exchange. Like the compression cycle chillers, there is an approach temperature between the boiling refrigerant and the leaving chilled water. This approach is likely to be 2°F or 3°F for an absorption chiller because this heat exchange is good, **Figure 48–65.**

The absorber heat exchanger exchanges heat between the absorber solution and the water returning from the cooling tower. The design cooling tower water under full load in hot weather is 85°F. The water from this heat exchange continues to the condenser where the refrigerant vapor is condensed for reuse in the evaporator. The leaving water is likely to be from 95°F to 103°F on a hot day at full load, **Figure 48–66.**

The other heat exchange for standard equipment is between the heating medium: steam, hot water, or flue gases and the refrigerant. This tube bundle may have some high-temperature differences because it may be at room temperature until the system is started and hot water or steam is turned into the bundle. A great deal of stress in the form of tube expansion occurs at this time. The manufacturers claim that a tube can grow in length up to 1/4 in. during this process. Different manufacturers deal with this stress in different ways.

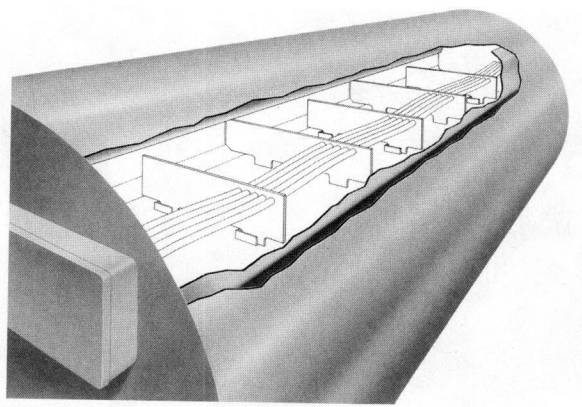

Figure 48–67 The tubes in the absorption chiller are subject to great expansion due to significant temperature differences. Special methods are used to allow for tube expansion. *Courtesy Trane Company*

Figure 48–67 shows a tube bundle that has the capability of floating to minimize problems.

Absorption equipment manufacturers provide thermometer wells located at strategic spots to check the solution temperatures for assessing machine performance and to determine if the tubes are dirty. The manufacturers provide the correct temperatures and differences for their machines.

48.34 DIRECT-FIRED SYSTEMS

Some manufacturers furnish direct-fired equipment. These use gas or oil for the heat source. The system can be furnished as a dual-fuel machine for applications where gas demand may require a second fuel.

These machines range in capacities from approximately 100 to 1500 tons. They also may have the ability to either heat or cool by furnishing either hot or chilled water. In some cases they may be used in an old installation to replace a boiler and an old absorption machine as one piece of equipment. **Figure 48–68** shows direct-fired equipment manufactured by two different companies.

48.35 MOTORS AND DRIVES FOR COMPRESSION CYCLE CHILLERS

All of the motors that drive high-pressure and low-pressure chillers are highly efficient three-phase motors. These motors all give off some heat while in operation. This heat must be removed from the motor or it will become overheated and burn. 🌐Some manufacturers choose to keep the motor in the atmosphere for cooling.🌐 This reduces the load on the refrigeration system. These motors are air cooled and give off heat to the equipment room, **Figure 48–69.** This heat must be removed or the room will overheat. The heat is typically removed using an exhaust fan system. This compressor must have a shaft seal to prevent refrigerant from leaking to the atmosphere, **Figure 48–70.** In the past this has been a source of

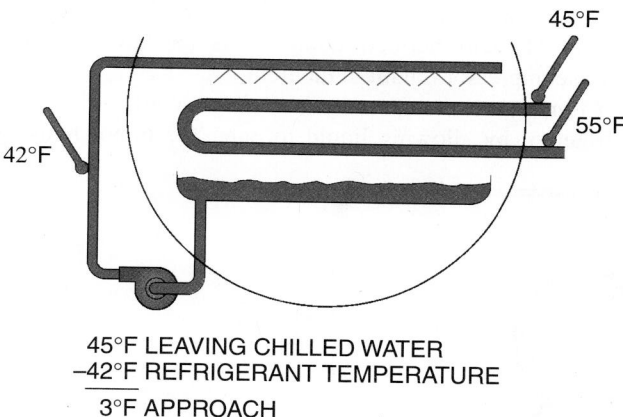

45°F

55°F

42°F

45°F LEAVING CHILLED WATER
−42°F REFRIGERANT TEMPERATURE

3°F APPROACH

Figure 48–65 The approach temperatures for an absorption chiller are very close.

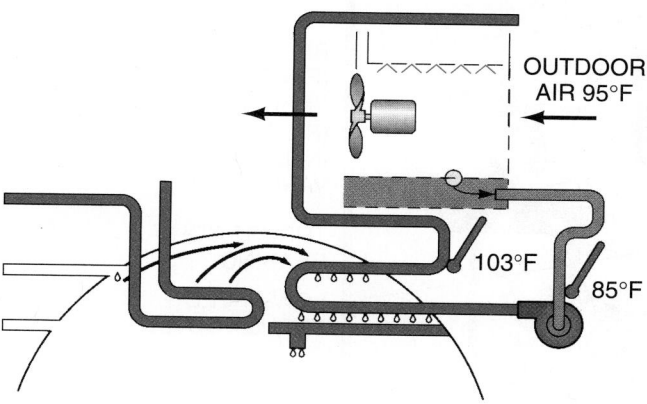

OUTDOOR AIR 95°F

103°F

85°F

Figure 48–66 A cooling tower circuit.

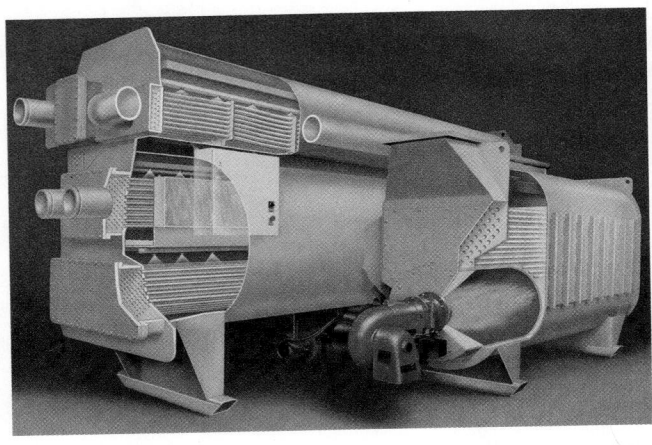

(A)

(B)

Figure 48–68   Direct-fired absorption chillers. *(A) Courtesy Trane Company. (B) Courtesy York International*

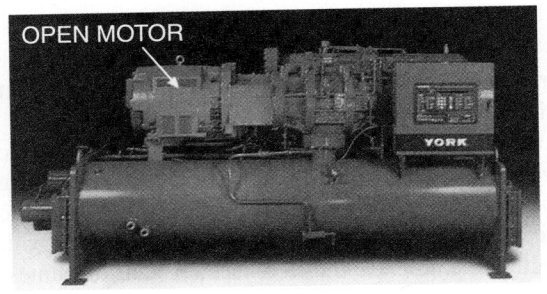

Figure 48–69   An air-cooled, direct-drive compressor motor. *Courtesy York International*

leaks, but modern seals and correct shaft alignment have improved the success with open-drive compressors.

Compressor motors may also be cooled with the vapor refrigerant leaving the evaporator. These compressors are called suction-cooled compressors and are either hermetic or semi-hermetic compressors. The fully hermetic compressors are in the smaller sizes and may be returned to the factory for rebuild. The semi-hermetic compressors may have the motor or the compressor rebuilt in the field.

Compressor motors may also be cooled with liquid refrigerant by allowing liquid to enter the motor housing

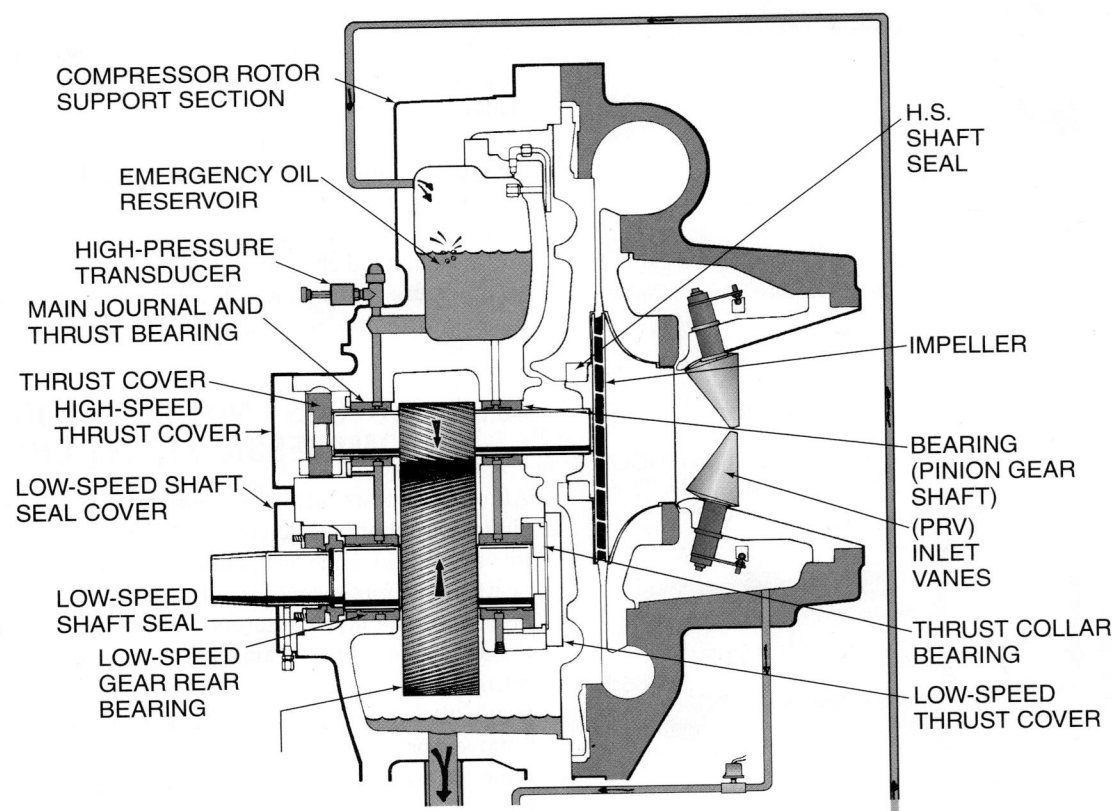

Figure 48–70   The shaft seal for a gear-drive compressor. *Courtesy York International*

around the motor. The liquid does not actually touch the motor or windings; it is contained in a jacket around the motor, **Figure 48–14.**

Large amounts of electrical power must be furnished to the motor through a leak-free connection. This power is often furnished through a nonconducting terminal block made of some form of phenolic or plastic with the motor terminals protruding through both sides. The motor terminals are fastened on one side and the field connections on the other with O rings used for the seal on the terminals, **Figure 48–71.** The terminal board has a gasket to seal it to the compressor housing. This terminal block is normally checked periodically for tightness of the field connections. **A loose connection may melt the board and cause a leak to develop.**

The motors used in all types of chillers are expensive and the manufacturer goes to great lengths to protect the motor by designing it to operate in a protective atmosphere. Part of getting long motor life is to use a good start-up procedure for the motor.

A large motor is usually started with a group of components called a *starter*. There are several types of starters. A motor draws about five times as much amperage at locked rotor on start-up as at full-load amperage. If a motor draws 200 A at full load, it would draw approximately 1000 A at start-up. This inrush of current can cause problems in the electrical service so manufacturers use several different methods to start motors to minimize the inrush of current and power-line fluctuations. The common ones are part-winding, autotransformer, wye-delta (often called star-delta), and electronic start.

48.36 PART-WINDING START

When compressor motors reach about 25 hp, manufacturers often use part-winding start motors. These motors are versatile and normally have nine leads. The same motor can be used for two different voltages. The compressor manufacturer can use the same compressor for 208/230-V and 460-V applications by changing the motor terminal arrangement, **Figure 48–72.** This motor is actually two motors in one. For example, a 100-hp compressor has two motors inside that are 50 hp each. When connected for 208/230-V application, the motors are wired in parallel and each motor is started separately, first one then the other. This is done with two motor starters and a time delay of up to about one second between them, **Figure 48–73.** When the first motor is started, the motor shaft starts turning. The second motor then starts to bring the compressor up to full speed, about 1800 or 3600 rpm depending on the motor's rated speed. 🜁 This only imposes the inrush current of a 50-hp motor on the line because when the second motor is energized, the shaft is turning, and the inrush current is less. 🜁

When the motor is used for 460-V application, the motors are wired in series and are started as one motor across the line, **Figure 48–74.** The higher voltage has much less inrush amperage on start-up. These motors are found on compressors of up to about 150 hp.

48.37 AUTOTRANSFORMER START

An autotransformer start installation is actually a reduced voltage start. A transformer-like coil is placed between the motor and the starter contacts, and the voltage to the motor is supplied through the transformer during start-up. This reduces the voltage to the motor until the motor is up to speed, **Figure 48–75.** When the motor is up to speed, a set of contacts close that short around the transformer to run the

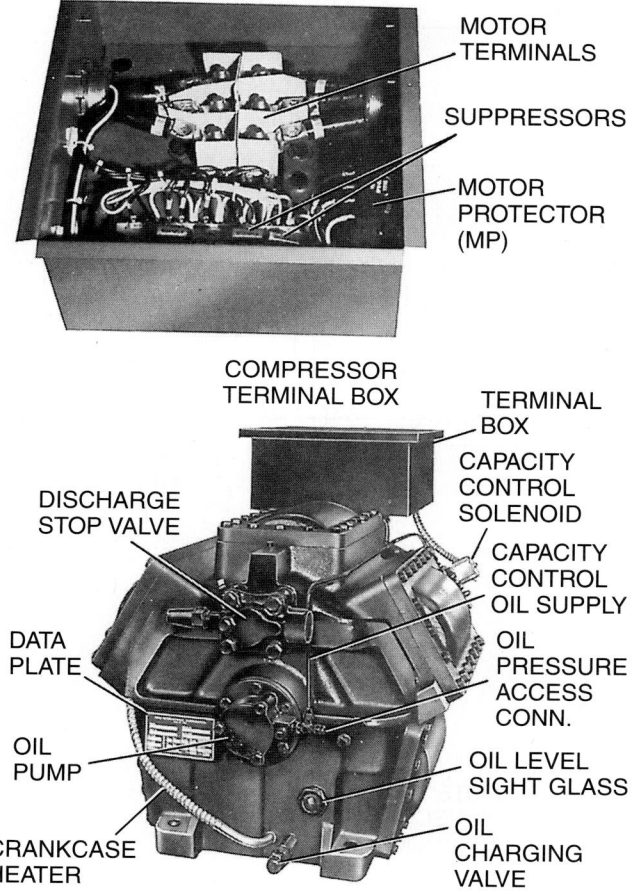

MOTOR TERMINALS

SUPPRESSORS

MOTOR PROTECTOR (MP)

COMPRESSOR TERMINAL BOX

TERMINAL BOX

CAPACITY CONTROL SOLENOID

DISCHARGE STOP VALVE

CAPACITY CONTROL OIL SUPPLY

DATA PLATE

OIL PRESSURE ACCESS CONN.

OIL PUMP

OIL LEVEL SIGHT GLASS

CRANKCASE HEATER

OIL CHARGING VALVE

Figure 48–71 The terminal block for a hermetic compressor motor. *Courtesy York International*

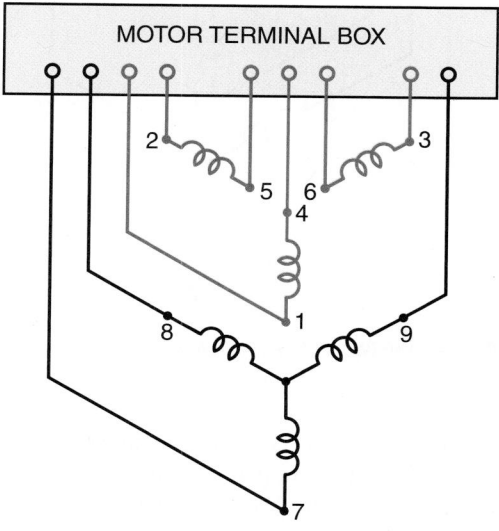

Figure 48–72 Nine-lead, dual-voltage motor connections.

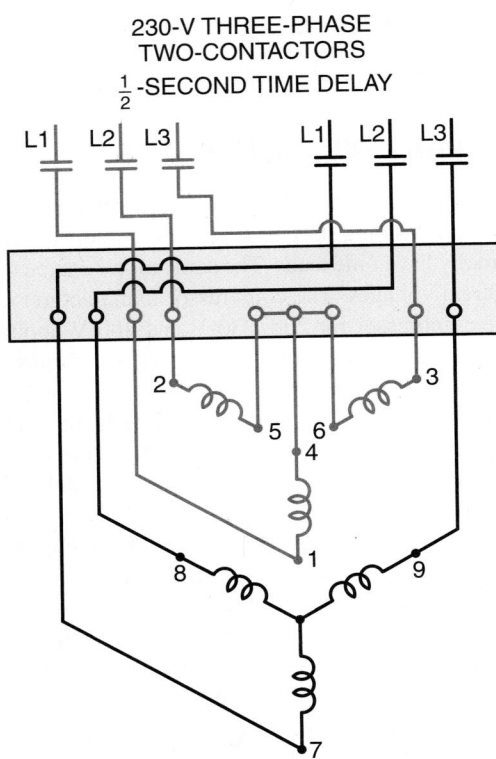

Figure 48–73 The wiring of a part-winding start motor wired for 208/230-V application.

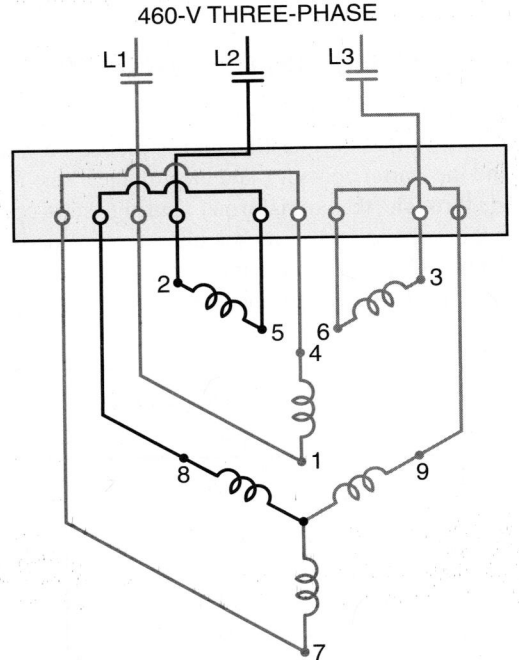

Figure 48–74 The two motors are now in series for the 460-V application.

motor at full voltage. The contacts that direct power through the transformer will then open so that current does not go through the transformer when the motor is running. This minimizes heat in the area of the transformer.

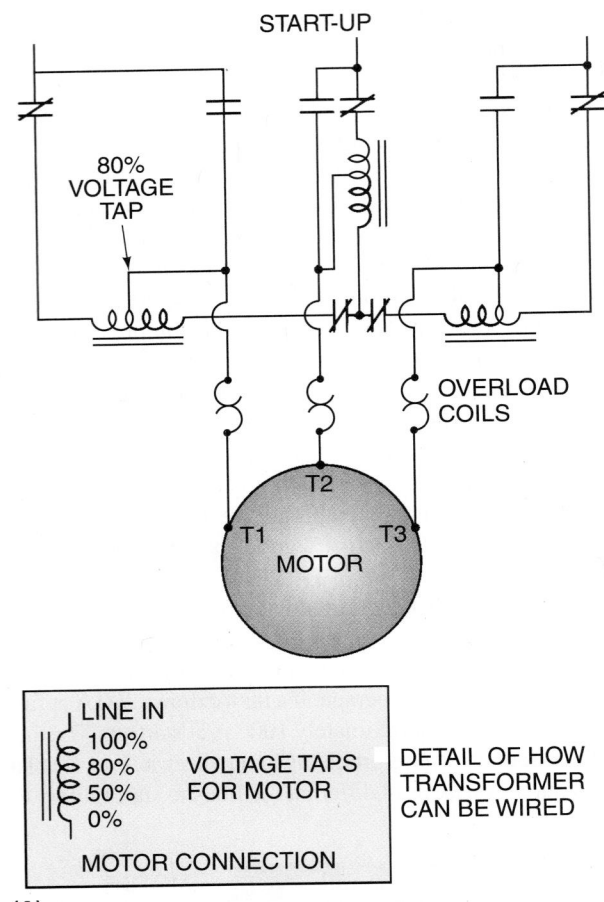

(A)

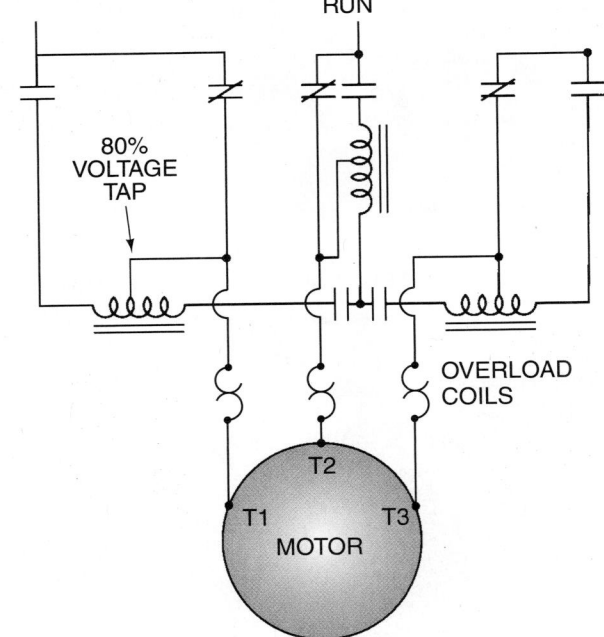

NOTE THAT THE TRANSFORMER IS DISCONNECTED DURING RUN TO MINIMIZE HEAT.

(B)

Figure 48–75 An autotransformer start (motor application).

The motor has very little starting torque because it is starting at reduced voltage so autotransformer start is used only in special applications. This type of motor start-up is common among large compressors that need little starting torque, such as centrifugal compressors where the pressures are completely equalized during the off cycle and compression does not start until the compressor is up to speed and the vanes start to open.

48.38   WYE-DELTA

Wye-delta start is often called star-delta start and is used with large motors with six leads and a single voltage. 🌐When the motor is starting in the wye configuration, it typically draws only about 33% of the current that it would draw starting in the delta configuration, **Figure 48–76.**🌐 After the motor is up to speed, a transition to a delta connection is made where the motor pulls the load and does the work more efficiently, **Figure 48–77.** The transition from star to delta is accomplished with a starting sequence that has three different contactors that are **electrically** and **mechanically interlocked;** two are energized in wye for start-up, then one is dropped out and the other engaged for delta operation. The time the motor operates in the wye connection depends on how long it takes the compressor to get up to speed. Large centrifugal compressors may take a minute or more to get up to speed in wye, then the transition is made to delta.

When the motor starter switches from wye to delta, the wye connection is actually disconnected electrically by means of a

contactor. This disconnect is proven both electrically by the timing of the auxiliary contacts and mechanically by means of a set of levers that interlock the contactors. Then the delta connection is made. The electrical and mechanical interlock feature is necessary because if the wye connection and the delta connection were to be made at the same time, a dead short from phase to phase would occur and likely destroy the starter components, **Figure 48–78.**

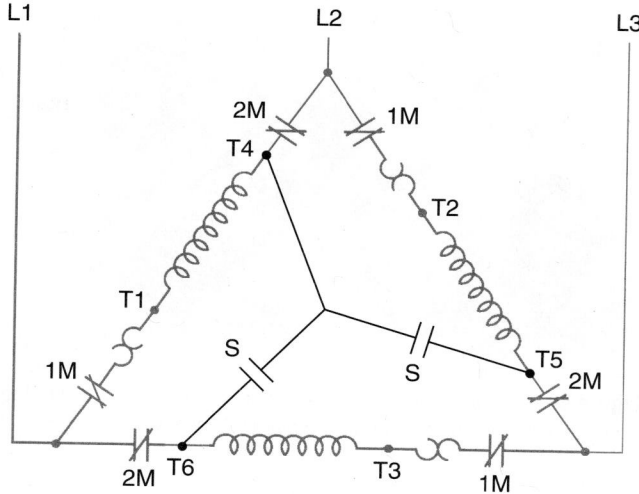

DELTA CONNECTION FOR RUN

NOTE: 1M AND 2M CONTACTS ARE CLOSED. S CONTACTS MUST BE OPEN.

Figure 48–77   Once the motor is up to speed, it is changed over to the delta connection shown in this diagram.

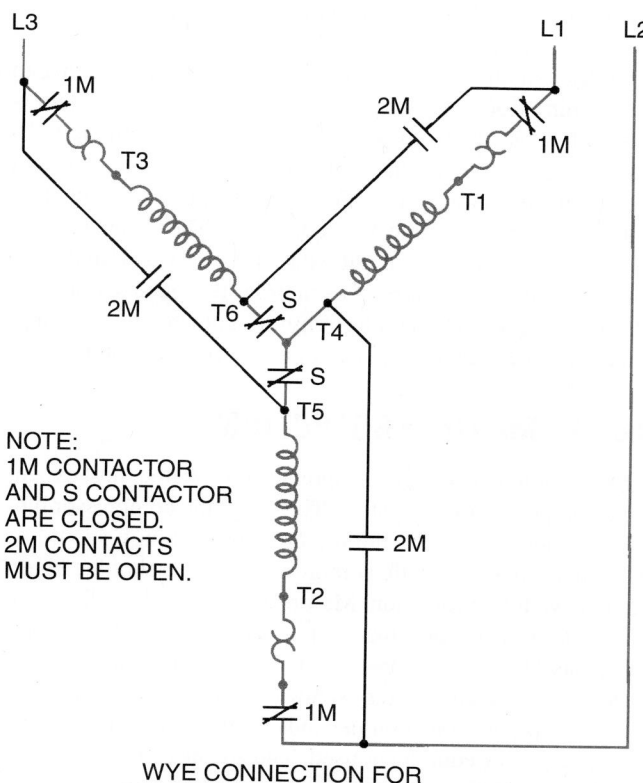

WYE CONNECTION FOR STARTING THE MOTOR

NOTE: 1M CONTACTOR AND S CONTACTOR ARE CLOSED. 2M CONTACTS MUST BE OPEN.

Figure 48–76   Starting a motor using a wye-delta circuit. The motor in this diagram is connected in the wye configuration.

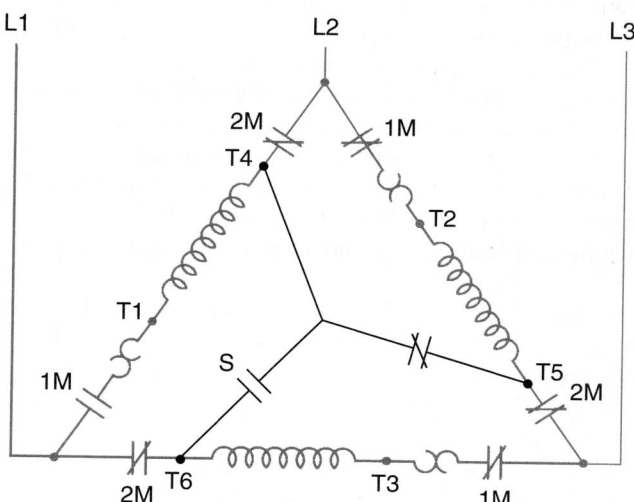

2M AND S CONTACTS CANNOT BE CLOSED AT THE SAME TIME. THESE CONTACTS ARE KEPT SEPARATED BY ELECTRICAL INTERLOCKS AND BY LEVERS CALLED MECHANICAL INTERLOCKS.

Figure 48–78   When too many of the contacts are closed, a short occurs. This diagram shows a short from line 2 to line 3 because the wye contacts did not open correctly.

Figure 48–79 New and old sets of contacts. The old ones are pitted from many motor starts. *Courtesy Square D Company*

When the delta connection is made, large amperage is drawn as the motor is totally disconnected and then reconnected. This spike of amperage could cause voltage problems in the vicinity of the motor, for example, a computer room located on the same service and close by. Many manufacturers furnish a motor starter with a set of resistors that are electrically connected and acting as the load while the transition is being made from wye to delta. This is known as a wye-delta closed transition starter. The starter would be a wye-delta open transition without the resistors. The larger the motor to be started, the more likely it is to have a closed transition starter.

All of the above starters use open contactors to start and stop the motors. When contacts of this nature are brought together to start a motor, they are under a great deal of stress with the inrush current. When the motor is disconnected from the line, an electrical arc tries to maintain the electrical path. This is much like an electrical welding arc, and damage to the contactor is caused each time the load is disconnected. The contacts become pitted due to this arc. The contacts will only be able to interrupt the load of a motor a limited number of times until replacement is required. **Figure 48–79** shows a new and an old set of contacts for a medium-sized contactor.

The contacts in the starter will also have an arc shield to prevent the arc when opening the contacts from spreading to the next phase of contacts. This arc shield must be in place when starting and stopping the motor or severe damage may occur.

SAFETY PRECAUTION: Starting and stopping a large motor using a starter requires a lot of electrical energy to be expended inside the starter cabinet. The operator should NEVER start and stop a motor with the starter door open. If something should happen inside the starter, particularly during start-up, most of this energy may be converted to heat energy, and the result may be molten metal blown toward the door. Do not take the chance—shut the door.

48.39 ELECTRONIC STARTERS

As mentioned before for other starters, the inrush current can be five times the full-load current for a motor. 🌎 This inrush current can be greatly reduced by using electronic starters. They are also called soft starters. These starters use electronic circuits to reduce the voltage and vary the frequency to

the motor at start-up. The result can be a motor that would typically draw 1000 A at locked rotor to be able to start up with 100 A at locked rotor.🌎 The motor speed is then accelerated up to full speed. This start-up is much easier on all components and reduces power line-voltage fluctuations.

Electronic starters are much smaller than conventional starters. For example, when an electronic starter is used to replace either an autotransformer or a wye-delta starter, it only takes one component to replace several. Electronic starters are much more compact, **Figure 48–80.**

The electronic starter uses electronic switching devices inside—typically SCRs to do the switching without electrical open contacts; see Unit 17 for an explanation of how these devices are switched on and off. Once the motor is up to speed and leveled out, these devices need to have the load taken off of them so they will not generate local heat. This is accomplished with a set of open contacts that are switched on very quickly after the motor is up to speed. These contacts will close, which effectively provides a circuit around the SCRs, **Figure 48–80.**

When it is time to stop the motor, the SCRs can be used to shut the motor down with a soft coast-down. It is also hard on other electrical circuits for a large electrical load to be running at full load and just shut off. The power companies' regulation system will allow a voltage spike with an abrupt shutdown. The SCRs can be used for shutdown by opening the bypass contactor contacts and then ramping the SCRs down with the electronics in the circuit.

In the event of a power failure, the system just shuts down by means of the bypass contactor.

The particular manufacturer's procedures must be used to troubleshoot all electronic equipment. There is a sequence to performing troubleshooting that must be used. These starters typically have an error code that is linked to a blinking LED (light-emitting diode) that will tell the technician that a fault has occurred and what it is. For example, the light may blink three times, then one time, for a code of 31. The technician then looks up that code to find out what the fault was. Usually the only tools the technician needs in order to troubleshoot these devices is a good VOM (volt-ohm-milliammeter) and a clamp-on ammeter for checking line amperage for the motor.

48.40 MOTOR PROTECTION

The motor that drives the compressor is often the most expensive component in the system. The larger the compressor motor the more expensive and the more protection it should be afforded. Small hermetic compressors in household refrigerators have little protection. Motors used in small chillers will have motor protection like that covered in Unit 19, "Motor Controls." Here we discuss only the motor protection used for large chillers such as rotary screw or centrifugals.

The type of protection depends on the size of the system and the type of equipment used. The advances made in electronic devices for monitoring voltage, heat, and amperage have improved the protection offered today compared with the protection offered several years ago. However, older motor protectors are still in use and will be for many years to come.

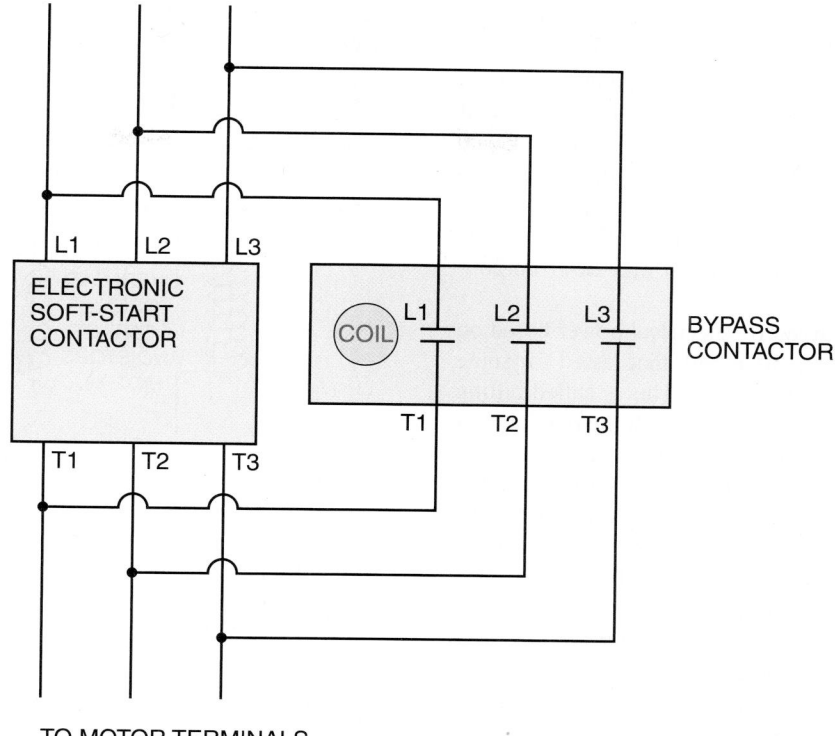

Figure 48–80 🌐 The electronic contactor is used to stop and start the motor because it offers a soft start and a slow stopping of the motor. This provides less inrush current and prevents a voltage spike when the motor is stopped. A set of bypass contacts (conventional contacts) are energized during the running cycle to take the load off of the electronic contactor. 🌐

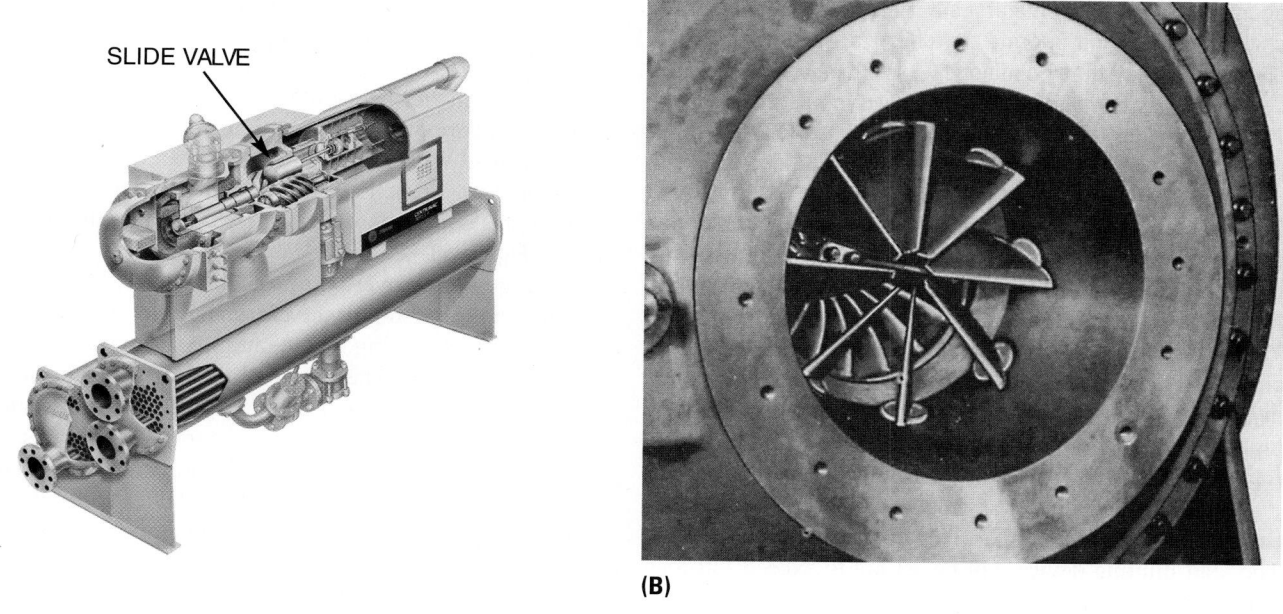

Figure 48–81   (A) A slide valve capacity control. **(B)** Prerotation guide vanes used for capacity control. **(A)** *Courtesy Trane Company.* **(B)** *Courtesy York International*

48.41 LOAD-LIMITING DEVICES

Motors used on rotary screw and centrifugal chillers use load-limiting devices to control the motor amperage. The load-limiting device monitors the current the motor is drawing while operating and throttles the refrigerant to the suction inlet of the compressor to prevent the motor from operating at an amperage higher than the full-load amperage. The load-limiting device controls the slide valve on a rotary screw or the prerotation vanes on a centrifugal, **Figure 48–81**. This device is a precision device that is adjusted at start-up and should not give any problems in the future. It is the first line of defense to prevent motor overload and is set at exactly the full-load amperage. NOTE: Full-load amperage may be derated to a lower value for a system that has high voltage so you cannot always go by the nameplate amperage. Look at the start-up log for the derated amperage.

The load limiter may also have a feature that allows the operator to operate the chiller at reduced load manually. Typically for rotary screw chillers this load may be varied from 10% to 100%. For centrifugal chillers it may be from 20% to 100%.

Most buildings are charged for electrical power based on the highest current draw for the month that lasted for some period of time, typically 15 to 30 min. This is called billing demand power charge with most power companies and is measured by a demand meter. Operating the equipment below the billing demand charge would be desirable whenever possible. For example, if an office building did not require cooling during the month until the last day and the operator started the chiller and allowed it to run at full load to reduce the building temperature quickly, this would take more than the required time to drive up the demand meter for the power company and the building would be charged for the whole month at a rate as though they had operated at full load all month. This excess could have been saved by an operator who started the system and kept it operating at part load and took longer to reduce the building temperature. For example, the chiller may have the feature that would allow it to run at 40% and take longer to accomplish the task.

🌐 Many buildings have computer-controlled power management. This type of management may take over from the chiller's control and operate the chiller at reduced capacity during high peak-load conditions.🌐 For example, if the peak demand of power consumption is reached for the month on the 15th, and on the 30th of the month a high air-conditioning load is needed, the system is allowed to operate above the peak demand set on the 15th, and a higher power bill will be paid for the entire month. 🌐The computer may reduce the power consumption on some portion of the building rather than allow this to happen. It is not uncommon to reduce the capacity of the air-conditioning chiller or chillers to accomplish this.🌐 This reduction may be easily accomplished through electronic circuits. It may be better to allow a temperature rise of a few degrees in the entire building rather than pay a considerably higher power bill for the month. This is called *load shedding,* for reducing the load and the demand charge.

The load limiter is set at motor full-load amperage. For example, assume a motor has a full-load amperage of 200 A. The load-limiting device will be set not to exceed full-load amperage, or 200 A.

48.42 MECHANICAL-ELECTRICAL MOTOR OVERLOAD PROTECTION

All motors must have some overload protection. A motor must not be allowed to operate at above the full-load amperage for long or damage will occur to the motor windings due

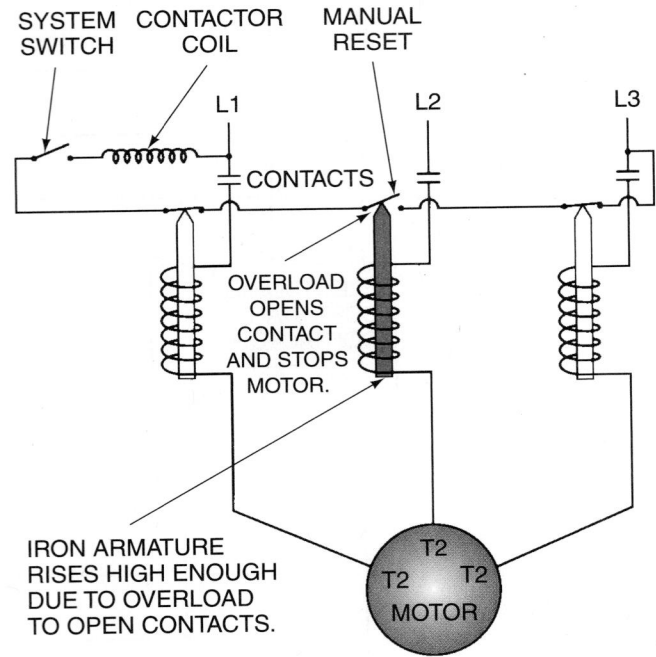

Figure 48–82 An iron core and current flow in a compressor overload.

to heat. Different motors have different types of protection. Overload protection for smaller motors, typically up to about 100 hp, was discussed in Unit 19, "Motor Controls."

Mechanical-electrical overload protection for large rotary screw or centrifugal compressors may be a simple dash-pot type of overload device. This device operates on the electromagnetic theory that when a coil of wire is wound around a core of iron, the core of iron will move when current flows. With the dash-pot, either the current from the motor or a branch of the current of the motor passing through a coil around an iron core will cause the iron core to rise, **Figure 48–82.**

Large motor overload protectors are typically rated to stop the motor should the current rise to 105% of the motor's full-load rating. For example, assume a motor has a rating of 200 A. The overload device would be set up to trip at 210 A (200 × 1.05 = 210). The motor should be allowed to run at this amperage for a few minutes to prevent unwanted overload trips before stopping the motor. These motors have a load-limiting device to limit the motor amperage, so if this device is functioning correctly, the amperage should not be excessive for longer than it takes this control to react. At this point, the amperage should be no more than full load.

The reason this overload device is called a dash-pot is because it contains a time delay to allow the motor to be able to be started. Without the time delay, the inrush current would trip the device. The time delay in the dash-pot is accomplished with a piston and a thick fluid that the piston must rise up through before tripping, **Figure 48–83.** The overload mechanism is wired into the main control circuit to stop

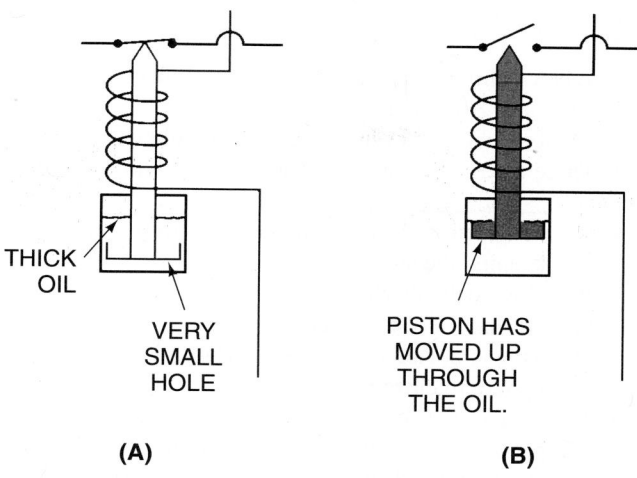

THICK OIL

VERY SMALL HOLE

PISTON HAS MOVED UP THROUGH THE OIL.

(A)   **(B)**

Figure 48–83   A dash-pot of oil gives the overload the time delay needed for motor start-up.

the motor when it is tripped. Overload protection for large motors is a manual reset type so the operator will be aware of any problem.

48.43 ELECTRONIC SOLID-STATE OVERLOAD DEVICE PROTECTION

These protective devices are wired into the control circuit like the dash-pot but act and react as an electronic control. They have the capability to allow the motor to be started and also monitor the full-load motor amperage closely. They typically are located in the starter cabinet between the starter and the motor. They are the last line of defense. They are much more compact than the dash-pot overloads and have many more features for close motor overcurrent control. These electronic overloads often have features such as the following:

- Motor overload protection
- Phase reversal protection
- Phase failure (single-phase) protection
- Voltage unbalance protection
- Manual and automatic reset, often from a remote location
- Ambient temperature compensation
- Trouble codes, much like the electronic contactors

Some of these features are discussed in more detail in the next portion of the text.

48.44 ANTI-RECYCLE CONTROL

All large motors should be protected from starting too often over a specified period of time. The anti-recycle timer is a device that prevents the motor from restarting unless it has had enough running time or off time to dissipate the heat from the last start-up. Manufacturers have different ideas as to how much this time should be. Typically the larger the motor, the more time required. Many centrifugals have a 30-min time. If the motor has not been started or tried to start for 30 min, it is ready for a start. Check the manufacturer's literature for the recycle time for a specific chiller.

48.45 PHASE FAILURE PROTECTION

Large motors all use three-phase power. The typical voltages are 208, 230, 460, and 575 V. Higher voltages are used for some applications, such as 4160 V or 13,000 V. Whatever the voltage, all three phases must be furnished or the motor will overload immediately. Electronic phase protection monitors the power and ensures that all three phases are present.

48.46 VOLTAGE UNBALANCE

The voltage to a compressor must be balanced within certain limits. Usually the manufacturers use 2% voltage unbalance as the maximum.

$$\text{Voltage unbalance} = \frac{\text{Maximum deviation from average voltage}}{\text{Average voltage}}$$

For example, a technician measures the voltage on a nominal 460-V system to be

Phase 1 to phase 2   475 V
Phase 1 to phase 3   448 V
Phase 2 to phase 3   461 V

The average voltage is

$$\frac{475\text{ V} + 448\text{ V} + 461\text{ V}}{3} = 461.3\text{ V}$$

The maximum deviation from average is

$$475\text{ V} - 461.3\text{ V} = 13.7\text{ V}$$

The voltage unbalance is

$$\frac{13.7}{461.3\text{ V}} = 0.0297 \text{ or } 2.97\%$$

This is more than some manufacturers' maximum allowable. The operating technician should keep an eye on voltage unbalance because it causes the motor to overheat. Some sophisticated electronic systems have voltage unbalance protection as part of their features; otherwise, it is up to the equipment operator to monitor.

Phase unbalance is caused by the electrical load on the building being unbalanced or the power company supply voltage being out of balance.

48.47 PHASE REVERSAL

Three-phase motors turn in the direction in which they are wired to run. If the phases are reversed, the motor rotation will reverse. *This can be detrimental for many compressors.* Reciprocating compressors will normally perform in either direction because of bidirectional oil pumps. Scroll, rotary screw, and centrifugal compressors must turn in the correct direction. Phase protection is often part of the control package for these compressors. Any compressor with a separate oil pump, such as the centrifugal, will not start under these conditions because the oil pump will not pump when reversed. It is not uncommon for the power company to reverse the phases of the power servicing a building or for an electrician to reverse the phases of a circuit within a building. It is very good practice for a building technician to be on the alert for any phase reversal after the building electrical system has been serviced by anyone. A phase meter may be used to determine the proper phasing of the power supply. This meter is a good tool for the heavy commercial or industrial technician to carry.

SUMMARY

■ A chiller refrigerates circulating water.

■ The compression cycle chiller has the same four basic components as other refrigeration systems discussed previously in this text: the compressor, evaporator, condenser, and metering device.

■ R-22, R-134a, and other refrigerants which are environmentally friendly alternatives are used in reciprocating compressor chillers.

■ Cylinder unloading is used to control the capacity of a reciprocating compressor.

■ Rotary screw compressors are used for larger-capacity chillers using high-pressure refrigerants.

■ These compressors are lubricated with an oil pump and separate motor.

■ The evaporators used for high-pressure chillers are either direct-expansion or flooded evaporators.

■ The condenser for high-pressure chillers may be either water or air cooled.

■ Many condensers have subcooling circuits that reduce the liquid temperature.

■ The condenser, like the evaporator, has an approach relationship between the refrigerant condensing temperature and the leaving-water temperature.

■ Air-cooled condensers eliminate the need for using water towers.

■ Centrifugal compressors may be manufactured so that they can be operated in series; this is called multistage operation.

■ For prelubrication of the bearings, the oil pump on a centrifugal compressor is energized before the compressor is started.

■ The orifice or the high- or low-side float metering device is typically used for low-pressure chillers.

■ When a centrifugal uses a low-pressure refrigerant, the low-pressure side is always in a vacuum. If there is a leak, air will enter the system.

■ Absorption refrigeration is a process that uses heat as the driving force rather than a compressor.

■ The absorption chiller uses water as the refrigerant.

■ The absorption chiller uses a salt solution consisting of lithium-bromide (Li-Br) as the attractant in the refrigeration process.

■ These solutions are corrosive, and air must be kept from them.

■ The absorption chillers also have a purge system.

■ Many motor controls include motor overload protection devices, load-limiting devices, anti-recycle control, phase failure protection, voltage unbalance, and phase reversal protection.

REVIEW QUESTIONS

1. When a chiller is used, the secondary refrigerant that circulates in the building is _____.
2. What are the two basic categories of chillers?
3. Three types of compressors are used for chillers: _____, _____, and _____.
4. Blocked suction cylinder unloading is accomplished by
   A. keeping the suction valve closed.
   B. closing the discharge.
   C. closing the suction service valve.
   D. using a device in the low-pressure side of the system to close it off.
5. True or False: The capacity control for a screw compressor is accomplished with cylinder unloading.
6. Which of the following methods are used to control the capacity of a centrifugal compressor?
   A. Cylinder unloading
   B. Prerotation vane control
   C. Blocked discharge
   D. High- to low-side bypass
7. Describe a direct-expansion versus a flooded evaporator.
8. What is the purpose of multiple passes in an evaporator in a chiller?
9. The approach temperature for the evaporator is
   A. the difference between the suction and head pressure converted to temperature.
   B. the difference between the refrigerant boiling temperature and the suction-line temperature.
   C. the difference between the refrigerant boiling temperature and the inlet water temperature.
   D. the difference between the refrigerant boiling temperature and the leaving-water temperature.
10. What type of water-cooled condenser is used in large chillers?
11. The compressor used in low-pressure chillers is the
    A. centrifugal.
    B. screw.
    C. rotary.
    D. reciprocating.
12. List the types of metering devices used with high-pressure chillers.
13. The subcooling temperature in a condenser can be measured by taking the difference between the
    A. suction pressure converted to temperature and the boiling refrigerant.
    B. boiling temperature and the condensing temperature.
    C. condensing temperature and the leaving-refrigerant temperature.
    D. condensing temperature and the entering-refrigerant temperature.
14. What causes a surge in a centrifugal chiller?
15. Why is it important to keep an operating log on a chiller?

16. Which of the following metering devices is used on low-pressure chillers?
   A.   A low- or high-side float
   B.   A capillary tube
   C.   An automatic expansion valve
   D.   A thermostatic expansion valve

17. The purge unit on a low-pressure chiller removes
   A.   overcharge of refrigerant.
   B.   excess oil.
   C.   condensable refrigerant.
   D.   noncondensables.

18. In an absorption chiller, _____ is used to create the difference in pressures.

19. The types of energy used to power a typical absorption refrigeration machine are _____ and _____.

20. True or False: Some absorption machines are directly fired with gas or oil.

21. What is the common refrigerant used in absorption refrigeration machines?

22. How is capacity control accomplished with a steam-driven absorption refrigeration machine?

23. True or False: Single-phase induction run-repulsion start motors are used with centrifugal machines.

24. Describe how a wye-delta motor is started and runs.

25. Why is it important to have phase protection for large motors?

UNIT 49 Cooling Towers and Pumps

OBJECTIVES

After studying this unit, you should be able to

- describe the purpose of water cooling towers used with chilled-water systems.
- state the relationship of the cooling capacity of the water tower and the wet-bulb temperature of the outside air.
- state the means by which the cooling tower reduces water temperature.
- describe three types of water cooling towers.
- explain the various uses of fill material in water cooling towers.
- list the two types of fan drives.
- state the two types of fans used in water cooling towers.
- explain the purpose of the water tower sump.
- explain the purpose of makeup water.
- describe a centrifugal pump.
- describe water vortexing.
- explain two types of motor-pump alignment.

SAFETY CHECKLIST

- ✔ Ensure that all electrical safety precautions are observed when servicing pump motors.
- ✔ Ensure that all guards are in place when working in the area near a turning fan.
- ✔ Do not approach a turning fan that is not properly guarded.
- ✔ Be careful of your footing and balance when climbing up to or down from a cooling tower location. Do not move in a way that will cause you to lose your balance when you are working around an elevated tower.
- ✔ When lifting a motor or pump, lift with your back straight. Wear a back belt support if recommended by your employer or insurance carrier.

49.1 COOLING TOWER FUNCTION

The heat that is absorbed into the boiling refrigerant in the chiller's evaporator is pumped to the condenser as hot gas via the compressor. The hot refrigerant gas is then condensed with water from the *cooling tower.* The water from the condenser now contains the heat absorbed from the building. It is pumped to the cooling tower, where atmospheric air is passed over the water to remove the heat from the water so that most of it can be reused, **Figure 49–1.**

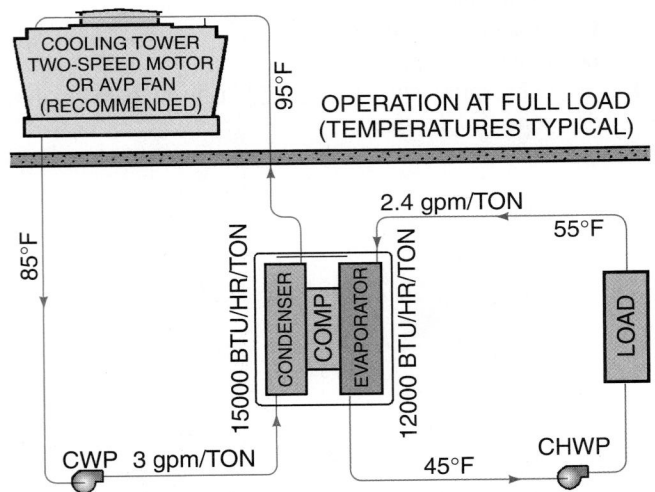

Figure 49–1 An example of a condenser water circuit with a cooling tower. *Courtesy Marley Cooling Tower Company*

In compression systems the cooling tower must reject more heat than the chiller absorbs from the structure. The chiller absorbs the heat from the chilled water circuit, and in compression systems the compressor adds heat of compression to the hot gas pumped to the condenser, **Figure 49–2.** The compressor adds about 25% more heat, so the cooling tower must reject about 25% more heat than the capacity of the chiller. For example, a 1000-ton chiller would need a cooling tower that could reject about 1250 tons of heat.

The condenser must be furnished water within the design limits of the system, or the system will not perform adequately. The design temperature for the water leaving the cooling tower for most refrigeration systems, including the absorption

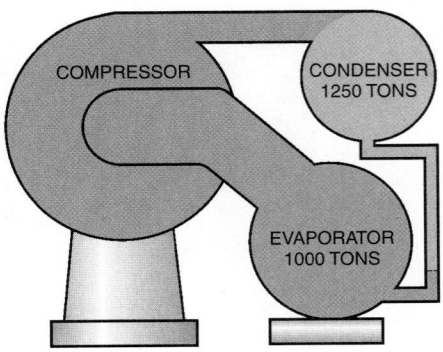

Figure 49–2 Heat of compression and the condenser.

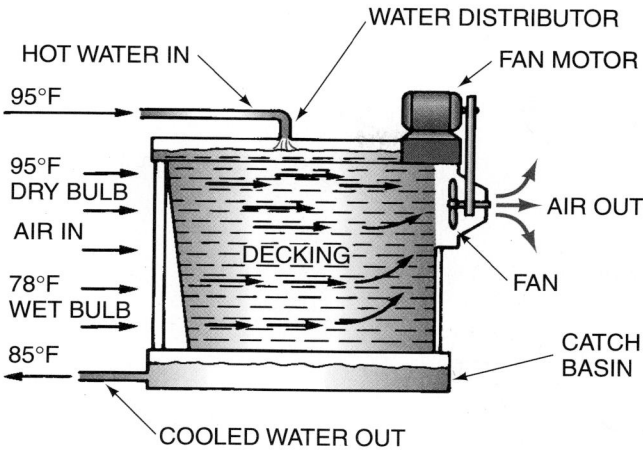

Figure 49–3   The typical cooling tower approach temperature.

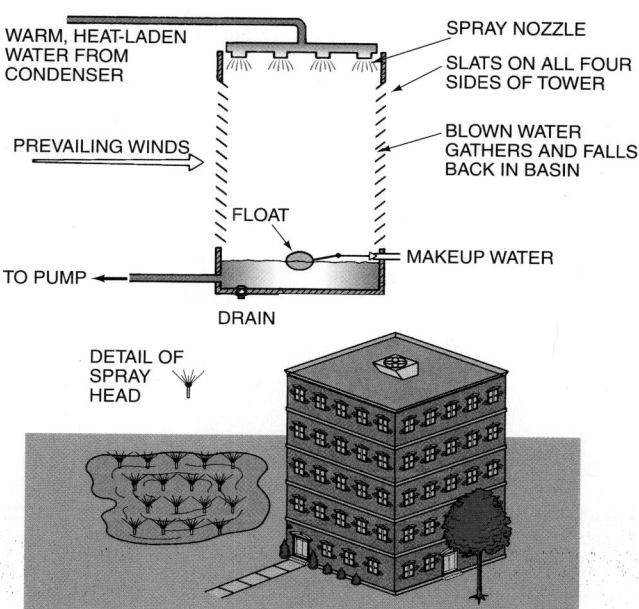

Figure 49–4   Natural-draft tower types.

chiller, is 85°F. Using a design temperature of 95°F dry-bulb and 78°F wet-bulb for a typical southern section of the country, the cooling tower can lower the water temperature through the tower to within 7°F of the wet-bulb temperature of the outside air. When the wet-bulb temperature is 78°F, the water temperature leaving the tower would be 85°F, **Figure 49–3.** This occurs even though the air dry-bulb temperature is 95°F.

Most cooling towers work in a similar manner; they reduce the temperature of the water in the tower by means of evaporation. As the water moves through the tower, the surface area of the water is increased to enhance the evaporation. The different methods used to increase the surface area of the water is part of what makes one water tower different from another. When the water is spread out to create more surface area, the water will evaporate faster and be cooled closer to the wet-bulb temperature of the air. The water cannot be cooled below the temperature of the cooling medium, which is the wet-bulb temperature. A nearly perfect cooling tower could reduce the temperature to the entering wet-bulb temperature; however, this tower would be tremendous in size. Manufacturers make towers that approach the wet-bulb temperature by 7°F for typical applications. Towers are manufactured to a 5°F approach, but they are not typically used for air-conditioning applications.

49.2   TYPES OF COOLING TOWERS

Three types of cooling towers are in common use: the *natural-draft* and *forced- or induced-draft tower* and another type utilizing three different operating modes. ⬤ This is a closed-circuit hybrid cooling tower. ⬤ The larger the installation, the more elaborate the tower.

The natural-draft tower may be anything from a spray pond in front of a building to a tower on top of a building, **Figure 49–4.** The spray pond and natural-draft cooling tower rely on the prevailing winds and will cool the water to a lower temperature when the prevailing winds are greater. The natural-draft tower or spray pond application usually has an approach temperature of about 10°F because of relying only on the winds.

The spray pond can be located at ground level or it may be on the roof of a building, but wherever it is, the spray must be contained in the area of the pond. The wind can blow the spray water out of the pond and annoy people or damage property. Spray ponds on rooftops have made a comeback in the last few years because they cool the roof and reduce the solar load. If the rooftop can be cooled to 85°F, a great reduction in air-conditioning load can be realized.

The spray pond and natural-draft cooling tower use pump pressure to increase the area of the water by atomizing it into droplets. Spray heads are located at strategic locations within the pond or tower and provide a spray pattern for the water. The spray nozzles are a source of restriction and must be kept clean or they will not atomize the water. Atomizing the water requires pump horsepower that may be considered an unacceptable expense because of the large volume of water used in these applications.

Forced-draft and induced-draft towers use a fan to move the air through the tower, **Figure 49–5.** These are the most popular applications used today because the efficiency is reliable. Centrifugal fans are used to move the air in some towers. When the tower air must be ducted to the outside of the building, the centrifugal fan must be used because it can overcome the static pressure in the ductwork. Towers with centrifugal fans can be more compact and therefore more desirable for some applications. These towers are available up to about 500 tons of capacity and heat rejection.

Larger towers use propeller-type fans. These fans may be either belt driven or gear driven. Belt-driven fans will require belt maintenance on a regular basis. Gear-driven fans have a transmission that will require only lubrication.

⬤ The closed-circuit tower, **Figure 49–6,** has a dry/wet mode, an adiabatic mode, and a dry mode. This tower has a finned coil, **Figure 49–7;** a prime surface coil, **Figure 49–8;** and a wet deck surface, **Figure 49–9.** ⬤

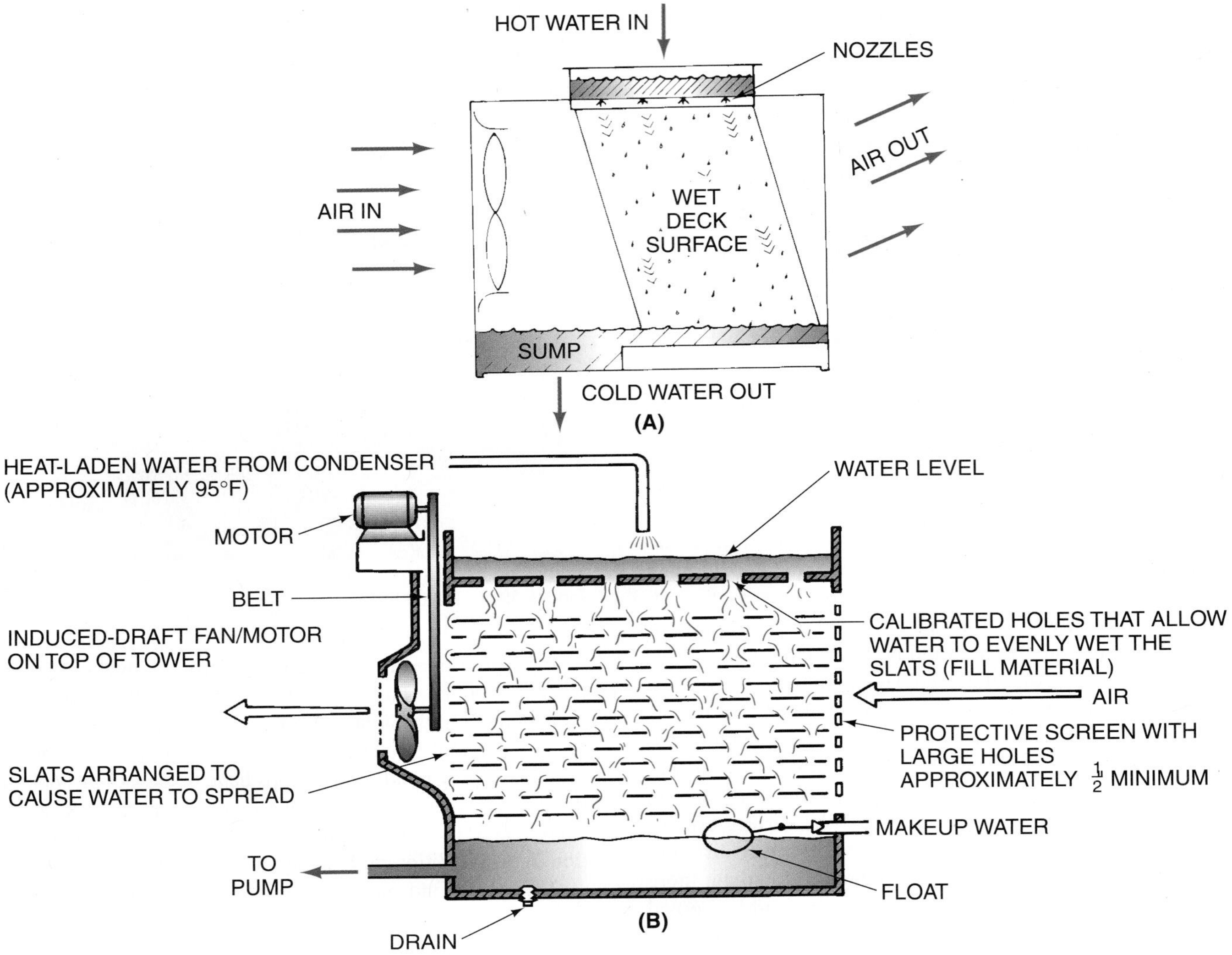

HOT WATER IN

NOZZLES

AIR OUT

AIR IN

WET DECK SURFACE

SUMP

COLD WATER OUT

(A)

HEAT-LADEN WATER FROM CONDENSER (APPROXIMATELY 95°F)

WATER LEVEL

MOTOR

BELT

INDUCED-DRAFT FAN/MOTOR ON TOP OF TOWER

CALIBRATED HOLES THAT ALLOW WATER TO EVENLY WET THE SLATS (FILL MATERIAL)

AIR

PROTECTIVE SCREEN WITH LARGE HOLES APPROXIMATELY $\frac{1}{2}$ MINIMUM

SLATS ARRANGED TO CAUSE WATER TO SPREAD

MAKEUP WATER

TO PUMP

FLOAT

DRAIN

(B)

Figure 49–5 **(A)** A forced-draft cooling tower. **(B)** An induced-draft cooling tower.

Figure 49–6 A closed-circuit cooling tower. *Copyright, Baltimore Air Coil Company, 2003*

FINNED COIL

Figure 49–7 A finned coil located at the top of the closed-circuit cooling tower. *Copyright, Baltimore Air Coil Company, 2003*

Figure 49–8   A prime surface coil located below the water distribution system in the closed-circuit cooling tower. *Copyright, Baltimore Air Coil Company, 2003*

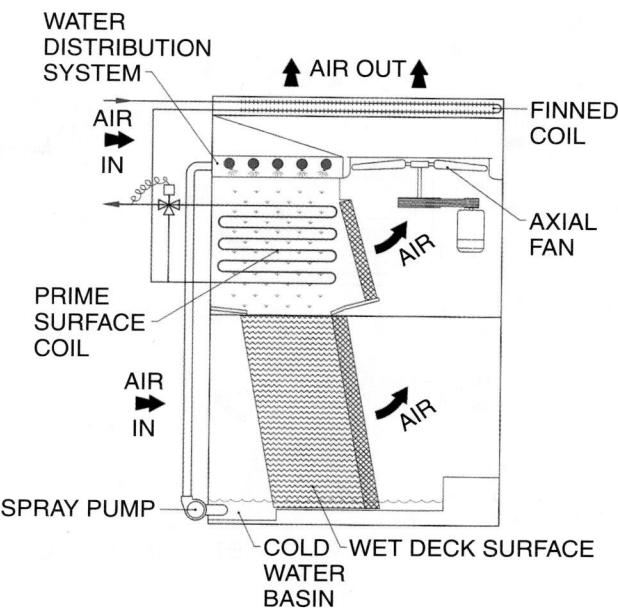

DRY/WET MODE

Figure 49–10   🌐An illustration of the closed-circuit cooling tower in the dry/wet mode.🌐 *Copyright, Baltimore Air Coil Company, 2003*

Figure 49–9   A PVC wet deck surface located below the prime surface coil in the closed-circuit cooling tower. *Copyright, Baltimore Air Coil Company, 2003*

In the dry/wet mode the fluid to be cooled is fed to the dry finned coil and then to the prime surface coil, **Figure 49–10.** The fluid leaves the tower and is pumped back to the system, probably a condenser, where its cooled effect is used. Spray water from the cold water sump is pumped to a distribution system above the prime surface coil and allowed to fall over the prime surface coil, thus providing evaporative cooling. The spray water then falls over the wet deck surface, where it is cooled as more evaporative heat transfer takes place. Air is drawn through the prime surface coil and the wet deck surface, leaving after passing through the finned coil, where more cooling occurs. When heat load or ambient temperatures are reduced, the water usage over the evaporative cooling portion can be reduced. A valve which controls the outlet fluid temperature can modulate this water usage.

The term *adiabatic* refers to the process of cooling the circulating condenser water by evaporating the tower water. It

means that there is no heat added or removed from the process from external means and that the sensible heat transferred from the condenser circulating water is equal to the latent heat transferred or gained into the air from the evaporating water in the cooling tower.

In the adiabatic mode the fluid to be cooled bypasses the prime surface coil and is pumped through the finned coil only, **Figure 49–11.** Spray water is used to help cool the air passing through the tower, providing sensible heat transfer through the finned coil.

Plume, which is the saturated discharge air, can be reduced (unlike conventional evaporative cooling towers)—some with the wet/dry mode and others in the adiabatic mode. Plume can be a problem in some areas when falling on parked cars or other highly finished objects. It can also be a safety factor when falling on areas where cars travel and at airports.

🌐 In the dry mode the fluid to be cooled is passed through both the finned coil and the prime surface coil, **Figure 49–12.** No spray water is used and consequently there is no plume. The fluid is cooled by the forced air being passed over the coils.🌐

The closed-circuit system protects the system fluid from contamination and helps to ensure maximum efficiency over time, including reduced water usage. The closed-loop feature of this tower not only protects the circulating fluid, it protects the heat transfer surface in the condenser. As mentioned earlier, the condenser tubes frequently have a slightly fouled surface that reduces the heat exchange. With the closed-loop system the tubes can be maintained with practically no fouling. Cooling towers may be selected for 86°F entering water instead of 85°F because of this condenser efficiency. This can be a savings in tower size and first cost.

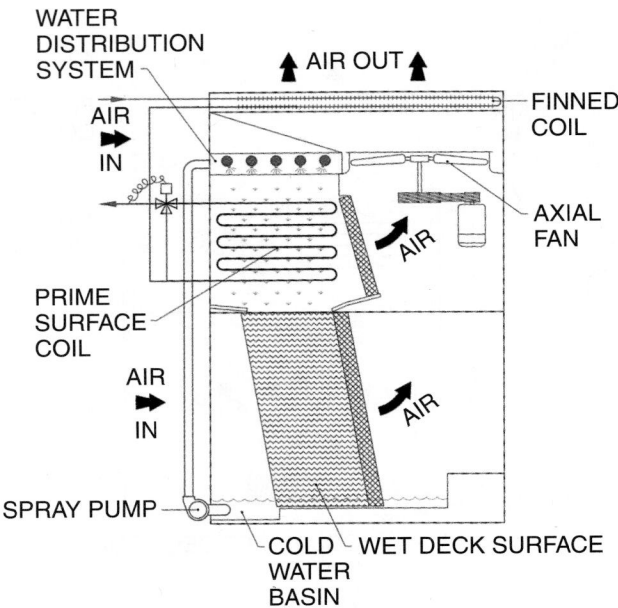

ADIABATIC MODE

Figure 49–11 🌐 An illustration of the closed-circuit cooling tower in the adiabatic mode. 🌐 *Copyright, Baltimore Air Coil Company, 2003*

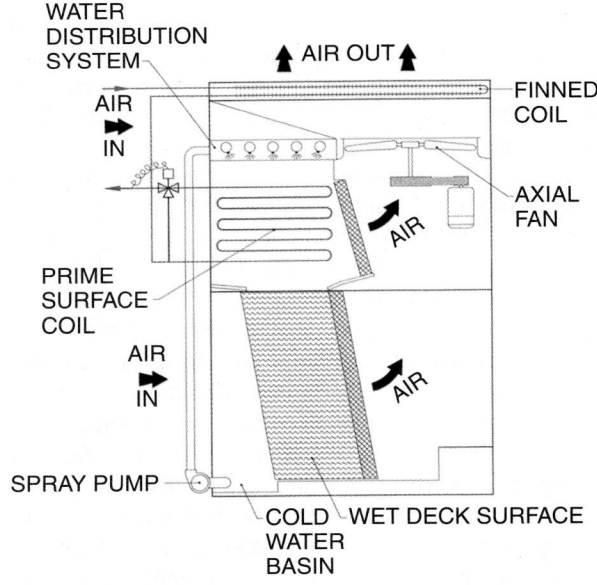

DRY MODE

Figure 49–12 🌐 An illustration of the closed-circuit cooling tower in the dry mode. 🌐 *Copyright, Baltimore Air Coil Company, 2003*

49.3 FIRE PROTECTION

The open type cooling tower is a place where a fire may occur during the off season when the tower components are dry. The tower may contain materials that are flammable, such as wood or some plastics. Some fire codes and insurance companies may require that the tower be manufactured of all fire-proof materials or they may require that a tower wetting system be in place. A tower wetting system may consist of sprinkler heads such as these required in buildings. Some towers are controlled in such a manner that the system pump starts up on a timed cycle to wet the tower down occasionally during the off cycle. Another wetting system may be an auxiliary pumping system that keeps the tower wet all the time whenever the weather is above freezing. This type of system also may prevent expansion and contraction of any wood construction in the tower because it is wet all the time. Local codes and insurance requirements will dictate what method should be used to protect the tower from fire.

49.4 FILL MATERIAL

The cooling tower is designed to keep the water and the air in contact for as long as possible. Manufacturers use various methods to slow the water as it trickles down through the tower with the air moving up through the tower. Two methods are used to evaporate the water in the forced- and induced-draft tower. These are the splash method and the fill, or wetted-surface, method. Both methods use a material in the tower often called the fill material. This material may be wood slats where the water drips down through layers of wood. Other types of materials are also used in the splash method, such as polyvinyl chloride (PVC) plastic or fiber-reinforced polyester (FRP) plastic. These have a slow burn rate and should be acceptable for use. Towers that use the splash system have a framework that supports the slats and is designed to keep them at the correct angle for proper water wetting of all slats with water flow from the top of the tower to the bottom.

The fill or wetted-surface type of tower uses a fill material that may be some form of plastic or fiberglass. The water is spread out on the surface while air is passed over it, **Figure 49–13**. This type of fill may have smaller passages for the air to travel through and is not used where particles from the surroundings can contaminate the tower and restrict the passages.

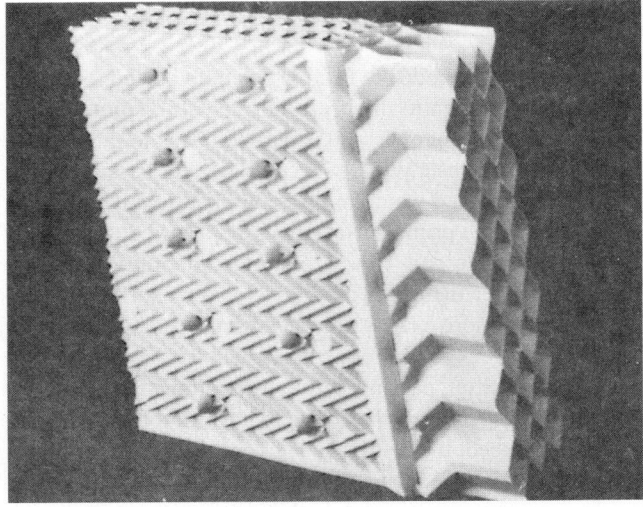

Figure 49–13 The wetted surface tower material. *Courtesy Marley Cooling Tower Company*

Both types of fill material rely on water to run down across the fill material, which should be kept at the angle the manufacturer recommends or the water will not take the proper path. Water running to one side or other will cause the capacity of the tower to be reduced. If the fill material is removed for maintenance, care should be taken to replace it in the correct manner for correct water movement through the tower.

49.5 FLOW PATTERNS

There are two distinct airflow patterns for cooling towers: *crossflow* and *counterflow,* **Figure 49–14.** The crossflow tower introduces the air from the side and usually pulls the air to the top of the tower for exhaust. In smaller crossflow towers the fan is on the side of the tower and the exhaust is out the side, **Figure 49–5(B).** When the air is exhausted out the side of the tower, care must be taken that the moisture-laden air is not exhausted to a place where it will cause a problem, such as in a walkway or parking lot where cars may be spotted with the water and chemicals used in treatment, which may be corrosive to car finishes. In the tower, the water is moving downward and the air is moving at right angles to the water.

The counterflow tower introduces the air at the bottom of the tower and exhausts the air out the top. The water is moving downward as the air moves up through it.

The water in most towers, specifically the spray-type towers, has many small particles of water suspended in air. These particles are subject to being blown out of the tower by the prevailing winds. This loss of water is called drift and can be expensive and a nuisance for any surrounding areas with spray that contains chemicals. This drift of water is minimized with eliminators. Eliminators cause the spray to change direction and rub against a solid surface where the water should deposit and run back to the tower basin. The eliminators may be louvres on the side of the tower or they may be part of the fill material in some newer towers.

49.6 TOWER MATERIALS

The materials from which cooling towers are constructed must be able to withstand the environment in which the tower operates. There are towers all over the world and in different chemical environments so the materials are varied. The typical tower must be able to withstand wind, weight of tower components and related water, sun, cold, freezing weather (including any ice that may accumulate), and vibration of the fan and drive mechanisms. Towers must be carefully designed and many different materials may be used depending on the type and location of the tower. Typically, the smaller packaged towers are made of galvanized steel (for rust protection), fiberglass, or FRP. These towers are manufactured as a complete assembly and shipped to the job, **Figure 49–15.**

Larger towers may have a concrete base and sump to hold the water and the sides made of other materials such as corrugated asbestos-cement panels, wood (either treated or redwood), fiberglass, or corrugated FRP. Fire prevention must be considered in the selection of materials in many

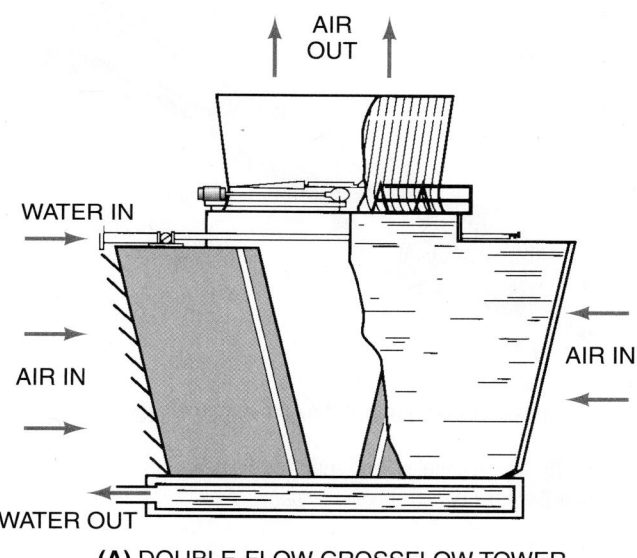

(A) DOUBLE-FLOW CROSSFLOW TOWER

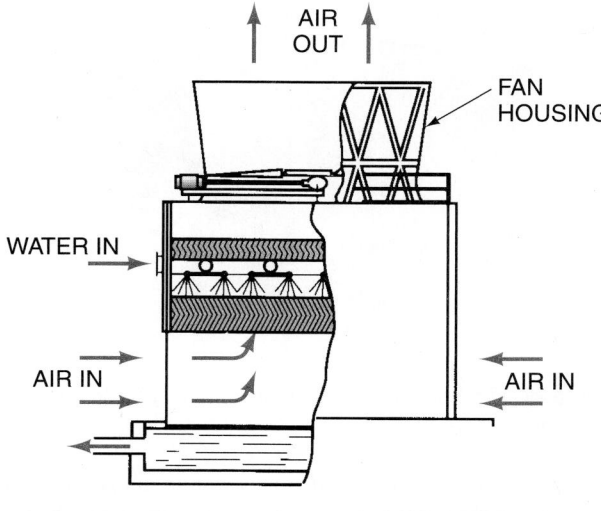

(B) INDUCED-DRAFT COUNTERFLOW TOWER

Figure 49–14 Crossflow and counterflow cooling towers. *Courtesy Marley Cooling Tower Company*

Figure 49–15 A packaged cooling tower. *Courtesy Baltimore Aircoil Company*

cases. Asbestos-cement panels are no longer being used. They may be found in older towers, and technicians should handle them with care. Contact the manufacturer for handling instructions.

49.7 FAN SECTION

The motor must have a method for turning the fan for all forced-draft towers. There are two different types of drive, the belt and gearbox (transmission) drives. The belt drive is normally used for the smaller towers and includes an adjustable motor mount. This mechanism will have the greatest wear in the cooling tower and require the most maintenance so it should be located where preventive maintenance and service can be performed with as little effort as possible.

Fans for larger towers have the motor mounted out to the side and a gearbox or transmission to change the direction of the motor drive shaft 90°. These motor/fan units may also be designed to change the motor-to-fan shaft speed relationship. The motor will turn at 1800 or 3600 rpm for a typical motor and the speed of the fan shaft will turn considerably slower, depending on the gear reduction. The motor is connected to the gearbox using a coupling and shaft, **Figure 49–16**. The motor, gearbox, and fan bearings must be accessible for service purposes.

The propeller fan blade is enclosed in a fan housing that improves the efficiency of the fan blades, **Figure 49–14**. This fan blade location is critical in most towers. The fan must be located at the correct distance from the top and sides for best performance.

🌎VFDs (variable-frequency drives) for motors are also being used for cooling tower fans. The capability of reducing the fan speed to correspond with system conditions will help control the system more closely and improve efficiency by reducing the energy used by the system.🌎 In the past, a thermostat in the water was used to turn the cooling tower fans on or off as needed to control water temperature. This way a minimum tower basin water temperature could be maintained. This system has a fairly wide temperature swing—causing the water temperature to vary considerably. As we will see later in this unit, motorized diverting valves may be used to maintain conditions closer to the condenser when water temperature swings occur. The VFD tower fan motor can be used for this function.

🌎 The VFDs also allow the motor to soft start and to coast down without just having to be shut off. Both of these features improve system performance and prevent power fluctuations.🌎 Some of these cooling towers have large motors.

49.8 TOWER ACCESS

All cooling towers will need service on a regular basis. The tower must have access doors to the fill material for cleaning and possible removal, **Figure 49–17**. The tower basin must also be accessible for cleaning because large amounts of sludge will accumulate. The cooling tower becomes a large filter for whatever may be airborne. Dirt, pollution, feathers, birds, plastic

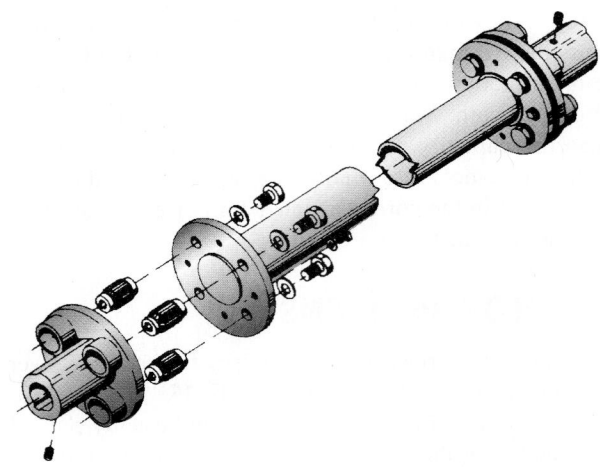

GROMMET-TYPE DRIVESHAFT COUPLING

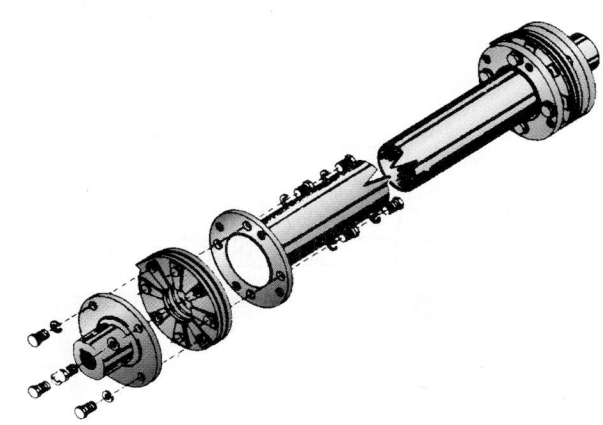

DISC-TYPE DRIVESHAFT COUPLING

Figure 49–16   A coupling and shaft for the gear reduction box. *Courtesy Marley Cooling Tower Company*

ACCESS DOOR

Figure 49–17   Tower access for service. *Courtesy Baltimore Aircoil Company*

wrappers, and cups are among the common items. These particles will gather in the sump and must be removed. There is a screen at the water outlet to the cooling tower that will stop objects from entering the pump and piping circuit. There

Figure 49–18   The stairway and guardrails for a large tower. *Courtesy Marley Cooling Tower Company*

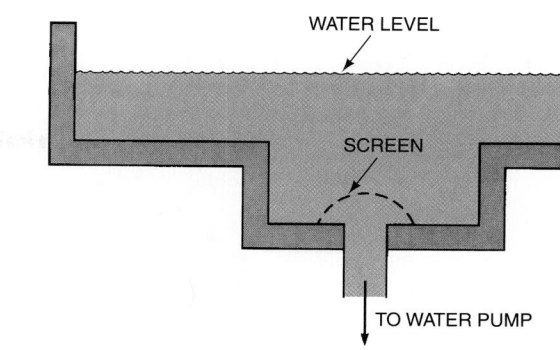

Figure 49–20   The filter system for a cooling tower sump.

should be an adequate water supply in the vicinity of the tower for the connection of a water hose for flushing the tower sump.

When the tower is tall, it will likely have a stairway or ladder to the top for servicing the fan and drive components, **Figure 49–18.** Provisions should be made in the vicinity of the motor and fan to lift any component parts that may need to be removed. These parts may need to be lowered to the ground for servicing. A stair or ladder installed for servicing must meet the local codes for safety and include proper handrails and barriers to prevent stepping over the edge.

49.9   TOWER SUMP

All cooling towers must have some sort of *sump* for the water to collect in. The sump on small cooling towers may consist of a metal pan that drains to a lower point to gather the water. When this sump contains water and is located outside, it must have some means to prevent freezing such as a sump heater that is thermostatically operated, **Figure 49–19.** These sump heaters are used for small installations. Larger installations that would require more heat may have a circulating hot water coil or a method of using low-pressure steam to heat the tower basin water in cold weather.

Many cooling towers use underground sumps made of concrete. These sumps will not freeze in the southern climates

but must be protected from the cold in northern climates unless the refrigeration system is operating and adding heat to the sump. Some sumps are located in the heated space of a building to prevent freezing. Antifreeze products also may be added to some systems to prevent freezing.

Wherever the sump is located, it is the catchall for all sediment and should be accessible for cleaning. A bypass filter system is often used to sweep the bottom of the sump and carry a portion of the water through the filter system for cleaning. The sump usually contains a coarse screen strainer to protect the pump, **Figure 49–20.** The return line may also have an in-line removable strainer.

49.10   MAKEUP WATER

Because the cooling tower operates using the principle of evaporation, water is continuously lost from the system. Several systems are used to make up water, the float valve being one of the most common. When the water level drops, the float ball drops with it and opens the valve to the makeup water supply. Usually this makeup water is from the normal municipal or other similar supply water. Another method is to use float switches that fall with the water level and energize a solenoid valve to allow water to fill the sump. When the water level reaches the proper level, the float rises and shuts off the solenoid valve. Still another method uses electrodes protruding into the water that sense the water level.

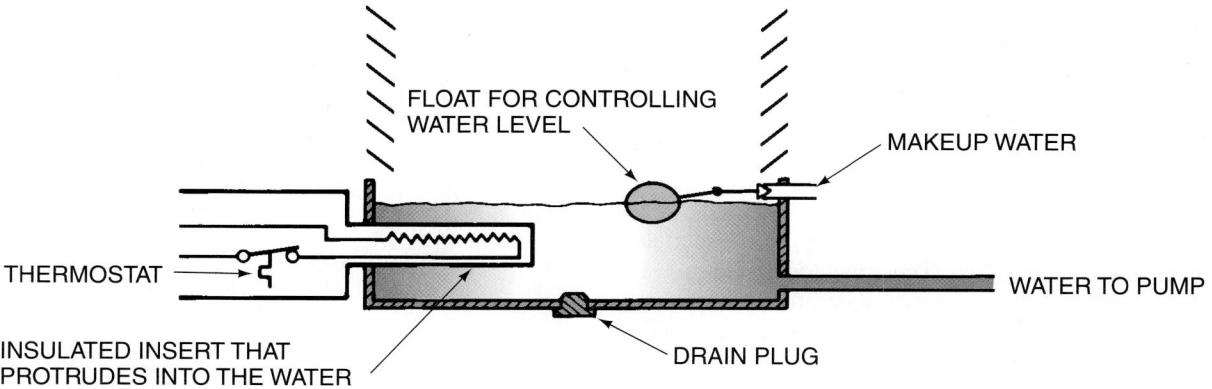

Figure 49–19   Freeze protection for a cooling tower sump.

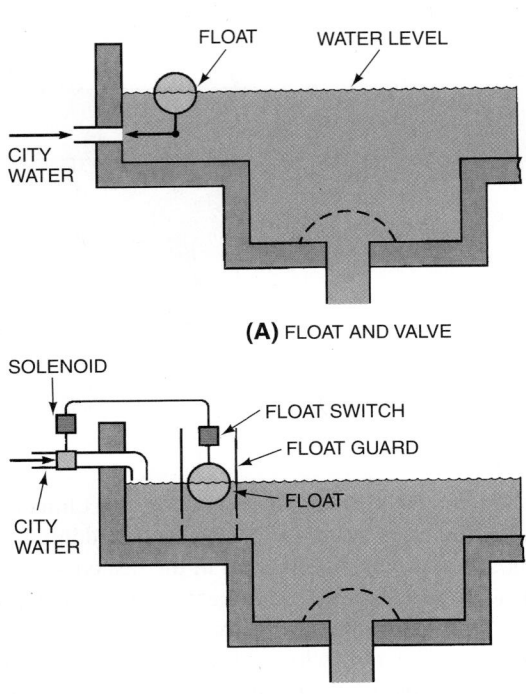

(A) FLOAT AND VALVE

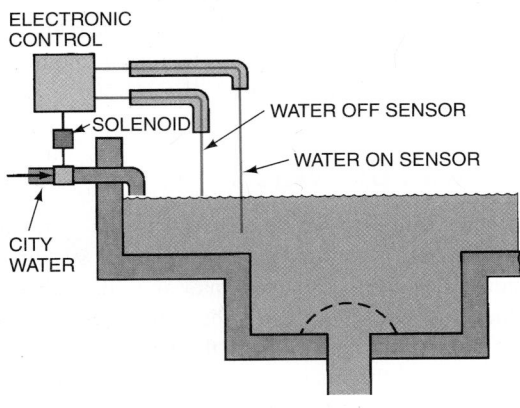

(B) FLOAT SWITCH AND SOLENOID

(C) ELECTRONIC WATER LEVEL CONTROL

Figure 49–21 Three types of cooling tower fill methods. **(A)** Float and valve. **(B)** Float and switches. **(C)** Electronic sensors.

When the electrodes sense that the water level is too low, a solenoid valve is opened, and makeup water is allowed to enter. As the water level rises, an upper electrode senses the water level and shuts off the water. **Figure 49–21** shows all three level-control methods.

49.11 BLOWDOWN

The water that evaporates leaves behind any solid materials that may have been in the water. These include dust particles, minerals, and algae, a lower form of plant life. As the water evaporates, the water left behind becomes more and more concentrated with these particles. If this continues, the particles will drop out of the water and deposit on the surfaces of the cooling tower. These will be hard to remove because they

turn to a substance much like cement. Even more important than what is happening to the cooling tower is what will happen to the condenser. Some of these particles will deposit at the hottest part of the condenser, the *outlet*. These deposits will act like an insulation to the heat exchanger. This will cause the head pressure to rise. The condenser approach will begin to spread. As mentioned in the section on condensers, the original operating log will give the technician a clue as to when this is happening.

Blowdown is the cure for part of these problems. It is merely a bleed off of the portion of the water that is being circulated. When new water is added, less sediment is present as some has gone with the water that was bled off. Fresh water is added to make up for water lost during blowdown. This fresh water dilutes the remaining water and reduces the mineral content. It is generally recognized that 3 gal/min/ton is the amount of water circulated in a water-cooled condenser used for air conditioning with a 10°F temperature rise through the condenser. **Figure 49–22** shows a 30-ton application and what would happen if the cooling tower water did not have blowdown as part of the piping system. **Figure 49–23** shows four methods that may be commonly used for obtaining the correct blowdown. **Figure 49–23(A)** shows a constant bleed with a flow control device. Often this is just a hand valve. **Figure 49–23(B)** shows two methods of taking a sample of water in a measured area. **Figure 49–23(C)** is a method of measuring

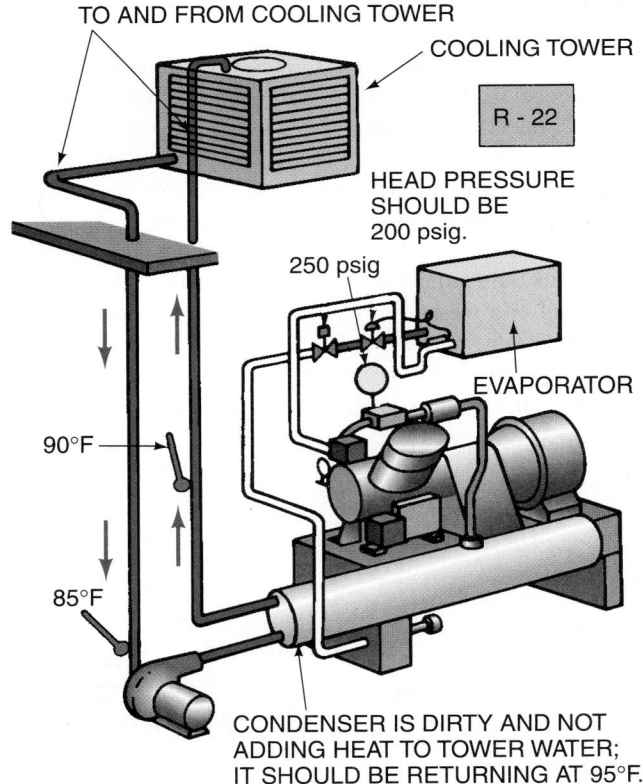

Figure 49–22 The flow rate of water is correct, but only a 5°F temperature rise occurs. The condenser is not removing enough heat because the tubes are fouled (dirty). High head pressure is the result.

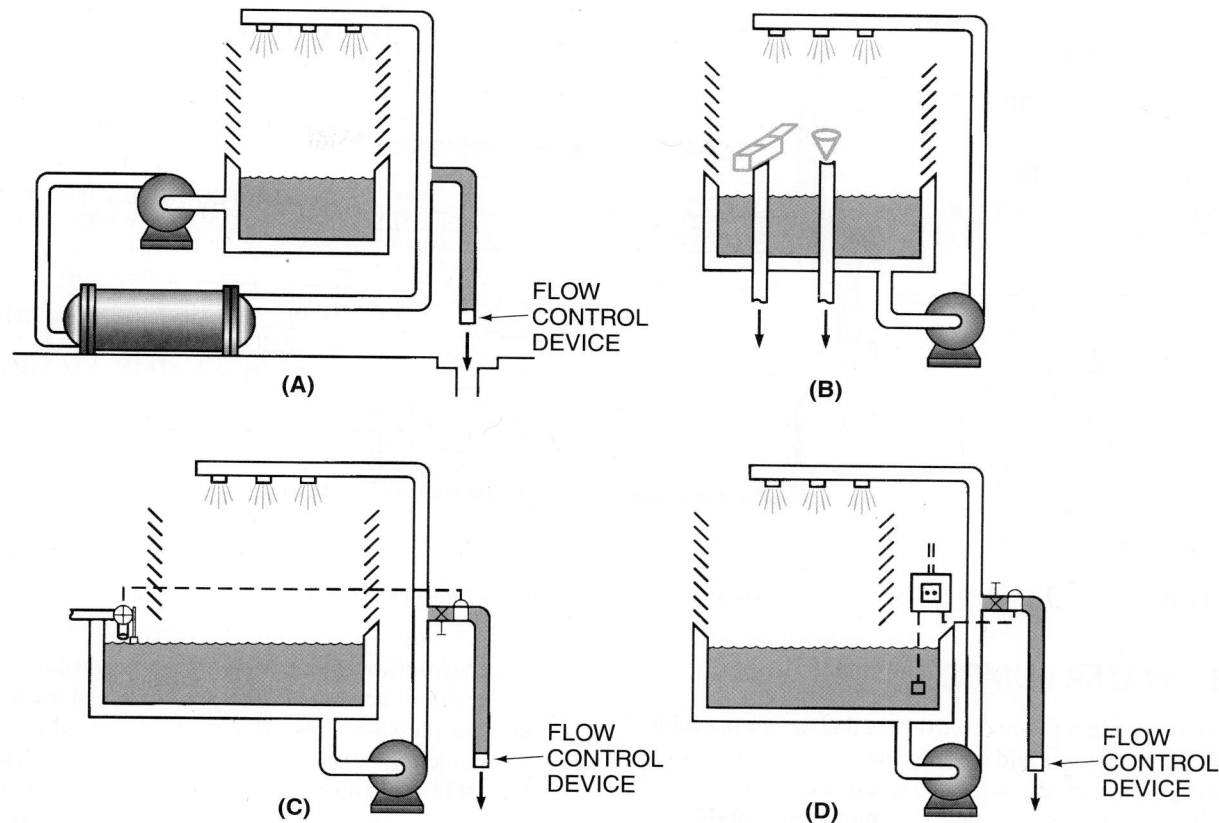

Figure 49–23   Four methods by which proper blowdown may be achieved. *Courtesy Nu-Calgon Wholesaler, Inc.*

the amount of water flow and bleeding off water on the basis of flow rate. **Figure 49–23(D)** shows an electronic sensor that senses the mineral content of the water and bleeds off water when the minerals have become too concentrated in the water. Often, building management personnel have a difficult time understanding the purpose of blowdown. They do not understand why water is being allowed to run down the drain. **Blowdown must be managed correctly because expensive water treatment chemicals are going down the drain with the water.**

49.12   BALANCING THE WATER FLOW FOR A TOWER

The water flow for even distribution must be correct for many towers. If a tower has two or more cells where the water is distributed, the same amount of water must be fed to each cell or the tower will not perform as designed. Many towers use a distribution pan at the top to distribute the water over the fill. These pans are often a series of calibrated holes drilled into the pan. Warm return water from the condenser is poured out in the pan. This discharge may be split to each side of the tower. A balancing valve or valves must be set to obtain the correct flow, **Figure 49–24**. The calibrated holes in the top of the tower must also be clean of foreign material and be of the correct size. **Figure 49–25** shows a pan where the holes are

rusted out and most of the water is flowing to one side of the tower and not wetting the entire fill deck. The tower will not perform correctly when this happens. For example, the tower may have a design water temperature of 85°F and it may be returning 90°F. This will cause high head pressure for the chiller and high operating costs.

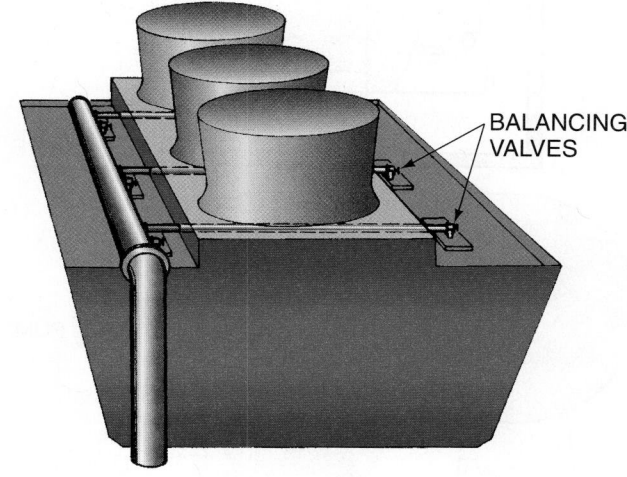

Figure 49–24   Balancing valves for balancing the water flow at the cooling tower. *Courtesy Marley Cooling Tower Company*

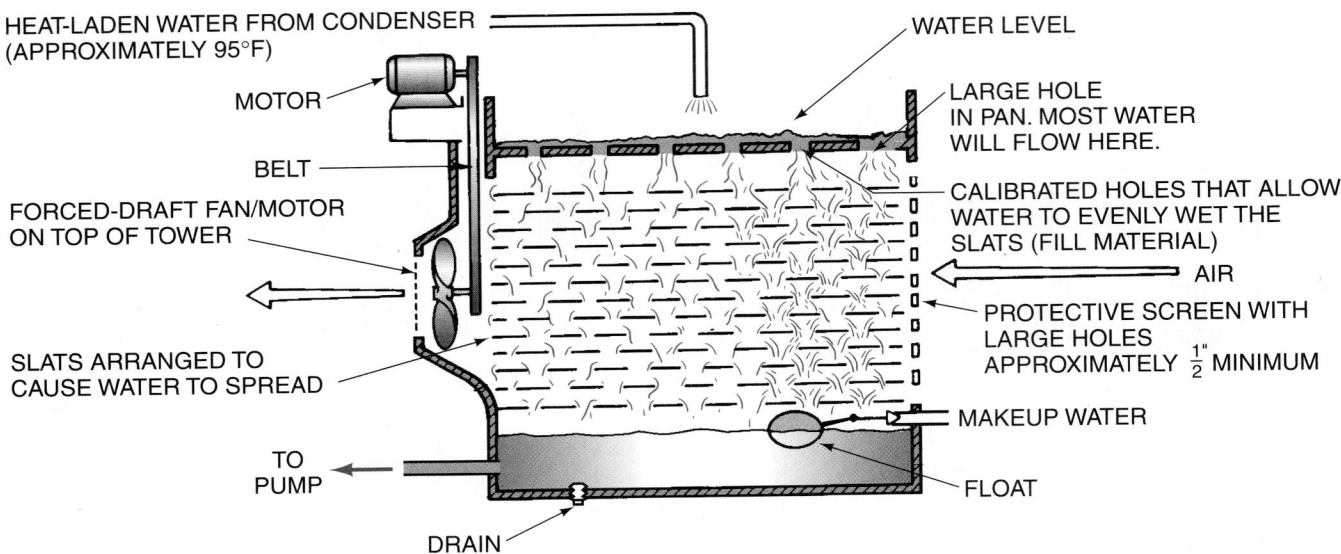

HEAT-LADEN WATER FROM CONDENSER (APPROXIMATELY 95°F)

MOTOR

BELT

FORCED-DRAFT FAN/MOTOR ON TOP OF TOWER

SLATS ARRANGED TO CAUSE WATER TO SPREAD

TO PUMP

DRAIN

WATER LEVEL

LARGE HOLE IN PAN. MOST WATER WILL FLOW HERE.

CALIBRATED HOLES THAT ALLOW WATER TO EVENLY WET THE SLATS (FILL MATERIAL)

AIR

PROTECTIVE SCREEN WITH LARGE HOLES APPROXIMATELY $\frac{1}{2}$" MINIMUM

MAKEUP WATER

FLOAT

Figure 49–25 Calibration holes for distributing water in the tower are enlarged due to rust. Most of the water is running down the right side of the tower.

49.13 WATER PUMPS

The condenser water pump is the device that moves the water through the condenser and cooling tower circuit. The cooling tower pump normally is located where it takes water from the cooling tower sump into the pump inlet, **Figure 49–26.** The water pump is the heart of the cooling tower system because it pumps the heat-laden water. This water must be pumped at the correct rate and delivered to the cooling tower at the correct pressure.

The condenser water pump is normally a *centrifugal pump.* It uses centrifugal action to impart velocity to the water that is converted to pressure. Pressure may be expressed in pounds per square inch (psig) or feet of head. One foot of head of water is 0.433 psig—that is, a column of water 1 ft high will cause a pressure gage to read 0.433 psig. A column of water 2.31 ft high (27.7 in.) will cause a pressure gage to read 1 psig, **Figure 49–27.** Pump capacities are discussed in feet

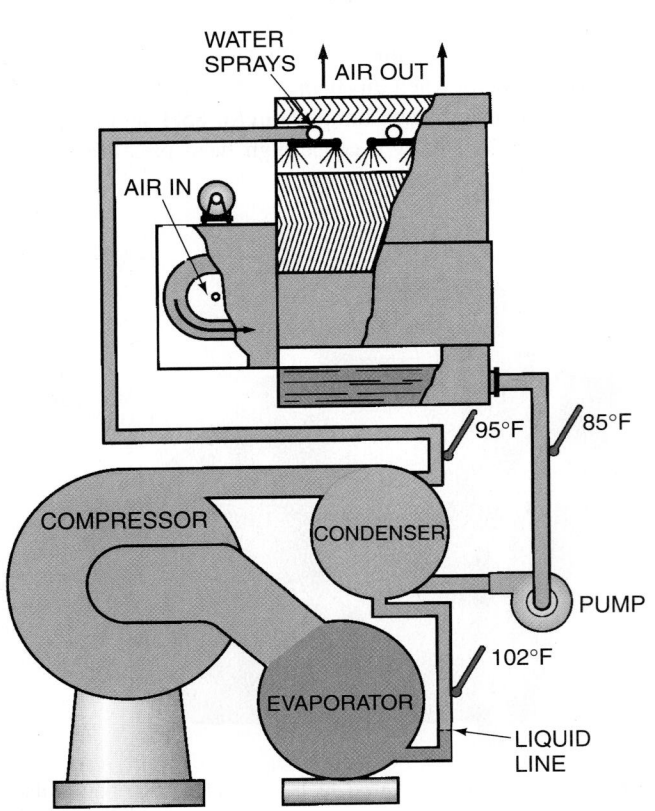

WATER SPRAYS

AIR OUT

AIR IN

95°F

85°F

COMPRESSOR

CONDENSER

PUMP

102°F

EVAPORATOR

LIQUID LINE

Figure 49–26 The cooling tower pump and system.

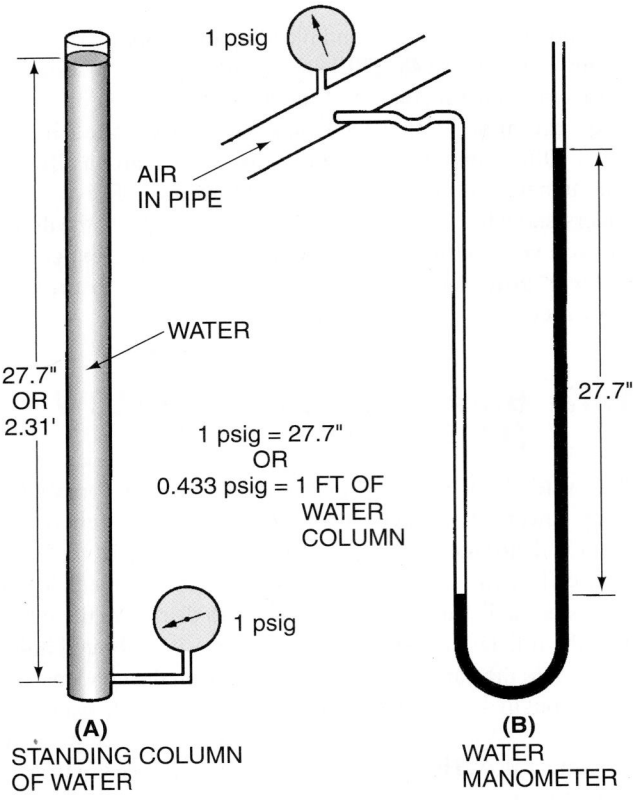

1 psig

AIR IN PIPE

WATER

27.7" OR 2.31'

1 psig = 27.7" OR 0.433 psig = 1 FT OF WATER COLUMN

1 psig

27.7"

(A) STANDING COLUMN OF WATER

(B) WATER MANOMETER

Figure 49–27 **(A)** A standing column of water and **(B)** the column as it would be measured with a water manometer.

Figure 49–28   A close-coupled pump assembly. *Courtesy Bell and Gossett, ITT Industries*

of head. Centrifugal pump action is discussed in Unit 33, "Hydronic Heat," and a study of that unit should be helpful.

Several types of condenser water pumps are available for large condenser water systems. In a *close-coupled pump* the pump is located very close to the motor with the pump impeller actually mounted on the end of the motor shaft, **Figure 49–28.** These pumps are used for small applications. Note that all of the water enters the side of the pump housing. The pump shaft must have a seal to prevent water from leaking to the atmosphere or to prevent atmosphere from leaking in should this portion of the piping be in a vacuum.

The *base-mounted* pump comes as an assembly with the pump mounted at the end of the pump shaft and a flexible coupling between the water and the pump, **Figure 49–29.** This pump may have a single- or double-sided impeller. The base is usually made of steel or cast iron and holds the pump and motor steady during operation. The pump base is usually fastened to the floor using cement as a filler in the open portion of the pump base. This is called grouting the pump in. The purpose is to make a more firm foundation that will last for many years. The pump and motor can still be removed from the steel or cast-iron base. One of the advantages of this pump is that the motor and pump shafts are aligned at the factory and alignment is not required in the field. These pumps are used in larger applications.

As pumps become larger, the need for different designs becomes apparent. Some pumps have single-inlet impellers with all of the water entering the pump impeller on one side, **Figure 49–30.** The double-inlet impeller allows water to flow into both sides, **Figure 49–31.** The pump housing must be designed differently to allow water to enter this pump. The

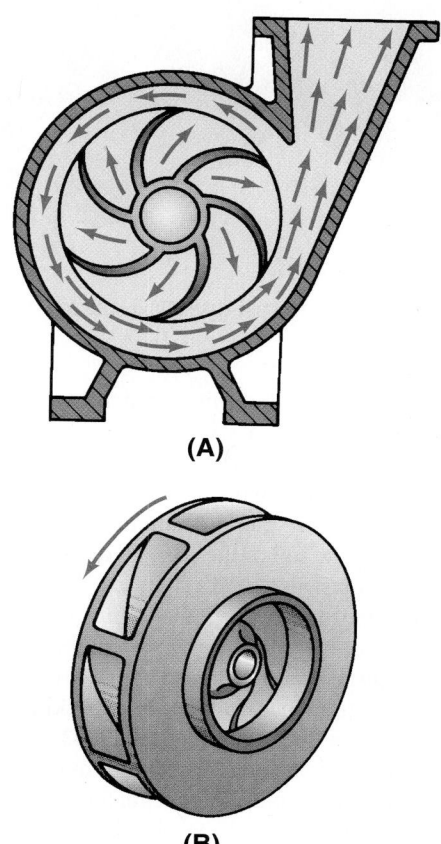

(A)

(B)

Figure 49–30   A single-inlet impeller.

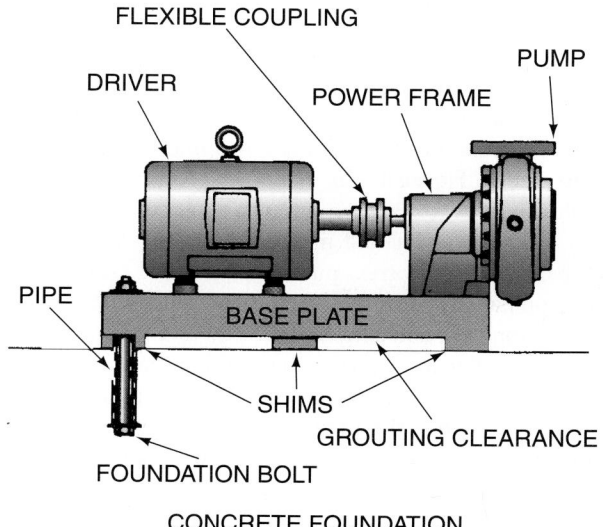

Figure 49–29   A flexible coupling between the pump and motor. *Courtesy Amtrol, Inc.*

INLET

Figure 49–31   A double-inlet impeller pump. *Courtesy Bell and Gossett, ITT Industries*

Figure 49–32   A horizontal split-case pump. *Courtesy Bell and Gossett, ITT Industries*

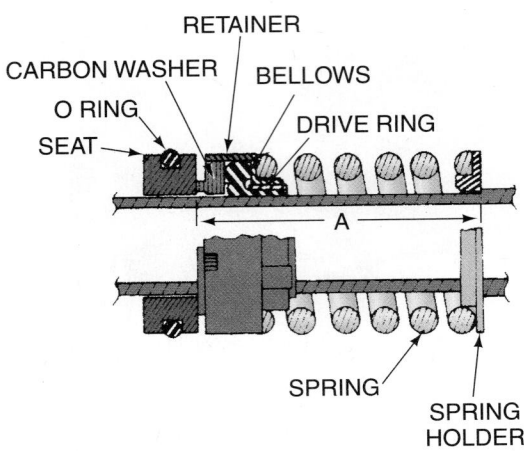

Figure 49–34   The mechanical shaft seal. *Courtesy Amtrol, Inc.*

housing at the inlet must be split so water can flow into both sides of the impeller. Two styles of *double-inlet impeller pumps* exist. With one type the pump is disassembled from the ends of the pump as with the base-mounted pump; the other type is a split-case pump where the pump is disassembled by removing the top of the pump, **Figure 49–32**.

These pumps must also have a shaft seal. Two kinds of seal may be used for the pumps; one is the *stuffing box* and the other the *mechanical type.* The stuffing-box seal is a packing-gland type seal that must be hand tightened with a wrench. When the nut on the seal is tightened, it squeezes a packing around the shaft to minimize leakage, **Figure 49–33**. Stuffing-box types of seals are typically used for pump pressures up to about 150 psig. These seals require some regular maintenance.

The mechanical-type seal may be used for higher pressures, up to about 300 psig, and usually has a carbon ring mounted on the end of a bellows that turns with the shaft. The bellows is sealed to the shaft with an O ring seal. The carbon ring rubs against a stationary ceramic ring while the shaft turns and seals water in the pump housing, **Figure 49–34**. Both of these seals are lubricated with the circulating water. The seals may have a special piping system that injects pump discharge water to the seal for split-case pumps, **Figure 49–35**.

Figure 49–35   Lubricating a shaft seal. *Courtesy Bell and Gossett, ITT Industries*

Pumps may also be manufactured in a vertical configuration with the motor on top. These pumps are mounted on top of the water sump and protrude down into the sump for the pump pickup, **Figure 49–36**.

Most pumps for cooling towers are manufactured of cast iron. They are heavy and may be used for pressures up to 300 psig with the correct piping flanges and arrangements. Cast iron has the capacity to last many years without deterioration from rust and corrosion.

The impeller in the typical centrifugal pump is made of brass and is often mounted on a stainless-steel shaft. This can be important when the time comes to remove the impeller from the shaft. This makes it easier to remove after many years of operation.

Impellers are furnished from the factory at a specified diameter for each pump to furnish a specific water flow against a specified pressure. Often the pump may furnish too

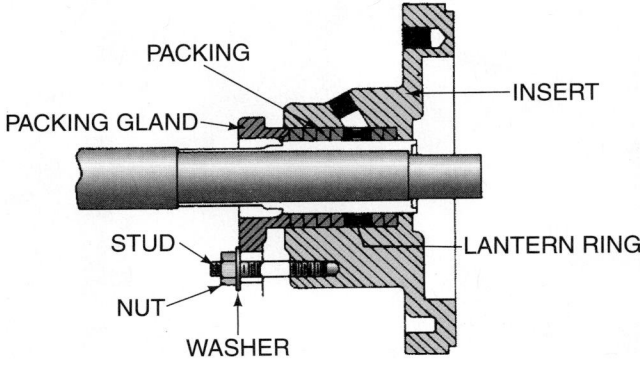

Figure 49–33   The stuffing-box shaft seal. *Courtesy Amtrol, Inc.*

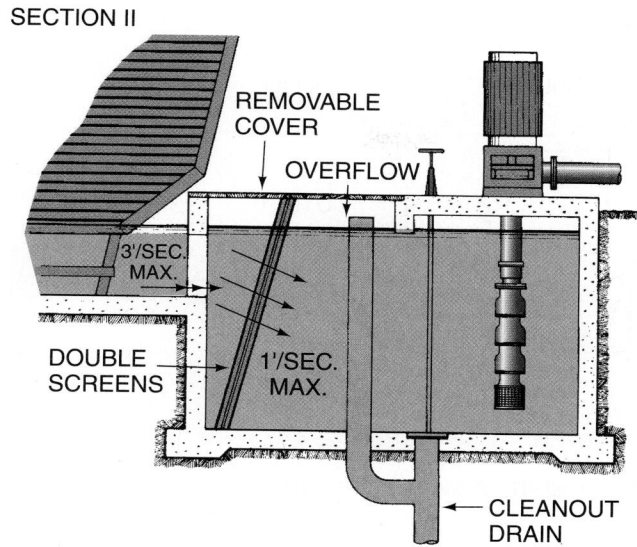

SECTION II

REMOVABLE COVER

OVERFLOW

3'/SEC. MAX.

DOUBLE SCREENS

1'/SEC. MAX.

CLEANOUT DRAIN

Figure 49–36 A vertical pump application. *Courtesy Marley Cooling Tower Company*

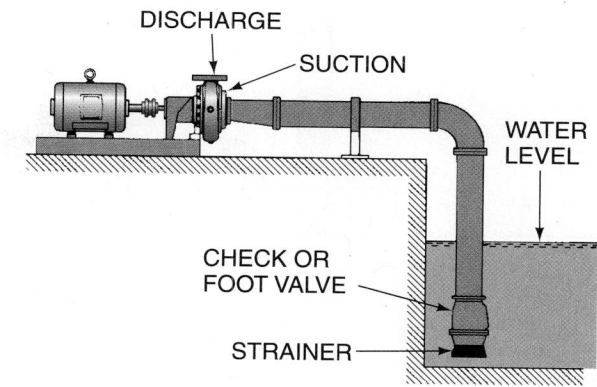

DISCHARGE

SUCTION

WATER LEVEL

CHECK OR FOOT VALVE

STRAINER

Figure 49–37 The cooling tower sump is located below the pump.

much water flow to the point that a smaller impeller is needed. These impellers are often trimmed in the field to new specifications for the actual water flow and pressure requirements. The manufacturer can be contacted for any modifications for a specific pump.

The location of the condenser water pump is important because these pumps are normally centrifugal pumps and they must be furnished water in such a manner that the eye of the impeller is under water during start-up. Otherwise, the pump will not move any water. They *do not* have the capacity to pump air to pull water into the impeller. The pump should be located below the tower water with nothing in the pump inlet piping that will impede flow. A free flow of water to the centrifugal pump inlet is always good practice. Several things can interfere with water flow to the pump inlet, such as vortexing in the tower, fine mesh strainers, and the cooling tower bypass.

There are times when the sump to the cooling tower is located below the pump. When this is the case, special care should be taken during start-up because only air will be in the pump casing. The pump should never be started with only air inside because the pump seal will have no water for lubrication. When the sump is below the pump, a foot or check valve must be located in the line to prevent the water from flowing back into the sump during shutdown, **Figure 49–37**. Water must be added to the pump inlet side until the pump casing is completely full before the pump is started. This can be done by connecting a water hose to the pump inlet side and filling it until water escapes the bleed port on top of the pump. If the pump inlet cannot be filled, the foot or check valve is leaking and must be repaired.

Vortexing in the tower is actually a whirlpool action. This vortex interferes because it introduces air to the pump inlet. The pump is designed to pump water, and air will cause problems with the pump and with the condenser at the chiller. Poor tower construction or piping practice are the cause of

vortexing. The sump may not be deep enough or antivortexing design may not be in place. Vortexing can often be eliminated by means of a device placed in the sump outlet that breaks up the vortex. This may be a cross type of configuration with a plate on top. The cross configuration creates four outlets from the sump, and the plate causes the water to be pulled from a further distance and from the side, **Figure 49–38**. Vortexing has been found in multistory buildings with the condenser water pump well below the tower and has been addressed using the above methods.

There must be some sort of strainer located between the tower water and the pump inlet to prevent trash from the tower from entering the pumping circuit. Typically a screen-type strainer is in the cooling tower exit to the sump. This may be a coarse screen. Often a fine mesh screen is located before the pump inlet. If this mesh screen is too fine, problems from pressure drop will occur as the screen becomes restricted. Pressure drop causes the pump inlet to operate in a vacuum. It may pull in air while operating in a vacuum if the pump seals leak. Pressure drop can also cause pump cavitation—

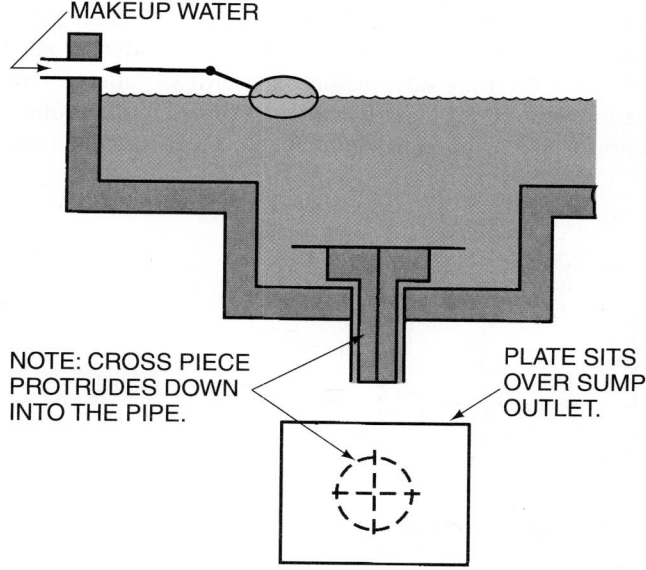

MAKEUP WATER

NOTE: CROSS PIECE PROTRUDES DOWN INTO THE PIPE.

PLATE SITS OVER SUMP OUTLET.

Figure 49–38 Vortex protection for a cooling tower.

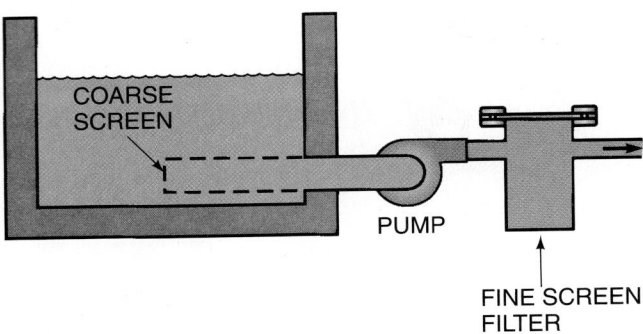

Figure 49–39 The best application for a fine screen filter is in the pump outlet.

water turning to a vapor if the cooling tower water is warm enough and the pressure is low enough. Cavitation or air at the pump inlet would be evident from noise at the pump. It will often sound like it is pumping small rocks. If a fine screen must be used, it may often be better applied at the pump outlet, **Figure 49–39.** It does less harm to throttle the pump discharge with pressure drop than the pump inlet.

The tower bypass valve helps to maintain the correct tower water temperature at the condenser during start-up and during low ambient operation. The tower bypass circuit allows water from the pump outlet to be recirculated to the pump inlet so that the cold cooling tower water will not reach the condenser, **Figure 49–40.** Condenser water that is too cold in the compression cycle systems will reduce the head pressure to the point that the condenser will be so efficient that too much refrigerant will be held in the condenser and starve the evaporator. This will also cause the oil to migrate in some machines. Cooling tower water that is too cold for an absorption machine will often cause the salt solution to crystallize. Two types of three-way bypass valves may be used for tower bypass. These are the mixing valve and the diverting valve. The mixing valve has two inlets and one outlet and would need to be located in the pump suction line between the tower and the pump inlet, **Figure 49–40.** This valve can cause pressure drop in the pump suction and should be avoided if possible. The diverting valve has one inlet and two outlets; it can be located in the pump discharge, and piped to the cooling tower or the pump inlet, **Figure 49–41.** These valves do not

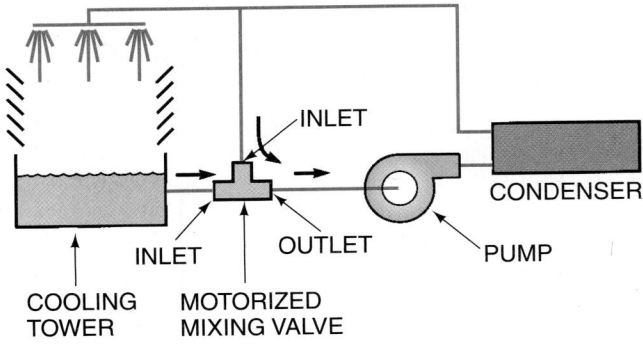

Figure 49–40 A mixing valve application.

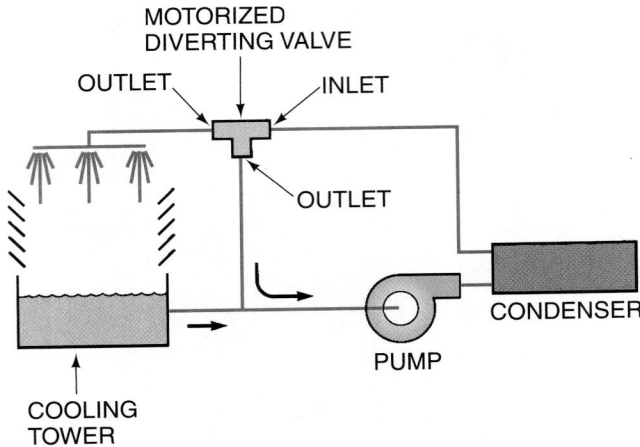

Figure 49–41 A diverting valve application.

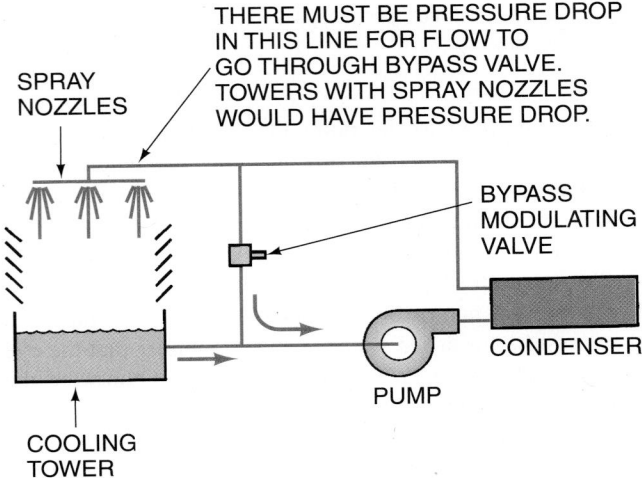

Figure 49–42 A straight-through valve for tower bypass.

come in large sizes, normally a maximum of 4 in. so other arrangements may have to be made if a larger size is needed. A straight-through valve can be used in the configuration shown in **Figure 49–42.** Both the diverting valve and the straight-through valve allow some cooling tower water from the basin to be introduced to the pump inlet but may cause problems at start-up until the tower water becomes the correct temperature.

All pumps must have bearings that support the turning shaft while under load. Sleeve bearings are used for small pumps, and ball or roller bearings are used for larger pumps. These bearings must be lubricated on a regular basis for satisfactory service. Notice that the bearings are outside the pump on the split-case pump, **Figure 49–43.**

The pump must be fastened to the motor shaft in most of the above pumps in such a manner that they are within alignment tolerances. A flexible coupling may be installed between the pump and motor shaft to take out small misalignment, **Figure 49–44.** Because this coupling can only handle slight misalignment, the shafts must be aligned.

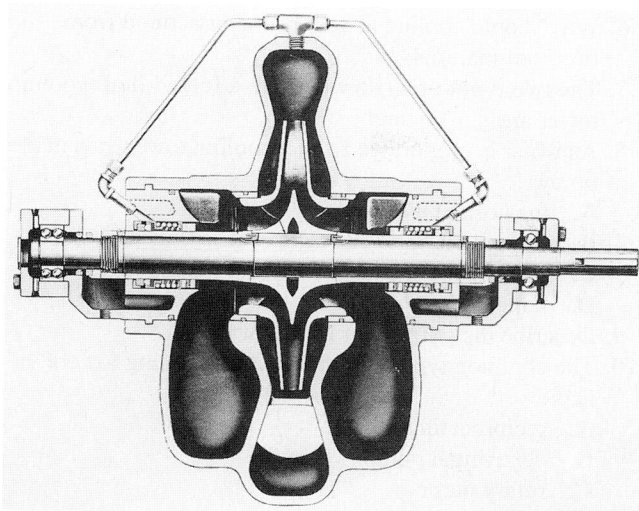

Figure 49–43   The bearings on a horizontal split-case pump. *Courtesy Bell and Gossett, ITT Industries*

Figure 49–44   A flexible coupling. *Courtesy TB Woods & Sons*

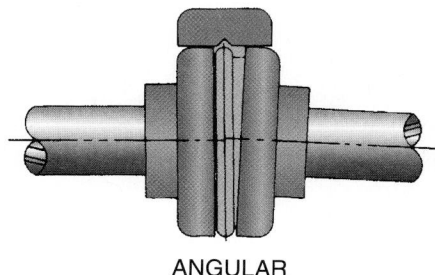

ANGULAR

Figure 49–45   Angular misalignment for a coupling. *Courtesy Amtrol, Inc.*

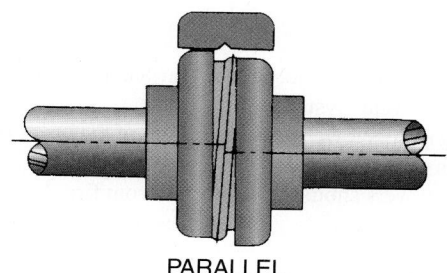

PARALLEL

Figure 49–46   Parallel misalignment for a coupling. *Courtesy Amtrol, Inc.*

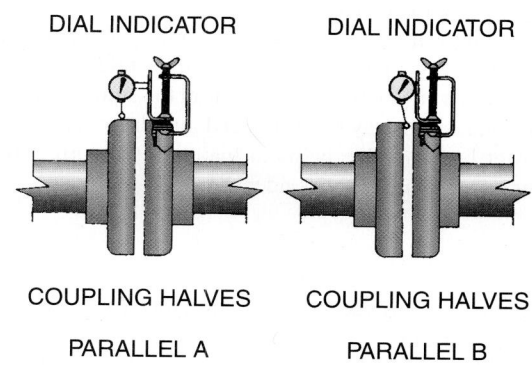

DIAL INDICATOR   DIAL INDICATOR

COUPLING HALVES   COUPLING HALVES

PARALLEL A   PARALLEL B

Figure 49–47   Dial indicators used to align coupling and shafts. *Courtesy Amtrol, Inc.*

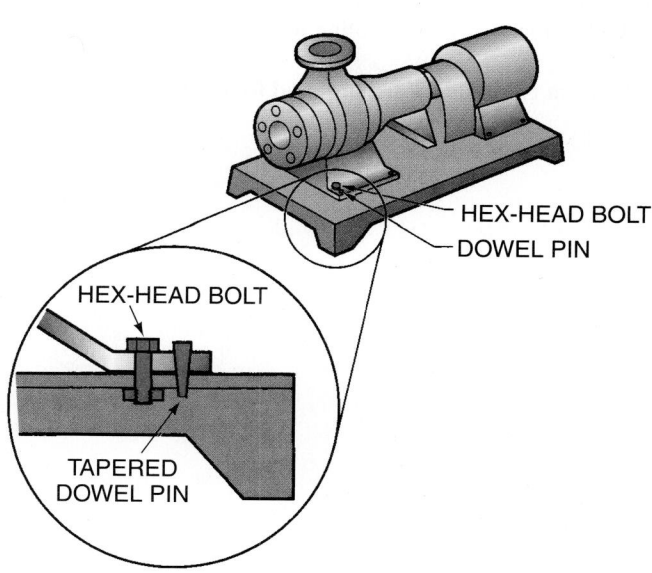

HEX-HEAD BOLT
DOWEL PIN

HEX-HEAD BOLT

TAPERED DOWEL PIN

Figure 49–48   The pump and motor are fastened to the base using tapered dowel pins.

Two planes of alignment must be considered, angular and parallel. Angular alignment ensures that both shafts are at the same angle with each other, **Figure 49–45.** Parallel alignment ensures that the shafts are end to end, **Figure 49–46.** Correct alignment is accomplished with a dial indicator that reads to 1/1000 of an inch. It is mounted on the shaft and the shafts are rotated through the full rotation, **Figure 49–47.** Shims (thin sheets of steel) are placed under the pump and motor until both angular and parallel alignment (sometimes called radial and axial alignment) are achieved to within the manufacturer's specifications. After alignment is attained, the pump and motor are tightened down and alignment is checked again. Then the base of both are drilled and tapered dowel pins are driven in the base, **Figure 49–48.** When done correctly, the motor and pump will give years of good service.

SUMMARY

- The cooling tower rejects heat that has been absorbed into the chilled water system.
- Cooling towers lower the temperature of the water in a tower by means of evaporation.
- Cooling towers should be protected from fire.
- The design of a cooling tower provides for distinct airflow patterns.
- Fan drives for forced-draft towers may be belt or gear type.
- The cooling tower sump needs to be flushed periodically.
- Sumps may have heaters to protect from freezing.
- Cooling towers must have makeup water systems.
- The water flow through the tower must follow the pattern designed by the manufacturer.
- Larger pumps may have single- or double-inlet impellers.
- Most pumps for cooling towers are manufactured of cast iron.
- Vortexing (whirlpooling) may occur in the sump. This is not good because air may be introduced into the pump.
- A tower bypass valve helps to maintain the correct tower water temperature at the condenser.

REVIEW QUESTIONS

1. What is the purpose of the cooling tower for a water-cooled system?
2. The typical difference in temperature between the cooling tower water supplied to the condenser and the outdoor wet-bulb temperature is _____ °F.
3. Spreading the water out as it passes through the cooling tower helps the water _____ faster.
4. Two types of water cooling tower are the _____ and the _____.
5. What two types of fan are used for cooling towers?
6. Why should cooling towers be manufactured from fireproof materials?
7. The two types of airflow through a forced-draft cooling tower are _____ and _____.
8. As water is evaporated from a cooling tower, it is made up by
   A. the normal water supply.
   B. dripping it back into the tower.
   C. condensing it and saving it for makeup.
   D. none of the above.
9. Describe the purpose of blowdown.
10. The common type of pump used for cooling towers is the
   A. reciprocating pump.
   B. centrifugal pump.
   C. rotary pump.
   D. all of the above.
11. Vortexing in a cooling tower occurs in the
   A. inlet piping.
   B. basin of the tower.
   C. outlet piping.
   D. pump discharge.
12. Describe the purpose of the tower bypass.
13. The result of operating a system when the cooling tower water is too cold is
   A. high head pressure.
   B. low head pressure.
   C. that the compressor motor will overheat.
   D. that the pump will cavitate.
14. What is the component that is installed between the pump shaft and the motor shaft?
15. What substance is used to lubricate the seals in a water pump?

UNIT 50

Operation, Maintenance, and Troubleshooting of Chilled-Water Air-Conditioning Systems

OBJECTIVES

After studying this unit, you should be able to

- ■ discuss the general start-up procedures for a chilled-water air-conditioning system.
- ■ describe specific chiller start-up procedures for chillers having a scroll, a reciprocating, a rotary screw, or a centrifugal compressor.
- ■ describe operating and monitoring procedures for scroll and reciprocating chilled-water systems.
- ■ discuss preventive and other electrical maintenance and service that should be performed at least annually on chillers.
- ■ list maintenance procedures that should be performed periodically on water-cooled chiller systems.
- ■ describe start-up procedures for an absorption chilled-water system.
- ■ list preventive and routine maintenance procedures for an absorption chiller.
- ■ state preventive and periodic maintenance procedures for the purge system on an absorption system.

SAFETY CHECKLIST

- ✔ Wear gloves and use caution when working around hot steam pipes and other heated components.
- ✔ Use caution when working around high-pressure systems. Do not attempt to tighten or loosen fittings or connections when a system is pressurized. Follow recommended procedures when using nitrogen.
- ✔ Follow all recommended safety procedures when working around electrical circuits.
- ✔ When working on an electrical system, lock disconnect box or panel and tag it with your name when power is turned off. There should be only one key for this lock, and it should be in the possession of the technician.
- ✔ Never start a motor with the door to the starter components open.
- ✔ To avoid excessive pressures, check pressures regularly as indicated by the manufacturer of the system.

The operation of these chillers involves starting, running, and stopping chiller systems in an orderly manner. The chiller operator should be familiar with the system and the machine before operating or monitoring any system. Most machines and systems are almost foolproof, but there is no need to take chances with this expensive equipment. It is good practice to become familiar with one piece of equipment at a time.

There should be literature at the site for any major equipment installation. The building blueprint for every major installation shows the piping diagram, an electrical diagram, a duct diagram, and a mechanical equipment layout. This information is invaluable for helping the technician locate wiring control points and valves for controlling fluid flows. The duct diagram will help the technician find dampers and air control points. All duct and pipe sizes and routes are listed on the prints. Electrical junction boxes and control points should be on the electrical print. If a job is old and has been modified, the prints should have been kept up to date with the changes. This is often overlooked.

Manufacturers' literature should be on site for the technician to become familiar with the equipment. These prints or literature should not be allowed to leave the job site. If for some reason the prints are lost, they can sometimes be copied from the original engineering firm that drew them up. They save copies for any future use. If the prints were created on a computer, there should be files where they can be re-created.

50.1 CHILLER START-UP

The first step in starting up a chilled-water system is to establish chilled-water flow. This is true if it is for the first time each season or the first time each day. If the system is water cooled, the next step (if the condenser is water cooled) is to establish condenser water flow. Air-cooled chillers have fans at the condensers and do not need the attention that water-cooled chillers do. Both the chilled-water and the condenser water circuit will have some means, such as a flow switch, of proving that water is flowing. The flow switch is often a paddle that protrudes into the water and when water passes, it moves and operates a switch. Some chiller manufacturers use pressure drop controls to establish water flow. If this is the case, these are built into the water circuit by the manufacturer. The technician should know the checkpoints to make sure that these switches are functioning. Because the paddle of a flow switch is in the water stream, it is not unusual for them to break off. If so, water flow may be established and not be proven with the flow switch. Water flow may be verified by the technician by means of the pressure gages. Never bypass a chilled-water flow switch or freeze conditions may occur. The starting point to finding out the correct water flow through a heat exchanger is to know what the pressure drop should be.

There should be a pressure test point at the inlet and outlet of any water heat exchanger, chiller or condenser, **Figure 50–1.** Note that two gages are used in the figure to check pressure.

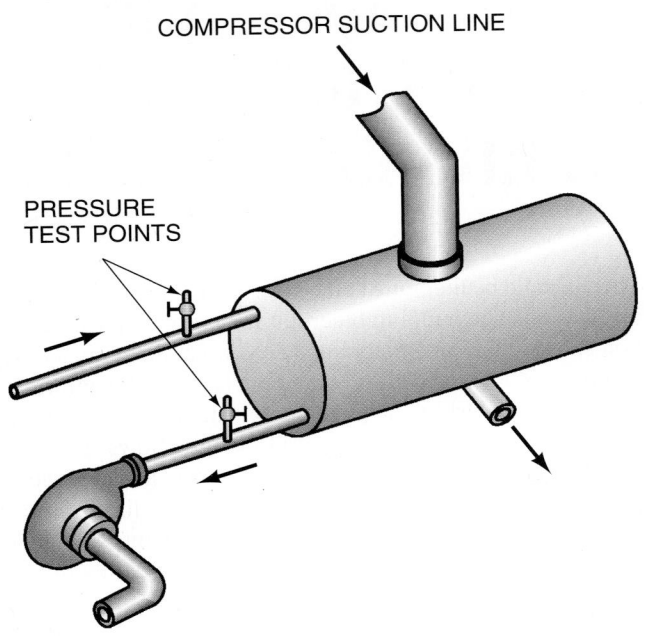

Figure 50–1   A condenser with inlet and leaving-water gage ports.

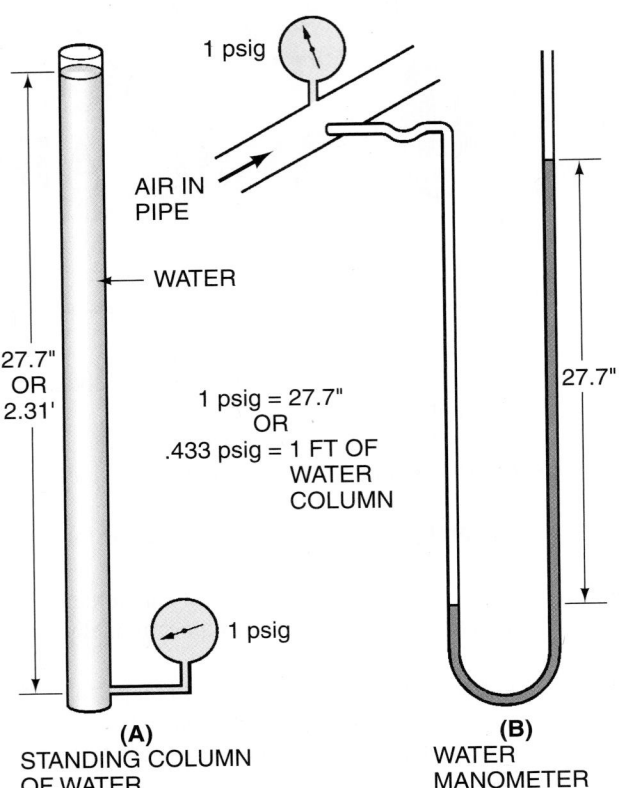

Figure 50–3   (A) Pressure at the bottom of a standing column of water, and **(B)** as it would be measured with a water manometer.

A better arrangement is to use one gage and two pressure connections, **Figure 50–2**. If two gages are used and they are at different heights, an error is automatically built in. For example, if one gage is 2.31 ft lower than the other, there is an error of 1 psig because of the difference in height. A standing column of water that is 2.31 ft high will have a pressure of 1 psig at the bottom, **Figure 50–3**. If the gages have a built-in error because they are not calibrated, they cannot be expected to give correct readings. If one gage reads 1 psig when open to the atmosphere while the other gage reads 0, there is an error of 1 psig in the pressure difference. Actually, all you want to know is the pressure difference at the two points. Using one gage that is not accurate at 0 may still yield the correct pressure difference.

The technician needs to know from the pressure readings that there is a pressure drop across the heat exchanger. As water flows through the heat exchanger, the pressure drops a specific amount. The heat exchanger is a calibrated pressure drop monitoring device that may be used to determine water flow. A pressure drop chart for the heat exchanger will tell you the gallons per minute (gpm) of water flow, **Figure 50–4**. The original operating log sheet for the installation will show the pressure drops at start-up. This is very good information for future use. If the original log cannot be found, the manufacturer may be contacted for a copy of the log sheet or a pressure drop chart for the chiller.

The technician should also be aware of the interlock circuit through the contactors that must be satisfied before start-up. The starting sequence for most chiller systems is to start the chilled-water pump first. A set of auxiliary contacts in the chilled-water starter then starts the condenser water pump.

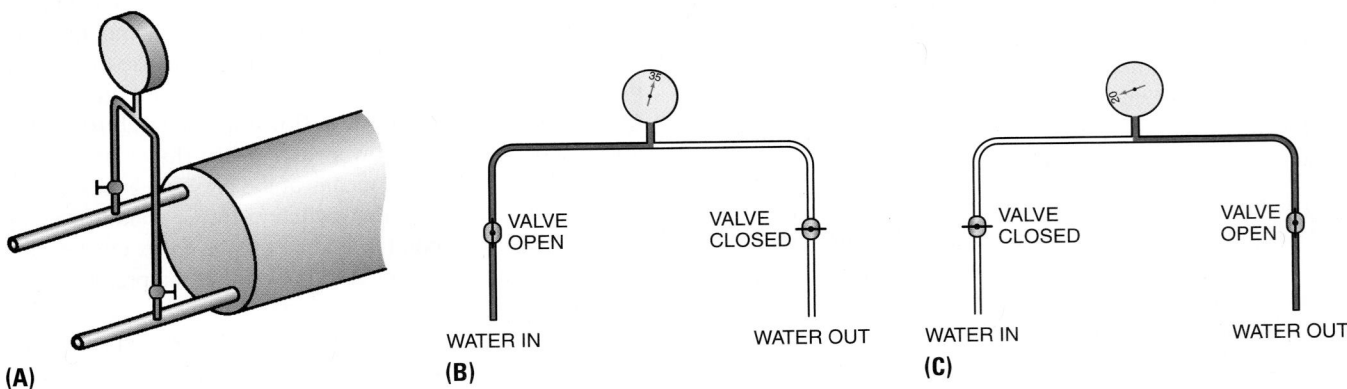

Figure 50–2   (A) One gage being used to check both ports. **(B)** Inlet water pressure. **(C)** Outlet water pressure. The difference between **(B)** and **(C)** is the actual pressure drop through the condenser, 15 psig.

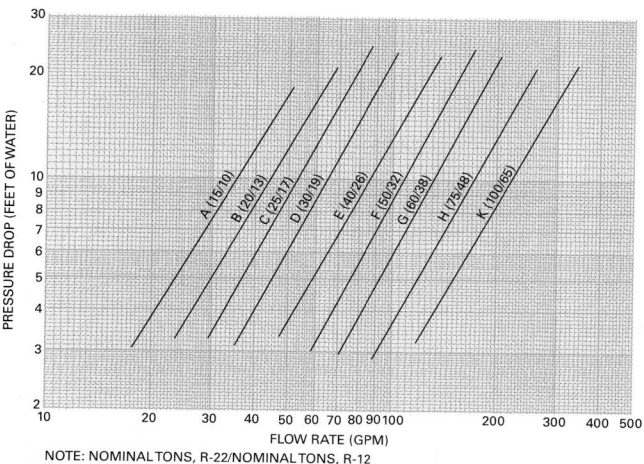

Figure 50–4 A pressure drop chart for a heat exchanger. *Courtesy Trane Company*

When the condenser water pump starts, a set of auxiliary contacts make and pass power to the cooling tower fan and the chiller circuit, **Figure 50–5.** When the signal is received at the chiller control circuit, the compressor should start. This signal is often called the field control circuit because it is the circuit furnished by the contractor to the manufacturer's control circuit. If a chiller should not start, the first thing the technician should do is check the field circuit to see if it is

satisfied. A technician should know where the field circuit is wired to the chiller circuit for each chiller on a job where there are several chillers. A quick check at this point will tell whether there is a field circuit problem (flow switches, pump interlock circuit, outdoor thermostat, and often the main controller) or a chiller problem. If there is no field control circuit power, there is no need to look at the chiller until there is power. Many chillers have a ready light that shows that the field control circuit is energized.

Different chillers have different starting sequences, usually depending on the type of lubrication the compressor has. When the chiller has a positive displacement compressor (scroll, reciprocating, or rotary screw), the compressor will start soon after the field control circuits are satisfied. Some manufacturers may have a time delay before the compressor starts after the field circuit is satisfied, but the compressor should start soon. Look for a time-delay circuit and make note of it and what the time delay is for any system, so you will not be waiting and wondering what the problem is. The technician may think there is a problem only to have the chiller start up unexpectedly after a planned time delay. The reciprocating, scroll, and rotary screw compressors are lubricated from within and do not have a separate oil pump that must be started first. The reciprocating and the rotary screw compressors will start up unloaded and will begin to load when oil pressure is developed. The scroll compressor does not have compressor unloading capabilities and starts up under load.

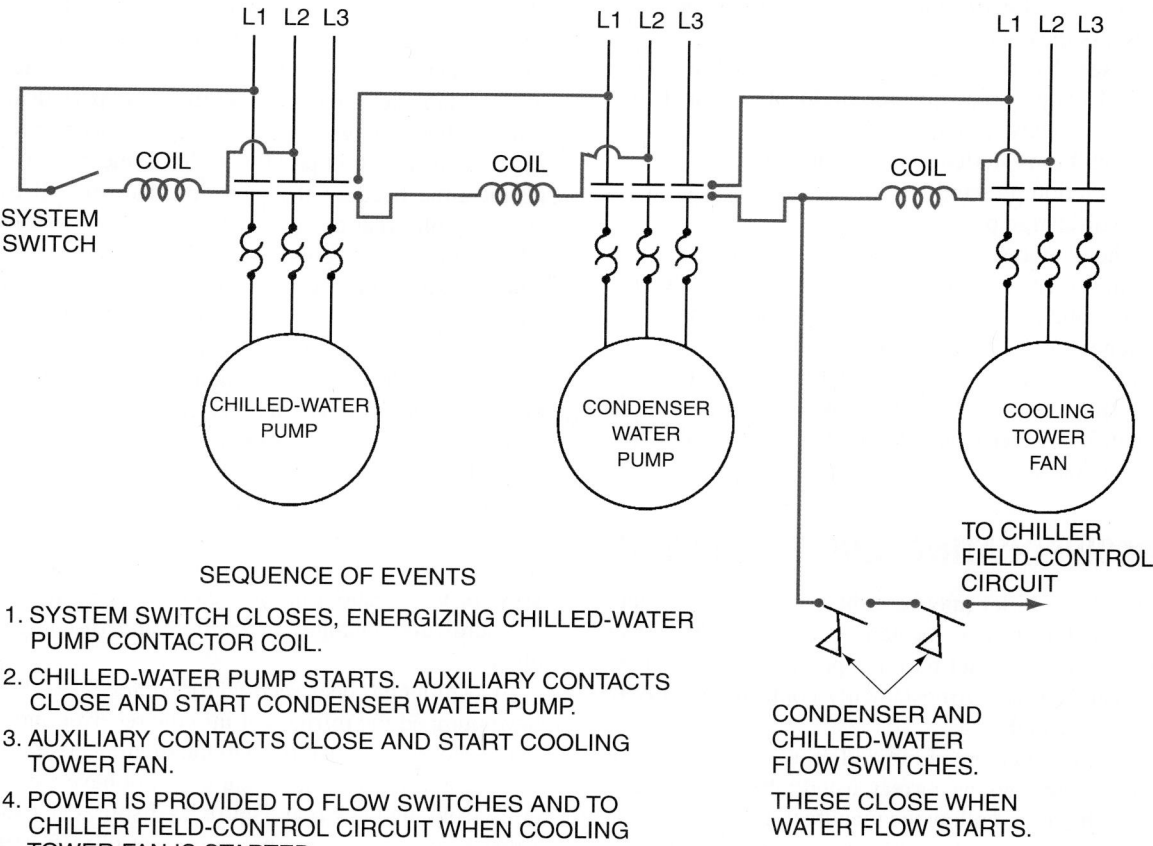

SEQUENCE OF EVENTS

1. SYSTEM SWITCH CLOSES, ENERGIZING CHILLED-WATER PUMP CONTACTOR COIL.

2. CHILLED-WATER PUMP STARTS. AUXILIARY CONTACTS CLOSE AND START CONDENSER WATER PUMP.

3. AUXILIARY CONTACTS CLOSE AND START COOLING TOWER FAN.

4. POWER IS PROVIDED TO FLOW SWITCHES AND TO CHILLER FIELD-CONTROL CIRCUIT WHEN COOLING TOWER FAN IS STARTED.

CONDENSER AND CHILLED-WATER FLOW SWITCHES.

THESE CLOSE WHEN WATER FLOW STARTS.

Figure 50–5 The control circuit for a typical chiller.

50.2 SCROLL CHILLER START-UP

Scroll chillers are either air or water cooled. Regardless of the type, before attempting to start the chiller make sure the water for the chilled water circuit is at the correct level. The system must be full. Then operate the chilled-water pump and verify the water flow using pressure drop across the chiller. Make sure that all valves in the refrigerant circuit are in the correct position, usually back seated. Visually check the system for leaks. Oil on the external portion of a fitting or valve is a sure sign of a leak as oil is entrained in the refrigerant. When refrigerant leaks out, small amounts of oil will be present. The compressor will have crankcase heat and it must be energized for the length of time the manufacturer requires, usually 24 hours. Do not start the compressor without adequate crankcase heat, or compressor damage may occur.

When the chiller is air cooled, it will be located outside. The chiller barrel will have a heater to prevent it from freezing in the winter. Be sure this heater is wired and operable. If you do not do this at start-up, you may forget it when winter arrives. This heat strip is thermostatically controlled and will shut off when not needed.

When air cooled, make sure that the condenser fans are free to turn.

When the chiller is water cooled, the water-cooled condenser portion must be in operation. The cooling tower must be full of fresh water and the condenser water pump must be started. Make sure that water flow is established. You can look at the tower and verify this or you can check the pressure drop across the condenser.

Check to see that the field control circuit is calling for cooling. If all of the preceding requirements are met, the chiller should be ready to start.

When the chiller is started, do the following:
1. Observe suction pressure.
2. Observe discharge pressure.
3. Check the compressor for liquid floodback.
4. Check the entering- and leaving-water temperature on the chiller and the condenser entering- and leaving-water temperature for water-cooled units.

When the chiller is operating normally, it can usually be left unattended except for the water tower when the unit is water cooled. The water tower should have regular observation and maintenance.

50.3 RECIPROCATING CHILLER START-UP

The reciprocating chiller may be either air or water cooled. When air cooled, it may be located outside and a heat strip will be applied to the chiller barrel to prevent freezing in the cold. The heat strip is thermostatically controlled and only operates when needed.

Before starting a reciprocating chiller, make sure that the oil level in the sight glass is correct, usually from 1/4 to 1/2 level at the sight glass. Follow the manufacturer's instruction on oil level. When the level is high, refrigerant may be in the oil; check for crankcase heat. Reciprocating chillers will

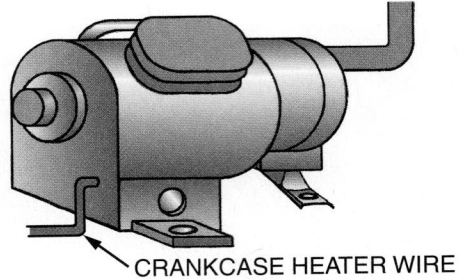

Figure 50–6 Heat from a crankcase heater.

have crankcase heaters to prevent refrigerant migration to the crankcase during the off cycle. When the crankcase heat is on, the compressor will be warm to the touch, **Figure 50–6.** This crankcase heat must be energized for the prescribed length of time before start-up to prevent damage to the compressor. Refrigerant in the oil will dilute the lubrication quality of the oil and can easily cause bearing damage that does not show up immediately. The safe time for crankcase heat to be energized is 24 hours unless the manufacturer states otherwise. It is not good practice to ever turn the crankcase heat off, even during winter shutdown because refrigerant will migrate to the crankcase, and it is not easy to boil it out to the point where the oil is at the correct consistency for proper lubrication.

When the compressor is started, the pressures can be observed when there are panel gages. Typically, chillers over 25 tons have gages permanently installed by the manufacturer. These gages may be isolated by means of a service valve to prevent damage to the gages during long periods of running time so these valves will need to be opened for reading the gages. It is not good practice to leave the gage valves open all the time because the gage mechanism will experience considerable wear during normal operation from reciprocating compressor gas pulsations. These gages are for intermittent checking purposes. The technician should check these gages for accuracy from time to time because they cannot be relied on during the entire life of the chiller.

Many chillers that are manufactured today have light-emitting diode (LED) readouts that show the pressures and temperatures. These pressures are registered with transducers and changed to an electrical or electronic signal for use in the electronic control system. The control system may use these signals for trouble and analysis and may have a history of operating conditions stored in memory. This can be valuable for troubleshooting as it can show any trends, such as pressure drop changes or suction and discharge pressure changes.

If a chiller has been secured for winter, the technician may have pumped the refrigerant into the receiver and the valves may need to be repositioned to the running position, which is back seated for any valve that does not have a control operating from the back seat. **Figure 50–7** shows an example of a valve in which the low-pressure control is fastened to the gage port on the back seat of the valve. If the valve is back

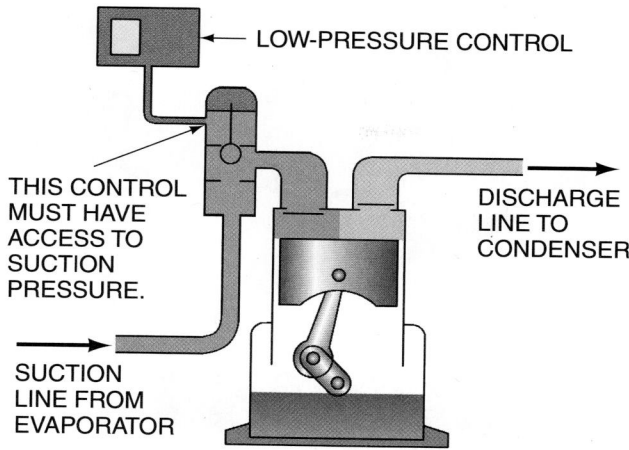

Figure 50–7 This service valve is positioned (cracked) just off-the-back-seat so that the low-pressure control can sense the pressure.

seated, the low-pressure control will not be functional because it is isolated.

When the compressor is started, it will start up unloaded until the oil pressure builds up and loads the compressor. An ammeter can be applied to the motor leads to determine what the level of load is on the compressor. When it is fully loaded, the amperage should be close to full load.

As the system temperature begins to pull down, the technician may notice that the oil level in the compressor rises and foams. This is normal for many systems, but the oil level should reach normal level after a short period of operation, normally within 15 min. The compressor is suction cooled and the technician should ensure that liquid refrigerant is not flooding back into the compressor during the running cycle. This can be determined by feeling the compressor housing; the compressor motor should be cold where the suction line enters, then it should gradually become warm where the motor housing reaches the compressor housing. The compressor crankcase should never be cool to the touch after 30 min of running time or liquid refrigerant should be suspected to be flooding back, **Figure 50–8**.

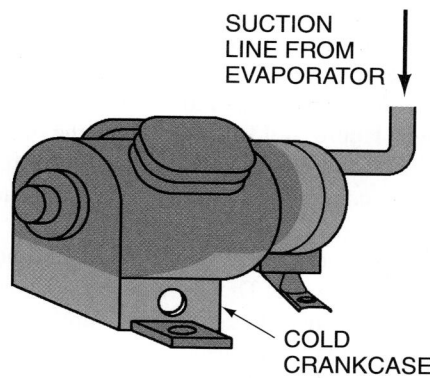

Figure 50–8 A compressor crankcase that is cold because of liquid refrigerant returning to the compressor.

50.4 ROTARY SCREW CHILLER START-UP

The rotary screw chiller may be either air or water cooled. If it is air cooled, it will be located outdoors and the heat for the chiller barrel should be verified. Again, it is thermostatically controlled and will only operate when needed.

The compressor must have crankcase heat before start-up. Again, most manufacturers require that this heat be energized for 24 hours before start-up. Start-up without crankcase heat can cause compressor damage.

The water flow must be verified through the chiller and the condenser if water cooled. The best way to verify this is to use pressure gages.

All valves must be in the correct position before start-up. Check the manufacturer's literature for valve positions. If the chiller was pumped down for winter, the correct procedures must be followed to let the refrigerant back into the evaporator before start-up.

Check the field control circuit to make sure that it is calling for cooling.

When all of the preceding is complete, the chiller is ready to start.

Start the compressor and watch for it to load up and start cooling the water. Observe the following:
1. Suction pressure
2. Discharge pressure
3. Water temperature (for both chiller and condenser when water cooled)
4. Liquid flooding back to the compressor

When the chiller is operating normally, it can usually be left unattended except for the water tower when the unit is water cooled. The water tower should be observed and maintained regularly.

50.5 VALVES FOR LARGE SYSTEMS

Refrigerant valves on large systems require proper use and maintenance in many systems. Valves increase in size as the systems become larger and they have some mechanical features that the technician must pay attention to. These valves are basically used in the refrigerant circuits.

The valves used for the service of refrigerant circuits are either diaphragm, ball, or service types of valves with back and front seats. These valves must be manufactured to high levels of leak-free performance. Since these valves control refrigerant, they must be reliable.

Diaphragm valves are used for small lines, **Figure 50–9**. The name refers to the diaphragm that is inside the valve. It is made from stainless steel and is under the valve cap. This valve should not be disassembled unless necessary. The valve may be fastened to the piping by a flare fitting, or by solder or brazing. Some technicians will disassemble the valve for brazing purposes to prevent overheating the valve. This is not necessary if the valve body is kept cool with a wet rag or heat-sink paste material during the brazing process. When heated for brazing, the valve should not be seated—back the valve off of the seat a few turns. Normally the technician has no

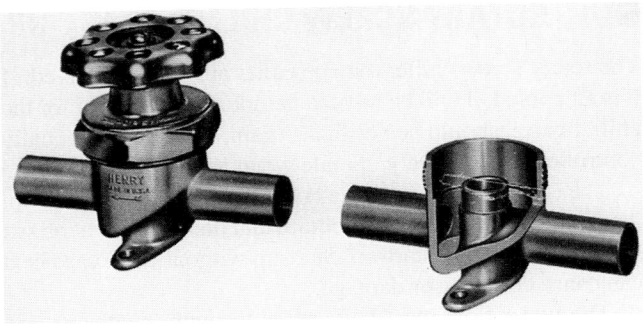

Figure 50–9 This hand valve is called a diaphragm valve. The actual diaphragm is under the valve cover and can't be seen. *Courtesy Henry Valve Company*

reason to take this valve apart. Once in service, it usually does not require service for the life of the system. The valve is not used for everyday service—usually only to valve the refrigerant off for isolation purposes. The diaphragm valve is normally used in the liquid line because it does not come in large sizes—only up to about 7/8-in. piping.

Ball valves for refrigerant piping come in sizes of up to about 2 1/8 in. and are used for larger applications, **Figure 50–10.** These ball valves usually do not have any serviceable parts inside and are not normally disassembled. Since they are used for refrigerant, they have an enclosed valve stem. It has a cover that must be replaced after use. The ball valve only needs to be turned one-quarter turn to be fully open or fully closed. It is typically brazed in the line and the manufacturer may require it to be disassembled for that purpose. The ball valve usually has a stainless-steel ball and stem that will not rust or corrode.

Service valves are valves that are used on the compressor and in some cases in the liquid line. These valves are brazed into the line when used in the piping and are bolted to the compressor and have a brazed flange for the piping when used on the compressor, **Figure 50–11.** These valves are used on many compressors of up to about 150 tons and that have large piping connections. These valves have a front seat, a back seat, and can be used in the midseated position for different purposes, **Figure 50–12.** Notice that the valve in **Figure 50–11** has a protective cap. This is the second line of defense from refrigerant leaks. The first line of defense is the backseat; the valve cap provides additional protection. The valve cap is also used to protect the valve stem from damage. This is particularly true with the suction-line valve, which often operates below

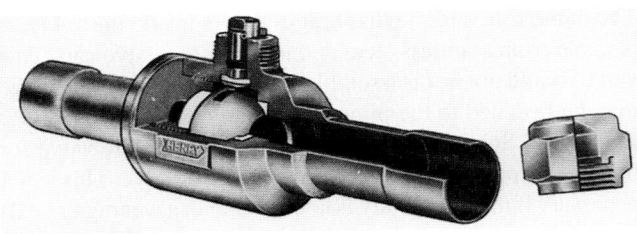

Figure 50–10 This is a ball valve, used for refrigerant. Notice that it has a protective cap to protect the valve stem and that it could also prevent a leak if the stem leaked. *Courtesy Henry Valve Company*

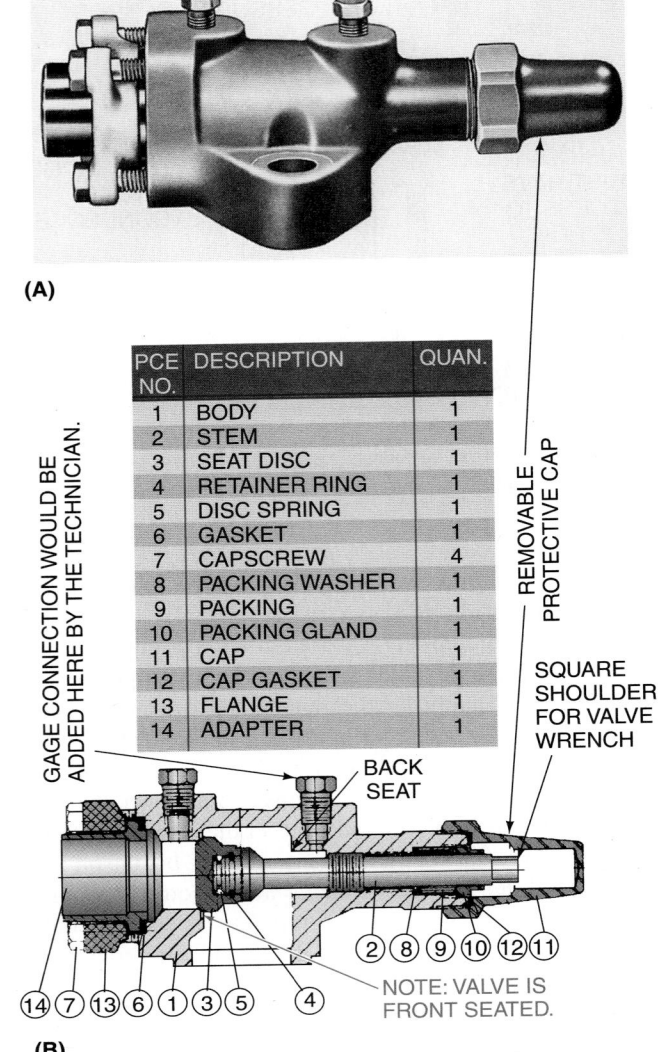

PCE NO.	DESCRIPTION	QUAN.
1	BODY	1
2	STEM	1
3	SEAT DISC	1
4	RETAINER RING	1
5	DISC SPRING	1
6	GASKET	1
7	CAPSCREW	4
8	PACKING WASHER	1
9	PACKING	1
10	PACKING GLAND	1
11	CAP	1
12	CAP GASKET	1
13	FLANGE	1
14	ADAPTER	1

Figure 50–11 **(A)** A large service valve like those used on many compressors for suction or discharge service. **(B)** Cutaway of the service valve. *Courtesy Henry Valve Company*

the surrounding dew point temperature and sweats. Often, the valve stem is made of ferrous metals (iron or steel) and the valve stem will become rusted. **If the technician turns the rusty valve stem down through the packing gland, damage to the packing gland will occur.** The proper practice is to use fine sand cloth (400 grit) to clean the high spots off of the valve stem, **Figure 50–13,** before turning the stem. It is also good practice to loosen the packing nut before turning the stem. This may cause refrigerant seepage, but the technician must preserve the packing.

The valve stem packing can often be removed and replaced by back seating the valve. The packing portion of the valve is isolated. You would need to talk to the compressor manufacturer to determine the correct packing if you cannot identify it when removing the old packing. It is often graphite or Teflon rope, which is available at plumbing supply houses. When large valves are used, it is much easier to replace the packing than to replace the valve. Another alternative is to leave the

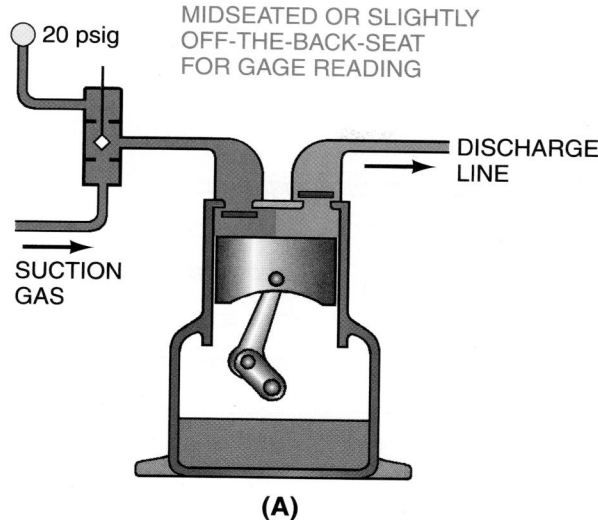

MIDSEATED OR SLIGHTLY OFF-THE-BACK-SEAT FOR GAGE READING

20 psig

DISCHARGE LINE

SUCTION GAS

(A)

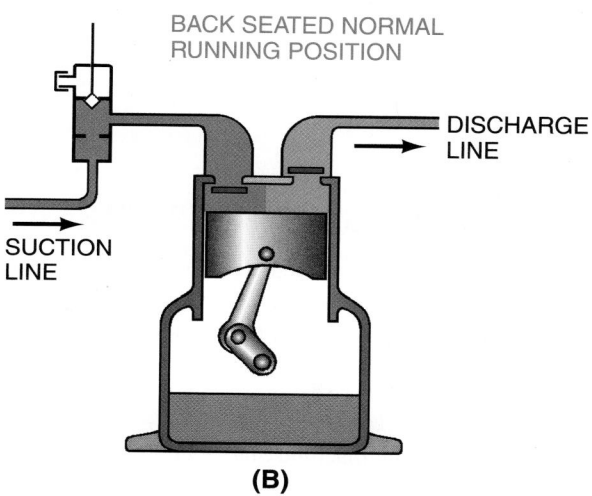

BACK SEATED NORMAL RUNNING POSITION

DISCHARGE LINE

SUCTION LINE

(B)

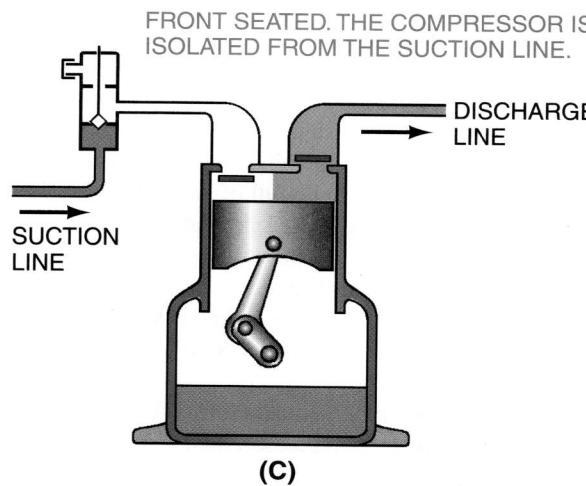

FRONT SEATED. THE COMPRESSOR IS ISOLATED FROM THE SUCTION LINE.

DISCHARGE LINE

SUCTION LINE

(C)

Figure 50–12   Three positions for a service valve **(A)–(C).**

piping sleeve on the piping and just replace the valve body, when it is available.

When the technician is through servicing the unit, the in-side of the valve cap should be cleaned up and the stem

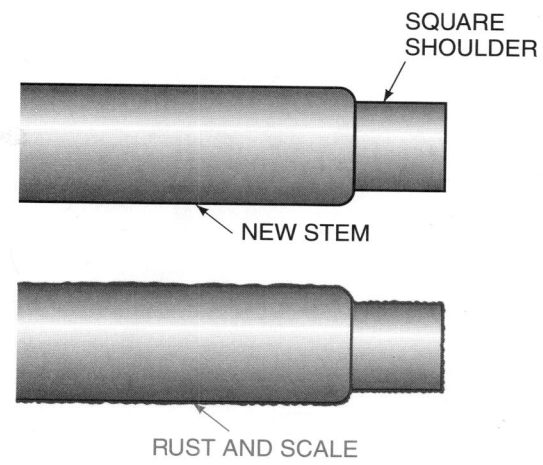

SQUARE SHOULDER

NEW STEM

RUST AND SCALE

Figure 50–13   A clean stem and a rusty stem.

should be wiped clear of any moisture. Many technicians then coat the valve stem with silicone spray or oil to protect it just before replacing the valve cap.

The valve cap is often hard to remove when it has been on for a long time as rust may have accumulated on the threads. Some of these valve caps are quite large. Many technicians use a pipe wrench to take the valve cap off. Another way to remove it without leaving scars on the valve cap is to lightly tap each flat on the nut with a hammer to break the rust loose, then use a hex-jaw wrench to remove it, **Figure 50–14.** You can actually tap the handle of this wrench with a hammer to dislodge the valve cap.

Figure 50–12 shows ways that the compressor service valves are used. They can be back seated to remove the gages; this is the normal running position for the valves. Often the low-pressure control is fastened to the service valve, and in such a case the valve should be slightly cracked (turned toward the front seat) in order for the control to be able to function. The valves can be midseated or cracked-off-the-back-seat for gage readings.

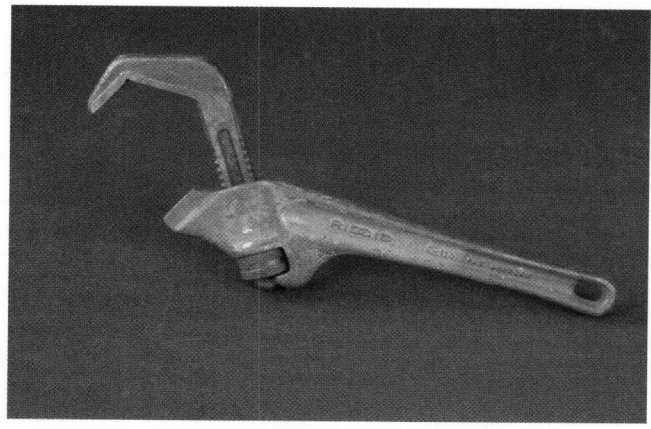

Figure 50–14   This wrench can be used to remove valve stem caps.
Photo by Bill Johnson

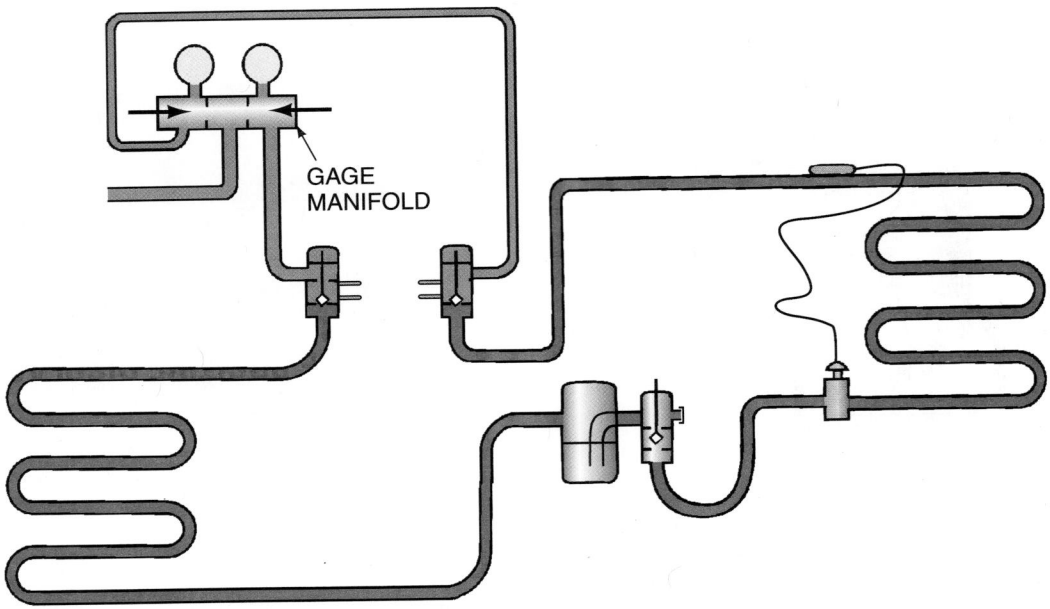

Figure 50–15 When the suction and discharge valves are front seated, the compressor can be removed. The valves are fastened to the compressor with bolts. The compressor becomes a separate vessel.

The valves can be completely front seated for compressor service. Notice that the compressor can be removed and replaced using the front seats by unbolting the compressor flange bolts, **Figure 50–15**. Only the refrigerant in the compressor may be recovered and the new compressor can be evacuated using the front-seated position.

The liquid-line king valve is a type of service valve that is usually found at the outlet of the refrigerant receiver. This valve is used to isolate the refrigerant in the condenser by pumping the refrigerant from the outlet of the receiver into the condenser and receiver, **Figure 50–16**. A technician can fasten a high-side gage to the king valve, front seat it, and start the compressor to pump the refrigerant to the condenser receiver. The gage reading will tell the technician what the pressure is in the liquid line.

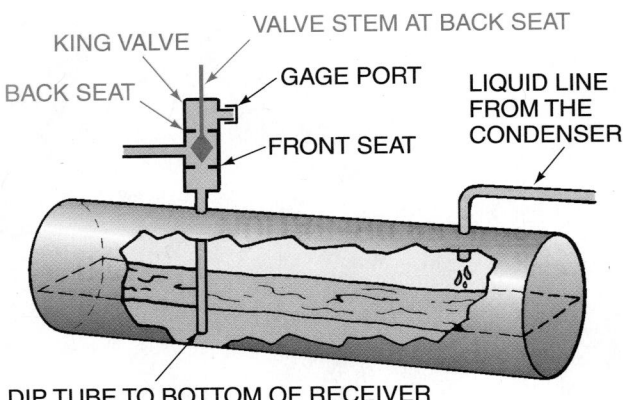

Figure 50–16 The king valve in the liquid line is a little different. The back seat is open to the gage port when the valve is front seated. This valve has a back seat so that the gages can be removed.

50.6 CENTRIFUGAL CHILLER START-UP

Centrifugal chillers often need to be started and watched for the first few minutes because of the separate oil sump system. When the chiller is centrifugal, the oil sump should be checked before a start-up is tried. You should look for the following:

1. Correct oil sump temperature, from 135°F to 165°F, depending on the machine manufacturer. NOTE: Do not attempt to start a compressor unless the oil sump temperature is within the range the compressor manufacturer recommends or serious problems may occur. After the machine has been operating for some time, check the bearing oil temperature to be sure that it is not overheating. If overheating occurs, check the oil cooling medium, usually water.

2. Correct level of oil in the oil sump. If the oil level is above the glass, it may be full of liquid refrigerant. Ensure that it is the correct temperature. You may start the oil pump in manual and observe what happens. If in doubt, call the manufacturer for recommendations. Unless you have experience as to what to do, do not try to start the compressor; marginal oil pressure may cause bearing damage.

3. Start the oil pump in manual and verify the oil pressure before starting the compressor, then turn it to automatic before starting the compressor.

When the chiller is started, the compressor oil pump will start and build oil pressure first. When satisfactory oil pressure is established, the compressor will start and run up to speed then change over to the run-winding configuration. This could be autotransformer, wye-delta, or electronic starter. The centrifugal compressor starts unloaded and only begins to load up after the motor is up to speed. This is accomplished with the prerotation guide vanes mentioned earlier. When the chiller is up to speed, the prerotation vanes begin to open and will open to full load unless the machine demand limit control stops the vanes at a lower percent of full load.

When the chiller compressor starts to work, pull full- or part-load current, the technician should observe the pressure gages. The technician should look for problems such as the following:

1. The suction pressure operating too low or too high. The technician should mark on the gage front or on a notepad the expected operating suction pressure for normal operation, which would be at design water temperature. Typically this would be 45°F leaving water.

2. The net oil pressure should be correct. This should be noted nearby for ready reference. If the machine has been off for a long period of time, observe the oil level in the oil sump as the compressor starts to accept the load and look for oil foaming. If the oil starts to foam, reduce the load on the compressor. This will raise the oil sump pressure and boil the refrigerant out of the oil at a slower pace. If you do not reduce the load, it is likely that the oil pressure will start to drop and the machine will shut off because of low oil pressure. If the compressor has an anti-recycle timer, you will not be able to restart the compressor until the timer completes its cycle and that may be 30 min. It is better to avoid allowing the compressor to shut off by watching the oil pressure.

3. Discharge pressure should be correct. This should be noted nearby for ready reference or marked on the pressure gage for typical operating conditions.

4. Compressor current should be correct and marked on the ammeter on the starter. The ammeter should be marked for all operating percentages, typically 40%, 60%, 80%, and 100%, so the technician will be aware at all times at what load the machine is operating. The operating arm for the prerotation vanes should be marked so the travel can be measured; this is closely correlated with the current.

Chillers manufactured in the last few years all have electronic controls, and a sequence of the starting events should be studied for the particular machine you are interested in. Typically they use everything from default lights on the control panel to electronic readout for what is happening. For example, a unit may have a sequence of LED lights on the circuit board that will be lit if a particular problem has occurred. The technician must use the manufacturer's trouble chart to discover what the nature of the problem may be. Some manufacturers use the flash sequence of LED lights to describe a problem. Some manufacturers use an electronic readout that explains the problem in words or code numbers. These control sequences can be lengthy, and it is beyond the scope of this text to describe them. Again, nothing beats an understanding of the equipment that is based on the manufacturer's own description. One of the intents of these electronic controls is to make troubleshooting easier for the technician.

50.7 SCROLL AND RECIPROCATING CHILLER OPERATION

Once the chiller is on and operating, the operator should observe the chiller from time to time to make sure it is operating correctly. Small air-cooled chillers may be located in remote locations and not observed on a regular basis. These chillers are usually operated as unattended. There is not much that can go wrong if the chilled-water supply is maintained.

All water-cooled chillers require more attention because of the water circuit and cooling tower. It is good practice to check any piece of operating equipment on a periodic basis, but it is expensive so most close observation is reserved for larger systems. It is good practice to check the cooling tower water several times per day to be sure that the water level is correct. Often, a partial loss of water pressure will cause the tower water to drop due to the fact that the water is evaporating faster than the water supply can make it up. With careful observation, the technician can catch this problem before it shuts the unit off and a complaint occurs. If there is no regular technician on the job, for example, at a small office building, someone can be assigned to make this check with very little instruction. These chillers are often on top of buildings where no one goes on a regular basis. Someone should be assigned to check these installations every week.

Water-cooled chillers must have water treatment to prevent minerals and algae from forming in the water circuit. A qualified water treatment specialist is necessary for all water tower applications, or problems will occur. Review the text on water towers for types of treatment needed. All towers must have blowdown, which is a percentage of the circulated water passing down the drain to prevent the tower water from becoming overconcentrated with minerals due to evaporation. Blowdown for the cooling tower is necessary but often management has a difficult time watching water that looks clean flow down the drain. The technician should know how to explain this to them.

50.8 LARGE POSITIVE DISPLACEMENT CHILLER OPERATION

Large chillers that use reciprocating or rotary screw compressors may range in size to more than 1000 tons of capacity. Chillers of this size are extremely expensive and must be observed on a regular basis or problems may occur. These problems can often be expensive to correct. These large chillers are reliable but observation can be important. The technician should look for any potential problem, as explained in the previous paragraphs. These chillers may also have pressure gages that are active all the time in that they are not valved off. If so, these may be observed as much as once per hour on some critical applications. Often an operating log is maintained on a regular basis. The frequency of the recording in the operating log depends on the importance of the job. If it is a chiller used for critical manufacturing, the entries in the log may be required hourly. If it is an office building, the log may be maintained on a daily or weekly basis. See **Figure 48–22** for an example of an operating log form.

50.9 CENTRIFUGAL CHILLER OPERATION

Centrifugal chillers were the largest chillers made for many years. Typically when the requirements were greater than 100 tons, a centrifugal chiller was considered or multiple reciprocating chillers were used. Centrifugal chillers range in size from 100 tons to about 10,000 tons for a single chiller. These were the most expensive of all chillers and also the most reliable so much attention has been given to the observation of the operation and maintenance of these chillers. For years, it has been customary for the operation of these chillers to be observed and an operating log maintained on a regular basis, usually daily. With all operating logs, slowly occurring problems can be plotted. For example, if the cooling tower water treatment has not been working and the condenser water tubes are beginning to have a mineral buildup, the condenser water approach will begin to spread. An alert operator will notice this and take corrective action. This action may be to change the water treatment and certainly to clean the water tubes because the heat exchange has been reduced.

50.10 AIR-COOLED CHILLER MAINTENANCE

Air-cooled chillers require little maintenance, which is one reason why they are so popular. The fan section of these chillers may require lubrication of the fan motors, or the motors may have permanently lubricated bearings. It is typical for these chillers to have multiple fans and motors. The motor horsepower can be reduced in this manner and the fans can be direct drive. This eliminates belts and the need for routine lubrication in many cases. If one fan fails, the chiller can continue to operate.

The following electrical maintenance should be performed annually for a typical system:

1. Inspect the complete power wiring circuit because this is where the most current is drawn. Look for places where hot spots have occurred. For example, the wires on the compressor contactor should show no signs of heat. The insulation should not appear to have been hot. If so, this must be repaired. This can be done by cutting the wire back to clean copper. If the lead is not long enough, a splice of new wire may be needed or the wire may need to be replaced back to the next junction.

2. Inspect the motor terminal connections at the compressor. These should not show any discoloration or a repair should be made. A hot terminal block can cause refrigerant leaks so the wiring should be inspected carefully.

3. The contacts in all contactors should be inspected and replaced if pitting is excessive. If excess pitting begins to occur, it is likely that the contacts will weld shut in the future and the motor cannot be stopped using the contactor. This can lead to single phasing of a wye-wound motor if any two contacts weld shut. Motor burn will then occur because there is nothing to shut the motor off except the breaker, **Figure 50–17.**

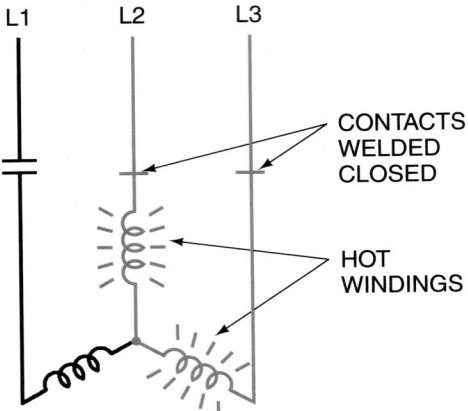

THIS MOTOR IS OPERATING IN A SINGLE-PHASE CONDITION.

Figure 50–17 The contacts are welded shut on a wye-wound motor, which will cause the motor to try to run on L2 and L3. This is called single phasing and will burn the motor.

4. The compressor motor should be checked for internal ground using an ohmmeter that can check ohm readings in the millions of ohms. This meter is called a megohmmeter. **Figure 50–18** shows an example of a basic megohmmeter. This instrument uses about 50 to 500 V DC to check for leaking circuits; others may use much higher voltages and have a crank-type generator to achieve the high voltage. The correct motor manufacturer's recommendations should be used for the allowable leakage of the circuit. Typically, a motor

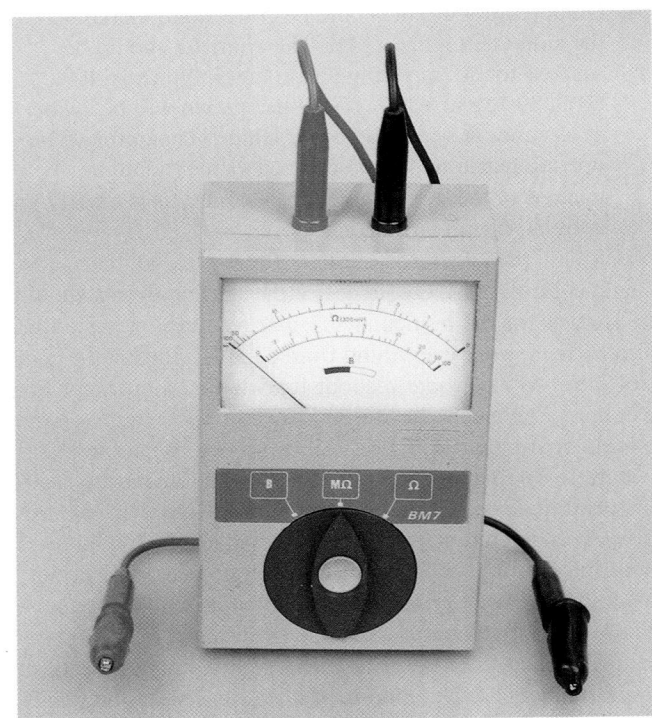

Figure 50–18 A megohmmeter for checking ground circuits in motors. *Photo by Bill Johnson*

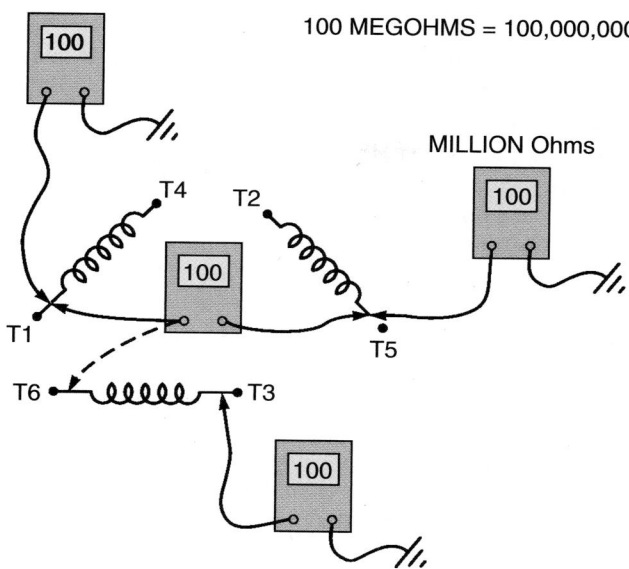

100 MEGOHMS = 100,000,000

MILLION Ohms

Figure 50–19 Using a megohmmeter to check a motor. Notice that it is checked from all windings to ground and from winding to winding.

should not have less than a 100-megohm circuit to ground or to another winding in the case of a delta-type motor or dual-voltage motor, **Figure 50–19.** The required reading depends on the motor temperature because the temperature changes the requirements. You should consult the manufacturer if proper guidelines are not known. A reliable local motor shop can also give you guidelines to use. It is important to start a process of megging a motor when it is new and keep good records. If the motor megohm value begins to reduce, it is a sign that moisture or some other foreign matter may be getting in the system.

5. An oil sample from the compressor crankcase can be sent to an oil laboratory to determine the condition of the compressor. This test, like the megohm test, should be started when the chiller is new and performed every year. If certain elements start appearing in greater quantities in the oil, it is evident that problems are occurring. For example, if the bearings are made of babbitt and the babbitt content increases each year, at some point it is evident that the bearings need to be inspected. If small amounts of water appear, the tubes need to be inspected for leaks. This will only occur if the water pressure is greater than the refrigerant pressure. For high-pressure chillers this can occur if the chiller is at the bottom of a high column of water.

6. Inspect the condenser for deposits such as lint, dust, or dirt on the coil surface. These must be removed because they block airflow. It is good practice to clean all air-cooled condensers once a year or more often if needed. If the operating log shows an increase in head pressure compared with the entering-air temperature, the coil is becoming dirty. Air-cooled coils can be deceiving in that they can be dirty and you may not

notice it because the dirt may be imbedded in the interior of the coil. This cannot be seen until the coil is cleaned. A coil should be cleaned by saturating the coil with an approved detergent and then washing the coil backward, in reverse to the direction of the airflow; this is called backwashing. An apparently clean coil may yield a large volume of dirt from the interior of the coil.

50.11 WATER-COOLED CHILLER MAINTENANCE

Routine maintenance may involve cleaning the equipment room and keeping all pipes and pumps in good working order. The technician should ensure that the room and equipment are kept clean and in good order. It is always easier to keep a room clean than to have to clean it when the chiller breaks down and needs to be disassembled. The cooling tower should be checked to be sure the strainer is not restricted. Look at the water for signs of rust, dirt, and floating debris and clean if necessary. The water treatment should be checked on a regular basis. Some systems call for the operating technician to perform water analysis on a daily basis to check for mineral content. When the water treatment gets out of balance, the technician should either make adjustments or call the water treatment company. Someone must be in charge of keeping the water chemicals at the correct level for best performance.

Annual maintenance should involve checking the complete system and chiller to prepare it for a season of routine maintenance. Some chillers operate year round and are only shut down for annual maintenance. During this maintenance, it is often wise to bring in the factory maintenance representatives. These technicians are privileged to know what is happening across the nation or around the world with regard to any service or failure problems with their equipment. They will provide the best technical knowledge for your equipment. Often, these technicians will perform the complete start-up procedure for your equipment just as though it is for the first time. All controls may be checked at this time to be sure they will perform up to standard. It is not uncommon for a system to operate for years without a control checkup and then fail—only to have the technician discover that a control has failed that should have saved the system but did not. A control checkup once a year can be good insurance against many failures.

During annual maintenance, all electrical connections should be checked; refer to the electrical maintenance for air-cooled systems stated previously. Water-cooled equipment has the same electrical symptoms.

The water-cooled condenser should be inspected on the inside every year by draining the water from the condenser and removing the heads. The tubes should be clean. You can tell when a tube is clean when you take a light and shine inside the tube, and all you see is copper tube with a dull copper finish. A penlight is very good for this. The penlight can be inserted inside the tube. Any film on the inside of the tube must be removed for proper heat exchange. Many technicians believe that if the tube is open enough to allow a good water flow it is clean. This is not true; the tube must be clean down to bare copper for proper efficiency.

When it is discovered that the tubes are dirty, several approaches may be used. The tubes may be brushed with a nylon brush that is the size of the inside of the tube. This is customarily done with a machine that turns the brush while flushing the tube with water. Some manufacturers recommend using a brush with fine brass bristles if the nylon will not remove the scale from the tubes. Check with the manufacturer because some do not recommend this practice. Always keep the tubes wet until it is time to brush them or the scale will become more hardened. It is good practice to be prepared to brush the tubes before removing the heads so the tubes will stay wet until they are brushed.

If brushing the tubes will not work, the tubes may be chemically cleaned using the recommended acid for the application, **Figure 50–20**. NOTE: This is a very delicate process because the tubes may be damaged. Consult the chemical and chiller manufacturer for this procedure.

It may be good practice to let the chemical manufacturer clean the tubes; then if damage occurs, they are responsible. However the process is accomplished, make sure that the chemicals are neutralized or damage to the tubes may occur. The damage that may occur would be severe because the chemicals may eat the tube material, and leaks between the water and refrigerant may occur.

After the tubes have been chemically treated, they may be clean, or the chemicals may just soften the scale and the tubes may need brushing to remove the scale.

The water pump should be checked to make sure the coupling is in good condition. SAFETY PRECAUTION: This can be done by turning off the power, locking and tagging the disconnect, and removing the cover to the pump coupling. If materials such as filings are found inside the coupling housing, deterioration of the coupling should be suspected. Some couplings are made of rubber and some are steel. Look for the flexible material that takes up the slack to see whether it is wearing.

After the condenser has been cleaned, the condenser tubes may be checked for defects such as stress cracks, erosion, corrosion, and wear on the outside by means of an eddy current check. This test should be performed on the evaporator tubes as well. This check determines if there are irregularities in the tube. One common problem is wear on the tube's outer surface where it is supported by the tube support sheets. Each set of tubes has a different type of stress applied to the outside. In the condenser, the hot discharge gas is pulsating as it enters the condenser and has a tendency to shake the tubes. In the evaporator, the boiling refrigerant shakes the tubes. The tube support sheets are sheets of steel that the tubes pass through to help prevent this action and are evenly spaced between the end sheets. They are located in the condenser and the evaporator. When the evaporator or condenser shell is tubed, the tubes are guided in place through the support sheets. When the tube is in place, a roll mechanism is inserted into the tube and expanded to hold the tube tight in the support sheet, **Figure 50–21**. The tube roller may miss a tube and this tube can shake or vibrate until the tube sheet wears the tube. If this wear continues, the tube will rupture and a leak between the water and refrigerant circuit will occur. The worst case can flood the chiller on the refrigerant side with water. This may happen in particular with low-pressure chillers. The eddy current test can be used to find potential tube failures and these tubes may be pulled and replaced or plugged if there are not too many.

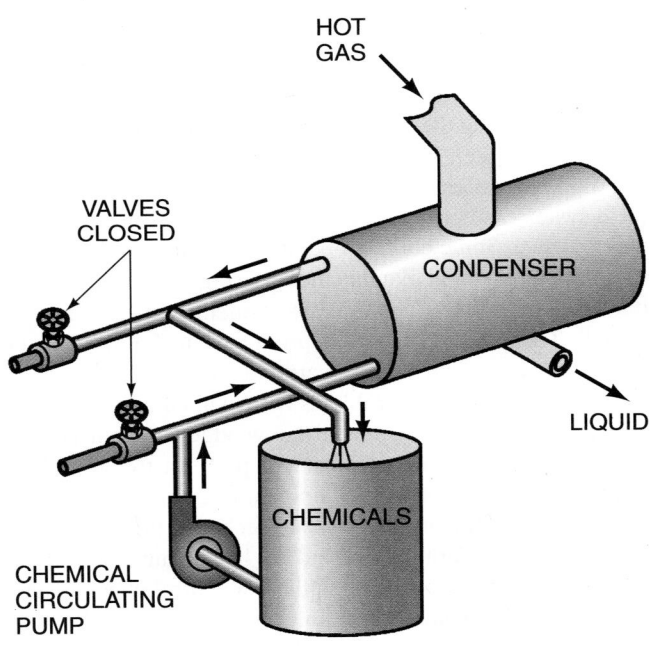

Figure 50–20 Chemically cleaning condenser tubes.

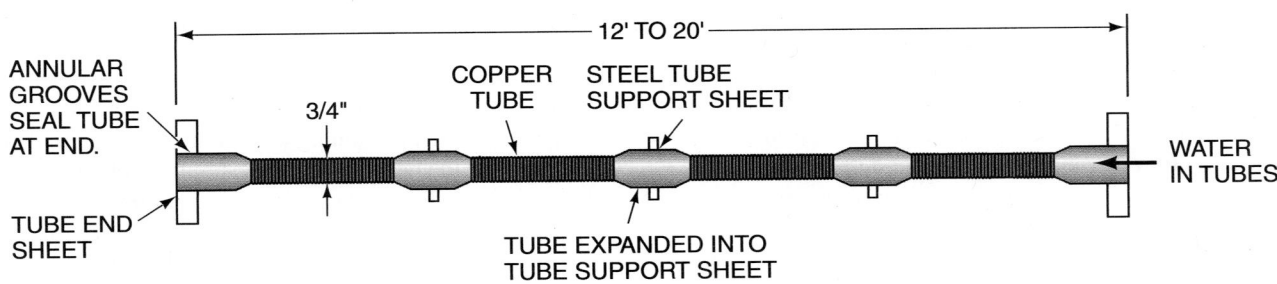

Figure 50–21 The tube assembly for a heat exchanger. Notice the tube support sheets that support the tubes.

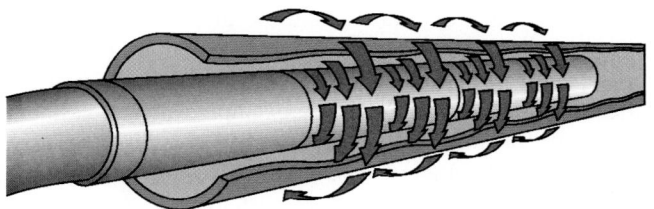

Figure 50–22 A probe for checking tubes that uses an eddy current tester can determine irregularities in the tubes.

The eddy current test instrument has a probe that can be pushed through the clean tube while an operator watches a screen monitor, **Figure 50–22.** As the probe passes through the tube, the probe sends out a magnetic current signal that reacts with the tube and shows a profile on the screen monitor. When unusual profiles are noticed, the tube is marked for further study, which may involve comparing the profile to that of a known good tube. A decision may need to be made to pull or plug the tube. A tube failure in the middle of the cooling season can take days to repair, and it is good practice to find these problems in advance.

Water and steam valves can be maintained as leak free as refrigerant valves, but are not as intricate. The *globe valve* is the most common valve used in water and steam circuits, **Figure 50–23.** It has a packing gland like the service valves on a compressor. This valve can normally be back seated to facilitate the removal of the existing packing and the addition of new packing. Note: This is the only time the valve should be back seated. The globe valve can be disassembled and the

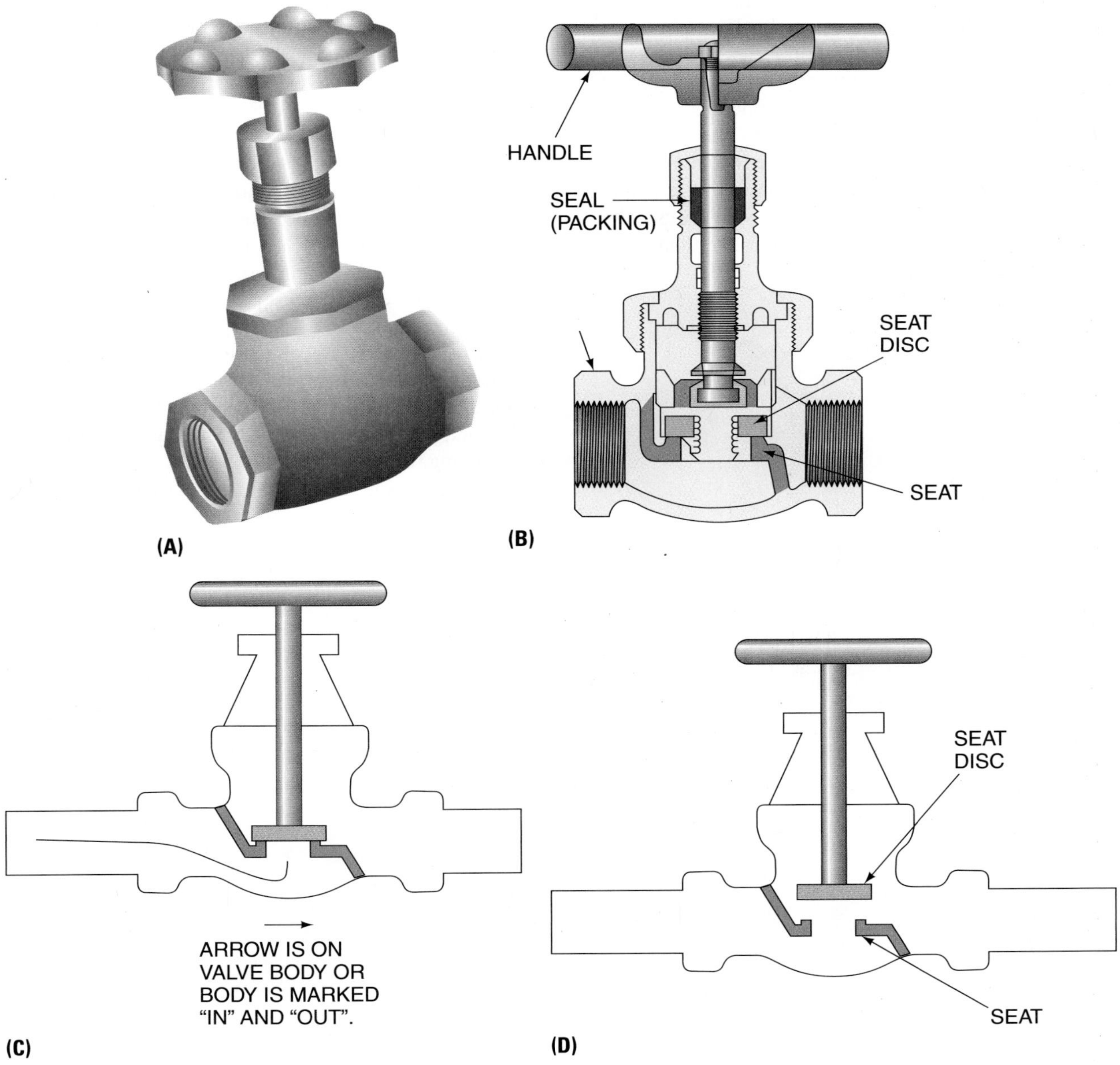

HANDLE

SEAL (PACKING)

SEAT DISC

SEAT

(A)

(B)

ARROW IS ON VALVE BODY OR BODY IS MARKED "IN" AND "OUT".

SEAT DISC

SEAT

(C)

(D)

Figure 50–23 **(A)** A globe valve. **(B)** Cutaway of a globe valve. **(C)** Globe valve shut. **(D)** Globe valve open. Notice that the fluid must change directions. **(A)** *Photo by Bill Johnson.* **(B)** *Courtesy Nibco*

valve seat can be replaced. The system must be drained in order to be disassembled. Notice that the valve has a directional arrow for flow. The stream to be controlled should be piped to the underside of the globe valve. If it is piped up backwards, the valve seat will often chatter or make a buzzing sound as the seat vibrates. The globe valve has a great deal of pressure drop because the fluid has to make several turns as it goes through the valve.

The *gate valve* is another common valve used for these applications, **Figure 50–24.** The gate valve creates much less pressure drop through the valve than the globe valve and is a much better choice where excess pressure drop is a concern. Notice that the gate valve has a packing gland and a back seat for packing-gland repair. The gate valve can also be disassembled for seat repair. It does not have a replaceable seat; the gate is finely ground to fit into the finely ground seat. For large valves, this seat may be resurfaced if needed. NOTE: The globe valve and gate valve should not be back seated when in operation. The reason for this is that if you back seat the valve and then later forget the position, you might do damage to the valve if you try to open it. It is good practice to open the valve all the way to the back seat and then turn it inward about one turn. Then, when you or someone else tries to use the valve, they will know the position of the valve. It is not uncommon for a technician to mistake the valve for closed when it is on the back seat and then try to open it and do damage.

To keep the valve loose, both the globe valve and gate valve should be run from all the way closed to all the way open from time to time. This must be done when it will not affect the system's operation, for example, when the system is shut off for some reason. These valves come in very large sizes—up to about 24 in. The larger valves have flanges for fastening them with bolts into the system.

The *butterfly valve* is used in water circuits and is a quick-shutoff valve—as it only needs to be turned one-quarter turn from on to off, like a ball valve.

Strainers are used in water circuits to remove particles that may be in suspension and circulating. The cooling tower can be particularly dirty because it uses outside air that can have many contaminants in it. A cooling tower that has water flowing and air being forced through it is like a large filter for the air. The particles are gathered in the water and then the tower sump and onto any strainers. There are several types of strainers; consult the strainer manufacturer if there is doubt about how to clean them. SAFETY PRECAUTION: When water towers are cleaned, the technician should be careful with the contaminants in the tower. If the tower is dry, do not breathe the dust. Rubber gloves and boots should be worn, along with a face mask.

Water pumps are just as important as the chiller itself. If the water pumps will not operate, there will be no chiller operation. Water pumps and their motors often require lubrication. When grease fittings are used, pay attention to the relief plug that allows excess grease to escape, **Figure 50–25.** If the relief plug is not removed, excess grease will be forced down the shaft and may harm the grease seal or enter the motor.

The pump and motor will have a flexible coupling between them and it should be inspected. Turn off the power, lock it out, remove the protective cover, and visually inspect the coupling for wear.

If the chiller and water pumps are to stand idle during the winter months, all precautions should be taken to prevent freezing of any components such as piping, sumps, pumps, or chiller if it is located where it may freeze. Even if a chiller

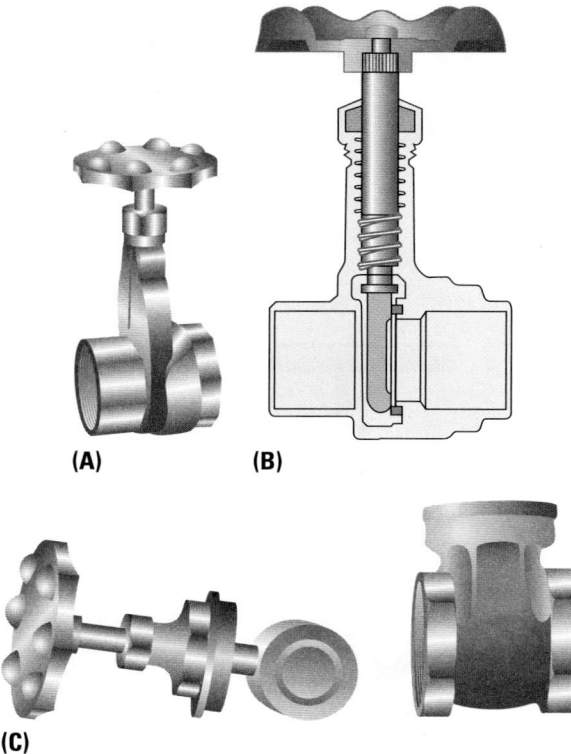

(A)   **(B)**

(C)

Figure 50–24   (A) A gate valve. **(B)** Cutaway of a gate valve. Notice that when the gate is open, the valve goes straight through, and there is very little pressure drop. **(C)** A gate valve disassembled. *(A) and (C) Photos by Bill Johnson. (B) Courtesy Nibco*

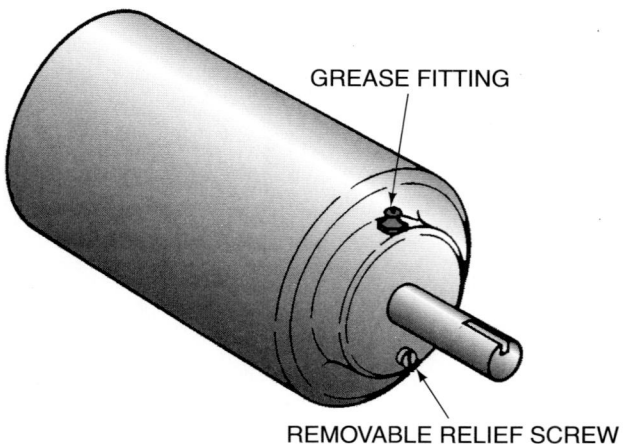

GREASE FITTING

REMOVABLE RELIEF SCREW

Figure 50–25   A motor bearing with a grease fitting. Notice the relief plug that prevents excess pressure when grease is added. The plug must be removed while grease is applied.

is in a building, it is good practice to drain it during winter because if the building loses power and the interior freezes, it would be expensive if the chiller tubes were to freeze. It is easier to just drain it and not take a chance.

50.12 ABSORPTION CHILLED-WATER SYSTEM START-UP

The absorption start-up is similar to any chiller; chilled water flow and condenser water flow must be established before the chiller is started. Ensure that the cooling tower water temperature is within the range of the chiller manufacturer's recommendations. If the absorption machine is started with the cooling tower water too cold, the lithium-bromide (Li-Br) may crystallize, causing a serious problem. In addition to this, the heat source must be verified whether it be steam, hot water, natural gas, or oil. Most absorption chillers are steam operated so the steam pressure must be available. If the chiller is operated from a boiler dedicated to it, the boiler must be started and operated until it is up to pressure.

The purge discharge may be monitored by placing the vacuum pump exhaust into a glass of water and looking for bubbles, **Figure 50–26.** Be sure to remove the purge exhaust from the water when the purge is not operating. If there are bubbles,

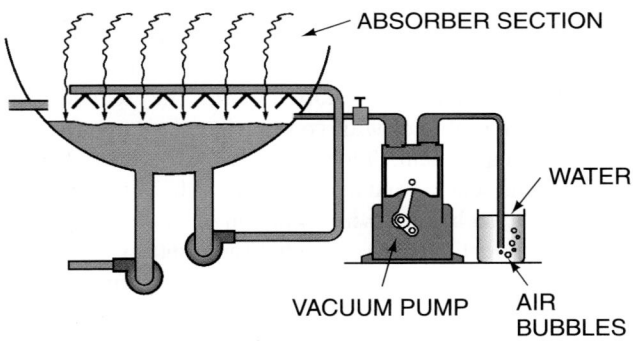

Figure 50–26 The vacuum pump exhaust is placed in a glass of water to determine noncondensable gas is being removed; bubbles are moving out of the exhaust. Note: Remove the glass before turning off the vacuum pump.

it is a sure sign that noncondensables are in the chiller. It is pointless to start the chiller until the bubbles stop. Let the vacuum pump run even when the chiller is started (unless the manufacturer recommends not to) because often noncondensables will migrate to the purge pickup point after start-up.

When all systems are ready, the chiller may be started. The fluid pumps will start to circulate the Li-Br and the refrigerant and the heat source will start. The chiller will begin to cool in a very few minutes and this can be verified by the temperature drop across the chilled water circuit. When absorption chillers are refrigerating, they make a sound like ice cracking. When this sound begins to come from the machine, the chilled-water temperature should begin to drop.

50.13 ABSORPTION CHILLER OPERATION AND MAINTENANCE

Absorption chillers must be observed for proper operation on a more regular basis than compression cycle chillers because they can be more intricate. Chiller manufacturers have different checkpoints to determine the chiller operation. The refrigerant temperature and the absorption fluid temperatures measured against the leaving chilled-water temperatures are among a few. The manufacturer's literature must be consulted for the checkpoints and procedures. The maintenance of an operating log is highly recommended for absorption installations.

When the chiller is operating, it is good practice to pay close attention to the purge operation. If the chiller is requiring excessive purge operation, the machine may have leaks or it may be manufacturing excess hydrogen internally and need additives. The machine must be kept free of noncondensables.

The heat source may need maintenance. If it is steam, the steam valve should be checked for leaks and operating problems. When steam is used, a condensate trap will also need checking to be sure it is operating properly. The condensate trap ensures that only water (condensed steam) returns to the boiler. A typical condensate trap may contain a float. When the water level rises, the float rises and only allows water to move into the condensate return line, **Figure 50–27.** There are several other different types of trap; this one is shown as an example.

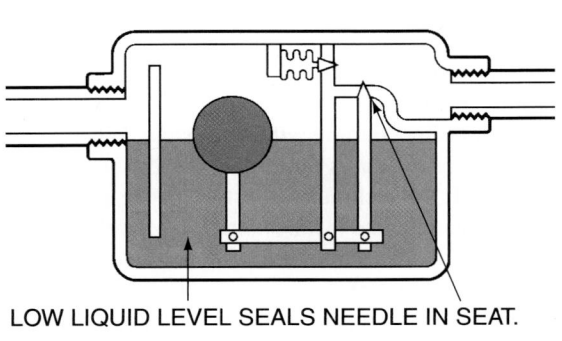

LOW LIQUID LEVEL SEALS NEEDLE IN SEAT.

(A)

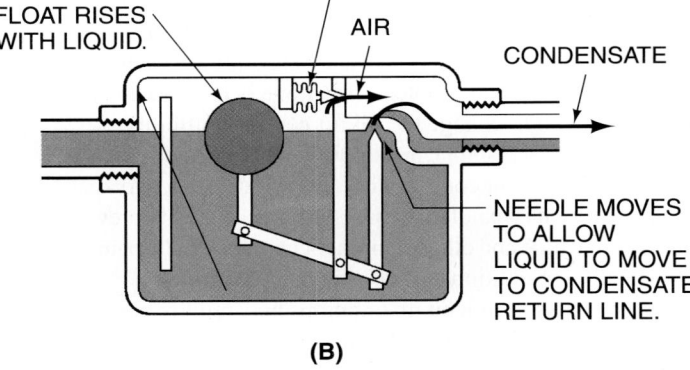

THERMOSTATIC ELEMENT CONTRACTS. THE CONTRACTION ALLOWS AIR TO ESCAPE.

FLOAT RISES WITH LIQUID.

AIR

CONDENSATE

NEEDLE MOVES TO ALLOW LIQUID TO MOVE TO CONDENSATE RETURN LINE.

(B)

Figure 50–27 The condensate float trap for a steam coil.

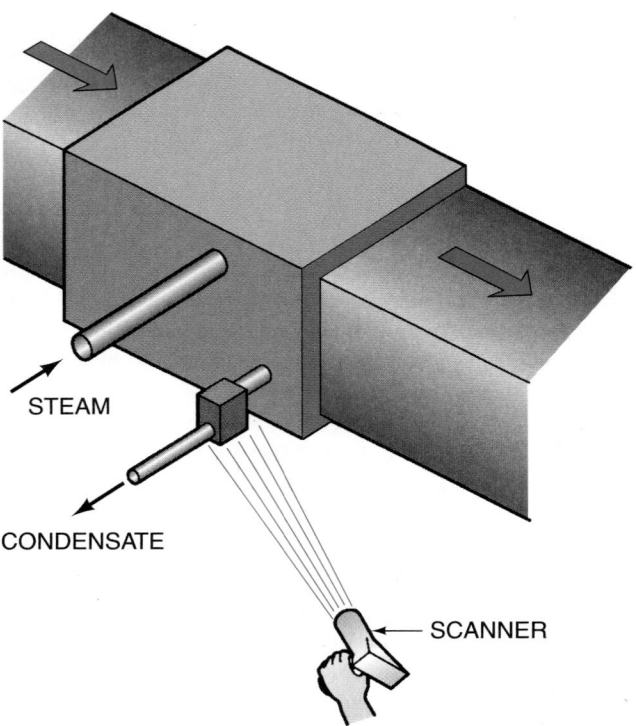

STEAM

CONDENSATE

SCANNER

Figure 50–28 Checking a condensate trap using an infrared temperature tester. If the trap and the entering and leaving piping are the same temperature, the trap is passing steam. When the entering piping from the coil is warmer than the leaving piping, the trap is passing only liquid condensate.

🌐 Condensate traps are often checked with an infrared checking device that measures the temperature on both sides of the trap, **Figure 50–28.**🌐

When the chiller is operated using hot water, the hot water valve should be observed for proper operation and to ensure that there are no leaks.

Gas- or oil-operated machines will need the typical maintenance for these fuels. Oil in particular requires maintenance of the filter systems and nozzles.

The purge system may require the most maintenance with the absorption chiller. If there is a vacuum pump with the chiller, regular oil changes will help to make it last longer. Li-Br is salt and will corrode the vacuum pump on the inside. It is not recommended that a vacuum pump used for a compression cycle system be used on an absorption system because of the corrosion problem, but if one is used, extra care should be taken to change the oil immediately after use.

Any Li-Br that is spilled in the equipment room will rust any Ferrous metal it is in contact with. It is good practice to keep the area where it is handled washed with fresh water to prevent this. The machine and other equipment may need to be painted periodically for additional protection in these areas.

The system solution pump or pumps may require periodic maintenance. A check with the manufacturer will detail the type and frequency of this maintenance.

The machine room must be kept above freezing or the refrigerant (water) inside the machine will freeze and cause

tube rupture problems. Unlike the other chillers, this water cannot be drained and must be protected.

The water pumps need regular maintenance as with any other system.

The steam or hot water valve and condensate trap should be inspected for any needed maintenance. A strainer will be located in the vicinity of the condensate trap and should be cleaned during the off season.

The cooling water must be maintained by cleaning and water treatment. Do not forget blowdown water. Absorption chillers operate at higher temperature than compression cycle chillers and good water treatment maintenance is vital because the tubes will become fouled much quicker.

Absorption chillers may need the tubes probed with an eddy current probe on a more regular basis than compression cycle chillers because of the higher temperatures in the condenser section. The tubes are stressed when the heat source is applied to these tubes and they expand and then contract when cooled during the off cycle. Checking with the manufacturer is a good idea.

50.14 GENERAL MAINTENANCE FOR ALL CHILLERS

Technicians who maintain chillers must be well qualified and should stay in contact with the manufacturers for the latest training schools and advice. Manufacturers receive feedback from all over the world on their equipment and know what is happening, such as premature failures. They should share any potential problems with you to prevent future problems with your equipment. Technicians should attend factory seminars and schools and make friends with the factory technicians. The technician who works on or observes work on large equipment should learn what to do during all service procedures. Because factory technicians perform much of the service on large system chillers, the operating technician will be present to observe the service procedures and assist the factory technician. These procedures may be anything from routine checking of the controls to major overhaul of the equipment. The operating technician should be able to recognize the procedures. When the factory technician leaves, the operating technician must operate the equipment.

Refrigerant management is a big factor for any chiller application because chillers contain large amounts of refrigerant. The job site often has a supply of refrigerant on hand for emergencies. If the job site is out of town it may take several days to obtain refrigerant, and this may not be satisfactory. The operating technician should be fully aware of the paperwork that must be kept current with the system. When a comfort cooling chiller contains more than 50 lb of refrigerant, which most will, the technician must be aware that adding refrigerant to the machine just because it loses refrigerant is not satisfactory. If a machine loses more than 15% of its refrigerant in a year or leaks at a rate of 15% per year the law requires that the leak be repaired within 30 days of discovery. The requirements are different for commercial and industrial process equipment; the allowable leak rate is 35% per year. The technician must keep up with all refrigerant purchased and where it is used and have a recordkeeping system.

50.15   LOW-PRESSURE CHILLERS

Low-pressure chillers are probably the most difficult to understand because they operate in both a positive pressure in the condenser and a vacuum in the evaporator while running. When they are shut down, both the condenser and the evaporator will be in a vacuum. The largest portion of the refrigerant is in the flooded evaporator section, and the hot gas from the condenser quickly condenses to the mass of liquid in the evaporator when shut down. Low-pressure chillers operate in a vacuum and will stay in a vacuum if the equipment room is not warm enough to keep the refrigerant warm. As mentioned earlier, a leak in a vacuum will pull air into the machine. The purge unit will operate excessively. Unless the machine has a high-efficiency purge, it will purge too much refrigerant out with the air that was pulled in. For example, R-11 boils at about 74°F at 0 psig (atmospheric pressure), and R-123 boils at about 83°F. This means that in the winter when the equipment is not operating, it may sit in a vacuum. Many equipment rooms are maintained at about 60°F during winter. If the refrigerant in the R-11 chiller is below 74°F, the chiller will be in a vacuum. The R-123 chiller will be in a vacuum at any temperature below 83°F. The owners will not want to keep the equipment room up to these temperatures, so chiller manufacturers provide ways to keep a slight positive pressure using heat in the chiller evaporator section, where the liquid refrigerant will accumulate. This heat can be thermostatically controlled with either warm water or electric heaters under the insulation of the chiller evaporator shell. It may be used to keep the chiller shell in a positive pressure to prevent air from being pulled into the shell, or it may be used to pressurize the chiller for service purposes.

Chillers may be pressurized using nitrogen pressure or heat added to the evaporator shell. Regulated nitrogen pressure can be used to quickly build pressure in the chiller shell, but it must be removed after the service call is completed. If the service involves recycling the refrigerant, nitrogen may be used in the interest of saving time. If the same refrigerant is going to be put back in the chiller, heat may be the best choice so the nitrogen will not have to be removed from the refrigerant.

Care must be used when the chiller shell is pressurized for any reason. The pressure must not exceed the stress level of the rupture disc. The rupture disc is a one-time use device, and it is not inexpensive. If its pressure rating is exceeded, the disc will rupture (blow out), refrigerant will be lost, and the rupture disc will have to be replaced. The rupture disc rating is 15 psig, so most technicians will not pressurize a system beyond 10 psig. The head pressure on a system may operate at a condensing temperature of 105°F using cooling tower water, which would equate to about 12 psig for R-11 and 8 psig for R-123. Since the rupture disc is located in the low-pressure part of the system, there is no chance that high head pressure will rupture the disc, **Figure 48–39**.

50.16   RECOVERING REFRIGERANT FROM A LOW-PRESSURE CHILLER

Once the chiller is pressurized to the point that the pressure inside the chiller is greater than the pressure in the recovery system, liquid is allowed to transfer to the recovery system. If the technician does not have a permanent storage cylinder on the job, a portable cylinder will need to be moved in for the recovery. A push-pull recovery machine may also be used to move the refrigerant. If the technician only pushed the liquid out of the machine, the vapor would still remain. When the liquid only is removed from a 350-ton R-11 machine, there are still about 100 lb of R-11 vapor left in the machine. This does not meet the EPA requirement of removing the refrigerant to the required pressure level. A typical R-11 chiller will operate on about 3 lb of refrigerant per ton. A 350-ton machine with a flooded evaporator is estimated to hold 1050 lb (350 × 3 = 1050).

Vapor from the top of the recovery cylinder is pulled out of the top and pushed into the chiller to push liquid to the recovery cylinder, **Figure 50–29**. Notice that the water is not connected to the recovery unit condenser, so no condensation takes place. When all the liquid is removed, the system is

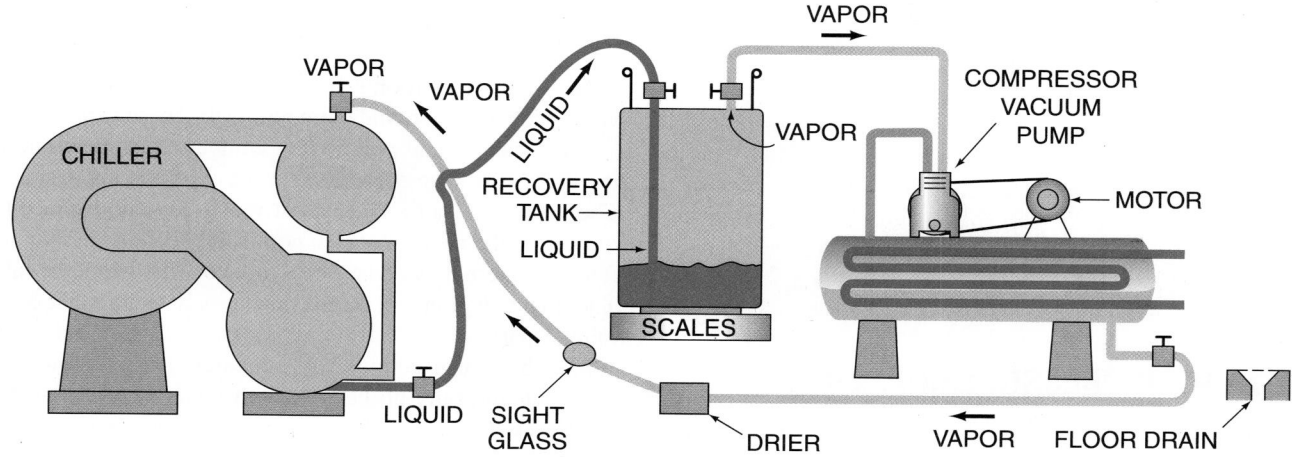

Figure 50–29   The recovery machine pulls vapor from the top of the recovery cylinder and puts it back into the chiller. This creates a low-pressure area in the recovery cylinder in relation to the chiller, and liquid moves into the recovery cylinder. Notice that there is no condensing taking place.

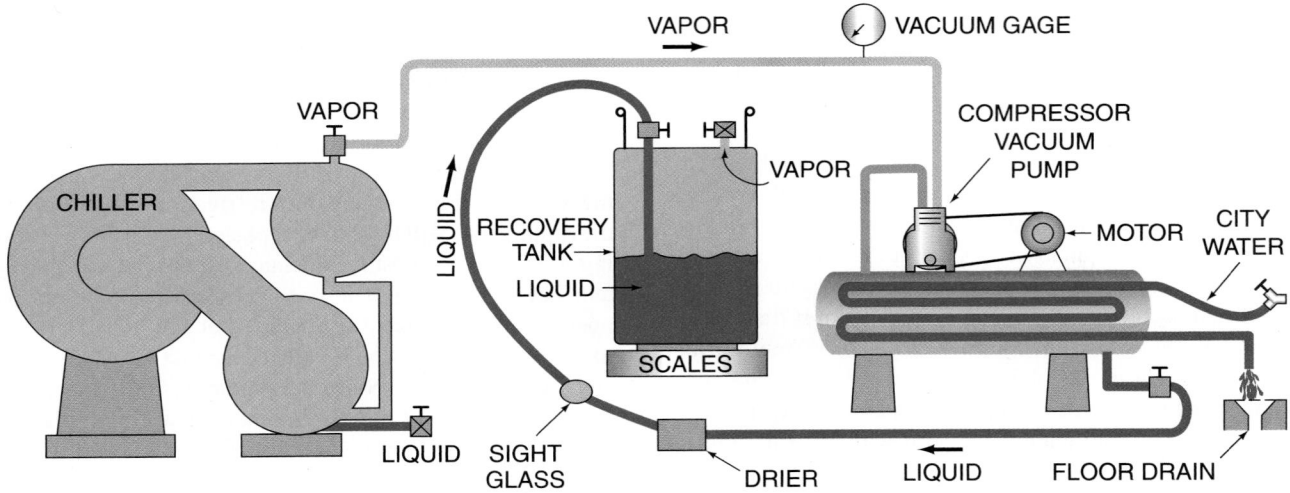

Figure 50–30 The flexible hoses are reconnected only to remove vapor from the cylinder. The vapor is condensed using water and moves to the recovery cylinder. The scales tell the technician when to stop adding refrigerant to the cylinder.

reversed, the liquid connection is shut off, and vapor is pulled out of the vapor space of the chiller, normally at the top of the condenser, **Figure 50–30.** The city water is now connected to the recovery machine condenser and condenses the liquid, and it moves to the recovery cylinder. The required evacuation level in the chiller is 25 in. Hg vacuum with recovery equipment manufactured before November 15, 1993. If the recovery equipment was manufactured after November 15, 1993, the system must be evacuated to 29 in. Hg vacuum (the same as 25 in. Hg absolute). After the service is completed, the system can be reversed and the refrigerant can be pushed back into the chiller. The push-pull equipment used is a recovery machine that is designed to handle large volumes of refrigerant. Some chillers can have upwards of 3000 lb of refrigerant. The recovery machines will have all the piping connections to fasten to the chiller and a water cooling circuit to cool the condenser that is used for condensing vapor while recovering. Water from the public water supply is normally used in this condenser and allowed to be dumped down the drain. This is not a procedure that is used except for service, so a permanent water connection is not needed. SAFETY PRECAUTION: When charging, vapor should be added only until the pressure inside the chiller is well above the pressure that corresponds to freezing. The freezing pressure is 17 in. Hg for R-11 and 20.3 in. Hg for R-123. The pressure in the machine must be above these two values before liquid refrigerant is introduced. Some technicians will circulate water through the evaporator water tubes and allow liquid to enter from the very beginning. This is taking a chance. If there is one tube in the shell that does not have water flowing, it is likely to freeze and split. Water will enter the shell when it thaws if it is split.

50.17 HIGH-PRESSURE CHILLERS

High-pressure chillers must be watched closely for leaks. A visual inspection is something that you can do any time you are at the machine. Where refrigerant leaks out, oil will leak

out also. Any time a fresh oil spot appears, you should suspect a leak. You can use an electronic leak detector or soap bubbles to check the place out. Be sure to clean the soap bubbles off with water, or dust will collect on the spot, giving the impression that oil has leaked and dust has collected on the oil. Some systems are designed so the refrigerant may be pumped into the condenser for repairs on the low-pressure side of the system. Otherwise, the refrigerant may have to be recovered for major repairs. A recovery system and procedures similar to the push-pull system described above is used.

The recovery of high-pressure refrigerants is much like recovering low-pressure refrigerant by means of a push-pull system that will handle high-pressure refrigerants. The same rules apply for charging refrigerant into a high-pressure machine. Make sure the pressure in the machine is above the freezing point of the refrigerant before allowing liquid to enter the machine, or a chiller tube may be frozen and broken.

50.18 REFRIGERANT SAFETY

Large equipment rooms become places where refrigerant vapors can gather should there be a leak. These vapors are all heavier than air and will collect without a technician noticing them. They are colorless, almost odorless, and tasteless. The vapors are not all toxic, but will gradually displace the oxygen in the room. Without notice, the technician may just faint and there may be no one to notice. This would be fatal. These large equipment rooms now have detectors that detect minute parts of the refrigerant vapor and sound an alarm. When the alarm sounds, the technician should evacuate the room and only go back in with breathing gear on. **Figure 50–31** shows two of these breathing devices in a yellow case where they can be readily seen—one is open and the other is closed.

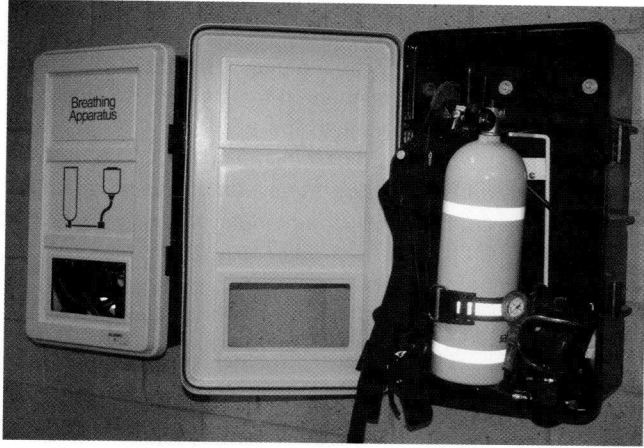

Figure 50–31   This breathing apparatus is mounted where a technician can easily reach it and put it on in case of a refrigerant leak alarm. *Photo by John Tomczyk. Courtesy Ferris State University*

50.19   SERVICE TECHNICIAN CALLS

SERVICE CALL 1

A building manager calls and reports that the building is becoming hot. *The problem is that the cooling tower fan has a broken belt and the fan is off. The chiller is off because of high head pressure.*

The technician arrives and goes to the basement where the chiller is located. It is a reciprocating water-cooled chiller. The chiller is off, and the technician looks around for the problem. The control panel is opened, and it is noticed that the reset button on the high-pressure control is out, signaling that the control needs resetting. The pressure drop across the condenser is normal, and the water temperature is normal (because the chiller has been off, the tower is able to lower the temperature without the fan). The technician resets the button, and the compressor starts. As the technician watches, the compressor starts to load up. Everything runs normally for about 10 min, and then the compressor sounds as though it is straining. The technician feels the condenser water line entering the condenser. It is warmer than hand temperature. The cooling tower water is too hot. The technician shuts the compressor off rather than letting it shut off with the high-pressure control.

The technician goes to the roof to look at the cooling tower, fully expecting to see it full of algae because the management does not require regular maintenance. When on the roof, it is noticed that the fan belt to the cooling tower fan is broken. The disconnect to the fan is shut off, locked out, and tagged for safety. The numbers from the old belt are written down, and the technician goes to the truck for a new one. While at the truck, the technician gets a grease gun for greasing the fan and motor bearings at the cooling tower. The belt is replaced, and the tower fan is turned on. It starts to turn, and the technician can see from the moist air leaving the tower that the water is loaded with heat.

The technician goes back to the basement and touches the water line leading from the tower to the condenser and notices that the water has cooled back to below hand temperature. The compressor is restarted. The technician stays with the compressor for about 30 min to make sure that it will stay on, and it does. This time is used for lubricating pump and fan motors and in general looking for any other potential problems.

SERVICE CALL 2

A building owner reports that the building is hot, and the chiller is off. The chiller is a low-pressure R-11 chiller and has an air leak. *The building maintenance personnel have been operating the purge for several days to remove air that is leaking and are waiting for the weekend to leak test the machine. The purge water has been valved off and the purge condenser is not operable, causing high purge pressure and releasing refrigerant.*

The technician arrives and goes to the equipment room where the chiller is located. The signal light indicates that the chiller is off because of freeze protection. The technician looks around and notices the purge pressure is very high as the purge is still running on manual. On inspection, it is noticed that the purge relief valve is seeping refrigerant. The technician then checks the water circuit to the purge and discovers that there is no water cooling the condenser in the purge. The valve is opened to the purge water condenser. This system uses chilled water, which is now warmed up to 80°F, but will still condense the refrigerant in the purge.

The technician then starts the compressor in the 40% mode and watches the suction pressure; it is not too low, so the load is advanced to 60%. The suction pressure begins to drop. The technician looks in the sight glass at the evaporator for the refrigerant level, and it is low. There is some refrigerant at the sight glass, so the technician gets set up to add refrigerant. The refrigerant drum is piped to the charging valve and the valve is opened. Meanwhile the technician places a thermometer in the well in the evaporator to check the refrigerant temperature. The chilled-water temperature leaving the chiller is down to 55°F, and the thermometer in the refrigerant reads 43°F. This is a 12°F approach, and it should be 7°F according to the log sheet that was left at start-up. Two hundred pounds of refrigerant is added before the compressor is operated at full load, and the refrigerant approach is lowered to 7°F. The technician is satisfied with the system charge. The chilled-water temperature pulls down to 45°F and the compressor begins to unload in about 30 min. The purge pressure is running normally and not relieving continuously as it was. The technician leaves and will return for the leak check later.

SERVICE CALL 3

A building manager reports that the chiller is off, and the building is hot. *The problem is that the flow switch in the condenser water circuit has a broken paddle, and the field control circuit is not passing power to the chiller and telling it to start.*

The technician arrives and goes straight to the chiller. This chiller has a ready light on the control panel that indicates when the field circuit is calling for cooling. The light is off. The technician first looks to see what the chilled-water temperature is. It is 80°F; the controller is calling for cooling. The technician then checks the water gages that register the pressure drop across the chiller and the condenser. The correct pressure drop registers across both so there is water flow. The technician then goes to the wiring diagram for the system and looks to see what other controls are in this circuit. The flow switches and the pump interlocks are all in the circuit. It is a matter of tracing the wiring down to see what has caused the chiller to shut down.

The technician realizes that the flow switches are likely to be a problem because there is a lot of movement of these devices since they are located in a moving water stream. The compressor switch is turned off so the compressor will not try to start during the control testing. The technician removes the cover from the chilled-water flow switch and checks voltage across it. There is no voltage.

NOTE: If the switch were open, there would be voltage. The cover is replaced and the same test is performed on the condenser-water flow switch and there is voltage. This flow switch is open. The technician jumps the switch with a jumper and the light on the chiller panel lights up. This is the problem.

The technician leaves the jumper on the flow switch and starts the chiller. Flow in the condenser water circuit is not as critical as it would be in a chilled-water circuit. If the water flow were to stop in the condenser circuit, the compressor would shut down because of high pressure. NOTE: The chilled-water flow switch should not be jumped with the compressor running because you could freeze the chiller.

The technician explains to the building manager that the flow switch paddle must be replaced. The manager requests this be done after hours so the technician leaves and will return at 5:00 PM with the parts to repair the switch.

The technician returns to the equipment room to make the repair. The chiller is off for the day because of the energy management time clock. The condenser water pump is turned off and locked. The valves on either side of the condenser system are valved off and the water is drained from the condenser. The flow switch is removed, repaired, and replaced. The water valves are turned on, and the condenser system is allowed to fill. Air is vented out of the water system at the high places in the piping. The technician then turns the water pump on and verifies that the flow switch is passing power. By leaving the meter leads on the flow switch and shutting off the water, it is verified the flow switch will also shut the system off if the water flow is stopped. The system is made ready to run the next morning when the time clock calls for cooling, and the technician leaves.

SERVICE CALL 4

A building manager telephones and reports that the centrifugal chiller in the basement of the building is making a screaming noise. He went to the door and was scared to go in. *The problem is that the cooling tower strainer is stopped up and restricting the water flow to the condenser. The centrifugal has high head pressure causing a surge, which produces a loud screaming sound but is normally otherwise harmless.*

The technician arrives and from the parking lot can hear the compressor screaming. When the technician enters the room, the machine capacity control is turned down to 40% to reduce the load. The machine has high head pressure.

A look at the pressure gages across the condenser shows that there is reduced water flow, so the technician proceeds to the cooling tower on the roof.

A look at the cooling tower basin shows what has happened; there are tree leaves in the cooling tower basin. The technician removes the cover from the cooling tower and removes the leaves from the strainer. It can be seen immediately that the water is flowing faster.

The technician replaces the cover to the tower basin, goes to the basement, and turns the capacity control up to 100%.

The technician then talks to the manager about cleaning the cooling tower after hours when the building air conditioning is off and gets permission to do this.

That afternoon, the technician arrives, and the system is off because of the time clock in the energy management system. The tower is cleaned and washed out. There are many small leaves in the tower. The technician then goes to the basement and prepares to check the condenser for leaves. The condenser circuit is valved off and drained. Then the technician removes the marine water box cover on the piping inlet end of the condenser where any trash would accumulate. There are small leaves in the condenser that have passed through the cooling tower strainer. These are cleaned out. While the water-box cover is removed, the technician uses a small penlight and examines the tubes. There is no scale or fouling of the tubes, which is good news.

The water-box cover is replaced, and the valves are opened to the condenser. The technician waits a few minutes for the tower to refill with water after the condenser has been filled and then starts the condenser water pump in the manual mode and verifies there is water flow. This is a necessary step if the condenser is large enough to drain most of the water from the tower; the tower must have time to refill before starting the pump. Now the technician knows that when the machine starts in the morning there should be no cooling tower water problems.

SERVICE CALL 5

A building maintenance person calls and says that the chiller is off. A light is lit that says freeze control. He wants to know if it should be reset. It is a mild day without much need for air conditioning. The dispatcher tells him no; a technician will be over within 30 min. *The problem is that the main controller is not calibrated and is drifting, causing the water temperature to fall too low.*

When the technician arrives, the maintenance person leads the way to the chiller, which is on the roof in a penthouse. The technician checks the water flow through the chiller by comparing the pressure in and out to the log sheet and it is correct. The freeze control is reset, and the chiller compressor starts. When it starts to lower the water temperature, the technician watches the controller. It is a pneumatic controller and as the water temperature drops, the air pressure should drop from the 15 psig furnished. When the water temperature reaches 43°F, the pneumatic air pressure should be 7 psig. When the water temperature reaches 41°F, the air pressure should be 3 psig, and the chiller should shut off. This does not happen; the chiller continues to run and the water temperature reaches 39°F without shutting off. The technician then adjusts the controller to the correct calibration. The chiller shuts off.

The technician waits for the chiller to restart when the building water warms up; this takes about 30 min. When it is established that the chiller is operating correctly, the technician leaves.

SERVICE CALL 6

A building manager calls and reports that the chiller is off, and the building is getting hot. This is a centrifugal chiller. *The problem is that the building technician has been adjusting the controls and has adjusted the load-limiting control, trying to get the motor to operate at more capacity to reduce the building temperature faster. The compressor was pulling too much current, which tripped the overload device.*

The technician arrives and goes to the equipment room with the building technician. The building technician does not tell the technician that the load-limiting control has been adjusted. The technician checks the field control circuit and discovers that all field controls are calling for cooling. A further check shows that an overload device has tripped. The technician knows that something has happened and caution should be used.

The technician shuts off the power to the compressor starter and checks the motor windings for a ground or a short circuit from winding to winding. The motor seems fine so power is restored. There is not much else to do except to start the motor and observe.

The technician starts the compressor in the 40% mode. When the compressor starts to load up, the current is observed. Start-up data show that the compressor should draw 600 A at full load and 240 A at 40%. The compressor is pulling 264 A; this is too much so the technician turns the load limiter back to 240 A and then turns the limiter up to 100%. The compressor begins to load up and reaches 600 A and throttles to hold this current. The technician then turns to the building maintenance technician and asks whether anyone has adjusted the demand limiter. The building technician admits that it has been done. The technician is told that these controls should not be adjusted,

because they are reliable. The technician then uses nail polish to mark the setting on the control so it will be known if it is adjusted again.

The technician realizes that the controls belong to the building owner, but they should be adjusted only by a qualified technician.

SERVICE CALL 7

A customer calls and says the building is becoming warm. The customer says that water is still flowing over the cooling tower, but there is no cooling. The unit is a 300-ton centrifugal with a hermetic compressor. *The problem is the compressor motor is burned and will need to be replaced or rewound.*

The technician goes straight to the equipment room on the 10th floor of an office building. It is noticed that the building is hot so there is a problem with the chiller.

The technician checks the field control circuit and discovers that all signals are correct; it is calling for cooling. The technician then goes to the starter and discovers that there is no line voltage to the starter. The technician goes to the main power supply and discovers the breaker is tripped. There must be something very wrong.

The technician locks out and tags the breaker and uses a Megger (megohmmeter) to check the motor leads in the starter to ground and discovers a circuit to ground. There is always a possibility that the problem may be in the wiring between the starter and the motor terminals on the motor so the technician removes the leads from the motor terminals. The motor is checked at the terminals without the wiring and it shows a circuit to ground.

The technician goes to the building management for advice as to the next step. It is explained that the motor will have to be removed from the equipment room for replacement or rebuild. ♻ Because the motor is a hermetically sealed motor, the refrigerant will have to be recovered. ♻ The technician explains that there are two possibilities for making the repair. The motor can be replaced with a factory motor, which may take several days for delivery, or the motor may be rebuilt at a local rebuild shop that is approved to work with hermetic motors. Building management gives permission to proceed with the job of a rebuild to save time because the building must have air conditioning. Like many modern buildings, no windows open and the computers rely on the air conditioning.

The technician calls the home office and requests backup help and the recommended tools for the job. This would include a hoist to lift the motor and a dolly to transport it on to the elevator and out to the truck, which is equipped with a lift gate.

While the helper is getting the tools together, the technician prepares to pull the motor. ♻ The refrigerant is recovered from the machine using the recovery system brought by the helper. Water in the chilled-water circuit and condenser water are circulated through the machine during recovery to prevent the tubes from freezing and to aid in boiling the refrigerant out of the machine. The refrigerant is stored in refrigerant drums. The machine uses R-11 and the refrigerant will be checked for acid content and recycled for reuse in the machine if needed. When the refrigerant is recovered down to 28 in. of vacuum in the machine, the technician breaks the vacuum with dry nitrogen that has been brought by the helper. Breaking the vacuum with dry nitrogen will minimize oxidation inside the machine when it is opened to the atmosphere. The machine is made of steel inside and it will quickly rust.

The technician then removes the wires from the motor terminal box and starts to disconnect the compressor from the motor. When the motor is free to move, the hoist is set up and the motor is set on a rolling dolly to transport it to the truck. The motor is then rolled onto the lift gate, lifted to the back of the truck, and secured. The helper takes the motor to the motor shop for rebuild. This will take about 48 hours. It is now 2:30 on Tuesday.

The technician goes back to the equipment room and prepares the equipment for when the motor is returned. A sample of the refrigerant and a sample of the refrigerant oil from the compressor sump are pulled and shipped by next-day carrier to a chemical analysis firm. They will be analyzed and the report can be called in by Wednesday afternoon to determine what must be done about the refrigerant. The refrigerant and oil sample do not smell as though they have very much acid, but the analysis will reveal the facts.

The technician checks the contacts in the starter. There must have been quite a surge of current when the motor grounded. The contacts are good and do not require replacement.

The oil heater is turned off, locked, and tagged, and the oil is removed. The technician then cleans all gasket sealant from all compressor flanges and is now ready for the motor to be returned. Time has been saved by doing all of this before the motor is brought back.

Thursday morning the technician checks with the rebuild shop and is told the motor will be ready by 11:30 AM. The technician arranges for the helper to pick up the motor and a compressor gasket kit. They meet at the building at 1:00 PM and proceed to install the motor. The motor is moved to the equipment room and lifted back into place. All gaskets are installed and everything is tightened. This system has a history of being very tight so the only leak checking that needs to be performed is on the fittings, including the gasketed flanges that were removed and in the area where the work was done.

When all is ready, the machine is pressured with nitrogen to 10 psig with a trace of R-22 refrigerant that is used to detect leaks. The machine is leak checked and found to be good. The nitrogen and refrigerant are exhausted, and it is time to pull the vacuum. The motor leads are connected to the motor terminals.

The vacuum pump is installed and started. It will take about 5 hours to pull the vacuum down to 0 in. Hg on a mercury manometer and it is 5:30 PM. The technician

leaves the job and decides to come back at 10:30 PM to check the vacuum. If the vacuum can be verified this evening and the vacuum pump shut off for 8 hours, the machine can be started in the morning if all is well. The technician checks the vacuum at about 11:00 PM, and the manometer reads 0 in. Hg vacuum. The vacuum pump is valved off, and the technician leaves to return in the morning.

When the technician arrives at 7:30 AM, the mercury manometer is still reading 0 in. Hg vacuum so the machine is leak free, and it is time to charge the system. The refrigerant and oil report came back "OK" so no extra precautions need to be taken with the refrigerant. The technician starts the chilled-water pump and the condenser water pump in preparation to start-up. A drum of refrigerant is connected to the machine at the drum vapor valve. The valve is slowly opened to the machine and vapor is allowed to enter the machine until the pressure is above that corresponding to freezing. This is important because if liquid refrigerant is allowed to enter the machine at such a low pressure, the tubes may freeze. The chilled water is circulating, but if there is one tube that does not have circulation, this tube may freeze. The pressure in the machine must be above 17 in. Hg vacuum to prevent freezing while charging liquid.

When the machine pressure reaches 17 in. Hg vacuum, the technician then starts charging liquid refrigerant. While the technician is charging refrigerant, the helper pulls the oil in the oil sump using the machine vacuum. When the oil is charged, the oil heater is turned on to bring the oil up to temperature.

The technician is not able to charge all of the refrigerant into the machine using the vacuum because the chilled-water circuit warms the refrigerant, and pressure is soon raised. The technician then changes over and pushes the remainder of the refrigerant into the machine by pressuring the drum with nitrogen on top of the liquid and drawing liquid from the bottom of the drum, **Figure 50–32.**

The technician starts the purge pump to remove any nitrogen that may have entered the system while charging, even though the technician was careful and could see no nitrogen in the clear charging hose while charging. The purge pressure does not rise, so the machine should be free of nitrogen. The technician has closed the main breaker and is ready to start the machine. The machine is started and quickly turned up to 100% to reduce the building temperature as quickly as possible. There is no apparent reason that the motor developed a circuit to ground; the motor rebuild shop could not find any problem so the motor failure is accounted for as random failure.

The chiller is quickly pulling the building water down to 45°F leaving-water temperature. The technician stays with the chiller for 2 hours and completes entries (including the motor current) in an operating log. When it is determined that the system is functioning properly the technician leaves.

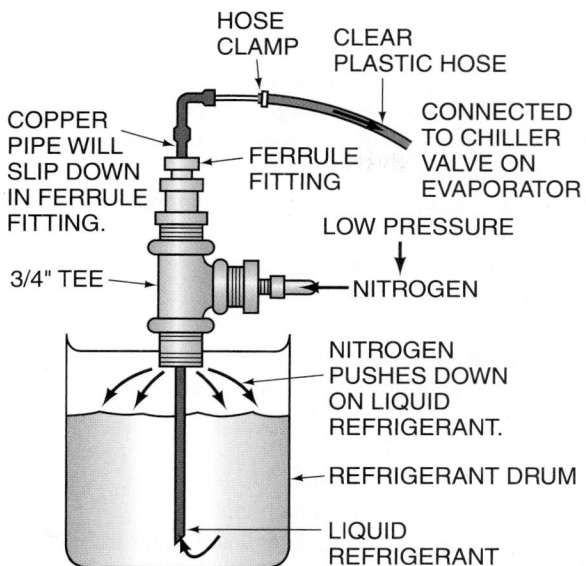

Figure 50–32   Using nitrogen to charge refrigerant into a low-pressure chiller. The clear plastic tube will tell the technician when any nitrogen is passing into the system.

SERVICE CALL 8

The technician is called to do routine maintenance on a reciprocating chiller. *This chiller has not had maintenance in several years. The last technician did not take proper care of the service valves and the suction service cap is rusted tight to the valve body.*

When the technician arrives, it is determined that the suction and discharge pressure should be taken to determine that the chiller is operating correctly. It is an air-cooled chiller using R-22. The technician removes the protective cap from the discharge service line and installs a high-pressure gage. The head pressure is 225 psig and the outside temperature is 80°F so the head pressure is good (80 + 30 = 110°F condensing temperature).

The technician then attempts to remove the cap from the suction service valve and discovers the cap is rusted tight. The technician then uses a soft face hammer and taps hard on all of the flats on the cap then uses a hex-jaw wrench, **Figure 50–14**, to loosen the cap. It still will not turn. Then a heavier steel hammer is used to tap the flat surfaces on the valve and it still will not loosen. The technician then uses the hammer to tap the handle of the wrench and sees the cap begin to turn. It is not good practice to hit a wrench handle with a hammer, but sometimes this is necessary. If the valve cap is struck too hard, it may split because it is cast iron. Cast iron can be brittle and breaks easily. Spraying a penetrating oil would help, but doing that will gather dust and lead the next technician to believe there is a leak.

When the valve cap is removed, the technician sees there has been moisture inside the cap for a long time. The valve stem is really rusty with scale buildup on it.

The valve stem is cleaned with a fine sand tape until there are no more high spots on the stem. The packing gland is then slightly loosened and the valve stem turned forward until a gage reading is obtained. The suction pressure reads 75 psig. The chilled water in is 50°F, and out, 45°F. The chiller would normally have a 10°F difference and the suction pressure would normally be 70 psig. The chiller has the correct water temperature so it is running at half capacity. It is operating at reduced capacity to fit the building's load. All is well.

The technician shuts the system off for about 30 minutes to give the chilled-water temperature a chance to warm up—at which point the technician will start the system up and run it at full load. Meanwhile, the pump motor is greased and the inlet to the condenser is examined to see if it needs cleaning. The fan motors are examined and they are permanently lubricated motors.

When the chiller is started back up, the suction pressure reduces to 70 psig, the compressor is now pulling full-load amperage, and the sight glass to the expansion valve is full. The chiller is running like it should.

The technician back seats both service valves and tightens the packing gland nut, then cleans the inside of the valve cap of the suction service valve to remove the rust. He then takes a small wire brush and cleans the threads on the valve body. The technician wipes the valve stem and sprays oil-free silicone spray and on it and on the inside of the valve cap and turns the cap onto the valve body. It is tightened with the square jaw wrench.

The technician then uses an electronic leak detector and leak checks all around the service valves where service has been performed.

SUMMARY

- The operation of chilled-water air-conditioning systems involves starting, running, and stopping the chiller systems in an orderly manner.
- When starting a chilled-water system, first check for sufficient chilled-water flow. If the system is water-cooled, check for condenser water flow. The interlock circuit through the contactors must be satisfied; the cooling tower fan should start, followed by the compressor. There may be a time delay before the compressor starts.
- The reciprocating, scroll, and rotary screw compressors may be either air or water cooled.
- The centrifugal chiller will have a separate oil lubrication system. It should be verified that the lubrication system is functioning satisfactorily before starting the compressor.
- After the chiller is on and operating, the technician should observe the operation for a period of time to ensure that it is all operating correctly.
- Water-cooled chillers must have water treatment to prevent minerals and algae from collecting and forming.

- Inspecting and cleaning the water tower should be done on a regular basis.
- Water-cooled condenser tubes should be checked at least annually. If they are found to be not clean or to have a scale buildup, they must be cleaned either with a brush or chemically.
- Condenser and evaporator tubes may be checked for defects with an eddy current test instrument.
- Absorption chilled-water system and compression cycle system start-up are similar in many respects. Condenser and chilled-water flow must be established before the chiller is started. The cooling tower water must not be too cold or the Li-Br may crystallize.
- Absorption chillers must be observed for proper operation on a more regular basis than compression cycle chillers because they can be more intricate.
- The purge system may require more maintenance than other systems on the absorption chiller. The vacuum pump oil should be changed regularly.
- Technicians who maintain chillers should stay in contact with the manufacturers of the equipment for the latest information advice, and the schedule for service schools.

REVIEW QUESTIONS

1. How does the compressor lubrication system differ for reciprocating and centrifugal compressors?
2. True or False: Most reciprocating compressors have crankcase heaters.
3. True or False: All reciprocating compressors start up fully loaded.
4. Describe how a centrifugal compressor starts up at part load and then goes to full load.
5. What system component must be started before the compressor in a chilled-water system?
6. When condenser tubes become fouled, they may be cleaned by
   A. using a steel brush.
   B. using a brass brush.
   C. using a chisel and hammer.
   D. running a mixture of sand and water through them.
7. Describe what happens when an absorption chiller is operated when the cooling tower water is too cold.

APPENDIX A—TEMPERATURE CONVERSION TABLE

	TEMPERATURE CONVERSION TABLE				
°F	Temperature to be Converted	°C	°F	Temperature to be Converted	°C
−76.0	−60	−51.1	23.0	− 5	−20.6
−74.2	−59	−50.6	24.8	− 4	−20.0
−72.4	−58	−50.0	26.6	− 3	−19.4
−70.6	−57	−49.4	28.4	− 2	−18.9
−68.8	−56	−48.9	30.2	− 1	−18.3
−67.0	−55	−48.3	32.0	0	−17.8
−65.2	−54	−47.8	33.8	1	−17.2
−63.4	−53	−47.2	35.6	2	−16.7
−61.6	−52	−46.7	37.4	3	−16.1
−59.8	−51	−46.1	39.2	4	−15.6
−58.0	−50	−45.6	41.0	5	−15.0
−56.2	−49	−45.0	42.8	6	−14.4
−54.4	−48	−44.4	44.6	7	−13.9
−52.6	−47	−43.9	46.4	8	−13.3
−50.8	−46	−43.3	48.2	9	−12.8
−49.0	−45	−42.8	50.0	10	−12.2
−47.2	−44	−42.2	51.8	11	−11.7
−45.4	−43	−41.7	53.6	12	−11.1
−43.6	−42	−41.1	55.4	13	−10.6
−41.8	−41	−40.6	57.2	14	−10.0
−40.0	−40	−40.0	59.0	15	− 9.4
−38.2	−39	−39.4	60.8	16	− 8.9
−36.4	−38	−38.9	62.6	17	− 8.3
−34.6	−37	−38.3	64.4	18	− 7.8
−32.8	−36	−37.8	66.2	19	− 7.2
−31.0	−35	−37.2	68.0	20	− 6.7
−29.2	−34	−36.7	69.8	21	− 6.1
−27.4	−33	−36.1	71.6	22	− 5.6
−25.6	−32	−35.6	73.4	23	− 5.0
−23.8	−31	−35.0	75.2	24	− 4.4
−22.0	−30	−34.4	77.0	25	− 3.9
−20.2	−29	−33.9	78.8	26	− 3.3
−18.4	−28	−33.3	80.6	27	− 2.8
−16.6	−27	−32.8	82.4	28	− 2.2
−14.8	−26	−32.2	84.2	29	− 1.7
−13.0	−25	−31.7	86.0	30	− 1.1
−11.2	−24	−31.1	87.8	31	− 0.6
− 9.4	−23	−30.6	89.6	32	0.0
− 7.6	−22	−30.0	91.4	33	0.6
− 5.8	−21	−29.4	93.2	34	1.1
− 4.0	−20	−28.9	95.0	35	1.7
− 2.2	−19	−28.3	96.8	36	2.2
− 0.4	−18	−27.8	98.6	37	2.8
1.4	−17	−27.2	100.4	38	3.3
3.2	−16	−26.7	102.2	39	3.9
5.0	−15	−26.1	104.0	40	4.4
6.8	−14	−25.6	105.8	41	5.0
8.6	−13	−25.0	107.6	42	5.6
10.4	−12	−24.4	109.4	43	6.1
12.2	−11	−23.9	111.2	44	6.7
14.0	−10	−23.3	113.0	45	7.2
15.8	− 9	−22.8	114.8	46	7.8
17.6	− 8	−22.2	116.6	47	8.3
19.4	− 7	−21.7	118.4	48	8.9
21.2	− 6	−21.1	120.2	49	9.4

°F	Temperature to be Converted	°C	°F	Temperature to be Converted	°C
122.0	50	10.0	208.4	98	36.7
123.8	51	10.6	210.2	99	37.2
125.6	52	11.1	212.0	100	37.8
127.4	53	11.7	213.8	101	38.3
129.2	54	12.2	215.6	102	38.9
131.0	55	12.8	217.4	103	39.4
132.8	56	13.3	219.2	104	40.0
134.6	57	13.9	221.0	105	40.6
136.4	58	14.4	222.8	106	41.1
138.2	59	15.0	224.6	107	41.7
140.0	60	15.6	226.4	108	42.2
141.8	61	16.1	228.2	109	42.8
143.6	62	16.7	230.0	110	43.3
145.4	63	17.2	231.8	111	43.9
147.2	64	17.8	233.6	112	44.4
149.0	65	18.3	235.4	113	45.0
150.8	66	18.9	237.2	114	45.6
152.6	67	19.4	239.0	115	46.1
154.4	68	20.0	240.8	116	46.6
156.2	69	20.6	242.6	117	47.2
158.0	70	21.1	244.4	118	47.7
159.8	71	21.7	246.2	119	48.3
161.8	72	22.2	248.0	120	48.8
163.4	73	22.8	257.0	125	51.7
165.2	74	23.3	266.0	130	54.4
167.0	75	23.9	275.0	135	57.2
168.8	76	24.4	284.0	140	60.0
170.6	77	25.0	293.0	145	62.8
172.4	78	25.6	302.0	150	65.6
174.2	79	26.1	311.0	155	68.3
176.0	80	26.7	320.0	160	71.1
177.8	81	27.2	329.0	165	73.9
179.6	82	27.8	338.0	170	76.7
181.4	83	28.3	347.0	175	79.4
183.2	84	28.9	356.0	180	82.2
185.0	85	29.4	365.0	185	85.0
186.8	86	30.0	374.0	190	87.8
188.6	87	30.6	383.0	195	90.6
190.4	88	31.1	392.0	200	93.3
192.2	89	31.7	401.0	205	96.1
194.0	90	32.2	410.0	210	98.9
195.8	91	32.8	413.6	212	100.0
197.6	92	33.3	428.0	220	104.4
199.4	93	33.9	446.0	230	110.0
201.2	94	34.4	464.0	240	115.6
203.0	95	35.0	482.0	250	121.1
204.8	96	35.6	500.0	260	126.7
206.6	97	36.1			

Example 1. To find 37°F as a Celsius equivalent, find 37 in the Temperature to be Converted column and read the value in the °C column, which is 2.8°C.

Example 2. To find 75°C as a Fahrenheit equivalent, find 75 in the Temperature to be Converted column and read the value in the °F column, which is 167.0°F.

APPENDIX B—ELECTRICAL SYMBOLS CHART

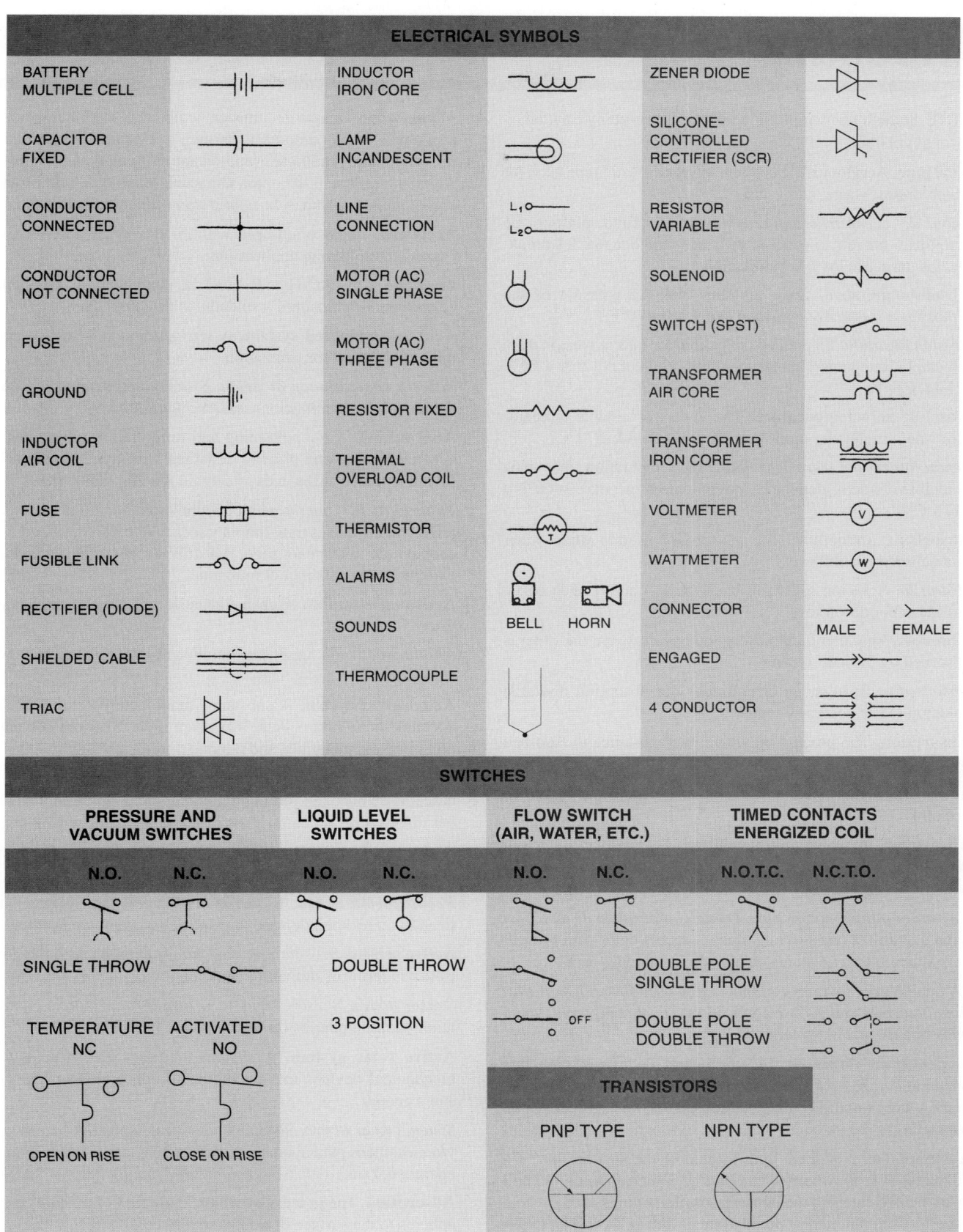

ELECTRICAL SYMBOLS

BATTERY MULTIPLE CELL		INDUCTOR IRON CORE	ZENER DIODE
CAPACITOR FIXED		LAMP INCANDESCENT	SILICONE-CONTROLLED RECTIFIER (SCR)
CONDUCTOR CONNECTED		LINE CONNECTION	RESISTOR VARIABLE
CONDUCTOR NOT CONNECTED		MOTOR (AC) SINGLE PHASE	SOLENOID
FUSE		MOTOR (AC) THREE PHASE	SWITCH (SPST)
GROUND		RESISTOR FIXED	TRANSFORMER AIR CORE
INDUCTOR AIR COIL		THERMAL OVERLOAD COIL	TRANSFORMER IRON CORE
FUSE		THERMISTOR	VOLTMETER
FUSIBLE LINK		ALARMS	WATTMETER
RECTIFIER (DIODE)		SOUNDS — BELL HORN	CONNECTOR — MALE FEMALE
SHIELDED CABLE		THERMOCOUPLE	ENGAGED
TRIAC			4 CONDUCTOR

SWITCHES

PRESSURE AND VACUUM SWITCHES

N.O.	N.C.

SINGLE THROW

TEMPERATURE NC — OPEN ON RISE

ACTIVATED NO — CLOSE ON RISE

LIQUID LEVEL SWITCHES

N.O.	N.C.

DOUBLE THROW

3 POSITION

FLOW SWITCH (AIR, WATER, ETC.)

N.O.	N.C.

OFF

TIMED CONTACTS ENERGIZED COIL

N.O.T.C.	N.C.T.O.

DOUBLE POLE SINGLE THROW

DOUBLE POLE DOUBLE THROW

TRANSISTORS

PNP TYPE   NPN TYPE

1335

Glossary

ABS pipe. Acrylonitrile-butadiene-styrene plastic pipe used for water, drains, waste, and venting.

Tubo de acronitrilo-butadieno-estireno. Tubo plástico de acronitrilo-butadieno-estireno utilizado para el agua, los drenajes, los desperdicios y la ventilación.

Absolute pressure. Gage pressure plus the pressure of the atmosphere, normally 14.696 at sea level at 70°F.

Presión absoluta. La presión del calibrador más la presión de la atmósfera, que generalmente es 14696 al nivel del mar a 70°F (21.11°C).

Absolute zero temperature. The lowest obtainable temperature where molecular motion stops, –460°F and –273°C.

Temperatura del cero absoluto. La temperatura más baja obtenible donde se detiene el movimiento molecular, –460°F y –273°C.

Absorbent (attractant). The salt solution used to attract water in an absorption chiller.

Hidrófilo. Solución salina utilizada para atraer el agua en un enfriador por absorción.

Absorber. That part of an absorption chiller where the water is absorbed by the salt solution.

Absorbedor. El lugar en el enfriador por absorción donde la solución salina absorbe el agua.

Absorption. The process by which one substance is absorbed by another.

Absorción. Proceso mediante el cual una sustancia es absorbida por otra.

Absorption air-conditioning chiller. A system using a salt substance, water, and heat to provide cooling for an air-conditioning system.

Enfriador por absorción para acondicionamiento de aire. Sistema que utiliza una sustancia salina, agua y calor para proveer enfriamiento en un sistema de acondicionamiento de aire.

"A" coil. An evaporator coil that can be used for upflow, downflow, and horizontal-flow applications. It actually consists of two coils shaped like a letter "A."

Serpentín en forma de "A." Serpentín de evaporación que puede utilizarse para aplicaciones de flujo ascendente, descendente y horizontal. En realidad, consiste en dos serpentines en forma de "A."

Accumulator. A storage tank located in the suction line of a compressor. It allows small amounts of liquid refrigerant to boil away before entering the compressor. Sometimes used to store excess refrigerant in heat pump systems during the winter cycle.

Acumulador. Tanque de almacenaje ubicado en el conducto de aspiración de un compresor. Permite que pequeñas cantidades de refrigerante líquido se evaporen antes de entrar al compresor. Algunas veces se utiliza para almacenar exceso de refrigerante en sistemas de bombas de calor durante el ciclo de invierno.

Acetylene. A gas often used with air or oxygen for welding, brazing, or soldering applications.

Acetileno. Gas usado con frecuencia, junto con aire u oxígeno, en trabajos de soldadura y soldaduras de cobre.

Acid-contaminated system. A refrigeration system that contains acid due to contamination.

Sistema contaminado de ácido. Sistema de refrigeración que, debido a la contaminación, contiene ácido.

ACR tubing. Air-conditioning and refrigeration tubing that is very clean, dry, and normally charged with dry nitrogen. The tubing is sealed at the ends to contain the nitrogen.

Tubería ACR. Tubería para el acondicionamiento de aire y la refrigeración que es muy limpia y seca, y que por lo general está cargada de nitrógeno seco. La tubería se sella en ambos extremos para contener el nitrógeno.

Activated alumina. A chemical desiccant used in refrigerant driers.

Alúmina activada. Disecante químico utilizado en secadores de refrigerantes.

Activated charcoal. A substance manufactured from coal or coconut shells into pellets. It is often used to adsorb solvents, other organic materials, and odors.

Carbón activado. Sustancia fabricada utilizando carbón o cáscaras de coco, en forma de gránulos. Se emplea para adsorber disolventes, así como otros materiales orgánicos y olores.

Active recovery. Recovering refrigerant with the use of a recovery machine that has its own built-in compressor.

Recuperación activa. El hecho de recuperar refrigerante utilizando un recuperador que dispone de su propio compresor.

Active sensor. Sensors that send information back to the controller in terms of milliamps (mA) or volts.

Sensor activo. Sensores que devuelven información al controlador en miliamperios (mA) o voltios.

Active solar system. A system that uses electrical and/or mechanical devices to help collect, store, and distribute the sun's energy.

Sistema solar activo. Sistema que utiliza dispositivos eléctricos y/o mecánicos para ayudar a acumular, almacenar y distribuir la energía del sol.

Adsorption. The process by which a thin film of a liquid or gas adheres to the surface of a solid substance.

Adsorción. Proceso mediante el cual una fina capa de líquido o gas se adhiere a la superficie de una sustancia sólida.

Air-acetylene. A mixture of air and acetylene gas that when ignited is used for soldering, brazing, and other applications.

Aire-acetilénico. Mezcla de aire y de gas acetileno que se utiliza en la soldadura, la broncesoldadura y otras aplicaciones al ser encendida.

Air conditioner. Equipment that conditions air by cleaning, cooling, heating, humidifying, or dehumidifying it. A term often applied to comfort cooling equipment.

Acondicionador de aire. Equipo que acondiciona el aire limpiándolo, enfriándolo, calentándolo, humidificándolo o deshumidificándolo. Término comúnmente aplicado al equípo de enfriamiento para comodidad.

Air conditioning. A process that maintains comfort conditions in a defined area.

Acondicionamiento de aire. Proceso que mantiene condiciones agradables en un área definida.

Air Conditioning and Refrigeration Institute (ARI). A nonprofit association that regulates equipment manufacturers and rates the capacity of equipment.

El Instituto del Aire Acondicionado (ARI en inglés). Una asociación sin fines lucrativos que regula los fabricantes del equipo y que establece la capacidad de dicho equipo.

Air-cooled condenser. One of the four main components of an air-cooled refrigeration system. It receives hot gas from the compressor and rejects heat to a place where it makes no difference.

Condensador enfriado por aire. Uno de los cuatro componentes principales de un sistema de refrigeración enfriado por aire. Recibe el gas caliente del compresor y dirige el calor a un lugar donde no afecte la temperatura.

Air friction chart. A chart used to determine proper round duct sizes.

Gráfica de fricción del aire. Gráfica utilizada para determinar los tamaños correctos de conductos redondos.

Air gap. The clearance between the rotating rotor and the stationary winding on an open motor. Known as a vapor gap in a hermetically sealed compressor motor.

Espacio de aire. Espacio libre entre el rotor giratorio y el devandado fijo en un motor abierto. Conocido como espacio de vapor en un motor de compresor sellado herméticamente.

Air handler. The device that moves the air across the heat exchanger in a forced air system—normally considered to be the fan and its housing.

Tratante de aire. Dispositivo que dirige el aire a través del intercambiador de calor en un sistema de aire forzado—considerado generalmente como el ventilador y su alojamiento.

Air heat exchanger. A device used to exchange heat between air and another medium at different temperature levels, such as air-to-air, air-to-water, or air-to-refrigerant.

Intercambiador de aire y calor. Dispositivo utilizado para intercambiar el calor entre el aire y otro medio, como por ejemplo aire y aire, aire y agua o aire y refrigerante, a diferentes niveles de temperatura.

Air loop. The heat pump's heating and cooling ducted air system, which exchanges heat with the refrigerant loop.

Circuito de aire. Sistema de tubería del aire de calentamiento y de refrigeración de la bomba de calor, que sirve para intercambiar el calor con el circuito de refrigeración.

Air pressure control (switch). Used to detect air pressure drop across the coil in a heat pump outdoor unit due to ice buildup.

Regulador de la presión de aire (conmutador). Utilizado para detectar una caída en la presión del aire a través de la bobina en una unidad de bomba de calor para exteriores debido a la acumulación de hielo.

Air sensor. A device that registers changes in air conditions such as pressure, velocity, temperature, or moisture content.

Sensor de aire. Dispositivo que registra los cambios en las condiciones del aire, como por ejemplo cambios en presión, velocidad, temperatura o contenido de humedad.

Air shutters. Devices that are placed in an airstream for the purpose of controlling airflow.

Contraventanas del aire. Aparatos que se ponen en un corriente de aire con el propósito de regular el fluir del aire.

Air, standard. Dry air at 70°F and 14.696 psi, at which it has a mass density of 0.075 lb/ft^3 and a specific volume of 13.33 ft^3/lb, ASHRAE.

Aire, estándar. Aire seco a 70°F (21.11°C) y 14696 psi (libra por pulgada cuadrada); a dicha temperatura tiene una densidad de masa de 0075 libra/pies^3 y un volumen específico de 1333 pies^3/libra, ASHRAE.

Air vent. A fitting used to vent air manually or automatically from a system.

Válvula de aire. Accesorio utilizado para darle al aire salida manual o automática de un sistema.

Alcohol. Antifreeze solution used in the water loop of geothermal heat pumps.

Alcohol. Líquido anticongelante que se utiliza en el circuito de agua en bombas de calor geotérmicas.

Algae. A form of green or black, slimy plant life that grows in water systems.

Alga. Tipo de planta legamosa de color verde o negro que crece en sistemas acuáticos.

Algorithm. A computer code or set of instructions to make specific calculations.

Algoritmo. Un código o juego de instrucciones computadora para hacer cálculos específicos.

Alkylbenzene. A popular synthetic lubricant that works best with HCFC-based refrigerant blends. This lubricant can be used with CFC and HCFC refrigerants.

Alquilbenceno. Lubricante sintético muy compatible con mezclas de refrigerantes basados en HCFC. También puede utilizarse con refrigerantes que usan CFC y HCFC.

Allen head. A recessed hex head in a fastener.

Cabeza allen. Cabeza de concavidad hexagonal en un asegurador.

All-weather system. System providing year-round conditioning of the air.

Sistema para todo el año. Sistema que proporciona una aclimatación ambiental todo el año.

Alternating current. An electric current that reverses its direction at regular intervals.

Corriente alterna. Corriente eléctrica que invierte su dirección a intervalos regulares.

Alternative refrigerant. One of the newer refrigerants that are replacing the traditional CFC refrigerants that have been used for many years. Many of these refrigerants have very low ozone depletion and global-warming indices. Some are completely chlorine-free.

Refrigerante alternativo. Cualquiera de los nuevos productos que sirve para sustituir a los refrigerantes basados en CFC que han sido utilizados durante muchos años. Muchos de estos nuevos refrigerantes tienen un índice muy bajo de desgaste de la capa de ozono y de calentamiento de la superficie terrestre. Algunos no utilizan cloro.

Altitude adjustment. An adjustment to a refrigerator thermostat to account for a lower-than-normal atmospheric pressure such as may be found at a high altitude.

Ajuste para elevación. Ajuste al termostato de un refrigerador para regular una presión atmosférica más baja que la normal, como la que se encuentra en elevaciones altas.

Ambient temperature. The surrounding air temperature.

Temperatura ambiente. Temperatura del aire circundante.

American standard pipe thread. Standard thread used on pipe to prevent leaks.

Rosca estándar estadounidense para tubos. Rosca estándar utilizada en tubos para evitar fugas.

Ammeter. A meter used to measure current flow in an electrical circuit.

Amperímetro. Instrumento utilizado para medir el flujo de corriente en un circuito eléctrico.

Amperage. Amount (quantity) of electron or current flow (the number of electrons passing a point in a given time) in an electrical circuit.

Amperaje. Cantidad de flujo de electrones o de corriente (el número de electrones que sobrepasa un punto específico en un tiempo fijo) en un circuito eléctrico.

Ampere. Unit of current flow.

Amperio. Unidad de flujo de corriente.

Analog electronic devices. Devices that generate continuous or modulating signals within a certain control range.

Aparatos electrónicos analógicos. Aparatos que generan señales continuas o modulares adentro de cierto registro de control.

Analog signal. A continuous signal with an infinite number of steps.

Señal analógico. Un señal continuo con un número infinitésimo de pasos.

Analog VOM. A volt-ohm-milliammeter constructed so that the meter indicator is a needle over a printed surface.

VOM analógico. Medidor de voltios-ohmios-miliamperímetros construido de forma que el indicador consiste en una aguja que se mueve encima de una superficie impresa.

Anemometer. An instrument used to measure the velocity of air.

Anemómetro. Instrumento utilizado para medir la velocidad del aire.

Aneroid barometer. An instrument used to measure atmospheric pressure. This barometer uses a closed bellows linked to a needle and can be easily carried from place to place.

Barómetro aneroide. Instrumento utilizado para medir la presión atmosférica. Este tipo de barómetro utiliza un fuelle cerrado unido a una aguja. Es muy fácil de trasladar de un lugar a otro.

Angle valve. Valve with one opening at a 90° angle from the other opening.

Válvua en ángulo. Válvula con una abertura a un ángulo de 90° con respecto a la otra abertura.

Annual Fuel Utilization Efficiency (AFUE). The U.S. Federal Trade Commission requires furnace manufacturers to provide this rating so consumers may compare furnace performances before purchasing.

Eficacia de uso de combustible anual (AFUE en inglés). La Comisión Federal de Comercio (Federal Trade Commission) de los EE.UU. exige que los fabricantes de hornos indiquen este valor con el fin de que los consumidores puedan comparar los rendimientos de los hornos antes de adquirirlos.

Anode. A terminal or connection point on a semiconductor.

Ánodo. Punto de conexión o terminal en un semiconductor.

ANSI. Abbreviation for the American National Standards Institute.

ANSI. Acrónimo de American National Standards Institute (Instituto Nacional Americano de Normas).

Anticipator. Resistance heaters in thermostats that lessen the effects of system lag and overshoot.

Anticipador. Calentador de resistencia en un termostato que reduce los efectos de subida y bajada del sistema.

Antifreeze solution. A solution that will mix with water that has a freezing point much lower than water. Different solutions are mixed with water to lower the freezing point to the desired level.

Solución anticongelante. Una solución que puede mezclarse con agua que tiene un punto de congelar mucho más bajo de lo del agua. Soluciones diferentes están mezcladas con agua para bajar el punto de congelar al nivel deseado.

Approach. The temperature difference between the water leaving the cooling tower and the wet-bulb temperature of the tower's entering air.

Acercamiento. La diferencia de temperatura entre el agua que sale de la torre de enfriar y la temperatura del bulbo mojado del aire que entra en la torre.

Approach temperature. The difference in temperature between the refrigerant and the leaving water in a chilled-water system.

Temperatura de acercamiento. Diferencia en temperatura entre el refrigerante y el agua de salida en un sistema de agua enfriada.

Aromatic oil. A refrigeration mineral oil in which the alkylbenzene lubricant group originates.

Aceite aromático. Aceite mineral de refrigeración del cual se deriva el grupo de lubricantes basados en alquilbenceno.

ASA. Abbreviation for the American Standards Association [now known as American National Standards Institute (ANSI)].

A.S.A. Abreviatura de Asociación Estadounidense de Normas [conocida ahora como Instituto Nacional Estadounidense de Normas (ANSI)].

ASHRAE. Abbreviation for the American Society of Heating, Refrigerating, and Air-Conditioning Engineers.

ASHRAE. Abreviatura de Sociedad Estadounidense de Ingenieros de Calefacción, Refrigeración y Acondicionamiento de Aire.

ASME. Abbreviation for the American Society of Mechanical Engineers.

ASME. Abreviatura de Sociedad Estadounidense de Ingenieros Mecánicos.

Aspect ratio. The ratio of the length to width of a component.

Coeficiente de alargamiento. Relación del largo al ancho de un componente.

Atmospheric pressure. The weight of the atmosphere's gases pressing down on the earth. Equal to 14.696 psi at sea level and 70°F.

Presión atmosférica. El peso de la presión ejercida por los gases de la atmósfera sobre la tierra, equivalente a 14696 psi al nivel del mar a 70°F.

Atom. The smallest particle of an element.

átomo. Partícula más pequeña de un elemento.

Atomization. The process of using pressure to change liquid to small particles of vapor.

Pulverización. El proceso de utilizar la presión para cambiar un líquido a partículas pequeñas de vapor.

Atomizing humidifier. A type of humidifier that discharges tiny water droplets into the air, which evaporate very rapidly into the duct airstream or directly into the conditioned space.

Humidificador por pulverización. Dispositivo que descarga al aire pequeñas gotas de agua que se evaporan rápidamente en el conducto de aire o en ambiente acondicionado.

Attractant. A type of salt solution, usually lithium bromide, used as an absorbent in absorption refrigeration.

Atrayente. Tipo de solución salina, normalmente bromuro de litio, utilizado como absorbente en la refrigeración por absorción.

Automatic changeover thermostat. A thermostat that changes from cool to heat automatically by room temperature.

Termostato de cambio automático. Un termostato que cambia de enfriar a calentar automáticamente con la temperatura del cuarto.

Automatic combination gas valve. A gas valve for gas furnaces that incorporates a manual control, gas supply for the pilot, adjustment and safety features for the pilot, pressure regulator, and the controls for and the main gas valve.

Válvula de gas de combinación automática. Válvula de gas para hornos de gas que incorpora un regulador manual, sumi-nistro de gas para la llama piloto, ajuste y dispositivos de seguridad, regulador de presión, la válvula de gas principal y los reguladores de la válvula.

Automatic control. Controls that react to a change in conditions to cause the condition to stabilize.

Regulador automático. Reguladores que reaccionan a un cambio en las condiciones para provocar la estabilidad de dicha condición.

Automatic defrost. Using automatic means to remove ice from a refrigeration coil at timed intervals.

Desempañador automático. La utilización de medios automáticos para remover el hielo de una bobina de refrigeración a intervalos programados.

Automatic expansion valve (AXV). A refrigerant control valve that maintains a constant pressure in an evaporator.

Válvula de expansión automática. Válvula de regulación del refrigerante que mantiene una presión constante en un evaporador.

Automatic pumpdown system. A control scheme in refrigeration consisting of a thermostat and liquid-line solenoid valve that clears refrigerant from the compressor's crankcase, evaporator, and suction line just before the compressor off cycle.

Sistema de bombeo hacia abajo automático. Un sistema de control en refrigeración que consiste de un termostato y una válvula de línea de líquido solenoide que vacía el refrigerante del cárter del cigüeñal del compresor, evaporador y línea de succión justo antes del ciclo de apagar del compresor.

Auxiliary drain pan. A separate drain pan that is placed under an air-conditioner evaporator to catch condensate in the event that the primary drain pan runs over.

Plato de drenaje auxiliar. Un plato separado de drenaje debajo del evaporador de un aire acondicionado para recoger la condensación en caso de que el plato principal se derrame.

Azeotropic blend. Two or more refrigerants mixed together that will have only one boiling and/or condensing point for each system pressure. Negligible fractionation or temperature glide will occur.

Mezcla azeotrópica. Mezcla de dos o más refrigerantes y que tiene un solo punto de ebullición y/o condensación para cada presión del sistema. Puede producirse un fraccionamiento o una variación térmica insignificantes.

Back electromotive force (BEMF). This is the voltage-generating effect of an electric motor's rotor turning within the motor.

Fuerza contraelectromotriz (BEMF en inglés). El efecto de generar tensión provocado por el rotor de un motor eléctrico al girar dentro del motor.

Back pressure. The pressure on the low-pressure side of a refrigeration system (also known as *suction pressure*).

Contrapresión. La presión en el lado de baja presión de un sistema de refrigeración (conocido también como *presión de aspiración*).

Back seat. The position of a refrigeration service valve when the stem is turned away from the valve body and seated, shutting off the service port.

Asiento trasero. Posición de una válvula de servicio de refrigeración cuando el vástago está orientado fuera del cuerpo de la válvula y aplicado sobre su asiento, cerrando así la apertura de servicio.

Baffle. A plate used to keep fluids from moving back and forth at will in a container.

Deflector. Placa utilizada para evitar el libre movimiento de líquidos en un recipiente.

Balance point. The temperature in a structure where the heat pump will run full-time and maintain the temperature in the structure. Auxiliary heat must be used to accompany the heat pump when the temperature is below the balance point.

Punto de balance. La temperatura en una estructura donde la bomba de calor corre todo el tiempo y mantiene la temperatura en la estructura. Calor auxiliar debe usarse para acompañar la bomba de calor cuando la temperatura está por debajo del punto de balance.

Balanced-port TXV. A valve that will meter refrigerant at the same rate when the condenser head pressure is low.

Válvula electrónica de expansión con conducto equilibrado. Válvula que medirá el refrigerante a la misma proporción cuando la presión en la cabeza del condensador sea baja.

Ball check valve. A valve with a ball-shaped internal assembly that only allows fluid flow in one direction.

Válvula de retención de bolas. Válvula con un conjunto interior en forma de bola que permite el flujo de fluido en una sola dirección.

Ball valve. Valve that uses a round ball in the flow chamber in the valve. The round ball is bored to the approximate inside diameter of the pipe and also offers very little resistance or pressure drop to the flow of water or fluid flow. The handle only has to be turned 90° to fully open or close the valve.

Válvula de bola. Válvula que utiliza una bola redonda en la cámara de flujo en la válvula. La bola redonda está calibrada aproximadamente al diámetro interior de la tubería y también presenta muy poca resistencia o caída de la presión al flujo del agua o fluido. El mango solamente se gira 90° para abrir o cerrar la válvula completamente.

Barometer. A device used to measure atmospheric pressure that is commonly calibrated in inches or millimeters of mercury. There are two types: mercury column and aneroid.

Barómetro. Dispositivo comúnmente calibrado en pulgadas o en milímetros de mercurio que se utiliza para medir la presión atmosférica. Existen dos tipos: columna de mercurio y aneroide.

Base. A terminal on a semiconductor.

Base. Punto terminal en un semiconductor.

Baseboard heating. Convection heaters providing whole-house, spot, or individual room heating. The heat is normally provided by electrical resistance or hot water.

Calefacción de zócalo. Calentadores por convección que proporcionan calefacción a toda la casa, en un punto específico o en una sola habitación. Normalmente, el calor se obtiene mediante una resistencia eléctrica o agua caliente.

Battery. A device that produces electricity from the interaction of metals and acid.

Pila. Dispositivo que genera electricidad de la interacción entre metales y el ácido.

Bearing. A device that surrounds a rotating shaft and provides a low-friction contact surface to reduce wear from the rotating shaft.

Cojinete. Dispositivo que rodea un árbol giratorio y provee una superficie de contacto de baja fricción para disminuir el desgaste de dicho árbol.

Bearing washout. A cleaning of the compressor's bearing surfaces, which causes lack of lubrication. It is usually caused by liquid refrigerant mixing with the compressor's crankcase oil due to liquid floodback or migration.

Derrubio del cojinete. Una limpieza de las superficies del cojinete del compresor, lo que lleva a falta de lubricación. Generalmente causado por refrigerante líquido mezclado con aceite del cárter del cigüeñal del compresor debido a inundación o migración del líquido.

Bellows. An accordion-like device that expands and contracts when internal pressure changes.

Fuelles. Dispositivo en forma de acordeón con pliegues que se expanden y contraen cuando la presión interna sufre cambios.

Bellows seal. A method of sealing a rotating shaft or valve stem that allows rotary movement of the shaft or stem without leaking.

Cierre hermético de fuelles. Método de sellar un árbol giratorio o el vástago de una válvula que permite el movimiento giratorio del árbol o del vástago sin producir fugas.

Belly-band-mount motor. An electric motor mounted with a strap around the motor secured with brackets on the strap.

Motor con barriguera. Motor eléctrico que se monta colocando una cincha a su alrededor y fijándola con abrazaderas.

Bending spring. A coil spring that can be fitted inside or outside a piece of tubing to prevent its walls from collapsing when being formed.

Muelle de flexión. Muelle helicoidal que puede acomodarse dentro o fuera de una pieza de tubería para evitar que sus paredes se doblen al ser formadas.

Bimetal. Two dissimilar metals fastened together to create a distortion of the assembly with temperature changes.

Bimetal. Dos metales distintos fijados entre sí para producir una distorción del conjunto al ocurrir cambios de temperatura.

Bimetal strip. Two dissimilar metal strips fastened back-to-back.

Banda bimetálica. Dos bandas de metales distintos fijadas entre sí en su parte posterior.

Binary. Consisting of 1s and 0s. The 1s and 0s represent numbers, words, or signals that can be stored in the computer's memory for future use. Calculators and computers are digital systems.

Binario. Consiste de 1s y 0s. Los 1s y 0s representan números, palabras o señales que pueden almacenarse en la memoria de la computadora para uso futuro. Las calculadoras y computadoras son sistemas digitales.

Bleeding. Allowing pressure to move from one pressure level to another very slowly.

Sangradura. Proceso a través del cual se permite el movimiento de presión de un nivel a otro de manera muy lenta.

Bleed valve. A valve with a small port usually used to bleed pressure from a vessel to the atmosphere.

Válvula de descarga. Válvula con un conducto pequeño utilizado normalmente para purgar la presión de un depósito a la atmósfera.

Blocked suction. A method of cylinder unloading. The suction line passage to a cylinder in a reciprocating compressor is blocked, thus causing that cylinder to stop pumping.

Aspiración obturada. Método de descarga de un cilindro. El paso del conducto de aspiración a un cilindro en un compresor alternativo se obtura, provocando así que el cilindro deje de bombear.

Blowdown. A system in a cooling tower whereby some of the circulating water is bled off and replaced with fresh water to dilute the sediment in the sump.

Vaciado. Sistema en una torre de refrigeración por medio del cual se purga parte del agua circulante y se reemplaza con agua fresca para diluir el sedimento en el sumidero.

Boiler. A container in which a liquid may be heated using any heat source. When the liquid is heated to the point that vapor forms and is used as the circulating medium, it is called a steam boiler.

Cardera. Recipiente en el que se puede calentar un líquido utilizando cualquier fuente de calor. Cuando se calienta el líquido al punto en que se produce vapor y se utiliza éste como el medio para la circulación, se llama caldera de vapor.

Boiling point. The temperature level of a liquid at which it begins to change to a vapor. The boiling temperature is controlled by the vapor pressure above the liquid.

Punto de ebullición. El nivel de temperatura de un líquido al que el líquido empieza a convertirse en vapor. La temperatura de ebullición se regula por medio de la presión del vapor sobre líquido.

Boiling temperature. The boiling temperature of the liquid can be controlled by controlling the pressure. The standard boiling pressure for water is an atmospheric pressure of 29.92 in. Hg (mercury) where water boils at 212°F.

Temperatura de ebullición. La temperatura de ebullición de un líquido puede controlarse controlando la presión. La presión de ebullición estándar para agua es una presión atmosférica de 29.92 pulgadas de mercurio donde el agua hierve a 212°F.

Booster compressor. The compressor on a parallel compressor system that is dedicated to the coldest evaporators.

Compresor elevador. Compresor montado en un sistema de compresión en paralelo, destinado a los evaporadores más fríos.

Booster pump. An additional pump that is used to build the pressure above what the primary pump can accomplish.

Bomba promotora. Una bomba adicional que se usa para aumentar la presión por encima de lo que la bomba primaria puede.

Boot. The connection between the branch line duct and the floor register. It transitions the branch duct to the register size. It may be from rectangular to another size of rectangle or from round to rectangular.

Manguito. La conexión entre el conducto de la línea ramal y el contador en el piso. Él hace la transición de la línea ramal al tamaño del contador. Puede conectar un rectángulo a otro rectángulo de tamaño diferente o círculo a un rectángulo.

Bore. The inside diameter of a cylinder.

Calibre. Diámetro interior de un cilindro.

Bourdon tube. C-shaped tube manufactured of thin metal and closed on one end. When pressure is increased inside, it tends to straighten. It is used in a gage to indicate pressure.

Tubo Bourdon. Tubo en forma de C fabricado de metal delgado y cerrado en uno de los extremos. Al aumentarse la presión en su interior, el tubo tiende a enderezarse. Se utiliza dentro de un calibrador para indicar la presión.

Brazing. High-temperature (above 800°F) melting of a filler metal for the joining of two metals.

Broncesoldadura. El derretir de un metal de relleno para fusionar dos metales a temperaturas altas (sobre los 800°F o 430°C).

Breaker. A heat-activated electrical device used to open an electrical circuit to protect it from excessive current flow.

Interruptor. Dispositivo eléctrico activado por el calor que se utiliza para abrir en circuito eléctrico a fin de protegerlo de un flujo excesivo de corriente.

British thermal unit (Btu). The amount (quantity) of heat required to raise the temperature of 1 lb of water 1°F.

Unidad térmica británica. Cantidad de calor necesario para elevar en 1°F (−17.56°C) la temperatura de una libra inglesa de agua.

Btu. Abbreviation for British thermal unit.

BTU. Abreviatura de unidad térmica británica.

Bubble point. The refrigerant temperature at which bubbles begin to appear in a saturated liquid.

Punto de burbujear. La temperatura refrigerada en que burbujas empiezan a aparecer en un liquído saturado.

Bulb, sensor. The part of a sealed automatic control used to sense temperature.

Bombilla sensora. Pieza de un regulador automático sellado que se utiliza para advertir la temperatura.

Burner. A device used to prepare and burn fuel.

Quemador. Dispositivo utilizado para la preparación y la quema de combustible.

Burr. Excess material squeezed into the end of tubing or pipe after a cut has been made. This burr must be removed.

Rebaba. Exceso de material introducido por fuerza en el extremo de una tubería después de hacerse un corte. Esta rebaba debe removerse.

Butane gas. A liquefied petroleum gas burned for heat.

Gas butano. Gas licuado derivado del petróleo que se quema para producir calor.

Butterfly valve. Valve constructed with a disk in the valve chamber that can be adjusted across the water or fluid stream. Imagine a coin that has a handle that can be turned crossways of the water or fluid flow to adjust or stop the flow. The valve handle only travels 90°, similar to the ball valve.

Válvula en forma de mariposa. Válvula construída con un disco en la cámara de la válvula que se puede ajustar a través del corriente del agua o fluido. Imagina una moneda con un mango que se puede girar transversal al flujo del agua o fluido para ajustar o parar el flujo. El mango de la válvula solamente pasa 90°, tal como la válvula de bola.

B-vent. A venting system for gas stoves or other gas appliances using approved piping through the roof or into a chimney.

Evacuación tipo B. Sistema de evacuación en estufas u otros dispositivos que funcionan con gas, utilizando conductos apropiados a través del techo o que se desembocan en una chimenea.

Cad cell. A device containing cadmium sulfide used to prove the flame in an oil-burning furnace.

Celda de cadmio. Dispositivo que contiene sulfuro de cadmio utilizado para probar la llama en un horno de aceite pesado.

Calibration. The adjustment of instruments or gages to the correct setting for known conditions.

Calibración. El ajuste de instrumentos o calibradores en posición correcta para su operación en condiciones conocidas.

Calorimeter. An instrument of laboratory-grade quality used to measure heat absorbed into a substance.

Calorímetro. Instrumento de calidad laboratorio que sirve para medir el calor absorbido por una sustancia.

Capacitance. The term used to describe the electrical storage ability of a capacitor.

Capacitancia. Término utilizado para describir la capacidad de almacenamiento eléctrico de un capacitador.

Capacitive circuit. When the current in a circuit leads the voltage by 90°.

Circuito capacitivo. Un circuito en que el corriente mueve el tensión por un 90°.

Capacitor. An electrical storage device used to start motors (start capacitor) and to improve the efficiency of motors (run capacitor).

Capacitador. Dispositivo de almacenamiento eléctrico utilizado para arrancar motores (capacitador de arranque) y para mejorar el rendimiento de motores (capacitador de funcionamiento).

Capacitor-start motor. A single-phase motor with a start and run winding that has a capacitor in series with the start winding, which remains in the circuit until the motor gets up to about 75% the run speed.

Motor de arranque capacitivo. Motor de fase sencilla con una bobina de arranque y marcha que tiene un condensador en serie con la bobina de arranque, el cual permanece en el circuito hasta que el motor alcanza alrededor de 75% de la velocidad de marcha.

Capacitor-start–capacitor-run motor. A single-phase motor that has a start capacitor in series with the start winding that is disconnected after start-up and a run capacitor that is also in parallel with the start windings that stays in the circuit while running. This capacitor is built for full-time duty and uses the potential voltage generated by the start winding to give the run winding more efficiency.

Motor de arranque capacitivo–marcha capacitiva. Motor de fase sencilla que tiene un condensador de arranque en serie con la bobina de arranque que se desconecta después del encendido, y un condensador de marcha que está en paralelo con las bobinas de arranque y que permanece en el circuito mientras está en marcha. Este condensador está hecho para servicio a tiempo

completo y usa el voltaje que genera la bobina de arranque para darle mayor eficiencia a la bobina de marcha.

Capacity. The rating system of equipment used to heat or cool substances.

Capacidad. Sistema de clasificación de equipo utilizado para calentar o enfriar sustancias.

Capillary attraction. The attraction of a liquid material between two pieces of material such as two pieces of copper or copper and brass. For instance, in a joint made up of copper tubing and a brass fitting, the solder filler material has a greater attraction to the copper and brass than to itself and is drawn into the space between them.

Atracción capilar. Atracción de un material líquido entre dos piezas de material, como por ejemplo dos piezas de cobre o cobre y latón. Por ejemplo, en una junta fabricada de tubería de cobre y un accesorio de latón, el material de relleno de la soldadura tiene mayor atracción al cobre y al latón que a sí mismo y es arrastrado hacia el espacio entre éstos.

Capillary tube. A fixed-bore metering device. This is a small-diameter tube that can vary in length from a few inches to several feet. The amount of refrigerant flow needed is predetermined and the length and diameter of the capillary tube is sized accordingly.

Tubo capilar. Dispositivo de medición de calibre fijo. Éste es un tubo de diámetro pequeño cuyo largo puede oscilar entre unas cuantas pulgadas y varios pies. La cantidad de flujo de refrigerante requerida es predeterminada y, de acuerdo a esto, se fijan el largo y el diámetro del tubo capilar.

Carbon dioxide. A by-product of natural gas combustion that is not harmful.

Bióxido de carbono. Subproducto de la combustión del gas natural que no es nocivo.

Carbon monoxide. A poisonous, colorless, odorless, tasteless gas generated by incomplete combustion.

Monóxido de carbono. Gas mortífero, inodoro, incoloro e insípido que se desprende en la combustión incompleta del carbono.

Catalytic combustor stove. A stove that contains a cell-like structure consisting of a substrate, washcoat, and catalyst that produces a chemical reaction causing pollutants to be burned at much lower temperatures.

Estufa de combustor catalítico. Estufa con una estructura en forma de celda compuesta de una subestructura, una capa brochada y un catalizador que produce una reacción química. Esta reacción provoca la quema de contaminantes a temperaturas muchas más bajas.

Cathode. A terminal or connection point on a semiconductor.

Cátodo. Punto de conexión o terminal en un semiconductor.

Cavitation. A vapor formed due to a drop in pressure in a pumping system. Vapor at a pump inlet may be caused at a cooling tower if the pressure is low and water is turned to vapor.

Cavitación. Vapor producido como consecuencia de una caída de presión en un sistema de bombeo. El vapor a la entrada de una bomba puede ser producido en una torre de refrigeración si la presión es baja y el agua se convierte en vapor.

Cellulose. A substance formed in wood plants from glucose or sugar.

Celulosa. Sustancia presente en plantas de madera y que se forma a partir de glucosa y azúcar.

Celsius scale. A temperature scale with 100-degree graduations between water freezing (0°C) and water boiling (100°C).

Escala Celsio. Escala dividida en cien grados, con el cero marcado a la temperatura de fusión del hielo (0°C) y el cien a la de ebullición del agua (100°C).

Centigrade scale. See Celsius scale.

Centígrado. Véase escala Celsio.

Centrifugal compressor. A compressor used for large refrigeration systems that uses centrifugal force to accomplish compression. It is not positive displacement, but it is similar to a blower.

Compresor centrífugo. Compresor utilizado en sistemas grandes de refrigeración que usa forza centrifuga para lograr compresión. No es desplazamiento positivo, pero es similar a un soplador.

Centrifugal pump. A pump that uses centrifugal force to move a fluid. An impeller is rotated rapidly within the pump, causing the fluid to fly away from the center, which forces the fluid through a piping system.

Bomba centrífuga. Bomba que utiliza la fuerza centrífuga para desplazar un fluido. Un propulsor dentro de la bomba gira a alta velocidad alejando el líquido del centro e impulsándolo a través de una tubería.

Centrifugal switch. A switch that uses a centrifugal action to disconnect the start windings from the circuit.

Conmutador centrífugo. Conmutador que utiliza una acción centrífuga para desconectar los devanados de arranque del circuito.

Change of state. The condition that occurs when a substance changes from one physical state to another, such as ice to water and water to steam.

Cambio de estado. Condición que ocurre cuando una sustancia cambia de un estado físico a otro, como por ejemplo el hielo a agua y el agua a vapor.

Charge of refrigerant. The quantity of refrigerant in a system.

Carga de refrigerante. Cantidad de refrigerante en un sistema.

Charging curve. A graphical method of assisting a service technician with charging an air-conditioning or heat pump system.

Curva de recarga. Un método gráfico para asistir al técnico de servicio con el recargar de un sistema de aire acondicionado o de bomba de calor.

Charging cylinder. A device that allows the technician to accurately charge a refrigeration system with refrigerant.

Cilindro cargador. Dispositivo que le permite al mecánico cargar correctamente un sistema de refrigeración con refrigerante.

Charging scale. A scale used to weigh refrigerant when charging a refrigeration or air-conditioning system.

Báscula de plancha. Báscula que se utiliza para pesar el refrigerante durante la carga de un sistema de refrigeración o de aire acondicionado.

Check valve. A device that permits fluid flow in one direction only.

Válvula de retención. Dispositivo que permite el flujo de fluido en una sola dirección.

Chill factor. A factor or number that is a combination of temperature, humidity, and wind velocity that is used to compare a relative condition to a known condition.

Factor de frío. Factor o número que es una combinación de la temperatura, la humedad y la velocidad del viento utilizado para comparar una condición relativa a una condición conocida.

Chilled-water system. An air-conditioning system that circulates refrigerated water to the area to be cooled. The refrigerated water picks up heat from the area, thus cooling the area.

Sistema de agua enfriada. Sistema de acondicionamiento de aire que hace circular agua refrigerada al área que será enfriada. El agua refrigerada atrapa el calor del área y la enfria.

Chiller purge unit. A system that removes air or noncondensables from a low-pressure chiller.

Unidad enfriadora de purga. Sistema que remueve el aire o sustancias no condensables de un enfriador de baja presión.

Chimney. A vertical shaft used to convey flue gases above the rooftop.

Chimenea. Cañón vertical utilizado para conducir los gases de combustión por encima del techo.

Chimney effect. A term used to describe air or gas when it expands and rises when heated.

Efecto de chimenea. Término utilizado para describir el aire o el gas cuando se expande y sube al calentarse.

Chlorofluorocarbons (CFCs). Those refrigerants thought to contribute to the depletion of the ozone layer.

Clorofluorocarburos (CFCs en inglés). Líquidos refrigerantes que, según algunos, han contribuido a la reducción de la capa de ozono.

Circuit. An electron or fluid-flow path that makes a complete loop.

Circuito. Electrón o trayectoria del flujo de fluido que hace un ciclo completo.

Circuit breaker. A device that opens an electric circuit when an overload occurs.

Interruptor para circuitos. Dispositivo que abre un circuito eléctrico cuando ocurre una sobrecarga.

Clamp-on ammeter. An instrument that can be clamped around one conductor in an electrical circuit and measure the current.

Amperímetro fijado con abrazadera. Instrumento que puede fijarse con una abrazadera a un conductor en un circuito eléctrico y medir la corriente.

Clearance volume. The volume at the top of the stroke in a reciprocating compressor cylinder between the top of the piston and the valve plate.

Volumen de holgura. Volumen en la parte superior de una carrera en el cilindro de un compresor recíproco entre la parte superior del pistón y la placa de una válvula.

Closed circuit. A complete path for electrons to flow on.

Circuito cerrado. Circuito de trayectoria ininterrumpida que permite un flujo continuo de electrones.

Closed-circuit cooling tower. May have a wet/dry mode, an adiabatic mode, and a dry mode.

Torre de enfriamiento de circuito cerrado. Puede tener un modo seco/mojado, un modo adiabático y un modo seco.

Closed loop. Piping circuit that is complete and not open to the atmosphere.

Ciclo cerrado. Circuito de tubería completo y no abierto a la atmósfera.

Closed-loop control configuration. A closed-loop control configuration can maintain the controlled variable, like the air in a duct, at its set point because the sensor is now located in the air being heated or cooled (controlled medium).

Configuración de control de circuito cerrado. Una configuración de control de circuito cerrado puede mantener el variable controlado, por ejemplo el aire en un conducto, en su punto de ajuste porque el sensor ya está ubicado en el aire que se calienta o se enfría (el medio controlado).

Closed-loop heat pump. Heat pump system that reuses the same heat transfer fluid, which is buried in plastic pipes within the earth or within a lake or pond for the heat source.

Bomba de calor de circuito cerrado. Sistema de bomba de calor que reutiliza el mismo líquido de transferencia térmica que se encuentra en tubos de plástico enterrados o sumergidos en un lago o estanque de los que obtiene el calor.

Coaxial heat exchanger. A tube-within-a-tube liquid heat exchanger. Typically it is used for water-source heat pumps and small water-cooled air conditioners.

Intercambiador de calor coaxial. Intercambiador de calor líquido con un tubo dentro de un tubo. Típicamente se usa para bombas de calor en fuentes de agua y acondicionadoras de aire pequeñas enfriadas por agua.

Code. The local, state, or national rules that govern safe installation and service of systems and equipment for the purpose of safety of the public and trade personnel.

Código. Reglamentos locales, estatales o federales que rigen la instalación segura y el servicio de sistemas y equipo con el propósito de garantizar la seguridad del personal público y profesional.

Coefficient of performance (COP). The ratio of usable output energy divided by input energy.

Coeficiente de rendimiento (COP en inglés). Relación de la de energía de salida utilizable dividida por la energía de entrada.

CO₂ indicator. An instrument used to detect the quantity of carbon dioxide in flue gas for efficiency purposes.

Indicador del CO₂. Instrumento utilizado para detectar la cantidad de bióxido de carbono en el gas de combustión a fin de lograr un mejor rendimiento.

Cold. The word used to describe heat at lower levels of intensity.

Frío. Término utilizado para describir el calor a niveles de intensidad más bajos.

Cold anticipator. A fixed resistor in a thermostat that is wired in parallel with the cooling contacts. This starts the cooling system before the thermostat calls for cooling, which allows the system to get up to capacity before the cooling is actually needed.

Anticipador de frío. Resistor fijo en un termostato, conectado en paralelo con los contactos de enfriamiento. Dicho resistor pone en marcha el sistema de enfriamiento antes de que lo haga el termostato, permitiendo así que el sistema alcance su plena capacidad antes de que realmente se necesite el enfriamiento.

Cold junction. The opposite junction to the hot junction in a thermocouple.

Empalme frío. El empalme opuesto al empalme caliente en un termopar.

Cold trap. A device to help trap moisture during the evacuation of a refrigeration system.

Trampa del frío. Dispositivo utilizado para ayudar a atrapar la humedad durante la evacuación de un sistema de refrigeración.

Cold wall. The term used in comfort heating to describe a cold outside wall and its effect on human comfort.

Pared fría. Término utilizado en la calefacción para comodidad que describe una pared exterior fría y sus efectos en la comodidad de una persona.

Collector. A terminal on a semiconductor.

Colector. Punto terminal en un semiconductor.

Combustion. A reaction called rapid oxidation or burning produced with the right combination of a fuel, oxygen, and heat.

Combustión. Reacción conocida como oxidación rápida o quema producida con la combinación correcta de combustible, oxígeno y calor.

Combustion analyzer. An instrument used to measure oxygen concentrations within flue gases. This analyzer can test smoke and test for carbon monoxide and other gases.

Analizador de combustión. Instrumento que se utiliza para medir la concentración del oxígeno en los gases de escape. Este tipo de analizador permite medir el contenido de monóxido de carbono y otros gases en los humos.

Comfort. People are said to be comfortable when they are not aware of the ambient air surrounding them. They do not feel cool or warm or sweaty.

Comodidad. Se dice que unas personas están cómodas cuandos no están conscientes del aire ambiental que les rodea. No sienten frío i calor, ni están sudadas.

Comfort chart. A chart used to compare the relative comfort of one temperature and humidity condition to another condition.

Esquema de comodidad. Esquema utilizado para comparar la comodidad relativa de un a condición de temperatura y humedad a otra condición.

Compound gage. A gage used to measure the pressure above and below the atmosphere's standard pressure. It is a Bourdon tube sensing device and can be found on all gage manifolds used for air-conditioning and refrigeration service work.

Calibrador compuesto. Calibrador utilizado para medir la presión mayor y menor que la presión estándar de la atmósfera. Es un dispositivo sensor de tubo Bourdon que puede encontrarse en todos los distribuidores de calibrador utilizados para el servicio de sistemas de acondicionamiento de aire y de refrigeración.

Compression. A term used to describe a vapor when pressure is applied and the molecules are compacted closer together.

Compresión. Término utilizado para describir un vapor cuando se aplica presión y se compactan las moléculas.

Compression ratio. A term used with compressors to describe the actual difference in the low- and high-pressure sides of the compression cycle. It is absolute discharge pressure divided by absolute suction pressure.

Relación de compresión. Término utilizado con compresores para describir la diferencia real en los lados de baja y alta presión del ciclo de compresión. Es la presión absoluta de descarga dividida por la presión absoluta de aspiración.

Compressor. A vapor pump that pumps vapor (refrigerant or air) from one pressure level to a higher pressure level.

Compresor. Bomba de vapor que bombea el vapor (refrigerante o aire) de un nivel de presión a un nivel de presión más alto.

Compressor crankcase. The internal part of the compressor that houses the crankshaft and lubricating oil.

Cárter del compresor. Parte interna del compresor donde se aloja el cigüeñal y el aceite de lubricación.

Compressor displacement. The internal volume of a compressor's cylinders, used to calculate the pumping capacity of the compressor.

Desplazamiento del compresor. Volumen interno de los cilindros de un compresor, utilizado para calcular la capacidad de bombeo del mismo.

Compressor head. The component that sits on top of the compressor cylinder and holds the components together.

Cabeza del compresor. El componente que está en la parte superior del cilindro y que mantiene a los componentes unidos.

Compressor oil cooler. One or more piping systems used for cooling the crankcase oil.

Enfriador del aceite del compresor. Uno o varios sistemas de tubería que sirven para enfriar el aceite del cárter.

Compressor shaft seal. The seal that prevents refrigerant inside the compressor from leaking around the rotating shaft.

Junta de estanqueidad del árbol del compresor. La junta de estanqueidad que evita la fuga, alrededor del árbol giratorio, del refrigerante en el interior del compresor.

Concentrator. That part of an absorption chiller where the dilute salt solution is boiled to release the water.

Concentrador. El lugar en el enfriador por absorción donde se hierve la solución salina diluida para liberar el agua.

Condensate. The moisture collected on an evaporator coil.

Condensado. Humedad acumulada en la bobina de un evaporador.

Condensate pump. A small pump used to pump condensate to a higher level.

Bomba para condensado. Bomba pequeña utilizada para bombear el condensado a un nivel más alto.

Condensation. Liquid formed when a vapor condenses.

Condensación. El líquido formado cuando se condensa un vapor.

Condense. Changing a vapor to a liquid.

Condensar. Convertir un vapor en líquido.

Condenser. The component in a refrigeration system that transfers heat from the system by condensing refrigerant.

Condensador. Componente en un sistema de refrigeración que transmite el calor del sistema al condensar el refrigerante.

Condenser flooding. An automatic method of maintaining the correct head pressure in mild weather by using refrigerant from an auxiliary receiver.

Inundación del condensador. Método automático de mantener una presión correcta en la cabeza en un tiempo utilizando refrigerante de un receptor auxiliar.

Condensing boiler. High-efficiency boiler designed to operate with low flue-gas temperatures. These boilers allow for the condensing of flue gas.

Caldera para condensación de gas. Una caldera muy eficiente diseñada para operar cuando la temperatura del gas de combustión está baja. Estas calderas permiten la condensación de los gases de combustión.

Condensing gas furnace. A furnace with a condensing heat exchanger that condenses moisture from the flue gases, resulting in greater efficiency.

Horno para condensación de gas. Horno con un intercambiador de calor para condensación que condensa la humedad de los gases de combustión. El resultado será un mayor rendimiento.

Condensing oil furnace. An oil furnace with a coil that causes moisture to condense from the flue gases, producing latent heat. This increases the efficiency of the furnace.

Horno de petróleo con serpentín de condensación. Horno de petróleo, con un serpentín que provoca la condensación de la humedad presente en los gases de combustión, produciendo un calor latente. Permite aumentar la eficacia del horno.

Condensing pressure. The pressure that corresponds to the condensing temperature in a refrigeration system.

Presión para condensación. La presión que corresponde a la temperatura de condensación en un sistema de refrigeración.

Condensing temperature. The temperature at which a vapor changes to a liquid.

Temperatura de condensación. Temperatura a la que un vapor se convierte en líquido.

Condensing unit. A complete unit that includes the compressor and the condensing coil.

Conjunto del condensador. Unidad completa que incluye el compresor y la bobina condensadora.

Conduction. Heat transfer from one molecule to another within a substance or from one substance to another.

Conducción. Transmisión de calor de una molécula a otra dentro de una sustancia o de una sustancia a otra.

Conductivity. The ability of a substance to conduct electricity or heat.

Conductividad. Capacidad de una sustancia de conducir electricidad o calor.

Conductor. A path for electrical energy to flow on.

Conductor. Trayectoria que permite un flujo continuo de energía eléctrica.

Connecting rod. A rod that connects the piston to the crankshaft.

Barra conectiva. Barra que conecta al pistón con el cigüeñal.

Contactor. A larger version of the relay. It can be repaired or rebuilt and has movable and stationary contacts.

Contactador. Versión más grande del relé. Puede ser reparado o reconstruido. Tiene contactos móviles y fijos.

Contaminant. Any substance in a refrigeration system that is foreign to the system, particularly if it causes damage.

Contaminante. Cualquier sustancia en un sistema de refrigeración extraña a éste, principalmente si causa averías.

Control. A device for stopping, starting, or modulating the flow of electricity or fluid to maintain a preset condition.

Regulador. Dispositivo para detener, poner en marcha o modular el flujo de electricidad o de fluido a fin de mantener una condición establecida con anticipación.

Control agent. The control agent is the fluid that transfers energy or mass to the controlled medium. It is the control agent (such as hot or cold water) that will flow through a final control device (water valves).

Agente de control. El agente de control es el fluido que transfiere energía o masa al medio controlado. Es el agente (como agua caliente o fría) que fluye por un dispositivo de regulación (válvulas de agua).

Control loop. A sensor, a controller, and a controlled device.

Circuito de control. Un sensor, un controlador y un aparato controlado.

Control point. The actual value of the controlled variable at a particular point in time.

Punto de control. El valor verdadero del variable controlado en un tiempo específico.

Control system. A network of controls to maintain desired conditions in a system or space.

Sistema de regulación. Red de reguladores que mantienen las condiciones deseadas en un sistema o un espacio.

Controlled device. May be any control that stops, starts, or modulates fuel, fluid flow, or air to provide expected conditions in the conditioned space.

Aparato controlado. Un aparato que puede ser cualquier control que detiene, enciende o modula el combustible, flujo de fluido o aire para proveer las condiciones esperadas en un espacio acondicionado.

Controlled environment. The conditioned space the HVACR system is trying to maintain.

Medio ambiente controlado. El espacio acondicionado que el sistema HVACR trata de conservar.

Controlled medium. The controlled medium is what absorbs or releases the energy or mass transferred to or from the HVACR process. Controlled mediums can be air in a duct or water in a pipe.

Medio controlado. El medio controlado absorbe o libera la energía o la masa transferidas al proceso HVACR o del mismo proceso. Un medio controlado puede ser el aire en un conducto o el agua en un tubo.

Controlled output device. Any device of an array of dampers, fans, variable-frequency drives (VFDs), cooling coil valves, heating coil valves, relays, and motors—all of which control the controlled environment.

Aparato de salida controlada. Una serie de desviadores, abanicos, propulsiones de frecuencia variable (VFDs en inglés), válvulas del serpentín de enfriamiento, válvulas del serpentín de calefacción, relés y motores—que controlan el medio ambiente controlado.

Controller. A device that provides the output to the controlled device.

Controlador. Aparato que provee la salida para el aparato controlado.

Convection. Heat transfer from one place to another using a fluid.

Convección. Transmisión de calor de un lugar a otro por medio de un fluido.

Conversion factor. A number used to convert from one equivalent value to another.

Factor de conversión. Número utilizado en la conversión de un valor equivalente a otro.

Cooler. A walk-in or reach-in refrigerated box.

Nevera. Caja refrigerada donde se puede entrar o introducir la mano.

Cooling tower. The final device in many water-cooled systems, which rejects heat from the system into the atmosphere by evaporation of water.

Torre de refrigeración. Dispositivo final en muchos sistemas enfriados por agua, que dirige el calor del sistema a la atmósfera por medio de la evaporación de agua.

Copper plating. Small amounts of copper are removed by electrolysis and deposited on the ferrous metal parts in a compressor.

Encobrado. Remoción de pequeñas cantidades de cobre por medio de electrólisis que luego se colocan en las piezas de metal férreo en un compresor.

Corrosion. A chemical action that eats into or wears away material from a substance.

Corrosión. Acción química que carcoma o desgasta el material de una sustancia.

Cotter pin. Used to secure a pin. The cotter pin is inserted through a hole in the pin, and the ends spread to retain it.

Pasador de chaveta. Se usa para asegurar una clavija. El pasador se inserta a través de un roto en la clavija y sus extremos se abren para asegurarla.

Counter EMF. Voltage generated or induced above the applied voltage in a single-phase motor.

Contra EMF. Tensión generada o inducida sobre la tensión aplicada en un motor unifásico.

Counterflow. Two fluids flowing in opposite directions.

Contraflujo. Dos fluidos que fluyen en direcciones opuestas.

Coupling. A device for joining two fluid-flow lines. Also the device connecting a motor driveshaft to the driven shaft in a direct-drive system.

Acoplamiento. Dispositivo utilizado para la conexión de dos conductos de flujo de fluido. Es también el dispositivo que conecta un árbol de mando del motor al árbol accionado en un sistema de mando directo.

CPVC (Chlorinated polyvinyl chloride). Plastic pipe similar to PVC except that it can be used with temperatures up to 180°F at 100 psig.

CPVC (Cloruro de polivinilo clorado). Tubo plástico similar al PVC, pero que puede utilizarse a temperaturas de hasta 180°F (82°C) a 100 psig [indicador de libras por pulgada cuadrada].

Crackage. Small spaces in a structure that allow air to infiltrate the structure.

Formación de grietas. Espacios pequeños en una estructura que permiten la infiltración del aire dentro de la misma.

Cradle-mount motor. A motor with a mounting cradle that fits the motor end housing on each end and is held down with a bracket.

Motor montado con cuña. Motor equipado de una cuña adaptada a la caja en los dos extremos y sujetado por fijaciones.

Crankcase heat. Heat provided to the compressor crankcase.

Calor para el cárter del cigüeñal. Calor suministrado al cárter del cigüeñal del compresor.

Crankcase pressure-regulating valve (CPR valve). A valve installed in the suction line, usually close to the compressor. It is used to keep a low-temperature compressor from overloading on a hot pulldown by limiting the pressure to the compressor.

Válvula reguladora de la presión del cárter del cigüeñal (CPR en inglés). Válvula instalada en el conducto de aspiración, normalmente cerca del compresor. Se utiliza para evitar la sobrecarga en un compresor de temperatura baja durante un arrastre caliente hacia abajo limitando la presión al compresor.

Crankshaft. In a reciprocating compressor, the crankshaft changes the round-and-round motion into the reciprocating back-and-forth motion of the pistons using off-center devices called throws.

Cigüeñal. En un compresor alternativo, el cigüeñal cambia el movimiento circular en un movimiento alternativo hacia delante y hacia atrás de los pistones usando aparatos descentrados llamados cigüeñas.

Crankshaft seal. Same as the compressor shaft seal.

Junta de estanqueidad del árbol del cigüeñal. Exactamente igual que la junta de estanqueidad del árbol del compresor.

Crankshaft throw. The off-center portion of a crankshaft that changes rotating motion to reciprocating motion.

Excentricidad del cigüeñal. Porción descentrada de un cigüeñal que cambia el movimiento giratorio a un movimiento alternativo.

Creosote. A mixture of unburned organic material found in the smoke from a wood-burning fire.

Creosota. Mezcla del material orgánico no quemado que se encuentra en el humo proveniente de un incendio de madera.

Crisper. A refrigerated compartment that maintains a high humidity and a low temperature.

Encrespador. Compartimiento refrigerado que mantiene una humedad alta y una temperatura baja.

Cross charge. A control with a sealed bulb that contains two different fluids that work together for a common specific condition.

Carga transversal. Regulador con una bombilla sellada compuesta de dos fluidos diferentes que pueden funcionar juntos para una condición común específica.

Cross liquid charge bulb. A type of charge in the sensing bulb of the TXV that has different characteristics from the system refrigerant. This is designed to help prevent liquid refrigerant from flooding to the compressor at start-up.

Bombilla de carga del líquido transversal. Tipo de carga en la bombilla sensora de la válvula electrónica de expansión que tiene características diferentes a las del refrigerante del sistema. La carga está diseñada para ayudar a evitar que el refrigerante líquido se derrame dentro del compresor durante la puesta en marcha.

Cross vapor charge bulb. Similar to the vapor charge bulb but contains a fluid different from the system refrigerant. This is a special-type charge and produces a different temperature/pressure relationship under different conditions.

Bombilla de carga del vapor transversal. Similar a la bombilla de carga del vapor pero contiene un fluido diferente al del refrigerante del sistema. Ésta es una carga de tipo especial y produce una relación diferente entre la temperatura y la presión bajo condiciones diferentes.

Crystallization. When a salt solution becomes too concentrated and part of the solution turns to salt.

Cristalización. Condición que ocurre cuando una solución salina se concentra demasiado y una parte de la solución se convierte en sal.

Cupronickel. A material used in heat exchangers. This material is made with an alloy of copper for a higher corrosion resistance for acid cleaning.

Cuproníquel. Material utilizado en intercambiadores térmicos. Se trata de una aleación de cobre y níquel con una elevada resistencia a la corrosión en procesos de limpieza con ácido.

Current, electrical. Electrons flowing along a conductor.

Corriente eléctrica. Electrones que fluyen a través de un conductor.

Current relay. An electrical device activated by a change in current flow.

Relé para corriente. Dispositivo eléctrico accionado por un cambio en el flujo de corriente.

Current sensing relay. An inductive relay coil usually located around a wire used to sense current flowing through the wire. Its action usually opens or closes a set of contacts.

Relé detector de corriente. Una bobina de relé inductiva que generalmente está ubicada cerca de un cable y se usa para detectar el flujo de corriente a través del cable. Su acción generalmente abre o cierra una serie de contactos.

Cut-in and cut-out. The two points at which a control opens or closes its contacts based on the condition it is supposed to maintain.

Puntos de conexión y desconexión. Los dos puntos en los que un regulador abre o cierra sus contactos según las condiciones que debe mantener.

Cycle. A complete sequence of events (from start to finish) in a system.

Ciclo. Secuencia completa de eventos, de comienzo a fin, que ocurre en un sistema.

Cylinder. A circular container with straight sides used to contain fluids or to contain the compression process (the piston movement) in a compressor.

Cilindro. Recipiente circular con lados rectos, utilizado para contener fluidos o el proceso de compresión (movimiento del pistón) en un compresor.

Cylinder, compressor. The part of the compressor that contains the piston and its travel.

Cilindro del compresor. Pieza del compresor que contiene el pistón y su movimiento.

Cylinder head, compressor. The top to the cylinder on the high-pressure side of the compressor.

Culata del cilindro del compresor. Tapa del cilindro en el lado de alta presión del compresor.

Cylinder, refrigerant. The container that holds refrigerant.

Cilindro del refrigerante. El recipiente que contiene el refrigerante.

Cylinder unloading. A method of providing capacity control by causing a cylinder in a reciprocating compressor to stop pumping.

Descarga del cilindro. Método de suministrar regulación de capacidad provocando que el cilindro en un compresor alternativo deje de bombear.

Damper. A component in an air-distribution system that restricts airflow for the purpose of air balance.

Desviador. Componente en un sistema de distribución de aire que limita el flujo de aire para mantener un equilibrio de aire.

Database. A collection of data or information stored in the memory of the controller to be used by the digital controller at a later point in time.

Base de datos. Una colección de datos o información almacenados en la memoria del controlador, para el uso del controlador digital más adelante.

DC converter. A type of rectifier that changes alternating current (AC) to direct current (DC).

Convertidor CD. Tipo de rectificador que cambia la corriente alterna (CA) a corriente directa (CD).

DC motor. A motor that operates on direct current (DC).

Motor CD. Un motor que opera con corriente directa (CD).

Declination angle. The angle of the tilt of the earth on its axis.

Ángulo de declinación. Ángulo de inclinación de la Tierra en su eje.

Decorative appliance. A term used by ANSI to indicate a gas appliance that is not designed for larger space-heating applications.

Aparato decorativo. Término usado por ANSI para indicar un aparato de gas que no está diseñado para calentar espacios grandes.

Deep vacuum. An attained vacuum that is below 250 microns.

Vacío profundo. Un vacío que se obtiene lo cual es menor de 250 micrones.

Defrost. Melting of ice.

Descongelar. Convertir hielo en líquido.

Defrost condensate. The condensate or water from a defrost application of a refrigeration system.

Condensado de descongelación. Condensado o agua causada por el dispositivo de descongelación en un sistema de refrigeración.

Defrost cycle. The portion of the refrigeration cycle that melts the ice off the evaporator.

Ciclo de descongelación. Parte del ciclo de refrigeración que derrite el hielo del evaporador.

Defrost termination switch. A temperature-activated switch that stops the defrost cycle and returns the equipment to the refrigeration cycle.

Interruptor de terminación de descongelación. Un interruptor activado por temperatura que detiene el ciclo de descongelación y regresa al equipo al ciclo de refrigeración.

Defrost timer. A timer used to start and stop the defrost cycle.

Temporizador de descongelación. Temporizador utilizado para poner en marcha y detener el ciclo de descongelación.

Degreaser. A cleaning solution used to remove grease from parts and coils.

Desengrasador. Solución limpiadora utilizada para remover la grasa de piezas y bobinas.

Dehumidify. To remove moisture from air.

Deshumidificar. Remover la humedad del aire.

Dehydration. Removing moisture from a sealed system or a product.

Deshidratación. Removiendo la humedad de un sistema sellado o un producto.

Delta transformer connection. A transformer connection that results in a voltage output of 115 V and 230 V.

Conexión de transformador delta. Una conexión de transformador que resulta en una salida de voltaje de 115 voltios y 230 voltios.

Delta-T. The temperature difference at two different points, such as the inlet and outlet temperature difference across a water chiller.

Delta-T. La diferencia en temperatura en dos puntos diferentes, tales como la diferencia de temperatura entre la entrada y la salida de un enfriador de agua.

Demand metering. In this system, the power company charges the customer based on the highest usage for a prescribed period of time during the billing period. The prescribed time for demand metering may be any 15- or 30-min period within the billing period.

Medición por demanda. Un sistema utilizado por la compañía de electricidad en el cual cobran al consumidor por el período de facturación basado en el uso más alto durante un período de tiempo prescrito durante el período de facturación. El tiempo prescrito para la medición por demanda puede ser cualquier período de 15 ó 30 minutos durante el período de facturación.

Density. The weight per unit of volume of a substance.

Densidad. Relación entre el peso de una sustancia y su volumen.

Department of Transportation (DOT). The governing body of the U.S. government that makes the rules for transporting items, such as volatile liquids.

Departamento de Transportación (DOT en inglés). El cuerpo regente del gobierno de los Estados Unidos que crea las reglas para transportar artículos tales como líquidos volátiles.

Derivative (differential) controller. A control mode where the controller will change its output signal according to the rate of change of the error or offset.

Controlador derivativo (diferencial). Un modo de control en el cual el controlador cambiará su salida de acuerdo a la razón de cambio del error o de la compensación.

Desiccant. Substance in a refrigeration system drier that collects moisture.

Disecante. Sustancia en el secador de un sistema de refrigeración que acumula la humedad.

Desiccant drier. A device that dehumidifies compressed air for use in controls or processing.

Secador desecante. Un aparato que se usa para deshumedecer el aire comprimido que se usa en los controles o procesos.

Design pressure. The pressure at which the system is designed to operate under normal conditions.

Presión de diseño. Presión a la que el sistema ha sido diseñado para funcionar bajo condiciones normales.

Desuperheating. Removing heat from the superheated hot refrigerant gas down to the condensing temperature.

Des-sobrecalentamiento. Reducir el calor del gas caliente del refrigerante sobrecalentado hasta alcanzar la temperatura de condensación.

Detector. A device used to search and find.

Detector. Dispositivo de búsqueda y detección.

Detent or snap action. The quick opening and closing of an electrical switch.

Acción de detén o de encaje. El abrir y cerrar rápido de un interruptor eléctrico.

Dew. Moisture droplets that form on a cool surface.

Rocío. Gotitas de humedad que se forman en una superficie fría.

Dew point temperature. The exact temperature at which moisture begins to form.

Temperatura del punto de rocío. Temperatura exacta a la que la humedad comienza a formarse.

Diac. A semiconductor often used as a voltage-sensitive switching device.

Diac. Semiconductor utilizado frecuentemente como dispositivo de conmutación sensible a la tensión.

Diagnostic thermostat. A thermostat that can receive information from various sensors and determine when a system is having a problem. It may give a fault signal in the form of a code indicating the problem.

Termostato diagnóstico. Un termostato que puede recibir información de varios sensores y determinar cuándo un sistema tiene problemas. Puede indicar una señal de fallo en forma de un código que indica el problema.

Diaphragm. A thin flexible material (metal, rubber, or plastic) that separates two pressure differences.

Diafragma. Material delgado y flexible, como por ejemplo el metal, el caucho o el plástico, que separa dos presiones diferentes.

Diaphragm valve. A valve that has a thin sheet of metal between the valve and the fluid flow. This thin sheet is called the diaphragm and contains the pressure in the system from the action of the valve.

Válvula de diafragma. Una válvula que tiene una lámina fina de metal entre la válvula y el flujo de fluido. La lámina fina se llama diafragma y separa la presión del sistema de la acción de la válvula.

Die. A tool used to make an external thread such as on the end of a piece of pipe.

Troquel. Herramienta utilizada para formar un filete externo, como por ejemplo en el extremo de un tubo.

Differential. The difference in the cut-in and cut-out points of a control, pressure, time, temperature, or level.

Diferencial. Diferencia entre los puntos de conexión y desconexión de un regulador, una presión, un intervalo de tiempo, una temperatura o un nivel.

Diffuse radiation. Radiation from the sun that reaches the earth after it is reflected from other substances, such as moisture or other particles in space.

Radiación difusa. Radiación solar que alcanza la Tierra después de haber sido reflejada por otras sustancias como gotas de agua u otras partículas presentes en el aire.

Diffuser. The terminal or end device in an air-distribution system that directs air in a specific direction using louvres.

Placa difusora. Punto o dispositivo terminal en un sistema de distribución de aire que dirige el aire a una dirección específica, utilizando aberturas tipo celosía.

Digital electronic devices. Devices that generate strings of data or groups of logic consisting of 1s and 0s.

Aparatos electrónicos digitales. Aparatos que generan cadenas de data o grupos de lógica que consisten de unos y ceros.

Digital electronic signal. An electrical signal, usually 0 to 10 V DC or 0 to 20 milliamps DC, that is used to control system conditions.

Señal electrónica digital. Una señal eléctrica, generalmente de 0 a 10 voltios CD o de 0 a 20 miliamperes CD, que se usa para controlar las condiciones del sistema.

Digital VOM. A volt-ohm-milliammeter that displays the reading in digits or numbers.

VOM digital. Medidor de voltios-ohmios-miliamperímetros que indica la lectura en dígitos o números.

Diode. A solid-state device composed of both P-type and N-type material. When connected in a circuit one way, current will flow. When the diode is reversed, current will not flow.

Diodo. Dispositivo de estado sólido compuesto de material P y de material N. Cuando se conecta a un circuito de una manera, la corriente fluye. Cuando la dirección del diodo cambia, la corriente deja de fluir.

DIP (dual inline pair) switch. A very small low-amperage, single-pole, double-throw switch used in electronic circuits to set up the program in the circuit.

Interruptor de doble paquete en línea (DIP en inglés). Un interruptor muy pequeño, de bajo amperaje, unipolar de doble tiro que se usa en los circuitos electrónicos para preparar el programa en un circuito.

Direct current. Electricity in which all electron flow goes continuously in one direction.

Corriente continua. Electricidad en la que todos los electrones fluyen continuamente en una sola dirección.

Direct digital control (DDC). Very low-voltage control signal, usually 0 to 10 V DC or 0 to 20 milliamps DC.

Control digital directo (DDC en inglés). Señal control de voltaje bien bajo generalmente 0 a 20 voltios CD o de 0 a 20 miliamperes CD.

Direct-drive compressor. A compressor that is connected directly to the end of the motor shaft. No pulleys are involved.

Compresor de conducción directa. Un compresor que está conectado directamente a un extremo del eje de un motor, sin usar poleas.

Direct-drive motor. A motor that is connected directly to the load, such as an oil burner motor or a furnace fan motor.

Motor de conducción directa. Un motor que está conectado directamente a la carga, tal como un motor de un quemador de aceite o el motor de un ventilador en un calefactor.

Direct expansion. The term used to describe an evaporator with an expansion device other than a low-side float type.

Expansión directa. Término utilizado para describir un evaporador con un dispositivo de expansión diferente al tipo de dispositivo flotador de lado bajo.

Direct-fired absorption system. An absorption refrigeration system that uses gas or oil as the heat source.

Sistema de absorción de inyección directa. Sistema de refrigeración por absorción que emplea gas o petróleo como fuente de calor.

Direct radiation. The energy from the sun that reaches the earth directly.

Radiación directa. Energía solar que alcanza la Tierra directamente.

Direct-spark ignition (DSI). A system that provides direct ignition to the main burner.

Encendido de chispa directa (DSI en inglés). Sistema que le provee un encendido directo al quemador principal.

Direct-vent. A venting system for a gas furnace, which pulls the outside air in for combustion and vents the flue gases to the outside.

Escape directo. Sistema de escape en un horno, con un conducto de doble pared. El conducto hace entrar el aire del exterior para la combustión, y evacúa los gases de combustión hacia afuera.

Discharge pressure. The pressure on the high-pressure side of a compressor.

Presión de descarga. La presión en el extremo de alta presión de un compresor.

Discharge valve. The valve at the top of a compressor cylinder that shuts on the downstroke to prevent high-pressure gas from reentering the refrigerant cylinder, allowing low-pressure gas to enter.

Válvula de descarga. La válvula que está en la parte superior del cilindro de un compresor que se cierra en el recorrido hacia abajo del pistón para evitar que el gas a alta presión regrese al cilindro del refrigerante y permitir que el gas a baja presión entre.

Discus compressor. A reciprocating compressor distinguished by its disc-type valve system.

Compresor de disco. Compresor alternativo caracterizado por su sistema de válvulas de tipo disco.

Discus valve. A reciprocating compressor valve design with a low clearance volume and larger bore.

Válvula de disco. Diseño de válvula de compresor alternativo con un volumen de holgura bajo y un calibre más grande.

Distributor. A component installed at the outlet of the expansion valve that distributes the refrigerant to each evaporator circuit.

Distribuidor. Componente instalado a la salida de la válvula de expansión que distribuye el refrigerante a cada circuito del evaporador.

Diverter tee. Special tee fitting used on one-pipe hydronic systems to facilitate water flow through remote terminal units.

Desviadores en forma de T. Un accesorio especial en forma de T usado en sistemas hidrónicos para facilitar la fluya de agua por las unidades de terminal remoto.

Domestic hot water loop. Domestic hot water heated by the refrigerant loop of the heat pump.

Circuito de agua caliente doméstica. Agua caliente doméstica calentada por el circuito de refrigeración de la bomba de calor.

Doping. Adding an impurity to a semiconductor to produce a desired charge.

Impurificación. La adición de una impureza para producir una carga deseada.

Double flare. A connection used on copper, aluminum, or steel tubing that folds tubing wall to a double thickness.

Abocinado doble. Conexión utilizada en tuberías de cobre, aluminio o acera que pliega la pared de la tubería y crea un espesor doble.

Dowel pin. A pin, which may or may not be tapered, used to align and fasten two parts.

Pasador de espiga. Pasador, que puede o no ser cónico, utilizado para alinear y fijar dos piezas.

Downflow furnace. This furnace sometimes is called a counterflow furnace. The air intake is at the top, and the discharge air is at the bottom.

Horno de corriente descendente. También conocido como horno de contracorriente. La entrada del aire está en la parte superior y la salida en la parte inferior.

Draft gage. A gage used to measure very small pressures (above and below atmospheric) and compare them with the atmosphere's pressure. Used to determine the flow of flue gas in a chimney or vent.

Calibrador de tiro. Calibrador utilizado para medir presiones sumamente pequeñas (mayores o menores que la atmosférica) y compararas con la presión de la atmósfera. Utilizado para determinar el flujo de gas de combustión en una chimenea o válvula.

Drier. A device used in a refrigerant line to remove moisture.

Secador. Dispositivo utilizado en un conducto de refrigerante para remover la humedad.

Drilled well. A well with a well casing, electric pump, and electric lines that is drilled into the ground into a water source.

Pozo perforado. Un pozo con recubrimiento, bomba eléctrica y cables de alimentación eléctrica perforado en el suelo para extraer agua.

Drip pan. A pan shaped to collect moisture condensing on an evaporator coil in an air-conditioning or refrigeration system.

Colector de goteo. Un colector formado para acumular la humedad que se condensa en la bobina de un evaporador en un sistema de acondicionamiento de aire o de refrigeración.

Drip-proof motor. A motor that can stand water dripping on it, such as a condenser fan motor in an outdoor condensing unit.

Motor a prueba de goteo. Un motor que puede tolerar el goteo de agua sobre él, tal como el motor del condensador en una unidad externa de condensación.

Drive clip. One of the fasteners that holds square or rectangular steel ducts together.

Grapa de conducto. Uno de los sujetadores que une los conductos cuadrados o rectangulares de acero.

Dry-bulb temperature. The temperature measured using a plain thermometer.

Temperatura de bombilla seca. Temperatura que se mide con un termómetro sencillo.

Dry well. A well used for the discharged water in an open-loop geothermal heat pump.

Pozo seco. Pozo que se utiliza para depositar agua de descarga en una bomba de calor geotérmica de circuito abierto.

Dual pressure control. Two controls that are mounted in the same housing, typically low- and high-pressure controls together.

Control de presión doble. Dos controles que se colocan en el mismo cárter, típicamente controles para alta y baja presión juntos.

Duct. A sealed channel used to convey air from the system to and from the point of utilization.

Conducto. Canal sellado que se emplea para dirigir el aire del sistema hacia y desde el punto de utilización.

Dust mites. Microscopic spiderlike insects. Dust mites and their remains are thought to be a primary irritant to some people.

Ácaros del polvo. Insectos microscópicos parecidos a arañas. Los ácaros del polvo y sus restos se consideran irritantes principales para algunas personas.

Earth-coupled heat pump. Another name for a closed-loop heat pump.

Bomba de calor conectada a tierra. Otro nombre para una bomba de calor de circuito cerrado.

Eccentric. An off-center device that rotates in a circle around a shaft.

Excéntrico. Dispositivo descentrado que gira en un círculo alrededor de un árbol.

ECM. An electronically commutated motor. This DC motor uses electronics to commutate the rotor instead of brushes. It is typically built for under 1 hp.

CEM (ECM en inglés). Un motor conmutado electrónicamente. Este motor CD usa electrónica para conmutar el rotor en lugar de escobillas. Típicamente están hechos para menos de 1 caballo de fuerza.

Eddy current test. A test with an instrument to find potential failures in evaporator or condenser tubes.

Prueba para la corriente de Foucault. Prueba que se realiza con un instrumento para detectar posibles fallas en los tubos del evaporador o del condensador.

EEPROM. An acronym for electrical erasable programmable read-only memory.

EEPROM. Un acrónimo para memoria electrónica, borrable, programable, de sólo-lectura (electrical erasable programmable read-only memory).

Effective temperature. Different combinations of temperature and humidity that provide the same comfort level.

Temperatura efectiva. Diferentes combinaciones de temperatura y humedad que proveen el mismo nivel de comodidad.

Electric forced-air furnace. An electrical resistance type of heating furnace used with a duct system to provide heat to more than one room.

Horno eléctrico de aire soplado. Horno con resistencia eléctrica que se utiliza con un sistema de conductos para proporcionar calefacción a varias habitaciones.

Electric heat. The process of converting electrical energy, using resistance, into heat.

Calor eléctrico. Proceso de convertir energía eléctrica en calor a través de la resistencia.

Electric hydronic boiler. A boiler using electrical resistance heat, which often has a closed-loop piping system to distribute heated water for space heating.

Caldera hidrónica eléctrica. Caldera que utiliza calor proporcionado por una resistencia eléctrica. A menudo cuenta con un sistema cerrado para distribuir agua caliente para usos de calefacción.

Electrical power. Electrical power is measured in watts. One watt is equal to one ampere flowing with a potential of one volt. Watts = Volts $\times$ Amperes ($P = E \times I$)

Potencia eléctrica. La potencia eléctrica se mide en watios. Un watio equivale a un amperio que fluye con una potencia de un voltio. Watios = voltios $\times$ amperios ($P = E \times I$)

Electrical shock. When an electrical current travels through a human body.

Sacudida eléctrica. Paso brusco de una corriente eléctrica a través del cuerpo humano.

Electrodes. Electrodes carry high voltage to the tips, where an arc is created for the purpose of ignition for oil or gas furnaces.

Electrodos. Los electrodos llevan alto voltaje hasta las puntas, donde se crea un arco con el propósito de encender los calefactores de aceite o de gas.

Electromagnet. A coil of wire wrapped around a soft iron core that creates a magnet.

Electroimán. Bobina de alambre devanado alrededor de un núcleo de hierro blando que crea un imán.

Electromechanical controls. Electromechanical controls convert some form of mechanical energy to operate an electrical function, such as a pressure-operated switch.

Controles electromecánicos. Los controles electromecánicos convierten algún tipo de energía mecánica para operar una función eléctrica, tal como un interruptor operado por presión.

Electromotive force. A term often used for voltage indicating the difference of potential in two charges.

Fuerza electromotriz. Término empleado a menudo para el voltaje, indicando la diferencia de potencia entre dos cargas.

Electron. The smallest portion of an atom that carries a negative charge and orbits around the nucleus of an atom.

Electrón. La parte más pequeña de un átomo, con carga negativa y que sigue una órbita alrededor del núcleo de un átomo.

Electronic air filter. A filter that charges dust particles using a high-voltage direct current and then collects these particles on a plate of an opposite charge.

Filtro de aire electrónico. Filtro que carga partículas de polvo utilizando una corriente continua de alta tensión y luego las acumula en una placa de carga opuesta.

Electronic charging scale. An electronically operated scale used to accurately charge refrigeration systems by weight.

Escala electrónica para carga. Escala accionada electrónicamente que se utiliza para cargar correctamente sistemas de refrigeración por peso.

Electronic circuit board. A phenolic type of plastic board that electronic components are mounted on. Typically, the circuits are routed on the back side of the board and the components are mounted on the front with prongs of wire that are soldered to the circuits on the back. These can be mass-produced and coated with a material that keeps the circuits separated if moisture and dust accumulate.

Tarjeta de circuitos electrónicos. Un tipo fenólico de tarjeta plástica en la cual se montan los componentes electrónicos. La tarjeta generalmente tiene los circuitos trazados en la parte de atrás de la tarjeta y los componentes se colocan en el frente con cables que se sueldan al circuito por detrás. Éstas pueden producirse en masa y pueden recubrirse con un material que mantiene separado a los circuitos si se acumula humedad y polvo.

Electronic controls. Controls that use solid-state semiconductors for electrical and electronic functions.

Controles electrónicos. Controles que usan semiconductores de estado sólido para las funciones eléctricas y electrónicas.

Electronic expansion valve (EXV). A metering valve that uses a thermistor as a temperature-sensing element that varies the voltage to a heat motor–operated valve.

Válvula electrónica de expansión (EXV en inglés). Válvula de medición que utiliza un termistor como elemento sensor de temperatura para variar la tensión a una válvula de calor accionada por motor.

Electronic leak detector. An instrument used to detect gases in very small portions by using electronic sensors and circuits.

Detector electrónico de fugas. Instrumento que se emplea para detectar cantidades de gases sumamente pequeñas utilizando sensores y circuitos electrónicos.

Electronic or programmable thermostat. A space thermostat that is electronic in nature with semiconductors that provide different timing programs for cycling the equipment.

Termostato electrónico o programable. Un termostato de espacio que es de naturaleza electrónica con semiconductores que proveen diferentes programas de cronometraje para ciclar el equipo.

Electronic relay. A solid-state relay with semiconductors used to stop, start, or modulate power in a circuit.

Relé electrónico. Un relé de estado sólido con semiconductores para detener, iniciar o modular la electricidad en un circuito.

Electronics. The use of electron flow in conductors, semiconductors, and other devices.

Electrónica. La utilización del flujo de electrones en conductores, semiconductores y otros dispositivos.

Electrostatic precipitator. Another term for an electronic air cleaner.

Precipitador electrostático. Otro término para un limpiador eléctrico del aire.

Emitter. A terminal on a semiconductor.

Emisor. Punto terminal en un semiconductor.

End bell. The end structure of an electric motor that normally contains the bearings and lubrication system.

Extremo acampanado. Estructura terminal de un motor eléctrico que generalmente contiene los cojinetes y el sistema de lubricación.

End-mount motor. An electric motor mounted with tabs or studs fastened to the motor housing end.

Motor con montaje en los extremos. Motor eléctrico con lengüetas o espigas de montaje en el extremo de su caja.

End play. The amount of lateral travel in a motor or pump shaft.

Holgadura. Amplitud de movimiento lateral en un motor o en el árbol de una bomba.

Energy. The capacity for doing work.

Energía. Capacida para realizar un trabajo.

Energy efficiency ratio (EER). An equipment efficiency rating that is determined by dividing the output in Btu/h by the input in watts. This does not take into account the start-up and shutdown for each cycle.

Relación del rendimiento de energía (EER en inglés). Clasificación del rendimiento de un equipo que se determina al dividir la salida en Btu/h por la entrada en watios. Esto no toma en cuenta la puesta en marcha y la parada de cada ciclo.

Energy management. The use of computerized or other methods to manage the power to a facility. This may include cycling off nonessential equipment, such as water fountain pumps or lighting, when it may not be needed. The air-conditioning and heating system is also operated at optimum times when needed instead of around the clock.

Manejo de energía. Cualquier facilidad que usa métodos computarizados o de otro tipo para manejar el consumo de energía en la facilidad. Esto puede incluir apagar los equipos no esenciales en ciclos, tales como las bombas de las fuentes de agua o las luces cuando no son necesarias. Los sistemas de aire acondicionado y de calefacción también se operan en tiempos óptimos cuando son necesarios en vez de todo el tiempo.

Energy recovery ventilator (ERV). In winter, this ventilation equipment recovers heat and moisture from exhaust air. In summer, heat and moisture are removed from incoming air and transferred to exhaust air.

Ventilador de recuperación de energía (ERV en inglés). En el invierno, este equipo de ventilación recupera el calor y la humedad del aire de escape. En el verano remueve el calor y la humedad del aire de entrada y lo transfiere al aire de salida.

Enthalpy. The amount of heat a substance contains from a predetermined base or point.

Entalpía. Cantidad de calor que contiene una sustancia, establecida desde una base o un punto predeterminado.

Entropy. Term to describe the change in heat content of a pound of refrigerant per degree Rankine. Expressed in units of Btu/lb/°R.

Entropía. Término utilizado para describir el cambio en el contenido de calor de una libra de refrigerante por grado Rankine. Expresado en unidades térmicas británicas (Btu)/libras/°Rankine.

Environment. Our surroundings, including the atmosphere.

Medio ambiente. Nuestros alrededores, incluyendo la atmósfera.

Environmental Protection Agency (EPA). A branch of the federal government dealing with the control of ozone-depleting refrigerants and other chemicals and the overall welfare of the environment.

Agencia de Protección Ambiental (EPA en inglés). Una rama del gobierno federal que trata con el control de los refrigerantes y otros químicos que repletan el ozono, y el bienestar completo del ambiente.

EPA. Abbreviation for the Environmental Protection Agency.

EPA. Acrónimo en inglés de Environmental Protection Agency (Agencia de protección ambiental).

Error. The signed or mathematical difference between the set point and the control point of a control process that tells whether the control point is above or below the control set point.

Error. La diferencia matemática o con signo entre el punto de ajuste y el punto de control en un proceso de control, y el cual dice si el punto de control está por encima o por debajo del punto de ajuste.

Error codes. Codes that may read out on an electronic panel that tell the technician what the problem may be.

Códigos de error. Códigos que pueden leerse en un panel electrónico y que le dicen al técnico cuál puede ser el problema.

Ester. A popular synthetic lubricant that performs best with HFCs and HFC-based blends.

Ester. Lubricante sintético de uso común que da óptimos resultados con HFC y mezclas basadas en HFC.

Ethane gas. The fossil fuel, natural gas, used for heat.

Gas etano. Combustible fósil, gas natural, utilizado para generar calor.

Eutectic solution. A mixture of two substances in such a ratio that it will provide for the lowest possible melting temperature for that solution.

Solución eutéctica. Una mezcla de dos sustancias a una razón que proveerá la temperatura de derretimiento más baja para esa solución.

Evacuation. The removal of any gases not characteristic to a system or vessel.

Evacuación. Remoción de los gases no característicos de un sistema o depósito.

Evaporation. The condition that occurs when heat is absorbed by liquid and it changes to vapor.

Evaporación. Condición que ocurre cuando un líquido absorbe calor y se convierte en vapor.

Evaporative condenser (cooling tower). A combination water cooling tower and condenser. The refrigerant from the compressor is routed to the cooling tower where the tower evaporates water to cool the refrigerant. In the evaporative condenser the refrigerant is routed to the tower and the water circulates only in the tower. In a cooling tower, the water is routed to the condenser at the compressor location.

Condensador evaporatorio (torre de enfriamiento). Una combinación de una torre de enfriamiento de agua y un condensador. El refrigerante del compresor se desvía a la torre de enfriamiento donde la torre evapora agua para enfriar al refrigerante. En el condensador evaporatorio el refrigerante se lleva a la torre y el agua circula sólo en la torre. En una torre de enfriamiento, el agua se pasa por el condensador donde está ubicado el compresor.

Evaporative cooling. Devices that provide this type of cooling use fiber mounted in a frame with water slowly running down the fiber. Fresh air is drawn in and through the water-soaked fiber and cooled by evaporation to a point close to the wet-bulb temperature of the air.

Enfriamiento por formación de vapor. Los dispositivos que proporcionan este tipo de enfriamiento utilizan agua que corre sobre fibra colocada en un marco. Se pasa aire nuevo a través de la fibra húmeda, el aire se enfría por evaporación a una temperatura próxima a la de una bombilla húmeda.

Evaporative humidifier. A humidifier that provides moisture on a media surface through which air is forced. The air picks up the moisture from the media as a vapor.

Humidificador por evaporación. Humidificador que proporciona humedad en una superficie a través de la cual pasa aire soplado. El aire recoge la humedad de la superficie, en forma de vapor.

Evaporator. The component in a refrigeration system that absorbs heat into the system and evaporates the liquid refrigerant.

Evaporador. El componente en un sistema de refrigeración que absorbe el calor hacia el sistema y evapora el refrigerante líquido.

Evaporator fan. A forced convector used to improve the efficiency of an evaporator by air movement over the coil.

Abanico del evaporador. Convector forzado que se utiliza para mejorar el rendimiento de un evaporador por medio del movimiento de aire a través de la bobina.

Evaporator pressure-regulating valve (EPR valve). A mechanical control installed in the suction line at the evaporator outlet that keeps the evaporator pressure from dropping below a certain point.

Válvula reguladora de presión del evaporador (EPR en inglés). Reguladora mecánica instalada en el conducto de aspiración de la salida del evaporador; evita que la presión del evaporador caiga hasta alcanzar un nivel por debajo del nivel específico.

Evaporator types. Flooded—an evaporator where the liquid refrigerant level is maintained to the top of the heat exchange coil. Dry type—an evaporator coil that achieves the heat exchange process with a minimum of refrigerant charge.

Clases de evaporadores. Inundado—un evaporador en el que se mantiene el nivel del refrigerante líquido en la parte superior de la bobina de intercambio de calor. Seco—una bobina de evaporador que logra el proceso de intercambio de calor con una mínima cantidad de carga de refrigerante.

Even parallel system. Parallel compressors of equal sizes mounted on a steel rack and controlled by a microprocessor.

Sistema paralelo homogéneo. Compresores de capacidades iguales, montados en paralelo en un bastidor de acero y controlados mediante un microprocesador.

Exhaust valve. The movable component in a refrigeration compressor that allows hot gas to flow to the condenser and prevents it from refilling the cylinder on the downstroke.

Válvula de escape. Componente móvil en un compresor de refrigeración que permite el flujo de gas caliente al condensador y evita que este gas rellene el cilindro durante la carrera descendente.

Expansion joint. A flexible portion of a piping system or building structure that allows for expansion of the materials due to temperature changes.

Junta de expansión. Parte flexible de un sistema de tubería o de la estructura de un edificio que permite la expansión de los materiales debido a cambios de temperatura.

Expansion (metering) device. The component between the high-pressure liquid line and the evaporator that feeds the liquid refrigerant into the evaporator.

Dispositivo de (medición) de expansión. Componente entre el conducto de líquido de alta presión y el evaporador que alimenta el refrigerante líquido hacia el evaporador.

Explosion-proof motor. A totally sealed motor and its connections that can be operated in an explosive atmosphere, such as in a natural gas plant.

Motor a prueba de explosión. Un motor totalmente sellado y con conexiones que puede operarse en una atmósfera explosiva, tal como dentro de una planta de gas natural.

External drive. An external type of compressor motor drive, as opposed to a hermetic compressor.

Motor externo. Motor tipo externo de un compresor, en comparación con un compresor hermético.

External equalizer. The connection from the evaporator outlet to the bottom of the diaphragm on a thermostatic expansion valve.

Equilibrador externo. Conexión de la salida del evaporador a la parte inferior del diafragma en una válvula de expansión termostática.

External heat defrost. A defrost system for a refrigeration system where the heat comes from some external source. It might be an electric strip heater in an air coil, or water in the case of an ice maker. *External* means other than hot gas defrost.

Descongelador de calor externo. Un sistema de descongelación para un sistema de refrigeración en el cual el calor viene de una fuente externa. La misma puede ser un calentador de tira eléctrico en un serpentín o agua en caso de una hielera. *Externo* se refiere a otro tipo de descongelación aparte del de gas caliente.

External motor protection. Motor overload protection that is mounted on the outside of the motor.

Protección externa para motor. Protección para un motor contra la sobrecarga y que está montada en el exterior del motor.

Fahrenheit scale. The temperature scale that places the boiling point of water at 212°F and the freezing point at 32°F.

Escala Fahrenheit. Escala de temperatura en la que el punto de ebullición del agua se encuentra a 212°F y el punto de fusión del hielo a 32°F.

Fail-safe. A programmed or elapsed time when the refrigeration system terminates its defrost cycle and begins another refrigeration cycle.

A prueba de fallo. En el sistema de refrigeración, el tiempo transcurrido programado después del cual el ciclo de descongelación termina y el ciclo de refrigeración comienza.

Fan. A device that produces a pressure difference in air to move it.

Abanico. Dispositivo que produce una diferencia de presión en el aire para moverlo.

Fan cycling. The use of a pressure control to turn a condenser fan on and off to maintain a correct pressure within the system.

Funcionamiento cíclico. La utilización de un regulador de presión para poner en marcha y detener el abanico de un condensador a fin de mantener una presión correcta dentro del sistema.

Fan relay coil. A magnetic coil that controls the starting and stopping of a fan.

Bobina de relé del ventilador. Bobina magnética que regula la puesta en marcha y la parada de un ventilador.

Farad. The unit of capacity of a capacitor. Capacitors in our industry are rated in microfarads.

Faradio. Unidad de capacidad de un capacitador. En nuestro medio, los capacitadores se clasifican en microfaradios.

Feedback loop. The circular data route in a control loop that usually travels from the control medium's sensor to the controller, then to the controlled device, and back into the controlled process to the sensor again as a change in the control point.

Circuito de retroalimentación. Ruta circular de la data en un circuito de control que generalmente va desde el sensor del medio de control al controlador y luego al aparato controlado, y de regreso al proceso controlado y al sensor nuevamente como un cambio en el punto de control.

Female thread. The internal thread in a fitting.

Filete hembra. Filete interno en un accesorio.

Fill or wetted-surface method. Water in a cooling tower is spread out over a wetted surface while air is passed over it to enhance evaporation.

Método de relleno o de superficie mojada. El agua en una torre de refrigeración se extiende sobra una superficie mojada mientras el aire se dirige por encima de la misma para facilitar la evaporación.

Film factor. The relationship between the medium giving up heat and the heat exchange surface (evaporator). This relates to the velocity of the medium passing over the evaporator. When the velocity is too slow, the film between the air and the evaporator becomes greater and becomes an insulator, which slows the heat exchange.

Factor de capa. Relación entre el medio que emite calor y la superficie del intercambiador de calor (evaporador). Esto se refiere a la velocidad del medio que pasa sobre el evaporador. Cuando la velocidad es demasiado lenta, la capa entre el aire y el evaporador se expande y se convierte en un aislador, disminuyendo así la velocidad del intercambio del calor.

Filter. A fine mesh or porous material that removes particles from passing fluids.

Filtro. Malla fina o material poroso que remueve partículas de los fluidos que pasan por él.

Filter drier. A type of refrigerant filter that includes a desiccant material that has an attraction for moisture. The filter drier will remove particles and moisture from refrigerant and oil.

Secador de filtro. Un tipo de filtro de refrigerante que incluye un material desecante el cual tiene una atracción a la humedad. El secador de filtro removerá las partículas y la humedad del aceite y del refrigerante.

Fin comb. A hand tool used to straighten the fins on an air-cooled condenser.

Herramienta para aletas. Herramienta manual utilizada para enderezar las aletas en un condensador enfriado por aire.

Finned-tube evaporator. A copper or aluminum tube that has fins, usually made of aluminum, pressed onto the copper lines to extend the surface area of the tubes.

Evaporador de tubo con aletas. Un tubo de cobre o aluminio que tiene aletas, generalmente de aluminio, colocadas a presión contra las líneas de cobre para extender el área de superficie de los tubos.

Fixed-bore device. An expansion device with a fixed diameter that does not adjust to varying load conditions.

Dispositivo de calibre fijo. Dispositivo de expansión con un diámetro fijo que no se ajusta a las condiciones de carga variables.

Fixed resistor. A nonadjustable resistor. The resistance cannot be changed.

Resistor fijo. Resistor no ajustable. La resistencia no se puede cambiar.

Flame impingement. When flame touches components where it is not supposed to. For example, the flame should not touch the heat exchanger in an oil or gas furnace. The flame should stay in the middle of the heat exchanger.

Impacto de llama. Cuando una llama toca componentes que no está supuesta a tocar. Por ejemplo, la llama no debería tocar el intercambiador de calor en un calefactor de aceite o gas. La llama debería mantenerse en el centro del intercambiador de calor.

Flame-proving device. A device in a gas or oil furnace to ensure there is a flame so that excess fuel will not be released, which may cause overheating in an oil furnace and explosion in a gas appliance.

Aparato probador de llama. Un aparato en un calefactor de gas o aceite que asegura que hay una llama presente para que no se libere combustible de más, lo cual puede causar sobrecalentamiento en un calefactor de aceite y una explosión en enseres de gas.

Flame rectification. In a gas furnace, the flame changes the normal current to direct current. The electronic components in the system will only energize and open the gas valve with a direct current.

Rectificación de llama. En un horno de gas, el calor generado por la llama convierte la corriente normal en corriente directa. Los componentes electrónicos del sistema sólo se activarán para abrir el paso del gas con corriente directa.

Flapper valve. See Reed valve.

Chapaleta. Véase válvula de lámina.

Flare. The angle that may be fashioned at the end of a piece of tubing to match a fitting and create a leak-free connection.

Abocinado. Ángulo que puede formarse en el extremo de una pieza de tubería para emparejar un accesorio y crear una conexión libre de fugas.

Flare nut. A threaded connector used in a flare assembly for tubing.

Tuerca abocinada. Conector de rosca utilizado en un conjunto abocinado para tuberías.

Flash gas. A term used to describe the pressure drop in an expansion device when some of the liquid passing through the valve is changed quickly to a gas and cools the remaining liquid to the corresponding temperature.

Gas instantáneo. Término utilizado para describir la caída de la presión en un dispositivo de expansión cuando una parte del líquido que pasa a través de la válvula se convierte rápidamente en gas y enfría el líquido restante a la temperatura correspondiente.

Float valve or switch. An assembly used to maintain or monitor a liquid level.

Válvula o conmutador de flotador. Conjunto utilizado para mantener o controlar el nivel de un líquido.

Floating head pressure. Letting the head pressure (condensing pressure) fluctuate with the ambient temperature from season to season for lower compression ratios and better efficiencies.

Fluctuar la presión de la carga. Hecho de dejar que la presión de la carga (presión de condensación) fluctúe con la temperatura del ambiente en cada temporada del año, para obtener relaciones de compresión más bajas y mejores rendimientos.

Flooded evaporator. A refrigeration system operated with the liquid refrigerant level very close to the outlet of the evaporator coil for improved heat exchange.

Sistema inundado. Sistema de refrigeración que funciona con el nivel del refrigerante líquido bastante próximo a la salida de la bobina del evaporador para mejorar el intercambio de calor.

Flooding. The term applied to a refrigeration system when the liquid refrigerant reaches the compressor.

Inundación. Término aplicado a un sistema de refrigeración cuando el nivel del refrigerante líquido llega al compresor.

Flue. The duct that carries the products of combustion out of a structure for a fossil- or a solid-fuel system.

Conducto de humo. Conducto que extraer los productos de combustión de una estructura en sistemas de combustible fósil o sólido.

Flue-gas analysis instruments. Instruments used to analyze the operation of fossil fuel–burning equipment such as oil and gas furnaces by analyzing the flue gases.

Instrumentos para el análisis del gas de combustión. Instrumentos utilizados para llevar a cabo un análisis del functionamiento de los quemadores de combustible fósil, como por ejemplo hornos de aceite pesado o gas, a través del estudio de los gases de combustión.

Fluid. The state of matter of liquids and gases.

Fluido. Estado de la materia de líquidos y gases.

Fluid expansion device. Using a bulb or sensor, tube, and diaphragm filled with fluid, this device will produce movement at the diaphragm when the fluid is heated or cooled. A bellows may be added to produce more movement. These devices may contain vapor and liquid.

Dispositivo para la expansión del fluido. Utilizando una bombilla o sensor, un tubo y un diafragma lleno de fluido, este dispositivo generará movimiento en el diafragma cuando se caliente o enfríe el fluido. Se le puede agregar un fuelle para generar aún más movimiento. Dichos dispositivos pueden contener vapor y líquido.

Flush. The process of using a fluid to push contaminants from a system.

Descarga. Proceso de utilizar un fluido para remover los contaminantes de un sistema.

Flux. A substance applied to soldered and brazed connections to prevent oxidation during the heating process.

Fundente. Sustancia aplicada a conexiones soldadas y broncesoldadas para evitar la oxidación durante el proceso de calentamiento.

Foaming. A term used to describe oil when it has liquid refrigerant boiling out of it.

Espumación. Término utilizado para describir el aceite cuando el refrigerante líquido se derrama del mismo.

Foot-pound. The amount of work accomplished by lifting 1 lb of weight 1 ft; a unit of energy.

Libra-pie. Medida de la cantidad de energía o fuerza que se requiere para levantar una libra a una distancia de un pie; unidad de energía.

Force. Energy exerted.

Fuerza. Energía ejercida sobre un objeto.

Forced convection. The movement of fluid by mechanical means.

Convección forzada. Movimiento de fluido por medios mecánicos.

Forced-draft cooling tower. A water cooling tower that has a fan on the side of the tower that pushes air through the tower, as opposed to an induced-draft tower, which has the fan on the side and draws air through the tower.

Torre de enfriamiento de ventilación forzada. Una torre de enfriamiento que tiene un ventilador en el lado de la torre y empuja aire a través de la torre, contrario a una torre de ventilación inducida que tiene un ventilador en el lado de la torre y jala el aire a través de la torre.

Forced-draft evaporator. An evaporator over which air is forced to spread the cooling more efficiently. This term usually refers to a domestic refrigerator or freezer.

Evaporador de tiro forzado. Evaporador encima del cual se envía aire soplado para obtener una distribución más eficaz del frío. Este término suele aplicarse a refrigeradores o congeladores domésticos.

Fossil fuels. Natural gas, oil, and coal formed millions of years ago from dead plants and animals.

Combustibles fósiles. El gas natural, el petróleo y el carbón que se formaron hace millones de años de plantas y animales muertos.

Four-way valve. The valve in a heat pump system that changes the direction of the refrigerant flow between the heating and cooling cycles.

Válvula con cuatro vías. Válvula en un sistema de bomba de calor que cambia la dirección del flujo de refrigerante entre los ciclos de calentamiento y enfriamiento.

Fractionation. When a zeotropic refrigerant blend phase changes, the different components in the blend all have different vapor pressures. This causes different vaporization and condensation rates and temperatures as they phase change.

Fraccionación. Cuando se produce un cambio en la fase de una mezcla zeotrópica de refrigerantes, los diferentes componentes que forman la mezcla tienen presiones de vapor diferentes. Esto provoca diferentes temperaturas y tasas de vaporización y de condensación según cambien de fase.

Freezer burn. The term applied to frozen food when it becomes dry and hard from dehydration due to poor packaging.

Quemadura del congelador. Término aplicado a la comida congelada cuando se seca y endurece debido a la deshidratación ocasionada por el empaque de calidad inferior.

Freeze-up. Excess ice or frost accumulation on an evaporator to the point that airflow may be affected.

Congelación. Acumulación excesiva de hielo o congelación en un evaporador a tal extremo que el flujo de aire puede ser afectado.

Freezing. The change of state of water from a liquid to a solid.

Congelamiento. Cambio de estado del agua de líquido a sólido.

Freon. The previous trade name for refrigerants manufactured by E. I. du Pont de Nemours & Co., Inc.

Freón. Marca registrada previa para refrigerantes fabricados por la compañía E. I. du Pont de Nemours, S.A.

Frequency. The cycles per second (cps) of the electrical current supplied by the power company. This is normally 60 cps in the United States.

Frecuencia. Ciclos por segundo (cps), generalmente 60 cps en los Estados Unidos, de la corriente eléctrica suministrada por la empresa de fuerza motriz.

Friction loss. The loss of pressure in a fluid flow system (air or water) due to the friction of the fluid rubbing on the sides. It is typically measured in feet of equivalent loss.

Pérdida por fricción. La pérdida de presión en un sistema de flujo de fluido (aire o agua) debido a la fricción del fluido al frotar contra los lados. Típicamente se mide en pies de pérdida equivalente.

Front seated. A position on a service valve that will not allow refrigerant flow in one direction.

Sentado delante. Posición en una válvula de servicio que no permite el flujo de refrigerante en una dirección.

Frost back. A condition of frost on the suction line and even the compressor body.

Obturación por congelación. Condición de congelación que ocurre en el conducto de aspiración e inclusive en el cuerpo del compresor.

Frostbite. When skin freezes.

Quemadura por frío. Congelación de la piel.

Frozen. The term used to describe water in the solid state; also used to describe a rotating shaft that will not turn.

Congelado. Término utilizado para describir el agua en un estado sólido; utilizado también para describir un árbol giratorio que no gira.

Fuel oil. The fossil fuel used for heating; a petroleum distillate.

Aceite pesado. Combustible fósil utilizado para calentar; un destilado de petróleo.

Full-load amperage (FLA). The current an electric motor draws while operating under a full-load condition. Also called the run-load amperage and rated-load amperage.

Amperaje de carga total (FLA en inglés). Corriente que un motor eléctrico consume mientras funciona en una condición de carga completa. Conocido también como amperaje de carga de funcionamiento y amperaja de carga standard.

Furnace. Equipment used to convert heating energy, such as fuel oil, gas, or electricity, to usable heat. It usually contains a heat exchanger, a blower, and the controls to operate the system.

Horno. Equipo utilizado para la conversión de energía calórica, como por ejemplo el aceite pesado, el gas o la electricidad, en calor utilizabe. Normalmente contiene un intercambiador de calor, un soplador y los reguladores para accionar el sistema.

Fuse. A safety device used in electrical circuits for the protection of the circuit conductor and components.

Fusible. Dispositivo de seguridad utilizado en circuitos eléctricos para la protección del conductor y los componentes del circuito.

Fusible link. An electrical safety device normally located in a furnace that burns and opens the circuit during an overheat situation.

Cartucho de fusible. Dispositivo eléctrico de seguridad ubicado por lo general en un horno, que quema y abre el circuito en caso de sobrecalentamiento.

Fusible plug. A device (made of low-melting temperature metal) used in pressure vessels that is sensitive to high temperatures and relieves the vessel contents in an overheating situation.

Tapón de fusible. Dispositivo utilizado en depósitos en presión, hecho de un metal que tiene una temperatura de fusión baja. Este dispositivo es sensible a temperaturas altas y alivia el contenido del depósito en caso de sobrecalentamiento.

Gage. An instrument used to indicate pressure.

Calibrador. Instrumento utilizado para indicar presión.

Gage manifold. A tool that may have more than one gage with a valve arrangement to control fluid flow.

Distribuidor de calibrador. Herramienta que puede tener más de un calibrador con las válvulas arregladas a fin de regular el flujo de fluido.

Gage port. The service port used to attach a gage for service procedures.

Orificio de calibrador. Orificio de servicio utilizado con el propósito de fijar un calibrador para procedimientos de servicio.

Gain. The term used to describe the sensitivity of a control, for example, how much change in output signal per degree of temperature change.

Ganancia. El término se usa para describir la sensibilidad de un control, por ejemplo, cuanto cambio en la señal de salida por cada grado de cambio en temperatura.

Gas. The vapor state of matter.

Gas. Estado de vapor de una materia.

Gas-pressure switch. Used to detect gas pressure before gas burners are allowed to ignite.

Conmutador de presión del gas. Utilizado para detectar la presión del gas antes de que los quemadores de gas puedan encenderse.

Gas valve. A valve used to stop, start, or modulate the flow of natural gas.

Válvula de gas. Válvula utilizada para detener, poner en marcha o modular el flujo de gas natural.

Gasket. A thin piece of flexible material used between two metal plates to prevent leakage.

Guarnición. Pieza delgada de material flexible utilizada entre dos piezas de metal para evitar fugas.

Gate. A terminal on a semiconductor.

Compuerta. Punto terminal en un semiconductor.

Gate valve. Valve that has a sliding wedge-shaped piece (gate) inside it that is connected to a handle. The handle can be turned to force the sliding gate into a fitted seat to stop or control water flow. Gate valves have very little pressure drop or resistance to water or fluid flow when wide open because the gate moves all of the way out of the water stream flow.

Válvula de compuerta. Válvula con una pieza (compuerta) deslizante en forma de una cuña que está conectada a un mango. Se gira el mango para forzar la compuerta sobre un asiento para parar o controlar el flujo del agua. Cuando están abiertas, las válvulas de compuerta tienen muy poca caída de la presión o resistencia al flujo del agua o fluido porque la compuerta se mueve fuera del corriente del flujo del agua.

Geothermal heat pump. A heat pump that uses the earth or water in the earth for its heat sources and sinks.

Bomba de calor geotérmica. Bomba de calor que utiliza el suelo o el agua de la tierra como fuente y depósito termal.

Geothermal well. A well dedicated to a geothermal heat pump that draws water from the top of the water column and returns the same water to the bottom of the water column.

Pozo geotermal. Pozo utilizado por una bomba de calor geotérmica que extrae agua de la parte superior de la columna de agua y la devuelve en la parte inferior de la misma columna.

Germanium. A substance from which many semiconductors are made.

Germanio. Sustancia de la que se fabrican muchos semiconductores.

Global warming. An earth-warming process caused by the atmosphere's absorption of the heat energy radiated from the earth's surface.

Calentamiento de la Tierra. Proceso de calentamiento de la Tierra provocado por la absorción de la energía radiada de la superficie de la Tierra, por la atmósfera, en forma de calor.

Global warming potential (GWP). An index that measures the direct effect of chemicals emitted into the atmosphere.

Potencial de calentamiento de la tierra (GWP en inglés). Índice que mide el efecto directo de los productos químicos que se emiten a la atmósfera.

Globe valve. Valve that uses a disk and a seat to stop or vary water or fluid flow. The handle raises or lowers the disk over the valve seat, which is built into the valve body. When the valve is fully open, the water must make two 90° turns to pass through the valve, creating pressure drop.

Válvula de globo. Válvula que utiliza un disco y un asiento para parar o variar el flujo del agua o fluido. El mango eleva o baja el disco sobre el asiento de la válvula construído en el cuerpo de la válvula. Cuando la válvula está completamente abierta, el agua debe girar 90° dos veces para pasar a través de la válvula, ocasionando así una caída de la presión.

Glow coil. A device that automatically reignites a pilot light if it goes out.

Bobina encendedora. Dispositivo que automáticamente vuelve a encender la llama piloto si ésta se apaga.

Glycol. Antifreeze solution used in the water loop of geothermal heat pumps.

Glicol. Líquido anticongelante que se emplea en el circuito de agua de las bombas de calor geotérmicas.

Graduated cylinder. A cylinder with a visible column of liquid refrigerant used to measure the refrigerant charged into a system. Refrigerant temperatures can be dialed on the graduated cylinder.

Cilindro graduado. Cilindro con una columna visible de refrigerante líquido utilizado para medir el refrigerante inyectado al sistema. Las temperaturas del refrigerante pueden marcarse en el cilindro graduado.

Grain. Unit of measure. One pound = 7000 grains.

Grano. Unidad de medida. Una libra equivale a 7.000 granos.

Gram (g). Metric measurement term used to express weight.

Gramo. Término utilizado para referirse a la unidad básica de peso en el sistema métrico.

Grille. A louvered, often decorative, component in an air system at the inlet or the outlet of the airflow.

Rejilla. Componente con celosías, comúnmente decorativo, en un sistema de aire que se encuentra a la entrada o a la salida del flujo de aire.

Grommet. A rubber, plastic, or metal protector usually used where wire or pipe goes through a metal panel.

Guardaojal. Protector de caucho, plástico o metal normalmente utilizado donde un alambre o un tubo pasa a través de una base de metal.

Ground, electrical. A circuit or path for electron flow to earth ground.

Tierra eléctrica. Circuito o trayectoria para el flujo de electrones a la puesta a tierra.

Ground fault circuit interrupter (GFCI). A circuit breaker that can detect very small leaks to ground, which, under certain circumstances, could cause an electrical shock. This small leak, which may not be detected by a conventional circuit breaker, will cause the GFCI circuit breaker to open the circuit.

Disyuntor por pérdidas a tierra (GFCI en inglés). Disyuntor capaz de detectar fugas muy pequeñas hacia tierra, que en determinadas circunstancias pueden provocar descargas eléctricas. Existe la posibilidad de que un disyuntor convencional no sea capaz de detectar las fugas pequeñas, en cuyo caso el disyuntor GFCI abre el circuito.

Ground loop. These loops of plastic pipe are buried in the ground in a closed-loop geothermal heat pump system and contain a heat transfer fluid.

Circuito de tierra. Circuitos de tubos de plástico enterrados y que forman parte de un sistema de bomba geotérmica de circuito cerrado. Dichos tubos contienen un fluido que permite la transferencia de calor.

Ground wire. A wire from the frame of an electrical device to be wired to the earth ground.

Alambre a tierra. Alambre que va desde el armazón de un dispositivo eléctrico para ser conectado a la puesta a tierra.

Grout. A cement-like material injected between the well casing and the drilled hole for a well. Grout adds rigidity and prevents contamination from entering the well.

Lechada. Pasta fina obtenida mezclando cemento y agua, que se inyecta entre la pared del pozo y la perforación para el mismo. Proporciona rigidez e impide la penetración de agentes contaminantes en el pozo.

Guide vanes. Vanes used to produce capacity control in a centrifugal compressor. Also called *prerotation guide vanes.*

Paletas directrices. Paletas utilizadas para producir la regulación de capacidad en un compresor centrífugo. Conocidas también como paletas directrices para prerotación.

Halide refrigerants. Refrigerants that contain halogen chemicals: R-12, R-22, R-500, and R-502 are among them.

Refrigerantes de hálido. Refrigerantes que contienen productos químicos de halógeno: entre ellos se encuentran el R-12, R-22, R-500 y R-502.

Halide torch. A torch-type leak detector used to detect the halogen refrigerants.

Soplete de hálido. Detector de fugas de tipo soplete utilizado para detectar los refrigerantes de halógeno.

Halogens. Chemical substances found in many refrigerants containing chlorine, bromine, iodine, and fluorine.

Halógenos. Sustancias químicas presentes en muchos refrigerantes que contienen cloro, bromo, yodo y flúor.

Hand truck. A two-wheeled piece of equipment that can be used for moving heavy objects.

Vagoneta para mano. Equipo con dos ruedas que puede utilizarse para transportar objetos pesados.

Hanger. A device used to support tubing, pipe, duct, or other components of a system.

Soporte. Dispositivo utilizado para apoyar tuberías, tubos, conductos u otros componentes de un sistema.

Head. Another term for pressure, usually referring to gas or liquid.

Carga. Otro término para presión, refiriéndose normalmente a gas o líquido.

Head pressure control. A control that regulates the head pressure in a refrigeration or air-conditioning system.

Regulador de la presión de la carga. Regulador que controla la presión de la carga en un sistema de refrigeración o de acondicionamiento de aire.

Header. A pipe or containment to which other pipe lines are connected.

Conductor principal. Tubo o conducto al que se conectan otras conexiones.

Heat. Energy that causes molecules to be in motion and to raise the temperature of a substance.

Calor. Energía que ocasiona el movimiento de las moléculas provocando un aumento de temperatura en una sustancia.

Heat anticipator. A device that anticipates the need for cutting off the heating system prematurely so the system does not overshoot the set point temperature.

Anticipador de calor. Dispositivo que anticipa la necesidad de detener la marcha del sistema de calentamiento para que el sistema no exceda la temperatura programada.

Heat coil. A device made of tubing or pipe designed to transfer heat to a cooler substance by using fluids.

Bobina de calor. Dispositivo hecho de tubos, diseñado para transmitir calor a una sustancia más fría por medio de fluidos.

Heat exchanger. A device that transfers heat from one substance to another.

Intercambiador de calor. Dispositivo que transmite calor de una sustancia a otra.

Heat fusion. A process that will permanently join sections of plastic pipe together.

Fusión térmica. Proceso mediante el cual se unen dos piezas de tubo de plástico permanentemente.

Heat of compression. That part of the energy from the pressurization of a gas or a liquid converted to heat.

Calor de compresión. La parte de la energía generada de la presurización de un gas o un líquido que se ha convertido en calor.

Heat of fusion. The heat released when a substance is changing from a liquid to a solid.

Calor de fusión. Calor liberado cuando una sustancia se convierte de líquido a sólido.

Heat of respiration. When oxygen and carbon hydrates are taken in by a substance or when carbon dioxide and water are given off. Associated with fresh fruits and vegetables during their aging process while stored.

Calor de respiración. Cuando se admiten oxígeno e hidratos de carbono en una sustancia o cuando se emiten bióxido de carbono y agua. Se asocia con el proceso de maduración de frutas y legumbres frescas durante su almacenamiento.

Heat pump. A refrigeration system used to supply heat or cooling using valves to reverse the refrigerant gas flow.

Bomba de calor. Sistema de refrigeración utilizado para suministrar calor o frío mediante válvulas que cambian la dirección del flujo de gas del refrigerante.

Heat reclaim. Using heat from a condenser for purposes such as space and domestic water heating.

Reclamación de calor. La utilización del calor de un condensador para propósitos tales como la calefacción de espacio y el calentamiento doméstico de agua.

Heat recovery ventilator. Units that recover heat only, used primarily in winter.

Ventilador de recuperación de calor. Unidades que recuperan calor solamente, usados principalmente en el invierno.

Heat sink. A low-temperature surface to which heat can transfer.

Fuente fría. Superficie de temperatura baja a la que puede transmitírsele calor.

Heat tape. Electric resistance wires embedded into a flexible housing usually wrapped around a pipe to keep it from freezing.

Cinta calefactora. Resistencia eléctrica incrustada en una cubierta flexible normalmente instalada alrededor de un tubo para impedir su congelación.

Heat transfer. The transfer of heat from a warmer to a colder substance.

Transmisión de calor. Cuando se transmite calor de una sustancia más caliente a una más fría.

Helix coil. A bimetal formed into a helix-shaped coil that provides longer travel when heated.

Bobina en forma de hélice. Bimetal encofrado en una bobina en forma de hélice que provee mayor movimiento al ser calentado.

HEPA filter. An abbreviation for high-efficiency particulate arrestor. These filters are used when a high degree of filtration is desired or required.

Filtro HEPA. Abreviatura para filtro de partículas de alto rendimiento. Este tipo de filtro se utiliza cuando se requiere un elevado grado de filtrado.

Hermetic compressor. A motor and compressor that are totally sealed by being welded in a container.

Compresor hermético. Un motor y un compresor que están totalmente sellados al ser soldados al contenedor.

Hermetic system. An enclosed refrigeration system where the motor and compressor are sealed within the same system with the refrigerant.

Sistema hermético. Sistema de refrigeracíon cerrado donde el motor y el compresor se obturan dentro del mismo sistema con el refrigerante.

Hertz. Cycles per second.

Hertz. Ciclos por segundo.

Hg. Abbreviation for the element mercury.

Hg. Abreviatura del elemento mercurio.

Hidden heat. See Latent heat.

Calor oculto. Véase Calor latente.

High-pressure control. A control that stops a boiler heating device or a compressor when the pressure becomes too high.

Regulador de alta presión. Regulador que detiene la marcha del dispositivo de calentamiento de una caldera o de un compresor cuando la presión alcanza un nivel demasiado alto.

High side. A term used to indicate the high-pressure or condensing side of the refrigeration system.

Lado de alta presión. Término utilizado para indicar el lado de alta presión o de condensación del sistema de refrigeración.

High-temperature refrigeration. A refrigeration temperature range starting with evaporator temperatures no lower than 35°F, a range usually used in air conditioning (cooling).

Refrigeración a temperatura alta. Margen de la temperatura de refrigeración que comienza con temperaturas de evaporadores no menores de 35°F (2°C). Este margen se utiliza normalmente en el acondicionamiento de aire (enfriamiento).

High-vacuum pump. A pump that can produce a vacuum in the low micron range.

Bomba de vacío alto. Bomba que puede generar un vacío dentro del margen de micrón bajo.

Hollow wall anchor. Can be used in plaster, wallboard, gypsum board, and similar materials. Once the anchor has been set, the screw may be removed as often as necessary.

Anclaje hueco para pared. Puede usarse en yeso, panel de yeso y materiales similares. Una vez se coloca el anclaje, el tornillo puede removerse tantas veces como sea necesario.

Horsepower (hp). A unit equal to 33000 ft-lb of work per minute.

Potencia en caballos (hp en inglés). Unidad equivalente a 33.000 libras-pies de trabajo por minuto.

Hot gas. The refrigerant vapor as it leaves the compressor. This is often used to defrost evaporators.

Gas caliente. El vapor del refrigerante al salir del compresor. Esto se utiliza con frecuencia para descongelar evaporadores.

Hot gas bypass. Piping that allows hot refrigerant gas into the cooler low-pressure side of a refrigeration system usually for system capacity control.

Desviación de gas caliente. Tubería que permite la entrada de gas caliente del refrigerante en el lado más frío de baja presión de un sistema de refrigeración, normalmente para la regulación de la capacidad del sistema.

Hot gas defrost. A system where the hot refrigerant gases are passed through the evaporator to defrost it.

Descongelación con gas caliente. Sistema en el que los gases calientes del refrigerante se pasan a través del evaporador para descongelarlo.

Hot gas line. The tubing between the compressor and condenser.

Conducto de gas caliente. Tubería entre el compresor y el condensador.

Hot junction. That part of a thermocouple or thermopile where heat is applied.

Empalme caliente. El lugar en un termopar o pila termoeléctrica donde se aplica el calor.

Hot pulldown. The process of lowering the refrigerated space to the design temperature after it has been allowed to warm up considerably over this temperature.

Descenso caliente. Proceso de bajar la temperatura del espacio refrigerado a la temperatura de diseño luego de habérsele permitido calentarse a un punto sumamente superior a esta temperatura.

Hot surface ignition. A silicon carbide or similar substance is placed in the gas stream of a gas furnace and allowed to get very hot. When the gas impinges on this surface, immediate ignition should occur.

Encendido por superficie caliente. Se coloca una sustancia de carburo de silicona o una semejante en el flujo de gas de un horno. Cuando se calienta la sustancia y el gas toca la superficie, se produce un encendido instantáneo.

Hot water heat. A heating system using hot water to distribute the heat.

Calor de agua caliente. Sistema de calefacción que utiliza agua caliente para la distribución del calor.

Hot wire. The wire in an electrical circuit that has a voltage potential between it and another electrical source or between it and ground.

Conductor electrizado. Conductor en un circuito eléctrico a través del cual fluye la tensión entre éste y otra fuente de electricidad o entre éste y la tierra.

Humidifier. A device used to add moisture to the air.

Humedecedor. Dispositivo utilizado para agregarle humedad al aire.

Humidistat. A control operated by a change in humidity.

Humidistato. Regulador activado por un cambio en la humedad.

Humidity. Moisture in the air.

Humedad. Vapor de agua existente en el ambiente.

Hunting. The open and close throttling of a valve that is searching for its set point.

Caza. El abrir y cerrar de una válvula que está buscando su punto de ajuste.

Hydraulics. Producing mechanical motion by using liquids under pressure.

Hidráulico. Generación de movimiento mecánico por medio de líquidos bajo presión.

Hydrocarbons (HCs). Organic compounds containing hydrogen and carbon found in many heating fuels.

Hidrocarburos (HCs en inglés). Compuestos orgánicos que contienen el hidrógeno y el carbón presentes en muchos combustibles de calentamiento.

Hydrochlorofluorocarbons (HCFCs). Refrigerants containing hydrogen, chlorine, fluorine, and carbon, thought to contribute to the depletion of the ozone layer, although not to the extent of chlorofluorocarbons.

Hidrocloroflurocarburos (HCFCs en inglés). Líquidos refrigerantes que contienen hidrógeno, cloro, flúor y carbono, y que, según algunos, han contribuido a la reducción de la capa de ozono aunque no en tal grado como los cloroflurocarburos.

Hydrofluorocarbon (HFC). A chlorine-free refrigerant containing hydrogen, fluorine, and carbon with zero ozone depletion potential.

Hidroflurocarbono (HFC en inglés). Refrigerante libre de cloro, compuesto de hidrógeno, fluoro y carbono que no tiene efectos perjudiciales sobre la capa de ozono.

Hydrofluoroether (HFE). A nontoxic, secondary fluid used in refrigeration systems. HFEs have a zero ozone depletion potential.

Hidrofluroéter (HFE en inglés). Fluido secundario no tóxico utilizado en sistemas de refrigeración. Los HFE no tienen efectos perjudiciales sobre la capa de ozono.

Hydrometer. An instrument used to measure the specific gravity of a liquid.

Hidrómetro. Instrumento utilizado para medir la gravedad específica de un líquido.

Hydronic. Usually refers to a hot water heating system.

Hidrónico. Normalmente se refiere a un sistema de calefacción de agua caliente.

Hygrometer. An instrument used to measure the amount of moisture in the air.

Higrómetro. Instrumento utilizado para medir la cantidad de humedad en el aire.

Idler. A pulley on which a belt rides. It does not transfer power but is used to provide tension or reduce vibration.

Polea tensora. Polea sobre la que se mueve una correa. No sirve para transmitir potencia, pero se utiliza para proveer tensión o disminuir la vibración.

Ignition or flame velocity. The rate at which the flame travels through the air and fuel mixture. Different gases have a different rate. The faster the rate of travel, the more explosive the gas.

Velocidad de ignición o de llama. La velocidad a la cual una llama viaja a través de la mezcla de aire y combustible. Diferentes gases tienen distintas velocidades. Mientras más rápido el velocidad del viaja, más explosivo en el gas.

Ignition transformer. Provides a high-voltage current, usually to produce a spark to ignite a furnace fuel, either gas or oil.

Transformador para encendido. Provee una corriente de alta tensión, normalmente para generar una chispa a fin de encender el combustible de uno horno, sea gas o aceite pesado.

Impedance. A form of resistance in an alternating current circuit.

Impedancia. Forma de resistencia en un circuito de corriente alterna.

Impeller. The rotating part of a pump that causes the centrifugal force to develop fluid flow and pressure difference.

Impulsor. Pieza giratoria de una bomba que hace que la fuerza centrífuga desarrolle flujo de fluido y una diferencia en presión.

Impingement. The condition in a gas or oil furnace when the flame strikes the sides of the combustion chamber, resulting in poor combustion efficiency.

Golpeo. Condición que occurre en un horno de gas o de aceite pesado cuando la llama golpea los lados de la cámara de combustión. Esta condición trae como resultado un rendimiento de combustión pobre.

Inclined water manometer. Indicates air pressures in very low-pressure systems.

Manómetro de agua inclinada. Señala las presiones de aire en sistemas de muy baja presión.

Indoor air quality (IAQ). This term generally refers to the study or research of air quality within buildings and the procedures used to improve air quality.

Calidad del aire en el interior (IAQ en inglés). Generalmente, este término hace referencia al estudio o investigación de la calidad del aire en el interior de los edificios, así como a los procesos empleados para su mejora.

Induced magnetism. Magnetism produced, usually in a metal, from another magnetic field.

Magnetismo inducido. Magnetismo generado, normalmente en un metal, desde otro campo magnético.

Inductance. An induced voltage producing a resistance in an alternating current circuit.

Inductancia. Tensión inducida que genera una resistencia en un circuito de corriente alterna.

Induction motor. An alternating current motor where the rotor turns from induced magnetism from the field windings.

Motor inductor. Motor de corriente alterna donde el rotor gira debido al magnetismo inducido desde los devanados inductores.

Inductive circuit. When the current in a circuit lags the voltage by 90°.

Circuito inductivo. Cuando la corriente en un circuito está atrasada al voltaje por 90°.

Inductive reactance. A resistance to the flow of an alternating current produced by an electromagnetic induction.

Reactancia inductiva. Resistencia al flujo de una corriente alterna generada por una inducción electromagnética.

Inefficient equipment. Equipment that is not operating at its design level of capacity because of some fault in the equipment, such as a cylinder not pumping in a multicylinder compressor.

Equipo ineficiente. Equipo que no está operando al nivel de capacidad al que fue diseñado debido a alguna falla en el equipo, tal como que un cilindro no esté bombeando en un compresor de múltiples cilindros.

Inert gas. A gas that will not support most chemical reactions, particularly oxidation.

Gas inerte. Gas incapaz de resistir la mayoría de las reacciones químicas, especialmente la oxidación.

Infiltration. Air that leaks into a structure through cracks, windows, doors, or other openings due to less pressure inside the structure than outside the structure.

Infiltración. Penetración de aire en una estructura a través de grietas, ventanas, puertas u otras aberturas debido a que la presión en el interior de la estructura es menor que en el exterior.

Infrared humidifier. A humidifier that has infrared lamps with reflectors to reflect the infrared energy onto the water. The water evaporates rapidly into the duct airstream and is carried throughout the conditioned space.

Humidificador por infrarrojos. Humidificador equipado con lámparas de infrarrojos cuyos reflectores reflejan la energía infrarroja sobre el agua haciendo que ésta se evapore rápidamente hacia el conducto de aire para ser transportada en el espacio acondicionado.

Infrared rays. The rays that transfer heat by radiation.

Rayos infrarrojos. Rayos que transmiten calor por medio de la radiación.

In. Hg vacuum. The atmosphere will support a column of mercury 29.92 in. high. To pull a complete vacuum in a refrigeration system, the pressure inside the system must be reduced to 29.92 in. Hg vacuum.

Vacío en mm Hg. La atmósfera soporta una columna de mercurio de 760 mm. Para poder crear un vacío completo en un sistema de refrigeración, la presión interna debe descender a 760 mm Hg.

Inherent motor protection. This is provided by internal protection such as a snap-disc or a thermistor.

Protección de motor inherente. Ésta es provista por una protección interna tal como un disco de encaje o un termisor.

Injection. Term used to describe any hydronic system that uses a circulator pump to push water into a secondary heating loop or circuit.

Inyección. Término utilizado para describir cualquier sistema hidrónico que usa una bomba de circulación para empujar el agua en un circuito de calefacción secundario.

In-phase. When two or more alternating current circuits have the same polarity at all times.

En fase. Cuando dos o más circuitos de corriente alterna tienen siempre la misma polaridad.

Input/output board. A solid-state electronic board that receives a signal from a microprocessor and initiates a response to cycle compressors or initiate defrost.

Tarjeta de entrada/salida. Placa electrónica de estado sólido que recibe una señal procedente de un microprocesador y activa o desactiva los compresores, o inicia el proceso de descongelación.

Insulation, electric. A substance that is a poor conductor of electricity.

Aislamiento eléctrico. Sustancia que es un conductor pobre de electricidad.

Insulation, thermal. A substance that is a poor conductor of the flow of heat.

Aislamiento térmico. Sustancia que es un conductor pobre de flujo de calor.

Insulator. A material with several electrons in the outer orbit of the atom making them poor conductors of electricity or good insulators. Examples are glass, rubber, and plastic.

Aislamiento. Material con varios electrones en la órbita exterior del átomo, que los convierte en malos conductores de electricidad o en buenos aislantes, por ejemplo vidrio, caucho y plástico.

Integral controller. A control mode where the controller will change the output signal according to the length of time the error or offset exists. It will calculate the amount of error that exists over a specific time interval.

Controlador integral. Un modo de control en cual el controlador cambia la señal de salida de acuerdo al tiempo de duración del error o de la compensación. El mismo calcula la cantidad de error que existe a través de un intervalo específico de tiempo.

Interlocking components. Mechanical and electrical interlocks that are used to prevent a piece of equipment from starting before it is safe to start. For example, the chilled-water pump and the condenser water pump must both be started before the compressor in a water-cooled chilled-water system.

Componentes con enclavamiento. Enclavamientos mecánicos y eléctricos se usan para evitar que una pieza de equipo se encienda antes de ser seguro que encienda. Por ejemplo, la bomba de agua enfriada y la bomba de agua del condensador deben encenderse antes que el compresor en un sistema de enfriamiento de agua enfriado por agua.

Intermittent ignition. Ignition system for a gas furnace that operates only when needed or when the furnace is operating.

Encendido interrumpido. Sistema de encendido para un horno de gas que funciona solamente cuando es necesario o cuando el horno está trabajando.

Internal heat defrost. Heat provided for defrost from inside the system, for example, hot gas defrost.

Descongelador por calor interno. El calor para descongelar se provee desde adentro del sistema, por ejemplo, descongelación de gas caliente.

Internal motor overload. An overload that is mounted inside the motor housing, such as a snap-disc or thermistor.

Sobrecarga interna del motor. Una sobrecarga que se monta dentro del cárter del motor tal como disco de encaje o un termisor.

Interrupted ignition. Ignition method that uses an electric spark only at the beginning of the burner's run cycle to ignite the atomized fuel.

Ignición interrumpida. Un método de ignición que utiliza una chispa eléctrica solamente al comienzo del ciclo de marcha del quemador para encender el combustible atomizado.

Inverter. A device that alters the frequency of an electronically altered sine wave, which will affect the speed of an alternating current motor.

Inversor. Dispositivo que hace alternar la frecuencia de una onda de signos alternados electrónicamente. Esta inversión tiene un efecto sobre la velocidad de un motor de corriente alterna.

Ion generator. A device that charges particles. These particles are then passed through positive- and negative-charged plates where the particles are repelled by the positive-charged plates and attracted to the negative-charged plates.

Generador de iones. Dispositivo cargador de partículas, las cuales se pasan entre placas con carga positiva y negativa, causando el rechazo de las partículas por las placas con carga positiva, y la atracción por las placas con carga negativa.

Isobars. Lines on a chart or graph that represent a constant pressure.

Isobarras. Líneas en una gráfica que representan una presión constante.

Isolation relays. Components used to prevent stray unwanted electrical feedback that can cause erratic operation.

Relés de aislación. Componentes utilizados para evitar la realimentación eléctrica dispersa no deseada que puede ocasionar un funcionamiento errático.

Isotherms. Lines on a chart or graph that represent a constant temperature.

Isotermas. Líneas en una gráfica que representan una temperatura constante.

Joule (J). Metric measurement term used to express the quantity of heat.

Joule (J en inglés). Término utilizado para referirse a la unidad básica de cantidad de calor en el sistema métrico.

Junction box. A metal or plastic box within which electrical connections are made.

Caja de empalme. Caja metálica o plástica dentro de la cual se nacen conexiones eléctricas.

Kelvin. A temperature scale where absolute 0 equals 0 or where molecular motion stops at 0. It has the same graduations per degree of change as the Celsius scale.

Escala absoluta. Escale de temperaturas donde el cero absoluto equivale a 0 ó donde el movimiento molecular se detiene en 0. Tiene las mismas graduaciones por grado de cambio que la escala Celsio.

Kilopascal. A metric unit of measurement for pressure used in the air-conditioning, heating, and refrigeration field. There are 6.89 kilopascals in 1 psi.

Kilopascal. Unided métrica de medida de presión utilizada en el ramo del acondicionamiento de aire, calefacción y refrigeración. 6.89 kilopascales equivalen a 1 psi.

Kilowatt. A unit of electrical power equal to 1000 watts.

Kilowatio. Unidad eléctrica de potencia equivalente a 1.000 watios.

Kilowatt-hour. 1 kilowatt (1000 watts) of energy used for 1 hour.

Kilowatio hora. Unidad de energía equivalente a la que produce un kilowatio durante una hora.

King valve. A service valve at the liquid receiver's outlet in a refrigeration system.

Válvula maestra. Válvula de servicio ubicada en el receptor del líquido.

Lag shield anchors. Used with lag screws to secure screws in masonry materials.

Anclajes de tornillos barraqueros. Se usan con tornillos barraqueros para asegurar tornillos en materiales de albañilería.

Latent heat (hidden heat). Heat energy absorbed or rejected when a substance is changing state and there is no change in temperature.

Calor latente (calor oculto). Energía calórica absorbida o rechazada cuando una sustancia cambia de estado y no se experimentan cambios de temperatura.

Latent heat of condensation. The latent heat given off when refrigerant condenses.

Calor latente de la condensación. Calor latente producido por la condensación del refrigerante.

Latent heat of vaporization. The latent heat absorbed when refrigerant evaporates.

Calor latente de vaporización. Calor latente absorbido por la evaporación del refrigerante.

Leads. Extended surfaces inside a heat exchanger used to enhance the heat transfer qualities of the heat exchanger.

Extensiones. Superficies extendidas dentro de un intercambiador de calor que se usan para mejorar las cualidades de transferencia de calor del intercambiador de calor.

Leak detector. Any device used to detect leaks in a pressurized system.

Detector de fugas. Cualquier dispositivo utilizado para detectar fugas en un sistema presurizado.

Lever truck. A long-handled, two-wheeled device that can be used to lift and assist in moving heavy objects.

Vagoneta con palanca. Dispositivo con dos ruedas y una manivela larga que puede utilizarse para levantar y ayudar a transportar objetos pesados.

Light-emitting diode (LED). A diode that emits light and often is used as a self-diagnostic tool within a microprocessor. The lights can spell out error codes or flash numbers.

Diodo emisor de luz (LED en inglés). Diodo que emite una luz. A menudo se utiliza como dispositivo de autodiagnóstico en un microprocesador. Las luces pueden indicar códigos de error o números.

Lignin. A fibrous material found in trees and that strengthens the cell wall.

Lignina. Tejido fibroso presente en la madera y que sirve para reforzar la pared celular.

Limit control. A control used to make a change in a system, usually to stop it when predetermined limits of pressure or temperature are reached.

Regulador de límite. Regulador utilizado para realizar un cambio en un sistema, normalmente para detener su marcha cuando se alcanzan niveles predeterminados de presión o de temperatura.

Limit switch. A switch that is designed to stop a piece of equipment before it does damage to itself or the surroundings, for example, a high limit on a furnace or an amperage limit on a motor.

Interruptor de límite. Interruptor que está diseñado para detener una pieza de equipo antes que ésta se haga daño o dañe sus alrededores, por ejemplo, un límite alto en un calefactor o un límite de amperaje en un motor.

Line set. A term used for tubing sets furnished by the manufacturer.

Juego de conductos. Término utilizado para referise a los juegos de tubería suministrados por el fabricante.

Line tap valve. A device that may be used for access to a refrigerant line.

Válvula de acceso en línea. Dispositivo que puede utilizarse para acceder al tubo del refrigerante.

Line-voltage thermostat. A thermostat that switches line voltage. For example, it is used for electric baseboard heat.

Termostato de voltaje de línea. Un termostato que interrumpe el voltaje de línea. Por ejemplo, para los calentadores eléctricos de rodapié.

Line wiring diagram. Sometimes called a ladder diagram, this type of diagram shows the power-consuming devices between the lines. Usually, the right side of the diagram consists of a common line.

Diagrama del cableado de línea. También conocido como diagrama de escalera, este tipo de diagrama muestra los dispositivos de consumo de corriente que hay entre las líneas. Generalmente, la línea común está en el lado derecho del diagrama.

Liquid. A substance where molecules push outward and downward and seek a uniform level.

Líquido. Sustancia donde las moléculas empujan hacia afuera y hacia abajo y buscan un nivel uniforme.

Liquid charge bulb. A type of charge in the sensing bulb of the thermostatic expansion valve. This charge is characteristic of the refrigerant in the system and contains enough liquid so that it will not totally boil away.

Bombilla de carga líquida. Tipo de carga en la bombilla sensora de la válvula de expansión termostática. Esta carga es característica del refrigerante en el sistema y contiene suficiente líquido para que el mismo no se evapore completamente.

Liquid collector design. A solar collector design in which either water or an antifreeze solution is passed through the collector and heated by the sun.

Colector de líquido. Colector a través del cual pasa el agua o el anticongelante para ser calentados por la energía solar.

Liquid-filled remote bulb. A remote-bulb thermostat that is completely liquid filled, such as the mercury bulb on some gas furnace pilot safety devices.

Bombillo remoto lleno de líquido. Un termostato de bombillo remoto que está completamente lleno con líquido, tal como el bulbo de mercurio en algunos dispositivos de seguridad de los calefactores de gas.

Liquid floodback. Liquid refrigerant returning to the compressor's crankcase during the running cycle.

Regreso de líquido. El regresar del refrigerante líquido al cárter del cigüeñal del compresor durante el ciclo de marcha.

Liquid hammer. The momentum force of liquid causing a noise or a disturbance when hitting against an object.

Martillo líquido. La fuerza mecánica de un líquido que causa un ruido o un disturbio cuando choca con un objeto.

Liquid line. A term applied in the industry to refer to the tubing or piping from the condenser to the expansion device.

Conducto de líquido. Término aplicado en nuestro medio para referirse a la tubería que va del condensador al dispositivo de expansión.

Liquid nitrogen. Nitrogen in liquid form.

Nitrógeno líquido. Nitrógeno en forma líquida.

Liquid receiver. A container in the refrigeration system where liquid refrigerant is stored.

Receptor del líquido. Recipiente en el sistema de refrigeración donde se almacena el refrigerante líquido.

Liquid refrigerant charging. The process of allowing liquid refrigerant to enter the refrigeration system through the liquid line to the condenser and evaporator.

Carga para refrigerante líquido. Proceso de permitir la entrada del refrigerante líquido al condensador y al evaporador en el sistema de refrigeración a través del conducto de líquido.

Liquid refrigerant distributor. This device is used between the expansion valve and the evaporator on multiple circuit evaporators to evenly distribute the refrigerant to all circuits.

Distribuidor de refrigerante líquido. Este aparato se usa entre la válvula de expansión y el evaporador en los evaporadores de múltiples circuitos para distribuir el refrigerante a todos los circuitos.

Liquified petroleum. Liquified propane, butane, or a combination of these gases. The gas is kept as a liquid under pressure until ready to use.

Petróleo licuado. Propano o butano licuados, o una combinación de estos gases. El gas se mantiene en estado líquido bajo presión hasta que se encuentre listo para usar.

Liquid slugging. A large amount of liquid refrigerant in the compressor cylinder, usually causing immediate damage.

Relleno de líquido. Acumulación de una gran cantidad de refrigerante líquido en el cilindro del compresor, que normalmente provoca una avería inmediata.

Lithium-bromide. A type of salt solution used in an absorption chiller.

Bromuro de litio. Tipo de solución salina utilizada en un enfriador por absorción.

Load matching. Trying to always match the capacity of the refrigeration or air-conditioning system with that of the heat load put on the evaporators.

Adaptación de carga. Hecho de intentar adaptar la capacidad del sistema de refrigeración o aire acondicionado a la carga térmica que deben soportar los evaporadores.

Load shed. Part of an energy management system where various systems in a structure may be cycled off to conserve energy.

Despojo de carga. Parte de un sistema de manejo de energía en el cual varios sistemas en una estructura pueden apagarse para conservar energía.

Locked-rotor amperage (LRA). The current an electric motor draws when it is first turned on. This is normally five times the full-load amperage.

Amperaje de rotor bloqueado (LRA en inglés). Corriente que un motor eléctrico consume al ser encendido, la cual generalmente es cinco veces mayor que el amperaje de carga completa.

Low-ambient control. Various types of controls that are used to control head pressure in air-cooled air-conditioning and refrigeration systems that must operate year-round or in cold weather.

Control de ambiente bajo. Varios tipos de controles que se usan para controlar la presión en los sistemas de aires acondicionados y refrigeración, enfriados por aire, y que tienen que operar todo el año o en climas fríos.

Low-boy furnace. This furnace is approximately 4 ft high, and the air intake and discharge are both at the top.

Horno bajo. Este tipo de horno tiene una altura de aproximadamente 1,2 metros, con la toma y evacuación del aire situadas en la parte de arriba.

Low-loss fitting. A fitting that is fastened to the end of a gage manifold that allows the technician to connect and disconnect gage lines with a minimum of refrigerant loss.

Acoplamiento de poca pérdida. Un tipo de acoplamiento que se conecta en un extremo de un colector de calibración y que le permite al técnico conectar y desconectar las líneas de calibración con una pérdida mínima de refrigerante.

Low-pressure control. A pressure switch that can provide low charge protection by shutting down the system on low pressure. It can also be used to control space temperature.

Regulador de baja presión. Commutador de presión que puede proveer protección contra una carga baja al detener el sistema si éste alcanza una presión demasiado baja. Puede utilizarse también para regular la temperatura de un espacio.

Low side. A term used to refer to that part of the refrigeration system that operates at the lowest pressure, between the expansion device and the compressor.

Lado bajo. Término utilizado para referirse a la parte del sistema de refrigeración que funciona a niveles de presión más baja, entre el dispositivo de expansión y el compresor.

Low-temperature refrigeration. A refrigeration temperature range starting with evaporator temperatures no higher than 0°F for storing frozen food.

Refrigeración a temperatura baja. Margen de la temperatura de refrigeración que comienza con temperaturas de evaporadores no mayores de 0°F (−18°C) para almacenar comida congelada.

Low-voltage thermostat. The typical thermostat used for residential and commercial air-conditioning and heating equipment to control space temperature. The supplied voltage is 24 V.

Termostato de bajo voltaje. El termostato típico que se usa en el equipo de aire acondicionado y de calefacción comercial y residencial para controlar la temperatura de un espacio. El voltaje suplido es de 24 voltios.

LP fuel. Liquefied petroleum, propane or butane. A substance used as a gas for fuel. It is transported and stored in the liquid state.

Combustible PL. Petróleo licuado, propano o butano. Sustancia utilizada como gas para combustible. El petróleo licuado se transporta y almacena en estado líquido.

Magnetic field. A field or space where magnetic lines of force exist.

Campó magnético. Campo o espacio donde existen líneas de fuerza magnética.

Magnetic overload protection. This protection reads the actual current draw of the motor and is able to shut it off based on actual current, versus the heat-operated thermal overloads, which are sensitive to the ambient heat of a hot cabinet.

Protección de sobrecarga magnética. Esta protección lee la toma de corriente actual del motor y es capaz de apagarlo basado en la corriente actual; esto es contrario a las sobrecargas termales operadas por calor, las cuales son sensibles al calor ambiental de un gabinete caliente.

Magnetism. A force causing a magnetic field to attract ferrous metals, or where like poles of a magnet repel and unlike poles attract each other.

Magnetismo. Fuerza que hace que un campo magnético atraiga metales férreos, o cuando los polos iguales de un imán se rechazan y los opuestos se atraen.

Makeup air. Air, usually from outdoors, provided to make up for the air used in combustion.

Aire de compensación. Aire, normalmente procedente del exterior, que se utiliza para compensar aquél utilizado en la combustión.

Makeup water. Water that is added back into any circulating water system due to loss of water. Makeup water in a cooling tower may be quite a large volume.

Agua de compensación. Agua que se añade a cualquier sistema de circulación de agua debido a la pérdida de agua. El agua de compensación en una torre de enfriamiento puede ser un volumen bastante grande.

Male thread. A thread on the outside of a pipe, fitting, or cylinder; an external thread.

Filete macho. Filete en la parte exterior de un tubo, accesorio o cilindro; filete externo.

Mandrel (clinching). Part of the pin rivet assembly that is pulled to the breaking point to swell the rivet into the drilled hole.

Mandrel (remachando). En la instalación con remaches, la parte que se tira hasta el límite para hinchar el remache en el hueco perforado.

Manifold. A device where multiple outlets or inlets can be controlled with valves or other devices. Our industry typically uses a gas manifold with orifices for gas-burning appliances and gage manifolds used by technicians.

Colector. Aparato desde el cual se pueden controlar varias entradas y salidas con válvulas u otros aparatos. Nuestra industria generalmente usa un colector de gas con orificios para los enseres que queman gas y colectores de calibración que los usan los técnicos.

Manometer. An instrument used to check low vapor pressures. The pressures may be checked against a column of mercury or water.

Manómetro. Instrumento utilizado para revisar las presiones bajas de vapor. Las presiones pueden revisarse comparándolas con una columna de mercurio o de agua.

Manual reset. A safety control that must be reset by a person, as opposed to automatically reset, to call attention to the problem. An electrical breaker is a manual reset device.

Reinicio manual. Un control de seguridad que una persona tiene que reiniciar, contrario a un reinicio automático, para llamar atención al problema. Un cortacircuito eléctrico es un aparato de reinicio manual.

Mapp gas. A composite gas similar to propane that may be used with air.

Gas Mapp. Gas compuesto similar al propano que puede utilizarse con aire.

Marine water box. A water box on a chiller or condenser with a removable cover.

Caja marina para agua. Caja para agua en un enfriador o condensador con un capón desmontable.

Mass. Matter held together to the extent that it is considered one body.

Masa. Materia compacta que se considera un solo cuerpo.

Mass spectrum analysis. An absorption machine factory leak test performed using helium.

Análisis del límite de masa. Prueba para fugas y absorción llevada a cabo en la fábrica utilizando helio.

Matter. A substance that takes up space and has weight.

Materia. Sustancia que ocupa espacio y tiene peso.

Mechanical controls. A control that has no connection to power, such as a water-regulating valve or a pressure relief valve.

Controles mecánicos. Un control que no tiene conexión a corriente, tales como una válvula reguladora de agua o una válvula de alivio de presión.

Medium-temperature refrigeration. Refrigeration where evaporator temperatures are 32°F or below, normally used for preserving fresh food.

Refrigeración a temperatura media. Refrigeración, donde las temperaturas del evaporador son 32°F (0°C) o menos, utilizada generalmente para preservar comida fresca.

Megohmmeter. An instrument that can detect very high resistances, in millions of ohms. A megohm is equal to 1,000,000 ohms.

Megaohmnímetro. Un instrumento que puede detectar resistencias muy altas, de millones de ohmios. Un megaohmio es equivalente a 1.000.000 de ohmios.

Melting point. The temperature at which a substance will change from a solid to a liquid.

Punto de fusión. Temperatura a la que una sustancia se convierte de sólido a líquido.

Memory. Electronic storage space located internally in a controller.

Memoria. Almacenes internos electrónicos en la computadora.

Mercury bulb. A glass bulb containing a small amount of mercury and electrical contacts used to make and break the electrical circuit in a low-voltage thermostat.

Bombilla de mercurio. Bombilla de cristal que contiene una pequeña cantidad de mercurio y que funciona como contacto eléctrico, utilizada para conectar y desconectar el circuito eléctrico en un termostato de baja tensión.

Metering device. A valve or small fixed-size tubing or orifice that meters liquid refrigerant into the evaporator.

Dispositivo de medida. Válvula o tubería pequeña u orificio que mide la cantidad de refrigerante líquido que entra en el evaporador.

Methane. Natural gas composed of 90% to 95% methane, a combustible hydrocarbon.

Metano. El gas natural se compone de un 90% a un 95% de metano, un hidrocarburo combustible.

Metric system. System International (SI); system of measurement used by most countries in the world.

Sistema métrico. Sistema internacional; el sistema de medida utilizado por la mayoría de los países del mundo.

Micro. A prefix meaning 1/1000000.

Micro. Prefijo que significa una parte de un millón.

Microfarad. Capacitor capacity equal to 1/1000000 of a farad.

Microfaradio. Capacidad de un capacitador equivalente a 1/1.000.000 de un faradio.

Micrometer. A precision measuring instrument.

Micrómetro. Instrumento de precisión utilizado para medir.

Micron. A unit of length equal to 1/1000 of a millimeter, 1/1,000,000 of a meter.

Micrón. Unidad de largo equivalente a 1/1000 de un milímetro, o 1/1.000.000 de un metro.

Micron gage. A gage used when it is necessary to measure pressure close to a perfect vacuum.

Calibrador de micrón. Calibrador utilizado cuando es necesario medir la presión de un vacío casi perfecto.

Microprocessor. A small, preprogrammed, solid-state microcomputer that acts as a main controller.

Microprocesador. Un microordenador de estado sólido preprogramado que actúa de controlador principal.

Midseated (cracked). A position on a service valve that allows refrigerant flow in all directions.

Sentado en el medio (agrietado). Posición en una válvula de servicio que permite el flujo de refrigerante en cualquier dirección.

Migration of oil or refrigerant. When the refrigerant moves to some place in the system where it is not supposed to be, such as when oil migrates to an evaporator or when refrigerant migrates to a compressor crankcase.

Migración de aceite o refrigerante. Cuando el refrigerante se mueve a cualquier lugar en el sistema donde no debe estar, como cuando aceite migra a un evaporador o cuando refrigerante se transplanta al cárter del cigüeñal del compresor.

Milli. A prefix meaning 1/1000.

Mili. Prefijo que significa una parte de mil.

Mineral oil. A traditional refrigeration lubricant used in CFC and HCFC systems.

Aceite mineral. Lubricante de refrigeración utilizado tradicionalmente en los sistemas de CFC y HCFC.

Minimum efficiency reporting value (MERV). Air filter rating system ranging from 1 to 20, with upper levels providing the most filtering.

Valor mínimo de reporte de eficacia (MERV en inglés). Sistema de clasificación de filtros de aire que va desde el 1 al 20 con los niveles mayores indicando más filtración.

Modulating flow. Controlling the flow between maximum or no flow. For example, the accelerator on a car provides modulating flow.

Flujo modulante. Controlado el flujo entre flujo que no es flujo máximo o ningún flujo. Por ejemplo, el acelerador de un carro provee un flujo modulante.

Modulator. A device that adjusts by small increments or changes.

Modulador. Dispositivo que se ajusta por medio de incrementos o cambios pequeños.

Moisture indicator. A device for determining moisture.

Indicador de humedad. Dispositivo utilizado para determinar la humedad.

Mold. A fungus found where there is moisture that develops and releases spores. Can be harmful to humans.

Moho. Un hongo en con trado donde hay humedad que se desarrolla y libera esporas. Puede ser nocivo a los humanos.

Molecular motion. The movement of molecules within a substance.

Movimiento molecular. Movimiento de moléculas dentro de una sustancia.

Molecule. The smallest particle that a substance can be broken into and still retain its chemical identity.

Molécula. La particula más pequeña en la que una sustancia puede dividirse y aún conservar sus propias características.

Monochlorodifluoromethane. The refrigerant R-22.

Monoclorodiflorometano. El refrigerante R-22.

Montreal Protocol. An agreement signed in 1987 by the United States and other countries to control the release of ozone-depleting gases.

Protocolo de Montreal. Un acuerdo firmado en 1987 por los Estados Unidos y otros países para controlar la liberación de gases que destruyen el ozono.

Motor service factor. A factor above an electric motor's normal operating design parameters, indicated on the nameplate, under which it can operate.

Factor de servicio del motor. Factor superior a los parametros de diseño normales de funcionamiento de un motor eléctrico, indicados en el marbete; este factor indica su nivel de funcionamiento.

Motor starter. Electromagnetic contactors that contain motor protection and are used for switching electric motors on and off.

Arrancador de motor. Contactadores electromagnéticos que contienen protección para el motor y se utilizan para arrancar y detener motores eléctricos.

Motor temperature-sensing thermostat. A thermostat that monitors the motor temperature and shuts it off for the motor's protection.

Termostato que detecta la temperatura del motor. Un termostato que vigila la temperatura del motor y lo apaga para la protección del motor.

Muffler, compressor. Sound absorber at the compressor.

Silenciador del compresor. Absorbedor de sonido ubicado en el compresor.

Mullion. Stationary frame between two doors.

Parteluz. Armazón fijo entre dos puertas.

Mullion heater. Heating element mounted in the mullion of a refrigerator to keep moisture from forming on it.

Calentador del parteluz. Elemento de calentamiento montado en el parteluz de un refrigerador para evitar la formación de humedad en el mismo.

Multimeter. An instrument that will measure voltage, resistance, and milliamperes.

Multímetro. Instrumento que mide la tensión, la resistencia y los miliamperios.

Multiple circuit coil. An evaporator or condenser coil that has more than one circuit because of the coil length. When the coil is too long, there will be an unacceptable pressure drop and loss of efficiency.

Serpentín de circuito múltiple. Un serpentín de evaporador o condensador que tiene más de un circuito por causa de la longitud del serpentín. Cuando el serpentín es demasiado largo, habrá una pérdida de presión y eficacia inaceptable.

Multiple evacuation. A procedure for evacuating a system. A vacuum is pulled, a small amount of refrigerant allowed into the system, and the procedure duplicated. This is often done three times.

Evacuación múltiple. Procedimiento para evacuar o vaciar un sistema. Se crea un vacío, se permite la entrada de una pequeña cantidad de refrigerante al sistema, y se repite el procedimiento. Con frecuencia esto se lleva a cabo tres veces.

Naphthenic oil. A refrigeration mineral oil refined from California and Texas crude oil.

Aceite nafténico. Aceite mineral refinado utilizado en la refrigeración, obtenido del petróleo crudo de California y Texas.

National Electrical Code® (NEC®). A publication that sets the standards for all electrical installations, including motor overload protection.

Código estadounidense de electridad. Publicación que establece las normas para todas las instalaciones eléctricas, incluyendo la protección contra la sobrecarga de un motor.

National Fire Protection Association (NFPA). An association organized to prevent fires through establishing standards, providing research, and providing public education.

National Fire Protection Association (NFPA en inglés) (Asociación nacional para la protección contra incendios). Asociación cuyo objetivo es prevenir incendios estableciendo normativas y facilitando la investigación y concienciación del público.

National pipe taper (NPT). The standard designation for a standard tapered pipe thread.

Cono estadounidense para tubos (NPT en inglés). Designación estándar para una rosca cónica para tubos estándar.

Natural convection. The natural movement of a gas or fluid caused by differences in temperature.

Convección natural. Movimiento natural de un gas o fluido ocasionado por diferencias en temperatura.

Natural-draft tower. A water cooling tower that does not have a fan to force air over the water. It relies on the natural breeze or airflow.

Torre de corriente de aire natural. Una torre de enfriamiento de agua que no tiene un ventilador para forzar el aire sobre el agua; depende de la brisa o flujo natural del aire.

Natural gas. A fossil fuel formed over millions of years from dead vegetation and animals that were deposited or washed deep into the earth.

Gas natural. Combustible fósil formado a través de millones de años de la vegetación y los animales muertos que fueron depositados o arrastrados a un gran profundidad dentro la tierra.

Near-azeotropic blend. Two or more refrigerants mixed together that will have a small range of boiling and/or condensing points for each system pressure. Small fractionation and temperature glides will occur but are often negligible.

Mezcla casi-azeotrópica. Mezcla de dos o más refrigerantes, que tiene un bajo rango de punto de ebullición y/o condensación para cada presión del sistema. Puede producirse una cierta fraccionación y variación de temperatura, aunque suelen ser insignificantes.

Needlepoint valve. A device having a needle and a very small orifice for controlling the flow of a fluid.

Válvula de aguja. Dispositivo que tiene una aguja y un orificio bastante pequeño para regular el flujo de un fluido.

Negative electrical charge. An atom or component that has an excess of electrons.

Carga eléctrica negativa. Átomo o componente que tiene un exceso de electrones.

Neoprene. Synthetic flexible material used for gaskets and seals.

Neopreno. Material sintético flexible utilizado en guarniciones y juntas de estanqueidad.

Net oil pressure. Difference in the suction pressure and the compressor oil pump outlet pressure.

Presión neta del aceite. Diferencia en la presión de aspiración y la presión a la salida de la bomba de aceite del compresor.

Net refrigeration effect (NRE). The quantity of heat in Btu/lb that the refrigerant absorbs from the refrigerated space to produce useful cooling.

Efecto neto de refrigeración (NRE en inglés). La cantidad de calor expresado en Btu/lb que el refrigerante absorbe del espacio refrigerado para producir refrigeración útil.

Net stack temperature. The temperature difference between the ambient temperature and the flue-gas temperature, typically for oil- and gas-burning equipment.

Temperatura neta de chimenea. La diferencia en temperatura entre la temperatura ambiental y la del conducto de gas, normalmente para equipo que quema aceite y gas.

Neutralizer. A substance used to counteract acids.

Neutralizador. Sustancia utilizada para contrarrestar ácidos.

Neutron. Neutrons and protons are located at the center of the nucleus of an atom. Neutrons have no charge.

Neutrón. Los neutrones y protones están situados en le centro del núcleo del átomo. Los neutrones carecen de carga.

Newton/meter². Metric unit of measurement for pressure. Also called a pascal.

Metro-Newton². Unidad métrica de medida de presión. Conocido también como pascal.

Nichrome. A metal made of nickel chromium that when formed into a wire is used as a resistance heating element in electric heaters and furnaces.

Níquel-cromio. Metal fabricado de níquel-cromio que al ser convertido en alambre, se utiliza como un elemento de calentamiento de resistencia en calentadores y hornos eléctricos.

Nitrogen. An inert gas often used to "sweep" a refrigeration system to help ensure that all refrigerant and contaminants have been removed.

Nitrógeno. Gas inerte utilizado con frecuencia para purgar un sistema de refrigeración. Esta gas ayuda a asegurar la remoción de todo el refrigerante y los contaminantes del sistema.

Nitrogen dioxide. A combustion pollutant that causes irritation of the respiratory tract and can cause shortness of breath.

Bióxido de nitrógeno. Un contaminante por combustión que causa irritación a las vías respiratorias y puede causar falta de aliento.

Nominal. A rounded-off stated size. The nominal size is the closest rounded-off size.

Nominal. Tamaño redondeado establecido. El tamaño nominal es el tamaño redondeado más cercano.

Noncatalytic stove. Wood-burning stoves without catalytic combustors and that were manufactured after July 1, 1990, are allowed to emit no more than 7.5 g/hr of particulates.

Estufa no catalítica. Estufa de madera sin quemadores catalíticos. Las estufas fabricadas después del 1 de julio de 1990 no pueden emitir más de 7,5 g/hr de partículas.

Noncondensable gas. A gas that does not change into a liquid under normal operating conditions.

Gas no condensable. Gas que no se convierte en líquido bajo condiciones de funcionamiento normales.

Nonferrous. Metals containing no iron.

No férreos. Metales que no contienen hierro.

Nonpolycyclic organic matter. Air pollutants from wood-burning stoves considered to be health hazards.

Materia orgánica no policíclica. Contaminantes del aire procedentes de estufas de madera. Se considera que son peligrosos para la salud.

Non-power-consuming devices. Devices that do not consume power, such as switches, that pass power to power-consuming devices.

Aparatos que no consumen corriente. Aparatos que no consumen corriente, tales como interruptores que trasmiten corriente a aparatos que consumen corriente.

North Pole, magnetic. One end of a magnet or the magnetic north pole of the earth.

Polo norte magnético. El extremo de un imán o el polo norte magnético del mundo.

Nozzle. A drilled opening that measures liquid flow, such as an oil burner nozzle.

Tobera. Una apertura taladrada que mide el flujo de líquido, tal como la tobera de un quemador de aceite.

N-type material. Semiconductor material with an excess of electrons that move from negative to positive when a voltage is applied.

Material tipo N. Material semiconductor con un exceso de electrones que pasan de negativo a positivo al aplicarles una corriente.

Nut driver. These tools have a socket head used primarily to turn hex head screws on air-conditioning, heating, and refrigeration cabinets.

Extractor de tuercas. Estas herramientas tienen una cabeza hueca hexagonal usadas principalmente para darle vuelta a tuercas de cabeza hexagonal en gabinetes de acondicionamiento de aire, de calefacción y de refrigeración.

Off cycle. A period when a system is not operating.

Ciclo de apagado. Período de tiempo cuando un sistema no está en funcionamiento.

Off-cycle defrost. Used for medium-temperature refrigeration where the evaporator coil operates below freezing but the air in the cooler is above freezing. The coil is defrosted by the air inside the cooler while the compressor is off cycle.

Descongelación de período de reposo. Se usa para refrigeración de temperatura media en la cual el serpentín del evaporador funciona por debajo del punto de congelación, pero el aire en el enfriador está por encima del punto de congelación. El serpentín se descongela por el aire dentro del enfriador mientras el compresor está en reposo.

Offset. The absolute (not signed + or −) difference between the set point and the control point of a control process.

Compensación. La diferencia absoluta (sin signo + o −) entre el punto de ajuste y el punto de control de un proceso de control.

Offset. The position of ductwork that must be rerouted around an obstacle.

Desviación. El posición de un conducto que tiene que desviarse alrededor de un obstáculo.

Ohm. A unit of measurement of electrical resistance.

Ohmio. Unidad de medida de la resistencia eléctrica.

Ohmmeter. A meter that measures electrical resistance.

Ohmiómetro. Instrumento que mide la resistencia eléctrica.

Ohm's law. A law involving electrical relationships discovered by Georg Ohm: $E = I \times R$.

Ley de Ohm. Ley que define las relaciones eléctricas, descubierta por Georg Ohm: $E = I \times R$.

Oil level regulator. A needle valve and float system located on each compressor of a parallel compressor system. It senses the oil level in the compressor's crankcase and adds oil if necessary. It receives its oil from the oil reservoir.

Regulador del nivel de aceite. Sistema de válvula de aguja y flotador que se encuentra en cada compresor de un sistema de compresores en paralelo. Detecta el nivel del aceite en el cárter del compresor y permite la entrada de más aceite procedente de un depósito, en caso necesario.

Oil-pressure safety control (switch). A control used to ensure that a compressor has adequate oil lubricating pressure.

Regulador de seguridad para la presión de aceite (conmutador). Regulador utilizado para asegurar que un compresor tenga la presión de lubrificación de aceita adecuada.

Oil, refrigeration. Oil used in refrigeration systems.

Aceite de refrigeración. Aceite utilizado en sistemas de refrigeración.

Oil reservoir. A storage cylinder for oil usually used on parallel compressor systems. It is located between the oil separator and the oil level regulators. It receives its oil from the oil separator.

Depósito de aceite. Cilindro en el que se almacena el aceite utilizado en los sistemas de compresores en paralelo. Está ubicado entre el separador de aceite y el regulador del nivel. Recibe el aceite del separador.

Oil separator. Apparatus that removes oil from a gaseous refrigerant.

Separador de aceite. Aparato que remueve el aceite de un refrigerante gaseoso.

Onetime relief valve. A pressure relief valve that has a diaphragm that blows out due to excess pressure. It is set at a higher pressure than the spring-loaded relief valve in case it fails.

Válvula de alivio de una vez. Una válvula de alivio de presión que tiene un diafragma que revienta debido a la presión excesiva. La misma se fija a una presión más alta que la válvula de alivio de resorte para en caso de que ésta falle.

Open compressor. A compressor with an external drive.

Compresor abierto. Compresor con un motor externo.

Open-loop control configuration. In an open-loop control configuration, the sensor is located upstream of the location where the control agent is causing a change in the controlled medium.

Configuración de control de circuito abierto. En una configuración de control de circuito abierto, el sensor está ubicado a contracorriente de donde el agente de control está causando un cambio en el medio controlado.

Open-loop heat pump. Heat pump system that uses the water in the earth as the heat transfer medium and then expels the water back to the earth in some manner.

Bomba de calor de circuito abierto. Sistema de bomba de calor que utiliza el agua de la tierra como medio de transferencia del calor y luego devuelve el agua a la tierra de cierta manera.

Open winding. The condition that exists when there is a break and no continuity in an electric motor winding.

Devanado abierto. Condición que se presenta cuando hay una interrupción en la continuidad del devanado de un motor.

Operating pressure. The actual pressure under operating conditions.

Presión de funcionamiento. La presión real bajo las condiciones de funcionamiento.

Organic. Materials formed from living organisms.

Orgánico. Materiales formados de organismos vivos.

Orifice. A small opening through which fluid flows.

Orificio. Pequeña abertura a través de la cual fluye un fluido.

Outward clinch tacker. A stapler or tacker that will anchor staples outward and can be used with soft materials.

Grapadora de agarre hacia fuera. Grapadora o tachueladora que ancla las grapas hacia fuera y que puede usarse con materiales suaves.

Overload protection. A system or device that will shut down a system if an overcurrent condition exists.

Protección contra sobrecarga. Sistema o dispositivo que detendrá la marcha de un sistema si existe una condición de sobreintensidad.

Oxidation. The combining of a material with oxygen to form a different substance. This results in the deterioration of the original substance. Rust is oxidation.

Oxidación. La combinación de un material con oxígeno para formar una sustancia diferente, lo que ocasiona el deterioro de la sustancia original. Herrumbre es oxidación.

Ozone. A form of oxygen (O_3). A layer of ozone is in the stratosphere that protects the earth from certain of the sun's ultraviolet wavelengths.

Ozono. Forma de oxígeno (O_3). Una capa de ozono en la estratosfera protege la Tierra de ciertos rayos ultravioletas del sol.

Ozone depletion. The breaking up of the ozone molecule by the chlorine atom in the stratosphere. Stratosphere ozone protects us from ultraviolet radiation emitted by the sun.

Reducción del ozono. Descomposición de la molécula de ozono por el átomo de cloro en la estratosfera. El ozono presente en la estratosfera nos protege de las radiaciones ultravioletas del sol.

Ozone depletion potential (ODP). A scale used to measure how much a substance will deplete stratospheric ozone.

Potencial de depleción de ozono (ODP en inglés). Una escala que se usa para medir cuánta depleción del ozono de la estratosfera una sustancia va a causar.

Package equipment (unit). A refrigerating system where all major components are located in one cabinet.

Equipo (unidad) completo. Sistema de refrigeración donde todos los componentes principales se encuentran en un solo gabinete.

Packing. A soft material that can be shaped and compressed to provide a seal. It is commonly applied around valve stems.

Empaquetadura. Material blando que puede formarse y comprimirse para proveer una junta de estanqueidad. Comúnmente se aplica alrededor de los vástagos de válvulas.

Paraffinic oil. A refrigeration mineral oil containing some paraffin wax, which is refined from eastern U.S. crude oil.

Aceite de parafina. Aceite mineral que contiene parafina y que se utiliza en sistemas de refrigeración. El aceite se obtiene mediante el refinado de petróleo extraído en los EE.UU del este.

Parallel circuit. An electrical or fluid circuit where the current or fluid takes more than one path at a junction.

Circuito paralelo. Corriente eléctrica o fluida donde la corriente o el fluido siguen más de una trayectoria en un empalme.

Parallel compressor. Many compressors piped in parallel and mounted on a steel rack. The compressors are usually cycled by a microprocessor.

Compresor en paralelo. Varios compresores conectados en paralelo y montados en un bastidor de acero. Normalmente, se sirve de un microprocesador para activarlos y desactivarlos.

Parallel flow. A flow path in which many paths exist for the fluid to flow.

Flujo paralelo. Vía de flujo que consta de varias vías que permiten el flujo de un fluido.

Part-winding start. A large motor that is actually two motors in one housing. It starts on one and then the other is energized. This is to reduce inrush current at start-up. For example, a 100-hp motor may have two 50-hp motors built into the same winding. It will start using one motor followed by the start of the other one. They will both run under the load.

Arranque de bobina parcial. Un motor grande que es, en efecto, dos motores bajo un mismo cárter. El mismo arranca con uno y luego se activa el segundo. El propósito de esto es reducir la corriente interna al arrancar. Por ejemplo, un motor de 100 caballos de fuerza puede tener dos motores de 50 caballos de fuerza construidos con la misma bobina. El motor arrancará usando un motor, seguido por el arranque del otro motor. Los dos motores correrán bajo carga.

Pascal. A metric unit of measurement of pressure.

Pascal. Unidad métrica de medida de presión.

Passive recovery. Recovering refrigerant with the use of the refrigeration system's compressor or internal vapor pressure.

Recuperación pasiva. Recuperación de un refrigerante utilizando el compresor del sistema de refrigeración o la presión del vapor interno.

Passive sensor. Sensors that send information back to the controller in terms of resistance (ohms).

Sensor pasivo. Sensores que devuelven información al controlador a través de la resistencia (ohmios).

Passive solar design. The use of nonmoving parts of a building to provide heat or cooling, or to eliminate certain parts of a building that cause inefficient heating or cooling.

Diseño solar pasivo. La utilización de piezas fijas de un edificio para proveer calefacción o enfriamiento, o para eliminar ciertas piezas de un edificio que causan calefacción o enfriamiento ineficientes.

PE (polyethylene). Plastic pipe used for water, gas, and irrigation systems.

Polietileno. Tubo plástico utilizado en sistemas de agua, de gas y de irrigación.

Pellet stove. A stove that burns small compressed pellets made from sawdust, cardboard, or sunflower or cherry seeds.

Estufa de gránulos. Estufa que quema gránulos comprimidos fabricados con aserrín, cartón y semillas de girasol o de cereza.

Percent refrigerant quality. Percent vapor.

Calidad porcentual de refrigerante. Porcentaje de vapor.

Permanent magnet. An object that has its own permanent magnetic field.

Imán permanente. Objeto que tiene su propio campo magnético permanente.

Permanent split-capacitor motor (PSC). A split-phase motor with a run capacitor only. It has a very low starting torque.

Motor permanente de capacitador separado (PSC en inglés). Motor de fase separada que sólo tiene un capacitador de funcionamiento. Su par de arranque es sumamente bajo.

PEX tubing. PEX stands for polyethylene cross (X)-linked tubing, a common name for polyethylene tubing.

Tubería PEX. PEX significa tubería polietilena unida al través y es un nombre que se usa frecuentemente para tubería polietilena.

Phase. One distinct part of a cycle.

Fase. Una parte específica de un ciclo.

Phase-change loop. The loop of piping, usually in a geothermal heat pump system, where there is a change of phase of the heat transfer fluid from liquid to vapor or vapor to liquid.

Circuito de cambio de fase. Circuito de tubería, generalmente en un sistema de bombeo de calor geotermal, en el cual hay un cambio de fase del fluido de transferencia de calor de líquido a vapor o de vapor a líquido.

Phase failure protection. Used on three-phase equipment to interrupt the power source when one phase becomes deenergized. The motors cannot be allowed to run on the two remaining phases or damage will occur.

Protección de fallo de fase. Se usa en equipo trifásico para interrumpir la fuente de potencia cuando se energiza una fase. No se puede permitir que los motores corran con las dos fases restantes o podría ocurrir una avería.

Phase reversal. Phase reversal can occur if someone switches any two wires on a three-phase system. Any system with a three-phase motor will reverse, and this cannot be allowed on some equipment.

Inversión de fase. La inversión de fase puede ocurrir si por cualquier razón alguien intercambia cualquier par de cables en un sistema trifásico. Cualquier sistema con un motor trifásico irá en dirección contraria y esto no puede permitirse en algunos sistemas.

Pictorial wiring diagram. This type of diagram shows the location of each component as it appears to the person installing or servicing the equipment.

Diagrama representativo del cableado. Este tipo de diagrama indica la ubicación de cada componente tal y como lo verá el personal técnico.

Piercing valve. A device that is used to pierce a pipe or tube to obtain a pressure reading without interrupting the flow of fluid. Also called a line tap valve.

Válvula punzante. Aparato que se usa para punzar un tubo para obtener una lectura de la presión sin interrumpir el flujo del fluido. También se llama una válvula de toma.

Pilot duty relay. A small relay that is used in control circuits for switching purposes. It is small and cannot take a lot of current flow, such as to start a motor.

Relé de función piloto. Un pequeño relé que se usa en los circuitos de control con propósitos de interrupción. Es pequeño y no puede tolerar un flujo alto de corriente, como para arrancar un motor.

Pilot light. The flame that ignites the main burner on a gas furnace.

Llama piloto. Llama que enciende el quemador principal en un horno de gas.

Pilot positioner. Used in a pneumatic air system to control the large volume of air that must be used to operate large pneumatic devices. The pneumatic thermostat sends a very small air signal to the pilot positioner in a 1/4" line and the pilot positioner controls the airflow in a 3/8" line for the large volume of air to the controlled device.

Posicionador piloto. Se usa en un sistema neumático de aire para controlar el gran volumen de aire necesario para operar los aparatos neumáticos grandes. El termostato neumático envía una señal de aire pequeña al posicionador piloto por una línea de 1/4 de pulgada, y el posicionador piloto controla el flujo de aire, una línea de 3/8 de pulgada para el gran volumen de aire, al aparato controlado.

Pin rivet assembly. Also known as a *blind rivet,* this is a fastener assembly of hollow rivets assembled on a pin, often called a mandrel. The rivets are used to join two pieces of sheet metal and are inserted and set from only one side of the metal.

Instalación con remaches. También llamados *remaches ciegos,* estos son remaches huecos con clavija. Los remaches se utilizan para unir dos láminas metálicas, y se introducen y se colocan desde una de las caras de las láminas.

Piston. The part that moves up and down in a cylinder.

Pistón. La pieza que asciende y desciende dentro de un cilindro.

Piston displacement. The volume within the cylinder that is displaced with the movement of the piston from top to bottom.

Despiazamiento del pistón. Volumen dentro del cilindro que se desplaza de arriba a abajo con el movimiento del pistón.

Pitot tube. Part of an instrument for measuring air velocities.

Tubo Pitot. Pieza de un instrumento para medir velocidades de aire.

Planned defrost. Shutting the compressor off with a timer so that the space temperature can provide the defrost.

Descongelación proyectada. Detención de la marcha de un compresor con un temporizador para que la temperatura del espacio lleve a cabo la descongelación.

Plenum. A sealed chamber at the inlet or outlet of an air handler. The duct attaches to the plenum.

Plenum. Cámara sellada a la entrada o a la salida de un tratante de aire. El conducto se fija al plenum.

Pneumatic atomizing humidifier. Air pressure is used to break up the water into a mist of tiny droplets and to disperse them.

Humidificador por pulverización neumática. Se utiliza aire a presión para convertir el agua en neblina y dispersarla.

Pneumatic controls. Controls operated by low-pressure air, typically 20 psig.

Controles neumáticos. Controles que se operan por aire de baja presión, generalmente 20 libras por pulgada cuadrada de presión de manómetro (Puig).

Polyalkylene glycol (PAG). A popular synthetic glycol-based lubricant used with HFC refrigerants, mainly in automotive systems. This was the first generation of oil used with HFC refrigerants.

Glicol polialkilénico (PAG en inglés). Lubricante sintético de uso común basado en glicol, usado con refrigerantes HFC, principalmente en sistemas de automóviles. Ésta fue la primera generación de aceites usados con refrigerantes HFC.

Polybutylene. A material used for the buried piping in geothermal heat pumps.

Polibutileno. Material utilizado para la fabricación de tubos enterrados, en sistemas de bombas de calor geotérmicas.

Polycyclic organic matter. By-products of wood combustion found in smoke and considered to be health hazards.

Materna orgánica policíclica. Subproductos de la combustión de madera presentes en el humo y considerados nocivos para la salud.

Polyethylene. A material used for the buried piping in geothermal heat pumps.

Polietileno. Material utilizado para la fabricación de tubos enterrados, en sistemas de bombas de calor geotérmicas.

Polyol ester (POE). A very popular ester-based lubricant often used in HFC refrigerant systems.

Poliol éster (POE en inglés). Lubricante muy popular basado en éster usado frecuentemente en sistemas con refrigerantes HFC.

Polyphase. Three or more phases.

Polifase. Tres o más fases.

Polyphosphate. A scale inhibitor with many phosphate molecules.

Polifosfato. Un inhibidor de escama que contiene muchas moléculas de fosfato.

Porcelain. A ceramic material.

Porcelana. Material cerámico.

Portable dolly. A small platform with four wheels on which heavy objects can be placed and moved.

Carretilla portátil. Plataforma pequeña con cuatro ruedas sobre la que pueden colocarse y transportarse objetos pesados.

Positive displacement. A term used with a pumping device such as a compressor that is designed to move all matter from a volume such as a cylinder or it will stall, possibly causing failure of a part.

Desplazamiento positivo. Término utilizado con un dispositivo de bombeo, como por ejemplo un compresor, diseñado para mover toda la materia de un volumen, como un cilindro, o se bloqueará, posiblemente causándole fallas a una pieza.

Positive electrical charge. An atom or component that has a shortage of electrons.

Carga eléctrica positiva. Átomo o componente que tiene una insuficiencia de electrones.

Positive temperature coefficient (PTC) start device. A thermistor used to provide start assistance to a permanent split-capacitor motor.

Dispositivo de arranque de coeficiente de temperatura positiva (PTC en inglés). Termistor utilizado para ayudar a arrancar un motor permanente de capacitador separado.

Potential relay. A switching device used with hermetic motors that breaks the circuit to the start capacitor and/or start windings after the motor has reached approximately 75% of its running speed.

Relé de potencial. Dispositivo de conmutación utilizado con motores herméticos que interrumpe el circuito del capacitador y/o de los devandos de arranque antes de que el motor haya alcanzado aproximadamente un 75% de su velocidad de marcha.

Potential voltage. The voltage measured across the start winding in a single-phase motor while it is turning at full speed. This voltage is much greater than the applied voltage to the run winding. For example, the run winding may have 230 V applied to it, and a measured voltage across the start winding may be 300 V. This is created by the motor stator turning in the magnetic field of the run winding. *Voltage potential* is the difference in voltage between any two parts of a circuit.

Potencial de voltaje. El voltaje medido a través de la bobina de arranque en un motor de fase sencilla mientras está girando a velocidad completa. Este voltaje es mucho mayor que el voltaje aplicado a la bobina de marcha. Por ejemplo, a la bobina de marcha se le puede aplicar 230 V, y el voltaje medido a través de la bobina de arranque puede ser 300 V. Este voltaje lo crea el estator del motor al girar en el campo magnético de la bobina de marcha. *Potencial de voltaje* es la diferencia en voltaje entre cualesquiera dos partes del circuito.

Potentiometer. An instrument that controls electrical current.

Potenciómetro. Instrumento que regula corriente eléctrica.

Powder actuated tool (PAT). A tool with a powder load that forces a pin, threaded stud, or other fastener into masonry.

Herramienta activada por pólvora (PAT en inglés). Una herramienta con una carga de pólvora que inserta una clavija, una clavija con rosca u otro sujetador en la albañilería.

Power. The rate at which work is done.

Potencia. Velocidad a la que se realiza un trabajo.

Power-consuming devices. A power-consuming device is considered the electrical load. For example, in a lightbulb circuit, the switch is a power-passing device that passes power to the lightbulb that consumes the power and produces light.

Aparatos consumidores de potencia. Un aparato consumidor de potencia se considera la carga eléctrica. Por ejemplo, en un circuito de bombilla de luz, el interruptor es un dispositivo que pasa corriente que pasa la corriente a la bombilla, la cual consume electricidad y produce luz.

Pressure. Force per unit of area.

Presión. Fuerza por unidad de área.

Pressure access ports. Places in a system where pressure can be taken or registered.

Puerto de acceso a presión. Lugares en un sistema donde se puede tomar o registrar la presión.

Pressure differential valve. A valve that senses a pressure differential and opens when a specific pressure differential is reached.

Válvula de presión diferencial. Válvula que detecta diferencia de presiones y se abre cuando se alcanza una diferencia específica.

Pressure drop. The difference in pressure between two points.

Caída de presión. Diferencia en presión entre dos puntos.

Pressure/enthalpy diagram. A chart indicating the pressure and heat content of a refrigerant and the extent to which the refrigerant is a liquid and vapor.

Diagrama de presión y entalpía. Esquema que indica la presión y el contenido de calor de un refrigerante y el punto en que el refrigerante es líquido y vapor.

Pressure limiter. A device that opens when a certain pressure is reached.

Dispositivo limitador de presión. Dispositivo que se abre cuando se alcanza una presión específica.

Pressure-limiting TXV. A valve designed to allow the evaporator to build only to a predetermined pressure when the valve will shut off the flow of refrigerant.

Válvula electrónica de expansión limitadora de presión. Válvula diseñada para permitir que la temperatura del evaporador alcance una presión predeterminada cuando la válvula detenga el flujo de refrigerante.

Pressure regulator. A valve capable of maintaining a constant outlet pressure when a variable inlet pressure occurs. Used for regulating fluid flow such as natural gas, refrigerant, and water.

Regulador de presión. Válvula capaz de mantener una presión constante a la salida cuando ocurre una presión variable a la entrada. Utilizado para regular el flujo de fluidos, como por ejemplo el gas natural, el refrigerante y el agua.

Pressure switch. A switch operated by a change in pressure.

Commutador accionado por presión. Commutador accionado por un cambio en presión.

Pressure tank. A pressurized tank for water storage located in the water piping of an open-loop geothermal heat pump system. It prevents short cycling of the well pump.

Depósito de presión. Depósito presurizado que sirve para almacenar el agua contenida en la tubería de un sistema de bomba de calor geotérmica de circuito abierto. Impide el funcionamiento de la bomba del pozo en ciclos cortos.

Pressure/temperature relationship. This refers to the pressure/temperature relationship of a liquid and vapor in a closed container. If the temperature increases, the pressure will also increase. If the temperature is lowered, the pressure will decrease.

Relación entre presión y temperatura. Se refiere a la relación entre la presión y la temperatura de un líquido y un vapor en un recipiente cerrado. Si la temperatura aumenta, la presión también aumentará. Si la temperatura baja, habrá una caída de presión.

Pressure transducer. A pressure-sensitive device located in the piping of a refrigeration system that will transform a pressure signal to an electronic signal. The electronic signal will then feed a microprocessor.

Transductor de presión. Dispositivo sensible a la presión, situado en la tubería del sistema de refrigeración y que convierte una señal de presión en señal eléctrica. Seguidamente, la señal eléctrica se envía a un microprocesador.

Pressure vessels and piping. Piping, tubing, cylinders, drums, and other containers that have pressurized contents.

Depósitos y tubería con presión. Tubería, cilindros, tambores y otros recipientes que tienen un contenido presurizado.

Preventive maintenance. The action of performing regularly scheduled maintenance on a unit, including inspection, cleaning, and servicing.

Mantenimiento preventivo. La acción de dar mantenimiento regular a una unidad incluyendo inspección, limpieza y servicio.

Primary air. Air that is introduced to a furnace's burner before the combustion process has taken place.

Aire primario. Aire que se introduce en el quemador de un calefactor antes de que ocurra el proceso de combustión.

Primary control. Controlling device for an oil burner to ensure ignition within a specific time span, usually 90 seconds.

Regulador principal. Dispositivo de regulación para un quemador de aceite pesado. El regulador principal asegura el encendido dentro de un período de tiempo específico, normalmente 90 segundos.

Programmable thermostat. An electronic thermostat that can be set up to provide desired conditions at desired times.

Termostato programable. Un termostato electrónico que se puede programar para proveer las condiciones deseadas en tiempos deseados.

Propane. An LP (liquified petroleum) gas used for heat.

Propano. Gas de petróleo licuado que se utiliza para producir calor.

Propeller fan. This fan is used in exhaust fan and condenser fan applications. It will handle large volumes of air at low-pressure differentials.

Ventilador helicoidal. Se utiliza en ventiladores de evacuación y de condensador. Es capaz de mover grandes volúmenes de aire a bajas diferenciales de presión.

Propylene glycol. An antifreeze fluid cooled by a primary (phase-change) refrigerant, which then is circulated by pumps throughout the refrigeration system to absorb heat.

Glicol propílico. Líquido anticongelante enfriado por un refrigerante primario (cambio de fase). Seguidamente, una bomba lo hace circular por el sistema de refrigeración para que absorba el calor.

Proportional controller. A modulating control mode where the controller changes or modifies its output signal in proportion to the size of the change in the error.

Controlador proporcional. Un modo de control modulante donde el controlador cambia o modifica su señal de salida en proporción al tamaño del cambio en el error.

Proton. That part of an atom having a positive charge.

Protón. Parte de un átomo que tiene carga positiva.

Protozoa. A microscopic organism with a complex life cycle.

Protozoario. Un organismo microscópico con un ciclo de vida complejo.

PSC motor. See Permanent split-capacitor motor.

Motor PSC. Véase motor permanente de capacitador separado.

psi. Abbreviation for pounds per square inch.

psi. Abreviatura de libras por pulgada cuadrada.

psia. Abbreviation for pounds per square inch absolute.

psia. Abreviatura de libras por pulgada cuadrada absoluta.

psig. Abbreviation for pounds per square inch gage.

psig. Abreviatura de indicador de libras por pulgada cuadrada.

Psychrometer. An instrument for determining relative humidity.

Sicrómetro. Instrumento para medir la humedad relativa.

Psychrometric chart. A chart that shows the relationship of temperature, pressure, and humidity in the air.

Esquema psicrométrico. Esquema que indica la relación entre la temperatura, la presión y la humedad en el aire.

Psychrometrics. The study of air and its properties, particularly the moisture content.

Psicrometría. El estudio del aire y sus propiedades, particularmente el contenido de humedad.

P-type material. Semiconductor material with a positive charge.

Material tipo P. Material con carga positiva utilizado en semiconductores.

Pulse furnace. This furnace ignites minute quantities of gas 60 to 70 times per second. Small amounts of natural gas and air enter the combustion chamber and are ignited with a spark igniter that forces the combustion materials down a tailpipe to an exhaust decoupler. The pulse is reflected back to the combustion chamber, igniting another gas and air mixture.

Horno de impulsos. Este tipo de horno enciende pequeñas cantidades de gas, 60 a 70 veces por segundo. Pequeñas cantidades de gas natural y aire entran en la cámara de combustión y son encendidas mediante chispa. A continuación son forzadas por un tubo hacia un desconectador de evacuación. El impulso se devuelve a la cámara de combustión para encender otra mezcla de gas y aire.

Pulse-width modulator (PWM). An electronic device in a motor circuit that is used to control motor speed for variable-speed motors.

Modulador de ancho de pulso (PWM en inglés). Un aparato electrónico en un circuito de motor que se usa para controlar la velocidad del motor para motores de velocidad variable.

Pump. A device that forces fluids through a system.

Bomba. Dispositivo que introduce fluidos por fuerza a través de un sistema.

Pump down. To use a compressor to pump the refrigerant charge into the condenser and/or receiver.

Extraer con bomba. Utilizar un compresor para bombear la carga del refrigerante dentro del condensador y/o receptor.

Pure compound. A substance formed in definite proportions by weight with only one molecule present.

Componente puro. Una sustancia formada en proporciones por peso definidas con sólo una molécula presente.

Purge. To remove or release fluid from a system.

Purga. Remover o liberar el fluido de un sistema.

PVC (Polyvinyl chloride). Plastic pipe used in pressure applications for water and gas as well as for sewage and certain industrial applications.

Cloruro de polivinilo (PVC en inglés). Tubo plástico utilizado tanto en aplicaciones de presión para agua y gas, como en ciertas aplicaciones industriales y de aguas negras.

Quench. To submerge a hot object in a fluid for cooling.

Entriamiento por inmersión. Sumersión de un objeto caliente en un fluido para enfriarlo.

Quick-connect coupling. A device designed for easy connecting or disconnecting of fluid lines.

Acoplamiento de conexión rápida. Dispositivo diseñado para facilitar la conexión o desconexión de conductos de fluido.

R-12. Dichlorodifluoromethane, once a popular refrigerant for refrigeration systems. It can no longer be manufactured in the United States and many other countries.

R-12. Diclorodiflorometano, que fue una vez un refrigerante muy utilizado en sistemas de refrigeración. Ya no se puede fabricar ni en los Estados Unidos ni en muchos otros países.

R-22. Monochlorodifluoromethane, a popular HCFC refrigerant for air-conditioning systems.

R-22. Monoclorodiflorometano, refrigerante HCFC muy utilizado en sistemas de acondicionamiento de aire.

R-123. Dichlorotrifluoroethane, an HCFC refrigerant developed for low-pressure application.

R-123. Diclorotrifloroetano, refrigerante HCFC elaborado para aplicaciones de baja presión.

R-134a. Tetrafluoroethane, an HFC refrigerant developed for refrigeration systems and as a replacement for R-12.

R-134a. Tetrafloroetano, refrigerante HFC elaborado para sistemas de refrigeración y como sustituto del R-12.

R-502. An azeotropic mixture of R-22 and R-115, a once popular refrigerant for low-temperature refrigeration systems. It can no longer be manufactured in the United States and many other countries.

R-502. Mezcla azeotrópica de R-22 y R-115, que fue una vez un refrigerante muy utilizado en sistemas de refrigeración de temperatura baja. Ya no se puede fabricar ni en los Estados Unidos ni en muchos otros países.

Rack system. Many compressors piped in parallel and mounted on a steel rack. The compressors are usually cycled by a microprocessor.

Sistema de bastidor. Varios compresores conectados en paralelo y montados en un bastidor de acero. Normalmente, un microprocesador se encarga de activar y desactivar los compresores.

Radiant heat. Heat that passes through air, heating solid objects that in turn heat the surrounding area.

Calor radiante. Calor que pasa a través del aire y calienta objetos sólidos que a su vez calientan el ambiente.

Radiation. Heat transfer. See Radiant heat.

Radiación. Transferencia de calor. Véase Calor radiante.

Radon. A colorless, odorless, and radioactive gas. Radon can enter buildings through cracks in concrete floors and walls, floor drains, and sumps.

Radón. Gas incoloro, inodoro y radioactivo. El radón puede penetrar en los edificios a través de las grietas en el hormigón y suelos, desagües y sumideros.

Random or off-cycle defrost. Defrost provided by the space temperature during the normal off cycle.

Descongelación variable o de ciclo apagado. Descongelación llevada a cabo por la temperatura del espacio durante el ciclo normal de apagado.

Range. The pressure or temperature settings of a control defining certain boundaries of temperature or pressure.

Rango. Los valores de presión o temperatura para un control que definen los límites de la temperatura o presión.

Rankine. The absolute Fahrenheit scale with 0 at the point where all molecular motion stops.

Rankine. Escala absoluta de Fahrenheit con el 0 al punto donde se detiene todo movimiento molecular.

Rapid oxidation. A reaction between the fuel, oxygen, and heat that is known as rapid oxidation or the process of burning.

Oxidación rápida. Reacción producida entre el combustible y el oxígeno. El calor producido se conoce como oxidación rápida o proceso de quemado.

Reactance. A type of resistance in an alternating current circuit.

Reactancia. Tipo de resistencia en un circuito de corriente alterna.

Reamer. Tool to remove burrs from inside a pipe after it has been cut.

Escariador. Herramienta utilizada para remover las rebabas de un tubo después de haber sido cortado.

Receiver-drier. A component in a refrigeration system for storing and drying refrigerant.

Receptor-secador. Componente en un sistema de refrigeración que almacena y seca el refrigerante.

Reciprocating. Back-and-forth motion.

Movimiento alternativa. Movimiento de atrás para adelante.

Reciprocating compressor. A compressor that uses a piston in a cylinder and a back-and-forth motion to compress vapor.

Compresor alternativo. Compresor que utiliza un pistón en un cilindro y un movimiento de atrás para adelante a fin de comprir el vapor.

Recirculated water system. A system where water is used over and over, such as a chilled water or cooling tower system.

Sistema de agua recirculada. Sistema donde el agua se usa una y otra vez, tal como en un sistema de agua enfriada o de torre enfriamiento.

Recovery cylinder. A cylinder into which refrigerant is transferred; should be approved by the Department of Transportation as a recovery cylinder. The color code for these cylinders is a yellow top with a gray body.

Cilindro de recuperación. Cilindro al que se transfiere el refrigerante y que debe ser homologado por el departamento de transporte como cilindro de recuperación. Este tipo de cilindro se identifica pintando su parte superior en amarillo y el resto del cuerpo en gris.

Rectifier. A device for changing alternating current to direct current.

Rectificador. Dispositivo utilizado para convertir corriente alterna en corriente continua.

Reed valve. A thin steel plate used as a valve in a compressor.

Válvula de lámina. Placa delgada de acero utilizada como una válvula en un compresor.

Refrigerant. The fluid in a refrigeration system that changes from a liquid to a vapor and back to a liquid at practical pressures.

Refrigerante. Fluido en un sistema de refrigeración que se convierte de líquido en vapor y nuevamente en líquido a presiones prácticas.

Refrigerant blend. Two or more refrigerants blended or mixed together to make another refrigerant. Blends can combine as either azeotropic or zeotropic blends.

Mezcla de refrigerante. Dos o más refrigerantes mezclados para crear otro. Las mezclas pueden ser de tipo azeotrópico o zeotrópico.

Refrigerant loop. The heat pump's refrigeration system, which exchanges energy with the fluid in the ground loop and the air side of the system.

Circuito de refrigeración. Sistema de refrigeración de la bomba de calor que sirve para intercambiar energía entre el fluido en el circuito de tierra y la parte del sistema que contiene el aire.

Refrigerant receiver. A storage tank in a refrigeration system where the excess refrigerant is stored. Since many systems use different amounts of refrigerant during the season, the excess is stored in the receiver tank when not needed. The refrigerant can also be pumped to the receiver when repairs on the low-pressure side of the system are made.

Recibidor de refrigerante. Tanque de almacenamiento en un sistema de refrigeración donde se almacena el exceso de refrigerante. Como muchos sistemas usan diferentes cantidades de refrigerantes durante la temporada, el exceso se almacena en el tanque cuando no se necesita. El refrigerante también puede bombearse al recibidor cuando se hacen reparaciones al extremo de baja presión del sistema.

Refrigerant reclaim. "To process refrigerant to new product specifications by means which may include distillation. It will require chemical analysis of the refrigerant to determine that appropriate product specifications are met. This term usually implies the use of processes or procedures available only at a reprocessing or manufacturing facility."

Recuperación del refrigerante. "Procesar refrigerante según nuevas especificaciones para productos a través de métodos que pueden incluir la destilación. Se requiere un análisis químico del refrigerante para asegurar el cumplimiento de las especificaciones para productos a través de métodos que pueden incluir la destilación. Se requiere un análisis químico del refrigerante para asegurar el cumplimiento de las especificaciones para productos adecuadas. Por lo general este término supone la utilización de procesos o de procedimientos disponibles solamente en fábricas de reprocesamiento o manufactura."

Refrigerant recovery. "To remove refrigerant in any condition from a system and store it in an external container without necessarily testing or processing it in any way."

Recobrar refrigerante líquido. "Remover refrigerante en cualquier estado de un sistema y almacenarlo en un recipiente externo sin ponerlo a prueba o elaborarlo de ninguna manera."

Refrigerant recycling. "To clean the refrigerant by oil separation and single or multiple passes through devices, such as replaceable core filter driers, which reduce moisture, acidity, and particulate matter. This term usually applies to procedures implemented at the job site or at a local service shop."

Recirculación de refrigerante. "Limpieza del refrigerante por medio de la separación del aceite y pasadas sencillas o múltiples a traves de dispositivos, como por ejemplo secadores filtros con núcleos reemplazables que disminuyen la humedad, la acidez y las partículas. Por lo general este término se aplica a los procedimientos utilizados en el lugar del trabajo o en un taller de servicio local."

Refrigerated air driers. A device that removes the excess moisture from compressed air.

Secadores de aire refrigerados. Aparato que remueve el exceso de humedad del aire comprimido.

Refrigeration. The process of removing heat from a place where it is not wanted and transferring that heat to a place where it makes little or no difference.

Refrigeración. Proceso de remover el calor de un lugar donde no es deseado y transferirlo a un lugar donde no afecte la temperatura.

Register. A terminal device on an air-distribution system that directs air but also has a damper to adjust airflow.

Registro. Dispositivo de terminal en un sistema de distribución de aire que dirige el aire y además tiene un desviador para ajustar su flujo.

Regulator. A valve used to control the pressure in liquid systems to some value. Many households have a water pressure regulator to reduce the pressure from the main to a more usable pressure in the house. Gas systems all have pressure regulators to stabilize the pressure to the burners.

Regulador. Una válvula que se usa para controlar y fijar la presión en los sistemas líquidos a algún valor. Muchas casas tienen un regulador de presión de agua para reducir la presión de la tubería principal a una presión más útil en la casa. Todos los sistemas de gas tienen reguladores de presión para estabilizar la presión en el quemador.

Relative humidity. The amount of moisture contained in the air as compared to the amount the air could hold at that temperature.

Humedad relativa. Cantidad de humedad presente en el aire, comparada con la cantidad de humedad que el aire pueda contener a dicha temperatura.

Relay. A small electromagnetic device to control a switch, motor, or valve.

Relé. Pequeño dispositivo electromagnético utilizado para regular un conmutador, un motor o una válvula.

Relief valve. A valve designed to open and release vapors at a certain pressure.

Válvula para alivio. Válvula diseñada para abrir y liberar vapores a una presión específica.

Remote system. Often called a split system where the condenser is located away from the evaporator and/or other parts of the system.

Sistema remoto. Llamado muchas veces sistema separado donde el condensador se coloca lejos del evaporador y/o otras piezas del sistema.

Resilient-mount motor. Electric motor that uses various materials to isolate the motor noise from metal framework. This type of motor requires a ground strap.

Motor con montaje antivibratorio. Motor eléctrico que utiliza varios materiales para aislar el ruido del bastidor metálico. Este tipo de motor requiere conexión a tierra.

Resistance. The opposition to the flow of an electrical current or a fluid.

Resistencia. Oposición al flujo de una corriente eléctrica o de un fluido.

Resistor. An electrical or electronic component with a specific opposition to electron flow. It is used to create voltage drop or heat.

Resistor. Componente eléctrico o electrónico con una oposición específica al flujo de electrones; se utiliza para producir una caída de tensión o calor.

Restrictor. A device used to create a planned resistance to fluid flow.

Limitador. Dispositivo utilizado para producir una resistencia proyectada al flujo de fluido.

Retrofit guidelines. Guidelines intended to make the transition from a CFC/mineral oil system to a system containing an alternative refrigerant and its appropriate oil.

Directrices de reconversión. Directrices destinadas a facilitar la transición entre un sistema de aceite CFC/mineral y otro con refrigerante alternativo y su aceite correspondiente.

Return well. A well for return water after it has experienced the heat exchanger of the geothermal heat pump.

Pozo de retorno. Pozo donde se acumula el agua de retorno una vez ha pasado por el intercambiador térmico de la bomba de calor geotérmica.

Reverse cycle. The ability to direct the hot gas flow into the indoor or the outdoor coil in a heat pump to control the system for heating or cooling purposes.

Ciclo invertido. Capacidad de dirigir el flujo de gas caliente dentro de la bobina interior o exterior en una bomba de calor a fin de regular el sistema para propósitos de calentamiento o enfriamiento.

Rigid-mount motor. Electric motor that is bolted metal-to-metal to a frame. This type of motor will transmit noise.

Motor con montaje rígido. Motor eléctrico que se encuentra sujeto directamente a un bastidor mediante pernos. Este tipo de motor genera ruido.

Rod and tube. The rod and tube are each made of a different metal. The tube has a high expansion rate and the rod a low expansion rate.

Varilla y tubo. La varilla y el tubo se fabrican de un metal diferente. El tubo tiene una tasa de expansión alta y la varilla una tasa de expansión baja.

Room heater. A gas stove or appliance considered by ANSI to be a heating appliance. This heater will have an efficiency rating.

Calentador de sala. Estufa de gas u otro dispositivo que, según ANSI, es un dispositivo de calefacción. Este tipo de calentador cuenta con una clasificación de eficacia.

Root mean square (RMS) voltage. The alternating current voltage effective value. This is the value measured by most voltmeters. The RMS voltage is $0.707 \times$ the peak voltage.

Voltaje de la raíz del valor medio cuadrado (RMS en inglés). El valor efectivo del voltaje de corriente alterna. Este valor es el que miden la mayoría de los voltímetros. El voltaje RMS es 0.707 por el voltaje pico.

Rotary compressor. A compressor that uses rotary motion to pump fluids. It is a positive displacement pump.

Compresor giratorio. Compresor que utiliza un movimiento giratorio para bombear fluidos. Es una bomba de desplazamiento positivo.

Rotor. The rotating or moving component of a motor, including the shaft.

Rotor. Componente giratorio o en movimiento de un motor, incluyendo el arbol.

Run-load amperage (RLA). The amperage at which a motor can safely operate while under full load, unless it has a service (reserve) factor allowing more amperage.

Amperaje de operación con carga (RLA en inglés). Amperaje bajo el cual el motor puede operar seguramente bajo carga completa, a menos que tenga un factor de servicio (reserva) que permite más amperaje.

Running time. The time a unit operates. Also called the *on time.*

Período de funcionamiento. El período de tiempo en que funciona una unidad. Conocido también como *período de conexión.*

Run winding. The electrical winding in a motor that draws current during the entire running cycle.

Devanado de funcionamiento. Devanado eléctrico en un motor que consume corriente durante todo el ciclo de funcionamiento.

Rupture disk. Pressure safety device for a centrifugal low-pressure chiller.

Disco de ruptura. Dispositivo de seguridad para un enfriador centrífugo de baja presión.

S fastener. A device used to connect square and rectangular sheet metal duct together, in conjunction with an S clip.

Sujetador tipo "S". Aparato que se usa para unir conductos cuadrados y rectangulares de metal en lámina en conjunción con una grapa tipo "S".

Saddle valve. A valve that straddles a fluid line and is fastened by solder or screws. It normally contains a device to puncture the line for pressure readings.

Válvula de silleta. Válvula que está sentada a horcajadas en un conducto de fluido y se fija por medio de la soldadura o tornillos. Por lo general contiene un dispositivo para agujerear el conducto a fin de que se puedan tomar lecturas de presión.

Safety control. An electrical, electronic, mechanical, or electromechanical control to protect the equipment or public from harm.

Regulador de seguridad. Regulador eléctrico, electrónico, mecánico o electromecánico para proteger al equipo de posibles averías o al público de sufrir alguna lesión.

Safety plug. A fusible plug that blows out when high temperature occurs.

Tapón de seguridad. Tapón fusible que se sale cuando se presentan temperaturas altas.

Sail switch. A safety switch with a lightweight, sensitive sail that operates by sensing an airflow.

Conmutador con vela. Conmutador de seguridad con una vela liviana sensible que funciona al advertir el flujo de aire.

Salt solution. Antifreeze solution used in a closed water loop of geothermal heat pumps.

Solucion de sal. Líquido anticongelante utilizado en el circuito cerrado de agua de una bomba de calor geotérmica.

Satellite compressor. The compressor on a parallel compressor system that is dedicated to the coldest evaporators.

Compresor auxiliar. Compresor montado en un sistema paralelo y que está dedicado a los evaporadores más fríos.

Saturated vapor. The refrigerant when all of the liquid has just changed to a vapor.

Vapor saturada. El refrigerante cuando todo el líquido acaba de convertirse en vapor.

Saturation. A term used to describe a substance when it contains all of another substance it can hold.

Saturación. Término utilizado para describir una sustancia cuando contiene lo más que puede de otra sustancia.

Scavenger pump. A pump used to remove the fluid from a sump.

Bomba de barrido. Bomba utilizada para remover el fluido de un sumidero.

Schematic wiring diagram. Sometimes called a line or ladder diagram, this type of diagram shows the electrical current path to the various components.

Diagrama de cableado esquematizado. También conocido como de línea o de escalera, este tipo de diagrama muestra la ruta actual que sigue la electricidad para llegar a los diferentes componentes.

Schrader valve. A valve similar to the valve on an auto tire that allows refrigerant to be charged or discharged from the system.

Válvula Schrader. Válvula similar a la válvula del neumático de un automóvil que permite la entrada o la salida de refrigerante del sistema.

Scotch yoke. A mechanism used to create reciprocating motion from the electric motor drive in very small compressors.

Yugo escocés. Mecanismo utilizado para producir movimiento alternativo del accionador del motor eléctrico en compresores bastante pequeños.

Screw compressor. A form of positive displacement compressor that squeezes fluid from a low-pressure area to a high-pressure area, using screw-type mechanisms.

Compresor de tornillo. Forma de compresor de desplazamiento positivo que introduce por fuerza el fluido de un área de baja presión a un área de alta presión, a través de mecanismos de tipo de tornillo.

Scroll compressor. A compressor that uses two scroll-type components, one stationary and one orbiting, to compress vapor.

Compresor espiral. Compresor que utiliza dos componentes de tipo espiral para comprimir el vapor.

Sealed unit. The term used to describe a refrigeration system, including the compressor, that is completely welded closed. The pressures can be accessed by saddle valves.

Unidad sellada. Término utilizado para describir un sistema de refrigeración, incluyendo el compresor, que es soldado completamente cerrado. Las presiones son accesibles por medio de válvulas de dilleta.

Seasonal energy efficiency ratio (SEER). An equipment efficiency rating that takes into account the start-up and shutdown for each cycle.

Relación del rendimiento de energía temporal (SEER en inglés). Clasificación del rendimiento de un equipo que toma en cuenta la puesta en marcha y la parada de cada ciclo.

Seat. The stationary part of a valve that the moving part of the valve presses against for shutoff.

Asiento. Pieza fija de una válvula contra la que la pieza en movimiento de la válvula presiona para cerrarla.

Secondary air. Air that is introduced to a furnace after combustion takes place and that supports combustion.

Aire secundario. Aire que se introduce en un calefactor después que occure la combustión y que ayuda la combustión.

Secondary fluid. An antifreeze fluid cooled by a primary (phase-change) refrigerant, which is then circulated by pumps throughout the refrigeration system to absorb heat.

Fluido secundario. Líquido anticongelante refrigerado por otro primario (cambio de fase), que luego es propulsado por las bombas a través del sistema de refrigeración para absorber el calor.

Semiconductor. A component in an electronic system that is considered neither an insulator nor a conductor but a partial conductor. It conducts current in a controlled and predictable manner.

Semiconductor. Componente en un sistema eléctrico que no se considera ni aislante ni conductor, sino conductor parcial. Conduce la corriente de una manera controlada y predecible.

Semi-hermetic compressor. A motor compressor that can be opened or disassembled by removing bolts and flanges. Also known as a *serviceable hermetic*.

Compresor semihermético. Compresor de un motor que puede abrirse o desmontarse al removerle los pernos y bridas. Conocido también como *hermético utilizable*.

Sensible heat. Heat that causes a change in temperature.

Calor sensible. Calor que produce un cambio en la temperatura.

Sensitivity. See Gain.

Sensibilidad. Vea Ganancia.

Sensors. Devices that can measure some type of environmental parameter or controlled variable and convert this parameter to a value that the controller can understand.

Sensores. Dispositivos que miden algún tipo de parámetro ambiental o variable controlado y convierten este parámetro a un valor que el controlador puede entender.

Sequencer. A control that causes a staging of events, such as a sequencer between stages of electrical heat.

Regulador de secuencia. Regulador que produce una sucesión de acontecimientos, como por ejemplo etapas sucesivas de calor eléctrico.

Series circuit. An electrical or piping circuit where all of the current or fluid flows through the entire circuit.

Circuito en serie. Cicuito eléctrico o de tubería donde toda la corriente o todo el fluido fluye a través de todo el circuito.

Series flow. A flow path in which only one path exists for fluid to flow.

Flujo en serie. Ruta de flujo única para el líquido.

Service valve. A manually operated valve in a refrigeration system used for various service procedures.

Válvula servicio. Válvula de un sistema de refrigeración accionada manualmente que se utiliza en varios procedimientos de servicio.

Serviceable hermetic. See Semi-hermetic compressor.

Compresor hermético utilizable. Véase Compresor semihermético.

Servo pressure regulator. A sensitive pressure regulator located inside a combination gas valve that senses the outlet or working pressure of the gas valve.

Regulador de presión por servomotor. Un regulador de presión sensible que está ubicado dentro de una válvula de gas de combinación que detecta la presión de salida o de trabajo de la válvula de presión.

Set point. The desired control point's magnitude in a control process.

Punto de ajuste. La magnitud deseada de un punto de control en un proceso de control.

Shaded-pole motor. An alternating current motor used for very light loads.

Motor polar en sombra. Motor de corriente alterna utilizado en cargas sumamente livianas.

Shell and coil. A vessel with a coil of tubing inside that is used as a heat exchanger.

Coraza y bobina. Depósito con una bobina de tubería en su interior que se utiliza como intercambiador de calor.

Shell and tube. A heat exchanger with straight tubes in a shell that can normally be mechanically cleaned.

Coraza y tubo. Intercambiador de calor con tubos rectos en una coraza que por lo general puede limpiarse mecánicamente.

Short circuit. A circuit that does not have the correct measurable resistance: too much current flows and will overload the conductors.

Cortocircuito. Corriente que no tiene la resistencia medible correcta: un exceso de corriente fluye a través del circuito provocando una sobrecarga de los conductores.

Short cycle. The term used to describe the running time (on time) of a unit when it is not running long enough.

Cico corto. Término utilizado para describir el período de funcionamiento (de encendido) de una unidad cuando no funciona por un período de tiempo suficiente.

Shorted motor winding. Part of an electric motor winding is shorted out because one part of the winding touches another part, where the insulation is worn or in some way defective.

Devanado de motor en cortocircuito. Debido a un aislamiento deficiente u otro defecto, una parte de los elementos del devanado en un motor eléctrico entran en contacto con otra, causando un cortocircuito.

Shroud. A fan housing that ensures maximum airflow through the coil.

Boveda. Alojamiento del abanico que asegura un flujo máximo de aire a través de la bobina.

Sight glass. A clear window in a fluid line.

Mirilla para observación. Ventana clara en un conducto de fluido.

Signal converter. An electronic device that converts one signal type to another.

Convertidor de señales. Un aparato electrónico que convierte un tipo de señal a otro.

Silica gel. A chemical compound often used in refrigerant driers to remove moisture from the refrigerant.

Gel silíceo. Compuesto químico utilizado a menudo en secadores de refrigerantes para remover la humedad del refrigerante.

Silicon. A substance from which many semiconductors are made.

Silicio. Sustancia de la cual se fabrican muchos semiconductores.

Silicon-controlled rectifier (SCR). A semiconductor control device.

Rectificador controlado por silicio. Dispositivo para regular un semiconductor.

Silver brazing. A high-temperature (above 800°F) brazing process for bonding metals.

Soldadura con plata. Soldadura a temperatura alta (sobre los 800°F ó 430°C) para unir metales.

Sine wave. The graph or curve used to describe the characteristics of alternating current and voltage.

Onda sinusoidal. Gráfica o curva utilizada para describir las características de tensión y de corriente alterna.

Single phase. The electrical power supplied to equipment or small motors, normally under 7 1/2 hp.

Monofásico. Potencia eléctrica suministrada a equipos o motores pequeños, por lo general menor de 7 1/2 hp.

Single-phase hermetic motor. A sealed motor, such as with a small compressor, that operates off single-phase power.

Motor de fase sencilla hermético. Un motor sellado, tal como un compresor pequeño, que opera con electricidad de fase sencilla.

Single phasing. The condition in a three-phase motor when one phase of the power supply is open.

Fasaje sencillo. Condición en un motor trifásico cuando una fase de la fuente de alimentación está abierta.

Sling psychrometer. A device with two thermometers, one a wet bulb and one a dry bulb, used for checking air conditions, wet-bulb and dry-bulb.

Sicrómetro con eslinga. Dispositivo con dos termómetros, uno con una bombilla húmeda y otro con una bombilla seca, utilizados para revisar las condiciones del aire, de la temperatura y de la humedad.

Slinger ring. A ring attached to the blade tips of a condenser fan. This ring throws condensate onto the condenser coil, where it is evaporated.

Anillo tubular. Anillo instalado en los extremos de las palas de un ventilador de condensación. Sirve para lanzar la condensación sobre el serpentín de refrigeración donde es evaporada.

Slinky loop. A flattened, circular coil of plastic pipe resembling a Slinky, which is used as a ground loop.

Espiral. Tubo plástico plano en forma helicoidal que se utiliza como circuito de tierra.

Slip. The difference in the rated rpm of a motor and the actual operating rpm when under a load.

Deslizamiento. Diferencia entre las rpm nominales de un motor y las rpm de funcionamiento reales.

Slugging. A term used to describe the condition when large amounts of liquid enter a pumping compressor cylinder.

Relleno. Término utilizado para describir la condición donde grandes cantidades de líquido entran en el cilindro de un compresor de bombeo.

Smart recovery. That start-up of a building air-conditioning or heating system whereby the system is started up earlier and allowed to run at a reduced capacity for the purpose of recording a low demand metering for the billing period.

Recuperación inteligente. El encendido del sistema de aire acondicionado o de calefacción de un edificio en el cual el sistema se enciende más temprano y se deja correr a una capacidad reducida con el propósito de registrar una medición por demanda baja por el período de facturación.

Smoke test. A test performed to determine the amount of unburned fuel in an oil burner flue-gas sample.

Prueba de humo. Prueba llevada a cabo para determinar la cantidad de combustible no quemado en una muestra de gas de combustión que se obtiene de un quemador de aceite pesado.

Snap-disc. An application of the bimetal. Two different metals fastened together in the form of a disc that provides a warping condition when heated. This also provides a snap action that is beneficial in controls that start and stop current flow in electrical circuits.

Disco de acción rápida. Aplicación del bimetal. Dos metales diferentes fijados entre sí en forma de un disco que provee un deformación al ser calentado. Esto provee también una acción rápida, ventajosa para reguladores que ponen en marcha y detienen el flujo de corriente en circuitos eléctricos.

Snap or detent action. The quick opening of a control that is used to minimize the arc when making and breaking an electrical circuit.

Acción de encaje o de detén. El abrir rápido de un control que se usa para minimizar el arco al abrir y cerrar un circuito eléctrico.

Soapstone. A stone used sometimes in the manufacture of woodstoves.

Esteatita. Piedra que se utiliza algunas veces en la fabricación de estufas de madera.

Software. Computer programs written to give specific instructions to computers.

Software. Programas de computadoras escritos para darles instrucciones específicas a las computadoras.

Solar collectors. Components of a solar system designed to collect the heat from the sun, using air, a liquid, or refrigerant as the medium.

Colectores solares. Componentes de un sistema solar diseñados para acumular el calor emitido por el sol, utilizando el aire, un líquido o un refrigerante como el medio.

Solar constant. The rate of solar energy reaching the outer limits of the earth's atmosphere has been determined to be 429 Btu/ft^2/h on a surface perpendicular (90°) to the sun's rays.

Constante solar. Se ha determinado que el índice de energía solar que alcanza los límites exteriores del atmósfera de la Tierra es de 429 Btu/pies^2/h, en una superficie perpendicular (90°) a los rayos solares.

Solar heat. Heat from the sun's rays.

Calor solar. Calor emitido por los rayos del sol.

Solar influence. The heat that the sun imposes on a structure.

Influencia solar. El calor que el sol impone en una estructura.

Solar radiant heat. Solar-heated water or an antifreeze solution is piped through heating coils embedded in concrete in the floor or in plaster in ceilings or walls.

Calor de radiación solar. El agua o líquido anticongelante calentado por energía solar se canaliza a través de serpentines de calefacción instalados en el hormigón del suelo o en el yeso de los techos o paredes.

Solder pot. A device using a low-melting solder and an overload heater sized for the amperage of the motor it is protecting. The solder will melt, opening the circuit when there is an overload. It can be reset.

Olla para soldadura. Dispositivo que utiliza una soldadura con un punto de fusión bajo y un calentador de sobrecarga diseñado para el amperaje del motor al que provee protección. La soldadura se fundirá, abriendo así el circuito cuando ocurra una sobrecarga. Puede ser reconectado.

Soldering. Fastening two base metals together by using a third, filler metal that melts at a temperature below 800°F.

Soldadura. La fijación entre sí de dos metales bases utilizando un tercer metal de relleno que se funde a una temperatura menor de 800°F (430°C).

Solderless terminals. Used to fasten stranded wire to various terminals or to connect two lengths of stranded wire together.

Terminales sin soldadura. Se usan para fijar cable trenzado a varios terminales o para unir dos pedazos de cable.

Solenoid. A coil of wire designed to carry an electrical current producing a magnetic field.

Solenoide. Bobina de alambre diseñada para conducir una corriente eléctrica generando un campo magnético.

Solid. Molecules of a solid are highly attracted to each other, forming a mass that exerts all of its weight downward.

Sólido. Las moléculas de un sólido se atraen entre sí y forman una masa que ejerce todo su peso hacia abajo.

Space cooling and heating thermostat. The device used to control the temperature of a space, such as a home thermostat that controls the temperature in a home or office.

Termostato de enfriamiento o calefacción de espacio. Aparato usado para controlar la temperatura de un espacio, tal como un termostato de hogar que controla la temperatura en un hogar u oficina.

Specific gravity. The weight of a substance compared to the weight of an equal volume of water.

Gravedad específica. El peso de una sustancia comparada con el peso de un volumen igual de agua.

Specific heat. The amount of heat required to raise the temperature of 1 lb of a substance 1°F.

Calor específico. La cantidad de calor requerida para elevar la temperatura de una libra de una sustancia 1°F (–17°C).

Specific volume. The volume occupied by 1 lb of a fluid.

Volumen específico. Volumen que ocupa una libra de fluido.

Splash lubrication system. A system of furnishing lubrication to a compressor by agitating the oil.

Sistema de lubrificación por salpicadura. Método de proveerle lubrificación a un compresor agitando el aceite.

Splash method. A method of water dropping from a higher level in a cooling tower and splashing on slats with air passing through for more efficient evaporation.

Método de salpicaduras. Método dé dejar caer agua desde un nivel más alto en una torre de refrigeración y salpicándola en listones, mientras el aire pasa a través de los mismos con el propósito de lograr una evaporación más eficaz.

Split-phase motor. A motor with run and start windings.

Motor de fase separada. Motor con devandos de functionamiento y de arranque.

Split suction. When the common suction line of a parallel compressor system has been valved in such a way as to provide for multiple temperature applications in one refrigeration package.

Succión dividida. Cuando la línea común de succión de un sistema de compresor paralelo ha sido dividida de tal manera que provee para aplicaciones de múltiples temperaturas en un empaque de refrigeración.

Split system. A refrigeration or air-conditioning system that has the condensing unit remote from the indoor (evaporator) coil.

Sistema separado. Sistema de refrigeración o de acondicionamiento de aire cuya unidad de condensación se encuentra en un sitio alejado de la bobina interior del evaporador.

Spray pond. A pond with spray heads used for cooling water in water-cooled air-conditioning or refrigeration systems.

Tanque de rociado. Tanque con una cabeza rociadora utilizada para enfriar el agua en sistemas de acondicionamiento de aire o de refrigeración enfriados por agua.

Spring-loaded relief valve. A fluid (refrigerant, air, water, or steam) relief valve that can function more than one time because a spring returns the valve to a seat.

Válvula de alivio de resorte. Una válvula de alivio de fluido (refrigerante, aire, agua o vapor de agua) que puede funcionar más de una vez porque el resorte regresa la válvula a su asiento.

Squirrel cage fan. A cylindrically shaped fan assembly used to move air.

Abanico con jaula de ardilla. Conjunto cilíndrico de abanico utilizado para mover el aire.

Squirrel cage rotor. Describes the construction of a motor rotor.

Rotor de jaula de ardilla. Describe la construcción del rotor de un motor.

Stack switch. A safety device placed in the flue of an oil furnace that proves combustion within a time frame.

Interruptor de chimenea. Un aparato de seguridad que se coloca en la chimenea de un calefactor de aceite para comprueba combustión dentro de un período de tiempo.

Stamped evaporator. An evaporator that has stamped refrigerant passages in sheet steel or aluminum.

Evaporador estampado. Un evaporador que tiene pasajes para el refrigerante estampados en lata o aluminio.

Standard atmosphere or standard conditions. Air at sea level at 70°F when the atmosphere's pressure is 14.696 psia (29.92 in. Hg). Air at this condition has a volume of 13.33 ft³/lb.

Atmósfera estándar o condiciones estándares. El aire al nivel del mar a una temperatura de 70°F (15°C) cuando la presión de la atmósfera es 14.696 psia (29.92 pulgadas Hg). Bajo esta condición, el aire tiene un volumen de 13.33 ft³/lb (pies³/libras).

Standing pilot. Pilot flame that remains burning continuously.

Piloto constante. Llama piloto que se quema de manera continua.

Start capacitor. A capacitor used to help an electric motor start.

Capacitador de arranque. Capacitador utilizado para ayudar en el arranque de un motor eléctrico.

Starting relay. An electrical relay used to disconnect the start capacitor and/or start winding in a hermetic compressor.

Relé de arranque. Relé eléctrico utilizado para desconectar el capacitador y/o el devanado de arranque en un compresor hermético.

Starting winding. The winding in a motor used primarily to give the motor extra starting torque.

Devanado de arranque. Devanado en un motor utilizado principalmente para proveerle al motor mayor para el arranque.

Starved coil. The condition in an evaporator when the metering device is not feeding enough refrigerant to the evaporator.

Bobina estrangulada. Condición que ocurre en un evaporador cuando el dispositivo de medida no le suministra suficiente refrigerante al evaporador.

Static pressure. The bursting pressure or outward force in a duct system.

Presión estática. La presión de estallido o la fuerza hacia fuera en un sistema de conductos.

Stator. The component in a motor that contains the windings: it does not turn.

Estátor. Componente en un motor que contiene los devanados y que no gira.

Steady-state condition. A stabilized condition of a piece of heating or cooling equipment where not much change is taking place.

Condición de régimen estable. Condición estabilizada de un dispositivo de calefacción o de refrigeración en el cual no hay muchos cambios.

Steam. The vapor state of water.

Vapor. Estado de vapor del agua.

Step motor. An electric motor that moves with very small increments or "steps," usually in either direction, and is usually controlled by a microprocessor with input and output controlling devices.

Motor a pasos. Un motor eléctrico que se mueve en incrementos muy pequeños o "pasos", generalmente en cualquier dirección, y generalmente son controlados por un microprocesador de aparatos de control con entradas y salidas.

Strainer. A fine-mesh device that allows fluid flow and holds back solid particles.

Colador. Dispositivo de malla fina que permite el flujo de fluido a través de él y atrapa partículas sólidas.

Stratification. The condition where a fluid appears in layers.

Estratificación. Condición que ocurre cuando un fluido aparece en capas.

Stratosphere. An atmospheric level that is located from 7 to 30 miles above the earth. Good ozone is found in the stratosphere.

Estratosfera. Capa del atmósfera que se encuentra a una altura entre 11 y 48 kilómetros encima de la Tierra. Contiene una buena capa de ozono.

Stress crack. A crack in piping or other component caused by age or abnormal conditions such as vibration.

Grieta por tensión. Grieta que aparece en una tubería u otro componente ocasionada por envejecimiento o condiciones anormales, como por ejemplo vibración.

Subbase. The part of a space temperature thermostat that is mounted on the wall and to which the interconnecting wiring is attached.

Subbase. Pieza de un termóstato que mide la temperatura de un espacio que se monta sobre la pared y a la que se fijan los conductores eléctricos interconectados.

Subcooled. The temperature of a liquid when it is cooled below its condensing temperature.

Subenfriado. La temperatura de un líquido cuando se enfría a una temperatura menor que su temperatura de condensación.

Sublimation. When a substance changes from the solid state to the vapor state without going through the liquid state.

Sublimación. Cuando una sustancia cambia de sólido a vapor sin covertirse primero en líquido.

Suction gas. The refrigerant vapor in an operating refrigeration system found in the tubing from the evaporator to the compressor and in the compressor shell.

Gas de aspiración. El vapor del refrigerante en un sistema de refrigeración en funcionamiento presente en la tubería que va del evaporador al compresor y en la coraza del compresor.

Suction line. The pipe that carries the heat-laden refrigerant gas from the evaporator to the compressor.

Conducto de aspiración. Tubo que conduce el gas de refrigerante lleno de calor del evaporador al compresor.

Suction-line accumulator. A reservoir in a refrigeration system suction line that protects the compressor from liquid floodback.

Acumulador de la línea de succión. Un estanque en la línea de succión de un sistema de refrigeración que protege al compresor de una inundación de líquido.

Suction pressure. The pressure created by the boiling refrigerant on the evaporator or low-pressure side of the system.

Presión de succión. La presión creada por el refrigerante hirviendo en el evaporador o en el lado de baja presión del sistema.

Suction service valve. A manually operated valve with front and back seats located at the compressor.

Válvula de aspiración para servicio. Válvula accionada manualmente que tiene asientos delanteros y traseros ubicados en el compresor.

Suction valve. The valve at the compressor cylinder that allows refrigerant from the evaporator to enter the compressor cylinder and prevents it from being pumped back out to the suction line.

Válvula de succión. La válvula en el cilindro de un compresor que permite que el refrigerante del evaporador entre al cilindro del compresor y evita que se bombee nuevamente a la línea de succión.

Suction-valve-lift unloading. The suction valve in a reciprocating compressor cylinder is lifted, causing that cylinder to stop pumping.

Descarga por levantamiento de la válvula de aspiración. La válvula de aspiración en el cilindro de un compresor alternativo se levanta, provocando que el cilindro deje de bombear.

Sulfur dioxide. A combustion pollutant that causes eye, nose, and respiratory tract irritation and possibly breathing problems.

Bióxido de azufre. Un contaminante por combustión que causa irritación en los ojos, la nariz y las vías respiratorias, y posiblemente problemas respiratorios.

Sump. A reservoir at the bottom of a cooling tower to collect the water that has passed through the tower.

Sumidero. Tanque que se encuentra en el fondo de una torre de refrigeración para acumular el agua que ha pasado a través de la torre.

Superheat. The temperature of vapor refrigerant above its saturation (change-of-state) temperature.

Sobrecalor. Temperatura del refrigerante de vapor mayor que su temperatura de cambio de estado de saturación.

Surge. When the head pressure becomes too great or the evaporator pressure too low, refrigerant will flow from the high- to the low-pressure side of a centrifugal compressor system, making a loud sound.

Movimiento repentino. Cuando la presión en la cabeza aumenta demasiado o la presión en el evaporador es demasiado baja, el refrigerante fluye del lado de alta presión al lado de baja presión de un sistema de compresor centrífugo. Este movimiento produce un sonido fuerte.

Swaged joint. The joining of two pieces of copper tubing by expanding or stretching the end of one piece of tubing to fit over the other piece.

Junta estampada. La conexión de dos piezas de tubería de cobre dilatando o alargando el extremo de una pieza de tubería para ajustarla sobre otra.

Swaging. See Swaged joint.

Estampar. Véase Junta estampada.

Swaging tool. A tool used to enlarge a piece of tubing for a solder or braze connection.

Herramienta de estampado. Herramienta utilizada para agrandar una pieza de tubería a utilizarse en una conexión soldada o broncesoldada.

Swamp cooler. A slang term used to describe an evaporative cooler.

Nevera pantanosa. Término del argot utilizado para describir una nevera de evaporación.

Sweating. A word used to describe moisture collection on a line or coil that is operating below the dew point temperature of the air.

Exudación. Término utilizado para describir la acumulación de humedad en un conducto o una bobina que está funcionando a una temperatura menor que la del punto de rocío de aire.

System charge. The refrigerant in a system, both liquid and vapor. The correct charge is a balance where the system will give the most efficiency.

Carga del sistema. El refrigerante en un sistema, tanto líquido y vapor. La carga correcta es un balance donde el sistema dará la mayor eficiencia.

System lag. The temperature drop of the controlled space below the set point of the thermostat.

Retardo del sistema. Caída de temperatura de un espacio controlado, por debajo del nivel programado en el termostato.

System overshoot. The temperature rise of the controlled space above the set point of the thermostat.

Exceso del sistema. Subida de la temperatura de un espacio controlado, por encima del nivel programado en el termostato.

System pressure-regulating valve (SPR valve). A valve located between the compressor and receiver of a refrigeration system. This valve controls the amount of liquid refrigerant that bypasses a parallel liquid receiver.

Válvula reguladora de presión del sistema (SPR en inglés). Válvula situada entre el compresor y el destino de un sistema de refrigeración. Esta válvula controla la cantidad de refrigerante que no entra en un receptor de líquido paralelo.

Tank. A closed vessel used to contain a fluid.

Tanque. Desposito cerrado utilizado para contener un fluido.

Tap. A tool used to cut internal threads in a fastener or fitting.

Macho de roscar. Herramienta utilizada para cortar filetes internos en un aparto fijador o en un accesorio.

Technician. A person who performs maintenance, service, testing, or repair to air-conditioning or refrigeration equipment. *Note:* the EPA defines this person as someone who could reasonably be expected to release CFCs or HCFCs into the atmosphere.

Técnicos. Una persona que lleva a cabo mantenimiento, servicio o reparaciones a equipos de aire acondicionado o refrigeración. *Nota:* Esta persona, según defunido por la EPA, es una persona del cual razonablemente se estaría esparado que libere CFC (clorofluorocarbonos) a la atmósfera.

Temperature. A word used to describe the level of heat or molecular activity, expressed in Fahrenheit, Rankine, Celsius, or Kelvin units.

Temperatura. Término utilizado para describir el nivel de calor o actividad molecular, expresado en unidades Fahrenheit, Rankine, Celsio o Kelvin.

Temperature difference (TD). The difference between the inlet temperature and outlet temperature of a heat exchanger. For example, an evaporator may have a 20°F TD: 75°F air in and 55°F air out.

Diferencia entre temperaturas. La diferencia existente entre la temperatura de entrada y la de salida de un intercambiador de calor. Por ejemplo, en un evaporador puede existir una diferencia de 20°F entre la temperatura de entrada y la de salida: 75°F (la toma del aire) y 55°F (la evacuacíon del aire).

Temperature glide. When a refrigerant blend has different temperatures when it evaporates and condenses at a single given pressure.

Variación de temperatura. Una mezcla de refrigerantes tiene varias temperaturas a las cuales se produce una evaporación a una presión determinada.

Temperature-measuring instruments. Devices that accurately measure the level of temperature.

Instrumentos que miden temperatura. Aparatos que miden el nivel de la temperatura con precisión.

Temperature/pressure relationship. This refers to the temperature/pressure relationship of a liquid and vapor in a closed container. If the temperature increases, the pressure will also increase. If the temperature is lowered, the pressure will decrease.

Relación entre temperatura y presión. Se refiere a la relación entre la presión y la temperatura de un líquido y un vapor en un recipiente cerrado. Si la temperatura aumenta, la presión también aumentará. Si la temperatura baja, habrá una caída de presión.

Temperature reference points. Various points that may be used to calibrate a temperature-measuring device, such as boiling or freezing water.

Puntos de referencia de temperatura. Varios puntos que pueden usarse para calibrar un aparato que mide temperatura, tales como agua congelada o hirvienda.

Temperature-sensing elements. Various devices in a system that are used to detect temperature.

Elementos que detectan temperatura. Varios aparatos en un sistema que se usan para detectar temperatura.

Temperature swing. The temperature difference between the low and high temperatures of the controlled space.

Oscilación de temperatura. Diferencia existente entre las temperaturas altas y bajas de un espacio controlado.

Test light. A lightbulb arrangement used to prove the presence of electrical power in a circuit.

Luz de prueba. Arreglo de bombillas utilizado para probar la presencia de fuerza eléctrica en un circuito.

Testing, Adjusting and Balancing Bureau (TABB). A certification bureau for individuals involved in working with ventilation.

Agencia de Prueba, Ajuste y Balanceo (TABB en inglés). Agencia de certificación para los individuos involucrados en el trabajo de ventilación.

Therm. Quantity of heat, 100,000 Btu.

Therm. Cantidad de calor, mil unidades térmicas inglesas.

Thermistor. A semiconductor electronic device that changes resistance with a change in temperature.

Termistor. Dispositivo eléctrico semiconductor que cambia su resistencia cuando se produce un cambio en temperatura.

Thermocouple. A device made of two unlike metals that generates electricity when there is a difference in temperature from one end to the other. Thermocouples have a hot and cold junction.

Thermopar. Dispositivo hecho de dos metales distintos que genera electricidad cuando hay una diferencia en temperatura de un extremo al otro. Los termopares tienen un empalme caliente y uno frío.

Thermometer. An instrument used to detect differences in the level of heat.

Termómetro. Instrumento utilizado para detectar diferencias en el nivel de calor.

Thermopile. A group of thermocouples connected in series to increase voltage output.

Pila termoeléctrica. Grupo de termopares conectados en serie para aumentar la salida de tensión.

Thermostat. A device that senses temperature change and changes some dimension or condition within to control an operating device.

Termostato. Dispositivo que advierte un cambio en temperatura y cambia alguna dimensión o condición dentro de sí para regular un dispositivo en funcionamiento.

Thermostatic expansion valve (TXV). A valve used in refrigeration systems to control the superheat in an evaporator by metering the correct refrigerant flow to the evaporator.

Válvula de gobierno termostático para expansión. Válvula utilizada en sistemas de refrigeración para regular el sobrecalor en un evaporador midiendo el flujo correcto de refrigerante al evaporador.

Three-phase power. A type of power supply usually used for operating heavy loads. It consists of three sine waves that are out of phase by 120° with each other.

Potencia trifásica. Tipo de fuente de alimentación normalmente utilizada en el funcionamiento de cargas pesadas. Consiste de tres ondas sinusoidales que no están en fase la una con la otra por 120°.

Throttling. Creating a planned or regulated restriction in a fluid line for the purpose of controlling fluid flow.

Estrangulamiento. Que ocasiona una restricción intencional o programada en un conducto de fluido, a fin de controlar el flujo del fluido.

Thrust surface. A term that usually applies to bearings that have a pushing pressure to the side and that therefore need an additional surface to absorb the push. Most motor shifts cradle in their bearings because they operate in a horizontal mode, like holding a stick in the palm of your hand. When a shaft is turned to the vertical mode, a thrust surface must support the weight of the shaft along with the load the shaft may impose on the thrust surface. The action of a vertical fan shaft that pushes air up is actually pushing the shaft downward.

Superficie de empuje. Un término que generalmente se aplica a cojinetes que sostienen una presión de empuje a un lado y que necesitan una superficie adicional para absorber este empuje. La mayoría de los ejes de motor están al abrigo en sus cojinetes, porque funcionen en una modalidad horizontal, como cuando uno sostiene una vara en la mano. Cuando un eje está sintonizado a la modalidad vertical, una superficie de empuje debe sostener el peso del eje junto con la carga que el eje puede imponer sobre la superficie de empuje. La moción de un eje vertical de un ventilador que empuja aire hacia arriba está en realidad empujando el eje hacia abajo.

Time delay. A device that prevents a component from starting for a prescribed time. For example, many systems start the fans and use a time delay relay to start the compressor at a later time to prevent too much inrush current.

Retraso de tiempo. Un aparato que evita que un componente se encienda por un período prescrito de tiempo. Por ejemplo, muchos sistemas encienden los ventiladores y, usando un relé de retraso, encienden el compresor un tiempo después para evitar mucha corriente interna.

Timers. Clock-operated devices used to time various sequences of events in circuits.

Temporizadores. Dispositivos accionados por un reloj utilizados para medir el tiempo de varias secuencias de eventos en circuitos.

Toggle bolt. Provides a secure anchoring in hollow tiles, building block, plaster over lath, and gypsum board. The toggle folds and can be inserted through a hole, where it opens.

Tornillo de fiador. Proveen un anclaje seguro en losetas huecas, bloques de construcción, yeso sobre listón y tablón de yeso. El fiador se dobla y puede insertarse a través de un roto y después se abre.

Ton of refrigeration. The amount of heat required to melt a ton (2000 lb) of ice at 32°F in 24 hours, 288,000 Btu/24 h, 12000 Btu/h, or 200 Btu/min.

Tonelada de refrigeración. Cantidad de calor necesario para fundir una tonelada (2.000 libras) de hielo a 32°F (0°C en 24 horas), 288.000 Btu/24 h 12.000 Btu/h o 200 Btu/min.

Torque. The twisting force often applied to the starting power of a motor.

Par de torsión. Fuerza de torsión aplicada con frecuencia a la fuerza de arranque de un motor.

Torque wrench. A wrench used to apply a prescribed amount of torque or tightening to a connector.

Llave de torsión. Llave utilizada para aplicar una cantidad específica de torsión o de apriete a un conector.

Total equivalent warming impact (TEWI). A global warming index that takes into account both the direct effects of chemicals emitted into the atmosphere and the indirect effects caused by system inefficiencies.

Impacto de calentamiento equivalente total (TEWI en inglés). Índice de calentamiento de la Tierra que tiene en cuenta los efectos directos de los productos químicos emitidos en la atmósfera, y los efectos indirectos causados por la ineficacia de un sistema.

Total heat. The total amount of sensible heat and latent heat contained in a substance from a reference point.

Calor total. Cantidad total de calor sensible o de calor latente presente en una sustancia desde un punto de referencia.

Total pressure. The sum of the velocity and the static pressure in an air duct system.

Presión total. La suma de la velocidad y la presión estática en un sistema de conducto de aire.

Transformer. A coil of wire wrapped around an iron core that induces a current to another coil of wire wrapped around the same iron core. Note: A transformer can have an air core.

Transformador. Bobina de alambre devanado alrededor de un núcleo de hierro que induce una corriente a otra bobina de alambre devanado alrededor del mismo núcleo de hierro. Nota: un transformador puede tener un núcleo de aire.

Transistor. A semiconductor often used as a switch or amplifier.

Transistor. Semiconductor que suele utilizarse como conmutador o amplificador.

TRIAC. A semiconductor switching device.

TRIAC. Dispositivo de conmutación para semiconductores.

Troposphere. The lower atmospheric level that extends upward from ground level to about 7 miles. Global warming takes place in the troposphere.

Troposfera. Capa más baja de la del atmósfera que va desde el nivel del suelo hasta una altura de aproximadamente 11 kilómetros. El efecto del calentamiento de la tierra se produce en la troposfera.

Tube-within-a-tube coil. A coil used for heat transfer that has a pipe in a pipe and is fastened together so that the outer tube becomes one circuit and the inner tube another.

Bobina de tubo dentro de un tubo. Bobina utilizada en la transferencia de calor que tiene un tubo dentro de otro y se sujeta de manera que el tubo exterior se convierte en un circuito y el tubo interior en otro circuito.

Tubing. Pipe with a thin wall used to carry fluids.

Tubería. Tubo que tiene una pared delgada utilizada para conducir fluidos.

Twinning. Two furnaces connected side by side and sharing a common ducting system.

Aparear. Dos calefactores conectados lado a lado o que comparten un sistema de conductos común.

Two-pole, split-phase motor. This motor runs at 3600 rpm when not loaded and at about 3450 rpm when loaded. A four-pole motor runs at about 1800 rpm not loaded and at about 1725 fully loaded.

Motor de dos polos, de fase dividida. Este motor corre a 3.600 revoluciones por minuto cuando no está cargado y cuando está cargado corre a 3.450 revoluciones por minuto. Un motor de 4 polos corre a 1.800 revoluciones por minuto si no está cargado y alrededor de 1.725 revoluciones con carga completa.

Two-speed compressor motor. Can be a four-pole motor that can be connected as a two-pole motor for high speed (3450 rpm) and connected as a four-pole motor for running at 1725 rpm for low speed. This is accomplished with relays outside the compressor.

Motor de compresor de dos velocidades. Puede ser un motor de 4 polos que puede conectarse como un motor de 2 polos para velocidades altas (3.450 revoluciones por minuto) y conectarse como un motor de 4 polos para correr a 1.725 revoluciones por minuto para velocidades bajas. Esto se hace con relés fuera del compresor.

Two-temperature valve. A valve used in systems with multiple evaporators to control the evaporator pressures and maintain different temperatures in each evaporator. Sometimes called a *holdback valve.*

Válvula de dos temperaturas. Válvula utilizada en sistemas con evaporadores múltiples para regular las presiones de los evaporadores y mantener temperaturas diferentes en cada uno de ellos. Conocida también como *válvula de retención.*

Ultrasound leak detector. Detectors that use sound from escaping refrigerant to detect leaks.

Detector de escapes ultrasónico. Detectores que usan sonido del refrigerante que está escapando para detectar escapes.

Ultraviolet. Light waves that can only be seen under a special lamp.

Ultravioleta. Ondas de luz que pueden observarse solamente utilizando una lámpara especial.

Ultraviolet light. Light frequency between 200 and 400 nanometers.

Luz ultravioleta. Luz con frecuencia entre 200 y 400 nanómetros.

Uneven parallel system. Parallel compressors of unequal sizes mounted on a steel rack and controlled by a microprocessor.

Sistema paralelo heterogéneo. Compresores de capacidades distintas montados en paralelo en un bastidor de acero y controlados mediante microprocesador.

Upflow furnace. This furnace takes in air from the bottom or from sides near the bottom and discharges hot air out the top.

Horno de flujo ascendente. La entrada del aire en este tipo de horno se hace desde abajo o en los laterales, cerca del suelo. La evacuación se realiza en la parte superior.

Urethane foam. A foam that can be applied between two walls for insulation.

Espuma de uretano. Espuma que puede aplicarse entre dos paredes para crear un aislamiento.

U-Tube mercury manometer. A U-tube containing mercury, which indicates the level of vacuum while evacuating a refrigeration system.

Manómetro de mercurio de tubo en U. Tubo en U que contiene mercurio y que indica el nivel del vacío mientras vacía un sistema de refrigeración.

U-Tube water manometer. Indicates natural gas and propane gas pressures. It is usually calibrated in inches of water.

Manómetro de agua de tubo en U. Indica las presiones del gas natural y del propano. Se calibra normalmente en pulgadas de agua.

Vacuum. The pressure range between the earth's atmospheric pressure and no pressure, normally expressed in inches of mercury (in. Hg) vacuum.

Vacío. Margen de presión entre la presión de la atmósfera de la Tierra y cero presión, por lo general expresado en pulgadas de mercurio (pulgadas Hg) en vacío.

Vacuum gage. An instrument that measures the vacuum when evacuating a refrigeration, air-conditioning, or heat pump system.

Medidor del vacío. Instrumento que se utiliza para medir el vacío al vaciar un sistema de refrigeración, de aire acondicionado, o de bomba de calor.

Vacuum pump. A pump used to remove some fluids such as air and moisture from a system at a pressure below the earth's atmosphere.

Bomba de vacío. Bomba utilizada para remover algunos fluidos, como por ejemplo aire y humedad de un sistema a una presión menor que la de la atmósfera de la Tierra.

Valve. A device used to control fluid flow.

Válvula. Dispositivo utilizado para regular el flujo de fluido.

Valve plate. A plate of steel bolted between the head and the body of a compressor that contains the suction and discharge reed or flapper valves.

Placa de válvula. Placa de acero empernado entre la cabeza y el cuerpo de un compresor que contiene la lámina de aspiración y de descarga o las chapaletas.

Valve seat. That part of a valve that is usually stationary. The movable part comes in contact with the valve seat to stop the flow of fluids.

Asiento de la válvula. Pieza de una válvula que es normalmente fija. La pieza móvil entra en contacto con el asiento de la válvula para detener el flujo de fluidos.

Valve stem depressor. A service tool used to access pressure at a Schrader valve connection.

Depresor de vástago de válvula. Una herramienta de servicio que se usa para acceder la presión en una conexión de válvula Schrader.

Vapor. The gaseous state of a substance.

Vapor. Estado gaseoso de una sustancia.

Vapor barrier. A thin film used in construction to keep moisture from migrating through building materials.

Película impermeable. Película delgada utilizada en construcciones para evitar que la humedad penetre a través de los materiales de construcción.

Vapor charge bulb. A charge in a thermostatic expansion valve bulb that boils to a complete vapor. When this point is reached, an increase in temperature will not produce an increase in pressure.

Válvula para la carga de vapor. Carga en la bombilla de una válvula de expansión termostática que hierve a un vapor completo. Al llegar a este punto, un aumento en temperatura no produce un aumento en presión.

Vapor lock. A condition where vapor is trapped in a liquid line and impedes liquid flow.

Bolsa de vapor. Condición que ocurre cuando el vapor queda atrapado en el conducto de líquido e impide el flujo de líquido.

Vapor pressure. The pressure exerted on top of a saturated liquid.

Presión del vapor. Presión que se ejerce en la superficie de un líquido saturado.

Vapor pump. Another term for compressor.

Bomba de vapor. Otro término para compresor.

Vapor refrigerant charging. Adding refrigerant to a system by allowing vapor to move out of the vapor space of a refrigerant cylinder and into the low-pressure side of the refrigeration system.

Carga del refrigerante de vapor. Agregarle refrigerante a un sistema permitiendo que el vapor salga del espacio de vapor de un cilindro de refrigerante y que entre en el lado de baja presión del sistema de refrigeración.

Vaporization. The changing of a liquid to a gas or vapor.

Vaporización. Cuando un líquido se convierte en gas o vapor.

Variable-frequency drive (VFD). An electrical device that varies the frequency (hertz) for the purpose of providing a variable speed.

Propulsión de frecuencia variable (VFD en inglés). Un aparato eléctrico que varía la frecuencia (hertz) con el propósito de proveer velocidad variable.

Variable pitch pulley. A pulley whose diameter can be adjusted.

Polea de paso variable. Polea cuyo diámetro puede ajustarse.

Variable resistor. A type of resistor where the resistance can be varied.

Resistor variable. Tipo de resistor donde la resistencia puede variarse.

Variable-speed motor. A motor that can be controlled, with an electronic system, to operate at more than one speed.

Motor de velocidad variable. Un motor que puede controlarse, con un sistema electrónico, para operar a más de una velocidad.

V belt. A belt that has a V-shaped contact surface and is used to drive compressors, fans, or pumps.

Correa en V. Correa que tiene una superficie de contacto en forma de V y se utiliza para accionar compresores, abanicos o bombas.

Velocity. The speed at which a substance passes a point.

Velocidad. Rapidez a la que una sustancia sobrepasa un punto.

Velocity meter. A meter used to detect the velocity of fluids, air, or water.

Velocímetro. Instrumento utilizado para medir la velocidad de fluidos, aire o agua.

Velometer. An instrument used to measure the air velocity in a duct system.

Velómetro. Instrumento utilizado para medir la velocidad del aire en un conducto.

Vent-free. Certain gas stoves or gas fireplaces are not required to be vented; therefore, they are called "vent-free."

Sin ventilación. Ciertas estufas de gas y chimeneas de gas no requieren una ventilación, por lo que se conocen como estufas "sin ventilación."

Ventilation. The process of supplying and removing air by natural or mechanical means to and from a particular space.

Ventilación. Proceso de suministrar y evacuar el aire de un espacio determinado, utilizando procesos naturales o mecánicos.

Venting products of combustion. Venting flue gases that are generated from the burning process of fossil fuels.

Descargar productos de combustión. Descargar los gases de la chimenea que se generan del proceso de combustión de los combustibles fósiles.

Venturi. A smoothly tapered device used in fluid flow to increase the velocity of the fluid.

Venturi. Un aparato que está ahusado lisamente y que se usa en el flujo de fluido para aumentar la velocidad del fluido.

Volt-ohm-milliammeter (VOM). A multimeter that measures voltage, resistance, and current in milliamperes.

Voltio-ohmio-miliamperimetro (VOM en inglés). Multímetro que mide tensión, resistencia y corriente en miliamperios.

Voltage. The potential electrical difference for electron flow from one line to another in an electrical circuit.

Tensión. Diferencia de potencial eléctrico del flujo de electrones de un conducto a otro en un circuito eléctrico.

Voltage feedback. Voltage potential that travels through a power-consuming device when it is not energized.

Retroalimentación de voltaje. El voltaje que viaja a través de un aparato de consumo de electricidad cuando no está energizado.

Voltmeter. An instrument used for checking electrical potential.

Voltímetro. Instrumento utilizado para revisar la potencia eléctrica.

Volumetric efficiency. The pumping efficiency of a compressor or vacuum pump that describes the pumping capacity in relationship to the actual volume of the pump.

Rendimiento volumétrico. Rendimiento de bombeo de un compresor o de una bomba de vacío que describe la capacidad de bombeo con relación al volumen real de la bomba.

Vortexing. A whirlpool action in the sump of a cooling tower.

Acción de vórtice. Torbellino en el sumidero de una torre de refrigeración.

Walk-in cooler. A large refrigerated space used for storage of refrigerated products.

Nevera con acceso al interior. Espacio refrigerado grande utilizado para almacenar productos refrigerados.

Wastewater system. A refrigeration system that uses water one time and then exhausts it to a waste system.

Sistema de aguas residuales. Sistema de refrigeración que usa agua una vez y luego la vacía en el sistema de aguas residuales.

Water box. A container or reservoir at the end of a chiller where water is introduced and contained.

Caja de agua. Recipiente o depósito al extremo de un enfriador por donde entra y se retiene el agua.

Water column (WC). The pressure it takes to push a column of water up vertically. One inch of water column is the amount of pressure it would take to push a column of water in a tube up one inch.

Columna de agua (WC en inglés). Presión necesaria para levantar una columna de agua verticalmente. Una pulgada de columna de agua es la cantidad de presión necesaria para levantar una columna de agua a una distancia de una pulgada en un tubo.

Water-cooled condenser. A condenser used to reject heat from a refrigeration system into water.

Condensador enfriado por agua. Condensador utilizado para dirigir el calor de un sistema de refrigeración al agua.

Water cooler. A small refrigeration machine that is typically used to refrigerate water for drinking purposes.

Enfriador de agua. Una máquina de refrigeración pequeña que típicamente se usa para refrigerar agua para beber.

Water hammer. A loud pressure pulse that occurs when valves close fast.

Martillo de agua. Impulso de presión fuerte que se produce cuando una válvula se cierra rápidamente.

Water loop. These loops of plastic pipe are buried in the ground in a closed-loop geothermal heat pump system and contain a heat transfer fluid.

Circuito de agua. Estos circuitos de tubos de plástico se encuentran enterrados en el suelo, en el caso de una bomba de calor geotérmica de circuito cerrado y contiene un líquido que transpaso lo caliente.

Water manometer. A device that uses a column of water to measure low pressures in air or gas systems.

Manómetro de agua. Un aparato que usa una columna de agua para medir bajas presiones en sistemas de aire o de gas.

Water pump. Used to pump water or other fluids from one pressure level to another to promote water flow in piping systems. See Centrifugal pump.

Bomba de agua. Bomba utilizada para bombear agua u otros fluidos de un nivel de presión a otro para mejorar el flujo de agua atravéz de los sistemas de tubería. Véase Bomba centrífuga.

Water-regulating valve. An operating control regulating the flow of water.

Válvula reguladora de agua. Regulador de mando que controla el flujo de agua.

Watt (W). A unit of power applied to electron flow. One watt equals 3.414 Btu.

Watio (W en inglés). Unidad de potencia eléctrica aplicada al flujo de electrones. Un watio equivale a 3414 Btu.

Watt-hour. The unit of power that takes into consideration the time of consumption. It is the equivalent of a 1-watt bulb burning for 1 hour.

Watio hora. Unidad de potencia eléctrica que toma en cuenta la duración de consumo. Es el equivalente de una bombilla de 1 watio encendida por espacio de una hora.

Weep holes. Holes that connect each cell in an ice machine's cell-type evaporator that allow air entering from the edges of the ice to travel along the entire ice slab to relieve the suction force and allow the ice to fall off of the evaporator.

Aberturas de exudación. Aberturas que conectan cada célula en el evaporador de tipo celular de una hielera, que permiten que el aire que entra de los bordes del hielo viaja a lo largo de todo el pedazo de hielo para aliviar la fuerza de succión y permitir que el hielo se caiga del evaporador.

Weight. The force that matter (solid, liquid, or gas) applies to a supporting surface when it is at rest.

Peso. La fuerza que la materia (sólido, líquido, o gas) en reposo aplica a una superficie de apoyo.

Welded hermetic compressor. A compressor that is completely sealed by welding, versus a semi-hermetic compressor that is sealed by bolts and flanges.

Compresor hermético soldado. Un compresor que está completamente sellado por soldadura, contrario a un compresor semi-hermético que está sellado con tornillos y bridas.

Wet-bulb depression. The difference between the wet-bulb and the dry-bulb reading when readings are taken in air.

Depresión de bulbo mojado. La diferencia entre la lectura de bulbo mojado y bulbo seco cuando las lecturas están tomadas en aire.

Wet-bulb temperature. A wet-bulb temperature of air is used to evaluate the humidity in the air. It is obtained with a wet thermometer bulb to record the evaporation rate with an airstream passing over the bulb to help in evaporation.

Temperatura de una bombilla húmeda. La temperatura de una bombilla húmeda se utiliza para evaluar la humedad presente en el aire. Se obtiene con la bombilla húmeda de un termómetro para registrar el margen de evaporación con un flujo de aire circulando sobre la bombilla para ayudar en evaporar el agua.

Wet compression. Saturated refrigerant vapors being compressed by the compressor, which will turn into liquid refrigerant droplets when compressed.

Compresión mojada. Vapores saturados de refrigerante que están siendo comprimidos por el compresor y que a su vez se convertirán en gotitas de refrigerantes al comprimirse.

Wet heat. A heating system using steam or hot water as the heating medium.

Calor húmedo. Sistema de calentamiento que utiliza vapor o agua caliente como medio de calentamiento.

Winding thermostat. A safety device used in electric motor windings to detect overtemperature conditions.

Termostato de bobina. Un aparato de seguridad usado en un motor eléctrico para detectar condiciones de exceso de temperatura.

Window unit. An air conditioner installed in a window that rejects the heat outside the structure.

Acondicionador de aire para la ventana. Acondicionador de aire instalado en una ventana que desvía el calor proveniente del exterior de la estructura.

Wire connectors (screw-on). Used to connect two or more wires together.

Conectores de cables (de rosca). Se usan para conectar dos o más cables.

Work. A force moving an object in the direction of the force. Work = Force × Distance.

Trabajo. Fuerza que mueve un objeto en la dirección de la fuerza. Trabajo = Fuerza × Distancia.

WYE transformer connection. Typically furnishes 208 V and 115 V to a customer.

Conexión de transformador WYE. Típicamente provee 208 voltios y 115 voltios a los consumidores.

Zeotropic blend. Two or more refrigerants mixed together that will have a range of boiling and/or condensing points for each system pressure. Noticeable fractionation and temperature glide will occur.

Mezcla zeotrópica. Mezcla de dos o más refrigerantes que tiene un rango de ebullición y/o punto de condensación para cada presión en el sistema. Se produce una fraccionación y una variación de temperatura notables.

Zone valve. Zone control valves are thermostatically controlled valves that control water flow in various zones in a hydronic heating system.

Válvula de sector. Se trata de válvulas controladas termostáticamente, que controlan el flujo del agua en varios sectores en un sistema de calefacción hidrónico.

Index

Note: Entries followed by an f *denote figures.*